Conversion Factors

Length

1 m = 39.37 in. = 3.281 ft

1 in. = 2.54 cm

1 km = 0.621 mi

1 mi = 5280 ft = 1.609 km

1 lightyear = 9.461×10^{15} m

1 angstrom (Å) = 10^{-10} m

Mass

1 kg = 10^3 g = 6.85×10^{-2} slug

1 slug = 14.59 kg

1 u = 1.66×10^{-27} kg

Time

1 min = 60 s

1 h = 3600 s

1 day = 8.64×10^4 s

1 year = 365.242 days = 3.156×10^7 s

Volume

1 liter = 1000 cm^3 = 3.531×10^{-2} ft^3

1 ft^3 = 2.832×10^{-2} m^3

1 gallon = 3.786 liter = 231 in.^{3}

Angle

180° = π rad

1 rad = 57.30°

1° = 60 min = 1.745×10^{-2} rad

Speed

1 km/h = 0.278 m/s = 0.621 mi/h

1 m/s = 2.237 mi/h = 3.281 ft/s

1 mi/h = 1.61 km/h = 0.447 m/s = 1.47 ft/s

Force

1 N = 0.2248 lb = 10^5 dynes

1 lb = 4.448 N

1 dyne = 10^{-5} N = 2.248×10^{-6} lb

Work and energy

1 J = 10^7 erg = 0.738 ft·lb = 0.239 cal

1 cal = 4.186 J

1 ft·lb = 1.356 J

1 Btu = 1.054×10^3 J = 252 cal

1 J = 6.24×10^{18} eV

1 eV = 1.602×10^{-19} J

1 kWh = 3.60×10^6 J

Pressure

1 atm = 1.013×10^5 N/m^2 (or Pa) = 14.70 lb/in.^{2}

1 Pa = 1 N/m^2 = 1.45×10^{-4} lb/in.^{2}

1 lb/in.^{2} = 6.895×10^3 N/m^2

Power

1 hp = 550 ft·lb/s = 0.746 kW

1 W = 1 J/s = 0.738 ft·lb/s

1 Btu/h = 0.293 W

COLLEGE PHYSICS

Fifth Edition

TECHNOLOGY VERSION

Raymond A. Serway

Jerry S. Faughn

SAUNDERS COLLEGE PUBLISHING

A Division of Harcourt College Publishers

FORT WORTH   PHILADELPHIA   SAN DIEGO   NEW YORK   ORLANDO   AUSTIN
SAN ANTONIO   TORONTO   MONTREAL   LONDON   SYDNEY   TOKYO

Publisher: Emily Barrosse
Publisher: John Vondeling
Product Managers: David Theisen and Pauline E. Mula
Developmental Editor: Susan Dust Pashos
Project Editors: Elizabeth Ahrens and Frank Messina
Production Manager: Charlene Catlett Squibb
Art Director: Carol Bleistine
Text and Cover Designer: Ruth A. Hoover

Cover Image and Credit: Mountaineering, Mt. Ranier, Washington, © Chris Noble/Tony Stone
Images
COLLEGE PHYSICS, TECHNOLOGY VERSION, Fifth Edition
ISBN: 0-03-031901-3
Library of Congress Catalog Card Number: 99-068658

2000 Version
Copyright © 1999, 1995, 1992, 1989, 1985 by Raymond A. Serway

Address for domestic orders:
Saunders College Publishing, 6277 Sea Harbor Drive, Orlando, FL 32887-6777
1-800-782-4479
e-mail: collegesales@ harcourt.com

Address for international orders:
International Customer Service, Harcourt, Inc.
6277 Sea Harbor Drive, Orlando, FL 32887-6777
(407) 345-3800
Fax: (407) 345-4060
e-mail: hbintl@ harcourt.com

Address for editorial correspondence:
Saunders College Publishing, Public Ledger Building, Suite 1250, 150 S. Independence Mall West,
Philadelphia, PA 19106-3412

Web Site Address
http://www.harcourtcollege.com

Printed in the United States of America

012345678   032   10 98765432

Preface

Welcome to the Technology Version of *College Physics,* 5th edition! This kit consists of:

- the paperback textbook you are now reading.
- the *Saunders Core Concepts in College Physics* CD-ROM, a set of three disks described in greater detail later in the Preface
- the Workbook to accompany the *Saunders Core Concepts in College Physics* CD-ROM
- a single CD-ROM containing the Textbook version of Interactive Physics™ software (developed by MSC Working Knowledge), as well as simulations files by Raymond A. Serway that run on this software.

Instructors interested in technology-based education should also consider the three options for on-line homework described on page xii.

College Physics is written for a one-year course in introductory physics. The course is usually taken by students majoring in biology; the health professions; environmental, earth, and social sciences; and technical fields such as architecture. The mathematical techniques used in the book include algebra, geometry, and trigonometry, but not calculus.

The main objectives of this introductory book are twofold: to provide the student with a clear and logical presentation of the basic concepts and principles of physics, and to strengthen an understanding of the concepts and principles through a broad range of interesting applications to the real world. In order to meet these objectives, emphasis is placed on sound physical arguments and discussions of everyday experiences and observations. At the same time, we have attempted to motivate the student through practical examples that demonstrate the role of physics in other disciplines.

This book, which covers the standard topics in classical physics and 20th century physics, is divided into six parts. Part 1 (Chapters 1–9) deals with Newtonian mechanics and the physics of fluids; Part 2 (Chapters 10–12) is concerned with heat and thermodynamics; Part 3 (Chapters 13–14) covers wave motion and sound; Part 4 (Chapters 15–21) is concerned with electricity and magnetism; Part 5 (Chapters 22–25) treats the properties of light and the field of geometric and wave optics; and Part 6 (Chapters 26–30) represents an introduction to special relativity, quantum physics, and atomic and nuclear physics.

The textbook for the Technology Version of *College Physics* has been designed to facilitate the use of the *Saunders Core Concepts in College Physics* CD-ROM as an additional instructional aid. An icon in the shape of a CD has been placed in the textbook margins at locations where it is helpful to use the CD-ROM. Other marginal notes and footnotes refer to the Workbook accompanying the CD-ROM or refer the reader to the textbook's Web site for links to additional information on the World Wide Web.

© *Index Stock Photography, Inc.*

CHANGES TO THE FIFTH EDITION

A number of changes and improvements have been made in preparing the fifth edition of this text. Many changes were in response to comments and suggestions submitted by reviewers of the manuscript and instructors using the fourth edition. The new features added to this edition are based on current trends in science education. The following describe the major changes in the fifth edition.

Conceptual Thinking

A concerted effort was made to place more emphasis on critical thinking and teaching physical concepts. This was accomplished through the use of approximately 200 conceptual examples, called **Thinking Physics,** and over 400 **Conceptual Questions.** Many of the **Thinking Physics** examples provide students with a means of reviewing concepts presented in that section. Some examples demonstrate the connection between the concepts presented in that chapter and other scientific disciplines. These examples also serve as models for students when assigned the task of responding to the **Conceptual Questions** presented at the end of each chapter. The answers to the odd-numbered **Conceptual Questions** are contained in the answer section at the end of the textbook.

Applications

We have expanded the number of applications of physics throughout the text. Some applications are developed within the text proper, while many are in the form of new examples called **Applying Physics.** Because the examples chosen should be familiar to most students, they should help motivate the student and show the relevance of physical concepts as they apply to real-world problems. Applications in the text proper are identified by special marginal notes of the form

APPLICATION

Connecting Your Stereo
Speakers.

A complete list of the applications can be found on pages xv–xvi.

QuickLabs

This new feature encourages students to perform simple experiments on their own so that they will be more engaged in the learning process. Most **QuickLab** experiments can be performed with low-cost items such as string, rubber bands, a ruler, tape, and balloons. In most cases, students are simply asked to observe the outcome of an experiment and explain their results in terms of what they have learned in that chapter. Where appropriate, students are also asked to obtain some data and graph their results.

Multiple-Choice Questions

We have included a number of multiple-choice questions at the end of each chapter, with answers provided in the answer section at the back of the book. These questions are intended to give students practice in this form of testing, commonly used in the MCAT exam and in many large enrollment courses.

Problems and Conceptual Questions

A substantial revision of the end-of-chapter problems and conceptual questions was made in this fifth edition. Most of the new problems that have been added are intermediate in level (as identified by their blue problem numbers). All problems have been carefully worded and have been checked for clarity and accuracy. Some problem sets begin with a Review Problem that requires the use of concepts learned in previous chapters. Solutions to approximately 20 percent of the end-of-chapter problems are included in the *Study Guide and Student Solutions Manual.* These problems are identified by boxes around their numbers. A smaller subset of solutions has been

QUICKLAB

Place a cup of water on a sheet of paper with one side of the paper overhanging the table that supports them. Remove the sheet of paper from underneath the cup by *quickly* pulling on the sheet. (Until you become good at this, you might want to use a plastic cup rather than a glass cup in case of an accident.) Explain why this is possible.

posted on the World Wide Web (**http://www.harcourtcollege.com**) at a site accessible to students and instructors using *College Physics;* these problems are identified with the label WEB. See below for a complete description of other features of the problem set.

Chapter-Opening Photos and Puzzlers

All chapter-opening photographs now include puzzlers in the form of a question to the student. The student should be able to answer the puzzler after completing that chapter.

Other Content Changes

The material entitled "For Further Study" in the fourth edition for Chapters 7, 9, and 15 was modified and moved from the back of the book to the main text in the form of optional sections. Some features found in the fourth edition were either deleted or reduced in length to allow for the new features described earlier. The biographical sketches of important scientists were replaced by short descriptions of their work and placed in margins. Most essays by guest authors were omitted from the text. However, those essays that pertain to the health sciences were retained in this edition. The lists of suggested additional readings were removed from the text but are included in the *Instructor's Manual* at the beginning of each chapter's solutions.

FEATURES

We have retained or added many pedagogical features in the textbook that should enhance its usefulness to both the student and instructor. These are as follows:

Style

We have attempted to write the book in a style that is clear, relaxed, and pleasing to the reader. New terms are carefully defined, and we have tried to avoid jargon. At the same time, we have attempted to keep the presentation accurate and precise.

Organization

The book is divided into six parts: mechanics, thermodynamics, vibrations and wave motion, electricity and magnetism, light and optics, and modern physics. Each part includes an overview of the subject matter to be covered in that part and some historical perspectives.

Introductory Chapter

The introductory chapter, which "sets the stage" for the text, discusses the building blocks of matter, the units of physical quantities, order-of-magnitude calculations, dimensional analysis, significant figures, and mathematical notation.

Units

The International System of Units (SI) is used throughout the book. The British engineering system of units is used only to a limited extent in the problem sets of the early chapters on mechanics.

Previews

Most chapters begin with a chapter preview, which includes a brief discussion of the chapter objectives and content.

Equations, Marginal Notes, and Important Statements

Important equations are highlighted with a light gold screen, and marginal notes are often used to describe their meaning. Marginal notes are also used to locate specific definitions and important statements in the text. Most important statements and definitions are highlighted with a light blue screen or are set in **boldface type** for added emphasis and ease of review.

Telegraph Colour Library/FPG

Problem-Solving Strategies
General strategies and suggestions are included for solving the types of problems featured in both the worked examples and end-of-chapter problems. This feature is intended to help students identify the essential steps in solving problems and increase their skills as problem solvers. This feature is highlighted by a light blue screen to help students locate the strategies quickly.

Physics in Action
This boxed material focuses on photographs of interesting demonstrations and phenomena in physics, accompanied by detailed explanations. The material can also serve as a source of information for initiating classroom discussions.

Topics Relevant to the Life Sciences and Other Disciplines
Many chapters include sections which are intended to expose the student to various practical and interesting applications of physical principles. Those topics dealing with applications of physics to the life sciences are identified by the DNA icon.

Worked Examples
A large number of worked examples, including many new ones, are presented as an aid in understanding and/or reinforcing physical concepts. In many cases, these examples serve as models for solving end-of-chapter problems. The examples are set off from the text for ease of location and are given titles to describe their content. Many examples include a **Reasoning** section to illustrate the underlying concepts and methodology used in arriving at a correct solution. This will help students understand the logic behind the solution and the advantage of using a particular approach to solve the problem. The solution answer is highlighted with a light blue screen. Many worked examples are followed immediately by exercises with answers. These exercises represent extensions of the worked examples and are intended to sharpen students' problem-solving skills and test their understanding of concepts. Students who work through these exercises on a regular basis should find the end-of-chapter problems less intimidating.

Illustrations and Photographs
The text material, worked examples, and end-of-chapter questions and problems are accompanied by numerous figures, photographs, and tables. Full color is used to add clarity to the figures and to make the visual presentation as realistic and pleasing as possible. Three-dimensional effects are rendered where appropriate. Vectors are color coded, and curves in xy-plots are drawn in color. Color photographs have been carefully selected, and their accompanying captions have been written to serve as an added instructional tool. A complete description of the pedagogical use of color appears on the inside front cover.

Summaries
Each chapter contains a summary which reviews the important concepts and equations discussed in that chapter.

Conceptual Questions
A set of conceptual questions is provided at the end of each chapter. The **Thinking Physics** and **Applying Physics** examples presented in the text should serve as models for students when conceptual questions are assigned or used in tests. The questions all provide the student with a means of self-testing the concepts presented in the chapter. Some conceptual questions are appropriate for initiating classroom discussions. Answers to all odd-numbered conceptual questions are located in the answer section at the end of the book.

Multiple-Choice Questions
A set of multiple-choice questions is included at the end of each chapter. The answers to these questions are given at the end of the book.

NASA

End-of-Chapter Problems An extensive set of problems is included at the end of each chapter. Answers to odd-numbered problems are given at the end of the book. For the convenience of both the student and instructor, about two thirds of the problems are keyed to specific sections of the chapter. The remaining problems, labeled "Additional Problems," are not keyed to specific sections. There are three levels of problems graded according to their difficulty. Straightforward problems are numbered in black, intermediate-level problems are numbered in blue, and the most challenging problems are numbered in magenta. Those problems that are accompanied by an Interactive Physics™ computer simulation (see below) are labeled with the Interactive Physics™ icon (⬛), and those with a focus on the life sciences are identified by the DNA icon (⬛). Solutions for problems marked web may be found on our Web Site: **http://www.harcourtcollege.com,** provided the proper instructor or student password is known.

Appendices Several appendices are provided at the end of the book. The appendix material includes a review of mathematical techniques used in the book, such as scientific notation, algebra, geometry, and trigonometry. Reference to these appendices is made as needed throughout the book. Most of the mathematical review sections include worked examples and exercises with answers. Some appendices contain useful tables that supplement textual information. For easy reference, the front endpapers contain a chart explaining the use of color throughout the book and a list of frequently used conversion factors.

SAUNDERS CORE CONCEPTS IN COLLEGE PHYSICS CD-ROM

The *Saunders Core Concepts in College Physics* CD-ROM set developed by Archipelago Productions applies the power of multimedia to the introductory physics course, offering full-motion animation and video, engaging interactive graphics, clear and concise text, and guiding narration. *Saunders Core Concepts in College Physics* focuses on those concepts students typically find most difficult in the course, drawing from topics in mechanics, thermodynamics, electric fields, magnetic fields, and optics. The animations and graphics are presented to aid the student in developing accurate conceptual models of difficult topics—topics often hard to explain with words or chalkboard illustrations. The CD-ROM set also presents a step-by-step exploration of problem-solving strategies and provides animations of problems in order to promote conceptual understanding and to sharpen problem-solving skills.

The material in the *Saunders Core Concepts in College Physics* CD-ROM is divided into modules, with Modules 1–5 on Disk 1, Modules 6–9 on Disk 2, and Modules 10–14 on Disk 3. Each module is organized into a series of main screens that address a single topic or a group of closely related topics. Additionally, many of the main screens are divided into two sections. The first section appears when you access the main screen. The second section, if available, appears when you select a colored bar found in the lower right-hand corner of the screen. Installation instructions and additional information concerning the CD-ROM set can be found in the Introduction and User's Guide in the accompanying Workbook.

Icons Within the Textbook When you see the icon ⬛ **3.2**, it suggests that it is appropriate to access Module 3, Screen 2, which is displayed at the top of the following page:

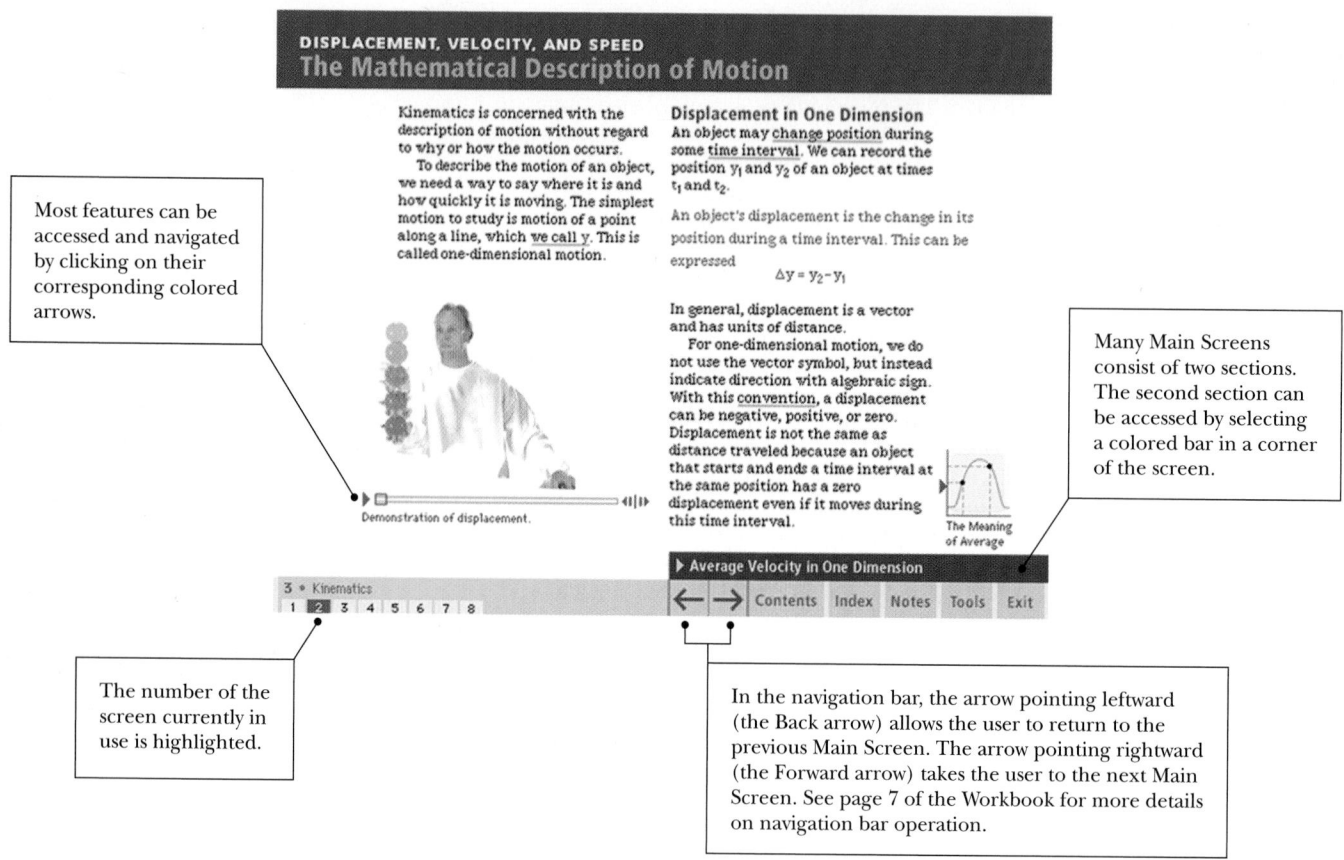

Most features can be accessed and navigated by clicking on their corresponding colored arrows.

DISPLACEMENT, VELOCITY, AND SPEED
The Mathematical Description of Motion

Kinematics is concerned with the description of motion without regard to why or how the motion occurs.

To describe the motion of an object, we need a way to say where it is and how quickly it is moving. The simplest motion to study is motion of a point along a line, which we call y. This is called one-dimensional motion.

Displacement in One Dimension
An object may change position during some time interval. We can record the position y_1 and y_2 of an object at times t_1 and t_2.

An object's displacement is the change in its position during a time interval. This can be expressed

$$\Delta y = y_2 - y_1$$

In general, displacement is a vector and has units of distance.

For one-dimensional motion, we do not use the vector symbol, but instead indicate direction with algebraic sign. With this convention, a displacement can be negative, positive, or zero. Displacement is not the same as distance traveled because an object that starts and ends a time interval at the same position has a zero displacement even if it moves during this time interval.

Demonstration of displacement.

The Meaning of Average

Many Main Screens consist of two sections. The second section can be accessed by selecting a colored bar in a corner of the screen.

▶ Average Velocity in One Dimension

3 ● Kinematics
1 2 3 4 5 6 7 8

← →   Contents   Index   Notes   Tools   Exit

The number of the screen currently in use is highlighted.

In the navigation bar, the arrow pointing leftward (the Back arrow) allows the user to return to the previous Main Screen. The arrow pointing rightward (the Forward arrow) takes the user to the next Main Screen. See page 7 of the Workbook for more details on navigation bar operation.

Workbook   The Workbook that accompanies *Saunders Core Concepts in College Physics* CD-ROM contains 93 example problems worked out in detail. Step-by-step guidance is provided, and full solutions are given at the end of the Workbook. The Workbook problems are referenced in the textbook by marginal notes or by footnotes within the end-of-chapter problems. The marginal notes and footnotes will assist in approaching a textbook problem in *College Physics* that is similar or occasionally identical to a problem in the Workbook.

INTERACTIVE PHYSICS™ PROGRAM AND SIMULATIONS

Many physical situations involving the motion of objects discussed in the textbook are brought to life by computer simulations created using the highly acclaimed Interactive Physics™ software developed by MSC Working Knowledge. Approximately 94 simulations, most of which are keyed to specific worked examples or end-of-chapter problems, are included on a separate CD-ROM that contains the Textbook version of the program. (This Textbook version enables students to print out their work but not to save it or to export their data.) Other simulations portray material that is difficult to display in a printed textbook (such as the tracking of a specific point on a rotating object) or complement concepts or applications discussed in the text. A complete list of Interactive Physics™ simulations is found on pages xvii–xviii.

Instructors who wish to take full advantage of the power of the Interactive Physics™ program are encouraged to contact MSC Working Knowledge at (800) 766-6615 or by logging on to their Web site at **http://www.krev.com.**

STUDENT ANCILLARIES

Saunders College Publishing, a division of Harcourt College Publishers, offers several items to supplement and enhance the classroom experience. The following ancillaries originally developed for *College Physics* will allow instructors to customize the textbook to their students' needs and their own style of instruction.

Study Guide and Student Solutions Manual by John R. Gordon, Charles Teague, and Raymond A. Serway. The manual features detailed solutions to 20 percent of the end-of-chapter problems from the text. These are indicated in the text with boxed problem numbers. The manual also features a skills section, important notes from key sections of the text, and a list of important equations and concepts.

Pocket Guide to Accompany College Physics by V. Gordon Lind. This 5″ by 7″ paperback is a section-by-section capsule of the textbook that provides a handy guide for looking up important concepts, equations, and problem-solving hints.

Physics Laboratory Manual, second edition by David Loyd. Updated and redesigned, this manual supplements the learning of basic physical principles while introducing laboratory procedures and equipment. Each chapter contains a pre-laboratory assignment, objectives, an equipment list, the theory behind the experiment, step-by-step experimental procedures, and questions. A laboratory report form is provided for each experiment so that students can record data and make calculations. Students are encouraged to apply statistical analysis to their data in order to develop their ability to judge the validity of their results.

So You Want to Take Physics: A Preparatory Course by Rodney Cole. This introductory-level book covers numerous physical principles and is ideal for strengthening mathematical skills essential to the study of physics.

Practice Problems with Solutions   After consulting with their instructor, students may wish to purchase an additional set of skill-building practice problems with solutions in print form. These problems will provide extra practice in unit conversions and other mathematical basics for students who need to build confidence in solving physics problems.

Student Web Site   Students will find an abundance of material at our Web site **(http://www.harcourtcollege.com/physics/cptech),** including a sampling of solutions from the Study Guide/Student Solutions Manual. They can access a Practice Exercises and Testing area, view sample chapters of the *Pocket Guide,* and link to other Web sites for supplemental information. Students are encouraged to visit the site frequently to check for new developments.

INSTRUCTOR'S ANCILLARIES

Except where noted, the following ancillaries were originally developed for *College Physics* and are appropriate for use with the Technology Version.

Instructor's Manual with Complete Solutions by Charles Teague and Jerry S. Faughn. This manual consists of complete solutions to all the problems in the text, as well

Richard Megna/Fundamental Photographs, NYC

© *R. Folwell/SPL/Photo Researchers, Inc.*

as answers to even-numbered problems and conceptual questions. Instructors will also find a list of suggested readings from journals and other resources.

Printed Test Bank by Robert Beichner. The test bank for *College Physics* has been newly revised, updated, and corrected and is available for use with both the Technology Version and the regular version of the textbook. The printed test bank contains approximately 2500 problems and questions (both open-ended and multiple-choice). The format allows instructors to duplicate pages for distribution to students.

Computerized Test Bank Available in both Windows™ and Macintosh® formats, the test bank contains approximately 2500 open-ended and multiple-choice problems and questions, representing every chapter of the text. The test bank allows instructors to customize tests by rearranging, editing, or adding questions. The questions are graded in level of difficulty for the instructor's convenience; the program places the answers to all questions on a separate grading key.

CAPA: A Computer-Assisted Personalized Approach CAPA is a network system for learning, teaching, assessment, and administration. It provides students with personalized problem sets, quizzes, and examinations consisting of qualitative conceptual problems and quantitative problems, including problems from *College Physics*. CAPA was developed through a collaborative effort of the Physics–Astronomy, Computer Science, and Chemistry Departments at Michigan State University. Students are given instant feedback and relevant hints via the Internet and may correct errors without penalty before an assignment's due date. The system records each student's participation and performance on assignments, quizzes, and examinations; and records are available on-line to both the individual student and to his or her instructor. For more information, visit the CAPA Web site at **http://www.pa.msu.edu/educ/CAPA/**

WebAssign: A Web-Based Homework System WebAssign is a Web-based homework delivery, collection, grading, and recording service developed at North Carolina State University. Instructors who sign up for WebAssign can assign homework to their students, using questions and problems taken directly from *College Physics*. WebAssign gives students immediate feedback on their homework that helps them to master information and skills, leading to greater competence and better grades. WebAssign can free instructors from the drudgery of grading homework and recording scores, allowing them to devote more time to meeting with students and preparing classroom presentations. Details about and a demonstration of WebAssign are available at **http://webassign.net/info.** For more information about ordering this service, contact WebAssign at **webassign@ncsu.edu.**

Homework Service With this service, instructors can reduce their grading workload by assigning thought-proving homework problems using the World Wide Web. Instructors browse problem banks that include problems from *College Physics,* select those they wish to assign to their students, and then let the Homework Service take over the delivery and grading. This system was developed and is maintained by Fred Moore at the University of Texas **(moore@physics.utexas.edu).** Students download their unique problems, submit their answers, and obtain immediate feedback; if students' answers are incorrect, they can resubmit them. This rapid grading feature facilitates effective learning. After the due date of their assignments, students can obtain the solutions to their problems. Minimal on-line connect time is required. The Homework Service uses algorithm-based problems: This means that each student solves sets of problems that are different from those given to other students. Details about this service and a demonstration of it are available at **http://hw10.ph.utexas.edu/instInst.html.**

Overhead Transparency Acetates The collection of transparencies consists of approximately 250 full-color figures from the text to enhance lectures; they feature large print for easy viewing in the classroom.

Practice Problems with Solutions   About 400 problems that do not appear in the text are available with full solutions in a printed version and on the Instructor's Resource CD-ROM. These can be used for examinations, homework assignments, or student practice and drill exercises. This problems booklet is also available at a low cost to the student.

Instructor's Manual for Physics Laboratory Manual, second edition by David Loyd. Each chapter contains a discussion of the experiment, teaching hints, answers to selected questions from the student laboratory manual, and a post-laboratory quiz with short answers and essay questions. The author has also included a list of the suppliers of scientific equipment and a summary of the equipment needed for all the experiments in the manual.

Physics Demonstration Videotapes   A unique collection of 70 physics demonstrations is provided on videotape to supplement classroom presentations and to help motivate students.

Instructor's Web Site   At the instructor's area for *College Physics* on the World Wide Web **(http://www.harcourtcollege.com),** instructors will find an overhead transparency listing; a chapter-by-chapter guide to relevant experiments in *Physics Laboratory Manual,* 2nd edition, by David Loyd; and a correlation guide between chapters in *College Physics,* 5th edition, and the *Saunders Core Concepts in College Physics* CD-ROM. Sample lesson plans will be available for guidance on use of the electronic media, and new developments may be expected periodically.

Instructor's Resource CD-ROM   Presentation CD-ROMs have been created to provide our textbook users with an exciting new tool for classroom use. The Instructor's Resource CD-ROM accompanying *College Physics,* 5th edition, contains a collection of graphics files of line art from the textbook. These files can be opened directly or can be imported into a variety of presentation packages. The labels for each piece of art have been enlarged and made bold for classroom viewing. The CD-ROM also contains electronic files of the Instructor's Manual and Test Bank, plus the Practice Problems with Solutions.

TEACHING OPTIONS

College Physics, 5th edition, contains more than enough material for a one-year course in introductory physics. This serves two purposes. First, it gives the instructor more flexibility in choosing topics for a specific course. Second, the book becomes more useful as a resource for students. On the average, it should be possible to cover about one chapter each week for a class that meets three hours per week. Many sections containing interesting applications are labeled as being optional material. Those optional sections dealing with applications of physics to the life sciences are identified with the DNA icon (🧬). Instructors are encouraged to cover those optional sections which best match their students' interests. We offer the following suggestions for shorter courses for those instructors who choose to move at a slower pace through the year:

Option A: If you choose to place more emphasis on contemporary topics in physics, you should consider omitting all or parts of Chapter 8 (Rotational Equilibrium and Rotational Dynamics), Chapter 21 (Alternating Current Circuits and Electromagnetic Waves), and Chapter 25 (Optical Instruments).

Option B: If you choose to place more emphasis on classical physics, you could omit all or parts of Part 6 of the textbook, which deals with special relativity and other topics in 20th century physics.

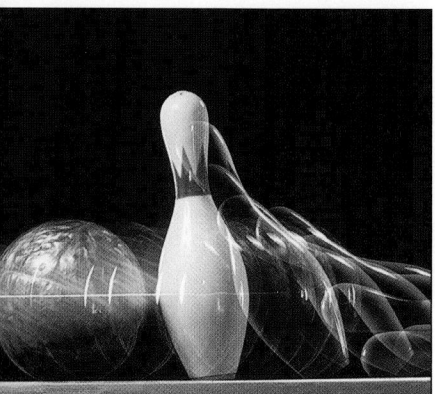

Ben Rose/The Image Bank

ACKNOWLEDGMENTS

In preparing the fifth edition of this textbook, we have been guided by the expertise of many people who have reviewed manuscript and/or provided pre-revision suggestions. We wish to acknowledge the following reviewers and express our sincere appreciation for their helpful suggestions, criticisms, and encouragement:

Paul D. Beale, *University of Colorado at Boulder*

David H. Bennum, *University of Nevada at Reno*

Neal M. Cason, *University of Notre Dame*

Steven D. Davis, *University of Arkansas at Little Rock*

John DeFord, *University of Utah*

Chris J. DeMarco, *Jackson Community College*

Tom French, *Montgomery County Community College*

Grant Hart, *Brigham Young University*

John Ho, *State University of New York at Buffalo*

Murshed Hossain, *Rowan University*

Ivan Kramer, *University of Maryland, Baltimore County*

David Markowitz, *University of Connecticut*

Steven Morris, *Los Angeles Harbor College*

Martin Nikolo, *Saint Louis University*

Ed Oberhofer, *University of North Carolina at Charlotte*

David G. Onn, *University of Delaware*

John Simon, *University of Toledo*

We thank William G. Buckman of Western Kentucky University, David Griffing of Miami University, Paul Davidovits of Boston College, and Isaac Abella of the University of Chicago for their interesting essays that follow Chapters 9, 11, 18, and 28.

We are especially grateful to John W. Jewett, Jr. of the California State Polytechnic University at Pomona for his exceptional contribution of chapter opener puzzlers, new applications, and many of the **Thinking Physics** and **Applying Physics** examples. We thank Ed Oberhofer for his contribution of new problems, including all of the Review Problems in this edition. Charles Teague carefully compiled the problems manuscript, in addition to selecting and editing solutions to appear in the *Study Guide and Student Solutions Manual.* We thank Ralph McGrew of Broome Community College and John Jewett for their painstaking accuracy checking of the text. John R. Gordon put together the many features of the *Study Guide and Student Solutions Manual* and oversaw its composition and layout by Michael Rudmin. We are grateful to our friend and colleague Robert J. Beichner for his suggestions and encouragement regarding the **QuickLabs** and other features of the text.

We thank Linda Miller for her painstaking work with a word processor in typing and checking the fifth edition of the manuscript. Dena Digilio-Betz is to be recognized for locating and suggesting many excellent photographs for this edition.

We are extremely grateful to the publishing team at Saunders College Publishing for their expertise and outstanding work in all aspects of this project. Susan Pashos carefully analyzed the reviews of the fifth edition, provided important guidance in all aspects of the project, and made many important improvements in the final manuscript. Beth Ahrens managed *College Physics* through all production stages and was invaluable in her help with the revised illustrations. We appreciate the fine work of Carol Bleistine for managing the overall art and design program. We also recognize important contributions by Frank Messina, Walter Neary, Sally Kusch, Pauline Mula, and Charlene Squibb. We thank John Vondeling for his continued friendship, enthusiasm, and sense of humor, even under a very tight schedule.

Finally, we dedicate this book to our wives and children, for their love, support, and long-term sacrifices.

Raymond A. Serway
Moneta, Virgina

Jerry S. Faughn
Richmond, Kentucky

Applications

Although physics is relevant to so much in our modern lives, this may not be obvious to students in an introductory course. In this fifth edition of *College Physics,* we have attempted to make the relevance of physics to everyday life more obvious by pointing out specific applications with a new design feature in the form of a marginal note. Additionally, some of these applications we have marked pertain to the life sciences and are identified with the DNA icon (🧬). The list below is not intended to be a complete listing of all the applications of the principles of physics found in the textbook. Many other applications are to be found within the text and especially in the worked examples, Conceptual Questions, and end-of-chapter problems.

◀ **APPLICATION**

Directs you to sections discussing applied principles of physics.

Interactive Physics™ Simulations

The following worked examples, end-of-chapter problems, and other items have simulations written for use with Interactive Physics™ by MSC Working Knowledge. A dual-platform (Windows™- and Macintosh®-compatible) CD-ROM is included in this kit, containing the Textbook version of the program and the simulations files.

To the Student

We feel it is appropriate to offer some words of advice that should be of benefit to you, the student. Before doing so, we shall assume that you have read the preface, which describes the various features of the text that will help you through the course.

HOW TO STUDY

Very often we are asked "How should I study physics and prepare for examinations?" There is no simple answer to this question, but we would like to offer some suggestions based on our own experiences in learning and teaching over the years.

First and foremost, maintain a positive attitude toward the subject matter, keeping in mind that physics is the most fundamental of all natural sciences. Other science courses that follow will use the same physical principles, so it is important that you understand and be able to apply the various concepts and theories discussed in the text.

CONCEPTS AND PRINCIPLES

It is essential that you understand the basic concepts and principles *before* attempting to solve assigned problems. This is best accomplished through a careful reading of the textbook before attending your lecture on that material. In the process, it is useful to jot down certain points that are not clear to you. Take careful notes in class, and then ask questions pertaining to those ideas that require clarification. Keep in mind that few people are able to absorb the full meaning of scientific material after one reading. Several readings of the text and notes may be necessary. Your lectures and laboratory work should supplement the text and clarify some of the more difficult material. You should reduce memorization of material to a minimum. Memorizing passages from a text, equations, and derivations does not necessarily mean you understand the material. Your understanding of the material will be enhanced through a combination of efficient study habits, discussions with other students and instructors, and your ability to solve the problems in the text. Ask questions whenever you feel it is necessary. If you are reluctant to ask questions in class, seek private consultation or initiate discussions with your classmates. Many individuals are able to speed up the learning process when the subject is discussed on a one-to-one basis.

STUDY SCHEDULE

It is important to set up a regular study schedule, preferably on a daily basis. Be sure to read the syllabus for the course and adhere to the schedule set by your instructor. The lectures will be much more meaningful if you have read the corresponding textual material *before* attending the lecture. As a general rule, you should devote about two hours of study time for every hour in class. If you are having trouble with the course, seek the advice of the instructor or students who have already taken the course. You may find it neces-

Gerard Vandystadt/All Sport

sary to seek further instruction from experienced students. Very often, instructors offer review sessions in addition to regular class periods. It is important that you avoid the practice of delaying study until a day or two before an exam. More often than not, this will lead to disastrous results. Rather than an all-night study session before an exam, it is better to briefly review the basic concepts and equations, followed by a good night's rest. The Pocket Guide that accompanies the text provides formulas and helpful hints and can be a useful review tool before exams. If you feel you need additional help in understanding the concepts, preparing for exams, or in problem solving, we suggest that you acquire a copy of the student study guide that accompanies the text; it should be available at your college bookstore.

USE THE FEATURES

You should make *full* use of the various features of the text discussed in the preface. For example, marginal notes are useful for locating and describing important equations, while important statements and definitions are highlighted with a light blue screen. Many useful tables are contained in the appendices, but most are incorporated into the text where they are used most often. Appendix A is a convenient review of mathematical techniques. Answers to odd-numbered problems are given at the end of the text, along with answers to the odd-numbered Conceptual Questions and all of the Multiple-Choice Questions. Exercises (with answers), which follow some worked examples, represent extensions of those examples, and in most cases you are expected to perform a simple calculation. Their purpose is to test your problem-solving skills as you read through the text. Problem-solving strategies are included in selected chapters throughout the text to give you additional information to help you solve the problems. An overview of the entire text is given in the table of contents, while the index will enable you to locate specific material quickly. Footnotes are sometimes used to supplement the discussion or to cite other references on the subject. Your instructor can suggest additional readings for you.

After reading a chapter, you should be able to define any new quantities introduced in that chapter and to discuss the principles and assumptions that were used to arrive at certain key relations. The chapter summaries and the review sections of the study guide should help you in this regard. In some cases, it will be necessary to refer to the index of the text to locate certain topics. You should be able to correctly associate with each physical quantity a symbol used to represent that quantity and the unit in which the quantity is specified. Furthermore, you should be able to express each important relation in a concise and accurate prose statement.

THINKING PHYSICS AND APPLYING PHYSICS

These new features are intended to develop your conceptual understanding of physics. Answers ("Explanation") are provided immediately for you because these non-mathematical examples often extend or further develop the chapter content. They may, for instance, illuminate the connection between physics and everyday life. You should attempt to answer the Conceptual Questions at the end of each chapter using the model for your answers provided by the Explanations in **Thinking Physics** and **Applying Physics.**

PROBLEM SOLVING

R. P. Feynman, Nobel laureate in physics, once said, "You do not know anything until you have practiced." In keeping with this statement, we strongly advise that you develop the skills necessary to solve a wide range of problems. Your ability to solve problems will be one

Heinz Fischer/The Image Bank

of the main tests of your knowledge of physics, and therefore you should try to solve as many problems as possible. It is essential that you understand basic concepts and principles before attempting to solve problems. It is good practice to try to find alternative solutions to the same problem. For example, problems in mechanics can be solved using Newton's laws, but very often an alternative method using energy considerations is more direct. You should not deceive yourself into thinking you understand the problem after seeing its solution in class. You must be able to solve the problem and similar problems on your own.

The method of solving problems should be carefully planned. A systematic plan is especially important when a problem involves several concepts. First, read the problem several times until you are confident you understand what is being asked. Look for any key words that will help you interpret the problem, and perhaps allow you to make certain assumptions. Your ability to interpret the question properly is an integral part of problem solving. You should acquire the habit of writing down the information given in a problem, and decide what quantities need to be found. You might want to construct a table listing quantities given and quantities to be found. This procedure is sometimes used in the worked examples of the text. After you have decided on the method you feel is appropriate for the situation, proceed with your solution. General problem-solving strategies of this type are included in the text and are highlighted by a light blue screen.

We often find that students fail to recognize the limitations of certain formulas or physical laws in a particular situation. It is very important that you understand and remember the assumptions underlying a particular theory or formalism. For example, we shall find that certain equations in kinematics apply only to an object moving with constant acceleration. They are not valid for situations in which the acceleration is not constant, as in the cases of the motion of an object connected to a spring and the motion of an object through a fluid.

Telegraph Colour Library/FPG

EXPERIMENTS AND "QUICKLABS"

Physics is a science based upon experimental observations. In view of this fact, we recommend that you make every effort to supplement the text through various types of "hands-on" experiments, either in the laboratory or at home, especially using the **QuickLabs** in this book. Many simple experiments can be used to test ideas and models discussed in class or in the text. For example, an object swinging on the end of a long string together with a wristwatch can be used to investigate pendulum motion; an object attached to the end of a vertical spring or a rubber band can be used to determine the nature of restoring forces and to investigate periodic motion; collisions between equal masses can be observed while playing billiards; an approximate value for the acceleration due to gravity can be obtained by dropping an object from a known height and simply measuring the time of its fall with a stopwatch; traveling waves can be investigated with the aid of a stretched rope or the common "Slinky" toy (a stretched spring); a pair of Polaroid sunglasses and some discarded lenses or a magnifying glass can be used to perform various experiments in optics. The list is endless. When physical models are not available, be imaginative and try to develop models of your own.

ELECTRONIC MEDIA

This textbook is part of a kit containing two multimedia products. The single CD-ROM containing the Textbook version of the Interactive Physics™ program by MSC Working Knowledge also includes simulations folders. The Interactive Physics™ simulations are keyed to selected worked examples and end-of-chapter problems dealing with mechanics and wave motion. The three-disk set comprising the *Saunders Core Concepts in College Physics* CD-ROM was developed by Archipelago Productions for a complete multimedia presentation of selected topics in mechanics, thermodynamics, electromagnetism, and optics. It is far easier to understand physics if you see it in action, and these electronic materials will enable you to be a part of that action.

You should also frequently visit the Student Web site (Student's home page) at **http://www.harcourtcollege.com/physics/cptech.** You can obtain the user ID and password from your instructor. Due to the fluid nature of the Web, the features and learning resources on this site may change and grow throughout the academic term.

CLOSING COMMENTS

Someone once said that there are only two professions in which people truly enjoy what they are doing: professional sports and physics. Although this statement is most likely an exaggeration, both professions are truly exciting and stretch your skills to the limit. It is our sincere hope that you too will find physics exciting and that you will benefit from this experience, regardless of your chosen profession.

Welcome to the exciting world of physics.

> *To see a World in a Grain of Sand*
> *And a Heaven in a Wild Flower,*
> *Hold infinity in the palm of your hand*
> *And Eternity in an hour.*

W. Blake, "Auguries of Innocence"

Contents Overview

Superstock

Contents

Jean Y. Rusniewski/Tony Stone Images

Stephen Krasemann/Tony Stone Images

Tom Mareschel/The Image Bank

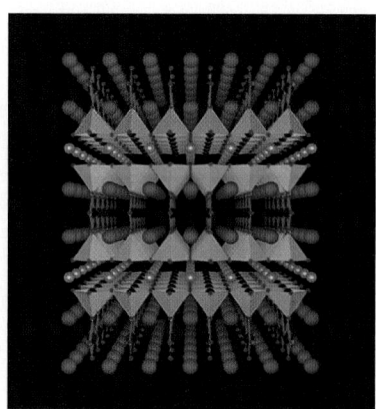

Courtesy of IBM Research

© Danny Lehman

5 Light and Optics   *721*

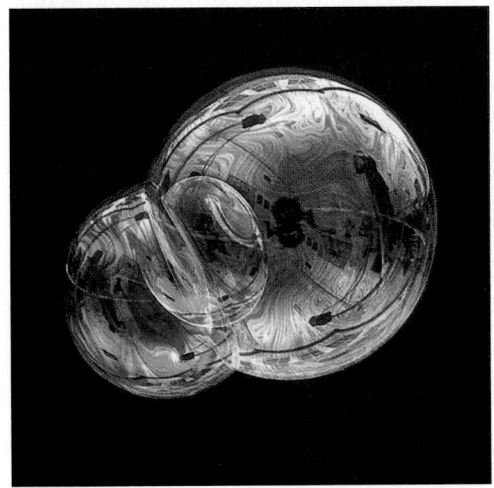

*Dr. Jeremy Burgess/Science Photo Library/Photo
Researchers, Inc.*

Dembinsky Photo Assoc.

PART 1

Mechanics

Physics, the most fundamental science, is concerned with the basic principles of the Universe. It is one of the foundations on which the other physical sciences — astronomy, chemistry, and geology — are based. The beauty of physics lies in the simplicity of its fundamental theories and in the way just a small number of basic concepts, equations, and assumptions can alter and expand our view of the world.

The myriad physical phenomena in our world are parts of one or more of the following five areas of physics:

- Mechanics, which is concerned with the effects of forces on material objects
- Thermodynamics, which deals with heat, temperature, and the behavior of large numbers of particles
- Electromagnetism, which deals with charges, currents, and electromagnetic fields
- Relativity, a theory that describes particles moving at any speed, and connects space and time
- Quantum mechanics, a theory dealing with the behavior of particles at the submicroscopic level as well as the macroscopic world

The first part of this textbook addresses mechanics, sometimes referred to as classical mechanics or Newtonian mechanics. This is an appropriate subject with which to begin an introductory book because many of the basic principles of mechanical systems can be used later to describe such natural phenomena as waves and heat transfer. Furthermore, the laws of conservation of energy and momentum to be introduced in our study of mechanics retain their importance in the fundamental theories that follow, including the theories of modern physics.

The first serious attempts to develop a theory of motion were made by the Greek astronomers and philosophers. Although they devised a complex model to describe the motions of heavenly bodies, their model lacked correlation between such motions and the motions of objects on Earth. The study of mechanics was enhanced by a number of careful astronomical investigations by Copernicus, Brahe, and Kepler in the 16th century. In the 16th and 17th centuries, Galileo attempted to relate the motions of falling bodies and projectiles to the motions of planetary

bodies, and Sevin and Hooke studied forces and their relationship to motion. A major development in the theory of mechanics was provided by Newton in 1687 when he published his *Principia*. Newton's elegant theory, which remained unchallenged for more than 200 years, was based on his hypothesis of universal gravitation together with contributions made by Galileo and others.

Today, mechanics is of vital importance to students from all disciplines. It is highly successful in describing the motions of material bodies, such as planets, rockets, and baseballs. In the first nine chapters of this book, we shall describe the laws of mechanics and examine a wide range of phenomena that can be understood through these laws.

Introduction

WEB

For Web links related to Stonehenge, visit the textbook Web Site at **http://www.harcourtcollege.com/physics/cptech**

▲ PHYSICS PUZZLER

Shown here is an eclipse over Stonehenge. Stonehenge is a circle of stone built by the ancient Britons on Salisbury Point, England. Its orientation marks the seasonal rising and setting points of the Sun. During the summer solstice, where would the builders of Stonehenge place a marker stone to locate the rising Sun from the center of the stone circle? *(Fred Espenak/Science Photo Library/ Photo Researchers, Inc.)*

The goal of physics is to provide an understanding of nature by developing theories based on experiments. The theories are usually expressed in mathematical form. Fortunately, it is possible to explain the behavior of a variety of physical systems with a limited number of fundamental laws.

Scientists continually work at improving our comprehension of fundamental laws, and new discoveries are made every day. In many research areas, a great deal of overlap occurs among physics, chemistry, and biology. The numerous recent technological advances are results of the efforts of many scientists, engineers, and technicians. Some of the most notable include unmanned space missions and manned Moon landings, microcircuitry and high-speed computers, and sophisticated imaging techniques used in scientific research and medicine. The impacts of such developments on our society have indeed been great, and it is likely that future discoveries will be exciting, challenging, and of great benefit to humanity.

Because following chapters will be concerned with the laws of physics, we must begin by clearly defining the basic quantities involved in these laws. For example, such physical quantities as force, velocity, volume, and acceleration can be described in terms of more fundamental quantities. In the next several chapters we shall encounter three basic quantities: **length** (L), **mass** (M), and **time** (T). In later chapters we will need to add two other standard units to our list, for temperature (the kelvin) and for electric current (the ampere). In our study of mechanics, however, we shall be concerned only with the units of length, mass, and time.

1.1 STANDARDS OF LENGTH, MASS, AND TIME

If we are to report the result of a measurement of a certain quantity to someone who wishes to reproduce this measurement, a unit for the quantity must be defined. For example, if someone familiar with our system of measurement and weights reports that a wall is 2.0 meters high and our fundamental unit of length is defined to be 1.0 meter, we know that the height of the wall is twice the fundamental unit of length. Likewise, if we are told that a person has a mass of 75 kilograms and our fundamental unit of mass is defined as 1.0 kilogram, then that person has a mass 75 times as great as the fundamental unit of mass. In 1960, an international committee agreed on a system of standards and designations for these fundamental quantities, called the **SI system** (Système International) of units. Its units of length, mass, and time are the meter, kilogram, and second.

Length

In 1799, the legal standard of length in France became the meter, defined as one ten-millionth of the distance from the Equator to the North Pole. As recently as 1960, the official length of the meter was the distance between two lines on a specific bar of platinum-iridium alloy stored under controlled conditions. This standard was abandoned for several reasons, a principal one being that the limited accuracy with which the separation between the lines can be determined does not meet the current requirements of science and technology. Then the meter was defined as 1 650 763.73 wavelengths of orange-red light emitted from a krypton-86 lamp. In October 1983, this definition was abandoned also and **the meter was redefined as the distance traveled by light in vacuum during a time interval**

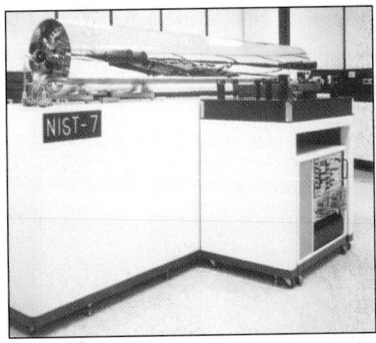

The primary frequency standard (an atomic clock) at the National Institute of Standards and Technology. This device keeps time with an accuracy of about one millionth of a second per year. *(Courtesy of National Institute of Standards and Technology, U.S. Department of Commerce)*

WEB

For Web links related to standards of length, mass, and time, visit the textbook Web Site at http://www.harcourtcollege.com/physics/cptech

◀ Definition of the meter

of 1/299 792 458 second. This latest definition establishes that the speed of light is 299 792 458 meters per second.

Mass

Definition of the kilogram ▶

Magnified view of the gears in a Swiss watch. *(Dale Boyer/Tony Stone Images)*

Definition of the second ▶

The SI unit of mass, the kilogram, is defined as the mass of a specific platinum-iridium alloy cylinder kept at the International Bureau of Weights and Measures at Sèvres, France. As we shall see in Chapter 4, mass is a quantity used to measure the resistance to a change in state of motion of an object. It is more difficult to cause a change in the state of motion of an object with a large mass than an object with a small mass.

Time

Before 1960, the time standard was defined in terms of the average length of a solar day in the year 1900. (A solar day is the time interval between successive appearances of the Sun at the highest point it reaches in the sky each day.) The basic unit of time, the second, was defined to be $(1/60)(1/60)(1/24) = 1/86\,400$ of the average solar day. In 1967, the second was redefined to take advantage of the high precision attainable with a device known as an atomic clock, which uses the characteristic frequency of the light emitted from the cesium-133 atom as its "reference clock." **The second is now defined as 9 192 631 700 times the period of oscillation of radiation from the cesium atom.**

Approximate Values for Length, Mass, and Time Intervals

Approximate values of some lengths, masses, and time intervals are presented in Tables 1.1, 1.2, and 1.3, respectively. Note the wide ranges of values. Study these tables and get a feel for what is meant by a kilogram of mass (this book has a mass

TABLE 1.2
Approximate Values of Some Masses

	Mass (kg)
Universe	10^{52}
Milky Way galaxy	7×10^{41}
Sun	2×10^{30}
Earth	6×10^{24}
Moon	7×10^{22}
Shark	1×10^{2}
Human	7×10^{1}
Frog	1×10^{-1}
Mosquito	1×10^{-5}
Bacterium	1×10^{-15}
Hydrogen atom	2×10^{-27}
Electron	9×10^{-31}

TABLE 1.1   Approximate Values of Some Measured Lengths

	Length (m)
Distance from Earth to most remote known quasar	1×10^{26}
Distance from Earth to most remote known normal galaxies	4×10^{25}
Distance from Earth to nearest large galaxy (M31 in Andromeda)	2×10^{22}
Distance from Earth to nearest star (Proxima Centauri)	4×10^{16}
One lightyear	9×10^{15}
Mean orbit radius of the Earth about the Sun	2×10^{11}
Mean distance from Earth to Moon	4×10^{8}
Mean radius of the Earth	6×10^{6}
Typical altitude of a satellite orbiting Earth	2×10^{5}
Length of a football field	9×10^{1}
Length of a housefly	5×10^{-3}
Size of smallest dust particles	1×10^{-4}
Size of cells of most living organisms	1×10^{-5}
Diameter of a hydrogen atom	1×10^{-10}
Diameter of an atomic nucleus	1×10^{-14}
Diameter of a proton	1×10^{-15}

TABLE 1.3 Approximate Values of Some Time Intervals

	Time Interval (s)
Age of the Universe	5×10^{17}
Age of the Earth	1×10^{17}
Average age of a college student	6×10^{8}
One year	3×10^{7}
One day (time required for one revolution of Earth about its axis)	9×10^{4}
Time between normal heartbeats	8×10^{-1}
Period[a] of audible sound waves	1×10^{-3}
Period of typical radio waves	1×10^{-6}
Period of vibration of an atom in a solid	1×10^{-13}
Period of visible light waves	2×10^{-15}
Duration of a nuclear collision	1×10^{-22}
Time required for light to cross a proton	3×10^{-24}

[a] A *period* is defined as the time required for one complete vibration.

TABLE 1.4
Some Prefixes for Powers of Ten Used with Metric Units

Power	Prefix	Abbreviation
10^{-18}	atto-	a
10^{-15}	femto-	f
10^{-12}	pico-	p
10^{-9}	nano-	n
10^{-6}	micro-	μ
10^{-3}	milli-	m
10^{-2}	centi-	c
10^{-1}	deci-	d
10^{1}	deka-	da
10^{3}	kilo-	k
10^{6}	mega-	M
10^{9}	giga-	G
10^{12}	tera-	T
10^{15}	peta-	P
10^{18}	exa-	E

of about 2 kilograms) or a time interval of 10^{10} seconds (one year is about 3×10^{7} seconds). Study Appendix A if you need to learn or review the notation for powers of 10, such as the expression of the number 50 000 in the form 5×10^{4}.

Systems of units commonly used are the SI system, in which the units of length, mass, and time are the meter (m), kilogram (kg), and second (s), respectively; the cgs or Gaussian system, in which the units of length, mass, and time are the centimeter (cm), gram (g), and second, respectively; and the British engineering system (sometimes called the conventional system), in which the units of length, mass, and time are the foot (ft), slug, and second, respectively. Throughout most of this book we shall use SI units, because they are almost universally accepted in science and industry. We shall make limited use of British engineering units in the study of mechanics.

Some of the most frequently used metric prefixes representing powers of 10 and their abbreviations are listed in Table 1.4. For example, 10^{-3} m is equivalent to 1 millimeter (mm), and 10^{3} m is 1 kilometer (km). Likewise, 1 kg is equal to 10^{3} g, and 1 megavolt (MV) is 10^{6} volts (V).

1.2 THE BUILDING BLOCKS OF MATTER

A 1-kg cube of solid gold has a length of about 3.73 cm on a side. Is this cube nothing but wall-to-wall gold, with no empty space? If the cube is cut in half, the two resulting pieces still retain their chemical identity as solid gold. But what if the pieces of the cube are cut again and again, indefinitely? Will the smaller and smaller pieces always be the same substance, gold? Questions such as these can be traced back to early Greek philosophers. Two of them—Leucippus and Democritus— could not accept the idea that such cutting could go on forever. They speculated that the process ultimately must end when it produces a particle that can no longer be cut. In Greek, *atomos* means "not sliceable." From this comes our English word *atom*, once believed to be the smallest, ultimate particle of matter. Elementary-particle physicists still engage in speculation and experimentation concerning the ultimate building blocks of matter.

Let us review briefly what is known about the ultimate structure of the world around us. It is useful to view the atom as a miniature Solar System with a dense, positively charged nucleus occupying the position of the Sun and negatively charged electrons orbiting like the planets. This model of the atom enables us to understand certain properties of the simpler atoms, such as hydrogen, but fails to explain many fine details of atomic structure.

Following the discovery of the nucleus in the early 1900s, the question arose: Does it have structure? That is, is the nucleus a single particle or a collection of particles? The exact composition of the nucleus has not been defined completely even today, but by the early 1930s a model evolved that helps us understand how the nucleus behaves. Scientists determined that occupying the nucleus are two basic entities, protons and neutrons. The *proton* is nature's fundamental carrier of positive charge, and the number of protons in a nucleus determines what element the material is. For instance, one proton in the nucleus means that the atom is an atom of hydrogen, regardless of how many neutrons may be present; two protons mean an atom of helium.

The existence of *neutrons* was verified conclusively in 1932. A neutron has no charge and a mass about equal to that of a proton. One of its primary purposes is to act as a "glue" to hold the nucleus together. If neutrons were not present, the repulsive electrical force between the positively charged particles would cause the nucleus to fly apart.

But is this where the breaking down stops? As we shall explore more carefully in Chapter 30, even more elementary building blocks than protons and neutrons exist. Protons, neutrons, and a zoo of other exotic particles are now thought to be composed of six particles called **quarks,** which have been given the names *up, down, strange, charmed, bottom,* and *top* (Fig. 1.1). The up, charmed, and top quarks have charges of $+2/3$ that of the proton, whereas the down, strange, and bottom quarks have charges of $-1/3$ that of the proton. The proton consists of two up quarks and one down quark, which one can easily show leads to the correct charge for the proton. Likewise, the neutron is composed of two down quarks and one up quark, giving a net charge of zero.

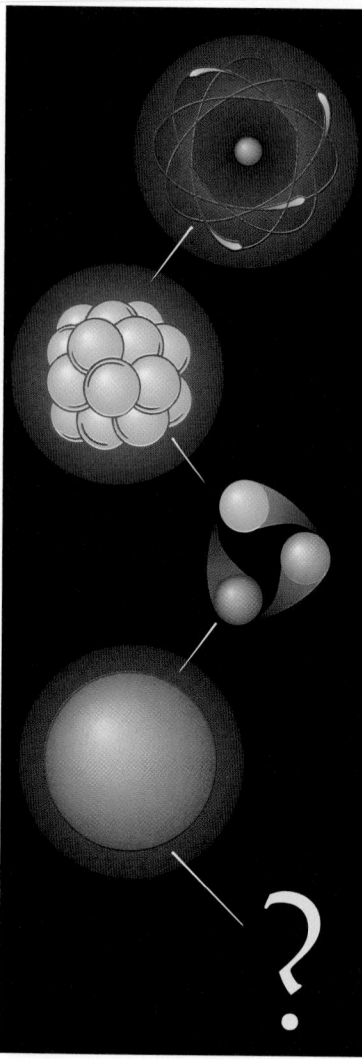

Figure 1.1 Distances at the frontier of nuclear physics are astonishingly short. An atom is so small that a single-file line of 250 000 of them would fit within the thickness of aluminum foil. The nucleus at the atom's center is a cluster of nucleons, each 100 000 times smaller than the atom itself. The three quarks inside each nucleon are smaller still. *(Courtesy of SURA, Inc.)*

1.3   DIMENSIONAL ANALYSIS

The word *dimension* has a special meaning in physics. It usually denotes the physical nature of a quantity. Whether the separation between two points is measured in units of feet or meters or furlongs, it is a distance. We say its dimension is *length*.

The symbols that will be used in this section to specify length, mass, and time are L, M, and T, respectively. We shall often use brackets [ ] to denote the dimensions of a physical quantity. For example, in this notation the dimensions of velocity, v, are written $[v] = L/T$, and the dimensions of area, A, are $[A] = L^2$. The dimensions of area, volume, velocity, and acceleration are listed in Table 1.5, along with their units in the three common systems. The dimensions of other quantities, such as force and energy, will be described later as they are introduced.

In many situations, you may have to derive or check a specific formula. Although you may have forgotten the details of the derivation, you can use a powerful procedure called **dimensional analysis.** Dimensional analysis makes use of the fact

TABLE 1.5 Dimensions and Some Units of Area,
                     Volume, Velocity, and Acceleration

System	Area (L^2)	Volume (L^3)	Velocity (L/T)	Acceleration (L/T^2)
SI	m^2	m^3	m/s	m/s^2
cgs	cm^2	cm^3	cm/s	cm/s^2
British engineering (conventional)	ft^2	ft^3	ft/s	ft/s^2

that **dimensions can be treated as algebraic quantities:** That is, quantities can be added or subtracted only if they have the same dimensions. Furthermore, the terms on both sides of an equation must have the same dimensions. By following these simple rules, you can use dimensional analysis to help determine whether or not an expression has the correct form. The relationship can be correct only if the dimensions on both sides of the equation are the same.

To illustrate this procedure, suppose you wish to derive a formula for the distance x traveled by a car in a time t if the car starts from rest and moves with constant acceleration a. In Chapter 2 we shall find that the correct expression for this special case is $x = \frac{1}{2}at^2$. Let us check the validity of this expression from a dimensional analysis approach.

The quantity x on the left side has the dimension length. In order for the equation to be dimensionally correct, the quantity on the right side must also have the dimension length. We can perform a dimensional check by substituting the basic units for acceleration, L/T^2, and time, T, into the equation. That is, the dimensional form of the equation $x = \frac{1}{2}at^2$ can be written

$$L = \frac{L}{T^2} \cdot T^2 = L$$

The units of time cancel as shown, leaving the unit of length. Note that the factor $\frac{1}{2}$ was ignored because it has no dimensions. Dimensional analysis cannot give us numerical factors; this is its characteristic limitation.

EXAMPLE 1.1 Analysis of an Equation

Show that the expression $v = v_0 + at$ is dimensionally correct, where v and v_0 represent velocities, a is acceleration, and t is a time interval.

Solution Because

$$[v] = [v_0] = \frac{L}{T}$$

and the dimensions of acceleration are L/T^2, the dimensions of at are

$$[at] = \frac{L}{T^2}(T) = \frac{L}{T}$$

and the expression is dimensionally correct. However, if the expression were given as $v = v_0 + at^2$, it would be dimensionally *incorrect*. Try it and see!

1.4   SIGNIFICANT FIGURES

When measurements are performed on certain quantities, the measured values are known only to within the limits of the experimental uncertainty. The value of the uncertainty can depend on factors such as the quality of the apparatus, the skill of the experimenter, and the number of measurements performed.

Suppose that in a laboratory experiment we are asked to measure the area of a rectangular plate with a meter stick. Let us assume that the accuracy to which we can measure a particular dimension of the plate is ± 0.1 cm. If the length of the plate is measured to be 16.3 cm, we can claim only that its length lies somewhere between 16.2 cm and 16.4 cm. In this case, we say that the measured value has three significant figures. Likewise, if the plate's width is measured to be 4.5 cm, the actual value lies between 4.4 cm and 4.6 cm. This measured value has only two significant figures. Note that the significant figures include the first estimated digit. Thus, we could write the measured values as 16.3 ± 0.1 cm and 4.5 ± 0.1 cm.

Suppose now that we would like to find the area of the plate by multiplying the two measured values together. If we were to claim that the area is $(16.3 \text{ cm})(4.5 \text{ cm}) = 73.35 \text{ cm}^2$, our answer would be unjustifiable, because it contains four significant figures, which is greater than the number of significant figures in either of the measured lengths. A good rule of thumb for determining the number of significant figures that can be claimed is as follows. **When multiplying several quantities, the number of significant figures in the final answer is the same as the number of significant figures in the *least accurate* of the quantities being multiplied, where *least accurate* means *having the lowest number of significant figures*. The same rule applies to division.**

Applying this rule to the preceding multiplication example, we see that the answer for the area can have only two significant figures because the dimension 4.5 cm has only two significant figures. Thus, we can only claim the area to be 73 cm^2, realizing that the value can range between $(16.2 \text{ cm})(4.4 \text{ cm}) = 71 \text{ cm}^2$ and $(16.4 \text{ cm})(4.6 \text{ cm}) = 75 \text{ cm}^2$.

Zeros may or may not be significant figures. Those used to position the decimal point in such numbers as 0.03 and 0.0075 are not significant. Thus there are one and two significant figures, respectively, in these two values. When the positioning of zeros comes after other digits, however, there is the possibility of misinterpretation. For example, suppose the mass of an object is given as 1500 g. This value is ambiguous because we do not know whether the last two zeros are being used to locate the decimal point or whether they represent significant figures in the measurement. In order to remove this ambiguity, it is common to use scientific notation to indicate the number of significant figures. In this case, we would express the mass as 1.5×10^3 g if there are two significant figures in the measured value, 1.50×10^3 g if there are three significant figures, and 1.500×10^3 g if there are four. Likewise, 0.000 15 should be expressed in scientific notation as 1.5×10^{-4} if it has two significant figures or as 1.50×10^{-4} if it has three significant figures. The three zeros between the decimal point and the digit 1 in the number 0.000 15 are not counted as significant figures because they are present only to locate the decimal point. In general, a **significant figure** is a reliably known digit (other than a zero used to locate the decimal point).

For addition and subtraction, the number of decimal places must be considered when you are determining how many significant figures to report. **When numbers**

This sundial is located in Majorca, Spain. The principle of the sundial is as follows. If the number 12 points to true north, the shadow gives the correct standard time. *(David R. Frazier/Photo Researchers, Inc.)*

are added or subtracted, the number of decimal places in the result should equal the smallest number of decimal places of any term in the sum. For example, if we wish to compute 123 + 5.35, the answer is 128 and not 128.35. If we compute the sum 1.0001 + 0.0003 = 1.0004, the result has the correct number of decimal places; consequently it has five significant figures even though one of the terms in the sum, 0.0003, has only one significant figure. Likewise, if we perform the subtraction 1.002 − 0.998 = 0.004, the result has only one significant figure even though one term has four significant figures and the other has three. In this book, **most of the numerical examples and end-of-chapter problems will yield answers having either two or three significant figures.**

EXAMPLE 1.2 Installing a Carpet

A carpet is to be installed in a room whose length is measured to be 12.71 m (four significant figures) and whose width is measured to be 3.46 m (three significant figures). Find the area of the room.

Solution If you multiply 12.71 m by 3.46 m on your calculator, you will get the answer 43.9766 m^2. How many of these numbers should you claim? Our rule of thumb for multiplication says that you can claim only the number of significant figures in the least accurate of the quantities being measured. In this example, we have three significant figures in our less accurate measurement, so we should express our final answer as 44.0 m^2. Note that in the answer given, we used a general rule for rounding off numbers, which states that the last digit retained is to be increased by 1 if the first digit dropped was equal to 5 or greater. Furthermore, if the last digit is 5, the result should be rounded to the nearest even number. (This helps avoid accumulation of errors.)

1.5 CONVERSION OF UNITS

Sometimes it is necessary to convert units from one system to another. Conversion factors between the SI and conventional systems for units of length are as follows:

$$1 \text{ mile} = 1609 \text{ m} = 1.609 \text{ km} \qquad 1 \text{ ft} = 0.3048 \text{ m} = 30.48 \text{ cm}$$

$$1 \text{ m} = 39.37 \text{ in.} = 3.281 \text{ ft} \qquad 1 \text{ in.} = 0.0254 \text{ m} = 2.54 \text{ cm}$$

A more extensive list of conversion factors can be found on the inside of the front cover.

Units can be treated as algebraic quantities that can cancel each other. For example, suppose we wish to convert 15.0 in. to centimeters. Because 1 in. = 2.54 cm, we find that

$$15.0 \text{ in.} = 15.0 \text{ in.} \times 2.54 \frac{\text{cm}}{\text{in.}} = 38.1 \text{ cm}$$

This road sign near Raleigh, North Carolina, shows distances in miles and kilometers. How accurate are the conversions? *(Billy E. Barnes/Stock Boston)*

EXAMPLE 1.3 Pull Over, Buddy!

If a car is traveling at a speed of 28.0 m/s, is it exceeding the speed limit of 55.0 mi/h?

Solution We can perform the conversion from meters per second to miles per hour in two steps—meters to miles and then seconds to hours. Converting the length units, we find that

$$28.0 \text{ m/s} = \left(28.0\, \frac{\text{m}}{\text{s}} \right) \left(\frac{1 \text{ mi}}{1609 \text{ m}} \right) = 1.74 \times 10^{-2} \text{ mi/s}$$

Now we finish by converting seconds to hours, as follows:

$$1.74 \times 10^{-2} \text{ mi/s} = \left(1.74 \times 10^{-2}\, \frac{\text{mi}}{\text{s}} \right) \left(60\, \frac{\text{s}}{\text{min}} \right) \left(60\, \frac{\text{min}}{\text{h}} \right) = 62.6 \text{ mi/h}$$

The car should slow down because it is exceeding the speed limit.

An alternative approach is to use the single conversion relationship 1 m/s = 2.237 mi/h:

$$28.0 \text{ m/s} = \left(28.0\, \frac{\text{m}}{\text{s}} \right) \left(\frac{2.237 \text{ mi/h}}{1 \text{ m/s}} \right) = 62.6 \text{ mi/h}$$

Exercise Use the speedometer pictured in Figure 1.2 to see whether the foregoing result is in agreement with the speedometer dial (calibrated in miles per hour and kilometers per hour).

Figure 1.2 (Example 1.3) This car speedometer gives readings in both miles per hour and kilometers per hour. See if you can confirm the conversions for a few readings on the dial. *(Paul Silverman, Fundamental Photographs)*

1.6   ORDER-OF-MAGNITUDE CALCULATIONS

Often it is useful to estimate an answer to a problem in which little information is given. This answer can then be used to determine whether or not a more precise calculation is necessary. Such an approximation is usually based on certain assumptions, which must be modified if greater precision is needed. Sometimes it is necessary to know a quantity only within a factor of 10. In such a case we refer to the **order of magnitude** of the quantity, by which we mean the power of ten that is closest to the actual value of the quantity. For example, the mass of a person might be 75 kg ≈ 10^2 kg. We would say that the person's mass is *on the order of* 10^2 kg. Usually, when an order-of-magnitude calculation is made, the results are reliable to within a factor of 10. If a quantity increases in value by three orders of magnitude, this means that its value increases by a factor of $10^3 = 1000$.

EXAMPLE 1.4   French Fries to the Moon?

McDonald's sells about 250 million packages of French fries per year. If these fries were placed end to end, how far would they reach?

Solution   This is a relatively simple estimation problem to give you some idea of how to approach more complicated problems. First, we must guess the length of an average French fry. A reasonable guess is three inches. Of course, some are longer and some are shorter, but three inches seems to be a reasonable estimate of the length of a typical fry. Next, we estimate that the number of fries in a typical package is 30. Multiplying the length of an average fry by the total number of fries gives us

$$(30 \text{ fries/package}) \ (250 \times 10^6 \text{ packages}) \ (3 \text{ in./fry}) \approx 2 \times 10^{10} \text{ in.}$$

which is about 5×10^8 m. Because the Earth-Moon distance is about 4×10^8 m, the chain of French fries would reach the Moon. Our estimate of 5×10^8 m should be accurate to within a factor of ten. That is, our stretched-out length of fries will be longer than about 5×10^7 m but shorter than 5×10^9 m.

EXAMPLE 1.5   How Much Gasoline Do We Use?

Estimate the number of gallons of gasoline used by all cars in the United States each year.

Solution   Because there are about 240 million people in the United States, an estimate of the number of cars in the country is 60 million (assuming one car and four people per family). We also estimate that the average distance traveled per car per year is 10 000 miles. If we assume a gas mileage of 20 mi/gal, corresponding to a gasoline consumption of 0.05 gal/mi, each car uses about 500 gal/year. Multiplying this by the total number of cars in the United States gives an estimated total consumption of 3×10^{10} gal. This corresponds to a yearly consumer expenditure of more than $30 billion! This is probably a low estimate because we have not accounted for commercial consumption and families with two or more vehicles.

EXAMPLE 1.6   The High Price of Gold

All the gold that has ever been mined could be melted and molded into a cube approximately 20 m on a side. Use the fact that 1 kg of gold occupies a cube 3.73 cm on a side and the fact that gold sells for about $12 000 per kg to estimate the value of all the gold existing on this planet.

Solution   The volume of a block 20 m on a side is $(20 \text{ m})(20 \text{ m})(20 \text{ m}) = 8\,000 \text{ m}^3$, and the volume of 1 kg of gold is

$$(3.73 \times 10^{-2} \text{ m})(3.73 \times 10^{-2} \text{ m})(3.73 \times 10^{-2} \text{ m}) \approx 5 \times 10^{-5} \text{ m}^3$$

Thus, our 8 000 m³ block of gold would have a mass of

$$\frac{8\,000 \text{ m}^3}{5 \times 10^{-5} \text{ m}^3/\text{kg}} = 1.5 \times 10^8 \text{ kg}$$

At \$12 000 per kg, this mass of gold would cost about 2×10^{12}.

1.7   MATHEMATICAL NOTATION

Many mathematical symbols will be used throughout this book. You are no doubt familiar with some, such as the symbol = to denote the equality of two quantities.

The symbol $\propto$ denotes a proportionality. For example, $y \propto x^2$ means that y is proportional to the square of x.

The symbol $<$ means *is less than,* and $>$ means *is greater than.* For example, $x > y$ means x is greater than y.

The symbol $\ll$ means *is much less than,* and $\gg$ means *is much greater than.*

The symbol $\approx$ indicates that two quantities are *approximately equal* to each other.

The symbol $\equiv$ means *is defined as.* This is a stronger statement than a simple =.

It is convenient to use the notation Δx (read as "delta x") to indicate the *change in the quantity* x. (Note that Δx does not mean "the product of Δ and x.") For example, suppose that a person out for a morning stroll starts measuring her distance away from home when she is 10 m from her doorway. The person then moves along a straight-line path and stops strolling 50 m from the door. Her change in position during the walk is $\Delta x = 50 \text{ m} - 10 \text{ m} = 40 \text{ m}$ or, in symbolic form,

$$\Delta x = x_f - x_i$$

In this equation x_f is the *final position* and x_i is the *initial position.*

We shall often have occasion to add several quantities. A useful abbreviation for representing such a sum is the Greek letter Σ (capital sigma). Suppose we wish to add a set of five numbers represented by $x_1, x_2, x_3, x_4,$ and x_5. In the abbreviated notation, we would write the sum as

$$x_1 + x_2 + x_3 + x_4 + x_5 = \sum_{i=1}^{5} x_i$$

where the subscript i on x represents any one of the numbers in the set. For example, if there are five masses in a system, $m_1, m_2, m_{3,}, m_4,$ and m_5, the total mass of the system $M = m_1 + m_2 + m_3 + m_4 + m_5$ could be expressed as

$$M = \sum_{i=1}^{5} m_i$$

Finally, the magnitude of a quantity x, written $|x|$, is simply the absolute value of that quantity. The sign of $|x|$ is always positive, regardless of the sign of x. For example, if $x = -5$, $|x| = 5$; if $x = 8$, $|x| = 8$.

2.2, SECTION 1

1.8 COORDINATE SYSTEMS AND FRAMES OF REFERENCE

Many aspects of physics deal in some way with locations in space. For example, the mathematical description of the motion of an object requires a method for describing the position of the object. Thus, it is fitting that we first discuss how to describe the position of a point in space. This is done by means of coordinates. A point on a line can be located with one coordinate; a point in a plane is located with two coordinates; and three coordinates are required to locate a point in space.

A coordinate system used to specify locations in space consists of

- A fixed reference point O, called the *origin*
- A set of specified axes or directions with an appropriate scale and labels on the axes
- Instructions that tell us how to label a point in space relative to the origin and axes

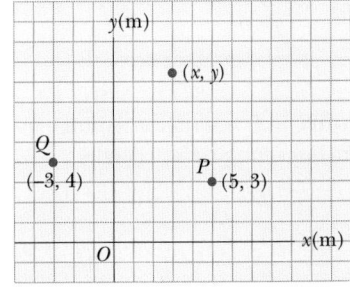

Figure 1.3 Designation of points in a two-dimensional Cartesian coordinate system. Every point is labeled with coordinates (x, y).

One convenient coordinate system that we shall use frequently is the **Cartesian coordinate system,** sometimes called the **rectangular coordinate system.** Such a system in two dimensions is illustrated in Figure 1.3. An arbitrary point in this system is labeled with the coordinates (x, y). For example, the point P in the figure has coordinates $(5, 3)$. This notation means that if we start at the origin (O), we can reach P by moving 5 meters to the right of the origin and then 3 meters above the origin. (Positive x is usually selected to the right of the origin and positive y upward from the origin. Negative x is usually to the left of the origin and negative y downward from the origin. This convention is not an absolute necessity, however. There may be instances in which you would like to take positive x to the left of the origin and negative x to the right. Feel free to do so.) In the same way, the point Q has coordinates $(-3, 4)$, which corresponds to going 3 meters to the left of the origin and 4 meters above the origin.

Sometimes it is more convenient to locate a point in space by its **plane polar coordinates** (r, θ), as in Figure 1.4. In this coordinate system, an origin and a reference line are selected, as shown. A point is then specified by the distance r from the origin to the point and by the angle θ between a line from the origin to the point and the reference line. (Frequently the reference line is selected to be the positive x axis of a Cartesian coordinate system, although this is not necessary.) The angle θ is considered positive when measured counterclockwise from the reference line and negative when measured clockwise. For example, if a point is specified by the polar coordinates 3 m and 60°, one locates this point by moving out 3 m from the origin at an angle of 60° above (or counterclockwise from) the reference line. A point specified by polar coordinates 3 m and −60° is located 3 m out from the origin and 60° below (or clockwise from) the reference line.

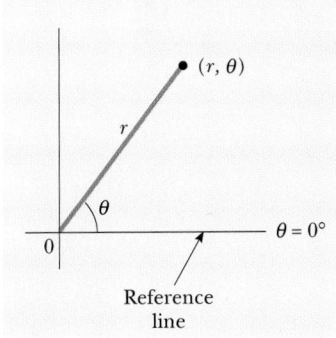

Figure 1.4 A polar coordinate system.

1.9 TRIGONOMETRY

The portion of mathematics that is based on the special properties of a right triangle is called trigonometry. Many of the concepts of this branch of mathematics are of utmost importance in the study of physics. We will now review

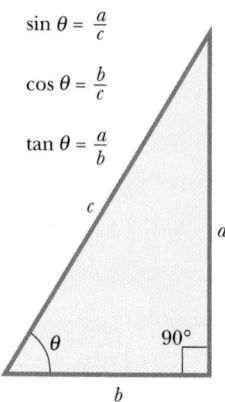

$$\sin \theta = \frac{a}{c}$$

$$\cos \theta = \frac{b}{c}$$

$$\tan \theta = \frac{a}{b}$$

Figure 1.5 Certain trigonometric functions for a right triangle.

some of the more inherent fundamental definitions and ideas that you will need to know.

Consider the right triangle shown in Figure 1.5, where side a is opposite the angle θ, side b is adjacent to the angle θ, and side c is the hypotenuse of the triangle. The basic trigonometric functions defined by such a triangle are the ratios of the lengths of certain sides of the triangle. Specifically, the most important relationships are called the sine (sin), cosine (cos), and tangent (tan) functions. In terms of the angle θ, these functions are[1]

$$\sin \theta = \frac{\text{side opposite } \theta}{\text{hypotenuse}} = \frac{a}{c}$$

$$\cos \theta = \frac{\text{side adjacent to } \theta}{\text{hypotenuse}} = \frac{b}{c} \qquad \text{[1.1]}$$

$$\tan \theta = \frac{\text{side opposite } \theta}{\text{side adjacent to } \theta} = \frac{a}{b}$$

For example, if the angle θ is equal to 30°, it is found that the ratio of a to c is always 0.50; that is, sin 30° = 0.50. Note that the sin, cos, and tan are quantities without units, because each represents the ratio of two lengths.

Another important relationship, called the **Pythagorean theorem,** exists between the lengths of the sides of a right triangle; it is

$$c^2 = a^2 + b^2 \qquad \text{[1.2]}$$

Finally, it will often be necessary to find the value of **inverse** relationships. For example, suppose you know the value of the sine of an angle, but you need to know the value of the angle itself. The inverse sine function may be expressed as $\sin^{-1}(0.866)$, which is a shorthand way of asking the question, "What angle has a sine of 0.866?" Punching a couple of buttons on your calculator reveals that this angle is 60°. Try it for yourself and show that $\tan^{-1}(0.400) = 21.8°$.

The following examples will give you a little practice in working with these trigonometric principles in practical situations.

EXAMPLE 1.7 How High Is the Building?

A person attempts to measure the height of a building by walking out a distance of 46.0 m from its base and shining a flashlight beam toward its top. He finds that when the beam is elevated at an angle of 39.0° with respect to the horizontal, as shown in Figure 1.6, the beam just strikes the top of the building. Find the height of the building and the distance the flashlight beam has to travel before it strikes the top of the building.

---

[1] In order to recall these trigonometric relationships, consider using the following memory device. Recall the mnemonic *SOHCAHTOA* whose letters stand for *S*ine = *O*pposite/*H*ypotenuse, *C*osine = *A*djacent/*H*ypotenuse, and *T*angent = *O*pposite/*A*djacent. (Thanks go to Professor Don Chodrow for pointing this out.)

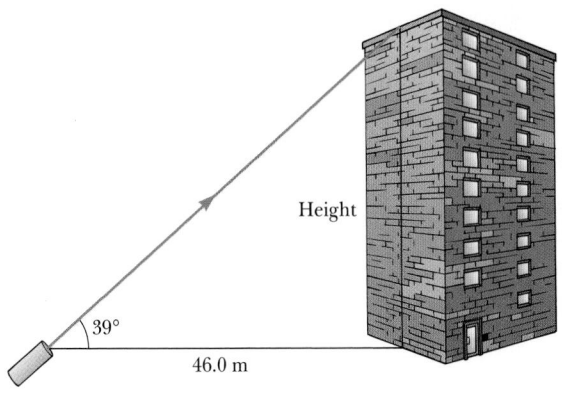

Figure 1.6 (Example 1.7)

Height

39°

46.0 m

Solution   We know the length of the adjacent side of the triangle shown in the figure, and we know the value of the angle. Thus, we can use the definition of the tangent function to find the height of the building, which is the opposite side of the triangle.

$$\tan 39.0° = \frac{\text{height of building}}{46.0 \text{ m}}$$

or

$$\text{Height of building} = (\tan 39.0°)(46.0 \text{ m}) = (0.810)(46.0 \text{ m}) = \boxed{37.3 \text{ m}}$$

Now that we know the lengths of both the adjacent side and the opposite side of the triangle, we can find the length of the hypotenuse, which is the distance, c, that the beam travels before it strikes the top of the building.

$$c = \sqrt{a^2 + b^2} = \sqrt{(37.3 \text{ m})^2 + (46.0 \text{ m})^2} = \boxed{59.2 \text{ m}}$$

Exercise   Try using a different method to find the distance traveled by the light. Would another trigonometric function, such as the sin or cos, give you the answer? Try it.

WEB

For Web links to biographical and other information on Pythagoras, visit the textbook Web site at **http://www.harcourtcollege.com/physics/cptech**

EXAMPLE 1.8   Things to Do When the CB Goes Bad

A truck driver moves up a straight mountain highway, as shown in Figure 1.7. Elevation markers at the beginning and ending points of the trip show that he has risen vertically 0.530 km, and the mileage indicator on the truck shows that he has traveled a total distance of 3.00 km during the ascent. Find the angle of incline of the hill.

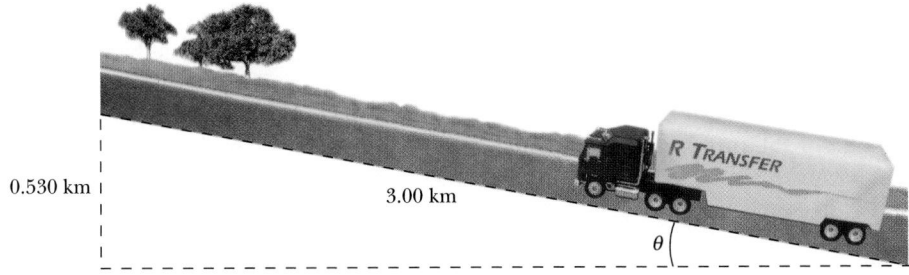

0.530 km

3.00 km

R TRANSFER

θ

Figure 1.7 (Example 1.8)

Solution   This is an example in which we must use an inverse trigonometric relationship. From the sine of the angle, defined as the opposite side over the hypotenuse, we find

$$\sin \theta = \frac{0.530}{3.00} = 0.177$$

To find θ, we use the inverse sine relationship:

$$\theta = \sin^{-1}(0.177) = \boxed{10.2°}$$

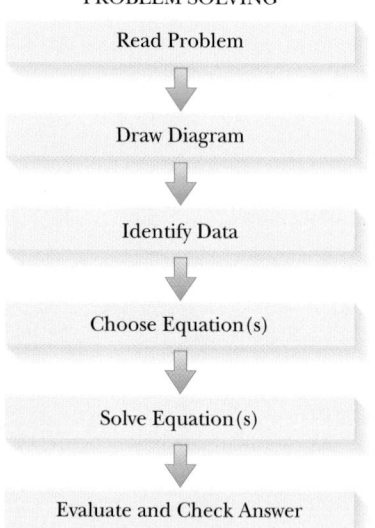

Complete all sections of module 1.

A GUIDE FOR
PROBLEM SOLVING

Read Problem

Draw Diagram

Identify Data

Choose Equation(s)

Solve Equation(s)

Evaluate and Check Answer

1.10   PROBLEM-SOLVING STRATEGY

Most courses in general physics require the student to learn the skills of problem solving, and examinations usually include problems that test such skills. This brief section presents some useful suggestions that will help increase your accuracy in solving problems, enhance your understanding of physical concepts, eliminate initial panic or lack of direction in approaching a problem, and organize your work. One way to help accomplish these goals is to adopt a problem-solving strategy. Many chapters in this book will include a section labeled "Problem-Solving Strategy" that should help you through the rough spots.

The following steps are commonly used to develop a problem-solving strategy:

1. Read the problem carefully at least twice. Be sure you understand the nature of the problem before proceeding further.
2. Draw a suitable diagram with appropriate labels and coordinate axes, if needed.
3. Imagine a movie, running in your mind, of what happens in the problem.
4. As you examine what is being asked in the problem, identify the basic physical principle or principles that are involved, listing the knowns and unknowns.
5. Select a basic relationship or derive an equation that can be used to find the unknown, and symbolically solve the equation for the unknown.
6. Substitute the given values with the appropriate units into the equation.
7. Obtain a numerical value with units for the unknown. You can have confidence in your result if the following questions can be properly answered: Do the units match? Is the answer reasonable? Is the plus or minus sign proper or meaningful?

One of the purposes of this strategy is to promote accuracy. Properly drawn diagrams can eliminate many sign errors. Diagrams also help to isolate the physical principles of the problem. Symbolic solutions and carefully labeled knowns and unknowns will help eliminate other careless errors. The use of symbolic solutions should help you think in terms of the physics of the problem. A check of units at the end of the problem can indicate a possible algebraic error. The physical layout and organization of your problem will make the final product more understandable and easier to follow. Once you have developed an organized system for examining problems and extracting relevant information, you will become a more confident problem solver.

The following example illustrates this procedure.

EXAMPLE 1.9 A Round Trip by Air

An airplane travels 450 km due east and then travels an unknown distance due north. Finally, it returns to its starting point by traveling a distance of 525 km. How far did the airplane travel in the northerly direction?

Solution After reading the problem carefully, draw a diagram of the situation, as in Figure 1.8. Because the path of the airplane forms a right triangle, you can use the Pythagorean theorem, which relates the three sides of the triangle:

$$c^2 = a^2 + b^2$$

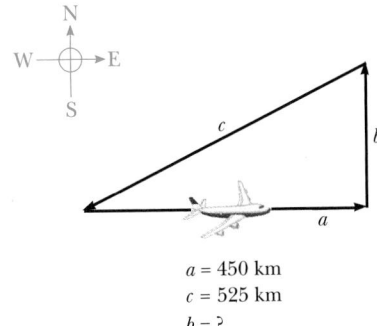

$a = 450$ km
$c = 525$ km
$b = ?$

Figure 1.8 (Example 1.9)

In this problem, the hypotenuse c and the side a are known quantities. We must solve for the unknown quantity, b. Subtracting a^2 from both sides of the preceding equation gives

$$b^2 = c^2 - a^2$$

We now take the square root of both sides to find the solution for b:

$$b = \sqrt{c^2 - a^2}$$

Note that the negative solution has been disregarded because it is not physically meaningful.

Finally, we substitute the given values for the quantities c and a to find a numerical value for b:

$$b = \sqrt{(525 \text{ km})^2 - (450 \text{ km})^2} = \boxed{270 \text{ km}}$$

SUMMARY

The physical quantities we shall encounter in our study of mechanics can be expressed in terms of three fundamental quantities, length, mass, and time—which have the units meters (m), kilograms (kg), and seconds (s), respectively, in the SI system. It is often helpful to use dimensional analysis to check equations and to assist in deriving equations.

When inserting numerical factors into equations, you must be sure that the dimensions of these factors are consistent throughout the equation. In many cases it will be necessary to use the table of conversion factors on the inside front cover to convert from one system of units to another.

Often it is useful to estimate an answer to a problem in which little information is given. Such estimates are called **order-of-magnitude calculations.**

The three most basic trigonometric functions for a right triangle are the sine, cosine, and tangent, defined as

$$\sin \theta = \frac{\text{side opposite } \theta}{\text{hypotenuse}}$$

$$\cos \theta = \frac{\text{side adjacent to } \theta}{\text{hypotenuse}}$$

$$\tan \theta = \frac{\text{side opposite } \theta}{\text{side adjacent to } \theta}$$

The **Pythagorean theorem** is an important relationship between the lengths of the sides of a right triangle:

$$c^2 = a^2 + b^2$$

where c is the hypotenuse of the triangle and a and b are its other two sides.

MULTIPLE-CHOICE QUESTIONS

1. A fathom is a unit of length about 6 ft long commonly used in measuring depths of water. Because a fathom is about the length of two outspread arms, it should not be surprising that it derives its name from the Viking word for *embrace*. Assuming that the distance to the Moon is about 4×10^8 m, the distance to the Moon in fathoms is about
   (a) 4×10^8 fathoms (b) 2×10^8 fathoms
   (c) 3×10^8 fathoms (d) 8×10^8 fathoms

2. Suppose that two quantities, A and B, have different dimensions. Determine which of the following arithmetic operations *could* be physically meaningful.
   (a) $A + B$ (b) $B - A$ (c) $A - B$ (d) A/B

3. Estimate the number of breaths taken during an average life span of 70 years. The answer is approximately
   (a) 3×10^7 breaths (b) 3×10^8 breaths
   (c) 3×10^9 breaths

4. Which of the following relationships is dimensionally consistent with an expression yielding a value for acceleration? Acceleration has the units of distance divided by time squared. In these equations, x is a distance, t is time, and v is velocity with units of distance divided by time.
   (a) v/t^2 (b) v/x^2 (c) v^2/t (d) v^2/x

5. Use the rules for using significant figures to find the answer to the following addition problem:
   $21.4 + 15 + 17.17 + 4.003$
   (a) 57.573 (b) 57.57 (c) 57.6 (d) 58

6. A sphere has a surface area of 100 m². A second sphere has a radius twice that of the first sphere. What is the surface area of the second sphere? (*Hint:* You do not need to find the radius of the first or the second sphere.)
   (a) 50 m² (b) 200 m² (c) 157 m²
   (d) 400 m² (e) 800 m²

7. A sphere has a volume of 100 m³. A second sphere has a radius twice that of the first sphere. The volume of the second sphere is
   (a) 33.3 m³ (b) 300 m³ (c) 400 m³
   (d) 800 m³ (e) No answer is correct.

CONCEPTUAL QUESTIONS

1. Estimate the order of magnitude of the length, in meters, of each of the following: (a) a mouse, (b) a pool cue, (c) a basketball court, (d) an elephant, (e) a city block.
2. Estimate the number of times your heart beats in a month.
3. Estimate your age in seconds.
4. Estimate the number of atoms in 1 cm³ of a solid. (Note that the diameter of an atom is about 10^{-10} m.)
5. The height of a horse is sometimes given in units of "hands." Why is this a poor standard of length?
6. How many of the lengths or time intervals given in Tables 1.2 and 1.3 could you verify using only equipment found in a typical dormitory room?
7. An ancient unit of length called the *cubit* was equal to six palms, where a palm was the width of the four fingers of an open hand. Noah's ark was 300 cubits long, 50 cubits wide, and 30 cubits high. Estimate the volume of the ark in cubic feet. Also, estimate the volume of a typical home and compare it to the volume of the ark in cubic meters.
8. If an equation is dimensionally correct, does this mean that the equation must be true?
9. An automobile tire is rated to last for 50 000 miles. Estimate the number of revolutions the tire will make in its lifetime.
10. Figure Q1.10 is a photograph showing unit conversions on the labels of some grocery-store items. Check the ac-

curacy of these conversions. Are the manufacturers using significant figures correctly?

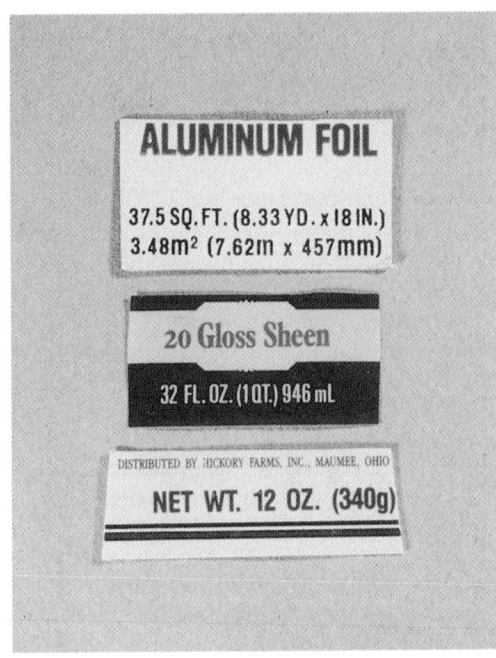

Figure Q1.10   *(Courtesy of Henry Leap and Jim Lehman)*

PROBLEMS

1, **2**, **3** = straightforward, intermediate, challenging   ☐ = full solution available in Study Guide/Student Solutions Manual   ◤ = Core Concepts Workbook
WEB = solution posted at **http://www.harcourtcollege.com/physics/cptech**   ◩ = biomedical application   ▥ = Interactive Physics

Section 1.3 Dimensional Analysis

1. A shape that covers an area A and has a uniform height h has a volume given by $V = Ah$. (a) Show that $V = Ah$ is dimensionally correct. (b) Show that the volumes of a cylinder and of a rectangular box can be written in the form $V = Ah$, identifying A in each case. (Note that A, sometimes called the "footprint" of the object, can have any shape and the height can be replaced by average thickness in general.)
2. In a desperate attempt to come up with an equation to use during an examination, a student tries $v^2 = ax$. Use dimensional analysis to determine whether this equation might be valid.

3. The period of a simple pendulum, defined as the time for one complete oscillation, is measured in time units and is given by

$$T = 2\pi\sqrt{\frac{\ell}{g}}$$

where ℓ is the length of the pendulum and g is the acceleration due to gravity, in units of length divided by time squared. Show that this equation is dimensionally consistent.
4. (a) One of the fundamental laws of motion states that the acceleration of an object is directly proportional to the resultant force on it and inversely proportional to its

mass. From this statement, determine the dimensions of force. **(b)** The newton is the SI unit of force. According to the results for **(a)**, how can you express a force having units of newtons using the fundamental units of mass, length, and time?

5. **(a)** Suppose that the displacement of an object is related to the time according to the expression $x = Bt^2$. What are the dimensions of B? **(b)** A displacement is related to the time as $x = A \sin(2\pi ft)$, where A and f are constants. Find the dimensions of A. (*Hint:* A trigonometric function appearing in an equation must be dimensionless.)

Section 1.4   Significant Figures

6. The value of the speed of light is now known to be $2.99\ 792\ 458 \times 10^8$ m/s. Express the speed of light to **(a)** three significant figures, **(b)** five significant figures, and **(c)** seven significant figures.

7. How many significant figures are there in **(a)** 78.9 ± 0.2, **(b)** 3.788×10^9, **(c)** 2.46×10^{-6}, **(d)** 0.0032?

8. Carry out the following arithmetic operations: **(a)** the sum of the numbers 756, 37.2, 0.83, and 2.5; **(b)** the product 3.2×3.563; **(c)** the product $5.67 \times \pi$.

9. **(a)** Using your calculator find, in scientific notation with appropriate rounding, the value of $(2.437 \times 10^4)(6.5211 \times 10^9)/(5.37 \times 10^4)$. **(b)** Find the value of $(3.14159 \times 10^2)(27.01 \times 10^4)/(1234 \times 10^6)$.

10. Calculate **(a)** the circumference of a circle of radius 3.5 cm and **(b)** the area of a circle of radius 4.65 cm.

11. A farmer measures the distance around a rectangular field. The length of each long side of the rectangle is found to be 38.44 m, and the length of each short side is found to be 19.5 m. What is the total distance around the field?

12. A fisherman catches two sturgeon. The smaller of the two has a measured length of 93.46 cm (two decimal places, four significant figures), and the larger fish has a measured length of 135.3 cm (one decimal place, four significant figures). What is the total length of fish caught for the day?

Section 1.5   Conversion of Units

13. Estimate the distance to the nearest star, in feet, using the data in Table 1.1 and the appropriate conversion factors.

14. Estimate the age of the Earth in years, using the data in Table 1.3 and the appropriate conversion factors.

15. The speed of light is about 3.00×10^8 m/s. Convert this to miles per hour.

16. **(a)** Find a conversion factor to convert from miles per hour to kilometers per hour. **(b)** For a while, federal law mandated that the maximum highway speed would be 55 mi/h. Use the conversion factor from part **(a)** to find the speed in kilometers per hour. **(c)** The maximum

highway speed has been raised to 65 mi/h in some places. In kilometers per hour, how much of an increase is this over the 55-mi/h limit?

17. A painter is to cover the walls in a room that is 8.0 ft high and 12.0 ft along a side. What surface area, in square meters, must she cover?

18. A house is 50.0 ft long and 26.0 ft wide, and has 8.0-ft-high ceilings. What is the volume of the interior of the house in cubic meters and in cubic centimeters?

19. Assume that 2.0 million stone blocks that average 2.5 tons each are used to build a pyramid. Find the weight of this pyramid in pounds and in newtons (see Problem 4).

20. A billionaire offers to give you $1 billion if you can count it out using only $1 bills. Should you accept her offer? Assume that you can count at an average rate of one bill every second, and be sure to allow for the fact that you need about 8 hours a day for sleeping and eating.

21. The amount of water in reservoirs is often measured in acre-ft. One acre-ft is a volume that covers an area of one acre to a depth of one foot. An acre is an area of $43\ 560$ ft^2. Find the volume in SI units of a reservoir containing 25.0 acre-ft of water.

22. The base of a pyramid covers an area of 13.0 acres (1 acre = $43\ 560$ ft^2) and has a height of 481 ft (Fig. P1.22). If the volume of a pyramid is given by the expression $V = (1/3)bh$, where b is the area of the base and h is the height, find the volume of this pyramid in cubic meters.

Figure P1.22

23. A quart-container of ice cream is to be made in the form of a cube. What should be the length of a side, in centimeters? (Use the conversion 1 gallon = 3.786 liter.)

24. One cubic meter (1.00 m^3) of aluminum has a mass of 2.70×10^3 kg, and 1.00 m^3 of iron has a mass of 7.86×10^3 kg. Find the radius of an aluminum sphere whose mass is the same as that of an iron sphere of radius 2.00 cm. (*Note:* Density is defined as the mass of an object divided by its volume.)

25. One cubic centimeter (1.0 cm³) of water has a mass of 1.0×10^{-3} kg. (a) Determine the mass of 1.0 m³ of water. (b) Assuming biological substances are 98% water, estimate the masses of a cell with a diameter of 1.0 μm, a human kidney, and a fly. Assume a kidney is roughly a sphere with a radius of 4.0 cm, and a fly is roughly a cylinder 4.0 mm long and 2.0 mm in diameter. (See the note in Problem 24.)

Section 1.6 Order-of-Magnitude Calculations

Note: In developing answers to the problems in this section, you should state your important assumptions, including the numerical values assigned to parameters used in the solution. Because only order-of-magnitude results are expected, do not be surprised if your results differ from those given in the answer section.

26. A hamburger chain advertises that it has sold more than 50 billion hamburgers. Estimate how many pounds of hamburger meat must have been used by the restaurant chain and how many head of cattle were required to furnish the meat.

27. Imagine that you are the equipment manager of a professional baseball team. One of your jobs is to keep baseballs on hand for games. Balls are sometimes lost when players hit them into the stands as either home runs or foul balls. Estimate how many baseballs you have to buy per season in order to make up for such losses. Assume your team plays an 81-game home schedule in a season.

28. Estimate the number of piano tuners living in New York City. This question was raised by the physicist Enrico Fermi, who was well known for making order-of-magnitude calculations.

Section 1.8 Coordinate Systems and Frames of Reference

29. A point is located in a polar coordinate system by the coordinates $r = 2.5$ m and $\theta = 35°$. Find the x and y coordinates of this point, assuming the two coordinate systems have the same origin.

30. A certain corner of a room is selected as the origin of a rectangular coordinate system. If a fly is crawling on an adjacent wall at a point having coordinates (2.0, 1.0), where the units are meters, what is the distance of the fly from the corner of the room?

31. Express the location of the fly in Problem 30 in polar coordinates.

32. Two points in a rectangular coordinate system have the coordinates (5.0, 3.0) and (−3.0, 4.0), where the units are centimeters. Determine the distance between these points.

Section 1.9 Trigonometry

33. For the triangle shown in Figure P1.33, what are (a) the length of the unknown side, (b) the tangent of θ, and (c) the sin of ϕ?

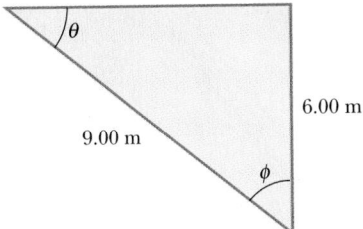

Figure P1.33

WEB 34. A right triangle has a hypotenuse of length 3.00 m, and one of its angles is 30.0°. What are the lengths of (a) the side opposite the 30.0° angle and (b) the side adjacent to the 30.0° angle?

35. In Figure P1.35, find (a) the side opposite θ, (b) the side adjacent to ϕ, (c) cos θ, (d) sin ϕ, and (e) tan ϕ.

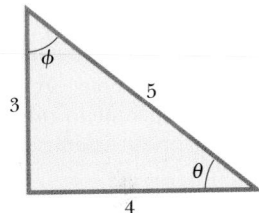

Figure P1.35

36. In a certain right triangle, the two sides that are perpendicular to each other are 5.00 m and 7.00 m long. What is the length of the third side of the triangle?

37. In Problem 36, what is the tangent of the angle for which 5.00 m is the opposite side?

ADDITIONAL PROBLEMS

38. Soft drinks are commonly sold in aluminum containers. Estimate the number of such containers thrown away each year by U.S. consumers. Approximately how many tons of aluminum does this represent?

39. The radius of the planet Saturn is 5.85×10^7 m, and its mass is 5.68×10^{26} kg (Fig. P1.39). (a) Find the density of Saturn (its mass divided by its volume) in grams per cubic centimeter. (The volume of a sphere is given by $(4/3)\pi r^3$.) (b) Find the surface area of Saturn in square feet. (The surface area of a sphere is given by $4\pi r^2$.)

Figure P1.39   A view of Saturn from *Voyager 2*.
(Courtesy of NASA)

40. The consumption of natural gas by a company satisfies the empirical equation $V = 1.5t + 0.0080t^2$, where V is the volume in millions of cubic feet and t the time in months. Express this equation in cubic feet and seconds. Put the proper units on the coefficients. Assume a month is 30 days.

WEB 41. The displacement of an object moving under uniform acceleration is some function of time and the acceleration. Suppose we write this displacement as $s = k\,a^m t^n$, where k is a dimensionless constant. Show by dimensional analysis that this expression is satisfied if $m = 1$ and $n = 2$. Can this analysis give the value of k?

42. Sphere 1 has surface area A_1 and volume V_1 and sphere 2 has surface area A_2 and volume V_2. If the radius of sphere 2 is double the radius of sphere 1, what is the ratio of (a) the areas A_2/A_1, and (b) the volumes V_2/V_1?

43. Compute the order of magnitude of the mass of (a) a bathtub filled with water, and (b) a bathtub filled with pennies. In your solution list the quantities you estimate and the value you estimate for each.

44. One gallon of paint (volume $= 3.78 \times 10^{-3}$ m^3) covers an area of 25.0 m^2. What is the thickness of the paint on the wall?

45. The radius r of a circle inscribed in any triangle whose sides are a, b, and c is given by

$$r = \left[\frac{(s - a)(s - b)(s - c)}{s}\right]^{1/2}$$

where s is an abbreviation for $(a + b + c)/2$. Check this formula for dimensional consistency.

46. Newton's law of universal gravitation is given by

$$F = G\frac{Mm}{r^2}$$

Here F is the force of gravity, M and m are masses, and r is a length. Force has the units kg·m/s^2. What are the SI units of the proportionality constant G?

47. (a) How many seconds are there in a year? (b) If one micrometeorite (a sphere with a diameter of 1.0×10^{-6} m) struck each square meter of the Moon each second, how many years would it take to cover the Moon to a depth of 1.0 m? (*Hint:* Consider a cubic box, 1.0 m on a side, on the Moon, and find how long it would take to fill the box.)

WEB 48. You can obtain a rough estimate of the size of a molecule by the following simple experiment. Let a droplet of oil spread out on a smooth water surface. The resulting oil slick will be approximately one molecule thick. Given an oil droplet of mass 9.00×10^{-7} kg and density 918 kg/m^3 that spreads out into a circle of radius 41.8 cm on the water surface, what is the diameter of an oil molecule?

49. At the time of this book's printing, the U.S. national debt is about \$4 trillion. (a) If this debt were paid at a rate of \$1000 per second, how long (in years) would it take to pay off the debt, assuming no interest were charged? (b) A dollar bill is about 15.5 cm long. If 4 trillion dollar bills were laid end to end around the Earth's Equator, how many times would they circle the Earth? Take the radius of the Earth to be about 6378 km. (*Note:* Before doing any of the calculations, try to guess at the answers. You may be surprised.)

Motion in One Dimension

▲ PHYSICS PUZZLER

Downhill skiers can reach speeds of greater than 100 km/h. If the skier starts from rest at the top of the slope and moves in a straight line, can you estimate the time it takes the skier to reach this speed? *(Jean Y. Ruszniewski/Tony Stone Images)*

3.1

People have always been concerned with some form of motion. For example, in today's world you consider motion when you describe how fast your new car will go or how much pickup it has. Likewise, prehistoric cave dwellers must have, in their own way, pondered motion as they devised methods for capturing a rapidly moving antelope. The branch of physics concerned with the study of the motion of an object and the relationship of this motion to such physical concepts as force and mass is called **dynamics.** The part of dynamics that describes motion without regard to its causes is called **kinematics.** In this chapter we shall focus on kinematics and on one-dimensional motion — that is, motion along a straight line. We shall start by discussing displacement, velocity, and acceleration. These concepts will enable us to study the motion of objects undergoing constant acceleration. In Chapter 3 we shall discuss the motions of objects in two dimensions.

The first recorded evidence of the study of mechanics can be traced to the peoples of ancient Sumerian and Egyptian civilizations, whose primary concern was to understand the motions of heavenly bodies. The most systematic and detailed early studies of the heavens were conducted by the Greeks from about 300 B.C. to A.D. 300. Ancient scientists and lay people regarded the Earth as the center of the Universe. This geocentric model was accepted by such notables as Aristotle (384–322 B.C.) and Claudius Ptolemy (about A.D. 140). Largely because of the authority of Aristotle, the geocentric model became the accepted theory of the Universe until the 17th century.

About 250 B.C. the Greek philosopher Aristarchus worked out the details of a model of the Solar System based on a moving Earth. He proposed that the sky appears to turn westward because the Earth is really turning eastward. This model was not given much consideration because it was believed that if the Earth turned, it would set up a great wind as it moved through the air. The critics could not see that the Earth was carrying the air and everything else with it as it rotated.

The Polish astronomer Nicolaus Copernicus (1473–1543) is credited with initiating the revolution that finally replaced the geocentric model. In his system, called the *heliocentric model,* the Earth and the other planets revolve in circular orbits around the Sun.

This early knowledge formed the foundation for the work of Galileo Galilei (1564–1642). Galileo stands out as perhaps the dominant facilitator of the entrance of physics into the modern era. In 1609 he became one of the first to make astronomical observations with a telescope. He observed mountains on the Moon, the larger satellites of Jupiter, the rings of Saturn, spots on the Sun, and the phases of Venus. His observations convinced him of the correctness of the Copernican theory. Galileo's work with motion is particularly well known, and because of his leadership, experimentation has become an important part of our search for knowledge.

3.2, SECTION 1

2.1 DISPLACEMENT

In order to describe the motion of an object, one must be able to specify its position at all times using some convenient coordinate system and a specified origin. For example, consider a sprinter moving along the x axis from an initial position, x_i, to some final position, x_f, as in Figure 2.1. **The displacement of an object, defined as its *change in position*, is given by the difference between its final and initial coordinates, or $x_f - x_i$.** As mentioned in Chapter 1, we use the Greek letter delta

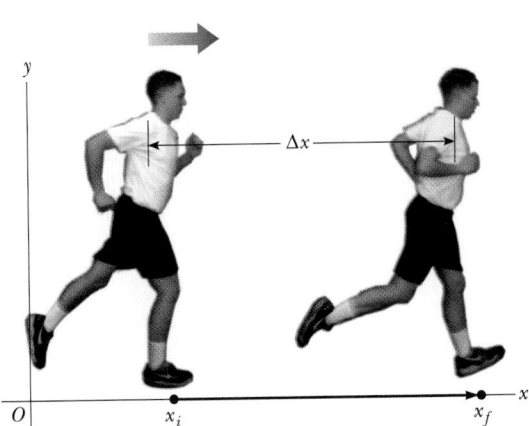

Figure 2.1 A sprinter moving along the x axis from x_i to x_f undergoes a displacement of $\Delta x = x_f - x_i$.

(Δ) to denote a change in a quantity. Therefore, we write the displacement, or change in position, of the object as

$$\Delta x \equiv x_f - x_i \qquad [2.1]$$

◀ Definition of displacement

From this definition, we see that Δx is positive if x_f is greater than x_i and negative if x_f is less than x_i. If the sprinter moves from an initial position of $x_i = 3$ m to a final position of $x_f = 15$ m, his displacement is $\Delta x = 15$ m $- 3$ m $= 12$ m.

Displacement is an example of a vector quantity. Many physical quantities in this book, including displacement, velocity, and acceleration, are vectors. In general, **a vector is a physical quantity that requires the specification of both direction and magnitude.** By contrast, **a scalar is a quantity that has magnitude and no direction.** Scalar quantities, such as mass and temperature, can be specified by numbers with appropriate units.

◀ A vector has both direction and magnitude.

We shall usually designate vector quantities with boldface type. For example, **v** denotes a velocity vector, and **a** denotes an acceleration vector. In this chapter, which deals with one-dimensional motion, it will not be necessary to use this notation. The reason is that in one-dimensional motion there are only two directions in which an object can move, and these two directions are easily specified by plus and minus signs. For example, if a truck moves from an initial position of 10 m to a final position of 80 m, as in Figure 2.2a, its displacement is

$$\Delta x = x_f - x_i = 80 \text{ m} - 10 \text{ m} = +70 \text{ m}$$

In this case, the displacement has a magnitude of 70 m and is directed in the positive x direction, as indicated by the plus sign of the result. (Sometimes the plus sign is omitted, but a result of 70 m for the displacement is understood to be the same as $+70$ m.) The displacement vector is usually represented by an arrow, as in Figure 2.2a. The length of the arrow represents the magnitude of the displacement, and the head of the arrow indicates its direction.

Now suppose the truck moves to the left from an initial position of 80 m to a final position of 20 m, as in Figure 2.2b. In this situation, the displacement is

$$\Delta x = x_f - x_i = 20 \text{ m} - 80 \text{ m} = -60 \text{ m}$$

The minus sign in this result indicates that the displacement is in the negative x direction. Likewise, the arrow representing the displacement vector is to the left, as in Figure 2.2b.

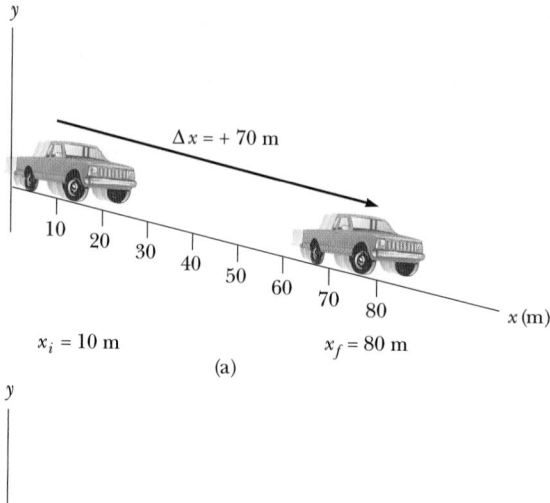

Figure 2.2 (a) A truck moving to the right from $x_i = 10$ m to $x_f = 80$ m undergoes a displacement $\Delta x = +70$ m.
(b) A truck moving to the left from $x_i = 80$ m to $x_f = 20$ m undergoes a displacement $\Delta x = -60$ m.

2.2   AVERAGE VELOCITY

3.2, SECTION 2

The motion of an object is completely known if its position is known at all times. Consider a truck moving along a highway (the x axis), as in Figure 2.2. Let the truck's position be x_i at some time t_i, and let its position be x_f at time t_f. (The indices i and f refer to the initial and final locations, respectively.) In the time interval $\Delta t = t_f - t_i$, the displacement of the truck is $\Delta x = x_f - x_i$.

▷ Definition of average velocity

   The average velocity, $\bar{v}$, is defined as the displacement, Δx, divided by the time interval during which the displacement occurred:

$$\bar{v} \equiv \frac{\Delta x}{\Delta t} = \frac{x_f - x_i}{t_f - t_i} \qquad [2.2]$$

The average velocity of an object can be either positive or negative, depending on the sign of the displacement. (The time interval, Δt, is always positive.) For example, if we select a coordinate system so that a car moves from a position 100 m from the origin to a position 50 m from that origin in a time interval of 2 s, the x component of the average velocity is -25 m/s. In this case, the minus sign indicates motion to the left. If an object moves in the negative x direction during some time interval, its velocity in the x direction during that interval is necessarily negative.

   In order to understand the vector nature of velocity, consider the following situation. Suppose a friend tells you that she will be taking a trip in her car and will travel at a constant rate of 55 mi/h in a straight line for 1 h. If she starts from her home, where will she be at the end of the trip? Obviously, you cannot answer this question because she did not specify the direction of her trip. All you can say is that she will be located 55 mi from her starting point. However, if she tells you that she will be driving at the rate of 55 mi/h directly northward, then her final location will be known exactly. The velocity of an object is known only if its direction and its magnitude (speed) are specified. In general, **any physical quantity that is a vector must be characterized by both a magnitude and a direction.**

As an example of the use of Equation 2.2, suppose that the truck in Figure 2.2a moves 100 m to the right in 5.0 s. The substitution of these values into Equation 2.2 gives the average velocity in this time interval as

$$\bar{v} = \frac{\Delta x}{\Delta t} = \frac{100 \text{ m}}{5.0 \text{ s}} = +20 \text{ m/s}$$

Note that the units of average velocity are units of length divided by units of time. These are meters per second (m/s) in SI. Other units for velocity might be feet per second (ft/s) in the conventional system or centimeters per second (cm/s) if we decided to measure distances in centimeters.

Let us assume we are watching a drag race from the Goodyear blimp. In one run we see a car follow the straight-line path from P to Q shown in Figure 2.3 during the time interval Δt, and in a second run a car follows the curved path during the same time interval. If you examine the definition of average velocity (Eq. 2.2) carefully, you will see that the two cars had the same average velocity. This is because they had the same displacement ($x_f - x_i$) during the same time interval (Δt), as indicated in Figure 2.3.

Figure 2.4 shows the unusual path of a confused football player. He receives a kickoff at his own goal, runs downfield to within inches of a touchdown, and then reverses direction to race backward until he is tackled at the exact location where he first caught the ball. What is his average velocity in the x direction (along the field) during this run? From the definition of average velocity, we see that the football player's average velocity is zero because his displacement is zero. In other words, because x_i and x_f have the same value, $\bar{v} = \Delta x/\Delta t = 0$. From this we see that displacement should not be confused with distance traveled. The confused football player clearly traveled a great distance, almost 200 yards; yet his displacement is zero.

Figure 2.3 A drag race viewed from a blimp. One car follows the red straight-line path from P to Q, and a second car follows the blue curved path.

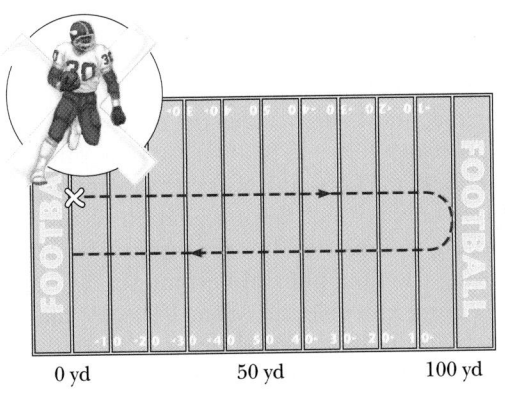

0 yd   50 yd   100 yd

Figure 2.4 The path followed by a confused football player.

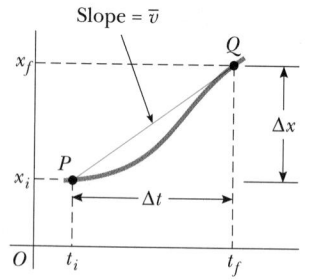

Figure 2.5 Position-time graph for an object moving along the x axis. The average velocity $\bar{v}$ in the time interval $\Delta t = t_f - t_i$ is the slope of the blue straight line connecting the points P and Q.

3.3, SECTION 1

Definition of instantaneous velocity ▶

Graphical Interpretation of Velocity

Figure 2.5 is a graph of the motion of an object moving along a straight-line path from the position x_i at time t_i to the position x_f at time t_f. (Note that the motion is along a straight line, yet the position-time graph is not a straight line. Why?) The straight line connecting points P and Q provides us with a geometric interpretation of average velocity. The slope of the line is Δx ($= x_f - x_i$) divided by the time interval for the motion, Δt ($= t_f - t_i$). Therefore, **the average velocity of an object during the time interval t_i to t_f is equal to the slope of the straight line joining the initial and final points on a graph of the position of the object plotted versus time.**

2.3   INSTANTANEOUS VELOCITY

Imagine that you take a trip in your car along a perfectly straight highway. At the end of your journey, it is a relatively simple task to calculate your average velocity. The car's odometer gives you the distance traveled, a compass can give you direction, and a watch can supply the time interval. However, such a calculation would omit a great deal of information about what actually occurred on your trip. For example, if your calculation indicated that your average velocity was 55 mi/h, this would not necessarily mean that your velocity at every instant was 55 mi/h. Instead, you might have traveled at 55 mi/h for a short distance, stopped for lunch and then made up some time by traveling at 70 mi/h, paused again as a police officer wrote out a ticket, and then raced forward again when the coast was clear.

More precisely, **the instantaneous velocity, v, is defined as the limit of the average velocity as the time interval Δt becomes infinitesimally short.** In mathematical language this is written

$$v \equiv \lim_{\Delta t \to 0} \frac{\Delta x}{\Delta t} \qquad [2.3]$$

The notation $\lim \Delta t \to 0$ means that the ratio $\Delta x/\Delta t$ is to be evaluated as the time interval Δt approaches zero.

To better understand the meaning of instantaneous velocity as expressed by Equation 2.3, consider the data in Table 2.1. Assume you have been observing a runner racing along a track. One second after starting into motion, the runner has moved to a position 1.00 m from the starting point; at $t = 1.50$ s, the runner is 2.25 m from the starting point; and so on. After collecting these data, suppose you wish to determine the velocity of the runner at the time $t = 1.00$ s. Table 2.2 presents some of the calculations you might perform to determine the velocity in question. Let us use $t = 1.00$ s as our initial time, and first consider the top row of the table. For the observed portion of the run (from 1.00 s to 3.00 s), the time interval $\Delta t = 2.00$ s and the displacement $\Delta x = +8.00$ m. Thus, the average velocity in this interval is $\Delta x/\Delta t = +4.00$ m/s. This gives only a rough approximation of the instantaneous velocity at $t = 1.00$ s. According to Equation 2.3, we can find increasingly more nearly correct answers by letting the time interval become smaller and smaller. Therefore, consider the second entry in Table 2.2, from 1.00 s to 2.00 s, corresponding to a time interval of $\Delta t = 1.00$ s. In this interval, $\Delta x = +3.00$ m and

TABLE 2.1
Positions of a Runner at Specific Instants of Time

t (s)	x (m)
1.00	1.00
1.01	1.02
1.10	1.21
1.20	1.44
1.50	2.25
2.00	4.00
3.00	9.00

TABLE 2.2 Calculated Values of the Time Intervals, Displacements, and Average Velocities for the Runner of Table 2.1

Time Interval (s)	Δt (s)	Δx (m)	$\bar{v}$ (m/s)
1 to 3.00	2.00	8.00	4.00
1 to 2.00	1.00	3.00	3.00
1 to 1.50	0.50	1.25	2.5
1 to 1.20	0.20	0.44	2.2
1 to 1.10	0.10	0.21	2.1
1 to 1.01	0.01	0.02	2

so the average velocity is $+3.00$ m/s. This is closer to the correct answer because the time interval is smaller. Now let us consider a very short time interval indicated by the last entry in Table 2.2. In this case, $\Delta t = 0.01$ s and the displacement $\Delta x = +0.02$ m. Thus, the average velocity in this interval is $+2$ m/s. We could improve the reliability of our calculation by allowing the time interval to become even smaller, but we can state with some degree of confidence that the instantaneous velocity of the runner was $+2$ m/s at the time $t = 1.00$ s.

In day-to-day usage, the terms *speed* and *velocity* are interchangeable. In physics, however, there is a clear distinction between these two quantities. **The instantaneous speed of an object, which is a scalar quantity, is defined as the magnitude of the instantaneous velocity.** Hence, by definition, speed can never be negative.

At appropriate places within each chapter we will include interludes that we hope will motivate you to think critically about the material you have been reading. These interesting interludes, called "Thinking Physics" or "Applying Physics," are conceptual rather than mathematical and should serve as models in responding to the conceptual questions at the end of each chapter.

Applying Physics 1

Does an automobile speedometer measure average or instantaneous speed?

Explanation Based on the fact that changes in the speed of the automobile are reflected in changes in the speedometer reading, we might be tempted to say that a speedometer measures instantaneous speed. We would be wrong, however. The reading on a speedometer is related to the rotation of the wheels. Often, a rotating magnet, whose rotation is related to that of the wheels, is used in a speedometer. The reading is based on the *time interval* between rotations of the magnet. Thus, the reading is an average speed over this time interval. This time interval is generally very short, so that the average speed is close to the instantaneous speed. As long as the speed changes are gradual, this approximation is very good. If the automobile brakes to a quick stop, however, the speedometer may not be able to keep up with the rapid changes in speed, and the measured average speed over the time interval may be very different than the instantaneous speed at the end of the interval.

APPLICATION

Automobile Speedometers.

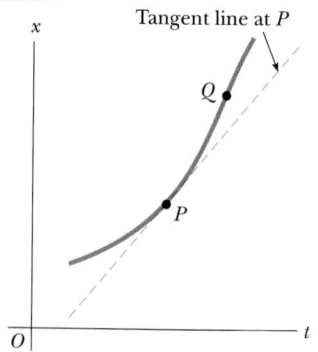

x

Tangent line at P

Q

P

O   t

Figure 2.6 Geometric construction for obtaining the instantaneous velocity from the *x* versus *t* curve. The instantaneous velocity at *P* is defined as the slope of the line tangent to the curve at *P*.

Graphical Interpretation of Instantaneous Velocity

Figure 2.6 is a graph, similar to Figure 2.5, for the position of an object versus time. To find the instantaneous velocity of the object at point *P*, we must find the average velocity during an infinitesimally short time interval. This means that point *Q* on the curve must be brought closer and closer to point *P* until the two points nearly overlap each other. From this construction we see that the line joining *P* and *Q* is approaching the line tangent to the curve at point *P*. **The slope of the line tangent to the position-time curve at *P* is defined to be the instantaneous velocity at that time.**

EXAMPLE 2.1 A Toy Train

A toy train moves slowly along a straight portion of track according to the graph of position versus time in Figure 2.7. Find (a) the average velocity for the total trip, (b) the average velocity during the first 4.0 s of motion, (c) the average velocity during the next 4.0 s of motion, (d) the instantaneous velocity at $t = 2.0$ s, and (e) the instantaneous velocity at $t = 5.0$ s.

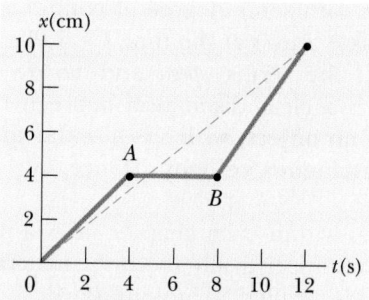

Figure 2.7 (Example 2.1)

Solution (a) The slope of the line joining the starting point and end point on the graph (the dashed line) provides the average velocity for the total trip. A measurement of the slope gives

$$\bar{v} = \frac{\Delta x}{\Delta t} = \frac{10 \text{ cm}}{12 \text{ s}} = \boxed{+0.83 \text{ cm/s}}$$

(b) The slope of the line joining the starting point to the point at $t = 4.0$ s on the curve gives us the average velocity during the first 4 s:

$$\bar{v} = \frac{\Delta x}{\Delta t} = \frac{4.0 \text{ cm}}{4.0 \text{ s}} = \boxed{+1.0 \text{ cm/s}}$$

(c) Following the same procedure for the next 4.0-s interval, we see that the slope of the line between points *A* and *B* is zero. During this time interval, the train has remained at the same location, 4.0 cm from the starting point.

(d) A line drawn tangent to the curve at the point corresponding to $t = 2.0$ s has the same slope as the line in part (b). Thus, the instantaneous velocity at this time is $+1.0$ cm/s. This has to be true because the graph indicates that during the first 4.0 s of motion the train covers equal distances in equal intervals of time. In other words, the train moves at a constant velocity during the first 4.0 s. Under these conditions the average velocity and the instantaneous velocity are identical at all times.

(e) At $t = 5.0$ s, the slope of the position-time curve is zero. Therefore, the instantaneous velocity is zero at this instant. In fact, the train is at rest during the entire time interval between 4.0 s and 8.0 s.

2.4 ACCELERATION

As you travel from place to place in your car, you normally do not travel long distances at a constant velocity. The velocity of the car increases when you step on the gas and decreases when you apply the brakes. Furthermore, the velocity changes when you round a curve, altering your direction of motion. When the velocity of an object changes with time, the object is said to undergo an *acceleration*. However, we need a more precise definition of acceleration.

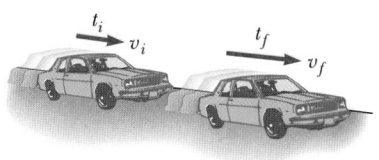

Figure 2.8 A car moving to the right accelerates from a velocity of v_i to a velocity of v_f in the time interval $\Delta t = t_f - t_i$.

Average Acceleration

Suppose a car moves along a straight highway as in Figure 2.8. At time t_i it has a velocity of v_i, and at time t_f its velocity is v_f. **The average acceleration during this time interval is defined as the change in velocity divided by the time interval during which this change occurs:**

◀ Definition of average acceleration

$$\bar{a} \equiv \frac{\Delta v}{\Delta t} = \frac{v_f - v_i}{t_f - t_i} \qquad [2.4]$$

 **3.3**, SECTION 2

For example, suppose the car shown in Figure 2.8 accelerates from an initial velocity of $v_i = +10$ m/s to a final velocity of $v_f = +30$ m/s in a time interval of 2.0 s. (Note that both velocities are toward the right, the direction selected as the positive direction.) These values can be inserted into Equation 2.4 to give the average acceleration:

$$\bar{a} = \frac{30 \text{ m/s} - 10 \text{ m/s}}{2.0 \text{ s}} = +10 \text{ m/s}^2$$

Acceleration is a vector quantity having dimensions of length divided by the square of time. Some common units of acceleration are meters per second per second [(m/s)/s, which is usually written m/s²] and feet per second per second (ft/s²). The acceleration calculated in the preceding case was $+10$ m/s². This notation means that, on the average, the car accelerates in the positive x direction such that its velocity increases at a rate of 10 m/s every second.

As a second example, consider the car pictured in Figure 2.9. In this case, the velocity of the car has changed from an initial value of $+30$ m/s to a final value of $+10$ m/s in a time interval of 2.0 s. The average acceleration of the car during this time interval is

$$\bar{a} = \frac{10 \text{ m/s} - 30 \text{ m/s}}{2.0 \text{ s}} = -10 \text{ m/s}^2$$

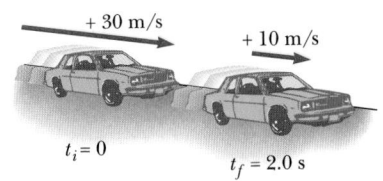

Figure 2.9 The velocity of the car decreases from $+30$ m/s to $+10$ m/s in a time interval of 2.0 s.

The minus sign indicates that the acceleration vector is in the negative x direction (to the left). For the case of motion in a straight line, the direction of the velocity of an object and the direction of its acceleration are related as follows. **When the object's velocity and acceleration are in the same direction, the speed of the object increases with time.** (The first case we cited demonstrates this situation.) **When the object's velocity and acceleration are in opposite directions, the speed of the object decreases with time.** (Decreases in speed are sometimes called *decelerations*.)

To clarify this point, consider the following situation. Suppose the velocity of a car changes from -10 m/s to -30 m/s in a time interval of 2.0 s. The minus signs indicate that the velocity of the car is in the negative x direction. The average acceleration of the car in this time interval is

$$\bar{a} = \frac{-30 \text{ m/s} - (-10 \text{ m/s})}{2.0 \text{ s}} = -10 \text{ m/s}^2$$

The minus sign indicates that the acceleration vector is also in the negative x direction. Because the velocity and acceleration vectors are in the same direction, the speed of the car must increase as the car moves to the left.

Instantaneous Acceleration

In some situations, the value of the average acceleration differs in different time intervals. It is useful, therefore, to define instantaneous acceleration. This concept is analogous to the definition of instantaneous velocity discussed in Section 2.3. In mathematical terms, **the instantaneous acceleration, a, is defined as the limit of the average acceleration as the time interval Δt goes to zero.** That is,

▶ Definition of instantaneous acceleration

$$a \equiv \lim_{\Delta t \to 0} \frac{\Delta v}{\Delta t} \qquad [2.5]$$

Here again, the notation $\Delta t \to 0$ means that the ratio $\Delta v/\Delta t$ is to be evaluated as the time interval Δt approaches zero.

Figure 2.10 is useful for understanding the concept of instantaneous acceleration. It plots the velocity of an object versus time. This could represent, for example, the motion of a car along a busy street. The average acceleration of the car between times t_i and t_f can be found by determining the slope of the line joining points P and Q. If we imagine that point Q is brought closer and closer to point P, the value that we find for the average acceleration between these points approaches the value of the acceleration of the car at point P. The instantaneous acceleration at point P, for example, is the slope of the graph at the point. That is, **the instantaneous acceleration of an object at a certain time equals the slope of the velocity-time graph at that instant of time.** From now on we shall use the term *acceleration* to mean "instantaneous acceleration."

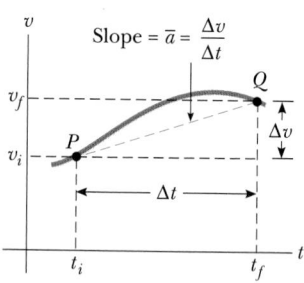

Figure 2.10 Velocity-time graph for an object moving in a straight line. The slope of the blue line connecting points P and Q is defined as the average acceleration in the time interval $\Delta t = t_f - t_i$.

EXAMPLE 2.2 A Fly Ball Is Caught

A baseball player moves in a straight-line path in order to catch a fly ball hit to the outfield. His velocity as a function of time is shown in Figure 2.11. Find his instantaneous acceleration at points *A*, *B*, and *C* on the curve.

Solution The instantaneous acceleration at any time is the slope of the velocity-time curve at that instant. The slope of the curve at point *A* is

$$a = \frac{\Delta v}{\Delta t} = \frac{4 \text{ m/s}}{2 \text{ s}} = \quad +2 \text{ m/s}^2$$

At point *B*, the slope of the line is zero and hence the instantaneous acceleration at this time is also zero. Note that even though the instantaneous acceleration has dropped to zero at this instant, the velocity of the player is not zero. (In general, if *a* = 0, the velocity is constant in time but *not necessarily* zero.) Instead, the player is running at a constant velocity of + 4 m/s. A value of zero for an instantaneous acceleration means that at a particular instant the velocity of the object is not changing.

Finally, at point *C* we calculate the slope of the velocity-time curve and find that the instantaneous acceleration is

$$a = \frac{\Delta v}{\Delta t} = \frac{-2 \text{ m/s}}{1 \text{ s}} = \quad -2 \text{ m/s}^2$$

For this situation, the minus sign and the fact that the velocities are positive indicate that the speed of the player is decreasing as he approaches the location where he will attempt to catch the baseball.

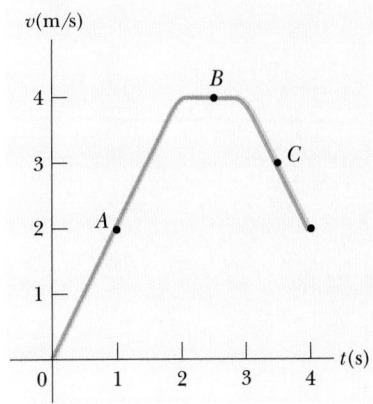

Figure 2.11 (Example 2.2)

 Workbook Problem 7 – (Workbook page 22) is similar to Example 2.2.

Thinking Physics 1

Parts (a), (b), and (c) of Figure 2.12 represent three graphs of the velocities of different objects moving in straight-line paths as functions of time. The possible accelerations of each object as functions of time are shown in parts (d), (e), and (f). Match each velocity-time graph with the acceleration-time graph that best describes the motion.

Explanation The velocity-time graph (a) has a constant slope, indicating a constant acceleration, which is represented by acceleration-time graph (e).

Graph (b) represents an object that has a constantly increasing speed, but the speed is not increasing at a uniform rate. Thus, the acceleration must be increasing, and the acceleration-time graph that best indicates this is (d).

Graph (c) depicts an object that first has a velocity that increases at a constant rate, which means its acceleration is constant. The motion then changes to one at constant speed, indicating that the acceleration of the object becomes zero. Thus, the best match to this situation is graph (f).

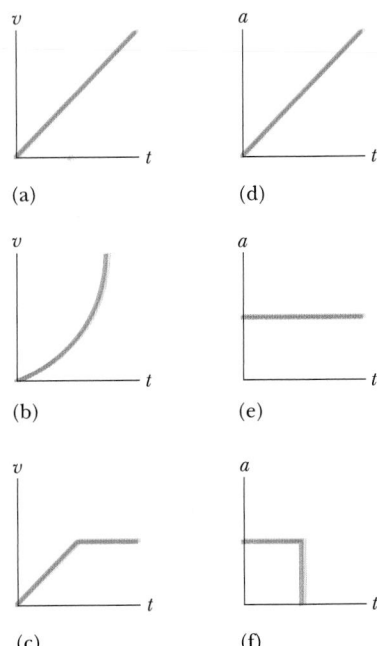

Figure 2.12 (Thinking Physics 1) Match each velocity-time graph to its corresponding acceleration-time graph.

2.5   MOTION DIAGRAMS

The concepts of velocity and acceleration are often confused with each other, but in fact they are quite different quantities. It is instructive to make use of motion diagrams to describe the velocity and acceleration vectors as time progresses while an object is in motion. In order not to confuse these two vector quantities, we use red for velocity vectors and violet for acceleration vectors as in Figure 2.13.[1] The vectors are sketched at several instants during the motion of the object, and the time intervals between adjacent positions are assumed to be equal.

A stroboscopic photograph of a moving object shows several images of the object, each image taken as the strobe light flashes. Figure 2.13 represents three sets of strobe photographs of cars moving along a straight roadway, from left to right. The time intervals between flashes of the stroboscope are equal in each diagram. Let us describe the motion of the car in each diagram.

In Figure 2.13a, the images of the car are equally spaced—the car moves the same distance in each time interval. Thus, the car moves with *constant positive velocity* and has zero acceleration.

In Figure 2.13b, the images of the car become farther apart as time progresses. In this case, the velocity vector increases in time because the car's displacement between adjacent positions increases as time progresses. Thus, the car is moving with a *positive velocity*, and a *positive acceleration*.

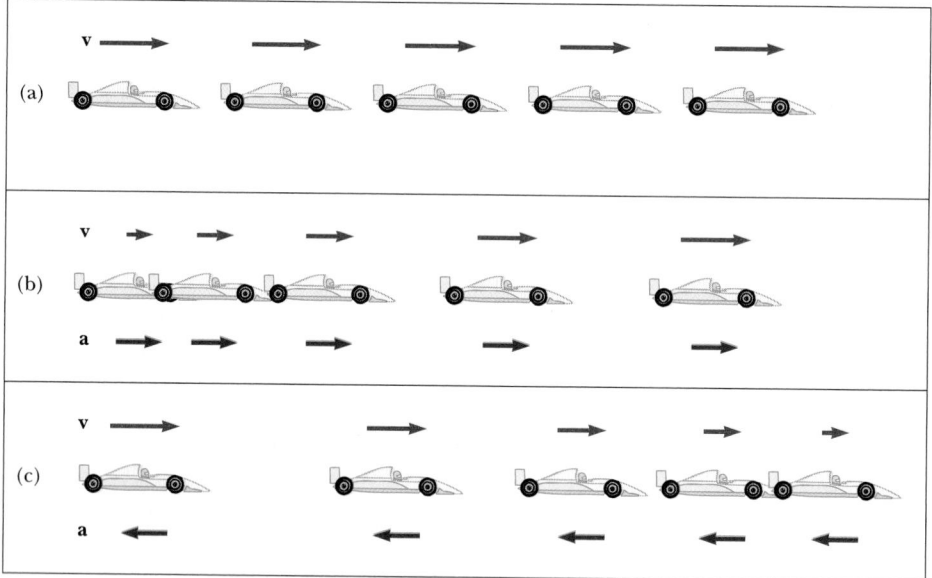

Figure 2.13   (a) Motion diagram for a car moving at constant velocity (zero acceleration). (b) Motion diagram for a car whose constant acceleration is in the direction of its velocity. The velocity vector at each instant is indicated by a red arrow, and the constant acceleration vector by a violet arrow. (c) Motion diagram for a car whose constant acceleration is in the direction *opposite* the velocity at each instant.

---

[1] Items marked with the icon are accompanied by Interactive Physics 4.0 simulations on disk.

In Figure 2.13c, the car slows as it moves to the right because its displacement between adjacent positions decreases as time progresses. In this case, the car moves initially to the right with a constant negative acceleration. The velocity vector decreases in time and eventually reaches zero. (This type of motion is exhibited by a car that skids to a stop after applying its brakes.) From this diagram we see that the acceleration and velocity vectors are *not* in the same direction. The car is moving with a *positive velocity*, but with a *negative acceleration*.

You should be able to construct motion diagrams for a car that moves initially to the left with a constant positive or negative acceleration. You should also construct appropriate motion diagrams after completing the mathematical solutions to kinematic problems, to see if your answers are consistent with the diagrams.

2.6 ONE-DIMENSIONAL MOTION WITH CONSTANT ACCELERATION

Most of the applications in this book that deal with mechanics will be concerned with objects moving with *constant acceleration*. This type of motion is important because it applies to many objects in nature. For example, an object in free fall near the Earth's surface moves in the vertical direction with constant acceleration, assuming that air resistance can be neglected. When an object moves with constant acceleration, *the average acceleration equals the instantaneous acceleration*. Consequently, the velocity increases or decreases at the same rate throughout the motion.

Because the average acceleration equals the instantaneous acceleration when a is constant, we can eliminate the bar used to denote average values from our defining equation for acceleration. That is, because $\bar{a} = a$, we can write Equation 2.4 as

$$a = \frac{v_f - v_i}{t_f - t_i}$$

For convenience, let $t_i = 0$ and t_f be any arbitrary time t. Also, we shall let $v_i = v_0$ (the initial velocity at $t = 0$) and $v_f = v$ (the velocity at any arbitrary time t). With this notation, we can express the acceleration as

$$a = \frac{v - v_0}{t}$$

or

$$v = v_0 + at \qquad \text{(for constant } a) \qquad \qquad \textbf{[2.6]}$$

One of the features of one-dimensional motion with constant acceleration is the manner in which the initial, final, and average velocities are related. Because the velocity is increasing or decreasing *uniformly* with time, we can express the **average velocity** in any time interval as the arithmetic average of the initial velocity, v_0, and the final velocity, v:

$$\bar{v} = \frac{v_0 + v}{2} \qquad \text{(for constant } a) \qquad \qquad \textbf{[2.7]}$$

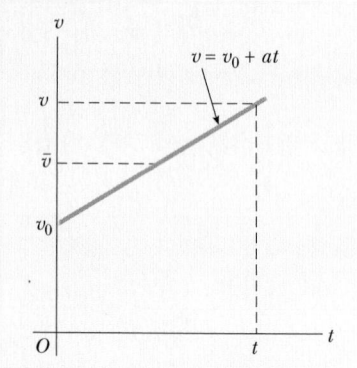

Figure 2.14   The velocity varies linearly with time for a particle moving with constant acceleration. The average velocity is the mean value of the initial and final velocities.

Note that this expression is valid only when the acceleration is constant — that is, when the velocity changes uniformly with time. The graphical interpretation of $\bar{v}$ is shown in Figure 2.14. As you can see, the velocity varies linearly with time according to Equation 2.6.

We can now use this result along with the defining equation for average velocity, Equation 2.2, to obtain an expression for the displacement of an object as a function of time. Again we choose $t_i = 0$, and for convenience we shall also assume that we have selected our coordinate system so that the initial position of the object under consideration is at the origin; that is, $x_i = 0$. This gives

$$x = \bar{v}t = \left(\frac{v_0 + v}{2}\right) t$$

$$x = \tfrac{1}{2}(v + v_0)t \tag{2.8}$$

We can obtain another useful expression for displacement by substituting the equation for v (Eq. 2.6) into Equation 2.8:

$$x = \tfrac{1}{2}(v_0 + at + v_0)t$$

$$x = v_0 t + \tfrac{1}{2}at^2 \qquad \text{(for constant } a) \tag{2.9}$$

Figure 2.15 shows a plot of x versus t for this equation. For example, you should be able to show that the area under the curve shown in Figure 2.14 is equal to $v_0 t + \tfrac{1}{2}at^2$, which is equal to the displacement. In fact, the area under *any* curve of v versus t can be shown to be equal to the displacement of the object.

Finally, we can obtain an expression that does not contain time by substituting the value of t from Equation 2.6 into Equation 2.8. This gives

$$x = \tfrac{1}{2}(v + v_0)\left(\frac{v - v_0}{a}\right) = \frac{v^2 - v_0^2}{2a}$$

$$v^2 = v_0^2 + 2ax \qquad \text{(for constant } a) \tag{2.10}$$

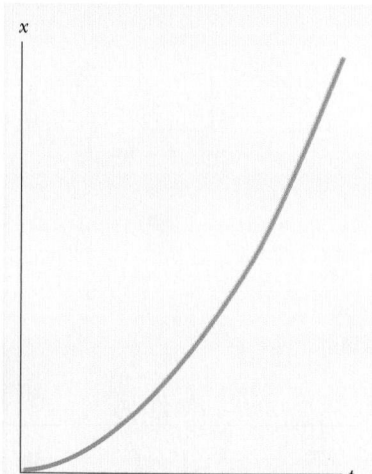

Figure 2.15   The displacement versus time for an object moving with constant acceleration. Note that the displacement varies as t^2. The constantly increasing slope indicates that the velocity increases with time.

TABLE 2.3   Equations for Motion in a Straight Line Under Constant Acceleration

Equation	Information Given by Equation
$v = v_0 + at$	Velocity as a function of time
$x = \tfrac{1}{2}(v_0 + v)t$	Displacement as a function of velocity
$x = v_0 t + \tfrac{1}{2}at^2$	Displacement as a function of time
$v^2 = v_0^2 + 2ax$	Velocity as a function of displacement

Note: Motion is along the x axis. At $t = 0$, the particle is at the origin ($x_0 = 0$) and its velocity is v_0.

Equations 2.6 through 2.10 may be used to solve any problem in one-dimensional motion with constant acceleration. The four equations that are used most often are listed in Table 2.3 for convenience.

The best way to gain confidence in the use of these equations is to work a number of problems. Many times you will discover that there is more than one method for solving a given problem.

Thinking Physics 2

Consider the following motions of an object in one dimension: (a) A ball is thrown directly upward, rises to a highest point and falls back into the thrower's hand. (b) A race car starts from rest and speeds up to 100 m/s. (c) A *Voyager* spacecraft drifts through space at constant velocity. Are there any points in the motion of these particles at which the average velocity (over the entire interval) and the instantaneous velocity (at an instant of time within the interval) are the same? If so, identify the point(s).

Explanation   (a) The average velocity for the thrown ball is zero—the ball returns to the starting point at the end of the time interval. There is one point at which the instantaneous velocity is zero—at the top of the motion. (b) The average velocity for the motion of the race car cannot be evaluated unambiguously with the information given, but it must be some value between 0 and 100 m/s. Because the car will have every instantaneous velocity between 0 and 100 m/s at some time during the interval, there must be some instant at which the instantaneous velocity is equal to the average velocity. (c) Because the instantaneous velocity of the spacecraft is constant, its instantaneous velocity over *any* time interval and its average velocity at *any* time are the same.

Problem-Solving Strategy

Accelerated Motion

The following procedure is recommended for solving problems involving accelerated motion:

1. Make sure all the units in the problem are consistent. That is, if distances are measured in meters, be sure that velocities have units of meters per second and accelerations have units of meters per second per second.
2. Chose a coordinate system.
3. Make a list of all the quantities given in the problem and a separate list of those to be determined.
4. Select from the list of kinematic equations the one or ones that will enable you to determine the unknowns.
5. Construct an appropriate motion diagram, and check to see if your answers are consistent with the diagram.

EXAMPLE 2.3   The Indianapolis 500

A race car starting from rest accelerates at a rate of 5.00 m/s². What is the velocity of the car after it has traveled 100 ft?

Reasoning and Solution   You should refer to the preceding Problem-Solving Strategy section to see how the steps indicated there are applied to this example. (Step 1) Be sure that the units you use are consistent. The units in this problem are *not* consistent. If we choose to leave the distance in feet, we must change the length dimension of the units of acceleration from meters to feet. Alternatively, we can leave the units of acceleration as they are and convert the distance traveled to meters. Let's do the latter. The table of conversion factors on the inside front cover gives 1 m = 3.281 ft, so we find that 1 ft = 0.305 m and, thus, 100 ft = 30.5 m.

   (Step 2) You must choose a coordinate system. A convenient one for this problem is shown in Figure 2.16. The origin of the coordinate system is at the initial location of the car, and the positive direction is to the right. Using this convention, we require that velocities, accelerations, and displacements to the right are positive, and vice versa.

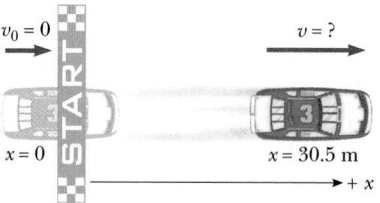

$v_0 = 0$ $\qquad\qquad\qquad$ $v = ?$

$x = 0$ $\qquad\qquad\qquad\qquad$ $x = 30.5$ m

$+x$

Figure 2.16   (Example 2.3)

   (Step 3) Make a list of the quantities given in the problem and a separate list of those to be determined:

Given	To Be Determined
$v_0 = 0$	v
$a = +5.00$ m/s²	
$x = +30.5$ m	

   (Step 4) Select from the kinematic equations (Table 2.3) those that will allow you to determine the unknowns. In the present case, the equation

$$v^2 = v_0^2 + 2ax$$

is our best choice because it will give us a value for v directly:

$$v^2 = (0)^2 + 2(5.00 \text{ m/s}^2)(30.5 \text{ m}) = 305 \text{ m}^2/\text{s}^2$$

from which

$$v = \sqrt{305 \text{ m}^2/\text{s}^2} = \pm 17.5 \text{ m/s}$$

Because the car is moving to the right, we choose +17.5 m/s as the correct solution for v. Alternatively, the problem can be solved by using $x = v_0 t + \frac{1}{2}at^2$ to find t and then using the expression $v = v_0 + at$ to find v. Try it!

EXAMPLE 2.4   Watch Out for the Speed Limit

A car traveling at a constant speed of 30.0 m/s ($\approx$67 mi/h) passes a trooper hidden behind a billboard. One second after the speeding car passes the billboard, the trooper sets off in chase with a constant acceleration of 3.00 m/s². How long does it take the trooper to overtake the speeding car?

Workbook Problem 9 – (Workbook page 26) is similar to Example 2.4.

Reasoning   To solve this problem algebraically, we write an expression for the position of each vehicle as a function of time. It is convenient to choose the origin at the position of the billboard and take $t = 0$ as the time the trooper begins moving. At that instant, the speeding car has already traveled a distance of 30.0 m, because it travels at a constant speed of 30.0 m/s. Thus, the initial position of the speeding car is $x_0 = 30.0$ m.

Solution   Because the car moves with constant speed, its acceleration is zero, and applying Equation 2.9 gives

$$x_C = 30.0 \text{ m} + (30.0 \text{ m/s})t$$

Note that at $t = 0$, this expression does give the car's correct initial position, $x_C = x_0 = 30.0$ m.

    For the trooper, who starts from the origin at $t = 0$, we have $x_0 = 0$, $v_0 = 0$, and $a = 3.00$ m/s². Hence, the position of the trooper as a function of time is

$$x_T = \tfrac{1}{2}at^2 = \tfrac{1}{2}(3.00 \text{ m/s}^2)t^2$$

The trooper overtakes the car at the instant that $x_T = x_C$, or

$$\tfrac{1}{2}(3.00 \text{ m/s}^2)t^2 = 30.0 \text{ m} + (30.0 \text{ m/s})t$$

This gives the quadratic equation

$$1.50t^2 - 30.0t - 30.0 = 0$$

whose positive solution is $t = 21.0$ s. Note that in this time interval, the trooper travels a distance of about 660 m.

    Frequently in this book, an example will be followed by an exercise whose purpose is to test your understanding of the example. The answer will be provided immediately after the exercise, when appropriate. Here is your first exercise, which relates to Example 2.4.

Exercise   This problem also can be easily solved graphically. On the *same* graph, plot the position versus time for each vehicle, and from the intersection of the two curves determine the time at which the trooper overtakes the speeding car.

2.7   FREELY FALLING OBJECTS

It is now well known that all objects dropped near the surface of the Earth in the absence of air resistance fall toward the Earth with the same constant acceleration. It was not until about 1600 that this conclusion was accepted. Prior to that time,

the teachings of the great philosopher Aristotle (384–322 B.C.) had held that heavier objects fell faster than lighter ones.

It was Galileo who originated our present-day ideas concerning falling objects. There is a legend that he discovered the law of falling objects by observing that two different weights dropped simultaneously from the Leaning Tower of Pisa hit the ground at approximately the same time. Although it is doubtful that this particular experiment was carried out, it is well established that Galileo performed many systematic experiments on objects moving on inclined planes. In his experiments he rolled balls down a slight incline and measured the distances they covered in successive time intervals. The purpose of the incline was to reduce the acceleration and enable Galileo to make accurate measurements of the time intervals. (Some people refer to this experiment as "diluting gravity.") By gradually increasing the slope of the incline, he was finally able to draw conclusions about freely falling objects, because a falling ball is equivalent to a ball falling down a vertical incline. Galileo's achievements in the science of mechanics paved the way for Newton in his development of the laws of motion, which we shall study in Chapter 4.

You might want to try the following experiment. Simultaneously drop a coin and a crumpled-up piece of paper from the same height. If the effects of air friction are negligible, both will have the same motion and hit the floor at the same time. In the idealized case, in which air resistance is absent, such motion is referred to as *free fall*. If this same experiment could be conducted in a vacuum, where air friction is zero, the paper and coin would fall with the same acceleration, regardless of the shape of the paper. (See the photographs in Physics in Action.) On August 2, 1971, such a demonstration was conducted on the Moon by astronaut David Scott. He simultaneously released a hammer and a feather, and they fell with the same acceleration to the lunar surface. This demonstration surely would have pleased Galileo!

We shall denote the free-fall acceleration with the symbol **g.** The magnitude of **g** decreases with increasing altitude. Furthermore, slight variations in the magnitude of **g** occur with latitude. However, at the surface of the Earth the magnitude of **g** is approximately 9.8 m/s², or 980 cm/s², or 32 ft/s². Unless stated otherwise, we shall use the value 9.80 m/s² when doing calculations. Furthermore, we shall assume that the vector **g** is directed downward toward the center of the Earth.

When we use the expression *freely falling object* we do not necessarily mean an object dropped from rest. **A freely falling object is an object moving under the influence of gravity only, regardless of its initial motion. Objects thrown upward or downward and those released from rest are all falling freely once they are released! Once they are in free fall, all objects have an acceleration downward, which is the free-fall acceleration g.**

If we neglect air resistance and assume that the free-fall acceleration does not vary with altitude over short vertical distances, then the motion of a freely falling object is equivalent to motion in one dimension under constant acceleration. Therefore, the equations developed in Section 2.6 for objects moving with constant acceleration can be applied. The only modification that we need to make in these equations for freely falling objects is to note that the motion is in the vertical direction (the y direction) rather than the horizontal (x) direction, and that the acceleration is downward and has a magnitude of 9.80 m/s². Thus, we always take $a = -g = -9.80$ m/s², where the minus sign means that the acceleration of a freely falling object is downward. In Chapter 7 we shall study how to deal with variations in g with altitude.

PHYSICS IN ACTION

Freely Falling Objects

On the left is a multiflash photograph of a falling billiard ball. As the ball falls, the spacing between successive images increases, indicating that the ball accelerates downward. The motion diagram shows that the ball's velocity (red arrows) increases with time, and its acceleration (violet arrows) remains constant.

The middle photograph shows a multiflash image of two freely falling balls released simultaneously. The ball on the left side is solid, and the ball on the right is a hollow Ping-Pong ball. The time interval between flashes is (1/30) s and the scale is in centimeters. Can you determine **g** from these data? Note that the effect of air resistance is greater for the smaller, hollow ball on the right, as indicated by the smaller separation between successive images.

In the photograph on the right, a feather and an apple are released from rest in a 4-ft vacuum chamber. The trap door that released the two objects was opened with an electronic switch at the same instant the camera shutter was opened. The two objects fell at approximately the same rate, as indicated by the horizontal alignment of the multiple images. When air resistance is negligible, all objects fall with the same acceleration, regardless of their masses.

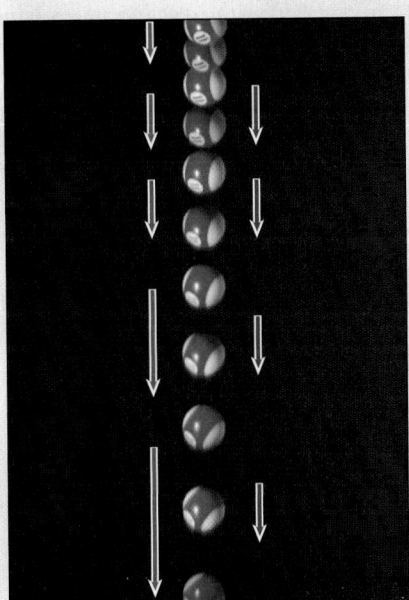

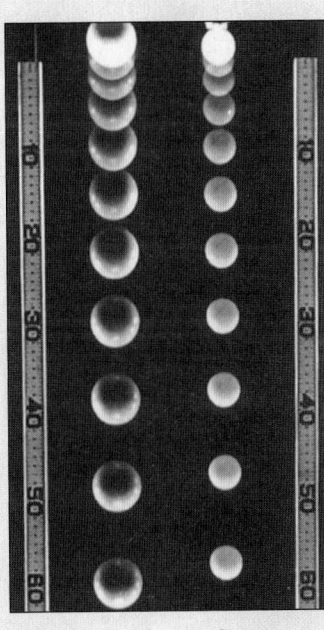

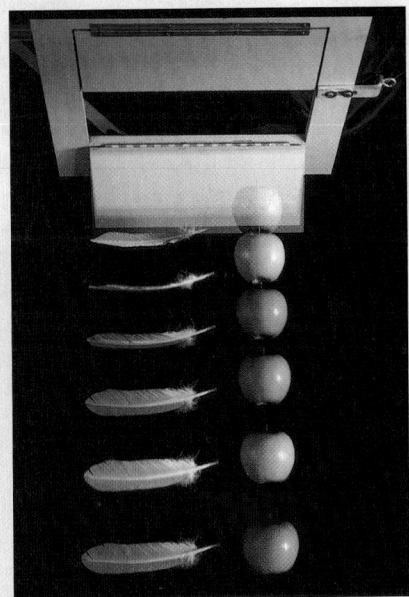

(Richard Megna 1990, Fundamental Photographs)

(Education Development Center, Newton, MA)

(© James Sugar, Black Star)

Applying Physics 2

A skydiver jumps out of a helicopter. A few seconds later, another skydiver jumps out, so that they both fall along the same vertical line. Ignore air resistance, so that both skydivers fall with the same acceleration. Does the vertical distance between them stay the same? Does the difference in their velocities stay the same? If they were connected

APPLICATION

Skydiving.

QUICKLAB

Reaction Time

To measure your reaction time, have a friend hold a ruler vertically between your index finger and thumb as shown. Note the position of the ruler with respect to your index finger. Your friend must release the ruler and you must catch it (without moving your hand downward) after it falls through some distance that you measure from the new position of the ruler. The ruler, a freely falling object, falls through a distance $d = \frac{1}{2}gt^2$, where t is the reaction time and $g = 9.80$ m/s². Repeat this measurement of d five times, average your results, and calculate an average value of t. Now measure your friend's reaction time using the same procedure. Compare your results. For most people, the reaction time is at best about 0.2 s.

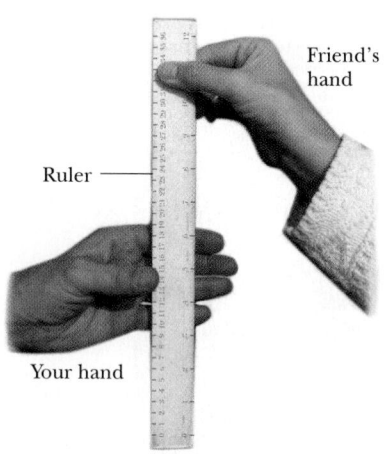

Friend's hand

Ruler

Your hand

by a long bungee cord, would the tension in the cord become larger, smaller, or stay the same?

Explanation   At any given instant of time, the velocities of the jumpers are definitely different, because one had a head start. In a time interval after this instant, however, each jumper increases his or her velocity by the same amount, because they have the same acceleration. Thus, the difference in velocities remains the same. The first jumper will always be moving with a higher velocity than the second. Thus, in a given time interval, the first jumper will cover more distance than the second. Thus, the separation distance between them increases. As a result, once the distance between the divers reaches the value at which the bungee cord becomes straight, the tension in the bungee cord will increase as the jumpers move apart. Of course, the bungee cord will pull downward on the second jumper and upward on the first, resulting in a force in addition to gravity that will affect their motion. If the bungee cord is robust enough, this will result in the separation distance of the divers becoming smaller as the cord pulls them together.

EXAMPLE 2.5   Look Out Below!

A golf ball is released from rest at the top of a very tall building. Neglecting air resistance, calculate the position and velocity of the ball after 1.00, 2.00, and 3.00 s.

Solution   We choose our coordinates so that the starting point of the ball is at the origin ($y_0 = 0$ at $t = 0$) and remember that we have defined y to be positive upward. Because $v_0 = 0$, and $a = -g = -9.80$ m/s², Equations 2.6 and 2.9 become

$$v = at = (-9.80 \text{ m/s}^2)t$$

$$y = \tfrac{1}{2}at^2 = \tfrac{1}{2}(-9.80 \text{ m/s}^2)t^2$$

where t is in seconds, v is in meters per second, and y is in meters. These expressions give the velocity and displacement at any time t after the ball is released. Therefore, at $t = 1.00$ s,

$$v = (-9.80 \text{ m/s}^2)(1.00 \text{ s}) = \boxed{-9.80 \text{ m/s}}$$

$$y = \tfrac{1}{2}(-9.80 \text{ m/s}^2)(1.00 \text{ s})^2 = \boxed{-4.90 \text{ m}}$$

Likewise, at $t = 2.00$ s, we find that $v = -19.6$ m/s and $y = -19.6$ m. Finally, at $t = 3.00$ s, $v = -29.4$ m/s and $y = -44.1$ m. The minus sign for v indicates that the velocity vector is directed downward, and the minus sign for y indicates a position below the original position.

Exercise   Calculate the position and velocity of the ball after 4.00 s.

Answer -78.4 m, -39.2 m/s.

EXAMPLE 2.6   Not a Bad Throw for a Rookie!

A stone is thrown from the top of a building with an initial velocity of 20.0 m/s straight upward. The building is 50.0 m high, and the stone just misses the edge of the roof on its way down, as in Figure 2.17. Determine (a) the time needed for the stone to reach its maximum height, (b) the maximum height, (c) the time needed for the stone to return to the level of the thrower, (d) the velocity of the stone at this instant, and (e) the velocity and position of the stone at $t = 5.00$ s.

Solution (a) To find the time necessary for the stone to reach the maximum height, use Equation 2.6, $v = v_0 + at$, noting that $v = 0$ at maximum height:

$$20.0 \text{ m/s} + (-9.80 \text{ m/s}^2)t = 0$$

$$t = \frac{20.0 \text{ m/s}}{9.80 \text{ m/s}^2} = \boxed{2.04 \text{ s}}$$

(b) This value of time can be substituted into Equation 2.9, $y = v_0 t + \frac{1}{2}at^2$, to give the maximum height as measured from the position of the thrower:

$$y_{max} = (20.0 \text{ m/s})(2.04 \text{ s}) + \tfrac{1}{2}(-9.80 \text{ m/s}^2)(2.04 \text{ s})^2 = \boxed{20.4 \text{ m}}$$

(c) When the stone is back at the height of the thrower, the y coordinate is zero. From the expression $y = v_0 t + \frac{1}{2}at^2$ (Eq. 2.9), with $y = 0$, we obtain the expression

$$0 = 20.0t - 4.90t^2$$

This is a quadratic equation and has two solutions for t. The equation can be factored to give

$$t(20.0 - 4.90t) = 0$$

One solution is $t = 0$, corresponding to the time at which the stone starts its motion. The other solution is $t = 4.08$ s, which is the solution we are after.

(d) The value for t found in (c) can be inserted into $v = v_0 + at$ (Eq. 2.6) to give

$$v = 20.0 \text{ m/s} + (-9.80 \text{ m/s}^2)(4.08 \text{ s}) = \boxed{-20.0 \text{ m/s}}$$

Note that the velocity of the stone when it arrives back at its original height is equal in magnitude to the stone's initial velocity but opposite in direction. This indicates that the motion is symmetric.

(e) From $v = v_0 + at$ (Eq. 2.6), the velocity after 5.00 s is

$$v = 20.0 \text{ m/s} + (-9.80 \text{ m/s}^2)(5.00 \text{ s}) = \boxed{-29.0 \text{ m/s}}$$

We can use $y = v_0 t + \frac{1}{2}at^2$ (Eq. 2.9) to find the position of the particle at $t = 5.00$ s:

$$y = (20.0 \text{ m/s})(5.00 \text{ s}) + \tfrac{1}{2}(-9.80 \text{ m/s}^2)(5.00 \text{ s})^2 = \boxed{-22.5 \text{ m}}$$

Exercise Find the velocity of the stone just before it hits the ground.

Answer -37.0 m/s.

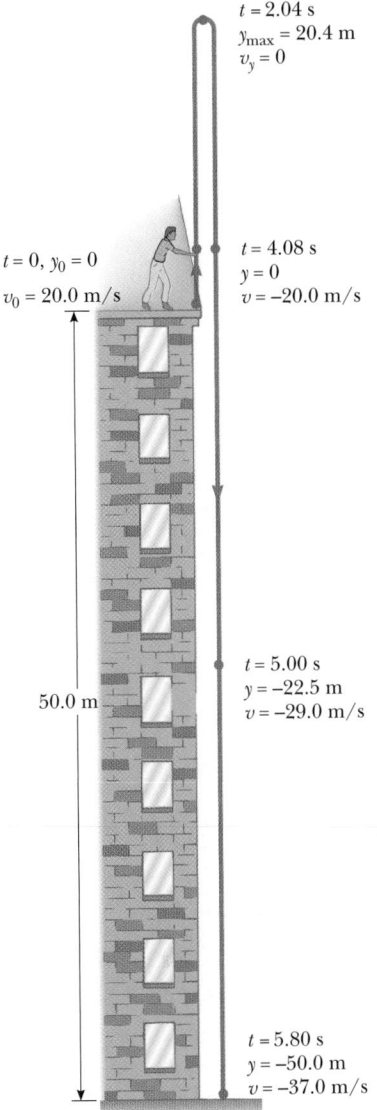

Figure 2.17 (Example 2.6) Position and velocity versus time for a freely falling object thrown initially upward with a velocity of $v_0 = +20.0$ m/s.

SUMMARY

The **displacement** of an object moving along the x axis is defined as the change in position of the object, and is given by

$$\Delta x \equiv x_f - x_i \qquad \text{[2.1]}$$

where x_i is the initial position of the object and x_f is its final position.

The **average velocity** of an object moving along the x axis during some time interval is equal to the displacement of the object, Δx, divided by the time interval, Δt, during which the displacement occurred.

3.8

$$\bar{v} \equiv \frac{\Delta x}{\Delta t} = \frac{x_f - x_i}{t_f - t_i} \qquad \text{[2.2]}$$

The average velocity is equal to the slope of the straight line joining the initial and final points on a graph of the position of the object versus time.

The slope of the line tangent to the position-time curve at some point is equal to the instantaneous velocity at that time. The **instantaneous speed** of an object is defined as the magnitude of the instantaneous velocity.

The **average acceleration** of an object during some time interval is defined as the change in velocity, Δv, divided by the time interval, Δt, during which the change occurred:

$$\bar{a} \equiv \frac{\Delta v}{\Delta t} = \frac{v_f - v_i}{t_f - t_i} \qquad \text{[2.4]}$$

The **instantaneous acceleration** of an object at a certain time equals the slope of a velocity-time graph at that instant.

A **vector** quantity is any quantity that is characterized by both a magnitude and a direction. In contrast, a **scalar** quantity such as mass has a magnitude only. Displacement, velocity, and acceleration are all vector quantities. In the case of motion along a straight line, we use algebraic signs to describe the directions of the vectors. For example, an object moving to the right with a speed of 8 m/s would have a velocity of $+8$ m/s, assuming the positive direction has been chosen to point to the right. If the object were moving to the left, its velocity would be -8 m/s.

The equations that describe the motion of an object moving with constant acceleration along the x axis are

$$v = v_0 + at \qquad \text{[2.6]}$$

$$x = v_0 t + \tfrac{1}{2} at^2 \qquad \text{[2.9]}$$

$$v^2 = v_0^2 + 2ax \qquad \text{[2.10]}$$

An object falling in the presence of the Earth's gravity experiences a free-fall acceleration directed toward the center of the Earth. If air friction is neglected and if the altitude of the falling object is small compared with the Earth's radius, then one can assume that the free-fall acceleration, **g,** is constant over the range of motion, where g is equal to 9.80 m/s^2, or 32.0 ft/s^2. Assuming that the positive direction for y is chosen to be upward, the acceleration is $-g$ (downward) and the equations describing the motion of the falling object are the same as the foregoing equations, with the substitutions $x \rightarrow y$ and $a \rightarrow -g$.

MULTIPLE-CHOICE QUESTIONS

1. People become uncomfortable in an elevator if it accelerates downward at a rate such that it attains a speed of about 6 m/s after ten stories (about 30 m). What is this acceleration?
   (a) 9.8 m/s^2 (b) 0.33 m/s^2 (c) 0.6 m/s^2
   (d) 1.2 m/s^2 (e) 2 m/s^2

2. Races are timed to 1/1000 of a second. What distance could a roller blader moving at a speed of 10 m/s travel in that period of time?
   (a) 10 mm   (b) 10 cm   (c) 10 m   (d) 1 mm   (e) 1 km

3. A rock is thrown downward from the top of a 40-m tower with an initial speed of 12 m/s. Assuming negligi-

ble air resistance, what is the speed of the rock just before hitting the ground?
(a) 28 m/s (b) 30 m/s (c) 56 m/s (d) 784 m/s

4. An x versus t graph is drawn for a ball moving in one direction. The graph starts at the origin, increases with a positive slope for five seconds, and then levels off to a zero slope for 2 s. We can be sure that
   (a) the speed is decreasing during the first 5 seconds.
   (b) the acceleration is constant throughout the complete motion.

(c) the speed is zero between $t = 5$ s and $t = 7$ s.
(d) the displacement is zero throughout the motion.
(e) more than one of the above is true.

5. A ball is thrown straight up. For which situation are both the instantaneous velocity and the acceleration zero?
   (a) on the way up
   (b) at the top of the flight path
   (c) on the way down
   (d) none of the above

CONCEPTUAL QUESTIONS

1. If a car is traveling eastward, can its acceleration be westward? Explain.

2. The speed of sound in air is 331 m/s. During the next thunderstorm, try to estimate your distance from a lightning bolt by measuring the time lag between the flash and the thunderclap. You can ignore the time it takes for the light flash to reach you. Why?

3. Can an object having constant acceleration ever stop and *stay* stopped?

4. If the average velocity of an object is zero in some time interval, what can you say about the displacement of the object for that interval?

5. A child throws a marble into the air with an initial speed v_0. Another child drops a ball at the same instant. Compare the accelerations of the two objects while they are in flight.

6. The strobe photographs in Figure Q2.6 were taken of a disk moving from left to right under different conditions, and the time interval between images is constant. Taking the direction to the right to be positive, describe the motion of the disk in each case. For which case is (a) the acceleration positive? (b) the acceleration negative? (c) the velocity constant?

7. Can the instantaneous velocity of an object at an instant of time ever be greater in magnitude than the average velocity over a time interval containing the instant? Can it ever be less?

8. Car A, traveling from New York to Miami, has a speed of 25 m/s. Car B, traveling from New York to Chicago, also has a speed of 25 m/s. Are their velocities equal? Explain.

9. A ball is thrown vertically upward. (a) What are its velocity and acceleration when it reaches its maximum altitude? (b) What is its acceleration just before it hits the ground?

10. A rule of thumb for driving is that a separation of one car length for each 10 mi/h of speed should be maintained between moving vehicles. Assuming a constant reaction time, discuss the relevance of this rule for (a) mo-

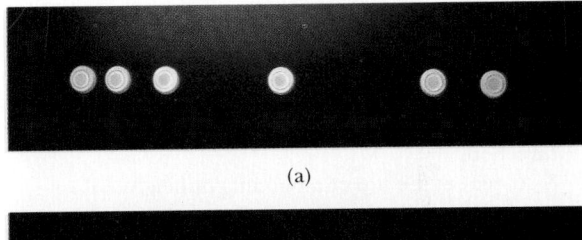

(a)

(b)

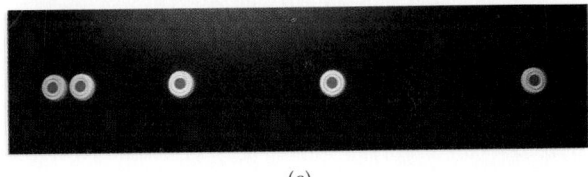

(c)

Figure Q2.6   *(Courtesy David Rogers)*

tion with constant velocity and (b) motion with constant acceleration.

11. Two cars are moving in the same direction in parallel lanes along a highway. At some instant, the velocity of car A exceeds the velocity of car B. Does this mean that the acceleration of A is greater than that of B? Explain.

12. Consider the following combinations of signs and values for velocity and acceleration of a particle with respect to a one-dimensional x-axis:

	Velocity	Acceleration		Velocity	Acceleration
a.	Positive	Positive	e.	Negative	Negative
b.	Positive	Negative	f.	Negative	Zero
c.	Positive	Zero	g.	Zero	Positive
d.	Negative	Positive	h.	Zero	Negative

Figure Q2.15

Describe what a particle is doing in each case, and give a real-life example for an automobile on an east–west one-dimensional axis, with east considered the positive direction.

13. A student at the top of a building of height h throws one ball upward with a speed of v_0 and then throws a second ball downward with the same initial speed, v_0. How do the final velocities of the balls compare when they reach the ground?

14. You drop a ball from a window on an upper floor of a building. It strikes the ground with velocity v. You now repeat the drop, but you have a friend down on the street who throws another ball upward at velocity v. Your friend throws the ball upward at exactly the same time that you drop yours from the window. At some location, the balls pass each other. Is this location *at the* halfway point between window and ground, *above* this point, or *below* this point?

15. A pebble is dropped into a water well, and the splash is heard 16 s later, as illustrated in the cartoon strip shown in Figure Q2.15. Estimate the distance from the rim of the well to the water's surface.

16. A ball rolls in a straight line along the horizontal direction. Using motion diagrams (or multiflash photographs) as in Figure 2.13, describe the velocity and acceleration of the ball for each of the following situations: (a) The ball moves to the right at a constant speed. (b) The ball moves from right to left and continually slows down. (c) The ball moves from right to left and continually speeds up. (d) The ball moves to the right, first speeding up at a constant rate, and then slowing down at a constant rate.

PROBLEMS

1, 2, 3 = straightforward, intermediate, challenging   ☐ = full solution available in Study Guide/Student Solutions Manual   ◤ = Core Concepts Workbook
WEB = solution posted at **http://www.harcourtcollege.com/physics/cptech**   🐛 = biomedical application   ⚙ = Interactive Physics

Section 2.2   Average Velocity

Section 2.3   Instantaneous Velocity

1. A person travels by car from one city to another with different constant speeds between pairs of cities. She drives for 30.0 min at 80.0 km/h, 12.0 min at 100 km/h, and 45.0 min at 40.0 km/h, and spends 15.0 min eating lunch and buying gas. (a) Determine the average speed for the trip. (b) Determine the distance between the initial and final cities along this route.

2. An athlete swims the length of a 50.0-m pool in 20.0 s and makes the return trip to the starting position in 22.0 s. Determine her average velocities in (a) the first half of the swim, (b) the second half of the swim, and (c) the round trip.

3. Two boats start together and race across a 60 km wide lake and back. Boat A goes across at 60 km/h and returns at 60 km/h. Boat B goes across at 30 km/h and its crew, realizing how far behind it is getting, returns at 90 km/h. Turn-around times are negligible, and the boat that completes the round-trip first wins. (a) Which boat wins and by how much? (Or is it a tie?) (b) What is the average velocity of the winning boat?

4. A certain bacterium swims with a speed of 3.5 μm/s. How long would it take this bacterium to swim across a petri dish having a diameter of 8.4 cm?

Problem 2 is the same as Workbook Problem 8 (Workbook page 24).

5. The Olympic record for the marathon is 2h, 9 min, 21 s. The marathon distance is 26 mi, 385 yd. Determine the average speed (in miles per hour) of this record.

6. A tennis player moves in a straight-line path as shown in Figure P2.6. Find her average velocities in the time intervals (a) 0 to 1.0 s, (b) 0 to 4.0 s, (c) 1.0 s to 5.0 s, and (d) 0 to 5.0 s.

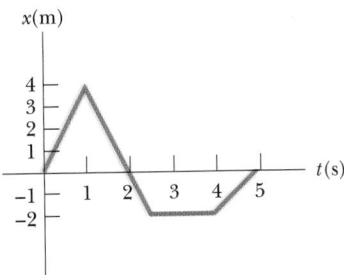

Figure P2.6   (Problems 6 and 15)

7. Two cars travel in the same direction along a straight highway, one at a constant speed of 55 mi/h and the other at 70 mi/h. (a) Assuming that they start at the same point, how much sooner does the faster car arrive at a destination 10 mi away? (b) How far must the faster car travel before it has a 15-min lead on the slower car?

8. If the average speed of an orbiting space shuttle is 19 800 mi/h, determine the time required for it to circle the Earth. Make sure you consider the fact that the shuttle is orbiting about 200 mi above the Earth's surface, and assume that the Earth's radius is 3963 miles.

9. A person takes a trip, driving with a constant speed of 89.5 km/h except for a 22.0-min rest stop. If the person's average speed is 77.8 km/h, how much time is spent on the trip and how far does the person travel?

10. A tortoise can run with a speed of 10.0 cm/s, and a hare can run 20 times as fast. In a race, they both start at the same time, but the hare stops to rest for 2.0 minutes. The tortoise wins by a shell (20 cm). (a) How long does the race take? (b) What is the length of the race?

WEB 11. In order to qualify for the finals in a racing event, a race car must achieve an average speed of 250 km/h on a track with a total length of 1600 m. If a particular car covers the first half of the track at an average speed of 230 km/h, what minimum average speed must it have in the second half of the event in order to qualify?

12. Runner A is initially 4.0 mi west of a flagpole and is running with a constant velocity of 6.0 mi/h due east. Runner B is initially 3.0 mi east of the flagpole and is running with a constant velocity of 5.0 mi/h due west. How far are the runners from the flagpole when they meet?

13. A runner moves so that his positions at certain times are given by the data in the following table. Use these data to construct a table like Table 2.2 in the text. From your

table, find (a) the average velocity during the complete interval and (b) the instantaneous velocity at $t = 2.00$ s.

t(s)	x(m)	t(s)	x(m)
2.00	5.66	2.50	6.32
2.01	5.674	3.00	6.92
2.20	5.93	4.00	8.00

14. A race car moves such that its position fits the relationship

$$x = (5.0 \text{ m/s})t + (0.75 \text{ m/s}^3)t^3$$

where x is measured in meters and t in seconds. (a) Plot a graph of position versus time. (b) Determine the instantaneous velocity at $t = 4.0$ s, using time intervals of 0.40 s, 0.20 s, and 0.10 s. (c) Compare the average velocity during the first 4.0 s with the results of (b).

15. Find the instantaneous velocities of the tennis player of Figure P2.6 at (a) 0.50 s, (b) 2.0 s, (c) 3.0 s, (d) 4.5 s.

Section 2.4 Acceleration

16. A car traveling initially at $+7.0$ m/s accelerates at the rate of $+0.80$ m/s^2 for an interval of 2.0 s. What is its velocity at the end of the acceleration?

17. A car traveling in a straight line has a velocity of $+5.0$ m/s at some instant. After 4.0 s, its velocity is $+8.0$ m/s. What is its average acceleration in this time interval?

18. A tennis ball with a speed of 10.0 m/s is thrown perpendicularly at a wall. After striking the wall, the ball rebounds in the opposite direction with a speed of 8.0 m/s. If the ball is in contact with the wall for 0.012 s, what is the average acceleration of the ball while it is in contact with the wall?

WEB 19. A certain car is capable of accelerating at a rate of $+0.60$ m/s^2. How long does it take for this car to go from a speed of 55 mi/h to a speed of 60 mi/h?

20. The velocity-versus-time graph for an object moving along a straight path is shown in Figure P2.20.

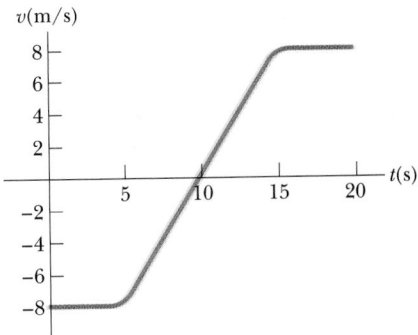

Figure P2.20

(a) Find the average accelerations of this object during the time intervals 0 to 5.0 s, 5.0 s to 15 s, and 0 to 20 s. (b) Find the instantaneous accelerations at 2.0 s, 10 s, and 18 s.

21. The engine of a model rocket accelerates the rocket vertically upward for 2.0 s as follows: at $t = 0$, its speed is zero; at $t = 1.0$ s, its speed is 5.0 m/s; at $t = 2.0$ s, its speed is 16 m/s. Plot a velocity-time graph for this motion, and from it determine (a) the average acceleration during the 2.0-s interval and (b) the instantaneous acceleration at $t = 1.5$ s.

Section 2.6   One-Dimensional Motion with Constant Acceleration

22. Jules Verne in 1865 proposed sending men to the Moon by firing a space capsule from a 220-m-long cannon with final velocity of 10.97 km/s. What would have been the unrealistically large acceleration experienced by the space travelers during launch? Compare your answer with the free-fall acceleration, 9.80 m/s².

23. A racing car reaches a speed of 40 m/s. At this instant, it begins a uniform negative acceleration, using a parachute and a braking system, and comes to rest 5.0 s later. (a) Determine the acceleration of the car. (b) How far does the car travel after acceleration starts?

24. A jet plane lands with a velocity of +100 m/s and can accelerate at a maximum rate of −5.0 m/s² as it comes to rest. (a) From the instant it touches the runway, what is the minimum time needed before it can come to rest? (b) Can this plane land on a small island airport where the runway is 0.80 km long?

25. A truck on a straight road starts from rest accelerating at 2.0 m/s² until it reaches a speed of 20 m/s. Then the truck travels for 20 s at constant speed until the brakes are applied, stopping the truck in a uniform manner in an additional 5.0 s. (a) How long is the truck in motion? (b) What is the average velocity of the truck for the motion described?

26. A Cessna aircraft has a lift-off speed of 120 km/h. (a) What minimum constant acceleration does this require if the aircraft is to be airborne after a takeoff run of 240 m? (b) How long does it take the aircraft to become airborne?

27. An electron moving in a straight line has an initial speed of 3.0×10^5 m/s. It undergoes an acceleration of 8.0×10^{14} m/s². (a) How long will it take to reach a speed of 5.4×10^5 m/s? (b) How far will it have traveled in this time?

28. A driver in a car traveling at a speed of 60 mi/h sees a deer 100 m away on the road. Calculate the minimum constant acceleration that is necessary for the car to stop without hitting the deer (assuming that the deer does not move in the meantime).

29. A speedboat increases in speed uniformly from 20 m/s to 30 m/s in a distance of 200 m. Find (a) the magnitude of its acceleration and (b) the time it takes the boat to travel this distance.

30. Two cars are traveling along a straight line in the same direction, the lead car at 25 m/s and the other car at 30 m/s. At the moment the cars are 40 m apart, the lead driver applies the brakes, causing her car to have an acceleration of −2.0 m/s². (a) How long does it take for the lead car to stop? (b) Assuming that the chasing car brakes at the same time as the lead car, what must be the chasing car's minimum negative acceleration so as not to hit the lead car? (c) How long does it take for the chasing car to stop?

31. A record of travel along a straight path is as follows.
    1. Start from rest with constant acceleration of 2.77 m/s² for 15.0 s
    2. Constant velocity for the next 2.05 min
    3. Constant negative acceleration −9.47 m/s² for 4.39 s
    (a) What was the total displacement for the complete trip?
    (b) What were the average speeds for legs 1, 2, and 3 of the trip as well as for the complete trip?

32. A train is traveling down a straight track at 20 m/s when the engineer applies the brakes, resulting in an acceleration of −1.0 m/s² as long as the train is in motion. How far does the train move during a 40 s time interval starting at the instant the brakes are applied?

33. A car accelerates uniformly from rest to a speed of 40.0 mi/h in 12.0 s. (a) Find the distance the car travels during this time and (b) the constant acceleration of the car.

34. A car starts from rest and travels for 5.0 s with a uniform acceleration of +1.5 m/s². The driver then applies the brakes, causing a uniform acceleration of −2.0 m/s². If the brakes are applied for 3.0 s, how fast is the car going at the end of the braking period, and how far has it gone?

35. A train 400 m long is moving on a straight track with a speed of 82.4 km/h. The engineer applies the brakes at a crossing, and later the last car passes the crossing with a speed of 16.4 km/h. Assuming constant acceleration, determine how long the train blocked the crossing. Disregard the width of the crossing.

36. A hockey player is standing on his skates on a frozen pond when an opposing player, moving with a uniform speed of 12 m/s, skates by with the puck. After 3.0 s, the first player makes up his mind to chase his opponent. If he accelerates uniformly at 4.0 m/s², (a) how long does it take him to catch his opponent, and (b) how far has he traveled in this time? (Assume the player with the puck remains in motion at constant speed.)

Problem 22 is similar to Workbook Problem 10 (Workbook page 28).

Section 2.7   Freely Falling Objects

37. A ball is thrown vertically upward with a speed of 25.0 m/s. (a) How high does it rise? (b) How long does it take to reach its highest point? (c) How long does it take to hit the ground after it reaches its highest point? (d) What is its speed when it returns to the level from which it started?

38. A peregrine falcon dives at a pigeon. The falcon starts downward from rest and falls with free-fall acceleration. If the pigeon is 76.0 m below the initial position of the falcon, how long does it take the falcon to reach the pigeon? Assume that the pigeon remains at rest.

39. A small mailbag is released from a helicopter that is descending steadily at 1.50 m/s. After 2.00 s, (a) what is the speed of the mailbag, and (b) how far is it below the helicopter? (c) What are your answers to parts (a) and (b) if the helicopter is rising steadily at 1.50 m/s?

40. A rocket moves upward, starting from rest with an acceleration of +29.4 m/s^2 for 4.00 s. It runs out of fuel at the end of this 4.00 s and continues to move upward for a while. How high does it rise above its original starting point?

41. A ball thrown vertically upward is caught by the thrower after 2.00 s. Find (a) the initial velocity of the ball and (b) the maximum height it reaches.

42. A model rocket is launched straight upward with an initial speed of 50.0 m/s. It accelerates with a constant upward acceleration of 2.00 m/s^2 until its engines stop at an altitude of 150 m. (a) What is the maximum height reached by the rocket? (b) How long after lift-off does the rocket reach its maximum height? (c) How long is the rocket in the air?

43. A parachutist with a camera, both descending at a speed of 10 m/s, releases that camera at an altitude of 50 m. (a) How long does it take the camera to reach the ground? (b) What is the velocity of the camera just before it hits the ground?

ADDITIONAL PROBLEMS

44. The total area under the velocity versus time curve shown in Figure P2.44 represents the distance traveled. Break this area into a rectangular area A_1 and a triangular area A_2. Compute the areas A_1 and A_2 and compare their sum to the result given in Equation 2.8.

45. A bullet is fired through a board 10.0 cm thick in such a way that the bullet's line of motion is perpendicular to the face of the board. If the initial speed of the bullet is 400 m/s and it emerges from the other side of the board with a speed of 300 m/s, find (a) the acceleration of the bullet as it passes through the board and (b) the total time the bullet is in contact with the board.

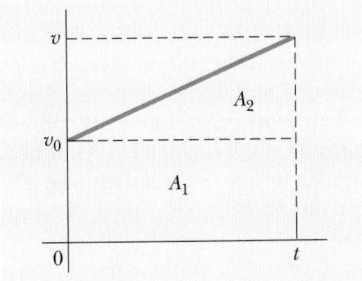

Figure P2.44

46. A ball is thrown upward from the ground with an initial speed of 25 m/s; at the same instant, a ball is dropped from a building 15 m high. After how long will the balls be at the same height?

47. A ranger in a national park is driving at 35.0 mi/h when a deer jumps into the road 200 ft ahead of the vehicle. After a reaction time of t, the ranger applies the brakes to produce an acceleration of $a = -9.00$ ft/s^2. What is the maximum reaction time allowed if she is to avoid hitting the deer?

48. Two students are on a balcony 19.6 m above the street. One student throws a ball vertically downward at 14.7 m/s; at the same instant the other student throws a ball vertically upward at the same speed. The second ball just misses the balcony on the way down. (a) What is the difference in their time in air? (b) What is the velocity of each ball as it strikes the ground? (c) How far apart are the balls 0.800 s after they are thrown?

49. The driver of a car slams on the brakes as a truck approaches the car head-on while slowing uniformly with acceleration -5.60 m/s^2. The driver of the car is frozen with horror for 4.20 s while the truck makes skid marks 62.4 m long. With what speed does the truck strike the car?

50. A young woman named Kathy Kool buys a sports car that can accelerate at the rate of 4.90 m/s^2. She decides to test the car by dragging with another speedster, Stan Speedy. Both start from rest, but experienced Stan leaves the starting line 1.00 s before Kathy. If Stan moves with a constant acceleration of 3.50 m/s^2 and Kathy maintains an acceleration of 4.90 m/s^2, find (a) the time it takes Kathy to overtake Stan, (b) the distance she travels before she catches him, and (c) the speeds of both cars at the instant she overtakes him.

51. A mountain climber stands at the top of a 50.0-m cliff that overhangs a calm pool of water. He throws two stones vertically downward 1.0 s apart and observes that they cause a single splash. The first stone has an initial velocity of -2.00 m/s. (a) How long after release of the first stone will the two stones hit the water?

(b) What initial velocity must the second stone have if they are to hit simultaneously? **(c)** What will the velocity of each stone be at the instant they hit the water?

52. In order to pass a physical education class at a university, a student must run 1.0 mi in 12 min. After running for 10 min, she still has 500 yd to go. If her maximum acceleration is 0.15 m/s², can she make it? If the answer is no, determine what acceleration she would need to be successful.

53. In Mostar, Bosnia, the ultimate test of a young man's courage once was to jump off a 400-year-old bridge (now destroyed) into the River Neretva, 23 m below the bridge. **(a)** How long did the jump last? **(b)** How fast was the diver traveling on impact with the river? **(c)** If the speed of sound in air is 340 m/s, how long after the diver took off did a spectator on the bridge hear the splash?

54. One swimmer in a relay race has a 0.50-s lead and is swimming at a constant speed of 4.0 m/s. He has 50 m to swim before reaching the end of the pool. A second swimmer moves in the same direction as the leader. What constant speed must the second swimmer have in order to catch up to the leader at the end of the pool?

55. A person sees a lightning bolt pass close to an airplane that is flying in the distance. The person hears thunder 5.0 s after seeing the bolt, and sees the airplane overhead 10 s after hearing the thunder. The speed of sound in air is 1100 ft/s. **(a)** Find the distance of the airplane from the person at the instant of the bolt. (Neglect the time it takes the light to travel from the bolt to the eye.) **(b)** Assuming that the plane travels with a constant speed toward the person, find the velocity of the airplane. **(c)** Look up the speed of light in air, and defend the approximation used in (a).

56. A person walks first at a constant speed of 5.00 m/s along a straight line from Point *A* to Point *B* and then back along the line from *B* to *A* at a constant speed of 3.00 m/s. What is **(a)** her average speed over the entire trip? **(b)** her average velocity over the entire trip?

57. A hard rubber ball, released at chest height, falls to the pavement and bounces back to nearly the same height. When it is in contact with the pavement, the lower side of the ball is temporarily flattened. Before this dent in the ball pops out, suppose that its maximum depth is on the order of one centimeter. Estimate the maximum acceleration of the ball. State your assumptions, the quantities you estimate, and the values you estimate for them.

58. A ball is thrown directly downward, with an initial speed of 8.00 m/s, from a height of 30.0 m. After what interval does the ball strike the ground?

59. A hot air balloon is traveling vertically upward at a constant speed of 5.00 m/s. When it is 21.0 m above the ground, a package is released from the balloon. **(a)** For how long after being released is the package in the air? **(b)** What is the velocity of the package just before impact with the ground? **(c)** Repeat (a) and (b) for the case of the balloon descending at 5.00 m/s.

WEB 60. A stunt woman sitting on a tree limb wishes to drop vertically onto a horse galloping under the tree. The speed of the horse is 10.0 m/s, and the woman is initially 3.00 m above the level of the saddle. **(a)** What must be the horizontal distance between the saddle and limb when the woman makes her move? **(b)** How long is she in the air?

The Wizard of Id   **by Parker and Hart**

By permission of John Hart and Field Enterprises, Inc.

Vectors and Two-Dimensional Motion

▲ **PHYSICS PUZZLER**

The circus stuntmen being shot out of cannons are human projectiles. Neglecting air resistance, they move in parabolic paths until they land in a stategically placed net. What initial condition(s) will determine where the catching net should be placed? *(Ringling Brothers Circus)*

2.1

I n our discussion of one-dimensional motion in Chapter 2, we used the concept of vectors only to a limited extent. As we progress in our study of motion, the ability to manipulate vector quantities will become increasingly important. As a result, much of this chapter will be devoted to techniques for adding vectors, subtracting them, and so forth. We will then apply our newfound skills to a special case of two-dimensional motion — projectiles. We shall also see that an understanding of vector manipulation is necessary in order to work with and understand relative motion.

2.3, BOTH SECTIONS

3.1   VECTORS AND SCALARS REVISITED

Each of the physical quantities we shall encounter in this book can be categorized as either a *scalar* or a *vector quantity*. A scalar is a quantity that can be completely specified by its magnitude with appropriate units; that is, a **scalar** has only magnitude and no direction. A **vector** is a physical quantity that requires the specification of both direction and magnitude.

Temperature is an example of a scalar quantity. If someone tells you that the temperature of an object is $-5°C$, that information completely specifies the temperature of the object; no direction is required. Other examples of scalars are masses, time intervals, and the number of pages in this textbook. Scalar quantities can be manipulated using the rules of ordinary arithmetic. For example, if you have 2 liters of water in a container (where 1 liter is defined to be 1000 cm^3) and you add 3 more liters, by ordinary arithmetic the amount of water in the container is then 5 liters.

An example of a vector quantity is force. If you are told that someone is going to exert a force of 10 lb on an object, that is not enough information to let you

(Left) The number of apples in the basket is one example of a scalar quantity. Can you think of other examples? *(Superstock)*   *(Right)* Jennifer pointing in the right direction tells us to travel 5 blocks to the north to reach the courthouse. A vector is a physical quantity that must be specified by both magnitude and direction. *(Raymond A. Serway)*

know what will happen to the object. The effect of a force of 10 lb exerted horizontally is different from the effect of a force of 10 lb exerted vertically upward or downward. In other words, you need to know the direction of the force as well as its magnitude.

Velocity is another example of a vector quantity. If we wish to describe the velocity of a moving vehicle, we must specify both its speed (say, 30 m/s) and the direction in which the vehicle is moving (say, northeast). Other examples of vector quantities include displacement and acceleration, which were defined in Chapter 2.

3.2 SOME PROPERTIES OF VECTORS

Equality of Two Vectors.   Two vectors, **A** and **B**, are defined as equal if they have the same magnitude and the same direction. This property allows us to translate a vector parallel to itself in a diagram without affecting the vector. In fact, for most purposes, any vector can be moved parallel to itself without being affected.

Adding Vectors.   When two or more vectors are added together, they must all have the same units. For example, it would be meaningless to add a velocity vector to a displacement vector, because they are different physical quantities. Scalars also obey this rule. For example, it would be meaningless to add temperatures and areas.

2.4, SECTION 1

When a vector quantity is handwritten, it is often represented with an arrow over the letter ($\vec{A}$). In this book, a vector quantity will be represented by boldface type (for example, **A**). The magnitude of a vector such as **A** will be represented by italic type such as A. Likewise, italic type will be used to represent scalars.

The procedures for adding vectors rely on geometric methods. (Later we shall develop an algebraic technique for adding vectors that is much more convenient and will be used throughout the remainder of the text.) To add vector **B** to vector **A,** first draw **A** on a piece of graph paper to some scale such as 1 cm = 1 m. Vector **A** must be drawn so that its direction is specified relative to a coordinate system. Then draw vector **B** to the same scale and with the tail of **B** starting at the tip of **A,** as in Figure 3.1a. Vector **B** must be drawn along the direction that makes the proper angle relative to vector **A.** The resultant vector, **R,** given by **R = A + B,** is

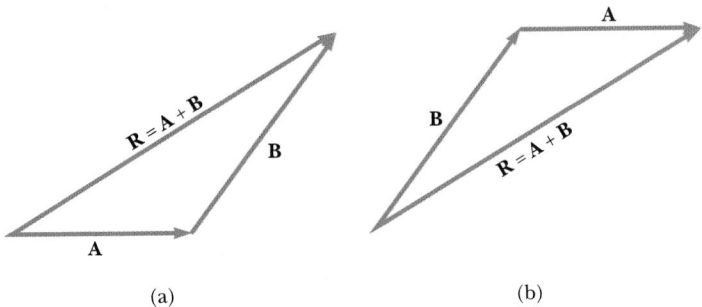

(a)   (b)

Figure 3.1   (a) When vector **B** is added to vector **A,** the vector sum, **R,** is the vector that runs from the tail of **A** to the tip of **B.** (b) Here the resultant runs from the tail of **B** to the tip of **A.** These constructions prove that **A + B = B + A.**

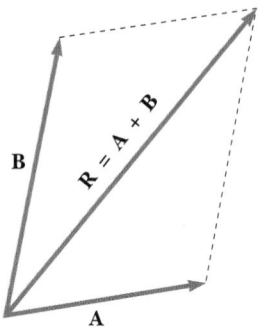

Figure 3.2 In this construction, the resultant **R** is the diagonal of a parallelogram with sides **A** and **B.**

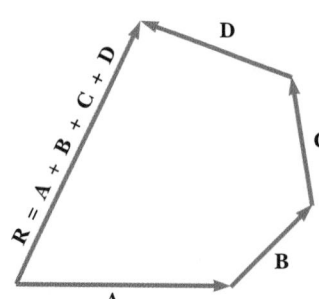

Figure 3.3 A geometric construction for summing four vectors. The resultant vector, **R,** is the vector that completes the polygon.

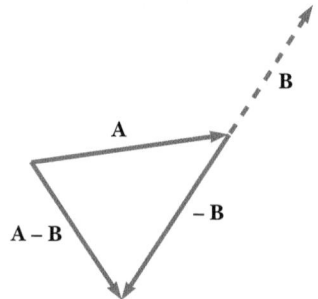

Figure 3.4 This construction shows how to subtract vector **B** from vector **A.** The vector $-$**B** has the same magnitude as the vector **B** but points in the opposite direction.

the vector drawn from the tail of **A** to the tip of **B.** This is known as the *triangle method of addition.*

When two vectors are added, the sum is independent of the order of the addition. That is, **A** + **B** = **B** + **A.** This can be seen from the geometric construction in Figure 3.1b.

An alternative graphical procedure for adding two vectors, known as the *parallelogram rule of addition,* is shown in Figure 3.2. In this construction, the tails of vectors **A** and **B** are joined together, and the resultant vector, **R,** is the diagonal of the parallelogram formed with **A** and **B** as its sides.

This same general approach can also be used to add more than two vectors, as is done in Figure 3.3 for four vectors. The resultant vector sum, **R** = **A** + **B** + **C** + **D,** is the vector drawn from the tail of the first vector to the tip of the last vector. Again, the order in which vectors are added is unimportant.

Negative of a Vector. The negative of the vector **A** is defined as the vector that when added to **A** gives zero for the vector sum. This means that **A** and $-$**A** have the same magnitude but opposite directions.

Subtraction of Vectors. Vector subtraction makes use of the definition of the negative of a vector. We define the operation **A** $-$ **B** as vector $-$**B** added to vector **A:**

$$\mathbf{A} - \mathbf{B} = \mathbf{A} + (-\mathbf{B}) \qquad [3.1]$$

Thus, vector subtraction is really a special case of vector addition. The geometric construction for subtracting two vectors is shown in Figure 3.4.

Multiplication and Division of Vectors by Scalars. The multiplication or division of a vector by a scalar gives a vector. For example, if a vector, **A,** is multiplied by the scalar number 3, the result, written 3**A,** is a vector with a magnitude three times that of the original vector **A,** pointing in the same direction as **A.** On the other hand, if we multiply vector **A** by the scalar $-$3, the result is a vector with a magnitude three times that of **A,** pointing in the direction opposite **A** (because of the negative sign).

EXAMPLE 3.1 Taking a Trip

A car travels 20.0 km due north and then 35.0 km in a direction 60.0° west of north, as in Figure 3.5. Find the magnitude and direction of the car's resultant displacement.

Solution The problem can be solved geometrically using graph paper and a protractor, as shown in Figure 3.5. The resultant displacement, **R,** is the sum of the two individual displacements, **A** and **B.**

The length of **R,** drawn to the same scale as **A** and **B,** indicates that the displacement of the car is 48.2 km, and a measurement of the angle β shows that the displacement is approximately 38.9° west of north.

Workbook Problem 2 — (Workbook page 14) is similar to Example 3.1.

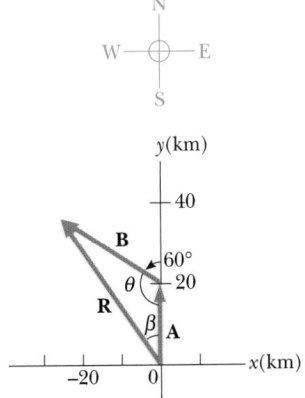

Figure 3.5 (Example 3.1) A graphical method for finding the resultant displacement vector **R = A + B.**

 2.5

3.3 COMPONENTS OF A VECTOR

One method of adding vectors makes use of the projections of a vector along the axes of a rectangular coordinate system. These projections are called **components.** Any vector can be completely described by its components.

Consider a vector, **A,** in a rectangular coordinate system, as shown in Figure 3.6. Note that **A** can be expressed as the sum of two vectors, $\mathbf{A}_x$ parallel to the x axis, and $\mathbf{A}_y$ parallel to the y axis. That is,

$$\mathbf{A} = \mathbf{A}_x + \mathbf{A}_y$$

$\mathbf{A}_x$ and $\mathbf{A}_y$ are the component vectors of **A.** The projection of **A** along the x axis, A_x, is called the x component of **A,** and the projection of **A** along the y axis, A_y, is called the y component of **A.** These components can be either positive or negative numbers with units. From the definitions of sine and cosine of an angle, we see that $\cos \theta = A_x/A$ and $\sin \theta = A_y/A$. Hence, the magnitudes of the components of **A** are

$$A_x = A \cos \theta$$
$$A_y = A \sin \theta \qquad [3.2]$$

These components form two sides of a right triangle, the hypotenuse of which has the magnitude A. Thus, it follows that **A**'s magnitude and direction are related to its components through the Pythagorean theorem and the definition of the tangent:

$$A = \sqrt{A_x^2 + A_y^2} \qquad [3.3]$$

$$\tan \theta = \frac{A_y}{A_x} \qquad [3.4]$$

To solve for the angle θ, we can write Equation 3.4 in the form

$$\theta = \tan^{-1}\left(\frac{A_y}{A_x}\right)$$

If a coordinate system other than the one shown in Figure 3.6 is chosen, the components of the vector must be modified accordingly. In many applications it is

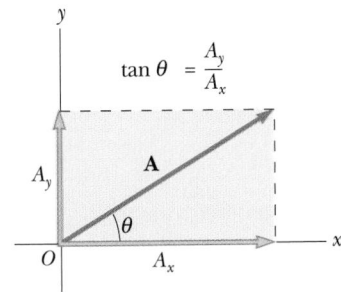

Figure 3.6 Any vector **A** lying in the xy plane can be represented by its rectangular components, A_x and A_y.

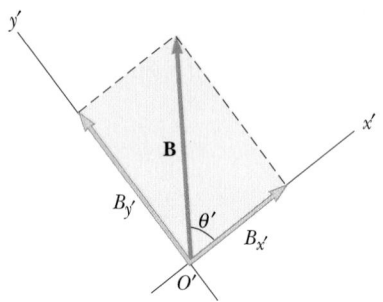

Figure 3.7 The components of vector **B** in a tilted coordinate system.

more convenient to express the components of a vector in a coordinate system having axes that are not horizontal and vertical but are still perpendicular to each other. Suppose a vector, **B,** makes an angle of θ' with the x' axis defined in Figure 3.7. The rectangular components of **B** along the axes of Figure 3.7 are given by $B_{x'} = B \cos \theta'$ and $B_{y'} = B \sin \theta'$, as in Equation 3.2. The magnitude and direction of **B** are obtained from expressions equivalent to Equations 3.3 and 3.4. Thus, we can express the components of a vector in any coordinate system that is convenient for the situation.

Thinking Physics 1

You may have asked someone directions to a destination in a city and have been told something like, "Walk 3 blocks east and then 5 blocks south." If so, are you experienced with vector components?

Explanation   Yes, you are! Although you may not have thought of vector component language when you heard these directions, this is exactly what the directions represent. The perpendicular streets of the city reflect an x–y coordinate system—we can assign the x axis to the east–west streets and the y axis to the north–south streets. Thus, the comment of the person giving you directions can be translated as, "Undergo a displacement vector that has an x component of $+3$ blocks and a y component of -5 blocks." You would arrive at the same destination by undergoing the y-component first, followed by the x component, demonstrating the commutative law of addition.

Thinking Physics 2

Consider your commute to work or school in the morning. Which is larger, the distance you traveled or the magnitude of the displacement vector?

Explanation   The distance traveled, unless you have a very unusual commute, *must* be larger than the magnitude of the displacement vector. The distance includes all of the twists and turns that you made in following the roads from home to work or school. However, the magnitude of the displacement vector is the length of a straight line from your home to work or school. This is often described informally as "the distance as the crow flies." The only way that the distance could be the same as the magnitude of the displacement vector is if your commute is a perfect straight line, which is unlikely! The distance could *never* be less than the magnitude of the displacement vector, because the shortest distance between two points is a straight line.

Adding Vectors

In order to add vectors algebraically rather than graphically, the following procedure is used. First, find the components of all the vectors in some coordinate system that is appropriate for the problem. Next, add all the x components to find the resultant component in the x direction. Similarly, add all the y components to find the resultant component in the y direction. Because the resultant x and y components are at right angles to each other, you can use the Pythagorean theorem to determine the magnitude of the resultant vector. Finally, use a suitable trigonometric function and the components of the resultant vector to find the angle that the resultant vector makes with the *x* axis. These steps are summarized as a Problem-Solving Strategy and are illustrated in the worked examples that follow.

Problem-Solving Strategy

Adding Vectors

When two or more vectors are to be added, the following steps are used.

1. Select a coordinate system.
2. Draw a sketch of the vectors to be added (or subtracted), with a label on each vector.
3. Find the *x* and *y* components of all vectors.
4. Find the resultant components (the algebraic sum of the components) in both the *x* and *y* directions.
5. Use the Pythagorean theorem to find the magnitude of the resultant vector.
6. Use a suitable trigonometric function to find the angle the resultant vector makes with the *x* axis.

EXAMPLE 3.2   Help Is on the Way!

Find the horizontal and vertical components of the 100-m displacement of a superhero who flies from the top of a tall building along the path shown in Figure 3.8a.

> **Workbook Problem 3 –** (Workbook page 15) is similar to Example 3.2.

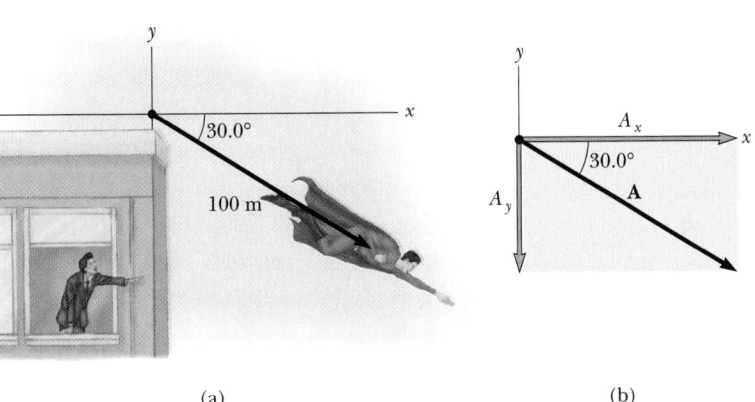

(a)   (b)

Figure 3.8   (Example 3.2)

Solution   The triangle formed by the displacement and its components is shown in Figure 3.8b. Because $A = 100$ m and $\theta = -30.0°$ (θ is negative because it is measured clockwise from the *x* axis), we have

$$A_y = A \sin \theta = (100 \text{ m}) \sin(-30.0°) = \boxed{-50.0 \text{ m}}$$

Note that $\sin(-\theta) = -\sin \theta$. The negative sign for A_y reflects the fact that displacement in the *y* direction is *downward* from the origin.

The x component of displacement is

$$A_x = A \cos \theta = (100 \text{ m}) \cos(-30.0°) = \boxed{+86.6 \text{ m}}$$

Note that $\cos(-\theta) = \cos \theta$. Also, from an inspection of the figure, you should be able to see that A_x is positive in this case.

EXAMPLE 3.3 Taking a Hike

A hiker begins a trip by first walking 25.0 km due southeast from her base camp. On the second day she walks 40.0 km in a direction 60.0° north of east, at which point she discovers a forest ranger's tower.
(a) Determine the components of the hiker's displacements in the first and second days.

Solution If we denote the displacement vectors on the first and second days by **A** and **B,** respectively, and use the camp as the origin of coordinates, we get the vectors shown in Figure 3.9. Displacement **A** has a magnitude of 25.0 km and is 45.0° south of east. Its components are

$$A_x = A \cos(-45.0°) = (25.0 \text{ km})(0.707) = \boxed{17.7 \text{ km}}$$

$$A_y = A \sin(-45.0°) = -(25.0 \text{ km})(0.707) = \boxed{-17.7 \text{ km}}$$

The negative value of A_y indicates that the y coordinate decreased in this displacement. The signs of A_x and A_y are also evident from Figure 3.9.

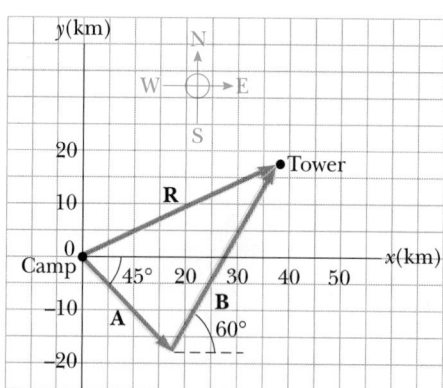

Figure 3.9 (Example 3.3)

The second displacement, **B,** has a magnitude of 40.0 km and is 60.0° north of east. Its components are

$$B_x = B \cos 60.0° = (40.0 \text{ km})(0.500) = \boxed{20.0 \text{ km}}$$

$$B_y = B \sin 60.0° = (40.0 \text{ km})(0.866) = \boxed{34.6 \text{ km}}$$

(b) Determine the components of the hiker's total displacement for the trip.

Solution The resultant displacement for the trip, $\mathbf{R} = \mathbf{A} + \mathbf{B}$, has components given by

$$R_x = A_x + B_x = 17.7 \text{ km} + 20.0 \text{ km} = \boxed{37.7 \text{ km}}$$

$$R_y = A_y + B_y = -17.7 \text{ km} + 34.6 \text{ km} = \boxed{16.9 \text{ km}}$$

Exercise Determine the magnitude and direction of the total displacement.

Answer 41.3 km, 24.1° north of east from the base camp.

3.4 VELOCITY AND ACCELERATION IN TWO DIMENSIONS

In our discussion of one-dimensional motion in Chapter 2, the vector nature of displacement, velocity, and acceleration was taken into account through the use of positive and negative signs. To completely describe the motion of an object in two or three dimensions, we must make use of vectors.

Consider an object moving through space as shown in Figure 3.10. When the object is at some point P at time t_i, its position is described by the position vector $\mathbf{r}_i$, drawn from the origin to P. Likewise, when the object has moved to some other point, Q, at time t_f, its position vector is $\mathbf{r}_f$. As you can see from the vector diagram in Figure 3.10, the final position vector is the sum of the initial position vector and $\Delta\mathbf{r}$. Because $\mathbf{r}_f = \mathbf{r}_i + \Delta\mathbf{r}$, the displacement of the object is defined as the change in the position vector:

$$\Delta\mathbf{r} \equiv \mathbf{r}_f - \mathbf{r}_i \qquad [3.5]$$

The average velocity of a particle during the time interval Δt is the ratio of the displacement to the time interval for this displacement.

$$\overline{\mathbf{v}} \equiv \frac{\Delta\mathbf{r}}{\Delta t} \qquad [3.6]$$

Because the displacement is a vector and the time interval is a scalar, we conclude that the average velocity is a *vector* quantity directed along $\Delta\mathbf{r}$.

The instantaneous velocity, v, is defined as the limit of the average velocity, $\Delta\mathbf{r}/\Delta t$, as Δt goes to zero:

$$\mathbf{v} \equiv \lim_{\Delta t \to 0} \frac{\Delta\mathbf{r}}{\Delta t} \qquad [3.7]$$

The direction of the instantaneous velocity vector is along a line that is tangent to the path of the particle and in the direction of motion.

The average acceleration of an object whose velocity changes by $\Delta\mathbf{v}$ in the time interval Δt is a vector defined as the ratio $\Delta\mathbf{v}/\Delta t$.

$$\overline{\mathbf{a}} \equiv \frac{\Delta\mathbf{v}}{\Delta t} \qquad [3.8]$$

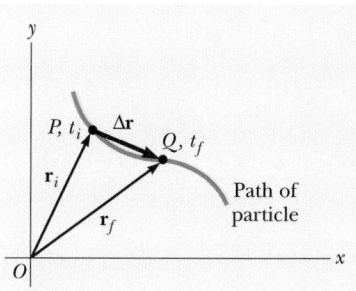

Figure 3.10 An object moving along some curved path between points P and Q. The displacement vector, $\Delta\mathbf{r}$, is the difference in the position vectors. That is, $\Delta\mathbf{r} = \mathbf{r}_f - \mathbf{r}_i$.

3.5, SECTION 1

◀ Average velocity

◀ Instantaneous velocity

◀ Average acceleration

Instantaneous acceleration ▷

The instantaneous acceleration vector, a, is defined as the limit of the average acceleration vector as Δt goes to zero.

It is important to recognize that a particle can accelerate in several ways. First, the magnitude of the velocity vector (the speed) may change with time. Second, a particle accelerates when the direction of the velocity vector changes with time (makes a curved path) even though the speed is constant. Third, acceleration may be due to changes in both the magnitude and the direction of the velocity vector.

Thinking Physics 3

The gas pedal in an automobile is called the *accelerator*. Are there any other controls that could also be considered to be an accelerator?

Explanation   The gas pedal is called the accelerator because the common usage of the word *acceleration* refers to an *increase in speed*. The scientific definition, however, is that **acceleration occurs whenever the velocity changes in any way.** Thus, the *brake pedal* can also be considered to be an accelerator, because it causes the car to slow down. The *steering wheel* is also an accelerator, because it changes the direction of the velocity vector.

3.5   PROJECTILE MOTION

3.5, SECTION 2

In the situations we considered in Chapter 2, objects moved along straight-line paths, such as the x axis. Now let us look at some cases in which an object moves in a plane. By this we mean that the object may move in both the x and y directions simultaneously or move in two dimensions. The particular form of two-dimensional

Projectile motion ▷

motion we shall concentrate on is called **projectile motion.** Anyone who has observed a baseball in motion (or, for that matter, any object thrown into the air) has observed projectile motion. It is surprisingly simple to analyze if the following three assumptions are made.

1. The free-fall acceleration, **g**, has a magnitude of 9.80 m/s², is constant over the range of motion, and is directed downward.
2. The effect of air resistance is negligible.
3. The rotation of the Earth does not affect the motion.

With these assumptions, we shall find that the path of a projectile is curved as shown in Figure 3.11. Such a curve is called a *parabola*.

Let us choose our coordinate system so that the y direction is vertical and positive upward. In this case, **the acceleration in the y direction is $-g$, just as in free fall, and the acceleration in the x direction is 0 (because air resistance is neglected).** Furthermore, let us assume that at $t = 0$, the projectile leaves the origin

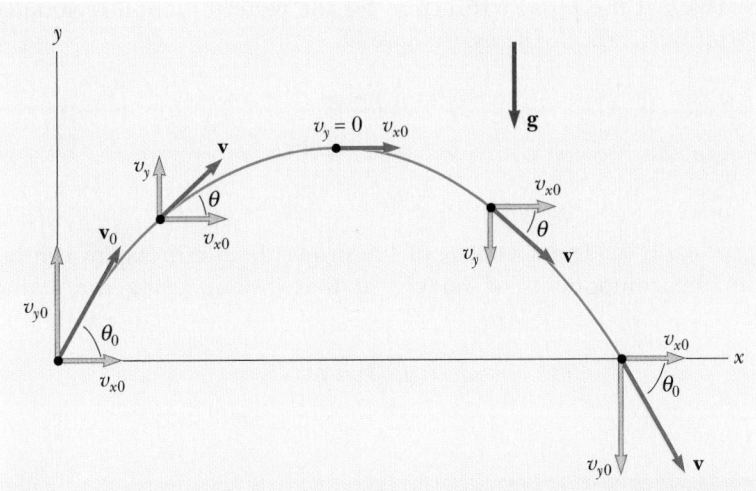

🔋**Figure 3.11** The parabolic trajectory of a particle that leaves the origin with a velocity of $\mathbf{v}_0$. Note that the velocity, $\mathbf{v}$, changes with time. However, the x component of the velocity, v_x, remains constant in time. Also, $v_y = 0$ at the peak, but the acceleration is always equal to the free-fall acceleration and acts vertically downward.

with a velocity of $\mathbf{v}_0$, as in Figure 3.11. If the velocity vector makes an angle of θ_0 with the horizontal where θ_0 is called the projection angle, then from the definitions of cosine and sine functions and Figure 3.11, we have

$$v_{x0} = v_0 \cos \theta_0 \quad \text{and} \quad v_{y0} = v_0 \sin \theta_0$$

In order to analyze projectile motion, we shall separate the motion into two parts, the x (or horizontal) motion and the y (or vertical) motion, and solve each part separately. We shall look first at the x motion. As noted before, motion in the x direction occurs with $a_x = 0$. This means that **the velocity component along the x direction remains constant.** Thus, if the initial value of the velocity component in the x direction is $v_{x0} = v_0 \cos \theta_0$, this is also the value of the velocity at any later time. That is,

$$v_x = v_{x0} = v_0 \cos \theta_0 = \text{constant} \qquad [3.9]$$

Equation 3.9 can be substituted into the defining equation for velocity (Eq. 2.2) to give us an expression for the horizontal position of the projectile as a function of time:

$$x = v_{x0}t = (v_0 \cos \theta_0)t \qquad [3.10]$$

These equations tell us all we need to know about the motion in the x direction. Let us now consider the y motion. Because it has constant acceleration, the equations developed in Section 2.6 can be used. In these equations, v_{y0} shall denote the initial velocity in the y direction and $-g$ the free-fall acceleration. The negative sign

An evening view of an illuminated water fountain in la Place de la Concorde, Paris. The individual water streams follow parabolic trajectories. The horizontal range and maximum height of a given stream of water depend on the elevation angle of its initial velocity as well as its initial speed. (© *The Telegraph Colour Library/ FPG*)

For this investigation, you need to be outdoors with a small ball such as a tennis ball and a wristwatch with a second hand. Throw the ball upward as hard as you can and determine the initial speed of your throw and the approximate maximum height of the ball using only your wristwatch.

APPLICATION

Trajectory of a Baseball.

for g indicates that the positive direction for the vertical motion is assumed to be upward. With this choice of signs, we have

$$v_y = v_{y0} - gt \qquad [3.11]$$

$$y = v_{y0}t - \tfrac{1}{2}gt^2 \qquad [3.12]$$

$$v_y^2 = v_{y0}^2 - 2gy \qquad [3.13]$$

where $v_{y0} = v_0 \sin \theta_0$. The speed, v, of the projectile at any instant can be calculated from the components of velocity at that instant, using the Pythagorean theorem:

$$v = \sqrt{v_x^2 + v_y^2}$$

Applying Physics 1

A home run is hit in a baseball game. The ball is hit from home plate into the center field stands, along a long parabolic path. What is the acceleration of the ball (a) while it is rising, (b) at the highest point of the trajectory, and (c) while it is descending after reaching the highest point? Ignore air resistance.

Explanation The answers to all three parts are the same—the acceleration is that due to gravity, 9.8 m/s², because the force of gravity is pulling downward on the ball during the entire motion. During the rising part of the trajectory, the downward acceleration results in the decreasing positive values of the vertical component of the velocity of the ball. During the falling part of the trajectory, the downward acceleration results in the increasing negative values of the vertical component of the velocity. Many people have trouble with the topmost point, claiming that the acceleration of the ball at the highest point is zero. This interpretation arises from a confusion between zero vertical velocity and zero acceleration. Even though the ball has momentarily come to rest (vertically; it is still moving horizontally), it is still accelerating, because the force of gravity is still acting. If the ball were to come to rest and experience zero acceleration, then the velocity would not change—the ball would remain suspended at the highest point! We do not see that happening, because the acceleration is indeed not equal to zero.

Before we examine some numerical examples dealing with projectile motion, let us pause to summarize what we have learned so far about this kind of motion:

- Provided air resistance is negligible, the horizontal component of velocity, v_x, remains constant because there is no horizontal component of acceleration.
- The vertical component of acceleration is equal to the free-fall acceleration, g.
- The vertical component of velocity, v_y, and the displacement in the y direction are identical to those of a freely falling body.
- Projectile motion can be described as a superposition of the two motions in the x and y directions.

Problem-Solving Strategy

Projectile Motion

We suggest that you use the following approach to solving projectile motion problems:

1. Select a coordinate system.
2. Resolve the initial velocity vector into x and y components.
3. Treat the horizontal motion and the vertical motion independently.
4. Follow the techniques for solving problems with constant velocity to analyze the horizontal motion of the projectile.
5. Follow the techniques for solving problems with constant acceleration to analyze the vertical motion of the projectile.

EXAMPLE 3.4 The Stranded Explorers

An Alaskan rescue plane drops a package of emergency rations to a stranded party of explorers, as shown in Figure 3.12. The plane is traveling horizontally at 40.0 m/s at a height of 100 m above the ground.
(a) Where does the package strike the ground relative to the point at which it was released?

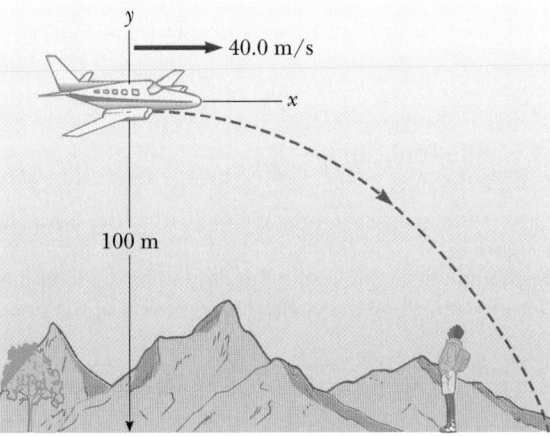

Figure 3.12 (Example 3.4) From the point of view of an observer on the ground, a package released from the rescue plane travels along the path shown.

Solution The coordinate system for this problem is selected as shown in Figure 3.12, with the positive x direction to the right and the positive y direction upward.

Consider first the horizontal motion of the package. The only equation available to us is

$$x = v_{x0}t$$

The initial x component of the package velocity is the same as the velocity of the plane when the package was released, 40.0 m/s. Thus, we have

$$x = (40.0 \text{ m/s})t$$

If we know t, the length of time the package is in the air, we can determine x, the distance traveled by the package along the horizontal. To find t, we move to the equations for the vertical motion of the package. We know that at the instant the package hits the ground, its y coordinate is -100 m. We also know that the initial velocity of the package in the vertical direction, v_{y0}, is zero because the package was released with only a horizontal component of velocity. From Equation 3.12 we have

$$y = -\tfrac{1}{2}gt^2$$

$$-100 \text{ m} = -\tfrac{1}{2}(9.80 \text{ m/s}^2)t^2$$

$$t^2 = 20.4 \text{ s}^2$$

$$t = 4.51 \text{ s}$$

This value for the time of flight substituted into the equation for the x coordinate gives

$$x = (40.0 \text{ m/s})(4.51 \text{ s}) = \boxed{180 \text{ m}}$$

(b) What are the horizontal and vertical components of the velocity of the package just before it hits the ground?

Solution   We already know the horizontal component of the velocity of the package just before it hits, because the velocity in the horizontal direction remains constant at 40.0 m/s throughout the flight.

The vertical component of the velocity just before the package hits the ground may be found by using Equation 3.11, with $v_{y0} = 0$:

$$v_y = v_{y0} - gt = 0 - (9.80 \text{ m/s}^2)(4.51 \text{ s}) = \boxed{-44.1 \text{ m/s}}$$

EXAMPLE 3.5   The Long Jump

A long jumper leaves the ground at an angle of 20.0° to the horizontal and at a speed of 11.0 m/s, as in Figure 3.13.
(a) How far does he jump? (Assume that the motion of the long jumper is equivalent to that of a particle—in other words, disregard the motion of the jumper's arms and legs).

Solution   His horizontal motion is described by using Equation 3.10:

$$x = (v_0 \cos \theta_0)t = (11.0 \text{ m/s})(\cos 20.0°)t$$

Figure 3.13   (Example 3.5) Carl Lewis performs a gold-medal-winning long jump of 8.50 m in the 1996 Olympics held in Atlanta, Georgia. (*J. O. Atlanta 96/Gamma*)

The value of x can be found if *t*, the total duration of the jump, is known. We can find *t* with $v_y = v_0 \sin \theta_0 - gt$ (Eq. 3.11) by noting that at the top of the jump the vertical component of velocity goes to zero:

$$v_y = v_0 \sin \theta_0 - gt$$

$$0 = (11.0 \text{ m/s}) \sin 20.0° - (9.80 \text{ m/s}^2)t_1$$

$$t_1 = 0.384 \text{ s}$$

Note that t_1 is the time interval to the *top* of the jump. Because of the symmetry of the vertical motion, an identical time interval passes before the jumper returns to the ground. Therefore, the *total time* in the air is $t = 2t_1 = 0.768$ s, and the distance jumped is

$$x = (11.0 \text{ m/s})(\cos 20.0°)(0.768 \text{ s}) = \boxed{7.94 \text{ m}}$$

(b) What is the maximum height reached?

Solution The maximum height reached is found using $y = (v_0 \sin \theta_0)t - \frac{1}{2}gt^2$ (Eq. 3.12) with $t = t_1 = 0.384$ s.

$$y_{max} = (11.0 \text{ m/s})(\sin 20.0°)(0.384 \text{ s}) - \frac{1}{2}(9.80 \text{ m/s}^2)(0.384 \text{ s})^2 = \boxed{0.722 \text{ m}}$$

Although the assumption that the motion of the long jumper is that of a projectile is an oversimplification of the situation, the values obtained are reasonable.

EXAMPLE 3.6 That's Quite an Arm

A stone is thrown upward from the top of a building at an angle of 30.0° to the horizontal and with an initial speed of 20.0 m/s, as in Figure 3.14. The height of the building is 45.0 m.

(a) How long is the stone "in flight"?

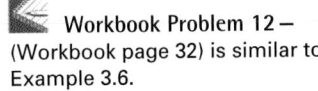

Workbook Problem 12 – (Workbook page 32) is similar to Example 3.6.

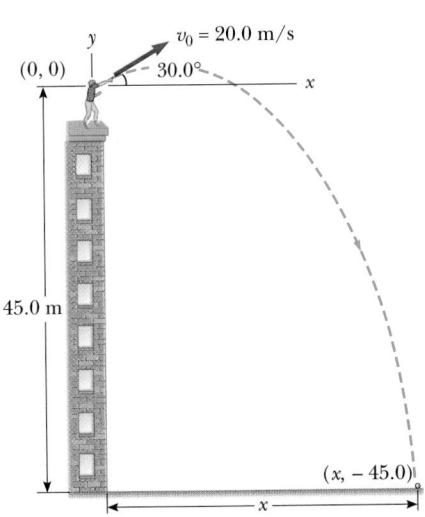

Figure 3.14   (Example 3.6)

Solution The initial *x* and *y* components of the velocity are

$$v_{x0} = v_0 \cos \theta_0 = (20.0 \text{ m/s})(\cos 30.0°) = +17.3 \text{ m/s}$$

$$v_{y0} = v_0 \sin \theta_0 = (20.0 \text{ m/s})(\sin 30.0°) = +10.0 \text{ m/s}$$

To find t, we can use $y = v_{y0}t - \frac{1}{2}gt^2$ (Eq. 3.12) with $y = -45.0$ m and $v_{y0} = 10.0$ m/s. (We have chosen the point of release as the origin, as shown in Fig. 3.14.)

$$-45.0 \text{ m} = (10.0 \text{ m/s})t - \frac{1}{2}(9.80 \text{ m/s}^2)t^2$$

Solving the quadratic equation for t (see Appendix A) gives, for the positive root, the value $t = 4.22$ s. Does the negative root have any physical meaning? (Why not think of another way of finding t from the information given?)

(b) What is the speed of the stone just before it strikes the ground?

Solution The y component of the velocity just before the stone strikes the ground can be obtained using Equation 3.11 with $t = 4.22$ s:

$$v_y = v_{y0} - gt = 10.0 \text{ m/s} - (9.80 \text{ m/s}^2)(4.22 \text{ s}) = -31.4 \text{ m/s}$$

Because $v_x = v_{x0} = 17.3$ m/s, the required speed is given by

$$v = \sqrt{v_x^2 + v_y^2} = \sqrt{(17.3)^2 + (-31.4)^2} \text{ m/s} = \boxed{35.9 \text{ m/s}}$$

Exercise Where does the stone strike the ground?

Answer 73.0 m from the base of the building.

3.6   RELATIVE VELOCITY

3.7

Observers in different frames of reference may measure different displacements or velocities for an object in motion. That is, two observers moving with respect to each other would generally not agree on the outcome of a measurement.

For example, if two cars were moving in the same direction with speeds of 50 mi/h and 60 mi/h, a passenger in the slower car would measure the speed of the faster car relative to the slower car as 10 mi/h. Of course, a stationary observer would measure the speed of the faster car as 60 mi/h. This simple example demonstrates that velocity measurements differ in different frames of reference.

We solved the previous problem with a minimum of thought and effort, but you will encounter many situations in which a more systematic method for attacking such problems is beneficial. To develop this method, let us write down all the information we are given and that which we want to know in the form of velocities with subscripts appended. We have

$\mathbf{v}_{se} = +50$ mi/h   (The subscript "se" means the velocity of the *slower* car with respect to the *Earth*.)

$\mathbf{v}_{fe} = +60$ mi/h   (The subscript "fe" means the velocity of the *faster* car with respect to the *Earth*.)

We want to know $\mathbf{v}_{fs}$, which is the velocity of the *faster* car with respect to the *slower* car. To find this, we write an equation for $\mathbf{v}_{fs}$ in terms of the other velocities. For this case, we have

$$\mathbf{v}_{\boxed{fs}} = \mathbf{v}_{\boxed{f}e} + \mathbf{v}_{e\boxed{s}} \qquad [3.14]$$

Note the pattern of the subscripts in this expression. The first subscript f on the left side of the equation is also the first subscript on the right side of the equation.

PHYSICS IN ACTION

Parabolic Paths

On the left is a multiflash photograph of a popular lecture demonstration in which a projectile is fired at a target that is being held by a magnet in the device at the top right of the photograph. The conditions of the experiment are that the gun is aimed at the target and the projectile leaves the gun at the instant the target is released from rest. Under these conditions the projectile will hit the target, independent of the initial speed of the projectile. The reason is that they both experience the same downward acceleration and hence the velocities of the projectile and target change by the same amount in the same time interval. Note that the velocity of the projectile (red arrows) changes in direction and magnitude, and its downward acceleration (violet arrows) remains constant.

On the right, a welder cuts holes through a heavy metal construction beam with a hot torch. The sparks generated in the process follow parabolic paths.

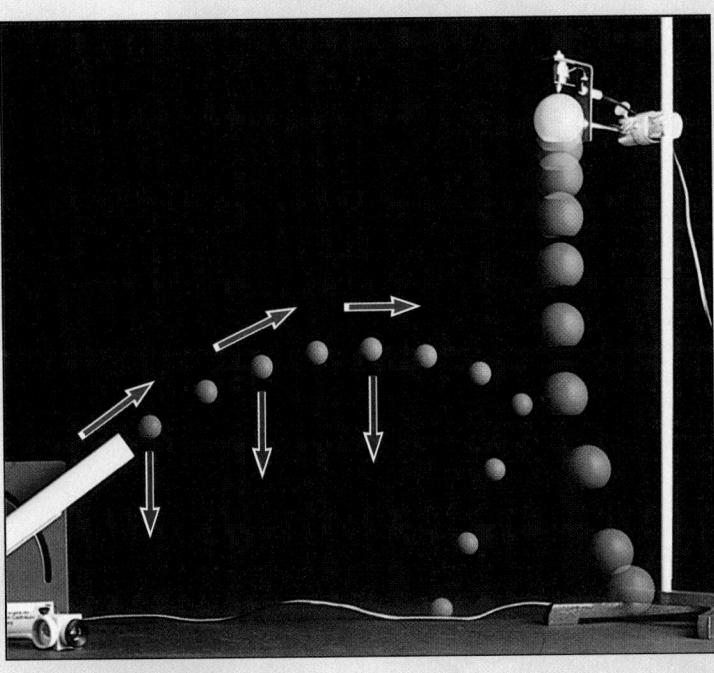

(Central Scientific Company)

(© The Telegraph Colour Library/FPG)

The second subscript s on the left side of the equation is also the last subscript on the right side of the equation. Boxes are used in the subscripts to emphasize this pattern. The boldface notation is, of course, indicative of the fact that velocity is a vector quantity. As we shall see in the following examples, this vector nature of the velocity is of paramount importance in certain instances.

We know that $\mathbf{v}_{es} = -\mathbf{v}_{se}$, so

$$\mathbf{v}_{fs} = +60 \text{ mi/h} - 50 \text{ mi/h} = +10 \text{ mi/h}$$

There is no general equation to memorize in order to work relative velocity problems; instead, you should develop the necessary equations on your own by following the technique already demonstrated for writing subscripts. We suggest that you practice this technique to work the examples that follow.

EXAMPLE 3.7   Where Does the Ball Land?

A passenger at the rear of a train, traveling at 15 m/s relative to the Earth, throws a baseball with a speed of 15 m/s in the direction opposite the motion of the train. What is the velocity of the baseball relative to the Earth?

Solution   We first write down our knowns and unknowns with appropriate subscripts:

$$\mathbf{v}_{te} = +15 \text{ m/s} \qquad \text{(velocity of the } \textit{train} \text{ relative to the } \textit{Earth})$$

$$\mathbf{v}_{bt} = -15 \text{ m/s} \qquad \text{(velocity of the } \textit{baseball} \text{ relative to the } \textit{train})$$

We need $\mathbf{v}_{be}$, the velocity of the *baseball* relative to the *Earth*. Following the strategy for watching the subscripts while we write down the equation for our unknown, we have

$$\mathbf{v}_{\boxed{be}} = \mathbf{v}_{\boxed{b}t} + \mathbf{v}_{t\boxed{e}}$$

$$\mathbf{v}_{\boxed{be}} = -15 \text{ m/s} + 15 \text{ m/s} = \boxed{0}$$

EXAMPLE 3.8   Crossing a River

A boat heading due north crosses a wide river with a velocity of 10.0 km/h relative to the water. The river has a uniform velocity of 5.00 km/h due east. Determine the velocity of the boat with respect to an observer on the riverbank.

Solution   We have

$$\mathbf{v}_{br} = 10.0 \text{ km/h due north} \qquad \text{(velocity of the } \textit{boat} \text{ with respect to the } \textit{river})$$

$$\mathbf{v}_{re} = 5.00 \text{ km/h due east} \qquad \text{(velocity of the } \textit{river} \text{ with respect to the } \textit{Earth})$$

and we want, $\mathbf{v}_{be}$, the velocity of the *boat* with respect to the *Earth*. Our equation becomes

$$\mathbf{v}_{\boxed{be}} = \mathbf{v}_{\boxed{b}r} + \mathbf{v}_{r\boxed{e}}$$

Because the velocities are not along the same direction as in preceding examples, the terms in the equation must be manipulated as vector quantities, which are shown in Figure 3.15. The quantity $\mathbf{v}_{br}$ is due north, $\mathbf{v}_{re}$ is due east, and the vector sum of the two, $\mathbf{v}_{be}$, is at the angle θ, as defined in Figure 3.15. Thus, we see that the velocity of the boat with respect to the Earth can be found from the Pythagorean theorem as

$$v_{be} = \sqrt{(v_{br})^2 + (v_{re})^2} = \sqrt{(10.0 \text{ km/h})^2 + (5.00 \text{ km/h})^2} = \boxed{11.2 \text{ km/h}}$$

and the direction of $\mathbf{v}_{be}$ is

$$\theta = \tan^{-1}\left(\frac{v_{re}}{v_{br}}\right) = \tan^{-1}\frac{5.00}{10.0} = 26.6°$$

Therefore, the boat travels at a speed of 11.2 km/h in the direction 63.4° north of east with respect to the Earth.

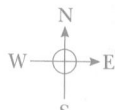

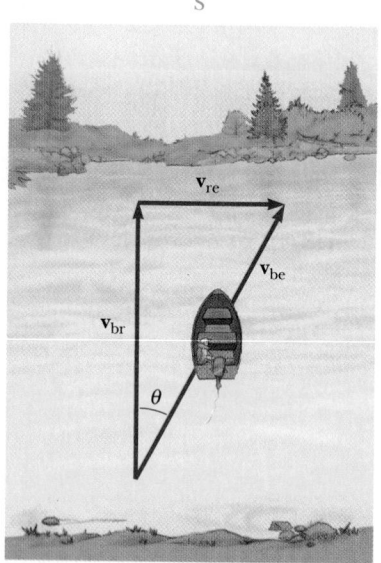

Figure 3.15   (Example 3.8)

EXAMPLE 3.9 Which Way Should the Boat Head?

If the boat of the preceding example travels with the same speed of 10.0 km/h relative to the water and is to travel due north, as in Figure 3.16, in what direction should it travel?

Solution We know

$$\mathbf{v}_{br} = \text{velocity of the } \textit{boat} \text{ with respect to the } \textit{river}$$

$$\mathbf{v}_{re} = \text{velocity of the } \textit{river} \text{ with respect to the } \textit{Earth}$$

and we need $\mathbf{v}_{be}$, the velocity of the *boat* with respect to the *Earth*. We have

$$\mathbf{v}_{\boxed{be}} = \mathbf{v}_{\boxed{b}r} + \mathbf{v}_{r\boxed{e}}$$

The relationship among these three vectors is shown in Figure 3.16; it agrees with our intuitive guess that the boat must head upstream in order to be pushed directly northward across the water. The speed v_{be} can be found from the Pythagorean theorem:

$$v_{be} = \sqrt{(v_{br})^2 - (v_{re})^2} = \sqrt{(10.0 \text{ km/h})^2 - (5.00 \text{ km/h})^2} = 8.66 \text{ km/h}$$

and the direction of v_{be} is

$$\theta = \tan^{-1} \frac{v_{re}}{v_{be}} = \tan^{-1}\left(\frac{5.00}{8.66}\right) = 30°$$

where θ is west of north.

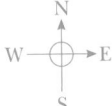

Workbook Problem 14 – (Workbook page 36) is similar to Example 3.9.

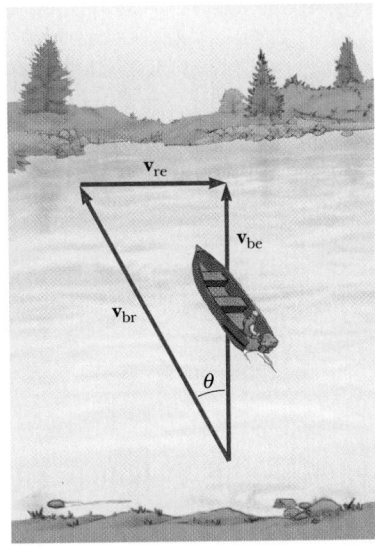

Figure 3.16 (Example 3.9)

SUMMARY

Two vectors, **A** and **B,** can be added geometrically by either the triangle method or the parallelogram rule. In the **triangle method,** the two vectors are drawn to scale, on graph paper, so that the tail of the second vector starts at the tip of the first. The resultant vector is the vector drawn from the tail of the first to the tip of the second. In the **parallelogram method,** the vectors are drawn to scale on graph paper with the tails of the two vectors joined. The resultant vector is the diagonal of the parallelogram formed with the two vectors as its sides.

The negative of a vector **A** is a vector with the same magnitude as **A** but pointing in the opposite direction.

The x component of a vector is equivalent to its projection along the x axis of a coordinate system. Likewise, the y component is the projection of the vector along the y axis of this coordinate system. The resultant of two or more vectors can be found mathematically by resolving all vectors into their x and y components, finding the resultant x and y components, and then using the Pythagorean theorem to find the resultant vector. The angle of the resultant vector with respect to the x axis can be found by use of a suitable trigonometric function.

An object moving above the surface of the Earth in both the x and y directions simultaneously is said to be undergoing **projectile motion.** The object moves along the horizontal (x) direction so that its velocity in this direction, v_x, is a constant, and the object also moves in the vertical (y) direction with a constant downward free-fall acceleration of magnitude $g = 9.80$ m/s^2. The equations describing the motion of a projectile are

$$x = v_{x0}t \qquad\qquad [3.10]$$

$$v_y = v_{y0} - gt \qquad\qquad [3.11]$$

$$y = v_{y0}t - \tfrac{1}{2}gt^2 \qquad\qquad [3.12]$$

$$v_y^2 = v_{y0}^2 - 2gy \qquad\qquad [3.13]$$

where $v_{y0} = v_0 \sin \theta_0$ is the initial vertical component of the velocity and $v_{x0} = v_0 \cos \theta_0$ is the initial horizontal component of the velocity.

Observations made in different frames of reference can be related to one another through the techniques of the transformation of relative velocities. There is no general equation for you to memorize in order to work relative velocity problems. Instead, you should follow the technique for writing subscripts on each velocity under consideration as described in the text.

MULTIPLE-CHOICE QUESTIONS

1. A 6-km displacement and an 8-km displacement are at right angles. The resultant displacement is found to be 10 km due east. Which of the following is a possible set of directions for the 6 km and 8 km displacements?
   (a) The 6 km is north of east and the 8 km is south of east.
   (b) The 6 km is north of west and the 8 km is south of west.
   (c) The 6 km is north of east and the 8 km is north of west.
   (d) The 6 km is south of west and the 8 km is south of east.

2. Ball A weighs twice as much as ball B. A is dropped straight down from the roof of a building and at the same time B is thrown off the building horizontally at a high speed. Neglect air resistance and determine which statement below is true.
   (a) A hits the ground before B.
   (b) B hits the ground before A.
   (c) They both hit at the same time.
   (d) There is insufficient information in this problem to enable one to determine an answer.

3. A tennis ball is given an initial horizontal velocity on the surface of a table. Which of the diagrams that follow best describes the motion of the ball as it rolls off the end of the table?

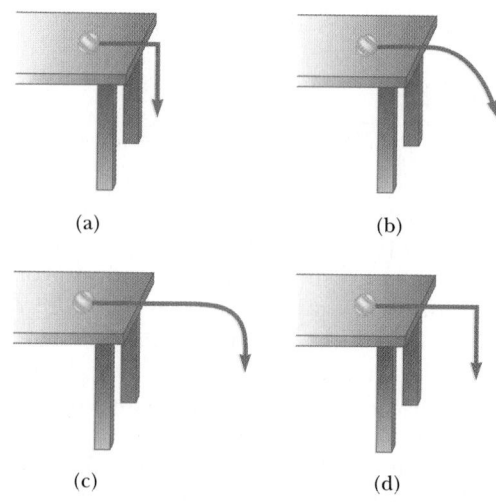

(a)   (b)

(c)   (d)

4. The record for a ski jump is 180 m set in 1989. Assume the jumper comes off the end of the ski jump horizontally and falls 90 m vertically before contacting the ground. What was the initial horizontal speed of the jumper?
   (a) 18.4 m/s   (b) 4.28 m/s   (c) 42 m/s
   (d) 84 m/s   (e) 360 m/s

5. It is said that in his youth George Washington threw a silver dollar across the Potomac River. Assuming the river was 75 m wide and the coin was thrown at a 45° angle, what minimum initial speed was required to get the coin across the river?
(a) 7 m/s (b) 14 m/s (c) 21 m/s (d) 27 m/s

CONCEPTUAL QUESTIONS

1. (a) Can an object accelerate if its speed is constant?
(b) Can an object accelerate if its velocity is constant?

2. If **B** is added to **A,** under what conditions does the resultant vector have a magnitude equal to $A + B$? Under what conditions is the resultant vector equal to zero?

3. (a) Can an object have a constant velocity and varying speed? (b) Can an object have a constant speed and varying velocity? Give examples.

4. Two vectors have unequal magnitudes. Can their sum be zero? Explain.

5. Suppose that you are running at constant speed. You wish to throw a ball such that you will catch it as it comes back down. How should you throw the ball?

6. Can a vector have a component greater than its magnitude?

7. Vector **A** lies in the xy plane. For what orientations will both of its rectangular components be negative? For what orientations will its components have opposite signs?

8. Under what circumstances would a vector have components that are equal in magnitude?

9. As a projectile moves in its path, is there any point along the path where the velocity and acceleration vectors are (a) perpendicular to each other? (b) parallel to each other?

10. A rock is dropped at the same instant that a ball, at the same initial elevation, is thrown horizontally. Which will have the greater speed when it reaches ground level?

11. Explain whether or not the following particles have an acceleration: (a) a particle moving in a straight line with constant speed and (b) a particle moving around a curve with constant speed.

12. Correct the following statement: "The racing car rounds the turn at a constant velocity of 90 miles per hour."

13. A spacecraft drifts through space at a constant velocity. Suddenly a gas leak in the side of the spacecraft causes a constant acceleration of the spacecraft in a direction perpendicular to the initial velocity. The orientation of the spacecraft does not change, so that the acceleration remains perpendicular to the original direction of the velocity. What is the shape of the path followed by the spacecraft in this situation?

14. A ball is projected horizontally from the top of a building. One second later another ball is projected horizontally from the same point with the same velocity. At what point in the motion will the balls be closest to each other? Will the first ball always be traveling faster than the second ball? What will be the time difference between when the balls hit the ground? Can the horizontal projection velocity of the second ball be changed so that the balls arrive at the ground at the same time?

15. Two projectiles are thrown with the same magnitude of initial velocity, one at an angle θ with respect to the level ground and the other at angle $90° - \theta$. Both projectiles will strike the ground at the same distance from the projection point. Will both projectiles be in the air for the same time interval?

16. A baseball is thrown such that its initial x and y components of velocity are known. Neglecting air resistance, describe how you would calculate, at the instant the ball reaches the top of its trajectory, (a) its coordinates, (b) its velocity, and (c) its acceleration. How would these results change if air resistance were taken into account?

17. A projectile is fired at some angle to the horizontal with some initial speed v_0, and air resistance is neglected. Is the projectile a freely falling body? What is its acceleration in the vertical direction? What is its acceleration in the horizontal direction?

18. The yellow ball in Figure Q3.18 is projected horizontally with some initial velocity. At the instant the yellow ball is projected, the red ball is released from the same height

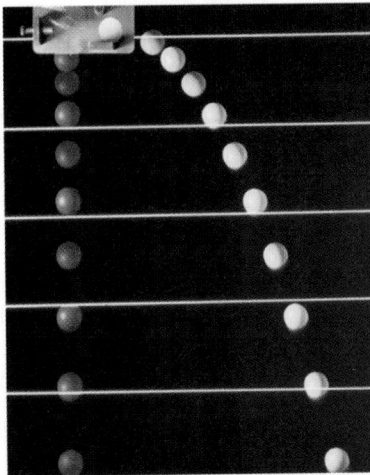

Figure Q3.18 (© *Richard Megna 1990, Fundamental Photographs*)

and falls to the floor. Explain why the two balls hit the floor simultaneously, even though the yellow ball has an initial velocity.

19. Describe how a driver can steer a car traveling at constant speed so that (a) the acceleration is zero or (b) the magnitude of the acceleration remains constant.

20. A ball is thrown upward in the air by a passenger on a train that is moving with constant velocity. (a) Describe the path of the ball as seen by the passenger. Describe the path as seen by a stationary observer outside the train. (b) How would these observations change if the train were accelerating along the track?

PROBLEMS

1, 2, 3 = straightforward, intermediate, challenging   ☐ = full solution available in Study Guide/Student Solutions Manual   ◿ = Core Concepts Workbook
WEB = solution posted at **http://www.harcourtcollege.com/physics/cptech**   ▧ = biomedical application   ▧ = Interactive Physics

Review Problem

When the triangular plot of land shown in the figure is surveyed, it is found that the side from *A* to *B* has a length of 311.4 ft and runs due east from point *A*. The side from *B* to *C* has a length of 260.8 ft and is in a direction 40.0° west of north. (a) When the side from *C* to *A* is surveyed, what is the expected length and direction for this side? (*Hint:* First draw a line from point *C* perpendicular to *AB*, giving you two right triangles to consider.) (b) Find the area of the plot of land in acres (1 acre = 43 560 ft²).

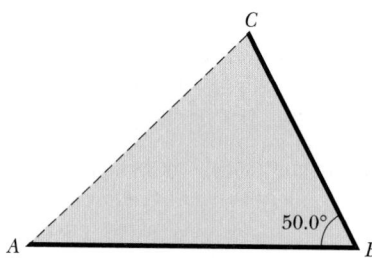

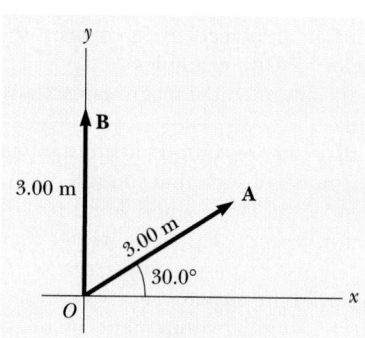

Figure P3.3

5. A man lost in a maze makes three consecutive displacements so that at the end of the walk he is right back where he started. The first displacement is 8.00 m westward, and the second is 13.0 m northward. Find the magnitude and direction of the third displacement, using the graphical method.

6. A jogger runs 100 m due west, then changes direction for the second leg of the run. At the end of the run, she is 175 m away from the starting point at an angle of 15.0° north of west. What were the direction and length of her second displacement? Use graphical techniques.

Section 3.2 Some Properties of Vectors

1. A dog searching for a bone walks 3.50 m south, then 8.20 m at an angle 30.0° north of east, and finally 15.0 m west. Find the dog's resultant displacement vector using graphical techniques.

2. A roller coaster moves 200 ft horizontally, then rises 135 ft at an angle of 30.0° above the horizontal. It then travels 135 ft at an angle of 40.0° below the horizontal. Use graphical techniques to find its displacement from the starting point at the end of this movement.

3. Each of the displacement vectors **A** and **B** shown in Figure P3.3 has a magnitude of 3.0 m. Graphically find (a) **A** + **B**, (b) **A** − **B**, (c) **B** − **A**, and (d) **A** − 2**B**.

4. Vector **A** is 3.00 units in length and points along the positive *x* axis. Vector **B** is 4.00 units in length and points along the negative *y* axis. Use graphical methods to find the magnitude and direction of the vectors (a) **A** + **B** and (b) **A** − **B**.

Section 3.3 Components of a Vector

7. A golfer takes two putts to get his ball into the hole once he is on the green. The first putt displaces the ball 6.00 m east, and the second, 5.40 m south. What displacement would have been needed to get the ball into the hole on the first putt?

8. A person walks 25.0° north of east for 3.10 km. How far would a person walk due north and due east to arrive at the same location?

WEB 9. A girl delivering newspapers covers her route by traveling 3.00 blocks west, 4.00 blocks north, then 6.00 blocks east. (a) What is her resultant displacement? (b) What is the total distance she travels?

10. A quarterback takes the ball from the line of scrimmage, runs backward for 10.0 yards, then runs sideways parallel to the line of scrimmage for 15.0 yards. At this point, he throws a 50.0-yard forward pass straight downfield, perpendicular to the line of scrimmage. What is the magnitude of the football's resultant displacement?

11. The eye of a hurricane passes over Grand Bahama Island. It is moving in a direction 60.0° north of west with a speed of 41.0 km/h. Three hours later, the course of the hurricane suddenly shifts due north, and its speed slows to 25.0 km/h. How far from Grand Bahama is the hurricane 4.50 h after it passes over the island?

12. A small map shows Atlanta to be 730 miles in a direction of 5° north of east from Dallas. The same map shows that Chicago is 560 miles in a direction of 21° west of north from Atlanta. Assume a flat Earth and use this information to find the displacement from Dallas to Chicago.

13. A commuter airplane starts from an airport and takes the route shown in Figure P3.13. It first flies to city A located at 175 km in a direction 30.0° north of east. Next, it flies 150 km 20.0° west of north to city B. Finally, it flies 190 km due west to city C. Find the location of city C relative to the location of the starting point.

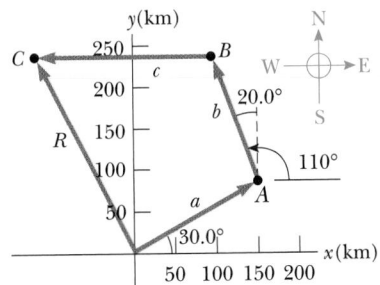

Figure P3.13

14. Two people pull on a stubborn mule, as seen from a helicopter in Figure P3.14. Find **(a)** the single force that is equivalent to the two forces shown, and **(b)** the force that a third person would have to exert on the mule to make the net force equal to zero.

15. A man pushing a mop across a floor causes it to undergo two displacements. The first has a magnitude of 150 cm and makes an angle of 120° with the positive x axis. The resultant displacement has a magnitude of 140 cm and is directed at an angle of 35.0° to the positive x axis. Find the magnitude and direction of the second displacement.

Section 3.4 Velocity and Acceleration in Two Dimensions

Section 3.5 Projectile Motion

16. The fastest recorded pitch in major-league baseball, thrown by Nolan Ryan in 1974, was clocked at

Problem 20 is the same as Workbook Problem 11 (Workbook page 30).

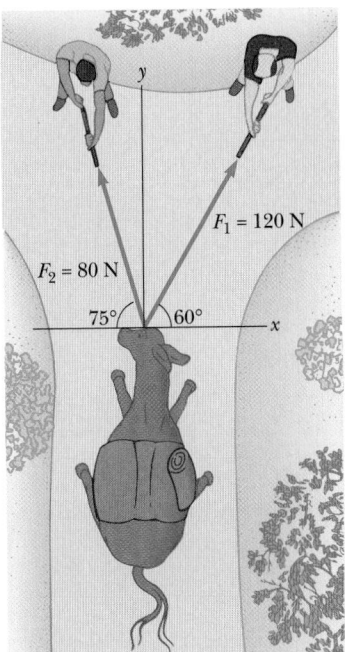

Figure P3.14

100.8 mi/h. If a pitch were thrown horizontally with this velocity, how far would the ball fall vertically by the time it reached home plate 60.0 ft away?

WEB 17. Tom the cat is chasing Jerry the mouse across a table surface 1.5 m above the floor. Jerry steps out of the way at the last second, and Tom slides off the edge of the table at a speed of 5.0 m/s. Where will Tom strike the floor, and what velocity components will he have just before he hits?

18. A student stands at the edge of a cliff and throws a stone horizontally over the edge with a speed of 18.0 m/s. The cliff is 50.0 m above a flat horizontal beach, as shown in Figure P3.18. How long after being released does the stone strike the beach below the cliff? With what speed and angle of impact does it land?

19. A brick is thrown upward from the top of a building at an angle of 25° to the horizontal and with an initial speed of 15 m/s. If the brick is in flight for 3.0 s, how tall is the building?

20. A place kicker must kick a football from a point 36.0 m (about 40.0 yd) from the goal, and the ball must clear the crossbar, which is 3.05 m high. When kicked, the ball leaves the ground with a speed of 20.0 m/s at an angle of 53.0° to the horizontal. **(a)** By how much does the ball clear or fall short of clearing the crossbar? **(b)** Does the ball approach the crossbar while still rising or while falling?

21. A car is parked on a cliff overlooking the ocean on an incline that makes an angle of 24.0° below the horizontal. The negligent driver leaves the car in neutral, and

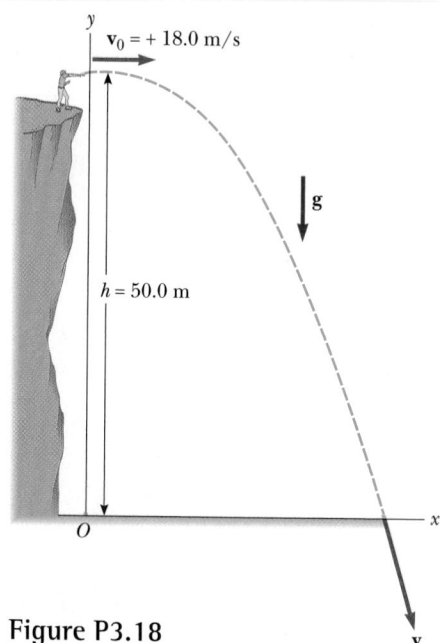

Figure P3.18

the emergency brakes are defective. The car rolls from rest down the incline with a constant acceleration of 4.00 m/s^2 for a distance of 50.0 m to the edge of the cliff. The cliff is 30.0 m above the ocean. Find (a) the car's position relative to the base of the cliff when the car lands in the ocean, and (b) the length of time the car is in the air.

22. A firefighter, 50.0 m away from a burning building, directs a stream of water from a ground level fire hose at an angle of 30.0° above the horizontal. If the speed of the stream as it leaves the hose is 40.0 m/s, at what height will the stream of water strike the building?

23. A projectile is launched with an initial speed of 60 m/s at an angle of 30° above the horizontal. The projectile lands on a hillside 4.0 s later. Neglect air friction. (a) What is the projectile's velocity at the highest point of its trajectory? (b) What is the straight-line distance from where the projectile was launched to where it hits?

Section 3.6   Relative Velocity

24. A jet airliner moving initially at 300 mph due east enters a region where the wind is blowing at 100 mph in a direction 30.0° north of east. What is the new velocity of the aircraft relative to the ground?

25. A boat moves through the water of a river at 10 m/s relative to the water, regardless of the boat's direction. If the water in the river is flowing at 1.5 m/s, how long does it take the boat to make a round trip consisting of a 300-m displacement downstream followed by a 300-m displacement upstream?

26. A river flows due east at 1.50 m/s. A boat crosses the river from the south shore to the north shore by maintaining a constant velocity of 10.0 m/s due north relative to the water. (a) What is the velocity of the boat relative to shore? (b) If the river is 300 m wide, how far downstream has the boat moved by the time it reaches the north shore?

WEB 27. A rowboat crosses a river with a velocity of 3.30 mi/h at an angle 62.5° north of west relative to the water. The river is 0.505 mi wide and carries an eastward current of 1.25 mi/h. How far upstream is the boat when it reaches the opposite shore?

28. The pilot of an aircraft wishes to fly due west in a 50.0-km/h wind blowing toward the south. If the speed of the aircraft relative to the air is 200 km/h, (a) in what direction should the aircraft head, and (b) what will be its speed relative to the ground?

29. How long does it take an automobile traveling in the left lane at 60 km/h to overtake (become even with) another car that is traveling in the right lane at 40 km/h, when the cars' front bumpers are initially 100 m apart?

30. A science student is riding on a flatcar of a train traveling along a straight horizontal track at a constant speed of 10.0 m/s. The student throws a ball along a path that she judges to make an initial angle of 60.0° with the horizontal and to be in line with the track. The student's professor, who is standing on the ground nearby, observes the ball to rise vertically. How high does the ball rise?

ADDITIONAL PROBLEMS

31. A person walks half way around the circumference of a circular path of radius 5.00 m. (a) Find the magnitude of the displacement vector. (b) How far did the person walk? (c) What is the magnitude of the displacement if the circle is completed?

32. A particle undergoes two displacements. The first has a magnitude of 150 cm and makes an angle of 120.0° with the positive x axis. The *resultant* of the two displacements is 140 cm directed at an angle of 35.0° to the positive x axis. Find the magnitude and direction of the second displacement.

33. Figure P3.33 is a multiflash photograph of two golf balls released simultaneously. The time interval between flashes is 0.033 s, and the white parallel lines were placed 15 cm apart. (a) Find the speed at which the right ball was projected, and (b) show that both balls should be expected to reach the floor simultaneously.

34. A ball is projected horizontally from the edge of a table that is 1.00 m high, and it strikes the floor at a point 1.20 m from the base of the table. (a) What is the initial speed of the ball? (b) How high is the ball above the floor when its velocity vector makes a 45.0° angle with the horizontal?

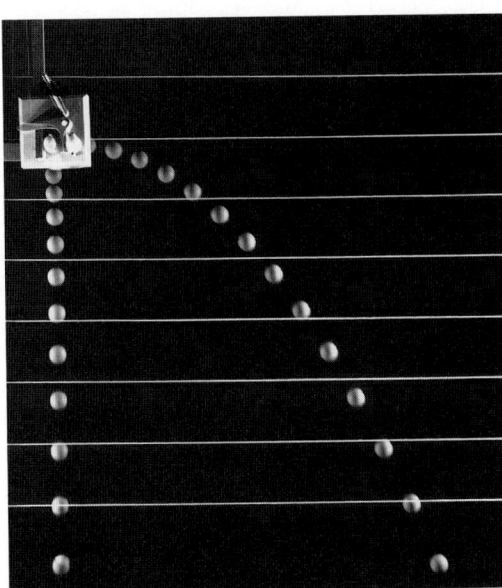

Figure P3.33   *(Educational Development Center)*

35. A car travels due east with a speed of 50.0 km/h. Rain is falling vertically with respect to the Earth. The traces of the rain on the side windows of the car make an angle of 60.0° with the vertical. Find the velocity of the rain with respect to **(a)** the car and **(b)** the Earth.

36. Towns A and B in Figure P3.36 are 80.0 km apart. A couple arranges to drive from town A and meet a couple driving from town B at the lake, L. The two couples leave simultaneously and drive for 2.50 h in the direc-

tions shown. Car 1 has a speed of 90.0 km/h. If the cars arrive simultaneously at the lake, what is the speed of car 2?

37. A rocket is launched at an angle of 53.0° above the horizontal with an initial speed of 100 m/s. It moves for 3.00 s along its initial line of motion with an acceleration of 30.0 m/s². At this time its engines fail and the rocket proceeds to move as a free body. Find **(a)** the maximum altitude reached by the rocket, **(b)** its total time of flight, and **(c)** its horizontal range.

38. Two paddlers in identical canoes exert the same effort paddling and hence maintain the same speed relative to the water. One paddles directly upstream (and moves upstream), whereas the other paddles directly downstream. If downstream is the positive direction, an observer on shore determines the velocities of the two canoes to be −1.2 m/s and +2.9 m/s, respectively. **(a)** What is the speed of the water relative to shore? **(b)** What is the speed of each canoe relative to the water?

39. If a person can jump a maximum horizontal distance (by using a 45° projection angle) of 3.0 m on the Earth, what would be his maximum range on the Moon, where the free-fall acceleration is $g/6$ and $g = 9.80$ m/s²? Repeat for Mars, where the acceleration due to gravity is $0.38g$.

WEB **40.** A daredevil decides to jump a canyon of width 10 m. To do so, he drives a motorcycle up an incline sloped at an angle of 15°. What minimum speed must she have in order to clear the canyon?

41. The determined coyote is out once more to try to capture the elusive roadrunner. The coyote wears a new pair of Acme power roller skates, which provide a constant horizontal acceleration of 15 m/s², as shown in Figure P3.41. The coyote starts off at rest 70 m from the edge of a cliff at the instant the roadrunner zips by in the direction of the cliff. **(a)** If the roadrunner moves with constant speed, determine the minimum speed he

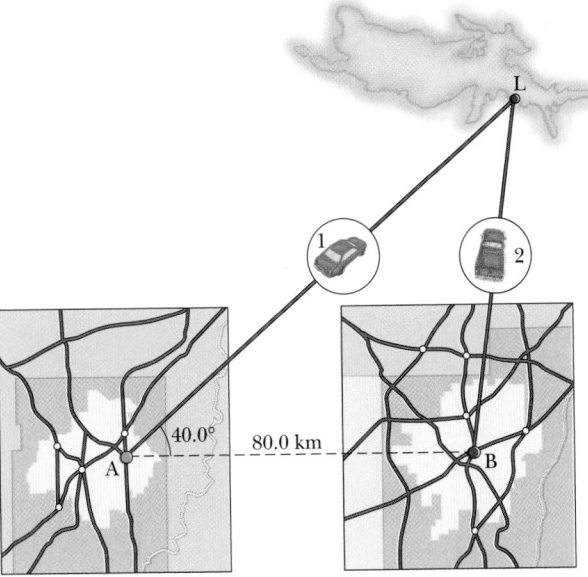

Figure P3.36

Figure P3.41

must have in order to reach the cliff before the coyote.
(b) If the cliff is 100 m above the base of a canyon, de-termine where the coyote lands in the canyon. (Assume that his skates are still in operation when he is in "flight" and that his horizontal component of accelera-tion remains constant at 15 m/s².)

42. An artillery shell is fired with an initial speed of 1.70×10^3 m/s (approximately five times the speed of sound) at an angle of 55.0° above the horizontal and re-turns to its original level before impact. Neglecting air resistance, find **(a)** the time it is in motion and **(b)** the horizontal distance traveled.

43. A home run is hit in such a way that the baseball just clears a wall 21 m high, located 130 m from home plate. The ball is hit at an angle of 35° to the horizontal, and air resistance is negligible. Find **(a)** the initial speed of the ball, **(b)** the time it takes the ball to reach the wall, and **(c)** the velocity components and the speed of the ball when it reaches the wall. (Assume the ball is hit at a height of 1.0 m above the ground.)

44. A ball is thrown straight upward and returns to the thrower's hand after 3.00 s in the air. A second ball is thrown at an angle of 30.0° with the horizontal. At what speed must the second ball be thrown so that it reaches the same height as the one thrown vertically?

45. A quarterback throws a football toward a receiver with an initial speed of 20 m/s, at an angle of 30° above the horizontal. At that instant, the receiver is 20 m from the quarterback. In what direction and with what constant speed should the receiver run in order to catch the foot-ball at the level at which it was thrown?

46. A 2.00-m-tall basketball player wants to make a goal from 10.0 m from the basket, as in Figure P3.46. If he shoots the ball at a 45.0° angle, at what initial speed must he throw the basketball so that it goes through the hoop without striking the backboard?

47. In a popular lecture demonstration, a projectile is fired at a falling target as in Figure P3.47. The projectile leaves the gun at the same instant that the target is dropped from rest. Assuming that the gun is initially aimed at the target, show that the projectile will hit the target. (One restriction of this experiment is that the projectile must reach the target before the target strikes the floor.)

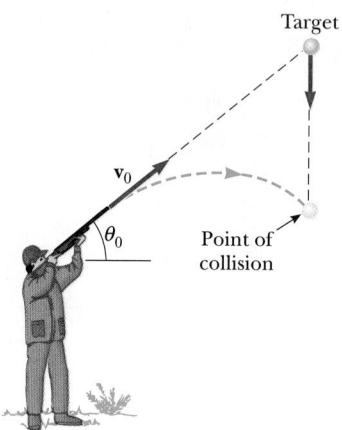

Figure P3.47

48. Figure P3.48 illustrates the difference in proportions be-tween the male and female anatomies. The displace-ments $\mathbf{d}_{1m}$ and $\mathbf{d}_{1f}$ from the bottom of the feet to the navel have magnitudes of 104 cm and 84.0 cm, respec-tively. The displacements $\mathbf{d}_{2m}$ and $\mathbf{d}_{2f}$ have magnitudes of 50.0 cm and 43.0 cm, respectively. **(a)** Find the vector sum of the displacements $\mathbf{d}_1$ and $\mathbf{d}_2$ in each case. **(b)** The male figure is 180 cm tall, the female 168 cm. Normalize the displacements of each figure to a com-mon height of 200 cm, and re-form the vector sums as in part **(a)**. Then find the vector difference between the two sums.

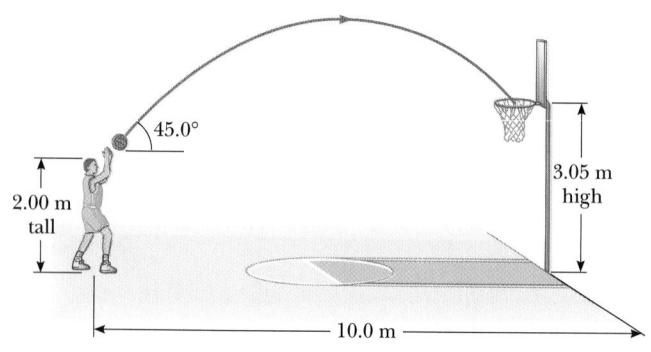

Figure P3.46

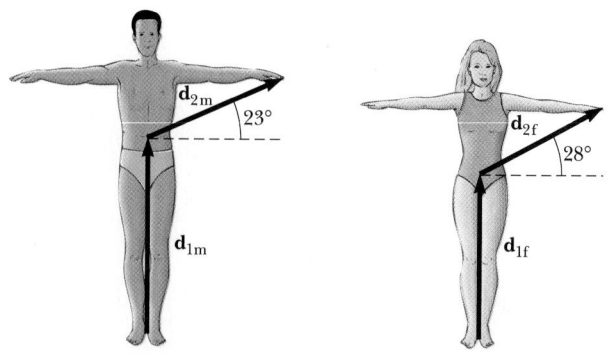

Figure P3.48

49. A girl can throw a ball a maximum horizontal distance of R on a level field. How far can she throw the same ball vertically upward? Assume that her muscles give the ball the same speed in each case. (Is this assumption valid?)

50. A projectile is fired with an initial speed of v_0 at an angle of θ_0 to the horizontal, as in Figure 3.11. When it reaches its peak, it has (x, y) coordinates given by $(R/2, h)$, and when it strikes the ground, its coordinates are $(R, 0)$, where R is called the horizontal range.
(a) Show that it reaches a maximum height, h, given by

$$h = \frac{v_0{}^2 \sin^2 \theta_0}{2g}$$

(b) Show that its horizontal range is given by

$$R = \frac{v_0{}^2 \sin 2\theta_0}{g}$$

51. A hunter wishes to cross a river that is 1.5 km wide and flows with a speed of 5.0 km/h parallel to its banks. The hunter uses a small powerboat that moves at a maximum speed of 12 km/h with respect to the water. What is the minimum time necessary for crossing?

52. A skater (water spider) maintains an average position on the surface of a stream by darting upstream against the current, then drifting downstream with the current to its original position. The current in the stream is 0.500 m/s relative to the shore, and the skater darts upstream 0.560 m (relative to a spot on shore) in 0.800 s during the first part of its motion. Take upstream as the positive direction. (a) Determine the velocity of the skater relative to the water (i) during its dash upstream and (ii) during its drift downstream. (b) How far upstream relative to the water does the skater move during one cycle of this motion? (c) What is the average velocity of the skater relative to the water?

53. A daredevil is shot out of a cannon at 45.0° to the horizontal with an initial speed of 25.0 m/s. A net is positioned a horizontal distance of 50.0 m from the cannon. At what height above the cannon should the net be placed in order to catch the daredevil?

54. A dart gun is fired while being held horizontally at a height of 1.00 m above ground level and at rest relative to the ground. The dart from the gun travels a horizontal distance of 5.00 m. A college student holds the same gun in a horizontal position while sliding down a 45.0° incline at a constant speed of 2.00 m/s. How far will the dart travel if the student fires the gun when it is 1.00 m above the ground?

55. A student decides to measure the muzzle velocity of a pellet from her gun. She points the gun horizontally. She places a target on a vertical wall a distance x away from the gun. The pellet hits the target a vertical dis-

tance y below the gun. (a) Show that the position of the pellet when traveling through the air is given by $y = Ax^2$, where A is a constant. (b) Express the constant A in terms of the initial velocity and the free-fall acceleration. (c) If $x = 3.00$ m and $y = 0.210$ m, what is the initial speed of the pellet?

56. Indiana Jones is trapped in a maze. To find his way out, he walks 10.0 m, makes a 90.0° right turn, walks 5.00 m, makes another 90.0° right turn, and walks 7.00 m. What is his displacement from his initial position?

57. Instructions for finding a buried treasure include the following: Go 75 paces at 240°, turn to 135° and walk 125 paces, then travel 100 paces at 160°. Determine the resultant displacement from the starting point.

58. A jet airliner moving initially at 300 mph due east enters a region where the wind is blowing at 100 mph in a direction 30.0° north of east. What are the new velocity and direction of the aircraft?

59. One strategy in a snowball fight is to throw a snowball at a high angle over level ground. While your opponent is watching the first one, you throw a second snowball at a low angle timed to arrive before or at the same time as the first one. Assume both snowballs are thrown with a speed of 25.0 m/s. The first one is thrown at an angle of 70.0° with respect to the horizontal. (a) At what angle should the second snowball be thrown to arrive at the same point as the first? (b) How many seconds later should the second snowball be thrown after the first to arrive at the same time?

60. When baseball outfielders throw the ball, they usually allow it to take one bounce on the theory that the ball arrives sooner this way. Suppose that after the bounce the ball rebounds at the same angle θ as it had when released (Fig. P3.60) but loses half its speed. (a) Assuming the ball is always thrown with the same initial speed, at what angle θ should the ball be thrown in order to go the same distance D with one bounce (blue path) as one thrown upward at 45.0° with no bounce (green path)? (b) Determine the ratio of the times for the one-bounce and no-bounce throws.

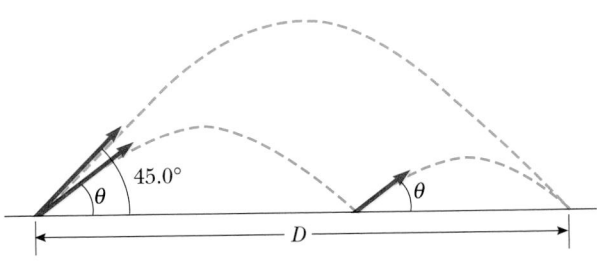

Figure P3.60

Problem 57 is the same as Workbook Problem 4 (Workbook page 16).

61. A projectile is fired up an incline (having an angle of ϕ) with an initial velocity of v_0 at an angle of θ_0 with respect to the horizontal ($\theta_0 > \phi$), as shown in Figure P3.61. Show that the projectile will travel a distance of d up the incline, where

$$d = \frac{2v_0^2 \cos(\theta_0) \sin(\theta_0 - \phi)}{g \cos^2(\phi)}$$

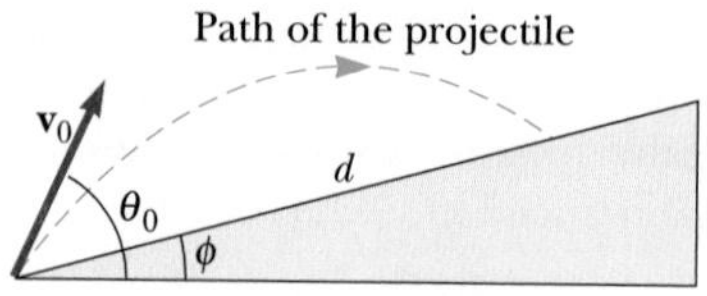

Figure P3.61

The Laws of Motion

▲ PHYSICS PUZZLER

A pole vaulter falling from a height of about 6 m is protected from serious injury by landing on a thick soft pad. Why does the padding reduce the average force exerted on the vaulter? What information would you need to estimate the acceleration of the vaulter when he is in contact with the pad? *(Gerard Vandystadt/All Sport)*

4.1

Classical mechanics describes the relationship between the motion of an object and the forces acting on it. There are conditions under which classical mechanics does not apply or applies in only a limited way. Most often these conditions are encountered when dealing with objects whose size is comparable to that of atoms or smaller (about 10^{-10} m) and/or which move at speeds near the speed of light (3×10^8 m/s). Our study of relativity and quantum mechanics in later chapters will enable us to handle such situations. However, we can make accurate calculations and investigations in physics with the laws of classical mechanics discussed in this chapter.

We learn in this chapter that an object remains in motion with constant velocity if no net external force acts on it. We will also see that if a net external force is exerted on an object, the object accelerates in response to this net force. The acceleration can be determined if the net force acting on the object is known and if the object's mass is known.

4.1   THE CONCEPT OF FORCE

When we think of a force, we usually imagine a push or a pull exerted on some object. For instance, you exert a force on a ball when you throw it or kick it, and you exert a force on a chair when you sit down on it. What happens to an object when it is acted on by a force depends on the magnitude and the direction of the force. Force is a vector quantity; thus, we denote it with a directed arrow, just as we do velocity and acceleration.

If you pull on a spring, as in Figure 4.1a, the spring stretches. If a child pulls hard enough on a wagon, as in Figure 4.1b, the wagon moves. When a football is kicked, as in Figure 4.1c, it is deformed and set in motion. These are all examples of **contact forces,** so named because they result from physical contact between two objects.

Another class of forces does not involve physical contact between two objects. Early scientists, including Newton, were uneasy with the concept of forces that act between two disconnected objects. To overcome this conceptual difficulty, Michael Faraday (1791–1867) introduced the concept of a *field*. The corresponding forces are called **field forces.** According to this approach, when a mass, m, is placed at some point P near a second mass, M, we say that m interacts with M by virtue of the gravitational field that exists at P. Thus, the force of gravitational attraction between two objects, illustrated in Figure 4.1d, is an example of a field force. This force keeps objects bound to the Earth and gives rise to what we commonly call the *weight* of the object. The planets of our Solar System move under the actions of gravitational forces.

Another common example of a field force is the electric force that one electric charge exerts on another, as in Figure 4.1e. A third example is the force exerted by a bar magnet on a piece of iron, as in Figure 4.1f. The known fundamental forces in nature are all field forces. These are, in order of decreasing strength, (1) strong nuclear forces between subatomic particles, (2) electromagnetic forces between electric charges at rest or in motion, (3) weak nuclear forces, which arise in certain radioactive decay processes, and (4) gravitational attractions between objects. In classical physics we are concerned only with gravitational and electromagnetic forces.

Whenever a force is exerted on an object, its shape can change. For example, when you squeeze a rubber ball or strike a punching bag with your fist, the object

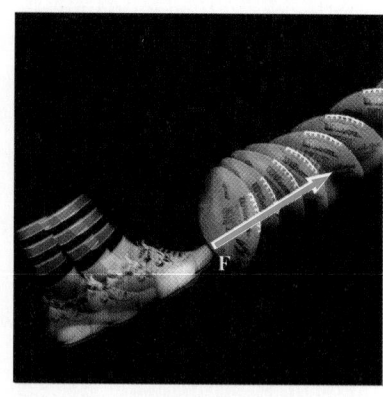

A football is set in motion by the contact force, **F,** on it due to the kicker's foot. The ball is deformed during the short time in contact with the foot. *(Ralph Cowan, Tony Stone Worldwide)*

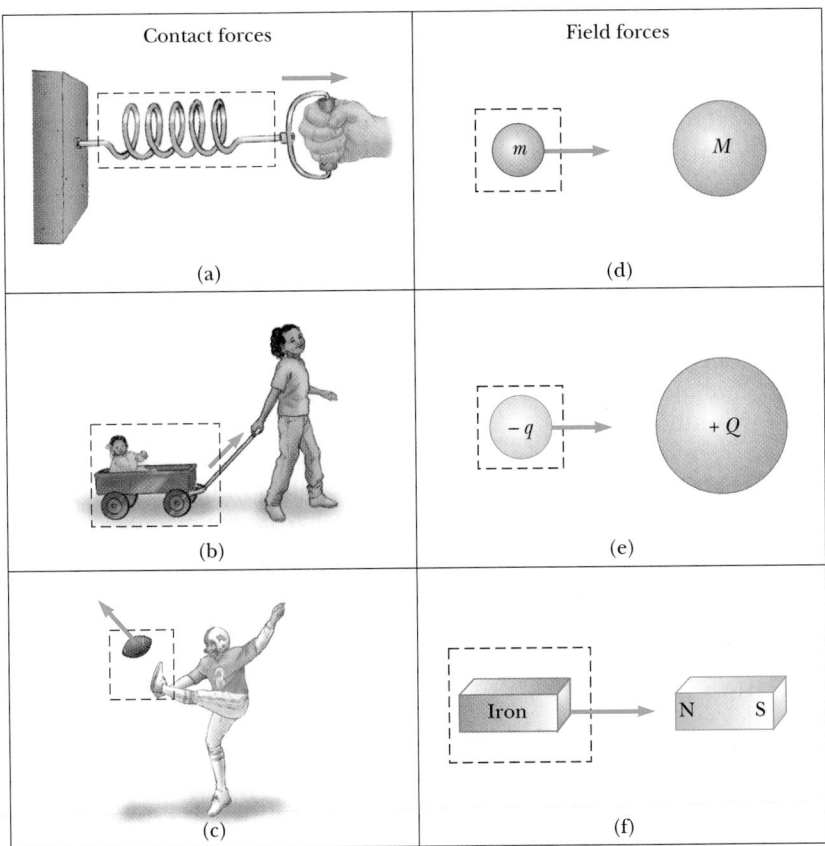

Contact forces	Field forces
(a)	(d)
(b)	(e)
(c)	(f)

Figure 4.1 Examples of forces applied to various objects. In each case a force acts on the object surrounded by the dashed lines. Something in the environment external to the boxed area exerts this force.

deforms to some extent. Even more rigid objects, such as automobiles, are deformed under the action of external forces. Often the deformations are permanent, as in the case of a collision between vehicles.

4.2 NEWTON'S FIRST LAW

Consider the following simple experiment. Suppose a book is lying on a table. Obviously, the book remains at rest if left alone. Now imagine that you push the book with a horizontal force great enough to overcome the force of friction between book and table, so that the book is set in motion. If the magnitude of your applied force is equal to the magnitude of the friction force, the book moves with constant velocity. If the magnitude of your applied force exceeds the magnitude of the friction force, the book accelerates. If you stop applying the force, the book stops sliding after traveling a short distance because the force of friction retards its motion. Now imagine pushing the book across a smooth, waxed floor. The book again comes to rest once the force is no longer applied, but not as soon as before. Finally, imagine that the book is moving on a horizontal, frictionless surface. In this situation, the book continues to move in a straight line with constant velocity until it hits a wall or some other object.

Before about 1600, scientists felt that the natural state of matter was the state of rest. Galileo was the first to take a different approach. He devised thought

A multiflash exposure of a tennis ball being struck by a racket. The ball experiences a large contact force as it is struck, and a correspondingly large acceleration. *(Ben Rose/The Image Bank)*

experiments—such as an object moving on a frictionless surface, as just described—and concluded that it is not the nature of an object to *stop* once set in motion; rather, it is an object's nature to *resist acceleration*. This approach to motion was later formalized by Newton in a form that has come to be known as **Newton's first law of motion:**

Newton's first law ▶

> **An object at rest remains at rest, and an object in motion continues in motion with constant velocity (that is, constant speed in a straight line), unless it experiences a net external force.**

4.2, BOTH SECTIONS

WEB

For Web links to more information on Newton, visit the textbook Web site at http://www.harcourtcollege.com/physics/cptech

By external force we mean any force that results from the interaction between the object and its environment, such as the force exerted on an object when it is lifted. In simpler terms, Newton's first law says that **when the net external force on an object is zero, its acceleration is zero.** That is, when $\Sigma\mathbf{F} = 0$, then $\mathbf{a} = 0$.

Finally, consider a spaceship traveling in space, far from any planets or other matter. The spaceship requires a propulsion system to change its velocity. However, if the propulsion system is turned off when a velocity, $\mathbf{v}$, is reached, the spaceship coasts in space with a constant velocity and the astronauts get a free ride.

Mass and Inertia

Imagine a bowling ball and a golf ball sitting side by side on the ground. Newton's first law tells us that both remain at rest as long as no net external force acts on them. Now imagine supplying a net force by striking each ball with a golf club.

An object at rest will remain at rest and an object in motion will continue in motion with constant velocity unless a net external force acts on the object. In this case, the wall of the building did not exert a large enough force on the moving train to stop it. (*Photo Roger Viollet, Mill Valley, CA, University Science Books, 1982*)

Both balls resist your attempt to change their states of motion. But you know from everyday experience that if the two are struck with equal force, the golf ball will travel much farther than the bowling ball. That is, the bowling ball is more successful in resisting your attempt to change the state of motion. The tendency of an object to resist any attempt to change its motion is called the **inertia** of the object.

Mass is a measurement of inertia, and the SI unit of mass is the kilogram. The greater the mass of a body, the less it accelerates under the action of an applied force. For example, if a given force acting on a 3-kg mass produces an acceleration of 4 m/s², the same force applied to a 6-kg mass will produce an acceleration of only 2 m/s². Mass is a scalar quantity that obeys the rules of ordinary arithmetic.

Inertia is the principle that underlies the operation of seat belts. In the event of an accident, the purpose of the seat belt is to hold the passenger firmly in place relative to the car, to prevent serious injury. Figure 4.2 illustrates how one type of shoulder harness operates. Under normal conditions, the ratchet turns freely to allow the harness to wind on or unwind from the pulleys as the passenger moves. If an accident occurs, the car undergoes a large negative acceleration and rapidly comes to rest. The large mass under the seat, because of its inertia, continues to slide forward along the tracks. The pin connection between the mass and the rod causes the rod to pivot about its center and engage the ratchet wheel. At this point the ratchet wheel locks in place, and the harness no longer unwinds.

A slight modification enables this device to also activate an air bag in the car. In this case, movement of the mass and pivoting of the rod open a valve on a cylinder that contains nitrogen under pressure. The nitrogen rushes into the air bag, causing it to expand rapidly so that it serves as a protective cushion for the passenger.

4.3

A P P L I C A T I O N

Seat Belts and Air Bags.

Isaac Newton, English physicist and mathematician (1642–1727)

Newton was one of the most brilliant scientists in history. Before the age of 30, he formulated the basic concepts and laws of mechanics, discovered the law of universal gravitation, and invented the mathematical methods of calculus. As a consequence of his theories, Newton was able to explain the motions of the planets, the ebb and flow of the tides, and many special features of the motions of the Moon and the Earth. He also interpreted many fundamental observations concerning the nature of light. His contributions to physical theories dominated scientific thought for two centuries and remain important today. *(Giraudon/Art Resource)*

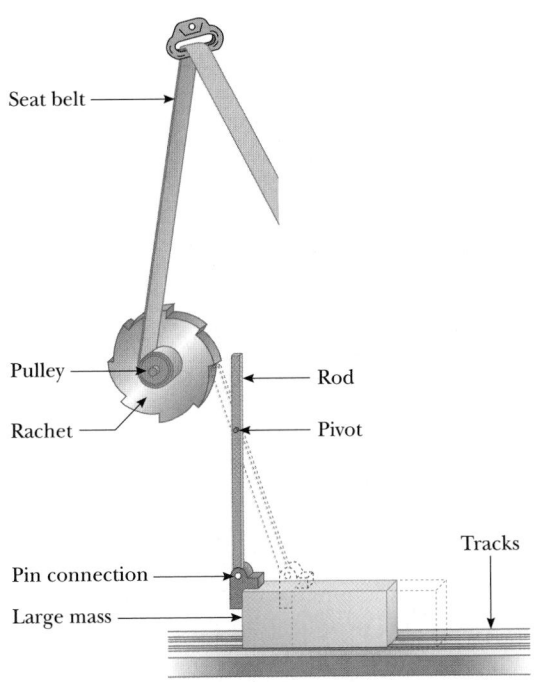

Figure 4.2 A mechanical arrangement for a safety belt.

4.3   NEWTON'S SECOND LAW

Newton's first law explains what happens to an object when the net force acting on it is zero: the object either remains at rest or moves in a straight line with constant speed. Newton's second law answers the question of what happens to an object that has a nonzero net force acting on it.

Imagine pushing a block of ice across a frictionless horizontal surface. When you exert some horizontal force on the block, it moves with an acceleration of, say, 2 m/s^2. If you apply a force twice as large, the acceleration doubles. Pushing three times as hard triples the acceleration, and so on. From such observations, we conclude that **the acceleration of an object is directly proportional to the net force acting on it.**

Common experience with pushing objects tells you that mass also affects acceleration. Imagine that you stack identical blocks of ice on top of each other while pushing the stack with constant force. If the force when you push one block produces an acceleration of 2 m/s^2, the acceleration will drop to half that value when two blocks are pushed, one third that initial value when three blocks are pushed, and so on. We conclude that **the acceleration of an object is inversely proportional to its mass.** These observations are summarized in **Newton's second law:**

Newton's second law ▶

> The acceleration of an object is directly proportional to the net force acting on it and inversely proportional to its mass.

In equation form, we can state Newton's second law as

$$\Sigma \mathbf{F} = m\mathbf{a} \qquad [4.1]$$

where $\mathbf{a}$ is the acceleration of the object, m is its mass, and $\Sigma \mathbf{F}$ represents the *vector sum of all external forces acting on the object.* (An external force is one that results from interaction between the object and its environment.) You should note that, because this is a vector equation, it is equivalent to the following three component equations:

$$\Sigma F_x = ma_x \qquad \Sigma F_y = ma_y \qquad \Sigma F_z = ma_z \qquad [4.2]$$

Units of Force and Mass

Definition of newton ▶

The SI unit of force is the **newton,** defined as the force that, when acting on a 1-kg mass, produces an acceleration of 1 m/s^2. From this definition and Newton's second law, we see that the newton can be expressed in terms of the fundamental units of mass, length, and time:

$$1 \text{ N} \equiv 1 \text{ kg} \cdot \text{m/s}^2 \qquad [4.3]$$

Definition of dyne ▶

The unit of force in the cgs system is called the **dyne** and is defined as the force that, when acting on a 1-g mass, produces an acceleration equal to 1 cm/s^2:

$$1 \text{ dyne} \equiv 1 \text{ g} \cdot \text{cm/s}^2 \qquad [4.4]$$

TABLE 4.1 Units of Mass, Acceleration, and Force

System	Mass	Acceleration	Force
SI	kg	m/s^2	$N = kg \cdot m/s^2$
cgs	g	cm/s^2	$dyne = g \cdot cm/s^2$

Note: $1 \text{ N} = 10^5 \text{ dyne} = 0.225 \text{ lb}$.

In the British engineering system the unit of force is the **pound.** The following conversion from pounds to newtons will be useful to you in many problems:

$$1 \text{ N} \equiv 0.225 \text{ lb} \qquad [4.5]$$

◀ Definition of pound

The units of mass, acceleration, and force in the SI and cgs system are summarized in Table 4.1.

Weight and the Gravitational Force

We are well aware that objects are attracted to the Earth. The force exerted by the Earth on an object is the gravitational force $\mathbf{F}_g$. This force is directed approximately toward the center of the Earth, and its magnitude varies with location. The magnitude of the gravitational force is called the **weight** of the object, $w = mg$, where m is the mass of the object.

We have seen that a freely falling object experiences an acceleration, $\mathbf{g}$, acting toward the center of the Earth. Applying Newton's second law to the freely falling object shown in Figure 4.3, we have $\mathbf{F} = m\mathbf{a}$. Because $\mathbf{F}_g = m\mathbf{g}$, it follows that $\mathbf{a} = \mathbf{g}$ and

$$w = mg \qquad [4.6]$$

Because it depends on g, weight varies with geographic location. Bodies weigh less at higher altitudes than at sea level because g decreases with increasing distance above the Earth's surface. Hence, weight, unlike mass, is not an inherent property of an object. For example, if an object has a mass of 70.0 kg, then its weight at a location where $g = 9.80 \text{ m/s}^2$ is $mg = 686$ N. At the top of a mountain, where g might be 9.76 m/s^2, the object's weight would be 683 N. Therefore, if you want to lose weight quickly without going on a diet, move to the top of a mountain or weigh yourself at an altitude of 30 000 ft during a flight on a jet airplane.

Because $w = mg$, we can compare the masses of two objects by measuring their weights with a spring scale or balance. At a given location the ratio of the weights of the two objects equals the ratio of their masses.

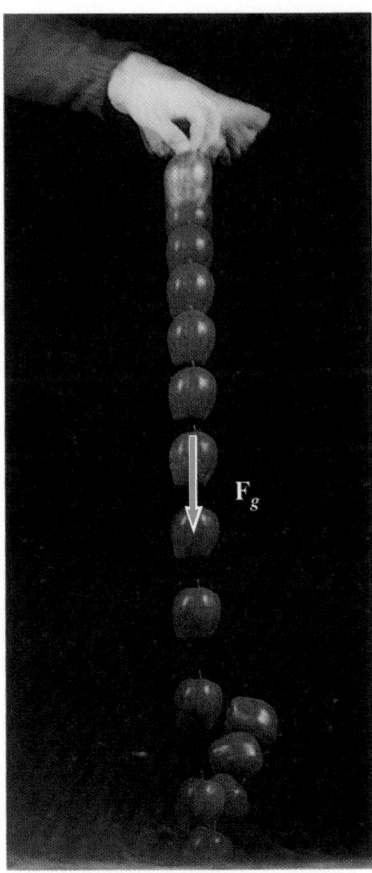

Figure 4.3 The only force acting on an object in free fall is the force of gravity, $\mathbf{F}_g$. The magnitude of the force of gravity is defined as the weight of the object, w. Newton's second law applied in the vertical direction gives $w = mg$.

Thinking Physics 1

Suppose you are talking by interplanetary telephone to your friend, who lives on the Moon. She tells you that she has just won 1 newton of gold in a contest. Excitedly, you tell her that you entered the Earth version of the same contest and also won 1 newton of gold! Who is richer?

Astronaut Edwin E. Aldrin, Jr., walking on the Moon after the Apollo 11 lunar landing. The weight of this astronaut on the Moon is less than it is on Earth, but his mass remains the same. *(NASA)*

Newton's third law ▶

4.5, SECTION 1

Explanation   Because the value of *g* on the Moon is smaller than on Earth by a factor of about 6, more mass of gold would be required on the Moon to represent a weight of 1 newton. Thus, your friend on the Moon is richer, by a factor of about 6!

4.4   NEWTON'S THIRD LAW

In Section 4.1 we found that a force is exerted on an object when that object comes into contact with some other object. For example, consider the task of driving a nail into a block of wood, as illustrated in Figure 4.4a. To accelerate the nail and drive it into the block, a net force must be supplied to the nail by the hammer. However, Newton recognized that a single isolated force (such as the force on the nail by the hammer) cannot exist. Instead, **forces in nature always exist in pairs.** According to Newton, the hammer exerts a force on the nail, and the nail exerts a force on the hammer. There is clearly a net force on the hammer, because it rapidly slows down after coming into contact with the nail.

Newton described this type of situation in terms of his **third law of motion:**

> If two objects interact, the force exerted on object 1 by object 2 is equal in magnitude but opposite in direction to the force exerted on object 2 by object 1.

This law, which is illustrated in Figure 4.4b, is equivalent to stating that **forces always occur in pairs,** or that **a single isolated force cannot exist.** The force that body 1 exerts on body 2 is sometimes called the *action force,* and the force of body 2 on body 1 is called the *reaction force.* In reality, either force can be labeled the action or reaction force. **The action force is equal in magnitude to the reaction force and opposite in direction. In all cases, the action and reaction forces act on different objects and must be of the same type.** For example, the force acting on a freely falling projectile is the force of the Earth on the projectile, $\mathbf{F}_g$, and the magnitude of this force is *mg*. The reaction to this force is the force of the projectile

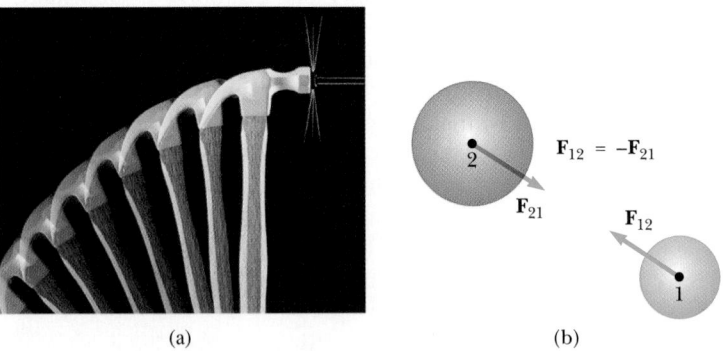

(a)   (b)

Figure 4.4   Newton's third law. (a) The force exerted by the hammer on the nail is equal to and opposite the force exerted by the nail on the hammer. (b) The force exerted by object 1 on object 2 is equal to and opposite the force exerted by object 2 on object 1. *(John Gillmoure, The Stock Market)*

on the Earth, $\mathbf{F}_g' = -\mathbf{F}_g$. The reaction force, $\mathbf{F}_g'$, must accelerate the Earth toward the projectile just as the action force, $\mathbf{F}_g$, accelerates the projectile toward the Earth. However, because the Earth has such a large mass, its acceleration due to this reaction force is negligibly small.

You directly experience the law if you slam your fist against a wall or kick a football with your bare foot. You should be able to identify the action and reaction forces in these cases.

As another application of Newton's third law, consider helicopters that you have seen. The most familiar helicopters have a large set of blades rotating in a horizontal plane above the body and another smaller set rotating in a vertical plane at the back. Other helicopters have two large sets of blades above the body rotating in opposite directions. Why do helicopters always have two sets of blades? The helicopter engine applies a force to the blades, causing them to change their rotational motion, but according to Newton's third law, the blades must exert an equal and opposite force on the helicopter. This force would cause the body of the helicopter to rotate in the direction opposite to the blades. A rotating helicopter would be impossible to control, so a second set of blades is used. The small blades in the back provide a force opposite to that tending to rotate the body of the helicopter, keeping the body oriented stably. In the case of two sets of large, counter-rotating blades, the engines apply forces in opposite directions, so that there is no net reaction force rotating the helicopter.

As we mentioned earlier, the Earth exerts a force $\mathbf{F}_g$ on any object. If the object is a TV at rest on a table, as in Figure 4.5a, the reaction force to $\mathbf{F}_g$ is the force the TV exerts on the Earth, $\mathbf{F}_g'$. The TV does not accelerate, because it is held up by the table. The table, therefore, exerts on the TV an upward action force, $\mathbf{n}$, called the **normal force.** This is the force that prevents the TV from falling through the

APPLICATION

Helicopter Flight.

 4.5, SECTION 2

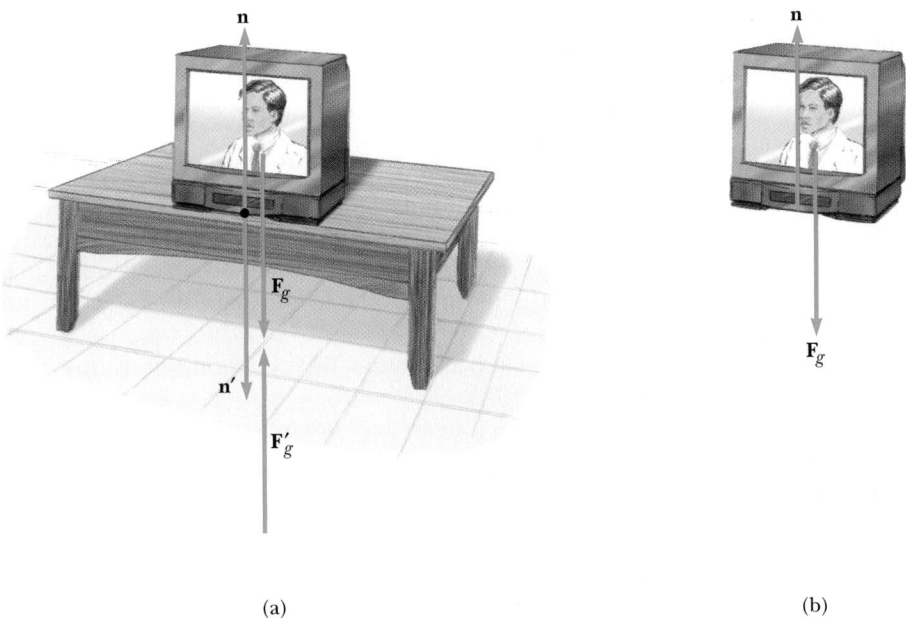

(a)   (b)

Figure 4.5 When a TV set is sitting on a table, the forces acting on the set are the normal force, **n,** and the force of gravity, $\mathbf{F}_g$, as illustrated in (b). The reaction to **n** is the force exerted by the TV set on the table, **n′.** The reaction to $\mathbf{F}_g$ is the force exerted by the TV set on the Earth, $\mathbf{F}_g'$.

PHYSICS IN ACTION

Forces and Motion

On the left is an athlete running at sunset. The external forces acting on the athlete, as described by the blue vectors, are (a) the force exerted by the Earth, $\mathbf{F}_g$, (b) the force exerted by the ground, $\mathbf{F}$, and (c) the force of air resistance, $\mathbf{R}$.

On the right is a multiflash exposure of a tennis player executing a backhand stroke. The parabolic path of the ball before it is struck is visible, as is the follow-through of the swing. The force exerted by the racket on the ball is equal to and opposite the force exerted by the ball on the racket.

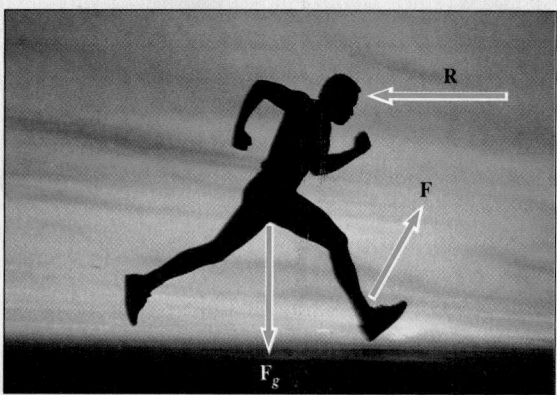

(Mitchell Funk/The Image Bank)

(© Zimmerman/FPG)

table; it can have any value needed, up to the point of breaking the table. The normal force balances the weight and provides equilibrium. The reaction to $\mathbf{n}$ is the force of the TV on the table, $\mathbf{n}'$. Therefore, we conclude that

$$\mathbf{F}_g = -\mathbf{F}_g' \qquad \text{and} \qquad \mathbf{n} = -\mathbf{n}'$$

The forces $\mathbf{n}$ and $\mathbf{n}'$ have the same magnitude, which is the same as $\mathbf{F}_g$ unless the table has broken. Note that the forces acting on the TV are $\mathbf{F}_g$ and $\mathbf{n}$, as shown in Figure 4.5b. The two reaction forces, $\mathbf{F}_g'$ and $\mathbf{n}'$ are exerted on objects other than the TV. Remember, the two forces in an action-reaction pair always act on two different objects.

From the second law, we see that, because the TV is in equilibrium ($\mathbf{a} = 0$), it follows that $F_g = n = mg$.

Applying Physics 1

APPLICATION

Colliding Vehicles.

If a small sports car collides head-on with a massive truck, which vehicle experiences the greater impact force? Which vehicle experiences the greater acceleration?

Explanation The car and truck experience forces that are equal in magnitude but in opposite directions. A calibrated spring scale placed between the colliding vehicles reads the same whichever way it faces. Because the car has the smaller mass, it experiences much greater acceleration.

Thinking Physics 2

Consider a horse attempting to pull a wagon from rest. A student makes the following argument: If the horse pulls forward on the wagon, then Newton's third law holds that the wagon pulls back equally hard on the horse. Thus, the forces cancel and nothing happens.

Explanation This is a common conundrum that has its basis in an incomplete understanding of Newton's third law. When applying the third law, it is important to recognize the objects on which the pair of forces act: "If object 1 exerts a force on object 2, then 2 exerts an equal and opposite force on 1." This form of the third law indicates clearly that the two forces act on *different* objects. In the statement of this example, we are equating the force of the horse *on the wagon* to the force of the wagon *on the horse*—the forces are acting on different objects. However, if you want to determine the motion of an object, you must consider *only the forces on that object*. The force that accelerates the horse and wagon (the object) is the force of the Earth on the horse's feet, which is the reaction to the force that the horse exerts backward on the Earth.

Thinking Physics 3

You have most likely had the experience of standing in an elevator that accelerates upward as it leaves to move toward a higher floor. In this case, you *feel* heavier. If you are standing on a bathroom scale at the time, it will *measure* a force larger than your weight. Thus, you have evidence that leads you to believe that you are heavier in this situation. *Are* you heavier?

Explanation No, you are not. Your weight is unchanged. In order to provide the acceleration upward, the floor or the bathroom scale must apply an upward normal force larger than your weight. It is this larger force that you feel, which you interpret as feeling heavier. A bathroom scale reads this upward force, not your weight, so its reading also increases.

4.5   SOME APPLICATIONS OF NEWTON'S LAWS

This section applies Newton's laws to objects moving under the actions of constant external forces. We assume that objects behave as particles, and so we need not worry about rotational motion. We also neglect any friction effects. Finally, we neglect the masses of any ropes or strings involved; in these approximations, the magnitude of the force exerted along a rope (the tension) is the same at all points in the rope. 4.6

When we apply Newton's law to an object, we are interested only in those *external* forces that act *on the object*. For example, in Figure 4.4b, the only external forces acting on the TV are $\mathbf{n}$ and $\mathbf{F}_g$. The reactions to these forces, $\mathbf{n}'$ and $\mathbf{F}_g{}'$, act on the table and on the Earth, respectively, and do not appear in Newton's second law as applied to the TV.

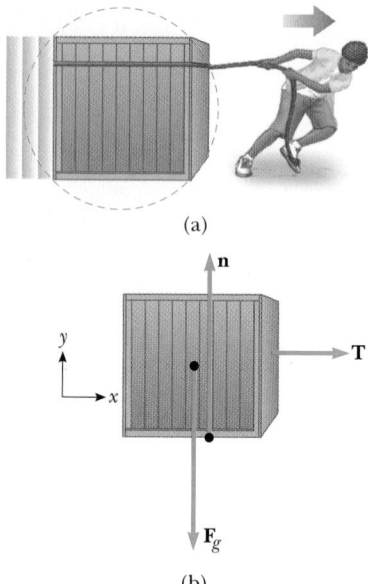

(a)

(b)

Figure 4.6   (a) A crate being pulled to the right on a frictionless surface. (b) The free-body diagram that represents the external forces on the crate.

Consider a crate being pulled to the right on a frictionless, horizontal surface, as in Figure 4.6a. Suppose you are asked to find the acceleration of the crate and the force the surface exerts on it. The horizontal force applied to the crate acts through the rope. The force that the rope exerts on the crate is denoted by **T**. The magnitude of **T** is equal to the tension in the rope. A dashed circle is drawn around the crate in Figure 4.6a to remind you to isolate it from its surroundings.

Because we are interested only in the motion of the crate, we must be able to identify all external forces acting on it. These are illustrated in Figure 4.6b. In addition to the force **T,** the force diagram for the crate includes the force of gravity, $\mathbf{F}_g$, and the normal force, **n,** exerted by the floor on the crate. Such a force diagram is referred to as a **free-body diagram.** The construction of a correct free-body diagram is an essential step in applying Newton's laws; its importance cannot be overemphasized. The *reactions* to the forces we have listed—namely, the force exerted by the rope on the hand doing the pulling, the force exerted by the crate on the Earth, and the force exerted by the crate on the floor—are not included in the free-body diagram because they act on other objects and not on the crate.

Now let us apply Newton's second law to the crate. First we must choose an appropriate coordinate system. In this case it is convenient to use the one shown in Figure 4.6b, with the x axis horizontal and the y axis vertical. We can apply Newton's second law in the x direction, y direction, or both, depending on what we are asked to find in a problem. In addition, we may be able to use the equations of motion for constant acceleration that are found in Chapter 2. However, you should use these equations only when the acceleration is constant.

Objects in Equilibrium and Newton's First Law

Objects that are either at rest or moving with constant velocity are said to be in equilibrium, and Newton's first law is a statement of one condition that must be true for equilibrium conditions to prevail. In equation form, this condition of equilibrium can be expressed as

$$\Sigma\mathbf{F} = 0 \tag{4.7}$$

This statement signifies that the *vector* sum of all the forces (the net force) acting on an object in equilibrium is zero.

Usually, equilibrium problems are solved more easily if we work with Equation 4.7 in terms of the components of the external forces acting on an object. By this we mean that, in a two-dimensional problem, the sum of all the external forces in the x and y directions must separately equal zero; that is,

$$\Sigma F_x = 0 \quad \text{and} \quad \Sigma F_y = 0 \tag{4.8}$$

This set of equations is often referred to as the **first condition for equilibrium.** We shall not consider three-dimensional problems in this book, but the extension of Equation 4.8 to a three-dimensional situation can be made by adding a third equation, $\Sigma F_z = 0$.

Problem-Solving Strategy

Objects in Equilibrium

The following procedure is recommended for problems involving objects in equilibrium:

1. Make a sketch of the situation described in the problem statement.
2. Draw a free-body diagram for the *isolated* object under consideration, and label all external forces acting on the object.
3. Resolve all forces into x and y components, choosing a convenient coordinate system.
4. Use the equations $\Sigma F_x = 0$ and $\Sigma F_y = 0$. Keep track of the signs of the various force components.
5. Application of Step 4 leads to a set of equations with several unknowns. Solve the simultaneous equations for the unknowns in terms of the known quantities.

EXAMPLE 4.1   A Traffic Light at Rest

A traffic light weighing 100 N hangs from a vertical cable tied to two other cables that are fastened to a support, as in Figure 4.7a. The upper cables make angles of 37.0° and 53.0° with the horizontal. Find the tension in each of the three cables.

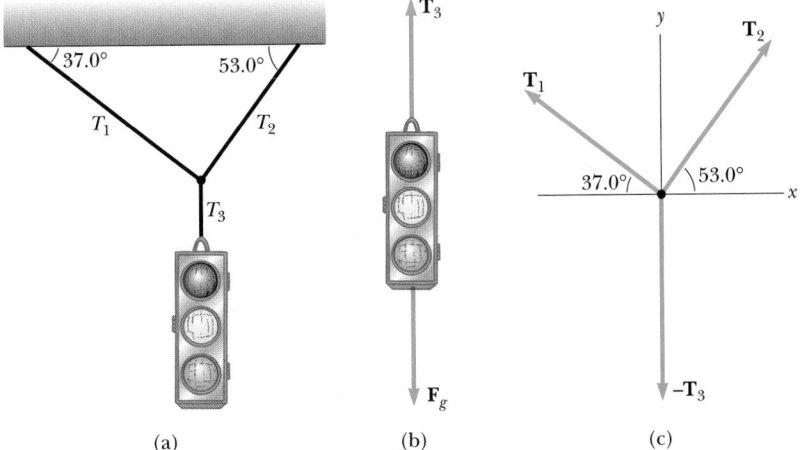

(a)   (b)   (c)

Workbook Problem 20 — (Workbook page 47) is similar to Example 4.1.

Figure 4.7   (Example 4.1) (a) A traffic light suspended by cables. (b) A free-body diagram for the traffic light. (c) A free-body diagram for the knot joining the cables.

Reasoning   We must construct two free-body diagrams in order to work this problem. The first of these is for the traffic light, shown in Figure 4.7b; the second is for the knot that holds the three cables together, as in Figure 4.7c. The knot is a convenient point to choose because all forces in question act at this point.

Solution   From the free-body diagram in Figure 4.7b, we see that $T_3 = w = 100$ N. Next, we choose the coordinate axes shown in Figure 4.7c and resolve the forces into their x and y components:

Force	x Component	y Component
$\mathbf{T}_1$	$-T_1 \cos 37.0°$	$T_1 \sin 37.0°$
$\mathbf{T}_2$	$T_2 \cos 53.0°$	$T_2 \sin 53.0°$
$\mathbf{T}_3$	0	-100 N

The first condition for equilibrium gives us the equations

$$(1) \qquad \Sigma F_x = T_2 \cos 53.0° - T_1 \cos 37.0° = 0$$

$$(2) \qquad \Sigma F_y = T_1 \sin 37.0° + T_2 \sin 53.0° - 100 \text{ N} = 0$$

From (1) we see that the horizontal components of $\mathbf{T}_1$ and $\mathbf{T}_2$ must be equal in magnitude, and from (2) we see that the sum of the vertical components of $\mathbf{T}_1$ and $\mathbf{T}_2$ must balance the force of gravity acting on the light. We can solve (1) for T_2 in terms of T_1 to give

$$T_2 = T_1 \left(\frac{\cos 37.0°}{\cos 53.0°}\right) = T_1 \left(\frac{0.799}{0.602}\right) = 1.33 T_1$$

This value for T_2 can be substituted into (2) to give

$$T_1 \sin 37.0° + (1.33 T_1)(\sin 53.0°) - 100 \text{ N} = 0$$

$$T_1 = \boxed{60.1 \text{ N}}$$

$$T_2 = 1.33 T_1 = 1.33(60.0 \text{ N}) = \boxed{79.9 \text{ N}}$$

Exercise   When will $T_1 = T_2$?

Answer   When the supporting cables make equal angles with the horizontal support.

EXAMPLE 4.2   Sled on a Frictionless Hill

A child holds a sled at rest on a frictionless, snow-covered hill, as shown in Figure 4.8a. If the sled weighs 77.0 N, find the force exerted on the rope by the child and the force exerted on the sled by the hill.

Reasoning   Figure 4.8b shows the forces acting on the sled and a convenient coordinate system to use for this type of problem. Note that **n,** the force exerted on the sled by the ground, is perpendicular to the hill. The ground can exert a component of force along the incline only if there is friction between the sled and the hill. Because the sled is at rest, we are able to apply the first condition for equilibrium as $\Sigma F_x = 0$ and $\Sigma F_y = 0$.

Solution   Applying the first condition for equilibrium to the sled, we find that

$$\Sigma F_x = T - (77.0 \text{ N})(\sin 30.0°) = 0$$

$$T = \boxed{38.5 \text{ N}}$$

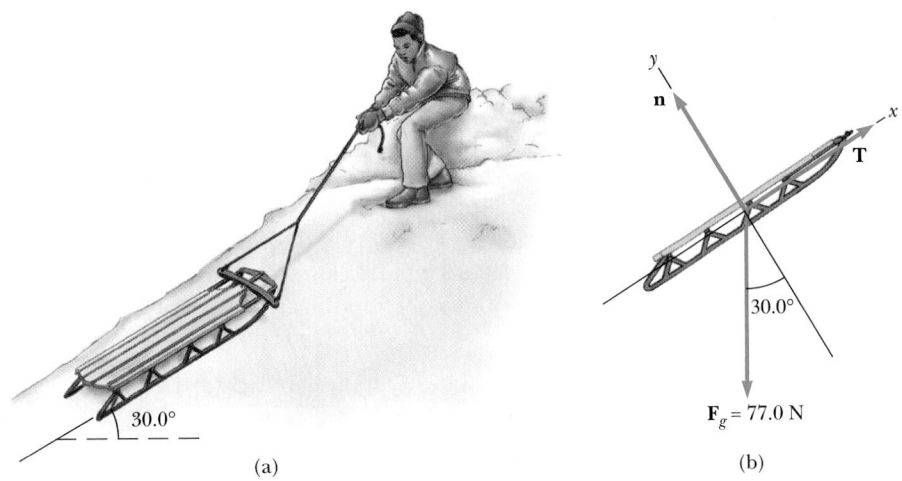

Figure 4.8 (Example 4.2) (a) A child holding a sled on a frictionless hill. (b) A free-body diagram for the sled.

$$\Sigma F_y = n - (77.0 \text{ N})(\cos 30.0°) = 0$$

$$n = \boxed{66.7 \text{ N}}$$

Note that n is *less* than the weight of the sled in this situation. This is so because the sled is on an incline and **n** is equal to and opposite the component of the force of gravity perpendicular to the incline.

Exercise What happens to the normal force as the angle of incline increases?

Answer It decreases.

Exercise When is the normal force equal to the weight of the sled?

Answer When the sled is on a horizontal surface and the applied force is either zero or along the horizontal.

Accelerating Objects and Newton's Second Law

In a situation in which a net force acts on an object, the object accelerates, and we use Newton's second law in order to analyze the motion. The representative examples and suggestions that follow should help you solve problems of this kind.

Problem-Solving Strategy

Newton's Second Law

The following procedure is recommended for working problems that involve the application of Newton's second law:

1. Draw a diagram of the system.
2. Isolate the object of interest whose motion is being analyzed. Draw a free-body diagram for this object, showing all external forces acting on it. For

systems containing more than one object, draw a *separate* diagram for each object.

3. Establish convenient coordinate axes for each object, and find the components of the forces along these axes. Apply Newton's second law in the x and y directions for each object.

4. Solve the component equations for the unknowns. Remember that, in order to obtain a complete solution, you must have as many independent equations as you have unknowns.

5. If necessary, use the equations of kinematics (motion with constant acceleration) from Chapter 2 to find all the unknowns.

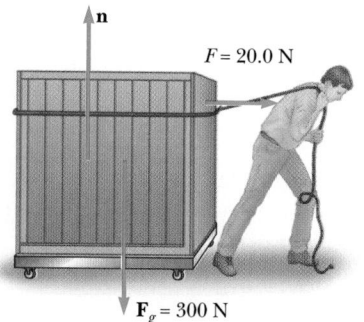

Figure 4.9   (Example 4.3)

  Workbook Problem 16 —
(Workbook page 41) is similar to
Example 4.3.

EXAMPLE 4.3   Moving a Crate

The combined weight of the crate and dolly in Figure 4.9 is 300 N. If the person pulls on the rope with a constant force of 20.0 N, what is the acceleration of the system (crate plus dolly), and how far will it move in 2.00 s? Assume that the system starts from rest and that there are no frictional forces opposing the motion of the system.

Reasoning   We can find the acceleration of the system from Newton's second law. Because the force exerted on the system is constant, its acceleration is constant. Therefore, we can apply the equations of motion with constant acceleration to find the distance traveled.

Solution   In order to apply Newton's second law to the system, we must first know the system's mass:

$$m = \frac{w}{g} = \frac{300 \text{ N}}{9.80 \text{ m/s}^2} = 30.6 \text{ kg}$$

Now we can find the acceleration of the system from the second law:

$$a_x = \frac{F_x}{m} = \frac{20.0 \text{ N}}{30.6 \text{ kg}} = \boxed{0.654 \text{ m/s}^2}$$

Because the acceleration is constant, we can find the distance the system moves in 2.00 s using the relation $x = v_0 t + \frac{1}{2}at^2$ with $v_0 = 0$:

$$x = \tfrac{1}{2}at^2 = \tfrac{1}{2}(0.654 \text{ m/s}^2)(2.00 \text{ s})^2 = \boxed{1.31 \text{ m}}$$

It is important to note that the constant applied force of 20.0 N is assumed to act on the system all during its motion. If the force were removed at some instant, the system would continue to move with constant velocity and hence zero acceleration.

EXAMPLE 4.4   The Run-Away Car

A car of mass m is on an icy driveway inclined at an angle $\theta = 20.0°$, as in Figure 4.10a. Determine the acceleration of the car, assuming the incline is frictionless.

Reasoning   The free-body diagram for the car is shown in Figure 4.10b. The only forces on the car are the normal force, **n,** acting perpendicular to the driveway surface, and the force of gravity, $\mathbf{F}_g$, acting vertically downward. It is convenient to choose the coordinate axes with the x axis along the incline and the y axis perpendicular to it. Then we replace

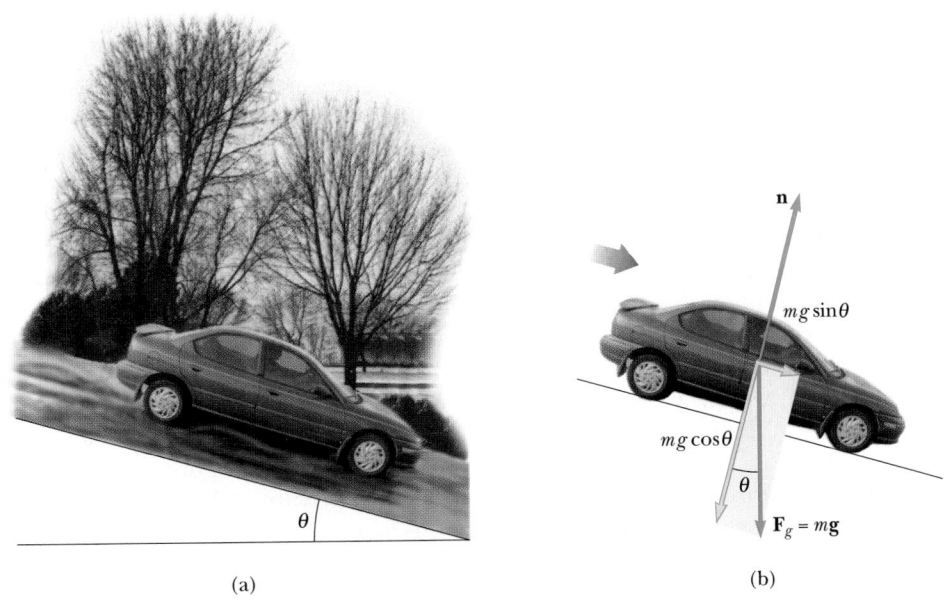

Figure 4.10 (Example 4.4)

the force of gravity with a component of magnitude $mg \sin \theta$ along the positive x axis and a component of magnitude $mg \cos \theta$ in the negative y direction.

Solution Applying Newton's second law in component form, with $a_y = 0$, gives

$$(1) \qquad \Sigma F_x = mg \sin \theta = ma_x$$

$$(2) \qquad \Sigma F_y = n - mg \cos \theta = 0$$

From (1) we see that the acceleration of the car along the driveway is provided by the component of the force of gravity directed down the incline:

$$(3) \qquad a_x = g \sin \theta$$

Note that the acceleration given by (3) is *constant* and *independent of the mass* of the car — it depends only on the angle of inclination and on g. In our example, $\theta = 20.0°$ and so, we find that

$$a_x = \boxed{3.35 \text{ m/s}^2}$$

Exercise A compact car and a large luxury sedan are at rest at the top of an ice-covered (frictionless) driveway. If they both slide down the driveway, which reaches the bottom first?

Answer Because the acceleration is independent of the mass, and thus is the same for both vehicles, they arrive at the bottom simultaneously.

Exercise If the length of the driveway is 25.0 m and a car starts from rest at the top, how long does it take to travel to the bottom? What is the car's speed at the bottom?

Answer 3.86 s; 12.9 m/s

4.6   FORCE OF FRICTION

When a body is in motion either on a surface or through a viscous medium such as air or water, there is resistance to the motion because the body interacts with its surroundings. We call such resistance a **force of friction.** Forces of friction are very important in our everyday lives. They allow us to walk or run and are necessary for the motion of wheeled vehicles.

Consider a block on a horizontal table, as in Figure 4.11a. If we apply an external horizontal force **F** to the block, acting to the right, the block remains stationary if **F** is not too large. The force that counteracts **F** and keeps the block from moving acts to the left and is called the force of static friction, f_s. As long as the block is not moving, $f_s = F$. Thus, if **F** is increased, f_s also increases. Likewise, if **F** decreases, f_s also decreases. Experiments show that the frictional force arises from the nature of the two surfaces: Because of their roughness, contact is made only at a few points, as shown in the magnified view of the surface in Figure 4.11a. Actually, the frictional force is much more complicated than presented, because it ultimately involves the electrostatic force between atoms or molecules.

If we increase the magnitude of **F,** as in Figure 4.11b, the block eventually slips. When the block is on the verge of slipping, f_s is a maximum as shown in Figure

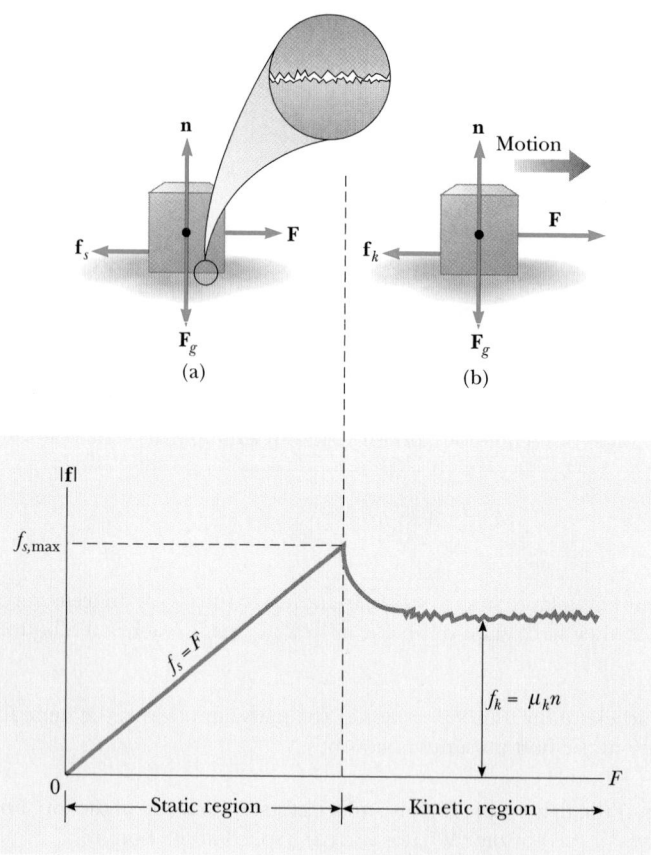

Figure 4.11   The direction of the force of friction, **f,** between a block and a horizontal surface is opposite the direction of an applied force, **F.** (a) The force of static friction equals the applied force. (b) When the magnitude of the applied force exceeds the maximum static friction force, $f_{s,max}$, the block accelerates to the right. (c) A graph of the magnitude of the frictional force versus the applied force. Note that $f_{s,max} > f_k$.

4.11c. When F exceeds $f_{s,max}$, the block moves and accelerates to the right. When the block is in motion, the frictional force becomes less than $f_{s,max}$ (Fig. 4.11c). We call the frictional force for an object in motion the **force of kinetic friction, f_k.** The unbalanced force in the x direction, $F - f_k$, produces an acceleration to the right. If $F = f_k$, the block moves to the right with constant speed. If the applied force is removed, then the frictional force acting to the left accelerates the block in the $-x$ direction and eventually brings it to rest.

Experimentally, one finds that, to a good approximation, both $f_{s,max}$ and f_k are proportional to the normal force acting on the block. The experimental observations can be summarized as follows:

- The magnitude of the force of static friction between any two surfaces in contact can have the values

$$f_s \leq \mu_s n \qquad [4.9]$$

where the dimensionless constant μ_s is called the **coefficient of static friction** and n is the magnitude of the normal force. The equality in Equation 4.9 holds when the block is on the verge of slipping, that is, when $f_s = f_{s,max} \equiv \mu_s n$. The inequality holds when the applied force is less than this value.
- The magnitude of the force of kinetic friction acting on an object is given by

$$f_k = \mu_k n \qquad [4.10]$$

where μ_k is the **coefficient of kinetic friction.**
- The values of μ_k and μ_s depend on the nature of the surfaces, but μ_k is generally less than μ_s. Table 4.2 lists some reported values.
- The coefficients of friction are nearly independent of the area of contact between the surfaces.

Finally, although the coefficient of kinetic friction varies with speed, we shall neglect any such variations. The approximate nature of the equations is easily demonstrated by trying to get an object to slide down an incline at constant speed.

TABLE 4.2 Coefficients of Friction [a]

	μ_s	μ_k
Steel on steel	0.74	0.57
Aluminum on steel	0.61	0.47
Copper on steel	0.53	0.36
Rubber on concrete	1.0	0.8
Wood on wood	0.25–0.5	0.2
Glass on glass	0.94	0.4
Waxed wood on wet snow	0.14	0.1
Waxed wood on dry snow	—	0.04
Metal on metal (lubricated)	0.15	0.06
Ice on ice	0.1	0.03
Teflon on Teflon	0.04	0.04
Synovial joints in humans	0.01	0.003

[a] All values are approximate.

QUICKLAB

Measuring μ

This is a simple method for measuring the coefficients of static and kinetic friction between an object and some surface. For this investigation, you will need a few coins, your book or some other flat surface that can be inclined, a protractor, and some double-stick tape. Place a coin at one edge of the book as it lies on a table, and lift that edge of the book until the coin just slips down the incline as in the figure. When this occurs, measure the inclined angle using your protractor. Repeat the measurement five times, and find the average value of this critical angle θ_C. The coefficient of static friction between the coin and book's surface is $\mu_s = \tan \theta_C$. (You should prove this as an exercise.) Calculate the average value of μ_s. To measure the coefficient of kinetic friction, find the angle θ_C' at which the coin moves down the incline with constant speed. This angle should be less than θ_C. Measure this new angle five times, and get its average value. Calculate the average value of μ_k using the fact that $\mu_k = \tan \theta_C'$, where $\theta_C' < \theta_C$. Repeat these measurement using two or three stacked coins, with double-stick tape between them. You should get the same results as with one coin. Why?

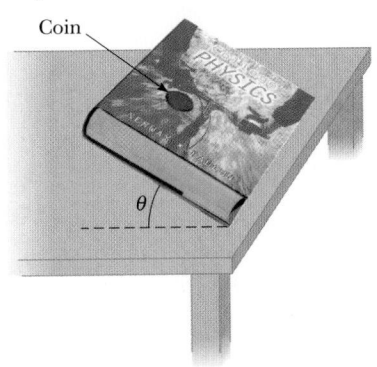
Coin

θ

Applying Physics 2

You are playing with your younger sister in the snow. She is sitting on a sled and asking you to slide her across a flat, horizontal field. You have a choice of pushing her from behind, by applying a force at 30° below the horizontal, or attaching a rope to the front of the sled and pulling with a force at 30° above the horizontal. Which would be easier for you and why?

Explanation   It is easier to attach the rope and pull. In this case, there is a component of your applied force that is upward. This reduces the normal force between the sled and the snow. In turn, this reduces the friction force between the sled and the snow, making it easier to move. If you push from behind, with a force with a downward component, the normal force is larger, the friction force is larger, and the sled is harder to move.

EXAMPLE 4.5   Moving into the Dormitory

At the beginning of a new school term, a student moves a box of books by attaching a rope to the box and pulling with a force of 90.0 N at an angle of 30.0°, as shown in Figure 4.12. The box of books has a mass of 20.0 kg, and the coefficient of kinetic friction between the bottom of the box and the floor is 0.500. Find the acceleration of the box.

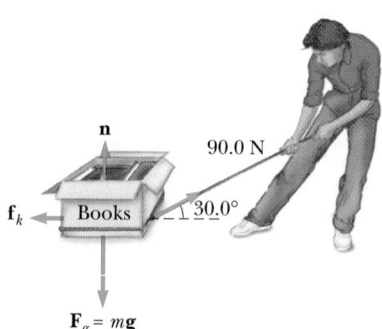

Figure 4.12   (Example 4.5) A box of books being pulled to the right at an angle of 30.0°.

Reasoning   There are basically three steps required for the solution to this problem. (1) First find the normal force $\mathbf{n}$ by applying the first condition of equilibrium, $\Sigma F_y = 0$, in the vertical direction. (2) Calculate the force of kinetic friction on the box from $f_k = \mu_k n$. (3) Apply Newton's second law along the horizontal direction to find the acceleration of the box.

Solution   (Step 1) The box is not accelerating in the vertical direction, and so we find the normal force from $\Sigma F_y = 0$. The forces in the y direction are the force of gravity $\mathbf{F}_g$, the normal force, $\mathbf{n},$ and the vertical component of the applied 90.0-N force. This vertical component has a magnitude equal to $(90.0 \text{ N})(\sin 30.0°)$. We find that

$$\Sigma F_y = n + (90.0 \text{ N})(\sin 30.0°) - (20.0 \text{ kg})(9.80 \text{ m/s}^2) = 0$$

$$n = 151 \text{ N}$$

The magnitude of the normal force is *not* equal to the weight of the box, because the vertical component of the 90.0-N force is helping to support some of that weight.

(Step 2) Knowing the normal force, we can find the force of kinetic friction:

$$f_k = \mu_k n = (0.500)(151 \text{ N}) = 75.5 \text{ N} \qquad \text{(to the left)}$$

(Step 3) Finally, we determine the horizontal acceleration using Newton's second law:

$$\Sigma F_x = (90.0 \text{ N})(\cos 30.0°) - 75.5 \text{ N} = (20.0 \text{ kg})(a_x)$$

$$a_x = \; +0.122 \text{ m/s}^2$$

Exercise If the initial speed of the box is zero, what is its speed after it has traveled 2.00 m? How long does it take to pull it this distance?

Answer 0.699 m/s; 5.73 s

EXAMPLE 4.6 The Sliding Hockey Puck

A hockey puck is given an initial speed of 20.0 m/s on a frozen pond, as in Figure 4.13. The puck remains on the ice and slides 120 m before coming to rest. Determine the coefficient of kinetic friction between puck and ice.

Reasoning The puck slides to rest with a constant acceleration along the horizontal. Thus, we can use the kinematic equation $v^2 = v_0^2 + 2ax$ to find a. Newton's second law applied in the horizontal direction is $-f_k = -\mu_k n = ma$. To find μ_k, we first find the normal force $\mathbf{n}$ by applying $\Sigma F_y = 0$ in the vertical direction.

Solution With the final speed, v, equal to zero; the initial speed, $v_0 = 20.0$ m/s; and the displacement, $x = 120$ m:

$$v^2 = v_0^2 + 2ax$$

$$0 = (20.0 \text{ m/s})^2 + 2a(120 \text{ m})$$

$$a = -1.67 \text{ m/s}^2$$

The negative sign means that the acceleration is to the left, *opposite* the direction of the velocity.

The magnitude of the force of kinetic friction is found from $f_k = \mu_k n$, and n is found from $\Sigma F_y = 0$ as follows:

$$\Sigma F_y = n - F_g = 0$$

$$n = F_g = mg$$

Thus,

$$f_k = \mu_k n = \mu_k mg$$

Now we apply Newton's second law along the horizontal direction, taking the positive direction toward the right:

$$\Sigma F_x = -f_k = ma$$

$$-\mu_k mg = m(-1.67 \text{ m/s}^2)$$

$$\mu_k = \frac{1.67 \text{ m/s}^2}{9.80 \text{ m/s}^2} = \; 0.170$$

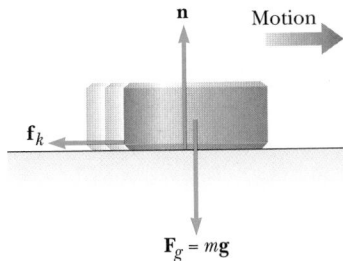

Figure 4.13 (Example 4.6) *After* the puck is given an initial velocity to the right, the external forces acting on it are the force of gravity, $\mathbf{F}_g$, the normal force, $\mathbf{n}$, and the force of kinetic friction, $\mathbf{f}_k$.

EXAMPLE 4.7   Connected Objects

Two objects are connected by a light string that passes over a frictionless pulley, as in Figure 4.14a. The coefficient of sliding friction between the 4.00-kg object and the surface is 0.300. Find the acceleration of the two objects and the tension in the string.

Workbook Problem 19 — (Workbook page 45) is similar to Example 4.7.

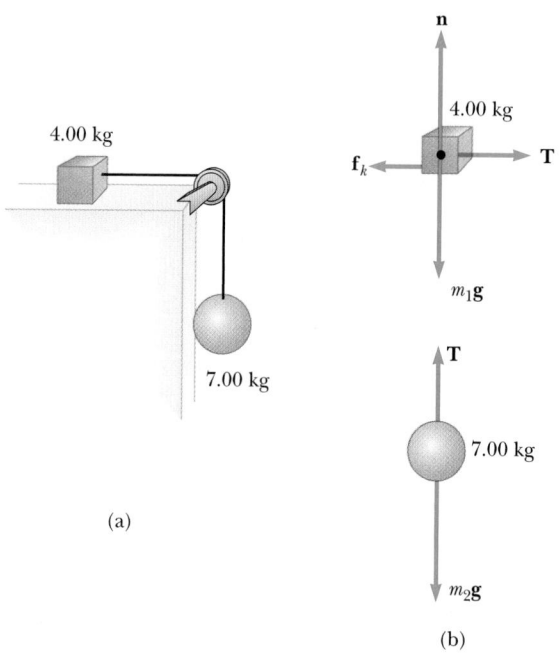

Figure 4.14   (Example 4.7) (a) Two objects connected by a light string that passes over a frictionless pulley. (b) Free-body diagrams for the objects.

Reasoning   Connected objects are handled by applying Newton's second law separately to each. The free-body diagrams for the block on the table and for the hanging block are shown in Figure 4.14b. We shall obtain two equations involving the unknowns T and a that can be solved simultaneously.

Solution   With the positive x direction to the right and the positive y direction upward, Newton's second law applied to the 4.00-kg object gives

$$\Sigma F_x = T - f_k = (4.00 \text{ kg})(a)$$
$$\Sigma F_y = n - (4.00 \text{ kg})(g) = 0$$

Because $f_k = \mu_k n$ and $n = (4.00 \text{ kg})g = 39.2$ N, we have $f_k = (0.300)(39.2 \text{ N}) = 11.8$ N. Therefore,

$$(1)\qquad T = f_k + (4.00 \text{ kg})(a) = 11.8 \text{ N} + (4.00 \text{ kg})(a)$$

Now we apply Newton's second law to the 7.00-kg object moving in the vertical direction, this time selecting downward as the positive direction:

$$\Sigma F_y = (7.00 \text{ kg})(g) - T = (7.00 \text{ kg})(a)$$

or

$$(2) \qquad T = 68.6 \text{ N} - (7.00 \text{ kg})(a)$$

Subtracting (1) from (2) eliminates T:

$$56.8 \text{ N} - (11.0 \text{ kg})(a) = 0$$

$$a = \boxed{5.16 \text{ m/s}^2}$$

When this value for the acceleration is substituted into (1), we get

$$T = \boxed{32.4 \text{ N}}$$

Friction and the Motion of a Car

Forces of friction are important in the analysis of the motion of cars and other wheeled vehicles. There are several types of friction forces to consider, the main ones being the force of friction between tires and road surface and the retarding force produced by air resistance.

As each of four wheels turns to propel a car forward, it exerts a backward force on the road through its tire. The reaction to this backward force is a forward force, **f,** exerted by the road on the tire (Fig. 4.15). If we assume that the same forward force **f** is exerted on each tire, the net forward force on the car is 4**f,** and the car's acceleration is therefore **a** = 4**f**/m.

When the car is in motion, we must also consider the force of air resistance, **R,** which acts in the direction opposite its velocity. The net force on the car is therefore 4**f** − **R,** and so the car's acceleration is **a** = (4**f** − **R**)/m. At normal driving speeds, the magnitude of **R** is proportional to the first power of the speed. That is, $R = bv$, where b is a constant. Thus, the force of air resistance increases with increasing speed. When R is equal to 4f the acceleration is zero, and the car moves at a constant speed.

A similar situation occurs when an object falls through air. When the upward force of air resistance balances the downward force of gravity, the net force on the object is zero and hence the object's acceleration is zero. Once this condition is reached, the object continues to move downward with some constant maximum speed called the **terminal speed.**

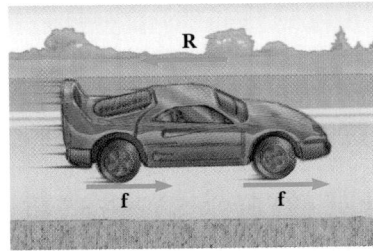

Figure 4.15 The horizontal forces acting on the car are the *forward* forces, **f,** exerted on each tire by the road and the force of air resistance, **R,** which acts *opposite* the car's velocity. (The wheels of the car exert a backward force on the road not shown in the diagram.)

Applying Physics 3

Consider a skydiver falling through air before reaching her terminal speed, as in Figure 4.16. As the speed of the skydiver increases, what happens to her acceleration? What is her acceleration once she reaches terminal speed?

Explanation   The forces exerted on the skydiver are the downward force of gravity, m**g,** and an upward force of air resistance, **R.** The magnitude of **R** is less than her weight before she reaches terminal speed. As her downward speed increases, the force of air resistance increases. The vector sum of the force of gravity and the force of air resistance gives a total force that decreases with time, so her acceleration decreases. Once the two forces balance each other so that the net force is zero, the acceleration is consequently zero, and she has reached terminal speed.

APPLICATION

Skydiving.

WEB

For Web links to more information about skydiving, visit the textbook Web site at **http://www.harcourtcollege.com/physics/cptech**

Figure 4.16   (Applying Physics 3) By spreading their arms and legs and by keeping their bodies parallel to the ground, skydivers experience maximum air resistance resulting in a minimum terminal speed. *(Guy Sauvage, Photo Researchers, Inc.)*

SUMMARY

Newton's first law states that an object at rest remains at rest, and an object in motion continues in motion with a constant velocity, unless it experiences a net external force.

The resistance of an object to a change in its state of motion is called **inertia.** Mass is the physical quantity used to measure inertia.

Newton's second law states that the resultant force acting on an object is equal to the product of the mass of the object and its acceleration:

$$\Sigma \mathbf{F} = m\mathbf{a} \qquad [4.1]$$

The magnitude of the force of gravity exerted on an object is called the **weight** of an object, denoted by w. The weight of an object of mass m is equal to the product mg:

$$w = mg \qquad [4.6]$$

Newton's third law states that, if two objects interact, the force exerted by object 1 on object 2 is equal in magnitude and opposite in direction to the force exerted by object 2 on object 1. Thus, an isolated force can never occur in nature.

An **object in equilibrium** has no net external force acting on it, and the first law, in component form, implies that $\Sigma F_x = 0$ and $\Sigma F_y = 0$.

The maximum force of static friction, $\mathbf{f}_{s,\max}$, between an object and a surface is proportional to the normal force acting on the object. This maximum force occurs when the object is on the verge of slipping. In general,

$$f_s \leq \mu_s n \qquad [4.9]$$

where μ_s is the **coefficient of static friction.** When an object slides over a surface, the direction of the force of kinetic friction, $\mathbf{f}_k$, is opposite the direction of the motion, and the magnitude is proportional to that of the normal force. The magnitude of $\mathbf{f}_k$ is

$$f_k = \mu_k n \qquad\qquad [4.10]$$

where μ_k is the **coefficient of kinetic friction.** In general, $\mu_k < \mu_s$.

MULTIPLE-CHOICE QUESTIONS

1. A block of mass 10 kg remains at rest on a hill inclined at an angle of 30° with the horizontal. Which one of the statements below is correct about the magnitude of the force of friction that acts on the mass?
   (a) It is larger than the weight of the block.
   (b) It is equal to the weight of the block.
   (c) It is greater than the component of the force of gravity down the plane.
   (d) It is less than the component of the force of gravity down the plane.
   (e) It is equal to the component of the force of gravity down the plane.
2. A hockey puck struck by a hockey stick is given an initial speed of 10 m/s. If the coefficient of kinetic friction between ice and puck is 0.10, how far will the puck slide before stopping?
   (a) 39 m   (b) 45 m   (c) 51 m   (d) 57 m
3. A thrown brick hits a window, breaking the glass, and ends up on the floor inside the room. Even though the brick broke the glass, we know that
   (a) the force of the brick on the glass was bigger than the force of the glass on the brick.
   (b) the force of the brick on the glass was the same size as the force of the glass on the brick.
   (c) the force of the brick on the glass was less than the force of the glass on the brick.
   (d) the brick didn't slow down as it broke the glass.

4. A manager of a restaurant pushes horizontally with a force of magnitude 200 N on a box of meat. The box moves across the floor with a constant forward acceleration. Which statement is most true about the magnitude of the force of kinetic friction acting on the box?
   (a) It is greater than 200 N.
   (b) It is less than 200 N.
   (c) It is equal to 200 N.
   (d) None of the above are necessarily true.
5. In the absence of air resistance, the force on a football at the top of its path after having been kicked is/are
   (a) the force due to the horizontal motion of the football.
   (b) the force of gravity.
   (c) the force exerted on it by the kicker and the force of gravity.
   (d) the force exerted on it by the Earth and the force of gravity.
6. Four forces act on an object, given by $\mathbf{A}$ = 40 N east, $\mathbf{B}$ = 50 N north, $\mathbf{C}$ = 70 N west, and $\mathbf{D}$ = 90 N south. What is the magnitude of the net force on the object?
   (a) 50 N   (b) 131 N   (c) 170 N   (d) 250 N
   (e) not enough information given

CONCEPTUAL QUESTIONS

1. (a) An object has only one force acting on it. Can it be at rest? Can it have an acceleration? (b) An object has zero acceleration. Does this mean that no forces act on it?
2. If an object is at rest, can we conclude that no external forces are acting on it?
3. Is it possible for an object to move if no net force acts on it?
4. If gold were sold by weight, would you rather buy it in Denver or in Death Valley? If it were sold by mass, in

which of the two locations would you prefer to buy it? Why?
5. A passenger sitting in the rear of a bus claims that she was injured as the driver slammed on the brakes, causing a suitcase to come flying toward her from the front of the bus. If you were the judge in this case, what disposition would you make? Why?
6. A space explorer is moving through space far from any planet or star. She notices a large rock, taken as a specimen from an alien planet, floating around the cabin of

the ship. Should she push it gently or kick it toward the storage compartment? Why?

7. What force causes an automobile to move? A propeller-driven airplane? A rowboat?

8. Analyze the motion of a rock dropped in water in terms of its speed and acceleration as it falls. Assume that a resistive force is acting on the rock that increases as the velocity increases.

9. In the motion picture *It Happened One Night* (Columbia Pictures, 1934), Clark Gable is standing inside a stationary bus in front of Claudette Colbert, who is seated. The bus suddenly starts moving forward and Clark falls into Claudette's lap. Why did this happen?

10. A weightlifter stands on a bathroom scale. He pumps a barbell up and down. What happens to the reading on the bathroom scale as this is done? Suppose he is strong enough to actually *throw* the barbell upward. How does the reading on the scale vary now?

11. In a tug-of-war between two athletes, each pulls on the rope with a force of 200 N. What is the tension in the rope? If the rope does not move, what horizontal force does each athlete exert against the ground?

12. As a rocket is fired from a launching pad, its speed and acceleration increase with time as its engines continue to operate. Explain why this occurs even though the thrust of the engines remains constant.

13. Identify the action-reaction pairs in the following situations: a man takes a step; a snowball hits a girl in the back; a baseball player catches a ball; a gust of wind strikes a window.

14. The driver of a speeding empty truck slams on the brakes and skids to a stop through a distance *d*. (a) If the truck carried a load that doubled its mass, what would be the truck's "skidding distance"? (b) If the initial speed of the truck were halved, what would be the truck's "skidding distance"?

15. Suppose you are driving a car at a high speed. Why should you avoid slamming on your brakes when you want to stop in the shortest possible distance?

16. Suppose a truck loaded with sand accelerates along a highway. If the driving force on the truck remains constant, what happens to the truck's acceleration if its trailer leaks sand at a constant rate through a hole in its bottom?

17. A large crate is placed on the bed of a truck but not tied down. (a) As the truck accelerates forward, the crate remains at rest relative to the truck. What force causes the crate to accelerate forward? (b) If the driver slammed on the brakes, what could happen to the crate?

18. Describe a few examples in which the force of friction exerted on an object is in the direction of motion of the object.

PROBLEMS

1, 2, 3 = straightforward, intermediate, challenging   ☐ = full solution available in Study Guide/Student Solutions Manual   ◤◢ = Core Concepts Workbook
WEB = solution posted at **http://www.harcourtcollege.com/physics/cptech**   ⬚ = biomedical application   ▨ = Interactive Physics

Section 4.1   The Concept of Force

Section 4.2   Newton's First Law

Section 4.3   Newton's Second Law

Section 4.4   Newton's Third Law

1. A 6.0-kg object undergoes an acceleration of 2.0 m/s². (a) What is the magnitude of the resultant force acting on it? (b) If this same force is applied to a 4.0-kg object, what acceleration is produced?

2. A football punter accelerates a football from rest to a speed of 10 m/s during the time in which his toe is in contact with the ball (about 0.20 s). If the football has a mass of 0.50 kg, what average force does the punter exert on the ball?

3. A bag of sugar weighs 5.00 lb on Earth. What should it weigh in newtons on the Moon, where the acceleration due to gravity is 1/6 that on Earth? Repeat for Jupiter,

where *g* is 2.64 times Earth gravity. Find the mass in kilograms at each of the three locations.

4. A freight train has a mass of 1.5×10^7 kg. If the locomotive can exert a constant pull of 7.5×10^5 N, how long does it take to increase the speed of the train from rest to 80 km/h?

5. The air exerts a forward force of 10 N on the propeller of a 0.20-kg model airplane. If the plane accelerates forward at 2.0 m/s², what is the magnitude of the resistive force exerted by the air on the airplane?

6. A 5.0-g bullet leaves the muzzle of a rifle with a speed of 320 m/s. What force (assumed constant) is exerted on the bullet while it is traveling down the 0.82-m-long barrel of the rifle?

7. A boat moves through the water with two forces acting on it. One is a 2000-N forward push by the motor, and the other is an 1800-N resistive force due to the water. (a) What is the acceleration of the 1000-kg boat? (b) If it starts from rest, how far will it move in 10.0 s? (c) What will its velocity be at the end of this time?

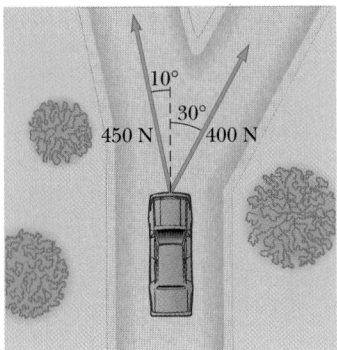

Figure P4.8

8. Two forces are applied to a car in an effort to move it, as shown in Figure P4.8. **(a)** What is the resultant of these two forces? **(b)** If the car has a mass of 3000 kg, what acceleration does it have?

9. The force of the wind on the sails of a sailboat is 390 N north. The water exerts a force of 180 N east. If the boat (including crew) has a mass of 270 kg, what are the magnitude and direction of its acceleration?

10. After falling from rest at a height of 30 m, a 0.50-kg ball rebounds upward, reaching a height of 20 m. If the contact between ball and ground lasted 2.0 ms, what average force was exerted on the ball?

Section 4.5 Some Applications of Newton's Laws

11. Find the tension in each cable supporting the 600-N cat burglar in Figure P4.11.

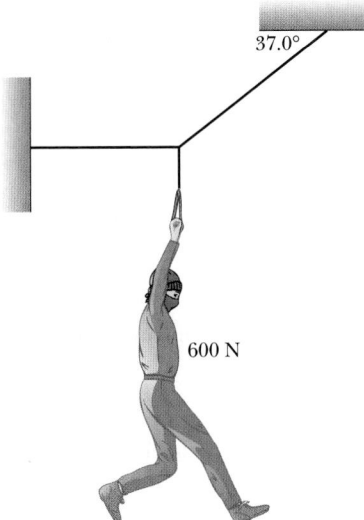

Figure P4.11

12. Find the tension in the two wires that support the 100-N light fixture in Figure P4.12.

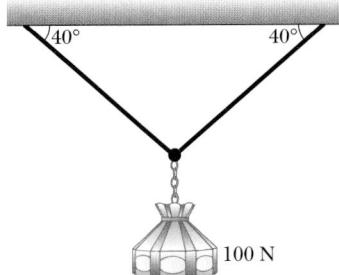

Figure P4.12

WEB 13. A 150-N bird feeder is supported by three cables, as shown in Figure P4.13. Find the tension in each cable.

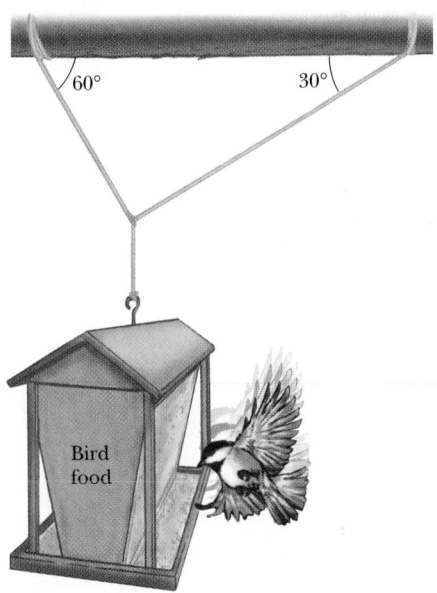

Figure P4.13

14. The leg and cast in Figure P4.14 weigh 220 N (w_1). Determine the weight w_2 and the angle α needed in order that there be no force exerted on the hip joint by the leg plus cast.

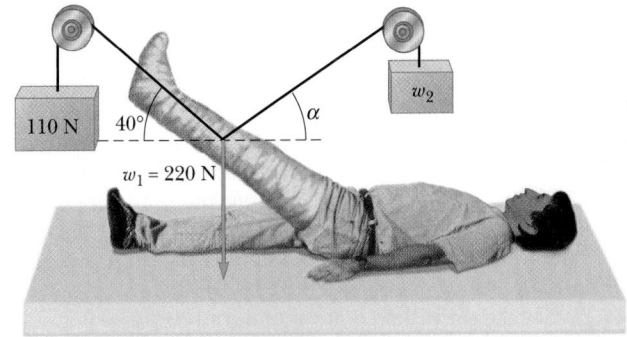

Figure P4.14

15. A block of mass $m = 2.0$ kg is held in equilibrium on an incline of angle $\theta = 60°$ by the horizontal force **F**, as shown in Figure P4.15. (a) Determine the value of F. (b) Determine the normal force exerted by the incline on the block (ignore friction).

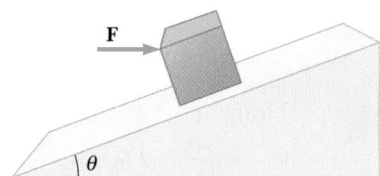

Figure P4.15 (Problems 15 and 42)

16. Two persons are pulling a boat through the water, as in Figure P4.16. Each exerts a force of 600 N directed at a 30.0° angle relative to the forward motion of the boat. If the boat moves with constant velocity, find the resistive force, **F**, exerted on the boat by the water.

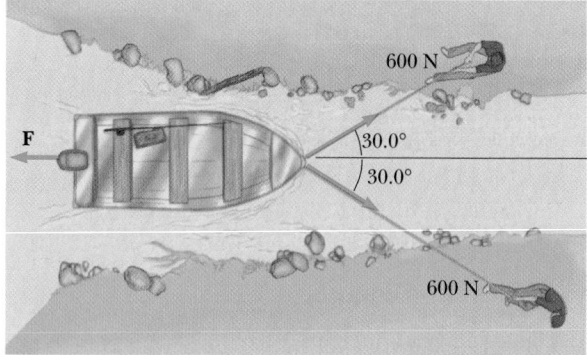

Figure P4.16 Two people pulling a boat.

17. A 5.0-kg bucket of water is raised from a well by a rope. If the upward acceleration of the bucket is 3.0 m/s^2, find the force exerted by the rope on the bucket.

18. A shopper in a supermarket pushes a loaded cart with a horizontal force of 10 N. The cart has a mass of 30 kg. (a) How far will it move in 3.0 s, starting from rest? (Ignore friction.) (b) How far will it move in 3.0 s if the shopper places her 30-N child in the cart before she begins to push it?

19. A 2000-kg car is slowed down uniformly from 20.0 m/s to 5.00 m/s in 4.00 s. What average force acted on the

car during this time, and how far did the car travel during the deceleration?

20. Two objects of masses 10.0 kg and 5.00 kg are connected by a light string that passes over a frictionless pulley as in Figure P4.20. The 5.00-kg object lies on a smooth incline of angle 40.0°. Find the acceleration of the 5.00 kg object and the tension in the string.

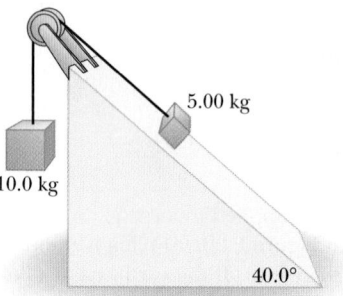

Figure P4.20

21. Assume that the three blocks in Figure P4.21 move on a frictionless surface and that a 42 N force acts as shown on the 3.0-kg block. Determine (a) the acceleration given this system, (b) the tension in the cord connecting the 3.0-kg and the 1.0-kg blocks, and (c) the force exerted on the 2.0-kg block by the 1.0-kg block.

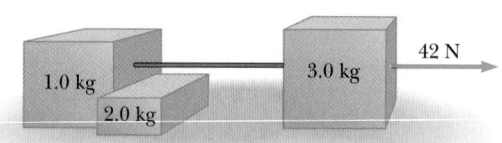

Figure P4.21

22. A train has a mass of 5.22×10^6 kg and is moving at 90.0 km/h. The engineer applies the brakes, which results in a net backward force of 1.87×10^6 N on the train. The brakes are held on for 30.0 s. (a) What is the new speed of the train? (b) How far does it travel during this period?

23. A 2.0-kg mass starts from rest and slides down an inclined plane 80 cm long in 0.50 s. What *net force* is acting on the mass along the incline?

24. A 40.0-kg wagon is towed up a hill inclined at 18.5° with respect to the horizontal. The tow rope is parallel to the incline and has a tension of 140 N in it. Assume that the wagon starts from rest at the bottom of the hill, and neglect friction. How fast is the wagon going after moving 80 m up the hill?

Problem 16 is the same as Workbook Problem 15 (Workbook page 40).
Problem 21 is similar to Workbook Problems 17 and 18 (Workbook pages 42 to 44).

25. A mass, $m_1 = 5.00$ kg, resting on a frictionless horizontal table is connected to a cable that passes over a pulley and then is fastened to a hanging mass, $m_2 = 10.0$ kg, as in Figure P4.25. Find the acceleration of each mass and the tension in the cable.

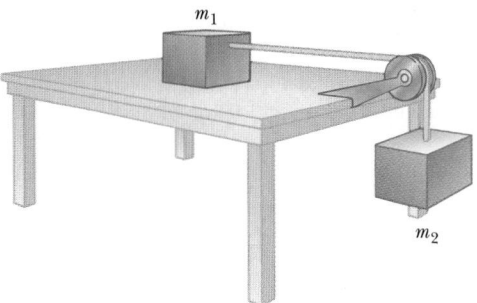

Figure P4.25 (Problems 25, 30, and 37)

26. Two blocks are fastened to the ceiling of an elevator, as in Figure P4.26. The elevator accelerates upward at 2.00 m/s². Find the tension in each rope.

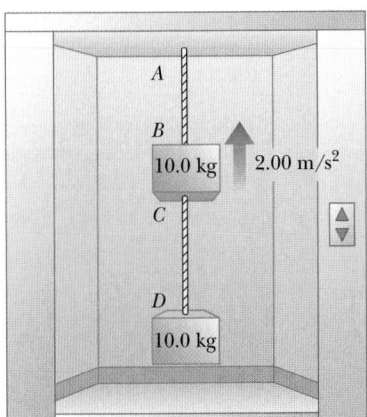

Figure P4.26

WEB **27.** A 1000-kg car is pulling a 300-kg trailer. Together the car and trailer have an acceleration of 2.15 m/s² in the forward direction. Neglecting frictional forces on the trailer, determine **(a)** the net force on the car; **(b)** the net force on the trailer; **(c)** the force exerted on the car by the trailer; **(d)** the resultant force exerted on the road by the car.

28. Two masses of 3.00 kg and 5.00 kg are connected by a light string that passes over a frictionless pulley as in Figure P4.28. Determine **(a)** the tension in the string,

(b) the acceleration of each mass, and **(c)** the distance each mass will move in the first second of motion if both masses start from rest.

Figure P4.28

Section 4.6 Force of Friction

29. A dockworker loading crates on a ship finds that a 20-kg crate, initially at rest on a horizontal surface, requires a 75-N horizontal force to set it in motion. However, after the crate is in motion, a horizontal force of 60 N is required to keep it moving with a constant speed. Find the coefficients of static and kinetic friction between crate and floor.

30. In Figure P4.25, $m_1 = 10$ kg and $m_2 = 4.0$ kg. The coefficient of static friction between m_1 and the horizontal surface is 0.50 while the coefficient of kinetic friction is 0.30. **(a)** If the system is released from rest, what will its acceleration be? **(b)** If the system is set in motion with m_2 moving downward, what will be the acceleration of the system?

31. A 1000-N crate is being pushed across a level floor at a constant speed by a force **F** of 300 N at an angle of 20.0° below the horizontal as shown in Figure P4.31a. **(a)** What is the coefficient of kinetic friction between the crate and the floor? **(b)** If the 300-N force is instead pulling the block at an angle of 20.0° above the horizontal as shown in Figure P4.31b, what will be the acceleration of the crate? Assume that the coefficient of friction is the same as found in **(a)**.

Figure P4.31

32. A hockey puck is hit on a frozen lake and starts moving with a speed of 12.0 m/s. Five seconds later, its speed is 6.00 m/s. (a) What is its average acceleration? (b) What is the average value of the coefficient of kinetic friction between puck and ice? (c) How far does the puck travel during this 5.00-s interval?

33. The coefficient of static friction between the 3.00-kg crate and the 35.0° incline of Figure P4.33 is 0.300. What minimum force **F** must be applied to the crate perpendicular to the incline to prevent the crate from sliding down the incline?

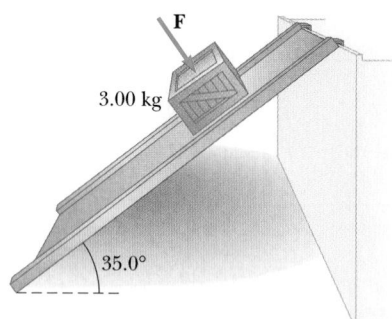

F

3.00 kg

35.0°

Figure P4.33

34. A box of books weighing 300 N is shoved across the floor of an apartment by a force of 400 N exerted downward at an angle of 35.2° below the horizontal. If the coefficient of kinetic friction between box and floor is 0.57, how long does it take to move the box 4.00 m, starting from rest?

35. An object falling under the pull of gravity experiences a frictional force of air resistance. The magnitude of this force is approximately proportional to the speed of the object, $f = bv$. Assume that $b = 15$ kg/s and $m = 50$ kg. (a) What is the terminal speed that the object reaches while falling? (b) Does your answer to part (a) depend on the initial speed of the object? Explain.

36. A student decides to move a box of books into her dormitory room by pulling on a rope attached to the box. She pulls with a force of 80.0 N at an angle of 25.0° above the horizontal. The box has a mass of 25.0 kg, and the coefficient of kinetic friction between box and floor is 0.300. (a) Find the acceleration of the box. (b) The student now starts moving the box up a 10.0° incline, keeping her 80.0 N force directed at 25.0° above the line of the incline. If the coefficient of friction is unchanged, what is the new acceleration of the box?

37. Masses $m_1 = 10.0$ kg and $m_2 = 5.00$ kg are connected by a light string that passes over a frictionless pulley as in Figure P4.25. If, when the system starts from rest, m_2 falls 1.00 m in 1.20 s, determine the coefficient of kinetic friction between m_1 and the table.

38. A car is traveling at 50.0 km/h on a flat highway. (a) If the coefficient of friction between road and tires on a rainy day is 0.100, what is the minimum distance in which the car will stop? (b) What is the stopping distance when the surface is dry and the coefficient of friction is 0.600?

39. A box slides down a 30.0° ramp with an acceleration of 1.20 m/s². Determine the coefficient of kinetic friction between the box and the ramp.

40. Masses $m_1 = 4.00$ kg and $m_2 = 9.00$ kg are connected by a light string that passes over a frictionless pulley. As shown in Figure P4.40, m_1 is held at rest on the floor and m_2 rests on a fixed incline of $\theta = 40.0°$. The masses are released from rest, and m_2 slides 1.00 m down the incline in 4.00 s. Determine (a) the acceleration of each mass, (b) the tension in the string, and (c) the coefficient of kinetic friction between m_2 and the incline.

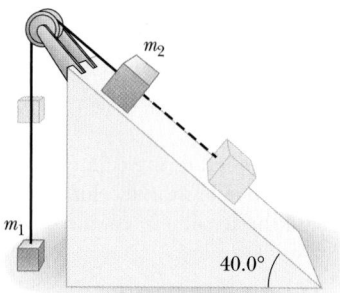

m_2

m_1

40.0°

Figure P4.40

41. Find the acceleration experienced by each of the two masses shown in Figure P4.41 if the coefficient of kinetic friction between the 7.00-kg mass and the plane is 0.250.

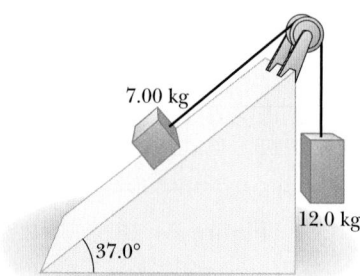

7.00 kg

12.0 kg

37.0°

Figure P4.41

42. A 2.00-kg block is held in equilibrium on an incline of angle $\theta = 60.0°$ by a horizontal force, **F**, applied in the direction shown in Figure P4.15. If the coefficient of static friction between block and incline is $\mu_s = 0.300$, determine (a) the minimum value of **F** and (b) the normal force of the incline on the block.

43. The person in Figure P4.43 weighs 170 lb. The crutches each make an angle of 22.0° with the vertical (as seen from the front). Half of the person's weight is sup-

ported by the crutches. The other half is supported by the vertical forces of the ground on his feet. Assuming the person is at rest and the force of the ground on the crutches acts along the crutches, determine (a) the smallest possible coefficient of friction between crutches and ground and (b) the magnitude of the compression force supported by each crutch.

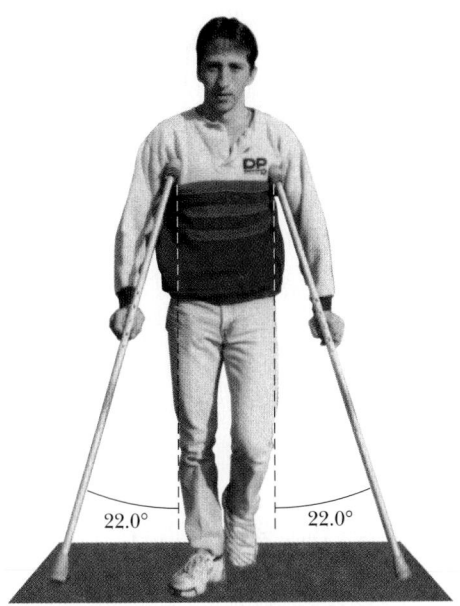

22.0°   22.0°

Figure P4.43

44. A block of mass $m = 2.00$ kg rests on the left edge of a block of length $L = 3.00$ m and mass $M = 8.00$ kg. The coefficient of kinetic friction between the two blocks is $\mu_k = 0.300$, and the surface on which the 8.00-kg block rests is frictionless. A constant horizontal force of magnitude $F = 10.0$ N is applied to the 2.00-kg block, setting it in motion as shown in Figure P4.44a. (a) How long will

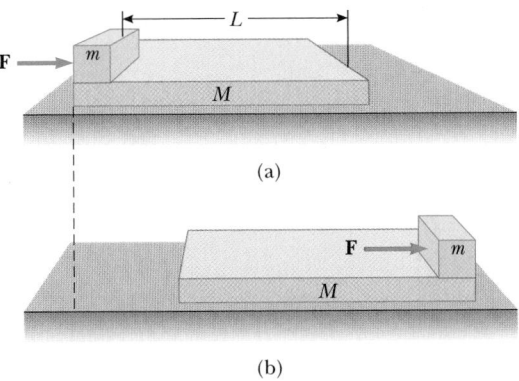

(a)

(b)

Figure P4.44

it take before this block makes it to the right side of the 8.00-kg block, as shown in Figure P4.44b? (*Note:* Both blocks are set in motion when **F** is applied.) (b) How far does the 8.00-kg block move in the process?

ADDITIONAL PROBLEMS

45. (a) What is the resultant force exerted by the two cables supporting the traffic light in Figure P4.45? (b) What is the weight of the light?

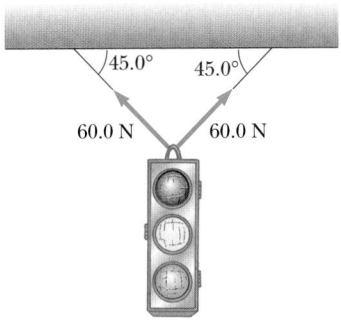

45.0°   45.0°

60.0 N   60.0 N

Figure P4.45

46. As a protest against the umpire's calls, a baseball pitcher throws a ball straight up into the air at a speed of 20.0 m/s. In the process, he moves his hand through a distance of 1.50 m. If the ball has a mass of 0.150 kg, find the average force he exerts on the ball to give it this upward speed.

47. A girl coasts down a hill on a sled, reaching a level surface at the bottom with a speed of 7.0 m/s. If the coefficient of friction between runners and snow is 0.050 and the girl and sled together weigh 600 N, how far does the sled travel on the level surface before coming to rest?

48. (a) What is the minimum force of friction required to hold the system of Figure P4.48 in equilibrium?

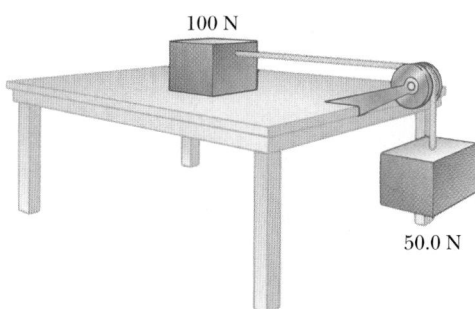

100 N

50.0 N

Figure P4.48

(b) What coefficient of static friction between the 100-N block and the table ensures equilibrium? **(c)** If the coefficient of kinetic friction between the 100-N block and the table is 0.250, what hanging weight should replace the 50.0-N weight to allow the system to move at a constant speed once it is set in motion?

WEB **49.** A box rests on the back of a truck. The coefficient of static friction between box and truck bed is 0.300. **(a)** When the truck accelerates forward, what force accelerates the box? **(b)** Find the maximum acceleration the truck can have before the box slides.

50. A 4.00-kg block is pushed along the ceiling with a constant applied force of 85.0 N that acts at an angle of 55.0° with the horizontal, as in Figure P4.50. The block accelerates to the right at 6.00 m/s². Determine the coefficient of kinetic friction between block and ceiling.

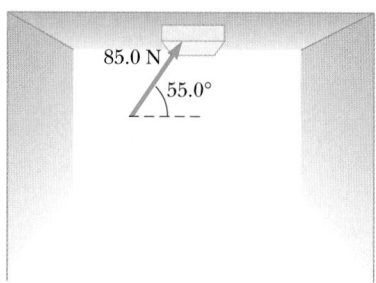

85.0 N

55.0°

Figure P4.50

51. A 2.00-kg aluminum block and a 6.00-kg copper block are connected by a light string over a frictionless pulley. They are allowed to move on a fixed steel block-wedge (of angle $\theta = 30.0°$) as shown in Figure P4.51. Making use of Table 4.2, determine **(a)** the acceleration of the two blocks and **(b)** the tension in the string.

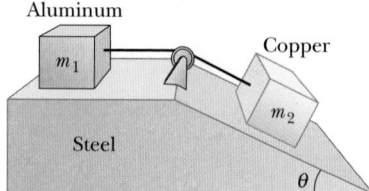

Aluminum

m_1

Copper

m_2

Steel

θ

Figure P4.51

52. Three masses are connected by light strings, as shown in Figure P4.52. The string connecting the 4.00-kg mass

and the 5.00-kg mass passes over a light frictionless pulley. Determine **(a)** the acceleration of each mass and **(b)** the tensions in the two strings.

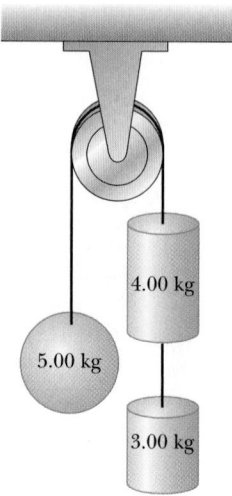

4.00 kg

5.00 kg

3.00 kg

Figure P4.52

53. A 3.00-kg block starts from rest at the top of a 30.0° incline and slides 2.00 m down the incline in 1.50 s. Find **(a)** the acceleration of the block, **(b)** the coefficient of kinetic friction between it and the incline, **(c)** the frictional force acting on the block, and **(d)** the speed of the block after it has slid 2.00 m.

54. A 5.0-kg penguin sits on a 10-kg sled, as in Figure P4.54. A horizontal force of 45 N is applied to the sled, but the penguin attempts to impede the motion by holding onto a cord attached to a tree. The coefficient of kinetic friction between the sled and snow as well as that between the sled and the penguin is 0.20. **(a)** Draw a free-body diagram for the penguin and one for the sled, and identify the reaction force for each force you include. Determine **(b)** the tension in the cord and, **(c)** the acceleration of the sled.

45 N

Figure P4.54

55. Two blocks on a frictionless horizontal surface are connected by a light string as in Figure P4.55, where $m_1 = 10$ kg and $m_2 = 20$ kg. A force of 50 N is applied to the 20-kg block. **(a)** Determine the acceleration of each block and the tension in the string. **(b)** Repeat the problem for the case where the coefficient of kinetic friction between each block and the surface is 0.10.

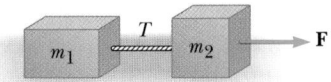

Figure P4.55

56. In Figure P4.56, the coefficient of kinetic friction between the two blocks is 0.30. The table surface and the pulleys are frictionless. **(a)** Draw a free-body diagram for each block. **(b)** Determine the acceleration of each block. **(c)** Find the tensions in the strings.

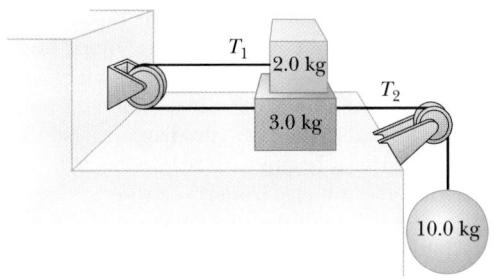

Figure P4.56

57. Two people pull as hard as they can on ropes attached to a 200-kg boat. If they pull in the same direction, the boat has an acceleration of 1.52 m/s² to the right. If they pull in opposite directions, the boat has an acceleration of 0.518 m/s² to the left. What is the force exerted by each person on the boat? (Disregard any other forces on the boat.)

58. A 3.0-kg mass hangs at one end of a rope that is attached to a support on a railroad car. When the car accelerates to the right, the rope makes an angle of 4.0° with the vertical, as shown in Figure P4.58. Find the acceleration of the car.

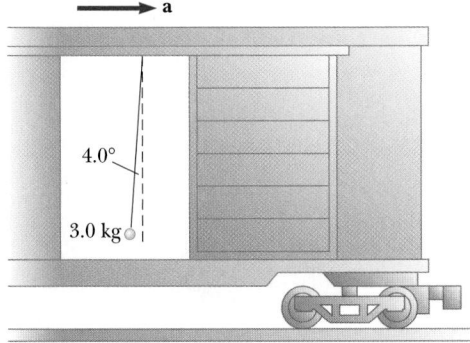

Figure P4.58

Problem 59 is similar to Workbook Problem 21 (Workbook page 48).

59. The three blocks of masses 10.0 kg, 5.00 kg, and 3.00 kg are connected by light strings that pass over frictionless pulleys as shown in Figure P4.59. The acceleration of the 5.00-kg block is 2.00 m/s² to the left, and the surfaces are rough. Find **(a)** the tension in each string and **(b)** the coefficient of kinetic friction between blocks and surfaces. (Assume the same μ_k for both blocks in contact with surfaces.)

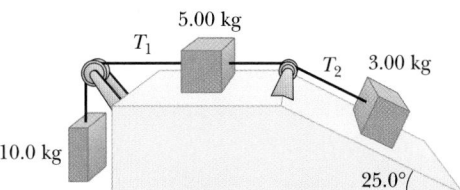

Figure P4.59

60. A sled weighing 60.0 N is pulled horizontally across snow and the coefficient of kinetic friction between sled and snow is 0.100. A penguin weighing 70.0 N rides on the sled. (See Fig. P4.60.) If the coefficient of static friction between penguin and sled is 0.700, find the maximum horizontal force that can be exerted on the sled before the penguin begins to slide off.

Figure P4.60

61. An 80-kg person escapes from a burning building by jumping from a window situated 30 m above a catching net. Assuming that air resistance exerts a 100-N force on the person as he falls, determine his velocity just before he hits the net.

62. The parachute on a race car of weight 8820 N opens at the end of a quarter-mile run when the car is traveling at 35 m/s. What total retarding force must be supplied by the parachute to stop the car in a distance of 1000 m?

WEB 63. On takeoff, the combined action of the engines and wings of an airplane exerts an 8000-N force on the plane, directed upward at an angle of 65.0° above the horizontal. The plane rises with constant velocity in the vertical direction while continuing to accelerate in the horizontal direction. (a) What is the weight of the plane? (b) What is its horizontal acceleration?

64. A van accelerates down a hill (Fig. P4.64), going from rest to 30.0 m/s in 6.00 s. During the acceleration, a toy ($m = 0.100$ kg) hangs by a string from the van's ceiling. The acceleration is such that the string remains perpendicular to the ceiling. Determine (a) the angle θ and (b) the tension in the string.

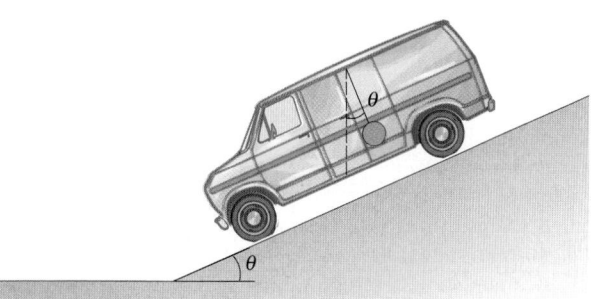

Figure P4.64

65. The board sandwiched between two other boards in Figure P4.65 weighs 95.5 N. If the coefficient of friction between the boards is 0.663, what must be the magnitude of the compression forces (assume horizontal) acting on both sides of the center board to keep it from slipping?

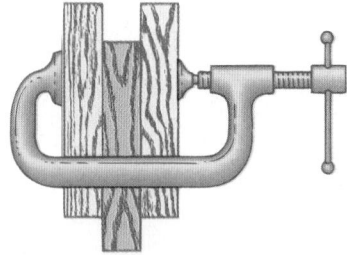

Figure P4.65

66. A 72-kg man stands on a spring scale in an elevator. Starting from rest, the elevator ascends, attaining its maximum speed of 1.2 m/s in 0.80 s. It travels with this constant speed for 5.0 s, undergoes a uniform *negative* acceleration for 1.5 s, and comes to rest. What does the spring scale register (a) before the elevator starts to

move? (b) during the first 0.80 s? (c) while the elevator is traveling at constant speed? (d) during the negative acceleration?

67. Bob and Kathy, two construction workers on the roof of a building, are about to raise a keg of nails from the ground by means of a light rope passing over a light frictionless pulley 10.0 m above the ground. Bob weighs 900 N, Kathy 600 N, the keg 300 N, and the nails 600 N. Both workers slip off the roof, and the following unfortunate sequence of events takes place. Hanging together on the rope, Bob and Kathy strike the ground just as the keg hits the pulley. Unnerved by his fall, Bob lets go of the rope, and the keg pulls Kathy up to the roof, where she cracks her head against the pulley but gamely hangs on. However, the nails spill out of the keg when it strikes the ground, and the empty keg rises as Kathy returns to the ground. Finally, she has had enough, lets go of the rope, and remains on the ground, only to be hit by the empty keg again. Ignoring the possible mid-air collisions that merely added insult to injury, how long did it take this industrial accident to run its course? Assume that all collisions with the ground and pulley serve to start each subsequent motion from rest, so that each trip up and down begins with zero velocity.

68. A bag of cement hangs from three wires, as shown in Figure P4.68. Two of the wires make angles θ_1 and θ_2 with the horizontal. If the system is in equilibrium, (a) show that

$$T_1 = \frac{w \cos \theta_2}{\sin(\theta_1 + \theta_2)}$$

(b) Given that $w = 325$ N, $\theta_1 = 10.0°$ and $\theta_2 = 25.0°$, find the tensions T_1, T_2, and T_3 in the wires.

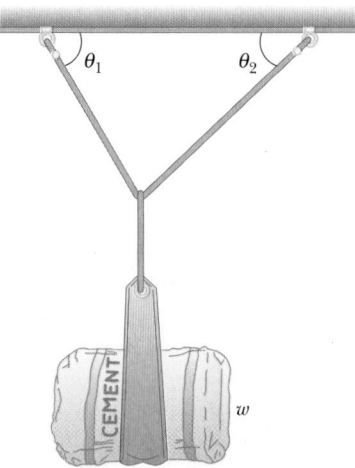

Figure P4.68

Problem 63 is the same as Workbook Problem 22 (Workbook page 49).

69. An inventive child named Pat wants to reach an apple in a tree without climbing the tree. Sitting in a chair connected to a rope that passes over a frictionless pulley (Fig. P4.69), Pat pulls on the loose end of the rope with such a force that the spring scale reads 250 N. Pat's true weight is 320 N, and the chair weighs 160 N. **(a)** Show that the acceleration of the system of Pat and the chair is *upward* and find its magnitude. **(b)** Find the force Pat exerts on the chair.

70. A fire helicopter carries a 620-kg bucket of water at the end of a 20.0-m long cable. Flying back from a fire at a constant speed of 40.0 m/s, the cable makes an angle of 40.0° with respect to the vertical. Determine the force of air resistance on the bucket.

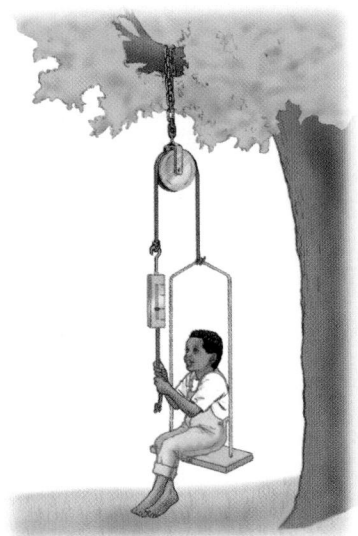

Figure P4.69

5

Work and Energy

▲ PHYSICS PUZZLER

Chum salmon leaping up a waterfall in the McNeil River in Alaska. The salmon must reach high speeds before leaping out of the water. Why are fish ladders often built around dams? Can you estimate how fast they must swim to leap a vertical distance of 2 m? *(Daniel J. Cox/Tony Stone Images)*

5.1

The concept of energy is one of the most important in the world of science. In everyday use, the term *energy* has to do with the cost of fuel for transportation and heating, electricity for lights and appliances, and the foods we consume. However, these ideas do not really define energy. They tell us only that fuels are needed to do a job and that those fuels provide us with something we call energy.

Energy is present in the Universe in a variety of forms, including mechanical energy, chemical energy, electromagnetic energy, and nuclear energy. Although energy can be transformed from one form to another, the total amount of energy in the Universe remains the same. If one form of energy in an isolated system decreases, then, by the principle of conservation of energy, another form of energy in the system must increase. For example, if the system consists of a motor connected to a battery, chemical energy is converted to electrical energy, which in turn is converted to mechanical energy. The transformation of energy from one form into another is essential in the study of physics, chemistry, biology, geology, and astronomy.

In this chapter we are concerned only with mechanical energy. We introduce the concept of *kinetic energy,* which is defined as the energy associated with motion, and the concept of *potential energy,* the energy associated with position. We shall see that the ideas of work and energy can be used in place of Newton's laws to solve certain problems.

We begin by defining *work,* a concept that provides a link between force and energy. With this as a foundation, we can then discuss the principle of conservation of energy and apply it to problems.

5.1 WORK

Almost all the terms we have used thus far have conveyed the same meaning in physics as in everyday life. Now, however, we encounter a term whose meaning in physics is distinctly different from its meaning in our day-to-day affairs. This new term is **work.** It can be defined with the help of Figure 5.1. Here we see an object that undergoes a displacement of **s** along a straight line while acted on by a constant force, **F,** that makes an angle of θ with **s.**

> The work W done by an agent exerting a constant force is defined as the product of the component of the force along the direction of displacement and the magnitude of the displacement:

$$W \equiv (F \cos \theta)\, s \qquad [5.1]$$

As an example of the distinction between this definition of *work* and our everyday understanding of the word, consider holding a heavy chair at arm's length for 10 min. At the end of this time interval, your tired arms may lead you to think that you have done a considerable amount of work. According to our definition, how-

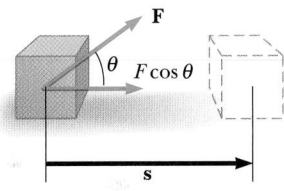

Figure 5.1 If an object undergoes a displacement of **s,** the work done by the force **F** is $(F \cos \theta)\, s$.

5.2

QUICKLAB

Slide a heavy book across your desk using just one finger. What is the best angle between your finger and the direction of motion to accomplish this task? Describe what happens to the component of the force along the direction of motion as you move your hand so that your finger is closer to vertical. What does this do to the work you are doing on the book?

Figure 5.2 No work is done when a bucket is moved horizontally, because the applied force, **F**, is perpendicular to the displacement.

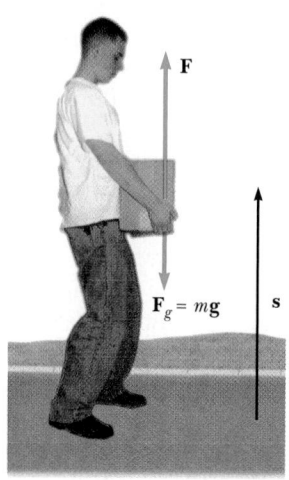

Figure 5.3 Positive work is done by this person when the box is lifted, because the applied force, **F**, is in the same direction as the displacement. When the box is lowered to the floor, the work done by the person is negative.

ever, you have done no work on the chair.[1] You have exerted a force in order to support the chair, but you have not moved it. A force does no work on an object if the object does not move. This can be seen by noting that if $s = 0$, Equation 5.1 gives $W = 0$.

Also note from Equation 5.1 that the work done by a force is zero when the force is perpendicular to the displacement. That is, if $\theta = 90°$, then $W = 0$ because $\cos 90° = 0$. For example, no work is done when a bucket of water is carried horizontally at constant velocity, because the upward force exerted to support the bucket is perpendicular to the displacement of the bucket, as shown in Figure 5.2. Likewise, the work done by the force of gravity during the horizontal displacement is also zero, for the same reason.

The sign of the work depends on the direction of **F** relative to **s.** Work is positive when the component $F \cos \theta$ is in the same direction as the displacement. For example, when you lift a box as in Figure 5.3, the work done by the force you exert on the box is positive because that force is upward, in the same direction as the displacement. Work is negative when **F** $\cos \theta$ is in the direction opposite the displacement. The negative sign comes from the fact that $\theta = 180°$ and $\cos 180° = -1$, which, from Equation 5.1, gives a negative value for W. Finally, if an

---

[1] Actually, you do burn calories while holding a chair at arm's length, because your muscles are continuously contracting and relaxing while the chair is being supported. Thus, work is being done on your body—but not on the chair.

TABLE 5.1   Units of Work in the Three Common Systems of Measurement

System	Unit of Work	Name of Combined Unit
SI	newton-meter (N · m)	joule (J)
cgs	dyne-centimeter (dyne · cm)	erg
British engineering (conventional)	foot-pound (ft · lb)	foot-pound

applied force acts along the direction of the displacement, then $\theta = 0°$. Because $\cos 0° = 1$, in this case Equation 5.1 becomes

$$W = Fs \qquad [5.2]$$

Work is a scalar quantity, and its units are force times length. Therefore, the SI unit of work is the **newton-meter (N·m).** Another name for the newton-meter is the **joule (J).** The unit of work in the cgs system is the **dyne-centimeter (dyne·cm),** which is also called the **erg,** and the unit in the conventional (British engineering) system is the **foot-pound (ft · lb).** These are summarized in Table 5.1. Note that $1 \text{ J} = 10^7$ ergs.

Thinking Physics 1

A person lifts a cement block of mass m a vertical height h, and then walks horizontally a distance d while holding the block, as in Figure 5.4. Determine the work done by the person and by the force of gravity in this process.

Explanation   Assuming that the person lifts the block with a force of magnitude equal to the weight of the block, mg, the work done by the person during the vertical displacement is mgh, because the force in this case is in the direction of the displacement. The work done by the person during the horizontal displacement of the block is zero because the applied force in this process is perpendicular to the displacement. Thus, the net work done by the person is mgh. The work done by the force of gravity during the vertical displacement of the block is $-mgh$, because this force is opposite the displacement. The work done by the force of gravity is zero during the horizontal displacement because this force is also perpendicular to the displacement. Hence, the net work done by the force of gravity is $-mgh$. The net work on the block is zero $(+mgh - mgh = 0)$.

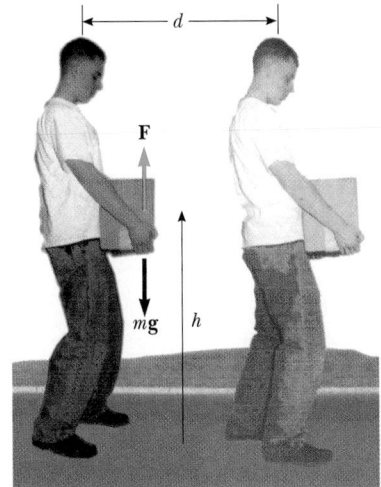

Figure 5.4   (Thinking Physics 1) A person lifts a cement block of mass m a vertical distance h and then walks horizontally a distance d.

EXAMPLE 5.1   Mr. Clean

A man cleaning his apartment pulls the canister of a vacuum cleaner with a force of magnitude $F = 50.0$ N at an angle of $30.0°$, as shown in Figure 5.5. He moves the vacuum cleaner a distance of 3.00 m. Calculate the work done by the 50.0-N force.

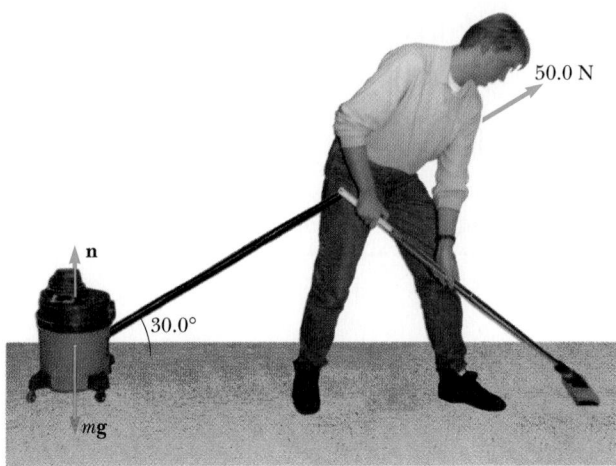

Figure 5.5   (Example 5.1) A vacuum cleaner being pulled at an angle of 30.0° with the horizontal.

Solution   We can use $W = (F \cos \theta) s$, with $F = 50.0$ N, $\theta = 30.0°$, and $s = 3.00$ m, to get

$$W = (50.0 \text{ N})(\cos 30.0°)(3.00 \text{ m}) = \boxed{130 \text{ J}}$$

Note that the normal force, **n,** the weight, $m\mathbf{g},$ and the upward component of the applied force $(50.0 \text{ N}) \cos 30.0°$ do *no* work because they are perpendicular to the displacement.

Exercise   Find the work done on the vacuum cleaner by the man if he pulls it 3.00 m with a horizontal force of $F = 50.0$ N.

Answer   150 J

5.2   KINETIC ENERGY AND THE WORK–KINETIC ENERGY THEOREM

Solving problems with Newton's second law can be difficult if the forces involved are complex. An alternative approach to such problems is to relate the speed of an object to its displacement under the influence of some net force. If the work done by the net force on the object can be calculated for a given displacement, the change in the object's speed is easy to evaluate.

Figure 5.6 shows a particle of mass m moving to the right under the action of a constant net force, **F.** Because the force is constant, we know from Newton's second law that the particle moves with a constant acceleration, **a.** If the particle is displaced a distance of s, the work done by **F** is

$$W_{\text{net}} = Fs = (ma)s \tag{5.3}$$

However, in Chapter 2 we found that the following relationship holds when an object undergoes constant acceleration:

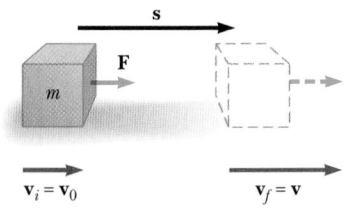

Figure 5.6   An object undergoing a displacement and a change in velocity under the action of a constant net force, **F.**

$$v^2 = v_0^2 + 2as \quad \text{or} \quad as = \frac{v^2 - v_0^2}{2}$$

5.4

This expression is now substituted into Equation 5.3 to give

$$W_{\text{net}} = m\left(\frac{v^2 - v_0^2}{2}\right)$$

[5.4]

$$W_{\text{net}} = \tfrac{1}{2}mv^2 - \tfrac{1}{2}mv_0^2$$

The quantity $mv^2/2$ has a special name in physics: **kinetic energy.** Any object of mass m and speed v is defined to have a kinetic energy, KE, of

$$KE \equiv \tfrac{1}{2}mv^2$$

[5.5] ◀ Kinetic energy

Kinetic energy is a scalar quantity and has the same units as work. For example, a 1.0-kg mass moving with a speed of 4.0 m/s has a kinetic energy of 8.0 J. We can think of kinetic energy as the energy associated with the motion of an object.

It is often convenient to write Equation 5.4 as

$$W_{\text{net}} = KE_f - KE_i$$

[5.6]

According to this result, the work done by a *net* force acting on an object is to change the kinetic energy of the object from some initial value, KE_i, to some final value, KE_f. Equation 5.6 is an important result known as the **work-kinetic energy theorem.** Thus, we conclude that

◀ Work-kinetic energy theorem

> **The net work done on an object by a net force acting on it is equal to the change in the kinetic energy of the object.**

From the work-kinetic energy theorem, we also find that the speed of the object increases if the net work done on it is positive, because the final kinetic energy is greater than the initial kinetic energy. The object's speed decreases if the net work is negative, because the final kinetic energy is less than the initial kinetic energy. Notice that the speed and kinetic energy of an object will change only if work is done on the object by some external force.

Consider the relationship between the work done on an object and the change in its kinetic energy as expressed by Equation 5.4. Because of this connection, we can also think of kinetic energy as the work the object can do in coming to rest. For example, suppose a hammer is on the verge of striking a nail, as in Figure 5.7. The moving hammer has kinetic energy and can do work on the nail. The work done on the nail is *Fs*, where *F* is the average force exerted on the nail by the hammer and *s* is the distance the nail is driven into the wall.

For convenience, Equation 5.4 was derived under the assumption that the net force acting on the object was constant. A more general derivation would show that this equation is valid under all circumstances, including that of a variable force.

Figure 5.7 The moving hammer has kinetic energy and thus can do work on the nail, driving it into the wall.

Thinking Physics 2

A car traveling at a speed v skids a distance d after its brakes lock. Estimate how far it will skid if its brakes lock when its initial speed is $2v$. What happens to the car's kinetic energy as it stops?

Explanation   Let us assume that the force of kinetic friction between car and road surface is constant and the same in both cases. The net force times the displacement of the car is equal its initial kinetic energy. If the speed is doubled, as in this example, the kinetic energy of the car is quadrupled. For a given applied force (in this case, the frictional force), the distance traveled is four times as great when the initial speed is doubled, so the estimated distance it skids is $4d$. The kinetic energy of the car is changed into internal energy associated with the tires, brake pads, and road as they increase in temperature.

Thinking Physics 3

More energy is expended by walking downstairs than by walking horizontally at the same speed. Why do you think this is so?

Explanation   When walking downstairs, a person has to stop the motion of her entire body at each step. This requires work to be done by the muscles. The work done by the muscles is negative, because the forces applied are opposite to the displacements representing the movement of the body. Negative work represents a transfer of energy out of the muscles (from the store of energy from the food eaten by the walker). When walking horizontally, the body can be kept in motion with an almost constant horizontal speed—only the feet and legs start and stop. Thus, more energy is transferred from the body in bringing the entire body to a stop in the trip downstairs than in walking.

Workbook Problem 27 — (Workbook page 59) will help you apply the concept of kinetic energy.

EXAMPLE 5.2   Towing a Car

A 1400-kg car has a net forward force of 4500 N applied to it. The car starts from rest and travels down a horizontal highway. What are its kinetic energy and speed after it has traveled 100 m? (Ignore losses in kinetic energy because of friction and air resistance.)

Solution   With the initial velocity given as zero, Equation 5.4 reduces to

$$(1) \qquad W_{\text{net}} = \tfrac{1}{2}mv^2$$

The work done by the net force on the car is

$$W_{\text{net}} = Fs = (4500 \text{ N})(100 \text{ m}) = 4.50 \times 10^5 \text{ J}$$

This work all goes into changing the kinetic energy of the car; thus the final value of the kinetic energy, from (1), is also 4.50×10^5 J.

The speed of the car can be found from (1) as follows:

$$\tfrac{1}{2}mv^2 = 4.50 \times 10^5 \, \text{J}$$

$$v^2 = \frac{2(4.50 \times 10^5 \, \text{J})}{1400 \text{ kg}} = 643 \text{ m}^2/\text{s}^2$$

$$v = \boxed{25.4 \text{ m/s}}$$

5.3 POTENTIAL ENERGY

In the preceding section we saw that an object that has kinetic energy can do work on another object, as illustrated by the moving hammer driving a nail into the wall. Now we introduce another form of energy associated with the position of an object.

Gravitational Potential Energy

As an object falls freely in a gravitational field, the field exerts a force on it, doing positive work on it and thereby increasing its kinetic energy. Consider Figure 5.8, which shows a brick of mass m held at a height of y_i above a nail in a board lying on the ground. When the brick is released, it falls toward the ground, gaining speed and thereby gaining kinetic energy. As a result of its position in space, the brick has potential energy (it has the *potential* to do work), and this potential energy is converted to kinetic energy as the brick falls. When the brick reaches the ground, it does work on the nail, driving it into the board. The energy that an object has as a result of its position in space near the surface of the Earth is called **gravitational potential energy.**

Let us now derive an expression for the gravitational potential energy of an object at a given location. Consider a block of mass m at an initial height of y_i above the ground, as in Figure 5.9, and let us neglect air resistance. As the block falls, the only force that does work on it is the gravitational force, $m\mathbf{g}$. The work done by the gravitational force as the block undergoes a downward displacement of $s = y_i - y_f$ is

$$W_g = mgs = mgy_i - mgy_f$$

We define the quantity mgy to be the gravitational potential energy, PE:

$$PE \equiv mgy \qquad [5.7]$$

Thus, the gravitational potential energy associated with any object at a point in space is the product of the object's weight and its vertical coordinate. It is important to note that this relationship is valid only for objects near the Earth's surface, where g is approximately constant.

If we substitute this expression for PE into the expression for W_g, we have

$$W_g = PE_i - PE_f \qquad [5.8]$$

The work done on any object by the force of gravity is equal to the object's initial potential energy minus its final potential energy.

The units of gravitational potential energy are the same as those of work: joules, ergs, or foot-pounds. Potential energy, like work and kinetic energy, is a scalar quantity.

Brick

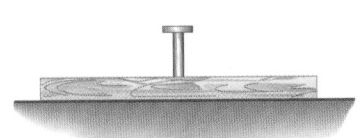

Figure 5.8 Because of its position in space, the brick can do work on the nail.

5.3, SECTION 1

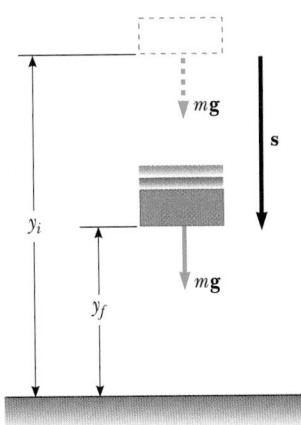

Figure 5.9 The work done by the gravitational force as the block falls from y_i to y_f is equal to $mgy_i - mgy_f$.

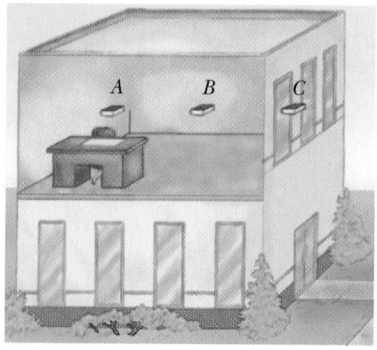

Figure 5.10 Any reference level could be used for measuring the gravitational potential energy of the book.

Reference Levels for Gravitational Potential Energy

In working problems involving the concept of gravitational potential energy, you must choose a location at which the gravitational potential energy is equal to zero. The location of this zero level is completely arbitrary. This is true because the quantity of importance is the *difference* in potential energy, and the difference in potential energy between two points is independent of the choice of zero level. It is often convenient to choose the surface of the Earth as the reference position for zero potential energy, but again this is not essential. Often the statement of a problem will suggest a convenient level to use. As an example, consider a book at several possible locations, as in Figure 5.10. When the book is at *A*, above the surface of a desk, a logical zero level for potential energy would be the surface of the desk. When the book is at *B*, however, the floor might be a more appropriate zero reference level. Finally, a location such as *C*, where the book is held out a window, would suggest choosing the surface of the Earth as the zero level of potential energy. The choice, however, makes no difference — any of the reference levels mentioned could be used as the zero level, regardless of whether the book is at *A*, *B*, or *C*. Example 5.3 illustrates this important point.

Applying Physics 1

APPLICATION

"Ringing the Bell" at a Carnival.

A common scene at a carnival is the Ring-the-Bell attraction, in which the player swings a hammer downward in an attempt to project a mass upward to ring a bell. The goal is to provide the mass with enough gravitational potential energy to lift it at least

Workbook Problem 28 — (Workbook page 60) will help you think about reference levels for gravitational potential energy.

"Step right up, folks!" In this carnival event, part of the kinetic energy of the heavy hammer is transferred to a pivoted target, which in turn lifts a weight and hopefully rings the bell. *(Robert E. Daemmrich/ Tony Stone Images)*

as high as the bell. This gravitational potential energy will be converted from kinetic energy of the mass as it moves upward from its lowest position near the ground. In turn, this kinetic energy is related to how much work was done by the player in striking the device with the hammer. To maximize this work, and ultimately the height of the mass, the player must strike with as much force as possible, and move the hammer with as much distance as possible while it is in contact with the device.

EXAMPLE 5.3 Wax Your Skis

A 60.0-kg skier is at the top of a slope, as shown in Figure 5.11a. At the initial point A, the skier is 10.0 m vertically above point B.
(a) Setting the zero level for gravitational potential energy at B, find the gravitational potential energy of the skier at A and B, and then find the difference in potential energy between these two points.

(a)   (b)

Figure 5.11 (Example 5.3) *(b, Gregg Adams/Tony Stone Worldwide)*

Solution The gravitational potential energy at B is zero by choice. Hence, the potential energy at A is

$$PE_i = mgy_i = (60.0 \text{ kg})(9.80 \text{ m/s}^2)(10.0 \text{ m}) = \boxed{5880 \text{ J}}$$

Because $PE_f = 0$, the difference in potential energy is

$$PE_i - PE_f = 5880 \text{ J} - 0 \text{ J} = \boxed{5880 \text{ J}}$$

(b) Repeat this problem with the zero level at point *A*.

Solution   In this case, the initial potential energy is zero because of the chosen reference level. The final potential energy is

$$PE_f = mgy_f = (60.0 \text{ kg})(9.80 \text{ m/s}^2)(-10.0 \text{ m}) = \boxed{-5880 \text{ J}}$$

The distance y_f is negative because the final point is 10.0 m below the zero reference level. The difference in potential energy is now

$$PE_i = PE_f = 0 \text{ J} - (-5880 \text{ J}) = \boxed{5880 \text{ J}}$$

These calculations show that the potential energy of the skier at the top of the slope is greater than the potential energy at the bottom by 5880 J, *regardless of the zero level selected.*

Exercise   If the zero level for gravitational potential energy is selected to be midway down the slope, at a height of 5.00 m, find the initial potential energy, the final potential energy, and the difference in potential energy between points *A* and *B*.

Answer   2940 J, −2940 J, 5880 J

5.4   CONSERVATIVE AND NONCONSERVATIVE FORCES

If you slide down a very slick hill on a sled, your speed, and hence your kinetic energy, increases dramatically. However, if the hill is not very icy, your speed and kinetic energy do not increase as rapidly. What has happened to this "lost" kinetic energy? To answer this question, we shall examine the properties of two categories of forces that exist in nature — conservative and nonconservative forces.

Conservative Forces

Definition of a conservative force ▶

5.6

A force is conservative if the work it does on an object moving between two points is independent of the path the object takes between the points. In other words, the work done on an object by a conservative force depends only on the initial and final positions of the object. Furthermore, **a force is conservative if the work it does on an object moving through any closed path is zero.**

   The force of gravity is conservative. As we learned in the preceding section, the work done by the gravitational force on an object moving between any two points near the Earth's surface is

$$W_g = mgy_i - mgy_f$$

From this we see that W_g depends only on the initial and final coordinates of the object and hence is independent of path. Furthermore, note that W_g is zero when the object moves over any closed path (where $y_i = y_f$).

PHYSICS IN ACTION

On the left, these cyclists are working hard and expending energy as they pedal up-hill in Marin County, California.

On the right, Twin Falls on the Island of Kauai, Hawaii. The potential energy of the water at the top of the falls is converted to kinetic energy at the bottom. In many locations, this mechanical energy is used to produce electrical energy.

Work and Energy in Sports and Nature

(David Madison, Tony Stone Images)

(Bruce Byers, FPG)

We can associate a potential energy function with any conservative force. In the preceding section, the potential energy function associated with the gravitational force was found to be

$$PE = mgy$$

Potential energy functions can be defined only for conservative forces. In general, the work, W_c, done on an object by a conservative force is equal to the initial potential energy of the object minus the final value:

$$W_c = PE_i - PE_f \qquad \text{[5.9]}$$

Nonconservative Forces

A force is nonconservative if it leads to a dissipation of mechanical energy. If you moved an object on a horizontal surface, returning it to the same location and same state of motion, but found it necessary to do net work on the object, then

◀ Definition of a nonconservative force

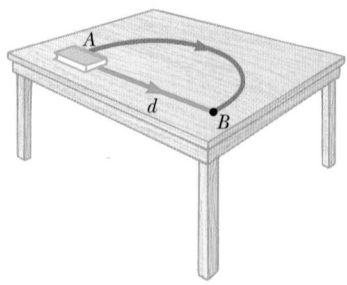

Figure 5.12 The loss in mechanical energy due to the force of friction depends on the path taken as the book is moved from *A* to *B*.

5.5, ALL SECTIONS

As this skillful rock climber moves cautiously up the mountain, her internal (biochemical) energy is transformed into gravitational potential energy. Can you identify other forms of energy transformations if she descends the mountain with the help of supporting ropes? *(Telegraph Colour Library/FPG)*

something must have dissipated the energy transferred to the object. That dissipative force is recognized as friction between object and surface. Friction is a dissipative, or nonconservative, force.

Suppose you displace a book between two points on a table. If you displace it in a straight line along the blue path shown in Figure 5.12, the change in mechanical energy due to friction is $-fd$, where d is the distance between the two points. However, if you move the book along any other path between the two points, the change in mechanical energy due to friction is greater (in absolute magnitude) than $-fd$. For example, the change in mechanical energy due to friction along the red semicircular path shown in Figure 5.12 is equal to $-f(\pi d/2)$, where d is the diameter of the circle.

5.5   CONSERVATION OF MECHANICAL ENERGY

Conservation principles play a very important role in physics, and you will encounter a number of them as you continue in this course. The first, conservation of energy, is one of the most important. Before we describe its mathematical details, let us pause briefly to examine what is meant by conservation. When we say that a physical quantity is *conserved*, we simply mean that the value of the quantity remains constant. Although the form of the quantity may change in some manner, its final value is the same as its initial value. For example, the energy in an isolated system may change from gravitational potential energy to kinetic energy or to one of a variety of other forms, but energy is never lost from the system.

In order to develop the principle of conservation of energy, let us return to the work–kinetic energy theorem, Equation 5.6, which states that the net work done on a system equals the change in the system's kinetic energy. Let us assume that the only force doing work on the system is conservative. In this case the net work on the system is equal to W_c, and from Equation 5.9 we have

$$W_{\text{net}} = W_c = PE_i - PE_f$$

We now substitute this expression into Equation 5.6 for W_{net} to find

$$PE_i - PE_f = KE_f - KE_i$$

or

$$KE_i + PE_i = KE_f + PE_f \qquad [5.10]$$

More formally, the principle of conservation of mechanical energy states that **the total mechanical energy in any isolated system of objects remains constant if the objects interact only through conservative forces.** It is important to note that Equation 5.10 is valid *provided* no energy is added to or removed from the system. Furthermore, there must be no nonconservative forces within the system. Therefore, if a conservative system (one subject only to conservative forces) has some initial combined amount of kinetic and potential energy, some of that energy may later transform from one kind into the other, but the total mechanical energy remains constant. This is equivalent to saying that, if the kinetic energy of a conservative system increases (or decreases) by some amount, the potential energy of the system must decrease (or increase) by the same amount.

If the force of gravity is the *only* force doing work on an object, then the total mechanical energy of the object remains constant, and the principle of conservation of mechanical energy takes the form

$$\tfrac{1}{2}mv_i^2 + mgy_i = \tfrac{1}{2}mv_f^2 + mgy_f \qquad [5.11]$$

Because there are other forms of energy besides kinetic and gravitational potential energy, we shall modify Equation 5.11 to include these new forms as we learn about them.

Applying Physics 2

You have graduated from college and are designing roller coasters for a living. Your boss asks you to design a roller coaster that is supposed to release a car from rest at the top of a hill of height h, and then roll freely down the hill and reach the peak of the next hill of height $1.1h$. What would you tell your boss in this situation?

Explanation You should politely tell your boss to study some physics. It is impossible to design such a roller coaster because it would violate the principle of conservation of mechanical energy. At the starting point, the roller coaster has no kinetic energy and gravitational potential energy mgh relative to ground level. If it were able to reach the top of the second hill, it would have gravitational potential energy $1.1mgh$, which is greater than its initial potential energy. If this roller coaster were actually built, the car would move up the second hill to a height h (neglecting friction), stop short of the peak, and then start rolling backward, becoming trapped between the two peaks.

Thinking Physics 4

Three identical balls are thrown from the top of a building, all with the same initial speed. Ball 1 is thrown horizontally, ball 2 is thrown at some angle above the horizontal, and ball 3 is thrown at some angle below the horizontal as in Figure 5.13. Neglecting air resistance, describe their motions and compare their speeds as they reach the ground.

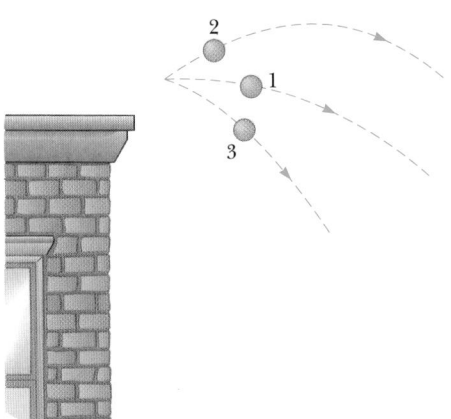

Figure 5.13 (Thinking Physics 4) Three identical balls are thrown with the same initial speed from the top of a building.

Explanation   Ball 1 and ball 3 speed up after they are thrown, while ball 2 first slows down and then speeds up after reaching its peak. The paths of all three are portions of parabolas. The three take different times to reach the ground. However, all have the same impact speed because all start with the same kinetic energy and undergo the same change in gravitational potential energy. That is, $E_{total} = \frac{1}{2}mv^2 + mgh$ is the same for all three balls.

Problem-Solving Strategy

Conservation of Energy

Take the following steps in applying the principle of conservation of energy:

1. Define your system, which may consist of more than one object.
2. Select a reference position for the zero point of gravitational potential energy.
3. Determine whether or not nonconservative forces are present.
4. If mechanical energy is conserved (that is, if only conservative forces are present), you can write the total initial energy at some point as the sum of the kinetic and potential energies at that point. Then, write an expression for the total final energy, $KE_f + PE_f$, at the final point of interest. Because mechanical energy is conserved, you can equate the two total energies and solve for the unknown.

EXAMPLE 5.4   The Daring Diver

A 755-N diver drops from a board 10.0 m above the water surface, as in Figure 5.14. (a) Use conservation of mechanical energy to find his speed 5.00 m above the water surface.

Reasoning   As the diver falls toward the water, only one force acts on him: the force of gravity. (This assumes that air resistance can be neglected.) Therefore, we can be assured that only the force of gravity does any work on him and that mechanical energy is conserved. In order to find the diver's speed at the 5-m mark, we will choose the water surface as the zero level for potential energy. Also, note that the diver drops from the board; that is, he leaves with zero velocity and therefore zero kinetic energy.

Solution   Conservation of mechanical energy, Equation 5.11, gives

$$\frac{1}{2}mv_i^2 + mgy_i = \frac{1}{2}mv_f^2 + mgy_f$$

$$0 + (755 \text{ N})(10.0 \text{ m}) = \frac{1}{2}(77.0 \text{ kg})v_f^2 + (755 \text{ N})(5.00 \text{ m})$$

$$v_f = \boxed{9.90 \text{ m/s}}$$

(b) Find the speed of the diver just before he strikes the water.

Solution   With the final position of the diver at the surface of the pool, Equation 5.11 gives

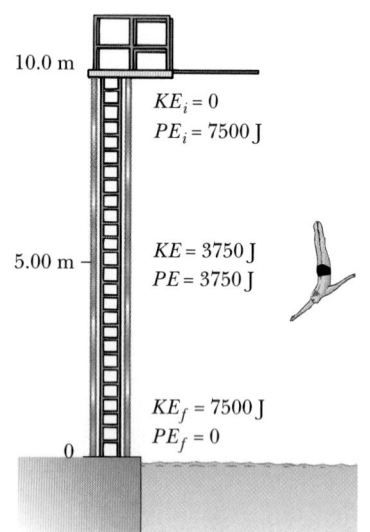

Figure 5.14   (Example 5.4) The kinetic energy and potential energy of a diver at various heights. The zero of potential energy is taken to be at the surface of the pool.

$$0 + (755 \text{ N})(10.0 \text{ m}) = \tfrac{1}{2}(77.0 \text{ kg})v_f^2 + 0$$

$$v_f = 14.0 \text{ m/s}$$

Exercise If the diver pushes off so that he leaves the board with an initial speed of 2.00 m/s, find his speed when he strikes the water. Use conservation of energy.

Answer 14.1 m/s

EXAMPLE 5.5 Sliding Down a Frictionless Hill

A sled and its rider together weigh 800 N. They move down a frictionless hill through a vertical distance of 10.0 m, as shown in Figure 5.15a. Use conservation of mechanical energy to find the speed of the sled–rider system at the bottom of the hill, assuming the rider pushes off with an initial speed of 5.00 m/s.

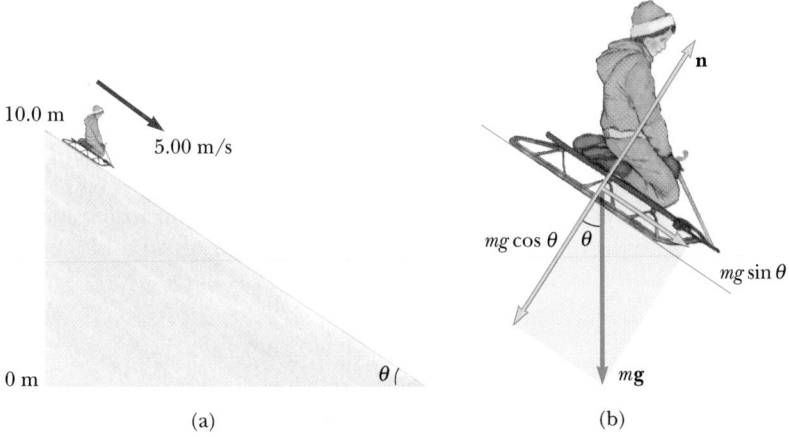

(a)   (b)

Figure 5.15 (Example 5.5) (a) A sled and rider start from the top of a frictionless hill with a speed of 5.00 m/s. (b) A free-body diagram for the system (sled plus rider).

Reasoning The forces acting on the sled and rider as they move down the hill are shown in Figure 5.15b. In the absence of a frictional force, the only forces acting are the normal force, **n**, and the gravitational force. At all points along the path, **n** is perpendicular to the direction of travel and hence does no work. Likewise, the component of weight perpendicular to the incline ($mg \cos \theta$) does no work. The only force that does any work is the component of the gravitational force, $mg \sin \theta$, along the slope of the hill. As a result, we are justified in using Equation 5.11.

Solution In this case the initial energy includes kinetic energy because of the initial speed:

$$\tfrac{1}{2}mv_i^2 + mgy_i = \tfrac{1}{2}mv_f^2 + mgy_f$$

or, after canceling m throughout the equation,

$$\tfrac{1}{2}v_i^2 + gy_i = \tfrac{1}{2}v_f^2 + gy_f$$

If we set the origin at the bottom of the incline, we see that the initial and final y coordinates are $y_i = 10.0$ m and $y_f = 0$. Thus we get

$$\tfrac{1}{2}(5.00 \text{ m/s})^2 + (9.80 \text{ m/s}^2)(10.0 \text{ m}) = \tfrac{1}{2}v_f^2 + (9.80 \text{ m/s}^2)(0)$$

$$v_f = \boxed{14.9 \text{ m/s}}$$

Notice that the solution of this problem does not depend on the shape of the hill. We only needed to know the rider's initial speed and the vertical distance traveled. If the hill had a varying curvature with bumps and troughs, the conservation of energy approach would yield the same result. If you try to solve the problem using Newton's laws, the varying slope of the hill would cause the acceleration to vary along the hill, making this approach very complicated. As you progress further with your study of physics, you will find many problems that are difficult or impossible to solve using Newton's laws, whereas conservation of energy provides a simple, yet powerful, alternative.

Exercise   If the sled and rider start at the bottom of the incline and are given an initial speed of 5.00 m/s up the incline, how high will they rise vertically, and what will their speed be when they return to the bottom of the hill?

Answer   1.28 m; 5.00 m/s

Potential Energy Stored in a Spring

5.3, SECTION 2

The concept of potential energy is of tremendous value in descriptions of certain types of mechanical motion. One of these, the motion of a mass attached to a stretched or compressed spring, is discussed more completely in Chapter 13, but it is instructive to examine it here.

Consider Figure 5.16a, which shows a spring in its equilibrium position — that is, the spring is neither compressed nor stretched. If we push a block against the

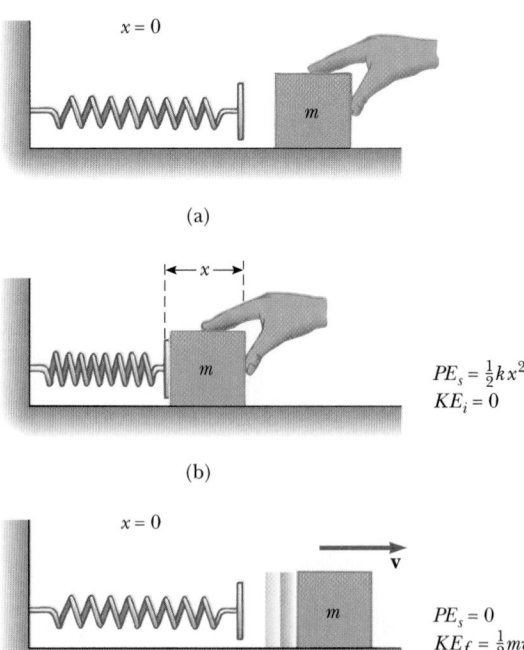

(a)

(b)

$$PE_s = \tfrac{1}{2}kx^2$$
$$KE_i = 0$$

(c)

$$PE_s = 0$$
$$KE_f = \tfrac{1}{2}mv^2$$

Figure 5.16   A block of mass m on a frictionless surface is pushed against a spring and released from rest. If x is the compression in the spring, as in (b), the potential energy stored in the spring is $\tfrac{1}{2}kx^2$. This energy is transferred to the block in the form of kinetic energy, as in (c).

spring, as in Figure 5.16b, the spring is compressed a distance of x. In order to compress the spring, we must exert on the block a force of magnitude $F = kx$, where k is a constant for a particular spring called the **spring constant.** For a flexible spring, k is a small number, whereas for a stiff spring, k is large. The equation $F = kx$ says that the more we compress the spring, the larger the force we must exert. Because the block is in equilibrium in Figure 5.16b, the spring must be exerting on it a force of $F_s = -kx$. This equation for F_s, describing the force exerted by a stretched or compressed spring, is often called **Hooke's law** after Sir Robert Hooke, who discovered the relationship between F_s and x.

To find an expression for the potential energy associated with the spring force, called the **elastic potential energy,** let us determine the work required to compress a spring from its equilibrium position to some final arbitrary position x. The force that must be applied to compress the spring varies from $F = 0$ to $F = kx$ at maximum compression. Because the force increases linearly with position (that is, $F \propto x$), the average force that must be applied is

$$\bar{F} = \frac{F_0 + F_x}{2} = \frac{0 + kx}{2} = \tfrac{1}{2}kx$$

Therefore, the work *done by the applied force* is

$$W = \bar{F}x = \tfrac{1}{2}kx^2$$

This work is stored in the compressed spring as elastic potential energy. Thus, we define the elastic potential energy, PE_s, as

$$PE_s \equiv \tfrac{1}{2}kx^2 \qquad\qquad [5.12]$$

Figure 5.16 shows how this stored elastic potential energy can be recovered. When the block is released, the spring snaps back to its original length and the stored elastic potential energy is converted to kinetic energy of the block. The elastic potential energy stored in the spring is zero when the spring is in equilibrium ($x = 0$). Note that energy is also stored in the spring when it is stretched and is given by Equation 5.12. Furthermore, the elastic potential energy is a maximum when the spring has reached its maximum compression or extension. Finally, the potential energy in the spring is always positive because it is proportional to x^2.

This new form of energy is included as another term in our equation for conservation of mechanical energy:

$$(KE + PE_g + PE_s)_i = (KE + PE_g + PE_s)_f$$

where PE_g is the gravitational potential energy.

Applying Physics 3

An old-fashioned manual phonograph, which you may have seen at an antique store, is a device that operates by use of stored elastic potential energy. A crank on the side of the record player was turned several times. This stored elastic potential energy in a spring in the motor mechanism. This energy was released through a series of gears to appear as kinetic energy of rotation of the turntable and record. Some of this energy ultimately left the phonograph by sound, allowing you to hear the information on the record. This same process is still used in nonelectric wind-up watches and wind-up toys.

APPLICATION

Wind-Up Devices.

5.9, SECTION 1

 EXAMPLE 5.6   A Block Projected on a Frictionless Surface

A 0.50-kg block rests on a horizontal, frictionless surface as in Figure 5.16. The block is pressed against a light spring having a spring constant of $k = 80.0$ N/m. The spring is compressed a distance of 2.0 cm and released. Consider the system to be the spring and block, and find the speed of the block when the block is at the $x = 0$ position.

Reasoning   Our expression for the conservation of mechanical energy is

$$(KE + PE_g + PE_s)_i = (KE + PE_g + PE_s)_f$$

The initial kinetic energy of the block is zero, and because the block remains at the same level throughout its motion, the gravitational potential energy terms on both sides of the equation are equal. Thus,

$$PE_{si} = KE_f$$

Solution   From the last expression we have

$$\tfrac{1}{2}kx_i^2 = \tfrac{1}{2}mv_f^2$$

$$\tfrac{1}{2}(80.0 \text{ N/m})(2.0 \times 10^{-2} \text{ m})^2 = \tfrac{1}{2}(0.50 \text{ kg})v_f^2$$

$$v_f = \boxed{0.25 \text{ m/s}}$$

5.6   NONCONSERVATIVE FORCES AND THE WORK–KINETIC ENERGY THEOREM

In realistic situations, nonconservative forces such as friction are usually present. In such situations, the total mechanical energy of the system is not constant, and one cannot apply Equation 5.10. In order to account for nonconservative forces, let us return to the work-kinetic energy theorem:

$$W_{\text{net}} = \tfrac{1}{2}mv_f^2 - \tfrac{1}{2}mv_i^2$$

Let us separate the net work into two parts, that done by the nonconservative forces, W_{nc}, and that done by the conservative forces, W_c. The work-kinetic energy relationship then becomes

$$W_{nc} + W_c = \tfrac{1}{2}mv_f^2 - \tfrac{1}{2}mv_i^2 \qquad [5.13]$$

Solving for W_{nc} and substituting the expression for W_c provided by Equation 5.9, we have

$$W_{nc} = \tfrac{1}{2}mv_f^2 - \tfrac{1}{2}mv_i^2 - (PE_i - PE_f)$$

$$W_{nc} = (KE_f + PE_f) - (KE_i + PE_i) \qquad [5.14]$$

The work done by all nonconservative forces equals the change in mechanical energy of the system.

5.7, SECTION 2

Thinking Physics 5

An automobile with kinetic energy, and potential energy carried within the gasoline, strikes a tree and comes to rest. The total energy in the system (the car) is now less than before. How did the energy leave the system, and in what form is energy left in the system?

Explanation   There are a number of ways that energy transferred out of the system. The automobile did some *work* on the tree during the collision, causing the tree to deform. There was a large crash during the collision, representing transfer of energy by *sound*. If the gasoline tank leaks after the collision, the automobile is losing energy by means of *mass transfer*. There will also be some *electromagnetic radiation*, because the car has a temperature, as well as the possibility of radiation from any sparks that occur during the collision. After the car has come to rest, there is no more kinetic energy. There is some *potential energy* left in any gas remaining within the tank. There is also more *internal energy* in the automobile, because its temperature is likely to be higher after the collision than before.

Problem-Solving Strategy

Conservation of Energy Revisited

If nonconservative forces such as friction act on a system, mechanical energy is not conserved. Hence, the problem-solving strategy given immediately before Example 5.4 must be modified as follows. First, write expressions for the total initial and total final mechanical energies. In this case, the difference between the two total energies is equal to the work done by the nonconservative force(s). These steps are summarized in Equation 5.14.

EXAMPLE 5.7   A Crate Sliding Down a Ramp

A 3.00-kg crate slides down a ramp at a loading dock. The ramp is 1.00 m long and inclined at an angle of 30.0°, as shown in Figure 5.17. The crate starts from rest at the top, experiences a constant frictional force of magnitude 5.00 N, and continues to move a short distance on the flat floor. Use energy methods to determine the speed of the crate when it reaches the bottom of the ramp.

Reasoning   The crate's initial energy is all in the form of gravitational potential energy, and its final energy is all kinetic. However, we cannot say that $PE_i = KE_f$, because there is an external nonconservative force that removes mechanical energy from the crate: the force of friction. Thus, we must use Equation 5.13 to find the final speed.

Solution   Because $v_i = 0$, the initial kinetic energy is zero. If the y coordinate is measured from the bottom of the incline, then $y_i = 0.500$ m. Therefore, the total mechanical energy of the crate at the top is all potential energy, given by

$$PE_i = mgy_i = (3.00 \text{ kg})(9.80 \text{ m/s}^2)(0.500 \text{ m}) = 14.7 \text{ J}$$

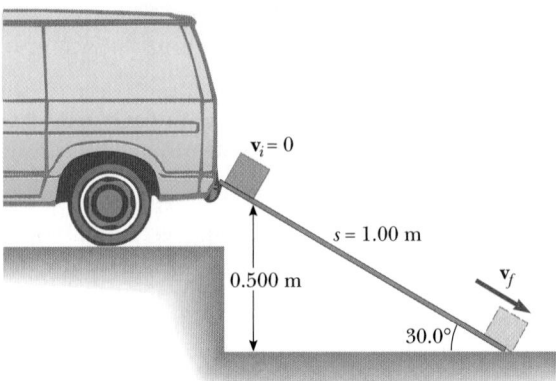

Figure 5.17   (Example 5.7) A crate slides down an incline under the influence of gravity. The potential energy of the crate decreases and its kinetic energy increases.

Workbook Problems 30 and 33 — (pages 65 and 71 of the Workbook) also deal with kinetic energy, potential energy, and nonconservative forces.

When the crate reaches the bottom, its potential energy is zero because its elevation is $y_f = 0$. Therefore, the total mechanical energy at the bottom is all kinetic energy:

$$KE_f = \tfrac{1}{2}mv_f^2$$

In this case, $W_{nc} = -fs$, where s is the displacement along the ramp. (Recall that the forces normal to the ramp do no work on the crate because they are perpendicular to the displacement.) With $f = 5.00$ N and $s = 1.00$ m, we have

$$W_{nc} = -fs = (-5.00\ \text{N})(1.00\ \text{m}) = -5.00\ \text{J}$$

This says that some mechanical energy is lost because of the presence of the retarding frictional force. Applying Equation 5.13 gives

$$-fs = \tfrac{1}{2}mv_f^2 - mgy_i$$

$$\tfrac{1}{2}mv_f^2 = 14.7\ \text{J} - 5.00\ \text{J} = 9.7\ \text{J}$$

$$v_f^2 = \frac{19.4\ \text{J}}{3.00\ \text{kg}} = 6.47\ \text{m}^2/\text{s}^2$$

$$v_f = \boxed{2.54\ \text{m/s}}$$

Exercise   Find the speed at the bottom of the ramp if the ramp surface is frictionless.

Answer $v_f = 3.13$ m/s

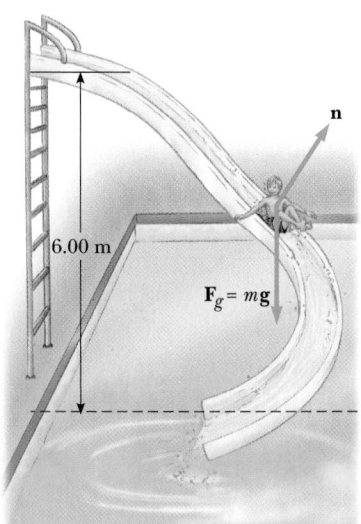

Figure 5.18   (Example 5.8) If the slide is frictionless, the speed of the child at the bottom depends on the height of the slide and is independent of its shape.

EXAMPLE 5.8   Fun on the Water Slide

A child of mass 20.0 kg takes a ride on an irregularly curved water slide of height 6.00 m, as in Figure 5.18. The child starts from rest at the top. Determine her speed at the bottom, assuming no friction is present.

Reasoning   The normal force does no work on the child because this force is always perpendicular to the displacement. Furthermore, because there is no friction, $W_{nc} = 0$ and we can apply the principle of conservation of mechanical energy.

Solution If we measure the y coordinate from the bottom of the slide, then $y_i = 6.00$ m, $y_f = 0$, and we get

$$KE_i + PE_i = KE_f + PE_f$$

$$0 + (20.0 \text{ kg})(9.80 \text{ m/s}^2)(6.00 \text{ m}) = \tfrac{1}{2}(20.0 \text{ kg})(v_f^2) + 0$$

$$v_f = \boxed{10.8 \text{ m/s}}$$

Note that this speed is the same as if the child had fallen vertically a distance of 6.00 m! From the free-fall equation, $v_f^2 = v_i^2 - 2gy$, we get the same result:

$$v_f = \sqrt{2(9.80 \text{ m/s}^2)(6.00 \text{ m})} = 10.8 \text{ m/s}$$

The effect of the slide is to direct the motion at the bottom horizontally rather than vertically.

Exercise If the child's speed at the bottom is 8.00 m/s rather than 10.8 m/s, how much mechanical energy is removed from the system?

Answer 536 J

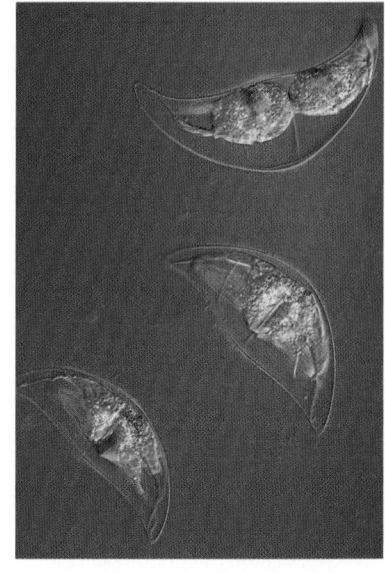

This small plant, found in warm Southern waters, exhibits bioluminescence, a process in which chemical energy is converted to light. The red areas are chlorophyll, which fluoresces when excited by blue light. (*Jan Hinsch/Science Photo Library/ Photo Researchers, Inc.*)

5.7 CONSERVATION OF ENERGY IN GENERAL

We can generalize the energy conservation principle to include all forces, both conservative and nonconservative, acting on a system. In the study of thermodynamics we shall find that mechanical energy can be transformed into internal energy of the system. For example, when a block slides over a rough surface, the mechanical energy lost is transformed into internal energy stored in the block and the surface, as evidenced by measurable increases in the temperatures of both. We shall see that, on an atomic scale, this internal energy is associated with the vibration of atoms about their equilibrium positions. Such internal atomic motion has kinetic and potential energy, and so one can say that frictional forces arise fundamentally from forces that are conservative at the atomic level. Therefore, if we include this increase in internal energy in our energy expression, total energy is conserved.

The principle of conservation of energy is not solely confined to physics. In biology, energy transformations take place in myriad ways inside bacteria. Examples include the transformation of chemical energy to mechanical energy that cause flagella to move and propel an organism. Some bacteria use chemical energy to produce light. Although the mechanisms that produce these light emissions are not well understood, living creatures often rely on this light for their existence. For example, certain fish have sacs beneath their eyes filled with light-emitting bacteria. The emitted light attracts other smaller creatures that become food for the fish.

If you analyze an isolated system, you will always find that its total energy does not change, as long as you account for all forms of energy. That is, **energy can never be created or destroyed. Energy may be transformed from one form into another, but the total energy of an isolated system is always constant.** From a universal point of view, we can say that **the total energy of the Universe is**

APPLICATION

Flagellar Movement; Bioluminescence.

◀ Total energy is always conserved.

constant: If one part of the Universe gains energy in some form, another part must lose an equal amount of energy. No violation of this principle has been found.

5.8   POWER

5.8

From a practical viewpoint, it is interesting to know not only the amount of energy transferred to or from a system, but also the rate at which the transfer occurred. **Power is defined as the time rate of energy transfer.**

Let us use our familiar concept of work as an example of energy transfer. If an external force is applied to an object and if the work done by this force is W in the time interval Δt, then the average power, $\bar{P}$, during this interval is defined as the ratio of the work done to the time interval:

$$\bar{P} = \frac{W}{\Delta t} \qquad [5.15]$$

It is sometimes useful to rewrite Equation 5.15 by substituting $W = F\Delta s$ and noting that $\Delta s / \Delta t$ is the average speed of the object during the time Δt:

Average power ▶

$$\bar{P} = \frac{W}{\Delta t} = \frac{F\Delta s}{\Delta t} = F\bar{v} \qquad . \qquad [5.16]$$

That is, the average power delivered either to or by an object is equal to the product of the force acting on the object during some time interval and the object's average speed during this time interval. In Equation 5.16, F is the component of force in the direction of the average velocity.

A result of the same form as Equation 5.16 is obtained for instantaneous values. That is, the instantaneous power delivered to an object is equal to the product of the force on the object at that instant and the instantaneous speed, or $P = Fv$.

The units of power in the SI system are joules per second, which are also called **watts** (W) after James Watt:

$$1\ \text{W} = 1\ \text{J/s} = 1\ \text{kg} \cdot \text{m}^2/\text{s}^3 \qquad [5.17]$$

The unit of power in the British engineering system is the horsepower (hp), in which

$$1\ \text{hp} \equiv 550\ \frac{\text{ft} \cdot \text{lb}}{\text{s}} = 746\ \text{W} \qquad [5.18]$$

The watt is commonly used in electrical applications, but it can be used in other scientific areas as well. For example, European sports-car engines are now rated in kilowatts.

We now define units of energy (or work) in terms of the unit of power, the watt. One kilowatt-hour (kWh) is the energy converted or consumed in 1 h at the constant rate of 1 kW = 1000 J/s. Therefore,

$$1\ \text{kWh} = (10^3\ \text{W})(3600\ \text{s}) = (10^3\ \text{J/s})(3600\ \text{s}) = 3.60 \times 10^6\ \text{J}$$

It is important to realize that a kilowatt-hour is a unit of energy, not power. When you pay your electric bill, you are buying energy, and the amount of electricity used by an appliance is usually expressed in multiples of kilowatt-hours. For example, an electric bulb rated at 100 W would "consume" 3.6×10^5 J of energy in 1 h.

Thinking Physics 6

A lightbulb is described by some individuals as "having 60 watts." What is wrong with this phrase?

Explanation The number given represents the power of the lightbulb, which is the rate at which energy passes through it, entering as electrical energy from the power source. It is not a value that the lightbulb "possesses," however. It is only the approximate power rating of the bulb when it is connected to a normal 120-volt household supply of electricity. If the bulb were connected to a 12-volt supply, energy would enter the bulb at a rate other than 60 watts. Thus, the 60 watts is not an intrinsic property of the bulb, such as density is for a given solid material.

EXAMPLE 5.9 Power Delivered by an Elevator Motor

A 1000-kg elevator carries a maximum load of 800 kg. A constant frictional force of 4000 N retards its motion upward, as in Figure 5.19. What minimum power, in kilowatts and in horsepower, must the motor deliver to lift the fully loaded elevator at a constant speed of 3.00 m/s?

Solution The motor must supply the force, **T,** that pulls the elevator upward. From Newton's second law and from the fact that **a** = 0 because **v** is constant, we get

$$T - f - Mg = 0$$

where M is the total mass (elevator plus load), equal to 1800 kg. Therefore,

$$
\begin{aligned}
T &= f + Mg \\
&= 4.00 \times 10^3 \text{ N} + (1.80 \times 10^3 \text{ kg})(9.80 \text{ m/s}^2) \\
&= 2.16 \times 10^4 \text{ N}
\end{aligned}
$$

From $P = Fv$ and the fact that **T** is in the same direction as **v**, we have

$$
\begin{aligned}
P &= Tv \\
&= (2.16 \times 10^4 \text{ N})(3.00 \text{ m/s}) = 6.48 \times 10^4 \text{ W} \\
&= \boxed{64.8 \text{ kW} = 86.9 \text{ hp}}
\end{aligned}
$$

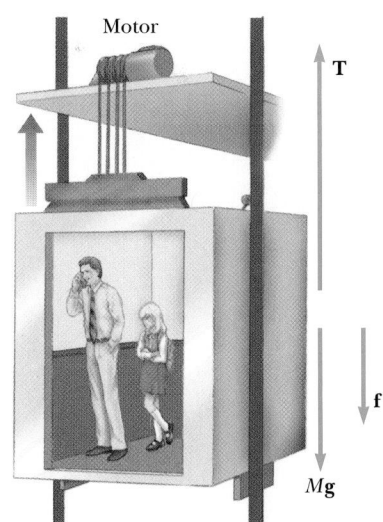

Figure 5.19 (Example 5.9) The motor provides an upward force, **T,** on the elevator. A frictional force, **f,** and the force of gravity, M**g,** act downward.

SUMMARY

The **work** done by a *constant* force, **F**, acting on an object is defined as the product of the component of the force in the direction of the object's displacement and the magnitude of the displacement. If the force makes an angle of θ with the displacement **s,** the work done by **F** is

$$W \equiv (F \cos \theta)s \qquad [5.1]$$

The **kinetic energy** of an object of mass m moving with speed v is defined as

$$KE \equiv \tfrac{1}{2}mv^2 \qquad [5.5]$$

The **work-kinetic energy theorem** states that the net work done on an object by external forces equals the change in kinetic energy of the object:

$$W_{\text{net}} = KE_f - KE_i \qquad [5.6]$$

The **gravitational potential energy** of an object of mass m that is elevated a distance of y above the Earth's surface is given by

$$PE \equiv mgy \qquad [5.7]$$

A force is **conservative** if the work it does on an object depends only on the initial and final positions of the object and not on the path taken between those positions. A force for which this is not true is said to be **nonconservative.**

The **principle of conservation of mechanical energy** states that if the only force acting on a system is conservative, total mechanical energy remains constant.

$$KE_i + PE_i = KE_f + PE_f \qquad [5.10]$$

The **elastic potential energy** stored in a spring is given by

$$PE_s \equiv \tfrac{1}{2}kx^2 \qquad [5.12]$$

where k is the spring constant and x is the distance the spring is compressed or extended from its unstretched (equilibrium) position.

Energy can never be created or destroyed. Energy may be transformed from one form into another, but the total energy of an isolated system is always constant. The work done by all nonconservative forces acting on a system equals the change in the total mechanical energy of the system:

$$W_{nc} = (KE_f + PE_f) - (KE_i + PE_i) \qquad [5.14]$$

Average power is the ratio of energy transfer to the time interval during which the transfer occurs. For work, this expression is

$$\overline{P} = \frac{W}{\Delta t} \qquad [5.15]$$

If a force, **F,** acts on an object moving with an average speed of $\bar{v}$, the average power delivered to the object is

$$\bar{P} = F\bar{v} \qquad\qquad [5.16]$$

MULTIPLE-CHOICE QUESTIONS

1. Bobby, of mass m, drops from a tree limb at the same time that Sally, also of mass m, begins her descent down a frictionless slide. The slide is in the shape of a quadrant of a circle. If they both start at the same height above the ground, which of the following is true about their kinetic energies as they reach the Earth?
   (a) The kinetic energy of Bobby is greater than that of Sally.
   (b) The kinetic energy of Sally is greater than that of Bobby.
   (c) Both have the same kinetic energy.
   (d) It is impossible to say from the information given.

2. A whale has a length of about 18 m and weighs about 40 000 N. Assume that its leap from the water carries him about half his height out of the water and that all the upward surge is achieved solely by its speed at the instant of leaving the water. How fast was it going as it left the water?
   (a) 9 m/s   (b) 13 m/s   (c) 18 m/s   (d) 176 m/s

(Multiple-Choice Question 2)
(Fred Felleman/Tony Stone Images)

vaulter associated with his muscles, tendons, and ligaments. For this problem, however, consider a simplified version of all these factors to get an estimate of how high a vaulter can ever hope to reach. To do this assume the vaulter's center of gravity is at a height of 1.1 m above the ground when he takes off. A reasonable estimate of the maximum takeoff speed is 11.0 m/s when carrying a pole. Use conservation of energy to estimate the maximum height. (*Hint:* The answer you will get is about a meter higher than the current record.)
   (a) 6.2 m   (b) 7.3 m   (c) 11.0 m   (d) 14.6 m

(Multiple-Choice Question 3)
(Estate of Harold Edgerton, courtesy of Palm Press, Inc.)

3. The analysis of a pole vault requires one to consider the kinetic energy of the runner, elastic potential energy of the pole, and the gravitational potential energy of the vaulter. Other factors include the internal energy of the

4. A 40.0-N crate starting at rest slides down a rough 6.0-m-long ramp, inclined at 30° with the horizontal. The force of friction between crate and ramp is 6.0 N.

What will be the speed of the crate at the bottom of the incline?

(a) 1.6 m/s   (b) 3.3 m/s   (c) 4.5 m/s   (d) 6.4 m/s

5. A worker pushes a wheelbarrow with a force of 50 N over a level distance of 5.0 m. If a frictional force of

43 N acts on the wheelbarrow in a direction opposite to that of the worker, what net work is done on the wheelbarrow?

(a) 250 J   (b) 215 J   (c) 35 J   (d) 10 J

CONCEPTUAL QUESTIONS

1. Consider a tug-of-war in which two teams pulling on a rope are evenly matched, so that no motion takes place. Is work done on the rope? On the pullers? On the ground? Is work done on anything?

2. Discuss whether any work is being done by each of the following agents and, if so, whether the work is positive or negative: (a) a chicken scratching the ground, (b) a person studying, (c) a crane lifting a bucket of concrete, (d) the force of gravity on the bucket in part (c), (e) the leg muscles of a person in the act of sitting down.

3. A team of furniture movers wishes to load a truck using a ramp from the ground to the rear of the truck. One of the movers claims that less work would be required to load the truck if the length of the ramp were increased, reducing the angle with respect to the horizontal. Is his claim valid? Explain.

4. (a) Can the kinetic energy of an object be negative? (b) Can the gravitational potential energy of an object be negative? Explain.

5. Roads going up mountains are formed into switchbacks with the road weaving back and forth along the face of the slope, such that there is only a gentle rise on any portion of the roadway. Does this require any less work to be done by an automobile climbing the mountain, compared to driving on a roadway that is straight up the slope? Why are switchbacks used?

6. (a) If the speed of a particle is doubled, what happens to its kinetic energy? (b) If the net work done on a particle is zero, what can be said about its speed?

7. As a simple pendulum swings back and forth, the forces acting on the suspended mass are the force of gravity, the tension in the supporting cord, and air resistance. (a) Which of these forces, if any, does no work on the pendulum? (b) Which of these forces does negative work at all times during its motion? (c) Describe the work done by the force of gravity while the pendulum is swinging.

8. A bowling ball is suspended from the ceiling of a lecture hall by a strong cord. The ball is drawn away from its equilibrium position and released from rest at the tip of the demonstrator's nose as in Figure Q5.8. If the dem-

onstrator remains stationary, explain why she is not struck by the ball on its return swing. Would this demonstrator be safe if the ball were given a slight push from its starting position at her nose?

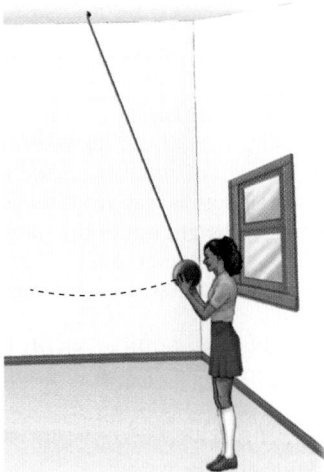

Figure Q5.8

9. An older model car accelerates from 0 to speed v in 10 seconds. A newer, more powerful sports car accelerates from 0 to $2v$ in the same time period. What is the ratio of powers expended by the two cars? Consider the energy coming from the engine to appear only as kinetic energy of the cars.

10. During a stress test of the cardiovascular system, a patient walks and runs on a treadmill. (a) Is the energy expended by the patient equivalent to the energy of walking and running on the ground? Explain. (b) If the treadmill is tilted upward, what effect does this have? Discuss.

11. In most situations we have encountered in this chapter, frictional forces tend to reduce the kinetic energy of an object. However, frictional forces can sometimes in-

crease an object's kinetic energy. Describe a few situations in which friction causes an increase in kinetic energy.

12. Discuss the energy transformations that occur during the pole vault event pictured in the multiflash photograph shown on page 139. Ignore rotational motion.

13. A weight is connected to a spring that is suspended vertically from the ceiling. If the weight is displaced downward from its equilibrium position and released, it will oscillate up and down. If air resistance is neglected, will the total energy of the system (weight plus spring) be conserved? How many forms of potential energy are there for this situation?

14. The driver of a car slams on her brakes to avoid colliding with a deer crossing the highway. What happens to the car's kinetic energy as it comes to rest?

15. You are reshelving books in a library. You lift a book from the floor to the top shelf. The kinetic energy of the book on the floor was zero, and the kinetic energy of the book on the top shelf is zero, so there is no change in kinetic energy. Yet you did some work in lifting the book. Is the work-kinetic energy theorem violated?

16. A ball is thrown straight up into the air. At what position is its kinetic energy a maximum? At what position is the gravitational potential energy a maximum?

17. An Earth satellite is in a circular orbit at an altitude of 500 km. Explain why the work done by the gravitational force acting on the satellite is zero. Using the work-kinetic energy theorem, what can you say about the speed of the satellite?

PROBLEMS

1, 2, 3 = straightforward, intermediate, challenging ☐ = full solution available in Study Guide/Student Solutions Manual ![]= Core Concepts Workbook
WEB = solution posted at **http://www.harcourtcollege.com/physics/cptech** ![]= biomedical application ![]= Interactive Physics

Review Problem

A 10-kg mass is being pulled down an incline by a force $F = 80$ N parallel to the incline as indicated in the figure. The mass starts from rest, and a frictional force of 20 N opposes this motion. When the mass has moved 4.0 m, find its speed by using (a) energy concepts and (b) Newton's second law of motion.

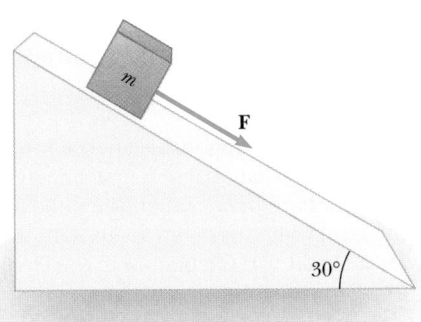

Figure P5.1 *(Gerard Vandystadt/Photo Researchers)*

2. If a woman lifts a 20.0-kg bucket from a well and does 6.00 kJ of work, how deep is the well? Assume that the speed of the bucket remains constant as it is lifted.

3. A tugboat exerts a constant force of 5.00×10^3 N on a ship moving at constant speed through a harbor. How much work does the tugboat do on the ship if each moves a distance of 3.00 km?

4. A shopper in a supermarket pushes a cart with a force of 35 N directed at an angle of 25° downward from the horizontal. Find the work done by the shopper as he moves down a 50-m length of aisle.

Section 5.1 Work

![] 1. A weight lifter lifts a 350-N set of weights from ground level to a position over his head, a vertical distance of 2.00 m. How much work does the weight lifter do, assuming he moves the weights at constant speed?

Problem 1 is similar to Workbook Problem 26 (Workbook page 58).

5. Starting from rest, a 5.0-kg block slides 2.5 m down a rough 30.0° incline. The coefficient of kinetic friction between the block and the incline is $\mu_k = 0.436$. Determine (a) the work done by the force of gravity, (b) the work done by the friction force between block and incline, and (c) the work done by the normal force.

6. A block of mass 2.50 kg is pushed 2.20 m along a frictionless horizontal table by a constant 16.0 N force directed 25.0° below the horizontal. Determine the work done by (a) the applied force, (b) the normal force exerted by the table, (c) the force of gravity, and (d) the net force on the block.

Section 5.2   Kinetic Energy and the Work–Kinetic Energy Theorem

7. A mechanic pushes a 2.50×10^3-kg car from rest to a speed of v, doing 5000 J of work in the process. During this time, the car moves 25.0 m. Neglecting friction between car and road, find (a) v and (b) the horizontal force exerted on the car.

8. A 7.00-kg bowling ball moves at 3.00 m/s. How fast must a 2.45-g Ping-Pong ball move so that the two balls have the same kinetic energy?

9. A 70.0-kg base runner begins his slide into second base when moving at a speed of 4.0 m/s. The coefficient of friction between his clothes and Earth is 0.70. He slides so that his speed is zero just as he reaches the base. (a) How much mechanical energy is lost due to friction acting on the runner? (b) How far does he slide?

10. A 0.60-kg particle has a speed of 2.0 m/s at point A and kinetic energy of 7.5 J at point B. What is (a) its kinetic energy at A? (b) its speed at point B? (c) the total work done on the particle as it moves from A to B?

WEB 11. A person doing a chin-up weighs 700 N exclusive of her arms. During the first 25.0 cm of the lift, each arm exerts an upward force of 355 N on the torso. If the upward movement starts from rest, what is her speed at this point?

12. An outfielder throws a 0.150-kg baseball at a speed of 40.0 m/s and an initial angle of 30.0°. What is the kinetic energy of the ball at the highest point of its motion?

13. A 2.0-g bullet leaves the barrel of a gun at a speed of 300 m/s. (a) Find its kinetic energy. (b) Find the average force exerted on the bullet by the expanding gases as the bullet moves the length of the 50-cm-long barrel.

14. A 10.0-kg crate is pulled up a rough 20° incline by a 100-N force parallel to the incline. The initial speed of the crate is 1.50 m/s, the coefficient of kinetic friction is 0.40, and the crate is pulled a distance of 5.00 m. Determine how much work is done by (a) the gravitational force, and (b) the 100-N force? (c) What is the change in kinetic energy of the crate? (d) What is the speed of the crate after it is pulled 5.00 m?

15. A 2000-kg car moves down a level highway under the actions of two forces. One is a 1000-N forward force exerted on the drive wheels by the road; the other is a 950-N resistive force. Use the work–kinetic energy theorem to find the speed of the car after it has moved a distance of 20 m, assuming it starts from rest.

16. On a frozen pond, a 10-kg sled is given a kick that imparts to it an initial speed of $v_0 = 2.0$ m/s. The coefficient of kinetic friction between sled and ice is $\mu_k = 0.10$. Use the work–kinetic energy theorem to find the distance the sled moves before coming to rest.

Section 5.3   Potential Energy

Section 5.4   Conservative and Nonconservative Forces

17. A 1.00×10^3-kg roller-coaster car is initially at the top of a rise, at point A. It then moves 50.0 m at an angle of 40.0° below the horizontal to a lower point, B. (a) Choosing point B as the zero level for gravitational potential energy, find the potential energy of the car at A and B and the difference in potential energy between these points. (b) Repeat part (a), choosing point A as the zero reference level.

18. A 2.00-kg ball is attached to a ceiling by a 1.00-m-long string. The height of the room is 3 m. What is the gravitational potential energy of the ball relative to (a) the ceiling? (b) the floor? (c) a point at the same elevation as the ball?

19. A 2.0-m-long pendulum is released from rest when the support string is at an angle of 25° with the vertical. What is the speed of the bob at the bottom of the swing?

20. A softball pitcher rotates a 0.250-kg ball around a vertical circular path of radius 0.6 m before releasing it. The pitcher exerts a 30.0-N force directed parallel to the motion of the ball around the complete circular path. The speed of the ball at the top of the circle is 15.0 m/s. If the ball is released at the bottom of the circle, what is its speed on release?

21. A 40-N child is in a swing that is attached to ropes 2.0 m long. Find the gravitational potential energy of the child relative to her lowest position (a) when the ropes are horizontal, (b) when the ropes make a 30° angle with the vertical, and (c) at the bottom of the circular arc.

Section 5.5   Conservation of Mechanical Energy

22. A 50-kg pole vaulter running at 10 m/s vaults over the bar. Her speed when she is over the bar is 1.0 m/s. Neglect air resistance, as well as any energy absorbed by the pole, and determine her altitude as she crosses the bar.

WEB 23. A child and sled with a combined mass of 50.0 kg slide down a frictionless hill. If the sled starts from rest and has a speed of 3.00 m/s at the bottom, what is the height of the hill?

Problem 5 is similar to Workbook Problem 25 (Workbook page 56).

24. A 0.400-kg bead slides on a curved wire, starting from rest at point A in Figure P5.24. If the wire is frictionless, find the speed of the bead **(a)** at B and **(b)** at C.

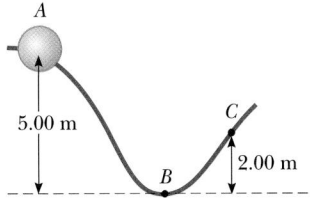

Figure P5.24

25. A 200-g particle is released from rest at point A on the inside of a smooth hemispherical bowl of radius $R = 30.0$ cm (Fig. P5.25). Calculate **(a)** its gravitational potential energy at A relative to B, **(b)** its kinetic energy at B, **(c)** its speed at B, and **(d)** its potential energy at C relative to B and its kinetic energy at C.

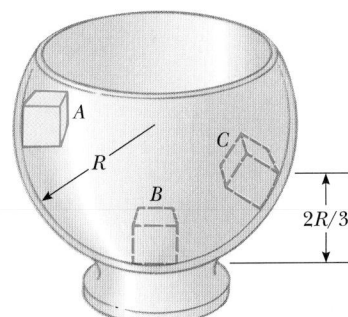

Figure P5.25

26. Tarzan swings on a 30.0-m-long vine initially inclined at an angle of 37.0° with the vertical. What is his speed at the bottom of the swing **(a)** if he starts from rest? **(b)** if he pushes off with a speed of 4.00 m/s?

27. Three masses, $m_1 = 5.0$ kg, $m_2 = 10.0$ kg, and $m_3 = 15.0$ kg, are attached by strings over frictionless pulleys

Figure P5.27 (Problems 27 and 38)

Problem 28 is the same as Workbook Problem 29 (Workbook page 62).

as indicated in Figure P5.27. The horizontal surface is frictionless and the system is released from rest. Using energy concepts, find the speed of m_3 after it moves down 4.0 m.

28. The launching mechanism of a toy gun consists of a spring of unknown spring constant, as shown in Figure P5.28a. If the spring is compressed a distance of 0.120 m and the gun fired vertically as shown, the gun can launch a 20.0-g projectile from rest to a maximum height of 20.0 m above the starting point of the projectile. Neglecting all resistive forces, determine **(a)** the spring constant and **(b)** the speed of the projectile as it moves through the equilibrium position of the spring (where $x = 0$), as shown in Figure P5.28b.

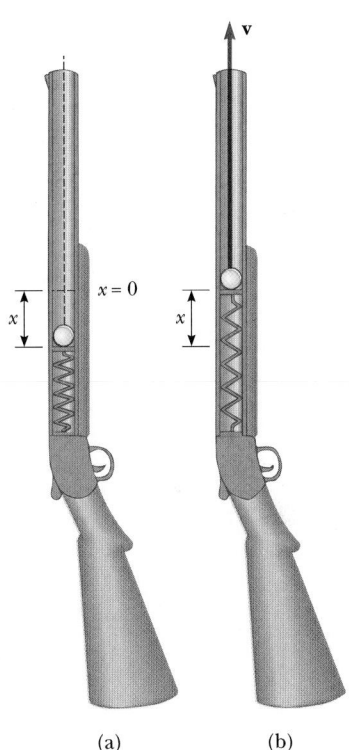

(a)   (b)

Figure P5.28

29. A projectile is launched with a speed of 40 m/s at an angle of 60° above the horizontal. Find the maximum height reached by the projectile during its flight by using conservation of energy.

30. A 0.250-kg block is placed on a vertical spring ($k = 5.00 \times 10^3$ N/m) and pushed downward, compressing the spring 0.100 m. As the block is released, it leaves the spring and continues to travel upward. What height above the point of release will the block reach if air resistance is negligible?

Section 5.6 Nonconservative Forces and the Work–Kinetic Energy Theorem

31. If the wire in Problem 24 (Fig. P5.24) is frictionless between points A and B and rough between B and C, and if the 0.400-kg bead starts from rest at A, **(a)** find its speed at B. **(b)** If the bead comes to rest at C, find the loss in mechanical energy as it goes from B to C.

32. An 80.0-N box is pulled 20.0 m up a 30° incline by an applied force of 100 N that points upward, parallel to the incline. If the coefficient of kinetic friction between box and incline is 0.220, calculate the change in the kinetic energy of the box.

WEB 33. A 70-kg diver steps off a 10-m tower and drops, from rest, straight down into the water. If he comes to rest 5.0 m beneath the surface, determine the average resistance force exerted on him by the water.

34. An airplane of mass 1.5×10^4 kg is moving at 60 m/s. The pilot then revs up the engine so that the forward thrust of the propeller becomes 7.5×10^4 N. If the force of air resistance has a magnitude of 4.0×10^4 N, find the speed of the airplane after it has traveled 500 m. Assume that the airplane is in level flight throughout this motion.

35. A 2.10×10^3-kg car starts from rest at the top of a 5.0-m-long driveway that is sloped at 20° with the horizontal. If an average friction force of 4.0×10^3 N impedes the motion, find the speed of the car at the bottom of the driveway.

36. A 25.0-kg child on a 2.00-m-long swing is released from rest when the swing supports make an angle of 30.0° with the vertical. **(a)** Neglecting friction, find the child's speed at the lowest position. **(b)** If the speed of the child at the lowest position is 2.00 m/s, what is the mechanical energy lost due to friction?

37. Starting from rest, a 10.0-kg block slides 3.00 m down a frictionless ramp (inclined at 30.0° from the floor) to the bottom. The block then slides an additional 5.00 m along the floor before coming to a stop. Determine **(a)** the speed of the block at the bottom of the ramp, **(b)** the coefficient of kinetic friction between block and floor, and **(c)** the mechanical energy lost due to friction.

38. Three masses, $m_1 = 5.0$ kg, $m_2 = 10.0$ kg, and $m_3 = 15.0$ kg, are attached by strings over frictionless pulleys, as indicated in Figure P5.27. The horizontal surface exerts a force of friction of 30 N on m_2. If the system is released from rest, use energy concepts to find the speed of m_3 after it moves 4.0 m.

39. A skier starts from rest at the top of a hill that is inclined at 10.5° with the horizontal. The hillside is 200 m long, and the coefficient of friction between snow and skis is 0.0750. At the bottom of the hill, the snow is level and the coefficient of friction is unchanged. How far does the skier move along the horizontal portion of the snow before coming to rest?

40. In a circus performance, a monkey is strapped to a sled and both are given an initial speed of 4.0 m/s up a 20° inclined track. The combined mass of monkey and sled is 20 kg, and the coefficient of kinetic friction between sled and incline is 0.20. How far up the incline do the monkey and sled move?

Section 5.8 Power

41. A skier of mass 70 kg is pulled up a slope by a motor-driven cable. **(a)** How much work is required to pull him 60 m up a 30° slope (assumed frictionless) at a constant speed of 2.0 m/s? **(b)** How many horsepower must a motor have to perform this task?

42. A 50.0-kg student climbs a 5.00-m-long rope and stops at the top. **(a)** What must her average speed be in order to match the power output of a 200-W lightbulb? **(b)** How much work does she do?

43. Water flows over a section of Niagara Falls at the rate of 1.2×10^6 kg/s and falls 50 m. How much power is generated by the falling water?

Figure P5.43 *(Ron Dorman/FPG)*

44. A 1.50×10^3-kg car accelerates uniformly from rest to 10.0 m/s in 3.00 s. Find **(a)** the work done on the car in this time interval, **(b)** the average horsepower delivered by the engine in this time interval, and **(c)** the instantaneous horsepower delivered by the engine at $t = 2.00$ s.

45. A 1.50×10^3-kg car starts from rest and accelerates uniformly to 18.0 m/s in 12.0 s. Assume that air resistance remains constant at 400 N during this time. Find **(a)** the average power developed by the engine and **(b)** the instantaneous power output of the engine at $t = 12.0$ s.

46. A 650-kg elevator starts from rest. It moves upward for 3.00 s with constant acceleration until it reaches its cruising speed, 1.75 m/s. **(a)** What is the average power of the elevator motor during this period? **(b)** How does this compare with its power during an upward cruise with constant speed?

Problem 45 is similar to Workbook Problem 32 (Workbook page 69).
Problem 46 is the same as Workbook Problem 31 (Workbook page 67).

47. While running, a person dissipates about 0.60 J of mechanical energy per step per kilogram of body mass. If a 60-kg person develops a power of 70 W during a race, how fast is the person running? Assume that a running step is 1.5 m long.

ADDITIONAL PROBLEMS

48. A car traveling at 50 km/h skids a distance of 35 m after its brakes lock. Estimate how far it will skid if its brakes lock when its initial speed is 100 km/h. What happens to the car's kinetic energy as it comes to rest?

49. A 98.0-N grocery cart is pushed 12.0 m by a shopper who exerts a constant horizontal force of 40.0 N. If all frictional forces are neglected and the cart starts from rest, what is its final speed?

50. (a) A 75-kg man steps out a window and falls (from rest) 1.0 m to a sidewalk. What is his speed just before his feet strike the pavement? (b) If the man falls with his knees and ankles locked, the only cushion for his fall is an approximately 0.50-cm give in the pads of his feet. Calculate the average force exerted on him by the ground in this situation. This average force is sufficient to cause cartilage damage in the joints or to break bones.

WEB **51.** A toy gun uses a spring to project horizontally a 5.3-g soft rubber sphere. The spring constant is 8.0 N/m, the barrel of the gun is 15 cm long, and a constant frictional force of 0.032 N exists between barrel and projectile. With what speed does the projectile leave the barrel if the spring was compressed 5.0 cm for this launch?

52. Two masses are connected by a light string passing over a light, frictionless pulley as in Figure P5.52. The 5.00-kg mass is released from rest at a point 4.00 m above the floor. (a) Determine the speed of each mass when the two pass each other. (b) Determine the speed of each mass at the moment the 5.00-kg mass hits the floor. (c) How much higher does the 3.00-kg mass travel after the 5.00-kg mass hits the floor?

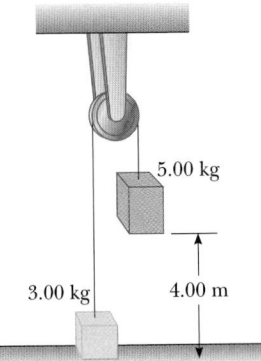

Figure P5.52

53. Two blocks, A and B (with mass 50 kg and 100 kg, respectively), are connected by a string, as shown in Figure P5.53. The pulley is frictionless and of negligible mass. The coefficient of kinetic friction between block A and the incline is $\mu_k = 0.25$. Determine the change in the kinetic energy of block A as it moves from C to D, a distance of 20 m up the incline if the system starts from rest.

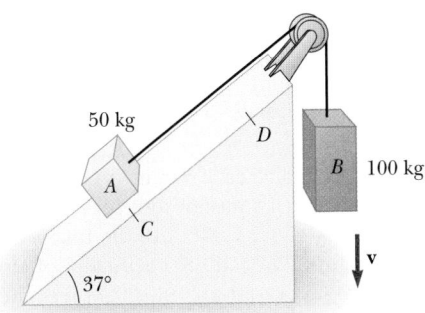

Figure P5.53

54. The light horizontal spring in Figure 5.16 has a force constant of $k = 100$ N/m. A 2.00-kg block is pressed against one end of the spring, compressing it 0.100 m. After the block is released, it starts from rest and moves 0.250 m to the right of its starting position before coming to rest. Show that the coefficient of kinetic friction between the horizontal surface and the block is 0.102.

55. A catcher "gives" with the ball when he catches a 0.15-kg baseball moving at 25 m/s. (a) If he moves his glove a distance of 2.0 cm, what is the average force acting on his hand? (b) Repeat for the case in which his glove and hand move 10 cm.

56. A 2.0×10^3-kg car starts from rest and accelerates along a horizontal roadway to $+20$ m/s in 15 s. Assume that air resistance remains constant at -500 N during this time. Find (a) the average power developed by the engine and (b) the instantaneous power developed at $t = 15$ s.

57. A ski jumper starts from rest 50.0 m above the ground on a frictionless track, and flies off the track at an angle of 45.0° above the horizontal and at a height of 10.0 m from the ground. Neglect air resistance. (a) What is his speed when he leaves the track? (b) What is the maximum altitude he attains after leaving the track? (c) Where does he land relative to the end of the track?

58. A spring of length 0.80 m rests along a frictionless 30° incline, as in Figure P5.58a. A 2.0-kg mass, at rest against the end of the spring, compresses the spring by 0.10 m (Fig. P5.58b). (a) Determine the spring constant k. (b) The mass is pushed down, compressing the spring an *additional* 0.60 m, and then released. If the incline is

2.0 m long, determine how far beyond the rightmost edge of the incline the mass lands.

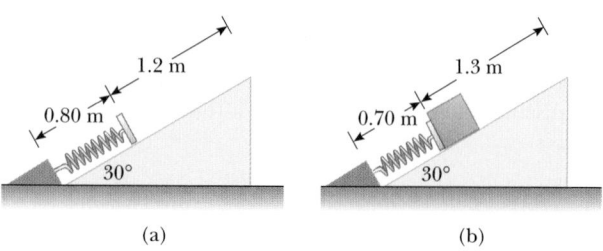

(a)   (b)

Figure P5.58

59. A 5.0-kg block is pushed 3.0 m up a vertical wall with constant speed by a constant force of magnitude F applied at an angle of $\theta = 30°$ with the horizontal, as shown in Figure P5.59. If the coefficient of kinetic friction between block and wall is 0.30, determine the work done by (a) **F**, (b) gravity, and (c) the normal force between block and wall. (d) By how much does the gravitational potential energy of the block increase?

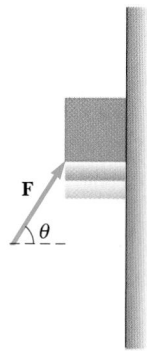

Figure P5.59

60. A 0.300-kg piece of putty is slowly placed on the top of a massless vertical spring of constant 19.6 N/m. (a) Determine the maximum compression of the spring. (b) What is the answer to part (a) if the putty is dropped onto the spring from a height of 0.500 m from the top of the spring?

61. The masses of the javelin, discus, and shot are 0.80 kg, 2.0 kg, and 7.2 kg, respectively, and record throws in the corresponding track events are about 89 m, 69 m, and 21 m, respectively. Neglecting air resistance, (a) calculate the minimum initial kinetic energies that would produce these throws, and (b) estimate the average force exerted on each object during the throw, assuming the force acts over a distance of 2.0 m. (c) Do your results suggest that air resistance is an important factor?

62. Two identical massless springs of constant $k = 200$ N/m are fixed at opposite ends of a level track, as shown in Figure P5.62. A 5.00-kg block is pressed against the left spring, compressing it by 0.150 m. The block (initially at rest) is then released, as shown in Figure P5.62a. The entire track is frictionless *except* for the section between A and B. Given that the coefficient of kinetic friction between block and track along AB is $\mu_k = 0.0800$, and given that the length AB is 0.250 m, (a) determine the maximum compression of the spring on the right (see Fig. P5.62b). (b) Determine where the block eventually comes to rest, as measured from A (see Fig. P5.62c).

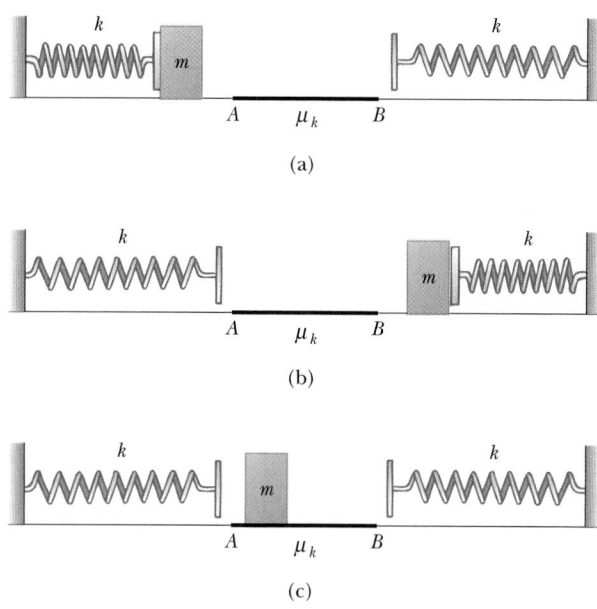

(a)

(b)

(c)

Figure P5.62

63. Jane, whose mass is 50.0 kg, needs to swing across a river filled with man-eating crocodiles in order to save Tarzan from danger. However, she must swing into a *constant* horizontal wind force, **F,** on a vine that is initially at an angle of θ with the vertical (see Fig. P5.63). $D = 50.0$ m, $F = 110$ N, $L = 40.0$ m, and $\theta = 50.0°$. (a) With what minimum speed must Jane begin her swing in order to just make it to the other side? (*Hint:* First determine the potential energy associated with the wind force. Because the wind force is constant, use an analogy with the constant gravitational force.) (b) Once the rescue is complete, Tarzan and Jane must swing back across the river. With what minimum speed must they begin their swing? Assume that Tarzan has a mass of 80.0 kg.

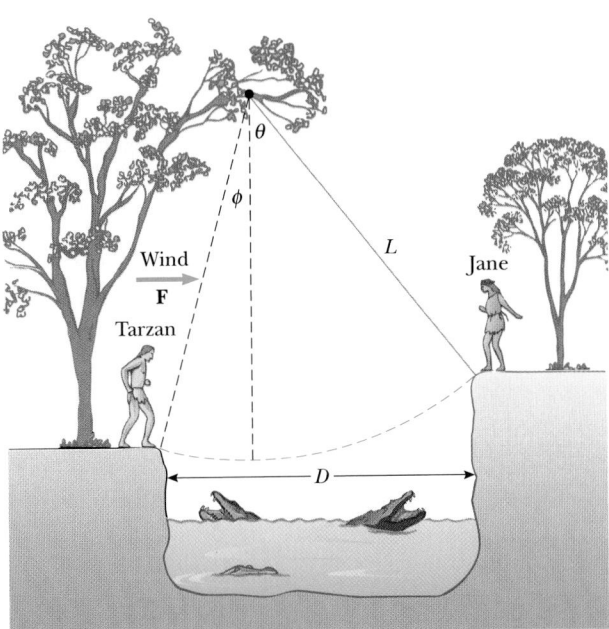

Figure P5.63

64. The ball launcher in a pinball machine has a spring that has a force constant of 1.20 N/cm (Fig. P5.64). The surface on which the ball moves is inclined 10.0° with respect to the horizontal. If the spring is initially compressed 5.00 cm, find the launching speed of a 0.100-kg ball when the plunger is released. Friction and the mass of the plunger are negligible.

10.0°

Figure P5.64

65. Suppose a car is modeled as a cylinder with cross-sectional area A moving with a speed v, as in Figure P5.65. In a time Δt, a column of air of mass Δm must be

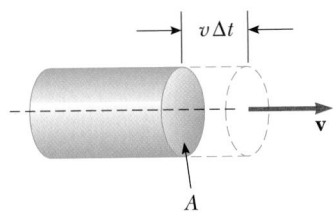

$v \Delta t$

$\mathbf{v}$

A

Figure P5.65

moved a distance $v \Delta t$ and hence must be given a kinetic energy $\frac{1}{2}(\Delta m) v^2$. Using this model, show that the power loss due to air resistance is $\frac{1}{2}\rho A v^3$ and the resistive force is $\frac{1}{2}\rho A v^2$, where ρ is the density of air.

66. An 80.0-kg skydiver jumps out of an airplane at an altitude of 1000 m and opens the parachute at an altitude of 200.0 m. (a) Assuming that the total retarding force on the diver is constant at 50.0 N with the parachute closed and constant at 3600 N with the parachute open, what is the speed of the diver when he lands on the ground? (b) Do you think the skydiver will get hurt? Explain. (c) At what height should the parachute be opened so that the final speed of the skydiver when he hits the ground is 5.00 m/s? (d) How realistic is the assumption that the total retarding force is constant? Explain.

67. A child's pogo stick (Fig. P5.67) stores energy in a spring ($k = 2.50 \times 10^4$ N/m). At position A ($x_1 = -0.100$ m), the spring compression is a maximum and the child is momentarily at rest. At position B ($x = 0$), the spring is relaxed and the child is moving upward. At position C, the child is again momentarily at rest at the top of the jump. Assuming that the combined mass of child and pogo stick is 25.0 kg, (a) calculate the total energy of the system if both potential energies are zero at $x = 0$, (b) determine x_2, (c) calculate the speed of the child at $x = 0$, (d) determine the value of x for which the kinetic energy of the system is a maximum, and (e) obtain the child's maximum upward speed.

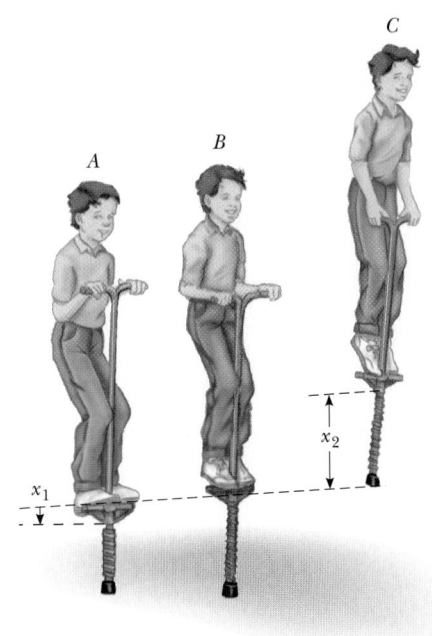

Figure P5.67

68. A 2.00-kg block situated on a rough incline is connected to a spring of negligible mass having a spring constant of 100 N/m (Fig. P5.68). The block is released from rest when the spring is unstretched, and the pulley is frictionless. The block moves 20.0 cm down the incline before coming to rest. Find the coefficient of kinetic friction between block and incline.

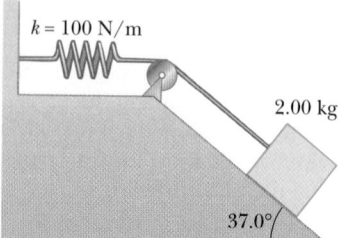

$k = 100$ N/m

2.00 kg

37.0°

Figure P5.68

69. In the dangerous "sport" of bungee jumping, a daring student jumps from a balloon with a specially designed elastic cord attached to his ankles, as shown in Figure P5.69. The unstretched length of the cord is 25.0 m, the student weighs 700 N, and the balloon is 36.0 m above the surface of a river below. Calculate the required force constant of the cord if the student is to stop safely 4.00 m above the river.

Figure P5.69 Bungee jumping. *(Gamma)*

70. An object of mass m is suspended from the top of a cart by a string of length L, as in Figure P5.70a. The cart and object are initially moving to the right at constant speed v_0. The cart comes to rest after colliding and sticking to a bumper, as in Figure P5.70b, and the suspended object swings through an angle θ. (a) Show that the initial speed $v_0 = \sqrt{2gL(1 - \cos\theta)}$. (b) If $L = 1.20$ m and $\theta = 35.0°$, find the initial speed of the cart. (*Hint:* The force exerted by the string on the object does no work on the object.)

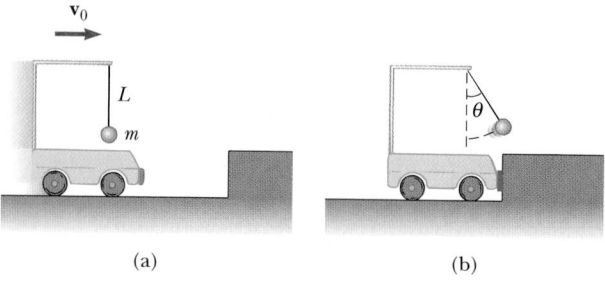

$\mathbf{v}_0$

L

m

θ

(a)   (b)

Figure P5.70

6

Momentum and Collisions

▲ PHYSICS PUZZLER

As these charging rams crash together head-on, they exert equal and opposite forces on each other and are nearly at rest after the collision. Both rams have kinetic energy before colliding, but their kinetic energy is almost zero after colliding. What happens to the kinetic energy of the rams? *(Stephen Krasemann/ Tony Stone Images)*

6.1

onsider what happens when a golf ball is struck by a club. As you well know, the ball receives a high velocity and goes a long distance because of the force exerted on it by the club. But why doesn't the club also receive a high velocity? It has the same force acting on it. Why does a good follow-through on the swing help the ball to go farther? Does the deformation of the ball on contact have any effect on its subsequent motion?

As a first step in answering such questions, we shall introduce the term *momentum*, a term that may have slightly different meaning in physics than it does in everyday life. For example, a candidate moving rapidly in the opinion polls is said to be gaining momentum, but from the standpoint of physics, is she? As another example, a very massive football player is said to have a great deal of momentum as he runs down the field. Can a much less massive player have the same momentum? What must he do to achieve this momentum? Can a common housefly have the same momentum as the charging rams in the chapter opener?

The concept of momentum will lead us to a second conservation law: conservation of momentum. This law is especially useful for treating problems that involve collisions between objects.

6.1   MOMENTUM AND IMPULSE

The linear momentum of an object of mass *m* moving with a velocity v is defined as the product of the mass and the velocity:

Linear momentum ▶

$$\mathbf{p} \equiv m\mathbf{v} \qquad [6.1]$$

As its definition shows, momentum is a vector quantity, with its direction matching that of the velocity. Momentum has dimensions ML/T, and its SI units are kilogram-meters per second (kg·m/s).

6.2, SECTION 1

Often we shall find it advantageous to work with the components of momentum. For two-dimensional motion, these are

$$p_x = mv_x \qquad p_y = mv_y$$

where p_x represents the momentum of an object in the *x* direction and p_y its momentum in the *y* direction.

6.3

The definition of momentum in Equation 6.1 coincides with our everyday use of the word. When we think of a massive object moving with a high velocity, we often say that the object has a large momentum, in accordance with Equation 6.1. Likewise, a small object moving slowly is said to have a small momentum. On the other hand, a small object moving with a high velocity can have a large momentum.

When Newton first expressed the second law in mathematical form, he wrote it not as $\mathbf{F} = m\mathbf{a}$ but as

Newton's second law ▶

$$\mathbf{F} = \frac{\text{change in momentum}}{\text{time interval}} = \frac{\Delta \mathbf{p}}{\Delta t} \qquad [6.2]$$

where Δt is the time interval during which the momentum changes by $\Delta \mathbf{p}$. This equation states that **the time rate of change of momentum of an object is equal to the net force acting on the object.** To see that this is equivalent to $\mathbf{F} = m\mathbf{a}$ for an object of constant mass, consider a constant force, **F,** acting on a particle and producing a constant acceleration. We can write Equation 6.2 as

$$\mathbf{F} = \frac{\Delta \mathbf{p}}{\Delta t} = \frac{m\mathbf{v}_f - m\mathbf{v}_i}{\Delta t} = \frac{m(\mathbf{v}_f - \mathbf{v}_i)}{\Delta t} \qquad [6.3]$$

Now recall that the velocity of an object moving with constant acceleration varies with time as

$$\mathbf{v}_f = \mathbf{v}_i + \mathbf{a}t$$

If we take $\Delta t = t$ and substitute for v_f in Equation 6.3, we see that **F** reduces to the familiar equation

$$\mathbf{F} = m\mathbf{a}$$

Note from Equation 6.2 that if the net force **F** is zero, the momentum of the object does not change. In other words, the linear momentum and velocity of a particle are conserved when **F** = 0. This property of momentum is important in the analysis of collisions, a subject we shall take up in a later section.

Equation 6.2 can be written as **F** $\Delta t = \Delta \mathbf{p}$ or

$$\mathbf{F} \, \Delta t = \Delta \mathbf{p} = m\mathbf{v}_f - m\mathbf{v}_i \qquad [6.4]$$

◀ Impulse-momentum theorem

 **6.4,** SECTION 1

This result is often called the **impulse-momentum theorem.** The term on the left side of the equation, **F** Δt, is called the **impulse** of the force **F** for the time interval Δt. According to this result, **the impulse of the force acting on an object equals the change in momentum of that object.**

This equation tells us that if we exert a force on an object for time interval Δt, the effect of this force is to change the momentum of the object from some initial value, $m\mathbf{v}_i$, to some final value, $m\mathbf{v}_f$. For example, suppose a pitcher throws a baseball with a velocity of $\mathbf{v}_i$, and a batter hits the ball head on so as to reverse the direction of its velocity. The force, **F,** that the bat exerts on the ball can change both the direction and the magnitude of the initial velocity to a higher value, $\mathbf{v}_f$.

A word of caution. If you tried to solve a problem such as this using Newton's second law, you would encounter some difficulty in measuring the value to be used for **F,** because the force exerted on the ball is large, of short duration, and not constant. Instead, it might be represented by a curve like that in Figure 6.1a. The

Q U I C K L A B

Place a cup of water on a sheet of paper with one side of the paper overhanging the table that supports them. Remove the sheet of paper from underneath the cup by *quickly* pulling on the sheet. (Until you become good at this, you might want to use a plastic cup rather than a glass cup in case of an accident.) Explain why this is possible.

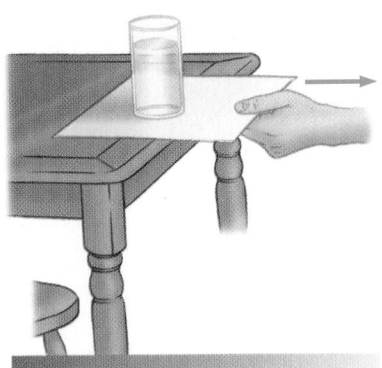

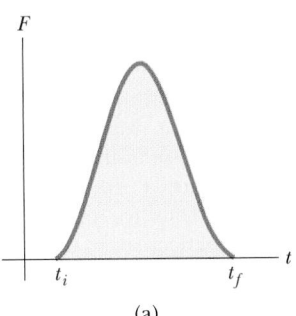
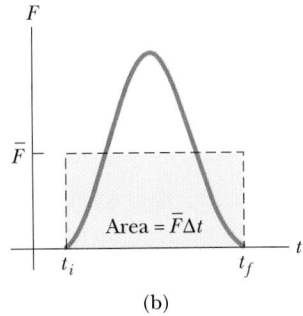

Figure 6.1 (a) A force acting on an object may vary in time. The impulse is the area under the force-time curve. (b) The average force (horizontal dashed line) gives the same impulse to the object in the time interval Δt as the real time-varying force described in (a).

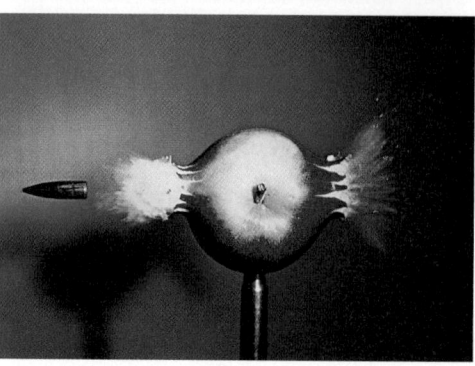

Figure 6.2   An apple being pierced by a 30-caliber bullet traveling at a supersonic speed of 900 m/s. This collision was photographed with a microflash stroboscope using an exposure time of 0.33 μs. Shortly after the photograph was taken, the apple disintegrated completely. Note that the points of entry and exit of the bullet are visually explosive. *(Harold E. Edgerton, Palm Press, Inc.)*

force starts out small as the bat comes in contact with the ball, rises to a maximum value when they are firmly in contact, and then drops off as the ball leaves the bat. In such instances, it is necessary to define an **average force, $\overline{\mathbf{F}}$,** shown as a dashed line in Figure 6.1b. This average force can be thought of as the constant force that would give the same impulse to the object in the time interval Δt as the actual time-varying force gives in this interval. The brief collision between a bullet and an apple is illustrated in Figure 6.2. **The impulse imparted by a force during the time interval Δt is equal to the area under the force-time graph from the beginning to the end of the time interval.**

APPLICATION

Cushioning Your Jump.

Think about what you do when you jump from a high position to the ground. As you strike the ground, you bend your knees. If you were to land on the ground with your legs locked, you would receive a painful shock in your legs as well as along your spine. The impulse-momentum theorem helps us understand why the landing is much less painful if you bend your knees. When you land, your momentum changes from its value just before you hit the ground to zero. By bending your knees, this change in momentum occurs over a longer time interval than with the knees locked. Thus, the force on the body is less than with the knees locked. For relatively short jumps, this force is low enough to do no damage to the bone structure.

Applying Physics 1

APPLICATION

Padded Gloves for Boxing.

In boxing matches of the nineteenth century, bare fists were used. In modern boxing, fighters wear padded gloves. How does this protect the brain of the boxer from injury? Boxers often "roll with the punch." How does this protect their health?

Explanation   The brain is immersed in a cushioning fluid inside the skull. If the head is struck suddenly by a bare fist, there is a rapid acceleration of the skull. The brain matches this acceleration only because of the large impulsive force of the skull on the brain. This large and sudden force can cause severe brain injury. Padded gloves extend the time over which the force is applied to the head. Thus, for a given impulse, the gloves provide a longer time interval than the bare fist, decreasing the average force. Because the average force is decreased, the acceleration of the skull is decreased, reducing (but not eliminating) the chance for brain injury. The same argument can be made for "rolling with the punch." If the head is held steady while being struck, the time interval over which the force is applied is relatively short and the average force is large. If the head is allowed to move in the same direction of the punch, the time interval is lengthened, and the average force reduced.

EXAMPLE 6.1 Teeing Off

A 50-g golf ball is struck with a club, as in Figure 6.3. The force on the ball varies from zero when contact is made up to some maximum value (when the ball is deformed) and then back to zero when the ball leaves the club. Thus, the force-time graph is somewhat like that in Figure 6.1. Assume that the ball leaves the club face with a velocity of $+44$ m/s.

(a) Estimate the impulse due to the collision.

Solution   Before the club hits the ball, the ball is at rest on the tee, and as a result its initial momentum is zero. The magnitude of the momentum immediately after the collision is

$$p_f = mv_f = (50 \times 10^{-3} \text{ kg})(44 \text{ m/s}) = +2.2 \text{ kg} \cdot \text{m/s}$$

Thus the impulse imparted to the ball, which equals its change in momentum, is

$$\Delta p = mv_f - mv_i = \boxed{+2.2 \text{ kg} \cdot \text{m/s}}$$

(b) Estimate the duration of the collision and the average force on the ball.

Solution   From Figure 6.3, it appears that a reasonable estimate of the distance the ball travels while in contact with the club is the radius of the ball, about 2.0 cm. The time it takes the club to move this distance (the contact time) is then

$$\Delta t = \frac{\Delta x}{\bar{v}_i} = \frac{2.0 \times 10^{-2} \text{ m}}{22 \text{ m/s}} = \boxed{9.1 \times 10^{-4} \text{ s}}$$

Finally, the magnitude of the average force is estimated to be

$$\bar{F} = \frac{\Delta p}{\Delta t} = \frac{2.2 \text{ kg} \cdot \text{m/s}}{9.1 \times 10^{-4} \text{ s}} = \boxed{2.4 \times 10^3 \text{ N}}$$

Figure 6.3 (Example 6.1) A golf ball being struck by a club. (© *Harold E. Edgerton. Courtesy of Palm Press, Inc.*)

EXAMPLE 6.2 How Good Are the Bumpers?

In a particular crash test, a 1.50×10^3 kg automobile collides with a wall, as in Figure 6.4a. The initial and final velocities of the automobile are $v_i = -15.0$ m/s and $v_f = +2.60$ m/s, respectively. If the collision lasts for 0.150 s, find the impulse due to the collision and the average force exerted on the automobile.

Solution   The initial and final momenta of the automobile are

$$p_i = mv_i = (1.50 \times 10^3 \text{ kg})(-15.0 \text{ m/s}) = -2.25 \times 10^4 \text{ kg} \cdot \text{m/s}$$

$$p_f = mv_f = (1.50 \times 10^3 \text{ kg})(2.60 \text{ m/s}) = +0.390 \times 10^4 \text{ kg} \cdot \text{m/s}$$

Hence, the impulse, which equals the change in momentum, is

$$\bar{F} \Delta t = \Delta p = p_f - p_i$$
$$= 0.390 \times 10^4 \text{ kg} \cdot \text{m/s} - (-2.25 \times 10^4 \text{ kg} \cdot \text{m/s})$$

$$\bar{F} \Delta t = \boxed{+2.64 \times 10^4 \text{ kg} \cdot \text{m/s}}$$

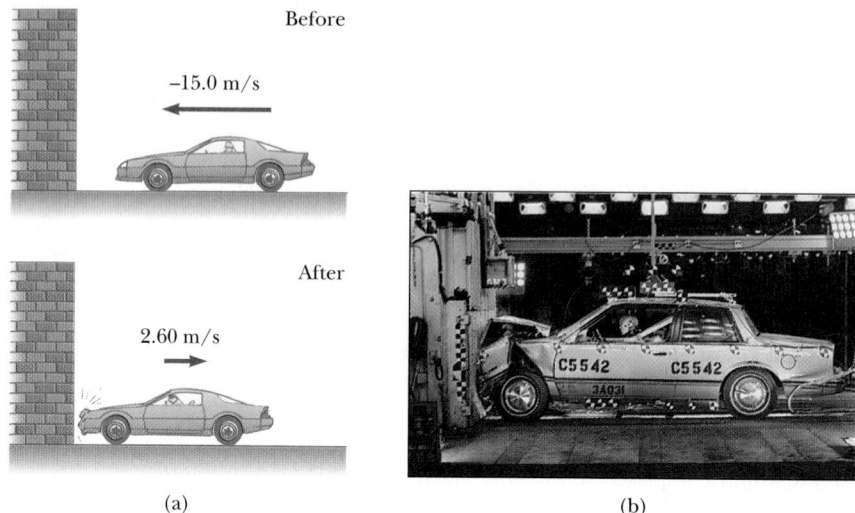

Before

−15.0 m/s

After

2.60 m/s

(a)

(b)

Figure 6.4 (Example 6.2) (a) This car's momentum changes as a result of its collision with the wall. (b) In a crash test (an inelastic collision), much of the car's initial kinetic energy is transformed into the energy it took to damage the vehicle. *(b, Courtesy of General Motors)*

The average force exerted on the automobile is

$$\bar{F} = \frac{\Delta p}{\Delta t} = \frac{2.64 \times 10^4 \text{ kg} \cdot \text{m/s}}{0.150 \text{ s}} = +1.76 \times 10^5 \text{ N}$$

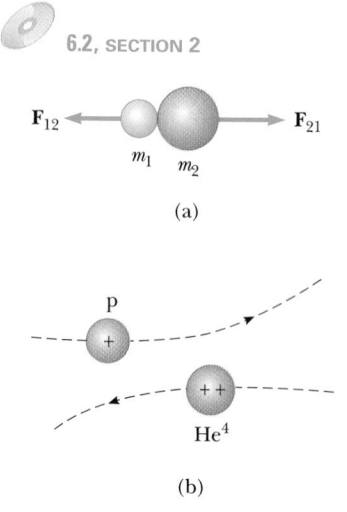

6.2, SECTION 2

$\mathbf{F}_{12}$ m_1 m_2 $\mathbf{F}_{21}$

(a)

p

He^4

(b)

Figure 6.5 (a) A collision between two objects resulting from direct contact. (b) A collision between two charged objects (in this case, a proton and a helium nucleus).

6.2   CONSERVATION OF MOMENTUM

An important feature of all isolated collisions is that momentum is conserved. In this section we shall see how the laws of motion lead us to this important conservation law.

A collision may be the result of physical contact between two objects, as illustrated in Figure 6.5a. This is a common observation when two macroscopic objects, such as two billiard balls or a baseball and a bat, strike each other. But "contact" on a submicroscopic scale is ill defined and hence meaningless, and so the notion of *collision* must be generalized. More accurately, forces between two bodies arise from the electrostatic interaction of the electrons in the surface atoms of the bodies. To understand the distinction between macroscopic and microscopic collisions, consider the collision between two positive charges as in Figure 6.5b. Because the two particles are positively charged, they repel each other and do not touch.

Figure 6.6 shows two isolated particles before and after they collide. By *isolated* we mean that no external forces, such as the gravitational force or friction, are present. Before the collision, the velocities of the two particles are $\mathbf{v}_{1i}$ and $\mathbf{v}_{2i}$; after the collision, the velocities are $\mathbf{v}_{1f}$ and $\mathbf{v}_{2f}$. The impulse-momentum theorem applied to m_1 becomes

$$\bar{\mathbf{F}}_1 \, \Delta t = m_1 \mathbf{v}_{1f} - m_1 \mathbf{v}_{1i}$$

Before collision

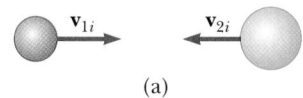

(a)

After collision

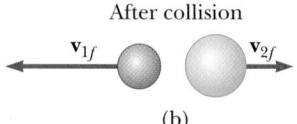

(b)

Figure 6.6 Before and after a head-on collision between two objects. The momentum of each object changes as a result of the collision, but the total momentum of the system remains constant.

During collision

F_1 F_2

m_1 m_2

Figure 6.7 When two objects collide, the force $\mathbf{F}_1$ exerted on m_1 is equal in magnitude and opposite in direction to the force $\mathbf{F}_2$ exerted on m_2.

The force from a nitrogen-propelled, hand-controlled device allows an astronaut to move about freely in space without restrictive tethers. *(Courtesy of NASA)*

Likewise, for m_2 we have

$$\overline{\mathbf{F}}_2 \, \Delta t = m_2\mathbf{v}_{2f} - m_2\mathbf{v}_{2i}$$

where $\overline{\mathbf{F}}_1$ is the force on m_1 due to m_2 during the collision and $\overline{\mathbf{F}}_2$ is the force on m_2 due to m_1 during the collision (Fig. 6.7).

We are using average values for $\overline{\mathbf{F}}_1$ and $\overline{\mathbf{F}}_2$ even though the actual forces may vary in time in a complicated way, as is the case in Figure 6.8. Newton's third law states that at all times these two forces are equal in magnitude and opposite in direction ($\mathbf{F}_1 = -\mathbf{F}_2$). In addition, the two forces act for the same time interval. Thus,

$$\overline{\mathbf{F}}_1 \, \Delta t = -\overline{\mathbf{F}}_2 \, \Delta t$$

or

$$m_1\mathbf{v}_{1f} - m_1\mathbf{v}_{1i} = -(m_2\mathbf{v}_{2f} - m_2\mathbf{v}_{2i})$$

from which we find

$$m_1\mathbf{v}_{1i} + m_2\mathbf{v}_{2i} = m_1\mathbf{v}_{1f} + m_2\mathbf{v}_{2f} \qquad [6.5]$$

This result is known as **conservation of momentum.**

The principle of conservation of momentum states that, when no external forces act on a system consisting of two objects that collide with each other, the total momentum of the system before the collision is equal to the total momentum of the system after the collision.

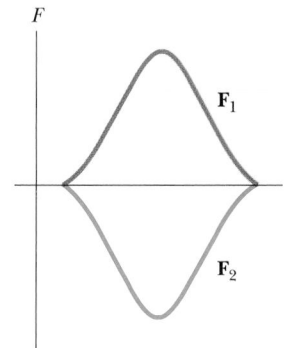

Figure 6.8 Force as a function of time for the two colliding particles in Figure 6.7. Note that $\mathbf{F}_1 = -\mathbf{F}_2$.

◀ Conservation of momentum

Note that momentum is conserved for a *system* of objects. In the example used to derive Equation 6.5, the system was taken to be two colliding objects. More

A squid propels itself by expelling water at a high velocity. *(Mike Severns/ Tony Stone Images)*

APPLICATION

Squid Propulsion.

generally, a system includes all the objects that are interacting with one another. In addition, we have assumed that the only forces acting during the collision are internal forces, meaning the forces that arise between the interacting objects of the system. For example, in the collision of two objects shown in Figure 6.7, the internal forces are $\mathbf{F}_1$ and $\mathbf{F}_2$. If a third object outside the system (consisting of m_1 and m_2) were to exert a force on either m_1 and m_2 (or both objects) during the collision, momentum would not be conserved for the system. In all the example problems that we shall consider, the system will be assumed to be isolated.

Our derivation has assumed that only two objects interact, but the result remains valid regardless of the number involved. In its most general form, conservation of momentum can be stated as follows: **The total momentum of an isolated system of objects is conserved regardless of the nature of the forces between the objects.**

To understand what this statement means, imagine that you initially stand at rest and then jump upward, leaving the ground with velocity **v**. Obviously, your momentum is not conserved, because before the jump it was zero and it became $m\mathbf{v}$ as you began to rise. However, the total momentum of the system is conserved if the system selected includes all objects that exert forces on one another. You must include the Earth as part of this system because you exert a downward force on the Earth when you jump. The Earth in turn exerts on you an upward force of the same magnitude, as required by Newton's third law. Momentum is conserved for the system consisting of you and the Earth. Thus, as you move upward with some momentum $m\mathbf{v}$, the Earth moves in the opposite direction with momentum of the same magnitude. The recoil velocity of the Earth due to this event is imperceptibly small, of course, because the Earth is so massive, but its momentum is not zero.

Conservation of momentum is in operation in the movement of squids through the water. To accelerate, a squid fills areas in its body walls with water and then squirts the water backward through a tube beneath its head. Because the squid is increasing the momentum of the water by exerting a force on it in the backward direction, the momentum of the squid increases in the forward direction. This conserves the momentum of the system of squid and ejected water. The same principle is used for rocket engines on spacecraft and jet engines on aircraft.

Applying Physics 2

A baseball is projected into the air at an angle to the ground. As it moves through its trajectory, its velocity and, therefore, its momentum constantly changes. Is this a violation of conservation of momentum?

Explanation   The principle of conservation of momentum states that the momentum of a particle or a system of particles is conserved *in the absence of external forces*. A ball projected through the air is subject to the external force of gravity, so we would not expect its momentum to be conserved. The gravitational force is in the vertical direction, so it is actually only the vertical component of the momentum that changes due to this force. In the horizontal direction, there is no force (ignoring air resistance), so the horizontal component of the ball's momentum is conserved. We used the same idea earlier in stating that the horizontal component of the *velocity* of a projectile remains constant.

If we consider the baseball and the Earth as a system of particles, then the gravitational force is an internal force to this system. The momentum of the ball-Earth system is conserved—as the ball rises and falls, the Earth sinks (due to the downward force from the thrower's feet) and rises (as the ball pulls it back upward, although imperceptibly!), so that the total momentum of the system remains unchanged.

EXAMPLE 6.3 The Recoiling Pitching Machine

A baseball player attempts to use a pitching machine to help him improve his batting average. He places the 50-kg machine on a frozen pond, as in Figure 6.9. The machine fires a 0.15-kg baseball horizontally with a speed of 36 m/s. What is the recoil velocity of the machine?

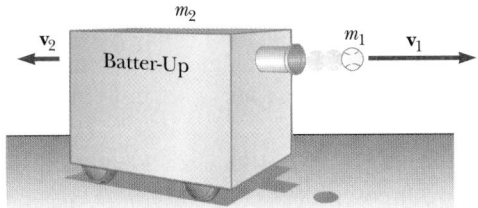

Figure 6.9 (Example 6.3) When the baseball is fired to the right, the pitching machine recoils to the left.

Solution We take the system to consist of the baseball and the pitching machine. Because of the force of gravity and the normal force, the system is not really isolated. However, both forces are directed perpendicularly to the motion of the system. Therefore, momentum is conserved in the x direction, because there are no external forces in this direction (assuming the surface is frictionless).

Because the baseball and pitching machine are at rest before firing, the total momentum of the system is zero. Therefore, the total momentum after firing must also be zero; that is,

$$m_1\mathbf{v}_1 + m_2\mathbf{v}_2 = 0$$

With $m_1 = 0.15$ kg, $\mathbf{v}_1 = +36$ m/s, and $m_2 = 50$ kg, we find the recoil velocity of the pitching machine to be

$$\mathbf{v}_2 = -\frac{m_1}{m_2}\mathbf{v}_1 = -\left(\frac{0.15\text{ kg}}{50\text{ kg}}\right)(36\text{ m/s}) = -0.11\text{ m/s}$$

The negative sign for $\mathbf{v}_2$ indicates that the pitching machine is moving to the left after firing, in the direction *opposite* the motion of the baseball.

6.3 COLLISIONS

We have seen that, for any type of collision, the total momentum of the system just before collision equals the total momentum just after collision. We can say that the total momentum is always conserved for any type of collision. However, the total kinetic energy is generally not conserved in a collision, because some of the kinetic

APPLICATION

Glaucoma Testing.

energy is converted to thermal energy and internal elastic potential energy when the objects deform.

We define an inelastic collision as a collision in which momentum is conserved but kinetic energy is not. The collision of a rubber ball with a hard surface is inelastic, because some of the kinetic energy is lost when the ball is deformed during contact with the surface. When two objects collide and stick together, the collision is called **perfectly inelastic.** For example, if two pieces of putty collide, they stick together and move with some common velocity after the collision. If a meteorite collides head on with the Earth, it becomes buried in the Earth and the collision is considered perfectly inelastic. Not all of the initial kinetic energy is necessarily lost in a perfectly inelastic collision.

An elastic collision is defined as one in which both momentum and kinetic energy are conserved. Billiard ball collisions and the collisions of air molecules with the walls of a container at ordinary temperatures are highly elastic. Macroscopic collisions such as those between billiard balls are only approximately elastic, because some permanent deformation, and hence some loss of kinetic energy, takes place. Perfectly elastic collisions do occur, however, between atomic and subatomic particles. Elastic and perfectly inelastic collisions are *limiting* cases; most actual collisions fall into a category between them.

A practical example of an inelastic collision is used to detect glaucoma, a disease in which the pressure inside the eye builds up and leads to blindness by damaging the cells of the retina. In this application, medical professionals use a device called a tonometer to measure the pressure inside the eye. This device releases a puff of air against the outer surface of the eye and measures the speed of the air after reflection from the eye. At normal pressure, the eye is slightly spongy, and the pulse is reflected at low speed. As the pressure inside the eye increases, the outer surface becomes more rigid, and the speed of the reflected pulse increases. Thus, the speed of the reflected puff of air is used to measure the internal pressure of the eye.

We can summarize the types of collisions as follows:

Elastic collision ▶

Inelastic collision ▶

- An elastic collision is one in which both momentum and kinetic energy are conserved.
- An inelastic collision is one in which momentum is conserved but kinetic energy is not.
- A perfectly inelastic collision is an inelastic collision in which the two objects stick together after the collision, so that their final velocities are the same and the momentum of the system is conserved.

In the remainder of this section, we shall treat perfectly inelastic and elastic collisions in one dimension.

Perfectly Inelastic Collisions

Consider two objects of masses m_1 and m_2 moving with initial velocities v_{1i} and v_{2i} along a straight line, as in Figure 6.10. We shall assume that the objects collide head on so that they move along the same line of motion after the collision. If the two objects stick together and move with some common velocity, v_f, after the collision,

6.5, BOTH SECTIONS

then only the momentum of the system is conserved, and we can say that the total momenta before and after the collision are equal:

$$m_1 v_{1i} + m_2 v_{2i} = (m_1 + m_2) v_f \qquad [6.6]$$

It is important to note that v_{1i}, v_{2i}, and v_f represent the x components of the vectors $\mathbf{v}_{1i}$, $\mathbf{v}_{2i}$, and $\mathbf{v}_f$, so one must be careful with signs. For example, in Figure 6.10, v_{1i} would have a positive value (m_1 moving to the right), whereas v_{2i} would have a negative value (m_2 moving to the left).

In a typical inelastic collision problem, only one quantity in the preceding equation is unknown, and so conservation of momentum is sufficient to tell us what we need to know.

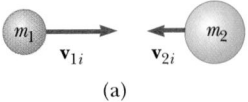

Before collision

EXAMPLE 6.4 The Cadillac Versus the Compact Car

An 1800-kg luxury sedan stopped at a traffic light is struck from the rear by a compact car with a mass of 900 kg. The two cars become entangled as a result of the collision. If the compact car were moving at a velocity of $+20.0$ m/s before the collision, what is the velocity of the entangled mass after the collision?

Solution The momentum before the collision is that of the compact car alone, because the large car is initially at rest. Thus, for the momentum before the collision we have

$$p_i = m_1 v_i = (900 \text{ kg})(20.0 \text{ m/s}) = +1.80 \times 10^4 \text{ kg·m/s}$$

After the collision, the mass that moves is the sum of the masses of the large car and the compact car, and the momentum of the combination is

$$p_f = (m_1 + m_2) v_f = (2700 \text{ kg})(v_f)$$

Equating the momentum before the collision to the momentum after the collision and solving for v_f, the velocity of the wreckage, we get

$$v_f = \frac{p_i}{m_1 + m_2} = \frac{1.80 \times 10^4 \text{ kg·m/s}}{2700 \text{ kg}} = +6.67 \text{ m/s}$$

Figure 6.10 (a) Before and (b) after a perfectly inelastic head-on collision between two objects.

WEB

For Web links to information on automobile safety, visit the textbook Web site at http://www.harcourtcollege.com/physics/cptech

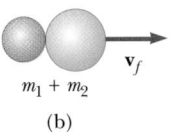 **Workbook Problems 36 and 37** — (pages 79 to 82 in the Workbook) also deal with perfectly inelastic collisions.

EXAMPLE 6.5 Here's Mud in Your Eye

Two balls of mud collide head on in a perfectly inelastic collision, as in Figure 6.10. Suppose $m_1 = 0.500$ kg, $m_2 = 0.250$ kg, $v_{1i} = +4.00$ m/s, and $v_{2i} = -3.00$ m/s. (a) Find the velocity of the composite ball of mud after the collision.

Solution Writing Equation 6.6 for conservation of momentum with the positive direction for velocity to the right, we can find the velocity of the combined mass after the collision:

$$m_1 v_{1i} + m_2 v_{2i} = (m_1 + m_2) v_f$$

$$(0.500 \text{ kg})(4.00 \text{ m/s}) + (0.250 \text{ kg})(-3.00 \text{ m/s}) = (0.750 \text{ kg})(v_f)$$

$$v_f = +1.67 \text{ m/s}$$

(b) How much kinetic energy is lost in the collision?

Solution   The kinetic energy of the system before the collision is

$$KE_i = KE_1 + KE_2 = \tfrac{1}{2}m_1v_{1i}^2 + \tfrac{1}{2}m_2v_{2i}^2$$
$$= \tfrac{1}{2}(0.500 \text{ kg})(4.00 \text{ m/s})^2 + \tfrac{1}{2}(0.250 \text{ kg})(-3.00 \text{ m/s})^2$$
$$= 5.13 \text{ J}$$

The kinetic energy of the system after the collision is

$$KE_f = \tfrac{1}{2}(m_1 + m_2)v_f^2 = \tfrac{1}{2}(0.750 \text{ kg})(1.67 \text{ m/s})^2 = 1.05 \text{ J}$$

Hence, the loss in kinetic energy is

$$KE_i - KE_f = \boxed{4.08 \text{ J}}$$

Most of this lost energy is converted to thermal energy and internal elastic potential energy as the two objects collide and deform.

EXAMPLE 6.6   The Ballistic Pendulum

The ballistic pendulum (Fig. 6.11) is a device used to measure the velocity of a fast-moving projectile such as a bullet. The bullet is fired into a large block of wood suspended from some light wires. The bullet is stopped by the block, and the entire system swings through

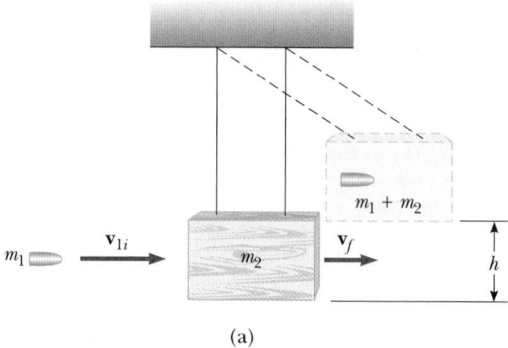

(a)

(b)

Figure 6.11   (Example 6.6) (a) Diagram of a ballistic pendulum. Note that $\mathbf{v}_f$ is the velocity of the system just after the perfectly inelastic collision. (b) Multiflash photograph of a laboratory ballistic pendulum. *(Courtesy of Central Scientific Company)*

the vertical distance h. It is possible to obtain the initial velocity of the bullet by measuring h and the two masses. As an example of the technique, assume that the mass of the bullet, m_1, is 5.00 g, the mass of the pendulum, m_2, is 1.00 kg, and h is 5.00 cm. Find the initial velocity of the bullet, v_{1i}.

Solution The collision between the bullet and the block (the system) is perfectly inelastic. Writing the conservation of momentum for the collision in the form of Equation 6.6, we have

$$m_1 v_{1i} = (m_1 + m_2) v_f$$

(1) $\qquad (5.00 \times 10^{-3} \text{ kg}) (v_{1i}) = (1.0050 \text{ kg}) (v_f)$

There are two unknowns in this equation, v_{1i} and v_f; the latter is the velocity of the block plus embedded bullet *immediately after the collision*. We must look for additional information if we are to complete the problem. Kinetic energy is not conserved during an inelastic collision. However, mechanical energy is conserved after the collision, and so the kinetic energy of the system at the bottom is transformed into the potential energy of the bullet plus block at height h:

$$\tfrac{1}{2}(m_1 + m_2) v_f^2 = (m_1 + m_2) gh$$

$$\tfrac{1}{2}(1.0050 \text{ kg}) (v_f^2) = (1.0050 \text{ kg}) (9.80 \text{ m/s}^2) (5.00 \times 10^{-2} \text{ m})$$

giving

$$v_f = 0.990 \text{ m/s}$$

With v_f now known, (1) yields v_{1i}:

$$v_{1i} = \frac{(1.0050 \text{ kg}) (0.990 \text{ m/s})}{5.00 \times 10^{-3} \text{ kg}} = \boxed{199 \text{ m/s}}$$

Exercise Explain why it would be incorrect to equate the initial kinetic energy of the incoming bullet to the final gravitational potential energy of the bullet-block combination.

Elastic Collisions

Now consider two objects that undergo an elastic head-on collision (Fig. 6.12). In this situation, *both momentum and kinetic energy are conserved;* we can write these conditions as

$$m_1 v_{1i} + m_2 v_{2i} = m_1 v_{1f} + m_2 v_{2f} \qquad \text{[6.7]}$$

$$\tfrac{1}{2} m_1 v_{1i}^2 + \tfrac{1}{2} m_2 v_{2i}^2 = \tfrac{1}{2} m_1 v_{1f}^2 + \tfrac{1}{2} m_2 v_{2f}^2 \qquad \text{[6.8]}$$

where v is positive if an object moves to the right and negative if it moves to the left.

In a typical problem involving elastic collisions, there are two unknown quantities, and Equations 6.7 and 6.8 can be solved simultaneously to find them. An alternative approach, employing a little mathematical manipulation of Equation 6.8, often simplifies this process. To see this, let's cancel the factor $\tfrac{1}{2}$ in Equation 6.8 and rewrite it as

$$m_1 (v_{1i}^2 - v_{1f}^2) = m_2 (v_{2f}^2 - v_{2i}^2)$$

6.6, SECTION 1

Before collision

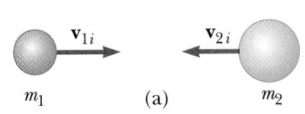

m_1 $\qquad$ (a) $\qquad$ m_2

After collision

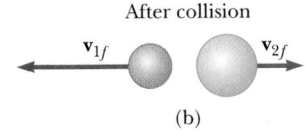

(b)

Figure 6.12 (a) Before and (b) after an elastic head-on collision between two hard spheres.

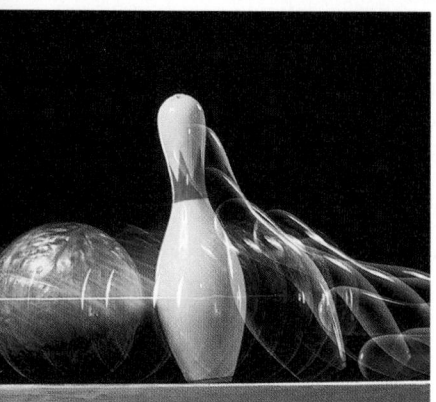

As a result of the collision between the bowling ball and pin, part of the ball's momentum is transferred to the pin. As a consequence, the pin acquires momentum and kinetic energy, and the ball loses momentum and kinetic energy. However, the total momentum of the system (ball and pin) remains constant. *(Ben Rose/ The Image Bank)*

Here we have moved the terms containing m_1 to one side of the equation and those containing m_2 to the other. Next, let us factor both sides of the equation:

$$m_1(v_{1i} - v_{1f})(v_{1i} + v_{1f}) = m_2(v_{2f} - v_{2i})(v_{2f} + v_{2i}) \qquad [6.9]$$

We now separate the terms containing m_1 and m_2 in the equation for the conservation of momentum (Eq. 6.7) to get

$$m_1(v_{1i} - v_{1f}) = m_2(v_{2f} - v_{2i}) \qquad [6.10]$$

To obtain our final result, we divide Equation 6.9 by Equation 6.10 and get

$$v_{1i} + v_{1f} = v_{2f} + v_{2i}$$

$$v_{1i} - v_{2i} = -(v_{1f} - v_{2f}) \qquad [6.11]$$

This equation, in combination with the condition for conservation of momentum, will be used to solve problems dealing with perfectly elastic, head-on collisions. According to Equation 6.11, the relative velocity of the two objects before the collision, $v_{1i} - v_{2i}$, equals the negative of the relative velocity of the two objects after the collision, $-(v_{1f} - v_{2f})$.

Applying Physics 3

A great deal of expense is involved in repairing automobiles after collisions. It seems that much of this money could be saved if cars were built from hard rubber, so that they made approximately elastic collisions and simply bounced off each other without denting. Why isn't this done?

Explanation   Making cars out of hard rubber would save on automobile repair bills, but the results for the drivers and passengers would be disastrous. Imagine two identical cars traveling toward each other at identical speeds collide head on. In the approximation that the collision is elastic, the cars bounce off each other and leave the collision with their velocities reversed in direction. Thus, the inhabitants of the cars would suffer a large change in momentum in a short period of time. As a result, the force on the inhabitants would become very large, causing significant bodily injury. For real cars in this situation, the cars would come to rest during the collision, as the front sections of the cars crumpled. Thus, the change in momentum is only half what it was in the elastic case, and the change occurs over the longer time period during which the cars crumpled. The resulting smaller force on the inhabitants will result in fewer injuries.

PROBLEM-SOLVING STRATEGY

Conservation of Momentum

The following procedure is recommended for problems involving collisions between two objects:

1. Set up a coordinate system and define your velocities with respect to that system. That is, objects moving in the direction selected as the positive di-

rection of the *x* axis are considered to have a positive velocity, and those moving in the negative *x* direction, a negative velocity. It is convenient to make the *x* axis coincide with one of the initial velocities.

2. In your sketch of the coordinate system, draw all velocity vectors with labels and include all the given information.

3. Write expressions for the momenta of each object before and after the collision. (In two-dimensional collision problems, write expressions for the *x* and *y* components of momentum before and after the collision.) Remember to include the appropriate signs for the velocity vectors.

4. Now write expressions for the *total* momentum *before* and *after* the collision and equate the two. (For two-dimensional collisions, this expression should be written for the momentum in both the *x* and *y* directions.) Remember, it is the momentum of the *system* (the two colliding objects) that is conserved, *not* the momenta of the individual objects.

5. If the collision is *inelastic*, kinetic energy is not conserved. Proceed to solve the momentum equations for the unknown quantities.

6. If the collision is *elastic*, kinetic energy is conserved, so you can equate the total kinetic energies before and after the collision. This gives an additional relationship between the velocities.

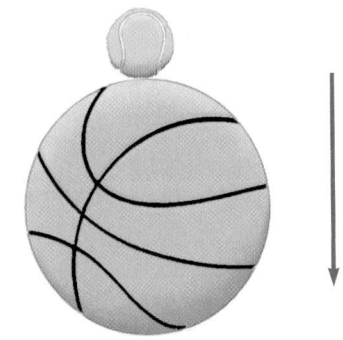

EXAMPLE 6.7 Let's Play Pool

Two billiard balls move toward one another, as in Figure 6.12. The balls have identical masses, and assume that the collision between them is elastic. If the initial velocities of the balls are $+30$ cm/s and -20 cm/s, what is the velocity of each ball after the collision?

Solution We turn first to Equation 6.7. The equal masses cancel on each side, and after substituting the appropriate values for the initial velocities, we have

$$30 \text{ cm/s} + (-20 \text{ cm/s}) = v_{1f} + v_{2f}$$

$$(1) \qquad 10 \text{ cm/s} = v_{1f} + v_{2f}$$

Because kinetic energy is also conserved, we can apply Equation 6.11, which gives

$$30 \text{ cm/s} - (-20 \text{ cm/s}) = v_{2f} - v_{1f}$$

$$(2) \qquad 50 \text{ cm/s} = v_{2f} - v_{1f}$$

Solving (1) and (2) simultaneously, we find

$$v_{1f} = \boxed{-20 \text{ cm/s}} \qquad v_{2f} = \boxed{+30 \text{ cm/s}}$$

That is, the balls *exchange velocities!* This is always the case when two objects of equal mass collide elastically head on.

Exercise Find the final velocity of the two balls if the ball with initial velocity -20 cm/s has a mass equal to half that of the ball with initial velocity $+30$ cm/s.

Answer $v_{1f} = -3.0$ cm/s; $v_{2f} = +47$ cm/s

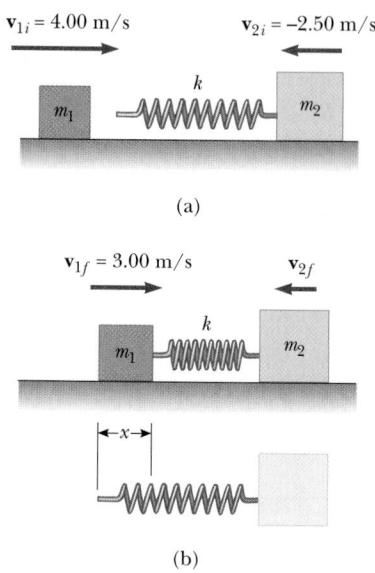

$v_{1i} = 4.00$ m/s $v_{2i} = -2.50$ m/s

(a)

$v_{1f} = 3.00$ m/s v_{2f}

(b)

Figure 6.13   (Example 6.8)

6.8, BOTH SECTIONS

EXAMPLE 6.8   A Two-Body Collision with Spring

A block of mass $m_1 = 1.60$ kg, moving to the right with a speed of 4.00 m/s on a friction-less, horizontal track, collides with a spring attached to a second block, of mass $m_2 = 2.10$ kg, that is moving to the left with a speed of 2.50 m/s (Fig. 6.13a). The spring constant is 600 N/m. For the instant when m_1 is moving to the right with a speed of 3.00 m/s, determine (a) the velocity of m_2 and (b) the distance, x, that the spring is compressed.

Solution   (a) First, note that the initial velocity of m_2 is $v_{2i} = -2.50$ m/s because it is moving to the left. Because the total momentum of the system is constant, we have

$$m_1 v_{1i} + m_2 v_{2i} = m_1 v_{1f} + m_2 v_{2f}$$

$$(1.60 \text{ kg})(4.00 \text{ m/s}) + (2.10 \text{ kg})(-2.50 \text{ m/s}) = (1.60 \text{ kg})(3.00 \text{ m/s}) + (2.10 \text{ kg}) v_{2f}$$

$$v_{2f} = -1.74 \text{ m/s}$$

This result shows that m_2 is still moving to the left at that instant.
(b) To determine the compression, x, in the spring in Figure 6.13b, we can make use of conservation of energy because no friction forces are acting on the system. Thus, we have

$$\tfrac{1}{2} m_1 v_{1i}^2 + \tfrac{1}{2} m_2 v_{2i}^2 = \tfrac{1}{2} m_1 v_{1f}^2 + \tfrac{1}{2} m_2 v_{2f}^2 + \tfrac{1}{2} k x^2$$

Substitution of the given values and the result of part (a) into this expression gives

$$x = 0.173 \text{ m}$$

Exercise   Find the velocity of m_1 and the compression in the spring at the instant m_2 is at rest.

Answer   0.719 m/s; 0.251 m

EXAMPLE 6.9   Executive Stress Reliever

An ingenious device that illustrates conservation of momentum and conservation of ki-netic energy is shown in Figure 6.14a. It consists of five identical hard balls supported by strings of equal lengths. When one ball is pulled out and released, after the almost elastic collision one ball moves out on the opposite side as in Figure 6.14b. If two balls are pulled out and released, two balls swing out on the opposite side, and so forth—but how do the balls know? For example, is it possible that on occasion when one ball is released, two will swing out on the opposite side traveling with half the speed of the incoming ball, as in Figure 6.14c? Assume the collision is elastic and use conservation of momentum and conservation of kinetic energy to show that the collision shown in Figure 6.14c can never occur.

Solution   First, let us apply conservation of momentum to the collision. The momentum of the system before the collision is mv, where m is the mass of the one ball released and v is its speed just before the collision. After the collision, we have two balls each of mass m, moving with speed $v/2$. Thus, the total momentum of the system after the collision is $m(v/2) + m(v/2) = mv$. Thus, momentum is conserved. However, let us now consider the kinetic energy of the system before and after the collision. The kinetic energy just before the collision is $KE_{\text{before}} = \tfrac{1}{2} mv^2$ and the kinetic energy just after the collision is

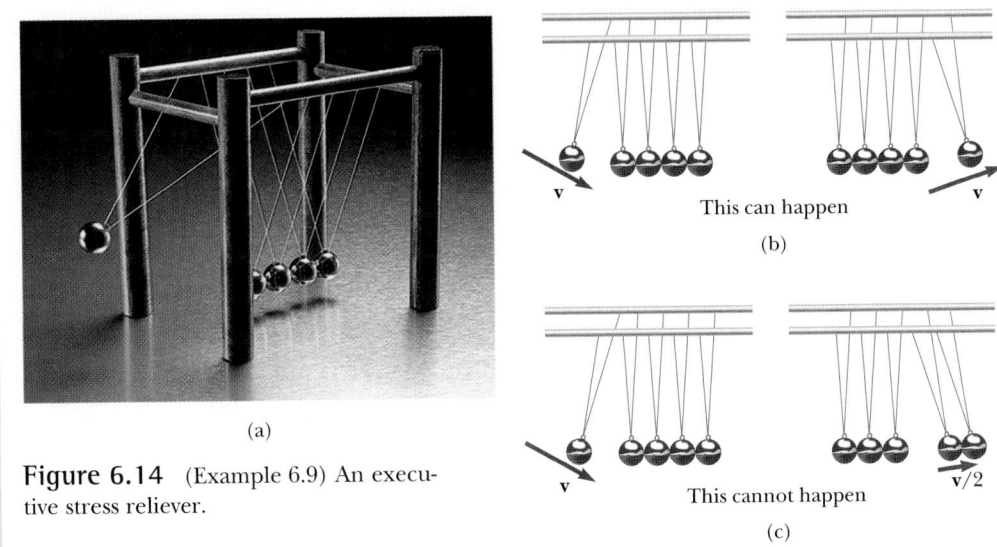

Figure 6.14 (Example 6.9) An executive stress reliever.

$KE_{after} = \frac{1}{2}m(v/2)^2 + \frac{1}{2}m(v/2)^2 = \frac{1}{4}mv^2$. Thus, kinetic energy is *not* conserved in this collision. The only way to have both momentum and kinetic energy conserved is for one ball to move out when one ball is released, or two balls to move out when two are released, and so on. Try other combinations for yourself to verify this statement.

6.4 GLANCING COLLISIONS

The collisions we have considered until now have been head-on collisions, in which the incident mass strikes a second mass head on and both rebound along a straight-line path that coincides with the line of motion of the incident mass. Anyone who has ever played billiards knows that such collisions are the exception rather than the rule. A more common type of collision is a *glancing collision*, in which the colliding masses rebound at some angle relative to the line of motion of the incident mass. Figure 6.15a shows a blue ball that travels with an initial speed of v_{1i} and strikes a red ball obliquely (off center). After the collision, the blue ball caroms off

6.6, SECTION 2

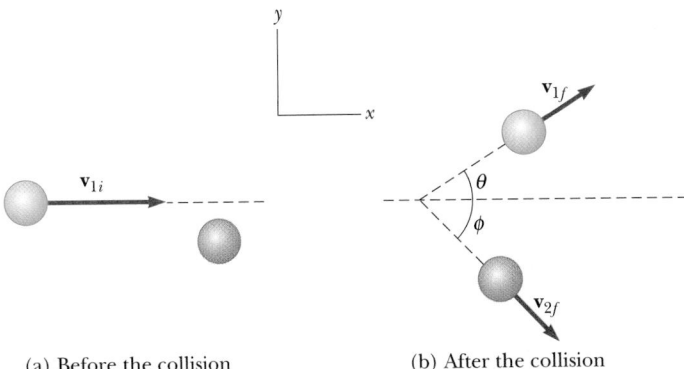

(a) Before the collision   (b) After the collision

Figure 6.15 (a) Before and (b) after a glancing collision between two balls.

at an angle of θ relative to its incident line of motion, and the red ball rebounds at an angle of ϕ.

As we emphasized earlier, **momentum is conserved in all collisions when no external forces are acting,** and glancing collisions are no exception. Because momentum is a vector quantity, the conservation of momentum principle must be written $\mathbf{p}_i = \mathbf{p}_f$. That is, the total initial momentum of the system (the two balls) must equal the total final momentum of the system. For a collision in two dimensions, such as that in Figure 6.16, this implies that the total momentum is conserved along the x direction *and* along the y direction. We can state this in equation form as

$$\Sigma p_{ix} = \Sigma p_{fx} \quad \text{and} \quad \Sigma p_{iy} = \Sigma p_{fy} \qquad \text{[6.12]}$$

The following example illustrates how to use the principle of conservation of momentum to treat glancing collisions.

EXAMPLE 6.10   Collision at an Intersection

At an intersection, a 1500-kg car traveling east at 25 m/s collides with a 2500-kg van traveling north at 20 m/s, as shown in Figure 6.16. Find the direction and magnitude of the velocity of the wreckage immediately after the collision, assuming that the vehicles undergo a perfectly inelastic collision (that is, they stick together).

Solution   Let us choose the positive x direction to be east and the positive y direction to be north, as in Figure 6.16. Before the collision, the only object having momentum in the x direction is the car. Thus, the total initial momentum of the system (car plus van) in the x direction is

$$\Sigma p_{ix} = (1500 \text{ kg})(25 \text{ m/s}) = 37\,500 \text{ kg} \cdot \text{m/s}$$

Now let us assume that the wreckage moves at an angle of θ and a speed of v after the collision, as in Figure 6.16. The total momentum in the x direction after the collision is

$$\Sigma p_{fx} = (4000 \text{ kg})(v \cos \theta)$$

Because momentum is conserved in the x direction, we have

$$\Sigma p_{ix} = \Sigma p_{fx}$$

(1) $\qquad 37\,500 \text{ kg} \cdot \text{m/s} = (4000 \text{ kg})(v \cos \theta)$

Similarly, the total initial momentum of the system in the y direction is that of the van, which equals $(2500 \text{ kg})(20 \text{ m/s})$. Applying conservation of momentum to the y direction, we have

$$\Sigma p_{iy} = \Sigma p_{fy}$$

$$(2500 \text{ kg})(20 \text{ m/s}) = (4000 \text{ kg})(v \sin \theta)$$

(2) $\qquad 50\,000 \text{ kg} \cdot \text{m/s} = (4000 \text{ kg})(v \sin \theta)$

If we divide (2) by (1), we get

$$\tan \theta = \frac{50\,000}{37\,500} = 1.33$$

$$\theta = \boxed{53°}$$

When this angle is substituted into (2) — or, alternatively, into (1) — the value of v is

$$v = \frac{50\,000 \text{ kg} \cdot \text{m/s}}{(4000 \text{ kg})(\sin 53°)} = \boxed{16 \text{ m/s}}$$

Workbook Problem 39 — (page 85 in the Workbook) deals with a perfectly elastic glancing collision.

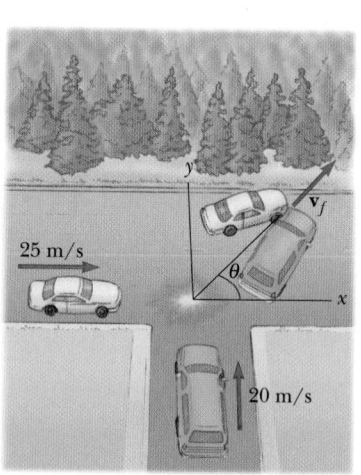

Figure 6.16 (Example 6.10) A top view of a perfectly inelastic collision between a car and a van.

PHYSICS IN ACTION

A Glancing Collision and Life on Mars

About 15 million years ago, a meteorite of unknown origin came hurtling into the atmosphere of Mars at a glancing angle. On impact with the Martian surface a lot of debris was tossed into space and out of the gravitational grasp of Mars. The energy of the collision was sufficiently great to melt a portion of one particular rock about the size of a potato that was a part of that debris. This melting created bubbles in the rock that captured traces of the Martian atmosphere. The makeup of this atmosphere is well-known because of samples taken by the Viking landers. This particular ejected rock eventually became a meteorite for Earth and landed on its surface. In 1984, this inconsequential appearing rock was collected by a team searching for meteorites in Antarctica and given the name ALH84001, a designation signifying that this was the first meteorite (001) collected from the Allen Hills Region of Antarctica (ALH). Analysis of the gas from these bubbles was later to provide convincing proof of the Martian origin of the rock. Between 1994 and 1996, David McKay and a team of scientists from the University of Georgia, Johnson Space Center, McGill University, and Stanford University subjected the meteorite to a wide array of tests in an effort to find evidence of life. In 1996, they announced that they had found four pieces of evidence to support the conclusion that tiny, cylindrical shaped organisms once lived in the interstices of the rock. Specifically, they found (1) traces of carbonate mineral. This mineral forms only in the presence of water. Thus, one of the presumed necessities of life, water, had at one time percolated through the microscopic channels of the rock. (2) Spectroscopic analysis of the materials in the rock showed the presence of a type of organic molecule called polycyclic aromatic hydrocarbons. This is a substance often formed when microbes die. (3) Minute specimens of magnetite, gregite, and pyrrhotite were revealed by a powerful electron microscope. All of these materials can be produced by various microscopic lifeforms. Finally, (4) this same microscope revealed cylindrical shapes that appear to be the fossilized remains of primitive lifeforms. The fact that all of these life signs increase with depth inside the rock indicate that they did not arise from contamination while the meteorite was on Earth.

The evidence is not conclusive to all scientists, but regardless of the opinion held, a glancing collision eons ago has certainly brought new interest in that age-old question, "Are we alone?"

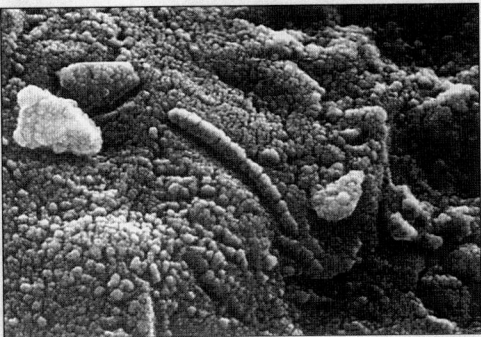

(NASA)

SUMMARY

6.9

The **linear momentum** of an object of mass m moving with a velocity of $\mathbf{v}$ is defined to be

$$\mathbf{p} \equiv m\mathbf{v} \qquad [6.1]$$

The **impulse** of a force, $\mathbf{F}$, acting on an object is equal to the product of the force and the time interval during which the force acts:

$$\text{Impulse} = \mathbf{F}\,\Delta t$$

The **impulse-momentum theorem** states that the impulse of a force on an object is equal to the change in momentum of the object:

$$\mathbf{F}\,\Delta t = \Delta \mathbf{p} = m\mathbf{v}_f - m\mathbf{v}_i \qquad [6.4]$$

Conservation of momentum applied to a system of two interacting objects states that, when no external forces act on the system, the total momentum of the system before the collision is equal to the total momentum of the system after the collision.

$$m_1\mathbf{v}_{1i} + m_2\mathbf{v}_{2i} = m_1\mathbf{v}_{1f} + m_2\mathbf{v}_{2f} \qquad [6.5]$$

An **inelastic collision** is one in which momentum is conserved but kinetic energy is not. A **perfectly inelastic collision** is one in which the colliding objects stick together after the collision. An **elastic collision** is one in which both momentum and kinetic energy are conserved.

In glancing collisions, conservation of momentum can be applied along two perpendicular directions — that is, along an x axis and a y axis.

MULTIPLE-CHOICE QUESTIONS

1. An object of mass m moves to the right with a speed v. It collides head-on in a perfectly inelastic collision with an object of twice the mass but half the speed moving in the opposite direction. What is the speed of the combined mass after the collision?
   (a) 0   (b) $v/2$ (c) v (d) $2v$

2. A baseball pitching machine of mass M (including the mass m of a baseball) rests on the pitcher's mound. What is the speed of recoil of the machine when it fires the baseball at a speed v?
   (a) 0   (b) $mv/(M - m)$
   (c) $mv/(M + m)$ (d) mv/M
   (e) $2mv/(M - m)$

3. A 0.10-kg object moving initially with a velocity of 0.20 m/s eastward makes an elastic head-on collision with a 0.15-kg object initially at rest. What is the final velocity of the 0.10-kg object after the collision?

(a) 0.16 m/s eastward
(b) 0.16 m/s westward
(c) 0.04 m/s eastward
(d) 0.04 m/s westward

4. When one catches a baseball, it hurts the hand less if one gives with the ball. The reason this helps the hand is that
   (a) this makes the kinetic energy change less.
   (b) this makes the momentum change less.
   (c) this makes the time interval for stopping greater.
   (d) this increases the impulse of the collision.

5. A 10.0-g bullet is fired into a 200-g block of wood at rest on a horizontal surface. After impact, the block slides 8.00 m before coming to rest. If the coefficient of friction is $\mu_k = 0.400$, find the speed of the bullet before impact.
   (a) 106 m/s   (b) 166 m/s   (c) 226 m/s   (d) 286 m/s

CONCEPTUAL QUESTIONS

1. In perfectly inelastic collisions between two objects, there are some instances in which all of the original kinetic energy is transformed to other forms than kinetic. Give an example of such a situation.

2. If two objects collide and one is initially at rest, is it possible for both to be at rest after the collision? Is it possible for one to be at rest after the collision? Explain.

3. A ball of clay of mass m is thrown with a speed v against a brick wall. The clay sticks to the wall and stops. Is the principle of conservation of momentum violated in this example?

4. A skater is standing still on a frictionless ice rink. Her friend throws a Frisbee straight at her. In which of the following cases is the largest momentum transferred to the skater? (a) The skater catches the Frisbee and holds onto it. (b) The skater catches the Frisbee momentarily but then drops it vertically downward. (c) The skater catches the Frisbee, holds it momentarily, and throws it back to her friend.

5. You are standing perfectly still and then you take a step forward. Before the step your momentum was zero, but afterward you have some momentum. Is the conservation of momentum violated in this case?

6. If two particles have equal kinetic energies, are their momenta necessarily equal? Explain.

7. A boy stands at one end of a floating raft that is stationary relative to the shore. He then walks to the opposite end of the raft, away from the shore. Does the raft move? Explain.

8. If two automobiles collide, they usually do not stick together. Does this mean the collision is perfectly elastic? Explain why a head-on collision is likely to be more dangerous than other types of collisions.

9. An open box slides across a frictionless, icy surface of a frozen lake. What happens to the speed of the box as water from a rain shower collects in it, assuming that the rain falls vertically downward into the box? Explain.

10. Consider a perfectly inelastic collision between a car and a large truck. Which vehicle loses more kinetic energy as a result of the collision?

11. Your physical education teacher, who knows something about physics, throws you a tennis ball at a certain velocity, and you catch it. You are now given the following choice: The teacher can throw you a medicine ball, which is much more massive than the tennis ball, with the same velocity as the tennis ball, the same momentum, or the same kinetic energy. Which choice would you make in order to make the easiest catch, and why?

12. You are watching a movie about a superhero and notice that the superhero hovers in the air and throws a piano at some bad guys while remaining stationary in the air. What's wrong with this scenario?

13. In golf, novice players are often advised to be sure to "follow through" with their swing. Why does this make the ball travel a longer distance? If a shot is taken near the green, little follow-through is required. Why?

14. An airbag is inflated when a collision occurs, which protects the passenger (the dummy, in this case) from serious injury (Fig. Q6.14). Why does the airbag soften the blow? Discuss the physics involved in this dramatic photograph.

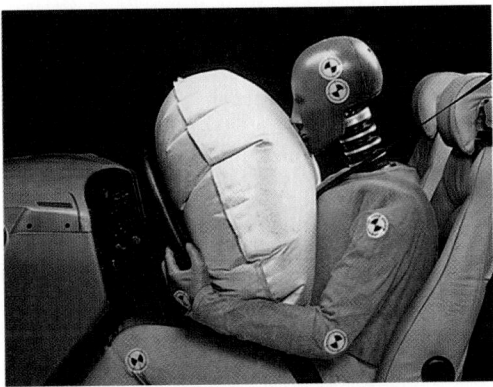

Figure Q6.14 *(Courtesy of Saab)*

15. A sharpshooter fires a rifle while standing with the butt of the gun against his shoulder. If the forward momentum of a bullet is the same as the backward momentum of the gun, why isn't it as dangerous to be hit by the gun as by the bullet?

16. A large bedsheet is held vertically by two students. A third student, who happens to be the star pitcher on the baseball team, throws a raw egg at the sheet. Explain why the egg does not break when it hits the sheet, regardless of its initial speed. (If you try this, make sure the pitcher hits the sheet near its center, and do not allow the egg to fall on the floor after being caught by the sheet.)

PROBLEMS

Section 6.1   Momentum and Impulse

1. Show that the kinetic energy of a particle of mass m is related to the magnitude of the momentum p of that particle by $KE = p^2/2m$. *Note:* This expression is invalid for very high speed particles.

2. Calculate the magnitude of the linear momentum for the following cases: **(a)** a proton with mass 1.67×10^{-27} kg, moving with a speed of 5.00×10^6 m/s; **(b)** a 15.0-g bullet moving with a speed of 300 m/s; **(c)** a 75.0-kg sprinter running with a speed of 10.0 m/s; **(d)** the Earth (mass = 5.98×10^{24} kg) moving with an orbital speed equal to 2.98×10^4 m/s.

3. A 0.10-kg ball is thrown straight up into the air with an initial speed of 15 m/s. Find the momentum of the ball **(a)** at its maximum height and **(b)** halfway to its maximum height.

4. A pitcher claims he can throw a 0.145-kg baseball with as much momentum as a 3.00-g bullet moving with a speed of 1.50×10^3 m/s. **(a)** What must the baseball's speed be if the pitcher's claim is valid? **(b)** Which has greater kinetic energy, the ball or the bullet?

5. A 1500-kg car moving with a speed of 15 m/s collides with a utility pole and is brought to rest in 0.30 s. Find the magnitude of the average force exerted on the car during the collision.

6. A car is stopped for a traffic signal. When the light turns green, the car accelerates, increasing its speed from 0 to 5.20 m/s in 0.832 s. What are the magnitudes of the linear impulse and the average force experienced by a 70.0-kg passenger in the car during this time?

7. A 0.500-kg football is thrown with a speed of 15.0 m/s. A stationary receiver catches the ball and brings it to rest in 0.020 s. **(a)** What is the impulse delivered to the ball? **(b)** What is the average force exerted on the receiver?

8. The force shown in the force-time diagram in Figure P6.8 acts on a 1.5-kg mass. Find **(a)** the impulse of the

force, **(b)** the final velocity of the mass if it is initially at rest, and **(c)** the final velocity of the mass if it is initially moving along the x axis with a velocity of -2.0 m/s.

9. The force, $\mathbf{F}_x$, acting on a 2.00-kg particle varies in time as shown in Figure P6.9. Find **(a)** the impulse of the force, **(b)** the final velocity of the particle if it is initially at rest, and **(c)** the final velocity of the particle if it is initially moving along the x axis with a velocity of -2.00 m/s.

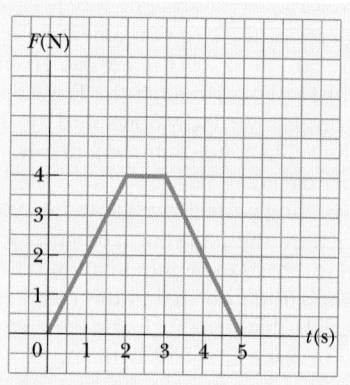

Figure P6.9   (Problems 9 and 38)

10. The forces shown in the force-time diagram in Figure P6.10 act on a 1.5-kg mass. Find **(a)** the impulse for the interval $t = 0$ to $t = 3.0$ s and **(b)** the impulse for the interval $t = 0$ to $t = 5.0$ s. **(c)** If the forces act on a 1.5-kg particle that is initially at rest, find the particle's speed at $t = 3.0$ s and at $t = 5.0$ s.

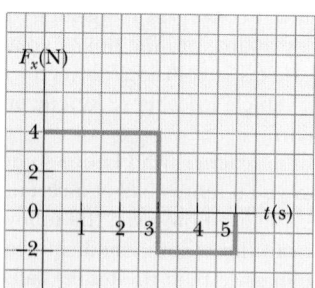

Figure P6.10

11. The front of a 1400-kg car is designed to absorb the shock of a collision by having a "crumple zone" in which the front 1.20 m of the car collapses in absorbing the shock of a collision. If a car traveling 25.0 m/s stops uni-

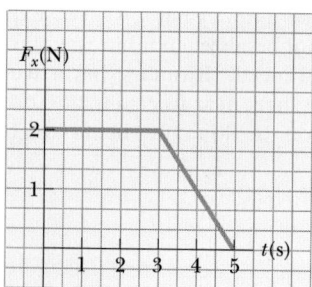

Figure P6.8

formly in 1.20 m, (a) how long does the collision last, (b) what is the magnitude of the average force on the car, and (c) what is the acceleration of the car in g's?

12. A 0.50-kg object is at rest at the origin of a coordinate system. A 3.0-N force in the $+x$ direction acts on the object for 1.50 s. (a) What is the velocity at the end of this interval? (b) At the end of this interval, a constant force of 4.0 N is applied in the $-x$ direction for 3.0 s. What is the velocity at the end of the 3.0 s?

WEB 13. A 0.15-kg baseball is thrown with a speed of 20 m/s. It is hit straight back at the pitcher with a final speed of 22 m/s. (a) What is the impulse delivered to the ball? (b) Find the average force exerted by the bat on the ball if the two are in contact for 2.0×10^{-3} s.

14. A 3.0-kg steel ball strikes a massive wall at 10 m/s at an angle of 60° with the plane of the wall. It bounces off with the same speed and angle (Fig. P6.14). If the ball is in contact with the wall for 0.20 s, what is the average force exerted on the ball by the wall?

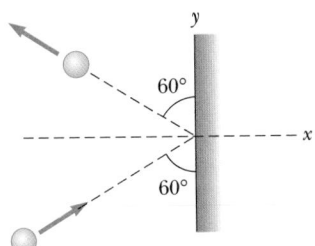

Figure P6.14

Section 6.2 Conservation of Momentum

15. A 730-N man stands in the middle of a frozen pond of radius 5.0 m. He is unable to get to the other side because of a lack of friction between his shoes and the ice. To overcome this difficulty, he throws his 1.2-kg physics textbook horizontally toward the north shore, at a speed of 5.0 m/s. How long does it take him to reach the south shore?

16. High-speed stroboscopic photographs show that the head of a 200-g golf club is traveling at 55 m/s just before it strikes a 46-g golf ball at rest on a tee. After the collision, the club head travels (in the same direction) at 40 m/s. Find the speed of the golf ball just after impact.

17. A rifle with a weight of 30 N fires a 5.0-g bullet with a speed of 300 m/s. (a) Find the recoil speed of the rifle. (b) If a 700-N man holds the rifle firmly against his shoulder, find the recoil speed of man and rifle.

18. A 45.0-kg girl is standing on a 150-kg plank. The plank, originally at rest, is free to slide on a frozen lake, which is a flat, frictionless surface. The girl begins to walk along the plank at a constant velocity of 1.50 m/s relative to the plank. (a) What is her velocity relative to the ice surface? (b) What is the velocity of the plank relative to the ice surface?

19. A 65.0-kg person throws a 0.0450-kg snowball forward with a ground speed of 30.0 m/s. A second person, with a mass of 60.0 kg, catches the snowball. Both people are on skates. The first person is initially moving forward with a speed of 2.50 m/s, and the second person is initially at rest. What are the velocities of the two people after the snowball is exchanged? Disregard the friction between the skates and the ice.

20. A 7.00-kg bowling ball is dropped from rest at an initial height of 3.00 m. (a) What is the speed of the Earth coming up to meet the ball just before the ball hits the ground? Use 5.98×10^{24} kg as the mass of the Earth. (b) Use your answer to part (a) to justify ignoring the motion of the Earth when dealing with the motions of terrestrial objects.

Section 6.3 Collisions

Section 6.4 Glancing Collisions

21. A 7.00-kg bowling ball collides head-on with a 2.00-kg bowling pin that was originally at rest. The pin flies forward with a speed of 3.00 m/s. If the ball continues forward with a speed of 1.80 m/s, what was the initial speed of the ball? Ignore rotation of the ball.

22. A 3.00-kg sphere makes a perfectly inelastic collision with a second sphere that is initially at rest. The composite system moves with a speed equal to one third the original speed of the 3.00-kg sphere. What is the mass of the second sphere?

23. A railroad car of mass 2.00×10^4 kg moving at 3.00 m/s collides and couples with two coupled railroad cars, each of the same mass as the single car and moving in the same direction at 1.20 m/s. (a) What is the speed of the three coupled cars after the collision? (b) How much kinetic energy is lost in the collision?

24. A 7.0-g bullet is fired into a 1.5-kg ballistic pendulum. The bullet emerges from the block with a speed of 200 m/s, and the block rises to a maximum height of 12 cm. Find the initial speed of the bullet.

WEB 25. A 0.030-kg bullet is fired vertically at 200 m/s into a 0.15-kg baseball that is initially at rest. How high does the combination rise after the collision, assuming the bullet embeds in the ball?

26. An 8.00-g bullet is fired into a 250-g block that is initially at rest at the edge of a table of 1.00-m height (Fig.

P6.26). The bullet remains in the block, and after the impact the block lands 2.00 m from the bottom of the table. Determine the initial speed of the bullet.

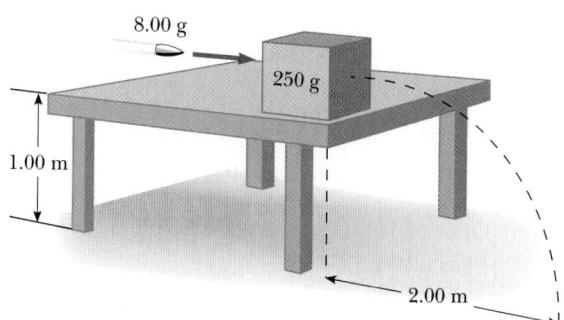

8.00 g

250 g

1.00 m

2.00 m

Figure P6.26

27. A 12.0-g bullet is fired horizontally into a 100-g wooden block that is initially at rest on a frictionless horizontal surface and connected to a spring of constant 150 N/m. If the bullet-block system compresses the spring by a maximum of 80.0 cm, what was the velocity of the bullet at impact with the block?

28. (a) Three carts of masses 4.0 kg, 10 kg, and 3.0 kg move on a frictionless horizontal track with speeds of 5.0 m/s, 3.0 m/s, and 4.0 m/s, as shown in Figure P6.28. The carts stick together after colliding. Find the final velocity of the three carts. (b) Does your answer require that all carts collide and stick together at the same time?

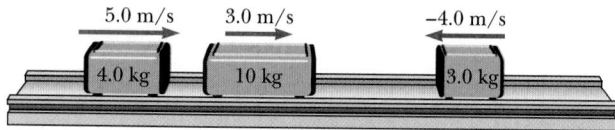

5.0 m/s   3.0 m/s   −4.0 m/s

4.0 kg   10 kg   3.0 kg

Figure P6.28

29. A 5.00-g object moving to the right at 20.0 cm/s makes an elastic head-on collision with a 10.0-g object that is initially at rest. Find (a) the velocity of each object after the collision and (b) the fraction of the initial kinetic energy transferred to the 10.0-g object.

30. A 10.0-g object moving to the right at 20.0 cm/s makes an elastic head-on collision with a 15.0-g object moving in the opposite direction at 30.0 cm/s. Find the velocity of each object after the collision.

Problem 40 is similar to Workbook Problem 34 (Workbook page 75).

31. A 25.0-g object moving to the right at 20.0 cm/s overtakes and collides elastically with a 10.0-g object moving in the same direction at 15.0 cm/s. Find the velocity of each object after the collision.

32. A billiard ball rolling across a table at 1.50 m/s makes a head-on elastic collision with an identical ball. Find the speed of each ball after the collision (a) when the second ball is initially at rest, (b) when the second ball is moving toward the first at a speed of 1.00 m/s, and (c) when the second ball is moving away from the first at a speed of 1.00 m/s.

33. A 90-kg fullback moving east with a speed of 5.0 m/s is tackled by a 95-kg opponent running north at 3.0 m/s. If the collision is perfectly inelastic, calculate (a) the velocity of the players just after the tackle and (b) the kinetic energy lost as a result of the collision. Can you account for the missing energy?

34. An 8.00-kg mass moving east at 15.0 m/s on a frictionless horizontal surface collides with a 10.0-kg mass that is initially at rest. After the collision, the 8.00-kg mass moves south at 4.00 m/s. (a) What is the velocity of the 10.0-kg mass after the collision? (b) What percentage of the initial kinetic energy is lost in the collision?

35. A 2000-kg car moving east at 10.0 m/s collides with a 3000-kg car moving north. The cars stick together and move as a unit after the collision, at an angle of 40.0° north of east and at a speed of 5.22 m/s. Find the velocity of the 3000-kg car before the collision.

36. A 0.30-kg puck, initially at rest on a frictionless horizontal surface, is struck by a 0.20-kg puck that is initially moving along the x axis with a velocity of 2.0 m/s. After the collision, the 0.20-kg puck has a speed of 1.0 m/s at an angle of $\theta = 53°$ to the positive x axis. (a) Determine the velocity of the 0.30-kg puck after the collision. (b) Find the fraction of kinetic energy lost in the collision.

WEB 37. A billiard ball moving at 5.00 m/s strikes a stationary ball of the same mass. After the collision, the first ball moves at 4.33 m/s at an angle of 30° with respect to the original line of motion. (a) Find the velocity (magnitude and direction) of the second ball after collision. (b) Was this an inelastic collision or an elastic collision?

ADDITIONAL PROBLEMS

38. Consult the force-time graph in Figure P6.9. Find the average force exerted on the particle for the time interval $t_i = 0$ to $t_f = 3.0$ s.

39. A 0.400-kg soccer ball approaches a player horizontally with a speed of 15.0 m/s. The player illegally strikes the ball with her hand and causes it to move in the opposite direction with a speed of 22.0 m/s. What impulse was delivered to the ball by the player?

40. An 80.0-kg astronaut is working on the engines of his spaceship, which is drifting through space with a con-

stant velocity. The astronaut, wishing to get a better view of the Universe, pushes against the ship and later finds himself 30.0 m behind the ship and moving so slowly that he can be considered to be at rest. Without a thruster, the only way to return to the ship is to throw his 0.500-kg wrench directly away from the ship. If he throws the wrench with a speed of 20.0 m/s, how long does it take him to reach the ship?

41. A 2.0-g particle moving at 8.0 m/s makes an elastic head-on collision with a resting 1.0-g object. (a) Find the speed of each after the collision. (b) If the stationary particle has a mass of 10 g, find the speed of each particle after the collision. (c) Find the final kinetic energy of the incident 2.0-g particle in the situations described in (a) and (b). In which case does the incident particle lose more kinetic energy?

42. Consider the ballistic pendulum described in Example 6.6 and shown in Figure 6.11. If the mass of the bullet is 8.00 g and the mass of the pendulum is 2.00 kg, find the ratio of the total kinetic energy after the collision to the kinetic energy before the collision. What accounts for the missing energy?

43. A 0.400-kg bead slides on a curved frictionless wire, starting from rest at point A in Figure P6.43. At point B the bead collides elastically with a 0.600-kg bead at rest. Find the distance the bead moves up the wire.

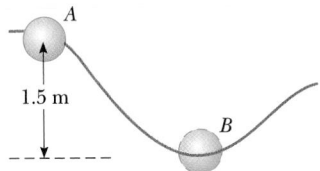

Figure P6.43

44. A standing 80-kg man steps off a 3.0-m high diving platform and begins to fall from rest. Two (2.0) seconds after reaching the water, the man comes to rest. What average force did the water exert on him?

WEB 45. A 12.0-g bullet is fired horizontally into a 100-g wooden block initially at rest on a horizontal surface. After impact, the block slides 7.5 m before coming to rest. If the coefficient of kinetic friction between block and surface is 0.650, what was the speed of the bullet immediately before impact?

46. Two blocks of masses $m_1 = 2.00$ kg and $m_2 = 4.00$ kg are each released from rest at a height of 5.00 m on a frictionless track, as shown in Figure P6.46, and undergo an elastic head-on collision. (a) Determine the velocity of each block just before the collision. (b) Determine the velocity of each block immediately after the collision.

Problem 46 is the same as Workbook Problem 38 (Workbook page 83).

(c) Determine the maximum heights to which m_1 and m_2 rise after collision.

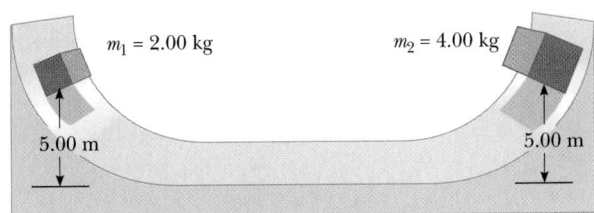

Figure P6.46

47. A 0.500-kg block is released from rest at the top of a frictionless track 2.50 m above the top of a table. It then collides elastically with a 1.00-kg mass that is initially at rest on the table, as shown in Figure P6.47. (a) Determine the speeds of the two masses just after the collision. (b) How high up the track does the 0.500-kg mass travel back after the collision? (c) How far away from the bottom of the table does the 1.00-kg mass land, given that the table is 2.00 m high? (d) How far away from the bottom of the table does the 0.500-kg mass eventually land?

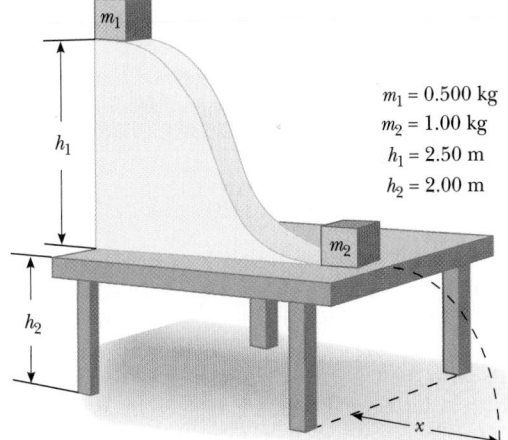

Figure P6.47

48. An 8.0-g bullet is fired into a 2.5-kg ballistic pendulum and becomes embedded in it. If the pendulum rises a vertical distance of 6.0 cm, calculate the initial speed of the bullet.

49. A small block of mass $m_1 = 0.500$ kg is released from rest at the top of a curved wedge of mass $m_2 = 3.00$ kg, which sits on a frictionless horizontal surface, as in Figure P6.49a. When the block leaves the wedge, its velocity

is measured to be 4.00 m/s to the right, as in Figure P6.49b. **(a)** What is the velocity of the wedge after the block reaches the horizontal surface? **(b)** What is the height, h, of the wedge?

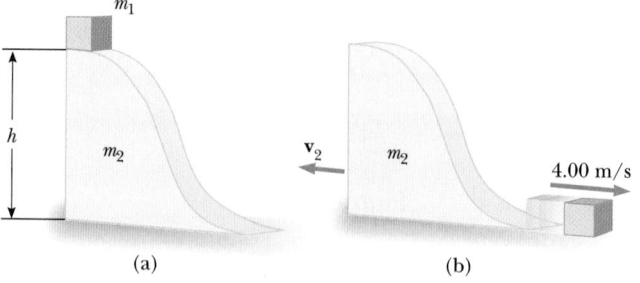

(a)   (b)

Figure P6.49

50. Two carts of equal mass, $m = 0.250$ kg, are placed on a frictionless track that has a light spring of force constant $k = 50.0$ N/m attached to one end of it, as in Figure P6.50. The red cart is given an initial velocity of $v_0 = 3.00$ m/s to the right, and the blue cart is initially at rest. If the carts collide elastically, find **(a)** the velocity of the carts just after the first collision and **(b)** the maximum compression in the spring.

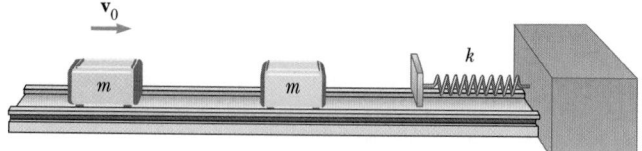

Figure P6.50

51. A cannon, initially resting on a frictionless surface, of mass $m_1 = 800$ kg (when unloaded) is loaded with a "shot" of mass $m_2 = 10.0$ kg. The cannon is aimed at mass $m_3 = 7990$ kg, which is connected to a massless spring of spring constant $k = 4500$ N/m, as in Figure P6.51a. The cannon is then fired, and the shot inelastically collides with mass m_3 and sticks in it, as shown in Figure P6.51b. The combined system then compresses the spring a maximum distance of $d = 0.5$ m, as in Figure P6.51c. **(a)** Determine the speed of m_2 just before it collides with m_3. (You may assume that m_2 travels in a straight line.) **(b)** Determine the recoil speed of the cannon. **(c)** The cannon recoils toward the right, and when it passes point A there is friction (with $\mu_k = 0.600$) between the cannon and the ground. How far to the right of point A does the cannon slide before coming to rest?

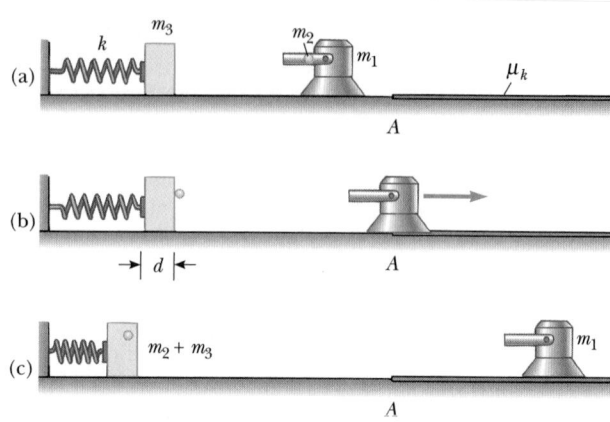

(a)

(b)

(c)

Figure P6.51

52. Water falls at the rate of 250 g/s from a height of 60 m into a 750-g bucket on a scale (without splashing). If the bucket is originally empty, what does the scale read after 3.0 s?

53. The bird perched on the swing in Figure P6.53 has a mass of 52.0 g, and the base of the swing has a mass of 153 g. Assume that the swing and bird are originally at rest and that the bird then takes off horizontally at 2.00 m/s. If the base can swing freely (i.e., without friction) around the pivot, how high will the base of the swing rise above its original level?

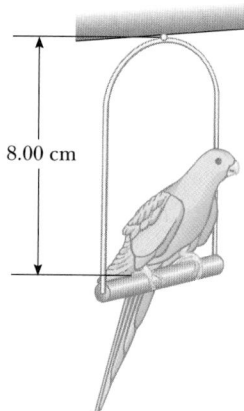

8.00 cm

Figure P6.53

54. A 2000-kg car traveling at 10.0 m/s collides with a 3000-kg car that is initially at rest at a stoplight. The cars stick together and move 2.00 m before friction causes them to stop. Determine the coefficient of kinetic friction between the cars and the road, assuming that the negative acceleration is constant and all wheels on both cars lock at the time of impact.

55. As shown in Figure P6.55, a bullet of mass m and speed v passes completely through a pendulum bob of mass M. The bullet emerges with a speed of $v/2$. The pendulum bob is suspended by a stiff rod of length ℓ and negligible mass. What is the minimum value of v such that the pendulum bob will barely swing through a complete vertical circle?

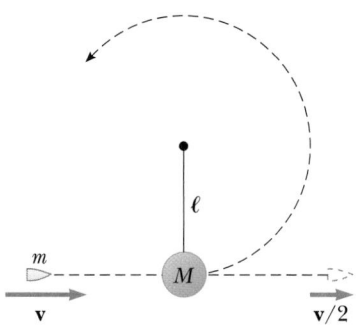

Figure P6.55

56. Two particles of masses m and $3m$ are moving toward each other along the x axis with the same initial speeds v_0. Mass m is traveling to the left, and mass $3m$ is traveling to the right. They undergo an elastic glancing collision such that mass m is moving downward after the collision at right angles from its initial direction. (a) Find the final speeds of the two masses. (b) What is the angle θ at which the mass $3m$ is scattered?

57. How fast can you set the Earth moving? In particular, when you jump straight up as high as you can, you give the Earth a maximum recoil speed of what order of magnitude? Visualize the Earth as a perfectly solid object. In your solution state the physical quantities you take as data and the values you measure or estimate for them.

58. A neutron in a reactor makes an elastic collision head on with a carbon atom that is initially at rest. (The mass of the carbon nucleus is about 12 times that of the neutron.) (a) What fraction of the neutron's kinetic energy is transferred to the carbon nucleus? (b) If the neutron's initial kinetic energy is 1.6×10^{-13} J, find its final kinetic energy and the kinetic energy of the carbon nucleus after the collision.

59. A cue ball traveling at 4.0 m/s makes a glancing, elastic collision with a target ball of equal mass that is initially at rest. The cue ball is deflected so that it makes an angle of 30° with its original direction of travel. Find (a) the angle between the velocity vectors of the two balls after the collision and (b) the speed of each ball after the collision.

60. A block of mass m lying on a rough horizontal surface is given an initial velocity of $\mathbf{v}_0$. After traveling a distance d, it makes a head-on elastic collision with a block of mass $2m$. How far does the second block move before coming to rest? (Assume the coefficient of friction μ is the same for both blocks.)

Circular Motion and the Law of Gravity

Chapter Outline

▲ PHYSICS PUZZLER

In a short race at a track event, such as a 200 m or 400 m sprint, the runners begin from staggered positions on the track. Why don't all the runners begin from the same line? (© Gerard Vandystadt/Photo Researchers, Inc.)

1n this chapter we shall investigate circular motion, a specific type of two-dimensional motion. We shall encounter such terms as *angular speed, angular acceleration,* and *centripetal acceleration.* The results derived will help you understand the motions of a diverse range of objects in our environment, from a car moving around a circular race track to clusters of galaxies orbiting a common center.

 7.1

We shall also introduce Newton's universal law of gravitation, one of the fundamental laws in nature, and show how this law, together with Newton's laws of motion, enables us to understand a variety of familiar phenomena.

Circular motion and the universal law of gravitation are related historically in that Newton discovered the law of gravity as a result of attempting to explain the circular motion of the Moon about the Earth and the motions of the planets about the Sun. Thus, it is appropriate to consider these two important physical topics together. A more general treatment of gravitational potential energy and the concept of escape velocity are provided in an optional section.

7.1 ANGULAR SPEED AND ANGULAR ACCELERATION

We began our study of linear motion by defining the terms *displacement, velocity,* and *acceleration.* We will take the same basic approach now as we turn to a study of rotational motion. Consider Figure 7.1a, a top view of a rotating compact disc. The axis of rotation is at the center of the disc, at O. A point P on the disc is at the distance r from the origin and moves about O in a circle of radius r. In fact, every point on the disc undergoes circular motion about O. To analyze such motion, it is convenient to set up a *fixed* reference line, as shown in Figure 7.1a. Let us assume that at time $t = 0$, the point P is on the reference line, as in Figure 7.1a, and a line is drawn on the disc from the origin out to P. After an interval of Δt has elapsed, P has advanced to a new position (Fig. 7.1b). In this time interval, the line OP has moved through the angle θ with respect to the reference line. Likewise, P has moved a distance of s measured along the circumference of the circle; s is called an *arc length.*

 7.2, SECTION 1

In situations we have encountered thus far, angles have been measured in degrees. However, in scientific work angles are often measured in *radians* (rad) rather than degrees, for the effect of simplifying certain equations. In fact, almost all of

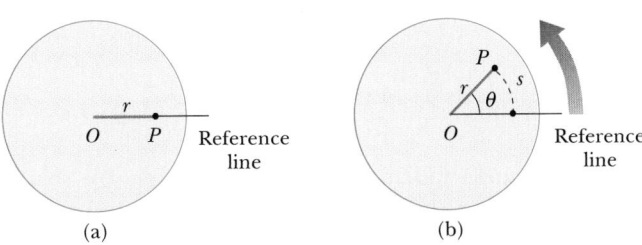

(a)   (b)

Figure 7.1   (a) The point P on a rotating compact disc at $t = 0$. (b) As the disc rotates, the point P moves through an arc length of s.

the equations derived in this chapter and the next require that angles be measured in radians. With reference to Figure 7.1b, when the arc length s is equal to the radius r, the angle θ swept out by r is equal to one radian. In general, any angle θ, measured in radians, is defined by the relation

$$\theta \equiv \frac{s}{r} \qquad [7.1]$$

The radian is a pure number with no dimensions. This can be seen from Equation 7.1, because θ is the ratio of an arc length (a distance) to the radius of the circle (also a distance).

To convert degrees to radians, note that when point P in Figure 7.1 moves through an angle of 360° (one revolution), the arc length s is equal to the circumference of the circle, $2\pi r$. From Equation 7.1 we see that the corresponding angle in radians is $2\pi r/r = 2\pi$ rad. Hence,

$$1 \text{ rad} \equiv \frac{360°}{2\pi} \approx 57.3°$$

From this definition, it follows that any angle in degrees can be converted to an angle in radians with the expression

$$\theta(\text{rad}) = \frac{\pi}{180°} \theta(\text{deg})$$

For example, 60° equals $\pi/3$ rad and 45° equals $\pi/4$ rad.

Returning to the compact disc, we see from Figure 7.2 that, as the disc rotates and a point on it moves from P to Q in a time of Δt, the angle through which the disc rotates is $\Delta\theta = \theta_2 - \theta_1$. We define $\Delta\theta$ as the **angular displacement. The average angular speed, $\overline{\omega}$ (ω is the Greek letter omega), of a rotating rigid object is the ratio of the angular displacement, $\Delta\theta$, to the time interval Δt:**

Average angular speed ▷

$$\overline{\omega} \equiv \frac{\theta_2 - \theta_1}{t_2 - t_1} = \frac{\Delta\theta}{\Delta t} \qquad [7.2]$$

By analogy with linear speed, **the instantaneous angular speed, ω, is defined as the limit of the average speed, $\Delta\theta/\Delta t$, as the time interval Δt approaches zero:**

Instantaneous angular speed ▷

$$\omega \equiv \lim_{\Delta t \to 0} \frac{\Delta\theta}{\Delta t} \qquad [7.3]$$

Angular speed has the units radians per second (rad/s). We shall take ω to be positive when θ is increasing (counterclockwise motion) and negative when θ is decreasing (clockwise motion). When the angular speed is constant, the instantaneous angular speed is equal to the average angular speed.

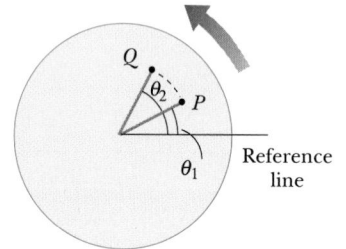

Figure 7.2 As a point on the compact disc moves from P to Q, the disc rotates through the angle $\Delta\theta = \theta_2 - \theta_1$.

EXAMPLE 7.1 Whirlybirds

The rotor on a helicopter turns at an angular speed of 320 revolutions per minute (in this book, we shall sometimes use the abbreviation rpm, but in most cases we shall use rev/min). Express this in radians per second.

Solution We shall use the conversion factors 1 rev = 2π rad and 60 s = 1 min, to give

$$320\,\frac{\text{rev}}{\text{min}} = 320\,\frac{\text{rev}}{\text{min}}\left(\frac{2\pi\,\text{rad}}{\text{rev}}\right)\left(\frac{1\,\text{min}}{60\,\text{s}}\right) = 10.7\pi\,\frac{\text{rad}}{\text{s}}$$

Figure 7.3 shows a bicycle turned upside down so that a repair person can work on the rear wheel. The bicycle pedals are turned so that at time t_1 the wheel has angular speed ω_1 (Fig. 7.3a), and at a later time, t_2, it has angular speed ω_2 (Fig. 7.3b). **The average angular acceleration, $\overline{\alpha}$ (α is the Greek letter alpha), of an object is defined as the ratio of the change in the angular speed to the time, Δt, it takes the object to undergo the change:**

$$\overline{\alpha} \equiv \frac{\text{change in angular speed}}{\text{time interval}} = \frac{\omega_2 - \omega_1}{t_2 - t_1} = \frac{\Delta\omega}{\Delta t} \qquad [7.4]$$

◀ Average angular acceleration

◀ Instantaneous angular acceleration

The instantaneous angular acceleration is defined as the limit of the ratio $\Delta\omega/\Delta t$ as Δt approaches zero. Angular acceleration has the units radians per second per second (rad/s^2). **When a rigid object rotates about a fixed axis, as does the bicycle wheel, every portion of the object has the same angular speed**

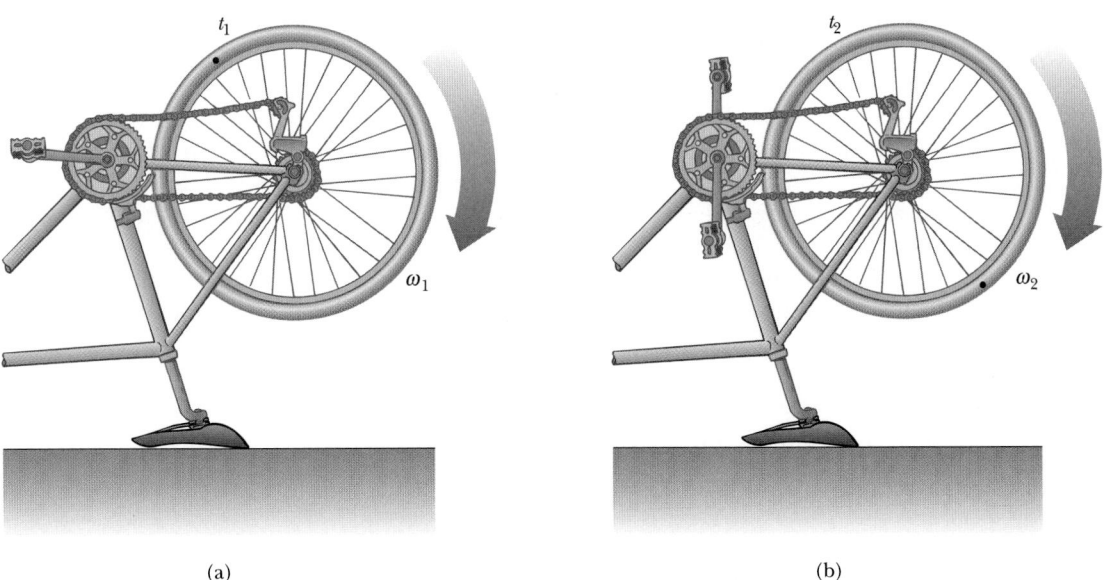

(a)   (b)

Figure 7.3 An accelerating bicycle wheel rotates with (a) angular speed ω_1 at time t_1 and (b) angular speed ω_2 at time t_2.

and the same angular acceleration. This, in fact, is precisely what makes these variables so useful for describing rotational motion.

The following argument should convince you that ω and α are the same for every point on the wheel. If a point on the rim of the wheel had a greater angular speed than a point nearer the center, the shape of the wheel would be changing. The wheel remains circular (symmetrically distributed about the axle) only if all points have the same angular speed and the same angular acceleration.

7.2   ROTATIONAL MOTION UNDER CONSTANT ANGULAR ACCELERATION

Let us pause for a moment to consider some similarities between the equations we have found thus far for rotational motion and those we found for linear motion in earlier chapters. For example, compare the defining equation for average angular speed,

$$\bar{\omega} \equiv \frac{\theta_f - \theta_i}{t_f - t_i} = \frac{\Delta\theta}{\Delta t}$$

with the defining equation for average linear speed,

$$\bar{v} \equiv \frac{x_f - x_i}{t_f - t_i} = \frac{\Delta x}{\Delta t}$$

The equations are similar in the sense that θ replaces x and ω replaces v. Take careful note of such similarities as you study rotational motion, because virtually every linear quantity we have encountered thus far has a corresponding "twin" in rotational motion. Once you are adept at recognizing such analogies, you will find it unnecessary to memorize many of the equations in this chapter. In addition, the techniques for solving rotational motion problems are quite similar to those you have already learned for linear motion. For example, problems concerned with objects that rotate with constant angular acceleration can be solved in much the same manner as those dealing with linear motion under constant acceleration. If you understand problems involving objects that move with constant linear acceleration, these rotational motion problems should be little more than a review for you.

An additional analogy between linear motion and rotational motion is revealed when we compare the defining equation for average angular acceleration,

$$\bar{\alpha} \equiv \frac{\omega_f - \omega_i}{t_f - t_i} = \frac{\Delta\omega}{\Delta t}$$

with the defining equation for average linear acceleration,

$$\bar{a} \equiv \frac{v_f - v_i}{t_f - t_i} = \frac{\Delta v}{\Delta t}$$

In light of the analogies between variables in linear motion and those in rotational motion, it should not surprise you that the equations of rotational motion involve the variables θ, ω, and α. In Chapter 2, Section 2.6, we developed a set of kinematic equations for linear motion under constant acceleration. The same procedure can be used to derive a similar set of equations for rotational motion under constant angular acceleration. The resulting equations of rotational kinematics, along with the corresponding equations for linear motion under constant acceleration, are as follows:

Rotational Motion About a Fixed Axis with α Constant (Variables: θ and ω)	Linear Motion with a Constant (Variables: x and v)	
$\omega = \omega_0 + \alpha t$	$v = v_0 + at$	[7.5]
$\theta = \omega_0 t + \frac{1}{2}\alpha t^2$	$x = v_0 t + \frac{1}{2}at^2$	[7.6]
$\omega^2 = \omega_0^2 + 2\alpha\theta$	$v^2 = v_0^2 + 2ax$	[7.7]

Again, note the one-to-one correspondence between the rotational equations involving the angular variables θ, ω, and α and the equations of linear motion involving the variables x, v, and a.

EXAMPLE 7.2 The Rotating Wheel

The bicycle wheel of Figure 7.3 rotates with a constant angular acceleration of 3.5 rad/s². If the initial angular speed of the wheel is 2.0 rad/s at $t_0 = 0$, (a) through what angle does the wheel rotate in 2.0 s?

Workbook Problem 42 – also deals with constant angular acceleration (see Workbook page 92).

Solution Because we are given $\omega_0 = 2.0$ rad/s and $\alpha = 3.5$ rad/s², we use

$$\theta = \omega_0 t + \frac{1}{2}\alpha t^2$$

$$\theta = (2.0 \text{ rad/s})(2.0 \text{ s}) + \frac{1}{2}(3.5 \text{ rad/s}^2)(2.0 \text{ s})^2$$

$$= 11 \text{ rad} = \boxed{630°}$$

(b) What is the angular speed at $t = 2.0$ s?

Solution Making use of Equation 7.5, we find that

$$\omega = \omega_0 + \alpha t = 2.0 \text{ rad/s} + (3.5 \text{ rad/s}^2)(2.0 \text{ s}) = \boxed{9.0 \text{ rad/s}}$$

Exercise Find the angular speed of the wheel at $t = 2.0$ s by making use of Equation 7.7 and the results of part (a).

7.3 RELATIONS BETWEEN ANGULAR AND LINEAR QUANTITIES

In this section we shall derive some useful relations between the angular speed and acceleration of a rotating object and the linear speed and acceleration of an arbitrary point in the object. Consider the arbitrarily shaped object in Figure 7.4, rotating about the z axis through the point O. Assume that the object rotates through

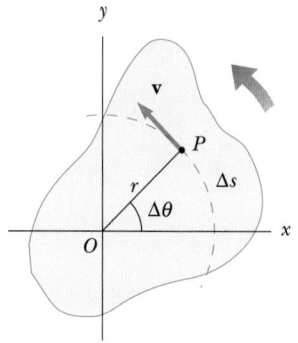

Figure 7.4 Rotation of an object about an axis through O that is perpendicular to the plane of the figure (the z axis). Note that a point, P, on the object rotates in a circle of radius r centered at O.

the angle $\Delta\theta$, and hence point P moves through the arc length Δs in the interval Δt. We know from the defining equation for radian measure that

$$\Delta\theta = \frac{\Delta s}{r}$$

Let us now divide both sides of this equation by Δt, the interval during which the rotation occurred:

$$\frac{\Delta\theta}{\Delta t} = \frac{1}{r}\frac{\Delta s}{\Delta t}$$

If Δt is very small, then the angle $\Delta\theta$ through which the object rotates is small and the ratio $\Delta\theta/\Delta t$ is the instantaneous angular speed, ω. Also, Δs is very small when Δt is very small, and the ratio $\Delta s/\Delta t$ equals the instantaneous linear speed, v, for small values of Δt. Hence, the preceding equation is equivalent to

$$\omega = \frac{v}{r}$$

Figure 7.4 allows us to interpret this equation. The distance Δs is traversed along an arc of the circular path followed by the point P as it moves during the time Δt. Thus, v must be the linear speed of a point lying along this arc, where the direction of v is *tangent to the circular path*. As a result, we often refer to this linear speed as the *tangential speed* of a particle moving in a circular path, and write

Tangential speed ▷

$$v_t = r\omega \qquad\qquad [7.8]$$

That is, **the tangential speed of a point on a rotating object equals the distance of that point from the axis of rotation multiplied by the angular speed.** Equation 7.8 shows that the linear speed of a point on the rotating object increases with movement outward from the center of rotation toward the rim, as one would intuitively expect. **Although every point on the rotating object has the same angular speed, not every point has the same linear (tangential) speed.**

Exercise caution when using Equation 7.8. It has been derived using the defining equation for radian measure and hence is valid only when ω is measured in radians per unit time. Other measures of angular speed, such as degrees per second and revolutions per second, are not to be used in Equation 7.8.

To find a second equation relating linear and angular quantities, imagine that an object rotating about a fixed axis (Fig. 7.4) changes its angular speed by $\Delta\omega$ in the interval Δt. At the end of this interval, the speed of a point on the object, such as P, has changed by the amount Δv_t. From Equation 7.8 we have

$$\Delta v_t = r\,\Delta\omega$$

Dividing by Δt gives

$$\frac{\Delta v_t}{\Delta t} = r\frac{\Delta\omega}{\Delta t}$$

If the time interval Δt is very small, then the ratio $\Delta v_t/\Delta t$ is the tangential acceleration of that point and $\Delta\omega/\Delta t$ is the angular acceleration. Therefore, we see that

Tangential acceleration ▷

$$a_t = r\alpha \qquad\qquad [7.9]$$

That is, **the tangential acceleration of a point on a rotating object equals the distance of that point from the axis of rotation multiplied by the angular acceleration.** Again, radian measure must be used for the angular acceleration term in this equation.

There is one more equation that relates linear quantities to angular quantities, but we shall defer its derivation to the next section.

Thinking Physics 1

The launch area for the European Space Agency is not in Europe—it is in South America. Why?

Explanation Placing a satellite in Earth orbit requires providing a large tangential speed to the satellite. This is the task of the rocket propulsion system. Anything that reduces the requirements on the rocket system is a welcome contribution. The surface of the Earth is already traveling toward the east at a high speed, due to the rotation of the Earth. Thus, if rockets are launched toward the east, the rotation of the Earth provides some initial tangential speed, reducing the requirements on the rocket. If rockets were launched from Europe, which is at a relatively large-angle latitude, the contribution of the Earth's rotation is relatively small, because the distance between Europe and the rotation axis of the Earth is relatively small. The ideal place for launching is at the Equator, which is as far as one can be from the rotation axis of the Earth and still be on the surface of the Earth. This results in the largest possible tangential speed due to the Earth's rotation. The European Space Agency exploits this advantage by launching from French Guiana, which is only a few degrees north of the equator.

Applying Physics 1

When a tall smokestack falls over, it often breaks somewhere along its length before it hits the ground, as in Figure 7.5. Why does this happen?

Explanation As the smokestack rotates around its base, each higher portion of the smokestack falls with an increasing tangential acceleration. The tangential acceleration of a given point on the smokestack is proportional to the distance of that portion from the base. As the acceleration increases, eventually higher portions of the smokestack will experience an acceleration greater than that which could result from gravity alone. This can only happen if these portions are being pulled downward by a force in addition to the gravitational force. The force that causes this to occur is the shear force from lower portions of the smokestack. Eventually, the shear force that provides this acceleration is larger than the smokestack can withstand, and it breaks.

Figure 7.5 (Applying Physics 1) A falling smokestack.

EXAMPLE 7.3 Computer Disks

A floppy disk in a computer rotates from rest up to an angular speed of 31.4 rad/s in a time of 0.892 s.
(a) What is the angular acceleration of the disk, assuming the angular acceleration is uniform?

Solution   If we use $\omega = \omega_0 + \alpha t$ and the fact that $\omega_0 = 0$ at $t = 0$, we get

$$\alpha = \frac{\omega}{t} = \frac{31.4 \text{ rad/s}}{0.892 \text{ s}} = \boxed{35.2 \text{ rad/s}^2}$$

(b) How many rotations does the disk make while coming up to speed?

Solution   Equation 7.6 enables us to find the angular displacement during the 0.892-s time interval. Taking $\omega_0 = 0$ gives

$$\theta = \omega_0 t + \tfrac{1}{2}\alpha t^2 = \tfrac{1}{2}(35.2 \text{ rad/s}^2)(0.892 \text{ s})^2 = 14.0 \text{ rad}$$

Because 2π rad = 1 rev, this angular displacement corresponds to $\boxed{2.23 \text{ rev.}}$

(c) If the radius of the disk is 4.45 cm, find the final linear speed of a microbe riding on the rim of the disk.

Solution   The relation $v_t = r\omega$ and the given values lead to

$$v_t = r\omega = (0.0445 \text{ m})(31.4 \text{ rad/s}) = \boxed{1.40 \text{ m/s}}$$

(d) What is the magnitude of the tangential acceleration of the microbe at this time?

Solution   We use $a_t = r\alpha$, which gives

$$a_t = r\alpha = (0.0445 \text{ m})(35.2 \text{ rad/s}^2) = \boxed{1.57 \text{ m/s}^2}$$

Exercise   What is the angular speed and angular displacement of the disk 0.300 s after it begins to rotate?

Answer   10.6 rad/s; 1.58 rad

APPLICATION

Phonograph Records and Compact Discs.

Before compact discs became the medium of choice for recorded music, phonograph records were popular. There are similarities between the rotational motion of phonograph records and compact discs, and there are some interesting differences. A phonograph record, for instance, rotates at a constant angular speed. Popular angular speeds were 33⅓ rev/min for long-playing albums (hence the nickname "LP"), 45 rev/min for "singles," and 78 rev/min used in very early recordings. At the outer edge of the record, the pickup needle (stylus) was moving over the vinyl material at a faster tangential speed than when the needle was close to the center of the record. As a result, the sound information was compressed into a smaller length of track near the center of the record than near the outer edge.

Compact discs, on the other hand, are designed such that the laser pickup moves over the information in the tracks at a constant tangential speed. Because the pickup moves radially as it follows the tracks of information, this requires that the angular speed of the compact disc must vary according to the radial position of the laser. Because the tangential speed is fixed, the information density (per length of track) anywhere on the disc is the same. Example 7.4 demonstrates numerical calculations for both compact discs and phonograph records.

EXAMPLE 7.4 Length of a Compact Disc Track

(a) A compact disc is designed such that as the read head moves out from the center of the disc, the angular speed of the disc changes so that the linear speed at the position of the head will always be at a constant value of about 1.3 m/s. Find the angular speed of the disc when the read head is at a distance of 5.0 cm from the center of the disc and at 8.0 cm.

Solution From $v_t = r\omega$, we have at $r = 5.0$ cm,

$$\omega = \frac{v_t}{r} = \frac{1.3 \text{ m/s}}{5.0 \times 10^{-2} \text{ m}} = \boxed{26.0 \text{ rad/s}}$$

and at 8.0 cm,

$$\omega = \frac{v_t}{r} = \frac{1.3 \text{ m/s}}{8.0 \times 10^{-2} \text{ m}} = \boxed{16.3 \text{ rad/s}}$$

(b) On the other hand, an old-fashioned record player rotated at a constant angular speed while the linear speed of the detector, a stylus, changed. Find the linear speed of a 45.0 rpm record at points 5.0 and 8.0 cm from the center.

Solution Because 45.0 rev/min = 4.7 rad/s, the linear speed at 5.0 cm is

$$v_t = r\omega = (5.0 \times 10^{-2} \text{ m})(4.7 \text{ rad/s}) = \boxed{0.24 \text{ m/s}}$$

At 8.0 cm, we have

$$v_t = r\omega = (8.0 \times 10^{-2} \text{ m})(4.7 \text{ rad/s}) = \boxed{0.38 \text{ m/s}}$$

(c) In both the CD and record, information is recorded in a continuous spiral track as one moves inward from the edge of the disc. Use the information above to calculate the total length of the track for a CD designed to play for 1 hour.

Solution The read head moves across the CD with a constant linear speed of 1.3 m/s. Thus, the total length of the track that it moves over in 3600 seconds is

$$x = v_t t = (1.3 \text{ m/s})(3600 \text{ s}) = \boxed{4680 \text{ m}}$$

7.4 CENTRIPETAL ACCELERATION

Figure 7.6a shows a car moving in a circular path with *constant linear speed v*. Students are often surprised to find that **even though the car moves at a constant speed, it still has an acceleration.** To see why this occurs, consider the defining equation for average acceleration:

 3.6

$$\mathbf{a} = \frac{\mathbf{v}_f - \mathbf{v}_i}{t_f - t_i} \qquad [7.10]$$

Note that the average acceleration depends on the *change in the velocity vector*. Because velocity is a vector, there are two ways in which an acceleration can be produced: by a change in the *magnitude* of the velocity and by a change in the *direction*

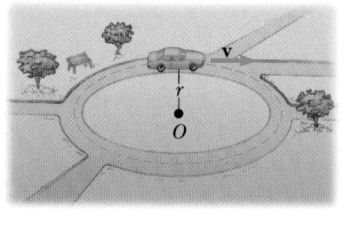

(a)

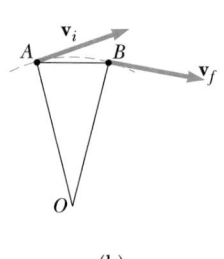

(b)

Figure 7.6   (a) Circular motion of a car moving with constant speed. (b) As the car moves along the circular path from A to B, the direction of its velocity vector changes, so the car undergoes a centripetal acceleration.

of the velocity. It is the latter situation that occurs for the car moving in a circular path with constant speed (see Fig. 7.6b). We shall show that the acceleration vector in this case is perpendicular to the path and always points toward the center of the circle. An acceleration of this nature is called a **centripetal** (center-seeking) **acceleration.** Its magnitude is given by

$$a_c = \frac{v^2}{r} \qquad [7.11]$$

To derive Equation 7.11, consider Figure 7.7a. Here an object is seen first at point A with velocity $\mathbf{v}_i$ at time t_i and then at point B with velocity $\mathbf{v}_f$ at a later time, t_f. Let us assume that here $\mathbf{v}_i$ and $\mathbf{v}_f$ differ only in direction; their magnitudes are the same (that is, $v_i = v_f = v$). To calculate the acceleration, we begin with Equation 7.10, which indicates that we must vectorially subtract $\mathbf{v}_i$ from $\mathbf{v}_f$:

$$\mathbf{a} = \frac{\mathbf{v}_f - \mathbf{v}_i}{t_f - t_i} = \frac{\Delta \mathbf{v}}{\Delta t} \qquad [7.12]$$

where $\Delta \mathbf{v} = \mathbf{v}_f - \mathbf{v}_i$ is the change in velocity. This can be accomplished graphically, as shown by the vector triangle in Figure 7.7b. Note that when Δt is very small, Δs and $\Delta \theta$ will also be very small. In this case, $\mathbf{v}_f$ will almost parallel $\mathbf{v}_i$, and the vector $\Delta \mathbf{v}$ will be approximately perpendicular to them, pointing toward the center of the circle. In the limiting case where Δt becomes vanishingly small, $\Delta \mathbf{v}$ will point exactly toward the center of the circle. Furthermore, in this limiting case the acceleration will also be directed toward the center of the circle because it is in the direction $\Delta \mathbf{v}$.

Now consider the triangle in Figure 7.7a, which has sides Δs and r. This triangle and the one formed by the vectors in Figure 7.7b are similar. This enables us to write a relationship between the lengths of the sides:

$$\frac{\Delta v}{v} = \frac{\Delta s}{r}$$

or

$$\Delta v = \frac{v}{r} \Delta s$$

This can be substituted into Equation 7.12 for Δv to give

$$a = \frac{v}{r} \frac{\Delta s}{\Delta t} \qquad [7.13]$$

In this situation, Δs is a small distance measured along the arc of the circle (a tangential distance), so $v = \Delta s / \Delta t$, where v is the tangential speed. Therefore, Equation 7.13 reduces to Equation 7.11:

$$a_c = \frac{v^2}{r}$$

Because the tangential speed is related to the angular speed through the relation $v_t = r\omega$ (Eq. 7.8), an alternative form of Equation 7.11 is

$$a_c = \frac{r^2 \omega^2}{r} = r\omega^2 \qquad [7.14]$$

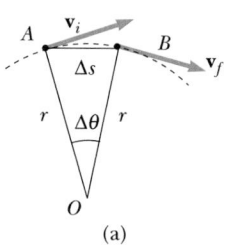

(a)

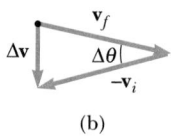

(b)

Figure 7.7   (a) As the particle moves from A to B, the direction of its velocity vector changes from $\mathbf{v}_i$ to $\mathbf{v}_f$. (b) The construction for determining the direction of the change in velocity, $\Delta \mathbf{v}$, which is toward the center of the circle.

Thus, we conclude that

> In circular motion the centripetal acceleration is directed inward toward the center of the circle and has a magnitude given by either v^2/r or $r\omega^2$.

◀ Centripetal acceleration

You should show that the dimensions of a_c are L/T^2, as required.

In order to clear up any misconceptions that might exist concerning centripetal and tangential acceleration, let us consider a car moving around a circular race track. Because the car is moving in a circular path, it always has a centripetal component of acceleration because its direction of travel, and hence the direction of its velocity, is continuously changing. If the car's speed is increasing or decreasing, it also has a tangential component of acceleration. To summarize, the tangential component of acceleration is due to changing speed; the centripetal component of acceleration is due to changing direction.

When both components of acceleration exist simultaneously, the tangential acceleration is tangent to the circular path and the centripetal acceleration points toward the center of the circular path. Because these components of acceleration are perpendicular to each other, we can find the magnitude of the **total acceleration** using the Pythagorean theorem:

$$a = \sqrt{a_t^2 + a_c^2}$$ [7.15]

◀ Total acceleration

EXAMPLE 7.5 Let's Go for a Spin

A test car moves at a constant speed of 10 m/s around a circular road of radius 50 m. Find the car's (a) centripetal acceleration and (b) angular speed.

Solution   (a) From Equation 7.11, the magnitude of the centripetal acceleration of the car is found to be

$$a_c = \frac{v^2}{r} = \frac{(10\ \text{m/s})^2}{50\ \text{m}} = \boxed{2.0\ \text{m/s}^2}$$

By definition, the direction of $\mathbf{a}_c$ is toward the center of curvature of the road.

(b) The angular speed of the car can be found using the expression $v_t = r\omega$, which gives

$$\omega = \frac{v_t}{r} = \frac{10\ \text{m/s}}{50\ \text{m}} = \boxed{0.20\ \text{rad/s}}$$

Note that we can also find the centripetal acceleration by using this value of ω, in an alternative method to part (a). From Equation 7.14 we have

$$a_c = r\omega^2 = (50\ \text{m})(0.20\ \text{rad/s})^2 = 2.0\ \text{m/s}^2$$

As expected, the two methods for finding the centripetal acceleration give the same answer.

Exercise   Find the car's tangential acceleration and total acceleration.

Answer   The tangential acceleration is zero because the speed of the car remains constant. Because $a_t = 0$, the total acceleration in this example equals the centripetal acceleration, 2.0 m/s², found in part (a).

Workbook Problem 13 – (Workbook page 34) extends Example 7.5 with an additional type of motion.

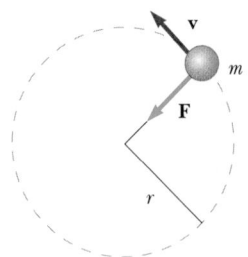

Figure 7.8 A ball attached to a string of length, r, rotating in a circular path at constant speed.

4.7, BOTH SECTIONS

7.2, SECTION 2

APPLICATION

Banked Roadways.

WEB

For Web links to information about amusement park rides (also see page 192), visit the textbook Web site at http://www.harcourtcollege.com/physics/cptech

7.5   FORCES CAUSING CENTRIPETAL ACCELERATION

Consider a ball of mass m tied to a string of length r and being whirled in a horizontal circular path, as in Figure 7.8. Let us assume that the ball moves with constant speed. Because the velocity vector, **v,** changes its direction continuously during the motion, the ball experiences a centripetal acceleration directed toward the center of motion, as described in Section 7.4, with magnitude

$$a_c = \frac{v_t^2}{r}$$

The inertia of the ball tends to maintain motion in a straight-line path; however, the string overcomes this by exerting a force on the ball that makes it instead follow a circular path. This force (equal to the force of tension) is directed along the length of the string toward the center of the circle in Figure 7.8. Thus, the equation for Newton's second law along the radial direction is

$$F = ma_c = m\frac{v_t^2}{r} \qquad \text{[7.16]}$$

Because it acts at right angles to the motion, a force that produces a centripetal acceleration causes a change in the direction of the velocity. Beyond this, such a force is no different from any of the other forces we have studied. For example, friction between tires and road provides the force that enables a race car to travel in a circular path on a flat road, and the gravitational force exerted on the Moon by the Earth provides the force necessary to keep the Moon in its orbit. (We shall discuss the gravitational force in more detail in Section 7.7.)

Regardless of the example used, if the force that produces a centripetal acceleration vanishes, the object does not continue to move in its circular path; instead, it moves along a straight-line path tangent to the circle. But as the race car in the previous paragraph travels faster and faster on the curved flat road, the centripetal acceleration will increase, as will the friction force needed to cause this acceleration. At some speed, the friction force needed to keep the car on a circular path will be greater than the maximum possible static friction force allowed by the static coefficient of friction. At this point, the car will skid off the road.

In order to prevent this from happening, roadways are often banked instead of being built with flat curves. By banking the roadway toward the inside of the circular path, there is a component of the normal force of the road on the car that is directed toward the center of the path. This component of the normal force adds to the centrally directed friction force. Thus, the car can negotiate a banked curve at a higher speed than it can an unbanked curve. The banking of roadways is very evident at high speed race tracks, in bicycle tracks in velodromes, and in bobsled tracks.

Thinking Physics 2

Suppose you are the only rider on a Ferris wheel. The drive mechanism breaks after the ride starts, so that you continue to rotate freely, without friction. As you pass through the lowest point (let us call it the 6 o'clock point), you will feel the heaviest—that is, the seat will exert the largest upward force on you. As you pass over the highest point (the 12 o'clock point), you will feel the lightest. At the 3 o'clock and 9 o'clock points, will you feel as if you have the "correct" weight? If not, where *do* you feel as if your weight is "correct"?

Explanation If the wheel were rotating at a constant speed, the answer to the question would be *yes*. However, because you are the only rider, the wheel is *unbalanced* and does not rotate with a constant speed. As you descend from the top point, your speed increases. After you pass through the lowest point, your speed decreases as you move back to the top. Thus, you will experience tangential acceleration as well as centripetal acceleration. At the 3 o'clock position, as you are moving downward, you have a downward tangential acceleration. Thus, just as in a downward accelerating elevator, you will feel lighter. At the 9 o'clock position, as you are moving upward, you are accelerating downward tangentially. Again, you will feel lighter at this point. In order to counteract the *downward* component of tangential acceleration near both positions, you need some *upward* component from the centripetal acceleration. This will be present if you are at positions *below* 3 o'clock or 9 o'clock, so that the centripetal acceleration has an upward component. The exact positions will depend on your velocity and the details of the Ferris wheel.

As these speed skaters round a curve, the force of friction exerted on their skates provides the centripetal acceleration. *(Bill Bachman/Photo Researchers, Inc.)*

Problem-Solving Strategy

Forces that Cause Centripetal Acceleration

Use the following steps when dealing with centripetal accelerations and forces that produce them:

1. Draw a free-body diagram of the object(s) under consideration, showing and labeling all forces that act on it.
2. Choose a coordinate system that has one axis perpendicular to the circular path followed by the object and one axis tangent to the circular path.
3. Find the net force toward the center of the circular path. This is the force that causes the centripetal acceleration.
4. From here on the steps are virtually identical to those encountered when solving Newton's second law problems with $\Sigma \mathbf{F} = m\mathbf{a}$. Also, note that the magnitude of the centripetal acceleration can always be written $a_c = v_t^2/r$.

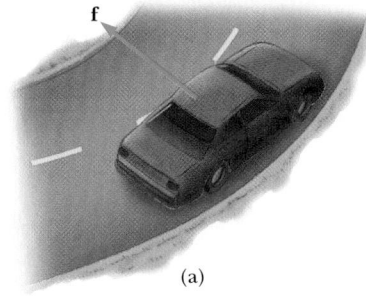

(a)

 EXAMPLE 7.6 Buckle Up for Safety

A car travels at a constant speed of 30.0 mi/h (13.4 m/s) on a level circular turn of radius 50.0 m, as shown in the bird's-eye view in Figure 7.9a. What is the minimum coefficient of static friction between the tires and roadway in order that the car makes the circular turn without sliding?

Reasoning The force in the radial direction acting on the car is the force of static friction directed toward the center of the circular path, as in Figure 7.9a. Thus, the objective in this example will be to find an expression for the maximum frictional force $f = \mu_s n$ and use this in Equation 7.16.

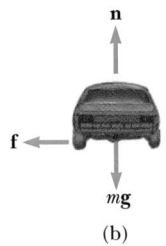

(b)

Figure 7.9 (Example 7.6) (a) Top view of a car on a curved path. (b) A free-body diagram for the car, showing an end view.

Solution   Thus, Equation 7.16 becomes

$$(1) \qquad f = m \frac{v_t^2}{r}$$

All the forces acting on the car are shown in Figure 7.9b. Because equilibrium is in the vertical direction, the normal force upward is balanced by the force of gravity downward, so that

$$n = mg$$

From this expression we can find the maximum force of static friction, as follows:

$$(2) \qquad f = \mu_s n = \mu_s mg$$

By setting the right-hand sides of (1) and (2) equal to each other, we find that

$$\mu_s \cancel{m} g = \cancel{m} \frac{v_t^2}{r}$$

or

$$(3) \qquad \mu_s = \frac{v_t^2}{rg}$$

Therefore, the minimum coefficient of static friction that is required for the car to make the turn without sliding outward is

$$\mu_s = \frac{v_t^2}{rg} = \frac{(13.4 \text{ m/s})^2}{(50.0 \text{ m})(9.80 \text{ m/s}^2)} = \boxed{0.366}$$

The value of μ_s for rubber on dry concrete is very close to 1; thus, the car can negotiate the curve with ease. If the road were wet or icy, however, the value for μ_s could be 0.2 or lower. Under such conditions, the radial force provided by static friction would not be great enough to enable the car to follow the circular path, and it would slide off the roadway.

EXAMPLE 7.7   Having Fun with a Yo-Yo

A child swings a yo-yo of weight mg in a horizontal circle so that the cord makes an angle of 30.0° with the vertical, as in Figure 7.10a. Find the centripetal acceleration of the yo-yo.

Reasoning   Two forces act on the yo-yo: the force of gravity and the force of tension in the cord, **T**, as in Figure 7.10b. The first condition for equilibrium applied in the y direction, $\Sigma F_y = 0$, will enable us to find T. The component of **T** toward the center of the circular path will be the force that produces the centripetal acceleration, and we can finally find a_c from $F = ma_c$.

Solution   Applying ΣF_y gives

$$T \cos 30.0° - mg = 0$$

$$T = \frac{mg}{\cos 30.0°}$$

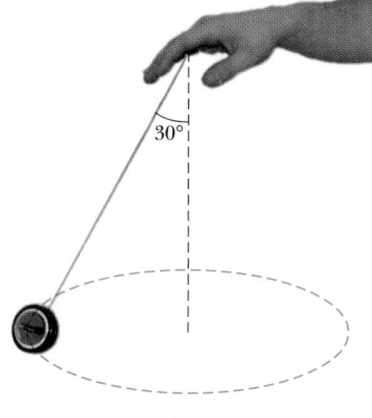

(a)

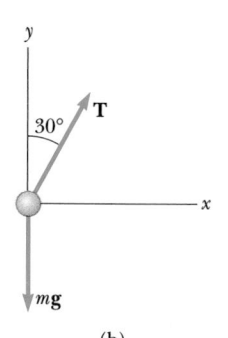

(b)

Figure 7.10   (Example 7.7) (a) A yo-yo swinging in a horizontal circle so that the cord makes a constant angle of 30° with the vertical. (b) A free-body diagram for the yo-yo.

The *horizontal component* of **T** causes the centripetal acceleration.

$$F = T \sin 30.0° = \frac{mg \sin 30.0°}{\cos 30.0°} = mg \tan 30.0°$$

The centripetal acceleration can now be found by applying Newton's second law along the horizontal direction:

$$F = ma_c$$

$$a_c = \frac{F}{m} = \frac{mg \tan 30.0°}{m} = g \tan 30.0° = (9.80 \text{ m/s}^2)(\tan 30.0°)$$

$$= \boxed{5.66 \text{ m/s}^2}$$

7.6 DESCRIBING MOTION OF A ROTATING SYSTEM

We have seen that an object moving in a circle of radius r with constant speed v_t has a centripetal acceleration whose magnitude is v_t^2/r and whose direction is toward the center of rotation. The force necessary to maintain this centripetal acceleration must also act toward the center of rotation. In the case of a ball rotating at the end of a string, the force exerted on the ball by the string (equal in magnitude to the tension in the string) is the force causing the centripetal acceleration; for a satellite in a circular orbit around the Earth, the force of gravity plays that role, and for a car rounding a curve on a level road, it is the force of friction between the tires and the pavement; and so forth.

In order to better understand the motion of a rotating system, consider a car approaching a curved exit ramp at high speed as in Figure 7.11. As the car takes the sharp left turn onto the ramp, a person sitting in the passenger seat slides to the right across the seat and hits the door. At that point, the force of the door keeps her from being ejected from the car. What causes the passenger to move toward the door? A popular but *erroneous* explanation is that some mysterious force pushes her outward. (This is sometimes erroneously called the "centrifugal" force.)

The phenomenon is correctly explained as follows. Before the car enters the ramp, the passenger is moving in a straight-line path. As the car enters the ramp and travels a curved path, the passenger, because of inertia, tends to move along the original straight-line path. This is in accordance with Newton's first law: The natural tendency of an object in motion is to continue moving in a straight line. However, if a sufficiently large force toward the center of curvature acts on the passenger, she moves in a curved path along with the car. The origin of this force is the force of friction between the passenger and the car seat. If this frictional force is not great enough, the passenger slides across the seat as the car turns under her. Because of the passenger's inertia, she continues to move in a straight-line path. Eventually the passenger encounters the door, which provides a large enough force to enable her to follow the same curved path as the car. The passenger slides toward the door not because of some mysterious outward force but because there is no force great enough to enable her to travel along the circular path followed by the car.

Figure 7.11 A car approaching a curved exit ramp.

APPLICATION

The Spin-Dry Cycle of a
Washing Machine.

As a second example, consider what happens when you run clothes through the rinse cycle of a washing machine. In the last phase of this cycle, the drum spins rapidly to remove water from the clothes. Why is the water thrown off? An *erroneous* explanation is that the rotating system creates some mysterious outward force on each drop of water, and this force causes the water to be hurled to the outer drum of the machine. The correct explanation is as follows. When the clothes are at rest in the machine, water is held to them by molecular forces between the water and the fabric. During the spin cycle, the clothes rotate and the molecular forces are not great enough to provide the necessary radial force to keep the water molecules moving in a circular path along with the clothes. Hence, the drops of water, because of their inertia, move in straight-line paths until they encounter the sides of the spinning drum.

In summary, in describing motion in an accelerating frame, one must be very careful to distinguish real forces from fictitious ones. An observer in a car rounding a curve is in an accelerating frame and invents a fictitious outward force to explain why he or she is thrown outward. A stationary observer outside the car, however, considers only real forces on the passenger. To the observer, the mysterious outward force *does not exist!* The only real external force on the passenger is the radial force due to friction or the contact force of the door.

EXAMPLE 7.8   Riding the Rails

Figure 7.12a shows a roller coaster moving around an almost circular loop of radius R. What speed must the car have at the bottom of the track so that it will just make it over the top of the loop?

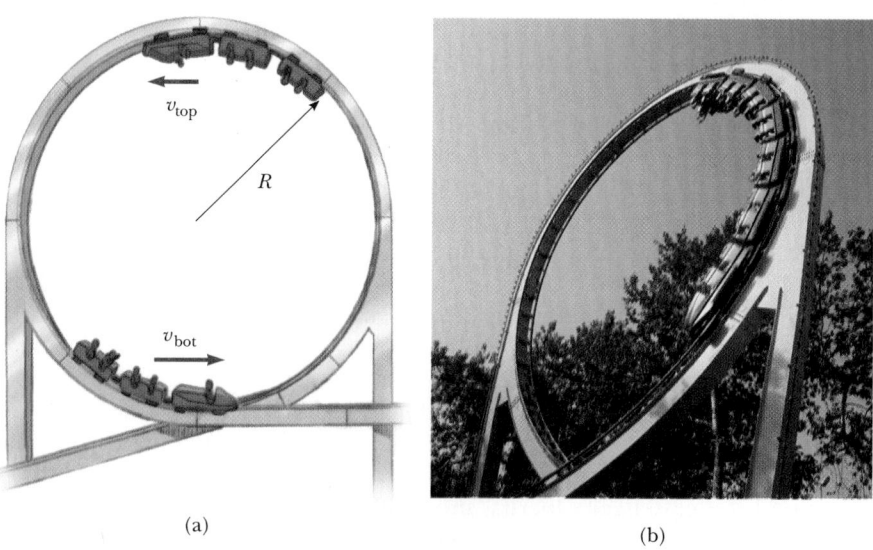

(a)   (b)

Figure 7.12   (Example 7.8) (a) A roller coaster traveling around a nearly circular track. (b) Passengers on a roller coaster at Knott's Berry Farm experience the thrill of a loop-the-loop. *(Superstock)*

Solution If the car is to just make it over the top and be able to complete the loop, the force of the track on the car must become zero at the top of the loop. Therefore, the only force exerted on the car at the top of the loop is the force of gravity, *mg*, and it is this force that provides the centripetal acceleration of the car. Thus, we have

$$mg = m\frac{v_{\text{top}}^2}{R}$$

where v_{top} is the minimum speed the car must have at the top to stay in contact with the tracks. Solving this equation for v_{top} gives

$$v_{\text{top}} = \sqrt{gR}$$

At the top of the loop, the car has both kinetic energy and potential energy. If we select the zero level for potential energy to be at the bottom of the loop, the mechanical energy of the car at the top is

$$E_{\text{top}} = \tfrac{1}{2}mv_{\text{top}}^2 + mg2R = \tfrac{1}{2}mgR + 2mgR = 2.5mgR$$

The mechanical energy of the car at the bottom of the loop is solely kinetic energy:

$$E_{\text{bot}} = \tfrac{1}{2}mv_{\text{bot}}^2$$

Because mechanical energy is conserved, the energy at the bottom must equal the energy at the top:

$$\tfrac{1}{2}mv_{\text{bot}}^2 = 2.5mgR$$

Solving for v_{bot} gives

$$v_{\text{bot}} = \sqrt{5gR}$$

For example, if the radius of the loop is 10.0 m, the velocity at the bottom must be at least

$$v_{\text{bot}} = \sqrt{5(9.8 \text{ m/s}^2)(10.0 \text{ m})} = 22.1 \text{ m/s}$$

Such a high speed can make the trip very uncomfortable for the passengers. The required speed can be reduced by changing the shape of the loop from a circular shape to one that closes down slightly at the top as shown in Figure 7.12b.

7.7 NEWTON'S UNIVERSAL LAW OF GRAVITATION

Prior to 1686 a great mass of data had been collected on the motions of the Moon and the planets, but a clear understanding of the forces that cause these celestial bodies to move the way they do was not available. In that year, Isaac Newton provided the key that unlocked the secrets of the heavens. He knew, from the first law, that a net force had to be acting on the Moon. If it were not, the Moon would move in a straight-line path rather than in its almost circular orbit around the Earth. Newton reasoned that this force arises as a result of an attractive field force between Moon and Earth, which we call the force of gravity. He also concluded that there could be nothing special about the Earth-Moon system or the Sun and its planets that would cause gravitational forces to act on them alone. In other words, he saw that the same force of attraction that causes the Moon to follow its path also causes an apple to fall to Earth from a tree. He wrote, "I deduced that the forces which keep the planets in their orbs must be reciprocally as the squares of their distances from the centers about which they revolve; and thereby compared the force requisite to keep the Moon in her orb with force of gravity at the surface of the Earth; and found them answer pretty nearly."

In 1687 Newton published his work on the universal law of gravitation, which states that

> Every particle in the Universe attracts every other particle with a force that is directly proportional to the product of their masses and inversely proportional to the square of the distance between them.

If the particles have masses m_1 and m_2 and are separated by the distance r, the magnitude of the gravitational force between them is

Universal law of gravitation ▶

$$F = G\frac{m_1 m_2}{r^2}$$ [7.17]

where G is a universal constant called the **constant of universal gravitation,** which has been measured experimentally. Its value in SI units is

$$G = 6.673 \times 10^{-11} \frac{\text{N} \cdot \text{m}^2}{\text{kg}^2}$$ [7.18]

This force law is an example of an **inverse-square law** in that it varies as the inverse square of the separation. The force acts so that the objects are always attracted to one another. From Newton's third law, we also know that the force on m_2 due to m_1, designated $\mathbf{F}_{21}$ in Figure 7.13, is equal in magnitude to the force on m_1 due to m_2, $\mathbf{F}_{12}$, but in the opposite direction. That is, these forces form an action-reaction pair.

Several features of the universal law of gravitation deserve some attention:

> 1. The gravitational force is a field force that always exists between two particles regardless of the medium that separates them.
> 2. The force varies as the inverse square of the distance between the particles and therefore decreases rapidly with increasing separation.
> 3. The force is proportional to the product of the particles' masses.

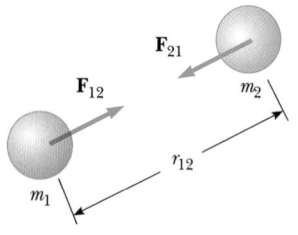

Figure 7.13 The gravitational force between two particles is attractive. Note that, according to Newton's third law, $\mathbf{F}_{12} = -\mathbf{F}_{21}$.

Another important fact is that **the gravitational force exerted by a spherical mass on a particle outside the sphere is the same as if the entire mass of the sphere were concentrated at its center.** For example, the force on a particle of mass m at the Earth's surface has the magnitude

$$F = G\frac{M_E m}{R_E^2}$$

where M_E is the Earth's mass and R_E is its radius. This force is directed toward the center of the Earth.

Thinking Physics 3

A student drops a ball and considers why the ball falls to the ground as opposed to the situation in which the ball stays stationary and the Earth moves up to meet it. The student comes up with the following explanation: "The Earth is much more massive than the ball, so the Earth pulls much harder on the ball than the ball pulls on the Earth. Thus, the ball falls while the Earth remains stationary." What do you think about this explanation?

Explanation According to Newton's universal law of gravity, the force between the ball and the Earth depends on the product of their masses, so both forces, that of the ball on the Earth, and that of the Earth on the ball, are equal in magnitude. This follows also, of course, from Newton's third law. The ball has large motion compared to the Earth because according to Newton's second law, the force gives a much greater acceleration to the small mass of the ball.

Measurement of the Gravitational Constant

The gravitational constant was first measured in an important experiment by Henry Cavendish in 1798. His apparatus consisted of two small spheres, each of mass m, fixed to the ends of a light horizontal rod suspended by a thin metal wire, as in Figure 7.14. Two large spheres, each of mass M, were placed near the smaller spheres. The attractive force between the smaller and larger spheres caused the rod to rotate in a horizontal plane and the wire to twist. The angle through which the suspended rod rotated was measured with a light beam reflected from a mirror attached to the vertical suspension. (Such a moving spot of light is an effective technique for amplifying the motion.) The experiment was carefully repeated with

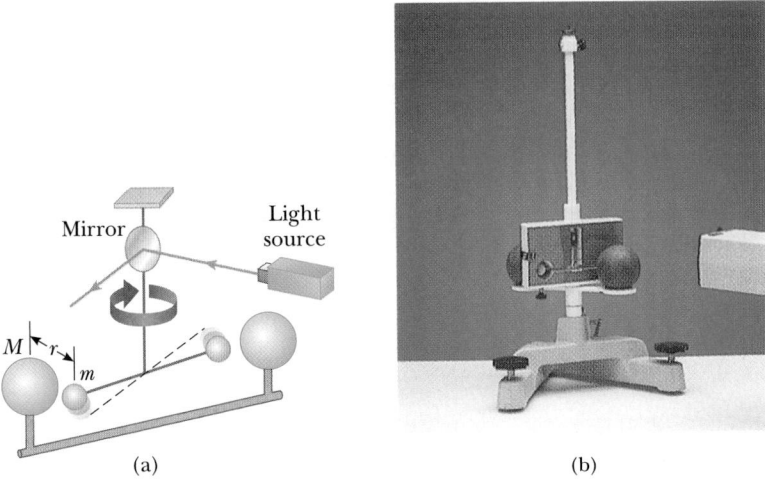

(a)   (b)

Figure 7.14 (a) A schematic diagram of the Cavendish apparatus for measuring G. The smaller spheres of mass m are attracted to the large spheres of mass M, and the rod rotates through a small angle. A light beam reflected from a mirror on the rotating apparatus measures the angle of rotation. (b) A student Cavendish apparatus. (*Courtesy of PASCO Scientific*)

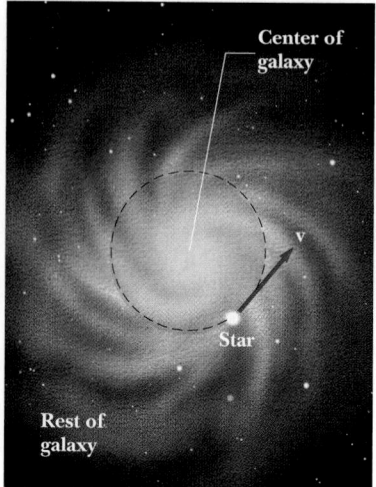

Figure 7.15 The star revolving at a speed v about the center of the galaxy is affected only by the mass inside its orbit (shaded region).

different masses at various separations. In addition to providing a value for G, the results showed that the force is attractive, proportional to the product mM, and inversely proportional to the square of the distance r.

Dark Matter

It is of interest to consider some facts from astronomy that can be understood based on the results of this section. In recent years astronomers have been interested in so-called missing mass in the Universe. To see why missing mass has been proposed, consider Fig. 7.15 which shows a star revolving at a speed v in a circular path of radius r about the center of a galaxy. We know that the force holding the star in its orbit is $G\dfrac{mM}{r^2}$ where m is the mass of the star and M is the total mass inside the circular orbit, light blue in the figure. From Newton's second law, we have

$$G\frac{mM}{r^2} = m\frac{v^2}{r}$$

or

$$v = \sqrt{\frac{GM}{r}}$$

From this equation we see that the greater the mass inside the orbit of the star, the faster it must move in order to remain at a given distance r from the center. However, when astronomers look at distant galaxies, they see stars moving at a speed v *larger* than expected compared to the amount of mass inside the orbit! The only explanation for this is that there is some form of mass present inside the orbit that cannot be seen by visible light. This material, known as **dark matter,** has been the source of intense experimental investigation in recent years. The interest in dark matter has come about in an effort to explain the future of the Universe — namely, will it continue to expand forever or will its present expansion slow, come to a halt, and begin to collapse? The determining factor will be the amount of mass present in the Universe. If the amount of mass is too low, the expansion of the Universe will continue forever, and the ultimate fate of the universe will be one in which all the stars die, cool, and move outward ad infinitum. On the other hand, if the amount of mass is large enough, the Universe will collapse on itself, perhaps to explode once again in another Big Bang to create yet another Universe. At present, even including dark matter, only about 10 percent of the amount of mass needed to produce a collapse has been found, but the search continues.

EXAMPLE 7.9 Billiards, Anyone?

Three 0.300-kg billiard balls are placed on a table at the corners of a right triangle, as shown from overhead in Figure 7.16. Find the net gravitational force on the ball designated as m_1 due to the forces exerted on it by the other two balls.

Solution To find the net gravitational force on m_1, we first calculate the force exerted on m_1 due to m_2. Then we find the force on m_1 due to m_3. Finally, we add these two forces *vectorially* to obtain the net force on m_1.

The force exerted on m_1 due to m_2, denoted by $\mathbf{F}_1$ in Figure 7.16, is upward. Its magnitude is calculated using Equation 7.17:

$$F_1 = G\frac{m_1 m_2}{r^2} = (6.67 \times 10^{-11} \text{ N·m}^2/\text{kg}^2)\frac{(0.300 \text{ kg})(0.300 \text{ kg})}{(0.400 \text{ m})^2}$$

$$= 3.75 \times 10^{-11} \text{ N}$$

This result shows that gravitational forces between the common objects that surround us have extremely small magnitudes.

Now let us calculate the gravitational force exerted on m_1 due to m_3, denoted by $\mathbf{F}_2$ in Figure 7.16. It is directed to the right, and its magnitude is

$$F_2 = G\frac{m_1 m_3}{r^2} = (6.67 \times 10^{-11} \text{ N·m}^2/\text{kg}^2)\frac{(0.300 \text{ kg})(0.300 \text{ kg})}{(0.300 \text{ m})^2}$$

$$= 6.67 \times 10^{-11} \text{ N}$$

The net gravitational force exerted on m_1 is found by adding $\mathbf{F}_1$ and $\mathbf{F}_2$ as *vectors*. The magnitude of this net force is given by

$$F = \sqrt{F_1{}^2 + F_2{}^2} = \sqrt{(3.75)^2 + (6.67)^2} \times 10^{-11} \text{ N} = \boxed{7.65 \times 10^{-11} \text{ N}}$$

Exercise Find the direction of the resultant force on m_1.

Answer The vector $\mathbf{F}$ makes an angle of 29.3° with the line joining m_1 and m_3.

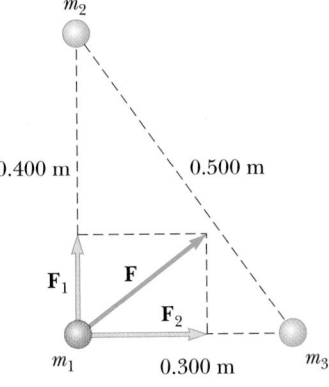

Figure 7.16 (Example 7.9)

EXAMPLE 7.10 The Mass of the Earth

Use the gravitational force law to find an approximate value for the mass of the Earth.

Solution Consider a baseball of mass m_b falling toward the Earth at a location where the free-fall acceleration is g. We know that the magnitude of the gravitational force exerted on the baseball by the Earth is the same as the weight of the ball and that this is given by $w = m_b g$. Because the magnitude of the gravitational force is given by Equation 7.17, we find that

$$m_b g = G\frac{M_E m_b}{R_E{}^2}$$

We can divide each side of this equation by m_b and solve for the mass of the Earth, M_E:

$$M_E = \frac{gR_E{}^2}{G}$$

The falling baseball is close enough to the Earth so that the distance of separation between the center of the ball and the center of the Earth can be taken as the radius of the Earth, 6.38×10^6 m. Thus, the mass of the Earth is

$$M_E = \frac{(9.80 \text{ m/s}^2)(6.38 \times 10^6 \text{ m})^2}{6.67 \times 10^{-11} \text{ N·m}^2/\text{kg}^2} = \boxed{5.98 \times 10^{24} \text{ kg}}$$

TABLE 7.1
Free-Fall Acceleration, g, at
Various Altitudes

Altitude (km) [a]	g (m/s^2)
1000	7.33
2000	5.68
3000	4.53
4000	3.70
5000	3.08
6000	2.60
7000	2.23
8000	1.93
9000	1.69
10 000	1.49
50 000	0.13

[a] All values are distances above the
Earth's surface.

EXAMPLE 7.11   Gravity and Altitude

Derive an expression that shows how the acceleration due to gravity varies with distance from the center of the Earth at an exterior point.

Solution   The falling baseball of Example 7.10 can be used here also. Now, however, assume that the ball is located at some arbitrary distance r from the Earth's center. The first equation in Example 7.10, with r replacing R_E and m_b removed from both sides, becomes

$$g = G \frac{M_E}{r^2}$$

This indicates that the free-fall acceleration at an exterior point decreases as the inverse square of the distance from the center of the Earth. Our assumption of Chapter 2 that objects fall with constant acceleration is obviously incorrect in light of the present example. For short falls, however, this change in g is so small that neglecting the variation does not introduce a significant error in the result.

Because the weight of an object is mg, we see that a change in the value of g produces a change in the weight of the object. For example, if you weigh 800 N at the surface of the Earth, you will weigh only 200 N at a height above the Earth equal to the radius of the Earth. Also, we see that if the distance of an object from the Earth becomes infinitely large, the weight approaches zero. Values of g at various altitudes are listed in Table 7.1.

Exercise   If an object weighs 270 N at the Earth's surface, what will it weigh at an altitude equal to twice the radius of the Earth?

Answer   30.0 N

Optional Section

7.8   GRAVITATIONAL POTENTIAL ENERGY REVISITED

In Chapter 5 we introduced the concept of gravitational potential energy and found that the potential energy of an object could be calculated using the equation $PE = mgh$, where h is the height of the object above or below some reference level. This equation is actually valid only when the object is near the Earth's surface. For objects high above the Earth's surface, such as a satellite, an alternative expression must be used to compute the gravitational potential energy. The general expression for the gravitational potential energy for an object of mass m at a distance r from the center of the Earth can be shown (using integral calculus) to be

General form of gravitational ▷
potential energy

$$PE = - G \frac{M_E m}{r} \qquad \text{[7.19]}$$

where M_E is the mass of the Earth.

This equation assumes that the zero level for potential energy is at an infinite distance from the center of the Earth. This point is necessary because the gravitational force goes to zero when r is set equal to infinity.

Thinking Physics 4

Why is the Sun hot?

Explanation The Sun was formed when a cloud of gas and dust coalesced, due to gravitational attraction, into a massive astronomical object. Before this occurred, the particles were widely scattered, representing a large amount of gravitational potential energy. As the particles came together to form the Sun, the gravitational potential energy decreased. According to the principle of conservation of energy, this potential energy must be transformed to another form. The form to which it transformed was internal energy, representing an increase in temperature. If enough particles come together, the temperature can rise to a point at which nuclear fusion occurs, and the object becomes a star. If there are not enough particles, the temperature rises, but not to a point at which fusion occurs—the object is a planet, if it is in orbit around a star. Jupiter is an example of a large planet that might have been a star if more particles were available to be collected.

EXAMPLE 7.12 Does the Potential Energy Reduce to *mgh*?

(a) Find expressions for the gravitational potential energy of an object at the surface of the Earth and for the same object when at a height h above the surface of the Earth (Fig. 7.17). (b) From the answers to part (a), show that the difference in potential energy between these two points reduces to the familiar expression $PE = mgh$ when h is small compared to the Earth's radius.

Solution Equation 7.19 gives the potential energy at the surface of the Earth as

$$PE_1 = -G\frac{M_E m}{R_E}$$

We can also use Equation 7.19 to find the potential energy of the object at the height h above the surface of the Earth. Taking $r = R_E + h$, we find

$$PE_2 = -G\frac{M_E m}{(R_E + h)}$$

(b) The difference in potential energy between these two points is found as follows:

$$PE_2 - PE_1 = -G\frac{M_E m}{(R_E + h)} - \left(-G\frac{M_E m}{R_E}\right)$$

$$= -GM_E m\left[\frac{1}{(R_E + h)} - \frac{1}{R_E}\right]$$

After finding a common denominator and applying some algebra to the equation above, we find

$$PE_2 - PE_1 = \frac{GM_E mh}{R_E(R_E + h)}$$

When the height h is very small compared to R_E, the denominator in the expression above is approximately equal to $R_E{}^2$. Thus, we have

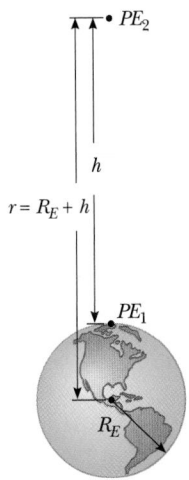

Figure 7.17 (Example 7.12)

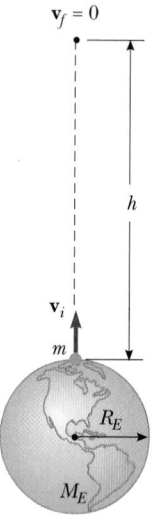

Figure 7.18 An object of mass m projected upward from the Earth's surface with an initial speed v_i reaches a maximum altitude h (where $M_E \gg m$).

  5.9, SECTION 2

WEB

For Web links related to space exploration, visit the textbook's Web site at http://www.harcourtcollege.com/physics/cptech

Escape speed ▶

TABLE 7.2
Escape Speeds for the Planets and the Moon

Planet	v_e(km/s)
Mercury	4.3
Venus	10.3
Earth	11.2
Moon	2.3
Mars	5.0
Jupiter	60.0
Saturn	36.0
Uranus	22.0
Neptune	24.0
Pluto	1.1

$$PE_2 - PE_1 \cong \frac{GM_E}{R_E^2}\, mh$$

Now note that the free-fall acceleration at the surface of the Earth is given by $g = GM_E/R_E^2$ (see Example 7.10). Thus,

$$PE_2 - PE_1 \cong mgh$$

Escape Speed

If an object is projected upward with a large enough speed it can escape the gravitational pull of the Earth and go soaring off into space. This particular speed is called the escape speed of an object from the Earth. We shall now proceed to derive an expression that will enable us to calculate the escape speed.

Suppose an object of mass m is projected vertically upward from the Earth's surface with an initial speed v_i, as in Figure 7.18. The initial mechanical energy (kinetic plus potential energy) is given by

$$KE_i + PE_i = \frac{1}{2}\, mv_i^2 - G\frac{M_E m}{R_E}$$

We neglect air resistance and assume that the initial speed is just large enough to allow the object to reach infinity with a speed of zero. We call this value of v_i the escape speed v_e. When the object is at an infinite distance from the Earth, its kinetic energy is zero, because $v_f = 0$, and the gravitational potential energy is also zero because our zero level of potential energy was selected at $r = \infty$. Hence, the total mechanical energy is zero, and the law of conservation of energy gives

$$\frac{1}{2}\, mv_e^2 - G\frac{M_E m}{R_E} = 0$$

and

$$v_e = \sqrt{\frac{2GM_E}{R_E}} \qquad\qquad [7.20]$$

From this equation, one finds that the escape speed for Earth is about 11.2 km/s, which corresponds to about 25 000 mi/h (see Example 7.13). Note that this expression for v_e is independent of the mass of the object projected from the Earth. For example, a spacecraft has the same escape speed as a molecule. A list of escape speeds for the planets and the Moon is given in Table 7.2. The data presented in this table help in understanding why some planets have atmospheres and others do not. For example, on the very hot planet Mercury, gas molecules often reach speeds that are greater than the escape speed of 4.3 km/s. Consequently, any gases that might have been present on the surface of the planet at its formation have long since wandered off into space. Likewise, in the atmosphere of our own Earth, hydrogen and helium molecules through collisions often gain speeds greater than 11.2 km/s, the escape speed on Earth. Therefore, these gases are not retained in the Earth's atmosphere. On the other hand, heavier gases such as oxygen and nitrogen in the Earth's atmosphere have average speeds of less than 11.2 km/s and do not escape.

EXAMPLE 7.13 Escape Speed of a Rocket

Calculate the escape speed from the Earth for a 5000-kg spacecraft.

Solution Using Equation 7.20 with $M_E = 5.98 \times 10^{24}$ kg and $R_E = 6.38 \times 10^6$ m gives

$$v_e = \sqrt{\frac{2GM_E}{R_E}} = \sqrt{\frac{2(6.68 \times 10^{-11})(5.98 \times 10^{24})}{6.37 \times 10^6}} = 1.12 \times 10^4 \text{ m/s}$$

This corresponds to about 25 000 mi/h or about 7 mi/s. Note that the mass of the spacecraft was not required for this calculation.

7.9 KEPLER'S LAWS

The movements of the planets, stars, and other celestial bodies have been observed for thousands of years. In early history, scientists regarded the Earth as the center of the Universe. This *geocentric model* was developed extensively by the Greek astronomer Claudius Ptolemy in the second century A.D. and was accepted for the next 1400 years. In 1543 the Polish astronomer Nicolaus Copernicus (1473–1543) showed that the Earth and the other planets revolve in circular orbits about the Sun (the *heliocentric hypothesis*).

The Danish astronomer Tycho Brahe (pronounced Brah or BRAH-huh; 1546–1601) made accurate astronomical measurements over a period of 20 years and provided the data for the currently accepted model of the Solar System. Brahe's precise observations of the planets and 777 stars were carried out with nothing more elaborate than a large sextant and compass; the telescope had not yet been invented.

The German astronomer Johannes Kepler, who was Brahe's assistant, acquired Brahe's astronomical data and spent about 16 years trying to deduce a mathematical model for the motions of the planets. After many laborious calculations, he found that Brahe's precise data on the motion of Mars about the Sun provided the answer. Kepler's analysis first showed that the concept of circular orbits about the Sun had to be abandoned. He eventually discovered that the orbit of Mars could be accurately described by an ellipse with the Sun at one focus. He then generalized this analysis to include the motions of all planets. The complete analysis is summarized in three statements known as **Kepler's laws:**

1. All planets move in elliptical orbits with the Sun at one of the focal points.
2. A line drawn from the Sun to any planet sweeps out equal areas in equal time intervals.
3. The square of the orbital period of any planet is proportional to the cube of the average distance from the planet to the Sun.

Newton later demonstrated that these laws are consequences of a simple force that exists between any two masses. Newton's universal law of gravity, together with

Johannes Kepler, German astronomer (1571–1630)

Kepler is best known for developing the laws of planetary motion based on the careful observations of Tycho Brahe. Throughout his life, Kepler was sidetracked by mystic ideas dating back to the ancient Greeks. For example, he believed in the notion of the "music of the spheres" proposed by Pythagoras, in which each planet in its motion sounds out an exact musical note. After spending several years trying to work out a "regular-solid theory" of the planets, he concluded that the Copernican view of circular planetary orbits had to be abandoned for the view that the planetary orbits are ellipses with the Sun always at one of the foci. *(Art Resource)*

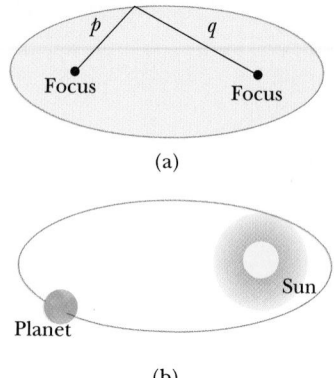

Figure 7.19 (a) The sum $p + q$ is the same for every point on the ellipse. (b) In the Solar System, the Sun is at one focus of the elliptical orbit of each planet and the other focus is empty.

his laws of motion, provides the basis for a full mathematical solution to the motions of planets and satellites. More important, Newton's universal law of gravity correctly describes the gravitational attractive force between *any* two masses.

Kepler's First Law

We shall not attempt to derive Kepler's first law. It can be shown that the first law arises as a natural consequence of the inverse-square nature of Newton's law of gravitation. That is, any object bound to another by a force that varies as $1/r^2$ will move in an elliptical orbit. As shown in Figure 7.19a, an ellipse is a curve drawn so that the sum of the distances from any point on the curve to two internal points called focal points or foci (singular, focus) is always the same. For the Sun-planet configuration (Fig. 7.19b), the Sun is at one focus and the other focus is empty. Because the orbit is an ellipse, the distance from the Sun to the planet continuously changes.

Kepler's Second Law

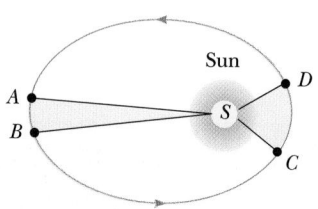

Figure 7.20 The two areas swept out by the planet in its elliptical orbit about the Sun are equal if the time interval between points A and B is equal to the time interval between points C and D.

Kepler's second law states that a line drawn from the Sun to any planet sweeps out equal areas in equal time intervals. Consider a planet in an elliptical orbit about the Sun, as in Figure 7.20. Imagine that at some instant of time we draw a line from the Sun to a planet, tracing out the line AS where S represents the position of the Sun. Exactly 30 days later, we repeat the process and draw the line BS. The area swept out is shown in color in the diagram. When the planet is at point C, we repeat the process, drawing the line CS from the planet to the Sun. We now wait 30 days, exactly the same amount of time we waited before, and draw the line DS. The area swept out in this interval is also shown in color in the figure. The two colored areas are equal.

Kepler's Third Law

The derivation of Kepler's third law is simple enough to carry out here. Consider a planet of mass M_p moving about the Sun, which has a mass of M_S, in a circular orbit (Fig. 7.21). (The assumption of a circular rather than an elliptical orbit will not introduce serious error into our approach, because the orbits of all planets except Mercury and Pluto are very nearly circular.) Because the gravitational force on the planet is the force needed to keep it moving in a circle,

$$\frac{GM_SM_p}{r^2} = \frac{M_p v^2}{r}$$

The speed, v, of the planet in its orbit is equal to the circumference of the orbit divided by the time required for one revolution, T, called the **period** of the planet. That is, $v = 2\pi r/T$, and the preceding expression becomes

$$\frac{GM_S}{r^2} = \frac{(2\pi r/T)^2}{r}$$

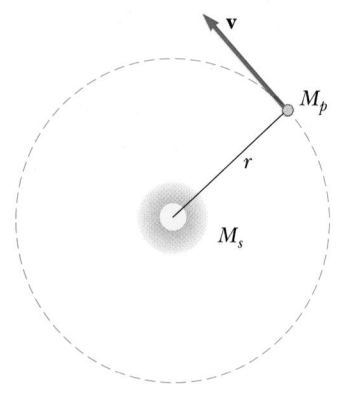

Figure 7.21 A planet of mass M_p moving in a circular orbit about the Sun. The orbits of all planets except Mercury and Pluto are nearly circular.

Kepler's third law ▶

$$T^2 = \left(\frac{4\pi^2}{GM_S}\right) r^3 = K_S r^3 \qquad [7.21]$$

PHYSICS IN ACTION

Views of the Planets

A **View of Saturn from Voyager 2.** In the photograph on the left, Saturn's rings are bright and its northern hemisphere defined by bright features as NASA's Voyager 2 approaches the planet, which it encountered on August 25, 1981. Three images, taken through ultraviolet, violet, and green filters on July 12, 1981, were combined to make this photograph. Voyager 2 was 43 million km (27 million miles) from Saturn when it took this photograph.

A View of Pluto and Charon. In the middle photograph, the Hubble Space Telescope's Faint Object Camera has obtained the clearest image ever of Pluto and its moon, Charon. Pluto is the bright object at the center of the frame; Charon is the fainter object in the lower left. Charon's orbit around Pluto is a circle seen nearly edge-on from Earth.

A Comet Colliding with Jupiter. On the right, a computer has put together separate views of Jupiter and of fragments of Comet Shoemaker-Levy 9, both taken with the Hubble Space Telescope about two months before Jupiter and the comet collided in July 1994. Their relative sizes and distance were altered. The black spot on Jupiter is the shadow of its moon Io.

(All photos courtesy of NASA)

where K_S is a constant given by

$$K_S = \frac{4\pi^2}{GM_S} = 2.97 \times 10^{-19} \ \text{s}^2/\text{m}^3$$

Equation 7.21 is Kepler's third law. It is also valid for elliptical orbits if r is replaced by a length equal to the semimajor axis of the ellipse. Note that K_S is independent of the mass of the planet. Therefore, Equation 7.21 is valid for any

TABLE 7.3   Useful Planetary Data

Body	Mass (kg)	Mean Radius (m)	Period (s)	Distance from Sun (m)	$\dfrac{T^2}{r^3}\left(\dfrac{s^2}{m^3}\right)$
Mercury	3.18×10^{23}	2.43×10^{6}	7.60×10^{6}	5.79×10^{10}	2.97×10^{-19}
Venus	4.88×10^{24}	6.06×10^{6}	1.94×10^{7}	1.08×10^{11}	2.99×10^{-19}
Earth	5.98×10^{24}	6.37×10^{6}	3.156×10^{7}	1.496×10^{11}	2.97×10^{-19}
Mars	6.42×10^{23}	3.37×10^{6}	5.94×10^{7}	2.28×10^{11}	2.98×10^{-19}
Jupiter	1.90×10^{27}	6.99×10^{7}	3.74×10^{8}	7.78×10^{11}	2.97×10^{-19}
Saturn	5.68×10^{26}	5.85×10^{7}	9.35×10^{8}	1.43×10^{12}	2.99×10^{-19}
Uranus	8.68×10^{25}	2.33×10^{7}	2.64×10^{9}	2.87×10^{12}	2.95×10^{-19}
Neptune	1.03×10^{26}	2.21×10^{7}	5.22×10^{9}	4.50×10^{12}	2.99×10^{-19}
Pluto	$\approx 1.4 \times 10^{22}$	$\approx 1.5 \times 10^{6}$	7.82×10^{9}	5.91×10^{12}	2.96×10^{-19}
Moon	7.36×10^{22}	1.74×10^{6}	—	—	—
Sun	1.991×10^{30}	6.96×10^{8}	—	—	—

A photomontage of the Earth and the surface of the Moon. The prominent colors are the blues of oceans and the white swirls of clouds. (© *Index Stock Photography, Inc.*)

planet. If we consider the orbit of a satellite, such as the Moon, about the Earth, then the constant has a different value, with the mass of the Sun replaced by the mass of the Earth. In this case, K_E equals $4\pi^2/GM_E$.

Kepler's third law gives us a method for measuring the mass of the Sun or of any celestial object around which at least one other object moves in orbit. Notice that the constant K_S in Equation 7.20 includes the mass of the Sun. The value of this constant can be found by substituting the values of the period and orbital radius and solving for K_S. The mass of the Sun is then

$$M_S = \frac{4\pi^2}{GK_S}$$

This same process can be used to calculate the mass of the Earth (by considering the period and orbital radius of the Moon) and the mass of other planets in the Solar System that possess moons.

A collection of useful planetary data is presented in Table 7.3, where the last column verifies that T^2/r^3 is a constant.

Thinking Physics 5

Some science fiction stories fantasize about a sister planet to the Earth, which is in the same orbit as the Earth but 180° ahead of us. Thus, we could never see this planet from Earth because it will always be behind the Sun, hidden from our point of view. How would you plan a visit to this proposed planet if you start from an Earth orbit?

Explanation   The quickest way to arrive at the planet is to fire your rocket so that you slow down in your orbit around the Sun. This will transfer you to an elliptical orbit around the Sun with a smaller major axis than that of the Earth's orbit. Thus, you will complete this orbit in a time less than one year. If you design your orbit so that the period is 0.5 year, then when you return to your original point, the sister planet will also be there.

A calculation using Kepler's third law shows that your spacecraft will dip inside the orbit of Venus, so you must be sure not to collide with Venus and that you have the appropriate thermal protection to deal with increased radiation from the Sun.

Another possibility is to fire your rockets so that you speed up in your orbit around the Sun, transferring to an elliptical orbit with a period of 1.5 years. This avoids the thermal problems of going near the Sun but takes three times as long.

Applying Physics 2

NASA employs a method called either "slingshotting" or "gravitational boosting" to add velocity to a spaceship as it passes a planet. To understand how this works, consider a spacecraft moving with a speed of 10.0 km/s as shown in Figure 7.22 approaching the planet Jupiter, which moves in its orbit at an average speed of about 13.1 km/s. Find the speed of the spacecraft after the "collision" with Jupiter assuming that it leaves traveling in a direction directly opposite to its incoming velocity.

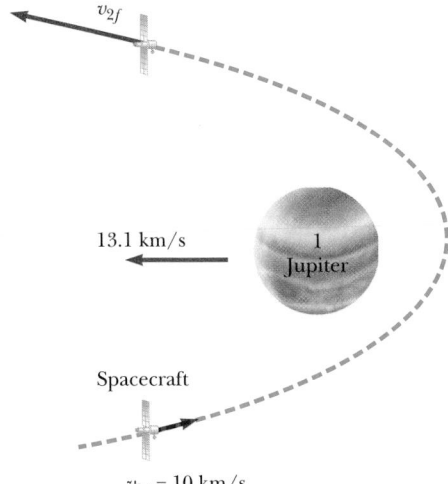

v_{2f}

13.1 km/s

1
Jupiter

Spacecraft

v_{2i} = 10 km/s

Figure 7.22   (Applying Physics 2)

Explanation   This is a collision in the pure sense of the word; the forces are not contact forces but a gravitational interaction. As the gravitational force is conservative, the collision is perfectly elastic. Because Jupiter is so massive, its velocity will be unaffected, so before and after the collision it will have a velocity of -13.1 km/s. To find the velocity of the spacecraft after the collision, we call Jupiter body 1 and the ship body 2 and apply Equation 6.11:

$$v_{1i} - v_{2i} = -(v_{1f} - v_{2f})$$

$$-13.1 \text{ km/s} - 10.0 \text{ km/s} = -(-13.1 \text{ km/s} - v_{2f})$$

$$v_{2f} = -36.2 \text{ km/s}$$

Thus, the spacecraft leaves the vicinity of Jupiter traveling more than three times faster than before its encounter with the planet.

Example 7.14   Geosynchronous Orbit

Satellite dishes do not have to change directions in order to stay focused on a signal from a satellite. This means that the satellite always has to be found at the same location with respect to the surface of the Earth. In order for this to occur, the satellite must be at a height such that its revolution period is the same as that of the Earth, 24 h. At what height must a satellite be to achieve this?

Reasoning and Solution   The force that produces the centripetal acceleration of the satellite is the gravitational force, so

$$(1) \qquad G\frac{M_E m}{r^2} = \frac{mv^2}{r}$$

where M_E is the mass of the Earth and r is the distance of the satellite from the center of the Earth.

Also, we find the velocity of the satellite to be

$$(2) \qquad v = \frac{d}{T} = \frac{2\pi r}{T}$$

where T is the orbital period of the satellite. Solving (1) and (2) simultaneously for r yields

$$r = \left(\frac{T^2}{4\pi^2} GM_E\right)^{1/3}$$

Exercise   Taking $T = 24$ h $= 86{,}400$ s and $M_E = 5.98 \times 10^{24}$ kg, show that the height above the Earth is approximately 22 000 miles. (Be careful; will $r = 22\,000$ miles?)

EXAMPLE 7.15   An Earth Satellite

A satellite of mass m moves in a circular orbit about the Earth with a constant speed of v and a height of $h = 1000$ km above the Earth's surface, as in Figure 7.23. (For clarity, this figure is not drawn to scale.) Find the orbital speed of the satellite. The radius of the Earth is 6.38×10^6 m, and its mass is 5.98×10^{24} kg.

Reasoning   The only force acting on the satellite is the gravitational force directed toward the center of the circular orbit. This can be equated to mv^2/r to find v. (The distance r is the distance from the center of the Earth to the satellite.)

Solution   The only external force on the satellite is that of the gravitational attraction exerted by the Earth. This force is directed toward the center of the satellite's circular path and is the force that produces the centripetal acceleration of the satellite. Because the force of gravity is $GM_E m/r^2$ we find that

$$F = G\frac{M_E m}{r^2} = m\frac{v^2}{r}$$

$$v^2 = \frac{GM_E}{r}$$

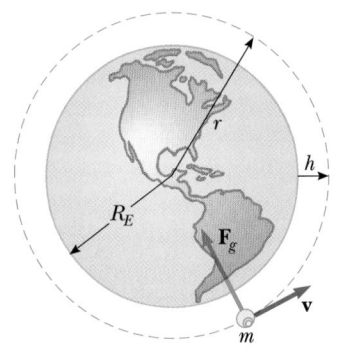

Figure 7.23   (Example 7.15) A satellite of mass m moving in a circular orbit of radius r and with constant speed v around the Earth (not drawn to scale). The force causing the centripetal acceleration is provided by the gravitational force acting on the satellite.

In this expression, the distance r is the Earth's radius plus the height of the satellite. That is, $r = R_E + h = 7.38 \times 10^6$ m, and so

$$v^2 = \frac{(6.67 \times 10^{-11} \text{ N} \cdot \text{m}^2/\text{kg}^2)(5.98 \times 10^{24} \text{ kg})}{7.38 \times 10^6 \text{ m}} = 5.40 \times 10^7 \text{ m}^2/\text{s}^2$$

Therefore,

$$v = \boxed{7.35 \times 10^3 \text{ m/s} \approx 16\ 400 \text{ mi/h}}$$

Note that v *is independent of the mass of the satellite!*

Exercise   Calculate the period of revolution (the time required for one revolution about the Earth), T, of the satellite.

Answer   105 min

7.10   THE VECTOR NATURE OF ANGULAR QUANTITIES   *Optional Section*

In this chapter we considered only the magnitudes of the angular velocity and angular acceleration. However, these are in fact vector quantities, having both magnitude and direction. Let us first describe how to find the direction of the angular velocity, $\boldsymbol{\omega}$.

When a rigid object such as a disk rotates about a fixed axis, the angular velocity, $\boldsymbol{\omega}$, points along this axis. A convenient right-hand rule, illustrated in Figure 7.24a, can be used to determine the direction of $\boldsymbol{\omega}$. Grasp the axis of rotation with your right hand so that your fingers wrap in the direction of rotation. Your extended thumb points in the direction of $\boldsymbol{\omega}$. Figure 7.24b illustrates that $\boldsymbol{\omega}$ is also in the direction of advance of a rotating right-handed screw.

Let us apply this rule to a rotating disk when viewed along the axis of rotation as in Figure 7.25. When the disk rotates clockwise, as in Figure 7.25a, applying the right-hand rule shows that the direction of $\boldsymbol{\omega}$ is into the page. When the disk rotates counterclockwise, as in Figure 7.25b, the direction of $\boldsymbol{\omega}$ is out of the page.

Finally, the direction of the angular acceleration vector, $\boldsymbol{\alpha}$, is in the direction of $\boldsymbol{\omega}$ if the angular speed (the magnitude of $\boldsymbol{\omega}$) is increasing in time and antiparallel to $\boldsymbol{\omega}$ if the angular speed is decreasing in time.

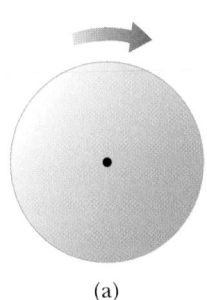

(a)

(b)

Figure 7.25   An end view of a disk rotating about an axis perpendicular to the page. (a) When the disk rotates clockwise, $\boldsymbol{\omega}$ is into the page. (b) When the disk rotates counterclockwise, $\boldsymbol{\omega}$ is out of the page.

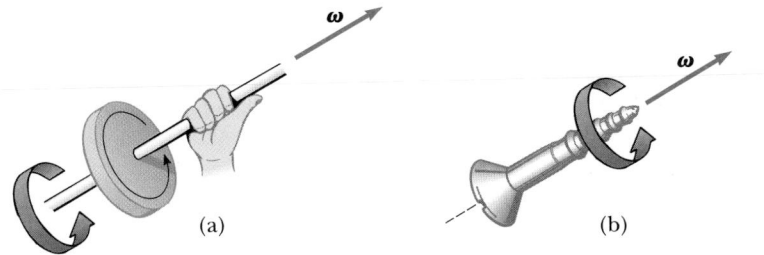

(a)   (b)

Figure 7.24   (a) The right-hand rule for determining the direction of the angular velocity vector. (b) The direction of $\boldsymbol{\omega}$ is in the direction of advance of a right-handed screw.

SUMMARY

The **average angular speed,** $\overline{\omega}$, of a rigid object is defined as the ratio of the angular displacement, $\Delta\theta$, to the time interval, Δt:

$$\overline{\omega} \equiv \frac{\theta_2 - \theta_1}{t_2 - t_1} = \frac{\Delta\theta}{\Delta t}$$ [7.2]

where $\overline{\omega}$ is in radians per second (rad/s).

The **average angular acceleration,** $\overline{\alpha}$, of a rotating object is defined as the ratio of the change in angular speed, $\Delta\omega$, to the time interval, Δt.

$$\overline{\alpha} \equiv \frac{\omega_2 - \omega_1}{t_2 - t_1} = \frac{\Delta\omega}{\Delta t}$$ [7.4]

where $\overline{\alpha}$ is in radians per second per second (rad/s²).

If an object undergoes rotational motion about a fixed axis under constant angular acceleration α, one can describe its motion with the following set of equations:

$$\omega = \omega_0 + \alpha t$$ [7.5]

$$\theta = \omega_0 t + \tfrac{1}{2}\alpha t^2$$ [7.6]

$$\omega^2 = \omega_0{}^2 + 2\alpha\theta$$ [7.7]

When an object rotates about a fixed axis, the angular speed and angular acceleration are related to the tangential speed and tangential acceleration through the relationships

$$v_t = r\omega$$ [7.8]

$$a_t = r\alpha$$ [7.9]

Any object moving in a circular path has an acceleration directed toward the center of the circular path, called a **centripetal acceleration.** Its magnitude is given by

$$a_c = \frac{v_t^2}{r} = r\omega^2$$ [7.11, 7.14]

Any object moving in a circular path must have a net force exerted on it that is directed toward the center of the circular path. From Newton's second law, this force and the centripetal acceleration are related by

$$F = m\frac{v_t^2}{r}$$ [7.16]

Some examples of forces that cause centripetal acceleration are the force of gravity (as in the motion of a satellite) and the force of tension in a string.

Newton's law of universal gravitation states that every particle in the Universe attracts every other particle with a force that is directly proportional to the product of their masses and inversely proportional to the square of the distance, r, between them:

$$F = G\frac{m_1 m_2}{r^2}$$ [7.17]

where $G = 6.673 \times 10^{-11}$ N·m²/kg² is the **constant of universal gravitation.**

Kepler's laws of planetary motion state that

1. All planets move in elliptical orbits with the Sun at one of the focal points.

2. A line drawn from the Sun to any planet sweeps out equal areas in equal time intervals.

3. The square of the orbital period of a planet is proportional to the cube of the mean distance from the planet to the Sun:

$$T^2 = \left(\frac{4\pi^2}{GM_S}\right) r^3 \qquad\qquad [7.21]$$

MULTIPLE-CHOICE QUESTIONS

A race track is constructed such that two semicircles of radius 90 m at A and 40 m at B are joined by two stretches of straight track, C and D. In a particular trial run, a driver traveled at a constant speed of 50 m/s for one complete lap. Answer the next three questions concerning this one lap.

1. The ratio of the tangential acceleration at A to that at B is
   (a) 1/2 (b) 1/4 (c) 2 (d) 4 (e) Undefined. The tangential acceleration is zero at both points.

2. The ratio of the centripetal acceleration at A to that at B is
   (a) 1/2 (b) 1/4 (c) 2 (d) 4 (e) Undefined. The centripetal acceleration is zero at both points.

3. The angular speed is greatest at
   (a) A (b) B (c) It is equal at both A and B.

4. Superman circles the Earth at a radius of $2R$, where R is the radius of the Earth. He then moves to a radius of $4R$. The gravitational force on him at this second orbit as compared to the first orbit is
   (a) the same.
   (b) twice as great.
   (c) four times as great.
   (d) half as great.
   (e) one fourth as great.

5. A 0.400-kg object is swung in a circular path and in a vertical plane on a 0.500-m length string. If a constant angular speed of 8.00 rad/s is maintained, what is the tension in the string when the object is at the top of the circle?
   (a) 8.88 N   (b) 10.5 N   (c) 12.8 N   (d) 19.6 N

CONCEPTUAL QUESTIONS

1. An object executes circular motion with a constant speed whenever a net force of constant magnitude acts perpendicular to the velocity. What happens to the speed if the force is not perpendicular to the velocity?

2. Explain why the Earth is not spherical in shape and bulges at the equator.

3. When a wheel of radius R rotates about a fixed axis do all points on the wheel have the same angular speed? Do they all have the same linear speed? If the angular speed is constant and equal to ω, describe the linear speeds and linear accelerations of the points located at $r = 0$, $r = R/2$, and $r = R$, where the points are measured from the center of the wheel.

4. If a car's wheels are replaced with wheels of greater diameter, will the reading of the speedometer change? Explain.

5. (1) At night, you are farther away from the Sun than during the day. (2) The force from the Sun on you is downward into the Earth at night and upward into the sky during the day. Based on these two facts, if you had a sensitive enough bathroom scale, would you appear to weigh more at night than during the day?

6. Correct the following statement: "The racing car rounds the turn at a constant velocity of 90 miles per hour."

7. A satellite in orbit is not truly traveling through a vacuum—it is moving through very thin air. Does the resulting air friction cause the satellite to slow down?

8. Why does an astronaut in a space capsule orbiting the Earth experience a feeling of weightlessness?

9. How would you explain the fact that Saturn and Jupiter have periods much greater than one year?

10. Because of the Earth's rotation about its axis, you would weigh slightly less at the Equator than at the poles. Why?

11. It has been suggested that rotating cylinders about 10 miles long and 5.0 miles in diameter be placed in space for colonies. The purpose of their rotations is to simulate gravity for the inhabitants. Explain the concept behind this proposal.

12. Describe the path of a moving body in the event that the acceleration is constant in magnitude at all times and (a) perpendicular to the velocity; (b) parallel to the velocity.

13. An object moves in a circular path with constant speed v. (a) Is the object's velocity constant? (b) Is its acceleration constant? Explain.

14. Centrifuges are often used in dairies to separate the cream from the milk. Which remains on the bottom?

15. Is it possible for a car to move in a circular path in such a way that it has a tangential acceleration but no centripetal acceleration?

16. Use Kepler's second law to convince yourself that the Earth must orbit faster during January, when it is closest to the Sun, than during July, when it is farthest from the Sun.

17. Why does a pilot sometimes black out when pulling out of a steep dive?

18. Consider the yo-yo in Figure 7.10. Is it possible to whirl it fast enough so that the cord makes an angle of 90° with the vertical? Explain.

19. A pail of water can be whirled in a vertical path such that no water is spilled. Why does the water remain in the pail, even when the pail is above your head?

PROBLEMS

1, 2, 3 = straightforward, intermediate, challenging   □ = full solution available in Study Guide/Student Solutions Manual   ◤ = Core Concepts Workbook
WEB = solution posted at **http://www.harcourtcollege.com/physics/cptech**   ▨ = biomedical application   ▨ = Interactive Physics

Review Problem

The Mars *Pathfinder* probe is designed to drop the instrument package from a height of 20 m above the surface, the package speed being brought to zero by a combination parachute–rocket system at that height. To cushion the landing, giant airbags surround the package. The mass of Mars is 0.1074 times that of Earth and the radius of Mars is 0.5282 that of Earth. Find (a) the acceleration due to gravity at the surface of Mars and (b) how long it takes for the instrument package to fall the last 20 m.

Section 7.1   Angular Speed and Angular Acceleration

1. The tires on a new compact car have a diameter of 2.0 ft and are warranted for 60 000 miles. (a) Determine the angle (in radians) through which one of these tires will rotate during the warranty period. (b) How many revolutions of the tire are equivalent to your answer in (a)?

2. A wheel has a radius of 4.1 m. How far (path length) does a point on the circumference travel if the wheel is rotated through angles of 30°, 30 rad, and 30 rev, respectively?

3. A rotating body has a constant angular speed of 33 rev/min. (a) What is its angular speed in rad/s? (b) Through what angle, in radians, does it rotate in 1.5 s?

4. Find the angular speed of the Earth about the Sun in radians per second and degrees per day.

5. A potter's wheel moves from rest to an angular speed of 0.20 rev/s in 30 s. Find its angular acceleration in radians per second per second.

Section 7.2   Rotational Motion Under Constant Angular Acceleration

Section 7.3   Relations Between Angular and Linear Quantities

6. A tire placed on a balancing machine in a service station starts from rest and turns through 4.7 revolutions in

1.2 s before reaching its final angular speed. Calculate its angular acceleration.

7. A machine part rotates at an angular speed of 0.60 rad/s; its speed is then increased to 2.2 rad/s at an angular acceleration of 0.70 rad/s². Find the angle through which the part rotates before reaching this final speed.

8. A dentist's drill starts from rest. After 3.20 s of constant angular acceleration it turns at a rate of 2.51×10^4 rev/min. (a) Find the drill's angular acceleration. (b) Determine the angle (in radians) through which the drill rotates during this period.

WEB 9. An electric motor rotating a workshop grinding wheel at a rate of 100 rev/min is switched off. Assume constant negative angular acceleration of magnitude 2.00 rad/s². (a) How long does it take for the grinding wheel to stop? (b) Through how many radians has the wheel turned during the interval found in (a)?

10. The tub of a washer goes into its spin-dry cycle, starting from rest and reaching an angular speed of 5.0 rev/s in 8.0 s. At this point the person doing the laundry opens the lid, and a safety switch turns off the washer. The tub slows to rest in 12.0 s. Through how many revolutions does the tub turn during this 20.0-s interval? Assume constant angular acceleration while it is starting and stopping.

11. A rotating wheel requires 3.00 s to rotate 37.0 revolutions. Its angular velocity at the end of the 3.00-s interval is 98.0 rad/s. What is the constant angular acceleration of the wheel?

12. A mass attached to a 50.0-cm-long string starts from rest and is rotated 40 times in 1.00 min before reaching a final angular speed. (a) Determine the angular acceleration of the mass, assuming that it is constant. (b) What is the angular speed of the mass after 1.00 min?

13. A coin with a diameter of 2.40 cm is dropped onto a horizontal surface. The coin starts out with an initial angular speed of 18.0 rad/s and rolls in a straight line without slipping. If the rotation slows with an angular

Problem 8 is the same as Workbook Problem 41 (Workbook page 90).

acceleration of magnitude 1.90 rad/s², how far does the coin roll before coming to rest?

14. A car is traveling at 17.0 m/s on a straight horizontal highway. The wheels of the car have radii of 48.0 cm. If the car speeds up with an acceleration of 2.00 m/s² for 5.00 s, find the number of revolutions of the wheels during this period.

15. The turntable of a record player rotates initially at 33 rev/min and takes 20 s to come to rest. (a) What is the angular acceleration of the turntable, assuming it is uniform? (b) How many rotations does the turntable make before coming to rest? (c) If the radius of the turntable is 0.14 m, what is the initial linear speed of a bug riding on the rim?

Section 7.4 Centripetal Acceleration

16. Find the centripetal accelerations of (a) a point on the Equator of the Earth and (b) a point at the North Pole of the Earth.

17. It has been suggested that rotating cylinders about 10 mi long and 5.0 mi in diameter be placed in space and used as colonies. What angular speed must such a cylinder have so that the centripetal acceleration at its surface equals Earth's gravity?

18. A tire of 2.0 ft diameter is placed on a balancing machine where it is spun so that its tread is moving at a constant speed of 60 mph. A small stone is stuck in the tread of the tire. What is the acceleration of the stone as the tire is being balanced?

19. (a) What is the tangential acceleration of a bug on the rim of a 10.0-in. diameter disk if the disk moves from rest to an angular speed of 78 revolutions per minute in 3.0 s? (b) When the disk is at its final speed, what is the tangential velocity of the bug? (c) One second after the bug starts from rest, what are its tangential acceleration, radial acceleration, and total acceleration?

20. A race car starts from rest on a circular track of radius 400 m. The car's speed increases at the constant rate of 0.500 m/s². At the point at which the magnitudes of the centripetal and tangential accelerations are equal, determine (a) the speed of the race car, (b) the distance traveled, and (c) the elapsed time.

Section 7.5 Forces Causing Centripetal Acceleration

21. A sample of blood is placed in a centrifuge of radius 15.0 cm. The mass of a red corpuscle is 3.0×10^{-16} kg, and the magnitude of the force required to make it settle out of the plasma is 4.0×10^{-11} N. At how many revolutions per second should the centrifuge be operated?

22. A 2000-kg car rounds a circular turn of radius 20 m. If the road is flat and the coefficient of friction between tires and road is 0.70, how fast can the car go without skidding?

23. A 50.0-kg child stands at the rim of a merry-go-round of radius 2.00 m, rotating with an angular speed of 3.00 rad/s. (a) What is the child's centripetal acceleration? (b) What is the minimum force between her feet and the floor of the merry-go-round that is required to keep her in the circular path? (c) What minimum coefficient of static friction is required? Is the answer you found reasonable? In other words, is she likely to be able to stay on the merry-go-round?

24. An airplane is flying in a horizontal circle at a speed of 100 m/s. The 80.0-kg pilot does not want her radial acceleration to exceed 7g. (a) What is the minimum radius of the circular path? (b) At this radius, what is the force causing the centripetal acceleration of the pilot?

25. An engineer wishes to design a curved exit ramp for a toll road in such a way that a car will not have to rely on friction to round the curve without skidding. He does so by banking the road in such a way that the necessary force causing the centripetal acceleration will be supplied by the component of the normal force toward the center of the circular path. (a) Show that for a given speed of v and a radius of r, the curve must be banked at the angle θ such that $\tan \theta = v^2/rg$. (b) Find the angle at which the curve should be banked if a typical car rounds it at a 50.0-m radius and a speed of 13.4 m/s.

26. An air puck of mass 0.25 kg is tied to a string and allowed to revolve in a circle of radius 1.0 m on a frictionless horizontal table. The other end of the string passes through a hole in the center of the table, and a mass of 1.0 kg is tied to it (Fig. P7.26). The suspended mass remains in equilibrium while the puck on the tabletop revolves. (a) What is the tension in the string? (b) What is the force causing the centripetal acceleration of the puck? (c) What is the speed of the puck?

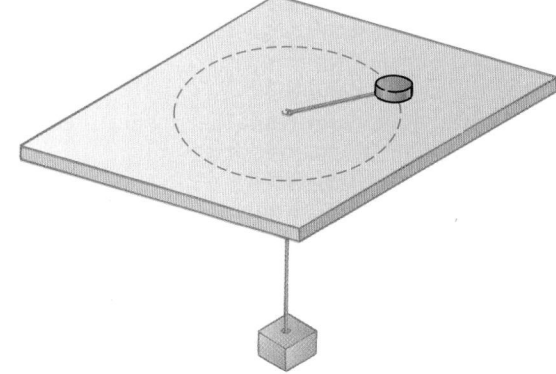

Figure P7.26

WEB

27. Tarzan ($m = 85$ kg) tries to cross a river by swinging from a 10-m-long vine. His speed at the bottom of the swing (as he just clears the water) is 8.0 m/s. Tarzan

doesn't know that the vine has a breaking strength of 1000 N. Does he make it safely across the river? Justify your answer.

28. A 40.0-kg child takes a ride on a Ferris wheel that rotates four times each minute and has a diameter of 18.0 m. (a) What is the centripetal acceleration of the child? (b) What force (magnitude and direction) does the seat exert on the child at the lowest point of the ride? (c) What force does the seat exert on the child at the highest point of the ride? (d) What force does the seat exert on the child when the child is halfway between the top and bottom?

29. A roller-coaster car has a mass of 500 kg when fully loaded with passengers (Fig. P7.29). (a) If the vehicle has a speed of 20.0 m/s at point A, what is the force of the track on the vehicle at this point? (b) What is the maximum speed the vehicle can have at B in order for gravity to hold it on the track?

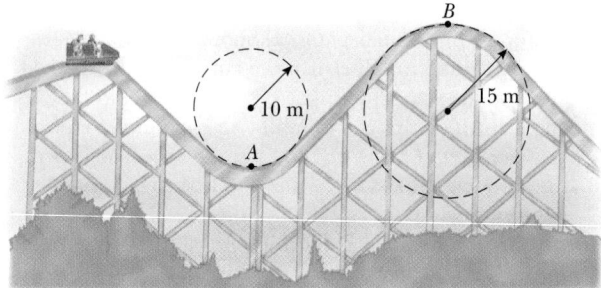

Figure P7.29

30. A pail of water is rotated in a vertical circle of radius 1.00 m (the approximate length of a person's arm). What must be the minimum speed of the pail at the top of the circle if no water is to spill out?

Section 7.7   Newton's Universal Law of Gravitation

Section 7.8   Gravitational Potential Energy Revisited (Optional)

Section 7.9   Kepler's Laws

31. The average distance separating the Earth and the Moon is 384 000 km. Use the data in Table 7.3 to find the net gravitational force the Earth and the Moon exerts on a 3.00×10^4-kg spaceship located halfway between them.

32. During a solar eclipse, the Moon, Earth, and Sun all lie on the same line, with the Moon between the Earth and the Sun. (a) What force is exerted on the Moon by the Sun? (b) What force is exerted on the Moon by the Earth? (c) What force is exerted on the Earth by the Sun? (See Table 7.3 and Problem 31 above.)

33. A coordinate system (in meters) is constructed on the surface of a pool table, and three masses are placed on the coordinate system as follows: a 2.0-kg mass at the origin, a 3.0-kg mass at (0, 2.0), and a 4.0-kg mass at (4.0, 0). Find the resultant gravitational force exerted on the mass at the origin by the other two masses.

34. Use the data of Table 7.3 to find the point between the Earth and the Sun at which an object can be placed so that the net gravitational force exerted on it by these two objects is zero.

35. Given that the Moon's period about the Earth is 27.32 days and the distance from the Earth to the Moon is 3.84×10^8 m, estimate the mass of the Earth. Assume the orbit is circular. Why do you suppose your estimate is high?

36. A satellite moves in a circular orbit around the Earth at a speed of 5000 m/s. Determine (a) the satellite's altitude above the surface of the Earth and (b) the period of the satellite's orbit.

WEB 37. A 600-kg satellite is in a circular orbit about the Earth at a height above the Earth equal to the Earth's mean radius. Find (a) the satellite's orbital speed, (b) the period of its revolution, and (c) the gravitational force acting on it.

38. A satellite is in a circular orbit just above the surface of the Moon. (See Table 7.3.) What are the satellite's (a) acceleration and (b) speed? (c) What is the period of the satellite orbit?

39. A satellite has a mass of 100 kg and is in an orbit at 2.00×10^6 m above the surface of the Earth. (a) What is the potential energy of the satellite at this location? (b) What is the magnitude of the gravitational force on the satellite?

40. Use data from Table 7.3 to determine the escape speed from (a) the Moon, (b) Mercury, and (c) Jupiter.

41. A satellite of mass 200 kg is launched from a site on the Earth's Equator into an orbit at 200 km above the surface of Earth. (a) Assuming a circular orbit, what is the orbital period of this satellite? (b) What is the satellite's speed in orbit? (c) What is the minimum energy necessary to place this satellite in orbit, assuming no air friction?

ADDITIONAL PROBLEMS

42. A rotating bicycle wheel has an angular speed of 3.00 rad/s at some instant of time. It is then given an angular acceleration of 1.50 rad/s². A chalk line drawn on the wheel is horizontal at $t = 0$. (a) What angle does this line make with its original direction at $t = 2.00$ s? (b) What is the angular speed of the wheel at $t = 2.00$ s?

43. Three masses are aligned along the x axis of a rectangular coordinate system so that a 2.0-kg mass is at the origin, a 3.0-kg mass is at (2.0, 0) m, and a 4.0-kg mass is at (4.0, 0) m. Find (a) the gravitational force exerted on

the 4.0-kg mass by the other two masses and (b) the magnitude and direction of the gravitational force exerted on the 3.0-kg mass by the other two.

44. An athlete swings a 5.00-kg ball horizontally on the end of a rope. The ball moves in a circle of radius 0.800 m at an angular speed of 0.500 rev/s. What are (a) the tangential speed of the ball and (b) its centripetal acceleration? (c) If the maximum tension the rope can withstand before breaking is 100 N, what is the maximum tangential speed the ball can have?

45. Io, a small moon of the giant planet Jupiter, has an orbital period of 1.77 days and an orbital radius equal to 4.22×10^5 km. From these data, determine the mass of Jupiter.

46. A high-speed sander has a disk 6.00 cm in radius that rotates about its axis at a constant rate of 1200 rev/min. Determine (a) the angular speed of the disk in radians per second, (b) the linear speed of a point 2.00 cm from the disk's center, (c) the centripetal acceleration of a point on the rim, and (d) the total distance traveled by a point on the rim in 2.00 s.

47. The Solar Maximum Mission Satellite was placed in a circular orbit about 150 mi above the Earth. Determine (a) the orbital speed of the satellite and (b) the time required for one complete revolution.

48. Geosynchronous satellites have an angular velocity that matches the rotation of the Earth and follow circular orbits in the equatorial plane of the Earth. (Almost all communications satellites are geosynchronous and appear to be stationary above a point on the Equator.) (a) What must be the radius of the orbit of a geosynchronous satellite? (b) How high (in miles) above the surface of the Earth are geosynchronous satellites located?

WEB **49.** A car moves at speed v across a bridge made in the shape of a circular arc of radius r. (a) Find an expression for the normal force acting on the car when it is at the top of the arc. (b) At what minimum speed will the normal force become zero (causing occupants of the car to seem weightless) if $r = 30.0$ m?

50. A 0.400-kg pendulum bob passes through the lowest part of its path at a speed of 3.00 m/s. (a) What is the tension in the pendulum cable at this point if the pendulum is 80.0 cm long? (b) When the pendulum reaches its highest point, what angle does the cable make with the vertical? (c) What is the tension in the pendulum cable when the pendulum reaches its highest point?

51. (a) Find the acceleration of gravity at the surface of a neutron star of mass 1.5 solar masses and having a radius of 10.0 km. (b) Find the weight of a 0.120-kg baseball on this star. (c) Assume the equation $PE = mgh$ applies and calculate the energy that a 70.0-kg person would expend climbing a 1.00-cm-tall mountain on this star.

52. Because of the Earth's rotation about its axis, a point on the Equator experiences a centripetal acceleration of 0.0340 m/s² while a point at the poles experiences no centripetal acceleration. (a) Show that at the Equator the gravitational force on an object (the true weight) must exceed the object's apparent weight. (b) What are the apparent weights at the Equator and at the poles of a 75.0-kg person? (Assume the Earth is a uniform sphere, and take $g = 9.800$ m/s².)

53. In a popular amusement park ride, a rotating cylinder of radius 3.00 m is set in rotation at an angular speed of 5.00 rad/s, as in Figure P7.53. The floor then drops away, leaving the riders suspended against the wall in a vertical position. What minimum coefficient of friction between a rider's clothing and the wall is needed to keep the rider from slipping? (*Hint:* Recall that the magnitude of the maximum force of static friction is equal to μn, where n is the normal force—in this case, the force causing the centripetal acceleration.)

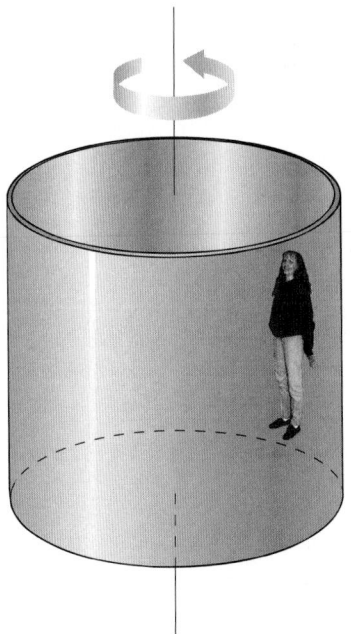

Figure P7.53

54. A stunt man whose mass is 70 kg swings from the end of a 4.0-m-long rope along the arc of a vertical circle. Assuming he starts from rest when the rope is horizontal, find the tensions in the rope that are required to make him follow his circular path, (a) at the beginning of his motion, (b) at a height of 1.5 m above the bottom of the circular arc, and (c) at the bottom of his arc.

55. A 0.50-kg ball that is tied to the end of a 1.5-m light cord is revolved in a horizontal plane with the cord making a 30° angle with the vertical (see Fig. 7.10).

(a) Determine the ball's speed. (b) If the ball is revolved so that its speed is 4.0 m/s, what angle does the cord make with the vertical? (c) If the cord can withstand a maximum tension of 9.8 N, what is the highest speed at which the ball can move?

56. A skier starts at rest at the top of a large hemispherical hill (Fig. P7.56). Neglecting friction, show that the skier will leave the hill and become airborne at a distance of $h = R/3$ below the top of the hill. (*Hint:* At this point, the normal force goes to zero.)

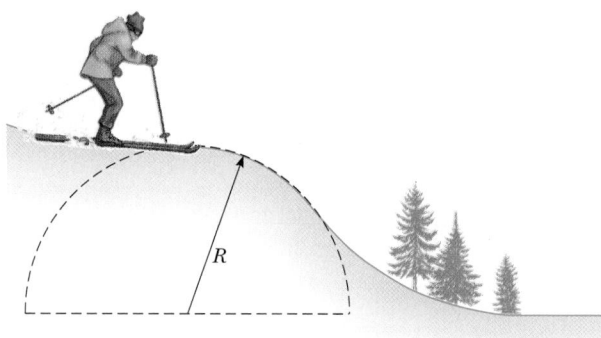

Figure P7.56

57. Halley's comet approaches the Sun to within 0.570 A.U., and its orbital period is 75.6 years. (A.U. is the abbreviation for astronomical unit, where 1 A.U. = 1.50×10^{11} m is the mean Earth–Sun distance.) How far from the Sun will Halley's comet travel before it starts its return journey? (See Fig. P.7.57.)

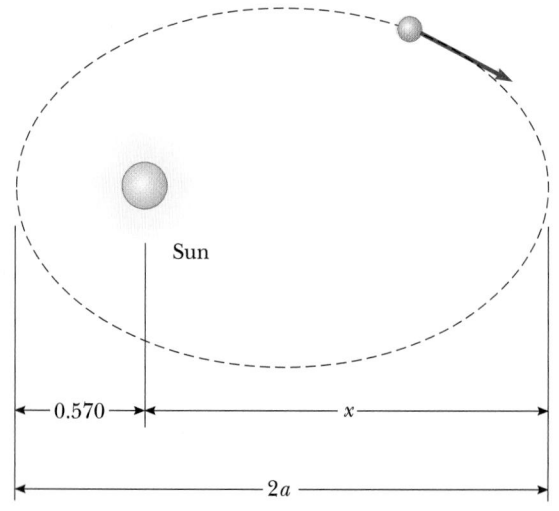

Figure P7.57

58. Two planets X and Y travel counterclockwise in circular orbits about a star, as in Figure P7.58. The radii of their orbits is in the ratio 3:1. At some time, they are aligned, as in Figure P7.58a, making a straight line with the star. Five years later, planet X has rotated through 90°, as in Figure P7.58b. Where is planet Y at this time?

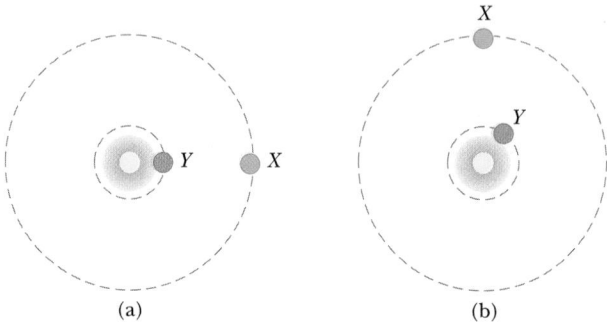

Figure P7.58

59. A spacecraft in the shape of a long cylinder has a length of 100 m and its mass with occupants is 1000 kg. It has strayed too close to a 1.0-m radius black hole having a mass 100 times that of the Sun (Fig. P7.59). If the nose of the spacecraft points toward the center of the black hole, and if distance between the nose of the spacecraft and the black hole's center is 10 km, (a) determine the total force on the spacecraft. (b) What is the difference in the force per kilogram of mass felt by the occupants in the nose of the ship and those in the rear of the ship farthest from the black hole?

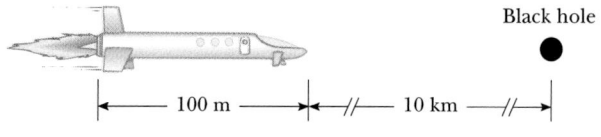

Figure P7.59

60. In Robert Heinlein's *The Moon Is a Harsh Mistress,* the colonial inhabitants of the moon threaten to launch rocks down onto the Earth if they are not given independence (or at least representation). Assuming that a rail gun could launch a rock of mass m at twice the lunar escape speed, calculate the speed of the rock as it enters the Earth's atmosphere.

61. Show that the escape speed from the surface of a planet of uniform density is directly proportional to the radius of the planet.

62. A massless spring of constant $k = 78.4$ N/m is fixed on the left side of a level track. A block of mass $m = 0.50$ kg is pressed against the spring and compresses it a distance of d, as in Figure P7.62. The block (initially at rest) is then released and travels toward a circular loop-the-loop of radius $R = 1.5$ m. The entire track and the loop-the-loop are frictionless except for the section of track between points A and B. Given that the coefficient of kinetic friction between the block and the track along AB is $\mu_k = 0.30$, and that the length of AB is 2.5 m, determine the minimum compression, d, of the spring that enables the block to just make it through the loop-the-loop at point C. (*Hint:* The force of the track on the block will be zero if the block barely makes it through the loop-the-loop.)

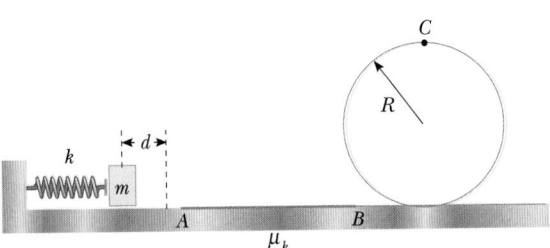

Figure P7.62

63. A small block of mass $m = 0.50$ kg is fired with an initial speed of $v_0 = 4.0$ m/s along a horizontal section of frictionless track, as shown in the top portion of Figure P7.63. The mass then moves along the frictionless semicircular, *vertical* tracks of radius $R = 1.5$ m. **(a)** Determine the force of the track on the block at points A and B. **(b)** The bottom of the track consists of a section ($L = 0.40$ m) with friction. Determine the coefficient of kinetic friction between the block and that portion of the bottom track if the mass just makes it to point C on the first trip. (*Hint:* If the block just makes it to point C, the force of contact exerted on the block by the track at that point should be zero.)

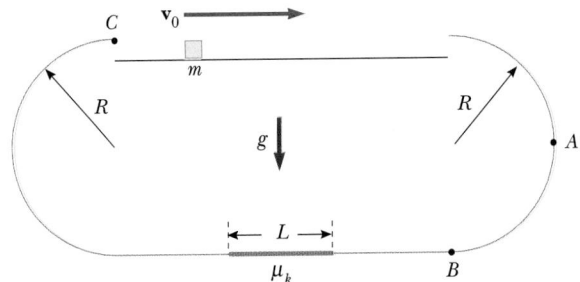

Figure P7.63

64. A frictionless roller coaster is given an initial velocity of v_0 at height h, as in Figure P7.64. The radius of curvature of the track at point A is R. **(a)** Find the maximum value of v_0 so that the roller coaster stays on the track at A solely because of gravity. **(b)** Using the value of v_0 calculated in (a), determine the value of h' that is necessary if the roller coaster is to just make it to point B.

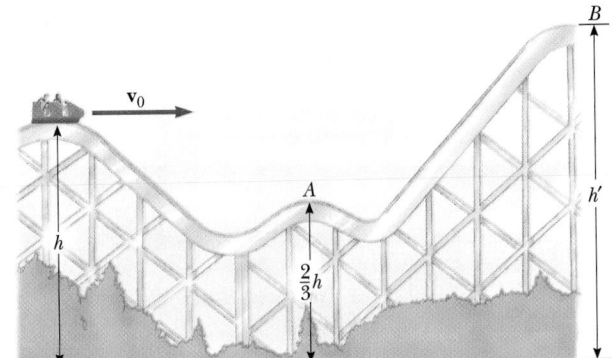

Figure P7.64

65. A car rounds a banked curve where the radius of curvature of the road is R, the banking angle is θ, and the coefficient of static friction is μ. **(a)** Determine the range of speeds the car can have without slipping up or down the road. **(b)** What is the range of speeds possible if $R = 100$ m, $\theta = 10°$, and $\mu = 0.10$ (slippery conditions)?

8

Rotational Equilibrium and Rotational Dynamics

Chapter Outline

▲ PHYSICS PUZZLER

This student is holding the axle of a spinning bicycle wheel while seated on a pivoted stool. The student and stool are initially at rest while the wheel spins in a horizontal plane. When the wheel is inverted about its center by 180°, the student and stool begin to rotate. If the wheel is inverted back to its original orientation, the student and stool stop rotating. Can you explain this set of events? *(Courtesy of Central Scientific Company)*

This chapter completes our study of equilibrium begun in Chapter 4. An understanding of equilibrium problems is important in a variety of fields. For example, students of architecture or industrial technology benefit from an understanding of the forces that act on buildings or large machines, and biology students should understand the forces at work in muscles and bones. Newton's first law, as discussed in Chapter 4, is the basis for much of our work in this chapter. In addition, in order to fully understand objects in equilibrium, we must consider torque. This concept will also play a key role in our discussion of rotational motion.

In this chapter we will also complete our study of rotational motion. We shall build on the definitions of angular speed and angular acceleration encountered in Chapter 7 by examining the relationship between these concepts and the forces that produce rotational motion. Specifically, we shall find the rotational analog of Newton's second law and define a term that needs to be added to our equation for conservation of mechanical energy: rotational kinetic energy. One of the central purposes of this chapter is to develop the concept of angular momentum, a quantity that plays a key role in rotational motion. Finally, just as we found that linear momentum is conserved, we shall also find that the angular momentum of any isolated system is always conserved.

8.1 TORQUE

Consider Figure 8.1, an overhead view of a door hinged at point O. From this viewpoint, the door is free to rotate about a line perpendicular to the page and passing through O. When the force $\mathbf{F}$ is applied at the outer edge, as shown, the door can easily be caused to rotate counterclockwise; that is, the rotational effect of the force is quite large. On the other hand, the same force applied at a point nearer the hinges produces a smaller rotational effect on the door.

The tendency of a force to rotate a body about some axis is measured by a quantity called the torque, τ. The torque due to a force of $\mathbf{F}$ has the magnitude

$$\tau = Fd \qquad [8.1]$$

7.5

In this equation, τ (the Greek letter tau) is the torque, and the distance d is the **lever arm** (or moment arm) of the force $\mathbf{F}$. **The lever arm is the perpendicular distance from the axis of rotation to a line drawn along the direction of the force.**

As an example, consider the wrench pivoted about the axis O in Figure 8.2a. In this case, the applied force $\mathbf{F}$ acts at an angle of ϕ with the horizontal. If you examine the definition of lever arm just given, you will see that in this case the lever arm is the distance d shown in the figure and not L, the length of the wrench. That is, d is the perpendicular distance from the axis of rotation to the line along which the applied force acts. In this case, d is related to L by the expression $d = L \sin \phi$. Thus, the net torque produced by the force $\mathbf{F}$ is given by $\tau = FL \sin \phi$. Actually, computing torque in situations such as this is usually best accomplished by resolving the force into components, as shown in Figure 8.2b. The torque about the axis of rotation produced by the component $F \cos \phi$ is zero because the lever arm of this component is zero. That is, the distance from the pivot to the line along which the

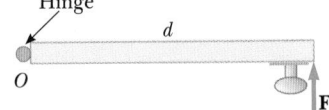

Figure 8.1 A bird's-eye view of a door hinged at O, with a force applied perpendicularly to the door.

Figure 8.2 (a) A force, **F**, acting at an angle of ϕ with the horizontal produces a torque of magnitude Fd about the pivot O. (b) The component $F \sin \phi$ tends to rotate the system about O. The component $F \cos \phi$ produces no torque about O.

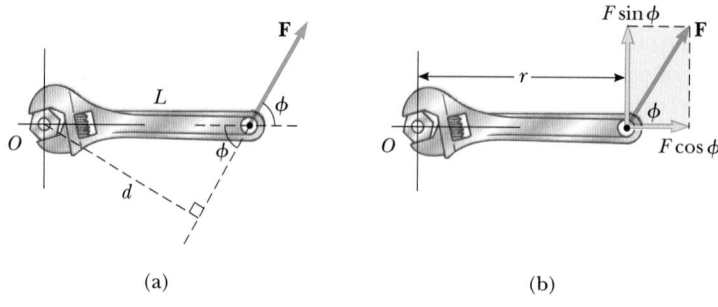

(a)   (b)

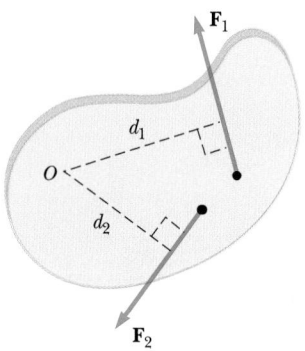

Figure 8.3 The force $\mathbf{F}_1$ tends to rotate the body counterclockwise about O, and $\mathbf{F}_2$ tends to rotate the body clockwise.

force acts is zero, because the line along which the force acts passes through the axis of rotation. The component $F \sin \phi$ has a lever arm of L, and the torque produced by this component is, as before, $\tau = FL \sin \phi$.

If two or more forces are acting on an object, as in Figure 8.3, then each has a tendency to produce a rotation about the pivot O. For example, $\mathbf{F}_2$ tends to rotate the object clockwise and $\mathbf{F}_1$ tends to rotate the object counterclockwise. We shall use the convention that the sign of the torque resulting from a force is positive if its turning tendency is counterclockwise and negative if its turning tendency is clockwise. In Figure 8.3, then, the torque resulting from $\mathbf{F}_1$, which has a moment arm of d_1, is positive and equal to $F_1 d_1$; the torque associated with $\mathbf{F}_2$ is negative and equal to $-F_2 d_2$. The *net torque* acting on the object about O is found by summing the torques:

$$\Sigma \tau = \tau_1 + \tau_2 = F_1 d_1 - F_2 d_2$$

Notice that the units of torque are units of force times length, such as the newton-meter (N·m) or the pound-foot (lb·ft).

EXAMPLE 8.1   The Spinning Crate

Figure 8.4 is a top view of a packing crate being pushed by two equal and opposite forces acting as shown. Find the net torque exerted on the crate if its width is 1.0 m. Assume an axis of rotation through the center of the crate.

Solution   The torque produced by $\mathbf{F}_1$ is

$$\tau_1 = F_1 d_1 = -(500 \text{ N})(0.50 \text{ m}) = -250 \text{ N·m}$$

and the torque produced by $\mathbf{F}_2$ is

$$\tau_2 = F_2 d_2 = -(500 \text{ N})(0.50 \text{ m}) = -250 \text{ N·m}$$

Because each force produces clockwise rotation, the torques are both negative. Thus, the net torque is -500 N·m.

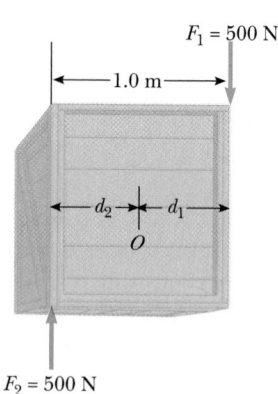

Figure 8.4 (Example 8.1) A top view of a packing crate being pushed with equal but opposite forces.

EXAMPLE 8.2 The Swinging Door

Find the torque produced by the 300-N force applied at an angle of 60° to the door of Figure 8.5a.

Reasoning   In Figure 8.5b, the 300-N force has been replaced by its horizontal and vertical components:

$$F_x = (300 \text{ N})\cos 60° = 150 \text{ N}$$

$$F_y = (300 \text{ N})\sin 60° = 260 \text{ N}$$

A convenient and obvious location for the axis of rotation is the hinge of the door.

Solution   In this case, the 150-N force produces zero torque about the axis of rotation, because the line along which the force acts passes through the axis of rotation and hence the lever arm is zero. The 260-N force has a lever arm of 2.0 m and thus produces a torque of

$$\tau = (260 \text{ N})(2.0 \text{ m}) = \boxed{520 \text{ N} \cdot \text{m}}$$

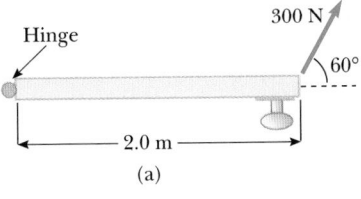

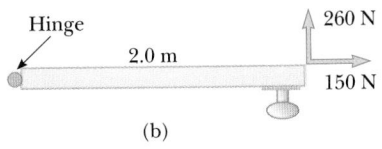

Figure 8.5   (Example 8.2) (a) A top view of a door being pulled by a 300-N force. (b) The components of the 300-N force.

8.2	## TORQUE AND THE SECOND CONDITION FOR EQUILIBRIUM

In Chapter 4 we examined some situations, under the heading of the first condition of equilibrium, in which an object has no net external force acting on it; that is, $\Sigma F_x = 0$ and $\Sigma F_y = 0$. We found there that such an object either remains at rest if it is at rest or moves with a constant velocity if it is in motion. However, the first condition is not sufficient to ensure that an object is in complete equilibrium. This can be understood by considering the situation illustrated in Figure 8.4. In this situation, the two applied forces acting on the crate are equal in magnitude and opposite in direction. The first condition for equilibrium is satisfied because the two external forces balance each other, and yet the crate can still move—it will rotate clockwise because the forces do not act through a common point.

This case illustrates that, to fully understand the effect of a force or group of forces on an object, one must know not only the magnitude and direction of the force(s) but the point of application. That is, the net torque acting on an object must be considered. An object in rotational equilibrium has no angular acceleration.

The second condition for equilibrium asserts that if an object is in rotational equilibrium, the net torque acting on it about any axis must be zero. That is,

$$\Sigma \tau = 0 \qquad\qquad [8.2]$$

You can see that a body in static equilibrium must satisfy two conditions:

1. **The resultant external force must be zero.** $\Sigma \mathbf{F} = 0$

2. **The resultant external torque must be zero.** $\Sigma \tau = 0$

◀ The two conditions for static equilibrium

The Chinese acrobats in this difficult formation represent a balanced
mechanical system. The external
forces acting on the system, as
shown by the blue vectors, are the
weights of the acrobats, **w**₁ and **w**₂,
and the upward force of the support, **n,** on the lower acrobat. The
vector sum of these external forces
must be zero in such a balanced system. The net external torque acting
on such a balanced system must also
be zero. (*J. P. Lafont, Sygma*)

The first condition is a statement of translational equilibrium; the second is a statement of rotational equilibrium. We shall discuss torque and its relation to rotational
motion in more detail later in the chapter.

Position of the Axis of Rotation

In the cases we have been describing so far, the axes of rotation for calculating
torques have been selected without explanation. Often the nature of a problem
suggests a convenient location for the axis, but just as often no single location stands
out as being preferable. You might ask, "What axis should I choose in calculating
the net torque?" The answer is as follows: **If the object is in equilibrium, it does
not matter where you put the axis of rotation for calculating the net torque;
the location of the axis is completely arbitrary.**

Applying Physics 1

When an automobile driver steps on the accelerator, the nose of the car moves upward. When the driver brakes, the nose moves downward. Why do these effects occur?

Explanation When the driver steps on the accelerator, there is an increased force
on the tires from the roadway. This force is parallel to the roadway and directed
toward the front of the automobile, as suggested in Figure 8.6a. This force provides a

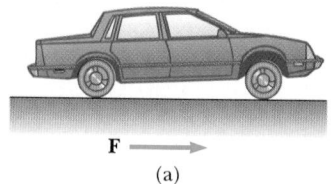

F ⟶

(a)

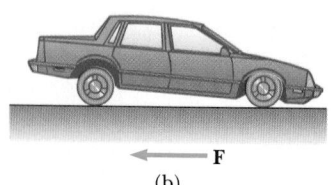

⟵ F

(b)

Figure 8.6 (Applying Physics 1)

torque that tends to cause the car to rotate in the counterclockwise direction in the diagram. The result of this rotation is a "nosing up" of the car. When the driver steps on the brake, there is an increased force on the tires from the roadway, directed toward the rear of the automobile, as suggested in Figure 8.6b. This force results in a torque that causes a clockwise rotation and the subsequent "nosing down" of the automobile.

8.3 THE CENTER OF GRAVITY

One of the forces that must be considered in dealing with a rigid object is the force of gravity acting on the object. To compute the torque due to the weight force, all of the weight can be thought of as concentrated at a single point called either the center of gravity or the center of mass.

Consider an object of arbitrary shape lying in the xy plane, as in Figure 8.7. The object is divided into a large number of very small particles of weight m_1g, m_2g, m_3g, . . . having coordinates (x_1, y_1), (x_2, y_2), (x_3, y_3), ... Each particle contributes a torque about the origin that is equal to its weight multiplied by its lever arm. For example, the torque due to the weight m_1g is m_1gx_1, and so forth.

We wish to locate the one position of the single force of magnitude w—the total weight of the object—whose effect on the rotation of the object is the same as that of the individual particles. This point is called the **center of gravity** of the object. Equating the torque exerted by w at the center of gravity to the sum of the torques acting on the individual particles gives

$$(m_1g + m_2g + m_3g + \ldots)x_{cg} = m_1gx_1 + m_2gx_2 + m_3gx_3 + \ldots$$

If we assume that g is uniform over the object (which will be the case in all the situations we will examine), then the g terms in the preceding equation cancel and we get

$$x_{cg} = \frac{m_1x_1 + m_2x_2 + m_3x_3 + \ldots}{m_1 + m_2 + m_3 + \ldots} = \frac{\Sigma m_i x_i}{\Sigma m_i} \qquad [8.3]$$

Similarly, the y coordinate of the center of gravity of the system can be found from

$$y_{cg} = \frac{\Sigma m_i y_i}{\Sigma m_i} \qquad [8.4]$$

The center of gravity of a homogeneous, symmetric body must lie on the axis of symmetry. For example, the center of gravity of a homogeneous rod must lie midway between the ends of the rod. The center of gravity of a homogeneous sphere or a homogeneous cube must lie at the geometric center of the object. One can determine the center of gravity of an irregularly shaped object, such as a wrench, experimentally by suspending the wrench from two different points (Fig. 8.8). The wrench is first hung from point A, and a vertical line, AB (which can be established with a plumb bob), is drawn when the wrench is in equilibrium. The wrench is then hung from point C, and a second vertical line, CD, is drawn. The

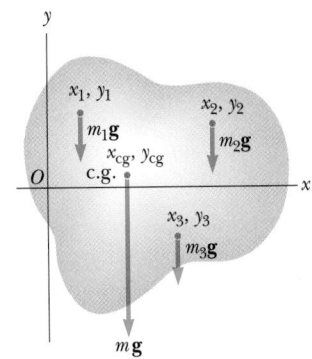

Figure 8.7 The center of gravity of an object is the point where all the weight of the object can be considered to be concentrated.

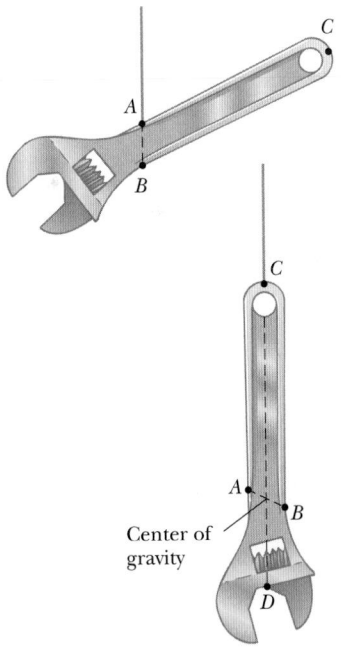

Figure 8.8 An experimental technique for determining the center of gravity of a wrench. The wrench is hung freely from two different pivots, A and C. The intersection of the two vertical lines, AB and CD, locates the center of gravity.

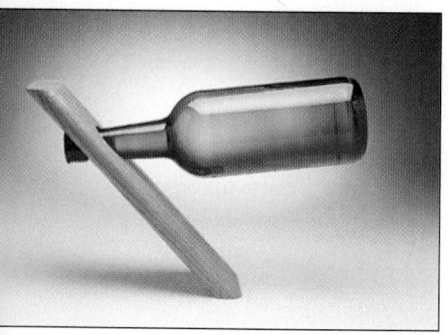

This one-bottle wine holder is an interesting example of a balanced mechanical system, which seems to defy gravity. The system (wine holder + bottle) is balanced when its center of gravity is directly over the lowest support point so that the net torque acting on the system is zero. *(Courtesy of Charles Winters)*

center of gravity coincides with the intersection of these two lines. In fact, if the wrench is hung freely from any point, the vertical line through that point must pass through the center of gravity.

In several examples in Section 8.4, we shall be concerned with homogeneous, symmetric objects whose centers of gravity coincide with their geometric centers. A rigid object in a uniform gravitational field can be balanced by a single force equal in magnitude to the weight of the object, as long as the force is directed upward through the object's center of gravity.

Thinking Physics 1

Why can't you put the back of your heels firmly against a wall and then bend over without falling?

Explanation   In order for you to remain in equilibrium, your center of gravity must always be over your point of support, the feet. When you bend over with your heels against the wall, your center of gravity shifts forward in front of your feet. This produces a torque on your body rotating you forward.

6.7, BOTH SECTIONS

EXAMPLE 8.3   Where Is the Center of Gravity?

Three particles are located in a coordinate system as shown in Figure 8.9. Find the center of gravity.

Reasoning   The y coordinate of the center of gravity is zero because all the particles are on the x axis. To find the x coordinate of the center of gravity, we use Equation 8.3:

$$x_{cg} = \frac{\sum m_i x_i}{\sum m_i}$$

Solution   For the numerator, we find

$$\sum m_i x_i = m_1 x_1 + m_2 x_2 + m_3 x_3$$
$$= (5.00 \text{ kg})(-0.500 \text{ m}) + (2.00 \text{ kg})(0 \text{ m}) + (4.00 \text{ kg})(1.00 \text{ m})$$
$$= 1.50 \text{ kg} \cdot \text{m}$$

The denominator is $\sum m_i = 11.0$ kg; therefore,

$$x_{cg} = \frac{1.50 \text{ kg} \cdot \text{m}}{11.0 \text{ kg}} = 0.136 \text{ m}$$

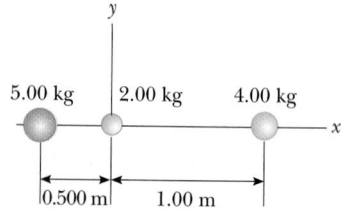

Figure 8.9   (Example 8.3) Locating the center of gravity for a system of three particles.

Exercise   If a fourth particle of mass 2.00 kg is placed at $x = 0$, $y = 0.250$ m, find the x and y coordinates of the center of gravity for this system of four particles.

Answer $x_{cg} = 0.115$ m; $y_{cg} = 0.0380$ m

8.4 EXAMPLES OF OBJECTS IN EQUILIBRIUM

In Chapter 4 we discussed some techniques for solving problems concerned with objects in equilibrium. Recall that when the objects were treated as geometric points, it was sufficient to simply apply the condition that the net force on the object must be zero. In this chapter we have shown that for objects of finite dimensions, a second condition for equilibrium must be satisfied—namely that the net torque on the object must also be zero. The following general procedure is recommended for solving problems that involve objects in equilibrium.

Problem-Solving Strategy

Objects in Equilibrium

1. Draw a simple, neat diagram of the system.
2. Isolate the object that is being analyzed. Draw a free-body diagram showing all external forces that are acting on this object. For systems that contain more than one object, draw a *separate* diagram for each object. Do not include forces that the object exerts on its surroundings.
3. Establish convenient coordinate axes for each body and find the components of the forces along these axes. Now apply the first condition of equilibrium (the net force on the object in the x and y direction must be zero) for each object under consideration.
4. Choose a convenient origin for calculating the net torque on the object. Now apply the second condition of equilibrium (the net torque on the object about any origin must be zero). Remember that the choice of the origin for the torque equation is arbitrary: therefore, choose an origin that will simplify your calculation as much as possible. Note that a force that acts along a line passing through the point chosen as the axis of rotation gives zero contribution to the torque.
5. The first and second conditions for equilibrium will give a set of simultaneous equations with several unknowns. All that is left to complete your solution is to solve for the unknowns in terms of the known quantities.

 ### EXAMPLE 8.4 The Seesaw

A uniform 40.0-N board supports two children weighing 500 N and 350 N (Fig. 8.10a). The support (often called the *fulcrum*) is under the center of gravity of the board, and the 500-N child is 1.50 m from the center.
(a) Determine the upward force, **n,** exerted on the board by the support.

Reasoning First note in Figure 8.10b that, in addition to **n,** the external forces acting on the board are the forces of gravity acting on the children and the board, all of which act downward. We can assume that the board's center of gravity is at its geometric center because we were told that the board is uniform. Because the system is in equilibrium, the upward force **n** must balance all the downward forces.

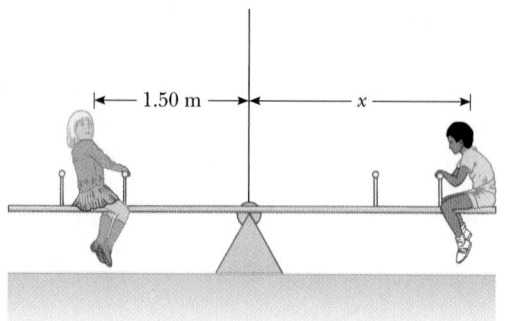

Figure 8.10 (Example 8.4) Two children balanced on a seesaw.

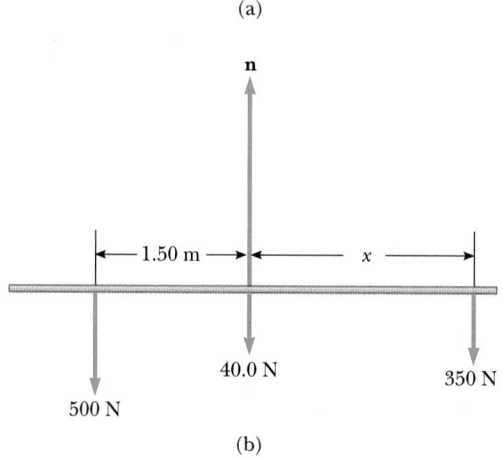

(a)

(b)

Solution   From $\Sigma F_y = 0$, we have

$$n - 500 \text{ N} - 350 \text{ N} - 40.0 \text{ N} = 0 \qquad \text{or} \qquad n = \boxed{890 \text{ N}}$$

Although the equation $\Sigma F_x = 0$ also applies to this situation, it is unnecessary to consider it because no forces are acting horizontally on the board.

(b) Determine where the 350-N child should sit to balance the system.

Reasoning   To find this position, we must invoke the second condition for equilibrium. We take the center of gravity of the board as the axis for the torque equation. This choice simplifies the problem, because the torques produced by both **n** and the 40.0-N weight are zero about this axis (because the lever arm of each is zero).

Solution   We apply $\Sigma \tau = 0$ to find

$$(500 \text{ N})(1.50 \text{ m}) - (350 \text{ N})(x) = 0$$

$$x = \boxed{2.14 \text{ m}}$$

(c) Repeat part (b), using another axis for the torque computations.

Reasoning   We stated that when an object is in equilibrium, the choice for the axis about which to compute torques is completely arbitrary. To illustrate this point, let us choose an axis perpendicular to the page and passing through the location of the 500-N child.

Solution In this case $\Sigma\tau = 0$ yields

$$n(1.50 \text{ m}) - (40.0 \text{ N})(1.50 \text{ m}) - (350 \text{ N})(1.50 + x) = 0$$

From part (a) we know that $n = 890$ N. Thus, we can solve for x to find $x = 2.14$ m, in agreement with the result of part (b).

EXAMPLE 8.5　A Weighted Forearm

A 50.0-N weight is held in a person's hand with the forearm horizontal, as in Figure 8.11a. The biceps muscle is attached 0.0300 m from the joint, and the weight is 0.350 m from the joint. Find the upward force exerted on the forearm (the ulna) by the biceps and the downward force on the forearm (the humerus) acting at the joint. Neglect the weight of the forearm.

Solution The forces acting on the forearm are equivalent to those acting on a bar of length 0.350 m, as shown in Figure 8.11b, where **F** is the upward force of the biceps and **R** is the downward force at the joint. From the first condition for equilibrium, we have

$$(1) \qquad \Sigma F_y = F - R - 50.0 \text{ N} = 0$$

From the second condition for equilibrium, we know that the sum of the torques about any point must be zero. With the joint O as the axis, we have

$$F(0.0300 \text{ m}) - (50.0 \text{ N})(0.350 \text{ m}) = 0$$

$$F = \boxed{583 \text{ N}}$$

This value for F can be substituted into (1) to give $\boxed{R = 533 \text{ N}}$. The two values correspond to $F = 131$ lb and $R = 119$ lb. Clearly, the forces at joints and in muscles can be extremely large.

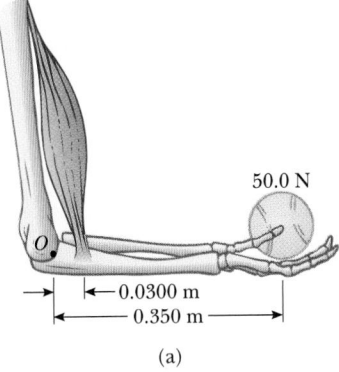

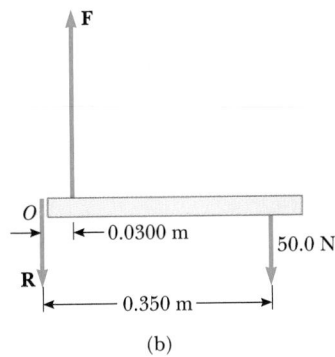

Figure 8.11 (Example 8.5) (a) A weight held with the forearm horizontal. (b) The mechanical model for the system.

EXAMPLE 8.6　Walking a Horizontal Beam

A uniform, horizontal 300-N beam, 5.00 m long, is attached to a wall by a pin connection that allows the beam to rotate. Its far end is supported by a cable that makes an angle of 53.0° with the horizontal (Fig. 8.12a). If a 600-N person stands 1.50 m from the wall, find the tension in the cable and the force exerted on the beam by the wall.

Reasoning First we must identify all the external forces acting on the beam and sketch them on a free-body diagram. This is shown in Figure 8.12b. The forces on the beam consist of the downward force of gravity acting on the beam, which has a magnitude of 300 N; the downward force exerted on the beam by the man, which is equal in magnitude to his weight, 600 N; the tension force, **T,** in the cable; and the force of the wall on the beam, **R.** We now resolve the forces **T** and **R** into their horizontal and vertical components, as shown in Figure 8.12c. Note that the x component of the tension force ($T\cos 53.0°$) is to the left, whereas the y component ($T\sin 53.0°$) is upward. The horizontal and vertical components of **R** are denoted by R_x and R_y, respectively. The first condition

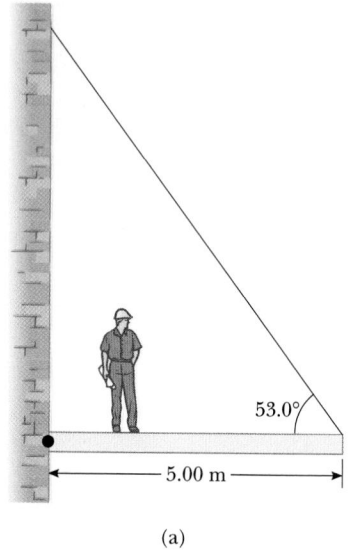

(a)

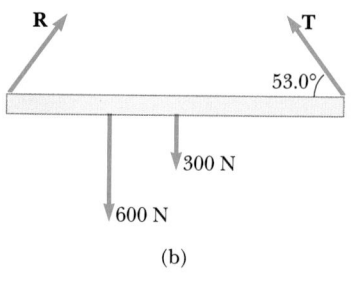

(b)

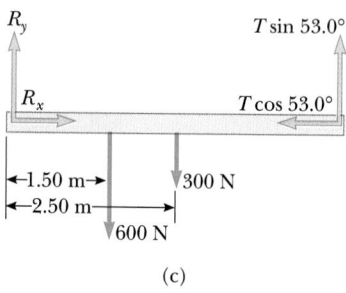

(c)

Figure 8.12 (Example 8.6) (a) A uniform beam attached to a wall and supported by a cable. (b) A free-body diagram for the beam. (c) The component form of the free-body diagram.

for equilibrium can now be applied in the x and y direction to give us two equations in terms of our unknowns R_x, R_y, and T. The necessary third equation can be found from the second condition of equilibrium.

Solution From the first condition for equilibrium, we find

(1) $F_x = R_x - T\cos 53.0° = 0$

(2) $F_y = R_y + T\sin 53.0° - 600\text{ N} - 300\text{ N} = 0$

The unknowns are R_x, R_y, and T. Because there are three unknowns and only two equations, we cannot find a solution from just the first condition of equilibrium.

   Now let us use the second condition of equilibrium. The axis that passes through the pivot at the wall is a convenient one to choose for the torque equation because the forces R_x, R_y, and $T\cos 53.0°$ all have lever arms of zero and hence have zero torque about this pivot. Recalling our sign convention for the torque about an axis and noting that the lever arms of the 600-N, 300-N, and $T\sin 53.0°$ forces are 1.50 m, 2.50 m, and 5.00 m, respectively, we get

(3) $\Sigma\tau_0 = (T\sin 53.0°)(5.00\text{ m}) - (300\text{ N})(2.50\text{ m}) - (600\text{ N})(1.50\text{ m}) = 0$

$T = \boxed{413\text{ N}}$

Thus, the torque equation using this axis gives us one of the unknowns immediately! This value for T is then substituted into (1) and (2) to give

$R_x = \boxed{249\text{ N}}$ $R_y = \boxed{570\text{ N}}$

If we selected some other axis for the torque equation, the solution would be the same. For example, if the axis were to pass through the center of gravity of the beam, the torque equation would involve both T and R_y; together with (1) and (2), however, it could still be solved for the unknowns. Try it!

Exercise Repeat this problem, but with the direction of R_x opposite that shown in Figure 8.12c. What answers do you get for T, R_x, and R_y?

Answer $T = 413$ N, $R_x = -249$ N, and $R_y = 570$ N. The negative sign for R_x means its direction was chosen incorrectly. The direction of R_x must be to the right, as shown in Figure 8.12c.

EXAMPLE 8.7 Don't Climb the Ladder

A uniform 10-m-long, 50-N ladder rests against a smooth vertical wall, as in Figure 8.13a. If the ladder is just on the verge of slipping when the angle it makes with the ground is 50°, find the coefficient of static friction between the ladder and ground.

Reasoning Figure 8.13b is the free-body diagram for the ladder, showing all external forces acting on it. At the base of the ladder, the Earth exerts an upward normal force, **n**, and a force of static friction, **f**, acts to the right. The wall exerts the force **P** to the left. Note that **P** is horizontal because the wall is smooth. (If the wall were rough, an upward frictional force would be exerted on the ladder.) The first condition for equilibrium can now be applied in the x and y directions to give us two equations in terms of our unknowns f, P, and n. The necessary third equation can be found from the second condition of equilibrium.

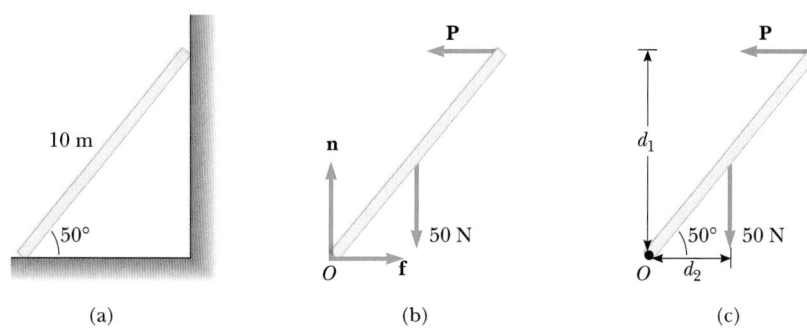

Figure 8.13 (Example 8.7) (a) A ladder leaning against a frictionless wall. (b) A free-body diagram for the ladder. (c) Lever arms for the force of gravity and **P**.

Solution From the first condition for equilibrium applied to the ladder, we have

$$(1) \quad \Sigma F_x = f - P = 0$$

$$(2) \quad \Sigma F_y = n - 50 \text{ N} = 0$$

From (2) we see that $n = 50$ N. Furthermore, when the ladder is on the verge of slipping, the force of static friction must be maximum and given by the relation $f_{s,\text{max}} = \mu_s n = \mu_s(50 \text{ N})$. Thus, (1) reduces to

$$(3) \quad \mu_s(50 \text{ N}) = P$$

Let us now apply the second condition of equilibrium and take the torques about the axis O at the bottom of the ladder, as in Figure 8.13c. The force **P** and the force of gravity acting on the ladder are the only forces that contribute to the torque about this axis, and their lever arms are shown in Figure 8.13c. Note that because the length of the ladder is 10 m, the lever arm for **P** is $d_1 = 10$ m sin 50°. Likewise, the lever arm for the 50-N force of gravity is $d_2 = 5.0$ m cos 50°, where the force of gravity acts through the center because the ladder is uniform. Thus, we find that

$$\Sigma \tau_0 = P(10 \text{ m sin } 50°) - (50 \text{ N})(5.0 \text{ m cos } 50°) = 0$$

$$P = 21 \text{ N}$$

Now that P is known, we can substitute its value into (3) to find μ_s:

$$\mu_s = \frac{21 \text{ N}}{50 \text{ N}} = 0.42$$

8.5 RELATIONSHIP BETWEEN TORQUE AND ANGULAR ACCELERATION

Earlier in this chapter we considered the situation in which both the net force and the net torque acting on an object are zero. Such objects are said to be in equilibrium. We shall now examine the behavior of an object when the net torque acting on it is not zero. As you shall see, when a rigid object is acted on by a net torque, it undergoes an angular acceleration. Furthermore, the angular acceleration is directly proportional to the net torque. The end result of our investigation will be an expression that is analogous to $\mathbf{F} = m\mathbf{a}$ in translational motion.

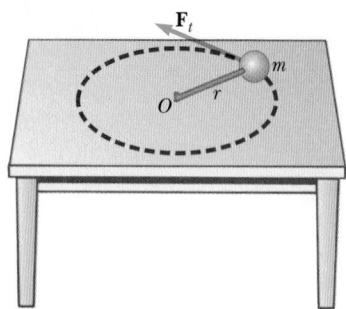

Figure 8.14 A mass, m, attached to a light rod of length r moves in a circular path on a frictionless horizontal surface while the tangential force $\mathbf{F}_t$ acts on it.

Let us begin by considering the system shown in Figure 8.14, which consists of a mass m connected to a very light rod of length r. The rod is pivoted at the point O, and its movement is confined to rotation on a frictionless *horizontal* table. Now let us assume that a force, $\mathbf{F}_t$, perpendicular to the rod and hence tangent to the circular orbit, is acting on m. Because there is no force to oppose this tangential force, the mass undergoes a tangential acceleration according to Newton's second law:

$$F_t = ma_t$$

Multiplying the left and right sides of this equation by r gives

$$F_t r = mra_t$$

In Chapter 7 we found that the tangential acceleration and angular acceleration for a particle rotating in a circular path are related by the expression

$$a_t = r\alpha$$

so now we find that

$$F_t r = mr^2\alpha \qquad [8.5]$$

The left side of Equation 8.5, which should be familiar to you, is the torque acting on the mass about its axis of rotation. That is, the torque is equal in magnitude to the force on m multiplied by the perpendicular distance from the pivot to the line along which the force acts, or $\tau = F_t r$. Hence, we can write Equation 8.5 as

$$\tau = mr^2\alpha \qquad [8.6]$$

Equation 8.6 shows that the torque on the system is proportional to angular acceleration, where the constant of proportionality, mr^2, is called the **moment of inertia** of the mass m. (Because the rod is very light, its moment of inertia can be neglected.)

Torque on a Rotating Object

Now consider a solid disk rotating about its axis as in Figure 8.15a. The disk consists of many particles at various distances from the axis of rotation, as in Figure 8.15b. The torque on each one of these particles is given by Equation 8.6. The *total* torque on the disk is given by the sum of the individual torques on all the particles:

$$\Sigma\tau = (\Sigma mr^2)\alpha \qquad [8.7]$$

Note that, because the disk is rigid, all particles have the *same* angular acceleration, so α is not involved in the sum. If the masses and distances of the particles are labeled with subscripts as in Figure 8.15b, then

$$\Sigma mr^2 = m_1 r_1^2 + m_2 r_2^2 + m_3 r_3^2 + \dots$$

This quantity is called the **moment of inertia** of the whole body and is given the symbol I:

Moment of inertia ▶

$$I \equiv \Sigma mr^2 \qquad [8.8]$$

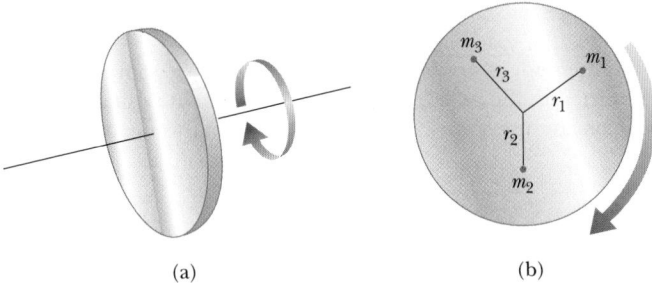

(a)   (b)

Figure 8.15 (a) A solid disk rotating about its axis. (b) The disk consists of many particles, all with the same angular acceleration.

The moment of inertia has the SI units $kg \cdot m^2$. Using this result in Equation 8.7, we see that the total torque on a rigid body rotating about a fixed axis is given by

$$\Sigma \tau = I\alpha \qquad [8.9]$$

◀ Relationship between net torque and angular acceleration

The angular acceleration of an object is proportional to the net torque acting on it. The proportionality constant, I, between the net torque and angular acceleration is the moment of inertia.

It is important to note that the equation $\Sigma \tau = I\alpha$ (Eq. 8.9) is the rotational counterpart to Newton's second law, $\Sigma F = ma$. Thus, the correspondence between rotational motion and linear motion continues. Recall from Chapter 7 that the linear variables x, v, and a are replaced in rotational motion by the variables θ, ω, and α. Likewise, we now see that **the force and mass in linear motion correspond to torque and moment of inertia in rotational motion.** In this chapter we shall develop other equations for the rotational kinetic energy and the angular momentum of a body rotating about a fixed axis. Based on the analogies already presented, you should be able to predict the form of these equations.

Applying Physics 2

The gear system on a bicycle provides an easily visible example of the relationship between torque and angular acceleration. Consider first a five-speed gear system in which the drive-chain can be adjusted to wrap around any of five gears attached to the back wheel (Figure 8.16). The gears are concentric with the wheel hub and are of different radii. When the cyclist begins from rest, the chain is attached to the largest gear. Because it has the largest radius, this gear provides the largest torque to the drive wheel. Large torque is required for starting, because the bicycle must change its motion from rest to some speed. As the bicycle rolls faster, the tangential speed of the chain increases and becomes too fast for the cyclist to maintain by pushing the pedals. The chain is then moved to the gear with the next smallest radius, and the chain now has a smaller tangential speed, which the cyclist can maintain by pushing the pedals. This gear does not provide as much torque as the first, but the cyclist does not need as much—he or she only needs to accelerate from some speed to a somewhat higher speed. This process continues as the bicycle moves faster and faster, and the cyclist shifts through all five gears. The fifth gear supplies the lowest torque, but now the main function of the supplied torque is to counter the frictional torque from the rolling tires, which will tend to slow the bicycle. The small radius of this gear allows the

Figure 8.16 (Applying Physics 2) The drive wheel and gears of a bicycle. *(Semple Design & Photography)*

QUICKLAB

Find a nearby door. After turning the doorknob, open the door by pushing on the doorknob. Estimate the force and, hence, the amount of torque required to open it. Now open the door by pushing on it half-way between the hinge line and the doorknob. Compare the force and torque required to that found when you pushed on the doorknob.

7.4, SECTION 1

cyclist to keep up with the chain's movement by pushing the pedals, despite the rapid movement of the bicycle.

A ten-speed bicycle has the same gear structure on the drive wheel, but has two gears on the sprocket connected to the pedals. By combining different positions of the chain on the rear gears and the sprocket gears, ten different torques are available.

More on the Moment of Inertia

Because the moment of inertia of a body—as defined by $I = \Sigma mr^2$ (Eq. 8.8)—will be used throughout the remainder of this chapter, it will be useful to examine it in more detail before discussing other aspects of rotational motion. As seen earlier, a small object (or particle) has a moment of inertia equal to mr^2 about some axis. Now consider a somewhat more complicated system, the baton being twirled by a majorette in Figure 8.17. Let us assume that the baton can be modeled as a very light rod of length 2ℓ with a heavy mass at each end. (The rod of a real baton has significant mass relative to its ends.) Because we are neglecting the mass of the rod, the moment of inertia of the baton about an axis through its center and perpendicular to its length is given by Equation 8.8:

$$I = \Sigma mr^2$$

Because in this system there are two equal masses equidistant from the axis of rotation, we see that $r = \ell$ for each mass, and the sum is

$$I = \Sigma mr^2 = m\ell^2 + m\ell^2 = 2m\ell^2$$

We pointed out earlier that I is the rotational counterpart of m. However, there are some important distinctions between the two. For example, mass is an intrinsic property of an object that does not change, whereas **the moment of inertia of a**

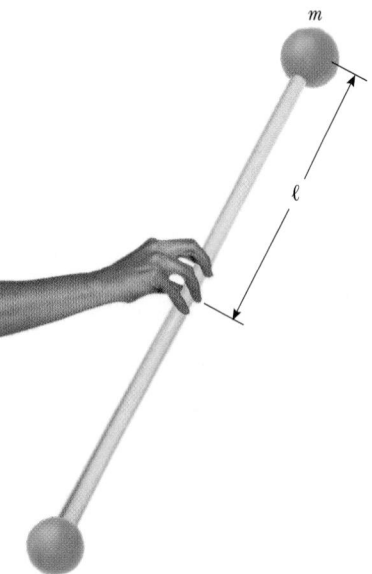

Figure 8.17 A baton of length 2ℓ and mass $2m$ (the mass of the connecting rod is neglected). The moment of inertia about the axis through the baton's center and perpendicular to its length is $2m\ell^2$.

system depends on the axis of rotation and on the manner in which the mass is distributed. Examples 8.8 and 8.9 illustrate this point.

EXAMPLE 8.8 The Baton Twirler

In an effort to be the star of the half-time show, a majorette twirls a highly unusual baton made up of four masses fastened to the ends of light rods (Fig. 8.18). Each rod is 1.0 m long. Find the moment of inertia of the system about an axis perpendicular to the page and passing through the point where the rods cross.

Solution   Applying Equation 8.8, we get

$$I = \Sigma mr^2 = m_1 r_1^2 + m_2 r_2^2 + m_3 r_3^2 + m_4 r_4^2$$

$$= (0.20 \text{ kg})(0.50 \text{ m})^2 + (0.30 \text{ kg})(0.50 \text{ m})^2 + (0.20 \text{ kg})(0.50 \text{ m})^2$$

$$+ (0.30 \text{ kg})(0.50 \text{ m})^2$$

$$= \boxed{0.25 \text{ kg} \cdot \text{m}^2}$$

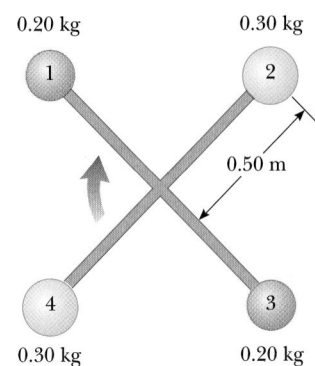

Figure 8.18   (Example 8.8) Four masses connected to light rods rotating in the plane of the page.

EXAMPLE 8.9 The Baton Twirler: Second Act

Not satisfied with the crowd reaction from the baton twirling of Example 8.8, the majorette tries spinning her strange baton about the axis $00'$, as shown in Figure 8.19. Calculate the moment of inertia about this axis.

Solution   Again applying $I = \Sigma mr^2$, we have

$$I = (0.20 \text{ kg})(0)^2 + (0.30 \text{ kg})(0.50 \text{ m})^2 + (0.20 \text{ kg})(0)^2 + (0.30 \text{ kg})(0.50 \text{ m})^2$$

$$= \boxed{0.15 \text{ kg} \cdot \text{m}^2}$$

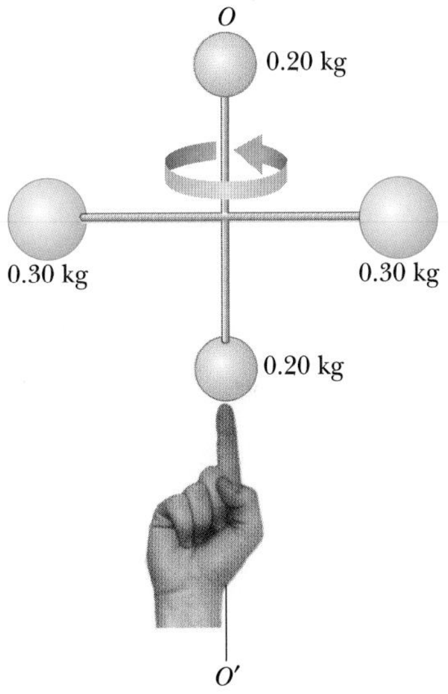

Figure 8.19   (Example 8.9) A double baton rotating about the axis OO'.

Calculation of Moments of Inertia for Extended Objects

The method used for calculating moments of inertia in Examples 8.8 and 8.9 is simple enough when you have only a few small masses rotating about an axis. The situation becomes much more complex when the object is an extended mass, such as a sphere, a cylinder, or a cone. One type of extended object that is amenable to a simple solution is a hoop rotating about an axis perpendicular to its plane and passing through its center, as shown in Figure 8.20. A bicycle tire, for example, would fit in this category.

To evaluate the moment of inertia of the hoop, we can still use the equation $I = \Sigma mr^2$ (Eq. 8.8) and imagine that the hoop is divided into a number of small segments having masses $m_1, m_2, m_3, \ldots$ as in Figure 8.20. This approach is just an extension of the baton problem described in the preceding examples, except that now we have a large number of small masses in rotation instead of only four.

We can express the sum for I as

$$I = \Sigma mr^2 = m_1 r_1^2 + m_2 r_2^2 + m_3 r_3^2 + \ldots$$

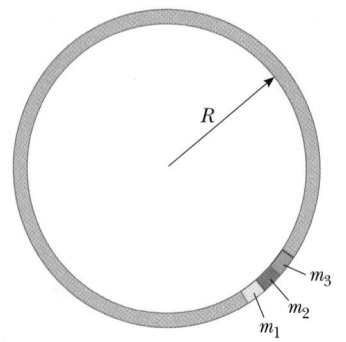

Figure 8.20 A uniform hoop can be divided into a large number of small segments that are equidistant from the center of the hoop.

All of the segments around the hoop are at the *same distance, R,* from the axis of rotation; thus, we can drop the subscripts on the distances and factor out the common factor R^2:

$$I = (m_1 + m_2 + m_3 + \ldots)R^2$$

We know, however, that the sum of the masses of all the segments must equal the total mass of the hoop, M:

$$M = m_1 + m_2 + m_3 + \ldots$$

and so we can express I as

$$I = MR^2 \qquad\qquad\qquad [8.10]$$

This expression can be used for the moment of inertia of any ring-shaped object rotating about an axis through its center and perpendicular to its plane. Note that the result is strictly valid only if the thickness of the ring is small relative to its inner radius.

The hoop we selected as an example is unique in that we were able to find an expression for its moment of inertia by using only simple algebra. Unfortunately, most extended objects are more difficult to work with, and the methods of integral calculus are required. Such methods are beyond the scope of this text. The moments of inertia for some common shapes are given without proof in Table 8.1.

TABLE 8.1 Moments of Inertia for Various Rigid Bodies of Uniform Composition

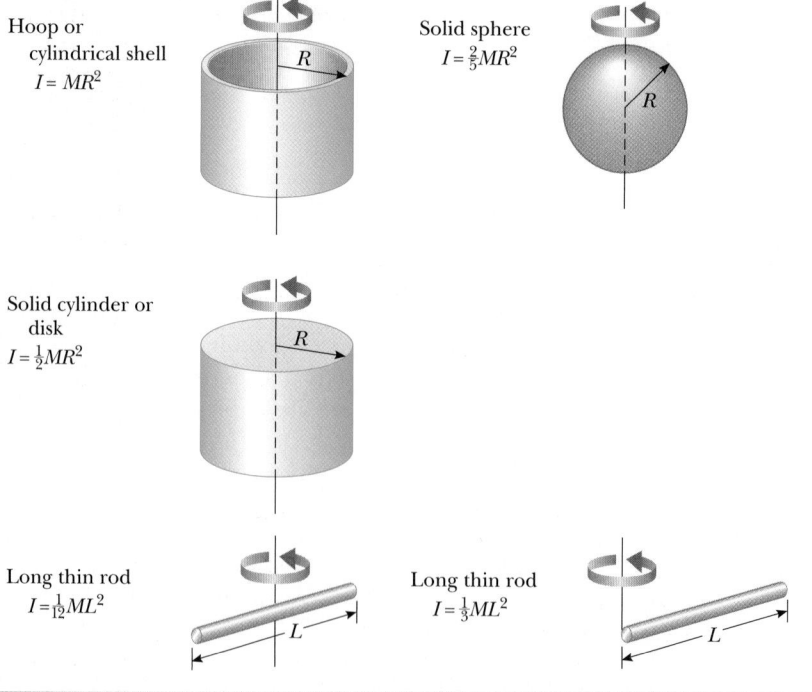

Hoop or cylindrical shell
$I = MR^2$

Solid sphere
$I = \frac{2}{5}MR^2$

Solid cylinder or disk
$I = \frac{1}{2}MR^2$

Long thin rod
$I = \frac{1}{12}ML^2$

Long thin rod
$I = \frac{1}{3}ML^2$

When the need arises, you can use this table to determine the moment of inertia of a body having any one of the listed shapes.

 ## EXAMPLE 8.10 Warming Up

A baseball player loosening up his arm before a game tosses a 0.150-kg baseball using only the rotation of his forearm to accelerate the ball (Fig. 8.21). The ball starts at rest and is released with a speed of 30.0 m/s in 0.300 s.
(a) Find the constant angular acceleration of the arm and ball.

Solution During its acceleration, the ball moves through an arc of a circle with a radius of 0.350 m. We can determine the angular acceleration using $\omega = \omega_0 + \alpha t$. Because $\omega_0 = 0$, however, $\omega = \alpha t$, or

$$\alpha = \frac{\omega}{t}$$

We also know that $v = r\omega$, and so we get

$$\alpha = \frac{\omega}{t} = \frac{v}{rt} = \frac{30.0 \text{ m/s}}{(0.350 \text{ m})(0.300 \text{ s})} = \boxed{286 \text{ rad/s}^2}$$

(b) Find the torque exerted on the ball to give it this angular acceleration.

Solution The moment of inertia of the ball about an axis that passes through the elbow, perpendicularly to the arm, is

$$I = mr^2 = (0.150 \text{ kg})(0.350 \text{ m})^2 = 1.84 \times 10^{-2} \text{ kg} \cdot \text{m}^2$$

Thus, the required torque is

$$\tau = I\alpha = (1.84 \times 10^{-2} \text{ kg} \cdot \text{m}^2)(286 \text{ rad/s}^2) = \boxed{5.26 \text{ N} \cdot \text{m}}$$

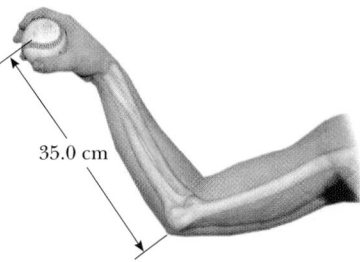

Figure 8.21 (Example 8.10) A ball being tossed by a pitcher. The forearm is being used to accelerate the ball.

Workbook Problem 44 – (Workbook page 96) also deals with the relationship between torque and angular acceleration.

EXAMPLE 8.11 The Falling Bucket

A solid, frictionless cylindrical pulley of mass $M = 3.00$ kg and radius $R = 0.400$ m is used to draw water from a well (Fig. 8.22a). A bucket of mass $m = 2.00$ kg is attached to a cord that is wrapped around the cylinder. If the bucket starts from rest at the top of the well and falls for 3.00 s before hitting the water, how far does it fall?

Reasoning Figure 8.22b shows the two forces on the bucket as it falls: **T** is the tension in the cord, and $m\mathbf{g}$ is the force of gravity acting on the bucket. We shall choose downward as the positive direction and write Newton's second law for the bucket as

$$mg - T = ma$$

When the given quantities are substituted into this equation, we have

$$(1) \qquad (2.00 \text{ kg})(9.80 \text{ m/s}^2) - T = (2.00 \text{ kg})a$$

With one equation and two unknowns, we must develop an additional equation to complete the problem. To obtain this second equation, let us consider the cylinder's rotational motion. Equation 8.9 applied to the cylinder gives the necessary expression:

$$\tau = I\alpha = \tfrac{1}{2}MR^2\alpha$$

(a)

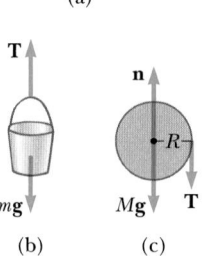

(b)   (c)

Figure 8.22 (Example 8.11) (a) A water bucket attached to a rope passing over a frictionless pulley. (b) A free-body diagram for the bucket. (c) The tension produces a torque on the cylinder about its axis of rotation.

Figure 8.22c shows that the only force producing a torque on the cylinder as it rotates about an axis through its center is **T,** the force due to the tension in the cord. Actually, two other forces act on the cylinder—the force of gravity and the upward force of the axle; but we do not have to consider them here because the lever arm of each about the axis of rotation is zero. Thus, we have

$$T(0.400 \text{ m}) = \tfrac{1}{2}(3.00 \text{ kg})(0.400 \text{ m})^2(\alpha)$$

$$(2) \qquad T = (0.600 \text{ kg} \cdot \text{m})\alpha$$

At this point, it is important to recognize that the downward acceleration of the bucket is equal to the tangential acceleration of a point on the rim of the cylinder. Therefore, the angular acceleration of the cylinder and the linear acceleration of the bucket are related by $a_t = r\alpha$. When this relation is used in (2), we get

$$(3) \qquad T = (1.50 \text{ kg})a_t$$

Solution Equations (1) and (3) can now be solved simultaneously to find a_t and T. This procedure gives

$$a_t = 5.60 \text{ m/s}^2 \qquad T = 8.40 \text{ N}$$

Finally, we turn to the equations for motion with constant linear acceleration to find the distance, d, that the bucket falls in 3.00 s. Because $v_0 = 0$, we get

$$d = v_0 t + \tfrac{1}{2}at^2 = \tfrac{1}{2}(5.60 \text{ m/s}^2)(3.00 \text{ s})^2 = \boxed{25.2 \text{ m}}$$

8.6 ROTATIONAL KINETIC ENERGY

In Chapter 5 we defined the kinetic energy of a particle moving through space with a speed v as the quantity $\frac{1}{2}mv^2$. Analogously, **a body rotating about some axis with an angular speed ω has rotational kinetic energy given by $\frac{1}{2}I\omega^2$.** To prove this, consider a rigid plane body rotating about some axis perpendicular to its plane, as in Figure 8.23. The body consists of many small particles, each of mass m. All these particles rotate in circular paths about the axis. If r is the distance of one of the particles from the axis of rotation, the speed of this particle is $v = r\omega$. Because the *total* kinetic energy of the body is the sum of all the kinetic energies associated with all the particles making up the body, we have

$$KE_r = \Sigma(\tfrac{1}{2}mv^2) = \Sigma(\tfrac{1}{2}mr^2\omega^2) = \tfrac{1}{2}(\Sigma mr^2)\omega^2$$

$$KE_r = \tfrac{1}{2}I\omega^2 \qquad\qquad [8.11]$$

where $I = \Sigma mr^2$ is the moment of inertia of the body. Note that the ω^2 term is factored out because it is the same for every particle.

We found the energy concept to be extremely useful for describing the linear motion of a system. It can be equally useful for simplifying the analysis of rotational motion. Consider a system that is described by three types of energy: **gravitational potential energy, PE_g, translational kinetic energy, KE_t,** and **rotational kinetic energy, KE_r.** We must include all these forms of energy in our equation for conservation of mechanical energy.

$$(KE_t + KE_r + PE_g)_i = (KE_t + KE_r + PE_g)_f \qquad [8.12]$$

where i and f refer to initial and final values, respectively. This relation is *only* true if we ignore dissipative forces such as friction.

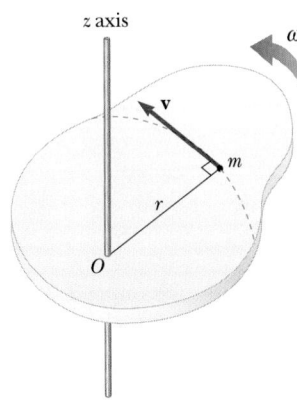

Figure 8.23 A rigid plane body rotating about the z axis with angular speed ω. The kinetic energy of a particle of mass m is $\frac{1}{2}mv^2$. The total kinetic energy of the body is $\frac{1}{2}I\omega^2$.

 7.3

◀ Conservation of mechanical energy

Thinking Physics 2

Two children are playing on a hill, rolling automobile tires down the slope. One child claims that the tire will roll faster if one of them curls up inside the tire and rides down inside of it. The other child claims that this will cause the tire to roll more slowly. Which child is correct?

Explanation The first child is correct. The original gravitational potential energy of the object turns into kinetic energy of translation plus kinetic energy of rotation plus work against friction. Dividing the mass out of the work-energy theorem shows that the final energy is measured by the expression

$$v^2\left(1 + \frac{I}{mR^2}\right)$$

where I is the moment of inertia of the object of mass m, and R is the radius on which it rolls. To make its center-of-mass speed v large, the children want to channel more energy into translation and less into rotation. That is, they want to make the fraction I/mR^2 small. Filling in the hole in the tire has this effect. It decreases the average distance from the axis out to bits of matter.

QUICKLAB

Compare the motion of an empty soup can and a filled soup can down the same incline such as a tilted table. If they are released from rest at the same height on the incline, which one reaches the bottom first? Repeat your observations with different kinds of soup (tomato, chicken noodle, etc.) and a can of beans. Compare their motions, and try to explain your observations. Finally, compare the motion of a filled soup can with a tennis ball, and explain your results to a friend.

EXAMPLE 8.12 A Ball Rolling Down an Incline

A ball of mass M and radius R starts from rest at a height of 2.00 m and rolls down a 30.0° slope, as shown in Figure 8.24. What is the linear speed of the ball when it leaves the incline? Assume that the ball rolls without slipping.

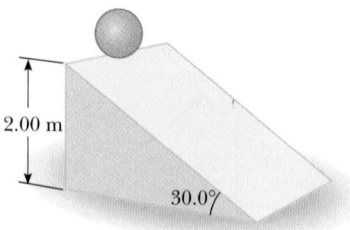

Figure 8.24 (Example 8.12) A ball starts from rest at the top of an incline and rolls to the bottom without slipping.

Workbook Problems 45 and 46 — (pages 98 to 101 of the Workbook) also deal with rotational kinetic energy and rolling motion.

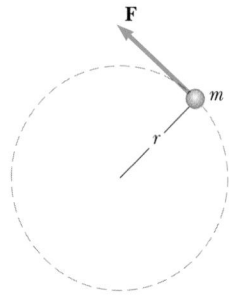

Figure 8.25 An object of mass m rotating in a circular path under the action of a constant torque.

Definition of angular momentum ▷

Reasoning The initial energy of the ball is gravitational potential energy, and when the ball reaches the bottom of the ramp, this potential energy has been converted to translational and rotational kinetic energy. The conservation of mechanical energy equation becomes

$$(PE_g)_i = (KE_t + KE_r)_f$$

$$Mgh = \tfrac{1}{2}Mv^2 + \tfrac{1}{2}(\tfrac{2}{5}MR^2)\omega^2$$

where h is the distance through which the ball's center of gravity falls and where we have used $I = \tfrac{2}{5}MR^2$ from Table 8.1 as the moment of inertia of the ball. If the ball rolls without slipping, a point on its surface must have the same instantaneous speed as its center of gravity has relative to the incline.[1] Thus, we can relate the ball's linear speed to its rotational speed:

$$v = R\omega$$

This expression can be used to eliminate ω from the equation for conservation of mechanical energy:

$$Mgh = \tfrac{1}{2}Mv^2 + \tfrac{1}{5}Mv^2$$

Solution Solving for v gives

$$v = \sqrt{\frac{10gh}{7}} = \sqrt{\frac{10(9.80 \text{ m/s}^2)(2.00 \text{ m})}{7}} = \boxed{5.29 \text{ m/s}}$$

Exercise Repeat this example for a solid cylinder of the same mass and radius as the ball and released from the same height. In a race between the two objects on the incline, which one would win?

Answer $v = \sqrt{4gh/3} = 5.11$ m/s; the ball would win.

8.7 ANGULAR MOMENTUM

In Figure 8.25, an object of mass m is positioned a distance of r away from a center of rotation. Under the action of a constant torque on the object, its angular speed will increase from the value ω_0 to the value ω in a time of Δt. Thus, we can write

$$\tau = I\alpha = I\left(\frac{\omega - \omega_0}{\Delta t}\right) = \frac{I\omega - I\omega_0}{\Delta t}$$

If we define the product

$$L \equiv I\omega \qquad [8.13]$$

as the angular momentum of the object, then we can write

$$\tau = \frac{\text{change in angular momentum}}{\text{time interval}} = \frac{\Delta L}{\Delta t} \qquad [8.14]$$

---

[1] Note that a point on the surface of the ball travels a distance of $2\pi R$ (relative to the center of the ball) during one rotation. If no slippage occurs, this is also the distance traveled by the center of gravity in the same time interval. Thus, the center of gravity has the same speed relative to the incline as a point on the surface of the ball has relative to the center.

Equation 8.14 is the rotational analog of Newton's second law, $F = \Delta p/\Delta t$, and states that **the torque acting on an object is equal to the time rate of change of the object's angular momentum.**

When the net external torque ($\Sigma \tau$) acting on the system is zero, we see from Equation 8.14 that $\Delta L/\Delta T = 0$. In this case, the rate of change of the system's angular momentum is zero. Therefore, **the product $I\omega$ remains constant in time.** That is, $L_i = L_f$, or

$$I_i \omega_i = I_f \omega_f \qquad \text{if} \qquad \Sigma \tau = 0 \qquad \qquad [8.15]$$

7.9

The angular momentum of a system is conserved when the net external torque acting on the system is zero. That is, when $\Sigma \tau = 0$, the initial angular momentum equals the final angular momentum.

◄ Conservation of angular momentum

In Equation 8.15 we have a third conservation law to add to our list: **conservation of angular momentum.** We can now state that **the energy, linear momentum, and angular momentum of an isolated system all remain constant.**

There are many examples of conservation of angular momentum, some of which should be familiar to you. You may have observed a figure skater spinning in the finale of her act. The skater's angular speed increases when she pulls her hands and feet close to the trunk of her body, as in Figure 8.26. That is, $\omega_2 > \omega_1$.

APPLICATION

Figure Skating.

(a)   (b)

Figure 8.26 (a) The angular speed of this skater increases when she pulls her arms in close to her body, demonstrating that angular momentum is conserved. (b) Photograph of Nancy Kerrigan in her final spinning routine. *(Reuters/Corbis-Bettmann)*

As the acrobat somersaults through the air, his angular momentum remains constant. Note that the center of mass of the acrobat follows a parabolic trajectory. *(Globus Bros. Studios/The Stock Market)*

Neglecting friction between skater and ice, we see that there are no external torques on the skater. The moment of inertia of her body decreases as her hands and feet are brought in. The resulting change in angular speed is accounted for as follows. Because angular momentum must be conserved, the product $I\omega$ has to remain constant, and a decrease of the moment of inertia of the skater must be compensated by a corresponding increase in the angular speed.

APPLICATION

Aerial Somersaults.

Similarly, when a diver or acrobat wishes to make several somersaults, he pulls his hands and feet close to the trunk of his body in order to rotate at a greater angular speed. In this case, the external force due to gravity acts through the center of gravity and hence exerts no torque about the axis of rotation, and the angular momentum about the center of gravity is conserved. For example, when a diver wishes to double his angular speed, he must reduce his moment of inertia to half its initial value.

Derek Swinson, professor of physics at the University of New Mexico, demonstrating the "gyro-ski" technique. The skier initiates a turn by lifting the axle of the rotating bicycle wheel. The direction of the turn depends on whether the left or right hand is used to lift the axle from the horizontal. Ignoring friction and gravity, the angular momentum of the system (the skier and the bicycle wheel) is conserved. *(Courtesy of Derek Swinson)*

Applying Physics 3

When a star exhausts its fuel, it tends to collapse to its center due to the mutual gravitational attraction of the mass of its material. For some massive stars, the gravitational force can be so strong as to squeeze protons and electrons together to form neutrons. The result is a neutron star—a star composed entirely of neutrons. On gravitational collapse, the star's radius becomes very small, on the order of a few kilometers. In general, a star rotates before it exhausts its fuel. As its radius becomes very small, its moment of inertia decreases even more drastically from its precollapse value. (Remember that moment of inertia is proportional to the square of the radius.) Because the star's angular momentum is conserved, its angular speed increases to an enormous value. And because stars possess magnetic fields (see Chapters 19–21), a rapidly rotating neutron star sends out an electromagnetic beacon that pulses at a very high frequency —hence the name *pulsar*. For example, the pulsar at the center of the Crab Nebula (which resulted from a supernova observed in 1054 A.D.) has a period of 33 ms, corresponding to a rotational frequency of 30 s^{-1}. Imagine a star rotating 30 times every second!

APPLICATION

Pulsars.

Thinking Physics 3

A common classroom demonstration is to spin on a rotating stool with heavy weights in your outstretched hands, and bring the weights close to your body (Fig. 8.27). The result is an increase in the angular velocity, in agreement with the conservation of angular momentum. As discussed previously, ice skaters use this effect to produce high-speed spins when they bring their arms close to the rotation axis. Let us imagine we

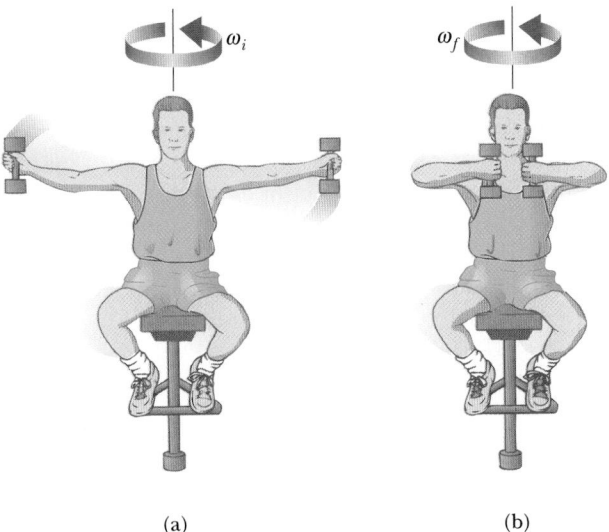

(a)   (b)

Figure 8.27   (Thinking Physics 3) (a) This student is given an initial angular speed while holding two masses as shown. (b) When the masses are pulled in close to the body, the angular speed of the system increases.

perform such a demonstration, and moving the weights inward results in halving the moment of inertia, and, therefore, doubling the angular velocity. If we consider the rotational kinetic energy, we see that the energy is *doubled* in this situation. Thus, angular momentum is conserved, but kinetic energy is not. Where does this extra energy come from?

Explanation As you spin with the weights in your hands, there is a tension force in your arms that causes the weights to move in a circular path. As you bring the weights in, this force from your arms does work on the weights, because it is moving them through a radial displacement. What's more, as the angular velocity increases in response to the weights moving inward, you must apply more and more force to keep the weights moving in the circular path of smaller radius. Thus, the work that you do in bringing the weights in increases as they are brought in. This work that you do with your arms appears as the increase in rotational kinetic energy of the system. You can perform this work due to the energy stored in your body from previous meals, so you are transforming potential energy stored in your body into rotational kinetic energy.

Problem-Solving Strategy

Rotational Motion

Keep the following facts and procedures in mind when solving rotational motion problems.

1. Very few new techniques must be learned in order to solve rotational motion problems. For example, problems involving the equation $\Sigma \tau = I\alpha$ are very similar to those encountered in Newton's second law problems, $\Sigma \mathbf{F} = m\mathbf{a}$. Note the correspondences between linear and rotational quantities in that $\mathbf{F}$ is replaced by τ, m by I, and $\mathbf{a}$ by α.
2. Other analogs between rotational quantities and linear quantities include the replacement of x by θ and v by ω. These are helpful as memory devices for such rotational motion quantities as rotational kinetic energy $KE_r = \frac{1}{2}I\omega^2$, and angular momentum, $L = I\omega$.
3. With the analogs mentioned in the previous step, techniques for conservation of energy are the same as those examined in Chapter 5 except for the fact that the object's rotational kinetic energy must be included in the expression for the conservation of energy.
4. Likewise, the techniques for solving problems in conservation of angular momentum are essentially the same as those for solving problems in conservation of linear momentum, except that total initial angular momentum is equated to total final angular momentum as $I_i\omega_i = I_f\omega_f$.

EXAMPLE 8.13 The Spinning Stool

A student sits on a pivoted stool while holding a pair of masses (see Fig. 8.27). The stool is free to rotate about a vertical axis with negligible friction. The moment of inertia of student, masses, and stool is 2.25 kg·m². The student is set in rotation with an initial angular speed of 5.00 rad/s, with masses outstretched. As he rotates, he pulls the masses inward so that the new moment of inertia of the system (student, masses, and stool) becomes 1.80 kg·m². What is the new angular speed of the system?

Reasoning We shall apply the principle of conservation of angular momentum to find the new angular speed. The initial angular momentum, $I_i\omega_i$, will be equated to the final angular momentum, $I_f\omega_f$.

Solution The initial angular momentum of the system is

$$L_i = I_i\omega_i = (2.25 \text{ kg}\cdot\text{m}^2)(5.00 \text{ rad/s}) = 11.3 \text{ kg}\cdot\text{m}^2/\text{s}$$

When the masses are pulled in, they are closer to the axis of rotation, and as a result the moment of inertia of the system is reduced. The new angular momentum is

$$L_f = I_f\omega_f = (1.80 \text{ kg}\cdot\text{m}^2)\omega_f$$

Because the net external torque on the system is zero, angular momentum is conserved. Thus, we find that

$$(11.3 \text{ kg}\cdot\text{m}^2/\text{s}) = (1.80 \text{ kg}\cdot\text{m}^2)\omega_f$$

$$\omega_f = 6.28 \text{ rad/s}$$

EXAMPLE 8.14 The Merry-Go-Round

A student stands at the edge of a circular platform that rotates in a horizontal plane about a frictionless vertical axle (Fig. 8.28). The platform has a mass of $M = 100$ kg and a radius of $R = 2.00$ m. The student, whose mass is $m = 60.0$ kg, walks slowly from the rim of the disk toward the center. If the angular speed of the system is 2.00 rad/s when the student is at the rim, calculate the angular speed when the student reaches a point 0.500 m from the center.

*Workbook Problem 47 –
(Workbook page 102) is similar to
Example 8.14.*

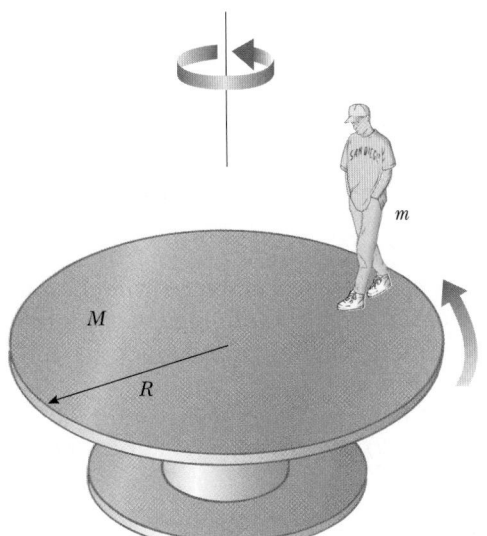

Figure 8.28 (Example 8.14) As this student walks toward the center of the rotating platform, the angular speed of the system increases because the angular momentum of the system (student + platform) must remain constant.

Reasoning We shall use the principle of conservation of angular momentum. The initial angular momentum of the system is the sum of the angular momentum of the platform plus that of the student when he is at the rim of the merry-go-round. The final angular momentum is the sum of the angular momentum of the platform plus that of the student when he is 0.500 m from the center.

Solution   The moment of inertia of the platform, I_p, is

$$I_p = \tfrac{1}{2}MR^2 = \tfrac{1}{2}(100 \text{ kg})(2.00 \text{ m})^2 = 200 \text{ kg·m}^2$$

Treating the student as a point mass, his initial moment of inertia is

$$I_s = mR^2 = (60.0 \text{ kg})(2.00 \text{ m})^2 = 240 \text{ kg·m}^2$$

Thus, the initial angular momentum of the platform plus student is

$$L_i = (I_p + I_s)(\omega_i) = (200 \text{ kg·m}^2 + 240 \text{ kg·m}^2)(2.00 \text{ rad/s}) = 880 \text{ kg·m}^2/\text{s}$$

When the student has walked to the position 0.500 m from the center, his moment of inertia is

$$I_s' = mr_f^2 = (60.0 \text{ kg})(0.500 \text{ m})^2 = 15.0 \text{ kg·m}^2$$

No change occurs in the moment of inertia of the platform. Because there are no external torques on the *system* (student plus platform) about the axis of rotation, we can apply the law of conservation of angular momentum:

$$L_i = L_f$$

$$880 \text{ kg·m}^2/\text{s} = 200\omega_f + 15.0\omega_f = 215\omega_f$$

$$\omega_f = \;\; 4.09 \text{ rad/s}$$

Exercise   Calculate the change in kinetic energy of the system (student plus platform) for this situation. What accounts for this change in energy?

Answer $KE_f - KE_i = 918$ J. The student must perform positive work in order to walk toward the center of the platform.

SUMMARY

7.10

The tendency of a force to rotate an object about some axis is measured by a quantity called **torque, τ. The magnitude of the torque is given by**

$$\tau = Fd \tag{8.1}$$

In this equation, d is the **lever arm**— the perpendicular distance from the axis of rotation to a line drawn along the direction of the force. The sign of the torque is negative if the turning tendency of the corresponding force is clockwise and positive if the turning tendency is counterclockwise.

     An object is in **equilibrium** when the following conditions are satisfied: (1) the net external force must be zero, and (2) the net external torque must be zero about any origin. That is,

$$\Sigma \mathbf{F} = 0 \qquad \Sigma \tau = 0$$

The first equation is called the **first condition for equilibrium.** When this equation holds, an object is either at rest or moving with a constant velocity. The second equation is called the **second condition for equilibrium.** When it holds, an object is said to be in rotational equilibrium.

     The **moment of inertia** of a group of particles is

$$I \equiv \Sigma mr^2 \tag{8.8}$$

If a rigid body free to rotate about a fixed axis has a net external torque acting on it, the body undergoes an angular acceleration, α, where

$$\Sigma\tau = I\alpha \qquad \text{[8.9]}$$

If a rigid object rotates about a fixed axis with angular speed ω, its **rotational kinetic energy** is

$$KE_r \equiv \tfrac{1}{2}I\omega^2 \qquad \text{[8.11]}$$

where I is the moment of inertia about the axis of rotation.

The **angular momentum** of a rotating object is

$$L \equiv I\omega \qquad \text{[8.13]}$$

If the net external torque acting on a system is zero, the total angular momentum of the system is constant. Applying the law of conservation of angular momentum to an object whose moment of inertia changes with time gives

$$I_i\omega_i = I_f\omega_f \qquad \text{[8.15]}$$

MULTIPLE-CHOICE QUESTIONS

1. A constant net torque is applied to an object. One of the following will definitely *not* be a constant. It is the object's
   (a) angular acceleration
   (b) angular velocity
   (c) moment of inertia
   (d) center of mass

2. A rod 7.0 m long is pivoted at a point 2.0 m from the left end. A downward force of 50 N acts at the left end, and a downward force of 200 N acts at the right end. At what distance from the pivot can a third force of 300 N acting upward be placed to produce rotational equilibrium? Neglect the weight of the rod.
   (a) 1.0 m   (b) 2.0 m   (c) 3.0 m   (d) 4.0 m
   (e) none of the above answers is correct

3. A horizontal disk with moment of inertia I_1 rotates with angular velocity ω_0 about a vertical frictionless axle. A second horizontal disk, with moment of inertia I_2 and initially not rotating, drops onto the first. Because the surfaces are rough, the two eventually reach the same angular velocity ω. The ratio ω/ω_0 is
   (a) I_1/I_2 (b) I_2/I_1 (c) $I_1/(I_1 + I_2)$ (d) $I_2/(I_1 + I_2)$

4. An 80.0 kg man is one fourth of the way up a 10.0 m ladder that is resting against a smooth, frictionless wall. If the ladder has a mass of 20.0 kg and makes an angle of 60.0° with the ground, find the force of friction of the ground on the foot of the ladder.
   (a) 784 N   (b) 196 N   (c) 50 N   (d) 170 N

5. What must be the angular velocity of a solid cylinder rolling on the ground at the bottom of a hill so that it will be able to roll to the top of the hill if the hill is 10.0 m long and 3.00 m high? The mass of the cylinder is 2.00 kg and its radius is 0.400 m.
   (a) 15.7 rad/s   (b) 27.1 rad/s   (c) 19.2 rad/s
   (d) 28.6 rad/s

CONCEPTUAL QUESTIONS

1. Both torque and work are products of force and distance. How are they different? Do they have the same units?

2. Is it possible to calculate the torque acting on a rigid body without specifying the origin? Is the torque independent of the location of the origin?

3. Suppose a pencil is balanced on its end on a perfectly frictionless table. If it falls over, what is the path followed by the center of mass of the pencil?

4. When a high diver wants to turn a flip in midair, she will often draw her legs up against her chest. Why does this make her rotate faster? What should she do when she wants to come out of her flip?

5. In some motorcycle races, the riders drive over small hills, and the motorcycle becomes airborne for a short time. If the motorcycle racer keeps the throttle open while leaving the hill and going into the air, the motorcycle tends to nose upward. Why does this happen?

6. Why does holding a long pole help a tightrope walker stay balanced?

7. Suppose you are designing a car for a coasting race—the cars in these races have no engines; they simply coast downhill. Do you want large wheels or small wheels? Do you want solid, disk-like wheels or hoop-like wheels? Should the wheels be heavy or light?

8. Three homogeneous rigid bodies—a solid sphere, a solid cylinder, and a hollow cylinder—are placed at the top of an incline (Fig. Q8.8). If all are released from rest at the same elevation and roll without slipping, which reaches the bottom first? Which reaches last? You should try this at home and note that the result is *independent* of the masses and radii.

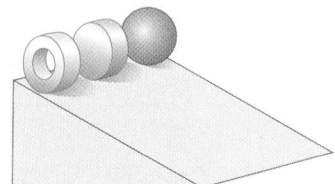

Figure Q8.8   Which object wins the race?

9. Stars originate as large bodies of slowly rotating gas. Because of gravity, these clumps of gas slowly decrease in size. What happens to the angular speed of a star as it shrinks? Explain.

10. Two solid spheres are rolled down a hill—a large, massive sphere and a small sphere with low mass. Which one reaches the bottom of the hill first? Next, we roll a large, low-density sphere, and a small, high-density sphere, and both spheres have the same mass. Which one wins the race?

11. If global warming occurs over the next century, it is likely that the polar ice caps of the Earth will melt and the water will be distributed closer to the Equator. How would this change the moment of inertia of the Earth? Would the length of the day (one revolution) increase or decrease?

12. If angular momentum is conserved for a propeller-driven airplane, it seems that if the propeller is turning clockwise, the airplane should be turning counterclockwise. Why doesn't this occur?

13. In a tape recorder, the tape is pulled past the read-and-write heads at a constant speed by the drive mechanism. Consider the reel from which the tape is pulled. As the tape is pulled off it, the radius of the roll of remaining tape decreases. How does the torque on the reel change

with time? How does the angular velocity of the reel change with time? If the tape mechanism is suddenly turned on so that the tape is quickly pulled with a large force, is the tape more likely to break when pulled from a nearly full reel or a nearly empty reel?

14. (a) Give an example in which the net force acting on an object is zero and yet the net torque is nonzero. (b) Give an example in which the net torque acting on an object is zero and yet the net force is nonzero.

15. A mouse is initially at rest on a horizontal turntable mounted on a frictionless vertical axle. If the mouse begins to walk clockwise around the perimeter, what happens to the turntable? Explain.

16. A cat usually lands on its feet regardless of the position from which it is dropped (though they can be hurt or even killed when they fail to land on their feet when dropped from excessive heights—so don't try it!). A slow-motion film of a cat falling (Fig. Q8.16) shows that the upper half of its body twists in one direction while the lower half twists in the opposite direction. Why does this type of rotation occur?

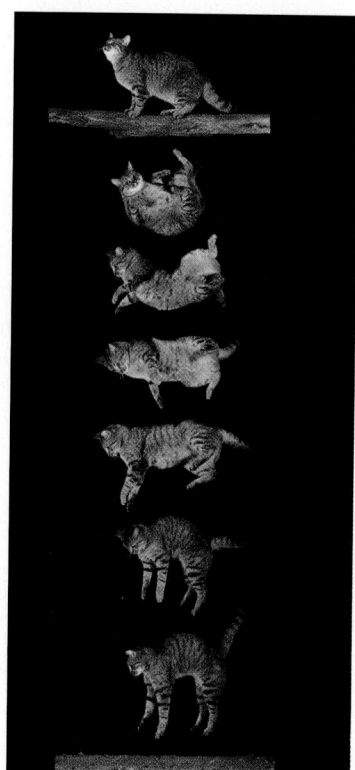

Figure Q8.16   A falling, twisting cat. (*© Gerard Lacz/NHPA*)

17. A ladder rests inclined against a wall. Would you feel safer climbing up the ladder if you were told that the floor is frictionless but the wall is rough or that the wall is frictionless but the floor is rough? Justify your answer.

PROBLEMS

1, **2**, **3** = straightforward, intermediate, challenging   ☐ = full solution available in Study Guide/Student Solutions Manual   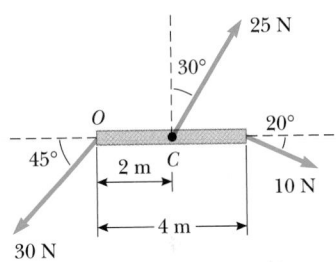 = Core Concepts Workbook
WEB = solution posted at **http://www.harcourtcollege.com/physics/cptech**   = biomedical application   = Interactive Physics

Review Problem

The mass of the Earth, M_E, is 81 times the mass of the Moon, M_M, the radius of the Earth, R_E, is 6400 km, and the distance, d_c, between the centers of the Earth and the Moon is 3.84×10^5 km. **(a)** Find the location of the point at which the gravitational forces exerted on an object by the Earth and the Moon will cancel each other. **(b)** Where is the location of the center of gravity (more properly called *center of mass* in this case) of the Earth-Moon system? Is this the same as your answer to **(a)**?

Section 8.1 Torque

1. If the torque required to loosen a nut that is holding a flat tire in place on a car has a magnitude of 40.0 N·m, what *minimum* force must be exerted by the mechanic at the end of a 30.0-cm lug wrench to accomplish the task?

2. A steel band exerts a horizontal force of 80.0 N on a tooth at point *B* in Figure P8.2. What is the torque on the root of the tooth about point *A*?

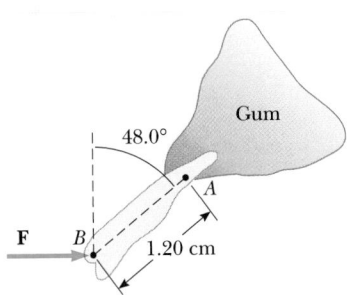

Figure P8.2

3. A simple pendulum consists of a 3.0-kg point mass hanging at the end of a 2.0-m-long light string that is connected to a pivot point. Calculate the magnitude of the torque (due to the force of gravity) about this pivot point when the string makes a 5.0° angle with the vertical.

4. A fishing pole is 2.00 m long and inclined to the horizontal at an angle of 20.0° (Fig. P8.4). What is the torque exerted by the fish about an axis perpendicular to the page and passing through the hand of the person holding the pole?

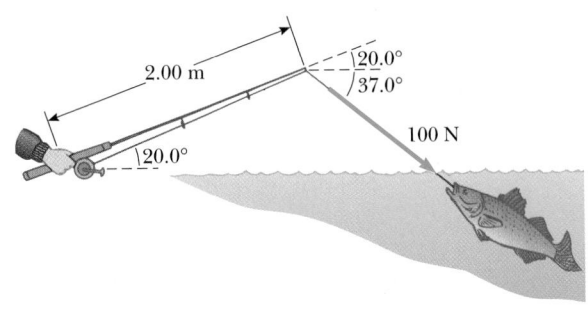

Figure P8.4

5. Calculate the net torque (magnitude and direction) on the beam in Figure P8.5 about **(a)** an axis through *O*, perpendicular to the page, and **(b)** an axis through *C*, perpendicular to the page.

Figure P8.5

Section 8.2 Torque and the Second Condition for Equilibrium

Section 8.3 The Center of Gravity

Section 8.4 Examples of Objects in Equilibrium

6. Four objects are situated along the *y* axis as follows: a 2.00-kg object is at $+3.00$ m, a 3.00-kg object is at $+2.50$ m, a 2.50-kg object is at the origin, and a 4.00-kg object is at -0.500 m. Where is the center of gravity of this system?

7. A water molecule consists of an oxygen atom with two hydrogen atoms bound to it, as shown in Figure P8.7. The bonds are 0.100 nm in length and the angle between the two bonds is 106°. Use the *x*-*y* axis shown and determine the location of the center of gravity of the molecule. Consider the mass of an oxygen atom to be 16 times the mass of a hydrogen atom.

Problem 7 is the same as Workbook Problem 40 (Workbook page 87).

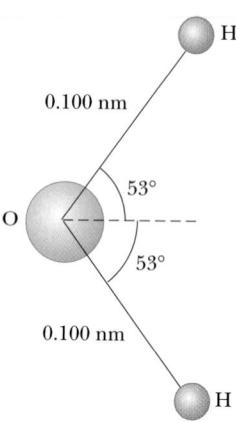

Figure P8.7

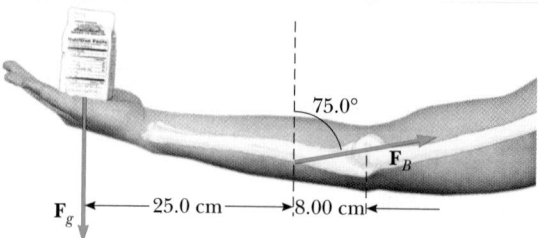

Figure P8.11

8. Consider the following mass distribution where the x-y coordinates are given in meters: 5 kg at (0, 0) m, 3 kg at (0, 4) m, and 4 kg at (3, 0) m. Where should a fourth mass of 8 kg be placed so the center of gravity of the four-mass arrangement will be at (0, 0) m?

9. Find the x- and y-coordinates of the center of gravity of a 4.00 ft by 8.00 ft uniform sheet of plywood with the upper right quadrant removed as shown in Figure P8.9.

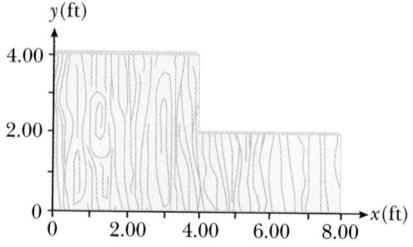

Figure P8.9

10. A meter stick is found to balance at the 49.7 cm mark when placed on a fulcrum. When a 50.0-g mass is attached at the 10.0-cm mark, the fulcrum must be moved to the 39.2-cm mark for balance. What is the mass of the meter stick?

11. A cook holds a 2.00-kg carton of milk at arm's length (Fig. P8.11). What force $\mathbf{F}_B$ must be exerted by the biceps muscle? (Ignore the weight of the forearm.)

12. A window washer is standing on a scaffold supported by a vertical rope at each end. The scaffold weighs 200 N and is 3.00 m long. What is the tension in each rope when the 700-N worker stands 1.00 m from one end?

13. The chewing muscle, the masseter, is one of the strongest in the human body. It is attached to the mandible (lower jawbone), as shown in Figure P8.13a. The jawbone is pivoted about a socket just in front of the auditory canal. The forces acting on the jawbone are equivalent to those acting on the curved bar in Figure P8.13b: $\mathbf{C}$ is the force exerted against the jawbone by the food being chewed, $\mathbf{T}$ is the tension in the masseter, and $\mathbf{R}$ is the force exerted on the mandible by the socket. Find $\mathbf{T}$ and $\mathbf{R}$ if you bite down on a piece of steak with a force of 50.0 N.

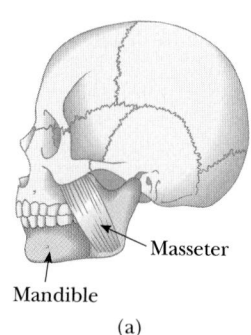

(a)

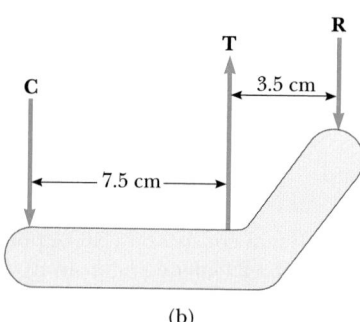

(b)

Figure P8.13

14. A hungry 700-N bear walks out on a beam in an attempt to retrieve some goodies hanging at the end (Fig. P8.14). The beam is uniform, weighs 200 N, and is 6.00 m long; the goodies weigh 80.0 N. (a) Draw a free-

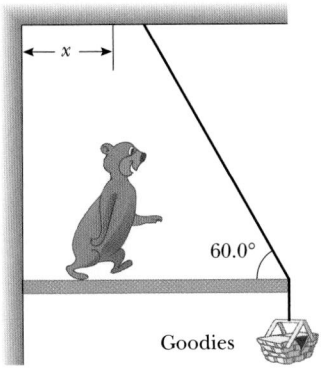

Figure P8.14

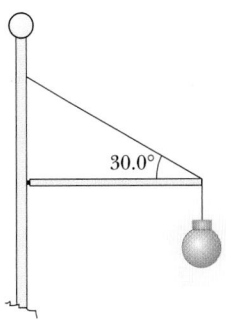

Figure P8.16

body diagram for the beam. **(b)** When the bear is at
$x = 1.00$ m, find the tension in the wire and the components of the reaction force at the hinge. **(c)** If the wire
can withstand a maximum tension of 900 N, what is the
maximum distance the bear can walk before the wire
breaks?

15. A uniform semicircular sign 1.00 m in diameter and of
weight W is supported by two wires, as shown in Figure
P8.15. What is the tension in each of the wires supporting the sign?

17. A 500-N uniform rectangular sign 4.00 m wide and
3.00 m high is suspended from a horizontal,
6.00-m-long, uniform, 100-N rod, as indicated in Figure
P8.17. The left end the rod is supported by a hinge and
the right end is supported by a thin cable making a
30.0° angle with the vertical. **(a)** Find the tension, T, in
the cable. **(b)** Find the horizontal and vertical components of force exerted on the left end of the rod by the
hinge.

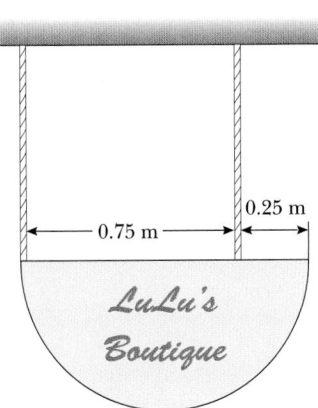

Figure P8.15

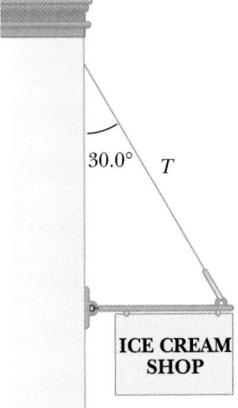

Figure P8.17

16. A 20.0-kg floodlight in a park is supported at the end
of a horizontal beam of negligible mass that is hinged to
a pole, as shown in Figure P8.16. A cable at an angle
of 30.0° with the beam helps to support the light.
Find **(a)** the tension in the cable and **(b)** the horizontal and vertical forces exerted on the beam by the
pole.

18. The arm in Figure P8.18 weighs 41.5 N. The force of
gravity acting on the arm acts through point A. Determine the magnitudes of the tension force $\mathbf{F}_t$ in the del-
toid muscle and the force $\mathbf{F}_s$ of the shoulder on the humerus (upper-arm bone) to hold the arm in the
position shown.

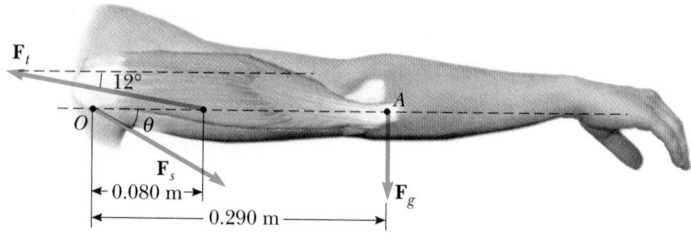

Figure P8.18

19. A uniform plank of length 2.00 m and mass 30.0 kg is supported by three ropes, as indicated by the blue vectors in Figure P8.19. Find the tension in each rope when a 700-N person is 0.500 m from the left end.

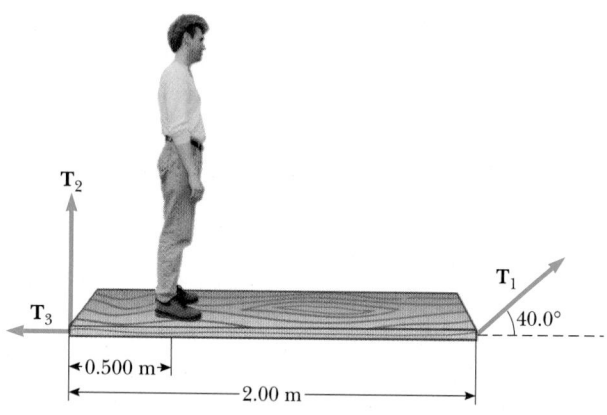

Figure P8.19

20. A 15.0-m, 500-N uniform ladder rests against a frictionless wall, making an angle of 60.0° with the horizontal. **(a)** Find the horizontal and vertical forces exerted on the base of the ladder by the Earth when an 800-N fire fighter is 4.00 m from the bottom. **(b)** If the ladder is just on the verge of slipping when the fire fighter is 9.00 m up, what is the coefficient of static friction between ladder and ground?

WEB **21.** An 8.0-m, 200-N uniform ladder rests against a smooth wall. The coefficient of static friction between the ladder and the ground is 0.60, and the ladder makes a 50.0° angle with the ground. How far up the ladder can an 800-N person climb before the ladder begins to slip?

22. A 1200-N uniform boom is supported by a cable perpendicular to the boom, as in Figure P8.22. The boom is hinged at the bottom, and a 2000-N weight hangs from its top. Find the tension in the supporting cable and the components of the reaction force exerted on the boom by the hinge.

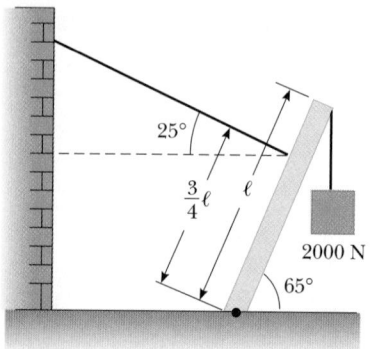

Figure P8.22

23. The large quadriceps muscle in the upper leg terminates at its lower end in a tendon attached to the upper end of the tibia (Fig. P8.23a). The forces on the lower leg when the leg is extended are modeled as in Figure P8.23b, where **T** is the tension in the tendon, **C** is the force of gravity acting on the lower leg, and **F** is the force of gravity acting on the foot. Find **T** when the tendon is at an angle of 25.0° with the tibia, assuming that $C = 30.0$ N, $F = 12.5$ N, and the leg is extended at an angle of 40.0° with the vertical ($\theta = 40.0°$). Assume that the center of gravity of the lower leg is at its center, and that the tendon attaches to the lower leg at a point one fifth of the way down the leg.

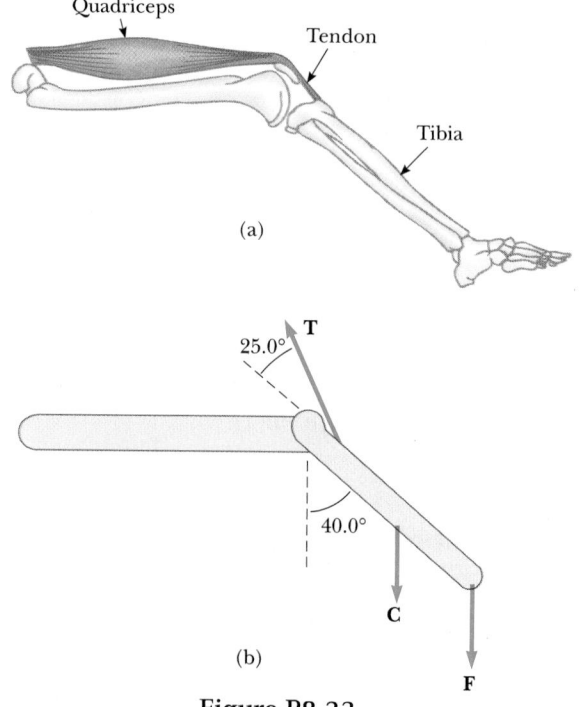

Figure P8.23

24. One end of a uniform 4.0-m long rod of weight w is supported by a cable. The other end rests against the wall, where it is held by friction (see Fig. P8.24). The coefficient of static friction between the wall and the rod is $\mu_s = 0.50$. Determine the minimum distance, x, from point A at which an additional weight w (same as the weight of the rod) can be hung without causing the rod to slip at point A.

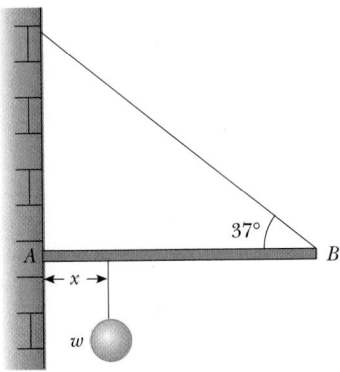

Figure P8.24

Section 8.5 Relationship Between Torque and Angular Acceleration

25. Four masses are held in position at the corners of a rectangle by light rods, as shown in Figure P8.25. Find the moment of inertia of the system about (a) the x axis, (b) the y axis, and (c) an axis through O and perpendicular to the page.

26. If the system shown in Figure P8.25 is set in rotation about each of the axes mentioned in Problem 25, find

the torque that will produce an angular acceleration of 1.50 rad/s^2 in each case.

27. A potter's wheel having a radius of 0.50 m and a moment of inertia of 12.5 kg·m^2 is rotating freely at 50 rev/min. The potter can stop the wheel in 6.0 s by pressing a wet rag against the rim and exerting a radially inward force of 70 N. Find the effective coefficient of kinetic friction between the wheel and the wet rag.

28. A cylindrical fishing reel has a moment of inertia of $I = 6.8 \times 10^{-4}$ kg·m^2 and a radius of 4.0 cm. A friction clutch in the reel exerts a restraining torque of 1.3 N·m if a fish pulls on the line. The fisherman gets a bite, and the reel begins to spin with an angular acceleration of 66 rad/s^2. **(a)** What is the force of the fish on the line? **(b)** How much line unwinds in 0.50 s?

29. A 150-kg merry-go-round in the shape of a uniform, solid, horizontal disk of radius 1.50 m is set in motion by wrapping a rope about the rim of the disk and pulling on the rope. What constant force would have to be exerted on the rope to bring the merry-go-round from rest to an angular speed of 0.500 rev/s in 2.00 s?

30. A cable passes over a pulley. Because of friction, the tension in the cable is not the same on opposite sides of the pulley. The force on one side is 120 N, and the force on the other side is 100 N. Assuming that the pulley is a uniform disk of mass 2.1 kg and radius 0.81 m, determine its angular acceleration.

WEB **31.** A cylindrical 5.00-kg pulley with a radius of 0.600 m is used to lower a 3.00-kg bucket into a well (Fig. P8.31). The bucket starts from rest and falls for 4.00 s. **(a)** What is the linear acceleration of the falling bucket? **(b)** How far does it drop? **(c)** What is the angular acceleration of the cylinder?

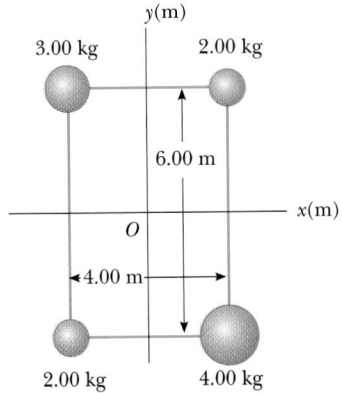

Figure P8.25 (Problems 25 and 26.)

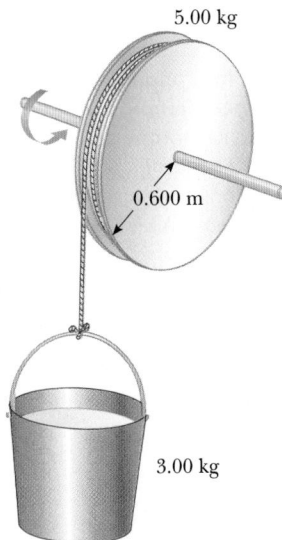

Figure P8.31

32. An airliner lands with a speed of 50.0 m/s. Each wheel of the plane has a radius of 1.25 m and a moment of inertia of 110 kg·m². At touchdown the wheels begin to spin under the action of friction. Each wheel supports a weight of 1.40×10^4 N, and the wheels attain the angular speed of rolling without slipping in 0.480 s. What is the coefficient of kinetic friction between the wheels and the runway? Assume that the speed of the plane is constant.

33. A light string is wrapped around a solid cylindrical spool of radius 0.500 m and mass 0.500 kg. A 5.00-kg mass is hung from the string, causing the spool to rotate and the string to unwind (Fig. P8.33). Assume that the system starts from rest and no slippage takes place between the string and the spool. By direct application of Newton's second law, determine the angular speed of the spool after the mass has dropped 4.00 m.

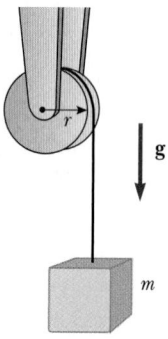

Figure P8.33 (Problems 33 and 41.)

34. A string is wrapped around a uniform cylinder of mass m and radius r (Fig. P8.34). One end of the string is attached to the ceiling, and the cylinder is allowed to fall from rest. (a) Write Newton's second law for the cylinder. (b) Find the net torque about the center of the cylinder and equate this to $I\alpha$. (c) Use the equations found in (a) and (b) along with $a = r\alpha$ to show that the linear acceleration of the cylinder is given by $a = (2/3)g$.

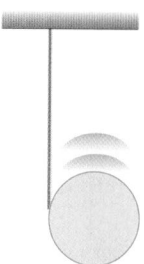

Figure P8.34

Section 8.6 Rotational Kinetic Energy

35. A 10.0-kg cylinder rolls without slipping on a rough surface. At the instant its center of mass has a speed of 10.0 m/s, determine (a) the translational kinetic energy of its center of mass, (b) the rotational kinetic energy about its center of mass, and (c) its total kinetic energy.

36. The net work done in accelerating a propeller from rest to an angular speed of 200 rad/s is 3000 J. What is the moment of inertia of the propeller?

37. A horizontal 800-N merry-go-round of radius 1.50 m is started from rest by a constant horizontal force of 50.0 N applied tangentially to the merry-go-round. Find the kinetic energy of the merry-go-round after 3.00 s. (Assume it is a solid cylinder.)

38. A car is designed to get its energy from a rotating flywheel with a radius of 2.00 m and a mass of 500 kg. Before a trip, the flywheel is attached to an electric motor, which brings the flywheel's rotational speed up to 5000 rev/min. (a) Find the kinetic energy stored in the flywheel. (b) If the flywheel is to supply energy to the car as would a 10.0-hp motor, find the length of time the car could run before the flywheel would have to be brought back up to speed.

39. The top in Figure P8.39 has a moment of inertia of 4.00×10^{-4} kg·m² and is initially at rest. It is free to rotate about a stationary axis, AA'. A string, wrapped around a peg along the axis of the top, is pulled in such a manner as to maintain a constant tension of 5.57 N in the string. If the string does not slip while wound around the peg, what is the angular speed of the top after 80.0 cm of string has been pulled off the peg? (*Hint:* Consider the work done.)

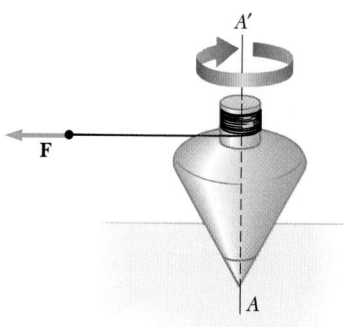

Figure P8.39

40. A 240-N sphere 0.20 m in radius rolls, without slipping, 6.0 m down a ramp that is inclined at 37° with the horizontal. What is the angular speed of the sphere at the bottom of the hill if it starts from rest?

WEB **41.** Use conservation of energy to determine the angular speed of the spool shown in Figure P8.33 after the mass m has fallen 4.00 m, starting from rest. The light string attached to this mass is wrapped around the spool and does not slip as it unwinds. Assume that the spool is a solid cylinder of radius 0.500 m and mass 0.500 kg, and that $m = 5.00$ kg.

Section 8.7 Angular Momentum

42. (a) Calculate the angular momentum of the Earth that arises from its spinning motion on its axis and (b) the angular momentum of the Earth that arises from its orbital motion about the Sun.

43. A light rigid rod 1.00 m in length rotates about an axis perpendicular to its length and through its center as shown in Figure P8.43. Two particles of masses 4.00 kg and 3.00 kg are connected to the ends of the rod. What is the angular momentum of the system if the speed of each particle is 5.00 m/s?

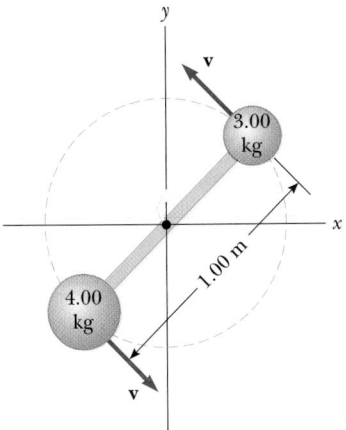

Figure P8.43

44. Halley's comet moves about the Sun in an elliptical orbit, with its closest approach to the Sun being 0.59 A.U. and its greatest distance being 35 A.U. (1 A.U. = Earth–Sun distance). If the comet's speed at closest approach is 54 km/s, what is its speed when it is farthest from the Sun? You may neglect any change in the comet's mass and assume that its angular momentum about the Sun is conserved.

45. The system of point masses shown in Figure P8.45 is rotating at an angular speed of 2.0 rev/s. The masses are connected by light, flexible spokes that can be lengthened or shortened. What is the new angular speed if the spokes are shortened to 0.50 m? (An effect similar to that illustrated in this problem occurred in the early stages of the formation of our Galaxy. As the massive cloud of dust and gas that was the source of the stars

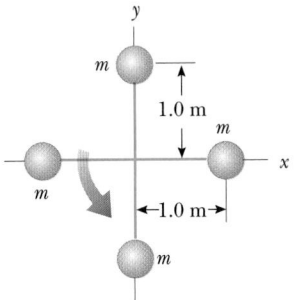

Figure P8.45

and planets contracted, an initially small rotation increased with time.)

46. A solid, horizontal cylinder of mass 10.0 kg and radius 1.00 m rotates with an angular speed of 7.00 rad/s about a fixed vertical axis through its center. A 0.250-kg piece of putty is dropped vertically onto the cylinder at a point 0.900 m from the center of rotation, and sticks to the cylinder. Determine the final angular speed of the system.

47. A student sits on a rotating stool holding two 3.0-kg masses. When his arms are extended horizontally, the masses are 1.0 m from the axis of rotation, and he rotates with an angular speed of 0.75 rad/s. The moment of inertia of the student plus stool is 3.0 kg·m² and is assumed to be constant. The student then pulls the masses horizontally to 0.30 m from the rotation axis. (a) Find the new angular speed of the student. (b) Find the kinetic energy of the student before and after the masses are pulled in.

48. The puck in Figure P8.48 has a mass of 0.120 kg. Its original distance from the center of rotation is 40.0 cm, and the puck is moving with a speed of 80.0 cm/s. The string is pulled downward 15.0 cm through the hole in the frictionless table. Determine the work done on the

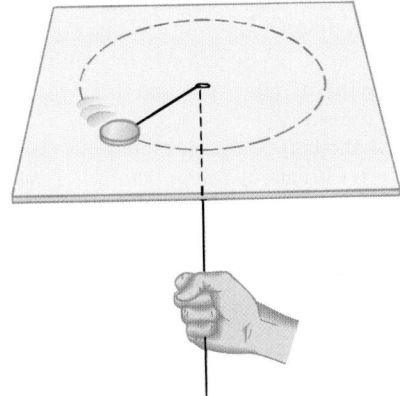

Figure P8.48

puck. (*Hint:* Consider the change of kinetic energy of the puck.)

49. A merry-go-round rotates at the rate of 0.20 rev/s with an 80-kg man standing at a point 2.0 m from the axis of rotation. **(a)** What is the new angular speed when the man walks to a point 1.0 m from the center? Assume that the merry-go-round is a solid 25-kg cylinder of radius 2.0 m **(b)** Calculate the change in kinetic energy due to this movement. How do you account for this change in kinetic energy?

ADDITIONAL PROBLEMS

50. A cylinder with moment of inertia I_1 rotates with angular velocity ω_0 about a frictionless vertical axle. A second cylinder, with moment of inertia I_2, initially not rotating, drops onto the first cylinder (Fig. P8.50). Because the surfaces are rough, the two eventually reach the same angular velocity, ω. **(a)** Calculate ω. **(b)** Show that kinetic energy is lost in this situation, and calculate the ratio of the final to the initial kinetic energy.

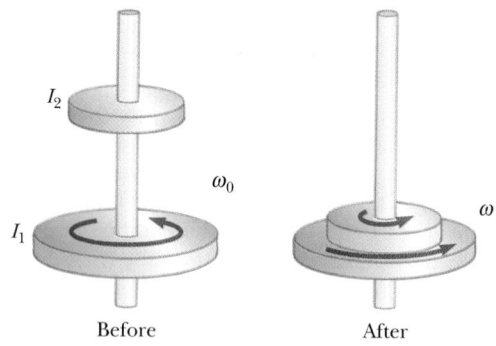

Before   After

Figure P8.50

51. A 0.100-kg meter stick is supported at its 40.0-cm mark by a string attached to the ceiling. A 0.700-kg mass hangs vertically from the 5.00-cm mark. A mass, *m*, is attached somewhere on the meter stick to keep it horizontal and in rotational and translational equilibrium. If the tension in the string attached to the ceiling is 19.6 N, determine **(a)** the value of *m* and **(b)** its point of attachment on the stick.

52. Show that the kinetic energy of an object rotating about a fixed axis with angular momentum $L = I\omega$ can be written as $KE = L^2/2I$.

53. A person bends over and lifts a 200-N weight, as in Figure P8.53a, with his back horizontal. The muscle that attaches two thirds of the way up the spine maintains the position of the back; the angle between the spine and this muscle is 12°. Using the mechanical model in Figure P8.53b and taking the weight of the upper body to be 350 N, find the tension in the back muscle and the compressional force in the spine.

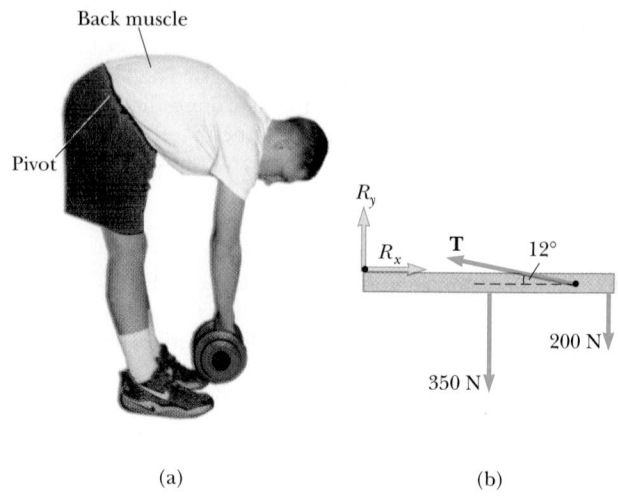

(a)   (b)

Figure P8.53

54. A 12.0-kg mass is attached to a cord that is wrapped around a wheel of radius $r = 10.0$ cm (Fig. P8.54). The acceleration of the mass down the frictionless incline is measured to be 2.00 m/s^2. Assuming the axle of the wheel to be frictionless, determine **(a)** the tension in the rope, **(b)** the moment of inertia of the wheel, and **(c)** the angular speed of the wheel 2.00 s after it begins rotating, starting from rest.

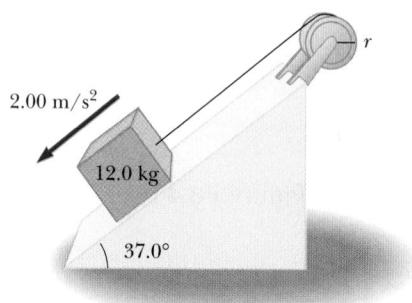

Figure P8.54

55. A uniform ladder of length L and weight W is leaning against a vertical wall. The coefficient of static friction between the ladder and the floor is the same as that between the ladder and the wall. If this coefficient of static friction is $\mu_s = 0.500$, determine the smallest angle the ladder can make with the floor without slipping.

56. A uniform 10.0-N picture frame is supported as shown in Figure P8.56. Find the tension in the cords and the magnitude of the horizontal force at P that are required to hold the frame in the position shown.

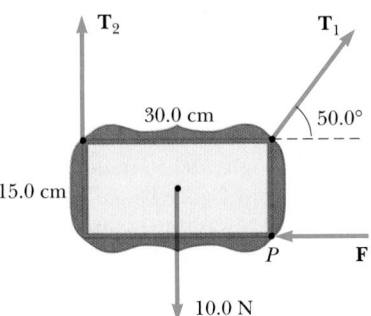

Figure P8.56

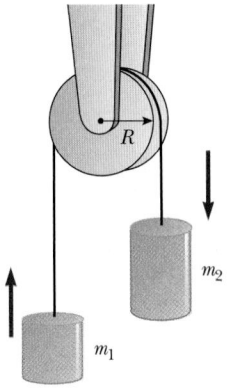

Figure P8.59

57. A solid 2.0-kg ball of radius 0.50 m starts at a height of 3.0 m above the surface of the Earth and *rolls* down a 20° slope. A solid disk and a ring start at the same time and the same height. The ring and disk each have the same mass and radius as the ball. Which of the three wins the race to the bottom if all roll without slipping?

58. A 40.0-kg child stands at one end of a 70.0-kg boat that is 4.00 m long (Fig. P8.58). The boat is initially 3.00 m from the pier. The child notices a turtle on a rock beyond the far end of the boat and proceeds to walk to that end to catch the turtle. (a) Neglecting friction between the boat and water, describe the motion of the system (child + boat). (b) Where will the child be relative to the pier when he reaches the far end of the boat? (c) Will he catch the turtle? (Assume that he can reach out 1.00 m from the end of the boat.)

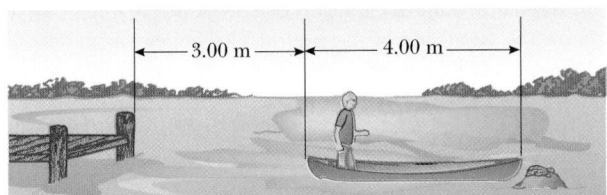

Figure P8.58

59. The pulley in Figure P8.59 has a moment of inertia of 5.0 kg·m² and a radius of 0.50 m. The cord supporting the masses m_1 and m_2 does not slip, and the axle is frictionless. (a) Find the acceleration of each mass when $m_1 = 2.0$ kg and $m_2 = 5.0$ kg. (b) Find the tension in the cable supporting m_1 and the tension in the cable supporting m_2. (Note: They are different.)

60. A 4.00-kg mass is connected by a light cord to a 3.00-kg mass on a smooth surface (Fig. P8.60). The pulley rotates about a frictionless axle and has a moment of inertia of 0.500 kg·m² and a radius of 0.300 m. Assuming that the cord does not slip on the pulley, find (a) the acceleration of the two masses and (b) the tensions T_1 and T_2.

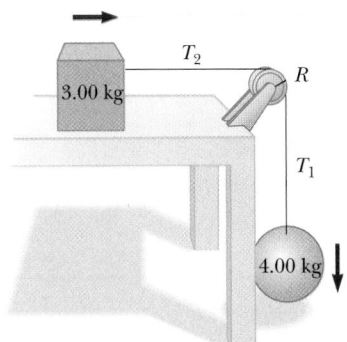

Figure P8.60

61. A model airplane whose mass is 0.750 kg is tethered by a wire so that it flies in a circle 30.0 m in radius. The airplane engine provides a net thrust of 0.800 N perpendicular to the tethering wire. (a) Find the torque the net thrust produces about the center of the circle. (b) Find the angular acceleration of the airplane when it is in level flight. (c) Find the linear acceleration of the airplane tangent to its flight path.

62. A puck of mass 80.0 g and radius 4.00 cm slides along an air table at a speed of 1.50 m/s, as shown in Figure P8.62a. It makes a glancing collision with a second puck

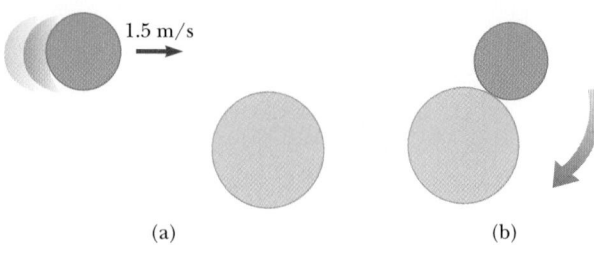

(a)   (b)

Figure P8.62

of radius 6.00 cm and mass 120 g (initially at rest) such that their rims just touch. The pucks stick together and spin after the collision (Fig. P8.62b). **(a)** What is the angular momentum of the system relative to the center of mass? **(b)** What is the angular velocity about the center of mass?

WEB 63. Two astronauts (Fig. P8.63), each having a mass of 75.0 kg, are connected by a 10.0-m rope of negligible mass. They are isolated in space, orbiting their center of mass at speeds of 5.00 m/s. Calculate **(a)** the magnitude of the angular momentum of the system by treating the astronauts as particles and **(b)** the rotational energy of the system. By pulling on the rope, the astronauts shorten the distance between them to 5.00 m. **(c)** What is the new angular momentum of the system? **(d)** What are their new speeds? **(e)** What is the new rotational energy of the system? **(f)** How much work is done by the astronauts in shortening the rope?

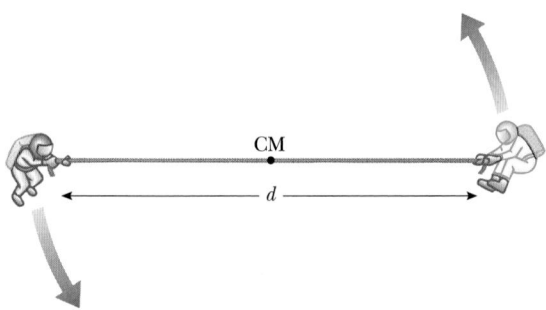

Figure P8.63   (Problems 63 and 64.)

64. Two astronauts (Fig. P8.63), each having a mass M, are connected by a rope of length d having negligible mass. They are isolated in space, orbiting their center of mass at speeds v. Calculate **(a)** the magnitude of the angular momentum of the system by treating the astronauts as particles and **(b)** the rotational energy of the system. By pulling on the rope, the astronauts shorten the distance between them to $d/2$. **(c)** What is the new angular momentum of the system? **(d)** What are their new speeds?

(e) What is the new rotational energy of the system?
(f) How much work is done by the astronauts in shortening the rope?

65. Two window washers Bob and Joe, are on a 3.00-m-long, 345-N scaffold supported by two cables attached to its ends. Bob, who weighs 750 N, stands 1.00 m from the left end, as shown in Figure P8.65. Two meters from the left end is the 500-N washing equipment. Joe is 0.500 m from the right end and weighs 1000 N. Given that the scaffold is in rotational and translational equilibrium, what are the forces on each cable?

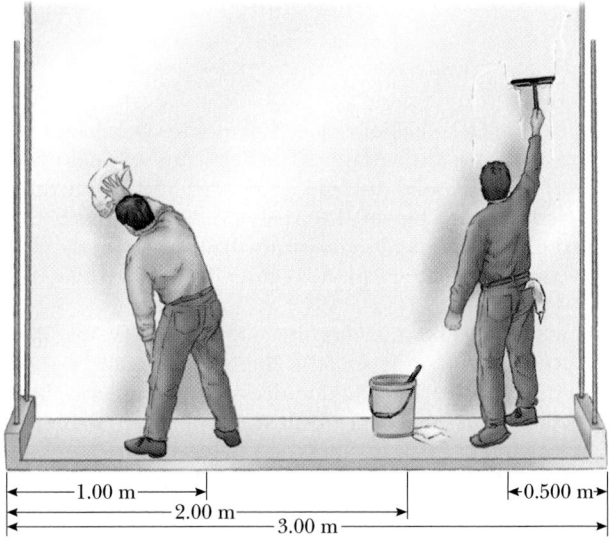

|←1.00 m→| |←0.500 m→|
|←2.00 m—→|
|—3.00 m—→|

Figure P8.65

66. When the motor in Figure P8.66 raises the 1000-kg mass, it produces a tension of 1.14×10^4 N in the cable on the right side of the pulley. The pulley has a mo-

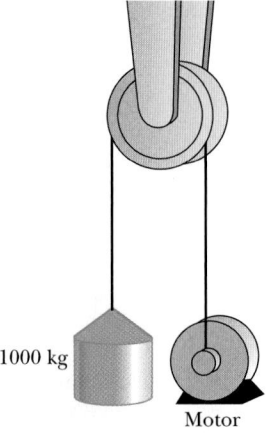

1000 kg

Motor

Figure P8.66

ment of inertia of 79.8 kg·m² and a radius of 0.762 m. The cable rides over the pulley without slipping. Determine the acceleration of the 1000-kg mass. (*Hint:* Draw free-body diagrams of the mass and the pulley. Do not assume that the tension in the cable is the same on both sides of the pulley.)

67. A 60-kg woman stands at the rim of a horizontal turntable having a moment of inertia of 500 kg·m² and a radius of 2.0 m. The system is initially at rest, and the turntable is free to rotate about a frictionless vertical axle through its center. The woman then starts walking clockwise (looking downward) around the rim at a constant speed of 1.5 m/s relative to the Earth. (a) In what direction and with what angular speed does the turntable rotate? (b) How much work does the woman do to set the system in motion?

68. We have all complained that there aren't enough hours in a day. In an attempt to change that, suppose that all the people in the world lined up at the Equator and all started running east at 2.5 m/s relative to the surface of the Earth. By how much would the length of a day increase? (Assume that there are 5.5×10^9 people in the world with an average mass of 70 kg each and that the Earth is a solid homogeneous sphere. In addition, you may use the result $1/(1 - x) \approx 1 + x$ for x small.)

69. In a circus performance, a large 5.0-kg hoop of radius 3.0 m rolls without slipping. If the hoop is given an angular speed of 3.0 rad/s while rolling on the horizontal and allowed to roll up a ramp inclined at 20° with the horizontal, how far (measured along the incline) does the hoop roll?

70. A uniform, solid cylinder of mass M and radius R rotates on a frictionless horizontal axle (Fig. P8.70). Two equal masses hang from light cords wrapped around the cylinder. If the system is released from rest, find (a) the tension in each cord and (b) the acceleration of each mass after the masses have descended a distance of h.

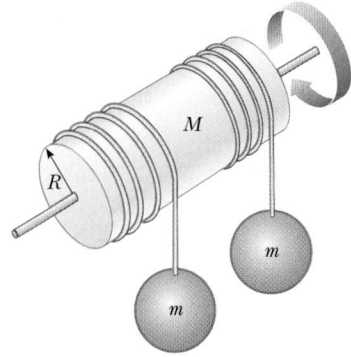

Figure P8.70

71. Figure P8.71 shows a vertical force applied tangentially to a uniform cylinder of weight w. The coefficient of static friction between the cylinder and all surfaces is 0.500. Find, in terms of w, the maximum force **F** that can be applied without causing the cylinder to rotate. (*Hint:* When the cylinder is on the verge of slipping, both friction forces are at their maximum values. Why?)

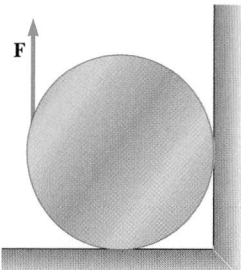

Figure P8.71

72. A common physics demonstration (Figure P8.72) consists of a ball resting at the end of a board of length ℓ that is elevated at an angle θ with the horizontal. A light cup is attached to the board at r_c so that it will catch the ball when the support stick is suddenly removed. (a) Show that the ball will lag behind the falling board when $\theta < 35.3°$, and (b) the ball will fall into the cup when the board is supported at this limiting angle and the cup is placed at

$$r_c = \frac{2\ell}{3 \cos \theta}$$

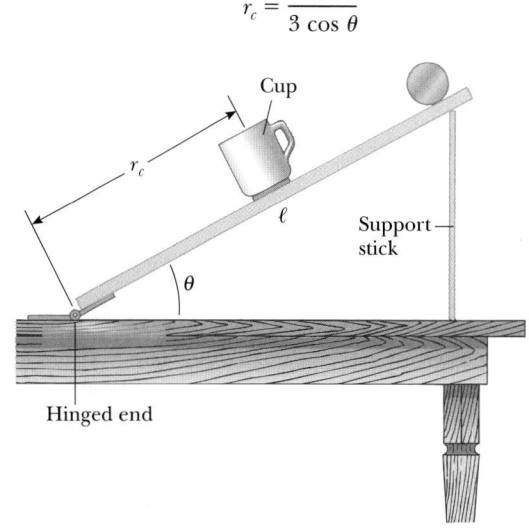

Figure P8.72

9

Solids and Fluids

▲ PHYSICS PUZZLER

A scuba diver plays with an octopus in Poor Knights Island, New Zealand. As the diver descends to greater depths, the water pressure increases above atmospheric pressure. Scuba divers have been able to swim at depths of more than 300 m. Can you estimate the pressure exerted on the diver at such great depths? How is the body able to withstand such pressure? Why is it necessary to breathe in air from pressurized tanks? *(Darryl Torckler/Tony Stone Images)*

I n this chapter we consider some properties of solids and fluids (both liquids and gases). We spend some time looking at properties that are peculiar to solids, but much of our emphasis is on the properties of fluids. We take this approach because an understanding of the behavior of fluids is fundamentally important to students in the life sciences. We open the fluids part of the chapter with a study of fluids at rest and finish with a discussion of the properties of fluids in motion. Additional topics discussed in this chapter include surface tension, viscosity, and transport phenomena.

Crystals of natural quartz (SiO_2), one of the most common minerals on Earth. Quartz crystals are used to make special lenses and prisms and in certain electronic applications. *(Charles Winters)*

9.1 STATES OF MATTER

Matter is normally classified as being in one of three states—solid, liquid, or gaseous. Often this classification system is extended to include a fourth state, referred to as a *plasma*. When matter is heated to high temperatures, many of the electrons surrounding each atom are freed from the nucleus. The resulting substance is a collection of free, electrically charged particles—negatively charged electrons and positively charged ions. Such a highly ionized substance containing equal amounts of positive and negative charges is a **plasma.** Plasmas exist inside stars, for example. If we were to take a grand tour of our Universe, we would find that there is far more matter in the plasma state than in the more familiar solid, liquid, and gaseous states because there are far more stars around than any other form of celestial matter. In this chapter, however, we ignore plasmas and concentrate on the more familiar solid, liquid, and gaseous forms that make up the environment on our planet.

Everyday experience tells us that a solid has definite volume and shape. A brick, for example, maintains its familiar shape and size day in and day out. We also know that a liquid has a definite volume but no definite shape. For instance, when you fill the tank on a lawn mower, the gasoline changes its shape from that of the original container to that of the tank on the mower. If there is a gallon of gasoline before you pour, however, there still is a gallon after. Finally, a gas has neither definite volume nor definite shape.

All matter consists of some distribution of atoms or molecules. The atoms in a solid are held, by forces that are mainly electrical, at specific positions with respect to one another and vibrate about these equilibrium positions. At low temperatures, however, the vibrating motion is slight and the atoms can be considered to be essentially fixed. As energy is added to the material, the amplitude of the vibrations increases. A vibrating atom can be viewed as being bound in its equilibrium position by springs attached to neighboring atoms. A collection of such atoms and imaginary springs is shown in Figure 9.1. If a solid is compressed by external forces, we can picture the forces as compressing these tiny internal springs. When the external forces are removed, the solid tends to return to its original shape and size. Consequently, a solid is said to have *elasticity*.

Solids can be classified as being either crystalline or amorphous. A **crystalline solid** is one in which the atoms have an ordered structure. For example, in the sodium chloride crystal (common table salt), sodium and chlorine atoms occupy alternate corners of a cube, as in Figure 9.2a. In an **amorphous solid,** such as glass, the atoms are arranged randomly, as in Figure 9.2b.

For any given substance, the liquid state exists at a higher temperature than the solid state. The intermolecular forces in a liquid are not strong enough to keep

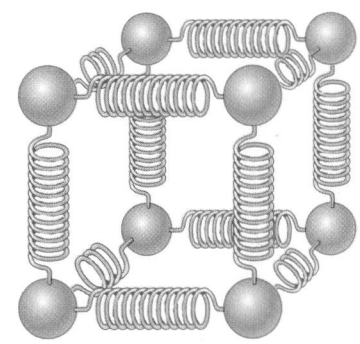

Figure 9.1 A model of a portion of a solid. The atoms (spheres) are imagined as being attached to each other by springs, which represent the elastic nature of the interatomic forces. A solid consists of trillions of segments like this with springs connecting all of them.

Figure 9.2   (a) The NaCl structure, with the Na$^+$ (red) and Cl$^-$
(blue) ions at alternate corners of a
cube. (b) In an amorphous solid,
the atoms are arranged randomly.
(c) Erratic motion of a molecule in
a liquid.

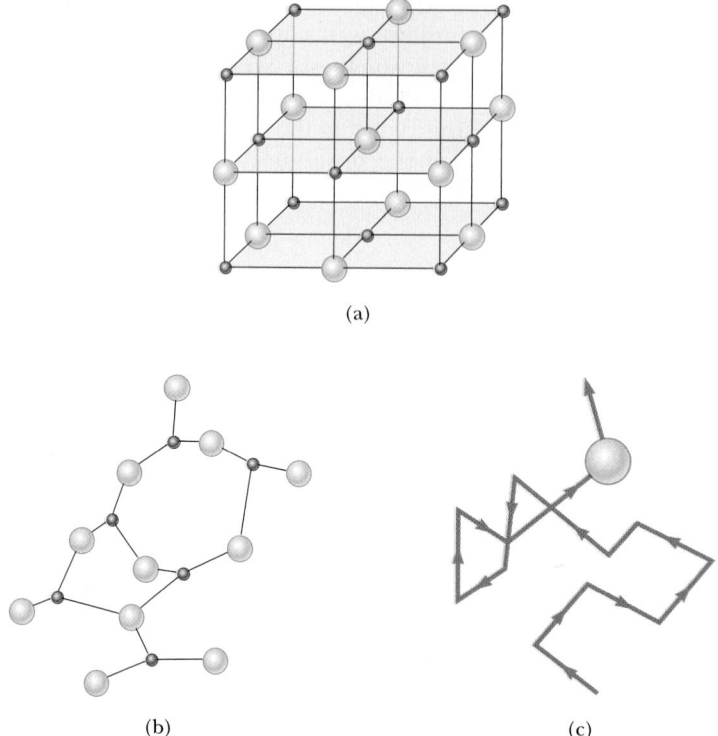

(a)

(b)   (c)

the molecules in fixed positions, and they wander through the liquid in a random
fashion (Fig. 9.2c). Solids and liquids have the following property in common:
When an attempt is made to compress them, strong repulsive atomic forces act
internally to resist compression.

   In the gaseous state, the molecules are in constant random motion and exert
only weak forces on each other. The average separation distances between the molecules of a gas are quite large compared with the size of the molecules. Occasionally
the molecules collide with each other, but most of the time they move as nearly
free, noninteracting particles. We shall say more about gases in subsequent
chapters.

9.2   THE DEFORMATION OF SOLIDS

We usually think of a solid as an object having definite shape and volume. In our
study of mechanics, to keep things simple, we assumed that objects remain unde-
formed when external forces act on them. In reality, all objects are deformable.
That is, it is possible to change the shape or size of an object (or both) through
the application of external forces. When the forces are removed, the object tends
to return to its original shape and size, which means that the deformation exhibits
an elastic behavior.

The elastic properties of solids are discussed in terms of stress and strain. **Stress** is related to the force causing a deformation; **strain** is a measure of the degree of deformation. It is found that, for sufficiently small stresses, stress is proportional to strain, and the constant of proportionality depends on the material being deformed and on the nature of the deformation. We call this proportionality constant the **elastic modulus:**

$$\text{Elastic modulus} \equiv \frac{\text{stress}}{\text{strain}} \qquad [9.1]$$

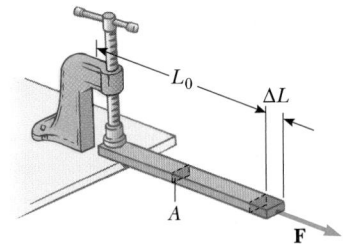

Figure 9.3 A long bar clamped at one end is stretched by the amount ΔL under the action of a force, **F.**

Young's Modulus: Elasticity in Length

Consider a long bar of cross-sectional area A and length L_0, clamped at one end (Fig. 9.3). When an external force, **F,** is applied along the bar, perpendicularly to the cross-section, internal forces in the bar resist the distortion ("stretching") that **F** tends to produce, but the bar nevertheless attains an equilibrium in which (1) its length is greater than L_0 and (2) the external force is balanced by internal forces. In such a situation the bar is said to be stressed. We define the **tensile stress** as the ratio of the magnitude of the external force, F, to the cross-sectional area, A. The SI units of stress are newtons per square meter (N/m^2). One N/m^2 is also given a special name, the **pascal** (Pa):

◁ The pascal

$$1 \text{ Pa} \equiv 1 \text{ N/m}^2 \qquad [9.2]$$

The **tensile strain** in this case is defined as the ratio of the change in length, ΔL, to the original length, L_0, and is therefore a dimensionless quantity. Thus, we can use Equation 9.1 to define **Young's modulus, Y:**

$$Y \equiv \frac{\text{tensile stress}}{\text{tensile strain}} = \frac{F/A}{\Delta L/L_0} = \frac{FL_0}{A \, \Delta L} \qquad [9.3]$$

This quantity is typically used to characterize a rod or wire stressed under *either tension or compression*. Note that because the strain is a dimensionless quantity, Y is in pascals. Typical values are given in Table 9.1. Experiments show that (1) the change in length for a fixed external force is proportional to the original length

TABLE 9.1 Typical Values for Elastic Modulus

Substance	Young's Modulus (Pa)	Shear Modulus (Pa)	Bulk Modulus (Pa)
Aluminum	7.0×10^{10}	2.5×10^{10}	7.0×10^{10}
Brass	9.1×10^{10}	3.5×10^{10}	6.1×10^{10}
Copper	11×10^{10}	4.2×10^{10}	14×10^{10}
Steel	20×10^{10}	8.4×10^{10}	16×10^{10}
Tungsten	35×10^{10}	14×10^{10}	20×10^{10}
Glass	$6.5–7.8 \times 10^{10}$	$2.6–3.2 \times 10^{10}$	$5.0–5.5 \times 10^{10}$
Quartz	5.6×10^{10}	2.6×10^{10}	2.7×10^{10}
Water	—	—	0.21×10^{10}
Mercury	—	—	2.8×10^{10}

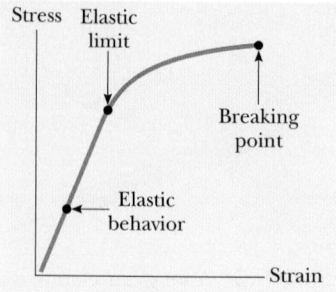

Figure 9.4   A stress-strain curve for a solid.

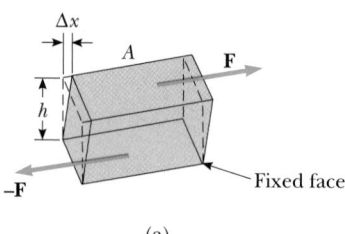

(a)

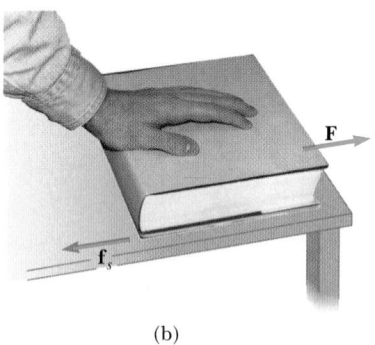

(b)

Figure 9.5   (a) A shear deformation in which a rectangular block is distorted by a force applied tangent to one of its faces. (b) A book under shear stress.

Bulk modulus ▶

and (2) the force necessary to produce a given strain is proportional to the cross-sectional area.

It is possible to exceed the *elastic limit* of a substance by applying a sufficiently great stress (Fig. 9.4). At the *elastic limit,* the stress-strain curve departs from a straight line. A material subjected to a stress beyond this level ordinarily does not return to its original length when the external force is removed. As the stress is increased further, the material ultimately breaks.

Shear Modulus: Elasticity of Shape

Another type of deformation occurs when a body is subjected to a force, **F,** tangential to one of its faces while the opposite face is held fixed (Fig. 9.5a). If the object is originally a rectangular block, such a tangential force results in a shape whose cross section is a parallelogram. In this situation, the stress is called a shear stress. A book pushed sideways, as in Figure 9.5b, is under a shear stress. There is no change in volume with this deformation. We define the **shear stress** as F/A, the ratio of the magnitude of the tangential force to the area, A, of the face being sheared. The **shear strain** is the ratio $\Delta x/h$, where Δx is the horizontal distance the sheared face moves and h is the height of the object. In terms of these quantities, the **shear modulus,** S, is defined as

$$S \equiv \frac{\text{shear stress}}{\text{shear strain}} = \frac{F/A}{\Delta x/h} \qquad \text{[9.4]}$$

Shear moduli for some representative materials are listed in Table 9.1. Note that the units of shear modulus are force per unit area (Pa).

Bulk Modulus: Volume Elasticity

Bulk modulus characterizes the response of a substance to uniform squeezing. Suppose that the external forces acting on an object are at right angles to all of its faces (Fig. 9.6) and distributed uniformly over all the faces. As we shall see later, this occurs when an object is immersed in a fluid. A body subject to this type of deformation undergoes a change in volume but no change in shape. The **volume stress,** ΔP, is defined as the ratio of the magnitude of the change in the normal force, ΔF, to the area, A. (When dealing with fluids, we shall refer to the quantity F/A as the **pressure.**) The volume strain is equal to the change in volume, ΔV, divided by the original volume, V. Thus, from Equation 9.1 we can characterize a volume compression in terms of the **bulk modulus,** B, defined as

$$B \equiv \frac{\text{volume stress}}{\text{volume strain}} = -\frac{\Delta F/A}{\Delta V/V} = -\frac{\Delta P}{\Delta V/V} \qquad \text{[9.5]}$$

Note that a negative sign is included in this defining equation so that B is always positive. An increase in pressure (positive ΔP) causes a decrease in volume (negative ΔV) and vice versa.

Table 9.1 lists bulk modulus values for some materials. If you look up such values in a different source, you will often find that the reciprocal of the bulk modulus, called the **compressibility** of the material, is listed. Note from Table 9.1 that both solids and liquids have bulk moduli. However, there is no shear modulus and no Young's modulus for liquids because a liquid will not sustain a shearing stress or a tensile stress (it flows instead).

EXAMPLE 9.1 Built to Last

A vertical steel beam in a building supports a load of 6.0×10^4 N. If the length of the beam is 4.0 m and its cross-sectional area is 8.0×10^{-3} m^2, find the distance it is compressed along its length.

Solution Because the beam is under compression, we can use Equation 9.3. Taking Young's modulus for steel, $Y = 20 \times 10^{10}$ Pa, from Table 9.1, we have

$$Y = \frac{FL_0}{A\,\Delta L}$$

or

$$\Delta L = \frac{FL_0}{YA} = \frac{(6.0 \times 10^4 \text{ N})(4.0 \text{ m})}{(20 \times 10^{10} \text{ Pa})(8.0 \times 10^{-3} \text{ m}^2)} = \boxed{1.5 \times 10^{-4} \text{ m}}$$

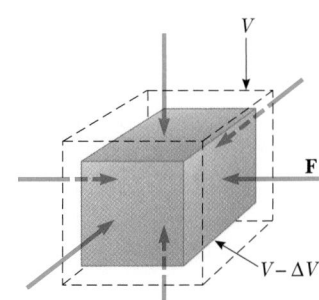

Figure 9.6 When a solid is under uniform pressure, it undergoes a change in volume but no change in shape. This cube is compressed on all sides by forces normal to its six faces.

EXAMPLE 9.2 Squeezing a Lead Sphere

A solid lead sphere of volume 0.50 m^3 is dropped in the ocean to a depth where the pressure increases by 2.0×10^7 Pa. Lead has a bulk modulus of 7.7×10^9 Pa. What is the change in volume of the sphere?

Solution From the definition of bulk modulus (Eq. 9.5), we have

$$B = -\frac{\Delta P}{\Delta V/V}$$

$$\Delta V = -\frac{V\,\Delta P}{B}$$

In this case, when the sphere is at the surface, where its volume is 0.50 m^3, the pressure exerted on it is atmospheric pressure. The increase in pressure, ΔP, when the sphere is submerged is 2.0×10^7 Pa. Therefore, the change in its volume when it is submerged is

$$\Delta V = -\frac{(0.50 \text{ m}^3)(2.0 \times 10^7 \text{ Pa})}{7.7 \times 10^9 \text{ Pa}} = \boxed{-1.3 \times 10^{-3} \text{ m}^3}$$

The negative sign indicates a *decrease* in volume.

9.3 DENSITY AND PRESSURE

The **density** of a substance of uniform composition is defined as its mass per unit volume.

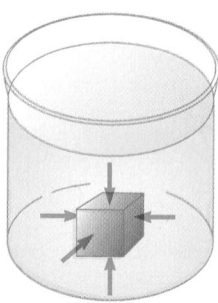

Figure 9.7 The force exerted by a fluid on a submerged object at any point is perpendicular to the surface of the object. The force exerted by the fluid on the walls of the container is perpendicular to the walls at all points and increases with depth.

TABLE 9.2   Density of Some Common Substances

Substance	$\rho(kg/m^3)^a$	Substance	$\rho(kg/m^3)^a$
Ice	0.917×10^3	Water	1.00×10^3
Aluminum	2.70×10^3	Glycerin	1.26×10^3
Iron	7.86×10^3	Ethyl alcohol	0.806×10^3
Copper	8.92×10^3	Benzene	0.879×10^3
Silver	10.5×10^3	Mercury	13.6×10^3
Lead	11.3×10^3	Air	1.29
Gold	19.3×10^3	Oxygen	1.43
Platinum	21.4×10^3	Hydrogen	8.99×10^{-2}
Uranium	18.7×10^3	Helium	1.79×10^{-1}

^{a} All values are at standard atmospheric pressure and temperature (STP), defined as 0°C (273 K) and 1 atm (1.01×10^5 Pa). To convert to grams per cubic centimeter, multiply by 10^{-3}.

In symbolic form, a substance of mass M and volume V has a density, ρ (Greek rho), given by

 Density

$$\rho \equiv \frac{M}{V} \qquad\qquad [9.6]$$

The units of density are kilograms per cubic meter in the SI system and grams per cubic centimeter in the cgs system. Table 9.2 lists the densities of some substances. The densities of most liquids and solids vary slightly with changes in temperature and pressure; the densities of gases vary greatly with such changes. Note that under normal conditions the densities of solids and liquids are about 1000 times greater than the densities of gases. This difference implies that the average spacing between molecules in a gas under these conditions is about ten times greater than that in a solid or liquid.

The **specific gravity** of a substance is the ratio of its density to the density of water at 4°C, which is 1.0×10^3 kg/m^3. By definition, specific gravity is a dimensionless quantity. For example, if the specific gravity of a substance is 3.0, its density is $3.0(1.0 \times 10^3$ kg/m^3) $= 3.0 \times 10^3$ kg/m^3.

We have seen that fluids do not sustain shearing stresses, and thus the only stress that can exist on an object submerged in a fluid is one that tends to compress the object. The force exerted by the fluid on the object is always perpendicular to the surfaces of the object, as shown in Figure 9.7.

The pressure at a specific point in a fluid can be measured with the device pictured in Figure 9.8—an evacuated cylinder enclosing a light piston connected to a spring. As the device is submerged in a fluid, the fluid presses down on the top of the piston and compresses the spring until the inward force of the fluid is balanced by the outward force of the spring. The fluid pressure can be measured directly if the spring is calibrated in advance. This is accomplished by applying a known force to the spring to compress it a given distance.

If F is the magnitude of the force exerted by the fluid on the piston and A is the area of the piston, then **the average pressure, P, of the fluid at the level to which the device has been submerged is defined as the ratio of force to area:**

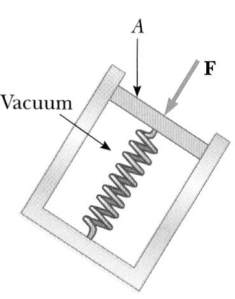

Figure 9.8 A simple device for measuring pressure in a fluid.

$$P \equiv \frac{F}{A} \qquad [9.7]$$

Because pressure is defined as force per unit area, it has units of pascals (newtons per square meter).

As you can see from this definition, one needs to know the magnitude of the force exerted on a surface *and* the area over which that force is applied. For example, a 700-N man can stand on a vinyl-covered floor in regular street shoes without damaging the surface. If he wears golf shoes with numerous metal cleats protruding from each sole, he does considerable damage to the floor. In both cases the net force applied to the floor is 700 N. However, when the man wears ordinary shoes, the area of his contact with the floor is considerably larger than when he wears golf shoes. (In the latter case, the only area in contact with the floor is the sum of the small cross-sectional areas of the metal cleats.) Hence, the *pressure* on the floor is much smaller when he wears ordinary shoes.

Snowshoes use this principle. The weight of the body—the force—is distributed over the very large areas of the snowshoes so that the pressure at any given point is relatively low and the person does not penetrate very deeply into the snow.

Snowshoes prevent the person from sinking into the soft snow because the person's weight is spread over a larger area, which reduces the pressure on the snow's surface. *(Earl Young/FPG)*

Thinking Physics 1

The daring physics professor, after a long lecture, stretches out for a nap on a bed of nails, as in Figure 9.9. How is this possible?

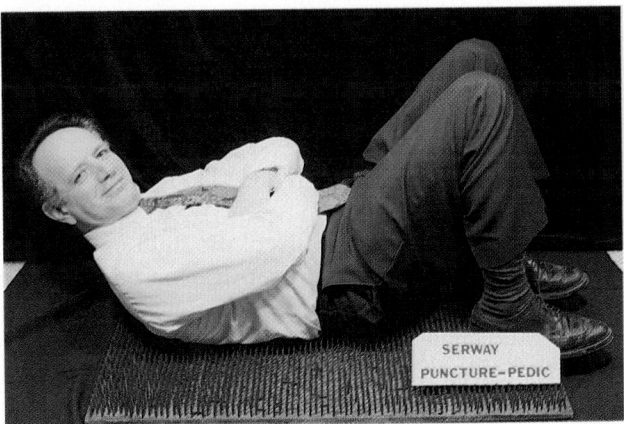

Figure 9.9 (Thinking Physics 1) Does anyone have a pillow?

Explanation If you try to support your entire weight on a single nail, the pressure on your body is your weight divided by the very small area of the nail. This pressure is sufficiently large to penetrate the skin. However, if you distribute your weight over several hundred nails, as the professor is doing, the pressure is considerably reduced because the area that supports your weight is the total area of all nails in contact with your body. (Note that lying on a bed of nails is much more comfortable than sitting on the bed. Why? Extend the logic to show that it would be even more uncomfortable to stand on a bed of nails without shoes.)

QUICKLAB

Place a tack between your thumb and index finger, as in the figure. Now squeeze the tack and note the sensation. The sharp end of the tack causes pain, and the blunt end does not. According to Newton's third law, the force exerted on the thumb is equal and opposite the force exerted on the index finger. However, the pressure at the sharp end of the tack is much greater than the pressure at the blunt end. (Remember that pressure is a measure of force per unit area.)

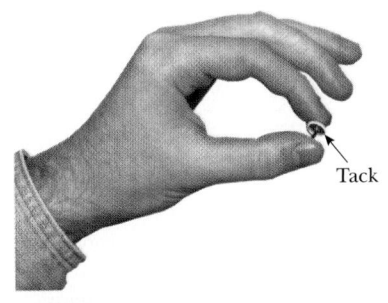

Tack

EXAMPLE 9.3   The Water Bed

A water bed is 2.00 m on a side and 30.0 cm deep.
(a) Find its weight.

Solution   Because the density of water is 1000 kg/m³ (Table 9.2) and the bed's volume is $(2.00 \times 2.00 \times 0.300)$ m³ $= 1.20$ m³, the mass of the bed is

$$M = \rho V = (1000 \text{ kg/m}^3)(1.20 \text{ m}^3) = 1.20 \times 10^3 \text{ kg}$$

and its weight is

$$w = Mg = (1.20 \times 10^3 \text{ kg})(9.80 \text{ m/s}^2) = \boxed{1.18 \times 10^4 \text{ N}}$$

This is equivalent to approximately 2640 lb. In order to support such a heavy load, you would be well advised to keep your water bed in the basement or on a sturdy, well-supported floor.
(b) Find the pressure that the water bed exerts on the floor. Assume that the entire lower surface of the bed makes contact with the floor.

Solution   The weight of the water bed is 1.18×10^4 N. Its cross-sectional area is 4.00 m². The pressure exerted on the floor is therefore

$$P = \frac{1.18 \times 10^4 \text{ N}}{4.00 \text{ m}^2} = \boxed{2.95 \times 10^3 \text{ Pa}}$$

Exercise   Calculate the pressure exerted by the bed on the floor if the bed rests on its side.

Answer   Because the area of the bed's side is 0.600 m², the pressure is 1.96×10^4 Pa.

9.4   VARIATION OF PRESSURE WITH DEPTH

If a fluid is at rest in a container, *all* portions of the fluid must be in static equilibrium. Furthermore, **all points at the same depth must be at the same pressure.** If this were not the case, a given portion of the fluid would not be in equilibrium. For example, consider the small block of fluid shown in Figure 9.10a. If the pressure were greater on the left side of the block than on the right, $\mathbf{F}_1$ would be greater than $\mathbf{F}_2$, and the block would accelerate and thus would not be in equilibrium.

Now let us examine the portion of the fluid contained within the volume indicated by the darker region in Figure 9.10b. This region has a cross-sectional area A and extends to a depth h below the surface of the water. Three external forces act on this volume of fluid: the force of gravity, Mg; the upward force, PA, exerted by the fluid below it; and a downward force, P_0A, exerted by the atmosphere, where P_0 is atmospheric pressure. Because this volume of fluid is in equilibrium, these forces must add to zero, and so we get

$$PA - Mg - P_0A = 0 \qquad \text{[9.8]}$$

From the relation $M = \rho V = \rho Ah$, the weight of the fluid in the volume is

$$w = Mg = \rho gAh \qquad \text{[9.9]}$$

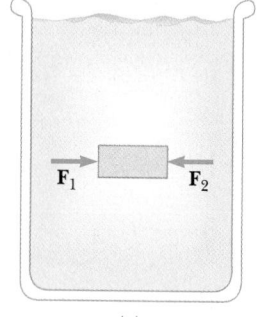

(a)

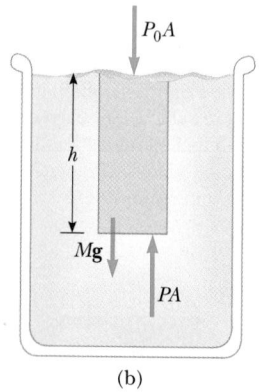

(b)

Figure 9.10   (a) If the block of fluid is to be in equilibrium, the force $\mathbf{F}_1$ must balance the force $\mathbf{F}_2$. (b) The net force on the volume of water within the darker region must be zero.

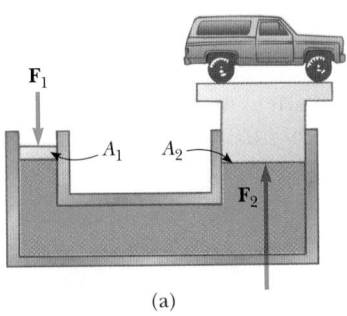

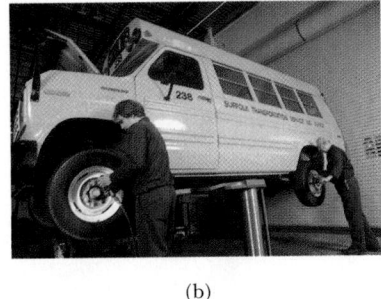

(a)   (b)

Figure 9.11 (a) Diagram of a hydraulic press. Because the pressure is the same at the left and right sides, a small force, $\mathbf{F}_1$, at the left produces a much larger force, $\mathbf{F}_2$, at the right. (b) A bus under repair is supported by a hydraulic lift in a garage. *(Superstock)*

When Equation 9.9 is substituted into Equation 9.8, we get

$$P = P_0 + \rho g h \qquad\qquad [9.10]$$

where normal atmospheric pressure at sea level is $P_0 = 1.01 \times 10^5$ Pa (equivalent to 14.7 lb/in.^{2}). According to Equation 9.10, **the pressure, P, at a depth of h below the surface of a liquid open to the atmosphere is greater than atmospheric pressure by the amount $\rho g h$.** Moreover, the pressure is not affected by the shape of the vessel.

Pressure in a fluid depends only on depth. Any increase in pressure at the surface is transmitted to every point in the fluid. This was first recognized by the French scientist Blaise Pascal (1623–1662) and is called **Pascal's principle:**

> **Pressure applied to an enclosed fluid is transmitted undiminished to every point of the fluid and to the walls of the containing vessel.**

An important application of Pascal's principle is the hydraulic press (Fig. 9.11a). A downward force, $\mathbf{F}_1$, is applied to a small piston of area A_1. The pressure is transmitted through a fluid to a larger piston of area A_2. Because the pressure is the same on both sides, we see that $P = F_1/A_1 = F_2/A_2$. Therefore, the magnitude of the force $\mathbf{F}_2$ is larger than the magnitude of $\mathbf{F}_1$ by the factor A_2/A_1. That is why a large load, such as a car, can be supported on the large piston by a much smaller force on the smaller piston. Hydraulic brakes, car lifts, hydraulic jacks, forklifts, and other machines make use of this principle.

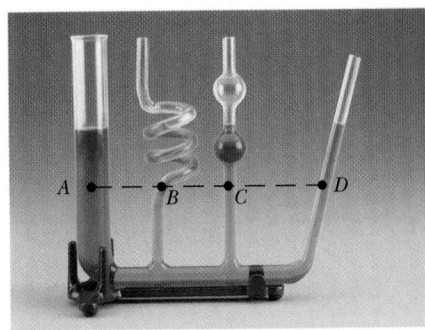

This photograph illustrates that the pressure in a liquid is the same at all points having the same elevation. For example, the pressure is the same at points A, B, C and D. Note that the shape of the vessel does not affect the pressure. *(Courtesy of Central Scientific Company)*

APPLICATION

Hydraulic Lifts.

Applying Physics 1

A corollary to the statement that pressure in a fluid increases with depth is that water always seeks its own level. This means that regardless of the shape of the vessel, when filled with water, the surface of the water is perfectly flat and at the same height at all points. The ancient Egyptians used this fact to level the pyramids. Devise a scheme showing how this could be done.

Explanation   There are many ways this could have been done, but Figure 9.12 shows the scheme used by the Egyptians. The builders cut grooves in the base of the pyramid as in (a) and partially filled the grooves with water. The height of the water was marked as in (b) and the rock was chiseled down to this line (c). Finally, the groove was filled with crushed rock and gravel, as in (d).

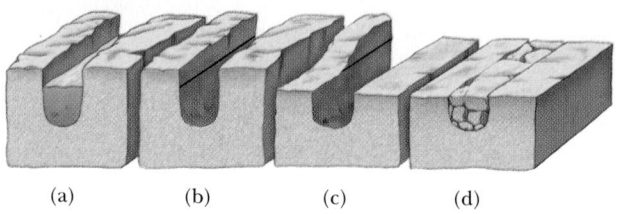

(a)   (b)   (c)   (d)

Figure 9.12   (Applying Physics 1)

9.5   PRESSURE MEASUREMENTS

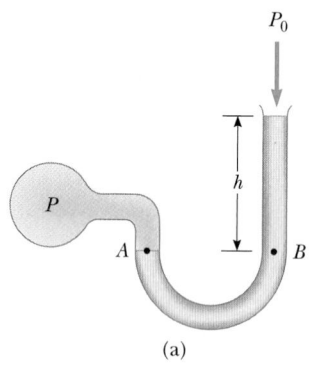

(a)

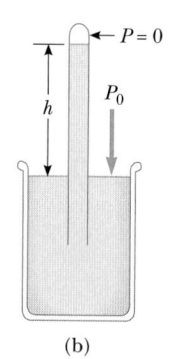

(b)

Figure 9.13   Two devices for measuring pressure: (a) an open-tube manometer and (b) a mercury barometer.

Another simple device for measuring pressure is the open-tube manometer (Fig. 9.13a). One end of a U-shaped tube containing a liquid is open to the atmosphere, and the other end is connected to a system of unknown pressure P. The pressure at point B equals $P_0 + \rho g h$, where ρ is the density of the fluid. The pressure at B, however, equals the pressure at A, which is also the unknown pressure P. Therefore, we conclude that

$$P = P_0 + \rho g h$$

The pressure P is called the *absolute pressure*, and $P - P_0$ is called the *gauge pressure*. Thus, if P in the system is greater than atmospheric pressure, h is positive. If P is less than atmospheric pressure (a partial vacuum), h is negative.

A third instrument that is used to measure pressure is the *barometer* (Fig. 9.13b), invented by Evangelista Torricelli (1608–1647). A long tube that is closed at one end is filled with mercury and then inverted into a dish of mercury. The closed end of the tube is nearly a vacuum, and so its pressure can be taken as zero. It follows that $P_0 = \rho g h$, where ρ is the density of the mercury and h is the height of the mercury column.

One atmosphere of pressure is defined to be the pressure equivalent of a column of mercury that is exactly 0.76 m in height at 0°C with $g = 9.806\ 65$ m/s². At this temperature mercury has a density of 13.595×10^3 kg/m³; therefore,

$$P_0 = \rho g h = (13.595 \times 10^3 \text{ kg/m}^3)(9.806\ 65 \text{ m/s}^2)(0.7600 \text{ m})$$
$$= 1.013 \times 10^5 \text{ Pa}$$

It is interesting to note that the force of the atmosphere on our bodies (assuming a body area of 2000 in.²) is extremely large, on the order of 30 000 lb! A natural question to raise is, "How can we exist under such great forces attempting to collapse our bodies?" The answer is that our body cavities and tissues are permeated with fluids and gases that are pushing outward with this same atmospheric pressure.

Consequently, our bodies are in equilibrium under the force of the atmosphere pushing in and an equal internal force pushing out.

Blood Pressure Measurements

A specialized manometer (called a sphygmomanometer) is often used to measure blood pressure. In this application, a rubber bulb forces air into a cuff wrapped tightly around the upper arm and simultaneously into a manometer, as in Figure 9.14. The pressure in the cuff is increased until the flow of blood through the brachial artery in the arm is stopped. A valve on the bulb is then opened and the measurer listens with a stethoscope to the artery at a point just below the cuff. When the pressure in the cuff and brachial artery is just below the maximum value produced by the heart (the systolic pressure) the artery opens momentarily on each beat of the heart. At this point, the velocity of the blood is high and turbulent, and the flow is noisy. The manometer is calibrated to read the pressure in millimeters of mercury, and the value obtained is about 120 mm for a normal heart. Values of 140 mm or above are considered to be high and medication to lower blood pressure is often prescribed. As the pressure in the cuff is lowered further, intermittent sounds are still heard until the pressure falls just below the minimum heart pressure (the diastolic pressure). At this point, continuous sounds are heard. In the normal heart, this transition occurs at about 80 mm of mercury, and values above 90 are considered to need medical intervention. Blood pressure readings are usually expressed as the ratio systolic/diastolic or 120/80 for a healthy heart.

APPLICATION

Measuring Blood Pressure.

Applying Physics 2

Blood pressure is normally measured with the cuff of the sphygmomanometer around the arm. Suppose that the blood pressure were measured with the cuff around the calf of the leg of a standing person. Would the reading of the blood pressure be the same here as it was for the arm?

Explanation The blood pressure measured at the calf would be larger than that measured at the arm. If we imagine the vascular system of the body to be a vessel containing a liquid (the blood), the pressure in the liquid will increase with depth. The blood at the calf is deeper in the liquid than that at the arm and is at a higher pressure.

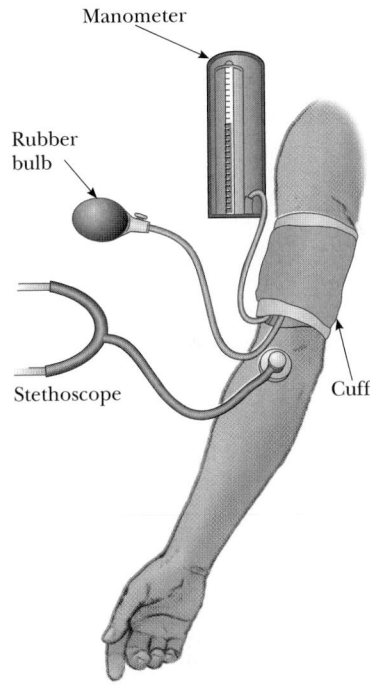

Figure 9.14 A sphygmomanometer can be used to measure blood pressure.

Applying Physics 3

In a ball-point pen, ink moves down a tube to the tip where it is spread on a sheet of paper by a rolling stainless steel ball. Near the top of the ink cartridge, there is a small hole open to the atmosphere. If you seal this hole, you will find that the pen no longer works. Use your knowledge of how a barometer works to explain this behavior.

Explanation If the hole is sealed, or if it were not present, the pressure of the air above the ink would decrease as some ink is used. Consequently, atmospheric pressure exerted against the ink at the bottom of the cartridge would prevent some of the ink from flowing out. The hole allows the pressure above the ink to remain at atmospheric pressure. Why does a ball-point pen seem to run out of ink when you write on a vertical surface?

APPLICATION

Ball-Point Pens.

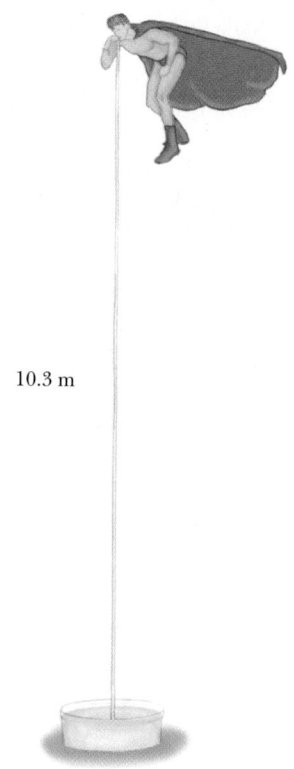

10.3 m

Figure 9.15 Even the man of steel is defeated if the straw is too long.

Applying Physics 4

Figure 9.15 shows Superman attempting to drink through a very long straw. Unbelievably, he finds that he is unable to lift the fluid to his lips if the straw is longer than about 10.3 m. Explain why this is the case, and then extend this reasoning to determine what would happen if Superman tried to drink through a straw on the Moon where there is no atmosphere.

Explanation Without a doubt, Superman is able to remove all the air from above the surface of the fluid. At this point, the pressure at the top of the straw (P_0) is zero. $P = P_0 + \rho g h$ becomes, with $P_0 = 0$, $P = P_a$ and with ρ being the density of the fluid (assumed to be water), $P_a = \rho g h$ or

$$h = \frac{P_a}{\rho g} = \frac{1.01 \times 10^5 \text{ Pa}}{(1.00 \times 10^3 \text{ kg/m}^3)(9.8 \text{ m/s}^2)} = 10.3 \text{ m}$$

By the same reasoning, if P_a goes to zero as it would on the Moon, Superman would not be able to lift any height of fluid into the straw.

EXAMPLE 9.4 The Car Lift

In a car lift, compressed air exerts a force on a piston with a radius of 5.00 cm. This pressure is transmitted to a second piston, of radius 15.0 cm. What force must the compressed air exert in order to lift a car weighing 1.33×10^4 N? What air pressure produces this force? Neglect the weights of the pistons.

Solution Because the pressure exerted by the compressed air is transmitted undiminished throughout the fluid, we have

$$F_1 = \left(\frac{A_1}{A_2}\right) F_2 = \frac{\pi(5.00 \times 10^{-2} \text{ m})^2}{\pi(15.0 \times 10^{-2} \text{ m})^2} (1.33 \times 10^4 \text{ N}) = \boxed{1.48 \times 10^3 \text{ N}}$$

The air pressure that produces this force is

$$P = \frac{F_1}{A_1} = \frac{1.48 \times 10^3 \text{ N}}{\pi(5.00 \times 10^{-2} \text{ m})^2} = \boxed{1.88 \times 10^5 \text{ Pa}}$$

This pressure is approximately twice atmospheric pressure. (Note that pressure here means gauge pressure.)

EXAMPLE 9.5 Pressure in the Ocean

Calculate the absolute pressure at an ocean depth of 1000 m. Assume that the density of water is 1.0×10^3 kg/m^3 and that $P_0 = 1.01 \times 10^5$ Pa.

Solution

$$P = P_0 + \rho g h$$
$$= 1.01 \times 10^5 \text{ Pa} + (1.0 \times 10^3 \text{ kg/m}^3)(9.80 \text{ m/s}^2)(1.00 \times 10^3 \text{ m})$$
$$P \approx \boxed{9.9 \times 10^6 \text{ Pa}}$$

This is approximately 100 times greater than atmospheric pressure! Obviously, the design and construction of vessels that can withstand such enormous pressures are not trivial matters.

Exercise Calculate the total force exerted on the outside of a 30-cm-diameter circular submarine window at this depth.

Answer 7.0×10^5 N

9.6 BUOYANT FORCES AND ARCHIMEDES'S PRINCIPLE

A fundamental principle affecting objects submerged in fluids was discovered by the Greek mathematician Archimedes. **Archimedes's principle** can be stated as follows:

> Any body completely or partially submerged in a fluid is buoyed up by a force whose magnitude is equal to the weight of the fluid displaced by the body.

Everyone has experienced Archimedes's principle. For example, recall that it is relatively easy to lift someone if you are both standing in a swimming pool, whereas lifting that same individual on dry land would be a difficult task. Water provides partial support to any object placed in it. We say that an object placed in a fluid is buoyed up by the fluid, and we call this upward force the **buoyant force.** According to Archimedes's principle, **the magnitude of the buoyant force always equals the weight of the fluid displaced by the object.** The buoyant force acts vertically upward through what was the center of gravity of the fluid before the fluid was displaced.

Archimedes's principle can be verified in the following manner. Suppose we focus our attention on the cube of water that is colored red in the container of Figure 9.16. This cube of water is in equilibrium under the action of the forces on it. One of those forces is the force of gravity. What cancels that downward force? Apparently, the water beneath the cube is buoying it up and holding it in equilibrium. Thus, the buoyant force, **B,** on the cube of water must be exactly equal in magnitude to the weight of the water inside the cube: $B = w$.

Now imagine that the cube of water is replaced by a cube of steel of the same dimensions. What is the buoyant force on the steel? The water surrounding a cube behaves in the same way whether the cube is made of water or steel; therefore, **the buoyant force acting on the steel is the same as the buoyant force acting on a cube of water of the same dimensions.** This result applies for a totally submerged object of any shape, size, or density.

Let us show explicitly that the buoyant force is equal in magnitude to the weight of the displaced fluid. The pressure at the bottom of the cube in Figure 9.16 is greater than the pressure at the top by the amount $\rho_f gh$, where ρ_f is the density of the fluid and h is the height of the cube. Because the pressure difference, ΔP, is equal to the buoyant force per unit area, or, $\Delta P = B/A$, we see that $B = (\Delta P)(A) = (\rho_f gh)(A) = \rho_f gV$, where V is the volume of the cube. Because the mass of the fluid in the cube is $M = \rho_f V$, we see that

$$B = \rho_f Vg = Mg = w_f \qquad [9.11]$$

where w_f is the weight of the displaced fluid.

Archimedes, Greek mathematician, physicist, and engineer (287–212 B.C.)

Archimedes was perhaps the greatest scientist of antiquity. He is well known for discovering the nature of the buoyant force and was a gifted inventor. According to legend, Archimedes was asked by King Hieron to determine whether the king's crown was made of pure gold or merely a gold alloy. The task was to be performed without damaging the crown. Archimedes allegedly arrived at a solution while taking a bath, noting a partial loss of weight after submerging his arms and legs in the water. As the story goes, he was so excited about his great discovery that he ran naked through the streets of Syracuse shouting, "Eureka!" which is Greek for "I have found it."

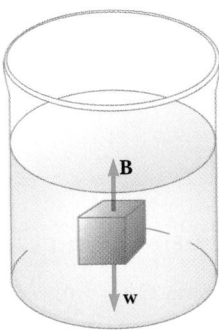

Figure 9.16 The external forces on a cube of water are the force of gravity, **w,** and the buoyant force, **B.** Under equilibrium conditions, $B = w$.

Hot-air balloons. Because hot air is less dense than cold air, there is a net upward buoyant force on the balloons. *(Richard Megna, Fundamental Photographs, NYC)*

The weight of the submerged object is $w_0 = mg = \rho_0 Vg$, where ρ_0 is the density of the object. Because $w_f = \rho_f Vg$ and $w_0 = \rho_0 Vg$, we see that if the density of the object is greater than the density of the fluid, the unsupported object sinks. If the density of the object is less than that of the fluid, the unsupported submerged object accelerates upward and ultimately floats. When a *floating* object is in equilibrium, only part of it is submerged. In this case, the magnitude of the buoyant force equals the weight of the object.

It is instructive to compare the forces on a totally submerged object with those on a floating object.

Case I: A Totally Submerged Object.   When an object is *totally* submerged in a fluid of density ρ_f, the upward buoyant force has a magnitude of $B = \rho_f V_0 g$, where V_0 is the volume of the object. If the object has density ρ_0, the downward force is equal to $mg = \rho_0 V_0 g$, and the net force on it is $B - w = (\rho_f - \rho_0) V_0 g$. Hence, if the density of the object is less than the density of the fluid, as in Figure 9.17a, the net force is positive (upward) and the unsupported object accelerates upward. If the density of the object is greater than the density of the fluid, as in Figure 9.17b, the net force is negative and the unsupported object sinks.

Case II: A Floating Object.   Now consider an object in static equilibrium floating on a fluid — that is, a partially submerged object. In this case, the upward buoyant force is balanced by the downward force of gravity acting on the object. If V_f is the volume of the fluid displaced by the object (which corresponds to the volume of the part of the object beneath the fluid level), then the magnitude of the buoyant force is given by $B = \rho_f V_f g$. Because the weight of the object is $w = mg = \rho_0 V_0 g$, and $w = B$, we see that $\rho_f V_f g = \rho_0 V_0 g$, or

$$\frac{\rho_0}{\rho_f} = \frac{V_f}{V_0} \qquad\qquad [9.12]$$

▶ **A P P L I C A T I O N**

Swim Bladders in Fish.

▶ **A P P L I C A T I O N**

Cerebrospinal Fluid.

A P P L I C A T I O N ▶

Testing Your Car's Antifreeze.

Under normal conditions, the average density of a fish is slightly greater than the density of water. This being the case, a fish would sink if it did not have a mechanism for adjusting its density: the internal regulation of the size of the swim bladder. In this manner, fish maintain neutral buoyancy as they swim to various depths.

The human brain is immersed in a fluid (the cerebrospinal fluid) of density 1007 kg/m^3, which is slightly less than the average density of the brain, 1040 kg/m^3. Consequently, most of the weight of the brain is supported by the buoyant force of the surrounding fluid. In some clinical procedures, it is necessary to remove a portion of this fluid for diagnostic purposes. During such procedures, the nerves and blood vessels in the brain are placed under great strain, which in turn can cause extreme discomfort and pain. Great care must be exercised with such patients until the initial brain fluid volume has been restored by the body.

When service station attendants check the antifreeze in your car or the condition of your battery, they often use devices that apply Archimedes's principle. Figure 9.18 shows a common device that is used to check the antifreeze in a car radiator. The small balls in the enclosed tube vary in density so that all of them float when the tube is filled with pure water, none float in pure antifreeze, one floats in a 5% mixture, two in a 10% mixture, and so forth. The number of balls that float thus serves as a measure of the percentage of antifreeze in the mixture, which in turn

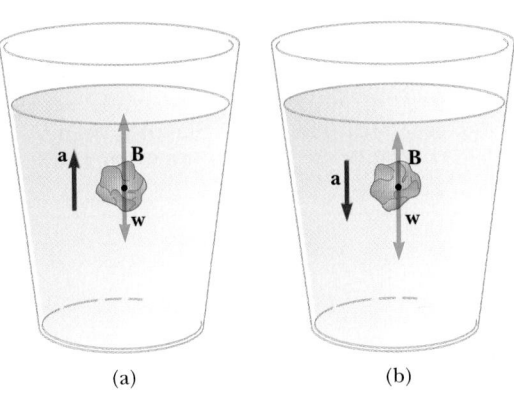

(a)   (b)

Figure 9.17 (a) A totally submerged object that is less dense than the fluid in which it is submerged experiences a net upward force. (b) A totally submerged object that is denser than the fluid sinks.

is used to determine the lowest temperature the mixture can withstand without freezing.

Similarly, the charge of some newer car batteries can be determined with a so-called "magic-dot" process that is built into the battery (Fig. 9.19). When one looks down into a viewing port in the top of the battery, a red dot indicates that the battery is sufficiently charged; a black dot indicates that the battery has lost its charge. This is because, if the battery has sufficient charge, the density of the battery fluid is high enough to cause the red ball to float. As the battery loses its charge, the density of the battery fluid decreases and the ball sinks beneath the surface of the fluid, where the dot appears black.

APPLICATION

Checking the Battery Charge.

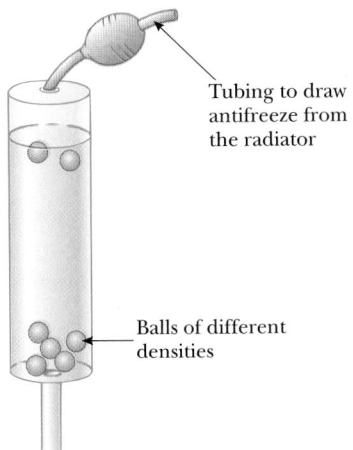

Figure 9.18 The number of balls that float in this device is a measure of the density of the antifreeze solution in a vehicle's radiator, and consequently a measure of the temperature at which freezing will occur.

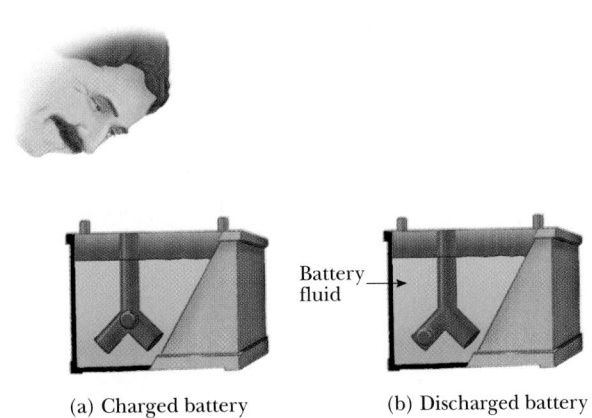

(a) Charged battery   (b) Discharged battery

Figure 9.19 The red ball in the plastic tube inside the battery serves as an indicator of whether the battery is charged or discharged. As the battery loses its charge, the density of the battery fluid decreases, and the ball sinks out of sight.

EXAMPLE 9.6 A Red-Tag Special on Crowns

A bargain hunter purchases a "gold" crown at a flea market. After she gets home, she hangs it from a scale and finds its weight to be 7.84 N (Fig. 9.20a). She then weighs the crown while it is immersed in water of density 1000 kg/m³, as in Figure 9.20b, and now the scale reads 6.86 N. Is the crown made of pure gold?

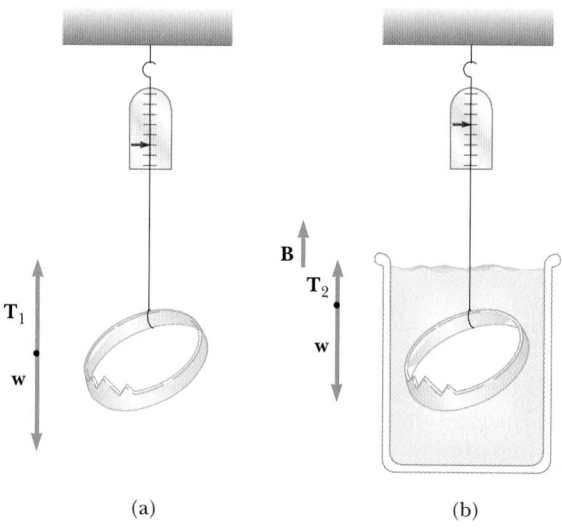

(a)   (b)

Figure 9.20 (Example 9.6) (a) When the crown is suspended in air, the scale reads the true weight, **w.** (b) When the crown is immersed in water, the buoyant force, **B,** reduces the scale reading to the apparent weight, $T_2 = w - B$.

Reasoning When the crown is suspended in air, the scale reads the true weight, w (neglecting the buoyancy of air). When it is immersed in water, the buoyant force, **B,** reduces the scale reading to an apparent weight of $T_2 = w - B$. Hence, the buoyant force on the crown is the difference between its weight in air and its weight in water:

$$B = w - T_2 = 7.84 \text{ N} - 6.86 \text{ N} = 0.98 \text{ N}$$

Because the buoyant force is equal in magnitude to the weight of the displaced fluid, w_f, we have

$$w_f = \rho_f g V_f = 0.98 \text{ N}$$

where V_f is the displaced fluid's volume, and ρ_f is its density (1000 kg/m³). Also, the volume of the crown, V_c, is equal to the volume of the displaced fluid (because the crown is completely submerged).

Solution We find that

$$V_c = V_f = \frac{0.98 \text{ N}}{g\rho_f} = \frac{0.98 \text{ N}}{(9.8 \text{ m/s}^2)(1000 \text{ kg/m}^3)} = 1.0 \times 10^{-4} \text{ m}^3$$

Finally, the density of the crown is

$$\rho_c = \frac{m_c}{V_c} = \frac{w_c}{g V_c} = \frac{7.84 \text{ N}}{(9.8 \text{ m/s}^2)(1.0 \times 10^{-4} \text{ m}^3)} = 8.0 \times 10^3 \text{ kg/m}^3$$

From Table 9.2 we see that the density of gold is 19.3×10^3 kg/m³. Thus, either the crown is hollow or it is made of an alloy. This was not a good day for the bargain hunter!

EXAMPLE 9.7 Floating Down the River

A raft is constructed of wood having a density of 600 kg/m³. Its surface area is 5.7 m² and its volume is 0.60 m³. When the raft is placed in fresh water of density 1000 kg/m³, as in Figure 9.21, how much of it is below water level?

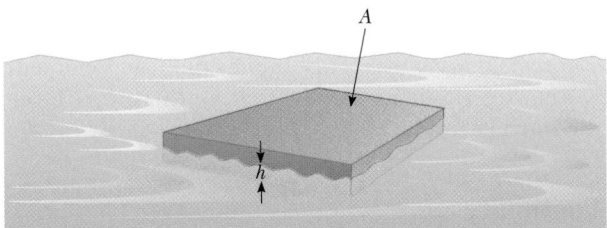

Figure 9.21 (Example 9.7) A raft partially submerged in water.

Reasoning The magnitude of the upward buoyant force acting on the raft must equal the weight of the raft if the raft is to float. In addition, from Archimedes's principle the magnitude of the buoyant force is equal to the weight of the displaced water.

Solution The weight of the raft is

$$w_r = \rho_r g V_r = (600 \text{ kg/m}^3)(9.8 \text{ m/s}^2)(0.60 \text{ m}^3) = 3.5 \times 10^3 \text{ N}$$

The magnitude of the upward buoyant force acting on the raft equals the weight of the displaced water, which in turn must equal the weight of the raft:

$$B = \rho_w g V_w = \rho_w g A h = 3.5 \times 10^3 \text{ N}$$

Because the area A and density ρ_w are known, we can find the depth, h, at which the raft sinks into the water:

$$h = \frac{B}{\rho_w g A} = \frac{3.5 \times 10^3 \text{ N}}{(1000 \text{ kg/m}^3)(9.8 \text{ m/s}^2)(5.7 \text{ m}^2)} = \boxed{0.06 \text{ m}}$$

9.7 FLUIDS IN MOTION

When a fluid is in motion, its flow can be characterized in one of two ways. The flow is said to be **streamline,** or **laminar,** if every particle that passes a particular point moves exactly along the smooth path followed by particles that passed that point earlier. The path is called a *streamline* (Fig. 9.22a). Different streamlines cannot cross each other under this steady-flow condition, and the streamline at any point coincides with the direction of fluid velocity at that point.

In contrast, the flow of a fluid becomes irregular, or **turbulent,** above a certain velocity or under any conditions that can cause abrupt changes in velocity. Irregular motions of the fluid, called *eddy currents,* are characteristic in turbulent flow, as shown in Figure 9.22b.

In discussions of fluid flow, the term *viscosity* is used for the degree of internal friction in the fluid. This internal friction is associated with the resistance between two adjacent layers of the fluid moving relative to each other. A fluid such as kerosene has a lower viscosity than crude oil or molasses.

Most of the volume of this iceberg is beneath the water. Can you determine what fraction of the total volume is underwater? *(Geraldine Prentice/ Tony Stone Images)*

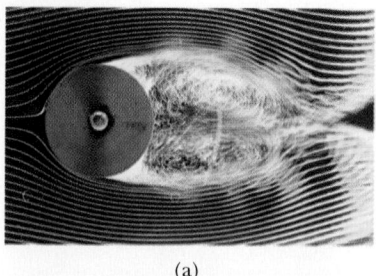

(a)

(b)

Figure 9.22 (a) This photograph was taken in a water tunnel using hydrogen bubbles to visualize the flow pattern around a cylinder. The flow was started from rest, and at this instant the pattern shows the development of a complex wake structure on the downstream side of the cylinder. *(Courtesy of Dr. Kenneth W. McAlister, Department of the Army)* (b) Hot gases from a cigarette made visible by smoke particles. The smoke first moves in streamline flow at the bottom and then in turbulent flow above. *(Werner Wolff/Black Star)*

Many features of fluid motion can be understood by considering the behavior of an **ideal fluid,** which satisfies the following conditions:

1. *The fluid is nonviscous;* that is, there is no internal friction force between adjacent layers.
2. *The fluid is incompressible,* which means that its density is constant.
3. *The fluid motion is steady,* meaning that the velocity, density, and pressure at each point in the fluid do not change in time.
4. *The fluid moves without turbulence.* This implies that each element of the fluid has zero angular velocity about its center; that is, there can be no eddy currents present in the moving fluid.

Equation of Continuity

Figure 9.23 represents a fluid flowing through a pipe of nonuniform size. The particles in the fluid move along the streamlines in steady-state flow. In a small time interval, Δt, the fluid entering the bottom end of the pipe moves a distance of $\Delta x_1 = v_1 \Delta t$, where v_1 is the speed of the fluid at this location. If A_1 is the cross-sectional area in this region, then the mass contained in the bottom blue region is $\Delta M_1 = \rho A_1 \Delta x_1 = \rho A_1 v_1 \Delta t$, where ρ is the density of the fluid. Similarly, the fluid that moves out of the upper end of the pipe in the same interval, Δt, has a mass of $\Delta M_2 = \rho A_2 v_2 \Delta t$. However, because mass is conserved and because the flow is steady, the mass that flows into the bottom of the pipe through A_1 in the time Δt must equal the mass that flows out through A_2 in the same interval. Therefore, $\Delta M_1 = \Delta M_2$, or

$$\rho A_1 v_1 = \rho A_2 v_2 \qquad [9.13]$$

Equation 9.13 reduces to

$$A_1 v_1 = A_2 v_2 \qquad [9.14]$$

This expression is called the **equation of continuity.** From this result, we see that **the product of any cross-sectional area of the pipe and the fluid speed at that cross-section is a constant.** Therefore, the speed is high where the tube is constricted and low where the tube has a larger diameter. The product Av, which has dimensions of volume per unit time, is called the **flow rate. The condition Av = constant is equivalent to the fact that the amount of fluid that enters one end of the tube in a given time interval equals the amount of fluid leaving the tube in the same interval, assuming the absence of leaks.**

There are many instances in your everyday experiences when you have seen the equation of continuity in action. For example, you place your thumb over the open end of a garden hose to make the water spray our farther and with greater speed. As you reduce the cross-sectional area of the nozzle, the water drops leave the nozzle at a higher speed, causing them to move a longer distance. Similar reasoning explains why rising smoke from a smoldering piece of wood behaves as it does. The smoke first rises in a streamline pattern, getting thinner as it rises, and eventually breaks up into a swirling, turbulent pattern. The smoke rises because it is less dense than air, and the buoyant force of the air accelerates it upward. As the speed of the smoke stream increases, the cross-sectional area of the stream decreases, as seen from the equation of continuity. However, the stream soon reaches a speed so great that streamline flow is not possible. We shall study the relationship between speed of fluid flow and turbulence in a later discussion on the Reynolds number.

EXAMPLE 9.8 Filling a Water Bucket

A water hose 1.00 cm in radius is used to fill a 20.0-liter bucket. If it takes 1.00 min to fill the bucket, what is the speed, v, at which the water leaves the hose? (1 liter = 10^3 cm^3.)

Solution The cross-sectional area of the hose is

$$A = \pi r^2 = \pi(1.00 \text{ cm})^2 = \pi \text{ cm}^2$$

The flow rate is equal to the product Av. Thus,

$$Av = 20.0 \frac{\text{liters}}{\text{min}} = \frac{20.0 \times 10^3 \text{ cm}^3}{60.0 \text{ s}}$$

$$v = \frac{20.0 \times 10^3 \text{ cm}^3}{(\pi \text{ cm}^2)(60.0 \text{ s})} = \boxed{106 \text{ cm/s}}$$

Exercise If the radius of the hose is reduced to 0.500 cm, what is the speed of the water as it leaves the hose, assuming the same flow rate?

Answer 424 cm/s

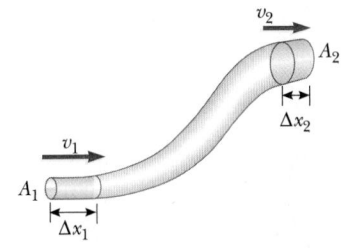

Figure 9.23 A fluid moving with streamline flow through a pipe of varying cross-sectional area. The volume of fluid flowing through A_1 in a time interval of Δt must equal the volume flowing through A_2 in the same time interval. Therefore, $A_1 v_1 = A_2 v_2$.

Bernoulli's Equation

As a fluid moves through a pipe of varying cross-section and elevation, the pressure changes along the pipe. In 1738 the Swiss physicist Daniel Bernoulli (1700–1782) derived a fundamental expression that relates pressure to fluid speed and elevation. Bernoulli's equation is not a freestanding law of physics. It is, instead, a consequence of energy conservation as applied to the ideal fluid.

In deriving Bernoulli's equation, we again assume that the fluid is incompressible and nonviscous and flows in a nonturbulent, steady-state manner. Consider the flow through a nonuniform pipe in the time Δt, as illustrated in Figure 9.24. The force on the lower end of the fluid is $P_1 A_1$, where P_1 is the pressure at the lower end. The work done on the lower end of the fluid by the fluid behind it is

$$W_1 = F_1 \Delta x_1 = P_1 A_1 \Delta x_1 = P_1 V$$

where V is the volume of the lower blue region in Figure 9.24. In a similar manner, the work done on the fluid on the upper portion in the time Δt is

$$W_2 = -P_2 A_2 \Delta x_2 = -P_2 V$$

(Remember that the volume of fluid that passes through A_1 in the time Δt equals the volume that passes through A_2 in the same interval.) The work W_2 is negative because the force on the fluid at the top is opposite its displacement. Thus, the net work done by these forces in the time Δt is

$$W = P_1 V - P_2 V$$

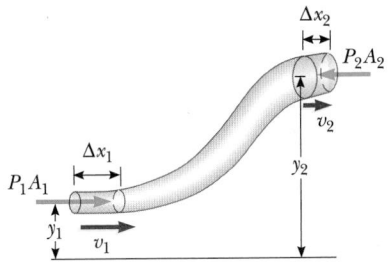

Figure 9.24 A fluid flowing through a constricted pipe with streamline flow. The fluid in the section with a length of Δx_1 moves to the section with a length of Δx_2. The volumes of fluid in the two sections are equal.

Daniel Bernoulli, Swiss physicist and mathematician (1700–1782)

Bernoulli made important discoveries in hydrodynamics. In his most famous work, *Hydrodynamica*, Bernoulli showed that, as the velocity of fluid flow increases, its pressure decreases. In this same publication, Bernoulli also attempted the first explanation of the behavior of gases with changing pressure and temperature; this was the beginning of kinetic theory of gases. *(North Wind Picture Archives)*

Part of this work goes into changing the fluid's kinetic energy, and part goes into changing its gravitational potential energy. If m is the mass of the fluid passing through the pipe in the time interval Δt, then the change in kinetic energy of the volume of fluid is

$$\Delta KE = \tfrac{1}{2}mv_2^2 - \tfrac{1}{2}mv_1^2$$

The change in its gravitational potential energy is

$$\Delta PE = mgy_2 - mgy_1$$

We can apply the work–energy theorem in the form $W = \Delta KE + \Delta PE$ (Chapter 5) to this volume of fluid to give

$$P_1V - P_2V = \tfrac{1}{2}mv_2^2 - \tfrac{1}{2}mv_1^2 + mgy_2 - mgy_1$$

If we divide each term by V and recall that $\rho = m/V$, this expression reduces to

$$P_1 - P_2 = \tfrac{1}{2}\rho v_2^2 - \tfrac{1}{2}\rho v_1^2 + \rho gy_2 - \rho gy_1$$

Let us move those terms that refer to point 1 to one side of the equation and those that refer to point 2 to the other side:

$$P_1 + \tfrac{1}{2}\rho v_1^2 + \rho gy_1 = P_2 + \tfrac{1}{2}\rho v_2^2 + \rho gy_2 \qquad [9.15]$$

This is **Bernoulli's equation.** It is often expressed as

$$P + \tfrac{1}{2}\rho v^2 + \rho gy = \text{constant} \qquad [9.16]$$

> Bernoulli's equation states that the sum of the pressure (P), the kinetic energy per unit volume ($\tfrac{1}{2}\rho v^2$), and the potential energy per unit volume (ρgy) has the same value at all points along a streamline.

An important consequence of Bernoulli's equation can be demonstrated by considering Figure 9.25, which shows water flowing through a horizontal constricted pipe from a region of large cross-sectional area into a region of smaller cross-sectional area. This device, called a *Venturi tube*, can be used to measure the speed of fluid flow. Let us compare the pressure at point 1 to the pressure at point 2. Because the pipe is horizontal, $y_1 = y_2$ and Equation 9.15 applied to points 1 and 2 gives

$$P_1 + \tfrac{1}{2}\rho v_1^2 = P_2 + \tfrac{1}{2}\rho v_2^2 \qquad [9.17]$$

Because the water is not backing up in the pipe, its speed in the constriction, v_2, must be greater than the speed v_1. From Equation 9.17, $v_2 > v_1$ means that P_2 must be less than P_1. This result is often expressed by the statement that **swiftly moving fluids exert less pressure than do slowly moving fluids.** As we shall see in the next section, this important result enables us to understand a wide range of everyday phenomena.

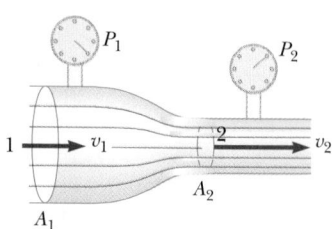

Figure 9.25 The pressure P_1 is greater than the pressure P_2, because $v_1 < v_2$. This device can be used to measure the speed of fluid flow.

EXAMPLE 9.9 Shoot-Out at the Old Water Tank

A nearsighted sheriff fires at a cattle rustler with his trusty six-shooter. Fortunately for the cattle rustler, the bullet misses him and penetrates the town water tank and causes a leak (Fig. 9.26). If the top of the tank is open to the atmosphere, determine the speed at which the water leaves the hole when the water level is 0.500 m above the hole.

Reasoning If we assume that the cross-sectional area of the tank is large relative to that of the hole ($A_2 \gg A_1$), then the water level drops very slowly and we can assume $v_2 \approx 0$. Let us apply Bernoulli's equation to points 1 and 2. If we note that $P_1 = P_0$ at the hole, we get

$$P_0 + \tfrac{1}{2}\rho v_1{}^2 + \rho g y_1 = P_0 + \rho g y_2$$

$$v_1 = \sqrt{2g(y_2 - y_1)} = \sqrt{2gh}$$

Solution This says that the speed of the water emerging from the hole is equal to the speed acquired by a body falling freely through the vertical distance h. This is known as **Torricelli's law.** If the height h is 0.500 m, for example, the speed of the stream is

$$v = \sqrt{2(9.80 \text{ m/s}^2)(0.500 \text{ m})} = \boxed{3.13 \text{ m/s}}$$

Exercise If the head of the cattle rustler is 3.00 m below the level of the hole in the tank, where must he stand to get doused with water?

Answer 2.45 m from the base of the tank.

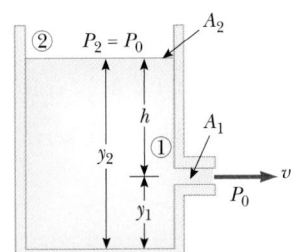

Figure 9.26 (Example 9.9) The water speed, v_1, from the hole in the side of the container is given by $v_1 = \sqrt{2gh}$.

9.8 OTHER APPLICATIONS OF BERNOULLI'S EQUATION

In this section we describe some common phenomena that can be explained, at least in part, by Bernoulli's equation.

First, consider the circulation of air around a thrown baseball. If the ball is not spinning (Fig. 9.27a), the motion of the airstream past the ball is nearly streamline.

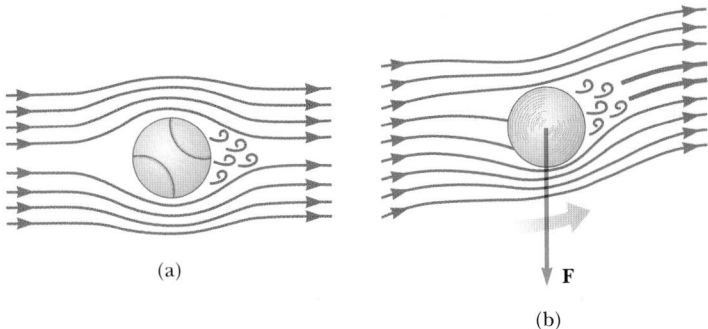

(a)

(b)

Figure 9.27 (a) The airstream around a nonrotating baseball moving from right to left. The streamlines represent the flow relative to the baseball. Note the symmetric region of turbulence behind the ball. (b) The airstream around a spinning baseball. The ball experiences a deflecting force because of the Bernoulli effect.

In this figure the ball is moving from right to left. Hence, from its point of view, the airstream is moving from left to right. A symmetric region of turbulence occurs behind the ball as shown. When the ball is spinning counterclockwise (Fig. 9.27b), layers of air near its surface are carried in the direction of spin because of viscosity. The combined effect of the steady flow of air and the air dragged along due to the spinning motion produces the streamlines shown in 9.24b and a corresponding turbulence pattern. The speed of the air below the ball is greater than the speed above the ball. Thus, from Bernoulli's equation, the air pressure above the ball is greater than the air pressure below the ball, and the ball experiences a downward deflecting force. When a pitcher wishes to throw a curve ball that deflects sideways, the spin axis should be vertical (perpendicular to the page in Fig. 9.27b). In contrast, if he wishes to throw a "sinker," the spin axis should be horizontal. Tennis balls and golf balls with spin also exhibit dynamic lift.

Many devices operate in the manner illustrated in Figure 9.28. A stream of air passing over an open tube reduces the pressure above the tube. This causes the liquid to rise into the airstream. The liquid is then dispersed into a fine spray of droplets. You might recognize that this so-called atomizer is used in perfume bottles and paint sprayers. The same principle is used in the carburetor of a gasoline engine. In that case, the low-pressure region in the carburetor is produced by air drawn in by the piston through the air filter. The gasoline vaporizes, mixes with the air, and enters the cylinder of the engine for combustion.

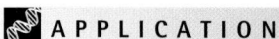
In a person with advanced arteriosclerosis, the Bernoulli effect produces a symptom called *vascular flutter*. In this situation, the artery is constricted as a result of accumulated plaque on its inner walls, as in Figure 9.29. To maintain a constant flow rate, the blood must travel faster than normal through the constriction. If the blood speed is sufficiently high in the constricted region, the artery may collapse under external pressure, causing a momentary interruption in blood flow. At this moment there is no Bernoulli effect, and the vessel reopens under arterial pressure. As the blood rushes through the constricted artery, the internal pressure drops and again the artery closes. Such variations in blood flow can be heard with a stethoscope. If the plaque becomes dislodged and ends up in a smaller vessel that delivers blood to the heart, the person can suffer a heart attack.

An aneurysm is a weakened spot on an artery where the artery walls have ballooned outward. Blood flows more slowly though this region, as can be seen from the equation of continuity, resulting in an increase in pressure in the vicinity of the aneurysm relative to the pressure in other parts of the artery. This condition is dangerous because the excess pressure can cause the artery to rupture.

The lift on an aircraft wing can also be explained, in part, by the Bernoulli effect. Airplane wings are designed so that the air speed above the wing is greater than that below. As a result, the air pressure above the wing is less than the pressure below, and there is a net upward force on the wing, called the "lift." Another factor influencing the lift on a wing is shown in Figure 9.30. The wing has a slight upward tilt that causes air molecules striking the bottom to be deflected downward. The air molecules bouncing off the wing at the bottom produce an upward force on the wing and a significant lift on the aircraft. Finally, turbulence also has an effect. If the wing is tilted too much, the flow of air across the upper surface becomes turbulent, and the pressure difference across the wing is not as great as that predicted by Bernoulli's equation. In an extreme case, this turbulence may cause the aircraft to stall.

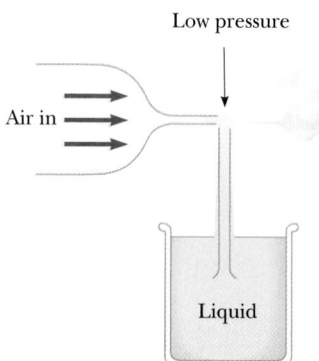

Figure 9.28 A stream of air passing over a tube dipped in a liquid causes the liquid to rise in the tube as shown. This effect is used in perfume bottles and paint sprayers.

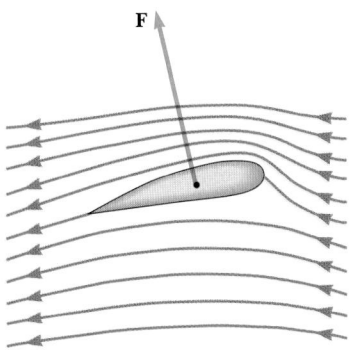

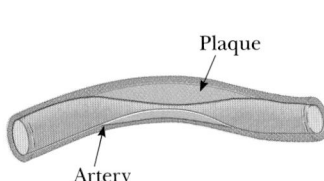

Figure 9.29 The flow of blood through a constricted artery.

Figure 9.30 Streamline flow around an airplane wing. The pressure above is less than the pressure below, and there is a dynamic lift force upward.

Q U I C K L A B

Place a coin and an empty drinking glass on a table. Challenge your friends to make the coin jump into the glass without touching either object. After their failed attempts, show how to do this by blowing on the coin from the side while the glass is set on its side as in the figure. You should practice this on your own to ensure success after a few tries. Explain the role of the Bernoulli principle in this demonstration.

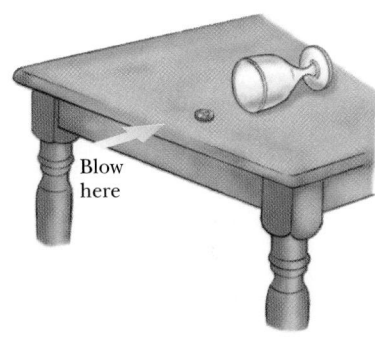

Applying Physics 5

The Bernoulli effect can be used to partially explain how a sailboat can accomplish the seemingly impossible task of sailing into the wind. How can this be done?

Explanation As shown in Figure 9.31, the wind blowing in the direction of the arrow causes the sail to billow out, taking on a shape similar to that of an airplane wing. The sail is oriented at an angle that is approximately halfway between the boat's axis (the keel axis) and the wind direction. The force of the wind on the sail, $\mathbf{F}_{wind}$, is due to the combined effect of the change in momentum of the wind as it bounces off the sail and the Bernoulli effect. The direction of this force is approximately perpendicular to the sail and tends to cause the boat to move sideways in the water. However, the keel (a broad vertical projection beneath the water) prevents this from happening because the water exerts a second force on the keel, $\mathbf{F}_{water}$, that is perpendicular to the keel axis. The resultant of the two forces, $\mathbf{F}_R$, is approximately in the forward direction as shown. The word almost is used here because a sailboat can move forward only when the wind direction is about 10 to 15° with respect to the forward direction. This means that in order to sail directly against the wind, a boat must follow a zigzag path, a procedure called tacking, so that the wind is always at some angle with respect to the direction of travel.

A P P L I C A T I O N

Sailing into the Wind.

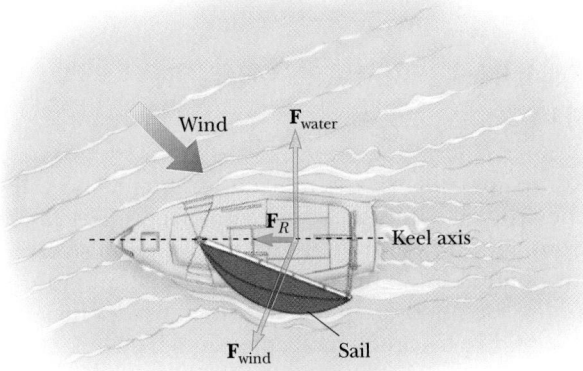

Figure 9.31
(Applying Physics 5)

Applying Physics 6

Consider the portion of a home plumbing system shown Figure 9.32. The water trap in the pipe below the sink captures a plug of water that prevents sewer gas from finding its way from the sewer pipe, up the sink drain, and into the home. Suppose the dishwasher is draining, so that water is moving to the right in the sewer pipe. What is the purpose of the vent, which is open to the air above the roof of the house? In which direction is air moving at the opening of the vent, upward or downward?

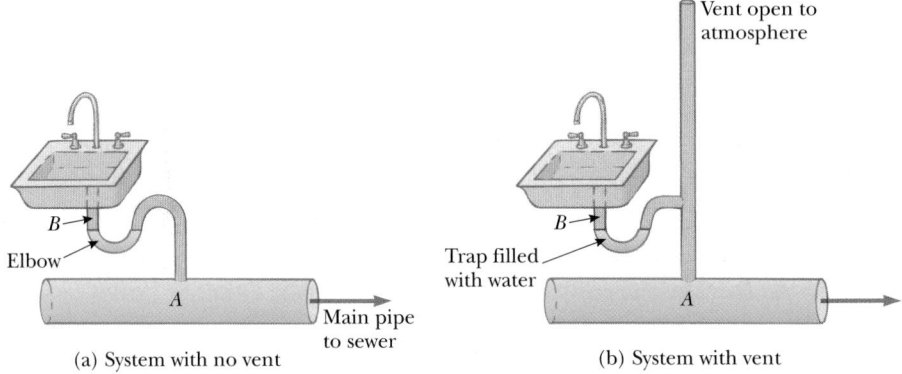

Figure 9.32   (Applying Physics 6)

Explanation   Let us imagine that the vent is not present, so that the drain pipe for the sink is simply connected through the trap to the sewer pipe. As water from the dishwasher moves to the right in the sewer pipe, the pressure in the sewer pipe is reduced below atmospheric pressure, according to Bernoulli's principle. The pressure at the drain in the sink is still at atmospheric pressure. Thus, this pressure differential can push the plug of water in the water trap of the sink down the drain pipe into the sewer pipe, removing it as a barrier to sewer gas. With the addition of the vent to the roof, the reduced pressure of the dishwasher water will result in air entering the vent pipe at the roof. This will keep the pressure in the vent pipe and the right-hand side of the sink drain pipe at a pressure close to atmospheric, so that the plug of water in the water trap will remain in the trap.

Optional Section

9.9   SURFACE TENSION, CAPILLARY ACTION, AND VISCOSITY

If you look closely at a dewdrop sparkling in the morning sunlight, you will find that the drop is spherical. The drop takes this shape because of a property of liquid surfaces called **surface tension.** In order to understand the origin of surface tension, consider a molecule at point A in a container of water, as in Figure 9.33. Although nearby molecules exert forces on this molecule, the net force on it is zero because it is completely surrounded by other molecules and hence is attracted equally in all directions. The molecule at B, however, is not attracted equally in all directions. Because there are no molecules above it to exert upward forces, the molecule at B is pulled toward the interior of the liquid. The contraction at the surface of the liquid ceases when the inward pull exerted on the surface molecules is balanced by the outward repulsive forces that arise from collisions with molecules in the interior of the liquid. **The net effect of this pull on all the surface molecules is to make the surface of the liquid contract and consequently to make the surface area of the liquid as small as possible.** Drops of water take on a spherical shape because a sphere has the smallest surface area for a given volume.

If you place a sewing needle very carefully on the surface of a bowl of water, you will find that the needle floats even though the density of steel is about eight times that of water. This also can be explained by surface tension. A close examination of the needle shows that it actually rests in a depression in the liquid surface, as shown in Figure 9.34. The water surface acts like an elastic membrane under tension. The weight of the needle produces a depression, thus increasing the surface area of the film. Molecular forces now act at all points along the depression in an attempt to restore the surface to its original horizontal position. The vertical components of these forces act to balance the force of gravity acting on the needle.

The **surface tension,** γ, in a film of liquid is defined as the ratio of the magnitude of the surface tension force, **F,** to the length along which the force acts:

$$\gamma \equiv \frac{F}{L} \qquad [9.18]$$

The SI units of surface tension are newtons per meter, and values for a few representative materials are given in Table 9.3.

The concept of surface tension can be thought of as the energy content of the fluid at its surface per unit surface area. To see that this is reasonable, we can manipulate the units of surface tension as

$$[\gamma] = \frac{N}{m} = \frac{N \cdot m}{m^2} = \frac{J}{m^2}$$

In general, **any equilibrium configuration of an object is one in which the energy is a minimum.** Consequently, a fluid will take on a shape such that its surface area is small as possible. For a given volume, a spherical shape is the one that has the smallest surface area; therefore, a drop of water takes on a spherical shape.

An apparatus used to measure the surface tension of liquids is shown in Figure 9.35. A circular wire with a circumference L is lifted from a body of liquid. The surface film clings to the inside and outside edges of the wire, holding back the wire and causing the spring to stretch. If the spring is calibrated, one can measure the force required to overcome the surface tension of the liquid. In this case, the surface tension is given by

$$\gamma = \frac{F}{2L}$$

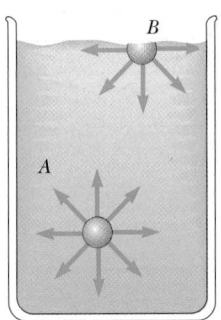

Figure 9.33 The net force on a molecule at A is zero because such a molecule is completely surrounded by other molecules. The net force on a surface molecule at B is downward because it is not completely surrounded by other molecules.

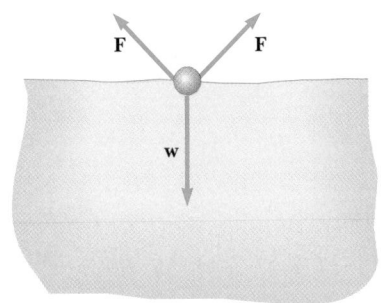

Figure 9.34 End view of a needle resting on the surface of water. The components of surface tension balance the weight force.

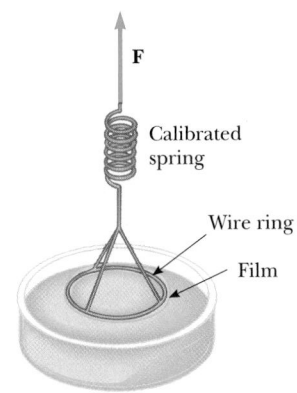

Figure 9.35 An apparatus for measuring the surface tension of liquids. The force on the wire ring is measured just before it breaks free of the liquid.

TABLE 9.3 Surface Tensions for Various Liquids

Liquid	$T(°C)$	Surface Tension (N/m)
Ethyl alcohol	20	0.022
Mercury	20	0.465
Soapy water	20	0.025
Water	20	0.073
Water	100	0.059

We must use $2L$ for the length because the surface film exerts forces on the inside and outside of the ring.

The surface tension of liquids decreases with increasing temperature. This occurs because the faster moving molecules of a hot liquid are not bound together as strongly as are those in a cooler liquid. Furthermore, certain ingredients added to liquids decrease surface tension. For example, soap or detergent decreases the surface tension of water. This reduction in surface tension makes it easier for soapy water to penetrate the cracks and crevices of your clothes to clean them better than plain water. An effect similar to this occurs in the lungs. The surface tissue of the air sacs in the lungs contains a fluid that has a surface tension of about 0.050 N/m. A liquid with a surface tension this high would make it very difficult for the lungs to expand as one inhales. However, as the area of the lungs increases with inhalation, the body secretes into the tissue a substance that gradually reduces the surface tension of the liquid. At full expansion, the surface tension of the lung fluid can drop to as low as 0.005 N/m.

APPLICATION

Surfactants.

EXAMPLE 9.10   Walking on Water

In this example, we shall illustrate how an insect is supported on the surface of water by surface tension. Let us assume that the insect's "foot" is spherical. When the insect steps onto the water with all six legs, a depression is formed in the water around each foot, as shown in Figure 9.36a. The surface tension of the water produces upward forces on the water that tend to restore the water surface to its normally flat shape. If the insect has a mass of 2.0×10^{-5} kg and if the radius of each foot is 1.5×10^{-4} m, find the angle θ.

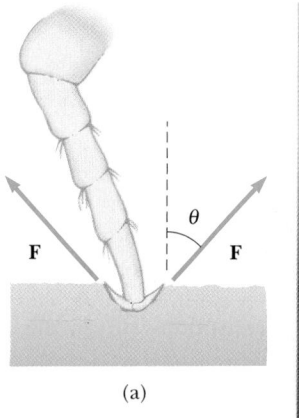

(a)

(b)

Figure 9.36   (Example 9.10) (a) One foot of an insect resting on the surface of water. (b) This water-strider resting on the surface of a lake is able to remain on the surface, rather than sink, because an upward surface tension force acts on each leg, which balances the weight of the insect. *(Hermann Eisenbeiss/Photo Researchers, Inc.)*

Solution   From the definition of surface tension, we can find the net force F directed tangential to the depressed part of the water surface:

$$F = \gamma L$$

The length L along which this force acts is equal to the distance around the insect's foot, $2\pi r$. (It is assumed that the insect depresses the water surface such that the radius of the depression is equal to the radius of the foot.) Thus,

$$F = \gamma 2\pi r$$

and the net vertical force is

$$F_v = \gamma 2\pi r \cos \theta$$

Because the insect has six legs, this upward force must equal one sixth the weight of the insect, assuming its weight is equally distributed on all six feet. Thus,

$$\gamma 2\pi r \cos \theta = \tfrac{1}{6} w = \tfrac{1}{6} mg$$

$$(1) \qquad \cos \theta = \frac{mg}{12\pi r\gamma} = \frac{(2.0 \times 10^{-5} \text{ kg})(9.80 \text{ m/s}^2)}{12\pi(1.5 \times 10^{-4} \text{ m})(0.073 \text{ N/m})}$$

$$\theta = \boxed{62°}$$

Note that if the weight of the insect were great enough to make the right side of (1) greater than unity, a solution for θ would be impossible because the cosine can never be greater than unity. Under these conditions, the insect would sink.

The Surface of Liquids

If you have ever closely examined the surface of water in a glass container, you may have noticed that the surface of the liquid near the walls of the glass curves upward as you move from the center to the edge, as shown in Figure 9.37a. However, if mercury is placed in a glass container, the mercury surface curves downward, as in Figure 9.37b. These surface effects can be explained by considering the forces between molecules. In particular, we must consider the forces that the molecules of the liquid exert on one another and the forces that the molecules of the glass surface exert on those of the liquid. In general terms, forces between like molecules, such as the forces between water molecules, are called *cohesive forces* and forces between unlike molecules, such as those of glass on water, are called *adhesive forces*.

Water tends to cling to the walls of the glass because the adhesive forces between the liquid molecules and the glass molecules are *greater* than the cohesive forces between the liquid molecules. In effect, the liquid molecules cling to the

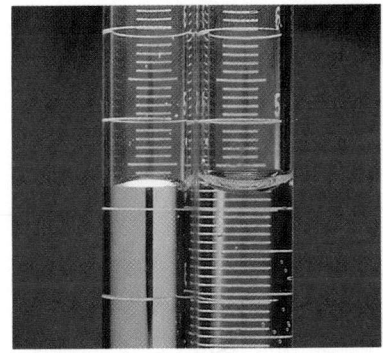

The surface of mercury *(left)* curves downward in a glass container, while the surface of water *(right)* curves upward as you move from the center to the edge. *(Charles D. Winters)*

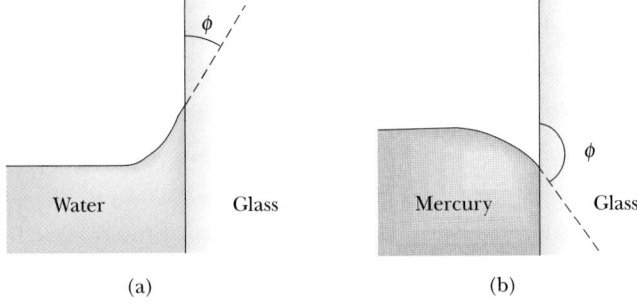

(a)   (b)

Figure 9.37 A liquid in contact with a solid surface. (a) For water, the adhesive force is greater than the cohesive force. (b) For mercury, the adhesive force is less than the cohesive force.

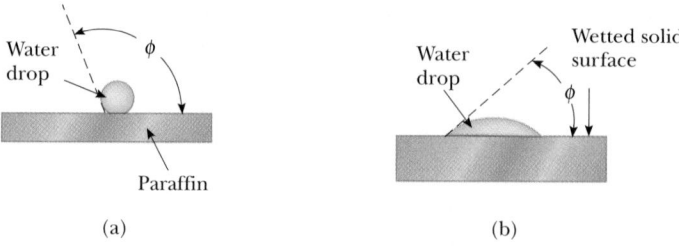

Figure 9.38 (a) The contact angle between water and paraffin is about 107°. In this case, the cohesive force is greater than the adhesive force. (b) When a chemical called a wetting agent is added to the water, it wets the paraffin surface, and $\phi < 90°$. In this case the adhesive force is greater than the cohesive force.

surface of the glass rather than fall back into the bulk of the liquid. When this condition prevails, the liquid is said to wet the glass surface. The surface of the mercury curves downward near the walls of the container because the cohesive forces between the mercury atoms are greater than the adhesive forces between mercury and glass. That is, a mercury atom near the surface is pulled more strongly toward other mercury atoms than toward the glass surface; hence mercury does not wet the glass surface.

The angle ϕ between the solid surface and a line drawn tangent to the liquid at the surface is called *the angle of contact* (Fig. 9.38a and 9.38b). Note that ϕ is less than 90° for any substance in which adhesive forces are stronger than cohesive forces and greater than 90° if cohesive forces predominate. For example, if a drop of water is placed on paraffin, the contact angle is approximately 107° (Fig. 9.38a). If certain chemicals, called wetting agents or detergents, are added to the water, the contact angle becomes less than 90°, as shown in Figure 9.38b. The addition of such substances to water is of value when one wants to ensure that water makes intimate contact with a surface and penetrates it. For this reason, detergents are added to water to wash clothes or dishes. On the other hand, it is often necessary to keep water from making intimate contact with a surface, as in waterproofing clothing, where a situation somewhat the reverse of that shown in Figure 9.38 is called for. The clothing is sprayed with a waterproofing agent, which changes ϕ from less than 90° to greater than 90°. Thus, the water beads up on the surface and does not easily penetrate the clothing.

Capillary Action

Capillary tubes are tubes in which the diameter of the opening is very small. In fact, the word *capillary* means "hair-like." If such a tube is inserted into a fluid for which adhesive forces dominate over cohesive forces, the liquid will rise into the tube, as shown in Figure 9.39. The rising of the liquid in the tube can be explained in terms of the shape of the surface of the liquid and in terms of the surface tension effects in the liquid. At the point of contact between liquid and solid, the upward force of surface tension is directed as shown in Figure 9.39. From Equation 9.18, the magnitude of this force is

$$F = \gamma L = \gamma(2\pi r)$$

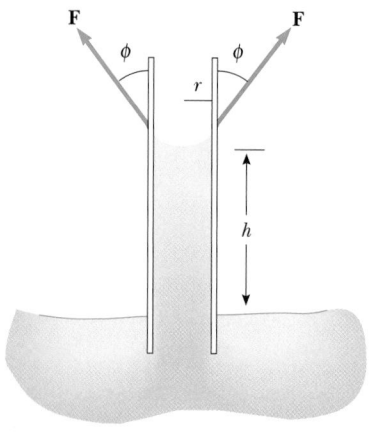

Figure 9.39 A liquid rises in a narrow tube because of capillary action, a result of surface tension and adhesive forces.

We use $L = 2\pi r$ here because the liquid is in contact with the surface of the tube at all points around its circumference. The vertical component of this force due to surface tension is

$$F_v = \gamma(2\pi r)(\cos \phi) \qquad [9.19]$$

In order for the liquid in the capillary tube to be in equilibrium, this upward force must be equal to the weight of the cylinder of water of height h inside the capillary tube. The weight of this water is

$$w = Mg = \rho Vg = \rho g \pi r^2 h \qquad [9.20]$$

Equating F_v in Equation 9.19 to w in Equation 9.20, we have

$$\gamma(2\pi r)(\cos \phi) = \rho g \pi r^2 h$$

Thus, the height to which water is drawn into the tube is

$$h = \frac{2\gamma}{\rho g r} \cos \phi \qquad [9.21]$$

If a capillary tube is inserted into a liquid in which cohesive forces dominate over adhesive forces, the level of the liquid in the capillary tube will be below the surface of the surrounding fluid, as shown in Figure 9.40. An analysis similar to that done above would show that the distance h the surface is depressed is given by Equation 9.21.

Capillary tubes are often used to draw small samples of blood from a needle prick in the skin. Capillary action must also be considered in the construction of concrete-block buildings because water seepage through capillary pores in the blocks or the mortar may cause damage to the inside of the building. To prevent this, the blocks are usually coated with a waterproofing agent either outside or inside the building. Water seepage through a wall is an undesirable effect of capillary action, but paper towels employ capillary action in a useful manner to absorb spilled fluids.

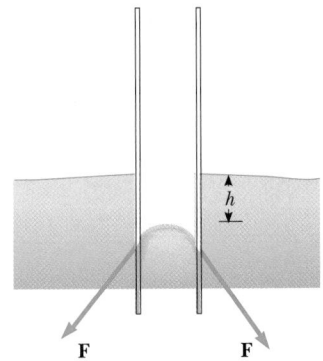

Figure 9.40   When cohesive forces between molecules of the liquid exceed adhesive forces, the liquid level in the capillary tube is below the surface of the surrounding fluid.

EXAMPLE 9.11   How High Does the Water Rise?

Find the height to which water would rise in a capillary tube with a radius equal to 5.0×10^{-5} m. Assume that the angle of contact between the water and the material of the tube is small enough to be considered zero.

Solution   The surface tension of water is 0.073 N/m. For a contact angle of 0°, we have $\cos \phi = \cos 0° = 1$, so that Equation 9.21 gives

$$h = \frac{2\gamma}{\rho g r} = \frac{2(0.073 \text{ N/m})}{(1.00 \times 10^3 \text{ kg/m}^3)(9.80 \text{ m/s}^2)(5.0 \times 10^{-5} \text{ m})} = \boxed{0.29 \text{ m}}$$

Viscosity

It is considerably easier to pour water out of a container than to pour syrup. This is because syrup has a higher viscosity than water. In a general sense, *viscosity refers to the internal friction of a fluid.* It is very difficult for layers of a viscous fluid to slide

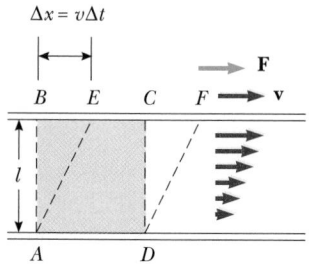

$\Delta x = v\Delta t$

Figure 9.41 A layer of liquid between two solid surfaces in which the lower surface is fixed and the upper surface moves to the right with a velocity **v.**

Coefficient of viscosity ▶

past one another. Likewise, it is difficult for one solid surface to slide past another if there is a highly viscous fluid, such as soft tar, between them.

To better understand the concept of viscosity, consider a liquid layer placed between two solid surfaces, as in Figure 9.41. The lower surface is fixed in position, and the top surface moves to the right with a velocity **v** under the action of an external force **F.** Because of this motion, a portion of the liquid is distorted from its original shape, *ABCD,* at one instant to the shape *AEFD* a moment later. You will recognize that the liquid has undergone a constantly increasing shear strain. Ideal fluids that have no internal friction forces between adjacent layers could not have their shape distorted. However, in viscous fluids, there are cohesive forces between molecules in the various layers that can lead to strains that change with time. By definition, the shear stress on the liquid is

$$\text{Shear stress} \equiv \frac{F}{A}$$

where A is the area of the top plate. Furthermore, the shear strain is defined as

$$\text{Shear strain} \equiv \frac{\Delta x}{L}$$

The velocity of the fluid changes from zero at the lower plate to **v** at the upper. Thus, in a time Δt, the fluid at the upper plate moves a distance $\Delta x = v \, \Delta t$. Therefore,

$$\frac{\text{Shear strain}}{\Delta t} = \frac{\Delta x/L}{\Delta t} = \frac{v}{L}$$

This equation states that the rate of change of the shearing strain is v/L.

The **coefficient of viscosity,** η (lowercase Greek letter eta), for the fluid is defined as the ratio of the shearing stress to the rate of change of the shear strain:

$$\eta \equiv \frac{FL}{Av} \qquad\qquad \text{[9.22]}$$

The SI units of viscosity are $N \cdot s/m^2$. You should note that the units of viscosity in many reference sources are often expressed in $dyne \cdot s/cm^2$, called 1 **poise** in honor of the French scientist J. L. Poiseuille (1799–1869). The relationship between the SI unit of viscosity and the poise is

$$1 \text{ poise} = 10^{-1} \, N \cdot s/m^2 \qquad\qquad \text{[9.23]}$$

Small viscosities are often expressed in centipoise (cp), where $1 \text{ cp} = 10^{-2}$ poise. The coefficients of viscosity for some common substances are listed in Table 9.4.

TABLE 9.4 The Viscosities of Various Fluids

Fluid	$T(°C)$	Viscosity $\eta(N \cdot s/m^2)$
Water	20	1.0×10^{-3}
Water	100	0.3×10^{-3}
Whole blood	37	2.7×10^{-3}
Glycerin	20	1500×10^{-3}
10-wt motor oil	30	250×10^{-3}

Poiseuille's Law

Figure 9.42 shows a section of a tube containing a fluid under a pressure P_1 at the left end and a pressure P_2 at the right. Because of this pressure difference, the fluid will flow through the tube. The rate of flow (volume per unit time) depends on the pressure difference $(P_1 - P_2)$, the dimensions of the tube, and the viscosity of the fluid. The result, known as **Poiseuille's law,** is

$$\text{Rate of flow} = \frac{\Delta V}{\Delta t} = \frac{(P_1 - P_2)(\pi R^4)}{8L\eta} \qquad [9.24]$$

 ◀ Poiseuille's law

where R is the radius of the tube, L is its length, and η is the coefficient of viscosity. We shall not attempt to derive this equation here because the methods of integral calculus are required. However, you should note that the equation does agree with common sense. That is, it is reasonable that the rate of flow should increase if the pressure difference across the tube or the tube radius increases. Likewise, the flow rate should decrease if the viscosity of the fluid or the length of the tube increases. Thus, the presence of R and the pressure difference in the numerator of Equation 9.24 and of L and η in the denominator makes sense.

From Poiseuille's law, we see that in order to maintain a constant flow rate, the pressure difference across the tube has to increase if the viscosity of the fluid increases. This is important when one considers the flow of blood through the circulatory system. The viscosity of blood increases as the number of red blood cells rises. Thus, blood with a high concentration of red blood cells requires greater pumping pressure from the heart to keep it circulating than does blood of lower red blood cell concentration.

Note that the flow rate varies as the radius of the tube raised to the fourth power. Consequently, if a constriction occurs in a vein or artery, the heart will have to work considerably harder in order to produce a higher pressure drop and hence to maintain the required flow rate.

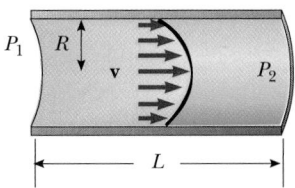

Figure 9.42 Velocity profile of a fluid flowing through a uniform pipe of circular cross-section. The rate of flow is given by Poiseuille's law. Note that the fluid velocity is greatest at the middle of the pipe.

EXAMPLE 9.12 A Blood Transfusion

A patient receives a blood transfusion through a needle of radius 0.20 mm and length 2.0 cm. The density of blood is 1050 kg/m³. The bottle supplying the blood is 0.50 m above the patient's arm. What is the rate of flow through the needle?

Solution The pressure differential between the level of the blood and the patient's arm is

$$P_1 - P_2 = \rho g h = (1050 \text{ kg/m}^3)(9.80 \text{ m/s}^2)(0.50 \text{ m}) = 5.15 \times 10^3 \text{ Pa}$$

Thus, the rate of flow, from Poiseuille's law, is

$$\frac{\Delta V}{\Delta t} = \frac{(P_1 - P_2)(\pi R^4)}{8L\eta}$$

$$= \frac{(5.15 \times 10^3 \text{ Pa})(\pi)(2.0 \times 10^{-4} \text{ m})^4}{8(2.0 \times 10^{-2} \text{ m})(2.7 \times 10^{-3} \text{ N} \cdot \text{s/m}^2)} = 6.0 \times 10^{-8} \text{ m}^3/\text{s}$$

Exercise How long will it take to inject 1 pint (500 cm³) of blood into the patient?

Answer 140 min

Reynolds Number

As we mentioned earlier, at sufficiently high velocities, fluid flow changes from simple streamline flow to turbulent flow—that is, flow characterized by a highly irregular motion of the fluid. It is found experimentally that the onset of turbulence in a tube is determined by a dimensionless factor called the **Reynolds number,** given by

Reynolds number ▶

$$RN = \frac{\rho v d}{\eta} \qquad [9.25]$$

where ρ is the density of the fluid, v is the average speed of the fluid along the direction of flow, d is the diameter of the tube, and η is the viscosity of the fluid. If RN is below about 2000, the flow of fluid through a tube is streamline; turbulence occurs if RN is above 3000. In the region between 2000 and 3000, the flow is unstable, meaning that the fluid can move in streamline flow but any small disturbance will cause its motion to change to turbulent flow.

EXAMPLE 9.13   Turbulent Flow of Blood

Determine the speed at which blood flowing through an artery of diameter 0.20 cm would become turbulent. Assume that the density of blood is 1.05×10^3 kg/m³ and that its viscosity is 2.7×10^{-3} N·s/m².

Solution   At the onset of turbulence, the Reynolds number is 3000. Thus, the speed of the blood would have to be

$$v = \frac{\eta(RN)}{\rho d} = \frac{(2.7 \times 10^{-3}\text{ N·s/m}^2)(3000)}{(1.05 \times 10^3\text{ kg/m}^3)(0.20 \times 10^{-2}\text{ m})} = \boxed{3.9\text{ m/s}}$$

Optional Section

9.10   TRANSPORT PHENOMENA

When a fluid flows through a tube, the basic mechanism that produces the flow is a difference in pressure across the ends of the tube. This pressure difference is responsible for the transport of a mass of fluid from one location to another. The fluid may also move from place to place because of a second mechanism, one that depends on a concentration difference between two points in the fluid, as opposed to a pressure difference. When the concentration (the number of molecules per unit volume) is higher at one location than at another, molecules will flow from the point where the concentration is high to the point where it is lower. The two fundamental processes involved in fluid transport resulting from concentration differences are called *diffusion* and *osmosis*. The following sections examine the nature and importance of these processes.

Diffusion

You can imagine what happens when someone wearing a strong shaving lotion or perfume strolls into a crowded room. All eyes turn to seek out the source of the delightful smell. The aroma spreads through the room by a process called diffusion.

In a diffusion process, molecules move from a region where their concentration is high to a region where their concentration is lower.

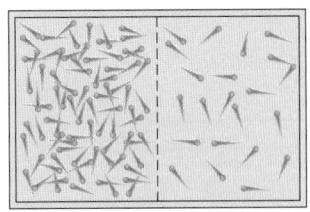

Figure 9.43 When the concentration of gas molecules on the left side of the container exceeds the concentration on the right side, there will be a net motion (diffusion) of molecules from left to right.

That is, the molecules of the lotion or perfume move from the source (near the person's face), where there are many molecules per unit volume, throughout the room, to regions where the concentration of these molecules is lower. Although the example used is one of diffusion in air, the process also occurs in liquids and, to a lesser extent, in solids. For example, if a drop of food coloring is placed in a glass of water, the coloring soon spreads throughout the liquid by diffusion. In either case, diffusion ceases when there is a uniform concentration at all locations in the fluid.

To understand why diffusion occurs, consider Figure 9.43, which represents a container in which a high concentration of molecules has been introduced into the left side. For example, this could be accomplished by releasing a few drops of perfume into the left side of the container. The dashed line in Figure 9.43 represents an imaginary barrier separating the region of high concentration from the region of lower concentration. Because the molecules are moving with high speeds in random directions, many of them will cross the imaginary barrier moving from left to right. Very few molecules of perfume will pass through this area moving from right to left simply because there are very few of them on the right side of the container at any instant. Thus, there will always be a *net* movement from the region where there are many molecules to the region where there are fewer molecules. For this reason, the concentration on the left side of the container will decrease in time and that on the right side will increase. Once a concentration equilibrium has been reached, there will be no *net* movement across the cross-sectional area. That is, when the concentration is the same on both sides, the number of molecules diffusing from right to left in a given time interval will equal the number moving from left to right in the same time interval.

The basic equation for diffusion is **Fick's law,** which in equation form is

$$\text{Diffusion rate} = \frac{\text{mass}}{\text{time}} = \frac{\Delta M}{\Delta t} = DA\left(\frac{C_2 - C_1}{L}\right) \qquad \text{[9.26]}$$

◀ Fick's law

where D is a constant of proportionality. The left side of this equation is called the diffusion rate and is a measure of the mass being transported per unit time. This equation says that

the rate of diffusion is proportional to the cross-sectional area A and to the change in concentration per unit distance, $(C_2 - C_1)/L$, which is called the concentration gradient.

The concentrations C_1 and C_2 are measured in kilograms per cubic meter. The proportionality constant D is called the **diffusion coefficient** and has units of square meters per second. Table 9.5 lists diffusion coefficients for a few substances.

TABLE 9.5	Diffusion Coefficients for Various Substances at 20°C
Substance	$D(m^2/s)$
Oxygen through air	6.4×10^{-5}
Oxygen through tissue	1×10^{-11}
Oxygen through water	1×10^{-9}
Sucrose through water	5×10^{-10}
Hemoglobin through water	76×10^{-11}

The Size of Cells and Osmosis

Diffusion through cell membranes is extremely vital in carrying oxygen to the cells of the body and in removing carbon dioxide and other waste products from them. Oxygen is required by the cells for those metabolic processes in which substances are either synthesized or broken down. In such metabolic processes, the cell uses up oxygen and produces carbon dioxide as a by-product. A fresh supply of oxygen diffuses from the blood, where its concentration is high, into the cell, where its concentration is low. Likewise, carbon dioxide diffuses from the cell into the blood, where it is in lower concentration. Water, ions, and other nutrients also pass into and out of cells by diffusion.

A common characteristic of cells in all plants and animals is their extremely small size. The adult human body contains literally trillions of cells. In order to understand why cells are so small, we must consider the relationship between the surface area of an object and its volume.

Let us consider a cube 2 cm on a side. The area of one of its faces is 2 cm $\times$ 2 cm = 4 cm^2, and because a cube has six sides, the total surface area is 24 cm^2. Its volume is 2 cm $\times$ 2 cm $\times$ 2 cm = 8 cm^3. Hence, the ratio of surface area to volume is 24/8 = 3. Now consider a larger cube, one measuring 3 cm on a side. Repeating the calculations gives us a surface area of 54 cm^2 and a volume of 27 cm^3. In this case, the ratio of surface area to volume is 54/27 = 2. Thus, we see that as the size of an object decreases, the ratio of its surface area to its volume increases. This, of course, says that a small cell has a larger surface-area-to-volume ratio than a large cell. But how does this pertain to the operation of a cell?

A cell can function properly only if it can (a) rapidly receive vital substances such as oxygen and (b) rapidly eliminate waste products. If such substances are to readily move into and out of cells, the cells should have a large surface area. However, if the volume of the cell is too large, it could take a considerable period of time for the nutrients to diffuse into the interior of the cell where they are needed. Under optimum conditions, the surface area of the cell should be large enough so that the exposed membrane area can exchange materials effectively while at the same time the volume should be small enough so that materials can reach or leave particular locations rapidly. To reach these optimum conditions, a small cell with its high surface-area-to-volume ratio is necessary.

As we have seen, the movement of material through cell membranes is necessary for the efficient functioning of cells. The diffusion of material through a membrane is partially determined by the size of the pores (holes) in the membrane wall. That is, small molecules, such as water, may pass through the pores easily and larger molecules, such as sugar, may pass through only with difficulty or not at all. A membrane that allows passage of some molecules but not others is called a selectively permeable membrane.

> **Osmosis** is defined as the movement of water from a region where its concentration is high, across a selectively permeable membrane, into a region where its concentration is lower.

As in the case of diffusion, osmosis continues until the concentrations on the two sides of the membrane are equal. Osmosis is often described simply as the diffusion of water across a membrane.

To understand the effect of osmosis on living cells, let us consider a particular cell in the body that contains a sugar concentration of 1%. (That is, 1 g of sugar is dissolved in enough water to make 100 ml of solution.) Now assume that this cell is immersed in a 5% sugar solution (5 g of sugar dissolved in enough water to make 100 ml). In such a situation, water would diffuse from inside the cell, where its concentration is higher, across the cell wall membrane, to the outside solution, where the concentration of water is lower. This loss of water from the cell would cause it to shrink and perhaps become damaged through dehydration. If the concentrations were reversed, water would diffuse into the cell, causing it to swell and perhaps burst. It should be obvious from this description that normal osmotic relationships must be maintained in the body. If solutions are introduced into the body intravenously, care must be taken to ensure that these solutions do not disturb the osmotic balance of the body because such a disturbance could lead to cell damage. For example, if 9% saline solution surrounds a red blood cell, the cell will shrink. On the other hand, if the saline solution is about 1%, the cell will eventually burst.

In the body, blood is cleansed of impurities by osmosis as it flows through the kidneys (see Fig. 9.44a). Arterial blood first passes through a bundle of capillaries known as a glomerulus where most of the waste products along with some essential salts and minerals are removed from the blood. From the glomerulus, a narrow tube emerges that is in intimate contact with other capillaries throughout its length. In its passage through the tubules, most of the essential elements are returned to the blood, and the waste products are not allowed to re-enter and are eventually removed as urine.

If the kidneys fail, an artificial kidney or a dialysis machine can filter the blood. Figure 9.44b shows how this is done. Blood from an artery in the arm is mixed with heparin, a blood thinner, and allowed to pass through a tube covered with a semipermeable membrane. This tubing is immersed in a bath of a dialysate fluid with the same chemical composition as purified blood. Waste products from the blood enter the dialysate by diffusion through the membrane walls. The filtered blood is then returned to a vein.

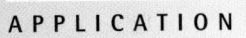

APPLICATION

Kidney Function and Dialysis.

Motion Through a Viscous Medium

When an object falls through air, its motion is impeded by the force of air resistance. In general, this force is dependent on the shape of the falling object and on its velocity. This viscous drag acts on all falling objects, but the exact details of the

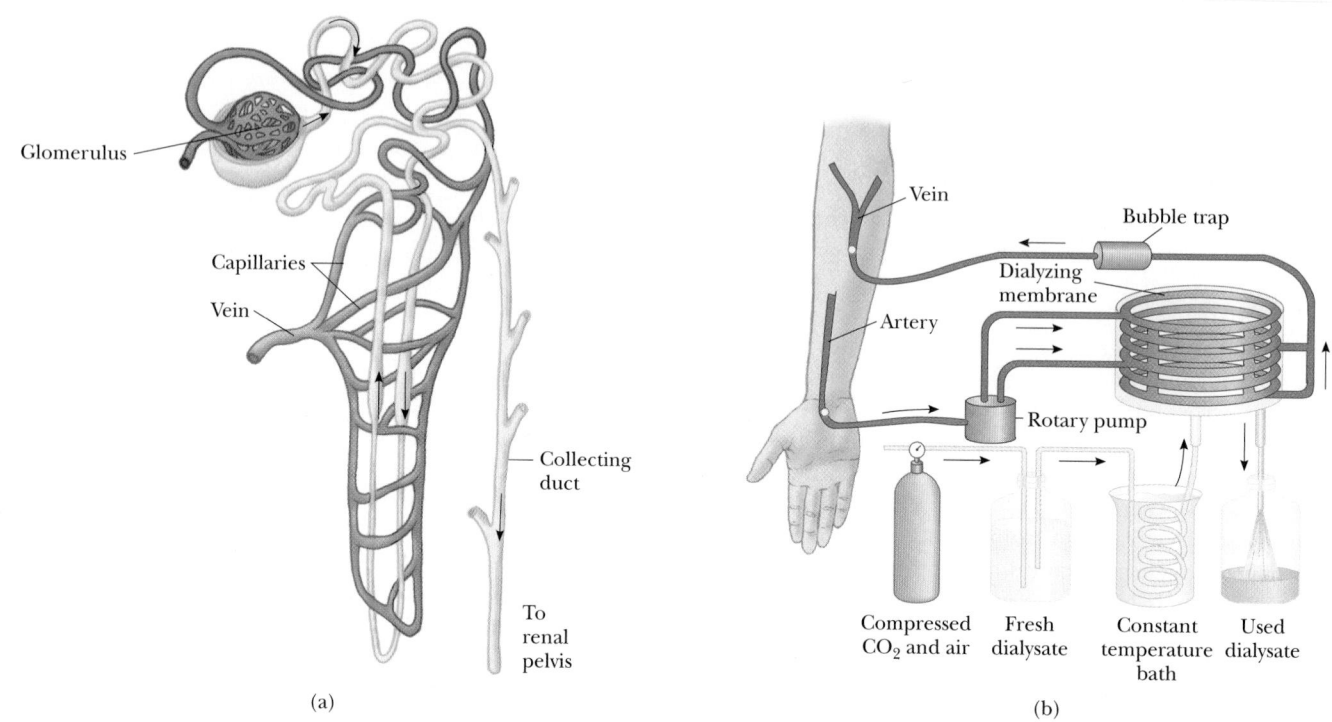

(a)

(b)

Figure 9.44 (a) Diagram of a single nephron in the human kidney system. (b) An artificial kidney.

motion can be calculated only for a few cases in which the object has a simple shape, such as a sphere. In this section, we shall examine the motion of a tiny spherical object falling slowly through a viscous medium.

In 1845 a scientist named George Stokes found that the magnitude of the resistive force on a very small spherical object of radius r falling slowly through a fluid of viscosity η with speed v is given by

$$F_r = 6\pi\eta r v \qquad [9.27]$$

This equation, called **Stokes's law,** has many important applications. For example, it describes the sedimentation of particulate matter in blood samples. It was used by Robert Millikan (1886–1953) to calculate the radius of charged oil droplets falling through air. From this, Millikan was ultimately able to determine the smallest known unit of electric charge. Millikan was awarded the Nobel prize in 1923 for this pioneering work on elemental charge.

As a sphere falls through a viscous medium, three forces act on it, as shown in Figure 9.45: $\mathbf{F}_r$ is the force of frictional resistance, $\mathbf{B}$ is the buoyant force of the fluid, and $\mathbf{w}$ is the force of gravity acting on the sphere, whose magnitude is given by

$$w = \rho g V = \rho g(\tfrac{4}{3}\pi r^3)$$

where ρ is the density of the sphere and $\tfrac{4}{3}\pi r^3$ is its volume. According to Archimedes's principle, the magnitude of the buoyant force is equal to the weight of the fluid displaced by the sphere:

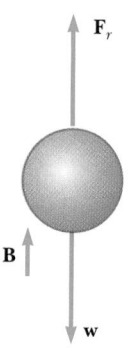

Figure 9.45 A sphere falling through a viscous medium. The forces acting on the sphere are the resistive frictional force $\mathbf{F}_r$, the buoyant force $\mathbf{B}$, and the weight of the sphere $\mathbf{w}$.

$$B = \rho_f g V = \rho_f g(\tfrac{4}{3}\pi r^3)$$

where ρ_f is the density of the fluid.

At the instant the sphere begins to fall, the force of frictional resistance is zero because the speed of the sphere is zero. As it accelerates, the speed increases and so does $\mathbf{F}_r$. Finally, at a speed called the **terminal speed** v_t, *the net force goes to zero.* This occurs when the net upward force balances the downward force of gravity. Hence, the sphere reaches terminal speed when ◀ Terminal speed

$$F_r + B = w$$

or

$$6\pi\eta r v_t + \rho_f g(\tfrac{4}{3}\pi r^3) = \rho g(\tfrac{4}{3}\pi r^3)$$

When this is solved for v_t, we get

$$v_t = \frac{2r^2 g}{9\eta}(\rho - \rho_f) \qquad\qquad [9.28]$$

EXAMPLE 9.14 A Falling Pearl

A pearl of density 2.0×10^3 kg/m^3 and radius 2.0 mm falls through a liquid shampoo of density 1.4×10^3 km/m^3 and viscosity 0.50×10^{-3} N·s/m^2. Find the terminal speed of the pearl.

Solution   Substituting the given values into Equation 9.28, we have

$$v_t = \frac{2r^2 g}{9\eta}(\rho - \rho_f)$$

$$= \frac{2(2.0 \times 10^{-3}\ \text{m})^2 (9.80\ \text{m/s}^2)}{9(0.50 \times 10^{-3}\ \text{N·s/m}^2)}(2.0 \times 10^3\ \text{kg/m}^3 - 1.4 \times 10^3\ \text{kg/m}^3)$$

$$= \boxed{1.1 \times 10^{-2}\ \text{m/s}}$$

By spreading their arms and legs out from their bodies while keeping the planes of their bodies parallel to the ground, skydivers experience maximum air drag. The forces acting on this skydiver are his weight (the force of gravity) and the force of air resistance. When these two forces balance each other, the skydiver has zero acceleration and reaches a terminal speed of about 60 m/s. *(Heinz Fischer/The Image Bank)*

Sedimentation and Centrifugation

If an object is not spherical, we can still use the basic approach just described to determine its terminal speed. The only difference will be that we shall not be able to use Stokes's law for the resistive force. Instead, let us assume that the resistive force has a magnitude given by $F_r = kv$, where k is a coefficient of frictional resistance that must be determined experimentally. As we discussed previously, the object reaches its terminal speed when the downward force of gravity is balanced by the net upward force or

$$w = B + F_r \qquad \text{[9.29]}$$

where B is the buoyant force, given by $B = \rho_f g V$.

We can use the fact that volume, V, of the displaced fluid is related to the density of the falling object, ρ, by $V = m/\rho$. Hence, we can express the buoyant force as

$$B = \frac{\rho_f}{\rho} mg$$

Let us substitute this expression for B and $F_r = kv_t$ into Equation 9.29 (terminal speed condition):

$$mg = \frac{\rho_f}{\rho} mg + kv_t$$

or

$$v_t = \frac{mg}{k}\left(1 - \frac{\rho_f}{\rho}\right) \qquad \text{[9.30]}$$

The terminal speed for particles in biological samples is usually quite small. For example, the terminal speed for blood cells falling through plasma is about 5 cm/h in the gravitational field of the Earth. The terminal speeds for the molecules that make up a cell are many orders of magnitude smaller than this because of their much smaller mass. The speed at which materials fall through a fluid is called the *sedimentation rate*. This number is often important in clinical analysis.

It is often desired to increase the sedimentation rate in a fluid. A common method used to accomplish this is to increase the effective acceleration g that appears in Equation 9.30. A fluid containing various biological molecules is placed in a centrifuge and whirled at very high angular speeds (Fig. 9.46). Under these conditions, the particles experience a large radial acceleration, $a_c = v^2/r = \omega^2 r$, which is much greater than the free-fall acceleration, and so we can replace g in Equation 9.30 by $\omega^2 r$:

$$v_t = \frac{m\omega^2 r}{k}\left(1 - \frac{\rho_f}{\rho}\right) \qquad \text{[9.31]}$$

This equation indicates that those particles having the greatest mass will have the largest terminal speed. Therefore, the most massive particles will settle out on the bottom of a test tube first.

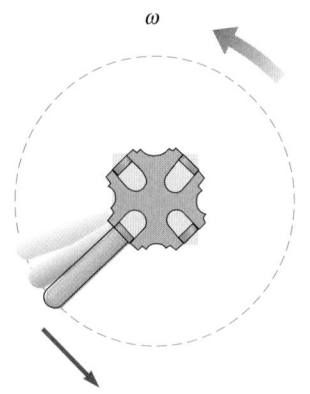

Figure 9.46 Simplified diagram of a centrifuge (top view).

APPLICATION

Centrifugation.

EXAMPLE 9.15 The Spinning Test Tube

A centrifuge rotates at 50 000 rev/min, which corresponds to an angular frequency of 5240 rad/s (a typical speed). A test tube placed in this device has its top 5.0 cm from the axis of rotation and its bottom 13 cm from this axis. Find the effective value of g at the midpoint of the test tube, which corresponds to a distance 9.0 cm from the axis of rotation.

Solution The acceleration experienced by the particles of the tube at a distance $r = 9.0$ cm from the axis of rotation is given by

$$a_c = \omega^2 r = \left(5240\,\frac{\text{rad}}{\text{s}}\right)^2 (9.0 \times 10^{-2}\,\text{m}) = 2.5 \times 10^6\,\text{m/s}^2$$

Exercise If the mass of the contents of the test tube is 15 g, find the force that the bottom of the tube must exert on the contents of the tube to provide this centripetal acceleration. Assume a centripetal acceleration equal to that found at the midpoint of the tube.

Answer 3.7×10^4 N, or about 8000 lb! (Because of such large forces, the base of the tube in a centrifuge must be rigidly supported to keep the glass from shattering.)

SUMMARY

Matter is normally classified as being in one of three states: solid, liquid, or gaseous.

The elastic properties of a solid can be described using the concepts of stress and strain. **Stress** is related to the force producing a deformation; **strain** is a measure of the degree of deformation. Stress is proportional to strain, and the constant of proportionality is the **elastic modulus:**

$$\text{Elastic modulus} \equiv \frac{\text{stress}}{\text{strain}} \qquad [9.1]$$

Three common types of deformation are, (1) the resistance of a solid to elongation or compression, characterized by **Young's modulus, Y;** (2) the resistance to displacement of the faces of a solid sliding past each other, characterized by the **shear modulus, S;** (3) the resistance of a solid or liquid to a volume change, characterized by the **bulk modulus, B.**

In the SI system, pressure is expressed in pascals (Pa), where $1\ \text{Pa} \equiv 1\ \text{N/m}^2$.

The **density, ρ,** of a substance of uniform composition is its mass per unit volume — kilograms per cubic meter (kg/m^3) in the SI system:

$$\rho \equiv \frac{M}{V} \qquad [9.6]$$

The **pressure, P,** in a fluid is the force per unit area that the fluid exerts on an object immersed in it:

$$P \equiv \frac{F}{A} \qquad [9.7]$$

The pressure in a fluid varies with depth, h, according to the expression

$$P = P_0 + \rho g h \qquad [9.10]$$

where P_0 is atmospheric pressure (1.01×10^5 Pa) and ρ is the density of the fluid.

Pascal's principle states that, when pressure is applied to an enclosed fluid, the pressure is transmitted undiminished to every point of the fluid and to the walls of the containing vessel.

When an object is partially or fully submerged in a fluid, the fluid exerts an upward force, called the **buoyant force,** on the object. According to **Archimedes's principle,** the magnitude of the buoyant force is equal to the weight of the fluid displaced by the object.

Certain aspects of a fluid in motion can be understood by assuming that the fluid is nonviscous and incompressible and that its motion is in a steady state with no turbulence.

1. The flow rate through the pipe is a constant, which is equivalent to stating that the product of the cross-sectional area, A, and the speed, v, at any point is constant.

$$A_1 v_1 = A_2 v_2 \qquad [9.14]$$

This relation is referred to as the **equation of continuity.**

2. The sum of the pressure, the kinetic energy per unit volume, and the potential energy per unit volume has the same value at all points along a streamline:

$$P + \tfrac{1}{2}\rho v^2 + \rho g y = \text{constant} \qquad [9.16]$$

This is known as **Bernoulli's equation.**

MULTIPLE-CHOICE QUESTIONS

1. A test is performed on copper wire of radius 1.00 mm to determine how it will withstand the buildup of ice on it when used as a power line. It is found that a force of 300 N will cause it to stretch from a length of 100.00 m to a length of 100.11 m, and that when the weight is removed, the wire snaps back to its original length. The stress applied to the wire is _____, the strain is _____ and the elastic modulus is _____, in SI units.
   (a) 8.68×10^{10}, 0.11×10^{-2}, 9.55×10^7
   (b) 9.55×10^7, 0.11×10^{-2}, 8.68×10^{10}
   (c) 9.55×10^7, 8.68×10^{10}, 0.11×10^{-2}
   (d) 0.11×10^{-2}, 8.68×10^{10}, 9.55×10^7
   (e) 8.68×10^{10}, 9.55×10^7, 0.11×10^{-2}

2. As the experiment of Question 1 continues, it is found out that when a certain weight is added to the wire, it ceases to snap back to its original length. In this case, one has exceeded the _____.
   (a) breaking point   (b) stress limit   (c) strain limit
   (d) elastic limit   (e) elasticity

3. According to Poiseuille's law, the most effective way to increase the rate of flow of a transfusing fluid into a patient is to
   (a) increase the pressure at the level of the arm by raising the height of the bag.
   (b) dissolve the fluid in a less viscous material.
   (c) increase the radius of the needle.
   (d) increase the length of tubing from bag to needle.

4. The distance from the feet to the heart for an individual is 1.20 m, and the density of blood is 1.06×10^3 kg/m^3. Find the difference in blood pressure between the level of the heart and the level of the feet.
   (a) 1270 Pa   (b) 1.06×10^3 Pa   (c) 1.25×10^4 Pa
   (d) 1.00 atm

5. A solid rock, suspended in air by a spring scale, has a measured mass of 0.80 kg. When the rock is submerged in water, the scale reads 0.70 kg. What is the density of the rock?
   (a) 4.5×10^3 kg/m^3 (b) 3.5×10^3 kg/m^3
   (c) 1.2×10^3 kg/m^3 (d) 2.7×10^3 kg/m^3
   (e) 8.0×10^3 kg/m^3

CONCEPTUAL QUESTIONS

1. A woman wearing high-heeled shoes is invited into a home in which the kitchen has vinyl floor covering. Why should the homeowner be concerned?

2. Figure Q9.2 shows aerial views from directly above two dams. Both dams are equally long (the vertical dimension in the diagram) and equally deep (into the page in the diagram). The dam on the left holds back a very large lake, and the dam on the right holds back a narrow river. Which dam has to be built more strongly?

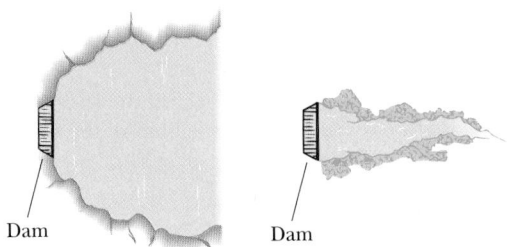

Dam   Dam

Figure Q9.2

3. A typical silo on a farm has many bands wrapped around its perimeter, as shown in Figure Q9.3. Why is the spacing between successive bands smaller at the lower portions of the silo?

Figure Q9.3

4. During inhalation, the pressure in the lungs is slightly less than external pressure and the muscles controlling exhalation are relaxed. Underwater, the body equalizes internal and external pressures. Discuss the condition of the muscles if a person underwater is breathing through a snorkel. Would a snorkel work in deep water?

5. Atmospheric pressure varies from day to day. Does a ship float higher in the water on a high pressure day compared to a low pressure day?

6. Lead has a greater density than iron, and both are denser than water. Is the buoyant force on a solid lead object greater than, less than, or equal to the buoyant force on a solid iron object of the same dimensions?

7. Suppose a damaged ship just barely floats in the ocean after a hole in its hull has been sealed. It is pulled by a tugboat toward shore and into a river, heading toward a drydock for repair. As it is pulled up the river, it sinks. Why?

8. Will an ice cube float higher in water or in an alcoholic beverage?

9. A pound of styrofoam and a pound of lead have the same weight. If they are placed on an equal arm balance, will it balance?

10. An ice cube is placed in a glass of water. What happens to the level of the water as the ice melts?

11. A person in a boat floating in a small pond throws an anchor overboard. Does the level of the pond rise, fall, or remain the same?

12. Prairie dogs live in underground burrows with at least two entrances. They ventilate their burrows by building a mound over one entrance, which is open to a stream of air, as in the photograph (Figure Q9.12). A second entrance at ground level is open to almost stagnant air. How does this construction create an air flow through the burrow?

Figure Q9.12 *(Pamela Zilly/The Image Bank)*

13. Will a ship ride higher in an inland lake or in the ocean? Why?

14. A barge is carrying a load of gravel along a river. It approaches a low bridge and the captain realizes that the top of the pile of gravel is not going to make it under. The captain orders the crew to quickly shovel gravel from the pile into the water. Is this a good decision?

15. When you are driving a small car on the freeway and a truck passes you at high speed, you feel pulled toward the truck. Why?

16. Tornadoes and hurricanes often lift the roofs of houses. Use the Bernoulli effect to explain why. Why should you keep your windows open under these conditions?

PROBLEMS

1, 2, 3 = straightforward, intermediate, challenging   □ = full solution available in Study Guide/Student Solutions Manual   ![] = Core Concepts Workbook
WEB = solution posted at **http://www.harcourtcollege.com/physics/cptech**   ![] = biomedical application   ![] = Interactive Physics

Section 9.2 The Deformation of Solids

1. The heels on a pair of women's shoes have radii of 0.50 cm at the bottom. If 30% of the weight of a woman weighing 480 N is supported by each heel, find the stress on each heel.

2. Find the minimum diameter of an 18.0-m-long steel wire that will stretch no more than 9.00 mm when a load of 380 kg is hung on the lower end.

3. For safety in climbing, a mountaineer uses a nylon rope that is 50 m long and 1.0 cm in diameter. When supporting a 90-kg climber, the rope elongates 1.6 m. Find its Young's modulus.

4. If the elastic limit of steel is 5.0×10^8 Pa, determine the minimum diameter a steel wire can have if it is to support a 70-kg circus performer without its elastic limit being exceeded.

5. Bone has a Young's modulus of about 14.5×10^9 Pa. Under compression, it can withstand a stress of about 160×10^6 Pa before breaking. Assume that a femur (thigh bone) is 0.50 m long and calculate the amount of compression this bone can withstand before breaking.

6. The distortion of the Earth's crustal plates is an example of shear on a large scale. A particular crustal rock has a shear modulus of 1.5×10^{10} Pa. What shear stress is involved when a 10-km layer of this rock is sheared through a distance of 5.0 m?

7. A child slides across a floor in a pair of rubber-soled shoes. The friction force acting on each foot is 20 N, the cross-sectional area of each foot is 14 cm², and the height of the soles is 5.0 mm. Find the horizontal distance traveled by the sheared face of the sole. The shear modulus of the rubber is 3.0×10^6 Pa.

8. A high-speed lifting mechanism supports an 800-kg mass with a steel cable 25.0 m long and 4.00 cm² in cross-sectional area. (a) Determine the elongation of the cable. (b) By what additional amount does the cable increase in length if the mass is accelerated upward at a rate of 3.0 m/s²? (c) What is the greatest mass that can be accelerated upward at 3.0 m/s² if the stress in the cable is not to exceed the elastic limit of the cable, 2.2×10^8 Pa?

Section 9.3 Density and Pressure

9. A 50.0-kg ballet dancer stands on her toes during a performance with four square inches (26.0 cm²) in contact with the floor. What is the pressure exerted by the floor over the area of contact (a) if the dancer is stationary, and (b) if the dancer is leaping upwards with an acceleration of 4.00 m/s²?

10. The four tires of an automobile are inflated to a gauge pressure of 2.0×10^5 Pa. Each tire has an area of 0.024 m² in contact with the ground. Determine the weight of the automobile.

11. A 70-kg man in a 5.0-kg chair tilts back so that all the weight is balanced on two legs of the chair. Assume that each leg makes contact with the floor over a circular area with a radius of 1.0 cm, and find the pressure exerted on the floor by each leg.

12. If 1.0 m³ of concrete weighs 5.0×10^4 N, what is the height of the tallest cylindrical concrete pillar that will not collapse under its own weight? The compression strength of concrete (the maximum pressure that can be exerted on the base of the structure) is 1.7×10^7 Pa.

Section 9.4 Variation of Pressure with Depth

Section 9.5 Pressure Measurements

13. Water is to be pumped to the top of the Empire State Building, which is 1200 ft high. What gauge pressure is needed in the water line at the base of the building to raise the water to this height?

14. A collapsible plastic bag (Fig. P9.14) contains a glucose solution. If the average gauge pressure in the artery is 1.33×10^4 Pa, what must be the minimum height, h, of the bag in order to infuse glucose into the artery? Assume that the specific gravity of the solution is 1.02.

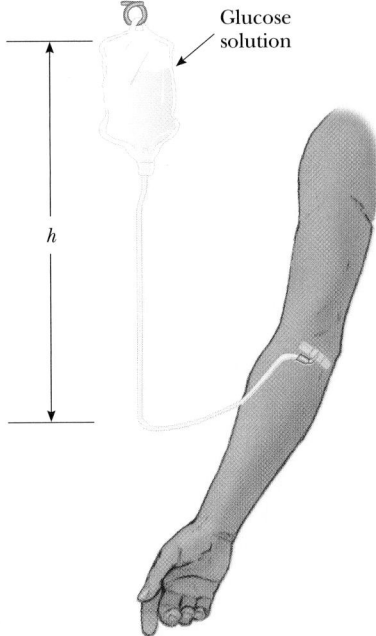

Figure P9.14

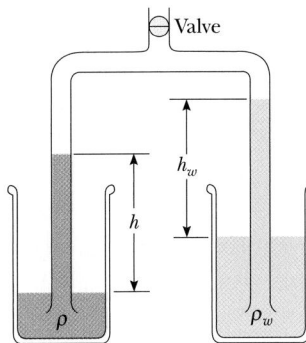

Figure P9.16

15. Air is trapped above liquid ethanol in a rigid container, as shown in Figure P9.15. If the air pressure above the liquid is 1.10 atm, determine the pressure inside a bubble 4.0 m below the surface of the ethanol.

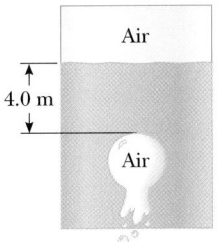

Figure P9.15

16. One method of measuring the density of a liquid is illustrated in Figure P9.16. One side of the U-tube is in the liquid being tested; the other side is in water of density ρ_w. When the air is partially removed at the upper part of the tube, show that the density of the liquid on the left is given by $\rho = (h_w/h)\rho_w$.

WEB **17.** A container is filled to a depth of 20.0 cm with water. On top of the water floats a 30.0-cm-thick layer of oil with specific gravity 0.700. What is the absolute pressure at the bottom of the container?

18. Piston 1 in Figure P9.18 has a diameter of 0.25 in.; piston 2 has a diameter of 1.5 in. In the absence of friction, determine the force, **F,** necessary to support the 500-lb weight.

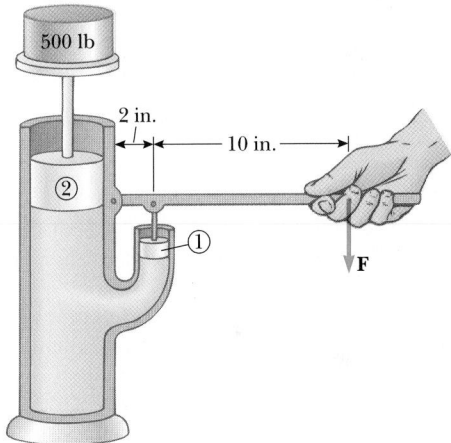

Figure P9.18

19. Figure P9.19 shows the essential parts of a hydraulic brake system. The area of the piston in the master cylin-

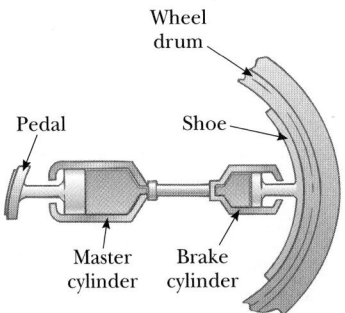

Figure P9.19

der is 6.4 cm² and that of the piston in the brake cylinder is 1.75 cm². The coefficient of friction between shoe and wheel drum is 0.50. If the wheel has a radius of 34 cm, determine the frictional torque about the axle when a force of 44 N is exerted on the brake pedal.

Section 9.6 Buoyant Forces and Archimedes's Principle

20. A frog in a hemispherical pod finds that he just floats without sinking in a fluid of density 1.35 g/cm³. If the pod has a radius of 6.00 cm and negligible mass, what is the mass of the frog? (See Fig. P9.20.)

Figure P9.20

21. A small ferry boat is 4.00 m wide and 6.00 m long. When a loaded truck pulls onto it, the boat sinks an additional 4.00 cm into the water. What is the weight of the truck?

22. The density of ice is 920 kg/m³, and that of seawater is 1030 kg/m³. What fraction of the total volume of an iceberg is exposed?

23. An empty rubber balloon has a mass of 0.0120 kg. The balloon is filled with helium at a density of 0.181 kg/m³. At this density the balloon is spherical with a radius of 0.500 m. If the filled balloon is fastened to a vertical line, what is the tension in the line?

24. A light spring of constant $k = 160$ N/m rests vertically on the bottom of a large beaker of water (Fig. P9.24a). A 5.00-kg block of wood (density = 650 kg/m³) is connected to the spring and the mass-spring system is allowed to come to static equilibrium (Fig. P9.24b). What is the elongation, ΔL, of the spring?

25. A sample of an unknown material weighs 300 N in air and 200 N when immersed in alcohol of specific gravity 0.700. What are (a) the volume and (b) the density of the material?

26. An object weighing 300 N in air is immersed in water after being tied to a string connected to a balance. The scale now reads 265 N. Immersed in oil, the object weighs 275 N. Find (a) the density of the object and (b) the density of the oil.

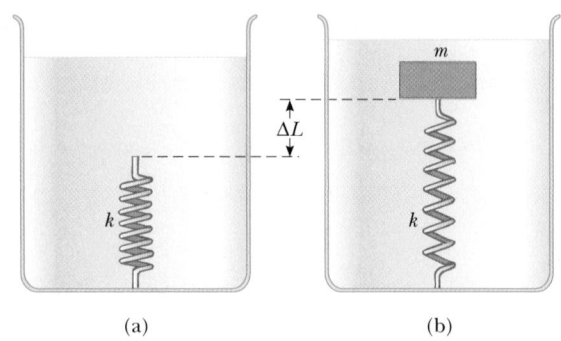

Figure P9.24

27. A thin spherical shell of mass 0.400 kg and diameter 0.200 m is filled with alcohol ($\rho = 806$ kg/m³). It is then released from rest on the bottom of a pool of water. Find the acceleration of the alcohol filled shell as it rises toward the surface of the water.

28. A hollow brass tube (diam. = 4.00 cm) is sealed at one end and loaded with lead shot to give a total mass of 0.200 kg. When the tube is floated in pure water, what is the depth, z, of its bottom end?

29. A rectangular air mattress is 2.0 m long, 0.50 m wide, and 0.08 m thick. If it has a mass of 2.0 kg, what additional mass can it support in water?

30. A 1.00-kg beaker containing 2.00 kg of oil (density = 916 kg/m³) rests on a scale. A 2.00-kg block of iron is suspended from a spring scale and completely submerged in the oil (Fig. P9.30). Find the equilibrium readings of both scales.

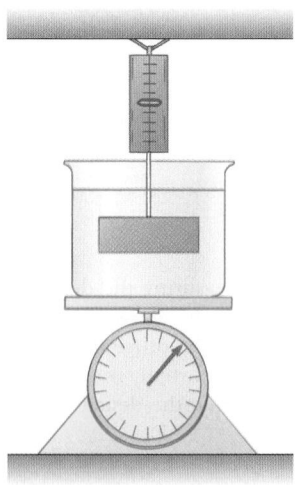

Figure P9.30

Section 9.7   Fluids in Motion

Section 9.8   Other Applications of Bernoulli's Equation

31. Water is pumped into a storage tank from a well delivering 20.0 gallons of water in 30.0 seconds through a pipe of 1.00 in.² cross-sectional area. What is the average velocity of the water in the pipe as the water is pumped from the well?

32. A cowboy at a dude ranch fills a horse trough that is 1.5 m long, 60 cm wide, and 40 cm deep. He uses a 2.0-cm-diameter hose from which water emerges at 1.5 m/s. How long does it take him to fill the trough?

33. (a) Calculate the mass flow rate (in grams per second) of blood ($\rho = 1.0$ g/cm³) in an aorta with a cross-sectional area of 2.0 cm² if the flow speed is 40 cm/s. (b) Assume that the aorta branches to form a large number of capillaries with a combined cross-sectional area of 3.0×10^3 cm². What is the flow speed in the capillaries?

34. What is the net upward force on an airplane wing of area 20.0 m² if the speed of air flow is 300 m/s across the top of the wing and 280 m/s across the bottom?

35. A liquid ($\rho = 1.65$ g/cm³) flows through two horizontal sections of tubing joined end to end. In the first section the cross-sectional area is 10.0 cm², the flow speed is 275 cm/s, and the pressure is 1.20×10^5 Pa. In the second section the cross-sectional area is 2.50 cm². Calculate the smaller section's (a) flow speed and (b) pressure.

36. A hypodermic syringe contains a medicine with the density of water (Fig. P9.36). The barrel of the syringe has a cross-sectional area of 2.50×10^{-5} m². In the absence of a force on the plunger, the pressure everywhere is 1.00 atm. A force, **F,** of magnitude 2.00 N is exerted on the plunger, making medicine squirt from the needle. Determine the medicine's flow speed through the needle. Assume that the pressure in the needle remains equal to 1.00 atm and that the syringe is horizontal.

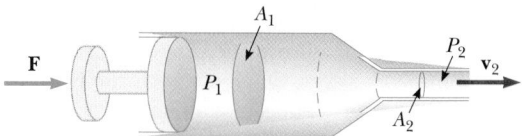

Figure P9.36

37. When a person inhales, air moves down the bronchus (windpipe) at 15 cm/s. The average flow speed of the air doubles through a constriction in the bronchus. Assuming incompressible flow, determine the pressure drop in the constriction.

38. A jet of water squirts out horizontally from a hole near the bottom of the tank in Figure P9.38. If the hole has a diameter of 3.50 mm, what is the height, h, of the water level in the tank?

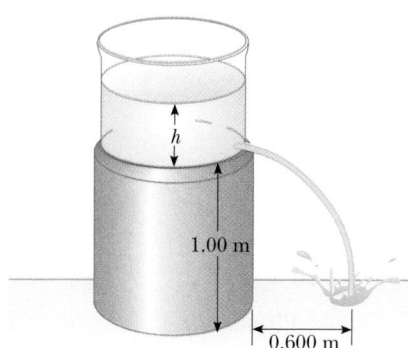

Figure P9.38

WEB **39.** A large storage tank, open to the atmosphere at the top and filled with water, develops a small hole in its side at a point 16.0 m below the water level. If the rate of flow from the leak is 2.50×10^{-3} m³/min, determine (a) the speed at which the water leaves the hole and (b) the diameter of the hole.

40. The inside diameters of the larger portions of the horizontal pipe in Figure P9.40 are 2.50 cm. Water flows to the right at a rate of 1.80×10^{-4} m³/s. Determine the inside diameter of the constriction.

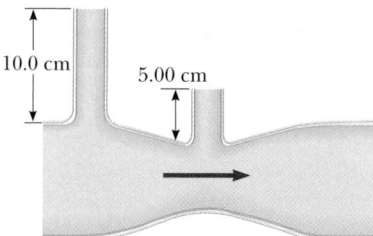

Figure P9.40

41. The water supply of a building is fed through a main entrance pipe 6.0 cm in diameter. A 2.0-cm-diameter faucet tap positioned 2.0 m above the main pipe fills a 25-liter container in 30 s. (a) What is the speed at which the water leaves the faucet? (b) What is the gauge pressure in the main pipe? (Assume that the faucet is the only outlet in the system.)

42. A siphon is a device that allows a fluid to seemingly defy gravity (Fig. P9.42). The flow must be initiated by a partial vacuum in the tube, as in a drinking straw. (a) Show that the speed at which the water emerges from the siphon is given by $v = \sqrt{2gh}$. (b) For what values of y will the siphon work?

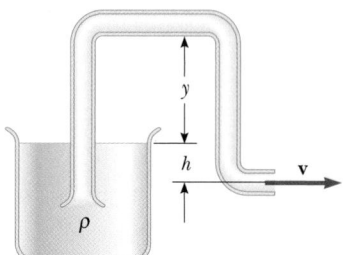

Figure P9.42

Section 9.9 Surface Tension, Capillary Action, and Viscosity (Optional)

43. In order to lift a wire ring of radius 1.75 cm from the surface of a container of blood plasma, a vertical force of 1.61×10^{-2} N greater than the weight of the ring is required. Calculate the surface tension of blood plasma from this information.

44. A square metal sheet 3.0 cm on a side and of negligible thickness is attached to a balance and inserted into a container of fluid. The contact angle is found to be zero, as shown in Figure P9.44a, and the balance to which the metal sheet is attached reads 0.40 N. A thin veneer of oil is then spread over the metal sheet and the contact angle becomes 180°, as shown in Figure P9.44b. The balance now reads 0.39 N. What is the surface tension of the fluid?

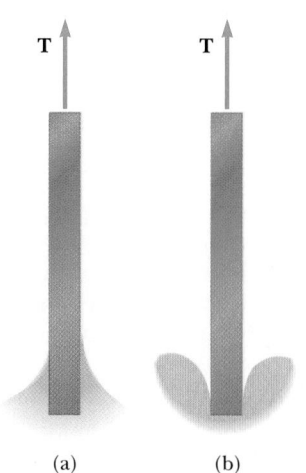

Figure P9.44

45. Whole blood has a surface tension of 0.058 N/m and a density of 1050 kg/m³. To what height can whole blood rise in a capillary blood vessel that has a radius of 2.0×10^{-6} m if the contact angle is zero?

46. A certain fluid has a density of 1080 kg/m³ and is observed to rise to a height of 2.1 cm in a 1.0-mm-diameter tube. The contact angle between the wall and the fluid is zero. Calculate the surface tension of the fluid.

47. A staining solution used in a microbiology laboratory has a surface tension of 0.088 N/m and a density 1.035 times the density of water. What must the diameter of a capillary tube be so that this solution will rise to a height of 5 cm? (Assume a contact angle of zero.)

48. The block of ice (temperature 0°C) shown in Figure P9.48 is drawn over a level surface lubricated by a layer of water 0.10 mm thick. Determine the magnitude of force **F** needed to pull the block with a constant speed of 0.50 m/s. At 0°C, the viscosity of water is equal to $\eta = 1.79 \times 10^{-3}$ N·s/m².

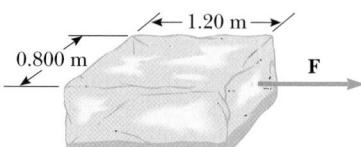

Figure P9.48

49. A 1.5-mm coating of glycerine has been placed between two microscope slides of width 1.0 cm and length 4.0 cm. Find the force required to pull one of the microscope slides at a constant speed of 0.30 m/s relative to the other.

50. A straight horizontal pipe with a diameter of 1.0 cm and a length of 50 m carries oil with a coefficient of viscosity of 0.12 Pa·s. At the output of the pipe, the flow rate is 8.6×10^{-5} m³/s and the pressure is 1.0 atmosphere. Find the gauge pressure at the pipe input.

51. The pulmonary artery, which connects the heart to the lungs, has an inner radius of 2.6 mm and is 8.4 cm long. If the pressure drop between the heart and lungs is 400 Pa, what is the average speed of blood in the pulmonary artery?

52. A hypodermic needle is 3.0 cm in length and 0.30 mm in diameter. What excess pressure is required along the needle so that the flow rate of water through it will be 1 g/s? (Use 1.0×10^{-3} Pa·s as the viscosity of water.)

WEB **53.** What diameter needle should be used to inject a volume of 500 cm³ of a solution into a patient in 30 min? Assume a needle length of 2.5 cm and that the solution is elevated 1.0 m above the point of injection. Furthermore, assume the viscosity and density of the solution are those of pure water and assume that the pressure inside the vein is atmospheric.

54. The aorta in humans has a diameter of about 2.0 cm and, at certain times, the blood speed through it is about 55 cm/s. Is the blood flow turbulent? The density of whole blood is 1050 kg/m^3, and its coefficient of viscosity is 2.7×10^{-3} N·s/m^2.

55. A pipe carrying 20°C water has a diameter of 2.5 cm. Estimate the maximum flow speed if the flow is to be laminar.

Section 9.10   Transport Phenomena (Optional)

56. Sucrose is allowed to diffuse along a 10-cm length of tubing filled with water. The tube is 6.0 cm^2 in cross-sectional area. The diffusion coefficient is equal to 5.0×10^{-10} m^2/s, and 8.0×10^{-14} kg is transported along the tube in 15 s. What is the difference in the concentration levels of sucrose at the two ends of the tube?

57. Glycerine in water diffuses along a horizontal column that has a cross-sectional area of 2.0 cm^2. The concentration gradient is 3.0×10^{-2} kg/m^4, and the diffusion rate is found to be 5.7×10^{-15} kg/s. Determine the diffusion coefficient.

58. The viscous force on an oil drop is measured to be 3×10^{-13} N when the drop is falling through air with a speed of 4.5×10^{-4} m/s. If the radius of the drop is 2.5×10^{-6} m, what is the viscosity of air?

59. Small spheres of diameter 1.00 mm fall through 20°C water with a terminal speed of 1.10 cm/s. Calculate the density of the spheres.

60. Spherical particles of a protein of density 1.8 g/cm^3 are shaken up in a solution of 20°C water. The solution is allowed to stand for 1.0 h. If the depth of water in the tube is 5.0 cm, find the radius of the largest particles that remain in solution at the end of the hour.

ADDITIONAL PROBLEMS

61. The approximate inside diameter of the aorta is 0.50 cm; that of a capillary is 10 μm. The approximate average blood flow speed is 1.0 m/s in the aorta and 1.0 cm/s in the capillaries. If all the blood in the aorta eventually flows through the capillaries, estimate the number of capillaries in the circulatory system.

62. Water at a pressure of 3.00×10^5 Pa flows through a horizontal pipe at a speed of 1.00 m/s. If the pipe narrows to one fourth its original diameter, find **(a)** the flow speed in the narrow section and **(b)** the pressure in the narrow section.

63. A block of wood weighs 50.0 N when weighed in air. A sinker is attached to the block, and the weight of the wood-sinker combination is 200 N when the sinker

alone is immersed in water. Finally, the wood-sinker combination is completely immersed and the weight is 140 N. Find the density of the block.

64. A 600-kg weather balloon is designed to lift a 4000-kg package. What volume should the balloon have after being inflated with helium at standard temperature and pressure in order that the total load can be lifted?

65. A helium-filled balloon at atmospheric pressure is tied to a 2.0-m-long, 0.050-kg string. The balloon is spherical with a radius of 0.40 m. When released, it lifts a length (h) of the string and then remains in equilibrium, as in Figure P9.65. Determine the value of h. When deflated, the balloon has a mass of 0.25 kg. (*Hint:* Only that part of the string above the floor contributes to the weight of the system in equilibrium.)

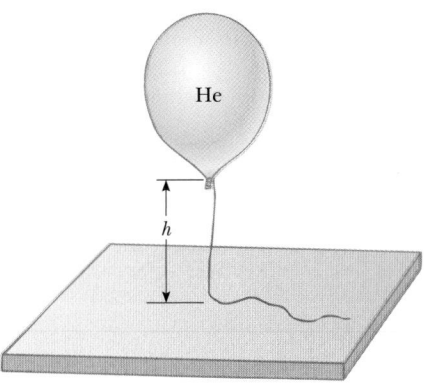

He

h

Figure P9.65

66. A stainless-steel orthodontic wire is applied to a tooth, as in Figure P9.66. The wire has an unstretched length of 3.1 cm and a diameter of 0.22 mm. If the wire is stretched 0.10 mm, find the magnitude and direction of the force on the tooth. Disregard the width of the tooth and assume that Young's modulus for stainless steel is 18×10^{10} Pa.

30°   30°

Figure P9.66

67. A small sphere 0.60 times as dense as water is dropped from a height of 10 m above the surface of a smooth lake. Determine the maximum depth to which the sphere will sink. Neglect any energy transferred to the water during impact and sinking.

68. A 2.0-cm-thick bar of soap is floating on a water surface so that 1.5 cm of the bar is underwater. Bath oil of specific gravity 0.60 is poured into the water and floats on top of the water. What is the depth of the oil layer when the top of the soap is just level with the upper surface of the oil?

69. A cube of ice whose edge is 20.0 mm is floating in a glass of ice-cold water with one of its faces parallel to the water surface. (a) How far below the water surface is the bottom face of the block? (b) Ice-cold ethyl alcohol is gently poured onto the water surface to form a layer 5.00 mm thick above the water. When the ice cube attains hydrostatic equilibrium again, what will be the distance from the top of the water to the bottom face of the block? (c) Additional cold ethyl alcohol is poured onto the water surface until the top surface of the alcohol coincides with the top surface of the ice cube (in hydrostatic equilibrium). How thick is the required layer of ethyl alcohol?

70. Water falls over a dam of height h meters at a rate of R kg/s. (a) Show that the power available from the water is given by

$$P = Rgh$$

where g is the acceleration of gravity. (b) Each hydroelectric unit at the Grand Coulee Dam discharges water at a rate of 8.50×10^5 kg/s from a height of 87.0 m. The power developed by the falling water is converted to electric power with an efficiency of 85.0%. How much electric power is produced by each hydroelectric unit?

71. A U-tube open at both ends is partially filled with water (Fig. 9.71a). Oil ($\rho = 750$ kg/m^3) is then poured into the right arm and forms a column $L = 5.00$ cm high (Fig. 9.71b). (a) Determine the difference, h, in the heights of the two liquid surfaces. Assume that the density of air is 1.29 kg/m^3, but be sure to include differences in the atmospheric pressure due to changes in al-

titude. (b) The right arm is then shielded from any air motion while air is blown across the top of the left arm until the surfaces of the two liquids are at the same height (Fig. 9.71c). Determine the speed of the air being blown across the left arm.

72. Figure P9.72 shows a water tank with a valve at the bottom. If this valve is opened, what is the maximum height attained by the water stream coming out of the right side of the tank? Assume that $h = 10$ m, $L = 2.0$ m, and $\theta = 30°$ and that the cross-sectional area at A is very large compared with that at B.

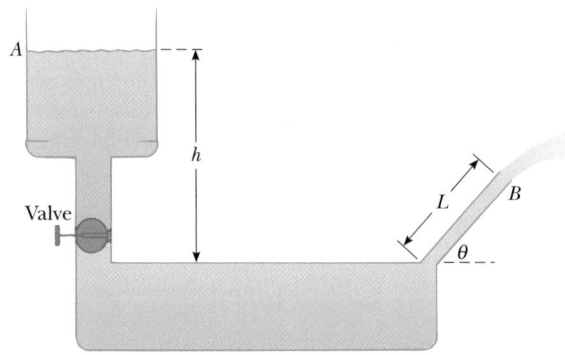

Figure P9.72

73. A solid copper ball with a diameter of 3.00 m at sea level is placed at the bottom of the ocean, at a depth of 10.0 km. If the density of the seawater is 1030 kg/m^3, how much does the diameter of the ball decrease when it reaches bottom?

74. A 1.0-kg hollow ball with a radius of 0.10 m, filled with air, is released from rest at the bottom of a 2.0-m-deep pool of water. How high above the water does the ball shoot upward? Neglect all frictional effects, and neglect the ball's motion when it is only partially submerged.

75. In 1657 Otto von Guericke, inventor of the air pump, evacuated a sphere made of two brass hemispheres (Fig. P9.75). Two teams of eight horses each could pull the hemispheres apart only on some trials, and then with the greatest difficulty. (a) Show that the force required to pull the evacuated hemispheres apart is $\pi R^2 (P_0 - P)$, where R is the radius of the hemispheres and P is the pressure inside the sphere, which is much less than atmospheric pressure, P_0. (b) Determine the required force if $P = 0.10 \, P_0$ and $R = 0.30$ m.

76. A water tank open to the atmosphere at the top has two holes punched in its side, one above the other. The holes are 5.00 cm and 12.0 cm above the floor. How high does water stand in the tank if the two streams of water hit the floor at the same place?

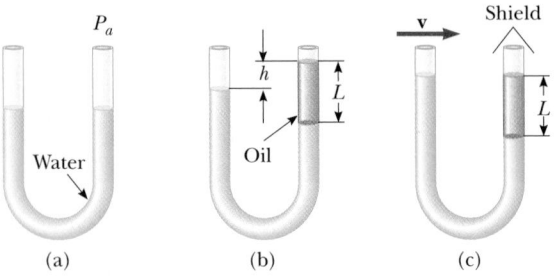

Figure P9.71

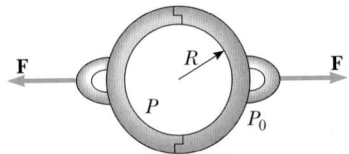

Figure P9.75   *(Courtesy of Henry Leap and Jim Lehman)*

WEB **77.** Oil having a density of 930 kg/m^3 floats on water. A rectangular block of wood 4.00 cm high and with a density of 960 kg/m^3 floats partly in the oil and partly in the water. The oil completely covers the block. How far below the interface between the two liquids is the bottom of the block?

78. A hollow object with an average density of 900 kg/m^3 floats in a pan containing 500 cm^3 of water. Ethanol is added to the water until the object is just on the verge of sinking. What volume of ethanol has been added? Disregard the loss of volume caused by mixing.

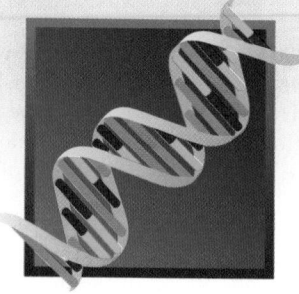

Physics of the Human Circulatory System

WILLIAM G. BUCKMAN
Western Kentucky University

The circulatory system is an extremely complex and vital part of the human body. The blood supplies food and oxygen to the tissues, carries away waste products from the cells, distributes the heat generated by the cells to equalize body temperature, carries hormones that stimulate and coordinate the activity of organs, distributes antibodies to fight infection, and performs numerous other functions.

William Harvey (1579–1657), an English physician and physiologist, studied blood flow and the action of the heart. He established the essential mechanics of the heart and found that the blood flows from the arterial system through capillary beds and into the veins to be returned to the heart.

The Physical Properties of Blood

Blood is a liquid tissue consisting of two principal parts: the plasma, which is the intercellular fluid, and the cells, which are suspended in the plasma. Plasma is about 90% water, 9% proteins, and 1.0% salts, sugar, and traces of other materials. Blood contains white blood cells and red blood cells. The individual red blood cells are biconcave and have an average diameter of 7.5 μm. There are about 5×10^6 red blood cells per cubic millimeter of blood. The five types of white blood cells found in the blood have an average concentration of 8000 per cubic millimeter, with the concentration normally varying between 4500 and 11 000 per cubic millimeter. The density of blood is about 1.05×10^3 kg/m^3, and its viscosity varies from 2.5 to 4 times that of water.

The Heart as a Pump

The heart can be considered a double pump, with each side consisting of an atrium and a ventricle (Fig. 1a). Oxygen-poor blood enters the right atrium, flows into the right ventricle, is pumped by the right ventricle to the lungs, and returns through the left atrium to the left ventricle. The left ventricle then pumps the oxygenated blood out through the aorta to the rest of the body. The heart has a system of one-way valves to ensure that the blood flows in the proper direction. The heart's pumping cycle has the two ventricles pumping at the same time, as shown in Figure 1b.

The pressure generated by the right ventricle is quite low (about 25 mm Hg), and the lungs offer a low resistance to blood flow. The left ventricle generates a larger pressure, typically greater than 120 mm Hg, at the peak (systole) of the pressure. During the resting stage (diastole) of the heartbeat, the pressure is typically about 80 mm Hg.

We shall now calculate the mechanical work done by the heart. Consider the fluid in the vessel shown in Figure 2. The net force on the fluid is equal to the product of the pressure drop across the fluid, ΔP, and the cross-sectional area, A. The power expended is equal to the net force times the average velocity: $(\Delta P A)(\overline{v})$. Because $A\overline{v} = AL/t = $ volume/time, which is the flow rate, we may now write for the power expended by the heart

$$\text{Power} = (\text{flow rate})(\Delta P)$$

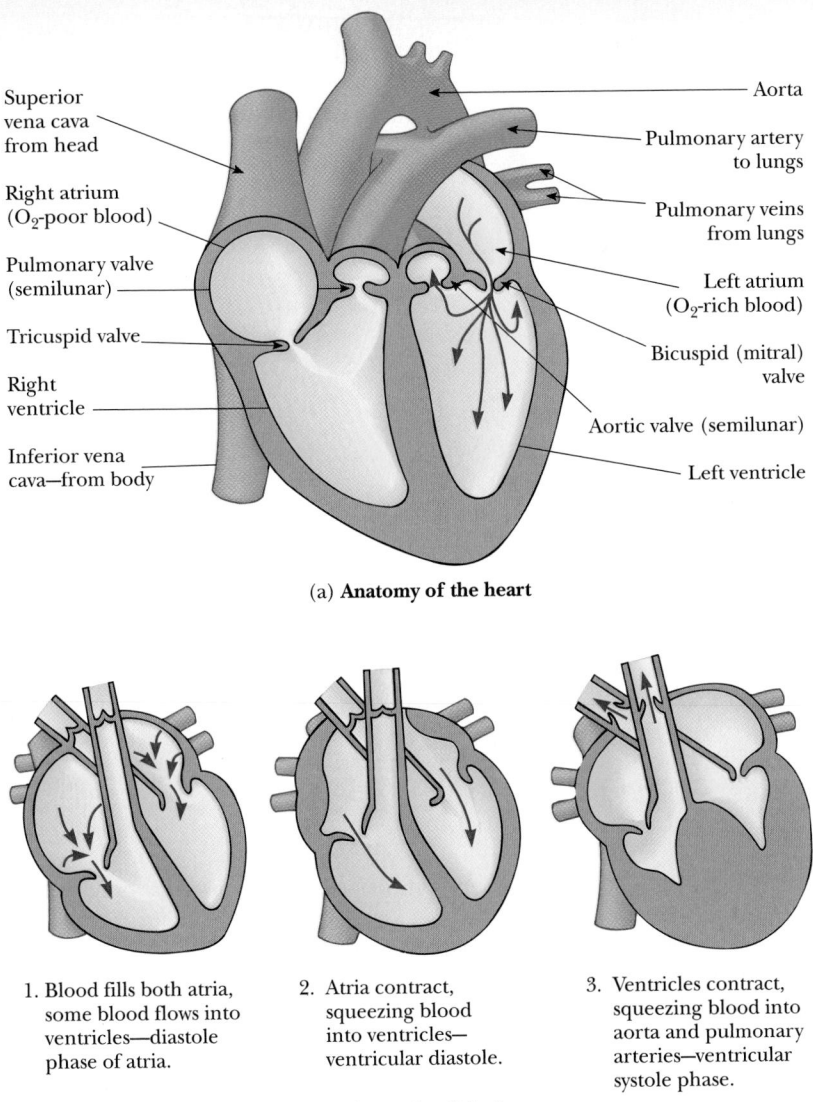

Superior
vena cava
from head

Right atrium
(O_2-poor blood)

Pulmonary valve
(semilunar)

Tricuspid valve

Right
ventricle

Inferior vena
cava—from body

Aorta

Pulmonary artery
to lungs

Pulmonary veins
from lungs

Left atrium
(O_2-rich blood)

Bicuspid (mitral)
valve

Aortic valve (semilunar)

Left ventricle

(a) **Anatomy of the heart**

1. Blood fills both atria,
   some blood flows into
   ventricles—diastole
   phase of atria.

2. Atria contract,
   squeezing blood
   into ventricles—
   ventricular diastole.

3. Ventricles contract,
   squeezing blood into
   aorta and pulmonary
   arteries—ventricular
   systole phase.

(b) **Pumping cycle of the heart**

Figure 1   (a) Anatomy of the heart. (b) Pumping cycle of the heart.

If a normal heart pumps blood at the rate of 97 cm³/s and the pressure drop from the arterial system to the venous system is 1.17×10^4 Pa, we then have

$$\text{Power} = (97 \text{ cm}^3/\text{s})(10^6 \text{ m}^3/\text{cm}^3)(1.17 \times 10^4 \text{ Pa}) = 1.1 \text{ W}$$

By measuring oxygen consumption, it is found that the heart of a 70-kg man at rest consumes about 10 W. In the previous calculation, it was determined that 1.1 W is required to do the mechanical work of pumping blood; hence, the heart is typically about 10% efficient. During strenuous exercise, the blood pressure may increase by 50% and the blood volume pumped may increase by a factor of 5 to yield an increase of 7.5 times in the power generated by the left ventricle. Because the right ventricle has a systolic pressure about

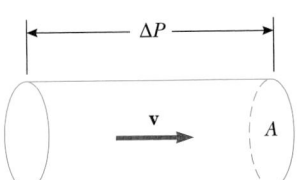

Figure 2   The power required to maintain blood flow against viscous forces.

307

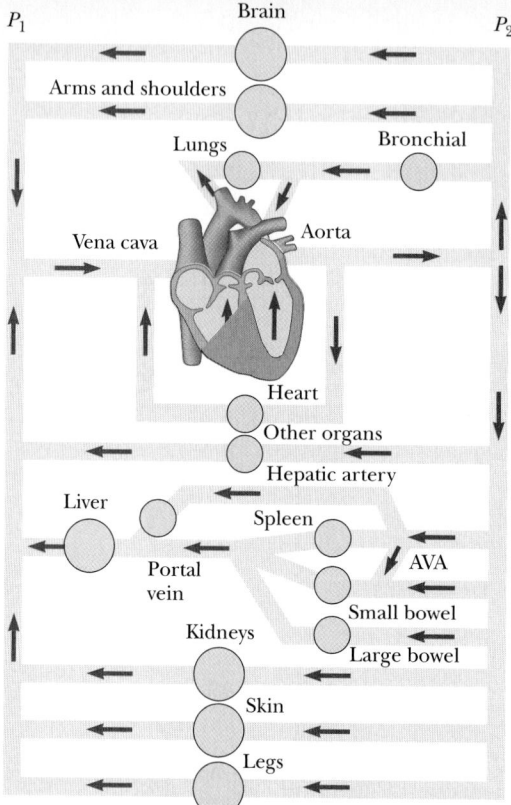

Figure 3 A diagram of the mammalian circulatory system. Pressure P_2 is that in the arterial system, and P_1 is that in the venous system. Arrows indicate the direction of blood flow.

one fifth that of the left ventricle, its power requirement is about one fifth that of the left ventricle.

When we listen to a heart with a stethoscope, we hear two sharp sounds. The first corresponds with the closing of the tricuspid and mitral valves, and the second corresponds with the closing of the aortic and pulmonary valves. Other sounds that are heard are those associated with the flow and turbulence of the blood.

The Cardiovascular System

The cardiovascular system includes the heart to pump the blood; arteries to carry the blood to the organs, muscle, and skin; and veins to return the blood to the heart (Fig. 3). The aorta branches to form smaller arteries, which in turn branch down to even smaller arteries, until finally the blood reaches the very small capillaries of the vascular bed. These capillaries are so small that the red blood cells must pass single file through them. After passing through the capillaries, where materials being carried by the blood are exchanged with the surrounding tissues, the blood flows to the veins and is returned to the heart.

The flow rate of the blood changes as the blood goes through this system. The cross-sectional area of the vascular bed, which is the product of the cross-sectional area and the number of capillaries, is much greater than the cross-sectional area of the aorta. Because the volume of the blood passing through a cross-sectional area per unit of time is Av, where v is the speed of the blood, we may express the volume flow rate of the blood as

$$\text{Flow rate} = A_{\text{aorta}} v_{\text{aorta}}$$

Furthermore, because the total average flow rate through the aorta and the capillaries must be the same, we have

$$\text{Flow rate} = A_{\text{aorta}}v_{\text{aorta}} = A_{\text{cap}}v_{\text{cap}}$$

EXAMPLE Flow of Blood in the Aorta and Capillaries

The speed of blood in the aorta is 50 cm/s and this vessel has a radius of 1.0 cm. (a) What is the rate of flow of blood through this aorta? (b) If the capillaries have a total cross-sectional area of 3000 cm^2, what is the speed of the blood in them?

Solution
(a) The cross-sectional area of the aorta is

$$A = \pi r^2 = \pi(1.0 \text{ cm})^2 = 3.14 \text{ cm}^2$$

$$\text{Flow rate} = Av = (3.14 \text{ cm}^2)(50 \text{ cm/s}) = \boxed{160 \text{ cm}^3/\text{s}}$$

(b) The flow rate in the capillaries = 160 cm^3/s = $A_c v_c$,

$$v_c = \frac{\text{flow rate}}{A_c} = \frac{160 \text{ cm}^3/\text{s}}{3000 \text{ cm}^2} = \boxed{0.053 \text{ cm/s}}$$

This low blood speed in the capillaries is necessary to enable the blood to exchange oxygen, carbon dioxide, and other nutrients with the surrounding tissues.

Questions and Problems

1. Explain why some individuals tend to black out when they stand up rapidly.
2. If a person is standing at rest, what is the relation between the blood pressure in the left arm and the left leg? What is the relation if the person is in a horizontal position?
3. At what upward acceleration would you expect the blood pressure in the brain to be zero? (Assume that no body mechanisms are operating to compensate for this condition.)
4. When a sphygmomanometer is used to measure blood pressure, will the blood pressure readings depend on the atmospheric pressure? If the atmospheric pressure decreases rapidly, how will this affect the blood pressure readings?
5. Why is it impractical to measure the pulse rate using a vein?
6. Assuming that an artery is clogged such that the effective radius is one half its normal radius, by what factor must the pressure differential be increased to obtain the normal flow rate through the clogged artery?
7. Determine the average speed of the blood in the aorta if it has a radius of 1.2 cm and the flow rate is 20 liters/min.
8. If the mean blood pressure in the aorta is 100 mm Hg, (a) determine the blood pressure in the artery located 6.5 cm above the heart. (b) One cannot apply, without significant error, Bernoulli's principle in the smaller arteries and the capillaries. Why not?
9. When the flow rate is 5.0 liters/min, the blood speed in the capillaries is 0.33 mm/s. Assuming the average diameter of a capillary to be 0.0080 mm, calculate the number of capillaries in the circulatory system.
10. An artery has a length of 20 cm and a radius of 0.50 cm, and blood is flowing at a rate of 6.0 liters/min. What is the difference in the pressure between the ends of the artery?

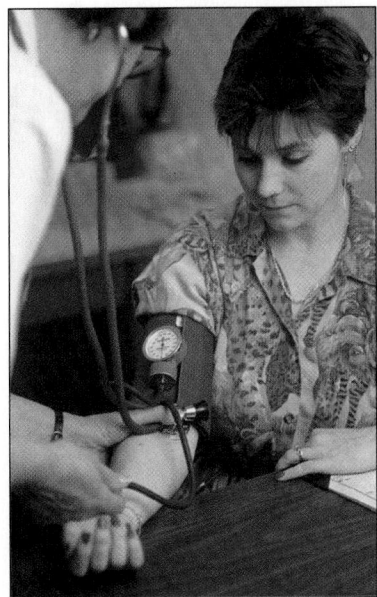

Figure 4 (Question 4) (*Charles Winters*)

PART 2

Thermodynamics

As we saw in the first part of this book, Newtonian mechanics explains a wide range of phenomena, such as the motions of baseballs, rockets, and planets. We now turn to the study of thermodynamics, which is concerned with the concepts of heat and temperature. As we shall see, thermodynamics is very successful in explaining the bulk properties of matter and the correlation between those properties and the mechanics of atoms and molecules.

Historically, the development of thermodynamics paralleled the development of the atomic theory of matter. By the middle of the nineteenth century, chemical experiments provided solid evidence for the existence of atoms. At that time, scientists recognized that there must be a connection between the heat and temperature theory and the structure of matter. In 1827 the botanist Robert Brown reported that grains of pollen suspended in a liquid moved erratically from one place to another, as if under constant agitation. In 1905 Albert Einstein developed a theory about the cause of this erratic motion, which today is called *Brownian motion*. Einstein explained the phenomenon by assuming that the grains of pollen are under constant bombardment by "invisible" molecules in the liquid, which themselves are moving erratically. Einstein's insight gave scientists a means of discovering vital information concerning molecular motion. It also gave reality to the concept of the atomic constituents of matter.

Have you ever wondered how a refrigerator cools, what types of transformations occur in an automobile engine, or what happens to the kinetic energy of a falling object once the object comes to rest? The laws of thermodynamics and the concepts of heat and temperature enable us to answer such practical questions.

Many things can happen to an object when it is heated. Its size generally changes slightly, but it may also melt, boil, ignite, or even explode. The outcome depends on the composition of the object and the degree to which it is heated. In general, thermodynamics must concern itself with the physical and chemical transformations of matter in all of its forms: solid, liquid, and gas.

◀ *Paul Chesley/Tony Stone Images*

Thermal Physics

▲ PHYSICS PUZZLER

After the bottle is shaken, the cork is blown off. Contrary to common belief, shaking the champagne bottle does not increase the CO_2 pressure inside. Because the temperature of the bottle and its contents remain constant, the equilibrium pressure does not change, as can be shown by replacing the cork with a pressure gauge. Why, then, is champagne expelled when the cork pops? *(Steve Niedorf, The IMAGE Bank)*

Our study thus far has focused exclusively on mechanics. Such concepts as mass, force, and kinetic energy have been carefully defined in order to make the subject quantitative. We now move to a new branch of physics, **thermal physics.** Here we shall find that quantitative descriptions of thermal phenomena require careful definitions of the concepts of temperature, heat, and internal energy. We will seek answers to such questions as: What happens to an object when energy is added to it or removed from it by heat? What physical changes occur in an object when its temperature increases or decreases? One familiar outcome of a temperature increase or decrease for an object is a change in size. We shall examine the details of this process in our discussion of linear expansion.

This chapter concludes with a study of ideal gases. We approach this study on two levels. The first examines ideal gases on the macroscopic scale. Here we are concerned with the relationships among such quantities as pressure, volume, and temperature. On the second level we examine gases on a microscopic scale, using a model that pictures the components of a gas as small particles. This latter approach, called the kinetic theory of gases, helps us understand what happens on the atomic level to affect such macroscopic properties as pressure and temperature.

10.1 TEMPERATURE AND THE ZEROTH LAW OF THERMODYNAMICS

We often associate the concept of temperature with how hot or cold an object feels when we touch it. Thus, our senses provide us with qualitative indications of temperature. However, our senses are unreliable and often misleading. For example, if we remove a metal ice tray and a package of frozen vegetables from the freezer, the ice tray feels colder to the hand than the vegetables even though the two are at the same temperature. This is because metal is a better heat conductor than cardboard, and so the ice tray conducts energy from our hand more efficiently than does the cardboard package. What is needed is a reliable and reproducible method of making quantitative measurements to establish the relative "hotness" or "coldness" of objects. Scientists have developed a variety of thermometers to fulfill this purpose.

We are all familiar with the fact that two objects at different initial temperatures may eventually reach some intermediate temperature when placed in contact with each other. For example, if two soft drinks, one warm and the other cold, are placed in an insulated container, the two eventually reach an equilibrium temperature once the cold one warms up and the warm one cools off. Likewise, if a cup of hot coffee is cooled with an ice cube, the ice eventually melts and the coffee's temperature decreases.

In order to understand the concept of temperature, it is useful to first define two often-used phrases, *thermal contact* and *thermal equilibrium.* To grasp the meaning of thermal contact, imagine two objects placed in an insulated container so that they interact with each other but not with the rest of the world. If the objects are at different temperatures, energy is exchanged between them. The energy exchanged between objects because of a temperature difference between them is called **heat.** We shall examine the concept of heat in more detail in Chapter 11. For purposes of the current discussion, we shall assume that two objects are in **thermal contact** with each other if energy can be exchanged between them. **Ther-**

Molten lava flowing down a mountain in Kilauea, Hawaii. In this case, the hot lava flows smoothly out of a central crater until it cools and solidifies to form the mountains. However, violent eruptions sometimes occur, as in the case of Mount St. Helens in 1980, which can cause both local and global (atmospheric) damage. *(Ken Sakomoto/Black Star)*

10.4

mal equilibrium is the situation in which two objects in thermal contact with each other cease to have any exchange of energy.

Now consider two objects, A and B, that are not in thermal contact with each other, and a third object, C, that acts as a thermometer. We wish to determine whether or not A and B would be in thermal equilibrium with each other, once placed in thermal contact. The thermometer (object C) is first placed in thermal contact with A until thermal equilibrium is reached. At that point, the thermometer's reading remains constant, and we record it. The thermometer is then placed in thermal contact with B, and its reading is recorded after thermal equilibrium is reached. If the two readings are the same, then A and B will also be in thermal equilibrium with each other when they are placed in thermal contact.

We can summarize these results in a statement known as the **zeroth law of thermodynamics (the law of equilibrium):**

Zeroth law of thermodynamics ▶

> If bodies A and B are separately in thermal equilibrium with a third body, C, then A and B will be in thermal equilibrium with each other if placed in thermal contact.

This statement, insignificant and obvious as it may seem, is easily proved experimentally and is very important because it makes it possible to define temperature. We can think of **temperature** as the property that determines whether or not an object will be in thermal equilibrium with other objects. **Two objects in thermal equilibrium with each other are at the same temperature.**

10.2   THERMOMETERS AND TEMPERATURE SCALES

Thermometers are devices used to measure the temperature of a system. All thermometers make use of a change in some physical property with temperature. One common thermometer in everyday use consists of a mass of liquid—usually mercury or alcohol—that expands into a glass capillary tube when heated (Fig. 10.1). In this case the physical property is the change in volume of a liquid, and one can define any temperature change to be proportional to the change in length of the liquid column in the capillary. The thermometer can be calibrated by placing it in thermal contact with some natural systems that remain at constant temperature. One such system is a mixture of water and ice in thermal equilibrium at atmospheric pressure. It is defined to have a temperature of zero degrees Celsius, written 0°C; this temperature is called **the ice point of water.** Another commonly used system is a mixture of water and steam in thermal equilibrium at atmospheric pressure; its temperature is 100°C, **the steam point of water.** Once the liquid levels in the thermometer at these two temperatures have been established, the column is divided into 100 equal segments, each corresponding to a change in temperature of one Celsius degree.

Thermometers calibrated in this way present problems when extremely accurate readings are needed. Because mercury and alcohol have different thermal expansion properties, when one indicates a temperature of 50°C, say, the other may indicate a slightly different value. In fact, an alcohol thermometer calibrated at the

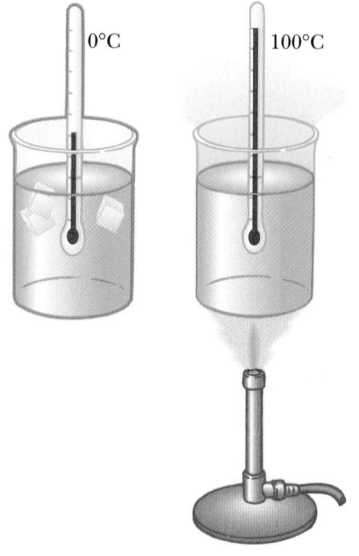

Figure 10.1 Schematic diagram of a mercury thermometer. Because of thermal expansion, the level of the mercury rises as the mercury is heated from 0°C (the ice point) to 100°C (the steam point).

ice and steam points of water might agree with a mercury thermometer only at the calibration points. The discrepancies between different types of thermometers are especially large when the temperatures to be measured are far from the calibration points.

An additional practical disadvantage of any thermometer is its limited temperature range. A mercury thermometer, for example, cannot be used below the freezing point of mercury, $-39°C$. To surmount such problems, we need a universal thermometer whose readings are independent of the substance used. The gas thermometer approaches this requirement.

The Constant-Volume Gas Thermometer and the Kelvin Scale

In a **gas thermometer,** the temperature readings are nearly independent of the substance used in the thermometer. One type of gas thermometer is the constant-volume unit shown in Figure 10.2. The physical property used in this device is the pressure variation with temperature of a fixed volume of gas. When the constant-volume gas thermometer was developed, it was calibrated using the ice and steam points of water. (A different calibration procedure, to be discussed shortly, is used now.) The gas flask was inserted into an ice bath, and mercury reservoir B was raised or lowered until the volume of the confined gas was at some value, indicated by the zero point on the scale. The height h, the difference between the levels in the reservoir and column A, indicated the pressure in the flask at $0°C$. The flask was inserted into water at the steam point, and reservoir B was readjusted until the height in column A was again brought to zero on the scale, ensuring that the gas volume was the same as it had been in the ice bath (hence the designation "constant-volume"). This gave a value for the pressure at $100°C$.

The pressure and temperature values were then plotted on a graph, as in Figure 10.3. The line connecting the two points serves as a calibration curve for measuring unknown temperatures. If we wanted to measure the temperature of a substance, we would place the gas flask in thermal contact with the substance and adjust the column of mercury until the gas took on its specified volume. The height of the mercury column would tell us the pressure of the gas, and we could then find the temperature of the substance from the graph.

Experiments show that the thermometer readings are nearly independent of the type of gas used, so long as the gas pressure is low and the temperature is well above the point at which the gas liquifies (Fig. 10.4). The agreement among thermometers using different gases improves as the pressure is reduced.

If the curves in Figure 10.4 are extended back toward negative temperatures, in every case the pressure is zero when the temperature is $-273.15°C$. This significant temperature is used as the basis for the Kelvin temperature scale, which sets $-273.15°C$ as its zero point (0 K). The size of a Kelvin unit (called a kelvin) is identical to the size of a degree on the Celsius scale. Thus, the relationship of conversion between these temperatures is simply

$$T_C = T - 273.15 \qquad [10.1]$$

where T_C is the **Celsius temperature** and T is the **Kelvin temperature.**

Early gas thermometers made use of ice and steam points according to the procedure just described. However, these points are experimentally difficult to duplicate because they are pressure sensitive. Consequently, a new procedure based

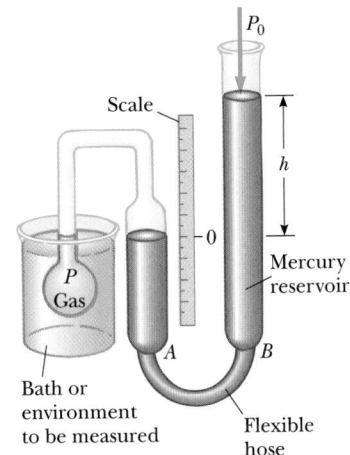

Figure 10.2 A constant-volume gas thermometer measures the pressure of the gas contained in the flask on the left. The volume of gas in the flask is kept constant by raising or lowering the column of mercury such that the mercury level remains constant.

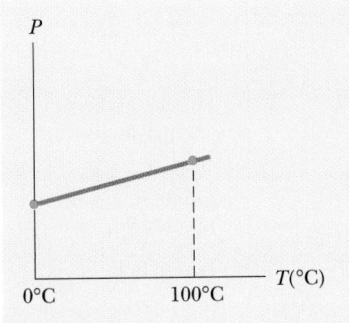

Figure 10.3 A typical graph of pressure versus temperature taken with a constant-volume gas thermometer. The dots represent known reference temperatures (the ice point and the steam point).

WEB

For Web links related to temperature and the temperature standard, visit the textbook Web site at http://www.harcourtcollege.com/physics/cptech

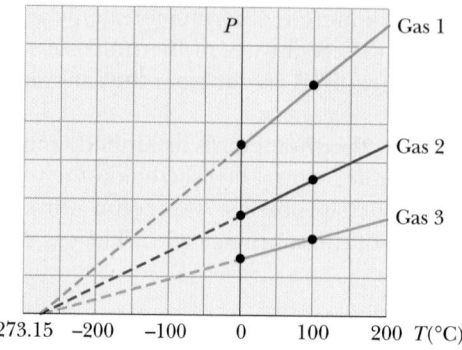

Figure 10.4 Pressure versus temperature for dilute gases. Note that, for all gases, the pressure extrapolates to zero at the unique temperature of $-273.15°C$.

on a single fixed point was adopted in 1954 by the International Committee on Weights and Measures. The **triple point of water, which is the single temperature and pressure at which water, water vapor, and ice can coexist in equilibrium,** was chosen as a convenient and reproducible reference temperature for the Kelvin scale. It occurs at a temperature of 0.01°C and a pressure of 4.58 mm of mercury. The temperature at the triple point of water on the Kelvin scale has been assigned a value of 273.16 kelvins (K). Thus, **the SI unit of temperature, the kelvin, is defined as 1/273.16 of the temperature of the triple point of water.**

⊳ The kelvin

Figure 10.5 shows the Kelvin temperatures for various physical processes and structures. The temperature 0 K is often referred to as **absolute zero,** and, as Figure 10.5 shows, this temperature has never been achieved, although laboratory experiments have come close.

What would happen to a substance if its temperature could reach 0 K? As Figure 10.4 indicates, the pressure the substance exerted on the walls of its container would be zero (assuming that it remains a gas without liquefying or solidifying all the way to absolute zero, which, of course, is not the case). In Section 10.6 we shall show that the pressure of a gas is proportional to the kinetic energy of the molecules of that gas. Thus, according to classical physics, the kinetic energy of the gas would go to zero, and there would be no motion at all of the individual components of the gas; hence, the molecules would settle out on the bottom of the container. Quantum theory, to be discussed in Chapter 27, modifies this statement to indicate that there would be some residual energy, called the *zero-point energy,* at this low temperature.

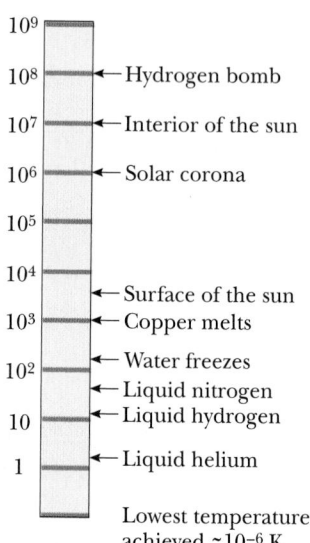

Temperature (K)

Figure 10.5 Absolute temperature at which various selected physical processes take place. Note that the scale is logarithmic.

The Celsius, Kelvin, and Fahrenheit Temperature Scales

Equation 10.1 shows that the Celsius temperature, T_C, is shifted from the absolute (Kelvin) temperature, T, by 273.15. Because the size of a degree is the same on the two scales, a temperature difference of 5°C is equal to a temperature difference of 5 K. The two scales differ only in the choice of zero point. Thus, the ice point (273.15 K) corresponds to 0.00°C, and the steam point (373.15 K) is equivalent to 100.00°C.

The most common temperature scale in use in the United States is the Fahrenheit scale. It sets the temperature of the ice point at 32°F and the temperature

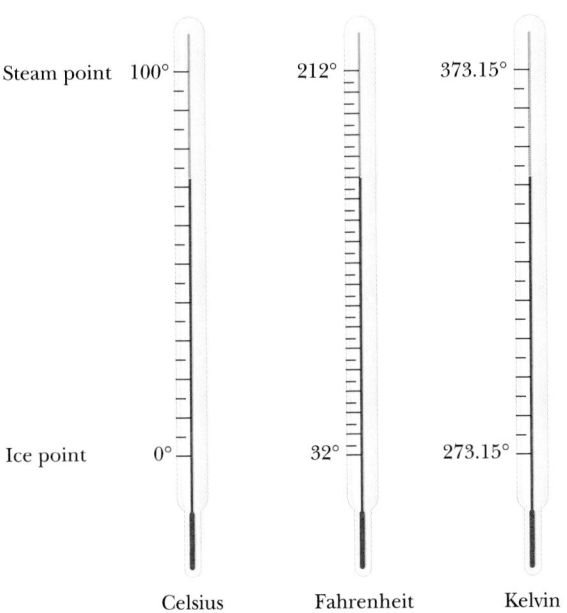

Figure 10.6 A comparison of the Celsius, Fahrenheit, and Kelvin temperature scales.

of the steam point at 212°F. The relationship between the Celsius and Fahrenheit temperature scales is

$$T_F = \tfrac{9}{5} T_C + 32 \qquad\qquad [10.2]$$

Equation 10.2 can easily be used to find a relationship between changes in temperature on the Celsius and Fahrenheit scales. In an end-of-chapter problem you will be asked to show that if the Celsius temperature changes by ΔT_C, the Fahrenheit temperature changes by the amount ΔT_F, given by

$$\Delta T_F = \tfrac{9}{5} \Delta T_C \qquad\qquad [10.3]$$

Figure 10.6 compares the Celsius, Fahrenheit and Kelvin temperature scales.

Thinking Physics 1

A group of future astronauts lands on an inhabited planet. They strike up a conversation with the life forms about temperature scales. It turns out that the inhabitants of this planet have a temperature scale based on the freezing and boiling points of water. Would these two temperatures on this planet be the same as those on Earth? Would the size of the planet inhabitants' degrees be the same as ours? Suppose that the inhabitants have also devised a scale similar to the Kelvin scale. Would their absolute zero be the same as ours?

Explanation The values of 0°C and 100°C for the freezing and boiling points of water are defined at atmospheric pressure. On another planet, it is unlikely that atmospheric pressure would be exactly the same as that on Earth. Thus, water would freeze and boil at different temperatures on the planet. The inhabitants may call these temperatures 0° and 100°, but their degrees would not be the same size as our Celsius degrees (unless their atmospheric pressure was the same as ours). For a version of the

Kelvin scale from this other planet, the absolute zero would be the same as ours, because it is based on a natural, universal definition, rather than being associated with a particular substance or a given atmospheric pressure.

EXAMPLE 10.1   Converting Temperatures

What is the temperature 50.0°F in degrees Celsius and in kelvins?

Solution   Let us solve Equation 10.2 for T_C and substitute $T_F = 50.0°F$:

$$T_C = \tfrac{5}{9}(T_F - 32.0) = \tfrac{5}{9}(50.0 - 32.0) = \boxed{10.0°C}$$

From Equation 10.1 we find that

$$T = T_C + 273.15 = \boxed{283 \text{ K}}$$

Exercise   On a hot summer day, the temperature is reported as 30.0°C. What is this temperature in Fahrenheit degrees and in kelvins?

Answer   86.0°F; 303 K

EXAMPLE 10.2   Heating a Pan of Water

A pan of water is heated from 25°C to 80°C. What is the change in its temperature on the Kelvin scale and on the Fahrenheit scale?

Solution   From Equation 10.1 we see that the change in temperature on the Celsius scale equals the change on the Kelvin scale. Therefore,

$$\Delta T = \Delta T_C = 80 - 25 = 55°C = \boxed{55 \text{ K}}$$

From Equation 10.3 we find that the change in temperature on the Fahrenheit scale is $\tfrac{9}{5}$ as great as the change on the Celsius scale. That is,

$$\Delta T_F = \tfrac{9}{5}\Delta T_C = \tfrac{9}{5}(80 - 25) = \boxed{99°F}$$

10.3   THERMAL EXPANSION OF SOLIDS AND LIQUIDS

Our discussion of the liquid thermometer made use of one of the best-known changes that occurs in a substance: As its temperature increases, its volume increases. (As we shall see, however, in some materials the volume *decreases* when the temperature increases.) This phenomenon, known as **thermal expansion,** plays an important role in numerous applications. For example, thermal expansion joints must be included in buildings, concrete highways, and bridges to compensate for changes in dimensions with temperature variations.

The overall thermal expansion of an object is a consequence of the change in the average separation between its constituent atoms or molecules. To understand this, consider how the atoms in a solid substance behave. At ordinary temperatures,

Thermal expansion joints are used to separate sections of roadways on bridges. Without these joints, the surfaces would buckle due to thermal expansion on very hot days or crack due to contraction on very cold days. (© *Frank Siteman, Stock/Boston*)

the atoms vibrate about their equilibrium positions with an amplitude of about 10^{-11} m, and the average spacing between the atoms is about 10^{-10} m. As the temperature of the solid increases, the atoms vibrate with greater amplitudes and the average separation between them increases. Consequently, the solid as a whole expands. If the thermal expansion of an object is sufficiently small compared with the object's initial dimensions, then the change in any dimension is, to a good approximation, proportional to the first power of the temperature change.

Suppose an object has an initial length of L_0 along some direction at some temperature. The length increases by ΔL for the change in temperature ΔT. Experiments show that when ΔT is small enough, ΔL is proportional to ΔT and to L_0:

$$\Delta L = \alpha L_0 \Delta T \qquad \text{[10.4]}$$

or

$$L - L_0 = \alpha L_0 (T - T_0)$$

where L is the final length, T is the final temperature, and the proportionality constant α is called the **average coefficient of linear expansion** for a given material and has units of $(°C)^{-1}$.

It may be helpful to picture a thermal expansion as a magnification or a photographic enlargement. For example, as a metal washer is heated (Fig. 10.7), all dimensions, including the radius of the hole, increase according to Equation 10.4. One practical application of thermal expansion is a common technique used to loosen a metal lid that is stuck on a glass jar. As you are aware, running hot water over the lid causes the lid to expand, thus enabling you to remove it from the jar. Table 10.1 lists the average coefficients of linear expansion for various materials. Note that for these materials α is positive, indicating an increase in length with increasing temperature.

Because the linear dimensions of an object change with temperature, it follows that surface area and volume also change with temperature. Consider a square

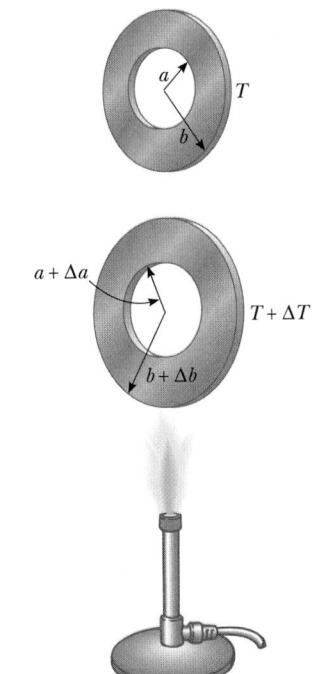

Figure 10.7 Thermal expansion of a homogeneous metal washer. As the washer is heated, all dimensions increase. (Note that the expansion is exaggerated in this figure.)

TABLE 10.1 Average Coefficients of Expansion for Some Materials Near Room Temperature

Material	Average Coefficient of Linear Expansion $[(°C)^{-1}]$	Material	Average Coefficient of Volume Expansion $[(°C)^{-1}]$
Aluminum	24×10^{-6}	Ethyl alcohol	1.12×10^{-4}
Brass and bronze	19×10^{-6}	Benzene	1.24×10^{-4}
Copper	17×10^{-6}	Acetone	1.5×10^{-4}
Glass (ordinary)	9×10^{-6}	Glycerin	4.85×10^{-4}
Glass (Pyrex®)	3.2×10^{-6}	Mercury	1.82×10^{-4}
Lead	29×10^{-6}	Turpentine	9.0×10^{-4}
Steel	11×10^{-6}	Gasoline	9.6×10^{-4}
Invar (Ni-Fe alloy)	0.9×10^{-6}	Air	3.67×10^{-3}
Concrete	12×10^{-6}	Helium	3.665×10^{-3}

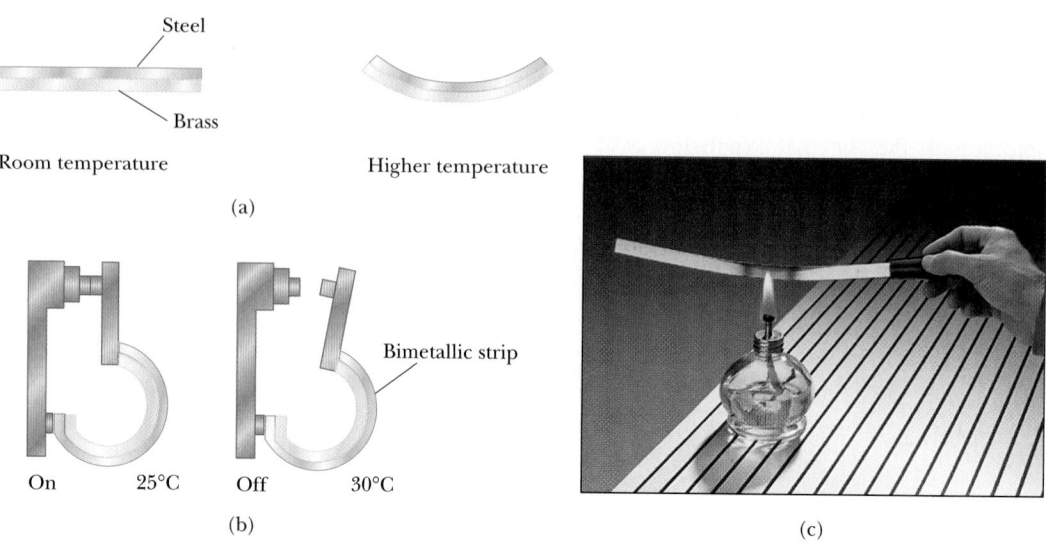

Figure 10.8 (a) A bimetallic strip bends as the temperature changes because the two metals have different expansion coefficients. (b) A bimetallic strip used in a thermostat to break or make electrical contact. (c) The blade being heated in the photograph is a bimetallic strip that was straight before being heated. Note that the blade bends when heated. Which way would it bend if it were cooled? *(Courtesy of Central Scientific Company)*

having an initial length L_0 on a side and therefore an initial area of $A_0 = L_0^2$. As the temperature is increased, the length of each side increases to

$$L = L_0 + \alpha L_0 \, \Delta T$$

We are now able to calculate the change in the area of the square object as follows. The new area $A = L^2$ is

$$L^2 = (L_0 + \alpha L_0 \, \Delta T)(L_0 + \alpha L_0 \, \Delta T) = L_0^2 + 2\alpha L_0^2 \, \Delta T + \alpha^2 L_0^2 (\Delta T)^2$$

The last term in this expression contains the quantity $\alpha \, \Delta T$ raised to the second power. Because $\alpha \, \Delta T$ is much less than unity, squaring it makes it even smaller. Therefore, we can neglect this term to get a simpler expression:

$$A = L^2 = L_0^2 + 2\alpha L_0^2 \, \Delta T$$

$$A = A_0 + 2\alpha A_0 \, \Delta T$$

or

$$\Delta A = A - A_0 = \gamma A_0 \, \Delta T \qquad [10.5]$$

where $\gamma = 2\alpha$. The quantity γ (Greek letter gamma) is called the **average coefficient of area expansion.**

By a similar procedure we can show that the *increase in volume* of an object accompanying a change in temperature is

$$\Delta V = \beta V_0 \, \Delta T \qquad [10.6]$$

where β, the **average coefficient of volume expansion,** is equal to 3α.

As Table 10.1 indicates, each substance has its own characteristic coefficients of expansion. For example, when the temperatures of a brass rod and a steel rod of equal length are raised by the same amount from some common initial value, the brass rod expands more than the steel rod because brass has a larger coefficient of expansion than steel. A simple device that utilizes this principle, called a bimetallic strip, is found in practical devices such as thermostats. The strip is made by securely bonding two different metals together. As the temperature of the strip increases, the two metals expand by different amounts and the strip bends, as in Figure 10.8.

◄ **APPLICATION**

Thermostats.

Applying Physics 1

A homeowner is painting the ceiling, and a drop of paint from the brush falls onto an operating incandescent light bulb. The bulb breaks. Why?

Explanation   The glass envelope of an incandescent light bulb receives energy on the inside surface by radiation from the very hot filament and convection of the gas filling the bulb. Thus, the glass can become very hot. If a drop of paint falls onto the glass, that portion of the glass envelope will suddenly become cold, and the contraction of this region could cause thermal stresses that could cause the glass to shatter.

Thermal expansion affects the choice of glassware used in kitchens and laboratories. If hot liquid is poured into a cold container made of ordinary glass, the container may well break due to thermal stress. The inside surface of the glass becomes hot and expands, while the outside surface is at room temperature, and ordinary glass may not withstand the differential expansion without breaking. Pyrex glass has a coefficient of linear expansion of about one third that of ordinary glass, so that the thermal stresses are smaller. Kitchen measuring cups and laboratory beakers are often made of Pyrex because they are likely to contain hot liquids.

◄ **APPLICATION**

Pyrex Glass.

EXAMPLE 10.3   Expansion of a Railroad Track

A steel railroad track has a length of 30.000 m when the temperature is 0°C. What is its length on a hot day when the temperature is 40°C?

Solution   If we use Table 10.1 and Equation 10.4, and note that the change in temperature is 40°C, we find that the *increase* in length is

$$\Delta L = \alpha L_0 \, \Delta T = [11 \times 10^{-6} \, (°C)^{-1}](30.000 \text{ m})(40°C) = 0.013 \text{ m}$$

Therefore, the track's length at 40°C is

$$L = L_0 + \Delta L = \boxed{30.013 \text{ m}}$$

Exercise   What is the length of the same railroad track on a cold winter day when the temperature is 0°F?

Answer   29.994 m

Thermal expansion: The extreme heat of a July day in Asbury Park, New Jersey, caused these railroad tracks to buckle. *(Wide World Photos)*

EXAMPLE 10.4 Does the Hole Get Bigger or Smaller?

A hole of cross-sectional area 100.00 cm² is cut in a piece of steel at 20°C. What is the area of the hole if the steel is heated from 20°C to 100°C?

Solution A hole in a substance expands in exactly the same way as would a piece of the substance having the same shape as the hole (see Fig. 10.7). The change in area of the hole can be found by using Equation 10.5.

$$\Delta A = \gamma A_0 \, \Delta T = [22 \times 10^{-6}(°C)^{-1}](100.00 \text{ cm}^2)(80°C) = 0.18 \text{ cm}^2$$

Therefore, the area of the hole at 100°C is

$$A = A_0 + \Delta A = \boxed{100.18 \text{ cm}^2}$$

Exercise In the derivation of Eq. 10.5, the term $\alpha^2 L_0^2(\Delta T)^2 = \alpha^2 A_0(\Delta T)^2$ was neglected because it was said to be negligibly small. Use the values found in Example 10.4 to verify that this term is, indeed, extremely small.

The Unusual Behavior of Water

Liquids generally increase in volume with increasing temperature and have volume expansion coefficients about ten times greater than those of solids. Water is an exception to this rule, as we can see from its density-versus-temperature curve in Figure 10.9. As the temperature increases from 0°C to 4°C, water contracts and thus its density increases. Above 4°C, water expands with increasing temperature. In other words, the density of water reaches its maximum value (1000 kg/m³) 4 degrees above the freezing point.

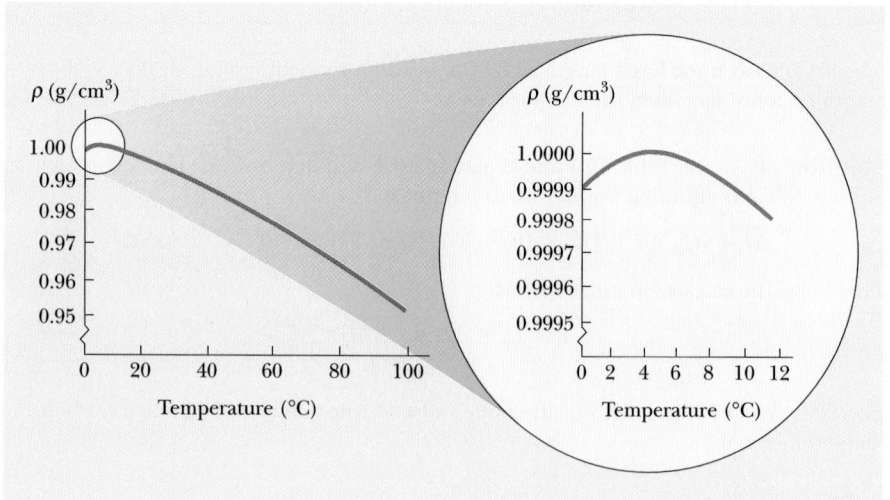

Figure 10.9 The density of water as a function of temperature. The inset at the right shows that the maximum density of water occurs at 4°C.

To understand why water behaves as it does, let us compare the volume of a given mass of water to the volume of an equal mass of ice. At a temperature of 0°C the ice occupies a larger volume than it does when in the form of liquid water at the same temperature. This is because solid ice has an open hexagonal crystal structure that results in relatively large amounts of empty space within its boundaries. On a microscopic level, the breakdown of all these intermolecular bonds is not complete until about 4°C. As liquid water is heated from 0°C to 4°C, its volume slowly decreases and, hence, its density slowly increases. Finally, at and above 4°C, random thermal motions lead to an increase in the volume of a given mass of water and a subsequent decrease in its density.

We can use this unusual thermal expansion behavior of water to explain why a pond freezes slowly from the top down. When the atmospheric temperature drops from, say, 7°C to 6°C, the water at the surface of the pond also cools and consequently decreases in volume. This means that the surface water is denser then the water below it, which has not cooled and decreased in volume. As a result, the surface water sinks and warmer water from below is buoyed up to the surface to be cooled. When the atmospheric temperature is between 4°C and 0°C, however, the surface water expands as it cools, becoming less dense than the water below it. The mixing process stops, and eventually the surface water freezes. As the water freezes, the ice remains on the surface because ice is less dense than water. The ice continues to build up on the surface, and water near the bottom of the pool remains at 4°C. If this did not happen, fish and other forms of marine life would not survive.

The same peculiar thermal expansion properties sometimes cause water pipes to burst in winter. As energy leaves the water through the pipe and is transferred to the outside cold air, the outer layers of water in the pipe freeze first. Continuing heat transfer causes ice formation to move ever closer to the center of the pipe. As long as there is still an opening through the ice, the water can expand as its temperature approaches 0°C or as it freezes into more ice, pushing itself into another part of the pipe. Eventually, however, the ice will freeze to the center somewhere along the pipe's length, forming a plug of ice there. If there is still liquid water between this plug and some other obstruction, such as another ice plug or a spigot, then there is no additional volume available for further expansion and freezing. It is possible for the pressure buildup due to this frustrated expansion to burst the pipe.

APPLICATION

Bursting Pipes in Winter.

Thinking Physics 2

Suppose a solid object is at the bottom of a container of water. Can the temperature of the system be changed in any way to make the object float to the surface?

Explanation Let us imagine lowering the temperature. Liquids tend to have larger coefficients of volume expansion than solids, so the density of water will increase more rapidly with the dropping temperature than that of the solid. If the density of the solid were initially only slightly larger than that of water, it would in principle be possible to make it float by lowering the temperature, although one faces the possible problem of reaching 4°C, the maximum density point, before this happens.

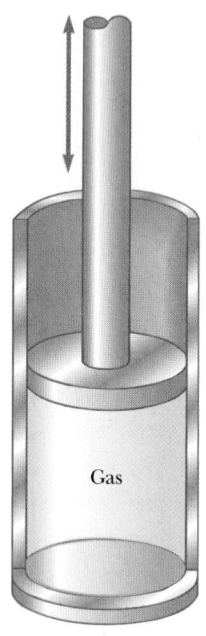

Gas

Figure 10.10   A gas confined to a cylinder whose volume can be varied with a movable piston.

10.4   MACROSCOPIC DESCRIPTION OF AN IDEAL GAS

In this section we are concerned with the properties of a gas of mass m confined to a container of volume V, pressure P, and temperature T. It is useful to know how these quantities are related. In general, the equation that interrelates them, called the *equation of state,* is very complicated. However, if the gas is maintained at a very low pressure (or low density), the equation of state is found experimentally to be quite simple. Such a low-density gas approximates what is called an **ideal gas.** Most gases at room temperature and atmospheric pressure behave as if they were ideal gases.

It is convenient to express the amount of gas in a given volume in terms of the number of moles, n. Recall that one mole of any substance is that mass of the substance that contains Avogadro's number, 6.022×10^{23}, of molecules. The number of moles, n, of a substance is related to its mass, m, by the expression

$$n = \frac{m}{M} \qquad\qquad [10.7]$$

where M is the molar mass, usually expressed in kilograms per mole. For example, the molar mass of molecular oxygen, O_2, is 32.0×10^{-3} kg/mol. Therefore, the mass of one mole of oxygen is 32.0×10^{-3} kg.

Now suppose an ideal gas is confined to a cylindrical container whose volume can be varied by means of a movable piston, as in Figure 10.10. We shall assume that the cylinder does not leak, and so the mass (or the number of moles) remains constant. For such a system, experiments provide the following information. First, when the gas is kept at a constant temperature, its pressure is inversely proportional to the volume (Boyle's law). Second, when the pressure of the gas is kept constant, the volume is directly proportional to the temperature (the law of Charles and Gay-Lussac). These observations can be summarized by the following **equation of state for an ideal gas:**

Equation of state for an ideal gas ▶

$$PV = nRT \qquad\qquad [10.8]$$

In this expression, called the **ideal gas law,** R is a constant for a specific gas that can be determined from experiments, and T is the temperature in kelvins. Experiments on several gases show that, as the pressure approaches zero, the quantity PV/nT approaches the same value of R for all gases. For this reason R is called the **universal gas constant.** In the SI system, where pressure is expressed in pascals and volume in cubic meters, the product PV has units of newton-meters, or joules, and R has the value

The universal gas constant ▶

$$R = 8.31 \ \text{J/mol·K} \qquad\qquad [10.9]$$

If the pressure is expressed in atmospheres and the volume is given in liters ($1 \ \text{L} = 10^3 \ \text{cm}^3 = 10^{-3} \ \text{m}^3$), then R has the value

$$R = 0.0821 \ \text{L·atm/mol·K}$$

Using this value of R and Equation 10.8, one finds that the volume occupied by 1 mol of any gas at atmospheric pressure and 0°C (273 K) is 22.4 L (see Example 10.5).

Applying Physics 2

On some canned foods, you can test the seal by observing the lid of the container when it is opened. When sealed, the lid is indented, but when opened, the lid springs back out. How does this work?

Explanation The food is sealed inside the container while it is still warm. As the food, and the air above it, cools, the pressure of the trapped air is lowered. Thus, the atmospheric pressure pressing against the outside of the lid causes it to indent inward. When you open the container, breaking the seal allows outside air to rush in, and the lid snaps back to its original shape.

EXAMPLE 10.5 The Volume of One Mole of Gas

Verify that one mole of oxygen occupies a volume of 22.4 L at 1 atm and 0°C.

Solution Let us solve the ideal gas equation for V:

$$V = \frac{nRT}{P}$$

In this problem the mass of the gas, m, is assumed to be that of one mole, M. Thus,

$$n = \frac{m}{M} = 1 \text{ mol}$$

Now let us convert the temperature to kelvins and substitute into the ideal gas equation:

$$V = \frac{nRT}{P} = \frac{(1 \text{ mol})(0.0821 \text{ L} \cdot \text{atm/mol} \cdot \text{K})(273 \text{ K})}{1 \text{ atm}} = \boxed{22.4 \text{ L}}$$

This answer has general validity. **One mole of any gas at standard temperature and pressure (STP) occupies a volume of 22.4 L.**

EXAMPLE 10.6 How Many Gas Molecules Are in a Container?

An ideal gas occupies a volume of 100 cm³ at 20°C and a pressure of 100 Pa. Determine the number of moles of gas in the container.

Solution The quantities given are volume, pressure, and temperature: $V = 100 \text{ cm}^3 = 1.00 \times 10^{-4} \text{ m}^3$, $P = 100$ Pa, and $T = 20°C = 293$ K. Using Equation 10.8, we get

$$n = \frac{PV}{RT} = \frac{(100 \text{ Pa})(10^{-4} \text{ m}^3)}{(8.31 \text{ J/mol} \cdot \text{K})(293 \text{ K})} = \boxed{4.11 \times 10^{-6} \text{ mol}}$$

QUICKLAB

Tie and tape two inverted empty paper bags to the ends of a rod as in the figure. Balance the setup. Then place a candle under one of the bags and note what happens. Why does this system become unbalanced? What do your results tell you concerning the density of warm air versus the density of cold air?

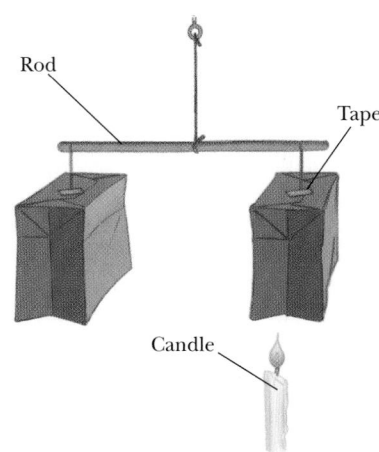

Rod

Tape

Candle

Exercise Show that the universal gas constant $R = 0.0821$ L·atm/mol·K is also equal to the value $R = 8.31$ J/mol·K.

EXAMPLE 10.7 Squeezing a Tank of Gas

Pure helium gas is admitted into a leakproof cylinder containing a movable piston. The initial volume, pressure, and temperature of the gas are 15 L, 2.0 atm, and 300 K. If the volume is decreased to 12 L and the pressure increased to 3.5 atm, find the final temperature of the gas. (Assume that helium behaves as an ideal gas.)

Solution Because no gas escapes from the cylinder, the number of moles remains constant; therefore, use of $PV = nRT$ at the initial and final points gives

$$\frac{P_i V_i}{T_i} = \frac{P_f V_f}{T_f}$$

where i and f refer to the initial and final values. Solving for T_f, we get

$$T_f = \left(\frac{P_f V_f}{P_i V_i}\right)(T_i) = \frac{(3.5 \text{ atm})(12 \text{ L})}{(2.0 \text{ atm})(15 \text{ L})}(300 \text{ K}) = \boxed{420 \text{ K}}$$

Workbook Problem 64 – (Workbook page 137) is similar to Example 10.8.

EXAMPLE 10.8 Heating a Bottle of Air

A sealed glass bottle at 27°C contains air at atmospheric pressure and has a volume of 30 cm³. The bottle is then tossed into an open fire. When the temperature of the air in the bottle reaches 200°C, what is the pressure inside the bottle? Assume that any volume changes of the bottle are small enough to be negligible.

Solution We start with the expression

$$\frac{P_i V_i}{T_i} = \frac{P_f V_f}{T_f}$$

Because the initial and final volumes of the gas are assumed equal, this expression reduces to

$$\frac{P_i}{T_i} = \frac{P_f}{T_f}$$

Before evaluating the final pressure, we must convert the given temperatures to kelvins: $T_i = 27°C = 300$ K and $T_f = 200°C = 473$ K. Thus,

$$P_f = \left(\frac{T_f}{T_i}\right)(P_i) = \left(\frac{473 \text{ K}}{300 \text{ K}}\right)(1.0 \text{ atm}) = \boxed{1.6 \text{ atm}}$$

Obviously, the higher the temperature, the higher the pressure exerted by the trapped air. Of course, if the pressure rises high enough, the bottle will shatter.

Exercise In this example we neglected the change in volume of the bottle. If the coefficient of volume expansion for glass is 27×10^{-6} (°C)$^{-1}$, find the magnitude of this volume change.

Answer 0.14 cm³

10.5 AVOGADRO'S NUMBER AND THE IDEAL GAS LAW

In the early 1800s an important field of experimental investigation was created to determine the relative masses of molecules. The equipment used in one such investigation is shown in Figure 10.11. The lower section of the glass vessel contains two electrodes that are connected to a battery so that an electric current can be passed through lightly salted water. (The purpose of the salt is to improve the electrical conductivity of the water.) Bubbles of gas are produced at each electrode and become trapped in the column above it. An analytical examination of the bubbles reveals that the column above the positive electrode contains oxygen gas and the column above the negative electrode contains hydrogen gas. Obviously, the current decomposes the water into its constituent parts.

Experimentally it is found that the volume of hydrogen gas collected is always exactly twice the volume of oxygen gas collected, as we now know in light of the chemical composition of water, H_2O. In addition, it is found that if 9 g of water are decomposed, 8 g of oxygen and 1 g of hydrogen are collected. From this information it is possible to determine the relative masses of oxygen and hydrogen molecules.

Following such an experimental investigation, Amedeo Avogadro in 1811 stated the following hypothesis:

> **Equal volumes of gas at the same temperature and pressure contain the same numbers of molecules.**

Based on this hypothesis and the fact that the ratio of hydrogen to oxygen molecules collected in the experiment is $2:1$, the mass of an oxygen molecule must be 16 times that of a hydrogen molecule.

In Example 10.8, we noted that single moles of all gases occupy the same volume at 1 atm and 0°C. Thus, a corollary to Avogadro's hypothesis is as follows:

> **One mole quantities of all gases at standard temperature and pressure contain the same numbers of molecules.**

Specifically, the number of molecules contained in one mole of any gas is Avogadro's number, 6.02×10^{23} molecules/mol, given the symbol N_A:

$$N_A = 6.02 \times 10^{23} \text{ molecules/mol} \qquad [10.10]$$

◀ Avogadro's number

It is of historical interest that Avogadro never knew, even approximately, the value for this number.

An alternative method for calculating the number of moles of any gas in a container is to divide the total number of molecules present, N, by Avogadro's number:

$$n = \frac{N}{N_A} \qquad [10.11]$$

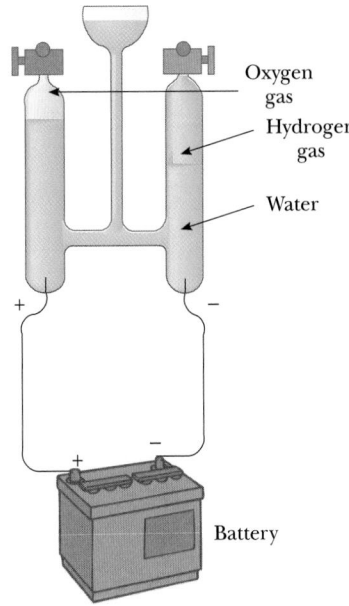

Figure 10.11 When water is decomposed by an electric current, the volume of hydrogen gas collected is twice that of oxygen.

With this expression we can rewrite the ideal gas law in the alternative form

$$PV = nRT = \frac{N}{N_A} RT$$

or

◀ Ideal gas law

$$PV = Nk_B T \qquad [10.12]$$

where k_B is **Boltzmann's constant** and has the value

◀ Boltzmann's constant

$$k_B = \frac{R}{N_A} = 1.38 \times 10^{-23} \, \text{J/K} \qquad [10.13]$$

EXAMPLE 10.9 What Is Avogadro's Number?

One mole of hydrogen atoms has a mass of 1.0078×10^{-3} kg. In the early twentieth century it was found that the mass of a hydrogen atom is approximately 1.673×10^{-27} kg. Use these values to find Avogadro's number.

Solution   The number of molecules in 1 mol of hydrogen can be found by dividing the mass of 1 mol by the mass per atom. Hence, we find that

$$N_A = \frac{1.0078 \times 10^{-3} \, \text{kg/mol}}{1.673 \times 10^{-27} \, \text{kg/atom}} = 6.024 \times 10^{23} \, \text{atoms/mol}$$

10.6   THE KINETIC THEORY OF GASES

10.5, SECTION 1

In Section 10.4 we discussed the properties of an ideal gas, using such quantities as pressure, volume, number of moles, and temperature. In this section we find that pressure and temperature can be understood on the basis of what is happening on the atomic scale. In addition, we reexamine the ideal gas law in terms of the behavior of the individual molecules that make up the gas. Our discussion is restricted to the behavior of gases because the molecular interactions in a gas are much weaker than those in solids and liquids.

Molecular Model for the Pressure of an Ideal Gas

Let us first use the kinetic theory of gases to show that the pressure a gas exerts on the walls of its container is a consequence of the collisions of the gas molecules with the walls. We make the following assumptions:

1. **The number of molecules is large, and the average separation between them is large compared with their dimensions.** This means that the molecules occupy a negligible volume in the container.

◀ Assumptions of kinetic theory for an ideal gas

2. **The molecules obey Newton's laws of motion, but as a whole they move randomly.** By "randomly" we mean that any molecule can move equally in any direction.

3. **The molecules undergo elastic collisions with each other and with the walls of the container.** Thus, in the collisions kinetic energy is constant.
4. **The forces between molecules are negligible except during a collision.** The forces between molecules are short-range, so the molecules interact with each other only during collisions.
5. **The gas under consideration is a pure substance;** that is, all molecules are identical.

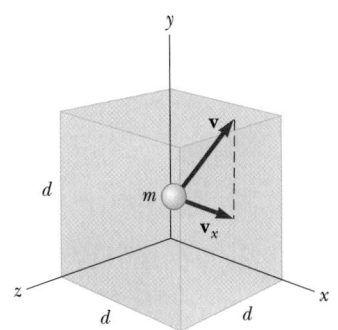

Figure 10.12 A cubical box with sides of length d containing an ideal gas. The molecule shown moves with velocity **v.**

Although we often picture an ideal gas as consisting of single atoms, molecular gases at low pressures provide approximations equally as good for ideal gases. Molecular rotations or vibrations have no average effect on the motions considered.

Now let us derive an expression for the pressure of N molecules of an ideal gas in a container of volume V. The container is a cube with edges of length d (Fig. 10.12). We shall focus our attention on one of these molecules, of mass m and assumed to be moving so that its component of velocity in the x direction is v_x (Fig. 10.13). As the molecule collides elastically with any wall, its velocity is reversed. Because the momentum, p, of the molecule is mv_x before the collision and $-mv_x$ after the collision, the change in momentum of the molecule is

$$\Delta p_x = mv_f - mv_i = -mv_x - mv_x = -2mv_x$$

Thus, the change in momentum of the wall is $2mv_x$. Applying the impulse-momentum theorem (Eq. 6.4) to the wall gives

$$F_1 \Delta t = \Delta p = 2mv_x$$

where F_1 is the average force on the molecule (or wall) and Δt is the interval between collisions.

In order for the molecule to make another collision with the same wall, it must travel a distance of $2d$ in the x direction. Therefore, the time interval between two collisions with the same wall is

$$\Delta t = \frac{2d}{v_x}$$

The substitution of this result into the impulse-momentum equation enables us to obtain the average force exerted by a molecule on the wall:

$$F_1 = \frac{2mv_x}{\Delta t} = \frac{2mv_x^2}{2d} = \frac{mv_x^2}{d}$$

The total force exerted on the wall by all the molecules is found by adding the average forces exerted by all the individual molecules:

$$F = \frac{m}{d} (v_{x1}^2 + v_{x2}^2 + \cdots + v_{xN}^2)$$

In this equation, v_{x1} is the x component of velocity of molecule 1, v_{x2} is the x component of velocity of molecule 2, and so on. The summation terminates when we reach molecule N because there are N molecules in the container.

To proceed further, note that the average value of the square of the velocity in the x direction for the N molecules is

$$\overline{v_x^2} = \frac{v_{x1}^2 + v_{x2}^2 + \cdots + v_{xN}^2}{N}$$

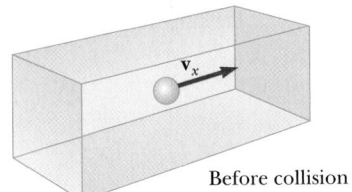

Before collision

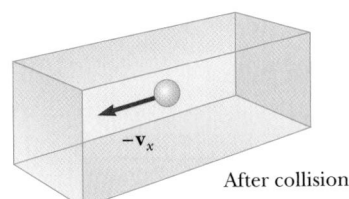

After collision

Figure 10.13 A molecule makes an elastic collision with the wall of the container. Its x component of momentum is reversed, and momentum is imparted to the wall.

Hence, the total force on the wall can be written

$$F = \frac{m}{d} N \overline{v_x^2}$$

Now let us focus on one molecule in the container and say that this molecule has velocity components v_x, v_y, and v_z. The Pythagorean theorem relates the square of the velocity to the square of these components:

$$v^2 = v_x^2 + v_y^2 + v_z^2$$

Hence, the average value of v^2 for all the molecules in the container is related to the average values of v_x^2, v_y^2, and v_z^2 according to the expression

$$\overline{v^2} = \overline{v_x^2} + \overline{v_y^2} + \overline{v_z^2}$$

Because all directions of motion are equivalent, the average velocity is the same in any direction. Therefore,

$$\overline{v_x^2} = \overline{v_y^2} = \overline{v_z^2}$$

and we have

$$\overline{v^2} = 3\overline{v_x^2}$$

Thus, the total force on the wall is

$$F = \frac{N}{3}\left(\frac{m\overline{v^2}}{d}\right)$$

This expression allows us to find the pressure exerted on the wall:

$$P = \frac{F}{A} = \frac{F}{d^2} = \frac{1}{3}\left(\frac{N}{d^3} \, m\overline{v^2}\right) = \frac{1}{3}\left(\frac{N}{V}\right)(m\overline{v^2})$$

[10.14]

$$P = \frac{2}{3}\left(\frac{N}{V}\right)(\tfrac{1}{2}m\overline{v^2})$$

◀ Pressure of an ideal gas

The glass vessel contains dry ice (solid carbon dioxide). The white cloud is carbon dioxide vapor, which is denser than air and hence falls from the vessel as shown.
(© R. Folwell/Science Photo Library)

This result shows that **the pressure is proportional to the number of molecules per unit volume and to the average translational kinetic energy of the molecules, $\frac{1}{2}m\overline{v^2}$.** With this simplified model of an ideal gas, we have arrived at an important result that relates the macroscopic quantity of pressure to an atomic quantity, the average value of the square of the molecular speed. Thus, we have a key link between the atomic world and the large-scale world.

Note that Equation 10.14 verifies some features of pressure that are probably familiar to you. One way to increase the pressure inside a container is to increase the number of molecules per unit volume in the container. You do this when you add air to a tire. The pressure in the tire can also be raised by increasing the average translational kinetic energy of the molecules in the tire. As we shall see shortly, this can be accomplished by increasing the temperature of the gas inside the tire. That is why the pressure inside a tire increases as the tire heats up during long trips. The continuous flexing of the tire as it moves along the road surface represents work being done which increases internal energy in the rubber. By heat, energy is partially transferred to the air inside, increasing the air's temperature, which in turn produces an increase in pressure.

Molecular Interpretation of Temperature

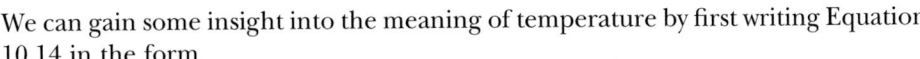

10.5, SECTION 2

We can gain some insight into the meaning of temperature by first writing Equation 10.14 in the form

$$PV = \tfrac{2}{3}N(\tfrac{1}{2}m\overline{v^2})$$

Let us compare this with the equation of state for an ideal gas:

$$PV = Nk_B T$$

Recall that the equation of state is based on experimental facts concerning the macroscopic behavior of gases. Equating the right sides of these expressions, we find that

$$T = \frac{2}{3k_B}(\tfrac{1}{2}m\overline{v^2}) \qquad [10.15]$$

◀ Temperature is proportional to average kinetic energy.

That is, **temperature is a direct measure of average molecular kinetic energy.** This result, together with Equation 10.14, implies that the pressure exerted by an ideal gas depends only on the number of molecules per unit volume and the temperature.

By rearranging Equation 10.15, we can relate the translational molecular kinetic energy to the temperature:

$$\tfrac{1}{2}m\overline{v^2} = \tfrac{3}{2}k_B T \qquad [10.16]$$

◀ Average kinetic energy per molecule

That is, the average translational kinetic energy per molecule is $\tfrac{3}{2}k_B T$. Because $\overline{v_x^2} = \tfrac{1}{3}\overline{v^2}$, it follows that

$$\tfrac{1}{2}m\overline{v_x^2} = \tfrac{1}{2}k_B T \qquad [10.17]$$

In a similar manner, for the y and z motions it follows that

$$\tfrac{1}{2}m\overline{v_y^2} = \tfrac{1}{2}k_B T \qquad \text{and} \qquad \tfrac{1}{2}m\overline{v_z^2} = \tfrac{1}{2}k_B T$$

Thus, each translational degree of freedom contributes an equal amount of energy to the gas, namely $\tfrac{1}{2}k_B T$. (In general, "degrees of freedom" refers to the number of independent means by which a molecule can possess energy.) A generalization of this result, known as the **theorem of equipartition of energy,** says that the energy of a system in thermal equilibrium is equally divided among all degrees of freedom.

◀ Theorem of equipartition of energy

The total translational kinetic energy of N molecules of gas is simply N times the average energy per molecule, which is given by Equation 10.16:

$$E = N(\tfrac{1}{2}m\overline{v^2}) = \tfrac{3}{2}Nk_B T = \tfrac{3}{2}nRT \qquad [10.18]$$

◀ Total kinetic energy of N molecules

where we have used $k_B = R/N_A$ for Boltzmann's constant and $n = N/N_A$ for the number of moles of gas. Equation 10.18 is an expression for the internal energy of the gas, and shows that the internal energy depends on the temperature.

The square root of $\overline{v^2}$ is called the **root-mean-square (rms) speed** of the molecules. From Equation 10.16 we get, for the rms speed,

$$v_{\text{rms}} = \sqrt{\overline{v^2}} = \sqrt{\frac{3k_B T}{m}} = \sqrt{\frac{3RT}{M}} \qquad [10.19]$$

◀ Root-mean-square speed

where M is the molar mass in kilograms per mole. This expression shows that, at a given temperature, lighter molecules move faster, on the average, than heavier

TABLE 10.2   Some rms Speeds

Gas	Molar Mass (kg/mol)	v_{rms} at 20°C (m/s)
H_2	2.02×10^{-3}	1902
He	4.0×10^{-3}	1352
H_2O	18×10^{-3}	637
Ne	20.1×10^{-3}	603
N_2 and CO	28×10^{-3}	511
NO	30×10^{-3}	494
CO_2	44×10^{-3}	408
SO_2	48×10^{-3}	390

molecules. For example, hydrogen, with a molar mass of 2×10^{-3} kg/mol, has molecules moving four times as fast as oxygen, whose molar mass is 32×10^{-3} kg/mol. The rms speed is not the speed at which a gas molecule moves across a room, because such a molecule undergoes several billion collisions per second with other molecules under standard conditions.

Table 10.2 lists the rms speeds for various molecules at 20°C.

Thinking Physics 3

Imagine a gas in an insulated cylinder with a movable piston. The piston has been pushed inward, compressing the gas, and is now released. As the molecules of the gas strike the piston, they move it outward. From the point of view of the kinetic theory, explain how the expansion of this gas causes its temperature to drop.

Explanation   From the point of view of kinetic theory, a molecule colliding with the piston causes the piston to rebound with some velocity. According to the conservation of momentum then, the molecule must rebound with less velocity than it had before the collision. Thus, as these collisions occur, the average velocity of the collection of molecules is reduced. Because temperature is related to the average velocity of the molecules, the temperature of the gas drops.

EXAMPLE 10.10   A Tank of Helium

A tank contains 2.0 mol of helium gas at 20°C. Assume that the helium behaves as an ideal gas.
(a) Find the total internal energy of the system.

Solution   Using Equation 10.18 with $n = 2.0$ and $T = 293$ K, we get

$$E = \tfrac{3}{2}nRT = \tfrac{3}{2}(2.0 \text{ mol})(8.31 \text{ J/mol·K})(293 \text{ K}) = \boxed{7.3 \times 10^3 \text{ J}}$$

(b) What is the average kinetic energy per molecule?

Solution   From Equation 10.16 we see that the average kinetic energy per molecule is

$$\tfrac{1}{2}m\overline{v^2} = \tfrac{3}{2}k_BT = \tfrac{3}{2}(1.38 \times 10^{-23} \text{ J/K})(293 \text{ K}) = \boxed{6.1 \times 10^{-21} \text{ J}}$$

Exercise Using the fact that the molar mass of helium is 4.0×10^{-3} kg/mol, determine the rms speed of the atoms at 20°C.

Answer 1.4×10^3 m/s

SUMMARY

The **zeroth law of thermodynamics** states that if two objects, A and B, are separately in thermal equilibrium with a third object, then A and B are in thermal equilibrium with each other.

The relationship between T_C, the **Celsius** temperature, and T, the **Kelvin (absolute)** temperature, is

$$T_C = T - 273.15 \qquad [10.1]$$

The relationship between the **Fahrenheit** and **Celsius** temperatures is

$$T_F = \tfrac{9}{5}T_C + 32 \qquad [10.2]$$

Generally, when a substance is heated, it expands. If an object has an initial length of L_0 at some temperature and undergoes a change in temperature of ΔT, this linear dimension changes by the amount ΔL, which is proportional to the object's initial length and the temperature change:

$$\Delta L = \alpha L_0 \, \Delta T \qquad [10.4]$$

The parameter α is called the **average coefficient of linear expansion.**

The change in area of a substance is given by

$$\Delta A = \gamma A_0 \, \Delta T \qquad [10.5]$$

where γ is the **average coefficient of area expansion** and is equal to 2α.

The change in volume of most substances is proportional to the initial volume, V_0, and the temperature change, ΔT:

$$\Delta V = \beta V_0 \, \Delta T \qquad [10.6]$$

where β is the **average coefficient of volume expansion** and is equal to 3α.

An **ideal gas** is one that obeys the equation

$$PV = nRT \qquad [10.8]$$

where P is the pressure of the gas, V is its volume, n is the number of moles of gas, R is the universal gas constant (8.31 J/mol·K), and T is the absolute temperature in kelvins. A real gas at very low pressures behaves approximately as an ideal gas.

The **pressure** of N molecules of an ideal gas contained in a volume V is given by

$$P = \tfrac{2}{3}\left(\frac{N}{V}\right)\left(\tfrac{1}{2}m\overline{v^2}\right) \qquad [10.14]$$

where $\tfrac{1}{2}m\overline{v^2}$ is the **average kinetic energy per molecule.**

The average kinetic energy of the molecules of a gas is directly proportional to the absolute temperature of the gas:

$$\tfrac{1}{2} m \overline{v^2} = \tfrac{3}{2} k_B T \qquad [10.16]$$

where k_B is **Boltzmann's constant** (1.38×10^{-23} J/K).
The **root-mean-square (rms) speed** of the molecules of gas is

$$v_{rms} = \sqrt{\frac{3 k_B T}{m}} = \sqrt{\frac{3RT}{M}} \qquad [10.19]$$

MULTIPLE-CHOICE QUESTIONS

1. Given that the molar mass of He is 4×10^{-3} kg/mol and for water vapor it is 18×10^{-3} kg/mol, at room temperature the ratio of the rms speed of He to water vapor molecules is approximately
   (a) 1 to 2   (b) 2 to 1   (c) 4.5 to 1
   (d) 1 to 4.5

2. One way to cool a gas is to let it expand. If a gas under 50 atm of pressure at 25°C is allowed to expand to 15 times its original volume, it reaches 1 atm pressure. What is its new temperature?
   (a) 177°C   (b) 0°C   (c) 200.7°C   (d) 89.4 K

3. The Statue of Liberty is 93.000 m tall and made of copper plates ($\alpha = 17 \times 10^{-6}$ C°$^{-1}$). If this is its height on a 20°C day, what is its height when the temperature is 35°C?
   (a) 92.764 m   (b) 93.024 m   (c) 95.46 m   (d) 108 m

4. An assumption below that is not a part of those made for the kinetic theory of gases is
   (a) The number of molecules is small.
   (b) The molecules obey Newton's laws of motion.
   (c) Collisions between molecules are elastic.
   (d) The gas is a pure substance.
   (e) The average separation between molecules is large compared to their dimensions.

5. Suppose for a brief moment that the gas molecules hitting a wall stuck to the wall instead of bouncing off the wall. How would the pressure on the wall be affected during that brief time?
   (a) the pressure would be zero.
   (b) the pressure would be halved.
   (c) the pressure would remain unchanged.
   (d) the pressure would be doubled.

CONCEPTUAL QUESTIONS

1. In an astronomy class, the temperature at the core of a star is given by the teacher as 1.5×10^7 degrees. A student asks if this is Kelvin or Celsius. How would you respond?

2. A piece of copper is dropped into a beaker of water. If the water's temperature rises, what happens to the temperature of the copper? When will the water and copper be in thermal equilibrium?

3. Common thermometers are made of a mercury column in a glass tube. Based on the operation of these common thermometers, which has the larger coefficient of linear expansion—glass or mercury? (Don't answer this by looking in a table.)

4. A steel wheel bearing is 1 mm smaller in diameter than an axle. How can the bearing be fit onto the axle without removing any material?

5. The boiling point of water is said to be 100°C in the text, but with a little luck and a very clean glass, you can often heat water in a microwave oven to 110°C for a few moments before it begins to boil. How might this occur?

6. Why is a power line more likely to break in winter than in summer even if it is loaded with the same weight?

7. Two spheres are made of the same metal and have the same radius, but one is hollow and the other is solid. The spheres are taken through the same temperature increase. Which sphere expands more?

8. After food is cooked in a pressure cooker, why is it very important to cool the container with cold water before attempting to remove the lid?

9. Some picnickers stop at a store to buy food, including bags of potato chips. They drive up into the mountains to their picnic site. When they unload the food, they no-

tice that the chip bags are puffed up like balloons. Why did this happen?

10. During normal breathing the lungs expand on inhalation and contract on exhalation. Does the ideal gas law apply to the volume of the lungs under these circumstances? Explain.

11. A common material for cushioning objects in packages is made by trapping small bubbles of air between sheets of plastic. Many individuals find satisfaction in popping these bubbles after receiving a package. (Other individuals who are nearby and listening to the sound may not be so satisfied.) Is this material more effective at cushioning the package contents on a hot day or a cold day?

12. Moving upward in the atmosphere, should you expect to find the ratio of nitrogen molecules relative to oxygen molecules increasing or decreasing? Explain.

13. Objects deep beneath the surface of the ocean are subjected to extremely high pressures, as we saw in Chapter 9. For example, some bacteria in these environments have adapted to pressures as much as a thousand times atmospheric. How might these bacteria be affected if they were rapidly moved to the surface of the ocean?

14. Two identical cylinders at the same temperature contain the same kind of gas. If cylinder *A* contains three times as much gas as cylinder *B*, what can you say about the relative pressures in the cylinders?

15. Small planets tend to have little or no atmosphere. Why is this?

16. Why do vapor bubbles in a pot of boiling water get larger as they approach the surface?

17. Markings to indicate length are placed on a steel tape in a room that is at a temperature of 22°C. Measurements are then made with the same tape on a day when the temperature is 27°C. Are the measurements too long, too short, or accurate?

18. Although the average speed of gas molecules in thermal equilibrium at some temperature is greater than zero, the average velocity is zero. Explain.

19. Explain why a column of mercury in a thermometer first descends slightly and then rises when placed in hot water.

20. One container is filled with helium gas and another with argon gas. If both containers are at the same temperature, which molecules have the higher rms speed?

21. What happens to a helium-filled balloon released into the air? Will it expand or contract? Will it stop rising at some height?

PROBLEMS

1, 2, 3 = straightforward, intermediate, challenging ☐ = full solution available in Study Guide/Student Solutions Manual ◣ = Core Concepts Workbook
WEB = solution posted at **http://www.harcourtcollege.com/physics/cptech** ▨ = biomedical application ◉ = Interactive Physics

Section 10.2 Thermometers and Temperature Scales

1. For each of the following temperatures, find the equivalent temperature on the indicated temperature scale: **(a)** $-273.15°C$ on the Fahrenheit scale, **(b)** 98.6°F on the Celsius scale, and **(c)** 100 K on the Fahrenheit scale.

2. The highest recorded temperature on Earth was 136°F, at Azizia, Libya, in 1922. The lowest recorded temperature was $-127°F$, at Vostok Station, Antarctica, in 1960. Express these temperature extremes in degrees Celsius.

3. Convert the following temperatures to Fahrenheit and kelvins: **(a)** the boiling point of liquid hydrogen $-252.87°C$; **(b)** the temperature of a room at 20°C.

4. Show that the temperature $-40°$ is unique in that it has the same numerical value on the Celsius and Fahrenheit scales.

WEB 5. A constant-volume gas thermometer is calibrated in dry ice ($-80.0°C$) and in boiling ethyl alcohol (78.0°C). The two pressures are 0.900 atm and 1.635 atm. **(a)** What value of absolute zero does the calibration yield? **(b)** What pressures would be found at the freezing and boiling points of water?

6. The pressure in a constant-volume gas thermometer is 0.700 atm at 100°C and 0.512 atm at 0°C. **(a)** What is the temperature when the pressure is 0.0400 atm? **(b)** What is the pressure at 450°C?

7. Show that if the temperature on the Celsius scale changes by ΔT_C, the Fahrenheit temperature changes by the amount $\Delta T_F = (9/5)\Delta T_C$.

Section 10.3 Thermal Expansion of Solids and Liquids

8. The New River Gorge bridge in West Virginia is a 518-m-long steel arch. How much will its length change between temperature extremes $-20°C$ and 35°C?

9. A gold ring has an inner diameter of 2.168 cm at a temperature of 15.0°C. Determine its inner diameter at 100°C. ($\alpha_{gold} = 1.42 \times 10^{-5}°C^{-1}$.)

10. A grandfather clock is controlled by a swinging brass pendulum that is 1.3000 m long at a temperature of 20.0°C. **(a)** What is the length of the pendulum rod when the temperature drops to 0.0°C? **(b)** If a pendulum's period is given by $T = 2\pi\sqrt{L/g}$, where L is its

length, does the change in length of the rod cause the clock to run fast or slow?

11. A pair of eyeglass frames are made of epoxy plastic (coefficient of linear expansion $= 130 \times 10^{-6}°C^{-1}$). At room temperature (assume 20.0°C) the frames have circular lens holes 2.20 cm in radius. To what temperature must the frames be heated in order to insert lenses 2.21 cm in radius?

12. A cube of solid aluminum has a volume of 1.00 m³ at 20°C. What temperature change is required to produce a 100 cm³ increase in the volume of the cube?

13. A cylindrical brass sleeve is to be shrink-fitted over a brass shaft whose diameter is 3.212 cm at 0°C. The diameter of the sleeve is 3.196 cm at 0°C. **(a)** To what temperature must the sleeve be heated before it will slip over the shaft? **(b)** Alternatively, to what temperature must the shaft be cooled before it will slip into the sleeve?

14. Show that the coefficient of volume expansion, β, is related to the coefficient of linear expansion, α, through the expression $\beta = 3\alpha$.

15. At 20.000°C, an aluminum ring has an inner diameter of 5.000 cm, and a brass rod has a diameter of 5.050 cm. **(a)** To what temperature must the aluminum ring be heated so that it will just slip over the brass rod? **(b)** To what temperature must *both* be heated so the aluminum ring will slip off the brass rod? Would this work?

16. A construction worker uses a steel tape to measure the length of an aluminum support column. If the measured length is 18.7 m when the temperature is 21.2°C, what is the measured length when the temperature rises to 29.4°C? (*Note:* Do not neglect the expansion of the steel tape.)

17. The band in Figure P10.17 is stainless steel (coefficient of linear expansion $= 17.3 \times 10^{-6}°C^{-1}$, Young's modulus $= 18 \times 10^{10}$ N/m²). It is essentially circular with an initial mean radius of 5.0 mm, a height of 4.0 mm, and a thickness of 0.50 mm. If the band just fits snugly over the tooth when heated to a temperature of 80°C, what is the tension in the band when it cools to a temperature of 37°C?

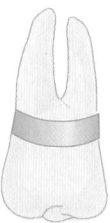

Figure P10.17

18. An automobile fuel tank is filled to the brim with 45 L (12 gal) of gasoline at 10°C. Immediately afterward, the vehicle is parked in the Sun, where the temperature is 35°C. How much gasoline overflows from the tank as a result of the expansion? (Neglect the expansion of the tank.)

19. An underground gasoline tank at 54°F can hold 1000 gallons of gasoline. If the driver of a tanker truck fills the underground tank on a day when the temperature is 90°F, how many gallons, according to his measure on the truck, can he pour in? Assume that the temperature of the gasoline cools to 54°F on entering the tank.

Section 10.4 Macroscopic Description of an Ideal Gas

20. The ideal gas constant in the SI system is 8.315 J/mol·K. In chemistry classes, the mixed units L·atm/mol·K are used for this constant. Find the ideal gas constant in the mixed units to 4 significant figures.

21. One mole of oxygen gas is at a pressure of 6.00 atm and a temperature of 27.0°C. **(a)** If the gas is heated at constant volume until the pressure triples, what is the final temperature? **(b)** If the gas is heated so that both the pressure and volume are doubled, what is the final temperature?

22. Gas is contained in an 8.0-L vessel at a temperature of 20°C and a pressure of 9.0 atm. **(a)** Determine the number of moles of gas in the vessel. **(b)** How many molecules are in the vessel?

23. **(a)** An ideal gas occupies a volume of 1.0 cm³ at 20°C and atmospheric pressure. Determine the number of molecules of gas in the container. **(b)** If the pressure of the 1.0 cm³ volume is reduced to 1.0×10^{-11} Pa (an extremely good vacuum) while the temperature remains constant, how many moles of gas remain in the container?

24. The pressure on an ideal gas is cut in half, resulting in a decrease in temperature to three fourths of the original value. Calculate the ratio of the final volume to the original volume of the gas.

25. A cylinder with a movable piston contains gas at a temperature of 27.0°C, a volume of 1.50 m³, and an absolute pressure of 0.200×10^5 Pa. What will be its final temperature if the gas is compressed to 0.700 m³ and the absolute pressure increases to 0.800×10^5 Pa?

26. Gas is confined in a tank at a pressure of 10.0 atm and a temperature of 15.0°C. If half of the gas is withdrawn and the temperature is raised to 65.0°C, what is the new pressure in the tank?

27. A rigid tank contains 0.40 moles of oxygen (O_2). Determine the mass (in kg) of oxygen that must be withdrawn from the tank to lower the pressure of the gas from 40 atm to 25 atm. Assume that the volume of the tank and the temperature of the oxygen are constant during this operation.

28. A weather balloon is designed to expand to a maximum radius of 20 m when in flight at its working altitude, where the air pressure is 0.030 atm and the temperature is 200 K. If the balloon is filled at atmospheric pressure and 300 K, what is its radius at lift-off?

WEB **29.** A cylindrical diving bell, 3.00 m in diameter and 4.00 m tall with an open bottom, is submerged to a depth of 220 m in the ocean. The surface temperature is 25.0°C, and the temperature 220 m down is 5.00°C. The density of seawater is 1025 kg/m³. How high does the seawater rise in the bell when it is submerged?

30. An air bubble has a volume of 1.50 cm³ when it is released by a submarine 100 m below the surface of a lake. What is the volume of the bubble when it reaches the surface? Assume that the temperature of the air in the bubble remains constant during ascent.

31. The density of helium gas at $T = 0°C$ and atmospheric pressure is $\rho_0 = 0.179$ kg/m³. The temperature is then raised to $T = 100°C$, but the pressure is kept constant. Assuming that helium is ideal, calculate the new density, ρ_f, of the gas.

Section 10.5 Avogadro's Number and the Ideal Gas Law

Section 10.6 The Kinetic Theory of Gases

32. A sealed cubical container 20.0 cm on a side contains three times Avogadro's number of molecules at a temperature of 20.0°C. Find the force exerted by the gas on one of the walls of the container.

33. What is the average kinetic energy of oxygen molecules at a temperature of 300 K?

34. (a) What is the total random kinetic energy of all the molecules in one mole of hydrogen at a temperature of 300 K? (b) With what speed would a mole of hydrogen have to move so that the kinetic energy of the mass as a whole would be equal to the total random kinetic energy of its molecules?

35. Use Avogadro's number to find the mass of a helium atom.

36. Three moles of nitrogen gas, N_2, at 27.0°C are contained in a 22.4-L cylinder. Find the pressure the gas exerts on the cylinder walls.

WEB **37.** The temperature near the top of the atmosphere on Venus is 240 K. (a) Find the rms speed of hydrogen (H_2) at this point in the atmosphere. (b) Repeat for carbon dioxide (CO_2). (c) It has been found that if the rms speed exceeds one sixth of the planet's escape velocity, the gas eventually leaks out of the atmosphere and into outer space. If the escape velocity on Venus is 10.3 km/s, does hydrogen escape? Does carbon dioxide?

38. A cylinder contains a mixture of helium and argon gas in equilibrium at a temperature of 150°C. (a) What is the average kinetic energy of each type of molecule? (b) What is the rms speed of each type of molecule?

39. Superman leaps in front of Lois Lane to save her from a volley of bullets. In a one-minute interval, an automatic weapon fires 150 bullets, each of mass 8.0 g, at 400 m/s. The bullets strike his mighty chest, which has an area of 0.75 m². Find the average force exerted on Superman's chest if the bullets bounce back after an elastic, head-on collision.

40. In a period of 1.0 s, 5.0×10^{23} nitrogen molecules strike a wall of area 8.0 cm². If the molecules move at 300 m/s and strike the wall head on in a perfectly elastic collision, find the pressure exerted on the wall. (The mass of one N_2 molecule is 4.68×10^{-26} kg.)

41. If 2.0 mol of a gas are confined to a 5.0-L vessel at a pressure of 8.0 atm, what is the average kinetic energy of a gas molecule?

ADDITIONAL PROBLEMS

42. The active element of a certain laser is an ordinary glass rod 20 cm long and 1.0 cm in diameter. If the temperature of the rod increases by 75°C, find its increases in (a) length, (b) diameter, and (c) volume.

43. A 1.5-m-long glass tube, closed at one end, is weighted and lowered to the bottom of a fresh-water lake. When the tube is recovered, an indicator mark shows that water rose to within 0.40 m of the closed end. Determine the depth of the lake. Assume constant temperature.

44. The density of gasoline is 730 kg/m³ at 0°C. Its volume expansion coefficient is $9.6 \times 10^{-4}°C^{-1}$. If one gallon of gasoline occupies 0.0038 m³, how many extra kilograms of gasoline are obtained when 10 gallons of gasoline are bought at 0°C rather than at 20°C?

45. A vertical cylinder of cross-sectional area 0.050 m² is fitted with a tight-fitting, frictionless piston of mass 5.0 kg (Fig. P10.45). If there are 3.0 mol of an ideal gas in the cylinder at 500 K, determine the height, h, at which the piston will be in equilibrium under its own weight.

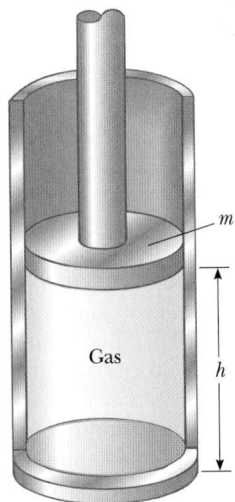

Figure P10.45

46. A liquid with coefficient of volume expansion β just fills a spherical container of volume V_0 at temperature T

(Fig. P10.46). The container is made of a material that has a coefficient of linear expansion of α. The liquid is free to expand into a capillary of cross-sectional area A at the top. (a) If the temperature increases by ΔT, show that the liquid rises in the capillary by the amount $\Delta h = (V_0/A)(\beta - 3\alpha)\Delta T$. (b) For a typical system, such as a mercury thermometer, why is it a good approximation to neglect the expansion of the container?

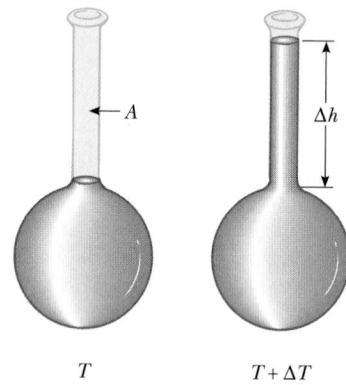

$$T \qquad\qquad T + \Delta T$$

Figure P10.46   (Problems 46 and 57)

47. A hollow aluminum cylinder is to be fitted over a steel piston. At 20°C the inside diameter of the cylinder is 99% of the outside diameter of the piston. To what common temperature should the two pieces be heated in order that the cylinder just fits over the piston?

48. A steel measuring tape was designed to read correctly at 20°C. A parent uses the tape to measure the height of a 1.1-m-tall child. If the measurement is made on a day when the temperature is 25°C, is the tape reading longer or shorter than the actual height, and by how much?

49. Before beginning a long trip on a hot day, a driver inflates an automobile tire to a gauge pressure of 1.80 atm at 300 K. At the end of the trip the gauge pressure has increased to 2.20 atm. (a) Assuming the volume has remained constant, what is the temperature of the air inside the tire? (b) What percentage of the original mass of air in the tire should be released so the pressure returns to the original value? Assume the temperature remains at the value found in (a), and the volume of the tire remains constant as air is released.

50. Two small containers of equal volume, 100 cm³, each contain helium gas at 0°C and 1.00 atm pressure. The two containers are joined by a small open tube of negligible volume, allowing gas to flow from one container to the other. What common pressure will exist in the two containers if the temperature of one container is raised to 100°C while the other container is kept at 0°C?

WEB 51. A copper rod and steel rod are heated. At 0°C the copper rod has a length of L_C, the steel one has a length

L_S. When the rods are being heated or cooled, a difference of 5.00 cm is maintained between their lengths. Determine the values of L_C and L_S.

52. Nine grams of water are placed in a 2.00-L pressure cooker and heated to 500°C. What is the pressure inside the container?

53. An expandable cylinder has its top connected to a spring of constant 2.0×10^3 N/m (see Fig. P10.53). The cylinder is filled with 5.00 L of gas with the spring relaxed at a pressure of 1.00 atm and a temperature of 20.0°C. (a) If the lid has a cross-sectional area of 0.0100 m² and negligible mass, how high will the lid rise when the temperature is raised to 250°C? (b) What is the pressure of the gas at 250°C?

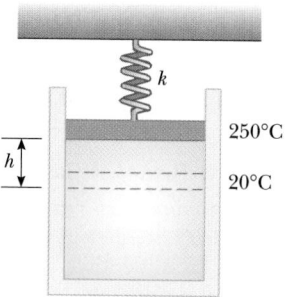

Figure P10.53

54. A 250-m-long bridge is improperly designed so that it cannot expand with temperature. It is made of concrete with $\alpha = 12 \times 10^{-6}$°C^{-1}. (a) Assuming that the maximum change in temperature at the site is expected to be 20°C, find the change in length the span would undergo if it were free to expand. (b) Show that the stress on an object with Young's modulus Y when raised by ΔT with its ends firmly fixed is given by $\alpha Y \Delta T$. (c) If the maximum stress the bridge can withstand without crumbling is 2.0×10^7 Pa, will it crumble because of this temperature increase? Young's modulus for concrete is about 2.0×10^{10} Pa.

55. A swimmer has 0.820 L of dry air in her lungs when she dives into a lake. Assuming the pressure of the dry air is 95% of the external pressure at all times, what is the volume of the dry air at a depth of 10.0 m? Assume that atmospheric pressure at the surface is 1.013×10^5 Pa.

56. Use the ideal gas equation and the relationship between moles and the mass of a gas to find an expression for the density of a gas.

57. A mercury thermometer is constructed as in Figure P10.46. The capillary tube has a diameter of 0.0050 cm, and the bulb has a diameter of 0.30 cm. Neglecting the expansion of the glass, find the change in height of the mercury column for a temperature change of 25°C.

58. A bimetallic bar is made of two thin strips of dissimilar metals bonded together. As they are heated, the one with the larger average coefficient of expansion expands more than the other, forcing the bar into an arc, with the outer radius having a larger circumference (see Fig. P10.58). **(a)** Derive an expression for the angle of bending θ as a function of the initial length of the strips, their average coefficients of linear expansion, the change in temperature, and the separation of the centers of the strips ($\Delta r = r_2 - r_1$). **(b)** Show that the angle of bending goes to zero when ΔT goes to zero or when the two coefficients of expansion become equal. **(c)** What happens if the bar is cooled?

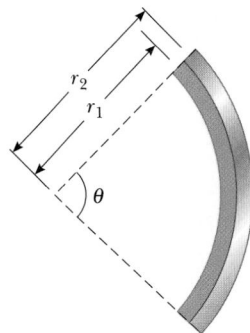

Figure P10.58

Heat

Chapter Outline

▲ PHYSICS PUZZLER

In regions where the temperature drops below freezing in winter, you will often see a road sign when approaching a bridge, warning "Bridge Freezes Before Roadway." Why does the bridge become icy before the roadway? *(© David R. Frazier Photolibrary)*

It is an experimentally established fact that when two objects at different temperatures are placed in thermal contact with each other, the temperature of the warmer object decreases and the temperature of the cooler object increases. If the two are left in contact for some time, they eventually reach a common equilibrium temperature that is intermediate between the two initial temperatures. When such a process occurs, we say that heat is transferred from the object at the higher temperature to the one at the lower temperature.

Up until about 1850, the subjects of heat and mechanics were considered to be two distinct branches of science, and the principle of conservation of energy seemed to be a rather specialized result that could be used to describe certain kinds of mechanical systems. After the two disciplines were shown to be related, the principle of conservation of energy emerged as a universal principle of nature. From this new perspective, heat and work are both treated as modes of energy transfer. Experiments performed by the Englishman James Joule (1818–1889) and his contemporaries demonstrated that whenever heat is gained or lost by a system during some process, the gain or loss can be accounted for by an equivalent quantity of mechanical work done on the system. Thus, with the broadening of the concept of energy to include heat as a path of energy transfer, the principle of energy conservation was extended.

The focus of this chapter is to introduce the concept of heat and some of the processes that enable heat to be transferred between a system and its surroundings.

James Prescott Joule, British physicist (1818–1889)

Joule received some formal education in mathematics, philosophy, and chemistry from John Dalton but was in large part self-educated. Joule's most active research period, from 1837 through 1847, led to the establishment of the principle of conservation of energy and the equivalence of heat and other forms of energy. His study of the quantitative relationship among electrical, mechanical, and chemical effects of heat culminated in his announcement in 1843 of the amount of work required to produce a unit of heat, called the mechanical equivalent of heat. *(By kind permission of the President and Council of the Royal Society)*

11.1 THE MECHANICAL EQUIVALENT OF HEAT

Heat (or **thermal energy**) is now defined as energy that is transferred between a system and its environment because of a temperature difference between them. Before scientists arrived at a correct understanding of heat, the units in which heat was measured had already been developed. They were chosen because of early misunderstandings about heat. These unusual units are still widely used in many applications, and so we shall discuss them briefly here. One of the most widely used is the **calorie (cal),** defined as **the heat required to raise the temperature of 1 g of water from 14.5°C to 15.5°C.** A related unit is the **kilocalorie (kcal),** 1 kcal = 10^3 cal. The Calorie with a capital C, used in describing the energy equivalent of foods, is equal to 1 kcal. The unit of heat in the British engineering system is the **British thermal unit** (Btu), defined as **the heat required to raise the temperature of 1 lb of water from 63°F to 64°F.**

◀ Definition of the calorie

Because heat is now recognized as energy being transferred, scientists are increasingly using the SI unit of energy, the **joule** (J), for quantities of heat. In this book, heat is most often measured in joules.

When the concept of mechanical energy was introduced in Chapter 5, we found that whenever friction is present in a mechanical system, some mechanical energy is lost. Experiments of various sorts show that this lost mechanical energy does not simply disappear; instead, it is transformed into thermal energy. Joule (1818–1889) was the first to establish the equivalence of the two forms of energy. He found that

$$1 \text{ cal} = 4.186 \text{ J} \qquad [11.1]$$

The ratio 4.186 J/cal is known, for purely historical reasons, as the **mechanical equivalent of heat.**

EXAMPLE 11.1 Losing Weight the Hard Way

A student eats a dinner rated at 2000 (food) Calories. He wishes to do an equivalent amount of work in the gymnasium by lifting a 50.0-kg mass. How many times must he raise the weight to expend this much energy? Assume that he raises the weight a distance of 2.00 m each time and that no work is done when the weight is dropped to the floor.

Solution Because 1 Calorie = 10^3 cal, the work required is 2.00×10^6 cal. Converting this to joules, we have, for the total work required,

$$W = (2.00 \times 10^6 \text{ cal})(4.186 \text{ J/cal}) = 8.37 \times 10^6 \text{ J}$$

The work done in lifting the weight once through the distance h is equal to mgh, and the work done in lifting the weight n times is $nmgh$. If we set $nmgh$ equal to the total work required, we have

$$W = nmgh = 8.37 \times 10^6 \text{ J}$$

Because $m = 50.0$ kg and $h = 2.00$ m, we get

$$n = \frac{8.37 \times 10^6 \text{ J}}{(50.0 \text{ kg})(9.80 \text{ m/s}^2)(2.00 \text{ m})} = 8.54 \times 10^3 \text{ times}$$

If the student is in good shape and lifts the weight, say, once every 5.0 s, it will take him about 12 h to perform this feat. Clearly, it is much easier to lose weight by dieting.

This problem is somewhat misleading in that it assumes perfect conversion of chemical energy into mechanical energy. In actual practice, a more realistic assessment of the number of repetitions can be found by dividing the given answer by 6. Thus, about 1400 lifts are required to burn off the calories.

TABLE 11.1
Specific Heats of Some Materials at Atmospheric Pressure

Substance	J/kg·°C	cal/g·°C
Aluminum	900	0.215
Beryllium	1820	0.436
Cadmium	230	0.055
Copper	387	0.0924
Germanium	322	0.077
Glass	837	0.200
Gold	129	0.0308
Ice	2090	0.500
Iron	448	0.107
Lead	128	0.0305
Mercury	138	0.033
Silicon	703	0.168
Silver	234	0.056
Steam	2010	0.480
Water	4186	1.00

11.2   SPECIFIC HEAT

The quantity of heat energy required to raise the temperature of a given mass of a substance by some amount varies from one substance to another. For example, the heat required to raise the temperature of 1 kg of water by 1°C is 4186 J, but the heat required to raise the temperature of 1 kg of copper by 1°C is only 387 J. Every substance has a unique value for the amount of heat required to change the temperature of 1 kg of it by 1°C, and this number is referred to as the specific heat of the substance. Table 11.1 lists specific heats for a few substances. Note that specific heats do vary with temperature. The values given in Table 11.1 represent average values in the range of "ordinary" temperatures where they are approximately constant.

Suppose that a quantity, Q, of heat is transferred to a substance of mass m, thereby changing its temperature by ΔT. The **specific heat,** c, of the substance is defined as

Specific heat ▶

$$c \equiv \frac{Q}{m \, \Delta T} \qquad \qquad [11.2]$$

From this definition we can express the heat transferred between a system of mass m and its surroundings for a temperature change of ΔT as

$$Q = mc\,\Delta T \qquad\qquad [11.3]$$

For example, the heat required to raise the temperature of 0.500 kg of water by 3.00°C is equal to (0.500 kg)(4186 J/kg·°C)(3.00°C) = 6280 J. Note that when the temperature increases, ΔT and Q are taken to be *positive*, corresponding to heat flowing *into* the system. Likewise, when the temperature decreases, ΔT and Q are *negative* and heat flows *out* of the system.

Note from Table 11.1 that water has the highest specific heat of the substances we are likely to encounter on a routine basis. This high specific heat is responsible for the moderate temperatures found in regions near large bodies of water. As the temperature of a body of water decreases during winter, the water gives off heat to the air, which carries the heat landward when prevailing winds are favorable. For example, the prevailing winds off the western coast of the United States are toward the land, and the heat liberated by the Pacific Ocean as it cools keeps coastal areas much warmer than they would otherwise be. This explains why the western coastal states generally have more favorable winter weather than the eastern coastal states, where the winds do not carry the heat toward land. The same effect also keeps summers cooler in San Francisco than in a similar region on the East coast.

The fact that the specific heat of water is higher than that of land is responsible for the pattern of air flow at a beach. During the day, the Sun adds roughly equal amounts of energy to beach and water, but the lower specific heat of sand causes the beach to reach a higher temperature than the water. As a consequence, the air above the land reaches a higher temperature than that over the water, and cooler air from above the water is drawn in to displace this rising hot air, resulting in a breeze from ocean to land during the day. Because the hot air gradually cools as it rises, it subsequently sinks, setting up the circulating pattern shown in Figure 11.1a. During the night, the land cools more quickly than the water, and the circulating pattern reverses itself because the hotter air is now over the water (Fig. 11.1b). The offshore and onshore breezes are certainly well known to sailors.

APPLICATION

Sea Breezes and Thermals.

Day

Night

(a)

(b)

Figure 11.1 Circulation of air at the beach. (a) On a hot day, the air above the sand warms faster than the air above the cooler water. The cooler air over the water moves toward the beach, displacing the rising warmer air. (b) At night the sand cools more rapidly than the water, and hence the air currents reverse direction.

A similar effect produces rising layers of air, called *thermals,* that can help eagles to soar higher and hang gliders to stay in flight longer. A thermal is created when a portion of the Earth reaches a higher temperature than neighboring regions. This often happens to plowed fields, which are heated by the Sun to higher temperatures than nearby fields shaded by vegetation. The cooler, denser air over the vegetation-covered fields pushes the expanding air over the plowed field upward, and a thermal is formed.

11.3   CONSERVATION OF ENERGY: CALORIMETRY

Situations in which mechanical energy is converted to thermal energy occur frequently. We shall look at some of them in examples and in the problems at the end of the chapter, but most of our attention here is directed toward a particular kind of conservation-of-energy situation. In problems using the procedure called *calorimetry,* only the transfer of thermal energy between the system and its surroundings is considered.

One technique for measuring the specific heat of a solid or liquid is simply to heat the substance to some known temperature, place it in a vessel containing water of known mass and temperature, and measure the temperature of the water after equilibrium is reached. Because a negligible amount of mechanical work is done in the process, the law of conservation of energy requires that the heat that leaves the warmer substance (of unknown specific heat) equal the heat that enters the water. Devices in which this heat transfer occurs are called **calorimeters.**

Suppose that m_x is the mass of a substance whose specific heat we wish to measure, c_x its specific heat, and T_x its initial temperature. Let m_w, c_w, and T_w represent the corresponding values for the water. If T is the final equilibrium temperature after everything is mixed, then from Equation 11.2 we find that the heat gained by the water is $m_w c_w(T - T_w)$ and the heat lost by the substance of unknown specific heat, c_x, is $m_x c_x(T_x - T)$. Assuming that the system (water + unknown) neither loses nor gains heat, and if the cup has negligible mass, it follows that the heat gained by the water must equal the heat lost by the unknown (conservation of energy):

$$m_w c_w(T - T_w) = m_x c_x(T_x - T)$$

Solving for c_x gives

$$c_x = \frac{m_w c_w(T - T_w)}{m_x(T_x - T)}$$

Do not attempt to memorize this equation. Instead, always start from first principles in solving calorimetry problems. That is, determine which substances lose heat and which gain heat, and then equate heat loss to heat gain.

A word about sign conventions: In a later chapter we shall find it necessary to use a sign convention in which a positive sign for Q indicates heat gained by a substance and a negative sign indicates heat lost. However, for calorimetry problems it is less confusing if you ignore the positive and negative signs and instead equate heat loss to heat gain. That is, you should always write ΔT as a positive quantity. For example, if T_f is greater than T_i, let $\Delta T = T_f - T_i$. If T_f is less than T_i, let $\Delta T = T_i - T_f$.

EXAMPLE 11.2   Cooling a Hot Ingot

A 0.0500-kg ingot of metal is heated to 200.0°C and then dropped into a beaker containing 0.400 kg of water that is initially at 20.0°C. If the final equilibrium temperature of the mixed system is 22.4°C, find the specific heat of the metal.

Solution   Because the heat lost by the ingot equals the heat gained by the water, we can write

$$m_x c_x (T_{ix} - T_{fx}) = m_w c_w (T_{fw} - T_{iw})$$

$$(0.0500 \text{ kg})(c_x)(200.0°C - 22.4°C) = (0.400 \text{ kg})(4186 \text{ J/kg·°C})(22.4°C - 20.0°C)$$

from which we find that

$$c_x = \boxed{453 \text{ J/kg·°C}}$$

The ingot is most likely iron, as can be seen by comparing this result with the data in Table 11.1.

EXAMPLE 11.3   Fun Time for a Cowboy

A cowboy fires a 2.00-g silver bullet at a muzzle speed of 200 m/s into the pine wall of a saloon. Assume that all the internal energy generated by the impact remains with the bullet. What is the temperature change of the bullet?

Solution   The kinetic energy of the bullet is

$$\tfrac{1}{2}mv^2 = \tfrac{1}{2}(2.00 \times 10^{-3} \text{ kg})(200 \text{ m/s})^2 = 40.0 \text{ J}$$

Nothing in the environment is hotter than the bullet, so the bullet gains no heat. Its temperature increases because the 40.0 J of kinetic energy becomes 40.0 J of extra internal energy. The temperature change would be the same as if 40.0 J of heat were transferred from a stove to the bullet, and we imagine this process to compute ΔT from

$$Q = mc\,\Delta T$$

Because the specific heat of silver is 234 J/kg·°C (Table 11.1), we get

$$\Delta T = \frac{Q}{mc} = \frac{40.0 \text{ J}}{(2.00 \times 10^{-3} \text{ kg})(234 \text{ J/kg·°C})} = \boxed{85.5°C}$$

Exercise   Suppose the cowboy runs out of silver bullets and fires a lead bullet of the same mass and velocity into the wall. What is the temperature change of the bullet?

Answer   157°C

11.4   LATENT HEAT AND PHASE CHANGES

A substance usually undergoes a change in temperature when heat is transferred between it and its surroundings. There are situations, however, in which the flow of heat does not result in a change in temperature. This is the case whenever the substance undergoes a physical alteration from one form to another, referred to as

TABLE 11.2   Latent Heats of Fusion and Vaporization

Substance	Melting Point (°C)	Latent Heat of Fusion J/kg	(cal/g)	Boiling Point (°C)	Latent Heat of Vaporization J/kg	(cal/g)
Helium	−269.65	5.23×10^3	(1.25)	−268.93	2.09×10^4	(4.99)
Nitrogen	−209.97	2.55×10^4	(6.09)	−195.81	2.01×10^5	(48.0)
Oxygen	−218.79	1.38×10^4	(3.30)	−182.97	2.13×10^5	(50.9)
Ethyl alcohol	−114	1.04×10^5	(24.9)	78	8.54×10^5	(204)
Water	0.00	3.33×10^5	(79.7)	100.00	2.26×10^6	(540)
Sulfur	119	3.81×10^4	(9.10)	444.60	3.26×10^5	(77.9)
Lead	327.3	2.45×10^4	(5.85)	1750	8.70×10^5	(208)
Aluminum	660	3.97×10^5	(94.8)	2450	1.14×10^7	(2720)
Silver	960.80	8.82×10^4	(21.1)	2193	2.33×10^6	(558)
Gold	1063.00	6.44×10^4	(15.4)	2660	1.58×10^6	(377)
Copper	1083	1.34×10^5	(32.0)	1187	5.06×10^6	(1210)

a **phase change.** Some common phase changes are solid to liquid (melting), liquid to gas (boiling), and a change in crystalline structure of a solid. Every phase change involves a change in internal energy.

The heat required to change the phase of a given mass, m, of a pure substance is

$$Q = mL \qquad [11.4]$$

where L is called the **latent heat** ("hidden" heat) of the substance and depends on the nature of the phase change as well as on the properties of the substance. **Latent heat of fusion, L_f,** is the term used when the phase change is from solid to liquid, and **latent heat of vaporization, L_v,** is the term used when the phase change is from liquid to gas.[1] For example, the latent heat of fusion for water at atmospheric pressure is 3.33×10^5 J/kg, and the latent heat of vaporization for water is 2.26×10^6 J/kg. The latent heats of different substances vary considerably, as Table 11.2 shows.

Consider, for example, the heat required to convert a 1.00-g block of ice at −30.0°C to steam (water vapor) at 120.0°C. Figure 11.2 indicates the experimental results obtained when heat is gradually added to the ice. Let us examine each portion of the curve separately.

Part A.   On this portion of the curve the temperature of the ice is changing from −30.0°C to 0.0°C. Because the specific heat of ice is 2090 J/kg · °C, we can calculate the amount of heat added from Equation 11.3:

$$Q = m_i c_i \, \Delta T = (1.00 \times 10^{-3} \text{ kg}) (2090 \text{ J/kg} \cdot {}^\circ\text{C}) (30.0 {}^\circ\text{C}) = 62.7 \text{ J}$$

[1] When a gas gives up heat, it eventually returns to the liquid phase, or *condenses*. The heat per unit mass given up during the process is called the *heat of condensation,* and it equals the heat of vaporization. When a liquid loses heat, it eventually solidifies, and the *heat of solidification* equals the heat of fusion.

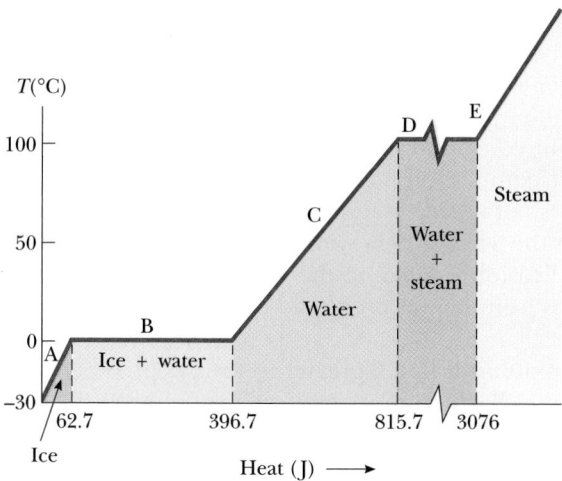

Figure 11.2 A plot of temperature versus heat added when 1.00 g of ice, initially at −30.0°C, is converted to steam at 120°C.

Part B. When the ice reaches 0°C, the ice–water mixture remains at this temperature — even though heat is being added — until all the ice melts to become water at 0°C. The heat required to melt 1.00 g of ice at 0°C is, from Equation 11.4,

$$Q = mL_f = (1.00 \times 10^{-3} \text{ kg})(3.33 \times 10^5 \text{ J/kg}) = 333 \text{ J}$$

Part C. Between 0°C and 100°C, nothing surprising happens. No phase change occurs in this region. The heat added to the water is being used to increase its temperature. The amount of heat necessary to increase the temperature from 0°C to 100°C is

$$Q = m_w c_w \Delta T = (1.00 \times 10^{-3} \text{ kg})(4.19 \times 10^3 \text{ J/kg} \cdot °C)(100°C)$$
$$= 4.19 \times 10^2 \text{ J}$$

Part D. At 100°C, another phase change occurs as the water changes from water at 100°C to steam at 100°C. Just as in Part B, the water–steam mixture remains at 100°C — even though heat is being added — until all the liquid has been converted to steam. The heat required to convert 1.00 g of water to steam at 100°C is

$$Q = mL_v = (1.00 \times 10^{-3} \text{ kg})(2.26 \times 10^6 \text{ J/kg}) = 2.26 \times 10^3 \text{ J}$$

Part E. On this portion of the curve, heat is being added to the steam with no phase change occurring. The heat that must be added to raise the temperature of the steam to 120°C is

$$Q = m_s c_s \Delta T = (1.00 \times 10^{-3} \text{ kg})(2.01 \times 10^3 \text{ J/kg} \cdot °C)(20°C) = 40.2 \text{ J}$$

The *total amount of heat* that must be added to change one gram of ice at −30°C to steam at 120°C is therefore about 3.11×10^3 J. Conversely, to cool one gram of steam at 120°C down to the point at which we have ice at −30°C, we must remove 3.11×10^3 J of heat.

Phase changes can be described in terms of rearrangements of molecules when heat is added to or removed from a substance. Consider first the liquid-to-gas phase change. The molecules in a liquid are close together, and the forces between them are stronger than those between the more widely separated molecules of a gas. Therefore, work must be done on the liquid against these attractive molecular forces in order to separate the molecules. The latent heat of vaporization is the amount of energy that must be added to the liquid to accomplish this.

Similarly, at the melting point of a solid, we imagine that the amplitude of vibration of the atoms about their equilibrium positions becomes great enough to allow the atoms to pass the barriers of adjacent atoms and move to their new positions. The new locations are, on the average, less symmetrical and therefore have higher energy. The latent heat of fusion is equal to the work required at the molecular level to transform the mass from the ordered solid phase to the disordered liquid phase.

The average distance between atoms is much greater in the gas phase than in either the liquid or the solid phase. Each atom or molecule is removed from its neighbors, without the compensation of attractive forces to new neighbors. Therefore, it is not surprising that more work is required at the molecular level to vaporize a given mass of substance than to melt it; thus the latent heat of vaporization is much greater than the latent heat of fusion (Table 11.2).

Thinking Physics 1

The intuitive result of adding energy to a substance is to observe an increase in its temperature. But if energy is added to ice at 0°C, its temperature does not increase. Where does the energy go?

Explanation When energy is added to ice at, say, −10°C, much of the energy goes into increased molecular vibration, which has a physical manifestation of an increased temperature. When energy is added to ice at 0°C, the temperature of the ice remains constant and the ice melts. Much of the energy goes into increasing the electrical potential energy and breaking the bonds between molecules. This is similar to adding energy to a rock–Earth system by throwing the rock so fast that it escapes Earth—the gravitational bond between the rock and the Earth has been broken. Breaking the bonds between the water molecules in the ice allows the molecules to move with respect to each other, which we recognize as the behavior of the liquid that results from the melting.

Problem Solving Strategy

Calorimetry Problems

If you are having difficulty with calorimetry problems, make the following considerations.

1. Be sure your units are consistent throughout. That is, if you are using specific heats measured in cal/g·°C, be sure that masses are in grams and temperatures are in Celsius units throughout.
2. Losses and gains in heat are found by using $Q = mc\,\Delta T$ only for those intervals in which no phase changes occur. Likewise, the equations

$Q = mL_f$ and $Q = mL_v$ are to be used only when phase changes are taking place.

3. Often sign errors occur in heat loss = heat gain equations. One way to determine whether your equation is correct is to examine the signs of all ΔT's that appear in your equation. Every one should be a positive number.

EXAMPLE 11.4   Cooling the Steam

What mass of steam that is initially at 130°C is needed to warm 200 g of water in a 100-g glass container from 20.0°C to 50.0°C?

Solution   This is a heat transfer problem in which we must equate the heat lost by the steam to the heat gained by the water and glass container. The steam loses heat in three stages. In the first, the steam is cooled to 100°C. The heat liberated in the process is

$$Q_1 = m_x c_s \, \Delta T = m_x (2.01 \times 10^3 \, \text{J/kg·°C}) (30.0°C) = m_x (6.03 \times 10^4 \, \text{J/kg})$$

In the second stage, the steam is converted to water. In this case, to find the heat removed, we use the heat of vaporization and $Q = mL_v$:

$$Q_2 = m_x (2.26 \times 10^6 \, \text{J/kg})$$

In the last stage, the temperature of the water is reduced to 50.0°C. This liberates heat in the amount of

$$Q_3 = m_x c_w \, \Delta T = m_x (4.19 \times 10^3 \, \text{J/kg·°C}) (50.0°C) = m_x (2.09 \times 10^5 \, \text{J/kg})$$

If we equate the heat lost by the steam to the heat gained by the water and glass, and use the given information, we find that

$$m_x (6.03 \times 10^4 \, \text{J/kg}) + m_x (2.26 \times 10^6 \, \text{J/kg}) + m_x (2.09 \times 10^5 \, \text{J/kg}) =$$
$$(0.200 \, \text{kg}) (4.19 \times 10^3 \, \text{J/kg·°C}) (30.0°C) + (0.100 \, \text{kg}) (837 \, \text{J/kg·°C}) (30.0°C)$$
$$m_x = 1.09 \times 10^{-2} \, \text{kg} = \boxed{10.9 \, \text{g}}$$

Workbook Problem 63 – Deals with heat transfer and change of phase. (See page 135 of the Workbook.)

WEB

For Web links to more information on Joule (see textbook page 341) and on global warming (see textbook page 359), visit the textbook Web site at http://www.harcourtcollege.com/physics/cptech

EXAMPLE 11.5   Boiling Liquid Helium

Liquid helium has a very low boiling point, 4.2 K, and a very low latent heat of vaporization, $2.09 \times 10^4 \, \text{J/kg}$ (Table 11.2). A constant power of 10.0 W (1 W = 1 J/s) is transferred to a container of liquid helium from an immersed electric heater. At this rate, how long does it take to boil away 1.00 kg of liquid helium?

Solution   Because $L_v = 2.09 \times 10^4 \, \text{J/kg}$, we must supply $2.09 \times 10^4 \, \text{J}$ of energy to boil away 1.00 kg. The power supplied to the helium is 10 W = 10 J/s. That is, in 1.00 s, 10.0 J of energy is transferred to the helium. Therefore, the time it takes to transfer $2.09 \times 10^4 \, \text{J}$ is

$$t = \frac{Q}{P} = \frac{2.09 \times 10^4 \, \text{J}}{10.0 \, \text{J/s}} = 2.09 \times 10^3 \, \text{s} \approx \boxed{35 \, \text{min}}$$

In contrast, 1.00 kg of liquid nitrogen ($L_v = 2.01 \times 10^5$ J/kg) would boil away in about 5.6 h with the same power input.

Exercise   If 10.0 W of power is supplied to 1.00 kg of water at 100°C, how long will it take for the water to completely boil away?

Answer   62.8 h

11.5   HEAT TRANSFER BY CONDUCTION

There are three ways in which heat energy can be transferred from one location to another: conduction, convection, and radiation. Regardless of the process, however, no net heat transfer takes place between a system and its surroundings when the two are at the same temperature. In this section we discuss heat transfer by conduction. Convection and radiation will be discussed in Sections 11.6 and 11.7.

Each of the methods of heat transfer can be examined by considering the ways in which you can warm your hands over an open fire. If you insert a copper rod into the flame, as in Figure 11.3, the temperature of the metal in your hand increases rapidly. **Conduction,** the process by which heat is transferred from the flame through the copper rod to your hand, can be understood by examining what is happening to the atoms of the metal. Initially, before the rod is inserted into the flame, the copper atoms are vibrating about their equilibrium positions. As the flame heats the rod, the copper atoms near the flame begin to vibrate with greater and greater amplitudes. These wildly vibrating atoms collide with their neighbors and transfer some of their energy in the collisions. Gradually, copper atoms progressively farther up the rod increase their amplitudes of vibration until those at the held end are reached. This increased vibration results in an increase in temperature of the metal, and possibly a burned hand.

Although the transfer of heat through a metal can be partially explained by atomic collisions, the rate of heat conduction also depends on the properties of the substance being heated. For example, it is possible to hold a piece of asbestos in a flame indefinitely. This implies that very little heat is being conducted through the asbestos. In general, metals are good conductors of heat, and materials such as asbestos, cork, paper, and fiber glass are poor conductors; gases also are poor heat conductors because of their dilute nature. Metals are good conductors of heat because they contain large numbers of electrons that are relatively free to move through the metal and transport energy from one region to another. Thus, in a good conductor, such as copper, heat conduction takes place both via the vibration of atoms and via the motions of free electrons.

In this section we consider the rate at which heat is transferred from one location to another. If Q is the amount of heat transferred from one location on an object to another in the time Δt, the **heat transfer rate, H** (sometimes called the *heat current*), is defined as

$$H \equiv \frac{Q}{\Delta t} \tag{11.5}$$

Figure 11.3   Heat reaches the hand by conduction through the copper rod.

The ceramic material being heated by a flame is a poor conductor of heat but is able to withstand a large temperature gradient; its left side is buried in ice (0°C) and its right side is extremely hot. (*Courtesy of Corning Glass Works*)

Note that H is expressed in watts when Q is in joules and Δt is in seconds (1 W = 1 J/s).

The conduction of heat occurs only if a difference in temperature exists between two parts of the conducting medium. Consider a slab of thickness L and cross-sectional area A, as in Figure 11.4. Suppose that one face is maintained at a temperature of T_2 and the other face is held at a lower temperature, T_1. Experimentally, one finds that the rate of heat flow—that is, the heat flow, Q, per unit time, Δt—is proportional to the difference in temperature $T_2 - T_1$ and the area A, and inversely proportional to the thickness of the slab. Specifically, the rate of flow of heat (or heat current) is given by

$$H = \frac{Q}{\Delta t} = kA\left(\frac{T_2 - T_1}{L}\right) \qquad \text{[11.6]}$$

where k is a constant called the **thermal conductivity** of the material. This constant is a property of the material. Table 11.3 lists some values of k for metals, gases, and nonmetals. Note that k is large for metals, which are good heat conductors, and small for gases and nonmetals, which are poor heat conductors (good insulators).

The fact that different materials have different k values should help you understand the following phenomenon, first mentioned in Chapter 10. If you remove a metal ice tray and a package of frozen food from the freezer, which feels colder? Experience tells you that the metal tray feels colder even though it is at the same initial temperature as the cardboard package. This is explained by the fact that metal has a much higher thermal conductivity than cardboard, and so it conducts heat more rapidly and hence removes heat from your hand at a higher rate. Consequently, the metal tray *feels* colder than the carton even though it isn't. By use of a similar argument, you should be able to explain why a tile floor feels colder to bare feet than a carpeted floor does.

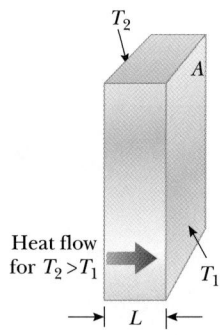

Figure 11.4 Heat transfer through a conducting slab of cross-sectional area A and thickness L. The opposite faces are at different temperatures, T_1 and T_2.

EXAMPLE 11.6 Heat Transfer Through a Concrete Wall

Find the amount of heat transferred in 1.00 h by conduction through a concrete wall 2.0 m high, 3.65 m long, and 0.20 m thick if one side of the wall is held at 20°C and the other side is at 5°C.

Solution Equation 11.6 gives the rate of heat transfer in joules per second. To find the amount of heat transferred in 1.00 h, we rewrite the equation as

$$Q = kA\,\Delta t\left(\frac{T_2 - T_1}{L}\right)$$

If we substitute the values given and consult Table 11.3, we find that

$$Q = (1.3\ \text{J/s·m·°C})(7.3\ \text{m}^2)(3600\ \text{s})\left(\frac{15°\text{C}}{0.20\ \text{m}}\right) = \boxed{2.6 \times 10^6\ \text{J}}$$

Early houses were insulated by the material of which their walls were constructed—thick masonry blocks. Masonry restricts heat loss by conduction because its k is relatively low. The large thickness, L, also decreases heat loss, as shown by Equation 11.6.

TABLE 11.3
Thermal Conductivities

Substance	Thermal Conductivity (J/s·m·°C)
Metals (at 25°C)	
Aluminum	238
Copper	397
Gold	314
Iron	79.5
Lead	34.7
Silver	427
Gases (at 20°C)	
Air	0.0234
Helium	0.138
Hydrogen	0.172
Nitrogen	0.0234
Oxygen	0.0238
Nonmetals	
Asbestos	0.25
Concrete	1.3
Glass	0.84
Ice	1.6
Rubber	0.2
Water	0.60
Wood	0.10

Heat is conducted from the inside to the exterior more rapidly where the snow has melted. The dormer in this house appears to have been added—and insulated. The main roof does not appear to be well insulated. *(Courtesy of A. A. Bartlett, University of Colorado, Boulder)*

Home Insulation

If you would like to do some calculating to determine whether or not to add insulation to a ceiling or to some other portion of a building, you need to slightly modify what you have just learned about conduction, for two reasons.

1. The insulating properties of materials used in buildings are usually expressed in engineering rather than SI units. For example, measurements stamped on a package of fiber glass insulating board will be in units such as British thermal units, feet, and degrees Fahrenheit.
2. In dealing with the insulation of a building, we must consider heat conduction through a compound slab, with each portion of the slab having a different thickness and a different thermal conductivity. For example, a typical wall in a house consists of an array of materials, such as wood paneling, dry wall, insulation, sheathing, and wood siding.

It is found that the rate of heat transfer through a compound slab is

$$\frac{Q}{\Delta t} = \frac{A(T_2 - T_1)}{\sum_i L_i/k_i} \qquad [11.7]$$

where T_1 and T_2 are the temperatures of the *outer extremities* of the slab and the summation is over all portions of the slab. For example, if the slab consists of three different materials, the denominator is the sum of three terms. In engineering practice, the term L/k for a particular substance is referred to as the **R value** of the material. Thus, Equation 11.7 reduces to

TABLE 11.4   R Values for Some Common Building Materials

Material	R value ($ft^2 \cdot °F \cdot h/Btu$)
Hardwood siding (1.0 in. thick)	0.91
Wood shingles (lapped)	0.87
Brick (4.0 in. thick)	4.00
Concrete block (filled cores)	1.93
Styrofoam (1.0 in. thick)	5.0
Fiber-glass batting (3.5 in. thick)	10.90
Fiber-glass batting (6.0 in. thick)	18.80
Fiber-glass board (1.0 in. thick)	4.35
Cellulose fiber (1.0 in. thick)	3.70
Flat glass (0.125 in. thick)	0.89
Insulating glass (0.25-in. space)	1.54
Vertical air space (3.5 in. thick)	1.01
Air film	0.17
Dry wall (0.50 in. thick)	0.45
Sheathing (0.50 in. thick)	1.32

$$\frac{Q}{\Delta t} = \frac{A(T_2 - T_1)}{\sum_i R_i}$$ [11.8]

The R values for a few common building materials are listed in Table 11.4 (note the units).

Next to any vertical outside surface is a very thin, stagnant layer of air that must be considered when the total R value for a wall is figured. The thickness of this stagnant layer depends on the speed of the wind. As a result, heat loss from a house on a day when the wind is blowing hard is greater than heat loss on a day when the wind speed is zero. A representative R value for the stagnant layer of air is given in Table 11.4.

EXAMPLE 11.7   The R Value of a Typical Wall

Calculate the total R value for a wall constructed as shown in Figure 11.5a. Starting outside the house (to the left in the figure) and moving inward, the wall consists of brick, 0.50 in. of sheathing, a vertical air space 3.5 in. thick, and 0.50 in. of dry wall. Do not forget the dead-air layers inside and outside the house.

Solution   Referring to Table 11.4, we find the total R value for the wall as follows:

$$R_1\text{(outside air film)} = 0.17 \text{ ft}^2 \cdot {}^\circ\text{F} \cdot \text{h/Btu}$$

$$R_2\text{(brick)} \qquad = 4.00$$

$$R_3\text{(sheathing)} \qquad = 1.32$$

$$R_4\text{(air space)} \qquad = 1.01$$

$$R_5\text{(dry wall)} \qquad = 0.45$$

$$R_6\text{(inside air film)} \quad = 0.17$$

$$R_{\text{total}} \qquad\qquad = \quad 7.12 \text{ ft}^2 \cdot {}^\circ\text{F} \cdot \text{h/Btu}$$

Exercise   If a layer of fiber glass insulation 3.5 in. thick is placed inside the wall to replace the air space, as in Figure 11.5b, what is the total R value of the wall? By what factor is the heat loss reduced?

Answer $R = 17 \text{ ft}^2 \cdot {}^\circ\text{F} \cdot \text{h/Btu}$; a factor of 2.5.

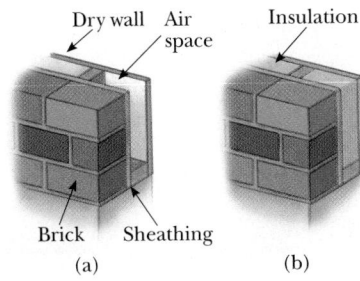

Figure 11.5   (Example 11.7) A cross-sectional view of an exterior wall containing (a) an air space and (b) insulation.

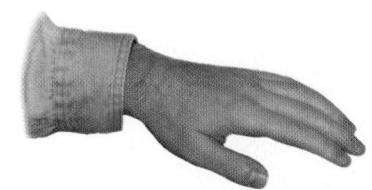

11.6   CONVECTION

You have probably warmed your hands by holding them over an open flame, as illustrated in Figure 11.6. In this situation, the air directly above the flame is heated and expands. As a result, the density of the air decreases and the air rises. This warmed mass of air heats your hands as it flows by. **Heat transferred by the movement of a heated substance is said to have been transferred by convection.** When the movement results from differences in density, as it does in air around a fire, it is referred to as *natural convection*. When the heated substance is forced to move by a fan or pump, as in some hot-air and hot-water heating systems, the process is called *forced convection*.

Figure 11.6   Heating a hand by convection.

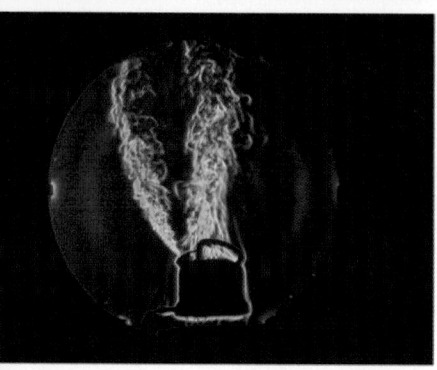

Schlieren photograph of a teakettle showing steam and turbulent convection air currents. *(Gary Settles/Science Source/Photo Researchers)*

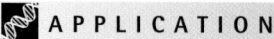

APPLICATION

Cooling Automobile Engines.

APPLICATION

Algal Blooms in Ponds and Lakes.

The circulating pattern of air flow at a beach (see Fig. 11.1) is an example of convection. Likewise, the mixing that occurs as water is cooled and eventually freezes at its surface (Chapter 10) is an example of convection in nature. Recall that mixing by convection currents ceases when the water temperature reaches 4°C. Because the water in the pool cannot be cooled by convection below 4°C, and because water is a relatively poor conductor of heat (see Table 11.3), the water near the bottom remains near 4°C for a long time. As a result, fish live in water of a comfortable temperature even in periods of prolonged cold weather.

If it were not for convection currents, it would be very difficult to boil water. The lower layers of water in a teakettle are warmed first. These heated regions expand and are buoyed up to the top because their density is lowered. Meanwhile, denser cool water replaces the warm water at the bottom of the kettle so that it can be heated.

The same process occurs when a room is heated by a radiator. The hot radiator warms the air in the lower regions of the room. The warm air expands and rises to the ceiling because of its lower density. The denser regions of cooler air from above replace the warm air, setting up the continuous air current pattern shown in Figure 11.7.

Your automobile engine is maintained at a safe operating temperature by a combination of conduction and forced convection. Water (actually, a mixture of water and antifreeze) circulates in the interior of the engine. As the metal of the engine block increases in temperature, heat passes from the hot metal to the cooler water by conduction. The water pump forces water out of the engine and into the radiator, carrying heat along with it (forced convection). In the radiator, the hot water passes through metal pipes that are in contact with the cooler outside air, and heat passes into the air by conduction. The cooled water is then returned to the engine by the water pump to absorb more energy.

The algal blooms often seen in temperate lakes and ponds during spring or fall are caused by convection currents in the water. To understand this process, consider Figure 11.8. During the summer, bodies of water develop temperature gradients such that an upper warm layer of water is separated from a lower cold layer by a buffer zone called a thermocline. In the spring or fall, the temperature changes in the water break down this thermocline, setting up convection currents that mix the water. This mixing process transports nutrients from the bottom to the surface. The nutrient-rich water forming at the surface can cause a rapid, temporary increase in the population of algae.

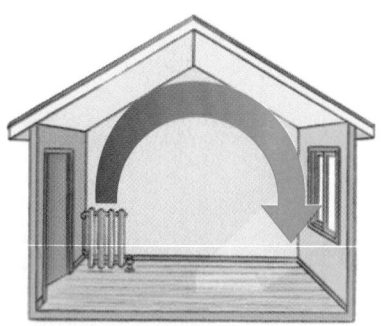

Figure 11.7 Convection currents are set up in a room heated by a radiator.

Applying Physics 1

 The body temperature of mammals ranges from about 35°C to 38°C, a fairly narrow range, while that of birds ranges from about 40°C to 43°C. To maintain this temperature in cold weather, it is necessary to provide insulation to prevent heat loss from the surface of the skin. How is this accomplished?

Explanation A natural method of accomplishing this is via layers of fat beneath the skin. Fat protects against both conduction and convection in that it has a low thermal conductivity, and there are few blood vessels in fat to carry blood to the surface where convective heat losses can occur. Birds ruffle their feathers in cold weather such that a layer of air with a low thermal conductivity is trapped between the feathers and the skin. Bristling the fur produces the same effect on fur-bearing animals.

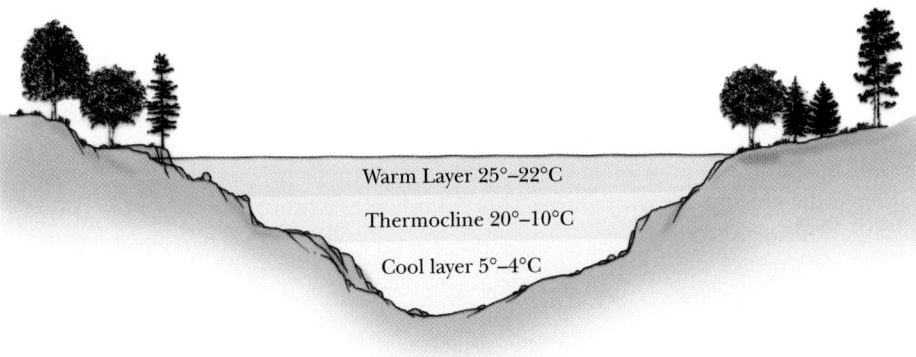

Warm Layer 25°–22°C

Thermocline 20°–10°C

Cool layer 5°–4°C

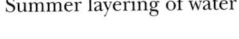

Summer layering of water

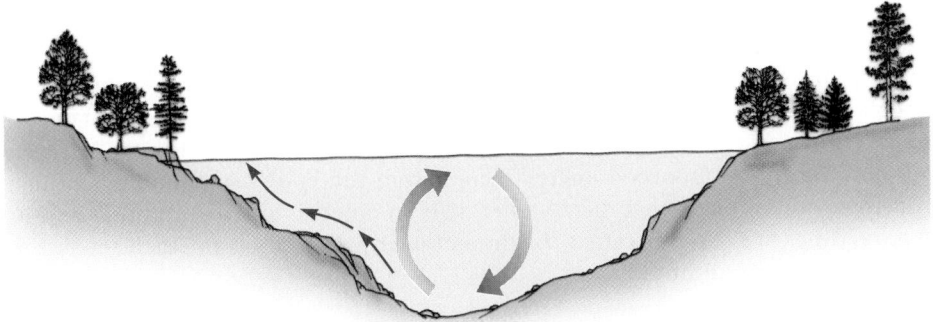

Fall and spring upwelling

Figure 11.8 (a) During the summer, the upper warm layer of water is separated from a cooler lower layer by a thermocline. (b) Convections currents during the spring or fall mix the water and can cause algal blooms.

11.7 RADIATION

The third way of transferring heat is through **radiation.** You have most likely experienced radiant heat when sitting in front of a fireplace. Figure 11.9 shows how you can warm your hands at an open flame by means of radiant heat. The hands are placed to one side of the flame; they are not in physical contact with the flame either directly or indirectly, and therefore conduction cannot account for the heat transfer. Furthermore, convection is not important in this situation because the hands are not above the flame in the path of convection currents. The important process in this case is the radiation of heat energy.

All objects continuously radiate energy in the form of electromagnetic waves, which we shall discuss in Chapter 21. Electromagnetic radiation associated with the loss of heat energy from an object at a temperature of a few hundred kelvins is referred to as *infrared* radiation.

The surface of the Sun is at a few thousand kelvins and most strongly radiates visible light. Approximately 1340 J of sunlight energy strikes 1 m^2 of the top of the Earth's atmosphere every second. Some of this energy is reflected back into space, and some is absorbed by the atmosphere, but enough arrives at the surface of the Earth each day to supply hundreds of times more energy than human technology requires—if it could be captured and used efficiently. The building of an increasing number of solar houses in this country is one attempt to make use of this free energy.

Radiant energy from the Sun affects our day-to-day existence in a number of ways. For example, consider what happens to the atmospheric temperature at night.

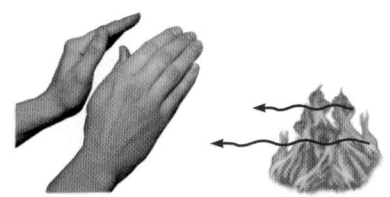

Figure 11.9 Warming hands by radiation.

If there is a cloud cover above the Earth, the water vapor in the clouds reflects back a part of the infrared radiation emitted by the Earth, and consequently the temperature remains moderate. In the absence of a cloud cover, however, there is nothing to prevent this radiation from escaping into space, and thus the temperature drops more on a clear night than when it is cloudy.

The rate at which an object emits radiant energy is proportional to the fourth power of its absolute temperature. This is known as **Stefan's law** and is expressed in equation form as

Stefan's law ▶

$$P = \sigma A e T^4 \qquad\qquad [11.9]$$

where P is the power radiated by the object in watts (or joules per second), σ is a constant equal to 5.6696×10^{-8} W/m^2·K^4, A is the surface area of the object in square meters, e is a constant called the **emissivity,** and T is the object's temperature in kelvins. The value of e can vary between zero and unity, depending on the properties of the surface.

An object radiates energy at a rate given by Equation 11.9. At the same time, the object also absorbs electromagnetic radiation. If the latter process did not occur, the object would eventually radiate all of its energy and its temperature would reach absolute zero. The absorbed energy comes from the body's surroundings, which consist of other objects that radiate energy. If an object is at temperature T and its surroundings are at temperature T_0, the net energy gained or lost each second by the object as a result of radiation is

$$P_{\text{net}} = \sigma A e (T^4 - T_0{}^4) \qquad\qquad [11.10]$$

When an object is in equilibrium with its surroundings, it radiates and absorbs energy at the same rate, and so its temperature remains constant. When an object is hotter than its surroundings, it radiates more energy than it absorbs, and so it cools. An *ideal absorber* is defined as an object that absorbs all of the energy incident on it; its emissivity is equal to unity. Such an object is often referred to as a **black body.** An ideal absorber is also an ideal radiator of energy. In contrast, an object with an emissivity equal to zero absorbs none of the energy incident on it. Such an object reflects all the incident energy and so is a *perfect reflector.*

APPLICATION

Light-Colored Summer Clothing.

White clothing is more comfortable to wear in the heat of the summer than black clothing. Black fabric acts as a good absorber of incoming sunlight and as a good emitter of this absorbed energy. However, about half of the emitted energy travels toward the body, causing the person wearing the garment to feel uncomfortably warm. In contrast, white or light-colored clothing reflects away much of the incoming energy.

APPLICATION

Thermography.

The amount of radiant energy emitted by a body can be measured with heat-sensitive recording equipment, using a technique called **thermography.** The radiation emitted by a body is greatest in the body's warmest regions. An image of the pattern formed by varying radiation levels, called a **thermogram,** is brightest in the warmest areas. Figure 11.10 reproduces a thermogram of a house. The center portions of the door and windows are yellow, signifying temperatures higher than those of surrounding areas. A higher temperature usually means that heat is escaping. Thermograms can be useful for purposes of energy conservation. For example, the owners of this house could conserve energy and reduce their heating costs by adding insulation to the attic area and by installing thermal draperies over the windows.

Figure 11.10 This thermogram of a home, made during cold weather, shows colors ranging from white and yellow (areas of greatest heat loss) to blue and purple (areas of least heat loss). *(Daedalus Enterprises, Inc./Peter Arnold, Inc.)*

Figure 11.11 shows a recently developed radiation thermometer that has removed most of the risk of taking the temperature of young children or the aged, risks such as bowel perforation or bacterial contamination. The instrument measures the intensity of the infrared radiation leaving the eardrum and surrounding tissues and converts this information to a standard numerical reading. The eardrum is a particularly good location to measure body temperature because it is near the hypothalamus—the body's temperature control center.

A P P L I C A T I O N

Radiation Thermometers for Measuring Body Temperature.

Applying Physics 2

If you sit in front of a fire in a fireplace with your eyes closed, you can feel significant warmth in your eyelids. If you now put on a pair of eyeglasses and repeat this activity, your eyelids will not feel as warm. Why?

Explanation Much of the warm feeling that one has while sitting in front of a fireplace is due to radiation from the fire. A large fraction of this radiation is in the infrared part of the spectrum. Your eyelids are particularly sensitive to infrared radiation. However, glass is partially opaque to infrared radiation. Thus, when you put on the glasses, you block much of the radiation from reaching your eyelids, and thus, they feel cooler.

Figure 11.11 A radiation thermometer measures a patient's temperature by monitoring the intensity of infrared radiation leaving the ear. *(© Blair Seitz/Photo Researchers, Inc.)*

EXAMPLE 11.8 Who Turned Down the Thermostat?

A student is trying to decide what to wear. The air in his bedroom is at 20.0°C. If the skin temperature of the unclothed student is 37.0°C, how much heat is lost from his body in 10.0 min? Assume that the emissivity of skin is 0.900 and that the surface area of the student is 1.50 m^2.

Solution Using Equation 11.10, the rate of heat loss from the skin is

$$P_{net} = \sigma A e (T^4 - T_0^4)$$
$$= (5.67 \times 10^{-8} \text{ W/m}^2 \cdot \text{K}^4)(1.50 \text{ m}^2)(0.90)[(310 \text{ K})^4 - (293 \text{ K})^4] = 143 \text{ J/s}$$

Note that it was necessary to change the temperature to kelvins. At this rate of heat loss, the total heat lost by the skin in 10.0 min is

$$Q = P_{net} \times \text{time} = (143 \text{ J/s})(600 \text{ s}) = \boxed{8.58 \times 10^4 \text{ J}}$$

11.8   HINDERING HEAT TRANSFER

APPLICATION

Thermos Bottles.

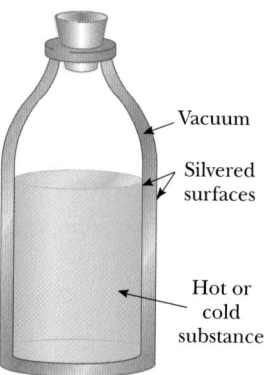

Figure 11.12 A cross-sectional view of a Dewar vessel designed to store hot or cold liquids.

APPLICATION

Dressing Warm in Winter.

The Thermos bottle, called a *Dewar flask* (after its inventor) in scientific applications, is designed to minimize heat transfer by conduction, convection, and radiation. It is used to store either cold or hot liquids for long periods of time. The standard vessel (Fig. 11.12) is double-walled Pyrex glass with a silvered inner wall. The space between the walls is evacuated to minimize heat transfer by conduction and convection. By reflecting most of the radiant heat, the silvered surface minimizes heat transfer by radiation. Very little heat is lost through the neck of the flask because Pyrex glass is a poor conductor. A further reduction in heat loss is achieved by reducing the size of the neck. A common scientific application of Dewar flasks is storage of liquid nitrogen (boiling point 77 K) and liquid oxygen (boiling point 90 K). For substances that have very low specific heats, such as liquid helium (boiling point 4.2 K), it is often necessary to use a double Dewar system in which the Dewar flask containing the liquid is surrounded by a second Dewar flask. The space between the two flasks is filled with liquid nitrogen.

Some of the principles of the Thermos bottle are used in the protection of sensitive electronic instruments in orbiting space satellites. In half of its orbit about the Earth, a satellite is exposed to intense radiation from the Sun, and in the other half it is in the Earth's cold shadow. Without protection, its interior would thus be subjected to tremendous extremes of heating and cooling. The interior of the satellite is wrapped with blankets of highly reflective aluminum foil. The foil's shiny surface reflects away much of the Sun's radiation while the satellite is in the unshaded part of the orbit, and helps retain interior heat while the satellite is in the Earth's shadow.

Wool sweaters and down jackets keep us warm by trapping the warmer air in regions close to our bodies and hence reducing heat loss by convection and conduction. In other words, what keeps us warm is not the clothing itself but the air trapped in the clothing.

11.9   METABOLISM AND LOSING WEIGHT

In order to maintain life, humans must receive an input of energy from their surroundings. This is accomplished by an intake of food that supplies chemical energy inside the cells of the body. The **basal metabolism** is defined as the rate of energy consumption while an individual is awake and resting. For an average 20-year-old man this is approximately 1.2 W/kg (2.87×10^{-4} Cal/s·kg) and for a 20-year-old woman 1.1 W/kg (2.63×10^{-4} Cal/s·kg). Do the conversion yourself to show that this is equivalent to about 1700 Cal per day for a 70 kg man and 1400 Cal per day for a 60 kg woman. For an average sedentary person the energy required for daily activity is about 2500 Cal per day. That is, if such an individual consumes 2500 Cal per day, weight will be maintained and eating more or less will lead to a gain or loss of body weight, respectively. Table 11.5 shows the metabolic rates in Cal/s·kg for various activities, and Table 11.6 shows the energy content per kg of food.

TABLE 11.5
Metabolic Rates for Various Activities

Activity	Cal/s·kg
Sleeping	2.63×10^{-4}
Awake but resting	2.87×10^{-4}
Sitting	3.58×10^{-4}
Standing	6.21×10^{-4}
Walking	1.00×10^{-3}
Biking	1.81×10^{-3}
Swimming	2.63×10^{-3}
Running	4.30×10^{-3}

Applying Physics 3

 A common "old wives' tale" is that a camel stores fat in the hump on its back and converts this fat to water when deprived in the desert. This cannot be the case, however, because converting fat to water would require a tremendous increase in oxygen intake by the camel. This would result in a tremendous loss of water from the lungs by respiration, more water in fact than would be converted. What other factor(s) could cause a camel to function so well in the desert?

Explanation The major factor that allows the camel to function so well in the desert is that its body temperature can vary from about 34°C during the cool night to about 41°C in the hottest part of the day. Because of this, the camel does not begin to sweat until its body temperature is about 41°C, and when it does sweat, very little of the water lost comes from the blood stream. In contrast, a human loses a considerable amount of water from blood plasma, making the blood much thicker and harder to pump than normal. The reduced circulation of blood makes it even more difficult for a human to dissipate heat from the body.

TABLE 11.6
Energy Content per Gram of Dry Food

Food	Cal/g
Carbohydrate	4.1
Protein	4.2
Fat	9.3

EXAMPLE 11.9 Trying to Burn the Weight Off

(a) How much energy is expended by a 70.00 kg man who exercises each morning by walking for one hour?

Solution From Table 11.5 we see that the energy consumed is

$$(1.00 \times 10^{-3} \text{ Cal/s·kg})(70.00 \text{ kg})(3600 \text{ s/h}) = \boxed{252.0 \text{ Cal/h}}$$

(b) If the man consumes body fat to produce this energy, how much mass will be lost?

Solution The energy equivalent of fat is 9.3 Cal/g from Table 11.6. Thus, the fat consumed is equal to 27.0 g. An obvious moral to this story is that the easiest way to lose weight is to limit food intake.

11.10 APPLICATION: GLOBAL WARMING AND GREENHOUSE GASES

Many of the principles of heat transfer, and its prevention, can be understood by studying the operation of a glass greenhouse. During the day, sunlight passes into the greenhouse and is absorbed by the walls, earth, plants, and so on. This absorbed visible light is subsequently re-radiated as infrared radiation, which causes the temperature of the interior to rise.

In addition, convection currents are inhibited in a greenhouse. As a result, heated air cannot rapidly pass over the surfaces of the greenhouse that are exposed to the outside air and thereby cause a heat loss through those surfaces. Most experts now consider this to be a more important warming effect than that of any trapped infrared radiation. In fact, experiments have shown that when the glass over a greenhouse is replaced by a special glass known to transmit infrared light, the temperature inside is lowered only slightly. Based on this evidence, the primary mechanism that heats a greenhouse is not the absorption of infrared radiation but the inhibition of air flow that occurs under any roof (in an attic, for example).

A phenomenon formerly known as the *greenhouse effect* can also play a major role in determining the Earth's temperature. First note that the Earth's atmosphere is a good transmitter (and hence a poor absorber) of visible radiation and a good absorber of infrared radiation. Carbon dioxide (CO_2) in the Earth's atmosphere allows incoming visible radiation from the Sun to pass through more easily than infrared radiation. The visible light that reaches the Earth's surface is absorbed and re-radiated as infrared light, which in turn is absorbed (trapped) by the Earth's atmosphere. An extreme case is the warmest planet, Venus, which has a carbon dioxide-rich atmosphere and temperatures approaching 850°F.

As fossil fuels (coal, oil, and natural gas) are burned, large amounts of carbon dioxide are released into the atmosphere, causing it to retain more heat. This is of great concern to scientists and governments throughout the world. Many scientists are convinced that the 10% increase in the amount of atmospheric carbon dioxide in the past 30 years could lead to drastic changes in world climate. According to one estimate, doubling the carbon dioxide content in the atmosphere will cause temperatures to increase by 2°C! In temperate regions, such as Europe and the United States, the temperature rise would save billions of dollars per year in fuel costs. Unfortunately, the temperature increase would also melt polar ice caps, which could cause flooding and destroy many coastal areas; increase the frequency of droughts; and consequently decrease already low crop yields in tropical and subtropical countries. Even slightly higher average temperatures might make it impossible for certain plants and animals to survive in their customary ranges.

At present, about 3.5×10^{11} tons of CO_2 are released into the atmosphere each year. Most of this gas results from human activities such as the burning of fossil fuels, the cutting of forests, and manufacturing processes. Other greenhouse gases are also increasing in concentration in the atmosphere. One of these is methane, CH_4, which is released in the digestive process of cows and other ruminants. This gas originates from that part of the animal's stomach called the *rumen* where cellulose is digested. Termites also are major producers of this gas. Finally, greenhouse gases such as nitrous oxide (N_2O) and sulfur dioxide (SO_2) are increasing due to automobile and industrial pollution.

Whether the increasing greenhouse gases are responsible or not, there is convincing evidence that global warming is certainly underway. The evidence comes from the melting of ice in Antarctica and the retreat of glaciers at widely scattered sites throughout the world. For example, satellite images of Antarctica show James Ross Island completely surrounded by water for the first time since maps of the area were made about 100 years ago. Previously the island was connected to the mainland by an ice bridge. In addition, at various places across the continent, ice shelves are retreating, some at a rapid rate.

Perhaps at no place in the world are glaciers monitored with greater interest than in Switzerland. There, it is found that the Alps have lost about 50% of their glacial ice compared to 130 years ago. As evidence of this, in 1818 the citizens of the town of Brig became concerned about the advance of the Aletsch glacier, the largest glacier in Switzerland. It was threatening their summer pastures and forests. A religious procession was held, and a cross was placed against the glacier as a symbolic blockade. Today, the glacier is about a mile farther back in the mountains away from the cross, and it continues to retreat.

The retreat of glaciers on high altitude peaks in the tropics is even more severe than in Switzerland. The Lewis glacier on Mount Kenya and the snows of Kilimanjaro are two examples. However, in certain regions of the planet where glaciers are near large bodies of water and are fed by large and frequent snows, glaciers continue to advance, so the overall picture of a catastrophic global warming scenario is premature. However, in about fifty years, the amount of carbon dioxide in the atmosphere is expected to be about twice what it was in the preindustrial era. Because of this, most scientists voice the concern that reductions in greenhouse gas emissions need to be made now.

SUMMARY

Heat is a transfer of energy that takes place as a consequence of a temperature difference between a system and its surroundings. The internal energy of a substance is a function of the state of the substance, and generally increases with increasing temperature.

The **calorie** is the amount of heat necessary to raise the temperature of 1 g of water from 14.5°C to 15.5°C. The **mechanical equivalent of heat** is 4.186 J/cal.

The heat required to change the temperature of a substance by ΔT is

$$Q = mc\,\Delta T \qquad\qquad [11.3]$$

where m is the mass of the substance and c is its **specific heat.**

The heat required to change the phase of a mass m of a pure substance is

$$Q = mL \qquad\qquad [11.4]$$

The parameter L is called the **latent heat** of the substance and depends on the nature of the phase change and the properties of the substance.

Heat may be transferred by three fundamentally distinct processes: *conduction, convection,* and *radiation.* **Conduction** can be viewed as an exchange of kinetic energy between colliding molecules or through the motion of electrons. The rate at which heat flows by conduction through a slab of area A and thickness L is

$$H = \frac{Q}{\Delta t} = kA\left(\frac{T_2 - T_1}{L}\right) \qquad\qquad [11.6]$$

where k is the **thermal conductivity** of the material making up the slab. Heat transferred by the movement of a heated substance is said to have been transferred by **convection.**

All objects **radiate** and absorb energy in the form of electromagnetic waves. An object that is hotter than its surroundings radiates more energy than it absorbs, whereas an object that is cooler than its surroundings absorbs more energy than it radiates. The rate at which an object emits radiant energy is given by **Stefan's law:**

$$P = \sigma A e T^4 \qquad\qquad [11.9]$$

where σ is a constant equal to 5.6696×10^{-8} W/m$^2 \cdot$K^4, and e is a constant called the **emissivity.**

MULTIPLE-CHOICE QUESTIONS

1. The specific heat of substance A is greater than that for substance B. If equal amounts of heat are added to both these substances, the one reaching the higher temperature, assuming no phase change occurs for either, will be
   (a) substance A.
   (b) substance B.
   (c) There will be no difference in the final temperatures.
   (d) Could be either A or B.

2. An amount of heat is added to ice, raising its temperature from $-10°C$ to $-5°C$. A larger amount of heat is added to the same mass of liquid water, raising its temperature from $15°C$ to $20°C$. From these results, we can conclude that
   (a) overcoming the latent heat of fusion of ice requires an input of energy.
   (b) the latent heat of fusion of ice delivers some energy to the system.
   (c) the specific heat of ice is less than that of water.
   (d) the specific heat of ice is greater than that of water.

3. Star A has twice the radius and twice the absolute temperature of star B. What is the ratio of the power output of star A to that of star B? The emissivity of both stars can assumed to be 1.
   (a) 4 (b) 8 (c) 16 (d) 32 (e) 64

4. How long would it take for a 1000 W heating element to melt 1 kg of ice at 0°C?
   (a) 4.19 s (b) 41.9 s (c) 5.55 min
   (d) 1 hour (e) 555 min

5. A 100-g piece of copper, initially at 95°C, is dropped into 200 g of water contained in a 280-g aluminum can; the water and can are initially at 15°C. What is the final temperature of the system? (Specific heats of copper and aluminum are 0.092 and 0.215 cal/g·°C, respectively.)
   (a) 16.8°C (b) 17.7°C (c) 23.7°C (d) 25.0°C

CONCEPTUAL QUESTIONS

1. Rub the palm of your hand on a metal surface for 30 to 45 seconds. Place the palm of your other hand on a unrubbed portion of the surface and then the rubbed portion. The rubbed portion will feel warmer. Now repeat this process on a wooden surface. You should notice that the difference in temperature between the rubbed and unrubbed portions of the wood surface seems larger than that of the metal surface. Why is this?

2. Pioneers stored fruits and vegetables in underground cellars. Discuss fully this choice for a storage site.

3. In usually warm climates that experience an occasional hard freeze, fruit growers will spray the fruit trees with water, hoping that a layer of ice will form on the fruit. Why is this advantageous?

4. In winter, the pioneers mentioned in Question 2 stored an open barrel of water alongside their produce. Why?

5. Cups of water for coffee or tea can be warmed with an immersion coil, which is immersed in the water and raised to a high temperature by means of electricity. The instructions warn the user not to operate the coils when they are out of water. Why? Can the immersion coil be used to warm up a cup of stew?

6. The U.S. penny is now made of copper-coated zinc. Can a calorimetric experiment be devised to test for the metal content in a collection of pennies? If so, describe the procedure you would use.

7. On a clear, cold night, why does frost tend to form on the tops of mailboxes and cars rather than the sides?

8. Consider two small balloons, one filled with air and the other with water. If a lighted match is placed within a few centimeters underneath the air-filled balloon, it bursts. However, if a lighted match is placed under the

water-filled balloon, the balloon does not burst even if the flame touches the balloon. Explain this observation in terms of heat conduction processes and the heat capacity of water.

9. A tile floor may feel uncomfortably cold to your bare feet, but a carpeted floor in an adjoining room at the same temperature feels warm. Why?

10. On a very hot day it is possible to cook an egg on the hood of a car. Would you select a black car or a white car on which to cook your egg? Why?

11. Concrete has a higher specific heat than does soil. Use this fact to explain (partially) why a city has a higher average temperature than the surrounding countryside. Would you expect breezes to blow from city to country or from country to city? Explain.

12. You need to pick up a very hot cooking pot in your kitchen. You have a pair of hot pads. Should you soak them in cold water or keep them dry in order to pick up the pot most comfortably?

13. In a daring lecture demonstration, a professor dips her wetted fingers into molten lead (327°C) and withdraws them quickly without getting burned. How is this possible?

14. The air temperature above coastal areas is profoundly influenced by the large specific heat of water. One reason is that the heat released when 1 cubic meter of water cools by 1.0°C will raise the temperature of an enormously larger volume of air by 1.0°C. Estimate the volume of air. The specific heat of air is approximately $1.0 \text{ kJ/kg} \cdot °C$. Take the density of air to be 1.3 kg/m^3.

15. Ethyl alcohol has about one half the specific heat of water. If equal masses of alcohol and water in separate beakers are supplied with the same amount of heat, compare the temperature increases of the two liquids.

PROBLEMS

1, 2, 3 = straightforward, intermediate, challenging   □ = full solution available in Study Guide/Student Solutions Manual   ◤ = Core Concepts Workbook
WEB = solution posted at **http://www.harcourtcollege.com/physics/cptech**   ◢ = biomedical application   ◢ = Interactive Physics

Review Problem

An aluminum rod is 20.000 cm long at 20°C and has a mass of 350 g. If 10,000 J of heat energy is added to the rod, what is its new length?

Section 11.1 The Mechanical Equivalent of Heat

Section 11.2 Specific Heat

1. As part of an exercise routine a 50.0-kg person climbs 10.0 meters up a vertical rope. How many (food) Calories are expended in a single climb up the rope? [1 (food) Calorie = 10^3 calories]

2. A 75.0-kg weight-watcher wishes to climb a mountain to work off the equivalent of a large piece of chocolate cake rated at 500 (food) Calories. How high must the person climb? [1 (food) Calorie = 10^3 calories]

3. How many joules of energy are required to raise the temperature of 100 g of gold from 20.0°C to 100°C?

4. A 50.0-g sample of copper is at 25°C. If 1200 J of heat energy is added to the copper, what is its final temperature?

5. A 5.00-g lead bullet traveling at 300 m/s is stopped by a large tree. If half the kinetic energy of the bullet is transformed into internal energy and remains with the bullet while the other half is transmitted to the tree, what is the increase in temperature of the bullet?

6. A 50.0-g piece of cadmium is at 20°C. If 400 cal of heat is added to the cadmium, what is its final temperature?

7. Water at the top of Niagara Falls has a temperature of 10.0°C. If it falls a distance of 50.0 m and all of its potential energy goes into heating the water, calculate the temperature of the water at the bottom of the falls.

8. A 1.5-kg copper block is given an initial speed of 3.0 m/s on a rough horizontal surface. Because of friction, the block finally comes to rest. (a) If the block absorbs 85% of its initial kinetic energy as internal energy, calculate its increase in temperature. (b) What happens to the remaining energy?

9. A 200-g aluminum cup contains 800 g of water in thermal equilibrium at 80°C. The combination of cup and water is cooled uniformly so that the temperature decreases by 1.5°C per minute. At what rate is heat energy being removed? Express your answer in watts.

10. The Btu is defined as the heat required to raise the temperature of 1 lbm (pound of mass) of water by 1°F. Determine the number of Joules in a Btu. [1 lbm = a quantity of material that weighs 1 lb = 1/32.174 slugs = 0.4536 kg]

Section 11.3 Conservation of Energy: Calorimetry

11. Lead pellets, each of mass 1.00 g, are heated to 200°C. How many pellets must be added to 500 g of water that is initially at 20.0°C to make the equilibrium temperature 25.0°C? Neglect any heat transfer to or from the container.

12. A 0.40-kg iron horseshoe that is initially at 500°C is dropped into a bucket containing 20 kg of water at

22°C. What is the final equilibrium temperature? Neglect any heat transfer to or from the surroundings.

13. What mass of water at 25.0°C must be allowed to come to thermal equilibrium with a 3.00-kg gold bar at 100°C in order to lower the temperature of the bar to 50.0°C?

14. An aluminum cup contains 225 g of water at 27°C. A 400-g sample of silver at an initial temperature of 87°C is placed in the water. A 40-g copper stirrer is used to stir the mixture until it reaches its final equilibrium temperature of 32°C. Calculate the mass of the aluminum cup.

WEB 15. If 200 g of water is contained in a 300-g aluminum vessel at 10°C and an additional 100 g of water at 100°C is poured into the container, what is the final equilibrium temperature of the mixture?

16. A 100-g aluminum calorimeter contains 250 g of water. The two substances are in thermal equilibrium at 10°C. Two metallic blocks are placed in the water. One is a 50-g piece of copper at 80°C. The other sample has a mass of 70 g and is originally at a temperature of 100°C. The entire system stabilizes at a final temperature of 20°C. Determine the specific heat of the unknown second sample.

17. It is desired to cool iron parts from 500°F to 100°F by dropping them into water that is initially at 75°F. Assuming that all the heat from the iron is transferred to the water and that none of the water evaporates, how many kilograms of water are needed per kilogram of iron?

18. A combination of 0.250 kg of water at 20.0°C, 0.400 kg of aluminum at 26.0°C, and 0.100 kg of copper at 100°C is mixed in an insulated container and allowed to come to thermal equilibrium. Neglect any heat transfer to or from the container and determine the final temperature of the mixture.

19. A student drops two metallic objects into a 120-g steel container holding 150 g of water at 25°C. One object is a 200-g cube of copper that is initially at 85°C, and the other is a chunk of aluminum that is initially at 5°C. To the surprise of the student, the water reaches a final temperature of 25°C, exactly where it started. What is the mass of the aluminum chunk?

Section 11.4 Latent Heat and Phase Changes

20. A 50-g ice cube at 0°C is heated until 45 g has become water at 100°C and 5.0 g has become steam at 100°C. How much heat was added to accomplish this?

21. A 100-g cube of ice at 0°C is dropped into 1.0 kg of water that is originally at 80°C. What is the final temperature of the water after the ice has melted?

22. How much heat is required to change a 40-g ice cube from ice at −10°C to steam at 110°C?

23. What mass of steam that is initially at 120°C is needed to warm 350 g of water and its 300-g aluminum container from 20°C to 50°C?

24. Assuming no heat loss, determine the mass of water that boils away when 1.94 kg of mercury at a temperature of 200°C is added to 0.050 kg of water at a temperature of 80.0°C.

25. A 75-kg cross-country skier moves across snow such that the coefficient of friction between skis and snow is 0.20. Assume all the snow beneath his skis is at 0°C and that all the internal energy generated by friction is added to snow which sticks to his skis until melted. How far would he have to ski to melt 1.0 kg of snow?

Figure P11.25 A cross-country skier.
(Nathan Bilow, Leo de Wys, Inc.)

26. A 100-g ice cube at 0°C is placed in 650 g of water at 25°C. What is the final temperature of the mixture?

WEB 27. A 40-g block of ice is cooled to −78°C. It is added to 560 g of water in an 80-g copper calorimeter at a temperature of 25°C. Determine the final temperature. (If not all the ice melts, determine how much ice is left.) Remember that the ice must first warm to 0°C, melt, and then continue warming as water. The specific heat of ice is 0.500 cal/g·°C = 2090 J/kg·°C.

28. A beaker of water sits in the sun until it reaches an equilibrium temperature of 30°C. The beaker is made of 100 g of aluminum and contains 180 g of water. In an attempt to cool this system, 100 g of ice at 0°C is added to the water. (a) Determine the final temperature. If $T_f = 0°C$, determine how much ice remains. (b) Repeat this for 50 g of ice.

29. Steam at 100°C is added to ice at 0°C. (a) Find the amount of ice melted and the final temperature when the mass of steam is 10 g and the mass of ice is 50 g. (b) Repeat with steam of mass 1.0 g and ice of mass 50 g.

Section 11.5 Heat Transfer by Conduction

30. (a) Find the rate of heat flow through a copper block of cross-sectional area 15 cm² and length 8.0 cm when a temperature difference of 30°C is established across the block. Repeat the calculation assuming the material is

(b) a block of air with these dimensions; (c) a block of wood with these dimensions.

31. A window has a glass surface of 1.6×10^3 cm^2 and a thickness of 3.0 mm. (a) Find the rate of heat transfer by conduction through this pane when the temperature of the inside surface of the glass is 70°F and the outside temperature is 90°F. (b) Repeat for the same inside temperature and an outside temperature of 0°F.

32. Determine the *R* value for a wall constructed as follows: The outside of the house consists of lapped wood shingles placed over 0.50-in.-thick sheathing, over 3.0 in. of cellulose fiber, over 0.50 in. of dry wall.

33. A Thermopane window consists of two glass panes, each 0.50 cm thick, with a 1.0-cm-thick sealed layer of air between. If the inside temperature is 23.0°C and the outside temperature is 0.0°C, determine the rate of heat transfer through 1.0 m^2 of the window. Compare this with the rate of heat transfer through 1.0 m^2 of a single 1.0-cm-thick pane of glass.

34. The average thermal conductivity of the walls (including windows) and roof of the house in Figure P11.34 is 4.8×10^{-4} kW/m·°C, and their average thickness is 21.0 cm. The house is heated with natural gas, with a heat of combustion (heat given off per cubic meter of gas burned) of 9300 kcal/m^3. How many cubic meters of gas must be burned each day to maintain an inside temperature of 25.0°C if the outside temperature is 0.0°C? Disregard radiation and heat loss through the ground.

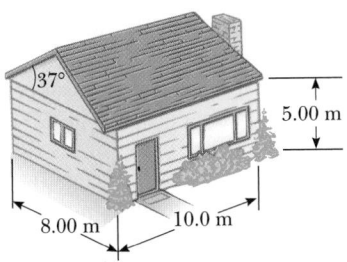

37°

5.00 m

8.00 m   10.0 m

Figure P11.34

35. A copper rod and an aluminum rod of equal diameter are joined end to end in good thermal contact. The temperature of the free end of the copper rod is held constant at 100°C, and that of the far end of the aluminum rod is held at 0°C. If the copper rod is 0.15 m long, what must be the length of the aluminum rod so that the temperature at the junction is 50°C?

36. A Styrofoam box has a surface area of 0.80 m^2 and a wall thickness of 2.0 cm. The temperature of the inner surface is 5°C, and that outside is 25°C. If it takes 8.0 h for 5.0 kg of ice to melt in the container, determine the thermal conductivity of the Styrofoam.

Section 11.7 Radiation

WEB 37. A sphere that is to be considered as a perfect black-body radiator has a radius of 0.060 m and is at 200°C in a room where the temperature is 22°C. Calculate the net rate at which the sphere radiates energy.

38. Two identical objects are in the same surroundings at 0°C. One is at a temperature of 1200 K, and the other is at 1100 K. Find the ratio of the net power emitted by the hotter object to the net power emitted by the cooler object.

39. Calculate the temperature at which a tungsten filament that has an emissivity of 0.25 and a surface area of 2.5×10^{-5} m^2 will radiate energy at the rate of 25 W in a room where the temperature is 22°C.

40. Measurements on two stars indicate that Star *X* has a surface temperature of 6000°C and Star *Y* has a surface temperature of 12000°C. If both stars have the same radius, what is the ratio of the luminosity (total power output) of Star *Y* to the luminosity of Star *X*? Both stars can be considered to have an emissivity of 1.

ADDITIONAL PROBLEMS

41. The bottom of a copper kettle has a 10.0-cm radius and is 2.0 mm thick. The temperature of the outside surface is 102°C, and the water inside the kettle is boiling at 1 atm of pressure. Find the rate at which heat is being transferred through the bottom of the kettle.

42. A 1.00-m^2 solar collector collects radiation from the Sun and focuses it on 250 g of water that is initially at 23°C. The average thermal energy arriving from the Sun at the surface of the Earth at this location is 550 W/m^2, and we assume that this is collected with 100% efficiency. Find the time required for the collector to raise the temperature of the water to 100°C.

43. In a showdown on the streets of Laredo, the good guy drops a 5.0-g silver bullet, at a temperature of 20°C, into a 100-cm^3 cup of water at 90°C. Simultaneously, the bad guy drops a 5.0-g copper bullet, at the same initial temperature, into an identical cup of water. Which one ends the showdown with the coolest cup of water in the West? Neglect any heat transfer into or away from the container.

44. A brass statue (60.0% copper, 40.0% zinc) has a mass of 50.0 kg. If its temperature increases by 20.0°C, what is the change of internal energy of the statue? (Specific heat of brass = 380 J/kg·°C = 0.092 cal/g·°C)

45. A 40-g ice cube floats in 200 g of water in a 100-g copper cup; all are at a temperature of 0°C. A piece of lead at 98°C is dropped into the cup, and the final equilibrium temperature is 12°C. What is the mass of the lead?

46. A family comes home from a long vacation with laundry to do and showers to take. The water heater has been turned off during the vacation. If the water heater has a capacity of 50.0 gallons and a 4800-W heating element,

how much time is required to raise the temperature of the water from 20.0°C to 60.0°C? Assume that the heater is well insulated and no water is withdrawn from the tank during this time.

47. At time $t = 0$, a vessel contains a mixture of 10 kg of water and an unknown mass of ice in equilibrium at 0°C. The temperature of the mixture is measured versus time, with the following results. During the first 50 min, the mixture remains at 0°C. From 50 min to 60 min, the temperature increases to 2°C. Neglecting the heat capacity of the vessel, determine the mass of ice that was initially placed in the vessel. Assume a constant power input to the container.

48. A 200-g block of copper at a temperature of 90°C is dropped into 400 g of water at 27°C. The water is contained in a 300-g glass container. What is the final temperature of the mixture?

49. A "solar cooker" consists of a curved reflecting mirror that focuses sunlight onto the object to be heated (Fig. P11.49). The solar power per unit area reaching the Earth at the location of a 0.50-m-diameter solar cooker is 600 W/m². Assuming that 50% of the incident energy is converted to heat energy, how long would it take to boil away 1.0 L of water initially at 20°C? (Neglect the specific heat of the container.)

Figure P11.49

50. A class of 10 students taking an exam has a power output per student of about 200 W. Assume that the initial temperature of the room is 20°C and that its dimensions are 6.0 m by 15.0 m by 3.0 m. What is the temperature of the room at the end of 1.0 h if all the heat remains in the air in the room and none is added by an outside source? The specific heat of air is 837 J/kg·°C, and its density is about 1.3×10^{-3} g/cm³.

WEB 51. An aluminum rod and an iron rod are joined end to end in a good thermal contact. The two rods have equal lengths and radii. The free end of the aluminum rod is maintained at a temperature of 100°C, and the free end

of the iron rod is maintained at 0°C. (a) Determine the temperature of the interface between the two rods. (b) If each rod is 15 cm long and each has a cross-sectional area of 5.0 cm², what quantity of heat energy is conducted across the combination in 30 min?

52. Water is being boiled in an open kettle that has a 0.500-cm-thick circular aluminum bottom with a radius of 12.0 cm. If the water boils away at rate of 0.500 kg/min, what is the temperature of the lower surface of the bottom of the kettle? Assume that the top surface of the bottom of the kettle is at 100°C.

53. A solar collector has an effective collecting area of 12 m². The collector is thermally insulated, so conduction is negligible in comparison with radiation. On a cold but sunny day the temperature outside is −20.0°C, and the Sun irradiates the collector with a power per unit area of 300 W/m². Treating the collector as a black body (i.e., emissivity = 1.0), determine its interior temperature after the collector has achieved a steady-state condition (radiating energy as fast as it is received).

54. A *flow calorimeter* is an apparatus used to measure the specific heat of a liquid. The technique is to measure the temperature difference between the input and output points of a flowing stream of the liquid while adding heat at a known rate. (a) Start with the equations $Q = mc(\Delta T)$ and $m = \rho V$, and show that the rate at which heat is added to the liquid is given by the expression $\Delta Q / \Delta t = \rho c (\Delta T)(\Delta V / \Delta t)$. (b) In a particular experiment, a liquid of density 0.72 g/cm³ flows through the calorimeter at the rate of 3.5 cm³/s. At steady state, a temperature difference of 5.8°C is established between the input and output points when heat is supplied at the rate of 40 J/s. What is the specific heat of the liquid?

55. Three liquids are at temperatures of 10°C, 20°C, and 30°C, respectively. Equal masses of the first two liquids are mixed, and the equilibrium temperature is 17°C. Equal masses of the second and third are then mixed, and the equilibrium temperature is 28°C. Find the equilibrium temperature when equal masses of the first and third are mixed.

56. Lake Erie contains roughly 4.00×10^{11} m³ of water. (a) How much heat is required to raise the temperature of that volume of water from 11.0°C to 12.0°C? (b) Approximately how many years would it take to supply this amount of heat by using the full output of a 1000-MW electric power plant?

57. A 3.00-g copper penny at 25.0°C drops 50.0 m to the ground. (a) If 60.0% of its initial potential energy goes into increasing the internal energy, determine its final temperature. (b) Does the result depend on the mass of the penny? Explain.

58. A 3.00-g lead bullet at 30.0°C is fired at a speed of 240 m/s into a large block of ice at 0°C, in which it embeds itself. What quantity of ice melts?

59. A bar of gold is in thermal contact with a bar of silver of the same length and area (Fig. P11.59). One end of the compound bar is maintained at 80.0°C while the opposite end is at 30.0°C. When the heat flow reaches steady state, find the temperature at the junction.

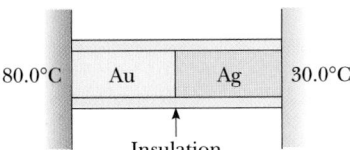

| 80.0°C | Au | Ag | 30.0°C |

Insulation

Figure P11.59

60. A box with a total surface area of 1.20 m² and a wall thickness of 4.00 cm is made of an insulating material. A 10.0-W electric heater inside the box maintains the inside temperature at 15.0°C above the outside temperature. Find the thermal conductivity k of the insulating material.

61. A 60-kg runner dissipates 300 W of power while running a marathon. Assuming that 10% of the runner's energy is dissipated in the muscle tissue and that the excess heat is removed from the body primarily by sweating, determine the volume of bodily fluid (assume it is water) lost per hour. (At 37°C the latent heat of vaporization of water is 575 kcal/kg.)

62. An iron plate is held against an iron wheel so that a sliding frictional force of 50 N acts between the two pieces of metal. The relative speed at which the two surfaces slide over each other is 40 m/s. (a) Calculate the rate at which mechanical energy is converted to internal energy. (b) The plate and the wheel have masses of 5.0 kg each, and each receives 50% of the frictional work. If the system is run as described for 10 s and each object is then allowed to reach a uniform internal temperature, what is the resultant temperature increase?

63. An automobile has a mass of 1500 kg, and its aluminum brakes have an overall mass of 60 kg. (a) Assuming that all of the frictional work produced when the car stops is deposited in the brakes, and neglecting heat transfer, how many times could the car be braked to rest starting from 25 m/s (56 mph) before the brakes would begin to melt? (Assume an initial temperature of 20°C.) (b) Identify some effects that are neglected in part (a) but are likely to be important in a more realistic assessment of the heating of brakes.

64. A 1.0-m-long aluminum rod of cross-sectional area 2.0 cm² is inserted vertically into a thermally insulated vessel containing liquid helium at 4.2 K. The rod is initially at 300 K. If half of the rod is inserted into the helium, how many liters of helium boil off in the very short time while the inserted half cools to 4.2 K?

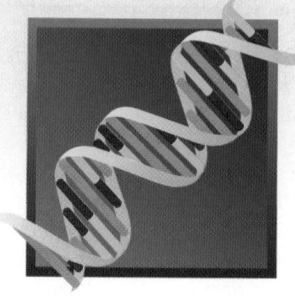

BIOLOGICAL PERSPECTIVE

Energy Management in the Human Body

DAVID GRIFFING
Miami University

Why can sprinters run at 10 m/s whereas distance runners can do only a little under 6 m/s? Why does a marathon runner's body temperature often increase by 3°F or more? Why is dieting a better strategy than exercise for losing weight? These and many other questions are readily answered if we examine energy management within the body.

Humans use energy to move, breathe, pump blood, and so forth. Chemical reactions within the body are responsible for storing, releasing, absorbing, and transferring this energy. Some of these reactions *require* energy (endothermic reactions), whereas others *release* energy (exothermic reactions).

The energy source our bodies use to produce endothermic reactions is the food we eat. Exothermic reactions release the energy needed to run, walk, build new cells, and so forth. As an example, locomotion in the human body depends on the cooperation of agonist/antagonist pairs of muscles that alternately contract and relax. Muscle fibers pull but cannot push. With skeletal support, the agonist pulls while the antagonist relaxes, to produce motion. Figure 1 shows the biceps and triceps muscles cooperating to bend the arm at the elbow.

The energy required for muscle contraction is made available in the muscle cells through a network of chemical reactions. Within this network, there are two input and two output ingredients. The input ingredients come from the air we breathe and the food we

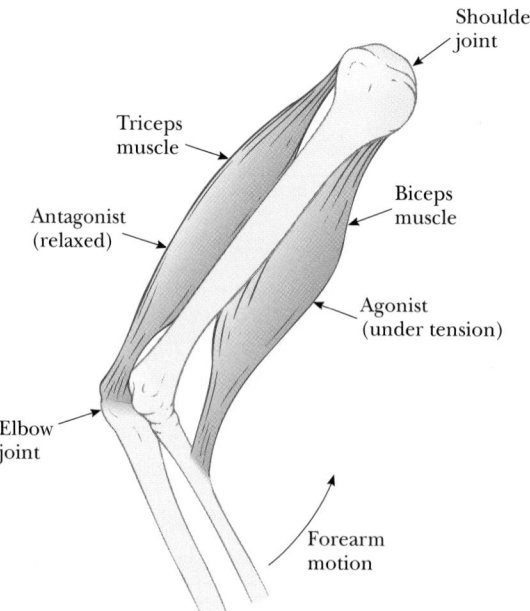

Figure 1 To bend the elbow, the biceps pulls (as agonist) while the triceps relaxes (as antagonist). To straighten the elbow, the roles of biceps and triceps are reversed. The bone is necessary because muscle fibers pull but cannot push.

368

eat, and the output ingredients are the carbon dioxide we exhale and the water produced as a by-product. In the lungs, the body removes oxygen from the inhaled air and transports it to muscle cells via hemoglobin in the bloodstream. In the mouth, stomach, and intestines, the body digests food, processing some of it into glucose ($C_6H_{12}O_6$). Some of this glucose is transported to the muscle cells. There, oxygen and glucose combine to form water and carbon dioxide in an exothermic reaction:

$$C_6H_{12}O_6 + 6O_2 \Rightarrow 6CO_2 + 6H_2O + E_{out}$$

Because this released energy is linked directly to the oxygen inhaled and carbon dioxide exhaled, it is theoretically simple to measure the energy a person generates. One measures either the net oxygen inhaled or the net carbon dioxide exhaled, and the energy released follows from the glucose oxidation reaction. Such measurements are routinely performed in hospitals and research laboratories.

Energy Production in Muscle Cells

The human body must take time to breathe, transport oxygen, and deliver glucose to the muscle cells. However, muscle fibers need not wait until glucose oxidation supplies the demanded energy because the body has the ability to store energy for future needs. Four processes are dedicated to energy production in a muscle cell; all use a high-energy molecule, adenosine triphosphate (ATP), to store and deliver energy in muscle cells. As long as ATP is available in the cell, immediate muscle contraction can occur without the benefit of breathing.

When the power demand in muscle fibers increases abruptly, the body needs time to adjust to the new level. Energy needed for muscle contraction during this transition time cannot be supplied by glucose oxidation because of the time needed to change to the new equilibrium state. To provide for this temporary energy gap, some "start-up" energy is stored in muscle cells in the form of existing ATP, and in the form of ingredients for anaerobic (without oxygen) production of ATP. After the transition time, respiration will gradually increase ATP production until a steady state is attained. During ordinary exercise, therefore, energy is supplied anaerobically for approximately the first two minutes and aerobically thereafter, via respiration.

Energy and Exercise

If either glucose or oxygen is absent, the production of energy in muscle cells stops. If either is in short supply, exercise is limited. Normally, enough glucose is stored in the muscle to supply ordinary energy needs, including vigorous exercise, for a few hours. Enough anaerobic energy is available for the body to function without oxygen for a couple of minutes. As an extreme example, Houdini, the celebrated escape artist of the 1920s, could function under water for 3 or 4 min without breathing.

When the body uses energy faster than respiration can support ATP production, ATP becomes scarce and the available energy declines. This condition is described as oxygen debt. At the end of vigorous exercise, when energy demand is back to normal, the debt is "repaid" as the exerciser continues to breathe rapidly, as when a sprinter gasps for air after the race.

Experiment has shown that the power, P, required to run a given distance is nearly directly proportional to the running speed v, that is, $P \propto v$. However, the power required to walk a given distance is $P \propto v^x$, with $x > 1$. So to walk a given distance the required energy increases with the walking speed, but to run the same distance the required energy is approximately independent of speed.[1] As a rule of thumb:

$$\text{Running energy/distance} \cong 1.5 \text{ kcal/mi} \cong 0.93 \text{ kcal/km}$$

Using this figure of merit, we can compare the energy expended running with the energy content of food. For example, an 80-kg person burns about 120 kcal running 1 mi at any speed. Burning 1 lb of fat generates about 4200 kcal. Thus, if the goal is to lose 1 lb of fat in 35 days, one could (1) run an extra mile a day, or (2) eat two fewer slices of bread a day.[2] Exercise probably stimulates the appetite enough so that the extra food one eats more than compensates for the extra energy needed for the exercise. Exercise is a poor substitute for dieting as a method to lose weight!

When the body is constructively stressed with exercise, it responds by improving its ability to cope with that stress. In weight lifting, for example, the muscles used to lift weights get stronger, while in training for distance running, the body's ability to ingest and process oxygen improves. In weight lifting, sudden bursts of energy are needed for short times. The exercise is largely anaerobic, and the body's ability to produce anaerobic bursts of energy improves. In distance running, however, a high level of steady-state energy production is needed. The exercise is largely aerobic, and the body responds by improving its capacity for aerobic energy production.

The primary goal of aerobic exercise is to attain good cardiovascular condition. Such exercise works constructively, provided the aerobic stress is regular and not harmfully intensive. Some forms of efficient aerobic activity are potentially harmful to the body in other ways. For example, running puts nonconstructive stress on the skeletal system, and joint problems are not uncommon among long-term distance runners and joggers. When a runner stops a regular exercise routine so that the stress is removed, the body quickly loses its high level of cardiovascular fitness. The time needed to achieve good fitness for distance running is several months or even several years. Other things being equal, runners who stop training during the summer cannot compete with those who train all year long.

To prevent glycogen (a polymer form of glucose) depletion during a marathon, glycogen loading has proven effective. Whereas long training times are needed for the body to develop excellent oxygen-processing in cells, times as short as a week can increase the glycogen stored in the muscle. The body quickly digests carbohydrates and converts them to glycogen. To increase the glycogen stored in muscle cells beyond the normal concentration, the athlete eats food containing no carbohydrates for a week or so and then eats food overconcentrated with carbohydrates for a few days before the race. In effect, the body is tricked into overstocking muscle cells with glycogen, which can provide extra fuel late in the race. Glycogen loading has been effective in helping marathon runners avoid "hitting the wall" after running about 20 mi.

Extreme energy demand, such as that required to swim the English Channel, to compete in the triathalon, or to cycle across the country, pushes the body to its limit both aerobically and anaerobically. Training for such high-stress events normally involves special high-energy diets or sleep deprivation.

Energy production in the body requires both oxygen and glucose. Oxygen cannot be stored, and so the steady-state rate of oxygen ingestion limits the long steady-state power production in muscles. Because glucose can be stored in the muscle cells and liver, the ultimate amount of energy that can be produced in the body without eating depends on

---

[1] Gilbert W. Fellingham et al., "Caloric Cost of Walking and Running," *Medicine and Science in Sports* 1978, 132–136.

[2] At 60 kcal/slice, 70 slices of bread correspond to 4200 kcal. At 2 slices a day, 35 days are needed.

how much glucose is stored in the form of glycogen. Beyond this point, the body will support itself by metabolizing body fat.

Work and Heat

A 70-kg person requires, depending on metabolism rate, about 2400 kcal a day for normal bodily functions. The power output of the body rises during exercise and falls during rest. What happens to the energy produced in the body? Some of this energy is used to perform work, and some appears as heat.

When a muscle contracts, a force is exerted through a distance, and so mechanical work is performed. Using only the upper body muscles, a skilled rope climber can climb 5 m in 2 s. For an 80-kg climber, about 2 kW of power is developed. Using leg muscles, up to 5–6 kW may be developed for similar short bursts of activity, such as in stair climbing or jumping. In extended activity, however, the power developed is considerably reduced. For example, in 1932 five men on the Polish Olympic Team climbed 362 m from the 5th to the 102nd floor of the Empire State Building in about 21 min. Again assuming a mass of 80 kg, this works out to a 225-W power output. The longer a person works, the lower the rate at which chemical energy is converted into mechanical work.

In a classic paper[3] the Nobel laureate A. V. Hill described measurements in which the work performed during muscle contraction is accompanied by a temperature increase of the muscle fiber. The energy delivered by ATP in the muscle is converted partly to thermal energy. To prevent the overall body temperature from rising when muscles contract, this thermal energy must be transferred from the body to the environment. A small amount is conducted away, and some is radiated away. Breathing also cools the body because of the water vapor exhaled as a by-product of respiration, and convection is an effective cooling process. Sweating is triggered to provide additional cooling when these cooling processes are saturated.

The body maintains a near-constant temperature by eliminating excess thermal energy as rapidly as possible. Blood vessels dilate so that blood flow increases, pores open so that sweating increases, and respiration increases. In hot weather, however, these mechanisms may be insufficient for extreme aerobic activities such as marathon running, and performance suffers. Athletes can pour water over themselves to increase cooling a little but may have to reduce their activity level to reduce their thermal energy production rate. It is not unusual for marathon runners to experience a rise in body temperature of 3 to 4°F during the race.

Joggers know that fewer clothes are needed to keep warm during exercise than when walking or standing. It is not unusual to observe joggers running in shorts and T-shirts in freezing weather. The heat developed during exercise keeps them warm. Surface blood vessels contract to inhibit heat loss. If any protection is needed against the cold, the most essential garment is a warm hat because up to 40% of the thermal energy lost by the body exits through the head, where a copious supply of blood vessels receives thermal energy from the rest of the body. (In bed in a cold room, the most efficient protection is a nightcap, as our ancestors well knew.)

The efficiency of converting energy to mechanical work is defined to be the ratio of the work output to the energy input. When viewed as an engine the human body's efficiency is probably less than 25%, depending on the basal metabolism rate of the individual. For comparison, this efficiency is about the same as that of an automobile engine.

[3] A. V. Hill, "The Heat of Shortening and the Dynamic Constants of Muscle," *Proc. Roy. Soc.* 1938, 136–195.

12

The Laws of Thermodynamics

▲ PHYSICS PUZZLER

Rockets, jet planes, gasoline automobiles, coal-fired and nuclear electric generating stations, and old-time steam locomotives all have the same basic limit to their efficiency. What is this limit? *(Phil Degginger/Tony Stone Images)*

The first law of thermodynamics is essentially the principle of conservation of energy generalized to include heat as a mode of energy transfer. According to the first law, the internal energy of an object (a concept to be discussed shortly) can be increased either by heat added to the object or by work done on it. The law places no restrictions on the types of energy conversions that occur. It tells us that the internal energy of a system can change as a result of two separate types of energy transfer across the boundary of the system.

The second law of thermodynamics, which can be stated in many equivalent ways, establishes which processes can occur in nature and which cannot. For example, the second law tells us that heat never flows spontaneously from a cold body to a hot body. One important application of this law is in the study of heat engines, such as the internal combustion engine, and the principles that limit their efficiency.

12.1 HEAT AND INTERNAL ENERGY

A major distinction must be made between internal energy and heat. **Internal energy** is all of the energy belonging to a system while it is stationary (neither translating nor rotating), including heat as well as nuclear energy, chemical energy, and strain energy (as in a compressed or stretched spring). **Thermal energy** is the portion of internal energy that changes when the temperature of the system changes. **Heat transfer** is caused by a temperature difference between the system and its surroundings.

In Chapter 10 we showed that the thermal energy of a monatomic ideal gas is associated with the motion of its atoms. In this special case, the thermal energy is simply kinetic energy on a microscopic scale; the higher the temperature of the gas, the greater the kinetic energy of the atoms and the greater the thermal energy of the gas. In general, however, thermal energy includes other forms of molecular energy, such as rotational energy and vibrational kinetic and potential energy (Fig. 12.1).

As an analogy, consider the distinction between work and energy that was discussed in Chapter 5. The work done on (or by) a system is a measure of the energy transferred between the system and its surroundings, whereas the system's mechanical energy (kinetic or potential) is a consequence of its motion and coordinates. Thus, when a person does work on a system, energy is transferred from the person to the system. It makes no sense to talk about the work *of* a system—one should refer only to the *work done on or by a system* when some process has occurred in which energy has been transferred to or from the system. Likewise, it makes no sense to use the term *heat* unless energy has been transferred as a result of a temperature difference.

It is also important to recognize that energy can be transferred between two systems even when no *thermal* energy transfer occurs. For example, when a piston quickly compresses a gas, the gas is warmed and its thermal energy increases, but no transfer of heat takes place; if the gas then expands rapidly, it cools and its thermal energy decreases, but there is no transfer of thermal energy to its surroundings. In each case, energy is transferred to or from the system as work but appears within the system as an increase or decrease in thermal energy. The change in

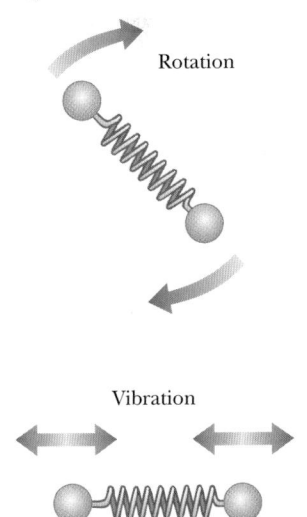

Figure 12.1 The rotational and vibrational energy of a diatomic gas contributes to the internal energy of the gas.

internal energy is equal to the change in thermal energy and is measured by a corresponding change in temperature.

12.2   WORK AND HEAT

10.6, SECTION 1

In the macroscopic approach to thermodynamics, we describe the *state* of a system with the use of such variables as pressure, volume, temperature, and internal energy. The number of macroscopic variables needed to characterize a system depends on the system's nature. For a homogeneous system, such as a gas containing only one type of molecule, usually only two variables are needed. It is important to note that a *macroscopic state* of an isolated system can be specified only if the system is in thermal equilibrium internally. In the case of a gas in a container, internal thermal equilibrium requires that every part of the container be at the same pressure and temperature.

Consider a gas contained by a cylinder fitted with a movable piston (Fig. 12.2). In equilibrium, the gas occupies a volume of V and exerts a uniform pressure, P, on the cylinder walls and piston. If the piston has cross-sectional area A, the force exerted by the gas on the piston is $F = PA$. Now let us assume that the gas expands slowly enough to allow the system to remain essentially in thermodynamic equilibrium at all times. As the piston moves up a distance of Δy, the work done on the piston by the gas is

$$W = F\,\Delta y = PA\,\Delta y$$

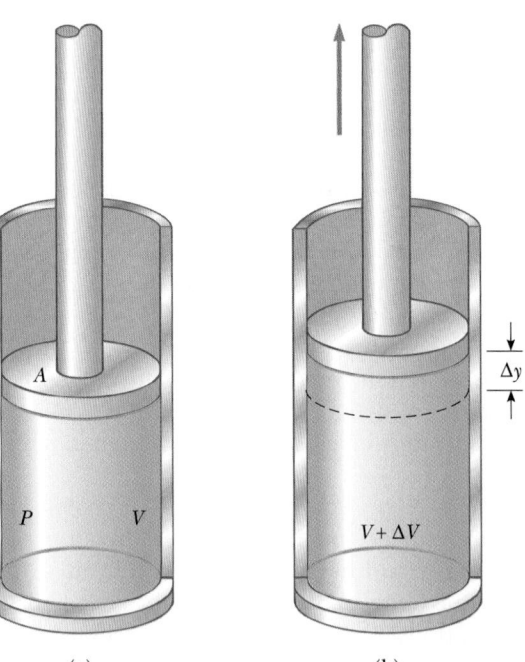

Figure 12.2 (a) A gas in a cylinder occupying a volume V at a pressure P. (b) As the gas expands at constant pressure and the volume increases by the amount ΔV, the work done by the gas is $P\,\Delta V$.

(a)   (b)

Because $A\,\Delta y$ is the increase in volume of the gas (that is, $\Delta V = A\,\Delta y$), we can express the work done as

$$W = P\,\Delta V \qquad\qquad [12.1]$$

If the gas expands, as in Figure 12.2b, ΔV is positive and the work done by the gas is positive. If the gas is compressed, ΔV is negative and the work done by the gas is negative. (In the latter case, negative work can be interpreted as work being done *on* the system.) Clearly, the work done by (or on) the system is zero when the volume remains constant.

Equation 12.1 can be used to calculate the work done on or by the system only when the pressure of the gas remains constant during the expansion or compression. If the pressure changes, calculus is required to determine the work done. We do not attempt such calculations here.

Consider the process represented by the pressure-volume diagram in Figure 12.3. The gas has expanded from an initial volume, V_i, to a final volume, V_f, at a constant pressure of P. From Equation 12.1 we see that the work done by the gas in this case is $P(V_f - V_i)$. Note that this is just the area under the pressure–volume curve. **In general, the work done in an expansion from some initial state to some final state is the area under the curve on a *PV* diagram.** This statement is true whether or not the pressure remains constant during the process.

As Figure 12.3 shows, the work done in the expansion from the initial state to the final state depends on the path taken between the two states. To illustrate this important point, consider several different paths connecting i and f (Fig. 12.4). In the process depicted in Figure 12.4a, the pressure of the gas is reduced from P_i to P_f by cooling at constant volume, V_i, and then the gas expands from V_i to V_f at constant pressure, P_f. The work done along this path is $P_f(V_f - V_i)$. In Figure 12.4b, the gas expands from V_i to V_f at constant pressure, P_i, and then its pressure is reduced to P_f at constant volume, V_f. The work done along this path is $P_i(V_f - V_i)$, which is greater than the work done in the process of Figure 12.4a. Finally, in the process depicted in Figure 12.4c, where both P and V change continuously, the work done has some value intermediate between the values obtained in the first two processes. To evaluate the work in this case, the shape of the PV curve must be known. Thus, the work done by a system depends on the process by which the system goes from the initial to the final state. In other words, the work done depends on the initial, final, and intermediate states of the system.

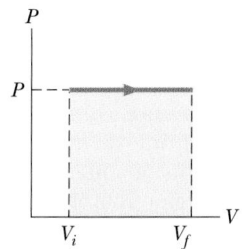

Figure 12.3 The *PV* diagram for a gas expanding at constant pressure. The shaded area represents the work done by the gas.

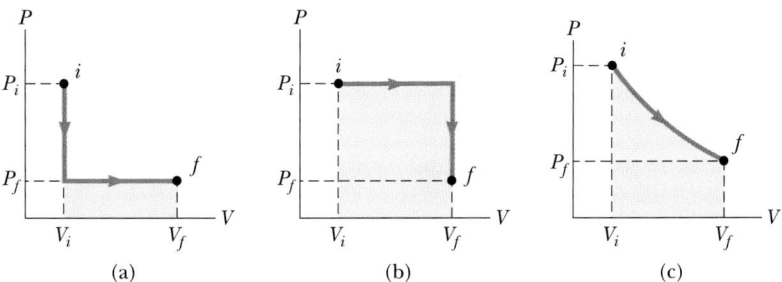

Figure 12.4 The work done by a gas as it is taken from an initial state to a final state depends on the path taken between these states.

Figure 12.5 (a) A gas at temperature T_i expands slowly by absorbing heat from a reservoir at the same temperature. (b) A gas expands rapidly into an evacuated region after a membrane separating it from that region is broken.

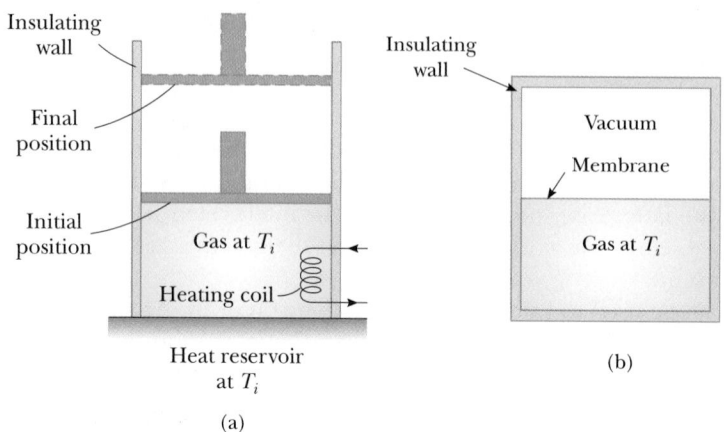

In a similar manner, the heat transferred into or out of a system is also found to depend on the process. This can be demonstrated by the situations depicted in Figure 12.5. In each case, the gas has the same initial volume, temperature, and pressure and is assumed to be ideal. In Figure 12.5a, the gas is in thermal contact

▶ **Heat reservoir**
with a heat reservoir, and a small heating coil allows energy to be transferred to the gas by heat from the surface of the coil. (A heat reservoir is a body whose heat capacity is so large that its temperature remains unchanged when heat is added or extracted from the body.) As an infinitesimal amount of energy enters the gas from the coil, the pressure of the gas becomes infinitesimally greater than atmospheric pressure, and the gas expands, causing the piston to rise. During this expansion to some final volume, V_f, and some final pressure, P_f, sufficient heat to maintain a constant temperature of T_i from the reservoir to the gas.

Now consider the thermally insulated system in Figure 12.5b. When the membrane is broken, the gas expands rapidly into the vacuum until it occupies a volume of V_f and is at a pressure of P_f. In this case the gas does no work, because there is no movable piston. Furthermore, no heat is transferred through the thermally in-

▶ **Free expansion of a gas**
sulated wall, which we call an *adiabatic wall*, and so the temperature remains at T_i. This process is often referred to as **adiabatic free expansion** or simply *free expansion*. In general, an adiabatic process is one in which no heat is transferred between the system and its surroundings.

The initial and final states of the ideal gas in Figure 12.5a are identical to the

▶ **Heat transfer and work done depend on the path between the initial and final states.**
initial and final states in Figure 12.5b, but the paths are different. In the first case, heat is transferred slowly to the gas, and the gas does work on the piston. In the second case, no heat is transferred, and the work done is zero. Therefore, we conclude that **heat transfer, like work, depends on the initial, final, and intermediate states of the system.**

Applying Physics 1

Warm air rises, so why is mountain air cold?

Explanation Consider what happens as warm air rises up a mountain. A certain volume of air expands as it rises because the pressure pushing against it is reduced. As the air expands, however, its temperature drops. By contrast, if prevailing winds drive air down a mountain, the falling air is compressed and, therefore, warmed. In Austria, a warm wind called the Foehn can blow from the Alps for several days at a time, proving very disconcerting for the Austrians. At one time, erratic behavior was excused, even by the courts, as being caused by the Foehn.

The usual situation, however, is for air to be pushed by prevailing winds from a valley to the top of a mountain. Air is a poor thermal conductor, so the process that occurs is one in which there is no heat transfer to or from the air. As the parcel of air rises, it moves into regions of lower atmospheric pressure. As a result, the parcel expands. By expanding, it is doing work on the surrounding air. This represents a transfer of energy out of the system, so that the internal energy drops and therefore the temperature drops. If the parcel of air contains water vapor, the drop in temperature as it rises up the mountain can result in condensation into raindrops. This is why the rain tends to fall on the windward side of mountains. On the leeward side, much of the water vapor has already condensed out as rain, and the air parcel is warming up again as it descends the mountain. Thus, there is little rain on the leeward side of the mountain.

An interesting example of this occurs in Hawaii. The prevailing winds are often from the northeast. The island of Kauai has a very tall mountain (Mount Waialeale, 1548 m), and it is one of the wettest places on Earth with an average rainfall of 1170 cm. Just to the southwest of Kauai is the island of Niihau, which is very dry and requires a system of irrigation for its farming. The leeward position of Niihau relative to Kauai is the reason.

APPLICATION

Rainfall Pattern Near Mountains.

Thinking Physics 1

Imagine a gas in an insulated cylinder with a movable piston. The piston has been pushed inward, compressing the gas, and is now released. As the molecules of the gas strike the piston, they move it outward. From the point of view of energy principles, explain how this expansion causes the temperature of the gas to drop.

Explanation From the point of view of energy principles, the molecules strike the piston and move it through a distance. Thus, the molecules do work on the piston, which represents a transfer of energy out of the gas. As a result, the internal energy of the gas drops. Because the temperature is related to internal energy, the temperature of the gas drops.

EXAMPLE 12.1   Work Done by an Expanding Gas

In the system shown in Figure 12.2, the gas in the cylinder is at a pressure of 8000 Pa and the piston has an area of 0.10 m^2. As heat is slowly added to the gas, the piston is pushed up a distance of 4.0 cm. Calculate the work done on the surroundings by the expanding gas. Assume that the pressure remains constant.

Solution The change in volume of the gas is

$$\Delta V = A\,\Delta y = (0.10 \text{ m}^2)(4.0 \times 10^{-2} \text{ m}) = 4.0 \times 10^{-3} \text{ m}^3$$

and from Equation 12.1, the work done is

$$W = P\,\Delta V = (8000 \text{ Pa})(4.0 \times 10^{-3} \text{ m}^3) = \boxed{32 \text{ J}}$$

10.6, SECTION 2

12.3 THE FIRST LAW OF THERMODYNAMICS

When the principle of conservation of energy was first introduced in Chapter 5, it was stated that the mechanical energy of a system is constant in the absence of nonconservative forces, such as friction. That mechanical model did not encompass changes in the internal energy of the system. We now broaden our scope to use the term *principle of conservation of energy* for a generalization encompassing possible changes in internal energy. This is a universally valid law that can be applied to all kinds of processes. Furthermore, it provides a connection between the microscopic and macroscopic worlds. The result will be the **first law of thermodynamics.**

▷ *Change in internal energy*

We have seen that energy can be transferred between a system and its surroundings in two ways: via work done by (or on) the system, which requires a macroscopic displacement of the point of application of a force (or pressure), and via heat transfer, which occurs through random molecular collisions. Each of these represents a change of energy of the system, and therefore usually results in measurable changes in the macroscopic variables of the system, such as the pressure, temperature, and volume of a gas.

To express these ideas more quantitatively, suppose a system undergoes a change from an initial state to a final state. During this change, positive Q is the heat transferred *to* the system, and positive W is the work done *by* the system. For example, suppose the system is a gas whose pressure and volume change from P_i and V_i to P_f and V_f. If the quantity $Q - W$ is measured for various paths connecting the initial and final equilibrium states (that is, for various *processes*), one finds that it is the same for *all* paths connecting the initial and final states. We conclude that **the quantity $Q - W$ is determined completely by the initial and final states of the system, and we call it the change in the internal energy of the system.** Although Q and W both depend on the path, $Q - W$ **is independent of the path.** If we represent the internal energy function with the letter U, then the *change* in internal energy, $\Delta U = U_f - U_i$, can be expressed as

▷ *First-law equation*

$$\Delta U = U_f - U_i = Q - W \qquad [12.2]$$

where all quantities must have the same energy units. Equation 12.2 is known as the first law of thermodynamics. This law applies universally to all systems.

Let us examine some special cases in which the only changes in energy are changes in internal energy. First consider an *isolated system*—that is, one that does not interact with its surroundings. In this case, no heat transfer takes place and the work done is zero; hence, the internal energy remains constant. That is, because

▷ *For isolated systems, U remains constant.*

$Q = W = 0$, $\Delta U = 0$, and so $U_i = U_f$. We conclude that **the internal energy of an isolated system remains constant.**

Next consider the case in which a system (not isolated from its surroundings) is taken through a **cyclic process**—that is, a process that originates and ends at the same state. In this case, the change in the internal energy must again be *zero*, and therefore the heat added to the system must equal the work done during the cycle. That is, in a cyclic process,

▷ *Cyclic process*

$$\Delta U = 0 \qquad \text{and} \qquad Q = W$$

Note that **the net work done per cycle equals the area enclosed by the path representing the process on a *PV* diagram.** As we shall see in the next section, cyclic processes are very important for describing the thermodynamics of *heat* *engines* — devices in which some part of the thermal energy input is converted into mechanical work.

If a process occurs in which the work done is zero, then the change in internal energy equals the heat entering or leaving the system. If heat enters the system, Q is positive and the internal energy increases. For a gas, we can associate this increase in internal energy with an increase in the kinetic energy of the molecules. If a process occurs in which the heat transferred is zero and work is done by the system, however, then the magnitude of the change in internal energy equals the negative of the work done by the system. That is, the internal energy of the system decreases. For example, if a gas is compressed with no heat transferred (by a moving piston, say), the work done by the gas is negative and the internal energy again increases. This is because kinetic energy is transferred from the moving piston to the gas molecules.

On a microscopic scale, no practical distinction exists between heat transfer and work. Each can produce a change in the internal energy of a system. Although the macroscopic quantities Q and W are *not* properties of a system, they are related to changes of the internal energy of a stationary system through the first law of thermodynamics. Once a process, or path, is defined, Q and W can be either calculated or measured, and the change the internal energy can be found from the first law of thermodynamics.

QUICKLAB

Bend a paper clip back and forth several times and touch it to your upper lip. Why does it warm up? Repeat this with a rubber band after it is stretched several times. You should notice that it also warms up after stretching. Now take the stretched rubber band from your lip and wait for it to come to equilibrium with the surrounding air. Touch the stretched band to your lip, let it relax to its unstretched length, and note that it becomes cooler. Explain how the first law of thermodynamics helps you understand these results.

EXAMPLE 12.2 An Isobaric Process

A gas is enclosed in a container fitted with a piston of cross-sectional area 0.10 m². The pressure of the gas is maintained at 8000 Pa while heat is slowly added; as a result, the piston is pushed up a distance of 4.0 cm. (Any process in which the pressure remains constant is called an **isobaric process.**) If 42 J of heat is added to the system during the expansion, what is the change in internal energy of the system?

Solution The work done by the gas is

$$W = P\Delta V = (8000 \text{ Pa})(0.10 \text{ m}^2)(4.0 \times 10^{-2} \text{ m}) = 32 \text{ N·m} = 32 \text{ J}$$

The change in internal energy is found from the first law of thermodynamics:

$$\Delta U = Q - W = 42 \text{ J} - 32 \text{ J} = \boxed{10 \text{ J}}$$

We see that **in an isobaric process the work done and the heat transferred are both nonzero.**

◀ An isobaric process is one that occurs at constant pressure.

Exercise If 42 J of heat is added to the system with the piston clamped in a *fixed* position, what is the work done by the gas? What is the change in its internal energy?

Answer No work is done; $\Delta U = 42$ J.

EXAMPLE 12.3 An Isovolumetric Process

Water with a mass of 2.0 kg is held at constant volume in a container while 10 000 J of heat is slowly added by a flame. The container is not well insulated, and as a result 2000 J of heat leaks out to the surroundings. What is the temperature increase of the water?

An isovolumetric process is one ▶
that takes place at constant
volume.

Solution   A process that takes place at constant volume is called an **isovolumetric process.** In such a process the work is equal to zero. Thus, the first law of thermodynamics gives

$$\Delta U = Q$$

This indicates that the net heat added to the water goes into increasing the internal energy of the water. The net heat added to the water is

$$Q = 10\ 000\ \text{J} - 2000\ \text{J} = 8000\ \text{J}$$

Because $Q = mc\,\Delta T$, the temperature increase of the water is

$$\Delta T = \frac{Q}{mc} = \frac{8000\ \text{J}}{(2.0\ \text{kg})(4.186 \times 10^3\ \text{J/kg·°C})} = \boxed{0.96°\text{C}}$$

EXAMPLE 12.4   Boiling Water

One gram of water at atmospheric pressure $(1.013 \times 10^5\ \text{Pa})$ occupies a volume of $1.0\ \text{cm}^3$. When this water is boiled, it becomes $1671\ \text{cm}^3$ of steam. Calculate the change in internal energy for this process.

Solution   Because the latent heat of vaporization of water is $2.26 \times 10^6\ \text{J/kg}$ at atmospheric pressure, the heat required to boil 1.0 g is

$$Q = mL_v = (1.0 \times 10^{-3}\ \text{kg})(2.26 \times 10^6\ \text{J/kg}) = 2260\ \text{J}$$

The work done by the system is positive and equal to

$$
\begin{aligned}
W &= P(V_{\text{steam}} - V_{\text{water}}) \\
&= (1.013 \times 10^5\ \text{Pa})[(1671 - 1.0) \times 10^{-6}\ \text{m}^3] = 169\ \text{J}
\end{aligned}
$$

Hence, the change in internal energy is

$$\Delta U = Q - W = 2260\ \text{J} - 169\ \text{J} = \boxed{2.1 \times 10^3\ \text{J}}$$

The positive ΔU tells us that the internal energy of the system has increased. We see that most of the heat (93%) transferred to the liquid goes into increasing the internal energy. Only a small fraction (7%) goes into external work.

EXAMPLE 12.5   Heat Transferred to a Solid

The internal energy of a solid also increases when heat is transferred to it from its surroundings. To illustrate this point, suppose a 1.0-kg bar of copper is heated at atmospheric pressure. Its temperature increases from 20°C to 50°C.
(a) Find the work done by the copper.

Solution   The change in volume of the copper can be calculated using Equation 10.6 and the linear expansion coefficient for copper taken from Table 10.1 (remembering that $\beta = 3\alpha$):

$$\Delta V = \beta V \Delta T = [5.1 \times 10^{-5}(°\text{C})^{-1}](50°\text{C} - 20°\text{C})\,V = (1.5 \times 10^{-3})\,V$$

The volume is equal to m/ρ, and the density of copper is 8.92×10^3 kg/m³. Hence,

$$\Delta V = (1.5 \times 10^{-3}) \left(\frac{m}{\rho}\right) = (1.5 \times 10^{-3}) \left(\frac{1.0 \text{ kg}}{8.92 \times 10^3 \text{ kg/m}^3}\right)$$

$$= 1.7 \times 10^{-7} \text{ m}^3$$

Because the expansion takes place at constant pressure (equal to normal atmospheric pressure), the work done is

$$W = P\Delta V = (1.013 \times 10^5 \text{ Pa})(1.7 \times 10^{-7} \text{ m}^3) = \boxed{1.9 \times 10^{-2} \text{ J}}$$

(b) What quantity of heat is transferred to the copper?

Solution   The specific heat of copper is given in Table 11.1, and from Equation 11.3 we find that the heat transferred is

$$Q = mc\,\Delta T = (1.0 \text{ kg})(387 \text{ J/kg} \cdot {}^\circ\text{C})(30{}^\circ\text{C}) = \boxed{1.2 \times 10^4 \text{ J}}$$

Exercise   What is the increase in internal energy of the copper?

Answer $\Delta U = 1.2 \times 10^4$ J

Applying Physics 2

In the late 1970s casino gambling was approved in Atlantic City, New Jersey, which can become quite cold in the winter. Energy projections that were performed on the design of the casinos showed that the air conditioning would need to operate in the casino even in the middle of a very cold January. Why?

Explanation   If we consider the air in the casino to be a gas to which we apply the first law, imagine that there is no air conditioning and no ventilation, so that this gas simply stays in the room. There is no work being done by the gas, so we focus on the heat input. A casino consists of a large number of people, many of whom are active (throwing dice, cheering, etc.). As a result, these people have large rates of energy flow by heat from their bodies. This energy goes into the air of the casino, resulting in an increase in internal energy of the air. With the large number of excited people in a casino (along with the large number of incandescent lights and machines), the temperature of the gas can rise quickly and to a high value. In order to keep the temperature at reasonable values, there must be an energy transfer out of the air. Calculations show that conduction through the walls even in a 10°F January are not sufficient to maintain the required energy transfer, so the air conditioning system must be in almost continuous use throughout the year.

12.4 HEAT ENGINES AND THE SECOND LAW OF THERMODYNAMICS

A heat engine is a device that converts thermal energy to other useful forms, such as electrical and mechanical energy. In a typical process for producing electricity in a power plant, for instance, coal or some other fuel is burned, and the thermal energy produced is used to convert water to steam. This steam is directed at the

 10.7, BOTH SECTIONS

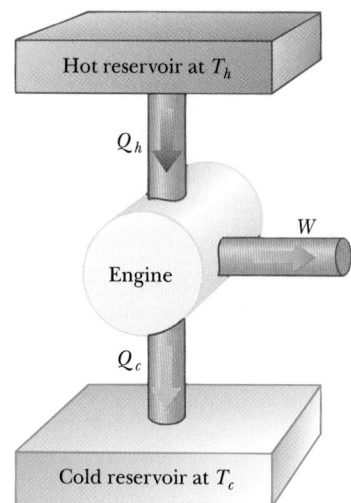

Figure 12.6 A schematic representation of a heat engine. The engine receives heat Q_h from the hot reservoir, expels heat Q_c to the cold reservoir, and does work W.

blades of a turbine, setting it in rotation. Finally, the mechanical energy associated with this rotation is used to drive an electric generator. Another heat engine, the automobile internal combustion engine, extracts heat from a burning fuel and converts a fraction of this energy to mechanical energy.

A heat engine carries some working substance through a cyclic process during which (1) heat is absorbed from a source at a high temperature, (2) work is done by the engine, and (3) heat is expelled by the engine to a reservoir at a lower temperature. As an example, consider the operation of a steam engine in which the working substance is water. The water is carried through a cycle in which it first evaporates into steam in a boiler and then expands against a piston. After the steam is condensed with cooling water, it is returned to the boiler, and the process is repeated.

It is useful to represent a heat engine schematically, as in Figure 12.6. The engine absorbs a quantity of heat, Q_h, from the hot reservoir, does work W, and then gives up heat Q_c to the cold reservoir. Because the working substance goes through a cycle, its initial and final internal energies are equal, so $\Delta U = 0$. Hence, from the first law of thermodynamics we see that **the net work, W, done by a heat engine equals the net heat flowing into it.** As we can see from Figure 12.6, $Q_{net} = Q_h - Q_c$; therefore,

$$W = Q_h - Q_c \qquad [12.3]$$

where Q_h and Q_c are taken to be positive quantities.

If the working substance is a gas, **the net work done for a cyclic process is the area enclosed by the curve representing the process on a *PV* diagram.** This is shown for an arbitrary cyclic process in Figure 12.7. **The thermal efficiency, *e*, of a heat engine is the ratio of the net work done to the heat absorbed at the higher temperature during one cycle:**

Thermal efficiency ▶

$$e \equiv \frac{W}{Q_h} = \frac{Q_h - Q_c}{Q_h} = 1 - \frac{Q_c}{Q_h} \qquad [12.4]$$

We can think of the efficiency as the ratio of what we get (mechanical energy) to what we give (thermal energy at the higher temperature). Equation 12.4 shows that a heat engine has 100% efficiency ($e = 1$) only if $Q_c = 0$ — that is, if no heat is expelled to the cold reservoir. In other words, a heat engine with perfect efficiency would have to convert all of the absorbed heat to mechanical work. One of the consequences of the second law of thermodynamics is that this is impossible.

Second law ▶

The **second law of thermodynamics** can be stated as follows: **It is impossible to construct a heat engine that, operating in a cycle, produces no other effect than the absorption of heat from a reservoir and the performance of an equal amount of work.**

This form of the second law is useful for understanding the operation of heat engines. With reference to Equation 12.4, the second law says that, during the operation of a heat engine, W can never be equal to Q_h or, alternatively, that some heat, Q_c, must be rejected to the environment. As a result, it is theoretically impossible to construct an engine that works with 100% efficiency.

Heat engines can, however, operate in reverse — we send in energy, modeled as work, W, in Figure 12.6, resulting in heat being extracted from the cold reservoir and transferred to the hot reservoir. This system is now operating as a heat pump. Your kitchen refrigerator is a common heat pump. Energy (heat) is extracted from

10.8

APPLICATION ➤

Heat Pumps and Refrigerators.

the cold food in the interior and delivered to the warmer air in the kitchen. The work is done in the compressor unit of the refrigerator, in which a piston compresses the refrigerant (freon, or other similar material) to a smaller volume, increasing its temperature. A household air conditioner is another example of a heat pump. Some homes are both heated and cooled by heat pumps. In the winter, the heat pump extracts energy from the outside air and delivers it to the inside, and the process is reversed for summer cooling. In a ground-water heat pump, energy is extracted in the winter from water deep in the ground rather than from the outside air, and energy is delivered to that water in the summer. This increases the year-round efficiency of the heating and cooling unit, because the ground water is at a higher temperature than the air in winter and is at a cooler temperature than the air in summer.

Our assessment of the first two laws of thermodynamics can be summed up as follows: the first law says **we cannot get a greater amount of energy out of a cyclic process than we put in,** and the second law says **we cannot break even.**

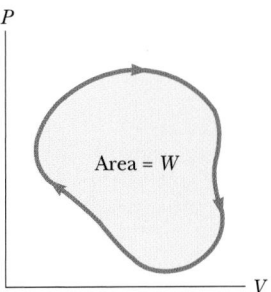

Figure 12.7 The *PV* diagram for an arbitrary cyclic process. The area enclosed by the curve equals the net work done.

EXAMPLE 12.6 The Efficiency of an Engine

Find the efficiency of an engine that introduces 2000 J of heat during the combustion phase and loses 1500 J at exhaust.

Solution The efficiency of the engine is given by Equation 12.4 as

$$e = 1 - \frac{Q_c}{Q_h} = 1 - \frac{1500\text{ J}}{2000\text{ J}} = \boxed{0.25 \text{ or } 25\%}$$

Exercise If an engine has an efficiency of 20% and loses 3000 J at exhaust and to the cooling water, how much work is done by the engine?

Answer 750 J

A model steam engine equipped with a built-in horizontal boiler. The water is heated electrically, generating steam that is used to power the electric generator at the left. (*Courtesy of Central Scientific Company*)

12.5 REVERSIBLE AND IRREVERSIBLE PROCESSES

In the next section we shall discuss a theoretical heat engine that is the most efficient engine possible. In order to understand its nature, we must first examine the meanings of reversible and irreversible processes. A **reversible** process is one in which every state between the initial and final states is an equilibrium state, and that can be reversed in order to be followed exactly from the final state back to the initial state. A process that does not satisfy these requirements is **irreversible.**

All natural processes are known to be irreversible. As an example, let us examine the free expansion of a gas (already discussed in Section 12.2) and show that it cannot be reversed. The gas is contained in an insulated container with a membrane separating it from a vacuum (Fig. 12.8). If the membrane is punctured, the gas expands freely into the vacuum. Because the gas does not exert a force on its surroundings through a distance, it does no work as it expands. In addition, no heat is transferred to or from the gas because the container is insulated from its surroundings. Thus, in this process the system has changed but the surroundings have not.

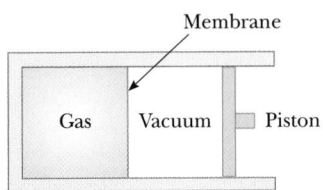

Figure 12.8 Free expansion of a gas.

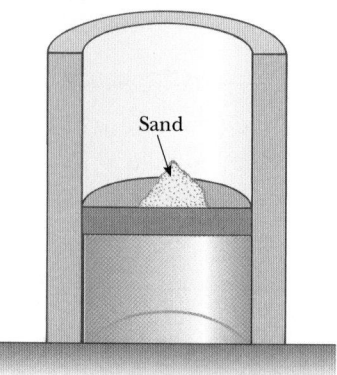

Figure 12.9 A gas in thermal contact with a heat reservoir is compressed slowly by grains of sand dropped onto the piston. The compression is isothermal and reversible.

Now imagine that we try to reverse the process by first compressing the gas to its original volume. Let's say an engine is being used to force the piston inward. Note, however, that this action is changing both the system and surroundings. The surroundings are changing because work is being done by an outside agent on the system, and the system is changing because the compression is increasing the temperature of the gas. We can lower the temperature of the gas by allowing it to come in contact with an external heat reservoir. Although this second procedure returns the gas to its original state, the surroundings are again affected because thermal energy is added to the surroundings. If this heat could somehow be used to drive the engine and compress the gas, the system and its surroundings could be returned to their initial states. However, our statement of the second law says that this extracted heat cannot be completely converted to mechanical energy isothermally. We must conclude that a reversible process has not occurred.

Although real processes are always irreversible, some are *almost* reversible. If a real process occurs very slowly so that the system is virtually always in equilibrium, the process can be considered reversible. For example, imagine compressing a gas very slowly by dropping some grains of sand onto a frictionless piston, as in Figure 12.9. The compression process can be reversed by the placement of the gas in thermal contact with a heat reservoir. The pressure, volume, and temperature of the gas are well defined during this isothermal compression. Each added grain of sand represents a change to a new equilibrium state. The process can be reversed by the slow removal of grains of sand from the piston.

A general characteristic of a reversible process is that no dissipative effects that convert mechanical energy to thermal energy, such as turbulence or friction, can be present. In reality, such effects are impossible to eliminate completely, and hence it is not surprising that real processes in nature are irreversible.

Sadi Carnot, French engineer (1796–1832)

Carnot is considered to be the founder of the science of thermodynamics. Some of his notes found after his death indicate that he was the first to recognize the relationship between work and heat.

12.6   THE CARNOT ENGINE

In 1824 a French engineer named Sadi Carnot (1796–1832) described a theoretical engine, now called a *Carnot engine,* that is of great importance from both practical and theoretical viewpoints. He showed that a heat engine operating in an ideal, reversible cycle — called a Carnot cycle — between two reservoirs is the most efficient engine possible. **Carnot's theorem** can be stated as follows:

> **No real engine operating between two heat reservoirs can be more efficient than a Carnot engine, operating between the same two reservoirs.**

To describe the Carnot cycle, we assume that the substance working between temperatures T_c and T_h is an ideal gas contained in a cylinder with a movable piston at one end. The cylinder walls and the piston are thermally nonconducting. Figure 12.10 shows four stages of the Carnot cycle, and Figure 12.11 is the *PV* diagram for the cycle. The cycle consists of two adiabatic and two isothermal processes, all reversible.

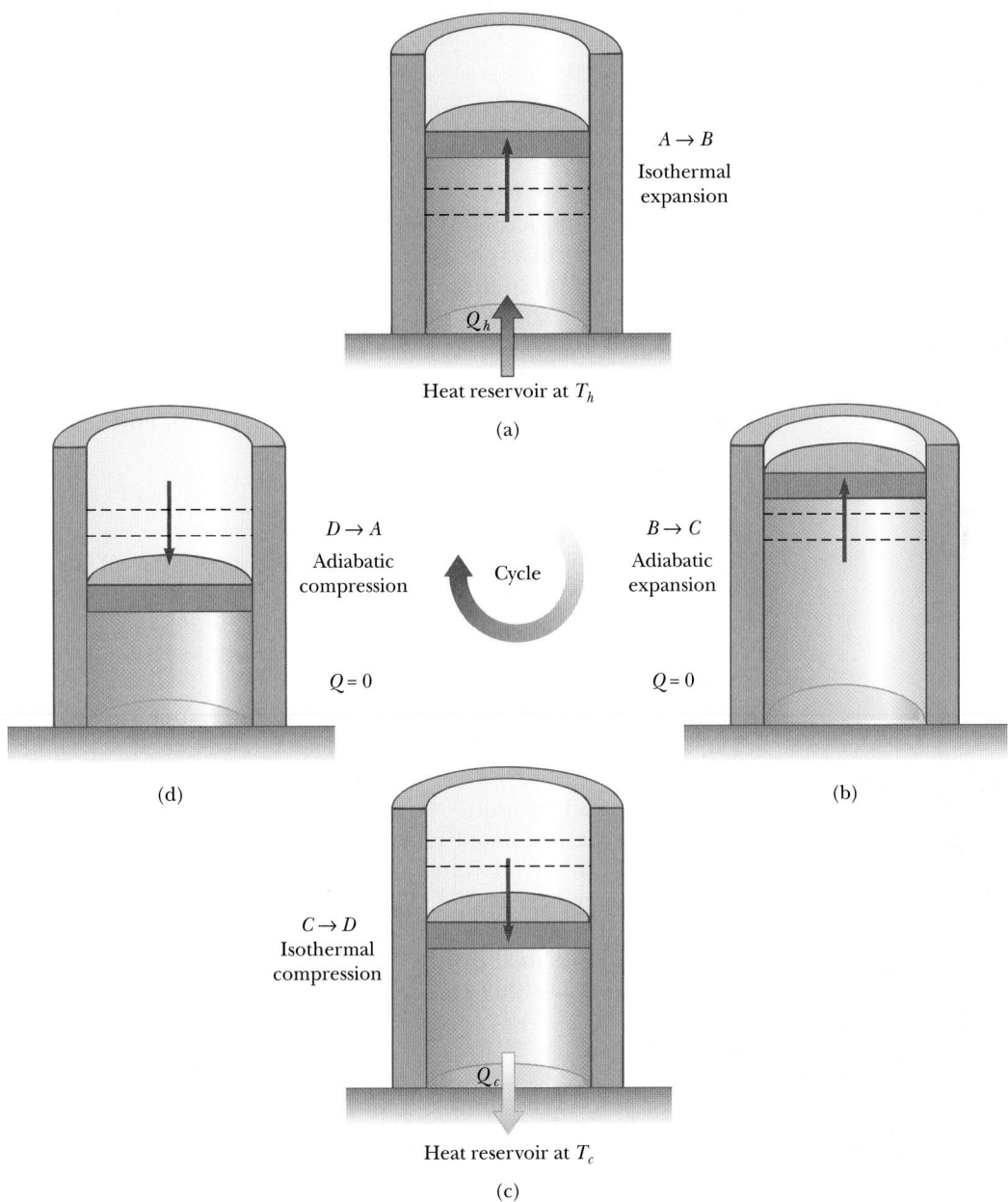

Figure 12.10 The Carnot cycle. In process $A \to B$, the gas expands isothermally while in contact with a reservoir at T_h. In process $B \to C$, the gas expands adiabatically ($Q = 0$). In process $C \to D$, the gas is compressed isothermally while in contact with a reservoir at $T_c < T_h$. In process $D \to A$, the gas is compressed adiabatically. The upward arrows on the piston indicate removal of sand during the expansions, and the downward arrows indicate addition of sand during the compressions.

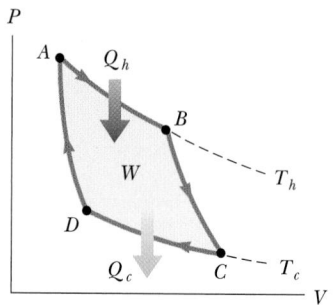

Figure 12.11 The *PV* diagram for the Carnot cycle. The net work done, *W*, equals the net heat received in one cycle, $Q_h - Q_c$.

10.9, BOTH SECTIONS

Lord Kelvin, British physicist and mathematician (1824–1907)

Born William Thomson in Belfast, Kelvin was the first to propose the use of an absolute scale of temperature. His study of Carnot's theory led to the idea that heat cannot pass spontaneously from a colder body to a hotter body; this is known as the second law of thermodynamics. *(J. L. Charmet/SPL/Photo Researchers)*

1. The process $A \rightarrow B$ is an isothermal expansion at temperature T_h, in which the gas is placed in thermal contact with a heat reservoir at temperature T_h (Fig. 12.10a). During the process, the gas absorbs heat Q_h from the reservoir and does work W_{AB} in raising the piston.
2. In the process $B \rightarrow C$, the base of the cylinder is replaced by a thermally nonconducting wall and the gas expands adiabatically; that is, no heat enters or leaves the system (Fig. 12.10b). During the process, the temperature falls from T_h to T_c and the gas does work W_{BC} in raising the piston.
3. In the process $C \rightarrow D$, the gas is placed in thermal contact with a heat reservoir at temperature T_C (Fig. 12.10c) and is compressed isothermally at temperature T_c. During this time, the gas expels heat Q_c to the reservoir, and the work done on the gas is W_{CD}.
4. In the final stage, $D \rightarrow A$, the base of the cylinder is again replaced by a thermally nonconducting wall (Fig. 12.10d) and the gas is compressed adiabatically. The temperature of the gas increases to T_h, and the work done on the gas is W_{DA}.

Carnot showed that the thermal efficiency of a Carnot engine is

$$e_c = \frac{T_h - T_c}{T_h} = 1 - \frac{T_c}{T_h} \qquad [12.5]$$

where T must be in kelvins. From this result, we see that all Carnot engines operating reversibly between the same two temperatures have the same efficiency. Furthermore, the efficiency of any reversible engine operating in a cycle between two temperatures is greater than the efficiency of any irreversible (real) engine operating between the same two temperatures.

Equation 12.5 can be applied to any working substance operating in a Carnot cycle between two heat reservoirs. According to this result, the efficiency is zero if $T_c = T_h$. The efficiency increases as T_c is lowered and as T_h is increased. However, the efficiency can be unity (100%) only if $T_c = 0$ K. Such reservoirs are not available, and so the maximum efficiency is always less than unity. In most practical cases, the cold reservoir is near room temperature, about 300 K. Therefore, one usually strives to increase the efficiency by raising the temperature of the hot reservoir. **All real engines are less efficient than the Carnot engine because they are subject to practical difficulties, including friction, but especially the need to operate irreversibly to complete a cycle in a brief time period.**

EXAMPLE 12.7 The Steam Engine

A steam engine has a boiler that operates at 500 K. The heat changes water to steam, which drives the piston. The temperature of the exhaust is that of the outside air, about 300 K. What is the maximum thermal efficiency of this steam engine?

Solution   From the expression for the efficiency of a Carnot engine, we find the maximum thermal efficiency for any engine operating between these temperatures:

$$e_c = 1 - \frac{T_c}{T_h} = 1 - \frac{300 \text{ K}}{500 \text{ K}} = \quad 0.4, \text{ or } 40\%$$

This is the highest theoretical efficiency of the engine. In practice, the efficiency is considerably lower.

PHYSICS IN ACTION

T he device in the photograph on the left, called *Hero's engine,* was invented around 150 B.C. by Hero in Alexandria. When water is boiled in the flask, which is suspended by a cord, steam exits through two tubes at the sides of the flask (in opposite directions), creating a torque that rotates the flask.

The device shown in the center and right-hand photographs, called a *thermoelectric converter,* uses a series of semiconductor cells to convert thermal energy to electrical energy. In the center photograph, the two "legs" of the device are at the same temperature, and no electrical energy is produced. However, when one leg is at a higher temperature than the other, as in the photograph on the right, electrical energy is produced as the device extracts energy from the hot reservoir and drives a small electric motor. How does this intriguing experiment demonstrate the second law of thermodynamics?

Devices That Convert Thermal Energy into Other Forms of Energy

(Courtesy of Central Scientific Company)

(Courtesy of PASCO Scientific Company)

Exercise Determine the maximum work the engine can perform in each cycle of operation if it absorbs 200 J of thermal energy from the hot reservoir during each cycle.

Answer 80 J

EXAMPLE 12.8 The Carnot Efficiency

The highest theoretical efficiency of a gasoline engine, based on the Carnot cycle, is 30%. If this engine expels its gases into the atmosphere, which has a temperature of 300 K, what is the temperature in the cylinder immediately after combustion?

Solution   The Carnot efficiency is used to find T_h:

$$e_c = 1 - \frac{T_c}{T_h}$$

$$T_h = \frac{T_c}{1 - e_c} = \frac{300 \text{ K}}{1 - 0.30} = \boxed{430 \text{ K}}$$

Actual gasoline engines operate on a cycle significantly different from the Carnot cycle and therefore have lower maximum possible efficiencies.

Exercise   If the heat engine absorbs 837 J of heat from the hot reservoir during each cycle, how much work can it perform in each cycle?

Answer   251 J

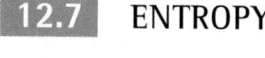 Workbook Problem 66 — (Workbook page 140) also deals with Carnot engines.

12.7   ENTROPY

The concept of temperature is involved in the zeroth law of thermodynamics (Chapter 10), and the concept of internal energy is involved in the first law. Temperature and internal energy are both state functions; that is, they can be used to describe the thermodynamic state of a system. Another state function related to the second law of thermodynamics is the **entropy function, S.**

Consider a reversible process between two equilibrium states. If ΔQ_r is the heat absorbed or expelled by the system during some small interval of the path, **the change in entropy, ΔS, between two equilibrium states is given by the heat transferred, ΔQ_r, divided by the absolute temperature, T, of the system in this interval.** That is,

$$\Delta S \equiv \frac{\Delta Q_r}{T} \qquad [12.6]$$

The subscript r on the term ΔQ_r emphasizes that the definition applies only to reversible processes. When heat is absorbed by the system, ΔQ_r is positive and hence the entropy increases. When heat is expelled by the system, ΔQ_r is negative and the entropy decreases. Note that Equation 12.6 defines not entropy but the change in entropy. This is consistent with the fact that a change in state always accompanies heat transfer. Hence, the meaningful quantity in a description of a process is the change in entropy.

The concept of entropy was introduced into the study of thermodynamics by Rudolf Clausius in 1865. One reason it became useful and gained wide acceptance is because it provides another variable to describe the state of a system, to go along with pressure, volume, and temperature. The concept of entropy reached a position of even more significance when it was found that **the entropy of the Universe increases in all natural processes.** This is yet another way of stating the second law of thermodynamics.

The statement of the second law just given must be interpreted with care. Although it says that the entropy of the Universe always increases in all natural processes, there are processes in which the entropy of a system decreases. That is, there

Rudolf Clausius (1822–1888)

"I propose . . . to call S the entropy of a body, after the Greek word 'transformation.' I have designedly coined the word 'entropy' to be similar to energy, for these two quantities are so analogous in their physical significance, that an analogy of denominations seems to be helpful." *(AIP Niels Bohr Library, Lande Collection)*

are situations in which the entropy of one system (system *A*) decreases but in correspondence with a net increase in entropy of some other system (system *B*). In all cases the change in entropy of system *B* is greater than the change in entropy of system *A*.

For centuries, individuals have attempted to build perpetual motion machines that would operate continuously without input of energy and without any net increase in entropy. Using the laws of thermodynamics, we can understand why a perpetual motion machine will *never* be invented. Perpetual motion machines of the first kind are those that violate the first law, putting out more energy than is put into the machine. Perpetual motion machines of the second kind are those that violate the second law. Imagine, for example, a waterfall in which water passes over a paddlewheel, which turns an electric generator providing electricity to a nearby motor. The temperature of the falling water has risen after it hits the paddlewheel and the pond below. The pond is used as the high temperature reservoir of a heat engine. The electric motor and the heat engine together are used to pump water from the pond back up to the top of the waterfall. This is a perpetual motion machine of both the first and second kinds. It violates the first law because it assumes all energy remains in the system so that it operates forever at the same rate. This would require no friction in any of the mechanisms, and it requires that the mechanisms make no sound, because sound would carry energy away from the system. The machine we have described also violates the second law because it assumes that there is no exhaust heat from the heat engine, which would result in energy loss from the system.

The concept of entropy is satisfying because it enables us to present the second law of thermodynamics in the form of a mathematical statement. In the next section we will find that entropy can also be interpreted in terms of probabilities, a relationship that has profound implications for our world.

APPLICATION

"Perpetual Motion" Machines.

EXAMPLE 12.9   Melting a Piece of Lead

Calculate the change in entropy when 300 g of lead melts at 327°C (600 K). Lead has a latent heat of fusion of 2.45×10^4 J/kg.

Solution   The amount of heat added to the lead to melt it is

$$Q = mL_f = (0.300 \text{ kg})(2.45 \times 10^4 \text{ J/kg}) = 7.35 \times 10^3 \text{ J}$$

From Equation 12.6, the entropy change of the lead is

$$\Delta S = \frac{Q}{T} = \frac{7.35 \times 10^3 \text{ J}}{600 \text{ K}} = \boxed{12.3 \text{ J/K}}$$

EXAMPLE 12.10   Which Way Does the Heat Flow?

A large, cold object is at 273 K, and a large, hot object is at 373 K. Show that it is impossible for a small amount of heat energy, say 8.00 J, to be transferred from the cold object to the hot object without decreasing the entropy of the Universe and hence violating the second law.

Solution　We assume here that during the heat transfer the two systems undergo no significant temperature change. This is not a necessary assumption; it is used to avoid a need for the techniques of integral calculus. The entropy change of the hot object is

$$\Delta S_h = \frac{Q}{T_h} = \frac{8.00\ \text{J}}{373\ \text{K}} = 0.0214\ \text{J/K}$$

The cold reservoir loses heat, and its entropy change is

$$\Delta S_c = \frac{Q}{T_c} = \frac{-8.00\ \text{J}}{273\ \text{K}} = -0.0293\ \text{J/K}$$

The net entropy change of the Universe is

$$\Delta S_U = \Delta S_c + \Delta S_h = -0.0079\ \text{J/K}$$

This is in violation of the law that the entropy of the Universe always increases in natural processes. That is, **the spontaneous transfer of heat from a cold object to a hot object cannot occur.**

Exercise　In the preceding example, suppose that 8.00 J of heat were transferred from the hot to the cold object. What would be the net change in entropy of the Universe?

Answer　+0.0079 J/K

12.8　ENTROPY AND DISORDER

10.10

As you look around at the beauty of nature, it is easy to recognize that the events of natural processes have in them a large element of chance. For example, the spacing between trees in a natural forest is quite random; if you were to discover a forest where all the trees were equally spaced, you would conclude that it was a planted forest. Likewise, leaves fall to the ground with random arrangements. It would be highly unlikely to find the leaves laid out in perfectly straight rows. We can express the results of such observations by saying that **a disorderly arrangement is much more probable than an orderly one if the laws of nature are allowed to act without interference.**

Entropy originally found its place in thermodynamics, but its importance grew tremendously as the field of statistical mechanics developed. This analytical approach employs an alternative interpretation of entropy. In statistical mechanics, the behavior of a substance is described in terms of the statistical behavior of the atoms and molecules contained in the substance. One of the main products of this approach is the conclusion that **isolated systems tend toward greater disorder, and entropy is a measure of that disorder.**

In light of this new view of entropy, Boltzmann found an alternative method for calculating entropy through use of the relation

$$S = k_B \ln W \qquad [12.7]$$

where k_B is Boltzmann's constant ($k_B = 1.38 \times 10^{-23}$ J/K) and W is a number proportional to the probability that the system has a particular configuration. The symbol "ln" in Equation 12.7 is an abbreviation for the natural logarithm discussed in Appendix A.

TABLE 12.1 Possible Results of Drawing Four Marbles from a Bag

End Result	Possible Draws	Total Number of Same Results
All R	RRRR	1
1G, 3R	RRRG, RRGR, RGRR, GRRR	4
2G, 2R	RRGG, RGRG, GRRG, RGGR, GRGR, GGRR	6
3G, 1R	GGGR, GGRG, GRGG, RGGG	4
All G	GGGG	1

Let us explore the meaning of this equation through a specific example. Imagine that you have a bag of 100 marbles, 50 red and 50 green. You are allowed to draw four marbles from the bag according to the following rules. Draw one marble, record its color, return it to the bag, and draw again. Continue this process until four marbles have been drawn. Note that because each marble is returned to the bag before the next one is drawn, the probability of drawing a red marble is always the same as the probability of drawing a green one.

The results of all possible drawing sequences are shown in Table 12.1. For example, the result RRGR means that a red marble was drawn first, a red one second, a green one third, and a red one fourth. This table indicates that there is only one possible way to draw four red marbles. There are four possible sequences that produce one green and three red marbles, six sequences that produce two green and two red, four sequences that produce three green and one red, and one sequence that produces all green. From Equation 12.7, we see that the state with the greatest disorder (two red and two green marbles) has the highest entropy because it is most probable. In contrast, the most ordered states (all red marbles and all green marbles) are least likely to occur and are states of lowest entropy.

The outcome of the draw can range between these highly ordered (lowest entropy) and highly disordered (highest entropy) states. Thus, entropy can be regarded as an index of how far a system has progressed from an ordered to a disordered state.

The second law of thermodynamics is really a statement of what is most probable rather than of what must be. Imagine placing an ice cube in contact with a hot piece of pizza. There is nothing in nature that absolutely forbids the transfer of heat from the ice to the much warmer pizza. Statistically, it is possible for a slow-moving molecule in the ice to collide with a faster-moving molecule in the pizza so that the slow one transfers some of its energy to the faster one. However, when the great number of molecules present in the ice and pizza are considered, the odds are overwhelmingly in favor of the transfer of energy from the faster-moving molecules to the slower-moving molecules. Furthermore, this example demonstrates that a system naturally tends to move from a state of order to a state of disorder. The initial state, in which all the pizza molecules have high kinetic energy and all the ice molecules have lower kinetic energy, is much more ordered than the final state after heat transfer has taken place and the ice has melted.

Even more generally, the second law of thermodynamics defines the direction of time for all events as the direction in which the entropy of the universe increases. Although conservation of energy is not violated if energy spontaneously flows from a cold object (the ice cube) to a hot object (the pizza slice), the event violates the

APPLICATION

The Direction of Time.

second law because it represents a spontaneous increase in order, and it also violates everyday experience. If the melting ice cube were filmed and the film speeded up, the difference between running the film in forward and reverse directions would be obvious to an audience. The same would be true of filming any event involving a large number of particles—such as a dish dropping to the floor and shattering.

Thinking Physics 2

Suppose you film the collision of two billiard balls on a pool table. Now suppose you run the film in reverse. Will the difference between forward and reverse directions be obvious? Why or why not?

Explanation The film shows nothing about the molecular arrangement of the billiard balls, so there are effectively only two particles in the event—the two billiard balls. During the collision, there is no change from order to disorder or vice versa, so the film appears to be authentic (describing a valid situation) when run in either direction.

As another example, suppose you were able to measure the velocities of all the air molecules in a room at some instant. It is very unlikely that you would find all molecules moving in the same direction with the same speed—this would be a highly ordered state, indeed. The most probable situation is a system of molecules moving haphazardly in all directions with a distribution of speeds—a highly disordered state. Let us compare this case to that of drawing marbles from a bag. If a container held 10^{23} molecules of a gas, the probability of finding all of the molecules moving in the same direction with the same speed at some instant would be similar to that of drawing a marble from the bag 10^{23} times and getting a red marble on every draw. This is clearly an unlikely set of events.

The tendency of nature to move toward a state of disorder affects the ability of a system to do work. Consider a ball thrown toward a wall. The ball has kinetic energy, and its state is an ordered one; that is, all of the atoms and molecules of the ball move in unison at the same speed and in the same direction (apart from their random thermal motions). When the ball hits the wall, however, part of the ball's kinetic energy is transformed into the random, disordered, thermal motion of the molecules in the ball and the wall and the temperatures of the ball and the wall both increase slightly. Before the collision, the ball was capable of doing work. It could drive a nail into the wall, for example. With the transformation of part of the ordered energy into disordered thermal energy, this capability of doing work is reduced. That is, the ball rebounds with less kinetic energy than it originally had, because the collision is inelastic.

Various forms of energy can be converted to thermal energy, as in the collision between the ball and the wall, but the reverse transformation is never complete. In general, if two kinds of energy, A and B, can be completely interconverted, we say that they are the *same grade*. However, if form A can be completely converted to form B and the reverse is never complete, then form A is a *higher grade* of energy than form B. In the case of a ball hitting a wall, the kinetic energy of the ball is of a higher grade than the thermal energy contained in the ball and the wall after the collision. Therefore, when high-grade energy is converted to thermal energy, it can never be fully recovered as high-grade energy.

A full house is a very good hand in the game of poker. Can you calculate the probability of being dealt a full house (a pair and three of a kind) from a standard deck of 52 cards? *(Tom Mareschel, The IMAGE Bank)*

This conversion of high-grade energy to thermal energy is referred to as **degradation of energy.** The energy is said to be degraded because it takes on a form that is less useful for doing work. In other words, **in all real processes where heat transfer occurs, the energy available for doing work decreases.**

Finally, note once again that the statement that entropy must increase in all natural processes is true only for isolated systems. There are instances in which the entropy of some system decreases, but with a corresponding net increase in entropy for some other system. When all systems are taken together to form the Universe, *the entropy of the Universe always increases.*

Ultimately, the entropy of the Universe should reach a maximum. At this time the Universe will be in a state of uniform temperature and density. All physical, chemical, and biological processes will have ceased, because a state of perfect disorder implies no available energy for doing work. This gloomy state of affairs is sometimes referred to as an ultimate "heat death" of the Universe.

An illustration from Flammarion's novel *La Fin du Monde*, depicting the heat death of the Universe.

Thinking Physics 3

If you shake a jar of jelly beans of two different sizes, the larger jelly beans tend to appear near the top, while the smaller ones tend to settle at the bottom. Why does this occur? Does this process violate the second law?

Explanation   Shaking opens up spaces between the jelly beans. The smaller ones have a chance of falling down into spaces below them. The accumulation of larger ones on top and smaller ones on the bottom implies an increase in order and a decrease in one contribution to the total entropy. However, the second law is not violated and the total entropy of the system increases. The increase in the internal energy of the system comes from the work required to shake the jar of beans (that is, work your muscles must do, with an increase in entropy accompanying the biological processes) and also from the small loss of gravitational potential energy as the beans settle together more compactly.

SUMMARY

The work done as a gas expands or contracts at a constant pressure is

$$W = P \Delta V \qquad [12.1]$$

The work done is negative if the gas is compressed and positive if the gas expands. In general, the work done in an expansion from some initial state to some final state is the area under the curve on a PV diagram.

 10.11

From the first law of thermodynamics, when a system undergoes a change from one state to another, the **change in its internal energy,** ΔU, is

$$\Delta U = U_f - U_i = Q - W \qquad [12.2]$$

where Q is the heat transferred into (or out of) the system and W is the work done by (or on) the system. Q is positive when heat enters the system and negative when the system loses heat.

The **first law of thermodynamics** is the generalization of the law of conservation of energy that includes heat transfer.

In a cyclic process (one in which the system returns to its initial state), $\Delta U = 0$ and therefore $Q = W$. That is, the heat transferred into the system equals the work done during the cycle.

An **adiabatic process** is one in which no heat is transferred between the system and its surroundings ($Q = 0$). In this case, the first law gives $\Delta U = -W$. That is, the internal energy changes as a consequence of work being done by (or on) the system.

An **isobaric process** is one that occurs at constant pressure. The work done in such a process is $P\,\Delta V$.

A **heat engine** is a device that converts thermal energy to other forms of energy, such as mechanical and electrical energy. The work done by a heat engine in carrying a substance through a cyclic process ($\Delta U = 0$) is

$$W = Q_h - Q_c \qquad [12.3]$$

where Q_h is the heat absorbed from a hot reservoir and Q_c is the heat expelled to a cold reservoir.

The **thermal efficiency** of a heat engine is defined as the ratio of the net work done to the heat absorbed per cycle:

$$e \equiv \frac{W}{Q_h} = 1 - \frac{Q_c}{Q_h} \qquad [12.4]$$

No real heat engine operating between the kelvin temperatures T_h and T_c can exceed the efficiency of an engine operating between the same two temperatures in a **Carnot cycle,** given by

$$e_c = 1 - \frac{T_c}{T_h} \qquad [12.5]$$

Real processes proceed in an order governed by the **second law of thermodynamics,** which can be stated in two ways:

1. Heat will not flow spontaneously from a cold object to a hot object.
2. No heat engine operating in a cycle can absorb thermal energy from a reservoir and perform an equal amount of work.

The second law can also be stated in terms of a quantity called **entropy.** The **change in entropy** of a system is equal to the heat flowing into (or out of) the system as the system changes from one state (A) to another (B), divided by the absolute temperature:

$$\Delta S \equiv \frac{\Delta Q_r}{T} \qquad [12.6]$$

One of the primary findings of statistical mechanics is that systems tend toward disorder and that entropy is a measure of this disorder. The entropy of the Universe increases in all natural processes; this is an alternative statement of the second law.

MULTIPLE-CHOICE QUESTIONS

1. If a process starts at (P_0, V_0) and doubles in volume, which of the following type involves the most work: **(a)** adiabatic **(b)** isobaric **(c)** isovolumetric **(d)** No answer is correct always.
2. An ideal gas is maintained at a constant pressure of 70 000 Pa during an isobaric process and its volume decreases by 0.2 m^3. What work is done by the system on its environment? **(a)** 14 000 J **(b)** 350 000 J **(c)** $-14\,000$ J **(d)** $-35\,000$ J
3. A 2-mole ideal-gas system is maintained at a constant volume of 4 liters; if 100 J of heat is added, what is the change in internal energy of the system? **(a)** zero **(b)** 50 J **(c)** 67 J **(d)** 100 J
4. A 1-kg block of ice at 0°C and 1 atm melts completely to water. What is its change in entropy? (For ice, $L_f = 3.34 \times 10^5$ J/kg.) **(a)** 3340 J/K **(b)** 2170 J/K **(c)** 613 J/K **(d)** 1220 J/K
5. An steam turbine operates at a boiler temperature of 450 K and an exhaust temperature of 300 K. What is the maximum theoretical efficiency of this system? **(a)** 24% **(b)** 50% **(c)** 33% **(d)** 67%

CONCEPTUAL QUESTIONS

1. For an ideal gas in an isothermal process, there is no change in internal energy. Suppose the gas does work W during such a process. How much energy was transferred by heat?
2. A designer of an electric heating unit describes her product as being 100% efficient. Is her claim correct?
3. Consider the human body performing a strenuous exercise, such as lifting weights or riding a bicycle. Work is being done by the body, and energy is leaving by conduction from the skin into the surrounding air. According to the first law of thermodynamics, the temperature of the body should be steadily decreasing during the exercise. This is not what happens, however. Is the first law invalid for this situation? Explain.
4. Clearly distinguish among temperature, thermal energy, heat, and internal energy.
5. What is wrong with the statement, "Given any two bodies, the one with the higher temperature contains more heat?"
6. A steam-driven turbine is one major component of an electric power plant. Why is it advantageous to increase the temperature of the steam as much as possible?
7. When a sealed Thermos bottle full of hot coffee is shaken, what changes, if any, take place in **(a)** the temperature of the coffee and **(b)** its internal energy?
8. In solar ponds constructed in Israel, the Sun's energy is concentrated near the bottom of a salty pond. With the proper layering of salt in the water, convection is prevented, and temperatures of 100°C may be reached. Can you guess the maximum efficiency with which useful energy can be extracted from the pond?
9. What are some factors that affect the efficiency of automobile engines?
10. Why does your automobile burn more gas in winter than in summer?
11. Is it possible to construct a heat engine that creates no thermal pollution?
12. Suppose your roommate is "Mr. Clean" and tidies up your messy room after a big party. Because more order is being created by your roommate, does this represent a violation of the second law of thermodynamics?
13. A thermodynamic process occurs in which the entropy of a system changes by -8.0 J/K. According to the second law of thermodynamics, what can you conclude about the entropy change of the environment?
14. If a supersaturated sugar solution is allowed to evaporate slowly, sugar crystals form in the container. Hence, sugar molecules go from a disordered form (in solution) to a highly ordered, crystalline form. Does this process violate the second law of thermodynamics? Explain.
15. The first law of thermodynamics says we cannot get more out of a process than we put in, but and the second law says that we cannot break even. Explain this statement.
16. Give some examples of irreversible processes that occur in nature. Give an example of a process in nature that is nearly reversible.
17. The thermoelectric converter shown on the right side photograph on page 387 uses semiconductor cells to convert thermal energy to electrical energy. **(a)** Why does a temperature difference between the two legs of the device produce electric energy? **(b)** In what sense does this intriguing device demonstrate the second law of thermodynamics?

PROBLEMS

1, 2, 3 = straightforward, intermediate, challenging   □ = full solution available in Study Guide/Student Solutions Manual   ![] = Core Concepts Workbook
WEB = solution posted at **http://www.harcourtcollege.com/physics/cptech**   ![] = biomedical application   ![] = Interactive Physics

Review Problem

A 40.0-gram projectile is launched by the expansion of hot gas in an arrangement shown in part (a) of the figure. The cross-sectional area of the launch tube is 1.0 cm², and the length that the projectile travels down the launch tube after starting from rest is 32 cm. As the gas expands, the pressure varies as shown in part (b) of the figure. The values for the initial pressure and volume are $P_0 = 11 \times 10^5$ Pa and $V_0 = 8.0$ cm³ and the final values are $P_f = 1.0 \times 10^5$ Pa and $V_f = 40.0$ cm³. Friction between the projectile and the launch tube is negligible. **(a)** If the projectile is launched into a vacuum, what is the speed of the projectile as it leaves the launch tube? **(b)** If instead the projectile is launched into air at a pressure of 1.0×10^5 Pa, what fraction of the work done by the expanding gas in the tube is spent by the projectile pushing air out of the way as it proceeds down the tube?

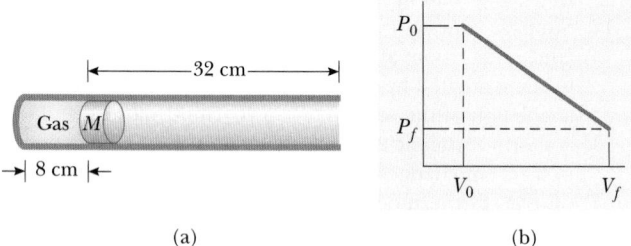

(a)   (b)

a volume of 1.0 L to 3.0 L at a constant pressure of 3.0 atm. **(b)** The gas is then cooled at constant volume until the pressure falls to 2.0 atm. **(c)** The gas is then compressed at a constant pressure of 2.0 atm from a volume of 3.0 L to 1.0 L. (*Note:* Be careful of signs.) **(d)** The gas is heated until its pressure increases from 2.0 atm to 3.0 atm at a constant volume. **(e)** Find the net work done during the complete cycle.

5. A gas expands from *I* to *F* along the three paths indicated in Figure P12.5. Calculate the work done by the gas along paths **(a)** *IAF*, **(b)** *IF*, and **(c)** *IBF*.

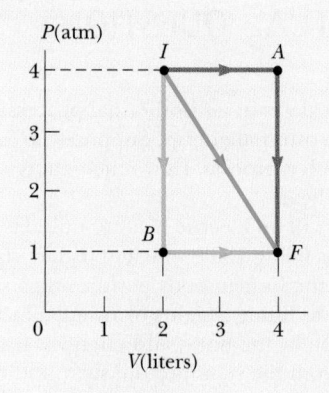

Figure P12.5

Section 12.1   Heat and Internal Energy

Section 12.2   Work and Heat

WEB **1.** The only form of energy possessed by molecules of a monatomic ideal gas is translational kinetic energy. Using the results from the discussion of kinetic theory in Section 10.6, show that the internal energy of a monatomic ideal gas at pressure *P* and occupying volume *V* may be written as $U = \frac{3}{2}PV$.

2. Steam moves into the cylinder of a steam engine at a constant pressure of 2.00×10^5 Pa. The diameter of the piston is 16.0 cm, and the piston travels 20.0 cm in one stroke. How much work is done during one stroke?

3. A container of volume 0.40 m³ contains 3.0 mol of argon gas at 30°C. Assuming argon behaves as an ideal gas, find the total internal energy of this gas. (*Hint:* See Problem 1.)

4. Sketch a *PV* diagram and find the work done by the gas during the following stages. **(a)** A gas is expanded from

6. Sketch a *PV* diagram of the following processes. **(a)** A gas expands at constant pressure P_1 from volume V_1 to volume V_2. It is then kept at constant volume while the pressure is reduced to P_2. **(b)** A gas is reduced in pressure from P_1 to P_2 while its volume is held constant at V_1. It is then expanded at constant pressure P_2 to a final volume, V_2. **(c)** In which of the processes is more work done? Why?

7. Gas in a container is at a pressure of 1.5 atm and a volume of 4.0 m³. What is the work done by the gas if **(a)** it expands at constant pressure to twice its initial volume? **(b)** it is compressed at constant pressure to one quarter its initial volume?

8. A movable piston having a mass of 8.00 kg and a cross-sectional area of 5.00 cm² traps 0.200 moles of an ideal gas in a vertical cylinder. If the piston slides without friction in the cylinder, how much work will the gas do when its temperature is increased from 20°C to 300°C?

9. One mole of an ideal gas initially at a temperature of $T_0 = 0°C$ undergoes an expansion, at a constant pressure of one atmosphere, to four times its original volume. (a) Calculate the new temperature of the gas, T_f. (b) Calculate the work done by the gas during the expansion.

Section 12.3   The First Law of Thermodynamics

10. A container is placed in a water bath and held at constant volume as a mixture of fuel and oxygen is burned inside it. The temperature of the water is observed to rise during the burning (the water is also held at constant volume). (a) Consider the burning mixture to be the system. What are the signs of Q, ΔU, and W? (b) What are the signs of these quantities if the water bath is considered to be the system?

11. A quantity of a monatomic ideal gas undergoes a process in which both its pressure and volume are doubled as shown in Figure P12.11. What is the heat absorbed by the gas during this process? (*Hint:* See Problem 1.)

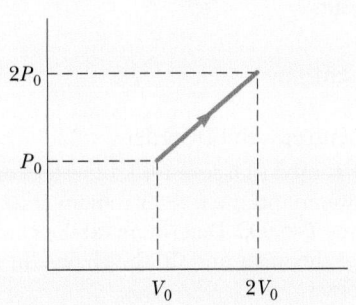

Figure P12.11

12. A monatomic ideal gas undergoes the thermodynamic process shown in the PV diagram of Figure P12.12. Determine whether each of the values ΔU, Q, and W for the gas are positive, negative, or zero. (*Hint:* See Problem 1.)

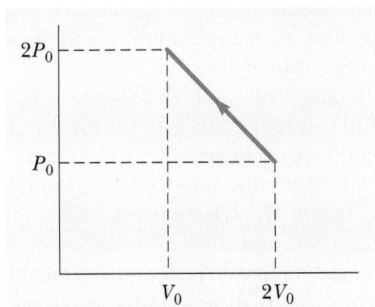

Figure P12.12

13. A gas is compressed at a constant pressure of 0.300 atm from a volume of 8.00 L to 3.00 L. In the process, 400 J of heat energy flows out of the gas. (a) What is the work done by the gas? (b) What is the change in its internal energy?

14. A gas expands from I to F in Figure P12.5. The heat added to the gas is 418 J when the gas goes from I to F along the diagonal path. (a) What is the change in internal energy of the gas? (b) How much heat must be added to the gas for the indirect path IAF to give the same change in internal energy?

15. A gas is taken through the cyclic process described by Figure P12.15. (a) Find the net heat transferred to the system during one complete cycle. (b) If the cycle is reversed—that is, the process follows the path $ACBA$—what is the net heat transferred per cycle?

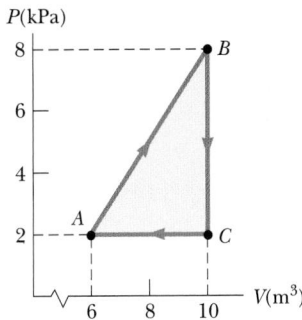

Figure P12.15   (Problems 15 and 18)

16. Two cm³ water is boiled at atmospheric pressure to become 3342 cm³ of steam, also at atmospheric pressure. (a) Calculate the work done by the gas during this process. (b) Find the amount of heat added to the water to accomplish this process. (c) From (a) and (b), find the change in internal energy.

17. A gas is enclosed in a container fitted with a piston of cross-sectional area 0.150 m². The pressure of the gas is maintained at 6000 Pa as the piston moves inward 20.0 cm. (a) Calculate the work done by the gas. (b) If the internal energy of the gas decreases by 8.00 J, find the amount of heat removed from the system during the compression.

18. Consider the cyclic process described by Figure P12.15. If Q is negative for the process BC and ΔU is negative for the process CA, determine the signs of Q, W, and ΔU associated with each process.

19. One gram of water changes from liquid to solid at a constant pressure of one atmosphere and a constant temperature of 0°C. In the process, the volume changes from 1.00 cm³ to 1.09 cm³. (a) Find the work done by the water, and (b) the change in the internal energy of the water.

20. A 5.0-kg block of aluminum is heated from 20°C to 90°C at atmospheric pressure. Find (a) the work done by the aluminum, (b) the amount of heat transferred to it, and (c) the increase in its internal energy.

21. One mole of gas is initially at a pressure of 2.00 atm, a volume of 0.300 L, and an internal energy equal to 91.0 J. In its final state the gas is at a pressure of 1.50 atm and a volume of 0.800 L, and its internal energy equals 180 J. For the paths IAF, IBF, and IF in Figure P12.21, calculate (a) the work done by the gas and (b) the net heat transferred to the gas in the process.

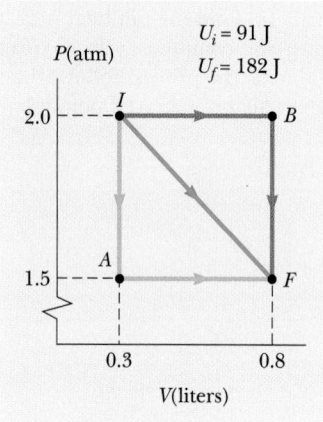

Figure P12.21

Section 12.4   Heat Engines and the Second Law of Thermodynamics

Section 12.5   Reversible and Irreversible Processes

Section 12.6   The Carnot Engine

22. A heat engine operates between two reservoirs at temperatures of 20°C and 300°C. What is the maximum efficiency possible for this engine?

WEB 23. The efficiency of a Carnot engine is 30%. The engine absorbs 800 J of heat per cycle from a hot reservoir at 500 K. Determine (a) the heat expelled per cycle and (b) the temperature of the cold reservoir.

24. The exhaust temperature of a Carnot heat engine is 300°C. What is the intake temperature if the efficiency of the engine is 30%?

25. A steam engine has a boiler that operates at 300°F, and the temperature of the exhaust is 150°F. Find the maximum efficiency of this engine.

26. A heat engine performs 200 J of work in each cycle and has an efficiency of 30%. For each cycle of operation, (a) how much heat is absorbed and (b) how much heat is expelled?

27. An engine absorbs 1700 J from a hot reservoir and expels 1200 J to a cold reservoir in each cycle. (a) What is the engine's efficiency? (b) How much work is done in

each cycle? (c) What is the power output of the engine if each cycle lasts for 0.300 s?

28. A particular engine has a power output of 5.00 kW and an efficiency of 25.0%. If the engine expels 8000 J of heat in each cycle, find (a) the heat absorbed in each cycle and (b) the time required for each cycle.

29. In one cycle, a heat engine absorbs 500 J from the high-temperature reservoir and expels 300 J to a low-temperature reservoir. If the efficiency of this engine is 60% of the efficiency of a Carnot engine, what is the ratio of the low temperature to the high temperature in the Carnot engine?

30. A heat engine operates in a Carnot cycle between 80.0°C and 350°C. It absorbs 21 000 J of heat per cycle from the hot reservoir. The duration of each cycle is 1.00 s. (a) What is the maximum power output of this engine? (b) How much heat does it expel in each cycle?

31. A nuclear power plant has a power output of 1000 MW and operates with an efficiency of 33%. If excess heat is carried away from the plant by a river with a flow rate of 1.0×10^6 kg/s, what is the rise in temperature of the flowing water?

Section 12.7   Entropy

Section 12.8   Entropy and Disorder

32. A freezer is used to freeze 1.0 L of water completely into ice. The water and the freezer remain at a constant temperature of $T = 0°C$. Determine (a) the change in the entropy of the water and (b) the change in the entropy of the freezer.

33. What is the change in entropy of 1.0 kg of liquid water at 100°C as it changes to steam at 100°C?

34. A 70-kg log falls from a height of 25 m into a lake. If the log, the lake, and the air are all at 300 K, find the change in entropy of the Universe for this process.

WEB 35. Two 2000-kg cars, both traveling at 20 m/s, undergo a head-on collision and stick together. Find the entropy change of the Universe during the collision if the temperature is 23°C.

36. Repeat the procedure used to construct Table 12.1 (a) for the case in which you draw three marbles rather than four from your bag and (b) for the case in which you draw five rather than four.

37. Prepare a table like Table 12.1 for the following occurrence. You toss four coins into the air simultaneously. Record all the possible results of the toss in terms of the numbers of heads and tails that can result. (For example, HHTH and HTHH are two possible ways in which three heads and one tail can be achieved.) (a) On the basis of your table, what is the most probable result of a toss? (b) In terms of entropy, what is the most ordered state and (c) what is the most disordered?

38. Consider a standard deck of 52 playing cards that has been thoroughly shuffled. (a) What is the probability of drawing the ace of spaces in one draw? **(b)** What is the probability of drawing any ace? **(c)** What is the probability of drawing any spade?

ADDITIONAL PROBLEMS

39. A student claims that she has constructed a heat engine that operates between the temperatures of 200 K and 100 K with 60% efficiency. The professor does not give her credit for the project. Why not?

40. A power plant that uses the temperature gradient of the ocean has been proposed. The system is to operate between 20°C (surface temperature) and 5.0°C (water temperature at a depth of about 1 km). **(a)** What is the maximum efficiency of such a system? **(b)** If the power output of the plant is 75 MW, what is the minimum thermal energy absorbed per hour? **(c)** In view of your answer to part (a), do you think such a system is worthwhile?

41. A heat engine extracts heat Q_h from a hot reservoir at constant temperature T_h and rejects heat Q_c to a cold reservoir at constant temperature T_c. Find the entropy changes of the **(a)** hot reservoir, **(b)** the cold reservoir, **(c)** the engine, and **(d)** the complete system.

42. One end of a copper rod is in thermal contact with a hot reservoir at $T = 500$ K, and the other end is in thermal contact with a cooler reservoir at $T = 300$ K. If 8000 J of thermal energy is transferred from one end to the other, with no change in the temperature distribution, find the entropy change of each reservoir and the total entropy change of the Universe.

43. Suppose that a nuclear power plant operates in a Carnot cycle. Estimate the temperature change in Celsius of a river due to the exhausted heat from the plant. Assume that the input power to the boiler in the plant is 25×10^8 W, the efficiency of the use of this power is 30%, and the river flow rate is 9.0×10^6 kg/min.

44. Every second at Niagara Falls, some 5000 m³ of water falls a distance of 50 m. What is the increase in entropy per second due to the falling water? (Assume a 20°C environment.)

Figure P12.44 Niagara Falls. (*Jan Kopec/Tony Stone Images*)

45. One object is at a temperature of T_h and another is at a lower temperature, T_c. Use the second law of thermodynamics to show that heat transfer can only occur from the hotter to the colder object. Assume a constant-temperature process.

46. When a gas follows path 123 on the *PV* diagram in Figure P12.46, 418 J of heat flows into the system and 167 J of work is done. **(a)** What is the change in the internal energy of the system? **(b)** How much heat flows into the system if the process follows path 143? The work done by the gas along this path is 63.0 J. What net work would be done on or by the system if the system followed **(c)** path 12341? **(d)** path 14321? **(e)** What is the change in internal energy of the system in the processes described in parts (c) and (d)?

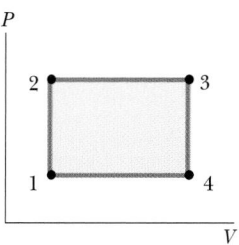

Figure P12.46

47. A substance undergoes the cyclic process shown in Figure P12.47. A work output occurs along path *AB*, while a work input is required along path *BC*, and no work is involved in the constant volume process *CA*. Heat transfers occur during each process involved in the cycle. **(a)** What is the work output during process *AB*? **(b)** How much work input is required during process *BC*? **(c)** What is the net heat input during this cycle?

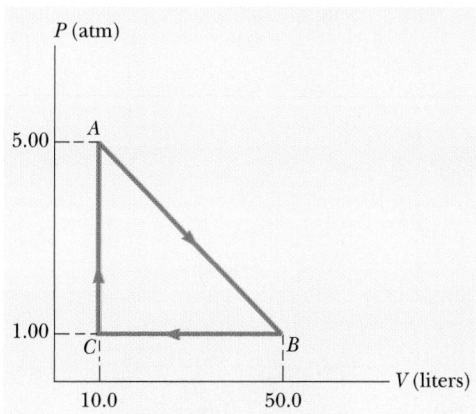

Figure P12.47

48. A 100-kg steel support rod in a building has a length of 2.0 m at a temperature of 20°C. The rod supports a load of 6000 kg. Find **(a)** the work done by the rod as the temperature increases to 40°C, **(b)** the heat added to the rod (assume the specific heat of steel is the same as that for iron), and **(c)** the change in internal energy of the rod.

49. A 1500-kW heat engine operates at 25% efficiency. The heat energy expelled at the low temperature is absorbed by a stream of water that enters the cooling coils at 20°C. If 60 L flows across the coils per second, determine the increase in temperature of the water.

50. An ideal gas initially at pressure P_0, volume V_0, and temperature T_0 is taken through the cycle described in Figure P12.50. **(a)** Find the net work done by the gas per cycle in terms of P_0 and V_0. **(b)** What is the net heat added to the system per cycle? **(c)** Obtain a numerical value for the net work done per cycle for 1.00 mol of gas initially at 0°C. (*Hint:* Recall that work equals the area under a *PV* curve.)

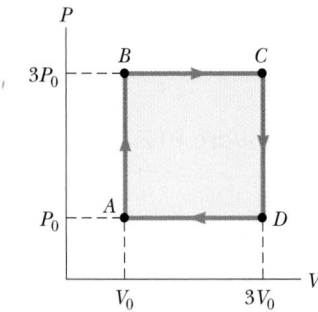

Figure P12.50

51. **(a)** Determine the work done by a fluid that expands from *i* to *f*, as indicated in Figure P12.51. **(b)** How much work is performed by the fluid if it is compressed from *f* to *i* along the same path?

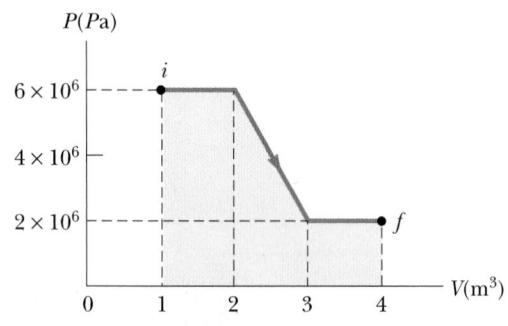

Figure P12.51

52. A 20.0%-efficient real engine is used to speed up a train from rest to 5.00 m/s. It is known that an ideal (Carnot) engine using the same cold and hot reservoirs would accelerate the same train from rest to a speed of 6.50 m/s using the same amount of fuel. If the engines use air at 300 K as a cold reservoir, find the temperature of the steam serving as the hot reservoir.

53. The surface of the Sun is approximately at 5700 K, and the temperature of the Earth's surface is approximately 290 K. What entropy change occurs when 1000 J of thermal energy is transferred from the Sun to the Earth?

54. A force, **F**, is applied to a metal wire, stretching it by the amount ΔL. Show that the work done on the wire is given by $W = \frac{1}{2}(\text{stress})(\text{strain})V$, where V is the volume of the wire.

55. A 1.0-kg block of aluminum is heated at atmospheric pressure so that its temperature increases from 22°C to 40°C. Find **(a)** the work done by the aluminum, **(b)** the heat added to the aluminum, and **(c)** the change in internal energy of the aluminum.

56. Suppose a heat engine is connected to two heat reservoirs, one a pool of molten aluminum (660°C) and the other a block of solid mercury ($-38.9°C$). The engine runs by freezing 1.00 g of aluminum and melting 15.0 g of mercury during each cycle. The latent heat of fusion of aluminum is 3.97×10^5 J/kg, and that of mercury is 1.18×10^4 J/kg. **(a)** What is the efficiency of this engine? **(b)** How does the efficiency compare with that of a Carnot engine?

WEB 57. One mole of neon gas is heated from 300 K to 420 K at constant pressure. Calculate **(a)** the heat energy transferred to the gas, **(b)** the change in the internal energy of the gas, and **(c)** the work done by the gas. Note that neon has a molar specific heat of $c = 20.79$ J/mol·K.

58. One mole of an ideal gas is taken through the reversible cycle shown in Figure P12.58. At point *A*, the pressure, volume, and temperature are P_0, V_0, and T_0. In terms of R and T_0, find **(a)** the total heat entering the system per cycle, **(b)** the total heat leaving the system per cycle,

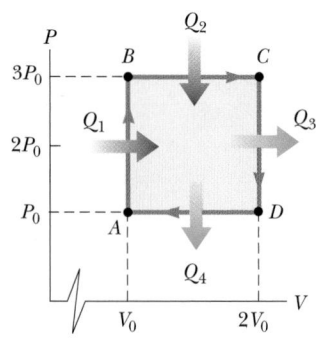

Figure P12.58

(c) the efficiency of an engine operating in this reversible cycle, and (d) the efficiency of an engine operating in a Carnot cycle between the temperature extremes for this process. (*Hint:* Recall that work equals the area under a *PV* curve.)

59. An electrical power plant has an overall efficiency of 15%. The plant is to deliver 150 MW of power to a city, and its turbines use coal as fuel. The burning coal produces steam at 190°C, which drives the turbines. This steam is then condensed into water at 25°C by passing through coils in contact with river water. (a) How many metric tons of coal does the plant consume each day (1 metric ton = 1×10^3 kg)? (b) What is the total cost of the fuel per year if the delivered price is $8 per metric ton? (c) If the river water is delivered at 20°C, at what minimum rate must it flow over the cooling coils in order that its temperature not exceed 25°C? (*Note:* The heat of combustion of coal is 7.8×10^3 cal/g.)

Vibrations and Wave Motion

As we look around us, we find many examples of objects that vibrate or oscillate: a pendulum, the strings of a guitar, an object suspended on a spring, the piston of an engine, the head of a drum, the reed of a saxophone. Most elastic objects vibrate when an impulse is applied to them; that is, once they are distorted, their shape tends to be restored to some equilibrium configuration. Even at the atomic level, the atoms in a solid vibrate about some position as if they were connected to their neighbors by imaginary springs.

Wave motion is closely related to the phenomenon of vibration. Sound waves, earthquake waves, waves on stretched strings, and water waves are all produced by vibrations. As a sound wave travels through a medium (such as air), the molecules of the medium vibrate back and forth; as a water wave travels across a pond, the water molecules vibrate up and down. When any wave travels through a medium, the particles of the medium move in repetitive cycles. Therefore, the motion of the particles bears a strong resemblance to the periodic motion of a vibrating pendulum or a mass attached to a spring.

Many other natural phenomena occur whose explanations require an understanding of vibrations and waves. Although many large structures, such as skyscrapers and bridges, appear to be rigid, they actually vibrate — a fact that must be taken into account by the architects and engineers who design and build them. To understand how radio and television work, we must comprehend the origin and nature of electromagnetic waves and how they propagate through space. Finally, much of what scientists have learned about atomic structure has come from information carried by waves; therefore, to understand the concepts and theories of atomic physics, we must first study waves and vibrations.

◀ *Giraudon/Art Resource, NY*

13

Vibrations and Waves

Chapter Outline

▲ PHYSICS PUZZLER

On musical instruments such as guitars, the frequency of the note played by a string can be changed in two ways—by turning the tuning pegs or by pressing the string to the fingerboard. What is being changed in each of these cases and why does this cause the frequency of vibration of the string to change? *(Michael Tamborrino/FPG International)*

This chapter constitutes a brief return to the subject of mechanics as we examine various forms of periodic motion. We concentrate especially on motion that occurs when the force on an object is proportional to the displacement of the object from its equilibrium position. When such a force acts only toward the equilibrium position, the result is a back-and-forth motion called *simple harmonic motion*—oscillation, or vibration, between two extreme positions for an indefinite period of time with no loss of energy. The terms *harmonic motion* and *periodic motion* are used interchangeably in this chapter. Both refer to back-and-forth motion. You may be familiar with several types of periodic motion, such as the oscillations of a mass on a spring, the motion of a pendulum, and the vibrations of a stringed musical instrument.

Because vibrations can move through a medium, we also study wave motion in this chapter. Many kinds of waves occur in nature, including sound waves, seismic waves, and electromagnetic waves. We end this chapter with a brief discussion of some terms and concepts that are common to all types of waves, and in later chapters we shall focus our attention on specific categories of waves.

8.1

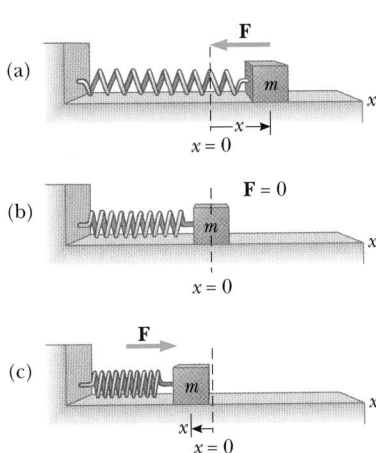

Figure 13.1 The force of a spring on a mass varies with the displacement of the mass from the equilibrium position, $x = 0$. (a) When x is positive (stretched spring), the spring force is to the left. (b) When x is zero (unstretched spring), the spring force is zero. (c) When x is negative (compressed spring), the spring force is to the right.

13.1 HOOKE'S LAW

One of the simplest types of vibrational motion is that of a mass attached to a spring, which we discussed briefly in Chapter 5. Let us assume that the mass moves on a frictionless horizontal surface. If the spring is stretched or compressed a small distance, x, from its unstretched, or equilibrium, position and then released, it exerts a force on the mass, as shown in Figure 13.1. From experiment this spring force is found to obey the equation

$$F_s = -kx \qquad [13.1]$$

where x is the displacement of the mass from its unstretched ($x = 0$) position and k is a positive constant called the **spring constant.** This force law for springs was discovered by Robert Hooke in 1678 and is known as **Hooke's Law.** The value of k is a measure of the stiffness of the spring. Stiff springs have large k values, and soft springs have small k values.

The negative sign in Equation 13.1 signifies that the force exerted by the spring is always directed *opposite* the displacement of the mass. When the mass is to the right of the equilibrium position, as in Figure 13.1a, x is positive and F_s is negative. This means that the force is in the negative direction, to the left. When the mass is to the left of the equilibrium position, as in Figure 13.1c, x is negative and F_s is positive, indicating that the direction of the force is to the right. Of course, when $x = 0$, as in Figure 13.1b, the spring is unstretched and $F_s = 0$. Because the spring force always acts toward the equilibrium position, it is sometimes called a *restoring force*. **The direction of the restoring force is such that the mass is being either pulled or pushed toward the equilibrium position.**

Let us examine the motion of the mass if it is initially pulled a distance of A to the right and released from rest. The force exerted on the mass by the spring pulls the mass back toward the equilibrium position. As the object moves toward $x = 0$, the magnitude of the force decreases (because x decreases) and reaches zero at $x = 0$. However, the mass gains speed as it moves toward the equilibrium position. In fact, the speed reaches its maximum value when $x = 0$. The momentum achieved

◀ Hooke's law

WEB

For Web links to more information on Robert Hooke, visit the textbook Web site at **http://www.harcourtcollege.com/physics/cptech**

Simple harmonic motion ▶

8.2, BOTH SECTIONS

8.10, SECTION 1

by the mass causes it to overshoot the equilibrium position and to compress the spring. As the object moves to the left of the equilibrium position (negative x values), the force acting on it begins to increase to the right and the speed of the mass begins to decrease. The mass finally comes to a stop at $x = -A$. The process is then repeated, and the mass continues to oscillate back and forth over the same path. This type of motion is called *simple harmonic motion*. **Simple harmonic motion occurs when the net force along the direction of motion is a Hooke's law type of force—that is, when the net force is proportional to the displacement and in the opposite direction.**

Not all repetitive motion over the same path can be classified as simple harmonic motion. For example, consider a ball being tossed back and forth between a parent and child. The ball moves repetitively, but the motion is not simple harmonic. Unless the force acting on an object along the direction of motion has the form of Equation 13.1, the object does not exhibit simple harmonic motion.

The motion of a mass suspended from a vertical spring is also simple harmonic. In this case, the force of gravity acting on the attached mass stretches the spring until equilibrium is reached. (The mass is suspended and at rest.) This position of equilibrium establishes the position of the mass for which $x = 0$. When the mass is stretched a distance of x beyond this equilibrium position and then released, a net force acts (in the form of Hooke's law) toward the equilibrium position. Because the net force is proportional to x, the motion is simple harmonic.

Before we can discuss simple harmonic motion in more detail, it is necessary to define a few terms.

- The **amplitude, A,** is the maximum distance traveled by an object away from its equilibrium position. In the absence of friction, an object continues in simple harmonic motion and reaches a maximum displacement equal to the amplitude on each side of the equilibrium position during each cycle.
- The **period, T,** is the time it takes the object to execute one complete cycle of motion.
- The **frequency, f,** is the number of cycles or vibrations per unit of time.

EXAMPLE 13.1 Measuring the Spring Constant

A common technique used to evaluate the spring constant is illustrated in Figure 13.2. The spring is hung vertically (Fig. 13.2a), and a body of mass m is attached to the lower end of the spring (Fig. 13.2b). The spring stretches a distance of d from its initial position under the action of the "load" mg. Because the spring force is upward, it must balance the weight, mg, downward *when the system is at rest*. In this case, we can apply Hooke's law to give

$$F_s = kd = mg$$

$$k = \frac{mg}{d}$$

For example, if a spring is stretched 2.0 cm by a mass of 0.55 kg, the force constant is

$$k = \frac{mg}{d} = \frac{(0.55 \text{ kg})(9.80 \text{ m/s}^2)}{2.0 \times 10^{-2} \text{ m}} = 2.7 \times 10^2 \text{ N/m}$$

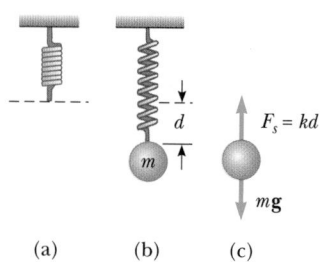

(a)   (b)   (c)

Figure 13.2 (Example 13.1) Determining the spring constant. The elongation, d, of the spring is due to the suspended weight, mg. Because the upward spring force balances the weight when the system is in equilibrium, it follows that $k = mg/d$.

EXAMPLE 13.2 Simple Harmonic Motion on a Frictionless Surface

A 0.35-kg mass attached to a spring of spring constant 130 N/m is free to move on a frictionless horizontal surface, as in Figure 13.1. If the mass is released from rest at $x = 0.10$ m, find the force on it and its acceleration at (a) $x = 0.10$ m, (b) $x = 0.050$ m, (c) $x = 0$ m, and (d) $x = -0.050$ m.

Solution (a) The point at which the mass is released defines the amplitude of the motion. In this case, $A = 0.10$ m. The mass moves continuously between the limits 0.10 m and -0.10 m. When x is a maximum ($x = A$), the force on the mass is a maximum and is calculated as follows:

$$F = -kx$$
$$F_{max} = -kA = -(130 \text{ N/m})(0.10 \text{ m}) = \boxed{-13 \text{ N}}$$

The negative sign indicates that the force acts to the left, in the negative x direction. We can use Newton's second law to calculate the acceleration at this position:

$$F_{max} = ma_{max}$$
$$a_{max} = \frac{F_{max}}{m} = -\frac{13 \text{ N}}{0.35 \text{ kg}} = \boxed{-37 \text{ m/s}^2}$$

Again, the negative sign indicates that the acceleration is to the left.

(b) We can use the same approach to find the force and acceleration at other positions. At $x = 0.05$ m, we have

$$F = -kx = -(130 \text{ N/m})(0.050 \text{ m}) = \boxed{-6.5 \text{ N}}$$
$$a = \frac{F}{m} = -\frac{6.5 \text{ N}}{0.35 \text{ kg}} = \boxed{-19 \text{ m/s}^2}$$

Note that the acceleration of an object moving with simple harmonic motion *is not constant* because F is not constant.

(c) At $x = 0$, the spring force is zero (because $F = -kx = 0$) and the acceleration is zero. In other words, when the spring is unstretched, it exerts no force on the mass attached to it.

(d) At $x = -0.050$ m, we have

$$F = -kx = -(130 \text{ N/m})(-0.050 \text{ m}) = \boxed{6.5 \text{ N}}$$
$$a = \frac{F}{m} = \frac{6.5 \text{ N}}{0.35 \text{ kg}} = \boxed{19 \text{ m/s}^2}$$

This result shows that the force and acceleration are positive when the mass is on the negative side of the equilibrium position. This indicates that the force of the spring on the mass is acting to the right as the spring is being compressed. At the same time, the mass is slowing down as it moves from $x = 0$ to $x = -0.050$ m.

Exercise Find the force and acceleration when $x = -0.10$ m.

Answer 13 N; 37 m/s^2

Q U I C K L A B

Firmly hold a ruler so that about half of
it is over the edge of your desk. Pull
down and release the free end, watching
how it vibrates. Now slide the ruler so
that only about a quarter of the ruler is
free to vibrate. How does its vibrational
frequency this time compare to what it
was before? Why?

As indicated in Example 13.2, the acceleration of an object moving with simple
harmonic motion can be found by using Hooke's law as the force in the equation
for Newton's second law, $F = ma$. This gives

$$-kx = ma$$

$$a = -\frac{k}{m}x \qquad\qquad [13.2]$$

Because the maximum value of x is defined to be the amplitude, A, we see that
the acceleration ranges over the values $-kA/m$ to $+kA/m$. Equation 13.2 enables
us to find the acceleration of the object as a function of its position. In subsequent
sections, we shall find equations for velocity as a function of position and position
as a function of time.

Earlier we stated that an object moves with simple harmonic motion when the
net force acting on it is proportional to its displacement from equilibrium and is
directed toward the equilibrium position. Equation 13.2 provides an alternative
definition of simple harmonic motion. An object moves with simple harmonic motion if its acceleration is proportional to its displacement and is in the opposite
direction to it.

13.2   ELASTIC POTENTIAL ENERGY

For the most part, we have worked thus far with three types of mechanical energy:
gravitational potential energy, translational kinetic energy, and rotational kinetic
energy. In Chapter 5 we briefly discussed a fourth type of mechanical energy, elastic
potential energy. We now consider this form of energy in more detail.

An object has potential energy by virtue of its shape or position. As we learned
in Chapter 5, an object of mass m at height h above the ground has gravitational
potential energy equal to mgh. This means that the object can do work after it is
released. Likewise, a compressed spring has potential energy by virtue of its shape.
In this case, the compressed spring, when allowed to expand, can move an object
and thus do work on it. As an example, Figure 13.3 shows a ball being projected
from a spring-loaded toy gun, where the spring is compressed a distance of x. As

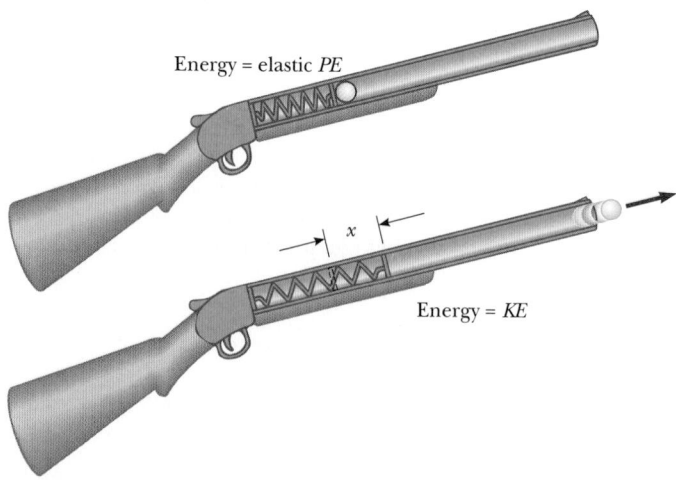

Figure 13.3 A ball projected
from a spring-loaded gun. The elastic potential energy stored in the
spring is transformed into the
kinetic energy of the ball.

the gun is fired, the compressed spring does work on the ball and imparts kinetic energy to it. **The energy stored in a stretched or compressed spring or other elastic material is called elastic potential energy, PE_s, given by**

$$PE_s \equiv \tfrac{1}{2}kx^2 \qquad\qquad [13.3]$$

◀ Elastic potential energy

Note that the elastic potential energy stored in a spring is zero when the spring is unstretched or uncompressed ($x = 0$). **Energy is stored in a spring only when it is either stretched or compressed. Furthermore, the elastic potential energy is a maximum when a spring has reached its maximum compression or extension.** Finally, the potential energy in a spring is always positive because it is proportional to x^2. We now include this new form of energy in our equation for conservation of mechanical energy:

$$(KE + PE_g + PE_s)_i = (KE + PE_g + PE_s)_f \qquad [13.4]$$

If nonconservative forces such as friction are present, then the final mechanical energy does not equal the initial mechanical energy. In this case, the difference in the two energies must equal the work done by the nonconservative force, W_{nc}. From the work-energy theorem,

$$W_{nc} = (KE + PE_g + PE_s)_f - (KE + PE_g + PE_s)_i \qquad [13.5]$$

As an example of the energy conversions that take place when a spring is included in the system, consider Figure 13.4. A block of mass m slides on a frictionless

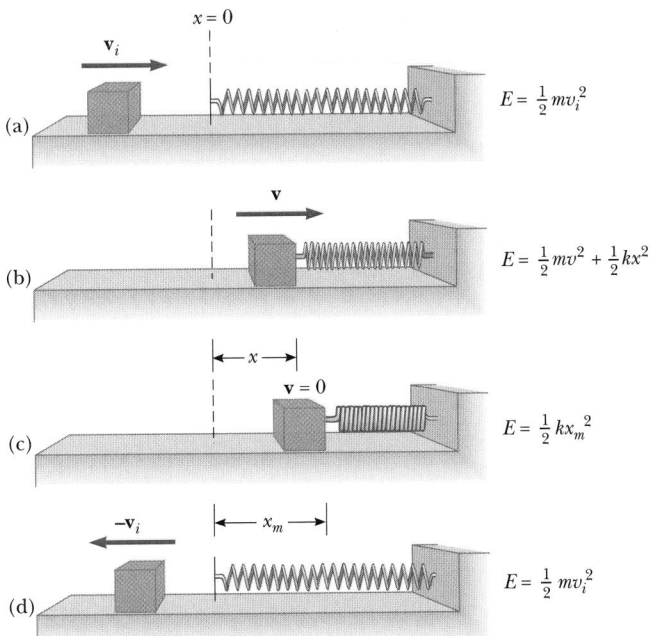

Figure 13.4 A block sliding on a frictionless horizontal surface collides with a light spring. (a) Initially, the mechanical energy is entirely the kinetic energy of the block. (b) The mechanical energy at some arbitrary position is the sum of the kinetic energy of the block and the elastic potential energy stored in the spring. (c) When the block comes to rest, the mechanical energy is entirely elastic potential energy stored in the compressed spring. (d) When the block leaves the spring, the mechanical energy is equal to the block's kinetic energy. The total energy remains constant.

Figure 13.5 Elastic potential energy is stored in this drawn bow. *(Eric Lars Baleke/Black Star)*

horizontal surface with constant velocity $\mathbf{v_i}$ and collides with a coiled spring. The description that follows is greatly simplified by assuming that the spring is very light and therefore has negligible kinetic energy. As the spring is compressed, it exerts a leftward force on the block. At maximum compression, the block momentarily stops (Fig. 13.4c). The initial total energy in the system before the collision (block plus spring) is the kinetic energy of the block. After the block collides with the spring and the spring is partially compressed, as in Figure 13.4b, the block has kinetic energy $\frac{1}{2}mv^2$ (where $v < v_i$) and the spring has potential energy $\frac{1}{2}kx^2$. When the block stops momentarily after colliding with the spring, the kinetic energy is zero. Because the spring force is conservative and because there are no external forces that can do work on the system, **the total mechanical energy of the system consisting of the block and spring remains constant.** Thus, energy is transformed from kinetic energy of the block into potential energy stored in the spring. As the spring expands, the block moves in the opposite direction and regains all of its initial kinetic energy, as in Figure 13.4d.

Elastic potential energy is stored in the bow of an archer when he or she pulls back the bowstring (Figure 13.5). This is a more complicated system than a single mass on a spring, but if we consider the essential features, there is a restoring force on the arrow as a result of the stretching of the string and the bending of the bow. The potential energy stored in the system is transformed into kinetic energy of the arrow as it is released. We could make similar statements about devices such as crossbows and slingshots.

APPLICATION

Archery.

EXAMPLE 13.3   Stop That Car!

A 13 000-N car starts at rest and rolls down a hill from a height of 10 m (Fig. 13.6). It then moves across a level surface and collides with a light spring-loaded guardrail. Neglecting any losses due to friction, find the maximum distance the spring is compressed. Assume a spring constant of 1.0×10^6 N/m.

Solution   The initial and final values of the kinetic energy of the car are zero. Thus, the initial potential energy of the car is completely converted to elastic potential energy in the spring at the end of the trip (assuming we neglect any energy losses due to friction during the collision). Conservation of energy gives

$$mgh = \tfrac{1}{2}kx^2$$

Figure 13.6 (Example 13.3) A car starts from rest on a hill at the position shown. When the car reaches the bottom of the hill, it collides with a spring-loaded guardrail.

Solving for x gives

$$x = \sqrt{\frac{2mgh}{k}} = \sqrt{\frac{2(13000 \text{ N})(10 \text{ m})}{1.0 \times 10^6 \text{ N/m}}} = 0.50 \text{ m}$$

Note that it was not necessary at any point to calculate the velocity of the car to obtain a solution. This demonstrates the power of the principle of conservation of energy. One works with the initial and final energy values only, without having to consider all the details in between.

Exercise What is the speed of the car just before it collides with the guardrail?

Answer 14 m/s

EXAMPLE 13.4 Motion With and Without Friction

A 1.6-kg block is attached to a spring with a spring constant of 1.0×10^3 N/m (Fig. 13.1). The spring is compressed a distance of 2.0 cm, and the block is released from rest.
(a) Calculate the speed of the block as it passes through the equilibrium position, $x = 0$, if the surface is frictionless.

Solution By using Equation 13.3, we can find the initial elastic potential energy of the spring when $x = -2.0$ cm $= -2.0 \times 10^{-2}$ m:

$$PE_s = \tfrac{1}{2}kx_i^2 = \tfrac{1}{2}(1.0 \times 10^3 \text{ N/m})(-2.0 \times 10^{-2} \text{ m})^2 = 0.20 \text{ J}$$

Because the block is always at the same height above the Earth's surface, its gravitational potential energy remains constant. Hence, the initial potential energy stored in the spring is converted to kinetic energy at $x = 0$. That is,

$$\tfrac{1}{2}kx_i^2 = \tfrac{1}{2}mv_f^2$$

$$0.20 \text{ J} = \tfrac{1}{2}(1.6 \text{ kg})(v_f^2)$$

$$v_f = 0.50 \text{ m/s}$$

(b) Calculate the speed of the block as it passes through the equilibrium position if a constant frictional force of 4.0 N retards its motion.

Solution Because sliding friction is present in this situation, we know that the final mechanical energy is less than the initial mechanical energy. The work done by the frictional force for a displacement of 2.0×10^{-2} m is

$$W_{nc} = W_f = -fs = -(4.0 \text{ N})(2.0 \times 10^{-2} \text{ m}) = -0.080 \text{ J}$$

Applying Equation 13.5 to this situation gives

$$-0.080 \text{ J} = \tfrac{1}{2}(1.6 \text{ kg})(v_f^2) - 0.20 \text{ J}$$

$$v_f = \boxed{0.39 \text{ m/s}}$$

Note that this value for v_f is less than that obtained in the frictionless case. Does the result make sense?

Exercise How far does the block travel before coming to rest? Assume a constant friction force of 4.0 N and the same initial conditions as before.

Answer 5.0 cm

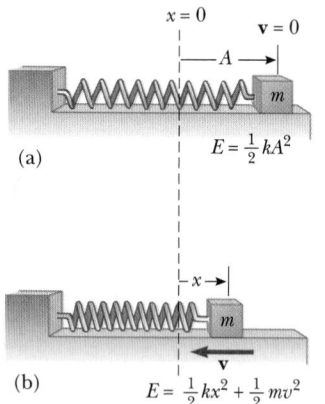

(a)

$x = 0$
$\mathbf{v} = 0$
A
m
$E = \tfrac{1}{2} kA^2$

(b)

x
m
$\mathbf{v}$
$E = \tfrac{1}{2} kx^2 + \tfrac{1}{2} mv^2$

Figure 13.7 (a) A mass attached to a spring on a frictionless surface is released from rest with the spring extended a distance of A. Just before release, the total energy is elastic potential energy, $kA^2/2$. (b) When the mass reaches position x, it has kinetic energy $mv^2/2$ and the elastic potential energy has decreased to $kx^2/2$.

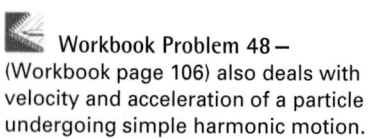

Workbook Problem 48 – (Workbook page 106) also deals with velocity and acceleration of a particle undergoing simple harmonic motion.

13.3   VELOCITY AS A FUNCTION OF POSITION

Conservation of energy provides a simple method of deriving an expression for the velocity of a mass undergoing periodic motion as a function of position. The mass in question is initially at its maximum extension, A (Fig. 13.7a), and is then released from rest. In this situation, the initial energy of the system is entirely elastic potential energy stored in the spring, $\tfrac{1}{2} kA^2$. As the mass moves toward the origin to some new position, x (Fig. 13.7b), part of this energy is transformed into kinetic energy, and the potential energy stored in the spring is reduced to $\tfrac{1}{2} kx^2$. Because the total energy of the system is equal to $\tfrac{1}{2} kA^2$ (the initial energy stored in the spring), we can equate this to the sum of the kinetic and potential energies at the final position:

$$\tfrac{1}{2} kA^2 = \tfrac{1}{2} mv^2 + \tfrac{1}{2} kx^2$$

Solving for v, we get

$$v = \pm \sqrt{\frac{k}{m}(A^2 - x^2)} \qquad [13.6]$$

This expression shows us that the speed is a maximum at $x = 0$ and zero at the extreme positions $x = \pm A$.

The right side of Equation 13.6 is preceded by the $\pm$ sign because the square root of a number can be either positive or negative. The sign of v that is selected depends on the circumstances of the motion. If the mass in Figure 13.7 is moving to the right, v must be positive; if the mass is moving to the left, v must be negative.

EXAMPLE 13.5 The Mass-Spring System Revisited

A 0.50-kg mass connected to a light spring with a spring constant of 20 N/m oscillates on a frictionless horizontal surface.
(a) Calculate the total energy of the system and the maximum speed of the mass if the amplitude of the motion is 3.0 cm.

Solution   From Equation 13.3, we have

$$E = PE_s = \tfrac{1}{2}kA^2 = \tfrac{1}{2}(20 \text{ N/m})(3.0 \times 10^{-2} \text{ m})^2 = \boxed{9.0 \times 10^{-3} \text{ J}}$$

When the mass is at $x = 0$, $PE_s = 0$ and $E = \tfrac{1}{2}mv_{max}^2$; therefore,

$$\tfrac{1}{2}mv_{max}^2 = 9.0 \times 10^{-3} \text{ J}$$

$$v_{max} = \sqrt{\frac{18 \times 10^{-3} \text{ J}}{0.50 \text{ kg}}} = \boxed{0.19 \text{ m/s}}$$

(b) What is the velocity of the mass when the displacement is 2.0 cm?

Solution   We can apply Equation 13.6 directly:

$$v = \pm\sqrt{\frac{k}{m}(A^2 - x^2)} = \pm\sqrt{\frac{20}{0.50}(3.0^2 - 2.0^2) \times 10^{-4}} = \boxed{\pm 0.14 \text{ m/s}}$$

The $\pm$ sign indicates that the mass could be moving to the right or to the left at this instant.

(c) Compute the kinetic and potential energies of the system when the displacement is 2.0 cm.

Solution   The results of (b) can be used to give

$$KE = \tfrac{1}{2}mv^2 = \tfrac{1}{2}(0.50 \text{ kg})(0.14 \text{ m/s})^2 = \boxed{5.0 \times 10^{-3} \text{ J}}$$

$$PE_s = \tfrac{1}{2}kx^2 = \tfrac{1}{2}(20 \text{ N/m})(2.0 \times 10^{-2} \text{ m})^2 = \boxed{4.0 \times 10^{-3} \text{ J}}$$

Note that the sum $KE + PE_s$ equals the total energy, E, found in part (a).

Exercise   For what values of x is the speed of the mass 0.10 m/s?

Answer ± 2.6 cm

13.4   COMPARING SIMPLE HARMONIC MOTION WITH UNIFORM CIRCULAR MOTION

We can better understand and visualize many aspects of simple harmonic motion along a straight line by looking at their relationships to uniform circular motion. Figure 13.8 is a top view of an experimental arrangement that is useful for this purpose. A ball is attached to the rim of a phonograph turntable of radius A, illuminated from the side by a lamp. Rather than concentrating on the ball, let us focus our attention on the shadow that the ball casts on the screen. We find that **as the turntable rotates with constant angular speed, the shadow of the ball moves back and forth with simple harmonic motion.**

In order to understand why, let us examine Equation 13.6 more closely. This equation says that the velocity of an object moving with simple harmonic motion is related to the displacement by

$$v = C\sqrt{A^2 - x^2}$$

where C is a constant. To see that the shadow also obeys this relation, consider Figure 13.9, which shows the ball moving with a constant speed, v_0, in a direction

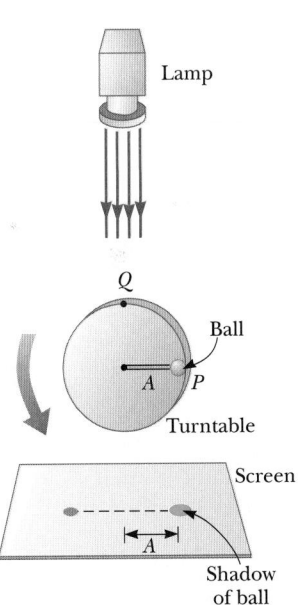

Figure 13.8   An experimental setup for demonstrating the connection between simple harmonic motion and uniform circular motion. As the ball rotates on the turntable with constant angular speed, its shadow on the screen moves back and forth with simple harmonic motion.

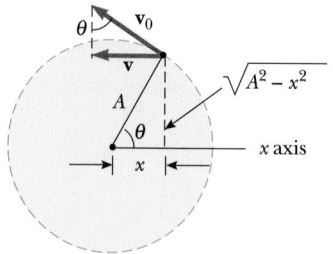

Figure 13.9 The ball rotates with constant speed v_0. The x component of the velocity of the ball equals the projection of $\mathbf{v}_0$ on the x axis.

tangent to the circular path. At this instant, the velocity of the ball in the x direction is given by $v = v_0 \sin \theta$, or

$$\sin \theta = \frac{v}{v_0}$$

Likewise, the larger triangle containing the angle θ in Figure 13.9 enables us to write

$$\sin \theta = \frac{\sqrt{A^2 - x^2}}{A}$$

Equating the right-hand sides of the last two expressions, we see that v is related to the displacement, x, as

$$\frac{v}{v_0} = \frac{\sqrt{A^2 - x^2}}{A}$$

or

$$v = \frac{v_0}{A}\sqrt{A^2 - x^2} = C\sqrt{A^2 - x^2}$$

The velocity of the ball in the x direction is related to the displacement, x, in exactly the same manner as the velocity of an object undergoing simple harmonic motion. Hence, the shadow moves with simple harmonic motion.

A valuable example of the relationship between simple harmonic motion and circular motion is seen in vehicles and machines that use the back-and-forth motion of a piston to create rotational motion of a wheel. For example, consider the drive wheels of a locomotive. In Figure 13.10, the black mechanism at the lower right contains a piston that moves back and forth in simple harmonic motion. The piston is connected to an arrangement of rods that transforms its back-and-forth motion into rotational motion of the wheels. A similar mechanism in an automobile engine transforms the back-and-forth motion of the pistons to rotational motion of the crankshaft.

APPLICATION

Pistons and Drive Wheels.

Figure 13.10 An old locomotive from the Strasburg Railroad, Pennsylvania. The piston housing is shown at the lower right. A piston moves back and forth with simple harmonic motion. *(Grant Heilman Photography)*

Period and Frequency

Note that the period, T, of the shadow in Figure 13.8, which represents the time required for one complete trip back and forth, is also the time it takes the ball to make one complete circular trip on the turntable. Because the ball moves through the distance $2\pi A$ (the circumference of the circle) in the time T, the speed, v_0, of the ball around the circular path is

$$v_0 = \frac{2\pi A}{T}$$

$$T = \frac{2\pi A}{v_0}$$

However, for our purposes, let us consider only a fraction of the complete trip. Imagine that the ball moves from P to Q, a quarter of a revolution, in Figure 13.8. This requires a time interval equal to one fourth of the period, and the distance traveled by the ball is $2\pi A/4$. Therefore,

8.10, SECTION 2

$$\frac{T}{4} = \frac{2\pi A}{4v_0} \qquad [13.7]$$

Now imagine that the motion of the shadow is equivalent to the horizontal motion of a mass on the end of a spring. During this quarter of a cycle, the shadow moves from a point where its energy is solely elastic potential energy to a point where its energy is solely kinetic energy. That is,

$$\tfrac{1}{2}kA^2 = \tfrac{1}{2}mv_0^2$$

$$\frac{A}{v_0} = \sqrt{\frac{m}{k}}$$

Substituting for A/v_0 in Equation 13.7, we find that the period is

$$T = 2\pi\sqrt{\frac{m}{k}} \qquad [13.8]$$

◀ The period of a mass-spring system moving with simple harmonic motion

This expression gives the time required for an object to make a complete cycle of its motion. Now recall that the frequency, f, is the number of cycles per unit of time. The symmetry in the units of period and frequency should lead you to see that the two must be related inversely as

$$f = \frac{1}{T} \qquad [13.9]$$

Therefore, the **frequency** of the periodic motion is

$$f = \frac{1}{2\pi}\sqrt{\frac{k}{m}} \qquad [13.10]$$

◀ Frequency

The units of frequency are s^{-1}, or hertz (Hz). The **angular frequency, ω,** is

$$\omega = 2\pi f = \sqrt{\frac{k}{m}} \qquad [13.11]$$

◀ Angular frequency

Thinking Physics 1

We know that the period of oscillation of a mass hung on a spring is proportional to the square root of the mass. Thus, if we perform an experiment in which we hang a range of masses on the end of a spring and measure the period of oscillation, a graph of the square of the period on the vertical axis versus the suspended mass m on the horizontal axis will result in a straight line. But we would find that the line does not go through the origin. That is, for $m = 0$, T^2 has a finite value. How do you explain this result?

Explanation The reason that the line does not go through the origin is that the spring itself has mass. Thus, the resistance to change in motion of the system is a combination of the mass hung on the end of the spring and the mass of the oscillating coils of the spring. The entire mass of the spring is not oscillating, however. The bottom-most coil is oscillating with the same amplitude as the mass, and the top-most coil is not oscillating at all. For a cylindrical spring, energy arguments can be used to show that the effective additional mass affecting the oscillations of a light spring is one third of the mass of the spring. The square of the period is proportional to the total oscillating mass. Therefore, a graph of period squared against total mass (mass hung on the spring plus the effective oscillating mass of the spring) would pass through the origin.

Thinking Physics 2

A mass oscillating on the end of a horizontal spring slides back and forth over a frictionless surface. During one oscillation, you set an identical mass on the first mass when the system is at the maximum displacement point where it is at rest. The two are connected by Velcro and continue their oscillation together. Does the period of the oscillation change? Does the amplitude of the oscillation change? Does the energy of the oscillation change?

Explanation   The period of the oscillation does change, because the period depends on the mass that is oscillating. The amplitude does not change. Because the new mass was added while the original mass was at rest, the combined masses are at rest at this point, also, defining the amplitude as the same as in the original oscillation. The energy does not change, either. At the maximum displacement point, the energy is all potential energy stored in the spring, which depends only on the spring constant and the amplitude, not the mass. The increased mass will pass through the equilibrium point with less velocity than in the original oscillation, but with the same kinetic energy. Another approach is to think about how energy could be transferred into the oscillating system—no work was done (nor was there any other form of energy transfer), so the energy in the system cannot change.

APPLICATION

Bungee Jumping.

Figure 13.11   (Applying Physics 1) Bungee jumping from a bridge. *(Telegraph Colour Library/FPG International)*

Applying Physics 1

A bungee cord can be modeled as a spring. If you go bungee jumping, you will bounce up and down at the end of the cord after your daring dive off a bridge (Fig. 13.11). Suppose you perform this dive and measure the frequency of your bouncing. You then move to another bridge. You discover that the bungee cord is too long for dives off this bridge; you will hit the ground. Thus, you fold the bungee cord in half and make the dive from the doubled bungee cord. How does the frequency of your bouncing at the end of this dive compare to the frequency after the dive from the first bridge?

Explanation   The force exerted by the bungee cord, modeled as a spring, is proportional to the separation of the coils as the spring is extended. Imagine that we extend a spring by a given distance and measure the distance between coils. We then cut the spring in half. If one of the half springs is now extended by the same distance, the coils will be twice as far apart as they were for the complete spring. Thus, it takes twice as much force to stretch the half-spring, from which we conclude that the half- spring has a spring constant that is twice that of the complete spring. Now consider the folded bungee cord that we model as two half-springs in parallel. Each half has a spring constant that is twice the original spring constant of the bungee cord. In addition, a given mass hanging on the folded bungee cord will experience two forces—one from each half-spring. As a result, the required force for a given extension will be four times as much as for the original bungee cord. The effective spring constant of the folded bungee cord is, therefore, four times as large as the original spring constant. Because the frequency of oscillation is proportional to the square root of the spring constant, your bouncing frequency on the folded cord will be twice that of the original cord.

WEB

For Web links to more information on bungee jumping, visit the textbook Web site at **http://www.harcourtcollege.com/physics/cptech**

EXAMPLE 13.6 That Car Needs a New Set of Shocks!

A 1300-kg car is constructed on a frame supported by four springs. Each spring has a spring constant of 20 000 N/m. If two people riding in the car have a combined mass of 160 kg, find the frequency of vibration of the car when it is driven over a pothole in the road.

Solution We assume that the weight is evenly distributed; thus each spring supports one fourth of the load. The total mass supported by the springs is 1460 kg, and therefore each spring supports 365 kg. Hence, the frequency of vibration is

$$f = \frac{1}{2\pi} \sqrt{\frac{k}{m}} = \frac{1}{2\pi} \sqrt{\frac{20\ 000\ \text{N/m}}{365\ \text{kg}}} = 1.18\ \text{Hz}$$

Exercise How long does it take the car to execute three complete vibrations?

Answer 2.54 s

13.5 POSITION, VELOCITY, AND ACCELERATION AS A FUNCTION OF TIME

We can obtain an expression for the position of an object moving with simple harmonic motion as a function of time by returning to the relationship between simple harmonic motion and uniform circular motion. Again, consider a ball on the rim of a rotating turntable of radius A, as in Figure 13.12. We shall refer to the circle made by the ball as the *reference circle* for the motion. Let us assume that the turntable revolves at a constant angular speed of ω. As the ball rotates on the reference circle, the angle, θ, made by the line OP with the x axis changes with time. Furthermore, as the ball rotates, the projection of P on the x axis, labeled point Q, moves back and forth along the axis with simple harmonic motion.

From the right triangle, OPQ, we see that $\cos\theta = x/A$. Therefore, the x coordinate of the ball is

$$x = A\cos\theta$$

Because the ball rotates with constant angular speed, it follows that $\theta = \omega t$ (see Chapter 7). Therefore,

$$x = A\cos(\omega t) \qquad \text{[13.12]}$$

In one complete revolution, the ball rotates through an angle of 2π rad in the period T. In other words, the motion repeats itself every T seconds. Therefore,

$$\omega = \frac{\Delta\theta}{\Delta t} = \frac{2\pi}{T} = 2\pi f \qquad \text{[13.13]}$$

8.3, BOTH SECTIONS

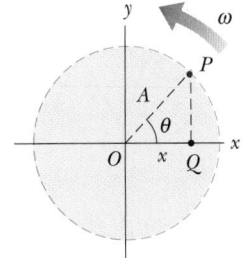

Figure 13.12 A reference circle. As the ball at P rotates in a circle with uniform angular speed, its projection, Q, along the x axis moves with simple harmonic motion.

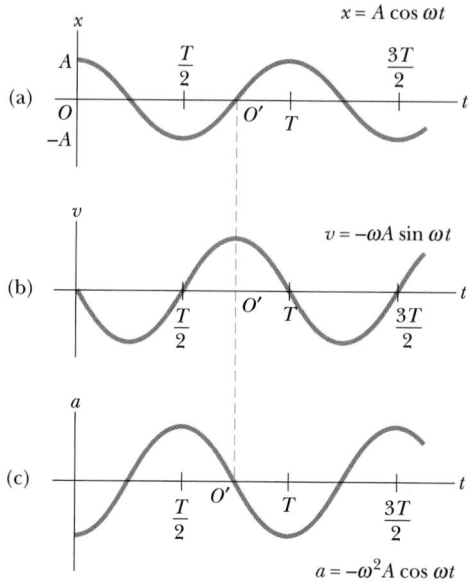

Figure 13.13 (a) Displacement, (b) velocity, and (c) acceleration versus time for an object moving with simple harmonic motion under the initial conditions that $x_0 = A$ and $v_0 = 0$ at $t = 0$.

where f is the frequency of the motion. The angular speed of the rotation on the reference circle is the same as the angular frequency of the projected simple harmonic motion. Consequently, Equation 13.12 can be written

$$x = A \cos(2\pi f t) \qquad [13.14]$$

This equation represents the position of an object moving with simple harmonic motion as a function of time; it is graphed in Figure 13.13a. The curve should be familiar to you from trigonometry. Its shape is called sinusoidal. Note that x varies between A and $-A$ because the cosine function varies between 1 and -1.

Figures 13.13b and 13.13c represent curves for velocity and acceleration as a function of time. An end-of-chapter problem will ask you to show that the velocity and acceleration are sinusoidal functions of time. Note that when x is a maximum or minimum, the velocity is zero, and when x is zero, the magnitude of the velocity is a maximum. Furthermore, when x has its maximum positive value, the acceleration is a maximum but in the negative x direction, and when x is at its maximum negative position, the acceleration has its maximum value in the positive direction. These curves are consistent with our earlier discussion of the points at which v and a reach their maximum, minimum, and zero values.

Figure 13.14 illustrates one experimental arrangement that demonstrates simple harmonic motion. A mass connected to a spring has a marking pen attached to it. While the mass vibrates vertically, a sheet of paper is moved horizontally with constant speed. The pen traces out a sinusoidal pattern.

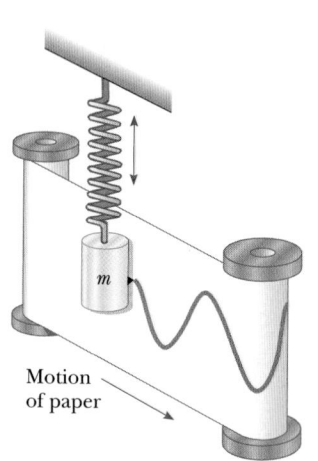

Figure 13.14 An experimental apparatus for demonstrating simple harmonic motion. A pen attached to the oscillating mass traces out a sinusoidal wave on the moving chart paper.

EXAMPLE 13.7   The Vibrating Mass-Spring System

(a) Find the amplitude, frequency, and period of motion for an object vibrating at the end of a spring if the equation for its position as a function of time is

$$x = (0.25 \text{ m}) \cos\left(\frac{\pi}{8.0} t\right)$$

Solution We can find two of our unknowns by comparing this equation with the general equation for such motion:

$$x = A \cos(2\pi f t)$$

By comparison, we see that

$$A = \boxed{0.25 \text{ m}}$$

$$2\pi f = \frac{\pi}{8.0} \text{ s}^{-1}$$

$$f = \boxed{\frac{1}{16} \text{ Hz}}$$

Because the period $T = 1/f$, if follows that $T = 1/f = \boxed{16 \text{ s}}$.

(b) What is the position of the object after 2.0 seconds have elapsed?

Solution Direct substitution for t in the expression for x gives

$$x = (0.25 \text{ m})\cos\left(\frac{\pi}{4}\right) = \boxed{0.18 \text{ m}}$$

Note that in evaluating the cosine function, the angle is in radians. Thus, you should either set your calculator to evaluate trigonometric functions based on radian measure or convert from radians to degrees. The angle is called the *phase* of the oscillation at that instant.

Workbook Problem 54—(Workbook page 114) is similar to Example 13.7.

8.11, BOTH SECTIONS

13.6 MOTION OF A PENDULUM

A simple pendulum is another mechanical system that exhibits periodic motion. It consists of a small bob of mass m suspended by a light string of length L fixed at its upper end, as in Figure 13.15. (By a light string, we mean that the string's mass is assumed to be very small compared to the mass of the bob and hence can be ignored.) When released, the mass swings to and fro over the same path; but is its motion simple harmonic? In order to answer this question, we must examine the force that acts as the restoring force on the pendulum. If this force is proportional to the displacement, s, then the force is of the Hooke's law form, $F = -ks$, and hence the motion is simple harmonic. Furthermore, because $s = L\theta$ in this case, we see that the motion is simple harmonic if F is proportional to the angle θ.

The component of the force of gravity tangential to the circular path is the force that acts to restore pendulum to its equilibrium position. Thus, the restoring force is

$$F_t = -mg \sin \theta$$

From this equation, we see that the restoring force is proportional to $\sin \theta$ rather than to θ. Thus, in general, the motion of a pendulum is *not* simple harmonic. However, for small angles, less than about 15 degrees, the angle θ measured in radians and the sine of the angle are approximately equal. Therefore, if we restrict the motion to small angles, the restoring force can be written as

$$F_t = -mg \theta$$

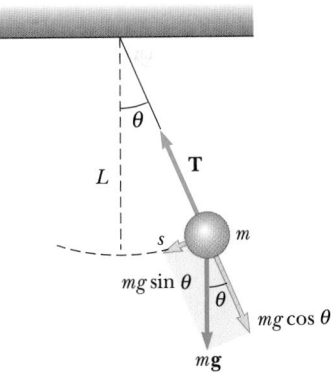

Figure 13.15 A simple pendulum consists of a bob of a mass m suspended by a light string of length L. (L is the distance from the pivot to the center of mass of the bob.) The restoring force that causes the pendulum to undergo simple harmonic motion is the component of weight tangent to the path of motion, $mg \sin \theta$.

A multiflash photograph of a swinging pendulum. Is the oscillating motion simple harmonic in this case?
(Richard Megna/Fundamental Photographs)

8.12

▶ The period of a simple pendulum depends only on *L* and *g*.

APPLICATION

Pendulum Clocks.

APPLICATION

Use of Pendulum in Prospecting.

Because $s = L\theta$, we have

$$F_t = -\left(\frac{mg}{L}\right) s$$

This equation follows the general form of the Hooke's law force, given by $F = -ks$, with $k = mg/L$. Thus, we are justified in saying that a pendulum undergoes simple harmonic motion only when it swings back and forth at very small amplitudes (or, in this case, small values of θ, so that $\sin \theta \cong \theta$).

We can find the period of a pendulum by first recalling that the period of a mass-spring system (Eq. 13.8) is

$$T = 2\pi\sqrt{\frac{m}{k}}$$

If we replace k with its equivalent, mg/L, we see that the period of a simple pendulum is

$$T = 2\pi\sqrt{\frac{m}{mg/L}}$$

$$T = 2\pi\sqrt{\frac{L}{g}} \qquad [13.15]$$

This equation reveals the somewhat surprising result that the period of a simple pendulum depends not on mass but only on the pendulum's length and on the free-fall acceleration. Furthermore, the amplitude of the motion is not a factor as long as it is relatively small. The analogy between the motion of a simple pendulum and the mass-spring system is illustrated in Figure 13.16.

It is of historical interest to point out that it was Galileo who first noted that the period of a pendulum was independent of its amplitude. He supposedly observed this while attending church services at a cathedral in Pisa. The pendulum he studied was a swinging chandelier that was set in motion when someone bumped it while lighting candles. Galileo was able to measure its frequency, and hence its period, by timing the swings with his pulse.

The dependence of the period of a pendulum on its length and on the free-fall acceleration allows us to use a pendulum as a timekeeper for a clock. A number of clock designs employ a pendulum, with the length adjusted so that its period serves as the basis for the rate at which the clock's hands turn. Of course, these clocks are used at different locations on the Earth, so there will be some variation of the free-fall acceleration. In order to compensate for this, the pendulum of a clock should have some movable mass so that the effective length can be adjusted for variations in *g*.

Geologists often make use of the simple pendulum and Equation 13.15 when prospecting for oil or minerals. Deposits beneath the Earth's surface can produce irregularities in the free-fall acceleration over the region being studied. A specially designed pendulum of known length is used to measure the period, which in turn is used to calculate *g*. Although such a measurement in itself is inconclusive, it is an important tool for geological surveys.

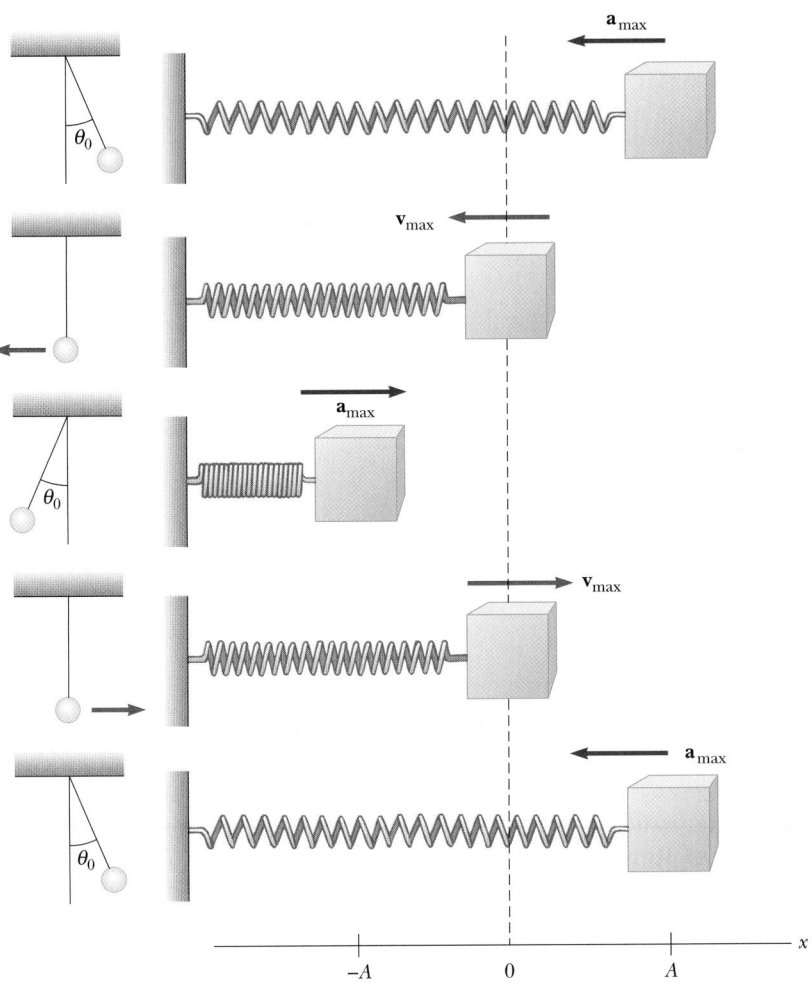

Figure 13.16 Simple harmonic motion for a mass-spring system and its analogy, the motion of a simple pendulum.

QUICKLAB

Construct a simple pendulum by tying a metal bolt to one end of a string of length 1.00 m, and measure the period of your pendulum while fixing the opposite end. You should obtain the period by timing 25 complete oscillations, making sure that the string always makes small angles with the vertical. Repeat the measurement for string lengths ranging from 0.40 m to 1.60 m in increments of 0.20 m. Plot the square of the period versus the length of the string, and measure the slope of the line through your data points. Does your slope agree with that predicted by Equation 13.15? What value of g do you obtain from your data?

EXAMPLE 13.8 What Is the Height of That Tower?

A man needs to know the height of a tower, but darkness obscures the ceiling. He knows, however, that a long pendulum extends from the ceiling almost to the floor and that its period is 12.0 s. How tall is the tower?

Solution If we use $T = 2\pi\sqrt{L/g}$ and solve for L, we get

$$L = \frac{gT^2}{4\pi^2} = \frac{(9.80 \text{ m/s}^2)(12.0 \text{ s})^2}{4\pi^2} = \boxed{35.8 \text{ m}}$$

Exercise If the length of the pendulum were halved, what would its period of vibration be?

Answer 8.49 s

Workbook Problem 55— (Workbook page 117) is similar to Example 13.8.

Figure 13.17 A cross-sectional view of a shock absorber connected to a spring in the suspension system of an automobile. The upper part of the shock absorber and the ends of the suspension spring are attached to the frame of the automobile (not shown).

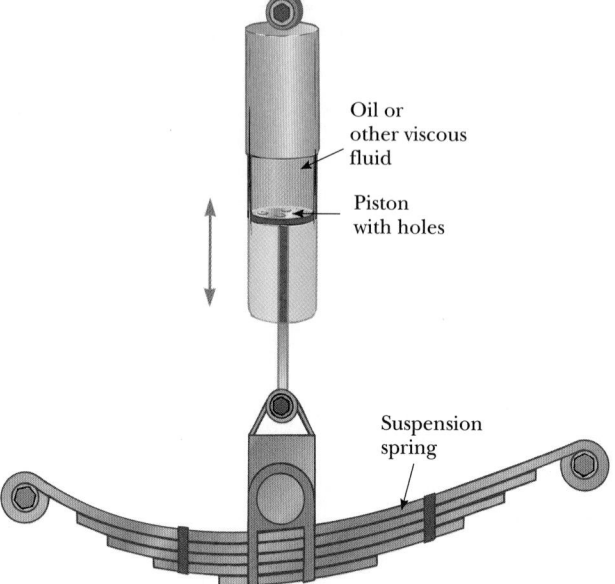

Oil or other viscous fluid

Piston with holes

Suspension spring

WEB

For Web links relating to shock absorbers, visit the textbook Web site at **http://www.harcourtcollege.com/ physics/cptech**

APPLICATION

Shock Absorbers.

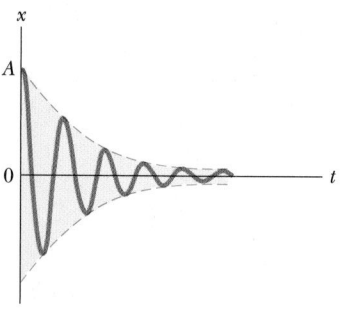

Figure 13.18 A graph of displacement versus time for an underdamped oscillator. Note the decrease in amplitude with time.

13.7   DAMPED OSCILLATIONS

The vibrating motions we have discussed so far have taken place in ideal systems —that is, systems that *oscillate indefinitely* under the action of a linear restoring force. In many real systems, forces of friction retard the motion, and consequently the systems do not oscillate indefinitely. The friction reduces the mechanical energy of the system as time passes, and the motion is said to be **damped.**

Shock absorbers in automobiles (Fig. 13.17) make practical application of damped motion. A shock absorber consists of a piston moving through a liquid such as oil. The upper part of the shock absorber is firmly attached to the body of the car, and the piston is attached to a leaf spring that, with the other springs, acts as the main suspension for the car. When the car travels over a bump in the road, holes in the piston allow it to move up and down in the fluid in a damped fashion.

Damped motion varies depending on the fluid used. For example, if the fluid has a relatively low viscosity, the vibrating motion is preserved but the amplitude of vibration decreases in time and the motion ultimately ceases. This is known as *underdamped* oscillation, and its position-time curve appears in Figure 13.18. If the fluid viscosity is increased, the mass returns rapidly to equilibrium after it is released and does not oscillate. In this case, the system is said to be *critically damped* (Fig. 13.19a), and the piston returns to the equilibrium position in the shortest time possible without once overshooting the equilibrium position. If the viscosity is made greater still, the system is said to be *overdamped.* In this case, the piston returns to equilibrium without ever passing through the equilibrium point, but the time required to reach equilibrium is greater than at critical damping, as shown by Figure 13.19b.

To make automobiles more comfortable to ride in, shock absorbers are designed to be slightly underdamped. This can be demonstrated by a sharp downward push on the hood of a car. After the applied force is removed, the body of the car oscillates a few times about the equilibrium position before returning to its fixed position.

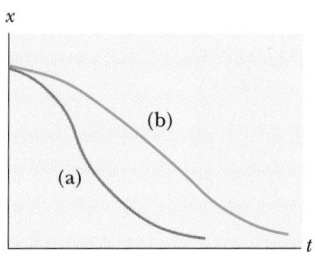

Figure 13.19 Plots of displacement versus time for (a) a critically damped oscillator and (b) an overdamped oscillator.

13.8 WAVE MOTION

Most of us first saw waves when, as children, we dropped a pebble into a pool of water. The disturbance created by the pebble generates water waves, which move outward until they finally reach the edge of the pool. There are a wide variety of physical phenomena that have wave-like characteristics. The world is full of waves: sound waves, waves on a string, earthquake waves, and electromagnetic waves, such as visible light, radio waves, television signals, and x-rays. All of these waves have as their source a vibrating object. Thus, we shall use the terminology and concepts of simple harmonic motion as we move into the study of wave motion.

In the case of sound waves, the vibrations that produce waves arise from such sources as a person's vocal cords or a plucked guitar string. The vibrations of electrons in an antenna produce radio or television waves, and the simple up-and-down motion of a hand can produce a wave on a string. Regardless of the type of wave under consideration, there are certain concepts common to all varieties. In the remainder of this chapter, we shall focus our attention on a general study of wave motion. In later chapters we shall study specific types of waves, such as sound and electromagnetic waves.

What Is a Wave?

As we said before, when you drop a pebble into a pool of water, the disturbance produced by the pebble excites water waves, which move away from the point at which the pebble entered the water. If you carefully examined the motion of a leaf floating near the disturbance, you would see that it moves up and down and back and forth about its original position but does not undergo any net displacement attributable to the disturbance. That is, the water wave (or disturbance) moves from one place to another *but the water is not carried with it.*

Einstein and Infeld made these remarks about wave phenomena:

> A bit of gossip starting in Washington reaches New York very quickly, even though not a single individual who takes part in spreading it travels between these two cities. There are two quite different motions involved, that of the rumor, Washington to New York, and that of the persons who spread the rumor. The wind, passing over a field of grain, sets up a wave which spreads out across the whole field. Here again we must distinguish between the motion of the wave and the motion of the separate plants, which undergo only small oscillations. . . . The particles constituting the medium perform only small vibrations, but the whole motion is that of a progressive wave. The essentially new thing here is that for the first time we consider the motion of something which is not matter, but energy propagated through matter.[1]

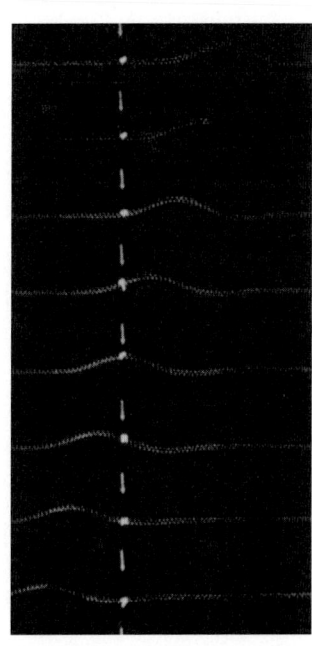

A disturbance traveling from right to left on a stretched spring. *(Education Development Center, Newton, MA)*

---

[1] A. Einstein and L. Infeld, *The Evolution of Physics,* New York, Simon and Schuster, 1961.

When we observe what is called a *water wave*, what we see is a rearrangement of the water's surface. Without the water there would be no wave. A wave traveling on a string would not exist without the string. Sound waves travel through air as a result of pressure variations from point to point. (We shall discuss sound waves in Chapter 14.) The wave motion we consider in this chapter corresponds to the disturbance of a body or medium. Therefore, we can consider a wave to be *the motion of a disturbance*. In a later chapter we shall discuss electromagnetic waves, which do not require a medium.

The mechanical waves discussed in this chapter require (1) some source of disturbance, (2) a medium that can be disturbed, and (3) some physical connection or mechanism through which adjacent portions of the medium can influence each other. All waves carry energy and momentum. The amount of energy transmitted through a medium and the mechanism responsible for the transport of energy differ from case to case. For instance, the energy carried by ocean waves during a storm is much greater than that carried by a sound wave generated by a single human voice.

Thinking Physics 3

We have shown that waves carry energy—not only waves on strings, but all waves. Devise a demonstration that shows that waves transfer momentum as well as energy.

Explanation   Imagine that we cover the ends of a cardboard tube with rubber sheets secured with rubber bands, to form a crude drum with two heads. We now lay the tube horizontally on the table and suspend a Ping-Pong ball from a string so that the Ping-Pong ball is just touching one of the rubber sheets. If we now thump the other rubber sheet with a finger, the sound will travel down the tube, strike the first rubber sheet, and the Ping-Pong ball will bounce away from the rubber sheet. This demonstrates that the sound wave is transferring momentum and energy to the Ping-Pong ball.

Applying Physics 2

Why is it important to be quiet in avalanche country? In a spy movie, the bad guys try to stop the hero, who is escaping on skis, by firing a gun and causing an avalanche. Why did the avalanche happen?

Explanation   The essence of a wave is the propagation of a disturbance through a medium. An impulsive sound, like the gunshot, can cause an acoustical disturbance that propagates through the air and can impact a ledge of snow that is just ready to break free to begin an avalanche. Such a disastrous event occurred in 1916 during World War I, when Austrian soldiers in the Alps were smothered by an avalanche caused by cannon fire.

13.9   TYPES OF WAVES

One of the simplest ways to demonstrate wave motion is to flip one end of a long rope that is under tension and has its opposite end fixed, as in Figure 13.20. The bump (called a *pulse*) travels to the right with a definite speed. A disturbance of

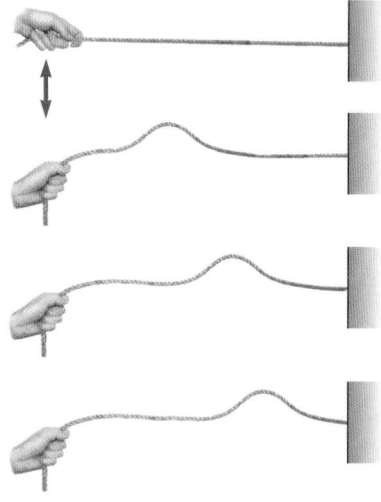

Figure 13.20 A wave pulse traveling along a stretched rope. The shape of the pulse is approximately unchanged as it travels.

this type is called a **traveling wave.** Figure 13.20 shows the shape of the rope at three closely spaced times.

As a traveling wave pulse travels along the rope, **each segment of the rope that is disturbed moves perpendicularly to the wave motion.** Figure 13.21 illustrates this point for a particular tiny segment, *P*. Never does the rope move in the direction of the wave. A traveling wave such as this, in which the particles of the disturbed medium move perpendicularly to the wave velocity, is called a **transverse wave.** Figure 13.22 illustrates the formation of transverse waves on a long spring.

In another class of waves, called **longitudinal waves, the particles of the medium undergo displacements parallel to the direction of wave motion.** Sound waves in air, for instance, are longitudinal. Their disturbance corresponds to a series of high- and low-pressure regions that may travel through air or through any material medium with a certain speed. A longitudinal pulse can be easily produced in a stretched spring, as in Figure 13.22b. The free end is pumped back and forth along the length of the spring. This action produces compressed and stretched regions of the coil that travel along the spring, parallel to the wave motion.

Picture of a Wave

Figure 13.23 shows the curved shape of a vibrating string. This pattern, sometimes called a *waveform*, should be familiar to you from our study of simple harmonic motion; it is a sinusoidal curve. The red curve can be thought of as a snapshot of a traveling wave taken at some instant of time, say *t* = 0; the blue curve, a snapshot of the same traveling wave at a later time. It is not difficult to imagine that this picture can as easily be used to represent a wave on water. In such a case, point *A* would correspond to the *crest* of the wave and point *B* to the low point, or *trough*, of the wave.

8.5

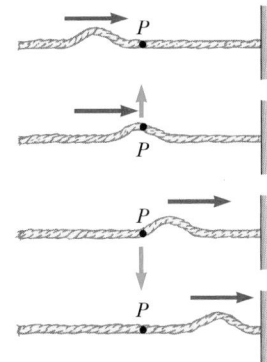

Figure 13.21 A pulse traveling on a stretched rope is a transverse wave. That is, any element *P* on the rope moves (blue arrows) perpendicularly to the wave motion (red arrows).

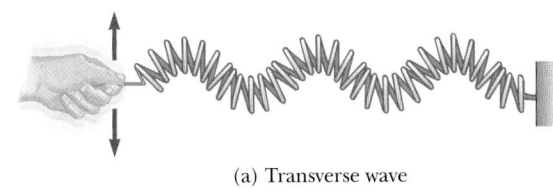

(a) Transverse wave

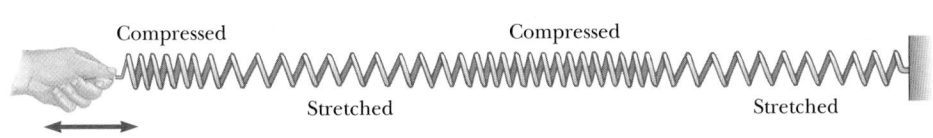

Compressed   Compressed

Stretched   Stretched

(b) Longitudinal wave

Figure 13.22 (a) A transverse wave is set up in a spring by moving one end of the spring perpendicularly to its length. (b) A longitudinal pulse along a stretched spring. The displacement of the coils is in the direction of the wave motion. For the starting motion described in the text, the compressed region is followed by a stretched region.

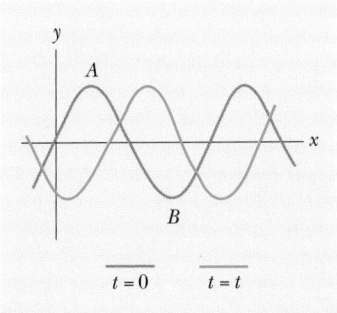

Figure 13.23 A one-dimensional sinusoidal wave traveling to the right with a speed of *v*. The red curve is a snapshot of the wave at *t* = 0, and the blue curve is another snapshot at some later time *t*.

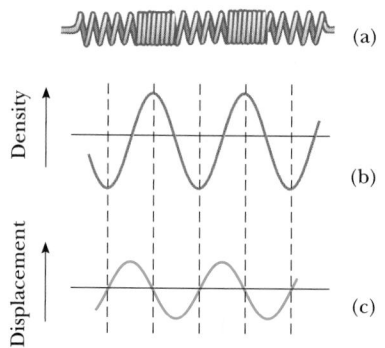

Density

Displacement

(a)

(b)

(c)

Figure 13.24 (a) A longitudinal wave on a spring. (b) The crests of the waveform correspond to compressed regions of the spring, and the troughs correspond to stretched regions of the spring. (c) The displacement wave.

8.6

This same waveform can be used to describe a longitudinal wave. To see how this is done, consider a longitudinal wave traveling on a spring. Figure 13.24a is a snapshot of this wave at some instant, and Figure 13.24b shows the sinusoidal curve that represents the wave. Points where the coils of the spring are compressed correspond to the crests of the waveform, and stretched regions correspond to troughs.

The type of wave represented by the curve in Figure 13.24b is often referred to as a *density* or *pressure wave*. This is because the crests, where the spring coils are compressed, are regions of high density, and the troughs, where the coils are stretched, are regions of low density.

An alternative method for representing wave motion along a spring is through the concept of a displacement wave, shown in Figure 13.24c. In this representation, coils that are displaced the greatest distance from equilibrium in one direction are indicated by crests, and coils that are displaced the greatest distance from equilibrium in the opposite direction are represented by troughs. Both of these wave representations will be used in future sections.

13.10 FREQUENCY, AMPLITUDE, AND WAVELENGTH

Figure 13.25 illustrates a method of producing a wave on a very long string. One end of the string is connected to a blade that is set in vibration. As the blade oscillates vertically with simple harmonic motion, a traveling wave moving to the right is set up in the string. Figure 13.25 consists of views of the wave at intervals of one quarter of a period. Note that **each particle of the string, such as P, oscillates vertically in the y direction with simple harmonic motion.** This must be the case, because each particle follows the simple harmonic motion of the blade. Therefore, every segment of the string can be treated as a simple harmonic oscillator vibrating with the same frequency as the blade that drives the string.

Figure 13.25 One method for producing traveling waves on a continuous string. The left end of the string is connected to a blade that is set in vibration. Note that every segment, such as *P*, oscillates vertically with simple harmonic motion.

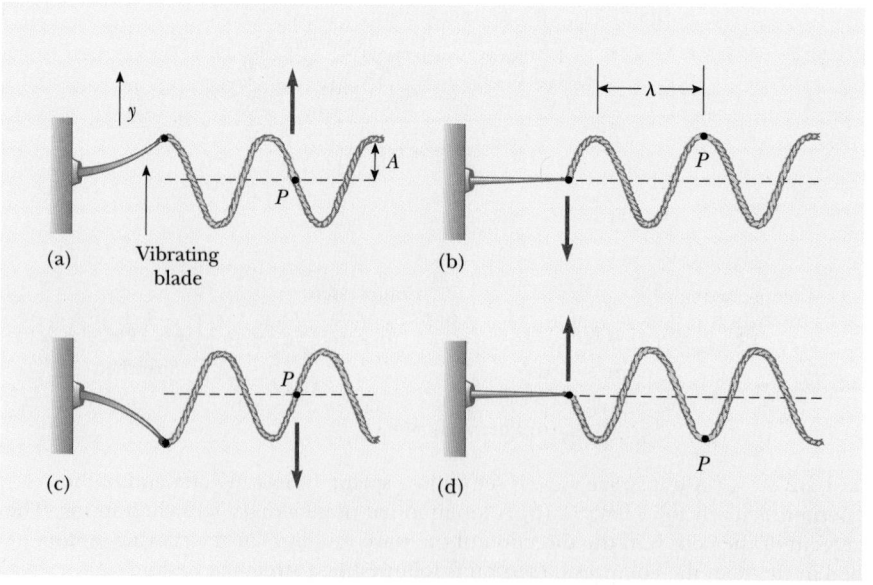

The frequencies of the waves we shall study will range from rather low values for waves on strings and waves on water to values between 20 Hz and 20 000 Hz (recall that $1 \text{ Hz} = 1 \text{ s}^{-1}$) for sound waves, and much higher frequencies for electromagnetic waves.

The horizontal dashed line in Figure 13.25 represents the position of the string if no wave were present. The maximum distance the string moves above or below this equilibrium value is called the **amplitude, A,** of the wave. For the waves we work with, the amplitudes at the crest and the trough will be identical.

Figure 13.25b illustrates another characteristic of a wave. The horizontal arrows show the distance between two successive points that behave identically. This distance is called the **wavelength, λ** (Greek letter lambda).

We can use these definitions to derive an expression for the velocity of a wave. We start with the defining equation for velocity:

$$v = \frac{\Delta x}{\Delta t}$$

A little reflection should convince you that a wave will advance a distance of one wavelength in a time interval equal to one period of the vibration. Thus

8.9, SECTION 1

$$v = \frac{\lambda}{T}$$

Because the frequency is equal to the reciprocal of the period, we have

$$v = f\lambda \qquad\qquad \text{[13.16]} \qquad \triangleleft \text{ Wave velocity}$$

We shall apply this equation to many types of waves. For example, we shall use it often in our study of sound and electromagnetic waves.

EXAMPLE 13.9 A Traveling Wave

A wave traveling in the positive x direction is pictured in Figure 13.26. Find the amplitude, wavelength, period, and speed of the wave if it has a frequency of 8.0 Hz.

Solution The amplitude and wavelength can be read directly from the figure:

$$A = \boxed{15 \text{ cm}} \qquad \lambda = 40 \text{ cm} = \boxed{0.40 \text{ m}}$$

The period of the wave is

$$T = \frac{1}{f} = \frac{1}{8.0} \text{ s} = \boxed{0.13 \text{ s}}$$

and the speed is

$$v = f\lambda = (8.0 \text{ s}^{-1})(0.40 \text{ m}) = \boxed{3.2 \text{ m/s}}$$

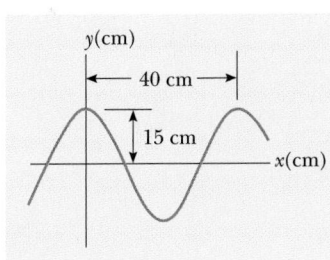

Figure 13.26 (Example 13.9) A harmonic wave of wavelength $\lambda = 40$ cm and amplitude $A = 15$ cm.

EXAMPLE 13.10 Give Me a "C" Note

The note middle C on a piano has a frequency of approximately 262 Hz and a wavelength in air of 1.31 m. Find the speed of sound in air.

Solution   By direct substitution into Equation 13.16, we find that

$$v = f\lambda = (262\ \text{s}^{-1})(1.31\ \text{m}) = \boxed{343\ \text{m/s}}$$

EXAMPLE 13.11   The Speed of Radio Waves

An FM station broadcasts at a frequency of 100 MHz (M = mega = 10^6), with a radio wave having a wavelength of 3.00 m. Find the speed of the radio wave.

Solution   As in the last example, we use Equation 13.16:

$$v = f\lambda = (100 \times 10^6\ \text{s}^{-1})(3.00\ \text{m}) = \boxed{3.00 \times 10^8\ \text{m/s}}$$

This, in fact, is the speed of *any* electromagnetic wave traveling through empty space.

Exercise   Find the wavelength of an electromagnetic wave whose frequency is 9.0 GHz = 9.0×10^9 Hz (G = giga = 10^9), which is the microwave range.

Answer   0.033 m

Workbook Problem 56 — (Workbook page 118) deals with the speeds of earthquake (seismic) waves.

9.2, BOTH SECTIONS

13.11   THE SPEED OF WAVES ON STRINGS

In this section we focus our attention on the speed of a wave on a stretched string. Rather than deriving the equation, we use dimensional analysis to verify that the expression can be valid.

It is easy to understand why the wave speed depends on the tension in the string. If a string under tension is pulled sideways and released, the tension is responsible for accelerating a particular segment back toward its equilibrium position. The acceleration and wave speed increase with increasing tension in the string. Likewise, the wave speed is inversely dependent on the mass per unit length of the string. This is because it is more difficult to accelerate (and impart a large wave speed to) a massive string than a light string. Thus, wave speed is directly dependent on the tension and inversely dependent on the mass per unit length of the string. The exact relationship of the wave speed, v, the tension, F, and the mass per length, μ, is

$$v = \sqrt{\frac{F}{\mu}} \qquad\qquad [13.17]$$

From this expression we see that the speed of a mechanical wave, such as a wave on a string, depends only on the properties of the medium through which the disturbance travels.

Now let us verify that this expression is dimensionally correct. The dimensions of F are ML/T^2, and the dimensions of μ are M/L. Therefore, the dimensions of F/μ are L^2/T^2, and those of $\sqrt{F/\mu}$ are L/T, which are indeed the dimensions of speed. No other combination of F and μ is dimensionally correct, assuming that these are the only variables relevant to the situation.

Equation 13.17 indicates that we can increase the speed of a wave on a stretched string by increasing the tension in the string. It also shows that if we wrap a string with a metallic winding, as is done to the bass strings of pianos and guitars, we decrease the speed of a transmitted wave because the mass per unit length is increased.

EXAMPLE 13.12   A Pulse Traveling on a String

A uniform string has a mass, M, of 0.300 kg and a length, L, of 6.00 m. Tension is maintained in the string by suspending a 2.00-kg block from one end (Fig. 13.27). Find the speed of a pulse on this string.

Solution   The tension, F, in the string is equal to the mass, m, of the block multiplied by the free-fall acceleration:

$$F = mg = (2.00 \text{ kg})(9.80 \text{ m/s}^2) = 19.6 \text{ N}$$

For the string, the mass per unit length, μ, is

$$\mu = \frac{M}{L} = \frac{0.300 \text{ kg}}{6.00 \text{ m}} = 0.0500 \text{ kg/m}$$

Therefore, the wave speed is

$$v = \sqrt{\frac{F}{\mu}} = \sqrt{\frac{19.6 \text{ N}}{0.0500 \text{ kg/m}}} = 19.8 \text{ m/s}$$

Exercise   Find the time it takes the pulse to travel from the wall to the pulley.

Answer   0.253 s

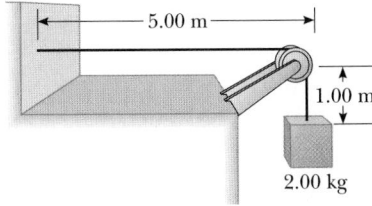

Figure 13.27   (Example 13.12) The tension, F, in the string is maintained by the suspended block. The wave speed is given by the expression $v = \sqrt{F/\mu}$.

Workbook Problem 57 — (Workbook page 122) extends Example 13.12 to a more complex case.

13.12   SUPERPOSITION AND INTERFERENCE OF WAVES

Many interesting wave phenomena in nature are impossible to describe with a single moving wave. Instead, one must analyze what happens when two or more waves pass through the same region of space. For such analyses one can use the **superposition principle:**

9.6, BOTH SECTIONS

> If two or more traveling waves are moving through a medium, the resultant wave is found by adding together the displacements of the individual waves point by point.

Experiments show that the superposition principle is valid only when the individual waves have small amplitudes of displacement—an assumption we make in all our examples.

One consequence of the superposition principle is that **two traveling waves can pass through each other without being destroyed or even altered.** For instance, when two pebbles are thrown into a pond, the expanding circular waves do

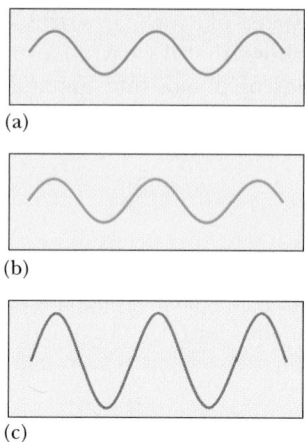

(a)

(b)

(c)

Figure 13.28 Constructive interference. If two waves having the same frequency and amplitude are in phase, as in (a) and (b), the resultant wave when they combine (c) has the same frequency as the individual waves but twice their amplitude.

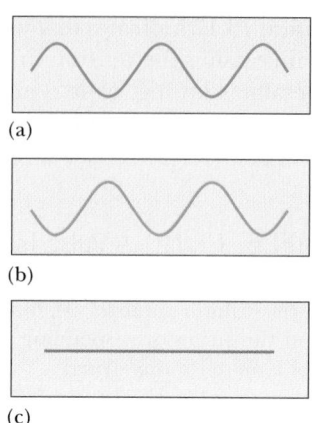

(a)

(b)

(c)

Figure 13.29 Destructive interference. When two waves with the same frequency and amplitude are 180° out of phase, as in (a) and (b), the result when they combine (c) is complete cancellation.

Figure 13.30 Interference patterns produced by outward-spreading waves from several drops of water falling into a pond. *(Martin Dohrn/SPL/Photo Researchers)*

not destroy each other. In fact, the ripples pass through each other. Likewise, when sound waves from two sources move through air, they pass through each other. The sound one hears at a given location is the result of both disturbances.

Figures 13.28a and 13.28b show two waves of the same amplitude and frequency. If at some instant of time these two waves attempted to travel through the same region of space, the resultant wave at that instant would have a shape like Figure 13.28c. For example, suppose these are water waves of amplitude 1 m. At the instant they overlap so that crest meets crest and trough meets trough, the resultant wave has an amplitude of 2 m. Waves coming together like this are said to be *in phase* and to undergo **constructive interference.**

Figures 13.29a and 13.29b show two similar waves. In this case, however, the crest of one coincides with the trough of the other; that is, one wave is *inverted* relative to the other. The resultant wave, shown in Figure 13.29c, is seen to be a state of complete cancellation. If these were water waves coming together, one of the waves would be trying to pull an individual drop of water upward at the same instant the other wave was trying to pull it downward. The result would be no motion of the water at all. In such a situation, the two waves are said to be 180° out of phase and to undergo **destructive interference.** Figure 13.30 illustrates the interference of water waves produced by drops of water falling into a pond.

Figure 13.31 shows constructive interference in two pulses moving toward each other along a stretched string; Figure 13.32 shows destructive interference in two pulses. Notice in each case that, when the two pulses separate, their shapes are unchanged, as if they had never met!

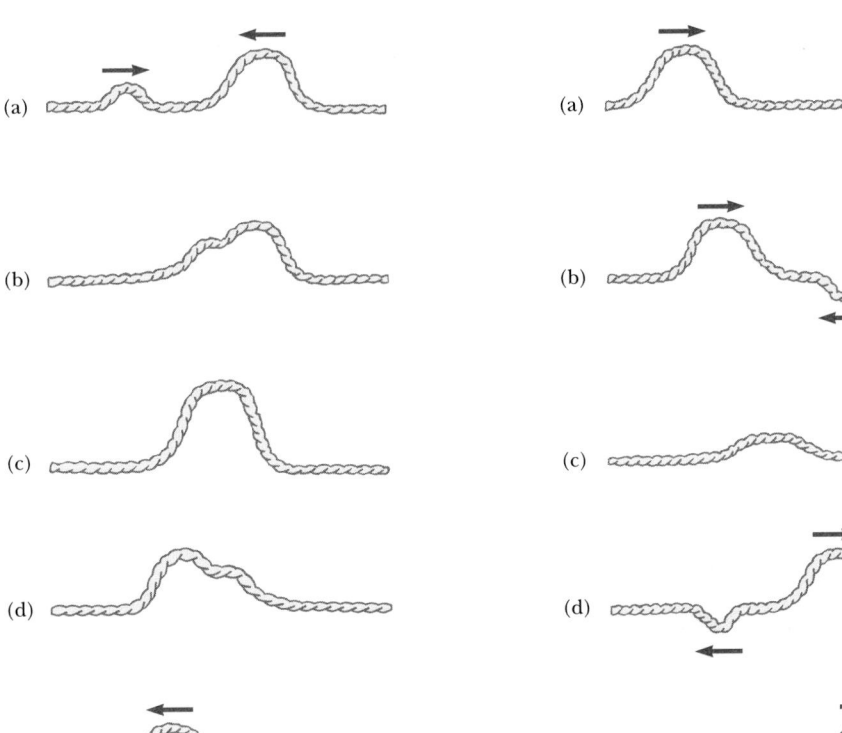

Figure 13.31 Two wave pulses traveling on a stretched string in opposite directions pass through each other. When the pulses overlap, as in (b), (c), and (d), the net displacement of the string equals the sum of the displacements produced by each pulse. Because each pulse produces positive displacements of the string, we refer to their superposition as *constructive interference*.

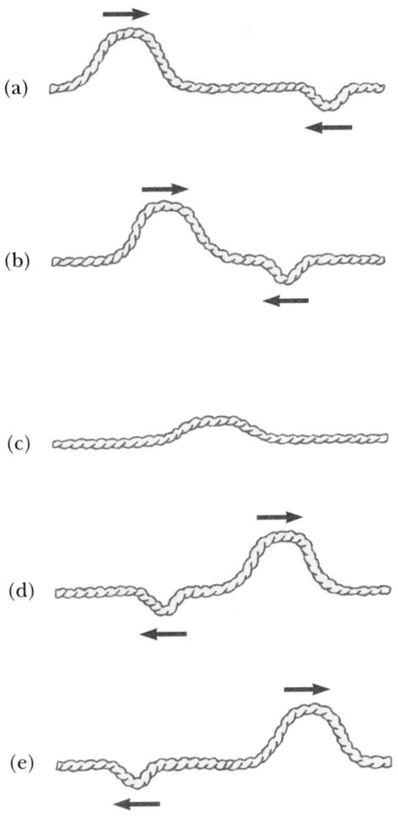

Figure 13.32 Two wave pulses traveling in opposite directions with displacements that are inverted relative to each other. When the two overlap as in (c), their displacements subtract from each other.

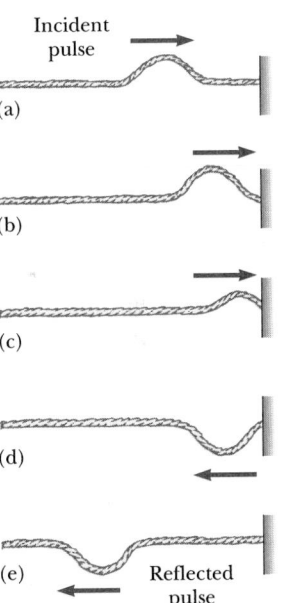

Figure 13.33 The reflection of a traveling wave at the fixed end of a stretched string. Note that the reflected pulse is inverted, but its shape remains the same.

9.3, BOTH SECTIONS

13.13 REFLECTION OF WAVES

In our discussion so far, we have assumed that the waves being analyzed could travel indefinitely without striking anything. Often, such conditions are not realized in practice. Whenever a traveling wave reaches a boundary, part or all of the wave is reflected. For example, consider a pulse traveling on a string that is fixed at one end (Fig. 13.33). When the pulse reaches the wall, it is reflected.

Note that the reflected pulse is inverted. This can be explained as follows. When the pulse meets the wall, the string exerts an upward force on the wall. According

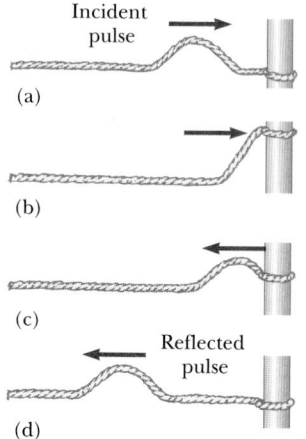

Figure 13.34 The reflection of a traveling wave at the free end of a stretched string. In this case, the reflected pulse is not inverted.

to Newton's third law, the wall must exert an equal and opposite (downward) reaction force on the string. This downward force causes the pulse to invert on reflection.

Now consider a case in which the pulse arrives at the string's end, which is attached to a ring of negligible mass that is free to slide along the post without friction (Fig. 13.34). Again the pulse is reflected, but this time it is not inverted. On reaching the post, the pulse exerts a force on the ring, causing it to accelerate upward. The ring is then returned to its original position by the downward component of the tension.

An alternative method of showing that a pulse is reflected without inversion when it strikes a free end of a string is to send a pulse down a string hanging vertically. When the pulse hits the free end, it is reflected without inversion, similarly to the pulse in Figure 13.34.

SUMMARY

Simple harmonic motion occurs when the net force along the direction of motion is a **Hooke's law** type of force — that is, when the net force is proportional to the displacement and in the opposite direction:

$$F_s = -kx \tag{13.1}$$

The time required for one complete vibration is called the **period** of the motion. The inverse of the period is the **frequency** of the motion, which is the number of oscillations per second.

When an object is moving with simple harmonic motion, its **acceleration** as a function of location is

$$a = -\frac{k}{m}x \tag{13.2}$$

The energy stored in a stretched or compressed spring or other elastic material is called **elastic potential energy:**

$$PE_s \equiv \tfrac{1}{2}kx^2 \tag{13.3}$$

The **velocity** of an object as a function of position, when the object is moving with simple harmonic motion, is

$$v = \pm\sqrt{\frac{k}{m}(A^2 - x^2)} \tag{13.6}$$

The **period** of an object of mass m moving with simple harmonic motion while attached to a spring of spring constant k is

$$T = 2\pi\sqrt{\frac{m}{k}} \tag{13.8}$$

Also, the **frequency** of a mass-spring system is $f = 1/T$.

The **position** of an object as a function of time, when the object is moving with simple harmonic motion, is

$$x = A\cos(2\pi ft) \tag{13.14}$$

A **simple pendulum** of length L moves with simple harmonic motion for small angular displacements from the vertical, with a period of

$$T = 2\pi\sqrt{\frac{L}{g}}$$ [13.15]

The period is independent of the suspended mass.

A **transverse wave** is one in which the particles of the medium move perpendicularly to the direction of the wave velocity. An example is a wave on a stretched string.

A **longitudinal wave** is one in which the particles of the medium move parallel to the direction of the wave velocity. An example is a sound wave.

The relationship of the velocity, wavelength, and frequency of a wave is

$$v = f\lambda$$ [13.16]

The speed of a wave traveling on a stretched string of mass per unit length μ and under tension F is

$$v = \sqrt{\frac{F}{\mu}}$$ [13.17]

The **superposition principle** states that if two or more traveling waves are moving through a medium, the resultant wave is found by adding the individual waves together point by point. When waves meet crest to crest and trough to trough, they undergo **constructive interference.** When crest meets trough, the waves undergo complete **destructive interference.**

When a wave pulse reflects from a rigid boundary, the pulse is inverted. When the boundary is free, the reflected pulse is not inverted.

MULTIPLE-CHOICE QUESTIONS

1. The distance between a crest of a sinusoidal water wave and the next trough is 2 m. If the frequency of this wave is 2 Hz, find the speed of the wave.
   (a) 4 m/s   (b) 1 m/s   (c) 8 m/s
   (d) Impossible to say from the information given.

2. Light waves and sound waves have the following in common.
   (a) Neither will travel in a vacuum.
   (b) Both are longitudinal waves.
   (c) They travel at the same speed.
   (d) None of the above is correct.

3. Replacing an object on a spring with one of nine times the mass will have the result of changing the frequency of the vibrating spring by what multiplying factor?
   (a) 1/9   (b) 1/3   (c) 3.0   (d) 9.0

4. A mass of 0.4 kg, hanging from a spring with a spring constant of 80 N/m, is set into an up-and-down simple harmonic motion. What is the speed of the mass when moving through the equilibrium point? The starting displacement is 0.1 m.
   (a) zero   (b) 1.4 m/s   (c) 2.0 m/s   (d) 3.4 m/s

5. A mass of 0.4 kg, hanging from a spring with a spring constant of 80 N/m, is set into an up-and-down simple harmonic motion. What is the acceleration of the mass when at its maximum displacement of 0.1 m?
   (a) zero   (b) 5 m/s^2 (c) 10 m/s^2 (d) 20 m/s^2

6. A runaway railroad car, with mass 30×10^4 kg, coasts across a level track at 2.0 m/s when it collides with a spring-loaded bumper at the end of the track. If the spring constant of the bumper is 2.0×10^6 N/m, what is the maximum compression of the spring during the collision? (Assume the collision is elastic.)
   (a) 0.77 m   (b) 0.58 m
   (c) 0.34 m   (d) 1.07 m

CONCEPTUAL QUESTIONS

1. A mass-spring system undergoes simple harmonic motion with an amplitude A. Does the total energy change if the mass is doubled but the amplitude is not changed? Are the kinetic and potential energies at a given point in its motion affected by the change in mass? Explain.

2. Does the acceleration of a simple harmonic oscillator remain constant during its motion? Is the acceleration ever zero? Explain.

3. A mass is hung on a spring and the frequency of the oscillation of the system, f, is measured. The mass, a second identical mass, and the spring are carried in the space shuttle to space. The two masses are attached to the ends of the spring, and the system is taken out on a space walk. The spring is extended, and the system is released to oscillate while floating in space. What is the frequency of oscillation for this system in terms of f?

4. If a mass-spring system is hung vertically and set into oscillation, why does the motion eventually stop?

5. Is a bouncing ball an example of simple harmonic motion? Is the daily movement of a student from home to school and back simple harmonic motion?

6. If a pendulum clock keeps perfect time at the base of a mountain, will it also keep perfect time when moved to the top of the mountain? Explain.

7. A simple pendulum is suspended from the ceiling of a stationary elevator, and the period is determined. Describe the changes, if any, in the period when the elevator (a) accelerates upward, (b) accelerates downward, and (c) moves with constant velocity.

8. If a grandfather clock were running slow, how could we adjust the length of the pendulum to correct the time?

9. A grandfather clock depends on the period of a pendulum to keep correct time. Suppose a grandfather clock is calibrated correctly and then the temperature of the room in which it resides increases. Does the grandfather clock run slow, fast, or correctly? (*Hint:* Remember that a metal expands when its temperature increases.)

10. The amplitude of a system moving in simple harmonic motion is doubled. Determine the change in (a) the total energy, (b) the maximum speed, (c) the maximum acceleration, and (d) the period.

11. In a long line of people waiting to buy tickets at a movie theater, when the first person leaves, a pulse of motion occurs as people step forward to fill in the gap. The gap moves through the line of people. What determines the speed of this pulse? Is it transverse or longitudinal? How about the "wave" at a baseball game, where people in the stands stand up and shout as the wave arrives at their location, and this pulse moves around the stadium —what determines the speed of this pulse? Is it transverse or longitudinal?

12. How would you create a longitudinal wave in a stretched spring? Would it be possible to create a transverse wave in a spring?

13. In mechanics, massless strings are often assumed. Why is this *not* a good assumption when discussing waves on strings?

14. What happens to the wavelength of a wave on a string when the frequency is doubled? Assume that the tension in the string remains the same.

15. Explain why the kinetic and potential energies of a mass-spring system can never be negative.

16. What happens to the speed of a wave on a string when the frequency is doubled? Assume that the tension in the string remains the same.

17. What is the total distance traveled by a body executing simple harmonic motion in a time equal to its period if its amplitude is A?

18. By what factor would you have to multiply the tension in a stretched spring in order to double the wave speed?

19. Determine whether or not the following quantities can be in the same direction for a simple harmonic oscillator: (a) displacement and velocity, (b) velocity and acceleration, (c) displacement and acceleration.

20. The motion of the Earth going around the Sun is periodic with a period of 1 year. Is this motion simple harmonic? Explain.

PROBLEMS

1, **2**, **3** = straightforward, intermediate, challenging   □ = full solution available in Study Guide/Student Solutions Manual   = Core Concepts Workbook

WEB = solution posted at **http://www.harcourtcollege.com/physics/cptech**   = biomedical application   = Interactive Physics

Section 13.1 Hooke's Law

1. A 0.40-kg mass is attached to a spring with a spring constant 160 N/m so that the mass is allowed to move on a horizontal frictionless surface. The mass is released from rest when the spring is compressed 0.15 m. Find (a) the force on the mass and (b) its acceleration at this instant.

2. A load of 50 N attached to a spring hanging vertically stretches the spring 5.0 cm. The spring is now placed

horizontally on a table and stretched 11 cm. **(a)** What force is required to stretch the spring by this amount? **(b)** Plot a graph of force (on the *y* axis) versus spring displacement from the equilibrium position along the *x* axis.

3. A ball dropped from a height of 4.00 m makes a perfectly elastic collision with the ground. Assuming no energy lost due to air resistance, **(a)** show that the motion is periodic and **(b)** determine the period of the motion. **(c)** Is the motion simple harmonic? Explain.

4. A small ball is set in horizontal motion by rolling it with a speed of 3.00 m/s across a room 12.0 m long, between two walls. Assume that the collisions made with each wall are perfectly elastic and that the motion is perpendicular to the two walls. **(a)** Show that the motion is periodic, and determine its period. **(b)** Is this motion simple harmonic? Explain.

Section 13.2 Elastic Potential Energy

5. A slingshot consists of a light leather cup containing a stone that is pulled back against two rubber bands. It takes a force of 30 N to stretch the bands 1.0 cm. **(a)** What is the potential energy stored in the bands when a 50-g stone is placed in the cup and pulled back 0.20 m from the equilibrium position? **(b)** With what speed does the stone leave the slingshot?

6. An archer pulls her bow string back 0.400 m by exerting a force that increases uniformly from zero to 230 N. **(a)** What is the equivalent spring constant of the bow? **(b)** How much work is done in pulling the bow?

WEB **7.** A child's toy consists of a piece of plastic attached to a spring (Fig. P13.7). The spring is compressed against the floor a distance of 2.00 cm, and the toy is released. If the toy has a mass of 100 g and rises to a maximum height of 60.0 cm, estimate the force constant of the spring.

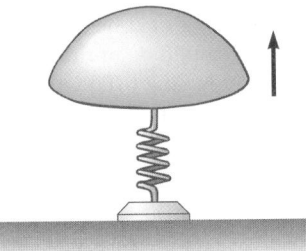

Figure P13.7

8. A child's pogo stick (Fig. P13.8) stores energy in a spring ($k = 2.50 \times 10^4$ N/m). At position *A* ($x_1 = -0.100$ m) the spring compression is a maximum, and the child momentarily stops. At position *B* ($x = 0$) the spring is relaxed, and the child is moving upward. At position *C* the spring remains relaxed and the child again stops at the top of the jump. Assume that the com-

bined mass of the child and pogo stick is 25.0 kg. **(a)** Calculate the total energy of the system if both the spring and the gravitational potential energies are zero at $x = 0$. **(b)** Determine x_2. **(c)** Calculate the speed of the child at $x = 0$. **(d)** Determine the acceleration of the child at $x = x_1$.

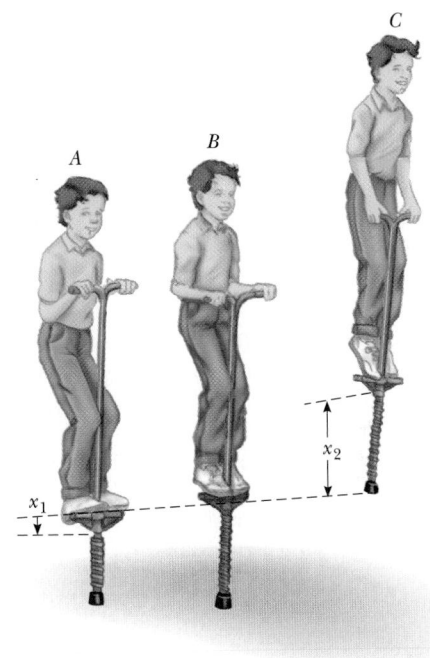

Figure P13.8

9. The spring constant of the spring in Figure P13.9 is 19.6 N/m, and the mass of the object is 1.5 kg. The spring is unstretched and the surface is frictionless. A constant 20-N force is applied horizontally to the object as shown. Find the speed of the object after it has moved a distance of 0.30 m.

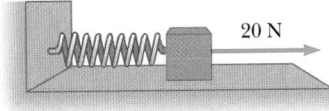

Figure P13.9

10. A 10.0-g bullet is fired into and embeds in a 2.00-kg block attached to a spring with a spring constant of 19.6 N/m and whose mass is negligible. How far is the spring compressed if the bullet has a speed of 300 m/s just before it strikes the block, and the block slides on a frictionless surface? (*Note:* You must use conservation of momentum in this problem. Why?)

11. A 1.5-kg block is attached to a spring with a spring constant of 2000 N/m. The spring is then stretched a distance of 0.30 cm and the block is released from rest. (a) Calculate the speed of the block as it passes through the equilibrium position if no friction is present. (b) Calculate the speed of the block as it passes through the equilibrium position if a constant frictional force of 2.0 N retards its motion. (c) What would be the strength of the frictional force if the block reached the equilibrium position the first time with zero velocity?

12. A simple harmonic oscillator has a total energy of E. (a) Determine the kinetic and potential energies when the displacement is one half the amplitude. (b) For what value of the displacement does the kinetic energy equal the potential energy?

Section 13.3   Velocity as a Function of Position

13. A mass of 0.40 kg connected to a light spring with a spring constant of 19.6 N/m oscillates on a frictionless horizontal surface. If the spring is compressed 4.0 cm and released from rest, determine (a) the maximum speed of the mass, (b) the speed of the mass when the spring is compressed 1.5 cm, and (c) the speed of the mass when the spring is stretched 1.5 cm. (d) For what value of x does the speed equal one half the maximum speed?

14. A mass-spring system oscillates with an amplitude of 3.5 cm. If the spring constant is 250 N/m and the mass is 0.50 kg, determine (a) the mechanical energy of the system, (b) the maximum speed of the mass, and (c) the maximum acceleration.

15. At an outdoor market, a bunch of bananas is set into oscillatory motion with an amplitude of 20.0 cm on a spring with a spring constant of 16.0 N/m. It is observed that the maximum speed of the bunch of bananas is 40.0 cm/s. What is the weight of the bananas in newtons?

16. A 50.0-g mass is attached to a horizontal spring with a spring constant of 10.0 N/m and released from rest with an amplitude of 25.0 cm. What is the velocity of the mass when it is halfway to the equilibrium position if the surface is frictionless?

Section 13.4   Comparing Simple Harmonic Motion with Uniform Circular Motion

17. An object moves uniformly around a circular path of radius 20.0 cm, making one complete revolution every 2.00 s. What is the (a) translational speed of the object, (b) the frequency of motion in hertz, and (c) the angular speed of the object?

18. Consider the simplified single-piston engine in Figure P13.18. If the wheel rotates at a constant angular speed of ω, explain why the piston rod oscillates in simple harmonic motion.

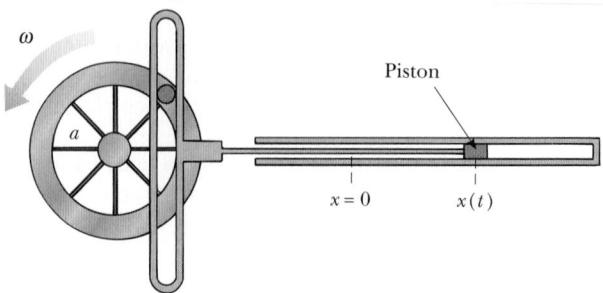

Figure P13.18

19. A 200-g mass is attached to a spring and executes simple harmonic motion with a period of 0.250 s. If the total energy of the system is 2.00 J, find (a) the force constant of the spring and (b) the amplitude of the motion.

20. The frequency of vibration of a mass-spring system is 5.00 Hz when a 4.00-g mass is attached to the spring. What is the force constant of the spring?

21. A spring stretches 3.9 cm when a 10-g mass is hung from it. If a total mass of 25 g attached to this spring oscillates in simple harmonic motion, calculate the period of motion.

22. When four people with a combined mass of 320 kg sit down in a car, they find that the car drops 0.80 cm lower on its springs. Then they get out of the car and bounce it up and down. What is the frequency of the car's vibration if its mass (empty) is 2.0×10^3 kg?

Section 13.5   Position, Velocity, and Acceleration as a Function of Time

23. The motion of an object is described by the equation

$$x = (0.30 \text{ m}) \cos(\pi t/3)$$

Find (a) the position of the object at $t = 0$ and $t = 0.60$ s, (b) the amplitude of the motion, (c) the frequency of the motion, and (d) the period of the motion.

24. A 2.00-kg mass on a frictionless horizontal track is attached to the end of a horizontal spring whose force constant is 5.00 N/m. The mass is displaced 3.00 m to the right from its equilibrium position and then released, which initiates simple harmonic motion. (a) What is the force (magnitude and direction) acting on the mass 3.50 s after it is released? (b) How many times does the mass oscillate in 3.50 s?

25. A spring of negligible mass stretches 3.00 cm from its relaxed length when a force of 7.50 N is applied. A 0.500-kg particle rests on a frictionless horizontal surface and is attached to the free end of the spring. The particle is pulled horizontally so that it stretches the spring 5.00 cm and is then released from rest at $t = 0$. (a) What is the force constant of the spring? (b) What are the period, frequency, and angular frequency (ω) of the mo-

tion? **(c)** What is the total energy of the system? **(d)** What is the amplitude of the motion? **(e)** What are the maximum velocity and the maximum acceleration of the particle? **(f)** Determine the displacement, x, of the particle from the equilibrium position at $t = 0.500$ s.

26. Given that $x = A \cos(\omega t)$ is a sinusoidal function of time, show that v (velocity) and a (acceleration) are also sinusoidal functions of time. (*Hint:* Use Equations 13.6 and 13.2.)

Section 13.6 Motion of a Pendulum

27. A simple 2.00-m-long pendulum oscillates in a location where $g = 9.80$ m/s². How many complete oscillations does it make in 5.00 min?

28. A visitor to a lighthouse wishes to determine the height of the tower. She ties a spool of thread to a small rock to make a simple pendulum, which she hangs down the center of a spiral staircase of the tower. The period of oscillation is 9.40 s. What is the height of the tower?

29. An aluminum clock pendulum having a period of 1.00 s keeps perfect time at 20.0°C. **(a)** When placed in a room at a temperature of -5.0°C, will it gain time or lose time? **(b)** How much time will it gain or lose every hour? (*Hint:* See Chapter 10.)

30. A pendulum clock that works perfectly on Earth is taken to the Moon. **(a)** Does it run fast or slow there? **(b)** If the clock is started at 12:00:00 A.M., what will it read after one Earth day (24.0 h)? Assume that the free-fall acceleration on the Moon is 1.63 m/s².

WEB 31. The acceleration due to gravity on Mars is 3.7 m/s². **(a)** What length pendulum has a period of one second on Earth? What length pendulum would have a one second period on Mars? **(b)** A mass is suspended from a spring with force constant 10 N/m. Find the mass suspended from this spring that would result in a one second period on Earth and on Mars.

Section 13.10 Frequency, Amplitude, and Wavelength

32. A wave traveling in the positive x direction has a frequency of 25.0 Hz, as in Figure P13.32. Find the **(a)** amplitude, **(b)** wavelength, **(c)** period, and **(d)** speed of the wave.

33. A bat can detect small objects such as an insect whose size is approximately equal to one wavelength of the sound the bat makes. If bats emit a chirp at a frequency of 60.0 kHz and if the speed of sound in air is 340 m/s, what is the smallest insect a bat can detect?

34. If the frequency of oscillation of the wave emitted by an FM radio station is 88.0 MHz, determine the wave's **(a)** period of vibration and **(b)** wavelength. (*Hint:* Radio waves travel at the speed of light, 3.00×10^8 m/s.)

35. The distance between two successive maxima of a transverse wave is 1.20 m. Eight crests, or maxima, pass a given point along the direction of travel every 12.0 s. Calculate the wave speed.

36. A piano emits frequencies that range from a low of about 28 Hz to a high of about 4200 Hz. Find the range of wavelengths spanned by this instrument. The speed of sound in air is approximately 343 m/s.

37. A harmonic wave is traveling along a rope. It is observed that the oscillator that generates the wave completes 40.0 vibrations in 30.0 s. Also, a given maximum travels 425 cm along the rope in 10.0 s. What is the wavelength?

38. Ocean waves are traveling to the east at 4 m/s with a distance of 20 m between crests. With what frequency do the waves hit the front of a boat **(a)** when the boat is at anchor and **(b)** when the boat is moving westward at 1 m/s?

Section 13.11 The Speed of Waves on Strings

39. A circus performer stretches a tightrope between two towers. He strikes one end of the rope and sends a wave along it toward the other tower. He notes that it takes the wave 0.800 s to reach the opposite tower, 20.0 m away. If one meter of the rope has a mass of 0.350 kg, find the tension in the tightrope.

40. Transverse waves with a speed of 50.0 m/s are to be produced on a stretched string. A 5.00-m length of string with a total mass of 0.0600 kg is used. **(a)** What is the required tension in the string? **(b)** Calculate the wave speed in the string if the tension is 8.00 N.

WEB 41. A string is 50.0 cm long and has a mass of 3.00 g. A wave travels at 5.00 m/s along this string. A second string has the same length but half the mass of the first. If the two strings are under the same tension, what is the speed of a wave along the second string?

42. Tension is maintained in a string, as in Figure P13.42.

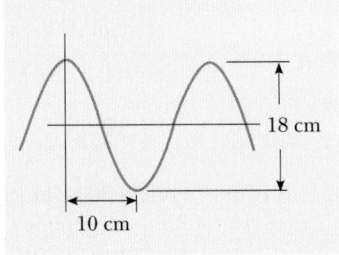

18 cm

10 cm

Figure P13.32

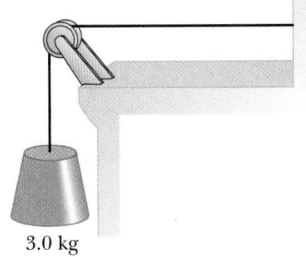

3.0 kg

Figure P13.42

Problems 33 and 36 are the same as Workbook Problems 50 and 51, respectively (Workbook pages 109 and 110).

The observed wave speed is 24 m/s when the suspended mass is 3.0 kg. (a) What is the mass per unit length of the string? (b) What is the wave speed when the suspended mass is 2.0 kg?

43. Transverse waves travel at 20.0 m/s on a string that is under a tension of 6.00 N. What tension is required for a wave speed of 30.0 m/s in the same string?

Section 13.12   Superposition and Interference of Waves

Section 13.13   Reflection of Waves

44. A series of pulses of amplitude 0.15 m are sent down a string that is attached to a post at one end. The pulses are reflected at the post and travel back along the string without loss of amplitude. What is the amplitude at a point on the string where two pulses are crossing, (a) if the string is rigidly attached to the post? (b) if the end at which reflection occurs is free to slide up and down?

45. A wave of amplitude 0.30 m interferes with a second wave of amplitude 0.20 m traveling in the same direction. What are the (a) largest and (b) smallest resultant amplitudes that can occur, and under what conditions will these maxima and minima occur?

ADDITIONAL PROBLEMS

46. The position of a 0.30-kg object attached to a spring is described by

$$x = (0.25 \text{ m}) \cos(0.4\pi t)$$

Find (a) the amplitude of the motion, (b) the spring constant, (c) the position at $t = 0.30$ s, and (d) the object's velocity at $t = 0.30$ s.

47. A spring with a spring constant of 30.0 N/m is stretched 0.200 m from its equilibrium position. How much work must be done to stretch it an additional 0.100 m?

48. A 500-g block is released from rest and slides down a frictionless track that begins 2.00 m above the horizontal, as shown in Figure P13.48. At the bottom of the track, where the surface is horizontal, the block strikes and sticks to a light spring with a spring constant of 20.0 N/m. Find the maximum distance the spring is compressed.

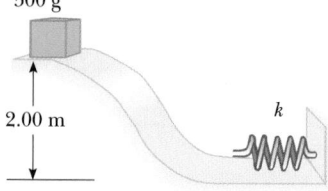

Figure P13.48

49. A 3.00-kg mass is fastened to a light spring that passes over a pulley (Fig. P13.49). The pulley is frictionless, and its inertia may be neglected. The mass is released from rest when the spring is unstretched. If the mass drops 10.0 cm before stopping, find (a) the spring constant of the spring and (b) the speed of the mass when it is 5.00 cm below its starting point.

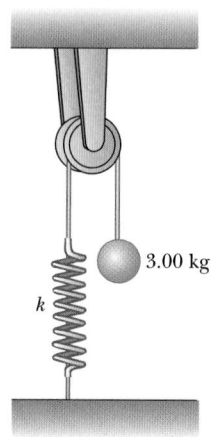

Figure P13.49

50. A 5.00-g bullet moving with an initial speed of 400 m/s is fired into and passes through a 1.00-kg block, as in Figure P13.50. The block, initially at rest on a frictionless horizontal surface, is connected to a spring with a spring constant of 900 N/m. If the block moves 5.00 cm to the right after impact, find (a) the speed at which the bullet emerges from the block and (b) the energy lost in the collision.

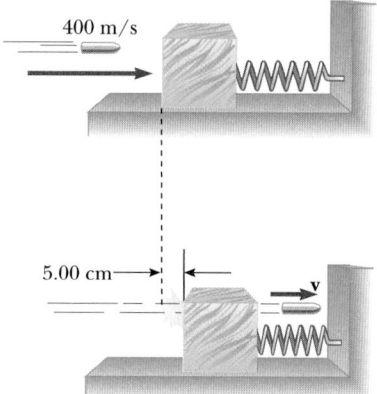

Figure P13.50

51. A 25-kg block is connected to a 30-kg block by a light string that passes over a frictionless pulley. The 30-kg block is connected to a light spring of force constant

200 N/m, as in Figure P13.51. The spring is unstretched when the system is as shown in the figure, and the incline is smooth. The 25-kg block is pulled 20 cm down the incline (so that the 30-kg block is 40 cm above the floor) and is released from rest. Find the speed of each block when the 30-kg block is 20 cm above the floor (that is, when the spring is unstretched).

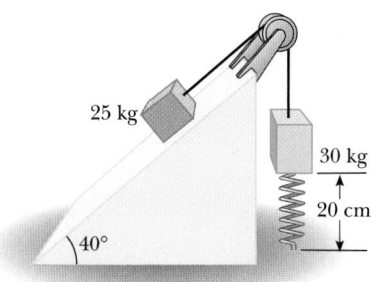

Figure P13.51

52. A spring in a toy gun has a spring constant of 9.80 N/m and can be compressed 20.0 cm beyond the equilibrium position. A 1.00-g pellet resting against the spring is propelled forward when the spring is released. (a) Find the muzzle speed of the pellet. (b) If the pellet is fired horizontally from a height of 1.00 m above the floor, what is its range?

53. A 2.0-kg block situated on a rough incline is connected to a light spring with a spring constant of 100 N/m (Fig. P13.53). The block is released from rest when the spring is unstretched, and the pulley is frictionless. The block moves 20 cm down the incline before stopping. Find the coefficient of kinetic friction between the block and the incline.

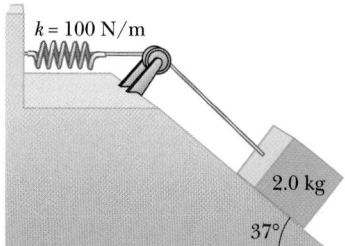

Figure P13.53

54. A 60.0-kg fire fighter slides down a pole while a constant frictional force of 300 N retards his motion. A horizontal 20.0-kg platform is supported by a spring at the bottom of the pole to cushion the fall. The fire fighter starts from rest 5.00 m above the platform, and the spring constant is 2500 N/m. Find (a) the fire fighter's speed just before he collides with the platform and

(b) the maximum distance the spring is compressed. Assume that the frictional force acts during the entire motion. (*Hint:* The collision between the fire fighter and the platform is perfectly inelastic.)

55. A mass, $m_1 = 9.0$ kg, is in equilibrium while connected to a light spring of constant $k = 100$ N/m that is fastened to a wall as in Figure P13.55a. A second mass, $m_2 = 7.0$ kg, is slowly pushed up against mass m_1, compressing the spring by the amount $A = 0.20$ m, as shown in Figure P13.55b. The system is then released, causing both masses to start moving to the right on the frictionless surface. (a) When m_1 reaches the equilibrium point, m_2 loses contact with m_1 (Fig. P13.55c) and moves to the right with speed v. Determine the value of v. (b) How far apart are the masses when the spring is fully stretched for the first time (Fig. P13.55d)? (*Hint:* First determine the period of oscillation and the amplitude of the m_1-spring system after m_2 loses contact with m_1.)

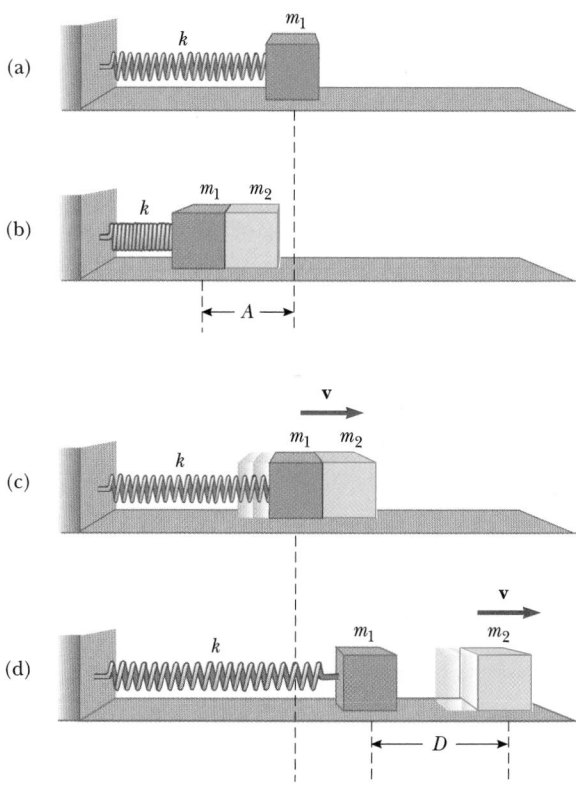

Figure P13.55

56. An astronaut on the Moon wishes to measure the local value of the free-fall acceleration g_{Moon} by timing pulses traveling down a wire that has a large mass suspended from it. Assume a wire of mass 4.00 g is 1.60 m long and has a 3.00 kg mass suspended from it. A pulse requires 36.1 ms to traverse the length of the wire. Calculate

g_{Moon} from these data. (You may neglect the mass of the wire when calculating the tension in it.)

57. A 2.00-kg block hangs without vibrating at the end of a spring ($k = 500$ N/m) that is attached to the ceiling of an elevator car. The car is rising with an upward acceleration of $g/3$ when the acceleration suddenly ceases (at $t = 0$). (a) What is the angular frequency of oscillation of the block after the acceleration ceases? (b) By what amount is the spring stretched during the time that the elevator car is accelerating?

58. A mass m is connected to two rubber bands of length L, each under tension F, as in Figure P13.58. The mass is displaced vertically by a small distance, y. Assuming the tension does not change, show that (a) the restoring force is $-(2F/L)y$ and (b) the system exhibits simple harmonic motion with an angular frequency $\omega = \sqrt{2F/mL}$.

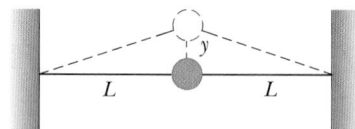

Figure P13.58

59. A light balloon filled with helium of density 0.180 kg/m³ is tied to a light string of length $L = 3.00$ m. The string is tied to the ground, forming an "inverted" simple pendulum (Fig. P13.59a). If the balloon is displaced slightly from equilibrium, as in Figure P13.59b, show that the motion is simple harmonic, and determine the period of the motion. Take the density of air to be 1.29 kg/m³. (*Hint:* Use an analogy with the simple pendulum discussed in the text, and see Chapter 9.)

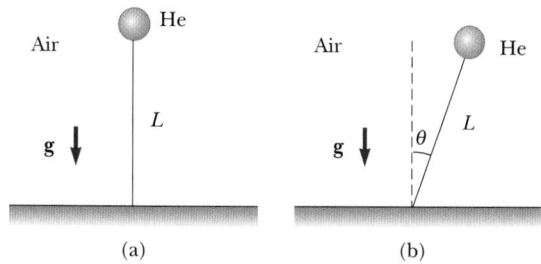

Figure P13.59

60. Two points, A and B, on the Earth are at the same longitude and 60.0° apart in latitude. An earthquake at point A sends two waves toward B. A transverse wave travels along the surface of the Earth at 4.50 km/s, and a longitudinal wave travels through the body of the Earth at 7.80 km/s. (a) Which wave arrives at B first? (b) What is the time difference between the arrivals of the two waves at B? Take the radius of the Earth to be 6.37×10^6 m.

WEB 61. A light string of mass 10.0 g and length $L = 3.00$ m has its ends tied to two walls that are separated by the distance $D = 2.00$ m. Two masses, each of mass $m = 2.00$ kg, are suspended from the string, as in Figure P13.61. If a wave pulse is sent from point A, how long does it take to travel to point B?

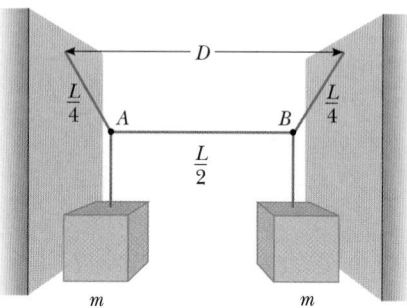

Figure P13.61

62. Figure P13.62 shows a crude model of an insect wing. The mass, m, represents the entire mass of the wing, which pivots about the fulcrum, F. The spring represents the surrounding connective tissue. Motion of the wing corresponds to vibration of the spring. Suppose the mass of the wing is 0.30 g and the effective spring constant of the tissue is 4.7×10^{-4} N/m. If the mass m moves up and down a distance of 2.0 mm from its position of equilibrium, what is the maximum speed of the outer tip of the wing?

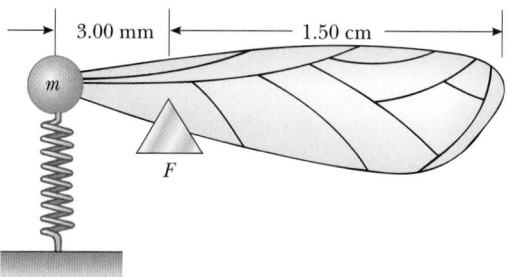

Figure P13.62

63. Assume that a hole is drilled through the center of the Earth. It can be shown that an object of mass m at a distance of r from the center of the Earth is pulled toward the center of the Earth only by the mass in the shaded portion of Figure P13.63. Write down Newton's law of gravitation for an object at distance r from the center of

the Earth, and show that the force on it is of Hooke's law form, $F = -kr$, with an effective force constant of $k = (4/3)\pi\rho Gm$, where ρ is the density of the Earth and G is the gravitational constant.

Earth

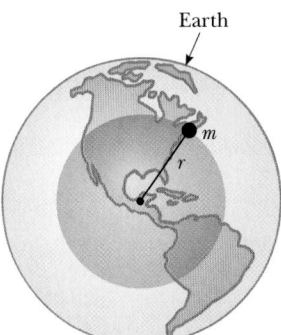

Figure P13.63

64. An 8.00-kg block travels on a rough horizontal surface and collides with a spring. The speed of the block just before the collision is 4.00 m/s. As it rebounds to the left with the spring uncompressed, the block travels at 3.00 m/s. If the coefficient of kinetic friction between the block and the surface is 0.400, determine (a) the loss in mechanical energy due to friction while the block is in contact with the spring and (b) the maximum distance the spring is compressed.

Sound

Chapter Outline

▲ PHYSICS PUZZLER

You may be familiar with a technique to determine how far away a lightning strike is. You count the seconds between the time when you see the lightning and when you hear the thunder. You divide that time interval by five, which gives you the distance to the lightning in miles. Why does this calculation give you this distance? *(Keith Kent/Peter Arnold, Inc.)*

Sound waves are the most important example of longitudinal waves. In this chapter we discuss the characteristics of sound waves—how they are produced, what they are, and how they travel through matter. We then investigate what happens when sound waves interfere with each other. The insights gained in this chapter will help you understand how we hear.

14.1 PRODUCING A SOUND WAVE

Whether it conveys the shrill whine of a jet engine or the soft melodies of a pop singer, any sound wave has its source in a vibrating object. Musical instruments produce sounds in a variety of ways. For example, the sound from a clarinet is produced by a vibrating reed, the sound from a drum by the vibration of the taut drum head, the sound from a piano by vibrating strings, and the sound from a singer by vibrating vocal folds.

Sound waves are longitudinal waves traveling through a medium, such as air. In order to investigate how sound waves are produced, we focus our attention on the tuning fork, a common device for producing pure musical notes. A tuning fork consists of two metal prongs, or tines, that vibrate when struck. Their vibration disturbs the air near them, as shown in Figure 14.1. (The amplitude of vibration of the tine in Figure 14.1 has been greatly exaggerated for clarity.) When a tine swings to the right, as in Figure 14.1a, the air molecules in front of its movement are forced closer together than normal. Such a region of high molecular density and high air pressure is called a **compression** or **condensation.** This compression moves away from the fork like a ripple on a pond. When the tine swings to the left, as in Figure 14.1b, the molecules to the right of the tine spread apart and the density and air pressure in this region are then lower than normal. Such a region of lower-than-normal density is called a **rarefaction.** Molecules to the right of the rarefaction in the figure move to the left. Hence, the rarefaction itself moves to the right, following the previously produced compression.

As the tuning fork continues to vibrate, a succession of condensations and rarefactions forms and spreads out from it. The resultant pattern in the air is somewhat like that pictured in Figure 14.2a. We can use a sinusoidal curve to represent a sound wave, as in Figure 14.2b. Notice that there are crests in the sinusoidal wave at the points where the sound wave has condensations and troughs where the sound

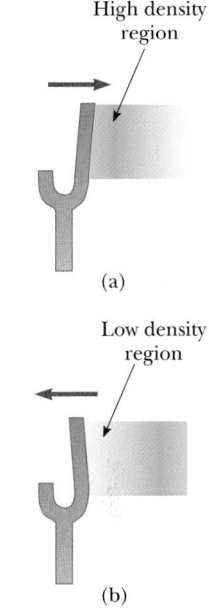

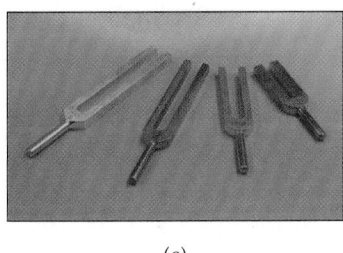

Figure 14.1 A vibrating tuning fork. (a) As the right tine of the fork moves to the right, a high-density region (condensation) of air is formed in front of its movement. (b) As the right tine moves to the left, a low-density region (rarefaction) of air is formed behind it. (c) A set of tuning forks. *(Courtesy of Henry Leap and Jim Lehman)*

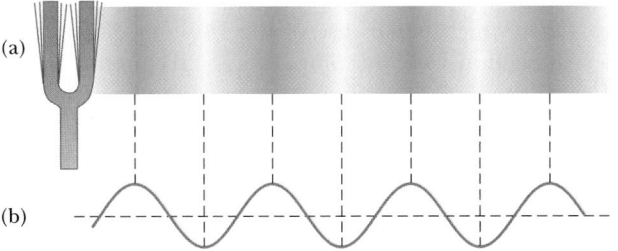

Figure 14.2 (a) As the tuning fork vibrates, a series of condensations and rarefactions moves outward, away from the fork. (b) The crests of the wave correspond to condensations, and the troughs correspond to rarefactions.

wave has rarefactions. The molecular motion of the sound waves is superposed on the random thermal motion of the atoms and molecules (discussed in Chapter 10).

14.2   CHARACTERISTICS OF SOUND WAVES

As already noted, the general motion of air molecules near a vibrating object is back and forth between regions of compression and rarefaction. Back-and-forth molecular motion in the direction of the disturbance is characteristic of a longitudinal wave. **The motion of the medium particles in a longitudinal sound wave is back and forth along the direction in which the wave travels.** In contrast, **in a transverse wave the vibrations of the medium are at right angles to the direction of travel of the wave.**

Categories of Sound Waves

Sound waves fall into three categories covering different ranges of frequencies. **Audible waves** are longitudinal waves that lie within the range of sensitivity of the human ear, approximately 20 Hz to 20 000 Hz. **Infrasonic waves** are longitudinal waves with frequencies below the audible range. Earthquake waves are an example. **Ultrasonic waves** are longitudinal waves with frequencies above the audible range for humans. For example, certain types of whistles produce ultrasonic waves. Some animals, such as dogs, can hear the waves emitted by these whistles, even though humans cannot.

Applications of Ultrasound

Ultrasonic waves are sound waves with frequencies from 20 kHz to 100 kHz, frequencies that are beyond the audible range. Because of their high frequency and corresponding short wavelengths, ultrasonic waves can be used to produce images of small objects and are currently in wide use in medical applications, both as a diagnostic tool and in certain treatments. Internal organs can be examined via the images produced by the reflection and absorption of ultrasonic waves. Although ultrasonic waves are far safer than x-rays, their images do not always have as much detail. Certain organs, however, such as the liver and the spleen, are invisible to x-rays but can be imaged with ultrasonic waves.

Medical workers can measure the speed of the blood flow in the body using a device called an *ultrasonic flow meter*, which makes use of the Doppler effect. The Doppler effect will be discussed in Section 14.6. By comparing the frequency of the waves scattered by the blood vessels with the incident frequency, one can obtain the flow speed.

Figure 14.3 illustrates the technique used to produce ultrasonic waves for clinical use. Electrical contacts are made to the opposite faces of a crystal, such as quartz or strontium titanate. If an alternating voltage of very high frequency is applied to these contacts, the crystal vibrates at the same frequency as the applied voltage, emitting a beam of ultrasonic waves. (At one time, this was how almost all headphones produced sound.) This method of transforming electrical energy into me-

APPLICATION

Medical Use of Ultrasound.

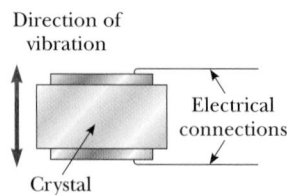

Figure 14.3   An alternating voltage applied to the faces of a piezoelectric crystal causes the crystal to vibrate.

chanical energy, called the **piezoelectric effect,** is also reversible: If some external source causes the crystal to vibrate, an alternating voltage is produced across the crystal. Hence, a single crystal can be used to both transmit and receive ultrasonic waves.

The primary physical principle that makes ultrasound imaging possible is the fact that a sound wave is partially reflected whenever it is incident on a boundary between two materials having different densities. If a sound wave is traveling in a material of density ρ_i, and strikes a material of density ρ_t, the percentage of the incident sound wave reflected, *PR*, is given by

$$PR = \left(\frac{\rho_i - \rho_t}{\rho_i + \rho_t}\right)^2 \times 100$$

This equation assumes that the incident sound wave travels perpendicularly to the boundary and that the speed of sound is approximately the same in the two materials. The latter assumption holds very well for the human body because the speed of sound does not vary much in the organs of the body.

Physicians commonly use ultrasonic waves to observe fetuses. This technique presents far less risk than x-rays, which can produce birth defects. First the abdomen of the mother is coated with a liquid, such as mineral oil. If this were not done, most of the incident ultrasonic waves from the piezoelectric source would be reflected at the boundary between the air and the mother's skin. Mineral oil has a density similar to that of skin, and a very small fraction of the incident ultrasonic wave is reflected when $\rho_i \approx \rho_t$. The ultrasound energy is emitted as pulses rather than as a continuous wave, so the same crystal can be used as a detector as well as a transmitter. The source-receiver is then passed over the mother's abdomen. The reflected sound waves picked up by the receiver are converted to an electric signal, which forms an image along a line on a fluorescent screen. The sound source is then moved a few centimeters on the mother's body, and the process is repeated. The reflected signal produces a second line on the fluorescent screen. In this fashion a complete scan of the fetus can be made. Difficulties such as the likelihood of spontaneous abortion or of breech birth are easily detected with this technique. Also, such fetal abnormalities as spina bifida and water on the brain are readily observable.

A relatively new medical application of ultrasonics is the cavitron ultrasonic surgical aspirator (CUSA). This device has made it possible to surgically remove brain tumors that were previously inoperable. It is a long needle that emits ultrasonic waves (about 23 kHz) at its tip. When the tip touches a tumor, the part of the tumor near the needle is shattered and the residue can be sucked up (aspirated) through the hollow needle.

Another interesting application of ultrasound is the ultrasonic ranging unit designed by the Polaroid Corporation and used in some of their cameras to provide an almost instantaneous measurement of the distance between the camera and object to be photographed. The principal component of this device is a crystal that acts as both a loudspeaker and a microphone. A pulse of ultrasonic waves is transmitted from the transducer to the object to be photographed. The object reflects part of the signal, producing an echo that is detected by the device. The time interval between the outgoing pulse and the detected echo is then electronically converted to a distance value, because the speed of sound is a known quantity.

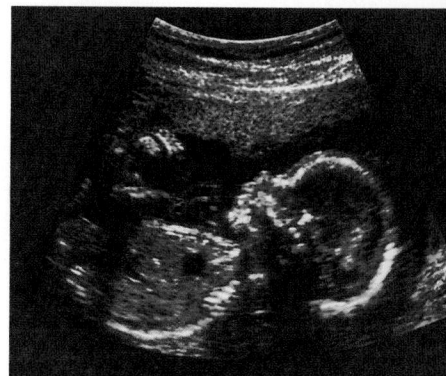

An ultrasound image of a human fetus in the womb after 20 weeks of development, showing the head, body, arms, and legs in profile. *(U.H.B. Trust/Tony Stone Images)*

APPLICATION

Ultrasonic Range Unit for Cameras.

14.3   THE SPEED OF SOUND

The speed of a sound wave in a liquid or gas depends on the medium's compressibility and inertia. If the fluid has a bulk modulus of B and an equilibrium density of ρ, the speed of sound is

◄ Speed of sound in a liquid

$$v = \sqrt{\frac{B}{\rho}} \qquad [14.1]$$

Recall from Chapter 9 that bulk modulus is defined as the ratio of the change in pressure, ΔP, to the resulting fractional change in volume, $\Delta V/V$:

9.4, BOTH SECTIONS

$$B \equiv -\frac{\Delta P}{\Delta V/V} \qquad [14.2]$$

Note that B is always positive because an increase in pressure (positive ΔP) results in a decrease in volume. Hence the ratio $\Delta P/\Delta V$ is always negative.

It is interesting to compare Equation 14.1 with Equation 13.17 for the speed of transverse waves on a string, $v = \sqrt{F/\mu}$, discussed in Chapter 13. In both cases, the wave speed depends on an elastic property of the medium (B or F) and on an inertial property of the medium (ρ or μ). In fact, the speed of all mechanical waves follows an expression of the general form

$$v = \sqrt{\frac{\text{elastic property}}{\text{inertial property}}}$$

Another example of this general form is the **speed of a longitudinal wave in a solid rod,** which is

$$v = \sqrt{\frac{Y}{\rho}} \qquad [14.3]$$

where Y is the Young's modulus of the solid, defined as the longitudinal stress divided by the longitudinal strain (Equation 9.3), and ρ is the density of the solid.

Table 14.1 lists the speeds of sound in various media. As you can see, the speed of sound is much higher in solids than in gases. This makes sense because the molecules in a solid are closer together than those in a gas and hence respond more rapidly to a disturbance. In general, sound travels more slowly in liquids than in solids because liquids are more compressible and hence have smaller bulk moduli. To show that sound travels faster in a solid than in air, suppose that a student places her ear against a long metal object, such as a long steel rail, and listens as her distant boyfriend strikes the rail with a hammer. The student will hear two sounds. The first is produced by the rapidly moving sound traveling through the rail, and the second is the sound taking the slower path through the air.

The speed of sound also depends on the temperature of the medium. For sound traveling through air, the relationship between the speed of sound and temperature is

$$v = (331 \text{ m/s})\sqrt{1 + \frac{T}{273}} \qquad [14.4]$$

TABLE 14.1
Speeds of Sound in Various Media

Medium	v (m/s)
Gases	
Air (0°C)	331
Air (100°C)	386
Hydrogen (0°C)	1290
Oxygen (0°C)	317
Helium (0°C)	972
Liquids at 25°C	
Water	1490
Methyl alcohol	1140
Seawater	1530
Solids	
Aluminum	5100
Copper	3560
Iron	5130
Lead	1320
Vulcanized rubber	54

where 331 m/s is the speed of sound in air at 0°C and T is the temperature in degrees Celsius. Using this equation, one finds that at 20°C the speed of sound in air is approximately 343 m/s.

Thinking Physics 1

Why does thunder produce an extended "rolling" sound? And how does lightning produce thunder in the first place?

Explanation   Let us assume that we are at ground level and neglect ground reflections. When lightning strikes, a channel of ionized air carries a very large electric current from a cloud to the ground. This results in a rapid temperature increase of this channel of air as the current moves through it. The temperature increase causes a sudden expansion of the air. This expansion is so sudden and so intense that a tremendous disturbance is produced in the air—thunder. The thunder rolls due to the fact that the lightning channel is a long extended source—the entire length of the channel produces the sound at essentially the same instant of time. Sound produced at the bottom of the channel reaches you first, if you are on the ground, because that is the point closest to you, and then sounds from progressively higher portions of the channel reach you. If the lightning channel were a perfectly straight line, the resulting sound might be a steady roar, but the zig-zagged shape of the path results in the rolling variation in loudness, with sound from some portions of the channel arriving at your location simultaneously. The result is a particularly loud sound, interspersed with other instants when the net sound reaching you is low in intensity.

EXAMPLE 14.1   Speed of Sound in a Liquid

Find the speed of sound in water, which has a bulk modulus of about 2.1×10^9 Pa and a density of about 1.0×10^3 kg/m^3.

Solution   From Equation 14.1, we find that

$$v_{\text{water}} = \sqrt{\frac{B}{\rho}} = \sqrt{\frac{2.1 \times 10^9 \text{ Pa}}{1.0 \times 10^3 \text{ kg/m}^3}} \approx 1500 \text{ m/s}$$

EXAMPLE 14.2   Sound Waves in a Solid Bar

If a solid bar is struck at one end with a hammer, a longitudinal pulse propagates down the bar. Find the speed of sound in a bar of aluminum, which has a Young's modulus of 7.0×10^{10} Pa and a density of 2.7×10^3 kg/m^3.

Solution   From Equation 14.3 we find that

$$v_{\text{Al}} = \sqrt{\frac{Y}{\rho}} = \sqrt{\frac{7.0 \times 10^{10} \text{ Pa}}{2.7 \times 10^3 \text{ kg/m}^3}} \approx 5100 \text{ m/s}$$

This is a typical value for the speed of sound in solids (see Table 14.1).

9.5

14.4   ENERGY AND INTENSITY OF SOUND WAVES

As the tines of a tuning fork move back and forth through the air, they exert a force on a layer of air and cause it to move. In other words, the tines do work on the layer of air. The fact that the fork pours energy into the air as sound is one of the reasons that the vibration of the fork slowly dies out. (Other factors, such as the energy lost to friction as the tines bend, also are responsible for the diminution of movement.)

> We define the **intensity, I,** of a wave to be the rate at which energy flows through a unit area, A, perpendicularly to the direction of travel of the wave.

In equation form this is

$$I \equiv \frac{1}{A}\frac{\Delta E}{\Delta t} \qquad\qquad [14.5]$$

Equation 14.5 can be written in an alternative form if you recall that the rate of transfer of energy is defined as power. Thus,

◀ Intensity of a wave

$$I \equiv \frac{\text{power}}{\text{area}} = \frac{P}{A} \qquad\qquad [14.6]$$

where P is the sound power passing through A, measured in watts, and the intensity has units of watts per square meter.

The faintest sounds the human ear can detect at a frequency of 1000 Hz have an intensity of about 1×10^{-12} W/m². This intensity is called the **threshold of hearing.** The loudest sounds the ear can tolerate have an intensity of about 1 W/m² (the **threshold of pain**). At the threshold of hearing, the increase in pressure in the ear is approximately 3×10^{-5} Pa over normal atmospheric pressure. Because atmospheric pressure is about 1×10^{5} Pa, this means the ear can detect pressure fluctuations as small as about 3 parts in 10^{10}! Also, at the threshold of hearing, the maximum displacement of an air molecule is about 1×10^{-11} m. This is a remarkably small number! If we compare this result with the diameter of a molecule (about 10^{-10} m), we see that the ear is an extremely sensitive detector of sound waves.

In a similar manner, one finds that the loudest sounds the human ear can tolerate correspond, at 1000 Hz, to a pressure variation of about 29 Pa away from normal atmospheric pressure, with a maximum displacement of air molecules of 1×10^{-5} m.

Intensity Levels in Decibels

As was just mentioned, the human ear can detect a wide range of intensities, with the loudest tolerable sounds having intensities about 1.0×10^{12} times greater than those of the faintest detectable sounds. However, the most intense sound is not perceived as being 1.0×10^{12} times louder than the faintest sound. This is because the sensation of loudness is approximately logarithmic in the human ear. The relative intensity of a sound is called the **intensity level** or **decibel level, β,** and is defined as

$$\beta \equiv 10 \log \left(\frac{I}{I_0} \right)$$ [14.7]   ◀ Intensity level

The constant I_0 is the reference intensity, taken to be the sound intensity at the threshold of normal acute hearing ($I_0 = 1.0 \times 10^{-12}$ W/m^2), I is any intensity, and β is the corresponding intensity level, measured in decibels (dB). (The word *decibel* comes from the name of the inventor of the telephone, Alexander Graham Bell, 1847–1922.) On this scale, the threshold of pain ($I = 1.0$ W/m^2) corresponds to an intensity level of $\beta = 10 \log(1/1 \times 10^{-12}) = 10 \log(10^{12}) = 120$ dB. Nearby jet airplanes can create intensity levels of 150 dB, and subways and riveting machines have levels of 90 to 100 dB. The electronically amplified sound heard at rock concerts can be at levels of up to 120 dB, the threshold of pain. Exposure to such high intensity levels can seriously damage the ear. Earplugs are recommended whenever prolonged intensity levels exceed 90 dB. Recent evidence suggests that noise pollution, which is common in most large cities and in some industrial environments, may be a contributing factor to high blood pressure, anxiety, and nervousness. Table 14.2 gives some idea of the intensity levels of various sounds.

TABLE 14.2
Intensity Levels in Decibels for Different Sources

Source of Sound	β (dB)
Nearby jet airplane	150
Jackhammer, machine gun	130
Siren, rock concert	120
Subway, power mower	100
Busy traffic	80
Vacuum cleaner	70
Normal conversation	50
Mosquito buzzing	40
Whisper	30
Rustling leaves	10
Threshold of hearing	0

EXAMPLE 14.3 Intensity Levels of Sound

Calculate the intensity level of a sound wave having an intensity of (a) 1.0×10^{-12} W/m^2; (b) 1.0×10^{-11} W/m^2; (c) 1.0×10^{-10} W/m^2.

Solution   (a) For an intensity of 1.0×10^{-12} W/m^2 the intensity level, in decibels, is

$$\beta = 10 \log \left(\frac{1.0 \times 10^{-12} \text{ W/m}^2}{1.0 \times 10^{-12} \text{ W/m}^2} \right) = 10 \log(1) = \boxed{0 \text{ dB}}$$

This answer should have been obvious without calculation, because an intensity of 1.0×10^{-12} W/m^2 corresponds to the threshold of hearing.

(b) In this case, the intensity is exactly ten times that in part (a). The intensity level is

$$\beta = 10 \log \left(\frac{1.0 \times 10^{-11} \text{ W/m}^2}{1.0 \times 10^{-12} \text{ W/m}^2} \right) = 10 \log(10) = \boxed{10 \text{ dB}}$$

(c) Here the intensity is 100 times greater than at the threshold of hearing, and the intensity level is

$$\beta = 10 \log \left(\frac{1.0 \times 10^{-10} \text{ W/m}^2}{1.0 \times 10^{-12} \text{ W/m}^2} \right) = 10 \log(100) = \boxed{20 \text{ dB}}$$

Note the pattern in these answers. A sound with an intensity level of 10 dB is ten times more intense than the 0-dB sound, and a sound with an intensity level of 20 dB is 100 times more intense than a 0-dB sound. This pattern is continued throughout the decibel scale. In short, on the decibel scale **an increase of 10 dB means that the intensity of the sound is multiplied by a factor of 10.** For example, a 50-dB sound is 10 times as intense as a 40-dB sound and a 60-dB sound is 100 times as intense as a 40-dB sound.

Exercise   Determine the intensity level of a sound wave with an intensity of 5.0×10^{-7} W/m^2.

Answer   57 dB

EXAMPLE 14.4   A Noisy Grinding Machine

A noisy grinding machine in a factory produces a sound intensity of 1.0×10^{-5} W/m². Find the decibel level of this machine, and calculate the new intensity level when a second, identical machine is added to the factory.

Solution   The intensity level of the single grinder is

$$\beta = 10 \log \left(\frac{1.0 \times 10^{-5} \text{ W/m}^2}{1.0 \times 10^{-12} \text{ W/m}^2} \right) = 10 \log(10^7) = \boxed{70 \text{ dB}}$$

Adding the second grinder doubles the energy input into sound and hence doubles the intensity. The new intensity level is

$$\beta = 10 \log \left(\frac{2.0 \times 10^{-5} \text{ W/m}^2}{1.0 \times 10^{-12} \text{ W/m}^2} \right) = \boxed{73 \text{ dB}}$$

APPLICATION

OSHA Noise Level
Regulations.

Federal regulations now demand that no office or factory worker be exposed to noise levels that average more than 90 dB over an 8-hour day. The results in this example read like one of the old jokes that start "There is some good news and some bad news." First the good news. Imagine that you are a manager analyzing the noise conditions in your factory. One machine in the factory produces a noise level of 70 dB. When you add a second machine, the noise level increases by only 3 dB. Because of the logarithmic nature of intensity levels, doubling the intensity does not double the intensity level; in fact, it alters it by a surprisingly small amount. This means that additional equipment can be added to the factory without appreciably altering the intensity level of the environment.

Now the bad news. The results also work in reverse. As you remove noisy machinery, the intensity level is not lowered appreciably. For example, consider a factory with 60 machines producing a noise level of 93 dB, which is 3 dB above the maximum allowed. In order to reduce the noise level by 3 dB, half the machines would have to be removed! That is, you would have to remove 30 machines to reduce the noise level to 90 dB. To reduce the level another 3 dB, you would have to remove half of the remaining machines, and so on.

14.5   SPHERICAL AND PLANE WAVES

If a small spherical object oscillates so that its radius changes periodically with time, a spherical sound wave is produced (Fig. 14.4). The wave moves outward from the source at a constant speed.

Because all points on the vibrating sphere behave in the same way, we conclude that the energy in a spherical wave propagates equally in all directions. That is, no one direction is preferred over any other. If P_{av} is the average power emitted by the source, then at any distance r from the source, this power must be distributed over a spherical surface of area $4\pi r^2$, assuming no absorption in the medium. (Recall that $4\pi r^2$ is the surface area of a sphere.) Hence, the **intensity** of the sound at a distance of r from the source is

$$I = \frac{\text{average power}}{\text{area}} = \frac{P_{av}}{A} = \frac{P_{av}}{4\pi r^2} \qquad \text{[14.8]}$$

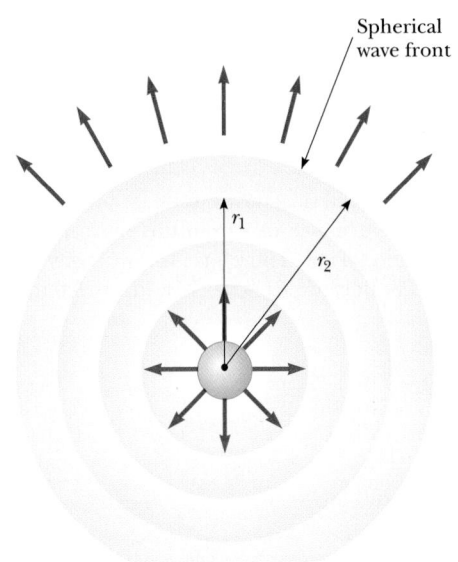

Figure 14.4 A spherical wave propagating radially outward from an oscillating sphere. The intensity of the spherical wave varies as $1/r^2$.

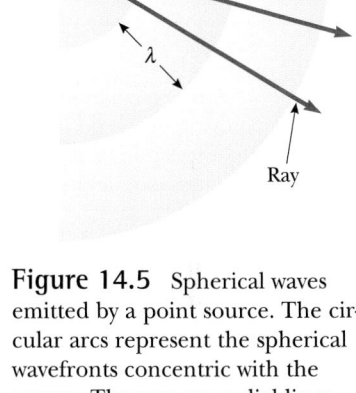

Figure 14.5 Spherical waves emitted by a point source. The circular arcs represent the spherical wavefronts concentric with the source. The rays are radial lines pointing outward from the source, perpendicular to the wavefronts.

This shows that the intensity of a wave decreases with increasing distance from its source, as you might expect. The fact that I varies as $1/r^2$ is a result of the assumption that the small source (sometimes called a **point source**) emits a spherical wave. (In fact, light waves also obey this so-called inverse square relationship.) Because the average power is the same through any spherical surface centered at the source, we see that the intensities at distances r_1 and r_2 (Fig. 14.4) from the center of the source are

$$I_1 = \frac{P_{av}}{4\pi r_1{}^2} \qquad I_2 = \frac{P_{av}}{4\pi r_2{}^2}$$

Therefore, the ratio of intensities at these two spherical surfaces is

$$\frac{I_1}{I_2} = \frac{r_2{}^2}{r_1{}^2}$$

It is useful to represent spherical waves graphically with a series of circular arcs (lines of maximum intensity) concentric with the source representing part of a spherical surface, as in Figure 14.5. We call such an arc a **wavefront.** The distance between adjacent wavefronts equals the wavelength, λ. The radial lines pointing outward from the source and cutting the arcs perpendicularly are called **rays.**

Now consider a small portion of a wavefront that is at a *great* distance (great relative to λ) from the source, as in Figure 14.6. In this case, the rays are nearly parallel to each other and the wavefronts are very close to being planes. Therefore, at distances from the source that are great relative to the wavelength, we can approximate the wavefront with parallel planes. We call such waves plane waves. Any small portion of a spherical wave that is far from the source can be considered a **plane wave.** Figure 14.7 illustrates a plane wave propagating along the x axis. If x is taken to be the direction of the wave motion (or ray) in this figure, then the wavefronts are parallel to the plane containing the y and z axes.

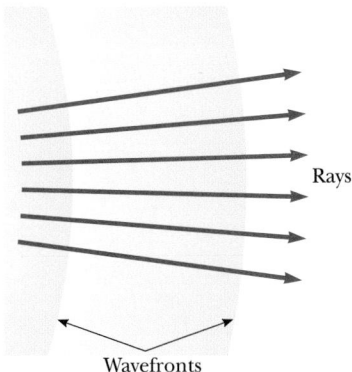

Figure 14.6 Far away from a point source, the wavefronts are nearly parallel planes and the rays are nearly parallel lines perpendicular to the planes. Hence, a small segment of a spherical wavefront is approximately a plane wave.

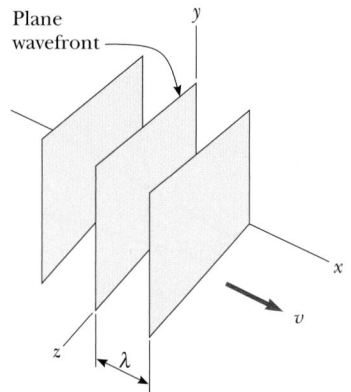

Plane
wavefront

Figure 14.7 A representation of a plane wave moving in the positive *x* direction with a speed of *v*. The wavefronts are planes parallel to the *yz* plane.

EXAMPLE 14.5   Intensity Variations of a Point Source

A small source emits sound waves with a power output of 80 W.
(a) Find the intensity 3.0 m from the source.

Solution   A small source emits energy in the form of spherical waves (see Fig. 14.4). Let P_{av} be the average power output of the source. At a distance of *r* from the source, the power is distributed over the surface area of a sphere, $4\pi r^2$. Therefore, the intensity at distance *r* from the source is given by Equation 14.8. Because $P_{av} = 80$ W and $r = 3.0$ m, we find that

$$I = \frac{P_{av}}{4\pi r^2} = \frac{80 \text{ W}}{4\pi (3.0 \text{ m})^2} = \boxed{0.71 \text{ W/m}^2}$$

which is close to the threshold of pain.

(b) Find the distance at which the sound level is 40 dB.

Solution   We can find the intensity at the 40-dB intensity level by using Equation 14.7 with $I_0 = 1.0 \times 10^{-12}$ W/m²:

$$40 = 10 \log(I/I_0)$$
$$4 = \log(I/I_0)$$
$$10^4 = I/I_0$$
$$I = 10^4 I_0 = 1.0 \times 10^{-8} \text{ W/m}^2$$

When this value for *I* is used in Equation 14.8, solving for *r* gives

$$r = \left(\frac{P_{av}}{4\pi I}\right)^{1/2} = \left(\frac{80 \text{ W}}{4\pi \times 10^{-8} \text{ W/m}^2}\right)^{1/2} = \boxed{2.5 \times 10^4 \text{ m}}$$

which is approximately 15 mi!

14.6   THE DOPPLER EFFECT

If a car or truck is moving while its horn is blowing, the frequency of the sound you hear is higher as the vehicle approaches you and lower as it moves away from you. This is one example of the Doppler effect, named for the Austrian physicist Christian Doppler (1803–1853), who discovered it.

In general, a **Doppler effect** is experienced whenever there is relative motion between a source of waves and an observer. When the source and observer are moving toward each other, the observer hears a frequency higher than the frequency of the source in the absence of relative motion. When the source and observer are moving away from each other, the observer hears a frequency lower than the source frequency.

Although the Doppler effect is most commonly experienced with sound waves, it is a phenomenon common to all waves. For example, the frequencies of light waves are also shifted by the relative motion of source and observer.

First let us consider the case in which the observer is moving and the sound source is stationary. For simplicity, we assume that the air is also stationary and that all velocity measurements are made relative to this stationary medium. Once a sound leaves its source, the speed of the wave is independent of the speed of the source. Figure 14.8 describes the situation in which the observer is moving with a speed of v_0 toward the source (considered a point source), which is at rest $(v_s = 0)$.

We shall take the frequency of the source to be f, the wavelength to be λ, and the speed of sound in air to be v. Clearly, if both observer and source were stationary, the observer would detect f wavefronts per second. (That is, when $v_0 = 0$ and $v_s = 0$, the observed frequency equals the source frequency.) When the observer is moving toward the source, he or she moves a distance of $v_0 t$ in t seconds. During this interval, **the observer detects an additional number of wavefronts.** The number of extra wavefronts detected is equal to the distance traveled, $v_0 t$, divided by the wavelength, λ. Thus,

$$\text{Additional wavefronts detected} = \frac{v_0 t}{\lambda}$$

The number of additional wavefronts detected *per second* is v_0/λ. Hence, the frequency f' heard by the observer is *increased* to

$$f' = f + \frac{v_0}{\lambda}$$

Using the fact that $\lambda = v/f$, we see that $v_0/\lambda = (v_0/v)f$. Hence, f' can be expressed as

$$f' = f\left(\frac{v + v_0}{v}\right) \tag{14.9}$$

An observer traveling *away* from the source, as in Figure 14.9, **detects fewer wavefronts per second.** Thus, it follows that the frequency heard by the observer in this case is *lowered* to

$$f' = f\left(\frac{v - v_0}{v}\right) \tag{14.10}$$

We can incorporate these two equations into one:

$$f' = f\left(\frac{v \pm v_0}{v}\right) \tag{14.11}$$

This general equation applies when an observer is moving with a speed of v_0 relative to a stationary source. **The positive sign is used when the observer is moving toward the source, and the negative sign is used when the observer is moving away from the source.**

Now consider the situation in which the source is in motion and the observer is at rest. If the source is moving directly toward observer A in Figure 14.10a, the

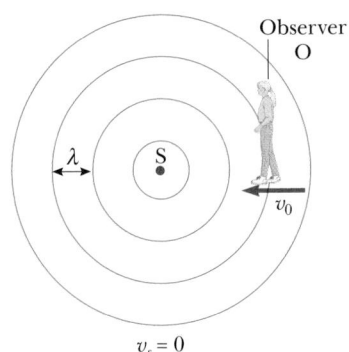

Figure 14.8 An observer moving with a speed of v_0 *toward* a stationary point source (S) hears a frequency, f', that is *greater* than the source frequency, f.

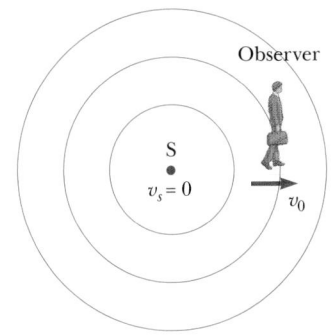

Figure 14.9 An observer moving with a speed of v_0 *away* from a stationary source hears a frequency, f', that is *lower* than the source frequency.

◄ Observed frequency — observer in motion

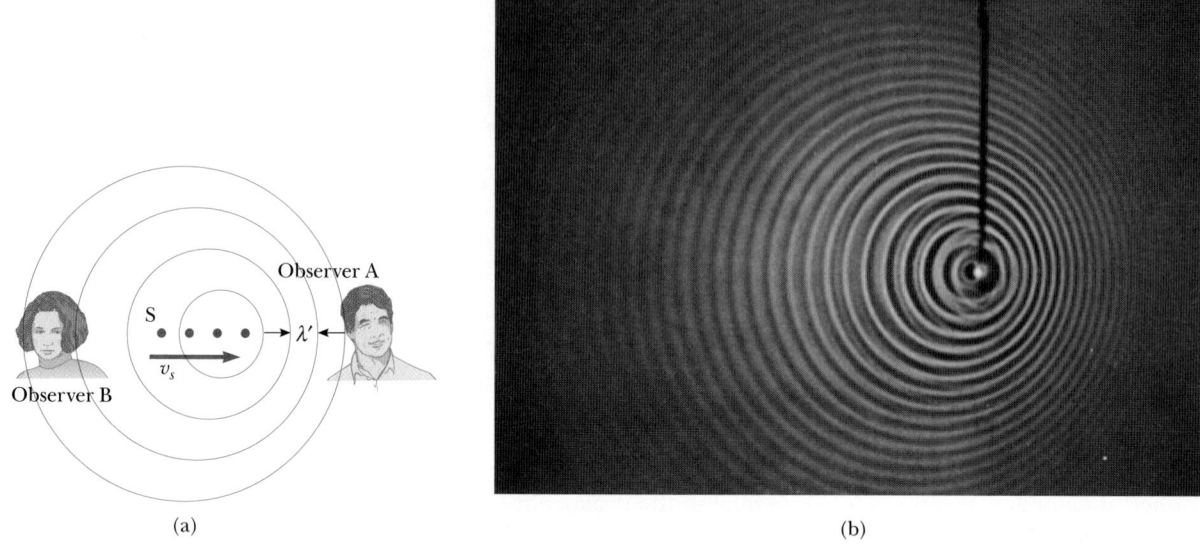

(a)   (b)

Figure 14.10 (a) A source, S, moving with speed v_s toward stationary observer A and away from stationary observer B. Observer A hears an *increased* frequency, and observer B hears a *decreased* frequency. (b) The Doppler effect in water, observed in a ripple tank. The source producing the water waves is moving to the right. *(Courtesy Educational Development Center, Newton, Mass.)*

wavefronts heard by A are closer together because the source is moving in the direction of the outgoing wave. As a result, the wavelength λ' measured by observer A is shorter than the true wavelength, λ, of the source. During each vibration, which lasts for an interval of T (the period), the source moves a distance of $v_s T = v_s/f$ and **the wavelength is shortened by this amount.** Therefore, the observed wavelength, λ', is given by

$$\lambda' = \lambda - \frac{v_s}{f}$$

Because $\lambda = v/f$, the frequency heard by observer A is

$$f' = \frac{v}{\lambda'} = \frac{v}{\lambda - \dfrac{v_s}{f}} = \frac{v}{\dfrac{v}{f} - \dfrac{v_s}{f}} = f\left(\frac{v}{v - v_s}\right) \qquad [14.12]$$

That is, **the observed frequency increases when the source is moving toward the observer.**

In Figure 14.10a the source is moving away from observer B, who is at rest to the left of the source. Thus, observer B measures a wavelength that is *greater* than λ and hears a *decreased* frequency of

$$f' = f\left(\frac{v}{v + v_s}\right) \qquad [14.13]$$

Combining Equations 14.12 and 14.13, we can express the general relationship for the observed frequency when the source is moving and the observer is at rest:

$$f' = f\left(\frac{v}{v \mp v_s}\right)$$ [14.14]

◀ Observed frequency — source in motion

The *negative* sign is used when the source is moving *toward* the observer, and the *positive* sign is used when the source is moving *away from* the observer.

Finally, if both the source and the observer are in motion, one finds the following general relationship for the observed frequency:

$$f' = f\left(\frac{v \pm v_0}{v \mp v_s}\right)$$ [14.15]

◀ Observed frequency — observer and source in motion

In this expression the *upper* signs ($+ v_0$ and $- v_s$) refer to motion of one *toward* the other, and the lower signs ($- v_0$ and $+ v_s$) refer to motion of one *away* from the other.

Thinking Physics 2

Suppose you place your stereo speakers far apart and run past them from right to left or left to right. If you run rapidly enough and have excellent pitch discrimination, you may notice that the music that is playing seems to be out of tune when you are between the speakers. Why?

Explanation When you are between the speakers, you are running away from one of them and toward the other. Thus, there is a Doppler shift downward for the sound from the speaker behind you and a Doppler shift upward for the sound from the speaker ahead of you. As a result, the sound from the two speakers will not be in tune. A calculation shows that a world-class sprint runner could run fast enough to generate about a semitone difference in the sound from the two speakers.

EXAMPLE 14.6 Listen, But Don't Stand on the Track

A train moving at a speed of 40.0 m/s sounds its whistle, which has a frequency of 500 Hz. Determine the frequency heard by a stationary observer as the train approaches the observer.

Solution We can use Equation 14.12 to get the apparent frequency as the train approaches the observer. Taking $v = 345$ m/s as the speed of sound in air, we have

$$f' = f\left(\frac{v}{v - v_s}\right) = (500 \text{ Hz})\left(\frac{345 \text{ m/s}}{345 \text{ m/s} - 40.0 \text{ m/s}}\right) = 566 \text{ Hz}$$

Exercise Determine the frequency heard by the stationary observer as the train recedes from the observer.

Answer 448 Hz

EXAMPLE 14.7   The Noisy Siren

An ambulance travels down a highway at a speed of 75.0 mi/h, its siren emitting sound at a frequency of 400 Hz. What frequency is heard by a passenger in a car traveling at 55.0 mi/h in the opposite direction as the car (a) approaches? (b) moves away from the ambulance?

Solution   Let us take the velocity of sound in air to be $v = 345$ m/s and use the conversion 1.00 mi/h = 0.447 m/s. Therefore, $v_s = 75.0$ mi/h = 33.5 m/s and $v_0 =$ 55.0 mi/h = 24.6 m/s. We can use Equation 14.15 in both cases.

(a) As the ambulance and car approach each other, the observed frequency is

$$f' = f\left(\frac{v + v_0}{v - v_s}\right) = (400 \text{ Hz})\left(\frac{345 \text{ m/s} + 24.6 \text{ m/s}}{345 \text{ m/s} - 33.5 \text{ m/s}}\right) = \boxed{475 \text{ Hz}}$$

(b) As the two vehicles recede from each other, the passenger in the car hears a frequency of

$$f' = f\left(\frac{v - v_0}{v + v_s}\right) = (400 \text{ Hz})\left(\frac{345 \text{ m/s} - 24.6 \text{ m/s}}{345 \text{ m/s} + 33.5 \text{ m/s}}\right) = \boxed{339 \text{ Hz}}$$

Shock Waves

Now let us consider what happens when the source speed, v_s, *exceeds* the wave velocity, v. Figure 14.11a describes this situation graphically. The circles represent spherical wavefronts emitted by the source at various times during its motion. At

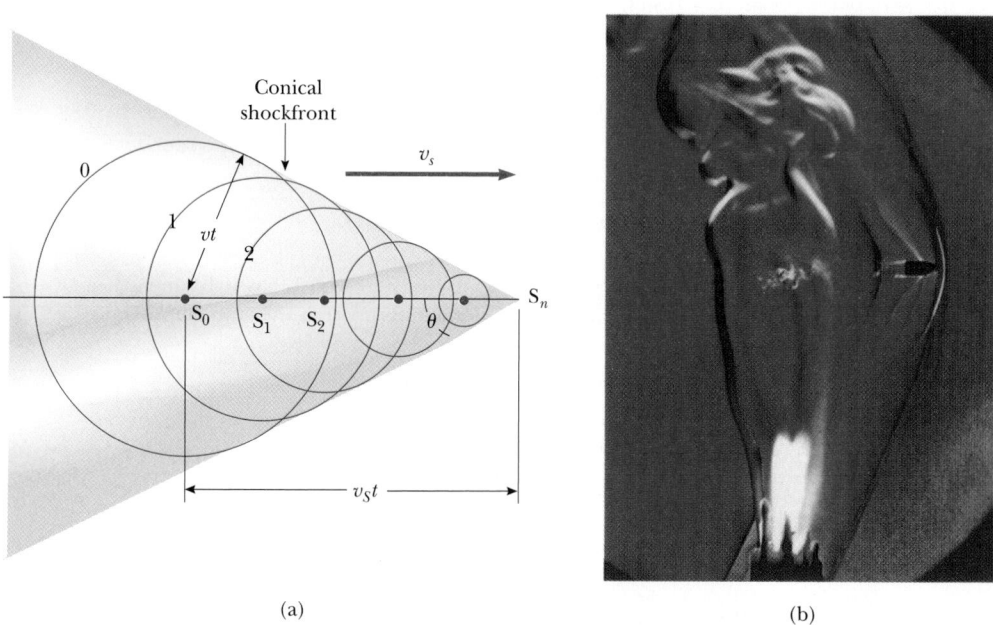

(a)   (b)

Figure 14.11   (a) A representation of a shock wave, produced when a source moves from S_0 to S_n with a speed, v_s, that is *greater* than the wave speed, v, in that medium. The envelope of the wavefronts forms a cone whose half-angle is $\sin \theta = v/v_s$. (b) A stroboscopic photograph of a bullet moving at supersonic speed through the hot air above a candle. *(Harold Edgerton, Courtesy of Palm Press, Inc.)*

Figure 14.12 The V-shaped bow wave of a boat is formed because the boat travels at a speed greater than the speed of water waves. A bow wave is analogous to a shock wave formed by an airplane traveling faster than sound. *(© 1994, Comstock)*

$t = 0$, the source is at point S_0, and at some later time t, the source is at point S_n. In the interval t, the wavefront centered at S_0 reaches a radius of vt. In this same interval, the source travels to S_n, a distance of $v_s t$. At the instant the source is at S_n, the waves just beginning to be generated at this point have wavefronts of zero radius. The line drawn from S_n to the wavefront centered on S_0 is tangent to all other wavefronts generated at intermediate times. All such tangent lines lie on the surface of a cone. The angle, θ, between one of these tangent lines and the direction of travel is given by

$$\sin \theta = \frac{v}{v_s}$$

The ratio v_s / v is referred to as the **Mach number.** The conical wavefront produced when $v_s > v$ (supersonic speeds) is known as a **shock wave.** Figure 14.11b is a photograph of a bullet traveling at supersonic speed through the hot air rising above a candle. Note the shock waves in the vicinity of the bullet. Another interesting example of shock waves is the V-shaped wavefront produced by a boat (the bow wave) when the boat's speed exceeds the speed of the water waves (Fig. 14.12).

Jet airplanes traveling at supersonic speeds produce shock waves, which are responsible for the loud explosion, or sonic boom, heard on the ground. A shock wave carries a great deal of energy concentrated on the surface of the cone, with correspondingly great pressure variations. Shock waves are unpleasant to hear and can damage buildings when aircraft fly supersonically at low altitudes. In fact, an airplane flying at supersonic speeds produces a double boom because two shock waves are formed, one from the nose of the plane and one from the tail (Fig. 14.13).

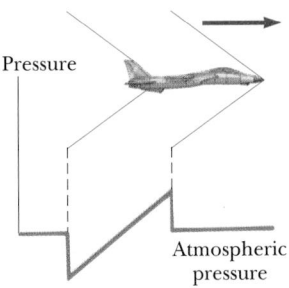

Pressure

Atmospheric pressure

Figure 14.13 The two shock waves produced by the nose and tail of a jet airplane traveling at supersonic speed.

A P P L I C A T I O N

Sonic Booms.

14.7 INTERFERENCE OF SOUND WAVES

Sound waves can be made to interfere with each other. This can be demonstrated with the device shown in Figure 14.14. Sound from a loudspeaker at S is sent into a tube at P, where there is a T-shaped junction. The sound splits and follows two

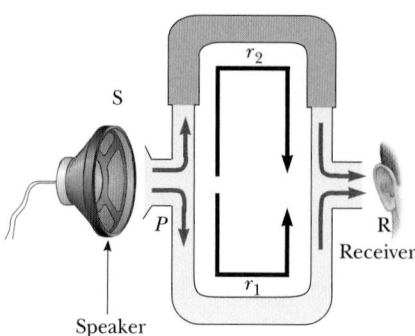

Figure 14.14 An acoustical system for demonstrating interference of sound waves. Sound from the speaker enters the tube and splits into two parts at *P*. The two waves combine at the opposite side and are detected at R. The upper path length is varied by the sliding section.

separate pathways indicated by the red arrows. Half of the sound travels upward, and half downward. Finally, the two sounds merge at an opening where a listener places her ear. If the two paths r_1 and r_2 are of the same length, a crest of the wave that enters the junction will separate into two halves, travel the two paths, and then combine again at the ear. This reuniting of the two waves produces constructive interference, and thus a loud sound is heard by the listener.

Suppose, however, that one of the path lengths (r_2) is adjusted by sliding the upper U-shaped tube upward so that the upper path is half a wavelength *longer* than the lower path r_1. In this case, an entering sound wave splits and travels the two paths as before, but now the wave along the upper path must travel a distance equivalent to half a wavelength farther than the wave traveling along the lower path. As a result, the crest of one wave meets the trough of the other when they merge at the receiver. Because this is the condition for destructive interference, no sound is detected at the receiver.

You should be able to predict what will be heard if the upper path is adjusted to be one full wavelength longer than the lower path. In this case, constructive interference of the two waves occurs, and a loud sound is detected at the receiver.

Nature provides many other examples of interference phenomena. In a later chapter we shall describe several interesting interference effects involving light waves.

APPLICATION

Connecting Your Stereo Speakers.

When connecting the wires between your stereo system and your loudspeakers, you may notice that the wires are color-coded and that the speakers have positive and negative signs on the connections. The reason for this is that the speakers need to be connected with the same "polarity." If they are not, then a sound that is fed to both speakers will result in one speaker cone moving outward at the same time that the other speaker cone is moving inward. In this case, the sounds leaving the two speakers will be 180° out of phase with each other. If you are sitting midway between the speakers, the sounds from both speakers travel the same distance and preserve the phase difference they had when they left. In an ideal situation, for a 180° phase difference you would get complete destructive interference and no sound! In reality, the cancellation is not complete and is much more significant for bass notes (which have a long wavelength) than for the shorter wavelength treble notes. Nevertheless, to avoid a significant reduction in the intensity of bass notes, the color-coded wires and the signs on the speaker connections should be noted carefully.

EXAMPLE 14.8 Interference from Two Loudspeakers

Two loudspeakers are placed as in Figure 14.15 and driven by the same source at a frequency of 2000 Hz. The top speaker is then moved left to position A. At this location, an observer at a great distance from the speakers and directly in front of them notices that the intensity of the sound from the two sources has decreased to a minimum. What is the minimum distance the top speaker has been moved? Assume that the speed of sound in air is 345 m/s.

Solution Initially, both speakers are at the same distance from the observer. Hence, the sound from each speaker must travel the same distance, and the observer hears a loud sound corresponding to constructive interference. When the top speaker is at position A, its sound must travel farther to reach the observer. Because the observer notices a minimum in the sound level when the top speaker is at A, destructive interference is taking place between the two separate sound signals. This means that the top speaker has been moved half a wavelength. With the speed of sound in air equal to 345 m/s, we can calculate the wavelength:

$$\lambda = \frac{v}{f} = \frac{345 \text{ m/s}}{2000 \text{ s}^{-1}} = 0.173 \text{ m}$$

Therefore, the distance moved is half this value, or $\boxed{0.0865 \text{ m.}}$

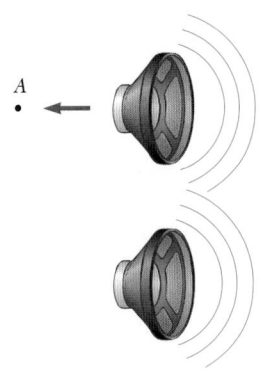

Figure 14.15 (Example 14.8) Two loudspeakers driven by the same source can produce interference.

14.8 STANDING WAVES

Standing waves can be set up in a stretched string by connecting one end of the string to a stationary clamp and connecting the other end to a vibrating object, such as the end of a tuning fork, or by shaking your hand up and down at a steady rate (Fig. 14.16). In this situation, traveling waves reflect from the ends, creating waves traveling in both directions on the string. The incident and reflected waves combine according to the superposition principle. If the string is vibrated at exactly the right frequency, the wave appears to stand still—hence its name, **standing wave.** A **node** occurs where the two traveling waves always have the same magnitude of displacement but of opposite sign, so that the net displacement is zero at that point. There is no motion in the string at the nodes but midway between two adjacent nodes, at an **antinode,** the string vibrates with the largest amplitude.

Figure 14.17 shows the oscillation of a standing wave during half of a cycle. Notice that **all points on the string oscillate vertically with the same frequency except for the node, which is stationary.** (The points at which the string is attached to the wall are also nodes, labeled N in Figure 14.17a.) Furthermore, different points have different amplitudes of motion. From the figure observe that the distance between adjacent nodes is one-half the wavelength of the wave:

$$d_{NN} = \tfrac{1}{2}\lambda$$

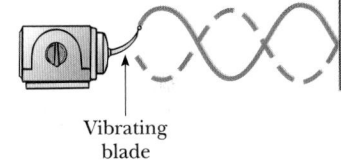

Vibrating blade

Figure 14.16 Standing waves can be set up in a stretched string by connecting one end of the string to a vibrating blade. When the blade vibrates at one of the natural frequencies of the string, large-amplitude standing waves are created.

9.8, BOTH SECTIONS

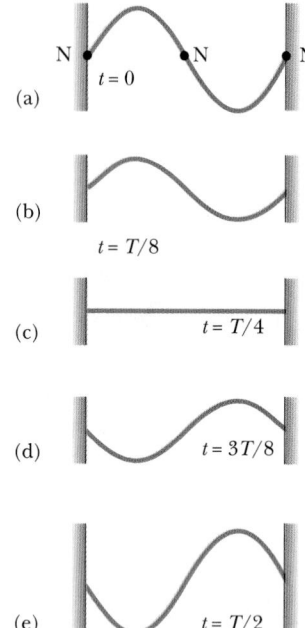

(a) $t = 0$

(b) $t = T/8$

(c) $t = T/4$

(d) $t = 3T/8$

(e) $t = T/2$

Figure 14.17   A standing wave pattern in a stretched string, shown by snapshots of the string during one half of a cycle.

Consider a string of length L that is fixed at both ends, as in Figure 14.18. The string has a number of natural patterns of vibration, called normal modes. Three of these are pictured in Figures 14.18b, 14.18c, and 14.18d. Each has a characteristic frequency, which we shall now calculate.

First, note that **the ends of the string must be nodes because these points are fixed.** If the string is displaced at its midpoint and released, the vibration shown in Figure 14.18b can be produced, in which case the center of the string is an antinode. For this normal mode, the length of the string equals $\lambda/2$ (the distance between nodes). Thus,

$$L = \frac{\lambda_1}{2} \quad \text{or} \quad \lambda_1 = 2L$$

and the frequency of this vibration is

$$f_1 = \frac{v}{\lambda_1} = \frac{v}{2L} \qquad \text{[14.16]}$$

In Chapter 13 (Equation 13.17), the speed of a wave on a string was given as $v = \sqrt{F/\mu}$, where F is the tension in the string and μ is its mass per unit length. Thus, we can express Equation 14.16 as

$$f_1 = \frac{1}{2L} \sqrt{\frac{F}{\mu}} \qquad \text{[14.17]}$$

This lowest frequency of vibration is called the **fundamental frequency** of the vibrating string.

The next normal mode, of wavelength λ_2 (Fig. 14.18c), occurs when the length of the string equals one wavelength — that is, when $\lambda_2 = L$. Hence,

$$f_2 = \frac{v}{L} = \frac{2v}{2L} = 2f_1 \qquad \text{[14.18]}$$

Note that this frequency is equal to *twice* the fundamental frequency. You should convince yourself that the next highest frequency of vibration, shown in Figure 14.18d, is

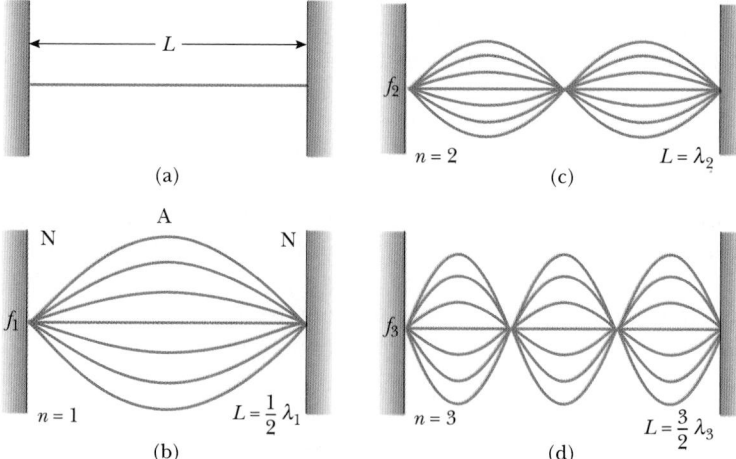

Figure 14.18   (a) Standing waves in a stretched string of length L, fixed at both ends. The normal frequencies of vibration form a harmonic series: (b) the fundamental frequency, or first harmonic; (c) the second harmonic; and (d) the third harmonic.

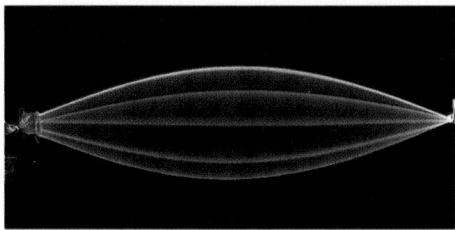
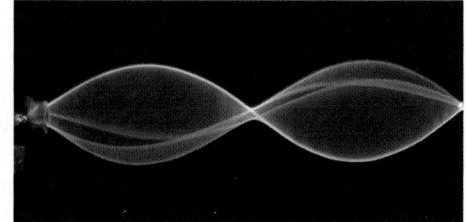
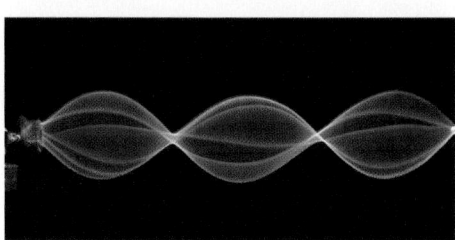

Multiflash photographs of standing wave patterns in a cord driven by a vibrator at the left end. The single loop pattern at the top left represents the fundamental ($n = 1$), the two-loop pattern at the right represents the second harmonic ($n = 2$), and the three-loop pattern at the lower left represents the third harmonic ($n = 3$). *(Richard Megna, Fundamental Photographs, NYC)*

$$f_3 = \frac{3v}{2L} = 3f_1 \qquad [14.19]$$

In general, the characteristic frequencies are given by

$$f_n = nf_1 = \frac{n}{2L}\sqrt{\frac{F}{\mu}} \qquad [14.20]$$

where $n = 1, 2, 3, \ldots$. In other words, the frequencies are integral multiples of the fundamental frequency. The frequencies f_1, $2f_1$, $3f_1$, and so on form a **harmonic series.** The fundamental f_1 corresponds to the **first harmonic;** the frequency $f_2 = 2f_1$, to the **second harmonic;** and so on.

When a stretched string is distorted to a shape that corresponds to any one of its harmonics, after being released it vibrates only at the frequency of that harmonic. If the string is struck or bowed, however, the resulting vibration includes frequencies of various harmonics, including the fundamental. Waves not in the harmonic series are quickly damped out on a string fixed at both ends. In effect, when disturbed, the string "selects" the normal-mode frequencies. As we shall see later, the presence of several harmonics on a string gives stringed instruments their characteristic sound, which enables us to distinguish one from another even when they are producing identical fundamental frequencies.

The frequency of a string on a musical instrument can be changed either by varying the tension or by changing the length. For example, the tension in guitar and violin strings is varied by turning pegs on the neck of the instrument. As the tension is increased, the frequency of the normal modes increases according to Equation 14.20. Once the instrument is tuned, the musician varies the frequency by pressing the strings against the neck at a variety of positions, thereby changing the effective lengths of the vibrating portions of the strings. As the length is reduced, the frequency increases, as Equation 14.20 indicates.

Finally, Equation 14.20 shows that a string of fixed length can be made to vibrate at a lower fundamental frequency by increasing its mass per unit length.

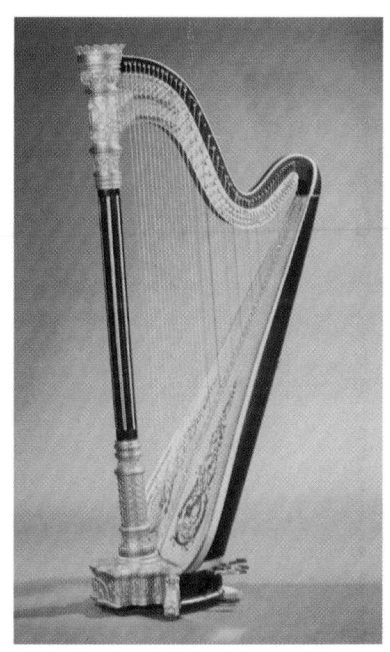

A concert-style harp. *(Lyon & Healy Harps, Chicago)*

APPLICATION

Tuning a Musical Instrument.

This is achieved in the bass strings of guitars and pianos by wrapping them with metal windings.

Workbook Problem 61 – (Workbook page 128) is similar to Example 14.9.

EXAMPLE 14.9   Harmonics of a Stretched String

Find the first four harmonics of a 1.0-m-long string if the string has a mass per unit length of 2.0×10^{-3} kg/m and is under a tension of 80 N.

Solution   The speed of the wave on the string is

$$v = \sqrt{\frac{F}{\mu}} = \sqrt{\frac{80 \text{ N}}{2.0 \times 10^{-3} \text{ kg/m}}} = 200 \text{ m/s}$$

The fundamental frequency can be found using Equation 14.16:

$$f_1 = \frac{v}{2L} = \frac{200 \text{ m/s}}{2(1.0 \text{ m})} = \boxed{100 \text{ Hz}}$$

The frequencies of the next three modes are $f_2 = 2f_1$, $f_3 = 3f_1$, and $f_4 = 4f_1$. Thus,

$$\boxed{f_2 = 200 \text{ Hz}, f_3 = 300 \text{ Hz, and } f_4 = 400 \text{ Hz.}}$$

Exercise   Find the tension in the string if the fundamental frequency is increased to 120 Hz.

Answer   120 N

14.9   FORCED VIBRATIONS AND RESONANCE

In Chapter 13 we learned that the energy of a damped oscillator decreases in time because of friction. It is possible to compensate for this energy loss by applying an external force that does positive work on the system.

For example, suppose a mass-spring system having some natural frequency of vibration, f_0, is pushed back and forth with a periodic force whose frequency is f. The system vibrates at the frequency, f, of the driving force. This type of motion is referred to as a **forced vibration.** Its amplitude reaches a maximum when the frequency of the driving force equals the natural frequency of the system, f_0, called the **resonant frequency** of the system. Under this condition, the system is said to be in **resonance.**

In Section 14.8 we learned that a stretched string can vibrate in one or more of its natural modes. Here again, if a periodic force is applied to the string, the amplitude of vibration increases as the frequency of the applied force approaches one of the natural frequencies of vibration.

Resonance vibrations occur in a wide variety of circumstances. Figure 14.19 illustrates one experiment that demonstrates a resonance condition. Several pendulums of different lengths are suspended from a flexible beam. If one of them, such as A, is set in motion, the others begin to oscillate because of vibrations in the flexible beam. Pendulum C, the same length as A, oscillates with the greatest amplitude because its natural frequency matches that of pendulum A (the driving force).

Another simple example of resonance is a child being pushed on a swing, which is essentially a pendulum with a natural frequency that depends on the length. The swing is kept in motion by a series of appropriately timed pushes. For its amplitude

9.10, BOTH SECTIONS

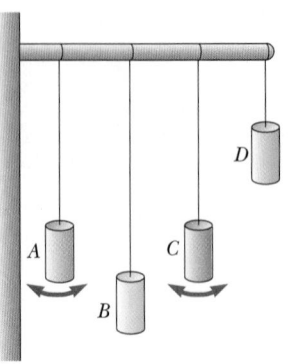

Figure 14.19 Resonance. If pendulum A is set in oscillation, only pendulum C, whose length matches that of A, will eventually oscillate with a large amplitude, or resonate. The arrows indicate motion perpendicular to the page.

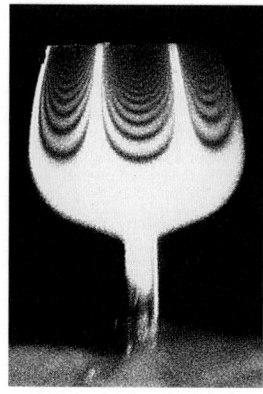 

Figure 14.20 (*Left*) Standing wave pattern in a vibrating wine glass. The wine glass will shatter if the amplitude of vibration becomes too large. *(Courtesy of Professor Thomas D. Rossing, Northern Illinois University)* (*Right*) A wine glass shattered by the amplified sound of a human voice. *(Ben Rose/The IMAGE Bank)*

to be increased, the swing must be pushed each time it returns to the person's hands. This corresponds to a frequency equal to the natural frequency of the swing. If the energy put into the system per cycle of motion exactly equals the energy lost due to friction, the amplitude remains constant.

Opera singers have been known to set crystal goblets in audible vibration with their powerful voices, as shown in Figure 14.20. This is yet another example of resonance: The sound waves emitted by the singer can set up large-amplitude vibrations in the glass. If a highly amplified sound wave has the right frequency, the amplitude of forced vibrations in the glass increases to the point where the glass becomes heavily strained and shatters.

A striking example of structural resonance occurred in 1940, when the Tacoma Narrows bridge in the state of Washington was set in oscillation by the wind (Fig. 14.21). The amplitude of the oscillations increased steadily until the bridge col-

A P P L I C A T I O N

Shattering Goblets with the Voice.

A P P L I C A T I O N

Structural Resonance in Bridges and Buildings.

 9.11

Figure 14.21 The collapse of the Tacoma Narrows suspension bridge in 1940 was a vivid demonstration of mechanical resonance. High winds set up standing waves in the bridge, causing it to oscillate at one of its natural frequencies. Once established, this resonance condition led to the bridge's collapse. *(United Press International Photo)*

lapsed. A more recent example of destruction by structural resonance occurred during the Loma Prieta earthquake near Oakland, California, in 1989. In one section—almost a mile long—of the double-decker Nimitz Freeway, the upper deck collapsed onto the lower deck, killing several people. The collapse of this particular section of roadway, while other sections escaped serious damage, has been traced to the fact that the earthquake waves had a frequency of approximately 1.5 Hz—very close to a natural resonant frequency of the section of roadway that gave way.

Thinking Physics 3

Passing ocean waves sometimes cause the water in a harbor to undergo very large oscillations, called *seiches*. Why would this happen?

Explanation   Water in a harbor is enclosed and possesses a natural frequency based on the size of the harbor. This is similar to the natural frequency of the enclosed air in a bottle, which can be excited by blowing across the edge of the opening. Ocean waves pass by the opening of the harbor at a certain frequency. If this frequency matches that of the enclosed harbor, then a large standing wave can be set up in the water by resonance. This can be simulated by carrying a fish tank with water. If your walking frequency matches the natural frequency of the water as it sloshes back and forth, a large standing wave in the fish tank can be established.

14.10   STANDING WAVES IN AIR COLUMNS

9.9

Standing longitudinal waves can be set up in a tube of air, such as an organ pipe, as the result of interference between sound waves traveling in opposite directions. The relationship between the incident wave and the reflected wave depends on whether the reflecting end of the tube is open or closed. A portion of the sound wave is reflected back into the tube at the open end. **If the reflecting end is closed, a node must exist at this end because the movement of air molecules is restricted. If the end is open, the air molecules have complete freedom of motion, and an antinode exists.**

Figure 14.22a shows the first three modes of vibration of a pipe open at both ends. When air is directed against an edge at the left, longitudinal standing waves are formed and the pipe vibrates at its natural frequencies. Note that for the fundamental frequency the wavelength is twice the length of the pipe and hence $f_1 = v/2L$. Similarly, one finds that the frequencies of the second and third harmonics are $2f_1, 3f_1, \ldots$. Thus, **in a pipe open at both ends, the natural frequencies of vibration form a series in which all harmonics are present. These harmonics are equal to integral multiples of the fundamental frequency.** We can express this harmonic series as

Pipe open at both ends; all ▷
harmonics present

$$f_n = n\,\frac{v}{2L} \qquad n = 1, 2, 3, \ldots \qquad [14.21]$$

where v is the speed of sound in air.

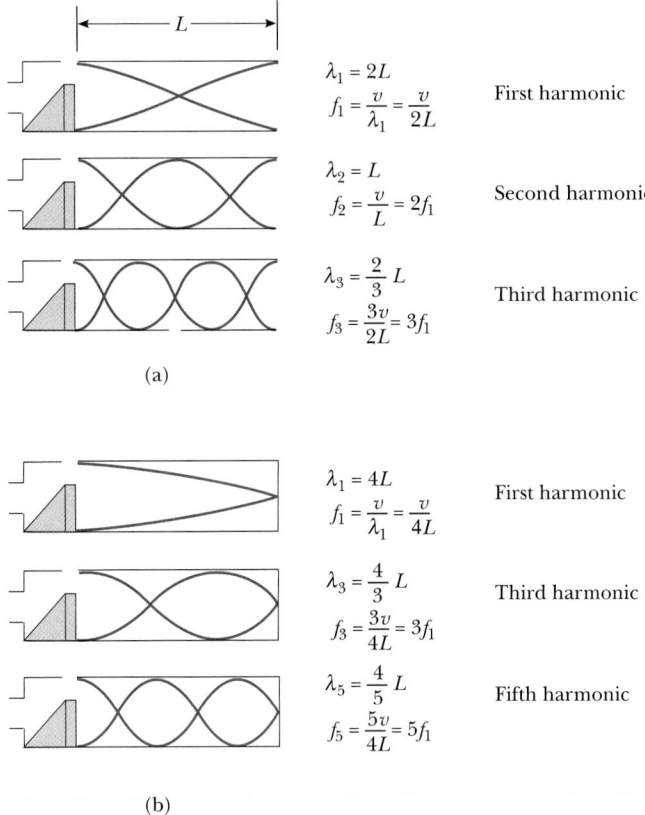

(a)

(b)

Figure 14.22 (a) Standing longitudinal waves in an organ pipe open at both ends. The natural frequencies f_1, $2f_1$, $3f_1$ . . . form a harmonic series. (b) Standing longitudinal waves in an organ pipe closed at one end. Only *odd* harmonics are present, and the natural frequencies are f_1, $3f_1$, $5f_1$, and so on.

If a pipe is closed at one end and open at the other, the closed end is a node (Fig. 14.22b). In this case, the wavelength of the fundamental mode is four times the length of the tube. Hence, $f_1 = v/4L$ and the frequencies of the third and fifth harmonics are $3f_1$, $5f_1$, ... That is, **in a pipe closed at one end and open at the other, only odd harmonics are present.** These are given by

$$f_n = n\frac{v}{4L} \qquad n = 1, 3, 5, \ldots \qquad [14.22]$$

◀ Pipe closed at one end; odd harmonics present

Applying Physics 1

If an orchestra doesn't warm up before a performance, the strings go flat and the wind instruments go sharp during the performance. Why?

Explanation Without warming up, all the instruments will be at room temperature at the beginning of the concert. As the wind instruments are played, they fill with warm air from the player's exhalation. The increase in temperature of the air in the instrument causes an increase in the speed of sound, which raises the resonance frequencies of the air columns. As a result, the instruments go sharp. The strings on the stringed instruments also increase in temperature due to the friction of rubbing with the bow. This results in thermal expansion, which causes a decrease in tension in the strings. With a decrease in tension, the wave speed on the strings drops, and the fundamental frequencies decrease. Thus, the stringed instruments go flat.

Applying Physics 2

A bugle has no valves, keys, slides, or finger holes. How can it play a song?

Explanation   Songs for the bugle are limited to harmonics of the fundamental frequency, because there is no control over frequencies without valves, keys, slides, or finger holes. The player obtains different notes by changing the tension in the lips as the bugle is played, in order to excite different harmonics. The normal playing range of a bugle is among the third, fourth, fifth, and sixth harmonics of the fundamental. For example "Reveille" is played with just the three notes G, C, and F. And "Taps" is played with these three notes and the G one octave above the lower G.

Workbook Problem 62 — also deals with standing waves in an air column. (See page 130 of the Workbook.)

EXAMPLE 14.10   Harmonics of a Pipe

A pipe is 2.46 m long.
(a) Determine the frequencies of the first three harmonics if the pipe is open at both ends. Take 345 m/s as the speed of sound in air.

Solution   The fundamental frequency of a pipe open at both ends can be found from Equation 14.21, with $n = 1$.

$$f_1 = \frac{v}{2L} = \frac{345 \text{ m/s}}{2(2.46 \text{ m})} = \boxed{70.0 \text{ Hz}}$$

Because all harmonics are present in a pipe open at both ends, the second and third harmonics have frequencies of $f_2 = 2f_1 = 140$ Hz and $f_3 = 3f_1 = 210$ Hz.

(b) What are the three lowest possible frequencies if the pipe is closed at one end and open at the other?

Solution   The fundamental frequency of a pipe closed at one end can be found from Equation 14.22, with $n = 1$.

$$f_1 = \frac{v}{4L} = \frac{345 \text{ m/s}}{4(2.46 \text{ m})} = \boxed{35.0 \text{ Hz}}$$

In this case, only odd harmonics are present, and so the third and fifth harmonics have frequencies of $f_3 = 3f_1 = 105$ Hz and $f_5 = 5f_1 = 175$ Hz.

Exercise   If the pipe is open at one end, how many harmonics are possible in the normal hearing range, 20 to 20 000 Hz?

Answer   286

QUICKLAB

Using an empty 1-liter soft drink container, blow over the open end and listen to the sound that is produced. Add some water to the container to change the height of the air column and repeat the experiment at various heights of the column. How does the frequency that you hear change with the height of the air column?

EXAMPLE 14.11   Resonance in a Tube of Variable Length

Figure 14.23 shows a simple apparatus for demonstrating resonance in a tube. A long tube open at both ends is partially submerged in a beaker of water, and a vibrating tuning fork of unknown frequency is placed near the top of the tube. The length of the air column, L, is adjusted by moving the tube vertically. The sound waves generated by the fork are reinforced when the length of the air column corresponds to one of the resonant frequencies of the tube. The smallest value of L for which a peak occurs in the sound intensity is 9.00 cm. For this measurement, determine the frequency of the tuning fork and the value of L for the next two resonant vibrations.

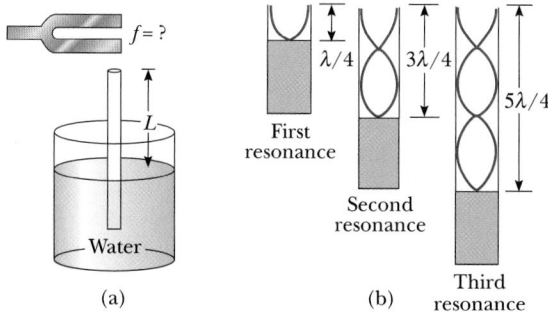

Figure 14.23 (Example 14.11) (a) Apparatus for demonstrating the resonance of sound waves in a tube closed at one end. The length, L, of the air column is varied by moving the tube vertically while it is partially submerged in water. (b) The first three normal frequencies of vibration for the system.

Reasoning and Solution   Once the tube is in the water, this setup represents a pipe closed at one end, and the fundamental has a frequency of $v/4L$ (Fig. 14.23b). If we take $v = 345$ m/s for the speed of sound in air, and $L = 0.0900$ m, we get

$$f_1 = \frac{v}{4L} = \frac{345 \text{ m/s}}{4(0.0900 \text{ m})} = \boxed{958 \text{ Hz}}$$

The fundamental wavelength of the pipe is given by $\lambda = 4L = 0.360$ m. Because the frequency of the source is constant, we see that the next resonance positions (Fig. 14.23b) correspond to lengths of $3\lambda/4 = 0.270$ m and $5\lambda/4 = 0.450$ m.

This arrangement is often used to measure the speed of sound, in which case the frequency of the tuning fork and the lengths at which resonance occurs must be known.

14.11   BEATS

The interference phenomena we have been discussing so far have involved the superposition of two or more waves with the same frequency, traveling in opposite directions. Let us now consider another type of interference effect that results from the superposition of two waves with slightly different frequencies. In this situation, the waves at some fixed point are periodically in and out of phase, corresponding to an alternation in time between constructive and destructive interference. In order to understand this phenomenon, consider Figure 14.24. The two waves in Figure 14.24a were emitted by two tuning forks having slightly different frequencies; Figure 14.24b shows the superposition of these two waves. At some time t_a, the two

Figure 14.24   Beats are formed by the combination of two waves of slightly different frequencies traveling in the same direction. (a) The individual waves. (b) The combined wave has an amplitude (dashed line) that oscillates in time.

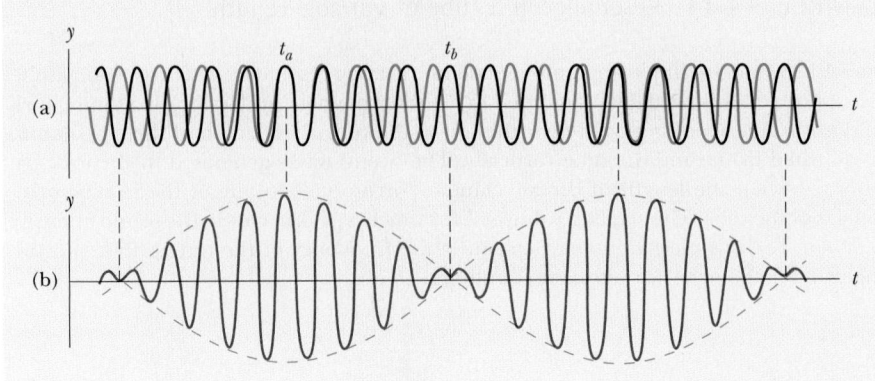

WEB

For a Web link relating to beats and the Doppler effect, visit the textbook Web site at http://www.harcourtcollege.com/physics/cptech

waves are in phase, and constructive interference occurs, as demonstrated by the resultant curve in Figure 14.24b. At some later time, however, the vibrations of the two forks move out of step with each other. At time t_b, one fork emits a compression while the other emits a rarefaction, and destructive interference occurs, as demonstrated by the curve in Figure 14.24b. As time passes, the vibrations of the two forks move out of phase, then into phase, and so on. As a consequence, a listener at some fixed point hears an alternation in loudness, known as **beats.** The number of beats per second, or beat frequency, equals the difference in frequency between the two sources. One can tune a stringed instrument, such as a piano, by beating a note on the instrument against a note of known frequency. The string can then be tuned to the desired frequency by adjusting the tension until no beats are heard.

APPLICATION

Another Way to Tune a Musical Instrument.

EXAMPLE 14.12   Sour Notes

A particular piano string is supposed to vibrate at a frequency of 440 Hz. In order to check its frequency, a tuning fork known to vibrate at a frequency of 440 Hz is sounded at the same time the piano key is struck, and a beat frequency of 4 beats per second is heard. Find the possible frequencies at which the string could be vibrating.

Solution   The number of beats per second is equal to the difference in frequency between the two sound sources. In this case, because one of the source frequencies is 440 Hz, 4 beats per second would be heard if the frequency of the string (the second source) were either 444 Hz or 436 Hz.

Optional Section

14.12   **QUALITY OF SOUND**

The sound-wave patterns produced by most musical instruments are very complex. Figure 14.25 shows characteristic waveforms produced by a tuning fork, a flute, and a clarinet, each playing the same steady tone. Although each instrument has its own

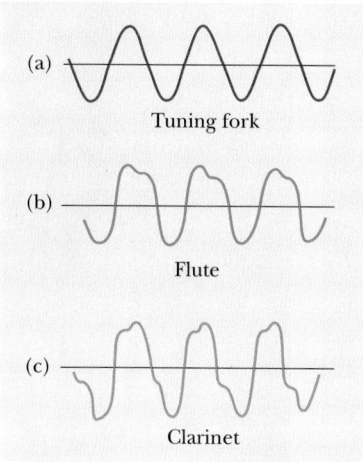

Figure 14.25 Waveforms produced by (a) a tuning fork, (b) a flute, and (c) a clarinet, all at approximately the same frequency. *(Adapted from C. A. Culver, Musical Acoustics, 4th ed., New York, McGraw-Hill, 1956)*

characteristic pattern, the figure reveals that each of the waveforms is periodic. Note that the tuning fork produces only one harmonic (the fundamental frequency), but the two instruments emit mixtures of harmonics. Figure 14.26 graphs the harmonics of the waveforms of Figure 14.25. When this note is played on the flute (Fig. 14.26b), part of the sound consists of a vibration at the fundamental frequency, an even higher intensity is contributed by the second harmonic, the fourth harmonic produces about the same intensity as the fundamental, and so on. These sounds add together according to the principle of superposition to give the complex waveform shown. The clarinet emits a certain intensity at a frequency of the first harmonic, about half as much intensity at the frequency of the second harmonic, and so forth. The resultant superposition of these frequencies produces the pattern shown in Figure 14.25c. The tuning fork (Figs. 14.25a and 14.26a) emits sound only at the frequency of the first harmonic.

In music, the mixture of harmonics that produces the characteristic sound of any instrument is referred to as the *quality*, or *timbre*, of the sound. We say that the

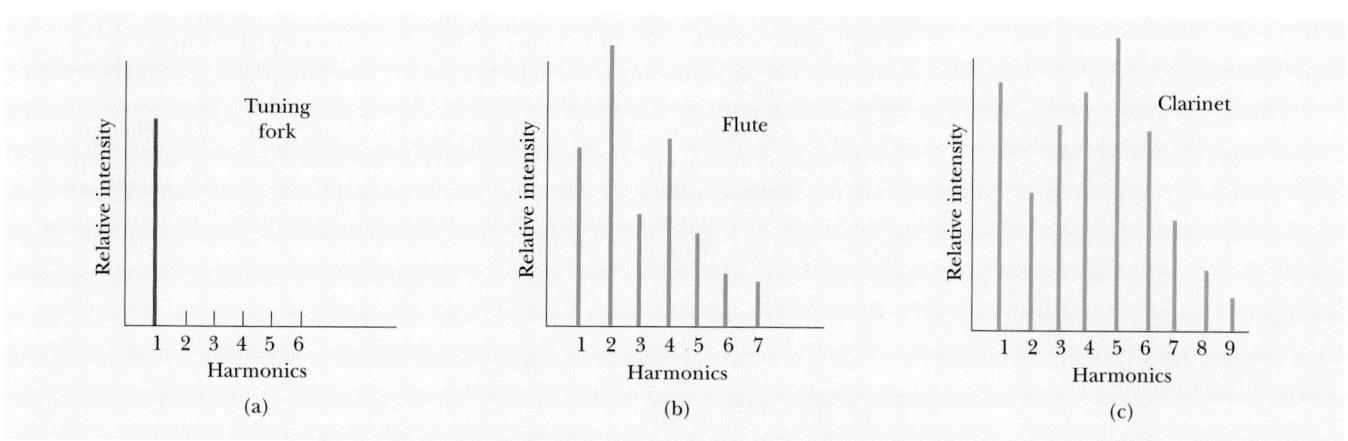

Figure 14.26 Harmonics of the waveforms in Figure 14.25. Note their variation in intensity. *(Adapted from C. A. Culver, Musical Acoustics, 4th ed., New York, McGraw-Hill, 1956)*

Each musical instrument has its own characteristic sound and mixture of harmonics. Instruments shown are (a) the violin, (b) the saxophone, and (c) the trumpet. *(Photographs courtesy of (a) © 1989 Gary Buss/FPG; (b) and (c) © 1989 Richard Laird/FPG)*

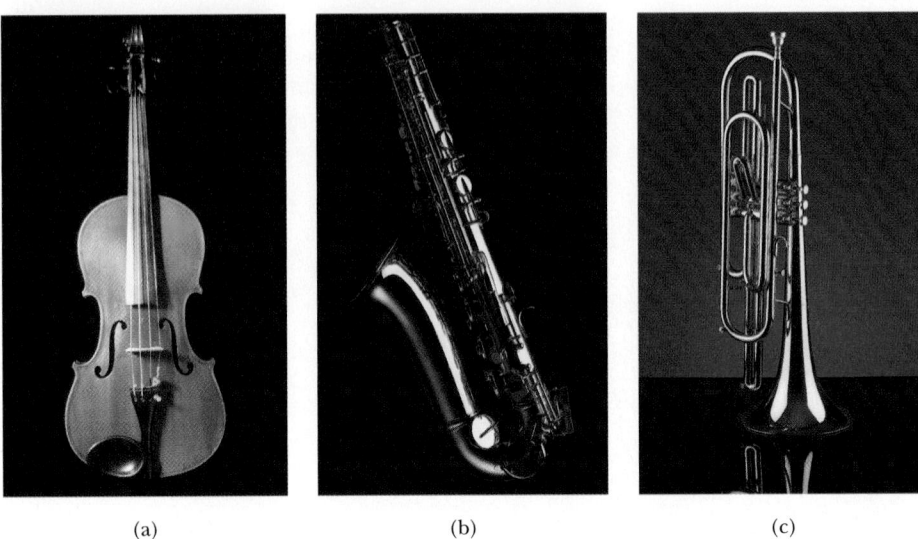

(a)   (b)   (c)

note C on a flute differs in quality from the same C on a clarinet. Instruments such as the bugle, trumpet, violin, and tuba are rich in harmonics. A musician playing a wind instrument can emphasize one or another of these harmonics by changing the configuration of his or her lips, and can thus play different musical notes with the same valve openings.

Thinking Physics 4

A professor performs a demonstration in which he breathes helium and then speaks with a comical voice. One student explains, "The velocity of sound in helium is higher than in air, so the fundamental frequency of the standing waves in the mouth is increased." Another student says, "No, the fundamental frequency is determined by the vocal folds and cannot be changed. Only the quality of the voice has changed." Which student is correct?

Explanation   The second student is correct. The fundamental frequency of the complex tone from the voice is determined by the vibration of the vocal folds and is not changed by substituting a different gas in the mouth. The introduction of the helium into the mouth results in harmonics of higher frequencies being excited more than in the normal voice, but the fundamental frequency of the voice is the same — only the quality has changed. The unusual inclusion of the higher frequency harmonics results in a common description of this effect as a "high-pitched" voice, but this is incorrect (it is really a quacky timbre).

14.13   THE EAR

The human ear is divided into three regions: the outer ear, the middle ear, and the inner ear (Fig. 14.27). The *outer ear* consists of the ear canal (open to the atmosphere), which terminates at the eardrum (tympanum). Sound waves travel

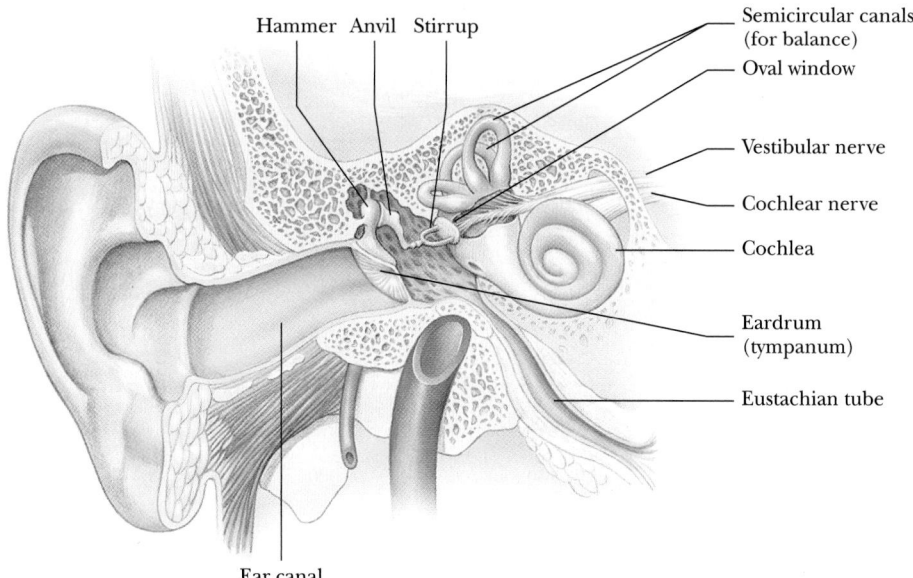

Hammer Anvil Stirrup

Semicircular canals
(for balance)

Oval window

Vestibular nerve

Cochlear nerve

Cochlea

Eardrum
(tympanum)

Eustachian tube

Ear canal

Figure 14.27 The structure of the human ear. The three tiny bones (ossicles) that connect the eardrum to the window of the cochlea act as a double-lever system to decrease the
amplitude of vibration and hence increase the pressure on the fluid in the cochlea.

down the ear canal to the eardrum, which vibrates in and out in phase with the
pushes and pulls caused by the alternating high and low pressures of the sound
wave. Behind the eardrum are three small bones of the *middle ear*, called the *hammer*,
the *anvil*, and the *stirrup* because of their shapes. These bones transmit the vibration
to the *inner ear*, which contains the *cochlea*, a snail-shaped tube about 2 cm long.
The cochlea makes contact with the stirrup at the oval window and is divided along
its length by the basilar membrane, which consists of small hairs and nerve fibers.
This membrane varies in mass per unit length and in tension along its length, and
different portions of it resonate at different frequencies. (Recall that the natural
frequency of a string depends on its mass per unit length and on the tension on
it.) Along the basilar membrane are numerous nerve endings, which sense the
vibration of the membrane and in turn transmit impulses to the brain. The brain
interprets the impulses as sounds of varying frequency, depending on the locations
along the basilar membrane of the impulse-transmitting nerves and on the rates at
which the impulses are transmitted. This is a simplified description of the complex
functioning of the ear.

Figure 14.28 shows the frequency response curves of an average human ear for
sounds of equal loudness, ranging from 0 to 120 dB. To interpret this series of
graphs, take the bottom curve as the threshold of hearing. Compare the intensity
level on the vertical axis for the two frequencies 100 Hz and 1000 Hz. The vertical
axis shows that the 100-Hz sound must be about 38 dB greater than the 1000-Hz
sound to be at the threshold of hearing. Thus, we see that the threshold of hearing
is very strongly dependent on frequency. The easiest frequencies to hear are around
3300 Hz, whereas frequencies above 12 000 Hz or below about 50 Hz must be relatively intense to be heard.

Figure 14.28 Curves of intensity level versus frequency for sounds that are perceived to be of equal loudness. Note that the ear is most sensitive at a frequency of about 3300 Hz. The lowest curve corresponds to the threshold of hearing for only about 1% of the population.

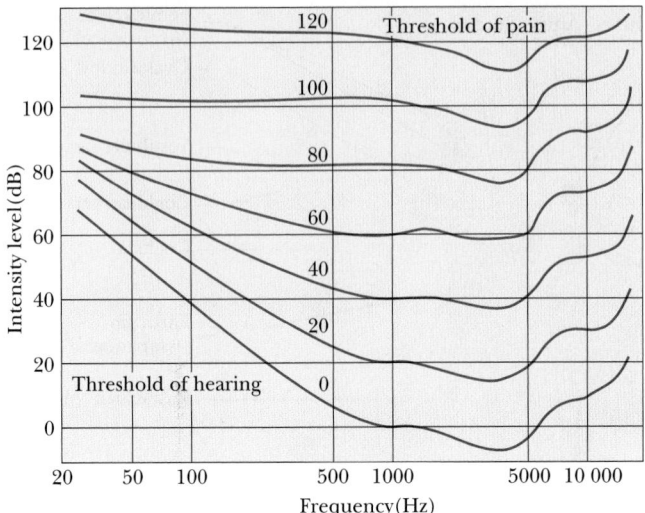

Now consider the curve labeled 80. This curve uses as its reference a 1000-Hz tone at an intensity level of 80 dB. The curve shows that a tone of frequency 100 Hz would have to be about 4 dB louder than the 80-dB, 1000-Hz tone in order to sound as loud. Notice that the curves flatten out as the intensity levels of the sounds increase. That is, when sounds are loud, all frequencies can be heard equally well.

The small bones in the middle ear represent an intricate lever system that increases the force on the oval window. The pressure is greatly magnified because the surface area of the eardrum is about 20 times that of the oval window (in analogy with a hydraulic press). The middle ear, together with the eardrum and oval window, in effect acts as a matching network between the air in the outer ear and the liquid in the inner ear. The overall energy transfer between the outer ear and inner ear is highly efficient, with pressure amplification factors of several thousand. In other words, pressure variations in the inner ear are much greater than those in the outer ear.

The ear has its own built-in protection against loud sounds. The muscles connecting the three middle-ear bones to the walls control the volume of the sound by changing the tension on the bones as sound builds up, thus hindering their ability to transmit vibrations. In addition, the eardrum becomes stiffer as the sound intensity increases. These two occurrences make the ear less sensitive to loud incoming sounds. There is a time delay between the onset of loud sound and the ear's protective reaction, however, so a very sudden loud sound can still damage the ear.

The complex structure of the human ear is believed to be related to the fact that mammals have evolved from sea-going creatures. In comparison, insect ears are considerably simpler in design because they have always been land residents. A typical insect ear consists of an eardrum exposed directly to the air on one side and to an air filled cavity on the other side. Nerve cells communicate directly with the cavity and the brain without the need for the complex intermediary of an inner and middle ear. This simple design allows the ear to be placed virtually anywhere on the body. For example, a grasshopper has its ears on its legs. One advantage of this location is that the distance and orientation of the ears can be varied so that sound location is easier to accomplish.

SUMMARY

Sound waves are longitudinal waves. **Audible waves** are sound waves with frequencies between 20 and 20 000 Hz. **Infrasonic waves** have frequencies below the audible range, and **ultrasonic waves** have frequencies above the audible range.

The speed of sound in a medium of bulk modulus B and density ρ is

$$v = \sqrt{\frac{B}{\rho}} \qquad [14.1]$$

The speed of sound also depends on the temperature of the medium. The relationship between temperature and speed for sound in air is

$$v = (331 \text{ m/s}) \sqrt{1 + \frac{T}{273}} \qquad [14.4]$$

where T is the temperature in degrees Celsius and 331 m/s is the speed of sound in air at 0°C.

The **intensity level** of a sound wave, in decibels, is given by

$$\beta \equiv 10 \log \left(\frac{I}{I_0} \right) \qquad [14.7]$$

The constant I_0 is a reference intensity, usually taken to be at the threshold of hearing ($I_0 = 1.0 \times 10^{-12}$ W/m²), and I is the intensity with level β, where β is measured in **decibels (dB).**

The **intensity** of a *spherical wave* produced by a point source is proportional to the average power emitted and inversely proportional to the square of the distance from the source:

$$I = \frac{P_{av}}{4\pi r^2} \qquad [14.8]$$

The change in frequency heard by an observer whenever there is relative motion between the source and observer is called the **Doppler effect.** If the observer is moving with a speed v_0 and the source is at rest, the observed frequency, f', is

$$f' = f \left(\frac{v \pm v_0}{v} \right) \qquad [14.11]$$

The positive sign is used when the observer is moving toward the source, and the negative sign when the observer is moving away from the source.

If the source is moving with a speed v_s and the observer is at rest, the observed frequency is

$$f' = f \left(\frac{v}{v \mp v_s} \right) \qquad [14.14]$$

where $-v_s$ refers to motion toward the observer and $+v_s$ refers to motion away from the observer.

When the observer and source are both moving, the observed frequency is

$$f' = f\left(\frac{v \pm v_0}{v \mp v_s}\right)$$ [14.15]

When waves interfere, the resultant wave is found by adding the individual waves together point by point. When crest meets crest and trough meets trough, the waves undergo **constructive interference.** When crest meets trough, **destructive interference** occurs.

Standing waves are formed when two waves having the same frequency, amplitude, and wavelength travel in opposite directions through a medium. One can set up standing waves of specific frequencies in a stretched string. The natural frequencies of vibration of a stretched string of length L, fixed at both ends, are

$$f_n = \frac{n}{2L}\sqrt{\frac{F}{\mu}} \qquad n = 1, 2, 3, \ldots$$ [14.20]

where F is the tension in the string and μ is its mass per unit length. The natural frequencies of vibration form a **harmonic series;** that is, the frequencies are integral multiples of the fundamental (lowest) frequency.

A system capable of oscillating is said to be in **resonance** with some driving force whenever the frequency of the driving force matches one of the natural frequencies of the system. When the system is resonating, it oscillates with maximum amplitude.

Standing waves can be produced in a tube of air. If the reflecting end of the tube is *open*, all harmonics are present and the natural frequencies of vibration are

$$f_n = n\frac{v}{2L} \qquad n = 1, 2, 3, \ldots$$ [14.21]

If the tube is *closed* at the reflecting end, only the odd harmonics are present, and the natural frequencies of vibration are

$$f_n = n\frac{v}{4L} \qquad n = 1, 3, 5, \ldots$$ [14.22]

The phenomenon of **beats** is an interference effect that occurs when two waves of slightly different frequencies travel in the same direction. For sound waves, the loudness of the resultant sound changes periodically with time.

MULTIPLE-CHOICE QUESTIONS

1. Two sirens are sounding so that the frequency from A is twice the frequency from B. Compared to the speed of sound from A, the speed of sound from B is
   (a) twice as fast   (b) half as fast   (c) four times as fast
   (d) one fourth as fast   (e) the same
2. A hollow pipe (such as an organ pipe open at both ends) is made to go into resonance at frequency f_{open}.

One end of the pipe is now covered and the pipe is again made to go into resonance, this time at frequency f_{closed}. Both resonances are first harmonics. How do these two resonances compare?
   (a) They are the same.   (b) $f_{open} = 2f_{closed}$
   (c) $f_{closed} = 2f_{open}$ (d) $f_{open} = f_{closed}$
   (e) $f_{closed} = \frac{3}{2}f_{open}$

3. When two tuning forks are sounded at the same time, a beat frequency of 5 Hz occurs. If one of the tuning forks has a frequency of 245 Hz, what is the frequency of the other tuning fork?
   (a) 240 Hz   (b) 242.5 Hz
   (c) 247.5 Hz   (d) 250 Hz
   (e) More than one answer could be correct.
4. Doubling the power output from a sound source emit-ting a single frequency will result in what increase in level?
   (a) 0.5 dB   (b) 2.0 dB   (c) 3.0 dB   (d) 20 dB
5. If a 1000-Hz sound source moves at a speed of 50.0 m/s toward a listener who moves at a speed of 30.0 m/s in a direction away from the source, what is the apparent frequency heard by the listener? (The velocity of sound is 340 m/s.)
   (a) 937 Hz   (b) 947 Hz   (c) 1060 Hz   (d) 1070 Hz

CONCEPTUAL QUESTIONS

1. An autofocus camera sends out a pulse of sound and measures the time for the pulse to reach an object, re-flect off of it and return to be detected. Can the temper-ature affect your focus for such a camera?
2. To keep animals away from their cars, some people mount short, thin pipes on the fenders. The pipes give out a high-pitched wail when the cars are moving. How do they create the sound?
3. Secret agents in the movies always want to get to a se-cure phone where a voice scrambler is in use. How do these devices work?
4. When a bell is rung, standing waves are set up around the bell's circumference. What boundary conditions must be satisfied by the resonant wavelengths? How does a crack in the bell, such as in the Liberty Bell, affect the satisfying of the boundary conditions and the sound em-anating from the bell?
5. Suppose the wind blows. Does this cause a Doppler ef-fect as you hear it for a sound moving through the air? Is it like a moving source or a moving observer?
6. Explain why your voice seems to sound richer than usual when you sing in the shower.
7. You are driving toward a cliff and you honk your horn. Is there a Doppler shift of the sound when you hear the echo? Is it like a moving source or moving observer? What if the reflection occurs not from a cliff but from the forward edge of a huge alien spacecraft that is mov-ing toward you as you drive?
8. Of the following sounds, state which is most likely to have an intensity level of 60 dB: a rock concert, the turning of a page in this text, a normal conversation, a cheering crowd at a football game, or background noise at a church?
9. Guitarists sometimes play a "harmonic" by lightly touch-ing a string at the exact center and plucking the string. The result is a clear note one octave higher than the fundamental of the string, even though the string is not pressed to the fingerboard. Why does this happen?

10. By listening to a band or orchestra, how can you deter-mine that the speed of sound is the same for all fre-quencies?
11. An archer shoots an arrow from a bow. Does the string of the bow exhibit standing waves after the arrow leaves? If so, and if the bow is perfectly symmetric so that the arrow leaves from the center of the string, what harmon-ics are excited?
12. The radar systems used by police to detect speeders are sensitive to the Doppler shift of a pulse of radio waves. Discuss how this sensitivity can be used to measure the speed of a car.
13. As oppositely moving pulses of the same shape (one up-ward, one downward) on a string pass through each other, there is one instant at which the string shows no displacement from the equilibrium position at any point. Has the energy carried by the pulses disappeared at this instant of time? If not, where is it?
14. A soft-drink bottle resonates as air is blown across its top. What happens to the resonant frequency as the level of fluid in the bottle decreases? (See QuickLab, p. 466.)
15. In Balboa Park in San Diego, California, there is a huge outdoor organ. Does the fundamental frequency of a particular pipe on this organ change on hot and cold days? How about on days with high and low atmospheric pressure?
16. Despite a reasonably steady hand, a person often spills his coffee when carrying it to his seat. Discuss resonance as a possible cause of this difficulty, and devise a means for solving the problem.
17. If you have a series of identical glass bottles, with varying amounts of water in them, you can play musical notes by either striking the bottles with a spoon or blowing across the open tops of the bottles. When hitting the bottles, the frequency of the note decreases as the water level rises. When blowing on the bottles, the frequency of the note increases as the water level rises. Why is the behav-ior of the frequency different in these two cases?

18. An airplane mechanic notices that the sound from a twin-engine aircraft rapidly varies in loudness when both engines are running. What could be causing this variation from loud to soft?

19. Why does a vibrating guitar string sound louder when placed on the instrument than it would if allowed to vibrate in the air while off the instrument?

PROBLEMS

1, 2, 3 = straightforward, intermediate, challenging   ☐ = full solution available in Study Guide/Student Solutions Manual   ◄ = Core Concepts Workbook
WEB = solution posted at **http://www.harcourtcollege.com/physics/cptech**   ◆ = biomedical application   ▨ = Interactive Physics

Section 14.2 Characteristics of Sound Waves

Section 14.3 The Speed of Sound

Note: Unless otherwise stated, use 345 m/s as the speed of sound in air.

1. A dolphin located in seawater at a temperature of 25°C emits a sound directed toward the bottom of the ocean 150 m below. How much time passes before it hears an echo?

2. A sound wave has a frequency of 700 Hz in air and a wavelength of 0.50 m. What is the temperature of the air?

3. The range of human hearing extends from approximately 20 Hz to 20 000 Hz. Find the wavelengths of these extremes at a temperature of 27°C.

4. A group of hikers hear an echo 3.00 s after they shout. If the temperature is 22.0°C, how far away is the mountain that reflected the sound wave?

5. The greatest value ever achieved for the speed of sound in air is about 1.0×10^4 m/s, and the highest frequency ever produced is about 2.0×10^{10} Hz. Find the wavelength of this wave.

6. A stone is dropped from rest into a well. The sound of the splash is heard exactly 2.00 s later. Find the depth of the well if the air temperature is 10.0°C.

Section 14.4 Energy and Intensity of Sound Waves

7. The area of a typical eardrum is about 5.0×10^{-5} m². Calculate the sound power (the energy per second) incident on an eardrum at **(a)** the threshold of hearing and **(b)** the threshold of pain.

8. A microphone in the ocean is sensitive to sounds emitted by porpoises. To produce a usable signal, sound waves striking the microphone must have an intensity of 10 dB. If porpoises emit sound waves with a power of 0.050 W, how far can a porpoise be from the microphone and still be heard? Disregard absorption of sound waves by the water.

WEB **9.** The intensity level of an orchestra is 85 dB. A single violin achieves a level of 70 dB. How does the intensity of the sound of the full orchestra compare with that of the violin's sound?

10. A noisy machine in a factory produces a decibel rating of 80 dB. How many identical machines could you add to the factory without exceeding the 90-dB limit?

11. Two sounds have measured intensities of $I_1 = 100$ W/m² and $I_2 = 200$ W/m². By how many decibels is the level of sound 1 lower than that of sound 2?

Section 14.5 Spherical and Plane Waves

12. A stereo speaker (considered a small source) emits sound waves with a power output of 100 W. **(a)** Find the intensity 10.0 m from the source. **(b)** Find the intensity level, in decibels, at this distance. **(c)** At what distance would you experience the sound at the threshold of pain, 120 dB?

13. A train sounds its horn as it approaches an intersection. The horn can just be heard at a level of 50 dB by an observer 10 km away. **(a)** What is the average power generated by the horn? **(b)** What intensity level of the horn's sound is observed by someone waiting at an intersection 50 m from the train? Treat the horn as a point source and neglect any absorption of sound by the air. Suppose the ground absorbs all the sound that is radiated downward.

14. A man shouting loudly produces a 70-dB sound at a distance of 5.0 m. How many watts of power does the man emit? (Treat the man as a point source.)

15. A skyrocket explodes 100 m above the ground (Fig. P14.15). Three observers are spaced 100 m apart, with the first (A) is directly under the point of the explosion.

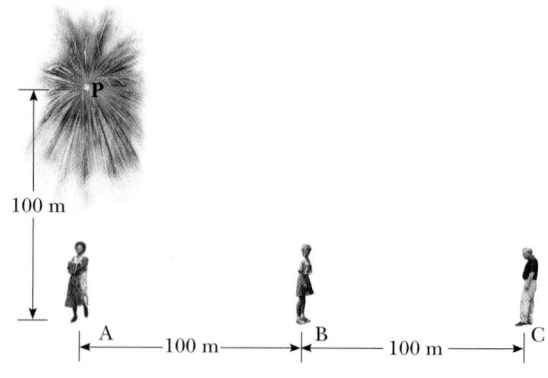

Figure P14.15

(a) What is the ratio of sound intensities heard by observers A and B? (b) What is the ratio of intensities heard by observers A and C?

Section 14.6 The Doppler Effect

16. A train at rest emits a sound at a frequency of 1000 Hz. An observer in a car travels away from the sound source at a speed of 30 m/s. What is the frequency heard by the observer?

17. An airplane traveling with half the speed of sound ($v = 172$ m/s) emits a sound of frequency 5.00 kHz. At what frequency does a stationary listener hear the sound (a) as the plane approaches? (b) after it passes?

18. At rest, a car's horn sounds the note A (440 Hz). The horn is sounded while the car is moving down the street. A bicyclist moving in the same direction with one third the car's speed hears a frequency of 415 Hz. What is the speed of the car? Is the cyclist ahead of or behind the car?

19. Two trains on separate tracks move toward one another. Train 1 has a speed of 130 km/h and train 2 a speed of 90.0 km/h. Train 2 blows its horn, emitting a frequency of 500 Hz. What is the frequency heard by the engineer on train 1?

20. A bat flying at 5.0 m/s emits a chirp at 40 kHz. If this sound pulse is reflected by a wall, what is the frequency of the echo received by the bat?

WEB **21.** An alert physics student stands beside the tracks as a train rolls slowly past. He notes that the frequency of the train whistle is 442 Hz when the train is approaching him and 441 Hz when the train is receding from him. From this he can find the speed of the train. What value does he find?

22. A supersonic jet traveling at Mach 3 (which means the speed of the jet is 3 times the speed of sound in air) at an altitude of 20 000 m is directly overhead at time $t = 0$, as in Figure P14.22. (a) How long will it be before the ground observer encounters the shock wave? (b) Where will the plane be when it is finally heard?

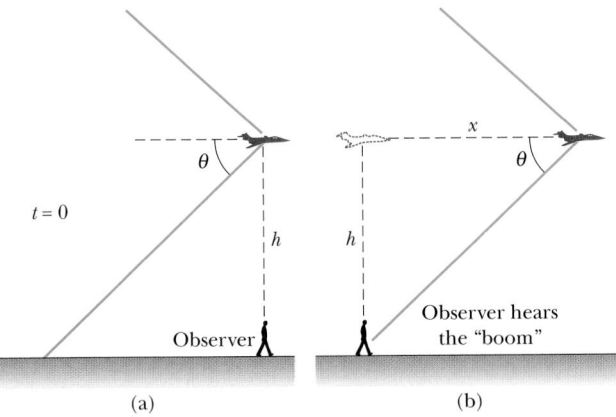

(a)

(b)

Figure P14.22

(Assume that the speed of sound in air is uniform at 345 m/s.)

23. The Concorde flies at Mach 1.5. What is the angle between the direction of propagation of the shock wave and the direction of the plane's velocity?

Section 14.7 Interference of Sound Waves

24. The sound interferometer shown in Figure 14.14 is driven by a speaker emitting a 400-Hz note. If *destructive* interference occurs at a particular instant, how much must the path length in the U-shaped tube be increased in order to hear (a) constructive interference and (b) destructive interference once again?

25. The ship in Figure P14.25 travels along a straight line parallel to the shore and 600 m from the shore. The ship's radio receives simultaneous signals of the same frequency from antennas A and B. The signals interfere constructively at point C, which is equidistant from A and B. The signal goes through the first minimum at point D. Determine the wavelength of the radio waves.

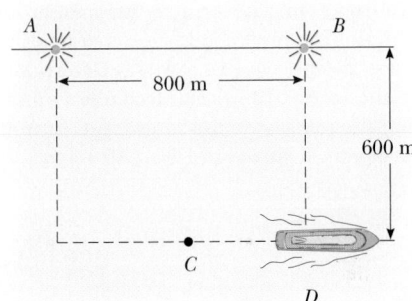

Figure P14.25

26. Two loudspeakers are placed above and below one another, as in Figure 14.15, and driven by the same source at a frequency of 500 Hz. (a) What minimum distance should the top speaker be moved back in order to create destructive interference between the two speakers? (b) If the top speaker is moved back twice the distance calculated in part (a), will constructive or destructive interference occur?

27. A pair of speakers separated by 0.700 m are driven by the same oscillator at a frequency of 690 Hz. An observer, originally positioned at one of the speakers, begins to walk along a line perpendicular to the line joining the two speakers. (a) How far must the observer walk before reaching a relative maximum in intensity? (b) How far will the observer be from the speaker when the first relative minimum is detected in the intensity?

Section 14.8   Standing Waves

28. A steel wire in a piano has a length of 0.7000 m and a mass of 4.300×10^{-3} kg. To what tension must this wire be stretched in order that the fundamental vibration correspond to middle C (f_C = 261.6 Hz on the chromatic musical scale)?

29. A stretched string fixed at each end has a mass of 40 g and a length of 8.0 m. The tension in the string is 49 N. (a) Determine the positions of the nodes and antinodes for the third harmonic. (b) What is the vibration frequency for this harmonic?

30. Resonance of sound waves can be produced within an aluminum rod by holding the rod at its midpoint and stroking it with an alcohol-saturated paper towel. In this resonance mode, the middle of the rod is a node and the ends are antinodes, there being no other nodes or antinodes present. What is the frequency of the resonance if the rod is 1 m long?

31. A 0.300-g wire is stretched between two points 70.0 cm apart. If the tension in the wire is 600 N, find the wire's first, second, and third harmonics.

32. A 12-kg mass hangs in equilibrium from a string of total length L = 5.0 m and linear mass density μ = 0.0010 kg/m. The string is wrapped around two light, frictionless pulleys that are separated by the distance d = 2.0 m (Fig. P14.32a). (a) Determine the tension in the string. (b) At what frequency must the string between the pulleys vibrate in order to form the standing wave pattern shown in Figure P14.32b?

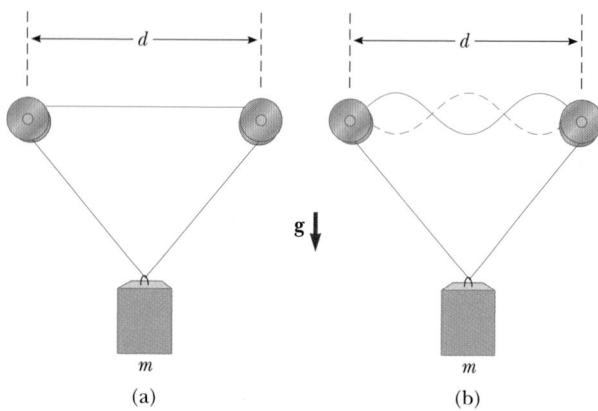

Figure P14.32

33. In the arrangement shown in Figure P14.33, a mass, m = 5.0 kg, hangs from a cord around a light pulley. The length of the cord between point P and the pulley is L = 2.0 m. (a) When the vibrator is set to a frequency of 150 Hz, a standing wave with six loops is formed. What must be the linear mass density of the cord? (b) How many loops (if any) will result if m is changed

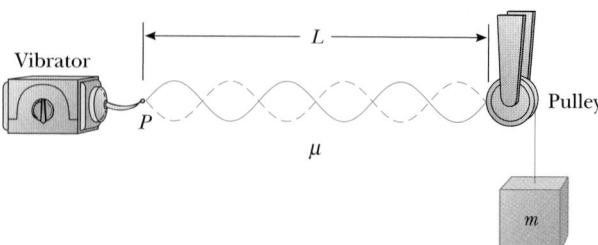

Figure P14.33

to 45 kg? (c) How many loops (if any) will result if m is changed to 10 kg?

34. Two pieces of steel wire with identical cross-sections have lengths of L and $2L$. Each of the wires is fixed at both ends and stretched so that the tension in the longer wire is four times greater than that in the shorter wire. If the fundamental frequency in the shorter wire is 60 Hz, what is the frequency of the second harmonic in the longer wire?

Section 14.9   Forced Vibrations and Resonance

35. A 5.0-kg mass connected to a spring is found to resonate when it is pushed at the frequency 2.4 Hz. Determine the spring constant for the spring.

36. The chains suspending a child's swing are 2.00 m long. At what frequency should a big brother push to make the child swing with largest amplitude?

Section 14.10   Standing Waves in Air Columns

37. A tuning fork is sounded above a resonating tube, as in Figure 14.23. The first resonant point is 0.080 m from the top of the tube, and the second is at 0.24 m. (a) Where is the third resonant point? (b) What is the frequency of the tuning fork?

38. The fundamental frequency of an open organ pipe corresponds to middle C (261.6 Hz on the chromatic musical scale). The third resonance of a closed organ pipe has the same frequency. What are the lengths of the two pipes?

39. The human ear canal is about 2.8 cm long. If it is regarded as a tube open at one end and closed at the eardrum, what is the fundamental frequency around which we would expect hearing to be best? Take the speed of sound to be 340 m/s.

40. A pipe open at each end has a fundamental frequency of 300 Hz when the speed of sound in air is 333 m/s. (a) How long is the pipe? (b) What is the frequency of the second harmonic when the temperature of the air is increased so that the speed of sound in the pipe is 344 m/s?

WEB **41.** A pipe open at both ends has a fundamental frequency of 300 Hz when the temperature is 0°C. (a) What is the length of the pipe? (b) What is the fundamental frequency at a temperature of 30°C?

42. A 2.00-m-long air column is open at both ends. The frequency of a certain harmonic is 410 Hz, and the frequency of the next higher harmonic is 492 Hz. Determine the speed of sound in the air column.

Section 14.11   Beats

43. Two identical mandolin strings under 200 N of tension are sounding tones with frequencies of 523 Hz. The peg of one string slips slightly, and the tension in it drops to 196 N. How often are beats heard?

44. The G string on a violin has a fundamental frequency of 196 Hz. It is 30.0 cm long and has a mass of 0.500 g. While this string is sounding, a nearby violinist effectively shortens (by sliding her finger down the string) the G string on her identical violin until a beat frequency of 2.00 Hz is heard between the two strings. When this occurs, what is the effective length of her string?

45. Two train whistles have identical frequencies of 180 Hz. When one train is at rest in the station, sounding its whistle, a beat frequency of 2 Hz is heard from a moving train. What two possible speeds and directions can the moving train have?

46. Two pipes, equal in length, are each open at one end. Each has a fundamental frequency of 480 Hz at 300 K. In one pipe the air temperature is increased to 305 K. If the two pipes are sounded together, what beat frequency results?

Section 14.13   The Ear

47. If a human ear canal can be thought of as resembling an organ pipe, closed at one end, that resonates at a fundamental frequency of 3000 Hz, what is the length of the canal? Use normal body temperature for your determination of the speed of sound in the canal.

48. Some studies indicate that the upper frequency limit of hearing is determined by the diameter of the eardrum. The wavelength of the sound wave and the diameter of the eardrum are approximately equal at this upper limit. If this is so, what is the diameter of the eardrum of a person capable of hearing 20 000 Hz? (Assume a body temperature of 37° C.)

ADDITIONAL PROBLEMS

49. Two cars are traveling in the same direction, both at a speed of 55 mph (80.7 ft/s). The driver of the trailing car sounds his horn, which has a frequency of 300 Hz. If the speed of sound is 1100 ft/s, what sound frequency is heard by the driver of the leading car? (*Hint:* Consider the relative motion between the source and observer in this case.)

50. A commuter train blows its horn as it passes a passenger platform at a constant speed of 40.0 m/s. The train horn sounds at a frequency of 320 Hz when the train is at rest. (a) What is the frequency observed by a person on the platform as the train approaches and (b) as the train recedes from him? (c) What wavelength does the observer find in each case?

51. On a workday the average decibel level of a busy street is 70 dB, with 100 cars passing a given point every minute. If the number of cars is reduced to 25 every minute on a weekend, what is the decibel level of the street?

52. A variable-length air column is placed just below a vibrating wire that is fixed at both ends. The length of the air column is gradually increased from zero until the first position of resonance is observed at $L = 34$ cm. The wire is 120 cm long and is vibrating in its third harmonic. If the speed of sound in air is 340 m/s, what is the speed of transverse waves in the wire?

53. A flute is designed so that it plays a frequency of 261.6 Hz, middle C, when all the holes are covered and the temperature is 20.0°C. (a) Consider the flute to be a pipe open at both ends, and find its length, assuming that the middle C frequency is the fundamental. (b) A second player, nearby in a colder room, also attempts to play middle C on an identical flute. A beat frequency of 3.00 beats/s is heard. What is the temperature of the room?

54. When at rest, two trains have sirens that emit a frequency of 300 Hz. The two trains travel toward one another and toward an observer stationed between them. One of the trains moves at 30 m/s, and the observer hears a beat frequency of 3 beats per second. What is the velocity of the second train, which travels faster than the first?

55. A speaker at the front of a room and an identical speaker at the rear of the room are being driven at 456 Hz by the same sound source. A student walks at a uniform rate of 1.50 m/s away from one speaker and toward the other. How many beats does the student hear per second?

56. A tree branch touches the surface of a smoothly gliding river. The wind makes the branch vibrate steadily to produce ripples that move outward with speed 0.500 m/s relative to the water. If the wavelength of the waves downstream is 50.0% longer than the wavelength of the ripples upstream from the branch, how fast is the river flowing?

57. A flower pot is knocked off a balcony 20.0 m above the sidewalk heading for a 1.75-m-tall man standing below. How high from the ground can the flower pot be after which it would be too late for a shouted warning to reach the man in time? Assume that the man below requires 0.300 s to respond to the warning.

58. A rescue plane flies horizontally at a constant speed searching for a disabled boat. When the plane is directly above the boat, the boat's crew blows a loud horn. By the time the plane's sound detector perceives the horn's sound, the plane has traveled a distance equal to one half its altitude above the ocean. If it takes the sound 2.00 s to reach the plane, determine (a) the speed of the plane and (b) its altitude. Take the speed of sound to be 343 m/s.

59. A tuning fork vibrating at 512 Hz falls from rest and accelerates at 9.80 m/s^2. How far below the point of release is the tuning fork when waves of frequency 485 Hz reach the release point? Take the speed of sound in air to be 340 m/s.

60. A block with a speaker bolted to it is connected to a spring having spring constant $k = 20.0 \text{ N/m}$, as in Figure P14.60. The total mass of the block and speaker is

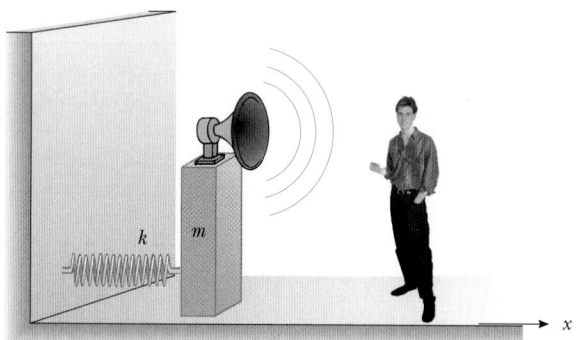

Figure P14.60

5.00 kg, and the amplitude of this unit's motion is 0.500 m. If the speaker emits sound waves of frequency 440 Hz, determine the range in frequencies heard by the person to the right of the speaker. Assume that the speed of sound is 343 m/s.

61. In order to be able to determine her speed, a skydiver carries a tone generator. A friend on the ground at the landing site has equipment for receiving and analyzing sound waves. While the skydiver is falling at terminal speed, her tone generator emits a steady tone of 1800 Hz. (Assume that the air is calm and that the sound speed is 343 m/s, independent of altitude.) (a) If her friend on the ground (directly beneath the skydiver) receives waves of frequency 2150 Hz, what is the skydiver's speed of descent? (b) If the skydiver were also carrying sound-receiving equipment sensitive enough to detect waves reflected from the ground, what frequency would she receive?

62. Two identical speakers separated by 10.0 m are driven by the same oscillator with a frequency of $f = 21.5 \text{ Hz}$ (Fig. P14.62). (a) Explain why a receiver at A records a

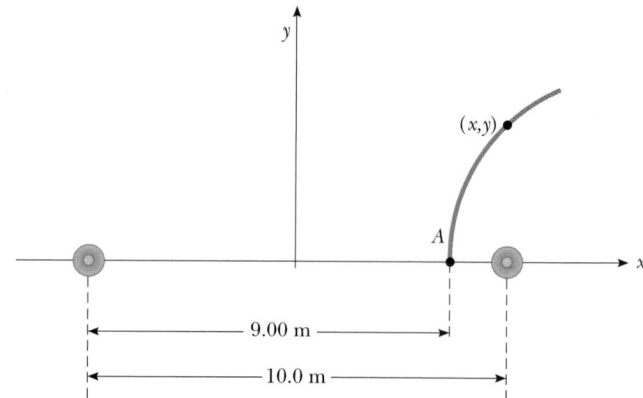

Figure P14.62

minimum in sound intensity from the two speakers. (b) If the receiver is moved in the plane of the speakers, what path should it take so that the intensity remains at a minimum? That is, determine the relationship between x and y (the coordinates of the receiver) that causes the receiver to record a minimum in sound intensity. Take the speed of sound to be 344 m/s.

WEB 63. By proper excitation, it is possible to produce both longitudinal and transverse waves in a long metal rod. In a particular case, the rod is 150 cm long and 0.200 cm in radius and has a mass of 50.9 g. Young's modulus for the material is 6.80×10^{10} Pa. Determine the required tension in the rod so that the ratio of the speed of longitudinal waves to the speed of transverse waves is 8.

64. A student stands several meters in front of a smooth reflecting wall, holding a board on which a wire is fixed at each end. The wire, vibrating in its third harmonic, is 75.0 cm long, has a mass of 2.25 g, and is under a tension of 400 N. A second student, moving toward the wall, hears 8.30 beats per second. What is the speed of the student approaching the wall? Use 340 m/s as the speed of sound in air.

65. Two ships are moving along a line due east. The trailing vessel has a speed of 64.0 km/h relative to a land-based observation point, and the leading ship has a speed of 45.0 km/h relative to the same station. The two ships are in a region of the ocean where the current is moving uniformly due west at 10.0 km/h. The trailing ship transmits a sonar signal at a frequency of 1200 Hz. What frequency is monitored by the leading ship? (Use 1520 m/s as the speed of sound in ocean water.)

66. The Doppler equation presented in the text is valid when the motion between the observer and the source occurs on a straight line, so that the source and observer are moving either directly toward or directly away from each other. If this restriction is relaxed, one must use the more general Doppler equation

$$f' = \left[\frac{v + v_0 \cos(\theta_0)}{v - v_s \cos(\theta_s)} \right] f$$

where θ_0 and θ_s are defined in Figure P14.66a. **(a)** If both observer and source are moving away from each other, show that the preceding equation reduces to the Doppler equations used in the text, with lower signs. **(b)** Use the preceding equation to solve the following problem. A train moves at a constant speed of 25.0 m/s toward the intersection shown in Figure P14.66b. A car is stopped near the intersection, 30.0 m from the tracks. If the train's horn emits a frequency of 500 Hz, what is the frequency heard by the passengers in the car when the train is 40 m from the intersection? Take the speed of sound to be 343 m/s.

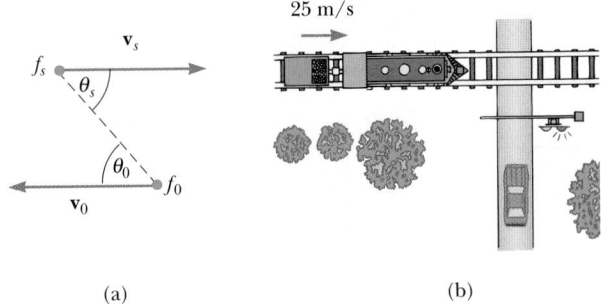

(a)   (b)

Figure P14.66

"I love hearing that lonesome wail of the train whistle as the magnitude of the frequency of the wave changes due to the Doppler effect."

PART 4

Electricity and Magnetism

We now begin a study of the branch of physics concerned with electric and magnetic phenomena. The laws of electricity and magnetism play central roles in the operation of many devices such as radios, televisions, electric motors, computers, high-energy accelerators, and a host of electronic devices used in medicine. More fundamentally, we now know that the interatomic and intermolecular forces that are responsible for the formation of solids and liquids are electric in origin. Furthermore, such forces as the pushes and pulls between objects and the elastic force in a spring arise from electric forces at the atomic level.

The ancient Greeks observed electric and magnetic phenomena as early as 700 B.C. They found that a piece of amber, when rubbed, became electrified and attracted pieces of straw or feathers. The existence of magnetic forces was known as a result of observations that a naturally occurring stone called *magnetite* (Fe_2O_3) was attracted to iron. (The word *electric* comes from the Greek word for amber, *elecktron*. The word *magnetic* comes from the name of the country where magnetite was found, *Magnesia*, now Turkey.)

In 1600 William Gilbert discovered that electrification was not limited to amber but is a general phenomenon. Experiments by Charles Coulomb confirmed the inverse-square force law for electricity.

It was not until the early part of the 19th century that scientists established that electricity and magnetism are, in fact, related phenomena. In 1820 Hans Oersted discovered that a compass needle is deflected when placed near a wire carrying an electric current. A few years later, Michael Faraday showed that when a wire is moved near a magnet (or, equivalently, when a magnet is moved near a wire), an electric current is observed in the wire. James Clerk Maxwell used these observations and other experimental facts as bases for formulating the laws of electromagnetism as we now know them. (*Electromagnetism* is a name given to the combined fields of electricity and magnetism.) Shortly thereafter, Heinrich Hertz verified Maxwell's predictions by producing electromagnetic waves in the laboratory. This was followed by such practical developments as radio and television.

Maxwell's contributions to the science of electromagnetism were especially significant because the laws he formulated are basic to *all* forms of electromagnetic phenomena. His work is comparable in importance to Newton's discovery of the laws of motion and the theory of gravitation.

◄ *Courtesy of Sandia National Laboratories*

Electric Forces and Electric Fields

▲ PHYSICS PUZZLER

This photo shows a variety of foods covered with plastic wrap for freshness. Why does plastic wrap stick readily to a dish and to itself but not to food?
(Semple Design and Photography)

11.1

The earliest known study of electricity was conducted by the Greeks about 700 B.C. By modern standards, their contributions to the field were modest. However, from those roots have sprung the enormous electrical distribution systems and sophisticated electronic instruments that are so much a part of our world today. It all began, apparently, when someone noticed that a fossil material called amber would attract small objects after being rubbed with wool. Since then we have learned that this phenomenon is not restricted to amber and wool but occurs (to some degree) when almost any two nonconducting substances are rubbed together.

In this chapter we use this effect, charging by friction, to begin an investigation of electric forces. We then discuss Coulomb's law, which is the fundamental law of force between any two stationary charged particles. The concept of an electric field associated with charges is then introduced and its effects on other charged particles described. We end with brief discussions of the Van de Graaff generator and the oscilloscope.

15.1   PROPERTIES OF ELECTRIC CHARGES

A number of simple experiments demonstrate the existence of electrostatic forces. For example, after running a plastic comb through your hair, you will find that the comb attracts bits of paper. The attractive force is often strong enough to suspend the paper from the comb. The same effect occurs with other rubbed materials, such as glass and hard rubber.

Another simple experiment is to rub an inflated balloon with wool (or across your hair). On a dry day, the rubbed balloon will then stick to the wall of a room, often for hours. When materials behave in this way, they are said to have become **electrically charged.** You can give your body an electric charge by vigorously rubbing your shoes on a wool rug or by sliding across a car seat. You can then feel, and remove, the charge on your body by lightly touching another person. Under the right conditions, a visible spark can be seen when you touch, and a slight tingle is felt by both parties. (Experiments such as these work best on a dry day because excessive moisture can provide a pathway for charge to leak off a charged object.)

Experiments also demonstrate that there are two kinds of electric charge, which Benjamin Franklin (1706–1790) named **positive** and **negative.** Figure 15.1 illustrates the interaction of the two charges. A hard rubber (or plastic) rod that has been rubbed with fur (or an acrylic material) is suspended by a piece of string. When a glass rod that has been rubbed with silk is brought near the rubber rod, the rubber rod is attracted toward the glass rod (Fig. 15.1a). If two charged rubber rods (or two charged glass rods) are brought near each other, as in Figure 15.1b, the force between them is repulsive. This observation demonstrates that the rubber and glass have different kinds of charge. We use the convention suggested by Franklin, wherein the electric charge on the glass rod is called positive and that on the rubber rod is called negative. On the basis of observations such as these, we conclude that **like charges repel one another and unlike charges attract one another.**

We now know that the origin of charge is the atom. As we shall see shortly, nature's basic carrier of positive electricity is the proton, a particle, along with

11.2, BOTH SECTIONS

◀ Like charges repel; unlike charges attract.

Figure 15.1 (a) A negatively charged rubber rod, suspended by a thread, is attracted to a positively charged glass rod. (b) A negatively charged rubber rod is repelled by another negatively charged rubber rod.

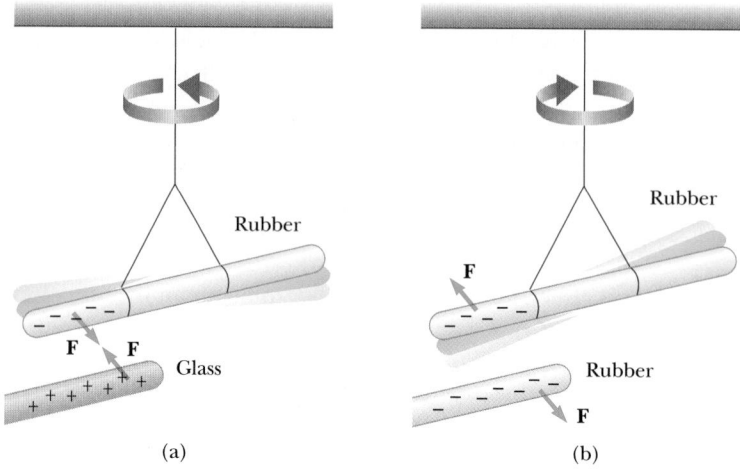

(a)   (b)

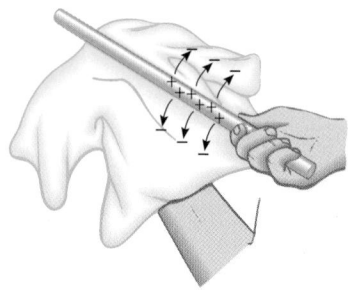

Figure 15.2 When a glass rod is rubbed with silk, electrons are transferred from the glass to the silk. Because of conservation of charge, each electron adds negative charge to the silk, and an equal positive charge is left behind on the rod. Also, because the charges are transferred in discrete bundles, the charges on the two objects are $\pm e$, or $\pm 2e$, or $\pm 3e$, and so on.

Charge is conserved; charge is ▶ quantized.

neutrons, located in the nucleus of an atom. Because the nucleus of an atom is held firmly in place inside a solid, protons are never moved from one material to another. Thus, when an object becomes charged, it does so because it has either gained or lost nature's basic carrier of negative electricity, the electron. It is convenient to picture the electron circling about the nucleus of an atom like a planet revolving about the Sun. The usual state of an atom is that for every positive charge on the nucleus there is an electron orbiting the nucleus to cancel the charge, leaving the atom as a neutral body. Any uncharged object in our large-scale world contains an enormous number of electrons and protons—about 10^{23} of each. However, for every negative electron a positively charged proton is present.

Charge has a natural tendency to be transferred between unlike materials. Rubbing the two materials together serves to increase the area of contact and thus to enhance the charge transfer process.

An important characteristic of charge is that **electric charge is always conserved.** That is, when two initially neutral objects are charged by being rubbed together, charge is not created in the process. The objects become charged because **negative charge is transferred from one object to the other.** One object gains some amount of negative charge while the other loses an equal amount of negative charge and hence is left with a positive charge. For example, when a glass rod is rubbed with silk, as in Figure 15.2, the silk obtains a negative charge that is equal in magnitude to the positive charge on the glass rod as negatively charged electrons are transferred from the glass to the silk in the rubbing process. Likewise, when rubber is rubbed with fur, electrons are transferred from the fur to the rubber.

In 1909 Robert Millikan (1886–1953) discovered that if an object is charged, its charge is always a multiple of a fundamental unit of charge, which we designate with the symbol e. In modern terms, the charge is said to be **quantized.** This means that charge occurs as discrete bundles in nature. Thus, an object may have a charge of $\pm e$, or $\pm 2e$, or $\pm 3e$, and so on, but never,[1] say, a fractional charge of $\pm 1.5e$.

---

[1] Recent developments have suggested the existence of fundamental particles called **quarks** that have charges of $\pm e/3$ or $\pm 2e/3$. A more complete discussion of quarks and their properties is presented in Chapter 30.

Other experiments in Millikan's time showed that the electron has a charge of $-e$ and the proton has an equal and opposite charge, $+e$. Some particles, such as a neutron, have no charge. A neutral atom (one with no net charge) contains as many protons as electrons. The value of e is now known to be $1.602\ 19 \times 10^{-19}$ C. (The unit of electric charge, the **coulomb** [C], will be defined more precisely in a later section.)

15.2 INSULATORS AND CONDUCTORS

It is convenient to classify substances in terms of their ability to conduct electric charge.

 11.3, SECTION 1

Conductors are materials in which electric charges move freely, and **insulators** are materials in which electric charges do not move freely.

Glass and rubber are insulators. When such materials are charged by rubbing, only the rubbed area becomes charged, and there is no tendency for the charge to move into other regions of the material. In contrast, materials such as copper, aluminum, and silver are good conductors. When such materials are charged in some small region, the charge readily distributes itself over the entire surface of the material. If you hold a copper rod in your hand and rub the rod with wool or fur, it will not attract a piece of paper. This might suggest that a metal cannot be charged. However, if you hold the copper rod with an insulator and then rub it with wool or fur, the rod remains charged and attracts the paper. In the first case, the electric charges produced by rubbing readily move from the copper through your body and finally to Earth. In the second case, the insulating handle prevents the flow of charge to Earth.

Semiconductors are a third class of materials, and their electrical properties are somewhere between those of insulators and those of conductors. Silicon and germanium are well-known semiconductors that are widely used in the fabrication of a variety of electronic devices. If controlled amounts of certain foreign atoms are added to semiconductors, their electrical properties can be changed by many orders of magnitude.

Charging by Conduction

Consider a negatively charged rubber rod brought into contact with a neutral conducting sphere that is insulated so that there is no conducting path for charges to leave the sphere. Some electrons on the rubber rod are now able to move onto the sphere, as in Figure 15.3. When the rubber rod is removed, the sphere is left with a negative charge. This process is referred to as charging by **conduction.** The object being charged in such a process (the sphere) is always left with a charge having the same sign as the object doing the charging (the rubber rod).

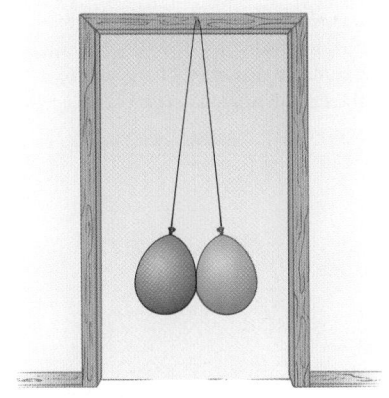

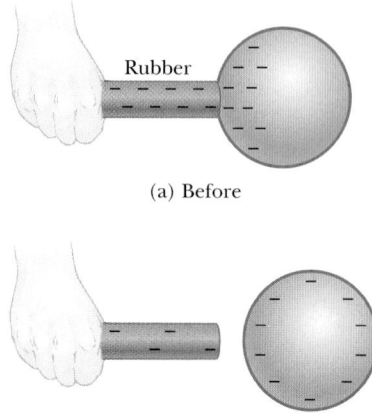

(a) Before

(b) After

Figure 15.3   Charging a metal object by conduction. (a) The charged rubber rod is placed in contact with the insulated metal sphere. Some electrons move from the rod onto the sphere. (b) When the rod is removed, the sphere is left with a negative charge.

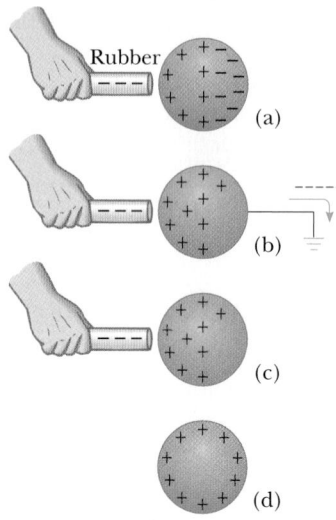

Figure 15.4   Charging a metal object by induction. (a) The charge on a neutral metal sphere is redistributed when a charged rubber rod is placed near the sphere. (b) The sphere is grounded, and some of the electrons leave the sphere through the ground wire. (c) The ground connection is removed, and the sphere is left with excess positive charge. (d) When the rubber rod is moved away, the sphere becomes uniformly charged.

Charging by Induction

When a conductor is connected to Earth by means of a conducting wire or copper pipe, it is said to be **grounded.** The Earth can be considered an infinite reservoir for electrons; this means that it can accept or supply an unlimited number of electrons. With this in mind, we can understand the charging of a conductor by a process known as **induction.**

Consider a negatively charged rubber rod brought near a neutral (uncharged) conducting sphere that is insulated so that there is no conducting path to ground (Fig. 15.4). The repulsive force between the electrons in the rod and those in the sphere causes a redistribution of charge on the sphere so that some electrons move to the side of the sphere farthest away from the rod (Fig. 15.4a). The region of the sphere nearest the negatively charged rod has an excess of positive charge because of the migration of electrons away from this location. If a grounded conducting wire is then connected to the sphere, as in Figure 15.4b, some of the electrons leave the sphere and travel to the Earth. If the wire to ground is then removed (Fig. 15.4c), the conducting sphere is left with an excess of induced positive charge. Finally, when the rubber rod is removed from the vicinity of the sphere (Fig. 15.4d), the induced positive charge remains on the ungrounded sphere. This excess positive charge becomes uniformly distributed over the surface of the ungrounded sphere because of the repulsive forces among the like charges and the high mobility of charge carriers in a metal.

In the process of inducing a charge on the sphere, the charged rubber rod loses none of its negative charge because it never came in contact with the sphere. Furthermore, the sphere is left with a charge opposite to that of the rubber rod. Note that **charging an object by induction requires no contact with the object inducing the charge.**

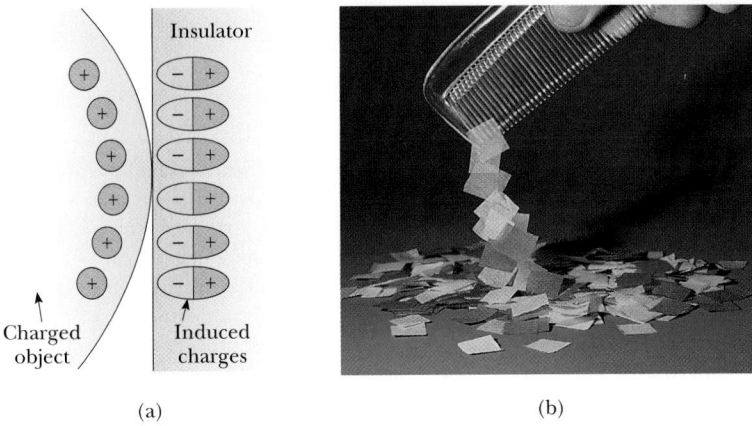

(a)   (b)

Figure 15.5 (a) The charged object on the left induces charges on the surface of an insulator. (b) A charged comb attracts bits of paper because charges are displaced in the paper. (*© 1968 Fundamental Photographs*)

A process very similar to charging by induction in conductors also takes place in insulators. In most neutral atoms or molecules, the center of positive charge coincides with the center of negative charge. However, in the presence of a charged object, these centers may shift slightly, resulting in more positive charge on one side of the molecule than on the other. This effect is known as **polarization.** The realignment of charge within individual molecules produces an induced charge on the surface of the insulator, as shown in Figure 15.5a. With these concepts, you should be able to explain why a comb that has been rubbed through hair will attract bits of neutral paper, or why a balloon that has been rubbed against your clothing can stick to a neutral wall.

Thinking Physics 1

A positively charged object hanging from a string is brought near a nonconducting object. Based on the behavior of the ball-string combination, the ball is seen to be attracted to the object. From this experiment, it is not possible to determine whether the object is negatively charged or neutral. Why not? What additional experiment would help you decide between these two possibilities?

Explanation The attraction between the ball and the object could be an attraction of unlike charges, or it could be an attraction between a charged object and a neutral object as a result of polarization of the molecules of the neutral object. Two additional experiments could help us determine if the object is charged. First, a known neutral ball could be brought near the object—if there is an attraction, the object is negatively charged. Another possibility is to bring a known negatively charged ball near the object—if there is a repulsion, then the object is negatively charged. If there is no attraction, then the object is neutral.

Applying Physics 1

You may have noticed that the vertical surface of your television screen or your computer monitor becomes very dusty. Why does this occur?

Explanation   The accumulation of dust on the vertical surface may surprise you, perhaps because it is more common to see dust on horizontal surfaces settling there under the action of gravity. The dust on the screen is attracted electrically. A television screen is constantly bombarded with electrons from the electron gun. As a result, it becomes negatively charged, and this charge will polarize dust particles in the air in front of the screen, just like a charged object polarizes molecules in a neutral object. This produces an attractive force on the dust particles, pulling them to the screen where they adhere.

15.3   COULOMB'S LAW

In 1785 Charles Coulomb (1736–1806) established the fundamental law of electric force between two stationary charged particles. Experiments show that

An **electric force** has the following properties:
1. It is inversely proportional to the square of the separation, r, between the two particles and is along the line joining them.
2. It is proportional to the product of the magnitudes of the charges, $|q_1|$ and $|q_2|$, on the two particles.
3. It is attractive if the charges are of opposite sign and repulsive if the charges have the same sign.

From these observations, we can express the magnitude of the electric force between two charges separated by a distance of r as

Coulomb's law ▶

$$F = k_e \frac{|q_1||q_2|}{r^2}$$ [15.1]

where k_e is a constant called the *Coulomb constant*.

The value of the Coulomb constant in Equation 15.1 depends on the choice of units. The SI unit of charge is the **coulomb (C),** which is defined in terms of a current unit called the **ampere (A),** where current is defined as the rate of flow of charge. (The ampere will be defined in Chapter 17.) When the current in a wire is 1 A, the amount of charge that flows past a given point in the wire in 1 s is 1 C. From experiment, we know that the **Coulomb constant** in SI units has the value

$$k_e = 8.9875 \times 10^9 \ \text{N} \cdot \text{m}^2/\text{C}^2$$ [15.2]

To simplify our calculations, we shall use the approximate value

$$k_e \approx 8.99 \times 10^9 \ \text{N} \cdot \text{m}^2/\text{C}^2$$ [15.3]

Charles Coulomb (1736–1806)

Coulomb's major contribution to science was in the field of electrostatics and magnetism. During his lifetime, he also investigated the strengths of materials and determined the forces that affect objects on beams, thereby contributing to the field of structural mechanics. In the field of ergonomics, his research provided a fundamental understanding of the ways in which people and animals can best do work. *(Photo courtesy of AIP Niels Bohr Library, E. Scott Barr Collection)*

TABLE 15.1 Charge and Mass of the Electron, Proton, and Neutron

Particle	Charge (C)	Mass (kg)
Electron	-1.60×10^{-19}	9.11×10^{-31}
Proton	$+1.60 \times 10^{-19}$	1.67×10^{-27}
Neutron	0	1.67×10^{-27}

The charge on the proton has a magnitude of $e = 1.6 \times 10^{-19}$ C. Therefore, it would take $1/e = 6.3 \times 10^{18}$ protons to create a total charge of $+1$ C. Likewise, 6.3×10^{18} electrons would have a total charge of -1 C. This can be compared with the number of free electrons in 1 cm^3 of copper, which is on the order of 10^{23}. Even so, 1 C is a substantial amount of charge. In typical electrostatic experiments, where a rubber or glass rod is charged by friction, a net charge on the order of 10^{-6} C $(= 1 \ \mu\text{C})$ is obtained. In other words, only a very small fraction of the total available charge is transferred between the rod and rubbing material. Table 15.1 lists the charges and masses of the electron, proton, and neutron.

When dealing with Coulomb's force law, remember that force is a vector quantity and must be treated accordingly. Furthermore, note that Coulomb's law applies exactly only to point charges or particles and to spherical distributions of charge. Figure 15.6a shows the electric force of repulsion between two positively charged particles. Electric forces obey Newton's third law, and hence the forces $\mathbf{F}_{12}$ and $\mathbf{F}_{21}$ are equal in magnitude but opposite in direction. (The notation $\mathbf{F}_{12}$ denotes the force on particle 1 exerted by particle 2. Likewise, $\mathbf{F}_{21}$ is the force on particle 2 exerted by particle 1.) Note that in this case, r is the distance between the centers of the spheres. It bears repeating that F_{12} and F_{21} are always equal regardless of whether q_1 and q_2 have the same magnitude.

The Coulomb force is the second example we have seen of a field force—a force exerted by one object on another even though there is no physical contact between them. Recall that another example of a field force is gravitational attraction. The mathematical form of the Coulomb force is the same as that of the gravitational force. That is, they are both inversely proportional to the square of the distance of separation. However, there are some important differences between electric and gravitational forces. Electric forces can be either attractive or repulsive, but gravitational forces are always attractive. Furthermore, gravitational forces are considerably weaker, as shown by the following example.

11.4, BOTH SECTIONS

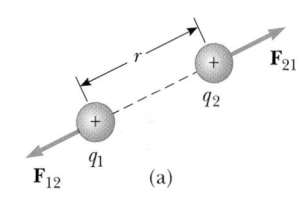

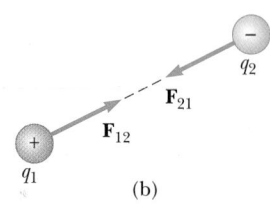

Figure 15.6 Two point charges separated by a distance of r exert a force on each other given by Coulomb's law. The force on q_1 is equal in magnitude and opposite in direction to the force on q_2. (a) When the charges are of the same sign, the force is repulsive. (b) When the charges are of opposite sign, the force is attractive.

EXAMPLE 15.1 The Electric Force and the Gravitational Force

The electron and proton of a hydrogen atom are separated (on the average) by a distance of about 5.3×10^{-11} m. Find the magnitudes of the electric force and the gravitational force that each particle exerts on the other.

Solution From Coulomb's law, we find that the attractive electric force has the magnitude

$$F_e = k_e \frac{|e|^2}{r^2} = \left(8.99 \times 10^9 \, \frac{\text{N} \cdot \text{m}^2}{\text{C}^2}\right) \frac{(1.6 \times 10^{-19} \, \text{C})^2}{(5.3 \times 10^{-11} \, \text{m})^2} = \boxed{8.2 \times 10^{-8} \, \text{N}}$$

From Newton's universal law of gravity and Table 15.1, we find that the gravitational force has the magnitude

$$F_g = G \frac{m_e m_p}{r^2} = \left(6.67 \times 10^{-11} \, \frac{\text{N} \cdot \text{m}^2}{\text{kg}^2}\right) \frac{(9.11 \times 10^{-31} \, \text{kg})(1.67 \times 10^{-27} \, \text{kg})}{(5.3 \times 10^{-11} \, \text{m})^2}$$

$$= \boxed{3.6 \times 10^{-47} \, \text{N}}$$

Because $F_e/F_g \approx 2 \times 10^{39}$, the gravitational force between the charged atomic particles is negligible compared with the electric force.

The Superposition Principle

Frequently, more than two charges are present and it is necessary to find the net electric force on one of them. This can be accomplished by noting that the electric force between any pair of charges is given by Equation 15.1. Therefore, the resultant force on any one charge equals the vector sum of the forces exerted by the other individual charges that are present. This is another example of the **superposition principle.** For example, if you have three charges and you want to find the force exerted on charge 1 by charges 2 and 3, you first find the force exerted on charge 1 by charge 2 and the force exerted on charge 1 by charge 3. You then add these two forces together vectorially to get the resultant force on charge 1. The following numerical example illustrates this procedure.

Workbook Problem 69 —
(Workbook page 148) is similar to
Example 15.2.

EXAMPLE 15.2   Using the Superposition Principle

Consider three point charges at the corners of a triangle, as in Figure 15.7, where $q_1 = 6.00 \times 10^{-9}$ C, $q_3 = 5.00 \times 10^{-9}$ C, $q_2 = -2.00 \times 10^{-9}$ C, and the distances of separation are shown in the figure. Find the resultant force on q_3.

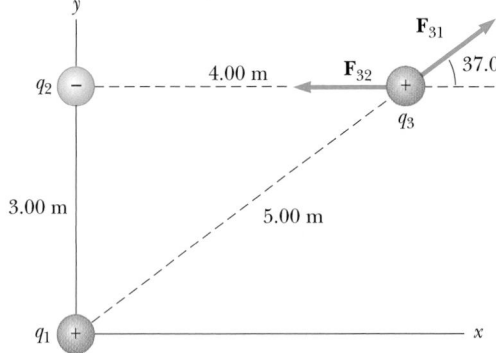

Figure 15.7 (Example 15.2) The force exerted on q_3 by q_1 is $\mathbf{F}_{31}$. The force exerted on q_3 by q_2 is $\mathbf{F}_{32}$. The *total force*, $\mathbf{F}_3$, on q_3 is the *vector sum* $\mathbf{F}_{31} + \mathbf{F}_{32}$.

Reasoning   It is first necessary to find the direction of the forces exerted on q_3 by q_1 and q_2. The force $\mathbf{F}_{32}$ exerted on q_3 by q_2 is attractive because q_2 and q_3 have opposite signs. The force $\mathbf{F}_{31}$ exerted on q_3 by q_1 is repulsive because both q_1 and q_3 are positive. To find the net force on q_3 it is necessary to find the magnitude of $\mathbf{F}_{32}$ and $\mathbf{F}_{31}$ by use of Coulomb's law and then add the two forces vectorially.

Solution The magnitude of the force exerted on q_3 by q_2 is

$$F_{32} = k_e \frac{|q_3||q_2|}{r^2} = (8.99 \times 10^9 \ \text{N} \cdot \text{m}^2/\text{C}^2) \frac{(5.00 \times 10^{-9} \ \text{C})(2.00 \times 10^{-9} \ \text{C})}{(4.00 \ \text{m})^2}$$

$$= 5.62 \times 10^{-9} \ \text{N}$$

The magnitude of the force exerted on q_3 by q_1 is

$$F_{31} = k_e \frac{|q_3||q_1|}{r^3} = (8.99 \times 10^9 \ \text{N} \cdot \text{m}^2/\text{C}^2) \frac{(5.00 \times 10^{-9} \ \text{C})(6.00 \times 10^{-9} \ \text{C})}{(5.00 \ \text{m})^2}$$

$$= 1.08 \times 10^{-8} \ \text{N}$$

The force $\mathbf{F}_{31}$ makes an angle of 37.0° with the x axis. Therefore, the x component of this force has the magnitude $F_{31} \cos 37.0° = 8.63 \times 10^{-9}$ N, and the y component has the magnitude $F_{31} \sin 37.0° = 6.50 \times 10^{-9}$ N. The force $\mathbf{F}_{32}$ is in the negative x direction. Hence, the x and y components of the resultant force on q_3 are

$$F_x = 8.63 \times 10^{-9} \ \text{N} - 5.62 \times 10^{-9} \ \text{N} = 3.01 \times 10^{-9} \ \text{N}$$

$$F_y = 6.50 \times 10^{-9} \ \text{N}$$

The magnitude of the resultant force on the charge q_3 is therefore

$$\sqrt{(3.01 \times 10^{-9} \ \text{N})^2 + (6.50 \times 10^{-9} \ \text{N})^2} = \ \ 7.16 \times 10^{-9} \ \text{N}$$

and the force vector makes an angle of 65.2° with the x axis.

EXAMPLE 15.3   Where Is the Resultant Force Zero?

Three charges lie along the x axis, as in Figure 15.8. The positive charge $q_1 = 15 \ \mu\text{C}$ is at $x = 2.0$ m, and the positive charge $q_2 = 6.0 \ \mu\text{C}$ is at the origin. Where must a *negative* charge, q_3, be placed on the x axis so that the resultant force on it is zero?

Reasoning The only location where the force exerted on q_3 by q_2 is opposite the force exerted on q_3 by q_1 lies on the x axis between q_1 and q_2, as in Figure 15.8. Because we require that the resultant force on q_3 be zero, then F_{32} must equal F_{31}.

Solution If we let x be the coordinate of q_3, then the forces $\mathbf{F}_{31}$ and $\mathbf{F}_{32}$ have the magnitudes

$$F_{31} = k_e \frac{|q_3|(15 \times 10^{-6} \ \text{C})}{(2.0 - x)^2} \quad \text{and} \quad F_{32} = k_e \frac{|q_3|(6.0 \times 10^{-6} \ \text{C})}{x^2}$$

If the resultant force on q_3 is zero, then $\mathbf{F}_{32}$ must be equal to and opposite $\mathbf{F}_{31}$, or

$$k_e \frac{|q_3|(15 \times 10^{-6} \ \text{C})}{(2.0 - x)^2} = k_e \frac{|q_3|(6.0 \times 10^{-6} \ \text{C})}{x^2}$$

Because k_e, 10^{-6}, and q_3 are common to both sides, they can be cancelled from the equation, and we have (after some reduction)

$$(2.0 - x)^2(6.0) = x^2(15)$$

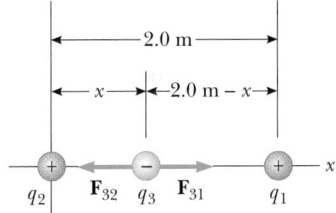

Figure 15.8 (Example 15.3) Three point charges are placed along the x axis. The charge q_3 is negative, whereas q_1 and q_2 are positive. If the net force on q_3 is zero, then the force on q_3 due to q_1 must be equal to and opposite the force on q_3 due to q_2.

This can be expanded to a quadratic equation, which then can be solved for x; but an easier approach is first to take the positive square root of both sides:

$$(2.0 - x)\sqrt{6.0} = x\sqrt{15}$$

$$(2.0 - x) = x(1.58)$$

$$x = \boxed{0.78 \text{ m}}$$

15.4 THE ELECTRIC FIELD

11.5, SECTION 1

Two different field forces have been introduced into our discussions so far: the gravitational force and the electrostatic force. As pointed out earlier, these forces are capable of acting through space, producing an effect even when there is no physical contact between the objects involved. Field forces can be discussed in a variety of ways, but an approach developed by Michael Faraday (1791–1867) is of such practical value that we shall devote much attention to it in the next several chapters. In this approach, an **electric field** is said to exist in the region of space around a charged object. When another charged object enters this electric field, forces of an electrical nature arise. As an example, consider Figure 15.9, which shows an object with a small positive charge, q_0, placed near a second object with a larger positive charge, Q.

We define the strength of the electric field at the location of the smaller charge to be the magnitude of the electric force acting on it, divided by the magnitude of its charge:

$$E \equiv \frac{|\mathbf{F}|}{|q_0|} \qquad [15.4]$$

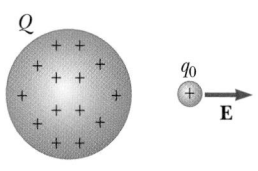

Figure 15.9 A small object with a positive charge, q_0, placed near an object with a larger positive charge, Q, experiences an electric field, **E,** directed as shown. The magnitude of the electric field is defined as the electric force on q_0 divided by the charge q_0.

Note that this is the electric field at the location of q_0 produced by the charge Q, not the field produced by q_0. The electric field is a vector quantity having the SI units newtons per coulomb (N/C). **The direction of E at a point is defined to be the direction of the electric force that would be exerted on a small positive**

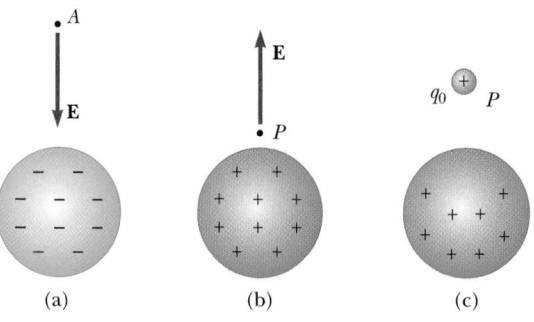

(a) (b) (c)

Figure 15.10 (a) The electric field at A due to the negatively charged sphere is downward, toward the negative charge. (b) The electric field at P due to the positively charged conducting sphere is upward, away from the positive charge. (c) A test charge, q_0, placed at P will cause a rearrangement of charge on the sphere unless q_0 is very small compared with the charge on the sphere.

charge placed at that point. Thus, in Figure 15.9, the direction of the electric field is horizontal and to the right. The electric field at point *A* in Figure 15.10a is vertical and downward because at this point a positive charge would experience a force of attraction toward the negatively charged sphere.

The definition we have provided for electric field has a serious difficulty. Consider the positively charged conducting sphere in Figure 15.10b. The field in the region surrounding the sphere could be explored by introducing a test charge, q_0, at a point such as *P*; finding the electric force on this charge; and then dividing this force by the magnitude of the charge on the test charge. Difficulties arise, however, when the magnitude of the test charge is great enough to influence the charge on the conducting sphere. For example, a large test charge can cause a rearrangement of the charges on the sphere, as in Figure 15.10c. As a result, the force exerted on the test charge is different from what it would be if the movement of charge on the sphere had not taken place. Furthermore, the strength of the measured electric field differs from what it would be in the absence of the test charge. To take care of this problem, we simply require that the test charge be small enough to have a negligible effect on the charges on the sphere.

Consider a point charge, *q*, located a distance of *r* from a test charge, q_0. According to Coulomb's law, the *magnitude* of the force on the test charge is

$$F = k_e \frac{|q||q_0|}{r^2}$$

Because the magnitude of the electric field at the position of the test charge is defined as $E = F/q_0$, we see that the *magnitude* of the electric field due to the charge *q* at the position of $|q_0|$ is

$$E = k_e \frac{|q|}{r^2} \qquad [15.5]$$

◀ Electric field due to a charge *q*

If *q* is *positive*, as in Figure 15.11a, the field at *P* due to this charge is *radially outward* from *q*. If *q* is *negative*, as in Figure 15.11b, the field at *P* is directed *toward q*. Equation 15.5 points out an important property of electric fields that makes them useful quantities for describing electrical phenomena. As the equation indicates, an electric field at a given point depends only on the charge, *q*, on the object setting up the field and the distance, *r*, from that object to a specific point in space. As a result, we can say that an electric field exists at point *P* in Figure 15.11 whether or not there is a charge at *P*.

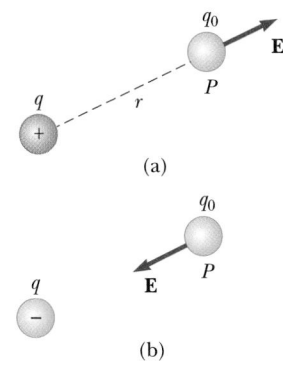

Figure 15.11 A test charge, q_0, at *P* is a distance of *r* from a point charge, *q*. (a) If *q* is positive, the electric field at *P* points radially *outward* from *q*. (b) If *q* is negative, the electric field at *P* points radially *inward* toward *q*.

Problem-Solving Strategy

Electric Forces and Fields

1. **Units.** When performing calculations that use the Coulomb constant, k_e, charges must be in coulombs and distances in meters. If they are given in other units, you must convert them.
2. **Applying Coulomb's law to point charges.** It is important to use the superposition principle properly when dealing with a collection of interacting point charges. If several charges are present, the resultant force on

any one of them is found by determining the individual force exerted on it by every other charge and then determining the vector sum of all these forces. The magnitude of the force that any charged object exerts on another is given by Coulomb's law, and the direction of the force is found by noting that the forces are repulsive between like charges and attractive between unlike charges.

3. **Calculating the electric field of point charges.** Remember that the superposition principle can be applied to electric fields, which are also vector quantities. To find the total electric field at a given point, first calculate the electric field at the point due to each individual charge. The vector sum of the fields due to all the individual charges is the resultant field at the point.

Thinking Physics 2

An electron moving horizontally passes between two horizontal plates, the upper charged negatively, the lower positively. A uniform, upward-directed electric field exists in this region, and this field exerts an electric force on the electron directed downward. Describe the movement of the electron in this region.

Explanation   The magnitude of the electric force on the electron of charge e due to a uniform electric field $\mathbf{E}$ is $F = eE$. Thus, the force is constant. Compare this to the force on a projectile of mass m moving in the gravitational field of the Earth. The magnitude of the gravitational force is mg. In both cases, the particle is subject to a constant force in the vertical direction and has an initial velocity in the horizontal direction. Thus, the path will be the same in each case—the electron will move as a projectile with an acceleration in the vertical direction and constant velocity in the horizontal direction. Once the electron leaves the region between the plates, the electric field disappears, and the electron continues moving in a straight line according to Newton's first law.

EXAMPLE 15.4   Electric Force on a Proton

Find the electric force on a proton placed in an electric field of 2.0×10^4 N/C that is directed along the positive x axis.

Solution   Because the charge on a proton is $+e = +1.6 \times 10^{-19}$ C, the electric force acting on the proton is

$$F = eE = (1.6 \times 10^{-19} \text{ C})(2.0 \times 10^4 \text{ N/C}) = \boxed{3.2 \times 10^{-15} \text{ N}}$$

where the force is in the positive x direction. The weight of the proton has the value $mg = (1.67 \times 10^{-27} \text{ kg})(9.80 \text{ m/s}^2) = 1.64 \times 10^{-26}$ N. Hence, the magnitude of the gravitational force is negligible compared with that of the electric force.

The principle of superposition holds when the electric field due to a group of point charges is calculated. We first use Equation 15.5 to calculate the electric field produced by each charge individually at a point, and then add these electric fields together as vectors.

EXAMPLE 15.5 Electric Field Due to Two Point Charges

Charge $q_1 = 7.00\ \mu C$ is at the origin, and charge $q_2 = -5.00\ \mu C$ is on the x axis, 0.300 m from the origin (Fig. 15.12). Find the electric field at point P, which has coordinates (0, 0.400) m.

Reasoning It is first necessary to find the direction of the field at P set up by each charge. The field $\mathbf{E}_1$ at P due to q_1 is vertically upward, as in Figure 15.12. Likewise, the field $\mathbf{E}_2$ at P due to q_2 is directed toward q_2, as in Figure 15.12. The magnitudes of the fields can be found from $E = k_e q/r^2$ and then added together vectorially.

Solution The magnitudes of $\mathbf{E}_1$ and $\mathbf{E}_2$ are

$$E_1 = k_e \frac{|q_1|}{r_1^2} = (8.99 \times 10^9\ \text{N}\cdot\text{m}^2/\text{C}^2)\frac{(7.00 \times 10^{-6}\ \text{C})}{(0.400\ \text{m})^2} = 3.93 \times 10^5\ \text{N/C}$$

$$E_2 = k_e \frac{|q_2|}{r_2^2} = (8.99 \times 10^9\ \text{N}\cdot\text{m}^2/\text{C}^2)\frac{(5.00 \times 10^{-6}\ \text{C})}{(0.500\ \text{m})^2} = 1.80 \times 10^5\ \text{N/C}$$

The vector $\mathbf{E}_1$ has an x component of zero. The vector $\mathbf{E}_2$ has an x component given by $E_2 \cos\theta = \frac{3}{5}E_2 = 1.08 \times 10^5\ \text{N/C}$ and a negative y component given by $-E_2 \sin\theta = -\frac{4}{5}E_2 = -1.44 \times 10^5\ \text{N/C}$. Hence, the resultant component in the x direction is

$$E_x = 1.08 \times 10^5\ \text{N/C}$$

and the resultant component in the y direction is

$$E_y = E_{y1} + E_{y2} = 3.93 \times 10^5\ \text{N/C} - 1.44 \times 10^5\ \text{N/C} = 2.49 \times 10^5\ \text{N/C}$$

From the Pythagorean theorem ($E = \sqrt{E_x^2 + E_y^2}$), we find that $\mathbf{E}$ has a magnitude of 2.72×10^5 N/C and makes an angle ϕ of 66.5° with the positive x axis.

Exercise Find the force on a positive test charge of 2.00×10^{-8} C placed at P.

Answer 5.44×10^{-3} N in the same direction as $\mathbf{E}$.

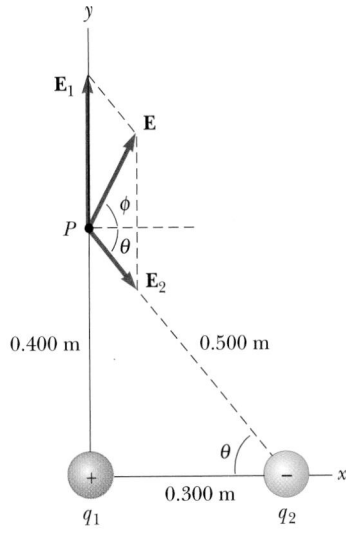

Figure 15.12 (Example 15.5) The total electric field $\mathbf{E}$ at P equals the vector sum $\mathbf{E}_1 + \mathbf{E}_2$, where $\mathbf{E}_1$ is the field due to the positive charge q_1 and $\mathbf{E}_2$ is the field due to the negative charge q_2.

 Workbook Problem 70 – (Workbook page 150) is similar to Example 15.5.

15.5 ELECTRIC FIELD LINES

A convenient aid for visualizing electric field patterns is to draw lines pointing in the direction of the electric field vector at any point. These lines, called **electric field lines,** are related to the electric field in any region of space in the following manner:

1. The electric field vector, $\mathbf{E}$, is tangent to the electric field lines at each point.
2. The number of lines per unit area through a surface perpendicular to the lines is proportional to the strength of the electric field in a given region.

Thus, $\mathbf{E}$ is large when the field lines are close together and small when they are far apart.

11.5, SECTION 2

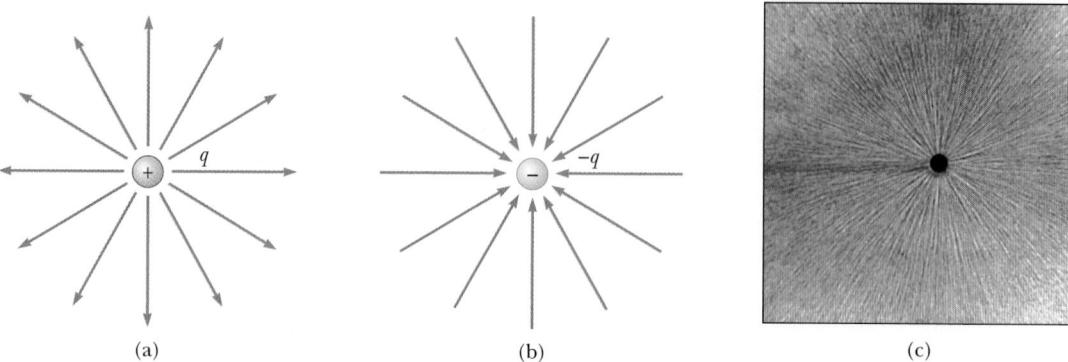

(a)　　　　　　(b)　　　　　　(c)

Figure 15.13 The electric field lines for a point charge. (a) For a positive point charge, the lines radiate outward. (b) For a negative point charge, the lines converge inward. Note that the figures show only those field lines that lie in the plane containing the charge. (c) The dark lines are small pieces of thread suspended in oil, which align with the electric field produced by a small charged conductor at the center. *(Photo courtesy of Harold M. Waage, Princeton University)*

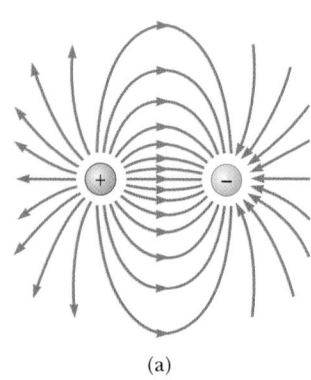

(a)

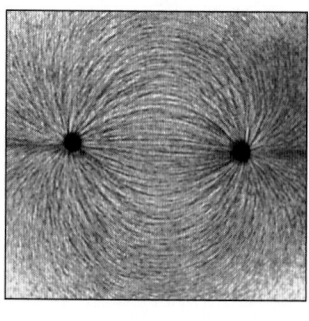

(b)

Figure 15.14 (a) The electric field lines for two equal and opposite point charges (an electric dipole). Note that the number of lines leaving the positive charge equals the number terminating at the negative charge. (b) The dark lines are small pieces of thread suspended in oil, which align with the electric field produced by two charged conductors. *(Photo courtesy of Harold M. Waage, Princeton University)*

Figure 15.13a shows some representative electric field lines for a single positive point charge. Note that this two-dimensional drawing contains only the field lines that lie in the plane containing the point charge. The lines are actually directed radially outward from the charge in *all* directions, somewhat like the quills of an angry porcupine. Because a positive test charge placed in this field would be repelled by the charge q, the lines are directed radially away from the positive charge. In a similar way, the electric field lines for a single negative point charge are directed toward the charge (Fig. 15.13b). In either case, the lines are radial and extend all the way to infinity. Note that the lines are closer together as they get near the charge, indicating that the strength of the field is increasing. Equation 15.5 verifies that this should indeed be the case.

The rules for drawing electric field lines for any charge distribution are as follows:

1. The lines must begin on positive charges (or at infinity) and must terminate on negative charges or, in the case of an excess of charge, at infinity.
2. The number of lines drawn leaving a positive charge or approaching a negative charge is proportional to the magnitude of the charge.
3. No two field lines can cross each other.

Figure 15.14 shows the electric field lines for two point charges of equal magnitude but opposite sign. This charge configuration is called an **electric dipole.** In this case the number of lines that begin at the positive charge must equal the number that terminate at the negative charge. At points very near either charge, the lines are nearly radial. The high density of lines between the charges indicates a strong electric field in this region.

Figure 15.15 shows the electric field lines in the vicinity of two equal positive point charges. Again, close to either charge the lines are nearly radial. The same number of lines emerges from each charge because the charges are equal in magnitude. At great distances from the charges, the field is approximately equal to that of a single point charge of magnitude $2q$. The bulging out of the electric field lines

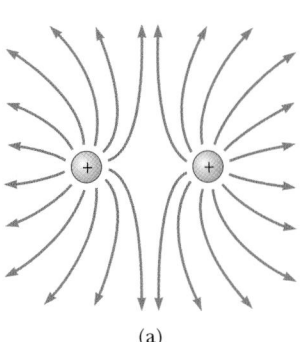

(a)

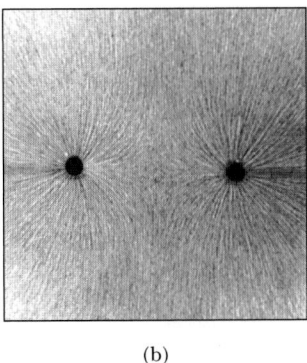

(b)

Figure 15.15 (a) The electric field lines for two positive point charges. (b) The dark lines are small pieces of thread suspended in oil, which align with the electric field produced by two charged conductors. *(Photo courtesy of Harold M. Waage, Princeton University)*

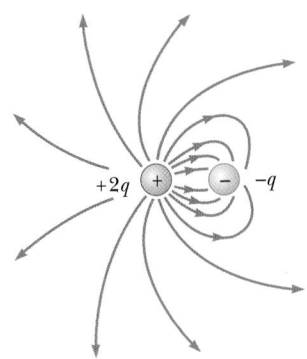

Figure 15.16 The electric field lines for a point charge of $+2q$ and a second point charge of $-q$. Note that two lines leave the charge $+2q$ for every line that terminates on $-q$.

between the charges indicates the repulsive nature of the electric force between like charges.

Finally, Figure 15.16 is a sketch of the electric field lines associated with the positive charge $+2q$ and the negative charge $-q$. In this case, the number of lines leaving charge $+2q$ is twice the number terminating on charge $-q$. Hence, only half of the lines that leave the positive charge end at the negative charge. The remaining half terminate on a negative charge that we assume to be located at infinity. At great distances from the charges (great compared with the charge separation), the electric field lines are equivalent to those of a single charge, $+q$.

Applying Physics 2

The electric field near the surface of the Earth in fair weather is about 100 N/C downward. Under a thundercloud, the electric field can be very large, on the order of 20 000 N/C. How are these electric fields measured?

Explanation A device for measuring these fields is called the *field mill.* Figure 15.17 shows the fundamental components of a field mill—two metal plates parallel to the ground. Each plate is connected to ground with a wire, with an ammeter (a device for measuring flow of charge, to be discussed in Section 19.6) in one path. Consider first just the lower plate. Because it is connected to ground, and the ground happens to carry a negative charge, the plate is negatively charged. Thus, electric field lines, directed downward, end on the plate, as in Figure 15.17a. Now, imagine that the upper plate is suddenly moved over the lower plate, as in Figure 15.17b. This plate is also connected to ground, and is also negatively charged, so the field lines now end on the upper plate. The negative charges in the lower plate are repelled by those on the upper plate and must pass through the ammeter, registering a flow of charge. The amount of charge that was on the lower plate is related to the strength of the electric field. Thus, the flow of charge through the ammeter can be calibrated to measure the electric field. The plates are normally designed like the blades of a fan, with the upper plate rotating so that the lower plate is alternately covered and uncovered. As a result, charges flow back and forth continually through the ammeter, and the reading can be related to the electric field strength.

A P P L I C A T I O N

Measuring Atmospheric Electric Fields.

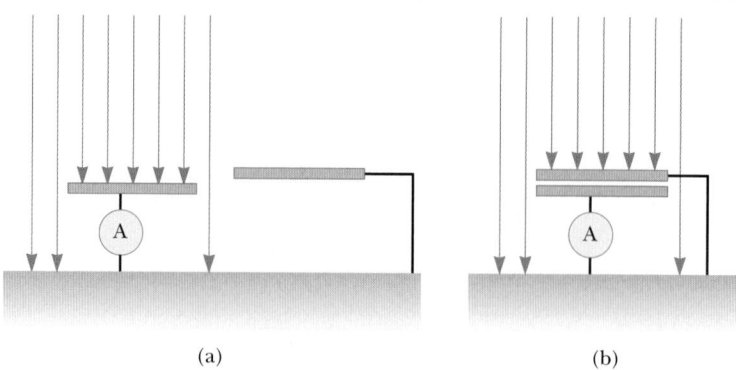

(a)   (b)

Figure 15.17 (Applying Physics 2) In (a), electric field lines end on negative charges on the lower plate. In (b), the second plate is moved above the lower plate. Electric field lines now end on the upper plate, and the negative charges in the lower plate are repelled through the ammeter.

15.6 CONDUCTORS IN ELECTROSTATIC EQUILIBRIUM

A good electric conductor, such as copper, contains charges (electrons) that are not bound to any atom and are free to move about within the material. When no net motion of charge occurs within a conductor, the conductor is said to be in **electrostatic equilibrium.** As we shall see, an isolated conductor (one that is insulated from ground) has the following properties:

Properties of an isolated conductor ▶

1. The electric field is zero everywhere inside the conductor.
2. Any excess charge on an isolated conductor resides entirely on its surface.
3. The electric field just outside a charged conductor is perpendicular to the conductor's surface.
4. On an irregularly shaped conductor, the charge tends to accumulate at locations where the radius of curvature of the surface is smallest—that is, at sharp points.

The first property can be understood by examining what would happen if it were *not* true. If there were an electric field inside a conductor, the free charge there would move and a flow of charge, or current, would be created. However, if there were a net movement of charge, there would no longer be electrostatic equilibrium.

Property 2 is a direct result of the $1/r^2$ repulsion between like charges described by Coulomb's law. If by some means an excess of charge is placed inside a conductor, the repulsive forces arising between the charges force them as far apart as possible, causing them to quickly migrate to the surface. (We shall not prove it here, but it is of interest to note that the excess charge resides on the surface due to the fact that Coulomb's law is an inverse-square law. With any other power law,

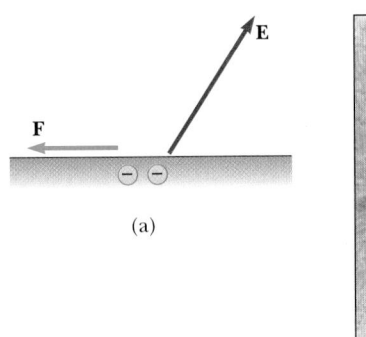

(a)

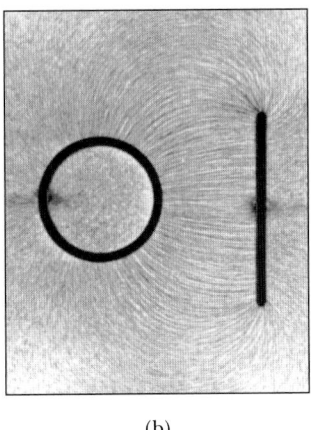

(b)

Figure 15.18 (a) Negative charges at the surface of a conductor. If the electric field were at an angle to the surface, as shown, an electric force would be exerted on the charges along the surface and they would move to the left. Because the conductor is assumed to be in electrostatic equilibrium, **E** cannot have a component along the surface and hence must be perpendicular to it. (b) The electric field pattern of a charged conducting plate near an oppositely charged conducting cylinder. Small pieces of thread suspended in oil align with the electric field lines. Note that (1) the electric field lines are perpendicular to the conductors and (2) there are no lines inside the cylinder (**E** = 0). *(Courtesy of Harold M. Waage, Princeton University)*

an excess of charge would exist on the surface, but there would be a distribution of charge, of either the same or opposite sign, inside the conductor.)

Property 3 can be understood by again considering what would happen if it were not true. If the electric field in Figure 15.18a were not perpendicular to the surface, the electric field would have a component along the surface, which would cause the free charges of the conductor to move (to the left in the figure). If the charges moved, however, a current would be created and there would no longer be electrostatic equilibrium. Hence, **E** must be perpendicular to the surface.

To see why property 4 must be true, consider Figure 15.19a, which shows a conductor that is fairly flat at one end and relatively pointed at the other. Any excess charge placed on the object moves to its surface. Figure 15.19b shows the forces between two such charges at the flatter end of the object. These forces are predominantly directed parallel to the surface. Thus, the charges move apart until repulsive forces from other nearby charges create an equilibrium situation. At the sharp end, however, the forces of repulsion between two charges are directed predominantly away from the surface, as in Figure 15.19c. As a result, there is less tendency for the charges to move apart along the surface here, and the amount of charge per unit area is greater than at the flat end. The cumulative effect of many such outward forces from nearby charges at the sharp end produces a large force directed away from the surface that can be great enough to cause charges to leap from the surface into the surrounding air.

Property 4 indicates that if a metal rod having sharp points is attached to a house, most of any charge on the house will pass through these points, thus eliminating the induced charge on the house produced by storm clouds. In addition, a lightning discharge striking the house can pass through the metal rod and be safely

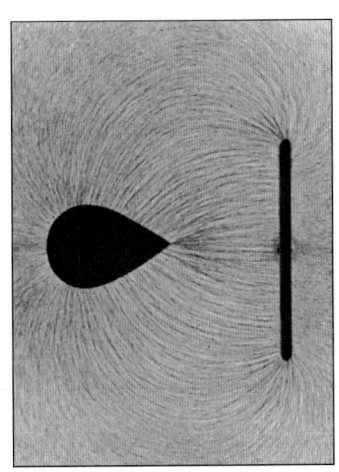

Electric field pattern of a charged conducting plate near an oppositely charged pointed conductor. Small pieces of thread suspended in oil align with the electric field lines. Note that the electric field is most intense near the pointed part of the conductor where the radius of curvature is the smallest. Also, the lines are perpendicular to the conductors. *(Courtesy of Harold M. Waage, Princeton University)*

APPLICATION

Lightning Rods.

Figure 15.19 (a) A conductor with a flatter end, *A*, and a relatively sharp end, *B*. Excess charge placed on a conductor resides entirely at its surface and is distributed so that (b) there is less charge per unit area on the flatter end and (c) there is a large charge per unit area on the sharper end.

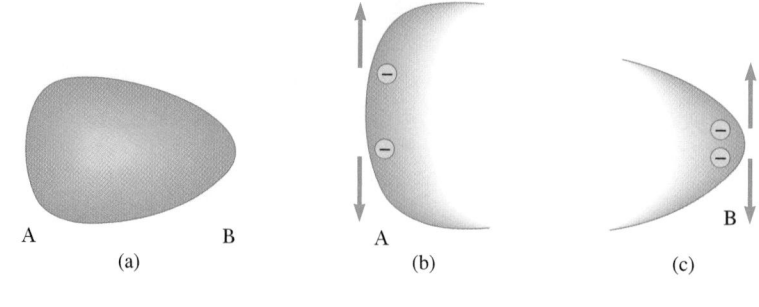

Figure 15.20 An experiment showing that any charge transferred to a conductor resides on its surface in electrostatic equilibrium. The hollow conductor is insulated from ground, and the small metal ball is supported by an insulating thread.

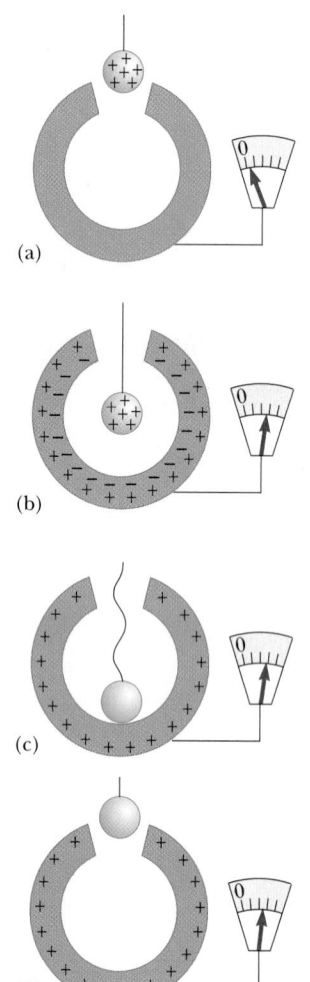

carried to the ground through wires leading from the rod to the Earth. Lightning rods using this principle were first developed by Benjamin Franklin. It is an interesting sidelight to American history to note that some European countries could not accept the fact that such a worthwhile idea could have originated in the New World. As a result, they "improved" the design by eliminating the sharp points. This modification in design drastically reduced the efficiency of their lightning rods, however.

Many experiments have shown that the net charge on a conductor resides on its surface. The experiment described here was first performed by Michael Faraday. A metal ball having a positive charge was lowered at the end of a silk thread into an uncharged hollow conductor insulated from ground, as in Figure 15.20a. (This experiment is referred to as **Faraday's ice-pail experiment** because he used a metal ice pail as the hollow conductor.) As the ball was lowered into the pail, the needle on an electrometer attached to the outer surface of the pail was observed to deflect. (An electrometer is a device used to measure charge.) The needle deflected because the charged ball induced a negative charge on the inner wall of the pail, which left an equal positive charge on the outer wall (Fig. 15.20b).

Faraday noted that the needle deflection did not change again, either when the ball touched the inner surface of the pail (Fig. 15.20c) or when it was removed (Fig. 15.20d). Furthermore, he found that the ball was now uncharged. Apparently, when the ball had touched the inside of the pail, the excess positive charge on the ball had been neutralized by the induced negative charge on the inner surface of the pail.

Faraday concluded that because the electrometer deflection did not change when the charged ball touched the inside of the pail, the negative charge induced on the inside surface of the pail was just enough to neutralize the positive charge on the ball. As a result of his investigations, he concluded that a charged object suspended inside a metal container causes a rearrangement of charge on the container in such a manner that the sign of the charge on the inside surface of the container is *opposite* the sign of the charge on the suspended object. This produces a charge on the outside surface of the container of the same sign as that on the suspended object.

Faraday also found that if the electrometer was connected to the inside surface of the pail after the experiment had been run, the needle showed no deflection. Thus, the *excess* charge acquired by the pail when contact was made between ball and pail appeared on the outer surface of the pail.

Thinking Physics 3

Suppose a point charge $+Q$ is in empty space. Wearing rubber gloves, we sneak up and surround the charge with a spherical conducting shell. What effect does this have on the field lines from the charge?

Explanation When the spherical shell is placed around the charge, the charges in the shell will adjust to satisfy the rules for a conductor in equilibrium. A net charge of $-Q$ will move to the interior surface of the conductor, so that the electric field inside the conductor will be zero. That is, the field lines originating on the $+Q$ charge will terminate on the $-Q$ charges. The movement of the $-Q$ charges to the inner surface of the sphere will leave a net charge of $+Q$ on the outer surface of the sphere. Thus, the only change in the field lines from the initial situation will be the absence of field lines within the conductor.

Applying Physics 3

Why is it safe to stay inside an automobile during a lightning storm?

Explanation Although many people believe that this is safe because of the insulating characteristics of the rubber tires, this is not true. Lightning is able to travel through several kilometers of air, so it can certainly penetrate a centimeter of rubber. The safety of remaining in the car is due to the fact that charges on the metal shell of the car will reside on the outer surface of the car, as noted in property 2 discussed earlier. Thus an occupant in the automobile touching the inner surfaces is not in danger.

APPLICATION

Driver Safety During
Electrical Storms.

15.7 THE MILLIKAN OIL-DROP EXPERIMENT

Optional Section

From 1909 to 1913, Robert Andrews Millikan (1868–1953) performed a brilliant set of experiments at the University of Chicago in which he measured the elementary charge, e, on an electron and demonstrated the quantized nature of the electronic charge. The apparatus he used, diagrammed in Figure 15.21, contains two parallel metal plates. Oil droplets that have been charged by friction in an atomizer are allowed to pass through a small hole in the upper plate. A horizontal light beam is used to illuminate the oil droplets, which are viewed by a telescope whose axis is at right angles to the beam. When the droplets are viewed in this manner, they appear as shining stars against a dark background, and the rate of fall of individual drops can be determined.

Let us assume that a single drop having a mass of m and carrying a charge of q is being viewed and that its charge is negative. If no electric field is present between the plates, the two forces acting on the charge are the force of gravity, $m\mathbf{g}$, acting downward, and an upward viscous drag force, $\mathbf{D}$ (Fig. 15.22a). The drag force is proportional to the speed of the drop. When the drop reaches its terminal speed, v, the two forces balance each other ($mg = D$).

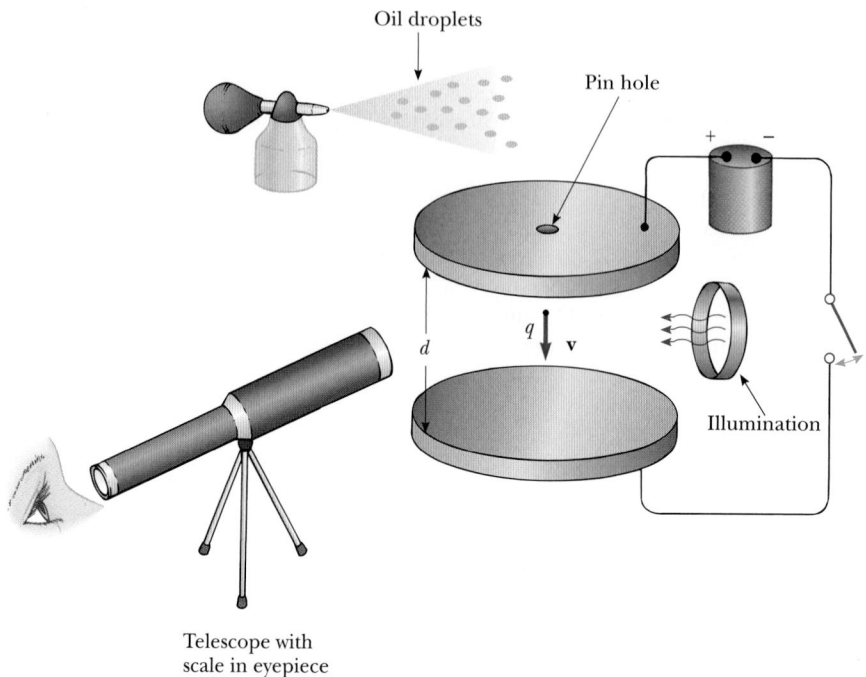

Figure 15.21   A schematic view of the Millikan oil-drop apparatus.

Now suppose that an electric field is set up between the plates by a battery connected so that the upper plate is positively charged. In this case, a third force, $q\mathbf{E}$, acts on the charged drop. Because q is negative and $\mathbf{E}$ is downward, the electric force is *upward,* as in Figure 15.22b. If this force is great enough, the drop moves upward and the drag force $\mathbf{D}'$ acts downward. When the upward electric force, $q\mathbf{E}$, balances the sum of the force of gravity and the drag force, both acting downward, the drop reaches a new terminal speed, v'.

With the field turned on, a drop moves slowly upward, typically at rates of *hundredths* of a centimeter per second. The rate of fall in the absence of a field is comparable. Hence, a single droplet with constant mass and radius can be fol-

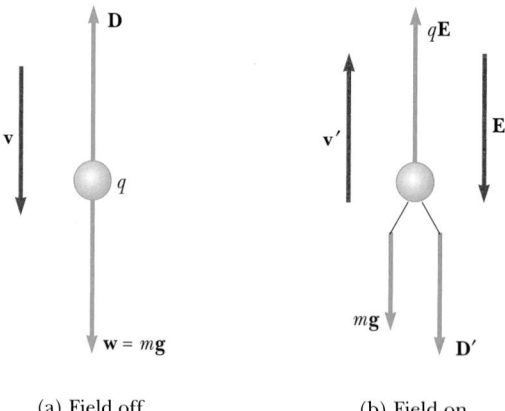

(a) Field off   (b) Field on

Figure 15.22   The forces on a charged oil droplet in the Millikan experiment.

lowed for hours as it alternately rises and falls by simply turning the electric field on and off.

After making measurements on thousands of droplets, Millikan and his co-workers found that every drop, to within about 1% precision, had a charge equal to some integer multiple of the elementary charge, *e*. That is,

$$q = ne \qquad n = 0, \pm 1, \pm 2, \pm 3, \ldots \qquad \text{[15.6]}$$

where $e = 1.60 \times 10^{-19}$ C. It was later established that a charge of $+e$ (for $n = +1$) would arise when an oil droplet had lost 1 electron. Likewise, a charge of $-e$ (for $n = -1$) would occur when a drop had gained 1 electron. Gains or losses in integral numbers provide conclusive evidence that charge is quantized. In 1923, Millikan was awarded the Nobel prize in physics for this work.

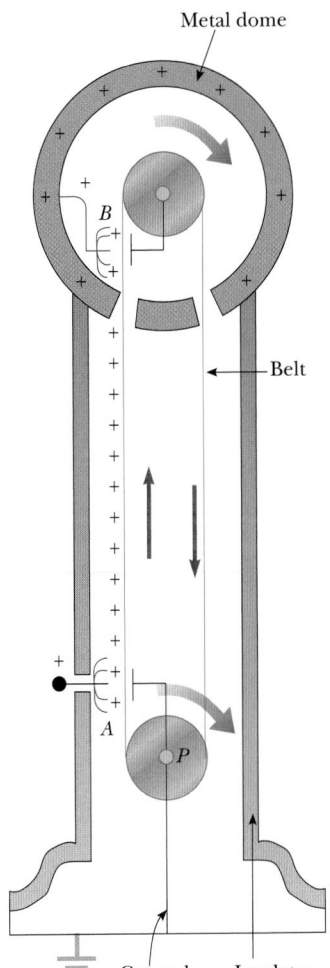

15.8 THE VAN DE GRAAFF GENERATOR   *Optional Section*

In 1929 Robert J. Van de Graaff designed and built an electrostatic generator that is used extensively in nuclear physics research. The principles of its operation can be understood with the help of the properties of electric fields and charges already presented in this chapter. Figure 15.23 shows the basic construction details of this device. A motor-driven pulley, *P*, moves a belt past positively charged comb-like metallic needles positioned at *A*. Negative charges are attracted to these needles from the belt, leaving the left side of the belt with a net positive charge. The positive charges attract electrons onto the belt as it moves past a second comb of needles at *B*, increasing the excess positive charge on the dome. Because the electric field inside the metal dome is negligible, the positive charge on it can easily be increased regardless of how much charge is already present. The result is that the dome is left with a large amount of positive charge.

This accumulation of charge on the dome cannot continue indefinitely, because eventually an electric discharge through the air takes place. To understand why, consider that, as more and more charge appears on the surface of the dome, the magnitude of the electric field at the surface of the dome is also increasing. Finally, the strength of the field becomes great enough to partially ionize the air near the surface, thus making the air partially conducting. Charges on the dome now have a pathway to leak off into the air, which can produce some spectacular "lightning bolts" as the discharge occurs. As noted earlier, charges find it easier to leap off a surface at points where the curvature is great. As a result, one way to inhibit the electric discharge, and to increase the amount of charge that can be stored on the dome, is to increase its radius. Another method for inhibiting discharge is to place the entire system in a container filled with a high-pressure gas, which is significantly more difficult to ionize than air at atmospheric pressure.

If protons (or other charged particles) are introduced into a tube attached to the dome, the large electric field of the dome exerts a repulsive force on the protons, causing them to accelerate to energies high enough to initiate nuclear reactions between the protons and various target nuclei.

Figure 15.23 A diagram of a Van de Graaff generator. Charge is transferred to the dome by means of a rotating belt. The charge is deposited on the belt at point *A* and transferred to the dome at point *B*.

11.10, BOTH SECTIONS

15.9   THE OSCILLOSCOPE

The oscilloscope is an electronic instrument widely used in making electrical measurements. Its main component is the cathode ray tube (CRT), shown in Figure 15.24. This tube is commonly used to create a visual display of electronic information for other applications, including radar systems, television receivers, and computers. The CRT is essentially a vacuum tube in which electrons are accelerated and deflected under the influence of electric fields.

The electron beam is produced by an assembly called an *electron gun* in the neck of the tube. The electron gun in Figure 15.24 consists of a heater (H), a cathode (C), and a positively charged anode (A). An electric current through the heater causes its temperature to rise, which in turn heats the cathode. The cathode reaches temperatures high enough to cause electrons to be "boiled off." Although they are not shown in the figure, the electron gun also includes an element that focuses the electron beam and one that controls the number of electrons reaching the anode (that is, a brightness control). The anode has a hole in its center that allows the electrons to pass through without striking the anode. If left undisturbed, the electrons travel in a straight-line path until they strike the face of the CRT. The screen at the front of the tube is coated with a fluorescent material that emits visible light when bombarded with electrons. This results in a visible spot on the screen of the CRT.

The electrons are moved in a variety of directions by two sets of deflection plates placed at right angles to each other in the neck of the tube (Fig. 15.24). In order to understand how the deflection plates operate, first consider the horizontal ones. External electric circuits can change the amount of charge present on these plates, with positive charge being placed on one plate and negative on the other. (In Chapter 16 we shall see that this can be accomplished by applying a voltage across the plates.) This increasing charge creates an increasing electric field between the two plates, which deflects the electron beam from its straight-line path. Slowly increasing the charge on the horizontal plates causes the electron beam to move gradually from the center toward the side of the screen. Because of the persistence of vision, however, one sees a horizontal line extending across the screen instead of the simple movement of a dot. The horizontal line can be maintained on the screen by rapid, repetitive tracing.

The vertical deflection plates act in exactly the same way as the horizontal plates, except that changing the charge on them causes a vertical line on the tube

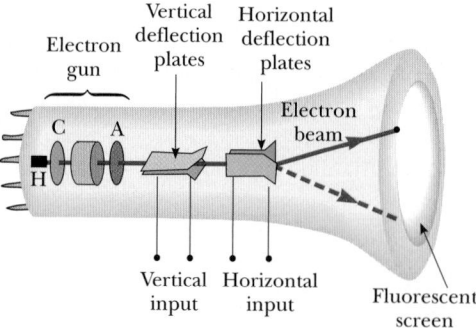

Figure 15.24 A diagram of a cathode ray tube. Electrons leaving the hot cathode, C, are accelerated to the anode, A, where some pass through a small hole. The electron gun is also used to focus the beam and the plates deflect the beam.

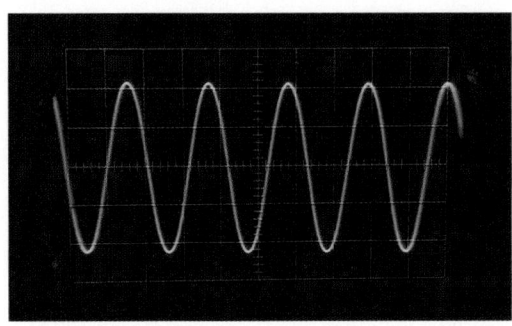

Figure 15.25 A sinusoidal vibration produced by a signal generator and displayed on the oscilloscope. *(Courtesy of Henry Leap and Jim Lehman)*

face. In practice, the horizontal and vertical deflection plates are used simultaneously.

To see how the oscilloscope can display visual information, let us examine how we can observe the sound wave from a tuning fork on the screen. For this purpose, the charge on the horizontal plates changes in such a manner that the beam sweeps across the face of the tube at a constant rate. The tuning fork is then sounded into a microphone, which changes the sound signal to an electric signal that is applied to the vertical plates. The combined effects of the horizontal and vertical plates cause the beam to sweep horizontally and up and down at the same time, with the vertical motion corresponding to the tuning fork signal. A pattern such as that in Figure 15.25 is formed on the screen.

15.10 ELECTRIC FLUX AND GAUSS'S LAW

Optional Section

This section describes an elegant technique for calculating electric fields that was developed by Karl Friedrich Gauss (1777–1855). Even though the procedure developed by Gauss is applicable only to situations in which the charge distribution is highly symmetric, it serves as a guide for understanding more complicated problems. In order to develop the law, we must first understand the concept of electric flux.

Consider an electric field that is uniform in both magnitude and direction, as in Figure 15.26. The electric field lines penetrate a surface of area A, which is perpendicular to the field. The technique used for drawing a figure such as Figure 15.26 is that the number of lines per unit area, Φ, is proportional to the magnitude of the electric field. Therefore,

$$\Phi = EA \qquad [15.7]$$

where Φ is the electric flux, which has units of $N \cdot m^2/C$ in SI units. Thus, if the area shown in Figure 15.26 is $1 \ m^2$ the magnitude of the electric field is $14 \ N/C$. (Count the lines in Fig. 15.26.) If the surface under consideration is not perpendicular to the field, the expression for the electric flux is

$$\Phi = EA \cos \theta \qquad [15.8]$$

Equation 15.8 can be easily understood by considering Figure 15.27, where the perpendicular to the area A is at an angle θ with respect to the field. The number

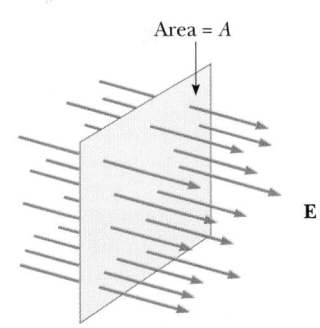

Figure 15.26 Field lines of a uniform electric field penetrating a plane of area A perpendicular to the field. The electric flux, Φ, through this area is equal to EA.

◀ Electric flux

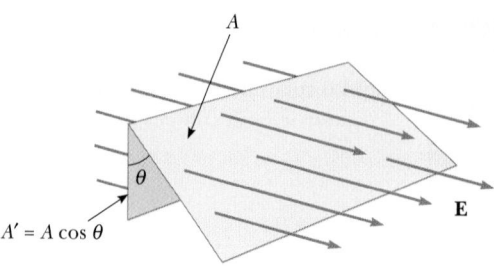

Figure 15.27   Field lines for a uniform electric field through an area A that is at an angle $(90° - \theta)$ to the field. Because the number of lines that go through the shaded area A' is the same as the number that go through A, we conclude that the flux through A' is equal to the flux through A and is given by $\Phi = EA \cos \theta$.

of lines that cross this area is equal to the number that cross the projected area A', which is perpendicular to the field. We see that the two areas are related by $A' = A \cos \theta$. From Equation 15.8 we see that the flux through a surface of fixed area has the maximum value, EA, when the surface is perpendicular to the field (when $\theta = 0°$) and that the flux is zero when the surface is parallel to the field (when $\theta = 90°$). **When the area is constructed such that a closed surface is formed, we shall adopt the convention that flux lines passing into the interior of the volume are negative and those passing out of the interior of the volume are positive.**

EXAMPLE 15.6   Flux Through a Cube

Consider a uniform electric field oriented in the x direction. Find the net electric flux through the surface of a cube of edges L oriented as shown in Figure 15.28.

Solution   The net flux can be evaluated by summing up the fluxes through each face of the cube. First, note that the flux through four of the faces is zero, because $\mathbf{E}$ is parallel to the area on these surfaces. These surfaces are those that are parallel to the xy and xz planes. For these surfaces, $\theta = 90°$, so $\Phi = EA \cos 90° = 0$. For surface 1 that lies in the yz plane in Figure 15.28, the flux lines pass into the interior of the cube, and the flux is taken to be negative. We have

$$\Phi_1 = -EA = -EL^2.$$

For surface 2 at the right face of the cube, the flux is positive and given by

$$\Phi_2 = EA = EL^2$$

The net flux through the surface of the cube is

$$\Phi_{net} = \Sigma EA = \Phi_1 + \Phi_2 = -EL^2 + EL^2 = \boxed{0}$$

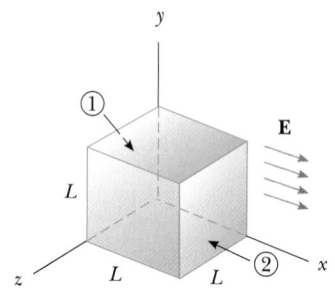

Figure 15.28   (Example 15.6) A hypothetical surface in the shape of a cube in a uniform electric field parallel to the x axis. The net flux through the surface is zero when the net charge inside the cube is zero.

Gauss's Law

We shall now describe a general relation between the net electric flux through a closed surface (often called a *gaussian surface*) and the charge *enclosed* by the surface. In our discussion of electric field lines we saw that the number of field lines originating on a positive charge or terminating on a negative charge is proportional to

Workbook Problem 71 — (Workbook page 152) deals with electric flux and Gauss's law.

the magnitude of the charge. That is, the net flux passing through a closed surface surrounding a charge Q is proportional to the magnitude of Q, or

$$\Phi_{\text{net}} = \Sigma EA \cos \theta \propto Q$$

11.6, BOTH SECTIONS

In free space, the constant of proportionality is $1/\epsilon_0$, where ϵ_0 is called the permittivity of free space and has the value

$$\epsilon_0 = \frac{1}{4\pi k_e} = \frac{1}{4\pi(8.99 \times 10^9 \text{ N} \cdot \text{m}^2/\text{C}^2)} = 8.85 \times 10^{-12} \text{ C}^2/\text{N} \cdot \text{m}^2$$

Thus, the net electric flux through a closed surface is

$$\Sigma EA \cos \theta = \frac{Q}{\epsilon_0} \qquad\qquad\text{[15.9]}$$

◀ Gauss's law

This result, known as **Gauss's law,** states that **the net electric flux through any closed gaussian surface is equal to the net charge inside the surface divided by ϵ_0.** It is the fundamental law describing how charges create electric fields. In principle, Gauss's law can always be used to calculate the electric field of a system of charges or a continuous distribution of charge. However, in practice the technique is useful only in a limited number of situations in which there is a high degree of symmetry.

To indicate the procedure for applying Gauss's law, consider the problem of calculating the electric field at a distance r from a positive point charge q. The first step is to construct a gaussian surface. To construct this surface, we must (1) consider the symmetry of the charge distribution, and (2) allow the gaussian surface to pass through the point at which we want to calculate the electric field. These two considerations lead us to the construction of a gaussian surface, which is a sphere with q at its center and having a radius of r, as shown in Figure 15.29.

Because there is only one surface to be considered, there is only one term in the sum on the left side of Gauss's law, and the left side of Equation 15.9 can be expressed as $EA \cos \theta$. Furthermore, the electric field is perpendicular to the area A at all points, so $\cos \theta = \cos 0° = 1$ at all points on the gaussian surface. Thus, the left side of Equation 15.9 reduces to

$$EA = E4\pi r^2$$

where $4\pi r^2$ is the surface area of the gaussian surface. Recall that the Q in Gauss's law is equal to the *net* charge enclosed by the gaussian surface, so the right side of Equation 15.9 reduces to

$$\frac{Q}{\epsilon_0} = \frac{q}{\epsilon_0}$$

and we have

$$E4\pi r^2 = \frac{q}{\epsilon_0}$$

or

$$E = \frac{q}{4\pi r^2 \epsilon_0} = k_e \frac{q}{r^2}$$

which is the equation used to calculate the magnitude of an electric field set up by a point charge q.

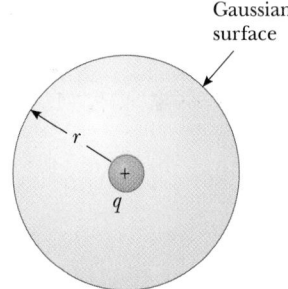

Gaussian surface

Figure 15.29 A spherical surface of radius r surrounding a point charge q. When the charge is at the center of the sphere, the electric field is normal to the surface and constant in magnitude everywhere on the surface.

Thinking Physics 4

A spherical gaussian surface surrounds a point charge q. Describe what happens to the total flux through the surface if (a) the charge is tripled, (b) the volume of the sphere is doubled, (c) the surface is changed to a cube, and (d) the charge is moved to another location inside the surface.

Explanation   (a) If the charge is tripled, the flux through the surface is also tripled, because the net flux is proportional to the charge inside the surface. (b) The flux remains constant when the volume changes because the surface surrounds the same amount of charge, regardless of its volume. (c) The total flux does not change when the shape of the closed surface changes. (d) The total flux through the closed surface remains unchanged as the charge inside the surface is moved to another location inside that surface. All of these conclusions are arrived at through an understanding of Gauss's law.

Thinking Physics 5

If the net flux through a surface is zero, which of the following statements are true? (a) There are no charges inside the surface. (b) The net charge inside the surface is zero. (c) The electric field is zero everywhere on the surface. (d) The number of electric field lines entering the surface equals the number leaving the surface.

Explanation   Statements (b) and (d) are true and follow from Gauss's law. Statement (a) is not necessarily true because Gauss's law says that the net flux through any closed surface equals the net charge inside the surface divided by ϵ_0. For example, a positive and a negative charge could be inside the surface. Statement (c) is not necessarily true. Although the net flux through the surface is zero, the electric field in that region may not be zero.

11.7, BOTH SECTIONS

EXAMPLE 15.7   The Electric Field of a Charged Thin Spherical Shell

A thin spherical shell of radius a has a total charge q distributed uniformly over its surface (Fig. 15.30). Find the electric field at points (a) outside and (b) inside the shell.

Solution   (a) The calculation of the field outside the shell is identical to that carried out in the body of the text for a point charge. Because the charge distribution is spherically symmetric, we select a spherical gaussian surface of radius r, concentric with the shell, as in Figure 15.30b. Following the same line of reasoning as that for a point charge, the left side of Gauss's law reduces to

$$\Sigma EA \cos \theta = EA\pi r^2$$

and the right side becomes

$$\frac{Q}{\epsilon_0} = \frac{q}{\epsilon_0}$$

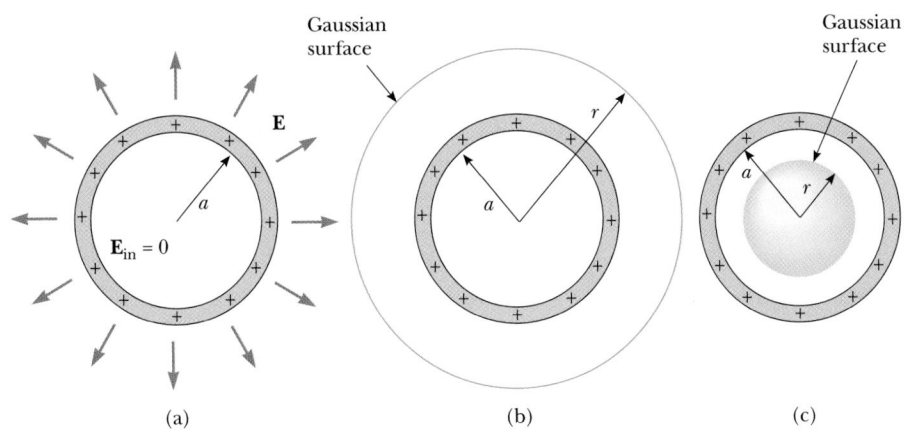

(a)   (b)   (c)

Figure 15.30 (Example 15.7) (a) The electric field inside a uniformly charged spherical shell is *zero*. The field outside is the same as that of a point charge having a total charge Q located at the center of the shell. (b) The construction of a gaussian surface for calculating the electric field *outside* a spherical shell. (c) The construction of a gaussian surface for calculating the electric field *inside* a spherical shell.

Thus, we have

$$E4\pi r^2 = \frac{q}{\epsilon_0}$$

or

$$E = \frac{q}{4\pi r^2 \epsilon_0} = k_e \frac{q}{r^2}$$

Therefore, the field at a point outside the shell is equivalent to that of a point charge located at the center of the shell.

(b) *The electric field inside the spherical shell is zero.* This follows from Gauss's law applied to a spherical gaussian surface of radius r placed inside the shell, as shown in Figure 15.30c. Because the net charge inside the surface is zero, $Q = 0$, we see that **E** = 0 in the region inside the shell.

Workbook Problem 72 — (Workbook page 153) shows another application of Gauss's law in calculating the magnitude of an electric field.

EXAMPLE 15.8 A Nonconducting Plane Sheet of Charge

Find the electric field due to a nonconducting infinite plane sheet of charge with uniform charge per unit area σ.

Solution The symmetry of the situation shows that the electric field must be perpendicular to the plane and that the direction of the field on one side of the plane must be opposite its direction on the other side, as shown in Figure 15.31. It is convenient to choose for our gaussian surface a small cylinder whose axis is perpendicular to the plane and whose ends each have an area A. The left side of Gauss's law, $\Sigma EA \cos \theta$, now is the sum of three terms, for each of the three surfaces labeled (1), (2), and (3) in Figure 15.31. Thus,

$$\Sigma EA \cos \theta = (EA \cos \theta)_1 + (EA \cos \theta)_2 + (EA \cos \theta)_3$$

But we see that because **E** is parallel to the cylindrical surface, surface 1, $\cos \theta = \cos 90° = 0$, so $(EA \cos \theta)_1 = 0$.

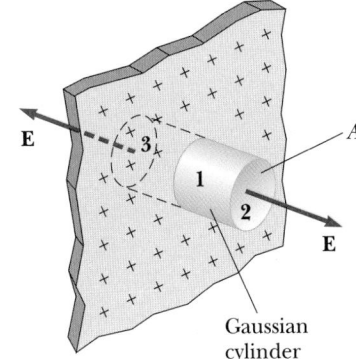

Figure 15.31 (Example 15.8) A cylindrical gaussian surface penetrating an infinite sheet of charge. The flux through each end of the gaussian surface is EA. There is no flux through the cylinder's curved surface.

Now consider surface (2). For this surface, we see that $\cos \theta = \cos 0° = 1$, and

$$(EA \cos \theta)_2 = EA$$

By the same reasoning, for surface (3),

$$(EA \cos \theta)_3 = EA$$

Therefore, the left side of Gauss's law becomes

$$(EA \cos \theta)_2 + (EA \cos \theta)_3 = 2EA$$

Because the total charge *inside* the gaussian surface is σA, the right side of Gauss's law becomes

$$\frac{Q}{\epsilon_0} = \frac{\sigma A}{\epsilon_0}$$

Applying Gauss's law to this situation gives

$$2EA = \frac{\sigma A}{\epsilon_0}$$

or,

$$E = \frac{\sigma}{2\epsilon_0}$$

Because the distance of the surfaces from the plane does not appear in our result, we conclude that $E = \sigma/2\epsilon_0$ at *any* distance from the plane. That is, the electric field is uniform everywhere.

SUMMARY

Electric charges have the following important properties:

1. Unlike charges attract one another and like charges repel one another.
2. Electric charge is always conserved.
3. Charge is quantized—that is, it exists in discrete packets that are integral multiples of the electronic charge.
4. The force between charged particles varies as the inverse square of their separation.

Conductors are materials in which charges move freely. **Insulators** are materials that do not readily transport charge.

Coulomb's law states that the electric force between two stationary charged particles separated by a distance of r has the magnitude

$$F = k_e \frac{|q_1||q_2|}{r^2} \qquad [15.1]$$

where $|q_1|$ and $|q_2|$ are the magnitudes of the charges on the particles in coulombs and k_e is the **Coulomb constant,** which has the approximate value

$$k_e \approx 8.99 \times 10^9 \text{ N} \cdot \text{m}^2/\text{C}^2 \qquad [15.3]$$

The magnitude of the **electric field, E,** at some point in space is defined as the magnitude of the electric force that acts on a small positive charge placed at that point, divided by the magnitude of its charge, $|q_0|$:

$$E \equiv \frac{|\mathbf{F}|}{|q_0|} \qquad [15.4]$$

The direction of the electric field at a point in space is defined to be the direction of the electric force that would be exerted on a small positive charge placed at that point.

The magnitude of the electric field due to a *point charge, q,* at distance r from the point charge is

$$E = k_e \frac{|q|}{r^2} \qquad [15.5]$$

Electric field lines are useful for describing the electric field in any region of space. The electric field vector, **E,** is tangent to the electric field lines at every point. Furthermore, the number of electric field lines per unit area through a surface perpendicular to the lines is proportional to the strength of the electric field in that region.

A **conductor in electrostatic equilibrium** has the following properties:

1. The electric field is zero everywhere inside the conductor.
2. Any excess charge on an isolated conductor must reside entirely on its surface.
3. The electric field just outside a charged conductor is perpendicular to the conductor's surface.
4. On an irregularly shaped conductor, charge tends to accumulate where the radius of curvature of the surface is smallest — that is, at sharp points.

MULTIPLE-CHOICE QUESTIONS

1. A circular ring of charge of radius b has a total charge q uniformly distributed around it. The magnitude of the electric field at the center of the ring is
   (a) 0   (b) $k_e q / b^2$ (c) $k_e q^2 / b^2$ (d) $k_e q^2 / b$
   (e) none of these.

2. The reason your hand will not penetrate into a solid wall is that your hand and the wall are covered on the surface with _____, which repel one another.
   (a) skin   (b) paint   (c) skin and paint   (d) atoms
   (e) electrons

3. A Styrofoam ball covered with a conducting paint has a mass of 5.0×10^{-3} kg and has a charge of 4.0 microcoulombs. What electric field directed upward will balance the weight of the ball?
   (a) 8.2×10^2 N/C   (b) 1.2×10^4 N/C
   (c) 2.0×10^{-2} N/C   (d) 5.1×10^6 N/C

4. An electron with a speed of 3.00×10^6 m/s moves into a uniform electric field of 1000 N/C. The field is parallel to the electron's motion. How far does the electron travel before it is brought to rest?
   ($m_e = 9.11 \times 10^{-31}$ kg, $q_e = 1.60 \times 10^{-19}$ C)
   (a) 2.56 cm   (b) 5.12 cm   (c) 11.2 cm   (d) 3.34 m

5. Charge A and charge B are 2.0 m apart, charge A is $+1.0$ C, and charge B is $+2.0$ C. Charge C (which is $+2.0$ C) is located between them at a certain point and the force on charge C is zero. How far from charge A is charge C?
   (a) 1.0 m   (b) 0.67 m   (c) 0.83 m   (d) 0.50 m

CONCEPTUAL QUESTIONS

1. Two insulated rods are oppositely charged on the ends. They are mounted at the centers so that they are free to rotate and then held in the position shown in a view from above (Fig. Q15.1). The plane of the rotation of the rods is in the plane of the paper. Will the rods stay in the positions when released? If not, into what position(s) will they move? Will the final configuration(s) be stable?

Figure Q15.1

2. Explain from an atomic viewpoint why charge is usually transferred by electrons.
3. If a suspended object A is attracted to object B, which is charged, can we conclude that object A is charged?
4. Operating-room personnel must wear special conducting shoes while working around oxygen. Why? Contrast this procedure with what might happen if personnel wore rubber shoes.
5. If a metal object receives a positive charge, does its mass increase, decrease, or stay the same? What happens to its mass if the object receives a negative charge?
6. When defining the electric field, why is it necessary to specify that the magnitude of the test charge be very small?
7. In fair weather, there is an electric field at the surface of the Earth, pointing down into the ground. What is the electric charge on the ground in this situation?
8. A student stands on a piece of insulating material, places her hand on top of a Van de Graaff generator, and then turns on the generator. Is she shocked?
9. An uncharged, metallic coated Styrofoam ball is suspended in the region between two vertical metal plates.

If the two plates are charged, one positive and one negative, describe the motion of the ball after it is brought into contact with one of the plates.
10. Is it possible for an electric field to exist in empty space? Explain.
11. There are great similarities between electric and gravitational fields. A room can be electrically shielded so that there are no electric fields in the room by surrounding it with a conductor. Can a room be gravitationally shielded? Why or why not?
12. A "free" electron and "free" proton are placed in an identical electric field. Compare the electric force on each particle. Compare their accelerations.
13. Explain why Gauss's law cannot be used to calculate the electric field of (a) a polar molecule consisting of a positive and a negative charge separated by a very small distance, (b) a charged disk, and (c) three point charges at the corner of a triangle.
14. Why should a ground wire be connected to the metal support rod for a television antenna?
15. A balloon is negatively charged by rubbing and then clings to a wall. Does this mean that the wall is positively charged? Why does the balloon eventually fall?
16. Why is it a bad idea to seek shelter under a tree during a lightning storm?
17. A charged comb will often attract small bits of dry paper that then fly away when they touch the comb. Explain.
18. Would life be different if the electron were positively charged and the proton were negatively charged? Does the choice of signs have any bearing on physical and chemical interactions? Explain.
19. If the total charge inside a closed surface is known but the distribution of charge is unspecified, can you use Gauss's law to find the electric field? Explain.

PROBLEMS

1, **2**, **3** = straightforward, intermediate, challenging   ☐ = full solution available in Study Guide/Student Solutions Manual   ◤ = Core Concepts Workbook
WEB = solution posted at **http://www.harcourtcollege.com/physics/cptech**   ◢ = biomedical application   ◖ = Interactive Physics

Review Problem

In the Bohr theory of the hydrogen atom, an electron moves in a circular orbit about a proton, where the radius of the orbit is 0.53×10^{-10} m. (a) Find the electrostatic force between the two. (b) If this force causes the centripetal acceleration of the electron, what is the speed of the electron?

Section 15.3 Coulomb's Law

1. A 27-g piece of aluminum that was originally electrically neutral is given a charge of $+1.6$ μC. (a) How many electrons were removed from the aluminum in the charging process? (b) What fraction of the electrons originally in the aluminum were involved in the charging process?

2. A 4.5×10^{-9} C charge is located 3.2 m from a -2.8×10^{-9} C charge. Find the electrostatic force exerted by one charge on the other.

3. Two identical conducting spheres are placed with their centers 0.30 m apart. One is given a charge of 12×10^{-9} C and the other a charge of -18×10^{-9} C. (a) Find the electrostatic force exerted on one sphere by the other. (b) The spheres are connected by a conducting wire. After equilibrium has occurred, find the electrostatic force between the two.

4. An alpha particle (charge $= +2.0e$) is sent at high speed toward a gold nucleus (charge $= +79e$). What is the electrical force acting on the alpha particle when it is 2.0×10^{-14} m from the gold nucleus?

5. The nucleus of ^{8}Be, which consists of four protons and four neutrons, is very unstable and spontaneously breaks into two alpha particles (helium nuclei, each consisting of two protons and two neutrons). (a) What is the force between the two alpha particles when they are 5.00×10^{-15} m apart, and (b) what will be the magnitude of the acceleration of the alpha particles because of this force?

6. Suppose that 1.00 g of hydrogen is separated into electrons and protons. Suppose also that the protons are placed at the Earth's North Pole and the electrons are placed at the South Pole. What is the resulting compressional force on the Earth?

7. An electron is released a short distance above the surface of the Earth. A second electron directly below it exerts an electrostatic force on the first electron just great enough to cancel the gravitational force on it. How far below the first electron is the second?

8. A 2.2×10^{-9} C charge is on the x axis at $x = -1.5$ m and a 5.4×10^{-9} C charge is on the x axis at $x = 2.0$ m. Find the net force exerted on a 3.5×10^{-9} C charge located at the origin.

9. Two small metallic spheres, each of mass 0.20 g, are suspended as pendulums by light strings from a common point, as shown in Figure P15.9. The spheres are given

the same electric charge, and it is found that the two come to equilibrium when each string is at an angle of $5.0°$ with the vertical. If each string is 30.0 cm long, what is the magnitude of the charge on each sphere?

10. Calculate the magnitude and direction of the Coulomb force on each of the three charges in Figure P15.10.

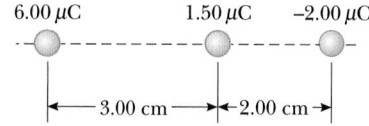

Figure P15.10 (Problems 10 and 16)

WEB 11. Three charges are arranged as shown in Figure P15.11. Find the magnitude and direction of the electrostatic force on the charge at the origin.

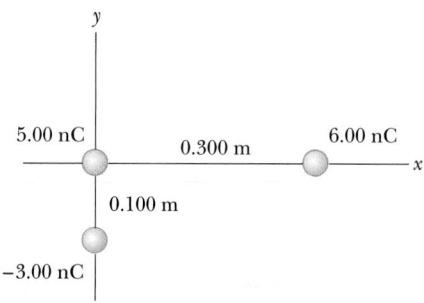

Figure P15.11

12. Three charges are arranged as shown in Figure P15.12. Find the magnitude and direction of the electrostatic force on the 6.00-nC charge.

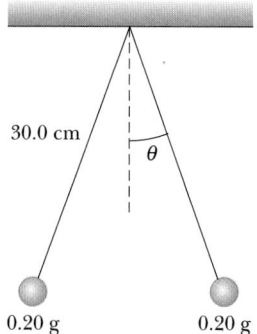

Figure P15.9

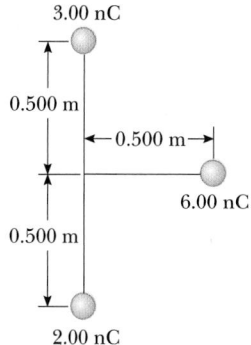

Figure P15.12

13. A charge of 6.00×10^{-9} C and a charge of -3.00×10^{-9} C are separated by a distance of 60.0 cm. Find the position at which a third charge, of 12.0×10^{-9} C, can be placed so that the net electrostatic force on it is zero.

14. A charge of 2.00×10^{-9} C is placed at the origin, and a charge of 4.00×10^{-9} C is placed at $x = 1.5$ m. Locate the point between the two charges at which a charge of 3.00×10^{-9} C should be placed so that the net electric force on it is zero.

Section 15.4 The Electric Field

15. In a hydrogen atom, what are the magnitude and direction of the electric field set up by the proton at the location of the electron (0.51×10^{-10} m away from the proton)?

16. (a) Determine the electric field strength at a point 1.00 cm to the left of the middle charge shown in Figure P15.10. (b) If a charge of -2.00 μC is placed at this point, what are the magnitude and direction of the force on it?

17. Find the electric field at a point midway between two charges of $+30.0 \times 10^{-9}$ C and (a) $+60.0 \times 10^{-9}$ C, separated by 30.0 cm; (b) -60.0×10^{-9} C, separated by 30.0 cm.

18. An electron is accelerated by a constant electric field of magnitude 300 N/C. (a) Find the acceleration of the electron. (b) Use the equations of motion with constant acceleration to find the electron's speed after 1.00×10^{-8} s, assuming it starts from rest.

19. A piece of aluminum foil of mass 5.00×10^{-2} kg is suspended by a string in an electric field directed vertically upward. If the charge on the foil is 3.00 μC, find the strength of the field that will reduce the tension in the string to zero.

20. A $+2.7$-μC point charge is on the x axis at $x = -3.0$ m, and a $+2.0$-μC point charge is on the x axis at $x = +1.0$ m. Determine the net electric field (magnitude and direction) on the y axis at $y = +2.0$ m.

21. A proton accelerates from rest in a uniform electric field of 640 N/C. At some later time, its speed is 1.20×10^6 m/s. (a) Find the magnitude of the acceleration of the proton. (b) How long does it take the proton to reach this speed? (c) How far has it moved in this interval? (d) What is its kinetic energy at the later time?

22. Each of the protons in a particle beam has a kinetic energy of 3.25×10^{-15} J. What are the magnitude and direction of the electric field that will stop these protons in a distance of 1.25 m?

23. Positive charges are situated at three corners of a rectangle, as shown in Figure P15.23. Find the electric field at the fourth corner.

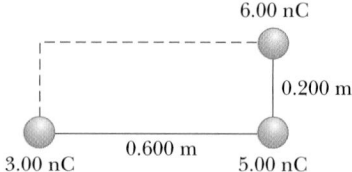

Figure P15.23

24. Three identical charges ($q = -5.0$ μC) are along a circle of 2.0-m radius at angles of 30°, 150°, and 270°, as shown in Figure P15.24. What is the resultant electric field at the center of the circle?

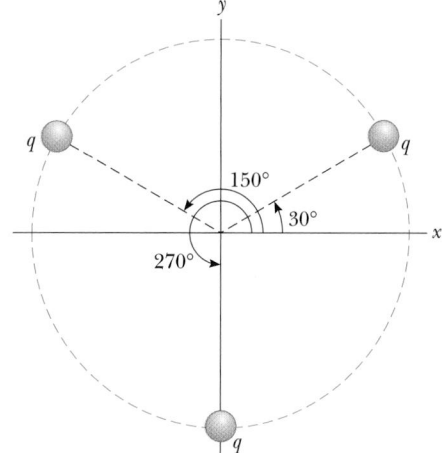

Figure P15.24

25. Each of the electrons in a particle beam has a kinetic energy of 1.60×10^{-17} J. (a) What is the magnitude of the uniform electric field (pointing in the direction of the electrons' movement) that will stop these electrons in a distance of 10.0 cm? (b) How long will it take to stop the electrons? (c) After the electrons stop, what will they do? Explain.

26. Three charges are at the corners of an equilateral triangle, as shown in Figure P15.26. Calculate the electric

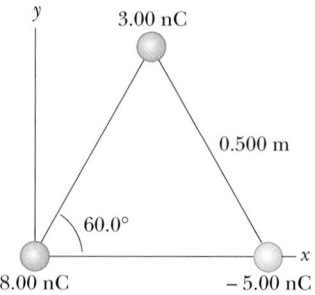

Figure P15.26

field at a point midway between the two charges on the x axis.

WEB 27. In Figure P15.27, determine the point (other than infinity) at which the total electric field is zero.

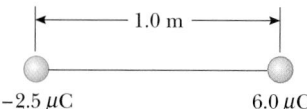

Figure P15.27   (Problems 27 and 49)

Section 15.5 Electric Field Lines

Section 15.6 Conductors in Electrostatic Equilibrium

28. Figure P15.28 shows the electric field lines for two point charges separated by a small distance. (a) Determine the ratio q_1/q_2. (b) What are the signs of q_1 and q_2?

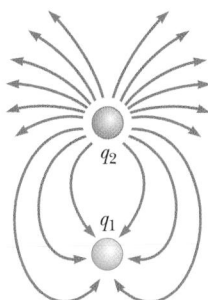

Figure P15.28

29. (a) Sketch the electric field lines around an isolated point charge, $q > 0$. (b) Sketch the electric field pattern around an isolated negative point charge of magnitude $-2q$.

30. (a) Sketch the electric field pattern around two positive point charges of magnitude 1 μC, placed close together. (b) Sketch the electric field pattern around two negative point charges of magnitude -2 μC, placed close together. (c) Sketch the pattern around a 1-μC positive point charge and a -2-μC charge placed close together.

31. Two point charges are a small distance apart. (a) Sketch the electric field lines for the two if one has a charge four times that of the other and both charges are positive. (b) Repeat for the case that both charges are negative.

32. (a) Sketch the electric field pattern set up by a positively charged hollow sphere. Include the lines both inside and outside the sphere. (b) A conducting cube is given a positive charge. Sketch the electric field pattern both inside and outside the cube.

33. Refer to Figure 15.20. The charge lowered into the center of the hollow conductor has a magnitude of 5.00 μC.

Find the magnitude and sign of the charge on the inside and outside of the hollow conductor when the charge is as shown in (a) Figure 15.20a, (b) Figure 15.20b, (c) Figure 15.20c, (d) Figure 15.20d.

Section 15.8 The Van de Graaff Generator (Optional)

34. The dome of a Van de Graaff generator receives a charge of 2.0×10^{-4} C. Find the strength of the electric field (a) inside the dome; (b) at the surface of the dome, assuming it has a radius of 1.0 m; (c) 4.0 m from the center of the dome. (*Hint:* See Section 15.6 to review properties of conductors in electrostatic equilibrium. Also use the fact that the points on the surface are outside a spherically symmetric charge distribution; the total charge may be considered as located at the center of the sphere.)

35. If the electric field strength in air exceeds 3.0×10^6 N/C, the air becomes a conductor. Using this fact, determine the maximum amount of charge that can be carried by a metal sphere 2.0 m in radius. (See the hint in Problem 34.)

36. Air breaks down (loses its insulating quality) and sparking results if the field strength is increased to about 3.0×10^6 N/C. (a) What acceleration does an electron experience in such a field? (b) If the electron starts from rest, in what distance does it acquire a speed equal to 10% of the speed of light?

37. A Van de Graaff generator is charged so that the electric field at its surface is 3.0×10^4 N/C. Find (a) the electric force exerted on a proton released at its surface and (b) the acceleration of the proton at this instant of time.

Section 15.10 Electric Flux and Gauss's Law (Optional)

38. A flat surface having an area of 3.2 m^2 is rotated in a uniform electric field of magnitude $E = 6.2 \times 10^5$ N/C. (a) Determine the electric flux through this area when the electric field is perpendicular to the surface and (b) when the electric field is parallel to the surface.

39. A 40-cm diameter loop is rotated in a uniform electric field until the position of maximum electric flux is found. The flux in this position is measured to be 5.2×10^5 N·m^2/C. Calculate the electric field strength in this region.

40. A point charge of $+5.00$ μC is located at the center of a sphere with a radius of 12.0 cm. Determine the electric flux through the surface of the sphere.

41. A point charge of magnitude q is located at the center of a spherical shell of radius a, which has a charge uniformly distributed on its surface of magnitude $-q$. Find the electric field (a) for all points outside the spherical shell and (b) for a point inside the shell a distance r from the center.

42. Use Gauss's law and the fact that the electric field inside any closed conductor in electrostatic equilibrium is zero

to show that any excess charge placed on the conductor must reside on its surface.

WEB 43. An infinite plane conductor has charge spread out on its surface as shown in Figure P15.43. Use Gauss's law to show that the electric field at any point outside the conductor is given by $E = \sigma/\epsilon_0$, where σ is the charge per unit area on the conductor. (*Hint:* Choose a gaussian surface in the shape of a cylinder with one end inside the conductor and one end outside the conductor.)

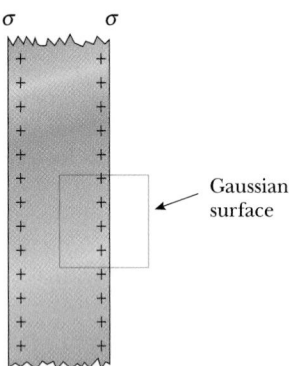

Figure P15.43

44. Show that the electric field just outside the surface of a good conductor of any shape is given by $E = \sigma/\epsilon_0$, where σ is the charge per unit area on the conductor.

ADDITIONAL PROBLEMS

45. Three point charges are aligned along the x axis, as shown in Figure P15.45. Find the electric field at the position $x = +2.0$ m, $y = 0$.

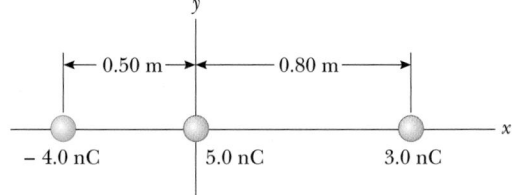

Figure P15.45

46. Three point charges lie along the y axis. A charge of $q_1 = -9.0\ \mu C$ is at $y = 6.0$ m, and a charge of $q_2 = -8.0\ \mu C$ is at $y = -4.0$ m. Where must the third positive charge, q_3, be placed so that the resultant force on it is zero?

47. A small 2.00-g plastic ball is suspended by a 20.0-cm-long string in a uniform electric field, as shown in Figure P15.47. If the ball is in equilibrium when the string

makes a 15.0° angle with the vertical as indicated, what is the net charge on the ball?

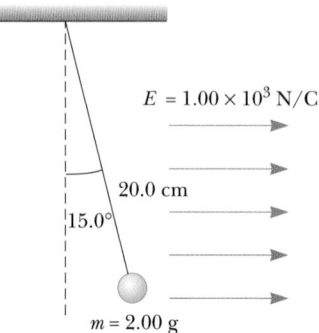

Figure P15.47

48. (a) Two identical point charges $+q$ are located on the y axis at $y = +a$ and $y = -a$. What is the electric field along the x axis at $x = b$? (b) A circular ring of charge of radius a has a total positive charge Q distributed uniformly around it. The ring is in the $x = 0$ plane with its center at the origin. What is the electric field along the x axis at $x = b$ due to the ring of charge? (*Hint:* Consider the charge Q to consist of a very many pairs of identical point charges positioned at ends of diameters of the ring.)

49. Consider the case in which the 6.0-μC charge in Figure P15.27 is replaced by a charge of $-6.0\ \mu C$. For the resulting charge distribution, determine the point (other than infinity) at which the total electric field is zero.

50. Two 2.0-g spheres are suspended by 10.0-cm-long light strings (Fig. P15.50). A uniform electric field is applied in the x direction. If the spheres have charges of -5.0×10^{-8} C and $+5.0 \times 10^{-8}$ C, determine the electric field intensity that enables the spheres to be in equilibrium at $\theta = 10°$.

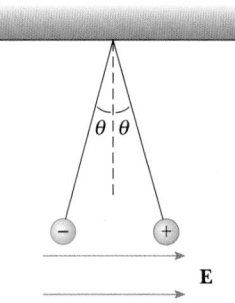

Figure P15.50

51. A solid conducting sphere of radius 2.00 cm has a charge 8.00 μC. A conducting spherical shell of inner

radius 4.00 cm and outer radius 5.00 cm is concentric with the solid sphere and has a charge $-4.00 \ \mu$C. Find the electric field at **(a)** $r = 1.00$ cm, **(b)** $r = 3.00$ cm, **(c)** $r = 4.50$ cm, and **(d)** $r = 7.00$ cm from the center of this charge configuration.

52. Two small silver spheres, each with a mass of 100 g, are separated by 1.00 m. Calculate the fraction of the electrons in one sphere that must be transferred to the other in order to produce an attractive force of 1.00×10^4 N (about a ton) between the spheres. (The number of electrons per atom of silver is 47, and the number of atoms per gram is Avogadro's number divided by the molar mass of silver, 107.87.)

53. The electrons in a particle beam each have a kinetic energy, K. What are the magnitude and direction of the electric field that will stop these electrons in a distance, d?

54. A molecule of DNA (deoxyribonucleic acid) is 2.17 μm long. The ends of the molecule become singly ionized —negative on one end, positive on the other. The helical molecule acts as a spring and compresses 1.00% on becoming charged. Determine the effective spring constant of the molecule.

55. Protons are projected with an initial speed of $v_0 = 9550$ m/s into a region in which a uniform electric field, $E = 720$ N/C, is present (Fig. P15.55). The protons are to hit a target that lies a horizontal distance of 1.27 mm from the point at which the protons are launched. Find **(a)** the two projection angles, θ, that will result in a hit and **(b)** the total duration of flight for each of these two trajectories.

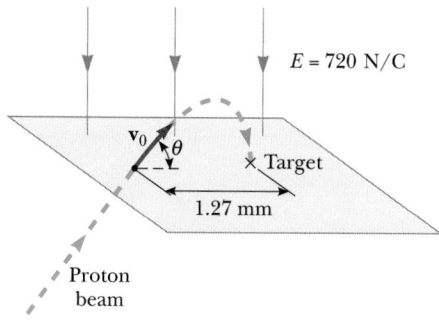

Figure P15.55

56. An electron traveling with an initial speed of 4.0×10^6 m/s enters a region with a uniform electric field of magnitude 2.5×10^4 N/C. The direction of travel of the electron is the direction of the field. **(a)** Find the acceleration of the electron. **(b)** Determine

the time it takes for the electron to stop after it enters the field. **(c)** How far does the electron move in the electric field before stopping?

57. A 2.00-μC charged 1.00-g cork ball is suspended vertically on a 0.500-m-long light string in the presence of a uniform downward-directed electric field of magnitude $E = 1.00 \times 10^5$ N/C. If the ball is displaced slightly from the vertical, it oscillates like a simple pendulum. **(a)** Determine the period of this oscillation. **(b)** Should gravity be included in the calculation for part (a)? Explain.

58. Two point charges such as those in Figure P15.58 are called an *electric dipole*. Show that the electric field at a distant point along the x axis is given by the expression $E_x = 4k_e qa/x^3$.

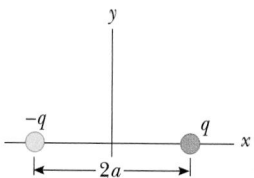

Figure P15.58

WEB 59. Two equal positive charges, q, are on the x axis at $x = a$ and $x = -a$. Show that the field along the positive y axis is in the y direction and is given by the relation $E_y = 2k_e qy(y^2 + a^2)^{-3/2}$.

60. A charged cork ball of mass 1.00 g is suspended on a light string in the presence of a uniform electric field, as in Figure P15.60. When the electric field has an x component of 3.00×10^5 N/C and a y component of 5.00×10^5 N/C, the ball is in equilibrium at $\theta = 37.0°$. Find **(a)** the charge on the ball and **(b)** the tension in the string.

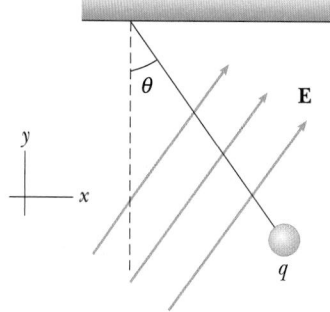

Figure P15.60

Electrical Energy and Capacitance

WEB

For Web links to sites with more information about defibrillators, visit the textbook Web site at **http://www.harcourtcollege.com/physics/cptech**

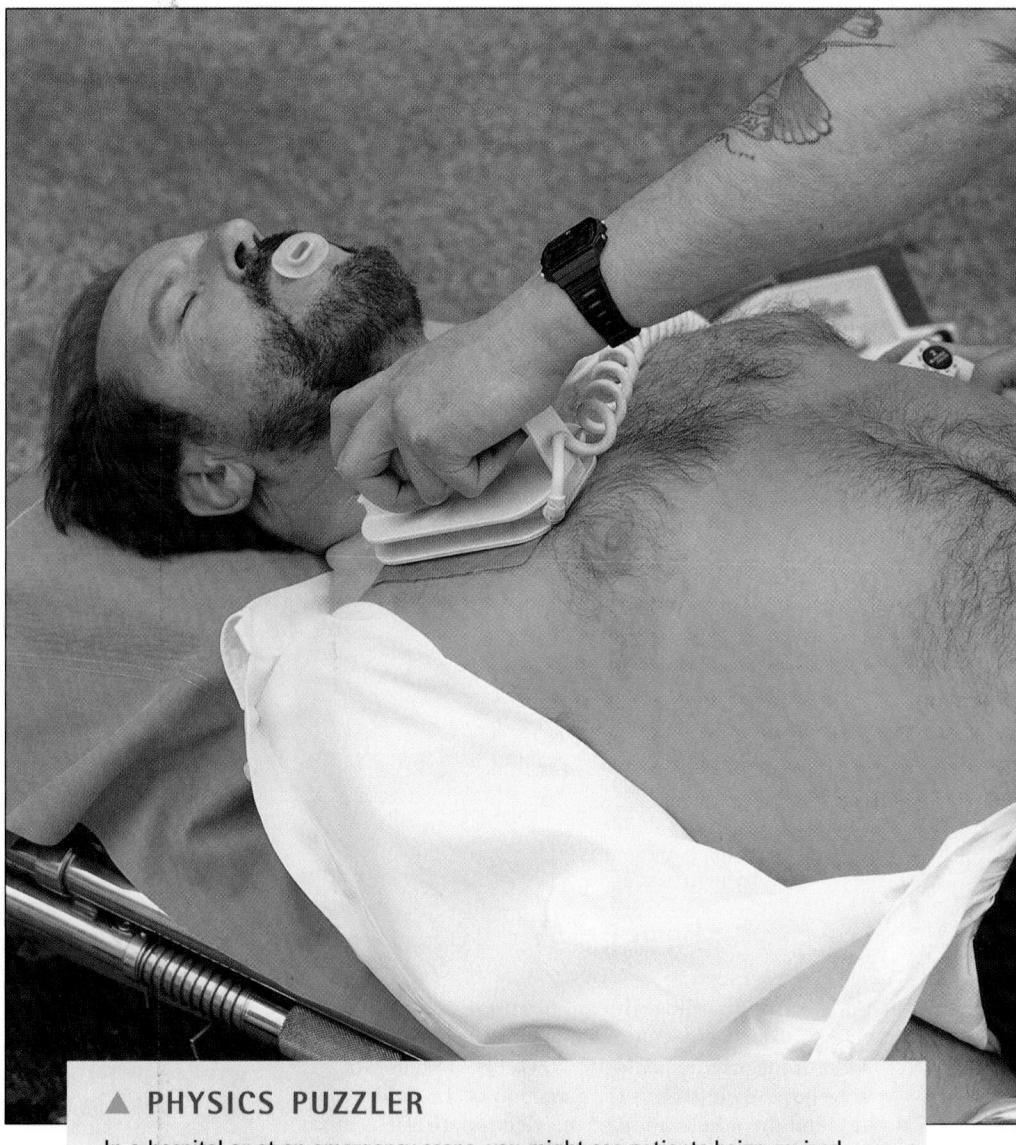

▲ **PHYSICS PUZZLER**

In a hospital or at an emergency scene, you might see patients being revived with a *defibrillator machine*. Paddles are applied to the patient's chest and a shock is sent through the patient. What does the electric shock do? Why do the caregivers have to wait before a subsequent shock can be applied? If the machine is a portable machine, how can batteries apply such a large voltage?

(Adam Hart-Davis/Science Photo Library/Photo Researchers, Inc.)

T he concept of potential energy was first introduced in Chapter 5. A potential energy function can be defined for any conservative force, such as the force of gravity. By using the principle of conservation of energy, we were often able to avoid working directly with forces when solving problems. In this chapter we discover that the potential energy concept is also useful in the study of electricity. Because the Coulomb force is conservative, we can define an electrical potential energy corresponding to the Coulomb force. This concept of potential energy is of value, but perhaps even more valuable is a quantity called *electric potential*, defined as potential energy per unit charge.

Electric potential is of great practical value for dealing with electric circuits. For example, when we speak of a voltage applied between two points, we are actually referring to an electric potential difference between those points. We take our first steps toward circuits with a discussion of electric potential, carried forward by an investigation of a common circuit element called a *capacitor*.

16.1 POTENTIAL DIFFERENCE AND ELECTRIC POTENTIAL

In Chapter 5 we showed that the gravitational force is a conservative force. As you may recall, this means that the work done on an object by this force depends only on the initial and final positions of the object and not on the path connecting the two positions. Furthermore, because the gravitational force is conservative, it is possible to define a potential energy function, which we call gravitational potential energy. Because the Coulomb force law is of the same form as the universal law of gravity, it follows that **the electrostatic force is also conservative.** Therefore, it is possible to define an electrical potential energy function associated with this force.

 11.8

Let us consider potential energy from the point of view of the particular situation shown in Figure 16.1. Imagine a small positive charge placed at point A in a uniform electric field of magnitude E. As the charge moves from point A to point B under the influence of the electric force exerted on it, qE, the work done on the charge by the electric force is

$$W = Fd = qEd$$

where d is the distance between A and B.

By definition, **the work done by a conservative force equals the negative of the change in potential energy, ΔPE.** The change in electrical potential energy is therefore

$$\Delta PE = -W = -qEd \qquad \text{[16.1]}$$

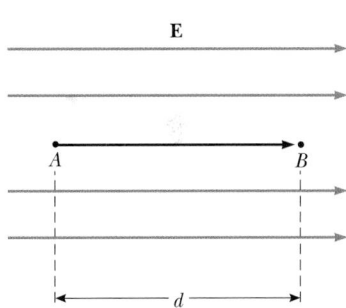

Figure 16.1 When a charge, q, moves in a uniform electric field, **E,** from point A to point B, the work done on the charge by the electric force is qEd.

◀ Change in electrical potential energy

Note that although potential energy can be defined for any electric field, **Equation 16.1 is valid only for the case of a uniform electric field.** In subsequent sections we shall examine situations in which the electric field is not uniform.

In the coming pages we shall often have occasion to use electrical potential energy, but of even more practical importance in the study of electricity is the concept of electric potential.

> The potential difference between points A and B, $V_B - V_A$, is defined as the change in potential energy (final value minus initial value) of a charge, q, moved from A to B, divided by the charge.

◀ Potential difference between two points

$$\Delta V \equiv V_B - V_A = \frac{\Delta PE}{q} \qquad [16.2]$$

Potential difference should not be confused with potential energy. The change in electric potential between two points is proportional to the change in electrical potential energy of a charge as it moves between the points, and we see from Equation 16.2 that the two are related as $\Delta PE = q\,\Delta V$. Because potential energy is a scalar quantity, **electric potential is also a scalar quantity.** From Equation 16.2 we see that electric potential difference is a measure of energy per unit charge. Alternatively, electrical potential difference is the work done to move a charge from a point A to a point B divided by the magnitude of the charge. Thus, the SI units of electric potential are joules per coulomb, called volts (V):[1]

$$1\text{ V} \equiv 1\text{ J/C} \qquad [16.3]$$

This says that 1 J of work must be done to move a 1-C charge between two points that are at a potential difference of 1 V. In the process of moving through a potential difference of 1 V, the 1-C charge gains (or loses) 1 J of energy. Dividing Equation 16.1 by q gives

$$\frac{\Delta PE}{q} = V_B - V_A = -Ed \qquad [16.4]$$

This equation shows that potential difference also has units of electric field times distance. From this, it follows that the SI units of electric field, newtons per coulomb, can also be expressed as volts per meter:

$$1\text{ N/C} = 1\text{ V/m}$$

Because Equation 16.4 is directly related to Equation 16.1, it, too, is valid only for the case of a uniform field.

Let us examine the changes in energy associated with movements of charge in the electric field pictured in Figure 16.2a. Because the positive charge q tends to move in the direction of the electric field, we must apply an upward external force on the charge to move it from B to A. Work is done on the charge, and this means that **a positive charge gains electrical potential energy when it is moved in a direction opposite the electric field.** This is analogous to a mass gaining gravitational potential energy when it rises to higher elevations in the presence of gravity, as in Figure 16.2b. If a positive charge is released from rest at point A, it experiences a force, qE, in the direction of the field (downward in Figure 16.2a). Therefore, it accelerates downward, gaining kinetic energy. **As it gains kinetic energy, it loses an equal amount of electrical potential energy.** Also, as Equation 16.4 shows, if a positive charge moves from A to B, its electric potential decreases.

---

[1] Note that the symbol V (italic) represents potential, whereas V (roman) is the symbol for the unit of this quantity—volts. Do not confuse these two symbols. On the *Core Concepts in College Physics* CD-ROM, the symbol V appears where we use ΔV in the textbook.

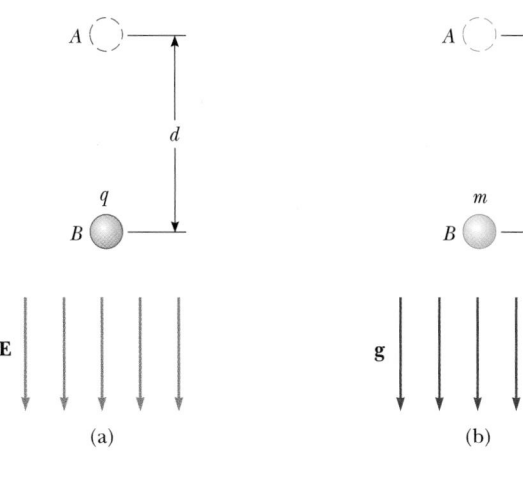

Figure 16.2 (a) When the electric field, **E,** is directed downward, point *B* is at a lower electric potential than point *A*. A positive test charge that moves from *A* to *B* loses electric potential energy. (b) A mass, *m*, moving downward in the direction of the gravitational field, **g,** loses gravitational potential energy.

By contrast, if the test charge *q* is negative, the situation is reversed. **A negative charge loses electrical potential energy when it moves in the direction opposite the electric field.** That is, a negative charge released from rest in the field **E** accelerates in a direction *opposite* the field.

Thus, when a positive charge is placed in an electric field, it moves in the direction of the field, from a point of high potential to a point of lower potential. In the process, its electrical potential energy decreases and its kinetic energy increases.

When a negative charge is placed in an electric field, it moves opposite to the direction of the field, from a point of low potential to a point of higher potential. In the process, it also undergoes a decrease in electrical potential energy and an increase in kinetic energy.

Let us pause briefly to discuss a situation that illustrates the concept of electric potential difference. Consider the common 12-V automobile battery. Such a battery maintains a potential difference across its terminals, where the positive terminal is 12 V higher in potential than the negative terminal. In practice, the negative terminal is usually connected to the metal body of the car, which can be considered at a potential of zero volts. The battery becomes a useful device when it is connected by conducting wires to such things as lightbulbs, a radio, power windows, motors, and so forth. Now consider a charge of + 1 C, to be moved around a circuit that contains the battery connected to some of these external devices. As the charge is moved inside the battery from the negative terminal (at 0 V) to the positive terminal (at 12 V), the work done on the charge by the battery is 12 J. Thus, every coulomb of positive charge that leaves the positive terminal of the battery carries an energy of 12 J. As this charge moves through the external circuit toward the negative terminal, it gives up its 12 J of electrical energy to the external devices. When the charge reaches the negative terminal, its electrical energy is zero. At this point, the battery takes over and restores 12 J of energy to the charge as it is moved from the negative to the positive terminal, enabling it to make another transit of the circuit. The actual amount of charge that leaves the battery and traverses the circuit depends on the properties of the external devices, as we shall see in the next chapter.

APPLICATION

Automobile Batteries.

QUICKLAB

Place a clean fluorescent tube on a dry towel lying on a table. In the dark, rub the tube with a sheet of plastic. What causes the tube to emit light?

EXAMPLE 16.1   The Field Between Two Parallel Plates of Opposite Charge

Figure 16.3 illustrates a situation in which a constant electric field can be set up. A 12-V battery is connected between two parallel metal plates separated by 0.30 cm. Find the strength of the electric field.

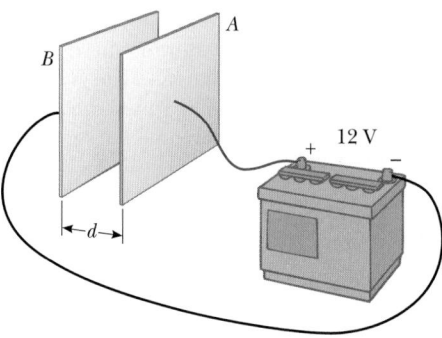

Figure 16.3   (Example 16.1) A 12-V battery connected to two parallel plates. The electric field between the plates has a magnitude given by the potential difference divided by the plate separation, *d*.

Reasoning   The electric field is uniform (except near the edges of the metal plates), and thus the relationship between potential difference and the magnitude of the field is given by Equation 16.4:

$$V_B - V_A = -Ed$$

As already noted, chemical forces inside a battery maintain one electrode, called the positive terminal, at a higher potential than a second electrode, the negative terminal. Thus, in Figure 16.3, plate *B*, which is connected to the negative terminal, must be at a lower potential than plate *A*, which is connected to the positive terminal.

Solution   We have

$$V_B - V_A = -12 \text{ V}$$

This gives a value for *E* of

$$E = -\frac{(V_B - V_A)}{d} = -\frac{(-12 \text{ V})}{0.30 \times 10^{-2} \text{ m}} = \boxed{4.0 \times 10^3 \text{ V/m}}$$

The direction of this field is from the positive plate to the negative plate. A device consisting of two plates separated by a small distance is called a *parallel-plate capacitor* (to be discussed later in this chapter).

EXAMPLE 16.2   Motion of a Proton in a Uniform Electric Field

A proton is released from *rest* in a uniform electric field of magnitude 8.0×10^4 V/m, directed along the positive *x* axis (Fig. 16.4). The proton undergoes a displacement of 0.50 m in the direction of the field.
(a) Find the *change* in electric potential of the proton as a result of this displacement.

Solution   From Equation 16.4, we have

$$\Delta V = V_B - V_A = -Ed = -(8.0 \times 10^4 \text{ V/m})(0.50 \text{ m}) = \boxed{-4.0 \times 10^4 \text{ V}}$$

Thus, the electric potential of the proton *decreases* as it moves from *A* to *B*.

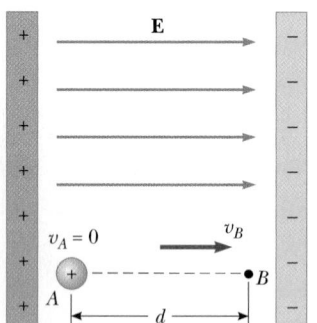

Figure 16.4   (Example 16.2) A proton accelerates from *A* to *B* in the direction of the uniform electric field.

(b) Find the change in electrical potential energy of the proton for this displacement.

Solution

$$\Delta PE = q\,\Delta V = e\,\Delta V = (1.6 \times 10^{-19}\ \text{C})(-4.0 \times 10^4\ \text{V}) = \boxed{-6.4 \times 10^{-15}\ \text{J}}$$

The negative sign here means that the electrical potential energy of the proton decreases as it moves in the direction of the electric field. This makes sense because, as the proton *accelerates* in the direction of the field, it gains kinetic energy and at the same time loses electrical potential energy (mechanical energy is conserved).

(c) Find the speed of the proton after it has moved 0.50 m, starting from rest.

Solution If no forces other than the conservative electrical force are acting on the proton, we can apply the principle of conservation of mechanical energy in the form

$$KE_i + PE_i = KE_f + PE_f$$

In our case, $KE_i = 0$; hence, the preceding expression gives

$$KE_f = PE_i - PE_f = -\Delta PE$$

With this equation and the results of part (b), we find that

$$\tfrac{1}{2}mv_f^2 = 6.4 \times 10^{-15}\ \text{J}$$

and

$$v_f^2 = \frac{2(6.4 \times 10^{-15}\ \text{J})}{1.67 \times 10^{-27}\ \text{kg}} = 7.66 \times 10^{12}\ \text{m}^2/\text{s}^2$$

$$v_f = \boxed{2.8 \times 10^6\ \text{m/s}}$$

16.2 ELECTRIC POTENTIAL AND POTENTIAL ENERGY DUE TO POINT CHARGES

In electric circuits, a point of zero electric potential is often defined by grounding (connecting to Earth) some point in the circuit. For example, if the negative plate in Example 16.1 were grounded, it would be considered to have a potential of zero and the positive plate to have a potential of 12 V. It is also possible to define the electric potential due to a point charge at a point in space. In this case, the point of zero electric potential is taken to be at an infinite distance from the charge. With this choice, the methods of calculus can be used to show that the electric potential created by a point charge, q, at any distance, r, from the charge is given by

$$V = k_e \frac{q}{r} \qquad\qquad [16.5]$$

◀ Electric potential created by a point charge

Equation 16.5 points out a significant property of electric potential that makes it an important quantity in the study of electricity: The potential at a given point depends only on the charge, q, on the object setting up the potential and the distance r from that object to a specific point in space. As a result, we can say that a potential exists at some point in space whether or not there is a charge at that point.

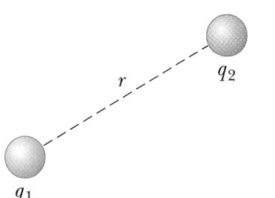

Figure 16.5 If two point charges are separated by the distance r, the potential energy of the pair is $k_e q_1 q_2 / r$.

Potential energy of a pair of ▶
charges

The electric potential of two or more charges is obtained by applying the **superposition principle.** That is, **the total electric potential at some point P due to several point charges is the algebraic sum of the electric potentials due to the individual charges.** This is similar to the method used in Chapter 15 to find the resultant electric field at a point in space. However, note that in the case of potentials, one must evaluate an *algebraic sum* of individual potentials to obtain the total, because **potentials are scalar quantities.** Thus, it is much easier to evaluate the electric potential at some point due to several charges than to evaluate the electric field, which is a vector quantity.

We now consider the electrical potential energy of interaction of a system of two charged particles. If V_1 is the electric potential due to charge q_1 at a point, P, then the work required to bring charge q_2 from infinity to P without acceleration is $q_2 V_1$. By definition, this work equals the potential energy, PE, of the two-particle system when the particles are separated by a distance of r (Fig. 16.5).

Therefore, we can express the electrical potential energy of the *pair* of charges as

$$PE = q_2 V_1 = k_e \frac{q_1 q_2}{r} \qquad \text{[16.6]}$$

Note that if the charges are of the *same* sign, PE is positive. This is consistent with the fact that like charges repel, and so positive work must be done on the system to bring two charges near one another. Conversely, if the charges are of *opposite* sign, the force is attractive and PE is negative. This means that negative work must be done to bring unlike charges close together.

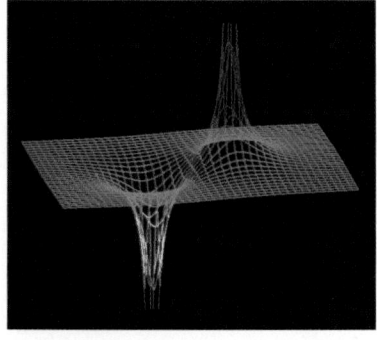

Computer-generated plot of the electric potential associated with an electric dipole. The charges lie in the horizontal plane, at the center of the potential spikes. The contour lines help visualize the size of the potential, the values of which are plotted vertically. *(Richard Megna, Fundamental Photographs)*

Thinking Physics 1

A spherical rubber balloon contains a charged object, which is mounted so that it always stays at the center of the balloon. As the balloon is inflated to a larger volume, what happens to the electric potential at the surface of the balloon? To the electric field at the surface of the balloon? Answer the same question for a metallic balloon being inflated.

Explanation The electric potential at the surface of the balloon will decrease in inverse proportion to the radius. The electric field will decrease as the reciprocal of the squared radius of the balloon. The answers will be exactly the same for the metallic balloon—the metal makes no difference, except for the fact that there will be no electric field within the skin of the balloon.

Problem-Solving Strategy

Electric Potential

1. When you work problems involving electric potential, remember that potential is a *scalar quantity* (rather than a vector quantity, like the electric field), so there are no components to worry about. Therefore, when using the superposition principle to evaluate the electric potential due to a system of point charges at a point, simply take the algebraic sum of the po-

tentials due to all charges. You must keep track of signs, however. The potential due to each positive charge is positive and the potential due to each negative charge is negative. Use the basic equation $V = k_e q/r$.

2. As in mechanics, only changes in electric potential are significant; hence, the point you choose for zero electric potential is arbitrary.

EXAMPLE 16.3 Finding the Electric Potential

A 5.0-μC point charge is at the origin, and a point charge of -2.0 μC is on the x axis at (3.0, 0) m, as in Figure 16.6.
(a) If the electric potential is taken to be zero at infinity, find the total electric potential due to these charges at point P, with coordinates (0, 4.0) m.

Reasoning The electric potential at P due to each charge can be calculated from $V = k_e q/r$. The total electric potential is the scalar sum of these two potentials.

Solution The electric potential at P due to the 5.0-μC charge is

$$V_1 = k_e \frac{q_1}{r_1} = \left(8.99 \times 10^9 \frac{\text{N} \cdot \text{m}^2}{\text{C}^2} \right) \left(\frac{5.0 \times 10^{-6} \text{ C}}{4.0 \text{ m}} \right) = 1.12 \times 10^4 \text{ V}$$

and the electric potential due to the -2.0-μC charge is

$$V_2 = k_e \frac{q_2}{r_2} = \left(8.99 \times 10^9 \frac{\text{N} \cdot \text{m}^2}{\text{C}^2} \right) \left(\frac{-2.0 \times 10^{-6} \text{ C}}{5.0 \text{ m}} \right) = -0.360 \times 10^4 \text{ V}$$

and

$$V_P = V_1 + V_2 = \boxed{7.6 \times 10^3 \text{ V}}$$

(b) How much work is required to bring a third point charge of 4.0 μC from infinity to P?

Solution

$$W = q_3 V_P = (4.0 \times 10^{-6} \text{ C})(7.6 \times 10^3 \text{ V})$$

Because 1 V = 1 J/C, W reduces to

$$W = \boxed{3.1 \times 10^{-2} \text{ J}}$$

Exercise Find the magnitude and direction of the electric field at point P.

Answer 2.3×10^3 N/C at an angle of 79° with the x axis.

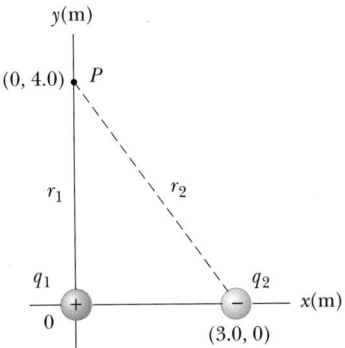

Figure 16.6 (Example 16.3) The electric potential at point P due to the point charges q_1 and q_2 is the algebraic sum of the potentials due to the individual charges.

16.3 POTENTIALS AND CHARGED CONDUCTORS

In order to determine the electric potential at all points on a charged conductor, let us combine Equations 16.1 and 16.2. From Equation 16.1 we see that the work done on a charge by electric forces is related to the change in electrical potential energy of the charge by

$$W = -\Delta PE$$

Furthermore, from Equation 16.2 we see that the change in electrical potential energy between two points, A and B, is related to the potential difference between these points by

$$\Delta PE = q(V_B - V_A)$$

Combining these two equations, we find that

$$W = -q(V_B - V_A) \qquad [16.7]$$

As we see from this result, **no work is required to move a charge between two points that are at the same electric potential. That is, $W = 0$ when $V_B = V_A$.**

In Chapter 15 we found that when a conductor is in electrostatic equilibrium, a net charge placed on it resides entirely on its surface. Furthermore, we showed that the electric field just outside the surface of a charged conductor in electrostatic equilibrium is perpendicular to the surface and that the field inside the conductor is zero. We shall now show that **all points on the surface of a charged conductor in electrostatic equilibrium are at the same potential.**

Consider a surface path connecting any points A and B on a charged conductor, as in Figure 16.7. The electric field, **E,** is always perpendicular to the displacement along this path; therefore, no work is done by the electric field if a charge is moved between these points. From Equation 16.7 we see that if the work done is zero, the difference in electric potential, $V_B - V_A$, is also zero. Therefore, **the electric potential is a constant everywhere on the surface of a charged conductor in equilibrium.** Furthermore, because the electric field inside a conductor is zero, no work is required to move a charge between two points inside the conductor. Again, Equation 16.7 shows that if the work done is zero, the difference in electric potential between any two points inside a conductor must also be zero. Thus, we conclude that the electric potential is constant everywhere inside a conductor.

Finally, because one of the points could be arbitrarily close to the surface of the conductor, we conclude that **the electric potential is constant everywhere inside a conductor and equal to its value at the surface.** As a consequence, no work is required to move a charge from the interior of a charged conductor to its surface. (Note that the electric potential inside a conductor is not necessarily zero even though the interior electric field is zero.)

▶ Properties of a charged conductor in equilibrium

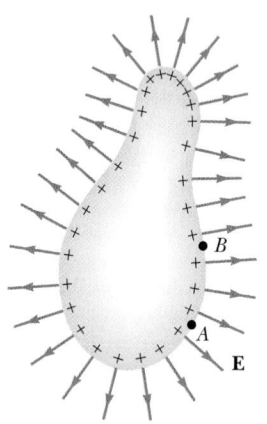

Figure 16.7 An arbitrarily shaped conductor with an excess positive charge. When the conductor is in electrostatic equilibrium, all of the charge resides at the surface, **E** = 0 inside the conductor, and the electric field just outside the conductor is perpendicular to the surface. The potential is constant inside the conductor and is equal to the potential at the surface.

The Electron Volt

A unit of energy commonly used in atomic and nuclear physics is the electron volt (eV).

▶ Definition of the electron volt

> The **electron volt** is defined as the energy that an electron (or proton) gains when accelerated through a potential difference of 1 V.

Because 1 V = 1 J/C and because the magnitude of charge on the electron or proton is 1.60×10^{-19} C, we see that the electron volt is related to the joule by

$$1 \text{ eV} = 1.60 \times 10^{-19} \text{ C} \cdot \text{V} = 1.60 \times 10^{-19} \text{ J} \qquad [16.8]$$

Thinking Physics 2

Suppose scientists had chosen to measure small energies in proton volts rather than electron volts. What difference would this make?

Explanation There would be no change at all. An electron volt is the kinetic energy gained by an electron in being accelerated through a potential difference of one volt. A proton accelerated through one volt would have the same kinetic energy, because it carries the same charge as the electron (except for sign). The proton would be moving in the opposite direction and more slowly after accelerating through 1 volt, due to its opposite charge and its larger mass, but it would still gain 1 electron volt, or 1 proton volt, of kinetic energy.

16.4 EQUIPOTENTIAL SURFACES

A surface on which all points are at the same potential is called an **equipotential surface.** The potential difference between any two points on an equipotential surface is zero. Hence, **no work is required to move a charge at constant speed on an equipotential surface.** Equipotential surfaces have a simple relationship to the electric field. **The electric field at every point of an equipotential surface is perpendicular to the surface.** If the electric field, **E,** had a component parallel to the surface, this component would produce an electric force on a charge placed on the surface. This force would do work on the charge as it moved from one point to another, in contradiction to the definition of an equipotential surface.

It is convenient to represent equipotential surfaces on a diagram by drawing **equipotential lines,** which are two-dimensional views of the intersections of the equipotential surfaces with the plane of the drawing. Figure 16.8a shows the equipotential lines (in blue) associated with a positive point charge. Note that the equi-

 11.9, SECTION 2

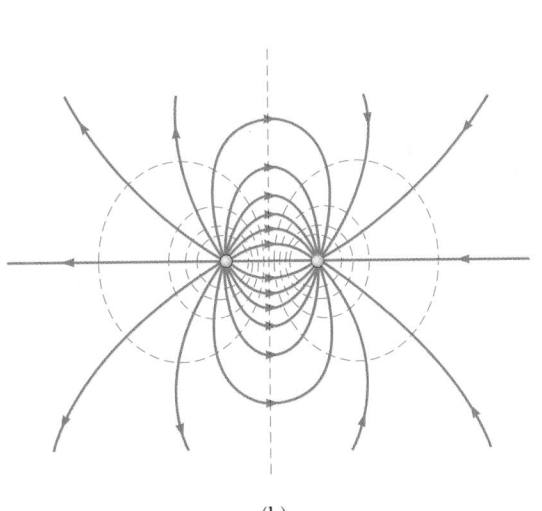

(a)   (b)

Figure 16.8 Equipotential surfaces (dashed blue lines) and electric field lines (red lines) for (a) a positive point charge and (b) two equal but opposite point charges. In all cases, the equipotential surfaces are *perpendicular* to the electric field lines at every point.

potential lines are perpendicular to the electric field lines (in red) at all points. Recall that the electric potential created by a point charge q is given by $V = k_e q / r$. This relation shows that, for a single point charge, the electric potential is constant on any surface in which r is constant. Therefore, the equipotential surfaces of a point charge are a family of spheres centered on the point charge. Figure 16.8b shows the equipotential lines associated with two charges of equal magnitude but opposite sign.

16.5   APPLICATIONS

The Electrostatic Precipitator

One important application of electric discharge in gases is a device called an *electrostatic precipitator*. It is used to remove particulate matter from combustion gases, thereby reducing air pollution. It is especially useful in coal-burning power plants and in industrial operations that generate large quantities of smoke. Systems currently in use can eliminate approximately 90% by mass of the ash and dust from the smoke. Unfortunately, a very high percentage of the lighter particles still escape, and these contribute significantly to smog and haze.

Figure 16.9 illustrates the basic idea of the electrostatic precipitator. A high voltage (typically 40 kV to 100 kV is maintained between a wire running down the center of a duct and the outer wall, which is grounded. The wire is maintained at a negative electric potential with respect to the wall, and so the electric field is directed toward the wire. The electric field near the wire reaches a high enough value to cause a discharge around the wire and the formation of positive ions, electrons, and negative ions, such as O_2^-. As the electrons and negative ions are accelerated toward the outer wall by the nonuniform electric field, the dirt particles in the streaming gas become charged by collisions and ion capture. Because most of the charged dirt particles are negative, they are also drawn to the outer wall by the electric field. When the duct is shaken, the particles fall loose and are collected at the bottom.

In addition to reducing the amounts of harmful gases and particulate matter in the atmosphere, the electrostatic precipitator recovers valuable metal oxides from the stack.

An *electrostatic air cleaner,* used in homes to relieve the discomfort of allergy sufferers, uses many of the same principles as the precipitator. Air laden with dust and pollen is drawn into the device across a positively charged mesh screen. The airborne particles become positively charged when they make intimate contact with the screen. The particles then pass through a second, negatively charged mesh screen. The electrostatic force of attraction between the positively charged particles in the air and the negatively charged screen causes the particles to precipitate out on the surface of the screen. In this fashion, a very high percentage of contaminants are removed from the air stream.

Xerography and Laser Printers

The process of xerography is widely used for making photocopies of printed materials. The basic idea behind the process was developed by Chester Carlson, who was granted a patent for his invention in 1940. In 1947 the Xerox Corporation

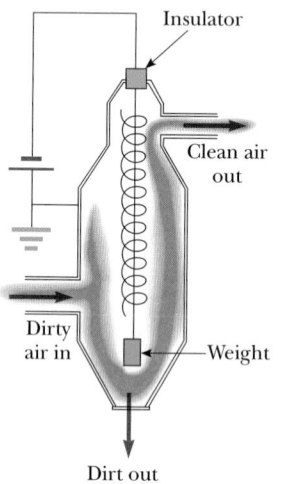

Figure 16.9 A schematic diagram of an electrostatic precipitator. The high voltage maintained on the central wires creates an electric discharge in the vicinity of the wire.

Q U I C K L A B

Sprinkle some salt and pepper in an open dish, and mix the two ingredients together. Now rub a comb through your hair several times and bring the comb to within about 1 cm of the ingredients. Why is the pepper attracted to the comb but not the salt? This is a simple demonstration of electrostatic separation.

launched a full-scale program to develop automated duplicating machines using Carlson's process. The huge success of that development is quite evident; today, practically all offices and libraries have one or more duplicating machines, and the capabilities of modern technology continues to evolve.

Some features of the xerographic process involve simple concepts from electrostatics and optics. However, the one idea that makes the process unique is the use of photoconductive material to form an image. (A photoconductor is a material that is a poor conductor of electricity in the dark but becomes a reasonably good electric conductor when exposed to light.)

Figure 16.10 illustrates the steps in the xerographic process. First, the surface of a plate or drum is coated with a thin film of the photoconductive material (usually selenium or some compound of selenium), and the photoconductive surface is given a positive electrostatic charge in the dark (Fig. 16.10a). The page to be copied is then projected onto the charged surface (Fig. 16.10b). The photoconducting surface becomes conducting only in areas where light strikes; there the light produces charge carriers in the photoconductor, which neutralize the positively charged surface. The charges remain on those areas of the photoconductor not exposed to light, however, leaving a hidden image of the object in the form of a positive surface-charge distribution.

APPLICATION

Photocopying.

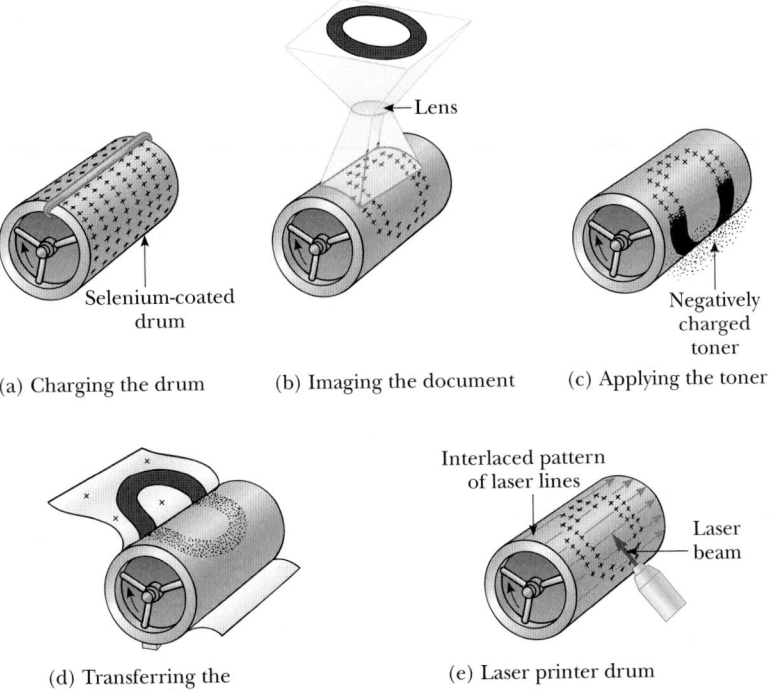

Figure 16.10 The xerographic process. (a) The photoconductive surface is positively charged. (b) Through the use of a light source and lens, a hidden image is formed on the charged surface in the form of positive charges. (c) The surface containing the image is covered with a negatively charged powder, which adheres only to the image area. (d) A piece of paper is placed over the surface and given a charge. This transfers the image to the paper, which is then heated to "fix" the powder to the paper. (e) The image on the drum of a laser printer is produced by turning a laser beam on and off as it sweeps across the selenium-coated drum.

Next, a powder called a *toner* is negatively charged and dusted onto the photoconducting surface (Fig. 16.10c). The charged powder adheres only to the areas that contain the positively charged image. At this point, the image becomes visible. It is then transferred to the surface of a sheet of positively charged paper (Fig. 16.10d). Finally, the toner is "fixed" to the surface of the paper by heat. This results in a permanent copy of the original.

APPLICATION

Laser Printers.

The steps for producing a document on a laser printer are similar to those used in a photocopy machine, in that parts (a), (c), and (d) of Figure 16.10 remain essentially the same. The difference between the two techniques lies in the way the image is formed on the selenium-coated drum. In a laser printer, the command to print the letter O, for instance, is sent to a laser from the memory of a computer. A rotating mirror inside the printer causes the beam of the laser to sweep across the selenium-coated drum in an interlaced pattern (Fig. 16.10e). Electrical signals generated by the printer turn the laser beam on and off in a pattern that traces out the letter "O" in the form of positive charges on the selenium. Toner is then applied to the drum, and the transfer to paper is accomplished as in a photocopy machine.

16.6 THE DEFINITION OF CAPACITANCE

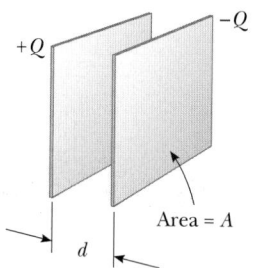

Figure 16.11 A parallel-plate capacitor consists of two parallel plates, each of area A, separated by a distance d. The plates carry equal and opposite charges.

A **capacitor** is a device used in a variety of electric circuits—for example, to tune the frequency of radio receivers, eliminate sparking in automobile ignition systems, or store short-term energy in electronic flash units. Figure 16.11 shows a typical design for a capacitor. It consists of two parallel metal plates separated by a distance of d. When used in an electric circuit, the plates are connected to the positive and negative terminals of a battery or some other voltage source. When this connection is made, electrons are pulled off one of the plates, leaving it with a charge of $+Q$, and transferred through the battery to the other plate, leaving it with a charge of $-Q$, as shown in the figure. This charge transfer stops when the potential difference across the plates equals the potential difference of the battery. Thus, a charged capacitor acts as a storehouse of charge and energy that can be reclaimed when needed for a specific application.

> The capacitance, C, of a capacitor is defined as the ratio of the magnitude of the charge on either conductor to the magnitude of the potential difference between the conductors:

Capacitance of a pair of conductors ▶

$$C \equiv \frac{Q}{\Delta V} \qquad [16.9]$$

From this equation we see that a large capacitance is needed to store a large amount of charge for a given applied voltage. Also, we see that capacitance has the SI units coulombs per volt, called **farads (F)** in honor of Michael Faraday. That is,

$$1 \text{ F} \equiv 1 \text{ C/V}$$

The farad is a very large unit of capacitance. In practice, most typical capacitors have capacitances ranging from microfarads ($1 \ \mu\text{F} = 1 \times 10^{-6} \ \text{F}$) to picofarads ($1 \ \text{pF} = 1 \times 10^{-12} \ \text{F}$).

EXAMPLE 16.4 The Charge on the Plates of a Capacitor

A 3.0-μF capacitor is connected to a 12-V battery. What is the magnitude of the charge on each plate of the capacitor?

Solution The definition of capacitance (Eq. 16.9) gives

$$Q = C \, \Delta V = (3.0 \times 10^{-6} \ \text{F})(12 \ \text{V}) = \boxed{36 \ \mu\text{C}}$$

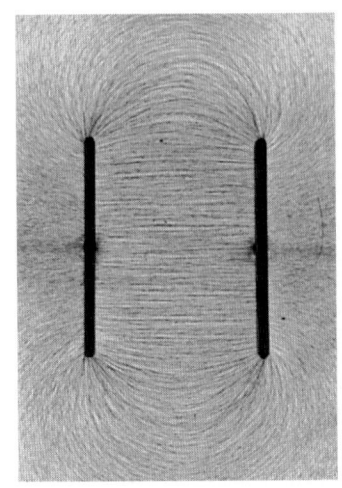

The electric field pattern of two oppositely charged conducting parallel plates. Note the nonuniform nature of the electric field at the ends of the plates. *(Courtesy of Harold M. Waage, Princeton University)*

16.7 THE PARALLEL-PLATE CAPACITOR

The capacitance of a device depends on the geometric arrangement of the conductors. For example, the capacitance of a parallel-plate capacitor whose plates are separated by air (see Fig. 16.11) is

$$C = \epsilon_0 \frac{A}{d} \qquad \text{[16.10]}$$

◀ Capacitance of a parallel-plate capacitor

where A is the area of one of the plates, d is the distance of separation of the plates, and ϵ_0 is a constant called the **permittivity of free space,** with the value

$$\epsilon_0 = 8.85 \times 10^{-12} \ \text{C}^2/\text{N} \cdot \text{m}^2$$

The permittivity of free space is related to the Coulomb constant, k_e, by

$$k_e = \frac{1}{4\pi\epsilon_0}$$

Although we shall not derive Equation 16.10, we shall attempt to make it seem plausible. As you can see from the definition of capacitance $C \equiv Q/\Delta V$, the amount of charge a given capacitor can store for a given potential difference across its plates increases as the capacitance increases. Therefore, it seems reasonable that a capacitor constructed from plates with large areas should be able to store a large charge. Furthermore, if the oppositely charged plates are close together, the attractive force between them will be large. In fact, for a given potential difference, the charge on the plates increases with decreasing plate separation.

One practical device that uses a capacitor is the flash attachment on a camera. A battery is used to charge the capacitor, and this stored charge is then released when the shutter-release button is pressed to take a picture. The stored charge is delivered to a flash tube very quickly, illuminating the subject at the instant more light is needed.

Computers make use of capacitors in many ways. For example, one type of computer keyboard has capacitors at the bases of its keys, as in Figure 16.12. Each key is connected to a movable plate, which represents one side of the capacitor; the fixed plate on the bottom of the keyboard represents the other side of the

APPLICATION

Camera Flash Attachments.

APPLICATION

Computer Keyboards.

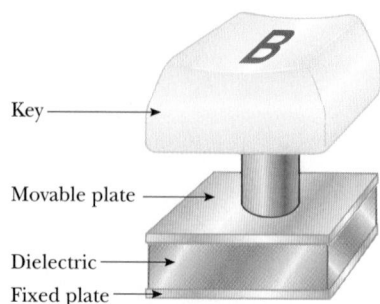

Key ——→

Movable plate ——→

Dielectric ——→
Fixed plate ——→

Figure 16.12 When the key of one type of keyboard is pressed, the capacitance of a parallel-plate capacitor increases as the plate spacing decreases.

capacitor. When a key is pressed, the capacitor spacing decreases, causing an increase in capacitance. External electronic circuits recognize each key by the *change* in its capacitance when it is pressed.

EXAMPLE 16.5   Calculating C for a Parallel-Plate Capacitor

A parallel-plate capacitor has an area of $A = 2.00$ cm^2 = 2.00×10^{-4} m^2 and a plate separation of $d = 1.00$ mm = 1.00×10^{-3} m. Find its capacitance.

Solution   From $C = \epsilon_0 A/d$ we find that

$$C = \epsilon_0 \frac{A}{d} = (8.85 \times 10^{-12} \text{ C}^2/\text{N} \cdot \text{m}^2) \left(\frac{2.00 \times 10^{-4} \text{ m}^2}{1.00 \times 10^{-3} \text{ m}} \right)$$

$$= 1.77 \times 10^{-12} \text{ F} = \boxed{1.77 \text{ pF}}$$

Exercise   Show that 1 C^2/N·m equals 1 F.

Symbols for Circuit Elements

The symbol that is commonly used to represent a capacitor in a circuit is ——│├——, or sometimes ——│├——. Do not confuse this with the circuit symbol ——┤├—— used to designate a battery (or any other direct current source). The positive terminal of the battery is at the higher potential and is represented by the longer vertical line in the battery symbol. In the next chapter we shall discuss another circuit element, called a resistor, represented by the symbol ——W——. The wires in a circuit that do not have appreciable resistance compared to other elements in the circuit will be represented by straight lines.

16.8   COMBINATIONS OF CAPACITORS

Two or more capacitors can be combined in circuits in several ways. The equivalent capacitances of certain combinations can be calculated with methods described in this section.

Parallel Combination

Two capacitors connected as shown in Figure 16.13a are known as a *parallel combination* of capacitors. The left plate of each capacitor is connected by a conducting wire to the positive terminal of the battery, and the left plates are therefore at the same potential. Likewise, the right plates are connected to the negative terminal of the battery. When the capacitors are first connected in the circuit, electrons are transferred from the left plates through the battery to the right plates, leaving the left plates positively charged and the right plates negatively charged. The energy source for this charge transfer is the internal chemical energy stored in the battery, which is converted to electrical energy. The flow of charge ceases when the voltage across the capacitors equals that of the battery. The capacitors reach their maximum

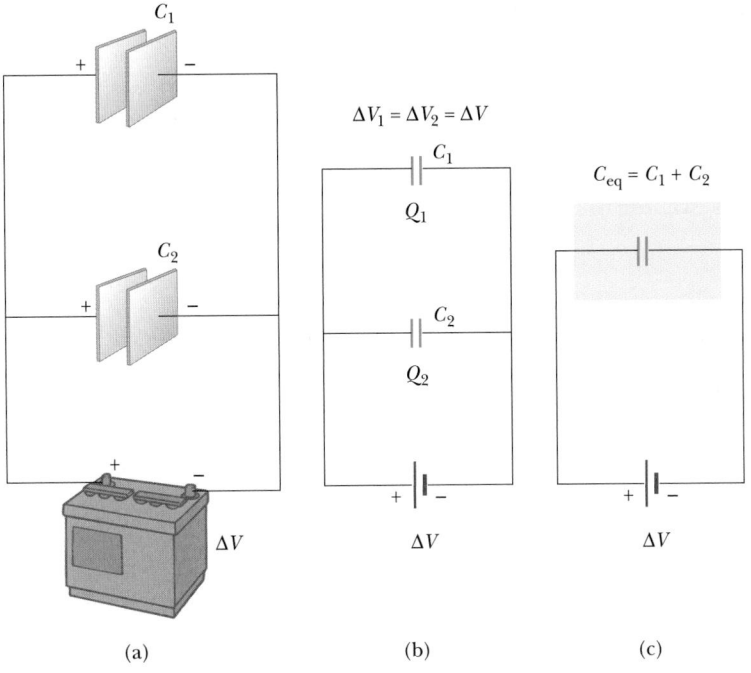

Figure 16.13 (a) A parallel connection of two capacitors. (b) The circuit diagram for the parallel combination. (c) The potential differences across the capacitors are the same, and the equivalent capacitance is $C_{eq} = C_1 + C_2$.

charge when the flow of charge ceases. Let us call the maximum charges on the two capacitors Q_1 and Q_2. Then the *total charge*, Q, stored by the two capacitors is

$$Q = Q_1 + Q_2 \qquad \text{[16.11]}$$

We can replace these two capacitors with one equivalent capacitor having a capacitance of C_{eq}. This equivalent capacitor must have exactly the same external effect on the circuit as the original two. That is, it must store Q units of charge. We also see from Figure 16.13b, that **the potential differences across the capacitors in a parallel circuit are the same; each is equal to the voltage of the battery, ΔV.** From Figure 16.13c, we see that the voltage across the equivalent capacitor is also ΔV. Thus, the charge on each capacitor is

◀ ΔV is the same across capacitors connected in parallel.

$$Q_1 = C_1 \, \Delta V \qquad \text{and} \qquad Q_2 = C_2 \, \Delta V$$

The charge on the equivalent capacitor is

$$Q = C_{eq} \, \Delta V$$

Substituting these relations into Equation 16.11 gives

$$C_{eq} \, \Delta V = C_1 \, \Delta V + C_2 \, \Delta V$$

or

$$C_{eq} = C_1 + C_2 \qquad \text{(parallel combination)} \qquad \text{[16.12]}$$

If we extend this treatment to three or more capacitors connected in parallel, the equivalent capacitance is found to be

$$C_{eq} = C_1 + C_2 + C_3 + \cdots \qquad \text{(parallel combination)} \qquad [16.13]$$

Thus, we see that **the equivalent capacitance of a parallel combination of capacitors is greater than any of the individual capacitances.**

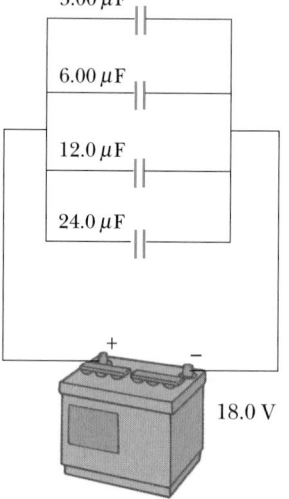

Figure 16.14 (Example 16.6) Four capacitors connected in parallel.

EXAMPLE 16.6   Four Capacitors Connected in Parallel

Determine the capacitance of the single capacitor that is equivalent to the parallel combination of capacitors shown in Figure 16.14, and find the charge on the 12.0-μF capacitor.

Solution   The equivalent capacitance is found by use of Equation 16.13:

$$C_{eq} = C_1 + C_2 + C_3 + C_4$$

$$= 3.00 \ \mu F + 6.00 \ \mu F + 12.0 \ \mu F + 24.0 \ \mu F = \boxed{45.0 \ \mu F}$$

The potential difference across the 12.0-μF capacitor (and all other capacitors in this case) is equal to the voltage of the battery, and so

$$Q = C \Delta V = (12.0 \times 10^{-6} \ \text{F})(18.0 \ \text{V}) = 216 \times 10^{-6} \ \text{C} = \boxed{216 \ \mu C}$$

Series Combination

Now consider two capacitors connected in *series*, as illustrated in Figure 16.15a. **For a series combination of capacitors, the magnitude of the charge must be the same on all the plates.** To see why this must be true, let us consider the charge

▶ *Q* is the same for all capacitors connected in series.

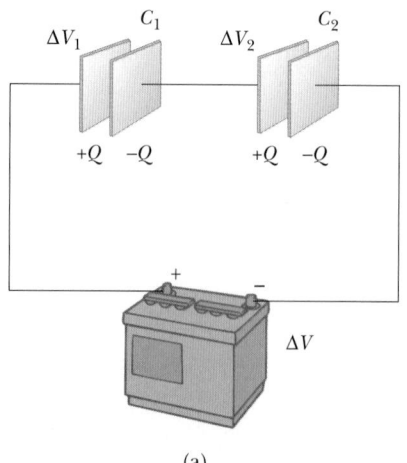

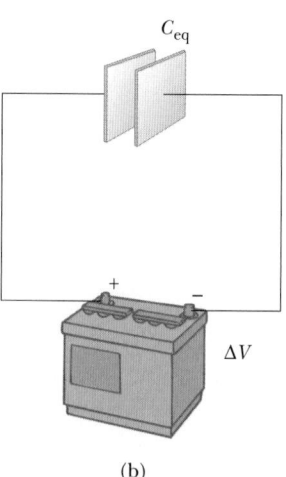

Figure 16.15 A series combination of two capacitors. The charges on the capacitors are the same, and the equivalent capacitance can be calculated from the reciprocal relationship $1/C_{eq} = (1/C_1) + (1/C_2)$.

(a)   (b)

transfer process in some detail. We start with uncharged capacitors. When a battery is connected to the circuit, electrons are transferred from the left plate of C_1 to the right plate of C_2 through the battery. As this negative charge accumulates on the right plate of C_2, an equivalent amount of negative charge is removed from the left plate of C_2, leaving it with an excess positive charge. The negative charge leaving the left plate of C_2 accumulates on the right plate of C_1, where again an equivalent amount of negative charge is removed from the left plate. The result of this is that **all of the right plates gain charges of $-Q$ and all the left plates have charges of $+Q$.** (This is a consequence of the conservation of charge.)

One can find an equivalent capacitor that performs the same function as the series combination. After it is fully charged, **the equivalent capacitor must end up with a charge of $-Q$ on its right plate and a charge of $+Q$ on its left plate.** By applying the definition of capacitance to the circuit in Figure 16.15b, we have

$$\Delta V = \frac{Q}{C_{eq}}$$

where ΔV is the potential difference between the terminals of the battery and C_{eq} is the equivalent capacitance. From Figure 16.15a we see that

$$\Delta V = \Delta V_1 + \Delta V_2 \qquad \text{[16.14]}$$

where ΔV_1 and ΔV_2 are the potential differences across capacitors C_1 and C_2. (This is a consequence of the conservation of energy.) **The potential difference across any number of capacitors (or other circuit elements) in series equals the sum of the potential differences across the individual capacitors.** Because $Q = C \Delta V$ can be applied to each capacitor, the potential differences across them are given by

$$\Delta V_1 = \frac{Q}{C_1} \qquad \Delta V_2 = \frac{Q}{C_2}$$

Substituting these expressions into Equation 16.14, and noting that $\Delta V = Q/C_{eq}$, we have

$$\frac{Q}{C_{eq}} = \frac{Q}{C_1} + \frac{Q}{C_2}$$

Cancelling Q, we arrive at the relationship

$$\frac{1}{C_{eq}} = \frac{1}{C_1} + \frac{1}{C_2} \qquad \text{(series combination)} \qquad \text{[16.15]}$$

If this analysis is applied to three or more capacitors connected in series, the equivalent capacitance is found to be

$$\frac{1}{C_{eq}} = \frac{1}{C_1} + \frac{1}{C_2} + \frac{1}{C_3} + \cdots \qquad \text{(series combination)} \qquad \text{[16.16]}$$

As we shall demonstrate in Example 16.7, this implies that **the equivalent capacitance of a series combination is always less than any individual capacitance in the combination.**

Problem-Solving Strategy

Capacitors

1. Be careful with your choice of units. To calculate the capacitance of a device in farads, make sure that distances are in meters and use the SI value of ϵ_0.
2. When two or more unequal capacitors are connected in *series*, they carry the same charge, but the potential differences across them are not the same. Their capacitances add as reciprocals, and the equivalent capacitance of the combination is always *less* than the smallest individual capacitor.
3. When two or more capacitors are connected in *parallel*, the potential differences across them are the same. The charge on each capacitor is proportional to its capacitance; hence, the capacitances add directly to give the equivalent capacitance of the parallel combination.
4. A complicated circuit consisting of capacitors can often be reduced to a simple circuit containing only one capacitor. To do this, examine your initial circuit and replace any capacitors in series or any in parallel with equivalent capacitors, using the rules in Steps 2 and 3. After making these changes, sketch your new circuit. Examine it and replace any series or parallel combinations again. Continue this process until a single, equivalent capacitor is found.
5. To find the charge on or the potential difference across one of the capacitors in the complicated circuit, start with the final circuit found in Step 4 and gradually work your way back through the circuits using $C = Q/\Delta V$ and the rules given in Steps 2 and 3.

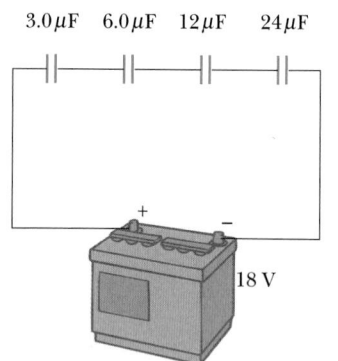

Figure 16.16 (Example 16.7) Four capacitors connected in series.

EXAMPLE 16.7   Four Capacitors Connected in Series

Four capacitors are connected in series with a battery, as in Figure 16.16. (a) Find the capacitance of the equivalent capacitor.

Solution   (a) The equivalent capacitance is found from Equation 16.16:

$$\frac{1}{C_{eq}} = \frac{1}{3.0 \ \mu F} + \frac{1}{6.0 \ \mu F} + \frac{1}{12 \ \mu F} + \frac{1}{24 \ \mu F}$$

$$C_{eq} = \boxed{1.6 \ \mu F}$$

Note that the equivalent capacitance is less than the capacitance of any of the individual capacitors in the combination.
(b) Find the charge on the 12-μF capacitor.

Solution   We find the charge on the equivalent capacitor:

$$Q = C_{eq} \ \Delta V = (1.6 \times 10^{-6} \ F)(18 \ V) = \boxed{29 \ \mu C}$$

This is also the charge on each of the capacitors it replaced. Thus, the charge on the 12-μF capacitor in the original circuit is 29 μC.

EXAMPLE 16.8 Equivalent Capacitance

Find the equivalent capacitance between a and b for the combination of capacitors shown in Figure 16.17a. All capacitances are in microfarads.

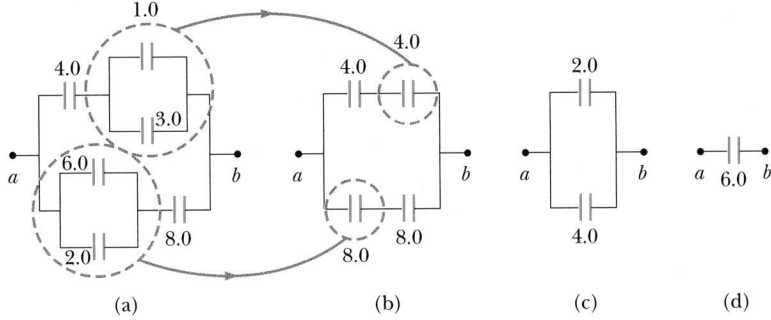

(a)   (b)   (c)   (d)

Figure 16.17 (Example 16.8) To find the equivalent capacitance of the circuit in (a), the circuit is reduced in steps—as indicated in (b), (c), and (d)—using the series and parallel rules described in the text.

Solution Using Equations 16.13 and 16.16, we reduce the combination step by step as indicated in the figure. The 1.0-μF and 3.0-μF capacitors are in *parallel* and combine according to $C_{eq} = C_1 + C_2$. Their equivalent capacitance is 4.0 μF. Likewise, the 2.0-μF and 6.0-μF capacitors are also in *parallel* and have an equivalent capacitance of 8.0 μF. The upper branch in Figure 16.17b now consists of two 4.0-μF capacitors in *series*, which combine according to

$$\frac{1}{C_{eq}} = \frac{1}{C_1} + \frac{1}{C_2} = \frac{1}{4.0 \ \mu F} + \frac{1}{4.0 \ \mu F} = \frac{1}{2.0 \ \mu F}$$

$$C_{eq} = 2.0 \ \mu F$$

Likewise, the lower branch in Figure 16.17b consists of two 8.0-μF capacitors in *series* with an equivalent capacitance of 4.0 μF. Finally, the 2.0-μF and 4.0-μF capacitors in Figure 16.17c are in *parallel* and have an equivalent capacitance of 6.0 μF. Hence, the equivalent capacitance of the circuit is 6.0 μF.

16.9 ENERGY STORED IN A CHARGED CAPACITOR

Almost everyone who works with electronic equipment has at some time verified that a capacitor can store energy. If the plates of a charged capacitor are connected by a conductor, such as a wire, charge transfers from one plate to the other until the two are uncharged. The discharge can often be observed as a visible spark. If you accidentally touched the opposite plates of a charged capacitor, your fingers would act as a pathway by which the capacitor could discharge, inflicting an electric shock. The degree of shock would depend on the capacitance and voltage applied to the capacitor. Where high voltages and large quantities of charge are present, as in the power supply of a television set, such a shock can be fatal.

    If a capacitor is initially uncharged (both plates neutral), so that the plates are at the same potential, almost no work is required to transfer a small amount of

13.5, SECTION 2

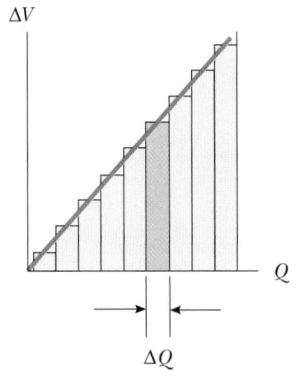

Figure 16.18 A plot of voltage versus charge for a capacitor is a straight line with the slope $1/C$. The work required to move a charge of ΔQ through a potential difference of ΔV across the capacitor plates is $\Delta W = \Delta V \Delta Q$, which equals the area of the blue rectangle. The *total work* required to charge the capacitor to a final charge of Q is the area under the straight line, which equals $Q \Delta V / 2$.

charge, ΔQ, from one plate to the other. However, once this charge has been transferred, a small potential difference, $\Delta V = \Delta Q / C$, appears between the plates. Therefore, work must be done to transfer additional charge through this potential difference. As more charge is transferred from one plate to the other, the potential difference increases in proportion. If the potential difference at any instant during the charging process is ΔV, the work required to move more charge, ΔQ, through this potential difference is $\Delta V \Delta Q$; that is,

$$\Delta W = \Delta V \Delta Q$$

We know that $\Delta V = Q/C$ for a capacitor that has a total charge of Q. Therefore, a plot of voltage versus charge gives a straight line with a slope of $1/C$, as shown in Figure 16.18. Because the work ΔW is the area of the shaded rectangle, the total work done in charging the capacitor to a final voltage, ΔV, is the area under the voltage-charge curve, which in this case equals the area under the straight line. Because the area under this line is the area of a triangle (which is one-half the product of the base and height), the total work done is

$$W = \tfrac{1}{2} Q \Delta V \qquad\qquad \text{[16.17]}$$

Note that this is also the energy stored in the capacitor, because the work required to charge the capacitor equals the energy stored in the capacitor after it is charged. From the definition of capacitance, we find $Q = C \Delta V$; hence, we can express the energy stored as

$$\text{Energy stored} = \tfrac{1}{2} Q \Delta V = \tfrac{1}{2} C (\Delta V)^2 = \frac{Q^2}{2C} \qquad\qquad \text{[16.18]}$$

This result applies to any capacitor. In practice, there is a limit to the maximum energy (or charge) that can be stored, because electrical breakdown ultimately occurs between the plates of the capacitor at a sufficiently large value of ΔV. For this reason, capacitors are usually labeled with a maximum operating voltage.

Large capacitors can store enough electrical energy to cause severe burns or even death if they are discharged so that the flow of charge can pass through the heart. Under the proper conditions, however, they can be used to sustain life by stopping cardiac fibrillation in heart attack victims. When fibrillation occurs, the heart produces a rapid, unregulated pattern of beats. A fast discharge of electrical energy through the heart can return the organ to its normal beat pattern. Emergency medical teams use defibrillators—batteries capable of charging a capacitor to a high voltage (see p. 520). (The circuit actually permits the capacitor to be charged to a much higher voltage than the battery.) The stored electrical energy is released through the heart by conducting electrodes, called *paddles*, that are placed on both sides of the victim's chest. The paramedics must wait between applications of the electrical energy due to the time necessary for the capacitors to become fully charged. The high voltage on the capacitor can be obtained from a low-voltage battery in a portable machine with the phenomenon of *electromagnetic induction*, to be studied in Chapter 20.

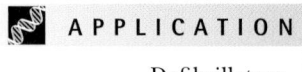

APPLICATION

Defibrillators.

Thinking Physics 3

You have three capacitors and two batteries. How should you connect the batteries to all three capacitors so that the capacitors will store the maximum possible energy?

Explanation   The energy stored in the capacitor is proportional to the capacitance and the square of the potential difference. Thus, we would like to maximize each of these quantities. We can do this by connecting the three capacitors in parallel (so that the capacitances add) across the two batteries in series (so that the potential differences add).

Thinking Physics 4

You charge a capacitor and then remove it from the battery. The capacitor consists of large movable plates with air between them. You pull the plates a bit farther apart. What happens to the charge on the capacitor? To the potential difference? To the energy stored? To the capacitance? To the electric field between the plates? Was work done in pulling the plates apart?

Explanation   Because the capacitor is removed from the battery, charges on the plates have nowhere to go. Thus, the charge on the capacitor plates remains the same as the plates are pulled apart. Because $\Delta V = Ed$, and because E is uniform between the plates, ΔV increases as d increases. Because energy stored is proportional to both charge and potential difference, the energy stored in the capacitor must increase. Because the same charge is stored at a higher potential difference, the capacitance has decreased. The electric field is uniform between the plates of the capacitor. The extra energy must have been transferred from somewhere, so work was done. This is consistent with the fact that the plates attract one another, so work must be done to pull them apart.

EXAMPLE 16.9   Energy Stored in a Charged Capacitor

Find the amount of energy stored in a 5.0-μF capacitor when it is connected across a 120-V battery.

Solution   Using Equation 16.18, we have

$$\text{Energy stored} = \tfrac{1}{2}C(\Delta V)^2 = \tfrac{1}{2}(5.0 \times 10^{-6}\,\text{F})(120\,\text{V})^2 = \boxed{3.6 \times 10^{-2}\,\text{J}}$$

Workbook Problem 84 — (Workbook page 180) deals with the energy stored in a network of capacitors.

16.10   CAPACITORS WITH DIELECTRICS

Optional Section

A **dielectric** is an insulating material, such as rubber, glass, or waxed paper. When a dielectric is inserted between the plates of a capacitor, the capacitance increases. If the dielectric completely fills the space between the plates, the capacitance is multiplied by the factor κ, called the **dielectric constant.**

The following experiment can be performed to illustrate the effect of a dielectric in a capacitor. Consider a parallel-plate capacitor of charge Q_0 and capacitance

Figure 16.19 (a) With air between the plates, the voltage across the capacitor is ΔV_0, the capacitance is C_0, and the charge is Q_0. (b) With a dielectric between the plates, the charge remains at Q_0, but the voltage and capacitance both change.

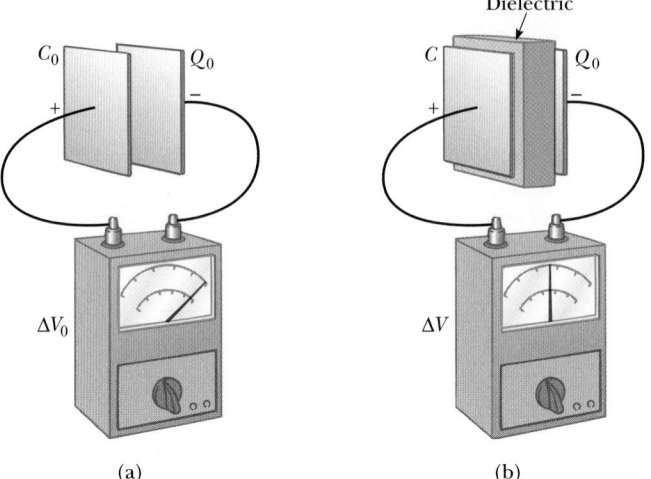

(a)   (b)

C_0 in the absence of a dielectric. The potential difference across the capacitor plates can be measured, and it is given by $\Delta V_0 = Q_0/C_0$ (Fig. 16.19a). Because the capacitor is not connected to an external circuit, there is no pathway for charge to leave or be added to the plates. If a dielectric is now inserted between the plates, as in Figure 16.19b, it is found that the voltage across the plates is *reduced* by the factor $\kappa(\,>\,1)$ to the value ΔV, where

$$\Delta V = \frac{\Delta V_0}{\kappa}$$

Because $\kappa > 1$, ΔV is less than ΔV_0. Because the charge Q_0 on the capacitor does not change, we conclude that the capacitance in the presence of the dielectric, C, must change to the value

$$C = \frac{Q_0}{\Delta V} = \frac{Q_0}{\Delta V_0/\kappa} = \frac{\kappa Q_0}{\Delta V_0}$$

or

$$C = \kappa C_0 \qquad\qquad [16.19]$$

According to this result, the capacitance is *multiplied* by the factor κ when the dielectric completely fills the region between the plates. For a parallel-plate capacitor, where the capacitance in the absence of a dielectric is $C_0 = \epsilon_0 A/d$, we can express the capacitance in the presence of a dielectric as

$$C = \kappa \epsilon_0 \frac{A}{d} \qquad\qquad [16.20]$$

From this result it appears that the capacitance could be made very large by decreasing d, the distance between the plates. In practice, the lowest value of d is limited by the electric discharge that can occur through the dielectric material separating the plates. For any given plate separation, there is a maximum electric field that can be produced in the dielectric before it breaks down and begins to conduct. This maximum electric field is called the **dielectric strength,** and for air

TABLE 16.1 Dielectric Constants and Dielectric Strengths of Various Materials at Room Temperature

Material	Dielectric Constant, κ	Dielectric Strength (V/m)
Vacuum	1.000 00	—
Air	1.000 59	3×10^6
Bakelite	4.9	24×10^6
Fused quartz	3.78	8×10^6
Pyrex glass	5.6	14×10^6
Polystyrene	2.56	24×10^6
Teflon	2.1	60×10^6
Neoprene rubber	6.7	12×10^6
Nylon	3.4	14×10^6
Paper	3.7	16×10^6
Strontium titanate	233	8×10^6
Water	80	—
Silicone oil	2.5	15×10^6

Dielectric breakdown in air. Sparks are produced when a large alternating voltage is applied across the electrodes using a high-voltage induction coil power supply. *(Courtesy of Central Scientific Company)*

its value is about 3×10^6 V/m. Most insulating materials have dielectric strengths greater than that of air, as indicated by the values in Table 16.1.

Commercial capacitors are often made using metal foil interlaced with thin sheets of paraffin-impregnated paper or mylar, which serves as the dielectric material. These alternate layers of metal foil and dielectric are then rolled into a small cylinder (Fig. 16.20a). A high-voltage capacitor commonly consists of a number of interwoven metal plates immersed in silicone oil (Fig. 16.20b). Small capacitors are often constructed from ceramic materials. Variable capacitors (typically 10 pF to 500 pF) usually consist of two interwoven sets of metal plates, one fixed and the other movable, with air as the dielectric.

An electrolytic capacitor (Fig. 16.20c) is often used to store large amounts of charge at relatively low voltages. It consists of a metal foil in contact with an

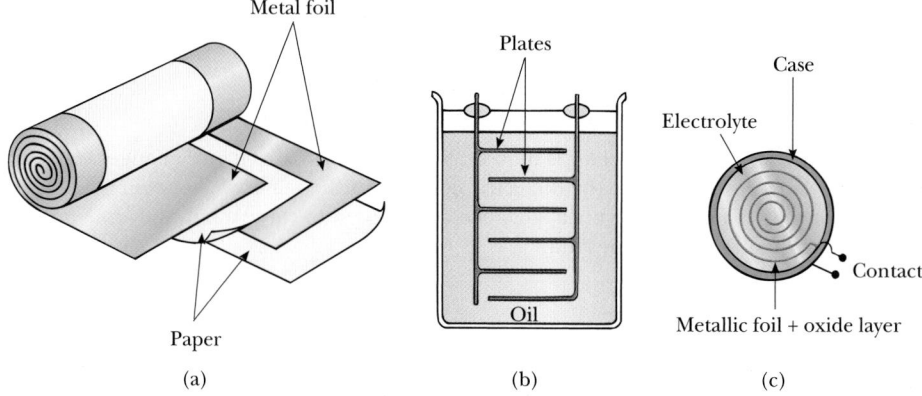

Figure 16.20 Three commercial capacitor designs: (a) a tubular capacitor whose plates are separated by paper and then rolled into a cylinder, (b) a high-voltage capacitor consisting of many parallel plates separated by oil, and (c) an electrolytic capacitor.

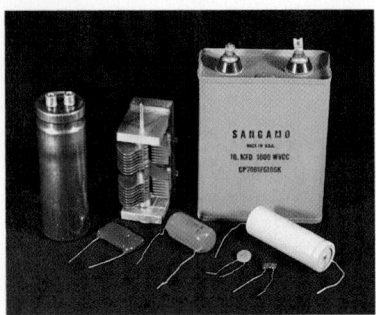

A collection of capacitors used in a variety of applications. *(Courtesy of Henry Leap and Jim Lehman)*

electrolyte—a solution that conducts charge by virtue of the motion of the ions contained in it. When a voltage is applied between the foil and the electrolyte, a thin layer of metal oxide (an insulator) is formed on the foil, and this layer serves as the dielectric. Enormous capacitances can be attained because the dielectric layer is very thin.

When electrolytic capacitors are used in circuits, the polarity (the plus and minus signs on the device) must be observed. If the polarity of the applied voltage is opposite that intended, the oxide layer will be removed and the capacitor will conduct rather than store charge. Furthermore, reversing the polarity can result in such a large current that the capacitor may either burn or produce steam and explode.

APPLICATION

Electronic Stud Finders.

Applying Physics 1

If you have ever tried to hang a picture on a wall securely, you know that it can be difficult to locate a wooden stud in which to anchor your nail or screw. The principles discussed in this section can be used to detect a stud electronically. The primary element of an electronic stud finder is a capacitor with its plates arranged side by side instead of facing one another, as in Figure 16.21. How does this device work?

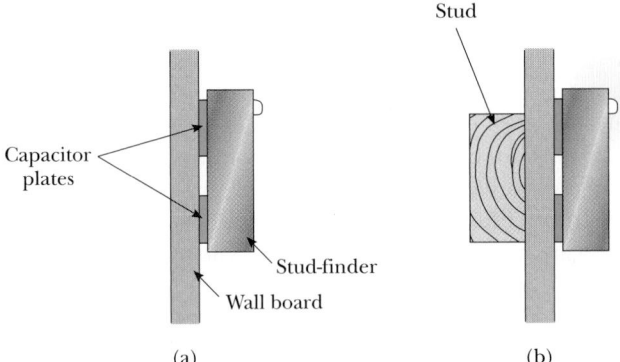

Figure 16.21 (Applying Physics 1) A stud finder. (a) The materials between the plates of the capacitor are the drywall and the air behind it. (b) The materials become drywall and wood when the detector moves across a stud in the wall. The change in the dielectric constant causes a signal light to illuminate.

Explanation As the detector is moved along a wall, its capacitance changes when it passes across a stud because the dielectric constant of the material "between" the plates changes. The change in capacitance can be used to cause a light to come on, signaling the presence of the stud.

EXAMPLE 16.10 A Paper-Filled Capacitor

A parallel-plate capacitor has plates 2.0 cm by 3.0 cm. The plates are separated by a 1.0-mm thickness of paper.
(a) Find the capacitance of this device.

Solution   (a) Because $\kappa = 3.7$ for paper (Table 16.1), we get

$$C = \kappa \epsilon_0 \frac{A}{d} = 3.7 \left(8.85 \times 10^{-12} \frac{C^2}{N \cdot m^2} \right) \left(\frac{6.0 \times 10^{-4} \text{ m}^2}{1.0 \times 10^{-3} \text{ m}} \right)$$

$$= 20 \times 10^{-12} \text{ F} = \boxed{20 \text{ pF}}$$

(b) Find the maximum charge that can be placed on the capacitor.

Solution   From Table 16.1 we see that the dielectric strength of paper is equal to 16×10^6 V/m. Because the paper thickness is 1.0 mm, the maximum voltage that can be applied before electrical breakdown occurs can be calculated using Equation 16.4:

$$\Delta V_{max} = E_{max} d = (16 \times 10^6 \text{ V/m})(1.0 \times 10^{-3} \text{ m}) = 16 \times 10^3 \text{ V}$$

Hence, the maximum charge that can be placed on the capacitor is

$$Q_{max} = C \Delta V_{max} = (20 \times 10^{-12} \text{ F})(16 \times 10^3 \text{ V}) = \boxed{0.32 \ \mu C}$$

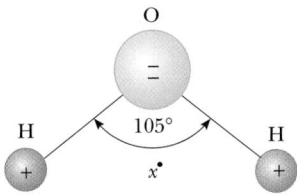

Figure 16.22   The water molecule, H_2O, has a permanent polarization resulting from its bent geometry. The point labeled x is the center of positive charge.

An Atomic Description of Dielectrics

The explanation of why a dielectric increases the capacitance of a capacitor is based on an atomic description of the material, which in turn involves a property of some molecules called **polarization.** A molecule is said to be polarized when there is a separation between the "centers of gravity" of its negative charge and its positive charge. In some molecules, such as water, this condition is always present. To see why, consider the geometry of a water molecule (Fig. 16.22). The molecule is arranged so that the negative oxygen atom is bonded to the positively charged hydrogen atoms with a 105° angle between the two bonds. The center of negative charge is at the oxygen atom, and the center of positive charge lies at a point midway along the line joining the hydrogen atoms (point x in the diagram). Materials composed of molecules that are permanently polarized in this fashion have large dielectric constants, and, indeed, Table 16.1 shows that the dielectric constant of water is quite large ($\kappa = 80$).

A symmetric molecule (Fig. 16.23a) can have no permanent polarization, but a polarization can be induced by an external electric field. A field directed to the left, as in Figure 16.23b, would cause the center of positive charge to shift to the left from its initial position, and the center of negative charge to shift to the right. This *induced polarization* is the effect that predominates in most materials used as dielectrics in capacitors.

To understand why the polarization of a dielectric can affect capacitance, consider Figure 16.24, which shows a slab of dielectric placed between the plates of a parallel-plate capacitor. The dielectric becomes polarized as shown because it is in the electric field that exists between the metal plates. Notice that a net positive charge appears on the dielectric surface adjacent to the negatively charged metal plate. The presence of this positive charge on the dielectric effectively reduces some of the negative charge on the metal, allowing more negative charge to be stored on the capacitor plates for a given applied voltage. From the definition of capacitance, $C = Q \Delta V$, we see that, because the plates can store more charge for a given voltage, the capacitance must increase.

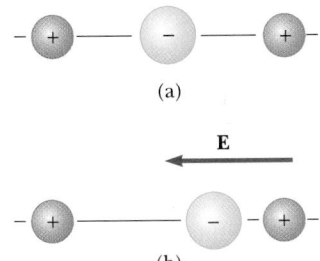

Figure 16.23   (a) A symmetric molecule has no permanent polarization. (b) An external electric field induces a polarization in the molecule.

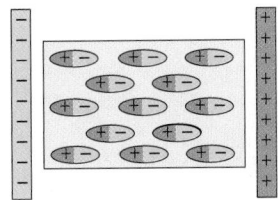

Figure 16.24   When a dielectric is placed between the plates of a charged parallel-plate capacitor, the dielectric becomes polarized. This creates a net positive induced charge on the left side of the dielectric and a net negative induced charge on the right side. As a result, the capacitance of the device is multiplied by the factor κ.

> ## Thinking Physics 5
>
> Consider a parallel-plate capacitor with a dielectric material between the plates. Is the capacitance higher on a cold day or a hot day?
>
> **Explanation**   The polarization of the molecules in the dielectric increases the capacitance when the dielectric is added. As the temperature increases, there is more vibrational motion of the polarized molecules. This disturbs the orderly arrangement of the polarized molecules, and the net polarization decreases. Thus, the capacitance must decrease as the temperature increases.

Optional Section

 **16.11**   APPLICATION: DNA AND FORENSIC SCIENCE

In 1868 biologists discovered the presence of giant molecules of **deoxyribonucleic acid (DNA)** in cells of the human body. However, it was not until the early 1950s when James Watson and Francis Crick determined the structure of DNA that a true understanding was achieved of how these molecules pass genetic information from one generation of cells to the next. More recently, DNA has found its way into courtrooms as a biological identifier used to identify suspects by way of DNA markers in their blood, semen, or other body fluids. The information we have learned in our study of electricity enables us to understand some of the fundamental principles of how DNA molecules are constructed, bonded, and used as a forensic tool.

Figure 16.25a shows the structure of a DNA molecule. It has three fundamental components, a sugar grouping, labeled S, a phosphate grouping, labeled P, and nucleotide bases, labeled either A, G, T, or C in the figure. The spine of the molecule is made up of alternating units of a sugar grouping bonded by electrical forces to a phosphate grouping. Extending from each sugar grouping is one of the four bases, adenine (A), thymine (T), cytosine (C), or guanine (G). Because of the structure of these bases, electrical forces of attraction sufficiently strong to bond the bases into pairs exist only between adenine and thymine and between cytosine and guanine, as shown in Figure 16.25b. Watson and Crick showed that the DNA

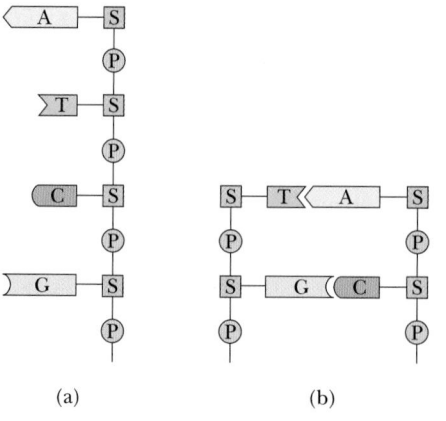

(a)   (b)

Figure 16.25   (a) A DNA strand consists of a sugar group, S, bound to a phosphate group, P. Extending from the sugar groups are nucleotide bases A, T, C, and G. (b) Strands of DNA are bound together by electrical forces between A and T and between C and G.

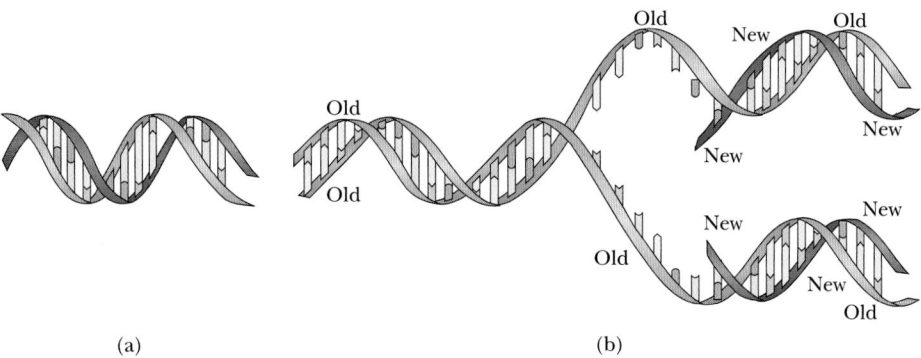

(a)   (b)

Figure 16.26   (a) The double helix structure of DNA. (b) As the DNA cell reproduces, the double helix separates at one end.

molecule consists of two strands coiled into a double helix, as shown in Figure 16.26a. When a cell reproduces, the double helix comes apart at one end as shown in Figure 16.26b. Each portion of the split double helix can now form a new double helix. The new helix must match the original exactly because all the A-T and C-G bonds must match perfectly throughout the strand.

The sequencing of bases along a strand of DNA provide instructions for producing amino acids and for how these should be linked together to form proteins. For example, the three-base sequence C-G-T along a strand instructs the body to manufacture the amino acid alanine. The next three bases provide information on another amino acid to be produced and connected to alanine. Finally, the amino acids are connected to form a protein.

Certainly, the instructions provided by the ordering of these bases is important to the well-being of the organism, but another feature of the DNA molecule makes it useful as a fingerprint to pinpoint the identity of individuals. For some unknown reason each DNA strand contains sequences of bases that repeat for a large number of times. That is, a sequence such as T-A-G may repeat for several thousand times along a strand, stop to allow information important to the body to be presented, then this sequence, or another, will repeat for perhaps several hundred times. These repeating sequences have no known purpose, but they are unique to each individual, thus making them important for DNA typing.

To type DNA, molecules are cut into fragments by use of **restriction enzymes.** These enzymes act like scissors that cut a strand whenever a specific sequence of bases is encountered. Within each strand may be one or several of these repeating sequences. The end result of these cuttings is that we have fragments of DNA all of different mass and net charge. By separating these fragments according to mass and charge, a profile of the individual from which they came can be formed. This separation is accomplished by a process called **electrophoresis,** shown in Figure 16.27. A voltage is applied across a support medium, such as a gel of starch or agar, and the body fluids to be DNA typed are inserted simultaneously at the origin. A known voltage across the gel is applied for a known period of time, causing the fragments to migrate through the gel. The fragments soon reach a terminal velocity that depends on their charge and mass. The result is a separation of fragments along the gel. In some instances the presence of the fragments can be indicated by staining, but more commonly radioactive tags can be applied that bind to the sep-

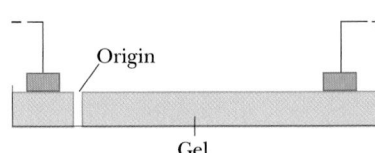

Figure 16.27 A voltage is applied across a gel in the electrophoresis separation of DNA fragments.

arated fragments. The emission from the tags can then be used to expose a sheet of x-ray film, producing a pattern that could be used to identify an individual.

The first crime solved by DNA typing occurred in Leicestershire, England, in 1987. Two young girls had been raped and brutally murdered, and the 5,500 citizens of the town agreed to be "blooded," DNA typed, to determine the guilty party. The guilty party actually was found when he attempted to avoid the tests. A happy adjunct to catching the true criminal was that an innocent suspect was freed from jail.

SUMMARY

The **difference in electric potential** between two points, A and B, is

$$V_B - V_A \equiv \frac{\Delta PE}{q}$$ [16.2]

where ΔPE is the *change* in electrical potential energy experienced by a charge, q, as it moves between A and B. The units of potential difference are joules per coulomb, or **volts**; $1\ \text{J/C} = 1\ \text{V}$.

The **electric potential difference** between two points, A and B, in a uniform electric field, E, is

$$V_B - V_A = -Ed$$ [16.4]

where d is the distance between A and B, and E is the strength of the electric field in that region.

The **electric potential** due to a point charge, q, at distance r from the point charge is

$$V = k_e \frac{q}{r}$$ [16.5]

The **electrical potential energy** of a pair of point charges separated by distance r is

$$PE = k_e \frac{q_1 q_2}{r}$$ [16.6]

Every point on the surface of a charged conductor in electrostatic equilibrium is at the same potential. Furthermore, the potential is constant everywhere inside the conductor and equals its value on the surface.

The **electron volt** is defined as the energy that an electron (or proton) gains when accelerated through a potential difference of 1 V. The conversion between electron volts and joules is

$$1\ \text{eV} = 1.60 \times 10^{-19}\ \text{J}$$ [16.8]

A **capacitor** consists of two metal plates with charges that are equal in magnitude but opposite in sign. The **capacitance** (C) of any capacitor is the ratio of the magnitude of the charge, Q, on either plate to the potential difference, ΔV, between them:

$$C \equiv \frac{Q}{\Delta V}$$ [16.9]

Capacitance has the units coulombs per volt, or **farads;** 1 C/V ≡ 1 F.

The capacitance of two *parallel metal plates* of area A separated by distance d is

$$C = \epsilon_0 \frac{A}{d} \qquad\qquad [16.10]$$

where ϵ_0 is a constant called the **permittivity of free space,** with the value

$$\epsilon_0 = 8.85 \times 10^{-12} \text{ C}^2/\text{N} \cdot \text{m}^2$$

The **equivalent capacitance of a parallel combination** of capacitors is

$$C_{eq} = C_1 + C_2 + C_3 + \ldots \qquad\qquad [16.13]$$

If two or more capacitors are connected in series, the **equivalent capacitance of the series combination** is

$$\frac{1}{C_{eq}} = \frac{1}{C_1} + \frac{1}{C_2} + \frac{1}{C_3} + \ldots \qquad\qquad [16.16]$$

Three equivalent expressions for calculating the **energy stored** in a charged capacitor are

$$\text{Energy stored} = \tfrac{1}{2} Q\,\Delta V = \tfrac{1}{2} C(\Delta V)^2 = \frac{Q^2}{2C} \qquad\qquad [16.18]$$

When a nonconducting material, called a **dielectric,** is placed between the plates of a capacitor, the capacitance is multiplied by the factor κ, which is called the **dielectric constant** and is a property of the dielectric material. The capacitance of a parallel-plate capacitor filled with a dielectric is

$$C = \kappa \epsilon_0 \frac{A}{d} \qquad\qquad [16.20]$$

MULTIPLE-CHOICE QUESTIONS

1. Using up to three of the following capacitors: 1 μF, 2 μF, and 3 μF, which of the following capacitances can be made?
   (a) 3 μF and 5 μF   (b) 6 μF and 3 μF
   (c) 3 μF and 7 mF   (d) Answers a and b
   (e) Answers a, b, and c

2. Using three 1-μF capacitors in combination, the number of different values of capacitance that can be obtained is
   (a) 2   (b) 3   (c) 4   (d) 5   (e) 6

3. Four point charges are positioned on the rim of a circle. The charge on each of the four (in microcoulombs) is $+0.5$, $+1.5$, -1.0, -0.5. If we are told that the electrical potential at the center of the circle due to the $+0.5$ charge alone is 4.5×10^4 V, what is the total potential at the center due to the four charges combined?

(a) 18.0×10^4 V   (b) 4.5×10^4 V   (c) zero
(d) -4.5×10^4 V

4. An electron in a television picture tube is accelerated through a potential difference of 10 kV before it hits the screen. What is the kinetic energy of the electron in electron volts?
   (a) 1.0×10^4 eV   (b) 1.6×10^{-15} eV   (c) 1.6×10^{-22} eV
   (d) 6.25×10^{22} eV

5. An electronics technician wishes to construct a parallel plate capacitor using rutile ($\kappa = 100$) as the dielectric. If the cross-sectional area of the plates is 1.00 cm^2, what is the capacitance if the rutile thickness is 1.00 mm?
   (a) 88.5 pF   (b) 177 pF   (c) 8.85 mF   (d) 100 mF

CONCEPTUAL QUESTIONS

1. If a proton is released from rest in a uniform electric field, does its electric potential increase or decrease?
2. Distinguish between electric potential and electrical potential energy.
3. Suppose you are sitting in a car and a 20,000-volt power line drops across the car. Should you stay in the car or get out? The power line potential is 20,000 volts compared to the potential of the ground.
4. Why is it important to avoid sharp edges or points on conductors used in high-voltage equipment?
5. If the electric potential at some point is zero, can you conclude that there are no charges in the vicinity of that point?
6. If you are given three different capacitors, C_1, C_2, C_3, how many different combinations of capacitance can you produce using all capacitors in your circuits?
7. Why is it dangerous to touch the terminals of a high-voltage capacitor even after the voltage source that charged the battery is disconnected from the capacitor? What can be done to make the capacitor safe to handle after the voltage source has been removed?
8. The plates of a capacitor are connected to a battery. What happens to the charge on the plates if the connecting wires are removed from the battery? What happens to the charge if the wires are removed from the battery and connected to each other?
9. Can electric field lines ever cross? Why or why not? Can equipotential surfaces ever cross? Why or why not?
10. A capacitor is designed so that one plate is large and the other is small. Do the plates have the same charge when connected to a battery?
11. Give a physical explanation of the fact that the potential energy of a pair of like charges is positive whereas the potential energy of a pair of unlike charges is negative.
12. Is it always possible to reduce a combination of capacitors to one equivalent capacitor with the rules developed in this chapter? Explain.
13. If you were asked to design a capacitor in which small size and large capacitance were required, what factors would be important in your design?
14. Explain why a dielectric increases the maximum operating voltage of a capacitor although the physical size of the capacitor does not change.
15. If the potential difference across a capacitor is doubled, by what factor is the energy stored by the capacitor multiplied?
16. Explain why, under static conditions, all points in a conductor must be at the same electric potential.
17. What happens to the charge on a capacitor if the potential difference between its plates is doubled?
18. A parallel-plate capacitor is charged by a battery, and the battery is then disconnected from the capacitor. Because the charges on the capacitor plates are equal and opposite, they attract each other. Hence, it takes positive work to increase the plate separation. What happens to the external work when the plate separation is increased?

PROBLEMS

Section 16.1   Potential Difference and Electric Potential

1. A proton moves 2.0 cm parallel to a uniform electric field of $E = 200$ N/C. (a) How much work is done on the proton by the field? (b) What change occurs in the potential energy of the proton? (c) What potential difference did the proton move through?
2. A uniform electric field of magnitude 250 V/m is directed in the positive x direction. A $+12$-μC charge moves from the origin to the point $(x, y) = (20$ cm, 50 cm$)$. (a) What was the change in the potential energy of this charge? (b) Through what potential difference did the charge move?
3. A potential difference of 90 mV exists between the inner and outer surfaces of the membrane of a cell. The inner surface is negative relative to the outer surface. How much work is required to eject a positive sodium ion (Na^+) from the interior of the cell?
4. An ion, when accelerated through a potential difference of 60.0 V, experiences a decrease in potential energy of 1.92×10^{-17} J. Calculate the charge on the ion.
5. The difference in potential between the accelerating plates of a television set is about 25 000 V. If the distance between these plates is 1.5 cm, find the magnitude of the uniform electric field in this region.
6. To recharge a 12-V battery, a battery charger must move 3.6×10^5 C of charge from the negative terminal to the positive terminal. How much work is done by the battery charger? Express your answer in joules.

WEB **7.** A pair of oppositely charged, parallel plates are separated by 5.33 mm. A potential difference of 600 V exists between the plates. **(a)** What is the magnitude of the electric field strength between the plates? **(b)** What is the magnitude of the force on an electron between the plates? **(c)** How much work must be done on the electron to move it to the negative plate if it is initially positioned 2.90 mm from the positive plate?

8. (a) Calculate the speed of a proton that is accelerated from rest through a potential difference of 120 V. **(b)** Calculate the speed of an electron that is accelerated through the same potential difference.

9. An electron moves from one plate to another across which there is a potential difference of 2000 V. **(a)** Find the speed with which the electron strikes the positive plate. **(b)** Repeat part (a) for a proton moving from the positive to the negative plate.

10. (a) Through what potential difference would an electron need to accelerate to achieve a speed of 60% of the speed of light, starting from rest? (The speed of light is 3.00×10^8 m/s.) **(b)** Repeat this calculation for a proton. (Do not consider relativistic effects.)

Section 16.2 Electric Potential and Potential Energy Due to Point Charges

Section 16.3 Potentials and Charged Conductors

Section 16.4 Equipotential Surfaces

11. (a) Find the potential 1.00 cm from a proton. **(b)** What is the potential difference between two points that are 1.00 cm and 2.00 cm from a proton?

12. Two point charges are on the y axis—one, of magnitude 3.0×10^{-9} C, is at the origin and a second, of magnitude 6.0×10^{-9} C, at the point $y = 30$ cm. Calculate the potential at $y = 60$ cm.

13. Find the electric potential, taking zero at infinity, at the upper right corner (the corner without a charge) of the rectangle in Figure P16.13.

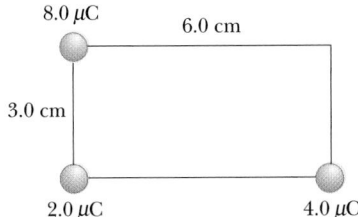

Figure P16.13   (Problems 13 and 14)

14. Three charges are situated at corners of a rectangle as in Figure P16.13. How much energy would be expended in moving the 8.0-μC charge to infinity?

Problem 7 is the same as Workbook Problem 74 (Workbook page 156).
Problem 15 is similar to Workbook Problem 73 (Workbook page 154).

15. Two point charges, $Q_1 = +5.00$ nC and $Q_2 = -3.00$ nC, are separated by 35.0 cm. **(a)** What is the electric potential at a point midway between the charges? **(b)** What is the potential energy of the pair of charges? What is the significance of the algebraic sign of your answer?

16. Calculate the speed of **(a)** an electron that has a kinetic energy of 1.00 eV and **(b)** a proton that has a kinetic energy of 1.00 eV.

17. A point charge of 9.00×10^{-9} C is located at the origin. How much work is required to bring a positive charge of 3.00×10^{-9} C from infinity to the location $x = 30.0$ cm?

18. An electron starts from rest 3.00 cm from the center of a uniformly charged sphere of radius 2.00 cm. If the sphere carries a total charge of 1.00×10^{-9} C, how fast will the electron be moving when it reaches the surface of the sphere?

WEB **19.** In Rutherford's famous scattering experiments that led to the planetary model of the atom, alpha particles (having charges of $+2e$ and masses of 6.6×10^{-27} kg) were fired toward a fixed gold nucleus with charge $+79e$. An alpha particle, initially very far from the gold nucleus, is fired at 2.0×10^7 m/s directly toward the gold nucleus, as in Figure P16.19. How close does the alpha particle get to the gold nucleus before turning around?.

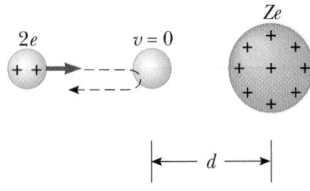

Figure P16.19

20. Starting with the definition of work, prove that at every point on an equipotential surface the surface must be perpendicular to the existing electric field.

Section 16.6 The Definition of Capacitance

Section 16.7 The Parallel–Plate Capacitor

21. (a) How much charge is on each plate of a 4.00-μF capacitor when it is connected to a 12.0-V battery? **(b)** If this same capacitor is connected to a 1.50-V battery, what charge is stored?

22. Consider the Earth and a cloud layer 800 m above the Earth to be the plates of a parallel-plate capacitor. **(a)** If the cloud layer has an area of 1.0 km$^2 = 1.0 \times 10^6$ m^2, what is the capacitance? **(b)** If an electric field strength greater than 3.0×10^6 N/C causes the air to break down and conduct charge (lightning), what is the maximum charge the cloud can hold?

23. The potential difference between a pair of oppositely charged parallel plates is 400 V. **(a)** If the spacing between the plates is doubled without altering the charge on the plates, what is the new potential difference between the plates? **(b)** If the plate spacing is doubled and the potential difference between the plates is kept constant, what is the ratio of the final charge on one of the plates to the original charge?

24. A parallel-plate capacitor has an area of 2.0 cm^2, and the plates are separated by 2.0 mm with air between them. How much charge does this capacitor store when connected to a 6.0-V battery?

25. The plates of a parallel-plate capacitor are separated by 0.100 mm. If the material between the plates is air, what plate area is required to provide a capacitance of 2.00 pF?

26. A parallel-plate capacitor has an area of 5.00 cm^2, and the plates are separated by 1.00 mm with air between them. It stores a charge of 400 pC. **(a)** What is the potential difference across the plates of the capacitor? **(b)** What is the magnitude of the uniform electric field in the region between the plates?

Section 16.8 Combinations of Capacitors

27. A series circuit consists of a 0.050-μF capacitor, a 0.100-μF capacitor, and a 400-V battery. Find the charge **(a)** on each of the capacitors; **(b)** on each of the capacitors if they are reconnected in parallel across the battery.

28. Three capacitors, $C_1 = 5.00\ \mu$F, $C_2 = 4.00\ \mu$F, and $C_3 = 9.00\ \mu$F, are connected together. **(a)** Find the effective capacitance of the group if they are all in parallel. **(b)** Find the effective capacitance of the group if they are all in series.

29. **(a)** Find the equivalent capacitance of the group of capacitors in Figure P16.29. **(b)** Find the charge on and the potential difference across each.

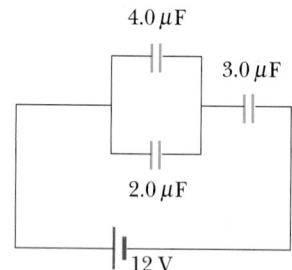

4.0 μF

3.0 μF

2.0 μF

12 V

Figure P16.29

30. Consider various combinations of three capacitors, each with a capacitance of 2.0 μF. **(a)** Sketch the arrangement that would give a circuit the largest equivalent capaci-

tance. **(b)** Sketch the arrangement that would give a circuit the smallest equivalent capacitance. **(c)** Sketch the arrangement that would give an equivalent capacitance of 3.0 μF.

31. Consider the combination of capacitors in Figure P16.31. **(a)** What is the equivalent capacitance of the group? **(b)** Determine the charge on each capacitor.

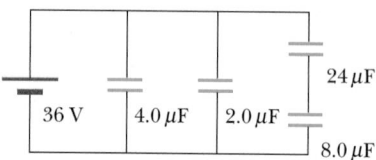

36 V   4.0 μF   2.0 μF

24 μF

8.0 μF

Figure P16.31

32. Find the charge on each of the capacitors in Figure P16.32.

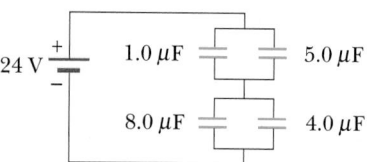

24 V   1.0 μF   5.0 μF

8.0 μF   4.0 μF

Figure P16.32

33. How should four 2.0-μF capacitors be connected to have a total capacitance of **(a)** 8.0 μF? **(b)** 2.0 μF? **(c)** 1.5 μF? **(d)** 0.50 μF?

34. To repair a power supply for a stereo amplifier, an electronics technician needs a 100-μF capacitor capable of withstanding a potential difference of 90 V between the plates. The only available supply is a box of five 100-μF capacitors, each having a maximum voltage capability of 50 V. Can the technician substitute a combination of these capacitors that has the proper electrical characteristics, and if so, what will be the maximum voltage across any of the capacitors used? (*Hint:* The technician may not have to use all the capacitors in the box.)

WEB **35.** A 25.0-μF capacitor and a 40.0-μF capacitor are charged by being connected across separate 50.0-V batteries. **(a)** Determine the resulting charge on each capacitor. **(b)** The capacitors are then disconnected from their batteries and connected to each other, with each negative plate connected to the other positive plate. What is the final charge of each capacitor, and what is the final potential difference across the 40.0-μF capacitor?

36. A 10.0-μF capacitor is fully charged across a 12.0-V battery. The capacitor is then disconnected from the battery and connected across an initially uncharged capacitor, *C*. The resulting voltage across each capacitor is 3.00 V. What is the capacitance *C*?

37. A 1.00-μF capacitor is first charged by being connected across a 10.0-V battery. It is then disconnected from the battery and connected across an uncharged 2.00-μF capacitor. Determine the resulting charge on each capacitor.

Section 16.9 Energy Stored in a Charged Capacitor

38. A parallel-plate capacitor has 2.00-cm^2 plates that are separated by 5.00 mm with air between them. If a 12.0-V battery is connected to this capacitor, how much energy does it store?

39. Two capacitors, $C_1 = 25\ \mu$F and $C_2 = 5.0\ \mu$F, are connected in parallel and charged with a 100-V power supply. (a) Calculate the total energy stored in the two capacitors. (b) What potential difference would be required across the same two capacitors connected in *series* in order that the combination store the same energy as in (a)?

40. Consider the parallel-plate capacitor formed by the Earth and a cloud layer as described in Problem 22. Assume this capacitor will discharge (i.e., lightning occurs) when the electric field strength between the plates reaches 3.0×10^6 N/C. What is the energy released if the capacitor discharges completely during a lighting strike?

Section 16.10 Capacitors with Dielectrics (Optional)

41. A capacitor with air between its plates is charged to 100 V and then disconnected from the battery. When a piece of glass is placed between the plates, the voltage across the capacitor drops to 25 V. What is the dielectric constant of this glass? (Assume the glass completely fills the space between the plates.)

42. Two parallel plates, each of area 2.00 cm^2, are separated by 2.00 mm with water between them. A voltage of 6.00 V is applied between the plates. Calculate (a) the magnitude of the electric field between the plates, (b) the charge stored on each plate, and (c) the charge stored on each plate if the water is removed and replaced with air.

43. Determine (a) the capacitance and (b) the maximum voltage that can be applied to a Teflon-filled parallel-plate capacitor having a plate area of 175 cm^2 and insulation thickness of 0.0400 mm.

44. When removing a wool sweater, sparks often occur. If a spark jumps a distance of 0.5 inch, what minimum potential difference could have caused this occurrence? (*Hint:* Assume that the spark occurred in air as a result of a uniform electric field.)

45. A model of a red blood cell portrays the cell as a spherical capacitor—a positively charged liquid sphere of surface area *A*, separated by a membrane of thickness *t*

from the surrounding, negatively charged fluid. Tiny electrodes introduced into the interior of the cell show a potential difference of 100 mV across the membrane. The membrane's thickness is estimated to be 100 nm and its dielectric constant to be 5.00. (a) If an average red blood cell has a mass of 1.00×10^{-12} kg, estimate the volume of the cell and thus find its surface area. The density of blood is 1100 kg/m^3. (b) Estimate the capacitance of the cell. (c) Calculate the charge on the surface of the membrane. How many electronic charges does this represent?

ADDITIONAL PROBLEMS

46. Three parallel plate capacitors are constructed, each having the same plate spacing *d*, and with C_1 having plate area A_1, C_2 having area A_2, and C_3 having area A_3. Show that the total capacitance *C* of these three capacitors connected in parallel is the same as a capacitor having plate spacing *d* and plate area $A = A_1 + A_2 + A_3$.

47. Three parallel plate capacitors are constructed, each having the same plate area *A*, and with C_1 having plate spacing d_1, C_2 having plate spacing d_2, and C_3 having plate spacing d_3. Show that the total capacitance *C* of these three capacitors connected in series is the same as a capacitor of plate area *A* and with plate spacing $d = d_1 + d_2 + d_3$.

48. Find the potential at point *P* for the rectangular grouping of charges shown in Figure P16.48.

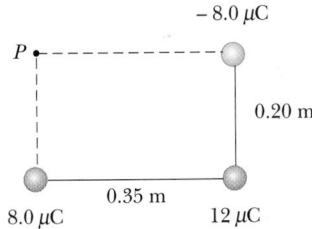

Figure P16.48

49. The three charges shown in Figure P16.49 are at the vertices of an isosceles triangle. If $q = -5.0 \times 10^{-9}$ C, calculate the electric potential at the midpoint of the base.

50. Find the equivalent capacitance of the group of capacitors in Figure P16.50.

51. When a certain air-filled parallel-plate capacitor is connected across a battery, it acquires a charge (on each plate) of 150 μC. While the battery connection is maintained, a dielectric slab is inserted into and fills the region between the plates. This results in the accumulation of an additional charge of 200 μC on each plate. What is the dielectric constant of the dielectric slab?

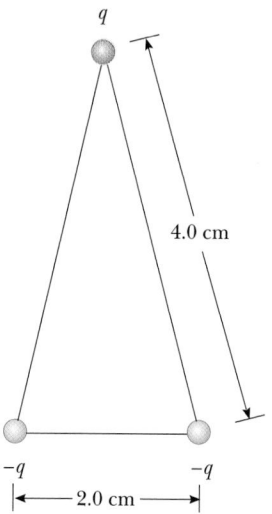

Figure P16.49

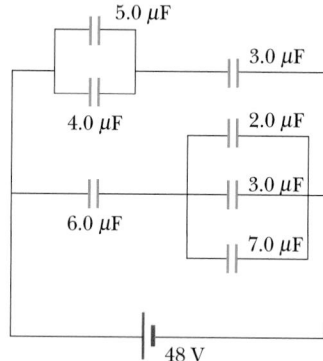

Figure P16.50

52. An isolated capacitor of unknown capacitance has been charged to a potential difference of 100 V. When the charged capacitor is disconnected from the battery and then connected in parallel to an uncharged 10.0-μF capacitor, the voltage across the combination is measured to be 30.0 V. Calculate the unknown capacitance.

53. A pair of oppositely charged, parallel plates are separated by a distance of 5.0 cm with a potential difference of 500 V between the plates. A proton is released from rest at the positive plate, and at the same time an electron is released from rest at the negative plate. Neglect any interaction between the proton and the electron. (a) After what interval (at what time) will their paths cross? (b) How fast will each particle be going when their paths cross? (c) At what time will the electron

reach the opposite plate? (d) At what time will the proton reach the opposite plate?

54. Capacitors $C_1 = 6.0$ μF and $C_2 = 2.0$ μF are charged as a parallel combination across a 250-V battery. The capacitors are disconnected from the battery and from each other. They are then connected positive plate to negative plate and negative plate to positive plate. Calculate the resulting charge on each capacitor.

55. Capacitors $C_1 = 4.0$ μF and $C_2 = 2.0$ μF are charged as a series combination across a 100-V battery. The two capacitors are disconnected from the battery and from each other. They are then connected positive plate to positive plate and negative plate to negative plate. Calculate the resulting charge on each capacitor.

56. On planet Tehar, the acceleration of gravity is the same as that on Earth but there is also a strong downward electric field with the field being uniform close to the planet's surface. A 2.00-kg ball having a charge of 5.00 μC is thrown upward at a speed of 20.1 m/s and it hits the ground after an interval of 4.10 s. What is the potential difference between the starting point and the top point of the trajectory?

57. Two capacitors when connected in parallel give an equivalent capacitance of 9.00 pF and an equivalent capacitance of 2.00 pF when connected in series. What is the capacitance of each capacitor?

58. Two capacitors when connected in parallel give an equivalent capacitance of C_p and an equivalent capacitance of C_s when connected in series. What is the capacitance of each capacitor?

59. Four capacitors are connected as shown in Figure P16.59. (a) Find the equivalent capacitance between points a and b. (b) Calculate the charge on each capacitor if $\Delta V_{ab} = 15.0$ V.

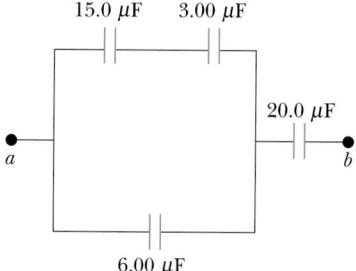

Figure P16.59

60. The charge distribution shown in Figure P16.60 is referred to as a *linear quadrupole*. (a) Show that the electric potential at a point on the x axis where $x > d$ is

$$V = \frac{2k_eQd^2}{x^3 - xd^2}$$

(b) Show that the expression obtained in (a) when $x \gg d$ reduces to

$$V = \frac{2k_e Q d^2}{x^3}$$

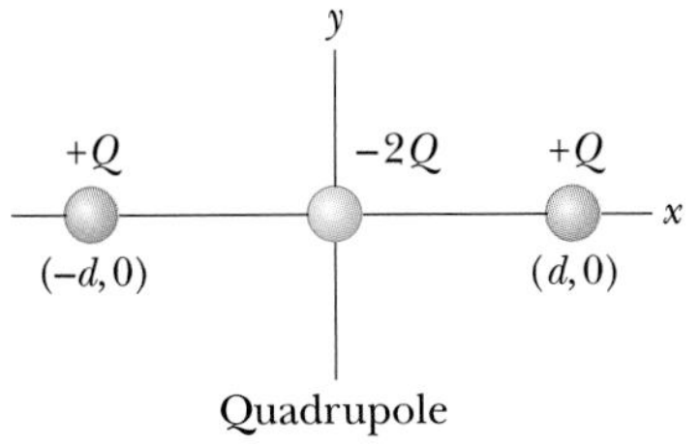

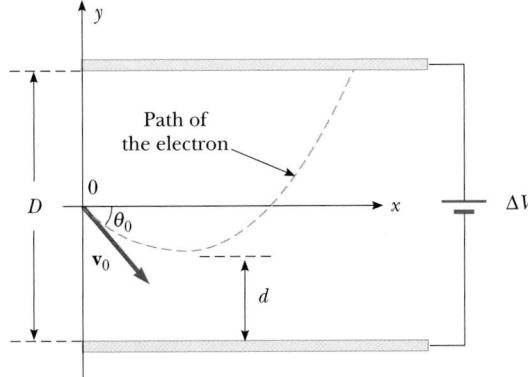

$+Q$ $-2Q$ $+Q$
$(-d,0)$ $(d,0)$

Quadrupole

Figure P16.60

61. The energy stored in a 52.0-μF capacitor is used to melt a 6.00-mg sample of lead. To what voltage must the capacitor be initially charged, assuming that the initial temperature of the lead is 20.0°C? Lead has a specific heat of 128 J/kg·°C, a melting point of 327.3°C, and a latent heat of fusion of 24.5 kJ/kg.

62. Consider a parallel-plate capacitor with charge Q and area A, filled with dielectric material having dielectric constant κ. It can be shown that the magnitude of the attractive force exerted on each plate by the other is given by $F = Q^2/(2\kappa\epsilon_0 A)$. When a potential difference of 100 V exists between the plates of an air-filled 20-μF parallel-plate capacitor, what force does each plate exert on the other if they are separated by 2.01 mm?

63. An electron is fired at a speed of $v_0 = 5.6 \times 10^6$ m/s and at an angle of $\theta_0 = -45°$, between two parallel conducting plates that are $D = 2.0$ mm apart, as in Figure P16.63. If the voltage difference between the plates is $\Delta V = 100$ V, determine **(a)** how close, d, the electron will get to the bottom plate and **(b)** where the electron will strike the top plate.

Figure P16.63

Current and Resistance

▲ PHYSICS PUZZLER

Electric workers at dusk, trying to restore power to the eastern Ontario town of St. Isadore that was without power for several days in January 1998 because of a severe ice storm. When power transmission lines fall because of lightning, ice storms, or earthquakes, it is very dangerous to touch them due to the high electric potential, possibly hundreds of thousands of volts. Why is such high voltage used in power transmission if it is so dangerous? *(AP/Wide World Photos/ Fred Chartrand)*

Many practical applications and devices are based on the principles of static electricity, but electricity truly became an inseparable part of our daily lives when scientists learned how to control the flow of electric charges. Electric currents power our lights, radios, television sets, air conditioners, and refrigerators; they ignite the gasoline in automobile engines, travel through miniature components making up the chips of microcomputers, and perform countless other invaluable tasks.

13.1

In this chapter we define current and discuss some of the factors that contribute to the resistance to flow of charge in conductors. We also discuss energy transformations in electric circuits. These topics will be the foundation for additional work with circuits in later chapters.

17.1 ELECTRIC CURRENT

Whenever electric charges of like signs move, a *current* is said to exist. To define current more precisely, suppose the charges are moving perpendicularly to a surface of area A, as in Figure 17.1. (This area could be the cross-sectional area of a wire, for example.) **The current is the rate at which charge flows through this surface.** If ΔQ is the amount of charge that passes through this area in a time interval of Δt, the current, I, is equal to the ratio of the charge to the time interval:

$$I \equiv \frac{\Delta Q}{\Delta t}$$ [17.1]

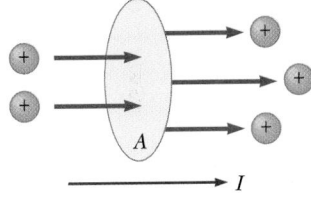

Figure 17.1 Charges in motion through an area of A. The time rate of flow of charge through the area is defined as the current I. The direction of the current is the direction of flow of positive charges.

The SI unit of current is the **ampere (A):**

$$1 \text{ A} = 1 \text{ C/s}$$ [17.2]

Thus, 1 A of current is equivalent to 1 C of charge passing through the cross-sectional area in a time interval of 1 s.

When charges flow through a surface as in Figure 17.1, they can be positive, negative, or both. **It is conventional to give the current the same direction as the flow of positive charge.** In a common conductor, such as copper, the current is due to the motion of the negatively charged electrons. Therefore, when we speak of current in such a conductor, the direction of the current is opposite the direction of flow of electrons. On the other hand, if one considers a beam of positively charged protons in an accelerator, the current is in the direction of motion of the protons. In some cases — gases and electrolytes, for example — the current is the result of the flows of both positive and negative charges. It is common to refer to a moving charge (whether it is positive or negative) as a mobile *charge carrier*. In a metal, for example, the charge carriers are electrons.

◀ Direction of current

13.2

EXAMPLE 17.1   The Current in a Lightbulb

The amount of charge that passes through the filament of a certain lightbulb in 2.00 s is 1.67 C. Find (a) the current in the lightbulb and (b) the number of electrons that pass through the filament in 1 second.

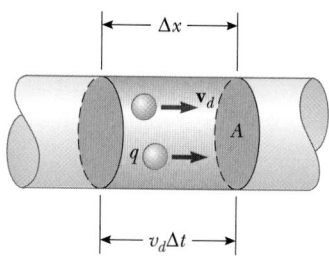

Figure 17.2 A section of a uniform conductor of cross-sectional area A. The charge carriers move with a speed of v_d, and the distance they travel in the time Δt is given by $\Delta x = v_d \, \Delta t$. The number of mobile charge carriers in the section of length Δx is given by $nAv_d \, \Delta t$, where n is the number of mobile carriers per unit volume.

Solution   (a) From Equation 17.1 we have

$$I = \frac{\Delta Q}{\Delta t} = \frac{1.67 \text{ C}}{2.00 \text{ s}} = \boxed{0.835 \text{ A}}$$

(b) In 1 second, 0.835 C of charge must pass the cross-sectional area of the filament. This total charge per second is equal to the number of electrons, N, times the charge on a single electron.

$$Nq = N(1.60 \times 10^{-19} \text{ C/electron}) = 0.835 \text{ C}$$

$$N = \boxed{5.22 \times 10^{18} \text{ electrons}}$$

17.2   CURRENT AND DRIFT SPEED

It is instructive to relate current to the motion of the charged particles. Consider the current in a conductor of cross-sectional area A (Fig. 17.2). The volume of an element of length Δx of the conductor is $A \, \Delta x$. If n represents the number of mobile charge carriers per unit volume, then the number of carriers in the volume element is $nA \, \Delta x$. Therefore, the charge, ΔQ, in this element is

$$\Delta Q = \text{number of carriers} \times \text{charge per carrier} = (nA \, \Delta x)q$$

where q is the charge on each carrier. If the carriers move with a speed of v_d, the distance they move in time Δt is $\Delta x = v_d \, \Delta t$. Therefore, we can write ΔQ as

$$\Delta Q = (nAv_d \, \Delta t)q$$

If we divide both sides of this equation by Δt, we see that the current in the conductor is

$$I = \frac{\Delta Q}{\Delta t} = nqv_d A \qquad [17.3]$$

▶ Current is proportional to the drift speed.

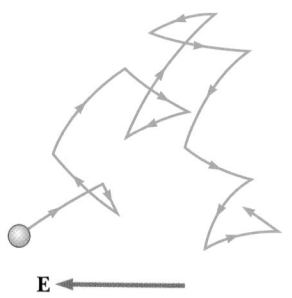

Figure 17.3 A schematic representation of the zigzag motion of a charge carrier in a conductor. The sharp changes in direction are due to collisions with atoms in the conductor. Note that the net motion of electrons is opposite the direction of the electric field.

The speed of the charge carriers, v_d, is an average speed called the **drift speed.** To understand its meaning, consider a conductor in which the charge carriers are free electrons. If the conductor is isolated, these electrons undergo random motion similar to that of gas molecules. When a potential difference is applied across the conductor (say, by means of a battery), an electric field is set up in the conductor, creating an electric force on the electrons and hence a current. In reality, the electrons do not simply move in straight lines along the conductor. Instead, they undergo repeated collisions with the metal atoms, and the result is a complicated zigzag motion (Fig. 17.3). The energy transferred from the electrons to the metal atoms during collision increases the vibrational energy of the atoms and causes a corresponding increase in the temperature of the conductor. However, despite the collisions, the electrons move slowly along the conductor (in a direction opposite $\mathbf{E}$) with the drift velocity, $\mathbf{v}_d$. The work done by the field on the electrons exceeds the average loss in energy due to collisions, and this work provides a steady current. One can think of the collisions within a conductor as being an effective internal friction (or drag) force similar to that experienced by the molecules of a liquid flowing through a pipe stuffed with steel wool.

EXAMPLE 17.2 The Drift Speed in a Copper Wire

A copper wire of cross-sectional area 3.00×10^{-6} m² carries a current of 10.0 A. Assuming that each copper atom contributes one free electron to the metal, find the drift speed of the electrons in this wire. The density of copper is 8.95 g/cm³.

Reasoning All the variables in Equation 17.3 are known except n, the number of free charge carriers per unit volume. We can find n by recalling that one mole of copper contains Avogadro's number (6.02×10^{23}) of atoms, and each atom contributes one charge carrier to the metal. The volume of one mole can be found from copper's known density and its atomic mass.

Solution From the periodic table of the elements, we find that the mass of one mole of copper is 63.5 g. Knowing the density of copper enables us to calculate the volume occupied by 63.5 g of copper:

$$V = \frac{m}{\rho} = \frac{63.5 \text{ g}}{8.95 \text{ g/cm}^3} = 7.09 \text{ cm}^3$$

If we now assume that each copper atom contributes one free electron to the body of the material, we have

$$n = \frac{6.02 \times 10^{23} \text{ electrons}}{7.09 \text{ cm}^3} = 8.48 \times 10^{22} \text{ electrons/cm}^3$$

$$= \left(8.48 \times 10^{22} \frac{\text{electrons}}{\text{cm}^3}\right)\left(10^6 \frac{\text{cm}^3}{\text{m}^3}\right) = 8.48 \times 10^{28} \text{ electrons/m}^3$$

From Equation 17.3, we find that the drift speed is

$$v_d = \frac{I}{nqA} = \frac{10.0 \text{ C/s}}{(8.48 \times 10^{28} \text{ electrons/m}^3)(1.60 \times 10^{-19} \text{ C})(3.00 \times 10^{-6} \text{ m}^2)}$$

$$= 2.46 \times 10^{-4} \text{ m/s}$$

Example 17.2 shows that drift speeds are typically very small. In fact, the drift speed is much smaller than the average speed between collisions; for instance, electrons traveling at 2.46×10^{-4} m/s would take about 68 min to travel 1 m! In view of this low speed, you might wonder why a light turns on almost instantaneously when a switch is thrown. Think of the flow of water through a pipe. If a drop of water is forced into one end of a pipe that is already filled with water, a drop must be pushed out the other end of the pipe. Although it may take an individual drop a long time to make it through the pipe, a flow initiated at one end produces a similar flow at the other end very quickly. Another familiar analogy is the motion of a bicycle chain. When the sprocket moves one link, the other links all move more or less immediately, even though it takes a given link some time to make a complete rotation. In a conductor, the electric field that drives the free electrons travels with a speed close to that of light. Thus, when you flip a light switch, the message for the electrons to start moving through the wire (the electric field) reaches them at a speed on the order of 10^8 m/s.

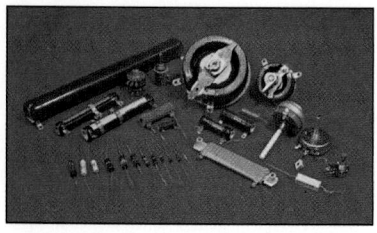

An assortment of resistors used for a variety of applications in electronic circuits. *(Courtesy of Henry Leap and Jim Lehman)*

Thinking Physics 1

Suppose a current-carrying wire has a cross-sectional area that gradually becomes smaller along the wire, so that the wire has the shape of a very long cone. How does the drift velocity vary along the wire?

Explanation   Every portion of the wire is carrying the same current. Thus, as the cross-sectional area decreases, the drift velocity must increase to maintain the constant value of the current. This increased drift velocity is a result of the electric field lines in the wire being compressed into a smaller area, thus increasing the strength of the field.

Thinking Physics 2

We have seen that an electric field must exist inside a conductor that carries a current. How is this possible in view of the fact that in electrostatics we concluded that the electric field is zero inside a conductor?

Explanation   In the electrostatic case in which charges are stationary, the internal electric field must be zero because a nonzero field would produce a current (by interacting with the free electrons in the conductor), which would violate the condition of static equilibrium. In this chapter we deal with conductors that carry current, a non-electrostatic situation. The current arises because of a potential difference applied between the ends of the conductor, which produces an internal electric field. So there is no paradox.

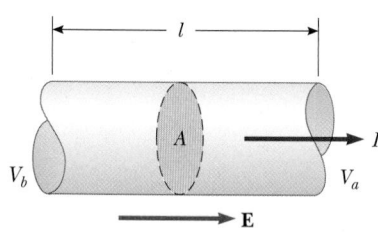

Figure 17.4   A uniform conductor of length l and cross-sectional area A. The current, I, in the conductor is proportional to the applied voltage, $\Delta V = V_b - V_a$. The electric field, **E**, set up in the conductor is also proportional to the current.

17.3   RESISTANCE AND OHM'S LAW

When a voltage (potential difference), ΔV, is applied across the ends of a metallic conductor as in Figure 17.4, the current in the conductor is found to be proportional to the applied voltage; that is, $I \propto \Delta V$. If the proportionality is exact, we can write $\Delta V = IR$, where the proportionality constant R is called the *resistance of the conductor*. In fact, we define this **resistance** as the ratio of the voltage across the conductor to the current it carries:

Resistance ▶

$$R \equiv \frac{\Delta V}{I}$$

[17.4]

13.3, SECTION 1

Resistance has the SI units volts per ampere, called **ohms (Ω).** Thus, if a potential difference of 1 V across a conductor produces a current of 1 A, the resistance of the conductor is 1 Ω. For example, if an electrical appliance connected to a 120-V source carries a current of 6 A, its resistance is 20 Ω.

It is useful to compare the concepts of electric current, voltage, and resistance with the flow of water in a river. As water flows downhill in a river of constant width and depth, the flow rate (water current) depends on the angle of flow and the effects of rocks, the river bank, and other obstructions. Based on this analogy, it seems reasonable that increasing the voltage applied to a circuit should increase

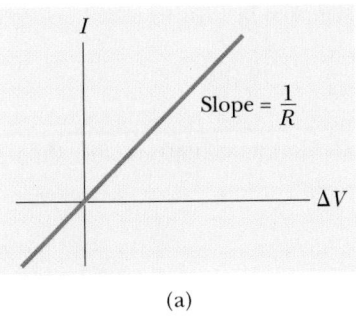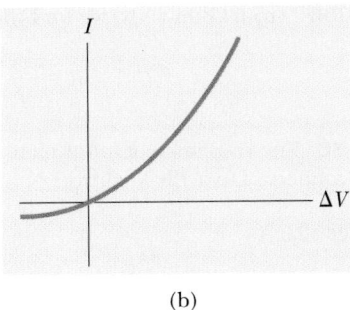

Figure 17.5 (a) The current-voltage curve for an ohmic material. The curve is linear, and the slope gives the resistance of the conductor. (b) A nonlinear current-voltage curve for a semiconducting diode. This device does not obey Ohm's law.

Georg Simon Ohm (1787–1854)

Ohm, a high school teacher in Cologne and later a professor at Munich, formulated the concept of resistance and discovered the proportionalities expressed in Equation 17.6. But Henry Cavendish had earllier determined the proportionality of current to voltage that we call Ohm's law. *(AIP Niels Bohr Library, E. Scott Barr Collection)*

the current in the circuit. For example, a 3-V battery should produce twice the current that would be produced by a 1.5 V battery. Resistance in a circuit arises due to collisions between the electrons carrying the current with fixed charges inside the conductor. These collisions inhibit the movement of charges in much the same way as would a force of friction. For many materials, including most metals, experiments show that **the resistance is constant over a wide range of applied voltages.** This statement is known as **Ohm's law** after Georg Simon Ohm (1789–1854), who was the first to conduct a systematic study of electrical resistance.

Ohm's law is an empirical relationship that is valid only for certain materials. Materials that obey Ohm's law, and hence have a constant resistance over a wide range of voltages, are said to be *ohmic*. Materials that do not obey Ohm's law are *nonohmic*. Ohmic materials have a linear current-voltage relationship over a large range of applied voltages (Fig. 17.5a). Nonohmic materials have a nonlinear current-voltage relationship (Fig. 17.5b). One common semiconducting device that is nonohmic is the diode. Its resistance is small for currents in one direction (positive ΔV) and large for currents in the reverse direction (negative ΔV). Most modern electronic devices, such as transistors, have nonlinear current-voltage relationships; their operation depends on the particular ways in which they violate Ohm's law.

It is common practice to express Ohm's law as

$$\Delta V = IR \qquad [17.5]$$

◄ Ohm's law

where R is understood to be independent of ΔV. We shall continue to use this traditional form of Ohm's law when discussing electrical circuits. A **resistor** is a conductor that provides a specified resistance in an electric circuit. The symbol for a resistor in circuit diagrams is a zigzag line, —⋀⋀⋀—.

EXAMPLE 17.3 The Resistance of a Steam Iron

All electric devices are required to have identifying plates that specify their electrical characteristics. The plate on a certain steam iron states that the iron carries a current of 6.4 A when connected to a 120-V source. What is the resistance of the steam iron?

Solution From Ohm's law, we find the resistance to be

$$R = \frac{\Delta V}{I} = \frac{120 \text{ V}}{6.4 \text{ A}} = \boxed{19 \; \Omega}$$

Exercise The resistance of a hot plate is 48 Ω. How much current does the plate carry when connected to a 120-V source?

Answer 2.5 A

17.4 RESISTIVITY

In an earlier section we pointed out that electrons do not move in straight-line paths through a conductor. Instead, they undergo repeated collisions with the metal atoms. Consider a conductor with a voltage applied between its ends. An electron gains speed as the electric force associated with the internal electric field accelerates it, giving it a velocity in the direction opposite that of the electric field. A collision with an atom randomizes the electron's velocity, thus reducing its velocity in the direction opposite the field. The process then repeats itself. Together these collisions affect the electron somewhat as a force of internal friction would. This is the origin of a material's resistance. The resistance of an ohmic conductor is proportional to its length, l, and inversely proportional to its cross-sectional area, A. That is,

$$R = \rho \frac{l}{A} \qquad \text{[17.6]}$$

where the constant of proportionality, ρ, is called the **resistivity** of the material.[1] Every material has a characteristic resistivity that depends on its electronic structure and on temperature. Good electric conductors have very low resistivities, and good insulators have very high resistivities. Table 17.1 lists the resistivities of a variety of materials at 20°C. Because resistance values are in ohms, resistivity values must be in ohm-meters.

Equation 17.6 shows that the resistance of a cylindrical conductor is proportional to its length and inversely proportional to its cross-sectional area. This is analogous to the flow of liquid through a pipe. As the length of the pipe is increased, the resistance to liquid flow increases because of a gain in friction between the fluid and the walls of the pipe. As its cross-sectional area is increased, the pipe can transport more fluid in a given time interval, so its resistance drops.

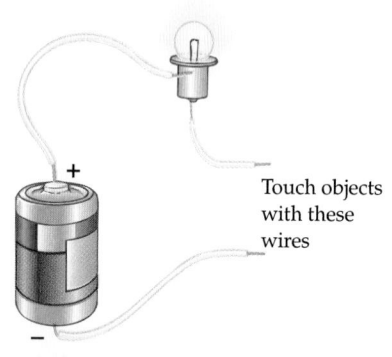
APPLICATION

Dimming of Aging Lightbulbs.

Applying Physics 1

It is a common observation that as a lightbulb ages, it gives off less light than when new. Why?

Explanation There are two reasons for this, one electrical and one optical, but both related to the same phenomenon occurring within the bulb. The filament of a

---

[1] The symbol ρ used for resistivity should not be confused with the same symbol used earlier in the book for density. Very often, a single symbol is used to represent different quantities.

lightbulb is made of a tungsten wire that, in an old lightbulb, has been kept at a high temperature for many hours. These high temperatures cause tungsten to be evaporated from the filament, thus decreasing its radius. From $R = \rho l/A$, we see that a decreased cross-sectional area leads to an increase in resistance of the filament. This increasing resistance with age means that the filament will carry less current for the same applied voltage. With less current in the filament, there is less light output, and the filament glows more dimly.

At the high operating temperature of the filament, tungsten atoms leave the surface of the filament, much as water molecules evaporate from a puddle of water. These atoms are carried away by convection currents in the gas in the bulb and are deposited on the inner surface of the glass. In time, the glass becomes less transparent because of this tungsten coating, which decreases the amount of light that passes through the glass.

TABLE 17.1 Resistivities and Temperature Coefficients of Resistivity for Various Materials [a]

Material	Resistivity $(\Omega \cdot m)$	Temperature Coefficient of Resistivity $[(°C)^{-1}]$
Silver	1.59×10^{-8}	3.8×10^{-3}
Copper	1.7×10^{-8}	3.9×10^{-3}
Gold	2.44×10^{-8}	3.4×10^{-3}
Aluminum	2.82×10^{-8}	3.9×10^{-3}
Tungsten	5.6×10^{-8}	4.5×10^{-3}
Iron	10.0×10^{-8}	5.0×10^{-3}
Platinum	11×10^{-8}	3.92×10^{-3}
Lead	22×10^{-8}	3.9×10^{-3}
Nichrome[b]	150×10^{-8}	0.4×10^{-3}
Carbon	3.5×10^{5}	-0.5×10^{-3}
Germanium	0.46	-48×10^{-3}
Silicon	640	-75×10^{-3}
Glass	$10^{10} - 10^{14}$	
Hard rubber	$\approx 10^{13}$	
Sulfur	10^{15}	
Quartz (fused)	75×10^{16}	

[a] All values are at 20°C.

[b] A nickel-chromium alloy commonly used in heating elements.

EXAMPLE 17.4 The Resistance of Nichrome Wire

(a) Calculate the resistance per unit length of a 22-gauge nichrome wire of radius 0.321 mm.

Solution The cross-sectional area of this wire is

$$A = \pi r^2 = \pi(0.321 \times 10^{-3} \text{ m})^2 = 3.24 \times 10^{-7} \text{ m}^2$$

The resistivity of nichrome is 1.5×10^{-6} $\Omega \cdot m$ (Table 17.1). Thus, we can use Equation

17.6 to find the resistance per unit length:

$$\frac{R}{l} = \frac{\rho}{A} = \frac{1.5 \times 10^{-6} \ \Omega \cdot m}{3.24 \times 10^{-7} \ m^2} = \boxed{4.6 \ \Omega/m}$$

(b) If a potential difference of 10.0 V is maintained across a 1.0-m length of the nichrome wire, what is the current in the wire?

Solution Because a 1.0-m length of this wire has a resistance of 4.6 Ω, Ohm's law gives

$$I = \frac{\Delta V}{R} = \frac{10.0 \ V}{4.6 \ \Omega} = \boxed{2.2 \ A}$$

Note from Table 17.1 that the resistivity of nichrome is about 100 times that of copper, a typical good conductor. Therefore, a copper wire of the same radius would have a resistance per unit length of only 0.052 Ω/m, and a 1.0-m length of copper wire of the same radius would carry the same current (2.2 A) with an applied voltage of only 0.11 V.

Because of its high resistivity and its resistance to oxidation, nichrome is often used for heating elements in toasters, irons, and electric heaters.

Exercise What is the resistance of a 6.0-m length of 22-gauge nichrome wire? How much current does it carry when connected to a 120-V source?

Answer 28 Ω; 4.3 A

17.5 TEMPERATURE VARIATION OF RESISTANCE

The resistivity, and hence the resistance, of a conductor depends on a number of factors. One of the most important is the temperature of the metal. For most metals, resistivity increases with increasing temperature. This correlation can be understood as follows. As the temperature of the material increases, its constituent atoms vibrate with increasingly greater amplitudes. Just as it is more difficult to weave one's way through a crowded room when the people are in motion than when they are standing still, so do the electrons find it more difficult to pass atoms moving with large amplitudes.

For most metals, resistivity increases approximately linearly with temperature over a limited temperature range, according to the expression

$$\rho = \rho_0[1 + \alpha(T - T_0)] \qquad [17.7]$$

where ρ is the resistivity at some temperature, T (in Celsius degrees); ρ_0 is the resistivity at some reference temperature, T_0 (usually taken to be 20°C); and α is a parameter called the **temperature coefficient of resistivity.** The temperature coefficients for various materials are provided in Table 17.1.

Because the resistance of a conductor with uniform cross section is proportional to the resistivity according to Equation 17.6 ($R = \rho l/A$), the temperature variation of resistance can be written

$$R = R_0[1 + \alpha(T - T_0)] \qquad [17.8]$$

Precise temperature measurements are often made using this property, as shown by the following example.

An old-fashioned carbon filament incandescent lamp. The resistance of such a lamp is typically 10 Ω, but it changes with temperature. *(Courtesy of Central Scientific Company)*

EXAMPLE 17.5 A Platinum Resistance Thermometer

A resistance thermometer, which measures temperature by measuring the change in resistance of a conductor, is made of platinum and has a resistance of 50.0 Ω at 20.0°C. When the device is immersed in a vessel containing melting indium, its resistance increases to 76.8 Ω. From this information, find the melting point of indium.

Solution If we solve Equation 17.8 for $T - T_0$ and get α for platinum from Table 17.1, we obtain

$$T - T_0 = \frac{R - R_0}{\alpha R_0} = \frac{76.8\ \Omega - 50.0\ \Omega}{[3.92 \times 10^{-3}\ (°C)^{-1}][50.0\ \Omega]} = 137°C$$

Because $T_0 = 20.0°C$, we find that the melting point of indium is

$$T = \boxed{157°C}$$

The Carbon Microphone

Figure 17.6 illustrates the construction of a carbon microphone, commonly used in the mouthpiece of a telephone. A flexible steel diaphragm is placed in contact with carbon granules inside a container. The carbon granules serve as the primary resistance medium in a circuit containing a source of current—here, a battery— and a transformer (described in Chapter 21).

The magnitude of the current in the circuit changes when a sound wave strikes the diaphragm. When a compression arrives at the microphone, the diaphragm flexes inward, causing the carbon granules to press together into a smaller than normal volume, corresponding to a decrease in the length of the resistive medium. This results in a lower circuit resistance and hence a greater current in the circuit. When a rarefaction arrives at the microphone, the diaphragm relaxes, and the carbon granules become more loosely packed, causing an increase in the circuit resistance and a corresponding decrease in current. These variations in current, following the changes of the sound wave, are sent through the transformer into the telephone company's transmission line. A speaker in the listener's earpiece then converts electric signals back to a sound wave.

The carbon microphone has a very poor frequency response. It adequately reproduces frequencies below 4000 Hz and is therefore suitable for speech trans-

APPLICATION

Telephone Mouthpieces.

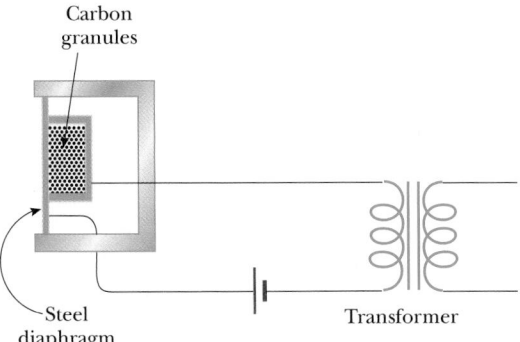

Carbon granules

Steel diaphragm

Transformer

Figure 17.6 A diagram of a carbon microphone.

mission, because the critical frequencies in normal conversation are usually below this value. However, its capabilities fall off dramatically at higher frequencies, rendering it useless for high-fidelity purposes, which require reliable sound reproduction at all frequencies between 20 Hz and 20 000 Hz.

17.6 SUPERCONDUCTORS

There is a class of metals and compounds whose resistances fall to virtually *zero* below a certain temperature, T_c, called the *critical temperature*. These materials are known as **superconductors.** The resistance-temperature graph for a superconductor follows that of a normal metal at temperatures above T_c (Fig. 17.7). When the temperature is at or below T_c, the resistance suddenly drops to zero. This phenomenon was discovered in 1911 by the Dutch physicist H. Kamerlingh Onnes as he and a graduate student worked with mercury, which is a superconductor below 4.1 K. Recent measurements have shown that the resistivities of superconductors below T_c are less than 4×10^{-25} $\Omega \cdot$m — around 10^{17} times smaller than the resistivity of copper and in practice considered to be zero.

Today thousands of superconductors are known, including such common metals as aluminum, tin, lead, zinc, and indium. Table 17.2 lists the critical temperatures of several superconductors. The value of T_c is sensitive to chemical composition, pressure, and crystalline structure. Interestingly, copper, silver, and gold, which are excellent conductors, do not exhibit superconductivity.

One of the truly remarkable features of superconductors is the fact that, once a current is set up in them, it persists *without any applied voltage* (because $R = 0$). In fact, steady currents in superconducting loops have been observed to persist for many years with no apparent decay!

An important development in physics that elicited much excitement in the scientific community was the discovery of high-temperature copper-oxide-based su-

TABLE 17.2
Critical Temperatures for Various Superconductors

Material	T_c (K)
Zn	0.88
Al	1.19
Sn	3.72
Hg	4.15
Pb	7.18
Nb	9.46
Nb_3Sn	18.05
Nb_3Ge	23.2
$YBa_2Cu_3O_7$	90
Bi-Sr-Ca-Cu-O	105
Tl-Ba-Ca-Cu-O	125

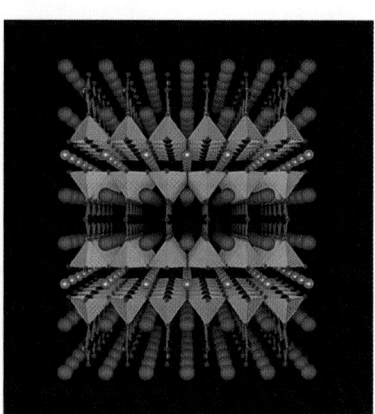

A computer-generated model of the high-temperature superconductor $YBa_2Cu_3O_7$, which is a member of a crystal structure family known as perovskites. *(Courtesy of IBM Research)*

Figure 17.7 Resistance versus temperature for a sample of mercury. The graph follows that of a normal metal above the critical temperature, T_c. The resistance drops to zero at the critical temperature, which is 4.1 K for mercury, and remains at zero for lower temperatures.

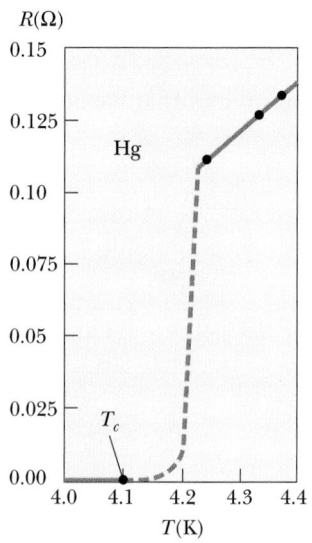

perconductors. The excitement began with a 1986 publication by J. Georg Bednorz and K. Alex Müller, scientists at the IBM Zurich Research Laboratory in Switzerland, in which they reported evidence for superconductivity at a temperature near 30 K in an oxide of barium, lanthanum, and copper. Bednorz and Müller were awarded the Nobel prize for physics in 1987 for their remarkable discovery. Shortly thereafter, a new family of compounds was open for investigation, and research activity in the field of superconductivity proceeded vigorously. In early 1987, groups at the University of Alabama at Huntsville and the University of Houston announced the discovery of superconductivity at about 92 K in an oxide of yttrium, barium, and copper ($YBa_2Cu_3O_7$). Late in 1987, teams of scientists from Japan and the United States reported superconductivity at 105 K in an oxide of bismuth, strontium, calcium, and copper. More recently, scientists have reported superconductivity at temperatures as high as 150 K in an oxide containing mercury. At this point one cannot rule out the possibility of room-temperature superconductivity, and the search for novel superconducting materials continues. It is an important search both for scientific reasons and because practical applications become more probable and widespread as the critical temperature is raised.

An important and useful application is superconducting magnets in which the magnetic field intensities are about ten times greater than those of the best normal electromagnets. Such magnets are being considered as a means of storing energy. The idea of using superconducting power lines for transmitting power efficiently is also receiving consideration. Modern superconducting electronic devices consisting of two thin-film superconductors separated by a thin insulator have been constructed. They include magnetometers (magnetic-field measuring devices) and various microwave devices.

A small permanent magnet floats freely above a ceramic disk of the superconductor $YBa_2Cu_3O_7$ cooled by liquid nitrogen at 77 K. The superconductor has zero electric resistance at temperatures below 92 K and expels any applied magnetic field. *(Courtesy of IBM Research Laboratory)*

WEB

For Web links to more information on superconductivity, visit the textbook Web site at **http://www.harcourtcollege.com/ physics/cptect**

17.7 ELECTRICAL ENERGY AND POWER

If a battery is used to establish an electric current in a conductor, chemical energy stored in the battery is continuously transformed into kinetic energy of the charge carriers. This kinetic energy is quickly lost as a result of collisions between the charge carriers and fixed atoms in the conductor, causing an increase in the temperature of the conductor. Thus, the chemical energy stored in the battery is continuously transformed into thermal energy.

In order to understand the process of energy transfer in a simple circuit, consider a battery whose terminals are connected to a resistor (Fig. 17.8). (Remember that the positive terminal of the battery is always at the higher potential.) Now imagine following a quantity of positive charge ΔQ around the circuit from point A through the battery and resistor and back to A. Point A is a reference point that is grounded (the ground symbol is ———), and its potential is taken to be zero. As the charge moves from A to B through the battery, its electrical potential energy increases by the amount $\Delta V \cdot \Delta Q$ and the chemical potential energy in the battery decreases by the same amount. (Recall from Chapter 16 that $\Delta PE = q \, \Delta V$.) However, as the charge moves from C to D through the resistor, it loses this electrical potential energy during collisions with atoms in the resistor, thereby producing thermal energy. Note that if we neglect the resistance of the interconnecting wires,

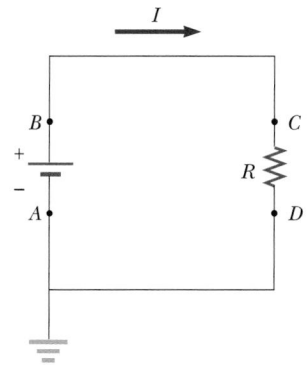

Figure 17.8 A circuit consisting of a battery and resistance R. Positive charge flows clockwise from the positive to the negative terminal of the battery. Point A is grounded.

13.3, SECTION 2

no loss in energy occurs for paths BC and DA. When the charge returns to point A, it must have the same potential energy (zero) as it had at the start.

The rate at which the charge ΔQ loses potential energy as it passes through the resistor is

$$\frac{\Delta Q}{\Delta t}\,\Delta V = I\,\Delta V$$

where I is the current in the circuit. Of course, the charge regains this energy when it passes through the battery. Because the rate at which the charge loses energy equals the power, P, dissipated in the resistor, we have

Power ▶
$$P = I\,\Delta V \qquad\qquad \text{[17.9]}$$

In this case, the power is supplied to a resistor by a battery. However, Equation 17.9 can be used to determine the power transferred from a battery to *any* device carrying a current, I, and having a potential difference, ΔV, between its terminals.

Using Equation 17.9 and the fact that $\Delta V = IR$ for a resistor, we can express the power dissipated by the resistor in the alternative form

Power dissipated by a resistor ▶
$$P = I^2 R = \frac{(\Delta V)^2}{R} \qquad\qquad \text{[17.10]}$$

When I is in amperes, ΔV in volts, and R in ohms, the SI unit of power is the watt (introduced in Chapter 5). The dissipation of power as heat in a conductor of resistance R is called *joule heating*. It is also often referred to as an I^2R loss.

Regardless of the ways in which you use electrical energy in your home, you ultimately must pay for it or risk having your power turned off. The unit of energy used by electric companies to calculate consumption, the **kilowatt-hour,** is defined

The kilowatt-hour is a unit of ▶
energy.

in terms of the unit of power. One kilowatt-hour (kWh) is the energy converted or consumed in 1 h at the constant rate of 1 kW. It has the numerical value

$$1 \text{ kWh} = (10^3 \text{ W})(3600 \text{ s}) = 3.60 \times 10^6 \text{ J} \qquad\qquad \text{[17.11]}$$

On an electric bill, the amount of electricity used in a given period is usually stated in multiples of kilowatt-hours.

Applying Physics 2

When is more power delivered to a lightbulb—just after it is turned on and the glow of the filament is increasing or after it has been on for a few seconds and the glow is steady?

Explanation   Once the switch is closed, the line voltage is applied across the lightbulb. As the voltage is applied across the cold filament when first turned on, the resistance of the filament is low, the current is high, and a relatively large amount of power is delivered to the bulb. As the filament warms, its resistance rises, and the current decreases. As a result, the power delivered to the bulb decreases. The large current spike at the beginning of operation is the reason that lightbulbs often fail just after they are turned on.

Applying Physics 3

Two lightbulbs, A and B, are connected across the same potential difference as in Figure 17.9. The resistance of A is twice that of B. Which lightbulb dissipates more power? Which carries the greater current?

Explanation Because the voltage across each lightbulb is the same, and the power dissipated by a conductor is $P = (\Delta V)^2/R$, the conductor with the lower resistance will dissipate more power. In this case, the power dissipated by B is twice that of A and provides twice as much illumination. Furthermore, because $P = I\Delta V$, we see that the current carried by B is twice that of A.

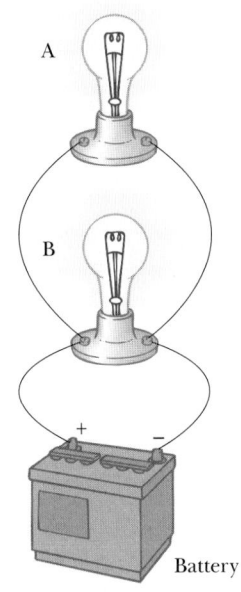

Figure 17.9 (Applying Physics 3)

Thinking Physics 3

When a lightning bolt strikes in a barnyard, it is more probable that a four-legged animal such as a cow will be killed than a two-legged creature such as a chicken. Why?

Explanation When a lightning bolt strikes the Earth, a large current moves outward away from the point of impact. Consider two points separated by about 1 m, about the distance between a cow's legs, along this radial direction. From $\Delta V = IR$, where I is the current from the bolt and R is the resistance of the Earth between these points, we see that the potential difference between these points can be very great. In fact, it is great enough to be lethal to the cow. A chicken does not stand a great chance of survival either, but because the distance between its feet is small, the voltage drop between these points would not be as great.

EXAMPLE 17.6 The Power Consumed by an Electric Heater

An electric heater is operated by applying a potential difference of 50.0 V to a nichrome wire of total resistance 8.00 Ω. Find the current carried by the wire and the power rating of the heater.

Solution Because $\Delta V = IR$, we have

$$I = \frac{\Delta V}{R} = \frac{50.0 \text{ V}}{8.00 \text{ } \Omega} = \boxed{6.25 \text{ A}}$$

We can find the power rating using $P = I^2 R$:

$$P = I^2 R = (6.25 \text{ A})^2 (8.00 \text{ } \Omega) = \boxed{313 \text{ W}}$$

Exercise If we doubled the applied voltage to the heater, what would happen to the current and power?

Answer The current would double, and the power would quadruple.

Workbook Problem 81 – (Workbook page 175) deals with the energy stored in a battery.

EXAMPLE 17.7   Electrical Rating of a Lightbulb

A lightbulb is rated at 120 V and 75 W. That is, at the intended operating voltage of 120 V, it converts 75 W of power. The bulb is powered by a 120-V direct current power supply. Find the current in the bulb and its resistance.

Solution   Because we know that the power rating of the bulb is 75 W and the operating voltage is 120 V, we can use $P = I \Delta V$ to find the current:

$$I = \frac{P}{\Delta V} = \frac{75 \text{ W}}{120 \text{ V}} = \boxed{0.63 \text{ A}}$$

Using Ohm's law, $\Delta V = IR$, the resistance is calculated to be

$$R = \frac{\Delta V}{I} = \frac{120 \text{ V}}{0.63 \text{ A}} = \boxed{190 \text{ }\Omega}$$

Exercise   What would the resistance be in a lamp rated at 120 V and 100 W?

Answer   140 Ω

EXAMPLE 17.8   The Cost of Operating a Lightbulb

How much does it cost to burn a 100-W lightbulb for 24 h if electric energy costs $0.080 per kilowatt-hour?

Solution   A 100-W lightbulb is equivalent to a 0.10-kW bulb. Because energy consumed equals power × time, the amount of energy you must pay for, expressed in kilowatt-hours, is

$$\text{Energy} = (0.10 \text{ kW})(24 \text{ h}) = 2.4 \text{ kWh}$$

If energy is purchased at $0.080 per kilowatt-hour, the 24-h cost is

$$\text{Cost} = (2.4 \text{ kWh})(\$0.080/\text{kWh}) = \boxed{\$0.19}$$

This is a small amount of money, but when larger and more complex electric devices are used, the cost goes up rapidly.

Exercise   If electric energy costs $0.080/kWh, what does it cost to operate an electric oven, which operates at 20.0 A and 220 V, for 5.0 h?

Answer $1.80

Optional Section

🔲 **17.8**   VOLTAGE MEASUREMENTS IN MEDICINE

Electrocardiograms

Every action involving the body's muscles is initiated by electrical activity. The voltages produced by muscular action in the heart are particularly important to physicians. Voltage pulses cause the heart to beat, and the waves of electrical excitation associated with the heartbeat are conducted through the body via the body fluids.

These voltage pulses are large enough to be detected by suitable monitoring equipment attached to the skin. Standard electric devices can be used to record these voltage pulses because the amplitude of a typical pulse associated with heart activity is of the order of 1 mV. These voltage pulses are recorded on an instrument called an **electrocardiograph,** and the pattern recorded by this instrument is called an **electrocardiogram** (EKG). In order to understand the information contained in an EKG pattern, it is useful to first describe the underlying principles concerning electrical activity in the heart.

The right atrium of the heart contains a specialized set of muscle fibers called the SA (sinoatrial) node, which initiate the heartbeat (Fig. 17.10). Electric impulses that originate in these fibers gradually spread from cell to cell throughout the right and left atrial muscles, causing them to contract. The pulse that passes through the muscle cells is often called a *depolarization wave* because of its effect on individual cells. If an individual muscle cell were examined, an electric charge distribution would be found on its surface, as shown in Figure 17.11a. The impulse generated by the SA node momentarily changes the cell's charge distribution to that shown in Figure 17.11b. The positively charged ions on the surface of the cell are temporarily able to diffuse through the membrane wall so that the cell attains an excess positive charge on its inside surface. As the depolarization wave travels from cell to cell throughout the atria, the cells recover to the charge distribution shown in Figure 17.11a. When the impulse reaches the AV (atrioventricular) node (Fig. 17.10), the muscles of the atria begin to relax, and the pulse is directed by the AV node to the ventricular muscles. The muscles of the ventricles contract as the depolarization wave spreads through the ventricles along a group of fibers called the *Purkinje fibers.* The ventricles then relax after the pulse has passed through. At this point, the SA node is again triggered and the cycle is repeated.

A sketch of the electrical activity registered on an EKG for one beat of a normal heart is shown in Figure 17.12. The pulse indicated by *P* occurs just before the atria begin to contract. The *QRS* pulse occurs in the ventricles just before they contract, and the *T* pulse occurs when the cells in the ventricles begin to recover. EKGs for an abnormal heart are shown in Figure 17.13. The *QRS* portion of the pattern shown in Figure 17.13a is wider than normal. This indicates that the patient may have an enlarged heart. Figure 17.13b indicates that there is no relationship between the *P* pulse and the *QRS* pulse. This suggests a blockage in the electrical conduction path between the SA and AV nodes. This can occur when the atria and ventricles beat independently. Finally, Figure 17.13c shows a situation in which there is no *P* pulse and an irregular spacing between the *QRS* pulses. This is symptomatic of irregular

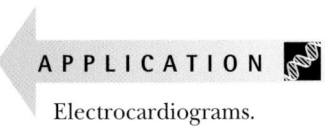

APPLICATION

Electrocardiograms.

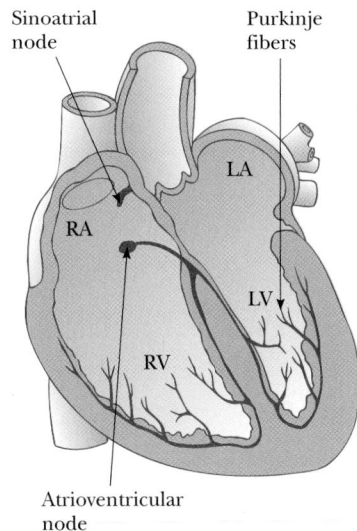

Figure 17.10 The electrical conduction system of the human heart. (RA: right atrium; LA: left atrium; RV: right ventricle; LV: left ventricle.)

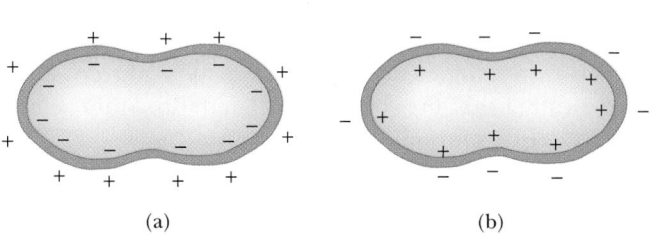

(a)   (b)

Figure 17.11 (a) Charge distribution of a muscle cell in the atrium before a depolarization wave has passed through the cell. (b) Charge distribution as the wave passes.

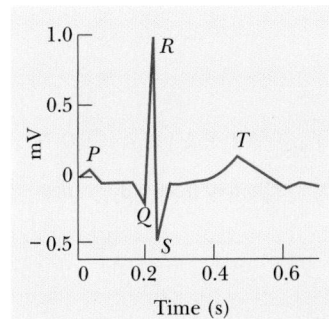

Figure 17.12 An EKG response for a normal heart.

Figure 17.13   Abnormal EKGs.

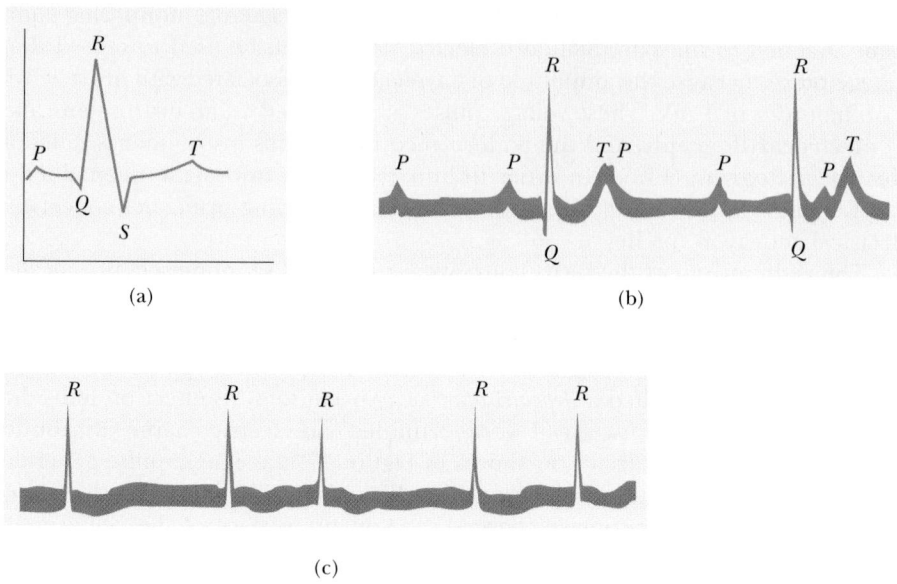

(a)

(b)

(c)

atrial contraction, which is called *fibrillation*. In this situation, the atrial and ventricular contractions are irregular.

As noted previously, the sinoatrial node directs the heart to beat at the appropriate rate, usually about 72 beats per minute. However, disease or the aging process can damage the heart so that the proper electrical activity cannot be maintained, and a medical assist may be necessary in the form of a pacemaker attached to the heart. This matchbox-sized electrical device has a lead that is connected to the wall of the right ventricle. Pulses from this lead stimulate the heart to maintain its proper rhythm. In general, a pacemaker is designed to produce pulses at a rate of about 60 per minute, slightly slower than the normal beats per minute but sufficient to maintain life. The circuitry consists of a capacitor charging up to a certain voltage and then discharging. The design of the circuit is so that if the heart is beating normally the capacitor is never allowed to charge completely and send pulses to the heart.

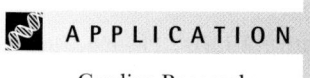

APPLICATION

Cardiac Pacemakers.

Electroencephalography

The electrical activity of the brain can be measured with an instrument called an **electroencephalograph** in much the same way that an electrocardiograph measures the electrical activity of the heart. The voltage pattern measured by an electroencephalograph is referred to as an **electroencephalogram (EEG).** An EEG pattern is recorded by placing electrodes on the patient's scalp. Although an EKG voltage pulse is typically 1 mV, the voltage associated with brain activity is only a few *micro*volts and is therefore more difficult to measure.

The EEGs in Figure 17.14 represent the brain wave pattern of a patient awake and then in various stages of sleep. In Figure 17.14b the patient begins to fall into a light sleep, and in Figure 17.14d the patient is in a deep sleep. Figure 17.14c is the EEG during a type of sleep called REM (rapid eye movement) sleep, which

APPLICATION

Electroencephalograms.

occurs approximately every 2 h. In this stage, the brain activity is quite similar to that of the patient while awake. It is interesting to note that a person in this stage of sleep is extremely difficult to awaken. Most researchers agree that REM sleep is necessary for psychological well-being. Anyone who is deprived of this stage of sleep for an extended period of time becomes extremely fatigued and irritable.

The EEG is an important diagnostic tool for detecting epilepsy, brain tumors, brain hemorrhages, meningitis, and so forth. For example, the EEG pattern of a patient suffering an epileptic seizure can show spike-shaped waves of amplitude greater than 100 μV.

Brain waves are often discussed in terms of their frequencies. Frequencies of about 10 Hz are referred to as *alpha waves,* frequencies between 10 Hz and 60 Hz are called *beta waves,* and those below 10 Hz are called *delta waves.* As a person falls asleep, the frequency of brain wave activity generally decreases. For example, the brain wave frequency for a person in a very deep sleep may drop as low as 1 or 2 Hz.

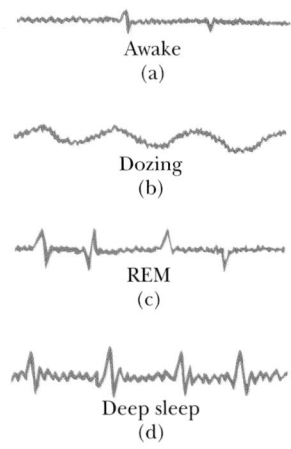

Figure 17.14 Brain waves from an individual in various stages of sleep.

SUMMARY

The **electric current,** *I,* in a conductor is defined as

$$I \equiv \frac{\Delta Q}{\Delta t} \qquad [17.1]$$

where ΔQ is the charge that passes through a cross section of the conductor in time Δt. The SI unit of current is the **ampere (A);** 1 A = 1 C/s. By convention, the direction of current is in the direction of flow of positive charge.

The current in a conductor is related to the motion of the charge carriers by

$$I = nqv_d A \qquad [17.3]$$

where n is the number of mobile charge carriers per unit volume, q is the charge on each carrier, v_d is the drift speed of the charges, and A is the cross-sectional area of the conductor.

The **resistance, *R,*** of a conductor is defined as the ratio of the potential difference across the conductor to the current:

$$R \equiv \frac{\Delta V}{I} \qquad [17.4]$$

The SI units of resistance are volts per ampere, or **ohms (Ω);** 1 Ω = 1 V/A.

Ohm's law describes many conductors, for which the applied voltage is directly proportional to the current it causes. The proportionality constant is the resistance:

$$\Delta V = IR \qquad [17.5]$$

If a conductor has length l and cross-sectional area A, its **resistance** is

$$R = \rho \frac{l}{A} \qquad [17.6]$$

where ρ is an intrinsic property of the conductor called the **electrical resistivity.** The SI unit of resistivity is the **ohm-meter** ($\Omega \cdot m$).

The resistivity of a conductor varies with temperature over a limited temperature range, according to the expression

$$\rho = \rho_0[1 + \alpha(T - T_0)] \qquad [17.7]$$

where α is the **temperature coefficient of resistivity** and ρ_0 is the resistivity at some reference temperature, T_0 (usually taken to be 20°C).

The resistance of a conductor varies with temperature according to the expression

$$R = R_0[1 + \alpha(T - T_0)] \qquad [17.8]$$

If a potential difference, ΔV, is maintained across a resistor, the **power,** or rate at which energy is supplied to the resistor, is

$$P = I\Delta V \qquad [17.9]$$

Because the potential difference across a resistor is $\Delta V = IR$, the power dissipated by a resistor can be expressed as

$$P = I^2R = \frac{(\Delta V)^2}{R} \qquad [17.10]$$

A **kilowatt-hour** is the amount of energy converted or consumed in 1 hour by a device supplied with power at the rate of 1 kW. This is equivalent to

$$1 \text{ kWh} = 3.60 \times 10^6 \text{ J} \qquad [17.11]$$

MULTIPLE-CHOICE QUESTIONS

1. What is the resistance of the electric device whose current versus voltage behavior is shown in Figure 17.15 when the potential difference across the device is 2 V? (a) 1 Ω (b) $\frac{3}{4}\Omega$ (c) $\frac{4}{3}\Omega$ (d) Undefined (e) None of these answers.

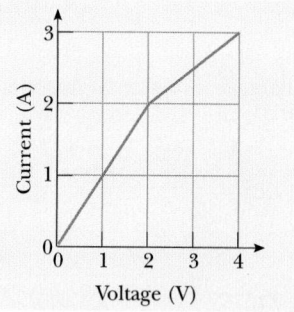

Figure 17.15   (Multiple-Choice Questions 1 and 2)

2. What is the resistance of the electrical device whose current versus voltage behavior is shown in Figure 17.15 when the potential difference across the device is 3 V? (a) $\frac{5}{6}\Omega$ (b) 1.2 Ω (c) 3 Ω (d) $\frac{4}{3}\Omega$ (e) None of these answers.

3. A metal wire has a resistance of 10.00 ohms at a temperature of 20°C. If the same wire has a resistance of 10.55 ohms at 90°C, what is the resistance of this same wire when its temperature is -20°C? (a) 0.7 Ω (b) 9.73 Ω (c) 10.31 Ω (d) 13.8 Ω

4. A color television set draws about 2.5 A when connected to 120 V. What is the cost (with electrical energy at 6 cents/kWh) of running the color television for 8 hours? (a) 1.4 cents (b) 3.0 cents (c) 14 cents (d) 30 cents

5. A 1.00-V potential difference is maintained across a 10.0-Ω resistor for a period of 20.0 s. What total charge passes through the wire in this time interval? (a) 200 C (b) 20.0 C (c) 2.00 C (d) 0.005 C

CONCEPTUAL QUESTIONS

1. Why don't the free electrons in a metal fall to the bottom of the metal due to gravity? And charges in a conductor are supposed to reside on the surface—why don't the free electrons all go to the surface?
2. In an analogy between traffic flow and electrical current, what would correspond to the charge, Q? What would correspond to the current, I?
3. Newspaper articles often have statements such as, "10,000 volts of electricity surged through the victim's body." What is wrong with this statement?
4. Two lightbulbs operate from 120 V, but one has a power rating of 25 W and the other has a power rating of 100 W. Which bulb has higher resistance? Which carries the greater current?
5. Aliens with strange powers visit Earth and double every linear dimension of every object on the surface of the planet. Does the electrical cord from the wall socket to your floor lamp now have more resistance than before, less resistance, or the same resistance? Does the lightbulb filament glow more brightly than before, less brightly, or the same? (Assume the resistivities of materials remain the same.)
6. When the voltage across a certain conductor is doubled, the current is observed to triple. What can you conclude about the conductor?
7. When incandescent lightbulbs burn out, they usually do so just after they are switched on. Why?
8. If electricity costs $0.080/kWh, estimate how much it costs a person to dry her hair with a 1500-W blow dryer during a year's time.
9. What factors affect the resistance of a conductor?
10. Some homes have light dimmers that are operated by rotation of a knob. What is being changed in the electric circuit when the knob is rotated?
11. Two wires A and B with circular cross sections are made of the same metal and have equal lengths, but the resistance of wire A is three times greater than that of wire B. What is the ratio of their cross-sectional areas? How do the radii compare?
12. What single experimental requirement makes superconducting devices expensive to operate? In principle, can this limitation be overcome?
13. Use the atomic theory of matter to explain why the resistance of a material should increase as its temperature increases.
14. Two conductors of the same length and radius are connected across the same voltage source. One conductor has twice the resistance of the other. Which conductor dissipates more power?
15. What could happen to the drift velocity of the electrons in a wire and to the current in the wire if the electrons could move freely without resistance through this metallic conductor?
16. Car batteries are often rated in ampere-hours. Does this designate the amount of current, power, energy, or charge that can be drawn from the battery?
17. If charges flow very slowly through a metal, why does it not require several hours for a light to come on when you throw a switch?

PROBLEMS

1, 2, 3 = straightforward, intermediate, challenging   ☐ = full solution available in Study Guide/Student Solutions Manual   ⬙ = Core Concepts Workbook
WEB = solution posted at **http://www.harcourtcollege.com/physics/cptech**   ⬙ = biomedical application   ⬙ = Interactive Physics

Section 17.1 Electric Current

Section 17.2 Current and Drift Speed

1. If a current of 80.0 mA exists in a metal wire, how many electrons flow past a given cross section of the wire in 10.0 min? Sketch the directions of the current and the electrons' motion.
2. The compressor on an air conditioner draws 90 A when it starts up. If the start-up time is about 0.50 s, how much charge passes a cross-sectional area of the circuit in this time?
3. A total charge of 6.0 mC passes through a cross-sectional area of a wire in 2.0 s. What is the current in the wire?

4. In a particular television picture tube, the measured beam current is 60.0 μA. How many electrons strike the screen every second?
5. In the Bohr model of the hydrogen atom, an electron in the lowest energy state moves at a speed of 2.19×10^6 m/s in a circular path having a radius of 5.29×10^{-11} m. What is the effective current associated with this orbiting electron?
6. If 3.25×10^{-3} kg of gold is deposited on the negative electrode of an electrolytic cell in a period of 2.78 h, what is the current through the cell in this period? Assume that the gold ions carry one elementary unit of positive charge.

7. A 200-km-long high-voltage transmission line 2.0 cm in diameter carries a steady current of 1000 A. If the conductor is copper with a free charge density of 8.5×10^{28} electrons per cubic meter, how long (in years) does it take one electron to travel the full length of the cable?

8. Calculate the number of free electrons per cubic meter for gold, assuming one free electron per atom. (Density of gold $= 19.3 \times 10^3$ kg/m^3.)

WEB 9. An aluminum wire with a cross-sectional area of 4.0×10^{-6} m^2 carries a current of 5.0 A. Find the drift speed of the electrons in the wire. The density of aluminum is 2.7 g/cm^3. (Assume that one electron is supplied by each atom.)

Section 17.3 Resistance and Ohm's Law

Section 17.4 Resistivity

10. A person notices a mild shock if the current along a path through the thumb and index finger exceeds 80 μA. Compare the maximum allowable voltage without shock across the thumb and index finger with a dry-skin resistance of 4.0×10^5 Ω and a wet-skin resistance of 2000 Ω.

11. When operating at 120 V, a resistor carries a current of 0.50 A. What current is carried if (a) the operating voltage is lowered to 90 V? (b) the voltage is raised to 130 V?

12. Calculate the diameter of a 2.0-cm length of tungsten filament in a small lightbulb if its resistance is 0.050 Ω.

13. Eighteen-gauge wire has a diameter of 1.024 mm. Calculate the resistance of 15 m of 18-gauge copper wire at 20°C.

14. A potential difference of 12 V is found to produce a current of 0.40 A in a 3.2-m length of wire with a uniform radius of 0.40 cm. What is (a) the resistance of the wire? (b) the resistivity of the wire?

15. A length, L_0, of copper wire has a resistance, R_0. The wire is cut into three pieces of equal length. The pieces are then connected as parallel lengths between points A and B. What resistance will this new wire of length $L_0/3$ have between points A and B?

16. A wire, 50.0 m long and 2.00 mm in diameter, is connected to a source with a potential difference of 9.11 V, and the current is found to be 36.0 A. Assume a temperature of 20°C and, using Table 17.1, identify the metal of the wire.

17. A rectangular block of copper has sides of length 10 cm, 20 cm, and 40 cm. If the block is connected to a 6.0-V source across opposite faces of the rectangular block, what are (a) the maximum current and (b) minimum current that can be carried?

Section 17.5 Temperature Variation of Resistance

18. (a) A 34.5-m length of copper wire at 20.0°C has a radius of 0.25 mm. If a potential difference of 9.0 V is applied across the length of the wire, determine the current in the wire. (b) If the wire is heated to 30.0°C and the 9.0-V potential difference is maintained, what is the resulting current in the wire?

WEB 19. A 1050-W toaster operates on a 120-V household circuit and has a 4.00-m length of nichrome wire as its heating element. The operating temperature of this element is 320°C. What is the cross-sectional area of the wire?

20. Suppose you wish to fabricate a uniform wire out of 1.0 g of copper. If the wire is to have a resistance of 0.50 Ω and all of the copper is to be used, what must be the (a) length and (b) radius of this wire?

21. A certain lightbulb has a tungsten filament with a resistance of 19.0 Ω when cold and 140 Ω when hot. Assume that Equation 17.8 can be used over the large temperature range involved here, and find the temperature of the filament when it is hot. Assume an initial temperature of 20°C.

22. If a silver wire has a resistance of 10.0 Ω at 20.0°C, what resistance does it have at 40.0°C? Neglect any change in length or cross-sectional area resulting from the change in temperature.

23. At 20°C the carbon resistor in an electric circuit, connected to a 5.0-V battery, has a resistance of 200 Ω. What is the current in the circuit when the temperature of the carbon rises to 80°C?

24. At 40.0°C, the resistance of a segment of gold wire is 100.0 Ω. When the wire is placed in a liquid bath, the resistance decreases to 97.0 Ω. What is the temperature of the bath? (*Hint:* First determine the resistance of the gold wire at room temperature, 20.0°C.)

25. The copper wire used in a house has a cross-sectional area of 3.00 mm^2. If 10.0 m of this wire is used to wire a circuit in the house at 20.0°C, find the resistance of the wire at temperatures of (a) 30.0°C and (b) 10.0°C.

26. A wire 3.00 m long and 0.450 mm^2 in cross-sectional area has a resistance of 41 Ω at 20°C. If its resistance increases to 41.4 Ω at 29.0°C, what is the temperature coefficient of resistivity?

27. A 100-cm-long copper wire 0.50 cm in radius has a potential difference across it sufficient to produce a current of 3.0 A at 20°C. (a) What is the potential difference? (b) If the temperature of the wire is increased to 200°C, what potential difference is now required to produce a current of 3.0 A?

28. In one form of plethysmograph (a device for measuring volume), a rubber capillary tube with an inside diameter of 1.00 mm is filled with mercury at 20°C. The resistance of the mercury is measured with the aid of electrodes sealed into the ends of the tube. If 100.00 cm of the

tube is wound in a spiral around a patient's upper arm, the blood flow during a heartbeat causes the arm to expand, stretching the tube to a length of 100.04 cm. From this observation (assuming cylindrical symmetry) you can find the change in volume of the arm, which gives an indication of blood flow. (a) Calculate the resistance of the mercury. (b) Calculate the fractional change in resistance during the heartbeat. (*Hint:* The fraction by which the cross-sectional area of the mercury thread decreases is the fraction by which the length increases, because the volume of mercury is constant.) Take $\rho_{Hg} = 9.4 \times 10^{-7}\ \Omega \cdot m$.

29. A platinum resistance thermometer has resistances of 200.0 Ω when placed in a 0°C ice bath and 253.8 Ω when immersed in a crucible containing melting potassium. What is the melting point of potassium? (*Hint:* First determine the resistance of the platinum resistance thermometer at room temperature, 20°C.)

Section 17.7   Electrical Energy and Power

30. Suppose that a voltage surge produces 140 V for a moment. By what percentage will the output of a 120-V, 100-W lightbulb increase, assuming its resistance does not change?

31. How many 100-W lightbulbs can you use in a 120-V circuit without tripping a 15-A circuit breaker? (The bulbs are connected in parallel.)

32. A high-voltage transmission line with a resistance of 0.31 Ω/km carries 1000 A, starting at 700 kV for a distance of 160 km. (a) What is the power loss due to resistance in the line? (b) What fraction of the transmitted power does this loss represent?

33. The power supplied to a typical black-and-white television set is 90 W when the set is connected to 120 V. (a) How much electric energy does this set consume in 1 hour? (b) A color television draws about 2.5 A when connected to 120 V. How much time is required for it to consume the same energy as the black-and-white model consumes in 1 hour?

34. The tungsten heating element in a 1500-W heater is 3.00 m long, and the resistor is to be connected to a 120-V source. What is the cross-sectional area of the wire? Assume a temperature of 20°C.

35. What is the required resistance of an immersion heater that will increase the temperature of 1.50 kg of water from 10.0°C to 50.0°C in 10.0 min while operating at 120 V?

36. A copper cable is designed to carry a current of 300 A with a power loss of 2.00 W/m. What is the required radius of this cable?

WEB 37. A small motor draws a current of 1.75 A from a 120-V line. The output power of the motor is 0.20 hp. (a) At a rate of $0.060/kWh, what is the cost of operating the motor for 4.0 h? (b) What is the efficiency of the motor?

38. How much does it cost to watch a complete 21-hour-long World Series on a 90.0-W black-and-white television set? Assume that electricity costs $0.0700/kWh.

39. A house is heated by a 24.0 kW electric furnace using resistance heating. The rate for electrical energy is $0.080/kWh. If the heating bill for January is $200, how long must the furnace have been running on an average January day?

40. A 110-V motor produces 2.50 hp of mechanical power. If this motor is 90.0% efficient in converting electrical power to mechanical power, find (a) the current drawn by the motor and (b) the total electrical energy (in both kWh and joules) used by this motor in running for one hour. (c) If electrical energy costs $0.080/kWh, what does it cost to run the motor for this hour?

41. An 11-W energy-efficient fluorescent lamp is designed to produce the same illumination as a conventional 40-W lamp. How much does the energy-efficient lamp save during 100 hours of use? Assume a cost of $0.080/kWh for electrical energy.

42. An electric resistance heater is to deliver 1500 kcal/h to a room using 110-V electricity. If fuses come in 10-A, 20-A, and 30-A sizes, what is the smallest fuse that can safely be used in the heater circuit?

43. The heating coil of a hot water heater has a resistance of 20 Ω and operates at 210 V. If electrical energy costs $0.080/kWh, what does it cost to raise the 200 kg of water in the tank from 15°C to 80°C? (See Chapter 11.)

ADDITIONAL PROBLEMS

44. A particular wire has a resistivity of $3.0 \times 10^{-8}\ \Omega \cdot m$ and a cross-sectional area of $4.0 \times 10^{-6}\ m^2$. A length of this wire is to be used as a resistor that will develop 48 W of power when connected across a 20-V battery. What length of wire is required?

45. A steam iron draws 6.0 A from a 120-V line. (a) How many joules of thermal energy are produced in 20 min? (b) How much does it cost, at $0.080/kWh, to run a steam iron for 20 min?

46. Storage batteries are often rated in terms of the amounts of charge they can deliver. How much charge can a 90-ampere-hour battery deliver?

47. Birds resting on high-voltage power lines are a common sight. The copper wire on which a bird stands is 2.2 cm in diameter and carries a current of 50 A. If the bird's feet are 4.0 cm apart, calculate the potential difference across its body.

48. A small sphere that carries a charge of 8.00 nC is whirled in a circle at the end of an insulating string. The angular speed is 100π rad/s. What average current does this rotating charge represent?

49. The current in a conductor varies in time as shown in Figure P17.49. (a) How many coulombs of charge pass through a cross section of the conductor in the interval $t = 0$ to $t = 5.0$ s? (b) What constant current would transport the same total charge during the 5.0-s interval as does the actual current?

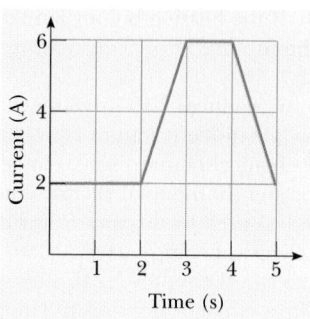

Figure P17.49

50. (a) A 115-g mass of aluminum is formed into a right circular cylinder, shaped so that its diameter equals its height. Calculate the resistance between the top and bottom faces of the cylinder at 20°C. (b) Calculate the resistance between opposite faces if the same mass of aluminum is formed into a cube.

51. A length of metal wire has a radius of 5.00×10^{-3} m and a resistance of 0.100 Ω. When the potential difference across the wire is 15.0 V, the electron drift speed is found to be 3.17×10^{-4} m/s. Based on these data, calculate the density of free electrons in the wire.

52. A 50.0-g sample of a conducting material is all that is available. The resistivity of the material is measured to be 11×10^{-8} Ω·m and the density is 7.86 g/cm³. The material is to be shaped into a solid cylindrical wire that has a total resistance of 1.5 Ω. (a) What length is required? (b) What must be the diameter of the wire?

53. A carbon wire and a nichrome wire are connected one after the other. If the combination has total resistance of 10.0 kΩ at 20°C, what is the resistance of each wire at 20°C so that the resistance of the combination does not change with temperature?

54. The temperature coefficients of resistivity in this chapter are at 20°C. What would they be at 0°C? (*Hint:* The temperature coefficient of resistivity at 20°C satisfies $R = R_0[1 + \alpha(T - T_0)]$ where R_0 is the resistance of the material at $T_0 = 20°C$. The temperature coefficient of resistivity, α', at 0°C must satisfy $R = R'(1 + \alpha'T)$, where R' is the resistance of the material at 0°C.)

55. (a) Determine the resistance of a lightbulb marked 100 W [at] 120 V. (b) Assuming that the filament is tungsten and has a cross-sectional area of 0.010 mm², determine the length of the wire inside the bulb when the bulb is turned on. (c) Why do you think the wire inside the bulb is tightly coiled? (d) If the temperature of the tungsten wire is 2600°C when the bulb is turned on, what is the length of the wire when the bulb is turned off and has cooled to 20°C? (See Chapter 10, and use 4.5×10^{-6}/°C as the coefficient of linear expansion for tungsten.)

56. In a certain stereo system, each speaker has a resistance of 4.00 Ω. The system is rated at 60.0 W in each channel. Each speaker circuit includes a fuse rated at a maximum current of 4.00 A. Is this system adequately protected against overload?

57. A resistor is constructed by forming a material of resistivity 3.5×10^5 Ω·m into the shape of a hollow cylinder of length 4.0 cm and inner and outer radii of 0.50 cm and 1.2 cm, respectively. In use, a potential difference is applied between the ends of the cylinder, producing a current parallel to the length of the cylinder. Find the resistance of the cylinder.

58. A wire of initial length L_0 and radius r_0 has a measured resistance of 1.0 Ω. The wire is drawn under tensile stress to a new uniform radius of $r = 0.25r_0$. What is the new resistance of the wire?

59. An x-ray tube used for cancer therapy operates at 4.0 MV, with a beam current of 25 mA striking the metal target. Nearly all the power in this beam is transferred to a stream of water flowing through holes drilled in the target. What rate of flow, in kilograms per second, is needed if the temperature rise (ΔT) of the water is not to exceed 50°C?

WEB 60. (a) A sheet of copper ($\rho = 1.7 \times 10^{-8}$ Ω·m) is 2.0 mm thick and has surface dimensions of 8.0 cm × 24 cm. If the long edges are joined to form a tube 24 cm in length, what is the resistance between the ends? (b) What mass of copper is required to manufacture a 1500-m-long spool of copper cable with a total resistance of 4.5 Ω?

61. When a straight wire is heated, its resistance changes according to the equation:

$$R = R_0[1 + \alpha(T - T_0)]$$

where α is the temperature coefficient of resistivity. (a) Show that a more precise result, which includes the fact that the length and area of a wire change when it is heated, is

$$R = \frac{R_0[1 + \alpha(T - T_0)][1 + \alpha'(T - T_0)]}{[1 + 2\alpha'(T - T_0)]}$$

where α' is the coefficient of linear expansion (see Chapter 10). (b) Compare these two results for a 2.00-m-long copper wire of radius 0.100 mm, starting at 20.0°C and heated to 100.0°C.

Direct Current Circuits

▲ PHYSICS PUZZLER

In a lightning storm, a lightning stroke often strikes the tallest object around. Thus, you may feel that it is safe to stand under a tree during a lightning storm, because the tree is taller than you are. This is *not* a good idea, however. Why not? *(Kent Wood/Peter Arnold, Inc.)*

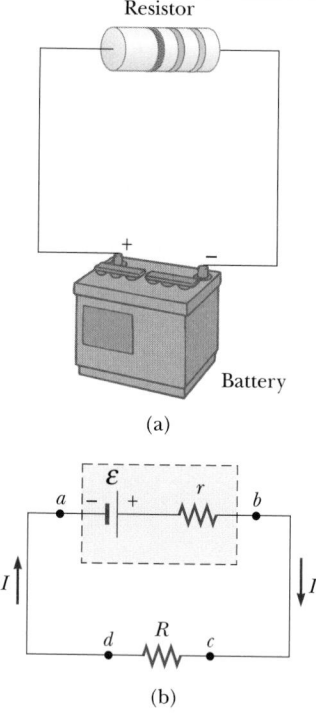

Resistor

+   −

Battery

(a)

$\mathcal{E}$

a −| |+ r b

I I

d R c

(b)

Figure 18.1   (a) A circuit consisting of a resistor connected to the terminals of a battery. (b) A circuit diagram of a source of emf, $\mathcal{E}$, of internal resistance r connected to an external resistor, R.

This chapter analyzes some simple circuits whose elements include batteries, resistors, and capacitors in varied combinations. Such analysis is simplified by the use of two rules known as Kirchhoff's rules, which follow from the principle of conservation of energy and the law of conservation of charge. Most of the circuits are assumed to be in *steady state,* which means that the currents are constant in magnitude and direction. We close the chapter with a discussion of circuits containing resistors and capacitors, in which current varies with time.

18.1   SOURCES OF emf

The source that maintains the constant current in a closed circuit is called a source of *emf.*[1] Any devices (such as batteries and generators) that increase the potential energy of charges circulating in circuits are sources of emf. One can think of such a source as a "charge pump" that forces electrons to move in a direction opposite the electrostatic field inside the source. The emf, $\mathcal{E}$, of a source is the work done per unit charge, and hence the SI unit of emf is the volt.

Consider the circuit in Figure 18.1a, consisting of a battery connected to a resistor. We assume that the connecting wires have no resistance. If we neglect the internal resistance of the battery, the potential drop across the battery (the terminal voltage) equals the emf of the battery. However, because a real battery always has some internal resistance, r, the terminal voltage is not equal to the emf. The circuit of Figure 18.1a can be described schematically by the diagram in Figure 18.1b. The battery, represented by the dashed rectangle, consists of a source of emf, $\mathcal{E}$, in series with an internal resistance, r. Now imagine a positive charge moving from a to b in Figure 18.1b. As the charge passes from the negative to the positive terminal of the battery, the potential of the charge increases by $\mathcal{E}$. However, as the charge moves through the resistance, r, its potential decreases by the amount Ir, where I is the current in the circuit. Thus, the terminal voltage of the battery, $\Delta V = V_b - V_a$, is

$$\Delta V = \mathcal{E} - Ir \qquad [18.1]$$

Note from this expression that **$\mathcal{E}$ is equal to the terminal voltage when the current is zero,** called the **open-circuit voltage.** By inspecting Figure 18.1b, we see that the terminal voltage, ΔV, must also equal the potential difference across the external resistance, R, often called the **load resistance;** that is, $\Delta V = IR$. Combining this with Equation 18.1, we see that

$$\mathcal{E} = IR + Ir \qquad [18.2]$$

Solving for the current gives

$$I = \frac{\mathcal{E}}{R + r}$$

This shows that the current in this simple circuit depends on both the resistance external to the battery and the internal resistance. If R is much greater than r, we can neglect r in our analysis and we do, for many circuits.

An assortment of batteries. *(Courtesy of Henry Leap and Jim Lehman)*

---

[1] The term was originally an abbreviation for *electromotive force,* but emf is not really a force, so the long form is discouraged.

If we multiply Equation 18.2 by the current, I, we get

$$I\mathcal{E} = I^2R + I^2r$$

This equation tells us that the total power output of the source of emf, $I\mathcal{E}$, is converted to power that is dissipated as joule heat in the load resistance, I^2R, *plus* power that is dissipated in the internal resistance, I^2r. Again, if $r \ll R$, most of the power delivered by the battery is transferred to the load resistance.

Unless otherwise stated, we will assume in our examples and end-of-chapter problems that the internal resistance of a battery in a circuit is negligible.

18.2   RESISTORS IN SERIES

The conservation laws are grand unifying principles that serve as bases for many outcomes in the study of physics. In particular, the next three sections of this chapter use two of these laws — the conservation of energy and the conservation of charge — as a framework. Rather than point out the places where these laws appear in our derivations, we will ask you, as an exercise, to identify their appearances yourself.

Figure 18.2 shows two resistors, R_1 and R_2, connected to a battery in a circuit called a series circuit, in which there is only one pathway for the current. Charges must pass through both resistors and the battery as they traverse the circuit. Hence, all charges in a series circuit must follow the same conducting path. **Note that the currents in all resistors in a series circuit are the same, because any charge that flows through R_1 must also flow through R_2.** This is analogous to water flowing through a pipe with two constrictions, corresponding to R_1 and R_2. Whatever volume of water flows in one end in a given time interval must exit the opposite end.

◀ For a series connection of resistors, the current is the same in all the resistors.

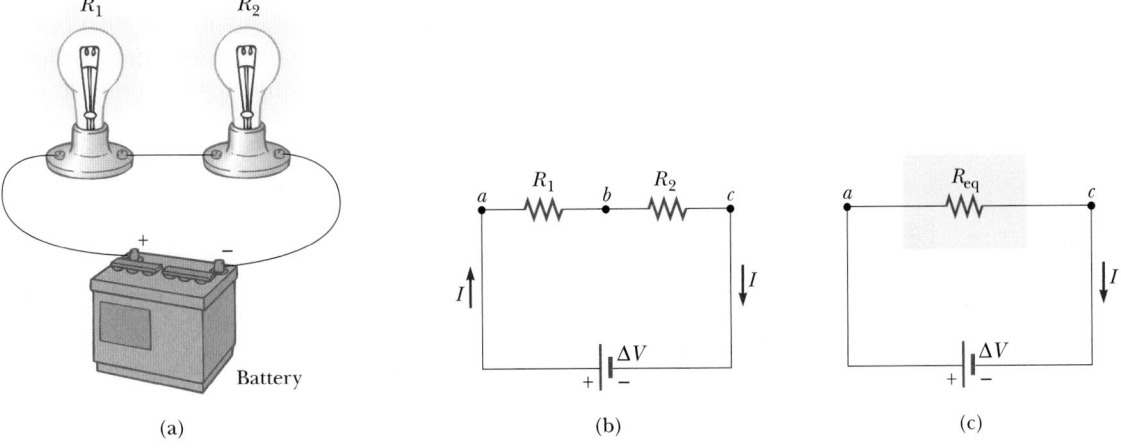

(a)   (b)   (c)

Figure 18.2 A series connection of two resistors, R_1 and R_2. The currents in the resistors are the same, and the equivalent resistance of the combination is given by $R_{eq} = R_1 + R_2$.

Because the potential drop from a to b in Figure 18.2b equals IR_1 and the potential drop from b to c equals IR_2, the potential drop from a to c is

$$\Delta V = IR_1 + IR_2$$

Regardless of how many resistors we have in series, the sum of the potential differences across the resistors is equal to the total potential difference across the combination. As we will show later, this is a consequence of the conservation of energy. Figure 18.2c shows an equivalent resistor, R_{eq}, that can replace the two resistors of the original circuit. Applying Ohm's law to this resistor, we have

$$\Delta V = IR_{eq}$$

Equating the preceding two expressions, we have

$$IR_{eq} = IR_1 + IR_2$$

or

$$R_{eq} = R_1 + R_2 \qquad \text{(series combination)} \qquad \text{[18.3]}$$

The resistance R_{eq} is equivalent to the series combination $R_1 + R_2$ in the sense that the circuit current is unchanged when R_{eq} replaces R_1 and R_2.

An extension of this analysis shows that the equivalent resistance of three or more resistors connected in series is

$$R_{eq} = R_1 + R_2 + R_3 + \ldots \qquad \text{(series combination)} \qquad \text{[18.4]}$$

From this, we see that **the equivalent resistance of a series combination of resistors is always greater than any individual resistance.**

Note that if the filament of one lightbulb in Figure 18.2 were to break, or "burn out," the circuit would no longer be complete (an open-circuit condition would exist) and the second bulb would also go out. Some Christmas-tree light sets (especially older ones) are connected in this way, and the task of determining which bulb is burned out is a tedious one.

In many circuits, fuses are used in series with other circuit elements for safety purposes. The conductor in the fuse is designed to melt and open the circuit at some maximum current, the value of which depends on the nature of the circuit. If a fuse were not used, excessive currents could damage circuit elements, overheat wires, and perhaps cause a fire. In modern home construction, circuit breakers are used in place of fuses. When the current in a circuit exceeds some value (typically 15 A), the circuit breaker acts as a switch and opens the circuit.

Applying Physics 1

APPLICATION

Christmas Lights in Series.

A new design for Christmas tree lights allows them to be connected in series. One might expect that a failed bulb in such a string would result in an open circuit, and all of the bulbs would go out. How can the bulbs be designed to prevent this from happening?

Explanation If the string of lights contained normal bulbs, a failed bulb would be hard to locate. Each bulb would have to be replaced with a good bulb, one by one, until the failed bulb is found. If there happened to be two failed bulbs in the string of lights, finding both of them would be a formidable task.

Christmas lights use specially designed bulbs that have an insulated loop of wire across the conducting supports to the bulb filaments. If the filament breaks and the bulb fails, the resistance of this bulb increases dramatically. As a result, most of the applied voltage appears across the loop of wire. This voltage causes the insulation around the loop of wire to burn, causing the metal wire to make electrical contact with the supports. This produces a conducting path through the bulb, enabling the other bulbs to remain lit.

EXAMPLE 18.1 Four Resistors in Series

Four resistors are arranged as shown in Figure 18.3a. Find (a) the equivalent resistance and (b) the current in the circuit if the emf of the battery is 6.0 V.

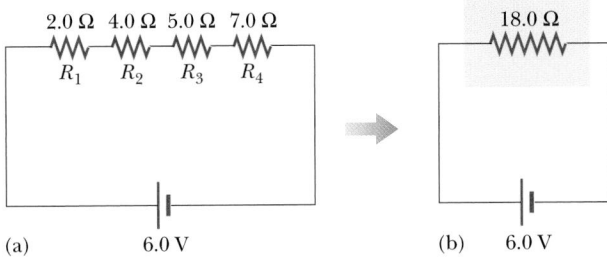

(a) 6.0 V   (b) 6.0 V

Figure 18.3 (Example 18.1) (a) Four resistors connected in series. (b) The equivalent resistance of the circuit in (a).

Solution (a) The equivalent resistance is found from Equation 18.4:

$$R_{eq} = R_1 + R_2 + R_3 + R_4 = 2.0\ \Omega + 4.0\ \Omega + 5.0\ \Omega + 7.0\ \Omega = 18.0\ \Omega$$

(b) If we apply Ohm's law to the equivalent resistor in Figure 18.3b, we find the current in the circuit to be

$$I = \frac{\Delta V}{R_{eq}} = \frac{6.0\ \text{V}}{18.0\ \Omega} = \tfrac{1}{3}\ \text{A}$$

Exercise Because the current in the equivalent resistor is $\tfrac{1}{3}$ A, this must also be the current in each resistor of the original circuit. Find the voltage drop across each resistor.

Answer $\Delta V_{2\Omega} = \tfrac{2}{3}$ V, $\Delta V_{4\Omega} = \tfrac{4}{3}$ V, $\Delta V_{5\Omega} = \tfrac{5}{3}$ V, $\Delta V_{7\Omega} = \tfrac{7}{3}$ V

18.3 RESISTORS IN PARALLEL

Now consider two resistors connected in parallel, as in Figure 18.4. **When resistors are connected in parallel, the potential differences across them are the same.** This must be true because the left sides of the resistors are connected to a common point, the positive side of the battery, point a in Figure 18.4b, and the right sides are connected to a common point, the negative terminal of the battery in Figure 18.4b.

13.4, SECTION 1

Figure 18.4   (a) A parallel connection of two resistors. (b) A circuit diagram for the parallel combination. (c) The voltages across the
resistors are the same, and the
equivalent resistance of the combination is given by the reciprocal relationship $1/R_{eq} = 1/R_1 + 1/R_2$.

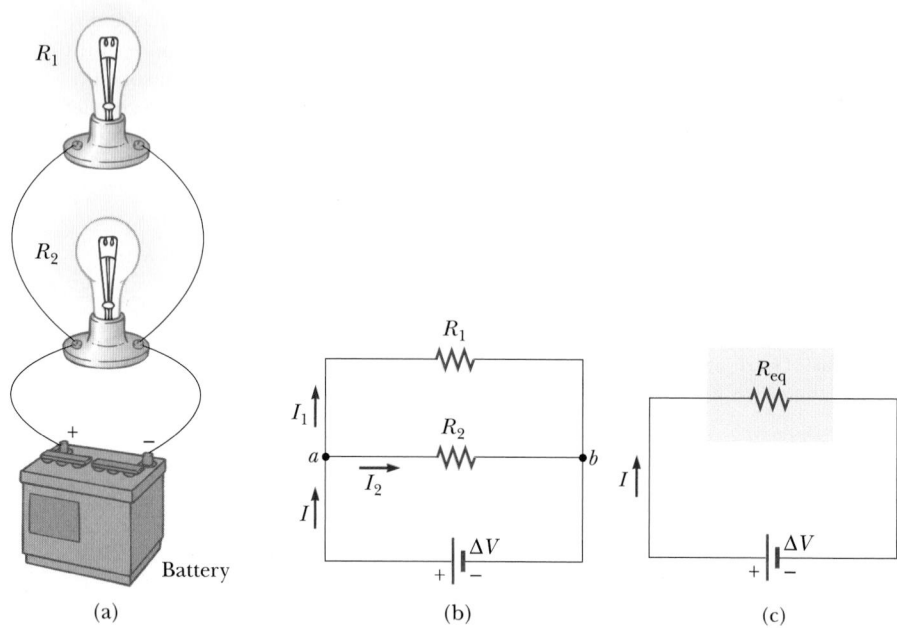

(a)   (b)   (c)

The currents, however, are generally not the same. They are the same only if
the resistors have the same resistance. When the current, I, reaches point a (called
a junction) in Figure 18.4b, it splits into two parts, I_1 going through R_1 and I_2 going
through R_2. If R_1 is greater than R_2, then I_1 is less than I_2. That is, the charge tends
to follow the path of least resistance. **Because charge is conserved, the current,
I, that enters point a must equal the total current leaving that point, $I_1 + I_2$.**
That is,

$$I = I_1 + I_2$$

The potential drop must be the same for the two resistors and must also equal the
potential drop across the battery. Ohm's law applied to each resistor gives

$$I_1 = \frac{\Delta V}{R_1} \qquad I_2 = \frac{\Delta V}{R_2}$$

Ohm's law applied to the equivalent resistor in Figure 18.4c gives

$$I = \frac{\Delta V}{R_{eq}}$$

When these expressions for the current are substituted into the equation
$I = I_1 + I_2$, and ΔV is cancelled, we obtain

$$\frac{1}{R_{eq}} = \frac{1}{R_1} + \frac{1}{R_2} \qquad \text{(parallel combination)} \qquad [18.5]$$

An extension of this analysis to three or more resistors in parallel produces the
following general expression for the equivalent resistance:

$$\frac{1}{R_{eq}} = \frac{1}{R_1} + \frac{1}{R_2} + \frac{1}{R_3} + \cdots \qquad \text{(parallel combination)} \qquad [18.6]$$

From this it can be shown that **the equivalent resistance of two or more resistors connected in parallel is always less than the smallest resistance in the group.**

Household circuits are always wired so that the lightbulbs (or appliances, or whatever) are connected in parallel, as in Figure 18.4a. In this manner, each device operates independently of the others, so that if one is switched off, the others remain on. Equally important, each device gets the same voltage.

Finally, it is interesting to note that parallel resistors combine in the same way series capacitors combine.

Problem-Solving Strategy

Resistors

1. When two or more unequal resistors are connected in *series,* they carry the same current, but the potential differences across them are not the same. The resistors add directly to give the equivalent resistance of the series combination.

2. When two or more unequal resistors are connected in *parallel,* the potential differences across them are the same. Because the current is inversely proportional to the resistance, the currents through them are not the same. The equivalent resistance of a parallel combination of resistors is found through reciprocal addition, and the equivalent resistance is always *less* than the smallest individual resistor in the combination.

3. A complicated circuit consisting of several resistors and batteries can often be reduced to a simple circuit with only one resistor. To do so, examine the initial circuit and replace any resistors in series or any in parallel using the procedures outlined in Steps 1 and 2. Sketch the new circuit after these changes have been made. Examine the new circuit and replace any series or parallel combinations. Continue this process until a single equivalent resistance is found.

4. If the current in or the potential difference across a resistor in the complicated circuit is to be identified, start with the final circuit found in Step 3 and gradually work back through the circuits, using $\Delta V = IR$ and the rules of Steps 1 and 2.

Thinking Physics 1

Predict the relative brightness of the four identical bulbs shown in Figure 18.5. What happens if bulb A "burns out," so that it cannot conduct current? What if C "burns out"? What if D "burns out"?

Explanation Bulbs A and B are connected in series across the emf of the battery, whereas bulb C is connected by itself across the emf. Thus, the emf is split between bulbs A and B. As a result, bulb C will be brighter than bulbs A and B, which should

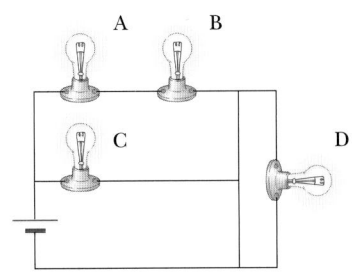

Figure 18.5 (Thinking Physics 1)

be equally as bright as each other. Bulb D has a resistanceless wire in parallel with it. There is no voltage drop across the wire, and because the wire is in parallel with bulb D, there also can be no voltage drop across it. Thus, bulb D does not glow at all. If bulb A "burns out," B goes out, but C stays lighted. If C "burns out," there is no effect on the other bulbs. If D "burns out" the event is undetectable, because D was not glowing anyway.

Applying Physics 2

Figure 18.6 illustrates how a three-way lightbulb is constructed to provide three levels of light intensity. The socket of the lamp is equipped with a three-way switch for selecting different light intensities. The bulb contains two filaments. Why are the filaments connected in parallel? Explain how the two filaments are used to provide three different light intensities.

Explanation If the filaments were connected in series and one of them were to burn out, no current could pass through the bulb, and the bulb would give no illumination, regardless of the switch position. However, when the filaments are connected in parallel and one of them (say the 75-W filament) burns out, the bulb will still operate in one of the switch positions as current passes through the other (100 W) filament. The three light intensities are made possible by selecting one of three values of filament resistance using a single value of 120 V for the applied voltage. The 75-W filament offers one value of resistance, the 100-W filament offers a second value, and the third resistance is obtained by combining the two filaments in parallel. When switch 1 is closed and switch 2 is opened, current passes only through the 75-W filament. When switch 1 is open and switch 2 is closed, current passes only through the 100-W filament. When both switches are closed, current passes through both filaments and a total illumination of 175 W is obtained.

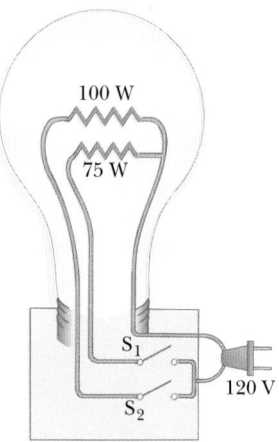

Figure 18.6 (Applying Physics 2)

EXAMPLE 18.2 Three Resistors in Parallel

Three resistors are connected in parallel, as in Figure 18.7. A potential difference of 18 V is maintained between points a and b.

(a) Find the current in each resistor.

Solution The resistors are in parallel, and so the potential difference across each is 18 V. Let us apply $\Delta V = IR$ to find the current in each resistor:

$$I_1 = \frac{\Delta V}{R_1} = \frac{18 \text{ V}}{3.0 \text{ }\Omega} = \boxed{6.0 \text{ A}}$$

$$I_2 = \frac{\Delta V}{R_2} = \frac{18 \text{ V}}{6.0 \text{ }\Omega} = \boxed{3.0 \text{ A}}$$

$$I_3 = \frac{\Delta V}{R_3} = \frac{18 \text{ V}}{9.0 \text{ }\Omega} = \boxed{2.0 \text{ A}}$$

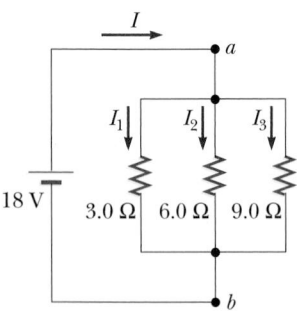

Figure 18.7 (Example 18.2) Three resistors connected in parallel. The voltage across each resistor is 18 V.

There is a saying that holds that current, like politicians, follows the path of least resistance. This is only partially correct, as our example demonstrates. Most of the current, 6.0 A, appears in the smallest resistance, the 3.0-Ω resistor, but still there is also current in the other resistors.

(b) Calculate the power dissipated by each resistor and the total power dissipated by the three resistors.

Solution Applying $P = I^2R$ to each resistor gives

$$3\ \Omega:\ P_1 = I_1^2R_1 = (6.0\ \text{A})^2(3.0\ \Omega) = \boxed{110\ \text{W}}$$

$$6\ \Omega:\ P_2 = I_2^2R_2 = (3.0\ \text{A})^2(6.0\ \Omega) = \boxed{54\ \text{W}}$$

$$9\ \Omega:\ P_3 = I_3^2R_3 = (2.0\ \text{A})^2(9.0\ \Omega) = \boxed{36\ \text{W}}$$

(Note that you can also use $P = (\Delta V)^2/R$ to find the power dissipated by each resistor.) Summing the three quantities gives a total power of 200 W.

Exercise Calculate the equivalent resistance of the three resistors, and from this result find the total power dissipated.

Answer $\frac{18}{11}$ Ω; 200 W

EXAMPLE 18.3 Equivalent Resistance

Four resistors are connected as shown in Figure 18.8a.
(a) Find the equivalent resistance between points a and c.

Solution The circuit can be reduced in steps, as shown in Figures 18.8b and 18.8c. The 8.0-Ω and 4.0-Ω resistors are in series, and so the equivalent resistance between a and b is 12 Ω (Eq. 18.4). The 6.0-Ω and 3.0-Ω resistors are in parallel, and so from Equation 18.6 we find that the equivalent resistance from b to c is 2.0 Ω. Hence, the equivalent resistance from a to c is $\boxed{14\ \Omega}$.

(b) What is the current in each resistor if a 42-V battery is placed between a and c?

Solution The current, I, is the same in the 8.0-Ω and 4.0-Ω resistors because they are in series. Using Ohm's law and the results of (a), we get

$$I = \frac{\Delta V_{ac}}{R_{eq}} = \frac{42\ \text{V}}{14\ \Omega} = 3.0\ \text{A}$$

When this current enters the junction at b, it splits; part of it passes through the 6.0-Ω resistor (I_1), and part passes through the 3.0-Ω resistor (I_2). Because the potential difference across these resistors, V_{bc}, is the *same* (they are in parallel), (6.0 Ω)I_1 = (3.0 Ω)I_2, or $I_2 = 2I_1$. Using this result and the fact that $I_1 + I_2 = 3.0$ A, we find that $I_1 = 1.0$ A and $I_2 = 2.0$ A. We could have guessed this from the start by noting that the current through the 3.0-Ω resistor has to be twice the current through the 6.0-Ω resistor in view of their relative resistances and the fact that the same voltage is applied to both.

As a final check, note that $\Delta V_{bc} = (6.0\ \Omega)I_1 = (3.0\ \Omega)I_2 = 6.0$ V and $\Delta V_{ab} = (12.0\ \Omega)I_1 = 36$ V; therefore, $\Delta V_{ac} = \Delta V_{ab} + \Delta V_{bc} = 42$ V, as expected.

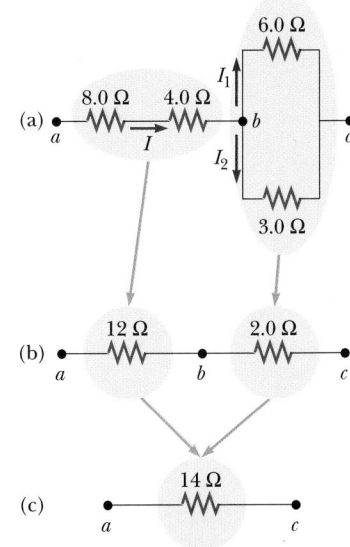

Figure 18.8 (Example 18.3) The four resistors shown in (a) can be reduced in steps to an equivalent 14-Ω resistor.

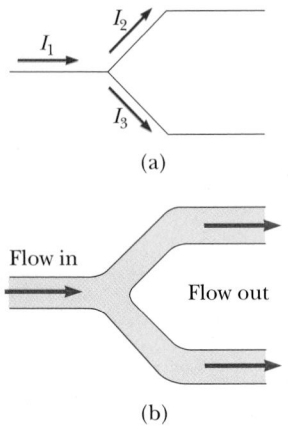

Figure 18.9 (a) A schematic diagram illustrating Kirchhoff's junction rule. Conservation of charge requires that whatever current enters a junction must leave that junction. Therefore, in this case, $I_1 = I_2 + I_3$. (b) A mechanical analog of the junction rule: The net flow out must equal the net flow in.

13.4, SECTION 2

Figure 18.10 Rules for determining the potential changes across a resistor and a battery, assuming the battery has no internal resistance.

18.4 KIRCHHOFF'S RULES AND COMPLEX DC CIRCUITS

As demonstrated in the preceding section, we can analyze simple circuits using Ohm's law and the rules for series and parallel combinations of resistors. However, there are many ways in which resistors can be connected so that the circuits formed cannot be reduced to a single equivalent resistor. The procedure for analyzing more complex circuits is greatly simplified by the use of two simple rules called **Kirchhoff's rules:**

> 1. The sum of the currents entering any junction must equal the sum of the currents leaving that junction. (This rule is often referred to as the **junction rule.**)
> 2. The sum of the potential differences across all the elements around any closed-circuit loop must be zero. (This rule is usually called the **loop rule.**)

The junction rule is a statement of *conservation of charge.* Whatever current enters a given point in a circuit must leave that point because charge cannot build up or disappear at a point. If we apply this rule to the junction in Figure 18.9a, we get

$$I_1 = I_2 + I_3$$

Figure 18.9b represents a mechanical analog to this situation, in which water flows through a branched pipe with no leaks. The flow rate into the pipe equals the total flow rate out of the two branches.

The loop rule is equivalent to the principle of *conservation of energy.* Any charge that moves around any closed loop in a circuit (starting and ending at the same point) must gain as much energy as it loses. It gains energy as it is pumped through a source of emf. Its energy may decrease in the form of a potential drop, $-IR$, across a resistor or as a result of flowing backward through a source of emf—that is, from the positive to the negative terminal inside the battery. In the latter case, electrical energy is converted to chemical energy as the battery is charged.

When applying Kirchhoff's rules, you must make two decisions at the beginning of the problem.

1. You must assign symbols and directions to the currents in all branches of the circuit. If you should happen to guess the wrong direction for a current, the end result for that current will be negative but its magnitude will be correct.
2. When applying the loop rule, you must choose a direction (clockwise or counterclockwise) for going around the loop. As you traverse the loop, record voltage drops and rises according to the rules stated below. They are summarized in Figure 18.10, where it is assumed that movement is from point a toward point b:
   (a) If a resistor is traversed in the direction of the current, the change in electric potential across the resistor is $-IR$ (Fig. 18.10a).
   (b) If a resistor is traversed in the direction opposite the current, the change in electric potential across the resistor is $+IR$ (Fig. 18.10b).
   (c) If a source of emf is traversed in the direction of the emf (from $-$ to $+$ on the terminals), the change in electric potential is $+\mathcal{E}$ (Fig. 18.10c).

(d) If a source of emf is traversed in the direction opposite the emf (from + to − on the terminals), the change in electric potential is $-\mathcal{E}$ (Fig. 18.10d).

There are limits to the numbers of times the junction rule and the loop rule can be used. You can use the junction rule as often as needed so long as, each time you write an equation, you include in it a current that has not been used in a previous junction-rule equation. (If this procedure is not followed, a new equation will not be produced.) In general, the number of times the junction rule can be used is one fewer than the number of junction points in the circuit. The loop rule can be used as often as needed so long as a new circuit element (resistor or battery) or a new current appears in each new equation. In general, **to solve a particular circuit problem, you need as many independent equations as you have unknowns.**

Gustav Kirchhoff, German physicist (1824–1887)

Kirchhoff, a professor at Heidelberg, and Robert Bunsen invented the spectroscope and founded the science of spectroscopy, which we study in Chapter 28. They discovered the elements cesium and rubidium and invented astronomical spectroscopy. Kirchhoff formulated another Kirchhoff's rule, namely, "a cool substance will absorb light of the same wavelengths that it emits when hot." *(AIP ESVA, W. F. Meggers Collection)*

Problem-Solving Strategy

Kirchhoff's Rules

1. First, draw the circuit diagram and assign labels and symbols to all the known and unknown quantities. You must assign *directions* to the currents in each part of the circuit. Do not be alarmed if you guess the direction of a current incorrectly; the resulting value will be negative, but *its magnitude will be correct.* Although the assignment of current directions is arbitrary, you must stick with it throughout as you apply Kirchhoff's rules.
2. Apply the junction rule to any junction in the circuit. The junction rule may be applied as many times as a new current (one not used in a previous application) appears in the resulting equation.
3. Now apply Kirchhoff's loop rule to as many loops in the circuit as are needed to solve for the unknowns. In order to apply this rule, you must correctly identify the change in electric potential as you cross each element in traversing the closed loop. Watch out for signs!
4. Solve the equations simultaneously for the unknown quantities. Be careful in your algebraic steps, and check your numerical answers for consistency.

EXAMPLE 18.4 Applying Kirchhoff's Rules

Find the currents in the circuit shown in Figure 18.11.

Reasoning There are three unknown currents in this circuit, and so we must obtain three independent equations. We can find the equations with one application of the junction rule and two applications of the loop rule.

Solution The first step is to assign a current to each branch of the circuit; these are our unknowns and are labeled I_1, I_2, and I_3 in Figure 18.11. It is also necessary to guess directions for the currents. Your experience with circuits such as this should tell you that the directions of all three have been chosen correctly. (However, recall that if a current

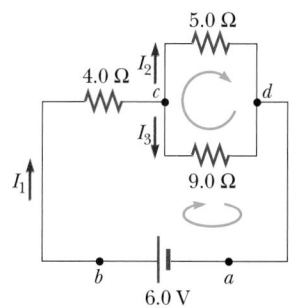

Figure 18.11 (Example 18.4) A multiloop circuit.

direction is chosen incorrectly, the numerical answer will turn out negative, but the magnitude will be correct. This point will be demonstrated in the next example.)

We now apply Kirchhoff's rules. First we can apply the junction rule using either c or d, the only two junctions in the circuit. Let us choose junction c. The net current into this junction is I_1, and the net current leaving it is $I_2 + I_3$. Thus, the junction rule applied to c gives

$$I_1 = I_2 + I_3$$

Recall that you may apply the junction rule over and over until you reach a situation in which no new currents appear in an equation. In this example we have reached that point with one application. If we apply the junction rule at d, we find that $I_1 = I_2 + I_3$—exactly the same equation.

We have three unknowns in our problem, I_1, I_2, and I_3; thus, we need two more independent equations before we can find a solution. We obtain these equations by applying the loop rule to the two loops indicated in the figure. Note that there are actually three loops in the circuit, but these two are sufficient to complete the problem. (Where is the loop that we do not use?)

When applying the loop rule, we must first choose the loops to be traversed, and then the directions in which to traverse them. We have selected the two loops indicated in the figure and have decided to traverse both of them clockwise. Other choices could be made, but the final result would be the same.

Starting at point a and moving clockwise around the large loop, we encounter the following voltage changes (see Fig. 18.10 for the basic rules):

- From a to b, we encounter a voltage change of 6.0 V.
- From b to c through the 4-Ω resistor, we encounter a voltage change of $-(4\,\Omega)I_1$.
- From c to d through the 9-Ω resistor, we encounter a voltage change of $-(9\,\Omega)I_3$.

No voltage change occurs from d back to a. Now that we have made a complete traversal of the loop, we can equate the sum of the voltage changes to zero:

$$6\text{ V} - (4\,\Omega)I_1 - (9\,\Omega)I_3 = 0$$

Moving clockwise around the small loop from point c, we encounter the following:

- From c to d through the 5.0-Ω resistor, a voltage change of $-(5\,\Omega)I_2$
- From d to c through the 9.0-Ω resistor, a voltage change of $+(9\,\Omega)I_3$

We find that

$$-(5\,\Omega)I_2 + (9\,\Omega)I_3 = 0$$

Thus, we have the following three equations to be solved for the three unknowns:

$$I_1 = I_2 + I_3$$

$$6\text{ V} - (4\,\Omega)I_1 - (9\,\Omega)I_3 = 0$$

$$-(5\,\Omega)I_2 + (9\,\Omega)I_3 = 0$$

If you need help in solving three equations with three unknowns, see Example 18.5. You should be able to obtain the following answers:

$$I_1 = \quad 0.83\text{ A} \qquad I_2 = \quad 0.53\text{ A} \qquad I_3 = \quad 0.30\text{ A}$$

Exercise   Solve this same problem by using the methods learned earlier for series and parallel combinations of resistors. First find the equivalent resistance of the circuit, which you can then use to obtain I_1.

QUICKLAB

Connect one terminal of a D-cell battery to the case of a flashlight bulb using insulated wire, tape a second wire to the other battery terminal, and tape a third wire to the center conductor of the bulb as in the top figure below. Make sure to remove about 1 cm of insulation from the ends of all wires before making connections. Connect the two open wires together to complete the circuit, and note the illumination of the bulb. Now add a second D-cell battery to the circuit as in the lower figure to give a total voltage of 3.0 V, connect the two open wires together to complete the circuit, and note the illumination of the bulb. Why does the bulb glow brighter in this case?

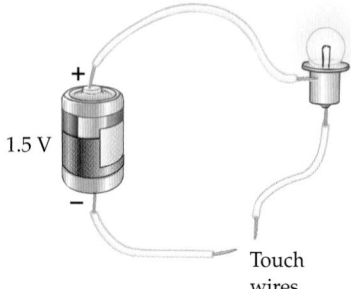

1.5 V

Touch wires

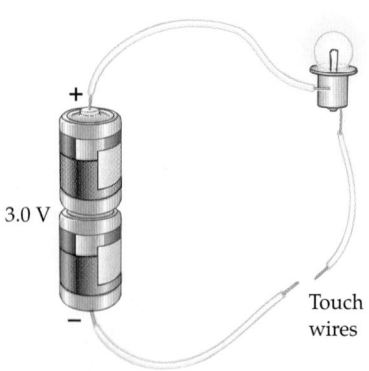

3.0 V

Touch wires

EXAMPLE 18.5 Another Application of Kirchhoff's Rules

Find I_1, I_2, and I_3 in Figure 18.12.

Reasoning To find the three unknown currents, we apply the junction rule once and the loop rule twice.

Solution We choose the directions of the currents as shown in the figure. Applying Kirchhoff's first rule to junction c gives

$$(1) \qquad I_1 + I_2 = I_3$$

The circuit has three loops: *abcda*, *befcb*, and *aefda*. We need only two loop equations to determine the unknown currents. The third loop equation would give no new information. Applying Kirchhoff's second rule to loops *abcda* and *befcb* and traversing these loops clockwise, we obtain the following expressions:

$$(2) \qquad \text{Loop } abcda: \qquad 10\text{ V} - (6\text{ }\Omega)I_1 - (2\text{ }\Omega)I_3 = 0$$

$$(3) \qquad \text{Loop } befcb: \qquad -14\text{ V} + (6\text{ }\Omega)I_1 - 10\text{ V} - (4\text{ }\Omega)I_2 = 0$$

Note that in loop *befcb*, a positive sign is obtained when the 6-Ω resistor is traversed, because the direction of the path is opposite the direction of the current I_1. A third loop equation for *aefda* gives $-14\text{ V} - (2\text{ }\Omega)I_3 - (4\text{ }\Omega)I_2 = 0$, which is just the sum of (2) and (3). Expressions (1), (2), and (3) represent three linear, independent equations with three unknowns.

We can solve the problem as follows: Substitution of (1) into (2) gives, with units ignored for the moment,

$$10 - 6I_1 - 2(I_1 + I_2) = 0$$

$$(4) \qquad 10 = 8I_1 + 2I_2$$

Dividing each term in (3) by 2 and rearranging the equation gives

$$(5) \qquad -12 = -3I_1 + 2I_2$$

Subtracting (5) from (4) eliminates I_2, giving

$$22 = 11I_1$$

$$I_1 = 2.0\text{ A}$$

Using this value of I_1 in (5) yields a value for I_2:

$$2I_2 = 3I_1 - 12 = 3(2) - 12 = -6$$

$$I_2 = -3.0\text{ A}$$

Finally, $I_3 = I_1 + I_2 = -1$ A. Hence, the currents have the values

$$I_1 = \boxed{2.0\text{ A}} \qquad I_2 = \boxed{-3.0\text{ A}} \qquad I_3 = \boxed{-1.0\text{ A}}$$

The fact that I_2 and I_3 are both negative indicates only that we chose the wrong directions for these currents. The numerical values are correct.

Exercise Find the potential difference between junctions b and c.

Answer $V_b - V_c = 2.0$ V

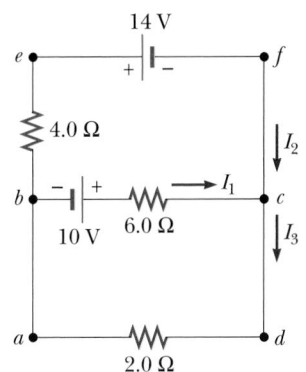

Figure 18.12 (Example 18.5) A circuit containing three loops.

Workbook Problem 83 – (Workbook page 179) provides additional practice at applying Kirchhoff's rules.

Voltages, currents, and resistances are frequently measured by digital multimeters like this one. *(Courtesy of Henry Leap and Jim Lehman)*

18.5   RC CIRCUITS

So far, we have been concerned with circuits with constant currents. We now consider direct-current circuits containing capacitors in which the currents vary with time. Consider the series circuit in Figure 18.13a. Let us assume that the capacitor is initially uncharged with the switch opened. After the switch is closed, the battery begins to charge the plates of the capacitor and a current passes through the resistor. The charging process continues until the capacitor is charged to its maximum equilibrium value, $Q = C\mathcal{E}$, where $\mathcal{E}$ is the maximum voltage across the capacitor. Once the capacitor is fully charged, the current in the circuit is zero. If we assume that the capacitor is uncharged before the switch is closed, and the switch is closed at $t = 0$, we find that the charge on the capacitor varies with time according to the expression

$$q = Q(1 - e^{-t/RC})$$ [18.7]

where $e = 2.718 \ldots$ is Euler's constant, the base of the natural logarithms. Figure 18.13b is a graph of this expression. Note that the charge is zero at $t = 0$ and approaches its maximum value, Q, as t approaches infinity. The voltage, ΔV, across the capacitor at any time is obtained by dividing the charge by the capacitance. That is, $\Delta V = q/C$.

As you can see from Equation 18.7, it takes an infinite amount of time for the capacitor to become fully charged. The term RC that appears in Equation 18.7, called the **time constant,** τ (Greek letter tau), is

$$\tau = RC$$ [18.8]

The time constant represents the time required for the charge to increase from zero to 63.2% of its maximum equilibrium value. That is, in one time constant, the charge on the capacitor increases from zero to $0.632Q$. This can be seen by substituting $t = \tau = RC$ in Equation 18.7 and solving for q. (Note that $e^{-1} = 1/e = 0.367$.) It is important to note that a capacitor charges very slowly in a circuit with a long time constant, whereas it charges very rapidly in a circuit with a short time constant.

Now consider the circuit in Figure 18.14a, consisting of a capacitor with an initial charge of Q, a resistor, and a switch. Before the switch is closed, the potential difference across the charged capacitor is Q/C. Once the switch is closed, the charge begins to flow through the resistor from one capacitor plate to the other until the capacitor is fully discharged. If the switch is closed at $t = 0$, it can

This versatile circuit enables the experimenter to examine the properties of circuit elements such as capacitors and resistors and their effects on circuit behavior. *(Courtesy of Central Scientific Company)*

Figure 18.13   (a) A capacitor in series with a resistor, a battery, and a switch. (b) A plot of the charge on the capacitor versus time after the switch for the circuit is closed. After one time constant, τ, the charge is 63% of the maximum value, $C\mathcal{E}$. The charge approaches its maximum value as t approaches infinity.

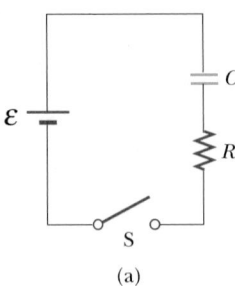

(a)

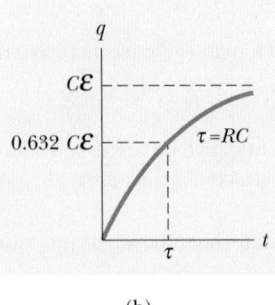

(b)

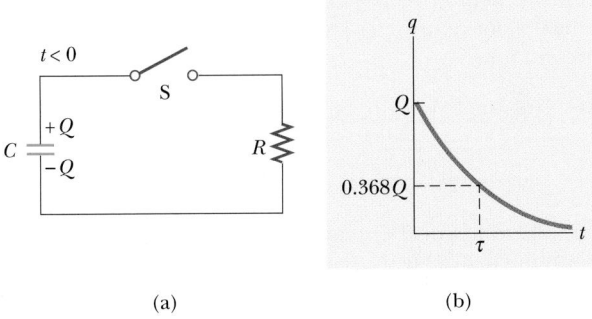

(a)   (b)

Figure 18.14 (a) A charged capacitor connected to a resistor and a switch. (b) A graph of the charge on the capacitor versus time after the switch is closed.

be shown that the charge, q, on the capacitor varies with time according to the expression

$$q = Qe^{-t/RC}$$ [18.9]

That is, the charge decreases exponentially with time, as shown in Figure 18.14b. In the interval $t = \tau = RC$, the charge decreases from its initial value, Q, to $0.368Q$. In other words, in one time constant, the capacitor loses 63.2% of its initial charge. Because $\Delta V = q/C$, we see that the voltage across the capacitor also decreases exponentially with time according to the expression $\Delta V = \mathcal{E}e^{-t/RC}$, where $\mathcal{E}$ (which equals Q/C) is the initial voltage across the fully charged capacitor.

Applying Physics 3

Many automobiles are equipped with windshield wipers that can be used intermittently during a light rainfall. How does the operation of this feature depend on the charging and discharging of a capacitor?

Explanation The wipers are part of an *RC* circuit, whose time constant can be varied by selecting different values of R through a multiposition switch. The brief time that the wipers remain on and the time they are off are determined by the value of the time constant of the circuit.

APPLICATION

Timed Windshield Wipers.

Applying Physics 4

In biological applications concerned with population growths, an equation is used that is similar to the exponential equations encountered in the analysis of *RC* circuits. It is

$$N_f = N_i 2^n$$

where N_f is the number of bacteria present at the beginning of an interval, N_i is the number present initially, and n is the number of growth cycles or doubling times. Doubling times vary according to the organism. The doubling time for the bacteria responsible for leprosy is about 30 days, and that for the salmonella bacteria responsible for food poisoning is about 20 minutes. For those of you with a biology inclination,

APPLICATION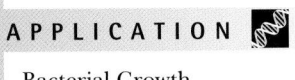

Bacterial Growth.

consider that only 10 salmonella bacteria find their way onto a turkey leg after your Thanksgiving meal. Four hours later you come back for a midnight snack. How many bacteria are present now?

Explanation   The number of doubling times is 240 min/20 min = 12. Thus, we have

$$N_f = N_i 2^n = (10 \text{ bacteria})(2^{12}) = 40\ 960 \text{ bacteria}$$

So, your system will have to deal with an invading host of about 41 000 bacteria, which are going to continue to double in a very promising environment.

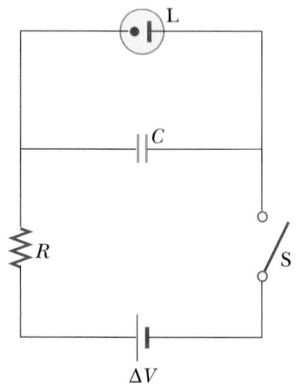

Figure 18.15
(Applying Physics 5)

Applying Physics 5

Many roadway construction sites have flashing yellow lights to warn motorists of possible dangers. A typical circuit for such a flasher is shown in Figure 18.15. The lamp L is a gas-filled lamp that acts as an open circuit until a large potential difference causes a discharge, which gives off a bright light. How does this arrangement produce a flashing signal?

Explanation   When the switch is closed, the battery charges up the capacitor. At first, the current is high and the charge on the capacitor is low, so that most of the potential difference appears across the resistor. As the capacitor charges, the potential difference across it increases, and the current and potential difference across the resistor are lowered. Eventually, the potential difference across the capacitor reaches a value at which the lamp will conduct electricity, causing a flash. The period between flashes can be adjusted by changing the time constant of the RC circuit.

EXAMPLE 18.6   Charging a Capacitor in an RC Circuit

An uncharged capacitor and a resistor are connected in series to a battery, as in Figure 18.13a. If $\mathcal{E} = 12$ V, $C = 5.0\ \mu$F, and $R = 8.0 \times 10^5\ \Omega$, find the time constant of the circuit, the maximum charge on the capacitor, and the charge on the capacitor after one time constant.

Solution   The time constant of the circuit is

$$\tau = RC = (8.0 \times 10^5\ \Omega)(5.0 \times 10^{-6}\ \text{F}) = \boxed{4.0 \text{ s}}$$

The maximum charge on the capacitor is

$$Q = C\mathcal{E} = (5.0 \times 10^{-6}\ \text{F})(12 \text{ V}) = \boxed{60\ \mu\text{C}}$$

After one time constant, the charge on the capacitor is 63.2% of its maximum value:

$$q = 0.632Q = 0.632(60 \times 10^{-6}\ \text{C}) = \boxed{38\ \mu\text{C}}$$

Exercise   Find the charge on the capacitor and the voltage across the capacitor after time t has elapsed.

Answer $q = (60\ \mu\text{C})(1 - e^{-t/4});\ \Delta V = (12 \text{ V})(1 - e^{-t/4})$

EXAMPLE 18.7 Discharging a Capacitor in an *RC* Circuit

Consider a capacitor, *C*, being discharged through a resistor, *R*, as in Figure 18.14a. After how many time constants does the charge on the capacitor drop to one fourth of its initial value?

Solution The charge on the capacitor varies with time according to Equation 18.9:

$$q(t) = Qe^{-t/RC}$$

where *Q* is the initial charge on the capacitor. To find the time it takes the charge *q* to drop to one fourth of its initial value, we substitute $q(t) = Q/4$ into this expression and solve for *t*:

$$\tfrac{1}{4}Q = Qe^{-t/RC}$$
$$\tfrac{1}{4} = e^{-t/RC}$$

Taking logarithms of both sides, we find that

$$-\ln 4 = -\frac{t}{RC}$$

$$t = RC \ln 4 = \boxed{1.39RC}$$

Exercise If $R = 8.0 \times 10^5 \ \Omega$, $C = 5.0 \ \mu F$, and the initial voltage across the capacitor is 6.0 V, what is the voltage across the capacitor after time *t* has elapsed?

Answer $\Delta V = (6.0 \ \text{V}) e^{-t/4}$

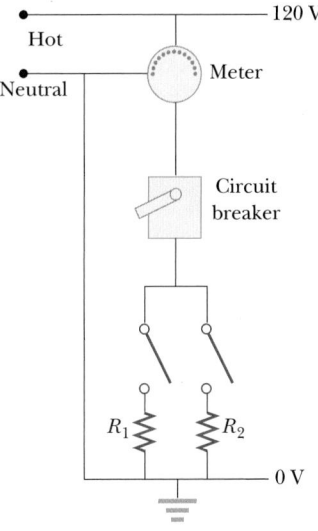

Figure 18.16 A wiring diagram for a household circuit. The resistances R_1 and R_2 represent appliances or other electrical devices that operate at an applied voltage of 120 V.

18.6 HOUSEHOLD CIRCUITS

Optional Section

Household circuits are a very practical application of some of the ideas presented in this chapter. In a typical installation, the utility company distributes electric power to individual houses with a pair of wires, or power lines. Electrical devices in a house are then connected in parallel to these lines, as shown in Figure 18.16. The potential drop between the two wires is about 120 V. (These are actually alternating currents and voltages, but for the present discussion we shall assume that they are direct currents and voltages.) One of the wires is connected to ground, and the other wire, sometimes called the "hot" wire, has a potential of 120 V. A meter and a circuit breaker or fuse are connected in series with the wire entering the house, as indicated in Figure 18.16. Figure 18.17 is a cutaway view of a fuse. The fuse is a small metallic strip that melts if the current exceeds a certain value. If a circuit did not include a fuse, excessive currents could damage circuit elements, overheat wires, and perhaps cause a fire.

In modern homes, circuit breakers are used in place of fuses. When the current in a circuit exceeds some value (typically 15 A), the circuit breaker acts as a switch and opens the circuit. Figure 18.18 is one design for a circuit breaker. Current passes through a bimetallic strip, the top of which bends to the left when excessive current heats it. If the strip bends far enough to the left, it settles into a groove in the spring-loaded metal bar. When this occurs, the bar drops enough to open the circuit at the contact point. The bar also flips a switch that indicates that the circuit

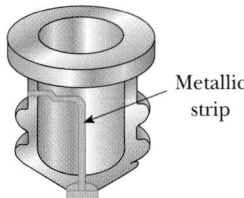

Figure 18.17 A cutaway view of a fuse.

APPLICATION

Fuses and Circuit Breakers.

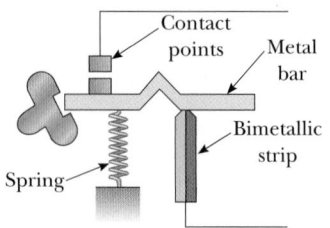

Figure 18.18   A circuit breaker that uses a bimetallic strip for its operation.

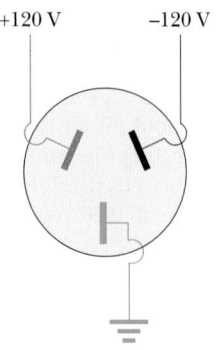

Figure 18.19   Power connections for a 240-V appliance.

breaker is not operational. (After the overload is removed, the switch can be flipped back on.) Circuit breakers based on this design have the disadvantage that some time is required for the heating of the strip, and thus the circuit is not opened rapidly enough when it is overloaded. As a consequence, many circuit breakers are now designed to use electromagnets, which we shall discuss in Chapter 19.

The wire and circuit breaker are carefully selected to meet the current demands of a circuit. If the circuit is to carry currents as large as 30 A, a heavy-duty wire and appropriate circuit breaker must be used. Household circuits that are normally used to power lamps and small appliances often require only 15 A. Each circuit has its own circuit breaker to accommodate its maximum safe load.

As an example, consider a circuit that powers a toaster, a microwave oven, and a heater (represented by R_1, R_2, . . . in Fig. 18.16). We can calculate the current through each appliance using the equation $P = I \Delta V$. The toaster, rated at 1000 W, draws a current of $1000/120 = 8.33$ A. The microwave oven, rated at 800 W, draws a current of 6.67 A, and the heater, rated at 1300 W, draws a current of 10.8 A. If the three appliances are operated simultaneously, they draw a total current of 25.8 A. Therefore, the breaker should be able to handle at least this much current, or else it will be tripped. As an alternative, one could operate the toaster and microwave oven on one 15-A circuit and the heater on a separate 15-A circuit.

Many heavy-duty appliances, such as electric ranges and clothes dryers, require 240 V to operate. The power company supplies this voltage by providing, in addition to a live wire that is 120 V above-ground potential, a wire, also considered live, that is 120 V below-ground potential (Fig. 18.19). Therefore, the potential drop across the two live wires is 240 V. An appliance operating from a 240-V line requires half the current of one operating from a 120-V line; therefore, smaller wires can be used in the higher voltage circuit without becoming overheated.

18.7   ELECTRICAL SAFETY

A person can be electrocuted by touching a live wire (which commonly is live because of a frayed cord and exposed conductors) while in contact with ground. The ground contact might be made by touching a water pipe (which is normally at ground potential) or by standing on the ground with wet feet, because impure water is a good conductor. Obviously such situations should be avoided at all costs.

Electric shock can result in fatal burns, or it can cause the muscles of vital organs, such as the heart, to malfunction. The degree of damage to the body depends on the magnitude of the current, the length of time it acts, and the part of the body through which it passes. Currents of 5 mA or less can cause a sensation of shock but ordinarily do little or no damage. If the current is larger than about 10 mA, the hand muscles contract and the person may be unable to let go of the live wire. If a current of about 100 mA passes through the body for just a few seconds, it can be fatal. Such large currents paralyze the respiratory muscles. In some cases, currents of about 1 A through the body produce serious (and sometimes fatal) burns.

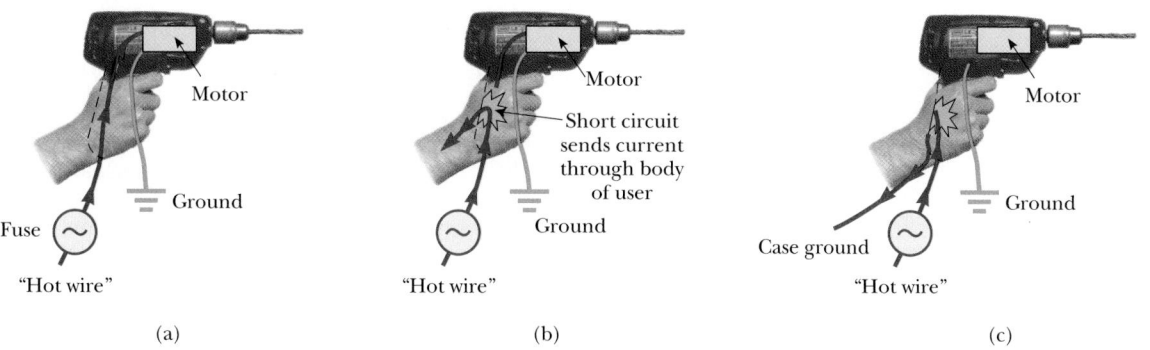

Figure 18.20 (a) When the drill is operated with two wires, the "hot wire," at 120 V, is always fused on this side of the circuit for safety. (b) A two-wire connection is potentially dangerous. If the high-voltage side comes in contact with the drill case, the person holding the drill receives an electrical shock. (c) Shock can be prevented by a third wire running from the drill case to the ground.

As an additional safety feature for consumers, electrical equipment manufacturers now use electrical cords that have a third wire, called a case ground. To understand how this works, consider the drill being used in Figure 18.20. Figure 18.20a shows a two-wire device that has one wire, called the "hot" wire, connected to the high-potential (120-V) side of the input power line, and the second wire is connected to ground (0 V). Under normal operating conditions, the path of the current through the drill is like that shown in Figure 18.20a. However, if the high-voltage wire comes in contact with the case of the drill (Fig. 18.20b), a "short circuit" can occur. In this undesirable circumstance, the pathway for the current is from the high-voltage wire through the person holding the drill and to Earth — a pathway that can kill. Protection is provided by a third wire, connected to the case of the drill (Fig. 18.20c). In this case, if a short occurs, the path of least resistance for the current is from the high-voltage wire through the case and back to ground through the third wire. The resulting high current produced will blow a fuse or trip a circuit breaker before the consumer is injured.

Special power outlets called ground-fault interrupters (GFIs) are now being used in kitchens, bathrooms, basements, and other hazardous areas of new homes. They are designed to protect people from electrical shock by sensing small currents — approximately 5 mA and greater — leaking to ground. When current above this level is detected, the device shuts off (interrupts) the current in less than a millisecond. Ground-fault interrupters will be discussed in Chapter 19.

APPLICATION

Third Wire on Consumer Appliances.

SUMMARY

A **source of emf** is any device that transforms nonelectrical energy into electrical energy.

The **equivalent resistance** of a set of resistors connected in **series** is

$$R_{eq} = R_1 + R_2 + R_3 + \ldots \qquad \text{[18.4]}$$

The **equivalent resistance** of a set of resistors connected in **parallel** is

$$\frac{1}{R_{eq}} = \frac{1}{R_1} + \frac{1}{R_2} + \frac{1}{R_3} + \dots \qquad [18.6]$$

Complex circuits are conveniently analyzed by using **Kirchhoff's rules:**

1. The sum of the currents entering any junction must equal the sum of the currents leaving that junction.
2. The sum of the potential differences across all the elements around any closed-circuit loop must be zero.

The first rule is a statement of **conservation of charge.** The second is a statement of **conservation of energy.**

As a capacitor is charged by a battery through a resistor, the current drops from a maximum value to zero. The **time constant,** $\tau = RC$, represents the time it takes the charge on the capacitor to increase from zero to 63% of its maximum value.

MULTIPLE-CHOICE QUESTIONS

1. What is the time constant of the circuit shown in Figure 18.21? Each of the five resistors has resistance R, and each of the five capacitors has capacitance C. The internal resistance of the battery is negligible.
   (a) RC (b) $5RC$ (c) $10RC$ (d) $25RC$
   (e) None of the answers is correct.

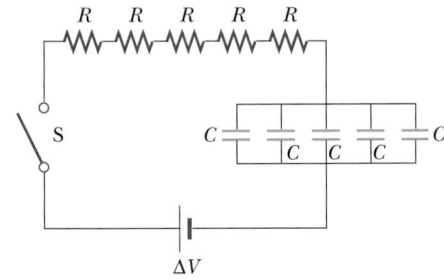

Figure 18.21   (Multiple-Choice Question 1)

2. If the time constant of the circuit shown in Figure 18.21 is actually $35RC$ rather than the answer found in Question 1, what is the internal resistance of the battery? Each of the five resistors has resistance R, and each of the five capacitors has capacitance C.
   (a) R (b) $2R$ (c) $35R$ (d) $\frac{5}{7}R$ (e) $\frac{7}{5}R$
3. What is the current in the 10.0-Ω resistance in the circuit shown in Figure 18.22?
   (a) 0.59 A   (b) 1.0 A   (c) 11 A   (d) 16 A

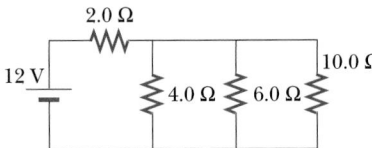

Figure 18.22   (Multiple-Choice Question 3)

4. What is the current in the 4.0-Ω resistor in Figure 18.23?
   (a) 1.0 A   (b) 0.50 A   (c) 1.5 A   (d) 2.0 A

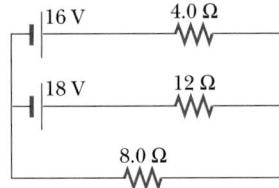

Figure 18.23   (Multiple-Choice Question 4)

5. An electric heater dissipates 1300 W, a toaster dissipates 1000 W, and an electric oven dissipates 1540 W. If all three of these appliances are operating in parallel on a 120-V circuit, what is the total current drawn from an external source?
   (a) 24 A   (b) 32 A   (c) 40 A   (d) 48 A

CONCEPTUAL QUESTIONS

1. If the energy transferred to a dead battery during charging is W, is the total energy transferred out of the battery to an external circuit during use in which it completely discharges also W?

2. How would you connect resistors so that the equivalent resistance is larger than the individual resistances? Give an example involving two or three resistors.

3. If you have your headlights on while you start your car, why do they dim while the car is starting?

4. How would you connect resistors so that the equivalent resistance is smaller than the individual resistances? Give an example involving two or three resistors.

5. Electrical devices are often rated with a voltage and a current—for example, 120 volts, 5 amperes. Batteries, however, are only rated with a voltage—for example, 1.5 volts. Why?

6. A "short circuit" is a circuit containing a path of very low resistance in parallel with some other part of the circuit. Discuss the effect of a short circuit on the portion of the circuit it parallels. Use a lamp with a frayed line cord as an example.

7. Connecting batteries in series increases the emf applied to a circuit. What advantage might there be to connecting them in parallel?

8. If electrical power is transmitted over long distances, the resistance of the wires becomes significant. Why? Which mode of transmission would result in less energy loss—high current and low voltage or low current and high voltage? Discuss.

9. You have a large supply of lightbulbs and a battery. You start with one lightbulb connected to the battery and notice its brightness. You then add one lightbulb at a time, each new bulb being added in series to the previous bulbs. As you add the lightbulbs, what happens to the brightness of the bulbs? To the current through the bulbs? To the power transferred from the battery? To the lifetime of the battery? Answer the same questions if the lightbulbs are added one by one in parallel with the first.

10. Two sets of Christmas tree lights are available. For set A, when one bulb is removed, the remaining bulbs remain illuminated. For set B, when one bulb is removed, the remaining bulbs do not operate. Explain the difference in wiring for the two sets.

11. Why is it possible for a bird to sit on a high-voltage wire without being electrocuted (Fig. Q18.11)?

12. Are the two headlights on a car wired in series or in parallel? How can you tell?

13. Embodied in Kirchhoff's rules are two conservation laws. What are they?

14. A ski resort consists of a few chair lifts and several interconnected downhill runs on the side of a mountain,

Figure Q18.11 Birds on a high-voltage wire. *(Superstock)*

with a lodge at the bottom. The lifts are analogous to batteries and the runs are analogous to resistors. Describe how two runs can be in series. Describe how three runs can be in parallel. Sketch a junction of one lift and two runs. One of the skiers is carrying an altimeter. State Kirchhoff's junction rule and Kirchhoff's loop rule for ski resorts.

15. Suppose you are flying a kite when it strikes a high-voltage wire (a very dangerous situation). What factors determine how great a shock you will receive?

16. Why is it dangerous to turn on a light when you are in a bathtub?

17. Suppose a parachutist lands on a high-voltage wire and grabs the wire as she prepares to be rescued. Will she be electrocuted? If the wire then breaks, should she continue to hold onto the wire as she falls to the ground?

18. Would a fuse work successfully if it were placed in parallel with the device it was supposed to protect?

19. A series circuit consists of three identical lamps connected to a battery, as in Figure Q18.19. When the

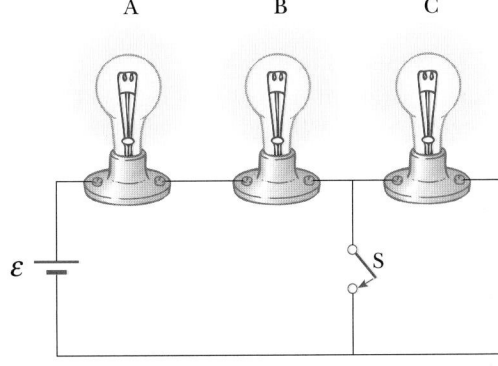

Figure Q18.19

switch S is closed, what happens **(a)** to the intensities of lamps A and B, **(b)** to the intensity of lamp C, **(c)** to the current in the circuit, and **(d)** to the voltage drop across the three lamps? **(e)** Does the power dissipated in the circuit increase, decrease, or remain the same?

20. Figure Q18.20 shows a series connection of three lamps, all rated at 120 V, with power ratings of 60 W, 75 W, and

200 W. Why do the intensities of the lamps differ? Which lamp has the greatest resistance? How would their intensities differ if they were connected in parallel?

21. In Figure Q18.21, describe what happens to the light-bulb after the switch is closed. Assume the capacitor has a large capacitance and is initially uncharged, and assume that the light illuminates when connected directly across the battery terminals.

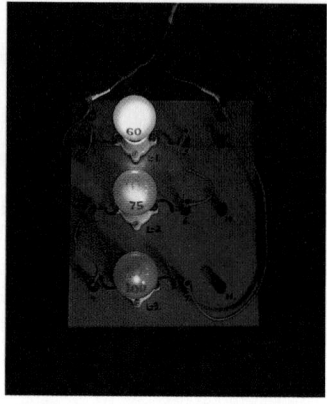

Figure Q18.20 *(Courtesy of Henry Leap and Jim Lehman)*

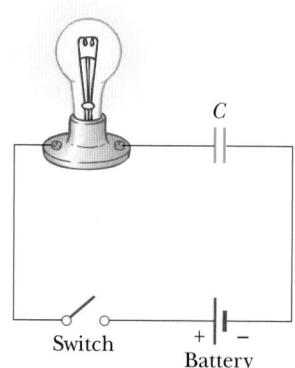

Figure Q18.21

PROBLEMS

1, 2, 3 = straightforward, intermediate, challenging   ☐ = full solution available in Study Guide/Student Solutions Manual   ◤ = Core Concepts Workbook
WEB = solution posted at **http://www.harcourtcollege.com/physics/cptech**   = biomedical application   = Interactive Physics

Section 18.2   Resistors in Series

Section 18.3   Resistors in Parallel

1. A battery having an emf of 9.00 V delivers 117 mA when connected to a 72.0-Ω load. Determine the internal resistance of the battery.

2. A 4.0-Ω resistor, an 8.0-Ω resistor, and a 12-Ω resistor are connected in series with a 24-V battery. What are **(a)** the equivalent resistance and **(b)** the current in each resistor?

3. An 18-Ω resistor and a 6.0-Ω resistor are connected in series across an 18-V battery. **(a)** Find the current through each resistor and the voltage drop across each

resistor. **(b)** Repeat part (a) for the situation in which the resistors are connected in parallel across the 18-V battery.

4. The resistors of Problem 2 are connected in parallel across a 24-V battery. Find **(a)** the equivalent resistance and **(b)** the current in each resistor.

5. A 9.0-Ω resistor and a 6.0-Ω resistor are connected in series with a power supply. **(a)** The voltage drop across the 6.0-Ω resistor is measured to be 12 V. Find the voltage output of the power supply. **(b)** The two resistors are connected in parallel across a power supply, and the current through the 9.0-Ω resistor is found to be 0.25 A. Find the voltage setting of the power supply.

6. Find the equivalent resistance of the circuit in Figure P18.6.

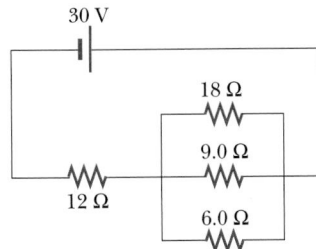

30 V
18 Ω
9.0 Ω
12 Ω
6.0 Ω

Figure P18.6

7. What is the equivalent resistance of the combination between points *a* and *b* in Figure P18.7? Note that one end of the vertical resistor is left free.

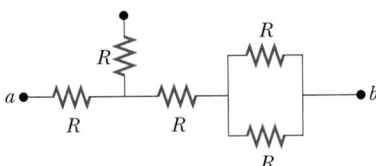

Figure P18.7

8. Find the equivalent resistance of the circuit in Figure P18.8.

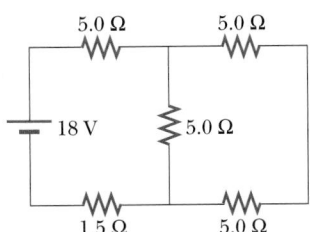

5.0 Ω   5.0 Ω
18 V   5.0 Ω
1.5 Ω   5.0 Ω

Figure P18.8

9. (a) Find the equivalent resistance of the circuit in Figure P18.9. (b) If the total power supplied to the circuit is 4.00 W, find the emf of the battery.

10. A technician has a box full of resistors, all with the same resistance, *R*. How many different values of effective resistance can the technician achieve using all the possible combinations of one to three separate resistors? Express the effective resistance of each combination in terms of *R*.

11. Two resistors, A and B, are connected in parallel across a 6.0-V battery. The current through B is found to be

Problem 13 is the same as Workbook Problem 82 (Workbook page 177).

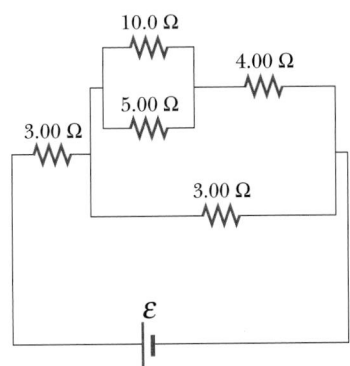

10.0 Ω
4.00 Ω
5.00 Ω
3.00 Ω
3.00 Ω
ε

Figure P18.9

2.0 A. When the two resistors are connected in series to the 6.0-V battery, a voltmeter connected across resistor A measures a voltage of 4.0 V. Find the resistances of A and B.

12. (a) You need a 45-Ω resistor but the stockroom has only 20-Ω and 50-Ω resistors. How can the desired resistance be achieved under these circumstances? (b) What can you do if you need a 35-Ω resistor?

13. The resistance between terminals *a* and *b* in Figure P18.13 is 75 Ω. If the resistors labeled *R* have the same value, determine *R*.

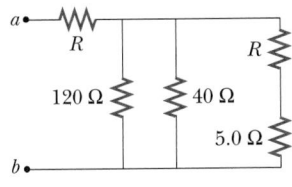

a
R   *R*
120 Ω   40 Ω
5.0 Ω
b

Figure P18.13

WEB **14.** Find the current in the 12-Ω resistor in Figure P18.14.

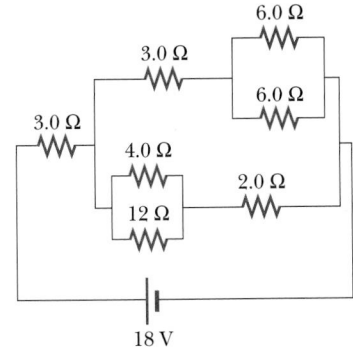

6.0 Ω
3.0 Ω
6.0 Ω
3.0 Ω   4.0 Ω
2.0 Ω
12 Ω
18 V

Figure P18.14

Section 18.4 Kirchhoff's Rules and Complex DC Circuits

15. Determine the potential difference, ΔV_{ab}, for the circuit in Figure P18.15.

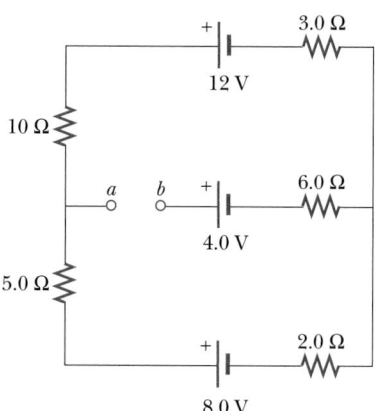

Figure P18.15

16. Figure P18.16 shows a circuit diagram. Determine **(a)** the current, **(b)** the potential of wire A relative to ground, and **(c)** the voltage drop across the 1500-Ω resistor.

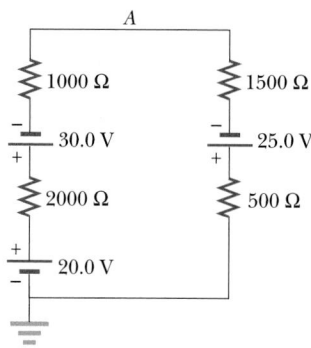

Figure P18.16

17. In the circuit of Figure P18.17, the current I_1 is 3 A and the value of $\mathcal{E}$ and R are unknown. What are the currents I_2 and I_3?

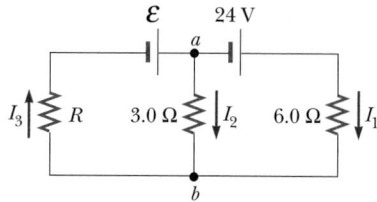

Figure P18.17   (Problems 17 and 19)

18. What is the emf, $\mathcal{E}$, of the battery in the circuit of Figure P18.18?

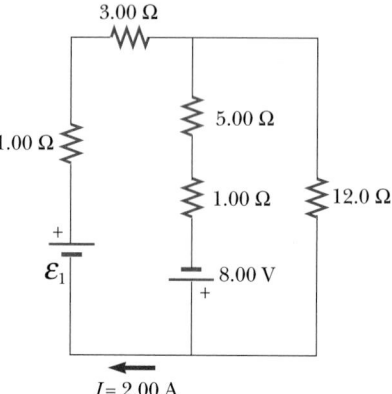

Figure P18.18

19. In the circuit shown in Figure P18.17, the current I_1 is 3 A and $R = 12\ \Omega$. What is the emf, $\mathcal{E}$, of the unknown battery?

20. Find the current through each of the three resistors of Figure P18.20 **(a)** by the rules for resistors in series and parallel and **(b)** by the use of Kirchhoff's rules.

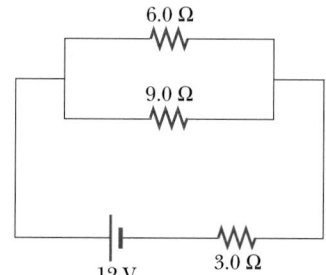

Figure P18.20

21. Two 1.50-V batteries—with their positive terminals in the same direction—are inserted in series into the barrel of a flashlight. One battery has an internal resistance of 0.255 Ω, the other an internal resistance of 0.153 Ω. When the switch is closed, a current of 0.600 A passes through the lamp. **(a)** What is the lamp's resistance? **(b)** What fraction of the power dissipated is dissipated in the batteries?

22. Four resistors are connected to a battery with a terminal voltage of 12 V, as shown in Figure P18.22. Determine the power lost in the 50-Ω resistor.

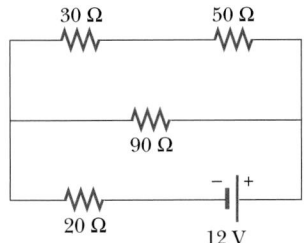

Figure P18.22

23. Calculate each of the unknown currents I_1, I_2, and I_3 for the circuit of Figure P18.23.

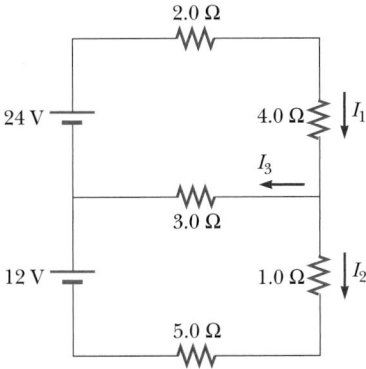

Figure P18.23

24. An unmarked battery has an unknown internal resistance. If the battery is connected to a fresh 5.60-V battery (negligible internal resistance) positive to positive and negative to negative, the current through the circuit is 10.0 mA. If the polarity of the unknown battery is reversed, the current increases to 25.0 mA. Determine the emf and internal resistance of the unknown battery. Assume that in each case the direction of the current is negative to positive within the 5.60-V battery.

25. Find the current in each resistor in Figure P18.25.

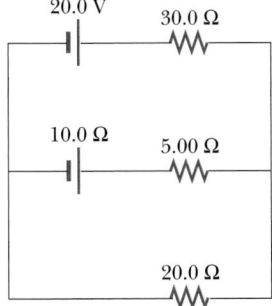

Figure P18.25

26. (a) Determine the potential difference, ΔV_{ab}, for the circuit in Figure P18.26. Note the each battery has an internal resistance, as indicated in the figure. (b) If points a and b are connected by a 7.0-Ω resistor, what is the current through this resistor?

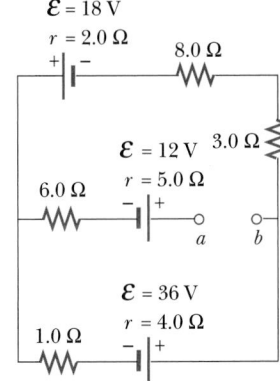

Figure P18.26

27. Find the potential difference across each resistor in Figure P18.27.

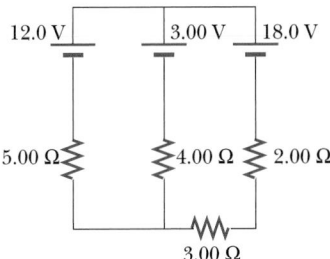

Figure P18.27

Section 18.5   *RC* Circuits (Optional)

28. Consider a series *RC* circuit for which $C = 6.0\ \mu\text{F}$, $R = 2.0 \times 10^6\ \Omega$, and $\mathcal{E} = 20$ V. Find (a) the time constant of the circuit and (b) the maximum charge on the capacitor after a switch in the circuit is closed.

29. Show that $\tau = RC$ has units of time.

30. An uncharged capacitor and a resistor are connected in series to a source of emf. If $\mathcal{E} = 9.00$ V, $C = 20.0\ \mu\text{F}$, and $R = 100\ \Omega$, find (a) the time constant of the circuit, (b) the maximum charge on the capacitor, and (c) the charge on the capacitor after one time constant.

WEB 31. Consider a series *RC* circuit (see Fig. 18.13) for which $R = 1.0\ \text{M}\Omega$, $C = 5.0\ \mu\text{F}$, and $\mathcal{E} = 30$ V. Find the charge on the capacitor 10 s after the switch is closed.

32. A series combination of a 12-kΩ resistor and an unknown capacitor is connected to a 12-V battery. One

second after the circuit is completed, the voltage across the capacitor is 10 V. Determine the capacitance of the capacitor.

33. A series RC circuit has a time constant of 0.960 s. The battery has an emf of 48.0 V, and the maximum current in the circuit is 500 mA. What are (a) the value of the capacitance and (b) the charge stored in the capacitor 1.92 s after the switch is closed?

Section 18.6 Household Circuits (Optional)

34. An electric heater is rated at 1300 W, a toaster is rated at 1000 W, and an electric grill is rated at 1500 W. The three appliances are connected in parallel to a common 120-V circuit. (a) How much current does each appliance draw? (b) Is a 30.0-A circuit breaker sufficient in this situation? Explain.

35. A lamp ($R = 150\ \Omega$), an electric heater ($R = 25\ \Omega$), and a fan ($R = 50\ \Omega$) are connected in parallel across a 120-V line. (a) What total current is supplied to the circuit? (b) What is the voltage across the fan? (c) What is the current in the lamp? (d) What power is expended in the heater?

36. A heating element in a stove is designed to dissipate 3000 W when connected to 240 V. (a) Assuming that the resistance is constant, calculate the current in this element if it is connected to 120 V. (b) Calculate the power it dissipates at this voltage.

37. Your toaster oven and coffee maker each dissipate 1200 W of power. Can you operate them together if the 120-V line that feeds them has a circuit breaker rated at 15 A? Explain.

ADDITIONAL PROBLEMS

38. Consider an RC circuit in which the capacitor is being charged by a battery connected in the circuit. After a time equal to two time constants, what percentage of the *final* charge is present on the capacitor?

39. Find the equivalent resistance between points a and b in Figure P18.39.

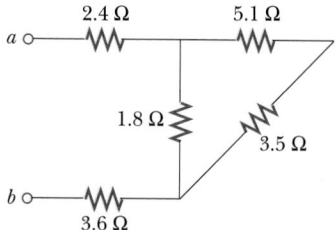

Figure P18.39

40. For the circuit in Figure P18.40, calculate (a) the equivalent resistance of the circuit and (b) the power dissipated by the entire circuit. (c) Find the current in the 5.0-Ω resistor.

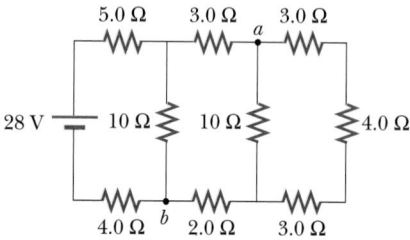

Figure P18.40

41. Find the equivalent resistance of the circuit in Figure P18.41.

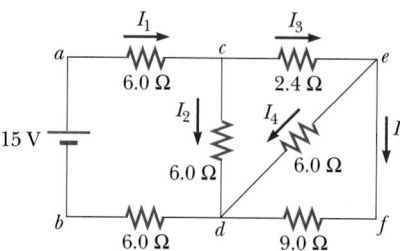

Figure P18.41   (Problems 41 and 42)

42. For the circuit in Figure P18.41, find (a) each current in the circuit, (b) the potential difference across each resistor, and (c) the power dissipated by each resistor.

WEB 43. An automobile battery has an emf of 12.60 V and an internal resistance of 0.080 Ω. The headlights have total resistance 5.00 Ω (assumed constant). What is the potential difference across the headlight bulbs (a) when they are the only load on the battery? (b) when the starter motor is operated, taking an additional 35.0 A from the battery?

44. Assume that you have five identical resistors, each having a resistance of 10 ohms. Show how all five of these resistors can be used in a combination to produce an equivalent resistance of 14 ohms.

45. Two resistors, R_1 and R_2, have an equivalent resistance of 690 Ω when they are connected in series and an equivalent resistance of 150 Ω when they are connected in parallel. What are R_1 and R_2?

46. Find the values of I_1, I_2, and I_3 for the circuit in Figure P18.46.

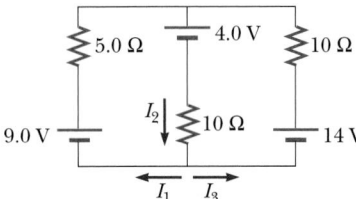

Figure P18.46

47. The resistance between points a and b in Figure P18.47 drops to one half its original value when switch S is closed. Determine the value of R.

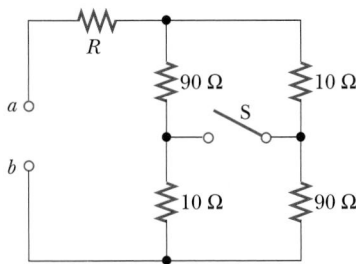

Figure P18.47

48. In Figure P18.48, $R_1 = 0.100\ \Omega$, $R_2 = 1.00\ \Omega$, and $R_3 = 10.0\ \Omega$. Find the equivalent resistance of the circuit and the current in each resistor when a 5.00-V power supply is connected between (a) points A and B, (b) points A and C, (c) points A and D.

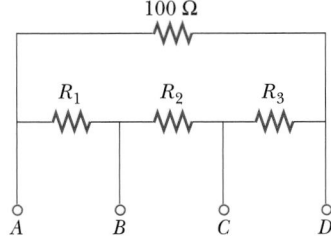

Figure P18.48

49. What are the expected readings of the ammeter and voltmeter for the circuit in Figure P18.49?

50. A generator has a terminal voltage of 110 V when it delivers 10.0 A, and 106 V when it delivers 30.0 A. Calculate the emf and the internal resistance of the generator.

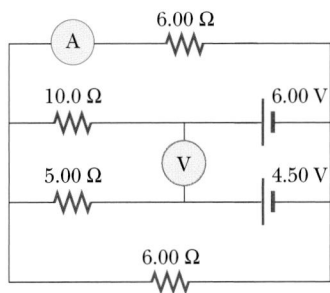

Figure P18.49

51. An emf of 10 V is connected to a series RC circuit consisting of a resistor of $2.0 \times 10^6\ \Omega$ and a capacitor of $3.0\ \mu F$. Find the time required for the charge on the capacitor to reach 90% of its final value.

52. (a) Apply Kirchhoff's loop rule to the RC circuit in Figure 18.13a and show that $\mathcal{E} - q/C - IR = 0$. Use this equation and Equation 18.7 to determine the current in the circuit as a function of time. (b) Apply Kirchhoff's loop rule to the RC circuit in Figure 18.14a and show that $q/C - IR = 0$. Use this equation and Equation 18.9 to determine the current in the circuit as a function of time.

53. The resistor R in Figure P18.53 dissipates 20.0 W of power. Determine the value of R.

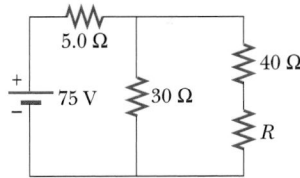

Figure P18.53

54. Consider the circuit shown in Figure P18.54. Find (a) the current in the 20.0-Ω resistor and (b) the potential difference between points a and b.

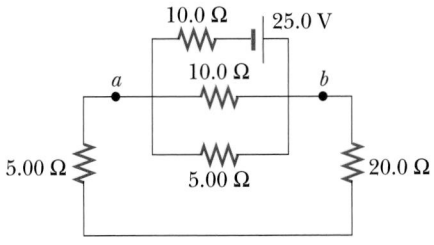

Figure P18.54

WEB 55. Determine the current in each branch of the circuit in Figure P18.55.

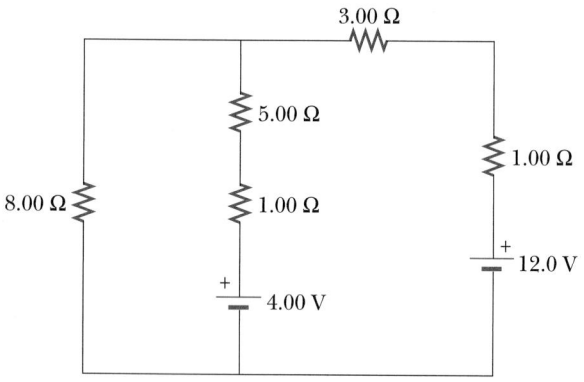

Figure P18.55

56. Using Kirchhoff's rules, **(a)** find the current in each resistor in Figure P18.56. **(b)** Find the potential difference between points c and f. Which point is at the higher potential?

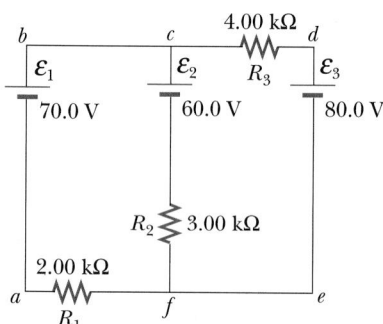

Figure P18.56

57. The student engineer of a campus radio station wishes to verify the effectiveness of the lightning rod on the antenna mast (Fig. P18.57). The unknown resistance R_x is between points C and E. Point E is a true ground but is inaccessible for direct measurement because this stratum is several meters below the Earth's surface. Two identical rods are driven into the ground at A and B, introducing an unknown resistance, R_y. The procedure is as follows: Measure resistance R_1 between points A and B, then connect A and B with a heavy conducting wire and measure resistance R_2 between points A and C. **(a)** Derive a formula for R_x in terms of the observable resistances, R_1 and R_2. **(b)** A satisfactory ground resis-

tance would be $R_x < 2.0\ \Omega$. Is the grounding of the station adequate if measurements give $R_1 = 13\ \Omega$ and $R_2 = 6.0\ \Omega$?

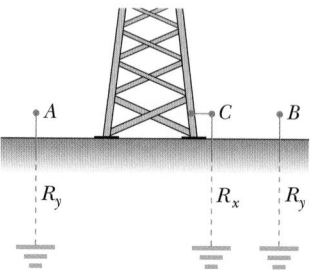

Figure P18.57

58. A voltage, ΔV, is applied to a series configuration of n resistors, each of value R. The circuit components are reconnected in a parallel configuration, and voltage ΔV is again applied. Show that the power consumed by the series configuration is $1/n^2$ times the power consumed by the parallel configuration.

59. For the network in Figure P18.59, show that the resistance between points a and b is $R_{ab} = \frac{27}{17}\ \Omega$. (*Hint*: Connect a battery with emf $\mathcal{E}$ across points a and b and determine $\mathcal{E}/I$, where I is the current through the battery.)

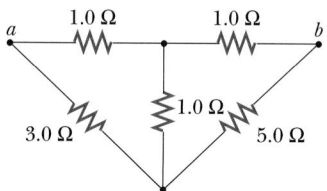

Figure P18.59

60. A battery with an internal resistance of $10.0\ \Omega$ produces an open-circuit voltage of 12.0 V. A variable load resistance with a range of 0 to $30.0\ \Omega$ is connected across the battery. (*Note:* A battery has a resistance that depends on the condition of its chemicals and increases as the battery ages. This so-called internal resistance can be represented in a simple circuit diagram as a resistor in series with the battery.) **(a)** Graph the power dissipated in the load resistor as a function of the load resistance. **(b)** With your graph, demonstrate the following important theorem: *The power delivered to a load is maximum if the load resistance equals the internal resistance of the source.*

61. The circuit in Figure P18.61 contains two resistors, $R_1 = 2.0$ kΩ and $R_2 = 3.0$ kΩ, and two capacitors, $C_1 = 2.0$ μF and $C_2 = 3.0$ μF, connected to a battery with emf $\mathcal{E} = 120$ V. If there are no charges on the capacitors before switch S is closed, determine the charges q_1 and q_2 on capacitors C_1 and C_2, respectively, after the switch is closed. (*Hint:* First reconstruct the circuit so that it becomes a simple *RC* circuit containing a single resistor and single capacitor in series, connected to the battery, and then determine the total charged, q, stored in the circuit.)

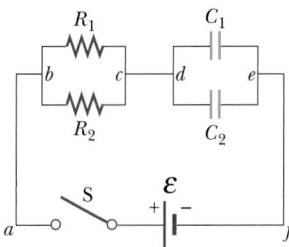

Figure P18.61

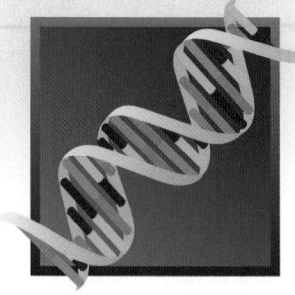

BIOLOGICAL PERSPECTIVE

Current in the Nervous System

PAUL DAVIDOVITS
Boston College

The most remarkable use of electrical phenomena in living organisms is found in the nervous system of animals. Specialized cells in the body called **neurons** form a complex network that receives, processes, and transmits information from one part of the body to another. The center of this network is located in the brain, which has the ability to store and analyze information. Based on this information, the nervous system controls parts of the body.

The nervous system is very complex: the human nervous system, for example, consists of about 10^{10} interconnected neurons. Some aspects of the nervous system are well known. Over the past 40 years, the method of signal propagation through the nervous system has been firmly established. The messages are electric pulses transmitted by neurons. When a neuron receives an appropriate stimulus, it produces electric pulses that are propagated along its cable-like structure. The strength of the stimulus is conveyed by the number of pulses produced. When the pulses reach the end of the "cable," they activate either muscle cells or other neurons.

The neurons, which are the basic units of the nervous system, can be divided into three classes: sensory neurons, motor neurons, and interneurons. The sensory neurons receive stimuli from sensory organs that monitor the external and internal environment of the body. Depending on their specialized functions, the sensory neurons convey messages about factors such as heat, light, pressure, muscle tension, and odor to higher centers in the nervous system. The motor neurons carry messages that control the muscle cells. The messages are based on the information provided by the sensory neurons and by the brain. The interneurons transmit information from one neuron to another.

Each neuron consists of a cell body to which are attached input ends called **dendrites** and a long tail called the **axon,** which propagates the signal away from the cell (Fig. 1). The far end of the axon branches into nerve endings that transmit the signal across small gaps to other neurons or to muscle cells. A simple sensory-motor neuron circuit is shown in Figure 2. A stimulus from a muscle produces nerve impulses that travel to the spine. Here the signal is transmitted to a motor neuron, which in turn sends impulses to control the muscle.

The axon, which is an extension of the neuron cell, conducts the electric impulses away from the cell body. Some axons are extremely long. In humans, for example, the axons connecting the spine with the fingers and toes are more than 1 m long. The neuron can transmit messages because of the special electrical characteristics of the axon. Most of the information about the electrical and chemical properties of the axon is obtained by inserting small needle-like probes into the axon. With such probes it is possible to measure currents in the axon and to sample its chemical composition. Such experiments are usually difficult to run because the diameter of most axons is very small. Even the largest axons in the human nervous system have a diameter of only about 20×10^{-4} cm. The giant squid, however, has an axon with a diameter of about 0.5 mm, which is large enough for the convenient insertion of probes. Much of the information about signal transmission in the nervous system has come from experiments with the squid axon, therefore.

In the aqueous environment of the body, salts and other molecules dissociate into positive and negative ions. As a result, body fluids are relatively good conductors of electricity.

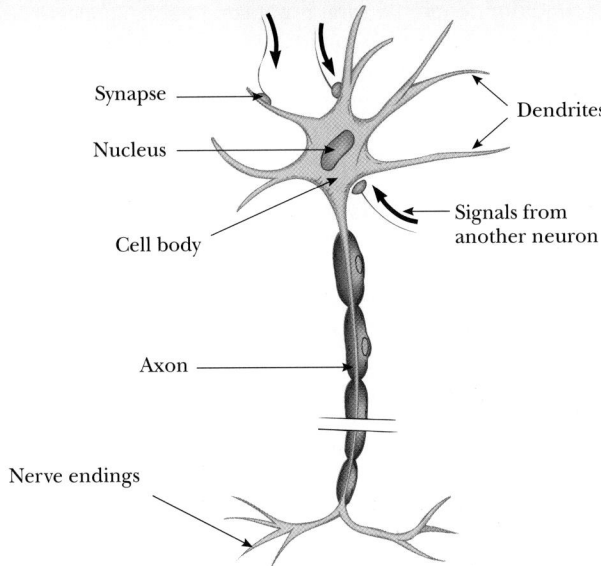

Figure 1   Diagram of a neuron.

The inside of the axon is filled with an ionic fluid that is separated from the surrounding body fluid by a thin membrane that is only about 5 nm to 10 nm thick.

The resistivities of the internal and external fluids are about the same, but their chemical compositions are substantially different. The external fluid is similar to seawater. Its ionic solutes are mostly positive sodium ions and negative chloride ions. Inside the axon, the positive ions are mostly potassium ions and the negative ions are mostly large organic ions.

Because there is a large concentration of sodium ions outside the axon and a large concentration of potassium ions inside, we may ask why the concentrations are not equalized by diffusion. In other words, why don't the sodium ions leak into the axon and the potassium ions leak out of it? The answer lies in the properties of the axon membrane.

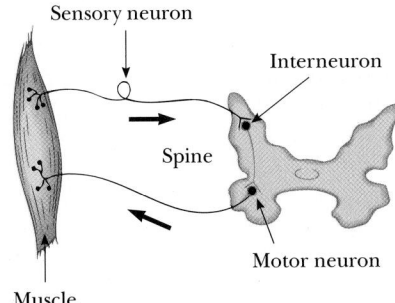

Figure 2   A simple neural circuit.

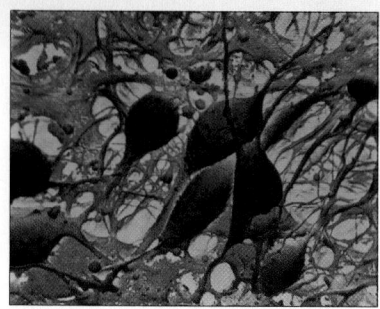

False-color scanning electron micrograph (SEM) of neurons from the human cerebral cortex (the outer, heavily folded gray matter of the brain). In the cerebral cortex, neurons of similar structure are arranged in a variable number of layers. *(CNRI/Science Photo Library/Photo Researchers, Inc.)*

In the resting condition, when the axon is not conducting an electric pulse, the axon membrane is highly permeable to potassium ions, slightly permeable to sodium ions, and impermeable to large organic ions. Thus, although sodium ions cannot easily leak into the axon, potassium ions can certainly leak out of it. As the potassium ions leak out of the axon, however, they leave behind the large negative organic ions, which cannot follow them through the membrane. As a result, a negative potential is produced inside the axon with respect to the outside. The negative potential, which has been measured at about 70 mV, holds back the outflow of potassium ions so that, at equilibrium, the concentration of ions is as we have stated.

The mechanism for the production of an electric signal by the neuron is conceptually remarkably simple. When a neuron receives an appropriate stimulus, which may be heat, pressure, or a signal from another neuron, the properties of its membrane change. As a result, sodium ions rush into the cell while potassium ions flow out of it. This flow of charged particle constitutes an electric current signal that propagates along the axon to its destination.

Although the axon is a highly complex structure, its main electrical properties can be represented by the standard electric circuit concepts of resistance and capacitance. The propagation of the signal along the axon is then well described by the techniques of electric circuit analysis discussed in the text.

Magnetism

▲ PHYSICS PUZZLER

Aurora borealis (the northern lights), photographed near Fairbanks, Alaska. Auroras occur when cosmic rays — electrically charged particles, mainly from the Sun — fall into the Earth's atmosphere over the magnetic poles and collide with other atoms, resulting in the emission of visible light. Why is it that only charged particles are trapped by the Earth's magnetic field? Why is the aurora most often observed around the magnetic poles of the Earth? *(George Lepp/Tony Stone Images)*

12.1

The list of important technological applications of magnetism is very long. For instance, large electromagnets are used to pick up heavy loads. Magnets are also used in such devices as meters, motors, and loudspeakers. Magnetic tapes are routinely used in sound and video recording equipment and for computer memory, and magnetic recording material is used on computer disks. Intense magnetic fields are currently being used in magnetic resonance imaging devices (MRI) to explore the human body with better resolution and greater safety than x-rays.

As we investigate magnetism in this chapter, you will find that the subject cannot be divorced from electricity. For example, magnetic fields affect moving charges and moving charges produce magnetic fields. The ultimate source of all magnetic fields is electric current, whether it be the current in a wire or the current produced by the motion of charges within atoms or molecules.

19.1 MAGNETS

Most people have had experience with some form of magnet. You are most likely familiar with the common iron horseshoe magnet that can pick up iron-containing objects such as paper clips and nails. In the discussion that follows, we shall assume that the magnet has the shape of a bar. Iron objects are most strongly attracted to the ends of such a bar magnet, called its **poles.** One end is called the **north pole** and the other the **south pole.** The names come from the behavior of a magnet in the presence of the Earth's magnetic field. If a bar magnet is suspended from its midpoint by a piece of string so that it can swing freely in a horizontal plane, it will rotate until its north pole points to the north of the Earth and its south pole points to the south of the Earth. The same idea is used to construct a simple compass. Magnetic poles also exert attractive or repulsive forces on each other similar to the electrical forces between charged objects. In fact, simple experiments with two bar magnets show that **like poles repel each other and unlike poles attract each other.** Although the force between two magnetic poles is similar to the force between two electric charges, there is an important difference. Electric charges can be isolated (witness the proton and the electron), but magnetic poles cannot be isolated. In fact, no matter how many times a permanent magnet is cut, each piece always has a north pole and a south pole. Thus, magnetic poles always occur in pairs. There is some theoretical basis for the speculation that magnetic monopoles (isolated north or south poles) may exist in nature, and attempts to detect them are currently an active experimental field of investigation. However, none of these attempts has proven successful.

There is yet another similarity between electric and magnetic effects, which concerns methods for making permanent magnets. In Chapter 15 we learned that when two materials such as rubber and wool are rubbed together, each becomes charged, one positively and the other negatively. In a somewhat analogous fashion, an unmagnetized piece of iron can be magnetized by stroking it with a magnet. Magnetism can be induced in iron (and other materials) by other means. For example, if a piece of unmagnetized iron is placed near a strong permanent magnet, the piece of iron eventually becomes magnetized. The process can be accelerated by either heating and cooling the iron or by hammering. Naturally occurring magnetic materials, such as magnetite, achieve their magnetism in this manner, because they have been subjected to the Earth's magnetic field over very long periods of time. The extent to which a piece of material retains its magnetism depends on

An assortment of commercially available magnets. The four red magnets and the large black magnet on the left are made of an alloy of iron, aluminum, and cobalt. The six horseshoe magnets on the right are made of different nickel alloy steels. The rectangular magnets on the lower right are ceramics made of iron, nickel, and beryllium oxides. *(Courtesy of Central Scientific Company)*

12.2, SECTION 1

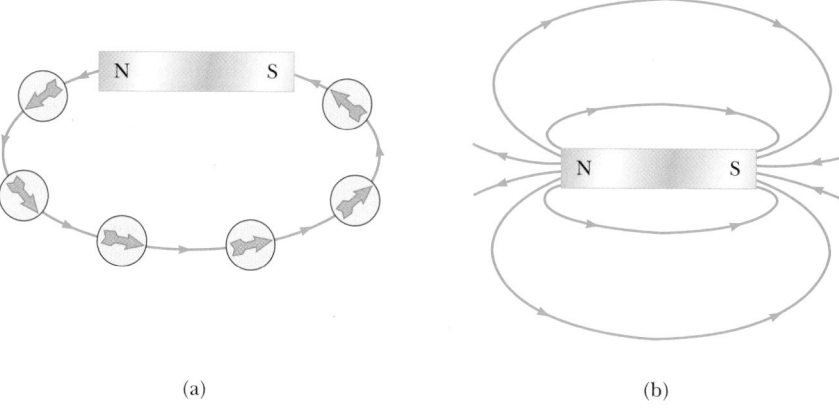

Figure 19.1 (a) Tracing the magnetic field of a bar magnet. (b) Several magnetic field lines of a bar magnet.

(a)   (b)

whether it is classified as being magnetically hard or soft. **Soft magnetic materials,** such as iron, are easily magnetized but also tend to lose their magnetism easily. In contrast, **hard magnetic materials** such as cobalt and nickel are difficult to magnetize but tend to retain their magnetism.

In earlier chapters we found it convenient to describe the interaction between charged objects in terms of electric fields. Recall that an electric field surrounds any electric charge. The region of space surrounding a *moving* charge also includes a magnetic field. In addition, a magnetic field surrounds any magnetized material.

To describe any type of field, we must define its magnitude, or strength, and its direction. The direction of the magnetic field, **B,** at any location is the direction in which the north pole of a compass needle points at that location. Figure 19.1a shows how the magnetic field of a bar magnet can be traced with the aid of a compass. Several magnetic field lines of a bar magnet traced out in this manner appear in Figure 19.1b. Magnetic field patterns can be displayed by small iron filings, as shown in Figure 19.2.

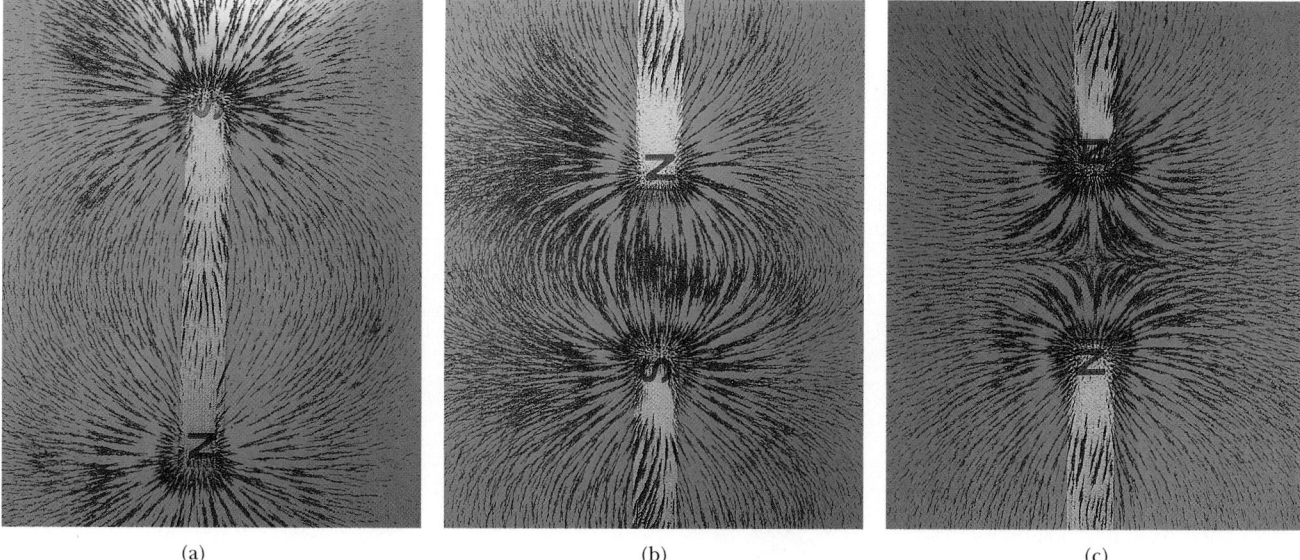

(a)   (b)   (c)

Figure 19.2 (a) The magnetic field pattern of a bar magnet, displayed by iron filings on a sheet of paper. (b) The magnetic field pattern between *unlike* poles of two bar magnets, displayed by iron filings. (c) The magnetic field pattern between two *like* poles. *(Courtesy of Henry Leap and Jim Lehman)*

APPLICATION

Dusting for Fingerprints.

A technique similar to that shown in Figure 19.2 is used by forensic scientists to find fingerprints at a crime scene. One way to find latent, or invisible, prints is by sprinkling a powder of iron dust on a surface. The iron adheres to perspiration or body oils present and can be spread around on the surface with a magnetic brush that never comes into contact with the powder or the surface.

19.2   MAGNETIC FIELD OF THE EARTH

When we speak of a small bar magnet as having north and south poles, we should more properly say that it has a "north-seeking" pole and a "south-seeking" pole. By this we mean that if such a magnet is used as a compass, one end will seek, or point to, the north geographic pole of the Earth. Thus, we conclude that **the geographic north pole corresponds to a magnetic south pole, and the geographic south pole corresponds to a magnetic north pole.** In fact, the configuration of the Earth's magnetic field, pictured in Figure 19.3, very much resembles what would be achieved by burying a bar magnet deep in the interior of the Earth.

If a compass needle is suspended in bearings that allow it to rotate in the vertical plane as well as in the horizontal plane, the needle is horizontal with respect to the Earth's surface only near the Equator. As the device is moved northward, the needle rotates so that it points more and more toward the surface of the Earth. The angle between the direction of the magnetic field and the horizontal is called the **dip angle.** Finally, at a point just north of Hudson Bay in Canada, the north pole of the needle points directly downward, and the dip angle is 90°. This site, first found in 1832, is considered to be the location of the south magnetic pole of the Earth. It is approximately 1300 mi from the Earth's geographic north pole and varies with time. Similarly, the magnetic north pole of the Earth is about 1200 miles from the geographic south pole. Thus, it is only approximately correct to say that a compass

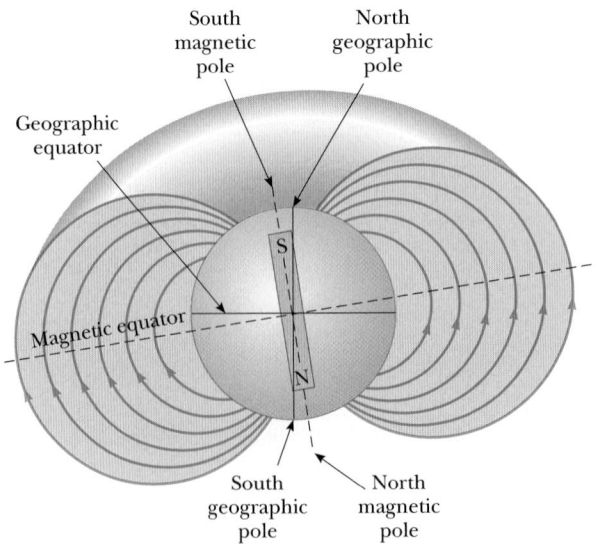

Figure 19.3 The Earth's magnetic field lines. Note that magnetic south is at the north geographic pole, and magnetic north is at the south geographic pole.

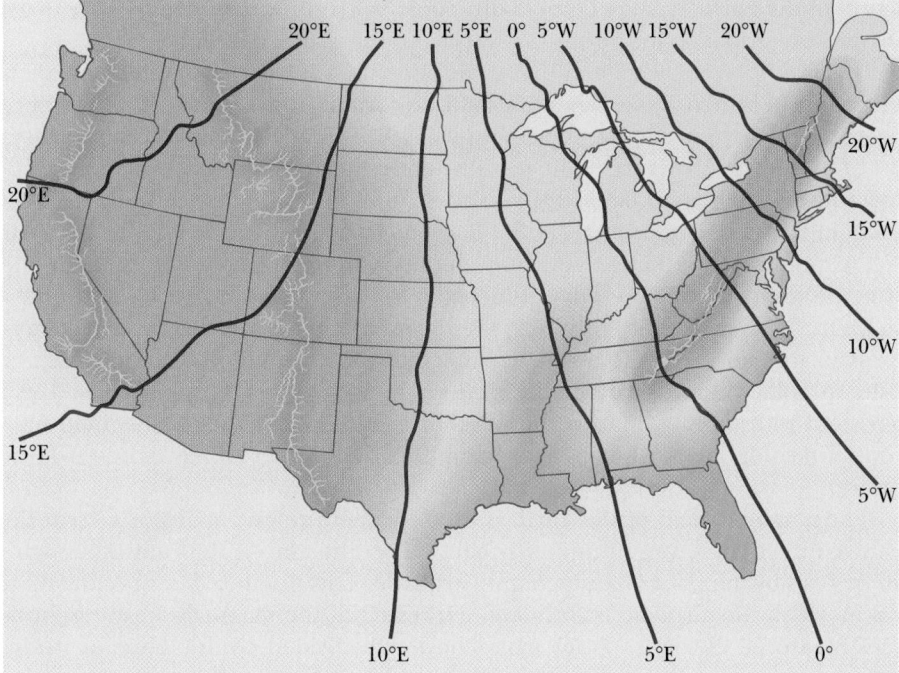

Figure 19.4 A map of the lower forty-eight United States showing the declination of a compass from true north.

needle points north. The difference between true north, defined as the geographic north pole, and north indicated by a compass varies from point to point on the Earth, and the difference is referred to as *magnetic declination*. For example, along a line through South Carolina and the Great Lakes, a compass indicates true north, whereas in Washington state it aligns 25° east of true north (Fig. 19.4).

Although the magnetic field pattern of the Earth is similar to the pattern that would be set up by a bar magnet deep in the Earth, it is easy to understand why the source of the Earth's field cannot be large masses of permanently magnetized material. The Earth does have large deposits of iron ore deep beneath its surface, but the high temperatures in the Earth's core prevent the iron from retaining any permanent magnetization. It is considered more likely that the true source of the Earth's magnetic field is charge-carrying convection currents in its core. Charged ions or electrons circling in the liquid interior could produce a magnetic field. There is also some evidence that the strength of a planet's magnetic field is related to the planet's rate of rotation. For example, Jupiter rotates faster than the Earth, and recent space probes indicate that Jupiter's magnetic field is stronger than ours. Venus, on the other hand, rotates more slowly than the Earth, and its magnetic field is found to be weaker. Investigation into the cause of the Earth's magnetism continues.

An interesting sidelight concerns the Earth's magnetic field. It has been found that the direction of the field reverses every few million years. Evidence for this is provided by basalt (an iron-containing rock) that is sometimes spewed forth by volcanic activity on the ocean floor. As the lava cools, it solidifies and retains a

QUICKLAB

For this experiment, you will need a small bar magnet, a small plastic container, and a bowl of water. Tape the magnet to the bottom of the small container, and float the container and magnet on the surface of the bowl of water as in the figure. The magnet and small container should rotate and come to equilibrium with the magnet pointing either north or south. The compass you have constructed is similar to the type used by early sailing vessels. How can you determine which direction is north and which is south?

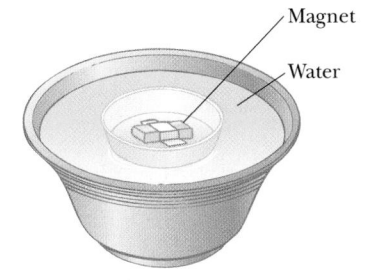

picture of the Earth's magnetic field direction. When the basalt deposits are dated, they provide evidence for periodic reversals of the magnetic field.

It has long been speculated that some animals, such as birds, use the magnetic field of the Earth to guide their migrations. Studies have shown that a type of anaerobic bacterium that lives in swamps has a magnetized chain of magnetite as part of its internal structure. (The term *anaerobic* means that these bacteria live and grow without oxygen; in fact, oxygen is toxic to them.) The magnetized chain acts as a compass needle that enables the bacteria to align with the Earth's magnetic field. When they find themselves out of the mud on the bottom of the swamp, they return to their oxygen-free environment by following the magnetic field lines of the Earth. Further evidence for their magnetic-sensing ability is the fact that bacteria found in the Northern Hemisphere have internal magnetite chains that are opposite in polarity to those of similar bacteria in the Southern Hemisphere. This is consistent with the fact that in the Northern Hemisphere the Earth's field has a downward component, whereas in the Southern Hemisphere it has an upward component.

The magnetic field of the Earth is used to label runways at airports according to their direction. A large number is painted on the end of the runway so that it can be read by the pilot of an incoming airplane. This number describes the direction in which the airplane is traveling, expressed as the magnetic heading, in degrees measured clockwise from magnetic north divided by 10. Thus, a runway marked 9 would be directed toward the east (90° divided by 10), and one marked 18 would be directed toward the magnetic south.

APPLICATION

Magnetotactic Bacteria.

APPLICATION

Labeling Airport Runways.

Applying Physics 1

On a business trip to Australia, you take along your American-made compass that you may have used on a camping trip. Does this compass work correctly in Australia?

Explanation   There is no problem with using the compass in Australia. The north pole of the magnet in the compass will be attracted to the south magnetic pole near the north geographic pole, just as it was in the United States. The only difference in the magnetic field lines is that they have an upward component in Australia, whereas they have a downward component in the United States. Your compass cannot detect this, however—it only displays the direction of the horizontal component of the magnetic field.

19.3   MAGNETIC FIELDS

12.2, SECTION 2

Experiments show that a stationary charged particle does not interact with a static magnetic field. However, **when moving through a magnetic field a charged particle experiences a magnetic force.** This force has its maximum value when the charge moves perpendicularly to the magnetic field lines, decreases in value at other angles, and becomes zero when the particle moves along the field lines. We shall make use of these observations in describing the magnetic field.

In our discussion of electricity, the electric field at some point in space was defined as the electric force per unit charge acting on some test charge placed at that point. In a similar manner, we can describe the properties of the magnetic field, **B,** at some point in terms of the magnetic force exerted on a test charge at that point. Our test object is assumed to be a charge, q, moving with velocity **v.** It is found experimentally that the strength of the magnetic force on the particle is proportional to the magnitude of the charge, q, the magnitude of the velocity, **v,** the strength of the external magnetic field, **B,** and the sine of the angle θ between the direction of **v** and the direction of **B.** These observations can be summarized by writing the magnitude of the magnetic force as

Magnetic force ▶ $$F = qvB \sin \theta \qquad [19.1]$$

This expression is used to define the magnitude of the magnetic field as

Magnetic field defined ▶ $$B \equiv \frac{F}{qv \sin \theta} \qquad [19.2]$$

If F is in newtons, q in coulombs, and v in meters per second, the SI unit of magnetic field is the **tesla (T),** also called the **weber (Wb) per square meter** (that is, $1\ \text{T} = 1\ \text{Wb/m}^2$). Thus, if a 1-C charge moves through a magnetic field of magnitude 1 T with a velocity of 1 m/s, perpendicularly to the field ($\sin \theta = 1$), the magnetic force exerted on the charge is 1 N. We can express the units of **B** as

$$[B] = T = \frac{\text{Wb}}{\text{m}^2} = \frac{\text{N}}{\text{C} \cdot \text{m/s}} = \frac{\text{N}}{\text{A} \cdot \text{m}} \qquad [19.3]$$

In practice, the cgs unit for magnetic field, the **gauss (G),** is often used. The gauss is related to the tesla through the conversion

$$1\ \text{T} = 10^4\ \text{G}$$

Conventional laboratory magnets can produce magnetic fields as large as about 25 000 G, or 2.5 T. Superconducting magnets that can generate magnetic fields as great as 3×10^5 G, or 30 T, have been constructed. These values can be compared with the value of the Earth's magnetic field near its surface, which is about 0.5 G, or 0.5×10^{-4} T.

From Equation 19.1 we see that the force on a charged particle moving in a magnetic field has its maximum value when the particle moves *perpendicularly* to the magnetic field, corresponding to $\theta = 90°$, so that $\sin \theta = 1$. The magnitude of this maximum force has the value

$$F_{\text{max}} = qvB \qquad [19.4]$$

Also, note from Equation 19.1 that F is zero when **v** is parallel to **B** (corresponding to $\theta = 0°$ or $180°$). Thus, no magnetic force is exerted on a charged particle when it moves in the direction of the magnetic field or opposite the field.

Experiments show that the direction of the magnetic force is always perpendicular to both **v** and **B,** as shown in Figure 19.5. To determine the direction of the force, we employ the following right-hand rule:

> **Hold your right hand open, as illustrated in Figure 19.6, and then place your fingers in the direction of B with your thumb pointing in the direction of v. The force, F, on a positive charge is directed *out* of the palm of your hand.**

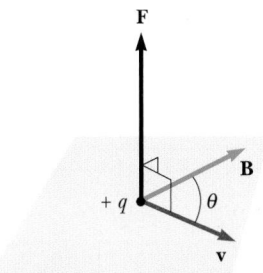

Figure 19.5 The direction of the magnetic force on a charged particle moving with a velocity of **v** in the presence of a magnetic field. When **v** is at an angle of θ with **B,** the magnetic force is perpendicular to both **v** and **B.**

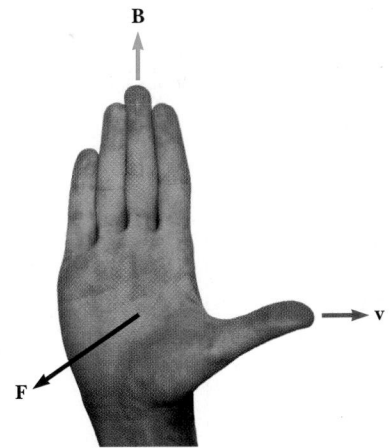

Figure 19.6 The right-hand rule for determining the direction of the magnetic force on a positive charge moving with a velocity of **v** in a magnetic field, **B.** With your thumb in the direction of **v** and your four fingers in the direction of **B,** the force is directed out of the palm of your hand.

If the charge is negative rather than positive, the force is directed *opposite* that shown in Figure 19.6. That is, if q is negative, simply use the right-hand rule to find the direction of **F** for positive q, and then reverse this direction for the negative charge.

EXAMPLE 19.1   A Proton Traveling in the Earth's Magnetic Field

A proton moves with a speed of 1.0×10^5 m/s through the Earth's magnetic field, which has a value of 55 μT at a particular location. When the proton moves eastward, the magnetic force acting on it is a maximum, and when it moves northward, no magnetic force acts on it. What is the strength of the magnetic force, and what is the direction of the magnetic field?

Solution   The magnitude of the force can be found from Equation 19.4:

$$F_{max} = qvB = (1.6 \times 10^{-19} \text{ C})(1.0 \times 10^5 \text{ m/s})(55 \times 10^{-6} \text{ T})$$

$$= \boxed{8.8 \times 10^{-19} \text{ N}}$$

The direction of the magnetic field cannot be determined precisely from the information given in the problem. Because no magnetic force acts on a charged particle when it is moving parallel to the field, all that we can say for sure is that the magnetic field is directed either northward or southward.

Exercise   Calculate the gravitational force on the proton and compare it with the magnetic force. Note that the mass of the proton is 1.67×10^{-27} kg.

Answer 1.6×10^{-26} N; $F_{grav}/F_{max} = 1.9 \times 10^{-8}$

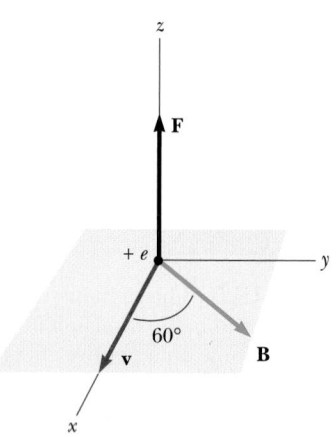

Workbook Problem 76— (Workbook page 163) is similar to Example 19.2.

Figure 19.7   (Example 19.2) The magnetic force, **F**, on a proton is in the positive z direction when **v** and **B** lie in the xy plane.

EXAMPLE 19.2   A Proton Moving in a Strong Magnetic Field

A proton moves at 8.0×10^6 m/s along the x axis. It enters a region in which there is a magnetic field of magnitude 2.5 T, directed at an angle of 60° with the x axis and lying in the xy plane (Fig. 19.7). Calculate the initial force on and acceleration of the proton.

Solution   From Equation 19.1 we get

$$F = qvB \sin \theta = (1.6 \times 10^{-19} \text{ C})(8.0 \times 10^6 \text{ m/s})(2.5 \text{ T})(\sin 60°)$$

$$= \boxed{2.8 \times 10^{-12} \text{ N}}$$

Use the right-hand rule, noting that the charge is positive, to see that the force is in the positive z direction. Verify that the units of F in the calculation reduce to newtons.

Because the mass of the proton is 1.67×10^{-27} kg, its initial acceleration is

$$a = \frac{F}{m} = \frac{2.8 \times 10^{-12} \text{ N}}{1.67 \times 10^{-27} \text{ kg}} = \boxed{1.7 \times 10^{15} \text{ m/s}^2}$$

in the positive z direction.

Exercise   Calculate the acceleration of an electron that moves through the same magnetic field at the same speed as the proton. The mass of an electron is 9.11×10^{-31} kg.

Answer 3.0×10^{18} m/s² in the negative z direction

19.4 MAGNETIC FORCE ON A CURRENT-CARRYING CONDUCTOR

If a force is exerted on a single charged particle when it moves through a magnetic field, it should be no surprise that a current-carrying wire also experiences a force when placed in a magnetic field. This follows from the fact that the current is a collection of many charged particles in motion; hence, the resultant force on the wire is due to the sum of the individual forces on the charged particles. The force on the particles is transmitted to the "bulk" of the wire through collisions with the atoms making up the wire.

Before we continue, some explanation is in order concerning notation in many of the figures. To indicate the direction of **B,** we use the following convention.

> If **B** is directed into the page, as in Figure 19.8, we use a series of blue crosses, representing the tails of arrows. If **B** is directed out of the page, we use a series of blue dots, representing the tips of arrows. If **B** lies in the plane of the page, we use a series of blue field lines with arrowheads.

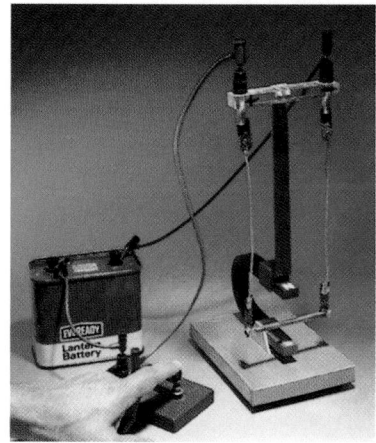

This apparatus demonstrates the force on a current-carrying conductor in an external magnetic field. Why does the bar swing *away* from the magnet after the switch is closed? *(Courtesy of Henry Leap and Jim Lehman)*

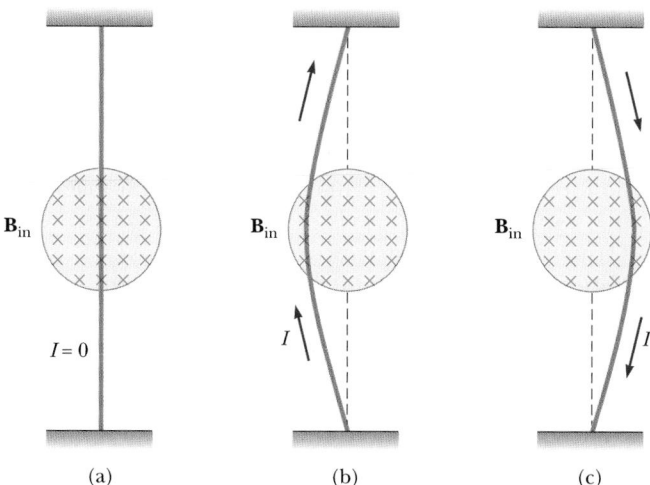

Figure 19.8 A segment of a flexible vertical wire partially stretched between the poles of a magnet, with the field (blue crosses) directed into the page. (a) When there is no current in the wire, it remains vertical. (b) When the current is upward, the wire deflects to the left. (c) When the current is downward, the wire deflects to the right.

The force on a current-carrying conductor can be demonstrated by hanging a wire between the poles of a magnet, as in Figure 19.8. In this figure, the magnetic field is directed into the page and covers the region within the shaded circle. The wire deflects to the right or left when a current is passed through it.

Let us quantify this discussion by considering a straight segment of wire of length ℓ and cross-sectional area A, carrying current I in a uniform external magnetic field, **B,** as in Figure 19.9. We assume that the magnetic field is perpendicular to the wire and is directed into the page. Each charge carrier in the wire experiences a force of magnitude $F_{max} = qv_dB$, where v_d is the drift velocity of the charge. To find the total force on the wire, we multiply the force on one charge carrier by the

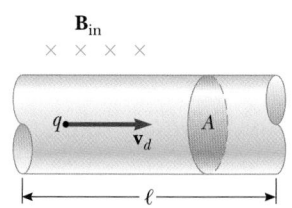

Figure 19.9 A section of a wire containing moving charges in an external magnetic field, **B.**

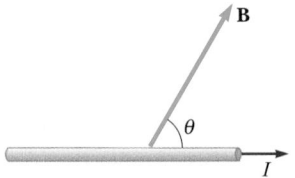

Figure 19.10 A wire carrying a current, I, in the presence of an external magnetic field, **B,** that makes an angle of θ with the wire.

number of carriers in the segment. Because the volume of the segment is $A\ell$, the number of carriers is $nA\ell$, where n is the number of carriers per unit volume. Hence, the magnitude of the total magnetic force on the wire of length ℓ is

Total force = force on each charge carrier × total number of carriers

$$F_{max} = (qv_d B)(nA\ell)$$

From Chapter 17, however, we know that the current in the wire is given by $I = nqv_d A$. Therefore, F_{max} can be expressed as

$$F_{max} = BI\ell \qquad [19.5]$$

This equation can be used only when the current and the magnetic field are at right angles to each other.

If the wire is not perpendicular to the field but is at some arbitrary angle, as in Figure 19.10, the magnitude of the magnetic force on the wire is

$$F = BI\ell \sin \theta \qquad [19.6]$$

where θ is the angle between **B** and the direction of the current. The direction of this force can be obtained by use of the right-hand rule. However, in this case you must place your thumb in the direction of the current rather than in the direction of **v.** In Figure 19.10, the direction of the magnetic force on the wire is out of the page.

Finally, when the current is either in the direction of the field or opposite the direction of the field, the magnetic force on the wire is zero.

APPLICATION

Loudspeaker Operation.

The fact that a magnetic force acts on a current-carrying wire in a magnetic field is the operating principle of most loudspeakers in sound systems. One speaker design, shown in Figure 19.11, consists of a coil of wire, called the *voice coil;* a flexible paper cone that acts as the speaker; and a permanent magnet. The coil of wire surrounding the north pole of the magnet is shaped so that the magnetic field lines are directed radially outward from the coil's axis. When an electrical signal is sent to the coil, producing a current in the direction shown in Figure 19.11, a magnetic force to the left acts on the coil. (This can be seen by applying the right-hand rule to each turn of wire.) When the current reverses direction, as it would for a sinusoidally varying current, the magnetic force on the coil also reverses direction, and the cone accelerates to the right. An alternating current through the coils causes an alternating force on the speaker, which results in vibrations of the cone. The vibrating cone creates sound waves as it pushes and pulls on the air in front of it. In this way an electrical signal is converted to a sound wave.

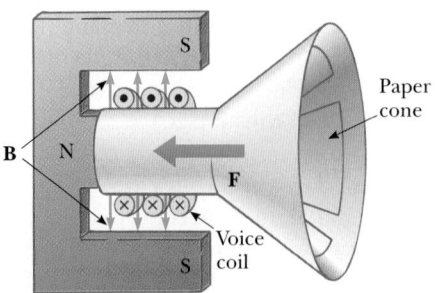

Figure 19.11 A diagram of a loudspeaker.

An unusual application of the force on a current-carrying conductor is illustrated by the electromagnetic pump shown in Figure 19.12. Artificial hearts require a pump that completely replaces the heart to keep the blood flowing, and kidney dialysis machines also require a pump to assist the heart in pumping blood that is to be cleansed. Ordinary mechanical pumps create difficulties because they damage the blood cells that move through. The mechanism shown in Figure 19.12 has demonstrated some promise in such applications. A magnetic field is established across a segment of the tube containing the blood, flowing in the direction of the velocity, **v.** An electric current passing through the fluid in the direction shown has a magnetic force acting on it in the direction of **v,** as application of the right-hand rule shows. This force helps to keep the blood in motion.

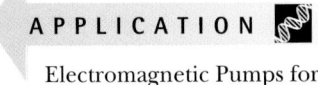

APPLICATION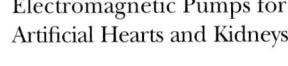

Electromagnetic Pumps for Artificial Hearts and Kidneys.

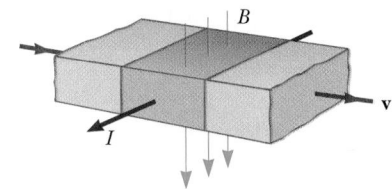

Figure 19.12 A simple electromagnetic pump has no moving parts to damage a conducting fluid, such as blood, passing through. Application of the right-hand rule shows that the force on the current-carrying segment of the fluid is in the direction of the velocity.

Thinking Physics 1

In a lightning strike, there is a rapid movement of negative charge from a cloud to the ground. In what direction is a lightning strike deflected by the Earth's magnetic field?

Explanation The downward flow of negative charge in a lightning strike is equivalent to an upward-moving current. Thus, we have an upward moving current in a northward-directed magnetic field. According to the right-hand rule, the lightning strike would be deflected toward the west.

EXAMPLE 19.3 A Current-Carrying Wire in the Earth's Magnetic Field

A wire carries a current of 22 A from east to west. Assume that at this location the magnetic field of the Earth is horizontal and directed from south to north, and that it has a magnitude of 0.50×10^{-4} T. Find the magnetic force on a 36-m length of wire. How does the force change if the current runs west to east?

Solution Because the directions of the current and magnetic field are at right angles, we can use Equation 19.5. The magnitude of the magnetic force is

$$F_{max} = BI\ell = (0.50 \times 10^{-4}\ \text{T})(22\ \text{A})(36\ \text{m}) = 4.0 \times 10^{-2}\ \text{N}$$

The right-hand rule shows that the force on the wire is directed toward the Earth.

If the current is directed from west to east, the force has the same magnitude but its direction is upward, away from the Earth.

Exercise If the current is directed north to south, what is the magnetic force on the wire?

Answer Zero

19.5 TORQUE ON A CURRENT LOOP

In the preceding section we showed how a force is exerted on a current-carrying conductor when the conductor is placed in an external magnetic field. With this as a starting point, we now show that a torque is exerted on a current loop placed

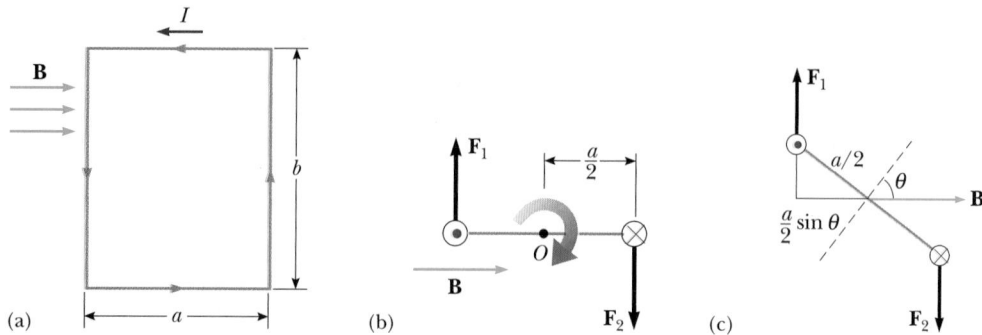

Figure 19.13 (a) Front view of a rectangular loop in a uniform magnetic field, **B.** There are no magnetic forces on the sides of length a parallel to **B,** but there are forces acting on the sides of length b. (b) Bottom view of the rectangular loop shows that the forces **F**$_1$ and **F**$_2$ on the sides of length b create a torque that tends to twist the loop clockwise. (c) If **B** is at an angle of θ with a line perpendicular to the plane of the loop, the torque is given by $BIA \sin \theta$.

in a magnetic field. The results of this analysis will be of great practical value when we discuss the galvanometer (in this chapter) and generators and motors (in Chapter 20).

Consider a rectangular loop carrying current I in the presence of an external uniform magnetic field in the plane of the loop, as shown in Figure 19.13a. The forces on the sides of length a are zero because these wires are parallel to the field. The magnitude of the magnetic forces on the sides of length b, however, is

$$F_1 = F_2 = BIb$$

The direction of **F**$_1$, the force on the left side of the loop, is out of the page, and that of **F**$_2$, the force on the right side of the loop, is into the page. If we view the loop from the bottom, as in Figure 19.13b, the forces are directed as shown. If we assume that the loop is pivoted so that it can rotate about point O, we see that these two forces produce a torque about O that rotates the loop clockwise. The magnitude of this torque, τ_{max}, is

$$\tau_{max} = F_1 \frac{a}{2} + F_2 \frac{a}{2} = (BIb) \frac{a}{2} + (BIb) \frac{a}{2} = BIab$$

where the moment arm about O is $a/2$ for both forces. Because the area of the loop is $A = ab$, the torque can be expressed as

$$\tau_{max} = BIA \qquad\qquad [19.7]$$

Note that this result is valid only when the magnetic field is parallel to the plane of the loop, as in the view in Figure 19.13b. If the field makes an angle of θ with a line perpendicular to the plane of the loop, as in Figure 19.13c, the moment arm for each force is given by $(a/2) \sin \theta$. An analysis such as the one just used produces, for the magnitude of the torque,

$$\tau = BIA \sin \theta \qquad\qquad [19.8]$$

This result shows that the torque has the *maximum* value, BIA, when the field is parallel to the plane of the loop ($\theta = 90°$) and is *zero* when the field is perpendicular to the plane of the loop ($\theta = 0$). As seen in Figure 19.13c, the loop tends to rotate to smaller values of θ (so that the normal to the plane of the loop rotates toward the direction of the magnetic field).

Although this analysis is for a rectangular loop, a more general derivation would indicate that Equation 19.8 applies regardless of the shape of the loop. Furthermore, the torque on a coil with N turns is

$$\tau = NBIA \sin \theta \qquad\qquad [19.9]$$

EXAMPLE 19.4   The Torque on a Circular Loop in a Magnetic Field

A circular wire loop of radius 50.0 cm is oriented at an angle of 30.0° to a magnetic field of 0.50 T, as shown in an edge view in Figure 19.14. The current in the loop is 2.0 A in the direction shown. Find the magnitude of the torque at this instant.

Solution   Regardless of the shape of the loop, Equation 19.8 is valid:

$$\tau = BIA \sin \theta = (0.50 \text{ T})(2.0 \text{ A})[\pi(0.50 \text{ m})^2](\sin 30.0°) = \boxed{0.39 \text{ N·m}}$$

Exercise   Find the torque on the loop if it has three turns rather than one.

Answer   The torque is three times that on the one-turn loop, or 1.2 N·m.

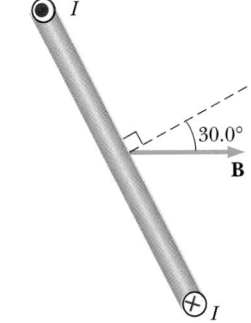

Figure 19.14   (Example 19.4) An edge view of a circular current loop in an external magnetic field, **B**.

19.6   THE GALVANOMETER AND ITS APPLICATIONS

The Galvanometer

A *galvanometer* is a device used in the construction of both ammeters and voltmeters. Its basic operation makes use of the fact that a torque acts on a current loop in the presence of a magnetic field. Figure 19.15a shows the main components of a gal-

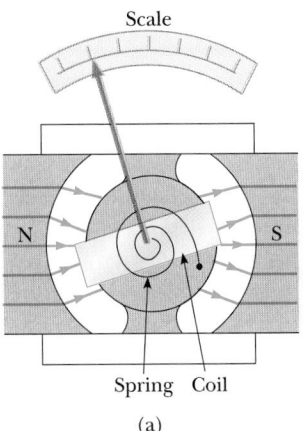

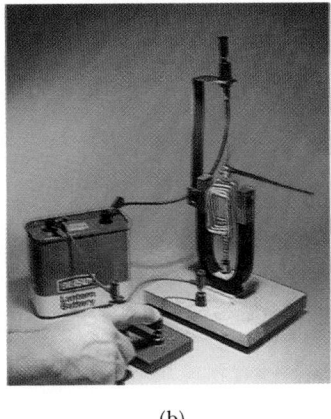

(a)   (b)

Figure 19.15   (a) The principal components of a galvanometer. When current passes through the coil, situated in a magnetic field, the magnetic torque causes the coil to twist. The angle through which the coil rotates is proportional to the current through it. (b) A large-scale demonstration model of a galvanometer movement. Why does the coil rotate about the vertical axis after the switch is closed? *(Courtesy of Jim Lehman)*

vanometer. It consists of a coil of wire mounted so that it is free to rotate on a pivot in a magnetic field provided by a permanent magnet. The torque experienced by the coil is proportional to the current. This means that the larger the current, the greater the torque and the more the coil will rotate before the spring tightens enough to stop the movement. Hence, the amount of deflection is proportional to the current. Once the instrument is properly calibrated, it can be used in conjunction with other circuit elements to measure either currents or potential differences. Figure 19.15b is a photograph of a large-scale galvanometer movement.

A Galvanometer Is the Basis of an Ammeter

A typical off-the-shelf galvanometer is usually not suitable for use as an ammeter (a current-measuring device). One of the main reasons is that a typical galvanometer has a resistance of about 60 Ω, and an ammeter resistance this large can considerably alter the current in the circuit in which it is placed. To understand this, consider the following case. Suppose you construct a simple series circuit containing a 3-V battery and a 3-Ω resistor. The current in such a circuit is 1 A. However, if you include a 60-Ω galvanometer in the circuit in an attempt to measure the current, the total resistance of the circuit is now 63 Ω and the current is reduced to 0.048 A.

A second factor that limits the use of a galvanometer as an ammeter is the fact that a typical galvanometer gives a full-scale deflection for very low currents, on the order of 1 mA and less. Consequently, such a galvanometer cannot be used directly to measure currents greater than 1 mA.

Now suppose we wish to convert a 60-Ω, 1-mA galvanometer to an ammeter that deflects full-scale when 2 A passes through it. In spite of the factors just described, this can be accomplished by simply placing a resistor, R_p, in *parallel* with the galvanometer, as in Figure 19.16. (The combination of the galvanometer and parallel resistor constitutes an ammeter.) The size of the resistor must be selected so that when 2 A passes through the ammeter, only 0.001 A passes through the galvanometer and the remaining 1.999 A passes through the resistor, R_p, sometimes called the *shunt resistor*. Because the galvanometer and shunt resistor are in parallel, the potential differences across them are the same. Thus, using Ohm's law, we get

$$(0.001 \text{ A})(60 \text{ }\Omega) = (1.999 \text{ A}) R_p$$

$$R_p = 0.030\ 02 \text{ }\Omega$$

Notice that the shunt resistance, R_p, is extremely small. Thus, the configuration in Figure 19.16 solves both problems associated with converting a galvanometer to an ammeter. The ammeter just described can measure a large current (2 A) and has a low resistance, on the order of 0.03 Ω. (Recall that the equivalent resistance of two resistors in parallel is always *less* than the value of either of the individual resistors.)

A Galvanometer Is the Basis of a Voltmeter

With the proper modification, the basic galvanometer can also be used to measure potential differences in a circuit. To understand how to accomplish this, let us first calculate the largest voltage that can be measured with a galvanometer. If the gal-

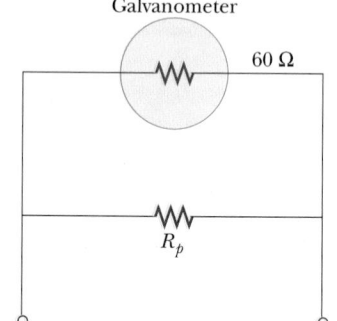

Figure 19.16 When a galvanometer is to be used as an ammeter, a resistor (R_p) is connected in parallel with the galvanometer.

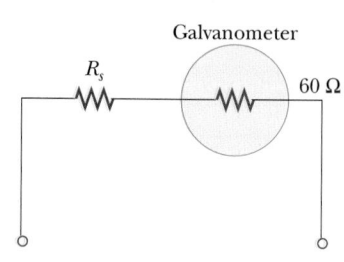

Figure 19.17 When a galvanometer is to be used as a voltmeter, a resistor (R_s) is connected in series with the galvanometer.

vanometer has a resistance of 60 Ω and gives a maximum deflection for a current of 1 mA, the largest voltage it can measure is

$$\Delta V_{max} = (0.001 \text{ A})(60 \text{ Ω}) = 0.06 \text{ V}$$

From this you can see that some modification is required to enable this device to measure larger voltages. Furthermore, a voltmeter must have a very high resistance to ensure that it will not disturb the circuit in which it is placed. The basic galvanometer, with a resistance of only 60 Ω, is not acceptable for direct voltage measurements.

The circuit in Figure 19.17 has the basic modification that is necessary to convert a galvanometer to a voltmeter. Suppose we want to construct a voltmeter capable of measuring a maximum voltage of 100 V. In this situation, a resistor, R_s, is placed in *series* with the galvanometer. The value of R_s is found by noting that a current of 1 mA must pass through the galvanometer when the voltmeter is connected across a potential difference of 100 V. Application of Ohm's law to this circuit gives

$$100 \text{ V} = (0.001 \text{ A})(R_s + 60 \text{ Ω})$$

$$R_s = 99\,940 \text{ Ω}$$

demonstrating that this voltmeter has a very high resistance.

When a voltmeter is constructed with several available ranges, values of R_s may be selected by use of a switch that can be connected to a preselected set of resistors. The required value of R_s increases as the maximum voltage to be measured increases.

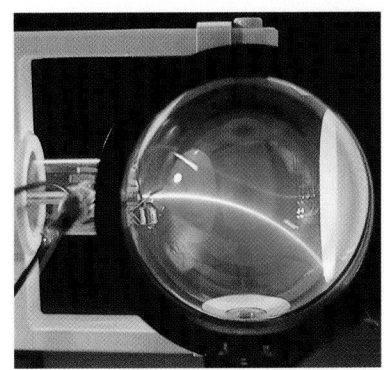

The bending of an electron beam in an external magnetic field. The tube contains gas at very low pressure. Electrons in the beam collide with gas atoms, which emit visible light. The beam is projected to the right and deflects downward in the presence of a magnetic field produced by a pair of current-carrying coils. What is the direction of the magnetic field? *(Courtesy of Central Scientific Company)*

12.3, BOTH SECTIONS

19.7 MOTION OF A CHARGED PARTICLE IN A MAGNETIC FIELD

Consider the case of a positively charged particle moving in a uniform magnetic field so that the direction of the particle's velocity is *perpendicular to the field*, as in Figure 19.18. The label $\mathbf{B}_{in}$ indicates that $\mathbf{B}$ is directed into the page. Application of the right-hand rule at point P shows that the direction of the magnetic force, $\mathbf{F}$, at this location is upward. This causes the particle to alter its direction of travel and to follow a curved path. Application of the right-hand rule at any point shows that **the magnetic force is always toward the center of the circular path**; therefore, the magnetic force causes the centripetal acceleration, which changes only the direction of $\mathbf{v}$ and not its magnitude. Because $\mathbf{F}$ produces the centripetal acceleration, we can equate its magnitude, qvB in this case, to the mass of the particle multiplied by the centripetal acceleration, v^2/r. From Newton's second law, we find that

$$F = qvB = \frac{mv^2}{r}$$

which gives

$$r = \frac{mv}{qB} \qquad\qquad [19.10]$$

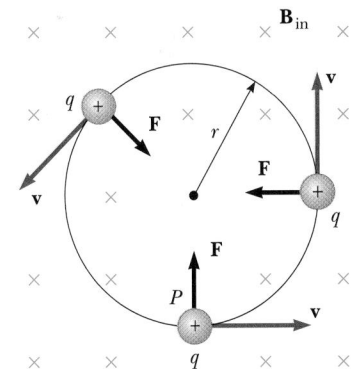

Figure 19.18 When the velocity of a charged particle is perpendicular to a uniform magnetic field, the particle moves in a circle whose plane is perpendicular to **B,** which is directed into the page. (The crosses represent the tails of the magnetic field vectors.) The magnetic force, **F,** on the charge is always directed toward the center of the circle.

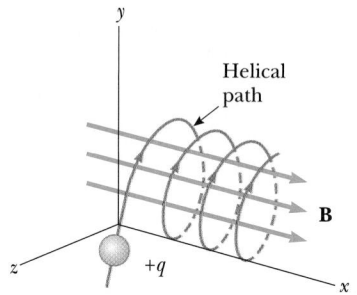

Figure 19.19 A charged particle that has a velocity vector with a component parallel to a uniform magnetic field moves in a helical path.

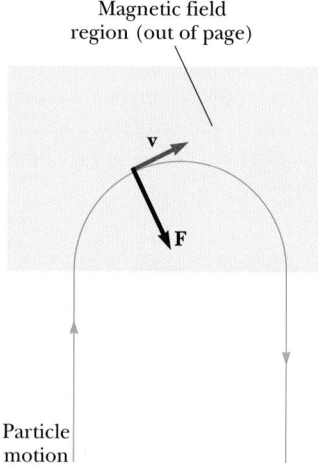

Magnetic field region (out of page)

Particle motion

Figure 19.20
(Thinking Physics 2)

Workbook Problem 75 — (Workbook page 161) is similar to Example 19.5.

APPLICATION

Mass Spectrometers.

This says that the radius of the path is proportional to the momentum, *mv*, of the particle and is inversely proportional to the magnetic field.

If the initial direction of the velocity of the charged particle is not perpendicular to the magnetic field but instead is directed at an angle to the field, as shown in Figure 19.19, the path followed by the particle is a spiral (called a *helix*) along the magnetic field lines.

Thinking Physics 2

Suppose a uniform magnetic field exists in a finite region of space. Can you inject a charged particle into this region from the outside and have it stay trapped in the region by the magnetic force?

Explanation   Let us consider separately the components of the particle velocity parallel and perpendicular to the field lines in the region. For the component parallel to the field lines, there will be no force on the particle—it will continue to move with the parallel component of the velocity. Now consider the component of velocity perpendicular to the field lines. This component will result in a magnetic force that is perpendicular to both the field lines and the velocity component. The path of a particle for which the force is always perpendicular to the velocity is a circle. Thus, the particle will follow a circular arc and exit the field on the other side of the circle, as shown in Figure 19.20 for a particle with constant kinetic energy. On the other hand, a particle can become trapped if it loses some kinetic energy in a collision after entering the region, as in Figure 19.19.

EXAMPLE 19.5   A Proton Moving Perpendicularly to a Uniform Magnetic Field

A proton is moving in a circular orbit of radius 14 cm in a uniform magnetic field of magnitude 0.35 T, directed perpendicularly to the velocity of the proton. Find the orbital speed of the proton.

Solution   From Equation 19.10, we get

$$v = \frac{qBr}{m} = \frac{(1.6 \times 10^{-19} \text{ C})(0.35 \text{ T})(14 \times 10^{-2} \text{ m})}{1.67 \times 10^{-27} \text{ kg}} = \boxed{4.7 \times 10^6 \text{ m/s}}$$

Exercise   If an electron moves perpendicularly to the same magnetic field with this speed, what is the radius of its circular orbit?

Answer 7.6×10^{-5} m

EXAMPLE 19.6   The Mass Spectrometer

Two singly ionized atoms move out of a slit at point *S* in Figure 19.21 and into a magnetic field of 0.10 T. Each has a speed of 1.0×10^6 m/s. The nucleus of the first atom contains one proton and has a mass of 1.67×10^{-27} kg, and the nucleus of the second atom contains a proton and a neutron and has a mass of 3.34×10^{-27} kg. Atoms with the same chemical properties but different masses are called isotopes. The two isotopes here are hydrogen and deuterium. Find their distance of separation when they strike a photographic plate at P.

Solution The radius of the circular path followed by the lighter isotope, hydrogen, is

$$r_1 = \frac{m_1 v}{qB} = \frac{(1.67 \times 10^{-27} \text{ kg})(1.0 \times 10^6 \text{ m/s})}{(1.6 \times 10^{-19} \text{ C})(0.10 \text{ T})} = 0.10 \text{ m}$$

The radius of the path of the heavier isotope, deuterium, is

$$r_2 = \frac{m_2 v}{qB} = \frac{(3.34 \times 10^{-27} \text{ kg})(1.0 \times 10^6 \text{ m/s})}{(1.6 \times 10^{-19} \text{ C})(0.10 \text{ T})} = 0.21 \text{ m}$$

The distance of separation is

$$x = 2r_2 - 2r_1 = \boxed{0.21 \text{ m}}$$

The concepts used in this example underlie the operation of a device called a **mass spectrometer,** which is sometimes used to separate isotopes according to their mass-to-charge ratios, but more often is used to measure masses.

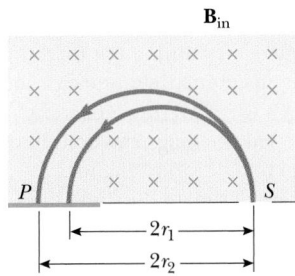

Figure 19.21 (Example 19.6) Two isotopes leave the slit at point S and travel in different circular paths before striking a photographic plate at P.

19.8   MAGNETIC FIELD OF A LONG, STRAIGHT WIRE AND AMPÈRE'S LAW

During a lecture demonstration in 1819, the Danish scientist Hans Oersted (1777–1851) found that an electric current in a wire deflected a nearby compass needle. This discovery, linking a magnetic field with an electric current, was the beginning of our understanding of the origin of magnetism.

A simple experiment first carried out by Oersted in 1820 clearly demonstrates that a current-carrying conductor produces a magnetic field. In this experiment, several compass needles are placed in a horizontal plane near a long vertical wire, as in Figure 19.22a. When there is no current in the wire, all needles point in the same direction (that of the Earth's field), as one would expect. However, when the wire carries a strong, steady current, the needles all deflect in directions tangent to

HANS CHRISTIAN OERSTED.

Hans Christian Oersted,
Danish physicist (1777–1851)

(North Wind Picture Archives)

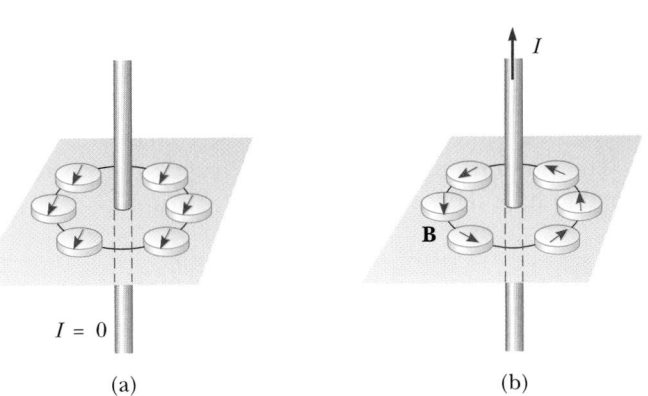

(a)   (b)

Figure 19.22 (a) When there is no current in the vertical wire, all compass needles point in the same direction. (b) When the wire carries a strong current, the compass needles deflect in directions tangent to the circle, pointing in the direction of **B** due to the current.

Figure 19.23   (a) The right-hand rule for determining the direction of the magnetic field due to a long, straight wire carrying a current. Note that the magnetic field lines form circles around the wire. (b) Circular magnetic field lines surrounding a current-carrying wire, displayed by iron filings. *(Courtesy of Henry Leap and Jim Lehman)*

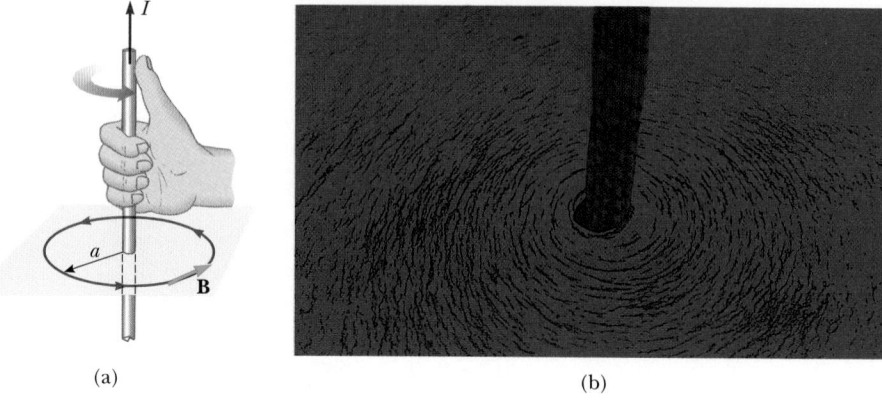

(a)   (b)

the circle, as in Figure 19.22b. These observations show that the direction of **B** is consistent with the following convenient rule:

> **If the wire is grasped in the right hand with the thumb in the direction of the current, as in Figure 19.23a, the fingers will curl in the direction of B.**

When the current is reversed, the needles in Figure 19.23b also reverse.

Because the needles point in the direction of **B,** we conclude that the lines of **B** form circles about the wire. By symmetry, the magnitude of **B** is the same everywhere on a circular path centered on the wire and lying in a plane perpendicular to the wire. By varying the current and distance from the wire, one finds that **B** is proportional to the current and inversely proportional to the distance from the wire.

Shortly after Oersted's discovery, scientists arrived at an expression for the strength of the magnetic field due to the current in a long, straight wire. The magnetic field strength at distance r from a wire carrying current I is

Magnetic field due to a ▶
long, straight wire
$$B = \frac{\mu_0 I}{2\pi r}$$
[19.11]

This result shows that the magnitude of the magnetic field is proportional to the current and decreases as the distance from the wire increases, as one might intuitively expect. The proportionality constant μ_0, called the **permeability of free space,** is defined to have the value

$$\mu_0 \equiv 4\pi \times 10^{-7} \text{ T} \cdot \text{m/A}$$
[19.12]

Ampère's Law and a Long, Straight Wire

Equation 19.11 enables us to calculate the magnetic field due to a long, straight wire carrying a current. A general procedure for deriving such equations was proposed by the French scientist André-Marie Ampère (1775–1836); it provides a relation between the current in an arbitrarily shaped wire and the magnetic field produced by the wire.

André-Marie Ampère
(1775–1836)

Ampère, a Frenchman, is credited with the discovery of electromagnetism — the relationship between electric currents and magnetic fields. Ampère's genius, particularly in mathematics, became evident by the age of 12, but his personal life was filled with tragedy. His father, a wealthy city official, was guillotined during the French Revolution, and his wife died young, in 1803. Ampère died at the age of 61 of pneumonia. His judgment of his life is clear from the epitaph he chose for his gravestone: *Tandem felix* (Happy at last). *(AIP Emilio Segre Visual Archive)*

Consider a circular path surrounding a current, as in Figure 19.23a. The path can be divided into many short segments, each of length $\Delta\ell$. Let us now multiply one of these lengths by the component of the magnetic field parallel to that segment, where the product is labeled $B_\parallel \Delta\ell$. According to Ampère, the sum of all such products over the closed path is equal to μ_0 times the net current, I, that passes through the surface bounded by the closed path. This statement, known as **Ampère's circuital law,** can be written

$$\sum B_\parallel \Delta\ell = \mu_0 I \qquad\qquad \text{[19.13]}$$

◀ Ampère's circuital law

where $\Sigma B_\parallel \Delta\ell$ means that we take the sum over all the products $B_\parallel \Delta\ell$ around the closed path. Ampère's law is the fundamental law describing how electric currents create magnetic fields in the surrounding empty space.

We can use Ampère's circuital law to derive the magnetic field due to a long, straight wire carrying a current, I. As discussed earlier, the magnetic field lines of this configuration form circles with the wire at their centers, as shown in Figure 19.23a. The magnetic field is tangent to this circle at every point and has the same value, $B_\parallel$, over the entire circumference of a circle of radius r. We now calculate the sum $\Sigma B_\parallel \Delta\ell$ over a circular path and note that $B_\parallel$ can be removed from the sum (because it has the same value for each element on the circle). Equation 19.13 then gives

$$\sum B_\parallel \Delta\ell = B_\parallel \sum \Delta\ell = B_\parallel (2\pi r) = \mu_0 I$$

Dividing both sides by $2\pi r$, we obtain

$$B = \frac{\mu_0 I}{2\pi r}$$

This is identical to Equation 19.11, which is the magnetic field of a long, straight current.

Ampère's circuital law provides an elegant and simple method for calculating the magnetic fields of highly symmetric current configurations. However, it cannot be used to calculate magnetic fields for complex current configurations that lack symmetry.

Thinking Physics 3

Consider a plastic ring encircling a long, straight wire, which is coming out of the page in Figure 19.24. On the plastic ring are fastened two bar magnets as shown. If a current flows in the wire in a direction out of the page, is there a net torque on the ring-magnet combination? If so, in which direction?

Explanation   Each of the poles on the bar magnets will experience a magnetic force due to the circular magnetic field of the wire. The magnetic field will fall off inversely with distance from the wire. Thus, the force on the north poles will be smaller than the force on the south poles. The torque, however, increases linearly with distance from the wire. Thus, the effects of the decrease in field strength and the increase in torque with distance cancel, and there is no net torque on the ring-magnet combination.

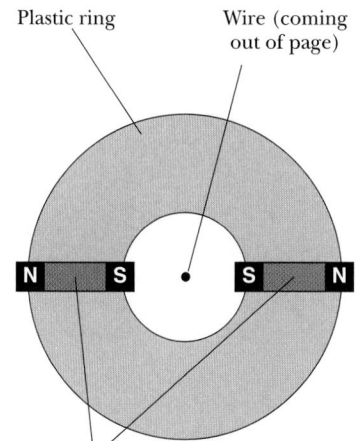

Figure 19.24
(Thinking Physics 3)

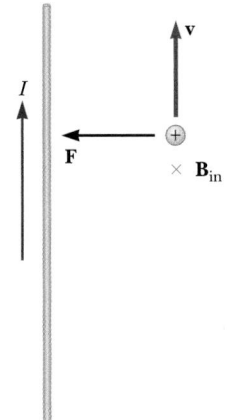

Figure 19.25 (Example 19.7) The magnetic field due to the current is into the page at the location of the proton, and the magnetic force on the proton is to the left.

EXAMPLE 19.7 The Magnetic Field of a Long Wire

A long, straight wire carries a current of 5.00 A. At one instant, a proton, 4.00 mm from the wire, travels at 1.50×10^3 m/s parallel to the wire and in the same direction as the current (Fig. 19.25). Find the magnitude and direction of the magnetic force that is acting on the proton because of the magnetic field produced by the wire.

Solution From Equation 19.11, the magnitude of the magnetic field produced by the current at a point 4.00 mm from the wire is

$$B = \frac{\mu_0 I}{2\pi r} = \frac{(4\pi \times 10^{-7} \ \text{T·m/A})(5.00 \ \text{A})}{2\pi(4.00 \times 10^{-3} \ \text{m})} = 2.50 \times 10^{-4} \ \text{T}$$

This field is directed into the page at the location of the proton, as shown by the right-hand rule for a long, straight wire (see Fig. 19.23a).

The magnetic force on the proton is

$$F = qvB = (1.60 \times 10^{-19} \ \text{C})(1.50 \times 10^3 \ \text{m/s})(2.50 \times 10^{-4} \ \text{T})$$

$$= \ 6.00 \times 10^{-20} \ \text{N}$$

The force is directed toward the wire, as shown by the right-hand rule for the force on a moving charge (see Fig. 19.6).

19.9   MAGNETIC FORCE BETWEEN TWO PARALLEL CONDUCTORS

As we have seen, a magnetic force acts on a current-carrying conductor when the conductor is placed in an external magnetic field. Because a current in a conductor creates its own magnetic field, it is easy to understand that two current-carrying wires placed close together exert magnetic forces on each other. Consider two long, straight, parallel wires separated by the distance d and carrying currents I_1 and I_2 in the same direction, as shown in Figure 19.26. Let us determine the magnetic force on one wire due to a magnetic field set up by the other wire.

Wire 2, which carries current I_2, sets up magnetic field $\mathbf{B}_2$ at wire 1. The direction of $\mathbf{B}_2$ is perpendicular to the wire, as shown in the figure. Using Equation 19.11, we see that the magnitude of this magnetic field is

$$B_2 = \frac{\mu_0 I_2}{2\pi d}$$

According to Equation 19.5, the magnetic force on wire 1 in the presence of field $\mathbf{B}_2$ due to I_2 is

$$F_1 = B_2 I_1 \ell = \left[\frac{\mu_0 I_2}{2\pi d}\right] I_1 \ell = \frac{\mu_0 I_1 I_2 \ell}{2\pi d}$$

We can rewrite this in terms of the force per unit length:

$$\frac{F_1}{\ell} = \frac{\mu_0 I_1 I_2}{2\pi d} \qquad\qquad [19.14]$$

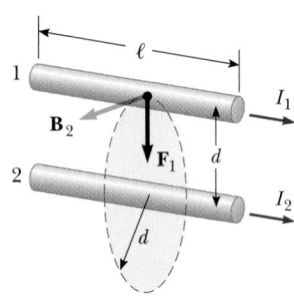

Figure 19.26 Two parallel wires, each carrying a steady current, exert forces on each other. The field at wire 1 due to wire 2, $\mathbf{B}_2$, produces a force on wire 1 given by $F_1 = B_2 I_1 \ell$. The force is attractive if the currents have the same direction, as shown, and repulsive if the two currents have opposite directions.

The direction of $\mathbf{F}_1$ is downward, toward wire 2, as indicated by the right-hand rule. If one considers the field set up at wire 2 due to wire 1, the force $\mathbf{F}_2$ on wire 2 is found to be equal to and opposite $\mathbf{F}_1$. This is what one would expect from Newton's third law of action-reaction.

We have shown that parallel conductors carrying currents in the same direction *attract* each other. You should use the approach indicated by Figure 19.26 and the steps leading to Equation 19.14 to show that parallel conductors carrying currents in opposite directions *repel* each other.

The force between two parallel wires carrying a current is used to define the SI unit of current, the **ampere (A),** as follows:

> If two long, parallel wires 1 m apart carry the same current, and the magnetic force per unit length on each wire is 2×10^{-7} N/m, then the current is defined to be 1 A.

The SI unit of charge, the **coulomb (C),** can now be defined in terms of the ampere as follows:

> If a conductor carries a steady current of 1 A, then the quantity of charge that flows through any cross section in 1 s is 1 C.

Workbook Problem 77 — (Workbook page 165) shows how to apply Ampère's law to the case of a packed bundle of wires.

EXAMPLE 19.8 Levitating a Wire

Two wires, each having a weight per unit length of 1.0×10^{-4} N/m, are strung parallel to one another above the surface of the Earth, one directly above the other. The wires are aligned in a north–south direction so that the Earth's magnetic field will not affect them. When their distance of separation is 0.10 m, what must be the current in each in order for the lower wire to levitate the upper wire? Assume that the wires carry the same currents, traveling in opposite directions.

Solution If the upper wire is to float, it must be in equilibrium under the action of two forces: the force of gravity and magnetic repulsion. The weight per unit length—here 1.0×10^{-4} N/m—must be equal and opposite the magnetic force per unit length given in Equation 19.14. Because the currents are the same, we have

$$\frac{F_1}{\ell} = \frac{mg}{\ell} = \frac{\mu_0 I^2}{2\pi d}$$

$$1.0 \times 10^{-4}\,\text{N/m} = \frac{(4\pi \times 10^{-7}\,\text{T·m/A})(I^2)}{(2\pi)(0.10\,\text{m})}$$

We solve for the current to find

$$I = \boxed{7.1\,\text{A}}$$

Exercise If the current in each wire is doubled, what is the equilibrium separation of the two wires?

Answer 0.40 m

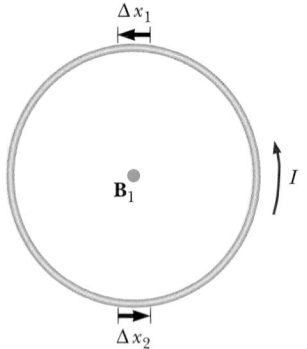

Figure 19.27   All segments of the current loop produce a magnetic field at the center of the loop, directed *out of the page*.

19.10   MAGNETIC FIELD OF A CURRENT LOOP

The strength of the magnetic field set up by a piece of wire carrying a current can be enhanced at a specific location if the wire is formed into a loop. You can understand this by considering the effect of several small segments of the current loop, as in Figure 19.27. The small segment at the top of the loop, labeled Δx_1, produces at the loop's center a magnetic field of magnitude B_1, directed out of the page. The direction of **B** can be verified using the right-hand rule for a long, straight wire. Imagine holding the wire with your right hand, with your thumb pointing in the direction of the current. Your fingers curl around in the direction of **B.**

A segment at the bottom of the loop, Δx_2, also contributes to the field at the center, thus increasing its strength. The field produced at the center of the current loop by the segment Δx_2 has the same magnitude as B_1 and is also directed out of the page. Similarly, all other such segments of the current loop contribute to the field. The net effect is a magnetic field for the current loop as pictured in Figure 19.28a.

Notice in Figure 19.28a that the magnetic field lines enter at the left side of the current loop and exit at the right. Thus, one side of the loop acts as though it were the north pole of a magnet, and the other acts as a south pole. The fact that the field set up by such a current loop bears a striking resemblance to the field of a bar magnet (Fig. 19.28c) will be of interest to us in a future section.

Applying Physics 2

In electrical circuits, it is often the case that wires carrying currents in opposite directions are twisted together. What is the advantage of doing this?

Explanation   If the wires are not twisted together, the combination of the two wires forms a current loop, which produces a relatively strong magnetic field. In fact, the magnetic field generated by the loop could be strong enough to affect adjacent circuits or components.

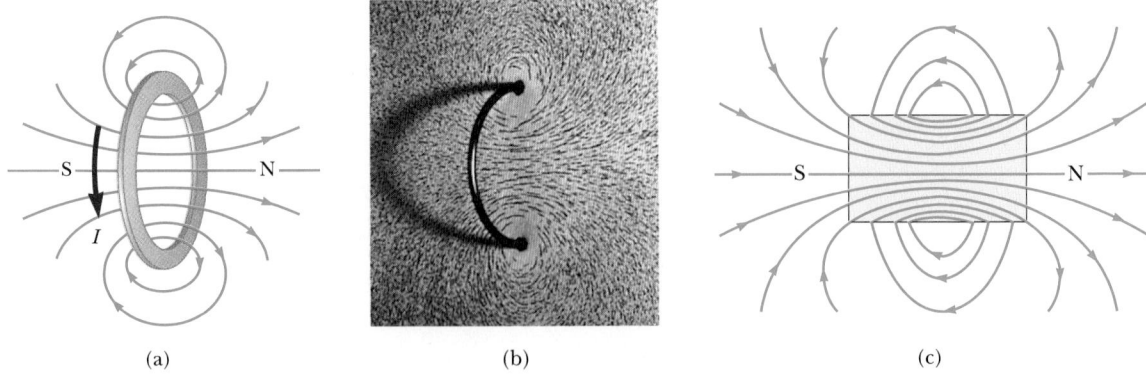

(a)   (b)   (c)

Figure 19.28   (a) Magnetic field lines for a current loop. Note that the magnetic field lines of the current loop resemble those of a bar magnet. (b) Field lines of a current loop, displayed by iron filings. *(Education Development Center, Newton, Mass.)* (c) The magnetic field of a bar magnet is similar to that of a current loop.

19.11 MAGNETIC FIELD OF A SOLENOID

If a long, straight wire is bent into a coil of several closely spaced loops, the resulting device is a **solenoid,** often called an **electromagnet.** This device is important in many applications because it acts as a magnet only when it carries a current. As we shall see, the magnetic field inside a solenoid increases with the current and is proportional to the number of coils per unit length.

Figure 19.29 shows the magnetic field lines of a loosely wound solenoid of length ℓ and total number of turns N. Note that the field lines inside the solenoid are nearly parallel, uniformly spaced, and close together. This indicates that the field inside the solenoid is nearly uniform and strong. The exterior field at the sides of the solenoid is nonuniform and is much weaker than the interior field.

If the turns are closely spaced, the field lines are as shown in Figure 19.30a, entering at one end of the solenoid and emerging at the other. This means that one end of the solenoid acts as a north pole and the other end acts as a south pole. If the length of the solenoid is much greater than its radius, the lines that leave the north end of the solenoid spread out over a wide region before returning to enter the south end. Hence, as you can see in Figure 19.30a, the magnetic field lines outside are widely separated, indicative of a weak field. This is in contrast to a much stronger field inside the solenoid, where the lines are close together. Also, the magnetic field inside the solenoid has a constant magnitude at all points far from its ends. The expression for the magnetic field inside the solenoid is

$$B = \mu_0 n I \qquad [19.15]$$

where $n = N/\ell$ is the number of turns per unit length.

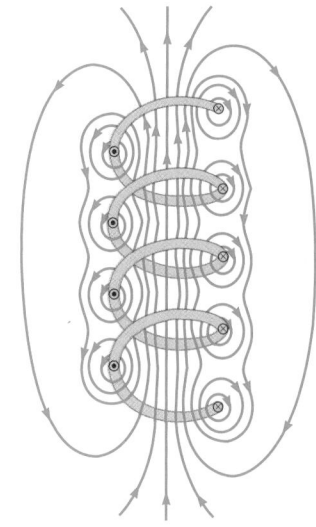

Figure 19.29 The magnetic field lines for a loosely wound solenoid.

◀ The magnetic field inside a solenoid

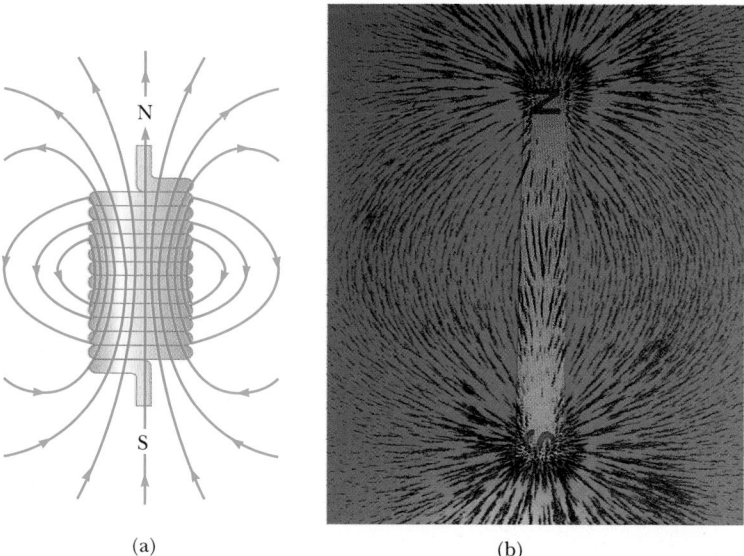

(a)   (b)

Figure 19.30 (a) Magnetic field lines for a tightly wound solenoid of finite length carrying a steady current. The magnetic field inside the solenoid is nearly uniform and strong. Note that the field lines resemble those of a bar magnet, so the solenoid effectively has north and south poles. (b) The magnetic field pattern of a bar magnet, displayed by small iron filings on a sheet of paper. *(Courtesy of Henry Leap and Jim Lehman)*

EXAMPLE 19.9   The Magnetic Field Inside a Solenoid

A certain solenoid consists of 100 turns of wire and has a length of 10.0 cm.
(a) Find the magnetic field inside the solenoid when it carries a current of 0.500 A.

Solution   The number of turns per unit length is

$$n = \frac{N}{L} = \frac{100 \text{ turns}}{0.10 \text{ m}} = 1000 \text{ turns/m}$$

so

$$B = \mu_0 n I = (4\pi \times 10^{-7} \text{ T·m/A})(1000 \text{ turns/m})(0.500 \text{ A}) = \boxed{6.28 \times 10^{-4} \text{ T}}$$

(b) Assume that the field has one half this value just outside the solenoid, like the point labeled N in Figure 19.30a. Find the magnitude and direction of the magnetic force acting on an electron that is moving from right to left in the figure, through point N, at 375 m/s.

Solution   The magnitude of the magnetic force on the electron is

$$F = qvB = (1.60 \times 10^{-19} \text{ C})(375 \text{ m/s})(3.14 \times 10^{-4} \text{ T}) = \boxed{1.88 \times 10^{-20} \text{ N}}$$

By use of the right-hand rule in Figure 19.6, the direction of this force is found to be out of the page. (Do not forget to change the direction of the magnetic force for the negatively charged electron.) This force will deflect the electron from its original direction of motion.

Exercise   How many turns should the solenoid have (assuming it carries the same current) if the field inside is to be five times as great?

Answer   500 turns

So-called steering magnets placed along the neck of the picture tube in a television set, as in Figure 19.31, are used to make the electron beam move to the desired locations on the screen, thus tracing out the images of your favorite program.

APPLICATION

Controlling the Electron
Beam in a Television Set.

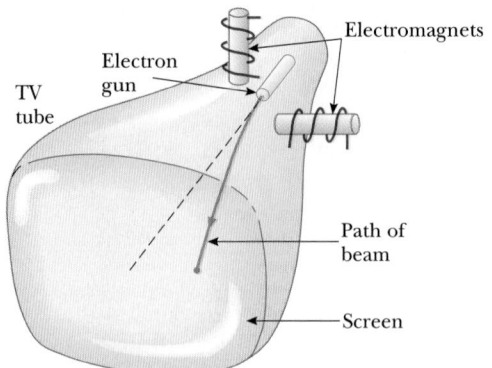

Figure 19.31   Electromagnets are used to deflect electrons to desired positions on the screen of a television tube.

Ampère's Law Applied to a Solenoid

We can use Ampère's law to obtain the expression for the magnetic field inside a solenoid carrying a current I. A cross section taken along the length of part of our solenoid is shown in Figure 19.32. **B** inside the solenoid is uniform and parallel to the axis, and **B** outside is zero. Consider a rectangular path of length L and width w, as shown in Figure 19.32. We can apply Ampère's law to this path by evaluating the sum of $B\,\Delta\ell$ over each side of the rectangle. The contribution along side 3 is clearly zero, because **B** = 0 in this region. The contributions from sides 2 and 4 are both zero, because **B** is perpendicular to $\Delta\ell$ along these paths. Side 1, the length of which is L, gives a contribution of BL to the sum, because **B** along this path is uniform and parallel to $\Delta\ell$. Therefore, the sum over the closed rectangular path has the value

$$\sum B\,\Delta\ell = BL$$

The right side of Ampère's law involves the total current that passes through the area bounded by the path chosen. In our case, the total current through the rectangular path equals the current through each turn of the solenoid multiplied by the number of turns. If N is the number of turns in the length L, then the total current through the rectangular path equals NI. Therefore, Ampère's law applied to this path gives

$$\sum B\,\Delta\ell = BL = \mu_0 NI$$

$$B = \mu_0 \frac{N}{L} I = \mu_0 n I$$

where $n = \dfrac{N}{L}$ is the number of turns per unit length.

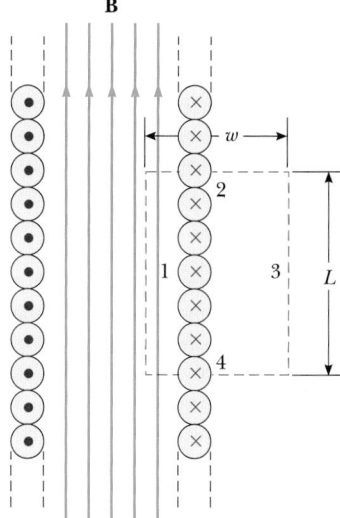

Figure 19.32 A cross-sectional view of a tightly wound solenoid. If the solenoid is long relative to its radius, we can assume that the magnetic field inside is uniform and the field outside is zero. Ampère's law applied to the red dashed rectangular path can then be used to calculate the field inside the solenoid.

19.12   MAGNETIC DOMAINS

The magnetic field produced by a current in a coil of wire gives us a hint as to what might cause certain materials to exhibit strong magnetic properties. A single coil like that in Figure 19.28 has a north pole and a south pole, but if this is true for a coil of wire, it should also be true for any current confined to a circular path. In particular, an individual atom should act as a magnet because of the motion of the electrons about the nucleus. Each electron, with its charge of 1.6×10^{-19} C, circles the atom once in about 10^{-16} s. If we divide the electronic charge by this time interval, we see that the orbiting electron is equivalent to a current of 1.6×10^{-3} A. Such a current produces a magnetic field on the order of 20 T at the center of the circular path. From this we see that a very strong magnetic field would be produced if several of these atomic magnets could be aligned inside a material. This does not occur, however, because the simple model we have described is not the complete story. A thorough analysis of atomic structure shows that the magnetic field produced by one electron in an atom is often canceled by an oppositely revolving electron in the same atom. The net result is that **the magnetic effect produced by the electrons orbiting the nucleus is either zero or very small for most materials.**

PHYSICS IN ACTION

The Motion of Charged Particles in Magnetic Fields

WEB

For a Web link to an experiment showing that grapes are repelled by magnets, visit the textbook Web site at http://www.harcourtcollege.com/physics/cptech

The blue-white arc in the photograph on the left indicates the circular path followed by an electron beam in a magnetic field. The vessel contains gas at very low pressure, and the beam is made visible as the electrons collide with the gas atoms, which emit visible light. The magnetic field is produced by two coils (not shown). The apparatus can be used to measure the ratio of e/m for the electron.

On the right, oxygen, a paramagnetic substance, is attracted to a magnetic field. The liquid oxygen in this photograph is suspended between the poles of a permanent magnet. Paramagnetic substances contain atoms (or ions) that have permanent magnetic dipole moments. These dipoles interact weakly with each other and are randomly oriented in the absence of an external magnetic field. When the substance is placed in an external magnetic field, its atomic dipoles tend to line up with the field.

(Courtesy of Central Scientific Company)

(Courtesy of Leon Lewandowski)

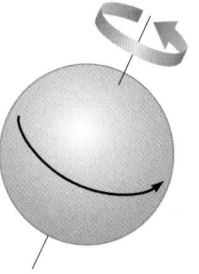

Figure 19.33 Classical model of a spinning electron.

The magnetic properties of many materials are explained by the fact that an electron not only circles in an orbit but also spins on its axis like a top (Fig. 19.33). (This classical description should not be taken literally. The property of *spin* can be understood only with the methods of quantum mechanics, which we shall not discuss here.) The spinning electron represents a charge in motion that produces a magnetic field. The field due to the spinning is generally stronger than the field due to the orbital motion. In atoms containing many electrons, the electrons usually pair up with their spins opposite each other, so that their fields cancel each other. That is why most substances are not magnets. However, in certain strongly magnetic materials such as iron, cobalt, and nickel, the magnetic fields produced by the electron spins do not cancel completely. Such materials are said to be **ferromagnetic.** In ferromagnetic materials, strong coupling occurs between neighboring atoms, to form large groups of atoms whose spins are aligned, called **domains.** Typically, the sizes of these groups range from about 10^{-4} cm to 0.1 cm. In an

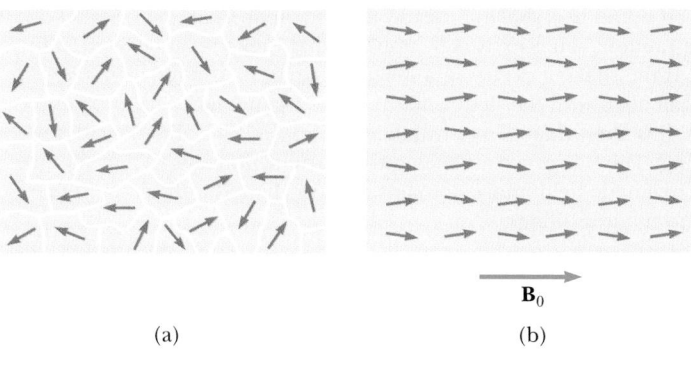

Figure 19.34 (a) Random orientation of domains in an unmagnetized substance. (b) When an external magnetic field, **B**$_0$, is applied, the domains tend to align with the magnetic field.

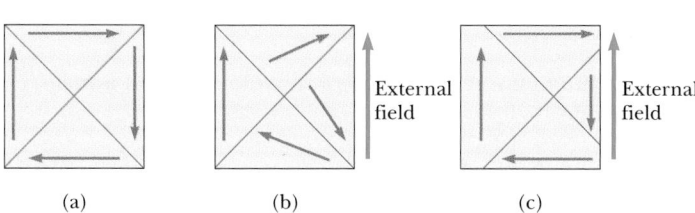

Figure 19.35 (a) An unmagnetized sample of four domains. (b) Magnetization due to rotation of the domains. (c) Magnetization due to a shift in domain boundaries.

unmagnetized substance the domains are randomly oriented, as shown in Figure 19.34a. When an external field is applied, as in Figures 19.34b, the magnetic field of each domain tends to come nearer alignment with the external field, resulting in magnetization.

In some substances a second effect occurs: Domains that are already aligned with the field tend to grow at the expense of the others (Fig. 19.35). Thus, two effects, both dependent on the domains, are responsible for a material's becoming magnetized.

In what are called hard magnetic materials, domain alignment persists after the external field is removed; the result is a **permanent magnet.** In soft magnetic materials, such as iron, once the external field is removed, thermal agitation produces domain motion and the material quickly returns to an unmagnetized state.

The alignment of domains explains why the strength of an electromagnet is increased dramatically by the insertion of an iron core into the magnet's center. The magnetic field produced by the current in the loops causes alignment of the domains, thus producing a large net external field. The use of iron as a core is also advantageous because it is a soft magnetic material and loses its magnetism almost instantaneously after the current in the coils is turned off.

Thinking Physics 4

Suppose a cardboard tube is filled with very fine iron filings. Each filing is in the form of a sliver—long and thin. If the tube of filings is shaken, what is the likelihood that all of them will wind up aligned in the same direction? Suppose the shaking is repeated while the tube is in a strong magnetic field. What is the likelihood now that the filings will be aligned? Is this a macroscopic analog to anything discussed in this section?

QUICKLAB

Construct an electromagnet by wrapping about 1 meter of small-diameter insulated wire around a steel nail. Tape the ends of the wires to a D-cell battery as in the figure. How many staples or paper clips can you pick up with your electromagnet? How would you increase the magnetization of the nail? Disconnect the wires from the battery, and test the magnetization of the nail by seeing how many staples it can pick up.

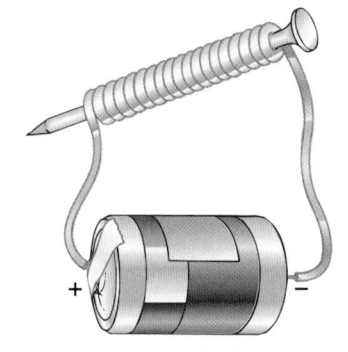

Explanation When the tube is shaken outside of a magnetic field, the orientation of the filings will be random—it is extremely unlikely that they will all be aligned in the same direction. If the shaking occurs in a magnetic field, the filings, which become magnetized, can align with the field while they are momentarily airborne and free to rotate. Thus, it is likely that a large number of filings will be aligned with the field after the shaking. The filings are an analog to magnetic domains. When the material is placed in a magnetic field, more domains end up aligned with the field than misaligned, similar to the iron filings.

SUMMARY

12.11

The **magnetic force** that acts on a charge, q, moving with velocity, **v**, in a magnetic field, **B**, has the magnitude

$$F = qvB \sin \theta \qquad [19.1]$$

where θ is the angle between **v** and **B**.

To find the direction of this force, you can use the **right-hand rule:** Place the fingers of your open right hand in the direction of **B** and point your thumb in the direction of the velocity, **v**. The force, **F**, on a positive charge is directed out of the palm of your hand.

If the charge is *negative* rather than positive, the force is directed opposite the force given by the right-hand rule.

The SI unit of magnetic field is the **tesla (T),** or weber per square meter (Wb/m^2). An additional commonly used unit for magnetic field is the **gauss (G);** $1 \text{ T} = 10^4 \text{ G}$.

If a straight conductor of length ℓ carries current I, the magnetic force on that conductor when it is placed in a uniform external magnetic field, **B,** is

$$F = BI\ell \sin \theta \qquad [19.6]$$

The right-hand rule also gives the direction of the magnetic force on the conductor. In this case, however, you must place your thumb in the direction of the current rather than in the direction of **v.**

The torque, τ, on a current-carrying loop of wire in a magnetic field, **B,** has the magnitude

$$\tau = BIA \sin \theta \qquad [19.8]$$

where I is the current in the loop and A is its cross-sectional area. The angle between **B** and a line drawn perpendicularly to the plane of the loop is θ.

The galvanometer can be used in the construction of both ammeters and voltmeters.

If a charged particle moves in a uniform magnetic field so that its initial velocity is perpendicular to the field, it will move in a circular path whose plane is perpendicular to the magnetic field. The radius, r, of the circular path is

$$r = \frac{mv}{qB} \qquad [19.10]$$

where m is the mass of the particle and q is its charge.

The magnetic field at distance r from a **long, straight wire** carrying current I has the magnitude

$$B = \frac{\mu_0 I}{2\pi r}$$ [19.11]

where $\mu_0 = 4\pi \times 10^{-7}$ T·m/A is the **permeability of free space.** The magnetic field lines around a long, straight wire are circles concentric with the wire.

Ampère's law can be used to find the magnetic field around certain simple current-carrying conductors. It can be written

$$\sum B_\parallel \, \Delta\ell = \mu_0 I$$ [19.13]

where $B_\parallel$ is the component of **B** tangent to a small current element of length $\Delta\ell$ that is part of a closed path, and I is the total current that penetrates the closed path.

The force per unit length on each of two parallel wires separated by the distance d and carrying currents I_1 and I_2 has the magnitude

$$\frac{F}{\ell} = \frac{\mu_0 I_1 I_2}{2\pi d}$$ [19.14]

The forces are attractive if the currents are in the same direction and repulsive if they are in opposite directions.

The magnetic field inside a solenoid has the magnitude

$$B = \mu_0 n I$$ [19.15]

where n is the number of turns of wire per unit length, $n = N/\ell$.

MULTIPLE-CHOICE QUESTIONS

1. A current in a long straight wire runs along the y axis in the positive direction, and another current in another long straight wire runs along the x axis in the positive direction, as in Figure 19.36. The axes divide the plane into 4 quadrants (I, II, III, and IV). In which quadrant is the contribution to the net magnetic field from both wires directed outward ($+z$ direction)?
   (a) I   (b) II   (c) III   (d) IV
   (e) It happens in more than one quadrant.

2. Once again, consider Figure 19.36. The quadrants in which one wire produces a field in the $+z$ direction and the second wire produces a field in the $-z$ direction are
   (a) I and II   (b) II and III   (c) I and III   (d) II and IV
   (e) not given as a choice

3. A thin copper rod 1.0 m long has a mass of 50 g. What is the minimum current in the rod that would allow it to "float" in a magnetic field of 0.1 T?
   (a) 1.2 A   (b) 2.5 A   (c) 4.9 A   (d) 9.8 A

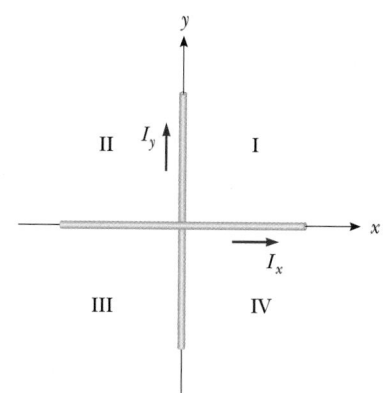

Figure 19.36 (Multiple-Choice Questions 1 and 2)

4. An electron moves across the Earth's Equator at a speed of 2.5×10^6 m/s and in a direction 35° N of E. At this point the Earth's magnetic field has a direction due north, is parallel to the surface, and has a value of 0.10×10^{-4} T. What is the magnitude of the force acting on the electron due to its interaction with the Earth's magnetic field?
(a) 5.1×10^{-18} N   (b) 4.0×10^{-18} N   (c) 3.3×10^{-18} N
(d) 2.3×10^{-18} N

5. What is the direction of the force on the electron in Question 4?
(a) due west   (b) into the Earth's surface
(c) out of the Earth's surface   (d) due south

6. A 10 Ω, 25-mA galvanometer is to be converted into an ammeter that reads 5.0 A at a full-scale deflection. What resistance should be placed in parallel with the galvanometer coil?
(a) 2.5 Ω   (b) 0.50 Ω   (c) 0.25 Ω   (d) 0.050 Ω

7. Repeat Question 6 except the conversion is to make the galvanometer into a 20-V voltmeter. The resistance required in series is
(a) 810 Ω   (b) 790 Ω   (c) 500 Ω   (d) 450 Ω

CONCEPTUAL QUESTIONS

1. If a charged particle moves in a straight line through some region of space, can you say that the magnetic field in that region is zero?

2. A current-carrying conductor experiences no magnetic force when placed in a certain manner in a uniform magnetic field. Explain.

3. How can the motion of a charged particle be used to distinguish between a magnetic field and an electric field in a certain region?

4. Which way would a compass point if you were at the north magnetic pole of the Earth?

5. Why does the picture on a television screen become distorted when a magnet is brought near the screen, as in Figure Q19.5? *Caution:* You should not do this at home on a color television set, because it may permanently affect the television picture quality.

Figure Q19.5 (*Courtesy of Henry Leap and Jim Lehman*)

6. A magnet attracts a piece of iron. The iron can then attract another piece of iron. On the basis of domain alignment, explain what happens in each piece of iron.

7. A charged particle moves in a circular path in the presence of a magnetic field applied perpendicular to the particle's velocity. Does the particle gain energy from the magnetic field?

8. Will a nail be attracted to either pole of a magnet? Explain what is happening inside the nail when placed near the magnet.

9. Suppose you move along a wire at the same speed as the drift speed of the electrons in the wire. Do you now measure a magnetic field of zero?

10. Why does hitting a magnet with a hammer cause the magnetism to be reduced?

11. Can you use a compass to detect the currents in wires in the walls near light switches in your home?

12. It is found that charged particles from outer space, called cosmic rays, strike the Earth more frequently at the poles than at the Equator. Why?

13. Two wires carry currents in opposite directions and are oriented parallel, with one above the other. The wires repel each other. Is the upper wire in a stable levitation over the lower wire? Suppose the current in one wire is reversed, so that the wires now attract. Is the lower wire hanging in a stable attraction to the upper wire?

14. How can a current loop be used to determine the presence of a magnetic field in a given region of space?

15. A hanging Slinky toy is attached to a powerful battery and a switch. When the switch is closed so that current suddenly flows through the Slinky, does the Slinky compress or expand?

16. Is it possible to orient a current loop in a uniform magnetic field such that the loop will not tend to rotate?

17. Two charged particles are projected into a region in which there is a magnetic field perpendicular to their velocities. If the charges are deflected in opposite directions, what can you say about them?

18. Can a constant magnetic field set into motion an electron at rest? Explain your answer.

19. Parallel wires exert magnetic forces on each other. What about perpendicular wires? Imagine two wires oriented perpendicular to each other and almost touching. Each wire carries a current. Is there a force between the wires?

20. A Hindu ruler once suggested that he be entombed in a magnetic coffin with the polarity arranged so that he could be forever suspended between heaven and Earth. Is such magnetic levitation possible? Discuss.

PROBLEMS

1, **2, 3** = straightforward, intermediate, challenging   ☐ = full solution available in Study Guide/Student Solutions Manual   🏃 = Core Concepts Workbook
WEB = solution posted at **http://www.harcourtcollege.com/physics/cptech**   📠 = biomedical application   🖥 = Interactive Physics

Review Problem

It is desired to construct a solenoid that will have a resistance of 5.00 Ω (at 20°C) and produce a magnetic field at its center of 4.00×10^{-2} T when a current of 4.00 A flows through it. The solenoid is to be constructed from copper wire having a diameter of 0.500 mm. If the radius of the solenoid is to be 1.00 cm, determine **(a)** the number of turns of wire needed and **(b)** the length the solenoid should have.

Section 19.3   Magnetic Fields

1. An electron gun fires electrons into a magnetic field that is directed straight downward. Find the direction of the force exerted on an electron by the field for each of the following directions of the electron's velocity: **(a)** horizontal and due north; **(b)** horizontal and 30° west of north; **(c)** due north, but at 30° below the horizontal; **(d)** straight upward. (Remember that an electron has a negative charge.)

2. **(a)** Find the direction of the force on a proton (a positively charged particle) moving through the magnetic fields in Figure P19.2, as shown. **(b)** Repeat part (a), assuming the moving particle is an electron.

3. Find the direction of the magnetic field on the positively charged particle moving in the various situations shown in Figure P19.3, if the direction of the magnetic force acting on it is as indicated.

4. A duck flying horizontally due north at 15 m/s passes over Atlanta, where the magnetic field of the Earth is 5.0×10^{-5} T in a direction 60° below a horizontal line running north and south. The duck has a positive charge of 4.0×10^{-8} C. What is the magnetic force acting on the duck?

5. A proton travels with a speed of 3.0×10^6 m/s at an angle of 37° with the direction of a magnetic field of 0.30 T in the $+y$ direction. What are **(a)** the magnitude of the magnetic force on the proton and **(b)** the proton's acceleration?

6. What speed would a proton need to circle the Earth 1000 km above the magnetic equator, where the Earth's

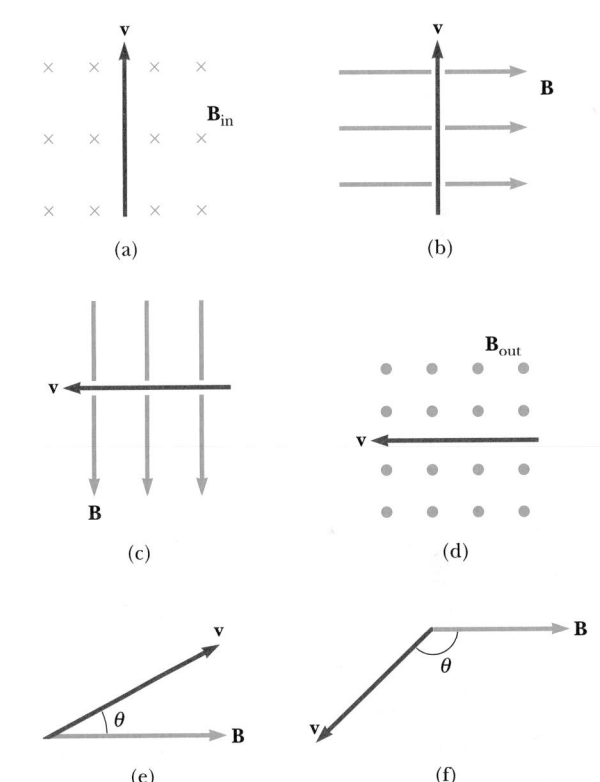

Figure P19.2   (Problems 2 and 13) For Problem 13, replace the velocity vector with a current in that direction.

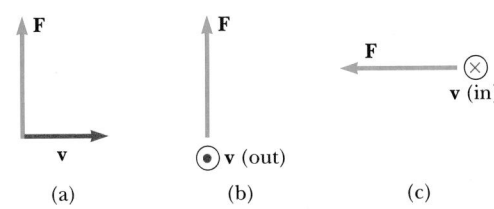

Figure P19.3   (Problems 3 and 12) For Problem 12, replace the velocity vector with a current in that direction.

magnetic field is directed on a line between magnetic north and south and has an intensity of 4.00×10^{-8} T?

WEB 7. An electron is accelerated through 2400 V from rest and then enters a region in which there is a uniform 1.70-T magnetic field. What are the (a) maximum and (b) minimum magnitudes of the magnetic force this charge can experience?

8. A proton moves perpendicularly to a uniform magnetic field, **B**, at 1.0×10^7 m/s and experiences an acceleration of 2.0×10^{13} m/s^2 in the $+x$ direction when its velocity is in the $+z$ direction. Determine the magnitude and direction of the field.

9. Sodium ions (Na$^+$) move at 0.851 m/s through a bloodstream in the arm of a person standing near a large magnet. The magnetic field has a strength of 0.254 T and makes an angle of 51.0° with the motion of the sodium ions. The arm contains 100 cm^3 of blood with 3.00×10^{20} Na$^+$ ions per cubic centimeter. If no other ions were present in the arm, what would be the magnetic force on the arm?

10. Show that the work done by the magnetic force on a charged particle moving in a uniform magnetic field is zero for any displacement of the particle.

Section 19.4 Magnetic Force on a Current-Carrying Conductor

11. A current, $I = 15$ A, is directed along the positive x axis and perpendicularly to a magnetic field. The conductor experiences a magnetic force per unit length of 0.12 N/m in the negative y direction. Calculate the magnitude and direction of the magnetic field in the region through which the current passes.

12. In Figure P19.3, assume that in each case the velocity vector shown is replaced with a wire carrying a current in the direction of the velocity vector. For each case, find the direction of the magnetic field that will produce the magnetic force shown.

13. In Figure P19.2, assume that in each case the velocity vector shown is replaced with a wire carrying a current in the direction of the velocity vector. For each case, find the direction of the magnetic force acting on the wire.

14. A wire carries a current of 10.0 A in a direction that makes an angle of 30.0° with the direction of a magnetic field of strength 0.300 T. Find the magnetic force on a 5.00-m length of the wire.

15. The Earth has a magnetic field of 0.60×10^{-4} T, pointing 75° below the horizontal in a north–south plane. A 10.0-m-long straight wire carries a 15-A current. (a) If the current is directed horizontally toward the east, what are the magnitude and direction of the magnetic force

on the wire? (b) What are the magnitude and direction of the force if the current is directed vertically upward?

16. A wire with a mass per unit length of 1.00 g/cm is placed on a horizontal surface with a coefficient of friction of 0.200. The wire carries a current of 1.50 A eastward and moves horizontally to the north. What are the magnitude and the direction of the *smallest* vertical magnetic field that enables the wire to move in this fashion?

17. An unusual message delivery system is pictured in Figure P19.17. A 15-cm length of conductor that is free to move is held in place between two thin conductors. When a 5.0-A current is directed as shown in the figure, the wire segment moves upward at a constant velocity. If the mass of the wire is 15 g, find the magnitude and direction of the minimum magnetic field that is required to move the wire. (The wire slides without friction on the two vertical conductors.)

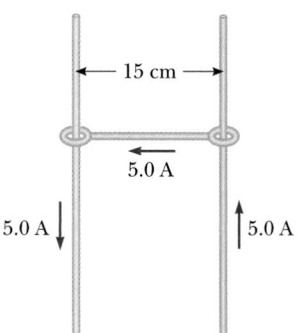

Figure P19.17

18. A thin, horizontal copper rod is 1.00 m long and has a mass of 50.0 g. What is the minimum current in the rod that can cause it to float in a horizontal magnetic field of 2.00 T?

Section 19.5 Torque on a Current Loop

19. A single circular wire loop of radius 50.0 cm, carrying a current of 2.00 A, is in a magnetic field of 0.400 T. (a) Find the maximum torque that acts on this loop. (b) Find the angle that the plane of the loop makes with the field when the torque is one half the value found in part (a).

20. An 8-turn coil encloses an elliptical area having a major axis of 40.0 cm and a minor axis of 30.0 cm (Fig. P19.20). The coil lies in the plane of the page and has a 6.00 A current flowing clockwise around it. If the coil is in a uniform magnetic field of 2.00×10^{-4} T, directed toward the left of the page, what is the magni-

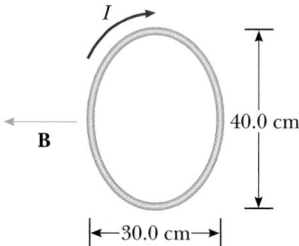

Figure P19.20

tude of the torque on the coil? (*Hint:* The area of an ellipse is $A = \pi ab$, where a and b are the semimajor and semiminor axes of the ellipse.)

21. A rectangular loop consists of 100 closely wrapped turns and has dimensions 0.40 m by 0.30 m. The loop is hinged along the y axis, and the plane of the coil makes an angle of 30.0° with the x axis (Fig. P19.21). What is the magnitude of the torque exerted on the loop by a uniform magnetic field of 0.80 T directed along the x axis, when the current in the windings has a value of 1.2 A in the direction shown? What is the expected direction of rotation of the loop?

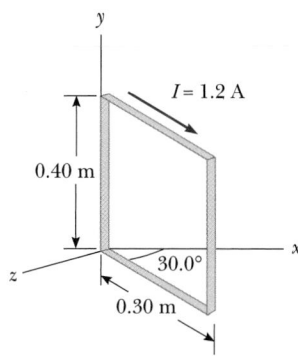

Figure P19.21

22. A 2.00-m-long wire carrying a current of 2.00 A forms a 1-turn loop in the shape of an equilateral triangle. If the loop is placed in a constant magnetic field of magnitude 0.500 T, determine the *maximum* torque that acts on it.

23. A copper wire is 8.00 m long and has a cross-sectional area of 1.00×10^{-4} m². This wire forms a 1-turn loop in the shape of a square and is then connected to a 0.100-V battery. If the loop is placed in a uniform magnetic field of magnitude 0.400 T, what is the maximum torque that can act on it? The resistivity of copper is 1.70×10^{-8} Ω·m.

Section 19.6 The Galvanometer and Its Applications

24. A 50.0-Ω, 10.0-mA galvanometer is to be converted to an ammeter that reads 3.00 A at full-scale deflection. What value of R_p should be placed in parallel with the coil?

WEB 25. A galvanometer has a resistance of 50.0 Ω and deflects full scale when the voltage across it is 50.0 mV. What is the magnitude of the shunt resistance needed to convert this galvanometer into an ammeter that reads 10.0 A at full-scale deflection?

26. Consider a galvanometer with an internal resistance of 60.0 Ω. If it deflects full-scale when it carries a current of 0.500 mA, what is the value of the series resistance that must be connected to it if this combination is to be used as a voltmeter having a full-scale deflection for a potential difference of 1.00 V?

27. A 40-Ω, 2.0-mA galvanometer is to be converted to a voltmeter that reads 150 V at full-scale deflection. What value of series resistance should be used with the galvanometer coil?

28. A galvanometer has an internal resistance of 100 Ω and deflects full-scale for a current of 100 μA. This galvanometer is to be converted to a multirange ammeter using the circuit shown in Figure P19.28. Find the values of R_1, R_2, and R_3 that will give the full-scale readings in the figure.

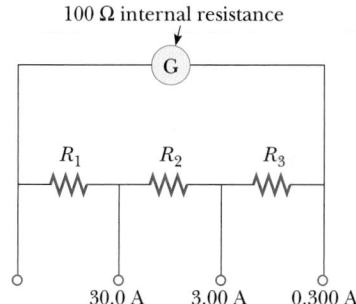

Figure P19.28

29. The galvanometer of Problem 28 is to be converted to a multirange voltmeter using the circuit shown in Figure P19.29. Find the values of R_1, R_2, and R_3 that will enable the meter to give the full-scale readings in the figure.

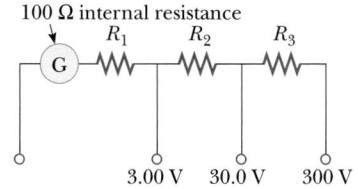

Figure P19.29

Section 19.7 Motion of a Charged Particle in a Magnetic Field

30. Figure P19.30a is a diagram of a device called a velocity selector, in which particles of a specific velocity pass through undeflected and those with greater or lesser velocities are deflected either upward or downward. An electric field is directed perpendicularly to a magnetic field. This produces on the charged particle an electric force and a magnetic force that are equal in magnitude and opposite in direction (Fig. P19.30b), and hence cancel. Show that particles with a speed of $v = E/B$ will pass through undeflected.

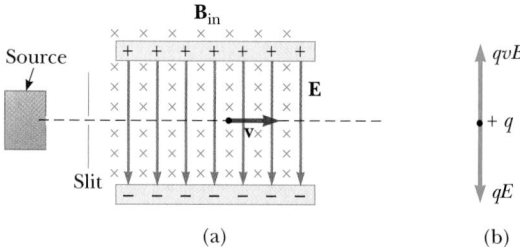

(a)   (b)

Figure P19.30

31. A $+2.0\text{-}\mu\text{C}$ charged particle with a kinetic energy of 0.090 J is fired into a uniform magnetic field of magnitude 0.10 T. If the particle moves in a circular path of radius 3.0 m, determine its mass.

32. Consider the mass spectrometer shown schematically in Figure P19.32. The electric field between the plates of the velocity selector is 950 V/m, and the magnetic fields in both the velocity selector and the deflection chamber have magnitudes of 0.930 T. Calculate the radius of the path in the system for a singly charged ion with mass $m = 2.18 \times 10^{-26}$ kg. (*Hint:* See Problem 30.)

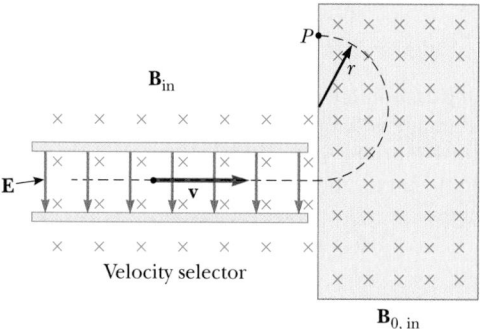

Figure P19.32 A mass spectrometer. Charged particles are first sent through a velocity selector. They then enter a region where a magnetic field, $\mathbf{B}_0$ (inward), causes positive ions to move in a semicircular path and strike a photographic film at P.

33. A singly charged positive ion has a mass of 2.50×10^{-26} kg. After being accelerated through a potential difference of 250 V, the ion enters a magnetic field of 0.500 T, in a direction perpendicular to the field. Calculate the radius of the path of the ion in the field.

34. A mass spectrometer is used to examine the isotopes of uranium. Ions in the beam emerge from the velocity selector at a speed of 3.00×10^5 m/s and enter a uniform magnetic field of 0.600 T directed perpendicularly to the velocity of the ions. What is the distance between the impact points formed on the photographic plate by singly charged ions of ^{235}U and ^{238}U?

35. A proton moves in a circular orbit perpendicularly to a uniform magnetic field of 0.758 T. Find the time it takes the proton to make one pass around the orbit.

Section 19.8 Magnetic Field of a Long, Straight Wire and Ampère's Law

36. Find the direction of the current in the wire in Figure P19.36 that would produce a magnetic field directed as shown, in each case.

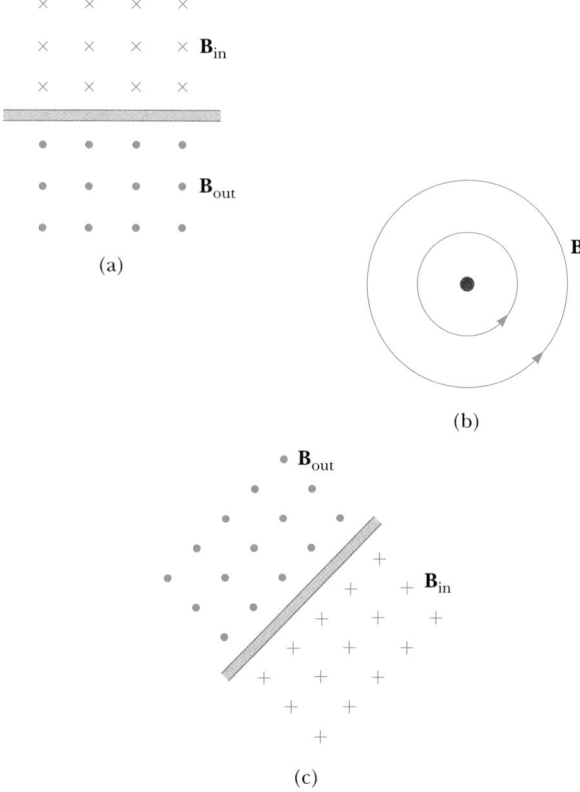

Figure P19.36

37. At what distance from a long, straight wire carrying a current of 5.0 A is the magnetic field due to the wire equal to the strength of the Earth's field, approximately 5.0×10^{-5} T?

38. The two wires shown in Figure P19.38 carry currents of 5.00 A in opposite directions and are separated by 10.0 cm. Find the direction and magnitude of the net magnetic field **(a)** at a point midway between the wires; **(b)** at point P_1—that is, 10.0 cm to the right of the wire on the right; and **(c)** at point P_2,—that is, 20.0 cm to the left of the wire on the left.

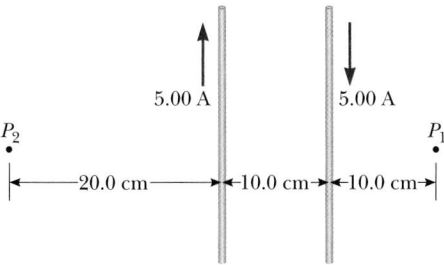

Figure P19.38

39. The two wires in Figure P19.39 carry currents of 3.00 A and 5.00 A in the direction indicated. **(a)** Find the direction and magnitude of the magnetic field at a point midway between the wires. **(b)** Find the magnitude and direction of the magnetic field at point P, located 20.0 cm above the wire carrying the 5.00-A current.

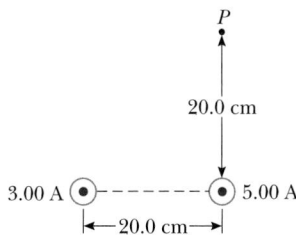

Figure P19.39

40. A very long wire carries a 7.00-A current along the x axis and another long wire carries a 6.00-A current along the y axis, as shown in Figure P19.40. What is the magnetic field at point P located at $x = 4.00$ m, $y = 3.00$ m?

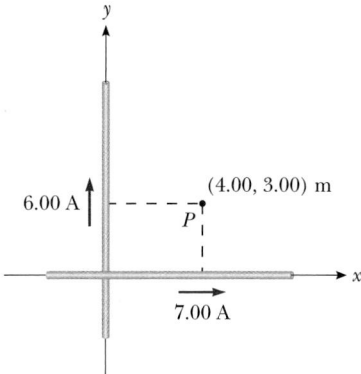

Figure P19.40

41. Figure P19.41 is a cross-sectional view of a coaxial cable (such as a VCR cable). The center conductor is surrounded by a rubber layer, which is surrounded by an outer conductor, which is surrounded by another rubber layer. The current in the inner conductor is 1.00 A out of the page, and the current in the outer conductor is 3.00 A into the page. Using Ampère's law, determine the magnitude and direction of the magnetic field at points A and B.

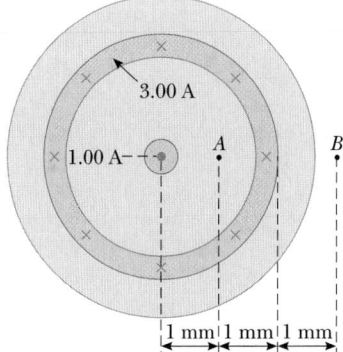

Figure P19.41

Section 19.9 Magnetic Force Between Two Parallel Conductors

42. Two parallel wires are 10.0 cm apart, and each carries a current of 10.0 A. **(a)** If the currents are in the same direction, find the force per unit length exerted on one of the wires by the other. Are the wires attracted or repelled? **(b)** Repeat the problem with the currents in opposite directions.

WEB **43.** A wire with a weight per unit length of 0.080 N/m is suspended directly above a second wire. The top wire carries a current of 30.0 A and the bottom wire carries a current of 60.0 A. Find the distance of separation between the wires so that the top wire will be held in place by magnetic repulsion.

44. Two long, parallel conductors are carrying currents in the same direction, as in Figure P19.44. Conductor A carries a current of 150 A and is held firmly in position;

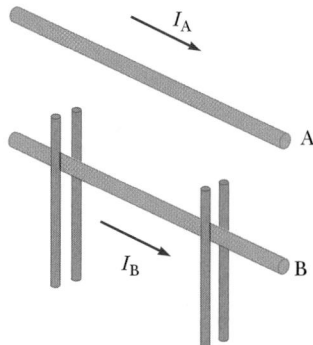

Figure P19.44

conductor B carries current I_B and is allowed to slide freely up and down (parallel to A) between a set of non-conducting guides. If the linear mass density of conductor B is 0.10 g/cm, what value of current I_B will result in equilibrium when the distance between the two conductors is 2.5 cm?

Section 19.11 Magnetic Field of a Solenoid

45. A single-turn square loop of wire, 2.00 cm on a side, carries a current of 0.200 A. The loop is inside a solenoid, with the plane of the loop perpendicular to the magnetic field of the solenoid. The solenoid has 30 turns per centimeter and carries a current of 15.0 A. Find the force on each side of the loop and the torque acting on it.

46. A solenoid of radius $R = 5.00$ cm is made of a long piece of wire of radius $r = 2.00$ mm, length $\ell = 10.0$ m ($\ell \gg R$), and resistivity $\rho = 1.70 \times 10^{-8}\ \Omega \cdot$m. Find the magnetic field at the center of the solenoid if the wire is connected to a battery having an emf $\mathcal{E} = 20.0$ V.

47. An electron moves at a speed of 1.0×10^4 m/s in a circular path of radius 2.0 cm inside a solenoid. The magnetic field of the solenoid is perpendicular to the plane of the electron's path. Find (a) the strength of the magnetic field inside the solenoid and (b) the current in the solenoid if it has 25 turns per centimeter.

ADDITIONAL PROBLEMS

48. A circular coil consisting of a single loop of wire has a radius of 30.0 cm and carries a current of 25 A. It is placed in an external magnetic field of 0.30 T. Find the torque on the wire when the plane of the coil makes an angle of 35° with the direction of the field.

49. Two long, straight wires cross each other at right angles, as shown in Figure P19.49. (a) Find the direction and

magnitude of the magnetic field at point P, which is in the same plane as the two wires. (b) Find the magnetic field at a point 30.0 cm above the point of intersection (30.0 cm out of the page, toward you).

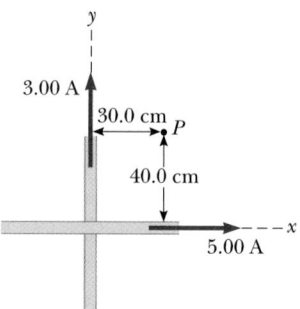

Figure P19.49

50. What magnetic field is required to constrain an electron with a kinetic energy of 400 eV to a circular path of radius 0.80 m?

51. Two species of singly charged positive ions of masses 20.0×10^{-27} kg and 23.4×10^{-27} kg enter a magnetic field at the same location with a speed of 1.00×10^5 m/s. If the strength of the field is 0.200 T, and the ions move perpendicularly to the field, find their distance of separation after they complete one half of their circular path.

52. Two parallel conductors carry currents in opposite directions, as shown in Figure P19.52. One conductor carries a current of 10.0 A. Point A is the midpoint between the wires, and point C is 5.00 cm to the right of the 10.0-A current. I is adjusted so that the magnetic field at C is zero. Find (a) the value of the current I and (b) the value of the magnetic field at A.

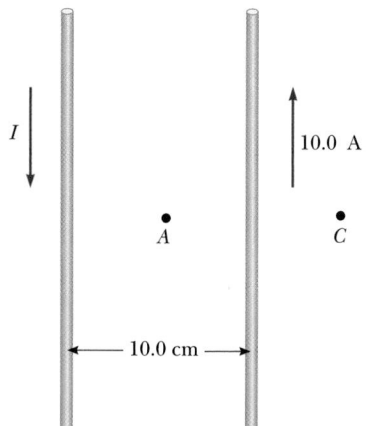

Figure P19.52

53. Four long, parallel conductors all carry currents of 4.00 A. Figure P19.53 is an end view of the conductors. The current direction is out of the page at points A and B (indicated by dots) and into the page at C and D (indicated by crosses). Calculate the magnitude and direction of the magnetic field at point P, at the center of the square of edge length of 0.200 m.

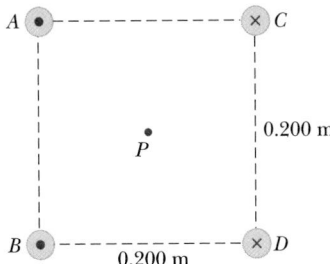

Figure P19.53

54. A singly charged heavy ion following a circular path in a uniform magnetic field of magnitude 0.050 T is observed to complete five revolutions in 1.50 ms. Calculate the mass of the ion.

55. A proton moves in a circular path perpendicularly to a constant magnetic field so that it takes 1.00 μs to complete one revolution. Determine the strength of the magnetic field.

56. For the arrangement shown in Figure P19.56, the current in the straight conductor has the value $I_1 = 5.0$ A and lies in the plane of the rectangular loop, which carries current $I_2 = 10.0$ A. The dimensions are $c = 0.10$ m, $a = 0.15$ m, and $\ell = 0.45$ m. Find the magnitude and direction of the *net force* exerted on the rectangle by the magnetic field of the straight current-carrying conductor.

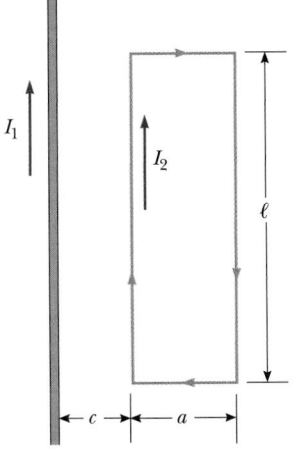

Figure P19.56

57. A straight wire of mass 10.0 g and length 5.0 cm is suspended from two identical springs that, in turn, form a closed circuit (Fig. P19.57). The springs stretch a distance of 0.50 cm under the weight of the wire. The circuit has a total resistance of 12 Ω. When a magnetic field is turned on, directed out of the page (indicated by the dots in Fig. P19.57), the springs are observed to stretch an additional 0.30 cm. What is the strength of the magnetic field? (The upper portion of the circuit is fixed.)

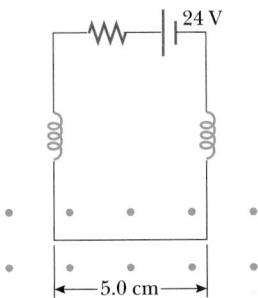

Figure P19.57

58. A metal ball having net charge $Q = 5.0$ μC is thrown out of a window horizontally at a speed $v = 20.0$ m/s. The window is at a height $h = 20.0$ m above the ground. A uniform horizontal magnetic field of magnitude $B = 0.0100$ T is perpendicular to the plane of the ball's trajectory. Find the magnetic force acting on the ball just before it hits the ground.

59. At the Fermilab accelerator in Batavia, Illinois, protons having momentum 4.80×10^{-16} kg·m/s are held in a circular orbit of radius 1.00 km by an upward magnetic field. What is the magnitude of this field?

60. Two circular loops are parallel, coaxial, and almost in contact, 1.00 mm apart (Fig. P19.60). Each loop is 10.0 cm in radius. The top loop carries a clockwise current of 140 A. The bottom loop carries a counterclockwise current of 140 A. **(a)** Calculate the magnetic force that the bottom loop exerts on the top loop. **(b)** The upper loop has a mass of 0.0210 kg. Calculate its acceleration, assuming that the only forces acting on it are the force in part (a) and its weight. (*Hint:* Think about how one loop looks to a bug perched on the other loop.)

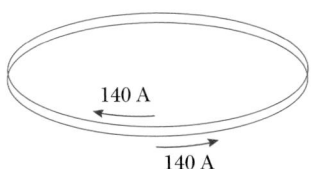

Figure P19.60

61. A strong magnet is placed under a horizontal conducting ring of radius $r = 2.0$ cm that carries a current of $I = 2.0$ A, as shown in Figure P19.61. (a) If the magnetic field lines have magnitude 0.010 T and make an angle of $\theta = 30°$ with the vertical at the ring's location, what are the magnitude and direction of the resultant force on the ring? (b) If this ring has a uniform linear mass density of 0.010 kg/m, what current in the ring will keep it levitated at its present location?

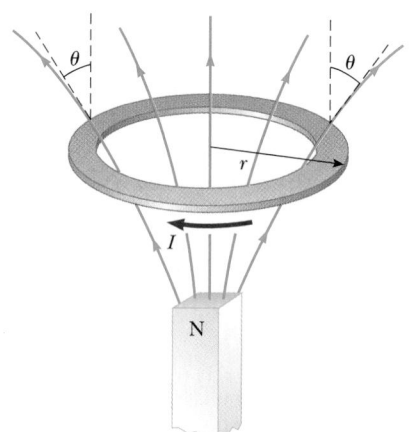

Figure P19.61

62. A uniform horizontal wire with a linear mass density of 0.50 g/m carries a 2.0-A current. It is placed in a constant magnetic field, with a strength of 4.0×10^{-3} T, that is horizontal and perpendicular to the wire. As the wire moves upward starting from rest, (a) what is its acceleration and (b) how long does it take to rise 50 cm? Neglect the magnetic field of the Earth.

63. Two ions with masses of 6.64×10^{-27} kg move out of the slit of a mass spectrometer and into a region in which the magnetic field is 0.20 T. Each has a speed of 1.0×10^6 m/s, but one ion is singly charged and the other is doubly charged. Find (a) the radius of the circular path followed by each in the field and (b) the distance of separation when they have moved through one half their circular path and strike a piece of photographic paper.

64. Three long, parallel conductors carry currents of $I = 2.0$ A. Figure P19.64 is an end view of the conductors, with each current coming out of the page. Given that $a = 1.0$ cm, determine the magnitude and direction of the magnetic field at points (a) A, (b) B, and (c) C.

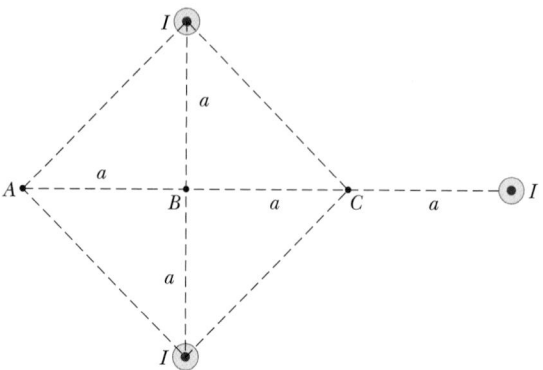

Figure P19.64

65. A heart surgeon monitors the flow rate of blood through an artery using an electromagnetic flowmeter (Fig. P19.65) in which electrodes A and B are attached to the outer surface of a blood vessel of inside diameter 3.0 mm. (a) For a magnetic field strength of 0.040 T, an emf of 160 μV is developed. Calculate the speed of the blood. (b) Verify that electrode A is positive, as shown. Does the sign of the emf depend on whether the ions in the blood are positively or negatively charged? Explain.

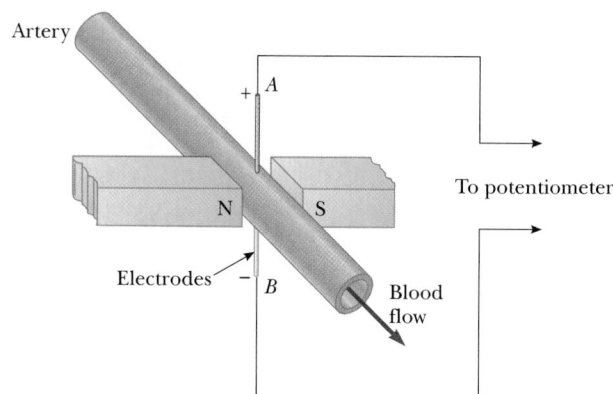

Figure P19.65

66. Two long, parallel wires, each with a mass per unit length of 40 g/m, are supported in a horizontal plane by 6.0-cm-long strings, as shown in Figure P19.66. Each wire carries the same current, I, causing the wires to repel each other so that the angle, θ, between the supporting strings is 16°. (a) Are the currents in the same or opposite directions? (b) Determine the magnitude of each current.

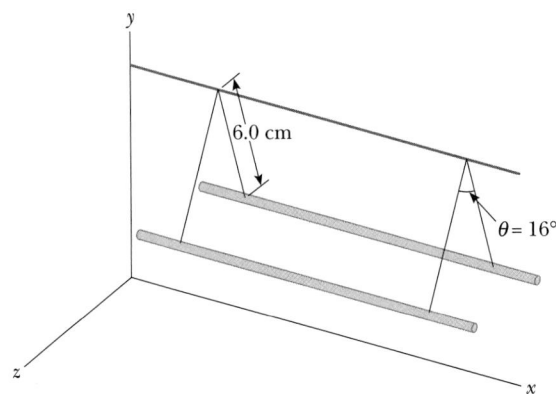

Figure P19.66

67. Protons having a kinetic energy of 5.00 MeV are moving in the positive x direction and enter a magnetic field having a magnitude of 0.0500 T directed out of the plane of the page, the y direction, and extending from $x = 0$ to $x = 1.00$ m, as in Figure P19.67. (a) Calculate the y component of the protons' momentum as they leave the magnetic field. (b) Find the angle α between the initial velocity vector of the proton beam and the velocity vector after the beam emerges from the field. (*Hint:* Neglect relativistic effects and note that $1 \text{ eV} = 1.60 \times 10^{-19}$ J.)

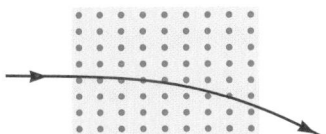

Figure P19.67

Induced Voltages and Inductance

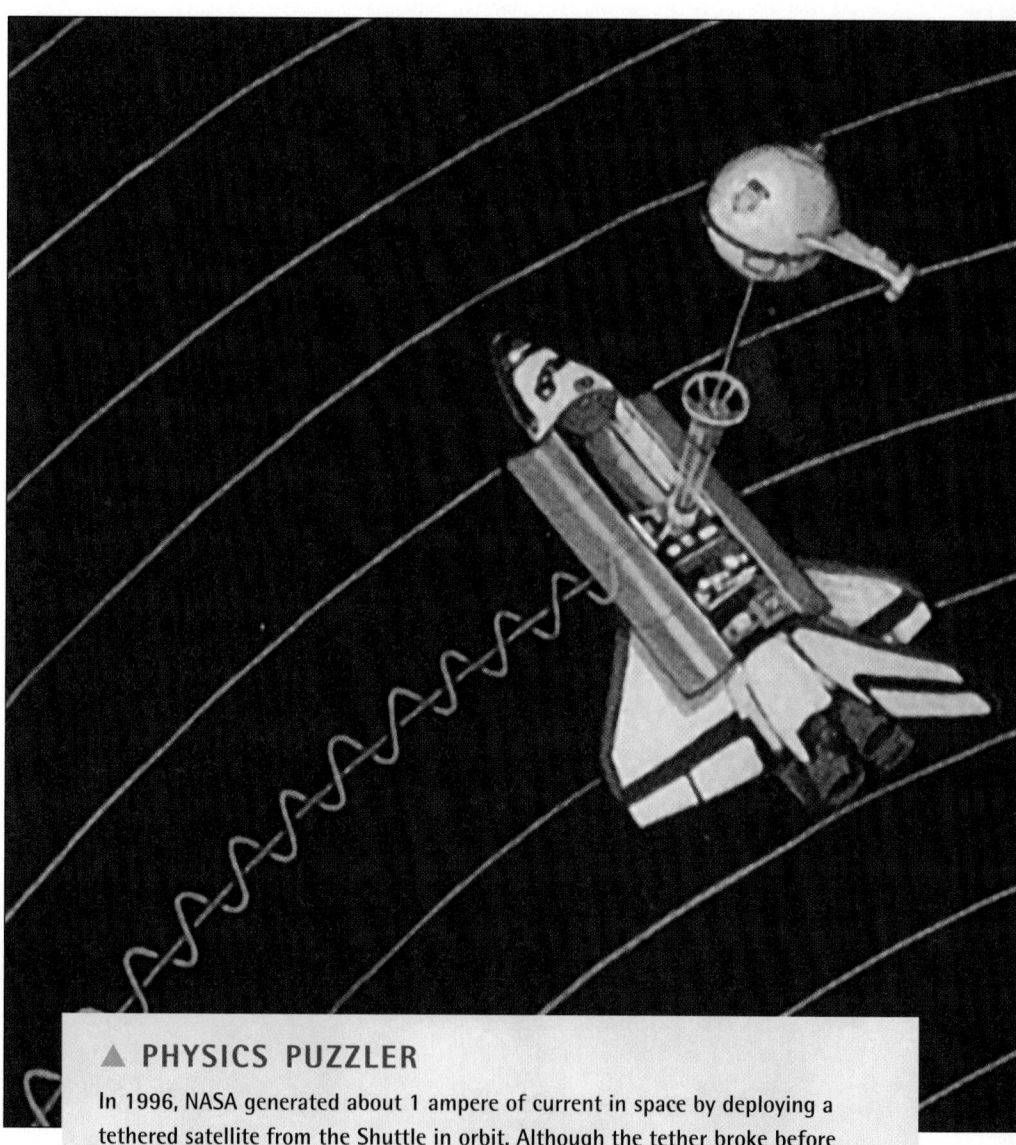

▲ PHYSICS PUZZLER

In 1996, NASA generated about 1 ampere of current in space by deploying a tethered satellite from the Shuttle in orbit. Although the tether broke before the experiment was completed, data collected from the experiment revised predictions of the efficiency of generating electricity in this way. How is it possible for a simple cable to generate electricity? *(NASA)*

I n 1819 Hans Christian Oersted discovered that a magnetic compass experiences a force in the vicinity of an electric current. Although there had long been speculation that such a relationship existed, this was the first evidence of a link between electricity and magnetism. Because nature is often symmetric, the discovery that electric currents produce magnetic fields led scientists to suspect that magnetic fields could produce electric currents. Indeed, experiments conducted by Michael Faraday in England and, independently, by Joseph Henry in the United States in 1831 showed that a changing magnetic field could induce an electric current in a circuit. The results of these experiments led to a very basic and important law known as Faraday's law. In this chapter we discuss several practical applications of Faraday's law, one of which is the production of electrical energy in power generation plants throughout the world.

Michael Faraday, British physicist and chemist (1791– 1867)

Faraday is often regarded as the greatest experimental scientist of the 1800s. His many contributions to the study of electricity include the invention of the electric motor, electric generator, and transformer, as well as the discovery of electromagnetic induction and the laws of electrolysis. Greatly influenced by religion, he refused to work on military poison gas for the British government. *(By kind permission of the President and Council of the Royal Society)*

20.1   INDUCED emf AND MAGNETIC FLUX

Induced emf

We begin this chapter by describing an experiment, first conducted by Faraday, that demonstrates that a current can be produced by a changing magnetic field. The apparatus shown in Figure 20.1 consists of a coil connected to a switch and a battery. We shall refer to this coil as the *primary coil* and to the corresponding circuit as the primary circuit. The coil is wrapped around an iron ring to intensify the magnetic field produced by the current through it. A second coil, at the right, is wrapped around the iron ring and is connected to a galvanometer. We shall refer to this as the *secondary coil* and to the corresponding circuit as the secondary circuit. Note that there is no battery in the secondary circuit. The only purpose of this circuit is to detect any current that might be produced by a magnetic field.

At first glance, you might guess that no current would ever be detected in the secondary circuit. However, when the switch in the primary circuit is suddenly closed or opened, something quite amazing happens. Just after the switch is closed, the galvanometer in the secondary circuit deflects in one direction and then returns to zero. When the switch is opened, the galvanometer deflects in the opposite direction and again returns to zero. Finally, when there is a steady current in the primary circuit, the galvanometer reads zero.

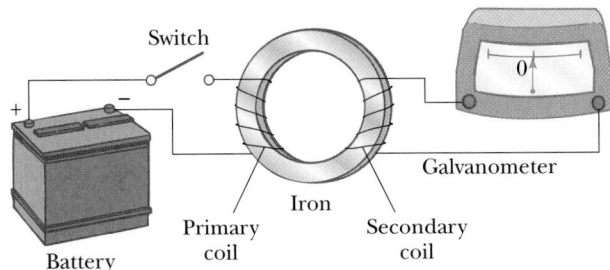

Figure 20.1 Faraday's experiment. When the switch in the primary circuit at the left is closed, the galvanometer in the secondary circuit at the right deflects momentarily. The emf in the secondary circuit is induced by the changing magnetic field through the coil in this circuit.

Figure 20.2   (a) A uniform magnetic field, **B,** making an angle of θ with the normal to the plane of a wire loop of area A. (b) An edge view of the loop.

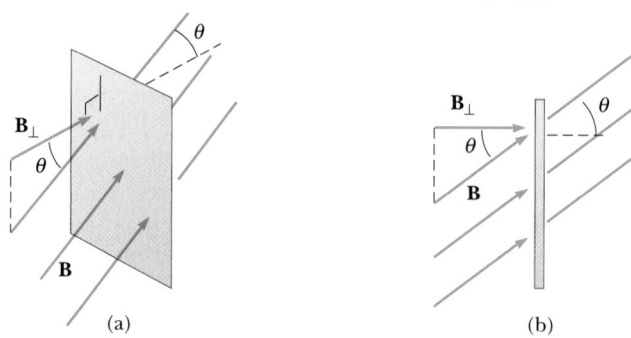

From observations such as these, Faraday concluded that an electric current can be produced by a changing magnetic field. (A steady magnetic field cannot produce a current.) The current produced in the secondary circuit occurs only for an instant while the magnetic field through the secondary coil is changing. In effect, the secondary circuit behaves as though a source of emf were connected to it for a short instant. It is customary to say that **an induced emf is produced in the secondary circuit by the changing magnetic field.**

Magnetic Flux

12.5, SECTION 1

In order to evaluate induced emfs quantitatively, it is first necessary to fully understand what factors affect the phenomenon. As you will see later, the emf is induced by a change in a quantity called the *magnetic flux* rather than simply by a change in the magnetic field.

Consider a loop of wire in the presence of a uniform magnetic field, **B.** If the loop has an area of A, the **magnetic flux, Φ,** through the loop is defined as

Magnetic flux ▶

$$\Phi \equiv B_\perp A = BA \cos \theta \qquad \text{[20.1]}$$

where $B_\perp$ is the component of **B** perpendicular to the plane of the loop, as in Figure 20.2a, and θ is the angle between **B** and the normal (perpendicular) to the plane of the loop. Figure 20.2b is an end view of the loop and the penetrating magnetic field lines. When the field is perpendicular to the plane of the loop, as in Figure 20.3a, $\theta = 0$ and Φ has a maximum value, $\Phi_{max} = BA$. When the plane of the loop is parallel to **B,** as in Figure 20.3b, $\theta = 90°$ and $\Phi = 0$. Because B is in teslas, or webers per square meter, the units of flux are $T \cdot m^2$, or webers (Wb).

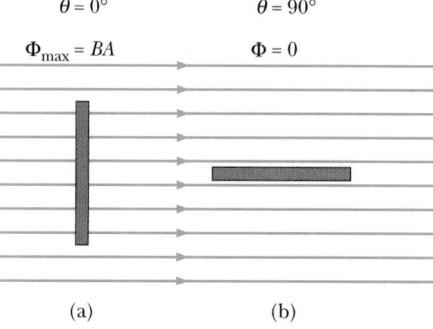

$\theta = 0°$

$\Phi_{max} = BA$

$\theta = 90°$

$\Phi = 0$

Figure 20.3   An edge view of a loop in a uniform magnetic field. (a) When the field lines are perpendicular to the plane of the loop, the magnetic flux through the loop is a maximum and equal to $\Phi_{max} = BA$. (a) When the field lines are parallel to the plane of the loop, the magnetic flux through the loop is zero.

We can emphasize the significance of Equation 20.1 by first drawing magnetic field lines, as in Figure 20.3. The number of lines per unit area increases as the field strength increases. **The value of the magnetic flux is proportional to the total number of lines passing through the loop.** Thus, we see that the most lines pass through the loop when its plane is perpendicular to the field, as in Figure 20.3a, and so the flux has its maximum value. As Figure 20.3b shows, no lines pass through the loop when its plane is parallel to the field, and so in this case $\Phi = 0$.

Thinking Physics 1

Argentina has more land area (2.8×10^6 km^2) than Greenland (2.2×10^6 km^2). Yet the magnetic flux of the Earth's magnetic field is larger through Greenland than through Argentina. Why?

Explanation There are two reasons for the larger magnetic flux through Greenland. Greenland (latitude 60° north to 80° north) is closer to a magnetic pole than is Argentina (latitude 20° south to 50° south). As a result, the angle that the magnetic field lines make with the vertical is smaller in Greenland than in Argentina. Thus, more field lines penetrate the surface in Greenland. The second reason is also a result of Greenland's proximity to a magnetic pole. The field strength is larger in the vicinity of a magnetic pole than it is farther from the pole. Thus, the magnitude of **B** in the definition of flux is larger for Greenland than for Argentina. These two influences will dominate over the slightly larger area of Argentina.

20.2 FARADAY'S LAW OF INDUCTION

The usefulness of the concept of magnetic flux can be made obvious by another simple experiment that demonstrates the basic idea of electromagnetic induction. Consider a wire loop connected to a galvanometer, as in Figure 20.4. If a magnet is moved toward the loop, the galvanometer needle deflects in one direction, as in Figure 20.4a. If the magnet is moved away from the loop, the galvanometer needle deflects in the opposite direction, as in Figure 20.4b. If the magnet is held stationary

12.6, SECTION 1

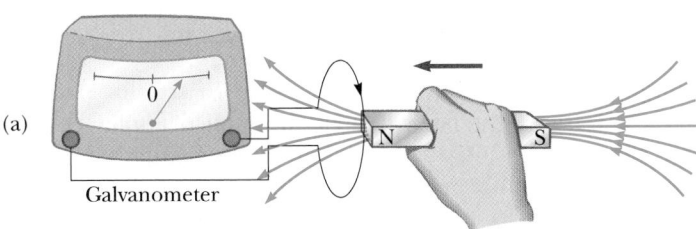

(a)

Galvanometer

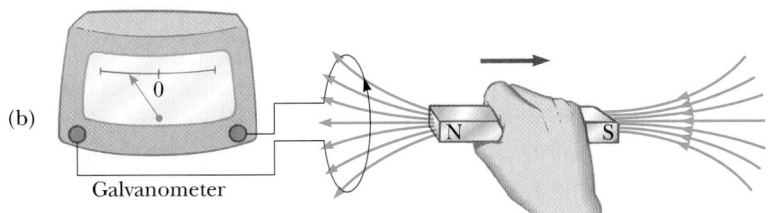

(b)

Galvanometer

Figure 20.4 (a) When a magnet is moved toward a wire loop connected to a galvanometer, the galvanometer deflects as shown. This shows that a current is induced in the loop. (b) When the magnet is moved away from the loop of wire, the galvanometer deflects in the opposite direction, indicating that the induced current is opposite that shown in (a).

and the loop is moved either toward or away from the magnet, the needle also deflects. From these observations, it can be concluded that **a current is set up in the circuit as long as there is relative motion between the magnet and the loop.** These results are quite remarkable in view of the fact that the circuit contains no batteries! We call such a current an **induced current** because it is produced by an induced emf.

This experiment has something in common with the Faraday experiment discussed in Section 20.1. In each case, an emf is induced in a current when the magnetic flux through the circuit changes with time. In fact, we can make the following general summary of such experiments involving induced currents and emfs:

> The instantaneous emf induced in a circuit equals the rate of change of magnetic flux through the circuit.

If a circuit contains N tightly wound loops and the flux through each loop changes by the amount $\Delta\Phi$ during the interval Δt, the average emf induced in the circuit during time Δt is

Faraday's law ▶

$$\mathcal{E} = -N\frac{\Delta\Phi}{\Delta t} \qquad [20.2]$$

This is a statement of **Faraday's law of magnetic induction.** The minus sign is included to indicate the polarity of the induced emf, which can be found by use of **Lenz's law:**

> The polarity of the induced emf is such that it produces a current whose magnetic field opposes the change in magnetic flux through the loop. That is, the induced current tends to maintain the original flux through the circuit.

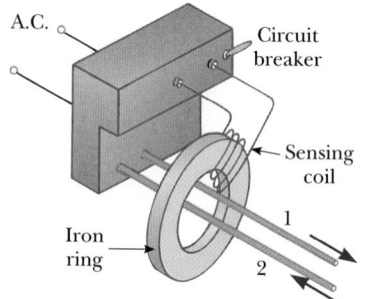

Figure 20.5 Essential components of a ground fault interrupter.

We shall consider several applications of Lenz's law in Section 20.4.

The ground fault interrupter (GFI) is an interesting safety device that protects users of electrical power against electric shock when they touch appliances. Its operation makes use of Faraday's law. Figure 20.5 shows the essential parts of a ground fault interrupter. Wire 1 leads from the wall outlet to the appliance to be protected, and wire 2 leads from the appliance back to the wall outlet. An iron ring surrounds the two wires to confine the magnetic field set up by each wire. A sensing coil, which can activate a circuit breaker when changes in magnetic flux occur, is wrapped around part of the iron ring. Because the currents in the wires are in opposite directions, the net magnetic field through the sensing coil due to the currents is zero. However, if a short circuit occurs in the appliance so that there is no returning current, the net magnetic field through the sensing coil is no longer zero. (This can happen, for example, if one of the wires loses its insulation and accidentally touches the metal case of the appliance, providing a direct path to ground.) Because the current is alternating, the magnetic flux through the sensing coil changes with time, producing an induced voltage in the coil. This induced voltage is used to trigger a circuit breaker, stopping the current before it reaches a level that might be harmful to the person using the appliance.

Another interesting application of Faraday's law is the production of sound in an electric guitar. A vibrating string induces an emf in a coil (Fig. 20.6). The pickup coil is placed near the vibrating guitar string, which is made of a metal that can be magnetized. The permanent magnet inside the coil magnetizes the portion of the string nearest the coil. When the guitar string vibrates at some frequency, its magnetized segment produces a changing magnetic flux through the pickup coil. The changing flux induces a voltage in the coil; the voltage is fed to an amplifier. The output of the amplifier is sent to the loudspeakers, producing the sound waves that we hear.

Sudden Infant Death Syndrome (SIDS) is a devastating affliction in which a baby suddenly stops breathing during sleep without an apparent cause. One type of monitoring device sometimes used to alert parents of the cessation of breathing uses induced currents, as shown in Fig. 20.7. A coil of wire attached to one side of the chest carries an alternating current. The varying magnetic field produced by this current threads through a pickup coil attached to the opposite side of the chest. Expansion and contraction of the chest caused by breathing or movement changes the strength of the voltage induced in the pickup coil. However, if breathing stops, the pattern of the induced voltage stabilizes, and external circuits monitoring the voltage sound an alarm to the caregivers after a momentary pause to ensure that a problem actually does exist.

APPLICATION

Electric Guitar Pickups.

APPLICATION

SIDS Monitors.

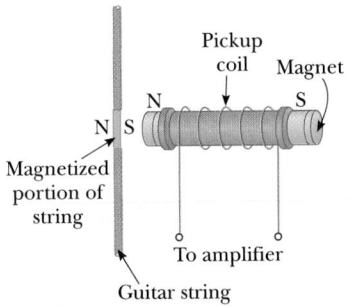

Figure 20.6 In an electric guitar, a vibrating string induces a voltage in the pickup coil.

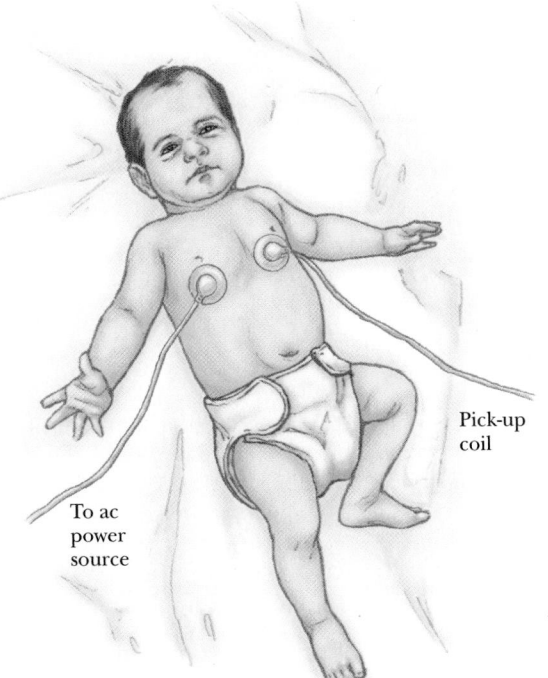

Figure 20.7 A monitor designed to detect sudden infant death syndrome (SIDS).

PHYSICS IN ACTION

Demonstrations of Electromagnetic Induction

On the left, when a strong magnet is moved toward or away from the coil attached to a galvanometer, a current is induced in the coil, indicated by the momentary deflection of the galvanometer during the movement of the magnet. What is the cause of this induced current? Would the galvanometer deflect if the coil were moved toward a stationary magnet?

On the right, to demonstrate electromagnetic induction, an ac voltage is applied to the lower coil in the apparatus. A voltage is induced in the upper coil, as indicated by the illuminated lamp connected to this coil. What do you think happens to the lamp's intensity as the upper coil is moved over the vertical tube? To answer this question, note that the magnetic field associated with the lower coil varies along the axis of the tube.

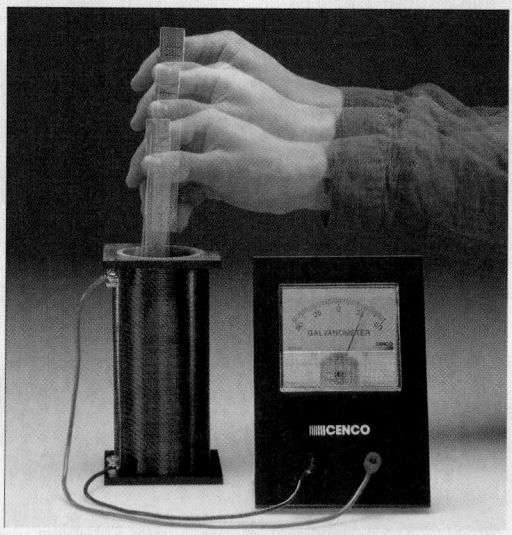

(Richard Megna, Fundamental Photographs)

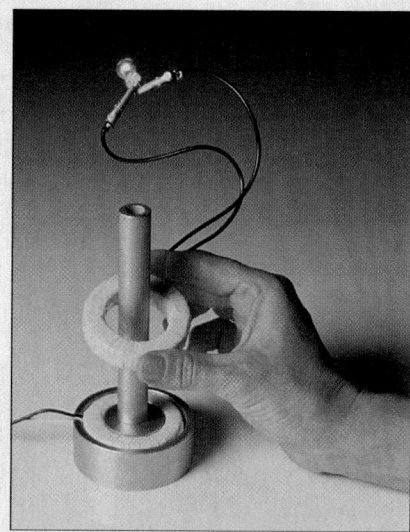

(Courtesy of Central Scientific Company)

Workbook Problem 80 – (Workbook page 170) contains a more complex application of Faraday's law.

EXAMPLE 20.1   Application of Faraday's Law

A coil with 200 turns of wire is wrapped on a square frame, 18.0 cm on a side. Each turn has the same area, equal to that of the frame, and the total resistance of the coil is 2.00 Ω. A uniform magnetic field is applied perpendicularly to the plane of the coil. If the field changes uniformly from 0 to 0.500 T in 0.800 s, find the magnitude of the induced emf in the coil while the field is changed.

Reasoning   The magnitude of the induced emf can be found from Faraday's law of induction, $\varepsilon = -N\dfrac{\Delta\Phi}{\Delta t}$. In this equation, $\Delta\Phi$ is the difference between the final and initial fluxes, where $\Phi = BA\cos\theta = BA\cos 0° = BA$.

Solution The area of the coil is $(0.180 \text{ m})^2 = 0.0324 \text{ m}^2$. The magnetic flux through the coil at $t = 0$ is zero because $B = 0$. At $t = 0.800$ s, the magnetic flux through the coil is

$$\Phi_f = BA = (0.500 \text{ T})(0.0324 \text{ m}^2) = 0.0162 \text{ T} \cdot \text{m}^2$$

Therefore, the *change* in flux through the coil during the 0.800-s interval is

$$\Delta\Phi = \Phi_f - \Phi_i = 0.0162 \text{ T} \cdot \text{m}^2$$

Faraday's law of induction enables us to find the magnitude of the induced emf:

$$|\mathcal{E}| = N\frac{\Delta\Phi}{\Delta t} = (200 \text{ turns})\left(\frac{0.0162 \text{ T} \cdot \text{m}^2}{0.800 \text{ s}}\right) = \boxed{4.05 \text{ V}}$$

(Note that $1 \text{ T} \cdot \text{m}^2/\text{s} = 1(\text{N} \cdot \text{s}/\text{C} \cdot \text{m})(\text{m}^2/\text{s}) = 1 \text{ N} \cdot \text{m}/\text{C} = 1 \text{ J}/\text{C} = 1 \text{ V}.$)

Exercise Find the magnitude of the induced current in the coil while the field is changing.

Answer 2.03 A

20.3 MOTIONAL emf

In Section 20.2, we considered a situation in which an emf is induced in a circuit when the magnetic field changes with time. In this section we describe a particular application of Faraday's law in which a so-called **motional emf** is produced. This is the emf induced in a conductor moving through a magnetic field.

12.6, SECTION 2

First consider a straight conductor of length ℓ moving with constant velocity through a uniform magnetic field directed into the paper, as in Figure 20.8. For simplicity, we assume that the conductor is moving perpendicularly to the field. The electrons in the conductor experience a force of magnitude $F = qvB$ directed downward along the conductor. Because of this magnetic force, the electrons move to the lower end and accumulate there, leaving a net positive charge at the upper end. As a result of this charge separation, an electric field is produced in the conductor. The charge at the ends builds up until the downward magnetic force, qvB, is balanced by the upward electric force, qE. At this point, charge stops flowing and the condition for equilibrium requires that

$$qE = qvB \qquad \text{or} \qquad E = vB$$

Because the electric field is constant, the field produced in the conductor is related to the potential difference across the ends by $\Delta V = E\ell$. Thus,

$$\Delta V = E\ell = B\ell v \qquad [20.3]$$

Because there is an excess of positive charge at the upper end of the conductor and an excess of negative charge at the lower end, the upper end is at a higher potential than the lower end. Thus,

---

A potential difference is maintained across the conductor as long as there is motion through the field. If the motion is reversed, the polarity of the potential difference is also reversed.

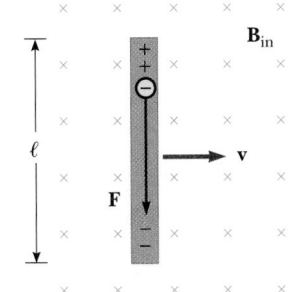

Figure 20.8 A straight conductor of length ℓ moving with velocity **v** through a uniform magnetic field, **B,** directed perpendicularly to **v.** The vector **F** is the force on an electron in the conductor. An emf of $B\ell v$ is induced between the ends of the bar.

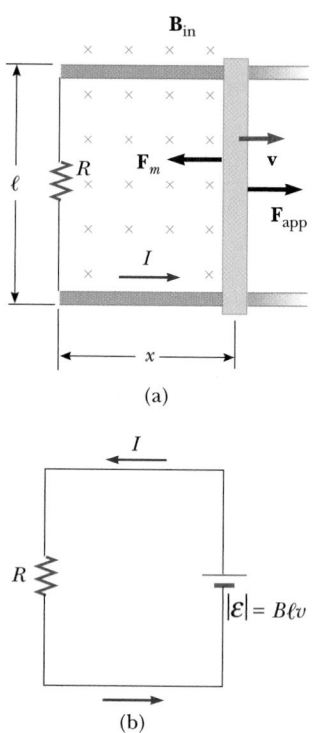

Figure 20.9   (a) A conducting bar sliding with velocity **v** along two conducting rails under the action of an applied force, $\mathbf{F}_{app}$. The magnetic force $\mathbf{F}_m$ opposes the motion, and a counterclockwise current is induced in the loop. (b) The equivalent circuit of (a).

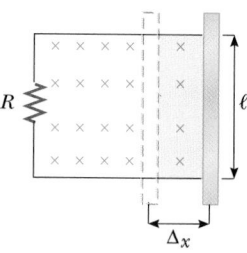

Figure 20.10   As the bar moves to the right, the area of the loop increases by the amount $\ell\,\Delta x$, and the magnetic flux through the loop increases by $B\ell\,\Delta x$.

APPLICATION

Railguns.

A more interesting situation occurs if the moving conductor is part of a closed conducting path. This situation is particularly useful for illustrating how a changing magnetic flux induces a current in a closed circuit. Consider a circuit consisting of a conducting bar of length ℓ, sliding along two fixed parallel conducting rails, as in Figure 20.9a. For simplicity, we assume that the moving bar has zero resistance and that the stationary part of the circuit has constant resistance R. A uniform and constant magnetic field, **B**, is applied perpendicularly to the plane of the circuit. As the bar is pulled to the right with velocity **v** under the influence of an applied force, $\mathbf{F}_{app}$, the free charges in the bar experience a magnetic force along the length of the bar. This force in turn sets up an induced current because the charges are free to move in a closed conducting path. In this case, the changing magnetic flux through the loop and the corresponding induced emf across the moving bar arise from the change in area of the loop as the bar moves through the magnetic field.

Let us assume that the bar moves a distance of Δx in time Δt, as shown in Figure 20.10. The increase in flux, $\Delta\Phi$, through the loop in that time is the amount of flux that now passes through the portion of the circuit that has area $\ell\,\Delta x$:

$$\Delta\Phi = BA = B\ell\,\Delta x$$

Using Faraday's law and noting that there is one loop ($N = 1$), we find that the induced emf has the magnitude

$$|\varepsilon| = \frac{\Delta\Phi}{\Delta t} = B\ell\,\frac{\Delta x}{\Delta t} = B\ell v \qquad [20.4]$$

This induced emf is often called a **motional emf** because it arises from the motion of a conductor through a magnetic field.

Furthermore, if the resistance of the circuit is R, the magnitude of the induced current is

$$I = \frac{\varepsilon}{R} = \frac{B\ell v}{R} \qquad [20.5]$$

Figure 20.9b is the equivalent circuit diagram for this example.

Applying Physics 1

We have discussed applying a force on the bar, which will result in an induced emf in the circuit shown in Figure 20.9a. Suppose we removed the external magnetic field in the diagram and replaced the resistor with a high-voltage source and a switch, as in Figure 20.11. What will happen when the switch is closed? Will the bar move, and does it matter which way we connect the high voltage source?

Explanation   Suppose the source is capable of establishing high current. The two horizontal conducting rods will create a strong magnetic field in the area between them, directed into the page. (The movable bar also creates a magnetic field, but this field cannot exert force on the bar itself.) As the current passes downward through the movable bar, it experiences a magnetic force to the right. Hence, it accelerates along the rails away from the power supply. If the polarity of the power were reversed, the magnetic field would be out of the page, the current in the bar would be upward, and the force on the bar would still be to the right. (This is the essence of a railgun. The $BI\ell$ force exerted by a magnetic field causes the bar to accelerate away from the voltage source. Studies have shown that it is possible to launch payloads into space with this technology. Very large accelerations can be obtained with currently available technology, with payloads being accelerated to a speed of several kilometers per second in a fraction of a second. This is a larger acceleration than humans can withstand.)

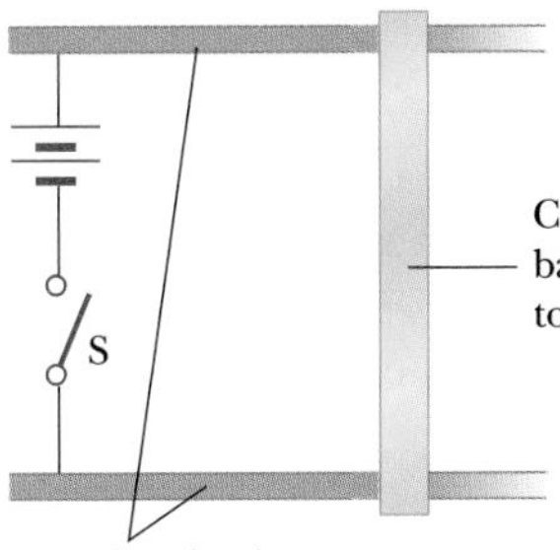

Thinking Physics 2

As an airplane flies from Los Angeles to Seattle, it cuts through the lines of the Earth's magnetic field. As a result, an emf is developed between the wingtips of the airplane. Which wingtip is positively charged and which is negatively charged? Would the wingtips be charged in the same way on an airplane traveling from Antarctica to Australia? Could this emf be a source of energy to operate the lighting system of the airplane?

Explanation The magnetic field of the Earth has a downward component in the northern hemisphere. As the airplane flies northward, the right-hand rule indicates that positive charge experiences a force to the left side of the airplane. Thus, the left wingtip becomes positively charged and the right wingtip negatively charged. In the southern hemisphere, the field has an upward component, so the charging of the wingtips would be in the opposite sense. The induced emf could not continuously operate a lighting system. If we tried to tap into the voltage generated by the airplane wing, we would have to add an insulated wire (with a little lightbulb) making a complete circuit between the wingtips. Then the extra wire would support its own motional emf and zero volts would appear across the bulb as the plane flies through a uniform field. The complete circuit is a loop with constant magnetic flux through it.

Figure 20.11 (Applying Physics 1)

EXAMPLE 20.2 The Electrified Airplane Wing

An airplane with a wing span of 30.0 m flies parallel to the Earth's surface at a location at which the downward component of the Earth's magnetic field is 0.60×10^{-4} T. Find the difference in potential between the wing tips when the speed of the plane is 250 m/s.

Solution Because the plane is flying horizontally, we do not have to concern ourselves with the horizontal component of the Earth's field. Thus, we find that

$$\mathcal{E} = B\ell v = (0.60 \times 10^{-4} \text{ T})(30.0 \text{ m})(250 \text{ m/s}) = \boxed{0.45 \text{ V}}$$

The chapter opener on page 650 shows how NASA plans to deploy a long conducting tether from a shuttle and use the motion through the Earth's nonuniform magnetic field to generate pulses of power.

EXAMPLE 20.3 Where Is the Energy Source?

(a) The sliding bar in Figure 20.9a has a length of 0.50 m and moves at 2.0 m/s in a magnetic field of magnitude 0.25 T. Find the induced voltage in the moving rod.

Solution   We use Equation 20.3 and find that

$$\mathcal{E} = B\ell v = (0.25 \text{ T})(0.50 \text{ m})(2.0 \text{ m/s}) = \boxed{0.25 \text{ V}}$$

(b) If the resistance in the circuit is 0.50 Ω, find the current in the circuit.

Solution   The current is found from Ohm's law to be

$$I = \frac{\mathcal{E}}{R} = \frac{0.25 \text{ V}}{0.50 \text{ }\Omega} = \boxed{0.50 \text{ A}}$$

(c) Find the amount of energy delivered to the 0.50-Ω resistor in one second.

Solution   The power dissipated by the resistor is

$$P = I\,\Delta V = (0.50 \text{ A})(0.25 \text{ V}) = \boxed{0.13 \text{ W}}$$

Because power is defined as the rate at which energy is converted in a device, the energy, W, dissipated in the resistor in one second is

$$W = Pt = (0.125 \text{ W})(1.0 \text{ s}) = \boxed{0.13 \text{ J}}$$

(d) The source of the energy calculated in part (c) is some external agent that keeps the bar moving at a constant speed of 2.0 m/s by exerting an applied force, F_{app}. Find the value of F_{app}.

Solution   From part (c), we know that the work done by the applied force in one second is 0.13 J. In one second, the bar moves a distance of

$$d = vt = (2.0 \text{ m/s})(1.0 \text{ s}) = 2.0 \text{ m}$$

Thus, from the definition of work, we find that $W = F_{app}\,d$, or

$$F_{app} = \frac{W}{d} = \frac{0.13 \text{ J}}{2.0 \text{ m}} = \boxed{0.063 \text{ N}}$$

Exercise   If the rod is to move at constant speed, the applied force must be equal in magnitude to the retarding magnetic force, $I\ell B$. Show that this approach also gives $F_{app} = 0.063$ N, as found in part (d).

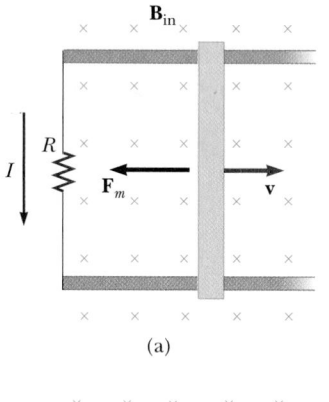

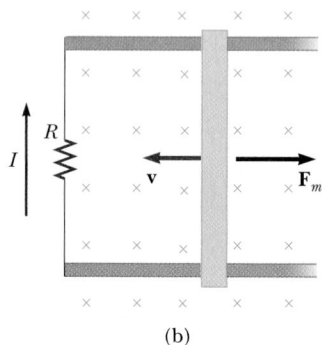

Figure 20.12   (a) As the conducting bar slides on the two fixed conducting rails, the magnetic flux through the loop increases in time. By Lenz's law, the induced current must be *counterclockwise* so as to produce a counteracting flux *out of the paper*. (b) When the bar moves to the left, the induced current must be *clockwise*. Why?

20.4   LENZ'S LAW REVISITED

To attain a better understanding of Lenz's law, let us return to the example of a bar moving to the right on two parallel rails in the presence of a uniform magnetic field directed into the paper (Fig. 20.12a). As the bar moves to the right, the magnetic flux through the circuit increases with time because the area of the loop increases. Lenz's law says that the induced current must be in a direction such that the flux *it* produces opposes the change in the external magnetic flux. Because the flux due to the external field is increasing *into* the paper, the induced current, to oppose the change, must produce a flux *out* of the paper. Hence, the induced current must be counterclockwise when the bar moves to the right. (Use the right-hand rule to verify this direction.) On the other hand, if the bar is moving to the left, as in Figure 20.12b, the magnetic flux through the loop decreases with time. Because the flux is into the paper, the induced current has to be clockwise to produce a flux into the paper. In either case, the induced current tends to maintain the original flux through the circuit.

12.7, BOTH SECTIONS

Let us examine this situation from the viewpoint of energy conservation. Suppose that the bar is given a slight push to the right. In the preceding analysis, we found that this motion led to a counterclockwise current in the loop. Let's see what would happen if we assumed that the current was clockwise. For a clockwise current I, the direction of the magnetic force, $BI\ell$, on the sliding bar would be to the right. This force would accelerate the rod and increase its velocity. This, in turn, would cause the area of the loop to increase more rapidly, thereby increasing the induced current, which would increase the force, which would increase the current, which would. . . . In effect, the system would acquire energy with zero input energy. This is clearly inconsistent with all experience and with the law of conservation of energy. We are forced to conclude that the current must be counterclockwise.

Consider another situation. A bar magnet is moved to the right toward a stationary loop of wire, as in Figure 20.13a. As the magnet moves, the magnetic flux through the loop increases with time. To counteract this, the induced current produces a flux to the left, as in Figure 20.13b; hence, the induced current is in the direction shown. Note that the magnetic field lines associated with the induced current oppose the motion of the magnet. Therefore, the left face of the current loop is a north pole and the right face is a south pole.

On the other hand, if the magnet were moving to the left, as in Figure 20.13c, its flux through the loop, which is toward the right, would decrease in time. Under these circumstances, the induced current in the loop would be in a direction to set up a field directed from left to right through the loop, in an effort to maintain a constant number of flux lines. Hence, the induced current in the loop would be as shown in Figure 20.13d. In this case, the left face of the loop would be a south pole and the right face would be a north pole.

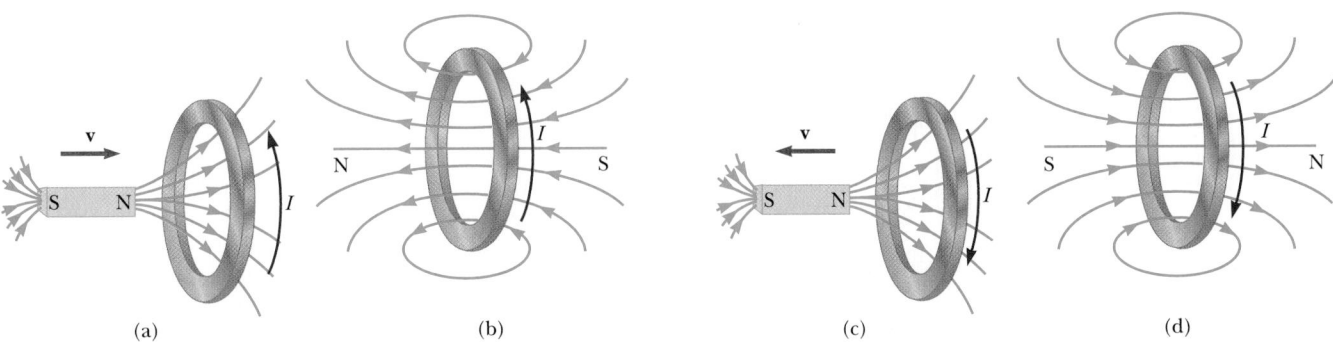

(a)   (b)   (c)   (d)

Figure 20.13 (a) When the magnet is moved toward the stationary conducting loop, a current is induced in the direction shown. (b) This induced current produces its own flux to the left to counteract the increasing external flux to the right. (c) When the magnet is moved away from the stationary conducting loop, a current is induced in the direction shown. (d) This induced current produces its own flux to the right to counteract the decreasing external flux to the right.

EXAMPLE 20.4 Application of Lenz's Law

A coil of wire is placed near an electromagnet, as in Figure 20.14a. Find the direction of the induced current in the coil (a) at the instant the switch is closed, (b) after the switch has been closed for several seconds, and (c) when the switch is opened.

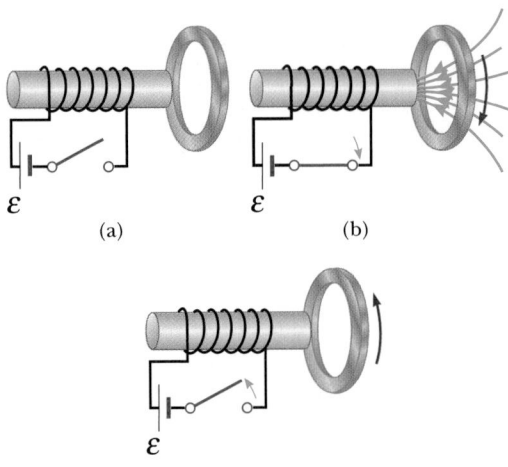

Figure 20.14 (Example 20.4)

Reasoning and Solution (a) When the switch is closed, the situation changes from a condition in which no lines of flux pass through the coil to one in which lines of flux pass through in the direction shown in Figure 20.14b. To counteract this change in the number of lines, the coil must set up a field from left to right in the figure. This requires a current directed as shown in Figure 20.14b.

(b) After the switch has been closed for several seconds, there is no change in the number of lines through the loop; hence, the induced current is zero.

(c) Opening the switch causes the magnetic field to change from a condition in which flux lines thread through the coil from right to left to a condition of zero flux. The induced current must then be as shown in Figure 20.14c, to set up its own field from right to left.

Application: Tape Recorders

APPLICATION

Magnetic Tape Recorders.

One common practical use of induced currents and emfs is in the tape recorder. Many different types of tape recorders are made, but the basic principles are the same for all. A magnetic tape moves past a recording head and a playback head, as in Figure 20.15a. The tape is a plastic ribbon coated with iron oxide or chromium oxide.

The recording process uses the fact that a current passing through an electromagnet produces a magnetic field. Figure 20.15b illustrates the steps in the process. A sound wave sent into a microphone is transformed into an electric current, amplified, and allowed to pass through a wire coiled around a doughnut-shaped piece of iron, which functions as the recording head. The iron ring and the wire constitute an electromagnet, in which the lines of the magnetic field are contained completely inside the iron except at the point where a slot is cut in the ring. Here the magnetic field fringes out of the iron and magnetizes the small pieces of iron oxide embedded in the tape. Thus, as the tape moves past the slot, it becomes magnetized in a pattern that reproduces both the frequency and the intensity of the sound signal entering the microphone.

To reconstruct the sound signal, the tape is allowed to pass through a recorder with the playback head in operation. This head is very similar to the recording head in that it consists of a wire-wound doughnut-shaped piece of iron with a slot in it. When the tape moves past this head, the varying magnetic fields on the tape produce changing field lines through the wire coil. The changing flux induces a current in the coil that corresponds to the current in the recording head that originally produced the tape. This changing electric current can be amplified and used to drive a speaker. Playback is thus an example of induction of a current by a moving magnet.

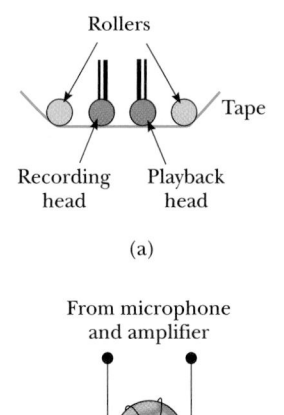

(a)

(b)

Figure 20.15 (a) The head of a magnetic tape recorder. (b) The fringing magnetic field magnetizes the tape during recording.

20.5 GENERATORS

Generators and motors are important practical devices that operate on the principle of electromagnetic induction. First, let us consider the **alternating current (ac) generator,** a device that converts mechanical energy to electrical energy. In its simplest form, the ac generator consists of a wire loop rotated in a magnetic field by some external means (Fig. 20.16a). In commercial power plants, the energy required to rotate the loop can be derived from a variety of sources. For example, in a hydroelectric plant, falling water directed against the blades of a turbine produces the rotary motion; in a coal-fired plant, heat produced by burning coal is used to convert water to steam, and this steam is directed against the turbine blades. As the loop rotates, the magnetic flux through it changes with time, inducing an emf and a current in an external circuit. The ends of the loop are connected to slip rings that rotate with the loop. Connections to the external circuit are made by stationary brushes in contact with the slip rings.

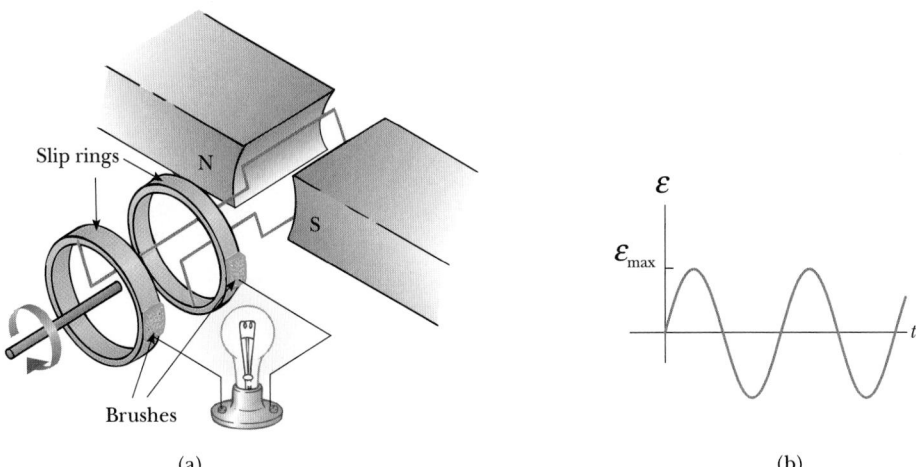

(a)

(b)

Figure 20.16 (a) A schematic diagram of an ac generator. An emf is induced in a coil, which rotates by some external means in a magnetic field. (b) A plot of the alternating emf induced in the loop versus time.

Figure 20.17 (a) A loop rotating at constant angular velocity in an external magnetic field. The emf induced in the loop varies sinusoidally with time. (b) An edge view of the rotating loop.

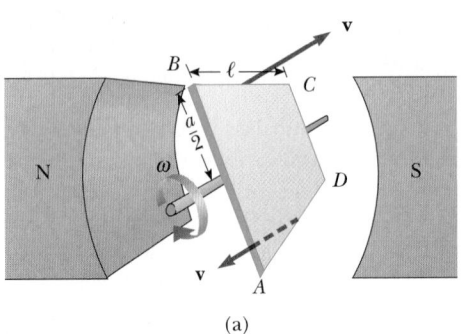

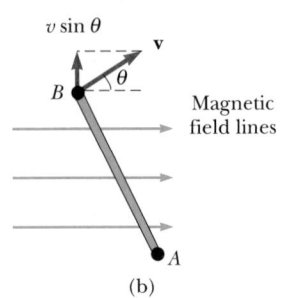

(a)   (b)

We can derive an expression for the emf generated in the rotating loop by making use of the expression for motional emf, $\mathcal{E} = B\ell v$. Figure 20.17a shows a loop of wire rotating clockwise in a uniform magnetic field directed to the right. The magnetic force (qvB) on the charges in wires AB and CD is not along the lengths of the wires. (The force on the electrons in these wires is perpendicular to the wires.) Hence, an emf is generated only in wires BC and AD. At any instant, wire BC has velocity **v** at an angle of θ with the magnetic field, as shown in Figure 20.17b. (Note that the component of velocity parallel to the field has no effect on the charges in the wire, whereas the component of velocity perpendicular to the field produces a magnetic force on the charges that moves electrons from C to B.) The emf generated in wire BC equals $B\ell v_\perp$, where ℓ is the length of the wire and $v_\perp$ is the component of velocity perpendicular to the field. An emf of $B\ell v_\perp$ is also generated in wire DA, and the sense of this emf is the same as in wire BC. Because $v_\perp = v \sin \theta$, the total emf is

$$\mathcal{E} = 2B\ell v_\perp = 2B\ell v \sin \theta \qquad [20.6]$$

If the loop rotates with a constant angular speed of ω, we can use the relation $\theta = \omega t$ in Equation 20.6. Furthermore, because every point on wires BC and DA rotates in a circle about the axis of rotation with the same angular speed, ω, we have $v = r\omega = (a/2)\omega$, where a is the length of sides AB and CD. Therefore, Equation 20.6 reduces to

$$\mathcal{E} = 2B\ell \left(\frac{a}{2} \right) \omega \sin \omega t = B\ell a\omega \sin \omega t$$

If a coil has N turns, the emf is N times as large because each loop has the same emf induced in it. Furthermore, because the area of the loop is $A = \ell a$, the total emf is

$$\mathcal{E} = NBA\omega \sin \omega t \qquad [20.7]$$

This result shows that the emf varies sinusoidally with time, as plotted in Figure 20.16b. Note that the maximum emf has the value

$$\mathcal{E}_{max} = NBA\omega \qquad [20.8]$$

which occurs when $\omega t = 90°$ or $270°$. In other words, $\mathcal{E} = \mathcal{E}_{max}$ when the plane of the loop is parallel to the magnetic field. Furthermore, the emf is zero when $\omega t = 0$ or $180°$—that is, when the magnetic field is perpendicular to the plane of

Electric generators driven by turbines at a hydroelectric power plant. *(Luis Castaneda/The IMAGE Bank)*

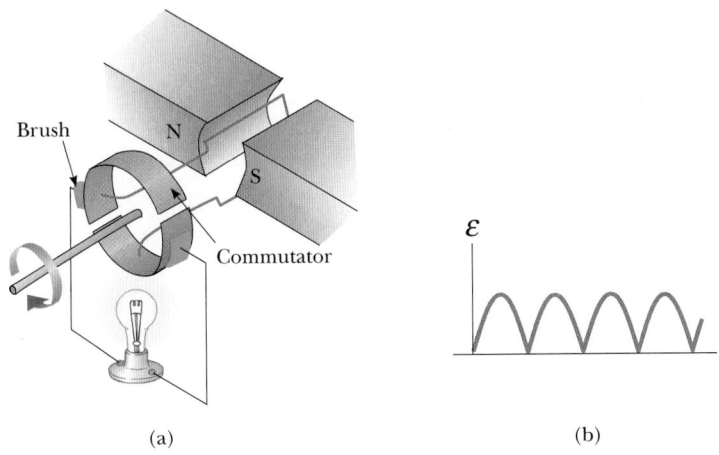

Figure 20.18 (a) A schematic diagram of a dc generator. (b) The emf fluctuates in magnitude but always has the same polarity.

the loop. In the United States and Canada, the frequency of rotation for commercial generators is 60 Hz, whereas in some European countries 50 Hz is used. (Recall that $\omega = 2\pi f$, where f is the frequency in hertz.)

The **direct current (dc) generator** is illustrated in Figure 20.18a. The components are essentially the same as those of the ac generator, except that the contacts to the rotating loop are made by a split ring, or commutator. In this design, the output voltage always has the same polarity and the current is a pulsating direct current, as in Figure 20.18b. This can be understood by noting that the contacts to the split ring reverse their roles every half cycle. At the same time, the polarity of the induced emf reverses. Hence, the polarity of the split ring remains the same.

A pulsating dc current is not suitable for most applications. To produce a steady dc current, commercial dc generators use many loops and commutators distributed around the axis of rotation so that the sinusoidal pulses from the loops overlap in phase. When these pulses are superimposed, the dc output is almost free of fluctuations.

EXAMPLE 20.5 Emf Induced in an ac Generator

An ac generator consists of eight turns of wire of area $A = 0.0900 \text{ m}^2$ with a total resistance of 12.0 Ω. The loop rotates in a magnetic field of 0.500 T at a constant frequency of 60.0 Hz.

(a) Find the maximum induced emf.

Solution First note that $\omega = 2\pi f = 2\pi(60.0 \text{ Hz}) = 377 \text{ rad/s}$. When we substitute the appropriate numerical values into Equation 20.8, we obtain

$$\mathcal{E}_{max} = NAB\omega = 8(0.0900 \text{ m}^2)(0.500 \text{ T})(377 \text{ rad/s}) = \boxed{136 \text{ V}}$$

(b) What is the maximum induced current?

Solution   From Ohm's law and the result of (a), we find that

$$I_{max} = \frac{\mathcal{E}_{max}}{R} = \frac{136 \text{ V}}{12.0 \text{ } \Omega} = \boxed{11.3 \text{ A}}$$

(c) Determine the induced emf as a function of time.

Solution   We can use Equation 20.7 to obtain the time variation of $\mathcal{E}$:

$$\mathcal{E} = \mathcal{E}_{max} \sin \omega t = \boxed{(136 \text{ V}) \sin 377t}$$

where t is in seconds.

Exercise   Determine the time variation of the induced current.

Answer $I = (11.3 \text{ A}) \sin 377t$

Motors and Back emf

Motors are devices that convert electrical energy to mechanical energy. Essentially, **a motor is a generator run in reverse.** Instead of a current being generated by a rotating loop, a current is supplied to the loop by a source of emf, and the magnetic torque on the current-carrying loop causes it to rotate.

A motor can perform useful mechanical work when a shaft connected to its rotating coil is attached to some external device. As the coil in the motor rotates, however, the changing magnetic flux through it induces an emf, which acts to reduce the current in the coil. If this were not the case, Lenz's law would be violated. The phrase **back emf** is used for an emf that tends to reduce the applied current. The back emf increases in magnitude as the rotational speed of the coil increases. We can picture this state of affairs as the equivalent circuit in Figure 20.19. For illustrative purposes, assume that the external power source attempting to drive current through the coil of the motor has a voltage of 120 V, that the coil has a resistance of 10 Ω, and that the back emf induced in the coil at this instant is 70 V. Thus, the voltage available to supply current equals the difference between the applied voltage and the back emf, 50 V in this case. It is clear that the current is limited by the back emf.

When a motor is turned on, there is no back emf initially and the current is very large because it is limited only by the resistance of the coil. As the coil begins to rotate, the induced back emf opposes the applied voltage and the current in the coil is reduced. If the mechanical load increases, the motor slows down, which decreases the back emf. This reduction in the back emf increases the current in the coil and therefore also increases the power needed from the external voltage source. As a consequence, the power requirements for starting a motor and for running it under heavy loads are greater than those for running the motor under average loads. If the motor is allowed to run under no mechanical load, the back emf reduces the current to a value just large enough to balance energy losses by heat and friction.

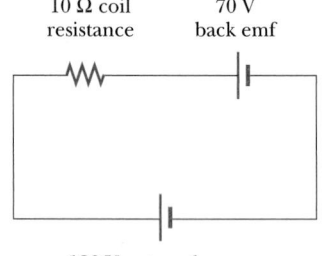

10 Ω coil resistance   70 V back emf

120 V external source

Figure 20.19   A motor can be represented as a resistance plus a back emf.

EXAMPLE 20.6   The Induced Current in a Motor

A motor has coils with a resistance of 10 Ω and is supplied by a voltage of 120 V. When the motor is running at its maximum speed, the back emf is 70 V. Find the current in the coils (a) when the motor is first turned on; (b) when the motor has reached maximum speed.

Solution   (a) When the motor is first turned on, the back emf is zero. (The coils are motionless.) Thus, the current in the coils is a maximum and is

$$I = \frac{\mathcal{E}}{R} = \frac{120 \text{ V}}{10 \text{ }\Omega} = \boxed{12 \text{ A}}$$

(b) At the maximum speed, the back emf has its maximum value. Thus, the effective supply voltage is now that of the external source minus the back emf, and the current is reduced to

$$I = \frac{\mathcal{E} - \mathcal{E}_{\text{back}}}{R} = \frac{120 \text{ V} - 70 \text{ V}}{10 \text{ }\Omega} = \frac{50 \text{ V}}{10 \text{ }\Omega} = \boxed{5.0 \text{ A}}$$

Exercise   If the current in the motor is 8.0 A at some instant, what is the back emf at this time?

Answer   40 V

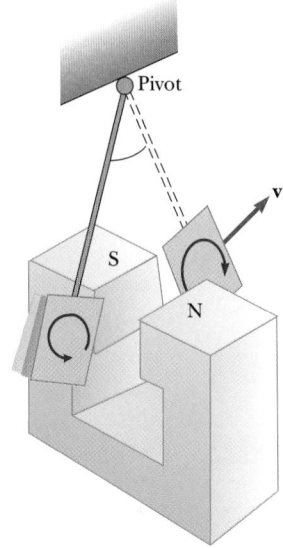

Figure 20.20   An apparatus that demonstrates the formation of eddy currents in a conductor moving through a magnetic field. As the plate enters or leaves the field, the changing magnetic flux sets up an induced emf, which causes the eddy currents in the plate.

20.6   EDDY CURRENTS   *Optional Section*

As we have seen, an emf and a current are induced in a circuit by a changing magnetic flux. In the same manner, circulating currents called **eddy currents** are set up in pieces of metal moving through a magnetic field. This can easily be demonstrated by allowing a flat metal plate at the end of a bar to swing through a magnetic field (Fig. 20.20). The metal should be a nonmagnetic material, such as aluminum or copper. As the plate enters the field, the changing flux creates an induced emf in the plate, which in turn causes the free electrons in the metal to move, producing swirling eddy currents. According to Lenz's law, the direction of the eddy currents must oppose the change that causes them. Thus, the eddy currents must produce effective magnetic poles on the plate, which are repelled by the poles of the magnet, giving rise to a repulsive force that opposes the swinging motion of the plate.

As indicated in Figure 20.21, when the magnetic field is into the paper, the eddy current is counterclockwise as the swinging plate enters the field at position 1. This is because the external flux into the paper is increasing, and hence, by Lenz's law, the induced current must provide a flux out of the paper. The opposite is true as the plate leaves the field at position 2, where the current is clockwise. Because the induced eddy current always produces a retarding force when the plate enters or leaves the field, the swinging plate quickly comes to rest.

If slots are cut in the metal plate, as in Figure 20.22, the eddy currents and the corresponding retarding force are greatly reduced. The cuts in the plate are open circuits for any large current loops that might otherwise be formed.

A task as simple as buying a candy bar from a coin-operated vending machine brings into use some of the principles of electricity and magnetism, as shown in Fig. 20.23. On entering the slot of the machine, the coin stops momentarily while its electrical resistance is checked. If its resistance falls within a broad range acceptable to the values stored in the memory of a computer chip inside the machine,

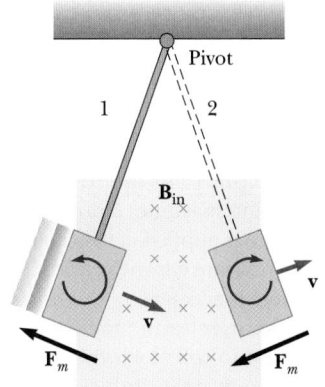

Figure 20.21   As the conducting plate enters the magnetic field in position 1, the eddy currents are counterclockwise. At position 2, however, the currents are clockwise. In either case, the magnetic force retards the motion of the plate.

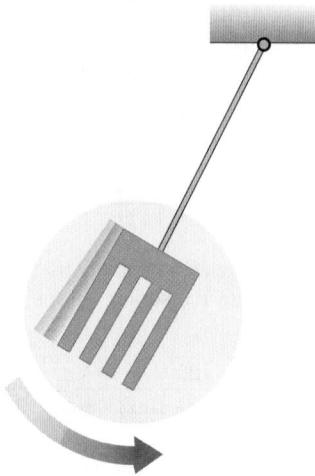

Figure 20.22 When slots are cut in the conducting plate, the eddy currents are reduced and the plate swings more freely through the magnetic field.

APPLICATION

Coin-Operated Vending Machines.

the coin is allowed to continue down a ramp and past a magnet. As it moves past the magnet, eddy currents are produced in the coin and magnetic forces act on it to slow it down slightly. The amount it is slowed depends on its metallic composition. Sensors measure its speed after moving past the magnet, and this speed is compared to values stored in the memory of the computer. If the computer approves, a gate is opened and the coin is accepted; otherwise, a second gate moves it into the reject runway.

Eddy currents are undesirable in motors and transformers because they dissipate energy in the form of heat. To reduce this energy loss, conducting parts are

Figure 20.23 As the coin enters the vending machine, a potential difference is applied across it at A, and its resistance is measured. If the resistance is acceptable, the holder drops down, releasing the coin and allowing it to roll along the inlet track. Two magnets induce eddy currents in the coin, and magnetic forces control its speed. If the speed sensors indicate that the coin has the correct speed, gate B swings up to allow the coin to be accepted—otherwise, gate C opens to allow the coin to follow the reject path.

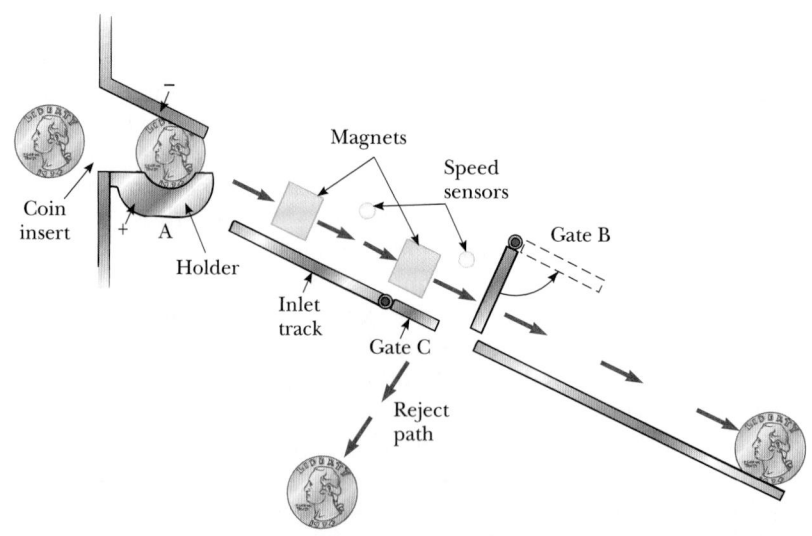

often laminated — that is, built up in thin layers separated by a nonconducting material, such as lacquer or metal oxide. This layered structure increases the resistance of the possible paths of the eddy currents and effectively confines the currents to individual layers. Lamination is used in the cores of transformers and motors to minimize eddy currents and thereby increase efficiency.

20.7 SELF-INDUCTANCE

Consider a circuit consisting of a switch, a resistor, and a source of emf, as in Figure 20.24. When the switch is closed, the current does not immediately change from zero to its maximum value, $\mathcal{E}/R$. The law of electromagnetic induction, Faraday's law, prevents this. What happens instead is the following. As the current increases with time, the magnetic flux through the loop due to this current also increases. The increasing flux induces an emf in the circuit that opposes the change in magnetic flux. By Lenz's law, the induced electric field in the loop must therefore be opposite the direction of the current. That is, the induced emf is in the direction indicated by the dashed battery in Figure 20.24. The net potential difference across the resistor is the emf of the battery minus the opposing induced emf. As the magnitude of the current increases, the *rate* of increase lessens and hence the induced emf decreases. This opposing emf results in a gradual increase in the current. For the same reason, when the switch is opened, the current gradually decreases to zero. This effect is called **self-induction** because the changing flux through the circuit arises from the circuit itself. The emf that is set up in this case is called a **self-induced emf.**

As a second example of self-inductance, consider Figure 20.25, which shows a coil wound on a cylindrical iron core. (A practical device would have several hundred turns.) Assume that the current changes with time. When the current is in the direction shown, a magnetic field is set up inside the coil, directed from right to left. As a result, some lines of magnetic flux pass through the cross-sectional area of the coil. As the current changes with time, the flux through the coil changes and induces an emf in the coil. Application of Lenz's law shows that this induced emf has a direction so as to oppose the change in the current. That is, if the current is increasing, the induced emf is as pictured in Figure 20.25b, and if the current is decreasing, the induced emf is as shown in Figure 20.25c.

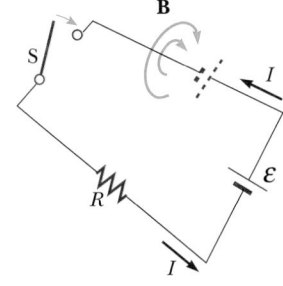

Figure 20.24 After the switch in the circuit is closed, the current produces its own magnetic flux through the loop. As the current increases toward its equilibrium value, the flux changes in time and induces an emf in the loop. The battery drawn with dashed lines is a symbol for the self-induced emf.

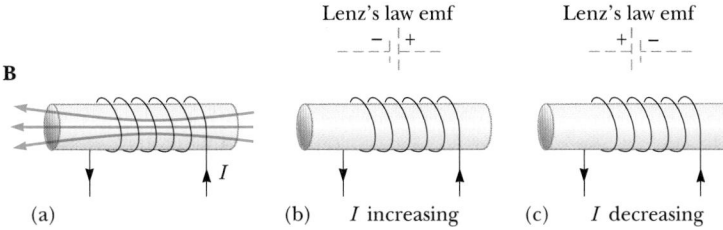

Figure 20.25 (a) A current in the coil produces a magnetic field directed to the left. (b) If the current increases, the coil acts as a source of emf directed as shown by the dashed battery. (c) The emf of the coil changes its polarity if the current decreases.

To evaluate self-inductance quantitatively, first note that, according to Faraday's law, the induced emf is as given by Equation 20.2:

$$\varepsilon = -N\frac{\Delta\Phi}{\Delta t}$$

The magnetic flux is proportional to the magnetic field, which is proportional to the current in the circuit. Thus, **the self-induced emf must be proportional to the time rate of change of the current:**

Self-induced emf ▶

$$\varepsilon \equiv -L\frac{\Delta I}{\Delta t} \qquad\qquad [20.9]$$

where L is a proportionality constant called the **inductance** of the device. The negative sign indicates that a changing current induces an emf in opposition to that change. This means that if the current is increasing (ΔI positive), the induced emf is negative to indicate opposition to the increase in current. Likewise, if the current is decreasing (ΔI negative), the sign of the induced emf is positive to indicate that the emf is acting to oppose the decrease.

The inductance of a coil depends on the cross-sectional area of the coil and other quantities, all of which can be grouped under the general heading of geometric factors. The SI unit of inductance is the **henry (H),** which, from Equation 20.9, is equal to 1 volt-second per ampere:

$$1\ \text{H} = 1\ \text{V}\cdot\text{s/A}$$

Examples 20.7 and 20.8 discuss simple situations for which self-inductances are easily evaluated. In the process, it is often convenient to equate Equations 20.2 and 20.9 to find an expression for L:

$$N\frac{\Delta\Phi}{\Delta t} = L\frac{\Delta I}{\Delta t}$$

Inductance ▶

$$L = N\frac{\Delta\Phi}{\Delta I} = \frac{N\Phi}{I} \qquad\qquad [20.10]$$

where we assume that $\Phi = 0$ when $I = 0$.

Joseph Henry, American physicist (1797–1878)

Henry became the first director of the Smithsonian Institution and first president of the Academy of Natural Science. He improved the design of the electromagnet and constructed one of the first motors. He also discovered the phenomenon of self-induction but failed to publish his findings. The unit of inductance, the henry, is named in his honor. *(North Wind Picture Archives)*

Thinking Physics 3

In some circuits, a spark occurs between the poles of a switch when the switch is opened. Yet, when the switch for this circuit is closed, there is no spark. Why is there this difference?

Explanation   According to Lenz's law, induced emfs are in a direction such as to attempt to maintain the original magnetic flux when a change occurs. When the switch is opened, the sudden drop in the magnetic field in the circuit induces an emf in a direction that attempts to maintain the original current. This can cause a spark as the current bridges the air gap between the poles of the switch. The spark does not occur when the switch is closed, because the original current is zero, and the induced emf attempts to maintain it at zero.

EXAMPLE 20.7   Inductance of a Solenoid

Find the inductance of a uniformly wound solenoid with N turns and length ℓ. Assume that ℓ is large compared with the radius and that the core of the solenoid is air.

Reasoning   The inductance can be found from $L = N\Phi/I$. The flux through each turn is $\Phi = BA$, and $B = \mu_0 nI$.

Solution   We take the interior field to be uniform and given by Equation 19.15:

$$B = \mu_0 nI = \mu_0 \frac{N}{\ell} I$$

where $n = N/\ell$ is the number of turns per unit length. The flux through each turn is

$$\Phi = BA = \mu_0 \frac{N}{\ell} AI$$

where A is the cross-sectional area of the solenoid. From this expression and Equation 20.10, we find that

$$L = \frac{N\Phi}{I} = \frac{\mu_0 N^2 A}{\ell} \qquad\qquad [20.11]$$

This shows that L depends on the geometric factors ℓ and A and on μ_0 and is proportional to the square of the number of turns. Because $N = n\ell$, we can also express the result in the form

$$L = \mu_0 \frac{(n\ell)^2}{\ell} A = \mu_0 n^2 A\ell = \mu_0 n^2 V \qquad\qquad [20.12]$$

where $V = A\ell$ is the volume of the solenoid.

EXAMPLE 20.8   Calculating Inductance and Self-Induced emf

(a) Calculate the inductance of a solenoid containing 300 turns if the length of the solenoid is 25.0 cm and its cross-sectional area is 4.00 cm² = 4.00×10^{-4} m².

Solution   Using Equation 20.11, we get

$$L = \frac{\mu_0 N^2 A}{\ell} = (4\pi \times 10^{-7}\ \text{T·m/A}) \frac{(300)^2 (4.00 \times 10^{-4}\ \text{m}^2)}{25.0 \times 10^{-2}\ \text{m}}$$

$$= 1.81 \times 10^{-4}\ \text{T·m}^2/\text{A} = \boxed{0.181\ \text{mH}}$$

(b) Calculate the self-induced emf in the solenoid described in (a) if the current through it is decreasing at the rate of 50.0 A/s.

Solution   Equation 20.9 can be combined with $\Delta I/\Delta t = -50.0$ A/s to give

$$\varepsilon = -L\frac{\Delta I}{\Delta t} = -(1.81 \times 10^{-4}\ \text{H})(-50.0\ \text{A/s}) = \boxed{9.05\ \text{mV}}$$

20.8   RL CIRCUITS

A circuit element that has a large inductance, such as a closely wrapped coil of many turns, is called an **inductor.** The circuit symbol for an inductor is ⟶‿‿‿⟶. We shall always assume that the self-inductance of the remainder of the circuit is negligible compared with that of the inductor in the circuit.

To gain some insight into the effect of an inductor in a circuit, consider the two circuits in Figure 20.26. Figure 20.26a shows a resistor connected to the terminals of a battery. For this circuit, Kirchhoff's loop rule is $\mathcal{E} - IR = 0$. The voltage drop across the resistor is

$$\Delta V = -IR \tag{20.13}$$

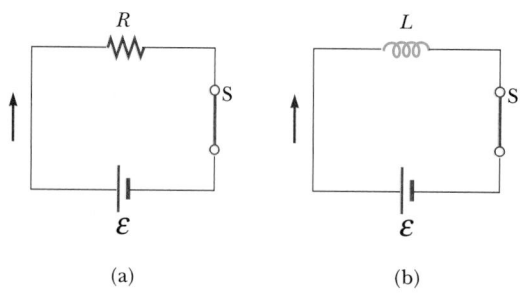

Figure 20.26   A comparison of the effect of a resistor to that of an inductor in a simple circuit.

(a)   (b)

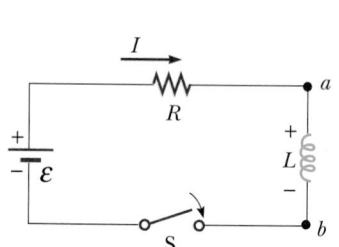

Figure 20.27   A series RL circuit. As the current increases toward its maximum value, the inductor produces an emf that opposes the increasing current.

In the past, we have interpreted resistance as a measure of opposition to the current. Now consider the circuit in Figure 20.26b, consisting of an inductor connected to the terminals of a battery. At the instant the switch in this circuit is closed, the emf of the battery equals the back emf generated in the coil. Thus, we have

$$\mathcal{E}_L = -L\frac{\Delta I}{\Delta t} \tag{20.14}$$

From this expression, **we can interpret L as a measure of opposition to the rate of change in current.**

Figure 20.27 shows a circuit consisting of a resistor, inductor, and battery. Suppose the switch is closed at $t = 0$. The current begins to increase, but the inductor produces an emf that opposes the increasing current. Thus, the current is unable to change from zero to its maximum value of $\mathcal{E}/R$ instantaneously. Equation 20.14 shows that the induced emf is a maximum when the current is changing most rapidly, which occurs when the switch is first closed. As the current approaches its steady-state value, the back emf of the coil falls off because the current is changing more slowly. Finally, when the current reaches its steady-state value, the rate of change is zero and the back emf is also zero. Figure 20.28 plots current in the circuit as a function of time.[1] This plot is very similar to that of the charge on a capacitor as a function of time, discussed in Chapter 18. In that case, we found it convenient to introduce a quantity called the *time constant of the circuit*, which told

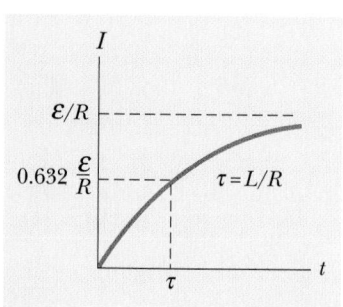

Figure 20.28   A plot of current versus time for the RL circuit shown in Figure 20.27. The switch is closed at $t = 0$, and the current increases toward its maximum value, $\mathcal{E}/R$. The time constant, τ, is the time it takes the current to reach 63.2% of its maximum value.

---

[1] The equation for the current in the circuit as a function of time is

$$I = \frac{\mathcal{E}}{R}(1 - e^{-Rt/L})$$

us something about the time required for the capacitor to approach its steady-state charge. In the same fashion, time constants are defined for circuits containing resistors and inductors. The **time constant** τ for an RL circuit is the time required for the current in the circuit to reach 63.2% of its final value, $\mathcal{E}/R$; the time constant of an RL circuit is given by

$$\tau = \frac{L}{R}$$ [20.15]

EXAMPLE 20.9 The Time Constant for an RL Circuit

The circuit shown in Figure 20.27 consists of a 30-mH inductor, a 6.0-Ω resistor, and a 12-V battery. The switch is closed at $t = 0$.

(a) Find the time constant of the circuit.

Solution The time constant is given by Equation 20.15:

$$\tau = \frac{L}{R} = \frac{30 \times 10^{-3} \text{ H}}{6.0 \ \Omega} = \boxed{5.0 \text{ ms}}$$

(b) Find the current after one time constant has elapsed.

Solution After one time constant, the current in the circuit has risen to 63.2% of its final value. Thus, the current is

$$I = 0.632 \frac{\mathcal{E}}{R} = (0.632)\left(\frac{12 \text{ V}}{6.0 \ \Omega}\right) = \boxed{1.3 \text{ A}}$$

Exercise What is the voltage drop across the resistor (a) at $t = 0$? (b) after one time constant?

Answer (a) 0; (b) 7.6 V

20.9 ENERGY STORED IN A MAGNETIC FIELD

The emf induced by an inductor prevents a battery from establishing a current in a circuit instantaneously. The battery has to do work to produce a current. We can think of this needed work as energy stored by the inductor in its magnetic field. In a manner quite similar to that used in Section 16.9 to find the energy stored by a capacitor, we find that the energy stored by an inductor is

$$PE_L = \tfrac{1}{2}LI^2$$ [20.16]

Note that the result is similar in form to the expression for the energy stored in a charged capacitor:

$$PE_C = \tfrac{1}{2}C(\Delta V)^2$$

13.6, SECTION 2

 Workbook Problem 85 – (Workbook page 182) asks you to calculate the energy stored in an inductor.

SUMMARY

The magnetic flux, Φ, through a closed loop is defined as

$$\Phi \equiv BA \cos \theta \qquad [20.1]$$

where B is the strength of the uniform magnetic field, A is the cross-sectional area of the loop, and θ is the angle between **B** and the direction perpendicular to the plane of the loop.

Faraday's law of induction states that the instantaneous emf induced in a circuit equals the rate of change of magnetic flux through the circuit:

$$\mathcal{E} = -N \frac{\Delta \Phi}{\Delta t} \qquad [20.2]$$

where N is the number of loops in the circuit.

Lenz's law states that the polarity of the induced emf is such that it produces a current whose magnetic field opposes the *change* in magnetic flux through a circuit.

If a conducting bar of length ℓ moves through a magnetic field with a speed, v, so that **B** is perpendicular to the bar, the emf induced in the bar, often called a **motional emf,** is

$$\mathcal{E} = B\ell v \qquad [20.4]$$

When a coil of wire with N turns, each of area A, rotates with constant angular speed ω in a uniform magnetic field **B** as in Figure 20.17, the emf induced in the coil is

$$\mathcal{E} = NAB\omega \sin \omega t \qquad [20.7]$$

When the current in a coil changes with time, an emf is induced in the coil according to Faraday's law. This **self-induced emf** is defined by the expression

$$\mathcal{E} \equiv -L \frac{\Delta I}{\Delta t} \qquad [20.9]$$

where L is the inductance of the coil. The SI unit for inductance is the henry (H); $1 \text{ H} = 1 \text{ V} \cdot \text{s/A}$.

The **inductance** of a coil can be found from the expression

$$L = \frac{N\Phi}{I} \qquad [20.10]$$

where N is the number of turns on the coil, I is the current in the coil, and Φ is the magnetic flux through the coil produced by that current.

If a resistor and inductor are connected in series to a battery and a switch is closed at $t = 0$, the current in the circuit does not rise instantly to its maximum value. After one **time constant,** $\tau = L/R$, the current in the circuit is 63.2% of its final value, $\mathcal{E}/R$.

The **energy stored** in the magnetic field of an inductor carrying current I is

$$PE_L = \tfrac{1}{2} LI^2 \qquad [20.16]$$

MULTIPLE-CHOICE QUESTIONS

1. What is the time constant for the circuit shown in Figure 20.29?

   (a) $\dfrac{L}{R}$ (b) $\dfrac{2L}{R}$ (c) $\dfrac{L}{2R}$ (d) $\dfrac{4L}{R}$ (e) $\dfrac{L}{R^2}$

   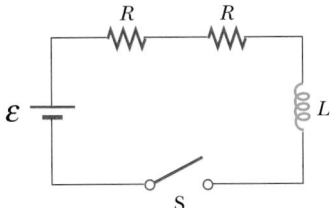

   Figure 20.29 (Multiple-Choice Question 1)

2. What is the time constant for the circuit shown in Figure 20.30?

   (a) $\dfrac{L}{R}$ (b) $\dfrac{2L}{R}$ (c) $\dfrac{L}{2R}$ (d) $\dfrac{4L}{R}$ (e) $\dfrac{L}{R^2}$

   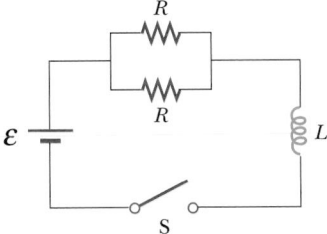

   Figure 20.30 (Multiple-Choice Question 2)

3. A circular coil with an area of 0.0500 m^2 is wrapped with 500 turns of wire and is placed in a uniform magnetic field perpendicular to its plane. If the field changes in value from -0.100 T to $+0.150$ T in an interval of 0.500 s, what average voltage is induced in the coil?
   (a) 12.5 V (b) 125 V (c) 188 V (d) 250 V

4. A bar magnet is falling through a loop of wire with constant velocity with the north pole entering first. Viewed from the same side of the coil as the magnet, as the north pole enters the wire, the induced current will be in what direction?
   (a) clockwise (b) counterclockwise (c) zero
   (d) along the length of the magnet

5. A small airplane with a wing span of 12 m flies horizontally and due north at a speed of 60 m/s in a region where the magnetic field of the Earth is 60 μT directed 60° below the horizontal. What is the magnitude of the induced emf between the ends of the wing?
   (a) 50 mV (b) 31 mV (c) 37 mV (d) 44 mV

CONCEPTUAL QUESTIONS

1. A circular loop is located in a uniform and constant magnetic field. Describe how an emf can be induced in the loop in this situation.

2. Does dropping a magnet down a copper tube produce a current in the tube? Explain.

3. A spacecraft orbiting the Earth has a coil of wire in it. An astronaut measures a small current in the coil, although there is no battery connected to it and there are no magnets on the spacecraft. What is causing the current?

4. Would you expect the tape from a tape recorder to be attracted to a magnet? (Try it, but not with a recording you wish to save.)

5. Suppose you would like to steal power for your home from the electric company by placing a loop of wire near a transmission cable in order to induce an emf in the loop (an illegal procedure). Should you orient the loop so that the transmission cable passes through your loop or simply place your loop near the transmission cable?

6. A bar magnet is dropped toward a conducting ring lying on the floor. As the magnet falls toward the ring, does it move as a freely falling body?

7. Is it possible to induce a constant emf for an infinite amount of time?

8. A piece of aluminum is dropped vertically between the

poles of an electromagnet. Does the magnetic field affect the velocity of the aluminum?

9. In equal arm balances used to weigh objects in the early twentieth century, it is sometimes observed that an aluminum sheet hangs from one of the arms and passes between the poles of a magnet. Why?

10. Why is the induced emf that appears in an inductor called a back (counter) emf?

11. A loop of wire is placed in a uniform magnetic field. For what orientation of the loop is the magnetic flux a maximum? For what orientation is the flux zero?

12. How is electrical energy produced in dams (that is, how is the energy of the motion of the water converted to ac electricity?)

13. If the current in an inductor is doubled, by what factor does the stored energy change?

14. As the conducting bar in Figure Q20.14 moves to the right, an electric field directed downward is set up in the conductor. If the bar were moving to the left, explain why the electric field would be upward.

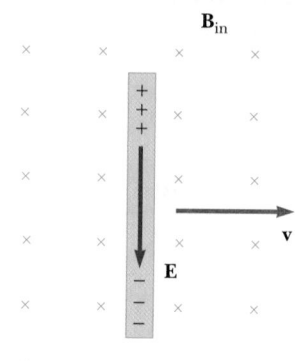

Figure Q20.14   (Conceptual Questions 14 and 15)

15. As the bar in Figure Q20.14 moves perpendicularly to the magnetic field, is an external force required to keep it moving with constant velocity? Explain.

PROBLEMS

1, 2, 3 = straightforward, intermediate, challenging   □ = full solution available in Study Guide/Student Solutions Manual   ◤ = Core Concepts Workbook
WEB = solution posted at **http://www.harcourtcollege.com/physics/cptech**   ▧ = biomedical application   ▨ = Interactive Physics

Section 20.1   Induced emf and Magnetic Flux

1. A magnetic field of strength 0.30 T is directed perpendicular to a plane circular loop of wire of radius 25 cm. Find the magnetic flux through the area enclosed by this loop.

2. A square loop 2.00 m on a side is placed in a magnetic field of strength 0.300 T. If the field makes an angle of 50.0° with the normal to the plane of the loop, as in Figure 20.2, determine the magnetic flux through the loop.

3. Find the flux of the Earth's magnetic field, of magnitude 5.00×10^{-5} T, through a square loop of area 20.0 cm², (a) when the field is perpendicular to the plane of the loop; (b) when the field makes a 30.0° angle with the normal to the plane of the loop; (c) when the field makes a 90.0° angle with the normal to the plane.

4. A long, straight wire carrying a current of 2.00 A is placed along the axis of a cylinder of radius 0.500 m and a length of 3.00 m. Determine the total magnetic flux through the cylinder.

5. A long, straight wire lies in the plane of a circular coil with a radius of 0.010 m. The wire carries a current of 2.0 A and is placed along a diameter of the coil. (a) What is the net flux through the coil? (b) If the wire passes through the center of the coil and is perpendicu-

lar to the plane of the coil, find the net flux through the coil.

6. A solenoid 4.00 cm in diameter and 20.0 cm long has 250 turns and carries a current of 15.0 A. Calculate the magnetic flux through the circular cross-sectional area of the solenoid.

7. A cube of edge length $\ell = 2.5$ cm is positioned as shown in Figure P20.7. There is a uniform magnetic field throughout the region with components of $B_x = +5.0$ T, $B_y = +4.0$ T, and $B_z = +3.0$ T. (a) Calculate the flux through the shaded face of the

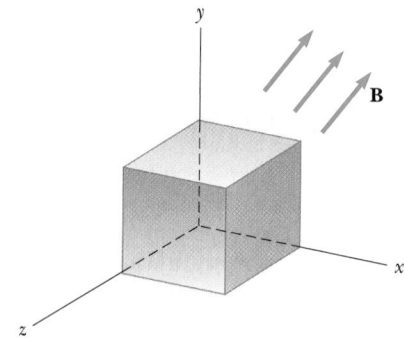

Figure P20.7

cube. **(b)** What is the total flux emerging from the volume enclosed by the cube (i.e., total flux through all six faces)?

Section 20.2 Faraday's Law of Induction

8. A circular loop of radius 20 cm is placed in an external magnetic field of strength 0.20 T so that the plane of the loop is perpendicular to the field. The loop is pulled out of the field in 0.30 s. Find the average induced emf during this interval.

9. A square, single-turn coil 0.20 m on a side is placed with its plane perpendicular to a constant magnetic field. An emf of 18 mV is induced in the winding when the area of the coil decreases at a rate of 0.10 m²/s. What is the magnitude of the magnetic field?

10. The flexible loop in Figure P20.10 has a radius of 12 cm and is in a magnetic field of strength 0.15 T. The loop is grasped at points A and B and stretched until it closes. If it takes 0.20 s to close the loop, find the magnitude of the average induced emf in it during this time.

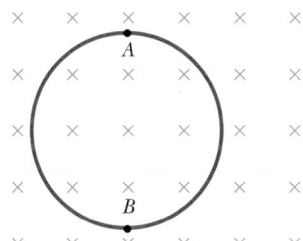

Figure P20.10

WEB **11.** The plane of a rectangular coil, 5.0 cm by 8.0 cm, is perpendicular to the direction of a magnetic field, **B**. If the coil has 75 turns and a total resistance of 8.0 Ω, at what rate must the magnitude of **B** change to induce a current of 0.10 A in the windings of the coil?

12. A 500-turn circular-loop coil 15.0 cm in diameter is initially aligned so that its axis is parallel to the Earth's magnetic field. In 2.77 ms the coil is flipped so that its axis is perpendicular to the Earth's magnetic field. If an average voltage of 0.166 V is thereby induced in the coil, what is the value of the Earth's magnetic field?

13. A wire loop of radius 0.30 m lies so that an external magnetic field of strength +0.30 T is perpendicular to the loop. The field changes to −0.20 T in 1.5 s. (The plus and minus signs here refer to opposite directions through the loop.) Find the magnitude of the average induced emf in the loop during this time.

14. A square, single-turn wire coil 1.00 cm on a side is placed inside a solenoid that has a circular cross section of radius 3.00 cm, as shown in Figure P20.14. The sole-

noid is 20.0 cm long and wound with 100 turns of wire. **(a)** If the current in the solenoid is 3.00 A, find the flux through the coil. **(b)** If the current in the solenoid is reduced to zero in 3.00 s, find the magnitude of the average induced emf in the coil.

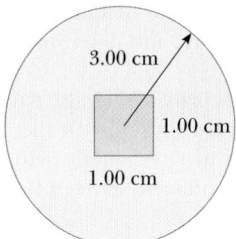

Figure P20.14

15. A 300-turn solenoid with a length of 20 cm and a radius of 1.5 cm carries a current of 2.0 A. A second coil of four turns is wrapped tightly about this solenoid so that it can be considered to have the same radius as the solenoid. Find **(a)** the change in the magnetic flux through the coil and **(b)** the magnitude of the average induced emf in the coil when the current in the solenoid increases to 5.0 A in a period of 0.90 s.

16. A circular coil, enclosing an area of 100 cm², is made of 200 turns of copper wire. The wire making up the coil has resistance of 5.0 Ω, and the ends of the wire are connected to form a closed loop. Initially, a 1.1-T uniform magnetic field points perpendicularly upward through the plane of the coil. The direction of the field then reverses so that the final magnetic field has a magnitude of 1.1 T and points downward through the coil. If the time required for the field to reverse directions is 0.10 s, what average current flows through the coil during this time?

Section 20.3 Motional emf

17. Consider the arrangement shown in Figure P20.17. Assume that $R = 6.0 \, \Omega$ and $\ell = 1.2$ m, and that a uniform 2.5-T magnetic field is directed *into* the page. At what

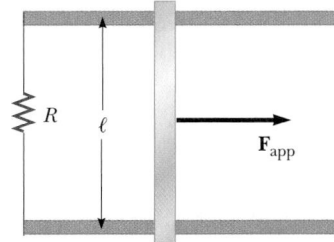

Figure P20.17 (Problems 17 and 56)

speed should the bar be moved to produce a current of 0.50 A in the resistor?

18. Over a region where the *vertical* component of the Earth's magnetic field is 40.0 μT directed downward, a 5.00-m length of wire is held in an east–west direction and moved horizontally to the north with a speed of 10.0 m/s. Calculate the potential difference between the ends of the wire, and determine which end is positive.

19. A helicopter has blades of length 3.0 m, rotating at 2.0 rev/s about a central hub. If the vertical component of the Earth's magnetic field is 5.0×10^{-5} T, what is the emf induced between the blade tip and the central hub?

20. A 12.0-m-long steel beam is accidentally dropped by a construction crane from a height of 9.00 m. The horizontal component of the Earth's magnetic field over the region is 18.0 μT. What is the induced emf in the beam just before impact with the Earth, assuming its long dimension remains in a horizontal plane, oriented perpendicularly to the horizontal component of the Earth's magnetic field?

Section 20.4 Lenz's Law Revisited

21. A bar magnet is held above the center of a wire loop in a horizontal plane, as shown in Figure P20.21. The south end of the magnet is toward the loop. The magnet is dropped. Find the direction of the current through the resistor **(a)** while the magnet is falling toward the loop and **(b)** after the magnet has passed through the loop and moves away from it.

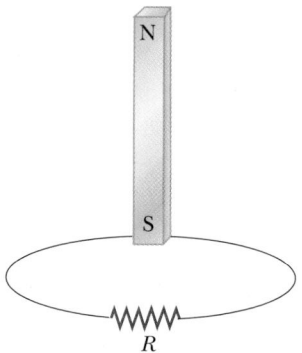

Figure P20.21

22. What is the direction of the current induced in the resistor when the current in the long, straight wire in Figure P20.22 decreases rapidly to zero?

Problem 19 is similar to Workbook Problem 79 (Workbook page 169).

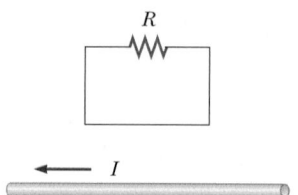

Figure P20.22

23. In Figure P20.23, what is the direction of the current induced in the resistor at the instant the switch is closed?

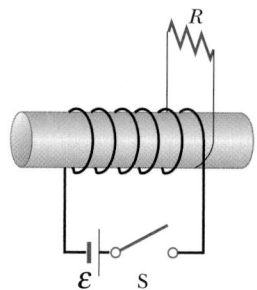

Figure P20.23

24. A bar magnet is positioned near a coil of wire, as shown in Figure P20.24. What is the direction of the current through the resistor when the magnet is moved **(a)** to the left? **(b)** to the right?

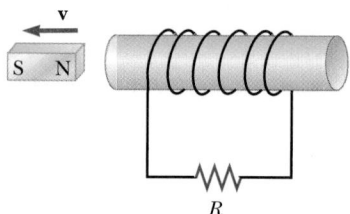

Figure P20.24

25. Find the direction of the current through the resistor in Figure P20.25 **(a)** at the instant the switch is closed,

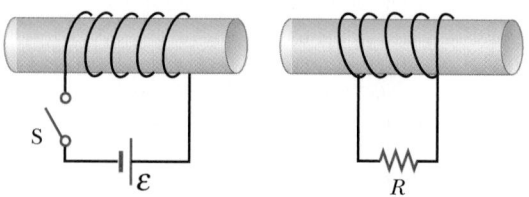

Figure P20.25

(b) after the switch has been closed for several minutes, and (c) at the instant the switch is opened.

26. Find the direction of the current in resistor R in Figure P20.26 after each of the following steps, (taken in the order given). (a) The switch is closed. (b) The variable resistance in series with the battery is decreased. (c) The circuit containing resistor R is moved to the left. (d) The switch is opened.

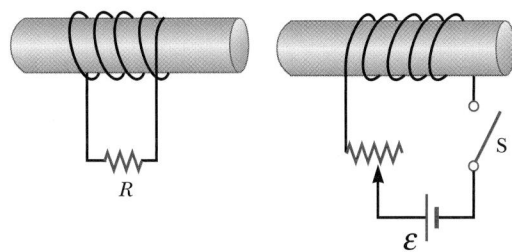

Figure P20.26

Section 20.5 Generators

27. A coil of 10.0 turns is in the shape of an ellipse having a major axis of 10.0 cm and a minor axis of 4.00 cm. The coil rotates at 100 rpm in a region in which the Earth's magnetic field is 55 μT. What is the maximum voltage induced in the coil if the axis of rotation of the coil is along its major axis and is aligned (a) perpendicular to the Earth's magnetic field; (b) parallel to the Earth's magnetic field? The area of an ellipse is given by $A = \pi ab$, where a is the *semi*-major axis and b is the *semi*-minor axis.

28. A 100-turn square wire coil of area 0.040 m² rotates about a vertical axis at 1500 rpm, as indicated in Figure P20.28. The horizontal component of the Earth's magnetic field at the location of the loop is 2.0×10^{-5} T. Calculate the maximum emf induced in the coil by the Earth's field.

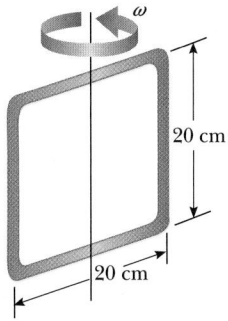

Figure P20.28

29. A motor has coils with a resistance of 30 Ω and operates from a voltage of 240 V. When the motor is operating at its maximum speed, the back emf is 145 V. Find the current in the coils (a) when the motor is first turned on and (b) when the motor has reached maximum speed. (c) If the current in the motor is 6.0 A at some instant, what is the back emf at that time?

30. When the coil of a motor is rotating at maximum speed, the current in the windings is 4.0 A. When the motor is first turned on, the current in the windings is 11 A. If the motor is operated at 120 V, find (a) the resistance of the windings and (b) the back emf in the coil at maximum speed.

31. A coil of area 0.10 m² is rotating at 60 rev/s with its axis of rotation perpendicular to a 0.20-T magnetic field. (a) If there are 1000 turns on the coil, what is the maximum voltage induced in the coil? (b) When the maximum induced voltage occurs, what is the orientation of the coil with respect to the magnetic field?

32. In a model ac generator, a 500-turn rectangular coil, 8.0 cm by 20 cm, rotates at 120 rev/min in a uniform magnetic field of 0.60 T. (a) What is the maximum emf induced in the coil? (b) What is the instantaneous value of the emf in the coil at $t = (\pi/32)$ s? Assume that the emf is zero at $t = 0$. (c) What is the smallest value of t for which the emf will have its maximum value?

Section 20.7 Self-Inductance

33. What is the inductance of a 510-turn solenoid that has a radius of 8.00 cm and an overall length of 1.40 m?

34. Show that the two expressions for inductance given by

$$L = \frac{N\Phi}{I} \quad \text{and} \quad L = \frac{-\mathcal{E}}{\Delta I/\Delta t}$$

have the same units.

35. A coil has an inductance of 3.0 mH, and the current through it changes from 0.20 A to 1.5 A in 0.20 s. Find the magnitude of the average induced emf in the coil during this period.

36. A solenoid of radius 2.5 cm has 400 turns and a length of 20 cm. Find (a) its inductance and (b) the rate at which current must change through it to produce an emf of 75 mV.

WEB 37. An emf of 24.0 mV is induced in a 500-turn coil when the current is changing at a rate of 10.0 A/s. What is the magnetic flux through each turn of the coil at an instant when the current is 4.00 A?

Section 20.8 *RL* Circuits (Optional)

38. Show that the SI units for the inductive time constant τ are seconds.

39. An *RL* circuit with $L = 3$ H and an *RC* circuit with $C = 3$ μF have the same time constant. If the two cir-

cuits have the same resistance, R, **(a)** what is the value of R, and **(b)** what is this common time constant?

40. A 6.0-V battery is connected in series with a resistor and an inductor. The series circuit has a time constant of 600 μs, and the maximum current is 300 mA. What is the value of the inductance?

41. A 25-mH inductor, an 8.0-Ω resistor, and a 6.0-V battery are connected in series. The switch is closed at $t = 0$. Find the voltage drop across the resistor **(a)** at $t = 0$ and **(b)** after one time constant has passed. Also, find the voltage drop across the inductor **(c)** at $t = 0$ and **(d)** after one time constant has elapsed.

42. The switch in a series RL circuit in which $R = 6.00\ \Omega$, $L = 3.00$ H, and $\mathcal{E} = 24.0$ V is closed at $t = 0$. **(a)** What is the maximum current in the circuit? **(b)** What is the current when $t = 0.500$ s?

Section 20.9 Energy Stored in a Magnetic Field

43. How much energy is stored in a 70.0-mH inductor at the instant when the current is 2.00 A?

44. A 300-turn solenoid has a radius of 5.00 cm and a length of 20.0 cm. Find the energy stored in it when the current is 0.500 A.

45. A 24-V battery is connected in series with a resistor and an inductor, where $R = 8.0\ \Omega$ and $L = 4.0$ H. Find the energy stored in the inductor **(a)** when the current reaches its maximum value and **(b)** one time constant after the switch is closed.

ADDITIONAL PROBLEMS

46. A tightly wound circular coil has 50 turns, each of radius 0.20 m. A uniform magnetic field is introduced perpendicularly to the plane of the coil. If the field increases in strength from 0 to 0.30 T in 0.40 s, what average emf is induced in the windings of the coil?

47. A coiled telephone cord has 70 turns, a cross-sectional diameter of 1.3 cm, and an unstretched length of 60 cm. Determine the self-inductance of the unstretched cord.

48. A 50-turn rectangular coil, 0.20 m by 0.30 m, is rotated at 90 rad/s in a magnetic field so that the axis of rotation is perpendicular to the direction of the field. The maximum emf induced in the coil is 0.50 V. What is the magnitude of the field?

49. In Figure P20.49, the bar magnet is being moved toward the loop. Is $V_a - V_b$ positive, negative, or zero during this motion? Explain.

50. A five-turn circular coil of radius 15 cm is oriented with its plane perpendicular to a uniform magnetic field of 0.15 T. During a 3.0-s time interval, this field increases to 0.20 T. If the resistance of the coil is 8.0 Ω, find the average current that flows in the coil during this interval.

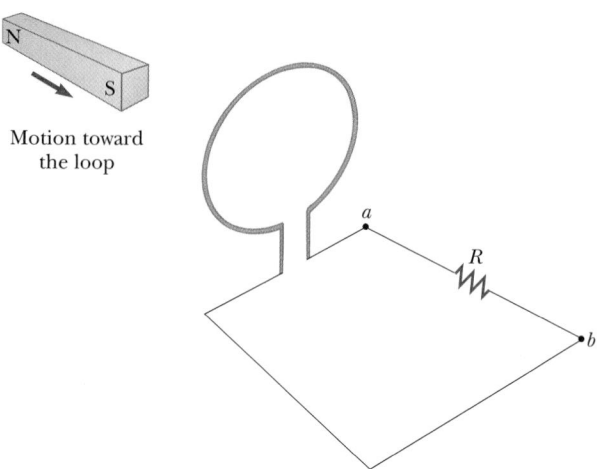

Figure P20.49

51. An 820-turn wire coil of resistance 24.0 Ω is placed on top of a 12 500-turn, 7.00-cm-long solenoid, as in Figure P20.51. Both coil and solenoid have cross-sectional areas of 1.00×10^{-4} m^2. **(a)** How long does it take the solenoid current to reach 0.632 times its maximum value? **(b)** Determine the average back emf caused by the self-inductance of the solenoid during this interval. **(c)** Determine the average rate of change in magnetic flux through each turn of the coil during this interval. **(d)** Find the magnitude of the average induced current in the coil.

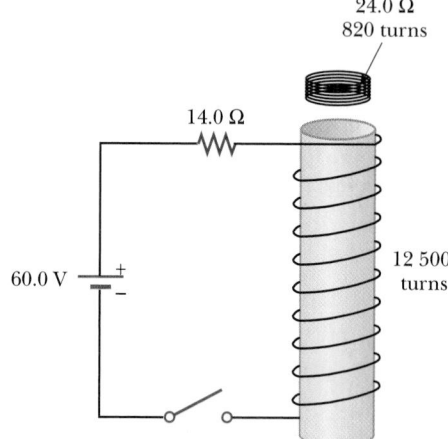

Figure P20.51

52. Figure P20.52 is a graph of induced emf versus time for a coil of N turns rotating with angular speed ω in a uniform magnetic field directed perpendicularly to the axis of rotation of the coil. Copy this sketch (increasing the scale), and on the same set of axes show the graph of

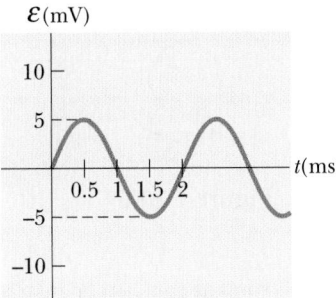

Figure P20.52

emf versus *t* when **(a)** the number of turns in the coil is doubled; **(b)** the angular speed is doubled; **(c)** the angular speed is doubled and the number of turns in the coil is halved.

53. An automobile starter motor draws a current of 3.5 A from a 12-V battery when operating at normal speed. A broken pulley locks the armature in position, and the current increases to 18 A. What was the back emf of the motor when operating normally?

54. A single-turn circular loop of radius 0.20 m is coaxial with a long 1600-turn solenoid of radius 0.050 m and length 0.80 m, as in Figure P20.54. The variable resistor is changed so that the solenoid current decreases linearly from 6.0 A to 1.5 A in 0.20 s. Calculate the induced emf in the circular loop. (The field just outside the solenoid is small enough to be negligible.)

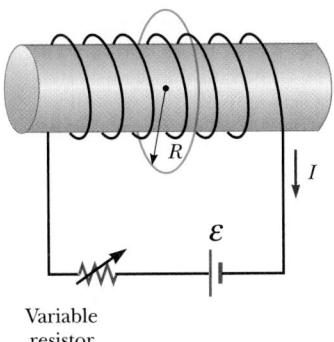

Variable
resistor

Figure P20.54

55. A horizontal wire is free to slide on the vertical rails of a conducting frame, as in Figure P20.55. The wire has

mass *m* and length ℓ, and the resistance of the circuit is *R*. If a uniform magnetic field is directed perpendicularly to the frame, what is the terminal speed of the wire as it falls under the force of gravity? (Neglect friction.)

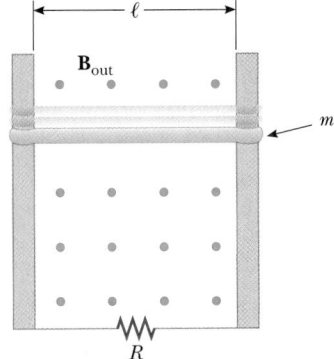

Figure P20.55

56. A conducting rod of length ℓ moves on two horizontal, frictionless rails, as in Figure P20.17. If a constant force of magnitude 1.00 N moves the bar at 2.00 m/s through a magnetic field **B** which is into the page, **(a)** what is the current through an 8.00-Ω resistor *R*? **(b)** What is the rate of energy dissipation in the resistor? **(c)** What is the mechanical power delivered by the force **F**?

57. The bolt of lightning depicted in Figure P20.57 passes 200 m from a 100-turn coil oriented as shown. If the current in the lightning bolt falls from 6.02×10^6 A to zero in 10.5 μs, what is the average voltage induced in the coil? Assume that the distance to the center of the coil determines the average magnetic field at the coil's position. Treat the lightning bolt as a long vertical wire.

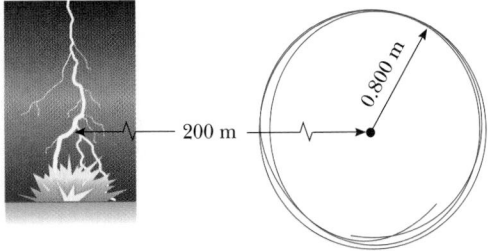

Figure P20.57

58. A two-turn circular wire coil of radius 0.500 m lies in a plane perpendicular to a uniform magnetic field of magnitude 0.40 T. If the wire is reshaped from a two-

turn circle to a one-turn circle in 0.10 s (while remaining in the same plane), what is the magnitude of the average induced emf in the wire during this time? (*Hint:* Use Faraday's law in the form $\mathcal{E} = -\Delta(N\Phi)/\Delta t$.)

59. The wire shown in Figure P20.59 is bent in the shape of a "tent," with $\theta = 60°$ and $L = 1.5$ m, and is placed in a uniform magnetic field of 0.30 T perpendicular to the tabletop. The wire is "hinged" at points a and b. If the tent is flattened out on the table in 0.10 s, what is the average induced emf in the wire during this time?

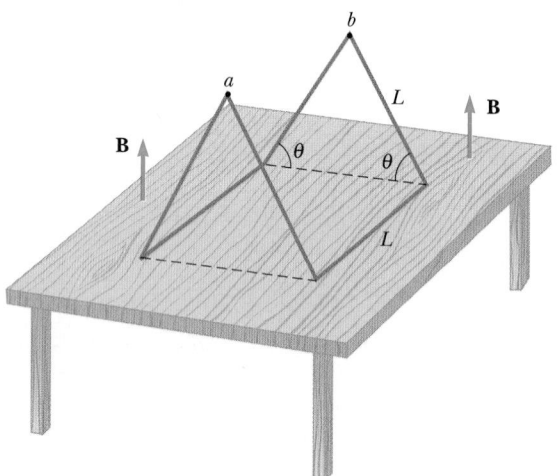

Figure P20.59

60. Over a region in which the vertical component of the Earth's magnetic field is 40.0 μT, a 0.500-m length of wire held in an east–west direction moves at 5.00 m/s along parallel rails, as shown in Figure 20.10. Find (a) the induced emf in the circuit and (b) the induced current in the circuit if there is a resistor of 5.00 Ω connected between the parallel rails. The resistance of the wire and track can be considered negligible. Find (c) the average power supplied to the circuit by the induced emf, (d) the power dissipated in the 5.00-Ω resistor, (e) the force required to keep the length of wire moving at a constant speed, and (f) the power supplied to the circuit by the agent exerting the force calculated in part (e).

61. The magnetic field shown in Figure P20.61 has a uniform magnitude of 25.0 mT directed into the paper. The initial diameter of the kink is 2.00 cm. (a) The wire is quickly pulled taut, and the kink shrinks to a diameter of zero in 50.0 ms. Determine the average voltage induced between endpoints A and B. Include the polarity. (b) Suppose the kink is undisturbed, but the magnetic field increases to 100 mT in 4.00×10^{-3} s. Determine

the average voltage across terminals A and B, including polarity, during this period.

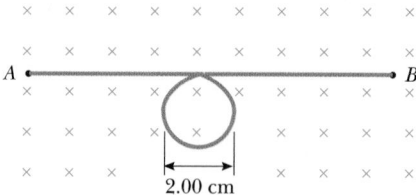

Figure P20.61

62. A bar magnet is spun at constant angular speed ω about an axis, as shown in Figure P20.62. A flat rectangular conducting loop surrounds the magnet, and at $t = 0$ the magnet is oriented as shown. Sketch the induced current in the loop as a function of time, plotting counter-clockwise currents as positive and clockwise currents as negative.

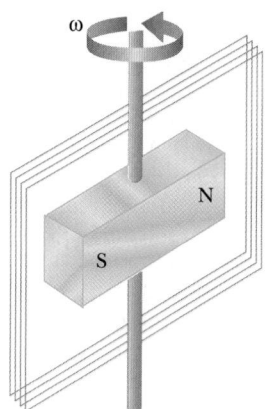

Figure P20.62

63. An aluminum ring of radius 5.0 cm and resistance 3.0×10^{-4} Ω is placed on top of a long air-core solenoid with 1000 turns per meter and radius 3.0 cm, as in Figure P20.63. The magnetic field due to the current in the solenoid at the location of the ring is one half that at the center of the solenoid. If the current in the solenoid is increasing at a constant rate of 270 A/s, what is the induced current in the ring?

64. Figure P20.64 shows a stationary conductor, whose shape is similar to the letter "e" ($a = 50.0$ cm), that is placed in a constant magnetic field of magnitude $B = 0.500$ T directed out of the page. A 50.0-cm-long, straight conducting rod pivoted about point O rotates with a constant angular speed of 2 rad/s. (a) Determine

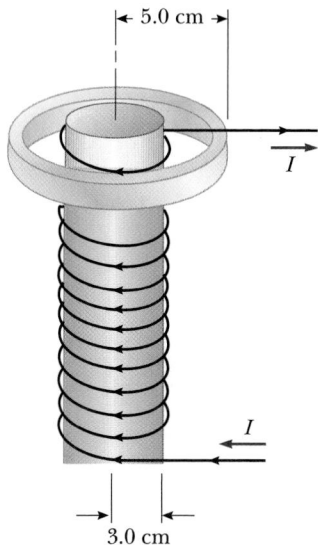

5.0 cm

I

I

3.0 cm

Figure P20.63

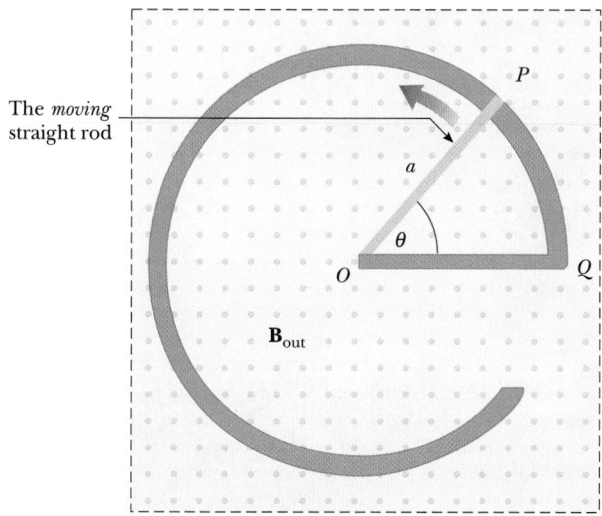

The *moving* straight rod

P

a

θ

O

Q

$\mathbf{B}_{out}$

Figure P20.64

ing one another on opposite sides of the river, a distance w apart and immersed entirely. The flow velocity of the river is $\mathbf{v}$, and the vertical component of the Earth's magnetic field is B. (a) Show that the current in the load resistor R is

$$I = \frac{abvB}{\rho + abR/w}$$

where ρ is the resistivity of the water. (b) Calculate the short-circuit current ($R = 0$) if $a = 100$ m, $b = 5.00$ m, $v = 3.00$ m/s, $B = 0.500\ \mu T$, and $\rho = 100\ \Omega \cdot m$.

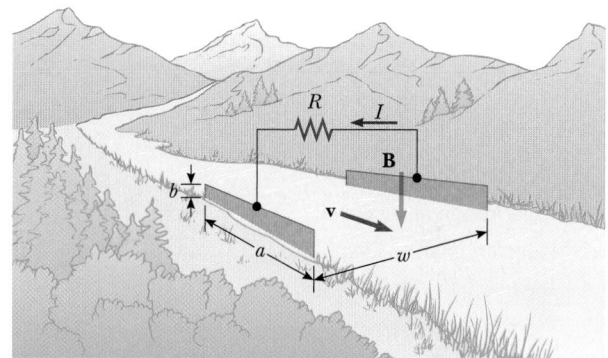

R

I

B

b

v

a

w

Figure P20.65

66. In Figure P20.66, the rolling axle, 1.50 m long, is pushed along horizontal rails at a constant speed $v = 3.00$ m/s. A resistor $R = 0.400\ \Omega$ is connected to the rails at points a and b, directly opposite each other. (The wheels make good electrical contact with the rails, and so the axle, rails, and R form a closed-loop circuit. The only significant resistance in the circuit is R.) There is a uniform magnetic field $B = 0.0800$ T vertically downward. (a) Find the induced current I in the resistor. (b) What horizontal force $\mathbf{F}$ is required to keep the axle rolling at constant speed? (c) Which end of the resistor, a or b, is at the higher electric potential? (d) After the axle rolls past the resistor, does the current in R reverse direction?

the induced emf in loop *POQ*. (*Hint:* The area of loop *POQ* is $A = \theta a^2/2$.) (b) If the conducting material has a resistance per unit length of 5.00 Ω/m, what is the induced current in loop *POQ* at 0.250 s? (*Hint:* The length of arc *PQ* is $a\theta$.)

65. In 1832 Faraday proposed that the apparatus shown in Figure P20.65 could be used to generate electric current from the flowing water in the Thames River. Two conducting planes of lengths a and widths b are placed fac-

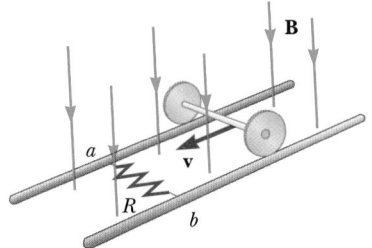

B

a

v

R

b

Figure P20.66

21

Alternating Current Circuits and Electromagnetic Waves

Chapter Outline

▲ PHYSICS PUZZLER

A rancher in St. Augustine Plains, New Mexico, rides past one of 27 radio telescopes that comprise the Very Large Array (VLA). Arranged in a Y-shaped configuration on a system of railroad tracks, the radio telescopes of the VLA capture and focus electromagnetic waves from space. What is an approximate wavelength for the radio waves that would be captured by one of these telescopes and how are these waves produced? *(© Danny Lehman)*

1t is important to understand the basic principles of alternating current (ac) circuits because they are so much a part of our everyday life. Every time we turn on a television set or a stereo, or any of a multitude of other electric appliances, we are calling on alternating currents to provide the power to operate them. We begin our study of ac circuits by examining the characteristics of a circuit containing a source of emf and one other circuit element: either a resistor, a capacitor, or an inductor. Then we examine what happens when these elements are connected in combination with each other. Our discussion is limited to situations in which the elements are arranged in simple series configurations.

We conclude this chapter with a discussion of **electromagnetic waves,** which are composed of fluctuating electric and magnetic fields. Electromagnetic waves in the form of visible light enable us to view the world around us; infrared waves warm our environment; radio-frequency waves carry our favorite television and radio programs; the list goes on and on.

21.1 RESISTORS IN AN ac CIRCUIT

An ac circuit consists of combinations of circuit elements and an ac generator, which provides the alternating current. We have seen that the output of an ac generator is sinusoidal and varies with time according to

$$\Delta v = \Delta V_m \sin 2\pi f t \qquad \text{[21.1]}$$

where Δv is the instantaneous voltage, ΔV_m is the maximum voltage of the ac generator, and f is the frequency at which the voltage changes, measured in hertz. We first consider a simple circuit consisting of a resistor and an ac generator (designated by the symbol —⊙—), as in Figure 21.1. The current and the voltage across the resistor are shown in Figure 21.2.

Let us briefly discuss the current-versus-time curve in Figure 21.2. At point a on the curve, the current has a maximum value in one direction, arbitrarily called the positive direction. Between points a and b, the current is decreasing in magnitude but is still in the positive direction. At point b, the current is momentarily zero; it then begins to increase in the opposite (negative) direction between points b and c. At point c, the current has reached its maximum value in the negative direction.

Note that the current and voltage are in step with each other because they vary identically with time. **Because the current and the voltage reach their maximum values at the same time, they are said to be in phase.** Note that **the average value of the current over one cycle is zero.** That is, the current is maintained in one direction (the positive direction) for the same amount of time and at the same magnitude as it is in the opposite direction (the negative direction). However, the direction of the current has no effect on the behavior of the resistor in the circuit. This can be understood by realizing that collisions between electrons and the fixed atoms of the resistor result in an increase in the temperature of the resistor. Although this temperature increase depends on the magnitude of the current, it is independent of its direction.

We can quantify this discussion by recalling that the rate at which electrical energy is converted to heat in a resistor, which is the power P, is

$$P = i^2 R$$

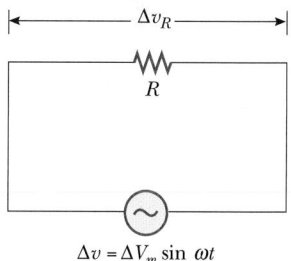

$$\Delta v = \Delta V_m \sin \omega t$$

Figure 21.1 A series circuit consisting of a resistor, R, connected to an ac generator, designated by the symbol —⊙—.

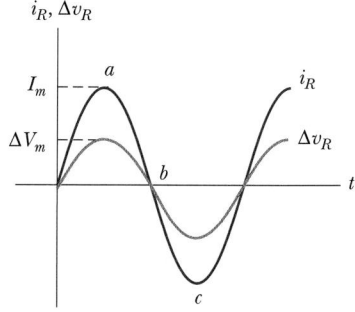

Figure 21.2 A plot of current and voltage across a resistor versus time.

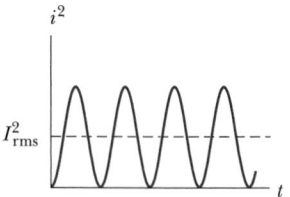

Figure 21.3   A plot of the square of the current in a resistor versus time. The rms current is the square root of the average of the square of the current.

rms current ▶

rms voltage ▶

where i is the instantaneous current in the resistor. Because the heating effect of a current is proportional to the *square* of the current, it makes no difference whether the current is direct or alternating — that is, whether the sign associated with the current is positive or negative. However, the heating effect produced by an alternating current with a maximum value of I_m is *not the same* as that produced by a direct current of the same value. This is because the alternating current is at this maximum value for only a very brief instant of time during a cycle. What is important in an ac circuit is an average value of current, referred to as the rms current. The **rms current** is the direct current that would dissipate the same amount of energy in a resistor as is dissipated by the actual alternating current. The term *rms* stands for *root mean square*, which simply means that the square root of the average value of the square of the current is taken. Because i^2 varies as $\sin^2 2\pi ft$, one can show that the average value of i^2 is $\frac{1}{2}I_m^2$ (Fig. 21.3).[1] Therefore, the rms current, I, is related to the maximum value of the alternating current, I_m, by

$$I = \frac{I_m}{\sqrt{2}} = 0.707 I_m \qquad [21.2]$$

This equation says that an alternating current with a maximum value of 3 A produces the same heating effect in a resistor as a direct current of $(3/\sqrt{2})$ A. Thus, we can say that the average power dissipated in a resistor that carries alternating current I is $P_{av} = I^2 R$, where I is the rms current.

Alternating voltages are also best discussed in terms of rms voltages, with the relationship being identical to the preceding one — that is, the rms voltage, ΔV, is related to the maximum value of the alternating voltage, ΔV_m, by

$$\Delta V = \frac{\Delta V_m}{\sqrt{2}} = 0.707 \, \Delta V_m \qquad [21.3]$$

When we speak of measuring an ac voltage of 120 V from an electric outlet, we really mean an *rms* voltage of 120 V. A quick calculation using Equation 21.3 shows that such an ac voltage actually has a peak value of about 170 V. In this chapter we use rms values when discussing alternating currents and voltages. One reason is that ac ammeters and voltmeters are designed to read rms values. Furthermore, if we use rms values, many of the equations we rely on will have the same form as those used in the study of direct current (dc) circuits. Table 21.1 summarizes the notations used in this chapter.

[1] The fact that the square root of the average value of the square of the current equals $I_m/\sqrt{2}$ can be shown as follows. The current in the circuit varies with time according to the expression $i = I_m \sin 2\pi ft$, and so $i^2 = I_m^2 \sin^2 2\pi ft$. Therefore, we can find the average value of i^2 by calculating the average value of $\sin^2 2\pi ft$. Note that a graph of $\cos^2 2\pi ft$ versus time is identical to a graph of $\sin^2 2\pi ft$ versus time, except that the points are shifted on the time axis. Thus, the time average of $\sin^2 2\pi ft$ is equal to the time average of $\cos^2 2\pi ft$ when taken over one or more cycles. That is,

$$(\sin^2 2\pi ft)_{av} = (\cos^2 2\pi ft)_{av}$$

With this fact and the trigonometric identity $\sin^2 \theta + \cos^2 \theta = 1$, we get

$$(\sin^2 2\pi ft)_{av} + (\cos^2 2\pi ft)_{av} = 2(\sin^2 2\pi ft)_{av} = 1$$

$$(\sin^2 2\pi ft)_{av} = \tfrac{1}{2}$$

When this result is substituted into the expression $i^2 = I_m^2 \sin^2 2\pi ft$, we get $(i^2)_{av} = I^2 = I_m^2/2$, or $I = I_m/\sqrt{2}$, where I is the rms current.

TABLE 21.1 Notation Used in This Chapter

	Voltage	Current
Instantaneous value	Δv	i
Maximum value	ΔV_m	I_m
rms value	ΔV	I

Consider the series circuit in Figure 21.1, consisting of a resistor connected to an ac generator. A resistor limits the current in an ac circuit just as it does in a dc circuit. Therefore, Ohm's law is valid for an ac circuit, and we have

$$\Delta V_R = IR \qquad [21.4]$$

That is, **the rms voltage across a resistor is equal to the rms current in the circuit times the resistance.** This equation also applies if maximum values of current and voltage are used. That is, the maximum voltage drop across a resistor equals the maximum current in the resistor times the resistance.

EXAMPLE 21.1 What Is the rms Current?

An ac voltage source has an output of $\Delta v = (200 \text{ V}) \sin 2\pi ft$. This source is connected to a 100-Ω resistor as in Figure 21.1. Find the rms current in the resistor.

Reasoning Compare the expression for the voltage output just given with the general form, $\Delta v = \Delta V_m \sin 2\pi ft$.

Solution By comparison, we see that the maximum output voltage of the device is 200 V. Thus, the rms voltage output of the source is

$$\Delta V = \frac{\Delta V_m}{\sqrt{2}} = \frac{200 \text{ V}}{\sqrt{2}} = 141 \text{ V}$$

Ohm's law can be used in resistive ac circuits as well as in dc circuits. The calculated rms voltage can be used with Ohm's law to find the rms current in the circuit:

$$I = \frac{\Delta V}{R} = \frac{141 \text{ V}}{100 \ \Omega} = \boxed{1.41 \text{ A}}$$

Exercise Find the maximum current in the circuit.

Answer 2.00 A

21.2 CAPACITORS IN AN ac CIRCUIT

To understand the effect of a capacitor on the behavior of a circuit containing an ac voltage source, let us first recall what happens when a capacitor is placed in a circuit containing a dc source, such as a battery. At the instant a switch is closed in

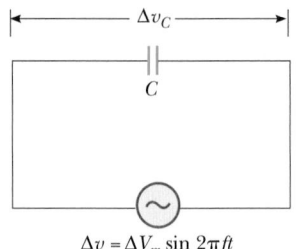

Figure 21.4 A series circuit consisting of a capacitor, C, connected to an ac generator.

a series circuit containing a battery, a resistor, and a capacitor, there is zero charge on the plates of the capacitor. Therefore, the motion of charge through the circuit is relatively free, and initially there is a large current in the circuit. As more charge accumulates on the capacitor, the voltage across it increases, opposing the current. After some time interval—which depends on the time constant, RC—has elapsed, the current approaches zero. From this, we see that a capacitor in a dc circuit limits, or impedes, the current so that it approaches zero after a brief time.

Now consider the simple series circuit in Figure 21.4, consisting of a capacitor connected to an ac generator. Let us sketch a curve of current versus time and one of voltage versus time, and then attempt to make the graphs seem reasonable. The curves are shown in Figure 21.5. First, note that the segment of the current curve from a to b indicates that the current starts out at a rather large value. This can be understood by recognizing that there is no charge on the capacitor at $t = 0$; as a consequence, there is nothing in the circuit except the resistance of the wires to hinder the flow of charge at this instant. However, the current decreases as the voltage across the capacitor increases from c to d on the voltage curve. When the voltage is at point d, the current reverses and begins to increase in the opposite direction (from b to e on the current curve). During this time, the voltage across the capacitor decreases from d to f because the plates are now losing the charge they accumulated earlier. The remainder of the cycle for both voltage and current is a repeat of what happened during the first half of the cycle. The current reaches a maximum value in the opposite direction at point e on the current curve and then decreases as the voltage across the capacitor builds up.

Note that the current and voltage are not in step with each other, as they are in a purely resistive circuit. The curves of Figure 21.5 indicate that, when an alternating voltage is applied across a capacitor, the voltage reaches its maximum value one quarter of a cycle after the current reaches its maximum value. In this situation, it is common to say that **the voltage always lags behind the current by 90°.**

▷ The voltage across a capacitor lags behind the current by 90°.

The impeding effect of a capacitor on the current in an ac circuit is expressed in terms of a factor called the **capacitive reactance,** X_C, defined as

▷ Capacitive reactance

$$X_C \equiv \frac{1}{2\pi f C} \qquad [21.5]$$

You will be asked in Problem 6 at the end of the chapter to show that when C is in farads and f is in hertz, the unit of X_C is the ohm.

Let us examine whether Equation 21.5 is reasonable. With a dc source (a dc source can be considered an ac source with zero frequency), X_C is infinitely large. This means that a capacitor impedes the direct current the same way a resistor of infinitely large resistance would. The current in such a circuit is zero. Indeed, we found that to be the case in Chapter 16. On the other hand, Equation 21.5 predicts that, as the frequency increases, the capacitive reactance decreases. This means that, before the charge on a capacitor has time to build up to the point where the current is zero, the direction of the current has reversed.

The analogy between capacitive reactance and resistance allows us to write an equation of the same form as Ohm's law to describe ac circuits containing capacitors. This equation relates the rms voltage and rms current in the circuit to the reactance as

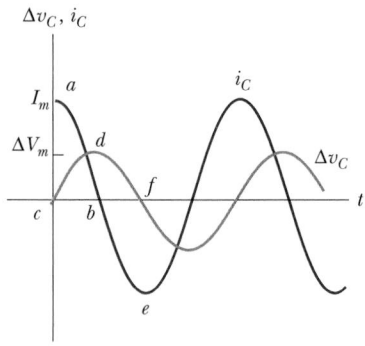

Figure 21.5 Plots of current and voltage across a capacitor versus time in an ac circuit. The voltage lags the current by 90°.

$$\Delta V_C = I X_C \qquad [21.6]$$

EXAMPLE 21.2   A Purely Capacitive ac Circuit

An 8.00-μF capacitor is connected to the terminals of an ac generator with an rms voltage of 150 V and a frequency of 60.0 Hz. Find the capacitive reactance and the rms current in the circuit.

Solution   From Equation 21.5 and the fact that $2\pi f = 377$ s^{-1}, we have

$$X_C = \frac{1}{2\pi fC} = \frac{1}{(377 \text{ s}^{-1})(8.00 \times 10^{-6} \text{ F})} = \boxed{332 \ \Omega}$$

If we substitute this result into Equation 21.6, we find that

$$I = \frac{\Delta V_C}{X_C} = \frac{150 \text{ V}}{332 \ \Omega} = \boxed{0.452 \text{ A}}$$

Exercise   If the frequency is doubled, what happens to the capacitive reactance and the current?

Answer X_C is halved, and I is doubled.

21.3   INDUCTORS IN AN ac CIRCUIT

Now consider an ac circuit consisting only of an inductor connected to the terminals of an ac generator, as in Figure 21.6. (In any real circuit, there is some resistance in the wire forming the inductive coil, but we ignore this for now.) The changing current output of the generator produces a back emf in the coil of magnitude

$$\Delta v_L = L\frac{\Delta I}{\Delta t} \qquad \text{[21.7]}$$

Thus, the current in the circuit is impeded by the back emf of the inductor. The effective resistance of the coil in an ac circuit is measured by a quantity called the **inductive reactance,** X_L:

$$X_L \equiv 2\pi fL \qquad \text{[21.8]}$$

◀ Inductive reactance

You will be asked in an end-of-chapter problem to show that when f is in hertz and L is in henries, the unit of X_L is the ohm. Note that the inductive reactance increases with increasing frequency and increasing inductance.

To understand the meaning of inductive reactance, let us compare this equation for X_L with Equation 21.7. First, note from Equation 21.8 that the inductive reactance depends on the inductance, L. This seems reasonable because the back emf (Eq. 21.7) is large for large values of L. Second, note that the inductive reactance depends on the frequency, f. This, too, seems reasonable because the back emf depends on $\Delta I/\Delta t$, a quantity that is large when the current changes rapidly, as it would for large frequencies.

With inductive reactance defined in this manner, we can write an equation of the same form as Ohm's law for the voltage across the coil or inductor:

$$\Delta V_L = IX_L \qquad \text{[21.9]}$$

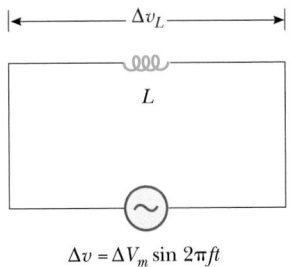

$\Delta v = \Delta V_m \sin 2\pi ft$

Figure 21.6   A series circuit consisting of an inductor, L, connected to an ac generator.

where ΔV_L is the rms voltage drop across the coil and I is the rms current in the coil.

Figure 21.7 shows the instantaneous voltage and instantaneous current across the coil as functions of time. When a sinusoidal voltage is applied across an inductor, the voltage reaches its maximum value one quarter of an oscillation period before the current reaches its maximum value. In this situation, we say that **the voltage always leads the current by 90°**.

To see why this phase relationship between voltage and current should exist, let us examine a few points on the curves of Figure 21.7. Note that at point a on the current curve, the current is beginning to increase in the positive direction. At this instant, the rate of change of current is at a maximum, and we see from Equation 21.7 that the voltage across the inductor is consequently also at a maximum at this time. As the current rises between points a and b on the curve, $\Delta I/\Delta t$ (the slope of the current curve) gradually decreases until it reaches zero at point b. As a result, the voltage across the inductor is decreasing during this same time interval, as the segment between c and d on the voltage curve indicates. Immediately after point b, the current begins to decrease, although it still has the same direction it had during the previous quarter cycle. As the current decreases to zero (from b to e on the curve), a voltage is again induced in the coil (d to f), but the sense of this voltage is opposite the sense of the voltage induced between c and d. This occurs because back emfs are always directed to oppose the change in the current.

We could continue to examine other segments of the curves, but no new information would be gained because the current and voltage variations are repetitive.

The voltage across an inductor leads ▶ the current by 90°.

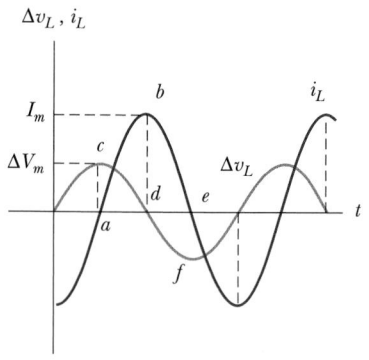

Figure 21.7 Plots of current and voltage across an inductor versus time in an ac circuit. The voltage leads the current by 90°.

Applying Physics 1

Imagine a circuit consisting of a series combination of an ac voltage source, a switch, an open coil inductor, and a lightbulb. The switch is closed and the circuit is allowed to come to equilibrium so that the lightbulb glows steadily. Now, an iron rod is inserted into the interior of the inductor. The glow of the lightbulb decreases as the iron rod is inserted. In older theatrical productions, this method was used to gradually dim the lights in the theater. Why does the lightbulb become dimmer?

Explanation   The reason for the dimming of the light bulb is related to the inductive reactance of the inductor. As the iron rod is inserted, the inductance of the coil increases, because the magnetic field inside the coil is increased. According to Equation 21.8, then, the inductive reactance of the coil increases. As a result, a larger fraction of the applied ac voltage appears across the inductor, leaving less voltage across the lightbulb. With less voltage across it, the lightbulb glows more dimly.

EXAMPLE 21.3   A Purely Inductive ac Circuit

In a purely inductive ac circuit (see Fig. 21.6), $L = 25.0$ mH and the rms voltage is 150 V. Find the inductive reactance and rms current in the circuit if the frequency is 60.0 Hz.

Solution First, note that $2\pi f = 2\pi(60.0) = 377 \text{ s}^{-1}$. Equation 21.8 then gives

$$X_L = 2\pi fL = (377 \text{ s}^{-1})(25.0 \times 10^{-3} \text{ H}) = \boxed{9.43 \text{ } \Omega}$$

Substituting this result into Equation 21.9 gives

$$I = \frac{\Delta V_L}{X_L} = \frac{150 \text{ V}}{9.43 \text{ } \Omega} = \boxed{15.9 \text{ A}}$$

Exercise Calculate the inductive reactance and rms current in the circuit if the frequency is 6 kHz.

Answer $X_L = 943 \text{ } \Omega$; $I = 0.159 \text{ A}$

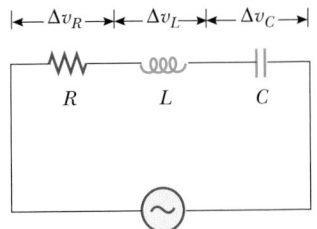

Figure 21.8 A series circuit consisting of a resistor, an inductor, and a capacitor connected to an ac generator.

21.4 THE *RLC* SERIES CIRCUIT

In the foregoing sections, we examined the effects of an inductor, a capacitor, and a resistor when they are connected separately across an ac voltage source. We now consider what happens when these devices are combined.

Figure 21.8 shows a circuit containing a resistor, an inductor, and a capacitor connected in series across an ac generator. The current in the circuit varies sinusoidally with time, as indicated in Figure 21.9a. Thus,

$$i = I_m \sin 2\pi ft$$

Earlier we learned that the voltage across each element may or may not be in phase with the current. The instantaneous voltages across the three elements, shown in Figure 21.9, have the following phase relations to the instantaneous current.

1. The instantaneous voltage across the resistor, Δv_R, is *in phase* with the instantaneous current. (See Fig. 21.9b.).
2. The instantaneous voltage across the inductor, Δv_L, *leads* the current by 90°. (See Fig. 21.9c.).
3. The instantaneous voltage across the capacitor, Δv_C, *lags behind* the current by 90°. (See Fig. 21.9d.)

The net instantaneous voltage, Δv, across all three elements is the sum of the instantaneous voltages across the separate elements: That is, $\Delta v = \Delta v_R + \Delta v_C + \Delta v_L$. However, it is simpler to use another technique involving vectors. We represent the voltage across each element with a rotating vector, as in Figure 21.10. The rotating vectors are referred to as **phasors,** and the diagram is called a **phasor diagram.** This particular diagram represents the circuit voltage given by the expression $\Delta v = \Delta V_m \sin(2\pi ft + \phi)$, where ΔV_m is the maximum voltage (the amplitude of the phasor) and ϕ is the angle between the phasor and the $+x$ axis. The phasor can be viewed as a vector of magnitude ΔV_m rotating at a constant frequency, f, so that its projection along the y axis is the instantaneous voltage in the circuit. Because ϕ is the phase angle between the voltage and current in the circuit, the phasor for the current (not shown in Figure 21.10) lies along the $+x$ axis and is expressed by the relation $i = I_m \sin(2\pi ft)$.

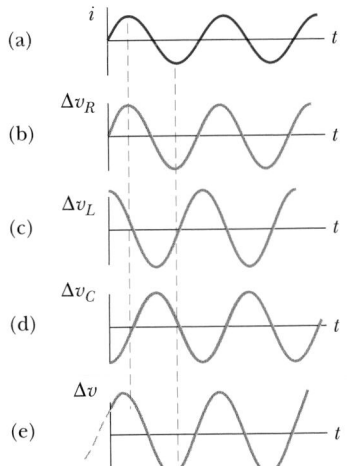

Figure 21.9 Phase relations in the series *RLC* circuit shown in Figure 21.8.

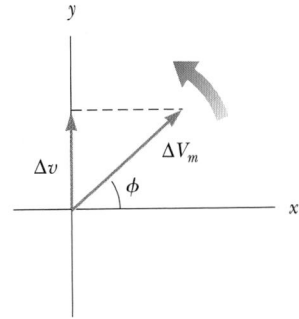

Figure 21.10 A phasor diagram for the voltage in an ac circuit, where ϕ is the phase angle between the voltage and current, and Δv is the instantaneous voltage.

Figure 21.11   (a) A phasor diagram for the *RLC* circuit. (b) Addition of the phasors as vectors gives $\Delta V = \sqrt{\Delta V_R^2 + (\Delta V_L - \Delta V_C)^2}$. (c) The reactance triangle that gives the impedance relation, $Z = \sqrt{R^2 + (X_L - X_C)^2}$.

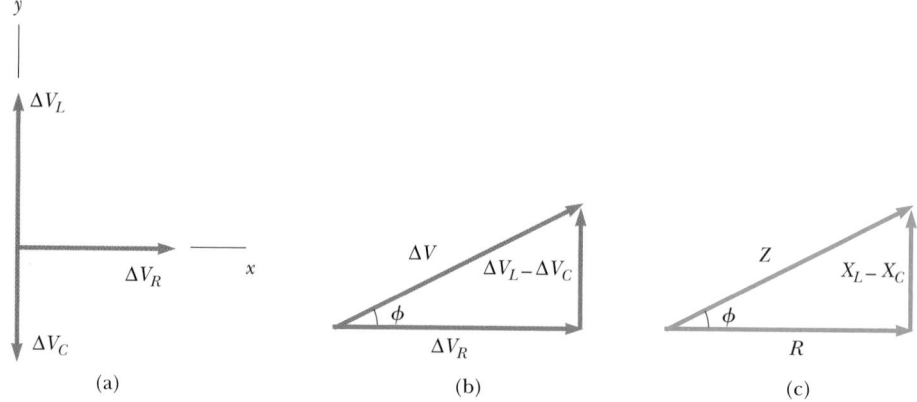

The phasor diagrams in Figure 21.11 are useful for analyzing the *series RLC* circuit. Voltages in phase with the current are represented by vectors along the $+x$ axis, and voltages out of phase with the current lie along other axes. Thus, ΔV_R is horizontal and to the right because it is in phase with the current. Likewise, ΔV_L is represented by a phasor along the $+y$ axis because it leads the current by $90°$. Finally, ΔV_C is along the $-y$ axis because it lags behind the current[2] by $90°$. If the phasors are added as vector quantities, Figure 21.11a shows that the only x component for the voltages is ΔV_R, and the net y component is $\Delta V_L - \Delta V_C$. It is now convenient to add the phasors vectorially (Fig. 21.11b), where ΔV is the total rms voltage in the circuit. The right triangle in Figure 21.11b gives the following equations for the total voltage and phase angle:

$$\Delta V = \sqrt{\Delta V_R^2 + (\Delta V_L - \Delta V_C)^2} \qquad [21.10]$$

$$\tan \phi = \frac{\Delta V_L - \Delta V_C}{\Delta V_R} \qquad [21.11]$$

where all voltages are rms values. Note that although we choose to use rms voltages in our analysis, the equations above apply equally well to peak voltages, because the two quantities are related to each other by the same factor for all circuit elements. The result for the total voltage, ΔV, as given by Equation 21.10, reinforces the fact that **the rms voltages across the resistor, capacitor, and inductor are not in phase, so one cannot simply add them to get the voltage across the combination of elements.**

We can write Equation 21.10 in the form of Ohm's law, using the relations $\Delta V_R = IR$, $\Delta V_L = IX_L$, and $\Delta V_C = IX_C$, where I is the rms current in the circuit:

$$\Delta V = I\sqrt{R^2 + (X_L - X_C)^2} \qquad [21.12]$$

---

[2] A mnemonic device to help you remember the phase relationships in *RLC* circuits is "*ELI* the *ICE* man." *E* represents the voltage, *I* the current, *L* the inductance, and *C* the capacitance. Thus, the name *ELI* means that in an inductive circuit, the voltage *E* leads the current *I*. In a capacitive circuit, *ICE* means that the current leads the voltage.

It is convenient to define a parameter called the **impedance, Z,** of the circuit as

$$Z \equiv \sqrt{R^2 + (X_L - X_C)^2}$$ [21.13] ◀ Impedance

so that Equation 21.12 becomes

$$\Delta V = IZ$$ [21.14]

Note that Equation 21.14 is in the form of Ohm's law, $\Delta V = IR$, where R is replaced by the impedance, in ohms. Equation 21.14 can be regarded as a generalized form of Ohm's law applied to a series ac circuit. Note that the current in the circuit depends on the resistance, the inductance, the capacitance, *and* the frequency because the reactances are frequency dependent.

It is useful to represent the impedance, Z, with a vector diagram such as the one depicted in Figure 21.11c. A right triangle is constructed; the right side is the quantity $X_L - X_C$, the base is R, and the hypotenuse is Z. Applying the Pythagorean theorem to this triangle, we see that

$$Z = \sqrt{R^2 + (X_L - X_C)^2}$$

which is consistent with Equation 21.13. Furthermore, we see from the vector diagram that the phase angle, ϕ, between the current and the voltage is given by

◀ Phase angle, ϕ

$$\tan \phi = \frac{X_L - X_C}{R}$$ [21.15]

The physical significance of the phase angle will become apparent in Section 21.5.

Figure 21.12 provides impedance values and phase angles for some series circuits containing different combinations of circuit elements.

Many parallel alternating current circuits are also useful in everyday applications. We shall not discuss them here, however, because their analysis is beyond the scope of this book.

Circuit elements	Impedance, Z	Phase angle, ϕ
R	R	0°
C	X_C	−90°
L	X_L	+90°
R C	$\sqrt{R^2 + X_C^2}$	Negative, between −90° and 0°
R L	$\sqrt{R^2 + X_L^2}$	Positive, between 0° and 90°
R L C	$\sqrt{R^2 + (X_L - X_C)^2}$	Negative if $X_C > X_L$ Positive if $X_C < X_L$

Figure 21.12 The impedance values and phase angles for various combinations of circuit elements. In each case, an ac voltage (not shown) is applied across the combination of elements (that is, across the dots).

Nikola Tesla (1856–1943)

Tesla was born in Croatia but spent most of his professional life as an inventor in the United States. He was a key figure in the development of alternating-current electricity, high-voltage transformers, and the transport of electrical power using ac transmission lines. Tesla's viewpoint was at odds with the ideas of Edison, who committed himself to the use of direct current in power transmission. Tesla's ac approach won out.
(UPI/Bettmann)

Problem-Solving Strategy

Alternating Current

The following procedures are recommended for solving alternating-current problems:

1. The first step in analyzing alternating-current circuits is to calculate as many of the unknown quantities, such as X_L and X_C, as possible. (When you calculate X_C, express the capacitance in farads rather than, say, microfarads.).

2. Apply the equation $\Delta V = IZ$ to the portion of the circuit that is of interest. For example, if you want to know the voltage drop across the combination of an inductor and a resistor, the equation for the voltage drop reduces to $\Delta V = I\sqrt{R^2 + X_L^2}$.

EXAMPLE 21.4 Analyzing a Series *RLC* ac Circuit

Analyze a series *RLC* ac circuit for which $R = 250\ \Omega$, $L = 0.600\ \text{H}$, $C = 3.50\ \mu\text{F}$, $f = 60\ \text{Hz}$, and $\Delta V = 150\ \text{V}$.

Solution The reactances are given by $X_L = 2\pi f L = 226\ \Omega$ and $X_C = 1/2\pi f C = 758\ \Omega$. Therefore, the impedance is

$$Z = \sqrt{R^2 + (X_L - X_C)^2} = \sqrt{(250\ \Omega)^2 + (226\ \Omega - 758\ \Omega)^2} = 588\ \Omega$$

The rms current is

$$I = \frac{\Delta V}{Z} = \frac{150\ \text{V}}{588\ \Omega} = 0.255\ \text{A}$$

The phase angle between the current and voltage is

$$\phi = \tan^{-1}\left(\frac{X_L - X_C}{R}\right) = \tan^{-1}\left(\frac{226\ \Omega - 758\ \Omega}{250\ \Omega}\right) = -64.8°$$

Because the circuit is more capacitive than inductive (that is, $X_C > X_L$), ϕ is negative. A negative phase angle means that the current leads the applied voltage.

The rms voltages across the elements are

$$\Delta V_R = IR = (0.255\ \text{A})(250\ \Omega) = 63.8\ \text{V}$$

$$\Delta V_L = IX_L = (0.255\ \text{A})(226\ \Omega) = 57.6\ \text{V}$$

$$\Delta V_C = IX_C = (0.255\ \text{A})(758\ \Omega) = 193\ \text{V}$$

Note that the sum of the three rms voltages, $\Delta V_R + \Delta V_L + \Delta V_C$, is 314 V, which is much greater than the rms voltage of the generator, 150 V. The sum 314 V is a meaningless quantity because, when alternating voltages are added, *both their amplitudes and their phases* must be taken into account. That is, the voltages must be added in a way that takes account of the different phases. The relationship among ΔV, ΔV_R, ΔV_L, and ΔV_C is given by Equation 21.10. You should use the values found above to verify this equation.

21.5   POWER IN AN ac CIRCUIT

No power losses are associated with capacitors and pure inductors in an ac circuit. (A *pure inductor* is defined as one with no resistance or capacitance.) Let us begin by analyzing the power dissipated in an ac circuit that contains only a generator and a capacitor.

When the current begins to increase in one direction in an ac circuit, charge begins to accumulate on the capacitor and a voltage drop appears across it. When this voltage reaches its maximum value, the energy stored in the capacitor is

$$PE_C = \tfrac{1}{2}C(\Delta V_m)^2$$

However, this energy storage is only momentary. When the current reverses direction, the charge leaves the capacitor plates and returns to the voltage source. Thus, during one half of each cycle the capacitor is being charged, and during the other half the charge is being returned to the voltage source. Therefore, the average power supplied by the source is zero. In other words, **a capacitor in an ac circuit does not dissipate energy.**

Similarly, the source must do work against the back emf of an inductor, which carries a current. When the current reaches its maximum value, the energy stored in the inductor is a maximum and is given by

$$PE_L = \tfrac{1}{2}LI_m^2$$

When the current begins to decrease in the circuit, this stored energy is returned to the source as the inductor attempts to maintain the current in the circuit. The only element in an *RLC* circuit that dissipates energy is the resistor. The average power lost in a resistor is

◀ The resistor is the only element in an *RLC* circuit that dissipates energy.

$$P_{\text{av}} = I^2R \qquad\qquad [21.16]$$

where *I* is the rms current in the circuit. An alternative equation for the average power dissipated in an ac circuit can be found by substituting (from Ohm's law) $R = \Delta V_R/I$ into Equation 21.17:

$$P_{\text{av}} = I\,\Delta V_R$$

It is convenient to refer to a voltage triangle that shows the relationship among ΔV, ΔV_R, and $\Delta V_L - \Delta V_C$, such as Figure 21.11b. From this figure, we see that the voltage drop across a resistor can be written in terms of the voltage of the source:

$$\Delta V_R = \Delta V \cos \phi$$

Hence, the average power dissipated in an ac circuit is

$$P_{\text{av}} = I\,\Delta V \cos \phi \qquad\qquad [21.17]$$

◀ Average power

where the quantity $\cos \phi$ is called the **power factor.**

EXAMPLE 21.5   Calculate the Average Power

Calculate the average power delivered to the series *RLC* circuit described in Example 21.4.

Reasoning and Solution   We are given 150 V for the rms voltage supplied to the circuit, and we have calculated that the rms current in the circuit is 0.255 A and the phase angle, ϕ, is $-64.8°$. Thus, the power factor, $\cos \phi$, is 0.426. From these values, we calculate the average power using Equation 21.17:

$$P_{av} = I\Delta V \cos \phi = (0.255 \text{ A})(150 \text{ V})(0.426) = \boxed{16.3 \text{ W}}$$

The same result can be obtained using Equation 21.16.

I $I = \dfrac{\Delta V}{Z}$

f_0 f

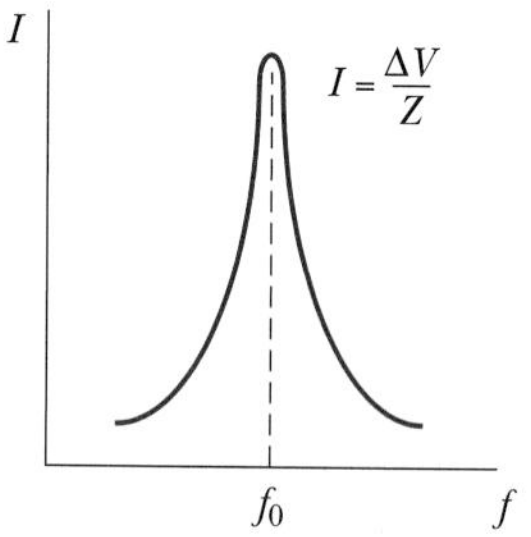

Figure 21.13   A plot of current amplitude in a series *RLC* circuit versus frequency of the generator voltage. Note that the current reaches its maximum value at the resonance frequency, f_0.

Resonance frequency ▶

13.7, SECTION 2

APPLICATION

Tuning Your Radio.

APPLICATION

Metal Detectors in Airports.

21.6   RESONANCE IN A SERIES *RLC* CIRCUIT

In general, the current in a series *RLC* circuit can be written

$$I = \frac{\Delta V}{Z} = \frac{\Delta V}{\sqrt{R^2 + (X_L - X_C)^2}} \qquad \text{[21.18]}$$

From this we see that the current has its *maximum* value when the impedance has its *minimum* value. This occurs when $X_L = X_C$. In such a circumstance, the impedance of the circuit reduces to $Z = R$. The frequency, f_0, at which this happens is called the **resonance frequency** of the circuit. To find f_0, we set $X_L = X_C$, which gives, from Equations 21.5 and 21.8,

$$2\pi f_0 L = \frac{1}{2\pi f_0 C}$$

$$f_0 = \frac{1}{2\pi \sqrt{LC}} \qquad \text{[21.19]}$$

Figure 21.13 is a plot of current as a function of frequency for a circuit containing a fixed capacitance and fixed inductance. From Equation 21.18 it must be concluded that the current would become infinite at resonance when $R = 0$. Although Equation 21.18 predicts this result, real circuits always have some resistance, which limits the value of the current.

The receiving circuit of a radio is an important application of a series resonance circuit. The radio is tuned to a particular station (which transmits a specific radio-frequency signal) by varying a capacitor, which changes the resonance frequency of the receiving circuit. When this resonance frequency matches that of the incoming radio wave, the current in the receiving circuit increases.

Applying Physics 2

When you walk through the doorway of an airport metal detector, you are really walking through a coil of many turns that is part of a resonant circuit. How might this work?

Explanation   The coil is connected to a capacitor tuned so that the circuit is in resonance. When you walk through with metal in your pocket, you change the inductance of the resonance circuit, resulting in a dramatic change in the current in the circuit. This change in current is detected, and electronic circuity causes a sound to be emitted as an alarm.

EXAMPLE 21.6   The Capacitance of a Circuit in Resonance

Consider a series *RLC* circuit for which $R = 150 \ \Omega$, $L = 20$ mH, $\Delta V = 20$ V, and $2\pi f = 5.0 \times 10^3 \ \text{s}^{-1}$. Determine the value of the capacitance for which the rms current is a maximum.

Reasoning   The current is a maximum at the resonance frequency, f_0, which should be made to match the driving frequency, $5.0 \times 10^3 \ \text{s}^{-1}$.

Solution   In this problem,

$$2\pi f_0 = 5.0 \times 10^3 \ \text{s}^{-1} = \frac{1}{\sqrt{LC}}$$

$$C = \frac{1}{(25 \times 10^6 \ \text{s}^{-2})L} = \frac{1}{(25 \times 10^6 \ \text{s}^{-2})(20.0 \times 10^{-3} \ \text{H})} = \boxed{2.0 \ \mu F}$$

Exercise   Calculate the maximum rms current in the circuit.

Answer   0.13 A

(Applying Physics 2) When you pass through a metal detector at an airport, you become part of a resonant circuit. Your body causes the inductance to change, causing the current in the circuit to change. *(Mason Morfit/FPG International, Inc.)*

Workbook Problem 86 — (Workbook page 183) considers an *LC* circuit (no resistance).

21.7   THE TRANSFORMER   *Optional Section*

In many situations it is desirable or necessary to change a small ac voltage to a larger one or vice versa. Before we examine a few such cases, let us consider the device that makes these conversions possible, the ac transformer.

In its simplest form, the **ac transformer** consists of two coils of wire wound around a core of soft iron, as in Figure 21.14. The coil on the left, which is connected to the input ac voltage source and has N_1 turns, is called the primary winding, or the *primary*. The coil on the right, which is connected to a resistor R and consists of N_2 turns, is the *secondary*. The purpose of the common iron core is to increase the magnetic flux and to provide a medium in which nearly all the flux through one coil passes through the other.

When an input ac voltage, ΔV_1, is applied to the primary, the induced voltage across it is given by

$$\Delta V_1 = -N_1 \frac{\Delta \Phi}{\Delta t} \qquad [21.20]$$

where Φ is the magnetic flux through each turn. If we assume that no flux leaks from the iron core, then the flux through each turn of the primary equals the

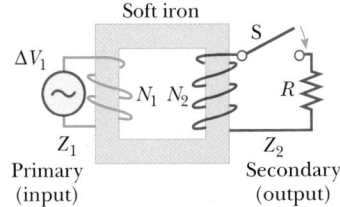

Figure 21.14   An ideal transformer consists of two coils wound on the same soft iron core. An ac voltage, ΔV_1, is applied to the primary coil, and the output voltage, ΔV_2, is observed across the load resistance, R.

flux through each turn of the secondary. Hence, the voltage across the secondary coil is

$$\Delta V_2 = -N_2 \frac{\Delta \Phi}{\Delta t} \qquad [21.21]$$

The term $\Delta \Phi / \Delta t$ is common to Equations 21.21 and 21.22. Therefore, we see that

$$\Delta V_2 = \frac{N_2}{N_1} \Delta V_1 \qquad [21.22]$$

When N_2 is greater than N_1, and thus ΔV_2 exceeds ΔV_1, the transformer is referred to as a *step-up transformer*. When N_2, is less than N_1, making ΔV_2 less than ΔV_1, we speak of a *step-down transformer*.

It should be clear that a voltage is generated across the secondary only when there is a *change* in the number of flux lines passing through the secondary. Thus, the input current in the primary must change with time, which is what happens when an alternating current is used. However, when the input at the primary is a direct current, a voltage output occurs at the secondary only at the instant a switch in the primary circuit is opened or closed. Once the current in the primary reaches a steady value, the output voltage at the secondary is zero.

It may seem that a transformer is a device in which it is possible to get something for nothing. For example, a step-up transformer can change an input voltage from, say, 10 V to 100 V. This means that each 1 coulomb of charge leaving the secondary has 100 J of energy, whereas each coulomb of charge entering the primary has only 10 J of energy. However, this is not an example of a breakdown in the principle of conservation of energy, because **the power input to the primary equals the power output at the secondary;** that is,

▷ In an ideal transformer, the input power equals the output power.

$$I_1 \Delta V_1 = I_2 \Delta V_2 \qquad [21.23]$$

Thus, if the voltage at the secondary is ten times that at the primary, the current at the secondary is reduced by a factor of 10. Equation 21.23 assumes an **ideal transformer,** in which there are no power losses between the primary and the secondary. Real transformers typically have power efficiencies ranging from 90 to 99%. Power losses occur because of such factors as eddy currents induced in the iron core of the transformer, which dissipate energy in the form of I^2R losses.

APPLICATION

Long Distance Electric Power Transmission.

When electric power is transmitted over large distances, it is economical to use a high voltage and a low current because the power lost via resistive heating in the transmission lines varies as I^2R. This means that if a utility company can reduce the current by a factor of 10, for example, the power loss is reduced by a factor of 100. In practice, the voltage is stepped up to around 230 000 V at the generating station, then stepped down to around 20 000 V at a distribution station, and finally stepped down to 120 V at the customer's utility pole.

APPLICATION

Automobile Ignition Systems.

In the electrical systems of automobiles, transformers are used with dc sources. In the engine, spark plugs produce a spark to ignite a gasoline-air mixture in the cylinders. For a plug to work, there must be a voltage high enough to cause the spark to jump across the plug's gap. The technique for creating this high voltage is indicated in Figure 21.15. The primary coil of the transformer is connected to the 12-V battery of the car. The secondary is connected to the spark plugs through the distributor. When in sequence a particular plug is supposed to fire, the distributor mechanism connects that spark plug to the secondary coil. A switch (called the

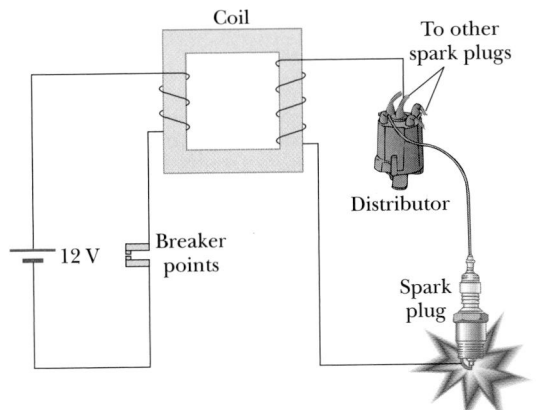

Figure 21.15 A circuit diagram of an automobile ignition system.

breaker points or simply the *points*) breaks the circuit between the battery and the primary. This interrupts the current in the primary and induces a voltage in the secondary. The transformer used is a step-up transformer, and the resulting voltage in the secondary is high enough (about 20 000 V) to cause a spark to jump when the voltage is applied across the gap of the spark plug.

EXAMPLE 21.7 Distributing Power to a City

A generator at a utility company produces 100 A of current at 4000 V. The voltage is stepped up to 240 000 V by a transformer before it is sent on a high-voltage transmission line across a rural area to a city. Assume that the effective resistance of the power line is 30.0 Ω.

(a) Determine the percentage of power lost.

Solution From Equation 21.23, the current in the transmission line is

$$I_2 = \frac{I_1 \Delta V_1}{\Delta V_2} = \frac{(100 \text{ A})(4000 \text{ V})}{2.40 \times 10^5 \text{ V}} = 1.67 \text{ A}$$

and the power lost in the transmission line is

$$P_{\text{lost}} = I_2^2 R = (1.67 \text{ A})^2 (30.0 \text{ Ω}) = 83.7 \text{ W}$$

The power output of the generator is

$$P = I\Delta V = (100 \text{ A})(4000 \text{ V}) = 4.00 \times 10^5 \text{ W}$$

From this we can find the percentage of power lost as

$$\% \text{ power lost} = \left(\frac{83.7 \text{ W}}{4.00 \times 10^5 \text{ W}} \right) \times 100 = \boxed{0.0209\%}$$

(b) What percentage of the original power would be lost in the transmission line if the voltage were not stepped up?

Solution If the voltage were not stepped up, the current in the transmission line would be 100 A and the power lost in the line would be

$$P_{\text{lost}} = I^2 R = (100 \text{ A})^2 (30.0 \text{ Ω}) = 3.00 \times 10^5 \text{ W}$$

In this case, the percentage of power lost would be

$$\% \text{ power lost} = \left(\frac{3.00 \times 10^5 \text{ W}}{4.00 \times 10^5 \text{ W}}\right) \times 100 = \boxed{75\%}$$

This example illustrates the advantage of high-voltage transmission lines. At the city, a transformer at a substation steps the voltage back down to about 4000 V, and this voltage is maintained across utility lines throughout the city. When the power is to be used at a home or business, a transformer on a utility pole near the establishment reduces the voltage to 240 V or 120 V.

Exercise If the transmission line is cooled so that the resistance is reduced to 5.0 Ω, how much power is lost in the line if it carries a current of 0.89 A?

Answer 4.0 W

James Clerk Maxwell
Scottish theoretical physicist
(1831–1879)

Maxwell developed the electromagnetic theory of light, the kinetic theory of gases, and explained the nature of Saturn's rings and color vision. Maxwell's successful interpretation of the electromagnetic field resulted in the field equations that bear his name. Formidable mathematical ability combined with great insight enabled him to lead the way in the study of electromagnetism and kinetic theory. He died of cancer before he was 50. *(North Wind Picture Archives)*

21.8 MAXWELL'S PREDICTIONS

During the early stages of their study and development, electric and magnetic phenomena were thought to be unrelated. In 1865, however, James Clerk Maxwell (1831–1879) provided a mathematical theory that showed a close relationship between all electric and magnetic phenomena. In addition, his theory predicted that electric and magnetic fields can move through space as waves. The theory he developed is based on the following four pieces of information:

1. Electric field lines originate on positive charges and terminate on negative charges. The electric field due to a point charge can be determined at a location by applying Coulomb's force law to a test charge placed at that location.
2. Magnetic field lines always form closed loops — that is, they do not begin or end anywhere.
3. A varying magnetic field induces an emf and hence an electric field. This is a statement of Faraday's law (Chapter 20).
4. Magnetic fields are generated by moving charges (or currents), as summarized in Ampère's law (Chapter 19).

Let us examine these statements further in order to understand their significance and Maxwell's contributions to the theory of electromagnetism. The first statement is a consequence of the nature of the electrostatic force between charged particles, given by Coulomb's law. It embodies a recognition of the fact that **free charges (electric monopoles) exist in nature.**

The second statement — that magnetic fields form continuous loops — is exemplified by the magnetic field lines around a long, straight wire, which are closed circles, and the magnetic field lines of a bar magnet, which form closed loops.

The third statement is equivalent to Faraday's law of induction; the fourth statement is equivalent to Ampère's law.

In one of the greatest theoretical developments of the 19th century, Maxwell used these four statements within a corresponding mathematical framework to prove that electric and magnetic fields play symmetric roles in nature. It was already known from experiments that a changing magnetic field produced an electric field

12.8

12.9, SECTION 1

according to Faraday's law. Maxwell believed that nature was symmetric, and he therefore hypothesized that a changing electric field should produce a magnetic field. This hypothesis could not be proven experimentally at the time it was developed, because the magnetic fields generated by changing electric fields are generally very weak and therefore difficult to detect.

◀ A changing electric field produces a magnetic field.

To justify his hypothesis, Maxwell searched for other phenomena that might be explained by it. He turned his attention to the motion of rapidly oscillating charges, such as those in a conducting rod connected to an alternating voltage. Such charges experience accelerations and, according to Maxwell's predictions, generate changing electric and magnetic fields. The changing fields cause electromagnetic disturbances that travel through space as waves, similar to the spreading water waves created by a pebble thrown into a pool. The waves sent out by the oscillating charges are fluctuating electric and magnetic fields, and so they are called *electromagnetic waves*. From Faraday's law and from his own generalization of Ampère's law, Maxwell calculated their speed to be equal to the speed of light, $c = 3 \times 10^8$ m/s. He concluded that light waves are electromagnetic in nature — that visible light and other electromagnetic waves consist of fluctuating electric and magnetic fields traveling through empty space with a speed of 3×10^8 m/s, by each field creating the other! This was truly one of the greatest discoveries of science. It had a profound influence on later scientific developments.

Heinrich Rudolf Hertz
German physicist (1857–1894)

Hertz made his most important discovery of radio waves in 1887. After finding that the speed of a radio wave was the same as that of light, Hertz showed that radio waves, like light waves, could be reflected, refracted, and diffracted. Hertz died of blood poisoning at the age of 36. During his short life, he made many contributions to science. The hertz, equal to one complete vibration or cycle per second, is named after him. *(The Bettmann Archive)*

21.9   HERTZ'S DISCOVERIES

In 1887, Heinrich Hertz (1857–1894) was the first to generate and detect electromagnetic waves in a laboratory setting. To appreciate the details of his experiment, let us reexamine the properties of an *LC* circuit. In such a circuit, a charged capacitor is connected to an inductor, as in Figure 21.16. When the switch is closed, oscillations occur in the current in the circuit and in the charge on the capacitor. If the resistance of the circuit is neglected, no energy is lost to heat, and the oscillations continue.

In the following analysis, we shall neglect the resistance in the circuit. Let us assume that the capacitor has an initial charge of Q_m and that the switch is closed at $t = 0$. It is convenient to describe what ensues from an energy viewpoint. When the capacitor is fully charged, the total energy in the circuit is stored in the electric field of the capacitor and is equal to $Q_m{}^2/2C$. At this time, the current is zero and so no energy is stored in the inductor. As the capacitor begins to discharge, the energy stored in its electric field decreases. At the same time, the current increases and energy equal to $LI^2/2$ is now stored in the magnetic field of the inductor. Thus, energy is transferred from the electric field of the capacitor to the magnetic field of the inductor. When the capacitor is fully discharged, it stores no energy. At this time, the current reaches its maximum value, and all of the energy is stored in the inductor. The process then repeats in the reverse direction. The energy continues to transfer between the inductor and the capacitor, corresponding to oscillations in the current and charge.

Figure 21.17 is a representation of this energy transfer. The circuit behavior is analogous to that of the oscillating mass–spring system studied in Chapter 13. The

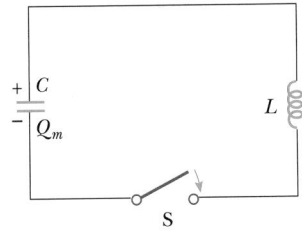

Figure 21.16 A simple *LC* circuit. The capacitor has an initial charge of Q_m, and the switch is closed at $t = 0$.

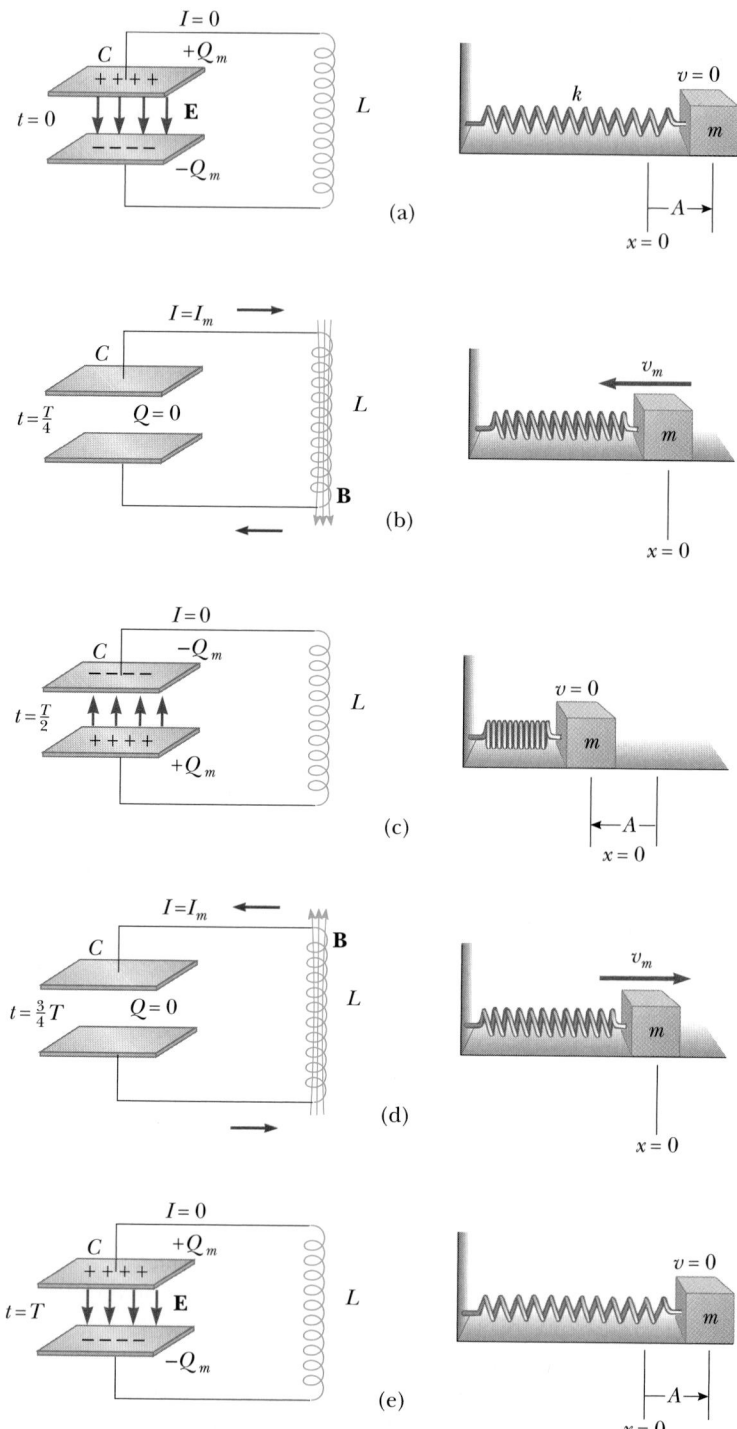

Figure 21.17 Stages of energy transfer in an *LC* circuit with zero resistance. The capacitor has a charge of Q_m at $t = 0$, when the switch is closed. The mechanical analog of this circuit, the mass-spring system, is shown at the right in each part of the figure.

potential energy stored in a stretched spring, $kx^2/2$, corresponds to the potential energy stored in the capacitor, $Q_m{}^2/2C$; the kinetic energy of the moving mass, $mv^2/2$, corresponds to the energy stored in the inductor, $LI^2/2$, which requires the presence of moving charges. In Figure 21.17a, all of the energy is stored as potential energy in the capacitor at $t = 0$ (because $I = 0$). In Figure 21.17b, all of the energy is stored as "kinetic" energy in the inductor, $LI_m{}^2/2$, where I_m is the maximum current. At intermediate points, part of the energy is potential energy and part is kinetic energy.

As we saw in Section 21.6, the frequency of oscillation of an LC circuit is called the *resonance frequency* of the circuit and is given by

$$f_0 = \frac{1}{2\pi\sqrt{LC}}$$

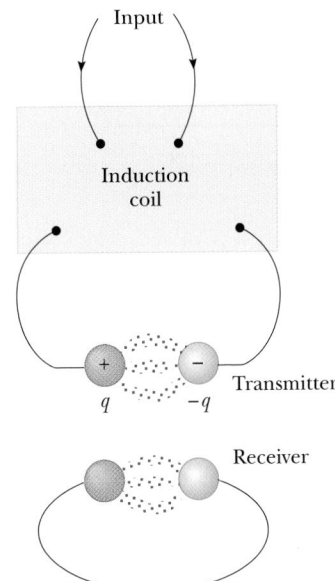

Figure 21.18 A schematic diagram of Hertz's apparatus for generating and detecting electromagnetic waves. The transmitter consists of two spherical electrodes connected to an induction coil, which provides short voltage surges to the spheres, setting up oscillations in the discharge. The receiver is a nearby single loop of wire containing a second spark gap.

The circuit Hertz used in his investigations of electromagnetic waves is similar to that just discussed and is shown schematically in Figure 21.18. An induction coil (a large coil of wire) is connected to two metal spheres with a narrow gap between them to form a capacitor. Oscillations are initiated in the circuit by short voltage pulses sent via the coil to the spheres, charging one positive, the other negative. Because L and C are quite small in this circuit, the frequency of oscillation is quite high, $f \approx 100$ MHz. This circuit is called a transmitter because it produces electromagnetic waves.

Several meters from the transmitter circuit, Hertz placed a second circuit, the receiver, which consisted of a single loop of wire connected to two spheres. It had its own effective inductance, capacitance, and natural frequency of oscillation. Hertz found that energy was being sent from the transmitter to the receiver when the resonance frequency of the receiver was adjusted to match that of the transmitter. The energy transfer was detected when the voltage across the spheres in the receiver circuit became high enough to produce ionization in the air, which caused sparks to appear in the air gap separating the spheres. Hertz's experiment is analogous to the mechanical phenomenon in which a tuning fork picks up the vibrations from another, identical tuning fork.

Hertz hypothesized that the energy transferred from the transmitter to the receiver is carried in the form of waves, which are now known to be electromagnetic waves. In a series of experiments, he also showed that the radiation generated by the transmitter exhibits wave properties: interference, diffraction, reflection, refraction, and polarization. As you will see shortly, all of these properties are exhibited by light. Thus, it became evident that these waves had properties similar to those of light waves and differed only in frequency and wavelength.

Perhaps the most convincing experiment Hertz performed was the measurement of the speed of waves from the transmitter, accomplished as follows. Waves of known frequency from the transmitter were reflected from a metal sheet so that an interference pattern was set up, much like the standing wave pattern on a stretched string. As we saw in our discussion of standing waves, the distance between nodes is $\lambda/2$, so Hertz was able to determine the wavelength, λ. Using the relationship $v = \lambda f$, he found that v was close to 3×10^8 m/s, the known speed of visible light. Hertz's experiments thus provided the first evidence in support of Maxwell's theory.

21.10 PRODUCTION OF ELECTROMAGNETIC WAVES BY AN ANTENNA

12.9, SECTION 2

An accelerating charge ▶
radiates energy.

APPLICATION

Radio Wave Transmission.

In the previous section, we found that the energy stored in an *LC* circuit is continually transferred between the electric field of the capacitor and the magnetic field of the inductor. However, this energy transfer continues for prolonged periods of time only when the changes occur slowly. If the current alternates rapidly, the circuit loses some of its energy in the form of electromagnetic waves. In fact, electromagnetic waves are radiated by *any* circuit carrying an alternating current. The fundamental mechanism responsible for this radiation is the acceleration of a charged particle. Whenever a charged particle undergoes an acceleration, it must radiate energy.

An alternating voltage applied to the wires of an antenna forces an electric charge in the antenna to oscillate. This is a common technique for accelerating charged particles and is the source of the radio waves emitted by the broadcast antenna of a radio station.

Figure 21.19 illustrates the production of an electromagnetic wave by oscillating electric charges in an antenna. Two metal rods are connected to an ac generator, which causes charges to oscillate between the two rods. The output voltage of the generator is sinusoidal. At $t = 0$, the upper rod is given a maximum positive charge and the bottom rod an equal negative charge, as in Figure 21.19a. The electric field near the antenna at this instant is also shown in Figure 21.19a. As the charges oscillate, the rods become less charged, the field near the rods decreases in strength, and the downward-directed maximum electric field produced at $t = 0$ moves away from the rod. When the charges are neutralized, as in Figure 21.19b, the electric field has dropped to zero. This occurs after an interval equal to one quarter of the period of oscillation. Continuing in this fashion, the upper rod soon

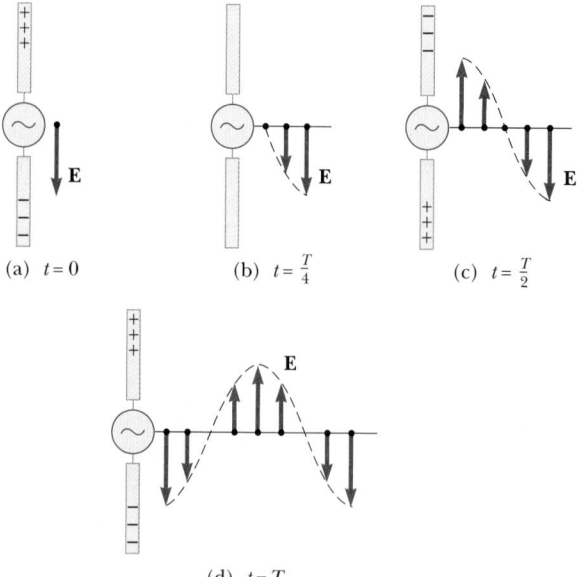

(a) $t = 0$ (b) $t = \frac{T}{4}$ (c) $t = \frac{T}{2}$

(d) $t = T$

Figure 21.19 An electric field set up by oscillating charges in an antenna. The field moves away from the antenna at the speed of light.

obtains a maximum negative charge and the lower rod becomes positive, as in Figure 21.19c, resulting in an electric field directed upward. This occurs after an interval equal to one half the period of oscillation. The oscillations continue as indicated in Figure 21.19d. Note that the electric field near the antenna oscillates in phase with the charge distribution. That is, the field points down when the upper rod is positive and up when the upper rod is negative. Furthermore, the magnitude of the field at any instant depends on the amount of charge on the rods at that instant.

As the charges continue to oscillate (and accelerate) between the rods, the electric field set up by the charges moves away from the antenna at the speed of light. Figure 21.19 shows the electric field pattern at certain times during the oscillation cycle. As you can see, one cycle of charge oscillation produces one full wavelength in the electric field pattern.

Because the oscillating charges create a current in the rods, a magnetic field is also generated when the current in the rods is upward, as shown in Figure 21.20. The magnetic field lines circle the antenna and are perpendicular to the electric field at all points. As the current changes with time, the magnetic field lines spread out from the antenna. At great distances from the antenna, the strengths of the electric and magnetic fields become very weak. However, at these distances it is necessary to take into account the facts that (1) a changing magnetic field produces an electric field, and (2) a changing electric field produces a magnetic field, as predicted by Maxwell. These induced electric and magnetic fields are in phase: at any point, the two fields reach their maximum values at the same instant. This is illustrated at one instant of time in Figure 21.21. Note that (1) these fields are perpendicular to each other, and (2) both fields are perpendicular to the direction of motion of the wave. This second property is characteristic of transverse waves. Hence, we see that **an electromagnetic wave is a transverse wave.**

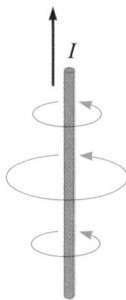

Figure 21.20 Magnetic field lines around an antenna carrying a changing current.

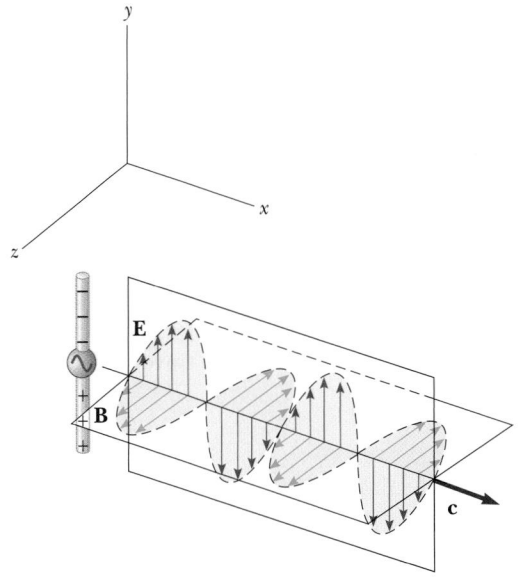

Figure 21.21 An electromagnetic wave sent out by oscillating charges in an antenna, represented at one instant of time. Note that the electric field is perpendicular to the magnetic field, and both are perpendicular to the direction of wave propagation.

We have seen that Maxwell's detailed analysis predicted the existence and properties of electromagnetic waves. We have already examined some of those properties. In this section we summarize what we know about electromagnetic waves thus far and consider some additional properties. In our discussion here and in future sections, we shall often make reference to a type of wave called a **plane wave.** A plane electromagnetic wave is a wave that travels in one direction only. Figure 21.21 pictures such a wave at a given instant of time. In this case, the oscillations of the electric and magnetic fields take place in planes perpendicular to the x axis and thus to the direction of travel for the wave. Because the electric and magnetic fields are perpendicular to the direction of travel of the wave, electromagnetic waves are transverse waves. In Figure 21.21, the electric field $\mathbf{E}$ is in the y direction and the magnetic field $\mathbf{B}$ is in the z direction.

Electromagnetic waves travel with the speed of light. In fact, it can be shown that the speed of an electromagnetic wave is related to the permeability and permittivity of the medium through which it travels. Maxwell found this relationship for free space to be

Speed of light ▶

$$c = \frac{1}{\sqrt{\mu_0 \epsilon_0}}$$ [21.24]

where c is the speed of light, $\mu_0 = 4\pi \times 10^{-7} \, \text{N} \cdot \text{s}^2/\text{C}^2$ is the permeability constant of vacuum, and $\epsilon_0 = 8.85419 \times 10^{-12} \, \text{C}^2/\text{N} \cdot \text{m}^2$ is the permittivity of free space. Substituting these values into Equation 21.25, we find that

$$c = 2.997\,92 \times 10^8 \, \text{m/s}$$ [21.25]

Because electromagnetic waves travel at a speed that is precisely the same as the speed of light in vacuum, one is led to believe (correctly) that **light is an electromagnetic wave.**

The ratio of the electric to the magnetic field in an electromagnetic wave equals the speed of light. That is,

$$\frac{E}{B} = c$$ [21.26]

Electromagnetic waves carry energy as they travel through space, and this energy can be transferred to objects placed in their paths. The average rate at which energy passes through an area perpendicular to the direction of travel of a wave, or the average power per unit area, is given by

$$\text{Average power per unit area} = \frac{E_m B_m}{2\mu_0}$$ [21.27]

As in Chapter 14, we call this quantity the intensity of the wave. Because $E = cB = B/\sqrt{\mu_0 \epsilon_0}$, this can also be expressed as

$$\text{Average power per unit area} = \frac{E_m{}^2}{2\mu_0 c} = \frac{c}{2\mu_0} B_m{}^2 \qquad \text{[21.28]}$$

Note that in these expressions we use the *average* power per unit area. Also note that the values to be used for E and B are the *maximum* values. It is interesting to note that a detailed analysis would show that the energy carried by an electromagnetic wave is shared equally by the electric and magnetic fields.

Electromagnetic waves transport linear momentum as well as energy. Hence it follows that pressure is exerted on a surface when an electromagnetic wave impinges on it. In what follows, we assume that the electromagnetic wave transports a total energy U to a surface in a time t. If the surface absorbs all the incident energy U in this time, Maxwell showed that the total momentum **p** delivered to this surface has a magnitude

◀ Light is an electromagnetic wave and transports energy and momentum.

$$p = \frac{U}{c} \qquad \text{(complete absorption)} \qquad \text{[21.29]}$$

If the surface is a perfect reflector, then the momentum delivered in a time t for normal incidence is twice that given by Equation 21.29. That is, a momentum U/c is delivered first by the incident wave and then again by the reflected wave, in analogy with a ball colliding elastically with a wall. Therefore,

$$p = \frac{2U}{c} \qquad \text{(complete reflection)} \qquad \text{[21.30]}$$

Although radiation pressures are very small (about $5 \times 10^{-6} \, \text{N/m}^2$ for direct sunlight), they have been measured with a device such as the one shown in Figure 21.22. Light is allowed to strike a mirror and a black disk that are connected to each other by a horizontal bar suspended from a fine fiber. Light striking the black disk is completely absorbed, so *all* of the momentum of the light is transferred to the disk. Light striking the mirror head-on is totally reflected; hence, the momentum transfer to the mirror is twice that transmitted to the disk. As a result, the horizontal bar supporting the disks twists counterclockwise as seen from above. The bar comes to equilibrium at some angle under the action of the torques caused by radiation pressure and the twisting of the fiber. The radiation pressure can be determined by measuring the angle at which equilibrium occurs. The apparatus must be placed in a high vacuum to eliminate the effects of air currents.

In summary, electromagnetic waves traveling through free space have the following properties:

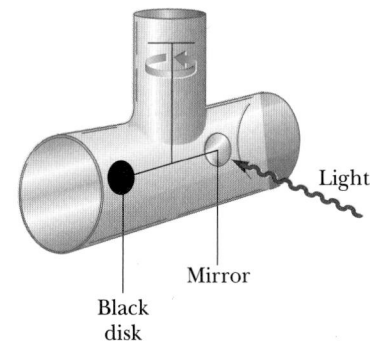

Figure 21.22 An apparatus for measuring the radiation pressure of light. In practice, the system is contained in a high vacuum.

1. Electromagnetic waves travel at the speed of light.
2. Electromagnetic waves are transverse waves, because the electric and magnetic fields are perpendicular to the direction of propagation of the wave and to each other.
3. The ratio of the electric field to the magnetic field in an electromagnetic wave equals the speed of light.
4. Electromagnetic waves carry both energy and momentum, which can be delivered to a surface.

Thinking Physics 1

In the interplanetary space in the solar system, there is a large amount of dust. Although interplanetary dust in theory can have a variety of sizes—from molecular size upward—there are very few dust particles smaller than about 0.2 μm in our solar system. Why? (*Hint:* The solar system originally contained dust particles of all sizes.)

Explanation   Dust particles in the solar system are subject to two forces—the gravitational force toward the Sun, and the force from radiation pressure, which is away from the Sun. The gravitational force is proportional to the cube of the radius of a spherical dust particle, because it is proportional to the mass (ρV) of the particle. The radiation pressure is proportional to the square of the radius, because it depends on the cross-sectional area of the particle. For large particles, the gravitational force is larger than the force of radiation pressure. For small particles, less than about 0.2 μm, the larger force from radiation pressure sweeps these particles out of the solar system.

EXAMPLE 21.8 Solar Energy

Assume that the Sun delivers an average power per unit area of about 1000 W/m² to the Earth's surface. Calculate the total power incident on a roof 8.00 m by 20.0 m. Assume that the radiation is incident *normal* to the roof (the Sun is directly overhead).

A solar home in Oregon.
(*John Neal/Photo Researchers, Inc.*)

Solution   The power per unit area, or light intensity, is 1000 W/m². For normal incidence we get

$$\text{Power} = (1000 \text{ W/m}^2)(8.00 \times 20.0 \text{ m}^2) = \boxed{1.60 \times 10^5 \text{ W}}$$

Note that if this power could *all* be converted to electric power, it would be more than enough for the average home. Unfortunately, solar energy is not easily harnessed, and the prospects for large-scale conversion are not as bright as they may appear from this simple calculation. For example, the conversion efficiency from solar to electrical energy is far less than 100%; 10% is typical for photovoltaic cells. Roof systems for converting solar energy to thermal energy with efficiencies of around 50% have been built. However, other practical problems must be considered, such as overcast days, geographic location, and energy storage.

Exercise How much solar energy (in joules) is incident on the roof in 1.00 h?

Answer 5.76×10^8 J

21.12 THE SPECTRUM OF ELECTROMAGNETIC WAVES

We have seen that all electromagnetic waves travel in a vacuum with the speed of light, c. These waves transport energy and momentum from some source to a receiver. In 1887 Hertz successfully generated and detected the radio-frequency electromagnetic waves predicted by Maxwell. Maxwell himself had recognized as electromagnetic waves both visible light and the infrared radiation discovered in 1800 by William Herschel. It is now known that other forms of electromagnetic waves exist that are distinguished by their frequencies and wavelengths.

Because all electromagnetic waves travel through vacuum with a speed of c, their frequency, f, and wavelength, λ, are related by the important expression

$$c = f\lambda \qquad \text{[21.31]}$$

The types of electromagnetic waves are presented in Figure 21.23. Note the wide range of frequencies and wavelengths. For instance, a radio wave with a frequency of 5.00 MHz (a typical value) has a wavelength of

$$\lambda = \frac{c}{f} = \frac{3.00 \times 10^8 \text{ m/s}}{5.00 \times 10^6 \text{ s}^{-1}} = 60.0 \text{ m}$$

The following abbreviations are often used to designate short wavelengths and distances:

$$1 \text{ micrometer } (\mu m) = 10^{-6} \text{ m}$$

$$1 \text{ nanometer } (nm) = 10^{-9} \text{ m}$$

$$1 \text{ angstrom } (\text{Å}) = 10^{-10} \text{ m}$$

The wavelengths of visible light, for example, range from 0.4 to 0.7 μm, or 400 to 700 nm, or 4000 to 7000 Å.

Brief descriptions of these wave types follow, in order of decreasing wavelength. There is no sharp division between one kind of wave and the next. Note that all forms of radiation are produced by accelerating charges.

Radio waves, which were discussed in Section 21.10, are the result of charges accelerating through conducting wires. They are, of course, used in radio and television communication systems.

Microwaves (short-wavelength radio waves) have wavelengths ranging between about 1 mm and 30 cm and are generated by electronic devices. Their short wavelengths make them well suited for the radar systems used in aircraft navigation and for the study of atomic and molecular properties of matter. Microwave ovens are an interesting domestic application of these waves. It has been suggested that solar energy might be harnessed by beaming microwaves to Earth from a solar collector in space.

Figure 21.23 The electromagnetic spectrum. Note the overlap between one type of wave and the next. There is no sharp division between the types.

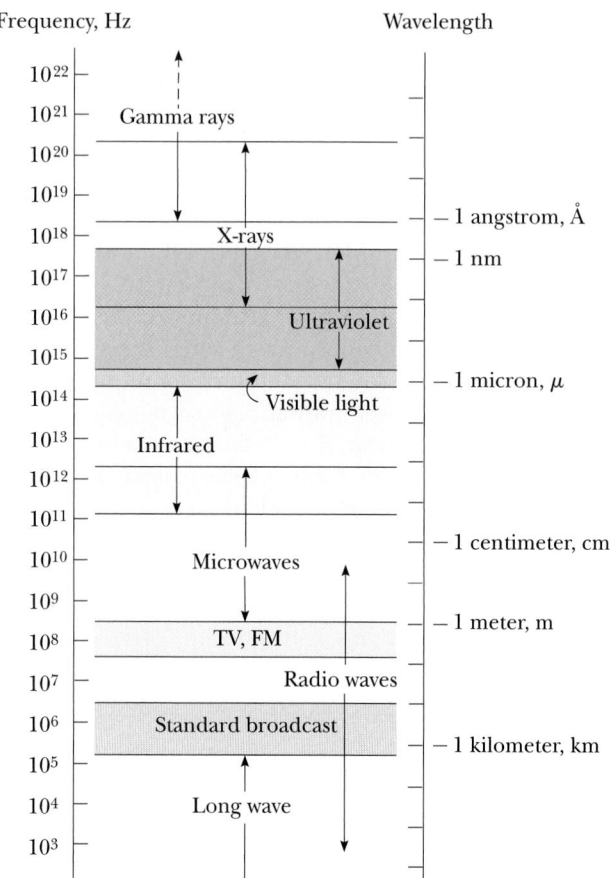

Infrared waves (sometimes called *heat waves*), produced by hot bodies and molecules, have wavelengths ranging from about 1 mm to the longest wavelength of visible light, 7×10^{-7} m. They are readily absorbed by most materials. The infrared energy absorbed by a substance appears as heat. This is because the energy agitates the atoms of the object, increasing their vibrational or translational motion, and the result is a temperature rise. Infrared radiation has many practical and scientific applications including physical therapy, infrared photography, and the study of the vibrations of atoms.

Visible light, the most familiar form of electromagnetic waves, may be defined as the part of the spectrum that is detected by the human eye. Light is produced by the rearrangement of electrons in atoms and molecules. The wavelengths of visible light are classified as colors ranging from violet ($\lambda \approx 4 \times 10^{-7}$ m) to red ($\lambda \approx 7 \times 10^{-7}$ m). The eye's sensitivity is a function of wavelength and is greatest at a wavelength of about 5.6×10^{-7} m (yellow-green).

Ultraviolet (uv) light covers wavelengths ranging from about 4×10^{-7} m (400 nm) down to 6×10^{-10} m (0.6 nm). The Sun is an important source of ultraviolet light (which is the main cause of suntans). Most of the ultraviolet light from the Sun is absorbed by atoms in the upper atmosphere, or stratosphere. This is fortunate, because uv light in large quantities has harmful effects on humans. One important constituent of the stratosphere is ozone (O_3) from reactions of oxygen with ultraviolet radiation. This ozone shield converts lethal high-energy ultraviolet radiation to heat, which warms the stratosphere.

X-rays are electromagnetic waves with wavelengths from about 10^{-8} m (10 nm) down to 10^{-13} m (10^{-4} nm). The most common source of x-rays is the acceleration of high-energy electrons bombarding a metal target. X-rays are used as a diagnostic tool in medicine and as a treatment for certain forms of cancer. Because x-rays damage or destroy living tissues and organisms, care must be taken to avoid unnecessary exposure and overexposure.

Gamma rays, electromagnetic waves emitted by radioactive nuclei, have wavelengths ranging from about 10^{-10} m to less than 10^{-14} m. They are highly penetrating and cause serious damage when absorbed by living tissues. As a consequence, those working near such radiation must be protected by garments containing heavily absorbing materials, such as layers of lead.

Thinking Physics 2

The center of sensitivity of our eyes coincides with the center of the wavelength distribution of the Sun. Is this an amazing coincidence?

Explanation   This is not a coincidence—it is the result of biological evolution. Humans have evolved with vision most sensitive to wavelengths that are strongest from the Sun. It is an interesting conjecture to imagine aliens from another planet, with a Sun of a different temperature, arriving at Earth. Their eyes would have the center of sensitivity at different wavelengths than ours. How would their vision of the Earth compare to ours?

21.13   THE DOPPLER EFFECT FOR ELECTROMAGNETIC WAVES

As we saw in Section 14.6, sound waves exhibit the Doppler effect when the observer of the wave, the source of the wave, or both are moving relative to the medium of propagation. Recall that in the Doppler effect, the observed frequency of the wave is larger or smaller than the frequency emitted by the source of the wave.

A Doppler effect can also occur for electromagnetic waves, but it differs from the Doppler effect for sound waves in two ways. First, in the Doppler effect for sound waves, motion relative to the medium is most important because sound waves require a medium in which to propagate. In contrast, the medium of propagation plays no role in the Doppler effect for electromagnetic waves because the waves require no medium in which to propagate. Second, the speed of sound which appears in the equation for the Doppler effect for sound depends on the reference frame in which it is measured. In contrast, as we shall see in Chapter 26, the speed of electromagnetic waves has the same value in all inertial frames.

The single equation that describes the Doppler effect for electromagnetic waves is given by the approximate expression

$$f' = f\left(1 \pm \frac{u}{c}\right) \qquad \text{if } u \ll c \qquad [21.32]$$

where f' is the observed frequency, f is the frequency emitted by the source, c is the speed of light in a vacuum, and u is the *relative* speed of the observer and source. Note that Equation 21.32 is valid only if u is much smaller than c. The positive sign in this equation must be used when the source and observer are moving toward one another, while the negative sign must be used when they are moving away from each other. Thus, we anticipate an increase in the observed frequency if the source and observer are approaching each other, and a decrease in observed frequency if the source and observer are receding from each other.

Astronomers have made important discoveries using Doppler observations on light reaching Earth from distant stars and galaxies. Such measurements have shown that distant galaxies are moving away from the Earth. Thus, the Universe is expanding. This is called a *red shift* because the observed wavelengths are shifted towards the red portion (longest wavelength) of the visible spectrum. Furthermore, measurements show that the speed of a galaxy increases with increasing distance from the Earth. More recent Doppler effect measurements made with the Hubble space telescope have shown that a galaxy labeled M87 is rotating. Its measured speed of rotation was used to identify a supermassive black hole located at the center of this galaxy.

SUMMARY

If an ac circuit consists of a generator and a resistor, the current in the circuit is in phase with the voltage. That is, the current and voltage reach their maximum values at the same time.

In discussions of voltages and currents in ac circuits, **rms values** of voltages are usually used. One reason is that ac ammeters and voltmeters are designed to read rms values. The rms values of currents and voltage (I and ΔV) are related to the maximum values of these quantities (I_m and ΔV_m) as follows:

$$I = \frac{I_m}{\sqrt{2}} \qquad \Delta V = \frac{\Delta V_m}{\sqrt{2}} \qquad [21.2, 21.3]$$

The rms voltage across a resistor is related to the rms current through the resistor by **Ohm's law:**

$$\Delta V_R = IR \qquad [21.4]$$

If an ac circuit consists of a generator and a capacitor, the voltage lags behind the current by 90°. That is, the voltage reaches its maximum value one quarter of a period after the current reaches its maximum value.

The impeding effect of a capacitor on current in an ac circuit is given by the **capacitive reactance, X_C,** defined as

$$X_C \equiv \frac{1}{2\pi f C} \qquad [21.5]$$

where f is the frequency of the ac generator.

The rms voltage across and the rms current through a capacitor are related by

$$\Delta V_C = I X_C \qquad \text{[21.6]}$$

If an ac circuit consists of a generator and an inductor, the voltage leads the current by 90°. That is, the voltage reaches its maximum value one quarter of a period before the current reaches its maximum value.

The effective impedance of a coil in an ac circuit is measured by a quantity called the **inductive reactance, X_L,** defined as

$$X_L \equiv 2\pi f L \qquad \text{[21.8]}$$

The rms voltage drop across a coil is related to the rms current through the coil by

$$\Delta V_L = I X_L \qquad \text{[21.9]}$$

In an *RLC* series ac circuit, the applied rms voltage, ΔV, is related to the rms voltages across the resistor (ΔV_R), capacitor (ΔV_C), and inductor (ΔV_L) by

$$\Delta V = \sqrt{\Delta V_R^2 + (\Delta V_L - \Delta V_C)^2} \qquad \text{[21.10]}$$

If an ac circuit contains a resistor, an inductor, and a capacitor, the limit they place on the current is described by the **impedance, Z,** of the circuit, defined as

$$Z \equiv \sqrt{R^2 + (X_L - X_C)^2} \qquad \text{[21.13]}$$

The relationship between the rms voltage supplied to an *RLC* circuit and the rms current in the circuit is

$$\Delta V = IZ \qquad \text{[21.14]}$$

In an *RLC* series ac circuit, the applied rms voltage and current are out of phase. The **phase angle, ϕ,** between the current and voltage is given by

$$\tan \phi = \frac{X_L - X_C}{R} \qquad \text{[21.15]}$$

The **average power** delivered by the generator in an *RLC* ac circuit is

$$P_{av} = I \Delta V \cos \phi \qquad \text{[21.17]}$$

where the constant $\cos \phi$ is called the **power factor.**

Electromagnetic waves were predicted by James Clerk Maxwell and later generated and detected by Heinrich Hertz. These waves have the following properties:

1. Electromagnetic waves are transverse waves, because the electric and magnetic fields are perpendicular to the direction of travel.
2. Electromagnetic waves travel with the speed of light.
3. The ratio of the electric field to the magnetic field in an electromagnetic wave equals the speed of light — that is,

$$\frac{E}{B} = c \qquad \text{[21.26]}$$

4. Electromagnetic waves carry energy as they travel through space. The average power per unit area is

$$\frac{E_m B_m}{2\mu_0} = \frac{E_m^2}{2\mu_0 c} = \frac{c}{2\mu_0} B_m^2 \qquad \text{[21.27, 21.28]}$$

where E_m and B_m are the maximum values of the electric and magnetic fields.

5. Electromagnetic waves transport momentum as well as energy. The speed, c, frequency, f, and wavelength, λ, of an electromagnetic wave are related by

$$c = f\lambda \qquad\qquad [21.29]$$

The **electromagnetic spectrum** includes waves covering a broad range of frequencies and wavelengths. These waves have a variety of applications and characteristics, depending on their frequencies or wavelengths.

MULTIPLE-CHOICE QUESTIONS

1. A satellite orbiting the Earth is powered by sunlight converted into electricity. The solar cell system efficiency, including temporary energy storage and a safety factor for possible cell failures, is 5%. Sunlight provides 1.4 kW/m² in the vicinity of the Earth. What is the area of the solar cell array if the power requirement is 1400 W?
   (a) 1 m²   (b) 10 m²   (c) 20 m²   (d) 200 m²
   (e) No answer is correct.

2. A radio transmitter broadcasts an average power of 100 W. At a distance of 1 km from the antenna, the peak magnetic field is measured by a sensor with 63 cm² area is B_m at a certain frequency. If the 100 W transmitter then broadcasts at double the original frequency, what will the peak magnetic field now be when measured again by the same sensor at a distance of 1 km from the antenna?
   (a) $0.5B_m$ (b) B_m (c) $2B_m$ (d) $4B_m$
   (e) No answer is correct.

3. An ac series circuit contains a resistor of 20 Ω, a capacitor of 0.75 μF, and an inductor of 120 mH. If an rms voltage of 120 V at $f = 500$ Hz is applied, what is the rms current in the circuit?
   (a) 2.3 A   (b) 6.0 A   (c) 10 A   (d) 17 A

4. Repeat the preceding problem to find the current when the circuit is in resonance.
   (a) 2.4 A   (b) 6.0 A   (c) 10 A   (d) 17 A

5. Which of the following is shown in the correct order, from lowest frequency to highest?
   (a) radio, visible, infrared, and x-ray
   (b) infrared, visible, x-ray, and gamma
   (c) visible, infrared, ultraviolet, and x-ray
   (d) infrared, ultraviolet, visible, and gamma

6. An electromagnetic wave with a peak magnetic field component of 1.5×10^{-7} T has an associated peak electric field component of what value?
   (a) 0.50×10^{-15} N/C   (b) 20×10^{-5} N/C
   (c) 2.2×10^4 N/C   (d) 45 N/C

CONCEPTUAL QUESTIONS

1. The switch in the circuit shown in Fig. Q21.1 is closed and the lightbulb glows steadily. The inductor is a simple air-core solenoid. An iron rod is inserted into the interior of the solenoid. What happens to the brightness of the lightbulb?

2. What is the impedance of an RLC circuit at the resonance frequency?

3. When a dc voltage is applied to a transformer, the primary coil sometimes will overheat and burn. Why?

4. Why are the primary and secondary coils of a transformer wrapped on an iron core that passes through both coils?

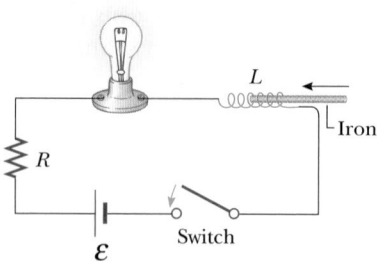

Figure Q21.1

5. Receiving radio antennas can be in the form of conducting lines or loops. What should the orientation of each of these types of antennas be, relative to a broadcasting antenna that is perpendicular to the Earth?

6. If the fundamental source of a sound wave is a vibrating object, what is the fundamental source of an electromagnetic wave?

7. In radio transmission, a radio wave serves as a carrier wave, and the sound signal is superimposed on the carrier wave. In amplitude modulation (AM) radio, the amplitude of the carrier wave varies according to the sound wave. In frequency modulation (FM) radio, the frequency of the carrier wave varies according to the sound wave. The navy sometimes uses flashing lights to send Morse code between neighboring ships, a process that has similarities to radio broadcasting. Is this AM or FM? What is the carrier frequency? What is the signal frequency? What is the broadcasting antenna? What is the receiving antenna?

8. When light (or other electromagnetic radiation) travels across a given region, what is it that moves?

9. In space sailing, which is a proposed alternative method to reach the planets, a spacecraft carries a very large sail that experiences a force due to radiation pressure from the Sun. Should the sail be absorptive or reflective to be most effective?

10. How can the average value of an alternating current be zero yet the square root of the average value squared not be zero?

11. Suppose a creature from another planet had eyes that were sensitive to infrared radiation. Describe what he would see if he looked around the room you are now in. That is, what would be bright and what would be dim?

12. Why should an infrared photograph of a person look different from a photograph taken with visible light?

13. Radio stations often advertise "instant news." If what they mean is that you hear the news at the instant they speak it, is their claim true? About how long would it take for a message to travel across this country by radio waves, assuming that these waves could travel this great distance and still be detected?

14. Would an inductor and a capacitor used together in a ac circuit dissipate any energy?

15. Does a wire connected to a battery emit an electromagnetic wave?

PROBLEMS

1, 2, 3 = straightforward, intermediate, challenging ☐ = full solution available in Study Guide/Student Solutions Manual ◤ = Core Concepts Workbook
WEB = solution posted at **http://www.harcourtcollege.com/physics/cptech** ▨ = biomedical application ▨ = Interactive Physics

Review Problem

An inductor with an air core is to be made of a 100-m length of 0.60-mm-diameter copper wire wound in a single layer on a cylindrical form of radius 2.0 cm. (a) Determine the number of turns of wire on the solenoid and its length. Find (b) the inductance and (c) the resistance of the completed solenoid. (d) Determine the power that will be dissipated when this device is connected to a 400-Hz, 5.0-V rms source.

Section 21.1 Resistors in an ac Circuit

1. An rms voltage of 100 V is applied to a purely resistive load of 5.00 Ω. Find (a) the maximum voltage applied, (b) the rms current supplied, (c) the maximum current supplied, and (d) the power dissipated.

2. An ac voltage source has an output of $\Delta v = 150 \sin 377t$. Find (a) the rms voltage output, (b) the frequency of the source, and (c) the the voltage at $t = 1/120$ s. (d) Find the maximum current in the circuit when the generator is connected to a 50.0-Ω resistor.

3. (a) What is the resistance of a lightbulb that uses an average power of 75 W when connected to a 60-Hz power source with an peak voltage of 170 V? (b) What is the resistance of a 100-W bulb?

4. An audio amplifier, represented by the ac source and the resistor R in Figure P21.4, delivers alternating voltages at audio frequencies to the speaker. If the source puts out an alternating voltage of 15.0 V (rms), resistance R is 8.20 Ω, and the speaker is equivalent to a resistance of 10.4 Ω, what is the time averaged power input to the speaker?

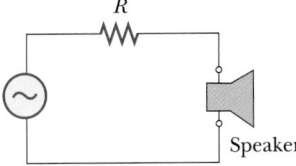

Figure P21.4

5. An ac power supply produces a peak voltage of $\Delta V_m = 100$ V. This power supply is connected to a 24-Ω resistor, and the current and resistor voltage are measured with an ideal ac ammeter and an ideal ac voltme-

ter, as shown in Figure P21.5. What does each meter read? Recall that an ideal ammeter has zero resistance and an ideal voltmeter has infinite resistance.

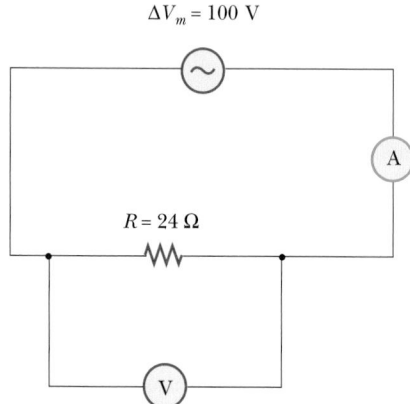

$\Delta V_m = 100$ V

$R = 24$ Ω

Figure P21.5

Section 21.2 Capacitors in an ac Circuit

6. Show that the SI unit of capacitive reactance, X_C is the ohm.

7. When a 4.0-μF capacitor is connected to a generator whose rms output is 30 V, the current in the circuit is observed to be 0.30 A. What is the frequency of the source?

8. A certain capacitor in a circuit has a capacitive reactance of 30.0 Ω when the frequency is 120 Hz. What capacitive reactance does the capacitor have at a frequency of 10 000 Hz?

WEB **9.** What value of capacitor must be inserted in a 60-Hz circuit in series with a generator of 170 V maximum output voltage to produce an rms current output of 0.75 A?

10. The generator in a purely capacitive ac circuit has an angular frequency of 120π rad/s. If $\Delta V_m = 140$ V and $C = 6.00$ μF, what is the rms current in the circuit?

Section 21.3 Inductors in an ac Circuit

11. Show that the inductive reactance, X_L, has SI units of ohms.

12. The generator in a purely inductive ac circuit has an angular frequency of 120π rad/s. If $\Delta V_m = 140$ V and $L = 0.100$ H, what is the rms current in the circuit?

13. An inductor has a 54.0-Ω reactance at 60.0 Hz. What will be the *peak* current if this inductor is connected to a 50.0-Hz source that produces a 100-V rms voltage?

14. An inductor is connected to a 20.0-Hz power supply that produces a 50.0-V rms voltage. What inductance is needed to keep the maximum current in the circuit below 80.0 mA?

15. A 2.40-μF capacitor is connected across an alternating voltage with an rms value of 9.00 V. The rms current in the capacitor is 25.0 mA. **(a)** What is the source fre-

quency? **(b)** If the capacitor is replaced by an ideal coil with an inductance of 0.160 H, what is the rms current in the coil?

Section 21.4 The *RLC* Series Circuit

16. A pure 20.0-mH inductor is connected in series with a 20.0-Ω resistor and a 60.0-Hz, 100-V rms source. Find **(a)** the rms current in the circuit, **(b)** the voltage drop across the inductor, **(c)** the voltage drop across the resistor, and **(d)** the phase angle for this circuit. **(e)** Sketch the phasor diagram for this circuit.

17. A 10.0-μF capacitor and a 2.00-H inductor are connected in series with a 60.0-Hz source whose rms output is 100 V. Find **(a)** the rms current in the circuit, **(b)** the voltage drop across the inductor, **(c)** the voltage drop across the capacitor, and **(d)** the phase angle for the circuit. **(e)** Sketch the phasor diagram for this circuit.

18. A 40.0-μF capacitor is connected to a 50.0-Ω resistor and a generator whose rms output is 30.0 V at 60.0 Hz. Find **(a)** the rms current in the circuit, **(b)** the voltage drop across the resistor, **(c)** the voltage drop across the capacitor, and **(d)** the phase angle for the circuit. **(e)** Sketch the phasor diagram for this circuit.

19. A resistor ($R = 900$ Ω), a capacitor ($C = 0.25$ μF), and an inductor ($L = 2.5$ H) are connected in series across a 240-Hz ac source for which $\Delta V_m = 140$ V. Calculate the **(a)** impedance of the circuit, **(b)** peak current delivered by the source, and **(c)** phase angle between the current and voltage. **(d)** Is the current leading or lagging behind the voltage?

20. A 50.0-Ω resistor, a 0.100-H inductor, and a 10.0-μF capacitor are connected in series to a 60.0-Hz source. The rms current in the circuit is 2.75 A. Find the rms voltages across **(a)** the resistor, **(b)** the inductor, **(c)** the capacitor, and **(d)** the *RLC* combination. **(e)** Sketch the phasor diagram for this circuit.

21. A 60.0-Ω resistor, a 3.00-μF capacitor, and a 0.400-H inductor are connected in series to a 90.0-V, 60.0-Hz source. Find **(a)** the voltage drop across the *LC* combination and **(b)** the voltage drop across the *RC* combination.

22. A 50.0-Ω resistor is connected in series with a 15.0-μF capacitor and a 60.0-Hz, 120-V source. **(a)** Find the current in the circuit. **(b)** What is the value of the inductor that must be inserted in the circuit to reduce the current to one half that found in **(a)**?

23. An ac source with a peak voltage of 150 V and $f = 50.0$ Hz is connected between points a and d in Figure P21.23. Calculate the rms voltages between points **(a)** a and b, **(b)** b and c, **(c)** c and d, **(d)** b and d.

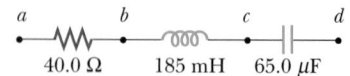

a b c d

40.0 Ω   185 mH   65.0 μF

Figure P21.23

Section 21.5 Power in an ac Circuit

24. A 50.0-Ω resistor is connected to a 30.0-μF capacitor and to a 60.0-Hz, 100-V rms source. (a) Find the power factor and the average power delivered to the circuit. (b) Repeat part (a) when the capacitor is replaced with a 0.300-H inductor.

25. A multimeter in an *RL* circuit records an rms current of 0.500 A and a 60.0-Hz rms generator voltage of 104 V. The average thermal power developed in the resistor is 10.0 W. Determine (a) the impedance in the circuit, (b) the resistance, R, and (c) the inductance, L.

26. In a certain *RLC* circuit, the rms current is 6.0 A, the rms voltage is 240 V, and the current leads the voltage by 53°. (a) What is the total resistance of the circuit? (b) Calculate the total reactance, $X_L - X_C$. (c) Find the average power dissipated in the circuit.

WEB **27.** An inductor and a resistor are connected in series. When connected to a 60-Hz, 90-V source, the voltage drop across the resistor is found to be 50 V and the power dissipated in the circuit is 14 W. Find (a) the value of the resistance and (b) the value of the inductance.

28. An ac voltage with an amplitude of 100 V is applied to a series combination of a 200-μF capacitor, a 100-mH inductor, and a 20.0-Ω resistor. Calculate the power dissipation and the power factor for frequencies of (a) 60.0 Hz and (b) 50.0 Hz.

Section 21.6 Resonance in a Series *RLC* Circuit

29. An *RLC* circuit is used to tune a radio to an FM station broadcasting at 88.9 MHz. The resistance in the circuit is 12.0 Ω and the capacitance is 1.40 pF. What inductance should be present in the circuit?

30. A resonant circuit in a radio receiver is tuned to a certain station when the inductor has a value of 0.200 mH and the capacitor has a value of 30.0 pF. Find the frequency of the radio station and the wavelength sent out by the station.

31. The AM band extends from approximately 500 kHz to 1600 kHz. If a 2.0-μH inductor is used in a tuning circuit for a radio, what are the extremes that a capacitor must reach in order to cover the complete band of frequencies?

32. A series circuit contains a 3.00-H inductor, a 3.00-μF capacitor, and a 30.0-Ω resistor connected to a 120-V rms source of variable frequency. Find the power delivered to the circuit when the frequency of the source is (a) the resonance frequency; (b) one half the resonance frequency; (c) one fourth the resonance frequency; (d) two times the resonance frequency; (e) four times the resonance frequency. From your calculations, can you draw a conclusion about the frequency at which the maximum power is delivered to the circuit?

33. The Q value of an *RLC* circuit is defined as the voltage drop across the inductor (or capacitor) at resonance, divided by the voltage drop across the resistor. The greater the Q value, the sharper, or narrower, is the curve of power versus frequency. Figure P21.33 shows such curves for small R and large R. (a) Show that the Q value is given by the expression $Q = 2\pi f_0 L / R$, where f_0 is the resonance frequency. (b) Calculate the Q value for the circuit of Problem 32.

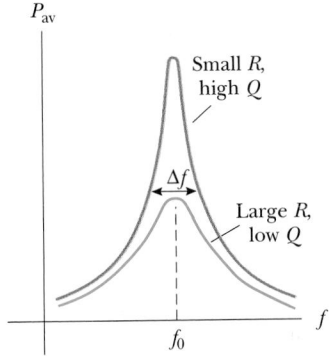

Figure P21.33   (Problems 33 and 34)

34. The curves of Figure P21.33 are characterized by the full width at half-maximum, Δf. This width is the difference in frequency between the two points on the curve where the power is one half the maximum value. It can be shown that this width is related to the Q value (when the Q value is large) by $\Delta f = f_0 / Q$. (a) Find Δf for the circuit in Problems 32 and 33. (b) Find the Q value and Δf for the circuit in Problem 32 if R is replaced by a 300-Ω resistor.

Section 21.7 The Transformer (Optional)

35. A step-up transformer is designed to have an output voltage of 2200 V (rms) when the primary is connected across a 110-V (rms) source. (a) If there are 80 turns on the primary winding, how many turns are required on the secondary? (b) If a load resistor across the secondary draws a current of 1.5 A, what is the current in the primary, assuming ideal conditions?

36. At a given moment, every inhabitant of a city of 20 000 people turns on a 100-W lightbulb. Assume no other power in the city is being used. (a) If the utility company furnishes this total power at 120 V, calculate the current in the power lines from the utility to the city. (b) Calculate this current if the power company first steps up the voltage to 200 000 V. (c) How much heat is lost in each 1.00-m length of the power lines if the resistance of the lines is 5.00×10^{-4} Ω/m? Repeat this calculation for situation (a) and for situation (b). (d) If an individual line from the utility can handle only 100 A, how many lines are required to handle the current in each situation described?

37. A transformer on a pole near a factory steps the voltage down from 3600 V to 120 V. The transformer is to deliver 1000 kW to the factory at 90% efficiency. Find (a) the power delivered to the primary, (b) the current in the primary, and (c) the current in the secondary.

38. An ac power generator produces 50 A (rms) at 3600 V. The voltage is stepped up to 100 000 V by an ideal transformer, and the energy is transmitted through a long-distance power line that has a resistance of 100 Ω. What percentage of the power delivered by the generator is dissipated as heat in the power line?

Section 21.10   Production of Electromagnetic Waves by an Antenna

Section 21.11   Properties of Electromagnetic Waves

39. The U.S. Navy has long proposed the construction of extremely low-frequency (ELF waves) communications systems; such waves could penetrate the oceans to reach distant submarines. Calculate the length of a quarter-wavelength antenna for a transmitter generating ELF waves of frequency 75 Hz. How practical is this?

40. Experimenters at the National Institute of Standards and Technology have made precise measurements of the speed of light using the fact that, in vacuum, the speed of electromagnetic waves is $c = 1/\sqrt{\mu_0\epsilon_0}$, where $\mu_0 = 4\pi \times 10^{-7}$ N·s^2/C^2 and $\epsilon_0 = 8.854 \times 10^{-12}$ C^2/N·m^2. What value (to four significant figures) does this give for the speed of light in vacuum?

WEB **41.** A particular electromagnetic wave traveling in vacuum has a magnetic field intensity of 1.5×10^{-7} T. Find (a) the electric field intensity and (b) the average power per unit area associated with the wave.

42. Assume that the solar radiation incident on the Earth is 1340 W/m^2 (at the top of the Earth's atmosphere). Calculate the total power radiated by the Sun, taking the average separation between the Earth and the Sun to be 1.49×10^{11} m.

43. The Sun delivers an average power of 1340 W/m^2 to the top of the Earth's atmosphere. Find the magnitudes of $\mathbf{E}_m$ and $\mathbf{B}_m$ for the electromagnetic waves at the top of the atmosphere.

Section 21.12   The Spectrum of Electromagnetic Waves

44. The human eye is sensitive to electromagnetic waves that have wavelengths in the range from 400 nm to 700 nm (1 nm = 10^{-9} m). What range of frequencies of electromagnetic radiation can the eye detect?

45. What are the wavelength ranges in the (a) AM radio band (540–1600 kHz) and (b) the FM radio band (88–108 MHz)?

46. A diathermy machine, used in physiotherapy, generates electromagnetic radiation that gives the effect of "deep heat" when absorbed in tissue. One assigned frequency for diathermy is 27.33 MHz. What is the wavelength of this radiation?

47. An important news announcement is transmitted by radio waves to people who are 100 km away, sitting next to their radios, and by sound waves to people sitting across the newsroom, 3.0 m from the newscaster. Who receives the news first? Explain. Take the speed of sound in air to be 343 m/s.

ADDITIONAL PROBLEMS

48. An ac adapter for a telephone answering unit uses a transformer to reduce the line voltage of 120 V to a voltage of 9.0 V. The rms current delivered to the answering system is 400 milliamps. (a) If the primary (input) coil in the transformer in the adapter has 240 turns, how many turns are there on the secondary (output) coil? (b) What is the rms power input to the transformer? Assume an ideal transformer.

49. The primary coil of a certain transformer has an inductance of 2.50 H and a resistance of 80.0 Ω (a) If the primary coil is connected to an ac source with a frequency of 60.0 Hz and a voltage of 110 V rms, what is the rms current through the primary? (b) If the primary is connected to 110 V dc, what is the current through the primary? Disregard initial effects. (c) In each case, compare the power dissipated in the resistance.

50. What value of inductance should be used in series with a capacitor of 1.50 pF to form an oscillating circuit that will radiate a wavelength of 5.25 m?

51. A small transformer is used to supply an ac voltage of 6.0 V to a model-railroad lighting circuit. The primary has 220 turns and is connected to a standard 110-V, 60-Hz line. Although the resistance of the primary may be neglected, it has an inductance of 150 mH. (a) How many turns are required on the secondary winding? (b) If the transformer is left plugged in, what current is drawn by the primary when the secondary is open? (c) What power is drawn by the primary when the secondary is open?

52. A 200-Ω resistor is connected in series with a 5.0-μF capacitor and a 60-Hz, 120-V rms line. If electrical energy costs $0.080 per kWh, how much does it cost to leave this circuit connected for 24 h?

WEB **53.** A series *RLC* circuit has a resonance frequency of $2000/\pi$ Hz. When it is operating at a frequency of $\omega > \omega_0$, $X_L = 12$ Ω and $X_C = 8.0$ Ω. Calculate the values of L and C for the circuit.

54. A 0.700-H inductor is connected in series with a fluorescent lamp to limit the current drawn by the lamp. If the combination is connected to a 60.0-Hz, 120-V line, and if the voltage across the lamp is to be 40.0 V, what is the current in the circuit? (*Hint:* The lamp is a pure resistive load.)

55. Two connections allow contact with two circuit elements in series inside a box, but it is not known whether the circuit elements are *R*, *L*, or *C*. In an attempt to find what is inside the box; you make some measurements, with the following results. When a 3.0-V dc power supply is connected across the terminals, there is a direct current of 300 mA in the circuit. When a 3.0-V, 60-Hz source is connected, the current becomes 200 mA. (a) What are the two elements in the box? (b) What are their values of *R*, *L*, or *C*?

56. (a) What capacitance will resonate with a one-turn loop of inductance 400 pH to give a radar wave of wavelength 3.0 cm? (b) If the capacitor has square parallel plates separated by 1.0 mm of air, what should the edge length of the plates be? (c) What is the common reactance of the loop and capacitor at resonance?

57. A dish antenna with a diameter of 20.0 m receives (at normal incidence) a radio signal from a distant source, as shown in Figure P21.57. The radio signal is a continuous sinusoidal wave with amplitude $E_m = 0.20$ μV/m. Assume the antenna absorbs all the radiation that falls on the dish. (a) What is the amplitude of the magnetic field in this wave? (b) What is the intensity of the radiation received by this antenna? (c) What is the power received by the antenna?

Figure P21.57

58. A microwave transmitter emits electromagnetic waves of a single wavelength. The maximum electric field 1.0 km from the transmitter is 6.0 V/m. Assuming that the transmitter is a point source, calculate (a) the maximum magnetic field at this distance and (b) the total power emitted by the transmitter.

59. What power must be radiated uniformly in all directions by a source if the amplitude of the magnetic field is 7.0×10^{-8} T at a distance of 2.0 m?

60. The electromagnetic power, *P*, radiated by a moving point charge, *q*, with acceleration *a* is

$$P = \frac{2k_e q^2 a^2}{3c^3}$$

where k_e is the Coulomb constant and *c* is the speed of light in vacuum. If an electron is placed in a constant electric field of 100 N/C, determine (a) the acceleration of the electron and (b) the power radiated by this electron. (c) Show that the right side of this equation is in watts.

61. A particular inductor has appreciable resistance. When the inductor is connected to a 12-V battery, the current through the inductor is 3.0 A. When it is connected to an ac source with an rms output of 12 V and a frequency of 60.0 Hz, the current drops to 2.0 A. What are (a) the impedance at 60.0 Hz and (b) the inductance of the inductor?

62. A transmission line with a resistance per unit length of 4.5×10^{-4} Ω/m is to be used to transmit 5000 kW of power over a distance of 400 miles (6.44×10^5 m). The output voltage of the generator is 4500 V. (a) What is the line loss if a transformer is used to step up the voltage to 500 kV? (b) What fraction of the input power is lost to the line under these circumstances? (c) What difficulties would be encountered in an attempt to transmit the 5000 kW of power at the generator voltage of 4500 V?

63. A possible means of space flight is to place a perfectly reflecting aluminized sheet into Earth's orbit and use the light from the Sun to push this solar sail. Suppose such a sail, of area 6.00×10^4 m² and mass 6000 kg, is placed in orbit facing the Sun. (a) What force is exerted on the sail? (b) What is the sail's acceleration? (c) How long does it take for this sail to reach the Moon, 3.84×10^8 m away? Ignore all gravitational effects, and assume a solar intensity of 1340 W/m². (*Hint:* The radiation pressure by a reflected wave is given by 2(average power per area)/*c*.)

64. Suppose you wish to use a transformer as an impedance-matching device between an audio amplifier that has an output impedance of 8000 Ω and a speaker that has an input impedance of 8.0 Ω. What should be the ratio of primary to secondary turns on the transformer?

65. Compute the average energy content of a liter of sunlight as it reaches the top of the Earth's atmosphere, where its intensity is 1340 W/m².

Light and Optics

Scientists have long been intrigued by the nature of light, and philosophers have argued endlessly concerning the proper definition and perception of light. It is important to understand the nature of this basic ingredient of life on Earth. Plants convert light energy from the Sun to chemical energy through photosynthesis. Light is the means by which we transmit and receive information from objects around us and throughout the Universe.

The Greeks believed that light consisted of tiny particles (corpuscles) that were emitted by a light source, then stimulated the perception of vision on striking the observer's eye. Newton used this corpuscular theory to explain the reflection and refraction of light. In 1670 one of Newton's contemporaries, the Dutch scientist Christian Huygens, succeeded in explaining many properties of light by proposing that light was wave-like. In 1801 Thomas Young gave strong support to the wave theory by showing that light beams can interfere with one

another. In 1865 Maxwell developed a brilliant theory that electromagnetic waves travel with the speed of light (Chapter 21). By that time, the wave theory of light seemed to be on firm ground.

However, at the beginning of the 20th century, Max Planck introduced the notion of quantization of electromagnetic radiation, and Albert Einstein returned to the corpuscular theory of light, in order to explain the radiation emitted by hot objects and the electrons emitted by a metal exposed to light (the photoelectric effect). We shall discuss those and other modern topics in the last part of this book.

Today scientists view light as having a dual nature. Experiments can be devised that will display either its particle-like or its wave-like nature. In this part of the book, we concentrate on the aspects of light that are best understood through the wave model. First we discuss the reflection of light at the boundary between two media and the refraction (bending) of light as it travels from one medium into another. We

use these ideas to study the refraction of light as it passes through lenses and the reflection of light from mirrored surfaces. Finally, we describe how lenses and mirrors can be used to view objects with telescopes and microscopes and how lenses are used in photography.

◀ *Erich Lessing/Magnum Photos, Inc.*

22

Reflection and Refraction of Light

Chapter Outline

▲ **PHYSICS PUZZLER**

When the Sun or Moon is low in the sky, its reflection in water may appear as a long line of light rather than as a circle. Why does this happen? And why does the Sun appear to be distorted when it is near the horizon? *(Stan Osolinski/ Dembinsky Photo Associates)*

22.1   THE NATURE OF LIGHT

Until the beginning of the 19th century, light was considered to be a stream of particles, emitted by a light source, that stimulated the sense of sight on entering the eye. The chief architect of the particle theory of light was Newton. With this theory he provided simple explanations of some known experimental facts concerning the nature of light—namely the laws of reflection and refraction.

Most scientists accepted Newton's particle theory of light. However, during Newton's lifetime another theory was proposed. In 1678 a Dutch physicist and astronomer, Christian Huygens (1629–1695), showed that a wave theory of light could also explain the laws of reflection and refraction. The wave theory did not receive immediate acceptance for several reasons. All the waves known at the time (sound, water, and so on) traveled through some sort of medium, but light from the Sun could travel to Earth through empty space. Furthermore, it was argued that if light were some form of wave, it would bend around obstacles; hence, we should be able to see around corners. It is now known that light does indeed bend around the edges of objects. This phenomenon, known as *diffraction*, is not easy to observe because light waves have such short wavelengths. Even though experimental evidence for the diffraction of light was discovered by Francesco Grimaldi (1618–1663) around 1660, for more than a century most scientists rejected the wave theory and adhered to Newton's particle theory. This was also due to Newton's great reputation as a scientist.

The first clear demonstration of the wave nature of light was provided by 1801 by Thomas Young (1773–1829), who showed that, under appropriate conditions, light exhibits interference behavior. That is, at certain points in the vicinity of two sources, light waves can combine and cancel each other by destructive interference. Such behavior could not be explained at that time by a particle theory.

The most important development concerning the theory of light was the work of Maxwell, who in 1865 predicted that light was a form of high-frequency electromagnetic wave (Chapter 21). His theory predicted that these waves should have a speed of 3×10^8 m/s, in agreement with the measured value.

Although the classical theory of electricity and magnetism explained most known properties of light, some subsequent experiments could not be explained by the assumption that light was a wave. The most striking of these was the *photoelectric effect* (which we shall examine more closely in Chapter 27), discovered by Hertz. Hertz found that clean metal surfaces emit charges when exposed to ultraviolet light.

In 1905 Einstein published a paper that formulated the theory of light quanta and explained the photoelectric effect. He reached the conclusion that light is composed of *corpuscles,* or discontinuous quanta of energy. Furthermore, he asserted that light interacting with matter also consists of quanta, and he brilliantly worked out the implications of the photoelectric process. More specifically, Einstein showed that the energy of a photon is proportional to the frequency of the electromagnetic wave:

$$E = hf \qquad\qquad [22.1]$$

◀ Energy of a photon

where $h = 6.63 \times 10^{-34}$ J · s is *Planck's constant*. This theory retains some features of both the wave and particle theories of light. As we shall discuss later, the photoelectric effect is the result of energy transfer from a single photon to an electron

Christian Huygens, Dutch physicist and astronomer (1629–1695)

Huygens is best known for his contributions to the fields of optics and dynamics. To Huygens, light was a type of vibratory motion, spreading out and producing the sensation of light when impinging on the eye. On the basis of this theory, he deduced the laws of reflection and refraction and explained the phenomenon of double refraction. *(Courtesy of Rijksmuseum voor de Geschiedenis der Natuurwetenschappen. Courtesy AIP Niels Bohr Library)*

in the metal. That is, the electron interacts with one photon of light as if the electron had been struck by a particle. Yet the photon has wave-like characteristics as implied by the fact that light exhibits interference phenomena.

In view of these developments, light must be regarded as having a *dual nature*. That is, **in some cases light acts as a wave and in others it acts as a particle.** Classical electromagnetic wave theory provides adequate explanations of light propagation and of the effects of interference, whereas the photoelectric effect and other experiments involving the interaction of light with matter are best explained by assuming that light is a particle. However, the question "Is light a wave or a particle?" is inappropriate; sometimes it acts as one, sometimes as the other. Fortunately, it never acts as both in the same experiment.

22.2   MEASUREMENTS OF THE SPEED OF LIGHT

Light travels so fast ($c \approx 3 \times 10^8$ m/s) that early attempts to measure its speed were unsuccessful. Galileo attempted to measure the speed of light by positioning two observers in towers separated by about 5 miles. Each observer carried a shuttered lantern. One observer opened his lantern first, and then the other opened his lantern at the moment he saw the light from the first lantern. In principle, the speed could then be obtained from the transit time of the light beams between lanterns. However, at a speed of 186 000 mi/s, light would travel the 10-mile round trip in approximately 54 μs. Because the transit time is so small compared with the reaction time of the observers, it is impossible to measure the speed of light in this manner.

Roemer's Method

The first successful estimate of the speed of light was made in 1675 by the Danish astronomer Ole Roemer (1644–1710). His technique involved astronomical observations of one of the moons of Jupiter, Io. It is of interest that Roemer did not set out to measure the speed of light. He was, instead, trying to explain a perplexing problem related to the prediction of eclipses of Io and a few other moons. After realizing that the underlying difficulty with predicting the eclipses had to do with the finite speed of light, he estimated that speed.

At the time of Roemer, only four of Jupiter's moons had been discovered, and the periods of their orbits were known. Io, the innermost moon, has a period of 42.5 h, and its orbit, the orbit of Jupiter, and the orbit of the Earth all lie in approximately the same plane. As a result, Io goes into eclipse behind Jupiter every 42.5 h, as seen from the Earth. Using the orbital motion of Io as a clock, Roemer expected to find a constant period over long time intervals. Instead, he observed a systematic variation in Io's period. He found that the periods were longer than average when Earth receded from Jupiter and shorter than average when Earth approached Jupiter. If Io had a constant period, Roemer should have been able to observe a particular eclipse and be able to predict when future eclipses would occur. But he found that his predictions often did not agree with the actual occurrences of the eclipses. For example, consider the situation diagrammed in Figure 22.1. Suppose an eclipse occurred when the Earth was at position E_1. Knowing the period

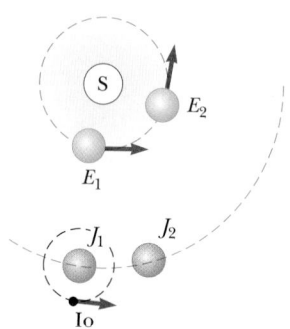

Figure 22.1   As the Earth moves from E_1 to E_2, Jupiter moves only from J_1 to J_2.

of Io, Roemer could predict when an eclipse should occur three months later, with the Earth at position E_2. However, the actual eclipse at E_2 occurred approximately 600 s later than the predicted time. Roemer attributed this discrepancy to the fact that the distance between the Earth and Jupiter was changing between the two observations. In three months (one quarter of the Earth's period), the Earth moved through one fourth of its orbit, from E_1 to E_2, as shown in Figure 22.1. In the same time interval, Jupiter, whose period is about 12 years, moved a much shorter distance, from J_1 to J_2. Therefore, as the Earth moved from E_1 to E_2, light from Jupiter had to travel an additional distance equal to the radius of the Earth's orbit.

Using the data available at that time, Roemer estimated the speed of light to be about 2.1×10^8 m/s. The large discrepancy between this value and the currently accepted value, 3×10^8 m/s, is due to a large error in the assumed radius of the Earth's orbit. Roemer's experiment is important historically because it demonstrated that light does have a finite speed and established a rough estimate of the magnitude of that speed.

Fizeau's Technique

The first successful method of measuring the speed of light using purely Earth-bound techniques was developed in 1849 by Armand H. L. Fizeau (1819–1896). Figure 22.2 is a simplified diagram of his apparatus. The basic idea is to measure the total time it takes light to travel from some point to a distant mirror and back. If d is the distance between the light source and the mirror and if the transmit time for one round trip is t, then the speed of light is $c = 2d/t$. To measure the transit time, Fizeau used a rotating toothed wheel, which converts an otherwise continuous beam of light to a series of light pulses. In addition, the rotation of the wheel controls what an observer at the light source sees. For example, if the light passing the opening at point A in Figure 22.2 returned at the instant tooth B had rotated into position to cover the return path, the light would not reach the observer. At a faster rate of rotation, the opening at point C could move into position to allow the reflected beam to pass and reach the observer. Knowing the distance, d, the number of teeth in the wheel, and the angular speed of the wheel, Fizeau arrived at a value of $c = 3.1 \times 10^8$ m/s. Similar measurements made by subsequent investigators yielded more accurate values for c, approximately 2.9977×10^8 m/s.

A variety of more accurate measurements have since been reported for c. A recent value, obtained using a laser technique, is

$$c = 2.997\ 924\ 574\ (12) \times 10^8 \text{ m/s}$$

where the (12) indicates the uncertainty in the last two digits. The number of significant figures here is certainly impressive. In fact, the speed of light has been determined with such high accuracy that it is now used to define the SI unit of length, the meter. As noted in Chapter 1, the meter is defined as the distance traveled by light in a vacuum during an interval of $1/299\ 792\ 458$ s.

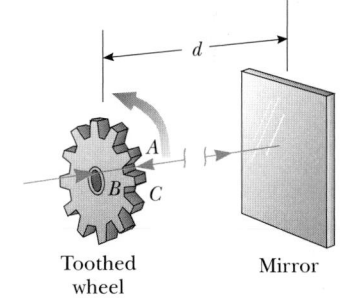

Figure 22.2   Fizeau's method for measuring the speed of light using a rotating toothed wheel.

EXAMPLE 22.1   Measuring the Speed of Light with Fizeau's Toothed Wheel

Assume that the toothed wheel of the Fizeau experiment has 360 teeth and is rotating at a speed of 27.5 rev/s when the light from the source is extinguished—that is, when a burst of light passing through opening A in Figure 22.2 is blocked by tooth B on return. If the distance to the mirror is 7500 m, find the speed of light.

Reasoning and Solution   If the wheel has 360 teeth, it turns through an angle of 1/720 rev in the time that it takes the light to make its round trip. From the definition of angular velocity, we see that the time is

$$t = \frac{\theta}{\omega} = \frac{(1/720) \text{ rev}}{27.5 \text{ rev/s}} = 5.05 \times 10^{-5} \text{ s}$$

Hence, the speed of light is

$$c = \frac{2d}{t} = \frac{2(7500 \text{ m})}{5.05 \times 10^{-5} \text{ s}} = \boxed{2.97 \times 10^8 \text{ m/s}}$$

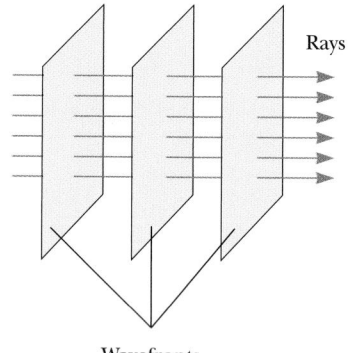

Figure 22.3 A plane wave traveling to the right. Note that the rays, corresponding to the direction of wave motion, are straight lines perpendicular to the wave fronts.

14.2, SECTION 1

14.3

22.3   THE RAY APPROXIMATION IN GEOMETRIC OPTICS

In studying geometric optics here and in future chapters, we shall make use of an important property of light that can be understood based on common experience. **Light travels in a straight line path until it encounters a boundary between two different materials.** As we will see, when light strikes a boundary it either is reflected from that boundary, passes into the material on the other side of the boundary, or partially does both.

Based on this observation, we will use what is called the *ray approximation* to represent our beams of light. As shown in Figure 22.3, a ray of light is an imaginary line drawn along the direction of travel of the light beam. For example, a beam of sunlight passing through a darkened room traces out the path of a light ray. We will also have occasion to refer to wavefronts of light. A wavefront is a surface passing through the points of a wave that have the same phase and amplitude. For instance, the wavefronts in Figure 22.3 could be surfaces passing through the crests of waves. You should note that the rays, corresponding to the direction of wave motion, are straight lines perpendicular to the wavefronts. Also, note that when light rays travel in parallel paths, the wavefronts are planes perpendicular to the rays.

22.4   REFLECTION AND REFRACTION

Reflection of Light

When a light ray traveling in a transparent medium encounters a boundary leading into a second medium, part of the incident ray is reflected back into the first medium. Figure 22.4a shows several rays of a beam of light incident on a smooth, mirror-like, reflecting surface. The reflected rays are parallel to each other, as indicated in the figure. Reflection of light from such a smooth surface is called *specular reflection*. On the other hand, if the reflecting surface is rough, as in Figure 22.4b, the surface reflects the rays in a variety of directions. Reflection from any rough surface is known as *diffuse reflection*. A surface behaves as a smooth surface as long as the surface variations are small compared with the wavelength of the incident light. Figures 22.4c and 22.4d are photographs of specular and diffuse reflection of laser light.

For instance, consider the two types of reflection from a road surface that one sees while driving at night. When the road is dry, light from oncoming vehicles is

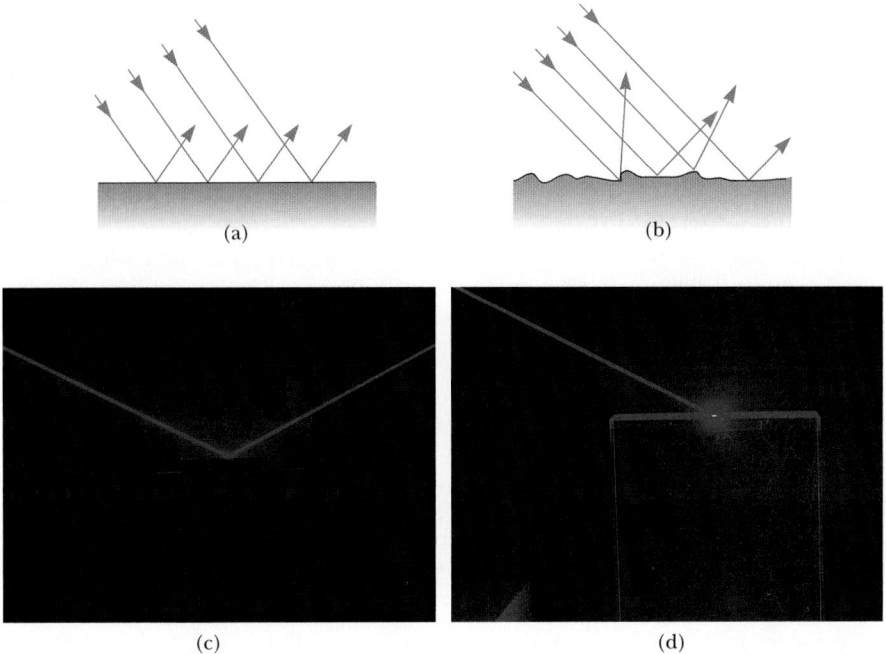

Figure 22.4 A schematic representation of (a) specular reflection, where the reflected rays are all parallel to each other, and (b) diffuse reflection, where the reflected rays travel in random directions. (c, d) Photographs of specular and diffuse reflection, using laser light. *(Photographs courtesy of Henry Leap and Jim Lehman)*

scattered off the road in different directions (diffuse reflection) and the road is quite visible. On a rainy night, when the road is wet, the road irregularities are filled with water. Because the wet surface is quite smooth, the light undergoes specular reflection. This means that the light is reflected straight ahead, and the driver of a car sees only what is directly in front of her. Light from the side never reaches her eyes. In this book we concern ourselves only with specular reflection, and we use the term *reflection* to mean specular reflection.

Consider a light ray traveling in air and incident at some angle on a flat, smooth surface, as in Figure 22.5. The incident and reflected rays make angles θ_1 and θ_1', respectively, with a line perpendicular to the surface at the point where the incident ray strikes the surface. We call this line the *normal* to the surface. Experiments show that **the angle of reflection equals the angle of incidence;** that is,

$$\theta_1' = \theta_1 \qquad [22.2]$$

You may have noticed a common occurrence in photographs of individuals — their eyes appear to be glowing red. This occurs when a photographic flash device is used and the flash unit is very close to the camera lens. Light from the flash unit enters the eye and is reflected back along its original path from the retina. This type of reflection back along the original direction is called *retroreflection*. If the flash unit and lens are close together, this retroreflected light can enter the lens. Most of the light reflected from the retina is red, due to the blood vessels at the back of the eye, giving the red-eye effect in the photograph.

APPLICATION

Seeing the Road on a Rainy Night.

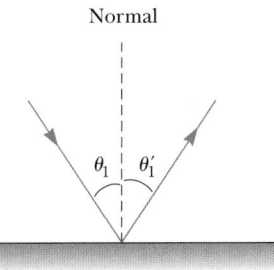

Figure 22.5 According to the law of reflection, $\theta_1 = \theta_1'$.

APPLICATION

Red Eyes in Flash Photographs.

Thinking Physics 1

An observer on the west-facing beach of a large lake is watching the beginning of a sunset. The water is very smooth except for some areas with small ripples. The observer notices that some areas of the water are blue and some are pink. Why does the water appear to be different colors in different areas?

Explanation The different colors arise from specular and diffuse reflection. The smooth areas of the water will specularly reflect the light from the west, which is the pink light from the sunset. The areas with small ripples will reflect the light diffusely. Thus, light from all parts of the sky will be reflected into the observers' eyes. Because most of the sky is still blue at the beginning of the sunset, these areas will appear to be blue.

Thinking Physics 2

When looking through a glass window to the outdoors at night, you sometimes see a double image of yourself. Why?

Explanation Reflection occurs whenever there is an interface between two different media. For the glass in the window, there are two such surfaces. The first is the inner surface of the glass, and the second is the outer surface. Each of these interfaces results in an image.

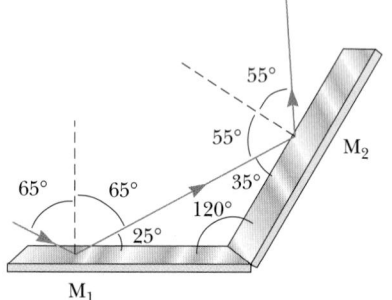

Figure 22.6 (Example 22.2) Mirrors M_1 and M_2 make an angle of 120° with each other.

14.4, SECTION 1

EXAMPLE 22.2 The Double-Reflecting Light Ray

Two mirrors make an angle of 120° with each other, as in Figure 22.6. A ray is incident on mirror M_1 at an angle of 65° to the normal. Find the direction of the ray after it is reflected from mirror M_2.

Reasoning and Solution From the law of reflection, we see that the first reflected ray also makes an angle of 65° with the normal. It follows that this same ray makes an angle of 90° − 65°, or 25°, with the horizontal. From the triangle made by the first reflected ray and the two mirrors, we see that the first reflected ray makes an angle of 35° with M_2 (because the sum of the interior angles of any triangle is 180°). This means that this ray makes an angle of 55° with the normal to M_2. From the law of reflection, it follows that the second reflected ray makes an angle of 55° with the normal to M_2.

Refraction of Light

When a ray of light traveling through a transparent medium encounters a boundary leading into another transparent medium, as in Figure 22.7a, part of the ray is reflected and part enters the second medium. The ray that enters the second medium is bent at the boundary and is said to be *refracted*. The incident ray, the reflected ray, the refracted ray, and the normal at the point of incidence all lie in the same plane. The **angle of refraction,** θ_2 in Figure 22.7a, depends on the properties

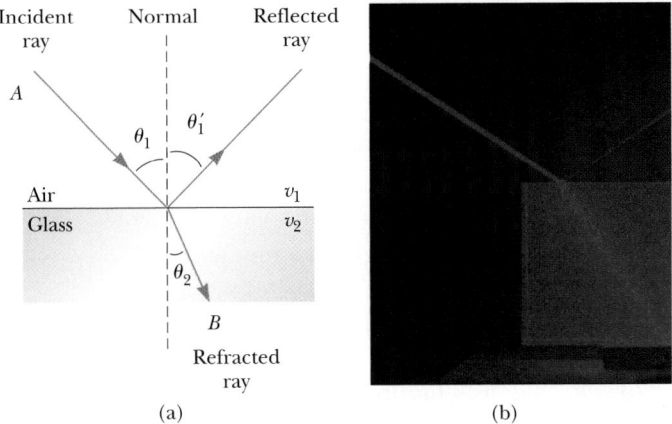

(a)   (b)

Figure 22.7 (a) A ray obliquely incident on an air-glass interface. The refracted ray is bent toward the normal because $v_2 < v_1$. (b) Light incident on the Lucite block bends when it enters the block. *(Courtesy of Henry Leap and Jim Lehman)*

of the two media and on the angle of incidence, through the relationship

$$\frac{\sin \theta_2}{\sin \theta_1} = \frac{v_2}{v_1} = \text{constant} \qquad [22.3]$$

where v_1 is the speed of light in medium 1 and v_2 is the speed of light in medium 2. Willebrord Snell (1591–1626) is usually credited with the experimental discovery of this relationship, which is therefore known as Snell's law. In Section 22.8 we shall derive the laws of reflection and refraction using Huygens's principle.

Experiment shows that **the path of a light ray through a refracting surface is reversible.** For example, the ray in Figure 22.7a travels from point A to point B. If the ray originated at B, it would follow the same path to reach point A, but the reflected ray would be in the glass.

When light moves from a material in which its speed is high to a material in which its speed is lower, the angle of refraction, θ_2, is less than the angle of incidence, as shown in Figure 22.8a. If the ray moves from a material in which its speed is low to a material in which its speed is higher, it is bent away from the normal, as in Figure 22.8b.

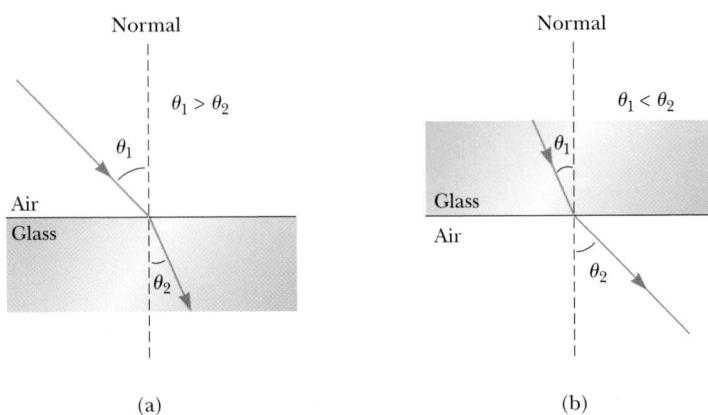

(a)   (b)

Figure 22.8 (a) When the light beam moves from air into glass, its path is bent toward the normal. (b) When the beam moves from glass into air, its path is bent away from the normal.

22.5   THE LAW OF REFRACTION

When light passes from one transparent medium to another, it is refracted because the speed of light is different in the two media. It is convenient to define the **index of refraction, _n_,** of a medium as the ratio

Index of refraction ▶

$$n = \frac{\text{speed of light in vacuum}}{\text{speed of light in a medium}} = \frac{c}{v}$$ [22.4]

From this definition, we see that the index of refraction is a dimensionless number that is greater than unity because v is always less than c. Furthermore, n equals unity for a vacuum. Table 22.1 lists the indices of refraction for some representative substances.

As light travels from one medium to another, its frequency does not change. To see why, consider Figure 22.9. Wave fronts pass an observer at point A in medium 1 with a certain frequency and are incident on the boundary between medium 1 and medium 2. The frequency with which the wave fronts pass an observer at point B in medium 2 must equal the frequency at which they arrive at point A in medium 1. If this were not the case, either wave fronts would pile up at the boundary or they would be destroyed or created at the boundary. Because there is no mechanism for this to occur, the frequency must be a constant as a light ray passes from one medium into another.

Therefore, because the relation $v = f\lambda$ must be valid in both media and because $f_1 = f_2 = f$, we see that

$$v_1 = f\lambda_1 \quad \text{and} \quad v_2 = f\lambda_2$$

where the subscripts refer to the two media. A relationship between index of refraction and wavelength can be obtained by dividing these two equations and mak-

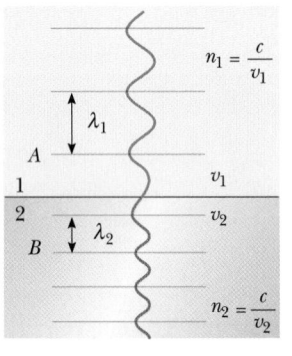

Figure 22.9 As the wave moves from medium 1 to medium 2, its wavelength changes but its frequency remains constant.

TABLE 22.1   Indices of Refraction for Various Substances, Measured with Light of Vacuum Wavelength $\lambda_0 = 589$ nm

Substance	Index of Refraction	Substance	Index of Refraction
Solids at 20°C		**Liquids at 20°C**	
Diamond (C)	2.419	Benzene	1.501
Fluorite (CaF$_2$)	1.434	Carbon disulfide	1.628
Fused quartz (SiO$_2$)	1.458	Carbon tetrachloride	1.461
Glass, crown	1.52	Ethyl alcohol	1.361
Glass, flint	1.66	Glycerine	1.473
Ice (H$_2$O) (at 0°C)	1.309	Water	1.333
Polystyrene	1.49		
Sodium chloride (NaCl)	1.544	**Gases at 0°C, 1 atm**	
Zircon	1.923	Air	1.000 293
		Carbon dioxide	1.000 45

ing use of the definition of index of refraction provided by Equation 22.4:

$$\frac{\lambda_1}{\lambda_2} = \frac{v_1}{v_2} = \frac{c/n_1}{c/n_2} = \frac{n_2}{n_1} \qquad [22.5]$$

which gives

$$\lambda_1 n_1 = \lambda_2 n_2 \qquad [22.6]$$

Let medium 1 be the vacuum so that $n_1 = 1$. From Equation 22.6, it follows that the index of refraction of any medium can be expressed as the ratio

$$n = \frac{\lambda_0}{\lambda_n} \qquad [22.7]$$

where λ_0 is the wavelength of light in vacuum and λ_n is the wavelength in a medium with index of refraction n. Figure 22.10 is a schematic representation of this reduction in wavelength when light passes from vacuum into a transparent medium.

We are now in a position to express Snell's law (Eq. 22.3) in an alternative form. If we substitute Equation 22.5 into Equation 22.3, we get

$$n_1 \sin \theta_1 = n_2 \sin \theta_2 \qquad [22.8]$$

◀ Snell's law

This is the most widely used and practical form of Snell's law.

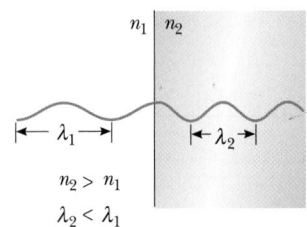

$n_2 > n_1$

$\lambda_2 < \lambda_1$

Figure 22.10 A schematic diagram of the *reduction* in wavelength when light travels from a medium with a low index of refraction to one with a higher index of refraction.

Applying Physics 1

When one looks at the stars on a clear night, they appear to be twinkling. Astronauts observing the stars from an orbiting space shuttle, however, will not see such twinkling. Why not?

Explanation The twinkling effect is due to refraction of light from a star as it passes through the Earth's atmosphere. Atmospheric turbulence causes rapid variation in local air density, and a change in air density results in a change in its index of refraction. As the light passes through regions of air whose index of refraction varies rapidly, it experiences many small shifts in its direction of travel. As an Earthbound observer views light from the star, the rapid changes in index of refraction cause the light to enter the eyes from slightly varying directions, resulting in the twinkling effect.

$Q\ U\ I\ C\ K\ L\ A\ B$

Fill a clear glass tumbler with water and place a pencil (or straw) into the tumbler as in the figure. Now observe the pencil from the side at an angle of about 45° to the surface, and note that the line of the portion of the pencil under water is not parallel with the line of the portion in air. That is, the pencil appears to be bent at the point where it enters the water. This is due to the refraction of light as it travels from water to air.

EXAMPLE 22.3 An Index of Refraction Measurement

A beam of light of wavelength 550 nm, traveling in air, is incident on a slab of transparent material. The incident beam makes an angle of 40.0° with the normal, and the refracted beam makes an angle of 26.0° with the normal. Find the index of refraction of the material.

Solution Snell's law of refraction (Eq. 22.8), together with the given data— $\theta_1 = 40.0°$, $n_1 = 1.00$ for air, and $\theta_2 = 26.0°$—gives

$$n_1 \sin \theta_1 = n_2 \sin \theta_2$$

$$n_2 = \frac{n_1 \sin \theta_1}{\sin \theta_2} = (1.00) \frac{\sin 40.0°}{\sin 26.0°} = \frac{0.643}{0.438} = 1.47$$

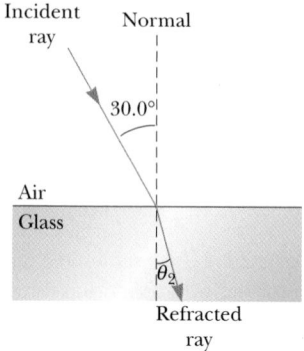

Figure 22.11 (Example 22.4) Refraction of light by glass.

Workbook Problem 89 — (Workbook page 189) is similar to Example 22.4.

Q U I C K L A B

Tape a coin to the bottom of a large opaque bowl as in figure (a) below. Look at the coin from the side and move backwards until you can no longer see the coin. Remain still at that position, and have a friend pour water into the bowl until it is full as in figure (b). Note that you are now able to see the coin because the light is refracted at the water-air interface.

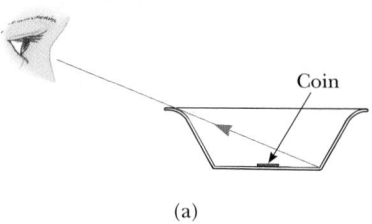

(a)

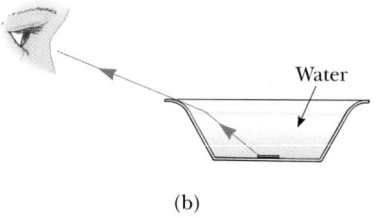

(b)

If we compare this value with the data in Table 22.1, we see that the material could be fused quartz.

Exercise   What is the wavelength of this light in the slab?

Answer   374 nm

EXAMPLE 22.4   Angle of Refraction for Glass

A light ray of wavelength 589 nm (produced by a sodium lamp) traveling through air is incident on a smooth, flat slab of crown glass at an angle of 30.0° to the normal, as sketched in Figure 22.11. Find the angle of refraction, θ_2.

Solution   Snell's law (Eq. 22.8) can be rearranged as

$$\sin \theta_2 = \frac{n_1}{n_2} \sin \theta_1$$

From Table 22.1, we find that $n_1 = 1.00$ for air and $n_2 = 1.52$ for crown glass. Therefore, the unknown refraction angle is determined by

$$\sin \theta_2 = \left(\frac{1.00}{1.52}\right)(\sin 30.0°) = 0.329$$

$$\theta_2 = \sin^{-1}(0.329) = \boxed{19.2°}$$

We see that the ray is bent *toward* the normal, as expected.

Exercise   If the light ray moves from inside the glass toward the glass-air interface at an angle of 30.0° to the normal, determine the angle of refraction.

Answer   49.5° *away* from the normal

EXAMPLE 22.5   The Speed of Light in Fused Quartz

Light of wavelength 589 nm in vacuum passes through a piece of fused quartz of index of refraction $n = 1.458$.

(a) Find the speed of light in fused quartz.

Solution   The speed of light in fused quartz can be obtained from Equation 22.4:

$$v = \frac{c}{n} = \frac{3.00 \times 10^8 \text{ m/s}}{1.458} = \boxed{2.06 \times 10^8 \text{ m/s}}$$

It is interesting to note that the speed of light in vacuum, 3.00×10^8 m/s, is an upper limit for the speed of material objects. In our treatment of relativity in Chapter 26, we shall find that this upper limit is consistent with experimental observations. However, it is possible for a particle moving in a medium to have a speed that exceeds the speed of light in that medium. For example, it is theoretically possible for a particle to travel through fused quartz at a speed greater than 2.06×10^8 m/s, but it must have a speed less than 3.00×10^8 m/s in a vacuum.

(b) What is the wavelength of this light in fused quartz?

Solution We can use $\lambda_n = \lambda_0/n$ (Eq. 22.7) to calculate the wavelength in fused quartz, noting that we are given $\lambda_0 = 589$ nm $= 589 \times 10^{-9}$ m:

$$\lambda_n = \frac{\lambda_0}{n} = \frac{589 \text{ nm}}{1.458} = \boxed{404 \text{ nm}}$$

Exercise Find the frequency of the light passing through the fused quartz.

Answer 5.09×10^{14} Hz

EXAMPLE 22.6 Light Passing Through a Slab

A light beam traveling through a transparent medium of index of refraction n_1 passes through a thick transparent slab with parallel faces and index of refraction n_2 (Fig. 22.12). Show that the emerging beam is parallel to the incident beam.

Reasoning To solve this problem, it is necessary to apply Snell's law twice, once at the upper surface and once at the lower surface. The two equations will be related because the angle of refraction at the upper surface equals the angle of incidence at the lower surface. The ray passing through the slab makes equal angles with the normals at the entry and exit points. This procedure will enable us to compare angles θ_1 and θ_3.

Solution First, let us apply Snell's law to the upper surface:

$$(1) \qquad \sin \theta_2 = \frac{n_1}{n_2} \sin \theta_1$$

Applying Snell's law to the lower surface gives

$$(2) \qquad \sin \theta_3 = \frac{n_2}{n_1} \sin \theta_2$$

Substituting (1) into (2) gives

$$\sin \theta_3 = \frac{n_2}{n_1} \left(\frac{n_1}{n_2} \sin \theta_1 \right) = \sin \theta_1$$

That is, $\theta_3 = \theta_1$, and so the slab does not alter the direction of the beam. It does, however, displace the beam. The same result is obtained when light passes through multiple layers of materials.

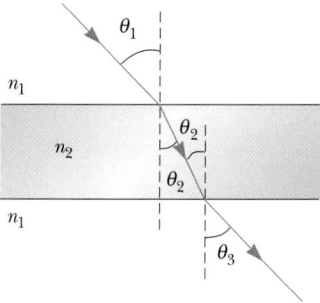

Figure 22.12 (Example 22.6) When light passes through a flat slab of material, the emerging beam is parallel to the incident beam, and therefore $\theta_1 = \theta_3$.

Workbook Problem 90— (Workbook page 190) also deals with the displacement of a light beam by a transparent block.

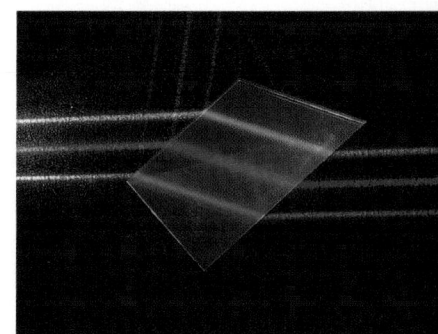

A transparent block has refracted three incident light beams, which exit in parallel but are offset. Note the partial reflection of both interfaces. (*©Leonard Lessin/Peter Arnold, Inc.*)

22.6 DISPERSION AND PRISMS

An important property of the index of refraction is that its value in anything but vacuum depends on the wavelength of light. This phenomenon is called **dispersion** (Fig. 22.13). Because n is a function of wavelength, Snell's law indicates that **light of different wavelengths is bent at different angles when incident on a refracting material.** As seen in Figure 22.13, the index of refraction generally decreases with increasing wavelength. This means that blue light ($\lambda \cong 470$ nm) bends more than red light ($\lambda \cong 650$ nm) when passing into a refracting material.

14.2, SECTION 2

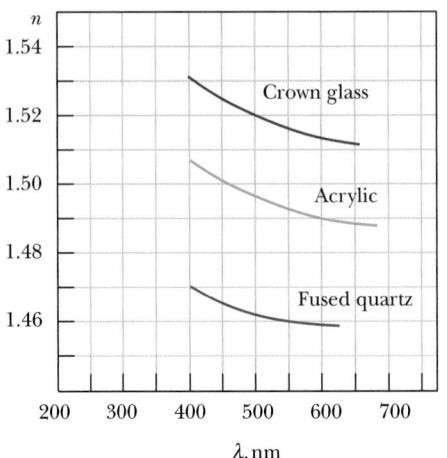

Figure 22.13   Variations of index of refraction in the visible spectrum with respect to vacuum wavelength for three materials.

To understand how dispersion can affect light, let us consider what happens when light strikes a prism, as in Figure 22.14a. A single ray of light that is incident on the prism from the left emerges bent away from its original direction of travel by an angle of δ, called the **angle of deviation.** Now suppose a beam of white light (a combination of all visible wavelengths) is incident on a prism, as in Figure 22.14b. Because of dispersion, the blue component of the incident beam is bent more than the red component, and the rays that emerge from the second face of the prism fan out in a series of colors known as a visible **spectrum,** as shown in Figure 22.15. These colors, in order of decreasing wavelength, are red, orange, yellow, green, blue, and violet. Clearly, the angle of deviation, δ, depends on wavelength. Violet light ($\lambda \cong 400$ nm) deviates the most, red light ($\lambda \cong 650$ nm) deviates the least, and the remaining colors in the visible spectrum fall between these extremes.

Prisms are often used in an instrument known as a **prism spectrometer,** the essential elements of which are shown in Figure 22.16a. This instrument is com-

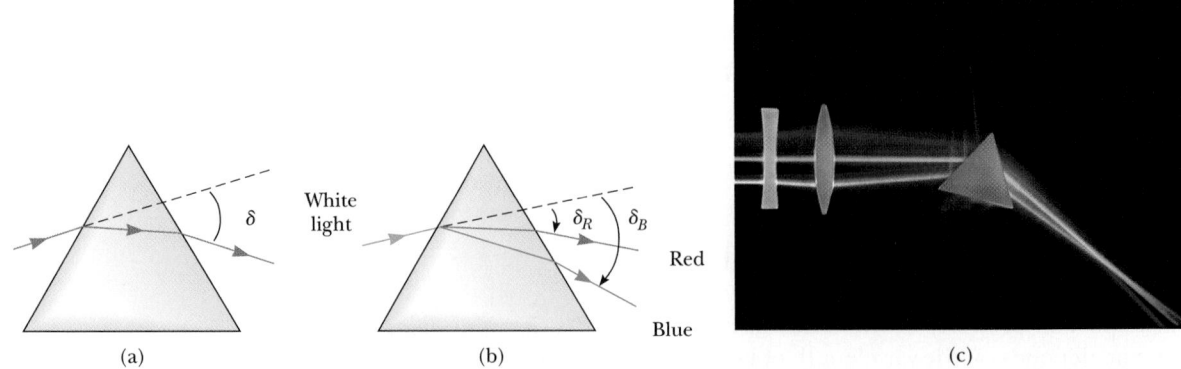

Figure 22.14   (a) A prism refracts a light ray and deviates the light through the angle δ. (b) When light is incident on a prism, the blue light is bent more than the red. (c) Light of different colors passes through a prism and two lenses. Note that as the light passes through the prism, different wavelengths are refracted at different angles. *(David Parker/ SPL/Photo Researchers)*

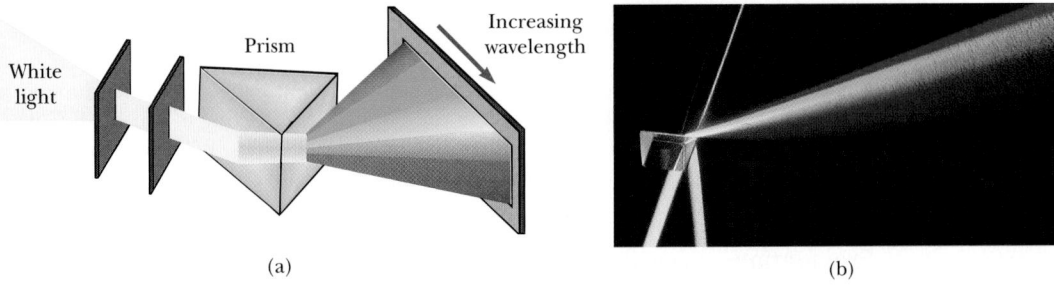

(a)   (b)

Figure 22.15 (a) Dispersion of white light by a prism. Because *n* varies with wavelength, the prism disperses the white light into its various spectral components. (b) Different colors of light that pass through a prism are refracted at different angles because the index of refraction of the glass depends on wavelength. The blue light bends the most; red light bends the least. *(Age/Peter Arnold, Inc.)*

monly used to study the wavelengths emitted by a light source, such as a sodium vapor lamp. Light from the source is sent through a narrow, adjustable slit and lens to produce a parallel, or collimated, beam. The light then passes through the prism and is dispersed into a spectrum. The refracted light is observed through a telescope. The experimenter sees an image of the slit through the eyepiece of the telescope. The telescope can be moved or the prism can be rotated in order to view the various wavelengths, which have different angles of deviation. Figure 22.16b shows the type of prism spectrometer used in undergraduate laboratories.

All hot, low-pressure gases emit their own characteristic spectra. Thus, one use of a prism spectrometer is to identify gases. For example, sodium emits only two wavelengths in the visible spectrum: two closely spaced yellow lines. A gas emitting these and only these colors can thus be identified as sodium. Likewise, mercury vapor has its own characteristic spectrum, consisting of four prominent wavelengths—orange, green, blue, and violet lines—along with some wavelengths of lower intensity. The particular wavelengths emitted by a gas serve as "fingerprints" of that gas.

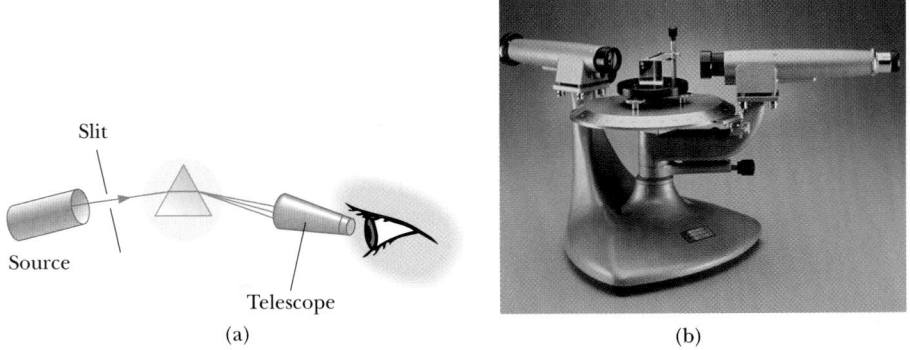

(a)   (b)

Figure 22.16 (a) A diagram of a prism spectrometer. The colors in the spectrum are viewed through a telescope. (b) A photograph of a prism spectrometer. *(Courtesy of Central Scientific Company)*

Tape a piece of black paper to the end of a flashlight, and cut a narrow slit in the middle of the paper as in the figure. Lean a flat mirror against one end of a tray partially filled with water. Shine your flashlight on that part of the mirror that is under water, and hold a sheet of white paper such that the reflected light shines on the paper. You should observe a spectrum of colors on the paper as the light is dispersed when it travels from air into water, and then from water into air. According to your observations, which color is bent the most? Which is bent the least?

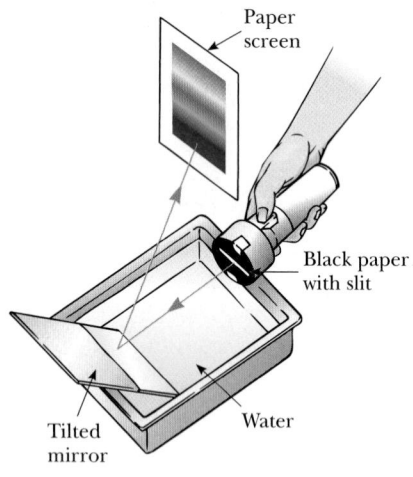

Paper screen

Black paper with slit

Tilted mirror

Water

Thinking Physics 3

When a beam of light enters a glass prism, which has nonparallel sides, the rainbow of color exiting the prism is a testimonial to the dispersion occurring in the glass. Suppose a beam of light enters a slab of material with parallel sides. When the beam exits the other side, traveling in the same direction as the original beam, is there any evidence of dispersion?

Explanation   Due to dispersion, light at the violet end of the spectrum will exhibit a larger angle of refraction on entering the glass than light at the red end. All colors of light will return to the original direction of propagation on refracting back out into the air. Thus, the outgoing beam will be white. But the net shift in the position of the violet light along the edge of the slab will be larger than the red light. Thus, one edge of the outgoing beam will have a bluish tinge to it (it will appear blue rather than violet, because the eye is not very sensitive to violet light), whereas the other edge will have a reddish tinge. This effect is indicated in Figure 22.17. The colored edges of the outgoing beam of white light are evidence of dispersion.

Figure 22.17   (Thinking Physics 3)

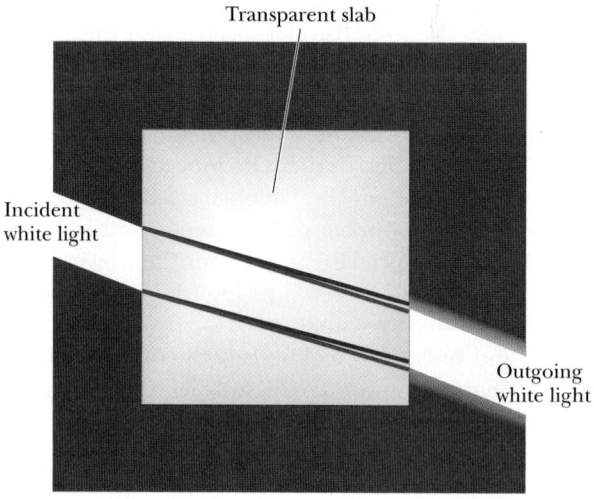

Transparent slab

Incident white light

Outgoing white light

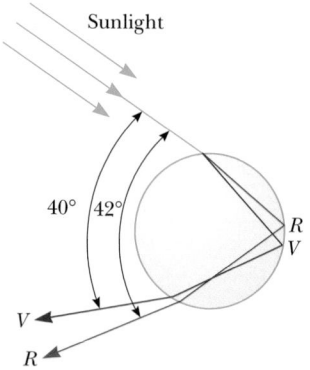

Sunlight

40° 42°

R
V

V

R

Figure 22.18   Refraction of sunlight by a spherical raindrop.

22.7   THE RAINBOW

The dispersion of light into a spectrum is demonstrated most vividly in nature through the formation of a rainbow, often seen by an observer positioned between the Sun and a rain shower. To understand how a rainbow is formed, consider Figure 22.18. A ray of light passing overhead strikes a drop of water in the atmosphere and is refracted and reflected as follows. It is first refracted at the front surface of the drop, with the violet light deviating the most and the red light the least. At the back surface of the drop, the light is reflected and returns to the front surface, where it again undergoes refraction as it moves from water into air. The rays leave the drop so that the angle between the incident white light and the returning violet ray is 40°, and the angle between the white light and the returning red ray is 42° (Fig.

(a)

(b)

Figure 22.19 (a) The formation of a rainbow. (b) Dramatic photograph of a rainbow over Niagara Falls in Ontario, Canada. (*John Edwards/Tony Stone Images*)

22.18). This small angular difference between the returning rays causes us to see the bow.

Now consider an observer viewing a rainbow, as in Figure 22.19a. If a raindrop high in the sky is being observed, the red light returning from the drop can reach the observer because it is deviated the most, but the violet light passes over the observer because it is deviated the least. Hence, the observer sees this drop as being red. Similarly, a drop lower in the sky would direct violet light toward the observer and appear to be violet. (The red light from this drop would strike the ground and not be seen.) The remaining colors of the spectrum would reach the observer from raindrops lying between these two extreme positions. Figure 22.19b shows a rainbow.

22.8   HUYGENS'S PRINCIPLE

The laws of reflection and refraction can be developed using a geometric method proposed by Huygens in 1678. Huygens assumed that light is some form of wave motion rather than a stream of particles. He had no knowledge of the nature of light or of its electromagnetic character. Nevertheless, his simplified wave model is adequate for understanding many practical aspects of the propagation of light.

Huygens's principle ▶

Huygens's principle is a geometric construction for determining at some instant the position of a new wave front from the knowledge of the wave front that preceded it. (A wave front is a surface passing through those points of a wave that have the same phase and amplitude. For instance, a wave front could be a surface passing through the crests of waves.) In Huygens's construction, **all points on a given wave front are taken as point sources for the production of spherical secondary waves, called wavelets, which propagate outward with speeds characteristic of waves in that medium. After some time has elapsed, the new position of the wave front is the surface tangent to the wavelets.**

Figure 22.20 illustrates two simple examples of Huygens's construction. First, consider a plane wave moving through free space, as in Figure 22.20a. At $t = 0$, the wave front is indicated by the plane labeled AA'. In Huygens's construction, each point on this wave front is considered a point source. For clarity, only a few points on AA' are shown. With these points as sources for the wavelets, we draw circles each of radius $c \Delta t$, where c is the speed of light in vacuum and Δt is the period of propagation from one wave front to the next. The surface drawn tangent to these wavelets is the plane BB', which is parallel to AA'. In a similar manner, Figure 22.20b shows Huygens's construction for an outgoing spherical wave.

A convincing demonstration of Huygens's principle is performed with water waves in a shallow tank (called a ripple tank), as in Figure 22.21. Plane waves at the

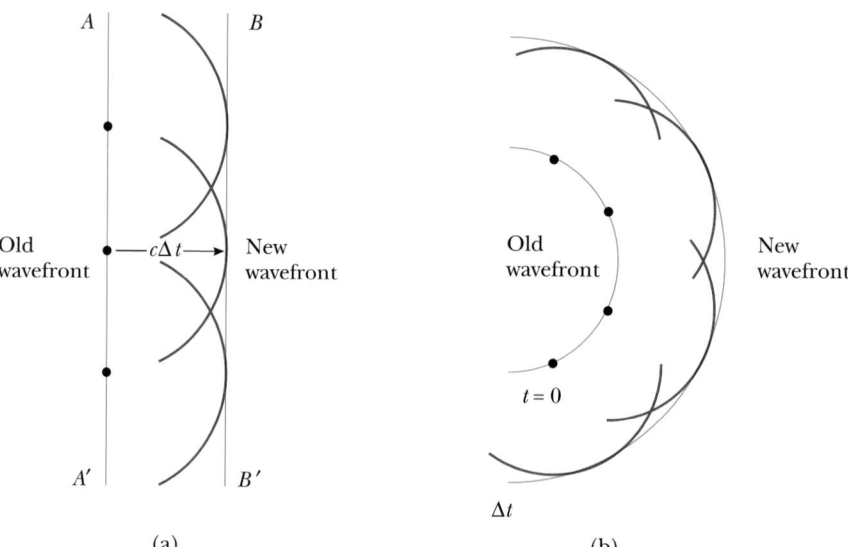

Figure 22.20 Huygens's constructions for (a) a plane wave propagating to the right and (b) a spherical wave.

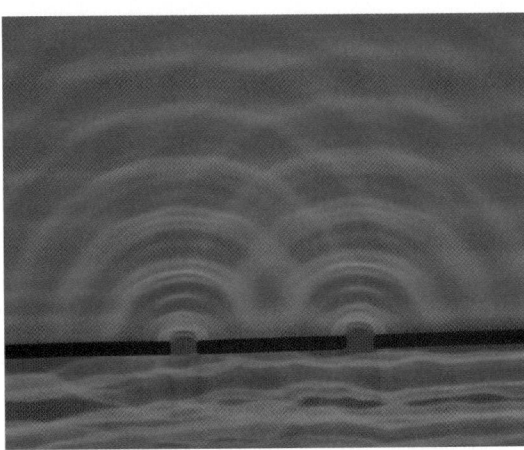

Figure 22.21 Water waves in a ripple tank demonstrate Huygens's wavelets. A plane wave at the bottom is incident on a barrier with two small openings. Each opening acts as a source of circular wavelets. *(Erich Schrempp/Photo Researchers)*

bottom are incident on a barrier that contains a small opening. The opening acts as a source of two-dimensional circular waves propagating outward.

Huygens's Principle Applied to Reflection and Refraction

The laws of reflection and refraction were stated earlier in this chapter without proof. We shall now derive these laws using Huygens's principle. For the law of reflection, refer to Figure 22.22a. The line AA' represents a wave front of the incident light. As ray 3 travels from A' to C, ray 1 reflects from A and produces a spherical wavelet of radius AD. (Recall that the radius of a Huygens wavelet is vt.) Because the two wavelets having radii $A'C$ and AD are in the same medium, they have the same speed, v, and thus $AD = A'C$. Meanwhile, the spherical wavelet centered at B has spread only half as far as the one centered at A, because ray 2 strikes the surface later than ray 1.

From Huygens's principle, we find that the reflected wave front is CD, a line tangent to all the outgoing spherical wavelets. The remainder of our analysis depends on geometry, as summarized in Figure 22.22b. Note that the right triangles ADC and $AA'C$ are congruent because they have the same hypotenuse, AC, and

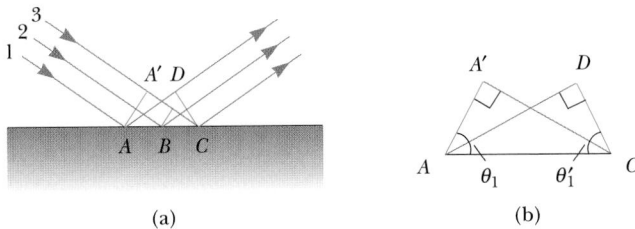

(a)   (b)

Figure 22.22 (a) Huygens's construction for proving the law of reflection. (b) Triangle ADC is identical to triangle $AA'C$.

because $AD = A'C$. From Figure 22.22b we have

$$\sin \theta_1 = \frac{A'C}{AC} \quad \text{and} \quad \sin \theta_1' = \frac{AD}{AC}$$

Thus,

$$\sin \theta = \sin \theta_1'$$

$$\theta_1 = \theta_1'$$

which is the law of reflection.

Now let us use Huygens's principle and Figure 22.23a to derive Snell's law of refraction. Note that in the time interval Δt, ray 1 moves from A to B and ray 2 moves from A' to C. The radius of the outgoing spherical wavelet centered at A is equal to $v_2 \Delta t$. The distance $A'C$ is equal to $v_1 \Delta t$. Geometric considerations show that angle $A'AC$ equals θ_1 and angle ACB equals θ_2. From triangles $AA'C$ and ACB, we find that

$$\sin \theta_1 = \frac{v_1 \Delta t}{AC} \quad \text{and} \quad \sin \theta_2 = \frac{v_2 \Delta t}{AC}$$

If we divide these two equations, we get

$$\frac{\sin \theta_1}{\sin \theta_2} = \frac{v_1}{v_2}$$

But from Equation 22.4 we know that $v_1 = c/n_1$ and $v_2 = c/n_2$. Therefore,

$$\frac{\sin \theta_1}{\sin \theta_2} = \frac{c/n_1}{c/n_2} = \frac{n_2}{n_1}$$

$$n_1 \sin \theta_1 = n_2 \sin \theta_2$$

which is the law of refraction.

A mechanical analog of refraction is shown in Figure 22.23b. The wheels change direction as they move from a concrete surface to a grass surface.

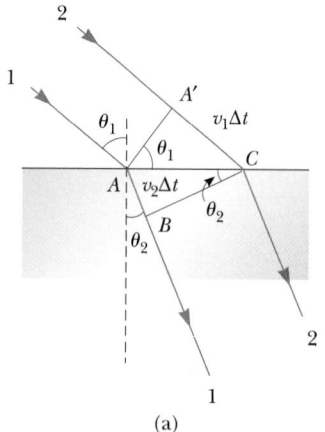
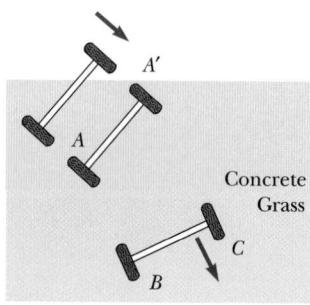

Figure 22.23 (a) Huygens's construction for proving the law of refraction. (b) A mechanical analog of refraction.

(a)

(b)

22.9 TOTAL INTERNAL REFLECTION

An interesting effect called *total internal reflection* can occur when light attempts to move from a medium with a *high* index of refraction to one with a *lower* index of refraction. Consider a light beam traveling in medium 1 and meeting the boundary between medium 1 and medium 2, where n_1 is greater than n_2 (Fig. 22.24). Possible directions of the beam are indicated by rays 1 through 5. Note that the refracted rays are bent away from the normal because n_1 is greater than n_2. At some particular angle of incidence, θ_c, called the **critical angle,** the refracted light ray moves parallel to the boundary so that $\theta_2 = 90°$ (Fig. 22.24b). *For angles of incidence greater than θ_c, the beam is entirely reflected at the boundary, as is ray 5 in Figure 22.24a. This ray is reflected at the boundary as though it had struck a perfectly reflecting surface. It and all rays like it obey the law of reflection; that is, the angle of incidence equals the angle of reflection.

We can use Snell's law to find the critical angle. When $\theta_1 = \theta_c$, $\theta_2 = 90°$ and Snell's law (Eq. 22.8) gives

$$n_1 \sin \theta_c = n_2 \sin 90° = n_2$$

$$\sin \theta_c = \frac{n_2}{n_1} \qquad \text{for } n_1 > n_2 \qquad \text{[22.9]}$$

Note that this equation can be used only when n_1 is greater than n_2. That is, **total internal reflection occurs only when light attempts to move from a medium of high index of refraction to a medium of lower index of refraction.** If n_1 were less than n_2, Equation 22.9 would give $\sin \theta_c > 1$, which is an absurd result because the sine of an angle can never be greater than unity.

When medium 2 is air, the critical angle is small for substances with large indices of refraction, such as diamond, where $n = 2.42$ and $\theta_c = 24.0°$. By comparison, for crown glass, $n = 1.52$ and $\theta_c = 41.0°$. This property, combined with proper faceting, causes diamonds to sparkle brilliantly.

This photograph shows nonparallel light rays entering a glass prism. The bottom two rays undergo total internal reflection at the longest side of the prism. The top three rays are refracted at the longest side as they leave the prism. *(Courtesy of Henry Leap and Jim Lehman)*

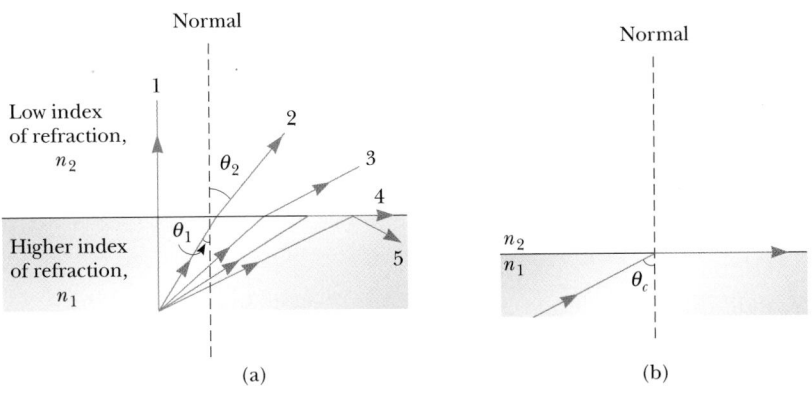

(a) (b)

Figure 22.24 (a) Rays from a medium with index of refraction n_1 travel to a medium with index of refraction n_2, where $n_1 > n_2$. As the angle of incidence increases, the angle of refraction θ_2 increases until θ_2 is 90° (ray 4). For even larger angles of incidence, total internal reflection occurs (ray 5). (b) The angle of incidence producing a 90° angle of refraction is often called the *critical angle, θ_c.*

Figure 22.25   Internal reflection in a prism. (a) The ray is deviated by 90°. (b) The direction of the ray is reversed. (c) Two prisms used as a periscope.

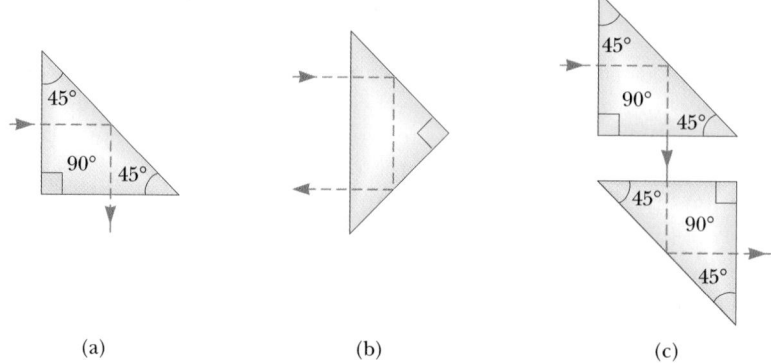

(a)   (b)   (c)

One can use a prism and the phenomenon of total internal reflection to alter the direction of travel of a light beam. Figure 22.25 illustrates two such possibilities. In one case the light beam is deflected by 90.0° (Fig. 22.25a), and in the second case the path of the beam is reversed (Fig. 22.25b). A common application of total internal reflection is a submarine periscope. In this device, two prisms are arranged, as in Figure 22.25c, so that an incident beam of light follows the path shown and the user can "see around corners."

APPLICATION

Submarine Periscopes.

Thinking Physics 4

A beam of white light is incident on the curved edge of a semicircular piece of glass, as shown in Figure 22.26. The light enters the curved surface along the normal, so it shows no refraction. It encounters the straight side at the center of curvature of the curved side and refracts into the air. The incoming beam is moved clockwise (so that the angle θ increases) such that the beam always enters along the normal to the curved side and encounters the straight side at the center of curvature of the curved side. As the refracted beam approaches a direction parallel to the straight side, it becomes redder. Why?

Explanation   When the outgoing beam approaches the direction parallel to the straight side, the incident angle is approaching the critical angle for total internal reflection. Dispersion occurs as the light passes out of the glass. The index of refraction for light at the violet end of the visible spectrum is larger than at the red end. Thus, as the outgoing beam approaches the straight side, the violet light experiences total internal reflection first, followed by the other colors. The red light is the last to experience total internal reflection, so just before the outgoing light disappears, it is composed of light from the red end of the visible spectrum.

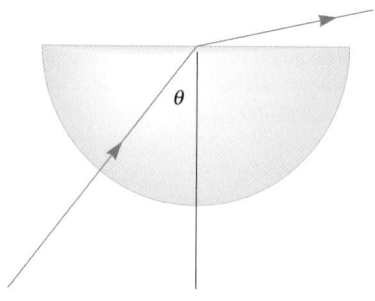

Figure 22.26   (Thinking Physics 4)

EXAMPLE 22.7   A View from the Fish's Eye

(a) Find the critical angle for a water-air boundary if the index of refraction of water is 1.33.

Solution   Applying Equation 22.9, we find the critical angle to be

$$\sin \theta_c = \frac{n_2}{n_1} = \frac{1.00}{1.33} = 0.752$$

$$\theta_c = \boxed{48.8°}$$

(b) Use the results of (a) to predict what a fish will see if it looks up toward the water surface at angles of 40.0°, 48.8°, and 60.0°.

Reasoning and Solution Because the path of a light ray is reversible, the fish can see out of the water if it looks toward the surface at an angle less than the critical angle. Thus, at 40.0°, the fish can see into the air above the water. At an angle of 48.8°, the critical angle for water, the light that reaches the fish has to skim along the water surface before being refracted to the fish's eye. At angles greater than the critical angle, the light reaching the fish comes via internal reflection at the surface. Thus, at 60.0°, the fish sees a reflection of some object on the bottom of the pool.

Strands of glass optical fibers are used to carry voice, video, and data signals in telecommunication networks. Typical fibers have diameters of 60 μm. (© *Richard Megna 1983, Fundamental Photographs*)

Application: Fiber Optics

Another interesting application of total internal reflection is the use of solid glass or transparent plastic rods to "pipe" light from one place to another. As indicated in Figure 22.27, light is confined to traveling within the rods, even around gentle curves, as a result of successive internal reflections. Such a light pipe can be flexible if thin fibers are used rather than thick rods. If a bundle of parallel fibers is used to construct an optical transmission line, images can be transferred from one point to another.

This technique is used in an industry known as *fiber optics*. Very little light intensity is lost in these fibers as a result of reflections on the sides. Any loss of intensity is due essentially to reflections from the two ends and absorption by the fiber material. Fiber-optic devices are particularly useful for viewing images produced at inaccessible locations. Physicians often use fiber-optic cables to aid in the diagnosis and correction of certain medical problems without the intrusion of major surgery. For example, a fiber-optic cable can be threaded through the esophagus and into the stomach to look for ulcers. In this application, the cable actually consists of two fiber-optic lines, one to transmit a beam of light into the stomach for illumination and the other to allow this light to be transmitted out of the stomach. The resulting image can, in some cases, be viewed directly by the physician but most often is displayed on a television monitor or captured on film. In a similar fashion, the cables can be used to examine the colon or to do repair work without the need for large incisions. Damaged knees and other joints can sometimes be repaired using a process called arthroscopic surgery. In this technique, a small incision is made into the joint. Repair is accomplished by inserting a small fiber-optic cable through the cut to provide illumination and then trimming cartilage or damaged tissue with a small knife at the end of a second cable. The field of fiber optics is also finding increasing use in telecommunications. The fibers can carry much higher volumes of telephone calls and other forms of communication than electrical wires because of the higher frequency of light.

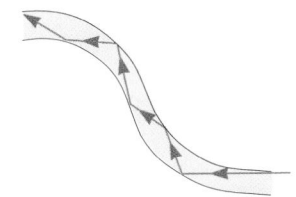

Figure 22.27 Light travels in a curved transparent rod by multiple internal reflections.

APPLICATION

Fiber Optics in Medical Diagnosis and Surgery.

APPLICATION

Fiber Optics in Telecommunications.

Workbook Problem 91 — (Workbook page 191) deals with total internal reflection within a fiber optic cable.

Applying Physics 2

An optical fiber consists of a transparent core, surrounded by cladding, which is a material with a lower index of refraction than the core. There is a cone of angles, called the acceptance cone, at the entrance to the fiber. Incoming light at angles within this cone will be transmitted through the fiber, whereas light entering the core from angles outside the cone will not be transmitted. Figure 22.28 shows a light ray entering the fiber just within the acceptance cone and experiencing total internal reflection at the interface between the core and the cladding. If it is technologically difficult to produce light entering the fiber from a small range of angles, how could you adjust the indices of refraction of the core and cladding to increase the size of the acceptance cone—would you design them to be farther apart or closer together?

Explanation   The acceptance cone would become larger if the critical angle (θ_c in the diagram) could be made smaller. This can be done by making the index of refraction of the cladding material smaller, so that the indices of refraction of the core and cladding material would be farther apart.

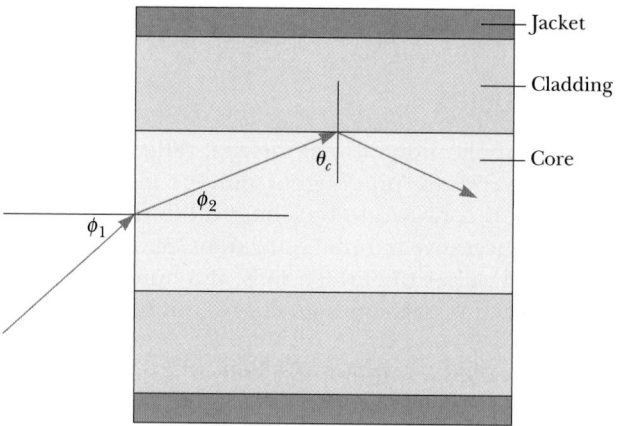

Figure 22.28   (Applying Physics 2)

SUMMARY

The **index of refraction** of a material, n, is defined as

$$n \equiv \frac{c}{v} \qquad [22.4]$$

where c is the speed of light in a vacuum and v is the speed of light in the material. The index of refraction of a material is also

$$n = \frac{\lambda_0}{\lambda_n} \qquad [22.7]$$

where λ_0 is the wavelength of the light in vacuum and λ_n is its wavelength in the material.

The **law of reflection** states that a wave reflects from a surface so that the *angle of reflection*, θ_1', equals the *angle of incidence*, θ_1.

The **law of refraction,** or **Snell's law,** states that

$$n_1 \sin \theta_1 = n_2 \sin \theta_2 \qquad \text{[22.8]}$$

Huygens's principle states that all points on a wave front are point sources for the production of spherical secondary waves called *wavelets*. These wavelets propagate outward at a speed characteristic of waves in a particular medium. After some time has elapsed, the new position of the wave front is the surface tangent to the wavelets.

Total internal reflection can occur when light attempts to move from a material with a high index of refraction to one with a lower index of refraction. The *maximum angle of incidence*, θ_c, for which light can move from a medium with index n_1 into a medium with index n_2, where n_1 is greater than n_2, is called the **critical angle** and is given by

$$\sin \theta_c = \frac{n_2}{n_1} \qquad \text{for } n_1 > n_2 \qquad \text{[22.9]}$$

MULTIPLE-CHOICE QUESTIONS

1. A retro-reflector sends a beam of light back in the direction from which it comes. At what angle should two touching mirrors be placed to form a retro-reflector? (a) 30° (b) 45° (c) 60° (d) 90° (e) 135°

2. How many 800-nm photons does it take to have the same total energy as four 200-nm photons? (a) 1 (b) 2 (c) 4 (d) 8 (e) 16

3. Carbon disulfide ($n = 1.63$) is poured into a container made of crown glass ($n = 1.52$). What is the critical angle for internal reflection of a ray in the liquid when it is incident on the liquid-to-glass surface? (a) 89° (b) 69° (c) 21° (d) 4.0°

4. A monochromatic light source emits a wavelength of 490 nm in air. When passing through a liquid, the wavelength reduces to 429 nm. What is the liquid's index of refraction? (a) 1.26 (b) 1.49 (c) 1.14 (d) 1.33

5. A ray of light is incident on the mid-point of a glass prism surface at an angle of 25.0° with the normal. For the glass, $n = 1.55$, and the prism apex angle is 30.0°. What is the angle of incidence at the glass-to-air surface on the side opposite where the ray entered the prism? (a) 14.2° (b) 22.1° (c) 28.2° (d) 46.0°

CONCEPTUAL QUESTIONS

1. The color of an object is said to depend on wavelength. So if you view colored objects under water, in which the wavelength of the light will be different, does the color change?

2. How is it possible that a complete circle of a rainbow can sometimes be seen from an airplane?

3. As light from the Sun passes through the atmosphere, it refracts, due to the small, but nonzero, index of refraction of air. The optical length of the day is defined as the time interval between the instant when the top of the Sun just appears above the horizon and the instant when the top of the Sun just disappears below the horizon. The geometric length of the day is defined as the time interval between the instant when a geometric straight line drawn from the observer to the top of the Sun just clears the horizon to the instant at which this line just dips below the horizon. Which is longer, the optical day or the geometric day?

4. Why does the arc of a rainbow appear with red on top and violet on the bottom?

5. If a beam of light with a given cross-section enters a new medium, the cross-section of the refracted beam is dif-

ferent from the incident beam. Is it larger or smaller, or is there no definite direction to the change?

6. Under what conditions is a mirage formed? On a hot day, what are we seeing when we observe a mirage water puddle on the road?

7. In dispersive materials, the angle of refraction for a light ray depends on the wavelength of the light. Does the angle of reflection from the surface of the material depend on the wavelength? Why or why not?

8. As light travels from vacuum ($n = 1$) to a medium such as glass ($n > 1$), does its wavelength change? Does its frequency change? Does its speed change?

9. Explain why a diamond loses most of its sparkle when submerged in carbon disulfide.

10. Suppose you are told that only two colors of light (X and Y) are sent through a glass prism and that X is bent more than Y. Which color travels more slowly in the prism?

11. The level of water in a clear, colorless glass is easily observed with the naked eye. The level of liquid helium in a clear glass vessel is extremely difficult to see with the naked eye. Explain. (*Hint:* The index of refraction of liquid helium is close to that of air.)

12. Is it possible to have total internal reflection for light incident from air on water? Explain.

13. Why does a diamond show flashes of color when observed under white light?

14. Explain why an oar partially in water appears to be bent.

15. Why do astronomers looking at distant galaxies talk about looking backward in time?

16. A solar eclipse occurs when the Moon gets between the Earth and the Sun. Use a diagram to show why some areas of the Earth see a total eclipse, other areas see a partial eclipse, and most areas see no eclipse.

17. Under certain circumstances, sound can be heard from extremely far away. This frequently happens over a body of water, where the air near the water surface is cooler than the air at higher altitudes. Explain how the refraction of sound waves could increase the distance over which sound can be heard.

PROBLEMS

1, **2**, **3** = straightforward, intermediate, challenging   ☐ = full solution available in Study Guide/Student Solutions Manual   ◤ = Core Concepts Workbook
WEB = solution posted at **http://www.harcourtcollege.com/physics/cptech**   ▩ = biomedical application   ▦ = Interactive Physics

Section 22.1   The Nature of Light

Section 22.2   Measurements of the Speed of Light

1. During the Apollo XI Moon landing, a highly reflecting screen was erected on the Moon's surface. When the Moon is directly overhead, the speed of light may be found by measuring the time it takes a laser beam to travel from Earth, reflect from the screen, and return to Earth. If this interval is measured to be 2.51 s, what is the measured speed of light? Take the center-to-center distance from Earth to Moon to be 3.84×10^8 m, and do not neglect the sizes of the Earth and Moon.

2. The Fizeau experiment is performed so that the round-trip distance for the light is 40 m. **(a)** Find the two lowest speeds of rotation that allow the light to pass through the notches. Assume that the wheel has 360 teeth and that the speed of light is 3.00×10^8 m/s. **(b)** Repeat for a round-trip distance of 4000 m.

WEB **3.** Albert A. Michelson very carefully measured the speed of light using an alternative version of the technique developed by Fizeau. Figure P22.3 shows the approach he used. Light was reflected from one face of a rotating eight-sided mirror toward a stationary mirror 35.0 km away. At certain rates of rotation, the returning beam of

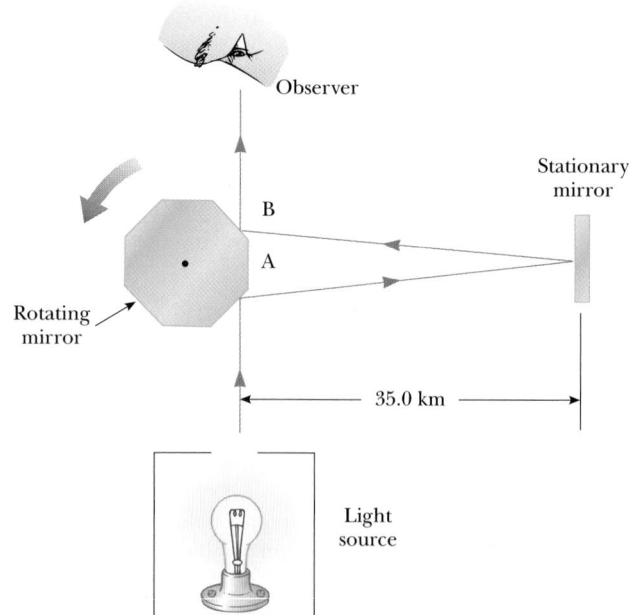

Figure P22.3

light was directed toward the eye of an observer as shown. **(a)** What minimum angular speed must the rotating mirror have in order that side A will have rotated to position B, causing the light to be reflected to the eye? **(b)** What is the next highest angular speed that will enable the source of light to be seen?

4. Figure P22.4 shows an apparatus used to measure the speed distribution of gas molecules. It consists of two slotted rotating disks separated by a distance s, with the slots displaced by the angle θ. Suppose the speed of light is measured by sending a light beam toward the right disk of this apparatus. **(a)** Show that a light beam will be seen in the detector (that is, will make it through both slots) only if its speed is given by $c = s\omega/\theta$, where ω is the angular speed of the disks and θ is measured in radians. **(b)** What is the measured speed of light if the distance between the two slotted rotating disks is 2.500 m, the slot in the second disk is displaced $\frac{1}{60}$ of one degree from the slot in the first disk, and the disks are rotating at 5555 rev/s?

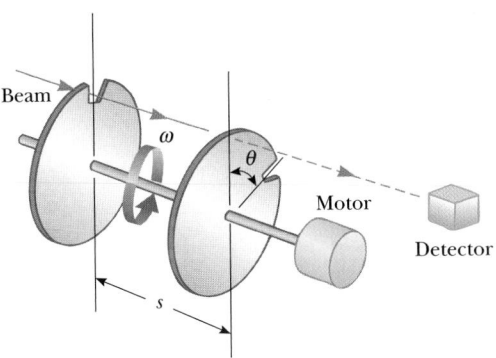

Figure P22.4

Section 22.4   Reflection and Refraction

Section 22.5   The Law of Refraction

5. The angle between the two mirrors in Figure P22.5 is a right angle. The beam of light in the vertical plane P strikes mirror 1 as shown. **(a)** Determine the distance the reflected light beam travels before striking mirror 2. **(b)** In what direction does the light beam travel after being reflected from mirror 2?

6. Light is incident normally on a 1.00-cm layer of water that lies on top of a flat Lucite plate with a thickness of 0.500 cm. How much more time is required for light to pass through this double layer than is required to traverse the same distance in air? ($n_{Lucite} = 1.59$)

7. Find the speeds of light in **(a)** flint glass, **(b)** water, and **(c)** zircon.

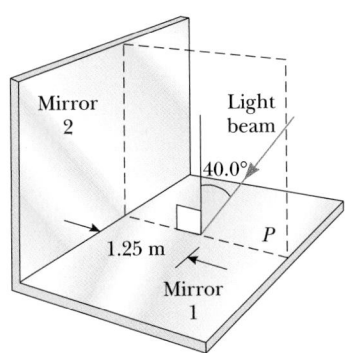

Figure P22.5

8. Light of wavelength λ_0 in a vacuum has a wavelength of 438 nm in water and a wavelength of 390 nm in benzene. **(a)** What is the wavelength λ_0 of this light in a vacuum? **(b)** Using only the given wavelengths, determine the ratio of the index of refraction of benzene to that of water.

9. Light of wavelength 436 nm in air enters a fishbowl filled with water, then exits through the crown-glass wall of the container. Find the wavelengths of the light **(a)** in the water and **(b)** in the glass.

10. A beam of light enters a layer of water at an angle of 36° with the vertical. What is the angle between the refracted ray and the vertical?

11. A ray of light is incident on the surface of a block of clear ice at an angle of 40.0° with the normal. Part of the light is reflected and part is refracted. Find the angle between the reflected and refracted light.

12. A narrow beam of sodium yellow light ($\lambda_0 = 589$ nm) is incident from air on a smooth surface of water at an angle of $\theta_1 = 35.0°$. Determine the angle of refraction, θ_2, and the wavelength of the light in water.

13. A beam of light, traveling in air, strikes the surface of mineral oil at an angle of 23.1° with the normal to the surface. If the light travels at 2.17×10^8 m/s through the oil, what is the angle of refraction?

14. A flashlight on the bottom of a 4.00-m-deep swimming pool sends a ray upward and at an angle so that the ray strikes the surface of the water 2.00 m from the point directly above the flashlight. What angle (in air) does the emerging ray make with the water's surface?

15. The laws of refraction and reflection are the same for sound as for light. The speed of sound is 340 m/s in air and 1510 m/s in water. If a sound wave traveling in air approaches a plane water surface at an angle of incidence of 12.0°, what is the angle of refraction?

16. A ray of light strikes a flat 2.00-cm-thick block of glass ($n = 1.50$) at an angle of 30.0° with the normal (Fig. P22.16). Trace the light beam through the glass and

find the angles of incidence and refraction at each surface.

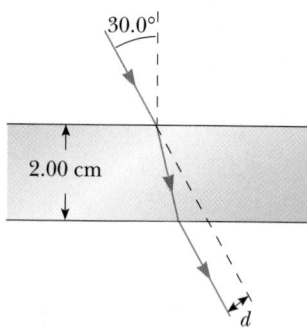

Figure P22.16   (Problems 16 and 17)

WEB **17.** When the light ray in Problem 16 passes through the glass block, it is shifted laterally by a distance d (Fig. P22.16). Find the value of d.

18. The light beam shown in Figure P22.18 makes an angle of 20.0° with the normal line NN' in the linseed oil. Determine the angles θ and θ'. (The refractive index for linseed oil is 1.48.)

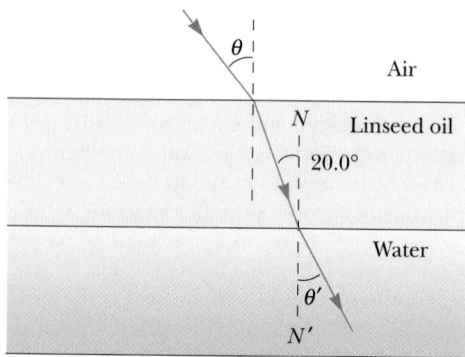

Figure P22.18

19. A submarine is 300 m horizontally out from the shore and 100 m beneath the surface of the water. A laser beam is sent from the sub so that it strikes the surface of the water at a point 210 m from the shore. If the beam just strikes the top of a building standing directly at the water's edge, find the height of the building.

20. A narrow beam of ultrasonic waves reflects off the liver tumor in Figure P22.20. If the speed of the wave is 10.0% less in the liver than in the surrounding medium, determine the depth of the tumor.

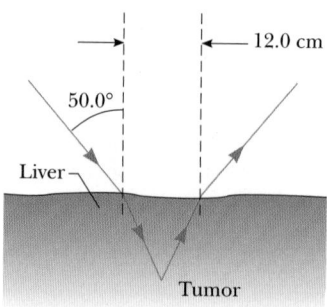

Figure P22.20

21. A beam of light both reflects and refracts at the surface between air and glass, as shown in Figure P22.21. If the index of refraction of the glass is n_g, find the angle of incidence, θ_1, in the air which would in result in the reflected ray and the refracted ray being perpendicular to each other. [Hint: Remember the identity $\sin(90° - \theta) = \cos\theta$.]

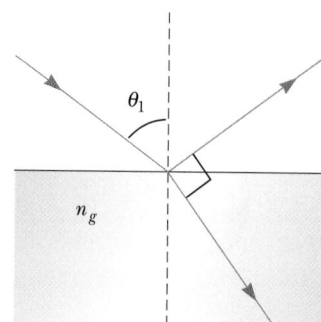

Figure P22.21

22. A ray of light strikes the midpoint of one face of an equiangular (60°-60°-60°) glass prism ($n = 1.5$) at an angle of incidence of 30°. (a) Trace the path of the light ray through the glass, and find the angles of incidence and refraction at each surface. (b) If a small fraction of light is also reflected at each surface, find the angles of incidence and reflection at these surfaces.

23. A cylindrical tank with an open top has a diameter of 3.00 m and is completely filled with water. When the setting Sun reaches an angle of 28.0° above the horizon, sunlight ceases to illuminate the bottom of the tank. How deep is the tank?

24. Three sheets of plastic have unknown indices of refraction. Sheet 1 is placed on top of sheet 2, and a laser beam is directed onto the sheets from above so that it strikes the interface at an angle of 26.5° with the normal. The refracted beam in sheet 2 makes an angle of 31.7° with the normal. The experiment is repeated with sheet 3 on top of sheet 2 and, with the same angle of incidence, the refracted beam makes an angle of 36.7° with the normal. If the experiment is repeated again with sheet 1 on top of sheet 3, what is the expected angle of refraction in sheet 3? Assume the same angle of incidence.

25. A cylindrical cistern, constructed below ground level, is 3.0 m in diameter and 2.0 m deep and is filled to the brim with a liquid whose index of refraction is 1.5. A small object rests on the bottom of the cistern at its center. How far from the edge of the cistern can a girl whose eyes are 1.2 m from the ground stand and still see the object?

Section 22.6 Dispersion and Prisms

26. Light of wavelength 400 nm is incident at an angle of 45° on acrylic and is refracted as it passes into the material. What wavelength of light incident on fused quartz at an angle of 45° would be refracted at exactly this same angle? (See Fig. 22.13.)

27. The index of refraction for red light in water is 1.331, and that for blue light is 1.340. If a ray of white light enters the water at an angle of incidence of 83.00°, what are the underwater angles of refraction for the blue and red components of the light?

28. A certain kind of glass has an index of refraction of 1.650 for blue light of wavelength 430 nm and an index of 1.615 for red light of wavelength 680 nm. If a beam containing these two colors is incident at an angle of 30.0° on a piece of this glass, what is the angle between the two beams inside the glass?

29. The index of refraction for violet light in silica flint glass is 1.66, and that for red light is 1.62. What is the angular dispersion of visible light passing through an equilateral prism of apex angle 60.0° if the angle of incidence is 50.0°? (See Fig. P22.29.)

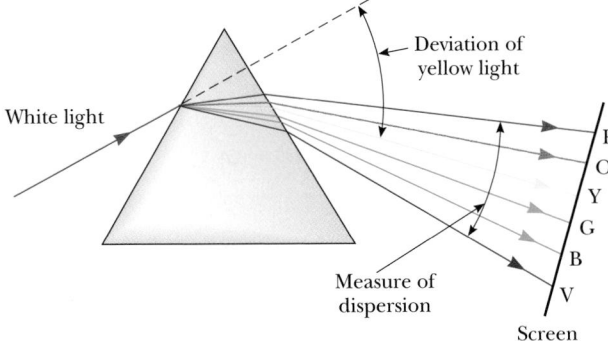

Figure P22.29

Section 22.9 Total Internal Reflection

30. Calculate the critical angles for the following materials when surrounded by air: (a) zircon, (b) fluorite, (c) ice. Assume that $\lambda = 589$ nm.

31. A beam of light is incident from air on the surface of a liquid. If the angle of incidence is 30.0° and the angle of refraction is 22.0°, find the critical angle for the liquid when surrounded by air.

32. From within a diamond, a light ray is incident on the interface between the diamond and air. What is the critical angle for total internal reflection? Use Table 22.1. (The smallness of θ_c for diamond means that light is easily "trapped" within a diamond and eventually emerges from the many cut faces after many internal reflections; this makes a diamond more brilliant than stones with smaller n and larger θ_c.)

33. A plastic light pipe has an index of refraction of 1.53. For total internal reflection, what is the minimum angle of incidence to the wall of the pipe if the pipe is in (a) air? (b) water?

34. A light pipe consists of a central strand of material surrounded by an outer coating. The interior portion of the pipe has an index of refraction of 1.60. If all rays striking the interior walls of the pipe with incident angles greater than 59.5° are subject to total internal reflection, what is the index of refraction of the coating?

WEB 35. Determine the maximum angle, θ, for which the light rays incident on the end of the pipe in Figure P22.35 are subject to total internal reflection along the walls of the pipe. Assume that the pipe has an index of refraction of 1.36 and that the outside medium is air.

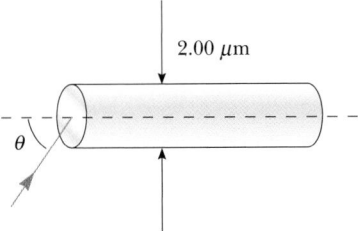

Figure P22.35

36. Three adjacent faces (that all share a corner) of a plastic cube of index of refraction n are painted black, with a clear spot at the painted corner serving as a source of diverging rays when light comes through the clear spot. Show that a ray from this corner to the center of a clear face is totally reflected if $n \geq \sqrt{3}$.

37. A jewel thief hides a diamond by placing it on the bottom of a public swimming pool. He places a circular raft

on the surface of the water directly above and centered on the diamond as shown in Figure P22.37. If the surface of the water is calm and the pool is 2.00 m deep, find the minimum diameter of the raft that would prevent the diamond from being seen.

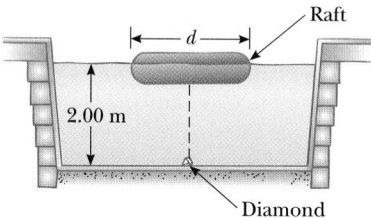

Figure P22.37

38. A light ray is incident normally to the long face (the hypotenuse) of a 45°-45°-90° prism surrounded by air, as shown in Figure 22.25b. Calculate the minimum index of refraction of the prism for which the ray will follow the path shown.

39. The light beam in Figure P22.39 strikes surface 2 at the critical angle. Determine the angle of incidence, θ_1.

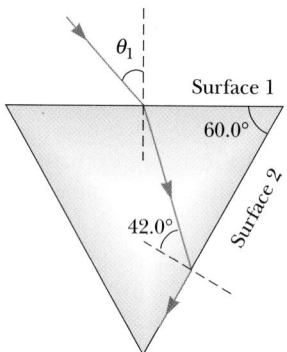

Figure P22.39

ADDITIONAL PROBLEMS

40. Repeat Example 22.2 for the case in which the two mirrors make a 90.0° angle with each other. Show that the ray of light is always reflected from the second mirror so that it travels opposite the original direction.

41. Light is incident on the surface of a prism, $n = 1.80$, as shown in Figure 22.25a. If the prism is surrounded by a fluid, what is the maximum index of refraction of the fluid that will still cause total internal reflection?

42. A layer of ice, having parallel sides, floats on water. If light is incident on the upper surface of the ice at an angle of incidence of 30.0°, what is the angle of refraction in the water?

43. A light ray of wavelength 589 nm is incident at an angle θ on the top surface of a block of polystyrene surrounded by air, as shown in Figure P22.43. (a) Find the maximum value of θ for which the refracted ray will undergo total internal reflection at the left vertical face of the block. (b) Repeat the calculation for the case in which the polystyrene block is immersed in water. (c) What happens if the block is immersed in carbon disulfide?

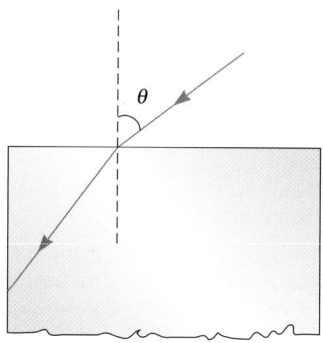

Figure P22.43

44. Figure P22.44 shows the path of a beam of light through several layers of different indices of refraction. (a) If $\theta_1 = 30.0°$, what is the angle, θ_2, of the emerging beam? (b) What must the incident angle, θ_1, be in order to have total internal reflection at the surface between the $n = 1.20$ medium and the $n = 1.00$ medium?

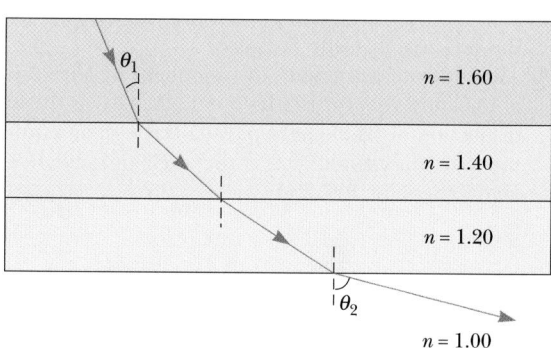

Figure P22.44

45. As shown in Figure P22.45, a light ray is incident normally on one face of a 30°-60°-90° block of dense flint glass (a prism) that is immersed in water. (a) Determine the exit angle, θ_4, of the ray. (b) A substance is dissolved in the water to increase the index of refraction. At what value of n_2 does total internal reflection cease at point P?

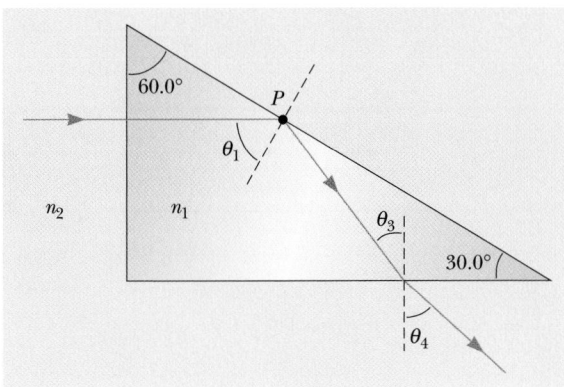

Figure P22.45

46. A thick plate of flint glass ($n = 1.66$) rests on top of a thick plate of transparent acrylic ($n = 1.50$), all surrounded by air. A beam of light is incident on the top surface of the flint glass at an angle θ_i. The beam passes through the glass and the acrylic and emerges from the acrylic at an angle of 40.0° with respect to the normal. Calculate the value of θ_i. A sketch of the light path through the two plates of refracting material would be helpful.

47. A narrow beam of light is incident from air onto a glass surface with index of refraction 1.56. Find the angle of incidence for which the corresponding angle of refraction is one half the angle of incidence. (*Hint:* You might want to use the trigonometric identity $\sin 2\theta = 2 \sin \theta \cos \theta$.)

48. One technique to measure the angle of a prism is shown in Figure P22.48. A parallel beam of light is directed on the apex of the prism so that the beam reflects from opposite faces of the prism. Show that the angular separation of the two reflected beams is given by $B = 2A$.

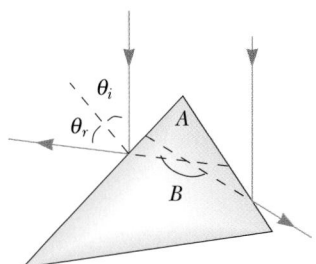

Figure P22.48

49. A shallow glass dish is 4.0 cm wide at the bottom, as shown in Figure P22.49. When an observer's eye is positioned as shown, the observer sees the edge of the bot-

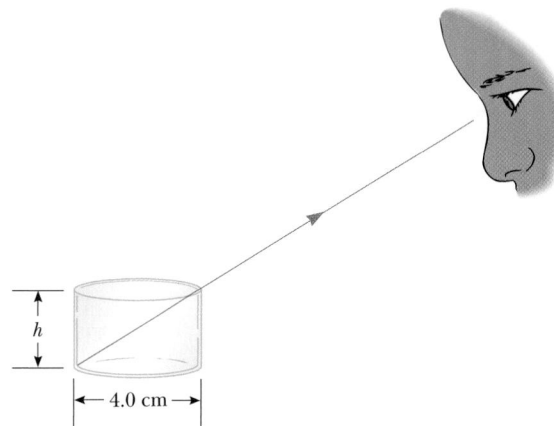

Figure P22.49

tom of the empty dish. When this dish is filled with water, the observer, with the eye positioned as before, sees the center of the bottom of the dish. Find the height of the dish.

50. For this problem, refer to Figure 22.14. For various angles of incidence, it can be shown that the angle δ is a minimum when the ray passes through the glass so that the ray is parallel to the base of the prism. A measurement of this minimum angle of deviation enables one to find the index of refraction of the prism material. Show that n is given by the expression

$$n = \frac{\sin[\frac{1}{2}(A + \delta_{min})]}{\sin\left(\dfrac{A}{2}\right)}$$

where A is the apex angle of the prism.

WEB 51. Two light pulses are emitted simultaneously from a source. Both pulses travel to a detector, but one first passes through 6.20 m of ice. Determine the difference in the pulses' times of arrival at the detector.

52. A piece of wire is bent through an angle θ. The bent wire is partially submerged in benzene (index of refraction = 1.50) so that looking along the dry part, the wire appears to be straight and makes an angle of 30.0° with the horizontal. Determine the value of θ.

53. When you look through a window, by how much time is the light you see delayed by having to go through glass instead of air? Make an order-of-magnitude estimate on the basis of data you specify. By how many wavelengths is it delayed?

54. A hiker stands on a mountain peak near sunset and observes a rainbow caused by water droplets in the air about 8.00 km away. The valley is 2.00 km below the mountain peak and entirely flat. What fraction of the complete circular arc of the rainbow is visible to the hiker?

55. A laser beam strikes one end of a slab of material, as in Figure P22.55. The index of refraction of the slab is 1.48. Determine the number of internal reflections of the beam before it emerges from the opposite end of the slab.

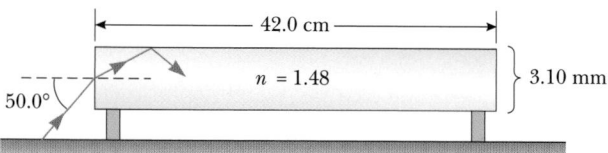

Figure P22.55

56. A cylindrical material of radius $R = 2.00$ m has a mirrored surface on its right half, as in Figure P22.56. A light ray traveling in air is incident on the left side of the cylinder. If the incident light ray and exiting light ray are parallel and $d = 2.00$ m, determine the index of refraction of the material.

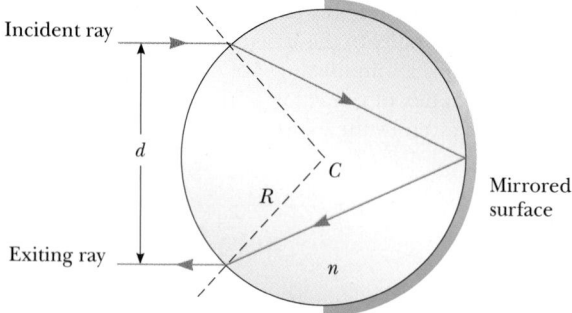

Figure P22.56

57. A. H. Pfund's method for measuring the index of refraction of glass is illustrated in Figure P22.57. One face of a slab of thickness t is painted white, and a small hole scraped clear at point P serves as a source of diverging rays when the slab is illuminated from below. Ray PBB'

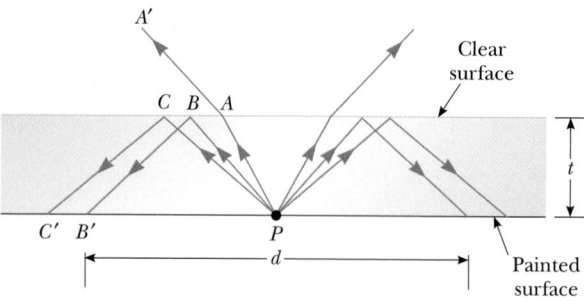

Figure P22.57

strikes the clear surface at the critical angle and is totally reflected, as are rays such as PCC'. Rays such as PAA' emerge from the clear surface. On the painted surface there appears a dark circle of diameter d, surrounded by an illuminated region, or halo. (a) Derive a formula for n in terms of the measured quantities d and t. (b) What is the diameter of the dark circle if $n = 1.52$ for a slab 0.600 cm thick? (c) If white light is used, the critical angle depends on color due to dispersion. Is the inner edge of the white halo tinged with red light or violet light? Explain.

58. A light ray is incident on a prism and refracted at the first surface, as shown in Figure P22.58. Let ϕ represent the apex angle of the prism and n its index of refraction. Find, in terms of n and ϕ, the smallest allowed value of the angle of incidence at the first surface for which the refracted ray will not undergo internal reflection at the second surface.

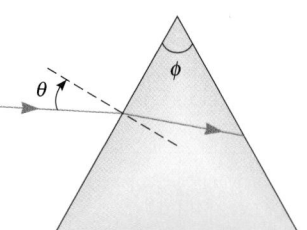

Figure P22.58

Mirrors and Lenses

▲ PHYSICS PUZZLER

This photo shows Mt. Hood reflected in Trillium Lake, Oregon. Why is the image inverted and the same size as the mountain? *(Raymond G. Barnes/Tony Stone Images)*

753

This chapter is concerned with the formation of images when plane and spherical waves fall on plane and spherical surfaces. Images can be formed by either reflection or refraction. Mirrors and lenses form images in both ways. In our study of mirrors and lenses, we continue to use the ray approximation and to assume that light travels in straight lines (in other words, we ignore diffraction).

23.1 FLAT MIRRORS

We begin our investigation by examining the simplest possible mirror, the flat mirror. Consider a point source of light placed at O in Figure 23.1, a distance of p in front of a flat mirror. The distance p is called the **object distance.** Light rays leave the source and are reflected from the mirror. After reflection, the rays diverge (spread apart), but they appear to the viewer to come from a point, I, behind the mirror. Point I is called the **image** of the object at O. Regardless of the system under study, **images are formed at the point at which rays of light actually intersect or at which they appear to originate.** Because the rays in Figure 23.1 appear to originate at I, which is a distance of q behind the mirror, this is the location of the image. The distance q is called the **image distance.**

Images are classified as real or virtual. A *real image* is one in which light actually intersects, or passes through, the image point; a *virtual image* is one in which the light does not pass through the image point but appears to come (diverge) from that point. The image formed by the flat mirror in Figure 23.1 is a virtual image. In fact, the images seen in flat mirrors are always virtual for real objects. Real images can be displayed on a screen (as at a movie), but virtual images cannot.

We shall examine some of the properties of the images formed by flat mirrors by using the simple geometric techniques shown in Figure 23.2. To find out where an image is formed, it is always necessary to follow at least two rays of light as they reflect from the mirror. One of those rays starts at P, follows a horizontal path, PQ, to the mirror, and reflects back on itself. The second ray follows the oblique path PR and reflects as shown. An observer to the left of the mirror would trace the two reflected rays back to the point from which they appear to have originated — that is, point P'. A continuation of this process for points other than P on the object

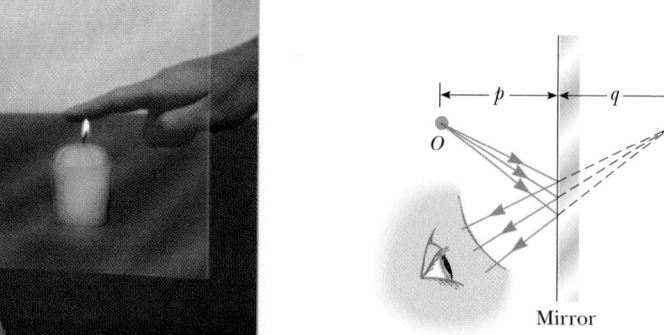

Figure 23.1 An image formed by reflection from a flat mirror. The image point, I, is behind the mirror at distance q, which is equal in magnitude to the object distance, p.

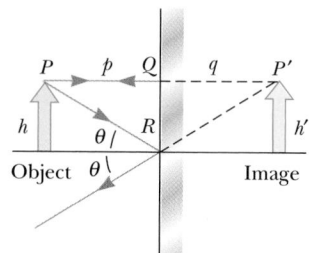

Figure 23.2 A geometric construction to locate the image of an object placed in front of a flat mirror. Because the triangles PQR and $P'QR$ are identical, $p = q$, and $h = h'$.

would result in a virtual image (drawn as a yellow arrow) to the right of the mirror. Because triangles *PQR* and *P'QR* are identical, *PQ* = *P'Q*. Hence, we conclude that **the image formed by an object placed in front of a flat mirror is as far behind the mirror as the object is in front of the mirror.** Geometry also shows that the object height, *h*, equals the image height, *h'*. **Lateral magnification, *M*,** is defined as follows:

$$M \equiv \frac{\text{image height}}{\text{object height}} = \frac{h'}{h} \qquad [23.1]$$

This is a general definition of the lateral magnification of any type of mirror. For a flat mirror, *M* = 1 because *h'* = *h*.

The observer sees that the image formed by a flat mirror has right-left reversal. This can be seen by standing in front of a mirror and raising your right hand. The image you see raises its left hand. Likewise, your hair appears to be parted on the opposite side and a mole on your right cheek appears to be on your left cheek.

In summary, the image formed by a flat mirror has the following properties:

1. **The image is as far behind the mirror as the object is in front.**
2. **The image is unmagnified, virtual, and upright.** (By *upright* we mean that if the object arrow points upward, as in Figure 23.2, so does the image arrow. The opposite of an upright image is an inverted image.)

Example 23.1 "I Can See Myself!"

A 1.80-m tall man stands in front of a mirror in hopes of seeing his full height, no more and no less. If his eyes are 0.10 m from the top of his head, what is the minimum height of the mirror?

Reasoning and Solution Figure 23.3 shows two rays of light originating at the extremes of the body, reflecting from the mirror, and entering the eye of the viewer. The ray from the feet just strikes the bottom of the mirror, so if the mirror were longer, it would be too long; if shorter, the ray would not be reflected. The angle of incidence and the angle of reflection are equal, labeled *θ*. The two triangles, *ABD* and *DBC* are identical because they are right triangles with a common side (*DB*) and two identical angles, *θ*. Thus, we have

$$AD = DC = \frac{1}{2} AC = \frac{1}{2} (1.80 \text{ m} - 0.10 \text{ m}) = 0.85 \text{ m}$$

Furthermore, because $\frac{1}{2} CF = \frac{1}{2} (0.10\text{m}) = 0.05$ m, we find that

$$d = FA - AD - \frac{1}{2} CF = 1.80 \text{ m} - 0.85 \text{ m} - 0.05 \text{ m} = \boxed{0.90 \text{ m}}$$

Thus, the mirror must be exactly equal to half the height of the man in order for him to see only his full height and nothing more or less.

Exercise Is the result of this problem still valid if the man moves farther away from the mirror? How does distance from the mirror affect the answer?

Workbook Problem 88 – (Workbook page 188) is similar to Example 23.1.

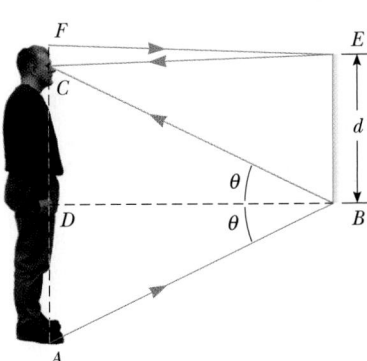

Figure 23.3 (Example 23.1)

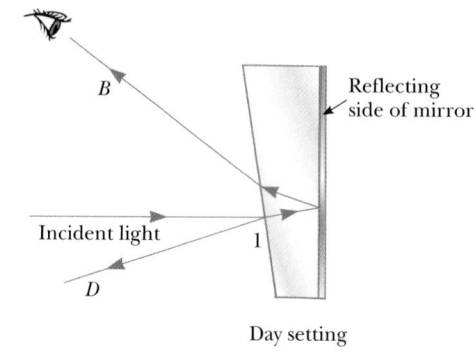

Figure 23.4   A cross-sectional view of a rearview mirror. (a) The day setting forms a bright image, *B*. (b) The night setting forms a dim image, *D*.

Reflecting side of mirror

Incident light

Day setting

(a)

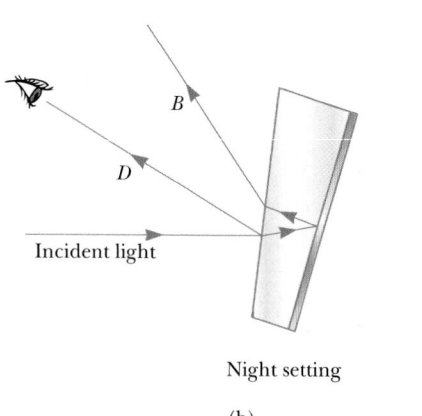

Incident light

Night setting

(b)

Figure 23.5 (Thinking Physics 1) *(Courtesy of Henry Leap and Jim Lehman)*

Most rearview mirrors in cars have a day setting and a night setting. The night setting greatly diminishes the intensity of the image so that lights from trailing cars will not blind the driver. To understand how such a mirror works, consider Figure 23.4. The mirror is a wedge of glass with a reflecting metallic coating on the back side. When the mirror is in the day setting, as in Figure 23.4a, light from an object behind the car strikes the mirror at point 1. Most of the light enters the wedge, is refracted, and reflects from the back of the mirror to return to the front surface, where it is refracted again as it re-enters the air as ray *B* (for *bright*). In addition, a small portion of the light is reflected at the front surface, as indicated by ray *D* (for *dim*). This dim reflected light is responsible for the image observed when the mirror is in the night setting, as in Figure 23.4b. In this case, the wedge is rotated so that the path followed by the bright light (ray *B*) does not lead to the eye. Instead, the dim light reflected from the front surface travels to the eye, and the brightness of trailing headlights does not become a hazard.

Thinking Physics 1

The professor in the box shown in Figure 23.5 appears to be balancing himself on a few fingers with both of his feet elevated from the floor. The professor can maintain this position for a long time, and he appears to defy gravity. How do you suppose this illusion was created?

Explanation This is one example of an optical illusion, used by magicians, that makes use of a mirror. The box that the professor is standing in is a cubical frame that contains a flat vertical mirror through a diagonal plane. The professor straddles the mirror so that the foot you see is in front of the mirror, and the other foot is behind the mirror where you cannot see it. When he raises the foot that you see in front of the mirror, the reflection of this foot also rises, so he appears to float in air.

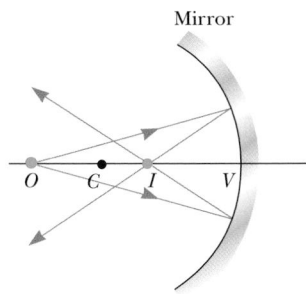

Figure 23.6 A point object placed at O, outside the center of curvature of a concave spherical mirror, forms a real image at I as shown. If the rays diverge from O at small angles, they all reflect through the same image point.

23.2 IMAGES FORMED BY SPHERICAL MIRRORS

Concave Mirrors

A **spherical mirror,** as its name implies, has the shape of a segment of a sphere. Figure 23.6 shows a spherical mirror with light reflecting from its silvered inner, concave surface; this is called a **concave mirror.** The mirror has radius of curvature R, and its center of curvature is at point C. Point V is the center of the spherical segment, and a line drawn from C to V is called the **principal axis** of the mirror.

Now consider a point source of light placed at point O in Figure 23.6, on the principal axis and outside point C. Several diverging rays originating at O are shown. After reflecting from the mirror, these rays converge to meet at I, called the **image point.** The rays then continue and diverge from I as if there were an object there. As a result, a real image is formed. **Whenever reflected light actually passes through a point, the image formed there is real.**

We assume that all rays that diverge from the object make small angles with the principal axis. All such rays reflect through the image point, as in Figure 23.6. Rays that are far from the principal axis, as in Figure 23.7, converge to other points on the principal axis, producing a blurred image. This effect, called **spherical aberration,** is present to some extent with any spherical mirror and will be discussed in Section 23.7.

We can use the geometry shown in Figure 23.8 to calculate the image distance, q, from the object distance, p, and radius of curvature, R. By convention, these distances are measured from point V. Figure 23.8 shows two rays of light leaving the tip of the object. One ray passes through the center of curvature, C, of the mirror, hitting the mirror head on (perpendicularly to the mirror surface) and

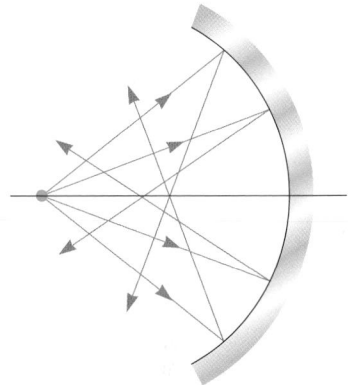

Figure 23.7 Rays at large angles from the horizontal axis reflect from a spherical, concave mirror to intersect the optic axis at different points, resulting in a blurred image. This is called *spherical aberration.*

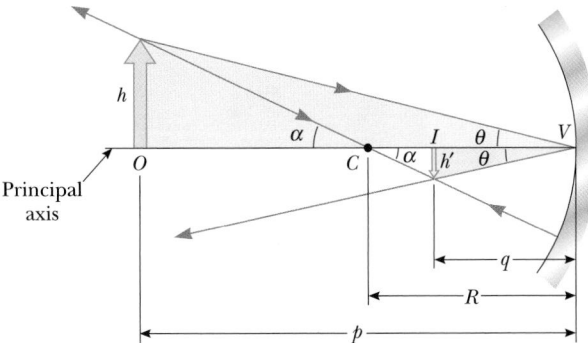

Figure 23.8 The image formed by a spherical concave mirror, where the object, at O, lies outside the center of curvature, C.

reflecting back on itself. The second ray strikes the mirror at point V and reflects as shown, obeying the law of reflection. The image of the tip of the arrow is at the point at which the two rays intersect. From the largest triangle in Figure 23.8 we see that $\tan \theta = h/p$; the light blue triangle gives $\tan \theta = -h'/q$. The negative sign signifies that the image is inverted, and so h' is negative. Thus, from Equation 23.1 and these results, we find that the magnification of the mirror is

$$M = \frac{h'}{h} = -\frac{q}{p}$$ [23.2]

We also note, from two other triangles in the figure, that

$$\tan \alpha = \frac{h}{p-R} \quad \text{and} \quad \tan \alpha = -\frac{h'}{R-q}$$

from which we find that

$$\frac{h'}{h} = -\frac{R-q}{p-R}$$ [23.3]

If we compare Equation 23.2 to Equation 23.3, we see that

$$\frac{R-q}{p-R} = \frac{q}{p}$$

Simple algebra reduces this to

Mirror equation ▶

$$\frac{1}{p} + \frac{1}{q} = \frac{2}{R}$$ [23.4]

This expression is called the **mirror equation.**

If the object is very far from the mirror — that is, if the object distance, p, is great enough compared with R that p can be said to approach infinity — then $1/p \approx 0$, and we see from Equation 23.4 that $q \approx R/2$. In other words, when the object is very far from the mirror, **the image point is halfway between the center of curvature and the center of the mirror,** as in Figure 23.9a. The incoming rays

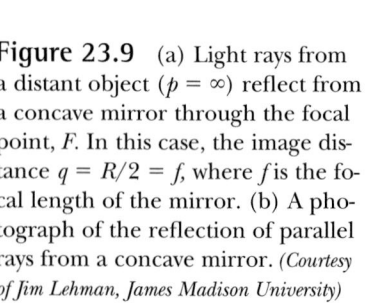

Figure 23.9 (a) Light rays from a distant object ($p = \infty$) reflect from a concave mirror through the focal point, F. In this case, the image distance $q = R/2 = f$, where f is the focal length of the mirror. (b) A photograph of the reflection of parallel rays from a concave mirror. *(Courtesy of Jim Lehman, James Madison University)*

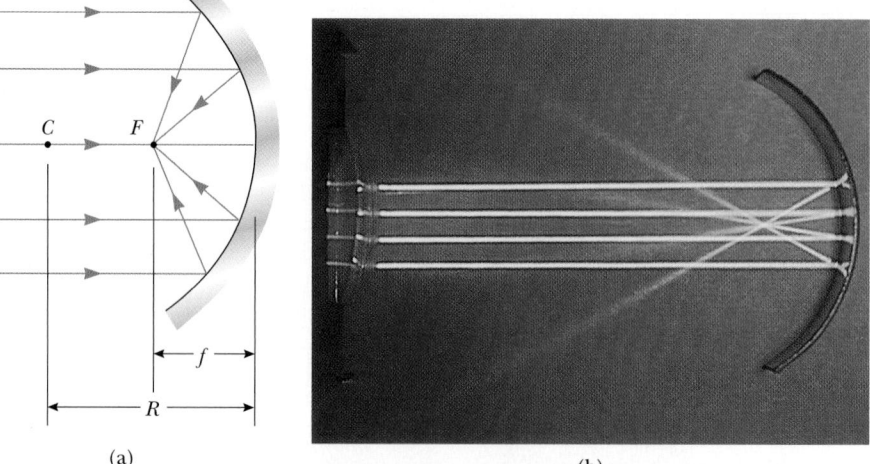

(a)

(b)

are essentially parallel in this figure because the source is assumed to be very far from the mirror. In this special case we call the image point the **focal point,** *F,* and the image distance the **focal length,** *f,* where

$$f = \frac{R}{2}$$ [23.5]   ◀ Focal length

The mirror equation can therefore be expressed in terms of the focal length:

$$\frac{1}{p} + \frac{1}{q} = \frac{1}{f}$$ [23.6]

Note that rays from objects at infinity are always focused at the focal point.

23.3   CONVEX MIRRORS AND SIGN CONVENTIONS

Figure 23.10 shows the formation of an image by a **convex mirror,** which is silvered so that light is reflected from the outer, convex surface. This is sometimes called a **diverging mirror** because the rays from any point on the object diverge after reflection as though they were coming from some point behind the mirror. The image in Figure 23.10 is virtual rather than real because it lies behind the mirror at the point at which the reflected rays appear to originate. In general, as shown in the figure, the image formed by a convex mirror is upright, virtual, and smaller than the object.

14.6, BOTH SECTIONS

We shall not derive any equations for convex spherical mirrors. If we did, we would find that the equations developed for concave mirrors can be used with convex mirrors if a particular sign convention is used. Let us call the region in which light rays move the *front side* of the mirror, and the other side, where virtual images are formed, the *back side.* For example, in Figure 23.8 and 23.10, the side to the left of the mirror is the front side, and the side to the right is the back side. Figure 23.11 is helpful for understanding the rules for object and image distances, and Table 23.1 summarizes the sign conventions for all the necessary quantities.

14.7, BOTH SECTIONS

Ray Diagrams for Mirrors

We can determine conveniently the positions and sizes of images formed by mirrors by constructing *ray diagrams* similar to the ones we have been using. This kind of graphical construction tells us the overall nature of the image and can be used to check parameters calculated from the mirror and magnification equations. To make a ray diagram, one needs to know the position of the object and the location

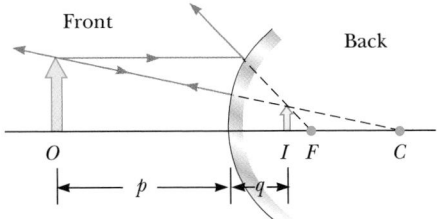

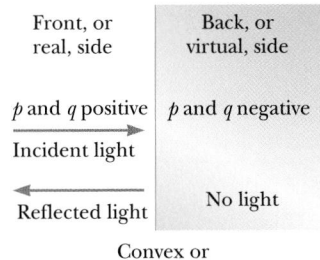

Figure 23.10 Formation of an image by a spherical convex mirror. Note that the image is virtual and upright.

Figure 23.11 A diagram describing the signs of *p* and *q* for convex and concave mirrors.

TABLE 23.1   Sign Conventions for Mirrors

p is $+$ if the object is in front of the mirror (real object).
p is $-$ if the object is in back of the mirror (virtual object).

q is $+$ if the image is in front of the mirror (real image).
q is $-$ if the image is in back of the mirror (virtual image).

Both f and R are $+$ if the center of curvature is in front of the mirror (concave mirror).
Both f and R are $-$ if the center of curvature is in back of the mirror (convex mirror).

If M is positive, the image is upright.
If M is negative, the image is inverted.

Note: p = object distance; q = image distance; f = focal length; R = radius of curvature; M = lateral magnification.

---

of the center of curvature. To locate the image, three rays are constructed (rather than just the two we have been constructing so far), as shown by the examples in Figure 23.12. All three rays start from the same object point; for these examples the tip of the arrow was chosen. For the concave mirrors in Figure 23.12a and b, the rays are drawn as follows:

1. Ray 1 is drawn parallel to the principal axis and is reflected back through the focal point, *F*.
2. Ray 2 is drawn through the focal point. Thus, it is reflected parallel to the principal axis.
3. Ray 3 is drawn through the center of curvature, *C*, and is reflected back on itself.

Note that rays actually go in all directions from the object; we choose to follow those moving in a direction that simplifies our drawing.

The intersection of any *two* of these rays at a point locates the image. The third ray serves as a check of construction. The image point obtained in this fashion must always agree with the value of q calculated from the mirror formula.

In the case of a concave mirror, note what happens as the object is moved closer to the mirror. The real, inverted image in Figure 23.12a moves to the left as the object approaches the focal point. When the object is at the focal point, the image is infinitely far to the left. However, when the object lies between the focal point and the mirror surface, as in Figure 23.12b, the image is virtual and upright.

With the convex mirror shown in Figure 23.12c, the image of a real object is always virtual and upright. As the object distance increases, the virtual image shrinks and approaches the focal point as p approaches infinity. You should construct a ray diagram to verify this.

The image-forming characteristics of curved mirrors obviously determine their uses. For example, suppose you want to design a mirror that will help people shave or apply cosmetics, such as the one in Figure 23.12b. That is, you need a concave mirror that puts the user inside the focal point. In such a situation, the image is upright and greatly enlarged. In contrast, suppose that the primary purpose of a

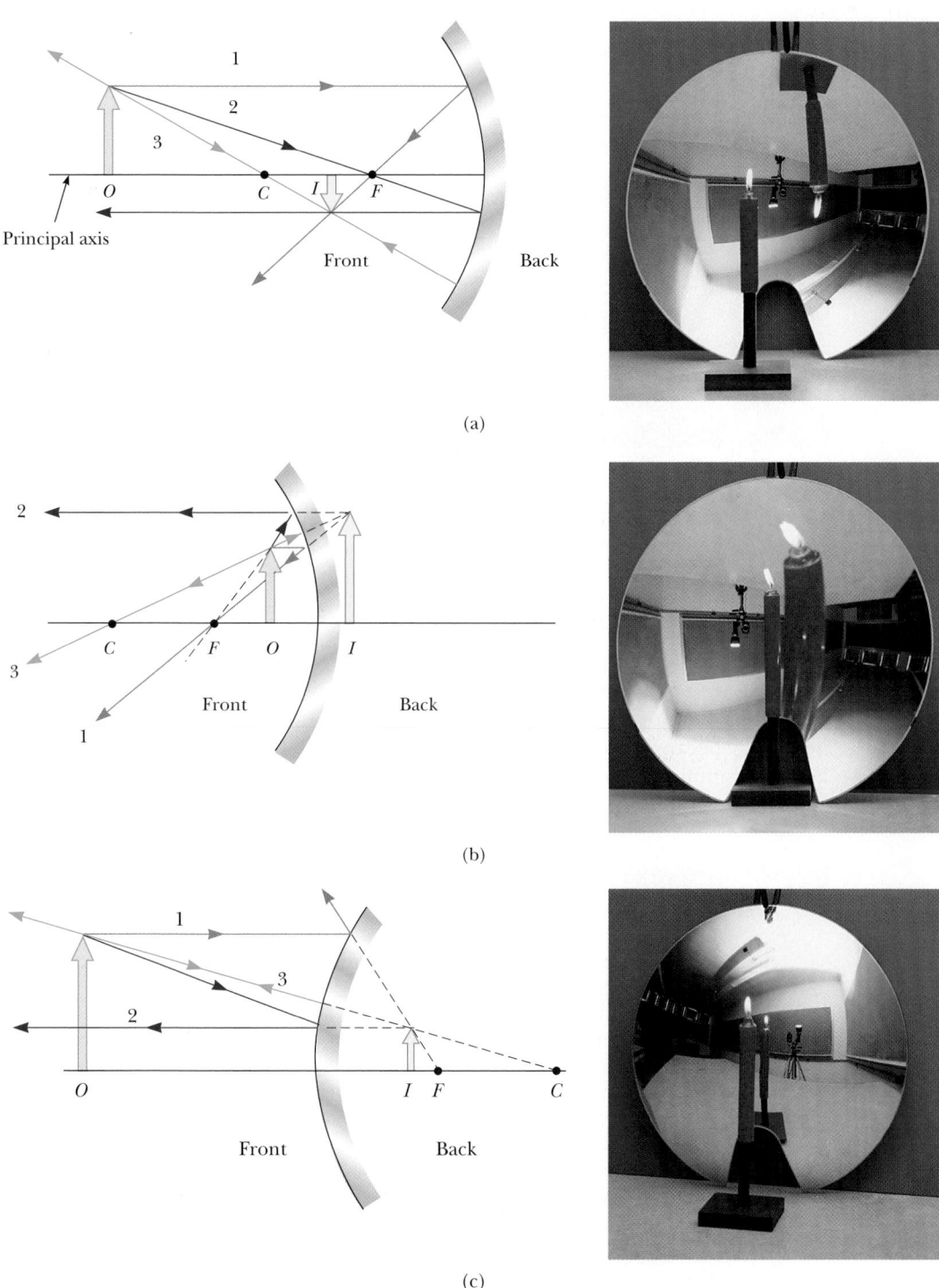

Figure 23.12 Ray diagrams for spherical mirrors, and corresponding photographs of the images of candles. (a) When an object is outside the center of curvature of a concave mirror, the image is real, inverted, and reduced in size. (b) When an object is between a concave mirror and the focal point, the image is virtual, upright, and magnified. (c) When an object is in front of a convex mirror, the image is virtual, upright, and reduced in size. *(Photos courtesy of David Rogers)*

Figure 23.13 A convex sideview mirror on a vehicle produces an upright image that is smaller than the object. *(© Junebug Clark 1988/Photo Researchers, Inc.)*

APPLICATION

Convex Sideview Mirrors on Cars.

mirror is to observe a large field of view, in which case you need a convex mirror such as the one in Figure 23.12c. The diminished size of the image means that a fairly large field of view is seen in the mirror. Mirrors such as this are often placed in stores to help employees watch for shoplifting. A second use is as a sideview mirror on a car (Fig. 23.13). This kind of mirror is usually placed on the passenger side of the car and carries the warning "Objects are closer than they appear." Without this warning, a driver might think she is looking into a flat mirror, which does not alter the size of the image. Thus, she could be fooled into believing that a truck is far away because it looks small, when it is actually a large semi very close behind her but diminished in size because of the image formation characteristics of the convex mirror.

Thinking Physics 2

For a concave mirror, a virtual image can be anywhere behind the mirror. For a convex mirror, however, there is a maximum distance at which the image can exist behind the mirror. Why?

Explanation Let us consider the concave mirror first and imagine two different light rays leaving a tiny object and striking the mirror. If the object is inside the focal point but infinitesimally close to it, the light rays reflecting from the mirror will be parallel to the mirror axis. They can be interpreted as forming a virtual image infinitely far away behind the mirror. As the object is brought closer to the mirror, the reflected rays will diverge through larger and larger angles, resulting in their extensions converging closer and closer to the back of the mirror. When the object is brought right up to the mirror, the image is right behind the mirror. We can argue that when the object is much closer to the mirror than the focal length, it looks like a flat mirror, so the image is just as far behind the mirror as the object is in front of it. Thus, the image can be anywhere from infinitely far away to right at the surface of the mirror. For the convex mirror, an object at infinity produces a virtual image at the focal point. As the object is brought closer, the reflected rays diverge more sharply and the image moves closer to the mirror. Thus, the virtual image is restricted to the region between the mirror and the focal point.

Applying Physics 1

Large trucks often have a sign on the back saying, "If you can't see my mirror, I can't see you." Explain this sign.

Explanation The trucking companies are making use of the principle of reversibility of light rays. In order for an image of you to be formed in the driver's mirror, there must be a pathway for rays of light to reach the mirror, allowing the driver to see your image. If you can't see the mirror, obviously there is no such pathway.

EXAMPLE 23.2 Images Formed by a Concave Mirror

Assume that a certain concave spherical mirror has a focal length of 10.0 cm. Locate the images for object distances of (a) 25.0 cm, (b) 10.0 cm, and (c) 5.00 cm. Describe the image in each case.

Solution (a) For an object distance of 25.0 cm, we find the image distance using the mirror equation:

$$\frac{1}{p} + \frac{1}{q} = \frac{1}{f}$$

$$\frac{1}{25.0 \text{ cm}} + \frac{1}{q} = \frac{1}{10.0 \text{ cm}}$$

$$q = \boxed{16.7 \text{ cm}}$$

The magnification is given by Equation 23.2:

$$M = -\frac{q}{p} = -\frac{16.7 \text{ cm}}{25.0 \text{ cm}} = \boxed{-0.667}$$

Thus, the image is smaller than the object. Furthermore, the image is inverted because M is negative. Finally, because q is positive, the image is on the front side of the mirror and is real. This situation is pictured in Figure 23.12a.

(b) When the object distance is 10.0 cm, the object is at the focal point. Substituting the values $p = 10.0$ cm and $f = 10.0$ cm into the mirror equation, we find that

$$\frac{1}{10.0 \text{ cm}} + \frac{1}{q} = \frac{1}{10.0 \text{ cm}}$$

$$q = \boxed{\infty}$$

Thus, we see that rays of light originating from an object at the focal point of a concave mirror are reflected so that the image is formed an infinite distance from the mirror—that is, the rays travel parallel to one another after reflection.

(c) When the object is at 5.00 cm, inside the focal point of the mirror, the mirror equation gives

$$\frac{1}{5.00 \text{ cm}} + \frac{1}{q} = \frac{1}{10.0 \text{ cm}}$$

$$q = \boxed{-10.0 \text{ cm}}$$

That is, the image is virtual because it is behind the mirror. The magnification is

$$M = -\frac{q}{p} = -\left(\frac{-10.0\ \text{cm}}{5.00\ \text{cm}}\right) = \boxed{2.00}$$

We see that the image height is magnified by a factor of 2, and the positive sign indicates that the image is upright (Fig. 23.12b).

Note the characteristics of an image formed by a concave spherical mirror. When the object is outside the focal point, the image is inverted and real; at the focal point, the image is formed at infinity; inside the focal point, the image is upright and virtual.

Exercise   If the object distance is 20.0 cm, find the image distance and the magnification of the mirror.

Answer $q = 20.0\ \text{cm}$, $M = -1.00$

EXAMPLE 23.3   Images Formed by a Convex Mirror

An object 3.00 cm high is placed 20.0 cm from a convex mirror with a focal length of 8.00 cm. Find (a) the position of the final image and (b) the magnification of the mirror.

Solution   (a) Because the mirror is convex, its focal length is negative. To find the image position, we use the mirror equation:

$$\frac{1}{p} + \frac{1}{q} = \frac{1}{f}$$

$$\frac{1}{20.0\ \text{cm}} + \frac{1}{q} = \frac{1}{-8.00\ \text{cm}}$$

$$q = \boxed{-5.71\ \text{cm}}$$

The negative value of q indicates that the image is virtual, or behind the mirror, as in Figure 23.12c.

(b) The magnification of the mirror is

$$M = -\frac{q}{p} = -\left(\frac{-5.71\ \text{cm}}{20.0\ \text{cm}}\right) = \boxed{0.286}$$

The image is upright because M is positive.

Exercise   Find the height of the image.

Answer   0.857 cm

EXAMPLE 23.4   An Enlarged Image

When a woman stands with her face 40.0 cm from a cosmetic mirror, the upright image is twice as tall as her face. What is the focal length of the mirror?

Reasoning   Most of the problems we have encountered so far have been simple applications of the mirror equation. However, to find f in this example, we must first find q, the image distance. Because the problem states that the image is upright, the magnification must be positive (in this case, $M = +2$), and because $M = -q/p$, we can determine q.

Solution The magnification equation gives us a relationship between the object and image distances:

$$M = -\frac{q}{p} = 2$$

$$q = -2p = -2(40.0 \text{ cm}) = -80.0 \text{ cm}$$

First, note that a virtual image is formed because the woman is able to see her upright image in the mirror. This explains why the image distance is negative. Substitute $q = -80.0$ cm into the mirror equation to obtain

$$\frac{1}{40.0 \text{ cm}} - \frac{1}{80.0 \text{ cm}} = \frac{1}{f}$$

$$f = \quad 80.0 \text{ cm}$$

The positive sign for the focal length indicates that the mirror is concave, a fact that we already knew because the mirror magnified the object. (A convex mirror would have produced a diminished image.)

23.4 IMAGES FORMED BY REFRACTION

In this section we describe how images are formed by refraction at a spherical surface. Consider two transparent media with indices of refraction n_1 and n_2, where the boundary between the two media is a spherical surface of radius R (Fig. 23.14). Let us assume that the medium to the right has a higher index of refraction than the one to the left; that is, $n_2 > n_1$. This would be the case for light entering a curved piece of glass from air or for light entering the water in a fishbowl from air. The rays originating at the object location, O, are refracted at the spherical surface and then converge to the image point, I. We can begin with Snell's law of refraction and use simple geometric techniques to show that the object distance, image distance, and radius of curvature are related by the equation

$$\frac{n_1}{p} + \frac{n_2}{q} = \frac{n_2 - n_1}{R} \qquad [23.7]$$

Furthermore, the magnification of a refracting surface is

$$M = \frac{h'}{h} = -\frac{n_1 q}{n_2 p} \qquad [23.8]$$

As with mirrors, we must use a sign convention if we are to apply these equations to a variety of circumstances. First note that real images are formed on the side of the surface *opposite* the side from which the light comes. This is in contrast with mirrors, where real images are formed on the *same* side of the reflecting surface. Therefore, **the sign convention for spherical refracting surfaces is the same as for mirrors, recognizing the change in sides of the surface for real and virtual images.** For example, in Figure 23.14, p, q, and R are all positive.

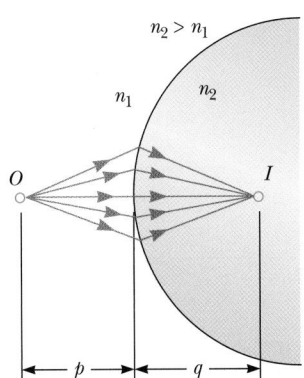

Figure 23.14 An image formed by refraction at a spherical surface. Rays making small angles with the optic axis diverge from a point object at O and pass through the image point, I.

TABLE 23.2	Sign Conventions for Refracting Surfaces

p is $+$ if the object is in front of the surface (real object).
p is $-$ if the object is in back of the surface (virtual object).

q is $+$ if the image is in back of the surface (real image).
q is $-$ if the image is in front of the surface (virtual image).

R is $+$ if the center of curvature is in back of the surface.
R is $-$ if the center of curvature is in front of the surface.

Note: p = object distance; q = image distance; R = radius of curvature.

The sign conventions for spherical refracting surfaces are summarized in Table 23.2. (The same conventions are used for thin lenses, which will be discussed in Section 23.6.) As with mirrors, we assume that the front of the refracting surface is the side from which the light approaches the surface.

Applying Physics 2

APPLICATION

Opening Your Eyes Underwater.

Why does a person with normal vision see a blurry image if the eyes are opened underwater with no goggles or diving mask in use?

Explanation   The eye presents a spherical refracting surface. The eye normally functions so that light entering from the air is refracted to form an image in the retina located at the back of the eyeball. The difference in index of refraction between water and the eye is smaller than the difference in index of refraction between air and the eye. Thus, light entering the eye from the water does not experience as much refraction as does light entering from the air, and the image is formed behind the retina. A diving mask or swimming goggles have no optical action of their own—they are simply flat pieces of glass or plastic in a rubber mount. However, they provide a region of air adjacent to the eyes, so that the correct refraction relationship is established, and images will be in focus.

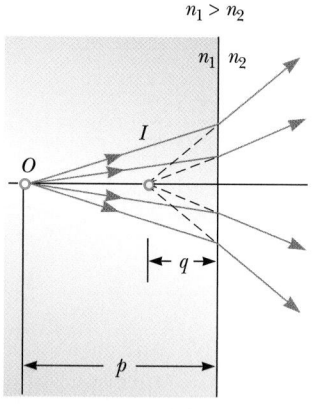

Figure 23.15 The image formed by a flat refracting surface is virtual; that is, it forms to the left of the refracting surface.

Flat Refracting Surfaces

If the refracting surface is flat, then R approaches infinity and Equation 23.7 reduces to

$$\frac{n_1}{p} = -\frac{n_2}{q}$$

$$q = -\frac{n_2}{n_1} p \qquad [23.9]$$

From Equation 23.9 we see that the sign of q is opposite that of p. Thus, **the image formed by a flat refracting surface is on the same side of the surface as the object.** This is illustrated in Figure 23.15 for the situation in which n_1 is greater than n_2, where a virtual image is formed between the object and the surface. Note that the refracted ray bends *away* from the normal in this case, because $n_1 > n_2$.

EXAMPLE 23.5   Gaze into the Crystal Ball

A coin 2.00 cm in diameter is embedded in a solid glass ball of radius 30.0 cm (Fig. 23.16). The index of refraction of the ball is 1.50, and the coin is 20.0 cm from the surface. Find the position and height of the image.

Solution   Because they are moving from a medium of high index of refraction to a medium of lower index of refraction, the rays originating at the object are refracted away from the normal at the surface and diverge outward. The image is formed in the glass and is virtual. Applying Equation 23.7 and taking $n_1 = 1.50$, $n_2 = 1.00$, $p = 20.0$ cm, and $R = -30.0$ cm, we get

$$\frac{n_1}{p} + \frac{n_2}{q} = \frac{n_2 - n_1}{R}$$

$$\frac{1.50}{20.0 \text{ cm}} + \frac{1.00}{q} = \frac{1.00 - 1.50}{-30.0 \text{ cm}}$$

$$q = \boxed{-17.1 \text{ cm}}$$

The negative sign indicates that the image is in the same medium as the object (the side of incident light), in agreement with our ray diagram, and therefore must be virtual.

To find the image height, we first use Equation 23.8 for the magnification:

$$M = -\frac{n_1 q}{n_2 p} = -\frac{1.50(-17.1 \text{ cm})}{1.00(20.0 \text{ cm})} = \frac{h'}{h} = 1.28$$

Therefore,

$$h' = 1.28h = (1.28)(2.00 \text{ cm}) = \boxed{2.56 \text{ cm}}$$

The positive value for M indicates an upright image.

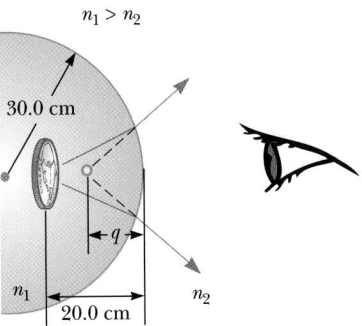

Figure 23.16   (Example 23.5) A coin embedded in a glass ball forms a virtual image between the coin and the glass surface.

EXAMPLE 23.6   The One That Got Away

A small fish is swimming at a depth of d below the surface of a pond (Fig. 23.17). What is the *apparent depth* of the fish as viewed from directly overhead?

Reasoning   In this example, the refracting surface is flat, and so R is infinite. Hence, we can use Equation 23.9 to determine the location of the image.

Solution   The facts that $n_1 = 1.33$ for water and $p = d$ give us

$$q = -\frac{n_2}{n_1} p = -\frac{1}{1.33} d = \boxed{-0.752d}$$

Again, because q is negative, the image is virtual, as indicated in Figure 23.17. The apparent depth is three fourths the actual depth. For instance, if $d = 4.0$ m, $q = -3.0$ m.

Exercise   If the fish is 12 cm long, how long is its image?

Answer   12 cm

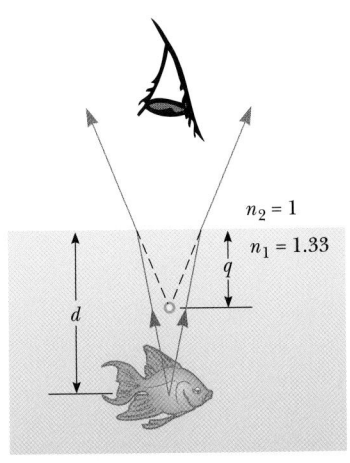

Figure 23.17   (Example 23.6) The apparent depth, q, of the fish is less than the true depth, d.

23.5   ATMOSPHERIC REFRACTION

Images formed by refraction in our atmosphere lead to some interesting results. In this section we look at two examples. A situation that occurs daily is the visibility of the Sun at dusk even though it has passed below the horizon. Figure 23.18 shows why this occurs. Rays of light from the Sun strike the Earth's atmosphere (represented by the shaded area around the Earth) and are bent as they pass into a medium that has an index of refraction different from that of the almost empty space in which they have been traveling. The bending in this situation differs somewhat from the bending we have considered previously in being gradual and continuous as the light moves through the atmosphere toward an observer at point *O*. This is because the light moves through layers of air that have a continuously changing index of refraction. When the rays reach the observer, the eye follows them back along the direction from which they appear to have come (indicated by the dashed path in the figure). The end result is that the Sun is seen to be above the horizon even after it has fallen below it.

The **mirage** is another phenomenon of nature produced by refraction in the atmosphere. A mirage can be observed when the ground is so hot that the air directly above it is warmer than the air at higher elevations. The desert is, of course, a region in which such circumstances prevail, but mirages are also seen on heated roadways during the summer. The layers of air at different heights above the Earth have different densities and different refractive indices. The effect this can have is pictured in Figure 23.19. In this situation the observer sees a tree in two different ways. One group of light rays reaches the observer by the straight-line path *A*, and the eye traces these rays back to see the tree in the normal fashion. In addition, a second group of rays travels along the curved path *B*. These rays are directed toward the ground and are then bent as a result of refraction. As a consequence, the observer also sees an inverted image of the tree as he traces these rays back to the point at which they appear to have originated. Because an upright image and an inverted image are seen when the image of a tree is observed in a reflecting pool of water, the observer unconsciously calls on this past experience and concludes that a pool of water must be in front of the tree.

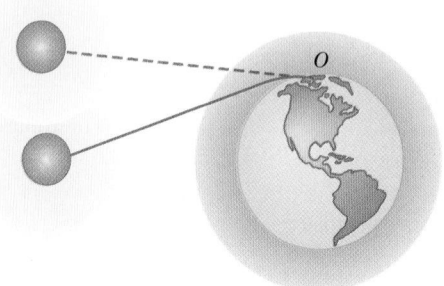

Figure 23.18   Because of refraction, an observer at *O* sees the Sun even though it has fallen below the horizon.

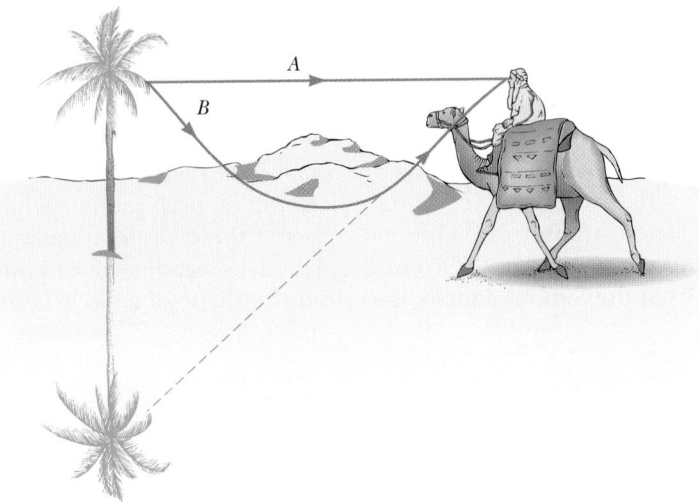

Figure 23.19 A mirage is produced by the bending of light rays in the atmosphere when there are large temperature differences between the ground and the air.

A common meteorological observance on cold winter days or nights is a halo around the Sun or Moon, as shown in Figure 23.20. An explanation of how these halos form depends on the details of refraction of light as the light passes through ice crystals high in the atmosphere. These halos are most commonly seen on winter days because an abundance of ice crystals in the sky is necessary for their production. The ice crystals form in a hexagonal pattern such as that shown in Figure 23.21a. When a light ray passes through, it is deflected by an angle δ known as the

Figure 23.20 Photograph of a halo observed around the Sun on a cold winter day. *(Ron Chapple/FPG)*

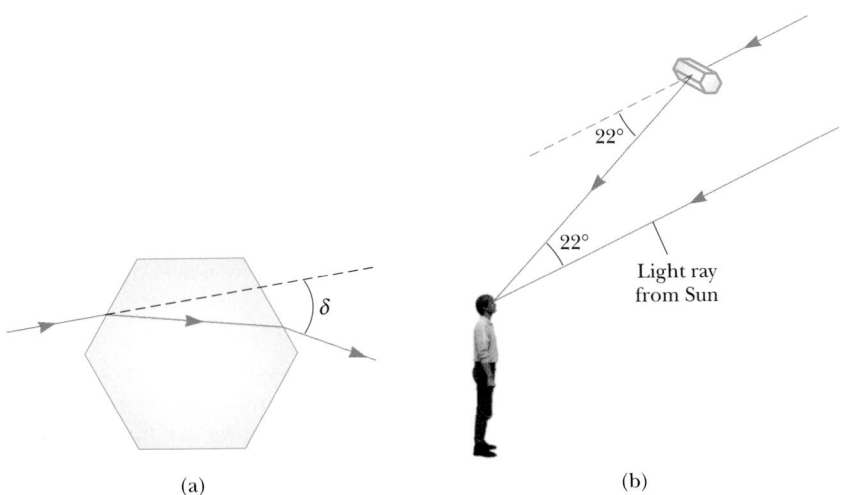

(a)   (b)

Figure 23.21 (a) Hexagonal ice crystals refract rays from the Sun or Moon through an angle δ. (b) The angle of deviation for a light ray passing through an ice crystal in the sky is 22°.

angle of deviation. When a ray of light passes through a hexagonal prism symmetrically, as shown in Figure 23.21b, the angle of deviation is a minimum and equal to 22°. Rotating the crystal in any direction either clockwise or counterclockwise increases this angle. However, there is a range of angles several degrees wide about this minimum deviation in which the light ray still passes through at an angle of deviation very close to 22°. This means that when light rays from many directions are incident on the face of the ice crystal, there will be preferential refraction for angles of deviation around 22°. One can observe these preferentially scattered rays by looking at an angle of 22° from the Sun. The location of all points at an angle of 22° from the Sun, of course, traces out the shape of a circle centered on the Sun.

23.6   THIN LENSES

A typical **thin lens** consists of a piece of glass or plastic, ground so that each of its two refracting surfaces is a segment of either a sphere or a plane. Lenses are commonly used to form images by refraction in optical instruments, such as cameras, telescopes, and microscopes. The equation that relates object and image distances for a lens is virtually identical to the mirror equation derived earlier, and the method used to derive it is also similar.

Figure 23.22 shows some representative shapes of lenses. Notice that we have placed these lenses in two groups. Those in Figure 23.22a are thicker at the center than at the rim, and those in Figure 23.22b are thinner at the center than at the rim. The lenses in the first group are examples of **converging lenses,** and those in the second group are **diverging lenses.** The reason for these names will become apparent shortly.

As we did for mirrors, it is convenient to define a point called the **focal point** for a lens. For example, in Figure 23.23a, a group of rays parallel to the axis passes through the focal point, F, after being converged by the lens. The distance from the focal point to the lens is called the **focal length, f. The focal length is the image distance that corresponds to an infinite object distance.** Recall that we are considering the lens to be very thin. As a result, it makes no difference whether we take the focal length to be the distance from the focal point to the surface of the lens or the distance from the focal point to the center of the lens, because the difference between these two lengths is negligible. A thin lens has *two* focal points, as illustrated in Figure 23.23, corresponding to parallel rays traveling from the left and from the right.

Rays parallel to the axis diverge after passing through a lens of biconcave shape in Figure 23.23b. In this case, the focal point is defined to be the point at which the diverged rays appear to originate, labeled F in the figure. Figures 23.23a and 23.23b indicate why the names *converging* and *diverging* are applied to these lenses.

Consider a ray of light passing through the center of a lens, labeled ray 1 in Figure 23.24. For a thin lens, a ray passing through the center is undeviated. Ray 2 in Figure 23.24 is parallel to the principal axis of the lens (the horizontal axis

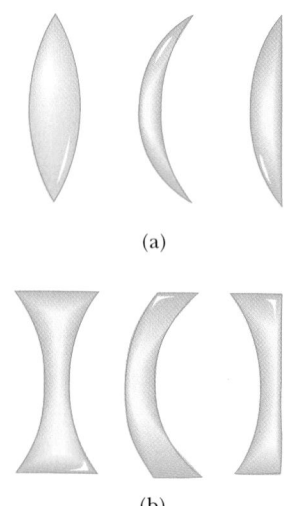

(a)

(b)

Figure 23.22 Lens shapes. (a) Converging lenses have positive focal lengths and are thickest at the middle. From left to right, these are biconvex, convex-concave, and plano-convex. (b) Diverging lenses have negative focal lengths and are thickest at the edges. From left to right are biconcave, convex-concave, and plano-convex lenses.

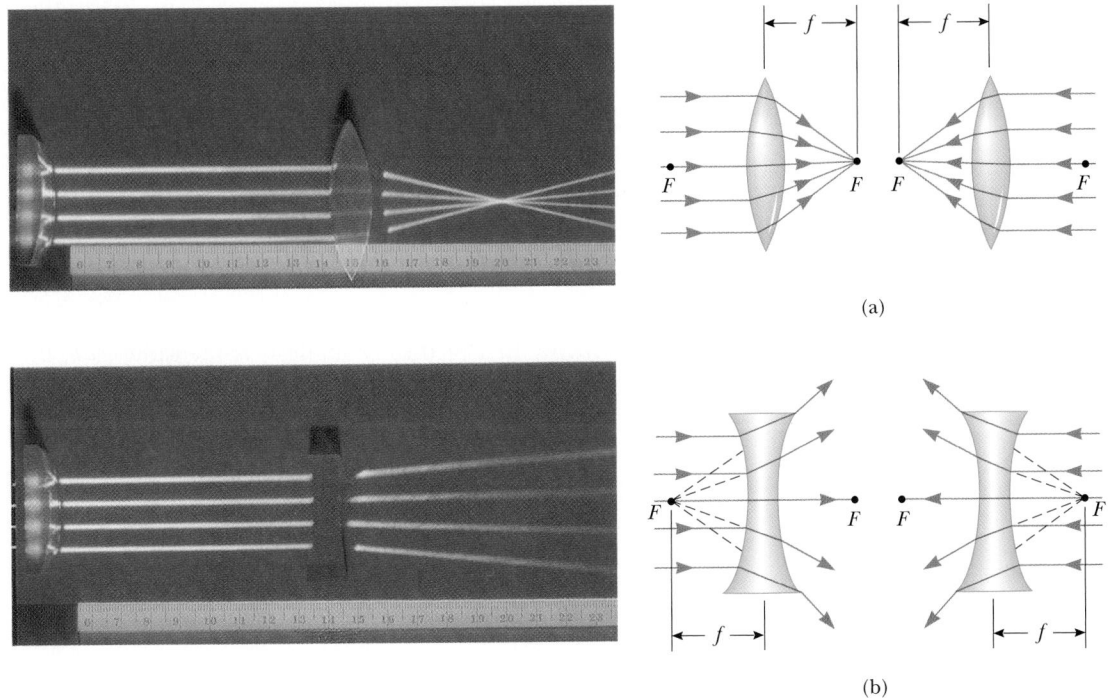

(a)

(b)

Figure 23.23 *(Left)* Photographs of the effects of converging and diverging lenses on parallel rays. *(Courtesy of Jim Lehman, James Madison University)* *(Right)* The focal points of (a) the biconvex lens and (b) the biconcave lens.

passing through *O*), and as a result it passes through the focal point, *F*, after re-  fraction. The point at which rays 1 and 2 intersect is the image point.

We first note that the tangent of the angle α can be found by using the shaded triangles in Figure 23.24:

14.8, BOTH SECTIONS

$$\tan \alpha = \frac{h}{p} \quad \text{or} \quad \tan \alpha = -\frac{h'}{q}$$

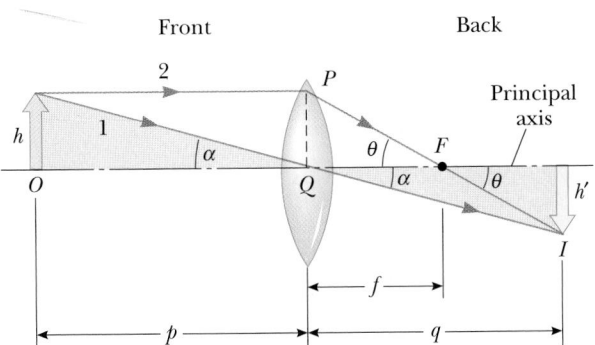

Figure 23.24 A geometric construction for developing the thin-lens equation.

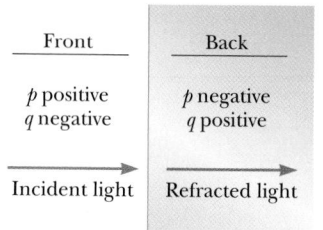

Front	Back
p positive q negative	p negative q positive
Incident light	Refracted light

Figure 23.25 A diagram for obtaining the signs of p and q for a thin lens or a refracting surface.

From this we find that

$$M = \frac{h'}{h} = -\frac{q}{p} \qquad [23.10]$$

Thus, the equation for magnification by a lens is the same as the equation for magnification by a mirror. We also note from Figure 23.24 that the tangent of θ is

$$\tan \theta = \frac{PQ}{f} \qquad \text{or} \qquad \tan \theta = -\frac{h'}{q-f}$$

However, the height PQ used in the first of these equations is the same as h, the height of the object. Therefore,

$$\frac{h}{f} = -\frac{h'}{q-f}$$

$$\frac{h'}{h} = -\frac{q-f}{f}$$

Using this in combination with Equation 23.10 gives

$$\frac{q}{p} = \frac{q-f}{f}$$

which reduces to

Thin lens equation ▶

$$\frac{1}{p} + \frac{1}{q} = \frac{1}{f} \qquad [23.11]$$

This equation, called the **thin-lens equation,** can be used with both converging and diverging lenses if we adhere to a set of sign conventions. Figure 23.25 is useful for obtaining the signs of p and q, and Table 23.3 gives the complete sign conventions for lenses. Note that **a converging lens has a positive focal length** under this convention, and **a diverging lens has a negative focal length.** Hence the names *positive* and *negative* are often given to these lenses.

TABLE 23.3 Sign Conventions for Thin Lenses

p is $+$ if the object is in front of the lens.
p is $-$ if the object is in back of the lens.

q is $+$ if the image is in back of the lens.
q is $-$ if the image is in front of the lens.

R_1 and R_2 are $+$ if the center of curvature for each surface is in back of the lens.
R_1 and R_2 are $-$ if the center of curvature for each surface is in front of the lens.

f is $+$ for a converging lens.
f is $-$ for a diverging lens.

Note: p = object distance; q = image distance; R_1 = radius of curvature of front surface; R_2 = radius of curvature of back surface; f = focal length.

The focal length for a lens in air is related to the curvatures of its front and back surfaces and to the index of refraction, n, of the lens material by

$$\frac{1}{f} = (n - 1)\left(\frac{1}{R_1} - \frac{1}{R_2}\right)$$ [23.12] ◀ Lens maker's equation

where R_1 is the radius of curvature of the front surface of the lens and R_2 is the radius of curvature of the back surface. (As with mirrors, we arbitrarily call the side from which the light approaches the *front* of the lens.) Equation 23.12 enables us to calculate the focal length from the known properties of the lens. It is called the **lens maker's equation.**

Ray Diagrams for Thin Lenses

Ray diagrams are convenient for determining the image formed by a thin lens or a system of lenses. They should also help clarify the sign conventions we have already discussed. Figure 23.26 illustrates this method for three single-lens situations. To locate the image formed by a converging lens (Fig. 23.26a and b), the following three rays are drawn from the top of the object:

1. The first ray is drawn parallel to the principal axis. After being refracted by the lens, this ray passes through (or appears to come from) one of the focal points.
2. The second ray is drawn through the center of the lens. This ray continues in a straight line.
3. The third ray is drawn through the other focal point and emerges from the lens parallel to the principal axis.

A similar construction is used to locate the image formed by a diverging lens, as shown in Figure 23.26c. The point of intersection of *any two* of the rays in these

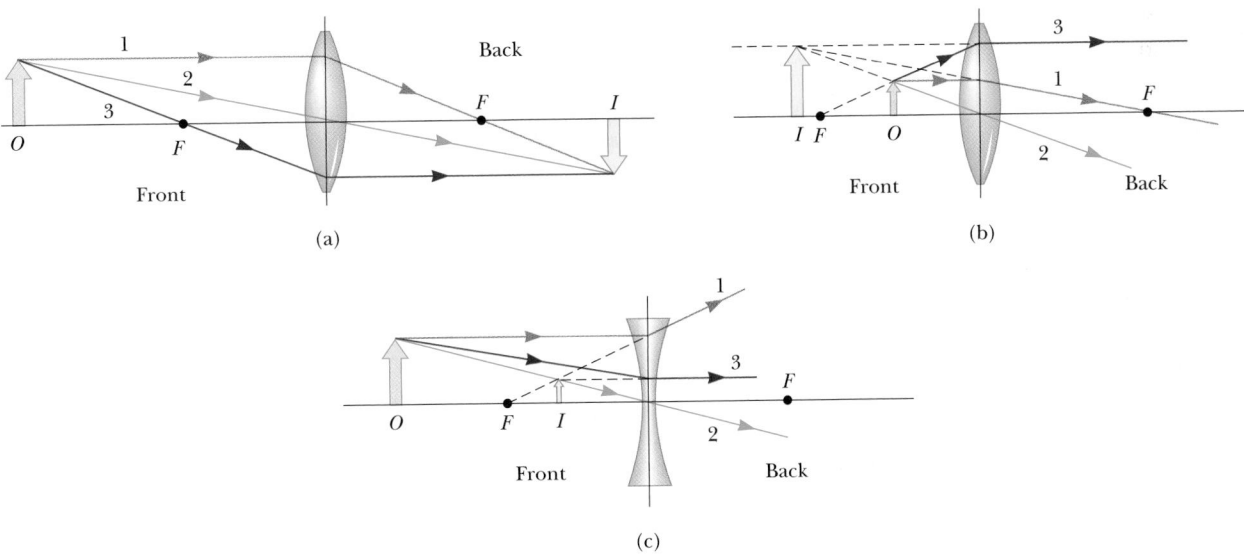

(a)

(b)

(c)

Figure 23.26 Ray diagrams for locating the image of an object. (a) The object is outside the focal point of a converging lens. (b) The object is inside the focal point of a converging lens. (c) The object is outside the focal point of a diverging lens.

diagrams can be used to locate the image. The third ray serves as a check on construction.

For the converging lens in Figure 23.26a, where the object is *outside* the front focal point ($p > f$), the image is real and inverted. When the real object is *inside* the front focal point ($p < f$), as in Figure 23.26b, the image is virtual and upright. For the diverging lens of Figure 23.26c, the image is virtual and upright.

APPLICATION

35-mm Slide Projectors.

You perform an exercise something like that shown in Figure 23.26a when you show 35-mm slides with a slide projector. The slide projector contains a lens that will project a real image onto a screen. The slide represents the object, which is close to but outside the focal point of the lens. Because the distance between the lens and the screen is much larger than the distance between the object and the lens, the magnification is large. This means that small 35-mm slides can be projected onto a screen to a size of several meters. If you have operated a slide projector, you probably noticed that you must insert the slides upside down. Because the image of the slide is inverted, the slide must be upside down in the projector in order to obtain an image on the screen that is right-side up for the audience.

Problem-Solving Strategy

Lenses and Mirrors

Your success or failure in working lens and mirror-problems will be determined largely by whether or not you make sign errors when substituting into the lens and mirror equations. The only way to ensure that you don't make sign errors is to become adept at using the sign conventions. The best way to do this is to work a multitude of problems on your own. Watching an instructor or reading the example problems is no substitute for practice.

Applying Physics 3

Diving masks often have a lens built into the glass for divers who do not have perfect vision. This allows the individual to dive without the necessity of glasses, because the lenses in the faceplate perform the necessary refraction to produce clear vision. Normal glasses have lenses that are curved on both the front and rear surfaces. The lenses in a diving mask faceplate often only have curved surfaces on the inside of the glass. Why is this design desirable?

Explanation   The main reason for curving only the inner surface of the lenses in the diving mask faceplate is so that the diver can see clearly while underwater and in the air. If there were curved surfaces on both the front and the back of the diving lens, there would be two refractions. The lens could be designed so that these two refractions would give clear vision while the diver is in air. When the diver goes underwater, however, the refraction between the water and the glass at the first interface is now different, because the index of refraction of water is different from that of air. Thus, the vision will not be clear underwater.

Thinking Physics 3

Consider a glass plano-convex lens, flat on one side and convex on the other. You project three laser beams through it, as shown in Figure 23.27a, and measure the focal length, f—that is, the distance from the lens to the point at which the three beams cross. Now you hold the flat side of the lens against the glass of an aquarium filled with water, as in Figure 23.27b. When you shine the laser beams through the lens from the outside of the aquarium, will the beams cross at a point closer to the lens than before, farther away, or at the same distance? What if you direct the laser beams through the lens from the other side of the aquarium, as in Figure 23.27c? What happens in this case?

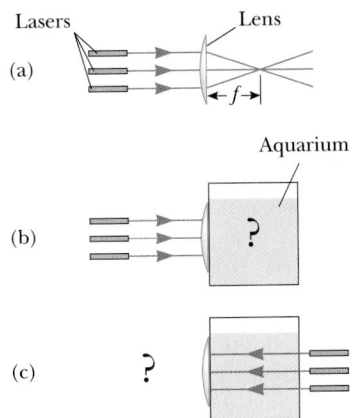

Figure 23.27 (Thinking Physics 3)

Explanation The laser beam going through the center of the lens is unaffected by the lens in all three cases. We need to look at what happens to the outer beams. In Figure 23.27a, these two laser beams will refract toward the normal as they pass from the air into the glass of the lens. Thus, they will be deviated *toward* the central beam. As they reach the other side, and pass from the glass back into the air, they will deviate away from the normal, which causes additional deviation toward the central beam. As a result, all three beams cross at the focal point. When the flat side of the lens is held against the glass side of the aquarium as in Figure 23.27b, the outgoing laser beams pass through the glass of the aquarium without additional refraction, but then enter *water*. The change in index of refraction in going from glass to water is much smaller than in the previous case, and the deviation toward the central beam is less. As a result, the crossing point for the three beams is farther from the lens. When the laser beams are directed through the water and then through the lens, as in Figure 23.27c, we must consider two factors. First, the focal length of a lens is independent of which way the light passes through it. Second, in the situation shown, the laser beams experience no refraction on entering the flat side of the lens, because they strike the glass at normal incidence. Thus, all of the refraction occurs at the interface of the air with the curved side of the lens. This is exactly the same situation as if the laser beams had struck the flat side of the lens while it was in air. Thus, there is no effect of the water in the aquarium. The laser beams cross at the same distance from the lens as they did in the first diagram. The third situation is very similar to the curvature of a lens on the inside of a diving mask, discussed previously.

EXAMPLE 23.7 The Lens Maker's Equation

The biconvex lens of Figure 23.28 has an index of refraction of 1.50. The radius of curvature of the front surface is $R_1 = 10.0$ cm, and that of the back surface is $R_2 = -15.0$ cm. Find the focal length of the lens.

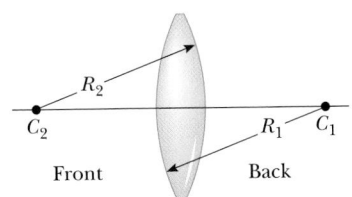

Figure 23.28 This converging lens has two curved surfaces with radii of curvature R_1 and R_2. The center of curvature R_1 lies to the right of the lens, and the center of R_2 lies to the left.

Solution From the sign conventions in Table 23.3 we find that $R_1 = +10.0$ cm and $R_2 = -15.0$ cm. Thus, using the lens maker's equation, we have

$$\frac{1}{f} = (n - 1)\left(\frac{1}{R_1} - \frac{1}{R_2}\right) = (1.50 - 1)\left(\frac{1}{10.0 \text{ cm}} - \frac{1}{-15.0 \text{ cm}}\right)$$

$$f = \boxed{12.0 \text{ cm}}$$

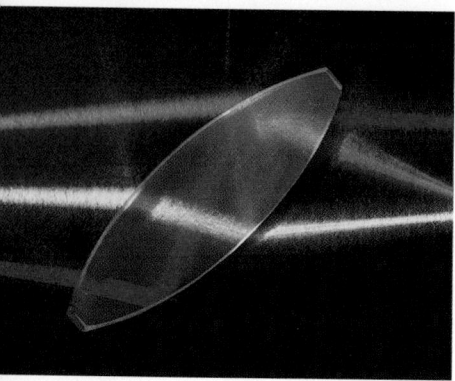

Three beams of light are incident on a biconvex lens. Part of each beam is reflected, and part refracted, by the glass lens at the left air-glass interface. The refracted beams entering the lens are then partially reflected and partially refracted at the opposite glass-air interface. How do you explain the fact that most of the red beam is reflected, and the yellow and blue beams are mostly refracted, at the second glass-air interface? (*Richard Megna, Fundamental Photographs*)

Workbook Problem 93 — (Workbook page 193) also deals with an image formed by a thin lens.

EXAMPLE 23.8   Images Formed by a Converging Lens

A converging lens of focal length 10.0 cm forms images of objects placed (a) 30.0 cm, (b) 10.0 cm, and (c) 5.00 cm from the lens. In each case, find the image distance and describe the image.

Solution   (a) The thin-lens equation, Equation 23.11, can be used to find the image distance:

$$\frac{1}{p} + \frac{1}{q} = \frac{1}{f}$$

$$\frac{1}{30.0 \text{ cm}} + \frac{1}{q} = \frac{1}{10.0 \text{ cm}}$$

$$q = 15.0 \text{ cm}$$

The positive sign for the image distance tells us that the image is real and on the back side of the lens (Fig. 23.25). The magnification of the lens is

$$M = -\frac{q}{p} = -\frac{15.0 \text{ cm}}{30.0 \text{ cm}} = -0.500$$

Thus, the image is reduced in height by one half, and the negative sign for M tells us that the image is inverted. The situation is like that in Figure 23.26a.

(b) No calculation is necessary for this case because we know that, when the object is placed at the focal point, the image is formed at infinity. This is readily verified by substituting $p = 10.0$ cm into the lens equation.

(c) We now move inside the focal point, to an object distance of 5.00 cm. In this case, the lens equation gives

$$\frac{1}{5.00 \text{ cm}} + \frac{1}{q} = \frac{1}{10.0 \text{ cm}}$$

$$q = -10.0 \text{ cm}$$

$$M = -\frac{q}{p} = -\left(\frac{-10.0 \text{ cm}}{5.00 \text{ cm}}\right) = 2.00$$

The negative image distance tells us that the image is virtual and formed on the side of the lens from which the light is incident, the front side (Fig. 23.25). The image is enlarged, and the positive sign for M tells us that the image is upright, as shown in Figure 23.26b.

EXAMPLE 23.9   The Case of a Diverging Lens

Repeat the problem of Example 23.8 for a *diverging* lens of focal length 10.0 cm.

Solution   (a) Let us apply the lens equation with an object distance of 30.0 cm:

$$\frac{1}{p} + \frac{1}{q} = \frac{1}{f}$$

$$\frac{1}{30.0 \text{ cm}} + \frac{1}{q} = -\frac{1}{10.0 \text{ cm}}$$

$$q = -7.50 \text{ cm}$$

The magnification is

$$M = -\frac{q}{p} = -\left(\frac{-7.50 \text{ cm}}{30.0 \text{ cm}}\right) = 0.250$$

Thus, the image is virtual, smaller than the object, and upright.

(b) When the object is at the focal point, $p = 10.0$ cm, we have

$$\frac{1}{10.0 \text{ cm}} + \frac{1}{q} = -\frac{1}{10.0 \text{ cm}}$$

$$q = -5.00 \text{ cm}$$

$$M = -\frac{q}{p} = -\left(\frac{-5.00 \text{ cm}}{10.0 \text{ cm}}\right) = 0.500$$

(c) When the object is inside the focal point, at 5.00 cm, we have

$$\frac{1}{5.00 \text{ cm}} + \frac{1}{q} = -\frac{1}{10.0 \text{ cm}}$$

$$q = -3.33 \text{ cm}$$

$$M = -\left(\frac{-3.33 \text{ cm}}{5.00 \text{ cm}}\right) = 0.667$$

Again, we have a virtual image that is smaller than the object and upright, as in Figure 23.26c.

Combination of Thin Lenses

If two thin lenses are used to form an image, the system can be treated in the following manner. First, the image of the first lens is calculated as though the second lens were not present. The light then approaches the second lens *as if* it had come from the image formed by the first lens. Hence, **the image formed by the first lens is treated as the object for the second lens.** The image formed by the second lens is the final image of the system. If the image formed by the first lens lies on the back side of the second lens, then the image is treated as a virtual object for the second lens (that is, p is negative). The same procedure can be extended to a system of three or more lenses. The overall magnification of a system of thin lenses is the *product* of the magnifications of the separate lenses.

EXAMPLE 23.10 Two Lenses in a Row

Two converging lenses are placed 20.0 cm apart, as shown in Figure 23.29. If the first lens has a focal length of 10.0 cm and the second has a focal length of 20.0 cm, locate the final image formed of an object 30.0 cm in front of the first lens. Find the magnification of the system.

Reasoning We apply the thin-lens equation to both lenses. The image formed by the first lens is treated as the object for the second lens. Also, we use the fact that the total magnification of the system is the product of the magnifications produced by the separate lenses.

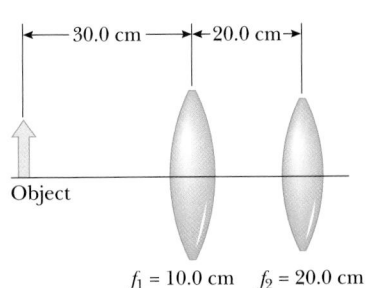

Figure 23.29 (Example 23.10)

Solution The location of the image formed by the first lens is found via the thin-lens equation:

$$\frac{1}{30.0 \text{ cm}} + \frac{1}{q} = \frac{1}{10.0 \text{ cm}}$$

$$q = 15.0 \text{ cm}$$

The magnification of this lens is

$$M_1 = -\frac{q}{p} = -\frac{15.0 \text{ cm}}{30.0 \text{ cm}} = -0.500$$

The image formed by this lens becomes the object for the second lens. Thus, the object distance for the second lens is 5.00 cm. We again apply the thin-lens equation to find the location of the final image.

$$\frac{1}{5.00 \text{ cm}} + \frac{1}{q} = \frac{1}{20.0 \text{ cm}}$$

$$q = -6.67 \text{ cm}$$

The magnification of the second lens is

$$M_2 = -\frac{q}{p} = -\frac{(-6.67 \text{ cm})}{5.00 \text{ cm}} = 1.33$$

Thus, the final image is 6.67 cm to the left of the second lens, and the overall magnification of the system is

$$M = M_1 M_2 = (-0.500)(1.33) = -0.667$$

The negative sign indicates that the final image is inverted with respect to the initial object.

Exercise If the two lenses in Figure 23.29 are separated by 10.0 cm, locate the final image and find the magnification of the system.

Answer 4.00 cm behind the second lens; $M = -0.400$

23.7 LENS ABERRATIONS

One of the basic problems of lenses and lens systems is the imperfect quality of the images, which is largely the result of defects in shape and form. The simple theory of mirrors and lenses assumes that rays make small angles with the principal axis and that all rays reaching the lens or mirror from a point source are focused at a single point, producing a sharp image. Clearly, this is not always true in the real world. Where the approximations used in this theory do not hold, imperfect images are formed.

If one wishes to analyze image formation precisely, it is necessary to trace each ray, using Snell's law, at each refracting surface. This procedure shows that there is no single point image; instead, the image is blurred. The departures of real (imperfect) images from the ideal predicted by the simple theory are called **aberrations.** Two common types of aberrations are spherical aberration and chromatic aberration. Photographs of three forms of lens aberrations are shown in Figure 23.30.

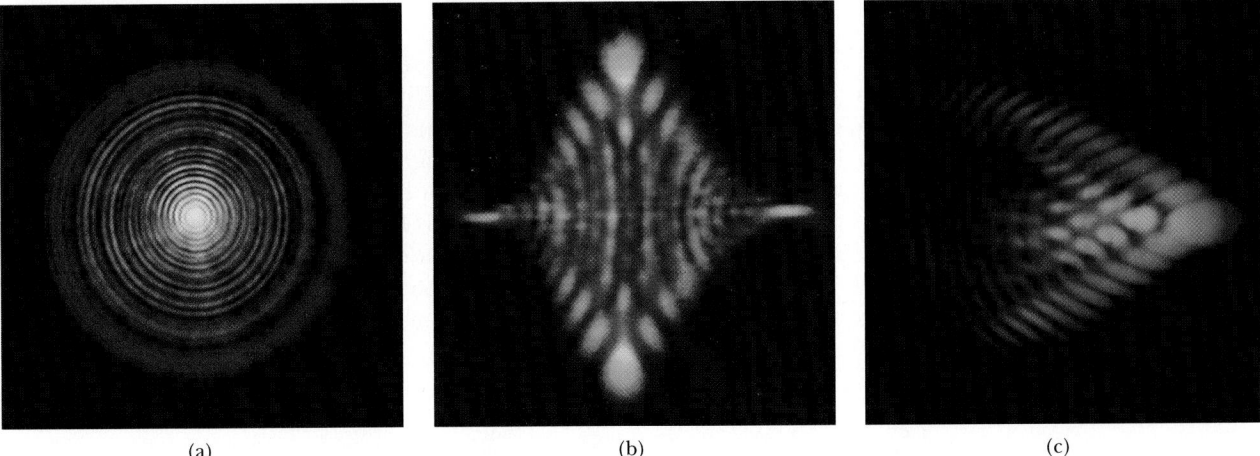

(a)   (b)   (c)

Figure 23.30 Lenses can produce varied forms of aberrations, as shown by these blurred photographic images of a point source. (a) Spherical aberration occurs when light passing through the lens at different distances from the optical axis is focused at different points. (b) Astigmatism is an aberration that occurs when the object is not on the optical axis of the lens. (c) Coma. This aberration occurs when light passing through the lens far from the optical axis focuses at a different part of the focal plane from light passing near the center of the lens. *(Photos by Norman Goldberg)*

Spherical Aberration

Spherical aberration results from the fact that the focal points of light rays far from the principal axis of a spherical lens (or mirror) are different from the focal points of rays with the same wavelength passing near the axis. Figure 23.31 illustrates spherical aberration for parallel rays passing through a converging lens. Rays near the middle of the lens are imaged farther from the lens than rays at the edges. Hence, there is no single focal length for a lens.

Most cameras are equipped with an adjustable aperture to control the light intensity and, when possible, reduce spherical aberration. (An aperture is an opening that controls the amount of light transmitted through the lens.) As the aperture size is reduced, sharper images are produced, because only the central portion of the lens is exposed to the incident light when the aperture is very small. At the same time, however, progressively less light is imaged. To compensate for this loss, a longer exposure time is used. An example of the results obtained with small apertures is the sharp image produced by a "pinhole" camera, with an aperture size of approximately 1 mm.

In the case of mirrors used for very distant objects, one can eliminate, or at least minimize, spherical aberration by employing a parabolic rather than spherical surface. Parabolic surfaces are not used in many applications, however, because they are very expensive to make with high-quality optics. Parallel light rays incident on such a surface focus at a common point. Parabolic reflecting surfaces are used in many astronomical telescopes to enhance the image quality. They are also used in searchlights, in which a nearly parallel light beam is produced from a small lamp placed at the focus of the reflecting surface.

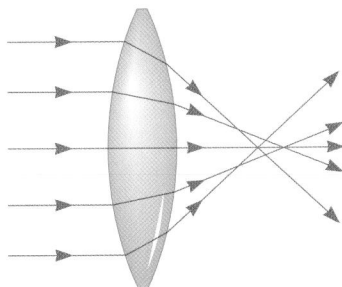

Figure 23.31 A spherical aberration produced by a converging lens. Does a diverging lens produce spherical aberration? (Angles are greatly exaggerated for clarity.)

APPLICATION

Adjustable Camera Apertures.

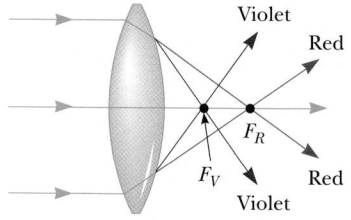

Figure 23.32 A chromatic aberration produced by a converging lens. Rays of different wavelengths focus at different points. (Angles are greatly exaggerated for clarity.)

Chromatic Aberration

The fact that different wavelengths of light refracted by a lens focus at different points gives rise to chromatic aberration. In Chapter 22 we described how the index of refraction of a material varies with wavelength. When white light passes through a lens, one finds, for example, that violet light rays are refracted more than red light rays (Fig. 23.32); thus, the focal length for red light is greater than that for violet light. Other wavelengths (not shown in Fig. 23.32) would have intermediate focal points. The chromatic aberration for a diverging lens is opposite that for a converging lens. Chromatic aberration can be greatly reduced by the use of a combination of converging and diverging lenses made from two different types of glass.

SUMMARY

Images are formed where rays of light intersect or where they appear to originate. A **real image** is one in which light intersects, or passes through, an image point. A **virtual image** is one in which the light does not pass through the image point but appears to diverge from that point.

The image formed by a flat mirror has the following properties:

1. The image is as far behind the mirror as the object is in front.
2. The image is unmagnified, virtual, and upright.

The **magnification, M,** of a mirror is defined as the ratio of **image height, h',** to **object height, h,** which is the negative of the ratio of image distance, q, to object distance, p:

$$M = \frac{h'}{h} = -\frac{q}{p} \qquad [23.2]$$

The **object distance** and **image distance** for a spherical mirror of radius R are related by the **mirror equation:**

$$\frac{1}{p} + \frac{1}{q} = \frac{1}{f} \qquad [23.6]$$

where $f = R/2$ is the **focal length** of the mirror.

An image can be formed by refraction at a spherical surface of radius R. The object and image distances for refraction from such a surface are related by

$$\frac{n_1}{p} + \frac{n_2}{q} = \frac{n_2 - n_1}{R} \qquad [23.7]$$

The **magnification of a refracting surface** is

$$M = \frac{h'}{h} = -\frac{n_1 q}{n_2 p} \qquad [23.8]$$

where the object is located in the medium with index of refraction n_1 and the image is formed in the medium with index of refraction n_2.

The **magnification for a thin lens** is

$$M = \frac{h'}{h} = -\frac{q}{p} \qquad \text{[23.10]}$$

and the object and image distances are related by the **thin-lens equation:**

$$\frac{1}{p} + \frac{1}{q} = \frac{1}{f} \qquad \text{[23.11]}$$

Aberrations are responsible for the formation of imperfect images by lenses and mirrors. **Spherical aberration** results from the fact that the focal points of light rays far from the principal axis of a spherical lens or mirror are different from those of rays passing through the center. **Chromatic aberration** arises from the fact that light rays of different wavelengths focus at different points when re-fracted by a lens.

MULTIPLE-CHOICE QUESTIONS

1. What is the closest an object and a screen can be placed so that an image can be formed on the screen using a diverging lens of focal length $-f$?
   (a) $f/2$ (b) f (c) $2f$ (d) $4f$ (e) No image will form.

2. Two thin lenses, one with focal length f and the other with focal length $-f$, are placed in contact along their optical axis. The focal length of the combination of these lenses is
   (a) 0   (b) $f/2$ (c) $2f$ (d) $-2f$ (e) infinite

3. If a man's face is 30.0 cm in front of a concave shaving mirror creating an upright image 1.50 times as large as the object, what is the mirror's focal length?
   (a) 12.0 cm   (b) 20.0 cm   (c) 70.0 cm   (d) 90.0 cm

4. Two thin lenses of focal lengths 15 and 10 cm, respec-tively, are separated by 35 cm along a common axis. The 15-cm lens is located to the left of the 10-cm lens. An object is now placed 50 cm to the left of the 15 cm lens. What is the magnification of the final image taken with respect to the object?
   (a) 0.6   (b) 1.2   (c) 2.4   (d) 3.6

5. A biconvex lens with two curved surfaces has a front sur-face with a radius of curvature of 10.0 cm, a back surface with a radius of curvature of 20.0 cm, and is made from material with an index of refraction of 2.50. What is the focal length of the lens?
   (a) 13.3 cm   (b) -13.3 cm   (c) 4.44 cm   (d) 0.250 cm

CONCEPTUAL QUESTIONS

1. Tape a picture of yourself on a bathroom mirror. Stand several centimeters away from the mirror. Can you focus your eyes on *both* the picture taped to the mirror *and* your image in the mirror *at the same time*? So, where is the image of yourself?

2. A person spear fishing from a boat sees a fish located 3 m from the boat at an apparent depth of 1 m. To spear the fish, should the person aim at, above, or below the image of the fish?

3. A flat mirror creates a virtual image of your face. Sup-pose the flat mirror is combined with another optical el-ement. Can a flat mirror form a real image in such a combination?

4. Explain why a mirror cannot give rise to chromatic aber-ration.

5. You are taking a picture of yourself with a camera that uses an ultrasonic range finder to measure the distance to the object. When you take a picture of yourself in a mirror with this camera, your image is out of focus. Why?

6. A solar furnace can be constructed by using a concave mirror to reflect and focus sunlight into a furnace enclosure. What factors in the design of the reflecting mirror will guarantee that very high temperatures can be achieved?

7. A virtual image is often described as one through which the light rays do not actually travel, as they do for a real image. Can a virtual image be photographed?

8. The rearview mirror on a late-model car warns the user that objects may be closer than they appear. What kind of mirror is used, and why was that type selected?

9. Suppose you want to use a converging lens to project the image of two trees onto a screen. One tree is a distance x from the lens, the other is at $2x$, as in Figure Q23.9. You adjust the screen, so that the near tree is in focus. If you now want to move the screen so that the far tree is in focus, do you move the screen toward, or away from, the lens?

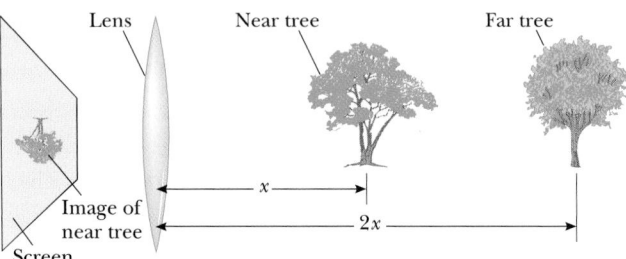

Figure Q23.9

10. Why does a clear stream always appear to be shallower than it actually is?

11. Can a converging lens be made to diverge light if placed in a liquid? How about a converging mirror?

12. A mirage is formed when the air gets gradually cooler as the height above the ground increases. What might happen if the air grows gradually warmer as the height is increased? This often happens over bodies of water or snow-covered ground: The effect is called *looming.*

13. A plastic sandwich bag filled with water can act as a crude converging lens in air. If the bag is filled with air and placed under water, is it a lens? If so, is it converging or diverging?

14. Lenses used in eyeglasses, whether converging or diverging, are always designed such that the middle of the lens curves away from the eye. Why?

15. Why does the focal length of a mirror not depend on the mirror material when the focal length of a lens does depend on the lens material?

16. If a cylinder of solid glass or clear plastic is placed above the words LEAD OXIDE and viewed from the side, as shown in Figure Q23.16, the word LEAD appears inverted but the word OXIDE does not. Explain.

Figure Q23.16

PROBLEMS

1, 2, 3 = straightforward, intermediate, challenging   ☐ = full solution available in Study Guide/Student Solutions Manual   = Core Concepts Workbook
WEB = solution posted at **http://www.harcourtcollege.com/physics/cptech**   = biomedical application   = Interactive Physics

Section 23.1 Flat Mirrors

1. A person walks into a room that has, on opposite walls, two flat mirrors producing multiple images. When the person is 5.00 ft from the mirror on the left wall and 10.0 ft from the mirror on the right wall, find the distances from the person to the first three images seen in the left-hand mirror.

2. Use Figure 23.2 to give a geometric proof that the virtual image formed by a flat mirror is the same distance behind the mirror as the object is in front of it.

Section 23.2 Images Formed by Spherical Mirrors

Section 23.3 Convex Mirrors and Sign Conventions

In the following problems, algebraic signs are not given. We leave it to you to determine the correct sign to use with each quantity, based on an analysis of the problem and the sign conventions in Table 23.1.

3. Under certain limiting conditions Equation 23.4 can be used for a flat mirror. (a) What value must be assumed for R if the mirror is flat? (b) What is the relationship

between the object and image distances, and what is the magnification, in this limiting case? **(c)** Do the results of **(b)** agree with the foregoing discussion of the flat mirror?

4. A spherical Christmas tree ornament is 6.00 cm in diameter. What is the magnification of an object placed 10.0 cm away from the ornament?

5. A concave spherical mirror has a radius of curvature of 20.0 cm. Locate the images for object distances of **(a)** 40.0 cm, **(b)** 20.0 cm, and **(c)** 10.0 cm. In each case, state whether the image is real or virtual and upright or inverted, and find the magnification.

6. A dentist uses a mirror to examine a tooth. The tooth is 1.00 cm in front of the mirror, and the image is formed 10.0 cm behind the mirror. Determine **(a)** the mirror's radius of curvature and **(b)** the magnification of the image.

7. A convex mirror with a radius of curvature of 0.550 m monitors the aisles in a store. Locate and describe the image of a customer 10.0 m from the mirror. Determine the magnification.

8. What type of mirror is required to form, on a wall 2.00 m from the mirror, an image of an object placed 10.0 cm in front of the mirror? What is the magnification of the image?

9. A convex mirror has a focal length of 20.0 cm. Determine the object location for which the image will be one half as tall as the object.

10. A 2.00-cm-high object is placed 10.0 cm in front of a mirror. What type of mirror and what radius of curvature are needed to create an upright image that is 4.00 cm high?

11. A 2.00-cm-high object is placed 3.0 cm in front of a concave mirror. If the image is 5.0 cm high and virtual, what is the focal length of the mirror?

12. A dedicated sports car enthusiast polishes the inside and outside surfaces of a hubcap that is a section of a sphere. When he looks into one side of the hubcap, he sees an image of his face 30.0 cm in back of the hubcap. He then turns the hubcap over, keeping it the same distance from his face. He now sees an image of his face 10.0 cm in back of the hubcap. **(a)** How far is his face from the hubcap? **(b)** What is the radius of curvature of the hubcap?

WEB 13. A concave makeup mirror is designed so that a person 25 cm in front of it sees an upright image magnified by a factor of two. What is the radius of curvature of the mirror?

14. A convex spherical mirror with a radius of curvature of 10.0 cm produces a virtual image one third the size of the real object. Where is the object?

15. A man standing 1.52 m in front of a shaving mirror produces an inverted image 18.0 cm in front of it. How close to the mirror should he stand if he wants to form

an upright image of his chin that is twice the chin's actual size?

16. A concave mirror has a focal length of 40.0 cm. Determine the object position that results in an upright image four times the height of the object.

17. A child holds a candy bar 10.0 cm in front of a convex mirror and notices that the image is only one half the size of the candy bar. What is the radius of curvature of the mirror?

18. It is observed that the size of a *real* image formed by a concave mirror is four times the size of the object when the object is 30.0 cm in front of the mirror. What is the radius of curvature of this mirror?

19. A spherical mirror is to be used to form an image, five times as tall as an object, on a screen positioned 5.0 m from the mirror. **(a)** Describe the type of mirror required. **(b)** Where should the mirror be positioned relative to the object?

Section 23.4 Images Formed by Refraction

20. A colored marble is dropped into a large tank completely filled with benzene ($n = 1.50$). What is the depth of the tank if the apparent depth of the marble, viewed from directly above the tank, is 35.0 cm?

21. A cubical block of ice 50.0 cm on an edge is placed on a level floor over a speck of dust. Locate the image of the speck, when viewed from directly above, if the index of refraction of ice is 1.309.

22. The top of a swimming pool is at ground level. If the pool is 2.00 meters deep, how far below ground level does the bottom of the pool appear to be located when **(a)** the pool is completely filled with water? **(b)** the pool is filled halfway with water?

WEB 23. A paperweight is made of a solid glass hemisphere of index of refraction 1.50. The radius of the circular cross section is 4.0 cm. The hemisphere is placed on its flat surface with the center directly over a 2.5-mm-long line drawn on a sheet of paper. What length of line is seen by someone looking vertically down on the hemisphere?

24. One end of a long glass rod ($n = 1.50$) is formed into the shape of a convex surface of radius 8.00 cm. An object is positioned in air along the axis of the rod. Find the image position that corresponds to each of the following object positions: **(a)** 20.0 cm, **(b)** 8.00 cm, **(c)** 4.00 cm, **(d)** 2.00 cm.

25. A transparent sphere of unknown composition is observed to form an image of the Sun on its surface opposite the Sun. What is the refractive index of the sphere material?

26. A flint glass block ($n = 1.66$) is 8.00 cm thick and rests on the bottom of an aquarium tank. The upper surface of the block is 12.0 cm below the surface of the water in the tank. Calculate the apparent thickness of the block

Problem 16 is the same as Workbook Problem 92 (Workbook page 192).

as viewed from above the water. (Assume nearly normal incidence.)

Section 23.6   Thin Lenses

27. A lens with radii of curvature of 52.5 cm and −61.9 cm has a focal length of +60.0 cm. Find its index of refraction.

28. A contact lens is made of plastic with an index of refraction of 1.58. The lens has a focal length of +25.0 cm, and its inner surface has a radius of curvature of +18.0 mm. What is the radius of curvature of the outer surface?

29. A glass converging lens ($n = 1.50$) is designed to look like the lens in Figure 23.28. The radius of the first surface is 15.0 cm, and the radius of the second surface is 10.0 cm. (a) Find the focal length of the lens when surrounded by air. Determine the positions of the images for object distances of (b) infinity, (c) $3f$, (d) f, and (e) $f/2$

30. A converging lens has a focal length of 20.0 cm. Locate the images for object distances of (a) 40.0 cm, (b) 20.0 cm, and (c) 10.0 cm. For each case, state whether the image is real or virtual and upright or inverted, and find the magnification.

31. A diverging lens ($n = 1.50$) is shaped like that in Figure 23.26c. The radius of the first surface is 15.0 cm, and that of the second surface is 10.0 cm. (a) Find the focal length of the lens. Determine the positions of the images for object distances of (b) infinity, (c) $3|f|$ (d) $|f|$, and (e) $|f|/2$.

32. A diverging lens has a focal length of 20.0 cm. Locate the images for object distances of (a) 40.0 cm, (b) 20.0 cm, and (c) 10.0 cm. For each case, state whether the image is real or virtual and upright or inverted, and find the magnification.

33. Where must an object be placed to have no magnification ($|M| = 1.00$) (a) for a converging lens of focal length 12.0 cm? (b) for a diverging lens of focal length 12.0 cm?

34. A convex lens of focal length 15.0 cm is used as a magnifying glass. At what distance from a postage stamp should you hold this lens to get a magnification of +2.00?

35. The nickel's image in Figure P23.35 has twice the diameter of the nickel when the lens is 2.84 cm from the nickel. Determine the focal length of the lens.

36. A slide projector is made by placing an illuminated slide slightly more than one focal length in front of a converging lens. If the lens has a 10.0-cm focal length and the object distance can be adjusted to any value between 10.2 cm and 11.0 cm, for what range of distances between projector and screen can a sharp image be obtained?

Figure P23.35

37. We want to form an image 30.0 cm from a diverging lens with a focal length of −40.0 cm. Where must we place the object? Determine the magnification.

38. An object's distance from a converging lens is ten times the focal length. How far is the image from the focal point? Express the answer as a fraction of the focal length.

39. A diverging lens is to be used to produce a virtual image one third as tall as the object. Where should the object be placed?

40. A diverging lens is used to form a virtual image of an object. The object is 80.0 cm to the left of the lens, and the image is 40.0 cm to the left of the lens. Determine the focal length of the lens.

41. A person uses a converging lens with a focal length of 12.5 cm to look at a gem. The lens forms a virtual image 30.0 cm from the lens. Determine the magnification. Is the image upright or inverted?

42. Two converging lenses, each of focal length 15.0 cm, are placed 40.0 cm apart, and an object is placed 30.0 cm in front of the first. Where is the final image formed, and what is the magnification of the system?

43. A converging lens is placed 30.0 cm to the right of a diverging lens of focal length 10.0 cm. A beam of parallel light enters the diverging lens from the left, and the beam is again parallel when it emerges from the converging lens. Calculate the focal length of the converging lens.

44. An object is placed 20 cm to the left of a converging lens of focal length 25 cm. A diverging lens of focal length 10 cm is 25 cm to the right of the converging lens. Find the position and magnification of the final image.

45. A microscope slide is placed in front of a converging lens with a focal length of 2.44 cm. The lens forms an image of the slide 12.9 cm from the slide. How far is the lens from the slide if the image is (a) real? (b) virtual?

46. An object is 5.00 m to the left of a flat screen. A converging lens with focal length $f = 0.800$ m is placed between the object and the screen. (a) Show that there are two positions for the lens that will cause an image to form on the screen, and determine how far these positions are from the object. (b) In what way would the two resultant images differ?

47. A 1.00-cm-high object is placed 4.00 cm to the left of a converging lens of focal length 8.00 cm. A diverging lens of focal length −16.00 cm is 6.00 cm to the right of the converging lens. Find the position and height of the final image. Is the image inverted or upright? Real or virtual?

48. Two converging lenses having focal lengths of 10.0 cm and 20.0 cm are placed 50.0 cm apart, as shown in Figure P23.48. The final image is to be located between the lenses, at the position indicated. (a) How far to the left of the first lens should the object be positioned? (b) What is the overall magnification? (c) Is the final image upright or inverted?

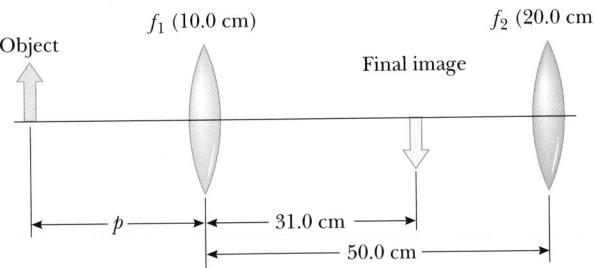

Figure P23.48

49. Lens L_1 in Figure P23.49 has a focal length of 15.0 cm and is located a fixed distance in front of the film plane of a camera. Lens L_2 has a focal length of 13.3 cm, and the distance, d, that it is from the film plane can be varied from 5.00 cm to 10.0 cm. Determine the range of distances for which objects can be focused on the film.

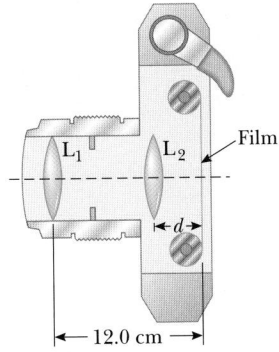

Figure P23.49

50. Consider two thin lenses, one of focal length f_1 and the other of focal length f_2, placed in contact with each other, as shown in Figure P23.50. Apply the thin lens equation to each of these lenses and combine these results to show that this combination of lenses behaves like a thin lens having a focal length f given by $1/f = 1/f_1 + 1/f_2$. Assume that the thicknesses of the lenses can be ignored in comparison to the other distances involved.

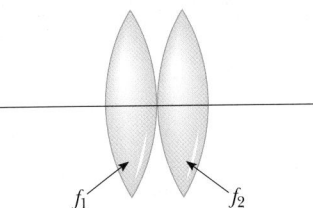

Figure P23.50

ADDITIONAL PROBLEMS

51. An object placed 10.0 cm from a concave spherical mirror produces a real image 8.00 cm from the mirror. If the object is moved to a new position 20.0 cm from the mirror, what is the position of the image? Is the final image real or virtual?

52. An object is placed 12 cm to the left of a diverging lens of focal length −6.0 cm. A converging lens of focal length 12 cm is placed a distance of d to the right of the diverging lens. Find the distance d that places the final image at infinity.

53. A converging lens with a 50.0-mm focal length is used to focus an image of a very distant scene onto a flat screen 35.0 mm wide. What is the angular width, α, of the scene included in the image on the screen?

54. The object in Figure P23.54 is midway between the lens and the mirror. The mirror's radius of curvature is 20.0 cm, and the lens has a focal length of −16.7 cm. Considering only the light that leaves the object and travels first toward the mirror, locate the final image formed by this system. Is this image real or virtual? Is it upright or inverted? What is the overall magnification?

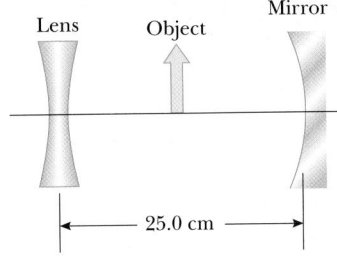

Figure P23.54

WEB 55. The lens and mirror in Figure P23.55 are separated by 1.00 m and have focal lengths of $+80.0$ cm and -50.0 cm, respectively. If an object is placed 1.00 m to the left of the lens as shown, locate the final image formed by light that has gone through the lens twice. State whether the image is upright or inverted, and determine the overall magnification.

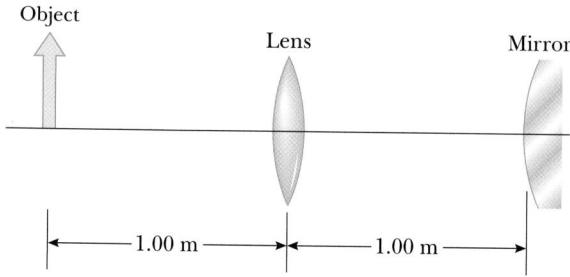

Figure P23.55

56. A converging lens of focal length 20.0 cm is separated by 50.0 cm from a converging lens of focal length 5.00 cm. (a) Find the position of the final image of an object placed 40.0 cm in front of the first lens. (b) If the height of the object is 2.00 cm, what is the height of the final image? Is the image real or virtual? (c) If the two lenses are now placed in contact with each other and the object is 5.00 cm in front of this combination, where will the image be located? (See Problem 50.)

57. To work this problem, use the fact that the image formed by the first surface becomes the object for the second surface. Figure P23.57 shows a piece of glass with index of refraction 1.50. The ends are hemispheres with radii 2.00 cm and 4.00 cm, and the centers of the hemispherical ends are separated by a distance of 8.00 cm. A point object is in air, 1.00 cm from the left end of the glass. Locate the image of the object due to refraction at the two spherical surfaces.

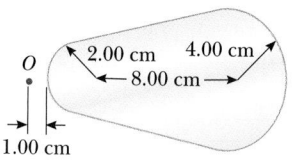

Figure P23.57

58. A "floating coin" illusion consists of two parabolic mirrors, each with a focal length of 7.5 cm, facing each other so that their centers are 7.5 cm apart (Fig. P23.58). If a few coins are placed on the lower mirror, an image of the coins forms at the small opening at the center of the top mirror. Show that the final image forms at that location, and describe its characteristics. (*Note:* A flashlight beam shone on these *images* has a very startling effect. Even at a glancing angle, the incoming light beam is seemingly reflected off the *images* of the coins! Do you understand why?)

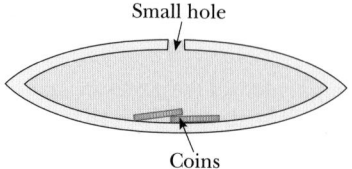

Figure P23.58

59. A ball is dropped from rest 3.00 m directly above the vertex of a concave mirror of radius 1.00 m. The mirror lies in a horizontal plane. (a) Describe the motion of the ball's image in the mirror. (b) At what times will the ball and its image coincide?

60. Figure P23.60 shows a converging lens with radii $R_1 = 9.00$ cm and $R_2 = -11.0$ cm, in front of a concave spherical mirror of radius $R = 8.00$ cm. The focal points (F_1 and F_2) for the thin lens and the center of curvature (C) of the mirror are also shown. (a) If the focal points F_1 and F_2 are 5.00 cm from the vertex of the thin lens, determine the index of refraction for the lens. (b) If the lens and mirror are 20.0 cm apart, and an object is placed 8.00 cm to the left of the lens, determine the position of the final image and its magnification as seen by the eye in the figure. (c) Is the final image inverted or upright? Explain.

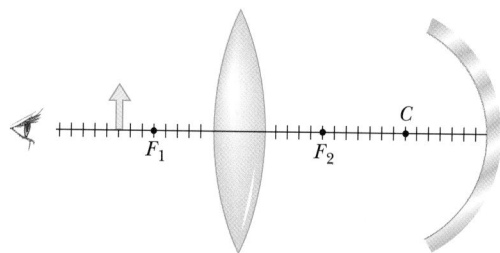

Figure P23.60

61. Object O_1 is 15.0 cm to the left of a converging lens of 10.0-cm focal length. A second lens is positioned 10.0 cm to the right of the first lens and is observed to form a final image at the position of the original object, O_1. (a) What is the focal length of the second lens?

(b) What is the overall magnification of this system?

(c) What is the nature (i.e., real or virtual, upright or inverted) of the final image?

62. The lens maker's equation for a lens with index n_1 immersed in a medium with index n_2 takes the form

$$\frac{1}{f} = \left(\frac{n_1}{n_2} - 1\right)\left(\frac{1}{R_1} - \frac{1}{R_2}\right)$$

A thin diverging glass (index = 1.5) lens with $R_1 = -3.0$ m and $R_2 = -6.0$ m is surrounded by air. An arrow is placed 10.0 m to the left of the lens. **(a)** Determine the position of the image. Repeat part **(a)** with the arrow and lens immersed in **(b)** water (index = 1.33); **(c)** a medium with an index of refraction of 2.0. **(d)** How can a lens that is diverging in air be changed into a converging lens?.

63. Find the object distances (in terms for f) for a thin converging lens of focal length f if **(a)** the image is real and the image distance is four times the focal length; **(b)** the image is virtual and the image distance is three times the focal length. **(c)** Calculate the magnification of the lens for cases **(a)** and **(b)**.

Wave Optics

▲ **PHYSICS PUZZLER**

Peacock feathers appear to be brilliant in color, especially in the blues and greens. However, those colors are not caused by pigments in the feathers. If not from pigments, how *are* these colors created? *(Terry Qing/FPG International, Inc.)*

O ur discussion of light has thus far been concerned with what happens when light passes through a lens or reflects from a mirror. Because explanations of such phenomena rely on a geometric analysis of light rays, that part of optics is often called *geometric optics*. We now expand our study of light into an area called *wave optics*. The three primary topics we examine in this chapter are interference, diffraction, and polarization. These phenomena cannot be adequately explained with ray optics, but the wave theory leads us to satisfying descriptions.

24.1 CONDITIONS FOR INTERFERENCE

In our discussion of interference of mechanical waves in Chapter 13, we found that two waves could add together either constructively or destructively. In constructive interference, the amplitude of the resultant wave is greater than that of either of the individual waves, whereas in destructive interference, the resultant amplitude is less than that of either individual wave. Electromagnetic waves also undergo interference. Fundamentally, all interference associated with electromagnetic waves arises from the combining of the electric and magnetic fields that constitute the individual waves.

Interference effects in light waves are not easy to observe because of the short wavelengths involved (about 4×10^{-7} m to about 7×10^{-7} m). For sustained interference between two sources of light to be observed, the following conditions must be met:

1. The sources must be **coherent;** that is, they must maintain a constant phase with respect to each other.
2. The sources must have identical wavelengths.
3. The superposition principle must apply.

◀ Conditions for interference

Let us examine the characteristics of coherent sources. Two sources (producing two traveling waves) are needed to create interference. To produce a stable interference pattern, the individual waves must maintain a constant phase with one another. When this situation prevails, the sources are said to be coherent. The sound waves emitted by two side-by-side loudspeakers driven by a single amplifier can produce interference because the two speakers respond to the amplifier in the same way at the same time—in other words, they are in phase.

If two light sources are placed side by side, however, no interference effects are observed, because the light waves from one source are emitted independently of the waves from the other source; hence, the emissions from the two sources do not maintain a constant phase relationship with each other during the time of observation. An ordinary light source undergoes random changes about once every 10^{-8} s. Therefore, the conditions for constructive interference, destructive interference, and intermediate states have durations on the order of 10^{-8} s. The result is that no interference effects are observed, because the eye cannot follow such short-term changes. Such light sources are said to be **noncoherent.**

A common method for producing two coherent light sources is to use a single-wavelength source to illuminate a screen containing two small slits. The light emerging from the two slits is coherent because a single source produces the original light beam and the slits serve only to separate the original beam into two parts (which is exactly what was done to the sound signal just mentioned). Any random change in the light emitted by the source will occur in the two separate beams at the same time, and interference effects can be observed.

24.2 YOUNG'S DOUBLE-SLIT INTERFERENCE

Interference in light waves from two sources was first demonstrated by Thomas Young in 1801. Figure 24.1a is a schematic diagram of the apparatus used in this experiment. (Young used pinholes rather than slits in his original experiments.) Light is incident on a screen in which there is a narrow slit, S_0. The light waves emerging from this slit arrive at a second screen that contains two narrow, parallel slits, S_1 and S_2. These slits serve as a pair of coherent light sources because waves emerging from them originate at the same wave front and therefore are always in phase. The light from the two slits produces on screen C a visible pattern consisting of a series of bright and dark parallel bands called **fringes** (Fig. 24.1b). When the light from slits S_1 and S_2 arrives at a point on the screen so that constructive interference occurs at that location, a bright line appears. When the light from the two slits combines destructively at any location on the screen, a dark line results. Figure 24.2 is a photograph of an interference pattern produced by two coherent vibrating sources in a water tank.

Figure 24.3 is a schematic diagram of some of the ways in which the two waves can combine at screen C. In Figure 24.3a, the two waves, which leave the two slits in phase, strike the screen at the central point, P. Because these waves travel equal distances, they arrive in phase at P, and as a result constructive interference occurs

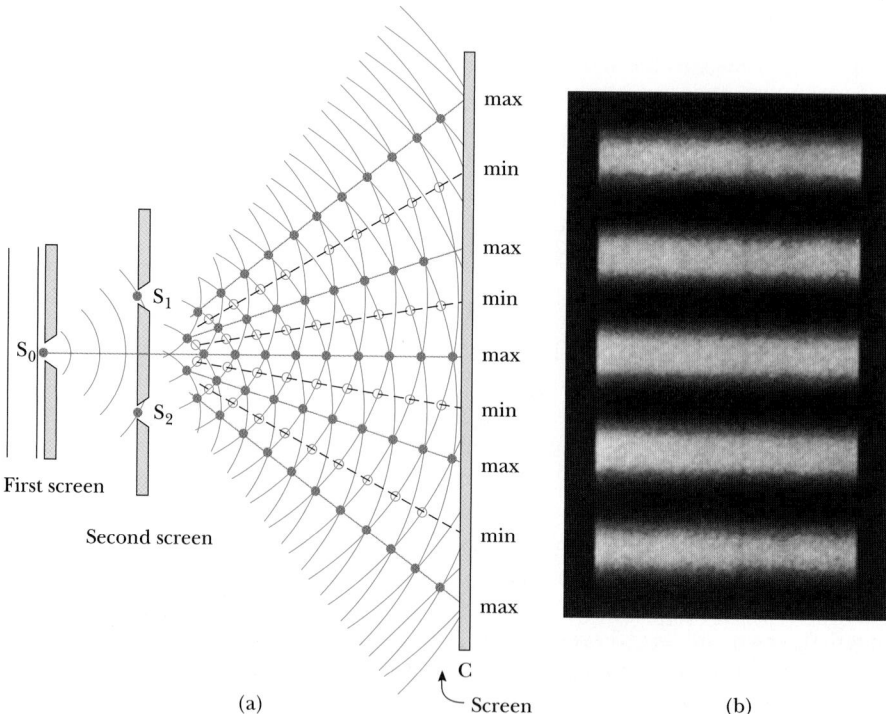

Figure 24.1 (a) A schematic diagram of Young's double-slit experiment. The narrow slits act as sources of waves. Slits S_1 and S_2 behave as coherent sources that produce an interference pattern on screen C. (This drawing is not to scale.) (b) The fringe pattern formed on screen C could look like this.

there and a bright fringe is observed. In Figure 24.3b, the two light waves again start in phase, but the upper wave has to travel one wavelength farther to reach point *Q* on the screen. Because the upper wave falls behind the lower one by exactly one wavelength, the two waves still arrive in phase at *Q*, and so a second bright fringe appears at that location. Now consider point *R*, midway between *P* and *Q* in Figure 24.3c. There the upper wave has fallen half a wavelength behind the lower wave. This means that the trough of the bottom wave overlaps the crest of the upper wave, giving rise to destructive interference at *R*. As a consequence, a dark region can be observed at *R*. Figure 24.3d shows the intensity distribution on the screen. Notice that the central fringe is most intense and that the intensity decreases for higher order fringes.

We can describe Young's experiment quantitatively with the help of Figure 24.4. Consider point *P* on the viewing screen; the screen is positioned a perpendicular distance of *L* from the screen containing slits S_1 and S_2, which are separated by distance *d*, and r_1 and r_2 are the distances the secondary waves travel from slit to screen. Let us assume that the waves emerging from S_1 and S_2 have the same single constant frequency, the same amplitude, and are in phase. The light intensity on

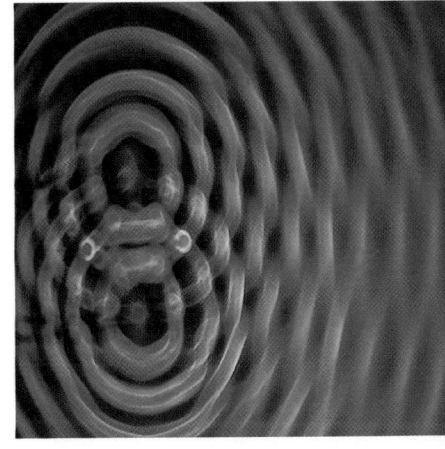

Figure 24.2 An interference pattern involving water waves is produced by two vibrating sources at the water's surface. The pattern is analogous to that observed in Young's double-slit experiment. Note the regions of constructive and destructive interference. *(Richard Megna, Fundamental Photographs)*

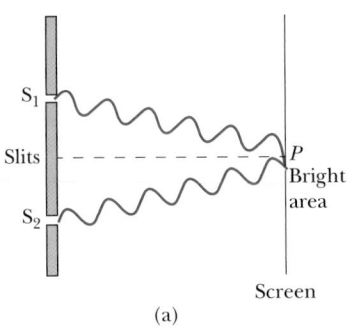

(a)

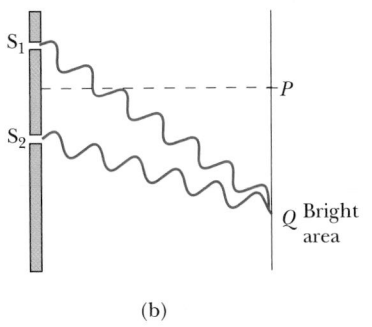

(b)

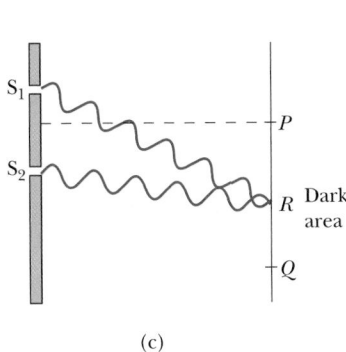

(c)

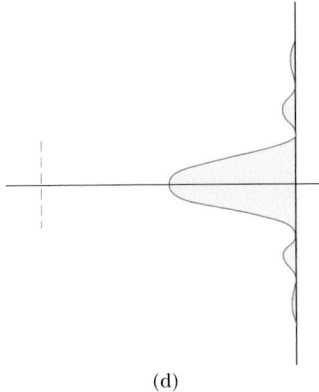

(d)

Figure 24.3 (a) Constructive interference occurs at *P* when the waves combine. (b) Constructive interference also occurs at *Q*. (c) Destructive interference occurs at *R* when the wave from the upper slit falls one half wavelength behind the wave from the lower slit. (d) The intensity of the fringes decreases with movement to higher orders. (These figures are not drawn to scale.)

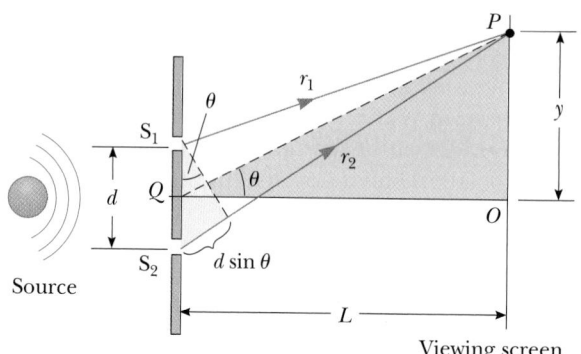

Figure 24.4 A geometric construction to describe Young's double-slit experiment. The path difference between the two rays is $r_2 - r_1 = d \sin \theta$. (This figure is not drawn to scale.)

the screen at P is the resultant of the light from both slits. Note that a wave from the lower slit travels farther than a wave from the upper slit by the amount $d \sin \theta$. This distance is called the **path difference,** δ (lowercase Greek delta), where

Path difference ▶

$$\delta = r_2 - r_1 = d \sin \theta \qquad\qquad \text{[24.1]}$$

This equation assumes that the two waves travel in parallel lines, which is approximately true, because L is much greater than d. As noted earlier, the value of this path difference determines whether the two waves are in phase when they arrive at P. If the path difference is either zero or some integral multiple of the wavelength, the two waves are in phase at P and constructive interference results. Therefore, the condition for bright fringes, or **constructive interference,** at P is

Condition for constructive ▶
interference (slits)

$$\delta = d \sin \theta = m\lambda \qquad m = 0, \pm 1, \pm 2, \dots \qquad \text{[24.2]}$$

The number m is called the **order number** of the fringe. The central bright fringe at $\theta = 0$ ($m = 0$) is called the *zeroth-order maximum;* the first maximum on either side, when $m = \pm 1$, is called the *first-order maximum;* and so forth.

Similarly, when the path difference is an odd multiple of $\lambda/2$, the two waves arriving at P are 180° out of phase and give rise to destructive interference. Therefore, the condition for dark fringes, or **destructive interference,** at P is

Condition for destructive ▶
interference (slits)

$$\delta = d \sin \theta = \left(m + \frac{1}{2} \right)\lambda \qquad m = 0, \pm 1, \pm 2, \dots \qquad \text{[24.3]}$$

If $m = 0$ in this equation, the path difference is $\delta = \lambda/2$, which is the condition for the location of the first dark line on either side of the central (bright) maximum. Likewise, if $m = 1$, $\delta = 3\lambda/2$, which is the condition for the second dark line on each side, and so forth.

It is useful to obtain expressions for the positions of the bright and dark fringes measured vertically from O to P. We assume that $L \gg d$ (Fig. 24.4) and that $d \gg \lambda$. The first assumption says that the distance from the slits to the screen is much greater than the distance between the two slits. The second says that the distance between the two slits is much greater than the wavelength. This situation can prevail in practice because L is often on the order of 1 m, whereas d is a fraction of a millimeter and λ is less than a micrometer for visible light. Under these conditions θ is small for the first several orders, and so we can use the approximation

$\sin \theta \cong \tan \theta$. From the triangle OPQ in Figure 24.4, we see that

$$\sin \theta \approx \tan \theta = \frac{y}{L} \qquad \text{[24.4]}$$

Using this result together with the substitution $\sin \theta = m\lambda / d$ from Equation 24.2, we see that the positions of the *bright fringes*, measured from O, are

$$y_{\text{bright}} = \frac{\lambda L}{d} m \qquad \text{[24.5]}$$

Similarly, using Equations 24.3 and 24.4, we find that the *dark fringes* are located at

$$y_{\text{dark}} = \frac{\lambda L}{d} \left(m + \frac{1}{2} \right) \qquad \text{[24.6]}$$

As we shall demonstrate in Example 24.1, Young's double-slit experiment provides a method of measuring the wavelength of light. In fact, Young used this technique to make the first measurement of the wavelength of light. In addition, his experiment gave the wave model of light a great deal of credibility. The phenomenon of interference explains many observations of wave-like behavior. For example, the observation of interference for electrons was essential for the development of quantum mechanics.

Reflection, interference, and diffraction can be seen in this aerial photograph of waves in the sea. As the waves pass through a narrow gap, they spread out (diffract), and the interference of two waveforms is manifested in cross-patterned areas. (*John S. Shelton*)

Thinking Physics 1

Consider a double-slit experiment in which a laser beam is passed through a pair of very closely spaced slits, and a clear interference pattern is displayed on a distant screen. Now suppose you place smoke particles between the double slit and the screen. With the presence of the smoke particles, will you see the effects of interference in the space between the slits and the screen, or will you only see the effects on the screen?

Explanation You will see the effects in the area filled with smoke. There will be bright lines directed toward the bright areas on the screen and dark lines directed toward the dark areas on the screen.

Applying Physics 1

If your stereo speakers are connected "out of phase"—that is, with one speaker connected correctly and the other with its wires reversed—the bass in the music tends to be weak. Why does this happen, and why is it a problem for the bass and not the treble notes?

Explanation This is an acoustic analog to double-slit interference. The two speakers act as sources of waves, just like the two slits in a Young's double-slit experiment. If the speakers are connected correctly, and the same sound signal is fed to each speaker, both speakers move inward and outward at the same time in response to the signal. Thus, the sound waves are in phase as they leave the speakers. If you are sitting at a point in front of the speakers and midway between them, you will be located at

APPLICATION

Acoustical Interference in Stereo Systems.

the zero-order maximum—the interference is constructive and the sound will be loud. If one speaker is wired backward, then one speaker will be moving outward while the other is moving inward. The sound leaves the two speakers half a wavelength out of phase. This is a particular problem for the bass due to the long wavelength of low frequency notes. This results in a very large region of destructive interference in front of the speakers, on the order of the size of the room. The much shorter wavelengths of the high-frequency notes result in closely spaced maxima and minima. The spacing can be on the order of the size of the head and smaller. Thus, if one ear is at a minimum, the other might be at a maximum. What's more, small movements of the head will result in a shift from the position of a minimum to that of a maximum.

APPLICATION

Television Signal Interference.

Applying Physics 2

Suppose you are watching television by means of an antenna rather than a cable system. If an airplane flies near your location, you may notice wavering ghost images in the television picture. What might cause this?

Explanation   Your television antenna receives two signals—the direct signal from the transmitting antenna and a signal reflected from the surface of the airplane. As the airplane changes position, there are some times when these two signals are in phase and other times when they are out of phase. As a result, there is a variation in the intensity of the combined signal received at your antenna. This variation is evidenced by the wavering of the ghost images of the picture.

EXAMPLE 24.1   Measuring the Wavelength of a Light Source

A screen is separated from a double-slit source by 1.2 m. The distance between the two slits is 0.030 mm. The second-order bright fringe ($m = 2$) is measured to be 4.5 cm from the centerline. Determine (a) the wavelength of the light and (b) the distance between adjacent bright fringes.

Solution   (a) We can use Equation 24.5 with $m = 2$, $y_2 = 4.5 \times 10^{-2}$ m, $L = 1.2$ m, and $d = 3.0 \times 10^{-5}$ m:

$$\lambda = \frac{y_2 d}{mL} = \frac{(4.5 \times 10^{-2} \text{ m})(3.0 \times 10^{-5} \text{ m})}{2(1.2 \text{ m})} = 5.6 \times 10^{-7} \text{ m} = \boxed{560 \text{ nm}}$$

Reasoning and Solution   (b) Because the positions of the bright fringes are given by Equation 24.5, we see that the distance between *any* adjacent bright fringes (say, those characterized by m and $m + 1$) is

$$\Delta y = y_{m+1} - y_m = \frac{\lambda L}{d}(m + 1) - \frac{\lambda L}{d}m = \frac{\lambda L}{d}$$

$$= \frac{(5.6 \times 10^{-7} \text{ m})(1.2 \text{ m})}{3.0 \times 10^{-5} \text{ m}} = \boxed{2.2 \text{ cm}}$$

24.3   CHANGE OF PHASE DUE TO REFLECTION

Young's method of producing two coherent light sources involves illuminating a pair of slits with a single source. Another simple, although ingenious, arrangement for producing an interference pattern with a single light source is known as *Lloyd's mirror*. A light source is placed at point *S*, close to a mirror, as illustrated in Figure 24.5. Waves can reach the viewing point, *P*, either by the direct path *SP* or by the path involving reflection from the mirror. The reflected ray can be treated as a ray originating at the source *S'*, behind the mirror. Source *S'*, which is the image of *S*, can be considered a virtual source.

At points far from the source, one would expect an interference pattern due to waves from *S* and *S'*, just as is observed for two real coherent sources. An interference pattern is indeed observed. However, the positions of the dark and bright fringes are *reversed* relative to the pattern of two real coherent sources (Young's experiment). This is because the coherent sources *S* and *S'* differ in phase by 180°. This 180° phase change is produced by reflection.

To illustrate this further, consider point *P'*, at which the mirror meets the screen. This point is equidistant from *S* and *S'*. If path difference alone were responsible for the phase difference, one would expect to see a bright fringe at *P'* (because the path difference is zero for this point), corresponding to the central fringe of the two-slit interference pattern. Instead, one observes a *dark* fringe at *P'* because of the 180° phase change produced by reflection. In general, an electromagnetic wave undergoes a phase change of 180° on reflection from a medium of higher index of refraction than the one in which it was traveling.

It is useful to draw an analogy between reflected light waves and the reflections of a transverse wave on a stretched string when the wave meets a boundary, as in Figure 24.6. The reflected pulse on a string undergoes a phase change of 180° when

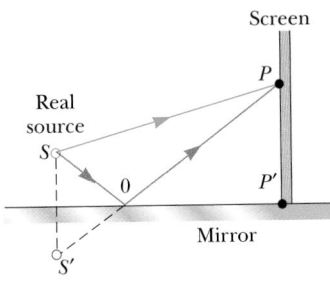

Figure 24.5 Lloyd's mirror. An interference pattern is produced on a screen at *P* as a result of the combination of the direct ray (blue) and the reflected ray (brown). The reflected ray undergoes a phase change of 180°.

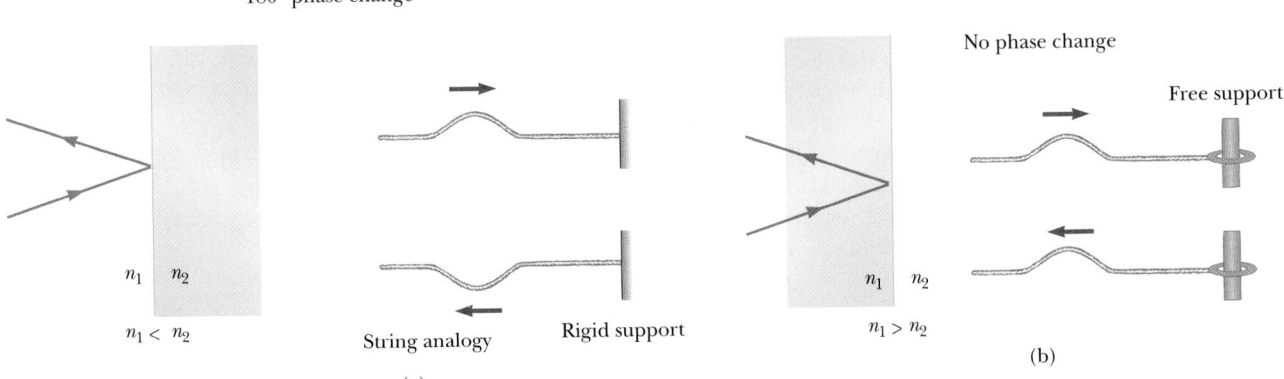

Figure 24.6 (a) A ray reflecting from a medium of higher refractive index undergoes a 180° phase change. The right side shows the analogy with a reflected pulse on a string. (b) A ray reflecting from a medium of lower refractive index undergoes no phase change.

it is reflected from the boundary of a denser medium, or from a rigid barrier, and no phase change when it is reflected from the boundary of a less dense medium. Similarly, an electromagnetic wave undergoes a 180° phase change when reflected from the boundary of a medium of higher index of refraction than the one in which it has been traveling. There is no phase change when the wave is reflected from a boundary leading to a medium of lower index of refraction. The part of the wave that crosses the boundary also undergoes no phase change.

24.4   INTERFERENCE IN THIN FILMS

Interference effects are commonly observed in thin films, such as soap bubbles and thin layers of oil on water. The varied colors observed with ordinary white light result from the interference of waves reflected from the opposite surfaces of the film.

Consider a film of uniform thickness t and index of refraction n, as in Figure 24.7. Let us assume that the light rays traveling in air are nearly normal to the two surfaces of the film. To determine whether the reflected rays interfere constructively or destructively, we must first note the following facts:

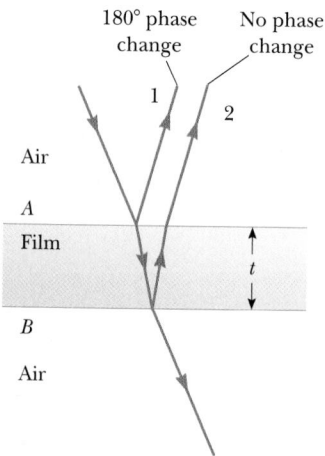

Figure 24.7 Interference observed in light reflected from a thin film is due to a combination of rays reflected from the upper and lower surfaces.

1. An electromagnetic wave traveling from a medium of index of refraction n_1 toward a medium of index of refraction n_2 undergoes a 180° phase change on reflection when $n_2 > n_1$. There is no phase change in the reflected wave if $n_2 < n_1$.
2. The wavelength of light, λ_n, in a medium with index of refraction n is

$$\lambda_n = \frac{\lambda}{n} \qquad [24.7]$$

where λ is the wavelength of light in vacuum.

We apply these rules to the film of Figure 24.7. According to the first rule, ray 1, which is reflected from the upper surface, A, undergoes a phase change of 180° with respect to the incident wave. Ray 2, which is reflected from the lower surface, B, undergoes no phase change with respect to the incident wave. Therefore, ray 1 is 180° out of phase with respect to ray 2, a situation that is equivalent to a path difference of $\lambda_n/2$. However, we must also consider the fact that ray 2 travels an extra distance of $2t$ before the waves recombine. For example, if $2t = \lambda_n/2$, rays 1 and 2 recombine in phase and constructive interference results. In general, the condition for constructive interference is

$$2t = (m + \tfrac{1}{2})\lambda_n \qquad m = 0, 1, 2, \ldots \qquad [24.8]$$

This condition takes into account two factors: (a) the difference in optical path length for the two rays (the term $m\lambda_n$) and (b) the 180° phase change on reflection (the term $\lambda_n/2$). Because $\lambda_n = \lambda/n$, we can write Equation 24.8 in the form

Condition for constructive ▷
interference (thin film)

$$2nt = (m + \tfrac{1}{2})\lambda \qquad m = 0, 1, 2, \ldots \qquad [24.9]$$

PHYSICS IN ACTION

Interference

On the left, a layer of soap-film bubbles on water. The colors, produced just before the bubbles burst, are due to interference between light rays reflected from the front and back of the thin film of soap making up the bubble. The color depends on the thickness of the film, ranging from black where the film is at its thinnest to magenta where it is thickest.

On the right, a thin film of oil on water displays interference, evidenced by the pattern of colors when white light is incident on the film. The film thickness varies in the vicinity of the blade, thereby producing the interesting color pattern.

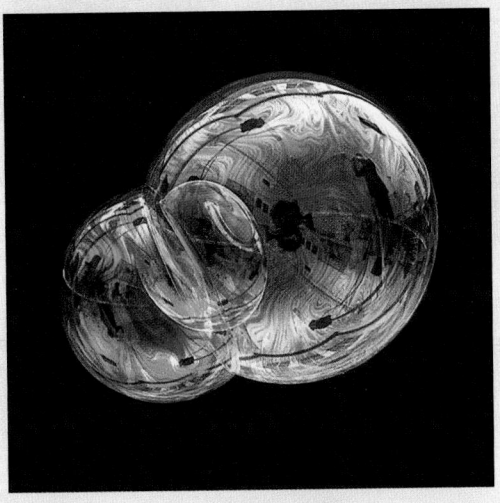

(Dr. Jeremy Burgess/Science Photo Library/Photo Researchers, Inc.)

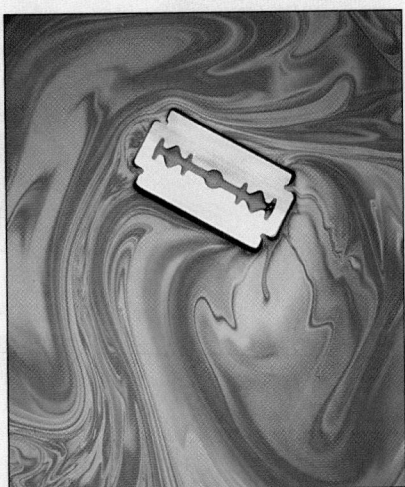

(Peter Aprahamian/Science Photo Library/Photo Researchers, Inc.)

If the extra distance $2t$ traveled by ray 2 is a multiple of λ_n, the two waves combine out of phase and destructive interference results. The general equation for destructive interference is

$$2nt = m\lambda \qquad m = 0, 1, 2, \ldots \qquad \text{[24.10]}$$

◀ Condition for destructive interference (thin film)

It is important to realize that two factors influence interference: (1) phase reversals on reflection and (2) differences in travel distance. The foregoing conditions for constructive and destructive interference are valid only when the medium above the top surface of the film is the same as the medium below the bottom surface. The surrounding medium may have a refractive index less than or greater than that of the film. In either case, the rays reflected from the two surfaces will be out of phase by 180°. If the film is placed between two *different* media, one of lower refractive index and one of higher refractive index, the conditions for constructive

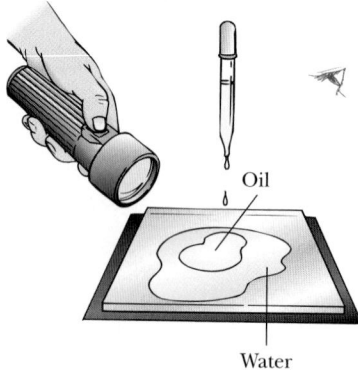

and destructive interference are reversed. In this case, either there is a phase change of 180° for both ray 1 reflecting from surface A and ray 2 reflecting from surface B, or there is no phase change for either ray; hence, the net change in relative phase due to the reflections is *zero*.

Newton's Rings

Another method for observing interference of light waves is to place a plano-convex lens on top of a flat glass surface, as in Figure 24.8a. With this arrangement, the air film between the glass surfaces varies in thickness from zero at the point of contact to some value t at P. If the radius of curvature of the lens, R, is very large compared with the distance r, and if the system is viewed from above using light of wavelength λ, a pattern of light and dark rings is observed (Fig. 24.8b). These circular fringes are called **Newton's rings,** after their discoverer. Newton's particle model of light could not explain the origin of the rings.

The interference is due to the combination of ray 1, reflected from the plate, with ray 2, reflected from the lower surface of the lens. Ray 1 undergoes a phase change of 180° on reflection, because it is reflected from a boundary leading into a medium of higher refractive index, whereas ray 2 undergoes no phase change. Hence, the conditions for constructive and destructive interference are given by Equations 24.9 and 24.10, respectively, with $n = 1$, because the "film" is air. Here again, one might guess that the contact point, O, would be bright, corresponding to constructive interference. Instead, it is dark, as seen in Figure 24.8b, because ray 1, reflected from the plate, undergoes a 180° phase change with respect to ray 2. Using the geometry shown in Figure 24.8a, one can obtain expressions for the radii

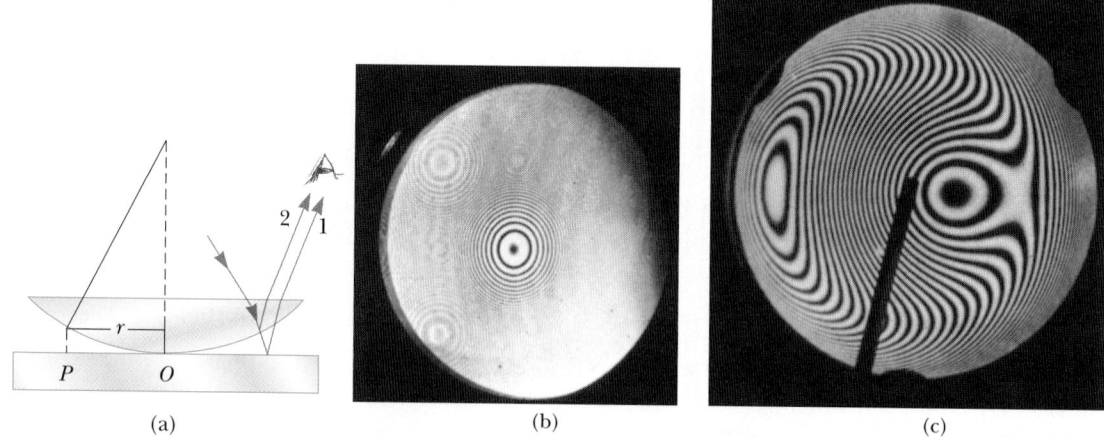

(a)   (b)   (c)

Figure 24.8   (a) The combination of rays reflected from the glass plate and the curved surface of the lens gives rise to an interference pattern known as Newton's rings. (b) A photograph of Newton's rings. *(Courtesy of Bausch & Lomb Optical Co.)* (c) This asymmetric interference pattern indicates imperfections in the lens. *(From Physical Science Study Committee,* College Physics, *Lexington, Mass., D. C. Heath and Co., 1968)*

of the bright and dark bands in terms of the radius of curvature, R, and vacuum wavelength, λ. For example, the dark rings have radii of $r \approx \sqrt{m\lambda R/n}$. In Problem 53 at the end of the chapter, you will be asked to supply the details.

One of the important uses of Newton's rings is in the testing of optical lenses. A circular pattern like that in Figure 24.8b is achieved only when the lens is ground to a perfectly spherical curvature. Variations from such symmetry might produce a pattern like that in Figure 24.8c. These variations give an indication of how the lens must be ground and polished to remove the imperfections.

APPLICATION

Checking for Imperfections in Optical Lenses.

Problem-Solving Strategy

Thin-Film Interference

The following features should be kept in mind when you work thin-film interference problems:

1. Identify the thin film causing the interference.
2. The type of interference that occurs is determined by the phase relationship between the portion of the wave reflected at the upper surface of the film and the portion reflected at the lower surface.
3. Phase differences between the two portions of the wave have two causes: (a) differences in the distances traveled by the two portions and (b) phase changes occurring on reflection. *Both* causes must be considered when you are determining which type of interference occurs.
4. When distance and phase changes on reflection are both taken into account, the interference is constructive if the path difference between the two waves is an integral multiple of λ, and destructive if the equivalent path difference is $\lambda/2$, $3\lambda/2$, $5\lambda/2$, and so forth.

EXAMPLE 24.2 Interference in a Soap Film

Calculate the minimum thickness of a soap-bubble film ($n = 1.33$) that will result in constructive interference in the reflected light if the film is illuminated by light with a wavelength in free space of 602 nm.

Reasoning The minimum film thickness for constructive interference corresponds to $m = 0$ in Equation 24.9. This gives $2nt = \lambda/2$.

Solution Because $2nt = \lambda/2$, we have

$$t = \frac{\lambda}{4n} = \frac{602 \text{ nm}}{4(1.33)} = \boxed{113 \text{ nm}}$$

Exercise What other film thicknesses will produce constructive interference?

Answer 338 nm, 564 nm, 789 nm, and so on

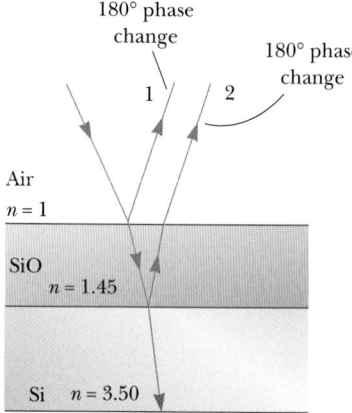

Figure 24.9 (Example 24.3) Reflective losses from a silicon solar cell are minimized by coating it with a thin film of silicon monoxide, SiO.

EXAMPLE 24.3 Nonreflective Coatings for Solar Cells

Semiconductors such as silicon are used to fabricate solar cells—devices that generate electric energy when exposed to sunlight. Solar cells are often coated with a transparent thin film, such as silicon monoxide (SiO; $n = 1.45$) to minimize reflective losses (Fig. 24.9). A silicon solar cell ($n = 3.50$) is coated with a thin film of silicon monoxide for this purpose. Assuming normal incidence, determine the minimum thickness of the film that will produce the least reflection at a wavelength of 552 nm.

Reasoning Reflection is least when rays 1 and 2 in Figure 24.9 meet the condition of destructive interference. Note that both rays undergo 180° phase changes on reflection. Hence the net change in phase due to reflection is zero, and the condition for a reflection *minimum* is a path difference of $\lambda_n/2$; therefore, $2t = \lambda/2n$.

Solution Because $2t = \lambda/2n$, the required thickness is

$$t = \frac{\lambda}{4n} = \frac{552 \text{ nm}}{4(1.45)} = \boxed{95.2 \text{ nm}}$$

Typically, such coatings reduce the reflective loss from 30% (with no coating) to 10% (with coating), thereby increasing the cell's efficiency, because more light is available to create charge carriers in the cell. In reality, the coating is never perfectly nonreflecting, because the required thickness is wavelength-dependent and the incident light covers a wide range of wavelengths.

Glass lenses used in cameras and other optical instruments are usually coated with a transparent thin film, such as magnesium fluoride (MgF_2) to reduce or eliminate unwanted reflection. More important, such coatings enhance the transmission of light through the lenses.

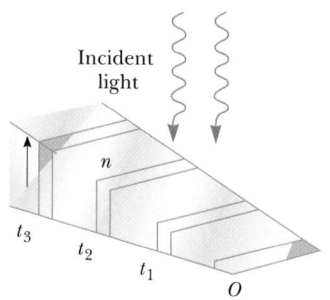

Figure 24.10 (Example 24.4) Interference bands in reflected light can be observed by illuminating a wedge-shaped film with monochromatic light. The dark areas correspond to positions of destructive interference.

EXAMPLE 24.4 Interference in a Wedge-Shaped Film

A thin, wedge-shaped film of refractive index n is illuminated with monochromatic light of wavelength λ, as illustrated in Figure 24.10. Describe the interference pattern observed in this case.

Reasoning and Solution The interference pattern is that of a thin film of variable thickness surrounded by air. Hence, the pattern is a series of alternating bright and dark parallel bands. A dark band corresponding to destructive interference appears at point O, the apex, because the upper reflected ray undergoes a 180° phase change and the lower one does not. According to Equation 24.10, other dark bands appear when $2nt = m\lambda$, so that $t_1 = \lambda/2n$, $t_2 = \lambda/n$, $t_3 = 3\lambda/2n$, and so on. Similarly, bright bands are observed when the thickness satisfies the condition $2nt = (m + \frac{1}{2})\lambda$, corresponding to thickness of $\lambda/4n$, $3\lambda/4n$, $5\lambda/4n$, and so on. If white light is used, bands of different colors are observed at different points, corresponding to the different wavelengths of light present.

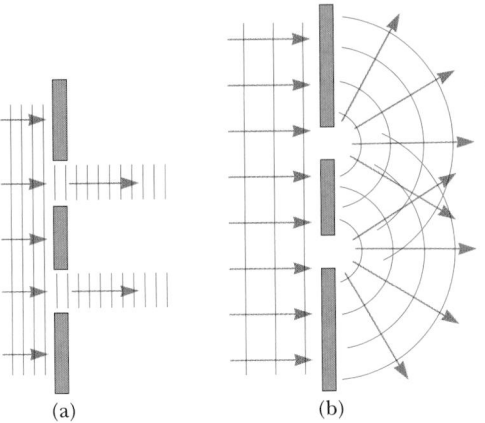

(a) (b)

Figure 24.11 (a) If light did not spread out after passing through the slits, no interference would occur. (b) The light from the two slits overlaps as it spreads out, filling the expected shadowed regions with light and producing interference fringes.

Figure 24.12 The diffraction pattern that appears on a screen when light passes through a narrow vertical slit. The pattern consists of a broad central band and a series of less intense and narrower side bands.

24.5 DIFFRACTION

Suppose a light beam is incident on two slits, as in Young's double-slit experiment. If the light truly traveled in straight-line paths after passing through the slits, as in Figure 24.11a, the waves would not overlap and no interference pattern would be seen. Instead, Huygens's principle requires that the waves spread out from the slits, as shown in Figure 24.11b. In other words, the light deviates from a straight-line path and enters the region that would otherwise be shadowed. This divergence of light from its initial line of travel is called **diffraction.**

In general, diffraction occurs when waves pass through small openings, around obstacles, or by sharp edges. For example, when a narrow slit is placed between a distant light source (or a laser beam) and a screen, the light produces a diffraction pattern like that in Figure 24.12. The pattern consists of a broad, intense central band, the **central maximum,** flanked by a series of narrower, less intense secondary bands (called **secondary maxima**) and a series of dark bands, or **minima.** This cannot be explained within the framework of geometric optics, which says that light rays traveling in straight lines should cast a sharp image of the slit on the screen.

Figure 24.13 shows the diffraction pattern and shadow of a penny. The pattern consists of the shadow, a bright spot at its center, and a series of bright and dark bands of light near the edge of the shadow. The bright spot at the center (called the *fresnel bright spot* after its discoverer, Augustin Fresnel) can be explained by the wave theory of light, which predicts constructive interference at this point. In contrast, from the viewpoint of geometric optics, the center of the pattern would be completely screened by the penny, and so one would never observe a central bright spot.

One type of diffraction, called **Fraunhofer diffraction,** occurs when the rays reaching the observing screen are approximately parallel. This can be achieved experimentally either by placing the observing screen far from the slit or by using a converging lens to focus the parallel rays on the screen, as in Figure 24.14a. A bright fringe is observed along the axis at $\theta = 0$, with alternating dark and bright

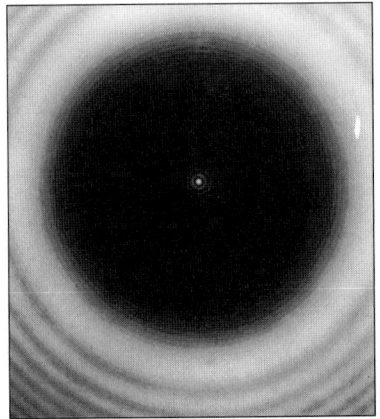

Figure 24.13 The diffraction pattern of a penny placed midway between the screen and the source. *(Courtesy of P. M. Rinard, from Am. J. Phys., 44:70, 1976)*

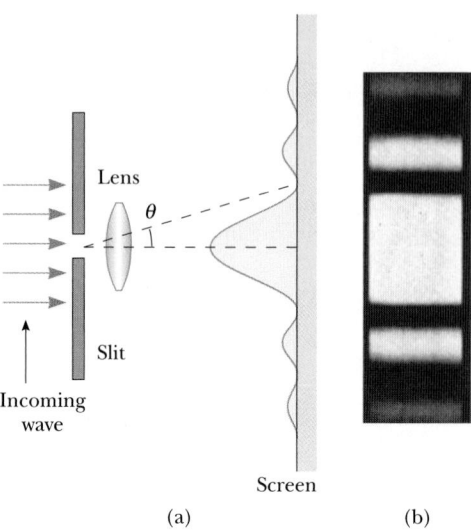

Figure 24.14 (a) The Fraunhofer diffraction pattern of a single slit. The parallel rays are brought into focus on the screen with a converging lens. The pattern consists of a central bright region flanked by much weaker maxima. (This drawing is not to scale.) (b) A photograph of a single-slit Fraunhofer diffraction pattern. *(From M. Cagnet, M. Francon, and J. C. Thierr,* Atlas of Optical Phenomena, *Berlin, Springer-Verlag, 1962, plate 18)*

fringes on each side of the central bright fringe. Figure 24.14b is a photograph of a single-slit Fraunhofer diffraction pattern.

24.6   SINGLE-SLIT DIFFRACTION

Until now we have assumed that slits are point sources of light. In this section we determine how their finite widths are the basis for understanding the nature of the Fraunhofer diffraction pattern produced by a single slit.

We can deduce some important features of this problem by examining waves coming from various portions of the slit, as shown in Figure 24.15. According to Huygens's principle, **each portion of the slit acts as a source of waves. Hence, light from one portion of the slit can interfere with light from another portion,** and the resultant intensity on the screen depends on the direction θ.

To analyze the diffraction pattern, it is convenient to divide the slit into halves, as in Figure 24.15. All the waves that originate at the slit are in phase. Consider waves 1 and 3, which originate at the bottom and center of the slit, respectively. Wave 1 travels farther than wave 3 by an amount equal to the path difference $(a/2)\sin\theta$, where a is the width of the slit. Similarly, the path difference between waves 3 and 5 is also $(a/2)\sin\theta$. If this path difference is exactly half of a wavelength (corresponding to a phase difference of 180°), the two waves cancel each other and destructive interference results. This is true, in fact, for any two waves that originate at points separated by half the slit width, because the phase difference between two such points is 180°. Therefore, waves from the upper half of the slit interfere *destructively* with waves from the lower half of the slit when

$$\frac{a}{2}\sin\theta = \frac{\lambda}{2}$$

or when

$$\sin\theta = \frac{\lambda}{a}$$

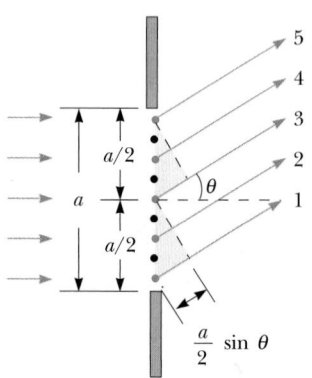

Figure 24.15 Diffraction of light by a narrow slit of width a. Each portion of the slit acts as a point source of waves. The path difference between rays 1 and 3 or between rays 2 and 4 is equal to $(a/2)\sin\theta$. (This drawing is not to scale, and the rays are assumed to converge at a distant point.)

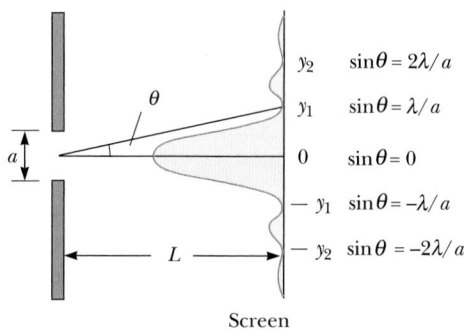

Screen

Figure 24.16 Positions of the minima for the Fraunhofer diffraction pattern of a single slit of width a. (This is not to scale.)

If we divide the slit into four parts rather than two, and use similar reasoning, we find that the screen is also dark when

$$\sin \theta = \frac{2\lambda}{a}$$

Likewise, we can divide the slit into six parts and show that darkness occurs on the screen when

$$\sin \theta = \frac{3\lambda}{a}$$

Therefore, the general condition for **destructive interference** is

$$\sin \theta = m \frac{\lambda}{a} \qquad m = \pm 1, \pm 2, \pm 3, \ldots \qquad \text{[24.11]}$$

Equation 24.11 gives the values of θ for which the diffraction pattern has zero intensity — that is, a dark fringe is formed. However, this equation tells us nothing about the variation in intensity along the screen. The general features of the intensity distribution along the screen are shown in Figure 24.16. A broad central bright fringe is observed, flanked by much weaker bright fringes alternating with dark fringes. The various dark fringes (points of zero intensity) occur at the values of θ that satisfy Equation 24.11. The points of constructive interference lie approximately halfway between the dark fringes. Note that the central bright fringe is twice as wide as the weaker maxima.

Thinking Physics 2

If a classroom door is open even just a small amount, you can hear sounds coming from the hallway. Yet you cannot see what is going on in the hallway. Why is there this difference?

Explanation The space between the slightly open door and the wall is acting as a single slit for waves. Sound waves have wavelengths larger than the slit width, so sound is effectively diffracted by the opening and the central maximum spreads throughout the room. Light wavelengths are much smaller than the slit width, so there is virtually no diffraction for the light. You must have a direct line of sight to detect the light waves.

EXAMPLE 24.5   Where Are the Dark Fringes?

Light of wavelength 580 nm is incident on a slit of width 0.30 mm. The observing screen is placed 2.0 m from the slit. Find the positions of the first dark fringes and the width of the central bright fringe.

Solution   The first dark fringes that flank the central bright fringe correspond to $m = \pm 1$ in Equation 24.11:

$$\sin \theta = \pm \frac{\lambda}{a} = \pm \frac{5.8 \times 10^{-7} \text{ m}}{0.30 \times 10^{-3} \text{ m}} = \pm 1.9 \times 10^{-3}$$

From the triangle in Figure 24.16, we see that $\tan \theta = y_1 / L$. Because θ is very small, we can use the approximation $\sin \theta \approx \tan \theta$, so that $\sin \theta \approx y_1 / L$. Therefore, the positions of the first minima, measured from central axis, are

$$y_1 \approx L \sin \theta = \pm L \frac{\lambda}{a} = \quad \pm 3.9 \times 10^{-3} \text{ m}$$

The positive and negative signs correspond to the first dark fringes on either side of the central bright fringe. Hence, the width of the central bright fringe is given by $2|y_1| = 7.8 \times 10^{-3}$ m $= 7.8$ mm. Note that this value is much greater than the width of the slit. However, as the width of the slit is *increased*, the diffraction pattern *narrows*, corresponding to smaller values of θ. In fact, for large values of a, the maxima and minima are so closely spaced that the only observable pattern is a large central bright area resembling the geometric image of the slit.

Exercise   Determine the width of the first-order bright fringe.

Answer   3.9 mm

24.7   POLARIZATION OF LIGHT WAVES

In Chapter 21 we described the transverse nature of electromagnetic waves. Figure 24.17 shows that the electric and magnetic field vectors associated with an electromagnetic wave are at right angles to each other and also to the direction of wave propagation. The phenomenon of polarization, described in this section, is firm evidence of the transverse nature of electromagnetic waves.

     An ordinary beam of light consists of a large number of waves emitted by the atoms or molecules of the light source. Each atom produces a wave with its own orientation of **E,** as in Figure 24.17, corresponding to the direction of atomic vi-

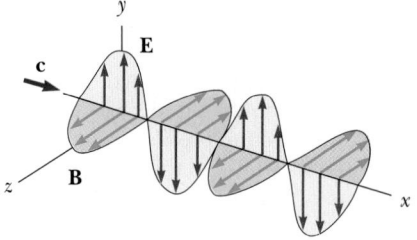

Figure 24.17   A schematic diagram of a polarized electromagnetic wave propagating in the x direction. The electric field vector, **E,** vibrates in the xy plane, and the magnetic field vector, **B,** vibrates in the xz plane.

bration. However, because all directions of vibration are possible, the resultant electromagnetic wave is a superposition of waves produced by the individual atomic sources. The result is an **unpolarized** light wave, represented schematically in Figure 24.18a. The direction of wave propagation in this figure is perpendicular to the page. Note that *all* directions of the electric field vector are equally probable and lie in a plane (such as the plane of this page) perpendicular to the direction of propagation. At any given point and at some instant of time, there is only one resultant electric field; do not be misled by Figure 24.18a.

A wave is said to be **linearly polarized** if **E** vibrates in the same direction *at all times* at a particular point, as in Figure 24.18b. (Sometimes such a wave is described as *plane-polarized* or simply *polarized*.) The wave in Figure 24.17 is an example of a wave linearly polarized in the y direction. As the wave propagates in the x direction, **E** is always in the y direction. The plane formed by **E** and the direction of propagation is called the *plane of polarization* of the wave. In Figure 24.17, the plane of polarization is the xy plane. It is possible to obtain a linearly polarized wave from an unpolarized wave by removing from the unpolarized wave all components except those whose electric field vectors oscillate in a single plane. We shall now discuss three processes for doing this: (1) selective absorption, (2) reflection, and (3) scattering.

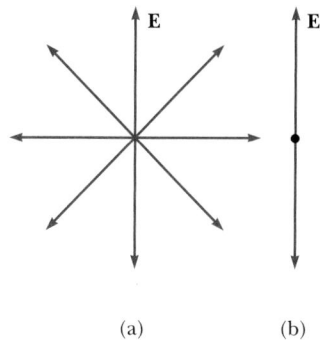

(a)   (b)

Figure 24.18 (a) An unpolarized light beam viewed along the direction of propagation (perpendicular to the page). The transverse electric field vector can vibrate in any direction with equal probability. (b) A linearly polarized light beam with the electric field vector vibrating in the vertical direction.

Polarization by Selective Absorption

The most common technique for polarizing light is to use a material that transmits waves whose electric field vectors vibrate in a plane parallel to a certain direction and absorbs those waves whose electric field vectors vibrate in directions perpendicular to that direction.

In 1932, E. H. Land discovered a material, which he called **polaroid,** that polarizes light through selective absorption by oriented molecules. This material is fabricated in thin sheets of long-chain hydrocarbons, which are stretched during manufacture so that the molecules align. After a sheet is dipped into a solution containing iodine, the molecules become good electrical conductors. However, the conduction takes place primarily along the hydrocarbon chains, because the valence electrons of the molecules can move easily only along the chains. (Recall that valence electrons are "free" electrons that can readily move through the conductor.) As a result, the molecules readily *absorb* light whose electric field vector is parallel to their lengths, and *transmit* light whose electric field vector is perpendicular to their lengths. It is common to refer to the direction perpendicular to the molecular chains as the **transmission axis.** In an ideal polarizer, all light with **E** parallel to the transmission axis is transmitted, and all light with **E** perpendicular to the transmission axis is absorbed.

Let us now describe the intensity of light that passes through a polarizing material. In Figure 24.19, an unpolarized light beam is incident on the first polarizing sheet, called the **polarizer,** where the transmission axis is as indicated. The light that is passing through this sheet is polarized vertically, and the transmitted electric field vector is $\mathbf{E}_0$. A second polarizing sheet, called the **analyzer,** intercepts this beam with its transmission axis at an angle of θ to the axis of the polarizer. The component of $\mathbf{E}_0$ that is perpendicular to the axis of the analyzer is completely absorbed, and the component parallel to the axis, $E_0 \cos \theta$, passes through. The

Figure 24.19 Two polarizing sheets whose transmission axes make an angle of θ with each other. Only a fraction of the polarized light incident on the analyzer is transmitted.

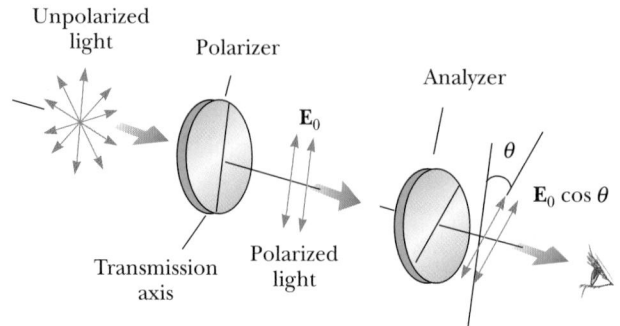

transmitted intensity varies as the *square* of the transmitted amplitude, and the intensity of the transmitted (polarized) light varies as

$$I = I_0 \cos^2 \theta \qquad [24.12]$$

where I_0 is the intensity of the polarized wave incident on the analyzer. This expression, known as **Malus's law,** applies to any two polarizing materials whose transmission axes are at an angle of θ to each other. From this expression, note that the transmitted intensity is a maximum when the transmission axes are parallel (θ = 0 or 180°) and zero (complete absorption by the analyzer) when the transmission axes are perpendicular to each other. This variation in transmitted intensity through a pair of polarizing sheets is illustrated in Figure 24.20.

When unpolarized light of intensity I_0 is sent through a single ideal polarizer, the transmitted linearly polarized light has intensity $\frac{1}{2} I_0$. This fact follows from Malus's law because the average value of $\cos^2 \theta$ is one half.

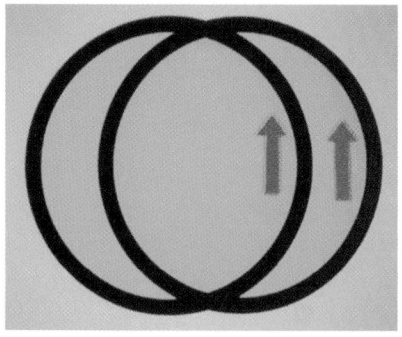
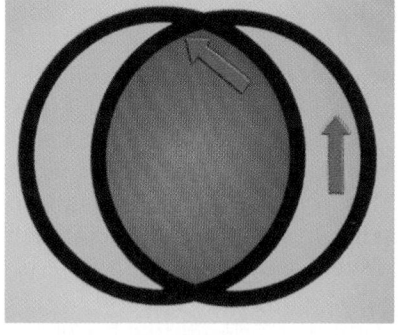
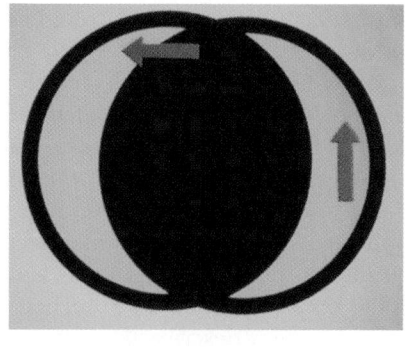

(a)   (b)   (c)

Figure 24.20 The intensity of light transmitted through two polarizers depends on the relative orientations of their transmission axes. (a) The transmitted light has *maximum* intensity when the transmission axes are *aligned* with each other. (b) The transmitted light intensity diminishes when the transmission axes are at an angle of 45° with each other. (c) The transmitted light intensity is a *minimum* when the transmission axes are at *right angles* to each other. *(Photos courtesy of Henry Leap)*

A polarizer for microwaves can be made as a grid of parallel metal wires about a centimeter apart. Is the electric field vector for microwaves transmitted through this polarizer parallel to, or perpendicular to, the metal wires?

Explanation Electric field vectors parallel to the metal wires will cause electrons in the metal to oscillate parallel to the wires. Thus, the energy from the waves with these electric field vectors will be transferred to the metal by accelerating these electrons and will eventually be transformed to internal energy through the resistance of the metal. Waves with electric field vectors perpendicular to the metal wires will not be able to accelerate electrons and will pass through. Thus, the electric field polarization will be perpendicular to the metal wires.

Polarization by Reflection

When an unpolarized light beam is reflected from a surface, the reflected light is completely polarized, partially polarized, or unpolarized, depending on the angle of incidence. If the angle of incidence is either 0° or 90° (a normal or grazing angle), the reflected beam is unpolarized. However, for angles of incidence between 0° and 90°, the reflected light is polarized to some extent. For one particular angle of incidence, the reflected beam is completely polarized. Let us now investigate that special angle.

Suppose an unpolarized light beam is incident on a surface, as in Figure 24.21a. The beam can be described by two electric field components, one parallel to the surface (represented by dots) and the other perpendicular to the first component

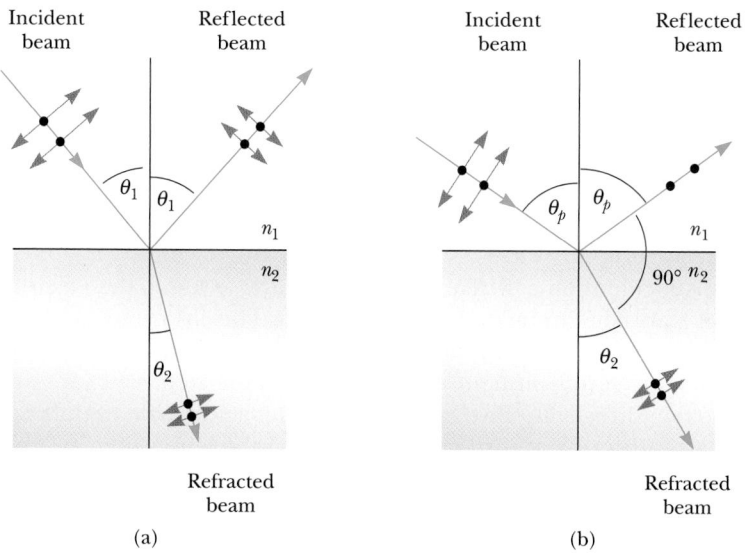

Figure 24.21 (a) When unpolarized light is incident on a reflecting surface, the reflected and refracted beams are partially polarized. (b) The reflected beam is completely polarized when the angle of incidence equals the polarizing angle, θ_p, satisfying the equation $n = \tan \theta_p$.

and to the direction of propagation (represented by brown arrows). It is found that the parallel component reflects more strongly than the other components, and this results in a partially polarized beam. Furthermore, the refracted beam is also partially polarized.

Now suppose that the angle of incidence, θ_1, is varied until the angle between the reflected and refracted beams is 90° (Fig. 24.21b). At this particular angle of incidence, the reflected beam is completely polarized, with its electric field vector parallel to the surface, while the refracted beam is partially polarized. The angle of incidence at which this occurs is called the **polarizing angle,** θ_p.

An expression relating the polarizing angle to the index of refraction of the reflecting surface can be obtained by use of Figure 24.21b. From this figure we see that, at the polarizing angle, $\theta_p + 90° + \theta_2 = 180°$, so that $\theta_2 = 90° - \theta_p$. Using Snell's law and taking $n_1 = n_{air} = 1.00$ and $n_2 = n$,

$$n = \frac{\sin \theta_1}{\sin \theta_2} = \frac{\sin \theta_p}{\sin \theta_2}$$

Because $\sin \theta_2 = \sin(90° - \theta_p) = \cos \theta_p$, the expression for n can be written

Brewster's law ▶

$$n = \frac{\sin \theta_p}{\cos \theta_p} = \tan \theta_p \qquad [24.13]$$

This expression is called **Brewster's law,** and the polarizing angle θ_p is sometimes called **Brewster's angle** after its discoverer, Sir David Brewster (1781–1868). For example, Brewster's angle for crown glass ($n = 1.52$) has the value $\theta_p = \tan^{-1}(1.52) = 56.7°$. Because n varies with wavelength for a given substance, Brewster's angle is also a function of wavelength.

Polarization by reflection is a common phenomenon. Sunlight reflected from water, glass, or snow is partially polarized. If the surface is horizontal, the electric field vector of the reflected light has a strong horizontal component. Sunglasses made of polarizing material reduce the glare of reflected light. The transmission axes of the lenses are oriented vertically to absorb the strong horizontal component of the reflected light.

A P P L I C A T I O N

Polaroid Sunglasses.

Polarization by Scattering

When light is incident on a system of particles, such as a gas, the electrons in the medium can absorb and reradiate part of the light. The absorption and reradiation of light by the medium, called **scattering,** is what causes sunlight reaching an observer on the Earth from straight overhead to be polarized. You can observe this effect by looking directly up through a pair of sunglasses made of polarizing glass. Less light passes through at certain orientations of the lenses than at others.

Figure 24.22 illustrates how the sunlight becomes polarized. The left side of the figure shows an incident unpolarized beam of sunlight on the verge of striking an air molecule. When the beam strikes the air molecule, it sets the electrons of the molecule into vibration. These vibrating charges act like those in an antenna except that they vibrate in a complicated pattern. The horizontal part of the electric field vector in the incident wave causes the charges to vibrate horizontally, and the vertical part of the vector simultaneously causes them to vibrate vertically. A horizontally polarized wave is emitted by the electrons as a result of their horizontal motion, and a vertically polarized wave is emitted parallel to the Earth as a result of their vertical motion.

Scientists have found that bees and homing pigeons use the polarization of sunlight as a navigational aid.

Optical Activity

Many important practical applications of polarized light involve the use of certain materials that display the property of **optical activity.** A substance is said to be optically active if it rotates the plane of polarization of transmitted light. To understand how this process occurs, suppose unpolarized light is incident on a polarizer from the left, as in Figure 24.23a. The transmitted light is polarized vertically, as shown. If this light is then incident on an analyzer with its axis perpendicular to that of the polarizer, no light emerges from it. If an optically active material is placed between the polarizer and analyzer, as in Figure 24.23b, the material causes the direction of the polarized beam to rotate through the angle θ. As a result, some light is able to pass through the analyzer. The angle through which the light is rotated by the material can be found by rotating the polarizer until the light is again extinguished. It is found that the angle of rotation depends on the length of the sample and, if the substance is in solution, on the concentration. One optically active material is a solution of common sugar, dextrose. A standard method for determining the concentration of sugar solutions is to measure the rotation produced by a fixed length of the solution.

Optical activity occurs in a material because of an asymmetry in the shape of its constituent molecules. For example, some proteins are optically active because of their spiral shapes. Other materials, such as glass and plastic, become optically active when placed under stress. If polarized light is passed through an unstressed piece of plastic and then through an analyzer with an axis perpendicular to that of

A P P L I C A T I O N

Finding the Concentrations of Solutions Using Their Optical Activity.

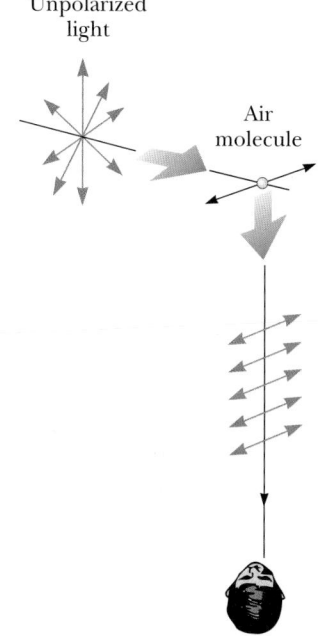

Figure 24.22 The scattering of unpolarized sunlight by air molecules. The light observed at right angles is linearly polarized because the vibrating molecule has a horizontal component of vibration.

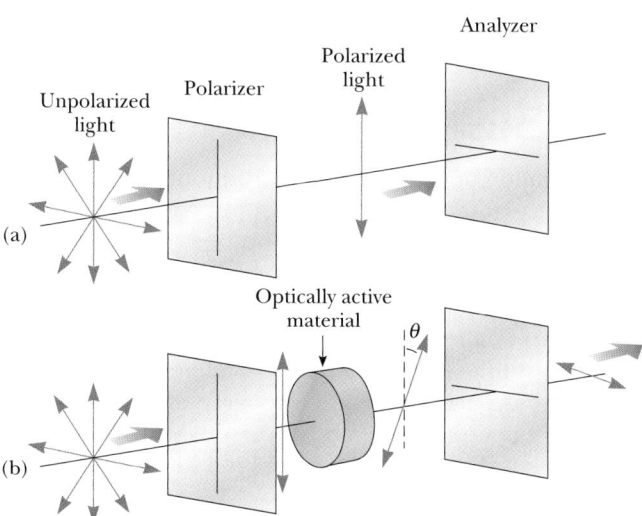

Figure 24.23 (a) When crossed polarizers are used, none of the polarized light can pass through the analyzer. (b) An optically active material rotates the direction of polarization through the angle θ, enabling some of this light to pass through the analyzer.

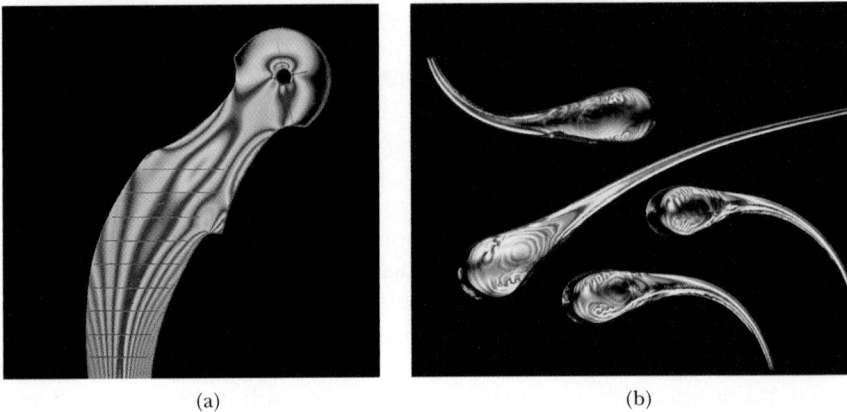

(a)

(b)

Figure 24.24   (a) Strain distribution in a plastic model of a hip replacement used in a medical research laboratory. The pattern is produced when the plastic model is placed between two crossed polarizers. *(Peter Aprahamian/Sharples Stress Engineers Ltd./SPL/Photo Researchers, Inc.)*   (b) These glass objects, called Prince Rupert drops, are made by dropping molten glass into water. The photograph was made by placing the objects between two crossed polarizers. The patterns observed represent the strain distribution in the glass. Studies of such patterns led to the development of tempered glass. *(James L. Amos, Peter Arnold, Inc.)*

the polarizer, none of the polarized light is transmitted. However, if the plastic is placed under stress, the regions of greatest stress produce the largest angles of rotation of polarized light. Hence, one observes a series of light and dark bands in the transmitted light. Engineers often use this procedure in the design of structures ranging from bridges to small tools. A plastic model is built and analyzed under different load conditions to determine positions of potential weakness and failure under stress. If the design is poor, patterns of light and dark bands will indicate the points of greatest weakness, and the design can be corrected at an early stage. Figure 24.24 shows examples of stress patterns in plastic and glass.

Application: Liquid Crystals

An effect similar to rotation of the plane of polarization is used to create the familiar displays on pocket calculators, wristwatches, laptop computers, and so forth. The properties of a unique type of substance called a liquid crystal make these displays (called LCDs for *liquid crystal displays*) possible. As its name implies, a **liquid crystal** is a substance with properties intermediate between those of a crystalline solid and those of a liquid—that is, the molecules of the substance are more orderly than those in a liquid but less than those in a pure crystalline solid. The forces that hold the molecules together in such a state are just barely strong enough to enable the substance to maintain a definite shape, so it is reasonable to call it a solid. However, small imputs of mechanical or electrical energy can disrupt these weak bonds and make the substance flow, rotate, or twist.

To see how liquid crystals can be used to create a display, consider Figure 24.25a. The liquid crystal is placed between two glass plates in the pattern shown, and electrical contacts, indicated by the thin lines, are made to the liquid crystal. When a voltage is applied across any segment in the display, that segment turns

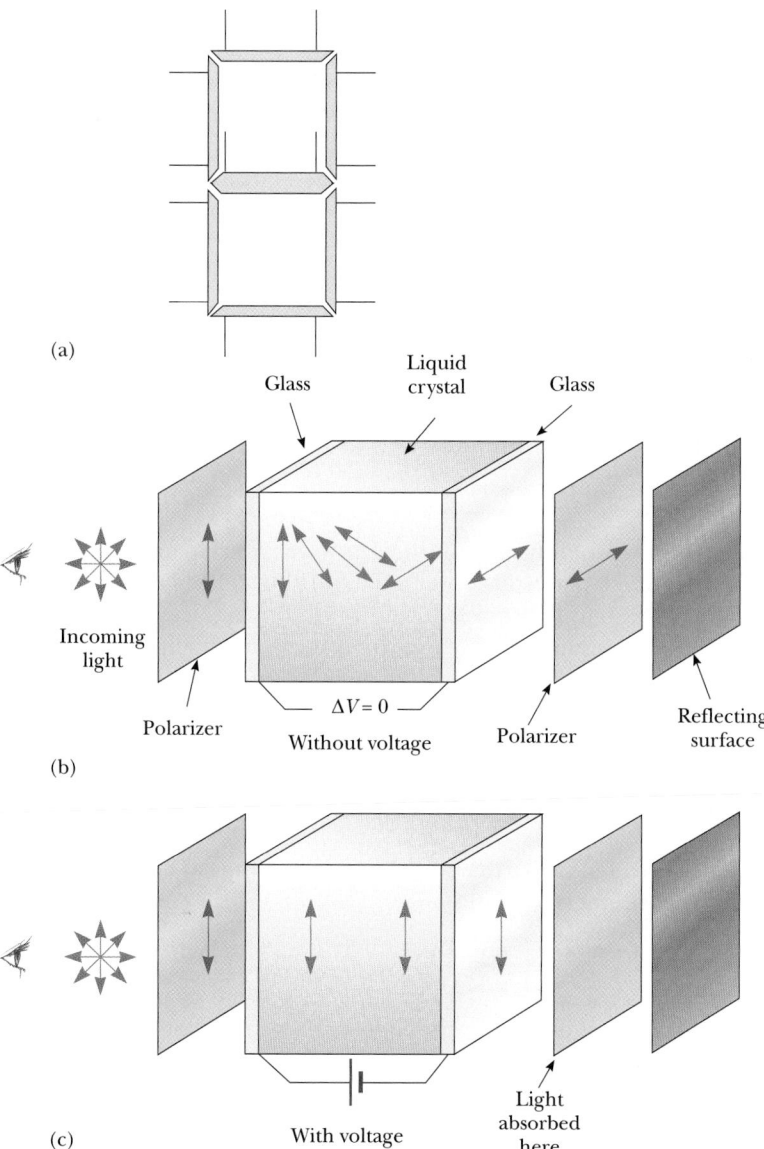

(a)

(b)

(c)

Figure 24.25 (a) The light-segment pattern of a liquid crystal display. (b) Rotation of a polarized light beam by a liquid crystal when the applied voltage is zero. (c) Molecules of the liquid crystal align with the electric field when a voltage is applied.

dark. In this fashion, any number between 0 and 9 can be formed by the pattern, depending on the voltages applied to the seven segments.

To see why a segment can be changed from dark to light by the application of a voltage, consider Figure 24.25b, which shows the basic construction of a portion of the display. The liquid crystal is placed between two glass substrates that are packaged between two pieces of Polaroid material with their transmission axes perpendicular. A reflecting surface is placed behind one of the pieces of Polaroid. First consider what happens when light falls on this package and no voltages are applied to the liquid crystal, as shown in Figure 24.25b. Incoming light is polarized by the polarizer on the left and then falls on the liquid crystal. As the light passes through the crystal, its plane of polarization is rotated by 90°, allowing it to pass through the polarizer on the right. It reflects from the reflecting surface and retraces its path through the crystal. Thus, an observer to the left of the crystal sees the segment as being bright. When a voltage is applied as in Figure 24.25c, the molecules of the liquid crystal do not rotate the plane of polarization of the light. In this case, the light is absorbed by the polarizer on the right and none is reflected back to the observer to the left of the crystal. Thus, the observer sees this segment as black. Changing the applied voltage to the crystal in a precise pattern and at precise times can make the pattern tick off the seconds on a watch, display a letter on a computer display, and so forth.

SUMMARY

Interference occurs when two or more light waves overlap at a given point. A sustained interference pattern is observed if (1) the sources are coherent (that is, they maintain a constant phase relationship with one another), (2) the sources have identical wavelengths, and (3) the superposition principle is applicable.

In **Young's double-slit experiment,** two slits separated by distance d are illuminated by a single-wavelength light source. An interference pattern consisting of bright and dark fringes is observed on a screen a distance of L from the slits. The condition for **bright fringes** (constructive interference) is

$$d \sin \theta = m\lambda \qquad m = 0, \pm 1, \pm 2, \ldots \qquad \text{[24.2]}$$

The number m is called the **order number** of the fringe. The condition for **dark fringes** (destructive interference) is

$$d \sin \theta = (m + \tfrac{1}{2})\lambda \qquad m = 0, \pm 1, \pm 2, \ldots \qquad \text{[24.3]}$$

An electromagnetic wave undergoes a phase change of 180° on reflection from a medium with an index of refraction higher than that of the medium in which the wave is traveling.

The wavelength of light, λ_n, in a medium with index of refraction n is

$$\lambda_n = \frac{\lambda}{n} \qquad \text{[24.7]}$$

where λ is the wavelength of the light in free space.

The **diffraction pattern** produced by a single slit on a distant screen consists of a central bright maximum flanked by less-bright fringes alternating with dark regions. The angles θ at which the diffraction pattern has zero intensity (regions of destructive interference) are described by

$$\sin \theta = \frac{m\lambda}{a} \qquad m = \pm 1, \pm 2, \pm 3, \ldots \qquad \text{[24.11]}$$

where a is the width of the slit and λ is the wavelength of the light incident on the slit.

Unpolarized light can be polarized by selective absorption, reflection, and scattering.

In general, light reflected from an amorphous material, such as glass, is partially polarized. Reflected light is completely polarized with its electric field parallel to the surface when the angle of incidence produces a 90° angle between the reflected and refracted beams. This angle of incidence, called the **polarizing angle,** θ_p, satisfies **Brewster's law,** given by

$$n = \tan \theta_p \qquad\qquad \text{[24.13]}$$

where n is the index of refraction of the reflecting medium.

MULTIPLE-CHOICE QUESTIONS

1. If linearly polarized light is sent through two polarizers, the first at 45° to the original plane of polarization and the second at 90° (45° more) to the original plane of polarization, what fraction of the original polarized intensity gets through the last polarizer?
   (a) 0 (b) 0.25 (c) 0.50 (d) 0.125 (e) No answer is correct.

2. The critical angle for sapphire surrounded by air is 34.4°. Calculate the polarizing angle for sapphire.
   (a) 60.5° (b) 59.7° (c) 58.6° (d) 56.3°

3. A Young's double slit has a slit separation of 3.00×10^{-5} m on which a monochromatic light beam is directed. The resultant bright fringe separation is 2.15×10^{-2} m on a screen 1.20 m from the double slit. What is the separation between the third-order bright fringe and the zeroth-order fringe?

   (a) 8.60×10^{-2} m (b) 6.45×10^{-2} m
   (c) 4.30×10^{-2} m (d) 2.15×10^{-2} m

4. A thin layer of oil ($n = 1.25$) is floating on water ($n = 1.33$). How thick is the oil in the region that reflects green light ($\lambda = 530$ nm)?
   (a) 212 nm (b) 313 nm (c) 404 nm (d) 500 nm

5. A Fraunhofer diffraction pattern is produced on a screen 1.30 m from a single slit. If a light source of 560 nm is used and the distance from the center of the central bright fringe to the first dark fringe is 5.30×10^{-3} m, what is the slit width?
   (a) 0.690×10^{-4} m (b) 1.37×10^{-4} m
   (c) 1.00×10^{-4} m (d) 0.810×10^{-4} m

CONCEPTUAL QUESTIONS

1. Consider a dark fringe in an interference pattern, at which almost no light energy is arriving. Light from both slits is arriving at this point, but the waves are canceling. Where does the energy go?

2. If Young's double-slit experiment were performed under water, how would the observed interference pattern be affected?

3. In a laboratory accident, you spill two liquids onto water, neither of which mixes with the water. They both form thin films on the water surface. You notice, as the films become very thin as they spread, that one film becomes bright and the other black in reflected light. Why might this be?

4. If white light is used in Young's double-slit experiment rather than monochromatic light, how does the interference pattern change?

5. In our discussion of thin film interference, we looked at light *reflecting* from a thin film. Consider one light ray, the direct ray, that transmits through the film without reflecting. Consider a second ray, the reflected ray, that transmits through the first surface, reflects back from the second, reflects again from the first, and then transmits out into the air, parallel to the direct ray. For normal incidence, how thick must the film be, in terms of the wavelength of the light, for the outgoing rays to interfere destructively? Is it the same thickness as for reflected destructive interference?

6. What is the necessary condition on path length difference between two waves that interfere (a) constructively and (b) destructively?

7. Astronomers often observe occultations, in which a star passes behind another object such as the Moon. When such an event occurs, it is found that the intensity of light from the star does not suddenly switch off as it passes behind the edge of the Moon. Instead the intensity fluctuates for a short time before dropping to zero. Why should this happen?

8. Describe the change in width of the central maximum of the single-slit diffraction pattern as the width of the slit is made smaller.

9. In everyday experience, radio waves are polarized, but light is not. Why?

10. Suppose reflected white light is used to observe a thin, transparent coating on glass as the coating material is gradually deposited by evaporation in a vacuum. Describe possible color changes that might occur during the process of building up the thickness of the coating.

11. Would it be possible to place a nonreflective coating on an airplane to cancel radar waves of wavelength 3 cm? Explain why or why not.

12. Certain sunglasses use a polarizing material to reduce the intensity of light reflected from shiny surfaces, such as water or the hood of a car. What orientation of the transmission axis should the material have to be most effective?

13. Why is it so much easier to perform interference experiments with a laser than with an ordinary light source?

14. A simple way of observing an interference pattern is to look at a distant light source through a stretched handkerchief or an open umbrella. Explain how this works.

15. Although we can hear around corners, we cannot see around corners. How can you explain this in view of the fact that sound and light are both waves?

16. Can a sound wave be polarized? Explain.

PROBLEMS

1, **2**, **3** = straightforward, intermediate, challenging   ☐ = full solution available in Study Guide/Student Solutions Manual   ![] = Core Concepts Workbook
WEB = solution posted at **http://www.harcourtcollege.com/physics/cptech**   ![] = biomedical application   ![] = Interactive Physics

Review Problem

A flat piece of glass is supported horizontally above the flat end of a 10.0-cm-long metal rod that has its lower end rigidly fixed. The thin film of air between the rod and glass is observed to be bright when illuminated by light of wavelength 500 nm. As the temperature is slowly increased by 25.0°C, the film changes from bright to dark and back to bright 200 times. What is the coefficient of linear expansion of the metal?

Section 24.2 Young's Double-Slit Interference

1. In a Young's double-slit experiment, a set of parallel slits with a separation of 0.100 mm is illuminated by light having a wavelength of 589 nm and the interference pattern observed on a screen 4.00 m from the slits. (a) What is the difference in path lengths from each of the slits to the screen location of a third-order bright fringe? (b) What is the difference in path lengths from the two slits to the screen location of the third dark fringe away from the center of the pattern?

2. A pair of narrow, parallel slits separated by 0.250 mm are illuminated by the green component from a mercury vapor lamp ($\lambda = 546.1$ nm). The interference pattern is observed on a screen 1.20 m from the plane of the parallel slits. Calculate the distance (a) from the central maximum to the first bright region on either side of the central maximum and (b) between the first and second dark bands in the interference pattern.

3. If the distance between two slits is 0.050 mm and the distance to a screen is 2.50 m, find the spacing between the first- and second-order bright fringes for yellow light of 600-nm wavelength.

4. A riverside warehouse has two open doors, as in Figure P24.4. A boat on the river sounds its horn. To person A the sound is loud and clear. To person B the sound is barely audible. The principal wavelength of the sound

waves is 3.00 m. Assuming person B is at the position of the first minimum, determine the distance between the doors, center to center.

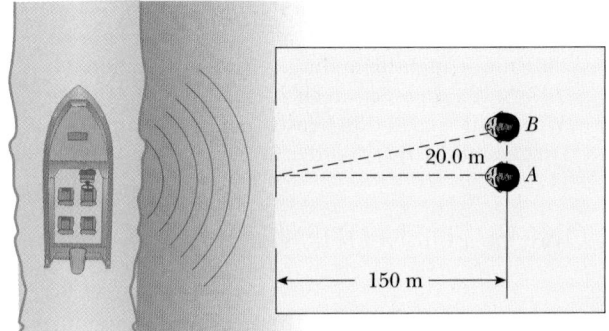

Figure P24.4

5. Light of wavelength 575 nm falls on a double slit, and the first bright fringe is seen at an angle of 16.5°. Find the distance between the two slits.

6. Light of wavelength 460 nm falls on two slits spaced 0.300 mm apart. What is the required distance from the slits to a screen if the spacing between the first and second dark fringes is to be 4.00 mm?

7. White light spans the wavelength range between about 400 nm and 700 nm. If white light passes through two slits 0.30 mm apart and falls on a screen 1.5 m from the slits, find the distance between the first-order violet and the first-order red fringes.

8. Two radio antennas separated by 300 m, as shown in Figure P24.8, simultaneously transmit identical signals of the same wavelength. A radio in a car traveling due north receives the signals. (a) If the car is at the position

of the second maximum, what is the wavelength of the signals? **(b)** How much farther must the car travel to encounter the next minimum in reception? (*Caution:* Avoid the small-angle approximation in this problem.)

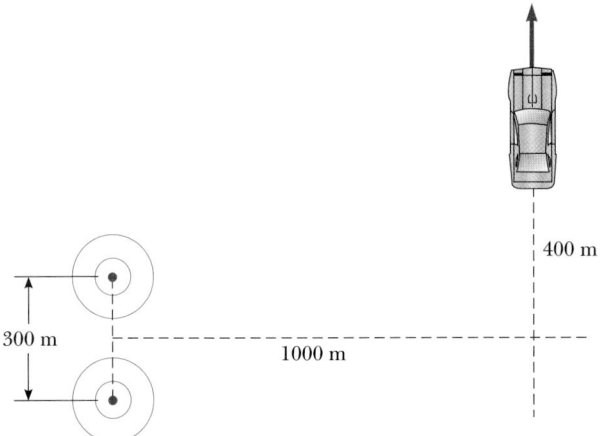

Figure P24.8

9. A Young's interference experiment is performed with blue-green argon laser light. The separation between the slits is 0.500 mm, and the interference pattern on a screen 3.30 m away shows the first maximum 3.40 mm from the center of the pattern. What is the wavelength of argon laser light?

10. Waves from a radio station have a wavelength of 300 m. They travel by two paths to a home receiver 20.0 km from the transmitter. One path is a direct path, and the other is by reflection from a mountain directly behind the home receiver. What is the minimum distance from the mountain to the receiver that produces destructive interference at the receiver? (Assume that no phase change occurs on reflection from the mountain.)

WEB 11. The waves from a radio station can reach a home receiver by two different paths. One is a straight-line path from the transmitter to the home, a distance of 30.0 km. The second path is by reflection from a storm cloud. Assume that this reflection takes place at a point midway between receiver and transmitter. If the wavelength broadcast by the radio station is 400 m, find the minimum height of the storm cloud that will produce destructive interference between the direct and reflected beams. (Assume no phase changes on reflection.)

12. Radio waves from a star, of wavelength 250.0 m, reach a radio telescope by two separate paths. One is a direct path to the receiver, which is situated on the edge of a cliff by the ocean. The second is by reflection off the water. The first minimum of destructive interference occurs when the star is 25.0° above the horizon. Find the height of the cliff. (Assume no phase change on reflection.)

Section 24.3 Change of Phase Due to Reflection

Section 24.4 Interference in Thin Films

13. Suppose the film shown in Figure 24.7 has an index of refraction of 1.36 and is surrounded by air on both sides. Find the minimum thickness, other than zero, that will produce constructive interference in the reflected light when the film is illuminated by light of wavelength 500 nm.

14. Two parallel glass plates are placed in contact and illuminated from above with light of wavelength 580 nm. As the plates are slowly moved apart and the reflected light is observed, darkness occurs at certain separations. What are the distances (other than, perhaps, zero) of the first three of these separations?

15. A thin layer of liquid methylene iodide ($n = 1.756$) is sandwiched between two flat parallel plates of glass ($n = 1.50$). What must be the thickness of the liquid layer if normally incident light with $\lambda = 600$ nm in air is to be strongly reflected?

16. A transparent oil of index of refraction 1.29 spills on the surface of water (index of refraction 1.33), producing a maximum of reflection with normally incident orange light (wavelength 600 nm in air). Assuming the maximum occurs in the first order, determine the thickness of the oil slick.

17. A coating is applied to a lens to minimize reflections. The index of refraction of the coating is 1.55, and that of the lens is 1.48. If the coating is 177.4 nm thick, what wavelength is minimally reflected for normal incidence in the lowest order?

18. Determine the minimum thickness of a soap film ($n = 1.330$) that will result in constructive interference of **(a)** the red H_α line ($\lambda = 656.3$ nm); **(b)** the blue H_γ line ($\lambda = 434.0$ nm).

19. A thin film of glass ($n = 1.50$) floats on a liquid of $n = 1.35$ and is illuminated by light of $\lambda = 580$ nm incident from air above it. Find the minimum thickness of the glass, other than zero, that will produce destructive interference in the reflected light.

20. A beam of light of wavelength 580 nm passes through two closely spaced glass plates, as shown in Figure P24.20. For what minimum nonzero value of the plate separation, *d*, will the transmitted light be bright?

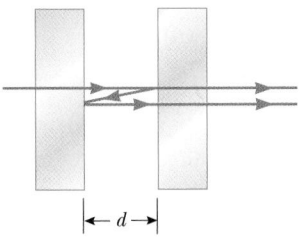

Figure P24.20

21. A planoconvex (flat on one side, convex on the other) lens rests with its curved side on a flat glass surface and is illuminated from above by light of wavelength 500 nm (see Fig. 24.8). A dark spot is observed at the center, surrounded by 19 concentric dark rings (with bright rings in between). How much thicker is the air wedge at the position of the 19th dark ring than at the center?

22. Two rectangular optically flat plates ($n = 1.52$) are in contact along one end and are separated along the other end by a spacer that is 2.00 μm thick (Fig. P24.22). The top plate is illuminated by monochromatic light of wavelength 546.1 nm. Calculate the number of dark parallel bands crossing the top plate (including the dark band at zero thickness along the edge of contact between the two plates).

Figure P24.22

23. A thin film of oil ($n = 1.38$) floats on water. What wavelength in the visible spectrum undergoes constructive interference and what is the color of the film at points at which the oil thickness is 300 nm? Assume that the film is viewed from above with white light.

WEB 24. A planoconvex lens (see Problem 21) with radius of curvature $R = 3.0$ m is in contact with a flat plate of glass. A light source and the observer's eye are both close to the normal, as shown in Figure 24.8. The radius of the 50th bright Newton's ring is found to be 9.8 mm. What is the wavelength of the light produced by the source?

25. Nonreflective coatings on camera lenses reduce the loss of light at the surfaces of multilens systems and prevent internal reflections that might mar the image. Find the minimum thickness of a layer of magnesium fluoride ($n = 1.38$) on flint glass ($n = 1.66$) that will cause destructive interference of reflected light of wavelength 550 nm near the middle of the visible spectrum.

26. A thin 1.00×10^{-5}-cm-thick film of MgF_2 ($n = 1.38$) is used to coat a camera lens. Are any wavelengths in the visible spectrum intensified in the reflected light?

Section 24.6 Single–Slit Diffraction

27. Light of wavelength 600 nm falls on a 0.40-mm-wide slit and forms a diffraction pattern on a screen 1.5 m away. (a) Find the position of the first dark band on each side of the central maximum. (b) Find the width of the central maximum.

28. Light of wavelength 587.5 nm illuminates a single 0.75-mm-wide slit. (a) At what distance from the slit should a screen be placed if the first minimum in the diffraction pattern is to be 0.85 mm from the central maximum? (b) Calculate the width of the central maximum.

29. Microwaves of wavelength 5.00 cm enter a long, narrow window in a building that is otherwise essentially opaque to the microwaves. If the window is 36.0 cm wide, what is the distance from the central maximum to the first-order minimum along a wall 6.50 m from the window?

30. Prove that if the wavelength of light is equal to or greater than the width of a slit, light striking the slit perpendicularly passes through without forming any dark interference bands.

31. A slit of width 0.50 mm is illuminated with light of wavelength 500 nm, and a screen is placed 120 cm in front of the slit. Find the widths of the first and second maxima on each side of the central maximum.

32. A screen is placed 50.0 cm from a single slit, which is illuminated with light of wavelength 680 nm. If the distance between the first and third minima in the diffraction pattern is 3.00 mm, what is the width of the slit?

Section 24.7 Polarization of Light Waves

33. The angle of incidence of a light beam in air onto a reflecting surface is continuously variable. The reflected ray is found to be completely polarized when the angle of incidence is 48.0°. (a) What is the index of refraction of the reflecting material? (b) If some of the incident light (at an angle of 48.0°) passes into the material below the surface, what is the angle of refraction?

34. The index of refraction of a glass plate is 1.52. What is Brewster's angle when the plate is (a) in air? (b) in water? (See Problem 38.)

35. At what angle above the horizon is the Sun if light from it is completely polarized on reflection from water?

36. A light beam is incident on heavy flint glass ($n = 1.65$) at the polarizing angle. Calculate the angle of refraction for the transmitted ray.

37. The critical angle for total internal reflection for sapphire surrounded by air is 34.4°. Calculate Brewster's angle for sapphire if the light is incident from the air.

38. Equation 24.13 assumes that the incident light is in air. If the light is incident from a medium of index n_1 on a medium of index n_2, follow the procedure used to derive Equation 24.13 to show that $\tan \theta_p = n_2/n_1$.

WEB 39. Light of intensity I_0 and polarized parallel to the transmission axis of a polarizer, is incident on an analyzer. (a) If the transmission axis of the analyzer makes an angle of 45° with the axis of the polarizer, what is the intensity of the transmitted light? (b) What should the angle between the transmission axes to be make $I/I_0 = \frac{1}{3}$?

40. Three polarizing plates whose planes are parallel are centered on a common axis. The directions of the transmission axes relative to the common vertical direction are shown in Figure P24.40. A linearly polarized beam

of light with the plane of polarization parallel to the vertical reference direction is incident from the left on the first disk with intensity $I_i = 10.0$ units (arbitrary). Calculate the transmitted intensity, I_f, when $\theta_1 = 20.0°$, $\theta_2 = 40.0°$, and $\theta_3 = 60.0°$. (*Hint:* Make repeated use of Malus's law.)

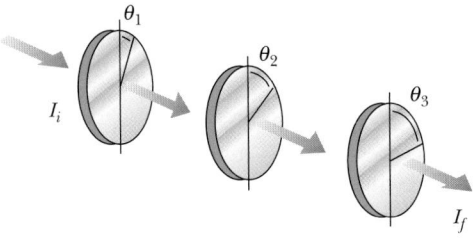

Figure P24.40   (Problems 40 and 47)

41. Light with a wavelength in vacuum of 546 nm falls perpendicularly on a biological specimen that is 1.00 μm thick. The light splits into two beams polarized at right angles, for which the indices of refraction are 1.32 and 1.33. (a) Calculate the wavelength of each component of the light while it is traversing the specimen. (b) Calculate the phase difference between the two beams when they emerge from the specimen.

ADDITIONAL PROBLEMS

42. A beam containing light of wavelengths λ_1 and λ_2 is incident on a set of parallel slits. In the interference pattern, the fourth bright line of the λ_1 light occurs at the same position as the fifth bright line of the λ_2 light. If λ_1 is known to be 540 nm, what is the value of λ_2?

43. Light of wavelength 546 nm (the intense green line from a mercury source) produces a Young's interference pattern in which the second minimum from the central maximum is along a direction that makes an angle of 18.0 min of arc with the axis through the central maximum. What is the distance between the parallel slits?

44. A thin layer of oil ($n = 1.25$) is floating on water. How thick is the oil in the region that strongly reflects green light ($\lambda = 525$ nm)?

45. Suppose that a slit 6.0 cm wide is placed in front of a microwave source operating at a frequency of 7.5 GHz. Calculate the angle (measured from the central maximum) where the first minimum in the diffraction pattern occurs.

46. When a monochromatic beam of light is incident from air at an angle of 37.0° with the normal on the surface of a glass block, it is observed that the refracted ray is directed at 22.0° with the normal. What angle of incidence from air would result in total polarization of the reflected beam?

47. Three polarizers, centered on a common axis and with their planes parallel to each other, have transmission axes oriented at angles of θ_1, θ_2 and θ_3 from the vertical as shown in Figure P24.40. Light of intensity I_i, polarized with its plane of polarization oriented vertically, is incident from the left on the first polarizer. What is the ratio I_f/I_i of the final transmitted intensity to the incident intensity if (a) $\theta_1 = 45°$, $\theta_2 = 90°$, and $\theta_3 = 0°$? (b) $\theta_1 = 0°$, $\theta_2 = 45°$, and $\theta_3 = 90°$?

48. Figure P24.48 shows a radio wave transmitter and a receiver, both $h = 50.0$ m above the ground and $d = 600$ m apart. The receiver can receive signals both directly from the transmitter and indirectly from signals that bounce off the ground. If that the ground is level between the transmitter and receiver and a $\lambda/2$ phase shift occurs on reflection, determine the longest wavelengths that interfere (a) constructively and (b) destructively.

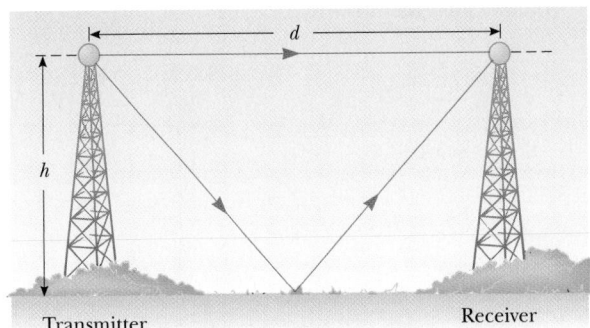

Figure P24.48

49. A pair of slits, separated by 0.150 mm, are illuminated by light having a wavelength of $\lambda = 643$ nm. An interference pattern is observed on a screen 140 cm from the slits. Consider a point on the screen located at $y = 1.80$ cm from the central maximum of this pattern. (a) What is the path difference, δ, for the two slits at this y location? (b) Express this path difference in terms of the wavelength. (c) Will the interference correspond to a maximum, a minimum, or an intermediate condition?

50. The condition for constructive interference by reflection from a thin film in air, as developed in Section 24.4, assumes nearly normal incidence. (a) Show that for large angles of incidence, the condition for constructive interference of light reflecting from a thin film of thickness t, index of refraction n, and surrounded by air may be written as

$$2nt \cos \theta_2 = \left(m + \frac{1}{2} \right) \lambda$$

where θ_2 is the angle of refraction. (b) Calculate the minimum thickness for constructive interference if sodium light ($\lambda = 590$ nm) is incident at an angle of $30.0°$ on a film with index of refraction 1.38.

51. Raise your hand and hold it flat. Think of the space between your index finger and your middle finger as one slit, and think of the space between middle finger and ring finger as a second slit. (a) Consider the interference resulting from sending coherent visible light perpendicularly through this pair of openings. Compute an order-of-magnitude estimate for the angle between adjacent zones of constructive interference. (b) To make the angles in the interference pattern easy to measure with a plastic protractor, you should use an electromagnetic wave with frequency of what order of magnitude? How is this wave classified on the electromagnetic spectrum?

52. Interference effects are produced at point P on a screen as a result of direct rays from a 500-nm source and reflected rays off the mirror, as in Figure P24.52. If the source is 100 m to the left of the screen, and 1.00 cm above the mirror, find the distance y (in millimeters) to the first dark band above the mirror.

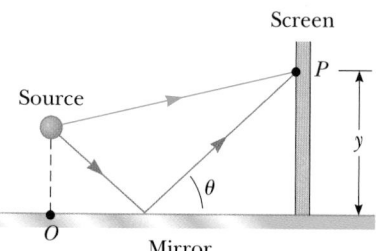

Figure P24.52

53. A planoconvex lens (flat on one side, convex on the other) with index of refraction n rests with its curved side (radius of curvature R) on a flat glass surface of the same index of refraction with a film of index n_{film} between them. The lens is illuminated from above by light of wavelength λ. Show that the dark Newton's rings have radii of

$$r \approx \sqrt{m\lambda R / n_{film}}$$

where m is an integer.

54. When a liquid is introduced into the air space between the lens and the plate in a Newton's-rings apparatus, the diameter of the tenth ring changes from 1.50 to 1.31 cm. Find the index of refraction of the liquid. (See Problem 53.)

WEB 55. (a) If light is incident at an angle of θ from a medium of index n_1 on a medium of index n_2 so that the angle between the reflected ray and refracted ray is β, show that

$$\tan \theta = \frac{n_2 \sin \beta}{n_1 - n_2 \cos \beta}$$

(*Hint:* Use the following identity.)

$$\sin(A + B) = \sin A \cos B + \cos A \sin B$$

(b) Show that the foregoing equation for $\tan \theta$ reduces to Brewster's law when $\beta = 90°$, $n_1 = 1$, and $n_2 = n$.

56. The transmitting antenna on a submarine is 5.00 m above the water when the ship surfaces. The captain wishes to transmit a message to a receiver on a 90.0-m-tall cliff at the ocean shore. If the signal is to be completely polarized by reflection off the ocean surface, how far must the ship be from the shore?

57. A diffraction pattern is produced on a screen 140 cm from a single slit, using monochromatic light of wavelength 500 nm. The distance from the center of the central maximum to the first-order maximum is 3.00 mm. Calculate the slit width. (*Hint:* Assume that the first-order maximum is halfway between the first- and second-order minima.)

58. A glass plate ($n = 1.61$) is covered with a thin, uniform layer of oil ($n = 1.20$). A light beam of variable wavelength from air is incident normally on the oil surface. Observation of the reflected beam shows destructive interference at 500 nm and constructive interference at 750 nm. From this information, calculate the thickness of the oil film.

59. A piece of transparent material with index of refraction n is cut into the shape of a wedge, as shown in Figure P24.59. The angle of the wedge is small, and monochromatic light of wavelength λ is normally incident from above. If the height of the wedge is h and its length is ℓ, show that bright fringes occur at the positions

$$x = \frac{\lambda\ell(m + \frac{1}{2})}{2hn}$$

and dark fringes occur at the positions $x = \lambda\ell m/2hn$, where $m = 0, 1, 2, \ldots$ and where x is measured as shown.

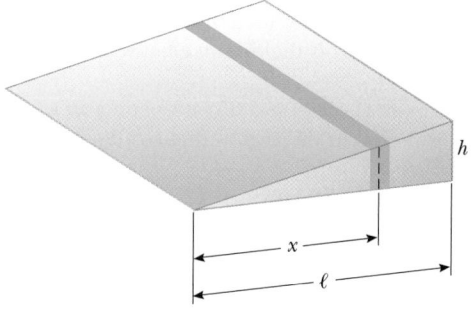

Figure P24.59

60. Figure P24.60 illustrates the formation of an interference pattern by the Lloyd's mirror method. Light from

source S reaches the screen via two different pathways. One is a direct path and the other is a reflection from a horizontal mirror. The interference functions as if light from two different sources, S and S', had interfered as in the Young's double-slit arrangement. Assume that the actual source, S, and the virtual source, S', are in a plane 25 cm to the left of the mirror, and the screen is a distance of $L = 120$ cm to the right of this plane. Source S is a distance of $d = 2.5$ mm above the top surface of the mirror, and the light is monochromatic with $\lambda = 620$ nm. Determine the distance of the first bright fringe above the surface of the mirror.

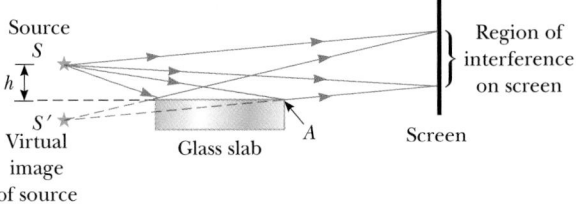

Figure P24.60

C H A P T E R

25

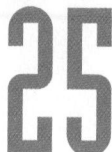

Optical Instruments

Chapter Outline

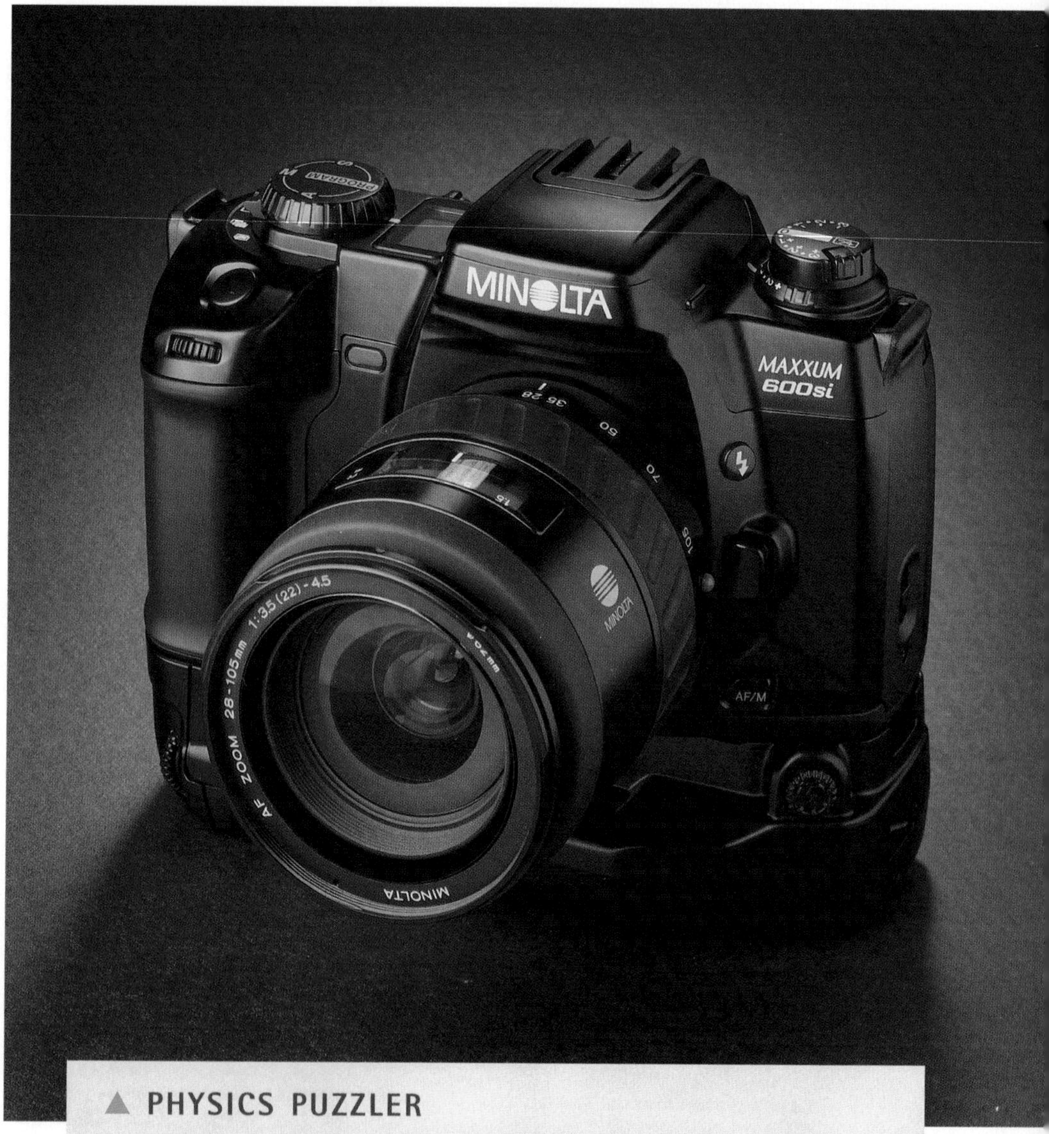

▲ PHYSICS PUZZLER

A normal lens for a camera might have a focal length of 50 mm. A telephoto lens might have a focal length of 200 mm. Why does the longer focal length lens result in a larger image on the film? *(Courtesy of Minolta Corporation)*

W e use devices made from lenses, mirrors, or other optical components every time we put on a pair of eyeglasses, take a photograph, look at the sky through a telescope, and so on. In this chapter we examine how these and other optical instruments work. For the most part, our analyses will involve the laws of reflection and refraction and the procedures of geometric optics. However, to explain certain phenomena we must use the wave nature of light.

25.1 THE CAMERA

The single-lens photographic **camera** is a simple optical instrument whose essential features are shown in Figure 25.1. It consists of a light-tight box, a converging lens that produces a real image, and a film behind the lens to receive the image. Focusing is accomplished by varying the distance between lens and film—with an adjustable bellows in old-style cameras and with other mechanical arrangements in newer models. For proper focusing, which leads to sharp images, the lens-to-film distance will depend on the object distance as well as on the focal length of the lens. The shutter, located behind the lens, is a mechanical device that is opened for selected time intervals. With this arrangement, moving objects can be photographed with the use of short exposure times, and dark scenes (low light levels) with the use of long exposure times. Without this control, it would be impossible to take stop-action photographs. For example, a speeding race car would move far enough while the shutter was open to produce a blurred image. Typical shutter "speeds" are 1/30, 1/60, 1/125, and 1/250 s. Stationary objects are often shot with a shutter speed of 1/60 s. More sophisticated cameras have a second adjustable aperture either behind or in front of the lens, to provide further control of the intensity of light reaching the film.

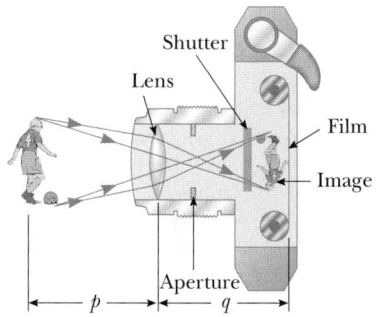

Figure 25.1 A cross-sectional view of a simple camera.

The brightness of the image focused on the film depends on the diameter and focal length of the lens. The amount of light reaching the film, and hence the brightness of the image formed on the film, increases with the size of the lens. The focal length of the lens also affects the brightness of the image. We can see this by considering the lateral magnification equation for a thin lens:

$$M = \frac{h'}{h} = -\frac{q}{p}$$

$$h' = -h\frac{q}{p}$$

where h and h' are the object and image heights, respectively, and p and q are the object and image distances. When p is large, q is approximately equal to the focal length, f. Thus, we have

$$h' \approx -h\frac{f}{p}$$

From this result, we see that a lens with a short focal length produces a small image, corresponding to a small value of h'.

A small image is brighter than a larger one because all of the incoming light is concentrated in a much smaller area. Because the brightness of the image depends

A Simple Pin–Hole Camera

Find a large cereal box and remove its paper lining. Cut one side of the box across the middle and two sides to form a flap, as in figure (a). Fold back the flap and tape a piece of waxed paper across the width and tape the paper about 2 inches from the bottom of the box. This serves as the viewing screen for your camera. Then tape the flap back to its original position using masking tape, and punch a hole in the bottom of the box with a pin or tack. Finally, remove the flaps from the top of the box and cut the top so that it conforms to the shape of your forehead and nose, as in figure (b). Now point the pinhole end of the box at a sunlit scene from your window and look into the box from the shaped end. You should see a small inverted image of the scene inside the box. Why is the image inverted? Draw a ray diagram that supports your observation.

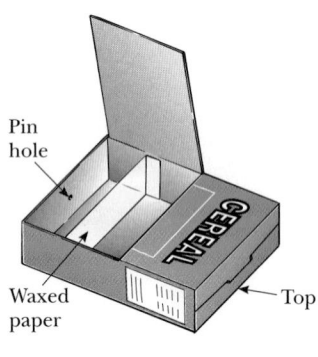

(a)

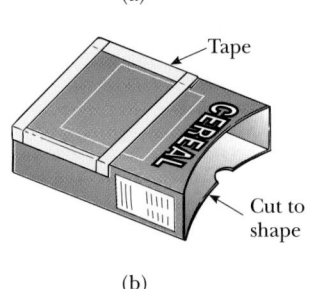

(b)

on f and on D, the diameter of the lens, a quantity called the **f-number** is defined as

$$f\text{-number} \equiv \frac{f}{D} \qquad [25.1]$$

For example, if a camera has a lens of focal length 52 mm and is set with f-number 4, the aperture diameter is $D = f/4 = (52 \text{ mm})/4 = 13 \text{ mm}$.

The f-number is a measure of the light-concentrating power of a lens and determines what is called the speed of the lens. A fast lens has a small f-number and usually a small focal length and large diameter. Camera lenses are often marked with a range of f-numbers such as 2.8, 4, 5.6, 8, 11, 16. They are selected by adjusting the aperture, which effectively changes D. When the f-number is increased by one position, or one "stop," the light admitted decreases by a factor of 2. Likewise, the shutter speed is changed in steps by a factor of 2. The smallest f-number corresponds to the case in which the aperture is wide open and as much of the lens area is in use as possible. Fast lenses, with f-numbers as low as 1.2, are relatively expensive because it is more difficult to keep aberrations acceptably small. A simple camera for routine snapshots usually has a fixed focal length and fixed aperture size, with an f-number of about 11.

EXAMPLE 25.1 Choosing the f-Number

Suppose you are using a single-lens 35-mm camera (35 mm is the width of the film strip) with only two f-stops, $f/2.8$ and $f/22$. Which f-number would you use on a cloudy day? Why?

Solution Substituting the given f-numbers into Equation 25.1, we have

$$2.8 = \frac{f_1}{D_1} \qquad \text{and} \qquad 22 = \frac{f_2}{D_2}$$

The focal length of the camera is fixed ($f_1 = f_2$ in the two equations), but the diameter of the aperture is not. On a cloudy day, you should make the shutter opening as large as possible. As these equations indicate, the largest value of D produces the smallest f-number. Thus, you should use the 2.8 setting.

25.2 THE EYE

The eye is a remarkable and extremely complex organ. Because of this complexity, defects sometimes arise that impair vision. To compensate for the defects, external aids, such as eyeglasses, are often used. In this section we describe the parts of the eye, their purposes, and some of the corrections that can be made when the eye does not function properly. You will find that the eye has much in common with the camera. Like the camera, the eye gathers light and produces a sharp image. However, the mechanisms by which the eye controls the amount of light admitted and adjusts itself to produce correctly focused images are far more complex, intricate, and effective than those in even the most sophisticated camera. In all respects, the eye is a wonder of design.

Figure 25.2 shows the essential parts of the eye. The front is covered by a transparent membrane called the *cornea*. Inward from the cornea are a clear liquid region (the *aqueous humor*), a variable aperture (the *iris* surrounding the *pupil*), and the *crystalline lens*. Most of the refraction occurs in the cornea, because the liquid medium surrounding the lens has an average index of refraction close to that of the lens. The iris, the colored portion of the eye, is a muscular diaphragm that regulates the amount of light entering the eye by dilating the pupil (increasing its diameter) in light of low intensity and contracting the pupil in high-intensity light. The pupil diameter can vary from about $f/2.8$ to $f/16$.

Light entering the eye is focused by the cornea-lens system onto the back surface of the eye, called the *retina*. The surface of the retina consists of millions of sensitive receptors called *rods* and *cones*. When stimulated by light, these structures send impulses via the optic nerve to the brain, where a distinct image of an object is perceived.

The eye focuses on a given object by varying the shape of the pliable crystalline lens through an amazing process called **accommodation.** An important component in accommodation is the *ciliary muscle*, which is attached to the lens. When the eye is focused on distant objects, the ciliary muscle is relaxed. For an object distance of infinity, the focal length of the eye (the distance between the lens and the retina) is about 1.7 cm. The eye focuses on nearby objects by tensing the ciliary muscle. This action effectively reduces the focal length by slightly decreasing the radius of curvature of the lens, which allows the image to be focused on the retina. This lens adjustment takes place so swiftly that we are not aware of the change. In this respect, as in others, even the finest electronic camera is a toy compared with the eye. It is evident that there is a limit to accommodation, because objects that are very close to the eye produce blurred images. **The near point is the smallest distance for which the lens will produce a sharp image on the retina.** This

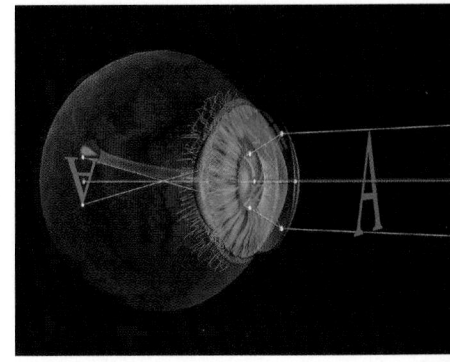

A model of a human eye. (*Douglas Struthers/Tony Stone Images*)

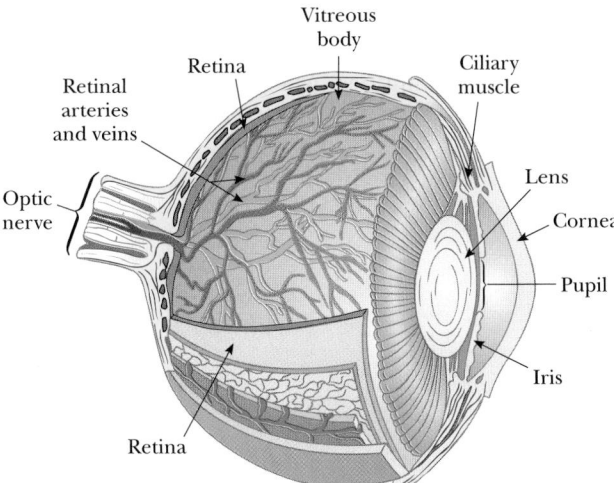

Figure 25.2 Essential parts of the eye. Can you correlate the essential parts of the eye with those of the simple camera in Figure 25.1?

distance usually increases with age. Typically, the near point of the eye is about 18 cm at age 10, about 25 cm at age 20, 50 cm at age 40, and 500 cm or greater at age 60.

Defects of the Eye

An eye can have several abnormalities that keep it from functioning properly. These can often be corrected with eyeglasses, contact lenses, or surgery.

When the relaxed eye produces an image of a nearby object *behind* the retina, as in Figure 25.3a, the abnormality is known as **hyperopia,** and the person is said to be *farsighted.* With this defect, distant objects are seen clearly but near objects are blurred. Either the hyperopic eye is too short or the ciliary muscle cannot change the shape of the lens enough to focus the image properly. The condition can be corrected with a converging lens, as shown in Figure 25.3b.

Another condition, known as **myopia,** or *nearsightedness,* occurs either when the eye is longer than normal or when the maximum focal length of the lens is insufficient to produce a clearly formed image on the retina. In this case, light from a distant object is focused in front of the retina (Fig. 25.4a). The distinguishing feature of this imperfection is that distant objects are not seen clearly. Nearsightedness can be corrected with a diverging lens, as in Figure 25.4b.

Beginning in middle age (around age 40) most people lose some of their accommodation power, usually as a result of hardening of the crystalline lens. This causes farsightedness, which can be corrected with converging lenses.

A common eye defect is **astigmatism,** in which light from a point source produces a line image on the retina. This occurs when the cornea or the crystalline lens or both are not perfectly spherical. Astigmatism can be corrected by lenses with different curvatures in two mutually perpendicular directions. A cylindrical lens (a segment of a cylinder) is typically used for this purpose.

The eye is also subject to several diseases. One, which usually occurs later in life, is the formation of **cataracts,** which make the lens partially or totally opaque. The common remedy for cataracts is surgical removal of the lens. Another disease, called **glaucoma,** arises from an abnormal increase in fluid pressure inside the eyeball. This pressure increase can cause a reduction in blood supply to the retina, which can eventually lead to blindness when the nerve fibers of the retina die. If the disease is discovered early enough, it can be treated with medicine or surgery.

Optometrists and ophthalmologists usually prescribe corrective lenses measured in diopters.

A P P L I C A T I O N

Correcting Eye Defects with Optical Lenses.

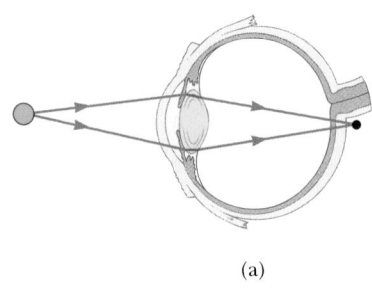

(a)

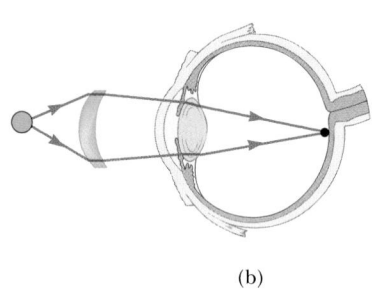

(b)

Figure 25.3   (a) A farsighted eye is slightly shorter than normal; hence, the image of a nearby object focuses *behind* the retina. (b) The condition can be corrected with a converging lens. (The object is assumed to be very small in these figures.)

The power, *P*, of a lens in diopters equals the inverse of the focal length in meters—that is, $P = 1/f$.

For example, a converging lens whose focal length is $+20$ cm has a power of $+5$ diopters, and a diverging lens whose focal length is -40 cm has a power of -2.5 diopters.

Thinking Physics 1

A classic science fiction story, *The Invisible Man,* tells of a person who becomes invisible by changing the index of refraction of his body to that of air. This story has been criticized by students who know how the eye works; they claim the invisible man would be unable to see. On the basis of your knowledge of the eye, could he see or not?

Explanation He would not be able to see. In order for the eye to "see" an object, incoming light must be refracted at the cornea and lens to form an image on the retina. If the cornea and lens have the same index of refraction as air, refraction cannot occur, and an image would not be formed.

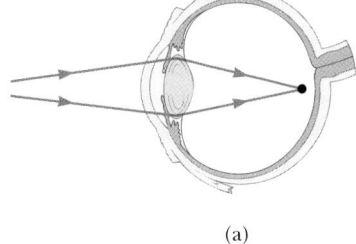

(a)

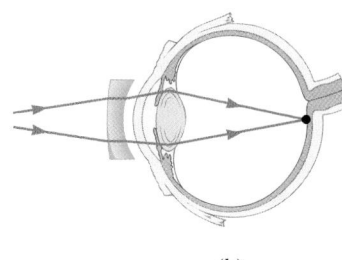

(b)

Figure 25.4 (a) A nearsighted eye is slightly longer than normal; hence, the image of a distant object focuses *in front* of the retina. (b) The condition can be corrected with a diverging lens. (The object is assumed to be very small in these figures.)

EXAMPLE 25.2 Prescribing a Lens

The near point of an eye is 50.0 cm. (a) What focal length must a corrective lens have to enable the eye to see clearly an object 25.0 cm away?

Reasoning The thin-lens equation (Eq. 23.11) enables us to solve this problem. We have placed an object at 25.0 cm, and we want the lens to form an image at the closest point that the eye can see clearly. This corresponds to the near point, 50.0 cm.

Solution Applying the thin-lens equation, we have

$$\frac{1}{25.0 \text{ cm}} + \frac{1}{(-50.0 \text{ cm})} = \frac{1}{f}$$

$$f = \boxed{50.0 \text{ cm}}$$

Why did we use a negative sign for the image distance? Notice that the focal length is positive, indicating the need for a converging lens to correct farsightedness such as this.

(b) What is the power of this lens?

Solution The power is the reciprocal of the focal length in meters:

$$P = \frac{1}{f} = \frac{1}{0.500 \text{ m}} = \boxed{2.00 \text{ diopters}}$$

EXAMPLE 25.3 A Case of Nearsightedness

A particular nearsighted person cannot see objects clearly when they are beyond 50.0 cm (the far point of the eye). What focal length should the prescribed lens have to correct this problem?

Reasoning For an object at infinity, the purpose of the lens in this instance is to place the image at a distance at which it can be seen clearly.

Solution From the thin-lens equation, we have

$$\frac{1}{p} + \frac{1}{q} = \frac{1}{\infty} + \frac{1}{(-50.0 \text{ cm})} = \frac{1}{f}$$

$$f = -50.0 \text{ cm}$$

Why did we use a negative sign for the image distance? As you should have suspected, the lens must be diverging (have a negative focal length) to correct nearsightedness.

Exercise What is the power of this lens?

Answer -2.00 diopters

25.3 THE SIMPLE MAGNIFIER

The **simple magnifier** is one of the most basic of all optical instruments because it consists only of a single converging lens. As the name implies, this device is used to increase the apparent size of an object. Suppose an object is viewed at some distance, p, from the eye, as in Figure 25.5. Clearly, the size of the image formed at the retina depends on the angle, θ, subtended by the object at the eye. As the object moves closer to the eye, θ increases and a larger image is observed. However, a normal eye cannot focus on an object closer than about 25 cm, the near point (Fig. 25.6a). (Try it!) Therefore, θ is maximum at the near point.

To further increase the apparent angular size of an object, a converging lens can be placed in front of the eye with the object positioned at point O, just inside the focal point of the lens, as in Figure 25.6b. At this location, the lens forms a virtual, upright, and enlarged image, as shown. The lens increases the angular size

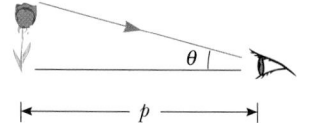

Figure 25.5 The size of the image formed on the retina depends on the angle, θ, subtended at the eye.

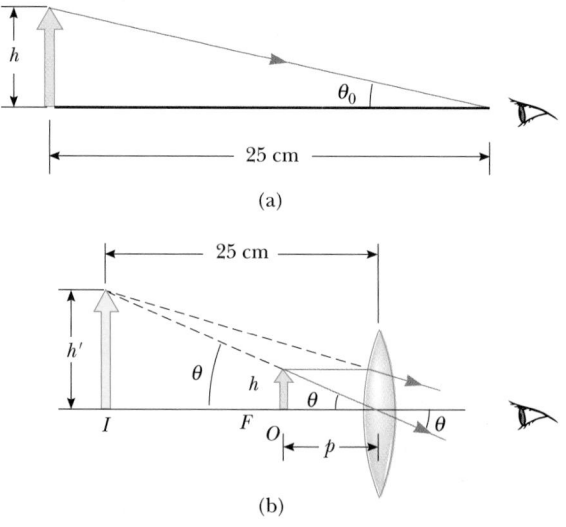

(a)

(b)

Figure 25.6 (a) An object placed at the near point ($p = 25$ cm) subtends an angle of $\theta_0 \approx h/25$ at the eye. (b) An object placed near the focal point of a converging lens produces a magnified image, which subtends an angle of $\theta \approx h'/25$ at the eye.

of the object. We define the **angular magnification, *m*,** as the ratio of the angle subtended by the object when the lens is in use (angle θ in Fig. 25.6b) to that subtended by the object when it is placed at the near point with no lens (angle θ_0 in Fig. 25.6a):

$$m \equiv \frac{\theta}{\theta_0} \qquad\qquad \text{[25.2]}$$

◀ Angular magnification

The angular magnification is a maximum when the image formed by the lens is at the near point of the eye—that is, when $q = -25$ cm (see Fig. 25.6b). The object distance that corresponds to this image distance can be calculated from the thin-lens equation:

$$\frac{1}{p} + \frac{1}{-25 \text{ cm}} = \frac{1}{f} \qquad\qquad \text{[25.3]}$$

$$p = \frac{25f}{25 + f}$$

where f is the focal length in centimeters. From Figures 25.6a and 25.6b, the small-angle approximation gives

$$\tan\theta_0 \approx \theta_0 \approx \frac{h}{25} \qquad \text{and} \qquad \tan\theta \approx \theta \approx \frac{h}{p} \qquad\qquad \text{[25.4]}$$

Thus, Equation 25.2 becomes

$$m = \frac{\theta}{\theta_0} = \frac{h/p}{h/25} = \frac{25}{p} = \frac{25}{25f/(25 + f)} \qquad\qquad \text{[25.5]}$$

$$m = 1 + \frac{25 \text{ cm}}{f}$$

The angular magnification given by Equation 25.5 is the ratio of the angular size seen with the lens to the angular size seen without the lens, with the object at the near point of the eye. The eye can actually focus on an image formed anywhere between the near point and infinity, and is most relaxed when the image is at infinity (Sec. 25.2). In order for the image formed by the magnifying lens to appear at infinity, the object must be placed at the focal point of the lens—that is, $p = f$. In this case, Equation 25.4 becomes

$$\theta_0 \approx \frac{h}{25} \qquad \text{and} \qquad \theta \approx \frac{h}{f}$$

and the angular magnification is

$$m = \frac{\theta}{\theta_0} = \frac{25 \text{ cm}}{f} \qquad\qquad \text{[25.6]}$$

With a single lens, it is possible to achieve angular magnifications up to about 4 without serious aberrations. Angular magnifications up to about 20 can be achieved by using a second lens to correct for aberrations.

QUICKLAB

Water-Drop Magnifier

Place a piece of waxed paper on a sheet of printed material such as a newspaper. Place a small drop of water on the waxed paper using a straw or medicine dropper, as in the figure. Note that the printed material you see through the drop is larger than the print on either side of the drop. Now place drops of various sizes on the waxed paper, and note that the smallest drop (having the greatest curvature on its upper surface) gives the largest magnification. Estimate the magnifications for the various drops (magnifying lenses) you have created.

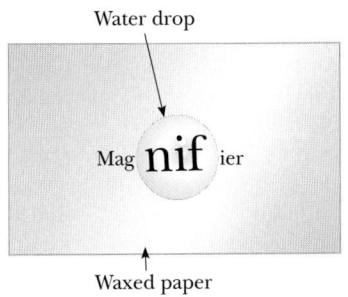

Water drop

Mag **nif** ier

Waxed paper

EXAMPLE 25.4   Maximum Angular Magnification of a Lens

What is the maximum angular magnification of a lens with a focal length of 10.0 cm, and what is the angular magnification of this lens when the eye is relaxed?

Reasoning   The maximum angular magnification occurs when the image formed by the lens is at the near point of the eye. Under these circumstances, Equation 25.5 gives us the maximum angular magnification.

Solution   Using Equation 25.5, we have

$$m = 1 + \frac{25 \text{ cm}}{f} = 1 + \frac{25 \text{ cm}}{10.0 \text{ cm}} = \boxed{3.5}$$

When the eye is relaxed, the image is at infinity. In this case, we use Equation 25.6:

$$m = \frac{25 \text{ cm}}{f} = \frac{25 \text{ cm}}{10.0 \text{ cm}} = \boxed{2.5}$$

25.4   THE COMPOUND MICROSCOPE

A simple magnifier provides only limited assistance with inspection of the minute details of an object. Greater magnification can be achieved by combining two lenses in a device called a *compound microscope,* a schematic diagram of which is shown in Figure 25.7a. It consists of two lenses: an objective with a very short focal length, f_o (where $f_o < 1$ cm), and an ocular lens, or eyepiece, with a focal length, f_e, of a few centimeters. The two lenses are separated by distance L, which is much greater than either f_o or f_e.

As you read the discussion that follows, note that the basic approach used to analyze the image formation properties of a microscope is that of two lenses in a row — that is, the image formed by the first becomes the object for the second. The object, O, placed just outside the focal length of the objective, forms a real, inverted image at I_1, which is at or just inside the focal point of the eyepiece. This image is

Figure 25.7   (a) A diagram of a compound microscope, which consists of an objective and an eyepiece, or ocular lens. (b) A compound microscope. The three-objective turret allows the user to switch to several different powers of magnification. Combinations of eyepieces with different focal lengths and different objectives can produce a wide range of magnifications. *(Henry Leap and Jim Lehman)*

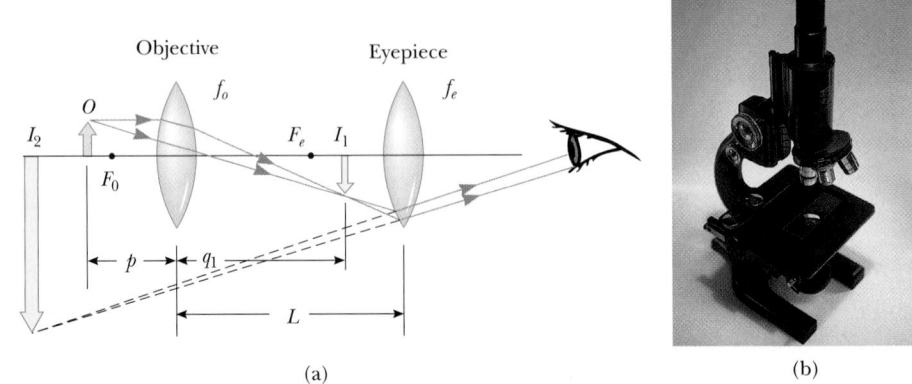

(a)

(b)

real and much enlarged. (For clarity, the enlargement of I_1 is not shown in Fig. 25.7a.) The eyepiece, which serves as a simple magnifier, uses the image at I_1 as its object and produces an image at I_2. The image seen by the eye at I_2 is virtual, inverted, and very much enlarged.

The lateral magnification, M_1, of the first image is $-q_1/p_1$. Note in Figure 25.7a that q_1 is approximately equal to L. This occurs because the object is placed close to the focal point of the objective lens, which ensures that the image formed will be far from the objective lens. Furthermore, because the object is very close to the focal point of the objective lens, $p_1 \approx f_o$. This gives a lateral magnification of

$$M_1 = -\frac{q_1}{p_1} \approx -\frac{L}{f_o}$$

for the objective. The angular magnification of the eyepiece for an object (corresponding to the image at I_1) placed at the focal point is found from Equation 25.6 to be

$$m_e = \frac{25 \text{ cm}}{f_e}$$

The overall magnification of the compound microscope is defined as the product of the lateral and angular magnifications:

$$M = M_1 m_e = -\frac{L}{f_o}\left(\frac{25 \text{ cm}}{f_e}\right) \qquad\qquad \text{[25.7]}$$

◀ Magnification of a microscope

The negative sign indicates that the image is inverted with respect to the object.

The microscope has extended our vision into the previously unknown realm of incredibly small objects, and the capabilities of this instrument have increased steadily with improved techniques in precision grinding of lenses. A question that is often asked about microscopes is, "With extreme patience and care, would it be possible to construct a microscope that would enable us to see an atom?" The answer to this question is no, as long as visible light is used to illuminate the object. The reason is that, in order to be seen, the object under a microscope must be at least as large as a wavelength of light. An atom is many times smaller than the wavelength of visible light, and so its mysteries must be probed via other techniques.

The wavelength dependence of the "seeing" ability of a wave can be illustrated by water waves set up in a bathtub in the following manner. Imagine that you vibrate your hand in the water until waves with a wavelength of about 6 in. are moving along the surface. If you fix a small object, such as a toothpick, in the path of the waves, you will find that the waves are not appreciably disturbed by the toothpick but continue along their path. Now suppose you fix a larger object, such as a toy sailboat, in the path of the waves. In this case, the waves are considerably "disturbed" by the object. The toothpick was much smaller than the wavelength of the waves, and as a result the waves did not "see" it. The toy sailboat, in contrast, is about the same size as the wavelength of the waves and hence creates a disturbance. Light waves behave in this same general way. The ability of an optical microscope to view an object depends on the size of the object relative to the wavelength of the light used to observe it. Hence, it will never be possible to observe atoms or molecules with such a microscope, because their dimensions are so small (≈ 0.1 nm) relative to the wavelength of the light (≈ 500 nm).

EXAMPLE 25.5 Magnifications of a Microscope

A certain microscope has two interchangeable objectives. One has a focal length of 20.0 mm, and the other has a focal length of 2.0 mm. Also available are two eyepieces of focal lengths 2.5 cm and 5.0 cm. If the length of the microscope is 18 cm, what magnifications are possible?

Reasoning and Solution The solution consists of applying Equation 25.7 to four different combinations of lenses. For the combination of the two long focal lengths, we have

$$M = -\frac{L}{f_o}\left(\frac{25 \text{ cm}}{f_e}\right) = -\frac{18}{2.0}\left(\frac{25}{5.0}\right) = \boxed{-45}$$

The combination of the 20.0-mm objective and the 2.5-cm eyepiece gives

$$M = -\frac{18}{2.0}\left(\frac{25}{2.5}\right) = \boxed{-90}$$

The 2.0-mm and 5.0-cm combination produces

$$M = -\frac{18}{0.20}\left(\frac{25}{5.0}\right) = \boxed{-450}$$

Finally, the two short focal lengths give

$$M = -\frac{18}{0.20}\left(\frac{25}{2.5}\right) = \boxed{-900}$$

25.5 THE TELESCOPE

There are two fundamentally different types of telescopes, both designed to help us view distant objects such as the planets in our Solar System. These two types are (1) the **refracting telescope,** which uses a combination of lenses to form an image, and (2) the **reflecting telescope,** which uses a curved mirror and a lens to form an image. Once again, we will be able to analyze the telescope by considering it to be a system of two optical elements in a row. The basic technique followed is that the image formed by the first element becomes the object for the second.

Let us first consider the refracting telescope. In this device, two lenses are arranged so that the objective forms a real, inverted image of the distant object very near the focal point of the eyepiece (Fig. 25.8a). Furthermore, the image at I_1 is formed at the focal point of the objective because the object is essentially at infinity. Hence, the two lenses are separated by the distance $f_o + f_e$, which corresponds to the length of the telescope's tube. The eyepiece finally forms, at I_2, an enlarged, inverted image of the image at I_1.

The angular magnification of the telescope is given by θ/θ_0, where θ_0 is the angle subtended by the object at the objective and θ is the angle subtended by the final image. From the triangles in Figure 25.8a, and for small angles, we have

$$\theta \approx \frac{h'}{f_e} \quad \text{and} \quad \theta_0 \approx \frac{h'}{f_o}$$

A large reflecting telescope used at Mount Laguna Observatory. (*Jerry Schad/Photo Researchers, Inc.*)

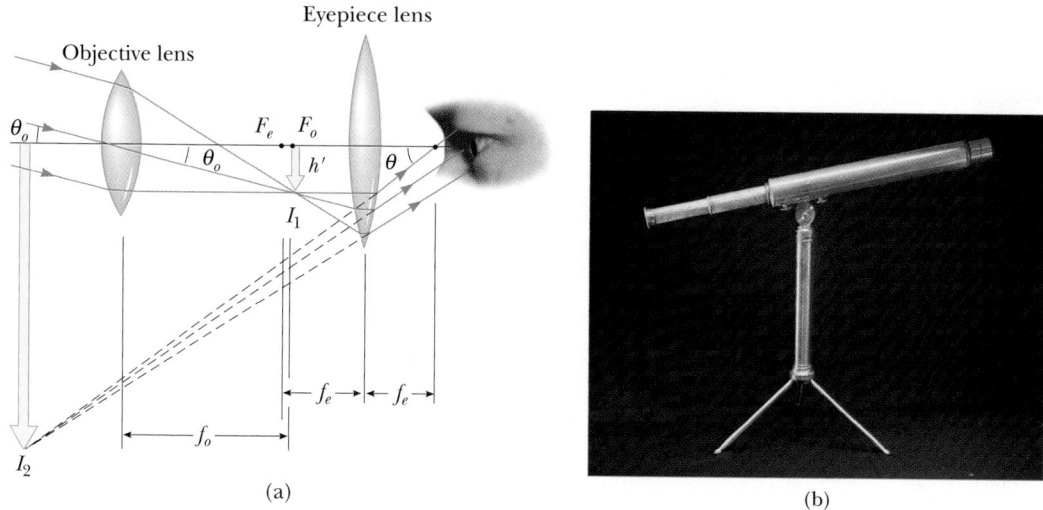

Figure 25.8 (a) A diagram of a refracting telescope, with the object at infinity. (b) A photograph of a refracting telescope. *(Photo courtesy of Henry Leap and Jim Lehman)*

Therefore, the angular magnification of the telescope can be expressed as

$$m = \frac{\theta}{\theta_0} = \frac{h'/f_e}{h'/f_o} = \frac{f_o}{f_e} \qquad [25.8]$$

This says that the angular magnification of a telescope equals the ratio of the objective focal length to the eyepiece focal length. Here again, the angular magnification is the ratio of the angular size seen with the telescope to the angular size seen with the unaided eye.

In some applications—for instance, the observation of relatively nearby objects such as the Sun, Moon, or planets—angular magnification is important. Stars, in contrast, are so far away that they always appear as small points of light regardless of how much angular magnification is used. The large research telescopes used to study very distant objects must have great diameters to gather as much light as possible. It is difficult and expensive to manufacture such large lenses for refracting telescopes. In addition, the heaviness of large lenses leads to sagging, which is another source of aberration.

These problems can be partially overcome by replacing the objective lens with a reflecting, concave mirror. Figure 25.9 shows the design for a typical reflecting telescope. Incoming light rays pass down the barrel of the telescope and are reflected by a parabolic mirror at the base. These rays converge toward point *A* in the figure, where an image would be formed on a photographic plate or another detector. However, before this image is formed, a small flat mirror at point *M* reflects the light toward an opening in the side of the tube that passes into an eyepiece. This design is said to have a *Newtonian focus,* after its developer. Note that in the reflecting telescope the light never passes through glass (except the small eyepiece). As a result, problems associated with chromatic aberration are virtually eliminated. Also, difficulties arising from spherical aberration are reduced by the parabolic shape of the mirror.

14.9, BOTH SECTIONS

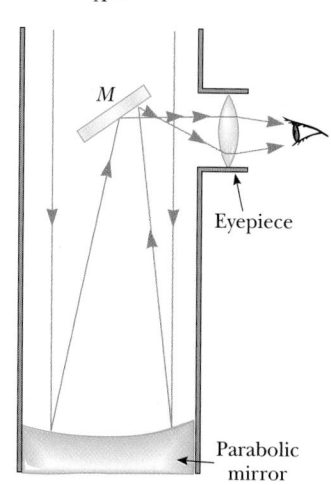

Figure 25.9 A reflecting telescope with a Newtonian focus.

Sunlight reflects off *Endeavour*'s aft windows and the shiny Hubble Space Telescope (HST) prior to its postservicing deployment near the end of the 11-day STS-61 mission. A hand-held Hasselblad camera was used inside *Endeavour*'s cabin to record the image. *(NASA)*

The largest telescopes in the world are the two 10-m-diameter Keck reflectors on Mauna Kea in Hawaii. The largest single-mirrored reflecting telescope in the United States is the 5-m-diameter instrument on Mount Palomar in California. In contrast, the largest refracting telescope in the world, at the Yerkes Observatory in Williams Bay, Wisconsin, has a diameter of only 1 m.

EXAMPLE 25.6 Angular Magnification of a Reflecting Telescope

A reflecting telescope has an 8-in.-diameter objective mirror with a focal length of 1500 mm. What is the angular magnification of this telescope when an eyepiece with an 18-mm focal length is used?

Solution The equation for finding the angular magnification of a reflector is the same as that for a refractor. Thus, Equation 25.8 gives

$$m = \frac{f_o}{f_e} = \frac{1500 \text{ mm}}{18 \text{ mm}} = \boxed{83}$$

25.6 RESOLUTION OF SINGLE-SLIT AND CIRCULAR APERTURES

The ability of an optical system such as a microscope or telescope to distinguish between closely spaced objects is limited because of the wave nature of light. To understand this difficulty, consider Figure 25.10, which shows two light sources far from a narrow slit of width a. The sources can be taken as two point sources, S_1 and S_2, that are *not* coherent. For example, they could be two distant stars. If no diffraction occurred, one would observe two distinct bright spots (or images) on the screen at the right in the figure. However, because of diffraction, each source is

Figure 25.10 Each of two point sources at some distance from a small aperture produces a diffraction pattern. (a) The angle subtended by the sources at the aperture is large enough so that the diffraction patterns are distinguishable. (b) The angle subtended by the sources is so small that the diffraction patterns are not distinguishable. (Note that the angles are greatly exaggerated.)

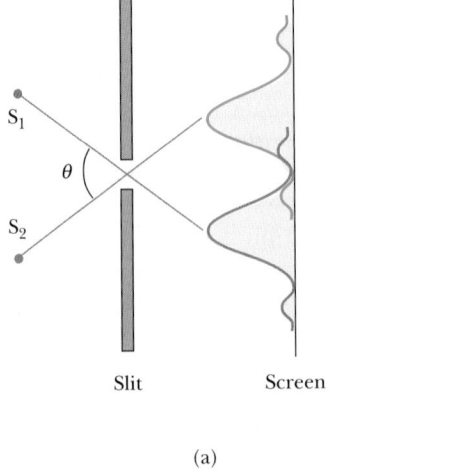

Slit   Screen

(a)

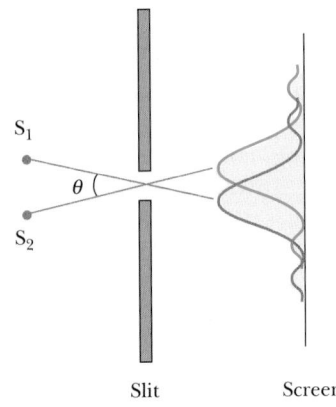

Slit   Screen

(b)

imaged as a bright central region flanked by weaker bright and dark rings. What is observed on the screen is the sum of two diffraction patterns, one from S_1 and the other from S_2.

If the two sources are separated so that their central maxima do not overlap, as in Figure 25.10a, their images can be distinguished and are said to be *resolved*. If the sources are close together, however, as in Figure 25.10b, the two central maxima may overlap and the images are *not resolved*. To decide whether two images are resolved, the following condition is often applied to their diffraction patterns:

> When the central maximum of one image falls on the first minimum of another image, the images are said to be just resolved. This limiting condition of resolution is known as **Rayleigh's criterion.**

Figure 25.11 shows diffraction patterns in three situations. When the objects are far apart, their images are well resolved (Fig. 25.11a). The images are just resolved when their angular separation satisfies Rayleigh's criterion (Fig. 25.11b). Finally, the images in Figure 25.11c are not resolved.

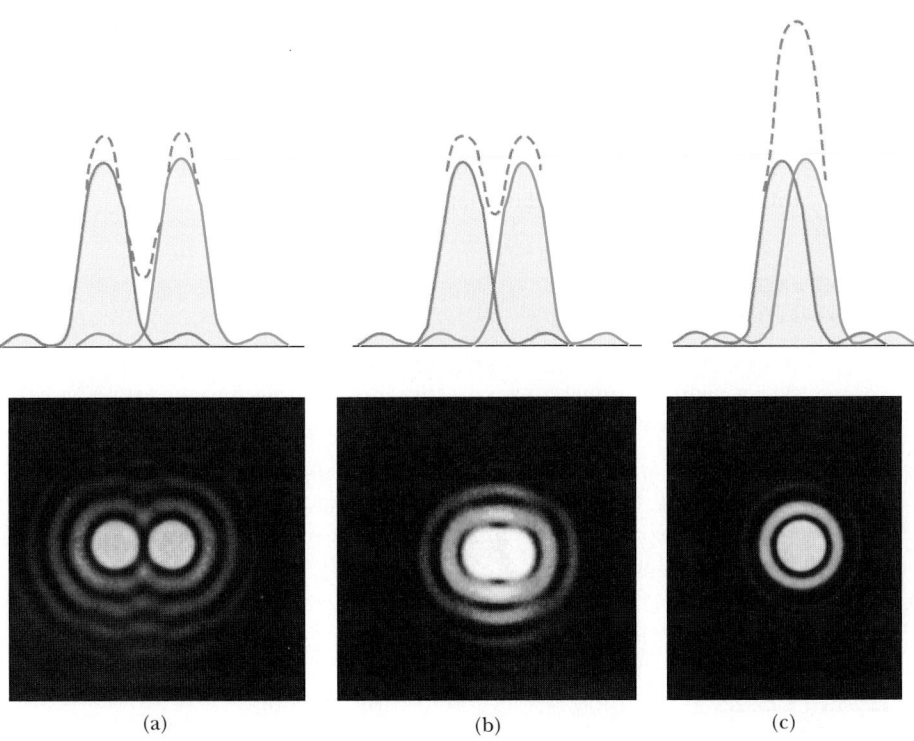

(a)   (b)   (c)

Figure 25.11 The diffraction patterns of two point sources (solid curves) and the resultant pattern (dashed curve) for three angular separations of the sources. (a) The sources are far apart, and their patterns are well resolved. (b) The sources are closer together, and their patterns are just resolved. (c) The sources are so close together that their patterns are not resolved. *(From M. Cagnet, M. Francon, and J. C. Thierr, Atlas of Optical Phenomena, Berlin, Springer-Verlag, 1962, plate 16)*

From Rayleigh's criterion, we can determine the minimum angular separation, θ_m, subtended by the source at the slit so that the images will be just resolved. In Chapter 24 we found that the first minimum in a single-slit diffraction pattern occurs at the angle that satisfies the relationship

$$\sin \theta = \frac{\lambda}{a}$$

where a is the width of the slit. According to Rayleigh's criterion, this expression gives the smallest angular separation for which two images can be resolved. Because $\lambda \ll a$ in most situations, $\sin \theta$ is small and we can use the approximation $\sin \theta \approx \theta$. Therefore, the limiting angle of resolution for a slit of width a is

◄ Limiting angle for a slit

$$\theta_m \approx \frac{\lambda}{a} \qquad\qquad [25.9]$$

where θ_m is in radians. Hence, the angle subtended by the two sources at the slit must be *greater* than λ/a if the images are to be resolved.

Many optical systems use circular apertures rather than slits. The diffraction pattern of a circular aperture (Fig. 25.12) consists of a central circular bright disk surrounded by progressively fainter rings. Analysis shows that the limiting angle of resolution of the circular aperture is

Limiting angle for a circular ► aperture

$$\theta_m = 1.22 \frac{\lambda}{D} \qquad\qquad [25.10]$$

where D is the diameter of the aperture. Note that Equation 25.10 is similar to Equation 25.9 except for the factor 1.22, which arises from a complex mathematical analysis of diffraction from a circular aperture.

Thinking Physics 2

Cats' eyes have vertical pupils when in dim light. Which would cats be most successful at resolving at night—headlights on a distant car or vertically separated running lights on a distant boat's mast?

Explanation The effective slit width in the vertical direction of the cat's eye is larger than that in the horizontal direction. Thus, it has more resolving power for lights separated in the vertical direction and would be more effective at resolving the mast lights on the boat.

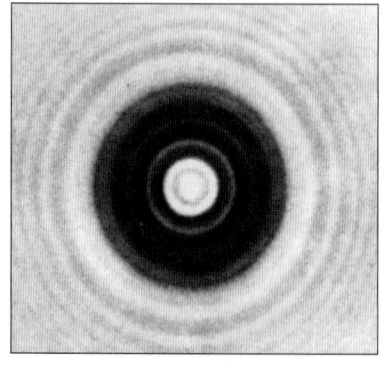

Figure 25.12 The Fresnel diffraction pattern of a circular aperture consists of a central bright disk surrounded by concentric bright and dark rings. *(From M. Cagnet, M. Francon, and J. C. Thierr,* Atlas of Optical Phenomena, *Berlin, Springer-Verlag, 1962, plate 34)*

EXAMPLE 25.7 Limiting Resolution of a Microscope

Sodium light of wavelength 589 nm is used to view an object under a microscope. The aperture of the objective has a diameter of 0.90 cm. (a) Find the limiting angle of resolution. (b) Using visible light of any wavelength you desire, what is the maximum limit of resolution for this microscope? (c) Suppose water of index of refraction 1.33 filled the space between the object and the objective. What effect would this have on the resolving power of the microscope?

Solution (a) From Equation 25.10, we find the limiting angle of resolution to be

$$\theta_m = 1.22 \left(\frac{589 \times 10^{-9} \text{ m}}{0.90 \times 10^{-2} \text{ m}} \right) = \boxed{8.0 \times 10^{-5} \text{ rad}}$$

This means that any two points on the object subtending an angle of less than 8.0×10^{-5} rad at the objective cannot be distinguished in the image.

(b) To obtain the maximum resolution, we have to use the shortest wavelength available in the visible spectrum. Violet light of wavelength 400 nm gives us a limiting angle of resolution of

$$\theta_m = 1.22 \left(\frac{400 \times 10^{-9} \text{ m}}{0.90 \times 10^{-2} \text{ m}} \right) = \boxed{5.4 \times 10^{-5} \text{ rad}}$$

(c) In this case, the wavelength of the sodium light in the water is found by $\lambda_w = \lambda_a / n$ (Chapter 22). Thus, we have

$$\lambda_w = \frac{\lambda_a}{n} = \frac{589 \text{ nm}}{1.33} = 443 \text{ nm}$$

The limiting angle of resolution at this wavelength is

$$\theta_m = 1.22 \left(\frac{443 \times 10^{-9} \text{ m}}{0.90 \times 10^{-2} \text{ m}} \right) = \boxed{6.0 \times 10^{-5} \text{ rad}}$$

EXAMPLE 25.8 Resolution of a Telescope

The Hale telescope at Mount Palomar has a diameter of 200 in. What is its limiting angle of resolution at a wavelength of 600 nm?

Solution Because $D = 200$ in. $= 5.08$ m and $\lambda = 6.00 \times 10^{-7}$ m, Equation 25.10 gives

$$\theta_m = 1.22 \frac{\lambda}{D} = 1.22 \left(\frac{6.00 \times 10^{-7} \text{ m}}{5.08 \text{ m}} \right) = \boxed{1.44 \times 10^{-7} \text{ rad}}$$

Therefore, any two stars that subtend an angle greater than or equal to this value could be resolved if the air above the telescope were perfectly steady.

It is interesting to compare this value with the resolution of a large radio telescope, such as the system at Arecibo, Puerto Rico, which has a diameter of 1000 ft (305 m). This telescope detects radio waves at a wavelength of 0.75 m. The corresponding minimum angle of resolution is calculated to be 3.0×10^{-3} rad (10 min 19 s of arc), which is more than 10 000 times larger than the calculated minimum angle for the Hale telescope.

EXAMPLE 25.9 Comparing Two Telescopes

Two telescopes have the following properties:

Telescope	Diameter of Objective (in.)	Focal Length of Objective (mm)	Focal Length of Eyepiece (mm)
A	6.00	1000	6.00
B	8.00	1250	25.0

(a) Which has the better resolving power? (b) Which has the greater light-gathering ability? (c) Which produces a greater magnification?

Solution (a) The telescope with the larger objective has the greater ability to discriminate between nearby objects, and this is telescope B.

(b) The telescope with the larger objective can collect more light. Hence, B is again the choice.

(c) The magnification of telescope A is

$$m = \frac{f_a}{f_e} = \frac{1000 \text{ mm}}{6.00 \text{ mm}} = 167$$

That of telescope B is

$$m = \frac{1250 \text{ mm}}{25.0 \text{ mm}} = 50.0$$

Thus, telescope A has the greater magnification.

25.7 THE MICHELSON INTERFEROMETER

The camera and the telescope are examples of commonly used optical instruments. In contrast, the Michelson interferometer is an optical instrument that is unfamiliar to most people. It has great scientific importance, however. Invented by the American physicist A. A. Michelson (1852–1931), it is an ingenious device that splits a light beam into two parts and then recombines them to form an interference pattern. The interferometer is used to make accurate length measurements.

Figure 25.13 is a schematic diagram of an interferometer. A beam of light provided by a monochromatic source is split into two rays by a partially silvered mirror, M, inclined at an angle of 45° relative to the incident light beam. One ray is reflected vertically upward to mirror M_1, and the other ray is transmitted horizontally through mirror M to mirror M_2. Hence, the two rays travel separate paths, L_1 and L_2. After reflecting from mirrors M_1 and M_2, the two rays eventually recombine to produce an interference pattern, which can be viewed through a telescope. The glass plate, P, equal in thickness to mirror M, is placed in the path of the horizontal ray to ensure that the two rays travel the same distance through glass.

The interference pattern for the two rays is determined by the difference in their path lengths. When the two rays are viewed as shown, the image of M_2 is at M_2' parallel to M_1. Hence, the space between M_2' and M_1 forms the equivalent of a parallel air film. The effective thickness of the air film is varied by using a finely threaded screw to move mirror M_1 in the direction indicated by the arrows in Figure 25.13. If one of the mirrors is tipped slightly with respect to the other, the thin film between the two is wedge-shaped, and an interference pattern consisting of parallel fringes is set up, as described in Example 24.4. Now suppose we focus on one of the dark lines with the crosshairs of a telescope. As the mirror M_1 is moved to lengthen the path L_1, the thickness of the wedge increases. When the thickness increases by $\lambda/4$, the destructive interference that initially produced the dark fringe has changed to constructive interference, and we now observe a bright fringe at

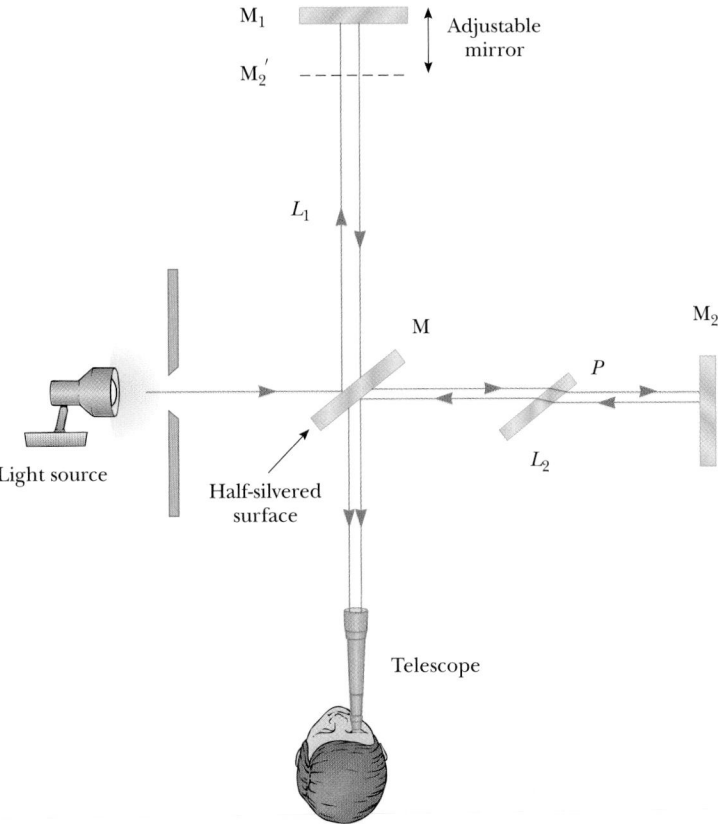

Figure 25.13 A diagram of the Michelson interferometer. A single beam is split into two rays by the half-silvered mirror, M. The path difference between the two rays is varied with the adjustable mirror, M_1.

the location of the crosshairs. The term *fringe shift* is used to describe the change in a fringe from dark to light or light to dark. Thus, successive light and dark fringes are formed each time M_1 is moved a distance of $\lambda/4$. The wavelength of light can be measured by counting the number of fringe shifts for a measured displacement of M_1. Conversely, if the wavelength is accurately known (as with a laser beam), the mirror displacement can be determined to within a fraction of the wavelength. Because the interferometer can measure displacements precisely, it is often used to make highly accurate measurements of the dimensions of mechanical components.

If the mirrors are perfectly aligned, rather than tipped with respect to one another, the path difference differs slightly for different angles of view. This results in an interference pattern that resembles Newton's rings. The pattern can be used in a fashion similar to that for tipped mirrors. One concentrates on the center spot in the interference pattern. For example, suppose the spot is initially dark, indicating that destructive interference is occurring. If M_1 is now moved by a distance of $\lambda/4$, this central spot changes to a light region, corresponding to a fringe shift.

Figure 25.14   A side view of a diffraction grating. The slit separation is d, and the path difference between adjacent slits is $d \sin \theta$.

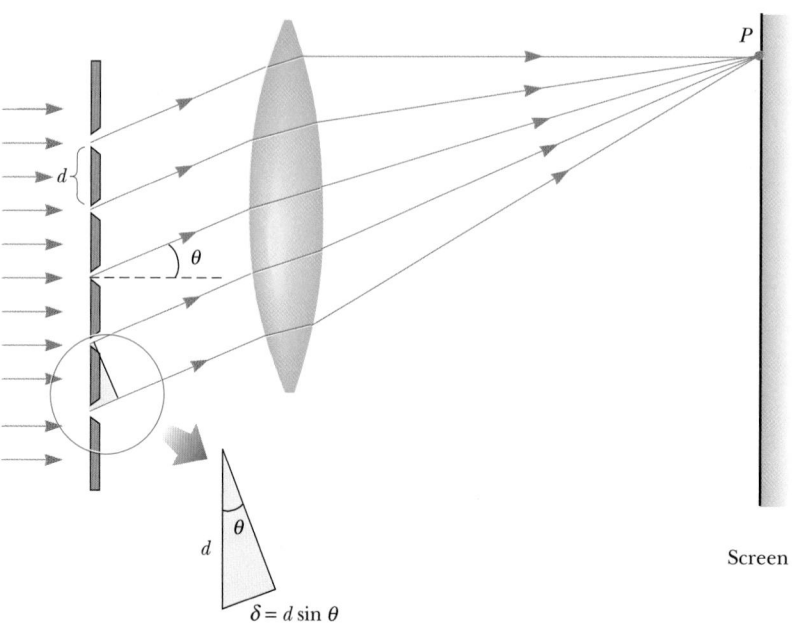

Screen

$\delta = d \sin \theta$

25.8   THE DIFFRACTION GRATING

The diffraction grating, a very useful device for analyzing light sources, consists of many equally spaced parallel slits. A grating can be made by scratching parallel lines on a glass plate with a precision machining technique. The spaces between the scratches are transparent to the light and hence act as separate slits. A typical grating contains several thousand lines per centimeter. For example, a grating ruled with 5000 lines/cm has a slit spacing, d, equal to the inverse of this number; hence, $d = (1/5000)$ cm $= 2 \times 10^{-4}$ cm.

Figure 25.14 is a schematic diagram of a section of a plane diffraction grating. A plane wave is incident from the left, normal to the plane of the grating. A converging lens can be used to bring the rays together at point P. The intensity of the pattern on the screen is the result of the combined effects of interference and diffraction. Each slit produces diffraction, and the diffracted beams in turn interfere with one another to produce the pattern. Moreover, each slit acts as a source of waves, and all waves start at the slits in phase. However, for some arbitrary direction θ measured from the horizontal, the waves must travel *different* path lengths before reaching a particular point P on the screen. From Figure 25.14, note that the path difference between waves from any two adjacent slits is $d \sin \theta$. If this path difference equals one wavelength or some integral multiple of a wavelength, waves from all slits will be in phase at P and a bright line will be observed. Therefore, the condition for **maxima** in the interference pattern at the angle θ is

$$d \sin \theta = m\lambda \qquad m = 0, 1, 2, \ldots \qquad [25.11]$$

This expression can be used to calculate the wavelength from the grating spacing and the angle of deviation, θ. The integer m is the **order number** of the diffraction pattern. If the incident radiation contains several wavelengths, each wavelength

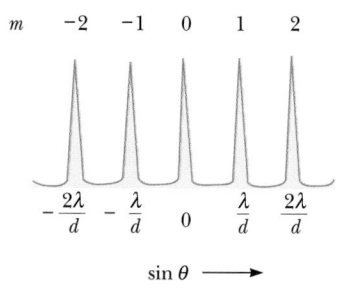

Figure 25.15   Intensity versus $\sin \theta$ for the diffraction grating. The zeroth-, first-, and second-order principal maxima are shown.

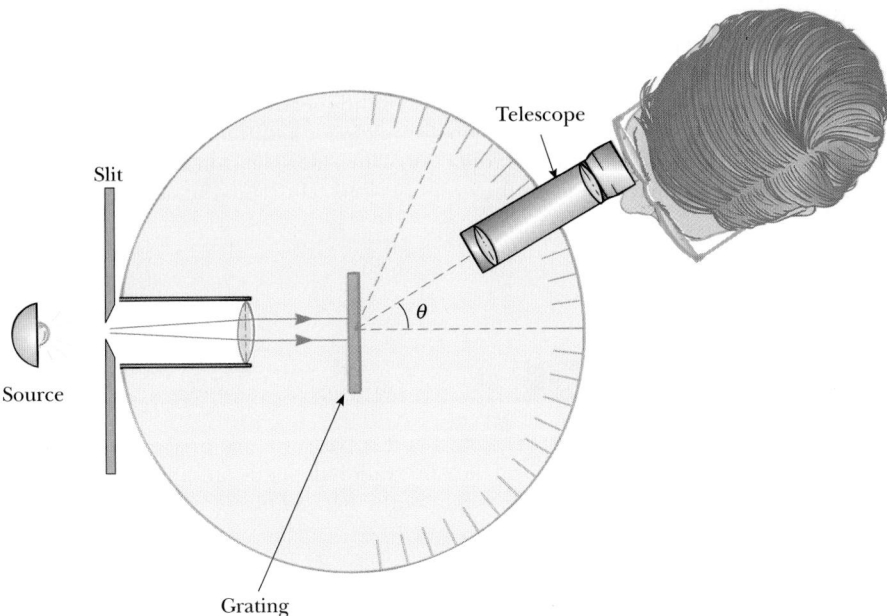

Figure 25.16 A diagram of a diffraction grating spectrometer. The collimated beam incident on the grating is diffracted into the various orders at the angles θ that satisfy the equation $d \sin \theta = m\lambda$, where $m = 0, 1, 2, \ldots$

deviates through a specific angle, which can be found from Equation 25.11. All wavelengths are focused at $\theta = 0$, corresponding to $m = 0$. This is called the *zeroth-order maximum*. The *first-order maximum*, corresponding to $m = 1$, is observed at an angle that satisfies the relationship $\sin \theta = \lambda/d$; the *second-order maximum*, corresponding to $m = 2$, is observed at a larger angle θ, and so on. Figure 25.15 is a sketch of the intensity distribution for some of the orders produced by a diffraction grating. Note the sharpness of the principal maxima and the broad range of the dark areas. This is in contrast to the broad bright fringes characteristic of the two-slit interference pattern.

A simple arrangement that can be used to measure the angles in a diffraction pattern is shown in Figure 25.16. This is a form of diffraction-grating spectrometer. The light to be analyzed passes through a slit and is formed into a parallel beam by a lens. The light then strikes the grating at a 90° angle. The diffracted light leaves the grating at angles that satisfy Equation 25.11. A telescope is used to view the image of the slit. The wavelength can be determined by measuring the angles at which the images of the slit appear for the various orders.

Thinking Physics 3

White light enters through an opening in an opaque box, exits through an opening on the other side of the box, and a spectrum of colors appears on the wall. How could you determine whether the box contains a prism or a diffraction grating?

Explanation   The determination could be made by noticing the order of the colors in the spectrum relative to the direction of the original beam of white light. For a prism, the separation of light is a result of dispersion, and the violet light will be refracted more than the red light. So the order of the spectrum from a prism will be from red, closest to the original direction, to violet. For a diffraction grating, the angle of diffraction increases with wavelength. Thus, the spectrum from the diffraction grating will have colors in the order of violet, closest to the original direction, to red.

Applying Physics 1

Light reflected from the surface of a compact disc has a multicolored appearance, as shown in Figure 25.17. Furthermore, the observation depends on the orientation of the disc relative to the eye and the position of the light source. Explain how this works.

Figure 25.17   (Applying Physics 1) Compact discs act as diffraction gratings when observed under white light. (© *Kristen Brochmann/Fundamental Photographs*)

Explanation   The surface of a compact disc has a spiral-shaped track (with a spacing of approximately 1.5 μm) that acts as a reflection grating. Musical information is encoded in a series of pits along this track. The light scattered by these closely spaced parallel tracks interferes constructively only in certain directions that depend on the wavelength and on the direction of the incident light. The pits within the tracks scatter light in all directions, and the flat reflective areas between the tracks specularly reflect the incident light. At a particular viewing angle, some regions of the disc will result in one color of light constructively interfering, and other regions will result in other colors undergoing constructive interference. As a result, all spectral colors are seen coming from various locations on the disc surface.

EXAMPLE 25.10   The Orders of a Diffraction Grating

Monochromatic light from a helium-neon laser ($\lambda = 632.8$ nm) is incident normally on a diffraction grating containing 6000 lines/cm. Find the angles at which one would observe the first-order maximum, the second-order maximum, and so forth.

Solution   First we must calculate the slit separation, which is the inverse of the number of lines per centimeter:

$$d = \frac{1}{6000} \text{ cm} = 1.667 \times 10^{-4} \text{ cm} = 1667 \text{ nm}$$

For the first-order maximum ($m = 1$), we get

$$\sin \theta_1 = \frac{\lambda}{d} = \frac{632.8 \text{ nm}}{1667 \text{ nm}} = 0.3796$$

$$\theta_1 = 22.31°$$

For $m = 2$ we find that

$$\sin \theta_2 = \frac{2\lambda}{d} = \frac{2(632.8 \text{ nm})}{1667 \text{ nm}} = 0.7592$$

$$\theta_2 = 49.39°$$

However, for $m = 3$ we find that $\sin \theta_3 = 1.139$. Because $\sin \theta$ cannot exceed unity, this is not a realistic solution. Hence, only zeroth-, first-, and second-order maxima would be observed in this situation.

Resolving Power of the Diffraction Grating

The diffraction grating is most useful for making accurate wavelength measurements. Like the prism, it can be used to disperse a spectrum into its components. Of the two devices, the grating is more precise if one wants to distinguish between two closely spaced wavelengths. We say that the grating spectrometer has a higher *resolution* than the prism spectrometer. If λ_1 and λ_2 are two nearly equal wavelengths between which the spectrometer can just barely distinguish, the **resolving power, R,** of the grating is defined as

$$R \equiv \frac{\lambda}{\lambda_2 - \lambda_1} = \frac{\lambda}{\Delta\lambda} \qquad [25.12]$$

where $\lambda \approx \lambda_1 \approx \lambda_2$ and $\Delta\lambda = \lambda_2 - \lambda_1$. Thus, we see that a grating with a high resolving power can distinguish small differences in wavelength. Furthermore, if N lines of the grating are illuminated, it can be shown that the resolving power in the mth-order diffraction equals the product Nm:

$$R = Nm \qquad [25.13]$$

◄ Resolving power of a grating

Thus, the resolving power increases with order number. Furthermore, R is large for a grating with a great number of illuminated slits. Note that for $m = 0$, $R = 0$, which signifies that *all wavelengths are indistinguishable* for the zeroth-order maximum (all wavelengths fall at the same point on the screen). However, consider the second-order diffraction pattern of a grating that has 5000 rulings illuminated by the light source. The resolving power of such a grating in second order is $R = 5000 \times 2 = 10\,000$. Therefore, the *minimum* wavelength separation between two spectral lines that can be just resolved, assuming a mean wavelength of 600 nm, is calculated from Equation 25.12 to be $\Delta\lambda = \lambda/R = 6 \times 10^{-2}$ nm. For the third-order principal maximum $R = 15\,000$ and $\Delta\lambda = 4 \times 10^{-2}$ nm, and so on.

EXAMPLE 25.11   Resolving the Sodium Spectral Lines

Two strong lines in the spectrum of sodium have wavelengths of 589.00 nm and 589.59 nm.

(a) What must the resolving power of a grating be in order to distinguish these wavelengths?

Solution   From Equation 25.12, we find that

$$R = \frac{\lambda}{\Delta\lambda} = \frac{589 \text{ nm}}{589.59 \text{ nm} - 589.00 \text{ nm}} = \frac{589}{0.59} = \boxed{998}$$

(b) To resolve these lines in the second-order spectrum, how many lines of the grating must be illuminated?

Solution   From Equation 25.13 and the result of (a), we find that

$$N = \frac{R}{m} = \frac{998}{2} = \boxed{499 \text{ lines}}$$

SUMMARY

The light-concentrating power of a lens of focal length f and diameter D is determined by the **f-number,** defined as

$$f\text{-number} \equiv \frac{f}{D} \qquad [25.1]$$

The smaller the f-number of a lens, the brighter the image formed.

    Hyperopia (farsightedness) is a defect of the eye that occurs either when the eyeball is too short or when the ciliary muscle cannot change the shape of the lens enough to form a properly focused image. **Myopia** (nearsightedness) occurs either when the eye is longer than normal or when the maximum focal length of the lens is insufficient to produce a clearly focused image on the retina.

    The **power** of a lens in **diopters** is the inverse of the focal length in meters.

    The **angular magnification of a lens** is defined as

$$m \equiv \frac{\theta}{\theta_0} \qquad [25.2]$$

where θ is the angle subtended by an object at the eye with a lens in use and θ_0 is the angle subtended by the object when it is placed at the near point of the eye and no lens is used. The **maximum angular magnification of a lens** is

$$m = 1 + \frac{25 \text{ cm}}{f} \qquad [25.5]$$

When the eye is relaxed, the angular magnification is

$$m = \frac{25 \text{ cm}}{f} \qquad [25.6]$$

The overall **magnification of a compound microscope** of length L is the product of the magnification produced by the objective, of focal length f_o, and the magnification produced by the eyepiece, of focal length f_e:

$$M = -\frac{L}{f_o}\left(\frac{25\text{ cm}}{f_e}\right) \qquad \text{[25.7]}$$

The **angular magnification of a telescope** is

$$m = \frac{f_o}{f_e} \qquad \text{[25.8]}$$

where f_o is the focal length of the objective and f_e is the focal length of the eyepiece.

Two images are said to be **just resolved** when the central maximum of the diffraction pattern for one image falls on the first minimum of the other image. This limiting condition of resolution is known as **Rayleigh's criterion.** The limiting angle of resolution for a **slit** of width a is

$$\theta_m \approx \frac{\lambda}{a} \qquad \text{[25.9]}$$

The limiting angle of resolution of a **circular aperture** is

$$\theta_m = 1.22\frac{\lambda}{D} \qquad \text{[25.10]}$$

where D is the diameter of the aperture.

A **diffraction grating** consists of many equally spaced, identical slits. The condition for **maximum intensity** in the interference pattern of a diffraction grating is

$$d \sin \theta = m\lambda \qquad m = 0, 1, 2, \ldots \qquad \text{[25.11]}$$

where d is the spacing between adjacent slits and m is the order number of the diffraction pattern. The **resolving power** of a diffraction grating in the mth order is

$$R = Nm \qquad \text{[25.13]}$$

where N is the number of illuminated rulings on the grating.

MULTIPLE-CHOICE QUESTIONS

1. A telescope can be used as a camera if the film is placed at the image position of the objective lens (or mirror) — in other words, the objective lens becomes the lens of the camera. If a 2000-mm focal length objective lens of a telescope is used to photograph the image of two stars that are 20 mm apart on the film, what is the angular separation of the two stars?
   (a) 0.010° (b) 0.10° (c) 0.29° (d) 0.57°
2. The visible spectrum ranges from 400 nm to 700 nm. What is the highest order continuous visible spectrum that can be viewed through a grating with no overlap of orders occurring? Assume that the grating has few enough lines/cm so that overlap will occur.
   (a) 1 (b) 2 (c) 3 (d) 4 (e) Depends on the grating.
3. A beam of monochromatic light is incident on a multiple slit diffraction grating with a slit separation of 2.50×10^{-6} m. If the angle between the normal and first-order bright fringe is 14.0°, what is the source wavelength?
   (a) 430 nm (b) 500 nm (c) 605 nm (d) 647 nm

4. The near point of a given individual's eye is 55 cm. What focal length corrective lens should be prescribed so that an object can be clearly seen when placed at 25 cm in front of the eye?
(a) −25 cm   (b) 17 cm   (c) 46 cm   (d) 30 cm

5. A compound microscope has objective and eyepiece lenses of focal lengths 0.80 and 4.0 cm, respectively. If the microscope length is 15 cm, what is the maximum magnification?
(a) 3.2   (b) 6.3   (c) 48   (d) 120

CONCEPTUAL QUESTIONS

1. Suppose you are observing a binary star system with a telescope and are having difficulty resolving the two stars. You decide to use a colored filter to help you. Should you use a blue or red filter?

2. Explain why two flashlights held close together do not produce an interference pattern on a distant screen.

3. Suppose you are observing the interference pattern formed by a Michelson interferometer in a laboratory and a joking colleague holds a lit match in the light path of one arm of the interferometer. Will this have an effect on the interference pattern?

4. Compare and contrast the eye and a camera. What parts of the camera correspond to the iris, the retina, and the cornea of the eye?

5. If laser light is reflected from a phonograph record or a compact disc, a diffraction pattern appears. This is because both devices contain parallel tracks of information that act as a reflection diffraction grating. Which device, record or compact disc, will result in diffraction maxima that are farther apart?

6. If you want to use a converging lens to set fire to a piece of paper, why should the light source be farther from the lens than its focal point?

7. When you receive a chest x-ray at a hospital, the x-rays pass through a series of parallel ribs in your chest. Do the ribs act as a diffraction grating for x-rays?

8. What difficulty would astronauts encounter in finding a landing site on the Moon if astronomers were not aware of the properties of the telescope?

9. Large telescopes are usually reflecting rather than refracting. List some reasons for this choice.

10. If you want to examine the fine detail of an object with a magnifying lens of focal length 15 cm, where should the object be placed in order to observe a magnified image of the object?

11. Explain why it is theoretically impossible to see an object as small as an atom regardless of the quality of the light microscope being used.

12. The optic nerve and the brain invert the image formed on the retina. Why do we not see everything upside down?

PROBLEMS

1, *2*, **3** = straightforward, intermediate, challenging   □ = full solution available in Study Guide/Student Solutions Manual   ◣ = Core Concepts Workbook
WEB = solution posted at **http://www.harcourtcollege.com/physics/cptech**   = biomedical application   = Interactive Physics

Section 25.1   The Camera

1. A camera used by a professional photographer to shoot portraits has a focal length of 25.0 cm. The photographer takes a portrait of a person 1.50 m in front of the camera. Where is the image formed, and what is the lateral magnification?

2. A photographic image of a building is 0.0920 m high. The image was made with a lens with a focal length of 52.0 mm. If the lens was 100 m from the building when the photograph was made, determine the height of the building.

3. The image area of a typical 35-mm slide is 23.5 mm by 35.0 mm. If a camera has a 55.0-mm focal length lens, will the full image this camera forms of the constellation Orion, which is 20° across, fit on a 35-mm slide?

4. The full Moon is photographed using a camera with a 120-mm focal length lens. Determine the diameter of the Moon's image on the film. (*Note:* The radius of the Moon is 1.74×10^6 m, and the Earth-Moon distance is 3.84×10^8 m.)

WEB 5. A camera is being used with the correct exposure at $f/4$ and a shutter speed of 1/32 s. In order to "stop" a fast-moving subject, the shutter speed is changed to 1/256 s. Find the new *f*-stop that should be used to maintain satisfactory exposure, assuming no change in lighting conditions occurs.

6. Assume that the camera in Figure 25.1 has a fixed focal length of 65.0 mm and is adjusted to focus the image of a distant object properly. How far and in what direction must the lens be moved to focus the image of an object that is 2.00 m away?

7. A certain type of film requires an exposure time of 0.010 s with an *f*/11 lens setting. Another type of film requires twice the light energy to produce the same level of exposure. What *f*-stop does the second type of film need with the 0.010-s exposure time?

Section 25.2 The Eye

8. A retired bank president can easily read the fine print of the financial page when the newspaper is held at arm's length, 60.0 cm from the eye. What should be the focal length of an eyeglass lens that will allow her to read at the more comfortable distance of 24.0 cm?

9. A person has far points 8.44 cm from the right eye and 12.2 cm from the left eye. Write a prescription for the powers of the corrective lenses.

10. The near point of an eye is 100 cm. A corrective lens is to be used to allow this eye to focus clearly on objects 25.0 cm in front of it. **(a)** What should be the focal length of this lens? **(b)** What is the power of the needed corrective lens?

11. An individual is nearsighted; his near point is 13.0 cm and his far point is 50.0 cm. **(a)** What lens power is needed to correct his nearsightedness? **(b)** When the lenses are in use, what is this person's near point?

12. An artificial lens is implanted in a person's eye to replace a diseased lens. The distance between the artificial lens and the retina is 2.80 cm. In the absence of the lens, an image of a distant object (formed by refraction at the cornea) falls 2.53 cm behind the retina. The lens is designed to put the image of the distant object on the retina. What is the power of the implanted lens? (*Hint:* Consider the image formed by the cornea as a virtual object.)

13. A person is to be fitted with bifocals. She can see clearly when the object is between 30 cm and 1.5 m from the eye. **(a)** The upper portions of the bifocals should be designed to enable her to see distant objects clearly. What power should they have? **(b)** The lower portions of the bifocals should enable her to see objects comfortably at 25 cm. What power should they have (Fig. P25.13)?

14. A certain child's near point is 10.0 cm; her far point (with eyes relaxed) is 125 cm. Each eye lens is 2.00 cm from the retina. **(a)** Between what limits, measured in di-

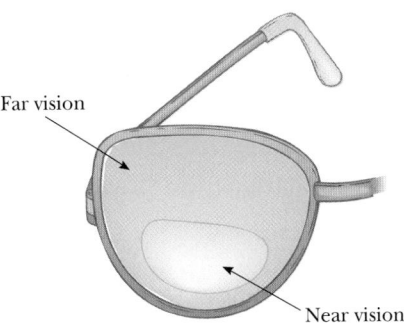

Figure P25.13

opters, does the power of this lens-cornea combination vary? **(b)** Calculate the power of the eyeglass lens this child should use for relaxed distance vision. Is the lens converging or diverging?

Section 25.3 The Simple Magnifier

15. A stamp collector uses a lens with 7.5-cm focal length as a simple magnifier. The virtual image is produced at the normal near point (25 cm). **(a)** How far from the lens should the stamp be placed? **(b)** What is the expected angular magnification?

16. A lens having a focal length of 25 cm is used as a simple magnifier. **(a)** What is the angular magnification obtained when the image is formed at the normal near point ($q = -25$ cm)? **(b)** What is the magnification produced by this lens when the eye is relaxed?

17. A biology student uses a simple magnifier to examine the structural features of the wing of an insect. The wing is held 3.50 cm in front of the lens, and the image is formed 25.0 cm from the eye. **(a)** What is the focal length of the lens? **(b)** What angular magnification is achieved?

18. A leaf of length h is positioned 71.0 cm in front of a converging lens with a focal length of 39.0 cm. An observer views the image of the leaf from a position 1.26 m behind the lens, as shown in Figure P25.18. **(a)** What is the magnitude of the lateral magnification (ratio of im-

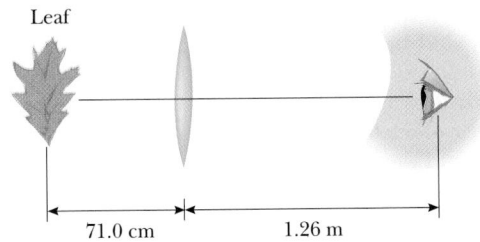

Figure P25.18

age size to object size) produced by the lens? **(b)** What angular magnification is achieved by viewing the image of the leaf rather than viewing the leaf directly?

Section 25.4 The Compound Microscope

Section 25.5 The Telescope

19. The objective lens in a microscope with a 20.0-cm-long tube has a magnification of 50.0 and the eyepiece has a magnification of 20.0. What are the focal lengths of **(a)** the objective and **(b)** the eyepiece? **(c)** What is the overall magnification of the microscope?

20. A microscope has an objective lens with a focal length of 16.22 mm and an eyepiece with a focal length of 9.50 mm. With the length of the barrel set at 29.0 cm, the diameter of a red blood cell's image subtends an angle of 1.43 mrad with the eye. If the final image distance is 29.0 cm from the eyepiece, what is the actual diameter of the red blood cell?

WEB **21.** The length of a microscope tube is 15.0 cm. The focal length of the objective is 1.00 cm, and the focal length of the eyepiece is 2.50 cm. What is the magnification of the microscope assuming it is adjusted so that the eye is relaxed?

22. An astronomical telescope has an objective with a focal length of 75 cm and an eyepiece with a focal length of 4.0 cm. What is the magnifying power of this instrument?

23. A certain telescope has a 5.00-in. diameter and an objective lens of focal length 1250 mm. **(a)** What is the *f*-number of this lens? **(b)** What is the angular magnification of this telescope when it is used with a 25-mm eyepiece?

24. A certain telescope has an objective of focal length 1500 cm. If the Moon is used as an object, a 1.0-cm length of the image formed by the objective corresponds to what distance, in miles, on the Moon? Assume 3.8×10^8 m for the Earth-Moon distance.

25. The lenses of an astronomical telescope are 92 cm apart when adjusted for viewing a distant object with minimum eyestrain. The angular magnification produced by the telescope is 45. Compute the focal length of each lens.

26. An elderly sailor is shipwrecked on a desert island but manages to save his eyeglasses. The lens for one eye has a power of +1.20 diopters, and the other lens has a power of +9.00 diopters. **(a)** What is the magnifying power of the telescope he can construct with these lenses? **(b)** How far apart are the lenses when the telescope is adjusted for minimum eyestrain?

27. A person decides to use an old pair of eyeglasses to make some optical instruments. He knows that the near

point in his left eye is 50.0 cm and the near point in his right eye is 100 cm. **(a)** What is the maximum angular magnification he can produce in a telescope? **(b)** If he places the lenses 10.0 cm apart, what is the maximum overall magnification he can produce in a microscope? [Go back to basics and use the thin-lens equation to solve part (b).]

Section 25.6 Resolution of Single-Slit and Circular Apertures

28. If the distance from the Earth to the Moon is 3.8×10^8 m, what diameter would be required for a telescope objective to resolve a Moon crater 300 m in diameter? Assume a wavelength of 500 nm.

29. A converging lens with a diameter of 30.0 cm forms an image of a satellite passing overhead. The satellite has two green lights (wavelength 500 nm) spaced 1.00 m apart. If the lights can just be resolved according to the Rayleigh criterion, what is the altitude of the satellite?

30. To increase the resolving power of a microscope, the object and the objective are immersed in oil ($n = 1.5$). If the limiting angle of resolution without the oil is 0.60 μrad, what is the limiting angle of resolution with the oil? (*Hint:* The oil changes the wavelength of the light.)

31. Two motorcycles, separated laterally by 2.0 m, are approaching an observer holding an infrared detector that is sensitive to radiation of wavelength 885 nm. What aperture diameter is required in the detector if the two headlights are to be resolved at a distance of 10.0 km?

32. **(a)** Calculate the limiting angle of resolution for the eye, assuming a pupil diameter of 2.00 mm, a wavelength of 500 nm *in air*, and an index of refraction for the eye of 1.33. **(b)** What is the maximum distance from the eye at which two points separated by 1.00 cm could be resolved?

33. Two stars in a binary system are 8.0 lightyears away from the observer and can just be resolved by a 20-in. telescope equipped with a filter that only allows light of wavelength 500 nm to pass. What is the distance between the two stars?

34. A spy satellite circles the Earth at an altitude of 200 km and carries out surveillance with a special high-resolution telescopic camera having a lens diameter of 35 cm. If the angular resolution of this camera is limited by diffraction, estimate the separation of two small objects on the Earth's surface that are just resolved in yellow-green light ($\lambda = 550$ nm).

35. Suppose a 5.00-m-diameter telescope were constructed on the Moon, where the absence of atmospheric distortion would permit excellent viewing. If observations were made using 500-nm light, what minimum separation between two objects could just be resolved on Mars at closest approach (when Mars is 8.0×10^7 km from the Moon)?

Section 25.7   The Michelson Interferometer

36. Light of wavelength 550 nm is used to calibrate a Michelson interferometer. By use of a micrometer screw, the platform on which one mirror is mounted is moved 0.180 mm. How many fringe shifts are counted?

37. An interferometer is used to measure the length of a bacterium. The wavelength of the light used is 650 nm. As one arm of the interferometer is moved from one end of the cell to the other, 310 fringe shifts are counted. How long is the bacterium?

38. The Michelson interferometer can be used to measure the index of refraction of a gas by placing an evacuated transparent tube in the light path along one arm of the device. Fringe shifts occur as the gas is slowly added to the tube. Assume that 600-nm light is used, the tube is 5.00 cm long, and 160 fringe shifts occur as the pressure of the gas in the tube increases to atmospheric pressure. What is the index of refraction of the gas? (*Hint:* The fringe shifts occur because the wavelength of the light changes inside the gas-filled tube.)

WEB 39. The light path in one arm of a Michelson interferometer includes a transparent cell that is 5.00 cm long. How many fringe shifts would be observed if all the air were evacuated from the cell? The wavelength of the light source is 590 nm and the refractive index of air is 1.000 29. (See the *Hint* in Problem 38.)

40. A thin sheet of transparent material has an index of refraction of 1.40 and is 15.0 μm thick. When it is inserted in the light path along one arm of an interferometer, how many fringe shifts occur in the pattern? Assume that the wavelength (in a vacuum) of the light used is 600 nm. (*Hint:* The wavelength will change within the material.)

Section 25.8   The Diffraction Grating

41. The 502-nm line in helium is observed at an angle of 30.0° in the second-order spectrum of a diffraction grating. Calculate the angular deviation of the 668-nm line in helium in the first-order spectrum for the same grating.

42. Three discrete spectral lines occur at angles of 10.1°, 13.7°, and 14.8° in the first-order spectrum of a diffraction grating spectrometer. (a) If the grating has 3660 slits/cm, what are the wavelengths of the light? (b) At what angles are these lines found in the second-order spectra?

43. Intense white light is incident on a diffraction grating that has 600 lines/mm. (a) What is the highest order in which the complete visible spectrum can be seen using this grating? (b) What is the angular separation between the violet edge (400 nm) and the red edge (700 nm) of the first-order spectrum produced by this grating?

44. A grating with 1500 slits per centimeter is illuminated with light of wavelength 500 nm. (a) What is the highest order number that can be observed with this grating? (b) Repeat for a grating of 15 000 slits per centimeter.

45. A diffraction grating with 2500 lines/cm is used to examine the sodium spectrum. Calculate the angular separation of the two closely spaced sodium-yellow lines (588.995 nm and 589.592 nm) in each of the first three orders.

46. A diffraction grating is calibrated by using the 546.1-nm line of mercury vapor. It is found that the first-order line is at an angle of 21.0°. Calculate the number of lines per millimeter on this grating.

47. A diffraction grating with 4000 lines/cm is illuminated by light from the Sun. The first-order solar spectrum is spread out on a white wall across the room. (a) At what angle from the centerline is violet light (400 nm)? (b) At what angle from the centerline does red light (650 nm) appear?

48. A light source emits two major spectral lines, an orange line of wavelength 610 nm and a blue-green line of wavelength 480 nm. If the spectrum is resolved by a diffraction grating having 5000 lines/cm and viewed on a screen 2.00 m from the grating, what is the distance (in centimeters) between the two spectral lines in the second-order spectrum?

49. The H_α line in hydrogen has a wavelength of 656.20 nm. This line differs in wavelength from the corresponding spectral line in deuterium (the heavy stable isotope of hydrogen) by 0.18 nm. (a) Determine the minimum number of lines a grating must have to resolve these two wavelengths in the first order. (b) Repeat part (a) for the second order.

50. A 15.0-cm-long grating has 6000 slits per centimeter. Can two lines of wavelengths 600.000 nm and 600.003 nm be separated using this grating? Explain.

ADDITIONAL PROBLEMS

WEB 51. The near point of an eye is 75.0 cm. (a) What should be the power of a corrective lens prescribed to enable the eye to see an object clearly at 25.0 cm? (b) If, using the corrective lens, the user can see an object clearly at 26.0 cm but not 25.0 cm, by how many diopters did the lens grinder miss the prescription?

52. If a typical eyeball is 2.00 cm long and has a pupil opening that can range from about 2.00 mm to 6.00 mm, what is (a) the focal length of the eye when it is focused on objects 1.00 m away, (b) the smallest *f*-number of the eye when it is focused on objects 1.00 m away, and (c) the largest *f*-number of the eye when it is focused on objects 1.00 m away?

53. A telescope has an objective of focal length 100 cm and an eyepiece of focal length 1.50 cm. (a) What is the distance between the two lenses when the telescope is used to view a distant object? (b) What is the angular magnification of the telescope?

54. The 546.1-nm line in mercury is measured at an angle of 81.0° in the third-order spectrum of a diffraction grating. Calculate the number of lines per millimeter for the grating.

55. The wavelengths of the sodium spectrum are $\lambda_1 = 589.00$ nm and $\lambda_2 = 589.59$ nm. Determine the minimum number of lines in a grating that will allow resolution of the sodium spectrum in (a) the first order and (b) the third order.

56. A fringe pattern is established in the field of view of a Michelson interferometer, using light of wavelength 580 nm. A parallel-faced sheet of transparent material 2.5 μm thick is placed in front of one of the mirrors, perpendicularly to the incident and reflected light beams. An observer counts 12 fringe shifts. What is the index of refraction of the sheet?

57. Sunlight is incident on a diffraction grating that has 2750 lines/cm. The second-order spectrum over the visible range (400–700 nm) is to be limited to 1.75 cm along a screen that is distance *L* from the grating. What is the required value of *L*?

58. The text discusses the astronomical telescope. Another type is the Galilean telescope, in which an objective lens gathers light (Fig. P25.58), and tends to form an image at point *A*. An eyepiece, consisting of a diverging lens, intercepts the light before it comes to a focus and forms a virtual image at point *B*. When adjusted for minimum eyestrain, point *B* is an infinite distance in front of the lens and parallel rays emerge from the lens, as in Figure P25.58b. An opera glass, which is a Galilean telescope, is used to view a 30.0-cm-tall singer's head that is 40.0 m from the objective lens. The focal length of the objective is +8.00 cm, and that of the eyepiece is −2.00 cm. The telescope is adjusted so parallel rays enter the eye. Compute (a) the size of the real image that would have been formed by the objective, (b) the virtual object distance for the diverging lens, (c) the distance between the lenses, and (d) the overall angular magnification.

59. A boy scout starts a fire by using a lens from his grandfather's eyeglasses to focus sunlight on kindling 5.0 cm from the lens. The boy scout has a near point of 15 cm. When he uses the lens as a simple magnifier, (a) what is the maximum magnification that can be achieved, and

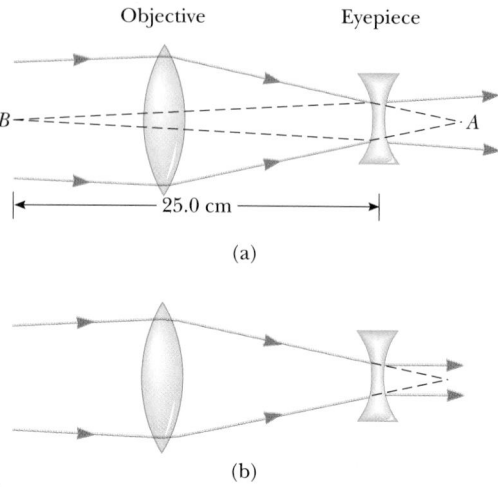

(a)

(b)

Figure P25.58

(b) what is the magnification when the eye is relaxed? (*Caution:* The equations derived in the text for a simple magnifier assume a "normal" eye.)

60. The separate stars of a double-star system are just resolved by a lens with a diameter of 10.0 cm and a focal length of 1.40 m. If the image is viewed from a distance of 30.0 cm, what is the angle subtended by the lines of sight of the two stars in the image? (Assume a 500-nm wavelength.)

61. Light containing two different wavelengths passes through a diffraction grating with 1200 slits/cm. On a screen 15.0 cm from the grating, the third-order maximum of the shorter wavelength falls on top of the first-order minimum of the longer wavelength. If the neighboring maxima of the longer wavelength are 8.44 mm apart on the screen, what are the wavelengths in the light? (*Hint:* Use the small angle approximation.)

62. A laboratory (astronomical) telescope is used to view a scale that is 300 cm from the objective, which has a focal length of 20.0 cm; the eyepiece has a focal length of 2.00 cm. Calculate the angular magnification when the telescope is adjusted for minimum eyestrain. (*Note:* The object is not at infinity, and so the simple expression $m = f_o/f_e$ is not sufficiently accurate for this problem. Also, assume small angles so that $\tan \theta \approx \theta$.)

63. If the aqueous humor of the eye has an index of refraction of 1.34 and the distance from the vertex of the cornea to the retina is 2.00 cm, what is the radius of curvature of the cornea for which distant objects will be focused on the retina? (For simplicity, assume that all refraction occurs in the aqueous humor.)

64. A source emits three lines: a violet line of wavelength 400 nm, a green line of wavelength 550 nm, and a red line of wavelength 700 nm. Images of these lines are formed by a diffraction grating with a line spacing of 2500 nm. Calculate $\sin \theta$ for all observable orders of each line, and arrange the images (13 in all) in sequence according to their $\sin \theta$ values—thus, V_1, G_1, R_1, V_2, ... The resulting sequence illustrates a phenomenon known as the *overlapping of orders* in a grating spectrum of visible light.

"JUST CHECKING."

(1) A. Piccard
(2) E. Henriot
(3) P. Ehrenfest
(4) E. Herzen
(5) Th. de Donder
(6) E. Schrödinger
(7) E. Verschaffelt
(8) W. Pauli
(9) W. Heisenberg
(10) R.H. Fowler

(11) L. Brillouin
(12) P. Debye
(13) M. Knudsen
(14) W.L. Bragg
(15) H.A. Kramers
(16) P.A.M. Dirac
(17) A.H. Compton
(18) L.V. de Broglie
(19) M. Born
(20) N. Bohr

(21) I. Langmuir
(22) M. Planck
(23) M. Curie
(24) H.A. Lorentz
(25) A. Einstein
(26) P. Langevin
(27) C.E. Guye
(28) C.T.R. Wilson
(29) O.W. Richardson

The "architects" of modern physics. This unique photograph shows many eminent scientists who participated in the Fifth International Congress of Physics held in 1927 by the Solvay Institute in Brussels. At this and similar conferences, held regularly from 1911 on, scientists were able to discuss and share the many dramatic developments in atomic and nuclear physics. This elite company of scientists includes fifteen Nobel prize winners in physics and three in chemistry.

(Photograph courtesy of AIP Niels Bohr Library)

PART 6

Modern Physics

At the end of the 19th century, scientists believed that they had learned most of what there was to know about physics. Newton's laws of motion and his universal theory of gravitation, Maxwell's theoretical work in unifying electricity and magnetism, and the laws of thermodynamics and kinetic theory were highly successful in explaining a wide variety of phenomena.

However, at the turn of the 20th century a major revolution shook the world of physics. In 1900 Planck provided the basic ideas that led to the formulation of the quantum theory, and in 1905 Einstein formulated his brilliant special theory of relativity. The excitement of the times is captured in Einstein's own words: "It was a marvelous time to be alive." Both theories were to have a profound effect on our understanding of nature, and within a few decades they inspired new developments and theories in the fields of atomic physics, nuclear physics, and condensed matter physics.

Our discussion of modern physics will begin with a treatment of the special theory of relativity in Chapter 26. Although its underlying concepts often violate our common sense, the theory provides us with a new and deeper view of physical laws. In Chapter 27 we shall discuss various developments in quantum theory, which provides us with a successful model for understanding electrons, atoms, and molecules. The last three chapters of the book are concerned with applications of quantum theory. Chapter 28 discusses the structure and properties of atoms using concepts from quantum mechanics. Chapter 29 is concerned with the structure and properties of the atomic nucleus. Chapter 30 discusses many practical applications of nuclear physics and concludes with a discussion of elementary particles.

Keep in mind that, although modern physics has been developed during the 20th century and has led to a multitude of important technological achievements, the story is still incomplete. New discoveries will be made during our lifetime, many of which will deepen or refine our understanding of nature and the world around us. It is *still* a marvelous time to be alive.

26

Relativity

▲ **PHYSICS PUZZLER**

A police officer can clock the speed of your automobile with a radar device. This device works using the Doppler effect for electromagnetic waves discussed in Chapter 21. Is it possible to accelerate an object such as a rocket to a speed greater than the speed of light? *(Trent Steffler/David R. Frazier Photolibrary)*

26.1 INTRODUCTION

Most of our everyday experiences and observations deal with objects that move at speeds much lower than the speed of light. Newtonian mechanics and the early ideas on space and time were formulated to describe the motion of such objects. As we saw in the chapters on mechanics, this formalism is very successful in describing a wide range of phenomena. Although Newtonian mechanics works very well at low speeds, it fails when applied to particles whose speeds approach that of light. The predictions of Newtonian theory at high speeds can be tested by accelerating an electron through a large electric potential difference. For example, it is possible to accelerate an electron to a speed of $0.99c$ by using a potential difference of several million volts. According to Newtonian mechanics, if the potential difference (as well as the corresponding energy) is increased by a factor of 4, then the speed of the electron should be doubled to $1.98c$. However, experiments show that the speed of the electron always remains *lower* than the speed of light, regardless of the size of the accelerating voltage. Because Newtonian mechanics places no upper limit on the speed that a particle can attain, it is contrary to modern experimental results and is clearly a limited theory.

In 1905, at the age of 26, Einstein published his special theory of relativity. Regarding the theory, Einstein wrote,

> The relativity theory arose from necessity, from serious and deep contradictions in the old theory from which there seemed no escape. The strength of the new theory lies in the consistency and simplicity with which it solves all these difficulties, using only a few very convincing assumptions.[1]

Although Einstein made many other important contributions to science, his theory of relativity alone represents one of the greatest intellectual achievements of the 20th century. With this theory, experimental observations over the range from $v = 0$ to speeds approaching the speed of light can be predicted. Newtonian mechanics, which was accepted for more than 200 years, is in fact a specialized case of Einstein's theory. This chapter introduces the special theory of relativity, with emphasis on some of the consequences of the theory. A discussion of general relativity and some of its consequences is presented in Section 26.10.

As we shall see, the special theory of relativity is based on two postulates:

1. The laws of physics are the same in all inertial reference systems.
2. The speed of light in a vacuum is always measured to be 3×10^8 m/s, and the measured value is independent of the motion of the observer or of the motion of the source of light.

Special relativity covers such phenomena as the slowing down of moving clocks and the contraction of moving rods as measured by a stationary observer. In addition to these topics, we also discuss the relativistic forms of momentum and energy, terminating the chapter with the famous mass-energy equivalence formula, $E = mc^2$.

---

[1] A. Einstein and L. Infeld, *The Evolution of Physics*, New York, Simon and Schuster, 1961.

Thinking Physics 1

Imagine a very powerful lighthouse with a rotating beacon. Imagine also drawing a horizontal circle around the lighthouse, with the lighthouse at the center. Along the circumference of the circle, the light beam lights up a portion of the circle and the lit portion of the circle moves around the circle at a certain tangential speed. If we now imagine a circle twice as big in radius, the tangential speed of the lit portion is faster, because it must travel a larger circumference in the time of one rotation of the light source. Imagine that we continue to make the circle larger and larger, eventually moving it out into space. The tangential speed of the lit portion will keep increasing. Is it possible that the tangential speed could become larger than the speed of light? Would this violate a principle of special relativity?

Explanation   For a large enough circle, it is possible that the tangential speed of the lit portion of the circle could be larger than the speed of light. This does not violate a principle of special relativity, however, because no matter or information is traveling faster than the speed of light.

26.2   THE PRINCIPLE OF RELATIVITY

In order to describe a physical event, it is necessary to choose a *frame of reference*. For example, when you perform an experiment in a laboratory, you select a coordinate system, or frame of reference, that is at rest with respect to the laboratory. However, suppose an observer in a passing car moving at a constant velocity with respect to the lab were to observe your experiment. Would the observations made by the moving observer differ dramatically from yours? That is, if you found Newton's first law to be valid in your frame of reference, would the moving observer agree with you? According to the principle of Newtonian relativity, **the laws of mechanics are the same in all inertial frames of reference.** Inertial frames of reference are those reference frames in which Newton's first law, the law of inertia, is valid. For the situation just described, the laboratory coordinate system and the coordinate system of the moving car are both inertial frames of reference. As a consequence, if the laws of mechanics are found to be true in the lab, the person in the car must also observe the same laws.

Let us describe a common observation to illustrate the equivalence of the laws of mechanics in different inertial frames. Consider an airplane in flight, moving with a constant velocity, as in Figure 26.1a. If a passenger in the airplane throws a ball straight up in the air, the passenger observes that the ball moves in a vertical path. The motion of the ball is precisely the same as it would be if the ball were thrown while at rest on Earth. The law of gravity and the equations of motion under constant acceleration are obeyed whether the airplane is at rest or in uniform motion. Now consider the same experiment when viewed by another observer at rest on the Earth. This stationary observer views the path of the ball to be a parabola, as in Figure 26.1b. Furthermore, according to this observer, the ball has a velocity to the right equal to the velocity of the plane. Although the two observers disagree on certain aspects of the experiment, both agree that the motion of the ball obeys

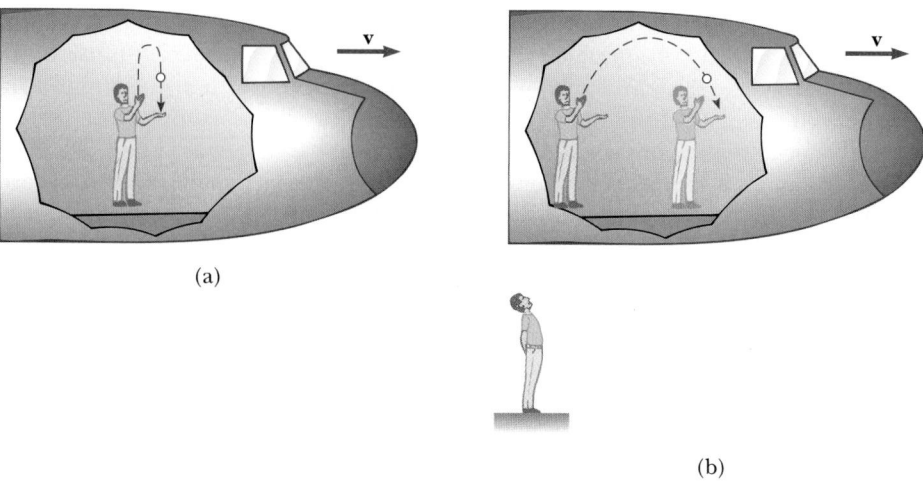

(a)

(b)

Figure 26.1 (a) The observer on the airplane sees the ball move in a vertical path when thrown upward. (b) The Earth observer views the path of the ball to be a parabola.

the law of gravity and Newton's laws of motion. Thus, we draw the following important conclusion: **There is no preferred frame of reference for describing the laws of mechanics.**

26.3 THE SPEED OF LIGHT

It is quite natural to ask whether the concept of Newtonian relativity in mechanics also applies to experiments in electricity, magnetism, optics, and other areas. For example, if we assume that the laws of electricity and magnetism are the same in all inertial frames, a paradox concerning the speed of light immediately arises. This can be understood by recalling that according to electromagnetic theory, the speed of light always has the fixed value of $2.997\ 924\ 58 \times 10^8$ m/s in free space. But this is in direct contradiction to common sense. For example, suppose a light pulse is sent out by an observer in a boxcar moving with a velocity $\mathbf{v}$ (Fig. 26.2). The light

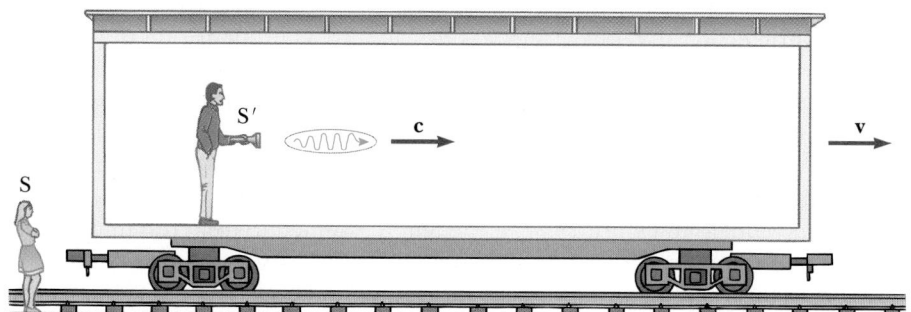

Figure 26.2 A pulse of light is sent out by a person in a moving boxcar. According to Newtonian relativity, the speed of the pulse should be $c + v$ relative to a stationary observer.

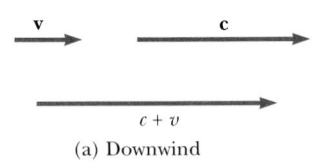

(a) Downwind

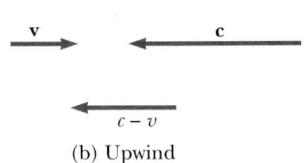

(b) Upwind

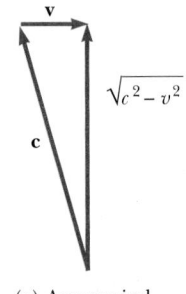

(c) Across wind

Figure 26.3 If the speed of the ether wind relative to the Earth is v, and c is the speed of light relative to the ether, the speed of light relative to the Earth is (a) $c + v$ in the downwind direction, (b) $c - v$ in the upwind direction, and (c) $(c^2 - v^2)^{1/2}$ in the direction perpendicular to the wind.

pulse has a velocity **c** relative to observer S′ in the boxcar. According to Newtonian relativity, the velocity of the pulse relative to the stationary observer S outside the boxcar should be **c** + **v.** This obviously contradicts Einstein's theory, which postulates that the velocity of the light pulse is the same for all observers.

In order to resolve this paradox, we must conclude either that (1) the addition law for velocities is incorrect or that (2) the laws of electricity and magnetism are not the same in all inertial frames. If the Newtonian addition law for velocities were incorrect, we would be forced to abandon the seemingly "obvious" notions of absolute time and absolute length that form the basis for this law.

If instead we assume that the second conclusion is true, then a preferred reference frame must exist in which the speed of light has the value c, whereas in any other reference frame the speed of light must have a value that is greater or less than c. It is useful to draw an analogy with sound waves, which propagate through a medium such as air. The speed of sound in air is about 330 m/s when measured in a reference frame in which the air is stationary. However, the speed of sound is greater or less than this value when measured from a reference frame that is moving with respect to the air.

In the case of light signals (electromagnetic waves), recall that electromagnetic theory predicted that such waves must propagate through free space with a speed equal to the speed of light. However, the theory does not require the presence of a medium for wave propagation. This is in contrast to other types of waves that we have studied, such as water and sound waves, that do require a medium to support the disturbances. In the 19th century, physicists thought that electromagnetic waves also required a medium in order to propagate. They proposed that such a medium existed, and they gave it the name **luminiferous ether.** The ether was assumed to be present everywhere, even in empty space, and light waves were viewed as ether oscillations. Furthermore, the ether would have to be a massless but rigid medium with no effect on the motion of planets or other objects. These are strange concepts indeed. In addition, it was found that the troublesome laws of electricity and magnetism would take on their simplest forms in a frame of reference at *rest* with respect to the ether. This frame was called the *absolute frame.* The laws of electricity and magnetism would be valid in this absolute frame, but they would have to be modified in any reference frame moving with respect to the absolute frame.

As a result of the importance attached to this absolute frame, it became of considerable interest in physics to prove by experiment that it existed. A direct method for detecting the ether wind was to measure its influence on the speed of light relative to a frame of reference on Earth. If **v** is the velocity of the ether relative to the Earth, then the speed of light should have its maximum value, $c + v$, when propagating downwind, as shown in Figure 26.3a. Likewise, the speed of light should have its minimum value, $c - v$, when propagating upwind, as in Figure 26.3b, and some intermediate value, $(c^2 - v^2)^{1/2}$, in the direction perpendicular to the ether wind, as in Figure 26.3c. If the Sun is assumed to be at rest in the ether, then the velocity of the ether wind would be equal to the orbital velocity of the Earth around the Sun, which has a magnitude of about 3×10^4 m/s. Because $c = 3 \times 10^8$ m/s, a change in speed of about 1 part in 10^4 m/s for measurements in the upwind or downwind directions should be detectable. However, as we shall see in the next section, all attempts to detect such changes and establish the existence of the ether (and hence the absolute frame) proved futile!

26.4 THE MICHELSON–MORLEY EXPERIMENT

The most famous experiment designed to detect small changes in the speed of light was performed in 1887 by A. A. Michelson (1852–1931) and E. W. Morley (1838–1923). We should state at the outset that the outcome of the experiment was *negative,* thus contradicting the ether hypothesis. The experiment was designed to determine the velocity of the Earth with respect to the hypothetical ether. The tool used was the Michelson interferometer, shown in Figure 26.4. When one of the arms of the interferometer was aligned along the direction of the Earth's motion through space, the motion of the Earth through the ether would have been equivalent to the ether flowing past the Earth in the opposite direction. This ether wind blowing in the opposite direction should have caused the speed of light as measured in the Earth's frame of reference to be $c - v$ as it approached the mirror M_2 in Figure 26.4 and $c + v$ after reflection. The speed v is the speed of the Earth through space, and hence the speed of the ether wind, and c is the speed of light in the absolute ether frame. In the experiment the two beams of light reflected from M_1 and M_2 recombined, and an interference pattern consisting of alternating dark and bright bands or fringes was formed. During the experiment, the interference pattern was observed while the interferometer was rotated through an angle of 90°. The effect of this rotation should have been to cause a slight but measurable shift in the fringe pattern. Measurements failed to show any change in the interference pattern! The Michelson–Morley experiment was repeated by other researchers under various conditions and at different locations, but the results were always the same: **No fringe shift of the magnitude required by the ether hypothesis was ever observed.**

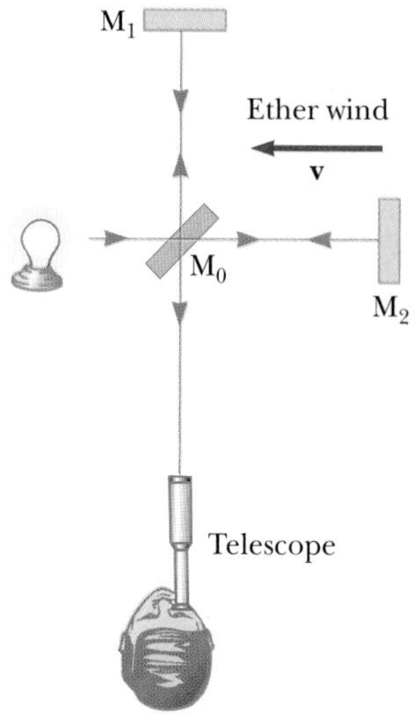

The negative results of the Michelson–Morley experiment meant that it was impossible to measure the absolute orbital velocity of the Earth with respect to the ether frame. However, as we shall see in the next section, Einstein developed a postulate for his theory of relativity that places quite a different interpretation on these results. In later years, when more was known about the nature of light, the idea of an ether that permeates all of space was relegated to the ash heap of worn-out concepts. Light is now understood to be an electromagnetic wave that requires no medium for its propagation. As a result, the idea of having an ether in which electromagnetic waves travel became unnecessary.

Figure 26.4 According to the ether-wind theory, the speed of light should be $c - v$ as the beam approaches mirror M_2 and $c + v$ after reflection.

Details of the Michelson–Morley Experiment

Optional Section

As we mentioned earlier, the Michelson–Morley experiment was designed to detect the motion of the Earth with respect to the ether. Before we examine the details of this important, historical experiment, it is instructive to first consider a race between two airplanes, as shown in Figure 26.5a. One airplane flies from point O to point A perpendicular to the direction of the wind, and the second airplane flies from point O to point B parallel to the wind. We shall assume that they start at O at the same time, travel the same distance L with the same cruising speed c with respect to the wind, and return to O. Which airplane will win the race? In order to answer this question, we shall first calculate the time of flight for both airplanes.

First, consider the airplane that moves along path I parallel to the wind. As it moves to the right, its speed is enhanced by the wind, and its velocity with respect

Figure 26.5 (a) If an airplane wishes to travel from O to A with a wind blowing to the right, it must head into the wind at some angle. (b) Vector diagram for determining the airplane's direction for the trip from O to A. (c) Vector diagram for determining its direction for the trip from A to O.

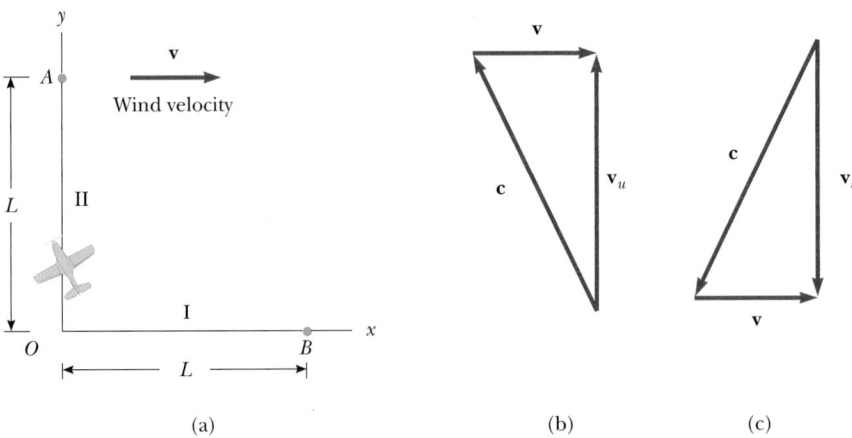

(a)   (b)   (c)

Albert A. Michelson,
German American physicist
(1852–1931)

Michelson spent much of his life making accurate measurements of the speed of light. In 1907 he was the first American to be awarded the Nobel prize, which he received for his work in optics. His most famous experiment, conducted with Edward Morley in 1887, implied that it was impossible to measure the absolute velocity of the Earth with respect to the ether. *(AIP Emilio Segré Visual Archives, Michelson Collection)*

to the Earth is $c + v$. As it moves to the left on its return journey, it must fly opposite the wind; hence its speed with respect to the Earth is $c - v$. Thus, the times of flight to the right and to the left are, respectively,

$$t_R = \frac{L}{c + v} \quad \text{and} \quad t_L = \frac{L}{c - v}$$

and the total time of flight for the airplane moving along path I is

$$t_1 = t_R + t_L = \frac{L}{c + v} + \frac{L}{c - v} = \frac{2Lc}{c^2 - v^2}$$

$$= \frac{2L}{c\left(1 - \dfrac{v^2}{c^2}\right)} \qquad\qquad \text{[26.1]}$$

Now consider the airplane flying along path II. If the pilot aims the airplane directly toward point A, it will be blown off course by the wind and will not reach its destination. To compensate for the wind, the pilot must point the airplane into the wind at some angle as shown in Figure 26.5a. This angle must be selected so that the vector sum of $\mathbf{c}$ and $\mathbf{v}$ leads to a velocity vector pointed directly toward A. The resultant vector diagram is shown in Figure 26.5b, where $\mathbf{v}_u$ is the velocity of the airplane with respect to the ground as it moves from O to A. From the Pythagorean theorem, the magnitude of the vector $\mathbf{v}_u$ is

$$v_u = \sqrt{c^2 - v^2} = c\sqrt{1 - \frac{v^2}{c^2}}$$

Likewise, on the return trip from A to O, the pilot must again head into the wind so that the airplane's velocity with respect to the Earth, $\mathbf{v}_d$, will be directed toward O, as shown in Figure 26.5c. From this figure, we see that

$$v_d = \sqrt{c^2 - v^2} = c\sqrt{1 - \frac{v^2}{c^2}}$$

Thus, the total time of flight for the trip along path II is

$$t_2 = \frac{L}{v_u} + \frac{L}{v_d} = \frac{L}{c\sqrt{1 - \dfrac{v^2}{c^2}}} + \frac{L}{c\sqrt{1 - \dfrac{v^2}{c^2}}}$$

$$= \frac{2L}{c\sqrt{1 - \dfrac{v^2}{c^2}}} \qquad \qquad \text{[26.2]}$$

Comparing Equations 26.1 and 26.2, we see that the airplane flying along path II wins the race. The difference in flight times is given by

$$\Delta t = t_1 - t_2 = \frac{2L}{c}\left[\frac{1}{\left(1 - \dfrac{v^2}{c^2}\right)} - \frac{1}{\sqrt{1 - \dfrac{v^2}{c^2}}}\right]$$

This expression can be simplified using the following binomial expansions in v/c (assumed to be much smaller than 1) after dropping all terms higher than second order:

$$\left(1 - \frac{v^2}{c^2}\right)^{-1} \approx 1 + \frac{v^2}{c^2}$$

and

$$\left(1 - \frac{v^2}{c^2}\right)^{-1/2} \approx 1 + \frac{1}{2}\frac{v^2}{c^2}$$

Therefore, the difference in flight times is

$$\Delta t = \frac{Lv^2}{c^3} \qquad \qquad \text{[26.3]}$$

The analogy between this airplane race and the Michelson–Morley experiment is shown in Figure 26.6a. Two beams of light travel along two arms of an interferometer. In this case, the "wind" is the ether blowing across the Earth from left to

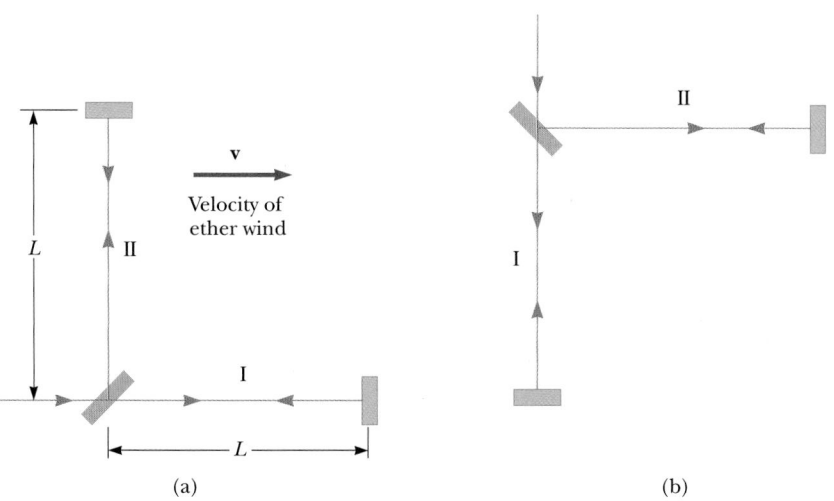

(a)

(b)

Figure 26.6 (a) Top view of the Michelson–Morley interferometer, where **v** is the velocity of the ether and L is the length of each arm. (b) When the interferometer is rotated by 90°, the role of each arm is reversed.

right as the Earth moves through the ether from right to left. Because the speed of the Earth in its orbital path is approximately equal to 3×10^4 m/s, the speed of the wind should be at least this great. The two light beams start out in phase and return to form an interference pattern. Let us assume that the interferometer is adjusted for parallel fringes and that a telescope is focused on one of these fringes. The time difference between the two light beams gives rise to a phase difference between the beams, producing an interference pattern when they combine at the position of the telescope. The difference in the pattern is detected by rotating the interferometer through 90° in a horizontal plane, so that the two beams exchange roles (Fig. 26.6b). This results in a net time shift of twice the time difference given by Equation 26.3. Thus, the net time difference is

$$\Delta t_{\text{net}} = 2\,\Delta t = \frac{2Lv^2}{c^3} \qquad\qquad \text{[26.4]}$$

The corresponding path difference is

$$\Delta d = c\,\Delta t_{\text{net}} = \frac{2Lv^2}{c^2} \qquad\qquad \text{[26.5]}$$

In the first experiments by Michelson and Morley, each light beam was reflected by the mirrors many times to give an increased effective path length L of about 11 meters. Using this value and taking v to be equal to 3×10^4 m/s gives a path difference of

$$\Delta d = \frac{2(11\text{ m})(3.0 \times 10^4\text{ m/s})^2}{(3.0 \times 10^8\text{ m/s})^2} = 2.2 \times 10^{-7}\text{ m}$$

This extra travel distance should produce a noticeable shift in the fringe pattern. Specifically, calculations show that if the pattern is viewed while the interferometer is rotated through 90°, a shift of about 0.4 fringes should be observed. The instrument used by Michelson and Morley was capable of detecting a shift in the fringe pattern as small as 0.01 fringes. However, *they detected no shift in the fringe pattern.* Since then, the experiment has been repeated many times by various scientists under various conditions and no fringe shift has ever been detected. Thus, it was concluded that the motion of the Earth with respect to the ether cannot be detected.

Many efforts were made to explain the null results of the Michelson–Morley experiment. For example, perhaps the Earth drags the ether with it in its motion through space. To test this assumption, interferometer measurements were made at various altitudes, but again no fringe shift was detected. In the 1890s G. F. Fitzgerald and H. A. Lorentz tried to explain the null results by making the following ad hoc assumption. They proposed that the length of an object moving along the direction of the ether wind would contract by a factor of $\sqrt{1 - v^2/c^2}$. The net result of this contraction would be a change in length of one of the arms of the interferometer such that no path difference would occur as the interferometer was rotated.

No other experiment in the history of physics has received such valiant efforts to explain the absence of an expected result as has the Michelson–Morley experiment. The stage was set for the brilliant Albert Einstein, who solved the problem in 1905 with his special theory of relativity.

26.5 EINSTEIN'S PRINCIPLE OF RELATIVITY

In the previous section we noted the serious contradiction between the Newtonian addition law for velocities and the fact that the speed of light is the same for all observers. In 1905 Albert Einstein proposed a theory that would resolve this contradiction but at the same time would completely alter our notions of space and time. Einstein based his special theory of relativity on the following general hypothesis, which is called the **principle of relativity:**

All the laws of physics are the same in all inertial frames.

An immediate consequence of the principle of relativity is that

The speed of light in a vacuum has the same value, $c = 2.997\ 924\ 58 \times 10^8$ m/s, in all inertial reference frames.

Albert Einstein (1879–1955)

Einstein, one of the greatest physicists of all times, was born in Ulm, Germany. In 1905, at the age of 26, he published four scientific papers that revolutionized physics. Two of these papers were concerned with what is now considered his most important contribution: the special theory of relativity. In 1916, Einstein published his work on the general theory of relativity. The most dramatic prediction of this theory is the degree to which light is deflected by a gravitational field. Measurements made by astronomers on bright stars in the vicinity of the eclipsed sun in 1919 confirmed Einstein's prediction, and as a result Einstein became a world celebrity. Einstein was deeply disturbed by the development of quantum mechanics in the 1920s despite his own role as a scientific revolutionary. In particular, he could never accept the probabilistic view of events in nature that is a central feature of quantum theory. The last few decades of his life were devoted to an unsuccessful search for a unified theory that would combine gravitation and electromagnetism. *(AIP Niels Bohr Library)*

In other words, anyone who measures the speed of light will get the same value, c. This implies that the ether does not exist. Together, the principle of relativity and its immediate consequence are often referred to as the two postulates of special relativity.

The null result of the Michelson–Morley experiment can be readily understood within the framework of Einstein's theory. According to his principle of relativity, the premises of the Michelson–Morley experiment were incorrect. In the process of trying to explain the expected results, we stated that when light traveled against the ether wind its speed was $c - v$. However, if the state of motion of the observer or of the source has no influence on the value found for the speed of light, the measured value will always be c. Likewise, the light makes the return trip after reflection from the mirror at a speed of c, not the speed of $c + v$. Thus, the motion of the Earth should not influence the fringe pattern observed in the Michelson–Morley experiment and a null result should be expected.

If we accept Einstein's theory of relativity, we must conclude that relative motion is unimportant when measuring the speed of light. At the same time, we must alter our common-sense notions of space and time and be prepared for some rather bizarre consequences.

26.6 CONSEQUENCES OF SPECIAL RELATIVITY

Almost everyone who has dabbled even superficially in science is aware of some of the startling predictions that arise because of Einstein's approach to relative motion. As we examine some of the consequences of relativity in this section, we shall find that they conflict with some of our basic notions of space and time. We shall restrict our discussion to the concepts of length, time, and simultaneity, which are quite different in relativistic mechanics from what they are in Newtonian mechanics. For example, we shall see that the distance between two points and the time interval

Absolute length and absolute time ▶
intervals are meaningless in
relativity.

between two events depend on the frame of reference in which they are measured. That is, in relativity, there is no such thing as absolute length or absolute time. Furthermore, events at different locations that occur simultaneously in one frame are not simultaneous in another frame.

Simultaneity and the Relativity of Time

A basic premise of Newtonian mechanics is that there is a universal time scale that is the same for all observers. In fact, Newton wrote, "Absolute, true, and mathematical time, of itself, and from its own nature, flows equably without relation to anything external." In his special theory of relativity, Einstein abandoned this assumption. According to Einstein, **time interval measurements depend on the reference frame in which they are made.**

Einstein devised the following thought experiment to illustrate this point. A boxcar moves with uniform velocity, and two lightning bolts strike its ends, as in Figure 26.7a, leaving marks on the boxcar and the ground. The marks left on the boxcar are labeled A' and B', and those on the ground are labeled A and B. An observer at O' moving with the boxcar is midway between A' and B', and an observer on the ground at O is midway between A and B. The events recorded by the observers are the light signals from the lightning bolts.

Let us assume that the two light signals reach the observer at O at the same time, as indicated in Figure 26.7b. This observer realizes that the light signals have traveled at the same speed over distances of equal length. Thus, the observer at O concludes that the events at A and B occurred simultaneously. Now consider the same events as viewed by the observer on the boxcar at O'. By the time the light has reached the observer at O, the observer at O' has moved, as indicated in Figure 26.7b. Thus, the light signal from B' has already swept past O', whereas the light from A' has not yet reached O'. According to Einstein's second postulate, the observer at O' must find that light travels at the same speed as that measured by the observer at O. Therefore, the observer at O' concludes that the lightning struck the front of the boxcar before it struck the back. This thought experiment clearly dem-

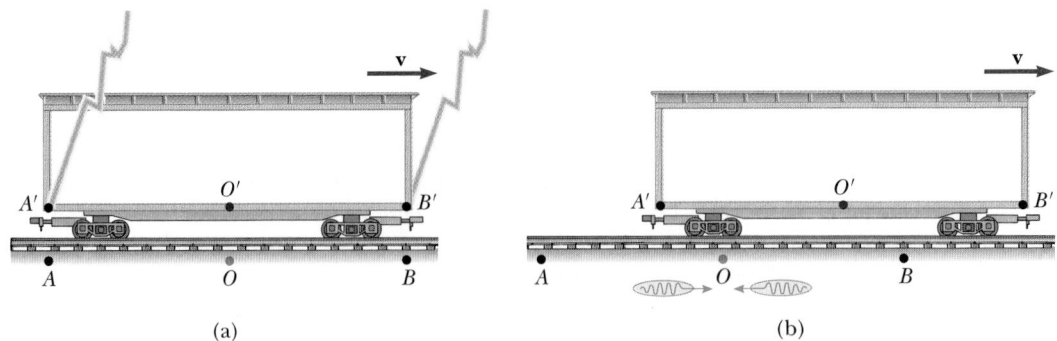

(a)   (b)

Figure 26.7   Two lightning bolts strike the ends of a moving boxcar. (a) The events appear to be simultaneous to the stationary observer at O, who is midway between A and B. (b) The events do not appear to be simultaneous to the observer at O', who claims that the front of the train is struck *before* the rear.

onstrates that the two events that appear to be simultaneous to the observer at O do not appear to be simultaneous to the observer at O'. In other words,

> Two events that are simultaneous in one reference frame are in general not simultaneous in a second frame moving with respect to the first. That is, simultaneity is not an absolute concept.

At this point, you might wonder which observer is right concerning the two events. The answer is that both are correct because the principle of relativity states that **there is no preferred inertial frame of reference.** Although the two observers reach different conclusions, both are correct in their own reference frames because the concept of simultaneity is not absolute.

Time Dilation

Consider a vehicle moving to the right with a speed v, as in Figure 26.8a. A perfectly reflecting mirror is fixed to the ceiling of the vehicle, and an observer at O' at rest in this system holds a flash gun a distance d below the mirror. At some instant, the flash gun goes off and a pulse of light is released. Because the light pulse has a speed c, the time it takes it to travel from the observer to the mirror and back again can be found from the definition of velocity,

$$\Delta t_p = \frac{\text{distance traveled}}{\text{velocity}} = \frac{2d}{c} \qquad [26.6]$$

where Δt_p is the time interval measured by O', the observer who is at rest in the moving vehicle.

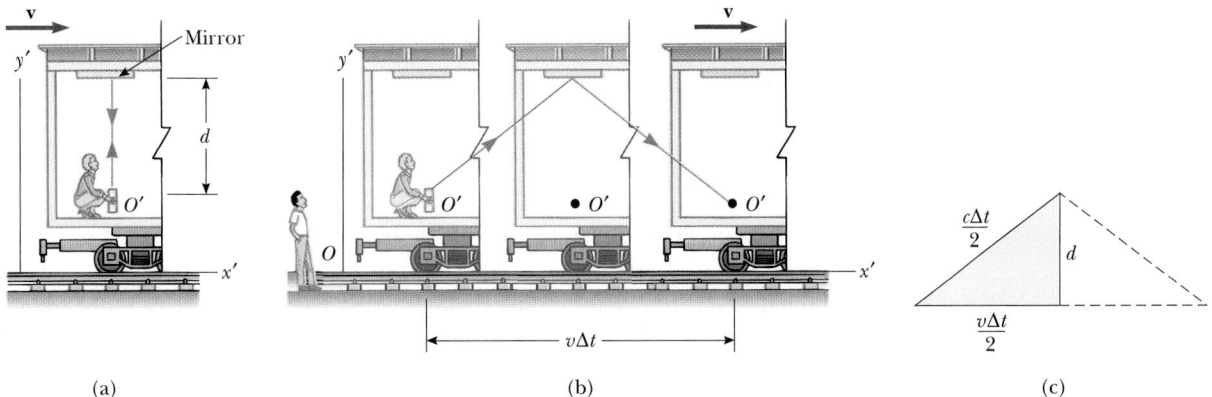

(a) (b) (c)

Figure 26.8 (a) A mirror is fixed to a moving vehicle, and a light pulse leaves O' at rest in the vehicle. (b) Relative to a stationary observer on Earth, the mirror and O' move with a speed v. Note that the distance the pulse travels is greater than $2d$ as measured by the stationary observer. (c) The right triangle for calculating the relationship between Δt and Δt_p.

Now consider the same set of events as viewed by an observer at O in a stationary frame (Fig. 26.8b). According to this observer, the mirror and flash gun are moving to the right with a speed of v. The sequence of events just described would appear entirely different to this stationary observer. By the time the light from the flash gun reaches the mirror, the mirror will have moved a distance of $v\,\Delta t/2$, where Δt is the time it takes the light pulse to travel from O' to the mirror and back, as measured by the stationary observer. In other words, the stationary observer concludes that, because of the motion of the system, the light, if it is to hit the mirror, will leave the flash gun at an angle with respect to the vertical. Comparing Figures 26.8a and 26.8b, we see that the light must travel farther in the stationary frame than in the moving frame.

Now, according to Einstein's second postulate, the speed of light must be c as measured by both observers. Therefore, it follows that the time interval Δt, measured by the observer in the stationary frame, is *longer* than the time interval Δt_p, measured by the observer in the moving frame. To obtain a relationship between Δt and Δt_p, it is convenient to use the right triangle shown in Figure 26.8c. The Pythagorean theorem applied to this triangle gives

$$\left(\frac{c\,\Delta t}{2}\right)^2 = \left(\frac{v\,\Delta t}{2}\right)^2 + d^2$$

Solving for Δt gives

$$\Delta t = \frac{2d}{\sqrt{c^2 - v^2}} = \frac{2d}{c\sqrt{1 - v^2/c^2}} \tag{26.7}$$

Because $\Delta t_p = 2d/c$, we can express Equation 26.7 as

Time dilation ▶

$$\Delta t = \frac{\Delta t_p}{\sqrt{1 - v^2/c^2}} = \gamma\,\Delta t_p \tag{26.8}$$

where $\gamma = 1\sqrt{1 - v^2/c^2}$. This result says that the time interval measured by the observer in the stationary frame is *longer* than that measured by the observer in the moving frame (γ is always greater than unity).

For example, suppose an observer in a moving vehicle has a clock that he uses to measure the time required for the light flash to leave the gun and return. Let us assume that the measured time interval in this frame of reference, Δt_p, is one second. (This would require a very tall vehicle.) Now let us find the time interval as measured by a stationary observer using an identical clock. If the vehicle is traveling at half the speed of light ($v = 0.500c$), then $\gamma = 1.15$, and according to Equation 26.8 $\Delta t = \gamma\,\Delta t_p = 1.15(1.00\text{ s}) = 1.15\text{ s}$. Thus, when the observer on the moving vehicle claims that 1.00 s has passed, a stationary observer claims that 1.15 s has passed. From this we may conclude that,

A clock in motion runs more slowly ▶
than an identical stationary clock.

> According to a stationary observer, a moving clock runs more slowly than an identical stationary clock by a factor of γ^{-1}. This effect is known as **time dilation.**

The time interval Δt_p in Equation 26.8 is called the *proper time*. In general, **proper time** is defined as **the time interval between two events as measured by an observer who sees the events occur at the same place.** In our case, the observer at O' measures the proper time. That is, **proper time is always the time interval measured with a single clock at rest in the frame in which the events take place at the same position.**

We have seen that moving clocks run slow by a factor of γ^{-1}. This is true for ordinary mechanical clocks as well as for the light clock just described. In fact, we can generalize these results by stating that **all physical processes, including chemical and biological reactions, slow down relative to a stationary clock when they occur in a moving frame.** For example, the heartbeat of an astronaut moving through space has to keep time with a clock inside the spaceship. Both the spaceship clock and the heartbeat are slowed down relative to a stationary clock. The astronaut would not, however, have any sensation of life slowing down in the spaceship.

Time dilation is a very real phenomenon that has been verified by various experiments. *Muons* are unstable elementary particles with a charge equal to that of the electron and a mass 207 times that of the electron. They can be produced by the absorption of cosmic radiation high in the atmosphere. These unstable particles have a lifetime of only 2.2 μs when measured in a reference frame at rest with respect to them. If we take 2.2 μs as the average lifetime of a muon and assume that their speed is close to the speed of light, we find that these particles can travel only about 600 m before they decay (Fig. 26.9a). Hence, they could never reach the Earth from the upper atmosphere where they are produced. However, experiments show that a large number of muons *do* reach the Earth, and the phenomenon of time dilation explains how. Relative to an observer on Earth, the muons have a lifetime equal to $\gamma\tau_p$, where $\tau_p = 2.2$ μs is the lifetime in a frame of reference traveling with the muons. For example, for $v = 0.99c$, $\gamma \approx 7.1$ and $\gamma\tau \approx 16$ μs. Hence, the average distance traveled as measured by an observer on Earth is $\gamma v\tau \approx 4800$ m, as indicated in Figure 26.9b.

In 1976 experiments with muons were conducted at the laboratory of the European Council for Nuclear Research (CERN) in Geneva. Muons were injected into a large storage ring, reaching speeds of about $0.9994c$. Electrons produced by the decaying muons were detected by counters around the ring, enabling scientists to measure the decay rate, and hence the lifetime of the muons. The lifetime of the moving muons was measured to be about 30 times as long as that of stationary muons to within two parts in a thousand, in agreement with the prediction of relativity.

The results of an experiment reported by Hafele and Keating provided direct evidence for the phenomenon of time dilation.[2] The experiment involved the use of very stable cesium-beam atomic clocks. Time intervals measured with four such clocks in jet flight were compared with time intervals measured by reference atomic clocks at the U.S. Naval Observatory. (Because of the Earth's rotation about its axis, a ground-based clock is not in a true inertial frame.) Time intervals measured with the flying clocks were compared to time intervals measured with the Earth-based reference clocks. In order to compare the results with the theory, many factors had to be considered, including periods of acceleration and deceleration relative to the

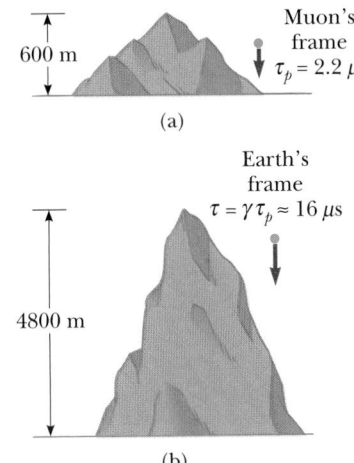

Figure 26.9 (a) The muons travel only about 600 m as measured in the muons' reference frame, in which their lifetime is about 2.2 μs. Because of time dilation, the muons' lifetime is longer as measured by the observer on Earth. (b) Muons traveling with a speed of $0.99c$ travel a distance of about 4800 m as measured by an observer on Earth.

[2] J. C. Hafele and R. E. Keating, "Around the World Atomic Clocks: Relativistic Time Gains Observed," *Science*, July 14, 1972, p. 168.

Earth, variations in direction of travel, and the weaker gravitational field experienced by the flying clocks. Their results were in good agreement with the predictions of the special theory of relativity. In their paper, Hafele and Keating report the following: "Relative to the atomic time scale of the U.S. Naval Observatory, the flying clocks lost 59 ± 10 ns during the eastward trip and gained 273 ± 7 ns during the westward trip. . . . These results provide an unambiguous empirical resolution of the famous clock paradox with macroscopic clocks."

Thinking Physics 2

Suppose a student explains time dilation with the following argument: If you start running at $0.99c$ away from a clock at 12:00, you would not see the time change, because the light from the clock representing 12:01 would never reach you. What is the flaw in this argument?

Explanation   The inference in this argument is that the velocity of light relative to the runner is approximately zero — "the light . . . would never reach you." This is a Newtonian relativity point of view, in which the relative velocity is a simple subtraction of running velocity from the light velocity. From the point of view of special relativity, one of the fundamental postulates is that the speed of light is the same for all observers, including one running away from the light source at the speed of light. Thus, the light from 12:01 will move toward the runner at the speed of light.

EXAMPLE 26.1   What Is the Period of the Pendulum?

The period of a pendulum is measured to be 3.0 s in the inertial frame of the pendulum. What is the period when measured by an observer moving at a speed of $0.95c$ with respect to the pendulum?

Reasoning and Solution   In this case, the proper time is 3.0 s. We can use Equation 26.8 to calculate the period measured by the moving observer:

$$T = \gamma\, T_p = \frac{1}{\sqrt{1 - \dfrac{(0.95c)^2}{c^2}}}\, T_p = (3.2)(3.0\text{ s}) = \boxed{9.6\text{ s}}$$

That is, the observer moving with a speed of $0.95c$ observes that the pendulum slows down.

The Twin Paradox

An interesting consequence of time dilation is the so-called twin paradox. Consider a controlled experiment involving 20-year-old twin brothers Speedo and Goslo (Fig. 26.10). Speedo, the more adventuresome twin, sets out on a journey toward a star located 30 lightyears from Earth. His spaceship is able to accelerate to a speed close to the speed of light. After reaching the star, Speedo becomes very homesick and immediately returns to Earth at the same high speed. On his return, he is shocked to find that many things have changed. Old cities have expanded and new cities have appeared. Lifestyles, fashions, and transportation systems have changed dramatically. Speedo's twin brother, Goslo, has aged to about 80 years old and is now

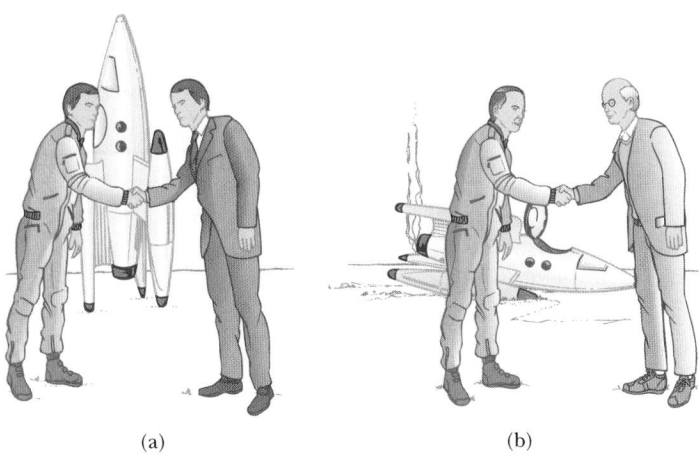

Figure 26.10 (a) As the twins depart, they are the same age. (b) When Speedo returns from his journey to Planet X, he is younger than his twin Goslo who remained on Earth.

wiser, feeble, and somewhat hard of hearing. Speedo, on the other hand, has aged only about 10 years. This is because his bodily processes slowed down during his travels in space.

It is quite natural to raise the question, "Which twin actually travels at a speed close to the speed of light, and therefore does not age as much?" Herein lies the paradox: From Goslo's frame of reference, he is at rest while his brother Speedo travels at a high velocity. On the other hand, according to the space traveler Speedo, it is he who is at rest while his brother zooms away from him on Earth and then returns. This leads to confusion about which twin actually ages more.

◀ The space traveler ages more slowly than his twin who remains on Earth.

In order to resolve this paradox, it should be pointed out that the trip is not as symmetrical as we may have led you to believe. Speedo, the space traveler, experiences a series of accelerations and decelerations during his journey to the star and back home, and therefore is not always in uniform motion. This means that Speedo is in a noninertial frame during part of his trip, so that predictions based on special relativity are not valid in his frame. On the other hand, the brother on Earth is in an inertial frame and can make reliable predictions based on the special theory. The situation is not symmetrical because Speedo experiences accelerations when his spaceship turns around, whereas Goslo is not subject to such accelerations. Therefore, the space traveler is indeed younger on returning to Earth.

Length Contraction

We have seen that measured time intervals are not absolute—that is, the time interval between two events depends on the frame of reference in which it is measured. Likewise, the measured distance between two points depends on the frame of reference. The **proper length** of an object is defined as **the length of the object measured in the reference frame in which the object is at rest.** The length of an object measured in a reference frame in which the object is moving is always less than the proper length. This effect is known as **relativistic length contraction.**

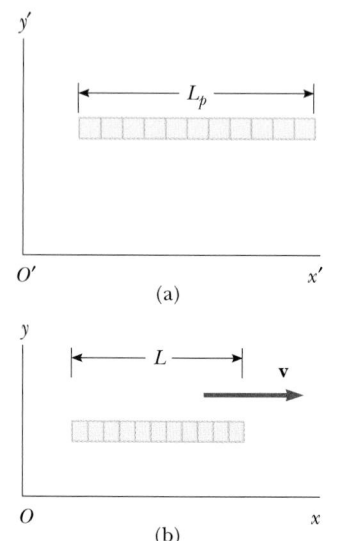

Figure 26.11   A meter stick moves to the right with a speed v. (a) The meter stick as viewed by an observer at rest with respect to the meter stick. (b) The meter stick as seen by an observer moving with a speed v with respect to the meter stick. The moving meter stick is always measured to be *shorter* than in its own rest frame by a factor of $\sqrt{1 - v^2/c^2}$.

APPLICATION

Passage of Time in Space Travel.

To understand relativistic length contraction quantitatively, let us consider a spaceship traveling with a speed v from one star to another, as seen by two observers. An observer at rest on Earth (and also assumed to be at rest with respect to the two stars) measures the distance between the stars to be L_p (where L_p is the proper length). According to this observer, it takes a time $\Delta t = L_p/v$ for the spaceship to complete the voyage. What does an observer in the spaceship measure? Because of time dilation, the space traveler measures a smaller time of travel: $\Delta t_p = \Delta t/\gamma$. The observer in the spaceship claims to be at rest and sees the destination star as moving toward the ship with speed v. Because the space traveler reaches the star in the time Δt_p, she concludes that the distance, L, between the stars is shorter than L_p. This distance is given by

$$L = v\,\Delta t_p = v\,\frac{\Delta t}{\gamma}$$

Because $L_p = v\,\Delta t$, we see that

$$L = \frac{L_p}{\gamma}$$

or,

Length contraction ▶ $$L = L_p \sqrt{1 - v^2/c^2}$$ [26.9]

According to this result, illustrated in Figure 26.11, if an observer at rest with respect to an object measures its length to be L_p, an observer moving at a relative speed v with respect to the object will find it to be shorter than its rest length by the factor $\sqrt{1 - v^2/c^2}$. You should note that **the length contraction takes place only along the direction of motion.**

Time dilation and length contraction effects have interesting applications for future space travel to distant stars. In order for the star to be reached in a reasonable fraction of a human lifetime, the trip must be taken at very high speeds. According to an Earth-bound observer, the time for a spacecraft to reach the destination star will be dilated compared to the time interval measured by the travelers. Thus, it will seem to the travelers to take less time to reach the star than for the Earth-bound observers, as was discussed in the treatment of the twin paradox. We can also argue this from length contraction. For the travelers, the distance from Earth to the star will appear to be contracted, and it will consequently take less time to cover this shorter distance. Thus, by the time the travelers reach the star, they have aged by some number of years, while their partners back on Earth will have aged a larger number of years, the exact ratio depending on the speed of the spacecraft. At a spacecraft speed of $0.94c$, this ratio is about $3:1$.

Another consideration for the space travelers is related to their paychecks. You are invited to explore this issue in Conceptual Question 3.

EXAMPLE 26.2   The Contraction of a Spaceship

A spaceship is measured to be 120 m long while it is at rest with respect to an observer. If this spaceship now flies past the observer with a speed of $0.99c$, what length will the observer measure for the spaceship?

Solution From Equation 26.9, the length measured by the observer is

$$L = L_p \sqrt{1 - v^2/c^2} = (120 \text{ m}) \sqrt{1 - \frac{(0.99c)^2}{c^2}} = \boxed{17 \text{ m}}$$

Exercise If the ship moves past the observer with a speed of $0.01000c$, what length will the observer measure?

Answer 119.994 m

EXAMPLE 26.3   How High Is the Spaceship?

An observer on Earth sees a spaceship at an altitude of 435 m moving downward toward the Earth with a speed of $0.970c$. What is the altitude of the spaceship as measured by an observer in the spaceship?

Solution The moving observer in the spaceship finds the altitude to be

$$L = L_p \sqrt{1 - v^2/c^2} = (435 \text{ m}) \sqrt{1 - \frac{(0.970c)^2}{c^2}} = \boxed{106 \text{ m}}$$

EXAMPLE 26.4   The Triangular Spaceship

A spaceship in the form of a triangle flies by an observer with a speed of $0.95c$. When the spaceship is at rest (Fig. 26.12a), the distances x and y are found to be 52 m and 25 m, respectively. What is the shape of the spaceship as seen by an observer at rest when the spaceship is in motion along the direction shown in Figure 26.12b?

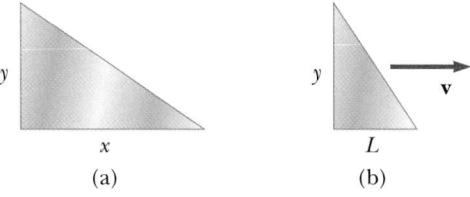

(a)   (b)

Figure 26.12   (Example 26.4) (a) When the spaceship is at rest, its shape is as shown. (b) The spaceship appears to look like this when it moves to the right with a speed v. Note that only its x dimension is contracted in this case.

Solution The observer sees the horizontal length of the spaceship to be contracted to a length of

$$L = L_p \sqrt{1 - v^2/c^2} = (52 \text{ m}) \sqrt{1 - \frac{(0.95c)^2}{c^2}} = \boxed{16 \text{ m}}$$

The 25-m vertical height is unchanged because it is perpendicular to the direction of relative motion between the observer and the spaceship. Figure 26.12b represents the shape of the spaceship as seen by the observer at rest.

26.7   RELATIVISTIC MOMENTUM

In order to describe properly the motion of particles within the framework of special relativity, we must generalize Newton's laws of motion and the definitions of momentum and energy. As we shall see, these generalized definitions reduce to the classical (nonrelativistic) definitions when v is much less than c.

First, recall that conservation of momentum states that when two objects collide, the total momentum of the system remains constant, assuming that the objects are isolated (that is, they interact only with each other). If such a collision is analyzed within the framework of Einstein's postulates of relativity, it is found that momentum is not conserved if the classical definition of momentum, $p = mv$, is used. However, according to the principle of relativity, momentum must be conserved in all reference systems. In view of this condition, it is necessary to modify the definition of momentum to satisfy the following conditions:

1. The relativistic momentum must be conserved in all collisions.
2. The relativistic momentum must approach the classical value mv as the quantity v/c approaches zero.

The correct relativistic equation for momentum that satisfies these conditions is

Momentum ▶

$$p \equiv \frac{mv}{\sqrt{1 - v^2/c^2}} = \gamma mv \qquad \text{[26.10]}$$

where v is the velocity of the particle. The theoretical derivation of this generalized expression for momentum is beyond the scope of this text. Note that when v is much less than c, the denominator of Equation 26.10 approaches unity, so that p approaches mv. Therefore, the relativistic equation for momentum reduces to the classical expression when v is small compared with c. When the speed of an object is less than $0.1c$, the classical expression mv will be equal to its actual (relativistic) momentum within 0.5% or better.

EXAMPLE 26.5   The Relativistic Momentum of an Electron

An electron, which has a mass of 9.11×10^{-31} kg, moves with a speed of $0.75c$. Find its relativistic momentum and compare this value to the momentum calculated from the classical expression.

Solution   From Equation 26.10, with $v = 0.75c$, we have

$$p = \frac{mv}{\sqrt{1 - v^2/c^2}}$$

$$= \frac{(9.11 \times 10^{-31} \text{ kg})(0.75 \times 3.00 \times 10^8 \text{ m/s})}{\sqrt{1 - (0.75c)^2/c^2}} = \boxed{3.1 \times 10^{-22} \text{ kg} \cdot \text{m/s}}$$

The classical expression gives

$$\text{Momentum} = mv = 2.1 \times 10^{-22} \text{ kg} \cdot \text{m/s}$$

The (correct) relativistic result is 50% greater than the classical result!

26.8   RELATIVISTIC ADDITION OF VELOCITIES

Imagine a motorcycle rider moving with a speed of $0.80c$ past a stationary observer, as shown in Figure 26.13. If the rider tosses a ball in the forward direction with a speed of $0.70c$ relative to himself, what is the speed of the ball as seen by the stationary observer at the side of the road? Common sense and the ideas of Newtonian relativity say that the speed should be the sum of the two speeds, or $1.50c$. This answer must be incorrect because it contradicts the assertion that no material object can travel faster than the speed of light.

Einstein resolved this dilemma by deriving an equation for the relativistic addition of velocities. For one-dimensional motion, this equation is

$$v_{ab} = \frac{v_{ad} + v_{db}}{1 + \dfrac{v_{ad}v_{db}}{c^2}}$$ **[26.11]**   ◀ Velocity addition

The left side of this equation and the numerator on the right are like the equations of Newtonian relativity discussed in Chapter 3, and the evaluation of subscripts is applied in the same fashion as discussed in Section 3.6. The denominator of Equation 26.11 is a correction to ordinary Newtonian relativity based on length contraction and time dilation. Let us apply this equation to the case of the speedy motorcycle rider and the stationary observer.

We have

v_{bm} = the velocity of the ball with respect to the motorcycle = $0.70c$,
v_{mo} = the velocity of the motorcycle with respect to the stationary observer = $0.80c$,

and we want to find

v_{bo} = the velocity of the ball with respect to the stationary observer.

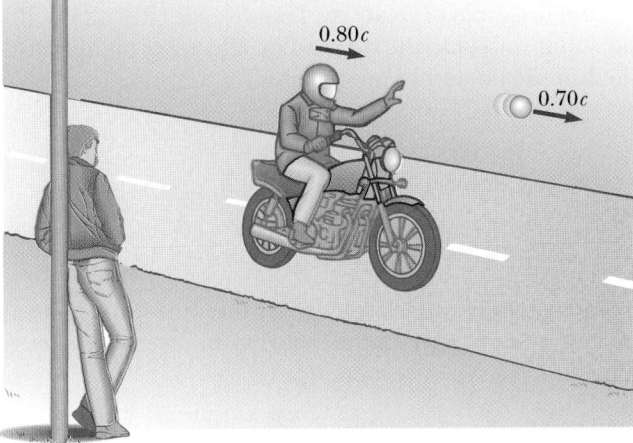

Figure 26.13 A motorcycle moves past a stationary observer with a speed of $0.80c$; the motorcyclist throws a ball in the direction of motion with a speed of $0.70c$ relative to himself.

The speed of light is the speed limit of the Universe.

Thus,

$$v_{bo} = \frac{v_{bm} + v_{mo}}{1 + \frac{v_{bm}v_{mo}}{c^2}} = \frac{0.70c + 0.80c}{1 + \frac{(0.70c)(0.80c)}{c^2}} = 0.96c$$

EXAMPLE 26.6 Measuring the Speed of a Light Beam

Suppose that the motorcyclist moving with a speed of $0.80c$ turns on beam of light that moves away from the motorcycle with a speed of c in the same direction as the moving motorcycle. What speed would the stationary observer measure for the beam of light?

Solution   In this case, we have

v_{lm} = the velocity of the light with respect to the motorcycle = c
v_{mo} = the velocity of the motorcycle with respect to the stationary observer = $0.80c$

and we want

v_{lo} = the velocity of the light with respect to the stationary observer

Thus,

$$v_{lo} = \frac{v_{lm} + v_{mo}}{1 + \frac{v_{lm}v_{mo}}{c^2}} = \frac{c + 0.80c}{1 + \frac{(c)(0.80c)}{c^2}} = c$$

This is consistent with the statement made earlier that **all observers measure the speed of light to be c regardless of the motion of the source of light.**

26.9   RELATIVISTIC ENERGY

We have seen that the definition of momentum required generalization to make it compatible with the principle of relativity. Likewise, the definition of kinetic energy requires modification in relativistic mechanics. Einstein found that the correct expression for the **kinetic energy** of an object is

Kinetic energy ▶

$$KE = \gamma mc^2 - mc^2 \tag{26.12}$$

The constant term mc^2 in Equation 26.12, which is independent of the speed of the object, is called the **rest energy** of the object, E_R.

Rest energy ▶

$$E_R = mc^2 \tag{26.13}$$

The term γmc^2 in Equation 26.12 depends on the object speed and is the sum of the kinetic and rest energies. We define γmc^2 to be the **total energy, E**—that is, total energy = kinetic energy + rest energy, or

$$E = \gamma mc^2 = KE + mc^2 \tag{26.14}$$

Because $\gamma = (1 - v^2/c^2)^{-1/2}$, we can express E as

Total energy ▶

$$E = \frac{mc^2}{\sqrt{1 - v^2/c^2}} \tag{26.15}$$

This, of course, is Einstein's famous mass-energy equivalence equation. The relation $E = \gamma mc^2 = \gamma E_R$ shows that **mass is one possible manifestation of energy.** Furthermore, this result shows that a small mass corresponds to an enormous amount of energy. This concept is fundamental to much of the field of nuclear physics.

In many situations, the momentum or energy of a particle is measured rather than its speed. It is therefore useful to have an expression relating the total energy E to the relativistic momentum p. This is accomplished by using the expressions $E = \gamma mc^2$ and $p = \gamma mv$. By squaring these equations and subtracting, we can eliminate v. The result, after some algebra, is

$$E^2 = p^2 c^2 + (mc^2)^2 \qquad [26.16]$$

When the particle is at rest, $p = 0$, and so $E = E_R = mc^2$. That is, the total energy equals the rest energy. For the case of particles that have zero mass, such as photons (massless, chargeless particles of light), we set $m = 0$ in Equation 26.16, and we see that

$$E = pc \qquad [26.17]$$

This equation is an exact expression relating energy and momentum for photons, which always travel at the speed of light.

Finally, note that because the mass m of a particle is independent of its motion, m must have the same value in all reference frames. For this reason, m is often called the **invariant mass.** On the other hand, the total energy and momentum of a particle depend on the reference frame in which they are measured, because they both depend on velocity. Because m is a constant, according to Equation 26.16 the quantity $E^2 - p^2 c^2$ must have the same value in all reference frames.

When dealing with subatomic particles, it is convenient to express their energy in electron volts (eV), because the particles are usually given this energy by acceleration through an electrostatic potential difference. The conversion factor is

$$1 \text{ eV} = 1.60 \times 10^{-19} \text{ J}$$

For example, the mass of an electron is 9.11×10^{-31} kg. Hence, the rest energy of the electron is

$$m_e c^2 = (9.11 \times 10^{-31} \text{ kg})(3.00 \times 10^8 \text{ m/s})^2 = 8.20 \times 10^{-14} \text{ J}$$

Converting this to eV, we have

$$m_e c^2 = (8.20 \times 10^{-14} \text{ J})(1 \text{ eV}/1.60 \times 10^{-19} \text{ J}) = 0.511 \text{ MeV}$$

Thinking Physics 3

A common principle learned in chemistry is conservation of mass. In practice, if the mass of the reactants is measured before a reaction and the mass of the products is measured afterward, the results will be the same. In light of special relativity, should we stop teaching the principle of conservation of mass in chemistry classes?

Explanation Consider a reaction that does not require energy input to occur. This type of reaction occurs because the products represent a lower overall rest energy than the reactants; the difference in rest energy is carried as kinetic energy of ejected particles or radiation. Because the rest energy of the reactants is smaller, according to relativity the mass of the reactants should be smaller than that of the products. Thus the law of conservation of mass is violated. The mass changes are so small, however, that in practice the law of conservation of mass is still useful.

EXAMPLE 26.7 The Energy Contained in a Baseball

If a 0.50-kg baseball could be converted completely to energy of forms other than mass, how much energy of other forms would be released?

Solution The energy equivalent of the baseball is found from Equation 26.14 (with $KE = 0$):

$$E = E_R = mc^2 = (0.50 \text{ kg})(3.0 \times 10^8 \text{ m/s})^2 = \boxed{4.5 \times 10^{16} \text{ J}}$$

This is enough energy to keep a 100-W lightbulb burning for approximately ten million years. However, it is generally impossible to achieve complete conversion from mass to energy of other forms. For example, mass is converted to energy in nuclear power plants, but only a small fraction of the mass actually undergoes conversion.

EXAMPLE 26.8 The Energy of a Speedy Electron

An electron moves with a speed of $v = 0.850c$. Find its total energy and kinetic energy in electron volts.

Solution The fact that the rest energy of an electron is 0.511 MeV, along with Equation 26.15, gives

$$E = \frac{m_e c^2}{\sqrt{1 - v^2/c^2}} = \frac{0.511 \text{ MeV}}{\sqrt{1 - \dfrac{(0.850c)^2}{c^2}}}$$

$$= 1.90(0.511 \text{ MeV}) = \boxed{0.970 \text{ MeV}}$$

The kinetic energy is obtained by subtracting the rest energy from the total energy:

$$KE = E - m_e c^2 = 0.970 \text{ MeV} - 0.511 \text{ MeV} = \boxed{0.459 \text{ MeV}}$$

EXAMPLE 26.9 The Energy of a Speedy Proton

The total energy of a proton is three times its rest energy.

(a) Find the proton's rest energy in electron volts.

Solution The rest energy is given by Equation 26.13:

$$E_R = m_p c^2 = (1.67 \times 10^{-27} \text{ kg})(3.00 \times 10^8 \text{ m/s})^2$$

$$= (1.50 \times 10^{-10} \text{ J}) \left(\frac{1 \text{ eV}}{1.60 \times 10^{-19} \text{ J}} \right) = \boxed{938 \text{ MeV}}$$

(b) With what speed is the proton moving?

Solution Because the total energy, E, is three times the rest energy, Equation 26.14 gives

$$E = \gamma m_p c^2 = 3m_p c^2 = \frac{m_p c^2}{\sqrt{1 - v^2/c^2}}$$

$$3 = \frac{1}{\sqrt{1 - v^2/c^2}}$$

Solving for v gives

$$1 - \frac{v^2}{c^2} = \frac{1}{9}$$

$$\frac{v^2}{c^2} = \frac{8}{9}$$

$$v = \frac{\sqrt{8}}{3} c = \boxed{2.83 \times 10^8 \text{ m/s}}$$

(c) Determine the kinetic energy of the proton in electron volts.

Solution

$$KE = E - m_p c^2 = 3m_p c^2 - m_p c^2 = 2m_p c^2$$

Because $m_p c^2 = 938$ MeV, $KE = \boxed{1880 \text{ MeV}}$

26.10 GENERAL RELATIVITY

Optional Section

Up to this point, we have sidestepped a curious puzzle. Mass has two seemingly different properties: a *gravitational attraction* for other masses and an *inertial property* that resists acceleration. To designate these two attributes, we use the subscripts g and i and write

Gravitational property $\quad F_g = m_g a$
Inertial property $\quad F_i = m_i a$

The value for the gravitational constant G was chosen to make the magnitudes of m_g and m_i numerically equal. Regardless of how G is chosen, however, the strict proportionality of m_g and m_i has been established experimentally to an extremely high degree: a few parts in 10^{12}. Thus, it appears that gravitational mass and inertial mass may indeed be exactly proportional.

But why? They seem to involve two entirely different concepts: a force of mutual gravitational attraction between two masses and the resistance of a single mass to being accelerated. This question, which puzzled Newton and many other physicists over the years, was answered when Einstein published his theory of gravitation, known as *general relativity*, in 1916. Because it is a mathematically complex theory, we merely offer a hint of its elegance and insight.

In Einstein's view, the remarkable coincidence that m_g and m_i seemed to be exactly proportional was evidence for a very intimate and basic connection between the two concepts. He pointed out that no mechanical experiment (such as dropping a mass) could distinguish between the two situations illustrated in Figures 26.14a and 26.14b. In each case, a mass released by the observer undergoes a downward acceleration of *g* relative to the floor.

Einstein carried this idea further and proposed that *no* experiment, mechanical or otherwise, could distinguish between the two cases. This extension to include all phenomena (not just mechanical ones) has interesting consequences. For example, suppose that a light pulse is sent horizontally across the box, as in Figure 26.14c. The trajectory of the light pulse bends downward as the box accelerates upward to meet it. Einstein proposed that a beam of light should also be bent downward by a gravitational field. (No such bending is predicted in Newton's theory of gravitation.)

The two postulates of Einstein's **general relativity** are as follows:

1. All the laws of nature have the same form for observers in any frame of reference, whether accelerated or not.
2. In the vicinity of any given point, a gravitational field is equivalent to an accelerated frame of reference in the absence of gravitational effects. (This is the *principle of equivalence*.)

The second postulate implies that gravitational mass and inertial mass are completely equivalent, not just proportional. What were thought to be two different types of mass are actually identical.

One interesting effect predicted by general relativity is that time scales are altered by gravity. A clock in the presence of gravity runs more slowly than one where gravity is negligible. As a consequence, the frequencies of radiation emitted by atoms in the presence of a strong gravitational field are shifted to lower frequencies when compared with the same emissions in a weak field. This gravitational shift has been detected in spectral lines emitted by atoms in massive stars. It has also been verified on the Earth by comparing the frequencies of gamma rays emitted from nuclei separated vertically by about 20 m.

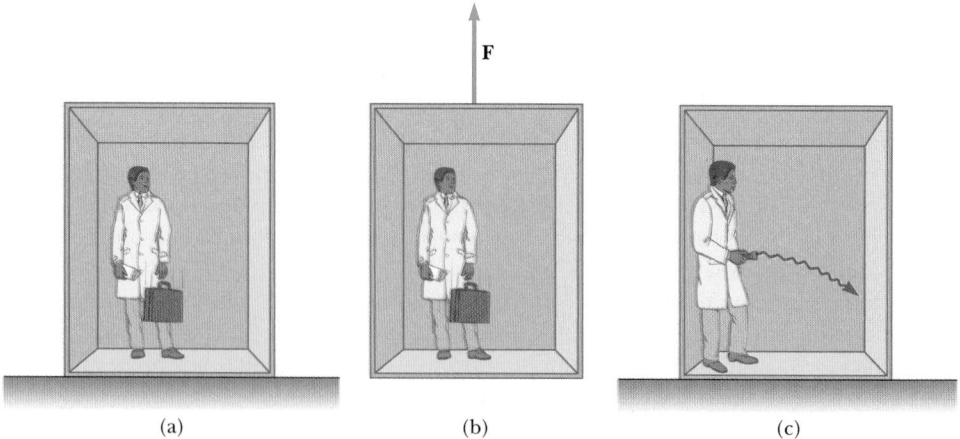

(a)   (b)   (c)

Figure 26.14   (a) The observer is at rest in a uniform gravitational field **g**. (b) The observer is in a region in which gravity is negligible, but the frame of reference is accelerated by an external force **F** that produces an acceleration **g**. According to Einstein, the frames of reference in parts (a) and (b) are equivalent in every way. No local experiment could distinguish any difference between the two frames. (c) If parts (a) and (b) are truly equivalent, as Einstein proposed, then a ray of light would bend in a gravitational field.

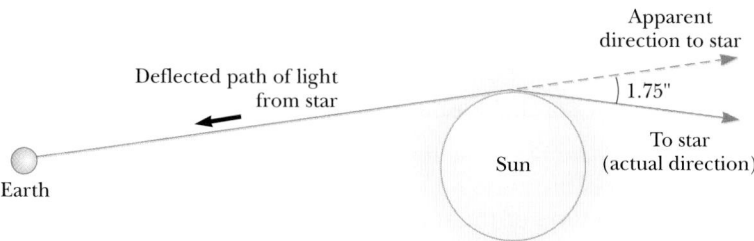

Figure 26.15 Deflection of starlight passing near the Sun. Because of this effect, the Sun and other remote objects can act as a *gravitational lens*. In his general theory of relativity, Einstein calculated that starlight just grazing the Sun's surface should be deflected by an angle of 1.75″.

The second postulate suggests that a gravitational field may be "transformed away" at any point if we choose an appropriate accelerated frame of reference—a freely falling one. Einstein developed an ingenious method of describing the acceleration necessary to make the gravitational field "disappear." He specified a certain quantity, the *curvature of space-time*, that describes the gravitational effect at every point. In fact, the curvature of space-time completely replaces Newton's gravitational theory. According to Einstein, there is no such thing as a gravitational force. Rather, the presence of a mass causes a curvature of space-time in the vicinity of the mass, and this curvature dictates the space-time path that all freely moving objects must follow. As one physicist says, "Mass one tells space-time how to curve; curved space-time tells mass two how to move." One important test of general relativity is the prediction that a light ray passing near the Sun should be deflected by some angle. This prediction was confirmed by astronomers as bending of starlight during a total solar eclipse shortly following World War I (Fig. 26.15).

If the concentration of mass becomes very great, as is believed to occur when a large star exhausts its nuclear fuel and collapses to a very small volume, a **black hole** may form. Here the curvature of space-time is so extreme that, within a certain distance from the center of the black hole, all matter and light become trapped.

Thinking Physics 4

Atomic clocks are extremely accurate; in fact an error of 1 second in 3 million years is typical. This error can be described as about one part in 10^{14}. On the other hand, the atomic clock in Boulder, Colorado, is often 15 ns faster than the one in Washington after only one day. This is an error of about one part in 6×10^{12}, which is about 17 times larger than the previously expressed error. If atomic clocks are so accurate, why does a clock in Boulder not remain in synchronism with one in Washington? (*Hint:* Denver, near Boulder, is known as the Mile High City.)

Explanation According to the general theory of relativity, the rate of passage of time depends on gravity—time runs more slowly in strong gravitational fields. Washington is at an elevation very close to sea level, whereas Boulder is about a mile higher in altitude. This will result in a weaker gravitational field at Boulder than at Washington. As a result, time runs more rapidly in Boulder than in Washington.

SUMMARY

The two basic postulates of the **special theory of relativity** are as follows:

1. The laws of physics are the same in all inertial frames of reference.
2. The speed of light is the same for all inertial observers, independent of their motion or of the motion of the source of light.

Some of the consequences of the special theory of relativity are as follows:

1. Clocks in motion relative to an observer slow down. This is known as **time dilation.** The relationship between time intervals in the moving and at-rest systems is

$$\Delta t = \gamma \, \Delta t_p \qquad [26.8]$$

where Δt is the time interval measured in the system in relative motion with respect to the clock, $\gamma = 1/\sqrt{1 - v^2/c^2}$, and Δt_p is the proper time interval measured in the system moving with the clock.

2. The length of an object in motion is *contracted* in the direction of motion. The equation for **length contraction** is

$$L = L_p\sqrt{1 - v^2/c^2} \qquad [26.9]$$

where L is the length measured in the system in motion relative to the object, and L_p is the proper length measured in the system in which the object is at rest.

3. Events that are simultaneous for one observer are not simultaneous for another observer in motion relative to the first.

The relativistic expression for the **momentum** of a particle moving with a velocity v is

$$p \equiv \frac{mv}{\sqrt{1 - v^2/c^2}} = \gamma mv \qquad [26.10]$$

The relativistic expression for the addition of velocities is

$$v_{ab} = \frac{v_{ad} + v_{db}}{1 + \dfrac{v_{ad}v_{db}}{c^2}} \qquad [26.11]$$

where v_{ab} is the velocity of object a with respect to object b, v_{ad} is the velocity of object a with respect to object d, and so forth.

The relativistic expression for the **kinetic energy** of an object is

$$KE = \gamma mc^2 - mc^2 \qquad [26.12]$$

where mc^2 is the **rest energy** of the object, E_R.

The **total energy** of a particle is

$$E = \frac{mc^2}{\sqrt{1 - v^2/c^2}} \qquad [26.15]$$

This is Einstein's famous mass-energy equivalence equation.

The relativistic momentum is related to the total energy through the equation

$$E^2 = p^2c^2 + (mc^2)^2 \qquad [26.16]$$

MULTIPLE-CHOICE QUESTIONS

1. Which has the greatest momentum: a 1 MeV photon, or a proton or an electron with kinetic energy 1 MeV?
(a) the photon　(b) the proton　(c) the electron
(d) the electron and the proton　(e) They are all the same.

2. An electron with a kinetic energy $2mc^2$ undergoes a head-on collision with another electron also with a kinetic energy of $2mc^2$. What is the kinetic energy of one of these electrons as viewed from the other electron just before the collision?
(a) mc^2　(b) $2mc^2$　(c) $4mc^2$　(d) $8mc^2$　(e) $16mc^2$

3. A mass-spring system moving with simple harmonic motion has a period T when measured by a ground observer. If the same system is then placed in an inertial frame of reference that moves past the ground observer at a speed of $0.50c$, by what factor should T be multiplied to give the system's period as measured by the ground observer?
(a) 0.50　(b) 0.87　(c) 1.0　(d) 1.2

4. A spacecraft was originally 100 m long. However, it is now moving toward a tunnel with a speed of $0.8c$. The lady living near the tunnel can control doors that open and shut at each end of the tunnel and she has found that the tunnel is 65 m long. The doors are open as the spacecraft approaches but, the moment that the back of the spacecraft is in the tunnel, she closes both doors and then opens the doors very quickly. According to the captain on the spacecraft,
(a) No door hit his spacecraft because the doors weren't closed simultaneously.
(b) No door hit his spacecraft because he finds that length contraction makes his spacecraft only 60 m long.
(c) No door hits the spacecraft because length contraction has made the tunnel 108.7 m long.
(d) A door hits his spacecraft.

5. The power output of the Sun is 3.7×10^{26} W. How much matter is converted into energy in the Sun every second?
(a) 4.1×10^9 kg/s　(b) 6.3×10^9 kg/s
(c) 7.4×10^9 kg/s　(d) 3.7×10^9 kg/s

CONCEPTUAL QUESTIONS

1. You are in a speedboat on a lake. You see ahead of you a wavefront, caused by the previous passage of another boat, moving away from you. You accelerate, catch up with, and pass the wavefront. Is this scenario possible if you are in a rocket and you detect a wavefront of light ahead of you?

2. What two speed measurements will two observers in relative motion always agree on?

3. Suppose astronauts were paid according to time spent traveling in space. After a long voyage traveling at a speed near that of light, astronauts return to Earth and open their pay envelopes. What will be their reaction?

4. Consider the incorrect statement, "Matter can neither be created nor destroyed." How would you correct this statement in view of the special theory of relativity?

5. You are packing for a trip to another star, to which you will be traveling at $0.99c$. Should you buy smaller sizes of your clothing, because you will be skinnier on the trip? Can you sleep in a smaller cabin than usual, because you will be shorter when you lie down?

6. It is said that Einstein, in his teenage years, asked the question, "What would I see in a mirror if I carried it in my hands and ran at a speed near that of light?" How would you answer this question?

7. You are observing a rocket moving away from you. You notice that it is measured to be shorter than when it was at rest on the ground next to you, and through the rocket window, you can see a clock. You observe that the passage of time on the clock is measured to be slower than that of the watch on your wrist. What if the rocket turns around and comes toward you? Will it appear to be longer and will the rocket-bound clock move faster?

8. Two identically constructed clocks are synchronized. One is put in orbit around the Earth and the other remains on Earth. Which clock runs more slowly? When the moving clock returns to Earth, will the two clocks still be synchronized?

9. A photon has a zero mass. If a photon is reflected from a surface, does it exert a force on the surface?

10. Imagine an astronaut on a trip to Sirius, which is 8 light-years from the Earth. On arrival at Sirius, the astronaut finds that the trip lasted 6 years. If the trip was made at a constant speed of $0.8c$, how can the 8-lightyear distance be reconciled with the 6-year duration?

11. Explain why it is necessary, when defining length, to specify that the positions of the ends of a rod are to be measured simultaneously.

12. The equation $E = mc^2$ is often given in popular descriptions of Einstein's theory of relativity. Is this expression strictly correct? For example, does it accurately account for the kinetic energy of a moving mass?

13. Give a physical argument that shows that it is impossible to accelerate an object m to the speed of light, even with a continuous force acting on it.

14. Some distant star-like objects, called *quasars*, are receding from us at half the speed of light (or greater). What is the speed of the light we receive from these quasars?

15. List some ways our day-to-day lives would change if the speed of light were only 50 m/s.

PROBLEMS

1, 2, 3 = straightforward, intermediate, challenging   ☐ = full solution available in Study Guide/Student Solutions Manual   ◩ = Core Concepts Workbook
WEB = solution posted at **http://www.harcourtcollege.com/physics/cptech**   ◪ = biomedical application   ◩ = Interactive Physics

Review Problem

If 3.00 moles of a monatomic ideal gas are heated at constant volume so that the temperature of the gas rises 900°F, how much does the mass of the gas increase?

Section 26.4 The Michelson–Morley Experiment

1. Two airplanes fly paths I and II, specified in Figure 26.5a. Both planes have air speeds of 100 m/s and fly a distance $L = 200$ km. The wind blows at 20.0 m/s in the direction shown in the figure. Find **(a)** the time of flight to each city, **(b)** the time to return, and **(c)** the difference in total flight times.

2. In one version of the Michelson–Morley experiment, the lengths L in Figure 26.6 were 28 m. Take v to be 3.0×10^4 m/s and find **(a)** the time difference caused by rotation of the interferometer and **(b)** the expected fringe shift, assuming that the light used has a wavelength of 550 nm.

Section 26.6 Consequences of Special Relativity

3. A deep-space probe moves away from Earth with a speed of $0.80c$. An antenna on the probe requires 3.0 s probe time to rotate through 1.0 rev. How much time is required for 1.0 rev according to an observer on Earth?

4. An astronaut at rest on Earth has a heartbeat rate of 70 beats/min. When the astronaut is traveling in a spaceship at $0.90c$, what will this rate be as measured by **(a)** an observer also in the ship and **(b)** an observer at rest on the Earth?

5. The average lifetime of a pi meson in its own frame of reference (i.e., the proper lifetime) is 2.6×10^{-8} s. If the meson moves with a speed of $0.98c$, what is **(a)** its mean lifetime as measured by an observer on Earth and **(b)** the average distance it travels before decaying as measured by an observer on Earth? **(c)** What distance would it travel if time dilation did not occur?

6. If astronauts could travel at $v = 0.950c$, we on Earth would say it takes $(4.20/0.950) = 4.42$ years to reach Alpha Centauri, 4.20 lightyears away. The astronauts dis-

agree. **(a)** How much time passes on the astronaut's clocks? **(b)** What is the distance to Alpha Centauri as measured by the astronauts?

WEB **7.** A muon formed high in the Earth's atmosphere travels at speed $v = 0.99c$ for a distance of 4.6 km before it decays into an electron, a neutrino, and an antineutrino $(\mu^- \rightarrow e^- + \nu + \bar{\nu})$. **(a)** How long does the muon live, as measured in its reference frame? **(b)** How far does the muon travel, as measured in its frame?

8. A friend in a spaceship travels past you at a high speed. He tells you that his ship is 20 m long and that the identical ship you are sitting in is 19 m long. According to your observations, **(a)** how long is your ship, **(b)** how long is his ship, and **(c)** what is the speed of your friend's ship?

9. A box is cubical with sides of proper lengths $L_1 = L_2 = L_3 = 2.0$ m, as shown in Figure P26.9, when viewed in its own rest frame. If this block moves parallel to one of its edges with a speed of $0.80c$ past an observer, **(a)** what shape does it appear to have to this observer, and **(b)** what is the length of each side as measured by this observer?

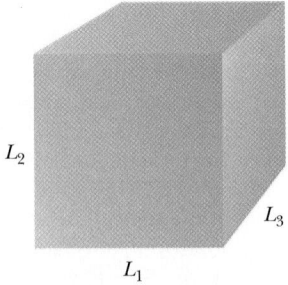

L_2
L_3
L_1

Figure P26.9

10. With what speed must a clock move in order to run at a rate that is one half the rate of a clock at rest?

11. The proper length of one spaceship is three times that of another. The two spaceships are traveling in the same direction and, while both are passing overhead, an Earth observer measures the two spaceships to have the same length. If the slower spaceship is moving with a speed of $0.35c$, determine the speed of the faster spaceship.

12. A supertrain of proper length 100 m travels at a speed of $0.95c$ as it passes through a tunnel having proper length 50 m. As seen by a trackside observer, is the train ever completely within the tunnel? If so, by how much?

13. An observer, moving at a speed of $0.995c$ relative to a rod (Fig. P26.13), measures its length to be 2.00 m and sees its length to be oriented at $30.0°$ with respect to the direction of motion. (a) What is the proper length of the rod? (b) What is the orientation angle in a reference frame moving with the rod?

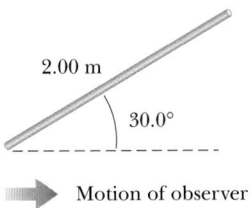

2.00 m

$30.0°$

Motion of observer

Figure P26.13   View of rod as seen by an observer moving to the right.

14. Observer A measures the length of two rods, one stationary, the other moving with a speed of $0.955c$. She finds that the rods have the same length, L_0. A second observer B travels along with the moving rod. What is the ratio of the length of A's rod to the length of B's rod according to observer B?

Section 26.7   Relativistic Momentum

15. An electron has a speed $v = 0.90c$. At what speed will a proton have a momentum equal to that of the electron?

16. Calculate the momentum of an electron moving with a speed of (a) $0.010c$, (b) $0.50c$, (c) $0.90c$.

17. Show that the speed of an object having momentum p and mass m is given by

$$v = \frac{c}{\sqrt{1 + (mc/p)^2}}$$

18. An unstable particle at rest breaks up into two fragments of *unequal mass*. The mass of the lighter fragment is 2.50×10^{-28} kg, and that of the heavier fragment is 1.67×10^{-27} kg. If the lighter fragment has a speed of $0.893c$ after the breakup, what is the speed of the heavier fragment?

WEB 19. The nonrelativistic expression for the momentum of a particle, $p = mv$, can be used if $v \ll c$. For what speed does the use of this formula give an error in the momentum of (a) 1.00% and (b) 10.0%?

Section 26.8   Relativistic Addition of Velocities

20. An electron moves to the right with a speed of $0.90c$ relative to the laboratory frame. A proton moves to the left with a speed of $0.70c$ relative to the electron. Find the speed of the proton relative to the laboratory frame.

21. Spaceship R is moving to the right at a speed of $0.70c$ with respect to the Earth. A second spaceship, L, moves to the left at the same speed with respect to the Earth. What is the speed of L with respect to R?

22. A space vehicle is moving at a speed of $0.75c$ with respect to an external observer. An atomic particle is projected at $0.90c$ in the same direction as the spaceship's velocity with respect to an observer inside the vehicle. What is the speed of the projectile as seen by the external observer?

23. A rocket moves with a velocity of $0.92c$ to the right with respect to a stationary observer A. An observer B moving relative to observer A finds that the rocket is moving with a velocity of $0.95c$ to the left. What is the velocity of observer B relative to observer A? (*Hint:* Consider observer B's velocity in the frame of reference of the rocket.)

24. A pulsar is a stellar object that emits light in short bursts. Suppose a pulsar with a speed of $0.950c$ approaches the Earth, and a rocket with a speed of $0.995c$ heads toward the pulsar (both speeds measured in the Earth's frame of reference). If the pulsar emits 10.0 pulses per second in its own frame of reference, at what rate are the pulses emitted in the rocket's frame of reference?

25. Spaceship I, which contains students taking a physics exam, approaches Earth with a speed of $0.60c$, while spaceship II, which contains instructors proctoring the exam, moves away from Earth at $0.28c$ as in Figure P26.25. If the instructors in spaceship II stop the exam after 50 min have passed on *their clock*, how long does the exam last as measured by (a) the students? (b) an observer on Earth?

I

II

$0.60c$

$0.28c$

Figure P26.25

Section 26.9   Relativistic Energy

26. A proton moves with a speed of $0.950c$. Calculate its (a) rest energy, (b) total energy, and (c) kinetic energy.

27. A mass of 0.50 kg is converted completely into energy of other forms. (a) How much energy of other forms is produced and (b) how long will this much energy keep a 100-W lightbulb burning?

28. The Sun radiates approximately 4.0×10^{26} J of energy into space each second. (a) How much mass is converted into energy of other forms each second? (b) If the mass of the Sun is 2.0×10^{30} kg, how long can the Sun survive if the energy transformation continues at the present rate?

29. What is the speed of a particle whose kinetic energy is equal to its own rest energy?

30. A proton in a high-energy accelerator is given a kinetic energy of 50.0 GeV. Determine (a) the momentum and (b) the speed of the proton.

31. In a color television tube, electrons are accelerated through a potential difference of 20 000 volts. With what speed do the electrons strike the screen?

32. What speed must a particle attain before its kinetic energy is double the value predicted by the nonrelativistic expression $KE = \frac{1}{2}mv^2$?

WEB 33. An unstable particle with a mass equal to 3.34×10^{-27} kg is initially at rest. The particle decays into two fragments that fly off with velocities of $0.987c$ and $-0.868c$. Find the masses of the fragments. (*Hint:* Conserve both mass-energy and momentum.)

34. If it takes 3750 MeV of work to accelerate a proton from rest to a speed of v, determine v.

ADDITIONAL PROBLEMS

35. Determine the energy required to accelerate an electron from (a) $0.500c$ to $0.750c$ and (b) $0.900c$ to $0.990c$.

36. How fast must a meter stick be moving if its length is observed to shrink to 0.500 m?

37. What is the speed of a proton that has been accelerated from rest through a difference of potential of (a) 500 V and (b) 5.00×10^8 V?

38. An electron has a total energy equal to five times its rest energy. (a) What is its momentum? (b) Repeat for a proton.

39. What is the momentum (in units of MeV/c) of an electron with a kinetic energy of 1.00 MeV?

40. An alarm clock is set to sound in 10 h. At $t = 0$ the clock is placed in a spaceship moving with a speed of $0.75c$ (relative to the Earth). What distance, as determined by an Earth observer, does the spaceship travel before the alarm clock sounds?

41. At what speed must an electron move for its energy to equal a proton's rest energy?

42. A radioactive nucleus moves with a speed of v relative to a laboratory observer. The nucleus emits an electron in the positive x direction with a speed of $0.70c$ relative to the decaying nucleus and a speed of $0.85c$ in the $+x$ direction relative to the laboratory observer. What is the value of v?

43. A certain quasar recedes from the Earth at $v = 0.870c$. A jet of material ejected from the quasar back toward the Earth moves at $0.550c$ relative to the quasar. Find the speed of the ejected material relative to the Earth.

44. A spaceship of proper length 300 m takes 0.75 μs to pass an Earth observer. Determine the speed of this spaceship as measured by the Earth observer.

45. Find the kinetic energy of a 78.0-kg spacecraft launched out of the Solar System with speed 106 km/s by using (a) the classical equation $KE = \frac{1}{2}mv^2$ and (b) the relativistic equation.

46. A physics professor on Earth gives an exam to her students who are on a rocketship traveling at speed of v with respect to Earth. The moment the ship passes the professor, she signals the start of the exam. If she wishes her students to have T_0 (rocket time) to complete the exam, show that she should wait a time of

$$T = T_0 \sqrt{\frac{1 - v/c}{1 + v/c}}$$

(Earth time) before sending a light signal telling them to stop. (*Hint:* Remember that it takes some time for the second light signal to travel from the professor to the students.)

47. Imagine that the entire Sun collapses to a sphere of radius R_g such that the work required to remove a small mass m from the surface would be equal to its rest energy mc^2. This radius is called the *gravitational radius* for the Sun. Find R_g. (It is believed that the ultimate fate of many stars is to collapse to their gravitational radii or smaller.)

48. A rod of length L_0 moves with a speed of v along the horizontal direction. The rod makes an angle of θ_0 with respect to the axis of a coordinate system moving with the rod. (a) Show that the length of the rod as measured by a stationary observer is given by

$$L = L_0 \left[1 - \left(\frac{v^2}{c^2} \right) \cos^2 \theta_0 \right]^{1/2}$$

(b) Show that the angle the rod makes with the axis as seen by the stationary observer is given by the expression $\tan \theta = \gamma \tan \theta_0$. These results show that the rod is both contracted and rotated. (Take the lower end of the rod to be at the origin of the moving coordinate system.)

49. Ted and Mary are playing a game of catch in frame S$'$, which is moving with a speed of $0.60c$; Jim in frame S is watching (Fig. P26.49). Ted throws the ball to Mary with a speed of $0.80c$ (according to Ted) and their separation (measured in S$'$) is 1.80×10^{12} m. (a) According to

Mary, how fast is the ball moving? **(b)** According to Mary, how long will it take the ball to reach her? **(c)** According to Jim, how far apart are Ted and Mary, and how fast is the ball moving?

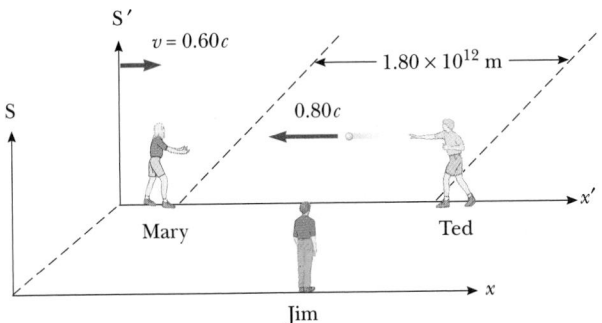

Figure P26.49

WEB 50. **(a)** Show that a potential difference of 1.02×10^6 V would be sufficient to give an electron a speed equal to twice the speed of light if Newtonian mechanics remained valid at high speeds. **(b)** What speed would an electron actually acquire in falling through a potential difference of 1.02×10^6 V?

51. Consider two inertial reference frames, S and S′, where S′ is moving to the right with constant speed $0.60c$ as measured by observers in S. Jennifer is located 1.80×10^{11} m to the right of the origin of S and is fixed in S (as measured by observers in S), and Matt is fixed

in S′ at the origin in S′ (as measured by observers in S′). At the instant their origins coincide, Matt throws a ball toward Jennifer at constant speed $0.80c$ as measured by Matt (Fig. P26.51). **(a)** What is the speed of the ball as measured by Jennifer? How long before Jennifer catches the ball, as measured by **(b)** Jennifer, **(c)** the ball, and **(d)** Matt?

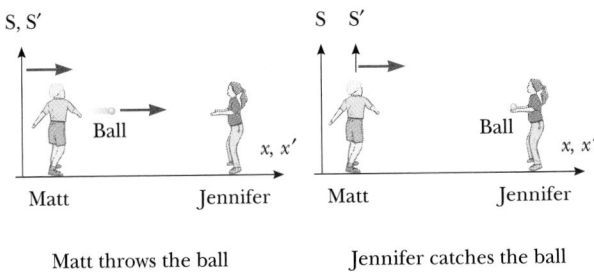

Matt throws the ball   Jennifer catches the ball

(a)   (b)

Figure P26.51

52. The muon is an unstable particle that spontaneously decays into an electron and two neutrinos. If the number of muons at $t = 0$ is N_0, the number at time t is given by $N = N_0 e^{-t/\tau}$ where τ is the mean lifetime, equal to 2.2 μs. Suppose that the muons move at a speed of $0.95c$ and that there are 5.0×10^4 muons at $t = 0$. **(a)** What is the observed lifetime of the muons? **(b)** How many muons remain after traveling a distance of 3.0 km?

Quantum Physics

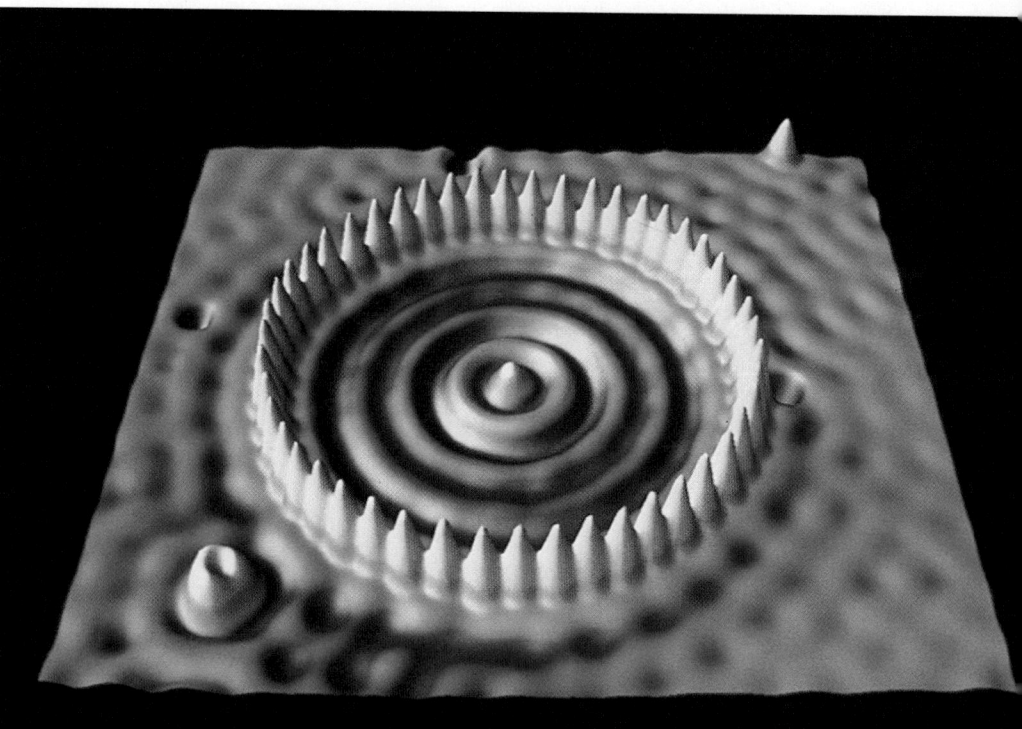

▲ **PHYSICS PUZZLER**

This is a photograph of a "quantum-corral" consisting of a ring of 48 iron atoms located on a copper surface. The diameter of the ring is 143 nm, and the photograph was obtained using a low-temperature scanning tunneling microscope (STM). Corrals and other structures are able to confine surface-electron waves. What role do you think such structures will play in the future development of small electronic devices? *(IBM Corporation Research Division)*

Although many problems were indeed resolved by the theory of relativity in the early part of the 20th century, many other problems remained unsolved. Attempts to explain the behavior of matter on the atomic level with the laws of classical physics were consistently unsuccessful. Various phenomena, such as blackbody radiation, the photoelectric effect, and the emission of sharp spectral lines by atoms in a gas discharge tube, could not be understood within the framework of classical physics. Between 1900 and 1930, however, a modern version of mechanics called *quantum mechanics* was highly successful in explaining the behavior of atoms, molecules, and nuclei. Moreover, the quantum theory reduces to classical physics when applied to macroscopic systems. As with relativity, the quantum theory requires a modification of our ideas concerning the physical world.

The earliest and most basic ideas of quantum theory were introduced by Planck, and most of the subsequent mathematical developments, interpretations, and improvements were made by a number of distinguished physicists, including Einstein, Bohr, Schrödinger, de Broglie, Heisenberg, Born, and Dirac. An extensive study of quantum theory is certainly beyond the scope of this book. This chapter is simply an introduction to the underlying ideas of quantum theory and the wave-particle nature of matter. We also discuss some simple applications of quantum theory, including the photoelectric effect, the Compton effect, and x-rays.

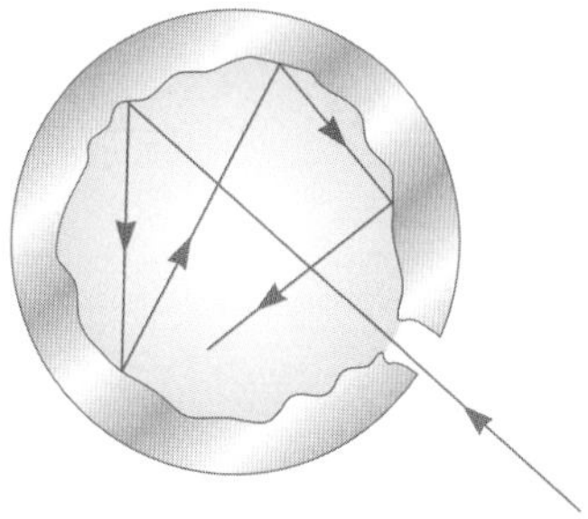

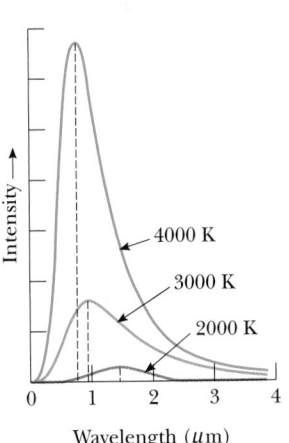

Figure 27.1 The opening in the cavity of a body is a good approximation of a blackbody. As light enters the cavity through the small opening, part is reflected and part is absorbed on each reflection from the interior walls. After many reflections, essentially all of the incident energy is absorbed.

27.1 BLACKBODY RADIATION AND PLANCK'S HYPOTHESIS

An object at any temperature is known to emit radiation that is sometimes referred to as **thermal radiation.** Stefan's law, which we discussed in Section 11.7, describes the total power radiated as electromagnetic waves. The spectrum of the radiation depends on the temperature and properties of the object. At low temperatures, the wavelengths of the thermal radiation are mainly in the infrared region and hence not observable by the eye. As the temperature of an object increases, the object eventually begins to glow red. At sufficiently high temperatures, it appears to be white, as in the glow of the hot tungsten filament of a lightbulb. A careful study of thermal radiation shows that it consists of a continuous distribution of wavelengths from the infrared, visible, and ultraviolet portions of the spectrum.

From a classical viewpoint, thermal radiation originates from accelerated charged particles near the surface of an object; those charges emit radiation much as small antennas do. The thermally agitated charges can have a distribution of accelerations, which accounts for the continuous spectrum of radiation emitted by the object. By the end of the 19th century, it had become apparent that the classical theory of thermal radiation was inadequate. The basic problem was in understanding the observed distribution of wavelengths in the radiation emitted by a blackbody. By definition, a blackbody is an ideal system that absorbs *all* radiation incident on it. A good approximation of a blackbody is a small hole leading to the inside of a hollow object, as shown in Figure 27.1. The nature of the radiation emitted through the small hole leading to the cavity depends only on the temperature of the cavity walls.

Experimental data for the distribution of energy in blackbody radiation at three temperatures are shown in Figure 27.2. The radiated energy varies with wavelength and temperature. As the temperature of the blackbody increases, the total amount

Figure 27.2 Intensity of blackbody radiation versus wavelength at three different temperatures. Note that the total radiation emitted (the area under a curve) increases with increasing temperature.

of energy it emits increases. Also, with increasing temperature, the peak of the distribution shifts to shorter wavelengths. This shift was found to obey the following relationship, called **Wien's displacement law:**

$$\lambda_{max} T = 0.2898 \times 10^{-2} \text{ m} \cdot \text{K} \qquad \text{[27.1]}$$

where λ_{max} is the wavelength at which the curve peaks and T is the absolute temperature of the object emitting the radiation.

Applying Physics 1

APPLICATION

Colors of the Stars.

If you look carefully at stars in the night sky, you can distinguish three main colors—red, white, and blue. What is the reason for these colors?

Explanation   These colors are a result of the thermal radiation from the surfaces of the stars. A relatively cool star, with a surface temperature of 3000 K, has a radiation curve like the middle curve in Figure 27.2. Notice that the peak in the curve is above the visible wavelengths, 0.4–0.7 μm. Thus, significantly more radiation is emitted within the visible range at the red end than the blue end. As a result, the star appears to be reddish in color, similar to the red glow from the burner of an electric stove.

A hotter star has a radiation curve more like the upper curve in Figure 27.2. In this case, the star emits significant radiation throughout the visible range and the combination of all colors causes the star to look white. This is the case with our own Sun, with a surface temperature of 5800 K. For very hot stars, the peak can be shifted so far below the visible range that significantly more blue radiation is emitted than red, so that the star appears bluish in color.

Early attempts to use classical ideas to explain the shape of the curves shown in Figure 27.2 failed. Figure 27.3 shows an experimental plot of the blackbody radiation spectrum (red) together with the theoretical picture of what this curve should look like based on classical theories (blue). At long wavelengths, classical theory is in good agreement with the experimental data. At short wavelengths, however, major disagreement exists between classical theory and experiment. As λ approaches zero, classical theory predicts that the amount of energy being radiated should increase. In fact, theory predicts that the intensity should be infinite. This is contrary to the experimental data, which show that as λ approaches zero, the amount of energy carried by short-wavelength radiation also approaches zero. This contradiction is often called the **ultraviolet catastrophe.**

In 1900 Planck developed a formula for blackbody radiation that was in complete agreement with experiments at all wavelengths. Planck's analysis led to a curve that is shown by the red line in Figure 27.3. Planck's original theoretical approach is rather abstract in that it involves arguments based on entropy and thermodynamics. We shall present arguments that are easier to visualize physically, while attempting to convey the spirit and revolutionary impact of Planck's original work.

Planck was convinced that blackbody radiation was produced by submicroscopic electric oscillators, which he called *resonators*. He assumed that the walls of a glowing cavity were composed of literally billions of these resonators (whose exact nature was unknown). In his theory, Planck made two bold and controversial as-

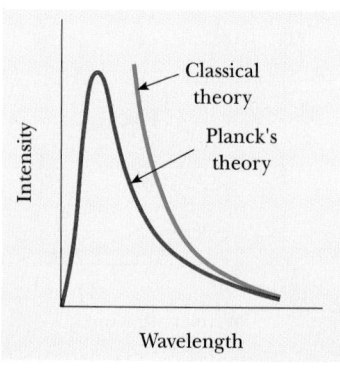

Figure 27.3   Comparison of the Planck theory with the classical theory for the distribution of blackbody radiation.

sumptions concerning the nature of the oscillating charges at the surface of the blackbody:

1. The resonators could have only certain discrete amounts of energy, E_n, given by

$$E_n = nhf \qquad [27.2]$$

where n is a positive integer called a **quantum number,** f is the frequency of vibration of a resonator, and h is a constant, known as **Planck's constant,** given by

$$h = 6.626 \times 10^{-34} \, \text{J} \cdot \text{s} \qquad [27.3]$$

Because the energy of each resonator can have only discrete values given by Equation 27.2, we say the energy is *quantized.* Each discrete energy value represents a different *quantum state,* with each value of n representing a specific quantum state. When the resonator is in the $n = 1$ quantum state, its energy is hf; when it is in the $n = 2$ quantum state, its energy is $2hf$; and so on.

2. The resonators emit discrete units of light energy that are called **photons.** The resonators emit or absorb photons by "jumping" from one quantum state to another. If the jump is from one state to an adjacent state — say, from the $n = 3$ state to the $n = 2$ state — Equation 27.2 shows that the amount of energy radiated by the resonator equals hf. Hence, the energy of one photon corresponds to the energy difference between two adjacent levels and is given by

$$E = hf \qquad [27.4]$$

The resonator will radiate or absorb energy only when it changes quantum states. If it remains in one quantum state, no energy is absorbed or emitted.

The key point in Planck's theory is the radical assumption of quantized energy states. This development marked the birth of the quantum theory. When Planck presented his theory, most scientists (including Planck!) did not consider the quantum concept to be realistic. Hence, Planck and others continued to search for a more rational explanation of blackbody radiation. However, subsequent developments showed that a theory based on the quantum concept (rather than on classical concepts) had to be used to explain a number of other phenomena at the atomic level.

Max Planck (1858–1947)

Planck introduced the concept of a "quantum of action" (Planck's constant, h) in an attempt to explain the spectral distribution of blackbody radiation, which laid the foundations for quantum theory. In 1918 he was awarded the Nobel Prize for this discovery of the quantized nature of energy. The work leading to the "lucky" blackbody radiation formula was described by Planck in his Nobel Prize acceptance speech (1920): "But even if the radiation formula proved to be perfectly correct, it would after all have been only an interpolation formula found by lucky guesswork and, thus, would have left us rather unsatisfied." *(Photo courtesy of AIP Niels Bohr Library, W. F. Meggers Collection)*

Thinking Physics 1

Suppose you are observing a photon while moving toward its source at a high speed. According to the Doppler effect, you should detect a higher frequency. According to the equation for the energy of the photon, $E = hf$, this should result in an increased energy. How can the energy of a photon be increased simply by moving toward it? Does this violate the conservation of energy principle?

Explanation There is no violation of the conservation of energy, because energy had to be added to the system of the photon and the observer to represent the motion of the observer. According to the observer, the increased energy appears in the form of a higher frequency of the photon. According to a second observer, external to the system, the increased energy appears as kinetic energy of the first observer.

EXAMPLE 27.1   Thermal Radiation from the Human Body

The temperature of the skin is approximately 35°C. At what wavelength does the radiation emitted from the skin reach its peak?

Solution   From Wien's displacement law (Eq. 27.1), we have

$$\lambda_{max}T = 0.2898 \times 10^{-2} \text{ m} \cdot \text{K}$$

Solving for λ_{max}, noting that 35°C corresponds to an absolute temperature of 308 K, we find

$$\lambda_{max} = \frac{0.2898 \times 10^{-2} \text{ m} \cdot \text{K}}{308 \text{ K}} = \boxed{940 \ \mu\text{m}}$$

This radiation is in the infrared region of the spectrum.

Exercise   (a) Find the wavelength corresponding to the peak of the radiation curve for the heating element of an electric oven at a temperature of 1.20×10^3 K. Note that although this radiation peak lies in the infrared, there is enough visible radiation at this temperature to give the element a red glow. (b) Calculate the wavelength corresponding to the peak of the radiation curve for an object whose temperature is 5.00×10^3 K, an approximate temperature for the surface of the Sun.

Answer   (a) 2.42 μm; (b) 580 nm in the visible region

EXAMPLE 27.2   The Quantized Oscillator

A 2.0-kg mass is attached to a massless spring of force constant $k = 25$ N/m. The spring is stretched 0.40 m from its equilibrium position and released. (a) Find the total energy and frequency of oscillation according to classical calculations.

Solution   The total energy of a simple harmonic oscillator having an amplitude A is $\frac{1}{2}kA^2$. Therefore,

$$E = \frac{1}{2}\ kA^2 = \frac{1}{2}\ (25 \text{ N/m})(0.40 \text{ m})^2 = \boxed{2.0 \text{ J}}$$

The frequency of oscillation is

$$f = \frac{1}{2\pi}\sqrt{\frac{k}{m}} = \frac{1}{2\pi}\sqrt{\frac{25 \text{ N/m}}{2.0 \text{ kg}}} = \boxed{0.56 \text{ Hz}}$$

(b) Assume that the energy is quantized and find the quantum number, n, for the system.

Solution   If the energy is quantized, we have $E_n = nhf$, and from the result of (a), we have

$$E_n = nhf = n(6.63 \times 10^{-34} \text{ J} \cdot \text{s})(0.56 \text{ Hz}) = 2.0 \text{ J}$$

Therefore,

$$n = \boxed{5.4 \times 10^{33}}$$

(c) How much energy would be carried away in a one-quantum change?

Solution The energy carried away in a one-quantum change of energy is

$$E = hf = (6.63 \times 10^{-34}\,\text{J} \cdot \text{s})(0.56\,\text{Hz}) = \boxed{3.7 \times 10^{-34}\,\text{J}}$$

The energy carried away by a one-quantum change is such a small fraction of the total energy of the oscillator that we could not expect to see it. Thus, even though the decrease in energy of a spring-mass system is quantized and does decrease by small quantum jumps, our senses perceive the decrease as continuous. Quantum effects become important and measurable only on the submicroscopic level of atoms and molecules.

EXAMPLE 27.3 The Energy of a "Yellow" Photon

Yellow light with a frequency of approximately 6.0×10^{14} Hz is the predominant frequency in sunlight. What is the energy carried by a quantum of this light?

Solution The energy carried by one quantum of light is given by Equation 27.4:

$$E = hf = (6.63 \times 10^{-34}\,\text{J} \cdot \text{s})(6.0 \times 10^{14}\,\text{Hz}) = 4.0 \times 10^{-19}\,\text{J} = \boxed{2.5\,\text{eV}}$$

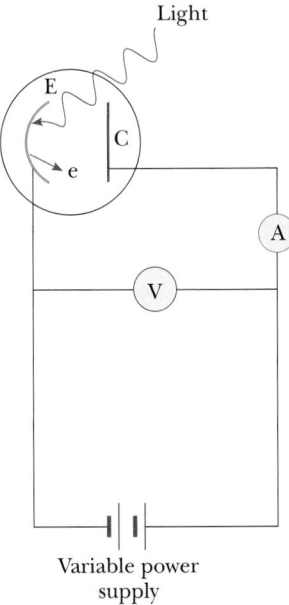

Figure 27.4 Circuit diagram for observing the photoelectric effect. When light strikes plate E, photoelectrons are ejected from the plate. Electrons collected at C and passing through the ammeter constitute a current in the circuit.

27.2 THE PHOTOELECTRIC EFFECT

In the latter part of the 19th century, experiments showed that, when light is incident on certain metallic surfaces, electrons are emitted from the surfaces. This phenomenon is known as the **photoelectric effect,** and the emitted electrons are called **photoelectrons.** The first discovery of this phenomenon was made by Hertz, who was also the first to produce the electromagnetic waves predicted by Maxwell.

Figure 27.4 is a schematic diagram of a photoelectric-effect apparatus. An evacuated glass tube contains a metal plate, E, connected to the negative terminal of a battery. Another metal plate, C, is maintained at a positive potential by the battery. When the tube is kept in the dark, the ammeter reads zero, indicating that there is no current in the circuit. However, when monochromatic light of an appropriate wavelength shines on plate E, a current is detected by the ammeter, indicating a flow of charges across the gap between E and C. The current associated with this process arises from electrons emitted from the negative plate (the emitter) and collected at the positive plate (the collector).

Figure 27.5 is a plot of the photoelectric current versus the potential difference, ΔV, between E and C for two light intensities. For large values of ΔV, the current reaches a maximum value. In addition, the current increases as the incident light intensity increases, as you might expect. Finally, when ΔV is negative — that is, when the battery in the circuit is reversed to make E positive and C negative — the current drops to a low value because photoelectrons are repelled by the negative collecting plate, C. Only those electrons having a kinetic energy greater than $e\,\Delta V$ will reach C, where e is the charge on the electron. When ΔV is less than or equal to ΔV_s, called the **stopping potential,** no electrons reach C and the current is zero. The stopping potential is *independent* of the radiation intensity. The maximum kinetic

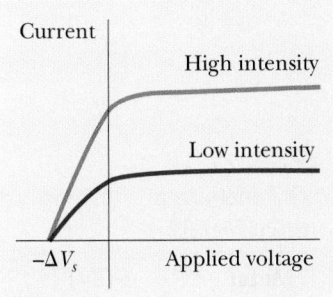

Figure 27.5 Photoelectric current versus applied voltage for two light intensities. The current increases with intensity but reaches a saturation level for large values of ΔV. At voltages equal to or less than $-\Delta V_s$, the current is zero.

energy of the photoelectrons is related to the stopping potential through the relation

$$KE_{max} = e \Delta V_s \qquad [27.5]$$

Several features of the photoelectric effect cannot be explained with classical physics or with the wave theory of light:

- No electrons are emitted if the incident light frequency falls below some **cutoff frequency,** f_c, which is characteristic of the material being illuminated. This is inconsistent with the wave theory, which predicts that the photoelectric effect should occur at any frequency, provided the light intensity is high enough.
- The maximum kinetic energy of the photoelectrons is independent of light intensity.
- The maximum kinetic energy of the photoelectrons increases with increasing light frequency.
- Electrons are emitted from the surface almost instantaneously (less than 10^{-9} s after the surface is illuminated), even at low light intensities. Classically, one would expect that the electrons would require some time to absorb the incident radiation before they acquired enough kinetic energy to escape from the metal.

A successful explanation of the photoelectric effect was given by Einstein in 1905, the same year he published his special theory of relativity. As part of a general paper on electromagnetic radiation, for which he received the Nobel prize in 1921, Einstein extended Planck's concept of quantization to electromagnetic waves. He assumed that light (or any other electromagnetic wave) of frequency f can be considered a stream of photons. Each photon has an energy E given by Equation 27.4.

Einstein's view was that a photon of the incident light can give *all* its energy, hf, to a single electron in the metal. Electrons emitted from the surface of the metal possess the maximum kinetic energy, KE_{max}. According to Einstein, the maximum kinetic energy for these liberated electrons is

Photoelectric effect equation ▶

$$KE_{max} = hf - \phi \qquad [27.6]$$

where ϕ is called the **work function** of the metal. **The work function represents the minimum energy with which an electron is bound in the metal,** and is on the order of a few electron volts. Table 27.1 lists selected values.

With the photon theory of light, one can explain the features of the photoelectric effect that cannot be understood using classical concepts:

- That the effect is not observed below a certain cutoff frequency follows from the fact that the photon energy must be $\geq \phi$. If the energy of the incoming photon is not $\geq \phi$, the electrons will never be ejected from the surface, regardless of light intensity.
- That KE_{max} is independent of the light intensity can be understood with the following argument. If the light intensity is doubled, the number of photons is doubled, which doubles the number of photoelectrons emitted. However, their kinetic energy, which equals $hf - \phi$, depends only on the light frequency and the work function, not on the light intensity.
- That KE_{max} increases with increasing frequency is easily understood with Equation 27.6.
- That the electrons are emitted almost instantaneously is consistent with the particle theory of light, in which the incident energy appears in small packets and

TABLE 27.1
Work Functions of
Selected Metals

Metal	ϕ (eV)
Na	2.46
Al	4.08
Cu	4.70
Zn	4.31
Ag	4.73
Pt	6.35
Pb	4.14
Fe	4.50

there is a one-to-one interaction between photons and electrons. This is in contrast to the energy of the light being spread thinly over a large area.

Experimental observation of a linear relationship between f and KE_{max} would be a final confirmation of Einstein's theory. Indeed, such a linear relationship is observed, as sketched in Figure 27.6. The slope of this curve is h. The intercept on the horizontal axis is the cutoff frequency, which is related to the work function through the relation $f_c = \phi/h$. This corresponds to a **cutoff wavelength** of

$$\lambda_c = \frac{c}{f_c} = \frac{c}{\phi/h} = \frac{hc}{\phi} \qquad [27.7]$$

where c is the speed of light. Wavelengths *greater* than λ_c incident on a material with a work function of ϕ do not result in the emission of photoelectrons.

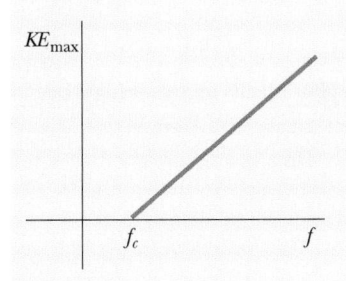

Figure 27.6 A sketch of KE_{max} versus frequency of incident light for photoelectrons in a typical photoelectric effect experiment. Photons with frequency less than f_c do not have sufficient energy to eject an electron from the metal.

EXAMPLE 27.4 The Photoelectric Effect for Sodium

A sodium surface is illuminated with light of wavelength 3.00×10^{-7} m. The work function for sodium is 2.46 eV. Find (a) the kinetic energy of the ejected photoelectrons and (b) the cutoff wavelength for sodium.

Solution (a) The energy of each photon of the illuminating light beam is

$$E = hf = \frac{hc}{\lambda} = \frac{(6.63 \times 10^{-34}\,\text{J} \cdot \text{s})(3.00 \times 10^8\,\text{m/s})}{3.00 \times 10^{-7}\,\text{m}}$$

$$= 6.63 \times 10^{-19}\,\text{J} = \frac{6.63 \times 10^{-19}\,\text{J}}{1.60 \times 10^{-19}\,\text{J/eV}} = 4.14\,\text{eV}$$

where we have used the conversion 1 eV = 1.6×10^{-19} J. Using Equation 27.6 gives

$$KE_{max} = hf - \phi = 4.14\,\text{eV} - 2.46\,\text{eV} = \boxed{1.68\,\text{eV}}$$

(b) The cutoff wavelength can be calculated from Equation 27.7 after we convert ϕ from electron volts to joules:

$$\phi = 2.46\,\text{eV} = (2.46\,\text{eV})(1.6 \times 10^{-19}\,\text{J/eV}) = 3.94 \times 10^{-19}\,\text{J}$$

Hence

$$\lambda_c = \frac{hc}{\phi} = \frac{(6.63 \times 10^{-34}\,\text{J} \cdot \text{s})(3.00 \times 10^8\,\text{m/s})}{3.94 \times 10^{-19}\,\text{J}}$$

$$= 5.05 \times 10^{-7}\,\text{m} = \boxed{505\,\text{nm}}$$

This wavelength is in the green region of the visible spectrum.

27.3 APPLICATIONS OF THE PHOTOELECTRIC EFFECT

The photoelectric cell shown in Figure 27.4 acts much like a switch in an electric circuit in that it produces a current in an external circuit when light of sufficiently high frequency falls on the cell, but it does not allow a current in the dark. Many practical devices in our everyday lives depend on the photoelectric effect. For ex-

APPLICATION

Photocells in Street Lights.

ample, a use familiar to everyone is that of turning street lights on at night and off in the morning. A photoelectric control unit in the base of the light activates a switch to turn off the streetlight when ambient light of the correct frequency falls on it.

Photocells are an integral part of a common law enforcement device called the Breathalyzer, used to measure the amount of alcohol present in the breath of an intoxicated individual. Figure 27.7 shows the essential elements of a typical Breathalyzer. When in use, an external switch is set to the TAKE position. This opens a valve allowing the breath from the suspect to enter a chamber and to raise a movable piston until two vent holes are exposed. When the breath is exhausted, the piston settles down into a fixed position below the vent holes, trapping a known volume of the breath inside the piston chamber.

APPLICATION

Photocells in the Breathalyzer.

When the switch is placed in the ANALYZE position, the trapped breath moves through a tube to a test vial filled with a fixed percentage mixture of potassium dichromate and silver nitrate dissolved in sulfuric acid. Any alcohol present in the breath immediately undergoes a chemical reaction with the potassium dichromate, converting it to acetic acid. The amount of potassium dichromate destroyed is used as a measure of the amount of alcohol originally present in the breath. (The silver nitrate present in the mixture serves as a catalyst to speed up the reaction.) The first step in the measurement process is to move a light source to a point between the test vial and a standard vial such that the intensity of the light reaching the photocells behind each vial is the same. The null meter reads zero when the source is positioned so that each photocell has the same output. The light from the source passes through filters that allows only light of wavelength 420 nm to reach the photocells. This particular wavelength is selected because it is known that potassium dichromate absorbs light very strongly at this wavelength. As alcohol bubbles into

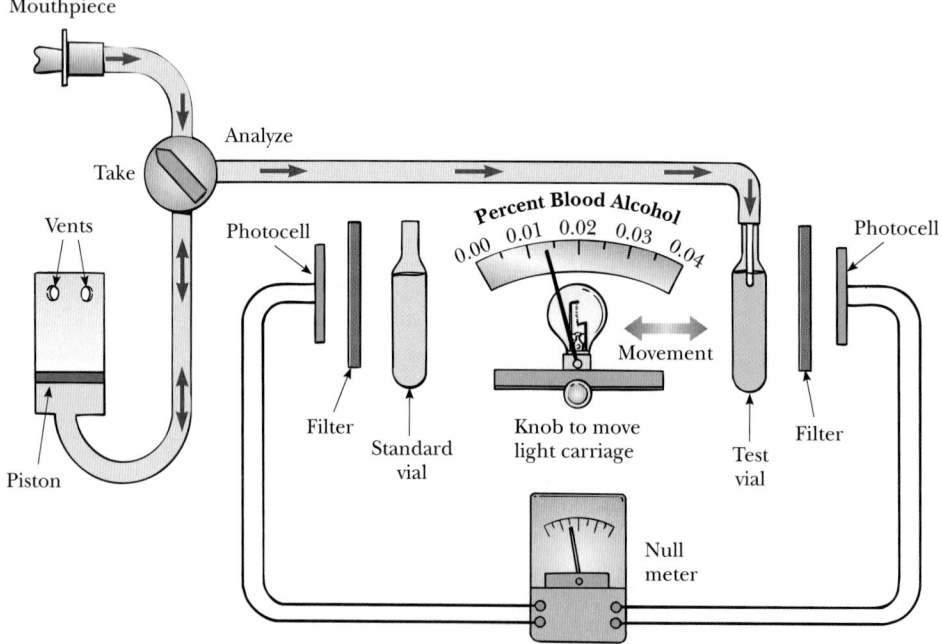

Figure 27.7 Diagram of a Breathalyzer.

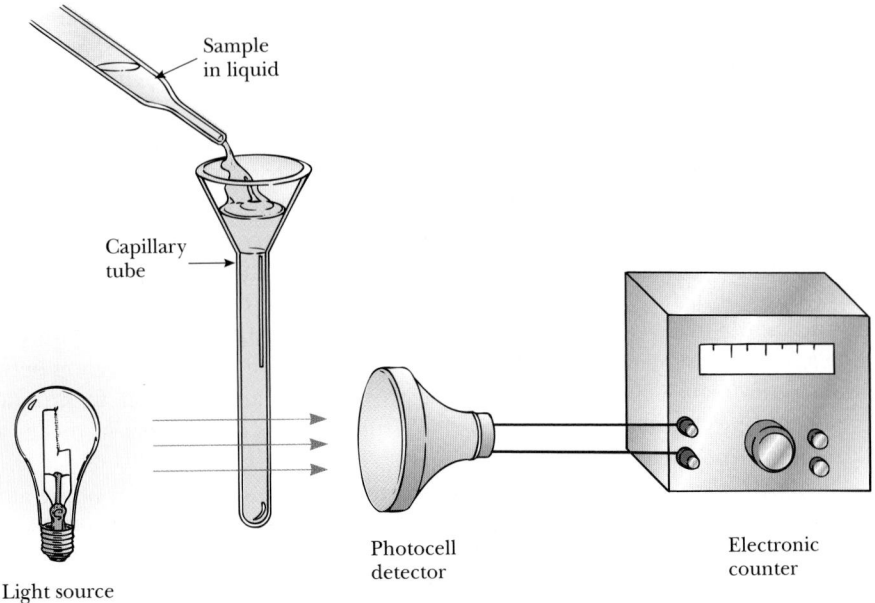

Figure 27.8 A Coulter counter for measuring the rate of bacterial growth.

the test vial, the chemical reaction proceeds, converting potassium dichromate to acetic acid, a substance that readily allows light of wavelength 420 nm to pass. As a result, the photocell behind the test vial receives more light than the standard vial and its current output goes up. To compensate for this, the examiner manually moves the light source toward the standard vial until the null meter again reads zero. The amount that the source has to be moved is a direct measure of the alcohol content in the breath.

A process similar to that used in the Breathalyzer is used by biologists to analyze bacterial growth quantitatively. A rough approximation to the rate of growth of bacteria is determined by measuring the turbidity (cloudiness) of a liquid containing the multiplying bacteria. A test tube containing a transparent growth medium is placed between a light source and a photocell. As bacterial growth occurs in the medium, the degree of turbidity of the sample is determined by the decline of the output of the photocell. This process produces only a gross approximation to the rate of growth. A more accurate count is achieved by use of a device called a Coulter counter, as shown in Figure 27.8. Bacteria in solution pass down a capillary tube located between a light source and a photocell. When the light is blocked by an organism, a counter is activated.

APPLICATION

Coulter Counters for Measuring Bacterial Growth.

27.4 X-RAYS

In 1895 at the University of Wurzburg, Wilhelm Roentgen (1845–1923) was studying electrical discharges in low-pressure gases when he noted that a fluorescent screen glowed even when placed several meters from the gas discharge tube and even when black cardboard was placed between the tube and the screen. He con-

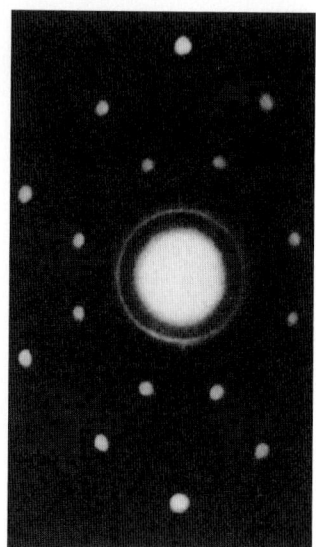

Figure 27.9 X-ray diffraction pattern of NaCl.

cluded that the effect was caused by a mysterious type of radiation, which he called **x-rays** because of their unknown nature. Subsequent study showed that these rays traveled at or near the speed of light and that they could not be deflected by either electric or magnetic fields. This last fact indicated that x-rays did not consist of beams of charged particles, although the possibility that they were beams of uncharged particles remained.

In 1912 Max von Laue (1879–1960) suggested that one should be able to diffract x-rays by using the regular atomic spacings of a crystal lattice as a diffraction grating, just as visible light is diffracted by a ruled grating. Shortly thereafter, researchers demonstrated that such a diffraction pattern could be observed, similar to that shown in Figure 27.9 for NaCl. The wavelengths of the x-rays were then determined from the diffraction data and the known values of the spacing between atoms in the crystal. X-ray diffraction has proved to be an invaluable technique for understanding the structure of matter. We shall discuss this subject in more detail in the next section.

Typical x-ray wavelengths are about 0.1 nm, which is on the order of the atomic spacing in a solid. We now know that x-rays are a part of the electromagnetic spectrum, characterized by frequencies higher than those of ultraviolet radiation and having the ability to penetrate most materials with relative ease.

X-rays are produced when high-speed electrons are suddenly decelerated, for example, when a metal target is struck by electrons that have been accelerated through a potential difference of several thousand volts. Figure 27.10a shows a schematic diagram of an x-ray tube. A current in the filament causes electrons to be boiled off, and these freed electrons are accelerated toward a dense metal target, such as tungsten, which is held at a higher potential than the filament.

Figure 27.11 represents a plot of x-ray intensity versus wavelength for the spectrum of radiation emitted by an x-ray tube. Note that there are two distinct patterns. One pattern is a continuous broad spectrum that depends on the voltage applied to the tube. Superimposed on this pattern is a series of sharp, intense lines that depend on the nature of the target material. The accelerating voltage must exceed a certain value, called the **threshold voltage,** in order to observe these sharp lines, which represent radiation emitted by the target atoms as their electrons undergo rearrangements. We shall discuss this further in Chapter 28. The continuous radiation is sometimes called **bremsstrahlung,** a German word meaning "braking radiation." The term arises from the nature of the mechanism responsible for the

Figure 27.10 (a) Diagram of an x-ray tube. (b) Photograph of an x-ray tube. *(Courtesy of GE Medical Systems)*

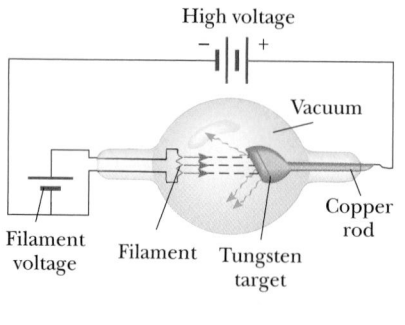

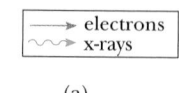

(a)

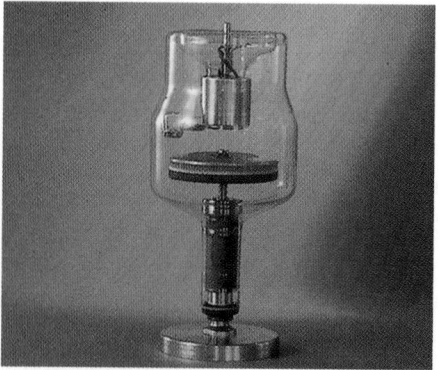

(b)

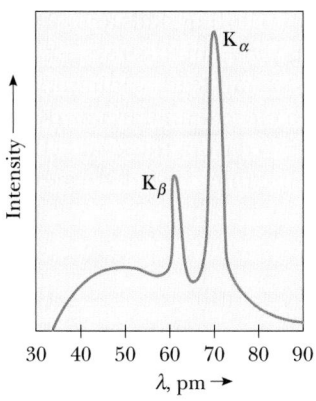

Figure 27.11 The x-ray spectrum of a metal target consists of a broad continuous spectrum plus a number of sharp lines, which are due to *characteristic x-rays*. The data shown were obtained when 35-keV electrons bombarded a molybdenum target. Note that $1 \text{ pm} = 10^{-12} \text{ m} = 10^{-3} \text{ nm}$.

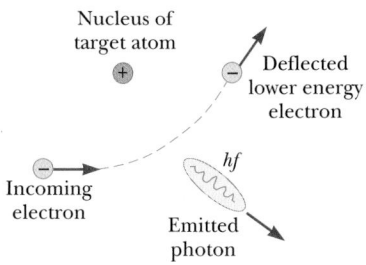

Figure 27.12 An electron passing near a charged target atom experiences an acceleration, and a photon is emitted in the process.

radiation. That is, electrons emit radiation when they undergo a deceleration inside the target.

The deceleration of the electrons by the target produces an effect similar to an inverse photoelectric effect. In the photoelectric process, a quantum of radiant energy is absorbed by an electron in a metal and the electron gains enough energy to escape the metal. In the case of x-ray production, the inverse of this process occurs, as shown in Figure 27.12. As an electron passes close to a positively charged nucleus contained in a target material, it is deflected from its path because of its electrical attraction to the nucleus, and hence experiences an acceleration. An analysis from classical physics shows that any charged particle will radiate energy in the form of electromagnetic radiation when it is accelerated. (An example of this is the production of electromagnetic waves by accelerated charges in a radio antenna, as described in Chapter 21.) According to quantum theory, this radiation must appear in the form of photons. Because the radiated photon shown in Figure 27.12 carries energy, the electron must lose kinetic energy because of its encounter with the target nucleus. Let us consider an extreme example in which the electron loses all of its energy in a single collision. In this case, the initial energy of the electron ($e \Delta V$) is transformed completely into the energy of the photon (hf_{max}). In equation form we have

$$e \Delta V = hf_{\text{max}} = \frac{hc}{\lambda_{\text{min}}} \qquad [27.8]$$

where $e \Delta V$ is the energy of the electron after it has been accelerated through a potential difference of ΔV volts and e is the charge on the electron. This says that the shortest wavelength radiation that can be produced is

$$\lambda_{\text{min}} = \frac{hc}{e \Delta V} \qquad [27.9]$$

The reason that all the radiation produced does not have this particular wavelength is because many of the electrons are not stopped in a single collision. This results in the production of the continuous spectrum of wavelengths.

As noted previously, x-rays are extremely penetrating and can produce burns or other complications if proper precautions are not taken. Between 1930 and 1950 an x-ray device called a fluoroscope was widely used in shoe stores to examine the bones of the foot. Such devices are no longer in use because they are now known to be health hazards, although physicians use similar devices to study the skeletal structures of their patients. An x-ray photograph of a human hand is shown in Figure 27.13.

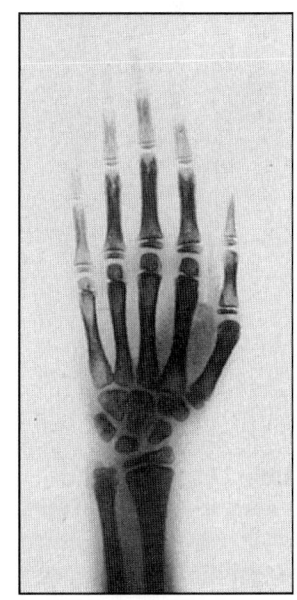

Figure 27.13 X-ray photograph of a human hand.

Interesting insights into the developments of painting skills and subject interest by old masters is being revealed by x-rays. Long wavelength x-rays are absorbed in varying degrees by some paints, such as those having lead, cadmium, chromium, or cobalt as a base. Also, thicker layers will absorb more than thin layers. To examine a painting, a film is placed behind it while it is x-rayed from the front. Ghost outlines of earlier paintings on the same canvas are sometimes revealed when the film is developed.

EXAMPLE 27.5 The Minimum X-Ray Wavelength

Calculate the minimum wavelength produced when electrons are accelerated through a potential difference of 100 000 V, a not-uncommon voltage for an x-ray tube.

Solution   From Equation 27.9, we have

$$\lambda_{min} = \frac{(6.63 \times 10^{-34}\,\text{J} \cdot \text{s})(3.00 \times 10^{8}\,\text{m/s})}{(1.60 \times 10^{-19}\,\text{C})(10^{5}\,\text{V})} = 1.24 \times 10^{-11}\,\text{m}$$

Optional Section

27.5   DIFFRACTION OF X-RAYS BY CRYSTALS

In Chapter 25 we described how a diffraction grating can be used to measure the wavelength of light. In principle, the wavelength of *any* electromagnetic wave can be measured if a grating having a suitable line spacing can be found. The spacing between lines must be approximately equal to the wavelength of the radiation to be measured. X-rays are electromagnetic waves with wavelengths on the order of 0.1 nm. It would be impossible to construct a grating with such a small spacing. As noted in the previous section, Max von Laue suggested that the regular array of atoms in a crystal could act as a three-dimensional grating for observing the diffraction of x-rays.

One experimental arrangement for observing x-ray diffraction is shown in Figure 27.14. A narrow beam of x-rays with a continuous wavelength range is incident on a crystal such as sodium chloride. The diffracted radiation is very intense in certain directions, corresponding to constructive interference from waves reflected from layers of atoms in the crystal. The diffracted radiation is detected by a photographic film and forms an array of spots known as a *Laue pattern*. The crystal structure is determined by analyzing the positions and intensities of the various spots in the pattern.

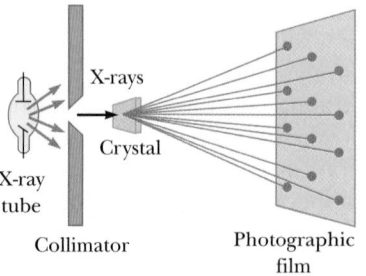

X-rays

Crystal

X-ray
tube

Collimator

Photographic
film

Figure 27.14   Schematic diagram of the technique used to observe the diffraction of x-rays by a single crystal. The array of spots formed on the film by the diffracted beams is called a Laue pattern.

The arrangement of atoms in a crystal of NaCl is shown in Figure 27.15. The smaller red spheres represent Na$^+$ ions and the larger blue spheres represent Cl$^-$ ions. The spacing between successive Na$^+$ (or Cl$^-$) ions in this cubic structure, denoted by the symbol a in Figure 27.14, is approximately 0.563 nm.

A careful examination of the NaCl structure shows that the ions lie in various planes. The shaded areas in Figure 27.15 represent one example in which the atoms lie in equally spaced planes. Now suppose an x-ray beam is incident at grazing angle θ on one of the planes, as in Figure 27.16. The beam can be reflected from both the upper and lower plane of atoms. However, the geometric construction in Figure 27.16 shows that the beam reflected from the lower surface travels farther than the beam reflected from the upper surface by a distance of $2d \sin \theta$. The two portions of the reflected beam will combine to produce constructive interference when this path difference equals some integral multiple of the wavelength λ. The same is true for reflection from the entire family of parallel planes. (Note the similarity between this analysis and that used to describe thin film interference.) The condition for constructive interference is given by

$$2d \sin \theta = m\lambda \qquad m = 1, 2, 3, \ldots \qquad \text{[27.10]}$$

This condition is known as **Bragg's law** after W. L. Bragg (1890–1971), who first derived the relationship. If the wavelength and diffraction angle are measured, Equation 27.10 can be used to calculate the spacing between atomic planes.

The method of x-ray diffraction to determine crystalline structures was thoroughly developed in England by W. H. Bragg and his son W. L. Bragg, who shared a Nobel prize in 1915 for their work. Since then, thousands of crystalline structures have been investigated. Recently, the technique of x-ray structural analysis has been used to unravel the mysteries of such complex organic systems as the important DNA molecule.

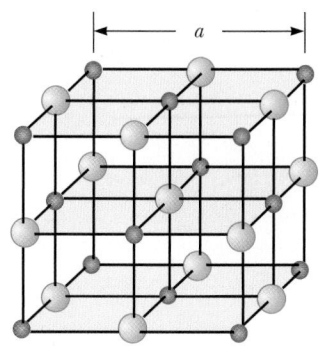

Figure 27.15 A model of the cubic crystalline structure of sodium chloride. The blue spheres represent the Cl$^-$ ions, and the red spheres represent the Na$^+$ ions. The length of the cube edge is $a = 0.563$ nm.

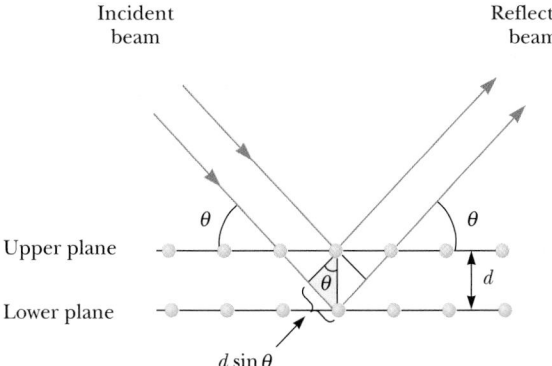

Incident beam

Reflected beam

Upper plane

Lower plane

θ θ

θ

d

$d \sin \theta$

Figure 27.16 A two-dimensional description of the reflection of an x-ray beam from two parallel crystalline planes separated by a distance d. The beam reflected from the lower plane travels farther than the one reflected from the upper plane by an amount equal to $2d \sin \theta$.

EXAMPLE 27.6 Reflection from Calcite

If the spacing between certain planes in a crystal of calcite is 0.314 nm, find the grazing angles at which first- and third-order interference will occur for x-rays of wavelength 0.070 nm.

Solution   For first-order interference, the value of m in Equation 27.10 is 1. Thus, the grazing angle corresponding to this order of interference is found as follows:

$$\sin \theta = \frac{m\lambda}{2d} = \frac{(0.0700 \text{ nm})}{2(0.314 \text{ nm})} = 0.111$$

$$\theta = \quad 6.37°$$

In third-order interference, $m = 3$, and we find

$$\sin \theta = \frac{m\lambda}{2d} = \frac{3(0.0700 \text{ nm})}{2(0.314 \text{ nm})} = 0.334$$

$$\theta = \quad 19.5°$$

27.6   THE COMPTON EFFECT

Further justification for the photon theory of light came from an experiment conducted by Arthur H. Compton in 1923. In his experiment, Compton directed an x-ray beam of wavelength λ_0 toward a block of graphite. He found that the scattered x-rays had a slightly longer wavelength, λ, than the incident x-rays, and hence the energies of the scattered rays were lower. The amount of energy reduction depended on the angle at which the x-rays were scattered. The change in wavelength, $\Delta\lambda$, between a scattered x-ray and an incident x-ray is called the **Compton shift.**

In order to explain this effect, Compton assumed that if a photon behaves like a particle, its collision with other particles is similar to that between two billiard balls. Hence, both energy and momentum must be conserved. If the incident photon collides with an electron initially at rest, as in Figure 27.17, the photon transfers some of its energy and momentum to the electron. As a consequence, the energy and frequency of the scattered photon are lowered and its wavelength increases.

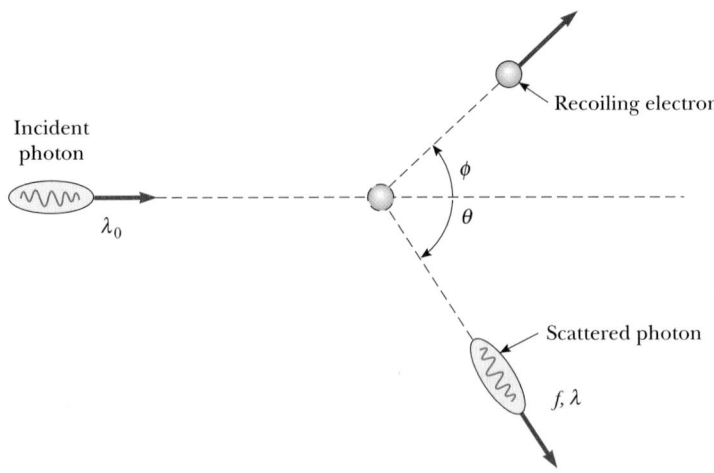

Figure 27.17   Diagram representing Compton scattering of a photon by an electron. The scattered photon has less energy (or longer wavelength) than the incident photon.

Applying relativistic energy and momentum conservation to the collision described in Figure 27.17, the shift in wavelength of the scattered photon is given by

$$\Delta\lambda = \lambda - \lambda_0 = \frac{h}{m_e c}(1 - \cos\theta)$$ [27.11]   ◀ The Compton shift formula

where m_e is the mass of the electron and θ is the angle between the directions of the scattered and incident photons. The quantity $h/m_e c$ is called the **Compton wavelength** and has a value of $h/m_e c = 0.002\ 43$ nm. Note that the Compton wavelength is very small relative to the wavelengths of visible light and hence would be difficult to detect if visible light were used. Furthermore, note that the Compton shift depends on the scattering angle, θ, and not on the wavelength. Experimental results for x-rays scattered from various targets strongly support the photon concept.

Arthur Holly Compton
American physicist (1892–1962)

(Courtesy of AIP Niels Bohr Library)

Thinking Physics 2

The Compton effect involves a change in wavelength as photons are scattered through different angles. Suppose we illuminate a piece of material with a beam of light and then view the material from different angles relative to the beam of light. Will we see a color change corresponding to the change in wavelength of the scattered light?

Explanation There will be a wavelength change for visible light scattered by the material, but the change will be far too small to detect as a color change. The largest possible wavelength change, at 180° scattering, will be twice the Compton wavelength, about 0.005 nm. This represents a change of less than 0.001% of the wavelength of red light. The Compton effect is only detectable for wavelengths that are very short to begin with, so that the Compton wavelength is an appreciable fraction of the incident wavelength. As a result, the usual radiation for observing the Compton effect is in the x-ray range of the electromagnetic spectrum.

EXAMPLE 27.7 Compton Scattering at 45°

X-rays of wavelength $\lambda_0 = 0.200\ 000$ nm are scattered from a block of material. The scattered x-rays are observed at an angle of 45.0° to the incident beam. Calculate the wavelength of the x-rays scattered at this angle.

Solution The shift in wavelength of the scattered x-rays is given by Equation 27.11. Taking $\theta = 45.0°$, we find that

$$\Delta\lambda = \frac{h}{m_e c}(1 - \cos\theta)$$

$$= \frac{6.626 \times 10^{-34}\ \text{J} \cdot \text{s}}{(9.11 \times 10^{-31}\ \text{kg})(3.00 \times 10^8\ \text{m/s})}(1 - \cos 45.0°)$$

$$= 7.10 \times 10^{-13}\ \text{m} = 0.000\ 710\ \text{nm}$$

Hence, the wavelength of the scattered x-ray at this angle is

$$\lambda = \Delta\lambda + \lambda_0 = \boxed{0.200\ 710\ \text{nm}}$$

Exercise Find the fraction of energy lost by the photon in this collision.

Answer Fraction $= \Delta E/E = 0.003\ 54$

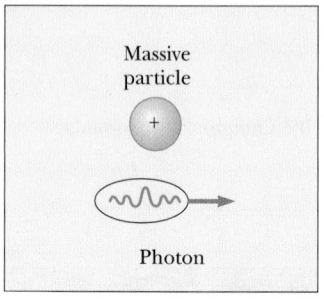

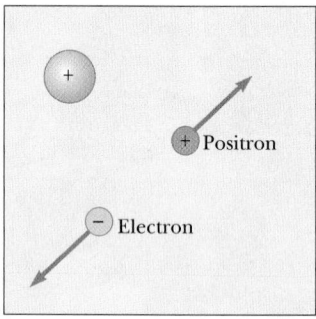

Figure 27.18 Representation of the process of pair production.

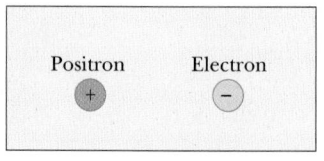

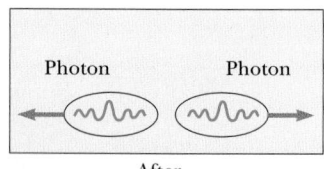

Figure 27.19 Representation of the process of pair annihilation.

27.7 **PAIR PRODUCTION AND ANNIHILATION**

In the photoelectric and Compton effects, the energy of a photon is transformed into the kinetic and potential energy of an electron. When a photon interacts with matter through the photoelectric effect, an electron is removed from an atom and the photon disappears. In the Compton effect, a photon is scattered off an electron (or a nucleus) and loses some energy in the process. We shall now describe a process in which the energy of a photon is converted completely into mass. This is a striking verification of the equivalence of mass and other forms of energy as predicted by Einstein's theory of relativity.

A common process in which a photon creates matter is called **pair production,** illustrated in Figure 27.18. In this process, an electron and a positron are simultaneously produced, while the photon disappears. (Note that the positron is a positively charged particle having the same mass as an electron. The positron is often called the *antiparticle* of the electron.) In order for pair production to occur, energy, momentum, and charge must all be conserved during the process. Note that it is impossible for a photon to produce a single electron because the photon has zero charge, and charge would not be conserved in the process. The *minimum* energy that a photon must have to produce an electron-positron pair can be found using conservation of energy by equating the photon energy to the total rest energy of the pair. That is,

$$hf_{min} = 2m_ec^2 \qquad [27.12]$$

Because the energy of an electron is $m_ec^2 = 0.51$ MeV, the minimum energy required for pair production is 1.02 MeV. The wavelength of a photon carrying this much energy is 0.0012 nm. Photons with such short wavelengths are in the gamma-ray (or very short x-ray) region of the spectrum.

Pair production cannot occur in a vacuum but can only take place in the presence of a massive particle such as an atomic nucleus. The massive particle must participate in the interaction in order that energy and momentum be conserved simultaneously.

Pair annihilation is a process in which an electron-positron pair produces two photons, the inverse of pair production. Figure 27.19 is one example of pair annihilation in which an electron and positron initially at rest combine with each other, disappear, and create two photons. Because the initial momentum of the pair is zero, it is impossible to produce a single photon. Momentum can be conserved only if two photons moving in opposite directions, both with the same energy and magnitude of momentum, are produced. We shall discuss particles and their antiparticles further in Chapter 30.

27.8 **PHOTONS AND ELECTROMAGNETIC WAVES**

An explanation of a phenomenon such as the photoelectric effect presents very convincing evidence in support of the photon (or particle) concept of light. An obvious question that arises at this point is, "How can light be considered a photon when it exhibits wave-like properties?" On the one hand, we describe light in terms of photons having energy and momentum. On the other hand, we must also recognize that light and other electromagnetic waves exhibit interference and diffraction effects that are consistent only with a wave interpretation. Which model is correct? Is light a wave or a particle? The answer depends on the specific phenomenon being observed. Some experiments can be better (or exclusively) explained on the basis of the photon concept, whereas others are best (or exclusively) described with a wave model. The end result is that **we must accept both models and admit that the true nature of light is not describable in terms of a single classical picture: Light has a dual nature.**

◀ Light has a dual nature.

We can perhaps understand why photons are compatible with electromagnetic waves in the following manner. We may suspect that long-wavelength radio waves do not exhibit particle characteristics. Consider, for instance, radio waves at a frequency of 2.5 MHz. The energy of a photon having this frequency is only about 10^{-8} eV. From a practical viewpoint, this energy is too small to be detected as a single photon. A sensitive radio receiver might require as many as 10^{10} of these photons to produce a detectable signal. With such a large number of photons reaching the detector every second, it would be unlikely for any graininess to appear in the detected signal; hence it would appear as a continuous wave. That is, we would not be able to detect the individual photons striking the antenna.

Now consider what happens as we go to higher frequencies or shorter wavelengths. In the visible region, it is possible to observe both the photon and the wave characteristics of light. As we mentioned earlier, a light beam shows interference phenomena and at the same time can produce photoelectrons, which can be understood best by using Einstein's photon concept. At even higher frequencies and correspondingly shorter wavelengths, the momentum and energy of the photons increase. As a consequence, the photon nature of light becomes more evident than its wave nature. For example, an x-ray photon is easily detected as a single event. However, as the wavelength decreases, wave effects, such as interference and diffraction, become more difficult to observe. Very indirect methods are required to detect the wave nature of very-high-frequency radiation, such as gamma rays.

All forms of electromagnetic radiation can be described from two points of view. At one extreme, the electromagnetic wave description suits the overall interference pattern formed by a large number of photons. At the other extreme, the photon description is natural when dealing with highly energetic photons of very short wavelength. Hence,

Louis de Broglie, French physicist (1892–1987)

De Broglie was awarded the Nobel prize in 1929 for his discovery of the wave nature of electrons. "It would seem that the basic idea of quantum theory is the impossibility of imaging an isolated quantity of energy without associating with it a certain frequency." *(AIP Niels Bohr Library)*

> Light has a dual nature: The photon, a quantum particle, exhibits characteristics of both a classical wave and a classical particle.

27.9 THE WAVE PROPERTIES OF PARTICLES

When students are first introduced to the dual nature of light, many often find the concept difficult to accept. Even more disconcerting is that *matter* has a dual nature as well!

In 1923, in his doctoral dissertation, Louis de Broglie postulated that **because photons have wave and particle characteristics, perhaps all forms of matter have wave as well as particle properties.** This was a highly revolutionary idea with no experimental confirmation at that time. According to de Broglie, electrons have a dual particle-wave nature. An electron in motion exhibits wave properties.

In Chapter 26 we found that the relationship between energy and momentum for a photon, which has a mass of zero, is $p = E/c$. We also know from Equation 27.4 that the energy of a photon is

Energy of a photon ▷

$$E = hf = \frac{hc}{\lambda}$$ [27.13]

Thus, the momentum of a photon can be expressed as

Momentum of a photon ▷

$$p = \frac{E}{c} = \frac{hc}{c\lambda} = \frac{h}{\lambda}$$ [27.14]

From this equation we see that the photon wavelength can be specified by its momentum, or $\lambda = h/p$. De Broglie suggested that **material particles, of momentum p, should also have wave properties and a corresponding wavelength.** Because the momentum of a particle of mass m and velocity v is $p = mv$, the **de Broglie wavelength** of a particle is

de Broglie's hypothesis ▷

$$\lambda = \frac{h}{p} = \frac{h}{mv}$$ [27.15]

Furthermore, in analogy with photons, de Broglie postulated that the frequencies of matter waves (that is, waves associated with real particles) obey the Einstein relation $E = hf$, so that

de Broglie wavelength ▷

$$f = \frac{E}{h}$$ [27.16]

The dual nature of matter is quite apparent in Equations 27.15 and 27.16 because each contains both particle concepts (mv and E) and wave concepts (λ and f). The fact that these relationships are established experimentally for photons makes the de Broglie hypothesis that much easier to accept.

The Davisson–Germer Experiment

De Broglie's proposal that any kind of particle exhibits both wave and particle properties was first regarded as pure speculation. If particles such as electrons had wave-like properties, then under the correct conditions they should exhibit diffraction effects. In 1927, three years after de Broglie published his work, C. J. Davisson and L. H. Germer of the United States succeeded in measuring the wavelength of electrons. Their important discovery provided the first experimental confirmation of the matter waves proposed by de Broglie.

It is interesting that the intent of the initial Davisson–Germer experiment was not to confirm the de Broglie hypothesis. In fact, their discovery was made by accident (as is often the case). The experiment involved the scattering of low-energy electrons (about 54 eV) shot at a nickel target in a vacuum. During one experiment, the nickel surface was badly oxidized because of an accidental break in the vacuum system. After the nickel target was heated in a flowing stream of hydrogen to remove the oxide coating, electrons scattered by it exhibited intensity maxima and minima at specific angles. The experimenters finally realized that the nickel had formed large crystal regions on heating and that the regularly spaced planes of atoms in the crystalline regions served as a diffraction grating for electron matter waves (see Section 27.5).

Shortly thereafter, Davisson and Germer performed more extensive diffraction measurements on electrons scattered from single-crystal targets. Their results showed conclusively the wave nature of electrons and confirmed the de Broglie relation $p = h/\lambda$. A year later in 1928, G. P. Thomson of Scotland observed electron diffraction patterns by passing electrons through very thin gold foils. Diffraction patterns have since been observed for helium atoms, hydrogen atoms, and neutrons. Hence, the universal nature of matter waves has been established in a variety of ways.

EXAMPLE 27.8 The Wavelength of an Electron

Calculate the de Broglie wavelength for an electron ($m_e = 9.11 \times 10^{-31}$ kg) moving at 1.00×10^7 m/s.

Solution Equation 27.15 gives

$$\lambda = \frac{h}{m_e v} = \frac{6.63 \times 10^{-34} \text{ J} \cdot \text{s}}{(9.11 \times 10^{-31} \text{ kg})(1.00 \times 10^7 \text{ m/s})} = \boxed{7.28 \times 10^{-11} \text{ m}}$$

This wavelength corresponds to that of x-rays in the electromagnetic spectrum.

Exercise Find the de Broglie wavelength of a proton ($m_p = 1.67 \times 10^{-27}$ kg) moving with a speed of 1.00×10^7 m/s.

Answer 3.97×10^{-14} m

EXAMPLE 27.9 The Wavelength of a Rock

A rock of mass 50.0 g is thrown with a speed of 40.0 m/s. What is the de Broglie wavelength of the rock?

Solution From Equation 27.15, we have

$$\lambda = \frac{h}{mv} = \frac{6.63 \times 10^{-34} \text{ J} \cdot \text{s}}{(50.0 \times 10^{-3} \text{ kg})(40.0 \text{ m/s})} = \boxed{3.32 \times 10^{-34} \text{ m}}$$

This wavelength is much smaller than any aperture through which the rock could possibly pass. This means that we could not observe diffraction effects, and as a result the wave properties of large-scale objects cannot be observed.

Application: The Electron Microscope

A practical device that relies on the wave characteristics of electrons is the **electron microscope,** which is in many respects similar to an ordinary compound microscope. One important difference is that the electron microscope has a much greater resolving power because electrons can be accelerated to high kinetic energies, giv-

APPLICATION

Electron Microscopes.

Figure 27.20   (a) Diagram of transmission electron microscope. The "lenses" that control the electron beam are magnetic deflection coils. (b) False-color transmission electron micrograph of a primary lysosome in a liver cell. The lysosome (shown here in red) is a membrane-bound structure that contains digestive enzymes. The cell nucleus is at left (yellow-green). *(CNRI/Science Photo Library/Photo Researchers, Inc.)*

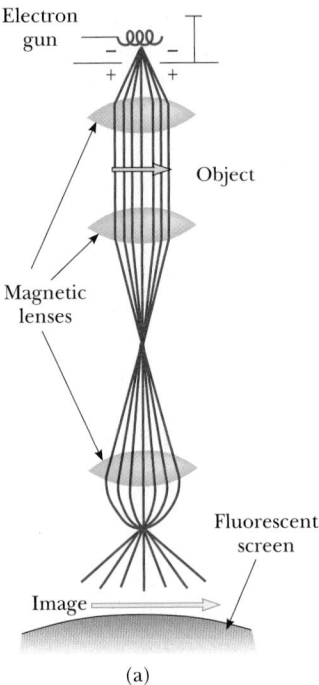

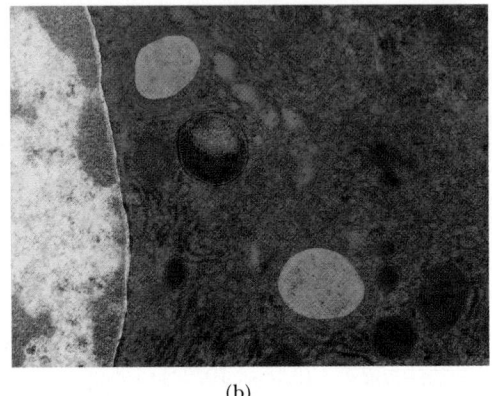

(a)

(b)

ing them a very short wavelength. Any microscope is capable of detecting details that are comparable in size to the wavelength of the radiation used to illuminate the object. The wavelengths of electrons typically are about 100 times shorter than those of the visible light used in optical microscopes. As a result, electron microscopes are able to distinguish details about 100 times smaller.

Figure 27.20a shows a diagram of one kind of electron microscope, the transmission electron microscope. In operation, a beam of electrons falls on a thin slice of the material to be examined. The sample must be very thin, typically a few hundred angstroms, in order to minimize undesirable effects such as absorption or scattering of the electrons. The electron beam is controlled by electrostatic or magnetic deflection, which acts on the charges to focus the beam to an image. Rather than examining the image through an eyepiece as in an ordinary microscope, a magnetic lens forms an image on a fluorescent screen. The fluorescent screen is necessary because the image produced would not otherwise be visible. An example of a photograph taken by an electron microscope is shown in Figure 27.20b.

Thinking Physics 3

Electron microscopes take advantage of the wave nature of particles. Electrons are accelerated to high speeds, giving them a short de Broglie wavelength. Imagine an electron microscope using electrons with a de Broglie wavelength of 0.2 nm. Why don't we design a microscope using 0.2 nm *photons* to do the same thing?

Explanation   Because electrons are charged particles, they will interact with the sample in the microscope and scatter according to the shape and density of various portions of the sample, providing a means of viewing the sample. Photons of wavelength 0.2 nm are uncharged and in the x-ray region of the spectrum. They will tend to simply pass through the sample without interacting. It is also much easier to image electrons because, unlike x-rays, electrons can be guided by electrostatic focusing fields.

27.10   THE WAVE FUNCTION

De Broglie's revolutionary idea that particles should have a wave nature soon moved out of the realm of skepticism to the point where it was viewed as a necessary concept in understanding the subatomic world. In 1926 the Austrian-German physicist Erwin Schrödinger proposed a wave equation that described the manner in which matter waves change in space and time. The Schrödinger wave equation represents a key element in the theory of quantum mechanics. It is as important in quantum mechanics as Newton's laws in classical mechanics. Schrödinger's equation has been successfully applied to the hydrogen atom and to many other microscopic systems. Its importance in most aspects of modern physics cannot be overemphasized.

We shall not go through a mathematical derivation of Schrödinger's wave equation, nor shall we even state the equation here because it involves mathematical operations beyond the scope of this textbook. When we attempt to solve the Schrödinger equation, the basic entity we seek to determine is a quantity, Ψ, called the **wave function.** Each particle is represented by a wave function Ψ that depends both on the position of the object and on time. Once Ψ is found, what information about the particle does it give? To answer this question, let us consider an analogy with light.

In Chapter 24 we discussed Young's double-slit experiment and explained experimental observations of the interference pattern solely in terms of the wave nature of light. Let us now discuss this same experiment in terms of both the wave and particle nature of light.

First, recall from Chapter 21 that the intensity of a light beam is proportional to the square of the electric field strength, E, associated with the beam. That is, $I \propto E^2$. According to the wave model of light, there are certain points on the viewing screen where the net electric field is zero as a result of destructive interference of waves from the two slits. Because E is zero at these points, the intensity is also zero, and the screen is dark at these locations. Likewise, at points on the screen at which constructive interference occurs, E is large, as is the intensity; hence these locations are bright.

Now consider the same experiment when light is viewed as having a particle nature. The number of photons reaching a point on the screen per second increases as the intensity (brightness) increases. Thus, the number of photons that strikes a unit area on the screen each second is proportional to the square of the electric field, or $N \propto E^2$. Now let us consider the behavior of a single photon. What will be the fate of the photon as it moves through the slits in Young's experiment? From a probabilistic point of view, a photon has a high probability of striking the screen at a point at which the intensity (and E^2) is high, and a low probability of striking the screen where the intensity is low.

When describing particles rather than photons, Ψ rather than E plays the role of the amplitude. Using an analogy with the description of light, we make the following interpretation of Ψ for particles: If Ψ is a wave function used to describe

Erwin Schrödinger, Austrian theoretical physicist (1887–1961)

Schrödinger is best known as the creator of wave mechanics. He demonstrated the mathematical equivalence between wave mechanics and the more abstract matrix mechanics developed by Heisenberg. In 1933 Schrödinger left Germany and eventually settled at the Dublin Institute of Advanced Study, where he spent 17 happy, creative years working on problems in general relativity, cosmology, and the application of quantum physics to biology. In 1956 he returned home to Austria and to his beloved Tirolean mountains, where he died in 1961.

a single particle, the value of Ψ^2 at some location at a given time is proportional to the probability of finding the particle at that location at that time.

27.11   THE UNCERTAINTY PRINCIPLE

If you were to measure the position and velocity of a particle at any instant, you would always be faced with reducing the experimental uncertainties in your measurements as much as possible. According to classical mechanics, there is no fundamental barrier to an ultimate refinement of the apparatus or experimental procedures. That is, in principle it would be possible to make such measurements with arbitrarily small uncertainty or with infinite accuracy. Quantum theory predicts, however, that **it is impossible to make simultaneous measurements of a particle's position and velocity with infinite accuracy.** This statement, known as the **uncertainty principle,** was first derived by Werner Heisenberg in 1927.

Consider a particle moving along the x axis and suppose that Δx and Δp_x represent the uncertainty in the measured values of the particle's position and momentum, respectively, at some instant. The uncertainty principle says that the product $\Delta x \, \Delta p_x$ is never less than a number of the order of Planck's constant. More specifically,

Uncertainty principle ▶
$$\Delta x \, \Delta p_x \geq \frac{h}{4\pi}$$
[27.17]

That is, **it is physically impossible to measure simultaneously the exact position and exact momentum of a particle.** If Δx is made very small, Δp_x will be large, and vice versa.

In order to understand the uncertainty principle, consider the following thought experiment. Suppose you wish to measure the position and momentum of an electron as accurately as possible. You might be able to do this by viewing the electron with a powerful microscope. In order for you to see the electron and thus determine its location, at least one photon of light must bounce off the electron and pass through the microscope into your eye. This incident photon is shown moving toward the electron in Figure 27.21a. When the photon strikes the electron, as in Figure 27.21b, it transfers some of its energy and momentum to the electron. Thus, in the process of attempting to locate the electron very accurately (that is, by making Δx very small), we have caused a rather large uncertainty in its momentum. In other words, **the measurement procedure itself limits the accuracy to which we can determine position and momentum simultaneously.**

Let us analyze the collision between the photon and the electron by first noting that the incoming photon has a momentum of h/λ. As a result of the collision, the photon transfers part or all of its momentum to the electron. Thus, the uncertainty in the electron's momentum after the collision is at least as great as the momentum of the incoming photon. That is, $\Delta p_x = h/\lambda$. Furthermore, because light also has wave properties, we would expect the uncertainty in the position of the electron to be on the order of one wavelength of the light being used to view it, because of diffraction effects. Thus, $\Delta x = \lambda$. Multiplying these two uncertainties gives

$$\Delta x \, \Delta p_x = \lambda \left(\frac{h}{\lambda} \right) = h$$

Werner Heisenberg, German theoretical physicist (1901–1976)

Heisenberg obtained his Ph.D. in 1923 at the University of Munich, where he studied under Arnold Sommerfeld. While physicists such as de Broglie and Schrödinger tried to develop physical models of the atom, Heisenberg developed an abstract mathematical model called *matrix mechanics* to explain the wavelengths of spectral lines. The more successful wave mechanics by Schrödinger, announced a few months later, was shown to be equivalent to Heisenberg's approach. Heisenberg made many other significant contributions to physics, including his famous uncertainty principle, for which he received the Nobel prize in 1932; the prediction of two forms of molecular hydrogen; and theoretical models of the nucleus. *(Courtesy of University of Hamburg)*

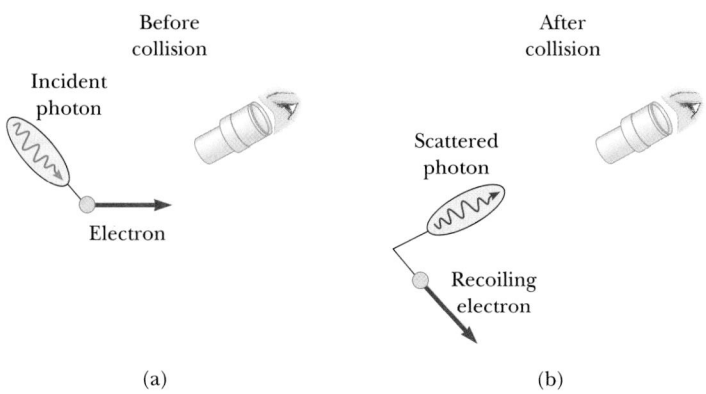

Figure 27.21   A thought experiment for viewing an electron with a powerful microscope. (a) The electron is viewed before colliding with the photon. (b) The electron recoils (is disturbed) as the result of the collision with the photon.

This represents the minimum in the products of the uncertainties. Because the uncertainty can always be greater than this minimum, we have

$$\Delta x \, \Delta p_x \geq h$$

This agrees with Equation 27.17 (apart from a numerical factor introduced by Heisenberg's more precise analysis).

Another form of the uncertainty principle that applies to the simultaneous measurement of energy and time is

$$\Delta E \, \Delta t \geq \frac{h}{4\pi} \qquad\qquad \textbf{[27.18]}$$

where ΔE is the uncertainty in a measurement of the energy and Δt is the time it takes to make the measurement. It can be inferred from this relationship that the energy of a particle cannot be measured with complete precision in an infinitely short interval of time. Thus, when an electron is viewed as a particle, the uncertainty principle tells us that (a) its position and velocity cannot both be known precisely at the same time and (b) its energy can be uncertain (or may not be conserved) for a period of time given by $\Delta t = h/(4\pi \, \Delta E)$.

We can arrive at the uncertainty principle in the form of Equation 27.18 by returning to Figure 27.21. Recall that the uncertainty in the position of the electron being viewed by the microscope is on the order of the wavelength of the light being used to detect the electron. The photon used to detect the electron travels with a speed c; therefore the time it takes the photon to travel a distance Δx is given by

$$\Delta t \approx \frac{\Delta x}{c} \approx \frac{\lambda}{c}$$

As the photon collides with the electron, it can transfer all or part of its energy to the electron. Thus, the uncertainty in the energy transferred is

$$\Delta E \approx hf \approx \frac{hc}{\lambda}$$

The product of these two uncertainties is

$$\Delta E \, \Delta t \approx h$$

Again, the result of our approximate analysis agrees with Heisenberg's more accurate and detailed analysis within a small numerical factor.

Heisenberg's uncertainty principle enables us to better understand the dual wave-particle nature of light and matter. We have seen that the wave description is quite different from the particle description. Therefore, if an experiment is designed to reveal the particle character of an electron (such as the photoelectric effect), its wave character will become fuzzy. Likewise, if the experiment is designed to measure the electron's wave properties accurately (such as diffraction from a crystal), its particle character will become fuzzy.

Thinking Physics 4

A common, but erroneous, description of the absolute zero of temperature is "that temperature at which all molecular motion ceases." How can the uncertainty principle be used to argue against this description?

Explanation   Let us imagine molecules in a piece of material. The molecules are confined within the material, so there is a fixed uncertainty in their position along one axis, Δx, which is the size of the piece of material along this axis. If the molecular motion were to cease at absolute zero, we would be claiming that the molecule velocity is zero, with no uncertainty, $\Delta v = 0$. The product of zero uncertainty in velocity and a nonzero uncertainty in position will be zero, violating the uncertainty principle. Thus, even at absolute zero, there must be some molecular motion.

EXAMPLE 27.10   Locating an Electron

The speed of an electron is measured to be 5.00×10^3 m/s to an accuracy of 0.003 00%. Find the uncertainty in determining the position of this electron.

Solution   The momentum of the electron is

$$p = mv = (9.11 \times 10^{-31} \text{ kg})(5.00 \times 10^3 \text{ m/s}) = 4.56 \times 10^{-27} \text{ kg} \cdot \text{m/s}$$

Because the uncertainty in p is 0.00 300% of this value, we get

$$\Delta p = 0.000\,0300p = (0.000\,0300)(4.56 \times 10^{-27} \text{ kg} \cdot \text{m/s})$$
$$= 1.37 \times 10^{-31} \text{ kg} \cdot \text{m/s}$$

The uncertainty in position can now be calculated by using this value of Δp and Equation 27.17:

$$\Delta x \, \Delta p_x \geq \frac{h}{4\pi}$$

$$\Delta x \geq \frac{h}{4\pi \, \Delta p_x} = \frac{6.626 \times 10^{-34} \text{ J} \cdot \text{s}}{4\pi (1.37 \times 10^{-31} \text{ kg} \cdot \text{m/s})}$$

$$= 0.384 \times 10^{-3} \text{ m} = \boxed{0.384 \text{ mm}}$$

EXAMPLE 27.11 Excited States of Atoms

As we shall see in the next chapter, electrons in atoms can be found in higher states of energy called *excited states* for short periods of time. If the average time that an electron exists in one of these states is 1.00×10^{-8} s, what is the minimum uncertainty in energy of the excited state?

Solution From the uncertainty principle in the form of Equation 27.18, we find that the minimum uncertainty in energy is

$$\Delta E = \frac{h}{4\pi \, \Delta t} = \frac{(6.63 \times 10^{-34} \, \text{J} \cdot \text{s})}{4\pi (1.00 \times 10^{-8} \, \text{s})} = 5.28 \times 10^{-27} \, \text{J} = \boxed{3.30 \times 10^{-8} \, \text{eV}}$$

27.12 THE SCANNING TUNNELING MICROSCOPE[1]

Optional Section

APPLICATION

Scanning Tunneling Microscopes.

One of the basic phenomena of quantum mechanics—tunneling—is at the heart of a very practical device, the scanning tunneling microscope, or STM, which enables us to get highly detailed images of surfaces with resolution comparable to the size of a *single atom.*

Figure 27.22, an image of the surface of a piece of graphite, shows what the STM can do. Note the high quality of the image and the recognizable rings of carbon atoms. What makes this image so remarkable is that its resolution (the size of the smallest detail that can be discerned) is about 0.2 nm. For an ordinary microscope, the resolution is limited by the wavelength of the waves used to make the image. Thus, an optical microscope has a resolution no better than 200 nm, about half the wavelength of visible light, and so could never show the detail displayed in Figure 27.22. Electron microscopes can have a resolution of 0.2 nm by using electron waves of this wavelength, given by the de Broglie formula $\lambda = h/p$. The electron momentum p required to give this wavelength is 10 000 eV/c, corresponding to an electron speed of 2% of the speed of light. Electrons traveling at this speed would penetrate into the interior of the sample in Figure 27.22 and so could not give us information about individual surface atoms.

The STM achieves its very fine resolution by using the basic idea shown in Figure 27.23. A conducting probe with a sharp tip is brought near the surface to be studied. Because it is attracted to the positive ions in the surface, an electron in the surface has a lower total energy than an electron in the empty space between surface and tip. The same thing is true for an electron in the probe tip, which is attracted to the positive ions in the tip. In Newtonian mechanics, this means that electrons cannot move between surface and tip because they lack the energy to escape either material. Because the electrons obey quantum mechanics, however, they can "tunnel" across the barrier of empty space. By applying a voltage between surface and tip, the electrons can be made to tunnel preferentially from surface to tip. In this way the tip samples the distribution of electrons just above the surface.

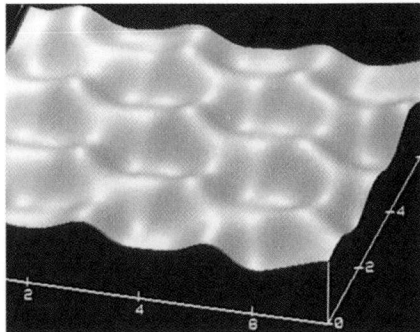

Figure 27.22 The surface of graphite as "viewed" with a scanning tunneling microscope. This technique enables scientists to see small details on surfaces with a lateral resolution of about 0.2 nm and a vertical resolution of 0.001 nm. The contours seen here represent the arrangement of individual carbon atoms on the crystal surface.

[1] This section was written by Roger A. Freedom and Paul K. Hansma, University of California, Santa Barbara.

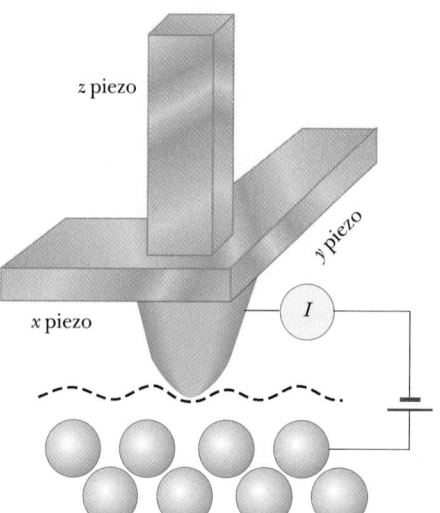

Figure 27.23 A schematic view of an STM. The tip, shown as a rounded cone, is mounted on a piezoelectric *x, y, z* scanner. A scan of the tip over the sample can reveal contours of the surface down to the atomic level. An STM image is composed of a series of scans displaced laterally from each other. *(Based on a drawing from P. K. Hansma, V. B. Elings, O. Marti, and C. Bracker,* Science *242:209, 1988. Copyright 1988 by the AAAS.)*

Because of the nature of tunneling, the STM is very sensitive to the distance z from tip to surface. The reason is that in the empty space between tip and surface, the electron wave function falls off exponentially with a decay length of order 0.1 nm—that is, the wave function decreases by $1/e$ over that distance. For distances z greater than 1 nm—that is, beyond a few atomic diameters—essentially no tunneling takes place. This exponential behavior causes the current of electrons tunneling from surface to tip to depend very strongly on z. This sensitivity is the basis of the operation of the STM: By monitoring the tunneling current as the tip is scanned over the surface, scientists obtain a sensitive measure of the topography of the electron distribution on the surface. The result of this scan is used to make images like that in Figure 27.22. In this way the STM can measure the height of surface features to within 0.001 nm, approximately 1/100 of an atomic diameter!

You can see just how sensitive the STM is by examining Figure 27.22. Of the six carbon atoms in each ring, three appear lower than the other three. In fact all six atoms are at the same level, but they all have slightly different electron distributions. The three atoms that appear lower are bonded to other carbon atoms directly beneath them in the underlying atomic layer, and so their electron distributions (which are responsible for the bonding) extend downward beneath the surface. The atoms in the surface layer that appear higher do not lie directly over subsurface atoms and hence are not bonded to carbon atoms beneath them. For these higher appearing atoms, the electron distribution extends upward into the space above the surface. This extra electron density is what makes these electrons appear higher in Figure 27.22, because what the STM maps is the topography of the electron distribution.

The STM has, however, one serious limitation: It depends on electrical conductivity of the sample and the tip. Unfortunately, most materials are not electrically conductive at their surface. Even metals such as aluminum are covered with nonconductive oxides. A newer microscope, the atomic force microscope, or AFM, overcomes this limitation. It measures the force between a tip and the sample rather than an electrical current. This force, which is typically a result of the exclusion principle, depends very strongly on the tip-sample separation just as the electron tunneling current does for the STM. Thus the AFM has comparable sensitivity for measuring topography and has become widely used for technological applications.

Perhaps the most remarkable thing about the STM is that its operation is based on a quantum-mechanical phenomenon — tunneling — that was well understood in the 1920s, even though the first STM was not built until the 1980s. What other applications of quantum mechanics may yet be waiting to be discovered?

SUMMARY

The characteristics of **blackbody radiation** cannot be explained using classical concepts. The peak of a blackbody radiation curve is given by **Wien's displacement law:**

$$\lambda_{max} T = 0.2898 \times 10^{-2} \text{ m} \cdot \text{K} \qquad [27.1]$$

where λ_{max} is the wavelength at which the curve peaks and T is the absolute temperature of the object emitting the radiation.

Planck first introduced the quantum concept when he assumed that the vibrating molecules responsible for blackbody radiation could have only discrete amounts of energy given by

$$E_n = nhf \qquad [27.2]$$

where n is a positive integer called a **quantum number** and f is the frequency of vibration of the molecule.

The **photoelectric effect** is a process whereby electrons are ejected from a metal surface when light is incident on that surface. Einstein provided a successful explanation of this effect by extending Planck's quantum hypothesis to electromagnetic waves. In this model, light is viewed as a stream of particles called photons, each with energy $E = hf$, where f is the frequency and h is **Planck's constant.** The maximum kinetic energy of the ejected photoelectrons is

$$KE_{max} = hf - \phi \qquad [27.6]$$

where ϕ is the **work function** of the metal.

X-rays are produced when high-speed electrons are suddenly decelerated. When electrons have been accelerated through a voltage ΔV, the shortest wavelength radiation that can be produced is

$$\lambda_{min} = \frac{hc}{e \, \Delta V} \qquad [27.9]$$

The regular array of atoms in a crystal can act as a diffraction grating for x-rays and for electrons. The condition for constructive interference of the diffracted rays is given by **Bragg's law:**

$$2d \sin \theta = m\lambda \qquad m = 1, 2, 3, \ldots \qquad [27.10]$$

X-rays from an incident beam are scattered at various angles by electrons in a target such as carbon. In such a scattering event, a shift in wavelength is observed for the scattered x-rays. This phenomenon is known as the **Compton shift.** Conservation of momentum applied to a photon-electron collision yields the following expression for the shift in wavelength of the scattered x-rays:

$$\Delta\lambda = \lambda - \lambda_0 = \frac{h}{m_e c}(1 - \cos\theta) \qquad [27.11]$$

where m_e is the mass of the electron, c is the speed of light, and θ is the scattering angle.

Pair production is a process in which the energy of a photon is converted into mass. In this process, the photon disappears as an electron-positron pair is created. Likewise, the energy of an electron-positron pair can be converted into electromagnetic radiation by the process of **pair annihilation.**

De Broglie proposed that all matter has both a particle and a wave nature. The **de Broglie wavelength** of any particle of mass m and speed v is

$$\lambda = \frac{h}{p} = \frac{h}{mv} \qquad [27.15]$$

De Broglie also proposed that the frequencies of the waves associated with particles obey the Einstein relationship, $E = hf.$

In the theory of **quantum mechanics,** each particle is described by a quantity Ψ called the **wave function.** The probability of finding the particle at a particular point at some instant is proportional to Ψ^2. Quantum mechanics is very successful in describing the behavior of atomic and molecular processes.

According to Heisenberg's **uncertainty principle,** it is impossible to measure simultaneously the exact position and exact momentum of a particle. If Δx is the uncertainty in the measured position and Δp_x the uncertainty in the momentum, the product $\Delta x \, \Delta p_x$ is given by

$$\Delta x \, \Delta p_x \geq \frac{h}{4\pi} \qquad [27.17]$$

Also

$$\Delta E \, \Delta t \geq \frac{h}{4\pi} \qquad [27.18]$$

where ΔE is the uncertainty in the energy of the particle and Δt is the uncertainty in the time it takes to measure the energy.

MULTIPLE-CHOICE QUESTIONS

1. A photon of energy E_0 strikes a free electron, with the scattered photon of energy E moving in the direction opposite that of the incident photon. In this Compton effect interaction, the resulting kinetic energy of the electron is
   (a) E_0 (b) E (c) $E_0 - E$ (d) $E_0 + E$ (e) Not given.

2. A photon of energy E_0 strikes a free electron, with the scattered photon of energy E moving in the direction opposite that of the incident photon. In this Compton effect interaction, the resulting momentum of the electron is
   (a) E_0/c (b) $<E_0/c$ (c) $>E_0/c$ (d) $(E_0 - E)/c$
   (e) $(E - E_0)/c$

3. A monochromatic light beam is incident on a barium target that has a work function of 2.50 eV. If a stopping potential of 1.00 V is required, what is the wavelength of the light beam?
   (a) 355 nm (b) 497 nm (c) 744 nm (d) 1.42 pm

4. The Sun's surface temperature is 5800 K and the peak wavelength in its radiation is 500 nm. What is the surface temperature of a distant star where the peak wavelength is 475 nm?
(a) 5510 K   (b) 5626 K   (c) 6105 K   (d) 6350 K

5. What is the de Broglie wavelength of an electron accelerated from rest through a potential difference of 50.0 V?
(a) 1.00×10^{-10} m   (b) 1.39×10^{-10} m
(c) 1.74×10^{-10} m   (d) 8.34×10^{-10} m

CONCEPTUAL QUESTIONS

1. If you observe objects inside a very hot kiln, it is difficult to discern the shapes of the objects. Why?
2. Why is an electron microscope more suitable than an optical microscope for "seeing" objects of an atomic size?
3. Are blackbodies really black?
4. Why is it impossible to measure simultaneously the position and velocity of a particle with infinite accuracy?
5. All objects radiate energy. Why, then, are we not able to see all objects in a dark room?
6. Is light a wave or a particle? Support your answer by citing specific experimental evidence.
7. A student claims that he is going to eject electrons from a piece of metal by placing a radio transmitter antenna adjacent to the metal and sending a strong AM radio signal into the antenna. The work function of a metal is typically a few electron volts. Will this work?
8. Does the process of pair production violate conservation of mass?
9. In the photoelectric effect, explain why the stopping po-

tential depends on the frequency of the light but not on the intensity.
10. Which has more energy, a photon of ultraviolet radiation or a photon of yellow light?
11. Why does the existence of a cutoff frequency in the photoelectric effect favor a particle theory of light rather than a wave theory?
12. What effect, if any, would you expect the temperature of a material to have on the ease with which electrons can be ejected from it in the photoelectric effect?
13. An x-ray photon is scattered by an electron. What happens to the frequency of the scattered photon relative to that of the incident photon?
14. The brightest star in the constellation Lyra is the bluish star Vega, whereas the brightest star in Bootes is the reddish star Arcturus. How do you account for the difference in color of the two stars?
15. If the photoelectric effect is observed for one metal, can you conclude that the effect will also be observed for another metal under the same conditions? Explain.

PROBLEMS

1, 2, 3 = straightforward, intermediate, challenging   □ = full solution available in Study Guide/Student Solutions Manual   ◤ = Core Concepts Workbook
WEB = solution posted at **http://www.harcourtcollege.com/physics/cptech**   ▨ = biomedical application   ◉ = Interactive Physics

Section 27.1 Blackbody Radiation and Planck's Hypothesis

1. (a) What is the surface temperature of Betelgeuse, a red giant star in the constellation of Orion, which radiates with a peak wavelength of about 970 nm? (b) Rigel, a bluish-white star in Orion, radiates with a peak wavelength of 145 nm. Find the temperature of Rigel's surface.
2. A certain light source is found to emit radiation whose peak value has a frequency of 1.00×10^{15} Hz. Find the temperature of the source assuming that it is a blackbody radiator.
3. If the surface temperature of the Sun is 5800 K, find the wavelength that corresponds to the maximum rate of energy emission from the Sun.
4. (a) Assuming that the tungsten filament of a lightbulb is a blackbody, determine its peak wavelength if its temperature is 2900 K. (b) Why does your answer to part

(a) suggest that more energy from a lightbulb goes into heat than into light?
5. Calculate the energy in electron volts of a photon having a wavelength in (a) the microwave range, 5.00 cm, (b) the visible light range, 500 nm, and (c) the x-ray range, 5.00 nm.
6. A quantum of electromagnetic radiation has an energy of 2.0 keV. What is its wavelength?
7. An FM radio transmitter has a power output of 150 kW and operates at a frequency of 99.7 MHz. How many photons per second does the transmitter emit?
8. The threshold of dark-adapted (scotopic) vision is 4.0×10^{-11} W/m² at a central wavelength of 500 nm. If light with this intensity and wavelength enters the eye when the pupil is open to its maximum diameter of 8.5 mm, how many photons per second enter the eye?

9. A 1.5-kg mass vibrates at an amplitude of 3.0 cm on the end of a spring of spring constant 20.0 N/m. (a) If the energy of the spring is quantized, find its quantum number. (b) If n changes by 1, find the fractional change in energy of the spring.

10. A 0.50-kg mass falls from a height of 3.0 m. If all of the energy of this mass could be converted to visible light of wavelength 5.0×10^{-7} m, how many photons would be produced?

Section 27.2   The Photoelectric Effect

11. From the scattering of sunlight, Thomson calculated that the classical radius of the electron has a value of 2.82×10^{-15} m. If sunlight having an intensity of 500 W/m² falls on a disk with this radius, estimate the time required to accumulate 1.00 eV of energy. Assume that light is a classical wave and that the light striking the disk is completely absorbed. How does your estimate compare with the observation that photoelectrons are promptly (within 10^{-9} s) emitted?

12. Electrons are ejected from a metallic surface with speeds ranging up to 4.6×10^5 m/s when light with a wavelength of $\lambda = 625$ nm is used. (a) What is the work function of the surface? (b) What is the cutoff frequency for this surface?

WEB 13. When light of wavelength 350 nm falls on a potassium surface, electrons are emitted that have a maximum kinetic energy of 1.31 eV. Find (a) the work function of potassium, (b) the cutoff wavelength, and (c) the frequency corresponding to the cutoff wavelength.

14. Consider the metals lithium, aluminum, and mercury, which have work functions of 2.3 eV, 4.1 eV, and 4.5 eV, respectively. If light of wavelength 3.0×10^{-7} m is incident on each of these metals, determine (a) which metals exhibit the photoelectric effect and (b) the maximum kinetic energy for the photoelectrons for those that exhibit the effect.

15. When a certain metal is illuminated with light of frequency 3.0×10^{15} Hz, a stopping potential of 7.0 V is required to stop the most energetic ejected electrons. What is the work function of this metal?

16. What wavelength light would have to fall on sodium (work function 2.46 eV) if it is to emit electrons with a maximum speed of 1.00×10^6 m/s?

17. When light of wavelength 254 nm falls on cesium, the required stopping potential is 3.00 V. If light of wavelength 436 nm is used, the stopping potential is 0.900 V. Use this information to plot a graph such as the one shown in Figure 27.6, and from the graph determine the cutoff frequency for cesium and its work function.

18. Ultraviolet light is incident normally on the surface of a certain substance. The binding energy of the electrons in this substance is 3.44 eV. The incident light has an intensity of 0.055 W/m². The electrons are photoelectri-

cally emitted with a maximum speed of 4.2×10^5 m/s. How many electrons are emitted from a square centimeter of the surface? Assume that 100% of the photons are absorbed.

19. The extremes of the x-ray portion of the electromagnetic spectrum range from approximately 1.0×10^{-8} m to 1.0×10^{-13} m. Find the minimum accelerating voltages required to produce wavelengths at these two extremes.

20. Calculate the minimum wavelength x-ray that can be produced when a target is struck by an electron that has been accelerated through a potential difference of (a) 15.0 kV; (b) 100 kV.

21. What minimum accelerating voltage would be required to produce an x-ray with a wavelength of 0.0300 nm?

Section 27.5   Diffraction of X-Rays by Crystals (Optional)

22. A monochromatic x-ray beam is incident on a NaCl crystal surface where $d = 0.353$ nm. The second-order maximum in the reflected beam is found when the angle between the incident beam and the surface is 20.5°. Determine the wavelength of the x-rays.

23. Potassium iodide has an interplanar spacing of $d = 0.296$ nm. A monochromatic x-ray beam shows a first-order diffraction maximum when the grazing angle is 7.6°. Calculate the x-ray wavelength.

24. The spacing between planes of nickel atoms in a nickel crystal is 0.352 nm. At what angle does a second-order Bragg reflection occur in nickel for 11.3-keV x-rays?

WEB 25. X-rays of wavelength 0.140 nm are reflected from a certain crystal, and the first-order maximum occurs at an angle of 14.4°. What value does this give for the interplanar spacing of this crystal?

Section 27.6   The Compton Effect

26. X-rays are scattered from electrons in a carbon target. The measured wavelength shift is 1.50×10^{-3} nm. Calculate the scattering angle.

27. A 0.0016-nm photon scatters from a free electron. For what (photon) scattering angle will the recoiling electron and scattered photon have the same kinetic energy?

28. A beam of 0.68-nm photons undergoes Compton scattering from free electrons. What are the energy and momentum of the photons that emerge at a 45° angle with respect to the incident beam?

29. X-rays with an energy of 300 keV undergo Compton scattering from a target. If the scattered rays are deflected at 37° relative to the direction of the incident rays, find (a) the Compton shift at this angle, (b) the energy of the scattered x-ray, and (c) the kinetic energy of the recoiling electron.

30. After a 0.800 nm x-ray photon scatters from a free electron, the electron recoils with a speed equal to 1.40×10^6 m/s. (a) What was the Compton shift in the

photon's wavelength? **(b)** Through what angle was the photon scattered?

31. A 0.45-nm x-ray photon is deflected through a 23° angle after scattering from a free electron. **(a)** What is the kinetic energy of the recoiling electron? **(b)** What is its speed?

Section 27.7 Pair Production and Annihilation

32. How much total kinetic energy will an electron-positron pair have if produced by a photon of energy 3.00 MeV?

33. If an electron-positron pair with a total kinetic energy of 2.50 MeV is produced, find **(a)** the energy of the photon that produced the pair and **(b)** its frequency.

34. Two photons are produced when a proton and an antiproton annihilate each other. What is the minimum frequency and corresponding wavelength of each photon?

35. An electron moving at a speed of $0.60c$ collides head on with a positron also moving at $0.60c$. Determine the energy and momentum of each photon produced in the process.

Section 27.9 The Wave Properties of Particles

36. A 0.200-kg ball is released from rest at the top of a 50.0-m tall building. Find the de Broglie wavelength of the ball just before it strikes the Earth.

37. **(a)** If the wavelength of an electron is equal to 5.00×10^{-7} m, how fast is it moving? **(b)** If the electron has a speed of 1.00×10^{7} m/s, what is its wavelength?

38. Through what potential difference would an electron have to be accelerated from rest to give it a de Broglie wavelength of 1.0×10^{-10} m?

39. Calculate the de Broglie wavelength of a proton moving at **(a)** 2.00×10^{4} m/s; **(b)** 2.00×10^{7} m/s.

40. A monoenergetic beam of electrons is incident on a single slit of width 0.500 nm. A diffraction pattern is formed on a screen 20.0 cm from the slit. If the distance between successive minima of the diffraction pattern is 2.10 cm, what is the energy of the incident electrons?

WEB 41. De Broglie postulated that the relationship $\lambda = h/p$ is valid for relativistic particles. What is the de Broglie wavelength for a (relativistic) electron whose kinetic energy is 3.00 MeV?

42. At what speed must an electron move so that its de Broglie wavelength equals its Compton wavelength? (*Hint:* This electron is relativistic.)

43. The resolving power of a microscope is proportional to the wavelength used. A resolution of approximately 1.0×10^{-11} m (0.010 nm) would be required in order to "see" an atom. **(a)** If electrons were used (electron microscope), what minimum kinetic energy would be required for the electrons? **(b)** If photons were used, what minimum photon energy would be needed to obtain 1.0×10^{-11} m resolution?

Section 27.11 The Uncertainty Principle

44. A 50.0-g ball moves at 30.0 m/s. If its speed is measured to an accuracy of 0.10%, what is the minimum uncertainty in its position?

45. A 0.50-kg block rests on the icy surface of a frozen pond, which we can assume to be frictionless. If the location of the block is measured to a precision of 0.50 cm, what speed must the block acquire because of the measurement process?

46. In the ground state of hydrogen, the uncertainty in the position of the electron is roughly 0.10 nm. If the speed of the electron is on the order of the uncertainty in the speed, how fast is the electron moving?

47. Suppose optical radiation ($\lambda = 5.00 \times 10^{-7}$ m) is used to determine the position of an electron to within the wavelength of the light. What will be the resulting uncertainty in the electron's velocity?

48. **(a)** Show that the kinetic energy of a nonrelativistic particle can be written in terms of its momentum as $KE = p^2/2m$. **(b)** Use the results of **(a)** to find the minimum kinetic energy of a proton confined within a nucleus having a diameter of 1.0×10^{-15} m.

ADDITIONAL PROBLEMS

49. Figure P27.49 shows the spectrum of light emitted by a firefly. Determine the temperature of a blackbody that would emit radiation peaked at the same frequency. Based on your result, would you say firefly radiation is blackbody radiation?

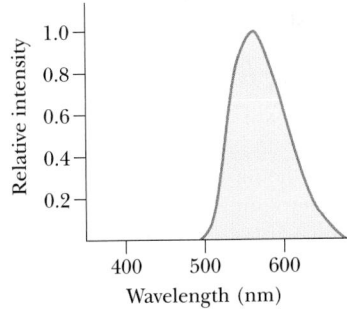

Figure P27.49

50. How many photons are emitted per second by a 100.0-W sodium lamp if the wavelength of sodium light is 589.3 nm?

51. A 70.0-kg gorilla swings at the end of a vine at a frequency of 0.50 Hz at 2.0 m/s as she moves through the lowest point on her arc. **(a)** Assume the energy is quantized and find the quantum number n for the system. **(b)** Find the energy carried away in a one-quantum change in her energy.

52. An x-ray tube is operated at 50 000 V. (a) Find the minimum wavelength of the radiation emitted by this tube. (b) If this radiation is directed at a crystal, the first-order maximum in the reflected radiation occurs when the angle of incidence is 2.5°. What is the spacing between reflecting planes in the crystal?

53. The spacing between certain planes in a crystal is known to be 0.30 nm. Find the smallest grazing angle at which constructive interference will occur for wavelength 0.070 nm.

54. Photons of wavelength 450 nm are incident on a metal. The most energetic electrons ejected from the metal are bent into a circular arc of radius 20.0 cm by a magnetic field whose strength is 2.00×10^{-5} T. What is the work function of the metal?

WEB 55. A light source of wavelength λ illuminates a metal and ejects photoelectrons with a maximum kinetic energy of 1.00 eV. A second light source of wavelength $\lambda/2$ ejects photoelectrons with a maximum kinetic energy of 4.00 eV. What is the work function of the metal?

56. Red light of wavelength 670.0 nm produces photoelectrons from a certain photoemissive material. Green light of wavelength 520.0 nm produces photoelectrons from the same material with 1.50 times the maximum kinetic energy. What is the material's work function?

57. How fast must an electron be moving if all its kinetic energy is lost to a single x-ray photon (a) at the high end of the x-ray electromagnetic spectrum with a wavelength of 1.00×10^{-8} m; (b) at the low end of the x-ray electromagnetic spectrum with a wavelength of 1.00×10^{-13} m?

58. After learning about de Broglie's hypothesis that particles of momentum p have wave characteristics with wavelength $\lambda = h/p$, an 80-kg student has grown concerned about being diffracted when passing through a 75-cm-wide doorway. Assuming that significant diffraction occurs when the width of the diffraction aperture is less that 10 times the wavelength of the wave being diffracted, (a) determine the maximum speed at which the student can pass through the doorway in order to

be significantly diffracted. (b) With that speed, how long will it take the student to pass through the doorway if it is 15 cm thick? Compare your result to the currently accepted age of the Universe, which is 4×10^{17} s. (c) Should this student worry about being diffracted?

59. Calculate the kinetic energy in MeV of electrons having a de Broglie wavelength of 1.0×10^{-3} nm.

60. Show that if an electron were confined inside an atomic nucleus of diameter 2.0×10^{-15} m, it would have to be moving relativistically, while a proton confined to the same nucleus can be moving at less than one tenth the speed of light.

61. A photon strikes a metal with a work function of ϕ and produces a photoelectron with a de Broglie wavelength equal to the wavelength of the original photon. (a) Show that the energy of this photon must have been given by
$$E = \frac{\phi(m_e c^2 - \phi/2)}{m_e c^2 - \phi},$$ where m_e is the mass of the electron. (*Hint:* Begin with conservation of energy, $E + m_e c^2 = \phi + \sqrt{(pc)^2 + (m_e c^2)^2}$.) (b) If one of these photons strikes platinum ($\phi = 6.35$ eV), determine the resulting maximum speed of the photoelectron.

62. In a Compton scattering event, the scattered photon has an energy of 120.0 keV and the recoiling electron has a kinetic energy of 40.0 keV. Find (a) the wavelength of the incident photon, (b) the angle θ at which the photon is scattered, and (c) the recoil angle of the electron. (*Hint:* Conserve both mass-energy and relativistic momentum.)

63. A woman on a ladder drops small pellets toward a spot on the floor. (a) Show that, according to the uncertainty principle, the average miss distance must be at least

$$\Delta x = \left(\frac{\hbar}{2m}\right)^{1/2} \left(\frac{H}{2g}\right)^{1/4}$$

where H is the initial height of each pellet above the floor and m is the mass of each pellet. (b) If $H = 2.00$ m and $m = 0.500$ g, what is Δx?

Atomic Physics

▲ **PHYSICS PUZZLER**

"Neon" lights are common in advertising signs and can appear in a variety of colors. How do these lights work, and what determines the color of the light?

(Dembinsky Photo Assoc.)

A large portion of this chapter is concerned with the study of the hydrogen atom. Although the hydrogen atom is the simplest atomic system, it is especially important for several reasons:

- The quantum numbers used to characterize the allowed states of hydrogen can also be used to describe (approximately) the allowed states of more complex atoms. This enables us to understand the periodic table of the elements, one of the greatest triumphs of quantum mechanics.
- The hydrogen atom is an ideal system for performing precise comparisons of theory and experiment and for improving our overall understanding of atomic structure.
- Much of what is learned about the hydrogen atom with its single electron can be extended to such single-electron ions as He^+ and Li^{2+}.

In this chapter we first discuss the Bohr model of hydrogen, which helps us understand many features of hydrogen but fails to explain many finer details of atomic structure. Next we examine the hydrogen atom from the viewpoint of quantum mechanics and the quantum numbers used to characterize various atomic states. In addition, we examine the physical significance of the quantum numbers and the effect of a magnetic field on certain quantum states. The Pauli exclusion principle is also presented. This physical principle is extremely important in understanding the properties of complex atoms and the arrangement of elements in the periodic table. Finally, we apply our knowledge of atomic structure to describe the mechanisms involved in the production of x-rays and the operation of a laser.

28.1   EARLY MODELS OF THE ATOM

The model of the atom in the days of Newton was a tiny, hard, indestructible sphere. Although this model was a good basis for the kinetic theory of gases, new models had to be devised when later experiments revealed the electrical nature of atoms. J. J. Thomson (1856–1940) suggested a model of the atom as a volume of positive charge with electrons embedded throughout the volume, much like the seeds in a watermelon (Fig. 28.1).

In 1911 Ernest Rutherford (1871–1937) and his students Hans Geiger and Ernest Marsden performed a critical experiment showing that Thomson's model could not be correct. In this experiment, a beam of positively charged **alpha particles** was projected against a thin metal foil, as in Figure 28.2a. The results of the experiment were astounding. Most of the alpha particles passed through the foil as if it were empty space. But many particles deflected from their original direction of travel were scattered through large angles. Some particles were even deflected backwards reversing their direction of travel. When Geiger informed Rutherford of these results, Rutherford wrote, "It was quite the most incredible event that has ever happened to me in my life. It was almost as incredible as if you fired a 15-inch shell at a piece of tissue paper and it came back and hit you."

Such large deflections were not expected on the basis of Thomson's model. According to this model, a positively charged alpha particle would never come close enough to a large positive charge to cause any large-angle deflections. Rutherford explained these astounding results by assuming that the positive charge in an atom was concentrated in a region that was small relative to the size of the atom. He

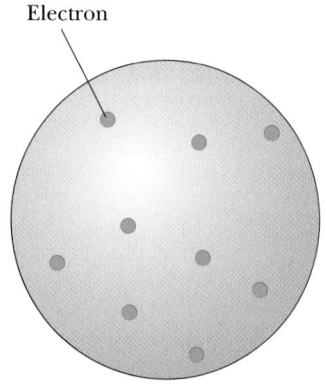

Electron

Figure 28.1 Thomson's model of the atom, with the electrons embedded inside the positive charge like seeds in a watermelon.

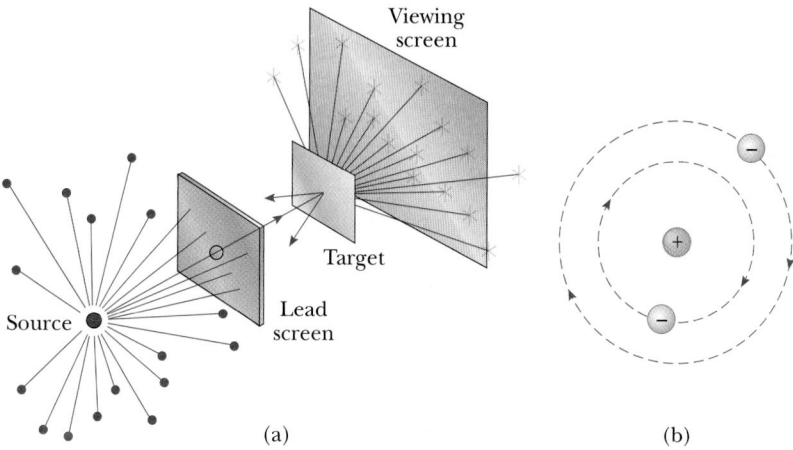

Figure 28.2 (a) Geiger and Marsden's technique for observing the scattering of alpha particles from a thin foil target. The source is a naturally occurring radioactive substance, such as radium. (b) Rutherford's planetary model of the atom.

called this concentration of positive charge the **nucleus** of the atom. Any electrons belonging to the atom were assumed to be in the relatively large volume outside the nucleus. In order to explain why electrons in this outer region of the atom were not pulled into the nucleus, Rutherford viewed them as moving in orbits about the positively charged nucleus in the same manner as the planets orbit the Sun, as shown in Figure 28.2b. Alpha particles themselves were later identified as the nuclei of helium atoms.

There are two basic difficulties with Rutherford's planetary model. As we shall see in the next section, an atom emits certain discrete characteristic frequencies of electromagnetic radiation and no others; the Rutherford model is unable to explain this phenomenon. A second difficulty is that Rutherford's electrons are undergoing a centripetal acceleration. According to Maxwell's theory of electromagnetism, centripetally accelerated charges revolving with frequency f should radiate electromagnetic waves of the same frequency. Unfortunately, this classical model leads to disaster when applied to the atom. As the electron radiates energy, the radius of its orbit steadily decreases and its frequency of revolution increases. This leads to an ever-increasing frequency of emitted radiation and a rapid collapse of the atom as the electron plunges into the nucleus.

28.2   ATOMIC SPECTRA

As you may have already learned in chemistry, the hydrogen atom is the simplest atomic system, and an especially important one to understand. Much of what we know about the hydrogen atom (which consists of one proton and one electron) can be extended directly to other single-electron ions such as He^+ and Li^{2+}. Furthermore, a thorough understanding of the physics underlying the hydrogen atom can then be used to describe more complex atoms and the periodic table of the elements.

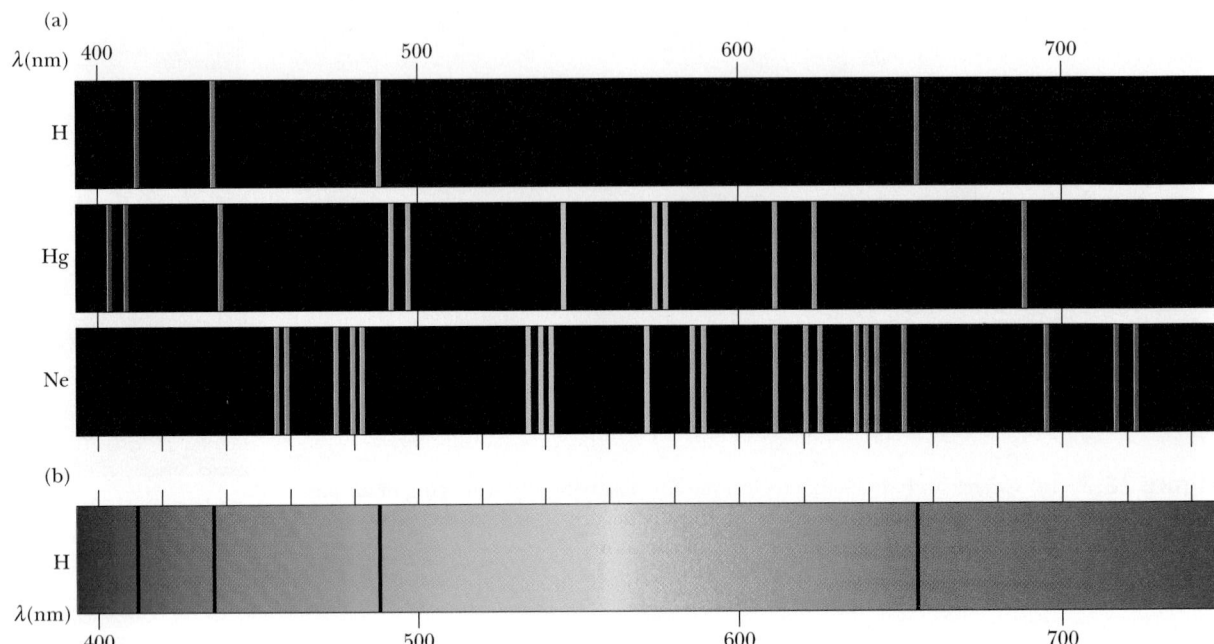

Figure 28.3 Visible spectra. (a) Line spectra produced by emission in the visible range for the elements hydrogen, mercury, and neon. (b) The absorption spectrum for hydrogen. The dark absorption lines occur at the same wavelengths as the emission lines for hydrogen shown in (a).

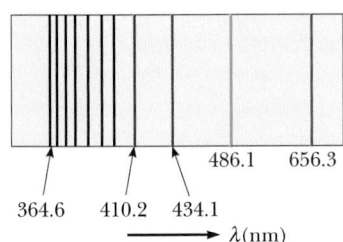

Figure 28.4 A series of spectral lines for atomic hydrogen. The prominent labeled lines are part of the Balmer series.

Suppose an evacuated glass tube is filled with hydrogen (or some other gas) at very low pressure. If a voltage applied between metal electrodes in the tube is great enough to produce an electric current in the gas, the tube emits light whose color is characteristic of the gas in the tube. (This is how a neon sign works.) When the emitted light is analyzed with a spectrometer, a series of discrete lines is observed, each line corresponding to a different wavelength, or color, of light. Such a series of spectral lines is commonly referred to as an **emission spectrum.** The wavelengths contained in a given line spectrum are characteristic of the element emitting the light (Fig. 28.3). Because no two elements emit the same line spectrum, this phenomenon represents a marvelous and reliable technique for identifying elements in a substance.

The emission spectrum of hydrogen shown in Figure 28.4 includes four prominent lines that occur at wavelengths of 656.3 nm, 486.1 nm, 434.1 nm, and 410.2 nm. In 1885 Johann Balmer (1825–1898) found that the wavelengths of these and less prominent lines can be described by the simple empirical equation:

Balmer series ▶

$$\frac{1}{\lambda} = R_{\mathrm{H}}\left(\frac{1}{2^2} - \frac{1}{n^2}\right)$$ [28.1]

where n may have integral values of 3, 4, 5, . . . , and R_{H} is a constant, called the **Rydberg constant.** If the wavelength is in meters, R_{H} has the value

Rydberg constant ▶

$$R_{\mathrm{H}} = 1.097\ 373\ 2 \times 10^7\ \mathrm{m}^{-1}$$ [28.2]

The first line in the Balmer series, at 656.3 nm, corresponds to $n = 3$ in Equation 28.1; the line of 486.1 nm corresponds to $n = 4$; and so on. In addition to this Balmer series of spectral lines, a Lyman series was subsequently discovered in the far ultraviolet, with the radiated wavelengths described by a similar equation.

In addition to emitting light at specific wavelengths, an element can also absorb light at specific wavelengths. The spectral lines corresponding to this process form what is known as an **absorption spectrum.** An absorption spectrum can be obtained by passing a continuous radiation spectrum (one containing all wavelengths) through a vapor of the element being analyzed. The absorption spectrum consists of a series of dark lines superimposed on the otherwise continuous spectrum. Each line in the absorption spectrum of a given element coincides with a line in the emission spectrum of the element. That is, if hydrogen is the absorbing vapor, dark lines will appear at the visible wavelengths 656.3 mn, 486.1 nm, 434.1 nm, and 410.2 nm, as shown in Figures 28.3b and 28.4.

The absorption spectrum of an element has many practical applications. For example, the continuous spectrum of radiation emitted by the Sun must pass through the cooler gases of the solar atmosphere and then through the Earth's atmosphere. The various absorption lines observed in the solar spectrum have been used to identify elements in the solar atmosphere. It is interesting to note that, when the solar spectrum was first being studied, some lines were found that did not correspond to any known element. A new element had been discovered! Because the Greek word for Sun is *helios,* the new element was named *helium.* It was later identified in underground gases on Earth. Scientists are able to examine the light from stars other than our Sun in this fashion, but elements other than those present on Earth have never been detected.

APPLICATION

Discovery of Helium.

Thinking Physics 1

You are observing a yellow candle flame, and your laboratory partner claims that the light from the flame is atomic in origin. You disagree, claiming that the candle flame is hot, so the radiation must be thermal in origin. Before the disagreement leads to fisticuffs, how could you determine who is correct?

Explanation A simple determination could be made by observing the light from the candle flame through a spectrometer, which is a slit and diffraction grating combination discussed in Chapter 25. If the spectrum of the light is continuous, then it is thermal in origin. If the spectrum shows discrete lines, it is atomic in origin. The results of the experiment show that the light is indeed thermal in origin and originates from random molecular motion in the candle flame.

Applying Physics 1

At extreme northern latitudes, the aurora borealis provides a beautiful and colorful display in the nighttime sky. A similar display occurs near the southern polar region and is called the aurora australis. What is the origin of the various colors of radiation seen in the auroras?

APPLICATION

The Northern and Southern Lights.

Explanation   The aurora is due to high-speed particles interacting with the Earth's magnetic field and entering the atmosphere, as discussed in Physics in Action in Chapter 19. When these particles collide with molecules in the atmosphere, they excite the atoms in a way similar to the voltage in the spectrum tubes discussed earlier in this section. In response, the molecules emit colors of light according to the characteristic spectrum of the atoms. For our atmosphere, the primary constituents are nitrogen and oxygen, which provide the red, blue, and green colors of the aurora.

28.3   THE BOHR THEORY OF HYDROGEN

Niels Bohr, Danish physicist (1885–1962)

Bohr was an active participant in the early development of quantum mechanics and provided much of its philosophical framework. During the 1920s and 1930s, Bohr headed the Institute for Advanced Studies in Copenhagen. The institute was a magnet for many of the world's best physicists and provided a forum for the exchange of ideas. When Bohr visited the United States in 1939 to attend a scientific conference, he brought news that the fission of uranium had been observed by Hahn and Strassman in Berlin. The results were the foundations of the atomic bomb developed in the United States during World War II. Bohr was awarded the 1922 Nobel prize for his investigation of the structure of atoms and of the radiation emanating from them.

At the beginning of the 20th century, scientists were perplexed by the failure of classical physics to explain the characteristics of spectra. Why did atoms of a given element emit only certain lines? Furthermore, why did the atoms absorb only those wavelengths that they emitted? In 1913 Bohr provided an explanation of atomic spectra that includes some features of the currently accepted theory. Using the simplest atom, hydrogen, Bohr developed a model of what he thought must be the atom's structure in an attempt to explain why the atom was stable. His model of the hydrogen atom contains some classical features as well as some revolutionary postulates that could not be justified within the framework of classical physics. The basic assumptions of the Bohr theory as it applies to the hydrogen atom are as follows:

1. The electron moves in circular orbits about the proton under the influence of the Coulomb force of attraction, as in Figure 28.5. In this case, the Coulomb force is the force that produces centripetal acceleration.
2. Only certain electron orbits are stable. These are orbits in which the hydrogen atom does not emit energy in the form of radiation. Hence, the total energy of the atom remains constant, and classical mechanics can be used to describe the electron's motion.
3. Radiation is emitted by the hydrogen atom when the electron "jumps" from a more energetic initial state to a lower state. The "jump" cannot be visualized or treated classically. In particular, the frequency, f, of the radiation emitted in the jump is related to the change in the atom's energy and is *independent of the frequency of the electron's orbital motion*. The frequency of the emitted radiation is

$$E_i - E_f = hf \qquad \text{[28.3]}$$

where E_i is the energy of the initial state, E_f is the energy of the final state, h is Planck's constant, and $E_i > E_f$.
4. The size of the allowed electron orbits is determined by a condition imposed on the electron's orbital angular momentum: The allowed orbits are those for which the electron's orbital angular momentum about the nucleus is an integral multiple of $\hbar = h/2\pi$.

$$m_e vr = n\hbar \qquad n = 1, 2, 3, \ldots \qquad \text{[28.4]}$$

With these assumptions, we can calculate the allowed energies and emission wavelengths of the hydrogen atom. We shall use the model pictured in Figure 28.5, in which the electron travels in a circular orbit of radius r with an orbit speed v.

The electrical potential energy of the atom is

$$PE = k_e \frac{q_1 q_2}{r} = k_e \frac{(-e)(e)}{r} = -k_e \frac{e^2}{r}$$

where k_e is the Coulomb constant. Assuming the nucleus is at rest the total energy, E, of the atom is the sum of the kinetic energy and the potential energy:

$$E = KE + PE = \tfrac{1}{2} m_e v^2 - k_e \frac{e^2}{r} \qquad \text{[28.5]}$$

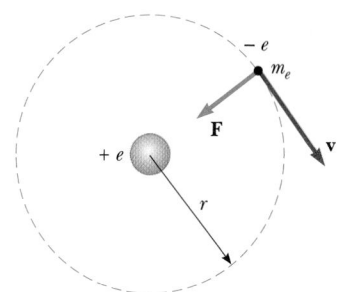

Figure 28.5 Diagram representing Bohr's model of the hydrogen atom. In this model, the orbiting electron is allowed only in specific orbits of discrete radius.

Let us apply Newton's second law to the electron. We know that the electric force of attraction on the electron, $k_e e^2 / r^2$, must equal $m_e a_r$, where $a_r = v^2 / r$ is the centripetal acceleration of the electron. Thus,

$$k_e \frac{e^2}{r^2} = m_e \frac{v^2}{r} \qquad \text{[28.6]}$$

From this equation, we see that the kinetic energy of the electron is

$$\tfrac{1}{2} m v^2 = \frac{k_e e^2}{2r} \qquad \text{[28.7]}$$

We can combine this result with Equation 28.5 and express the **total energy** of the atom as

$$E = -\frac{k_e e^2}{2r} \qquad \text{[28.8]}$$

◀ Total energy of the hydrogen atom

An expression for r is obtained by solving Equations 28.4 and 28.6 for v and equating the results:

$$v^2 = \frac{n^2 \hbar^2}{m_e^2 r^2} = \frac{k_e e^2}{m_e r}$$

$$r_n = \frac{n^2 \hbar^2}{m_e k_e e^2} \qquad n = 1, 2, 3, \dots \qquad \text{[28.9]}$$

◀ The radii of the Bohr orbits are quantized.

This equation is based on the assumption that the **electron can exist only in certain allowed orbits determined by the integer n.**

The orbit with the smallest radius, called the **Bohr radius,** a_0, corresponds to $n = 1$ and has the value

$$a_0 = \frac{\hbar^2}{m_e k_e e^2} = 0.0529 \text{ nm} \qquad \text{[28.10]}$$

A general expression for the radius of any orbit in the hydrogen atom is obtained by substituting Equation 28.10 into Equation 28.9:

$$r_n = n^2 a_0 = n^2 (0.0529 \text{ nm}) \qquad \text{[28.11]}$$

The first three Bohr orbits for hydrogen are shown in Figure 28.6.

Equation 28.9 may be substituted into Equation 28.8 to give the following expression for the energies of the quantum states:

$$E_n = -\frac{m_e k_e^2 e^4}{2\hbar^2} \left(\frac{1}{n^2} \right) \qquad n = 1, 2, 3, \dots \qquad \text{[28.12]}$$

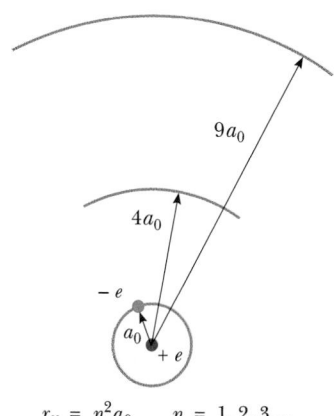

$$r_n = n^2 a_0 \qquad n = 1, 2, 3, \dots$$

Figure 28.6 The first three circular orbits predicted by the Bohr model of the hydrogen atom.

Allowed energies of the hydrogen ▶
atom

If we insert numerical values into Equation 28.12, we find

$$E_n = -\frac{13.6}{n^2}\ \text{eV} \qquad [28.13]$$

The lowest stationary energy state, or **ground state,** corresponds to $n = 1$ and has an energy $E_1 = -m_e k_e^2 e^4 / 2\hbar^2 = -13.6$ eV. The next state, corresponding to $n = 2$, has an energy $E_2 = E_1/4 = -3.40$ eV, and so on. An energy level diagram showing the energies of these stationary states and the corresponding quantum numbers is shown in Figure 28.7. The uppermost level shown, corresponding to $n \rightarrow \infty$, represents the state for which the electron is completely removed from the atom. In this case, $E = 0$ for $r = \infty$. The minimum energy required to ionize the atom — that is, to completely remove the electron — is called the **ionization energy.** The ionization energy for hydrogen is 13.6 eV.

Equations 28.3 and 28.12 and the third Bohr postulate show that if the electron jumps from one orbit, whose quantum number is n_i, to a second orbit, whose quantum number is n_f, it emits a photon of frequency f, given by

$$f = \frac{E_i - E_f}{h} = \frac{m_e k_e^2 e^4}{4\pi \hbar^3}\left(\frac{1}{n_f^2} - \frac{1}{n_i^2}\right) \qquad [28.14]$$

Finally, to compare this result with the empirical formulas for the various spectral series, we use the fact that, for light, $\lambda f = c$ and Equation 28.14 to get

$$\frac{1}{\lambda} = \frac{f}{c} = \frac{m_e k_e^2 e^4}{4\pi c\hbar^3}\left(\frac{1}{n_f^2} - \frac{1}{n_i^2}\right) \qquad [28.15]$$

A comparison of this result with Equation 28.1 gives the following expression for the Rydberg constant:

$$R_{\text{H}} = \frac{m_e k_e^2 e^4}{4\pi c\hbar^3} \qquad [28.16]$$

If we insert the known values of m_e, k_e, e, c, and $\hbar$ into this expression, the resulting theoretical value for R_{H} is found to be in excellent agreement with the value determined experimentally for the Rydberg constant. When Bohr demonstrated this agreement, it was recognized as a major accomplishment of his theory.

In order to compare Equation 28.15 with spectroscopic data, it is convenient to express it in the form

$$\frac{1}{\lambda} = R_{\text{H}}\left(\frac{1}{n_f^2} - \frac{1}{n_i^2}\right) \qquad [28.17]$$

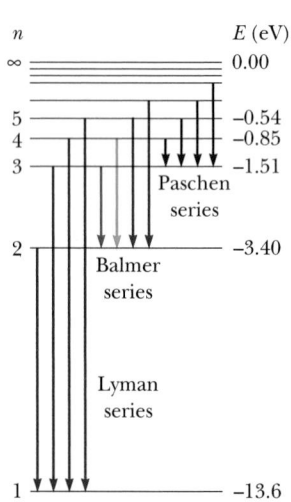

Figure 28.7 An energy level diagram for hydrogen. In such diagrams the discrete allowed energies are plotted on the vertical axis. Nothing is plotted on the horizontal axis, but the horizontal extent of the diagram is made large enough to show allowed transitions. Note that the quantum numbers are given on the left and the energies (in eV) are on the right.

We can use this expression to evaluate the wavelengths for the various series in the hydrogen spectrum. For example, in the Balmer series, $n_f = 2$ and $n_i = 3, 4, 5, \ldots$ (Eq. 28.1). For the Lyman series, we take $n_f = 1$ and $n_i = 2, 3, 4, \ldots$. The energy level diagram for hydrogen, shown in Figure 28.7, indicates the origin of the spectral lines described previously. The transitions between levels are represented by vertical arrows. Note that whenever a transition occurs between a state designated by n_i to one designated by n_f (where $n_i > n_f$), a photon with a frequency of $(E_i - E_f)/h$ is emitted. This can be interpreted as follows. The lines in the visible part of the hydrogen spectrum arise when the electron jumps from the third, fourth, or even higher orbit to the second orbit. Likewise, the lines of the Lyman series (in the ultraviolet) arise when the electron jumps from the second, third, or even higher orbit to the innermost ($n_f = 1$) orbit. Hence, the Bohr theory successfully predicts the wavelengths of all observed spectral lines of hydrogen.

Thinking Physics 2

According to the Bohr model of the hydrogen atom, the electron in the ground state moves in a circular orbit of radius 0.529×10^{-10} m, and the speed of the electron in this state is 2.2×10^6 m/s. How could Bohr's model be so successful initially when its specific orbital radii seem to contradict the "fuzziness" demanded by the Heisenberg uncertainty principle?

Explanation   We have no answer. The Bohr theory works much better than it should. The theory pictures an electron as having definite position and momentum, ignoring quantum uncertainty. The theory proceeds from Newton's second law, when really Schrödinger's equation describes the wave motion of the electron. Scattering experiments show that the electron does not lie on a flat circle, but fills a sphere around the nucleus.

EXAMPLE 28.1   An Electronic Transition in Hydrogen

The electron in the hydrogen atom makes a transition from the $n = 2$ energy state to the ground state (corresponding to $n = 1$). Find the wavelength and frequency of the emitted photon.

Solution   We can use Equation 28.17 directly to obtain λ, with $n_i = 2$ and $n_f = 1$:

$$\frac{1}{\lambda} = R_{\mathrm{H}} \left(\frac{1}{n_f^2} - \frac{1}{n_i^2} \right)$$

$$\frac{1}{\lambda} = R_{\mathrm{H}} \left(\frac{1}{1^2} - \frac{1}{2^2} \right) = \frac{3R_{\mathrm{H}}}{4}$$

$$\lambda = \frac{4}{3R_{\mathrm{H}}} = \frac{4}{3(1.097 \times 10^7 \ \mathrm{m}^{-1})} = 1.215 \times 10^{-7} \ \mathrm{m} = \boxed{121.5 \ \mathrm{nm}}$$

This wavelength lies in the ultraviolet region.
Because $c = f\lambda$, the frequency of the photon is

$$f = \frac{c}{\lambda} = \frac{3.00 \times 10^8 \ \mathrm{m/s}}{1.215 \times 10^{-7} \ \mathrm{m}} = \boxed{2.47 \times 10^{15} \ \mathrm{Hz}}$$

Exercise   What is the wavelength of the photon emitted by hydrogen when the electron makes a transition from the $n = 3$ state to the $n = 1$ state?

Answer $\dfrac{9}{8R_{\mathrm{H}}} = 102.6$ nm

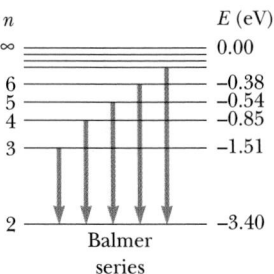

$$
\begin{array}{ll}
n & E \text{ (eV)} \\
\infty & 0.00 \\
6 & -0.38 \\
5 & -0.54 \\
4 & -0.85 \\
3 & -1.51 \\
2 & -3.40
\end{array}
$$

Balmer
series

Figure 28.8 (Example 28.2) Transitions responsible for the Balmer series for the hydrogen atom. All transitions terminate at the $n = 2$ level.

EXAMPLE 28.2 The Balmer Series for Hydrogen

The Balmer series for the hydrogen atom corresponds to electronic transitions that terminate in the state of quantum number $n = 2$, as shown in Figure 28.8.

(a) Find the longest wavelength photon emitted and determine its energy.

Solution The longest wavelength photon in the Balmer series results from the transition from $n = 3$ to $n = 2$. Using Equation 28.17 gives

$$
\frac{1}{\lambda} = R_H \left(\frac{1}{n_f^2} - \frac{1}{n_i^2} \right)
$$

$$
\frac{1}{\lambda_{\text{max}}} = R_H \left(\frac{1}{2^2} - \frac{1}{3^2} \right) = \frac{5}{36} R_H
$$

$$
\lambda_{\text{max}} = \frac{36}{5R_H} = \frac{36}{5(1.097 \times 10^7 \text{ m}^{-1})} = \boxed{656.3 \text{ nm}}
$$

This wavelength is in the red region of the visible spectrum.
The energy of this photon is

$$
E_{\text{photon}} = hf = \frac{hc}{\lambda_{\text{max}}}
$$

$$
= \frac{(6.626 \times 10^{-34} \text{ J} \cdot \text{s})(3.00 \times 10^8 \text{ m/s})}{656.3 \times 10^{-9} \text{ m}}
$$

$$
= 3.03 \times 10^{-19} \text{ J} = \boxed{1.89 \text{ eV}}
$$

We could also obtain the energy of the photon by using Equation 28.3 in the form $hf = E_3 - E_2$, where E_2 and E_3 are the energy levels of the hydrogen atom, calculated from Equation 28.13. Note that this is the lowest energy photon in this series, because it involves the smallest energy change.

(b) Find the shortest wavelength photon emitted in the Balmer series.

Solution The shortest wavelength photon in the Balmer series is emitted when the electron makes a transition from $n = \infty$ to $n = 2$. Therefore,

$$
\frac{1}{\lambda_{\text{min}}} = R_H \left(\frac{1}{2^2} - \frac{1}{\infty} \right) = \frac{R_H}{4}
$$

$$
\lambda_{\text{min}} = \frac{4}{R_H} = \frac{4}{1.097 \times 10^7 \text{ m}^{-1}} = \boxed{364.6 \text{ nm}}
$$

This wavelength is in the ultraviolet region and corresponds to the series limit.

Exercise Find the energy of the shortest wavelength photon emitted in the Balmer series for hydrogen.

Answer 3.40 eV

Bohr's Correspondence Principle

In our study of relativity in Chapter 26, we found that Newtonian mechanics cannot be used to describe phenomena that occur at speeds approaching the speed of light. Newtonian mechanics is a special case of relativistic mechanics and is usable only when v is much smaller than c. Similarly, **quantum mechanics is in agreement with classical physics when the energy differences between quantized levels are very small.** This principle, first set forth by Bohr, is called the **correspondence principle.**

For example, consider the hydrogen atom with $n > 10\ 000$. For such large values of n, the energy differences between adjacent levels approach zero and the levels are nearly continuous. As a consequence, the classical model is reasonably accurate in describing the system for large values of n. According to the classical model, the frequency of the light emitted by the atom is equal to the frequency of revolution of the electron in its orbit about the nucleus. Calculations show that for $n > 10\ 000$, this frequency is different from that predicted by quantum mechanics by less than 0.015%.

28.4 MODIFICATION OF THE BOHR THEORY

The Bohr theory of the hydrogen atom was a tremendous success in certain areas because it explained several features of the spectrum of hydrogen that had previously defied explanation. It accounted for the Balmer series and other series; it predicted a value for the Rydberg constant that is in excellent agreement with the experimental value; it gave an expression for the radius of the atom; and it predicted the energy levels of hydrogen. Although these successes were important to scientists, it is perhaps even more significant that the Bohr theory gave us a model of what the atom looks like and how it behaves. Once a basic model is constructed, refinements and modifications can be made to enlarge on the concept and to explain finer details.

The analysis used in the Bohr theory is also successful when applied to *hydrogen-like* atoms. An atom is said to be hydrogen-like when it contains only one electron. Examples are singly ionized helium, doubly ionized lithium, triply ionized beryllium, and so forth. The results of the Bohr theory for hydrogen can be extended to hydrogen-like atoms by substituting Ze^2 for e^2 in the hydrogen equations, where Z is the atomic number of the element. For example, Equations 28.12 and 28.15 become

$$E_n = -\frac{m_e k_e^2 Z^2 e^4}{2\hbar^2}\left(\frac{1}{n^2}\right) \qquad n = 1, 2, 3, \ldots \qquad \text{[28.18]}$$

and

$$\frac{1}{\lambda} = \frac{m_e k_e^2 Z^2 e^4}{4\pi c\hbar^3}\left(\frac{1}{n_f^2} - \frac{1}{n_i^2}\right) \qquad \text{[28.19]}$$

Although many attempts were made to extend the Bohr theory to more complex (multi-electron) atoms, the results were unsuccessful. Even today, only approximate methods are available for treating multi-electron atoms.

Within a few months following the publication of Bohr's theory, Arnold Sommerfeld (1868–1951) extended the results to include elliptical orbits. We shall examine his model briefly because much of the nomenclature used in this treatment is still in use today. Bohr's concept of quantization of angular momentum led to the **principal quantum number,** n, which determines the energy of the allowed

TABLE 28.1 Shell and Subshell Notations

n	Shell Symbol	ℓ	Subshell Symbol
1	K	0	s
2	L	1	p
3	M	2	d
4	N	3	f
5	O	4	g
6	P	5	h
...		...	

states of hydrogen. Sommerfeld's theory retained n, but also introduced a new quantum number, ℓ, called the **orbital quantum number,** where the value of ℓ ranges from 0 to $n - 1$ in integer steps. According to this model, an electron in any one of the allowed energy states of a hydrogen atom may move in any one of a number of orbits. For each value of n there are n possible orbits. Because $n = 1$ and $\ell = 0$ for the first energy level (ground state), there is only one possible orbit for this state. The second energy level, with $n = 2$, has two possible orbits corresponding to $\ell = 0$ and $\ell = 1$. The third energy level, with $n = 3$, has three possible orbits corresponding to $\ell = 0$, $\ell = 1$, and $\ell = 2$.

For historical reasons, **all states with the same principal quantum number are said to form a shell.** Shells are identified by the letters K, L, M, . . . , which designate the states for which $n = 1, 2, 3, ...$ Likewise, **the states with given values of n and ℓ are said to form a subshell.** The letters s, p, d, f, g, . . . are used to designate the states for which $\ell = 0, 1, 2, 3, 4, ...$ These notations are summarized in Table 28.1.

States that violate the rules given in Table 28.1 cannot exist. For instance, the $2d$ state, which would have $n = 2$ and $\ell = 2$, cannot exist because the highest allowed value of ℓ is $n - 1$, or 1 in this case. Thus, for $n = 2$, $2s$ and $2p$ are allowed states but $2d$, $2f$, . . . are not. For $n = 3$, the allowed subshells are $3s$, $3p$, and $3d$. The maximum number of electrons allowed in any given subshell is given by $2(2\ell + 1)$. For example, any p subshell ($\ell = 1$) is filled when it contains six electrons. This fact will be important to us later when we discuss the *Pauli exclusion principle.*

Another modification of the Bohr theory arose when it was discovered that the spectral lines of a gas are split into several closely spaced lines when the gas is placed in a strong magnetic field. (This is called the *Zeeman effect,* after its discoverer.) Figure 28.9 shows a single spectral line being split into three closely spaced lines. This observation indicates that the energy of an electron is slightly modified when the atom is immersed in a magnetic field. In order to explain this observation, a new quantum number, m_ℓ, called the **orbital magnetic quantum number,** was introduced. The theory is in accord with experimental results when m_ℓ is restricted to values ranging from $-\ell$ to $+\ell$, in integer steps.

Finally, very high resolution spectrometers revealed that spectral lines of gases are in fact two very closely spaced lines even in the absence of an external magnetic field. This splitting was referred to as **fine structure.** In 1925 Samuel Goudsmit and George Uhlenbeck introduced the idea of an electron spinning about its own axis

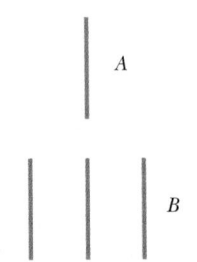

Figure 28.9 A single line (A) can split into three separate lines (B) in a magnetic field.

to explain the origin of fine structure. The results of their work introduced yet another quantum number, m_s, called the **spin magnetic quantum number.** We shall save further discussion of this quantum number for a later section. It is interesting to note that each of the new concepts introduced into the original Bohr theory added new quantum numbers and improved the original model. However, the most profound step forward in our current understanding of atomic structure came with the development of quantum mechanics.

EXAMPLE 28.3 Singly Ionized Helium

Singly ionized helium, He^+, a hydrogen-like system, has one electron in the $1s$ orbit when the atom is in its ground state. Find (a) the energy of the system in the ground state and (b) the radius of the ground-state orbit.

Solution (a) From Equation 28.18, the energy of a level whose principal quantum number is n is given by

$$E_n = -\frac{m_e k_e^2 Z^2 e^4}{2\hbar^2}\left(\frac{1}{n^2}\right)$$

This can be expressed in eV units as

$$E_n = -\frac{Z^2(13.6)}{n^2}\text{ eV}$$

Because $Z = 2$ for helium, and $n = 1$ in the ground state, we have

$$E_1 = -4(13.6)\text{ eV} = \boxed{-54.4\text{ eV}}$$

(b) The radius of the ground state orbit can be found with the help of Equation 28.9. This equation must be modified in the case of a hydrogen-like atom by substituting Ze^2 for e^2 to obtain

$$r_n = \frac{n^2\hbar^2}{m_e k_e Z e^2} = \frac{n^2}{Z}(0.0529\text{ nm})$$

For our case, $n = 1$ and $Z = 2$, and the result is

$$r_1 = \boxed{0.0265\text{ nm}}$$

28.5 DE BROGLIE WAVES AND THE HYDROGEN ATOM

One of the postulates made by Bohr in this theory of the hydrogen atom was that angular momentum of the electron is quantized in units of $\hbar = h/2\pi$, or

$$m_e vr = n\hbar$$

For more than a decade following Bohr's publication, no one was able to explain why the angular momentum of the electron was restricted to these discrete values. Finally, de Broglie recognized a connection between his theory of the wave character of material properties and the quantization condition given previously.

de Broglie assumed that an electron orbit would be stable (allowed) only if it contained an integral number of electron wavelengths. Figure 28.10a demonstrates this point when three complete wavelengths are contained in one circumference of the orbit. Similar patterns can be drawn for orbits containing one wavelength, two wavelengths, four wavelengths, five wavelengths, and so forth. This situation is analogous to that of standing waves on a string, discussed in Chapter 14. There we found that strings have preferred (resonant) frequencies of vibration. Figure 28.10b shows a standing wave pattern containing three wavelengths for a string fixed at each end. Now imagine that the vibrating string is removed from its supports at A and B and bent into a circular shape that brings points A and B together. The end result is a pattern such as the one shown in Figure 28.10a.

In general, the condition for a de Broglie standing wave in an electron orbit is that the circumference must contain an integral multiple of electron wavelengths. We can express this condition as

$$2\pi r = n\lambda \qquad n = 1, 2, 3, \ldots$$

Because the de Broglie wavelength of an electron is $\lambda = h/m_e v$, we can write the preceding equation as $2\pi r = nh/m_e v$, or

$$m_e v r = n\hbar$$

where $\hbar = h/2\pi$. This is precisely the quantization of angular momentum condition imposed by Bohr in his original theory of hydrogen.

The electron orbit shown in Figure 28.10a contains three complete wavelengths and corresponds to the case in which the principal quantum number n equals three. The orbit with one complete wavelength in its circumference corresponds to the first Bohr orbit, $n = 1$; the orbit with two complete wavelengths corresponds to the second Bohr orbit, $n = 2$, and so forth.

By applying the wave theory of matter to electrons in atoms, de Broglie was able to explain the appearance of integers in the Bohr theory as a natural consequence of interference. This was the first convincing argument that the wave nature of matter was at the heart of the behavior of atomic systems. Although the analysis provided by de Broglie was a promising first step, gigantic strides were made subsequently with the development of Schrödinger's wave equation and its application to atomic systems.

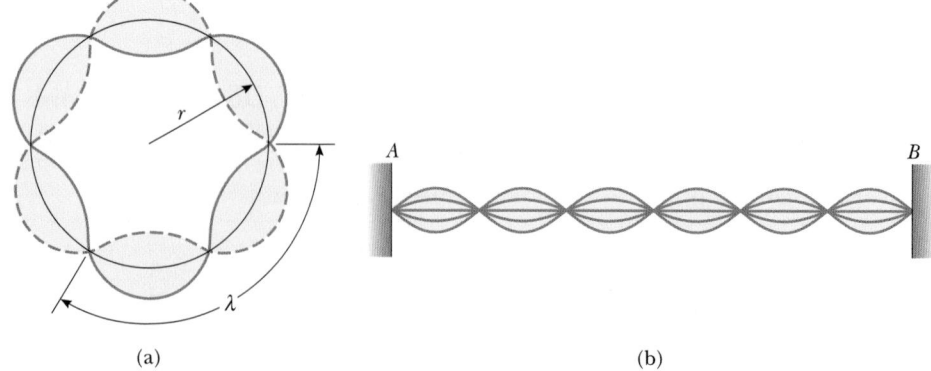

Figure 28.10 (a) Standing wave pattern for an electron wave in a stable orbit of hydrogen. There are three full wavelengths in this orbit. (b) Standing wave pattern for a vibrating stretched string fixed at its ends. This pattern has three full wavelengths.

(a)

(b)

28.6 QUANTUM MECHANICS AND THE HYDROGEN ATOM

One of the first great achievements of quantum mechanics was the solution of the wave equation for the hydrogen atom. We shall not attempt to carry out this solution. Rather, we shall simply describe its properties and some of its implications with regard to atomic structure.

According to quantum mechanics, the energies of the allowed states are in exact agreement with the values obtained by the Bohr theory (Eq. 28.12), when the allowed energies depend only on the principal quantum number, n.

In addition to the principal quantum number, two other quantum numbers emerged from the solution of the wave equation, ℓ and m_ℓ. The quantum number ℓ is called the **orbital quantum number**, and m_ℓ is called the **orbital magnetic quantum number.** As pointed out in Section 28.4, these quantum numbers had already appeared in modifications made to the Bohr theory. The significance of quantum mechanics is that these quantum numbers and the restrictions placed on their values arose directly from mathematics and not from any ad hoc assumptions to make the theory consistent with experimental observation. Because we shall need to make use of the various quantum numbers in the next several sections, the ranges of their values are repeated.

The values of n can range from 1 to ∞ in integer steps.
The values of ℓ can range from 0 to $n - 1$ in integer steps.
The values of m_ℓ can range from $-\ell$ to ℓ in integer steps.

For example, if $n = 1$, only $\ell = 0$ and $m_\ell = 0$ are permitted. If $n = 2$, the value of ℓ may be 0 or 1; if $\ell = 0$, then $m_\ell = 0$, but if $\ell = 1$, then m_ℓ may be 1, 0, or -1. Table 28.2 summarizes the rules for determining the allowed values of ℓ and m_ℓ for a given value of n.

States that violate the rules given in Table 28.2 cannot exist. For instance, one state that cannot exist is the $2d$ state, which would have $n = 2$ and $\ell = 2$. This state is not allowed because the highest allowed value of ℓ is $n - 1$, or 1 in this case. Thus, for $n = 2$, $2s$ and $2p$ are allowed states but $2d$, $2f$, . . . are not. For $n = 3$, the allowed states are $3s$, $3p$, and $3d$.

TABLE 28.2 Three Quantum Numbers for the Hydrogen Atom

Quantum Number	Name	Allowed Values	Number of Allowed States
n	Principal quantum number	1, 2, 3, $\cdots$	Any number
ℓ	Orbital quantum number	0, 1, 2, $\cdots$, $n - 1$	n
m_ℓ	Orbital magnetic quantum number	$-\ell$, $-\ell + 1$, $\cdots$, 0, $\cdots$, $\ell - 1$, ℓ	$2\ell + 1$

EXAMPLE 28.4 The $n = 2$ Level of Hydrogen

Determine the number of states in the hydrogen atom you expect for the principal quantum number $n = 2$ and calculate the energies of these states.

Solution For $n = 2$, ℓ can have the values 0 and 1. For $\ell = 0$, m_ℓ can only be 0; for $\ell = 1$, m_ℓ can be -1, 0, or 1. Hence we have one state designated as the $2s$ state associated with the quantum numbers $n = 2$, $\ell = 0$, and $m_\ell = 0$, and three states designated as $2p$ states, for which the quantum numbers are $n = 2$, $\ell = 1$, $m_\ell = -1$; $n = 2$, $\ell = 1$, $m_\ell = 0$; and $n = 2$, $\ell = 1$, $m_\ell = 1$.

Because all of these states have the same principal quantum number, $n = 2$, they also have the same energy, which can be calculated using the expression $E_n = -13.6/n^2$ (Eq. 28.13). For $n = 2$, this gives

$$E_2 = -\frac{13.6}{2^2}\ \text{eV} = -3.40\ \text{eV}$$

Exercise How many possible states are there for the $n = 3$ level of hydrogen? For the $n = 4$ level?

Answer 9 states with different values of ℓ or m_ℓ for $n = 3$, and 16 states for $n = 4$

28.7 THE SPIN MAGNETIC QUANTUM NUMBER

Example 28.4 was presented to give you some practice in manipulating quantum numbers, but as we shall see in this section, there actually are *eight* states corresponding to $n = 2$ for hydrogen, not four. This happens because it was later found that another quantum number, m_s, the **spin magnetic quantum number,** had to be introduced to explain the splitting of each level into two.

As pointed out in Section 28.4, the need for this new quantum number first came about because of an unusual feature in the spectra of certain gases, such as sodium vapor. Close examination of one of the prominent lines of sodium shows that it is, in fact, two very closely spaced lines. The wavelengths of these lines occur in the yellow region at 589.0 nm and 589.6 nm. In 1925, when this doublet was first noticed, atomic theory could not explain it. To resolve the dilemma, Samuel Goudsmit and George Uhlenbeck, following a suggestion by the Austrian physicist Wolfgang Pauli, proposed that a fourth quantum number, called the *spin quantum number,* be introduced to describe any atomic level.

In order to describe the spin quantum number, it is convenient (but incorrect) to think of the electron as spinning on its axis as it orbits the nucleus, just as the Earth spins on its axis as it orbits the Sun. There are only two ways in which the electron can spin as it orbits the nucleus, as shown in Figure 28.11. If the direction of spin is as shown in Figure 28.11a, the electron is said to have "spin up." If the direction of spin is reversed, as in Figure 28.11b, the electron is said to have "spin down." The energy of the electron is slightly different for the two spin directions, and this energy difference accounts for the sodium doublet. The quantum numbers associated with electron spin are $m_s = \frac{1}{2}$ for the spin-up state and $m_s = -\frac{1}{2}$ for the spin-down state. As we shall see in Example 28.5, this new quantum number doubles the number of allowed states specified by the quantum numbers n, ℓ, and m_ℓ.

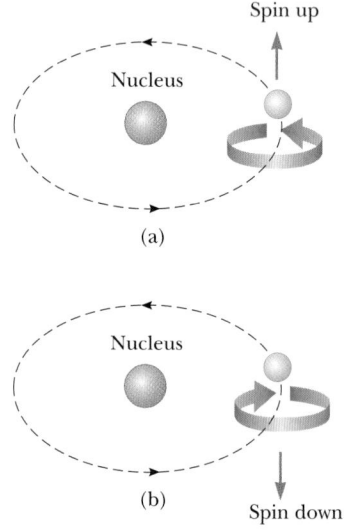

Figure 28.11 As an electron moves in its orbit about the nucleus, its spin can be either (a) up or (b) down.

Any classical description of electron spin is incorrect because quantum mechanics tells us that, because the electron cannot be precisely located in space, it cannot be considered to be a spinning solid object, as pictured in Figure 28.11. In spite of this conceptual difficulty, all experimental evidence supports the fact that an electron does have some intrinsic property that can be described by the spin magnetic quantum number.

Wolfgang Pauli and Niels Bohr watch a spinning top. *(Courtesy of AIP Niels Bohr Library, Margarethe Bohr Collection)*

EXAMPLE 28.5 Adding Electron Spin to Hydrogen

For a hydrogen atom, determine the quantum numbers associated with the possible states that correspond to the principal quantum number $n = 2$.

Solution With the addition of the spin quantum number, we have the following possibilities:

n	ℓ	m_ℓ	m_s	Subshell	Shell	Number of States in Subshell
2	0	0	$\frac{1}{2}$			
2	0	0	$-\frac{1}{2}$	$2s$	L	2
2	1	1	$\frac{1}{2}$			
2	1	1	$-\frac{1}{2}$			
2	1	0	$\frac{1}{2}$			
2	1	0	$-\frac{1}{2}$	$2p$	L	6
2	1	-1	$\frac{1}{2}$			
2	1	-1	$-\frac{1}{2}$			

Exercise Show that for $n = 3$, there are 18 possible states. (This follows from the restrictions that the maximum number of electrons in the $3s$ state is 2, the maximum number in the $3p$ state is 6, and the maximum number in the $3d$ state is 10.)

EXAMPLE 28.6 The Quantum Number for the 2p Subshell

List the quantum numbers for electrons in the $2p$ subshell.

Solution For this subshell, $n = 2$ and $\ell = 1$. The magnetic quantum number can have the values -1, 0, 1, and the spin quantum number is always $+\frac{1}{2}$ or $-\frac{1}{2}$. Thus, the six possibilities are

n	ℓ	m_ℓ	m_s
2	1	-1	$-\frac{1}{2}$
2	1	-1	$\frac{1}{2}$
2	1	0	$-\frac{1}{2}$
2	1	0	$\frac{1}{2}$
2	1	1	$-\frac{1}{2}$
2	1	1	$\frac{1}{2}$

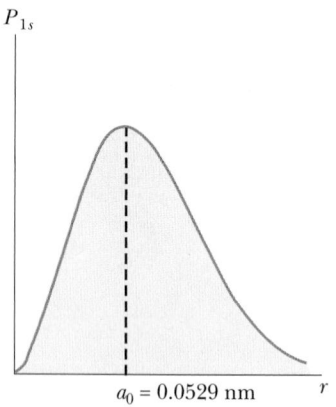

Figure 28.12 The probability of finding the electron versus distance from the nucleus for the hydrogen atom in the 1s (ground) state. Note that the probability has its maximum value when r equals the first Bohr radius, a_0.

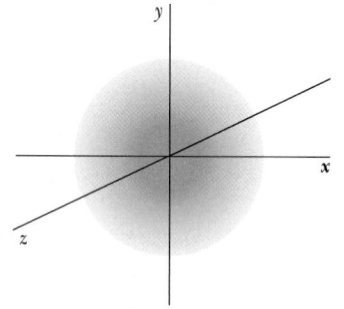

Figure 28.13 The spherical electron cloud for the hydrogen atom in its 1s state.

28.8   ELECTRON CLOUDS

The solution of the wave equation for the wave function Ψ, discussed in Section 27.10, yields a quantity dependent on the quantum numbers n, ℓ, and m_ℓ. Let us assume that we have found a value for Ψ and see what it may tell us about the hydrogen atom. Let us choose a value of $n = 1$ for the principal quantum number, which corresponds to the lowest energy state for hydrogen. For $n = 1$, the restrictions placed on the remaining, quantum numbers are that $\ell = 0$ and $m_\ell = 0$. We can now substitute these values into our expression for Ψ.

The quantity Ψ^2 has great physical significance because it is proportional to the probability of finding the electron at a given position. Figure 28.12 gives the probability of finding the electron at various distances from the nucleus in the 1s state of hydrogen. Some useful and surprising information can be extracted from this curve. First, the curve peaks at a value of $r = 0.0529$ nm, the Bohr value of the radius of the first electron orbit in hydrogen. This means that there is a probability of finding the electron at this distance from the nucleus. However, as the curve indicates, there is also a probability of finding the electron at some other distance from the nucleus. In other words, the electron is not confined to a particular orbital distance from the nucleus, as assumed in the Bohr model. The electron may be found at various distances from the nucleus, but **the probability of finding it at a distance corresponding to the first Bohr orbit is a maximum.** Quantum mechanics also predicts that the wave function for the hydrogen atom in the ground state is spherically symmetric; hence the electron can be found in a spherical region surrounding the nucleus. This is in contrast to the Bohr theory, which confines the position of the electron to points in a plane. This result is often interpreted by viewing the electron as a cloud surrounding the nucleus. An attempt at picturing this cloud-like behavior is shown in Figure 28.13. The densest regions of the cloud represent those locations where the electron is most likely to be found.

If a similar analysis is carried out for the $n = 2$, $\ell = 0$ state of hydrogen, a peak of the probability curve is found at $4a_0$. Likewise, for the $n = 3$, $\ell = 0$ state, the curve peaks at $9a_0$. Thus, quantum mechanics predicts a most probable electron location that is in agreement with the location predicted by the Bohr theory.

28.9   THE EXCLUSION PRINCIPLE AND THE PERIODIC TABLE

Earlier we found that the state of an electron in an atom is specified by four quantum numbers: n, ℓ, m_ℓ, and m_s. For example, an electron in the ground state of hydrogen could have quantum numbers of $n = 1$, $\ell = 0$, $m_\ell = 0$, $m_s = \frac{1}{2}$. As it turns out, the state of an electron in any other atom may also be specified by this same set of quantum numbers. In fact, these four quantum numbers can be used to describe all the electronic states of an atom regardless of the number of electrons in its structure.

An obvious question that arises here is, "How many electrons in an atom can have a particular set of quantum numbers?" This important question was answered by Pauli in 1925 in a powerful statement known as the **exclusion principle:**

> No two electrons in an atom can ever be in the same quantum state; that is, no two electrons in the same atom can have the same set of quantum numbers n, ℓ, m_ℓ, and m_s.

◀ The exclusion principle

It is interesting to note that if this principle were not valid, every electron would end up in the lowest energy state of the atom and the chemical behavior of the elements would be grossly different. Nature as we know it would not exist! In reality, we can view the electronic structure of complex atoms as a succession of filled levels increasing in energy, where the outermost electrons are primarily responsible for the chemical properties of the element.

As a general rule, the order that electrons fill an atom's subshell is as follows. Once one subshell is filled, the next electron goes into the vacant subshell that is lowest in energy. One can understand this principle by recognizing that if the atom were not in the lowest energy state available to it, it would radiate energy until it reached this state. A subshell is filled when it contains $2(2\ell + 1)$ electrons. This rule is based on the analysis of quantum numbers to be described later. Following this rule, shells and subshells can contain numbers of electrons according to the pattern given in Table 28.3.

The exclusion principle can be illustrated by an examination of the electronic arrangement in a few of the lighter atoms.

Hydrogen has only one electron, which, in its ground state, can be described by either of two sets of quantum numbers: $1, 0, 0, \frac{1}{2}$ or $1, 0, 0, -\frac{1}{2}$. The electronic configuration of this atom is often designated as $1s^1$. The notation $1s$ refers to a state for which $n = 1$ and $\ell = 0$, and the superscript indicates that one electron is present in this level.

Neutral *helium* has two electrons. In the ground state, the quantum numbers for these two electrons are $1, 0, 0, \frac{1}{2}$ and $1, 0, 0, -\frac{1}{2}$. No other possible combinations

TABLE 28.3 The Number of Electrons in Filled Subshells and Shells

Shell	Subshell	Number of Electrons in Filled Subshell	Number of Electrons in Filled Shell
K ($n = 1$)	s ($\ell = 0$)	2	2
L ($n = 2$)	s ($\ell = 0$) p ($\ell = 1$)	2 6	8
M ($n = 3$)	s ($\ell = 0$) p ($\ell = 1$) d ($\ell = 2$)	2 6 10	18
N ($n = 4$)	s ($\ell = 0$) p ($\ell = 1$) d ($\ell = 2$) f ($\ell = 3$)	2 6 10 14	32

Wolfgang Pauli (1900–1958)

An extremely talented Austrian theoretical physicist who made important contributions in many areas of modern physics, Pauli gained public recognition at the age of 21 with a masterful review article on relativity, which is still considered one of the finest and most comprehensive introductions to the subject. Other major contributions were the discovery of the exclusion principle, the explanation of the connection between particle spin and statistics, and theories of relativistic quantum electrodynamics, the neutrino hypothesis, and the hypothesis of nuclear spin. *(CERN, Courtesy of AIP Emilio Segre Visual Archive)*

of quantum numbers exist for this level, and we say that the K shell is filled. The helium electronic configuration is designated as $1s^2$.

Neutral *lithium* has three electrons. In the ground state, two of these are in the $1s$ subshell and the third is in the $2s$ subshell because it is lower in energy than the $2p$ subshell. Hence, the electronic configuration for lithium is $1s^2 2s^1$.

A list of electronic ground-state configurations for a number of atoms is provided in Table 28.4. In 1871 Dmitri Mendeleev (1834–1907), a Russian chemist, arranged the elements known at that time in a table according to their atomic masses and chemical similarities. The first table Mendeleev proposed contained many blank spaces, and he boldly stated that the gaps were there only because those elements had not yet been discovered. By noting the column in which these missing elements should be located, he was able to make rough predictions about their chemical properties. Within 20 years of this announcement, these elements were indeed discovered.

The elements in our current version of the periodic table are still arranged so that all those in a vertical column have similar chemical properties. For example, consider the elements in the last column: He (helium), Ne (neon), Ar (argon), Kr (krypton), Xe (xenon), and Rn (radon). The outstanding characteristic of these elements is that they do not normally take part in chemical reactions; that is, they do not join with other atoms to form molecules, and are therefore classified as inert. Because of this aloofness, they are referred to as the *noble gases*. We can partially understand their behavior by looking at the electronic configurations

TABLE 28.4 Electronic Configuration of Some Elements

Z	Symbol	Ground-State Configuration	Ionization Energy (eV)	Z	Symbol	Ground-State Configuration	Ionization Energy (eV)
1	H	$1s^1$	13.595	19	K	[Ar] $4s^1$	4.339
2	He	$1s^2$	24.581	20	Ca	$4s^2$	6.111
				21	Sc	$3d4s^2$	6.54
3	Li	[He] $2s^1$	5.390	22	Ti	$3d^2 4s^2$	6.83
4	Be	$2s^2$	9.320	23	V	$3d^3 4s^2$	6.74
5	B	$2s^2 2p^1$	8.296	24	Cr	$3d^5 4s^1$	6.76
6	C	$2s^2 2p^2$	11.256	25	Mn	$3d^5 4s^2$	7.432
7	N	$2s^2 2p^3$	14.545	26	Fe	$3d^6 4s^2$	7.87
8	O	$2s^2 2p^4$	13.614	27	Co	$3d^7 4s^2$	7.86
9	F	$2s^2 2p^5$	17.418	28	Ni	$3d^8 4s^2$	7.633
10	Ne	$2s^2 2p^6$	21.559	29	Cu	$3d^{10} 4s^1$	7.724
				30	Zn	$3d^{10} 4s^2$	9.391
11	Na	[Ne] $3s^1$	5.138	31	Ga	$3d^{10} 4s^2 4p^1$	6.00
12	Mg	$3s^2$	7.644	32	Ge	$3d^{10} 4s^2 4p^2$	7.88
13	Al	$3s^2 3p^1$	5.984	33	As	$3d^{10} 4s^2 4p^3$	9.81
14	Si	$3s^2 3p^2$	8.149	34	Se	$3d^{10} 4s^2 4p^4$	9.75
15	P	$3s^2 3p^3$	10.484	35	Br	$3d^{10} 4s^2 4p^5$	11.84
16	S	$3s^2 3p^4$	10.357	36	Kr	$3d^{10} 4s^2 4p^6$	13.996
17	Cl	$3s^2 3p^5$	13.01				
18	Ar	$3s^2 3p^6$	15.755				

Note: The bracket notation is used as a shorthand method to avoid repetition in indicating inner-shell electrons. Thus, [He] represents $1s^2$, [Ne] represents $1s^2 2s^2 2p^6$, [Ar] represents $1s^2 2s^2 2p^6 3s^2 3p^6$, and so on.

shown in Table 28.4. The element helium has the electronic configuration $1s^2$. In other words, one shell is filled. The electrons in this filled shell are considerably separated in energy from the next available level, the $2s$ level.

The electronic configuration for neon is $1s^2 2s^2 2p^6$. Again, the outer shell is filled and there is a large difference in energy between the $2p$ level and the $3s$ level. Argon has the configuration $1s^2 2s^2 2p^6 3s^2 3p^6$. Here, the $3p$ subshell is filled and there is a wide gap in energy between the $3p$ subshell and the $3d$ subshell. Through all the noble gases the pattern remains the same. A noble gas is formed when either a shell or a subshell is filled and there is a large gap in energy before the next possible level is encountered.

The elements in the first column of the periodic table are called the *alkali metals* and are very active chemically. Referring to Table 28.4, we can understand why these elements interact so strongly with other elements. All of these alkali metals have a single outer electron in an s subshell. This electron is shielded from the nucleus by all the electrons in the inner shells. Thus, it is only loosely bound to the atom and can readily be accepted by other atoms to form molecules.

All the elements in the seventh column of the periodic table (called the *halogens*) are also very active chemically. Note that all these elements are lacking one electron in a subshell. As a consequence, they readily accept electrons from other atoms to form molecules.

Thinking Physics 3

As one moves from left to right across one row of the periodic table, the effective size of the atoms first decreases and then increases. What would cause this behavior?

Explanation As one begins at the left side of the periodic table and moves toward the middle, the nuclear charge is increasing. As a result, there is an increasing Coulomb attraction between the nucleus and the electrons, and the electrons are pulled into an average position that is closer to the nucleus. Although the additional electrons repel one another, that repulsion is diffused throughout the volume defined by their orbits. From the middle of the row to the right side, the increasing number of electrons being placed in proximity to each other results in a mutual repulsion that increases the average distance from the nucleus and causes the atomic size to grow.

28.10 BUCKYBALLS

Diamond and graphite are well known as two forms of pure carbon, but now physicists have discovered a third form that goes under the strange name of **buckminsterfullerene**, or **buckyballs** for short. The description "ball" is appropriate for these molecules because they are the roundest molecule in existence. The simplest buckyball contains 60 carbon atoms bonded together in a molecular structure that looks much like a soccer ball (Fig. 28.14). The architecture that underlies such a surface is at the heart of the geodesic dome invented by visionary engineer R. Buckminster Fuller, hence the name for this new form of carbon. But the story does not stop with C_{60}. Another large molecule consisting of 70 carbon atoms and

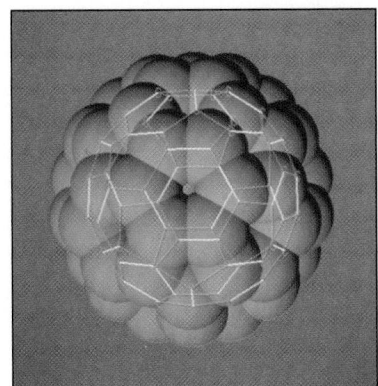

Figure 28.14 Structure of one of the recently discovered fullerene molecules, C_{60}, also known as the "buckyball." *(Charles Winters)*

looking like a rugby ball has also been produced in abundance. Buckyballs were discovered by two astrophysicists, W. Kratschmer and D. Huffman, who were interested in studying the carbon gas that exists between stars. Because this carbon has coalesced and formed under myriad conditions, the two scientists were investigating as many methods as possible for vaporizing and then condensing it. One of their techniques consisted of evaporating graphite in a helium atmosphere. They found that the residue had some peculiar properties, one of which was its strong absorption of high-frequency ultraviolet light. They later found that chemistry literature had predicted that pure carbon-60 would have the properties they were finding. Carbon-60 had been produced in the past under exacting conditions, but only in samples of a few hundred molecules. They were producing samples containing billions of molecules. The technique they used was so simple that scientists around the world became involved in related investigations, and within a short time buckyballs became a fascinating item to study in the world of materials science. Modest amounts of even more exotic buckyballs have been produced as a result of these investigations. Also, recently scientists have been able to stretch buckyballs into elongated shapes called *buckytubes* that are roughly the diameter of a DNA molecule. If these could be made into fibers, they could be molded into products that are very light and strong. Other hoped-for applications in the future include strong clothing, filters for purifying materials, faster microelectronic devices, lubricants, and integral parts of many medical devices. One of the underlying difficulties now is the inability to make buckyballs cheaper than gold.

28.11 CHARACTERISTIC X-RAYS

X-rays are emitted when a metal target is bombarded with high-energy electrons. The x-ray spectrum typically consists of a broad continuous band and a series of sharp lines that are dependent on the type of metal used for the target, as shown in Figure 28.15. These discrete lines, called **characteristic x-rays,** were discovered in 1908, but their origin remained unexplained until the details of atomic structure were developed.

The first step in the production of characteristic x-rays occurs when a bombarding electron collides with an electron in an inner shell of a target atom with sufficient energy to remove the electron from the atom. The vacancy created in the shell is filled when an electron in a higher level drops down into the lower energy level containing the vacancy. The time it takes for this to happen is very short, less than 10^{-9} s. This transition is accompanied by the emission of a photon whose energy equals the difference in energy between the two levels. Typically, the energy of such transitions is greater than 1000 eV, and the emitted x-ray photons have wavelengths in the range of 0.01 nm to 1 nm.

Let us assume that the incoming electron has dislodged an atomic electron from the innermost shell, the K shell. If the vacancy is filled by an electron dropping from the next higher shell, the L shell, the photon emitted in the process is referred to as the K_α line on the curve of Figure 28.15. If the vacancy is filled by an electron dropping from the M shell, the line produced is called the K_β line.

Other characteristic x-ray lines are formed when electrons drop from upper levels to vacancies other than those in the K shell. For example, L lines are produced when vacancies in the L shell are filled by electrons dropping from higher shells.

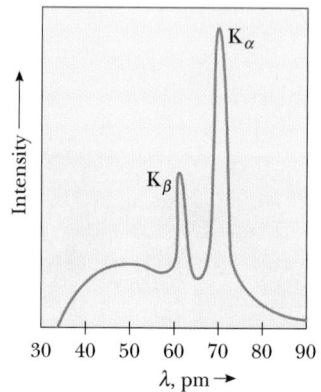

Figure 28.15 The x-ray spectrum of a metal target consists of broad continuous spectrum plus a number of sharp lines that are due to *characteristic x-rays*. The data shown were obtained when 35-keV electrons bombarded a molybdenum target. Note that 1 pm = 10^{-12} m = 0.001 nm.

An L_α line is produced as an electron drops from the M shell to the L shell, and an L_β line is produced by a transition from the N shell to the L shell.

We can estimate the energy of the emitted x-rays as follows. Consider two electrons in the K shell of an atom whose atomic number is Z. Each electron partially shields the other from the charge of the nucleus, Ze, and so each is subject to an effective nuclear charge of $Z_{eff} = (Z - 1)e$. We can now use a modified form of Equation 28.18 to estimate the energy of either electron in the K shell (with $n = 1$):

$$E_K = -m_e Z_{eff}^2 \frac{k_e^2 e^4}{2\hbar^2} = -Z_{eff}^2 E_0$$

where E_0 is the ground-state energy. Substituting $Z_{eff} = Z - 1$ gives

$$E_K = -(Z - 1)^2 (13.6 \text{ eV}) \qquad \text{[28.20]}$$

As Example 28.7 shows, we can estimate the energy of an electron in an L or M shell in a similar fashion. Taking the energy difference between these two levels, we can then calculate the energy and wavelength of the emitted photon.

In 1914 Henry G. J. Moseley plotted the Z values for a number of elements versus $\sqrt{1/\lambda}$, where λ is the wavelength of the K_α line for each element. He found that such a plot produced a straight line, as in Figure 28.16. This is consistent with our rough calculations of the energy levels based on Equation 28.20. From this plot, Moseley was able to determine the Z values of other elements, providing a periodic chart in excellent agreement with the known chemical properties of the elements.

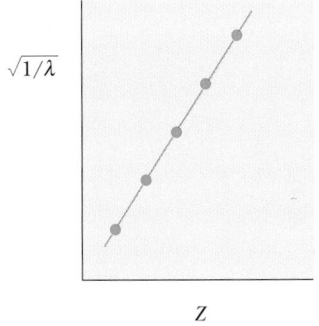

Figure 28.16 A Moseley plot. A straight line is obtained when $\sqrt{1/\lambda}$ is plotted versus Z for the K_α x-ray lines of a number of elements.

Thinking Physics 4

The Bohr theory gives good theoretical agreement with experimentally measured frequencies of spectral lines for the hydrogen atom, but it fails completely to predict the correct optical frequencies for heavier atoms. Moseley, however, used the Bohr theory to predict the frequencies of characteristic x-rays from heavy atoms. Why would the Bohr theory give good predictions for x-rays?

Explanation Characteristic x-rays have their origins in transitions between the lowest energy levels in the atom. An electron in a lower energy level is ejected from the atom by the accelerated electron in the x-ray tube, leaving a vacancy in the shell. Let us consider an electron ejected from the K shell and the resulting electrical interactions of an L shell electron with the other electrons. If we look at the electrons in higher shells than L, these electrons spend most of their time farther from the nucleus than the L electrons. What's more, when averaged over time, the charge of these outer electrons is uniform in all directions. The L electrons find themselves inside a set of perfectly symmetric shells of charge. Thus, the L electrons are not affected by these charges because they produce no electric field at the location of the L electrons. What remains to affect the L electrons is a nuclear charge shielded by the one remaining K electron and the other electrons in the L shell. If we ignore the other L electrons, we can model the L electron as belonging to a hydrogen atom, with a nuclear charge of $(Z - 1)$, where the nuclear charge Z is reduced by 1 due to the shielding effects of the remaining K electron. As a result, the Bohr theory works relatively well in describing the energy of the L electron, although it is not perfect, because we have ignored the other L electrons.

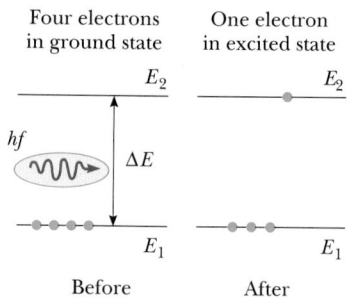

Figure 28.17 Energy level diagram of an atom with various allowed states. The lowest energy state, E_1, is the ground state. All others are excited states.

Figure 28.18 Diagram representing the process of *stimulated absorption* of a photon by an atom. The blue dots represent electrons in the various states. One electron is transferred from the ground state to the excited state when the atom absorbs one photon whose energy is $hf = E_2 - E_1$.

EXAMPLE 28.7   Estimating the Energy of an X-Ray

Estimate the energy of the characteristic x-ray emitted from a tungsten target when an electron drops from an M shell ($n = 3$ state) to a vacancy in the K shell ($n = 1$ state).

Solution   The atomic number for tungsten is $Z = 74$. Using Equation 28.20, we see that the energy of the electron in the K shell state is approximately

$$E_K = -(74 - 1)^2(13.6 \text{ eV}) = -72\,500 \text{ eV}$$

The electron in the M shell ($n = 3$) is subject to an effective nuclear charge that depends on the number of electrons in the $n = 1$ and $n = 2$ states, which shield the nucleus. Because there are eight electrons in the $n = 2$ state and one electron in the $n = 1$ state, roughly nine electrons shield the nucleus, and so $Z_{eff} = Z - 9$. Hence, the energy of an electron in the M shell ($n = 3$), following Equation 28.20, is equal to

$$E_M = -Z_{eff}^2 E_3 = -(Z - 9)^2 \frac{E_0}{3^2} = -(74 - 9)^2 \frac{(13.6 \text{ eV})}{9} = -6380 \text{ eV}$$

where E_3 is the energy of an electron in the $n = 3$ level of the hydrogen atom. Therefore, the emitted x-ray has an energy equal to $E_M - E_K = -6380 \text{ eV} - (-72\,500 \text{ eV}) =$ 66 100 eV.   Note that this energy difference is also equal to hf, where $hf = hc/\lambda$, and where λ is the wavelength of the emitted x-ray.

Exercise   Calculate the wavelength of the emitted x-ray for this transition.

Answer   0.0188 nm

28.12   ATOMIC TRANSITIONS

We have seen that an atom will emit radiation only at certain frequencies that correspond to the energy separation between the various allowed states. Consider an atom with many energy states, labeled $E_1, E_2, E_3, \ldots$, as in Figure 28.17. When light is incident on the atom, only those photons whose energy, hf, matches the energy separation, ΔE, between two levels can be absorbed by the atom. A schematic diagram representing this **stimulated absorption process** is shown in Figure 28.18. At ordinary temperatures, most of the atoms in a sample are in the ground state. If a vessel containing many atoms of a gaseous element is illuminated with a light beam containing all possible photon frequencies (that is, a continuous spectrum), only those photons of energies $E_2 - E_1, E_3 - E_1, E_4 - E_1$, and so on, can be absorbed. As a result of this absorption, some atoms are raised to various allowed higher energy levels, called **excited states.**

Once an atom is in an excited state, there is a constant probability that it will jump back to a lower level by emitting a photon, as shown in Figure 28.19. This process is known as **spontaneous emission.** Typically, an atom will remain in an excited state for only about 10^{-8} s.

A third process that is important in lasers, **stimulated emission,** was predicted by Einstein in 1917. Suppose an atom is in the excited state E_2, as in Figure 28.20, and a photon with energy $hf = E_2 - E_1$ is incident on it. The incoming photon increases the probability that the excited electron will return to the ground state and thereby emit a second photon having the same energy hf. Note that two identical photons result from stimulated emission—the incident photon and the emitted photon. The emitted photon is exactly in phase with the incident photon. These photons can stimulate other atoms to emit photons in a chain of similar processes. The many photons produced in this fashion are the source of the intense, coherent light in a laser.

Thinking Physics 5

A physics student is watching a meteor shower in the early morning hours. She notices that the streaks of light from the meteoroids entering the very high regions of the atmosphere last for as long as 2 or 3 seconds before fading. She also notices a lightning storm off in the distance. The streaks of light from the lightning fade away almost immediately after the flash, certainly in much less than 1 second. Both lightning and meteors cause the air to turn into a plasma because of the very high temperatures generated. The light is given off when the stripped electrons in the plasma recombine with the ionized atoms. Why would this light last longer for meteors than for lightning?

Explanation The answer lies in the subtle phrase in the description of the meteoroids—"entering the very high regions of the atmosphere." In the very high regions of the atmosphere, the pressure is very low. Thus, the density is very low, so that atoms of the gas are relatively far apart. After the air is ionized by the passing meteoroid, the probability of freed electrons finding an ionized atom with which to recombine is relatively low. As a result, the recombination process occurs over a relatively long time, measured in seconds. However, lightning occurs in the lower regions of the atmosphere (the troposphere) in which the pressure and density are relatively high. After the ionization by the lightning flash, the electrons and ionized atoms are much closer together than in the upper atmosphere. The probability of a recombination is much higher, and the time for the recombination to occur is much shorter.

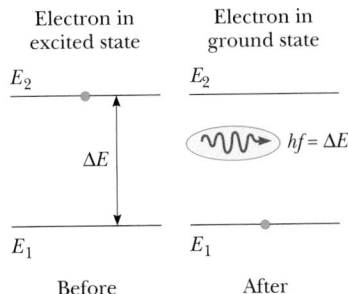

Figure 28.19 Diagram representing the process of *spontaneous emission* of a photon by an atom that is initially in the excited state E_2. When the electron falls to the ground state, the atom emits a photon whose energy is $hf = E_2 - E_1$.

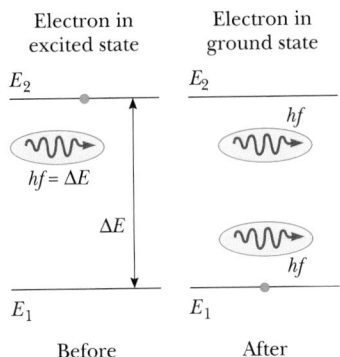

Figure 28.20 Diagram representing the process of *stimulated emission* of a photon by an incoming photon of energy hf. Initially, the atom is in the excited state. The incoming photon stimulates the atom to emit a second photon of energy $hf = E_2 - E_1$.

28.13 LASERS AND HOLOGRAPHY

We have described how an incident photon can cause atomic transitions either upward (stimulated absorption) or downward (stimulated emission). The two processes are equally probable. When light is incident on a system of atoms, there is usually a net absorption of energy because, when the system is in thermal equilibrium, there are many more atoms in the ground state than in excited states. However, if the situation can be inverted so that there are more atoms in an excited state than in the ground state, a net emission of photons can result. Such a condition is called **population inversion.** This is the fundamental principle involved in the operation of a laser, an acronym for **l**ight **a**mplification by **s**timulated **e**mission of **r**adiation. The amplification corresponds to a buildup of photons in the system as the result of a chain reaction of events. The following three conditions must be satisfied in order to achieve laser action:

1. The system must be in a state of population inversion (that is, more atoms in an excited state than in the ground state).

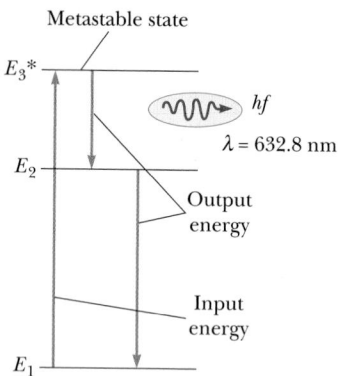

Figure 28.21 Energy-level diagram for the neon atom, which emits photons at a wavelength of 632.8 nm through stimulated emission. The photon at this wavelength arises from the transition $E_3^* \rightarrow E_2$. This is the source of coherent light in the helium-neon gas laser.

APPLICATION

Laser Technology.

2. The excited state of the system must be a *metastable state,* which means its lifetime must be long compared with the usually short lifetimes of excited states. When that is the case, stimulated emission will occur before spontaneous emission.
3. The emitted photons must be confined in the system long enough to allow them to stimulate further emission from other excited atoms. This is achieved by the use of reflecting mirrors at the ends of the system. One end is totally reflecting, and the other is slightly transparent to allow the laser beam to escape.

One device that exhibits stimulated emission of radiation is the helium-neon gas laser. Figure 28.21 is an energy-level diagram for the neon atom in this system. The mixture of helium and neon is confined to a glass tube sealed at the ends by mirrors. An oscillator connected to the tube causes electrons to sweep through the tube, colliding with the atoms of the gas and raising them into excited states. Neon atoms are excited to state E_3^* through this process and also as a result of collisions with excited helium atoms. When a neon atom makes a transition to state E_2, it stimulates emission by neighboring excited atoms. This results in the production of coherent light at a wavelength of 632.8 nm. Figure 28.22 summarizes the steps in the production of a laser beam.

Since the development of the first laser in 1960, laser technology has experienced tremendous growth. Lasers that cover wavelengths in the infrared, visible, and ultraviolet regions are now available. Applications include surgical "welding" of detached retinas, precision surveying and length measurement, a potential source for inducing nuclear fusion reactions, precision cutting of metals and other materials, and telephone communication along optical fibers. These and other applications are possible because of the unique characteristics of laser light. In addition to being highly monochromatic and coherent, laser light is also highly directional and can be sharply focused to produce regions of extremely intense light energy.

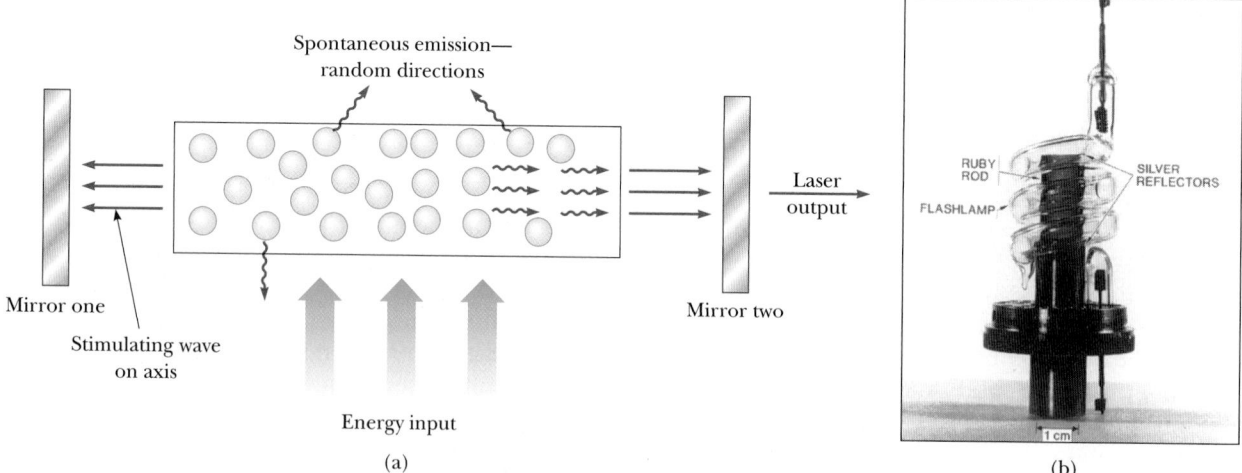

(a)   (b)

Figure 28.22 (a) A schematic of a laser design. The tube contains atoms, which represent the active medium. An external source of energy (optical, electrical, etc.) is needed to "pump" the atoms to excited energy states. The parallel end mirrors provide the feedback of the stimulating wave. (b) Photograph of the first ruby laser showing the flash lamp surrounding the ruby rod. *(Courtesy of Hughes Aircraft Company)*

Holography

One interesting application of the laser is holography, the production of three-dimensional images of objects. Figure 28.23a shows how a hologram is made. Light from the laser is split into two parts by a half-silvered mirror at *B*. One part of the beam reflects off the object to be photographed and strikes an ordinary photographic film. The other half of the beam is diverged by lens L_2, reflects from mirrors M_1 and M_2, and finally strikes the film. The two beams overlap to form an extremely complicated interference pattern on the film, one that can be produced only if the phase relationship of the two waves is constant throughout the exposure of the film. This condition is met through the use of light from a laser because such light is coherent. (All of the photons in the beam have the same phase.) The hologram records not only the intensity of the light scattered from the object (as in a conventional photograph) but also the phase difference between the reference beam and the beam scattered from the object. Because of this phase difference, an interference pattern is formed that produces an image with full three-dimensional perspective.

A hologram is best viewed by allowing coherent light to pass through the developed film while looking back along the direction from which the beam comes. Figure 28.23b is a photograph of a hologram made using a cylindrical film.

Application: Compact Disc Players

One of the most recent applications of the laser that has found its way into the life of college students is the compact disc (CD) player, a device that is rapidly replacing phonographs and tape players in ultra-high-fidelity sound reproduction. Figure 28.24 shows the important components of one of these devices. Light from a laser is sent through lens A toward a partially reflecting mirror. The portion of the light that passes through this mirror is redirected by another mirror through lens B toward a laser disc. The light reflected off this disc retraces the original path until it reaches the partially reflecting mirror, where a portion of it is reflected into a

Scientist checking the performance of an experimental laser-cutting device mounted on a robot arm. The laser is being used to cut through a metal plate. *(Philippe Plailly/SPL/Photo Researchers)*

APPLICATION

Use of Lasers in CD Players.

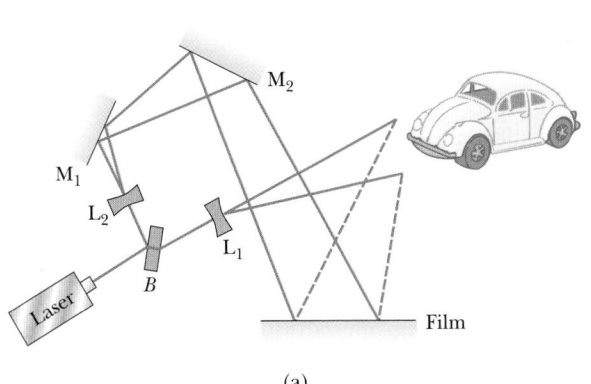

(a)

(b)

Figure 28.23 (a) Experimental arrangement for producing a hologram. (b) Photograph of a hologram made using a cylindrical film. Note the detail of the Volkswagen image. *(Courtesy of Central Scientific Company)*

Figure 28.24 A compact disc player.

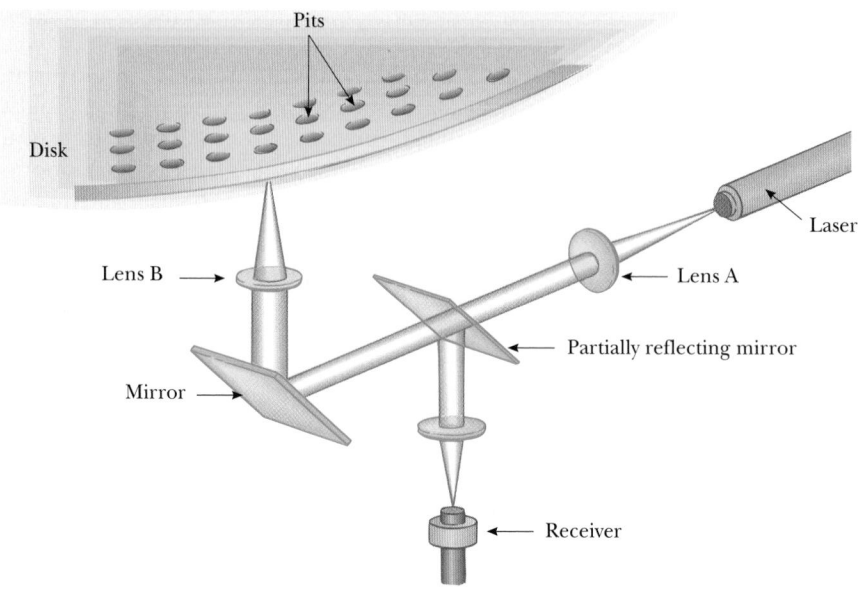

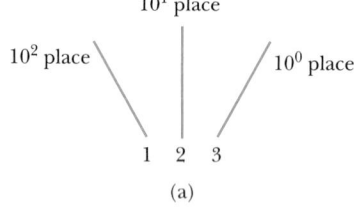

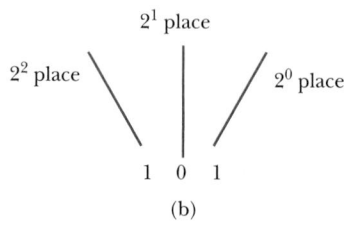

Figure 28.25 Weighting the digits in (a) the decimal system and (b) the binary system.

receiver. The receiver then uses this reflected signal to drive a loudspeaker and to recreate the information stored on the laser disc.

In order to explain how the light retrieves information from the disc, let us first describe the binary number system. Consider an ordinary decimal number such as 123, where each digit has a place value associated with it. In this case, the number 123 contains 1 hundred, 2 tens, and 3 ones. Figure 28.25a shows an alternative way of expressing this same information. Note that in the figure we say that the number 3 occupies the 10^0 location, which is equivalent to saying that it occupies the one's place because $10^0 = 1$. Also, the number 2 occupies the 10^1 location, and the number 1 occupies the 10^2 location. This procedure can be extended upward as high as you care to count, where the weight of each position would be determined by some power of the number system base, which is 10 for decimal numbers.

Each bit position of a binary number also carries a particular weight that determines its magnitude, except that the place value of each bit is determined by some power of the base number 2. For example, Figure 28.25b shows a binary number 101 with its corresponding weight positions. The 1 on the right is in the $2^0 = 1$ location, the 0 is in the 2^1 location, and the 1 on the left is in the 2^2 position. Each digit in a binary number is called a bit, and a collection of eight bits is a byte. Thus, we can represent a byte with a box constructed as shown in Figure 28.26, indicating the weight of each bit. The box says that the weight of the lowest order bit (the rightmost one) is 1 ($= 2^0$), and the weight of the highest order bit (the leftmost one) is 128 ($= 2^7$). Thus, we see that the binary number 101 is the decimal number 5. As an exercise, convert the following binary numbers to decimal numbers: 00101101, 11011100, and 11111111.[1]

---

[1] The answers are 45, 220, and 255, respectively.

Modern electronic devices such as computers use binary numbers in the form of a series of on and off signals. To understand this, consider the binary number 00000001, which is the binary representation of the decimal number 1. This binary number could be manipulated by an electronic circuit by turning a switch "off" for seven equal intervals of time, and then "on" for one interval.

The operation of a CD player is based on the procedure just described. The amplitude of a sound signal is sampled at regular intervals of time, as in Figure 28.27, and the loudness of the sound at each sampling is then converted to a binary number by an electronic circuit. For example, suppose the amplitude of the sound signal in Figure 28.27 at t_1, t_2, and t_3 is decimal 4, 5, 4, respectively. The circuitry will translate this to binary 00000100, 00000101, and 00000100. This pattern of binary numbers is then recorded by placing a series of pits and smooth places on a laser disc (each pit is smaller than the dot over an "i" in this textbook). When laser light is reflected from one of the smooth places on the disc, it is reflected back to the detector in Figure 28.24, which interprets the presence of the reflection as an "on" signal, or as a binary 1. On the other hand, when the light strikes a pit on the rotating disc, it reflects in some random direction and does not return to the detector, which interprets the absence of a reflection as an "off," or a binary 0. Electrical equipment connected to the detector then drives a loudspeaker according to the size of the binary number it receives, thus reconstructing the original sound that was used to etch the disc.

If a CD player is to reproduce sound faithfully, the laser beam must follow the spiral path of information perfectly. However, sometimes the laser beam can drift off track, and without a feedback procedure to let the player know this is happening, the fidelity of the music would be reduced greatly. Figure 28.28 shows how a diffraction grating is used to provide information to keep the beam on track. The central maximum of the diffraction pattern is used to read the information on the CD, and the two first-order maxima are used for steering. The grating is designed so that the first-order maxima fall on the smooth surface at each side of the information track. Both of these reflected beams have their own detectors, and because both are reflected from smooth, nonpitted surfaces, they should both have the same high intensity when they are detected. However, if the beam wanders off track, one of the beams will begin to strike pits on the information track and the amount of light reflected will diminish. This information can be used by suitable electronic circuits to guide the beam back to its desired location.

A new entrant in the high fidelity game is the Digital Audio Tape, DAT. The method of recording information is essentially the same as that used on compact discs in that the data is stored in the form of ones and zeros. The difference is that the medium is cassette tape. As noted in Chapter 20, information is recorded on cassette by magnetizing the tape, and the same process is used in DAT technology. Magnetizing a small portion of the tape in one direction is interpreted as the number one, and magnetizing it in the opposite direction is interpreted as a zero. The fidelity of CDs and the convenience of cassettes would seem to make DATs the perfect audio medium — except for the price.

Channel surfing with a television remote is another activity made possible by the use of binary arithmetic and a little physics. When you press a particular button on your remote, an infrared beam is sent out that is pulsed so that a binary digit is sent to the television. Electronic circuitry inside the television then decodes the message and directs that a particular task be performed. One remote will not in-

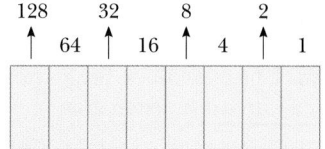

Figure 28.26 A byte consists of eight bits weighted as shown.

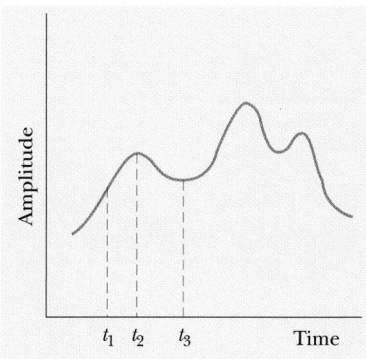

Figure 28.27 Sampling a sound signal whose amplitude varies with time.

APPLICATION

Use of Diffraction Grating in CD Player Tracking.

APPLICATION

Binary Codes in Television Remotes and Garage Door Openers.

Figure 28.28 The laser beam in a CD player is able to follow the spiral track by using three beams produced with a diffraction grating.

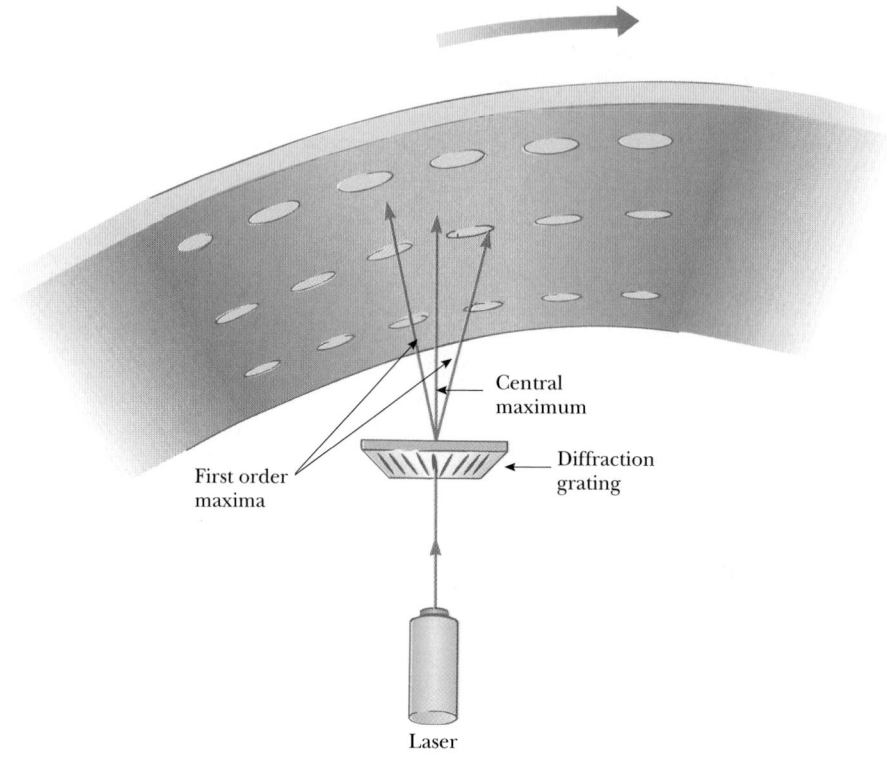

Central maximum

First order maxima

Diffraction grating

Laser

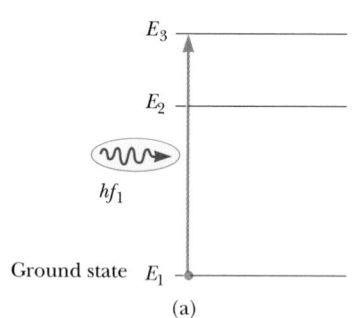

(a)

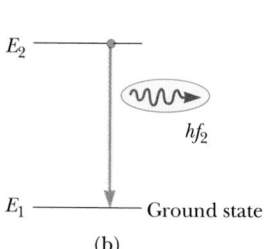

(b)

Figure 28.29 The process of fluorescence. (a) An atom absorbs a photon with energy hf_1 and ends up in an excited state, E_3. (b) The atom emits a photon of energy hf_2 when the electron moves from an intermediate state, E_2, back to the ground state.

terfere with another for, say, the VCR or a television from a different manufacturer because the binary digits for the completion of tasks are assigned differently. In a similar fashion, garage door openers perform their tasks by sending out a radio wave or high frequency sound wave modulated to carry a designated binary number recognized by your door's control mechanisms.

28.14 FLUORESCENCE AND PHOSPHORESCENCE

When an atom absorbs a photon and ends up in an excited state, it can return to the ground state via some intermediate states, as shown in Figure 28.29. The photons emitted by the atom will have lower energy, and therefore lower frequency, than the absorbed photon. The process of converting high frequency radiation to lower frequency radiation by this means is called **fluorescence.**

The common fluorescent light, which makes use of this principle, works as follows. Electrons are produced in the tube as a filament at the end is heated to sufficiently high temperatures. The electrons are accelerated by an applied voltage, and this causes them to collide with atoms of mercury vapor present in the tube. As a result of the collisions, many mercury atoms are raised to excited states. As the excited atoms drop to their normal levels, some ultraviolet photons are emitted, and these strike a phosphor coating on the inner surface of the tube. The coating absorbs these photons and emits visible light by means of fluorescence. Different phosphors emit light of different colors. "Cool white" fluorescent lights emit nearly all the visible colors and hence the light is very white. "Warm white" fluorescent lights have a phosphor that emits more red light and thereby produces a "warm" glow. It is interesting to note that the fluorescent lights above the meat counter in a grocery store are usually "warm white" to give the meat a redder color.

Recently, lasers entered the field of crime detection when it was found that a beam from an argon laser causes the perspiration and body oils of fingerprints to fluoresce. In this technique a beam from a laser is dispersed and shone around a darkened region where it is suspected that fingerprints are located. The forensic scientist wears goggles that absorb the laser light but allow the fluorescent color to pass through.

Another class of materials, called **phosphorescent** materials, continue to glow long after the illumination has been removed. An excited atom in a fluorescent material drops to its normal level in about 10^{-8} s, but an excited atom of a phosphorescent material may remain in an excited metastable state for periods ranging from a few seconds to several hours. Eventually, the atom will drop to its normal state and emit a visible photon. For this reason, phosphorescent materials emit light long after being placed in the dark. Paints made from these substances are often used to decorate the hands of watches and clocks, and to outline doors and stairways in large buildings so that these exits will be visible during power failures.

APPLICATION

Fluorescent Lights.

APPLICATION

Phosphorescent Paints.

SUMMARY

The **Bohr model** of the atom is successful in describing the spectra of atomic hydrogen and hydrogen-like ions. One of the basic assumptions of the model is that the electron can exist only in certain orbits such that its angular momentum, mvr, is an integral multiple of $h/2\pi = \hbar$, where h is Planck's constant. Assuming circular orbits and a Coulomb force of attraction between electron and proton, the energies of the quantum states for hydrogen are

$$E_n = -\frac{m_e k_e^2 e^4}{2\hbar^2}\left(\frac{1}{n^2}\right) \qquad n = 1, 2, 3, \ldots \qquad \text{[28.12]}$$

where k_e is the Coulomb constant, e is the charge on the electron, and n is an integer called a **quantum number.**

If the electron in the hydrogen atom jumps from an orbit whose quantum number is n_i to an orbit whose quantum number is n_f, it emits a photon of frequency f, given by

$$f = \frac{m_e k_e^2 e^4}{4\pi\hbar^3}\left(\frac{1}{n_f^2} - \frac{1}{n_i^2}\right) \qquad \text{[28.14]}$$

Bohr's **correspondence principle** states that quantum mechanics is in agreement with classical physics when the quantum numbers for a system are very large.

One of the many great successes of quantum mechanics is that the quantum numbers n, ℓ, and m_ℓ associated with atomic structure arise directly from the mathematics of the theory. The quantum number n is called the **principal quantum number,** ℓ is the **orbital quantum number,** and m_ℓ is the **orbital magnetic**

quantum number. In addition, a fourth quantum number, called the **spin magnetic quantum number,** m_s, is needed to explain certain features of atomic structure.

An understanding of the periodic table of the elements became possible when Pauli formulated the **exclusion principle,** which states that no two electrons in an atom can ever be in the same quantum state; that is, no two electrons in the same atom can have the same set of quantum numbers, n, ℓ, m_ℓ, and m_s.

Characteristic x-rays are produced when a bombarding electron collides with an electron in an inner shell of an atom with sufficient energy to remove the electron from the atom. The vacancy thus created is filled when an electron from a higher level drops down into the level containing the vacancy.

Lasers are monochromatic, coherent light sources that work on the principle of **stimulated emission** of radiation from a system of atoms.

MULTIPLE-CHOICE QUESTIONS

1. When the hydrogen atom absorbs a photon of energy E_{photon} resulting in the atom being in an excited state, the kinetic energy of the electron changes by
   (a) 0 (b) $E_{photon}/2$ (c) E_{photon}
   (d) $2E_{photon}$ (e) $-E_{photon}$

2. When the hydrogen atom absorbs a photon of energy E_{photon} resulting in the atom being in an excited state, the potential energy of the electron changes by
   (a) 0 (b) $-E_{photon}/2$ (c) $-E_{photon}$
   (d) $-2E_{photon}$ (e) $2E_{photon}$

3. Krypton (atomic number 36) has how many electrons in its next to outer shell ($n = 3$)?
   (a) 2 (b) 4 (c) 8 (d) 18

4. If an electron had a spin of 3/2, its spin quantum number, m_s, could have the following four values:
   $m_s = +3/2$, $+1/2$, $-1/2$, and $-3/2$. If this were true,

the first element with a filled shell would become the first of the noble gases, and it would be
   (a) He with 2 electrons (b) Li with 3 electrons
   (c) Be with 4 electrons (d) C with 5 electrons

5. In an analysis relating Bohr's theory to the de Broglie wavelength of electrons, when an electron moves from the $n = 1$ level to the $n = 3$ level, the circumference for its orbit becomes nine times greater. This occurs because
   (a) there are nine times as many wavelengths in the new orbit.
   (b) the wavelength of the electron becomes nine times as long.
   (c) there are three times as many wavelengths and each wavelength is three times as long.
   (d) the electron is moving nine times as fast.

CONCEPTUAL QUESTIONS

1. In the hydrogen atom, the quantum number n can increase without limit. Because of this, does the frequency of possible spectral lines from hydrogen also increase without limit?

2. Does the light emitted by a neon sign constitute a continuous spectrum or only a few colors? Defend your answer.

3. In an x-ray tube, if the energy with which the electrons strike the metal target is increased, the wavelengths of the characteristic x-rays do not change. Why not?

4. Must an atom first be ionized before it can emit light? Discuss.

5. Is it possible for a spectrum from an x-ray tube to show

the continuous spectrum of x-rays without the presence of the characteristic x-rays?

6. Suppose that the electron in the hydrogen atom obeyed classical mechanics rather than quantum mechanics. Why should such a hypothetical atom emit a continuous spectrum rather than the observed line spectrum?

7. When a hologram is produced, the system (including light source, object, beam splitter, and so on) must be held motionless within a quarter of a wavelength. Why?

8. If matter has a wave nature, why is this not observable in our daily experiences?

9. Discuss some consequences of the exclusion principle.

10. Can the electron in the ground state of hydrogen absorb a photon of energy less than 13.6 eV? Can it absorb a photon of energy greater than 13.6 eV? Explain.

11. Why do lithium, potassium, and sodium exhibit similar chemical properties?

12. List some ways in which quantum mechanics altered our view of the atom pictured by the Bohr theory.

13. It is easy to understand how two electrons (one spin up, one spin down) can fill the $1s$ shell for a helium atom. How is it possible that eight more electrons can fit into the $2s$, $2p$ level to complete the $1s^2 2s^2 2p^6$ shell for a neon atom?

14. The ionization energies for Li, Na, K, Rb, and Cs are 5.390, 5.138, 4.339, 4.176, and 3.893 eV, respectively. Explain why these values are to be expected in terms of the atomic structures.

PROBLEMS

1, 2, 3 = straightforward, intermediate, challenging   □ = full solution available in Study Guide/Student Solutions Manual   ◣ = Core Concepts Workbook
WEB = solution posted at **http://www.harcourtcollege.com/physics/cptech**   = biomedical application   = Interactive Physics

Review Problem

An electron in a hydrogen atom jumps from some initial Bohr orbit, n_i, to some final Bohr orbit, n_f, as the figure illustrates. **(a)** If the photon emitted in the process is capable of ejecting a photoelectron from tungsten (work function = 4.58 eV), determine n_f. **(b)** If a minimum stopping potential of $\Delta V_s = 7.51$ volts is required to prevent the photoelectron from hitting the anode, determine the value of n_i.

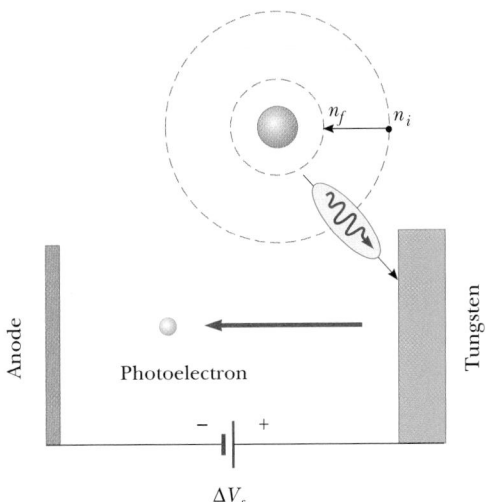

Section 28.1 Early Models of the Atom

Section 28.2 Atomic Spectra

1. Use Equation 28.1 to calculate the wavelength of the first three lines in the Balmer series for hydrogen.

2. **(a)** Suppose the Rydberg constant in Balmer's formula were given by $R_H = 2.00 \times 10^7$ m^{-1}. What part of the electromagnetic spectrum would the Balmer series correspond to? **(b)** Repeat for $R_H = 0.500 \times 10^7$ m^{-1}.

3. The "size" of the *atom* in Rutherford's model is about 1.0×10^{-10} m. **(a)** Determine the attractive electrostatic force between an electron and a proton separated by this distance. **(b)** Determine (in eV) the electrostatic potential energy of the atom.

4. The "size" of the *nucleus* in Rutherford's model of the atom is about 1.0 fm = 1.0×10^{-15} m. **(a)** Determine the repulsive electrostatic force between two protons separated by this distance. **(b)** Determine (in MeV) the electrostatic potential energy of the pair of protons.

WEB **5.** The "size" of the atom in Rutherford's model is about 1.0×10^{-10} m. **(a)** Determine the speed of an electron moving about the proton using the attractive electrostatic force between an electron and a proton separated by this distance. **(b)** Does this speed suggest that Einsteinian relativity must be considered when studying the atom? **(c)** Compute the de Broglie wavelength of the electron as it moves about the proton. **(d)** Does this wavelength suggest that wave effects, such as diffraction and interference, must be considered when studying the atom?

6. In a Rutherford scattering experiment, an α-particle (charge = $+2e$) heads directly toward a gold nucleus (charge = $+79e$). The α-particle had a kinetic energy of 5.0 MeV when very far ($r \rightarrow \infty$) from the nucleus. Assuming the gold nucleus to be fixed in space, determine the distance of closest approach. (*Hint:* Use conservation of energy with $PE = k_e q_1 q_2 / r$.)

Section 28.3 The Bohr Theory of Hydrogen

7. A hydrogen atom is in its first excited state ($n = 2$). Using the Bohr theory of the atom, calculate **(a)** the radius of the orbit, **(b)** the linear momentum of the electron, **(c)** the angular momentum of the electron, **(d)** the kinetic energy, **(e)** the potential energy, and **(f)** the total energy.

8. In the Bohr model of the hydrogen atom, what are the kinetic energy and the potential energy of the electron in the ground state? Express your answers in eV.

9. Show that the speed of the electron in the nth Bohr orbit in hydrogen is given by

$$v_n = \frac{k_e e^2}{n\hbar}$$

10. Show that the speed of the electron in the first (ground state) Bohr orbit of the hydrogen atom may be expressed as $v = (1/137)c$.

11. Calculate the Coulomb force of attraction on the electron when it is in the ground state of the hydrogen atom.

12. Four possible transitions for a hydrogen atom are listed below:

(A) $n_i = 2$; $n_f = 5$ (B) $n_i = 5$; $n_f = 3$
(C) $n_i = 7$; $n_f = 4$ (D) $n_i = 4$; $n_f = 7$

(a) Which transition will emit the shortest wavelength photon?
(b) For which transition will the atom gain the most energy?
(c) For which transition(s) does the atom lose energy?

13. What is the energy of the photon that, when absorbed by a hydrogen atom, could cause (a) an electronic transition from the $n = 3$ state to the $n = 5$ state and (b) an electronic transition from the $n = 5$ state to the $n = 7$ state?

14. A hydrogen atom initially in its ground state ($n = 1$) absorbs a photon and ends up in the state for which $n = 3$. (a) What is the energy of the absorbed photon? (b) If the atom eventually returns to the ground state, what photon energies could the atom emit?

15. A hydrogen atom emits a photon of wavelength 656 nm. From what energy orbit to what lower energy orbit did the electron jump?

16. How much energy is required to ionize hydrogen when it is in (a) the ground state and (b) the state for which $n = 3$?

17. Determine both the longest and the shortest wavelengths in (a) the Lyman series ($n_f = 1$) and (b) the Paschen series ($n_f = 3$) of hydrogen.

18. Consider a large number of hydrogen atoms, with electrons all initially in the $n = 4$ state. (a) How many different wavelengths would be observed in the emission spectrum of these atoms? (b) What is the longest wavelength that could be observed? To which series does it belong?

19. (a) If an electron makes a transition from the $n = 4$ Bohr orbit to the $n = 2$ orbit, determine the wavelength of the photon created in the process. (b) Assuming that the atom was initially at rest, determine the recoil speed of the hydrogen atom when this photon is emitted.

20. An electron is in the first Bohr orbit of hydrogen. Find (a) the speed of the electron, (b) the time required for the electron to circle the nucleus, and (c) the current in amperes corresponding to the motion of the electron.

WEB 21. A particle of charge q and mass m, moving with a constant speed, v, perpendicular to a constant magnetic field, B, follows a circular path. If the angular momentum about the center of this circle is quantized so that $mvr = n\hbar$, show that the allowed radii for the particle are

$$r_n = \sqrt{\frac{n\hbar}{qB}} \qquad \text{for} \qquad n = 1, 2, 3, \ldots$$

22. Two hydrogen atoms, both initially in the ground state, undergo a head-on collision. If both atoms are to be excited to the $n = 2$ level in this collision, what is the minimum speed each atom can have before the collision?

23. Analyze the Earth-Sun system following the Bohr model, where the gravitational force between Earth (mass m) and Sun (mass M) replaces the Coulomb force between the electron and proton (so that $F = GMm/r^2$ and $PE = -GMm/r$). Show that (a) the total energy of the Earth in an orbit of radius r is given by $E = -GMm/(2r)$, (b) the radius of the nth orbit is given by $r_n = r_0 n^2$, where $r_0 = \hbar^2/(GMm^2) = 2.32 \times 10^{-138}$ m, and (c) the energy of the nth orbit is given by $E_n = -E_0/n^2$, where $E_0 = G^2 M^2 m^3/(2\hbar^2) = 1.71 \times 10^{182}$ J. (d) Using the Earth-Sun orbit radius of $r = 1.49 \times 10^{11}$ m, determine the value of the quantum number n. (e) Should you expect to observe quantum effects in the Earth-Sun system?

24. (a) Calculate the angular momentum of the Moon as a result of its orbital motion about the Earth. In your calculation, use 3.84×10^8 m as the average Earth-Moon distance and 2.36×10^6 s as the period of the Moon in its orbit. (b) If the angular momentum of the moon obeys Bohr's quantization rule ($L = n\hbar$), determine the value of the quantum number, n. (c) By what fraction would the Earth-Moon radius have to be increased to increase the quantum number by 1?

25. Consider a hydrogen atom. (a) Calculate the frequency f of the $n = 2 \rightarrow n = 1$ transition and compare with the frequency f_{orb} of the electron orbital motion in the $n = 2$ state. (b) Make the same calculation for the $n = 10\,000 \rightarrow n = 9999$ transition. Comment on the results.

Section 28.4 Modification of the Bohr Theory

Section 28.5 de Broglie Waves and the Hydrogen Atom

26. (a) Find the energy of the electron in the ground state of doubly ionized lithium, which has an atomic number $Z = 3$. (b) Find the radius of its ground-state orbit.

27. Plot an energy-level diagram such as that in Figure 28.7 for singly ionized helium.

28. (a) Substitute numerical values into Equation 28.19 to find a value for the Rydberg constant for singly ionized helium, He⁺. (b) Use the result of part (a) to find the wavelength associated with a transition from the $n = 2$ state to the $n = 1$ state of He⁺. (c) Identify the region of the electromagnetic spectrum associated with this transition.

29. Construct an energy level diagram like that in Figure 28.7 for doubly ionized lithium (Li^{2+}), for which $Z = 3$.

30. Using the concept of standing waves, de Broglie was able to derive Bohr's stationary orbit postulate. He assumed that a confined electron could exist only in states where its de Broglie waves form standing wave patterns, as in Figure 28.10a. Consider a particle confined in a box of length L to be equivalent to a string of length L and fixed at both ends. Apply de Broglie's concept to show that (a) the linear momentum of this particle is quantized with $p = mv = nh/2L$ and (b) the allowed states correspond to particle energies of $E_n = n^2 E_0$, where $E_0 = h^2/(8mL^2)$.

31. Determine the wavelength of an electron in the third excited orbit of the hydrogen atom, with $n = 4$.

Section 28.6 Quantum Mechanics and the Hydrogen Atom

Section 28.7 The Spin Magnetic Quantum Number

32. When the principal quantum number is $n = 4$, how many different values of (a) ℓ and (b) m_ℓ are possible?

33. List the possible sets of quantum numbers for electrons in the $3p$ subshell.

34. The ρ-meson has a charge of $-e$, a spin quantum number of 1, and a mass 1507 times that of the electron. If the electrons in atoms were replaced by ρ-mesons, list the possible sets of quantum numbers for ρ-mesons in the $3d$ subshell.

Section 28.9 The Exclusion Principle and the Periodic Table

35. How many different sets of quantum numbers are possible for an electron for which (a) $n = 1$, (b) $n = 2$, (c) $n = 3$, (d) $n = 4$, and (e) $n = 5$? Check your results to show that they agree with the general rule that the number of different sets of quantum numbers is equal to $2n^2$.

36. (a) Write out the electronic configuration of the ground state for oxygen ($Z = 8$). (b) Write out the values for the set of quantum numbers n, ℓ, m_ℓ, and m_s for each of the electrons in oxygen.

WEB 37. Zirconium ($Z = 40$) has two electrons in an incomplete d subshell. (a) What are the values of n and ℓ for each electron? (b) What are all possible values of m_ℓ and m_s? (c) What is the electron configuration in the ground state of zirconium?

38. Suppose two electrons in the same system each have $\ell = 0$ and $n = 3$. (a) How many states would be possible if the exclusion principle were inoperative? (b) List the possible states, taking the exclusion principle into account.

Section 28.10 Characteristic X-Rays

39. The K_α x-ray is emitted when an electron undergoes a transition from the L shell ($n = 2$) to the K shell ($n = 1$). Use the method illustrated in Example 28.7 to calculate the wavelength of the K_α x-ray from a nickel target ($Z = 28$).

40. The K-shell ionization energy of copper is 8979 eV. The L-shell ionization energy is 951 eV. Determine the wavelength of the K_α emission line of copper. What must the minimum voltage be on an x-ray tube with a copper target in order to see the K_α line?

41. The K series of the discrete spectrum of tungsten contains wavelengths of 0.0185 nm, 0.0209 nm, and 0.0215 nm. The K-shell ionization energy is 69.5 keV. Determine the ionization energies of the L, M, and N shells. Sketch the transitions that produce these wavelengths.

42. When an electron drops from the M shell ($n = 3$) to a vacancy in the K shell ($n = 1$), the measured wavelength of the emitted x-ray is found to be 0.101 nm. Identify the element.

ADDITIONAL PROBLEMS

43. (a) How much energy is required to cause an electron in hydrogen to move from the $n = 1$ state to the $n = 2$ state? (b) If the electrons gain this energy by collision between hydrogen atoms in a high temperature gas, find the minimum temperature of the heated hydrogen gas. The thermal energy of the heated atoms is given by $3k_B T/2$, where k_B is the Boltzmann constant.

44. In a hydrogen atom, what is the principal quantum number of the electron orbit with a radius closest to $1.0\ \mu m$?

45. A laser used in a holography experiment has an average output power of 5.0 mW. The laser beam is actually a series of pulses of electromagnetic radiation at a wavelength of 632.8 nm, each having a duration of 25 ms. Calculate (a) the energy (in joules) radiated with each pulse and (b) the number of photons in each pulse.

46. When a muon with charge $-e$ is captured by a proton, the resulting bound system forms a "muonic atom," which is the same as hydrogen except with a muon (of mass 207 times that of an electron) replacing the electron. For this "muonic atom" determine (a) the Bohr radius and (b) the three lowest energy levels.

47. The Lyman series for a (new!?) one-electron atom is observed in a distant galaxy. The wavelengths of the first four lines and the short-wavelength limit of this Lyman series are given by the energy-level diagram in Figure P28.47. Based on this information, calculate (a) the energies of the ground state and first four excited states for this one-electron atom, and (b) the wavelengths of the longest-wavelength (alpha) line and of the short-

wavelength series limit in the Balmer series for this atom.

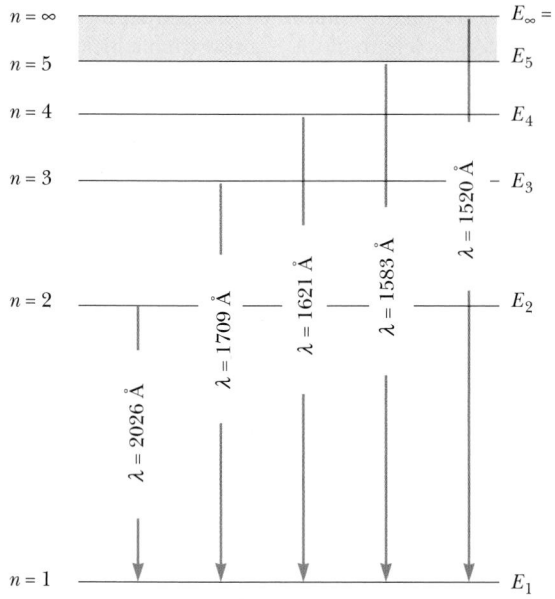

Figure P28.47

48. Mercury's ionization energy is 10.39 eV. The three longest wavelengths of the absorption spectrum of mercury are 253.7 nm, 185.0 nm, and 158.5 nm. (a) Construct an energy-level diagram for mercury. (b) Indicate all emission lines that can occur when an electron is raised to the third level above the ground state. (c) Disregarding recoil of the mercury atom, determine the minimum speed an electron must have in order to make an inelastic collision with a mercury atom.

49. Suppose the ionization energy of an atom is 4.100 eV. In this same atom, we observe emission lines with wavelengths 310.0 nm, 400.0 nm, and 1378 nm. Use this information to construct the energy-level diagram with the least number of levels. Assume the higher energy levels are closer together.

50. An electron is in the nth Bohr orbit of the hydrogen atom. (a) Show that the time it takes the electron to circle the nucleus once can be expressed as $T = \tau_0 n^3$, and determine the numerical value of τ_0. (b) On the average, an electron in the $n = 2$ orbit will exist in that orbit for about 1.0×10^{-8} s before it jumps down to the $n = 1$ (ground-state) orbit. How many revolutions about the nucleus are made by the electron before it jumps to the ground state? (c) If one revolution of the electron about the nucleus is defined as an "electron year" (analogous to an Earth year being one revolution of the Earth around the Sun), does the electron in the $n = 2$ orbit

"live" very long? Explain. (d) How does the above calculation support the "electron cloud" concept?

51. In order for an electron to be confined to a nucleus, its de Broglie wavelength would have to be less than 1.0×10^{-14} m. (a) What would be the kinetic energy of an electron confined to this region? (b) On the basis of this result, would you expect to find an electron in a nucleus? Explain. (*Hint:* This is a relativistic electron.)

52. An electron has a de Broglie wavelength equal to the diameter of a hydrogen atom in its ground state. (a) What is the kinetic energy of the electron? (b) How does this energy compare with the ground-state energy of the hydrogen atom?

53. A laser used in eye surgery emits a 3.00-mJ pulse in 1.00 ns, focused to a spot 30.0 μm in diameter on the retina. (a) Find (in SI units) the power per unit area at the retina. (This quantity is called the irradiance.) (b) What energy is delivered to an area of molecular size—say, a circular area 0.600 nm in diameter?

54. A pulsed laser emits light having wavelength λ. For a pulse of duration Δt having energy E, find (a) the physical length of the pulse as it travels through space and (b) the number of photons in it. (c) If the beam has a circular cross section having diameter d, find the number of photons per unit volume.

55. Light from a certain He-Ne laser has a power output of 1.00 mW and a cross-sectional area of 10.0 mm². The entire beam is incident on a metal target that requires 1.50 eV to remove an electron from its surface. (a) Perform a classical calculation to determine how long it takes one atom in the metal to absorb 1.50 eV from the incident beam. (*Hint:* Assume the face area of an atom is 1.00×10^{-29} m², and first calculate the energy incident on each atom per second.) (b) Compare the (wrong) answer obtained in (a) to the actual response time for photoelectric emission (about 10^{-9} s), and discuss the reasons for the large discrepancy.

WEB 56. A pi meson (π^-) of charge $-e$ and mass 273 times greater than that of the electron is captured by a helium nucleus ($Z = +2$) as shown in Figure P28.56. (a) Draw an energy level diagram (in units of eV) for this "Bohr-type" atom up to the first six energy levels. (b) When the pi meson makes a transition between two orbits, a photon is emitted that Compton scatters off a free electron initially at rest, producing a scattered photon of wavelength $\lambda' = 0.089\ 929\ 3$ nm at an angle of $\theta = 50.00°$, as shown on the right-hand side of Figure P28.56. Between which two orbits did the pi meson make a transition?

57. Use Bohr's model of the hydrogen atom to show that when the atom makes a transition from the state n to the state $n - 1$, the frequency of the emitted light is given by

$$f = \frac{2\pi^2 m k_e^2 e^4}{h^3} \left(\frac{2n - 1}{(n - 1)^2 n^2} \right)$$

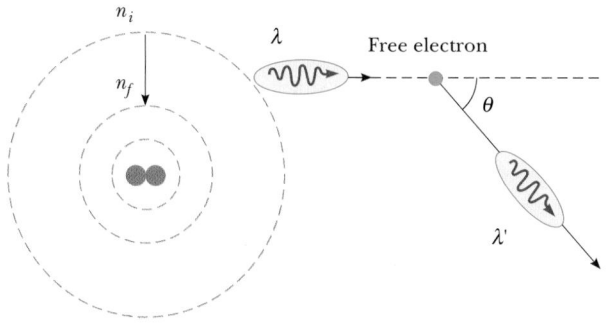

"Pi mesonic" He$^+$ atom
($Z = 2$, $m_\pi = 273m_e$)

Figure P28.56

58. Calculate the classical frequency for the light emitted by an atom. To do so, note that the frequency of revolution is $v/2\pi r$, where r is the Bohr radius. Show that as n approaches infinity in the equation of the preceding problem, the expression given there varies as $1/n^3$ and reduces to the classical frequency. (This is an example of the correspondence principle, which requires that the classical and quantum models agree for large values of n.)

59. A dimensionless number that often appears in atomic physics is the *fine-structure constant*, α, given by

$$\alpha = \frac{k_e e^2}{\hbar c}$$

where k_e is the Coulomb constant. **(a)** Obtain a numerical value for $1/\alpha$. **(b)** In terms of α, what is the ratio of the Bohr radius, a_0, to the Compton wavelength, $\lambda_C = h/m_e c$? **(c)** In terms of α, what is the ratio of the reciprocal of the Rydberg constant, $1/R_H$, to the Bohr radius?

60. In this problem you will estimate the classical lifetime of the hydrogen atom. An accelerating charge loses electromagnetic energy at a rate given by $P = -2k_e q^2 a^2/(3c^3)$, where k_e is the Coulomb constant, q is the charge of the particle, a is its acceleration, and c is the speed of light in a vacuum. Assuming that the electron is one Bohr radius (0.0529 nm) from the center of the hydrogen atom, **(a)** determine its acceleration. **(b)** Show that P has units of energy per unit time and determine the rate of energy loss. **(c)** Calculate the kinetic energy of the electron and determine how long it will take for all of this energy to be converted into electromagnetic waves, assuming that the rate calculated in part (b) remains constant throughout the electron's motion.

Lasers and Their Applications

ISAAC D. ABELLA
The University of Chicago

Laser Principles

We encounter lasers in many applications in the home and workplace. CD (compact disc) players in home audio systems incorporate a small laser disc reader. Laser printers use a laser beam to paint an image directly from the computer onto the xerographic surface, dispensing with the photographic lens of conventional copiers. The telephone company uses laser communications devices on the fiber-optic network. The supermarket checkout counter is where most people see the laser beam in action. This enormous growth of laser applications was stimulated by many scientific and engineering advances that exploit some of the unique properties of laser light.

These properties derive from the distinctive way laser light is produced in contrast to the generation of ordinary light (see Fig. 1). Laser light originates from *energized* ("excited") atoms, ions, or molecules through a process of *stimulated emission* of radiation. The suitably prepared *active* laser medium is contained in an enclosure or *cavity* that organizes the normally random emission process into an intense, directional, monochromatic, and coherent wave. The end *mirrors* provide the essential optical feedback that selectively builds up the stimulating wave along the tube axis. However, in an ordinary sodium vapor street lamp, for example, the energized atoms *spontaneously* emit in random directions and at irregular times, over a broad spectrum, resulting in isotropic illumination of incoherent light. Laser light has a well-defined phase, permitting a wide variety of applications based on interference or wave modulation.

Currently operating laser systems use a variety of atomic gases, solids, or liquids as the working laser substance. These devices are designed to emit either continuous or pulsed monochromatic beams, and operate over a broad range of the electromagnetic spectrum (ultraviolet, visible, infrared) with output powers from milliwatts (10^{-3}) to megawatts (10^6). The particular application determines the choice of laser system, wavelength, power level, or other relevant variables, because no one laser has all the desirable properties.

For a laser to operate successfully, several *atomic physics* conditions must be satisfied, as noted in Section 28.12. The requirement for population inversion—that is, more atoms in a particular *excited state* than in a lower state—essentially means that energy must be supplied from outside the system. Otherwise, atoms would eventually radiate and develop an increasing probability for absorbing light. Finally, they would fall to the lowest energy state and stop emitting altogether.

Therefore, all laser systems must be connected to external energy sources, usually electrical, as required by conservation of energy, to maintain this nonthermal equilibrium situation. For example, we can energize atoms by electron impact in gaseous discharges (so called "electrical pumping"). We can also supply energy to lamps whose light populates excited states by photon absorption ("optical pumping") for those solids or liquids that do not conduct electric charge. These pumping mechanisms tend to have low efficiency (ratio of laser energy output to electric energy supplied), typically a few percentages, with the balance discharged as heat into cooling water or circulating air.

Controlling the electrical input into the laser system allows a variable laser energy output, which may be important in many applications. Thus, the argon ion laser system can emit up to about 10 W in the green optical beam by adjustment of the electric current in the argon gas, which in turn controls the degree of population inversion. Chemical lasers, on the other hand, operate without direct electrical input. Several highly reactive gases are mixed in the laser chamber, with the energy released in the ensuing reaction populating

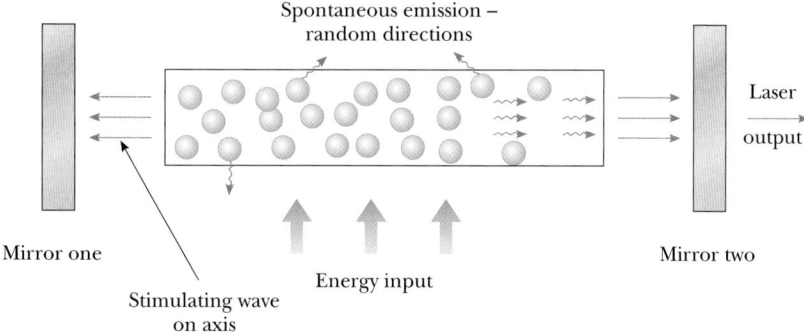

Figure 1 A schematic of a laser design. The tube contains atoms, which represent the active medium. An external source of energy (optical, electrical, etc.) is needed to "pump" the atoms to excited energy states. The parallel end mirrors provide the feedback of the stimulating wave.

the excited levels in the molecule. In this case, the reactants need to be resupplied for the laser to operate for any length of time.

Some laser systems have fluid media, containing dissolved dye molecules. The dye lasers are usually pumped to excited levels by an external laser. The advantage of this arrangement is that dye lasers can be continuously "tuned" over a wide range of wavelengths, using prisms or gratings, whereas the pump source has a fixed wavelength. Color variability is important for those cases in which the laser is directed at materials whose absorption depends on wavelength. Thus, the laser can be tuned into exact coincidence with selected energy states. For example, blood does not absorb red light to any extent, which excludes red light use for most surgical applications on blood-rich tissue.

Recent developments in tunable solid-state materials have permitted design of tunable lasers without the need for unstable dye molecules; most notable are sapphire crystals containing titanium ions. These materials are optically pumped by flashlamps or fixed wavelength lasers acting to populate the upper state directly.

A variety of laser systems are in general use today. They include the 1-mW helium-neon laser, usually appearing as a red beam at 632.8 nm (although yellow and green beams are available); the argon ion laser, which operates in the green or blue up to 10 W; the carbon dioxide gas laser, which emits in the infared at 10 mm and can produce several hundred watts; and the neodymium YAG laser, a powerful solid-state optically pumped system that emits at 1.06 mm either continuous or pulsed. The diode junction laser emits in the near infrared and operates by passage of current through the semiconductor material. The recombination radiation is essentially direct conversion of electrical energy to laser light and is a very efficient process. The diodes can emit up to 5 W and can be used to energize other laser materials.

Applications

We shall describe a few applications that should serve to illustrate the wide variety of laser uses. Other applications are discussed in this chapter. First, there is the use of lasers in precision long-range distance measurement (range finding). It has become important, for astronomical and geophysical purposes, to measure as precisely as possible the distance from various points on the surface of the Earth to a point on the Moon's surface. To facilitate this, the Apollo astronauts set up a compact array, a 0.5-m square of reflector prisms on the Moon, which allows laser pulses directed from an Earth station to be retroreflected

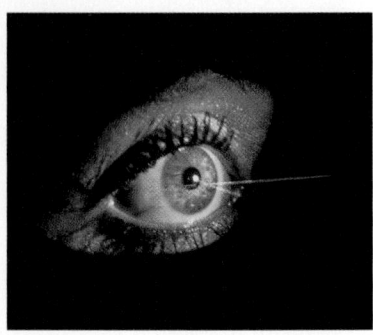

An argon laser passing through a cornea and lens during eye surgery. (© *Alexander Tsiaras, Science Source/ Photo Researchers, Inc.*)

to the same station. Using the known speed of light and the measured roundtrip travel time of a 1-ns pulse, one can determine the Earth-Moon distance, 380 000 km, to a precision of better than 10 cm. Such information would be useful, for example, in making more reliable earthquake predictions and for learning about the motions of the Earth-Moon system. This technique requires a high-power pulsed laser for its success, because a sufficient burst of photons must return to a collecting telescope on Earth and be detected. Variants of this method are also used to measure the distance to inaccessible points on the Earth as well as in military range finders.

Novel medical applications use the fact that the different laser wavelengths can be absorbed in specific biological tissues. A common eye condition, glaucoma, is manifested by a high fluid pressure in the eye, which can lead to destruction of the optic nerve. A simple laser operation (iridectomy) can "burn" open a tiny hole in a clogged membrane, relieving the destructive pressure. Along the same lines, a serious side effect of diabetes is neovascularization, the formation of weak blood vessels, which often leak blood into extremities. When this occurs in the eye, vision deteriorates (diabetic retinopathy) leading to blindness. It is now possible to direct the green light from the argon ion laser through the clear eye lens and eye fluid, focus on the retina edges, and photocoagulate the leaky vessels. These procedures have greatly reduced blindness in glaucoma and diabetes patients.

Laser surgery is now a practical reality. Infrared light at 10 mm from a carbon dioxide laser can cut through muscle tissue, primarily by heating and evaporating the water contained in cellular material. Laser power of about 100 W is required in this technique. The advantage of the laser knife over conventional methods is that laser radiation cuts and coagulates at the same time, leading to substantial reduction of blood loss. In addition, the technique virtually eliminates cell migration, which is very important in tumor removal. Furthermore, a laser beam can be trapped in fine glass-fiber light-guides (endoscopes) by means of total internal reflection (Section 22.9). The light fibers can be introduced through natural orifices, conducted around internal organs, and directed to specific interior body locations, eliminating the need for massive surgery. For example, bleeding in the gastrointestinal tract can be optically cauterized by fiber-optic endoscopes inserted through the mouth.

Finally, we describe an application to biological and medical research. It is often important to isolate and collect unusual cells for study and growth. A laser cell separator exploits the fact that specific cells can be tagged with fluorescent dyes. All cells are then dropped from a tiny charged nozzle and laser-scanned for the dye tag. If triggered by the correct light-emitting tag, a small voltage applied to parallel plates deflects the falling electrically charged cell into a collection beaker. This is an efficient method for extracting the proverbial needles from the haystack.

Nuclear Physics

▲ PHYSICS PUZZLER

In 1991, a tourist in the Italian Alps discovered the body of an ancient man who had been trapped in a glacier. The remains were dubbed the Iceman, and it was determined that the remains were more than 5000 years old. How were scientists able to determine the age of the remains and differentiate them from that of a recently deceased individual? *(Hanny Paul/Gamma Liaison)*

Ernest Rutherford, physicist from New Zealand (1871–1937)

Rutherford was awarded the Nobel prize in 1908 for discovering that atoms can be broken apart by alpha rays and for studying radioactivity. "On consideration, I realized that this scattering backward must be the result of a single collision, and when I made calculations I saw that it was impossible to get anything of that order of magnitude unless you took a system in which the greater part of the mass of the atom was concentrated in a minute nucleus. It was then that I had the idea of an atom with a minute massive center carrying a charge." *(North Wind Picture Archives)*

In 1896, the year that marks the birth of nuclear physics, Henri Becquerel (1852–1908) discovered radioactivity in uranium compounds. A great deal of activity followed this discovery as researchers attempted to understand and characterize the radiation that we now know to be emitted by radioactive nuclei. Pioneering work by Rutherford showed that the radiation was of three types, which he called *alpha, beta,* and *gamma rays.* These types are classified according to the nature of their electric charge and according to their ability to penetrate matter. Later experiments showed that alpha rays are helium nuclei, beta rays are electrons, and gamma rays are high-energy photons.

In 1911 Rutherford and his students Geiger and Marsden performed a number of important scattering experiments involving alpha particles. These experiments established that the nucleus of an atom can be regarded as essentially a point mass and point charge and that most of the atomic mass is contained in the nucleus. Furthermore, such studies demonstrated a wholly new type of force, the *nuclear force,* which is predominant at distances of less than about 10^{-14} m and zero at greater distances.

Other milestones in the development of nuclear physics include

- the observation of nuclear reactions by Cockcroft and Walton in 1930,
- the discovery of the neutron by Chadwick in 1932,
- the discovery of artificial radioactivity by Joliot and Irene Curie in 1933,
- the discovery of nuclear fission by Hahn and Strassman in 1938,
- the development of the first controlled fission reactor by Fermi and his collaborators in 1942.

In this chapter we discuss the properties and structure of the atomic nucleus. We start by describing the basic properties of nuclei and follow with a discussion of the phenomenon of radioactivity. Finally, we explore nuclear reactions and the various processes by which nuclei decay.

29.1   SOME PROPERTIES OF NUCLEI

All nuclei are composed of two types of particles: protons and neutrons. The only exception is the ordinary hydrogen nucleus, which is a single proton. In describing some of the properties of nuclei, such as their charge, mass, and radius, we make use of the following quantities:

- the **atomic number,** Z, which equals the number of protons in the nucleus,
- the **neutron number,** N, which equals the number of neutrons in the nucleus,
- the **mass number,** A, which equals the number of nucleons in the nucleus. (*Nucleon* is a generic term used to refer to either a proton or a neutron.)

The symbol we use to represent nuclei is $^A_Z X$, where X represents the chemical symbol for the element. For example, $^{27}_{13} Al$ has the mass number 27 and the atomic number 13; therefore, it contains 13 protons and 14 neutrons. When no confusion is likely to arise, we omit the subscript Z because the chemical symbol can always be used to determine Z.

The nuclei of all atoms of a particular element must contain the same number of protons, but they may contain different numbers of neutrons. Nuclei that are related in this way are called **isotopes. The isotopes of an element have the same Z value but different N and A values.** The natural abundances of isotopes can differ substantially. For example, $^{11}_{6}C$, $^{12}_{6}C$, $^{13}_{6}C$, and $^{14}_{6}C$ are four isotopes of carbon. The natural abundance of the $^{12}_{6}C$ isotope is about 98.9%, whereas that of the $^{13}_{6}C$ isotope is only about 1.1%. Some isotopes do not occur naturally but can be produced in the laboratory through nuclear reactions. Even the simplest element, hydrogen, has isotopes: $^{1}_{1}H$, hydrogen; $^{2}_{1}H$, deuterium; and $^{3}_{1}H$, tritium.

Charge and Mass

The proton carries a single positive charge, $+e$, the electron carries a single negative charge, $-e$, where $e = 1.6 \times 10^{-19}$ C, and the neutron is electrically neutral. Because the neutron has no charge, it is difficult to detect. The proton is about 1836 times as massive as the electron, and the masses of the proton and the neutron are almost equal. The masses of several particles are given in Table 29.1. The masses and some other properties of selected isotopes are provided in Appendix B.

It is convenient to define, for atomic masses, the **unified mass unit, u,** in such a way that the mass of one atom of the isotope ^{12}C is exactly 12 u, where 1 u = 1.660 540 $\times$ 10^{-27} kg. The proton and neutron each have a mass of about 1 u, and the electron has a mass that is only a small fraction of an atomic mass unit.

Because the rest energy of a particle is given by $E_R = mc^2$, it is often convenient to express the particle's mass in terms of its energy equivalent. For one atomic mass unit, we have

$$E_R = mc^2 = (1.660\ 540 \times 10^{-27}\ \text{kg})(2.997\ 92 \times 10^8\ \text{m/s})^2 = 931.494\ \text{MeV}$$

Nuclear physicists often express mass in terms of the unit MeV/c^2, where

$$1\ \text{u} = 931.494\ \text{MeV}/c^2$$

The Size of Nuclei

The size and structure of nuclei were first investigated in the scattering experiments of Rutherford, discussed in Section 28.1. Using the principle of conservation of energy, Rutherford found an expression for how close an alpha particle moving directly toward the nucleus can come to the nucleus before being turned around by Coulomb repulsion.

In such a head-on collision, the kinetic energy of the incoming alpha particle must be converted completely to electrical potential energy when the particle stops at the point of closest approach and turns around (Fig. 29.1). If we equate the initial kinetic energy of the alpha particle to the maximum electrical potential energy of the system (alpha particle plus target nucleus), we have

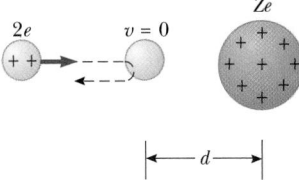

Figure 29.1 An alpha particle on a head-on collision course with a nucleus of charge Ze. Because of the Coulomb repulsion between the like charges, the alpha particle will stop instantaneously at a distance d from the nucleus.

TABLE 29.1 Mass of the Proton, Neutron, and Electron in Various Units

Particle	Mass		
	kg	u	MeV/c^2
Proton	1.6726 $\times$ 10^{-27}	1.007 276	938.28
Neutron	1.6750 $\times$ 10^{-27}	1.008 665	939.57
Electron	9.109 $\times$ 10^{-31}	5.486 $\times$ 10^{-4}	0.511

$$\tfrac{1}{2}mv^2 = k_e \frac{q_1 q_2}{r} = k_e \frac{(2e)(Ze)}{d}$$

where d is the distance of closest approach. Solving for d, we get

$$d = \frac{4k_e Ze^2}{mv^2}$$

From this expression, Rutherford found that alpha particles approached to within 3.2×10^{-14} m of a nucleus when the foil was made of gold. Thus, the radius of the gold nucleus must be less than this value. For silver atoms, the distance of closest approach was 2×10^{-14} m. From these results, Rutherford concluded that the positive charge in an atom is concentrated in a small sphere, which he called the nucleus, whose radius is no greater than about 10^{-14} m. Because such small lengths are common in nuclear physics, a convenient unit of length is the *femtometer* (fm), sometimes called the **fermi,** defined as

$$1 \text{ fm} \equiv 10^{-15} \text{ m}$$

Since the time of Rutherford's scattering experiments, a multitude of other experiments have shown that most nuclei are approximately spherical and have an average radius given by

$$r = r_0 A^{1/3} \qquad\qquad [29.1]$$

where A is the mass number and r_0 is a constant equal to 1.2×10^{-15} m. Because the volume of a sphere is proportional to the cube of its radius, it follows from Equation 29.1 that the volume of a nucleus (assumed to be spherical) is directly proportional to A, the total number of nucleons. This suggests that **all nuclei have nearly the same density.** Nucleons combine to form a nucleus *as though* they were tightly packed spheres (Fig. 29.2).

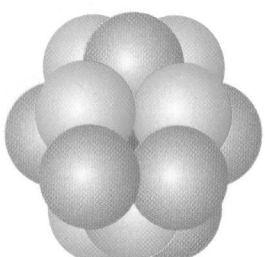

Figure 29.2 A nucleus can be visualized as a cluster of tightly packed spheres in which each sphere is a nucleon.

EXAMPLE 29.1 Nuclear Volume and Density

Find (a) an approximate expression for the mass of a nucleus of mass number A, (b) an expression for the volume of this nucleus in terms of the mass number, and (c) a numerical value for its density.

Solution   (a) The mass of the proton is approximately equal to that of the neutron. Thus, if the mass of one of these particles is m, the mass of the nucleus is approximately Am.

(b) Assuming the nucleus is spherical and using Equation 29.1, we find that the volume is

$$V = \tfrac{4}{3}\pi r^3 = \tfrac{4}{3}\pi r_0^3 A$$

(c) The nuclear density can be found as follows:

$$\rho_n = \frac{\text{mass}}{\text{volume}} = \frac{Am}{\tfrac{4}{3}\pi r_0^3 A} = \frac{3m}{4\pi r_0^3}$$

Taking $r_0 = 1.2 \times 10^{-15}$ m and $m = 1.67 \times 10^{-27}$ kg, we find that

$$\rho_n = \frac{3(1.67 \times 10^{-27} \text{ kg})}{4\pi(1.2 \times 10^{-15} \text{ m})^3} = 2.3 \times 10^{17} \text{ kg/m}^3$$

Note that the nuclear density is about 2.3×10^{14} times greater than the density of water $(1 \times 10^3 \text{ kg/m}^3)$!

Nuclear Stability

Given that the nucleus consists of a closely packed collection of protons and neutrons, you might be surprised that it can exist. The very large repulsive electrostatic forces between protons should cause the nucleus to fly apart. However, nuclei are stable because of the presence of another, short-range (about 2 fm) force, the **nuclear force.** This is an attractive force that acts between all nuclear particles. The protons attract each other via the nuclear force, and at the same time they repel each other through the Coulomb force. The nuclear force also acts between pairs of neutrons and between neutrons and protons.

The nuclear force dominates the Coulomb repulsive force within the nucleus (at short ranges). If this were not the case, stable nuclei would not exist. Moreover, the strong nuclear force is nearly independent of charge. In other words, the nuclear forces associated with the proton-proton, proton-neutron, and neutron-neutron interactions are approximately the same, apart from the additional repulsive Coulomb force for the proton-proton interaction.

There are about 260 stable nuclei; hundreds of others have been observed but are unstable. A plot of N versus Z for a number of stable nuclei is given in Figure 29.3. Note that light nuclei are most stable if they contain equal numbers of protons

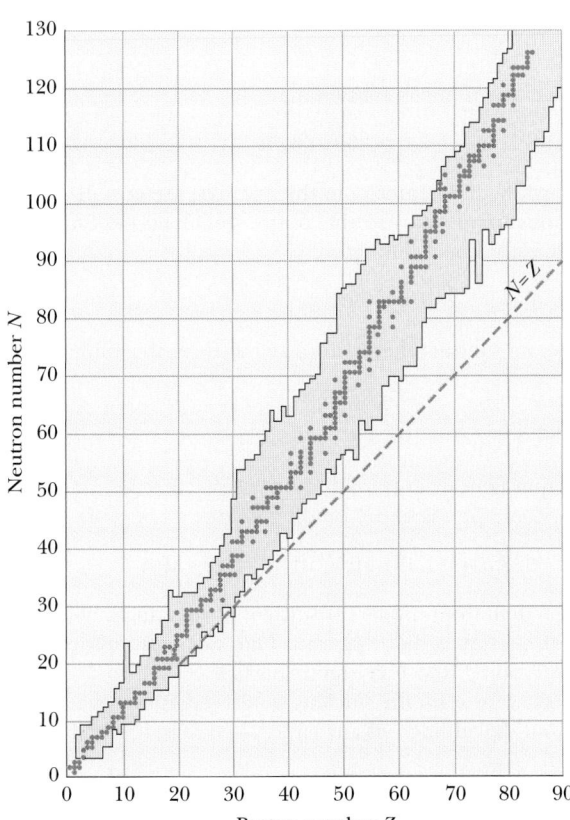

Figure 29.3 A plot of the neutron number N versus the proton number Z for the stable nuclei (solid points). The dashed straight line corresponds to the condition $N = Z$. The shaded area shows radioactive (unstable) nuclei.

Maria Goeppert-Mayer,
(1906–1972)

Goeppert-Mayer was born and educated in Germany. She is best known for her development of the shell model of the nucleus, published in 1950. A similar model was simultaneously developed by Hans Jensen, a German scientist. Maria Goeppert-Mayer and Hans Jensen were awarded the Nobel prize in physics in 1963 for their extraordinary work in understanding the structure of the nucleus. *(Courtesy of Louise Barker/AIP Niels Bohr Library)*

WEB

For Web links to topics in nuclear physics, including a detailed, "clickable" version of Figure 29.3, visit the textbook Web site at **http://www.harcourtcollege.com/physics/ cptech**

and neutrons—that is, if $N = Z$—but heavy nuclei are more stable if $N > Z$. This can be partially understood by recognizing that, as the number of protons increases, the strength of the Coulomb force increases, which tends to break the nucleus apart. As a result, more neutrons are needed to keep the nucleus stable, because neutrons experience only the attractive nuclear forces. In effect, the additional neutrons "dilute" the nuclear charge density. Eventually, when $Z = 83$, the repulsive forces between protons cannot be compensated by the addition of more neutrons. Elements that contain more than 83 protons do not have stable nuclei.

29.2   BINDING ENERGY

The total mass of a nucleus is always less than the sum of the masses of its nucleons. Because mass is another manifestation of energy, **the total energy of the bound system (the nucleus) is less than the combined energy of the separated nucleons.** This difference in energy is called the **binding energy** of the nucleus and can be thought of as the energy that must be added to a nucleus to break it apart into its components. Therefore, in order to separate a nucleus into protons and neutrons, energy must be put into the system.

EXAMPLE 29.2   The Binding Energy of the Deuteron

The nucleus of the deuterium atom, called the deuteron, consists of a proton and a neutron. Calculate the deuteron's binding energy, given that its mass is 2.014 102 u.

Solution   We know that the proton and neutron masses are

$$m_p = 1.007\ 825\ \text{u} \qquad m_n = 1.008\ 665\ \text{u}$$

Note that the masses used for the proton and deuteron in this example are actually those of the neutral atoms, as found in Appendix B. We are able to use atomic masses for these calculations because the electron masses cancel. Therefore,

$$m_p + m_n = 2.016\ 490\ \text{u}$$

To calculate the mass difference, we subtract the deuteron mass from this value:

$$\Delta m = (m_p + m_n) - m_d$$
$$= 2.016\ 490\ \text{u} - 2.014\ 102\ \text{u} = 0.002\ 388\ \text{u}$$

Because 1 u corresponds to an equivalent energy of 931.494 MeV (that is, $1\ \text{u} \cdot c^2 = 931.494$ MeV), the mass difference corresponds to the binding energy

$$E_b = (\Delta m)\,c^2 = (0.002\ 388\ \text{u})\,(931.494\ \text{MeV/u}) = \boxed{2.224\ \text{MeV}}$$

This result tells us that, to separate a deuteron into its constituent proton and a neutron, it is necessary to add 2.224 MeV of energy to the deuteron. One way of supplying the deuteron with this energy is by bombarding it with energetic particles.

If the binding energy of a nucleus were zero, the nucleus would separate into its constituent protons and neutrons without the addition of any energy—that is, it would spontaneously break apart.

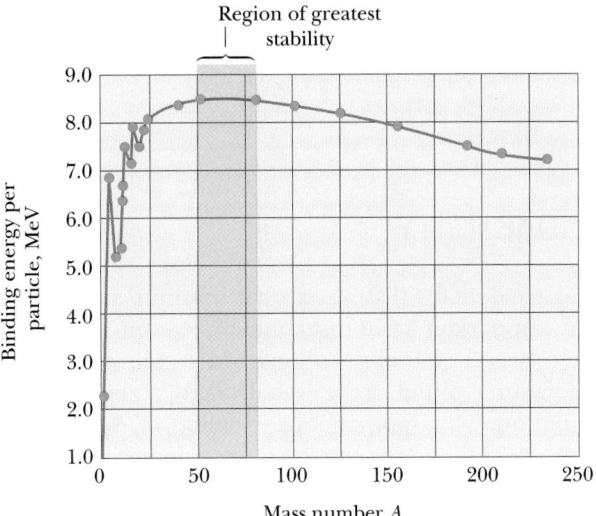

Figure 29.4 A plot of the binding energy per nucleon versus the mass number A for nuclei that are along the line of stability shown in Figure 29.3.

It is interesting to examine a plot of binding energy per nucleon, E_b/A, as a function of mass number for various stable nuclei (Fig. 29.4). Except for the lighter nuclei, the average binding energy per nucleon is about 8 MeV. Note that the curve peaks in the vicinity of $A = 60$. That is, nuclei with mass numbers greater or less than 60 are not as strongly bound as those near the middle of the periodic table. As we shall see later, this fact allows energy to be released in fission and fusion reactions. The curve is slowly varying for $A > 40$, which suggests that the nuclear force saturates. In other words, a particular nucleon can interact with only a limited number of other nucleons, which can be viewed as the "nearest neighbors" in the close-packed structure illustrated in Figure 29.2.

Thinking Physics 1

Figure 29.4 shows a graph of the amount of energy required to remove a nucleon from the nucleus. The figure indicates that an approximately constant amount of energy is necessary to remove a nucleon (above $A = 40$), whereas we saw in Chapter 28 that widely varying amounts of energy are required to remove an electron from the atom. Why does this difference occur?

Explanation In the case of Figure 29.4, the approximately constant value of the nuclear binding energy is a result of the short-range nature of the nuclear strong force. A given nucleon interacts only with its few nearest neighbors, rather than with all of the nucleons in the nucleus. Thus, no matter how many nucleons are present in the nucleus, pulling any one nucleon out involves separating it only from its nearest neighbors. The energy to do this, therefore, is approximately independent of how many nucleons are present. For clearest comparison with atomic electrons, think of averaging the energies required to strip all of the electrons out of a particular atom, from the outermost valence electron to the innermost K-shell electron. This average increases steeply with increasing atomic number. The electrical force binding the electrons to the nucleus in an atom is a long-range force. An electron in the atom interacts with all the protons in the nucleus. When the nuclear charge increases, there is a stronger attraction between the nucleus and the electrons. Therefore, as the nuclear charge increases, more energy is necessary to remove an average electron.

Marie Curie, Polish scientist (1867–1934)

In 1903 Marie Curie shared the Nobel prize in physics with her husband, Pierre, and with Becquerel for their studies of radioactive substances. In 1911 she was awarded a second Nobel prize in chemistry for the discovery of radium and polonium. Marie Curie died of leukemia caused by years of exposure to radioactive substances. "I persist in believing that the ideas that then guided us are the only ones which can lead to the true social progress. We cannot hope to build a better world without improving the individual. Toward this end, each of us must work toward his own highest development, accepting at the same time his share of responsibility in the general life of humanity." *(FPG)*

29.3   RADIOACTIVITY

In 1896 Becquerel accidentally discovered that uranium salt crystals emit an invisible radiation that can darken a photographic plate even if the plate is covered to exclude light. After several such observations under controlled conditions, he concluded that the radiation emitted by the crystals was of a new type, one that required no external stimulation. This spontaneous emission of radiation was soon called **radioactivity.** Subsequent experiments by other scientists showed that other substances were also radioactive.

The most significant investigations of this type were conducted by Marie and Pierre Curie. After several years of careful and laborious chemical separation processes on tons of pitchblende, a radioactive ore, the Curies reported the discovery of two previously unknown elements, both of which were radioactive. These were named polonium and radium. Subsequent experiments, including Rutherford's famous work on alpha-particle scattering, suggested that radioactivity was the result of the decay, or disintegration, of unstable nuclei.

Three types of radiation can be emitted by a radioactive substance: alpha (α) rays, in which the emitted particles are ^{4}He nuclei; beta (β) rays, in which the emitted particles are either electrons or positrons; and gamma (γ) rays, in which the emitted "rays" are high-energy photons. A **positron** is a particle similar to the electron in all respects except that it has a charge of $+e$. (The positron is said to be the **antiparticle** of the electron.) The symbol e^- is used to designate an electron and e^+ designates a positron.

It is possible to distinguish these three forms of radiation by using the scheme described in Figure 29.5. The radiation from a radioactive sample is directed into a region with a magnetic field, and the beam splits into three components, two bending in opposite directions and the third not changing direction. From this simple observation it can be concluded that the radiation of the undeflected beam

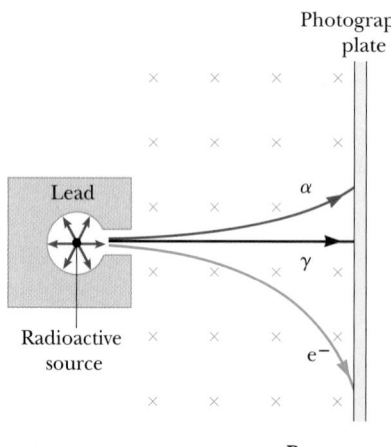

Figure 29.5 The radiation from a radioactive source, such as radium, can be separated into three components using a magnetic field to deflect the charged particles. The photographic plate at the right records the events.

carries no charge (the gamma ray), the component deflected upward contains positively charged particles (alpha particles), and the component deflected downward contains negatively charged particles (e⁻). If the beam includes a positron (e⁺), it is deflected upward.

The three types of radiation have quite different penetrating powers. Alpha particles barely penetrate a sheet of paper, beta particles can penetrate a few millimeters of aluminum, and gamma rays can penetrate several centimeters of lead.

The Decay Constant and Half-Life

If a radioactive sample contains N radioactive nuclei at some instant, it is found that the number of nuclei, ΔN, that decay in a small time interval Δt is proportional to N:

$$\Delta N = -\lambda N \, \Delta t \qquad \text{[29.2]}$$

where λ is a constant called the **decay constant.** The negative sign signifies that N decreases with time; that is, ΔN is negative. The value of λ for any isotope determines the rate at which that isotope will decay. **The decay rate, or activity, R, of a sample is defined as the number of decays per second.** From Equation 29.2, we see that the decay rate is

$$R = \left| \frac{\Delta N}{\Delta t} \right| = \lambda N \qquad \text{[29.3]}$$

◀ Decay rate

Thus we see that isotopes with a large λ value decay at a rapid rate and those with a small λ value decay slowly.

A general decay curve for a radioactive sample is shown in Figure 29.6. It can be shown from Equation 29.3 (using calculus) that the number of nuclei present varies with time according to the expression

$$N = N_0 e^{-\lambda t} \qquad \text{[29.4]}$$

where N is the number of radioactive nuclei present at time t, N_0 is the number present at time $t = 0$, and $e = 2.718 \ldots$ is the base of the natural logarithms. Processes that obey Equation 29.4 are sometimes said to undergo exponential decay.[1]

Another parameter that is useful for characterizing radioactive decay is the **half-life, $T_{1/2}$. The half-life of a radioactive substance is the time it takes half of a given number of radioactive nuclei to decay.** Setting $N = N_0/2$ and $t = T_{1/2}$ in Equation 29.4 gives

$$\frac{N_0}{2} = N_0 e^{-\lambda T_{1/2}}$$

Writing this in the form $e^{\lambda T_{1/2}} = 2$ and taking the natural logarithm of both sides, we get

$$T_{1/2} = \frac{\ln 2}{\lambda} = \frac{0.693}{\lambda} \qquad \text{[29.5]}$$

The hands and numbers of this luminous watch contain minute amounts of radium salt. The radioactive decay of radium causes the watch to glow in the dark. (© *Richard Megna 1990, Fundamental Photographs*)

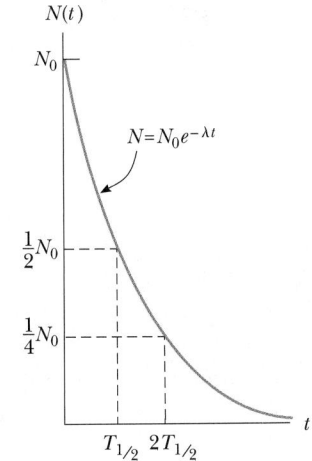

Figure 29.6 Plot of the exponential decay law for radioactive nuclei. The vertical axis represents the number of radioactive nuclei present at any time, t, and the horizontal axis is time. The parameter $T_{1/2}$ is the half-life of the sample.

---

[1] Other examples of exponential decays were discussed in Chapter 18 in connection with RC circuits, and in Chapter 20 in connection with RL circuits.

Chapter 29   Nuclear Physics

This is a convenient expression relating the half-life to the decay constant. Note that after an elapsed time of one half-life, $N_0/2$ radioactive nuclei remain (by definition); after two half-lives, half of these will have decayed and $N_0/4$ radioactive nuclei will be left; after three half-lives, $N_0/8$ will be left; and so on.

The unit of activity is the **curie (Ci),** defined as

$$1 \text{ Ci} \equiv 3.7 \times 10^{10} \text{ decays/s} \qquad \qquad \text{[29.6]}$$

This unit was selected as the original activity unit because it is the approximate activity of 1 g of radium. The SI unit of activity is the **becquerel (Bq):**

$$1 \text{ Bq} = 1 \text{ decay/s} \qquad \qquad \text{[29.7]}$$

Therefore, 1 Ci = 3.7×10^{10} Bq. The most commonly used units of activity are the millicurie (10^{-3} Ci) and the microcurie (10^{-6} Ci).

Thinking Physics 2

The isotope $^{14}_{6}\text{C}$ is radioactive and has a half-life of 5730 years. If you start with a sample of 1000 carbon-14 nuclei, how many will still be around in 17 190 years?

Explanation   In 5730 years, half the sample will have decayed, leaving 500 radioactive $^{14}_{6}\text{C}$ nuclei. In another 5730 years (for a total elapsed time of 11,460 years), the number will be reduced to 250 nuclei. After another 5730 years (total time 17 190 years), 125 remain.

These numbers represent ideal circumstances. Radioactive decay is an averaging process over a very large number of atoms, and the actual outcome depends on statistics. Our original sample in this example contained only 1000 nuclei, certainly not a very large number. Thus, if we were actually to count the number remaining after one half-life for this small sample, it probably would not be exactly 500.

EXAMPLE 29.3   The Activity of Radium

The half-life of the radioactive nucleus $^{226}_{86}\text{Ra}$ is 1.6×10^3 years. If a sample contains 3.0×10^{16} such nuclei, determine the activity at this time.

Solution   First, let us convert the half-life to seconds:

$$T_{1/2} = (1.6 \times 10^3 \text{ years})(3.16 \times 10^7 \text{ s/year}) = 5.0 \times 10^{10} \text{ s}$$

Now we can use this value in Equation 29.5 to get the decay constant:

$$\lambda = \frac{0.693}{T_{1/2}} = \frac{0.693}{5.0 \times 10^{10} \text{ s}} = 1.4 \times 10^{-11} \text{ s}^{-1}$$

We can calculate the activity of the sample at $t = 0$ using $R_0 = \lambda N_0$, where R_0 is the decay rate at $t = 0$ and N_0 is the number of radioactive nuclei present at $t = 0$:

$$R_0 = \lambda N_0 = (1.4 \times 10^{-11} \text{ s}^{-1})(3.0 \times 10^{16}) = 4.1 \times 10^5 \text{ decays/s}$$

Because 1 Ci = 3.7×10^{10} decays/s, the activity, or decay rate, at $t = 0$ is

$$R_0 = \boxed{11.1 \ \mu\text{Ci}}$$

EXAMPLE 29.4 The Activity of Radon Gas

Radon $^{222}_{86}$Rn is a radioactive gas that can be trapped in the basement of homes, and its presence in high concentrations is a known health hazard. Radon has a half-life of 3.83 days. A gas sample contains 4.0×10^8 radon atoms initially.

(a) How many atoms will remain after 12 days have passed?

Reasoning First note that 12 days corresponds to about 3.1 half-lives. In three half-lives, the number of radon atoms is reduced by a factor of $2^3 = 8$. Therefore, the number of remaining radon atoms is approximately $4.0 \times 10^8/8 = 5.0 \times 10^7$.

Solution A more precise answer is obtained by first finding the decay constant from Equation 29.5:

$$\lambda = \frac{0.693}{T_{1/2}} = \frac{0.693}{3.83 \text{ days}} = 0.181 \text{ days}^{-1}$$

We now use Equation 29.4, taking $N_0 = 4.0 \times 10^8$ and the value of λ just found to obtain the number N remaining after 12 days:

$$N = N_0 e^{-\lambda t} = (4.0 \times 10^8 \text{ atoms}) e^{-(0.181 \text{ days}^{-1})(12 \text{ days})} = \boxed{4.6 \times 10^7 \text{ atoms}}$$

This is very close to our original estimate of 5.0×10^7 atoms.

(b) What is the initial activity of the radon sample?

Solution First, we must express the decay constant in units of s^{-1}:

$$\lambda = \frac{0.693}{(3.83 \text{ days})(8.64 \times 10^4 \text{ s/day})} = 2.09 \times 10^{-6} \text{ s}^{-1}$$

From Equation 29.3 and the above value of λ, we find that the initial activity is

$$R = \lambda N_0 = (2.09 \times 10^{-6} \text{ s}^{-1})(4.0 \times 10^8) = 840 \text{ decays/s} = \boxed{840 \text{ Bq}}$$

Exercise Find the activity of the radon sample after 12 days have elapsed.

Answer 95 Bq

29.4 THE DECAY PROCESSES

As stated in the previous section, radioactive nuclei spontaneously decay via alpha, beta, and gamma decay. Let us discuss these in more detail.

Alpha Decay

If a nucleus emits an alpha particle ($^4_2$He), it loses two protons and two neutrons. Therefore, N (here meaning the neutron number of a single nucleus) decreases by 2, Z decreases by 2, and A decreases by 4. The decay can be written symbolically as

$$^A_Z X \longrightarrow \, ^{A-4}_{Z-2} Y + \, ^4_2 He \qquad\qquad [29.8]$$

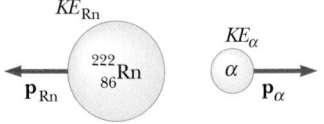

Figure 29.7 Alpha decay of radium. The radium nucleus is initially at rest. After the decay, the radon nucleus has kinetic energy KE_{Rn}, and momentum $\mathbf{p}_{Rn}$, and the alpha particle has kinetic energy KE_α and momentum $\mathbf{p}_\alpha$.

where X is called the **parent nucleus** and Y the **daughter nucleus.** As examples, ^{238}U and ^{226}Ra are both alpha emitters and decay according to the schemes

$$^{238}_{92}U \longrightarrow ^{234}_{90}Th + ^{4}_{2}He \qquad [29.9]$$

$$^{226}_{88}Ra \longrightarrow ^{222}_{86}Rn + ^{4}_{2}He \qquad [29.10]$$

The half-life for ^{238}U decay is 4.47×10^9 years, and the half-life for ^{226}Ra decay is 1.60×10^3 years. In both cases, note that the A of the daughter nucleus is 4 less than that of the parent nucleus. Likewise, Z is reduced by 2. The differences are accounted for in the emitted alpha particle (the ^{4}He nucleus).

The decay of ^{226}Ra is shown in Figure 29.7. When one element changes into another, as happens in alpha decay, the process is called **spontaneous decay,** or **transmutation.** As a general rule, (1) the sum of the mass numbers A must be the same on both sides of the equation, and (2) the sum of the atomic numbers Z must be the same on both sides of the equation.

In order for alpha emission to occur, the mass of the parent must be greater than the combined mass of the daughter and the alpha particle. In the decay process, this excess mass is converted into energy of other forms and appears in the form of kinetic energy in the daughter nucleus and the alpha particle. Most of the kinetic energy is carried away by the alpha particle because it is much less massive than the daughter nucleus. This can be understood by first noting that a particle's kinetic energy and momentum, p, are related as follows:

$$KE = \frac{p^2}{2m}$$

Because momentum is conserved, the two particles in a decay must have equal, but oppositely directed, momenta. Thus, the lighter particle has more kinetic energy than the more massive particle.

Thinking Physics 3

In comparing alpha decay energies from a number of radioactive nuclides, it is found that the half-life of the decay goes down as the energy of the decay goes up. Why is this?

Explanation It should seem reasonable that the higher the energy of the alpha particle, the more likely is it to escape the confines of the nucleus. The higher probability of escape translates to a faster rate of decay, which appears as a shorter half-life.

EXAMPLE 29.5   The Energy Liberated When Radium Decays

We showed that the $^{226}_{88}Ra$ nucleus undergoes alpha decay to $^{222}_{86}Rn$ (Eq. 29.10). Calculate the amount of energy liberated in this decay. Take the mass of $^{226}_{88}Ra$ to be 226.025 402 u, that of $^{222}_{86}Rn$ to be 222.017 571 u, and that of $^4_2$He to be 4.002 602 u, as found in Appendix B.

Solution After decay, the mass of the daughter, m_d, plus the mass of the alpha particle, m_α, is

$$m_d + m_\alpha = 222.017\ 571\ \text{u} + 4.002\ 602\ \text{u} = 226.020\ 173\ \text{u}$$

Thus, calling the mass of the parent nucleus M_p, we find that the mass lost during decay is

$$\Delta m = M_p - (m_d + m_\alpha) = 226.025\ 402\ \text{u} - 226.020\ 173\ \text{u} = 0.005\ 229\ \text{u}$$

Using the relationship $1\ \text{u} = 931.494\ \text{MeV}$, we find that the energy liberated is

$$E = (\Delta m)\,c^2 = (0.005\ 229\ \text{u})(931.494\ \text{MeV/u}) = \boxed{4.87\ \text{MeV}}$$

Beta Decay

When a radioactive nucleus undergoes beta decay, **the daughter nucleus has the same number of nucleons as the parent nucleus, but the atomic number is changed by 1:**

$$^A_Z X \longrightarrow\ ^{A}_{Z+1} Y + e^- \qquad \qquad \text{[29.11]}$$

$$^A_Z X \longrightarrow\ ^{A}_{Z-1} Y + e^+ \qquad \qquad \text{[29.12]}$$

Again, note that the nucleon number and total charge are both conserved in these decays. However, as we shall see shortly, these processes are not described completely by these expressions.

A typical beta decay event is

$$^{14}_6 C \longrightarrow\ ^{14}_7 N + e^- \qquad \qquad \text{[29.13]}$$

The emission of electrons from a *nucleus* is surprising because, in all our previous discussions, we stated that the nucleus is composed of protons and neutrons only. This apparent discrepancy can be explained by noting that the emitted electron is created in the nucleus by a process in which a neutron is transformed into a proton. This can be represented by the equation

$$^1_0 n \longrightarrow\ ^1_1 p + e^- \qquad \qquad \text{[29.14]}$$

Let us consider the energy of the system of Equation 29.13 before and after decay. As with alpha decay, energy must be conserved in beta decay. The following example illustrates how to calculate the amount of energy released in the beta decay of $^{14}_6 C$.

EXAMPLE 29.6 The Beta Decay of Carbon-14

Find the energy liberated in the beta decay of $^{14}_6 C$ to $^{14}_7 N$ as represented by Equation 29.13.

Solution Equation 29.13 refers to nuclei, while Appendix B shows masses of neutral atoms. Adding six electrons to both sides of Equation 29.13 gives

$$^{14}_6 C \text{ atom} \longrightarrow\ ^{14}_7 N \text{ atom}$$

We find from Appendix B that $^{14}_{6}C$ has a mass of 14.003 242 u and $^{14}_{7}N$ has a mass of 14.003 074 u. Here, the mass difference between the initial and final states is

$$\Delta m = 14.003\ 242\ u - 14.003\ 074\ u = 0.000\ 168\ u$$

This corresponds to an energy release of

$$E = (\Delta m)\,c^2 = (0.000\ 168\ u)\,(931.494\ MeV/u) = \boxed{0.156\ MeV}$$

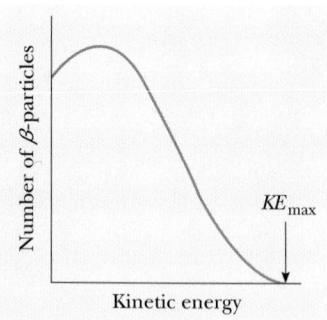

Figure 29.8   A typical beta-decay spectrum.

From Example 29.6, we see that the energy released in the beta decay of ^{14}C is approximately 0.16 MeV. As with alpha decay, we expect the electron to carry away virtually all of this as kinetic energy because apparently it is the lightest particle produced in the decay. However, as Figure 29.8 shows, only a small number of electrons have this maximum kinetic energy, represented as KE_{max} on the graph; most of the electrons emitted have kinetic energies lower than this predicted value. If the daughter nucleus and the electron are not carrying away this liberated energy, then the energy conservation requirement leads to the question, "What accounts for the missing energy?" As an additional complication, further analysis of beta decay shows that the principles of conservation of both angular momentum and linear momentum appear to be violated!

In 1930 Pauli proposed that a third particle must be present to carry away the "missing" energy and to conserve momentum. Enrico Fermi later named this particle the **neutrino** ("little neutral one") because it had to be electrically neutral and have little or no mass. Although it eluded detection for many years, the neutrino (ν) was finally detected experimentally in 1950. It has the following properties:

- Zero electric charge
▶ Properties of the neutrino
- A mass smaller than that of the electron, and possibly zero (Recent experiments suggest that any mass of the neutrino is less than 7 eV/c^2.)
- A spin of $\frac{1}{2}$, which satisfies the law of conservation of angular momentum
- Interaction that is very weak with matter, making the particle quite difficult to detect

Thus, with the introduction of the neutrino, we are now able to represent the beta decay process of Equation 29.13 in its correct form:

$$^{14}_{6}C \longrightarrow {}^{14}_{7}N + e^- + \bar{\nu} \qquad [29.15]$$

where the bar in the symbol $\bar{\nu}$ indicates an **antineutrino.** To explain what an antineutrino is, let us first consider the following decay:

$$^{12}_{7}N \longrightarrow {}^{12}_{6}C + e^+ + \nu \qquad [29.16]$$

Here we see that when ^{12}N decays into ^{12}C, a particle is produced that is identical to the electron except that it has a positive charge of $+e$. This particle is called a *positron*. Because it is like the electron in all respects except charge, the positron is said to be **antiparticle** to the electron. We shall discuss antiparticles further in Chapter 30. For now, it suffices to say that **in beta decay, an electron and an antineutrino are emitted or a positron and a neutrino are emitted.**

Gamma Decay

Very often a nucleus that undergoes radioactive decay is left in an excited energy state. The nucleus can then undergo a second decay to a lower energy state, perhaps to the ground state, by emitting one or more photons. The process is very similar to the emission of light by an atom. An atom emits radiation to release some extra energy when an electron "jumps" from a state of high energy to a state of lower energy. Likewise, the nucleus uses essentially the same method to release any extra energy it may have following a decay or some other nuclear event. In nuclear de-excitation, the "jumps" that release energy are made by protons or neutrons in the nucleus as they move from a higher energy level to a lower level. The photons emitted in such a de-excitation process are called **gamma rays,** which have very high energy relative to the energy of visible light.

A nucleus may reach an excited state as the result of a violent collision with another particle. However, it is more common for a nucleus to be in an excited state as a result of alpha or beta decay. The following sequence of events represents a typical situation in which gamma decay occurs:

$$^{12}_{5}\text{B} \longrightarrow {}^{12}_{6}\text{C}^* + e^- \qquad [29.17]$$

$$^{12}_{6}\text{C}^* \longrightarrow {}^{12}_{6}\text{C} + \gamma \qquad [29.18]$$

Equation 29.17 represents a beta decay in which ^{12}B decays to $^{12}\text{C}^*$, where the asterisk indicates that the carbon nucleus is left in an excited state following the decay. The excited carbon nucleus then decays to the ground state by emitting a gamma ray, as indicated by Equation 29.18. Note that gamma emission does not result in any change in either Z or A.

Practical Uses of Radioactivity

Carbon Dating

The beta decay of ^{14}C given by Equation 29.15 is commonly used to date organic samples. Cosmic rays (high-energy particles from outer space) in the upper atmosphere cause nuclear reactions that create ^{14}C from ^{14}N. In fact, the ratio of ^{14}C to ^{12}C (by numbers of nuclei) in the carbon dioxide molecules of our atmosphere has a constant value of about 1.3×10^{-12} as determined by measuring carbon ratios in tree rings. All living organisms have the same ratio of ^{14}C to ^{12}C because they continuously exchange carbon dioxide with their surroundings. When an organism dies, however, it no longer absorbs ^{14}C from the atmosphere, and so the ratio of ^{14}C to ^{12}C decreases as the result of the beta decay of ^{14}C. It is therefore possible to determine the age of a material by measuring its activity per unit mass as a result of the decay of ^{14}C. Using carbon dating, samples of wood, charcoal, bone, and shell have been identified as having lived from 1000 to 25 000 years ago. This knowledge has helped scientists and researchers to reconstruct the history of living organisms—including humans—during this time span.

A particularly interesting example is the dating of the Dead Sea Scrolls. This group of manuscripts was first discovered in 1947 by a young Bedouin boy in a cave at Qumran near the Dead Sea. Translation showed them to be religious documents, including most of the books of the Old Testament. Because of their historical and religious significance, scholars wanted to know their age. Carbon dating applied to fragments of the scrolls and to the material in which they were wrapped established that they were about 1950 years old. The scrolls are now stored at the Israel Museum in Jerusalem.

Enrico Fermi, Italian physicist (1901–1954)

Fermi was awarded the Nobel prize in 1938 for producing the transuranic elements by neutron irradiation and for his discovery of nuclear reactions bought about by slow neutrons. He made many other outstanding contributions to physics, including his theory of beta decay, the free electron theory of metals, and the development of the world's first fission reactor in 1942. Fermi was truly a gifted theoretical and experimental physicist. He was also well known for his ability to present physics in a clear and exciting manner. "Whatever Nature has in store for mankind, unpleasant as it may be, men must accept, for ignorance is never better than knowledge." *(National Accelerator Laboratory)*

APPLICATION

Carbon Dating of the Dead Sea Scrolls.

A P P L I C A T I O N

Smoke Detectors.

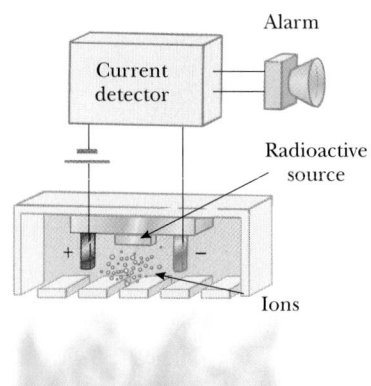

Figure 29.9 An ionization-type smoke detector. Smoke entering the chamber reduces the detected current, causing the alarm to sound.

A P P L I C A T I O N

Radon Pollution.

A P P L I C A T I O N

Radioactive Dating of Iceman.

Smoke Detectors

Smoke detectors are frequently used in homes and industry for fire protection. Most of the common ones are the ionization-type that use radioactive materials (see Fig. 29.9). A smoke detector consists of an ionization chamber, a sensitive current detector, and an alarm. A weak radioactive source ionizes the air in the chamber of the detector, which creates charged particles. A voltage is maintained between the plates inside the chamber, setting up a small but detectable current in the external circuit. As long as the current is maintained, the alarm is deactivated. However, if smoke drifts into the chamber, the ions become attached to the smoke particles. These heavier particles do not drift as readily as do the lighter ions, which causes a decrease in the detector current. The external circuit senses this decrease in current and sets off the alarm.

Radon Detecting

Radioactivity can also affect our daily lives in harmful ways. Soon after the discovery of radium by the Curies, it was found that the air in contact with radium compounds becomes radioactive. It was shown that this radioactivity came from the radium itself, and the product was therefore called "radium emanation." Rutherford and Soddy succeeded in condensing this "emanation," confirming that it is a real substance — the inert, gaseous element now called **radon, Rn.** We now know that the air in uranium mines is radioactive because of the presence of radon gas. The mines must therefore be well ventilated to help protect the miners. The fear of radon pollution has now moved from uranium mines into our own homes (see Example 29.4). Because certain types of rock, soil, brick, and concrete contain small quantities of radium, some of the resulting radon gas finds its way into our homes and other buildings. The most serious problems arise from leakage of radon from the ground into the structure. One practical remedy is to exhaust the air through a pipe just above the underlying soil or gravel directly to the outdoors by means of a small fan or blower.

Applying Physics 1

In 1991, a German tourist discovered the well-preserved remains of the Iceman trapped in a glacier in the Italian Alps (see photograph on page 957). Radioactive dating of a sample of Iceman revealed an age of 5300 years. Why did scientists date the sample using the isotope ^{14}C, rather than ^{11}C, a beta emitter with a half-life of 20.4 min?

Explanation ^{14}C has a long half-life of 5730 years, so the fraction of ^{14}C nuclei remaining after one half-life is high enough to measure accurate changes in the sample's activity. The ^{11}C isotope, which has a very short half-life, is not useful because its activity decreases to a vanishingly small value over the age of the sample, making it impossible to detect.

    If a sample to be dated is not very old, say about 50 years, then you should select the isotope of some other element whose half-life is comparable with the age of the sample. For example, if the sample contained hydrogen, you could measure the activity of ^{3}H (tritium), a beta emitter of half-life 12.3 years. As a general rule, the expected age of the sample should be long enough to measure a change in activity, but not so long that its activity cannot be detected.

Thinking Physics 4

A wooden coffin is found that contains a skeleton holding a gold statue. Which of the three objects (coffin, skeleton, and gold statue) can be carbon dated to find out how old it is?

Explanation Only the coffin and the skeleton can be carbon dated. These two items exchanged air with the environment during their lifetimes, resulting in a fixed ratio of carbon-14 to carbon-12. Once the human and the tree died, the amount of carbon-14 began to decrease due to radioactive decay. The gold in the statue was never alive and did not have an uptake of carbon; therefore, there is no carbon-14 to detect.

EXAMPLE 29.7 Should We Report This to Homicide?

A 50.0-g sample of carbon is taken from the pelvis bone of a skeleton and is found to have a carbon-14 decay rate of 200.0 decays/min. It is known that carbon from a living organism has a decay rate of 15.0 decays/min·g and that ^{14}C has a half-life of 5730 y $= 3.01 \times 10^9$ min. Find the age of the skeleton.

Solution Let us start with Equation 29.4

$$N = N_0 e^{-\lambda t}$$

and multiply both sides by λ to get

$$\lambda N = \lambda N_0 e^{-\lambda t}$$

But from Equation 29.3 we see that this is equivalent to

$$R = R_0 e^{-\lambda t}$$

where R is the present activity and R_0 was the activity when the skeleton was a part of a living organism. Because we are given the decay rate and mass of the sample, we can find R_0 as

$$R_0 = \left(15.0 \, \frac{\text{decays}}{\text{min} \cdot \text{g}} \right) (50.0 \text{ g}) = 750 \, \frac{\text{decays}}{\text{min}}$$

The decay constant is found from Equation 29.5 as

$$\lambda = \frac{0.693}{T_{1/2}} = \frac{0.693}{3.01 \times 10^9 \text{ min}} = 2.30 \times 10^{-10} \text{ min}^{-1}$$

Thus, we make the following substitutions:

$$R = R_0 e^{-\lambda t}$$

$$200.0 \, \frac{\text{decays}}{\text{min}} = \left(750 \, \frac{\text{decays}}{\text{min}} \right) e^{-(2.30 \times 10^{-10} \text{ min}^{-1}) t}$$

$$0.266 = e^{-(2.30 \times 10^{-10} \text{ min}^{-1}) t}$$

Now we take the natural log of both sides of the equation, to give

$$\ln(0.266) = -(2.30 \times 10^{-10} \text{ min}^{-1}) t$$

$$-1.32 = -(2.30 \times 10^{-10} \text{ min}^{-1}) t$$

$$t = 5.74 \times 10^9 \text{ min} = \boxed{10\,900 \text{ y}}$$

Application: Carbon–14 and the Shroud of Turin

Since the Middle Ages, many people have marveled at a 14-foot-long, yellowing piece of linen found in Turin, Italy, purported to be the burial shroud of Jesus Christ. The cloth bears a remarkable, full-size likeness of a crucified body, with wounds on the head that could have been caused by a crown of thorns, and another in the side that could have been the cause of death. Skepticism over the authenticity of the shroud has existed since its first public showing in 1354; in fact, a French bishop declared it to be a fraud at the time. Because of its controversial nature, religious bodies have taken a neutral stance on its authenticity.

In 1978 the bishop of Turin allowed the cloth to be subjected to scientific analysis, but notably missing from these tests was a carbon-14 dating. The reason for this omission was that, at the time, carbon-dating techniques required a piece of cloth about the size of a handkerchief. In 1988 the process had been refined to the point that pieces as small as one square inch were sufficient, and at that time permission was granted to allow the dating to proceed. Three labs were selected for the testing, and each was given four pieces of material. One of these was a piece of the shroud, and the other three pieces were control pieces similar in appearance to the shroud.

The testing procedure consisted of burning the cloth to produce carbon dioxide, which was then converted chemically to graphite. The graphite sample was subjected to carbon-14 analysis, and in the end all three labs agreed amazingly well on the age of the shroud. The average of their results gave a date for the cloth of A.D. 1320 ± 60 years, with an assurance that the cloth could not be older than A.D. 1200. Carbon-14 dating has thus unraveled the most important mystery concerning the shroud, but others remain. For example, investigators have not yet been able to explain how the image was imprinted.

APPLICATION

Carbon-14 Dating of the Shroud of Turin.

A portion of the Shroud of Turin as it appears in a photographic negative image. *(Santi Visali/The IMAGE Bank)*

29.5 NATURAL RADIOACTIVITY

Radioactive nuclei are generally classified into two groups: (1) unstable nuclei found in nature, which give rise to what is called **natural radioactivity,** and (2) nuclei produced in the laboratory through nuclear reactions, which exhibit **artificial radioactivity.**

Three series of naturally occurring radioactive nuclei exist (Table 29.2). Each series starts with a specific long-lived radioactive isotope whose half-life exceeds that of any of its descendants. The fourth series in Table 29.2 begins with ^{237}Np, a

TABLE 29.2 The Four Radioactive Series

Series	Starting Isotope	Half-Life (years)	Stable End Product
Uranium	$^{238}_{92}$U	4.47×10^9	$^{206}_{82}$Pb
Actinium	$^{235}_{92}$U	7.04×10^8	$^{207}_{82}$Pb
Thorium	$^{232}_{90}$Th	1.41×10^{10}	$^{208}_{82}$Pb
Neptunium	$^{237}_{93}$Np	2.14×10^6	$^{209}_{83}$Bi

transuranic element (one having an atomic number greater than that of uranium) not found in nature. This element has a half-life of "only" 2.14×10^6 years.

The two uranium series are somewhat more complex than the ^{232}Th series (Fig. 29.10). Also, there are several naturally occurring radioactive isotopes, such as ^{14}C and ^{40}K, that are not part of either decay series.

Natural radioactivity constantly resupplies our environment with radioactive elements that would otherwise have disappeared long ago. For example, because the Solar System is about 5×10^9 years old, the supply of ^{226}Ra (whose half-life is only 1600 years) would have been depleted by radioactive decay long ago if it were not for the decay series that starts with ^{238}U, with a half-life of 4.47×10^9 years.

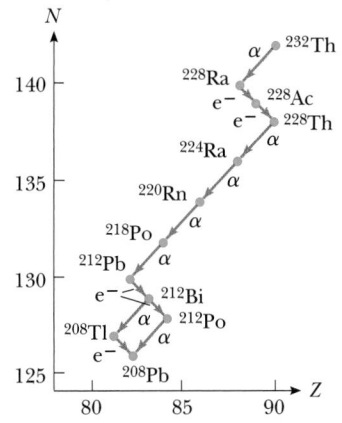

Figure 29.10 Decay series beginning with ^{232}Th.

29.6   NUCLEAR REACTIONS

It is possible to change the structure of nuclei by bombarding them with energetic particles. Such changes are called **nuclear reactions.** Rutherford was the first to observe nuclear reactions, using naturally occurring radioactive sources for the bombarding particles. He found that protons were released when alpha particles were allowed to collide with nitrogen atoms. The process can be represented symbolically as

$$\frac{4}{2}\text{He} + \frac{14}{7}\text{N} \longrightarrow \text{X} + \frac{1}{1}\text{H} \qquad [29.19]$$

This equation says that an alpha particle ($^4_2$He) strikes a nitrogen nucleus and produces an unknown product nucleus (X) and a proton ($^1_1$H). Balancing atomic numbers and mass numbers, as we did for radioactive decay, enables us to conclude that the unknown is characterized as $^{17}_8$X. Because the element with atomic number 8 is oxygen, we see that the reaction is

$$\frac{4}{2}\text{He} + \frac{14}{7}\text{N} \longrightarrow \frac{17}{8}\text{O} + \frac{1}{1}\text{H} \qquad [29.20]$$

This nuclear reaction starts with two stable isotopes, helium and nitrogen, and produces two different stable isotopes, hydrogen and oxygen.

Since the time of Rutherford, thousands of nuclear reactions have been observed, particularly following the development of charged-particle accelerators in the 1930s. With today's advanced technology in particle accelerators and particle detectors, it is possible to achieve particle energies of at least 1000 GeV = 1 TeV. These high-energy particles are used to create new particles whose properties are helping to solve the mysteries of the nucleus.

EXAMPLE 29.8   The Discovery of the Neutron

A nuclear reaction of significant historical note occurred in 1932 when Chadwick, in England, bombarded a beryllium target with alpha particles. Analysis of the experiment indicated that the following reaction occurred:

$$\frac{4}{2}\text{He} + \frac{9}{4}\text{Be} \longrightarrow \frac{12}{6}\text{C} + \text{X}$$

What is X in this reaction?

Solution Balancing mass numbers and atomic numbers, we see that the unknown particle must be represented as $_0^1X$, that is, with a mass of 1 and zero charge. Hence, the particle X is the neutron, $_0^1n$. This experiment was the first to provide positive proof of the existence of neutrons.

EXAMPLE 29.9 Synthetic Elements

(a) A beam of neutrons is directed at a target of $_{92}^{238}U$. The reaction products are a gamma ray and another isotope. What is the isotope? (b) This isotope is radioactive and emits a beta particle. Write the equation symbolizing this decay and identify the resulting isotope. (c) This isotope is also radioactive and decays by beta emission. What is the end product? (d) What is the significance of these reactions?

Solution

(a) Balancing input with output gives

$$_0^1n + {}_{92}^{238}U \longrightarrow {}_{92}^{239}U + \gamma$$

(b) The decay of ^{239}U by beta emission is

$$_{92}^{239}U \longrightarrow {}_{93}^{239}Np + e^- + \bar{\nu}$$

(c) The decay of $_{93}^{239}Np$ by beta emission gives

$$_{93}^{239}Np \longrightarrow {}_{94}^{239}Pu + e^- + \bar{\nu}$$

(d) The interesting feature of these reactions is the fact that uranium is the element with the greatest number of protons, 92, that exists in nature in any appreciable amount. The reactions in parts (a), (b), and (c) do occur occasionally in nature; hence minute traces of neptunium and plutonium are present. In 1940, however, researchers bombarded uranium with neutrons to produce plutonium and neptunium by the steps given previously. These two elements were thus the first elements made in the laboratory, and by bombarding them with neutrons and other particles, the list of synthetic elements has been extended to include those up to atomic number 112.

Q Values

We have just examined some nuclear reactions for which mass numbers and atomic numbers must be balanced in the equations. We shall now consider the energy involved in these reactions, because energy is another important quantity that must be conserved.

Let us illustrate this procedure by analyzing the following nuclear reaction:

$$_1^2H + {}_7^{14}N \longrightarrow {}_6^{12}C + {}_2^4He \qquad [29.21]$$

The total mass on the left side of the equation is the sum of the mass of $_1^2H$ (2.014 102 u) and the mass of $^{14}_7N$ (14.003 074 u), which equals 16.017 176 u. Similarly, the mass on the right side of the equation is the sum of the mass of $_6^{12}C$ (12.000 000 u) plus the mass of $_2^4He$ (4.002 602 u), for a total of 16.002 602 u. Thus, the total mass before the reaction is greater than the total mass after the reaction. The mass difference in this reaction is equal to 16.017 176 u −

16.002 602 u = 0.014 574 u. This "lost" mass is converted to the kinetic energy of the nuclei present after the reaction. In energy units, 0.014 574 u is equivalent to 13.576 MeV of kinetic energy carried away by the carbon and helium nuclei.

The energy required to balance the equation is called the Q value of the reaction. In Equation 29.21 the Q value is 13.576 MeV. Nuclear reactions in which there is a release of energy — that is, positive Q values — are said to be **exothermic reactions.**

The energy balance sheet is not complete, however. We must also consider the kinetic energy of the incident particle before the collision. As an example, let us assume that the deuteron in Equation 29.21 has a kinetic energy of 5 MeV. Adding this to our Q value, we find that the carbon and helium nuclei have a total kinetic energy of 18.576 MeV following the reaction.

Now consider the reaction

$$_2^4\text{He} + {}_7^{14}\text{N} \longrightarrow {}_8^{17}\text{O} + {}_1^1\text{H} \qquad \text{[29.22]}$$

Before the reaction, the total mass is the sum of the masses of the alpha particle and the nitrogen nucleus: 4.002 602 u + 14.003 074 u = 18.005 676 u. (See Table 29.4 and Appendix B.) After the reaction, the total mass is the sum of the masses of the oxygen nucleus and the proton: 16.999 133 u + 1.007 825 u = 18.006 958 u. In this case, the total mass after the reaction is *greater* than the total mass before the reaction. The mass deficit is 0.001 282 u, equivalent to an energy deficit of 1.194 MeV. This deficit is expressed by the negative Q value of the reaction, −1.194 MeV. Reactions with negative Q values are called **endothermic reactions.** Such reactions will not take place unless the incoming particle has at least enough kinetic energy to overcome the energy deficit.

At first it might appear that the reaction in Equation 29.22 could take place if the incoming alpha particle had a kinetic energy of 1.194 MeV. In practice, however, the alpha particle must have more energy than this. If it had an energy of only 1.194 MeV, energy would be conserved but careful analysis would show that momentum was not. This can easily be understood by recognizing that the incoming alpha particle has some momentum before the reaction. However, if its kinetic energy were only 1.194 MeV, the products (oxygen and a proton) would be created with zero kinetic energy and, thus, zero momentum. It can be shown that, in order to conserve both energy and momentum, the incoming particle must have a minimum kinetic energy given by

$$KE_{\text{min}} = \left(1 + \frac{m}{M}\right) |Q| \qquad \text{[29.23]}$$

where m is the mass of the incident particle, M is the mass of the target, and the absolute value of the Q value is used. For the reaction given by Equation 29.22, we find

$$KE_{\text{min}} = \left(1 + \frac{4.002\ 602}{14.003\ 074}\right) |-1.194\ \text{MeV}| = 1.535\ \text{MeV}$$

This minimum value of the kinetic energy of the incoming particle is called the **threshold energy.** The nuclear reaction shown in Equation 29.22 will not occur if the incoming alpha particle has an energy of less than 1.535 MeV, but can occur if the kinetic energy is equal to or greater than 1.535 MeV.

29.7 MEDICAL APPLICATIONS OF RADIATION

Radiation Damage in Matter

Radiation absorbed by matter can cause severe damage. The degree and type of damage depend on several factors, including the type and energy of the radiation and the properties of the absorbing material. Radiation damage in biological organisms is primarily due to ionization effects in cells. The normal function of a cell may be disrupted when highly reactive ions or radicals are formed as the result of ionizing radiation. For example, hydrogen and hydroxyl radicals produced from water molecules can induce chemical reactions that may break bonds in proteins and other vital molecules. Large acute doses of radiation are especially dangerous because damage to a great number of molecules in a cell may cause death of the cell. Also, cells that do survive the radiation may become defective, which can lead to cancer.

In biological systems, it is common to separate radiation damage into two categories: somatic damage and genetic damage. **Somatic damage** is radiation damage to any cells except the reproductive cells. Such damage can lead to cancer at high radiation levels or seriously alter the characteristics of specific organisms. **Genetic damage** affects only reproductive cells. Damage to the genes in reproductive cells can lead to defective offspring. Clearly, we must be concerned about the effect of diagnostic treatments, such as x-rays and other forms of radiation exposure.

Several units are used to quantify radiation exposure and dose. The **roentgen (R)** is defined as **that amount of ionizing radiation that will produce 2.08 × 10^9 ion pairs in 1 cm^3 of air under standard conditions.** Equivalently, the roentgen is **that amount of radiation that deposits 8.76 × 10^{-3} J of energy into 1 kg of air.**

For most applications, the roentgen has been replaced by the **rad** (which is an acronym for radiation absorbed dose), defined as follows: **one rad is that amount of radiation that deposits 10^{-2} J of energy into 1 kg of absorbing material.**

Although the rad is a perfectly good physical unit, it is not the best unit for measuring the degree of biological damage produced by radiation. This is because the degree of biological damage depends not only on the dose but also on the type of radiation. For example, a given dose of alpha particles causes about ten times more biological damage than an equal dose of x-rays. The **RBE** (relative biological effectiveness) factor is defined as **the number of rad of x-radiation or gamma radiation that produces the same biological damage as 1 rad of the radiation being used.** The RBE factors for different types of radiation are given in Table 29.3. Note that the values are only approximate because they vary with particle energy and form of damage.

Finally, the **rem** (roentgen equivalent in man) is defined as the product of the dose in rad and the RBE factor:

$$\text{Dose in rem} = \text{dose in rad} \times \text{RBE}$$

According to this definition, 1 rem of any two radiations will produce the same amount of biological damage. From Table 29.3, we see that a dose of 1 rad of fast neutrons represents an effective dose of 10 rem and that 1 rad of x-radiation is equivalent to a dose of 1 rem.

Low-level radiation from natural sources, such as cosmic rays and radioactive rocks and soil, delivers to each of us a dose of about 0.13 rem/year. The upper

TABLE 29.3
RBE Factors for Several Types of Radiation

Radiation	RBE Factor
X-rays and gamma rays	1.0
Beta particles	1.0–1.7
Alpha particles	10–20
Slow neutrons	4–5
Fast neutrons and protons	10
Heavy ions	20

limit of radiation dose recommended by the U.S. government (apart from background radiation and exposure related to medical procedures) is 0.5 rem/year. Many occupations involve higher levels of radiation exposure, and for individuals in these occupations an upper limit of 5 rem/year has been set for whole-body exposure. Higher upper limits are permissible for certain parts of the body, such as the hands and forearms. An acute whole-body dose of 400 to 500 rem results in a mortality rate of about 50%. The most dangerous form of exposure is ingestion or inhalation of radioactive isotopes, especially those elements the body retains and concentrates, such as ^{90}Sr. In some cases, a dose of 1000 rem can result from ingesting 1 mCi of radioactive material.

Sterilizing objects by exposing them to radiation has been going on for at least 25 years, but in recent years, the methods used have become safer to use and more economical. Most bacteria, worms, and insects are easily destroyed by exposure to radiation from radioactive cobalt. The process is very effective in destroying Trichinella worms in pork, salmonella bacteria in chickens, insect eggs in wheat, and surface bacteria on fruit and vegetables that can lead to rapid spoilage. Recently, this procedure has been expanded to include the sterilization of medical equipment while in its protective covering. Surgical gloves, sponges, sutures, and so forth are irradiated while packaged. Also, bone, cartilage, and skin used for grafting is often irradiated to reduce the chance for infection.

Tracing

Radioactive particles can be used to trace chemicals participating in various reactions. One of the most valuable uses of radioactive tracers is in medicine. For example, ^{131}I is an artificially produced isotope of iodine (the natural, nonradioactive isotope is ^{127}I). Iodine, which is a necessary nutrient for our bodies, is obtained largely through the intake of seafood and iodized salt. The thyroid gland plays a major role in the distribution of iodine throughout the body. In order to evaluate the performance of the thyroid, the patient drinks a very small amount of radioactive sodium iodide. Two hours later, the amount of iodine in the thyroid gland is determined by measuring the radiation intensity at the neck area.

A medical application of the use of radioactive tracers occurring in emergency situations is that of locating a hemorrhage inside the body. Often the location of the site cannot easily be determined, but radioactive Cr can identify the location with a high degree of precision. Chromium is taken up by red blood cells and carried uniformly throughout the body. However, at a hemorrhage site, the blood will be dumped, and the radioactivity of that region will increase markedly.

The tracer technique is also useful in agricultural research. Suppose the best method of fertilizing a plant is to be determined. A certain material in the fertilizer, such as nitrogen, can be tagged with one of its radioactive isotopes. The fertilizer is then sprayed on one group of plants, sprinkled on the ground for a second group, and raked into the soil for a third. A Geiger counter is then used to track the nitrogen through the three types of plants.

Tracing techniques are as wide-ranging as human ingenuity can devise. Present applications range from checking the absorption of fluorine by teeth to checking contamination of food-processing equipment by cleansers to monitoring deterioration inside an automobile engine. In the latter case, a radioactive material is used in the manufacture of the pistons, and the oil is checked for radioactivity to determine the amount of wear on the pistons.

APPLICATION

Occupational Radiation Exposure Limits.

APPLICATION

Irradiation of Food and Medical Equipment.

APPLICATION

Radioactive Tracers in Medicine.

APPLICATION

Radioactive Tracers in Agricultural Research.

Computed Axial Tomography (CAT Scans)

The normal x-ray of a human body has two primary disadvantages when used as a source of clinical diagnosis. First, it is difficult to distinguish between various types of tissue in the body because they all have similar x-ray absorption properties. Second, a conventional x-ray absorption picture is indicative of the average amount of absorption along a particular direction in the body, leading to somewhat obscured pictures. To overcome these problems, a device called a CAT scanner was developed in England in 1973; it is capable of producing pictures of much greater clarity and detail than were previously obtainable.

The operation of a CAT scanner can be understood by considering the following hypothetical experiment. Suppose a box consists of four compartments, labeled A, B, C, and D, as in Figure 29.11a. Each compartment has a different amount of absorbing material from any other compartment. What set of experimental procedures will enable us to determine the relative amounts of material in each compartment? The following steps outline one method that will provide this information. First, a beam of x-rays is passed through compartments A and C, as Figure 29.11b. The intensity of the exiting radiation is reduced by absorption by some number that we assign as 8. (The number 8 could mean, for example, that the intensity of the exiting beam is reduced by eight tenths of 1% from its initial value.) Because we do not know which of the compartments, A or C, was responsible for this reduction in intensity, half the loss is assigned to each compartment, as in Figure 29.11c. Next, a beam of x-rays is passed through compartments B and D, as in Figure 29.11b. The reduction in intensity for this beam is 10, and again we assign half the loss to each compartment. We now redirect the x-ray source so that it sends one beam through compartments A and B and another through compartments C and D, as in Figure 29.11d, and again measure the absorption. Suppose the ab-

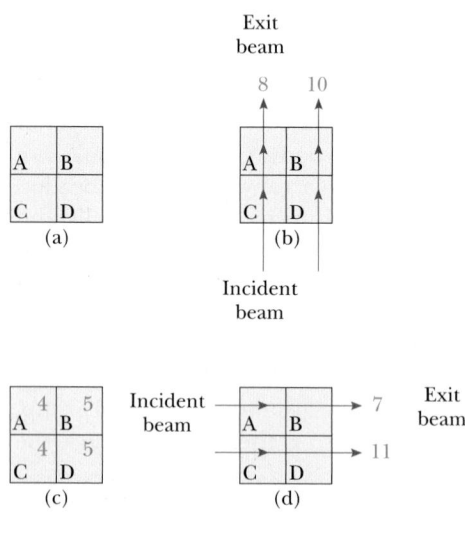

Figure 29.11 An experimental procedure for determining the relative amounts of x-ray absorption by four different compartments in a box.

sorption through compartments A and B in this experiment is measured to be 7 units. On the basis of our first experiment, we would have guessed it would be 9 units, 4 by compartment A, and 5 by compartment B. Thus, we have reduced the guessed absorption for each compartment by 1 unit so that the sum is 7 rather than 9, to give the numbers shown in Figure 29.11e. Likewise, when the beam is passed through compartments C and D, as in Figure 29.11d, we may find the total absorption to be 11 as compared to our first experiment of 9. In this case, we add 1 unit of absorption to each compartment to give a sum of 11, as in Figure 29.11e. This somewhat crude procedure could be improved by measuring the absorption along other paths. However, these simple measurements are sufficient to enable us to conclude that compartment D contains the most absorbing material and A the least. A visual representation of these results can be obtained by assigning to each compartment a shade of gray corresponding to the particular number associated with the absorption. In our example, compartment D would be very dark and compartment A would be very light.

The steps outlined previously are representative of how a CAT scanner produces images of the human body. A thin slice of the body is subdivided into perhaps 10 000 compartments, rather than 4 compartments as in our simple example. The function of the CAT scanner is to determine the relative absorption in each of these 10 000 compartments and to display a picture of its calculations in various shades of gray. Note that CAT stands for **computed axial tomography.** The term *axial* is used because the slice of the body to be analyzed corresponds to a plane perpendicular to the head-to-toe axis. *Tomos* is the Greek word for slice and *graph* is the Greek word for picture. In a typical diagnosis, the patient is placed in the position shown in Figure 29.12 and a narrow beam of x-rays is sent through the plane of interest. The emerging x-rays are detected and measured by photomultiplier tubes behind the patient. The x-ray tube is then rotated a few degrees, and the intensity is recorded again. An extensive amount of information is obtained by rotating the beam through 180° at intervals of about 1° per measurement, resulting in a set of

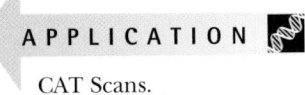

APPLICATION

CAT Scans.

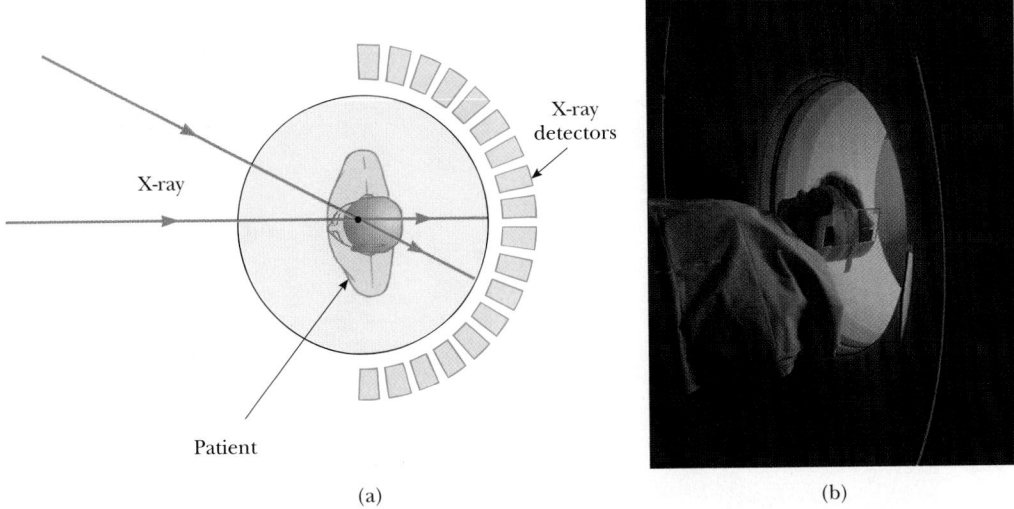

(a)   (b)

Figure 29.12   (a) CAT scanner detector assembly. (b) Photograph of a patient undergoing a CAT scan in a hospital. (*Jay Freis/The Image Bank*)

numbers assigned to each of the 10 000 "compartments" in the slice. These numbers are then converted by the computer to a photograph in various shades of gray for this segment of the body.

A brain scan of a patient can now be made in about 2 s, and a full-body scan requires about 6 s. The final result is a picture containing much greater quantitative information and clarity than a conventional x-ray photograph. Because CAT scanners use x-rays, which are an ionizing form of radiation, the technique presents a health risk to the patient being diagnosed.

Magnetic Resonance Imaging (MRI)

At the heart of magnetic resonance imaging (MRI) is the fact that when a nucleus having a magnetic moment is placed in an external magnetic field, its moment will precess about the magnetic field with a frequency that is proportional to the field. For example, a proton, whose spin is $\frac{1}{2}$, can occupy one of two energy states when placed in an external magnetic field. The lower energy state corresponds to the case in which the spin is aligned with the field, whereas the higher energy state corresponds to the case in which the spin is opposite the field. Transitions between these two states can be observed using a technique known as **nuclear magnetic resonance.** A dc magnetic field is applied to align the magnetic moments, and a second, weak oscillating magnetic field is applied perpendicular to the dc field. When the frequency of the oscillating field is adjusted to match the precessional frequency of the magnetic moments, the nuclei will "flip" between the two spin states. These transitions result in a net absorption of energy by the spin system, which can be detected electronically.

In MRI, image reconstruction is obtained using spatially varying magnetic fields and a procedure for encoding each point in the sample being imaged. Some MRI images taken on a human head are shown in Figure 29.13. In practice, a computer-controlled pulse sequencing technique is used to produce signals that are captured by a suitable processing device. This signal is then subjected to appropriate mathematical manipulations to provide data for the final image. The main advantage of MRI over other imaging techniques in medical diagnostics is that it causes minimal damage to cellular structures. Photons associated with the rf signals used in MRI

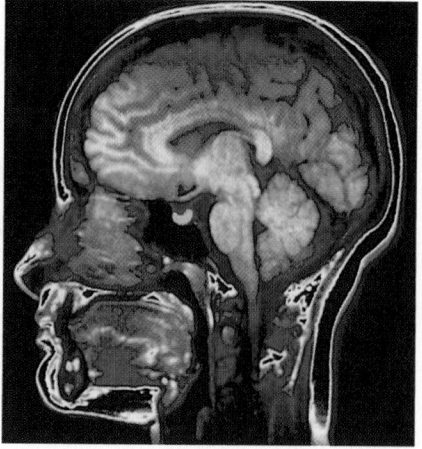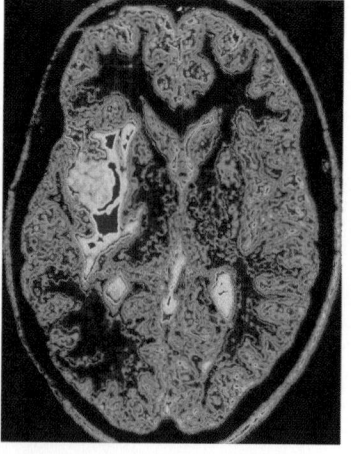

Figure 29.13 Computer-enhanced MRI images of (a) a normal human brain with the pituitary gland highlighted and (b) a human brain with a glioma tumor. *(Scott Camazine/Science Source/Photo Researchers, Inc.)*

(a)

(b)

have energies of only about 10^{-7} eV. Because molecular bond strengths are much larger (of the order of 1 eV), the rf photons cause little cellular damage. In comparison, x-rays or γ-rays have energies ranging from 10^4 to 10^6 eV and can cause considerable cellular damage.

SUMMARY

Nuclei are represented symbolically as $^A_Z X$, where X represents the chemical symbol for the element. The quantity A is the **mass number,** which equals the total number of nucleons (neutrons plus protons) in the nucleus. The quantity Z is the **atomic number,** which equals the number of protons in the nucleus. Nuclei that contain the same number of protons but different numbers of neutrons are called **isotopes.** In other words, isotopes have the same Z value but different A values.

Most nuclei are approximately spherical, with an average radius given by

$$r = r_0 A^{1/3} \qquad \text{[29.1]}$$

where A is the mass number and r_0 is a constant equal to 1.2×10^{-15} m.

The total mass of a nucleus is always less than the sum of the masses of its individual nucleons. This mass difference, Δm, multiplied by c^2 gives the **binding energy** of the nucleus.

The spontaneous emission of radiation by certain nuclei is called **radioactivity.** There are three processes by which a radioactive substance can decay: alpha (α) decay, in which the emitted particles are $^4_2$He nuclei; beta (β) decay, in which the emitted particles are electrons or positrons; and gamma (γ) decay, in which the emitted particles are high-energy photons.

The **decay rate,** or **activity, R,** of a sample is given by

$$R = \left| \frac{\Delta N}{\Delta t} \right| = \lambda N \qquad \text{[29.3]}$$

where N is the number of radioactive nuclei at some instant and λ is a constant for a given substance called the **decay constant.**

Nuclei in a radioactive substance decay in such a way that the number of nuclei present varies with time according to the expression

$$N = N_0 e^{-\lambda t} \qquad \text{[29.4]}$$

where N is the number of radioactive nuclei present at time t, N_0 is the number at time $t = 0$, and $e = 2.718. \ldots$

The **half-life, $T_{1/2}$,** of a radioactive substance is the time required for half of a given number of radioactive nuclei to decay. The half-life is related to the decay constant as

$$T_{1/2} = \frac{0.693}{\lambda} \qquad \text{[29.5]}$$

If a nucleus decays by alpha emission, it loses two protons and two neutrons. A typical alpha decay is

$$^{238}_{92}\text{U} \longrightarrow \ ^{234}_{90}\text{Th} + \ ^4_2\text{He} \qquad \text{[29.9]}$$

Note that in this decay, as in all radioactive decay processes, the sum of the Z values on the left equals the sum of the Z values on the right; the same is true for the A values.

A typical beta decay is

$$^{14}_{6}\text{C} \longrightarrow {}^{14}_{7}\text{N} + e^- + \bar{\nu} \qquad \textbf{[29.15]}$$

When a nucleus beta decays, an **antineutrino** ($\bar{\nu}$) is emitted along with an electron, or a **neutrino** (ν) along with a positron. A neutrino has zero electric charge and a small mass (which may be zero) and interacts weakly with matter.

Nuclei are often in an excited state following radioactive decay, and release their extra energy by emitting a high-energy photon called a **gamma ray (γ).** A typical gamma ray emission is

$$^{12}_{6}\text{C}^* \longrightarrow {}^{12}_{6}\text{C} + \gamma \qquad \textbf{[29.18]}$$

where the asterisk indicates that the carbon nucleus was in an excited state before gamma emission.

Nuclear reactions can occur when a bombarding particle strikes another nucleus. A typical nuclear reaction is

$$^{4}_{2}\text{He} + {}^{14}_{7}\text{N} \longrightarrow {}^{17}_{8}\text{O} + {}^{1}_{1}\text{H} \qquad \textbf{[29.20]}$$

In this reaction an alpha particle strikes a nitrogen nucleus, producing an oxygen nucleus and a proton. As in radioactive decay, atomic numbers and mass numbers balance on the two sides of the arrow.

Nuclear reactions in which energy is released are said to be **exothermic reactions** and are characterized by positive Q values. Reactions with negative Q values, called **endothermic reactions,** cannot occur unless the incoming particle has at least enough kinetic energy to overcome the energy deficit. In order to conserve both energy and momentum, the incoming particle must have a minimum kinetic energy, called the **threshold energy,** given by

$$KE_{min} = \left(1 + \frac{m}{M}\right)|Q| \qquad \textbf{[29.23]}$$

where m is the mass of the incident particle and M is the mass of the target atom.

MULTIPLE-CHOICE QUESTIONS

Use the following information in the next two problems: The activity per gram of carbon due to carbon-14 in living material is about 0.25 disintegrations per second, and carbon-14 has a half-life of 5730 years. The human body is composed of about 18% carbon.

1. The mass of ^{14}C in 1.0 gram of carbon in living material is
   (a) 4.8×10^{-20} g
   (b) 1.5×10^{-12} g
   (c) 1.3×10^{-12} g
   (d) 6.0×10^{-12} g

2. If a person has a mass of 60 kg, what is the radioactivity of that person due to ^{14}C?
   (a) 2.7×10^3 Bq   (b) 7.3×10^{-8} Ci
   (c) 73 nCi   (d) All answers are correct.

3. What is the Q-value for the reaction
   $^{9}\text{Be} + \alpha \rightarrow {}^{12}\text{C} + \text{n}$?
   (a) 8.4 MeV   (b) 7.3 MeV   (c) 6.2 MeV   (d) 5.7 MeV

4. The activity of a newly discovered radioactive isotope reduces to 96% of its original value in an interval of 2 hours. What is its half-life?
   (a) 10.2 h   (b) 34.0 h   (c) 44.0 h   (d) 68.6 h

5. An endothermic nuclear reaction occurs as a result of the collision of two reactant nuclei. If the Q value of this reaction is -2.17 MeV, which of the following describes the minimum kinetic energy needed in the reactant nuclei if the reaction is to occur?

(a) equal to 2.17 MeV
(b) greater than 2.17 MeV
(c) less than 2.17 MeV
(d) exactly half of 2.17 MeV

CONCEPTUAL QUESTIONS

1. Isotopes of a given element have many different properties, such as mass, but the same chemical properties. Why is this?
2. If a nucleus such as ^{226}Ra that is initially at rest undergoes alpha decay, which has more kinetic energy after the decay, the alpha particle or the daughter nucleus?
3. A student claims that a heavy form of hydrogen decays by alpha emission. How do you respond?
4. Explain the main differences between alpha, beta, and gamma rays.
5. In beta decay, the energy of the electron or positron emitted from the nucleus lies somewhere in a relatively large range of possibilities. In alpha decay, however, the alpha particle energy can only have discrete values. Why is there this difference?
6. If film is kept in a box, alpha particles from a radioactive source outside the box cannot expose the film, but beta particles can. Explain.
7. In positron decay, a proton in the nucleus becomes a neutron, and the positive charge is carried away by the positron. But a neutron has a larger rest energy than a proton. How is this possible?

8. An alpha particle has twice the charge of a beta particle. Why does the former deflect less than the latter when passing between electrically charged plates, assuming they both have the same speed?
9. Can carbon-14 dating be used to measure the age of a stone?
10. Pick any beta decay process and show that the neutrino must have zero charge.
11. Why do heavier elements require more neutrons in order to maintain stability?
12. Suppose it could be shown that cosmic ray intensity was much greater 10,000 years ago. How would this affect the ages we assign to ancient samples of once-living matter?
13. What fraction of a radioactive sample has decayed after two half-lives have elapsed?
14. Why is carbon dating unable to provide accurate estimates of ages on the order of one million years?
15. Two samples of the same radioactive nuclide are prepared. Sample A has twice the initial activity of sample B. How does the half-life of A compare with the half-life of B? After each has passed through five half-lives, what is the ratio of their activities?

PROBLEMS

1, **2**, **3** = straightforward, intermediate, challenging   □ = full solution available in Study Guide/Student Solutions Manual   ◤ = Core Concepts Workbook
WEB = solution posted at **http://www.harcourtcollege.com/physics/cptech**   ▩ = biomedical application   ▨ = Interactive Physics

Table 29.4 (page 986) will be useful for many of these problems. A more complete list of atomic masses is given in Appendix B.

Section 29.1 Some Properties of Nuclei

1. Compare the nuclear radii of the following nuclides: $^{2}_{1}$H, $^{60}_{27}$Co, $^{197}_{79}$Au, $^{239}_{94}$Pu.
2. Find the radius of a nucleus of (a) $^{4}_{2}$He and (b) $^{238}_{92}$U.
3. Using the result of Example 29.1, find the radius of a sphere of nuclear matter that would have a mass equal to that of the Earth. The Earth has a mass of 5.98×10^{24} kg and average radius of 6.37×10^{6} m.

4. Consider the hydrogen atom to be a sphere of radius equal to the Bohr radius, 0.53×10^{-10} m, and calculate the approximate value of the ratio of the nuclear density to the atomic density.
5. (a) Find the speed an alpha particle requires to come within 3.2×10^{-14} m of a stationary gold nucleus. (b) Find the energy of the alpha particle in MeV.
6. Use energy methods to calculate the distance of closest approach for a head-on collision between an alpha particle with an initial energy of 0.50 MeV and a gold nucleus (^{197}Au) at rest. Assume the gold nucleus remains at rest during the collision.

TABLE 29.4
Some Atomic Masses

Element	Atomic Mass (u)
$^{1}_{1}H$	1.007 825
$^{1}_{0}n$	1.008 665
$^{4}_{2}He$	4.002 602
$^{7}_{3}Li$	7.016 003
$^{9}_{4}Be$	9.012 174
$^{10}_{5}B$	10.012 936
$^{12}_{6}C$	12.000 000
$^{13}_{6}C$	13.003 355
$^{14}_{7}N$	14.003 074
$^{15}_{7}N$	15.000 108
$^{15}_{8}O$	15.003 065
$^{17}_{8}O$	16.999 133
$^{18}_{8}O$	17.999 160
$^{18}_{9}F$	18.000 937
$^{20}_{10}Ne$	19.992 435
$^{23}_{11}Na$	22.989 770
$^{23}_{12}Mg$	22.994 127
$^{27}_{13}Al$	26.981 538
$^{30}_{15}P$	29.978 310
$^{40}_{20}Ca$	39.962 591
$^{42}_{20}Ca$	41.958 63
$^{43}_{20}Ca$	42.958 770
$^{56}_{26}Fe$	55.934 940
$^{64}_{30}Zn$	63.929 144
$^{64}_{29}Cu$	63.929 599
$^{93}_{41}Nb$	92.906 3768
$^{197}_{79}Au$	196.966 543
$^{202}_{80}Hg$	201.970 617
$^{216}_{84}Po$	216.001 790
$^{220}_{86}Rn$	220.011 401
$^{234}_{90}Th$	234.043 583
$^{238}_{92}U$	238.050 784

7. An α particle ($Z = 2$, mass 6.64×10^{-27} kg) approaches to within 1.00×10^{-14} m of a carbon nucleus ($Z = 6$). What is the (a) maximum Coulomb force on the α particle, (b) acceleration of the α particle at this point, and (c) potential energy of the α particle at this point?

Section 29.2   Binding Energy

8. Compare the average binding energy per nucleon of $^{24}_{12}Mg$ and $^{85}_{37}Rb$.

9. Calculate the average binding energy per nucleon of $^{93}_{41}Nb$ and $^{197}_{79}Au$.

10. Two isotopes having the same mass number are known as isobars. Calculate the difference in binding energy per nucleon for the isobars $^{23}_{11}Na$ and $^{23}_{12}Mg$. How do you account for this difference?

WEB 11. A pair of nuclei for which $Z_1 = N_2$ and $Z_2 = N_1$ are called *mirror isobars* (the atomic and neutron numbers are interchanged). Binding energy measurements on such pairs can be used to obtain evidence of the charge independence of nuclear forces. Charge independence means that the proton-proton, proton-neutron, and neutron-neutron forces are approximately equal. Calculate the difference in binding energy for the two mirror nuclei, $^{15}_{8}O$ and $^{15}_{7}N$.

12. Calculate the binding energy of the last neutron in the $^{43}_{20}Ca$ nucleus. (*Hint:* You should compare the mass of $^{43}_{20}Ca$ with the mass of $^{42}_{20}Ca$ plus the mass of a neutron.)

Section 29.3   Radioactivity

13. The half-life of an isotope of phosphorus is 14 days. If a sample contains 3.0×10^{16} such nuclei, determine its activity. Express your answer in curies.

14. A drug tagged with $^{99}_{43}Tc$ (half-life = 6.05 h) is prepared for a patient. If the original activity of the sample was 1.1×10^4 Bq, what is its activity after it has sat on the shelf for 2.0 h?

15. The half-life of ^{131}I is 8.04 days. (a) Calculate the decay constant for this isotope. (b) Find the number of ^{131}I nuclei necessary to produce a sample with an activity of 0.50 μCi.

16. A radioactive sample contains 3.50 μg of pure ^{11}C, which has a half-life of 20.4 min. (a) How many moles of ^{11}C are present initially? (b) Determine the number of nuclei present initially. What is the activity of the sample (c) initially and (d) after 8.00 h?

17. After 2 days, the activity of a sample of an unknown type radioactive material has decreased to 84.2% of the initial activity. (a) What is the half-life of this material? (b) Can you identify it by using the table of isotopes in Appendix B?

18. Suppose that you start with 1.00×10^{-3} g of a pure radioactive substance and 2.0 h later determine that only 0.25×10^{-3} g of the substance remains. What is the half-life of this substance?

19. Radon gas has a half-life of 3.83 days. If 3.00 g of radon gas is present at time $t = 0$, what mass of radon will remain after 1.50 days have passed?

20. The ^{14}C content decreases after the death of a living system with a half-life of 5730 years. If an archaeologist finds an ancient firepit containing partially consumed

firewood, and the ^{14}C content of the wood is only 12.5% that of an equal carbon sample from a present-day tree, what is the age of the ancient site?

21. Many smoke detectors use small quantities of the isotope ^{241}Am in their operation. The half-life of ^{241}Am is 432 years. How long will it take for the activity of this material to decrease to 1.00×10^{-3} of the original activity?

22. A building has become accidentally contaminated with radioactivity. The longest lived material in the building is strontium-90 (the atomic mass of $^{90}_{38}$Sr is 89.9077). If the building initially contained 5.0 kg of this substance and the safe level is less than 10.0 counts/min, how long will the building be unsafe?

WEB 23. A freshly prepared sample of a certain radioactive isotope has an activity of 10.0 mCi. After 4.00 h, the activity is 8.00 mCi. (a) Find the decay constant and half-life of the isotope. (b) How many atoms of the isotope were contained in the freshly prepared sample? (c) What is the sample's activity 30 h after it is prepared?

Section 29.4 The Decay Processes

24. Complete the following radioactive decay formulas:

$$^{12}_{5}B \longrightarrow ? + e^{-} + \bar{\nu}$$

$$^{234}_{90}Th \longrightarrow ^{230}_{88}Ra + ?$$

$$? \longrightarrow ^{14}_{7}N + e^{-} + \bar{\nu}$$

25. Complete the following radioactive decay formulas:

$$^{212}_{83}Bi \longrightarrow ? + ^{4}_{2}He$$

$$^{95}_{36}Kr \longrightarrow ? + e^{-} + \bar{\nu}$$

$$? \longrightarrow ^{4}_{2}He + ^{140}_{58}Ce$$

26. Figure P29.26 shows the steps by which $^{235}_{92}U$ decays to $^{207}_{82}Pb$. Enter the correct isotope symbol in each square.

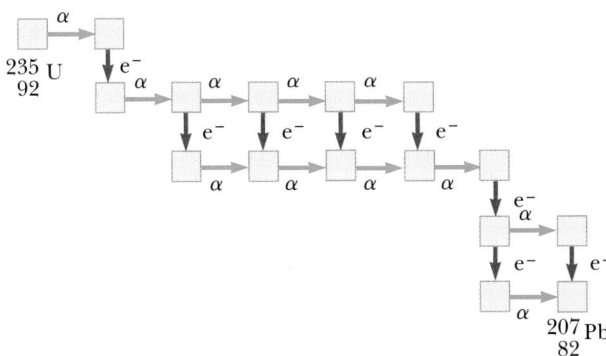

Figure P29.26

27. The mass of ^{56}Fe is 55.9349 u and the mass of ^{56}Co is 55.9399 u. Which isotope decays into the other, and by what process?

28. Find the energy released in the alpha decay of $^{238}_{92}U$. The following mass value will be useful: $^{234}_{90}Th$ has a mass of 234.043 583 u.

29. A nucleus of mass 228 u, initially at rest, undergoes alpha decay. If the alpha particle emitted has a kinetic energy of 4.00 MeV, what is the kinetic energy of the recoiling daughter nucleus?

30. Determine which of the following suggested decays can occur spontaneously: (a) $^{40}_{20}Ca \rightarrow e^{+} + ^{40}_{19}K$; (b) $^{144}_{60}Nd \rightarrow ^{4}_{2}He + ^{140}_{58}Ce$

31. An ^{3}H nucleus beta decays into ^{3}He by creating an electron and an antineutrino according to the reaction

$$^{3}_{1}H \longrightarrow ^{3}_{2}He + e^{-} + \bar{\nu}$$

Use Appendix B to determine the total energy released in this reaction.

32. $^{66}_{28}Ni$ (mass = 65.9291 u) decays by e^{-} emission to $^{66}_{29}Cu$ (mass = 65.9289 u). (a) Write the complete decay formula for this process. (b) Find the maximum kinetic energy of the emerging e^{-} particles.

33. A wooden artifact is found in an ancient tomb. Its carbon-14 ($^{14}_{6}C$) activity is measured to be 60.0% of that in a fresh sample of wood from the same region. Assuming the same amount of carbon-14 was initially present in the wood from which the artifact was made, determine the age of the artifact.

34. A piece of charcoal used for cooking is found at the remains of an ancient campsite. A 1.00-kg sample of carbon from the wood has an activity of 2.00×10^{3} decays per minute. Find the age of the charcoal. (*Hint:* Living material has an activity of 15.0 decays/minute per gram of carbon present.)

Section 29.6 Nuclear Reactions

35. (a) Suppose $^{10}_{5}B$ is struck by an alpha particle, releasing a proton and a product nucleus in the reaction. What is the product nucleus? (b) An alpha particle and a product nucleus are produced when $^{13}_{6}C$ is struck by a proton. What is the product nucleus?

36. Complete the following nuclear reactions:

$$? + ^{14}_{7}N \longrightarrow ^{1}_{1}H + ^{17}_{8}O$$

$$^{7}_{3}Li + ^{1}_{1}H \longrightarrow ^{4}_{2}He + ?$$

37. Natural gold has only one isotope, $^{197}_{79}Au$. If gold is bombarded with slow neutrons, e^{-} particles are emitted. (a) Write the appropriate reaction equation. (b) Calculate the maximum energy of the emitted beta particles. The mass of $^{198}_{80}Hg$ is 197.966 75 u.

38. The first nuclear reaction using particle accelerators was performed by Cockcroft and Walton. Accelerated pro-

tons were used to bombard lithium nuclei, producing the following reaction:

$$\,^1_1H + \,^7_3Li \longrightarrow \,^4_2He + \,^4_2He$$

Because the masses of the particles involved in the reaction were well known, these results were used to obtain an early proof of the Einstein mass-energy relation. Calculate the Q value of the reaction.

WEB **39.** The first known reaction in which the product nucleus was radioactive (achieved in 1934) was one in which $\,^{27}_{13}Al$ was bombarded with alpha particles. Produced in the reaction were a neutron and a product nucleus. (a) What was the product nucleus? (b) Find the Q value of the reaction.

40. (a) Determine the product of the reaction $\,^7_3Li + \,^4_2He \rightarrow$? + n. (b) What is the Q value of the reaction?

41. A beam of 6.61 MeV protons is incident on a target of $\,^{27}_{13}Al$. Those that collide produce the reaction

$$p + \,^{27}_{13}Al \longrightarrow \,^{27}_{14}Si + n$$

($\,^{27}_{14}Si$ has a mass of 26.986 721 u.) Neglect any recoil of the product nucleus and determine the kinetic energy of the emerging neutrons.

42. Find the threshold energy that the incident neutron must have to produce the reaction

$$\,^1_0n + \,^4_2He \longrightarrow \,^2_1H + \,^3_1H$$

43. When $\,^{18}O$ is struck by a proton, $\,^{18}F$ and another particle are produced. (a) What is the other particle? (b) This reaction has a Q value of -2.453 MeV, and the atomic mass of $\,^{18}O$ is 17.999 160 u. What is the atomic mass of $\,^{18}F$?

Section 29.7   Medical Applications of Radiation

44. In terms of biological damage, how many rad of heavy ions is equivalent to 100 rad of x-rays?

45. A person whose mass is 75.0 kg is exposed to a whole-body dose of 25.0 rad. How many joules of energy are deposited in the person's body?

46. A 200-rad dose of radiation is administered to a patient in an effort to combat a cancerous growth. Assuming all of the energy deposited is absorbed by the growth, (a) calculate the amount of energy delivered. (b) Assuming the growth has a mass of 0.25 kg and a specific heat equal to that of water, calculate its temperature rise.

47. A "clever" technician decides to heat some water for his coffee with an x-ray machine. If the machine produces 10 rad/s, how long will it take to raise the temperature of a cup of water by 50°C? Ignore heat losses during this time.

48. An x-ray technician works 5 days per week, 50 weeks per year. Assume that the technician takes an average of eight x-rays per day and receives a dose of 5.0 rem/year as a result. (a) Estimate the dose in rem per x-ray taken.

(b) How does this result compare with the amount of low-level background radiation the technician is exposed to?

WEB **49.** A patient swallows a radiopharmaceutical tagged with phosphorus-32 ($\,^{32}_{15}P$), an e^- emitter with a half-life of 14.3 days. The average kinetic energy of the emitted electrons is 700 keV. If the initial activity of the sample is 1.31 MBq, determine (a) the number of electrons emitted in a 10-day period, (b) the total energy deposited in the body during the 10 days, and (c) the absorbed dose if the electrons are completely absorbed in 100 g of tissue.

50. A particular radioactive source produces 100 mrad of 2-MeV gamma rays per hour at a distance of 1.0 m. (a) How long could a person stand at this distance before accumulating an intolerable dose of 1 rem? (b) Assuming the gamma radiation is emitted uniformly in all directions, at what distance would a person receive a dose of 10 mrad/h from this source?

ADDITIONAL PROBLEMS

51. A 200.0-mCi sample of a radioactive isotope is purchased by a medical supply house. If the sample has a half-life of 14 days, how long will it keep before its activity is reduced to 20.0 mCi?

52. If there are N_0 radioactive nuclei present at $t = 0$, the number remaining at time t is given by $N = N_0 e^{-\lambda t}$, where $\lambda = \dfrac{\ln 2}{T_{1/2}}$. Show that the expression $N = N_0 (\frac{1}{2})^{t/T_{1/2}}$ is equivalent to $N = N_0 e^{-\lambda t}$ by showing that both give the same value for the number of radioactive nuclei remaining at times of (a) $t = T_{1/2}$, (b) $t = 0.318T_{1/2}$, and (c) $t = 4.72T_{1/2}$.

53. A sample of organic material is found to contain 18 g of carbon. The investigators believe the material to be 20 000 years old based on samples of pottery found at the site. If so, what is the expected activity of the organic material? Take data from Example 29.7.

54. One method for producing neutrons for experimental use is to bombard $\,^7_3Li$ with protons. The neutrons are emitted according to the following reaction:

$$\,^1_1H + \,^7_3Li \longrightarrow \,^7_4Be + \,^1_0n$$

What is the minimum kinetic energy the incident proton must have if this reaction is to occur?

55. Deuterons that have been accelerated are used to bombard other deuterium nuclei, resulting in the reaction

$$\,^2_1H + \,^2_1H \longrightarrow \,^3_2He + \,^1_0n$$

Does this reaction require a threshold energy? If so, what is its value?

56. Many radioisotopes have important industrial, medical, and research applications. One of these is $\,^{60}Co$, which has a half-life of 5.2 years and decays by the emission of a beta particle (energy 0.31 MeV) and two gamma photons (energies 1.17 MeV and 1.33 MeV). A scientist wishes to prepare a ^{60}Co sealed source that will have an activity of at least 10 Ci after 30 months of use. What is the minimum initial mass of ^{60}Co required?

57. Free neutrons have a characteristic half-life of 12 min. What fraction of a group of free neutrons at thermal energy (0.040 eV) will decay before traveling a distance of 10.0 km?

58. A medical laboratory stock solution is prepared with an initial activity due to ^{24}Na of 2.5 mCi/ml, and 10.0 ml of the stock solution is diluted at $t_0 = 0$ to a working solution whose total volume is 250 ml. After 48 h, a 5.0-ml sample of the working solution is monitored with a counter. What is the measured activity? (Note that 1 ml = 1 milliliter.)

59. The theory of nuclear astrophysics proposes that all the heavy elements such as uranium are formed in explosions ending the lives of massive stars. These supernovas release the elements into space. If we assume that at the time of explosion there were equal amounts of ^{235}U and ^{238}U, how long ago were the elements that formed the Earth released, given that the present ^{235}U/^{238}U ratio is 0.007? (The half-lives of ^{235}U and ^{238}U are 0.70×10^9 y and 4.47×10^9 y, respectively.)

60. During the manufacture of a steel engine component, radioactive iron (^{59}Fe) is included in the total mass of 0.20 kg. The component is placed in a test engine when the activity due to the isotope is 20.0 μCi. After a 1000-h test period, oil is removed from the engine and found to contain enough ^{59}Fe to produce 800 disintegrations/min per liter of oil. The total volume of oil in the engine is 6.5 liters. Calculate the total mass worn from the engine component per hour of operation. (The half-life for ^{59}Fe is 45.1 days.)

61. In a piece of rock from the Moon, the ^{87}Rb content is assayed to be 1.82×10^{10} atoms per gram of material, and the ^{87}Sr content is found to be 1.07×10^9 atoms per gram. (The relevant decay is ^{87}Rb $\rightarrow$ ^{87}Sr + e$^-$. The half-life of the decay is 4.8×10^{10} years.) (a) Determine the age of the rock. (b) Could the material in the rock actually be much older? What assumptions are implicit in using the radioactive dating method?

62. After determining that the Sun has existed for hundreds of millions of years but before the discovery of nuclear physics, scientists could not explain why the Sun has continued to burn for such a long time. (If it used a nonnuclear burning process—e.g., coal—it would have burned up in 3000 years or so.) Assume that the Sun, whose mass is 1.99×10^{30} kg, consists entirely of hydrogen and that its total power output (or luminosity) is 3.9×10^{26} W. (a) If the energy-generating mechanism of the Sun is the "burning" or transforming of hydrogen into helium via the reaction,

$$4(^1_1\text{H}) \longrightarrow {}^4_2\text{He} + 2e^+ + 2\nu + \gamma$$

calculate the energy (in joules) given off by this reaction. (b) Determine how many hydrogen atoms are available for burning. Take the mass of one hydrogen atom (proton) to be 1.67×10^{-27} kg. (c) Assuming that the total power output remains constant, how long will it be before all the hydrogen is converted into helium and the Sun dies? (d) Why are your results larger than the accepted lifetime of about 10 billion years?

63. The energy required to construct a uniformly charged sphere of total charge Q and radius R is $E = 3k_eQ^2/5R$, where k_e is the Coulomb constant. Assume that a ^{40}Ca nucleus consists of 20 protons uniformly distributed in a spherical volume. (a) How much energy is required to counter the electrostatic repulsion given by the above equation? (*Hint:* First calculate the radius of a ^{40}Ca nucleus.) (b) Calculate the binding energy of ^{40}Ca and compare it to the result in part (a). (c) Explain why the result of part (b) is larger than that of part (a).

64. A by-product of some fission reactors is the isotope $^{239}_{94}$Pu, which is an alpha emitter with a half-life of 24 000 years:

$$^{239}_{94}\text{Pu} \longrightarrow {}^{235}_{92}\text{U} + {}^4_2\text{He}$$

Consider a sample of 1.0 kg of pure $^{239}_{94}$Pu at $t = 0$. Calculate (a) the number of $^{239}_{94}$Pu nuclei present at $t = 0$ and (b) the initial activity of the sample. (c) How long does the sample have to be stored if a "safe" activity level is 0.10 Bq?

65. A fission reactor is hit by a nuclear weapon, causing 5.0×10^6 Ci of ^{90}Sr ($T_{1/2} = 28.7$ years) to evaporate into the air. The ^{90}Sr falls out over an area of 10^4 km^2. How long will it take the activity of the ^{90}Sr to reach the agriculturally "safe" level of 2.0 μCi/m^2?

66. A 25-g piece of charcoal is known to be about 25 000 years old. (a) Determine the number of decays per minute expected from this sample. Take data from Example 29.7. (b) If the radioactive background in the counter without a sample is 20.0 counts/min and we assume 100% efficiency in counting, explain why 25 000 years is close to the limit of dating with this technique.

30

Nuclear Energy and Elementary Particles

Chapter Outline

▲ PHYSICS PUZZLER

A technician works on one of the particle detectors at CERN, the European center for particle physics near Geneva, Switzerland. In modern particle accelerators, it is preferable to have two colliding beams of particles hit each other rather than allowing a single beam to hit a stationary target. What is the advantage of the colliding beams? *(David Parker/Science Photo Library, Photo Researchers, Inc.)*

I n this concluding chapter we discuss the two means by which energy can be derived from nuclear reactions. These two techniques are fission, in which a nucleus of large mass-number splits, or fissions, into two smaller nuclei, and fusion, in which two light nuclei fuse to form a heavier nucleus. In either case, there is a release of large amounts of energy, which can be used destructively through bombs or constructively through the production of electric power.

We end our study of physics by examining the known subatomic particles and the fundamental interactions that govern their behavior. We also discuss the current theory of elementary particles, which states that all matter in nature is constructed from only two families of particles, quarks and leptons. Finally, we describe how clarifications of such models might help scientists understand the evolution of the Universe.

30.1 NUCLEAR FISSION

Nuclear fission occurs when a heavy nucleus, such as ^{235}U, splits, or fissions, into two smaller nuclei. In such a reaction, **the total mass of the products is less than the original mass of the heavy nucleus.**

Nuclear fission was first observed in 1939 by Otto Hahn and Fritz Strassman, following some basic studies by Fermi. After bombarding uranium ($Z = 92$) with neutrons, Hahn and Strassman discovered among the reaction products two medium-mass elements, barium and lanthanum. Shortly thereafter, Lisa Meitner and Otto Frisch explained what had happened. The uranium nucleus had split into two nearly equal fragments after absorbing a neutron. Such an occurrence was of considerable interest to physicists attempting to understand the nucleus, but it was to have even more far-reaching consequences. Measurements showed that about 200 MeV of energy is released in each fission event, and this fact was to affect the course of human history.

The fission of ^{235}U by slow (low energy) neutrons can be represented by the reaction

$$^{1}_{0}n + ^{235}_{92}U \longrightarrow ^{236}_{92}U^* \longrightarrow X + Y + neutrons \qquad [30.1]$$

where ^{236}U* is an intermediate state that lasts only for about 10^{-12} s before splitting into X and Y. The resulting nuclei, X and Y, are called **fission fragments.** There are many combinations of X and Y that satisfy the requirements of conservation of mass-energy and charge. In the fission of uranium, there are about 90 different daughter nuclei that can be formed. The process also results in the production of several (typically two or three) neutrons per fission event. On the average, 2.47 neutrons are released per event.

A typical reaction of this type is

$$^{1}_{0}n + ^{235}_{92}U \longrightarrow ^{141}_{56}Ba + ^{92}_{36}Kr + 3^{1}_{0}n \qquad [30.2]$$

The fission fragments, barium and krypton, and the released neutrons have a great deal of kinetic energy following the fission event.

The breakup of the uranium nucleus can be compared to what happens to a drop of water when excess energy is added to it. All of the atoms in the drop have energy, but not enough to break up the drop. However, if enough energy is added to set the drop vibrating, it will undergo elongation and compression until the amplitude of vibration becomes large enough to cause the drop to break apart. In

Figure 30.1 The stages involved in a nuclear fission event as described by the liquid-drop model of the nucleus.

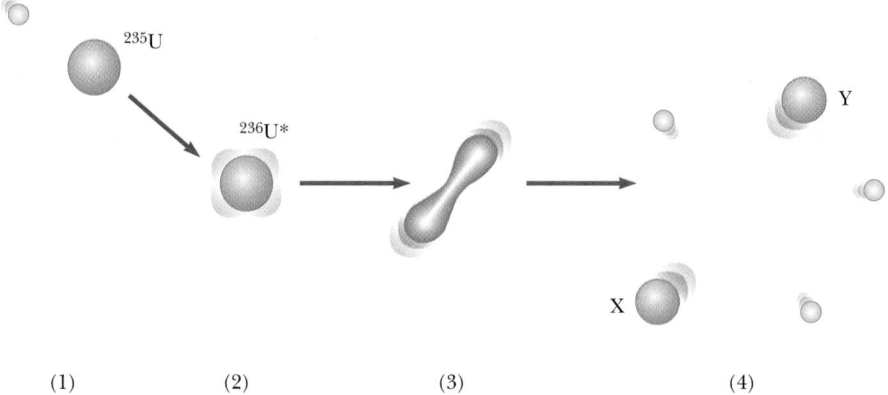

(1)   (2)   (3)   (4)

the uranium nucleus, a similar process occurs (Fig. 30.1). The sequence of events is as follows:

1. The ^{235}U nucleus captures a thermal (slow-moving) neutron.
2. This capture results in the formation of ^{236}U*, and the excess energy of this nucleus causes it to undergo violent oscillations.
3. The ^{236}U* nucleus becomes highly distorted, and the force of repulsion between protons in the two halves of the dumbbell shape tends to increase the distortion.
4. The nucleus splits into two fragments, emitting several neutrons in the process.

Let us estimate the disintegration energy, Q, released in a typical fission process. From Figure 29.4 we see that the binding energy per nucleon is about 7.2 MeV for heavy nuclei (those having a mass number of approximately 240) and about 8.2 MeV for nuclei of intermediate mass. This means that the nucleons in the fission fragments are more tightly bound and therefore have less mass than the nucleons in the original heavy nucleus. This decrease in mass per nucleon appears as released energy when fission occurs. The amount of energy released is (8.2 − 7.2) MeV per nucleon. Assuming a total of 240 nucleons, we find that the energy released per fission event is

$$Q = (240 \text{ nucleons}) \left(8.2 \, \frac{\text{MeV}}{\text{nucleon}} - 7.2 \, \frac{\text{MeV}}{\text{nucleon}} \right) = 240 \text{ MeV}$$

This is indeed a very large amount of energy relative to the amount released in chemical processes. For example, the energy released in the combustion of one molecule of the octane used in gasoline engines is about one hundred-millionth the energy released in a single fission event!

Thinking Physics 1

If a heavy nucleus were to fission into just two product nuclei, they would be very unstable. Why is this?

Explanation According to Figure 29.3, the ratio of the number of neutrons to the number of protons increases with Z. As a result, when a heavy nucleus splits in a fission reaction to two lighter nuclei, the lighter nuclei tend to have too many neutrons. This leads to instability, as the nucleus returns to the curve in Figure 29.3 by decay processes that reduce the number of neutrons.

EXAMPLE 30.1 The Fission of Uranium

Two other possible ways by which ^{235}U can undergo fission when bombarded with a neutron are (1) by the release of ^{140}Xe and ^{94}Sr as fission fragments and (2) by the release of ^{132}Sn and ^{101}Mo as fission fragments. In each case, neutrons are also released. Find the number of neutrons released in each of these events.

Solution By balancing mass numbers and atomic numbers, we find that these reactions can be written

$$_{0}^{1}n + _{92}^{235}U \longrightarrow _{54}^{140}Xe + _{38}^{94}Sr + 2_{0}^{1}n$$

$$_{0}^{1}n + _{92}^{235}U \longrightarrow _{50}^{132}Sn + _{42}^{101}Mo + 3_{0}^{1}n$$

Thus, two neutrons are released in the first event and three in the second.

EXAMPLE 30.2 The Energy Released in the Fission of ^{235}U

Calculate the total energy released if 1.00 kg of ^{235}U undergoes fission, taking the disintegration energy per event to be $Q = 208$ MeV (a more accurate value than the estimate given previously).

Solution We need to know the number of nuclei in 1.00 kg of uranium. Because $A = 235$, the number of nuclei is

$$N = \left(\frac{6.02 \times 10^{23} \text{ nuclei/mol}}{235 \text{ g/mol}} \right) (1.00 \times 10^{3} \text{ g}) = 2.56 \times 10^{24} \text{ nuclei}$$

Hence the disintegration energy

$$E = NQ = (2.56 \times 10^{24} \text{ nuclei}) \left(208 \frac{\text{MeV}}{\text{nucleus}} \right) = 5.32 \times 10^{26} \text{ MeV}$$

Because 1 MeV is equivalent to 4.45×10^{-20} kWh, $E = 2.37 \times 10^{7}$ kWh. This is enough energy to keep a 100-W lightbulb burning for about 30 000 years. Thus, 1.00 kg of ^{235}U is a relatively large amount of fissionable material.

30.2 NUCLEAR REACTORS

We have seen that when ^{235}U undergoes fission, an average of about 2.5 neutrons are emitted per event. These neutrons can in turn trigger other nuclei to undergo fission, with the possibility of a chain reaction (Fig. 30.2). Calculations show that if the chain reaction is not controlled (that is, if it does not proceed slowly), it could result in a violent explosion, with the release of an enormous amount of energy, even from only 1 g of ^{235}U. If the energy in 1 kg of ^{235}U were released, it would

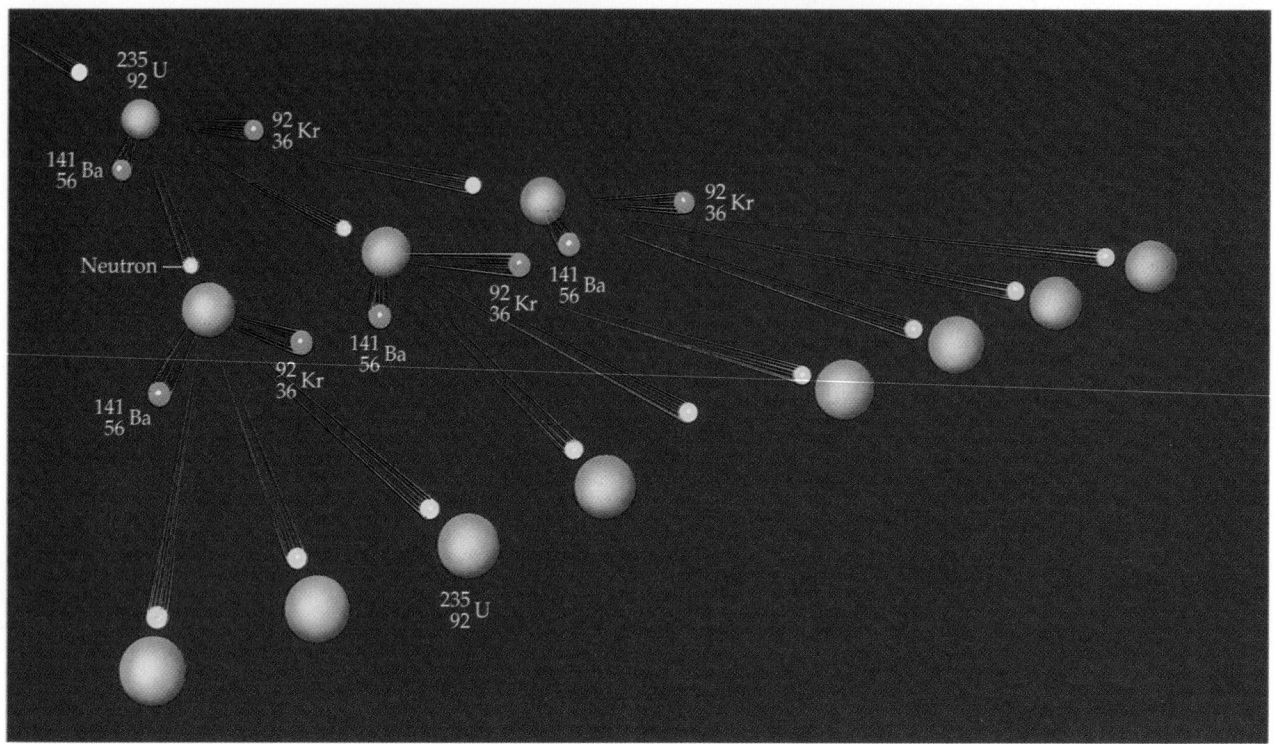

Figure 30.2 A nuclear chain reaction initiated by capture of a neutron. (Many pairs of different isotopes are produced, but only one pair is shown.)

equal that released by the detonation of about 20 000 tons of TNT! This, of course, is the principle behind the first nuclear bomb, an uncontrolled fission reaction.

A nuclear reactor is a system designed to maintain what is called a **self-sustained chain reaction.** This important process was first achieved in 1942 by Fermi at the University of Chicago, with natural uranium as the fuel. Most reactors in operation today also use uranium as fuel. Natural uranium contains only about 0.7% of the ^{235}U isotope, with the remaining 99.3% being the ^{238}U isotope. This is important to the operation of a reactor because ^{238}U almost never undergoes fission. Instead, it tends to absorb neutrons, producing neptunium and plutonium. For this reason, reactor fuels must be artificially enriched so that they end up with a few percentages of the ^{235}U isotope.

Earlier, we mentioned that an average of about 2.5 neutrons are emitted in each fission event of ^{235}U. In order to achieve a self-sustained chain reaction, one of these neutrons, on the average, must be captured by another ^{235}U nucleus and cause it to undergo fission. A useful parameter for describing the level of reactor operation is the **reproduction constant, K, defined as the average number of neutrons from each fission event that will cause another event.** As we have seen, K can have a maximum value of 2.5 in the fission of uranium. However, in practice K is less than this because of several factors, which we shall soon discuss.

A self-sustained chain reaction is achieved when $K = 1$. Under this condition, the reactor is said to be **critical.** When K is less than unity, the reactor is subcritical and the reaction dies out. When K is greater than unity, the reactor is said to be

Painting of the world's first reactor. Because of wartime secrecy, there are no photographs of the completed reactor. The reactor was composed of layers of graphite interspersed with uranium. A self-sustained chain reaction was first achieved on December 2, 1942. Word of the success was telephoned immediately to Washington with this message: "The Italian navigator has landed in the New World and found the natives very friendly." The historic event took place in an improvised laboratory in the racquet court under the west stands of the University of Chicago's Stagg Field; the Italian navigator was Fermi. *(Courtesy of Chicago Historical Society)*

supercritical, and a runaway reaction occurs. In a nuclear reactor used to furnish power to a utility company, it is necessary to maintain a K value close to unity.

The basic design of a nuclear reactor is shown in Figure 30.3. The fuel elements consist of enriched uranium. The function of the remaining parts of the reactor and some aspects of its design will now be described.

Neutron Leakage

In any reactor, a fraction of the neutrons produced in fission will leak out of the core before inducing other fission events. If the fraction leaking out is too large, the reactor will not operate. The percentage lost is large if the reactor is very small because leakage is a function of the ratio of surface area to volume. Therefore, a critical requirement of reactor design is choosing the correct surface-area-to-volume ratio so that a sustained reaction can be achieved.

Regulating Neutron Energies

The neutrons released in fission events are very energetic, with kinetic energies of about 2 MeV. It is found that slow neutrons are far more likely than fast neutrons to produce fission events in ^{235}U. Furthermore, ^{238}U does not absorb slow neutrons. Therefore, in order for the chain reaction to continue, the neutrons must be slowed down. This is accomplished by surrounding the fuel with a **moderator** substance.

In order to understand how neutrons are slowed down, consider a collision between a light object and a very massive one. In such an event, the light object rebounds from the collision with most of its original kinetic energy. However, if the

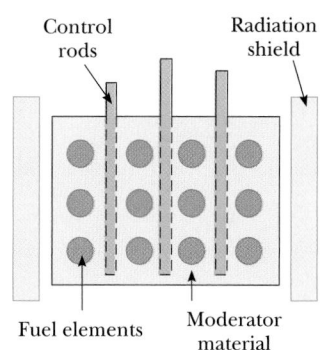

Figure 30.3 Cross section of a reactor core surrounded by a radiation shield.

collision is between objects whose masses are nearly the same, the incoming projectile will transfer a large percentage of its kinetic energy to the target. In the first nuclear reactor ever constructed, Fermi placed bricks of graphite (carbon) between the fuel elements. Carbon nuclei are about 12 times more massive than neutrons, but after about 100 collisions with carbon nuclei, a neutron is slowed sufficiently to increase its likelihood of fission with ^{235}U. In this design the carbon is the moderator; most modern reactors use heavy water (D$_2$O) as the moderator.

Neutron Capture

In the process of being slowed down, the neutrons may be captured by nuclei that do not undergo fission. The most common event of this type is neutron capture by ^{238}U. The probability of neutron capture by ^{238}U is very high when the neutrons have high kinetic energies and very low when they have low kinetic energies. Thus the slowing down of the neutrons by the moderator serves the dual purpose of making them available for reaction with ^{235}U and decreasing their chances of being captured by ^{238}U.

Control of Power Level

It is possible for a reactor to reach the critical stage ($K = 1$) after all the neutron losses described previously are minimized. However, a method of control is needed to adjust a K value near unity. If K were to rise above this value, the heat produced in the runaway reaction would melt the reactor. To control the power level, control rods are inserted into the reactor core (see Fig. 30.3). These rods are made of materials such as cadmium that are very efficient in absorbing neutrons. By adjusting the number and position of these control rods in the reactor core, the K value can be varied and any power level within the design range of the reactor can be achieved.

A diagram of a pressurized-water reactor is shown in Figure 30.4. This type of reactor is commonly used in electric power plants in the United States. Fission events in the reactor core supply heat to the water contained in the primary (closed) system, which is maintained at high pressure to keep it from boiling. This water also serves as the moderator. The hot water is pumped through a heat exchanger, and the heat is transferred to the water contained in the secondary system. There the hot water is converted to steam, which drives a turbine generator to create electric power. Note that the water in the secondary system is isolated from the water in the primary system in order to prevent contamination of the secondary water and steam by radioactive nuclei from the reactor core.

Reactor Safety[1]

The safety aspects of nuclear power reactors are often sensationalized by the media and misunderstood by the public. The 1979 near-disaster of Three Mile Island in Pennsylvania and the accident at the Chernobyl reactor in Ukraine rightfully focused attention on reactor safety. Yet the safety record in the United States is en-

---

[1] The authors are grateful to Professor Gene Skluzacek of the University of Nebraska at Omaha for rewriting this section on reactor safety.

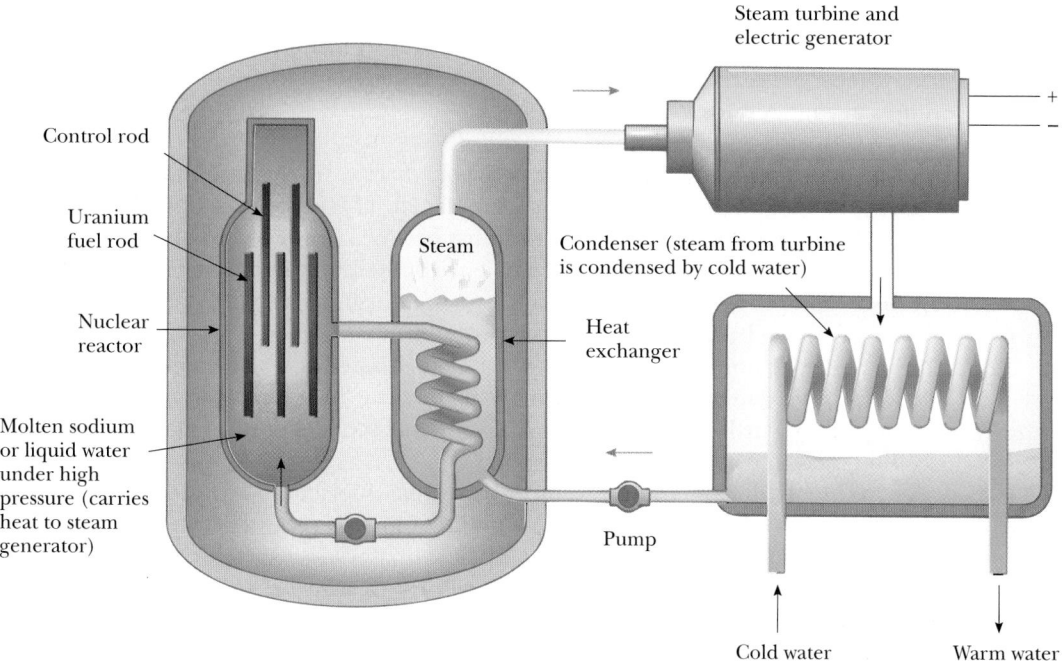

Figure 30.4   Main components of a pressurized-water reactor.

viable. The records show no fatalities attributed to commercial nuclear power gen-eration in the history of the United States nuclear industry.

Commercial reactors achieve safety through careful design and rigid operating procedures. Radiation exposure and the potential health risks associated with such exposure are controlled by three layers of containment. The fuel and radioactive fission products are contained inside the reactor vessel. Should this vessel rupture, the reactor building acts as a second containment structure to prevent radioactive material from contaminating the environment. Finally, the reactor facilities must be in a remote location to protect the general public from exposure should radi-ation escape the reactor building.

According to the Oak Ridge National Laboratory Review, "the health risk of living within 8 km (5 miles) of a nuclear reactor for 50 years is no greater than the risk of smoking 1.4 cigarettes, drinking 0.5 liters of wine, traveling 240 km by car, flying 9600 km by jet, or having one chest x-ray in a hospital. Each of these activities is estimated to increase a person's chances of dying in any given year by one in a million."

Another potential danger in nuclear reactor operations is the possibility that the water flow could be interrupted. Even if the nuclear fission chain reaction were stopped immediately, residual heat could build up in the reactor to the point of melting the fuel elements. The molten reactor core would melt to the bottom of the reactor vessel and conceivably melt its way into the ground below—the so-called "China syndrome." Although it might appear that this deep underground burial site would be an ideal safe haven for a radioactive blob, there would be danger of a steam explosion should the molten mass encounter water. This nonnuclear explosion could spread radioactive material to the areas surrounding the power plant. To prevent such an unlikely chain of events, nuclear reactors are designed with emergency core cooling systems requiring no power that automatically flood the reactor with water in the event of coolant loss. The emergency cooling water moderates heat build-up in the core, which in turn prevents the occurrence of melting.

A continuing concern in nuclear fission reactors is the safe disposal of radioactive material when the reactor core is replaced. This waste material contains long-lived, highly radioactive isotopes and must be stored over long periods of time in such a way that there is no chance of environmental contamination. At present, sealing radioactive wastes in waterproof containers and burying them in deep salt mines seems to be the most promising solution.

Transportation of reactor fuel and reactor wastes poses additional safety risks. However, the danger of theft during transport (say by a terrorist group) is greatly exaggerated. Furthermore, neither the waste nor the fuel of nuclear power reactors can be used to construct a nuclear bomb.

Accidents during transportation of nuclear fuel could expose the public to harmful levels of radiation. The Department of Energy requires stringent crash tests on all containers used to transport nuclear materials. Container manufacturers must demonstrate that their containers will not rupture even in high-speed collisions.

The safety issues associated with nuclear power reactors are complex and often emotional. All sources of energy have associated risks. In each case, one must weigh the risks against the benefits and the availability of the energy source.

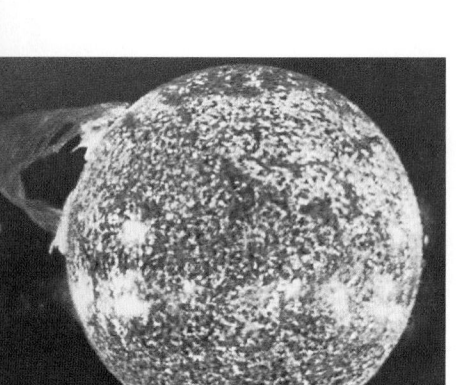

This photograph of the Sun, taken on December 19, 1973, during the third and final manned Skylab mission, shows one of the most spectacular solar flares ever recorded, spanning more than 588 000 km (365 000 mi) across the solar surface. The last picture, taken some 17 hours earlier, showed this feature as a large quiescent prominence on the eastern side of the Sun. The flare gives the distinct impression of a twisted sheet of gas in the process of unwinding itself. In this photograph the solar poles are distinguished by a relative absence of granulation and a much darker tone than the central portions of the disk. Several active regions are seen on the eastern side of the disk. The photograph was taken in the light of ionized helium by the extreme ultraviolet spectroheliograph instrument of the U.S. Naval Research Laboratory. *(NASA)*

30.3   NUCLEAR FUSION

Figure 29.4 shows that the binding energy for light nuclei (those having a mass number lower than 20) is much smaller than the binding energy for heavier nuclei. This suggests a possible process that is the reverse of fission. **When two light nuclei combine to form a heavier nucleus, the process is called nuclear fusion.** Because the mass of the final nucleus is less than the masses of the original nuclei, there is a loss of mass accompanied by a release of energy. Although fusion power plants have not yet been developed, a great worldwide effort is under way to harness the energy from fusion reactions in the laboratory. Later we shall discuss the possibilities and advantages of this process for generating electric power.

Fusion in the Sun

All stars generate their energy through fusion processes. About 90% of the stars, including the Sun, fuse hydrogen, whereas some older stars fuse helium or other heavier elements. Stars are born in regions of space containing vast clouds of dust and gas. Recent mathematical models of these clouds indicate that star formation is triggered by shock waves passing through a cloud. These shock waves are similar to sonic booms and are produced by events such as the explosion of a nearby star, called a *supernova explosion*. The shock wave compresses certain regions of the cloud, causing these regions to collapse under their own gravity. As the gas falls inward toward the center, the atoms gain speed, which causes the temperature of the gas to rise. Two conditions must be met before fusion reactions in the star can sustain its energy needs: (1) The temperature must be high enough (about 10^7 K for hydrogen) to allow the kinetic energy of the positively charged hydrogen nuclei to overcome their mutual Coulomb repulsion as they collide, and (2) the density of nuclei must be high enough to ensure a high rate of collision.

When fusion reactions occur at the core of a star, the energy liberated eventually becomes sufficient to prevent further collapse of the star under its own gravity. The star then continues to live out the remainder of its life under a balance between the inward force of gravity pulling it toward collapse and the outward force due to thermal effects and radiation pressure. The proton-proton cycle is a series of three nuclear reactions that are believed to be the stages in the liberation of energy in the Sun and other stars rich in hydrogen. An overall view of the proton-proton cycle is that four protons combine to form an alpha particle and two positrons, with the release of 25 MeV of energy in the process.

The three steps in the proton-proton cycle are

$$\begin{aligned} {}_1^1\text{H} + {}_1^1\text{H} &\longrightarrow {}_1^2\text{H} + \text{e}^+ + \nu \\ {}_1^1\text{H} + {}_1^2\text{H} &\longrightarrow {}_2^3\text{He} + \gamma \end{aligned}$$

[30.3]

This second reaction is followed by either

$$ {}_1^1\text{H} + {}_2^3\text{H} \longrightarrow {}_2^4\text{H} + \text{e}^+ + \nu $$

or

$$ {}_2^3\text{He} + {}_2^3\text{He} \longrightarrow {}_2^4\text{He} + {}_1^1\text{H} + {}_1^1\text{H} $$

The energy liberated is carried primarily by gamma rays, positrons, and neutrinos, as can be seen from the reactions. The gamma rays are soon absorbed by the dense gas, thus raising its temperature. The positrons combine with electrons to produce gamma rays, which in turn are also absorbed by the gas within a few centimeters. The neutrinos, however, almost never interact with matter; hence they escape from the star, carrying about 2% of the generated energy with them. These energy-liberating fusion reactions are called **thermonuclear fusion reactions.** The hydrogen (fusion) bomb, first exploded in 1952, is an example of an uncontrolled thermonuclear fusion reaction.

Fusion Reactors

The enormous amount of energy released in fusion reactions suggests the possibility of harnessing this energy for useful purposes on Earth. A great deal of effort is under way to develop a sustained and controllable thermonuclear reactor—a fusion power reactor. Controlled fusion is often called the ultimate energy source because of the availability of its fuel source: water. For example, if deuterium were used as the fuel, 0.06 g of it could be extracted from 1 gal of water at a cost of about four cents. Such rates would make the fuel costs of even an inefficient reactor almost insignificant. An additional advantage of fusion reactors is that comparatively few radioactive by-products are formed. As noted in Equation 30.3, the end product of the fusion of hydrogen nuclei is safe, nonradioactive helium. Unfortunately, a thermonuclear reactor that can deliver a net power output over a reasonable time interval is not yet a reality, and many difficulties must be solved before a successful device is constructed.

We have seen that the Sun's energy is based, in part, on a set of reactions in which ordinary hydrogen is converted to helium. Unfortunately, the proton-proton interaction is not suitable for use in a fusion reactor because the event requires very high pressures and densities. The process works in the Sun only because of the extremely high density of protons in the Sun's interior. In fact, even at the densities and temperatures that exist at the center of the Sun, the average proton takes 14 billion years to react.

The fusion reactions that appear most promising in the construction of a fusion power reactor involve deuterium and tritium, which are isotopes of hydrogen. These reactions are

$$\,^2_1\text{H} + \,^2_1\text{H} \longrightarrow \,^3_2\text{He} + \,^1_0\text{n} \qquad Q = 3.27 \text{ MeV}$$

$$\,^2_1\text{H} + \,^2_1\text{H} \longrightarrow \,^3_1\text{H} + \,^1_1\text{H} \qquad Q = 4.03 \text{ MeV} \qquad [30.4]$$

$$\,^2_1\text{H} + \,^3_1\text{H} \longrightarrow \,^4_2\text{He} + \,^1_0\text{n} \qquad Q = 17.59 \text{ MeV}$$

where the Q values refer to the amount of energy released per reaction. As noted earlier, deuterium is available in almost unlimited quantities from our lakes and oceans and is very inexpensive to extract. Tritium, however, is radioactive ($T_{1/2} = 12.3$ years) and undergoes beta decay to ^3He. For this reason, tritium does not occur naturally to any great extent and must be artificially produced.

One of the major problems in obtaining energy from nuclear fusion is the fact that the Coulomb repulsion force between two charged nuclei must be overcome before they can fuse. The fundamental challenge is to give the two nuclei enough kinetic energy to overcome this repulsive force. This can be accomplished by heating the fuel to extremely high temperatures (about 10^8 K, far greater than the interior temperature of the Sun). As you might expect, such high temperatures are not easy to obtain in a laboratory or a power plant. At these high temperatures the atoms are ionized, and the system consists of a collection of electrons and nuclei, commonly referred to as *plasma*.

In addition to the high temperature requirements, there are two other critical factors that determine whether or not a thermonuclear reactor will be successful: **plasma ion density, n,** and **plasma confinement time, τ,** the time the interacting ions are maintained at a temperature equal to or greater than that required for the reaction to proceed successfully. The density and confinement time must both be large enough to ensure that more fusion energy will be released than is required to heat the plasma.

Lawson's criterion states that a net power output in a fusion reactor is possible under the following conditions:

Lawson's criterion ▶

$$n\tau \geq 10^{14}\text{s/cm}^3 \quad \text{Deuterium-tritium interaction}$$
$$n\tau \geq 10^{16}\text{s/cm}^3 \quad \text{Deuterium-deuterium interaction} \qquad [30.5]$$

The problem of plasma confinement time has yet to be solved. How can a plasma be confined at a temperature of 10^8 K for times on the order of 1 s? The basic plasma-confinement technique under investigation is discussed following Example 30.3.

EXAMPLE 30.3 The Deuterium-Deuterium Reaction

Find the energy released in the deuterium-deuterium reaction

$$\,^2_1H + \,^2_1H \longrightarrow \,^3_1H + \,^1_1H$$

Solution The mass of the $\,^2_1H$ atom is 2.014 102 u. Thus, the total mass before the reaction is 4.028 204 u. After the reaction, the sum of the masses is equal to 3.016 049 u + 1.007 825 u = 4.023 874 u. Thus, the excess mass is 0.004 33 u. In energy units, this is equivalent to   4.03 MeV.

Magnetic Field Confinement

Most fusion experiments use magnetic field confinement to contain a plasma. One device, called a **tokamak,** has a doughnut-shaped geometry (a toroid), as shown in Figure 30.5a. This device, first developed in the former Soviet Union, uses a

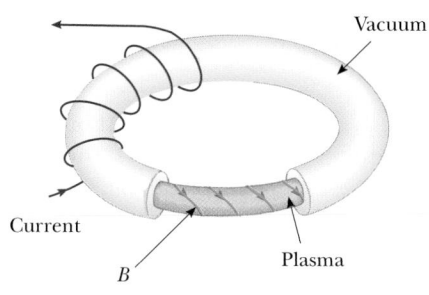

(a)

(b)

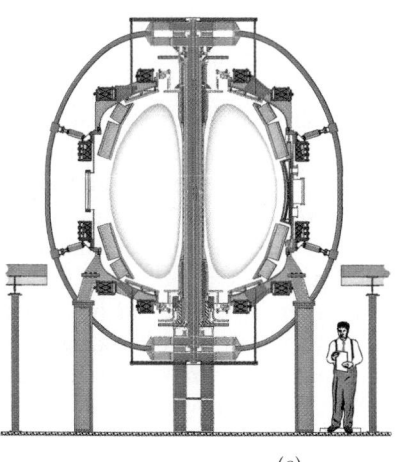

(c)

Figure 30.5 (a) Diagram of a tokamak used in the magnetic confinement scheme. The plasma is trapped within the spiraling magnetic field lines as shown. (b) Interior view of the recently closed Tokamak Fusion Test Reactor (TFTR) vacuum vessel at the Princeton Plasma Physics Laboratory, Princeton University, New Jersey. (c) A diagram of the National Spherical Torus Experiment (NSTX) now under construction at the Princeton Plasma Physics Laboratory. *(b, c, Courtesy of Princeton Plasma Physics Laboratory)*

WEB

For Web links to a "virtual tokamak" and the National Spherical Torus Experiment, visit the textbook Web site at http://www.harcourtcollege.com/physics/cptech

combination of two magnetic fields to confine the plasma inside the doughnut. A strong magnetic field is produced by the current in the windings, and a weaker magnetic field is produced by the current in the toroid. The resulting magnetic field lines are helical, as in Figure 30.5a. In this configuration, the field lines spiral around the plasma and prevent it from touching the walls of the vacuum chamber. In order for the plasma to reach ignition temperature, energetic neutral particles are injected into the plasma.

Figure 30.5b is a photograph of the Tokamak Fusion Test Reactor (TFTR) at the Princeton Plasma Physics Laboratory. Although recently shut down, this reactor produced a fusion output power of 6.1 MW with an input power of 29.4 MW. One of the new generation of fusion experiments under construction at the Princeton Plasma Physics Laboratory is the National Spherical Torus Experiment (NSTX) shown in Figure 30.5c. This experiment will produce a spherical plasma with a hole through its center, while the tokamak plasma is donut-shaped. The major advantage of the spherical plasma configuration is its ability to confine a higher pressure plasma in a given magnetic field. This alternative approach could lead to the development of smaller and more economical fusion reactors.

30.4   ELEMENTARY PARTICLES

The word "atom" is from the Greek word *atomos,* meaning "indivisible." At one time atoms were thought to be the indivisible constituents of matter; that is, they were regarded as elementary particles. Discoveries in the early part of the 20th century revealed that the atom is not elementary, but has as its constituents protons, neutrons, and electrons. Until 1932 physicists viewed these three constituent particles as elementary because, with the exception of the free neutron, they are very stable. The theory soon fell apart, however, and beginning in 1945, many new particles were discovered in experiments involving high-energy collisions between known particles. These new particles are characteristically unstable and have very short half-lives, ranging between 10^{-6} and 10^{-23} s. So far more than 300 of them have been cataloged.

Until the 1960s, physicists were bewildered by the large number and variety of subatomic particles being discovered. They wondered if the particles were like animals in a zoo or if a pattern was emerging that would provide a better understanding of the elaborate structure in the subnuclear world. In the past 30 years, physicists have made tremendous advances in our knowledge of the structure of matter by recognizing that all particles (with the exception of electrons, photons, and a few others) are made of smaller particles called *quarks.* Thus, protons and neutrons, for example, are not truly elementary but are systems of tightly bound quarks. The quark model has reduced the bewildering array of particles to a manageable number and has successfully predicted new quark combinations that were subsequently found in many experiments.

30.5   THE FUNDAMENTAL FORCES IN NATURE

The key to understanding the properties of elementary particles is to be able to describe the forces between them. All particles in nature are subject to four fundamental forces: strong, electromagnetic, weak, and gravitational.

TABLE 30.1 Particle Interactions

Interaction (Force)	Relative Strength[a]	Range of Force	Mediating Field Particle
Strong	1	Short (≈ 1 fm)	Gluon
Electromagnetic	10^{-2}	Long ($\propto 1/r^2$)	Photon
Weak	10^{-6}	Short ($\approx 10^{-3}$ fm)	$W^{\pm}$ and Z bosons
Gravitational	10^{-43}	Long ($\propto 1/r^2$)	Graviton

[a] For two quarks separated by 3×10^{-17} m

The **strong force** is responsible for the binding of neutrons and protons into nuclei. This force represents the "glue" that holds the nucleons together and is the strongest of all the fundamental forces. It is very short-ranged and is negligible for separations greater than about 10^{-15} m (the approximate size of the nucleus). The **electromagnetic force,** which is about 10^{-2} times the strength of the strong force, is responsible for the binding of atoms and molecules. It is a long-range force that decreases in strength as the inverse square of the separation between interacting particles. The **weak force** is a short-range nuclear force that tends to produce instability in certain nuclei. It is responsible for most radioactive decay processes such as beta decay, and its strength is only about 10^{-6} times that of the strong force. (As we shall discuss later, scientists now believe that the weak and electromagnetic forces are two manifestations of a single force called the *electroweak* force.) Finally, the **gravitational force** is a long-range force with a strength of only about 10^{-43} times that of the strong force. Although this familiar interaction is the force that holds the planets, stars, and galaxies together, its effect on elementary particles is negligible. Thus, the gravitational force is the weakest of all the fundamental forces.

Modern physics often describes the interactions between particles in terms of the actions of field particles or quanta. In the case of the familiar electromagnetic interaction, the field particles are photons. In the language of modern physics, it can be said that the electromagnetic force is *mediated* by photons, which are the quanta of the electromagnetic field. Likewise, the strong force is mediated by field particles called *gluons,* the weak force is mediated by particles called the W and Z *bosons,* and the gravitational force is mediated by quanta of the gravitational field called *gravitons.* All of these field quanta have been detected except for the graviton, which may never be found directly because of the weakness of the gravitational field. These interactions, their ranges, and their relative strengths are summarized in Table 30.1.

30.6 POSITRONS AND OTHER ANTIPARTICLES

In the 1920s, the theoretical physicist Paul Adrien Maurice Dirac (1902–1984) developed a version of quantum mechanics that incorporated special relativity. Dirac's theory successfully explained the origin of the electron's spin and its magnetic moment. But it had one major problem: Its relativistic wave equation required solutions corresponding to negative energy states, and if negative energy states

Paul Adrien Maurice Dirac (1902–1984)

Winner of the Nobel prize for physics in 1933. *(Courtesy AIP Emilio Segrè Visual Archives)*

existed, we would expect an electron in a state of positive energy to make a rapid transition to one of these states, emitting a photon in the process.

Dirac circumvented this difficulty by postulating that all negative energy states are filled. The electrons that occupy the negative energy states are said to be in the "Dirac sea" and not directly observable because the Pauli exclusion principle does not allow them to react to external forces. However, if one of these negative energy states is vacant, leaving a hole in the sea of filled states, the hole can react to external forces and therefore be observable. The profound implication of this theory is that **for every particle, there is an antiparticle.** The antiparticle has the same mass as the particle, but the opposite charge. For example, the electron's antiparticle (now called a *positron*) has a mass of 0.511 MeV/c^2 and a positive charge of 1.6×10^{-19} C. As noted in Chapter 29, we usually designate an antiparticle with a bar over the symbol for the particle. Thus, $\bar{p}$ denotes the antiproton and $\bar{\nu}$ the antineutrino. In this book, the notation e^+ is preferred for the positron.

The positron was discovered by Carl Anderson in 1932 and in 1936 he was awarded the Nobel prize for his achievement. Anderson discovered the positron while examining tracks created by electron-like particles of positive charge in a cloud chamber. (These early experiments used cosmic rays—mostly energetic protons passing through interstellar space—to initiate high-energy reactions on the order of several GeV.) In order to discriminate between positive and negative charges, the cloud chamber was placed in a magnetic field, causing moving charges to follow curved paths. Anderson noted that some of the electron-like tracks deflected in a direction corresponding to a positively charged particle.

Since Anderson's initial discovery, the positron has been observed in a number of experiments. Perhaps the most common process for producing positrons is **pair production.** In this process, a gamma ray with sufficiently high energy collides with a nucleus, creating an electron-positron pair. Because the total rest energy of the electron-positron pair is $2m_ec^2 = 1.02$ MeV (where m_e is the mass of the electron), the gamma ray must have at least this much energy to create an electron-positron pair. Figure 30.6 shows tracks of electron-positron pairs created by 300-MeV gamma rays striking a lead sheet.

The reverse process can also occur. Under the proper conditions, an electron and positron can annihilate each other and produce two photons with a combined

Figure 30.6 (a) Bubble-chamber tracks of electron-positron pairs produced by 300-MeV gamma rays striking a lead sheet. *(Courtesy of Lawrence Berkeley Laboratory, University of California)* (b) Sketch of the pertinent pair-production events. Note that the positrons deflect upward and the electrons deflect downward in an applied magnetic field directed into the diagram.

(a)

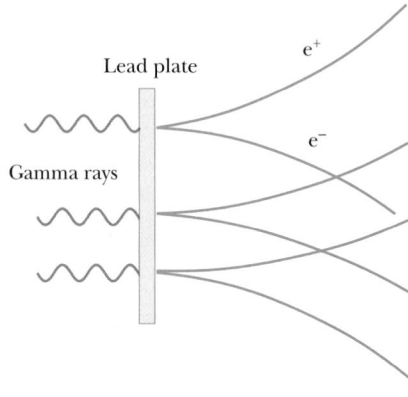

(b)

energy of at least 1.02 MeV:

$$e^- + e^+ \longrightarrow 2\gamma$$

Practically every known elementary particle has an antiparticle. Among the exceptions are the photon and the neutral pion (π^0). Following the construction of high-energy accelerators in the 1950s, many of these antiparticles were discovered. They included the antiproton, $\overline{p}$, discovered by Emilio Segrè and Owen Chamberlain in 1955 and the antineutron, $\overline{n}$, discovered shortly thereafter.

Thinking Physics 2

When an electron and a positron meet at low speeds in free space, why are two gamma rays produced rather than one gamma ray having twice the energy?

Explanation Gamma rays are photons, and photons carry momentum. If only one photon were produced, momentum would not be conserved because the total momentum of the electron-positron pair is approximately zero, whereas a single high-energy photon would have a large momentum. However, the two gamma-ray photons that are produced travel off in opposite directions, so their total momentum is zero.

The process of electron-positron annihilation is used in the medical diagnostic technique of positron emission tomography (PET). The patient is injected with a glucose solution containing a radioactive substance that decays by positron emission. Examples of such substances are oxygen-15, nitrogen-13, carbon-11, and fluorine-18. The radioactive material is carried to the brain. The emitted positron from a decay annihilates with an electron in the brain tissue, resulting in two gamma ray photons emitted in opposite directions. The gamma detector surrounding the patient locates the source of the gamma photons and, with the assistance of a computer, displays an image of the sites in the brain at which the glucose accumulates.

The images from a PET scan can indicate a wide variety of disorders in the brain, including Alzheimer's disease. In addition, because glucose metabolizes more rapidly in active areas of the brain, the PET scan can indicate which areas of the brain are involved in various processes such as language, music, and vision.

APPLICATION

Positron Emission Tomography (PET Scanning).

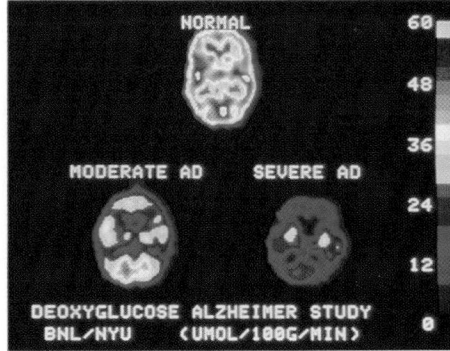

PET scans of the brains of a healthy older person and of patients suffering from Alzheimer's disease. Lighter colors indicate regions having higher concentrations of radioactive glucose and therefore more active metabolism. *(Dr. Monty de Leon/ New York University Medical Center and National Institute on Aging)*

30.7 MESONS AND THE BEGINNING OF PARTICLE PHYSICS

Physicists in the mid-1930s had a fairly simple view of the structure of matter. The building blocks were the proton, the electron, and the neutron. Three other particles were known or postulated at the time: the photon, the neutrino, and the positron. These six particles were considered the fundamental constituents of matter. Although the accepted picture of the world was marvelously simple, no one was able to provide an answer to the following important question: Because the many protons in proximity in any nucleus should strongly repel each other due to their like charges, what is the nature of the force that holds the nucleus together? Scientists recognized that this mysterious force must be much stronger than anything encountered up to that time.

Hideki Yukawa, Japanese physicist (1907–1981)

Yukawa was awarded the Nobel prize in 1949 for predicting the existence of mesons. This photograph of Yukawa at work was taken in 1950 in his office at Columbia University. *(UPI/Corbis-Bettman)*

The first theory to explain the nature of the strong force was proposed in 1935 by the Japanese physicist Hideki Yukawa (1907–1981), an effort that later earned him the Nobel prize. In order to understand Yukawa's theory, it is useful to first note that **two atoms can form a covalent chemical bond by the exchange of electrons.** Similarly, in the modern views of electromagnetic interactions, **charged particles interact by exchanging a photon.** Yukawa used this same idea to explain the strong force by proposing a new particle that is shared by nucleons in the nucleus to produce the strong force. Furthermore, he established that the range of the force is inversely proportional to the mass of this particle, and predicted that the mass would be about 200 times the mass of the electron. Because the new particle would have a mass between that of the electron and the proton, it was called a **meson** (from the Greek *meso,* meaning "middle").

In an effort to substantiate Yukawa's predictions, physicists began looking for the meson by studying cosmic rays that enter the Earth's atmosphere. In 1937 Carl Anderson and his collaborators discovered a particle whose mass was $106 \text{ MeV}/c^2$, about 207 times the mass of the electron. However, subsequent experiments showed that the particle interacted very weakly with matter, and hence could not be the carrier of the strong force. This puzzling situation inspired several theoreticians to propose that there are two mesons with slightly different masses, an idea that was confirmed in 1947 with the discovery of the pi meson (π), or simply *pion,* by Cecil Frank Powell (1903–1969) and Guiseppe P. S. Occhialini (1907–   ). The lighter meson discovered earlier by Anderson, now called a *muon* (μ), has only weak and electromagnetic interaction and plays no role in the strong interaction.

The pion comes in three varieties, corresponding to three charge states: π^+, π^-, and π^0. The π^+ and π^- particles have masses of $139.6 \text{ MeV}/c^2$, and the π^0 has a mass of $135.0 \text{ MeV}/c^2$. Pions and muons are very unstable particles. For example, the π^-, which has a lifetime of about 2.6×10^{-8} s, decays into a muon and an antineutrino. The muon, with a lifetime of 2.2 μs, then decays into an electron, a neutrino, and an antineutrino. The sequence of decays is

$$\pi^- \longrightarrow \mu^- + \bar{\nu}$$
$$\mu^- \longrightarrow e^- + \nu + \bar{\nu} \qquad \text{[30.6]}$$

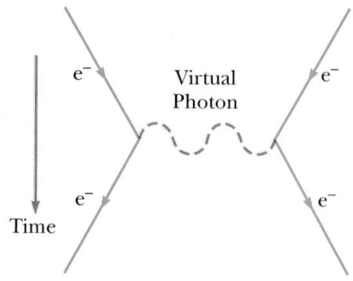

Figure 30.7 Feynman diagram representing a photon mediating the electromagnetic force between two electrons. The *t*-arrow shows the direction of increasing time.

The interaction between two particles can be understood in a simple diagram called a *Feynman diagram,* developed by Richard P. Feynman (1918–1988). Figure 30.7 is a Feynman diagram for the electromagnetic interaction between two electrons. In this simple case, a photon is the field particle that mediates the electromagnetic force between the electrons. The photon transfers energy and momentum from one electron to the other in this interaction. Such a photon, called a *virtual photon,* can never be detected directly because it is absorbed by the second electron very shortly after being emitted by the first electron. The existence of a virtual photon would violate the law of conservation of energy, but because of the uncertainty principle and its very short lifetime, Δt, the photon's excess energy is less than the uncertainty in its energy, given by $\Delta E \approx \hbar/\Delta t$.

Now consider the pion exchange between a proton and a neutron via the strong force (Fig. 30.8). One can reason that the energy, ΔE, needed to create a pion of mass m_π is given by $\Delta E = m_\pi c^2$. Again, the existence of the pion is allowed in spite of conservation of energy if this energy is surrendered in a short enough time, Δt, the time it takes the pion to transfer from one nucleon to the other. From the uncertainty principle, $\Delta E \, \Delta t \approx \hbar$, we get

$$\Delta t \approx \frac{\hbar}{\Delta E} = \frac{\hbar}{m_\pi c^2} \qquad \text{[30.7]}$$

Because the pion cannot travel faster than the speed of light, the maximum distance, d, it can travel in a time Δt is $c \, \Delta t$. Using Equation 30.7 and $d = c \, \Delta t$, we find this maximum distance to be

$$d \approx \frac{\hbar}{m_\pi c} \qquad \text{[30.8]}$$

The range of the strong force is about 1.5×10^{-15} m. Using this value for d in Equation 30.8, the rest energy of the pion is calculated to be

$$m_\pi c^2 \approx \frac{\hbar c}{d} = \frac{(1.05 \times 10^{-34} \, \text{J} \cdot \text{s})(3.00 \times 10^8 \, \text{m/s})}{1.5 \times 10^{-15} \, \text{m}}$$
$$= 2.1 \times 10^{-11} \, \text{J} \cong 130 \, \text{MeV}$$

This corresponds to a mass of 130 MeV/c^2 (about 250 times the mass of the electron), which is in good agreement with the observed mass of the pion.

The concept we have just described is quite revolutionary. In effect, it says that a proton can change into a proton plus a pion, as long as it returns to its original state in a very short time. High-energy physicists often say that a nucleon undergoes "fluctuations" as it emits and absorbs pions. As we have seen, these fluctuations are a consequence of a combination of quantum mechanics (through the uncertainty principle) and special relativity (through Einstein's energy-mass relation $E = mc^2$).

This section has dealt with the particles that mediate the strong force, namely the pions, and the mediators of the electromagnetic force, the photons. The graviton, which is the mediator of the gravitational force, has yet to be observed. The W and Z particles that mediate the weak force were discovered in 1983 by Carlo Rubbia and his associates at CERN.

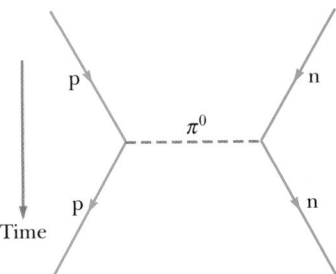

Figure 30.8 Feynman diagram representing a proton interacting with a neutron via the strong force. In this case, the pion mediates the nuclear force. The *t*-arrow shows the direction of increasing time.

Richard Feynman (1918–1988) with his son, Carl, in 1965

Feynman, together with Julian S. Schwinger and Shinichiro Tomonaga, won the 1965 Nobel prize for physics for fundamental work in the principles of quantum electrodynamics. His many important contributions to physics include the invention of simple diagrams to represent particle interactions graphically, the theory of the weak interaction of subatomic particles, a reformulation of quantum mechanics, the theory of superfluid helium, and his contribution to physics education through the magnificent three-volume text *The Feynman Lectures on Physics*. *(UPI Telephotos)*

30.8 CLASSIFICATION OF PARTICLES

Hadrons

All particles other than photons can be classified into two broad categories, hadrons and leptons, according to their interactions. Particles that interact through the strong force are called *hadrons*. There are two classes of hadrons, known as *mesons* and *baryons*, distinguished by their masses and spins. All mesons are known to decay finally into electrons, positrons, neutrinos, and photons. The pion is the lightest of known mesons, with a mass of about 140 MeV/c^2 and a spin of 0. Another is the K meson, with a mass of about 500 MeV/c^2 and spin 0.

Baryons have masses equal to or greater than the proton mass (the name *baryon* means "heavy" in Greek), and their spin is always a noninteger value ($\frac{1}{2}$ or $\frac{3}{2}$). Protons and neutrons are baryons, as are many other particles. With the exception of the proton, all baryons decay in such a way that the end products include a proton. For example, the baryon called the Ξ hyperon first decays to a Λ^0 in about 10^{-10} s. The Λ^0 then decays to a proton and a π^- in about 3×10^{-10} s.

Today it is believed that hadrons are composed of quarks. (Later we shall have more to say about the quark model.) Some of the important properties of hadrons are listed in Table 30.2.

Leptons

Leptons (from the Greek *leptos* meaning "small" or "light") are a group of particles that participate in the weak interaction. All leptons have a spin of $\frac{1}{2}$. Included in this group are electrons, muons, and neutrinos, which are less massive than the lightest hadron. Although hadrons have size and structure, leptons appear to be truly elementary, with no structure (that is, point-like).

Quite unlike hadrons, the number of known leptons is small. Currently, scientists believe there only are six leptons (each having an antiparticle) — the elec-

TABLE 30.2   Some Particles and Their Properties

Category	Particle Name	Symbol	Anti-particle	Mass (MeV/c^2)	B	L_e	L_μ	L_τ	S	Lifetime(s)	Principal Decay Modes[a]
Leptons	Electron	e^-	e^+	0.511	0	+1	0	0	0	Stable	
	Electron–Neutrino	ν_e	$\bar{\nu}_e$	$<7\ eV/c^2$	0	+1	0	0	0	Stable	
	Muon	μ^-	μ^+	105.7	0	0	+1	0	0	2.20×10^{-6}	$e^- \bar{\nu}_e \nu_\mu$
	Muon–Neutrino	ν_μ	$\bar{\nu}_\mu$	<0.3	0	0	+1	0	0	Stable	
	Tau	τ^-	τ^+	1784	0	0	0	+1	0	$<4 \times 10^{-13}$	$\mu^- \bar{\nu}_\mu \nu_\tau, e^- \bar{\nu}_e \nu_\tau$
	Tau–Neutrino	ν_τ	$\bar{\nu}_\tau$	<30	0	0	0	+1	0	Stable	
Hadrons											
Mesons	Pion	π^+	π^-	139.6	0	0	0	0	0	2.60×10^{-8}	$\mu^+ \nu_\mu$
		π^0	Self	135.0	0	0	0	0	0	0.83×10^{-16}	2γ
	Kaon	K^+	K^-	493.7	0	0	0	0	+1	1.24×10^{-8}	$\mu^+ \nu_\mu, \pi^+ \pi^0$
		K^0_S	$\bar{K}^0_S$	497.7	0	0	0	0	+1	0.89×10^{-10}	$\pi^+ \pi^-, 2\pi^0$
		K^0_L	$\bar{K}^0_L$	497.7	0	0	0	0	+1	5.2×10^{-8}	$\pi^\pm e^\mp \bar{\nu}_e, 3\pi^0$ $\pi^\pm \mu^\mp \bar{\nu}_\mu$
	Eta	η	Self	548.8	0	0	0	0	0	$<10^{-18}$	$2\gamma, 3\pi$
		η'	Self	958	0	0	0	0	0	2.2×10^{-21}	$\eta\pi^+ \pi^-$
Baryons	Proton	p	$\bar{p}$	938.3	+1	0	0	0	0	Stable	
	Neutron	n	$\bar{n}$	939.6	+1	0	0	0	0	920	$pe^- \bar{\nu}_e$
	Lambda	Λ^0	$\bar{\Lambda}^0$	1115.6	+1	0	0	0	-1	2.6×10^{-10}	$p\pi^-, n\pi^0$
	Sigma	Σ^+	$\bar{\Sigma}^-$	1189.4	+1	0	0	0	-1	0.80×10^{-10}	$p\pi^0, n\pi^+$
		Σ^0	$\bar{\Sigma}^0$	1192.5	+1	0	0	0	-1	6×10^{-20}	$\Lambda^0 \gamma$
		Σ^-	$\bar{\Sigma}^+$	1197.3	+1	0	0	0	-1	1.5×10^{-10}	$n\pi^-$
	Xi	Ξ^0	$\bar{\Xi}^0$	1315	+1	0	0	0	-2	2.9×10^{-10}	$\Lambda^0 \pi^0$
		Ξ^-	$\bar{\Xi}^+$	1321	+1	0	0	0	-2	1.64×10^{-10}	$\Lambda^0 \pi^-$
	Omega	Ω^-	Ω^+	1672	+1	0	0	0	-3	0.82×10^{-10}	$\Xi^0 \pi^-, \Lambda^0 K^-$

[a] Notations in this column such as $p\pi^-$, $n\pi^0$ mean two possible decay modes. In this case, the two possible decays are $\Lambda^0 \rightarrow p + \pi^-$ and $\Lambda^0 \rightarrow n + \pi^0$.

tron, the muon, the tau, and a neutrino associated with each:

$$\begin{pmatrix} e^- \\ \nu_e \end{pmatrix} \quad \begin{pmatrix} \mu^- \\ \nu_\mu \end{pmatrix} \quad \begin{pmatrix} \tau^- \\ \nu_\tau \end{pmatrix}$$

The tau lepton, discovered in 1975, has a mass about twice that of the proton.

Although neutrinos are thought to be massless, there is a possibility that they have some small nonzero mass. As we shall see later, a firm knowledge of the neutrino's mass could have great significance in cosmological models and the future of the Universe.

30.9 CONSERVATION LAWS

A number of conservation laws are important in the study of elementary particles. Although the two described here have no theoretical foundation, they are supported by abundant empirical evidence.

Baryon Number

The law of conservation of baryon number tells us that whenever a baryon is created in a reaction or decay, an antibaryon is also created. This can be quantified by assigning a baryon number: $B = +1$ for all baryons, $B = -1$ for all antibaryons, and $B = 0$ for all other particles. Thus, the **law of conservation of baryon number** states that whenever a nuclear reaction or decay occurs, the sum of the baryon numbers before the process must equal the sum of the baryon numbers after the process.

◀ Conservation of baryon number

Note that if baryon number is absolutely conserved, the proton must be absolutely stable. If it were not for the law of conservation of baryon number, the proton could decay into a positron and a neutral pion. However, such a decay has never been observed. At present, we can only say that the proton has a half-life of at least 10^{31} years (the estimated age of the Universe is about 10^{10} years). In one recent version of a so-called grand unified theory (GUT), physicists have predicted that the proton is actually unstable. According to this theory, the baryon number (sometimes called the *baryonic charge*) is not absolutely conserved, whereas electric charge is always conserved.

EXAMPLE 30.4 Checking Baryon Numbers

Determine whether or not each of the following reactions can occur based on the law of conservation of baryon number.

$$(1) \quad p + n \longrightarrow p + p + n + \bar{p}$$

$$(2) \quad p + n \longrightarrow p + p + \bar{p}$$

Solution Recall that $B = +1$ for baryons and $B = -1$ for antibaryons. Hence the left side of (1) gives a total baryon number of $1 + 1 = 2$. The right side gives a total baryon number of $1 + 1 + 1 + (-1) = 2$. Thus the reaction can occur provided the incoming proton has sufficient energy.

The left side of (2) gives a total baryon number of $1 + 1 = 2$. However, the right side gives $1 + 1 + (-1) = 1$. Because baryon number is not conserved, the reaction cannot occur.

Lepton Number

Conservation of lepton number ▷

There are three conservation laws involving lepton numbers, one for each variety of lepton. The **law of conservation of electron-lepton number** states that the sum of the electron-lepton numbers before a reaction or decay must equal the sum of the electron-lepton numbers after the reaction or decay. The electron and the electron neutrino are assigned a positive electron-lepton number, $L_e = +1$; the antileptons e^+ and $\bar{\nu}_e$ are assigned the electron-lepton number $L_e = -1$; and all other particles have $L_e = 0$. For example, consider the decay of the neutron

Neutron decay ▷

$$n \longrightarrow p^+ + e^- + \bar{\nu}_e$$

Before the decay, the electron-lepton number is $L_e = 0$; after the decay it is $0 + 1 + (-1) = 0$. Thus, the electron-lepton number is conserved. It is important to recognize that the baryon number must also be conserved. This can easily be seen by noting that before the decay $B = +1$, whereas after the decay $B = +1 + 0 + 0 = +1$.

Similarly, when a decay involves muons, the muon-lepton number, L_μ, is conserved. The μ^- and the ν_μ are assigned $L_\mu = +1$, the antimuons μ^+ and $\bar{\nu}_\mu$ are assigned $L_\mu = -1$, and all other particles have $L_\mu = 0$. Finally, the tau-lepton number, L_τ, is conserved, and similar assignments can be made for the τ lepton and its neutrino.

EXAMPLE 30.5   Checking Lepton Numbers

Determine which of the following decay schemes can occur on the basis of conservation of lepton number.

$$(1) \qquad \mu^- \longrightarrow e^- + \bar{\nu}_e + \nu_\mu$$

$$(2) \qquad \pi^+ \longrightarrow \mu^+ + \nu_\mu + \nu_e$$

Solution   Because decay 1 involves both a muon and an electron, L_μ and L_e must both be conserved. Before the decay, $L_\mu = +1$ and $L_e = 0$. After the decay, $L_\mu = 0 + 0 + 1 = +1$, and $L_e = +1 - 1 + 0 = 0$. Thus, both numbers are conserved, and on this basis the decay mode is possible.

Before decay 2 occurs, $L_\mu = 0$ and $L_e = 0$. After the decay, $L_\mu = -1 + 1 + 0 = 0$, but $L_e = +1$. Thus, the decay is not possible because the electron-lepton number is not conserved.

Exercise   Determine whether the decay $\mu^- \rightarrow e^- + \bar{\nu}_e$ can occur.

Answer   No. The muon-lepton number is $+1$ before the decay and 0 after.

A student claims to have observed a decay of an electron into two neutrinos, traveling in opposite directions. What conservation laws would be violated by this decay?

Explanation   Several conservation laws are violated. Conservation of electric charge is violated because the negative charge of the electron has disappeared. Conservation of electron lepton number is also violated, because there is one lepton before the decay and two afterward. If both neutrinos were electron-neutrinos, electron lepton number conservation would be violated in the final state. However, if one of the product neutrinos were other than an electron-neutrino, then another lepton conservation law would be violated, because there were no other leptons in the initial state. Other conservation laws are obeyed by this decay. Energy can be conserved—the rest energy of the electron appears as the kinetic energy (and possibly some small rest energy) of the neutrinos. The opposite directions of the velocities of the two neutrinos allows for conservation of momentum. Conservation of baryon number and conservation of other lepton numbers are also upheld in this decay.

30.10   STRANGE PARTICLES AND STRANGENESS

Many particles discovered in the 1950s were produced by the nuclear interaction of pions with protons and neutrons in the atmosphere. A group of these particles, namely the K, Λ, and Σ, were found to exhibit unusual properties in their production and decay, and hence were called *strange particles.*

One unusual property of strange particles is that they are always produced in pairs. For example, when a pion collides with a proton, two neutral strange particles are produced with high probability (Fig. 30.9) following the reaction

$$\pi^- + p \longrightarrow K^0 + \Lambda^0$$

On the other hand, the reaction $\pi^- + p \rightarrow K^0 + n$ never occurred, even though no known conservation laws were violated and the energy of the pion was sufficient to initiate the reaction.

The second peculiar feature of strange particles is that although they are produced by the strong interaction at a high rate, they do not decay into particles that interact via the strong force at a very high rate. Instead, they decay very slowly, which is characteristic of the weak interaction. Their half-lives are in the range 10^{-10} to 10^{-8} s; most other particles that interact via the strong force have lifetimes on the order of 10^{-23} s.

To explain these unusual properties of strange particles, a law called *conservation of strangeness* was introduced, together with a new quantum number, S, called **strangeness.** The strangeness numbers for some particles are given in Table 30.2. The production of strange particles in pairs is explained by assigning $S = +1$ to one of the particles and $S = -1$ to the other. All nonstrange particles are assigned strangeness $S = 0$. The **law of conservation of strangeness** states that whenever a nuclear reaction or decay occurs, the sum of the strangeness numbers before the process must equal the sum of the strangeness numbers after the process.

◀ Conservation of strangeness number

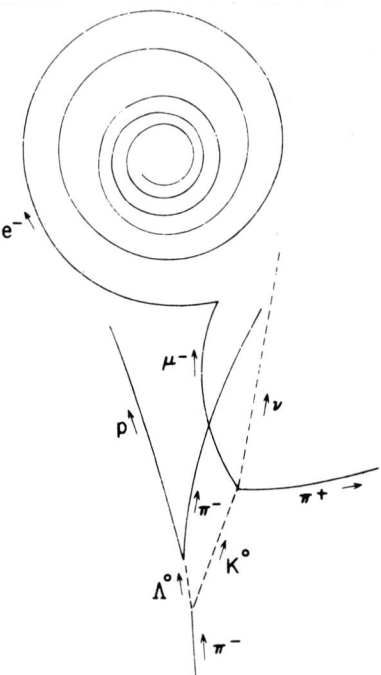

Figure 30.9   This drawing represents tracks of many events obtained by analyzing a bubble-chamber photograph. The strange particles Λ^0 and K^0 are formed (at the bottom) as the π^- interacts with a proton according to $\pi^- + p \rightarrow \Lambda^0 + K^0$. (Note that the neutral particles leave no tracks, as indicated by the dashed lines.) The Λ^0 and K^0 then decay according to $\Lambda^0 \rightarrow \pi^- + p$ and $K^0 \rightarrow \pi^+ + \mu^- + \bar{\nu}_\mu$. *(Courtesy Lawrence Berkeley Laboratory, University of California)*

One can explain the slow decay of strange particles by assuming that the strong and electromagnetic interactions obey the law of conservation of strangeness, whereas the weak interaction does not. Because the decay reaction involves the loss of one strange particle, it violates strangeness conservation and hence proceeds slowly via the weak interaction.

EXAMPLE 30.6   Is Strangeness Conserved?

(a) Determine whether the following reaction occurs on the basis of conservation of strangeness.

$$\pi^0 + n \longrightarrow K^+ + \Sigma^-$$

Solution   The initial state has a total strangeness of $S = 0 + 0 = 0$. Because the strangeness of the K^+ is $S = +1$, and the strangeness of the Σ^- is $S = -1$, the total strangeness of the final state is $+1 - 1 = 0$. Thus, strangeness is conserved and the reaction is allowed.

(b) Show that the following reaction does not conserve strangeness.

$$\pi^- + p \longrightarrow \pi^- + \Sigma^+$$

Solution   The initial state has strangeness $S = 0 + 0 = 0$, whereas the final state has strangeness $S = 0 + (-1) = -1$. Thus strangeness is not conserved.

Exercise   Show that the observed reaction $p + \pi^- \rightarrow K^0 + \Lambda^0$ obeys the law of conservation of strangeness.

30.11 THE EIGHTFOLD WAY

As we have seen, quantities such as spin, baryon number, lepton number, and strangeness are labels we associate with particles. Many classification schemes that group particles into families based on such labels have been proposed. First, consider the first eight baryons listed in Table 30.2, all having a spin of one half. The family consists of the proton, the neutron, and six other particles. If we plot their strangeness versus their charge using a sloping coordinate system, as in Figure 30.10a, a fascinating pattern emerges. Six of the baryons form a hexagon, and the remaining two are at the hexagon's center. (Particles with spin quantum number $\frac{1}{2}$ or $\frac{3}{2}$ are called fermions.)

Now consider the family of mesons listed in Table 30.2 with spins of zero. (Particles with spin quantum number 0 or 1 are called bosons.) If we count both particles and antiparticles, there are nine such mesons. Figure 30.10b is a plot of strangeness versus charge for this family. Again, a fascinating hexagonal pattern emerges. In this case, the particles on the perimeter of the hexagon lie opposite their antiparticles, and the remaining three (which form their own antiparticles) are at its center. These and related symmetric patterns, called the **eightfold way,** were proposed independently in 1961 by Murray Gell-Mann and Yuval Ne'eman.

The groups of baryons and mesons can be displayed in many other symmetric patterns within the framework of the eightfold way. For example, the family of spin-$\frac{3}{2}$ baryons contains ten particles arranged in a pattern such as the tenpins in a bowling alley. After the pattern was proposed, one of the particles was missing — it had yet to be discovered. Gell-Mann predicted that the missing particle, which he called the *omega minus* (Ω^-), should have a spin of $\frac{3}{2}$, a charge of -1, a strangeness of -3, and a mass of about 1680 MeV/c^2. Shortly thereafter, in 1964, scientists at the

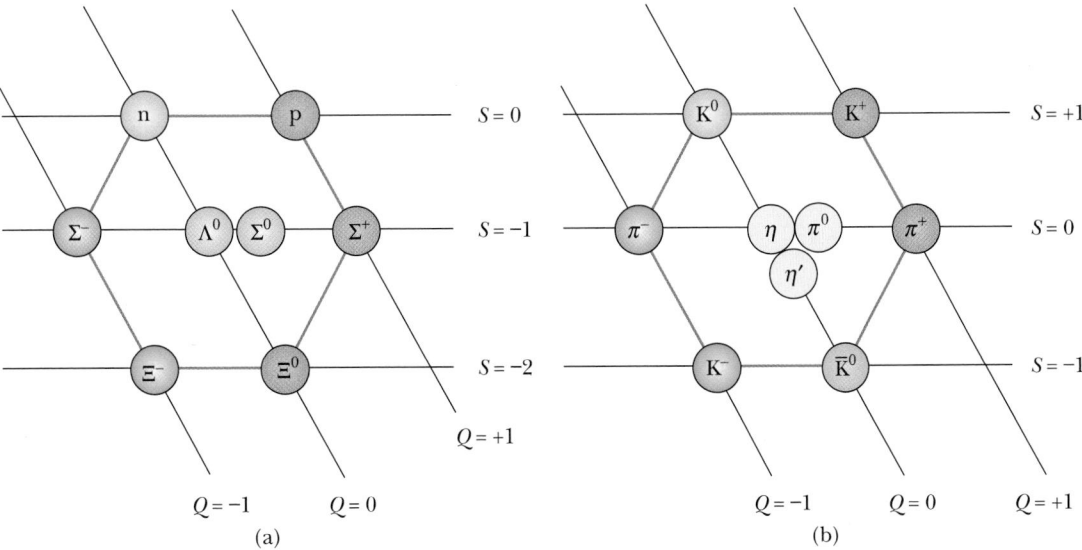

Figure 30.10 (a) The hexagonal eightfold-way pattern for the eight spin-$\frac{1}{2}$ baryons. This strangeness versus charge plot uses a horizontal axis for the strangeness values, S, but a sloping axis for the charge number Q. (b) The eightfold-way pattern for the nine spin-zero mesons.

Murray Gell-Mann,
American physicist (1929–   )

Gell-Mann was awarded the Nobel prize in 1969 for his theoretical studies dealing with subatomic particles. *(Photo courtesy of Michael R. Dressler)*

Brookhaven National Laboratory found the missing particle through careful analyses of bubble-chamber photographs, and confirmed all of its predicted properties.

The patterns of the eightfold way in the field of particle physics have much in common with the periodic table. Whenever a vacancy (a missing particle or element) occurs in the organized patterns, experimentalists have a guide for their investigations.

30.12   QUARKS

As we have noted, leptons appear to be truly elementary particles because they have no measurable size or internal structure, are limited in number, and do not seem to break down into smaller units. Hadrons, on the other hand, are complex particles with size and structure. Furthermore, we know that hadrons decay into other hadrons and are many in number. Table 30.2 lists only those hadrons that are stable against hadronic decay; hundreds of others have been discovered. These facts strongly suggest that hadrons cannot be truly elementary but have some substructure.

The Original Quark Model

In 1963 Gell-Mann and George Zweig independently proposed that hadrons have a more elementary substructure. According to their model, all hadrons are composite systems of two or three fundamental constituents called **quarks.** Gell-Mann

TABLE 30.3   Properties of Quarks and Antiquarks

				Quarks				
Name	Symbol	Spin	Charge	Baryon Number	Strangeness	Charm	Bottomness	Topness
Up	u	$\frac{1}{2}$	$+\frac{2}{3}e$	$\frac{1}{3}$	0	0	0	0
Down	d	$\frac{1}{2}$	$-\frac{1}{3}e$	$\frac{1}{3}$	0	0	0	0
Strange	s	$\frac{1}{2}$	$-\frac{1}{3}e$	$\frac{1}{3}$	-1	0	0	0
Charmed	c	$\frac{1}{2}$	$+\frac{2}{3}e$	$\frac{1}{3}$	0	$+1$	0	0
Bottom	b	$\frac{1}{2}$	$-\frac{1}{3}e$	$\frac{1}{3}$	0	0	$+1$	0
Top	t	$\frac{1}{2}$	$-\frac{2}{3}e$	$\frac{1}{3}$	0	0	0	$+1$

				Antiquarks				
Name	Symbol	Spin	Charge	Baryon Number	Strangeness	Charm	Bottomness	Topness
Anti-up	$\overline{u}$	$\frac{1}{2}$	$-\frac{2}{3}e$	$-\frac{1}{3}$	0	0	0	0
Anti-down	$\overline{d}$	$\frac{1}{2}$	$+\frac{1}{3}e$	$-\frac{1}{3}$	0	0	0	0
Anti-strange	$\overline{s}$	$\frac{1}{2}$	$+\frac{1}{3}e$	$-\frac{1}{3}$	$+1$	0	0	0
Anti-charmed	$\overline{c}$	$\frac{1}{2}$	$+\frac{2}{3}e$	$-\frac{1}{3}$	0	-1	0	0
Anti-bottom	$\overline{b}$	$\frac{1}{2}$	$+\frac{1}{3}e$	$-\frac{1}{3}$	0	0	-1	0
Anti-top	$\overline{t}$	$\frac{1}{2}$	$-\frac{2}{3}e$	$-\frac{1}{3}$	0	0	0	-1

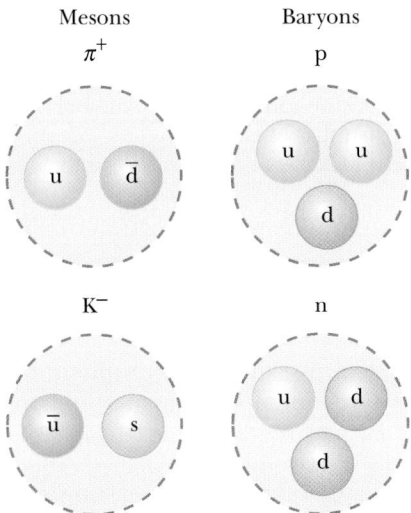

Mesons   Baryons

π^+ p

K^- n

Figure 30.11 Quark compositions of two mesons and two baryons. Note that the mesons on the left contain two quarks, and the baryons on the right contain three quarks.

borrowed the word *quark* from the passage "Three quarks for Muster Mark" in James Joyce's book *Finnegan's Wake*. In the original model, there were three types of quarks designated by the symbols u, d, and s. These were given the arbitrary names *up*, *down*, and *sideways* (or, now more commonly, *strange*).

An unusual property of quarks is that they have fractional electronic charges, as shown — along with other properties — in Table 30.3. Associated with each quark is an antiquark of opposite charge, baryon number, and strangeness. The compositions of all hadrons known when Gell-Mann and Zweig presented their models could be completely specified by three simple rules:

1. Mesons consist of one quark and one antiquark, giving them a baryon number of 0, as required.
2. Baryons consist of three quarks.
3. Antibaryons consist of three antiquarks.

Table 30.4 lists the quark compositions of several mesons and baryons. Note that just two of the quarks, u and d, are contained in all hadrons encountered in ordinary matter (protons and neutrons). The third quark, s, is needed only to construct strange particles with a strangeness of either $+1$ or -1. Figure 30.11 is a pictorial representation of the quark compositions of several particles.

TABLE 30.4
Quark Composition of Several Hadrons

Particle	Quark Composition
Mesons	
π^+	$u\bar{d}$
π^-	$\bar{u}d$
K^+	$u\bar{s}$
K^-	$\bar{u}s$
K^0	$d\bar{s}$
Baryons	
p	uud
n	udd
Λ^0	uds
Σ^+	uus
Σ^0	uds
Σ^-	dds
Ξ^0	uss
Ξ^-	dss
Ω^-	sss

Thinking Physics 4

We have seen a law of conservation of lepton number and a law of conservation of baryon number. Why isn't there a law of conservation of meson number?

Explanation We can argue this from the point of view of creating particle-antiparticle pairs from available energy. If energy is converted to rest energy of a lepton-antilepton pair, then there is no net change in lepton number, because the lepton has a lepton number of $+1$ and the antilepton -1. Energy could also be transformed into rest energy of a baryon-antibaryon pair. The baryon has baryon number $+1$, the antibaryon -1, and there is no net change in baryon number.

But now suppose energy is transformed into rest energy of a quark-antiquark pair. By definition in quark theory, a quark-antiquark pair is a meson. Thus, we have created a meson from energy—there was no meson before; now there is. Thus, there is no conservation of meson number. With more energy, we can create more mesons, with no restriction from a conservation law other than that of energy.

Charm and Other Recent Developments

Although the original quark model was highly successful in classifying particles into families, there were some discrepancies between predictions of the model and certain experimental decay rates. As a consequence, a fourth quark was proposed by several physicists in 1967. They argued that if there are four leptons (as was thought at the time), then there should also be four quarks because of an underlying symmetry in nature. The fourth quark, designated by c, was given a property called **charm.** A charmed quark would have the charge $+2e/3$, but its charm would distinguish it from the other three quarks. The new quark would have a charm of $C = +1$, its antiquark would have a charm of $C = -1$, and all other quarks would have $C = 0$, as indicated in Table 30.3. Charm, like strangeness, would be conserved in strong and electromagnetic interactions but not in weak interactions.

In 1974 a new heavy meson called the J/Ψ particle (or simply Ψ) was discovered independently by a group led by Burton Richter at the Stanford Linear Accelerator (SLAC) and another group led by Samuel Ting at the Brookhaven National Laboratory. Richter and Ting were awarded the Nobel prize in 1976 for this work. The J/Ψ particle did not fit into the three-quark model, but had the properties of a combination of a charmed quark and its antiquark ($c\bar{c}$). It was much heavier than the other known mesons (~ 3100 MeV/c^2) and its lifetime was much longer than those of other particles that decay via the strong force. In 1975 researchers at Stanford University reported strong evidence for the tau (τ) lepton, with a mass of 1784 MeV/c^2. Such discoveries led to more elaborate quark models and the proposal of two new quarks, named *top* (t) and *bottom* (b). (Some physicists prefer the whimsical names *truth* and *beauty*.) To distinguish these quarks from the old ones, quantum numbers called *topness* and *bottomness* were assigned to these new particles and are included in Table 30.3. In 1977 researchers at the Fermi National Laboratory, under the direction of Leon Lederman, reported the discovery of a very massive new meson, Υ, whose composition is considered to be $b\bar{b}$. In March of 1995, researchers at Fermilab announced the discovery of the top quark (supposedly the last of the quarks to be found) having mass 173 GeV/c^2.

You are probably wondering whether or not such discoveries will ever end. How many "building blocks" of matter really exist? At the present, physicists believe that the fundamental particles in nature include six quarks and six leptons (together with their antiparticles). Some of the properties of these particles are given in Table 30.5.

Despite many extensive experimental efforts, no isolated quark has ever been observed. Physicists now believe that quarks are permanently confined inside ordinary particles because of an exceptionally strong force that prevents them from escaping. This force, called the *color* force (which will be discussed in Section 30.13), increases with separation distance (similar to the force of a spring). The great strength of the force between quarks has been described by one author as follows:

TABLE 30.5
The Fundamental Particles and Some of Their Properties

Particle	Rest Energy	Charge
Quarks		
u	360 MeV	$+\frac{2}{3}e$
d	360 MeV	$-\frac{1}{3}e$
c	1500 MeV	$+\frac{2}{3}e$
s	540 MeV	$-\frac{1}{3}e$
t	173 GeV	$+\frac{2}{3}e$
b	5 GeV	$-\frac{1}{3}e$
Leptons		
e^-	511 keV	$-e$
μ^-	107 MeV	$-e$
τ^-	1784 MeV	$-e$
ν_e	<30 eV	0
ν_μ	<0.5 MeV	0
ν_τ	<250 MeV	0

Quarks are slaves of their own color charge, . . . bound like prisoners of a chain gang. . . . Any locksmith can break the chain between two prisoners, but no locksmith is expert enough to break the gluon chains between quarks. Quarks remain slaves forever.[2]

30.13 COLORED QUARKS

Shortly after the concept of quarks was proposed, scientists recognized that certain particles had quark compositions that were in violation of the Pauli exclusion principle. Because all quarks have spins of $\frac{1}{2}$, they are expected to follow the exclusion principle. One example of a particle that violates the exclusion principle is the Ω^- (sss) baryon that contains three s quarks having parallel spins, giving it a total spin of $\frac{3}{2}$. Other examples of baryons that have identical quarks with parallel spins are the Δ^{++} (uuu) and the Δ^- (ddd). To resolve this problem, Moo-Young Han and Yoichiro Nambu suggested in 1965 that quarks possess a new property called **color.** This property is similar in many respects to electric charge except that it occurs in three varieties labeled *red, green,* and *blue.* (The antiquarks are labeled *antired, antigreen,* and *antiblue.*) To satisfy the exclusion principle, all three quarks in a baryon must have different colors. Just as a combination of actual colors of light can produce the neutral color white, a combination of three quarks with different colors is also white, or colorless. A meson consists of a quark of one color and an antiquark of the corresponding anticolor. The result is that baryons and mesons are always colorless (or white).

Although the concept of color in the quark model was originally conceived to satisfy the exclusion principle, it also provided a better theory for explaining certain experimental results. For example, the modified theory correctly predicts the lifetime of the π^0 meson. The theory of how quarks interact with each other is called **quantum chromodynamics,** or QCD, to parallel quantum electrodynamics (the theory of interaction between electric charges). In QCD, the quark is said to carry a **color charge** in analogy to electric charge. The strong force between quarks is often called the **color force.**

As stated earlier, the strong interaction between hadrons is mediated by massless particles called **gluons** (analogous to photons for the electromagnetic force). According to QCD, there are eight gluons, all with color charge. When a quark emits or absorbs a gluon, its color changes. For example, a blue quark that emits a gluon may become a red quark, and the red quark that absorbs this gluon becomes a blue quark. The color force between quarks is analogous to the electric force between charges; like colors repel and opposite colors attract. Therefore, two red quarks repel each other, but a red quark will be attracted to an antired quark. The attraction between quarks of opposite color to form a meson ($q\bar{q}$) is indicated in Figure 30.12a. Differently colored quarks also attract each other, but with less intensity than opposite colors of quark and antiquark. For example, a cluster of red, blue, and green quarks all attract each other to form baryons as indicated in Figure 30.12b. Thus, every baryon contains three quarks of three different colors.

[2] H. Fritzsch, *Quarks, The Stuff of Matter,* London, Allen Lane, 1983.

An artist's version of a high-energy particle colliding with a nucleus. The quark structure of the nucleus is indicated by the small colored spheres inside the nucleus. *(Courtesy of Janie Martz/CEBAF)*

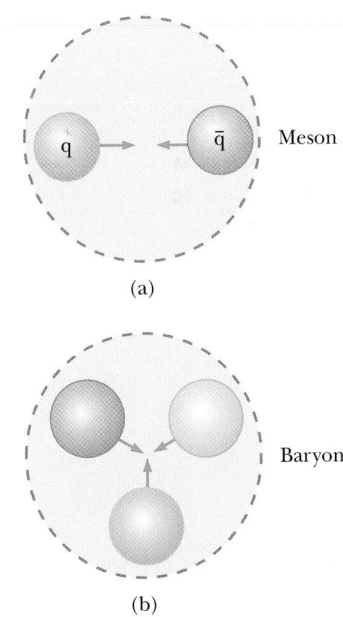

Figure 30.12 (a) A red quark is attracted to an antired quark. This forms a meson whose quark structure is ($q\bar{q}$). (b) Three different colored quarks attract each other to form a baryon.

Figure 30.13 (a) A nuclear interaction between a proton and a neutron explained in terms of Yukawa's pion exchange model. (b) The same interaction as in (a), explained in terms of quarks and gluons. Note that the exchanged ūd quark pair makes up a π^- meson.

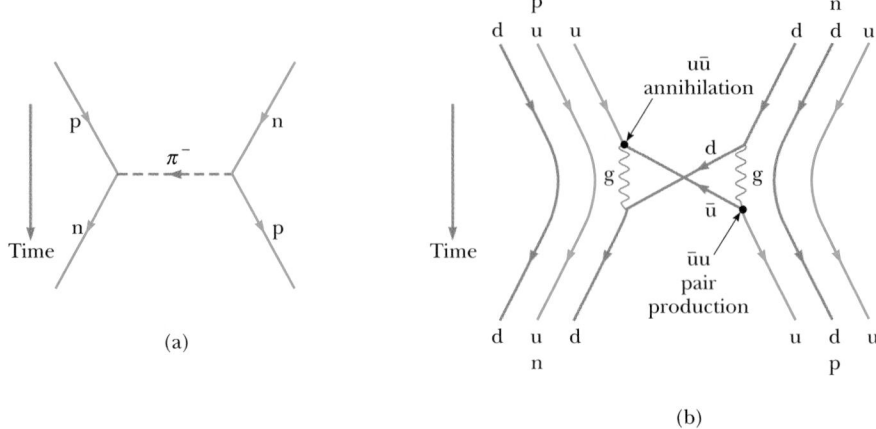

(a)

(b)

Although the color force between two color-neutral hadrons is negligible at large separations, the strong color force between their constituent quarks does not exactly cancel at small separations. This residual strong force is in fact the nuclear force that binds protons and neutrons to form nuclei. According to QCD, a more basic explanation of nuclear force can be given in terms of quarks and gluons, as shown in Figure 30.13, which shows contrasting Feynman diagrams of the same process. Each quark within the neutron and proton is continually emitting and absorbing virtual gluons and creating and annihilating virtual $(q\bar{q})$ pairs. When the neutron and proton approach within 1 fm of each other, these virtual gluons and quarks can be exchanged between the two nucleons, and such exchanges produce the nuclear force. Figure 30.13b depicts one likely possibility or contribution to the process shown in Figure 30.13a. A down quark emits a virtual gluon (represented by a wavy line g), which creates a u$\bar{\text{u}}$ pair. Both the recoiling d quark and the $\bar{\text{u}}$ are transmitted to the proton where the $\bar{\text{u}}$ annihilates a proton u quark (with the creation of a gluon) and the d is captured.

30.14 ELECTROWEAK THEORY AND THE STANDARD MODEL

Recall that the weak interaction is an extremely short range force having an interaction distance of approximately 10^{-18} m (Table 30.1). Such a short-range interaction implies that the quantized particles that carry the weak field (the spin one W^+, W^-, and Z^0 bosons) are quite massive as is indeed the case. These amazing bosons can be thought of as structureless, point-like particles as massive as krypton atoms. The weak interaction is, as mentioned earlier, responsible for neutron decay and the beta decay of other heavier baryons. More important, however, the weak interaction is responsible for the decay of the c, s, b, and t quarks into lighter, more stable u and d quarks as well as the decay of the massive μ and τ leptons into (lighter) electrons. Thus, **the weak interaction is very important because it governs the stability of the basic matter particles.**

A mysterious feature of the weak interaction is its lack of symmetry, especially compared to the high degree of symmetry shown by the strong, electromagnetic, and gravitational interactions. For example, the weak interaction, unlike the strong interaction, is not symmetric under mirror reflection or charge exchange. (*Mirror reflection* means that all the quantities in a given particle reaction are exchanged as in a mirror reflection—left for right, an inward motion toward the mirror for an outward motion. *Charge exchange* means that all the electric charges in a particle reaction are converted to their opposites—all positives to negatives and vice versa.) When we say that the weak interaction is not symmetric, we mean that the reaction with all quantities changed occurs less frequently than the direct reaction. For example, the decay of the K^0, which is governed by the weak interaction, is not symmetric under charge exchange because $K^0 \rightarrow \pi^- + e^+ + \nu_e$ occurs much more frequently than $K^0 \rightarrow \pi^+ + e^- + \bar{\nu}_e$.

In 1979, Sheldon Glashow, Abdus Salam, and Steven Weinberg won a Nobel prize for developing a theory that unified the electromagnetic and weak interactions. This **electroweak theory** postulates that the weak and electromagnetic interactions have the same strength at very high particle energies. Thus, the two interactions are viewed as two different manifestations of a single unifying electroweak interaction. The photon and the three massive bosons ($W^\pm$ and Z^0) play a key role in the electroweak theory. The theory makes many concrete predictions, but perhaps the most spectacular is the prediction of the masses of the W and Z particles at about 82 GeV/c^2 and 93 GeV/c^2, respectively. A 1984 Nobel prize was awarded to Carlo Rubbia and Simon van der Meer for their work leading to the discovery of these particles at just these energies at the CERN Laboratory in Geneva, Switzerland.

The combination of the electroweak theory and QCD for the strong interaction form what is referred to in high-energy physics as the **Standard Model.** Although the details of the Standard Model are complex, its essential ingredients can be summarized with the help of Figure 30.14. The strong force, mediated by gluons, holds quarks together to form composite particles such as protons, neutrons, and

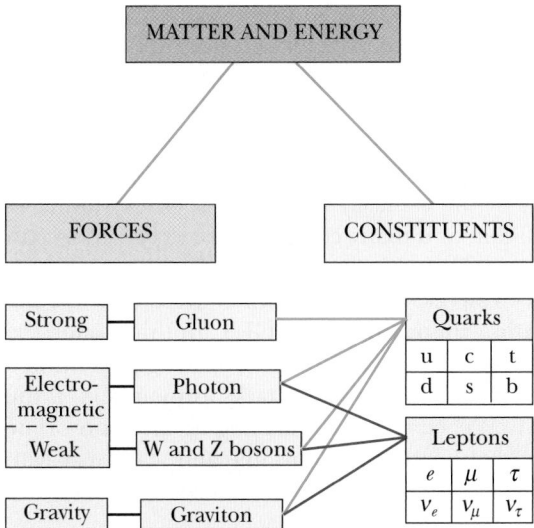

Figure 30.14 The Standard Model of particle physics.

WEB

For Web links related to particle physics, accelerators, colliders, and the invention of the World Wide Web by physicists at CERN, visit the textbook Web site at http://www.harcourtcollege.com/physics/cptech

mesons. Leptons participate only in the electromagnetic and weak interactions. The electromagnetic force is mediated by photons, and the weak force is mediated by W and Z bosons. Note that all fundamental forces are mediated by bosons (particles with spin 1) whose properties are given, to a large extent, by symmetries involved in the theories.

However, the Standard Model does not answer all questions. A major question is why the photon has no mass whereas the W and Z bosons do. Because of this mass difference, the electromagnetic and weak forces are quite distinct at low energies, but become similar in nature at very high energies, where the rest energies of the W and Z bosons are insignificant fractions of their total energies. This behavior as one goes from high to low energies, called **symmetry breaking,** leaves open the question of the origin of particle masses. To resolve this problem, a hypothetical particle called the **Higgs boson,** which provides a mechanism for breaking the electroweak symmetry, has been proposed. The Standard Model, including the Higgs mechanism, provides a logically consistent explanation of the massive nature of the W and Z bosons. Unfortunately, the Higgs boson has not yet been found, but physicists know that its mass should be less than $1 \text{ TeV}/c^2$ ($10^{12} \text{ eV}/c^2$).

In order to determine whether the Higgs boson exists, two quarks of at least 1 TeV of energy must collide, but calculations show that this requires injecting 40 TeV of energy within the volume of a proton. Scientists are convinced that because of the limited energy available in conventional accelerators using fixed targets, it is necessary to build colliding-beam accelerators called **colliders.** The concept of colliders is straightforward. Particles with equal masses and kinetic energies, traveling in opposite directions in an accelerator ring, collide head-on to produce the required reaction and the formation of new particles. Because the total momentum of the interacting particles is zero, all of their kinetic energy is available for the reaction. The Large Electron-Positron collider (LEP) at CERN near Geneva, Switzerland, and the Stanford Linear Collider in California collide both electrons and protons. The Super Proton Synchrotron at CERN accelerates protons and antiprotons to energies of 270 GeV, and the world's highest energy proton accelerator, the Tevatron at the Fermi National Laboratory in Illinois, produces protons at almost 1000 GeV (or 1 TeV). The Superconducting Super Collider (SSC), which was being built in Texas, was an accelerator designed to produce 20-TeV protons in a ring 52 mi in circumference. After much debate in Congress, and an investment of almost $2 billion, the SSC project was canceled by the U.S. Department of Energy in October 1993. CERN recently approved the development of the Large Hadron Collider (LHC), a proton-proton collider that will provide a center of mass energy of 14 TeV and allow an exploration of Higgs-boson physics. The accelerator will be constructed in the same 27-km circumference tunnel as CERN's Large Electron-Positron collider, and many countries are expected to participate in the project.

Following the success of the electroweak theory, scientists attempted to combine it with QCD in a **grand unification theory** known as GUT. In this model, the electroweak force was merged with the strong color force to form a grand unified force. One version of the theory considers leptons and quarks as members of the same family that are able to change into each other by exchanging an appropriate particle. Many GUT theories predict that protons are unstable and will decay with a life time of about 10^{31} years. This is far greater than the age of the Universe, and as yet proton decays have not been observed.

A view from inside the LEP (Large Electron-Positron Collider) tunnel, which is 27 km in circumference. *(Courtesy of CERN)*

Thinking Physics 5

Consider a car making a head-on collision with an identical car moving in the opposite direction at the same speed. Compare that collision to one in which one of the cars collides with the second car at rest. In which collision is there the larger transformation of kinetic energy to other forms during the collision? How does this relate to producing exotic particles in collisions?

Explanation In the head-on collision with both cars moving, conservation of momentum causes most if not all of the kinetic energy to be transformed to other forms. In the collision between a moving car and a stationary car, the cars are still moving after the collision, in the direction of the moving car, but with reduced speed. Thus, only part of the kinetic energy is transformed to other forms. This suggests the advantage of using colliding beams to produce exotic particles, as opposed to firing a beam into a stationary target. When particles moving in opposite directions collide, all of the kinetic energy is available for transformation into other forms—in this case, the creation of new particles. When a beam is fired into a stationary target, only part of the energy is available for transformation, so higher mass particles cannot be created.

30.15 THE COSMIC CONNECTION

As we have discussed, the world around us is dominated by protons, electrons, neutrons, and neutrinos. Other particles can be seen in cosmic rays. However, most of the new particles are produced using large, expensive machines that accelerate protons and electrons to energies in the GeV and TeV range. These energies are enormous when compared to the thermal energy in today's Universe. For example, $k_B T$ at the center of the Sun is only about 1 keV, but the temperature of the early Universe was high enough to reach energies of TeV and higher.

In this section we describe one of the most fascinating theories in all of science—the Big Bang theory of the creation of the Universe—and the experimental evidence that supports it. This theory of cosmology states that the Universe had a beginning, and that this beginning was so cataclysmic that it is impossible to look back beyond it. According to the theory, the Universe erupted from a point-like singularity about 15 to 20 billion years ago. The first few minutes after the Big Bang saw such extremes of energy that all four interactions of physics were unified and all matter melted down into an undifferentiated "quark soup."

The evolution of the four fundamental forces from the Big Bang to the present is shown in Figure 30.15. During the first 10^{-43} s (the ultra-hot epoch where $T \approx 10^{32}$ K), it is presumed that the strong, electroweak, and gravitational forces were joined to form a completely unified force. Between 10^{-43} s and 10^{-32} s following the Big Bang (the hot epoch where $T \approx 10^{29}$ K), gravity broke free of this unification and the strong and electroweak forces remained as one. (They are described by a grand unification theory.) This was a period when particle energies were so great ($> 10^{16}$ GeV) that very massive particles as well as quarks, leptons, and their antiparticles existed. Then the Universe rapidly expanded and cooled during the warm epoch when the temperatures ranged from 10^{29} to 10^{15} K, the strong and electroweak forces parted company, and the grand unification scheme was broken. As the Universe continued to cool, the electroweak force split into the weak force and the electromagnetic force about 10^{-10} s after the Big Bang.

Figure 30.15 A brief history of the Universe from the Big Bang to the present. The four fundamental forces became distinguishable during the first microsecond. Following this, all the quarks combined to form particles that interact via the strong force. The leptons, however, remained separate and exist as individually observable particles to this day.

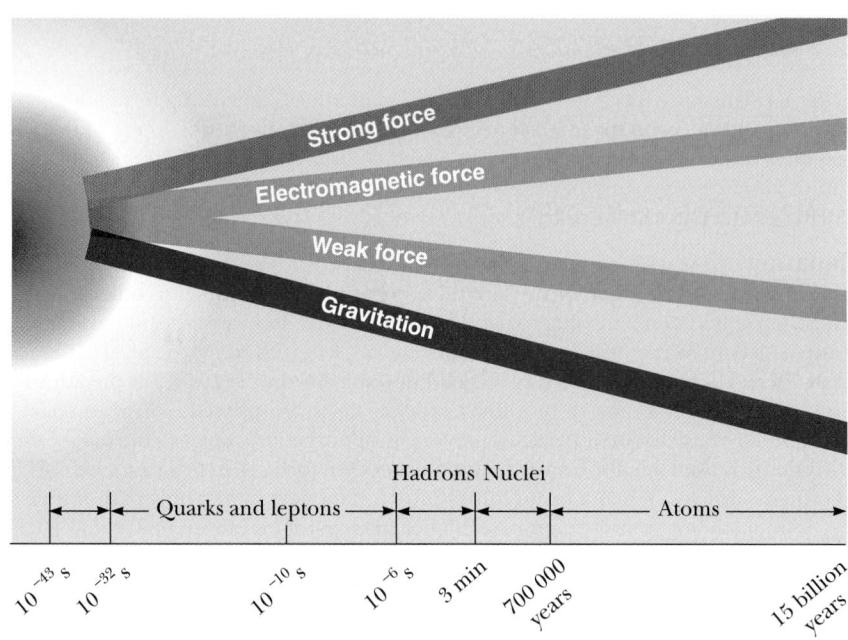

George Gamow (1904–1968)

Gamow and two of his students, Ralph Alpher and Robert Herman, were the first to take the first half-hour of the Universe seriously. In a mostly overlooked paper published in 1948, they made truly remarkable cosmological predictions. They correctly calculated the abundances of hydrogen and helium after the first half-hour (75% H and 25% He) and predicted that radiation from the Big Bang should still be present and have an apparent temperature of about 5 K.

After a few minutes, protons condensed out of the hot soup. For half an hour the Universe underwent thermonuclear detonation, exploding as a hydrogen bomb and producing most of the helium nuclei now present. Nevertheless, it continued to expand so fast that its temperature still dropped. Until about 700 000 years after the Big Bang, the Universe was dominated by radiation; ions absorbed and re-emitted photons, thereby ensuring thermal equilibrium of radiation and matter. Energetic radiation also prevented matter from forming clumps or even neutral hydrogen atoms. By the time the Universe was about 700 000 years old, it had expanded and cooled to about 3000 K, and protons could bind to electrons to form hydrogen atoms. Because neutral atoms do not appreciably scatter photons, the Universe suddenly became transparent to photons. Radiation no longer dominated the Universe, and clumps of neutral matter steadily grew—first atoms, followed by molecules, gas clouds, stars, and finally galaxies.

Observation of Radiation from the Primordial Fireball

In 1965 Arno A. Penzias and Robert W. Wilson of Bell Labs made an amazing discovery while testing a sensitive microwave receiver. A pesky signal producing a faint background hiss was interfering with their satellite communications experiments. In spite of their valiant efforts, the signal remained. Ultimately it became clear that they were observing microwave background radiation (at a wavelength of 7.35 cm) representing the leftover glow from the Big Bang.

The microwave horn that served as their receiving antenna is shown in Figure 30.16. The intensity of the detected signal remained unchanged as the antenna was pointed in different directions. The fact that the radiation has equal strengths in all directions suggested that the entire Universe was the source of this radiation. Evicting a flock of pigeons from the 20-foot horn and cooling the microwave detector both failed to remove the "spurious" signal. Through a casual conversation, Penzias and Wilson discovered that a group at Princeton had predicted the residual radiation from the Big Bang and were planning an experiment to confirm the theory. The excitement in the scientific community was high when Penzias and Wilson announced that they had already observed an excess microwave background compatible with a 3-K blackbody source.

Because the measurements of Penzias and Wilson were taken at a single wavelength, they did not completely confirm the radiation as 3-K blackbody radiation. Subsequent experiments by other groups added intensity data at different wavelengths as shown in Figure 30.17. The results confirm that the radiation is that of a blackbody at 2.9 K. This figure is perhaps the most clear-cut evidence for the Big Bang theory. The 1978 Nobel prize in physics was awarded to Penzias and Wilson for their important discovery.

The discovery of the cosmic background radiation provides strong confirmation of the Big Bang theory of the formation of the Universe and indeed is largely responsible for the wide acceptance of this theory. It also produces another problem, however. Scientists pondered the problem of the uniformity of the radiation. It was held that there would have to be slight fluctuations in this background for objects such as galaxies and quasars to form. In 1989, NASA launched a satellite called COBE (KOH-bee), for Cosmic Background Explorer, to study this radiation in great detail. In 1992, George Smoot of the Lawrence Berkeley Laboratory, based on the data collected, found that the background was not perfectly uniform. Instead, there were irregularities of only 0.0003 K in the background. It is these small temperature variations that provided nucleation sites for the formation of the galaxies and other objects we now see in the sky.

Figure 30.16 Robert W. Wilson (*left*) and Arno A. Penzias (*right*) with Bell Telephone Laboratories' horn-reflector antenna. (*AT&T Bell Laboratories*)

30.16 PROBLEMS AND PERSPECTIVES

Although particle physicists have been exploring the realm of the very small, cosmologists have been exploring cosmic history back to the first microsecond of the Big Bang. Observation of the events that occur when two particles collide in an accelerator is essential in reconstructing the early moments in cosmic history. Perhaps the key to understanding the early Universe is first to understand the world of elementary particles. Cosmologists and particle physicists find that they have many common goals and are joining efforts to attempt to study the physical world at its most fundamental level.

Our understanding of physics at short distances is far from complete. Particle physics is faced with many questions. Why is there so little antimatter in the Universe? Do neutrinos have a small mass, and if so, how do they contribute to the "dark matter" of the Universe? (Measurements on Supernova 1987A established an upper limit of 16 eV for the neutrino mass.) Is it possible to unify the strong and electroweak theories in a logical and consistent manner? Why do quarks and leptons form three similar but distinct families? Are muons the same as electrons (apart from their different masses), or do they have other subtle differences that have not been detected? Why are some particles charged and others neutral? Why do quarks carry a fractional charge? What determines the masses of the fundamental constituents? Can isolated quarks exist? The questions go on and on. Because of the rapid advances and new discoveries in the field of particle physics, by the time you read

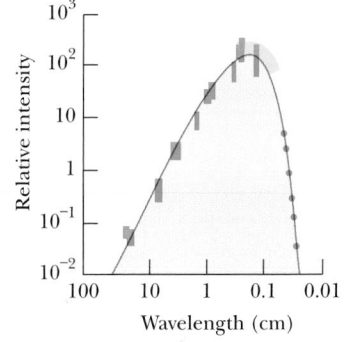

Figure 30.17 Radiation spectrum of the Big Bang. The blue areas are experimental results. The red line is the spectrum calculated for a blackbody at 2.9 K.

this book some of these questions will likely have been resolved and others may have emerged.

An important and obvious question that remains is whether leptons and quarks have a substructure. If they do, one could envision an infinite number of deeper structure levels. However, if leptons and quarks are indeed the ultimate constituents of matter, as physicists today tend to believe, we should be able to construct a final theory of the structure of matter as Einstein dreamed of doing. In the view of many physicists, the end of the road is in sight, but how long it will take to reach that goal is anyone's guess.

SUMMARY

In **nuclear fission** and **nuclear fusion,** the total mass of the products is always less than the original mass of the reactants. Nuclear fission occurs when a heavy nucleus splits, or fissions, into two smaller nuclei. In nuclear fusion, two light nuclei combine to form a heavier nucleus.

A **nuclear reactor** is a system designed to maintain a self-sustaining chain reaction. Nuclear reactors using controlled fission events are currently being used to generate electric power.

Controlled fusion events offer the hope of plentiful supplies of energy in the future. The nuclear fusion reactor is considered by many scientists to be the ultimate energy source because its fuel is water. **Lawson's criterion** states that a fusion reactor will provide a net output power if the product of the plasma ion density, n, and the plasma confinement time, τ, satisfies the following relationships:

$$n\tau \geq 10^{14} \text{ s/cm}^3 \qquad \text{Deuterium-tritium interaction}$$
$$n\tau \geq 10^{16} \text{ s/cm}^3 \qquad \text{Deuterium-deuterium interaction}$$

[30.5]

There are four fundamental forces in nature: **strong** (hadronic), **electromagnetic, weak,** and **gravitational.** The strong force is the force between nucleons that keeps the nucleus together. The weak force is responsible for beta decay. The electromagnetic and weak forces are now considered to be manifestations of a single force called the **electroweak** force.

Every fundamental interaction is said to be mediated by the exchange of field particles. The electromagnetic interaction is mediated by the photon, the weak interaction is mediated by the $W^{\pm}$ and Z^0 bosons, the gravitational interaction is mediated by gravitons, and the strong interaction is mediated by gluons.

An antiparticle and a particle have the same mass but opposite charge, and other properties may also have opposite values, such as lepton number and baryon number. It is possible to produce particle-antiparticle pairs in nuclear reactions if the available energy is greater than $2mc^2$, where m is the mass of the particle (or antiparticle).

Particles other than photons are classified as hadrons or leptons. **Hadrons** interact primarily through the strong force. They have size and structure and hence are not elementary particles. There are two types of hadrons, *baryons* and *mesons*. Mesons have a baryon number of zero and have either zero or integer spin. Baryons, which generally are the most massive particles, have nonzero baryon numbers and spins of $\frac{1}{2}$ or $\frac{3}{2}$. The neutron and proton are examples of baryons.

Leptons have no structure or size and are considered truly elementary particles. Leptons interact only through the weak and electromagnetic forces. There are six leptons: the electron, e^-; the muon, μ^-; the tau, τ^-; and their associated neutrinos, ν_e, ν_μ, and ν_τ.

In all reactions and decays, quantities such as energy, linear momentum, angular momentum, electric charge, baryon number, and lepton number are strictly conserved. Certain particles have properties called **strangeness** and **charm.** These unusual properties are conserved only in those reactions and decays that occur via the strong force.

Recent theories postulate that all hadrons are composed of smaller units known as **quarks,** which have fractional electric charges and baryon numbers of $\frac{1}{3}$, and come in six "flavors": up, down, strange, charmed, top, and bottom. Each baryon contains three quarks, and each meson contains one quark and one anti-quark.

According to the theory of **quantum chromodynamics,** quarks have a property called **color,** and the strong force between quarks is referred to as the **color force.**

Observation of background microwave radiation by Penzias and Wilson strongly confirmed that the Universe started with a Big Bang about 15 billion years ago. The background radiation is equivalent to that of a blackbody at a temperature of about 3 K.

MULTIPLE-CHOICE QUESTIONS

1. In the first atomic bomb, the energy released was equivalent to about 20 kilotons of TNT, where a ton of TNT equals 4.0×10^9 J. The amount of mass converted into energy in this event is nearest to
   (a) 1 μg (b) 1 mg (c) 1 g (d) 1 kg (e) 20 kilotons

2. If a reactor is fueled with a fuel of 3% enrichment and must be refueled when the enrichment reaches 1.5% in 1 year, how long can a similar reactor of the same power run without refueling if fuel of 91.5% enrichment is used until the enrichment reaches 1.5%?
   (a) 1 y (b) 6 y (c) 20 y (d) 30 y (e) 60 y

3. In a fission reaction, a ^{235}U nucleus captures a neutron. This results in the creation of the products ^{137}I and ^{96}Y along with how many neutrons?
   (a) 1 (b) 2 (c) 3 (d) 5

4. Which of the following particle reactions cannot occur?
   (a) $p + n \rightarrow p + p + \bar{p}$
   (b) $n \rightarrow p + e^- + \bar{\nu}_e$
   (c) $\mu^- \rightarrow e^- + \bar{\nu}_e + \nu_\mu$
   (d) $\pi^- \rightarrow \mu^- + \bar{\nu}_\mu$

5. Which of the following particle reactions cannot occur?
   (a) $p + \bar{p} \rightarrow 2\gamma$
   (b) $\gamma + p \rightarrow n + \pi^0$
   (c) $\pi^0 + n \rightarrow K^+ + \Sigma^-$
   (d) $\pi^+ + p \rightarrow K^+ + \Sigma^+$

CONCEPTUAL QUESTIONS

1. If high-energy electrons with the de Broglie wavelengths smaller than the size of the nucleus are scattered from nuclei, the behavior of the electrons is consistent with scattering from very massive structures much smaller in size than the nucleus—quarks. How is this similar to another classic experiment that detected small structures in an atom?

2. What factors make a fusion reaction difficult to achieve?

3. Doubly charged baryons are known to exist. Why are there no doubly charged mesons?

4. Why would a fusion reactor produce less radioactive waste than a fission reactor?

5. Atoms did not exist until hundreds of thousands of years after the Big Bang. Why?

6. Particles known as resonances have very short lifetimes, on the order of 10^{-23} s. Would you guess they are hadrons or leptons?

7. Describe the quark model of hadrons, including the properties of quarks.

8. In the theory of quantum chromodynamics, quarks come in three colors. How would you justify the statement that "all baryons and mesons are colorless"?

9. Describe the properties of baryons and mesons and the important differences between them.

10. Identify the particle decays in Table 30.2 that occur by the electromagnetic interaction. Justify your answer.

11. Kaons all decay into final states that contain no protons or neutrons. What is the baryon number of kaons?

12. How many quarks are there in (a) a baryon, (b) an antibaryon, (c) a meson, (d) an antimeson? How do you account for the fact that baryons have half-integral spins and mesons have spins of 0 or 1? (*Hint:* Quarks have spin $\frac{1}{2}$.)

13. Two protons in a nucleus interact via the strong interaction. Are they also subject to the weak interaction?

14. Why is a neutron stable inside the nucleus? (In free space, the neutron decays in 900 s.)

15. An antibaryon interacts with a meson. Can a baryon be produced in such an interaction? Explain.

16. Why is water a better shield against neutrons than lead or steel?

PROBLEMS

1, 2, 3 = straightforward, intermediate, challenging   ☐ = full solution available in Study Guide/Student Solutions Manual   ◥ = Core Concepts Workbook
WEB = solution posted at **http://www.harcourtcollege.com/physics/cptech**   ▨ = biomedical application   ◉ = Interactive Physics

Review Problem

Treat the cosmic background as blackbody radiation at a temperature of 3.0 K. (a) Determine the wavelength at which this blackbody radiation has its maximum value. (b) In which part of the electromagnetic spectrum does this peak wavelength lie?

Section 30.1   Nuclear Fission

Section 30.2   Nuclear Reactors

1. When ^{235}U absorbs a neutron, one possible fission reaction produces ^{141}Ba and ^{92}Kr as products. Write down this reaction. How many neutrons are released?

2. Find the energy released in the following fission reaction:

$$^{1}_{0}n + ^{235}_{92}U \longrightarrow ^{144}_{56}Ba + ^{89}_{36}Kr + 3^{1}_{0}n$$

3. Find the energy released in the following fission reaction:

$$^{1}_{0}n + ^{235}_{92}U \longrightarrow ^{88}_{38}Sr + ^{136}_{54}Xe + 12^{1}_{0}n$$

4. If the average energy released in a fission event is 208 MeV, find the total number of fission events required to operate a 100-W lightbulb for 1.0 h.

5. Assume that ordinary soil contains natural uranium in amounts of 1 part per million by mass. (a) How much uranium is in the top 1.00 meter of soil on a one acre(43 560 ft^2) plot of ground, assuming the specific gravity of soil is 4.00? (b) How much of the isotope ^{235}U,

appropriate for nuclear reactor fuel, is in this soil? (*Hint:* See Appendix B for the percentage of abundance of $^{235}_{92}$U.)

6. It has been estimated that the Earth contains 1.0×10^9 tons of natural uranium that can be mined economically. If all the world's energy needs (7.0×10^{12} J/s) were supplied by ^{235}U fission, how long would this supply last? (See the hint in Problem 5.)

WEB 7. An all-electric home uses approximately 2000 kWh of electric energy per month. How much ^{235}U would be required to provide this house with its energy needs for 1 year? (Assume 100% conversion efficiency and 208 MeV released per fission.)

8. What mass of ^{235}U must undergo fission to operate a 1000-MW power plant for one day if the conversion efficiency is 30.0%? (Assume 208 MeV released per fission event.)

WEB 9. Suppose that the water exerts an average frictional drag of 1.0×10^5 N on a nuclear-powered ship. How far can the ship travel per kilogram of fuel if the fuel consists of enriched uranium containing 1.7% of the fissionable isotope ^{235}U, and the ship's engine has a efficiency of 20%? (Assume 208 MeV released per fission event.)

Section 30.3   Nuclear Fusion

10. Find the energy released in the fusion reaction

$$^{1}_{1}H + ^{2}_{1}H \longrightarrow ^{3}_{2}He + \gamma$$

11. When a star has exhausted its hydrogen fuel, it may fuse other nuclear fuels. At temperatures above 1.0×10^8 K, helium fusion can occur. Write the equation for the processes described. (a) Two alpha particles fuse to produce a nucleus A and a gamma ray. What is nucleus A? (b) Nucleus A absorbs an alpha particle to produce a nucleus B and a gamma ray. What is nucleus B? (c) Find the total energy released in the reactions given in (a) and (b). (*Note:* The mass of $^8_4$Be $= 8.005\ 305$ u.)

12. Another series of nuclear reactions that can produce energy in the interior of stars is the cycle described next. This process is most efficient when the central temperature in a star is above 1.6×10^7 K. Because the temperature at the center of the Sun is only 1.5×10^7 K, the cycle below produces less than 10% of the Sun's energy. (a) A high-energy proton is absorbed by ^{12}C. Another nucleus, A, is produced in the reaction, along with a gamma ray. Identify nucleus A. (b) Nucleus A decays through positron emission to form nucleus B. Identify nucleus B. (c) Nucleus B absorbs a proton to produce nucleus C and a gamma ray. Identify nucleus C. (d) Nucleus C absorbs a proton to produce nucleus D and a gamma ray. Identify nucleus D. (e) Nucleus D decays through positron emission to produce nucleus E. Identify nucleus E. (f) Nucleus E absorbs a proton to produce nucleus F plus an alpha particle. What is nucleus F? (*Note:* If nucleus F is not ^{12}C, the nucleus you started with, you have made an error and should review the sequence of events.)

13. If an all-electric home uses approximately 2000 kWh of electric energy per month, how many fusion events described by the reaction $^2_1$H $+$ $^3_1$H $\rightarrow$ $^4_2$He $+$ $^1_0$n would be required to keep this house running for 1 year?

14. Of all the hydrogen in the ocean, 0.0156% of the mass is deuterium. (a) How many deuterium nuclei could be obtained from 1.0 gal of ordinary tap water? (b) If all of this deuterium could be converted to energy through reactions such as the first reaction in Equation 30.4, how much energy of forms other than mass could be obtained from the 1.0 gal of water? (c) The energy released through the burning of 1.0 gal of gasoline is approximately 2.0×10^8 J. How many gallons of gasoline would have to be burned in order to produce the same amount of energy as was obtained from 1.0 gal of water?

15. The oceans have a volume of 317 million cubic miles and contain 1.32×10^{21} kg of water. Of all the hydrogen nuclei in this water, 0.0156% of the mass is deuterium. (a) If all of these deuterium nuclei were fused to helium via the first reaction in Equation 30.4, determine the total amount of energy that could be released. (b) Present world electric power consumption is about 7.00×10^{12} W. If consumption were 100 times greater, how many years would the energy supply calculated in (a) last?

Section 30.6 Positrons and Other Antiparticles

16. Two photons are produced when a proton and an antiproton annihilate each other. What is the minimum frequency and corresponding wavelength of each photon?

17. A photon with an energy of 2.09 GeV creates a proton-antiproton pair in which the proton has a kinetic energy of 95.0 MeV. What is the kinetic energy of the antiproton?

Section 30.7 Mesons and the Beginning of Particle Physics

18. When a high-energy proton or pion traveling near the speed of light collides with a nucleus, it travels an average distance of 3.0×10^{-15} m before interacting. From this information, estimate the time for the strong interaction to occur.

WEB 19. One of the mediators of the weak interaction is the Z^0 boson, whose mass is 96 GeV/c^2. Use this information to find an approximate value for the range of the weak interaction.

20. If a π^0 at rest decays into two γ's, what is the energy of each of the γ's?

Section 30.9 Conservation Laws

Section 30.10 Strange Particles and Strangeness

21. Each of the following reactions is forbidden. Determine a conservation law that is violated for each reaction.
(a) $p + \bar{p} \rightarrow \mu^+ + e^-$
(b) $\pi^- + p \rightarrow p + \pi^+$
(c) $p + p \rightarrow p + \pi^+$
(d) $p + p \rightarrow p + p + n$
(e) $\gamma + p \rightarrow n + \pi^0$

22. For the following two reactions, the first may occur but the second cannot. Explain.

$$K^0 \longrightarrow \pi^+ + \pi^- \quad \text{(can occur)}$$

$$\Lambda^0 \longrightarrow \pi^+ + \pi^- \quad \text{(cannot occur)}$$

23. Identify the unknown particle on the left side of the reaction

$$? + p \longrightarrow n + \mu^+$$

24. Determine the type of neutrino or antineutrino involved in each of the following processes:
(a) $\pi^+ \rightarrow \pi^0 + e^+ + ?$
(b) $? + p \rightarrow \mu^- + p + \pi^+$
(c) $\Lambda^0 \rightarrow p + \mu^- + ?$
(d) $\tau^+ \rightarrow \mu^+ + ? + ?$

25. The following reactions or decays involve one or more neutrinos. Supply the missing neutrinos.
(a) $\pi^- \rightarrow \mu^- + ?$
(b) $K^+ \rightarrow \mu^+ + ?$
(c) $? + p \rightarrow n + e^+$

(d) $? + n \rightarrow p + e^-$

(e) $? + n \rightarrow p + \mu^-$

(f) $\mu^- \rightarrow e^- + ? + ?$

26. Determine which of the reactions can occur. For those that cannot occur, determine the conservation law (or laws) that each violates.

(a) $p \rightarrow \pi^+ + \pi^0$

(b) $p + p \rightarrow p + p + \pi^0$

(c) $p + p \rightarrow p + \pi^+$

(d) $\pi^+ \rightarrow \mu^+ + \nu_\mu$

(e) $n \rightarrow p + e^- + \bar{\nu}_e$

(f) $\pi^+ \rightarrow \pi^+ + n$

27. Which of the following processes are allowed by the strong interaction, the electromagnetic interaction, the weak interaction, or no interaction at all?

(a) $\pi^- + p \rightarrow 2\eta$

(b) $K^- + n \rightarrow \Lambda^0 + \pi^-$

(c) $K^- \rightarrow \pi^- + \pi^0$

(d) $\Omega^- \rightarrow \Xi^- + \pi^0$

(e) $\eta \rightarrow 2\gamma$

28. The neutral ρ meson decays by the strong interaction into two pions according to $\rho^0 \rightarrow \pi^+ + \pi^-$, with a half-life of about 10^{-23} s. The neutral K meson also decays into two pions according to $K^0 \rightarrow \pi^+ + \pi^-$, but with a much longer half-life of about 10^{-10} s. How do you explain these observations?

29. Determine whether or not strangeness is conserved in the following decays and reactions.

(a) $\Lambda^0 \rightarrow p + \pi^-$

(b) $\pi^- + p \rightarrow \Lambda^0 + K^0$

(c) $\bar{p} + p \rightarrow \bar{\Lambda}^0 + \Lambda^0$

(d) $\pi^- + p \rightarrow \pi^- + \Sigma^+$

(e) $\Xi^- \rightarrow \Lambda^0 + \pi^-$

(f) $\Xi^0 \rightarrow p + \pi^-$

30. (a) Show that baryon number and charge are conserved in the following reactions of a pion with a proton.

$$(1)\ \pi^+ + p \longrightarrow K^+ + \Sigma^+$$

$$(2)\ \pi^+ + p \longrightarrow \pi^+ + \Sigma^+$$

(b) The first reaction is observed, but the second never occurs. Explain these observations.

31. Identify the conserved quantities in the following processes:

(a) $\Xi^- \rightarrow \Lambda^0 + \mu^- + \nu_\mu$

(b) $K^0 \rightarrow 2\pi^0$

(c) $K^- + p \rightarrow \Sigma^0 + n$

(d) $\Sigma^0 \rightarrow \Lambda^0 + \gamma$

(e) $e^+ + e^- \rightarrow \mu^+ + \mu^-$

(f) $\bar{p} + n \rightarrow \bar{\Lambda}^0 + \Sigma^-$

32. Fill in the missing particle. Assume that (a) occurs via the strong interaction and (b) and (c) involve the weak interaction.

(a) $K^+ + p \rightarrow ? + p$

(b) $\Omega^- \rightarrow ? + \pi^-$

(c) $K^+ \rightarrow ? + \mu^+ + \nu_\mu$

Section 30.12 Quarks

Section 30.13 Colored Quarks

33. The quark composition of the proton is uud, whereas that of the neutron is udd. Show that the charge, baryon number, and strangeness of these particles equal the sums of these numbers for their quark constituents.

34. The quark compositions of the K^0 and Λ^0 particles are $d\bar{s}$ and uds, respectively. Show that the charge, baryon number, and strangeness of these particles equal the sums of these numbers for the quark constituents.

35. Identify the particles corresponding to the following quark states: (a) suu; (b) $\bar{u}$ d; (c) $\bar{s}$ d; (d) ssd.

36. What is the electrical charge of the baryons with the quark compositions (a) $\overline{uud}$ and (b) $\overline{udd}$? What are these baryons called?

37. Analyze the first three of the following reactions at the quark level and show that each conserves the net number of each type quark. In (d), identify the mystery particle.

(a) $\pi^- + p \rightarrow K^0 + \Lambda^0$

(b) $\pi^+ + p \rightarrow K^+ + \Sigma^+$

(c) $K^- + p \rightarrow K^+ + K^0 + \Omega^-$

(d) $p + p \rightarrow K^0 + p + \pi^+ + ?$

WEB 38. Neglect binding energies and estimate the mass of the u and d quarks from the mass of the proton and neutron.

ADDITIONAL PROBLEMS

39. It was stated in the text that the reaction $\pi^- + p \rightarrow K^0 + \Lambda^0$ occurs with high probability whereas the reaction $\pi^- + p \rightarrow K^0 + n$ never occurs. Analyze these reactions at the quark level and show that the first conserves the net number of each type of quark and the second reaction does not.

40. Two protons approach each other with equal and opposite velocities. What is the minimum kinetic energy of each of the protons if they are to produce a π^+ meson at rest in the reaction

$$p + p \longrightarrow p + n + \pi^+$$

41. A K^0 particle at rest decays into a π^+ and a π^-. What will be the speed of each of the pions? The mass of the K^0 is 497.7 MeV/c^2 and the mass of each pion is 139.6 MeV/c^2.

42. Calculate the range of the force that might be produced by the exchange of a virtual proton.

43. Find the energy released in the fusion reaction

$$^1_1H + {}^3_2He \longrightarrow {}^4_2He + e^- + \bar{\nu}_e$$

44. Occasionally, high-energy muons will collide with electrons and produce two neutrinos according to the reaction $\mu^+ + e^- \rightarrow 2\nu$. What kind of neutrinos are these?

45. Each of the following decays is forbidden. For each process, determine a conservation law that is violated.
(a) $\mu^- \rightarrow e^- + \gamma$
(b) $n \rightarrow p + e^- + \nu_e$
(c) $\Lambda^0 \rightarrow p + \pi^0$
(d) $p \rightarrow e^+ + \pi^0$
(e) $\Xi^0 \rightarrow n + \pi^0$

46. A Σ^0 particle at rest decays according to $\Sigma^0 \rightarrow \Lambda^0 + \gamma$. Find the gamma-ray energy. (*Hint:* Remember to conserve momentum.)

47. A 2.0-MeV neutron is emitted in a fission reactor. If it loses one half of its kinetic energy in each collision with a moderator atom, how many collisions must it undergo in order to achieve thermal energy (0.039 eV)?

48. Calculate the mass of ^{235}U required to provide the total energy requirements of a nuclear submarine during a 100-day patrol, assuming a constant power demand of 100 000 kW and a conversion efficiency of 30%. (Assume that the average energy released per fission is 208 MeV.)

49. If baryon number is not conserved, then one possible mechanism by which a proton can decay is

$$p \longrightarrow e^+ + \gamma$$

(a) Show that this reaction violates conservation of baryon number. (b) Assuming that this reaction occurs, and that the proton is initially at rest, determine the energy and momentum of the photon after the reaction. (*Hint:* Recall that energy and momentum must be conserved in the reaction.) (c) Determine the speed of the positron after the reaction.

50. (a) Show that about 1.0×10^{10} J would be released by the fusion of the deuterons in 1.0 gal of water. Note that 1 out of every 13 000 hydrogen atoms is a deuteron. (b) The average energy consumption rate of a person living in the United States is about 1.0×10^4 J/s (an average power of 10 kW). At this rate, how long would the energy needs of one person be supplied by the fusion of the deuterons in 1.0 gal of water? Assume that the energy released per deuteron is 1.64 MeV.

51. To understand why containment of a plasma is necessary, consider the rate at which a plasma would be lost if it were not contained. (a) Estimate the rms speed of deuterons in a plasma at 1.0×10^8 K. (b) Estimate the time such a plasma would remain in a cube 10 cm on an edge if no steps were taken to confine it.

52. Classical general relativity views the space-time manifold as a deterministic structure completely well defined down to arbitrarily small distances. On the other hand, quantum general relativity forbids distances smaller than the Planck length given by $L = (\hbar G / c^3)^{1/2}$. (a) Calculate the value of L. The answer suggests that after the Big Bang (when all the known Universe was reduced to a singularity), nothing could be observed until that singularity grew larger than the Planck length, L. Because the size of the singularity grew at the speed of light, we can infer that during the time it took for light to travel the Planck length, no observations were possible. (b) Determine this time (known as the Planck time, T) and compare it to the ultra-hot epoch discussed in the text. (c) Does this suggest that we may never know what happened between the time $t = 0$ and the time $t = T$?

"QUARKS. NEUTRINOS. MESONS. ALL THOSE DAMN PARTICLES YOU CAN'T SEE. THAT'S WHAT DROVE ME TO DRINK. BUT NOW I CAN SEE THEM."

Mathematical Review

A.1 SCIENTIFIC NOTATION

Many quantities that scientists deal with often have very large or very small values. For example, the speed of light is about 300 000 000 m/s and the ink required to make the dot over an *i* in this textbook has a mass of about 0.000 000 001 kg. Obviously, it is very cumbersome to read, write, and keep track of numbers such as these. We avoid this problem by using a method dealing with powers of the number 10:

$$10^0 = 1$$

$$10^1 = 10$$

$$10^2 = 10 \times 10 = 100$$

$$10^3 = 10 \times 10 \times 10 = 1000$$

$$10^4 = 10 \times 10 \times 10 \times 10 = 10\ 000$$

$$10^5 = 10 \times 10 \times 10 \times 10 \times 10 = 100\ 000$$

and so on. The number of zeros corresponds to the power to which 10 is raised, called the **exponent** of 10. For example, the speed of light, 300 000 000 m/s, can be expressed as 3×10^8 m/s.

For numbers less than one, we note the following:

$$10^{-1} = \frac{1}{10} = 0.1$$

$$10^{-2} = \frac{1}{10 \times 10} = 0.01$$

$$10^{-3} = \frac{1}{10 \times 10 \times 10} = 0.001$$

$$10^{-4} = \frac{1}{10 \times 10 \times 10 \times 10} = 0.0001$$

$$10^{-5} = \frac{1}{10 \times 10 \times 10 \times 10 \times 10} = 0.000\ 01$$

In these cases, the number of places the decimal point is to the left of the digit 1 equals the value of the (negative) exponent. Numbers that are expressed as some power of 10 multiplied by another number between 1 and 10 are said to be in

scientific notation. For example, the scientific notation for 5 943 000 000 is 5.943×10^9 and that for 0.000 083 2 is 8.32×10^{-5}.

When numbers expressed in scientific notation are being multiplied, the following general rule is very useful:

$$10^n \times 10^m = 10^{n+m} \qquad\qquad \text{[A.1]}$$

where n and m can be *any* numbers (not necessarily integers). For example, $10^2 \times 10^5 = 10^7$. The rule also applies if one of the exponents is negative. For example, $10^3 \times 10^{-8} = 10^{-5}$.

When dividing numbers expressed in scientific notation, note that

$$\frac{10^n}{10^m} = 10^n \times 10^{-m} = 10^{n-m} \qquad\qquad \text{[A.2]}$$

Exercises

With help from the above rules, verify the answers to the following:

1. $86\ 400 = 8.64 \times 10^4$
2. $9\ 816\ 762.5 = 9.816\ 762\ 5 \times 10^6$
3. $0.000\ 000\ 039\ 8 = 3.98 \times 10^{-8}$
4. $(4.0 \times 10^8)(9.0 \times 10^9) = 3.6 \times 10^{18}$
5. $(3.0 \times 10^7)(6.0 \times 10^{-12}) = 1.8 \times 10^{-4}$

6. $\dfrac{75 \times 10^{-11}}{5.0 \times 10^{-3}} = 1.5 \times 10^{-7}$

7. $\dfrac{(3 \times 10^6)(8 \times 10^{-2})}{(2 \times 10^{17})(6 \times 10^5)} = 2 \times 10^{-18}$

A.2   ALGEBRA

A. Some Basic Rules

When algebraic operations are performed, the laws of arithmetic apply. Symbols such as x, y, and z are usually used to represent quantities that are not specified, what are called the **unknowns.**

First, consider the equation

$$8x = 32$$

If we wish to solve for x, we can divide (or multiply) each side of the equation by the same factor without destroying the equality. In this case, if we divide both sides by 8, we have

$$\frac{8x}{8} = \frac{32}{8}$$

$$x = 4$$

Next consider the equation

$$x + 2 = 8$$

In this type of expression, we can add or subtract the same quantity from each side. If we subtract 2 from each side, we get

$$x + 2 - 2 = 8 - 2$$

$$x = 6$$

In general, if $x + a = b$, then $x = b - a$.

Now consider the equation

$$\frac{x}{5} = 9$$

If we multiply each side by 5, we are left with x on the left by itself and 45 on the right:

$$\left(\frac{x}{5}\right)(5) = 9 \times 5$$

$$x = 45$$

In all cases, **whatever operation is performed on the left side of the equality must also be performed on the right side.**

The following rules for multiplying, dividing, adding, and subtracting fractions should be recalled, where a, b, and c are three numbers:

	Rule	Example
Multiplying	$\left(\dfrac{a}{b}\right)\left(\dfrac{c}{d}\right) = \dfrac{ac}{bd}$	$\left(\dfrac{2}{3}\right)\left(\dfrac{4}{5}\right) = \dfrac{8}{15}$
Dividing	$\dfrac{(a/b)}{(c/d)} = \dfrac{ad}{bc}$	$\dfrac{2/3}{4/5} = \dfrac{(2)(5)}{(4)(3)} = \dfrac{10}{12}$
Adding	$\dfrac{a}{b} \pm \dfrac{c}{d} = \dfrac{ad \pm bc}{bd}$	$\dfrac{2}{3} - \dfrac{4}{5} = \dfrac{(2)(5) - (4)(3)}{(3)(5)} = -\dfrac{2}{15}$

Exercises

In the following exercises, solve for x:

Answers

1. $a = \dfrac{1}{1 + x}$ $x = \dfrac{1 - a}{a}$

2. $3x - 5 = 13$ $x = 6$

3. $ax - 5 = bx + 2$ $x = \dfrac{7}{a - b}$

4. $\dfrac{5}{2x + 6} = \dfrac{3}{4x + 8}$ $x = -\dfrac{11}{7}$

B. Powers

When powers of a given quantity x are multiplied, the following rule applies:

$$x^n x^m = x^{n+m} \qquad \text{[A.3]}$$

For example, $x^2 x^4 = x^{2+4} = x^6$.

When dividing the powers of a given quantity, note that

$$\frac{x^n}{x^m} = x^{n-m} \qquad \text{[A.4]}$$

For example, $x^8/x^2 = x^{8-2} = x^6$.

A power that is a fraction, such as $\frac{1}{3}$, corresponds to a root as follows:

$$x^{1/n} = \sqrt[n]{x} \qquad \text{[A.5]}$$

For example, $4^{1/3} = \sqrt[3]{4} = 1.5874$. (A scientific calculator is useful for such calculations.)

Finally, any quantity x^n that is raised to the mth power is

$$(x^n)^m = x^{nm} \qquad \text{[A.6]}$$

Table A.1 summarizes the rules of exponents.

TABLE A.1
Rules of Exponents

$$x^0 = 1$$
$$x^1 = x$$
$$x^n x^m = x^{n+m}$$
$$x^n/x^m = x^{n-m}$$
$$x^{1/n} = \sqrt[n]{x}$$
$$(x^n)^m = x^{nm}$$

Exercises

Verify the following:

1. $3^2 \times 3^3 = 243$
2. $x^5 x^{-8} = x^{-3}$
3. $x^{10}/x^{-5} = x^{15}$
4. $5^{1/3} = 1.709\,975$ (Use your calculator.)
5. $60^{1/4} = 2.783\,158$ (Use your calculator.)
6. $(x^4)^3 = x^{12}$

C. Factoring

Some useful formulas for factoring an equation are

$$ax + ay + az = a(x + y + z) \qquad \text{common factor}$$
$$a^2 + 2ab + b^2 = (a + b)^2 \qquad \text{perfect square}$$
$$a^2 - b^2 = (a + b)(a - b) \qquad \text{differences of squares}$$

D. Quadratic Equations

The general form of a quadratic equation is

$$ax^2 + bx + c = 0 \qquad \text{[A.7]}$$

where x is the unknown quantity and a, b, and c are numerical factors referred to as **coefficients** of the equation. This equation has two roots, given by

$$x = \frac{-b \pm \sqrt{b^2 - 4ac}}{2a} \qquad \text{[A.8]}$$

If $b^2 \geq 4ac$, the roots will be real.

EXAMPLE

The equation $x^2 + 5x + 4 = 0$ has the following roots corresponding to the two signs of the square-root term:

$$x = \frac{-5 \pm \sqrt{5^2 - (4)(1)(4)}}{2(1)} = \frac{-5 \pm \sqrt{9}}{2} = \frac{-5 \pm 3}{2}$$

that is,

$$x_+ = \frac{-5 + 3}{2} = -1 \qquad x_- = \frac{-5 - 3}{2} = -4$$

where x_+ refers to the root corresponding to the positive sign and x_- refers to the root corresponding to the negative sign.

Exercises

Solve the following quadratic equations:

Answers

1. $x^2 + 2x - 3 = 0$	$x_+ = 1$	$x_- = -3$
2. $2x^2 - 5x + 2 = 0$	$x_+ = 2$	$x_- = 1/2$
3. $2x^2 - 4x - 9 = 0$	$x_+ = 1 + \sqrt{22}/2$	$x_- = 1 - \sqrt{22}/2$

E. Linear Equations

A linear equation has the general form

$$y = ax + b \qquad \text{[A.9]}$$

where a and b are constants. This equation is referred to as being linear because the graph of y versus x is a straight line, as shown in Figure A.1. The constant b, called the **intercept,** represents the value of y at which the straight line intersects the y axis. The constant a is equal to the **slope** of the straight line and is also equal to the tangent of the angle that the line makes with the x axis. If any two points on the straight line are specified by the coordinates (x_1, y_1) and (x_2, y_2), as in Figure A.1, then the slope of the straight line can be expressed

$$\text{Slope} = \frac{y_2 - y_1}{x_2 - x_1} = \frac{\Delta y}{\Delta x} \qquad \text{[A.10]}$$

Note that a and b can have either positive or negative values. If $a > 0$, the straight line has a *positive* slope, as in Figure A.1. If $a < 0$, the straight line has a *negative* slope. In Figure A.1, both a and b are positive. Three other possible situations are shown in Figure A.2: $a > 0$, $b < 0$; $a < 0$, $b > 0$; and $a < 0$, $b < 0$.

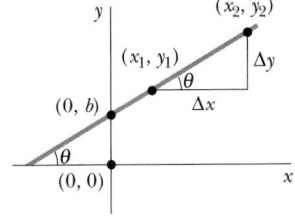

Figure A.1

Exercises

1. Draw graphs of the following straight lines:
   (a) $y = 5x + 3$ (b) $y = -2x + 4$ (c) $y = -3x - 6$
2. Find the slopes of the straight lines described in Exercise 1.
   Answers: (a) 5   (b) -2 (c) -3
3. Find the slopes of the straight lines that pass through the following sets of points:
   (a) $(0, -4)$ and $(4, 2)$, (b) $(0, 0)$ and $(2, -5)$, and (c) $(-5, 2)$ and $(4, -2)$
   Answers: (a) $3/2$ (b) $-5/2$ (c) $-4/9$

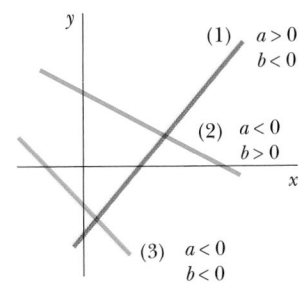

Figure A.2

F. Solving Simultaneous Linear Equations

Consider an equation such as $3x + 5y = 15$, which has two unknowns, x and y. Such an equation does not have a unique solution. That is, $(x = 0, y = 3)$, $(x = 5, y = 0)$, and $(x = 2, y = 9/5)$ are all solutions to this equation.

If a problem has two unknowns, a unique solution is possible only if we have *two* independent equations. In general, if a problem has n unknowns, its solution requires n independent equations. In order to solve two simultaneous equations involving two unknowns, x and y, we solve one of the equations for x in terms of y and substitute this expression into the other equation.

EXAMPLE

Solve the following two simultaneous equations:

$$(1) \quad 5x + y = -8 \qquad (2) \quad 2x - 2y = 4$$

Solution   From (2), we find that $x = y + 2$. Substitution of this into (1) gives

$$5(y + 2) + y = -8$$
$$6y = -18$$
$$y = -3$$
$$x = y + 2 = \;\; -1$$

Alternate solution: Multiply each term in (1) by the factor 2 and add the result to (2):

$$10x + 2y = -16$$
$$\underline{2x - 2y = 4}$$
$$12x = -12$$
$$x = \;\; -1$$
$$y = x - 2 = \;\; -3$$

Two linear equations with two unknowns can also be solved by a graphical method. If the straight lines corresponding to the two equations are plotted in a conventional coordinate system, the intersection of the two lines represents the solution. For example, consider the two equations

$$x - y = 2$$
$$x - 2y = -1$$

These are plotted in Figure A.3. The intersection of the two lines has the coordinates $x = 5$, $y = 3$. This represents the solution to the equations. You should check this solution by the analytical technique discussed above.

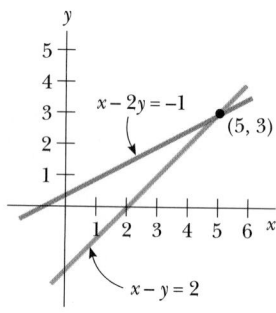

Figure A.3

Exercises

Solve the following pairs of simultaneous equations involving two unknowns:

                                      Answers

1. $x + y = 8$ $x = 5, y = 3$
    $x - y = 2$

2. $98 - T = 10a$ $T = 65, a = 3.27$
    $T - 49 = 5a$

3. $6x + 2y = 6$ $x = 2, y = -3$
    $8x - 4y = 28$

G. Logarithms

Suppose that a quantity x is expressed as a power of some quantity a:

$$x = a^y \qquad \text{[A.11]}$$

The number a is called the **base** number. The **logarithm** of x with respect to the base a is equal to the exponent to which the base must be raised in order to satisfy the expression $x = a^y$:

$$y = \log_a x \qquad \text{[A.12]}$$

Conversely, the **antilogarithm** of y is the number x:

$$x = \text{antilog}_a y \qquad \text{[A.13]}$$

In practice, the two bases most often used are base 10, called the *common* logarithm base, and base $e = 2.718 \ldots$, called the *natural* logarithm base. When common logarithms are used,

$$y = \log_{10} x \qquad (\text{or } x = 10^y) \qquad \text{[A.14]}$$

When natural logarithms are used,

$$y = \ln_e x \qquad (\text{or } x = e^y) \qquad \text{[A.15]}$$

For example, $\log_{10} 52 = 1.716$, so that $\text{antilog}_{10} 1.716 = 10^{1.716} = 52$. Likewise, $\ln_e 52 = 3.951$, so $\text{antiln}_e 3.951 = e^{3.951} = 52$.

In general, note that you can convert between base 10 and base e with the equality

$$\ln_e x = (2.302\ 585)\log_{10} x \qquad \text{[A.16]}$$

Finally, some useful properties of logarithms are

$$\log (ab) = \log a + \log b \qquad\qquad \ln e = 1$$

$$\log (a/b) = \log a - \log b \qquad\qquad \ln e^a = a$$

$$\log (a^n) = n \log a \qquad\qquad \ln\left(\frac{1}{a}\right) = -\ln a$$

A.3   GEOMETRY

Table A.2 gives the areas and volumes for several geometric shapes used throughout this text:

TABLE A.2   Useful Information for Geometry

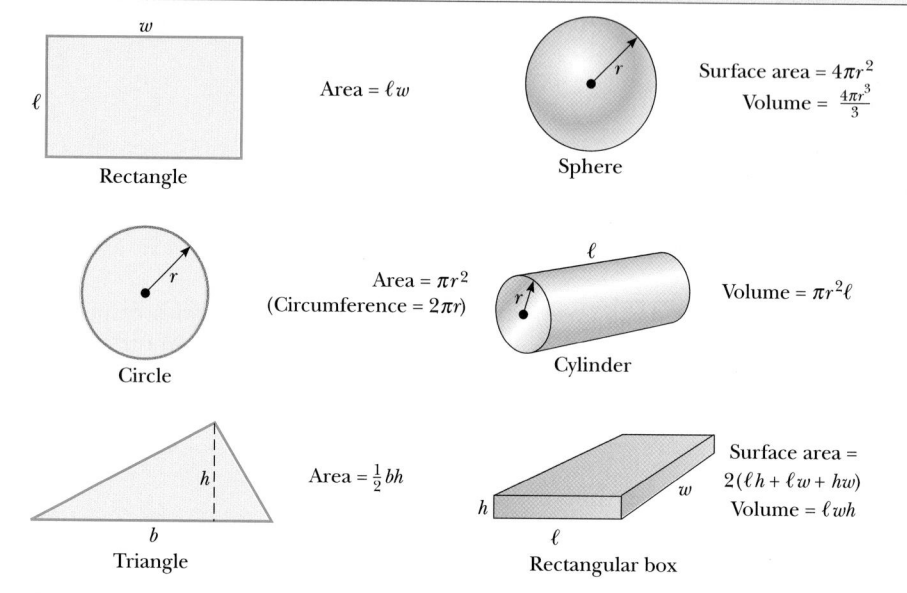

Area = ℓw

Rectangle

Surface area = $4\pi r^2$

Volume = $\frac{4\pi r^3}{3}$

Sphere

Area = πr^2

(Circumference = $2\pi r$)

Circle

Volume = $\pi r^2 \ell$

Cylinder

Area = $\frac{1}{2} bh$

Triangle

Surface area =

$2(\ell h + \ell w + hw)$

Volume = ℓwh

Rectangular box

A.4   TRIGONOMETRY

Some of the most basic facts concerning trigonometry are presented in Chapter 1, and we encourage you to study the material presented there if you are having trouble with this branch of mathematics. In addition to the discussion of Chapter 1, certain useful trig identities that can be of value to you follow.

$$\sin^2 \theta + \cos^2 \theta = 1$$

$$\sin \theta = \cos(90° - \theta)$$

$$\cos \theta = \sin(90° - \theta)$$

$$\sin 2\theta = 2 \sin \theta \cos \theta$$

$$\cos 2\theta = \cos^2 \theta - \sin^2 \theta$$

$$\sin(\theta \pm \phi) = \sin \theta \cos \phi \pm \cos \theta \sin \phi$$

$$\cos(\theta \pm \phi) = \cos \theta \cos \phi \mp \sin \theta \sin \phi$$

An Abbreviated Table of Isotopes

Atomic Number, Z	Element	Symbol	Chemical Atomic Mass (u)	Mass Number (* Indicates Radioactive) A	Atomic Mass (u) [a]	Percentage Abundance	Half-Life (if Radioactive) $T_{1/2}$
0	(Neutron)	n		1*	1.008 665		10.4 min
1	Hydrogen	H	1.0079	1	1.007 825	99.985	
	Deuterium	D		2	2.014 102	0.015	
	Tritium	T		3*	3.016 049		12.33 y
2	Helium	He	4.00260	3	3.016 029	0.00014	
				4	4.002 602	99.99986	
3	Lithium	Li	6.941	6	6.015 121	7.5	
				7	7.016 003	92.5	
4	Beryllium	Be	9.0122	7*	7.016 928		53.3 days
				9	9.012 174	100	
5	Boron	B	10.81	10	10.012 936	19.9	
				11	11.009 305	80.1	
6	Carbon	C	12.011	11*	11.011 433		20.4 min
				12	12.000 000	98.90	
				13	13.003 355	1.10	
				14*	14.003 242		5730 y
7	Nitrogen	N	14.0067	13*	13.005 738		9.96 min
				14	14.003 074	99.63	
				15	15.000 108	0.37	
8	Oxygen	O	15.9994	15*	15.003 065		122 s
				16	15.994 915	99.761	
				18	17.999 160	0.20	
9	Fluorine	F	18.99840	19	18.998 404	100	
10	Neon	Ne	20.180	20	19.992 435	90.48	
				22	21.991 383	9.25	
11	Sodium	Na	22.98987	22*	21.994 434		2.61 y
				23	22.989 770	100	
				24*	23.990 961		14.96 h
12	Magnesium	Mg	24.305	24	23.985 042	78.99	
				25	24.985 838	10.00	
				26	25.982 594	11.01	
13	Aluminum	Al	26.98154	27	26.981 538	100	
14	Silicon	Si	28.086	28	27.976 927	92.23	
15	Phosphorus	P	30.97376	31	30.973 762	100	
				32*	31.973 908		14.26 days

(Table continues)

[a] The masses in the sixth column are atomic masses, which include the mass of Z electrons. Data are from the National Nuclear Data Center, Brookhaven National Laboratory, prepared by Jagdish K. Tuli, July 1990. The data are based on experimental results reported in *Nuclear Data Sheets* and *Nuclear Physics* and also from *Chart of the Nuclides,* 14th ed. Atomic masses are based on those by A. H. Wapstra, G. Audi, and R. Hoekstra. Isotopic abundances are based on those by N. E. Holden.

Atomic Number, Z	Element	Symbol	Chemical Atomic Mass (u)	Mass Number (* Indicates Radioactive) A	Atomic Mass (u) [a]	Percentage Abundance	Half-Life (if Radioactive) $T_{1/2}$
16	Sulfur	S	32.066	32	31.972 071	95.02	
				35*	34.969 033		87.5 days
17	Chlorine	Cl	35.453	35	34.968 853	75.77	
				37	36.965 903	24.23	
18	Argon	Ar	39.948	40	39.962 384	99.600	
19	Potassium	K	39.0983	39	38.963 708	93.2581	
				40*	39.964 000	0.0117	1.28×10^9 y
20	Calcium	Ca	40.08	40	39.962 591	96.941	
21	Scandium	Sc	44.9559	45	44.955 911	100	
22	Titanium	Ti	47.88	48	47.947 947	73.8	
23	Vanadium	V	50.9415	51	50.943 962	99.75	
24	Chromium	Cr	51.996	52	51.940 511	83.79	
25	Manganese	Mn	54.93805	55	54.938 048	100	
26	Iron	Fe	55.847	56	55.934 940	91.72	
27	Cobalt	Co	58.93320	59	58.933 198	100	
				60*	59.933 820		5.27 y
28	Nickel	Ni	58.693	58	57.935 346	68.077	
				60	59.930 789	26.223	
29	Copper	Cu	63.54	63	62.929 599	69.17	
				65	64.927 791	30.83	
30	Zinc	Zn	65.39	64	63.929 144	48.6	
				66	65.926 035	27.9	
				68	67.924 845	18.8	
31	Gallium	Ga	69.723	69	68.925 580	60.108	
				71	70.924 703	39.892	
32	Germanium	Ge	72.61	70	69.924 250	21.23	
				72	71.922 079	27.66	
				74	73.921 177	35.94	
33	Arsenic	As	74.9216	75	74.921 594	100	
34	Selenium	Se	78.96	78	77.917 307	23.78	
				80	79.916 519	49.61	
35	Bromine	Br	79.904	79	78.918 336	50.69	
				81	80.916 287	49.31	
36	Krypton	Kr	83.80	82	81.913 481	11.6	
				83	82.914 136	11.5	
				84	83.911 508	57.0	
				86	85.910 615	17.3	
37	Rubidium	Rb	85.468	85	84.911 793	72.17	
				87*	86.909 186	27.83	4.75×10^{10} y
38	Strontium	Sr	87.62	86	85.909 266	9.86	
				88	87.905 618	82.58	
				90*	89.907 737		29.1 y
39	Yttrium	Y	88.9058	89	88.905 847	100	
40	Zirconium	Zr	91.224	90	89.904 702	51.45	
				91	90.905 643	11.22	
				92	91.905 038	17.15	
				94	93.906 314	17.38	

(Table continues)

Atomic Number, Z	Element	Symbol	Chemical Atomic Mass (u)	Mass Number (* Indicates Radioactive) A	Atomic Mass (u) [a]	Percentage Abundance	Half-Life (if Radioactive) $T_{1/2}$
41	Niobium	Nb	92.9064	93	92.906 376	100	
42	Molybdenum	Mo	95.94	92	91.906 807	14.84	
				95	94.905 841	15.92	
				96	95.904 678	16.68	
				98	97.905 407	24.13	
43	Technetium	Tc		98*	97.907 215		4.2×10^6 y
				99*	98.906 254		2.1×10^5 y
44	Ruthenium	Ru	101.07	99	98.905 939	12.7	
				100	99.904 219	12.6	
				101	100.905 558	17.1	
				102	101.904 348	31.6	
				104	103.905 428	18.6	
45	Rhodium	Rh	102.9055	103	102.905 502	100	
46	Palladium	Pd	106.42	104	103.904 033	11.14	
				105	104.905 082	22.33	
				106	105.903 481	27.33	
				108	107.903 893	26.46	
				110	109.905 158	11.72	
47	Silver	Ag	107.868	107	106.905 091	51.84	
				109	108.904 754	48.16	
48	Cadmium	Cd	112.41	110	109.903 004	12.49	
				111	110.904 182	12.80	
				112	111.902 760	24.13	
				113*	112.904 401	12.22	9.3×10^{15} y
				114	113.903 359	28.73	
49	Indium	In	114.82	115*	114.903 876	95.7	4.4×10^{14} y
50	Tin	Sn	118.71	116	115.901 743	14.53	
				118	117.901 605	24.22	
				120	119.902 197	32.59	
51	Antimony	Sb	121.76	121	120.903 820	57.36	
				123	122.904 215	42.64	
52	Tellurium	Te	127.60	126	125.903 309	18.93	
				128*	127.904 463	31.70	$>8 \times 10^{24}$ y
				130*	129.906 228	33.87	$\leqslant 1.25 \times 10^{21}$ y
53	Iodine	I	126.9045	127	126.904 474	100	
				129*	128.904 984		1.6×10^7 y
				131*	130.906 118		8.04 days
54	Xenon	Xe	131.29	129	128.904 779	26.4	
				131	130.905 069	21.2	
				132	131.904 141	26.9	
				134	133.905 394	10.4	
				136*	135.907 215	8.9	$\geqslant 2.36 \times 10^{21}$ y
55	Cesium	Cs	132.9054	133	132.905 436	100	
56	Barium	Ba	137.33	137	136.905 816	11.23	
				138	137.905 236	71.70	
				144*	143.922 673		11.9 s

(Table continues)

Atomic Number, Z	Element	Symbol	Chemical Atomic Mass (u)	Mass Number (* Indicates Radioactive) A	Atomic Mass (u) [a]	Percentage Abundance	Half-Life (if Radioactive) $T_{1/2}$
57	Lanthanum	La	138.905	139	138.906 346	99.9098	
58	Cerium	Ce	140.12	140	139.905 434	88.43	
				142*	141.909 241	11.13	$>5 \times 10^{16}$ y
59	Praseodymium	Pr	140.9076	141	140.907 647	100	
60	Neodymium	Nd	144.24	142	141.907 718	27.13	
				144*	143.910 082	23.80	2.3×10^{15} y
				146	145.913 113	17.19	
61	Promethium	Pm		145*	144.912 745		17.7 y
62	Samarium	Sm	150.36	147*	146.914 894	15.0	1.06×10^{11} y
				149*	148.917 180	13.8	$>2 \times 10^{15}$ y
				152	151.919 728	26.7	
				154	153.922 206	22.7	
63	Europium	Eu	151.96	151	150.919 846	47.8	
				153	152.921 226	52.2	
64	Gadolinium	Gd	157.25	156	155.922 119	20.47	
				158	157.924 099	24.84	
				160	159.927 050	21.86	
65	Terbium	Tb	158.9253	159	158.925 345	100	
66	Dysprosium	Dy	162.50	162	161.926 796	25.5	
				163	162.928 729	24.9	
				164	163.929 172	28.2	
67	Holmium	Ho	164.9303	165	164.930 316	100	
68	Erbium	Er	167.26	166	165.930 292	33.6	
				167	166.932 047	22.95	
				168	167.932 369	27.8	
69	Thulium	Tm	168.9342	169	168.934 213	100	
70	Ytterbium	Yb	173.04	172	171.936 380	21.9	
				173	172.938 209	16.12	
				174	173.938 861	31.8	
71	Lutecium	Lu	174.967	175	174.940 772	97.41	
72	Hafnium	Hf	178.49	177	176.943 218	18.606	
				178	177.943 697	27.297	
				179	178.945 813	13.629	
				180	179.946 547	35.100	
73	Tantalum	Ta	180.9479	181	180.947 993	99.988	
74	Tungsten	W	183.85	182	181.948 202	26.3	
				183	182.950 221	14.28	
				184	183.950 929	30.7	
				186	185.954 358	28.6	
75	Rhenium	Re	186.207	185	184.952 951	37.40	
				187*	186.955 746	62.60	4.4×10^{10} y
76	Osmium	Os	190.2	188	187.955 832	13.3	
				189	188.958 139	16.1	
				190	189.958 439	26.4	
				191*	190.960 94		15.4 days
				192	191.961 468	41.0	

(Table continues)

Atomic Number, Z	Element	Symbol	Chemical Atomic Mass (u)	Mass Number (* Indicates Radioactive) A	Atomic Mass (u) [a]	Percentage Abundance	Half-Life (if Radioactive) $T_{1/2}$
77	Iridium	Ir	192.2	191	190.960 585	37.3	
				193	192.962 916	62.7	
78	Platinum	Pt	195.08	194	193.962 655	32.9	
				195	194.964 765	33.8	
				196	195.964 926	25.3	
79	Gold	Au	196.9665	197	196.966 543	100	
80	Mercury	Hg	200.59	199	198.968 253	16.87	
				200	199.968 299	23.10	
				201	200.970 276	13.10	
				202	201.970 617	29.86	
81	Thallium	Tl	204.383	203	202.972 320	29.524	
				205	204.974 400	70.476	
		(Th C″)		208*	207.981 992		3.053 min
				210*	209.990 069		1.3 min
82	Lead	Pb	207.2	204*	203.973 020	1.4	$\geq 1.4 \times 10^{17}$ y
				206	205.974 440	24.1	
				207	206.975 871	22.1	
				208	207.976 627	52.4	
		(Ra D)		210*	209.984 163		22.3 y
		(Ac B)		211*	210.988 734		36.1 min
		(Th B)		212*	211.991 872		10.64 h
		(Ra B)		214*	213.999 798		26.8 min
83	Bismuth	Bi	208.9803	209	208.980 374	100	
		(Th C)		211*	210.987 254		2.14 min
84	Polonium	Po					
		(Ra F)		210*	209.982 848		138.38 days
		(Ra C′)		214*	213.995 177		164 μs
85	Astatine	At		218*	218.008 685		1.6 s
86	Radon	Rn		222*	222.017 571		3.823 days
87	Francium	Fr					
		(Ac K)		223*	223.019 733		22 min
88	Radium	Ra		226*	226.025 402		1600 y
		(Ms Th$_1$)		228*	228.031 064		5.75 y
89	Actinium	Ac		227*	227.027 749		21.77 y
90	Thorium	Th	232.0381				
		(Rd Th)		228*	228.028 716		1.913 y
				232*	232.038 051	100	1.40×10^{10} y
91	Protactinium	Pa		231*	231.035 880		32.760 y
92	Uranium	U	238.0289	232*	232.037 131		69 y
				233*	233.039 630		1.59×10^5 y
		(Ac U)		235*	235.043 924	0.720	7.04×10^8 y
				236*	236.045 562		2.34×10^7 y
		(UI)		238*	238.050 784	99.2745	4.47×10^9 y
93	Neptunium	Np		237*	237.048 168		2.14×10^6 y
94	Plutonium	Pu		239*	239.052 157		2.412×10^3 y
				242*	242.058 737		3.73×10^5 y
				244*	244.064 200		8.1×10^7 y

Appendix C

Some Useful Tables

TABLE C.1 Mathematical Symbols Used in the Text and Their Meaning

Symbol	Meaning
$=$	is equal to
$\neq$	is not equal to
$\equiv$	is defined as
$\propto$	is proportional to
$>$	is greater than
$<$	is less than
$\gg$	is much greater than
$\ll$	is much less than
$\approx$	is approximately equal to
Δx	change in x or uncertainty in x
Σx_i	sum of all quantities x_i
$\lvert x \rvert$	absolute value of x (always a positive quantity)

TABLE C.2 Standard Symbols for Units

Symbol	Unit	Symbol	Unit
A	ampere	K	kelvin
Å	angstrom	kcal	kilocalorie
atm	atmosphere	kg	kilogram
Btu	British thermal unit	km	kilometer
C	coulomb	kmol	kilomole
°C	degree Celsius	L	liter
cal	calorie	lb	pound
cm	centimeter	m	meter
deg	degree (angle)	min	minute
eV	electron volt	N	newton
°F	degree Fahrenheit	nm	nanometer
F	farad	rad	radian
ft	foot	rev	revolution
G	Gauss	s	second
g	gram	T	tesla
H	henry	u	atomic mass unit
h	hour	V	volt
hp	horsepower	W	watt
Hz	hertz	Wb	weber
in.	inch	μm	micrometer
J	joule	Ω	ohm

TABLE C.3 The Greek Alphabet

Alpha	A	α	Iota	I	ι	Rho	P	ρ
Beta	B	β	Kappa	K	κ	Sigma	Σ	σ
Gamma	Γ	γ	Lambda	Λ	λ	Tau	T	τ
Delta	Δ	δ	Mu	M	μ	Upsilon	Y	υ
Epsilon	E	ϵ	Nu	N	ν	Phi	Φ	ϕ
Zeta	Z	ζ	Xi	Ξ	ξ	Chi	X	χ
Eta	H	η	Omicron	O	o	Psi	Ψ	ψ
Theta	Θ	θ	Pi	Π	π	Omega	Ω	ω

TABLE C.4 Physical Data Often Used [a]

Average Earth-Moon distance	3.84×10^8 m
Average Earth-Sun distance	1.496×10^{11} m
Average radius of the Earth	6.37×10^6 m
Density of air (20°C and 1 atm)	1.20 kg/m^3
Density of water (20°C and 1 atm)	1.00×10^3 kg/m^3
Free-fall acceleration	9.80 m/s^2
Mass of the Earth	5.98×10^{24} kg
Mass of the Moon	7.36×10^{22} kg
Mass of the Sun	1.99×10^{30} kg
Standard atmospheric pressure	1.013×10^5 Pa

[a] These are the values of the constants as used in the text.

TABLE C.5 Some Fundamental Constants [a]

Quantity	Symbol	Value [b]	Quantity	Symbol	Value [b]
Atomic mass unit	u	$1.660\,540\,2(10) \times 10^{-27}$ kg $931.494\,32(28)$ MeV/c^2	Hydrogen ionization energy	$-E_1 = \dfrac{m_e e^4 k_e^2}{2\hbar^2} = \dfrac{e^2 k_e}{2a_0}$	$13.605\,698(40)$ eV
Avogadro's number	N_A	$6.022\,136\,7(36) \times 10^{23}$ (mol)$^{-1}$	Neutron mass	m_n	$1.674\,928\,6(10) \times 10^{-27}$ kg
Bohr radius	$a_0 = \dfrac{\hbar^2}{m_e e^2 k_e}$	$0.529\,177\,249(24) \times 10^{-10}$ m			$1.008\,664\,904(14)$ u $939.565\,63(28)$ MeV/c^2
Boltzmann's constant	$k_B = R/N_A$	$1.380\,658(12) \times 10^{-23}$ J/K	Permeability of free space	μ_0	$4\pi \times 10^{-7}$ N/A^2 (exact)
Compton wavelength	$\lambda_C = \dfrac{h}{m_e c}$	$2.426\,310\,58(22) \times 10^{-12}$ m	Permittivity of free space	$\epsilon_0 = 1/\mu_0 c^2$	$8.854\,187\,817 \times 10^{-12}$ C^2/N·m^2 (exact)
Coulomb's law force constant	$k_e = \dfrac{1}{4\pi\epsilon_0}$	$9.987\,551\,787 \times 10^9$ N·m^2/C^2 (exact)	Planck's constant	h $\hbar = h/2\pi$	$6.626\,075(40) \times 10^{-34}$ J·s $1.054\,572\,66(63) \times 10^{-34}$ J·s
Electron mass	m_e	$9.109\,389\,7(54) \times 10^{-31}$ kg $5.485\,799\,03(13) \times 10^{-4}$ u $0.510\,999\,06(15)$ MeV/c^2	Proton mass	m_p	$1.672\,623(10) \times 10^{-27}$ kg $1.007\,276\,470(12)$ u $938.272\,3(28)$ MeV/c^2
Electron-volt	eV	$1.602\,177\,33(49) \times 10^{-19}$ J	Rydberg constant	R_H	$1.097\,373\,153\,4(13) \times 10^7$ m^{-1}
Elementary charge	e	$1.602\,177\,33(49) \times 10^{-19}$ C	Speed of light in vacuum	c	$2.997\,924\,58 \times 10^8$ m/s (exact)
Gas constant	R	$8.314\,510(70)$ J/K·mol			
Gravitational constant	G	$6.672\,59(85) \times 10^{-11}$ N·m^2/kg^2			

[a] These constants are the values recommended in 1986 by CODATA, based on a least-squares adjustment of data from different measurements. For a more complete list, see Cohen, E. Richard, and Barry N. Taylor, *Rev. Mod. Phys.* **59**:1121, 1987.

[b] The numbers in parentheses for the values below represent the uncertainties in the last two digits.

Appendix D

SI Units

TABLE D.1 SI Base Units

Base Quantity	SI Base Unit Name	Symbol
Length	Meter	m
Mass	Kilogram	kg
Time	Second	s
Electric current	Ampere	A
Temperature	Kelvin	K
Amount of substance	Mole	mol
Luminous intensity	Candela	cd

TABLE D.2 Derived SI Units

Quantity	Name	Symbol	Expression in Terms of Base Units	Expression in Terms of Other SI Units
Plane angle	Radian	rad	m/m	
Frequency	Hertz	Hz	s^{-1}	
Force	Newton	N	$kg \cdot m/s^2$	J/m
Pressure	Pascal	Pa	$kg/m \cdot s^2$	N/m^2
Energy: work	Joule	J	$kg \cdot m^2/s^2$	$N \cdot m$
Power	Watt	W	$kg \cdot m^2/s^3$	J/s
Electric charge	Coulomb	C	$A \cdot s$	
Electric potential (emf)	Volt	V	$kg \cdot m^2/A \cdot s^3$	$W/A, J/C$
Capacitance	Farad	F	$A^2 \cdot s^4/kg \cdot m^2$	C/V
Electric resistance	Ohm	Ω	$kg \cdot m^2/A^2 \cdot s^3$	V/A
Magnetic flux	Weber	Wb	$kg \cdot m^2/A \cdot s^2$	$V \cdot s, T \cdot m^2$
Magnetic field intensity	Tesla	T	$kg/A \cdot s^2$	Wb/m^2
Inductance	Henry	H	$kg \cdot m^2/A^2 \cdot s^2$	Wb/A

Answers to Selected Questions and Problems

CHAPTER 1

Multiple-Choice Questions
1. b   2. d   3. b   4. d   5. d
6. d   7. d

Conceptual Questions
1. (a) ≈ 0.1 m   (b) ≈ 1 m   (c) ≈ 30 m   (d) ≈ 10 m
   (e) ≈ 100 m
3. $\approx 10^9$ s
5. The length of a hand varies from person to person, so it is not a useful standard of length.
7. The ark has an approximate volume of 4×10^4 m³, whereas a home has an approximate volume of 10^3 m³.
9. $\approx 10^8$ revolutions

Problems
1. (b) $A_{\text{cylinder}} = \pi R^2$, $A_{\text{rectangular solid}} = lw$.
5. (a) $\dfrac{L}{T^2}$ (b) L
7. (a) 3 significant figures   (b) 4 significant figures
   (c) 3 significant figures   (d) 2 significant figures.
9. (a) 2.96×10^9 (b) 6.876×10^{-2} (If you disagree with either the significant figures or the power of ten of either of these answers, please check your calculator manual.)
11. 115.9 m
13. 10^{17} ft
15. 6.71×10^8 mi/h
17. 36 m²
19. 10^{10} lb, 4.45×10^{10} N
21. 3.08×10^4 m³
23. 9.82 cm
25. (a) 1000 kg
    (b) $m_{\text{cell}} = 5.2 \times 10^{-16}$ kg
    $m_{\text{kidney}} = 0.27$ kg
    $m_{\text{fly}} = 1.3 \times 10^{-5}$ kg
27. 1800 balls (assumes 81 games per season, 9 innings per game, an average of 10 hitters per inning, and 1 ball lost for every four hitters)
29. (2.1, 1.4)
31. $r = 2.24$ m, $\theta = 26.6°$
33. (a) 6.71 m   (b) 0.894   (c) 0.746
35. (a) 3   (b) 3   (c) 4/5   (d) 4/5   (e) 4/3
37. $\frac{5}{7}$; the angle itself is 35.5°
39. (a) 0.677 g/cm³   (b) 4.63×10^{17} ft²

41. The value of k, a dimensionless constant, cannot be found by dimensional analysis.
43. (a) on the order of 100 kg   (b) on the order of 1000 kg
45. Both sides have the units of length, L.
47. (a) 3.16×10^7 s   (b) 6.0×10^{10} y
49. (a) 127 y   (b) 15 500 times

CHAPTER 2

Multiple-Choice Questions
1. c   2. a   3. b   4. c   5. d

Conceptual Questions
1. Yes. This occurs when a car is slowing down, so that the direction of its acceleration is opposite to its direction of motion.
3. No. For an accelerating particle to stop at all, the velocity and acceleration must have opposite signs, so that the speed is decreasing. Given that this is the case, the particle will eventually come to rest. If the acceleration remains constant, however, the particle must begin to move again, opposite to the direction of its original motion. If the particle comes to rest and then stays at rest, the acceleration has disappeared at the moment the motion stops. This is the case for a braking car—the acceleration is negative and goes to zero as the car comes to rest.
5. Once the objects leave the hand, both are freely falling objects, and both experience the same downward acceleration equal to the free-fall acceleration, $-g$.
7. Yes. Yes.
9. (a) The velocity of the ball changes continuously but its acceleration is always the free-fall acceleration, $g = 9.8$ m/s² downward. As the ball travels upward, its speed decreases by 9.8 m/s during each second of its motion, and its velocity vector is upward. When it reaches the peak of its motion, its speed becomes zero. (If the acceleration were zero at the peak when the velocity is zero, there would be no change in velocity thereafter, so the ball would stop at the peak and remain there, which is not what happens.) (b) The acceleration of the ball remains constant in magnitude and direction throughout its free flight, from the instant it leaves the hand until the

instant before it strikes the ground. Its magnitude is the free-fall acceleration, $g = 9.8$ m/s².

11. No. Car A might have greater acceleration than B, but they might both have zero acceleration, or otherwise equal accelerations: or the driver of B might have stepped very hard on the gas pedal in the recent past.

13. They are the same! You can see this for yourself by solving the kinematic equations for the two cases.

15. Ignoring air resistance, in 16 s the pebble would fall a distance of $\frac{1}{2}gt^2 = \frac{1}{2}(9.80$ m/s²$)(16.0$ s$) = 1250$ m $= 1.25$ km. Air resistance is an important force after the first few seconds, when the pebble has attained high speed. Also, part of the 16-s time interval must be occupied by the sound returning up the well. Thus the depth of the well is less than 1.25 km.

Problems

1. (a) 52.9 km/h   (b) 90.0 km
3. (a) Boat A by 60 km   (b) 0
5. 12.2 mph
7. (a) 2.34 min   (b) 64.2 miles
9. 2.81 h, 218 km
11. 76.1 m/s (274 km/h)
13. (a) 1.17 m/s   (b) 1.40 m/s
15. (a) 4.00 m/s   (b) -4.00 m/s   (c) 0
    (d) 2.00 m/s
17. 0.75 m/s²
19. 3.7 s
21. (a) 8.00 m/s²   (b) 11.0 m/s²
23. (a) -8.0 m/s²   (b) 100 m
25. (a) 35 s   (b) 16 m/s
27. (a) 3.0×10^{-10} s   (b) 1.26×10^{-4} m
29. (a) 1.3 m/s²   (b) 8.0 s
31. (a) 5.51 km
    (b) 20.8 m/s, 41.6 m/s, 20.8 m/s, 38.7 m/s
33. (a) 110 m   (b) 1.49 m/s²
35. 29.1 s
37. (a) 31.9 m   (b) 2.55 s   (c) 2.55 s
    (d) -25.0 m/s
39. (a) -21.1 m/s   (b) 19.6 m   (c) -18.1 m/s, 19.6 m
41. (a) 9.80 m/s   (b) 4.90 m
43. (a) 2.33 s   (b) -32.9 m/s
45. (a) -3.5×10^5 m/s²   (b) 2.86×10^{-4} s
47. 1.03 s
49. 3.10 m/s
51. (a) 3.0 s   (b) -15.2 m/s
    (c) -31.4 m/s, -34.8 m/s
53. (a) 2.17 s   (b) 21.3 m/s   (c) 2.24 s
55. (a) 5500 ft   (b) 367 ft/s   (c) The plane would travel 0.002 ft in the time it takes light from the bolt to reach the eye.
57. $a = 1500$ m/s²
59. (a) 2.64 s   (b) -20.9 m/s
    (c) 1.62 s, and 20.9 m/s down

CHAPTER 3

Multiple-Choice Questions
1. a   2. c   3. b   4. c   5. d

Conceptual Questions

1. (a) Yes. Although its speed may be constant, the direction of the motion may change, resulting in an acceleration. For example, an object moving in a circular path with constant speed has an acceleration because the direction of its velocity vector is changing. (b) No. An object that moves with constant velocity has zero acceleration. Note that constant velocity means that both the direction and magnitude of the velocity remain constant.

3. A constant velocity is interpreted as constant in both magnitude and direction. Thus, a constant velocity clearly implies a speed that does not change. On the other hand, a constant speed indicates that only the magnitude of the velocity vector is constant — the direction can change. Thus, as long as an object covers the same distance in equal time intervals, the speed is constant, and its direction can change continuously, giving a varying velocity. A familiar example is an object moving in a circular path at constant speed, such that the velocity is constantly changing direction.

5. You should throw it straight up in the air. Because the ball was moving along with you, it will follow the path of a projectile moving with a constant horizontal velocity that is the same as yours.

7. The components of **A** will both be negative when **A** lies in the third quadrant. The components of **A** will have opposite signs when **A** lies in either the second quadrant or the fourth quadrant.

9. (a) At the top of its flight, the velocity is horizontal and the acceleration is downward. This is the only point at which the velocity and acceleration vectors are perpendicular. (b) If the object is thrown straight up or down, then the velocity and acceleration will be parallel throughout the motion. For any other kind of projectile motion, the velocity and acceleration vectors are never parallel.

11. (a) The acceleration is zero, because |**v**| and its direction remain constant. (b) The particle has an acceleration because the direction of **v** changes.

13. The spacecraft will follow a parabolic path. This is equivalent to a projectile thrown off a cliff with a horizontal velocity. For the projectile, gravity provides an acceleration that is always perpendicular to the initial velocity, resulting in a parabolic path. For the spacecraft, the initial velocity plays the role of the horizontal velocity of the projectile. The acceleration provided by the leaking gas provides an acceleration that plays the role of gravity for the projectile. If the orientation of the spacecraft were to change in response to the gas leak (which is by far the more likely result), then the acceleration would change

direction and the motion could become quite complicated.

15. For angles $\theta < 45°$, the projectile thrown at angle θ will be in the air for a shorter time interval. For a small angle, the vertical component of the initial velocity is smaller than for the larger angle. Thus, the projectile thrown at the smaller angle will not go as high into the air and will spend less time in the air before landing.

17. Yes. The projectile is a freely falling body, because nothing counteracts the force of gravity. The vertical acceleration will be the local gravitational acceleration, g; the horizontal acceleration will be zero.

19. (a) The driver must drive in a straight line. (b) The driver must hold the steering wheel at a steady angle, so that the car will turn at a constant radius from the center of the turn.

Review Problem

(a) 246.2 ft at 35.7° west of south (from C to A)

(b) 0.714 acres

Problems

1. 7.92 m at 4.34° north of west
3. (a) 5.2 m at 60° above x axis
   (b) 3.0 m at 30° below x axis
   (c) 3.0 m at 150° with the x axis
   (d) 5.2 m at 60° below x axis
5. 15.3 m at 58° south of east
7. 8.07 m at 42° south of east
9. (a) 5 blocks at 53.1° north of east   (b) 13 blocks
11. 157 km
13. 245 km at 21.4° west of north
15. 196 cm at 14.7° below x axis
17. 2.8 m from base of table, $v_x = 5.0$ m/s, $v_y = -5.4$ m/s
19. 25 m
21. (a) 32.5 m horizontally from base of cliff   (b) 1.78 s
23. (a) 52 m/s directed horizontally   (b) 210 m
25. 61 s
27. 249 ft upstream
29. 18 s
31. (a) 10.0 m   (b) 15.7 m   (c) 0
33. (a) approximately 2.3 m/s horizontally
35. (a) 57.7 km/h at 60° west of vertical
    (b) 28.9 km/h downward
37. (a) 1.52×10^3 m   (b) 36.1 s   (c) 4.05×10^3 m
39. 18 m on the Moon, 7.9 m on Mars
41. (a) 23 m/s   (b) 360 m horizontally from base of cliff
43. (a) 42 m/s   (b) 3.8 s
    (c) $v_x = 34$ m/s, $v_y = -13$ m/s, $v = 37$ m/s
45. 7.5 m/s in the direction the ball was thrown
49. $R/2$
51. 7.5 min
53. 10.8 m above cannon

55. (a) and (b) $y = -\dfrac{1}{2} gt^2 = -\dfrac{g}{2}\left(\dfrac{x}{v_0}\right)^2 = Ax^2$

    where $A = -\dfrac{g}{2v_0^2}$ (c) 14.6 m/s

57. 227 paces at 165°

59. (a) 20.0°   (b) 3.05 s

CHAPTER 4

Multiple-Choice Questions

1. e   2. c   3. b   4. b   5. b   6. a

Conceptual Questions

1. (a) If a single force acts on it, the object must accelerate. If an object accelerates, at least one force must act on it. (b) If an object has no acceleration, you cannot conclude that no forces act on it. In this case, you can only say that the net force on the object is zero.

3. Motion can occur in the absence of a net force. Newton's first law holds that an object will continue to move with a constant speed and in a straight line if there is no net force acting on it.

5. The inertia of the suitcase would keep it moving forward as the bus stops. There would be no tendency for the suitcase to be thrown backward toward the passenger. The case should be dismissed.

7. The force causing an automobile to move is the force of friction between the tires and the roadway as the automobile attempts to push the roadway backward. The force driving a propeller airplane forward is the reaction force of the air on the propeller as the rotating propeller pushes the air backward (the action). In a rowboat, the rower pushes the water backward with the oars (the action). The water pushes forward on the oars and hence the boat (the reaction).

9. When the bus starts moving, the mass of Claudette is accelerated by the force of the back of the seat on her body. Clark is standing, however, and the only force on him is the friction between his shoes and the floor of the bus. Thus, when the bus starts moving, his feet start accelerating forward, but the rest of his body experiences almost no accelerating force (except that due to his being attached to his accelerating feet!). As a consequence, his body tends to stay almost at rest, according to Newton's first law, relative to the ground. Relative to Claudette, however, he is moving toward her and falls into her lap.

11. The tension in the rope is the maximum force that occurs in *both* directions. In this case, then, because both are pulling with a force of 200 N, the tension is 200 N. If the rope does not move, then the force on each athlete must equal zero. Therefore, each athlete exerts 200 N against the ground.

13. As a man takes a step, the action is the force his foot exerts on the Earth; the reaction is the force of the Earth on his foot. In the second case, the action is the force exerted on the girl's back by the snowball; the reaction is the force exerted on the snowball by the girl's back. The third action is the force of the glove on the ball; the reaction is the force of the ball on the glove. The fourth action is the force exerted on the window by the air molecules; the reaction is the force on the air molecules exerted by the window.

15. The brakes may lock and the car will slide farther than it would if the wheels continued to roll because the coefficient of kinetic friction is less than the coefficient of static friction. Hence, the force of kinetic friction is less than the maximum force of static friction.

17. (a) The crate accelerates because of a friction force exerted on it by the floor of the truck. (b) If the driver slams on the brakes, the inertia of the crate tends to keep it moving forward.

Problems

1. (a) 12 N   (b) 3.0 m/s^2
3. 3.71 N, 58.7 N, 2.27 kg
5. 9.6 N
7. (a) 0.20 m/s^2 (b) 10.0 m   (c) 2.00 m/s
9. 1.59 m/s^2 at 65.2° north of east
11. 796 N in the horizontal cable, 997 N in the inclined cable, 600 N in the vertical cable.
13. 75 N (right side cable), 130 N (left side cable), 150 N (vertical cable)
15. (a) 34 N   (b) 39 N
17. 64 N
19. 7500 N, 50.0 m
21. (a) 7.0 m/s^2 (b) 21 N   (c) 14 N toward the right
23. 13 N down the incline
25. 6.53 m/s^2, 32.7 N
27. (a) 2.15×10^3 N   (b) 645 N   (c) 645 N to rear
    (d) 1.02×10^4 N at 15.9° to rear of vertically downward
29. $\mu_S = 0.38$, $\mu_k = 0.31$
31. (a) 0.256   (b) 0.509 m/s^2
33. 32.1 N
35. (a) 32.7 m/s   (b) No. The object can speed up to 32.7 m/s from any higher speed or slow down to 32.7 m/s from any higher speed.
37. 0.288
39. 0.436
41. 3.32 m/s^2
43. (a) 0.404   (b) 45.8 lb
45. (a) 84.9 N vertically   (b) 84.9 N
47. 50 m
49. (a) friction between the box and truck   (b) 2.94 m/s^2
51. (a) 0.232 m/s^2 (b) 9.68 N
53. (a) 1.78 m/s^2 (b) 0.368   (c) 9.37 N
    (d) 2.67 m/s
55. (a) 1.7 m/s^2, 17 N   (b) 0.69 m/s^2, 17 N
57. 100 N, 204 N
59. (a) $T_1 = 78$ N, $T_2 = 35.9$ N   (b) $\mu_k = 0.66$
61. 23 m/s
63. (a) 7250 N   (b) 4.57 m/s^2
65. 72.0 N
67. 9.95 s
69. (a) 0.408 m/s^2 (b) 83.3 N

CHAPTER 5

Multiple-Choice Questions

1. c   2. b   3. b   4. d   5. c

Conceptual Questions

1. Because there is no motion taking place, the rope experiences no displacement. Thus, no work is done on it. For the same reason, no work is being done on the pullers or the ground. Work is only being done within the bodies of the pullers. For example, the heart of each puller is applying forces on the blood to move blood through the body.

3. Less force will be necessary with a longer ramp, but the force must act over a longer distance to do the same amount of work. Suppose a refrigerator is rolled up the ramp at constant speed. The normal force does no work because it acts at 90° to the motion. The work by gravity is just the weight of the refrigerator times the vertical height through which it is displaced ($W_g = -mgh$). Therefore, the movers must do work mgh on the refrigerator, however long the ramp.

5. If we ignore any effects due to rolling friction on the tires of the car, the same amount of work would be done in driving up the switchback and driving straight up the mountain, because the weight of the car is moved upward against gravity by the same vertical distance in each case. If we include friction, there is more work done in driving the switchback, because the distance over which the friction force acts is much longer. So why do we use switchbacks? The answer lies in the force required, not the work. The force from the engine required to follow a gentle rise is much smaller than that required to drive straight up the hill. Roadways running straight uphill would require redesigning engines in order to be able to apply much larger forces. This is similar to the relative ease with which heavy objects can be moved up ramps into trucks as in Conceptual Question 3.

7. (a) The tension in the supporting cord does no work, because the motion of the pendulum is always perpendicular to the cord, and therefore to the tension force. (b) The air resistance does negative work at all times, because the air resistance is always acting in a direction opposite to the motion. (c) The force of gravity always acts downward; therefore, the work done by gravity is positive on the downswing and negative on the upswing.

9. Because the time periods are the same for both cars, we need only to compare the work done. Because the sports car is moving twice as fast as the older car at the end of the time interval, it has four times the kinetic energy. Thus, according to the work-kinetic energy theorem, four times as much work was done, and the engine must have expended four times the power.

11. If a crate is located on the bed of a truck, and the truck accelerates, the friction force acting on the crate causes it to undergo the same acceleration as the truck, assuming the crate doesn't slip. Another example is a car that accelerates because of the frictional forces between the road surface and its tires. This force is in the direction of motion of the car and produces an increase in the car's kinetic energy.

13. Yes, the total mechanical energy of the system is conserved because the only forces acting are conservative—the force of gravity and the spring force. There are two forms of potential energy in this case, gravitational potential energy and elastic potential energy stored in the spring.

15. Let us assume you lift the book slowly. In this case there are two forces on the book that are almost equal in magnitude. They are the lifting force and the force of gravity acting on the book. Thus, the positive work done by you and the negative work done by gravity cancel. There is no net work performed, and no net change in the kinetic energy—the work-kinetic energy theorem is satisfied.

17. As the satellite moves in a circular orbit about the Earth, its displacement along the circular path during any small time interval is always perpendicular to the gravitational force, which always acts toward the center of the Earth. Therefore, the work done by the gravitational force during any displacement is zero. (Recall that the work done by a force is defined to be $Fs \cos \theta$, where θ is the angle between the force and the displacement. In this case, the angle is 90°, so the work done is zero.) Because the work-kinetic energy theorem says that the net work done on an object during any displacement is equal to the change in its kinetic energy, and the work done in this case is zero, the change in the satellite's kinetic energy is zero; hence its speed remains constant.

Review Problem
(a) 9.3 m/s   (b) 9.3 m/s

Problems
1. 700 J
3. 15.0 MJ
5. (a) 61 J   (b) −46 J   (c) 0
7. (a) 2.00 m/s   (b) 200 N
9. (a) 560 J   (b) 1.2 m
11. 0.265 m/s
13. (a) 90 J   (b) 180 N
15. 1.0 m/s

17. (a) 3.15×10^5 J, 0, 3.15×10^5 J
    (b) 0, -3.15×10^5 J, 3.15×10^5 J
19. 1.9 m/s
21. (a) 80 J   (b) 11 J   (c) 0
23. 0.459 m
25. (a) 0.588 J   (b) 0.588 J   (c) 2.42 m/s
    (d) $PE_C = 0.392$ J, $KE_C = 0.196$ J
27. 5.1 m/s
29. 61 m
31. (a) 9.90 m/s   (b) 11.8 J
33. 2.1×10^3 N
35. 3.8 m/s
37. (a) 5.42 m/s   (b) 0.30   (c) 147 J
39. 289 m
41. (a) 2.1×10^4 J   (b) 0.92 hp
43. 590 MW
45. (a) 2.39×10^4 W   (b) 4.77×10^4 W
47. 2.9 m/s
49. 9.80 m/s
51. 1.4 m/s
53. 3.9×10^3 J
55. (a) 2.3×10^3 N   (b) 4.7×10^2 N
57. (a) 28.0 m/s   (b) 30.0 m   (c) 89.0 m
59. (a) 310 J   (b) −150 J   (c) 0   (d) 150 J
61. (a) javelin requires 349 J, discus requires 676 J, and shot requires 741 J
    (b) 175 N on javelin, 338 N on discus, 371 N on shot
    (c) Yes
63. (a) 6.15 m/s   (b) 9.87 m/s
67. (a) 100 J   (b) 0.410 m   (c) 2.84 m/s
    (d) −9.80 mm   (e) 2.85 m/s
69. 914 N/m

CHAPTER 6

Multiple-Choice Questions
1. a   2. b   3. d   4. c   5. b

Conceptual Questions
1. If all the kinetic energy disappears, there must be no motion of either of the objects after the collision. If nothing is moving there is no momentum. Thus, if the final momentum of the system is zero, then the initial momentum of the system must also have been zero. A situation in which this could be true would be the head-on collision of two objects, each having equal but opposite momenta.

3. Initially the clay has momentum directed toward the wall. When it collides and sticks to the wall, nothing appears to have any momentum. Thus, it is tempting to conclude (wrongly) that momentum is not conserved. However, the "lost" momentum is actually imparted to the wall and Earth, causing both to move. Because of the enormous mass of the Earth, its recoil speed is too small to detect.

5. Before the step the momentum was zero, so afterward the net momentum must also be zero. Obviously, you have some momentum, so something must have momentum in the opposite direction. That something is the Earth. As noted in Conceptual Question 3, the enormous mass of the Earth ensures that its recoil speed will be too small to detect, but if you want to make the Earth move, it is as simple as taking a step.

7. The momentum of the system remains constant, so if the boy gains momentum away from the shore, something must have an equal and opposite momentum toward the shore so that the total momentum of the system remains zero. This something is the raft.

9. There are no external horizontal forces acting on the box, so its momentum cannot change as it moves along the horizontal surface. As the box slowly fills with water, its mass increases with time. Because the product mv must be a constant, and m is increasing, the speed of the box must decrease.

11. It will be easiest to catch the medicine ball when its speed (and kinetic energy) is lowest. The first option — throwing the medicine ball at the same velocity — will be the most difficult, because the speed will not be reduced at all. The second option, throwing the medicine ball with the same momentum, will reduce the velocity by the ratio of the masses. Because $m_t v_t = m_m v_m$,

$$v_m = v_t \left(\frac{m_t}{m_m} \right)$$

The third option, throwing the medicine ball with the same kinetic energy, will also reduce the velocity, but only by the square root of the ratio of the masses. Because $\frac{1}{2} m_t v_t^2 = \frac{1}{2} m_m v_m^2$,

$$v_m = v_t \sqrt{\frac{m_t}{m_m}}$$

Thus, the slowest — and easiest — throw will be made when the momentum is held constant. If you wish to check this, try substituting in values of $v_t = 1$ m/s, $m_t = 1$ kg, and $m_m = 100$ kg. The same-momentum throw will be caught at 1 cm/s, and the same-energy throw will be caught at 10 cm/s.

13. The follow-through keeps the club in contact with the ball as long as possible, maximizing the impulse. Thus, the ball experiences a larger change in momentum than without the follow-through, and it leaves the club with a higher velocity and it travels farther. With a short shot to the green, the primary factor is control, not distance. Thus, there is little or no follow-through, allowing for the golfer to have a better feel for how hard he or she is striking the ball.

15. It is the product mv that is the same for both the bullet and the gun. The bullet has a large velocity and a small mass, and the gun has a small velocity and a large mass. Furthermore, the bullet carries much more kinetic energy than the gun.

Problems

3. (a) 0   (b) 1.1 kg m/s
5. 7.5×10^4 N
7. (a) -7.50 kg m/s   (b) 375 N
9. (a) 12.0 N s   (b) 6.0 m/s   (c) 4.0 m/s
11. (a) 0.096 s   (b) 3.65×10^5 N   (c) 26.5 g
13. (a) 6.3 kg m/s
    (b) 3.2×10^3 N (both directed toward the pitcher)
15. 62 s
17. (a) 0.49 m/s   (b) 2.0×10^{-2} m/s
19. $v_{\text{thrower}} = 2.48$ m/s, $v_{\text{catcher}} = 2.25 \times 10^{-2}$ m/s
21. 2.66 m/s
23. (a) 1.80 m/s   (b) 2.16×10^4 J
25. 57 m
27. 273 m/s
29. (a) -6.67 cm/s, 13.3 cm/s   (b) 0.88
31. 17.1 cm/s (25-g object), 22.1 cm/s(10-g object)
33. (a) 2.9 m/s at 32° with respect to initial direction of travel of fullback   (b) 780 J
35. 5.59 m/s
37. (a) 2.5 m/s at $-60°$ with respect to the original line of motion of incident ball   (b) elastic
39. 14.8 kg·m/s in the direction of the final velocity of the ball
41. (a) $\frac{8}{3}$ m/s (incident particle), $\frac{32}{3}$ m/s (target)
    (b) $\frac{-16}{3}$ m/s (incident particle), $\frac{8}{3}$ m/s (target)
    (c) 7.1×10^{-3} J, in case (a), 2.8×10^{-3} J in case (b). The incident particle loses more energy in case (a).
43. 0.96 m above the level of point B
45. 91.2 m/s
47. (a) -2.33 m/s, 4.67 m/s   (b) 0.277 m   (c) 2.98 m
    (d) 1.49 m
49. (a) -0.67 m/s   (b) 0.953 m
51. (a) 300 m/s   (b) 3.75 m/s   (c) 1.20 m
53. 2.36 cm
55. $v = \frac{4M}{m} \sqrt{g\ell}$
57. about 10^{-23} m/s
59. (a) 90°   (b) $v_1 = 3.46$ m/s and $v_2 = 2.00$ m/s

CHAPTER 7

Multiple-Choice Questions

1. e   2. a   3. b   4. e   5. a

Conceptual Questions

1. An object can move in a circle even if the total force on it is not perpendicular to its velocity, but then its speed will change. Resolve the total force into an inward radial component and a perpendicular tangential component. If the tangential force is forward, the object will speed up, and if the tangential force acts backward, it will slow down.

3. Yes, all points on the wheel have the same angular speed. This is why we use angular quantities to describe rotational motion. Not all points on the wheel have the same linear speed. The point at $r = 0$ has zero linear speed and zero linear acceleration; a point at $r = R/2$ has a linear speed $v = R\omega/2$ and a linear acceleration equal to the centripetal acceleration $v^2/(R/2) = R\omega^2/2$. (The tangential acceleration is zero at all points because ω is a constant.) A point on the rim at $r = R$ has a linear speed $v = R\omega$ and a linear acceleration $R\omega^2$.

5. To a good first approximation, your bathroom scale reading is unaffected because you, Earth, and the scale are all in free fall in the Sun's gravitational field, in orbit around the Sun. To a precise second approximation, you weigh slightly less at noon and at midnight than you do at sunrise or sunset. The Sun's gravitational field is a little weaker at the center of the Earth than at the surface subsolar point, and a little weaker still on the far side of the planet. When the Sun is high in your sky, its gravity pulls up on you a little more strongly than on the Earth as a whole. At midnight the Sun pulls down on you a little less strongly than it does on the Earth below you. So you can have another doughnut with lunch, and your bedsprings will still last a little longer.

7. Air resistance causes a decrease in the energy of the satellite-Earth system. This reduces the diameter of the orbit, bringing the satellite closer to the surface of the Earth. A satellite in a smaller orbit, however, must travel faster. Thus, the effect of air resistance is to speed up the satellite!

9. Kepler's third law, which applies to all planets, tells us that the period of a planet is proportional to $r^{3/2}$. Because Saturn and Jupiter are farther from the Sun than Earth, they have longer periods. The Sun's gravitational field is much weaker at a distant Jovian planet. Thus, an outer planet experiences much smaller centripetal acceleration than Earth and a correspondingly longer period.

11. Consider an individual standing against the inside wall of the cylinder with her head pointed toward the axis of the cylinder. As the cylinder rotates, the person tends to move in a straight-line path tangent to the circular path followed by the cylinder wall. As a result, the person is forced against the wall, and the normal force exerted on her provides the radial force required to keep her moving in a circular path. If the rotational speed is adjusted such that this normal force is equal in magnitude to her weight on Earth, she would not be able to distinguish between the artificial gravity of the colony and ordinary gravity.

13. (a) As the object moves in its circular path with constant speed, the *direction* of the velocity vector changes. Thus, the velocity of the object is not constant. (b) The magnitude of its acceleration remains constant, and is equal to v_t^2/r. The acceleration vector is always directed toward the center of the circular path.

15. Any object that moves such that the *direction* of its velocity changes has an acceleration. A car moving in a circular path will always have a centripetal acceleration.

17. When a pilot is pulling out of a dive, blood leaves the head because there is not a great enough radial force to cause it to follow the circular path of the airplane. The loss of blood from the brain can cause the pilot to black out.

19. The tendency of the water is to move in a straight-line path tangent to the circular path followed by the container. As a result, at the top of the circular path, the water is forced against the bottom of the pail, and the normal force exerted on the water by the pail provides the radial force required to keep the water moving in its circular path.

Review Problem
(a) 3.77 m/s^2 (b) 3.26 s

Problems
1. (a) 3.2×10^8 rad   (b) 5.0×10^7 revolutions
3. (a) 3.5 rad/s   (b) 5.2 rad
5. 4.2×10^{-2} rad/s^2
7. 3.2 rad
9. (a) 5.24 s   (b) 27.4 rad
11. 13.7 rad/s^2
13. 1.02 m
15. (a) -0.17 rad/s^2 (b) 5.5 rev   (c) 0.48 m/s
17. 5.0×10^{-2} rad/s
19. (a) 3.5×10^{-1} m/s^2 (b) 1.0 m/s
    (c) 3.5×10^{-1} m/s^2, 0.94 m/s^2, 1.0 m/s^2 at 20° with respect to the direction of $\mathbf{a}_r$
21. 150 rev/s
23. (a) 18.0 m/s^2 (b) 900 N   (c) 1.84; a coefficient of friction greater than 1 is unreasonable. She will not be able to stay on the merry-go-round.
25. (b) 20.1°
27. The required tension in the vine is 1.4×10^3 N. He does not make it.
29. (a) 2.49×10^4 N   (b) 12.1 m/s
31. 321 N toward Earth
33. 1.1×10^{-10} N at 72° from $+x$ axis
35. 6.01×10^{24} kg. The estimate is high because the Moon actually orbits about the center of mass of the Earth–Moon system not about the center of the Earth.
37. (a) 5.58×10^3 m/s   (b) 239 min   (c) 1.47×10^3 N
39. (a) -4.76×10^9 J   (b) 5.68×10^2 N
41. (a) 88.5 min   (b) 7.79 km/s   (c) 6.43×10^9 J
43. (a) 2.3×10^{-10} N (in $-x$ direction)
    (b) 1.0×10^{-10} N (in $+x$ direction)
45. 1.90×10^{27} kg
47. (a) 7.8×10^3 m/s   (b) 5.4×10^3 s (89 min)
49. (a) $n = mg - \left(\dfrac{mv^2}{r}\right)$ (b) 17.1 m/s
51. (a) 2.0×10^{12} m/s^2 (b) 2.4×10^{11} N
    (c) 1.4×10^{12} J
53. 0.131
55. (a) 2.1 m/s   (b) 54°   (c) 4.7 m/s
57. 35.2 A.U.

59. (a) 1.31×10^{17} N   (b) 2.62×10^{12} N/kg
63. (a) 15.1 N at A, 29.8 N at B (b) 0.166
65. (a) $v_{min} = [Rg(\tan \theta - \mu)/(1 + \mu \tan \theta)]^{1/2}$,
    $v_{max} = [Rg(\tan \theta + \mu)/(1 - \mu \tan \theta)]^{1/2}$
    (b) 8.57 m/s to 16.6 m/s

CHAPTER 8

Multiple-Choice Questions

1. b   2. c   3. c   4. d   5. a

Conceptual Questions

1. There are two major differences between torque and work. The primary difference is that the displacement in the expression for work is directed along the force, whereas the important distance in the torque expression is perpendicular to the force. The second difference involves whether there is motion or not—in the case of work, there is work done only if the force succeeds in causing a displacement of the point of application of the force. However, a force applied at a perpendicular distance from a rotation axes results in a torque whether or not there is motion.

    As far as units are concerned, the mathematical expressions for both work and torque result in the product of newtons and meters, but this product is called a joule in the case of work and remains as a newton-meter in the case of torque.

3. On a frictionless table, the only forces on the pencil are the gravitational force, at the center of the mass, and the normal force on the pencil. These forces result in a torque on the pencil causing it to rotate. Both of these forces, however, are vertical—there are no horizontal forces on the pencil in the absence of friction. As a result, the center of mass of the pencil must fall straight downward. As the pencil falls, the tip slides to the side, so that the center of mass of the pencil follows a straight line downward to the floor.

5. As the motorcycle leaves the ground, the friction between the tire and the ground suddenly disappears. If the motorcycle driver keeps the throttle open while leaving the ground, the rear tire will increase its angular velocity and, hence, its angular momentum. The airborne motorcycle is now an isolated system, and its angular momentum must be conserved. The increase in angular momentum of the tire, say, directed clockwise must be compensated by an increase in angular momentum of the entire motorcycle counterclockwise. This rotation results in the nose of the motorcycle rising and the tail dropping.

7. In general, you want the rotational kinetic energy of the system to be as small a fraction of the total energy as possible—you want translation, not rotation. You want the wheels to have as little moment of inertia as possible, so that they represent the lowest resistance to changes in rotational motion. Disk-like wheels would have lower moments of inertia than hoop-like wheels, so disks are preferable. The lower the mass of the wheels, the less is the moment of inertia, so light wheels are preferable. The smaller the radius of the wheels, the less is the moment of inertia, so smaller wheels are preferable, within limits—you want the wheels to be large enough to be able to travel relatively smoothly over irregularities in the road.

9. The angular momentum of the gas cloud is conserved. Thus the product $I\omega$ remains constant. Hence, as the cloud shrinks in size, its moment of inertia decreases, so its angular speed ω must increase.

11. The Earth already bulges slightly at the Equator and is slightly flat at the poles. If more mass moved toward the Equator, it would essentially move the mass to a greater distance from the axis of rotation, and increase the moment of inertia. Because conservation of angular momentum requires that $I_z\omega_z$ = constant, an increase in the moment of inertia would decrease the angular velocity and slow down the spinning of the Earth. Thus, the length of each day would increase.

13. We can assume fairly accurately that the driving motor will run at a constant angular speed and at a constant torque. Therefore, as the radius of the takeup reel increases, the tension in the tape will decrease.

    (1) $T = \tau_{const}/R_{takeup}$

    (2) $\tau_{source} = FR_{source} = \tau_{const}R_{source}/R_{takeup}$

    As the radius of the source reel decreases, given a decreasing tension, the torque in the source reel will **decrease** even faster.

    This torque will be partly absorbed by friction in the feed heads (which we assume to be small); some will be absorbed by friction in the source reel. Another small amount of the torque will be absorbed by the increasing angular velocity of the source reel. However, in the case of a sudden jerk on the tape, the changing angular velocity of the source reel becomes important. If the source reel is full, then the moment of inertia will be large, and the tension in the tape will be large. If it is nearly empty, then the angular acceleration will be large instead. Thus, the tape will be more likely to break when the source reel is nearly **full.** One sees the same effect in the case of paper towels; it is easier to snap a towel free when the roll is new than when it is nearly empty.

15. The initial angular momentum of the system (mouse plus turntable) is zero. As the mouse begins to walk clockwise, its angular momentum increases, so the turntable must rotate in the counterclockwise direction with an angular momentum whose magnitude equals that of the mouse. This follows from the fact that the final angular momen-

tum of the system must equal the initial angular momentum, which is zero in this case.

17. When a ladder leans against a wall, both the wall and floor exert forces of friction on the ladder. If the floor is perfectly smooth, it can exert no frictional force in the horizontal direction to counterbalance the wall's normal force. Therefore a ladder on a smooth floor cannot stand in equilibrium. However, a smooth wall can still exert a normal force to hold the ladder in equilibrium against horizontal motion. The counterclockwise torque of this force prevents rotation about the foot of the ladder. So you should choose a rough floor.

Review Problem

(a) 3.46×10^5 km from center of Earth
(b) 4.68×10^3 km from center of Earth. No, the two points do not coincide.

Problems

1. 133 N
3. 5.1 N·m
5. (a) 29.6 N·m (counterclockwise)
   (b) 35.6 N·m (counterclockwise)
7. $x_{cg} = 6.69 \times 10^{-3}$ nm, $y_{cg} = 0$
9. $x_{cg} = 3.33$ ft, $y_{cg} = 1.67$ ft
11. 312 N
13. $T = 157$ N, $R = 107$ N
15. T(left wire) $= \frac{1}{3}W$, T(right wire) $= \frac{2}{3}W$
17. (a) 443 N (b) 222 N (to right) 216 N (upward)
19. $T_1 = 501$ N, $T_2 = 672$ N, $T_3 = 384$ N
21. 6.2 m
23. 209 N
25. (a) 99.0 kg·m^2 (b) 44.0 kg·m^2 (c) 140 kg·m^2
27. $\mu_k = 0.31$
29. 177 N
31. (a) 5.35 m/s^2 downward (b) 42.8 m
    (c) 8.91 rad/s^2
33. 17.3 rad/s
35. (a) 500 J (b) 250 J (c) 750 J
37. 276 J
39. 149 rad/s
41. 17.3 rad/s
43. 17.5 J·s counterclockwise
45. 8.0 rev/s
47. (a) 1.9 rad/s (b) $KE_i = 2.5$ J, $KE_f = 6.4$ J
49. (a) 3.6 rad/s (b) 540 J (difference results from work done by the man as he walks inward)
51. (a) 1.2 kg (b) at the 59.6 cm mark
53. $T = 2.70 \times 10^3$ N, $R_x = 2.65 \times 10^3$ N
55. 36.9 °
57. $a_{sphere} = \dfrac{g \sin \theta}{1.4}$, $a_{cylinder} = \dfrac{g \sin \theta}{1.5}$, $a_{ring} = \dfrac{g \sin \theta}{2.0}$
    Thus, sphere wins and ring comes in last.
59. (a) 1.1 m/s^2 (b) $T_1 = 22$ N $T_2 = 44$ N

61. (a) 24.0 N·m (b) 0.0356 rad/s^2 (c) 1.07 m/s^2
63. (a) 3750 kg·m^2/s (b) 1.88 kJ (c) 3750 kg·m^2/s
    (d) 10.0 m/s (e) 7.5 kJ (f) 5.62 kJ
65. $T_R = 1589$ N, $T_L = 1006$ N
67. (a) 0.36 rad/s, counterclockwise (b) 99.9 J
69. 24 m
71. $\frac{3}{8} w$

CHAPTER 9

Multiple-Choice Questions

1. b   2. d   3. c   4. c   5. e

Conceptual Questions

1. She can exert enough pressure on the floor to dent or puncture the floor covering. The large pressure is caused by the fact that her weight is distributed over the very small cross-sectional area of her high heels. If you are the homeowner, you might want to suggest that she remove her high heels and put on some slippers.

3. If you think of the grain stored in the silo as a fluid, the pressure the grain exerts on the walls of the silo increases with increasing depth just as water pressure in a lake increases with increasing depth. Thus, the spacing between bands is made smaller at the lower portions to overcome the larger outward forces on the walls in these regions.

5. The level of floating of a ship is unaffected by atmospheric pressure. The buoyant force results from the pressure differential in the fluid. On a high-pressure day, the pressure at all points in the water is higher than on a low-pressure day. Because water is almost incompressible, however, the rate of change of pressure with depth is the same, resulting in no change in the buoyant force.

7. In the ocean, the ship floats due to the buoyant force from *salt water*. Salt water is denser than fresh water. As the ship is pulled up the river, the buoyant force from the fresh water in the river is not sufficient to support the weight of the ship, and it sinks.

9. The balance will not be in equilibrium—the lead side will be lower. Despite the fact that the weights on both sides of the balance are the same, the styrofoam, due to its larger volume, will experience a larger buoyant force from the surrounding air. Thus, the net force of the weight and the buoyant force is larger, in the downward direction, for the lead than for the styrofoam.

11. The level of the pond falls. This is because the anchor displaces more water while in the boat. A floating object displaces a volume of water whose weight is equal to the weight of the object. A submerged object displaces a volume of water equal to the volume of the object. Because

the density of the anchor is greater than that of water, a volume of water that weighs the same as the anchor will be greater than the volume of the anchor.

13. According to Archimedes's principle, the magnitude of the buoyant force on the ship is equal to the weight of the water displaced by the ship. Because the density of salty ocean water is greater than fresh lake water, less ocean water needs to be displaced to enable the ship to float. Thus, the boat floats higher in the ocean than in the inland lake.

15. As the truck passes, the air between your car and the truck is compressed into the channel between you and the truck and moves at a higher speed than when your car is in the open. According to Bernoulli's principle, this high-speed air has a lower pressure than the air on the outer side of your car. The difference in pressure provides a net force on your car toward the truck.

Problems

1. 1.8×10^6 Pa
3. 3.5×10^8 Pa
5. 5.5 mm
7. 2.4×10^{-2} mm
9. (a) 1.88×10^5 Pa　　(b) 2.65×10^5 Pa
11. 1.2×10^6 Pa
13. 3.58×10^6 Pa
15. 1.4 atm
17. 1.05×10^5 Pa
19. 2.0 N·m
21. 9.41×10^3 N
23. 5.57 N
25. (a) 1.46×10^{-2} m^3　　(b) 2.10×10^3 kg/m^3
27. 1.07 m/s^2
29. 78 kg
31. 154 in/s
33. (a) 80 g/s　　(b) 2.7×10^{-2} cm/s
35. (a) 11.0 m/s　　(b) 2.64×10^4 Pa
37. 4.4×10^{-2} Pa
39. (a) 17.7 m/s　　(b) 1.73 mm
41. (a) 2.7 m/s　　(b) 2.3×10^4 Pa
43. 7.32×10^{-2} N/m
45. 5.6 m
47. 0.694 mm
49. 0.12 N
51. 1.5 m/s
53. 0.41 mm
55. 8.0 cm/s
57. 9.5×10^{-10} m^2/s
59. 1.02×10^3 kg/m^3
61. 2.5×10^7 capillaries
63. 833 kg/m^3
65. 1.9 m
67. 15 m

69. (a) 18.34 mm　　(b) 14.31 mm　　(c) 8.56 mm
71. (a) 1.25 cm　　(b) 13.8 m/s
73. 0.72 mm
75. (b) 2.6×10^4 N
77. 1.71 cm

CHAPTER 10

Multiple-Choice Questions

1. b　　2. d　　3. b　　4. a　　5. b

Conceptual Questions

1. The accurate answer is that it doesn't matter! Temperatures on the Kelvin and Celsius scales differ by only 273 degrees. This difference is insignificant for temperatures on the order of 10^7 degrees. If we imagine that the temperature is given in kelvin, and ignore any problems with significant figures, then the Celsius temperature is 1.4999727×10^7 °C.

3. Mercury must have the larger coefficient of expansion. As the temperature of a thermometer rises, both the mercury and the glass expand. If they both had the same coefficient of linear expansion, the mercury and the cavity in the glass would both expand by the same amount, and there would be no apparent movement of the end of the mercury column relative to the calibration scale on the glass. If the glass expanded more than the mercury, the reading would go down as the temperature went up! Now that we have argued this conceptually, we can look in a table and find that the coefficient for mercury is about 20 times as large as for glass, so that the expansion of the glass can sometimes be ignored.

5. This occurs because the water in the center of the glass is being heated, but the glass itself is not rising in temperature as rapidly as the water. The bubbles associated with boiling usually first appear at irregularities on the surface of the glass, but if the glass is not yet at 100°C, boiling does not take place. The superheated liquid is metastable and lacks nuclei for bubble formation.

7. A cavity in a material expands in exactly the same way as if the cavity were filled with material. Thus, both spheres will expand by the same amount.

9. The chip bags contain a sealed sample of air. When the bags are taken up the mountain, the external atmospheric pressure on the bags is reduced. As a result, the difference between the pressure of the air inside the bags and the reduced pressure outside results in a net force pushing the plastic of the bag outward.

11. On a cold day, the trapped air in the bubbles would be reduced in pressure, according to the ideal gas law. Thus, the volume of the bubbles may be smaller than on a hot day, and the material would not be as effective in cushioning the package contents.

13. One can think of each bacterium as being a small bag of liquid containing bubbles of gas at a very high pressure. If the bacterium is raised rapidly to the surface, the ideal gas law indicates that the volume must increase dramatically. In fact, this increase in volume is sufficient to rupture the cell.

15. The existence of an atmosphere on a planet is due to the gravitational force holding the gas of the atmosphere to the planet. On a small planet, the gravitational force is very small, and the escape speed is correspondingly small. If a small planet starts its existence with an atmosphere, the molecules of the gas will have a distribution of speeds, according to kinetic theory. Some of these molecules will have speeds higher than the escape speed of the planet and will leave the atmosphere. As the remaining atmosphere is warmed by radiation from the Sun, more molecules will attain speeds high enough to escape. As a result, the atmosphere bleeds off into space.

17. The measurements are too short. At 22°C the tape would read the width of the object accurately, but an increase in temperature causes the divisions ruled on the tape to be farther apart than they should be. This "too long" ruler will, then, measure objects to be shorter than they really are.

19. The glass surrounding the mercury expands before the mercury does, causing the level of the mercury to drop slightly. The mercury rises after it begins to heat up and approach the temperature of the hot water, because its temperature coefficient of expansion is greater than that for glass.

21. Imagine the balloon rising into air at uniform temperature. The air cannot be uniform in pressure because the lower layers support the weight of all the air above them. The rubber in a typical balloon stretches or contracts until interior and exterior pressures are nearly equal. So as the balloon rises, it expands; this can be considered as a constant temperature expansion with P decreasing as V increases by the same factor in $PV = nRT$. If the rubber wall is strong enough, it will eventually contain the helium at a pressure higher than the air outside but at the same density, so that the balloon will stop rising. It is more likely that the rubber will stretch and rupture, releasing the helium, which in turn will "boil out" of the Earth's atmosphere.

Problems

1. (a) $-459.67°F$ (b) $37.0°C$ (c) $-280°F$
3. (a) $-423°F$, 20.28 K (b) $68°F$, 293 K
5. (a) $-273.5°C$ (b) $P_f = 1.272$ atm, $P_b = 1.737$ atm
9. 2.171 cm
11. 55.0°C
13. (a) 263.5°C (b) $-262.2°C$
15. (a) 437°C (b) 2099°C; aluminum melts at 660°C
17. 270 N

19. 1020 gallons
21. (a) 627°C (b) 927°C
23. (a) 2.5×10^{19} molecules (b) 4.1×10^{-21} mol
25. 290°C
27. 4.8 g
29. 3.84 m
31. 0.131 kg/m^3
33. 6.21×10^{-21} J
35. 6.64×10^{-27} kg
37. (a) $v_{H_2} = 1.73$ km/s (b) $v_{CO_2} = 0.369$ km/s
(c) Hydrogen escapes; carbon dioxide does not.
39. 16 N
41. 5.1×10^{-21} J/molecule
43. 28 m
45. 2.4 m
47. 800°C
49. (a) 343 K (b) 12.5% of original mass
51. $L_C = 9.17$ cm, $L_S = 14.17$ cm
53. (a) 0.169 m (b) 1.35×10^5 Pa
55. 0.417 liters
57. 3.3 cm

CHAPTER 11

Multiple-Choice Questions
1. d 2. c 3. e 4. c 5. b

Conceptual Questions
1. When you rub the surface, you increase the temperature of the rubbed region. With the metal surface, some of this energy is transferred away from the rubbing site by conduction. Thus, the temperature in the rubbed area is not as high for the metal as it is for the wood, and it feels relatively cooler than the wood.

3. If the fruit were open to the cold air, it is possible that the temperature of the fruit could drop to a point several degrees below freezing, causing damage. By spraying with water, a thin film of ice can form on the fruit. This provides some insulation to the fruit so that its temperature can remain somewhat higher than that of the air.

5. The operation of the immersion coil depends on the convection of water to maintain a safe temperature. As the water near the coils warms up, it floats to the top due to Archimedes's principle. The temperature of the coils cannot go higher than the boiling temperature of water, 100°C. If the coils are operated in air, the convection process is reduced, and the upper limit of 100°C is removed. As a result, the coils can become hot enough to be damaged. If the coils are used in an attempt to warm a thick liquid like stew, the convection process cannot occur fast enough to carry energy away from the coils, so that they again become hot enough to be damaged.

7. One of the ways that objects transfer energy is by radiation. If we consider a mailbox, the top of the box is oriented toward the clear sky. Radiation emitted by the top of the mailbox goes upward and into space. There is little radiation coming down from space to the top of the mailbox. Radiation leaving the sides of the mailbox is absorbed by the environment. Radiation from the environment (tree, houses, cars, etc.), however, can enter the sides of the mailbox, keeping them warmer than the top. As a result, the top is the coldest portion and frost forms there first.

9. The tile is a better conductor of heat than carpet. Thus, energy is conducted away from your feet more rapidly by the tile than by the carpeted floor.

11. The large amount of energy stored in the concrete during the day as the sun falls on it is released at night resulting in an overall higher average temperature than the countryside. The heated air in a city rises to be replaced by cooler air drawn in from the countryside. Thus, evening breezes tend to blow from country to city.

13. The fingers are wetted to create a layer of steam between the fingers and the molten lead. The steam acts as an insulator and prevents serious burns. The molten lead demonstration is dangerous, and we do not recommend it.

15. The increase in temperature of the ethyl alcohol will be about twice that of the water.

Review Problem
20.02 cm

Problems
1. 1.17 Cal
3. 1.03×10^3 J
5. 176°C
7. 10.1°C
9. 88 W
11. 467 pellets
13. 185 g
15. 35°C
17. 1.7 kg
19. 0.26 kg
21. 66°C
23. 21 g
25. 2.3 km
27. 16°C
29. (a) all ice melts, $T_f = 40°C$ (b) 8.0 g, 0°C
31. (a) 470 J/s into house   (b) 1.7×10^3 J/s to outdoors
33. 52 J/s (Thermopane), 1.8×10^3 J/s (single pane)
35. 9.0 cm
37. 110 W
39. 2.6×10^{3}°C
41. 1.3×10^4 J/s
43. 89.8°C (silver), 89.7°C (copper); copper wins
45. 2.4 kg
47. 1.4 kg
49. 12 h
51. (a) 75°C   (b) 3.6×10^4 J
53. 38°C
55. 28°C
57. (a) 25.8°C   (b) no
59. 51.2°C
61. 400 cm^3/h
63. (a) 74 stops   (b) Answer assumes no heat loss to surroundings and that all internal energy generated remains with the brakes.

CHAPTER 12

Multiple-Choice Questions
1. b   2. c   3. d   4. d   5. c

Conceptual Questions
1. If there is no change in internal energy, then, according to the first law, the heat is equal to the work done by the gas, $Q = W$.

3. The energy that is leaving the body by work and heat is replaced by means of biological processes that transform chemical energy in the food that the individual ate into internal energy. Thus, the temperature of the body can be maintained.

5. The statement shows a misunderstanding of the concept of heat. Heat is energy in the process of being transferred, not a form of energy that is held or contained. If you wish to speak of energy that is "contained," you speak of *internal energy*, not heat. Correct statements would be, (1) "Given any two objects in thermal contact, the one with the higher temperature will transfer heat to the other." (2) "Given any two objects of equal mass, the one with the higher product of absolute temperature and specific heat contains more internal energy."

7. Although no heat is transferred into or out of the system, work is done on the system as the result of the agitation. As a consequence, both the temperature and the internal energy of the coffee increase.

9. The most basic limit of efficiency for automobile engines is the Carnot cycle; the efficiency can **never** exceed this ideal model. The efficiency is therefore affected by the maximum temperature that the engine block can stand and by the intake and exhaust pressures. However, the Carnot cycle assumes that the exhaust gases can be released over an infinite period of time. If you want your engine to act more quickly than "never," you must take additional losses. Other limits on efficiency are imposed by friction, the mass and acceleration of the engine parts, and the timing of the ignition.

11. Practically speaking, it is not possible to create a heat engine on Earth that creates no thermal pollution, because there must be both a hot heat source and a cold heat sink. The heat engine will warm the cold heat sink and

will cool down the heat source. If either of those two events is undesirable, then there will be thermal pollution.

There are some circumstances in which the thermal pollution would be negligible. For example, suppose a satellite in space were to run a heat pump between its sunny side and its dark side. The satellite would intercept some of the heat that gathered on one side and would dump it to the dark side. Because neither of those effects would be particularly undesirable, it could be said that such a heat pump produced no thermal pollution.

It is also possible to make the heat source and sink totally self-contained. In such a case, the engine would function until the sink became as hot as the source. However, building such an engine would probably be more wasteful than simply finding an acceptable form of thermal pollution.

13. The rest of the Universe must have an entropy change of +8.0 J/K or more.

15. The first law is a statement of conservation of energy that says that we cannot devise a process that produces more energy than we put into it. In addition, the second law says that during the operation of a heat engine, some heat must be rejected to the environment. As a result, it is theoretically impossible to construct an engine that will work with 100% efficiency.

17. (a) The semiconductor converter operates essentially like a thermocouple, which is a pair of wires of different metals and a junction at each end. When the junctions are at different temperatures, a small voltage appears around the loop, so that the device can be used to measure temperature or to drive a small motor. (b) The second law holds that it is impossible for a cycling engine simply to extract heat from one reservoir and convert it into work. This exactly describes the first situation, in which both legs are in contact with a single reservoir and the thermocouple fails to produce work. To convert heat to work, the device must transfer heat from a hot reservoir to a cold reservoir, as in the second situation.

Review Problem

(a) 31 m/s   (b) 1/6

Problems

3. 1.1×10^4 J
5. (a) 810 J   (b) 507 J   (c) 203 J
7. (a) 6.1×10^5 J   (b) -4.6×10^5 J
9. (a) 1093 K (820°C)   (b) 6.81×10^3 J
11. $6P_0V_0$
13. (a) -152 J   (b) -248 J
15. (a) 12.0 kJ   (b) -12.0 kJ
17. (a) -180 J   (b) -188 J
19. (a) 9.12×10^{-3} J   (b) -333 J
21. (a) $W_{IAF} = 76.0$ J, $W_{IBF} = 100$ J, $W_{IF} = 88.6$ J
    (b) $Q_{IAF} = 165$ J, $Q_{IBF} = 190$ J, $Q_{IF} = 178$ J

23. (a) 560 J   (b) 350 K
25. $e = 19.7\%$
27. (a) $e = 0.294$ (b) 500 J   (c) 1.67×10^3 W
29. $\dfrac{T_c}{T_h} = \dfrac{1}{3}$
31. 0.48°C
33. 6.1×10^3 J/K
35. 2.70×10^3 J/K
37. (a)

Result	Possible Combinations	Total
all heads	HHHH	1
3H, 1T	THHH, HTHH, HHTH, HHHT	4
2H, 2T	TTHH, THTH, THHT, HTTH, HTHT, HHTT	6
1H, 3T	HTTT, THTT, TTHT, TTTH	4
all tails	TTTT	1

Most probable result = 2H and 2T.
(b) All H or All T   (c) 2H and 2T

39. Maximum efficiency possible with these reservoirs = 50%; claim is invalid.
41. (a) $\dfrac{-Q_h}{T_h}$ (b) $\dfrac{Q_c}{T_c}$ (c) $\dfrac{Q_h}{T_h} - \dfrac{Q_c}{T_c}$ (d) 0
43. 2.8°C
47. (a) 1.22×10^4 J   (b) -4.05×10^3 J
    (c) 8.10×10^3 J
49. 18°C
51. (a) 12.0 MJ   (b) -12.0 MJ
53. 3.27 J/K
55. (a) 4.9×10^{-2} J   (b) 1.6×10^4 J   (c) 1.6×10^4 J
57. (a) 2.49×10^3 J   (b) 1.50×10^3 J   (c) 998 J
59. (a) $2.7 \times 10^3 \dfrac{\text{metric tons}}{\text{day}}$ (b) $\dfrac{\$7.7 \times 10^6}{\text{year}}$
    (c) 4.0×10^4 kg/s

CHAPTER 13

Multiple-Choice Questions

1. c   2. d   3. b   4. b   5. d   6. a

Conceptual Questions

1. No! Because the total energy $E = \frac{1}{2}kA^2$, changing the mass while keeping A constant has no effect on the total energy. When the mass is at displacement x from equilibrium, the potential energy is $\frac{1}{2}kx^2$, independent of mass, and the kinetic energy is $E - \frac{1}{2}kx^2$. The larger mass must move slower to have the same kinetic energy. At a particular instant in time, both kinetic and potential energy would change as the mass is increased.

3. When the spring with two masses is set into oscillation in space, the coil in the exact center of the spring does not move. Thus, we can imagine clamping the center coil in

place without affecting the motion. If we do this, we have two separate oscillating systems, one on each side of the clamp. The half-spring on each side of the clamp has twice the spring constant of the full spring, as shown by the following argument. The force exerted by a spring is proportional to the separation of the coils as the spring is extended. Imagine that we extend a spring by a given distance and measure the distance between coils. We then cut the spring in half. If one of the half-springs is now extended by the same distance, the coils will be twice as far apart as they were for the complete spring. Thus, it takes twice as much force to stretch the half-spring, from which we conclude that the half-spring has a spring constant that is twice that of the complete spring. Our clamped system of masses on two half-springs, therefore, will vibrate with a frequency that is higher than f by a factor of $\sqrt{2}$.

5. The bouncing ball is not an example of simple harmonic motion. The ball does not follow a sinusoidal function for its position as a function of time. The daily movement of a student is also not simple harmonic motion, because the student stays at a fixed location—school—for a long period of time. If this motion were sinusoidal, the student would move more and more slowly as she approached her desk, and as soon as she sat down at the desk, she would start to move back toward home again.

7. If it accelerates upward, the effective "g" is greater than the free-fall acceleration, so the period decreases. If it accelerates downward, the effective "g" is less than the free-fall acceleration, so the period increases. If it moves with constant velocity, the period does not change. (If the pendulum is in free-fall, it does not oscillate.)

9. As the temperature increases, the length of the pendulum will increase, due to thermal expansion. With a longer length, the period of the pendulum will increase. Thus, it will take longer to execute each swing, so that each second according to the clock will take longer than an actual second. Thus, the clock will run slow.

11. A pulse in a long line of people is longitudinal, because the movement of people is parallel to the direction of propagation of the pulse. The speed is determined by the reaction time of the people and the speed with which they can move once a space opens up. There is also a psychological factor, in that people will not want to fill a space that opens up in front of them too quickly, so as not to intimidate the person in front of them. The "wave" at a stadium is transverse, because the fans stand up vertically as the wave sweeps past them horizontally. The speed of this pulse depends on the limits of the fans' abilities to rise and sit rapidly and on psychological factors associated with the anticipation of seeing the pulse approach the observer's location.

13. A wave on a massless string would have an infinite speed of propagation because its linear mass density is zero.

15. The kinetic energy is proportional to the square of the speed, and the potential energy is proportional to the square of the displacement. Therefore, both must be positive quantities.

17. It travels a distance of $4A$.

19. In a simple harmonic oscillator, the velocity follows the displacement by $1/4$ of a cycle, and the acceleration follows the displacement by $1/2$ of a cycle. (a) There are times when both the displacement and the velocity are positive, and therefore in the same direction. (b) There are also times when both the velocity and the acceleration are positive, and therefore in the same direction. On the other hand, (c) when the displacement is positive, the acceleration is always negative, and therefore these two always act in opposite directions.

Problems

1. (a) 24 N   (b) 60 m/s^2
3. (b) 1.8 s   (c) No, the force is not of Hooke's law form.
5. (a) 60 J   (b) 49 m/s
7. 2.94×10^3 N/m
9. 2.6 m/s
11 (a) 11 cm/s   (b) 6.3 cm/s   (c) 3.0 N
13. (a) 28 cm/s   (b) 26 cm/s   (c) 26 cm/s
    (d) 3.5 cm
15. 39.2 N
17. (a) 0.628 m/s   (b) 0.500 Hz   (c) 3.14 rad/s
19. (a) 126 N/m   (b) 17.8 cm
21. 0.627 s
23. (a) at $t = 0$, $x = 0.30$ m; at $t = 0.60$ s, $x = 0.24$ m
    (b) 0.30 m   (c) $\frac{1}{6}$Hz   (d) 6.0 s
25. (a) 250 N/m   (b) 0.281 s, 3.56 Hz, 22.4 rad/s
    (c) 0.313 J   (d) 5.00 cm
    (e) $v_{max} = 1.12$ m/s, $a_{max} = 25.0$ m/s^2 (f) 1.01 cm
27. 105 complete oscillations
29. (a) gain   (b) 1.1 s
31. (a) $L_{Earth} = 25$ cm, and $L_{Mars} = 9.4$ cm,
    (b) $m_{Earth} = 0.25$ kg, $m_{Mars} = 0.25$ kg
33. 5.67 mm
35. 0.800 m/s
37. 31.9 cm
39. 219 N
41. 7.07 m/s
43. 13.5 N
45. (a) constructive interference yields $A = 0.50$ m
    (b) destructive interference yields $A = 0.10$ m
47. 0.750 J
49. (a) 588 N/m   (b) 0.700 m/s
51. 1.1 m/s
53. $\mu = 0.11$

55. (a) 0.50 m/s (b) 8.6 cm
57. (a) 15.8 rad/s (b) 5.23 cm
59. $F_{\text{tangential}} = -\left[(\rho_{\text{air}} - \rho_{\text{He}})\dfrac{Vg}{L}\right]s = -ks$, $T = 1.40$ s
61. 0.0329 s

CHAPTER 14

Multiple-Choice Questions
1. e   2. b   3. e   4. c   5. d

Conceptual Questions
1. The camera is designed such that it assumes the speed of sound is 345 m/s, the speed of sound at a room temperature of 23°C. If the temperature should decrease to, say, 0°C, the speed will also decrease, and the camera will note that it takes longer for the sound to make its round trip. Thus, it will think the object is farther away than it really is.
3. Sophisticated electronic devices break the frequency range of about 60 Hz to 4000 Hz used in telephone conversations into several frequency bands and then mix them in a predetermined pattern so that they become unintelligible. The descrambler, of course, moves the bands back into their proper order.
5. Wind can change a Doppler shift but cannot cause one. Both v_0 and v_s in our equations must be interpreted as speeds of observer and source relative to the air. If source and observer are moving relative to each other, the observer will hear one shifted frequency in still air and a different shifted frequency if wind is blowing. If the distance between source and observer is constant, there will never be a Doppler shift.
7. The echo is Doppler shifted, and the shift is like both a moving source and a moving observer. The sound that leaves your horn in the forward direction is Doppler shifted to a higher frequency, because it is coming from a moving source. As the sound reflects back and comes toward you, you are a moving observer, so there is a second Doppler shift to an even higher frequency. If the sound reflects from the spacecraft coming toward you, there is another moving source shift to an even higher frequency. The reflecting surface of the spacecraft acts as a moving source.
9. At the center of the string, there is a node for the second harmonic, as well as for every even-numbered harmonic. By placing the finger at the center and plucking, the guitarist is eliminating any harmonic that does not have a node at that point, which is all the odd harmonics. The even harmonics can vibrate relatively freely with the finger at the center because they exhibit no displacement at that point. The result is a sound with a mixture of frequencies that are integer multiples of the second harmonic, which is one octave higher than the fundamental.

11. The bow string is pulled away from equilibrium and released, similar to the way that a guitar string is pulled and released when it is plucked. Thus, standing waves will be excited in the bow string. If the arrow leaves from the exact center of the string, then a series of odd harmonics will be excited. Even harmonics will not be excited because they have a node at the point where the string exhibits its maximum displacement.
13. At the instant at which there is no displacement of the string, the string is moving at maximum speed. Thus, the energy at that instant is all kinetic energy of the string.
15. A change in temperature will result in a change in the speed of sound, and therefore, in the fundamental frequency. The speed of sound for audible frequencies in the open atmosphere is only a function of temperature and does not depend on the pressure. Thus, there will be no effect on the fundamental frequency caused by variations in atmospheric pressure.
17. When the bottles are struck, standing wave vibrations are established in the glass material of the bottles. The frequencies of these vibrations are determined by the tension in the glass and the mass of the glass material. As the water level rises, there is more mass, because the glass is in contact with the water. This increased mass decreases the frequency. On the other hand, blowing into a bottle establishes a standing wave vibration in the air cavity above the water. As the water level rises, the length of this cavity decreases, and the frequency rises.
19. A vibrating string is not able to set very much air into motion when vibrated alone. Thus it will not be very loud. If it is placed on the instrument, however, the string's vibration sets the sounding board of the guitar into vibration. A vibrating piece of wood is able to move a lot of air, and the note is louder.

Problems
1. 0.20 s
3. 1.7×10^{-2} m to 17 m
5. 5.0×10^{-7} m
7. (a) 5.0×10^{-17} W   (b) 5.0×10^{-5} W
9. $I_{\text{Full Orchestra}} = 32 I_{\text{Violin}}$
11. 3.01 dB
13. (a) 130 W   (b) 96 dB
15. (a) $\dfrac{I_A}{I_B} = 2$ (b) $\dfrac{I_A}{I_C} = 5$
17. (a) 10.0 kHz   (b) 3.33 kHz
19. 595 Hz
21. 0.391 m/s
23. $\theta = 42°$
25. 800 m
27. (a) 24.0 cm   (b) 85.5 cm
29. (a) Nodes at 0, 2.67 m, 5.33 m, and 8.00 m. Antinodes at 1.33 m, 4.00 m, and 6.67 m.
    (b) 18.6 Hz

31. 8.45×10^2 Hz, 1.69×10^3 Hz, 2.54×10^3 Hz
33. (a) 4.9×10^{-3} kg/m   (b) 2 loops   (c) No standing wave exists.
35. 1.1×10^3 N/m
37. (a) 0.40 m   (b) 1.1×10^3 Hz
39. 3.0×10^3 Hz
41. (a) 0.550 m   (b) 317 Hz
43. 5.26 beats/s
45. 3.8 m/s toward station, 3.9 m/s away from station
47. 2.94 cm
49. 300 Hz. There must be relative motion between the source and observer for a Doppler shift to occur.
51. 64 dB
53. (a) 0.655 m   (b) 13.4°C
55. 4.0 beats/s
57. 7.82 m
59. 19.3 m
61. (a) 55.8 m/s   (b) 2500 Hz
63. 1.34×10^4 N
65. 1.20×10^3 Hz

CHAPTER 15

Multiple-Choice Questions
1. a   2. e   3. b   4. a   5. c

Conceptual Questions
1. The configuration shown is inherently unstable. The negative charges are repelling each other. If there is any slight rotation of one of the rods, the repulsion can result in further rotation away from this configuration. There are three conceivable final configurations shown in Figure CQ15.1: Configuration (a) is stable—if the positive upper ends are pushed toward each other, they will repel and move the system back to the original configuration. Configuration (b) is an equilibrium configuration, but is unstable—if the lower ends are moved toward each other, the attraction of the lower ends will be larger than that of the upper ends and the configuration will shift to (c). Configuration (c) is another possible stable configuration.

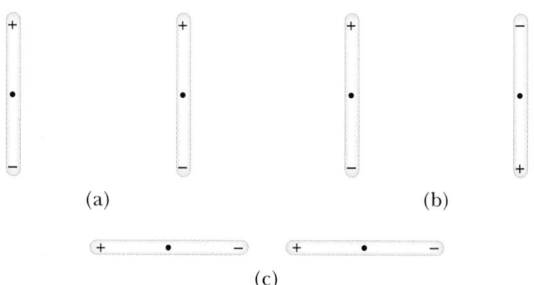

(a)   (b)

(c)

Figure CQ15.1

3. No. Object A might have a charge opposite in sign to that of B, but it also might be a neutral conductor. In this latter case, object B causes object A to be polarized, pulling charge of one sign to the near face of A and pushing an equal amount of charge of the opposite sign to the far face, as in Figure CQ15.3. Then the force of attraction exerted on B by the induced charge on the near side of A is slightly larger than the force of repulsion exerted on B by the induced charge on the far side of A. Therefore, the net force on A is toward B.

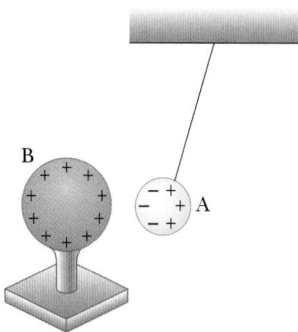

Figure CQ15.3

5. An object's mass decreases very slightly (immeasurably) when it is given a positive charge, because it loses electrons. When the object is given a negative charge, its mass increases slightly because it gains electrons.
7. Electric field lines start on positive charges and end on negative charges. Thus, if the fair weather field is directed into the ground, the ground must have a negative charge.
9. The two charged plates create a region of uniform electric field between them, directed from the positive toward the negative plate. Once the ball is disturbed so as to touch one plate, say the negative one, some negative charge will be transferred to the ball and it will experience an electric force that will accelerate it to the positive plate. Once the charge touches the positive plate, it will release its negative charge, acquire a positive charge, and accelerate back to the negative plate. The ball will continue to move back and forth between the plates until it has transferred all their net charge, thereby making both plates neutral.
11. The electric shielding effect of conductors depends on the fact that there are two kinds of charge—positive and negative. As a result, charges can move within the conductor so that the combination of positive and negative charges establishes an electric field that exactly cancels the external field within the conductor and any cavities inside the conductor. There is only one type of gravitation charge, however—there is no "negative mass." As a result, gravitational shielding is not possible.

13. The electric field patterns of each of these three configurations do not have sufficient symmetry to make the calculations practical. Gauss's law is only useful for calculating the electric field of highly symmetric charge distributions, such as uniformly charged spheres, cylinders, and sheets.

15. No. The balloon induces charge of opposite sign in the wall, causing it to be attracted. The balloon eventually falls because its charge slowly diminishes as it leaks to ground. Some of the charge could also be lost due to ions of opposite sign in the surrounding atmosphere, which would tend to neutralize the charge.

17. When the comb is nearby, charges separate on the paper, resulting in the paper being attracted. After contact, charges from the comb are transferred to the paper so that it has the same type charge as the comb. It is thus repelled.

19. No. This *can* be done in some cases—when the field is constant, for example. However, because we do not know the charge distribution, we cannot claim that the field is constant, and thus we cannot find the electric field.

    To illustrate this point, consider a sphere that contained a net charge of 100 μC. The charges could be located near the center, or they could all be grouped at the northernmost point within the sphere. In either case, the net electric flux would be the same, but the electric field would vary greatly.

Review Problem
(a) 8.2×10^{-8} N   (b) 2.2×10^6 m/s

Problems
1. (a) 1.0×10^{13} (b) 1.3×10^{-12}
3. (a) 2.2×10^{-5} N (attraction)
   (b) 9.0×10^{-7} N (repulsion)
5. (a) 36.8 N   (b) 5.51×10^{27} m/s^2
7. 5.08 m
9. 7.2×10^{-9} C
11. 1.38×10^{-5} N at 77.5° below $-x$ axis
13. 1.45 m beyond the -3.00×10^{-9} C charge
15. 5.53×10^{11} N/C (away from the proton)
17. (a) 1.20×10^4 N/C toward the 30 nC charge
    (b) 3.60×10^4 N/C toward the -60 nC charge
19. 1.63×10^5 N/C
21. (a) 6.11×10^{10} m/s^2 (b) 19.6 μs
    (c) 11.8 m   (d) 1.2×10^{-15} J
23. 756 N/C at 70.1° (CW from the $-x$ axis)
25. (a) 1.00×10^3 N/C   (b) 3.37×10^{-8} s
    (c) accelerate in the direction opposite to that of the field at 1.76×10^{14} m/s^2
27. 1.8 m to left of -2.5×10^{-6} C charge
33. (a) 0   (b) -5.00 μC inside, $+5.00$ μC outside
    (c) $+5.00$ μC outside, zero inside
    (d) $+5.00$ μC outside, zero inside
35. 1.3×10^{-3} C
37. (a) 4.8×10^{-15} N   (b) 2.9×10^{12} m/s^2

39. 4.1×10^6 N/C
41. (a) zero   (b) $E = k_e q/r^2$
45. 24 N/C in the $+x$ direction
47. 5.25 μC
49. 0.39 m to the right of the -2.5 μC charge
51. (a) 0   (b) 7.99×10^7 N/C   (c) 0
    (d) 7.35×10^6 N/C
53. $E = K/ed$ in the direction of motion
55. (a) 36.9° or 53.1°   (b) 1.66×10^{-7} s and 2.21×10^{-7} s
57. (a) 0.307 s   (b) Yes, the absence of gravity produces a 2.28% difference.

CHAPTER 16

Multiple-Choice Questions
1. d   2. c   3. b   4. a   5. a

Conceptual Questions
1. The proton is displaced in the direction of the electric field, so the electric potential and electric potential energy will *decrease*. The decrease in potential energy is accompanied by an equal increase in kinetic energy, as required by the law of conservation of energy.

3. The power line, if it makes electrical contact with the metal of the car, will raise the potential of the car to 20,000 volts. It will also raise the potential of your body to 20,000 volts, because you are in contact with the car. In itself, this is not a problem. If you step out of the car, your body at 20,000 volts will make contact with the ground, which is at zero volts. As a result, a current will pass through your body, and you will likely be injured. Thus, it is best to stay in the car until help arrives.

5. No. Suppose there are several charges in the vicinity of the point in question. If some charges are positive and some negative, their contributions to the potential at the point may cancel. For example, the electric potential halfway between two equal but opposite charges is zero.

7. The capacitor often remains charged long after the voltage source is disconnected. This residual charge can be lethal. The capacitor can be safely handled after discharging the plates by short circuiting the device with a conductor, such as a screwdriver with an insulating handle.

9. Field lines represent the direction of the electric force on a positive test charge. If electric field lines were to cross, then at the point of crossing there would be an ambiguity regarding the direction of the force on the test charge, because there would be two possible forces. Thus, electric field lines cannot cross. It is possible for equipotential surfaces to cross. (However, equipotential surfaces at different potentials cannot intersect.) For example, suppose two identical positive charges are at diagonally opposite corners of a square, and two negative charges of equal magnitude are at the other two corners. Then the planes perpendicular to the sides of the square at their mid-

points are equipotential surfaces. These two planes cross each other at the line perpendicular to the square at its center.

11. You may remember from the chapter on gravitational potential energy (Chapter 5) that potential energy of a system is defined to be positive when positive work must have been performed by an external agent. For example, a flag has a positive potential energy relative to the ground, because positive work must be done by an external force in order to raise it from the ground to the top of the pole.

When assembling like charges from an infinite separation, it takes work to move them closer together to some distance r; therefore energy is being stored, and the potential energy is positive. When assembling unlike charges from an infinite separation, the charges tend to accelerate toward each other, and thus energy is released as they approach a separation distance r. Therefore, the potential energy of a pair of unlike charges is negative.

13. You should use a dielectric-filled capacitor whose dielectric constant is very large. Furthermore, you should make the dielectric as thin as possible, keeping in mind that dielectric breakdown must be considered also.

15. Because the energy stored $= C(\Delta V)^2/2$, doubling ΔV will quadruple the stored energy.

17. Because the charge on the capacitor is given by $Q = C \Delta V$, and the capacitance is constant, doubling the potential difference would double the charge.

Problems

1. (a) 6.4×10^{-19} J   (b) -6.4×10^{-19} J   (c) -4.0 V
3. 1.44×10^{-20} J
5. 1.7×10^6 N/C
7. (a) 1.13×10^5 V/m   (b) 1.80×10^{-14} N
    (c) 4.38×10^{-17} J
9. (a) 2.65×10^7 m/s   (b) 6.19×10^5 m/s
11. (a) 1.44×10^{-7} V   (b) -7.19×10^{-8} V
13. 2.7×10^6 V
15. (a) 103 V   (b) -3.85×10^{-7} J Positive work must be done to separate the charges
17. 8.09×10^{-7} J
19. 2.8×10^{-14} m
21. (a) 48.0 μC   (b) 6.00 μC
23. (a) 800 V   (b) $Q_2 = \dfrac{Q_1}{2}$
25. 2.26×10^{-5} m^2
27. (a) 13.3 μC on each   (b) 20.0 μC, 40.0 μC
29. (a) 2.0 μF   (b) $Q_3 = 24$ μC, $Q_4 = 16$ μC, $Q_2 = 8.0$ μC, $\Delta V_2 = \Delta V_4 = 4.0$ V, $\Delta V_3 = 8.0$ V
31. (a) 12 μF
    (b) $Q_4 = 144$ μC, $Q_2 = 72$ μC, $Q_{24} = Q_8 = 216$ μC
33. (a) All four in parallel.   (b) Two in parallel followed by another group of two in parallel, or two in series that

are in parallel with another group of two in series.
    (c) One in series with a group of three in parallel.
    (d) All four in series.
35. (a) $Q_{25} = 1.25 \times 10^3$ μC, $Q_{40} = 2.00 \times 10^3$ μC
    (b) $Q_{25} = 288$ μC, $Q_{40} = 462$ μC, $\Delta V = 11.6$ V
37. $Q_1 = \dfrac{10}{3}$ μC, $Q_2 = \dfrac{20}{3}$ μC
39. (a) 0.15 J   (b) 270 V
41. $\kappa = 4.0$
43. (a) 8.13 nF   (b) 2.40 kV
45. (a) 4.54×10^{-10} m^2 (b) 2×10^{-13} F
    (c) 2.0×10^{-14} C, 1.25×10^5 charges
49. -7.8×10^3 V
51. $\kappa = 2.33$
53. (a) 7.54×10^{-9} s   (b) $v_e = 1.33 \times 10^7$ m/s and
    $v_p = 7.23$ km/s   (c) 7.55 ns   (d) 0.323 μs
55. 180 μC on C_1, 89 μC on C_2.
57. 6.00 pF and 3.00 pF
59. (a) 5.96 μF   (b) 89.5 μC on the 20-μF, 63.1 μC on
    6-μF, and 26.4 μC on both the 15-μF and the 3-μF
61. 121 V
63. (a) 0.11 mm   (b) 4.39 mm

CHAPTER 17

Multiple-Choice Questions
1. a   2. b   3. b   4. c   5. c

Conceptual Questions
1. The gravitational force pulling the electron to the bottom of a piece of metal is much smaller than the electrical repulsion pushing the electrons apart. Thus, they stay distributed throughout the metal. The concept of charges residing on the surface of a metal is true for a metal with an excess charge. The number of free electrons in a piece of metal is the same as the number of positive ions—the metal has zero net charge.

3. A voltage is not something that "surges through" a completed circuit. A voltage is a potential difference that is applied across a device or a circuit. What goes through the circuit is current. Thus, it would be more correct to say, "1 ampere of electricity surged through the victim's body." Although this current would have disastrous results on the human body, a value of 1 (ampere) doesn't sound as exciting for a newspaper article as 10,000 (volts). Another possibility is to write, "10,000 volts of electricity were applied across the victim's body," which still doesn't sound as exciting.

5. The length of the line cord will double in this event. This would tend to increase the resistance of the line cord. But the doubling of the radius of the line cord results in the cross-sectional area increasing by a factor of 4. This would reduce the resistance more than the doubling of the length increases it. The net result is a decrease in resis-

tance. The same effect will result in more current flowing through the filament, causing it to glow more brightly.

7. The bulb filaments are cold when the lamp is first switched on, hence they have a lower resistance and draw more current than when they are hot. The increased current can overheat the filament and destroy it.

9. The shape, dimensions, and the resistivity affect the resistance of a conductor. Because temperature and impurities also affect the conductor's resistivity, these factors also affect resistance.

11. The radius of wire B is $\sqrt{3}$ times the radius of wire A, to make its cross-sectional area three times larger.

13. The amplitude of atomic vibrations increases with temperature, thereby scattering conduction electrons more efficiently.

15. Drift velocity might increase steadily as time goes on, because collisions between electrons and atoms in the wire would be essentially nonexistent and the conduction electrons would move with constant acceleration. The current would rise steadily without bound also, because $I \propto v_d$ (Equation 17.3).

17. Because there are so many electrons in a conductor (approximately 10^{28} electrons/m^3), the average velocity of charges is very low. When you connect a wire to a potential difference, you establish an electric field everywhere in the wire nearly instantaneously to make electrons start drifting all at once.

Problems

1. 3.00×10^{20} electrons move past opposite to the direction of the current
3. 3.0 mA
5. 1.05 mA
7. 27 y
9. 1.3×10^{-4} m/s
11. (a) 0.38 A   (b) 0.54 A
13. 0.31 Ω
15. $R_0/9$
17. (a) 2.82×10^8 A   (b) 1.76×10^7 A
19. 4.9×10^{-7} m^2
21. 1300°C
23. 26 mA
25. (a) $5.89 \times 10^{-2}\ \Omega$ (b) $5.45 \times 10^{-2}\ \Omega$
27. (a) 0.65 mV   (b) 1.1 mV
29. 63.2°C
31. 18 bulbs
33. (a) 3.2×10^5 J   (b) 1.1×10^3 s (18 min)
35. 34.4 Ω
37. (a) 5.0 cents   (b) 71%
39. 3.36 h/day
41. 23 cents
43. $1.2
45. (a) 8.6×10^5 J   (b) 1.9 cents
47. 90 μV
49. (a) 18 C   (b) 3.6 A

51. 3.77×10^{28} electrons/m^3
53. $R_n = 5.6$ kΩ and $R_C = 4.4$ kΩ
55. (a) 140 Ω (b) 26 m   (c) To fit the required length into a small space   (d) 25 m
57. 37 MΩ
59. 0.48 kg/s
61. (b) 1.42 Ω is less precise than 1.418 Ω

CHAPTER 18

Multiple-Choice Questions
1. d   2. b   3. a   4. a   5. b

Conceptual Questions

1. The total amount of energy delivered by the battery will be less than W. Recall that a battery can be considered to be an ideal, resistanceless battery in series with the internal resistance. When charging, the energy delivered to the battery includes the energy necessary to charge the ideal battery, plus the energy that goes into raising the temperature of the battery due to I^2r heating in the internal resistance. This latter energy is not available during the discharge of the battery. During discharge, part of the reduced available energy again transforms into internal energy in the internal resistance, further reducing the available energy below W.

3. The starter in the automobile draws a relatively large current from the battery. This large current causes a significant voltage drop across the internal resistance of the battery. As a result, the terminal voltage of the battery is reduced, and the headlights dim accordingly.

5. An electrical appliance has a given resistance. Thus, when it is attached to a power source with a known potential difference, a definite current will be drawn. The device can be labeled with both the voltage and the current. Batteries, however, can be applied to a number of devices. Each device will have a different resistance, so the current from the battery will vary with the device. As a result, only the voltage of the battery can be specified.

7. Connecting batteries in parallel does not increase the emf. A high-current device connected to batteries in parallel can draw currents from both batteries. Thus, connecting the batteries in parallel does increase the possible current output and, therefore, the possible power outputs.

9. As you add more lightbulbs in series, the overall resistance of the circuit increases. Thus, the current in the bulbs and their brightness will decrease. This decrease in current will result in a decrease in power transferred from the battery. As a result, the battery lifetime will increase. As you add more lightbulbs in parallel, the overall resistance of the circuit decreases. The current in each bulb and their brightness remain the same (as long as the current capability of the battery is not surpassed). The brightness of each bulb is ideally constant. The current leaving

the battery will increase with the addition of each bulb. This increase in current will result in an increase in power transferred from the battery. As a result, the battery lifetime will decrease.

11. The bird is resting on a wire of a fixed potential. In order to be electrocuted, a potential difference is required. There is no potential difference between the bird's feet.

13. The junction rule is a statement of conservation of charge. It says that the amount of charge that enters the junction in some time interval must equal the charge that leaves the junction in that time interval. The loop rule is a statement of conservation of energy. It says that the potential increases and decreases around a closed loop in a circuit must add to zero.

15. A few of the factors involved are as follows: the conductivity of the string (is it wet or dry?); how well you are insulated from ground (are you wearing thick rubber or leather soled shoes?); the magnitude of the potential difference between you and the kite; the type and condition of the soil under your feet.

17. She will not be electrocuted if she holds on to only one high-voltage wire because she is not completing a circuit. There is no potential difference across her body as long as she clings to only one wire. However, she should immediately release the wire once it breaks, because she will become part of a closed circuit when she reaches the ground or comes into contact with another object.

19. (a) The intensity of each lamp increases because lamp C is shortcircuited and the current (which increases) passes only through lamps A and B. (b) The intensity of lamp C goes to zero because the current in this branch goes to zero. (c) The current in the circuit increases because the total resistance decreases from $3R$ (with the switch open) to $2R$ (after the switch is closed). (d) The voltage drop across lamps A and B increases, and the voltage drop across lamp C becomes zero. (e) The power dissipated increases from $\varepsilon^2/3R$ (with the switch open) to $\varepsilon^2/2R$ (after the switch is closed).

21. The lightbulb will glow for a very short while as the capacitor is being charged. Once the capacitor is almost totally charged, the current in the circuit will be nearly zero and the bulb will not glow.

Problems

1. 4.92 Ω

3. (a) 0.75 A, $\Delta V_{18} = 13.5$ V, $\Delta V_6 = 4.5$ V
   (b) $\Delta V_{18} = \Delta V_6 = 18$ V, $I_{18} = 1.0$ A, $I_6 = 3.0$ A

5. (a) 30 V   (b) 2.3 V

7. $2.5R$

9. (a) 5.13 Ω (b) 4.53 V

11. $R_A = 6.0$ Ω, $R_B = 3.0$ Ω

13. 55 Ω

15. 5.4 V (with a at higher potential than b)

17. $I_2 = -2$ A (flows from b toward a), $I_3 = 1$ A

19. 6 V (with the polarity shown)

21. (a) 4.59 Ω (b) fraction dissipated internally = 0.0816

23. $I_1 = 3.5$ A, $I_2 = 2.5$ A, $I_3 = 1.0$ A

25. $I_{30} = 0.353$ A, $I_5 = 0.118$ A, $I_{20} = 0.471$ A

27. $\Delta V_2 = 3.05$ V, $\Delta V_3 = 4.57$ V, $\Delta V_4 = 7.38$ V, $\Delta V_5 = 1.62$ V

31. 130 μC

33. (a) 1.00×10^{-2} F   (b) 0.415 C

35. (a) 8.0 A   (b) 120 V   (c) 0.80 A   (d) 580 W

37. No, a 20-A circuit breaker is needed.

39. 7.5 Ω

41. 15 Ω

43. (a) 12.4 V   (b) 9.65 V

45. 220 Ω, 470 Ω

47. 14 Ω

49. 0.390 A, 1.50 V

51. 14 s

53. $R = 20$ Ω or $R = 98.1$ Ω

55. $I_1 = \frac{11}{13}$ A (down in left branch), $I_2 = \frac{6}{13}$ A (down in center branch), $I_3 = \frac{17}{13}$ A (up in right branch)

57. (a) $R_x = R_2 - \frac{1}{4}R_1$
    (b) $R_x = 2.8$ Ω (inadequate grounding)

61. $q_1 = 0.4q = (240 \ \mu C)(1 - e^{-1000t/6})$ for 2-μF capacitor
    $q_2 = 0.6q = (360 \ \mu C)(1 - e^{-1000t/6})$ for 3-μF capacitor

CHAPTER 19

Multiple-Choice Questions

1. b   2. c   3. c   4. c   5. b   6. d   7. b

Conceptual Questions

1. No. There may be another field such as an electric field or gravitational field that produces another force on the charged particle that has the same magnitude but is opposite to the magnetic force, resulting in a net force of zero. Also a charged particle moving parallel or antiparallel to a magnetic field would experience no magnetic force.

3. The magnetic force on a moving charged particle is always perpendicular to the direction of motion. There is no magnetic force on the charge when it moves parallel to the direction of the magnetic field. However, the force on a charged particle moving in an electric field is never zero and is always parallel to the direction of the electric field. Therefore, by projecting the charged particle in different directions, it is possible to determine the nature of the field.

5. The magnetic field of the magnet produces a magnetic force on the electrons moving toward the screen that produce the image. This magnetic force deflects the electrons to regions on the screen other than the ones to which they are supposed to go. The result is a distorted image.

7. No. The magnetic field produces a magnetic force that is directed toward the center of the circular path of the particle, and this causes a change in direction of the parti-

cle's velocity vector. However, because the magnetic force is always perpendicular to the displacement of the particle, no work is done on the particle, and its speed remains constant. Hence, its kinetic energy also remains constant.

9. If you are moving along with the electrons, you will measure zero current for the electrons, so the electrons would not produce a magnetic field according to your observations. However, the fixed positive charges in the metal are now moving backward relative to you and creating a current equivalent to the forward motion of the electrons when you were stationary. Thus, you will measure the same magnetic field as when you were stationary, but it will be due to the positive charges presumed to be moving from your point of view.

11. A compass would not detect currents in wires near light switches for two reasons. Because the cable to the light switch contains two wires, with one carrying current to the switch and the other away from the switch, the net magnetic field would be very small and fall off rapidly. The second reason is that the current is alternating at 60 Hz. As a result, the magnetic field is oscillating at 60 Hz, also. This frequency would be too fast for the compass to follow, so the effect on the compass reading would average to zero.

13. The levitating wire is stable with respect to vertical motion — if it is displaced upward, the repulsive force weakens, and the wire drops back down. Conversely, if it drops lower, the repulsive force increases, and it moves back up. The wire is not stable, however, with respect to lateral movement. If it moves away from the vertical position directly over the lower wire, the repulsive force will have a sideways component, which will push the wire away. In the case of the attracting wires, the hanging wire is not stable for vertical movement. If it rises, the attractive force increases, and the wire moves even closer to the upper wire. If the hanging wire falls, the attractive force weakens, and the wire falls further. The hanging wire is also unstable to lateral movement. If the wire moves to the right, it moves farther from the upper wire and the attractive force decreases. Although there is a restoring force component pulling it back to the left, the vertical force component is not strong enough to hold it up and it falls.

15. Each coil of the Slinky will become a magnet, because a coil will act as a current loop. Because the sense of rotation of the current is the same in all coils, each coil becomes a magnet with the same orientation of poles. Thus, all of the coils attract, and the Slinky will compress.

17. If the magnetic field is uniform, and the velocity vectors are the same, the magnetic force $F = qvB \sin \theta$ should be the same for both particles. We can conclude that the only thing that could cause the force to be of opposite sign is if the charges were of opposite sign.

19. There will be no net force on the wires, but there will be a torque. To understand this, imagine a fixed vertical wire and a free horizontal wire (Figure CQ19.19). The vertical wire carries an upward current and creates a magnetic field that circles the wire, as shown in the figure. Each segment of the horizontal wire (of length ℓ) also carries a current that interacts with the magnetic field according to the equation $F = BI\ell \sin \theta$. Applying the right-hand rule, we see that the horizontal wire experiences an upward force on the left side, and an equal downward force on the right side. Although the forces cancel, they create a torque around the point at which the wires cross.

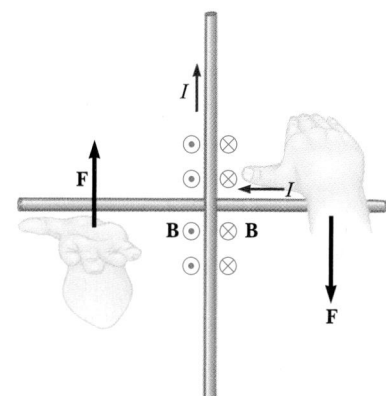

Figure CQ19.19

Review Problem
(a) 7.96×10^3 (b) 11.6 cm

Problems
1. (a) east   (b) 30° N of E   (c) east   (d) zero force
3. (a) into the page   (b) toward the right
   (c) toward the bottom of the page
5. (a) 8.7×10^{-14} N   (b) 5.2×10^{13} m/s^2
7. (a) 7.90×10^{-12} N   (b) 0
9. 806 N
11. 8.0×10^{-3} T in $+z$ direction
13. (a) to the left   (b) into the page
    (c) out of the page   (d) toward the top of the page
    (e) into the page   (f) out of the page
15. (a) 9.0×10^{-3} N at 15° above horizontal in northward direction   (b) 2.3×10^{-3} N horizontal and westward
17. 0.20 T out of the page
19. (a) 0.628 N·m   (b) 60.0°
21. 10 N·m clockwise (as viewed from above the loop)
23. 118 N·m
25. 5.00×10^{-3} Ω
27. 7.5×10^4 Ω
29. $R_1 = 2.99 \times 10^4$ Ω, $R_2 = 2.70 \times 10^5$ Ω,
    $R_3 = 2.70 \times 10^6$ Ω
31. 2.0×10^{-12} kg

33. 1.77 cm
35. 8.65×10^{-8} s
37. 2.0 cm
39. (a) 4.00×10^{-6} T downward
    (b) 6.67×10^{-6} T at 77.0° to the left of the vertical
41. At A, 2.00×10^{-4} T (counterclockwise)
    At B, 1.33×10^{-4} T (clockwise).
43. 4.50 mm
45. 2.26×10^{-4} N, torque $= 0$
47. (a) 2.8×10^{-6} T   (b) 0.89 mA
49. (a) 5.0×10^{-7} T out of the page   (b) 3.89×10^{-6} T in the plane parallel to the xy plane and at 31.0° counterclockwise from the $-y$ axis
51. 2.12 cm
53. 1.6×10^{-5} T toward the top of the page
55. 6.56×10^{-2} T
57. 0.59 T
59. 3.00 T
61. (a) 1.3×10^{-3} N upward   (b) 19.6 A
63. (a) 21 cm for singly charged and 10 cm for doubly charged   (b) 22 cm
65. (a) 1.33 m/s   (b) Does not depend on the charge.
67. (a) -8.00×10^{-21} kg·m/s   (b) 8.90°

CHAPTER 20

Multiple-Choice Questions
1. c   2. b   3. a   4. b   5. c

Conceptual Questions
1. According to Faraday's law, an emf is induced in a wire loop if the magnetic flux through the loop changes with time. In this situation, an emf can be induced by either rotating the loop around an arbitrary axis or by changing the shape of the loop.
3. As the spacecraft moves through space, it is apparently moving from a region of one magnetic field strength to a region of a different magnetic field strength. The changing magnetic field through the coil induces an emf and a corresponding current in the coil.
5. The magnetic field lines around the transmission cable will be circular. If you place your loop around the cable, there will be no field lines passing through the loop, so no emf will be induced. The loop needs to be placed next to the cable, with the plane of the loop containing the cable, to maximize the flux through its area.
7. A constant induced emf requires a magnetic field that is changing at a constant rate in one direction—for example, always increasing or always decreasing. It is impossible for a magnetic field to increase forever, both in terms of energy considerations and technological concerns. In the case of a decreasing field, once it reaches zero and then reverses direction, we again face the problem with the field increasing without bounds in the opposite direction.

9. The aluminum sheet and magnet supply magnetic damping to the system, so that the oscillations of the arms damp out rapidly. When the sheet moves in the magnetic field, eddy currents are induced in the aluminum. According to Lenz's law, these currents are in a direction to oppose the movement of the aluminum sheet, providing magnetic friction.
11. The flux is calculated as $\Phi \equiv B_\perp A = BA \cos \theta$. The flux is therefore maximum when the entire magnetic field vector is perpendicular to the loop of wire. We may also deduce that the flux is zero when there is no component of the magnetic field that is perpendicular to the loop.
13. The energy stored in an inductor carrying a current I is equal to $\frac{1}{2}LI^2$. Therefore, doubling the current will quadruple the energy stored in the inductor.
15. No. Once the bar is in motion and the charges are separated, no external force is necessary to maintain the motion. An applied external force in the x direction will cause the bar to accelerate in that direction.

Problems
1. 5.9×10^{-2} T·m²
3. (a) 1.00×10^{-7} T·m²   (b) 8.66×10^{-8} T·m²   (c) 0
5. (a) 0   (b) 0
7. (a) 3.1×10^{-2} T·m²   (b) $\Phi_{total} = 0$
9. 0.18 T
11. 2.7 T/s
13. 94 mV
15. (a) 4.0×10^{-6} T·m²   (b) 1.8×10^{-5} V
17. 1.0 m/s
19. 2.8 mV
21. (a) clockwise as viewed from above
    (b) counterclockwise as viewed from above
23. left to right
25. (a) left to right   (b) no current is present
    (c) right to left
27. (a) 18.1 μV   (b) 0
29. (a) 8.0 A   (b) 3.2 A   (c) 60 V
31. (a) 7.5×10^3 V   (b) when the plane of the coil is parallel to the field
33. 4.69 mH
35. 20 mV
37. 1.92×10^{-5} T·m²
39. (a) 1×10^3 Ω (b) 3 ms
41. (a) 0   (b) 3.8 V   (c) 6.0 V   (d) 2.2 V
43. 0.140 J
45. (a) 18 J   (b) 7.2 J
47. 1.4×10^{-6} H
49. negative ($V_a < V_b$)
51. (a) 0.0200 s   (b) 37.8 V   (c) 3.02×10^{-3} V
    (d) 0.103 A
53. 9.7 V
55. $v_t = \dfrac{mgR}{B^2\ell^2}$

57. 1.15×10^5 V

59. 6.8 V

61. (a) 0.157 mV (end B positive)

  (b) 5.89 mV (end A is positive)

63. 1.6 A

65. (b) 0.75 mA

CHAPTER 21

Multiple-Choice Questions

1. c   2. b   3. a   4. b   5. b   6. d

Conceptual Questions

1. When the iron rod is inserted into the solenoid, the inductance of the inductor increases. As a result, more potential difference appears across the inductor than before. As a consequence, less potential difference appears across the bulb, and its brightness decreases.

3. The primary coil of the transformer is an inductor. When an ac voltage is applied, the back emf due to the inductance will limit the current flow through the coil. If dc voltage is applied, there is no back emf, and the current can rise to a higher value. It is possible that this increased current will deliver so much energy to the resistance in the coil that its temperature rises to the point at which insulation on the wire can burn.

5. An antenna that is a conducting line responds to the electric field of the electromagnetic wave—the oscillating electric field causes an electric force on electrons in the wire along its length. The movement of electrons along the wire is detected as a current by the radio, demodulated, and amplified. Thus, a line antenna must have the same orientation as the broadcast antenna. A loop antenna responds to the magnetic field in the radio wave. The varying magnetic field induces a varying current in the loop (Faraday's law), and this signal is demodulated and amplified. The loop should be in the vertical plane containing the sight-line to the broadcast antenna, so that the magnetic field lines go through the area of the loop.

7. The flashing of the light according to Morse code is a drastic amplitude modulation—the amplitude is changing from a maximum to zero. In this sense, it is similar to the on-and-off binary code used in computers and compact disks. The carrier frequency is that of the light, on the order of 10^{14} Hz. The signal frequency depends on the skill of the signal operator, but it is on the order of a single hertz, as the light is flashed on and off. The broadcasting antenna for this modulated signal is the filament of the lightbulb in the signal source. The receiving antenna is the eye.

9. The sail should be as reflective as possible, so that the maximum momentum is transferred to the sail from the reflection of sunlight.

11. Suppose the extraterrestrial looks around your kitchen. Lightbulbs and the toaster glow brightly in the infrared. Somewhat fainter are the back of the refrigerator and the back of the television set, while the television screen is dark. The pipes under the sink show the same weak glow as the walls until you turn on the faucets. Then the pipe on the right gets darker and that on the left develops a gleam that quickly runs up along its length. The food on the plates shines, as does human skin, the same color for all races. Clothing is dark as a rule, but your seat and the chair seat glow alike after you stand up. Your face appears lit from within, like a jack-o-lantern; your nostrils and openings of your ear canals are bright; brighter still are the pupils of your eyes.

13. Radio waves move at the speed of light. They can travel around the curved surface of the Earth, bouncing between the ground and the ionosphere, which has an altitude that is small compared to the radius of the Earth. The distance across the lower forty-eight states is approximately 5000 km, requiring a travel time of $(5 \times 10^6$ m$)/(3 \times 10^8$ m/s$) \sim 10^{-2}$ s. Likewise, radio waves take only 0.07 s to travel halfway around the Earth. In other words, a speech can be heard on the other side of the world (in the form of radio waves) before it is heard at the back of the room (in the form of sound waves).

15. No. The wire will emit electromagnetic waves only if the current varies in time. The radiation is the result of accelerating charges, which can only occur when the current is not constant.

Review Problem

(a) 8.0×10^2 turns, 0.48 m   (b) 2.1 mH   (c) 6.0 Ω

(d) 2.4 W

Problems

1. (a) 141 V   (b) 20.0 A   (c) 28.3 A   (d) 2.00 kW

3. (a) 1.9×10^2 Ω (b) 1.4×10^2 Ω

5. 71 V, 3.0 A

7. 4.0×10^2 Hz

9. 17 μF

13. 3.14 A

15. (a) 184 Hz   (b) 48.6 mA

17. (a) 0.204 A   (b) 154 V   (c) 54.1 V   (d) 90.0°

19. (a) 1.4×10^3 Ω (b) 100 mA   (c) 51°

  (d) current lags voltage

21. (a) 89.4 V   (b) 108 V

23. (a) 104 V   (b) 150 V   (c) 127 V   (d) 23.6 V

25. (a) 208 Ω (b) 40.0 Ω (c) 0.541 H

27. (a) 1.8×10^2 Ω (b) 0.71 H

29. 2.29 μH

31. 4.9×10^{-9} F to 5.1×10^{-8} F

33. (b) $Q = 33.3$

35. (a) 1600   (b) 30 A

37. (a) 1.1×10^3 kW   (b) 3.1×10^2 A   (c) 8.3×10^3 A

39. 1000 km, there will always be a better use for tax money.

41. (a) 45 N/C   (b) 2.7 W/m²

43. 1.01×10^3 N/C, 3.35×10^{-6} T
45. (a) 188 m to 556 m   (b) 2.78 m to 3.4 m
47. People 100 km away receive the news 8.4 ms before people across the room because radio waves travel faster than sound waves.
49. (a) 0.116 A   (b) 1.38 A
    (c) Power for (a) = 1.08 W, for (b) 152 W
51. (a) 12 turns   (b) 1.9 A   (c) 0
53. 2.5 mH, 26 μF
55. (a) resistor and inductor   (b) $R = 10$ Ω, $L = 30$ mH
57. (a) 6.7×10^{-16} T   (b) 5.3×10^{-17} W/m^2
    (c) 1.7×10^{-14} W
59. 30 W
61. (a) 6.0 Ω (b) 12 mH
63. (a) 0.536 N   (b) 8.93×10^{-5} m/s^2 (c) 33.9 days
65. 4.47×10^{-9} J

CHAPTER 22

Multiple-Choice Questions
1. d   2. e   3. b   4. c   5. a

Conceptual Questions
1. The color will not change, for two reasons. First, despite the popular statement that color depends on wavelength, it actually depends on the *frequency* of the light, which does not change under water. Second, when the light enters the eye, it travels through the fluid within the eye. Thus, even if color did depend on wavelength, the important wavelength is that of the light in the ocular fluid, which does not depend on the medium through which the light traveled to reach the eye.
3. The optical day is longer than the geometric day. Due to the refraction of light by air, light rays from the Sun deviate slightly downward toward the surface of the Earth as the light enters the atmosphere. Thus, in the morning, light rays from the upper edge of the Sun will arrive at your eyes before the geometric line from your eyes to the top of the Sun clears the horizon. In the evening, light rays from the top of the Sun will continue to arrive at your eyes even after the geometric line from your eyes to the top of the Sun dips below the horizon.
5. The cross section can be visualized by considering just the two rays of light on the edges of the beam. If the beam of light enters a new medium with a higher index of refraction, the rays bend toward the normal, and the cross section of the refracted beam will be larger than that of the incident beam, as suggested by Figure CQ22.5a. If the new index of refraction is lower, the rays bend away from the normal, and the cross section of the beam is reduced, as shown in Figure CQ22.5b.
7. There is no dependence of the angle of reflection on wavelength, because the light does not enter deeply into the material during reflection—it reflects from the surface.

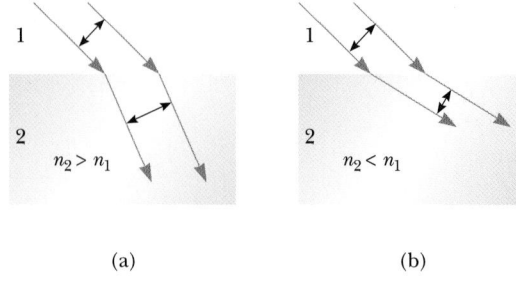

$n_2 > n_1$ $n_2 < n_1$

(a)   (b)

Figure CQ22.5

9. A ball covered with mirrors sparkles by reflecting light from its surface. On the other hand, a faceted diamond lets in light at the top, reflects it by total internal reflection in the bottom half, and sends the light out through the top again. Because of its high index of refraction, the critical angle for diamond in air for total internal reflection, $\theta_c = \sin^{-1}(n_{air}/n_{diamond})$, is small. Thus, light rays enter through a large area and exit through a very small area with a much higher intensity. When a diamond is immersed in carbon disulfide, the critical angle is increased to $\theta_c = \sin^{-1}(n_{CS_2}/n_{diamond})$. As a result, the light is emitted from the diamond over a larger area, and appears less intense.
11. The index of refraction of water is 1.333, quite different from that of air, which has an index of refraction of about 1. On the other hand, the index of refraction of liquid helium happens to be much closer to that of air. As a consequence, light undergoes less refraction in helium than it does in water.
13. The diamond acts like a prism in dispersing the light into its spectral components. Different colors are observed as a consequence of the manner in which the index of refraction varies with the wavelength.
15. Light travels through a vacuum at a speed of 3×10^8 m/s. Thus, an image we see from a distant star or galaxy must have been generated some time ago. For example, the star Altair is 16 lightyears away; if we look at an image of Altair today, we know only what Altair looked like 16 years ago. This initially may not seem significant; however, astronomers who look at other galaxies can get an idea of what galaxies looked like when they were much younger. Thus, it does make sense to speak of "looking backward in time."
17. A portion of the sound is bent back toward the Earth by refraction. This means that the sound can reach the listener by this path as well as by a direct path. Thus, the sound is louder than it would be if it followed the direct path only.

Problems
1. 2.995×10^8 m/s
3. (a) 536 rev/s   (b) 1.07×10^3 rev/s

5. (a) 1.94 m
   (b) 50.0° above horizontal (parallel to incident ray)
7. (a) 1.81×10^8 m/s   (b) 2.25×10^8 m/s
   (c) 1.56×10^8 m/s
9. (a) 327 nm   (b) 287 nm
11. 110.6°
13. 16.5°
15. 67.4°
17. 0.386 cm
19. 107 m
21. $\tan^{-1} n_g$
23. 3.39 m
25. 2.5 m
27. (a) $\theta_{red} = 48.2°$ (b) $\theta_{blue} = 47.8°$
29. 4.61°
31. 48.5°
33. (a) 40.8°   (b) 60.6°
35. 67.2°
37. 4.54 m
39. 27.5°
41. 1.27
43. (a) $\leq 90°$ (b) 30.3°
    (c) Not possible because $n_{carbon\ disulfide} < n_{water}$
45. (a) 38.5°   (b) $n_2 \geq 1.44$
47. 77.5°
49. 2.4 cm
51. 6.39 ns
53. about 10^{-11} s, 10^3 wavelengths
55. 82 reflections
57. (a) $n = \sqrt{1 + (4t/d)^2}$ (b) 2.10 cm
    (c) Violet light has a larger index of refraction, so the white halo is tinged with violet.

CHAPTER 23

Multiple-Choice Questions
1. e   2. e   3. d   4. b   5. c

Conceptual Questions
1. You will not be able to focus your eyes on both the picture and your image at the same time. To focus on the picture, you must adjust your eyes so that an object several centimeters away (the picture) is in focus. Thus, you are focusing on the mirror surface. But, your image in the mirror is as far behind the mirror as you are in front of it. Thus, you must focus your eyes beyond the mirror, twice as far away as the picture to bring the image into focus.
3. A single flat mirror forms a virtual image of an object due to two factors. First, the light rays from the object are necessarily diverging from the object, and second, the lack of curvature of the flat mirror cannot convert diverging rays to converging rays. If a second optical element is added that will cause light rays to converge, then the flat mirror can be placed in the region in which the converging rays

are present, and it will simply change the direction of the rays so that the real image is formed in a different location. For example, if a real image is formed by a convex lens, and the flat mirror is placed between the lens and the image position, the image formed by the mirror will be real.
5. The ultrasonic range finder sends out a sound wave and measures the time for the echo to return. Using this information, the camera calculates the distance to the subject and sets the camera lens. When facing a mirror, the ultrasonic signal reflects from the mirror surface and the camera adjusts its focus so that the mirror surface is at the correct focusing distance from the camera. But your image in the mirror is twice this distance from the camera, so it is blurry.
7. Light rays diverge from the position of a virtual image just as they do from an actual object. Thus, a virtual image can be as easily photographed as any object can. Of course, the camera would have to be placed near the axis of the lens or mirror in order to intercept the light rays.
9. We consider the two trees to be two separate objects. The far tree is an object that is farther from the lens than the near tree. Thus, the image of the far tree will be closer to the lens than the image of the near tree. The screen must be moved closer to the lens to put the far tree in focus.
11. If a converging lens is placed in a liquid having an index of refraction larger than that of the lens material, the direction of refractions at the lens surfaces will be reversed, and the lens will diverge light. A mirror depends only on reflection, which is independent of the surrounding material, so a converging mirror will be converging in any liquid.
13. The bag of air will act as a crude lens underwater. The direction of refraction at the bag surface as light goes from water to air will be the opposite of the directions for the light entering from air into the water-filled bag, so the air lens will diverge light underwater.
15. The focal length for a mirror is determined by the law of reflection from the mirror surface. The law of reflection is independent of the material of which the mirror is made and of the surrounding medium. Thus, the focal length depends only on the radius of curvature and not the material. The focal length of a lens depends on the law of refraction, and refraction is highly dependent on the indices of refraction of the lens material and surrounding medium. Thus, the focal length of a lens depends on the lens material.

Problems
1. 10.0 ft, 30.0 ft, 40.0 ft
3. (a) R is infinite   (b) $q = -p$, $M = +1$ (c) Yes
5. (a) 13.3 cm in front of mirror, real, inverted, $M = -0.333$

(b) 20.0 cm in front of mirror, real, inverted, $M = -1.00$

(c) No image is formed. Parallel rays leave the mirror.

7. 26.8 cm behind mirror, virtual, upright, $M = 0.0268$

9. 20.0 cm in front of mirror

11. 5.0 cm

13. 1.0 m

15. 8.05 cm

17. -20.0 cm

19. (a) It must be a concave mirror with focal length 83 cm.

(b) Object must be 1.0 m in front of mirror.

21. 38.2 cm below top surface of ice

23. 3.8 mm

25. $n = 2.00$

27. $n = 1.47$

29. (a) 12.0 cm   (b) 12.0 cm   (c) 18.0 cm

(d) no image is formed   (e) -12.0 cm

31. (a) -12.0 cm   (b) -12.0 cm   (c) -9.00 cm

(d) -6.00 cm   (e) -4.00 cm

33. (a) $M = -1.0$ for $p = +24$ cm, $M = +1.0$ only if $p = 0$ (object against lens)   (b) $M = -1.0$ for $p = -24$ cm, $M = +1.0$ only if $p = 0$ (object against lens)

35. 5.68 cm

37. 120 cm in front of lens, $M = +\frac{1}{4}$

39. At $p = -2f$

41. $M = 3.40$, upright

43. 40.0 cm

45. (a) May be either 3.27 cm or 9.63 cm   (b) 2.10 cm

47. 7.47 cm to left of second lens, 1.07 cm tall, upright, virtual

49. From 0.232 m to 225 m

51. Real image, 5.71 cm in front of mirror

53. 38.6°

55. 160 cm to left of lens, inverted, $M = -0.800$

57. 32.0 cm to the right of the second surface (real image)

59. (a) From $p = 3$ m to 0.5 m, the image is real and moves from 0.6 m to infinity. From 0.5 to contact, the image is virtual and moves from $-$infinity to 0.

(b) 0.639 s and 0.782 s

61. (a) -11.1 cm   (b) 2.50

(c) virtual, upright, and enlarged

63. (a) $4f/3$ (b) $3f/4$ (c) -3.00, 4.00

CHAPTER 24

Multiple-Choice Questions

1. b   2. a   3. b   4. a   5. b

Conceptual Questions

1. The result of the double slit is to redistribute the energy arriving at the screen. Although there is no energy at the location of a dark fringe, there is four times as much energy at the location of a bright fringe as there would be with only a single narrow slit. The total amount of energy arriving at the screen is twice as much as with a single slit, as it must be according to the conservation of energy.

3. One of the materials has a higher index of refraction than water, the other lower. The material with the higher index of refraction than water will appear black as it approaches zero thickness. There will be a 180° phase shift for the light reflected from the upper surface, but no such phase change from the lower surface, because the index of refraction for water on the other side is lower than that of the film. Thus, the two reflections will be out of phase and will interfere destructively. The material with index of refraction lower than that of water will have a phase change for the light reflected from both upper and lower surfaces, so that the reflections from the zero-thickness film will be back in phase, and the film will appear bright.

5. For normal incidence, the extra path length followed by the reflected ray is twice the thickness of the film. For destructive interference, this must be a distance of half a wavelength of the light in the material of the film. Because no 180° phase change will occur in these reflections, the thickness of the film must be one quarter wavelength, which is the same as the condition for constructive interference of reflected light.

7. As the edge of the Moon cuts across the light from the star, edge diffraction effects occur. Thus, as the edge of the Moon moves relative to the star, the observed light from the star proceeds through a series of maxima and minima.

9. For regional communication at the Earth's surface, radio waves are typically broadcast from currents oscillating in tall vertical towers. These waves have vertical planes of polarization. Light originates from the vibrations of atoms, or electronic transitions within atoms, which represents oscillations in all possible directions. Thus, light is not generally polarized.

11. Yes. In order to do this, first measure the radar-reflectivity of the metal of your airplane. Then choose a light, durable material that has approximately half the radar-reflectivity of the metal in your plane. Measure its index of refraction, and onto the metal plaster a coating equal in thickness to one quarter of 3 cm, divided by that index.

13. If you wish to perform an interference experiment, you need monochromatic coherent light. To obtain it, you must first pass light from an ordinary source through a prism or diffraction grating to disperse different colors into different directions. Using a single narrow slit, select a single color and make that light diffract to cover both slits for Young's experiment. The procedure is much simpler with a laser because its output is already monochromatic and coherent.

15. Audible sound has wavelengths on the order of meters or centimeters, and visible light has wavelengths on the order of half a micrometer. Sound, therefore, diffracts

around walls and doorways (roughly meter-sized apertures). Visible light diffracts only through very small angles as it passes ordinary sized objects or apertures, because sin $\theta = m\lambda/a$ by Equation 24.11, and λ/a is extremely small.

Review Problem

$20.0 \times 10^{-6} \,°C^{-1}$

Problems

1. (a) 1.77 μm   (b) 1.47 μm
3. 3.00 cm
5. 2.02 μm
7. 1.5 mm
9. 515 nm
11. 1.73 km
13. 91.9 nm
15. Any odd multiple of 85.4 nm
17. 550 nm
19. 190 nm
21. 4.75 μm
23. 552 nm; green
25. 99.6 nm
27. (a) 2.25 mm   (b) 4.50 mm
29. 91.2 cm
31. 1.2 mm, 1.2 mm
33. (a) $n = 1.11$ (b) 42.0°
35. 36.9°
37. 60.5°
39. (a) $I/I_0 = \frac{1}{2}$ (b) 54.7°
41. (a) 413.7 nm, 409.7 nm   (b) 8.64°
43. 0.156 mm
45. 42°
47. (a) 0   (b) 0.25
49. (a) 1.93 μm   (b) $\delta = 3\lambda$ (c) maximum
51. (a) 10^{-3} degrees   (b) about 10^{11} Hz, a microwave
57. 0.350 mm

CHAPTER 25

Multiple-Choice Questions

1. d   2. a   3. c   4. c   5. d

Conceptual Questions

1. We would like to reduce the minimum angular separation for two objects below the angle subtended by the two stars in the binary system. We can do that by reducing the wavelength of the light—this in essence makes the aperture larger, relative to the light wavelength, increasing the resolving power. Thus, we would choose a blue filter.
3. There will be an effect on the interference pattern—it will be distorted. The high temperature of the flame will change the index of refraction of air for the arm of the interferometer in which the match is held. As the index of refraction varies turbulently, the wavelength of the

light in that region will also vary turbulently. As a result, the effective difference in length between the two arms will vary, resulting in a wildly varying interference pattern.
5. The tracks of information on a compact disc are much closer together than on a phonograph record. As a result, the diffraction maxima from the compact disc will be farther apart than those from the record.
7. Strictly speaking, the ribs do act as a diffraction grating, but the separation distance of the ribs is so much larger than the wavelength of the x-rays that there are no observable effects.
9. Large lenses are difficult to manufacture and machine with accuracy. Their large weight leads to sagging, which produces a distorted image. In reflecting telescopes, light does not pass through glass; hence, problems associated with chromatic aberrations are eliminated. Large-diameter reflecting telescopes are also technically easier to construct. Some designs use a rotating pool of mercury as the reflecting surface.
11. In order to "see" an object, the wavelength of the light in the microscope must be comparable to the size of the object. An atom is much smaller than the wavelength of light in the visible spectrum, so an atom can never be seen using visible light.

Problems

1. 30.0 cm beyond the lens, $M = -\frac{1}{5}$
3. Yes, the maximum image dimension is only 19 mm.
5. $f/1.4$
7. $f/8$
9. For the right eye, $P = -11.8$ diopters; for the left eye, $P = -8.2$ diopters
11. (a) -2.00 diopters   (b) 17.6 cm
13. (a) -0.67 diopters   (b) $+0.67$ diopters
15. (a) 5.8 cm   (b) $m = 4.3$
17. (a) 4.07 cm   (b) $m = 7.14$
19. (a) 0.400 cm   (b) 1.25 cm   (c) $M = -1000$
21. $M = -115$
23. (a) 9.84   (b) $m = 50$
25. $f_e = +2.0$ cm, $f_o = +90$ cm
27. (a) $m = 1.50$ (b) $M = 1.91$
29. 493 km
31. 5.4 mm
33. 9.1×10^7 km
35. 9.8 km
37. 50.4 μm
39. 98 fringe shifts
41. 19.4°.
43. (a) 2 complete orders   (b) 10.9°.
45. $m = 1, \Delta\theta = 0.0087°$; $m = 2, \Delta\theta = 0.0179°$; $m = 3, \Delta\theta = 0.0286°$
47. (a) 9.21°   (b) 15.1°
49. (a) 3646 slits   (b) 1823 slits
51. (a) 2.67 diopters   (b) 0.16 diopters too low

53. (a) 101.5 cm   (b) $m = 67$
55. (a) 980 lines   (b) 330 lines
57. 10.1 cm
59. (a) $m = 4.0$ (b) $m = 3.0$
61. 469 nm and 78.2 nm
63. 5.07 mm

CHAPTER 26

Multiple-Choice Questions
1. b   2. e   3. d   4. a   5. a

Conceptual Questions
1. This scenario is not possible with light. Light waves are described by the principles of special relativity. As you detect the light wave ahead of you and moving away from you (which would be a pretty good trick—think about it!), its velocity relative to you is c. Thus, you will not be able to catch up to the light wave.
3. Assuming that their on-duty time was kept on Earth, they will be pleasantly surprised with a large paycheck. Less time will have passed for the astronauts in their frame of reference than for their employer back on Earth.
5. The answers to both of these questions is no. Both your clothing and your sleeping cabin are at rest in your reference frame, thus, they will have their proper length. There will be no change in measured lengths of objects within your spacecraft. Another observer, on a spacecraft traveling at a high speed relative to yours, will measure you as thinner (if your body is oriented in a direction perpendicular to the direction of motion relative to her) or will claim that you are able to fit into a shorter sleeping cabin (if your body is oriented in a direction parallel to your direction of travel relative to the other observer).
7. The incoming rocket will not appear to have a longer length and a faster clock. Length contraction and time dilation depend only on the magnitude of the relative velocity, not on the direction.
9. A reflected photon does exert a force on the surface. Although a photon has zero mass, a photon does carry momentum. When it reflects from a surface, there is a change in the momentum, just like the change in momentum of a ball bouncing off the floor. According to the momentum interpretation of Newton's second law, a change in momentum results in a force on the surface. This concept is used in theoretical studies of space sailing. These studies propose building nonpowered spacecraft with huge reflective sails oriented perpendicularly to the rays from the Sun. The large number of photons from the Sun reflecting from the surface of the sail will exert a force that, although small, will provide a continuous acceleration. This would allow the spacecraft to travel to other planets without fuel.
11. Your assignment: Measure the length of a rod as it slides past you. You mark the position of its front end on the floor and have an assistant mark the position of the back end. Then measure the distance between the two marks. This distance will represent the length of the rod only if the two marks were made simultaneously in your frame of reference.
13. As an object approaches the speed of light, its energy approaches infinity. Hence, it would take an infinite amount of work to accelerate the object to the speed of light under the action of a continuous force or it would take an infinitely large force.
15. For a wonderful fictional exploration of this question, get a "Mr. Tompkins" book by George Gamow. All of the relativity effects would be obvious in our lives. Time dilation and length contraction would both occur. Driving home in a hurry, you would push on the gas pedal not to increase your speed very much but to make the blocks shorter. Big Doppler shifts in wave frequencies would make red lights look green as you approached and make car horns and radios useless. High-speed transportation would be both very expensive, requiring huge fuel purchases, as well as dangerous, because a speeding car could knock down a building. When you got home, hungry for lunch, you would find that you had missed dinner; there would be a five-day delay in transit when you watched the Olympics in Australia on live television in the United States. Finally, we would not be able to see the Milky Way, because the fireball of the Big Bang would surround us at the distance of Rigel, or Deneb.

Review Problem
2.08×10^{-13} kg

Problems
1. (a) $t_{\text{I}} = 1.67 \times 10^3$ s, $t_{\text{II}} = 2.04 \times 10^3$ s
   (b) $t_{\text{I}}' = 2.50 \times 10^3$ s, $t_{\text{II}}' = 2.04 \times 10^3$ s   (c) $\Delta t = 90$ s
3. 5.0 s
5. (a) 1.3×10^{-7} s   (b) 38 m   (c) 7.6 m
7. (a) 2.2 μs   (b) 0.65 km
9. (a) a rectangular box   (b) sides perpendicular to velocity are 2.0 m long, and sides parallel to velocity are 1.2 m long.
11. $0.95c$
13. (a) 17.3 m   (b) 3.31° with respect to the direction of motion
15. 3.3×10^5 m/s
19. (a) $0.140c$ (b) $0.417c$
21. $-0.94c$
23. $0.998c$
25. (a) 76 min   (b) 52 min
27. (a) 4.5×10^{16} J   (b) 1.4×10^7 y
29. $0.866c$
31. $0.27c$
33. $m_{\text{faster}} = 2.51 \times 10^{-28}$ kg, $m_{\text{slower}} = 8.84 \times 10^{-28}$ kg
35. (a) 0.183 MeV   (b) 2.45 MeV

37. (a) 3.10×10^5 m/s
    (b) $0.757c$
39. 1.42 MeV/c
41. $0.995c$
43. $0.61c$ (away from Earth)
45. (a) 438 GJ   (b) 438 GJ
47. 1.47 km
49. (a) $0.80c$ (b) 7.5×10^3 s
    (c) 1.4×10^{12} m, $-0.39c$
51. (a) $0.95c$ (b) 630 s   (c) 210 s   (d) 600 s

CHAPTER 27

Multiple-Choice Questions
1. c   2. c   3. a   4. c   5. c

Conceptual Questions
1. The shape of an object is determined by observing the light reflecting from its surface. In a kiln, the objects will be very hot and will be glowing red. The emitted radiation is far stronger than the reflected radiation, and the thermal radiation emitted is only slightly dependent on the material from which the objects are made. Thus, we have a collection of objects glowing equally with emitted radiation, and only weak reflected light compared to the emitted light. This will result in indistinct outlines of the objects when viewed.
3. The "blackness" of a blackbody refers to its ideal property of absorbing all radiation incident on it. If an observed room temperature object in everyday life absorbs all radiation, we describe it as (visibly) black. The black appearance, however, is due to the fact that our eyes are sensitive only to visible light. If we could detect infrared light with our eyes, we would see the object emitting radiation. If the temperature of the blackbody is raised, Wien's law tells us that the emitted radiation will move into the visible. Thus, the blackbody could appear as red, white, or blue, depending on its temperature.
5. All objects do radiate energy, but at room temperature, this energy is primarily in the infrared region of the electromagnetic spectrum, which our eyes cannot detect. (Pit vipers have sensory organs that are sensitive to infrared radiation; thus they can seek out their warm-blooded prey in what we would consider absolute darkness.)
7. Most metals have cutoff frequencies corresponding to photons in or near the visible range of the electromagnetic spectrum. AM radio wave photons will have far too little energy to eject electrons from the metal.
9. We can picture higher frequency light as a stream of photons of higher energy. In a collision, one photon can give all of its energy to a single electron. The kinetic energy of such an electron is measured by the stopping potential. The reverse voltage (stopping voltage) required to stop

the current is proportional to the frequency of the incoming light. More intense light consists of more photons striking a unit area each second, but atoms are so small that one emitted electron never gets a "kick" from more than one photon. Increasing the light intensity will generally increase the size of the current but will not change the energy of the individual ejected electrons. Thus, the stopping potential remains constant.
11. Wave theory predicts that the photoelectric effect should occur at any frequency, provided that the light intensity is high enough. However, as seen in photoelectric experiments, the light must have sufficiently high frequency for the effect to occur.
13. The x-ray photon transfers some of its energy to the electron. Thus, its energy, and therefore its frequency, must be decreased.
15. No. Suppose that the incident light frequency at which you first observed the photoelectric effect is above the cutoff frequency of the first metal, but less than the cutoff frequency of the second metal. In that case, the photoelectric effect would not be observed at all in the second metal.

Problems
1. (a) ≈ 3000 K   (b) $\approx 20\,000$ K
3. 500 nm
5. (a) 2.49×10^{-5} eV   (b) 2.49 eV   (c) 249 eV
7. 2.27×10^{30} photons
9. (a) $n = 2.3 \times 10^{31}$ (b) $\dfrac{\Delta E}{E} = 4.2 \times 10^{-32}$
11. 148 days
13. (a) 2.24 eV   (b) 555 nm   (c) 5.41×10^{14} Hz
15. 5.4 eV
17. 4.77×10^{14} Hz, 2.03 eV
19. 1.2×10^2 V and 1.2×10^7 V, respectively
21. 4.14×10^4 V
23. 0.078 nm
25. 0.281 nm
27. $70.0°$
29. (a) 4.89×10^{-4} nm   (b) 268 keV   (c) 32 keV
31. (a) 2.0×10^{-19} J (1.25 eV)   (b) 6.6×10^5 m/s
33. (a) 3.52 MeV   (b) 8.50×10^{20} Hz
35. $E = 0.64$ MeV, $p = 0.64$ MeV/c
37. (a) 1.46×10^3 m/s   (b) 7.28×10^{-11} m
39. (a) 1.99×10^{-11} m   (b) 1.32×10^{-14} m
41. 3.58×10^{-13} m
43. (a) 15 keV   (b) 1.3×10^2 keV
45. 2.2×10^{-32} m/s
47. 116 m/s
49. ≈ 5200 K. Clearly, a firefly is not at this temperature, so this cannot be blackbody radiation.
51. (a) $n = 4.2 \times 10^{35}$ (b) 3.3×10^{-34} J
53. $6.9°$

55. 2.00 eV
57. (a) 0.0220c (b) 0.9992c
59. 0.83 MeV
61. (b) 3.69 × 10^3 m/s
63. (b) 1.84 × 10^{-16} m

CHAPTER 28

Multiple-Choice Questions
1. e   2. e   3. d   4. c   5. c

Conceptual Questions

1. If the energy of the hydrogen atom were proportional to n (or any positive power of n), then the energy would become infinite as n grew to infinity. But the energy of the atom is inversely proportional to n^2. Thus, as n grows to infinity, the energy of the atom approaches a value which is above the ground state by a finite amount, namely the ionization energy 13.6 eV. As the electron falls from one bound state to another, its energy loss is always less than the ionization energy. The energy and frequency of any emitted photon are finite.

3. The characteristic x-rays originate from transitions within the atoms of the target, such as an L shell electron making a transition to a vacancy in the K shell. This vacancy in the K shell is caused when an accelerated electron in the x-ray tube supplies energy to the K electron to eject it from the atom. If the energy of the bombarding electrons were to be increased, the K electron will be ejected from the atom with more remaining kinetic energy. But the energy difference between the K and L shell has not changed, so the emitted x-ray has exactly the same wavelength.

5. A continuous spectrum without characteristic x-rays is possible. At a low accelerating potential difference for the electron, the electron may not have enough energy to eject an electron from a target atom. As a result, there will be no characteristic x-rays. The change in speed of the electron as it enters the target will result in the continuous spectrum.

7. The hologram is an interference pattern between light scattered from the object and the reference beam. If anything moves by a distance comparable to the wavelength of the light (or more), the pattern will wash out. The effect is just like making the slits vibrate in Young's experiment, to make the interference fringes vibrate wildly so that a photograph of the screen displays only the average intensity everywhere.

9. If the Pauli exclusion principle were not valid, the elements and their chemical behavior would be grossly different because every electron would end up in the lowest energy level of the atom. All matter would therefore be nearly alike in its chemistry and composition, because the shell structures of each element would be identical. Most materials would have a much higher density, and the spectra of atoms and molecules would be very simple, resulting in the existence of less color in the world.

11. The three elements have similar electronic configurations, with filled inner shells, plus a single electron in an s orbital. Because atoms typically interact through their unfilled outer shells, and the outer shell of each of these atoms is similar, the chemical interactions of the three atoms are also similar.

13. Each of the eight electrons must have at least one quantum number different from each of the others. They can differ (in m_s) by being spin-up or spin-down. They can differ (in ℓ) in angular momentum and in the general shape of the wave function. Those electrons with $\ell = 1$ can differ (in m_ℓ) in orientation of angular momentum.

Review Problem
(a) $n_f = 1$ (b) $n_i = 3$

Problems

1. 656 nm, 486 nm, 434 nm
3. (a) 2.3 × 10^{-8} N   (b) −14 eV
5. (a) 1.6 × 10^6 m/s   (b) no, $v \ll c$ (c) 0.45 nm
   (d) Yes, the wavelength and the atom are roughly the same size.
7. (a) 0.212 nm   (b) 9.95 × 10^{-25} kg·m/s
   (c) 2.10 × 10^{-34} J·s   (d) 3.40 eV
   (e) −6.80 eV   (f) −3.40 eV
11. 8.22 × 10^{-8} N
13. (a) 0.967 eV   (b) 0.266 eV
15. $E = -1.51$ eV ($n = 3$) to $E = -3.40$ eV ($n = 2$)
17. (a) 122 nm, 91.4 nm   (b) 1.88 × 10^3 nm, 823 nm
19. (a) 486 nm   (b) 0.814 m/s
23. (d) $n = 2.53 \times 10^{74}$ (e) No. At such large quantum numbers, the allowed energies are essentially continuous.
25. (a) 2.46 × 10^{15} Hz, $f_{orb} = 8.29 \times 10^{14}$ Hz
    (b) 6.57 × 10^3 Hz, $f_{orb} = 6.56 \times 10^3$ Hz
    For large n, classical theory and quantum theory approach one another in their results.
27. Energy levels should be at −54.4 eV, −13.6 eV, −6.04 eV, −3.40 eV, −2.18 eV, and so on to 0.
29. Energy levels should be at −122.4 eV, −30.6 eV, −13.6 eV, −7.65 eV, −4.90 eV, and so on to 0.
31. 1.33 nm
33. $\begin{cases} n = 3 & \ell = 1 & m_\ell = 1 & m_s = \pm\frac{1}{2} \\ n = 3 & \ell = 1 & m_\ell = 0 & m_s = \pm\frac{1}{2} \\ n = 3 & \ell = 1 & m_\ell = -1 & m_s = \pm\frac{1}{2} \end{cases}$
35. (a) 2   (b) 8   (c) 18   (d) 32   (e) 50
37. (a) $n = 4$ and $\ell = 2$
    (b) $m_\ell = (0, \pm1, \pm2)$, $m_s = \pm\frac{1}{2}$
    (c) $1s^2 2s^2 2p^6 3s^2 3p^6 3d^{10} 4s^2 4p^6 4d^2 5s^2 = (\text{Kr})4d^2 5s^2$
39. 0.160 nm

41. L shell: 11.7 keV; M shell: 10.0 keV; N shell: 2.3 keV
43. (a) 10.2 eV (b) 7.88×10^4 K
45. (a) 1.3×10^{-4} J (b) 4.1×10^{14} photons
47. (a) -8.178 eV, -2.043 eV, -0.905 eV, -0.510 eV, -0.326 eV
    (b) 1092 nm and 608.4 nm
49. The simplest diagram has 4 levels with energies of -4.10 eV, -1.00 eV, -0.100 eV, and 0.
51. (a) 120 MeV (b) No, its energy is too large to allow it to be confined.
53. (a) 4.24×10^{15} W/m^2 (b) 1.20×10^{-12} J
55. (a) 0.240 s
    (b) Classical answer is too large by about a billion.
59. (a) 137.395 (b) $\dfrac{1}{2\pi\alpha}$ (c) $\dfrac{4\pi}{\alpha}$

CHAPTER 29

Multiple-Choice Questions
1. b   2. d   3. d   4. b   5. b

Conceptual Questions
1. Isotopes of a given element correspond to nuclei with different numbers of neutrons. This will result in a variety of different physical properties for the nuclei, including the obvious one of mass. The chemical behavior, however, is governed by the electrons. All isotopes of a given element have the same number of electrons and, therefore, the same chemical behavior.
3. An alpha particle contains two protons and two neutrons. Because a hydrogen nucleus only contains one proton, it cannot emit an alpha particle.
5. In alpha decay, there are only two final particles—the alpha particle and the daughter nucleus. There are also two conservation principles—energy and momentum. As a result, the alpha particle must be ejected with a discrete energy to satisfy both conservation principles. However, beta decay is a three-particle decay—the beta particle, the neutrino (or antineutrino), and the daughter nucleus. As a result, the energy and momentum can be shared in a variety of ways among the three particles while still satisfying the two conservation principles. This allows a continuous range of energies for the beta particle.
7. The larger rest energy of the neutron means that a free proton in space will not spontaneously decay into a neutron and a positron. When the proton is in the nucleus, however, the important question is that of the total rest energy of the nucleus. If it is energetically lower for the nucleus to have one less proton and one more neutron, then the decay process will occur to achieve this lower energy.
9. Carbon dating cannot generally be used to estimate the age of a stone, because the stone was not alive to take up carbon from the environment. Only the ages of artifacts that were once alive can be estimated with carbon dating.
11. The protons, although held together by the nuclear force, repel one another by the electrostatic force. If enough protons were placed together in a nucleus, the electrostatic force would overcome the nuclear force, which is based on the number of particles, and cause the nucleus to fission. The addition of neutrons prevents such fission. The neutron does not increase the electrostatic force, being electrically neutral, but does contribute to the nuclear force.
13. After the first half-life, half the original sample remains. After the second half-life, $(\frac{1}{2})(\frac{1}{2}) = \frac{1}{4}$, a quarter of the original sample remains, and three quarters of a radioactive sample has decayed after two half-lives.
15. Because the two samples are of the same radioactive nuclide, they have the same half-life; the $2:1$ difference in activity is due to a $2:1$ difference in the mass of each sample. After five half lives, each will have decreased in mass by a power of $2^5 = 32$. However, because this simply means that the mass of each is 32 times smaller, the ratio of the masses will still be $(2/32):(1/32)$, or $2:1$. Therefore, the ratio of their activities will *always* be $2:1$.

Problems
1. $A = 2$, $r = 1.5 \times 10^{-15}$ m; $A = 60$, $r = 4.7 \times 10^{-15}$ m; $A = 197$, $r = 7.0 \times 10^{-15}$ m; $A = 239$, $r = 7.4 \times 10^{-15}$ m
3. 1.8×10^2 m
5. (a) 1.9×10^7 m/s (b) 7.1 MeV
7. (a) 27.6 N (b) 4.16×10^{27} m/s^2 (c) 1.73 MeV
9. 8.66 MeV/nucleon for $^{93}_{41}$Nb, 7.92 MeV/nucleon for $^{197}_{79}$Au
11. 3.53 MeV
13. 0.47 curies
15. (a) 9.98×10^{-7} s^{-1} (b) 1.9×10^{10} nuclei
17. (a) 8.06 days
    (b) This could be $^{131}_{53}$I for which $T_{1/2} = 8.04$ days.
19. 2.29 g
21. 4.31×10^3 y
23. (a) 5.58×10^{-2} h^{-1}, 12.4 h (b) 2.39×10^{13} nuclei
    (c) 1.9 mCi
25. $^{208}_{81}$Tl, $^{95}_{37}$Rb, $^{144}_{60}$Nd
27. e^+ decay, $^{56}_{27}$Co $\rightarrow$ $^{56}_{26}$Fe $+ e^+ + \nu$
29. 71.4 keV
31. 18.6 keV
33. 4.22×10^3 y
35. (a) $^{13}_{6}$C (b) $^{10}_{5}$B
37. (a) $^{197}_{79}$Au $+ n \rightarrow$ $^{198}_{80}$Hg $+ e^- + \bar{\nu}$ (b) 7.88 MeV
39. (a) $^{30}_{15}$P (b) -2.64 MeV
41. 1.00 MeV
43. (a) $^1_0$n (b) Fluorine mass = 18.000 953 u
45. 18.8 J
47. 2.1×10^6 s ($\approx$ 24 days)
49. (a) 9.00×10^{11} electrons (b) 1.01×10^{-1} J
    (c) 101 rad
51. 46.5 days

53. 24 decays/min
55. $Q > 0$, no threshold energy is required
57. fraction decayed = 0.0035 (0.35%)
59. 5.9 billion years
61. 3.96×10^9 y   (b) It could be no older. The rock could be younger if some ^{87}Sr were initially present.
63. (a) 84.2 MeV   (b) 342 MeV
    (c) To be stable, the total binding energy must exceed the minimum energy needed to overcome electrostatic repulsion.
65. 230 y

CHAPTER 30

Multiple-Choice Questions

1. c   2. e   3. c   4. a   5. b

Conceptual Questions

1. The experiment described is a nice analogy to the Rutherford scattering experiment. In the Rutherford experiment, alpha particles were scattered from atoms, and the scattering was consistent with a small structure in the atom containing the positive charge.
3. The largest quark charge is $2e/3$, so a combination of only two particles, a quark and an antiquark forming a meson, could not possibly have electric charge up to $+2e$. Only particles containing three quarks, each with a charge of $2e/3$, can combine to produce a total charge of $2e$.
5. Until about 700 000 years after the Big Bang, the temperature of the Universe was high enough for any atoms that formed to be ionized by ambient radiation. Once the average radiation energy dropped below the hydrogen ionization energy of 13.6 eV, hydrogen atoms could form and remain as neutral atoms for a relatively long period of time.
7. In the quark model, all hadrons are composed of smaller units called quarks. Quarks have a fractional electric charge and a baryon number of $\frac{1}{3}$. There are six flavors of quarks: up (u), down (d), strange (s), charmed (c), top (t), and bottom (b). All baryons contain three quarks, and all mesons contain one quark and one antiquark. Section 30.12 has a more detailed discussion of the quark model.
9. Baryons and mesons are hadrons, interacting primarily through the strong force. They are not elementary particles, being composed of either three quarks (baryons), or a quark and an antiquark (mesons). Baryons have a non-zero baryon number with a spin of either $\frac{1}{2}$ or $\frac{3}{2}$. Mesons have a baryon number of zero and a spin of either 0 or 1.
11. All stable particles other than protons and neutrons have baryon number zero. Because the baryon number must be conserved, and the final states of the kaon decay contain no protons or neutrons, the baryon number of all kaons must be *zero*.

13. Yes, but the strong interaction predominates.
15. Unless the particles have enough kinetic energy to produce a baryon-antibaryon pair, the answer is *no*. Antibaryons have a baryon number of -1; baryons have a baryon number of $+1$; mesons have a baryon number of 0. If such an interaction were to occur and produce a baryon, the baryon number would not be conserved.

Review Problem

(a) $\lambda_{max} = 0.97$ mm (b) microwave region

Problems

1. $^1_0 \text{n} + ^{235}_{92}\text{U} \rightarrow ^{141}_{56}\text{Ba} + ^{92}_{36}\text{Kr} + 3^1_0\text{n}$, 3 neutrons released
3. 126.5 MeV
5. (a) 16.2 kg   (b) 0.117 kg
7. 1.01 g
9. 3.0×10^3 km ($\approx 1.9 \times 10^3$ mi)
11. (a) $^8_4$Be   (b) $^{12}_6$C   (c) 7.28 MeV
13. 3.07×10^{22} events/y
15. (a) 1.80×10^{30} J   (b) 81.4 million years
17. 118 MeV
19. 2.0×10^{-18} m
21. (a) conservation of electron-lepton number and conservation of muon-lepton number
    (b) conservation of charge
    (c) conservation of baryon number
    (d) conservation of baryon number
    (e) conservation of charge
23. $\bar{\nu}_\mu$
25. (a) $\bar{\nu}_\mu$ (b) ν_μ (c) $\bar{\nu}_e$ (d) ν_e (e) ν_μ
    (f) $\bar{\nu}_e$ and ν_μ
27. (a) not allowed; violates conservation of baryon number
    (b) strong interaction   (c) weak interaction
    (d) weak interaction   (e) electromagnetic interaction
29. (a) not conserved   (b) conserved   (c) conserved
    (d) not conserved   (e) not conserved
    (f) not conserved
31. (a) baryon number, charge, L_e, L_τ
    (b) baryon number, charge, L_e, L_μ, L_τ
    (c) strangeness, charge, L_e, L_μ, L_τ
    (d) baryon number, strangeness, charge, L_e, L_μ, L_τ
    (e) baryon number, strangeness, charge, L_e, L_μ, L_τ
    (f) baryon number, strangeness, charge, L_τ, L_μ, L_τ
33.

	Proton	u	u	d	Total
Strangeness	0	0	0	0	0
Baryon number	1	$\frac{1}{3}$	$\frac{1}{3}$	$\frac{1}{3}$	1
Charge	e	$2e/3$	$2e/3$	$-e/3$	e

	Neutron	u	d	d	Total
Strangeness	0	0	0	0	0
Baryon number	1	$\frac{1}{3}$	$\frac{1}{3}$	$\frac{1}{3}$	1
Charge	0	$2e/3$	$-e/3$	$-e/3$	0

35. (a) Σ^+ (b) π^- (c) K^0 (d) Ξ^-
37.

Reaction	At Quark Level	Net Quarks (Before and After)
(a)	$\bar{u}d + uud \rightarrow d\bar{s} + uds$	1 up, 2 down, 0 strange
(b)	$u\bar{d} + uud \rightarrow u\bar{s} + uus$	3 up, zero down, zero strange
(c)	$\bar{u}s + uud \rightarrow u\bar{s} + d\bar{s} + sss$	1 up, 1 down, 1 strange

(d) The mystery particle is a Λ^0.

39. First reaction: Net of 1 up and 2 down quarks both before and after. Second reaction: Net of 1 up and 2 down quarks before, but 1 up, 3 down, and 1 antistrange quark afterward.

41. $0.8279c$
43. 19.8 MeV
45. (a) electron-lepton and muon-lepton numbers not conserved
   (b) electron-lepton number not conserved
   (c) charge not conserved
   (d) baryon number not conserved
   (e) strangeness violated by 2 units
47. 26 collisions
49. (a) 1 baryon before decay, zero baryons after decay. Baryon number is not conserved.
   (b) $E_\gamma = 469$ MeV, $p_\gamma = 469$ MeV/c
   (c) $v = 0.9999994c$
51. (a) 1.1×10^6 m/s   (b) about 10^{-7} s

Index

Page numbers in *italics* indicate illustrations; page numbers followed by "n" indicate footnotes; page numbers followed by "t" indicate tables.

Physical Constants

Quantity	Symbol	Value	SI unit
Speed of light in vacuum	c	3.00×10^8	m/s
Permittivity of free space	ϵ_0	8.85×10^{-12}	$C^2/N \cdot m^2$
Coulomb constant, $1/4\pi\epsilon_0$	k_e	8.99×10^9	$N \cdot m^2/C^2$
Permeability of free space	μ_0	1.26×10^{-6} ($4\pi \times 10^{-7}$ exactly)	$T \cdot m/A$
Elementary charge	e	1.60×10^{-19}	C
Planck's constant	h	6.63×10^{-34}	J·s
	$\hbar = h/2\pi$	1.05×10^{-34}	J·s
Electron mass	m_e	9.11×10^{-31}	kg
		5.49×10^{-4}	u
Proton mass	m_p	$1.672\ 65 \times 10^{-27}$	kg
		$1.007\ 276$	u
Neutron mass	m_n	$1.674\ 95 \times 10^{-27}$	kg
		$1.008\ 665$	u
Avogadro's number	N_A	6.02×10^{23}	mol^{-1}
Universal gas constant	R	8.31	$J/mol \cdot K$
Boltzmann's constant	k_B	1.38×10^{-23}	J/K
Stefan-Boltzmann constant	σ	5.67×10^{-8}	$W/m^2 \cdot K^4$
Molar volume of ideal gas at STP	V	22.4	liters/mol
		2.24×10^{-2}	m^3/mol
Rydberg constant	R_H	1.10×10^7	m^{-1}
Bohr radius	a_0	5.29×10^{-11}	m
Electron Compton wavelength	$h/m_e c$	2.43×10^{-12}	m
Gravitational constant	G	6.67×10^{-11}	$N \cdot m^2/kg^2$
Standard free-fall acceleration	g	9.80	m/s^2
Radius of Earth (at equator)	R_E	6.38×10^6	m
Mass of Earth	M_E	5.98×10^{24}	kg
Radius of Moon	R_M	1.74×10^6	m
Mass of Moon	M_M	7.36×10^{22}	kg

The values presented in this table are those used in computations in the text. Generally, the physical constants are known to much better precision.